AA001015

International Symposium on Extreme Ultraviolet Lithography 2008

(2008 EUVL Symposium)

Lake Tahoe, California, USA
28 September – 1 October 2008

Volume 1 of 3

ISBN: 978-1-61567-661-3

Printed from e-media with permission by:

Curran Associates, Inc.
57 Morehouse Lane
Red Hook, NY 12571

Some format issues inherent in the e-media version may also appear in this print version.

Copyright© (2008) by SEMATECH
All rights reserved.

Printed by Curran Associates, Inc. (2009)

For permission requests, please contact SEMATECH
at the address below.

SEMATECH
2706 Montopolis Drive
Austin, Texas 78741

Phone: (512) 356-3500
Fax: (512) 356-7848

www.sematech.org

Additional copies of this publication are available from:

Curran Associates, Inc.
57 Morehouse Lane
Red Hook, NY 12571 USA
Phone: 845-758-0400
Fax: 845-758-2634
Email: curran@proceedings.com
Web: www.proceedings.com

International Symposium on Extreme Ultraviolet Lithography 2008

(2008 EUVL Symposium)

Lake Tahoe, California, USA
28 September – 1 October 2008

Volume 1 of 3

TABLE OF CONTENTS

Volume 1

OPENING PRESENTATIONS

Welcome to the 2008 EUVL Symposium! 1
Stefan Wurm, Ichiro Mori, Rob Hartman

Program Logistics & Overview 10
Obert Wood, Patrick Naulleau

EUV Lithography's Future 21
Harry J. Levinson

Samsung Lithography Strategy 71
Woosung Han

EUV Activities the EUVL Shop Future Plans 97
Rob Hartman

US Regional Update 124
Patrick Naulleau

IEUVI Update 136
Paolo Gargini

Asia Pacific Regional Update - Japan, Korea and Taiwan Regional Update 183
Ichiro Mori

TECHNICAL SESSION: EXPOSURE TOOL

EUV Alpha Demo Tools – Stepping Stones Towards Volume Production 199
H. Meiling, S. Lok, B. Hultermans, E. van Setten, B. Pierson, K. Cummings, C. Wagner, N. Harned

Nikon EUVL Development Progress Update 230
Takaharu Miura

Development Status of Canon's Full-Field EUVL Tool 264
Shigeyuki Uzawa, Tokuyuki Honda, Hideki Morishima, Takayuki Hasegawa

TECHNICAL SESSION: RESIST I

EUV Resist Performance on the ASML ADT and LBNL MET 291
Bill Pierson, Tom Wallow, Hiroyuki Mizuno, Anita Fumar-Pici, Linda Ohara, Karen Petrillo, Koen van Ingen-Schenau, Steve Hansen, Sang-In Han, Robert Watso, Lior Huli, Obert Wood, Joerg Mallmann, Bart Kessels, Robert Routh, Kevin Cummings

The SEMATECH Berkeley MET: Learning at the 22 nm Node 312
Patrick Naulleau, Chris Anderson, Paul Denham, Simi George, Ken Goldberg, Brian Hoef, Gideon Jones, Dimitra Niakoula, Ryan Miyakawa, John Roller, Chawon Koh, Warren Montgomery, Stefan Wurm, Bruno La Fontaine, Tom Wallow, Andy Ma, Joo-on Park

Evaluation of EUV Resists at Selete 335
Hiroaki Oizumi, Daisuke Kawamura, Koji Kaneyama, Shinji Kobayashi, Toshiro Itani

Advances in Resist Testing at PSI EUV-IL Exposure Tool 363
Vaida Auželyte, Pratap Sahoo, Menouer Saidani, Anja Weber, Harun H. Solak

Positive and Negative Tone Molecular Resists for 22-nm Node EUVL Patterning 376
Richard Lawson, Cheng-Tsung Lee, Laren M. Tolbert, Clifford L. Henderson

TECHNICAL SESSION: RESIST II

Sub-22nm Half-Pitch (HP) EUV Resist Imaging Results 398
Chawon Koh, Stefan Wurm, Joo-on Park, Andy Ma, Patrick Naulleau

Corner Rounding in Photoresists for Extreme Ultraviolet Lithography 426
Christopher N. Anderson, Patrick P. Naulleau, Thomas Wallow, Yunfei Deng

Reconciling Resist Resolution Metrics 487
Gregg M. Gallatin, Patrick P. Naulleau, Christopher N. Anderson
Feasibility Study on High-Sensitivity Chemically Amplified Resist by Polymer Absorption Enhancement in Extreme Ultraviolet Lithography 505
T. Kozawa, K. Okamoto, J. Nakamura, S. Tagawa

TECHNICAL SESSION: COST OF OWNERSHIP

Cost Implications of EUV Lithography Technology Decisions 520
Andrea F. Wüest, Andrew J. Hazelton, Greg Hughes, Lloyd C. Litt, Frank Goodwin

Volume 2

TECHNICAL SESSION: MASK I

An Overview of a Development Program for EUVL Mask Technologies in Selete 551
Osamu Suga, Tsuneo Terasawa, Hiroyuki Shigemura, Takao Taguchi, Iwao Nishiyama, Ichiro Mori
Development Status of EUVL Mask Blank and Substrate 563
Yoshiaki Ikuta
Multilayer Defect Compensation to Enable Quality Masks for EUVL Production 583
Ted Liang, Eric Gullikson
EUV Reticle Contamination and Cleaning 602
U. Okoroanyanwu, E. Langer, A. Fumar-Pici, T. Wallow, O. Wood, B. La Fontaine, C. Holfeld, J. H. Peters, M. Bender, M. Rossinger, S. Trogisch, F. Goodwin, A. Wüest, S. Huh, G. Denbeaux, Y. Fan, A. Antohe, L. Yankulin, R. Garg, K. Goldberg, P. Naulleau
High Transmission EUVL Pellicle Development 623
Yashesh A. Shroff, Pei-Yang Yan, Farhad Salamassi, Eric Gullikson
Applying Thinner Absorber to the EUVL Mask: EUV Printability and Integration Issues 646
Hwan-Seok Seo, Dong Gun Lee, Hoon Kim, Sungmin Huh, Byung-Sup Ahn, Hakseung Han, Dongwan Kim, Seong-Sue Kim, Han Ku Cho

TECHNICAL SESSION: MASK II

Study of Pit Defect Formation on EUV Blank Substrates 663
Abbas Rastegar, Sean Eichenlaub, Arun John Kadaksham, Matt House, Brian Cha, Henry Yun
Investigation of a Compensation Method for Pattern Placement Shifts of Chucked EUVL Masks 689
J. Sohn, K. Orvek, R. Engelstad, A. Lyons, J. Hartley
High Throughput Defect Mitigation 709
Patrick Kearney, C. C. Lin, T. Sugiyama, H. Yun, R. Randive, I. Reiss, P. Mirkarimi, E. Spiller, A. Hayes
Characterization of EUV Mask Defects: Printability and Repair Process 721
Hakseung Han, Donggun Lee, Hwan-Seok Seo, Kenneth A. Goldberg, Hoon Kim, Byung-Sub Ahn, In-Yong Kang, Wonil Cho, Sanghyeon Lee, Suyoung Lee, Geunbae Kim, Dongwan Kim, Seong-Sue Kim, HanKu Cho
Mask Defect Printability in Full Field EUV Lithography – Part 2 742
R. Jonckheere, F. Iwamoto, N. Stepanenko, A. M. Goethals, K. Ronse

TECHNICAL SESSION: METROLOGY/INSPECTION

A Practical Approach to EUV Reticle Inspection 774
Anna Tchikoulaeva, et.al.
Development of Actinic Mask Blank Inspection Technology at Selete 805
Tsuneo Terasawa, Takeshi Yamane, Teruo Iwasaki, Toshihiko Tanaka, Osamu Suga, Toshihisa Tomie
EUVL Blank Defect Inspection Capability at Intel 829
Andy Ma, Ted Liang, Seh-Jin Park, Guojing Zhang, Tomoya Tamura, Kazunori Omata
Analysis of Sub-22-nm Aerial Image Using Coherent Scattering Microscopy 848
Dong Gun Lee, Junki Kishimoto, Takeo Watanabe, Hiroo Kinoshita, Hwan-Seok Seo, Dongwan Kim, Seong-Sue Kim, HanKu Cho
Aerial Image Linewidth Measurement Capabilities of the Actinic Inspection Tool 867
Kenneth A. Goldberg, Iacopo Mochi, Patrick Naulleau, Bruno LaFontaine, Sungmin Huh

TECHNICAL SESSION: SOURCE I

Laser Produced Plasma Source System Development 890
David C. Brandt, et.al.
CO_2 Laser-produced Sn Plasma Source for EUV Lithography 916
Akira Endo, et.al.
EUV Sources based on DPP 935
Marc Corthout, Masaki Yoshioka
Time-Multiplexed Solid-State Laser-driven EUV Sources for Beta-Tools and HVM 956
K. Takenoshita, R. Bernath, R. Kamptaprasad, J. Szilagyi, S. A. George, J. Cunado, M. Richardson, B. Fulford, I. Henderson, N. Hay, S. Ellwi

TECHNICAL SESSION: SOURCE II

Guidelines and Promising Approach for LPP-EUV Light Source for HVM 975
K. Nishihara, et.al.
Technological Aspects of Sn RDE Source Development for HVM Lithography 996
V. M. Borisov, G. N. Borisova, A. S. Ivanov, Yu. B. Kirukhin, O. B. Khristoforov, V. A. Mishchenko, A. V. Prokofiev, A. Yu. Vinokhodov
Spectral Purity Filter Development for EUV HVM 1018
A. Yakunin, V. Banine, N. Salashchenko, E. Kluenkov, A. Lopatin, V. Luchin, N. Tsybin, L. Sjmaenok, W. Soer, M. Jak
Design and Fabrication Considerations of EUVL Collectors for HVM 1041
G. Bianucci, J. Kools, G. Salmaso, F. E. Zocchi

TECHNICAL SESSION: CONTAMINATION AND PARTICLES

Strategy for Minimizing EUV Optics Contamination During Exposure 1055
N. Harned, R. Moors, M. van Kampen, V. Banine, J. Huijbregtse, R. Vanneer, A. Kempen, D. Ehm, R. Verberk, E. te Sligte, A. Storm
Effective Debris Detection, Mitigation and Cleaning Methods for Source - Collector Optics 1077
David N. Ruzic, Ramasamy Raju, J. Sporre, H. Shin, W. M. Lytle, M. J. Neumann

Volume 3

TECHNICAL SESSION: CONTAMINATION AND PARTICLES (cont,)

Shielded Plasma's for Cleaning EUV Mirrors 1119
N. B. Koster, R. Koops, K. Agovic, F. P. J. de Groote, F. P. Wieringa, M. G. H. Meijerink
Carbon Accumulation on Model MLM Cap Layer: Interaction of Benzene Vapor with TiO2 Surface 1141
Boris Yakshinskiy, Shimon Zalkind, Theodore E. Madey
Progress on EUV Reticle Dual Pod Carriers for use in the Fab and Exposure Tools 1155
John Zimmerman, Long He

TECHNICAL SESSION: EXPOSURE TOOL EVALUATION

Lithographic Performance of Selete's Full Field EUV Exposure Tool 1174
Kazuo Tawarayama, Hajime Aoyama, Takashi Kamo, Shunko Magoshi, Yuusuke Tanaka, Seiichiro Shirai, Hiroyuki Tanaka
Full-field Patterning Test with ADT for 30-nm Node Device Application 1201
Doohoon Goo, Insung Kim, Joo-On Park, Jeonghoon Lee, Changmin Park, Jinhong Park, Jeong-Ho Yeo, Sungwoon Choi, Woosung Han
Flare Evaluation of an ASML Alpha Demo Tool 1227
Hiroyuki Mizuno, Martin Burkhardt, Chiew-seng Koay, Greg McIntyre, Tim Brunner, Bruno La Fontaine, Yunfei Deng, Obert Wood
Implementing Full field EUV Lithography using the ADT 1242
Anne-Marie Goethals, et.al.

TECHNICAL SESSION: OPTICS

Improvement of Optics for EUV Exposure Tools 1275
Katsuhiko Murakami, Tetsuya Oshino, Hiroyuki Kondo, Hiroshi Chiba, Kazushi Nomura

Optics for EUV Lithography 1300
Peter Kuerz, et.al.

Projection Optics for a 0.5-NA Microstepper Upgrade 1321
Michael Goldstein, Russ Hudyma, Patrick Naulleau

Projection Architectures for High NA EUVL 1341
Russ Hudyma, Mike Thomas, Mark Schwarz

CLOSING ADDRESS

2008 EUVL Symposium - Closing Address 1367
Stefan Wurm, Obert Wood, Patrick Naulleau

POSTERS: EXPOSURE TOOL

Ultra-High Resolution Extreme Ultraviolet Lithography by Incoherent to Coherent Conversion 1389
Christopher N. Anderson, Patrick P. Naulleau

MOSAIC - A New Way to Measure Optical Aberrations 1390
Christopher N. Anderson, Patrick P. Naulleau

First Italian EUV Micro Exposure Tool at 14.4 nm based on Kr DMS 1391
S. Bollanti, P. Di Lazzaro, F. Flora, L. Mezi, D. Murra, A. Torre

Lithographic Modeling of a 0.5-NA Microstepper Optic 1416
Patrick Naulleau, Michael Goldstein, Russ Hudyma

**Impact of Flare and Aberrations on Patterning Performance - Simulation with the EUV Full Field
Alpha Tool Conditions** 1417
Liping Ren, Frank Goodwin, Stefan Wurm, Chawon Koh, Sudharshanan Raghunathan, John Hartley

Multiple Catadioptric Simplified Extreme Ultraviolet Whole Lithography Machine and System 1418
Wynn L. Bear, Xiangwen Xiong

Reliability and Productivity Improvements on the Intel MET 1433
Roman Caudillo, Jeanette Roberts, Terence Bacuita, Gilroy Vandentop

**Composite Double Reflection Simplified Extreme Ultraviolet Whole Lithography Machine and
System** 1449
Wynn L. Bear, Xiangwen Xiong

POSTERS: SOURCE

Spectral Purity Filter Life-Time Testing on EQ-10 Source 1461
Chuck Schietinger, Dave Grove, Travis Ayers, Matthew Partlow, Debbie Gustafson

**Dependence of Laser Parameter on Conversion Efficiency in High-Repetition-Rate Laser-Ablation-
Discharge EUV Source** 1462
Takuma Yokoyama, Hiroshi Mizokoshi, Yusuke Teramoto, Daiki Yamatani, Hiroto Sato, Kazuaki Hotta

EUV Spectrometers for Source Development, Characterization and Optimization 1479
Scott Bergeson, Bryce Allred, Jershon Lopez, Jeffrey Kemp, Larry Knight, Alexander Shevelko

High Power from Low Etendue EUV Light Sources 1480
Michael Goldstein, Stefan Wurm, Frank Goodwin

The Characterization of the Ion Beam from a Sn-based DPP with Respect to the Ignition Parameters ... 1481
K. Gielissen, Y. Sidelnikov, W. Soer, M. van Herpen, D. Glushkov, V. Banine, J. van der Mullen

Present Status of Laser-produced Plasma EUV Light Source 1482
Hiroshi Komori, et.al.

Performance Evaluation of EUV SFET Source Collector Module 1483
Shunko Magoshi, Seiichiro Shirai, Hideto Mori, Kazuo Tawarayama, Yuusuke Tanaka, Hiroyuki Tanaka

POSTERS: RESISTS

Development of New Negative-tone Molecular Resists based on Alkylphenyl Callxarene for EUVL 1499
Masatoshi Echigo, Dai Oguro, Hiroaki Oizumi, Toshiro Itani

Survey and Comparison of Deprotection Blur Metrics for Extreme Ultraviolet Photoresists 1500
Christopher N. Anderson, Patrick P. Naulleau

Investigation of CA Resist Decomposition by EUV and EB Exposure 1501
Daiju Shiono, Taku Hirayama, Hideo Hada, Junichi Onodera, Takeo Watanabe, Hiroo Kinoshita

Novel Polyphenol Base Molecular Resist Having High Thermal Resistance 1513
Taku Hirayama, Takeyoshi Mimura, Jun Iwashita, Makiko Irie, Daiju Shiono, Hideo Hada, Takeshi Iwai

Electrons in EUV Resist Activation 1529
Theodore E. Madey, B. V. Yakshinskiy, E. Loginova, L. Sanche, P. Cloutier, R. Brainard, A. Wuest

DUV Source Integration into the 0.3 NA Berkeley SEMATECH MET for OOB Exposure Studies 1530
Simi A. George, Patrick P. Naulleau, Senajith Rekawa, Drew Kemp

Development of Novel Positive-tone Photoresists for EUVL 1531
Takanori Owada, Akinori Yomogita, Takashi Kashiwamura, Hiroaki Oizumi, Toshiro Itani

Relation between Acid Diffusion and Resolution in Chemically Amplified EUV Resists 1542
Y.uuki Hirai, Makoto Shimizu, Ken Maruyama, Toshiyuki Kai, Tsutomu Shimokawa, Toshiro Itani, Daisuke Kawamura

EUV Photoresists Twice as Fast as Previously Thought 1560
Patrick Naulleau, Eric Gullikson, Andrew Aquila, Paul Denham, Simi George, Drew Kemp, Dimitra Niakoula, Seno Rekawa

EUV Resist Outgassing Quantification and Qualification Analysis Methods 1561
Shinji Kobayashi, Julius Joseph Santillan, Hiroaki Oizumi, Toshiro Itani

Development of Partially Fluorinated EUV Resist Polymers for Sensitivity Improvement 1562
Takashi Sasaki, Osamu Yokokoji, Takeo Watanabe, Hiroo Kinoshita

POSTERS: MASK

Mask Effects on Line-Edge Roughness (LER) 1563
Patrick P. Naulleau, Kenneth A. Goldberg, Iacopo Mochi, Guojing Zhang

Experimental Study on Flatness of Electrostatically Chucked Reticle 1564
K. Ota, T. Taguchi, M. Amemiya, N. Nishimura, O. Suga

FIB Mask Repair Technology for EUV Lithography 1565
Tsuyoshi Amano, Yasushi Nishiyama, Hiroyuki Shigemura, Tsuneo Terasawa, Osamu Suga, Kensuke Shiina, Fumio Aramaki, Ryoji Hagiwara, Anto Yasaka

Analysis of Entrapped Object Size Effects on Out-of-Plane Distortion of the EUVL Mask in Electrostatic Chucking 1583
S. Lee, T. Yamamoto, K. Ota, N. Nishimura, T. Taguchi, I. Nishiyama, O. Suga, S. Warisawa, S. Ishihara

POSTERS: DEFECT INSPECTION

A Study of Optical Inspection on EUVL Mask for 32 nm Half Pitch Node Device and Beyond 1593
Yukiyasu Arisawa, Hiroyuki Shigemura, Tsuyoshi Amano, Hajime Aoyama, Toshihiko Tanaka, Osamu Suga

POSTERS: RETICLE CONTAMINATION

Characterization of EUV-Deposited Carboneous Contamination 1611
Toshihisa Anazawa, Yasushi Nishiyama, Hiroaki Oizumi, Osamu Suga, Iwao Nishiyama

Experimental Study of Particle-free Mask Handling Techniques using the MPE Tool 1631
Mitsuaki Amemiya, Kazuya Ota, Takao Taguchi, Osamu Suga

POSTERS: OPTICS AND ML COATINGS

Iterative Procedure for in situ Optical Testing using an Incoherent Source 1632
Ryan Miyakawa, Patrick P. Naulleau, Avideh Zakhor

Lateral Shearing Interferometry for EUV Optical Testing 1633
Ryan Miyakawa, Patrick P. Naulleau, Ken Goldberg

POSTERS: OPTICS CONTAMINATION

Resist Outgassing Measurements and Calibrations for High Volume Manufacturing 1634
Greg Denbeaux, Alin Antohe, Rashi Garg, Chimaobi Mbanaso, LeonidYankulin, Yu-Jen Fan, Kevin Orvek, Andrea Wüest

Experiment of Contamination Generation by EUV Irradiation with the Use of High-mass Hydrocarbon Gas 1648
Masahito Niibe, Keigo Koida

EUVL Optics Contamination from Resist Outgassing: Status Overview 1650
Kevin Orvek, Greg Denbeaux, Alin Antohe, Rashi Garg, Chimaobi Mbanaso

Moisture and Hydrocarbon Management for EUVL Tools: Ultra High Vacuum and Purge Gas Purification Solutions 1666
Andrea Conte, Cristian Landoni, Paolo Manini, Larry Rabellino, Sarah Riddle

POSTERS: DEVICE INTEGRATION

Characterization of New EUV Stable Silicon Photodiodes 1667
F. Scholze, C. Laubis, F. Sarubbi, Lis K. Nanver, S. N. Nihtianov

POSTERS: COST OF OWNERSHIP

The Comprehensive Cost for the Mainstream NGL and Simplified Extreme Ultraviolet Lithography Method 1668
Wynn L. Bear, Xiangwen Xiong

Author Index

Welcome to the 2008 EUVL Symposium!

Stefan Wurm (SEMATECH / AMD) - Symposium Chair

Ichiro Mori (Selete / Toshiba) – Symposium Co-Chair

Rob Hartman (ASML) - Symposium Co-Chair

Attendance by Geographic Region

**Asia / Pacific
117 (37%)**

**United States
129 (41%)**

**Europe
69 (22%)**

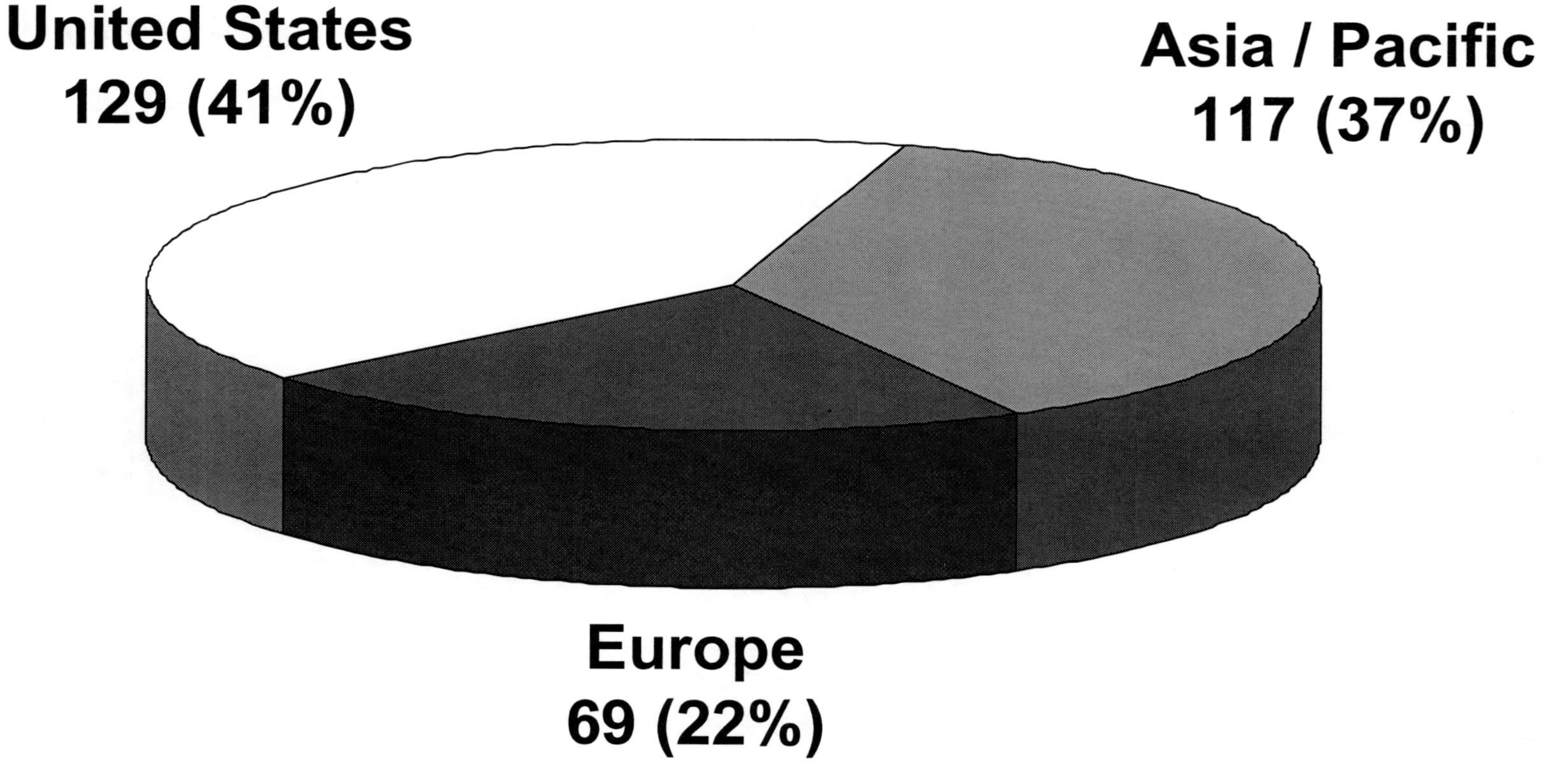

- **315 registered as of September 28**

Attendance by Year

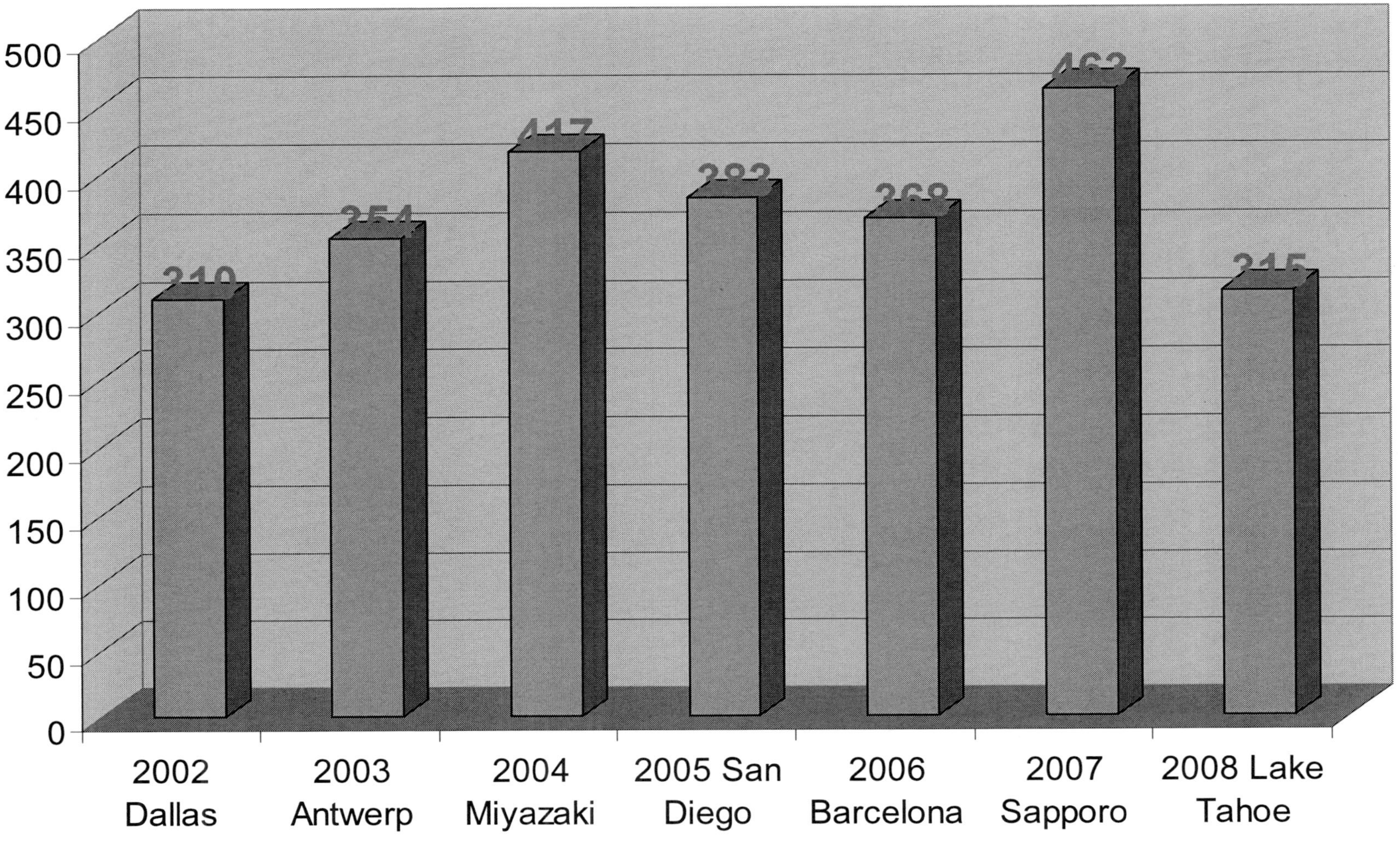

Year	Location	Attendance
2002	Dallas	210
2003	Antwerp	254
2004	Miyazaki	447
2005	San Diego	282
2006	Barcelona	269
2007	Sapporo	462
2008	Lake Tahoe	215

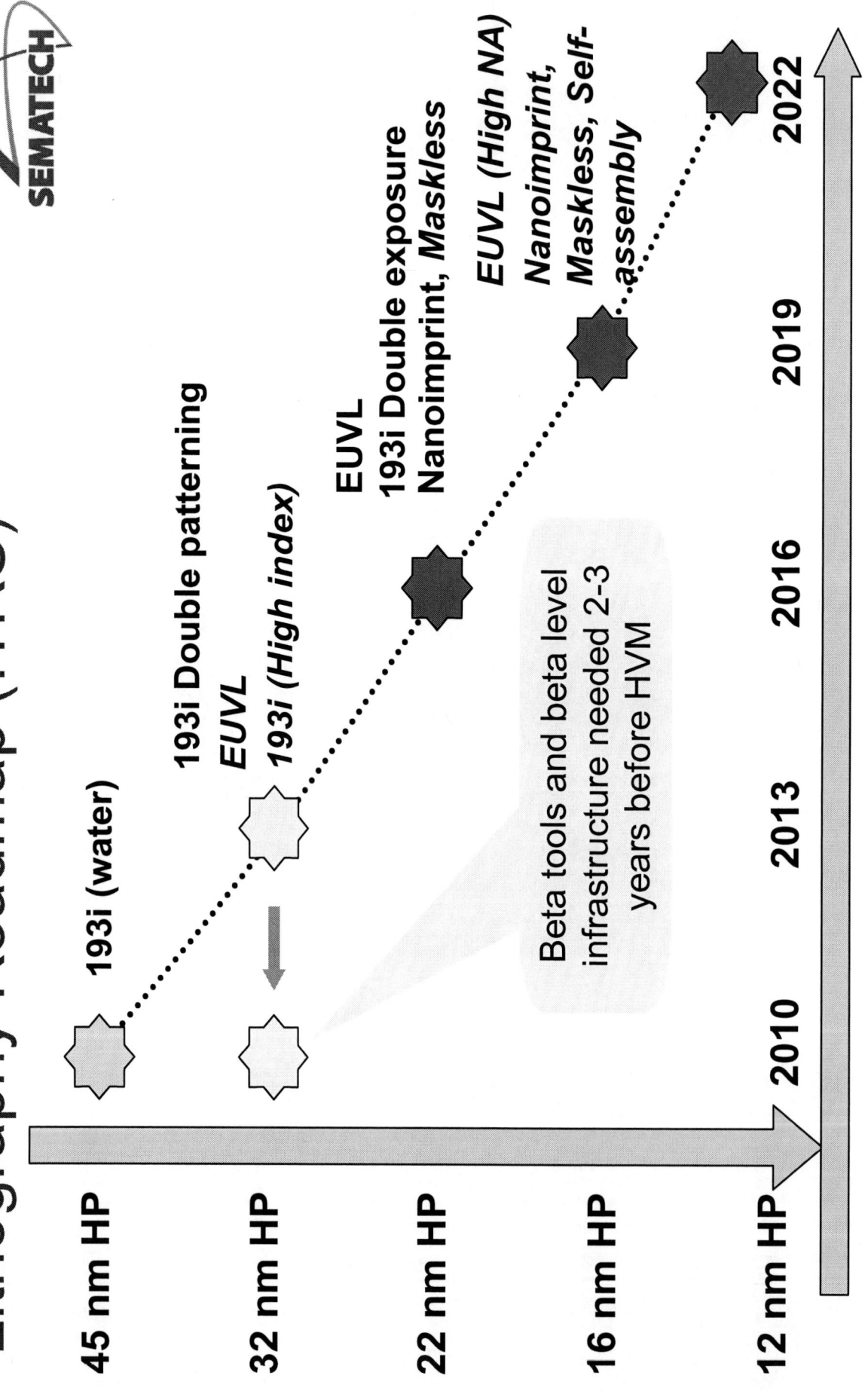

Litho Forum 2006 & 2008 Surveys: Roadmap Trends

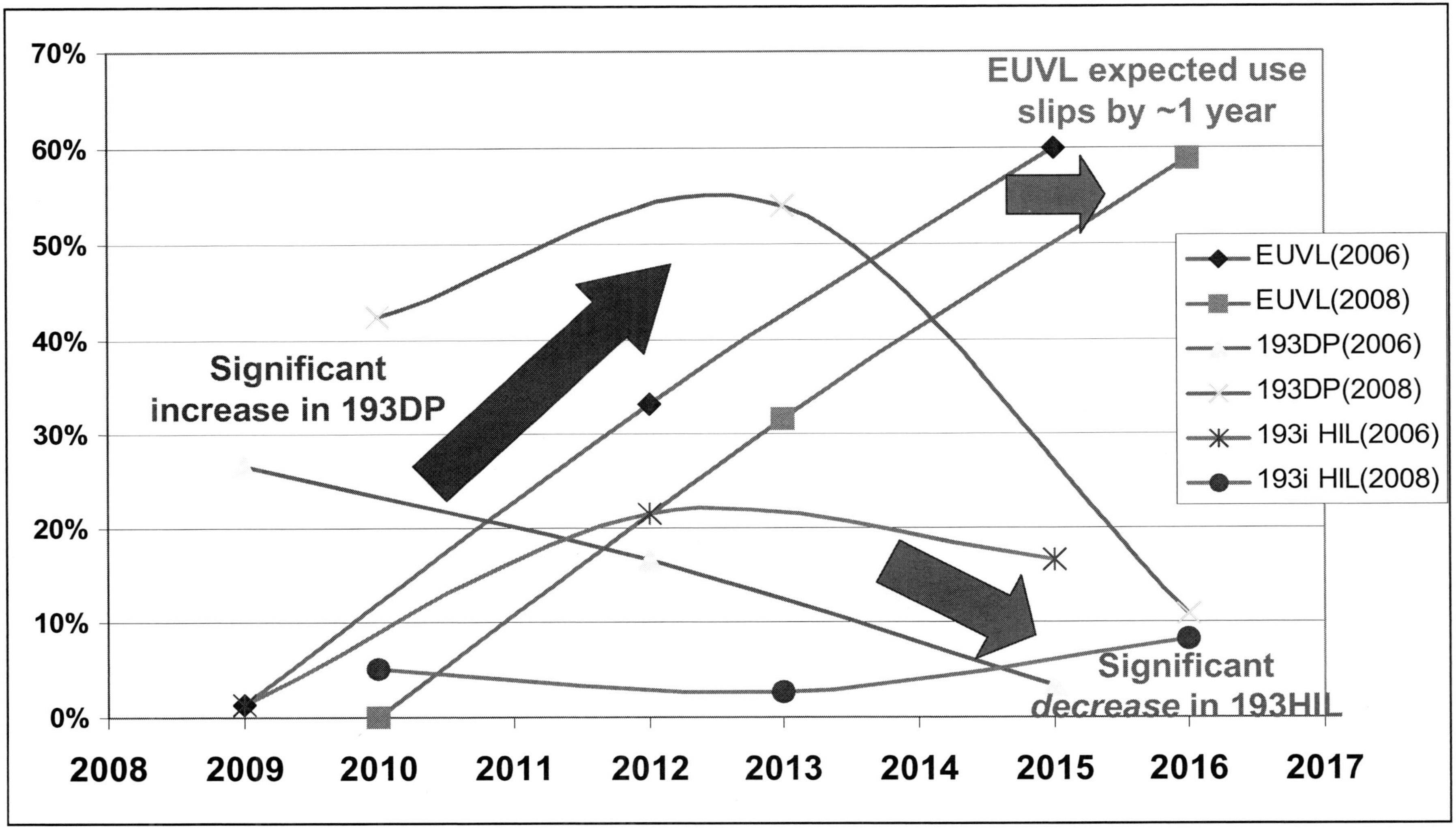

Legend:
- EUVL(2006)
- EUVL(2008)
- 193DP(2006)
- 193DP(2008)
- 193i HIL(2006)
- 193i HIL(2008)

EUVL expected use slips by ~1 year

Significant increase in 193DP

Significant decrease in 193HIL

2008 EUVL Symposium – Focus Questions:

- When can integrated EUV sources support beta tool source requirements (power, reliability, uptime)

- When can EUV masks support pilot line yield requirements

- When can EUV resists meet line edge roughness specifications for 22 nm half pitch at acceptable photospeed

- When can EUVL be ready for high volume manufacturing

The 2008 EUVL Program Committee

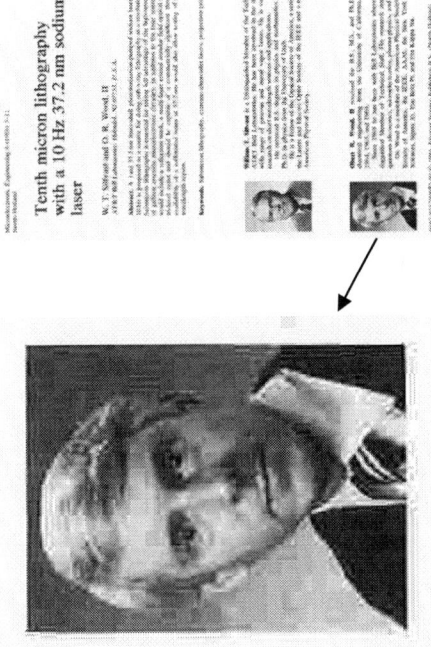

- Obert Wood (AMD) – Program Chair

- Patrick Naulleau (LBNL) – Program Co-Chair

- Iwao Nishiyama (Selete / NECEL) – Program Co-Chair

- Winfried Kaiser (Zeiss) – Program Co-Chair

2008 EUVL Symposium Sponsors

We express our sincere gratitude to the following organizations:

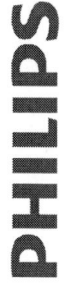
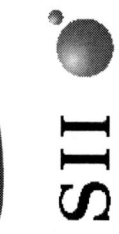

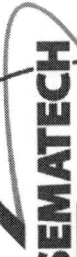

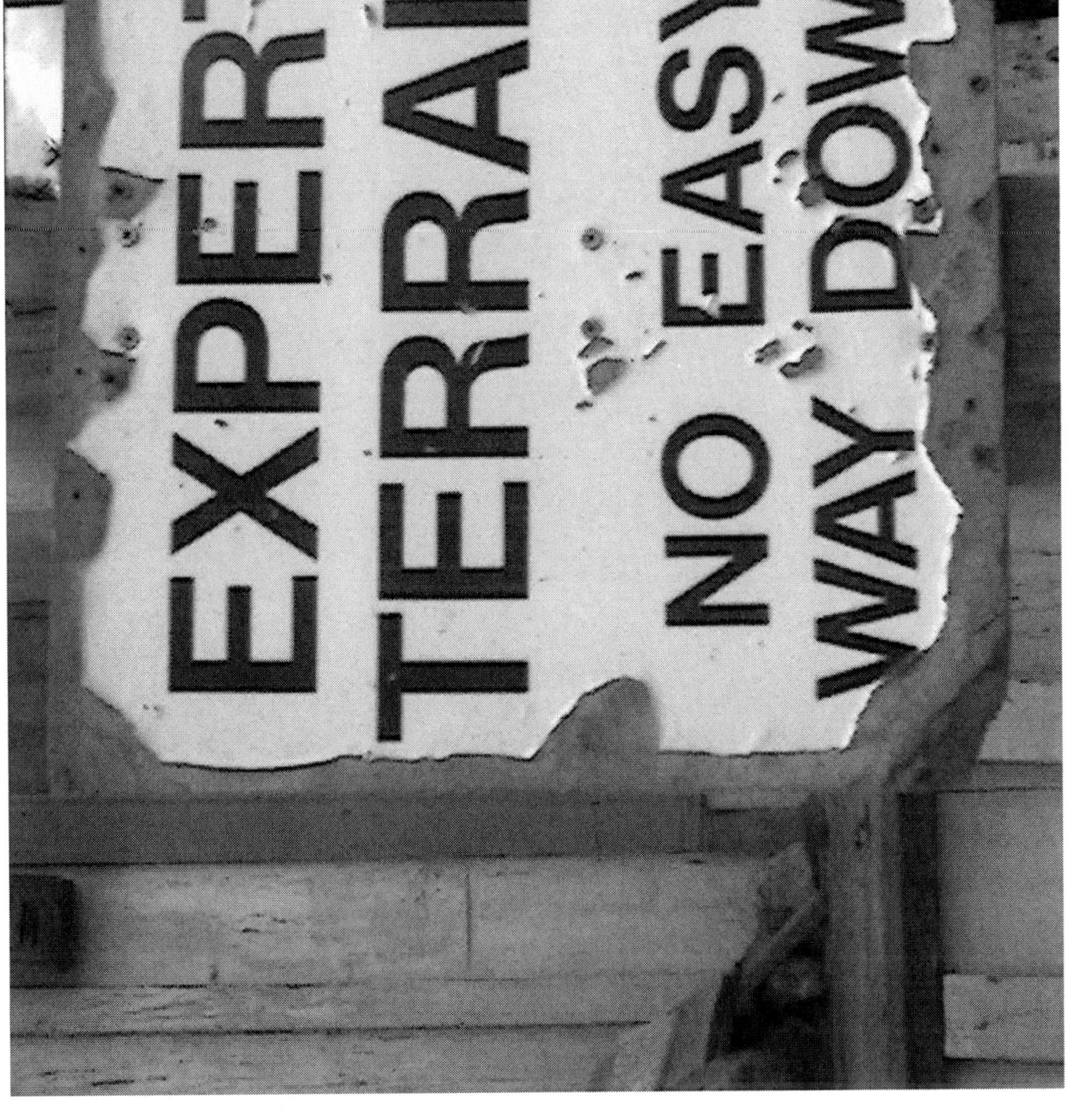
EXPERT
TERRAIN
NO EASY
WAY DOWN

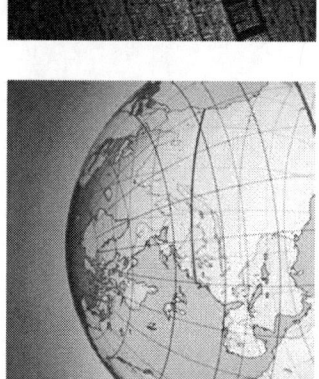
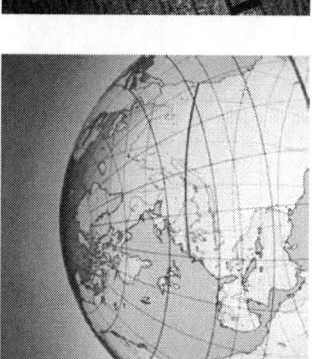

Accelerating the next technology revolution

2008 International EUVL Symposium
Program Logistics & Overview

Obert Wood, Program Chair
Patrick Naulleau, Program
Co-Chair

Presentations by Geographic Region

Asia /
Pacific
65 (34%)

United
States
76 (48%)

Europe
29 (18%)

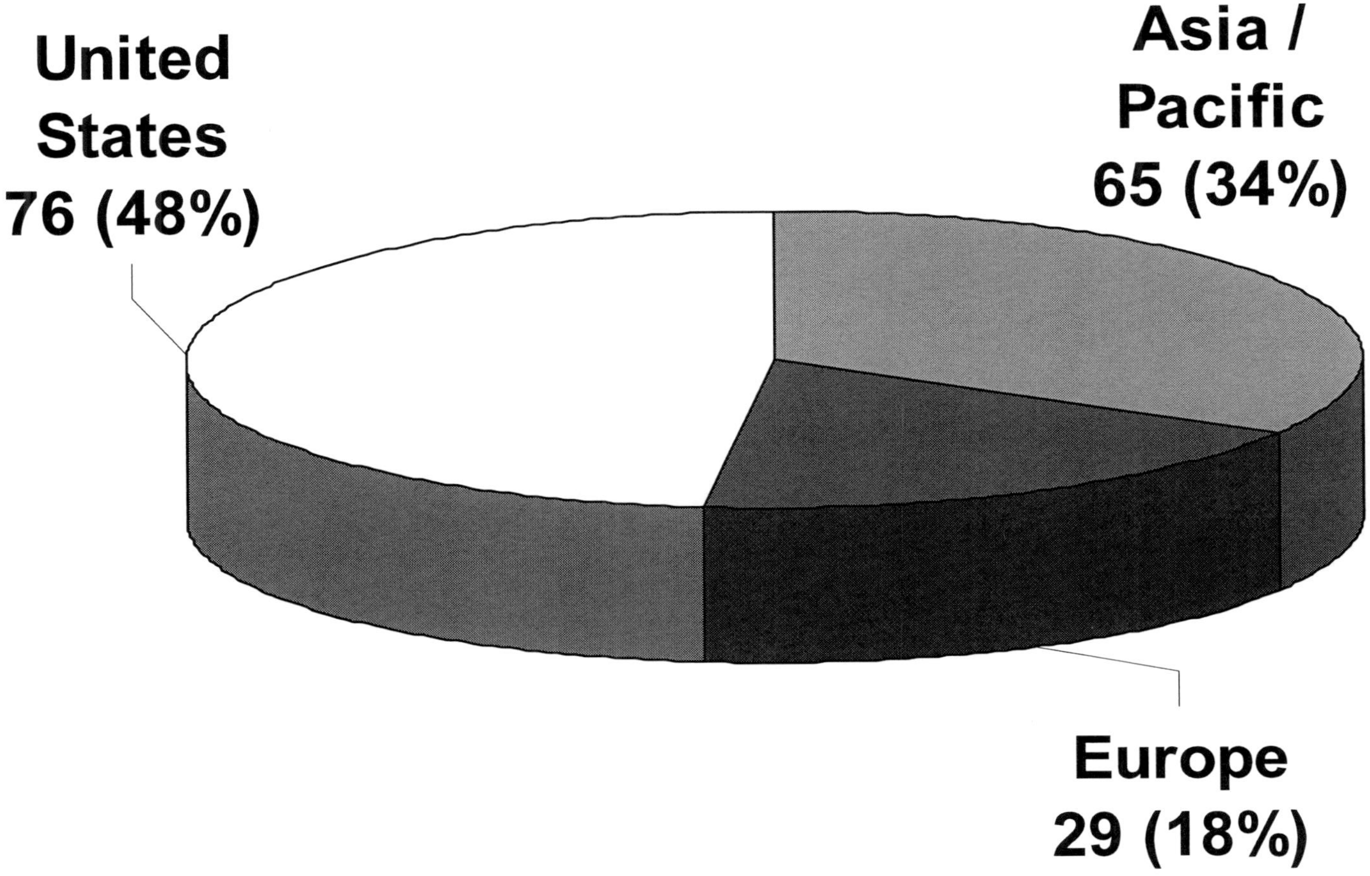

Oral: 49, Poster: 109, Total: 158

Presentations by Technical Topic

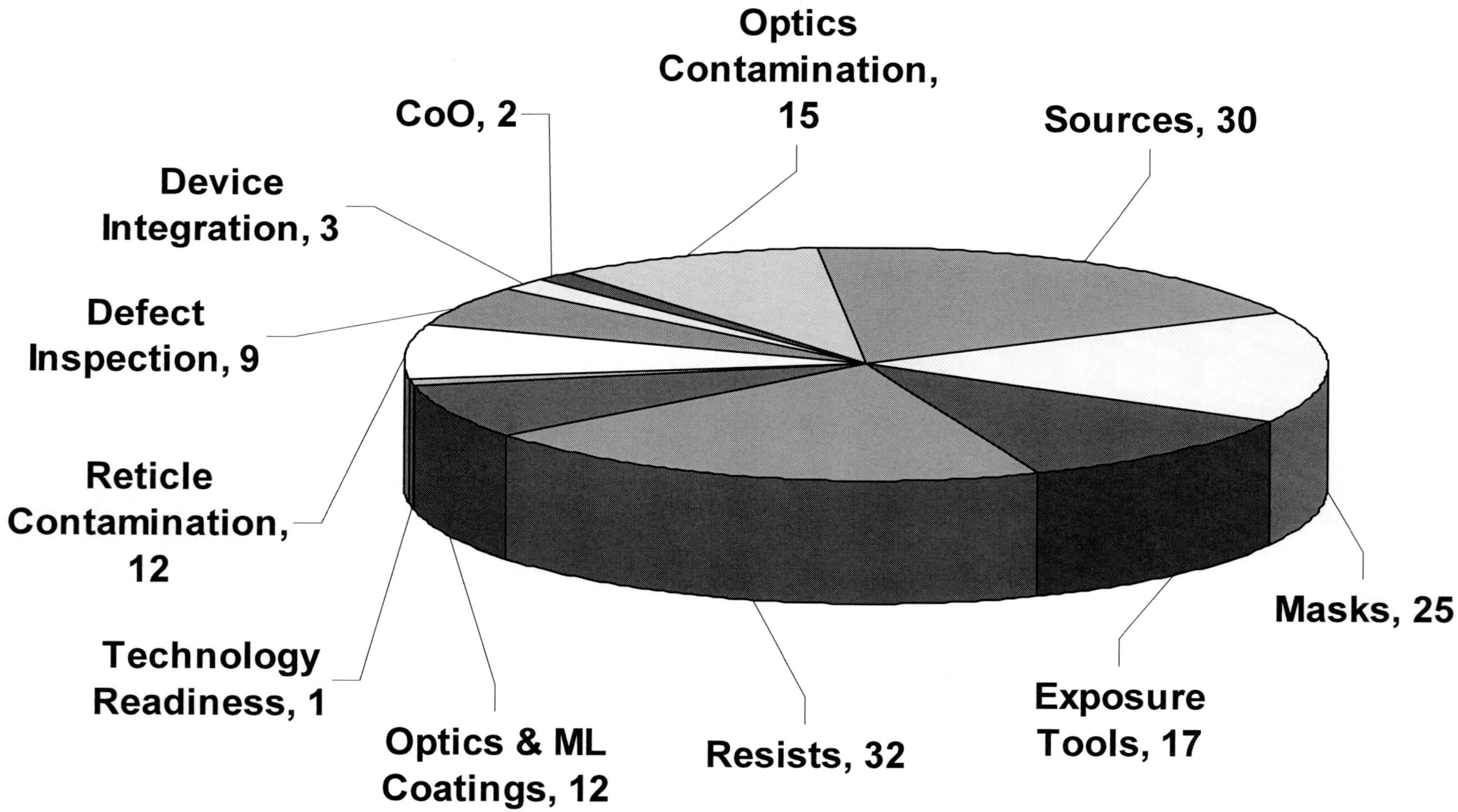

- Sources, 30
- Masks, 25
- Exposure Tools, 17
- Resists, 32
- Optics & ML Coatings, 12
- Technology Readiness, 1
- Reticle Contamination, 12
- Defect Inspection, 9
- Device Integration, 3
- CoO, 2
- Optics Contamination, 15

Presentations by Topic & Region

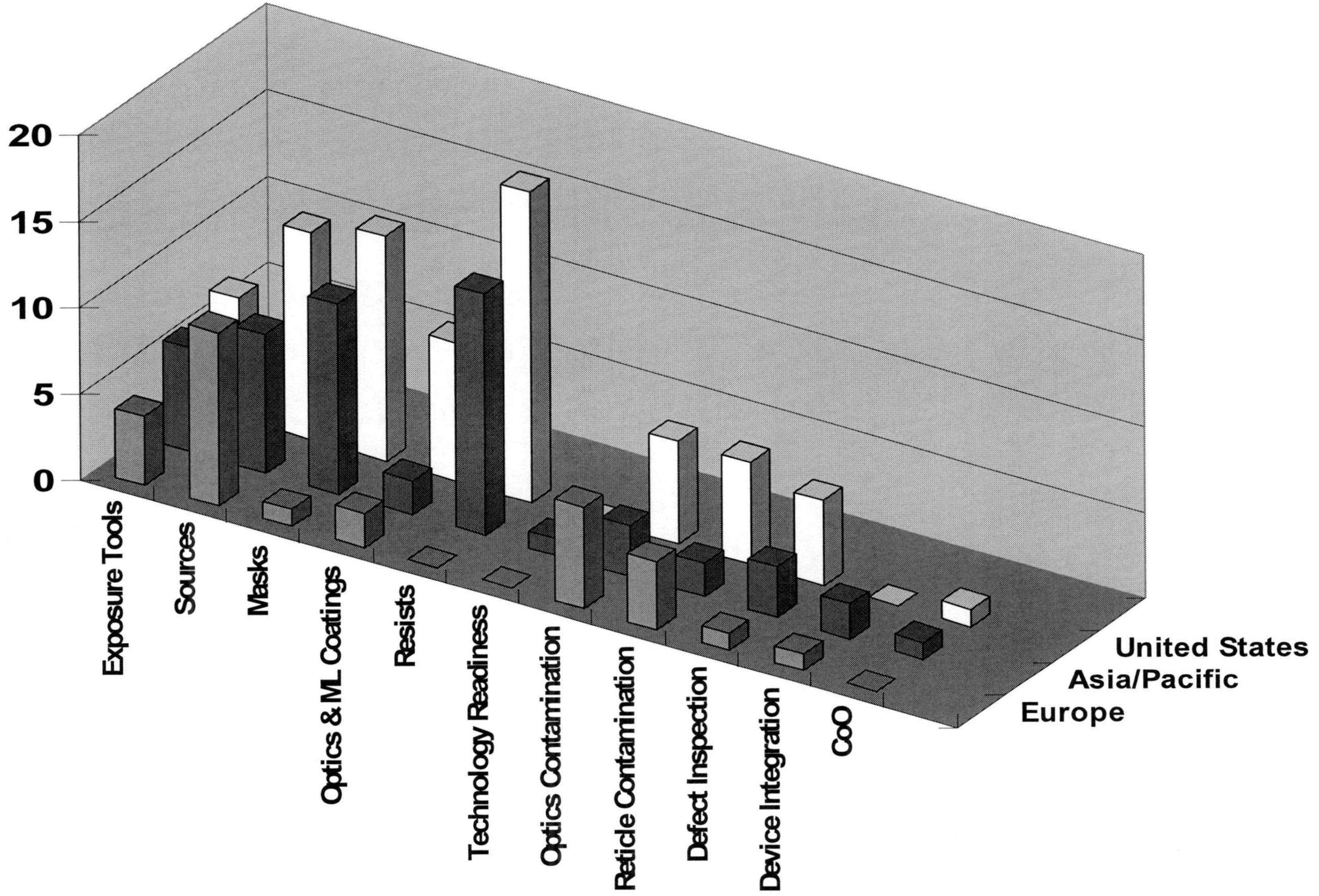

Housekeeping

- Please turn your mobile phones to silent or vibrate

- No photography, video or audio taping of the presentations is permitted

- The symposium proceedings (oral presentations and posters) will be available on a secure web site after the symposium

- Every registered attendee will receive an e-mail with access instructions and password

Meeting Logistics

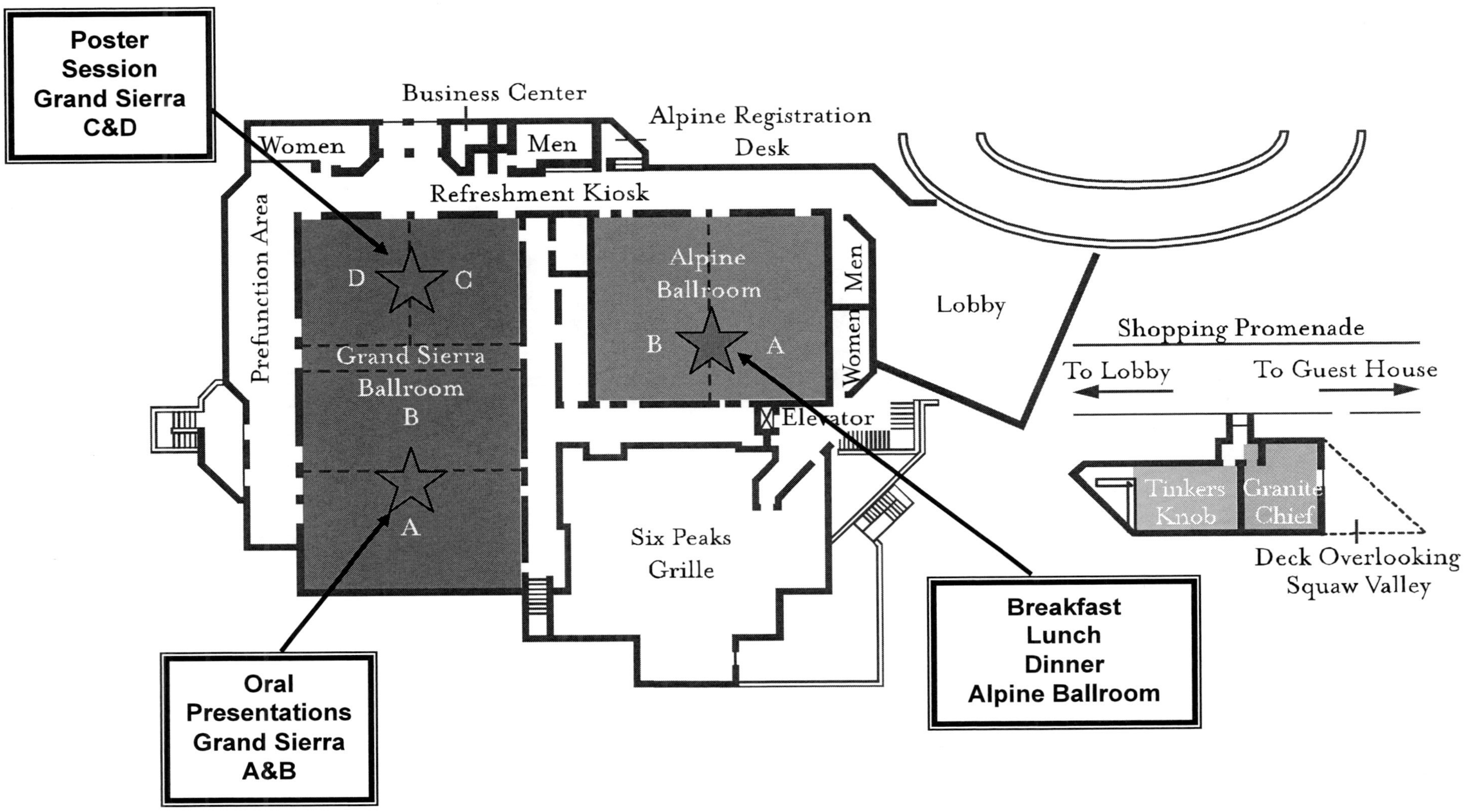

Poster Session Grand Sierra C&D

Oral Presentations Grand Sierra A&B

Breakfast Lunch Dinner Alpine Ballroom

Shopping Promenade

To Guest House

To Lobby

Deck Overlooking Squaw Valley

Tinkers Knob

Granite Chief

Lobby

Men

Women

Elevator

Six Peaks Grille

Alpine Ballroom

A

B

Alpine Registration Desk

Business Center

Men

Refreshment Kiosk

Women

Prefunction Area

Grand Sierra Ballroom

D C

B

A

Internet Access Location

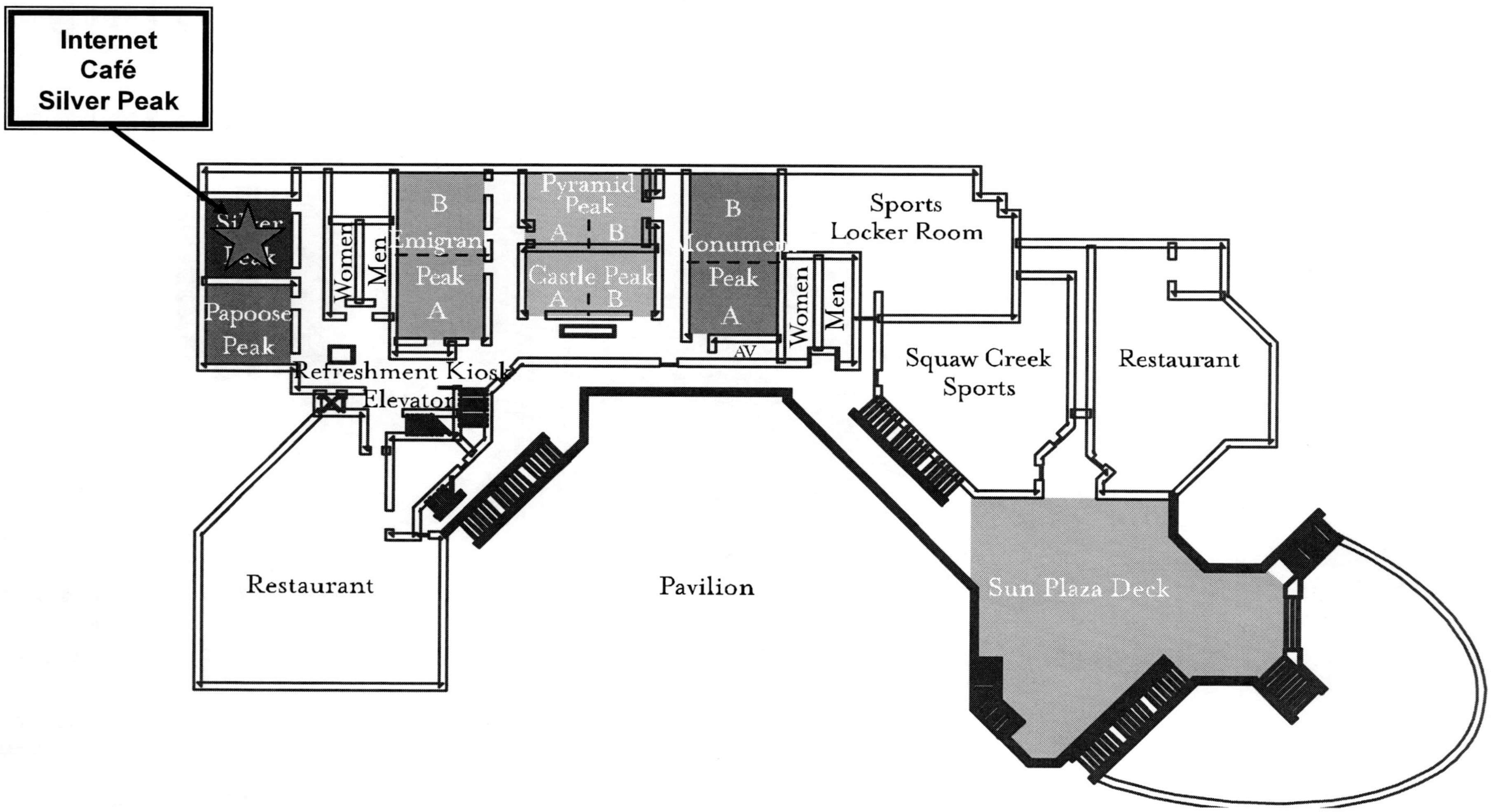

Internet Café Silver Peak

Silver Peak

Papoose Peak

Emigrant Peak B / A

Women / Men

Pyramid Peak A / B

Castle Peak A / B

Monument Peak B / A

AV

Women / Men

Sports Locker Room

Squaw Creek Sports

Restaurant

Refreshment Kiosk

Elevator

Pavilion

Sun Plaza Deck

Restaurant

Best Paper Award Voting

- **Awards for Best Paper and Best Poster will be presented on Day 3 of the Symposium**

- **Immediately following the final Session on Day 3, please vote for the Best Paper using the Voting Sheet illustrated below**

2008 EUVL Symposium Best Paper Voting Sheet

Presentation Title	Presenter	Presenting Organization	Best Paper Place X in
EUV Alpha Demo Tools – stepping stones towards volume production	H. Meiling	ASML	☐
Nikon EUVL development progress update	T. Miura	Nikon	☐
Development Status of Canon's Full-Field EUVL Tool	S. Uzawa	Canon	☐
EUV resist performance on the ASML ADT and LBNL MET	B. Pierson	ASML	☐
The SEMATECH Berkeley MET: learning at the 22-nm node	P. Naulleau	LBNL	☐
Evaluation of EUV resists at Selete	H. Oizumi	Selete	☐
Advances in resist testing at PSI EUV-IL exposure tool	V. Auzelyte	Paul Scherrer Institute	☐
Positive and negative tone molecular resists for 22 nm node EUVL patterning	C. Henderson	Georgia Institute of Technology	☐
Cost implications of EUV lithography technology decisions	A. Wüest	SEMATECH	☐

Best Poster Award Voting

Voting Stickers

Please vote for the poster with the best technical content or with the best quality of presentation by affixing one of the five blue stickers to the title of the poster. You may vote for the five best posters or you may attach all five of your stickers to one poster. The voting for Best Poster will close at 3:00 PM on Tuesday.

2008 EUVL Symposium Calendar

	Sunday 28-Sep	Monday 29-Sep	Tuesday 30-Sep	Wednesday 1-Oct	Thursday 2-Oct		Friday 3-Oct
8:00 – 9:00		EUVL Symposium Day 1	EUVL Symposium Day 2	EUVL Symposium Day 3	IEUVI Optics Cont. TWG	IEUVI Source TWG	IEUVI Board Meeting
9:00 – 10:00							
10:00 – 11:00							
11:00 – 12:00							
12:00 – 13:00							
13:00 – 14:00					IEUVI Resist TWG	IEUVI Mask TWG	
14:00 – 15:00	EUV Mask Workshop		Networking Time				
15:00 – 16:00							
16:00 – 17:00							
17:00 – 18:00							
18:00 – 19:00	Reception for Session Chairs & Organizers	Conference Reception	Dinner/Panel Discussion 7:00 PM	Dinner Cruise 6:30 PM			
19:00 – 20:00							
20:00 – 21:00							

Time Keeping for Oral Presentations

Operated by Session Chairs

Time for each Presentation:

Presentation: 17 minutes

Discussion: 3 minutes

Total: 20 minutes

Color indicator

1. Presentation start:

2. 2 minutes remaining (15 min past)

3. Presentation finished (17 min past)

EUV Lithography's Future

Harry J. Levinson

EUV optics

- To maintain a constant Strehl ratio, the wavefront error must scale with wavelength.

- 193 nm / 13.5 nm = 14.3

EUV optics

- To maintain a constant Strehl ratio, the wavefront error must scale with wavelength.

- 193 nm / 13.5 nm = 14.3

	Spatial frequency range	Parameter affected	Requirements (nm, rms)	Current best results (nm, rms)
Figure	∞ - 1 mm^{-1}	Aberrations	0.25	
Mid-spatial frequency roughness	1 mm^{-1} – 1 μm^{-1}	Flare	0.20	– – –
High-spatial frequency roughness	1 μm^{-1} – 50 μm^{-1}	Reflectivity	0.10	

Mask defects

Printability of substrate and absorber defects on extreme ultraviolet lithographic masks

K. B. Nguyen, A. K. Ray-Chaudhuri, R. H. Stulen, and K. Krenz
Sandia National Laboratories, Livermore, California 94551-0969

L. A. Fetter
AT&T Bell Laboratories, Holmdel, New Jersey 07733-3030

D. M. Tennant
AT&T Bell Laboratories, Murray Hill, New Jersey 07974-2070

D. L. Windt
AT&T Bell Laboratories, Holmdel, New Jersey 07733-3030

(Received 3 July 1995; accepted 4 August 1995)

Extreme ultraviolet lithography (EUVL) is a candidate for high-volume production of integrated circuits with 0.1 μm design rules. As a reduction imaging technique with robust mask substrates, EUVL reduces the mask contribution to the critical dimension (CD) error budget. However, the ability to manufacture EUVL mask blanks that are free of printable defects remains an important challenge. Electromagnetic simulations and imaging experiments have suggested that defects in the substrates and reflective coatings, in particular, may be highly printable and difficult to detect. A defect printability study using programmed defects was performed in order to determine the printing behavior of mask defects of different sizes and locations with respect to absorber features. Imaging was performed using a 10$\times$ Schwarzschild camera operating at 13.4 nm with a numerical aperture of 0.08, corresponding to a Rayleigh resolution of 0.1 μm. This system has an effective exposure field of 0.4 mm diam. Measurements of the defect-induced linewidth variations on the printed resist lines were performed with scanning electron microscopy and atomic force microscopy. Results show that defects located on the substrate and overcoated by the reflective coating are more printable compared to defects of the same sizes located above the reflective coating. In addition, defects centered in the clear region of lines-and-space (L/S) pattern are more printable compared to those located at the line edge programmed defects located in L/S patterns that are larger than 1/3 of the linewidth caused $\geq$10% CD variations, while defects that are $\leq$1/6 of the linewidth did not cause measurable CD variations. © *1995 American Vacuum Society.*

Sources

- Light sources for optical lithography

  - Mercury g-line
  - Mercury i-line
  - KrF excimer laser
  - ArF excimer laser

  All have high intensity within a narrow bandwidth

- Light sources for EUV lithography must match the wavelengths at which MoSi multilayers have high reflectivity.

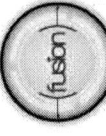

Sources

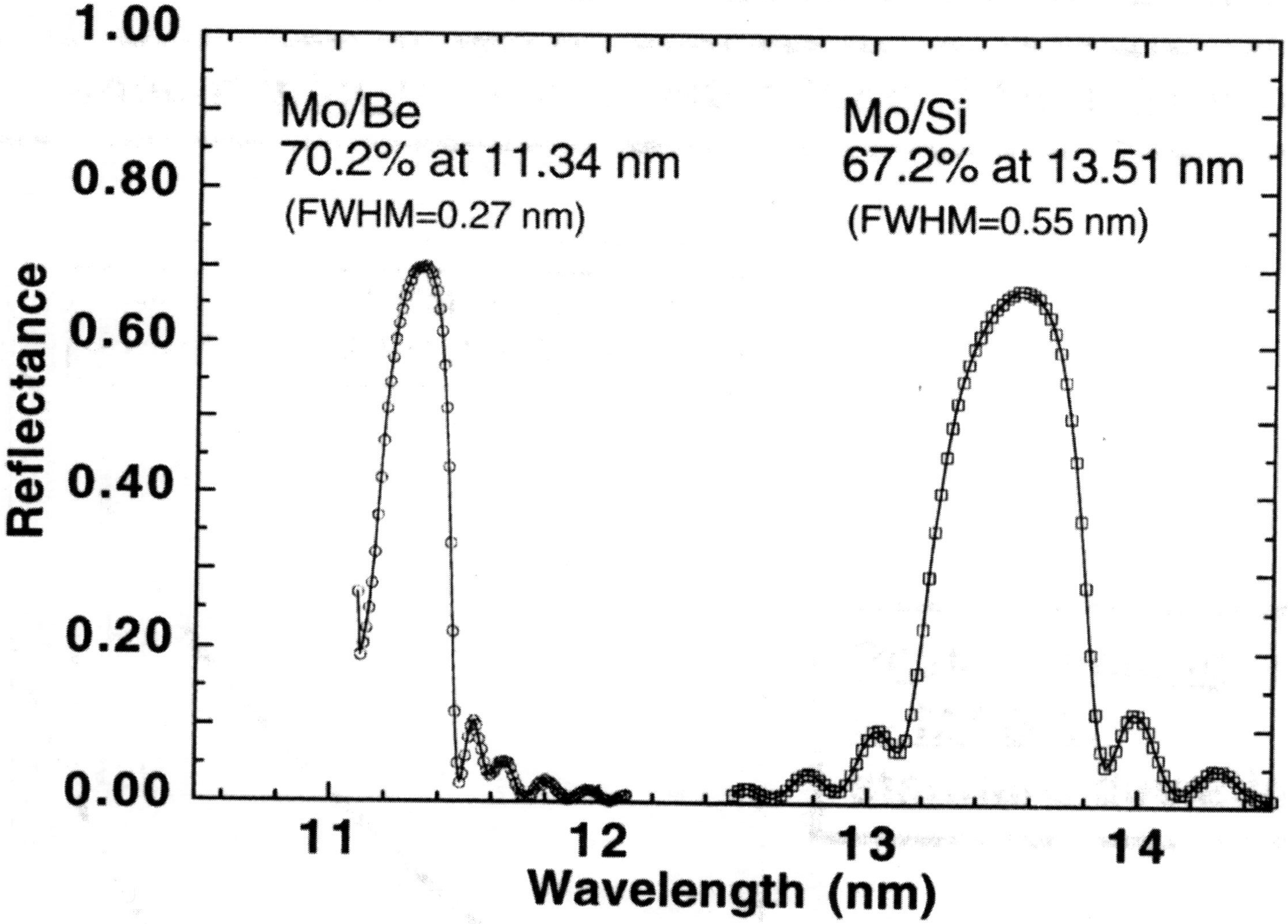

Mo/Be
70.2% at 11.34 nm
(FWHM=0.27 nm)

Mo/Si
67.2% at 13.51 nm
(FWHM=0.55 nm)

Reflectance

Wavelength (nm)

Resists

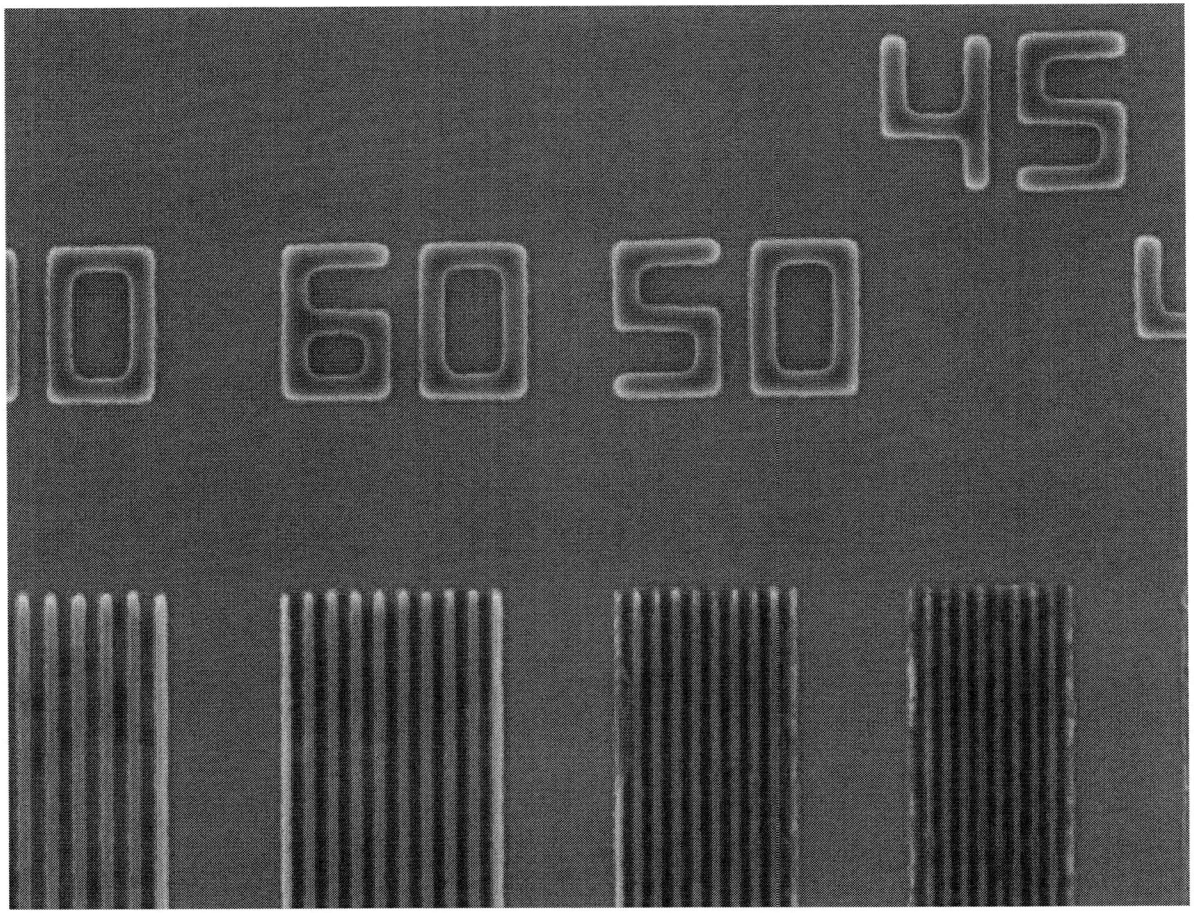

Tool costs

- EUV tools might be very expensive.

- E-mail replies from AMD executives when I included the estimated price for an EUV exposure tool in my weekly report:

  - My boss: "Wow!"

  - Director of integration: "Holy cow!"

  - Design vice-president: "Sticker shock!"

  - Vice-president who had assigned me to the EUV project: "I am glad that I am retiring."

"Typhoon" product demonstration

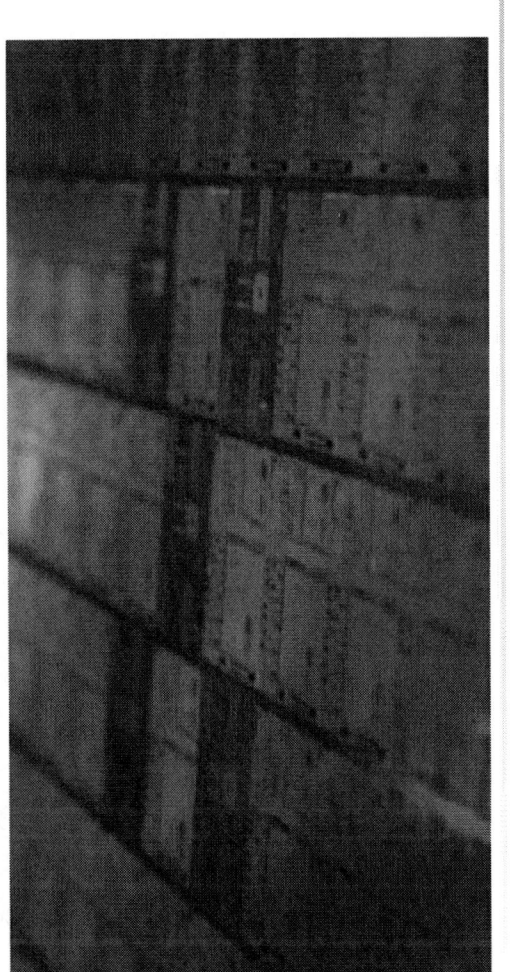

B. La Fontaine, et al., "The Use of EUV Lithography to Produce Demonstration Devices," SPIE 2008

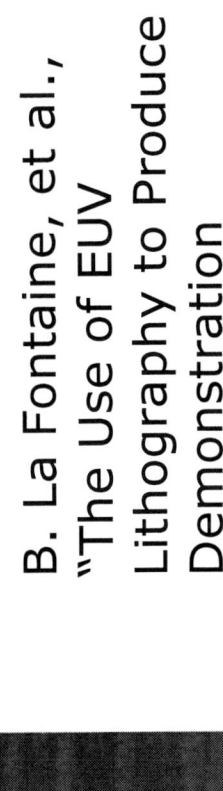

"Typhoon" product demonstration

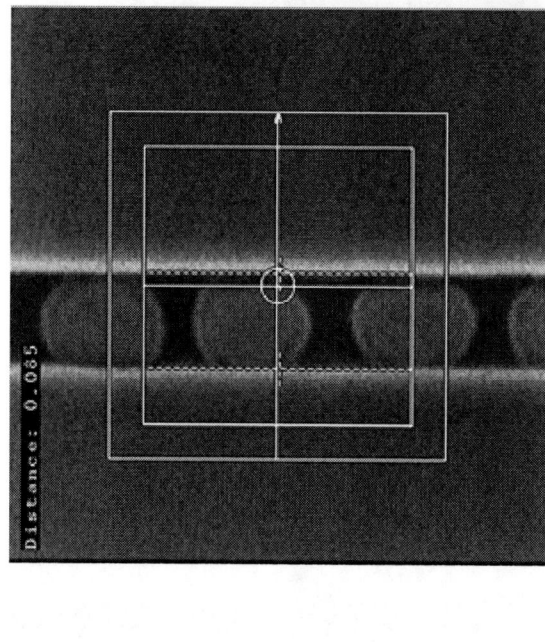

SRAM "Butterfly" curves

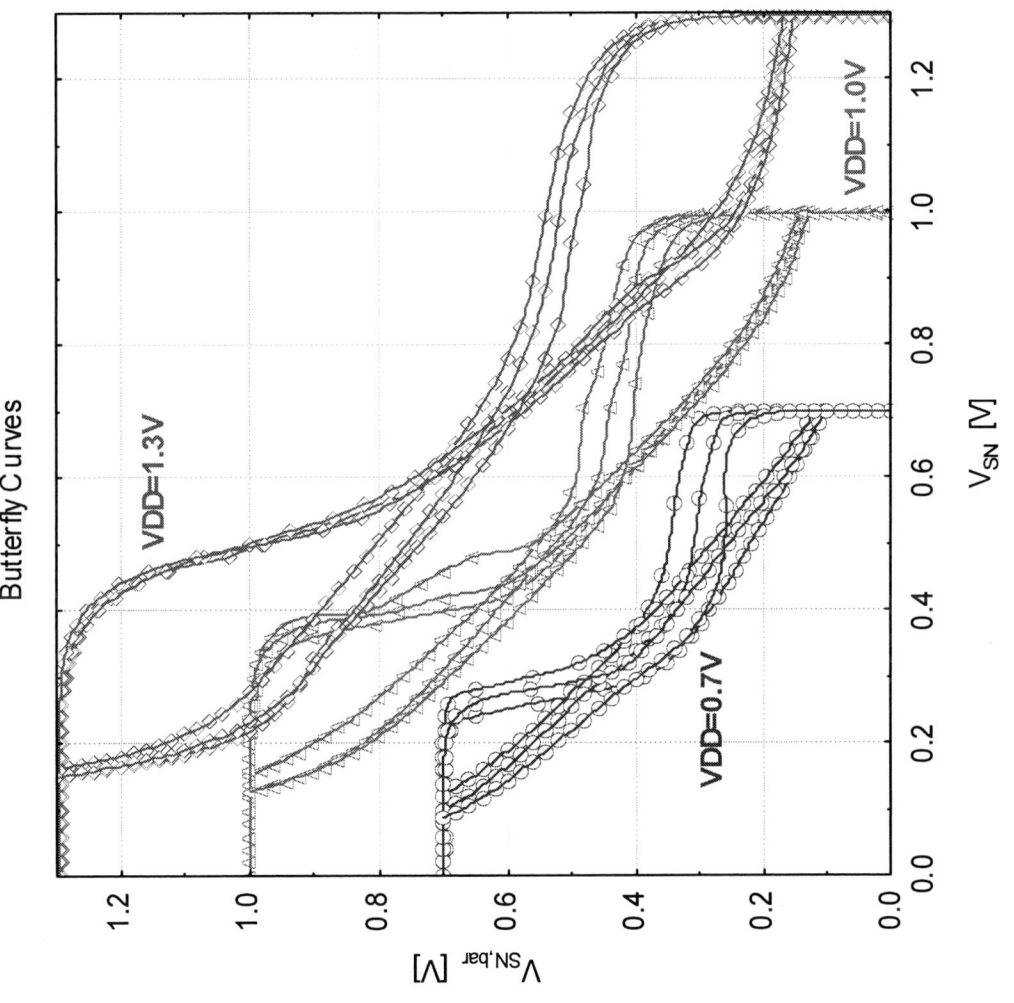

Things look clear where we are standing...

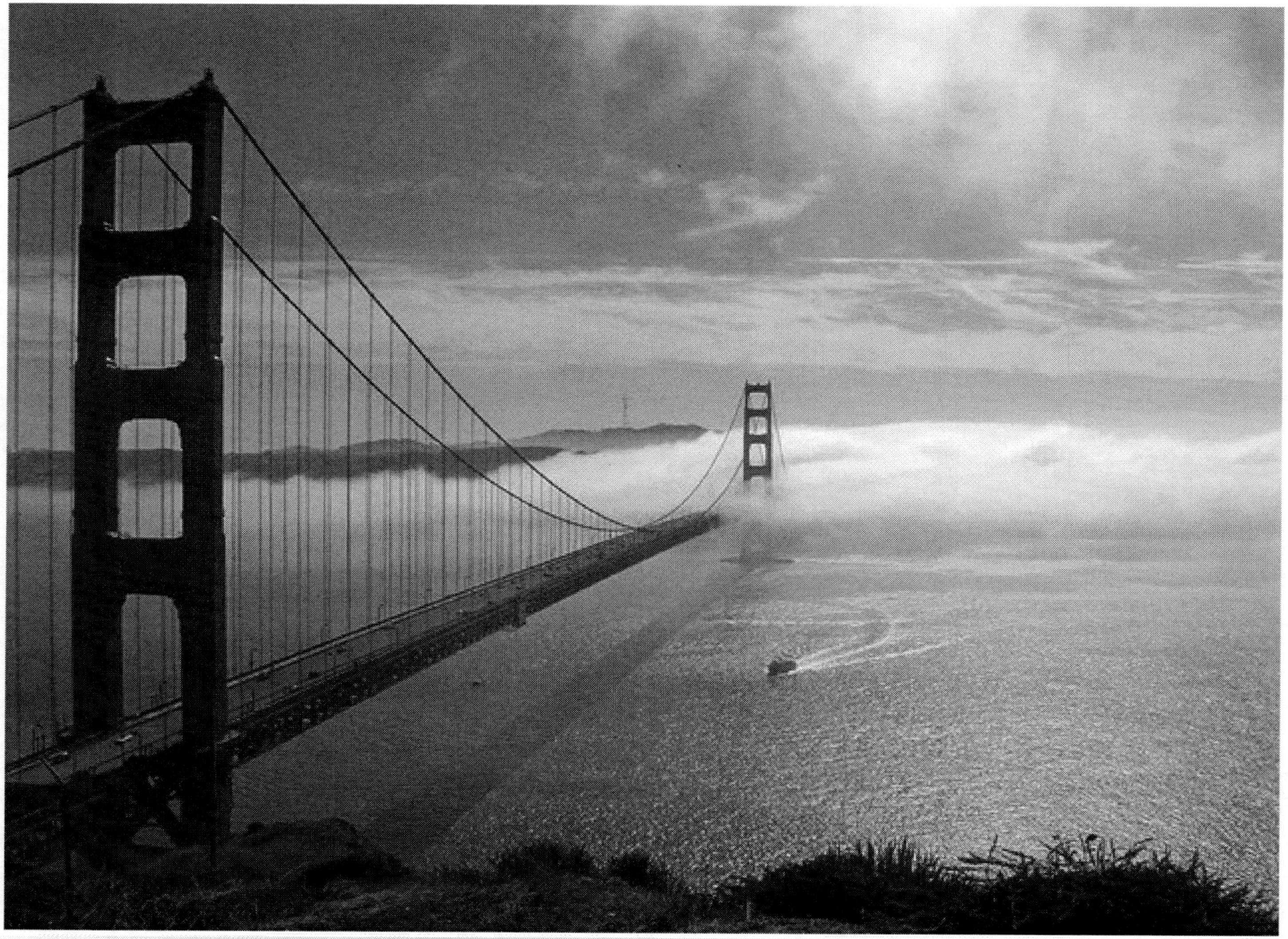

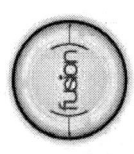

What about where we are going?

- Resists

- Mask defects

- Cost-effectiveness

- Timeliness

Resists

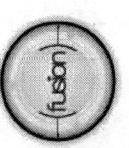

Good news!

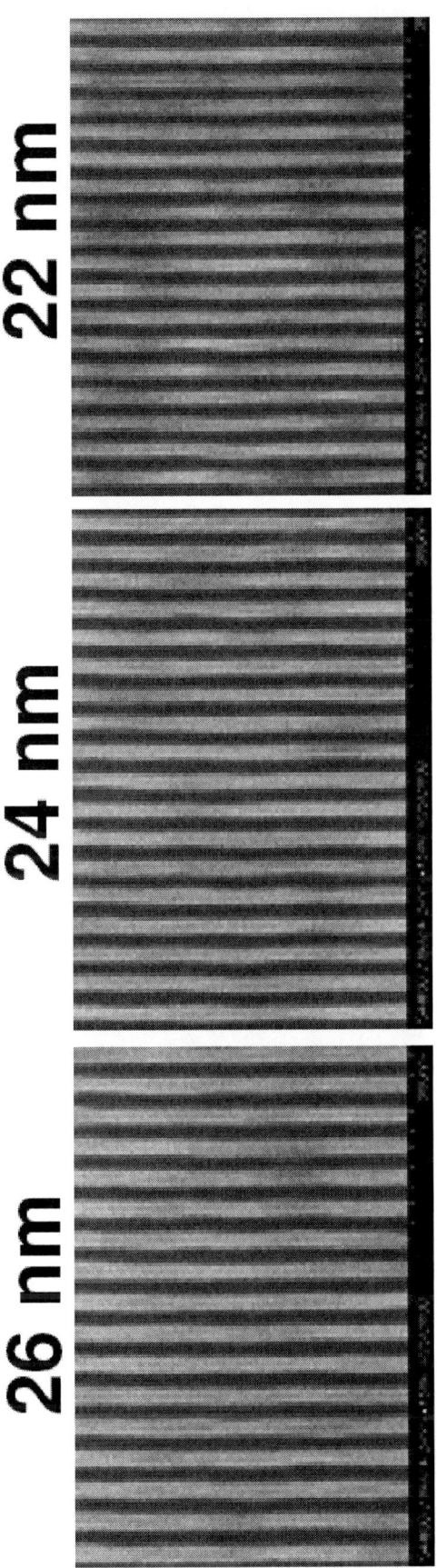

From P. Naulleau, et al., "Latest results from the Sematech Berkeley EUV microfield exposure tool," presented at EIPBN, 2008

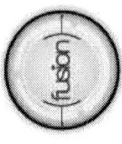

Masks

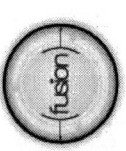

Masks

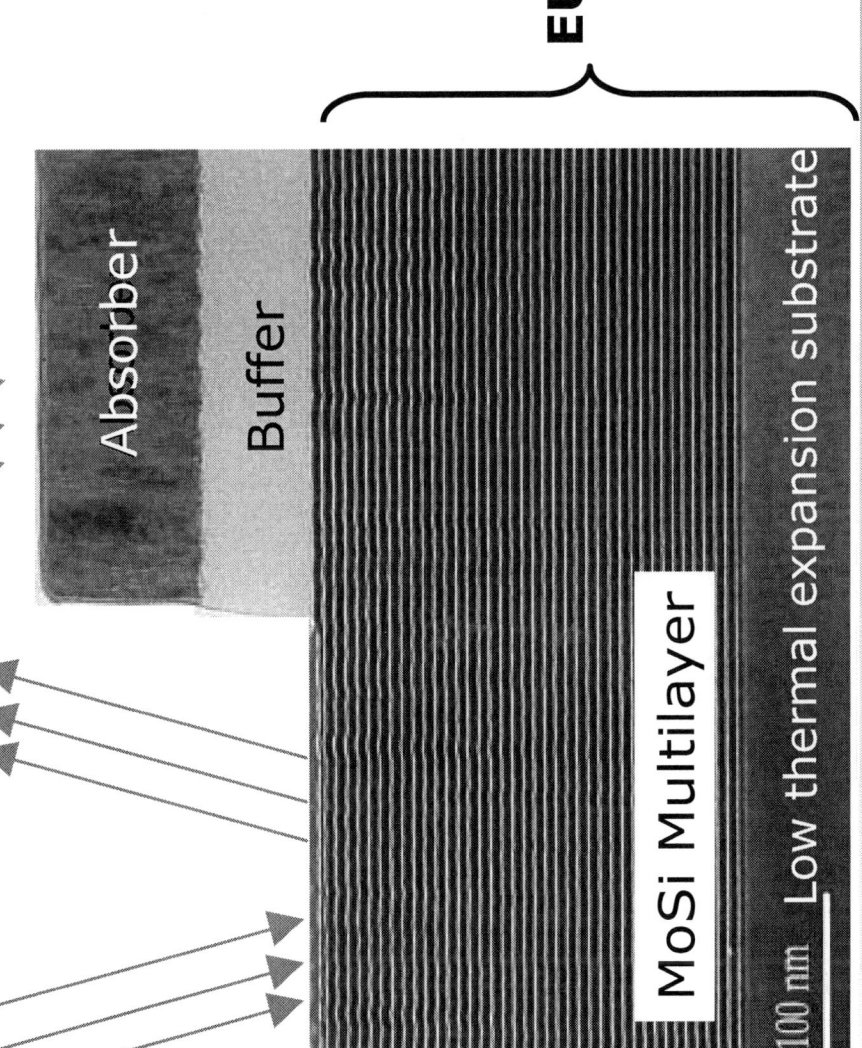

EUVL mask blank

Absorber

Buffer

MoSi Multilayer

Low thermal expansion substrate

100 nm

TEM: Janice Grey

The future is fusion

Masks

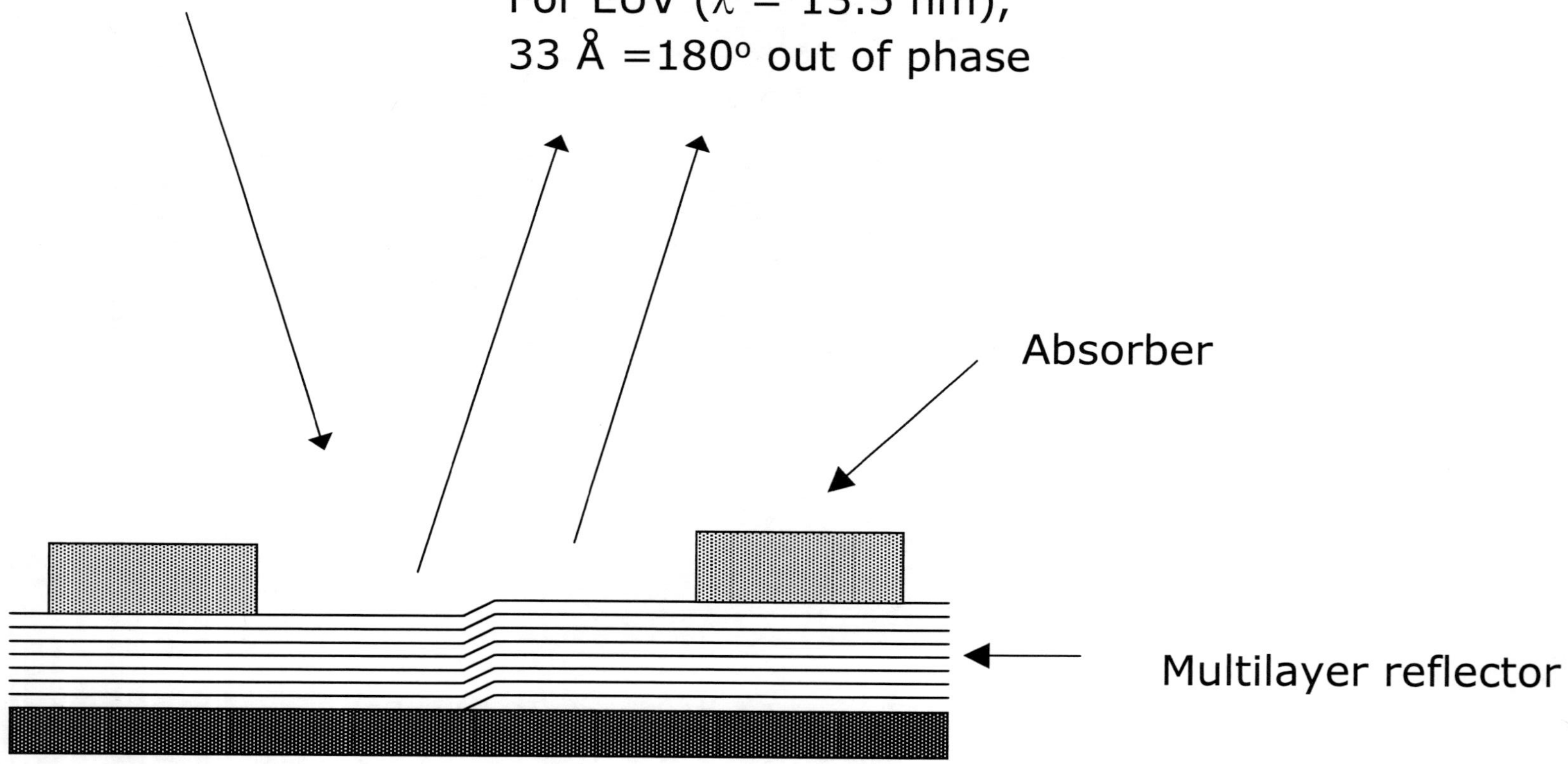

For EUV (λ = 13.5 nm),
33 Å =180° out of phase

Absorber

Multilayer reflector

Masks

EUV Resist Image

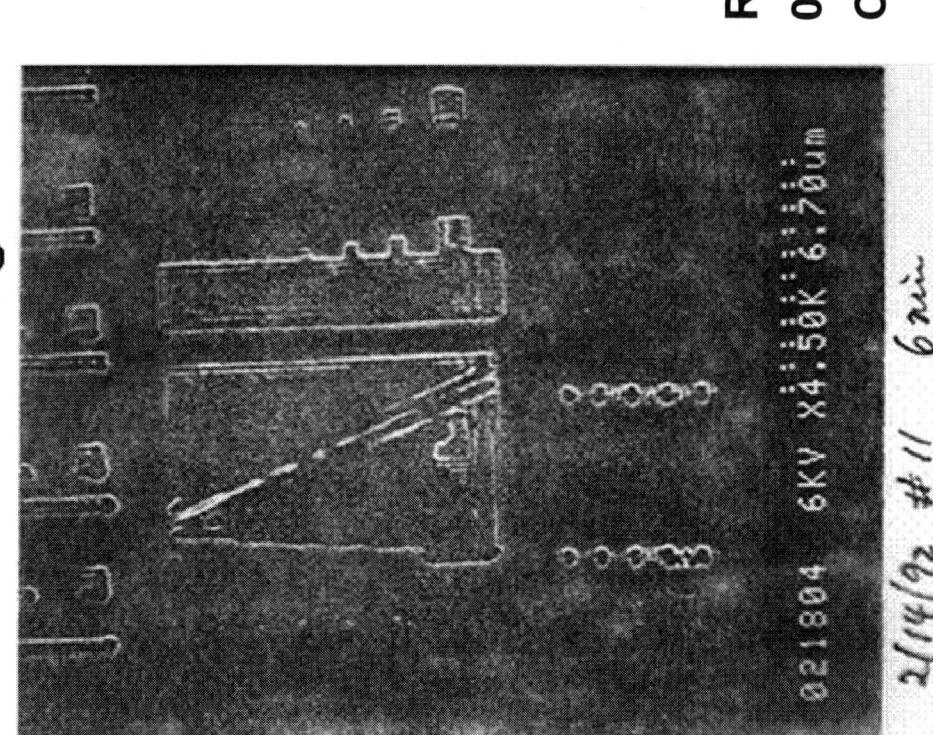

Interference Microscope Image

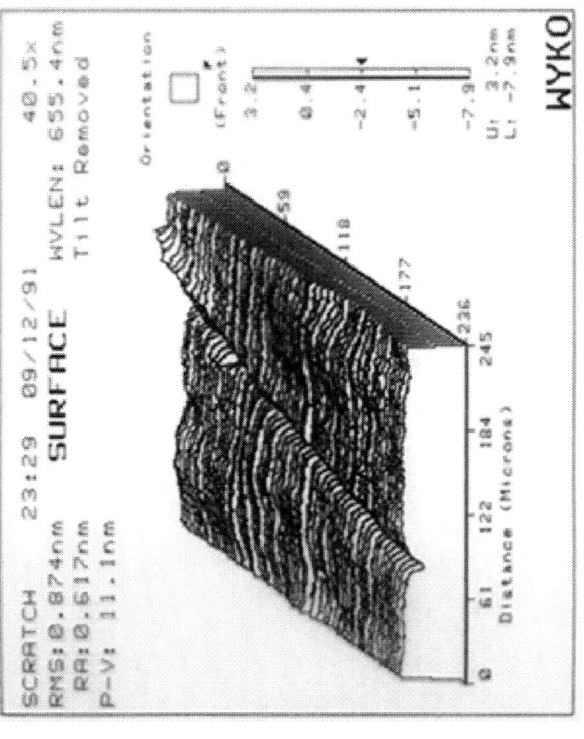

Depth of Scratch ~18 Å

Resist: 60-nm thick PMMA

0.08 NA Schwarzschild Optic

Coherent illumination @ 13.9 nm

Pictures courtesy of Dr. Obert Wood

The future is fusion

Multilayer deposited on small particle

TEM: Janice Grey

Masks

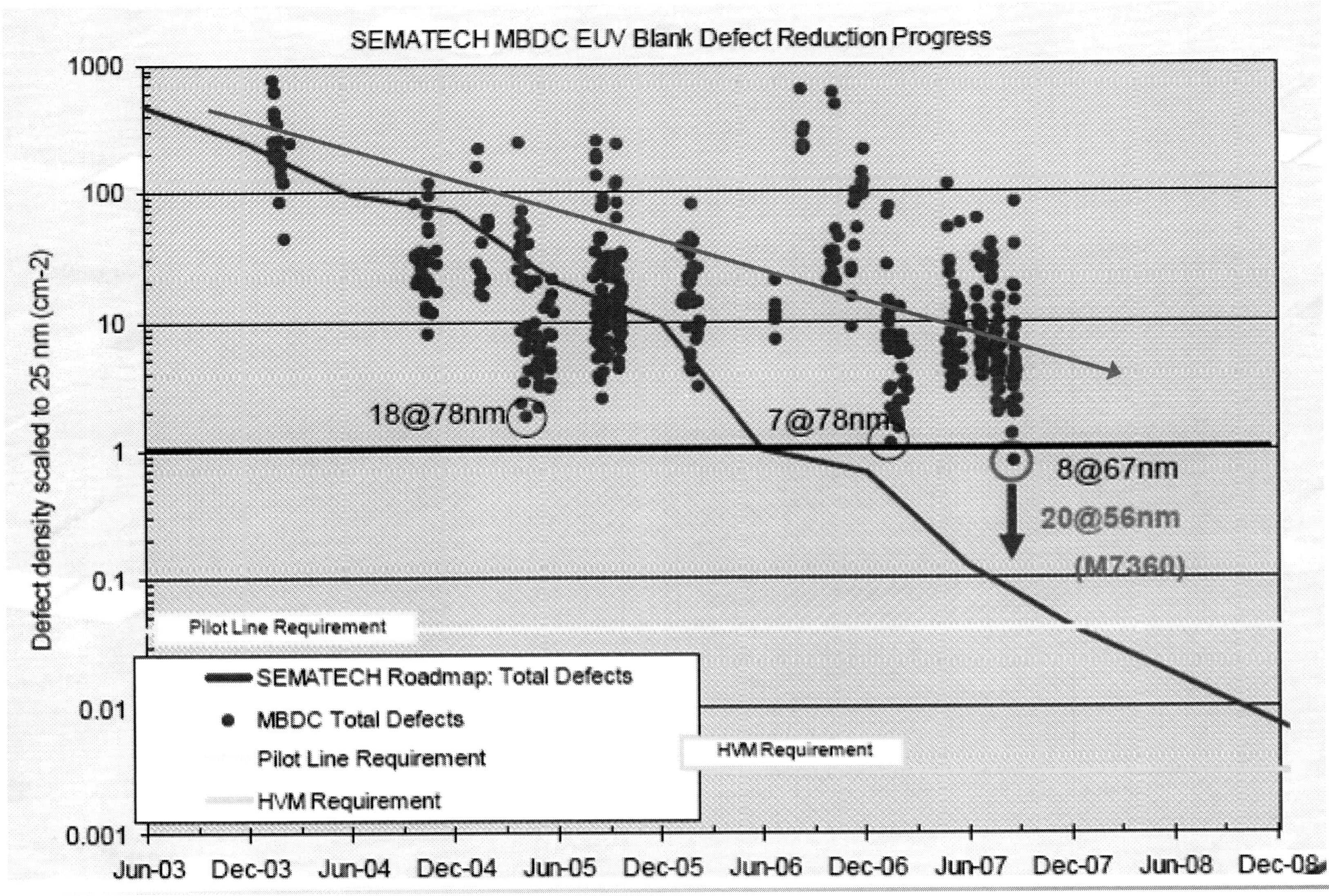

SEMATECH MBDC EUV Blank Defect Reduction Progress

Perfect mask blanks?

- It is hard to make something perfect.

- Is this necessary?

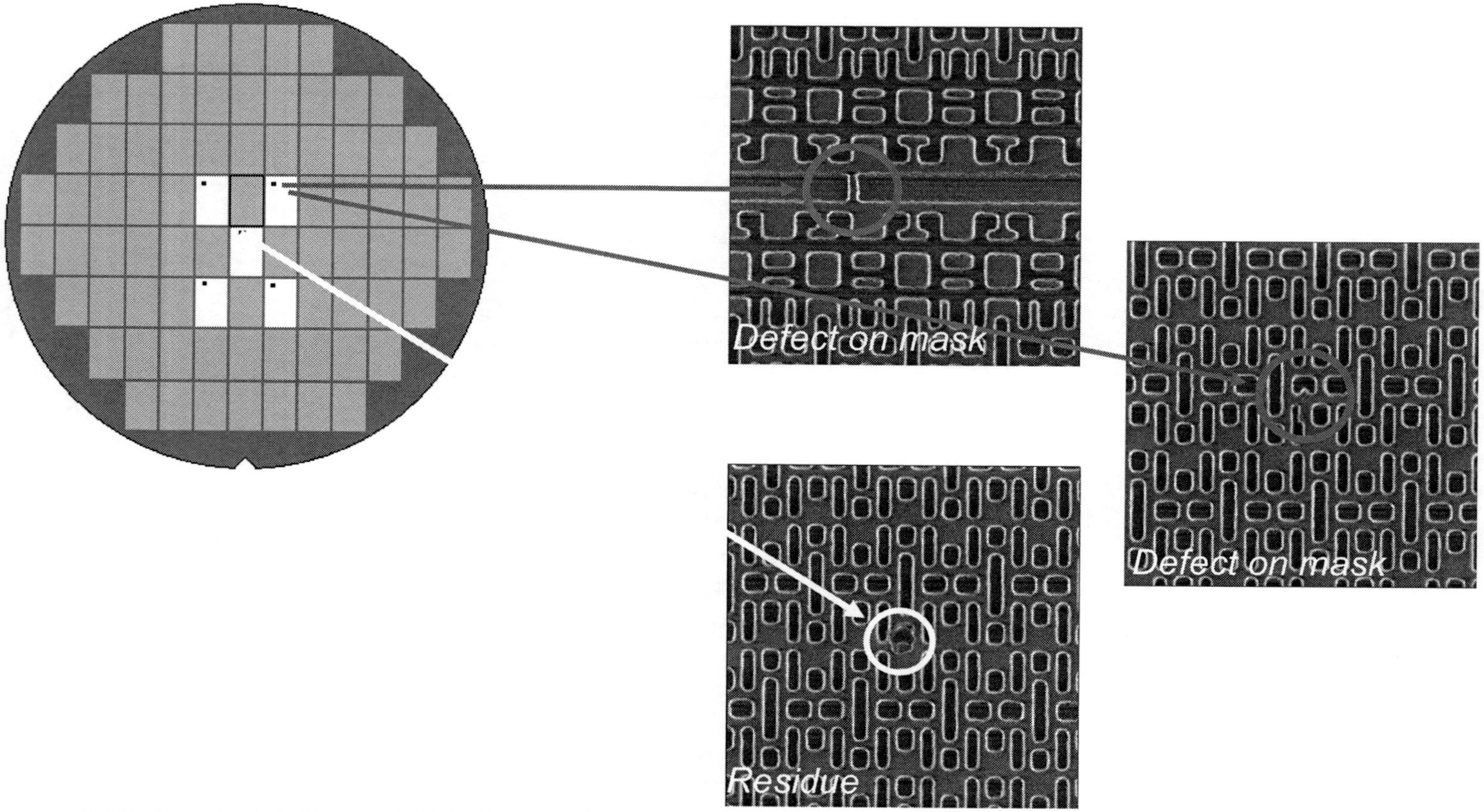

Defect on mask

Defect on mask

Residue

Mask defects

SEM image of mask

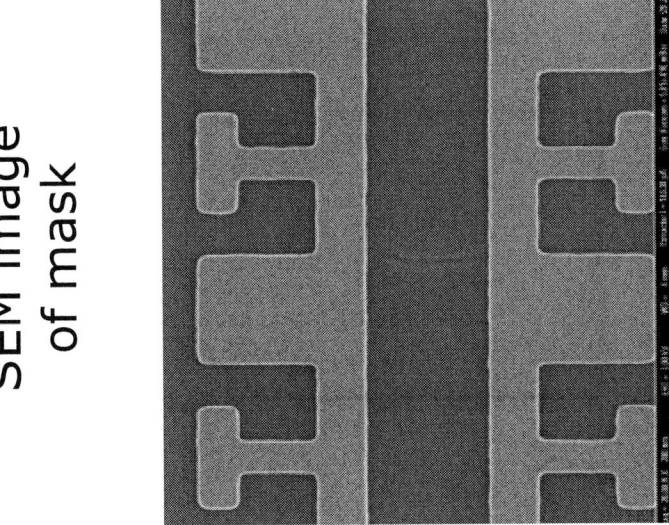

Image in mask defect inspection tool

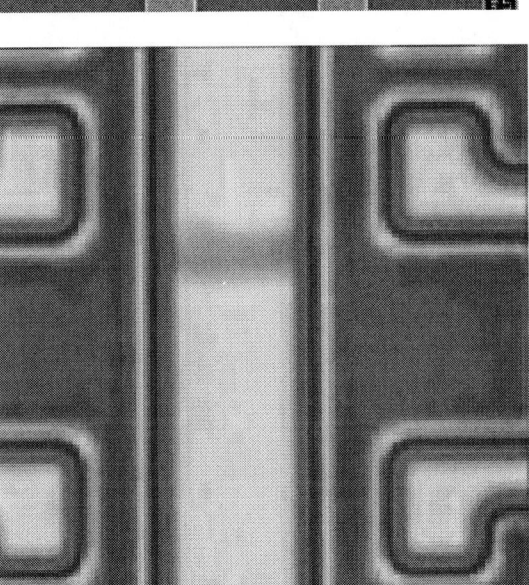

Resist pattern on wafer

The good news on mask defects

- Only a few percent of the substrate defects were observed to print.

- Too good to be true?

- More work is needed to understand mask substrate defect printability!

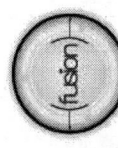

Pellicles, or lack thereof

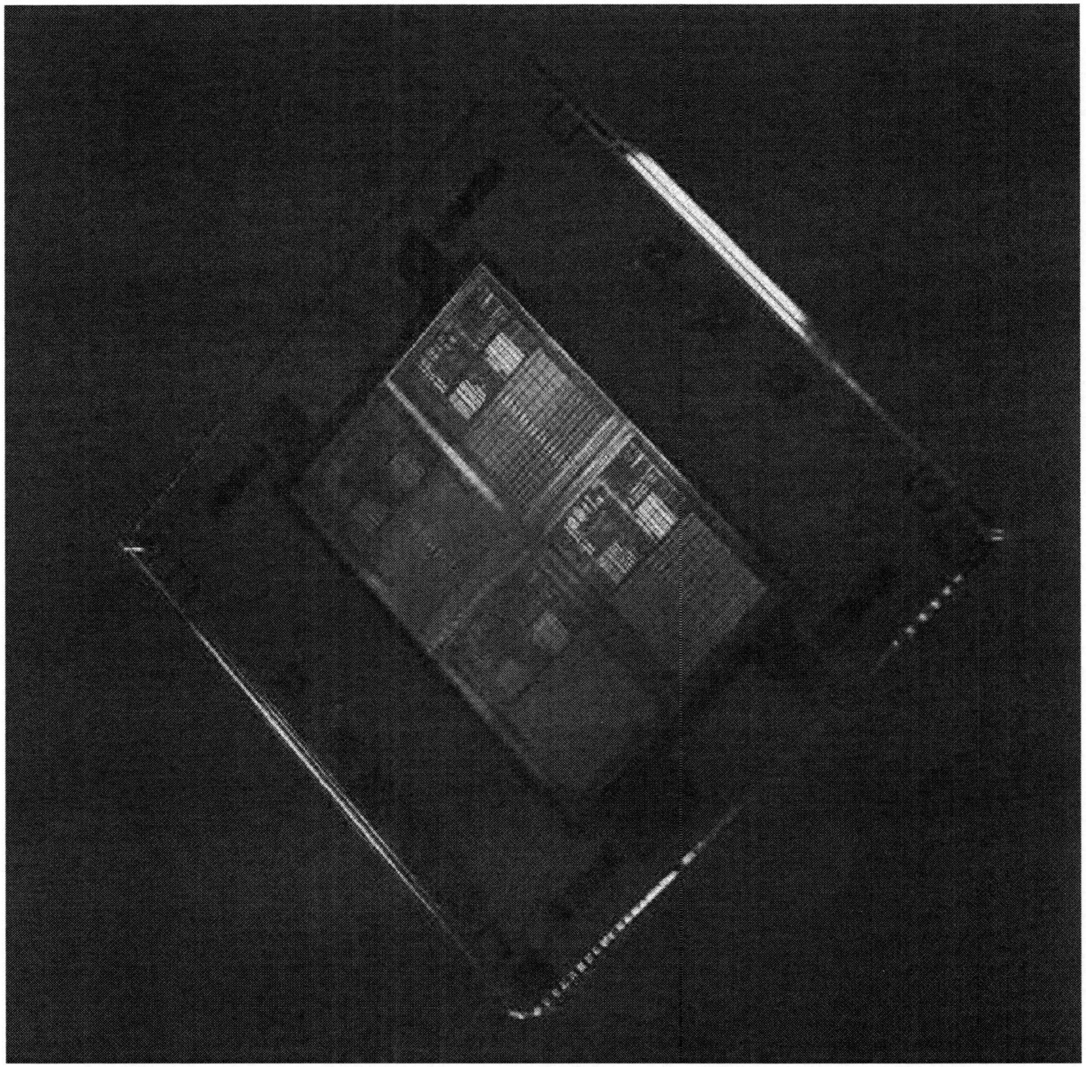

Particle-free mask handling

- sPod shows reticle protection down to 0.1 added particle/cycle @ 53 nm

sPod Carrier

with Inner Pod Exposed

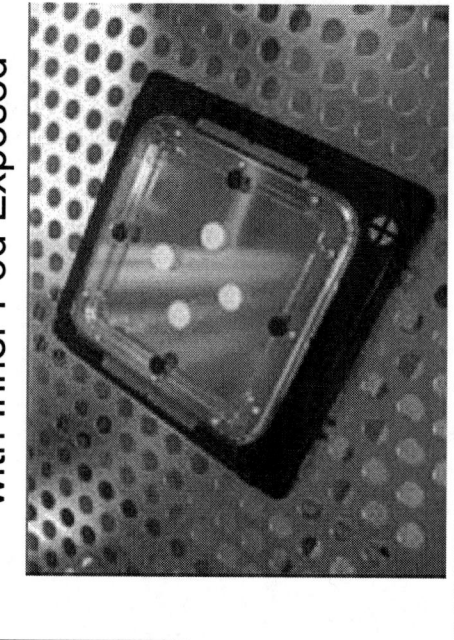

Ref: Long He, et al. – Proc. SPIE 6921, 69211Z (March 21, 2008)

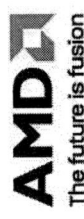

Cost-effectiveness

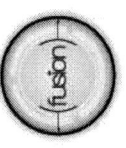

Lithography capital cost

Capital cost / wafer = Depreciation / throughput

1× to 2x double-patterning tools

Lithography capital cost

Capital cost / wafer = Depreciation / throughput

$$\sim \frac{1}{100} \times \text{double - patterning tools}$$

Light sources

- EUV light sources require highly ionized atoms.
  - Z^{10+}, etc.

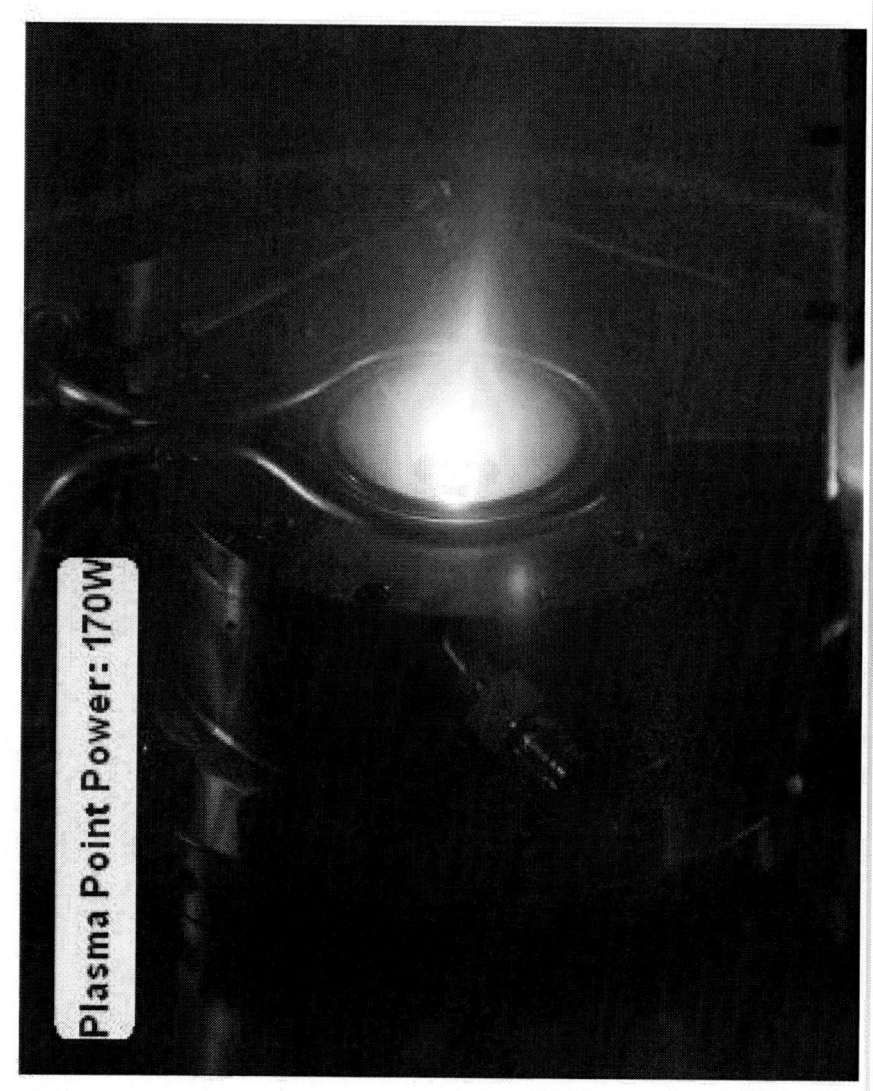

Plasma Point Power: 170W

Laser produced plasma (LPP) sources

-
- tin droplets

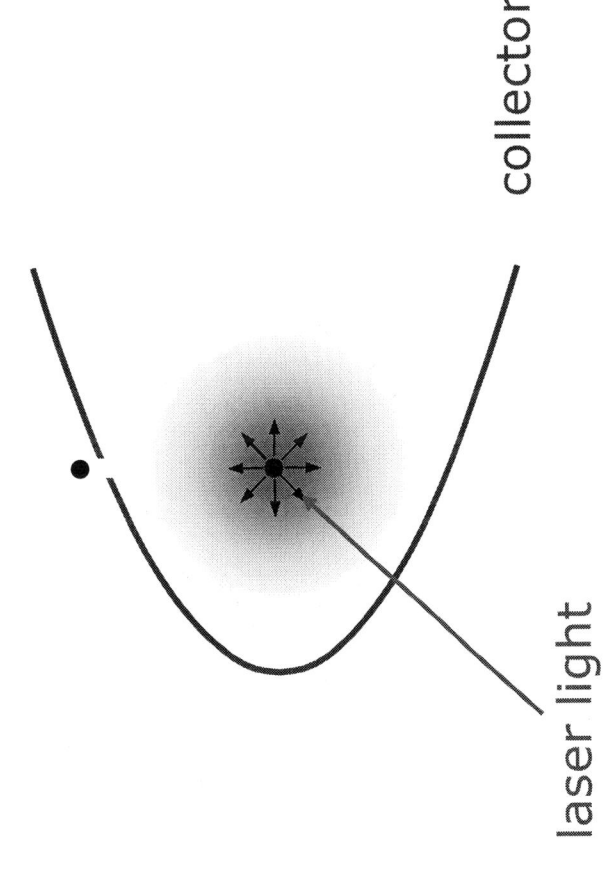

laser light

collector

EUV light sources

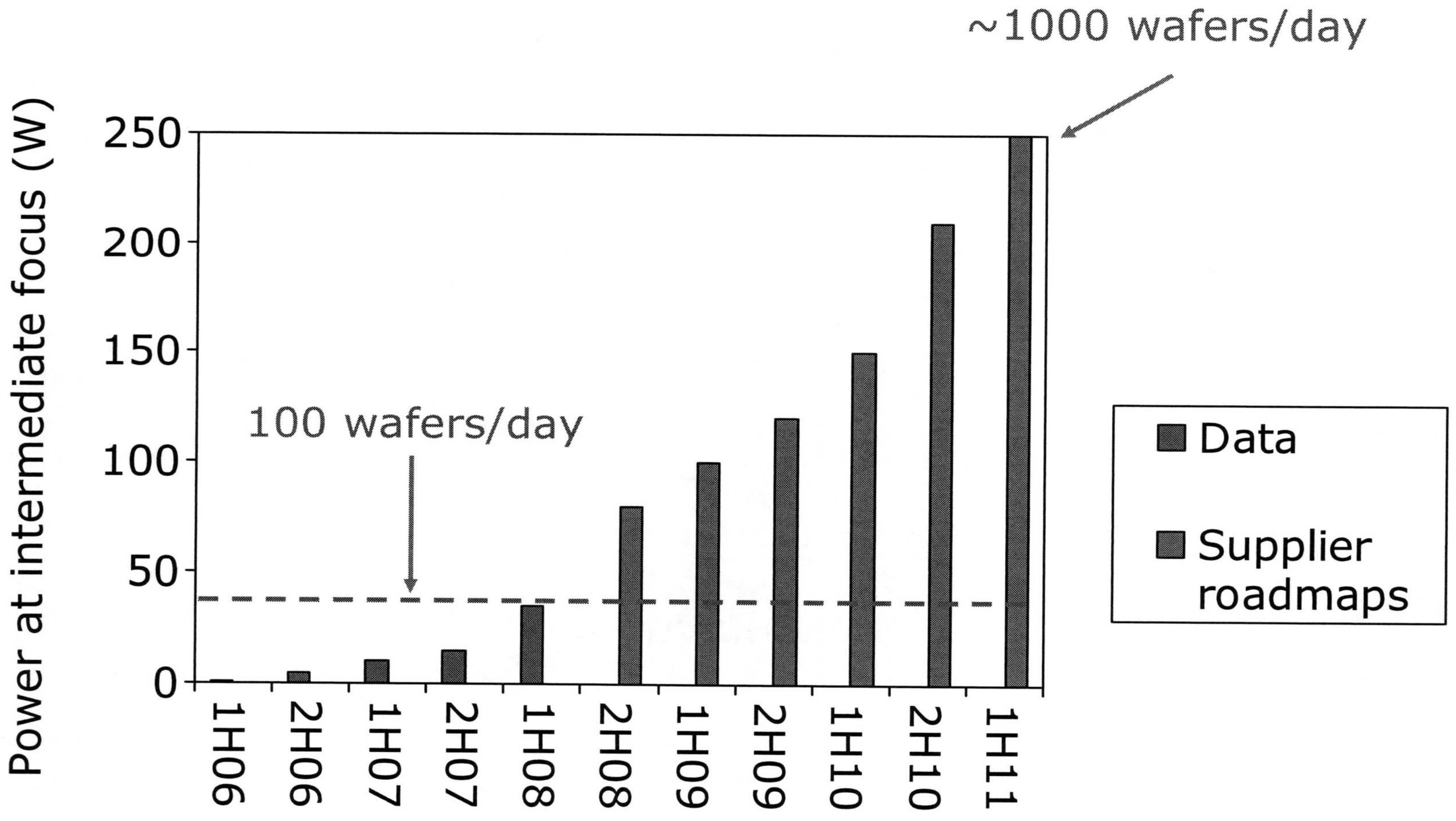

~1000 wafers/day

Data
Supplier roadmaps

100 wafers/day

Power at intermediate focus (W)

250 · 200 · 150 · 100 · 50 · 0

1H11 · 2H10 · 1H10 · 2H09 · 1H09 · 2H08 · 1H08 · 2H07 · 1H07 · 2H06 · 1H06

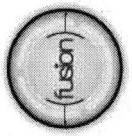

Electricity costs

- LPP generation of EUV light is not a highly efficient process.

- Infrared laser light → EUV light.   ≤ 4% efficiency.

- Electricity → infrared laser light.   ≤ 4% efficiency.

≤ 0.16% overall efficiency.

Fiber lasers

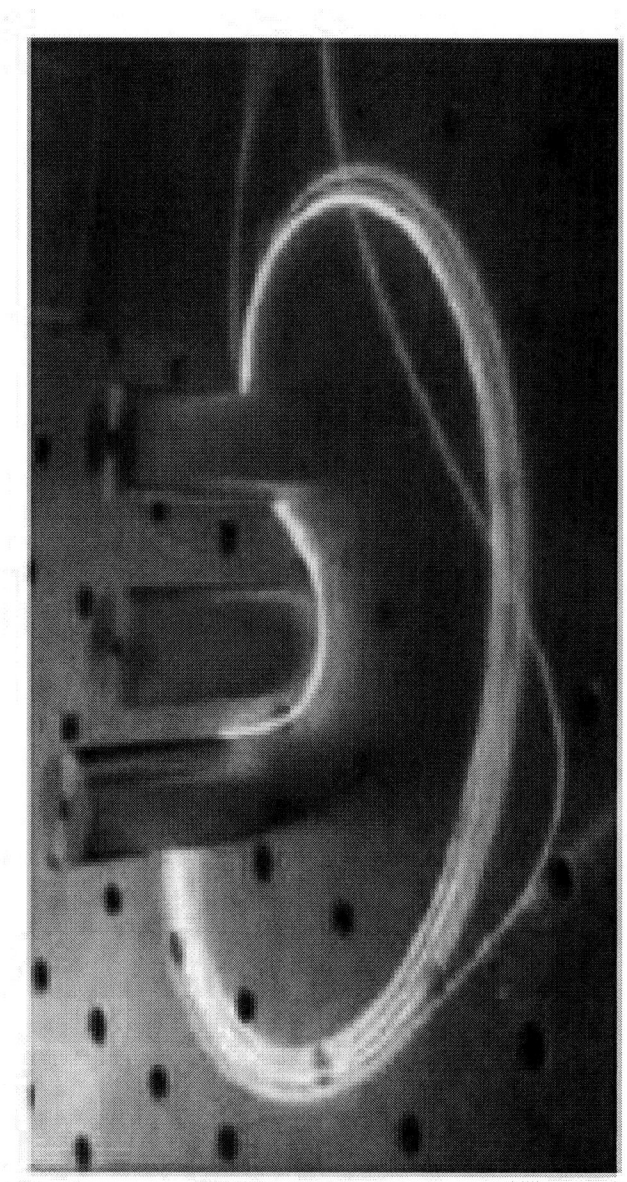

Electricity costs with fiber lasers

- Infrared laser light → EUV light.   ≤ 3% efficiency.

- Electricity → infrared laser light.   ≤ 40% efficiency.

        Up to 1.2% overall efficiency.

Critical issues: resist LER

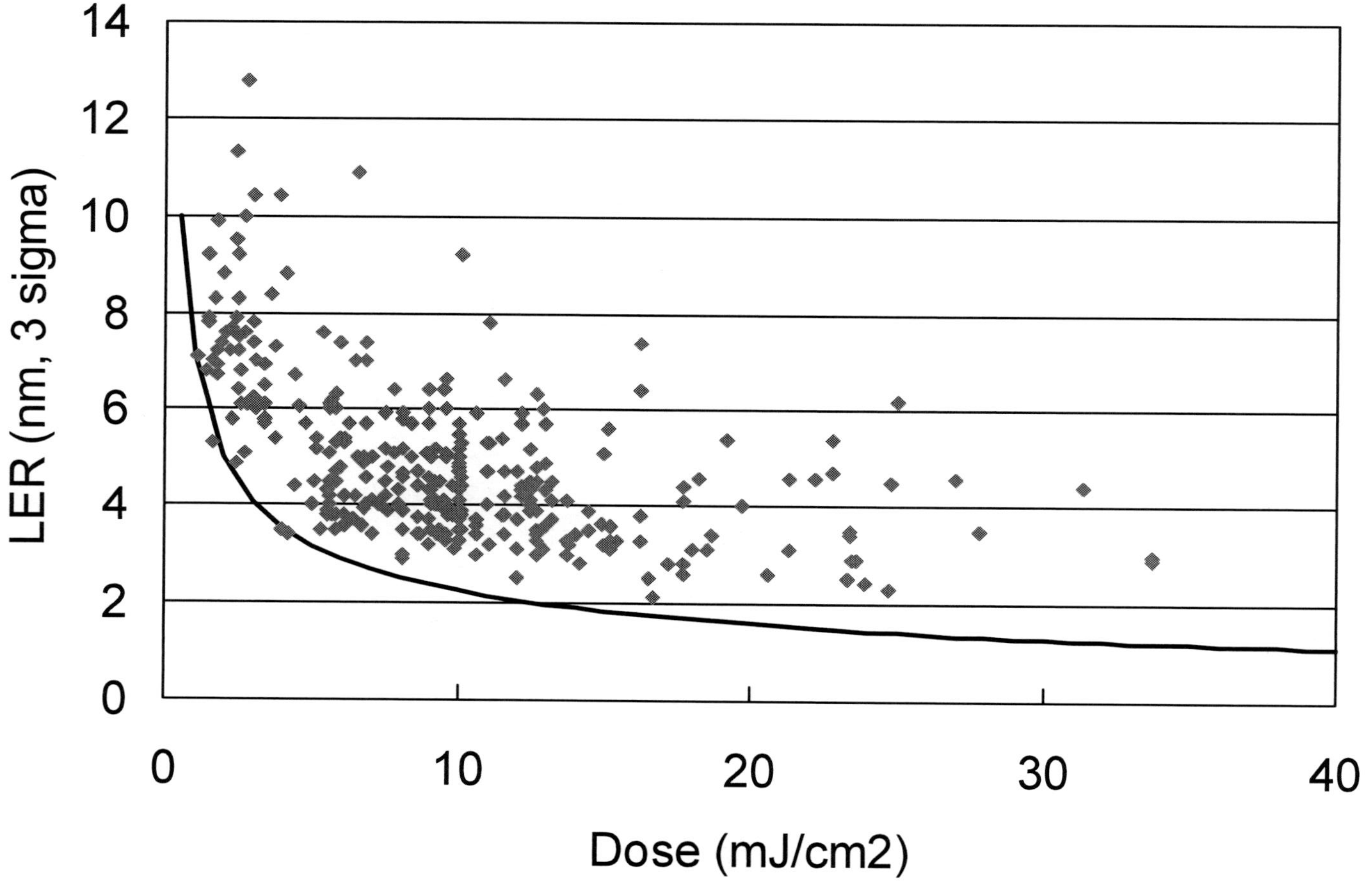

Simple shot-noise model predicts 1/√dose relationship between LER and dose

Data courtesy of Dr. P. Naulleau (LBNL) and Dr. T. Wallow (AMD)

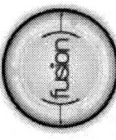

Need

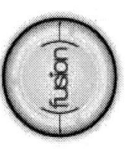

Need

- Why work so hard?

- Other options.

  - 193 nm $< \lambda \ll$ 13.5 nm?

    - Remember 157 nm?

  - Higher NA?

    - Can you spell "lutetium?"

  - Double patterning.

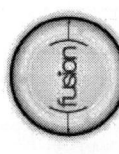

Need

- Memories have ~ 3 – 4 critical layers.

- Logic has ~ 2 – 7 more layers of dense metal than do memories.
  - Each metal layer involves two masking steps.

- Logic has ~ 30 critical layers.

Timeliness

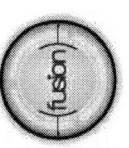

Timeliness

- The 22 nm logic node (40 – 45 nm ½-pitch) will be done with immersion lithography and some double patterning.

  - Manufacturing: 2011 – 2012

- Can EUV be ready in time for the 16 nm logic node?

  - Manufacturing: 2013 – 2014

- Technology development must start in 2012.

- EUV lithography must be moderately mature by then.

Cycles of learning

- It is important to exercise every part of the technology for EUV lithography to mature fully.
  - Pilot line.

- KrF lithography matured because it was being used to make a significant quantity of wafers.

- To be realistic, the pilot line must make something important.

- Importance → time pressure.

- The limitation for EUV pilot line operation today is the weak source.
  - < 100 wafer / week.

best sources are not yet integrated

Cycles of learning

Lessons from DUV pilot line

- Laser reliability improvements.

- Dose control calibration.

- Batch-to-batch resist control.

- Resist poisoning from the substrate.

- Diffusion effects in resist.

- Defect issues.

- Methods for productivity improvement.

Mirror contamination

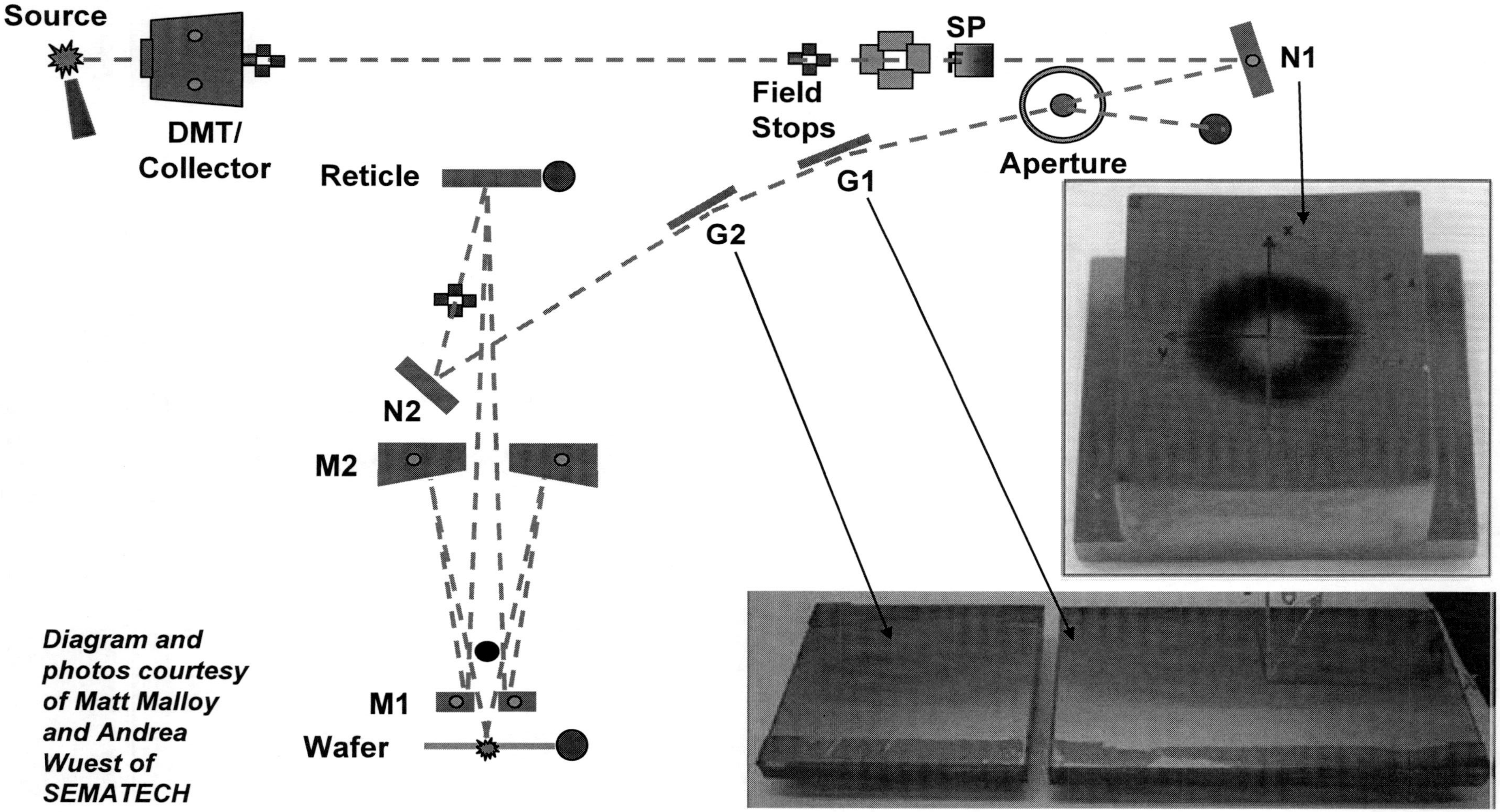

Source

DMT/
Collector

Reticle

Field
Stops

SP

Aperture

N1

G1

G2

N2

M2

M1

Wafer

Diagram and photos courtesy of Matt Malloy and Andrea Wuest of SEMATECH

Contamination

Un-contaminated sub-fields

Contaminated sub-fields

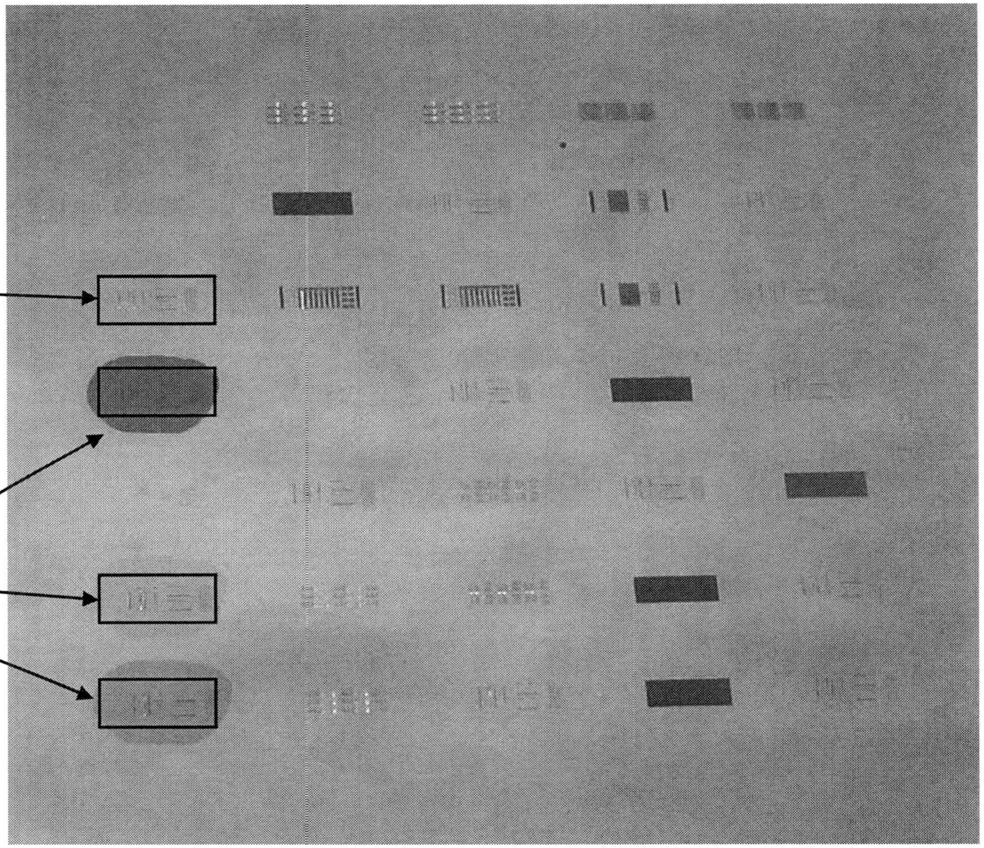

Cycles of learning

Lessons from DUV pilot line

- Laser reliability improvements.
- Dose control calibration.
- Batch-to-batch resist control.
- Resist poisoning from the substrate.
- Diffusion effects in resist.
- Defect issues.
- Methods for productivity improvement.

Lessons from EUV pilot line

- Contamination.
- ?
- ?
- ?
- ?
- ?
- ?

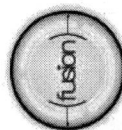

AMD
The future is fusion

EUV pilot lines

- EUV lithography is ready for use on pilot lines.

- It is essential for success to exercise the technology as much as possible and as soon as possible!

- Multiple pilot lines will help to accelerate technology development.

- All elements of the infrastructure need to be exercised.

- For a technology as immature as EUV, the year 2013 is not that far away.

Summary

Moving into manufacturing

- There are a few remaining known technical issues.

  - Mask defects: numbers and printability.

  - Source power.

- We need more cycles of learning!

- With enough progress in masks, equipment and materials, manufacturing using EUV lithography is possible by 2013/2014.

Trademark Attribution

AMD, the AMD Arrow logo and combinations thereof are trademarks of Advanced Micro Devices, Inc. in the United States and/or other jurisdictions. Other names used in this presentation are for identification purposes only and may be trademarks of their respective owners.

©2008 Advanced Micro Devices, Inc. All rights reserved.

Contents

 Memory scaling trend

 DPT vs. EUVL

 Other NGL options

 Status/Risks of EUVL

 Roadmap

SAMSUNG

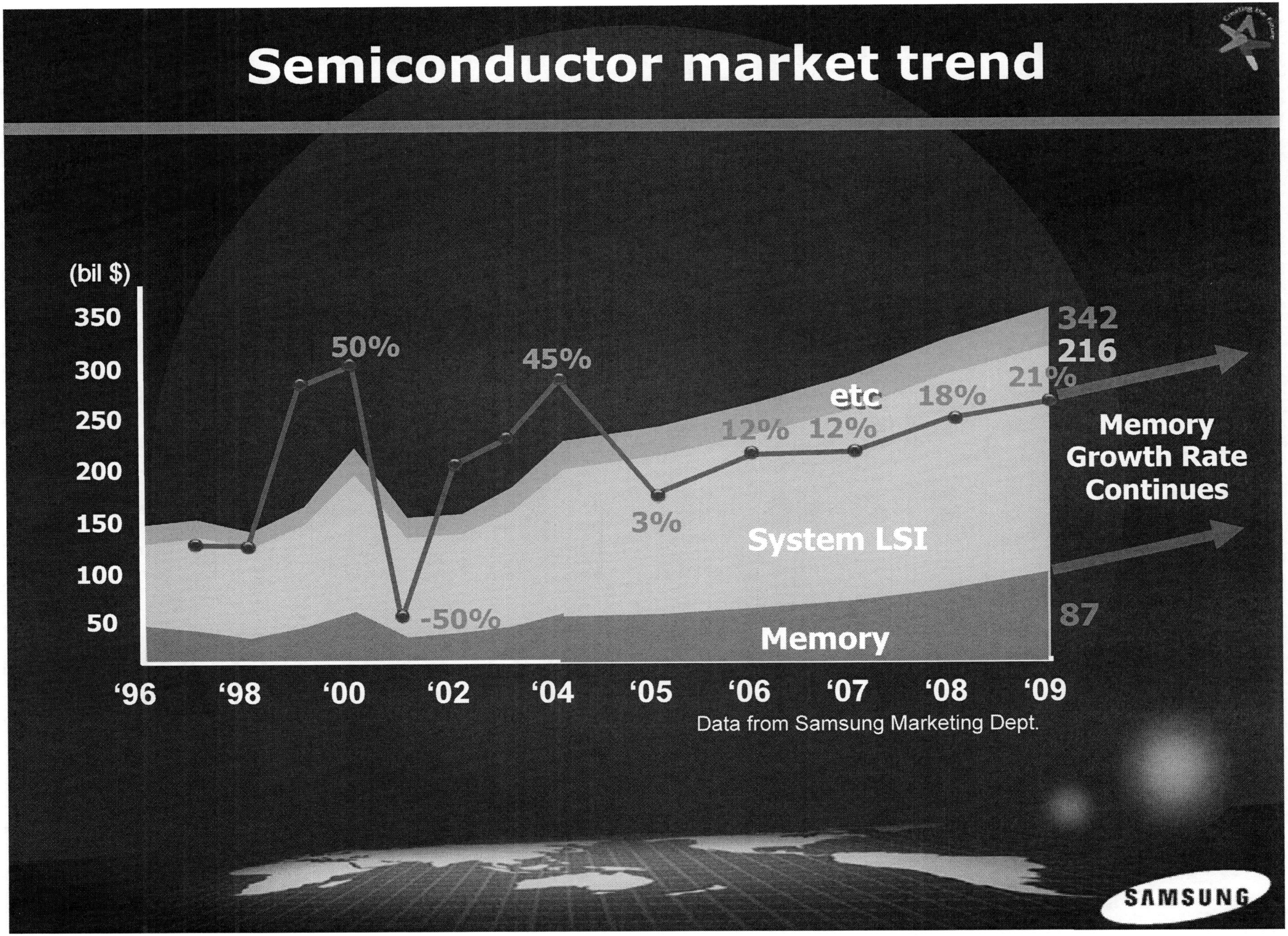

Semiconductor market trend

(bil $)

350
300
250
200
150
100
50

'96 '98 '00 '02 '04 '05 '06 '07 '08 '09

50%
45%
-50%
3%
12% 12%
etc 18%
21%
342
216
87

Memory Growth Rate Continues

System LSI

Memory

Data from Samsung Marketing Dept.

SAMSUNG

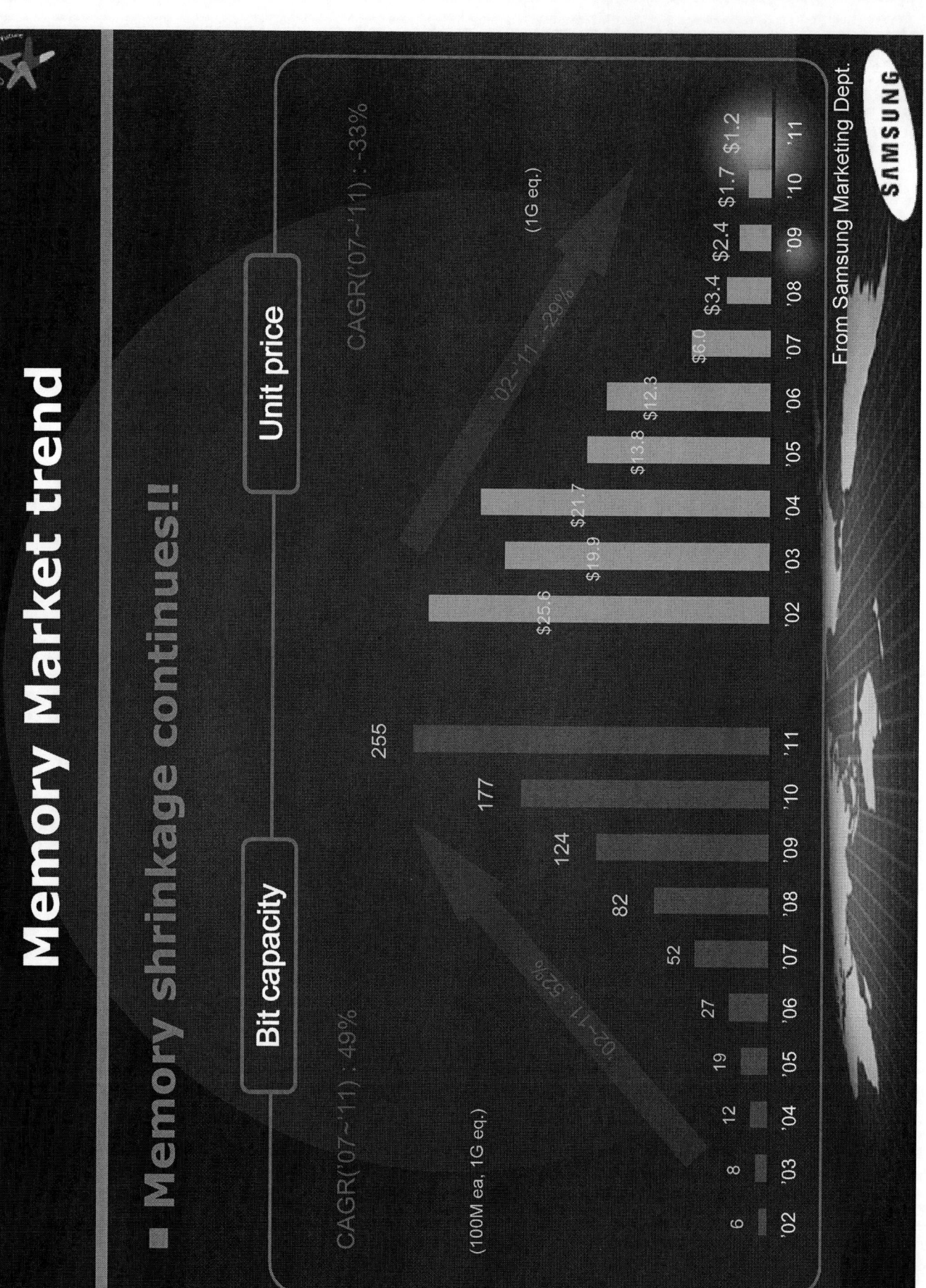

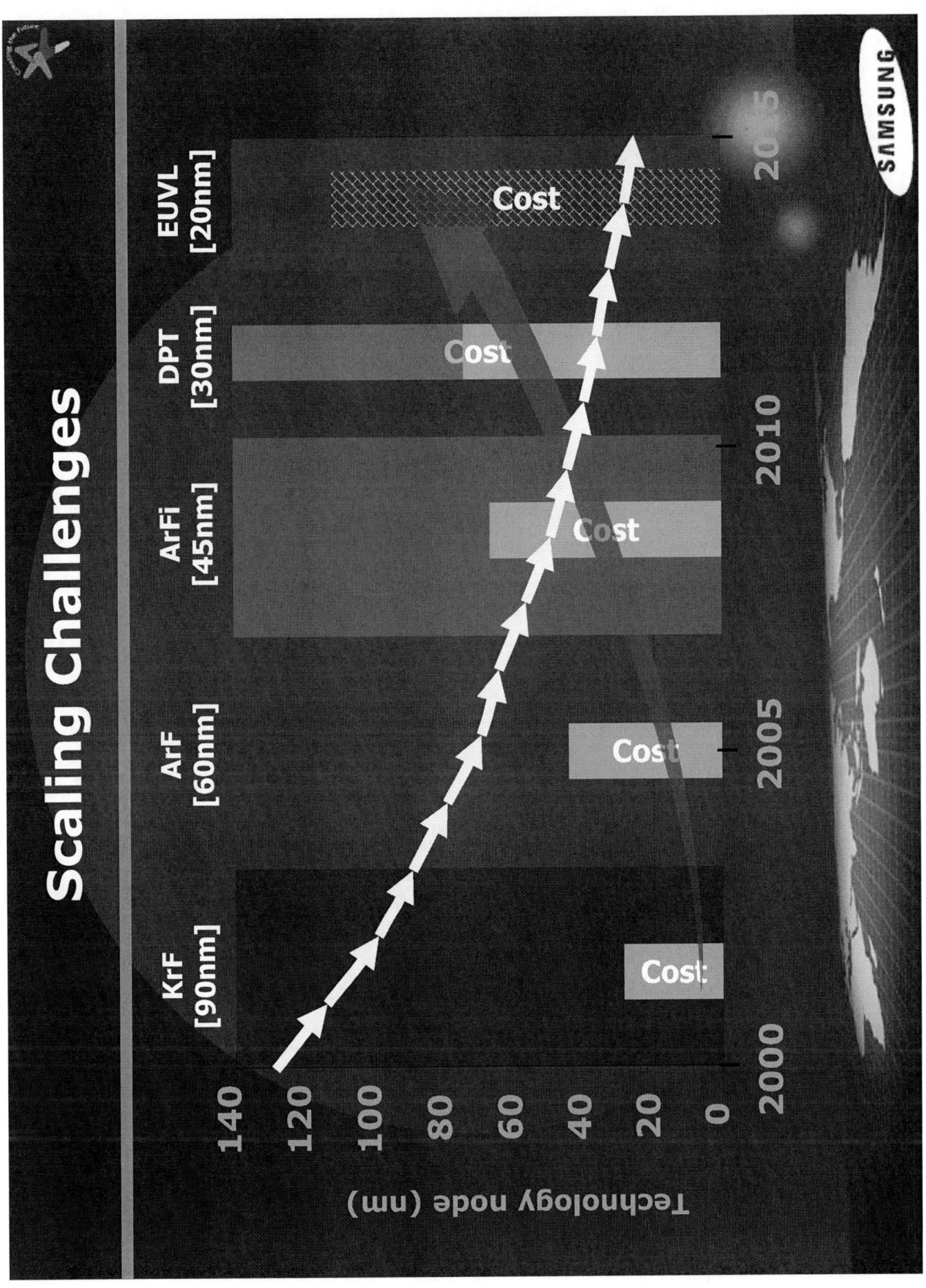

DPT vs. EUVL

❖ DPT and EUVL process

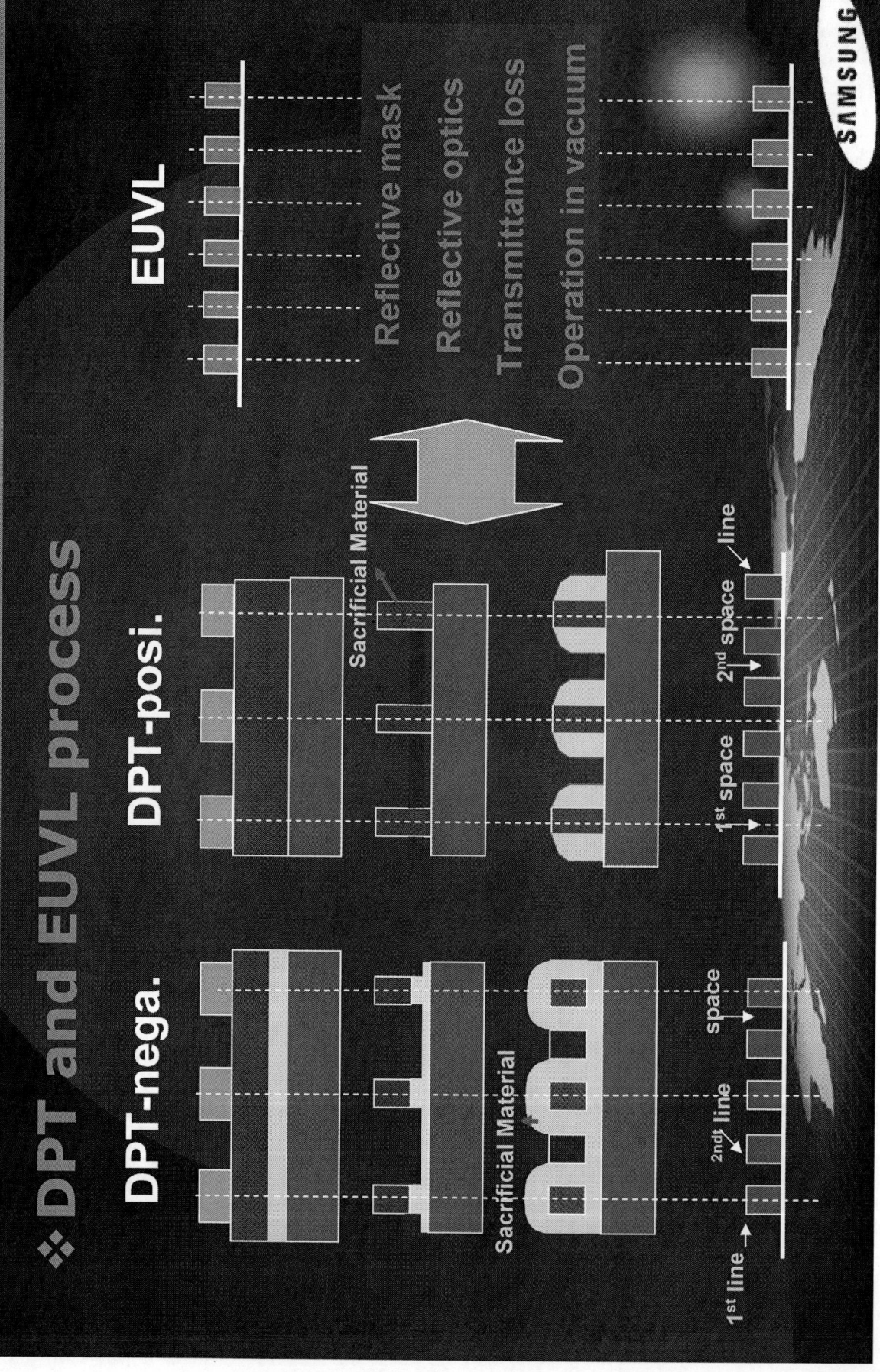

DPT cost vs. EVUL cost

	ArFi	DPT (spacer)	LELE	EUVL
Cap. Expense (a.u.) for new fab.	1	2.0	2.8	1.5*
Cap. Expense (a.u.) for existing fab.	0	1.0	1.8	1.5*
Complex Pattern	Line&Space Contacts Islands	Line&Space Only (1:1)	Contacts Islands	Line&Space Contacts Islands
Readiness	on Prod.	Ready	Ready	2012

* Dependent on EUVL tool cost & throughput

Requirements for the Viability

❖ **Technology alone may not be sufficient for NGL to become a finally viable solution**

Viable Lithographic Solution

Financial Support
Funding
Strong support

Stopper-less Technology
Prospect
Stopper-less

Market Needs
Cost
Timeliness

Service & Support
Satisfaction

SAMSUNG

Status of NGL options

	Resol. (current)	CoO (mass)	CoO (small)	T'put	OVL	Defect	Readiness
DPT	~30nm						
NIL	~30nm						
ML2	~40nm						
EUVL	< 20nm						

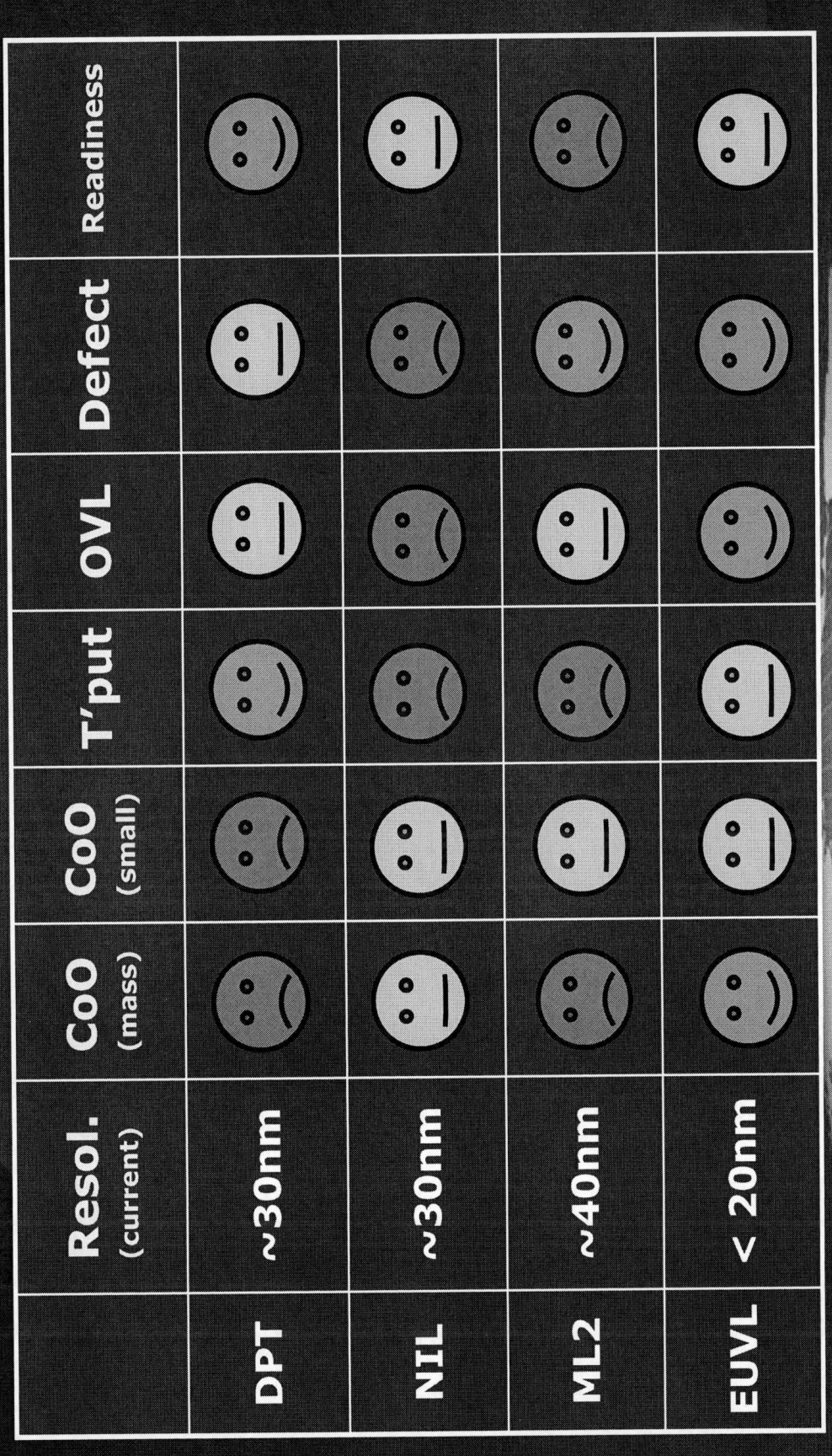

SAMSUNG

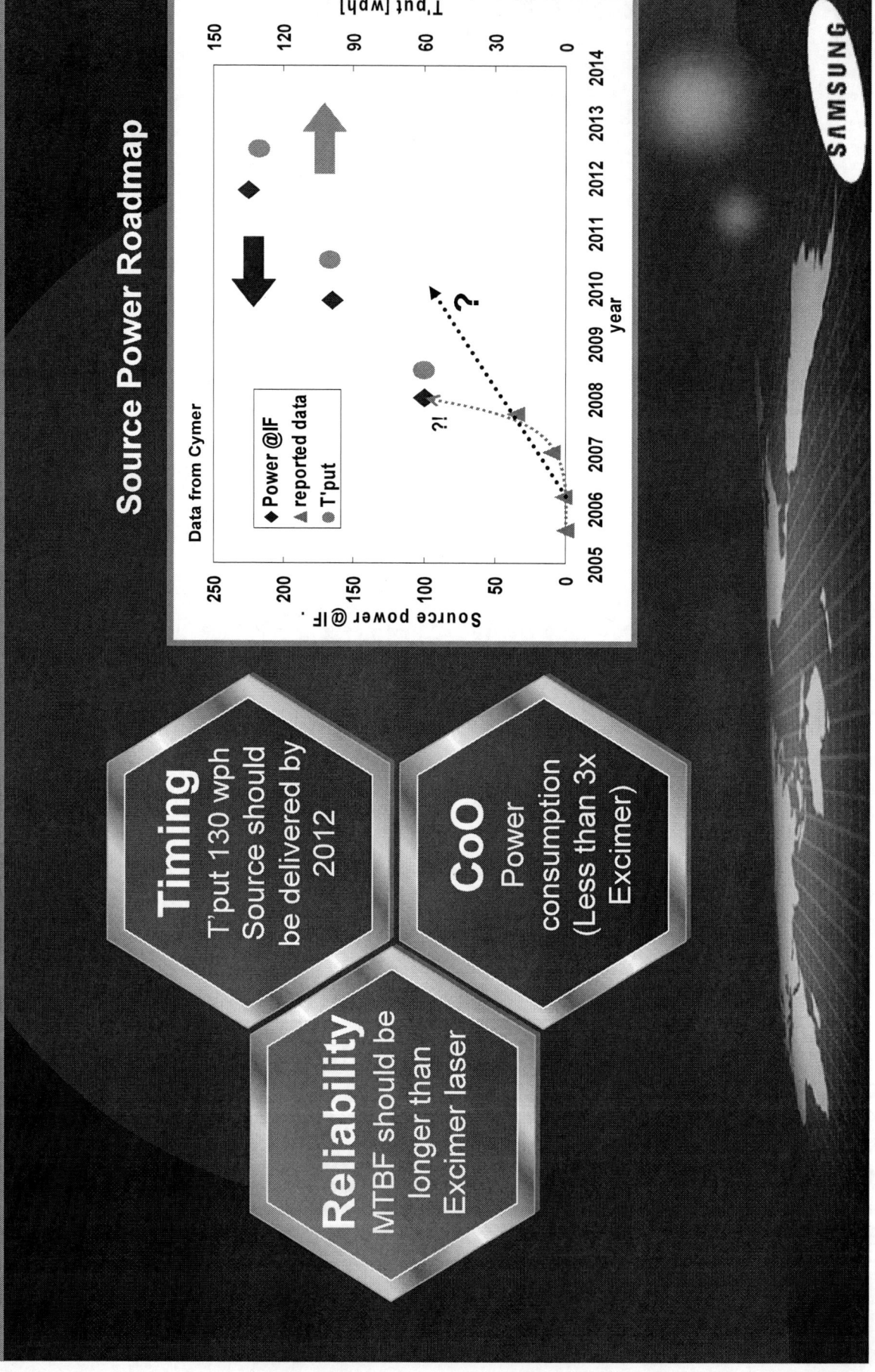

Challenges in EUVL Resist

Current status

Resolution
- reached (22nm)
LWR
- hard work needed (~4.0nm)
Sensitivity
- close (10mJ)

Hard to achieve each spec. simultaneously

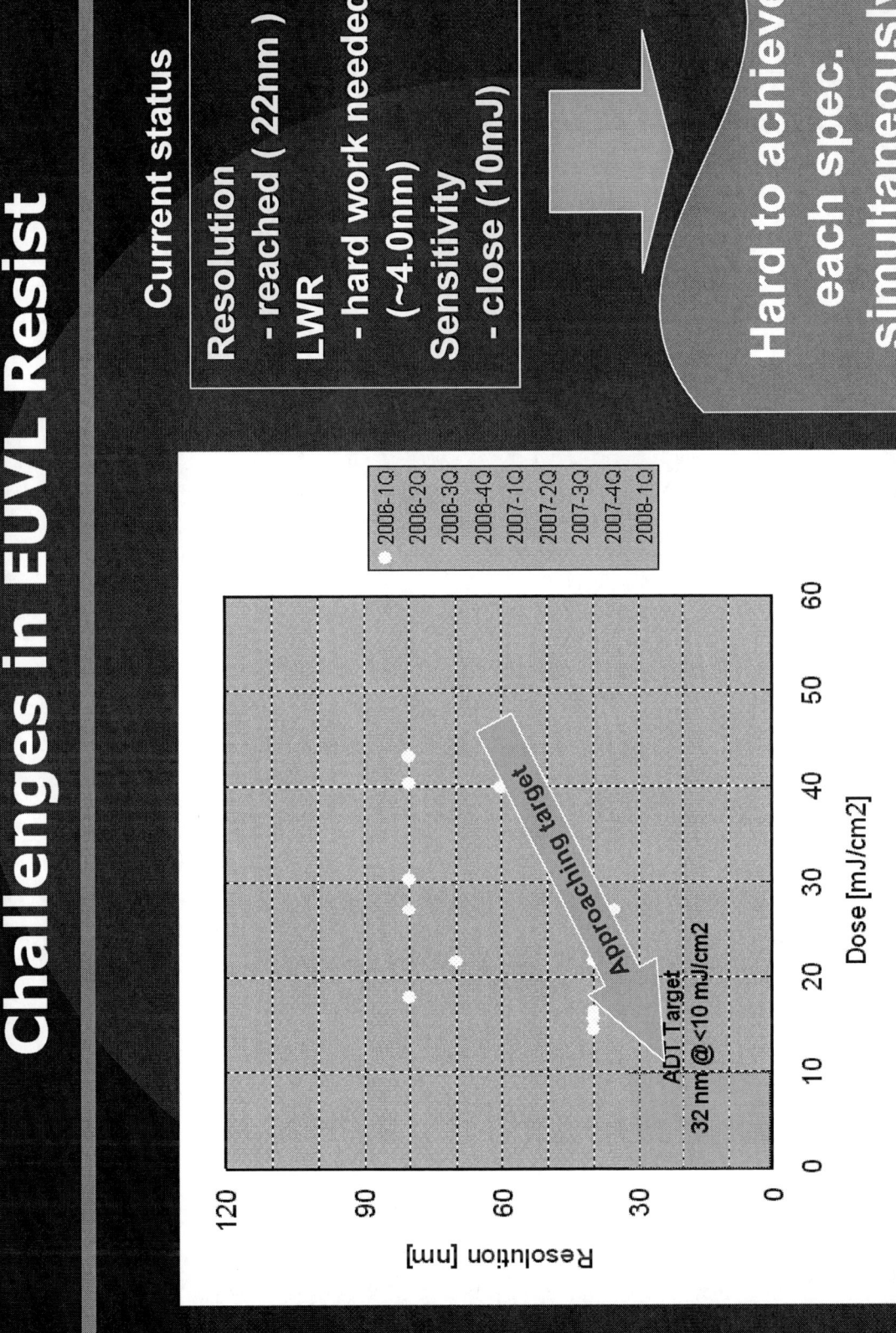

SAMSUNG

Challenges in EUVL Mask

Availability of Defect-Free Blank

- Achieving zero defect for EUVL masks in the next few years is extremely challenging.

Multilayer Defects

Substrate Defects

Bump

Pit

Blank's Defect Level required for HVM
< 0.003/cm² @25nm

⇕ Large GAP

Current Status (2008)
~ 0.3/cm² @60nm

Defect free blank
- Collaboration with supplier

Blank inspection
- DUV Inspection
  (M7360 → 2.5G)
- Actinic Blank Inspection

Tentative solution
- Layout Disposition
- Wafer inspection

SAMSUNG

New challenges in EUVL

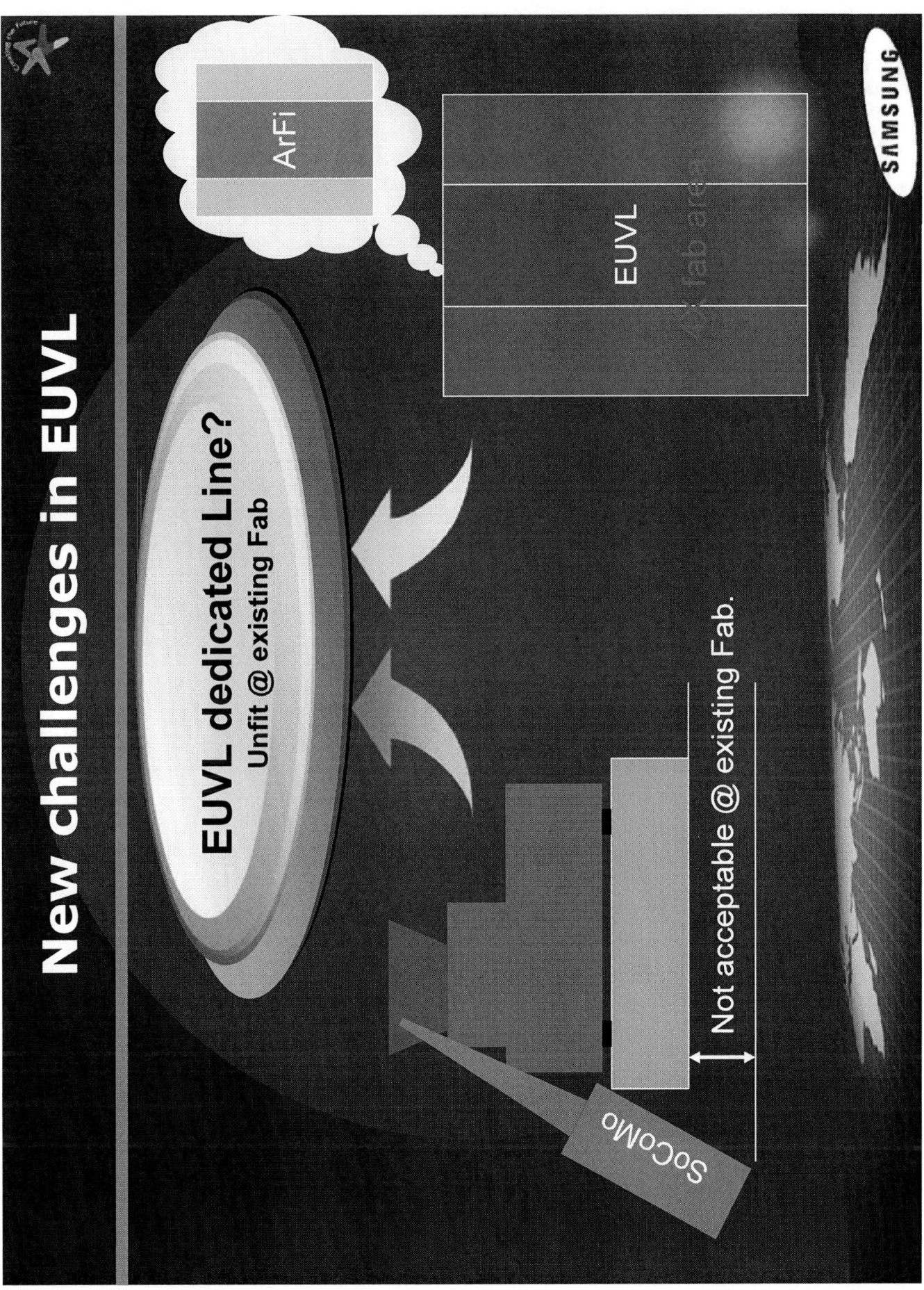

EUVL dedicated Line?
Unfit @ existing Fab

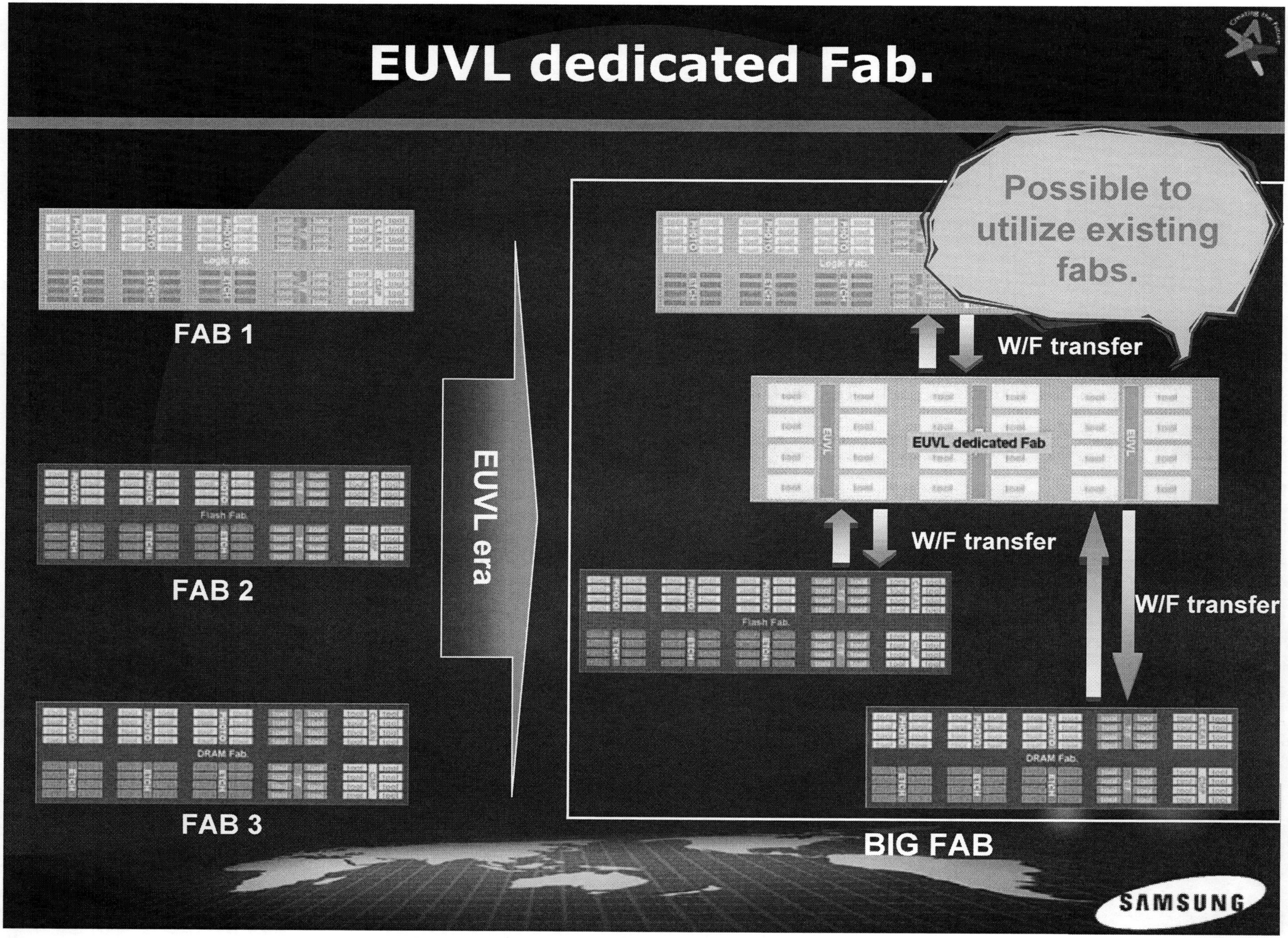
EUVL dedicated Fab.
FAB 1
Logic Fab.
FAB 2
Flash Fab.
FAB 3
DRAM Fab.
EUVL era
Possible to utilize existing fabs.
W/F transfer
EUVL dedicated Fab
EUVL
W/F transfer
W/F transfer
BIG FAB
SAMSUNG

Global Collaboration

SEMATECH
Blank Mask Process
Mask Inspection Infra
Standardization

Ministry of Knowledge Economy
mKE
Korean government Program

Supplier
Tool/Mask/Resist
Metrology/inspection

RAVE
HOYA Nikon
AGC
ROHM and HAAS
SUMITOMO CHEMICAL
ASML
tok
DONGJIN

SAMSUNG
Blank Evaluation
Process Development

IMEC
EUV ADT

LBNL
EUV MET
EUV AIT

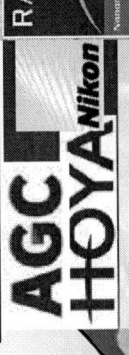

SAMSUNG

BERKELEY LAB

imec

EUVL for Memory litho

❖ Can EUVL be employed to 3xnm DRAM?

■ Test items

• Overlay (to immersion), Flare, Resolution..

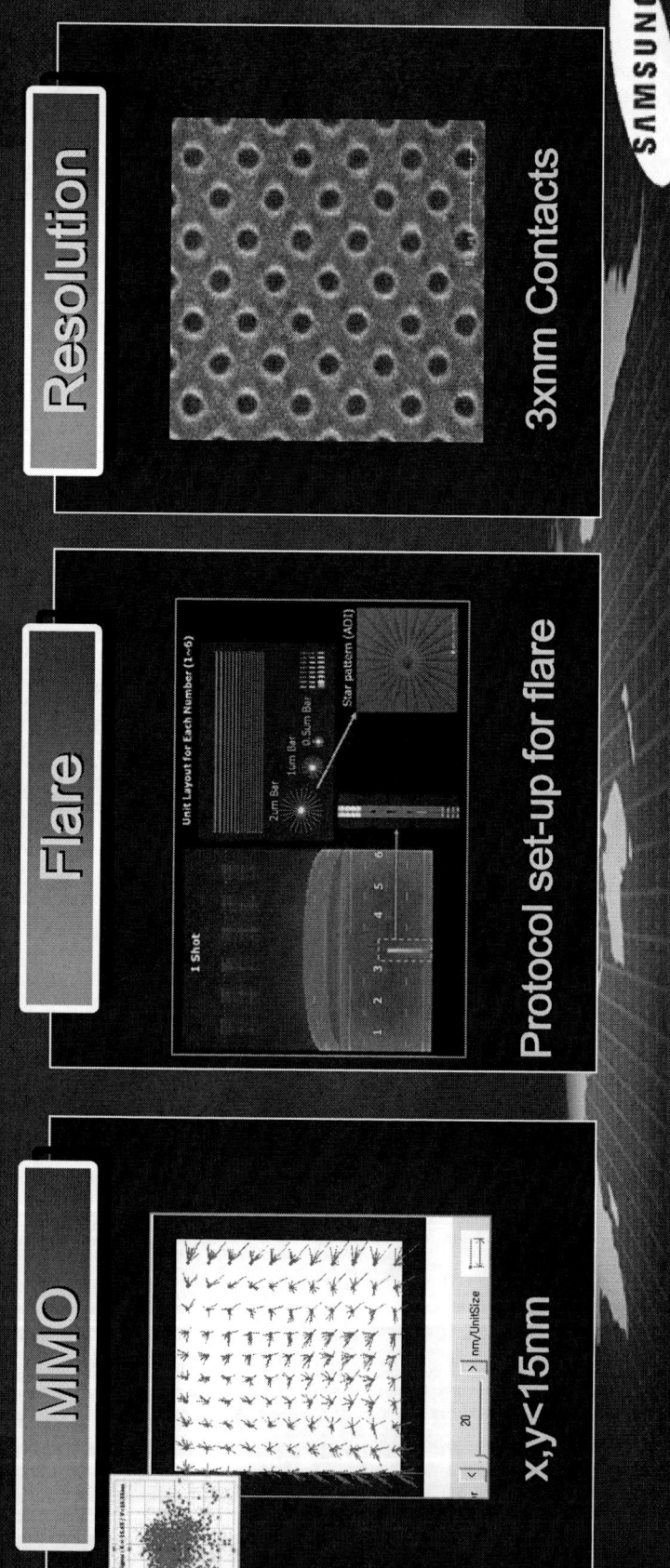

Resolution

3xnm Contacts

Flare

Protocol set-up for flare

MMO

x,y<15nm

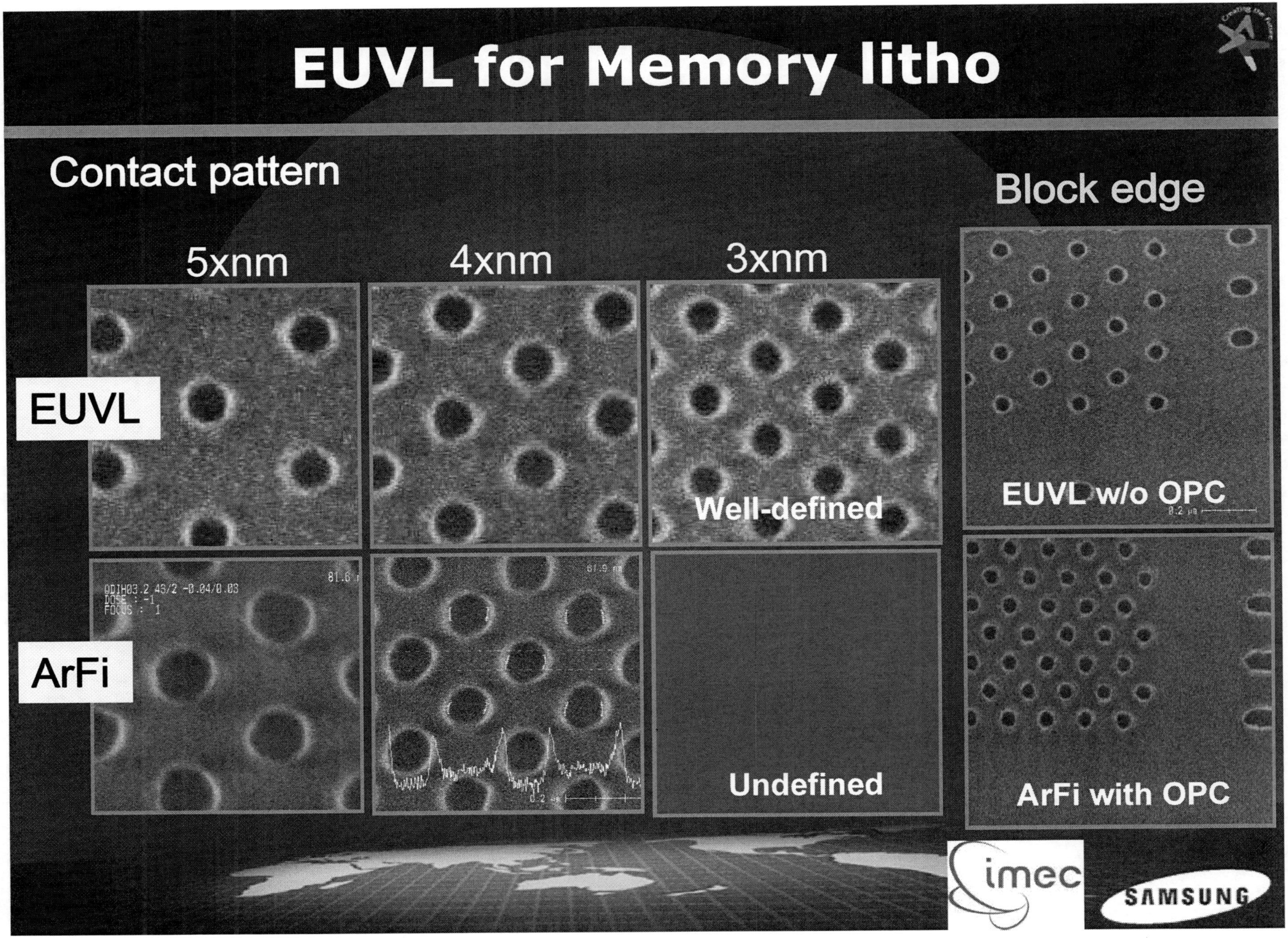

EUVL for Memory litho

Contact pattern

EUVL		
5xnm	4xnm	3xnm

EUVL — Well-defined

ArFi — Undefined

Block edge

EUVL w/o OPC

ArFi with OPC

EUVL for Memory litho

Line & space pattern

Peripheral patterns without OPC

EUVL

4xnm

EUVL

3xnm

EUVL

4xnm

EUVL

3xnm

ArFi

Poorly defined

Undefined

EUVL shows superior patterning performance to immersion ArF Lithography

Samsung DRAM Roadmap

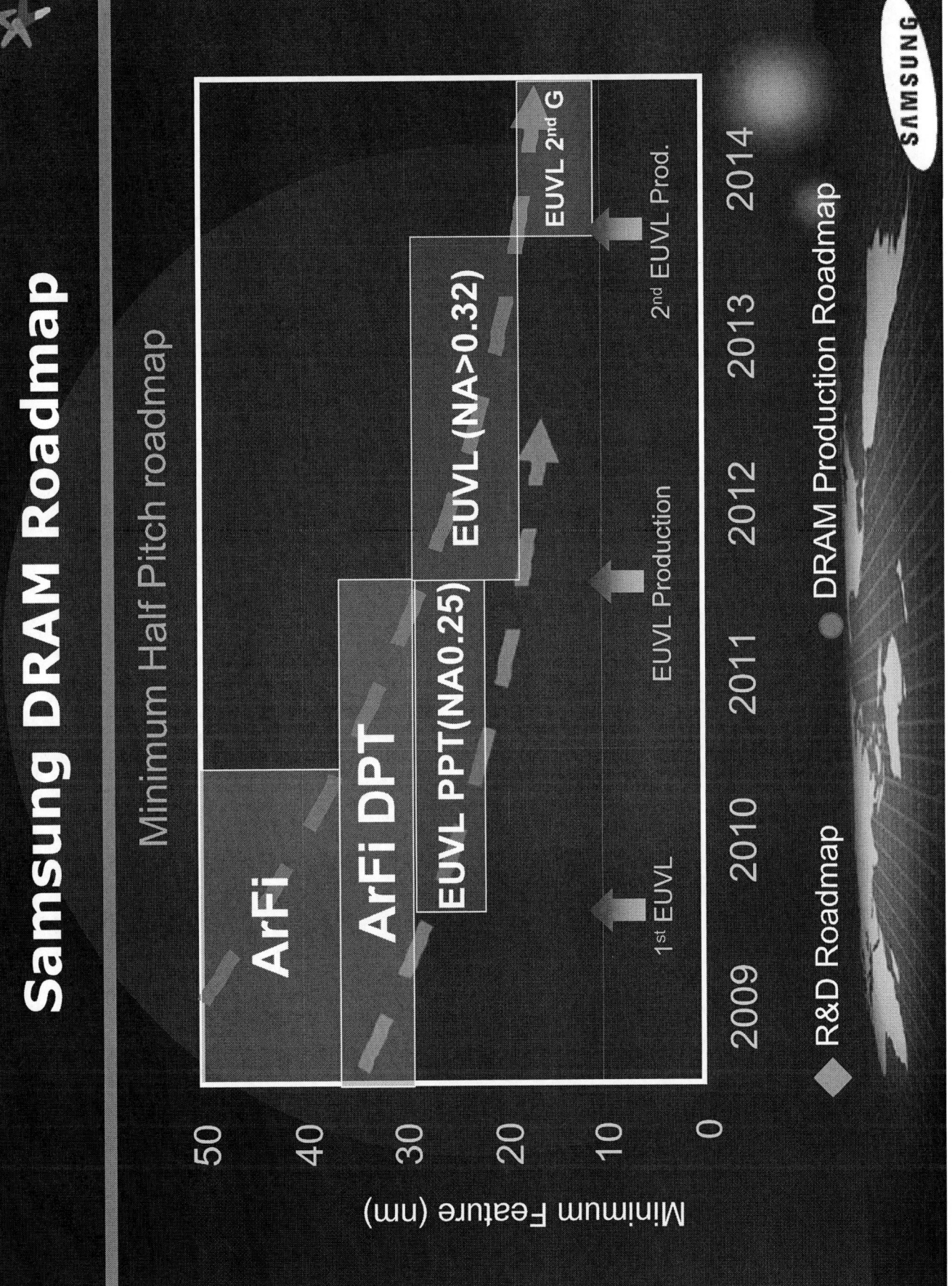

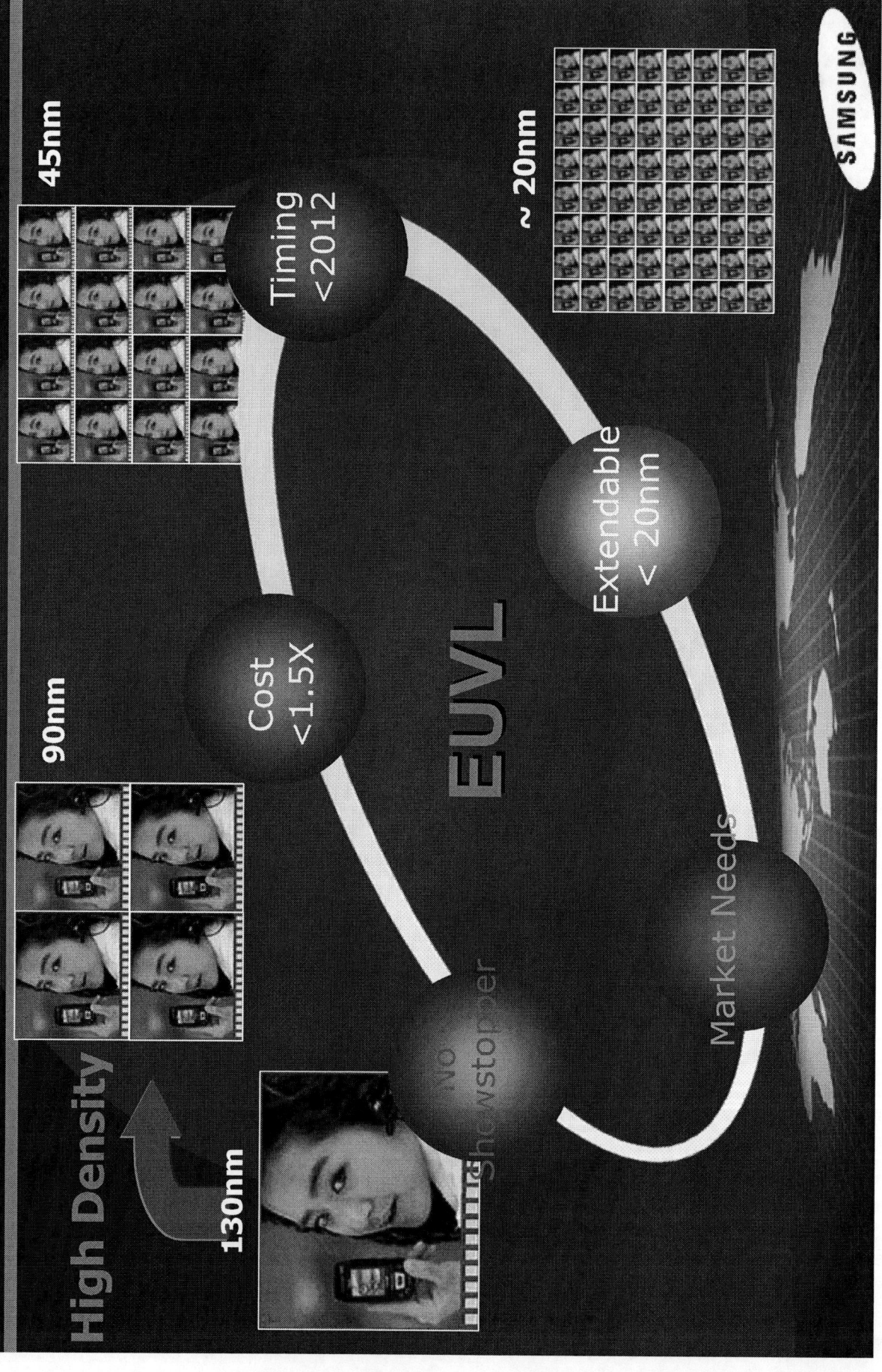

Thank You !

감사합니다
ありがとうございます
謝謝
Danke schön
Dank u
Grazie
Merci

SAMSUNG

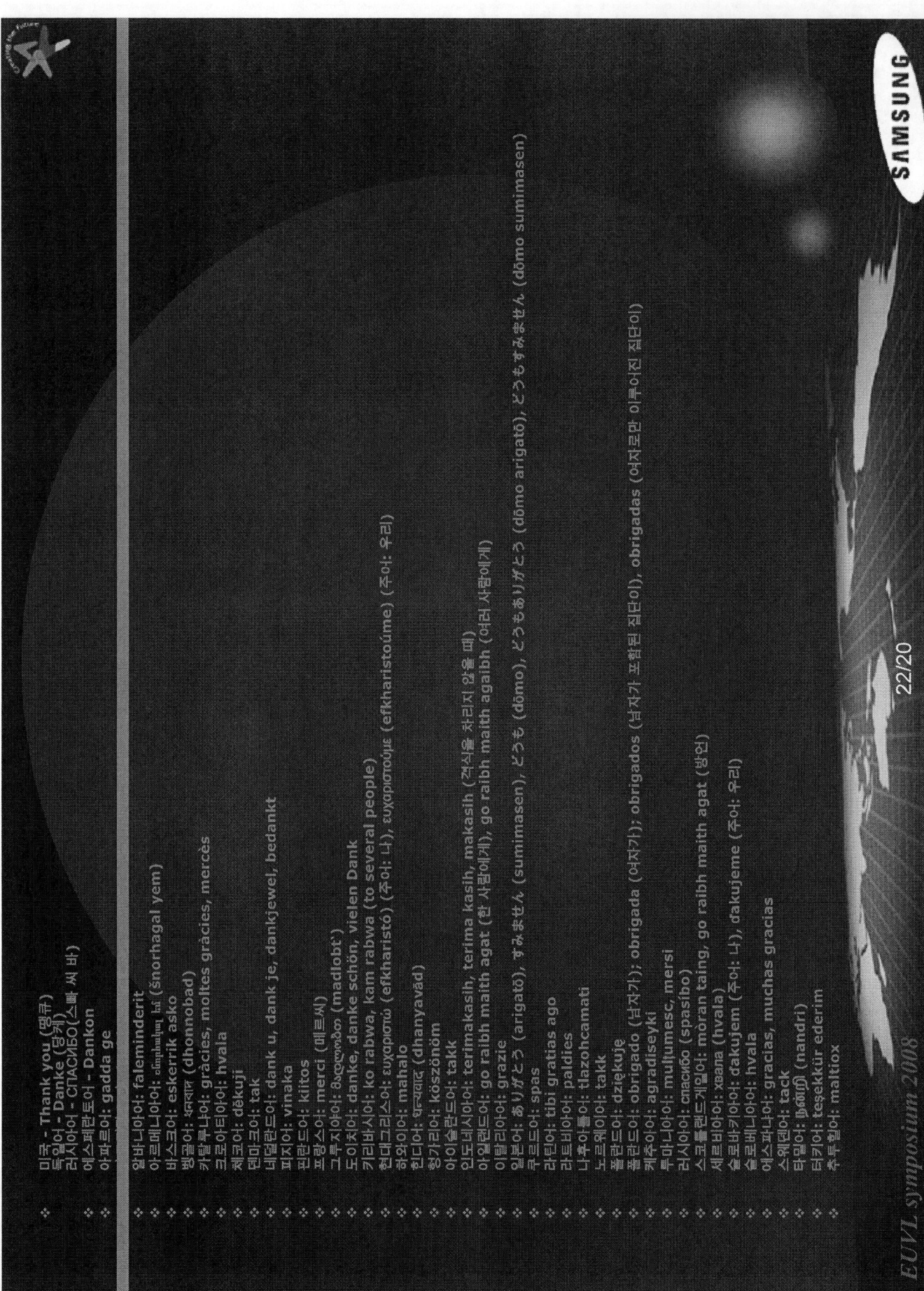

- 미국 – Thank you (땡큐)
- 독일어 – Danke (당케)
- 러시아어 – СПАСИБО (스빠 씨 바)
- 에스페란토어 – Dankon
- 이파르토어: gadda ge

- 알바니아어: faleminderit
- 아르메니아어: շնորհակալ եմ (šnorhagal yem)
- 바스크어: eskerrik asko
- 벵골어: ধন্যবাদ (dhonnobad)
- 카탈루냐어: gràcies, moltes gràcies, mercès
- 크로아티아어: hvala
- 체코어: děkuji
- 덴마크어: tak
- 네덜란드어: dank u, dank je, dankjewel, bedankt
- 피지어: vinaka
- 핀란드어: kiitos
- 프랑스어: merci (메르씨)
- 그루지아어: მადლობთ (madlobt')
- 도이치어: danke, danke schön, vielen Dank
- 기리바시어: ko rabwa, kam rabwa (to several people)
- 힌디어그리스어: ευχαριστώ (efkharistó) (주어: 나), ευχαριστούμε (efkharistoúme) (주어: 우리)
- 하와이어: mahalo
- 힌디어: धन्यवाद (dhanyavād)
- 헝가리어: köszönöm
- 아이슬란드어: takk
- 인도네시아어: terimakasih, terima kasih, makasih (격식을 차리지 않을 때)
- 아일랜드어: go raibh maith agat (한 사람에게), go raibh maith agaibh (여러 사람에게)
- 이탈리아어: grazie
- 일본어: ありがとう (arigató), ありがとうございます (dómo arigató), どうもありがとう (dómo arigató), どうもすみません (dómo sumimasen)
- 큐르드어: spas
- 라틴어: tibi gratias ago
- 라트비아어: paldies
- 나후아틀어: tlazohcamati
- 노르웨이어: takk
- 폴란드어: dziękuje
- 폴란드어: obrigado (남자가); obrigada (여자가); obrigados (남자가 포함된 집단이), obrigadas (여자로만 이루어진 집단이)
- 케추아어: agradiseyki
- 러시아어: спасибо (spasibo)
- 스코틀랜드게일어: mòran taing, go raibh maith agat (방언)
- 세르비아어: xвана (hvala)
- 슬로바키아어: dakujem (주어: 나), dakujeme (주어: 우리)
- 슬로베니아어: hvala
- 에스파니아어: gracias, muchas gracias
- 스웨덴어: tack
- 타밀어: நன்றி (nandri)
- 터키어: teşekkür ederim
- 주루얼어: maltiox

EUVL symposium 2008

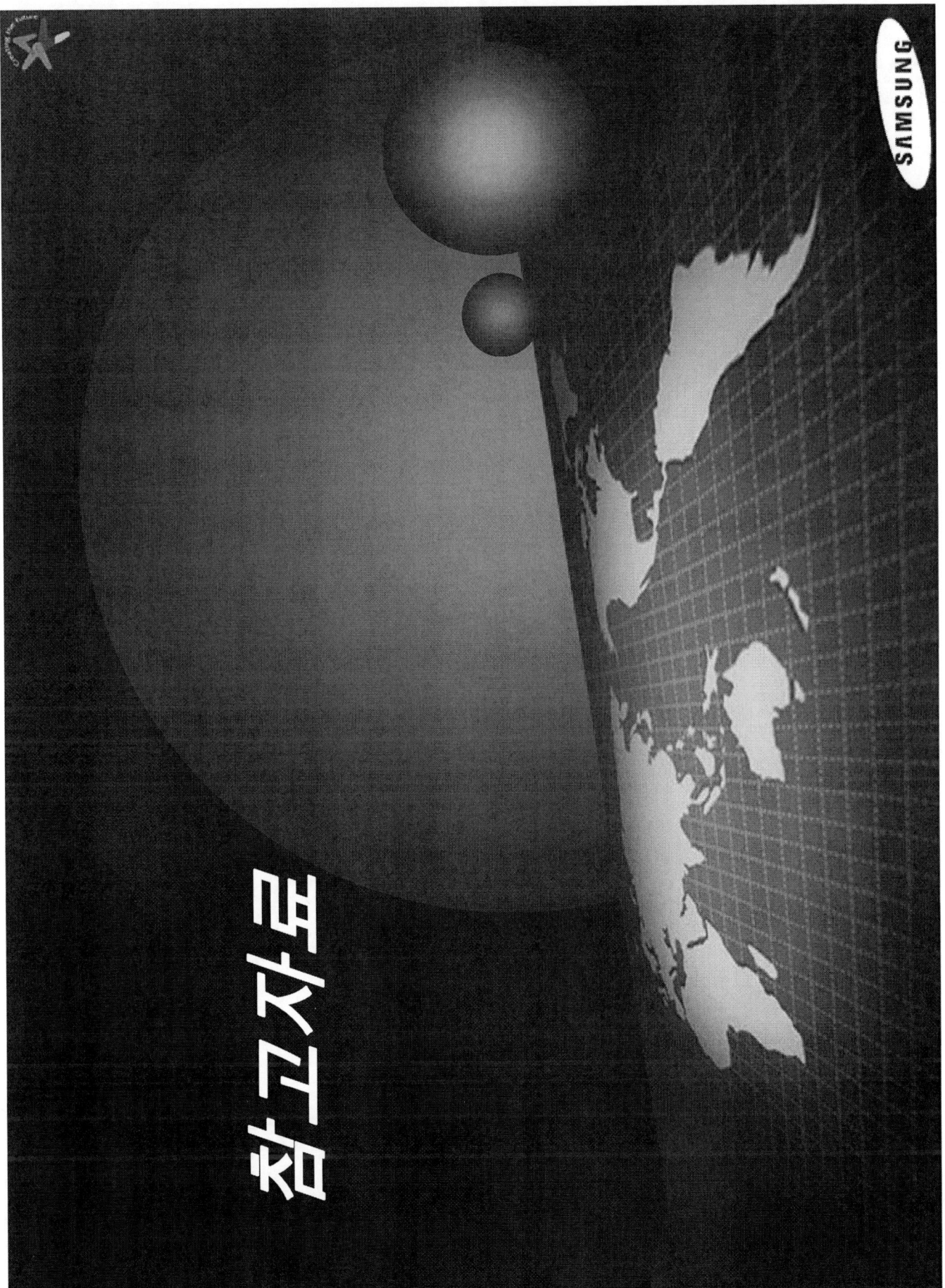

(Nano-?) Imprint for Memory ?

❖ **Defect Engineering and Placement control will be critical for NanoImprint to be adopted to Memory Production**

- Defect on Template
- Defect Inspection
- Defect Repair on Template
- Defect during Pattern Transfer
- Alignment & Overlay (Placement)

Recent Publications

IBM - EIPBN May 2007 →
Functional Storage Class memory 30nm

Samsung - SPIE Feb 2008 →
Full-Field Imprinting of Sub-40-nm Patterns

Toshiba - SPIE Feb 2008 →
Study of Nanoimprint Lithography for 22-nm node

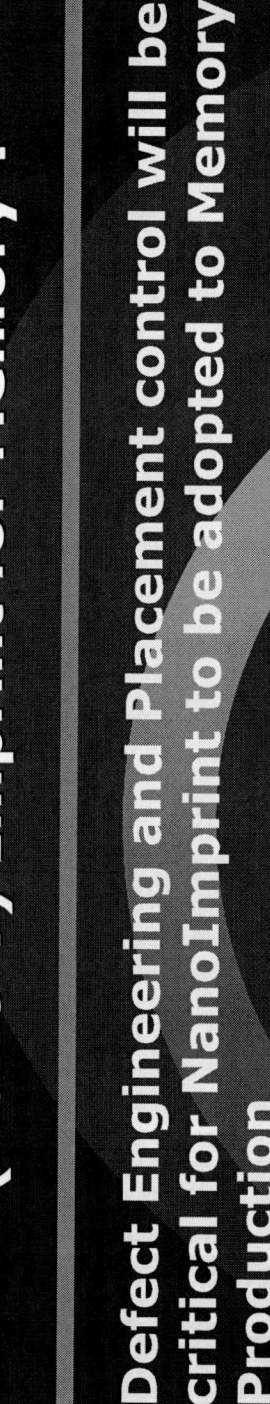

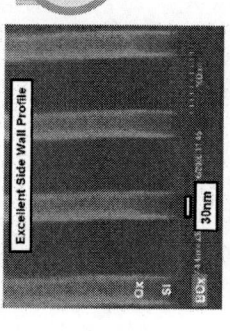

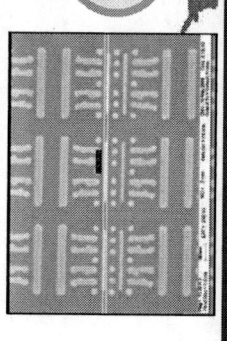

SAMSUNG

EUVL symposium 2008

24/20

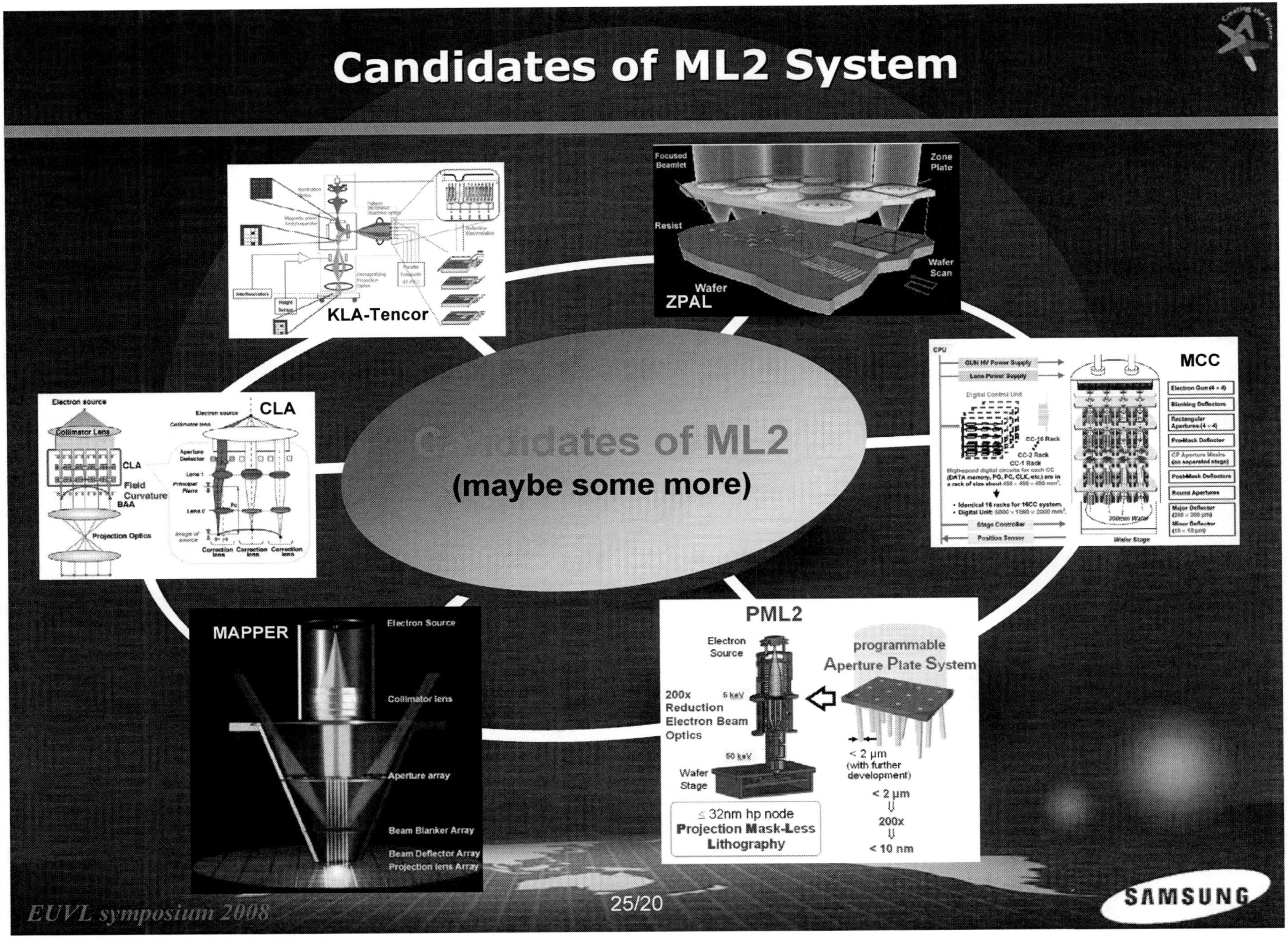
Candidates of ML2 System
KLA-Tencor
Focused Beamlet
Zone Plate
Resist
Wafer
Wafer Scan
ZPAL
Electron source
Collimator Lens
CLA
Field Curvature
BAA
Projection Optics
CLA
Candidates of ML2
(maybe some more)
MCC
CC-16 Rack
CC-2 Rack
CC-1 Rack
Stage Controller
Position Sensor
Electron Gun (4 × 4)
Blanking Deflectors
Rectangular Apertures (4 × 4)
Pre-Mask Deflector
CP Aperture Masks (on separated stage)
Post-Mask Deflectors
Round Apertures
Wafer Stage
MAPPER
Electron Source
Collimator lens
Aperture array
Beam Blanker Array
Beam Deflector Array
Projection lens Array
PML2
Electron Source
200x Reduction Electron Beam Optics
5 keV
50 keV
Wafer Stage
programmable Aperture Plate System
< 2 µm (with further development)
< 2 µm
200x
< 10 nm
≤ 32nm hp node
Projection Mask-Less Lithography
EUVL symposium 2008
25/20
SAMSUNG

Prospects of ML2

❖ Hurdles on its way to beat EUVL

Funding
Scalability & T/P
Data Management
Stitching & OVL

~ as of now

PML2, MAPPER,
REBL, ZPAL, MCC,
CLA, etc.

Pre-alpha Tool
POC proven @
Single Axis System

Researches based
on Funding
(Concept Proving)

~ 2010

Any Survivor ?

Alpha tool (Scaling)
Proving Multiple
Axis System

Driven by Demand
(More interest &
Funding support)

~ 2012

Final Survivor ?

Production tool
Cluster of Multiple
Axis System

Deliver for
Manufacturing
(Market Selection
based on COO)

Funding
Reliability & T/P
Stitching & OVL

SAMSUNG

EUVL symposium 2008

26/20

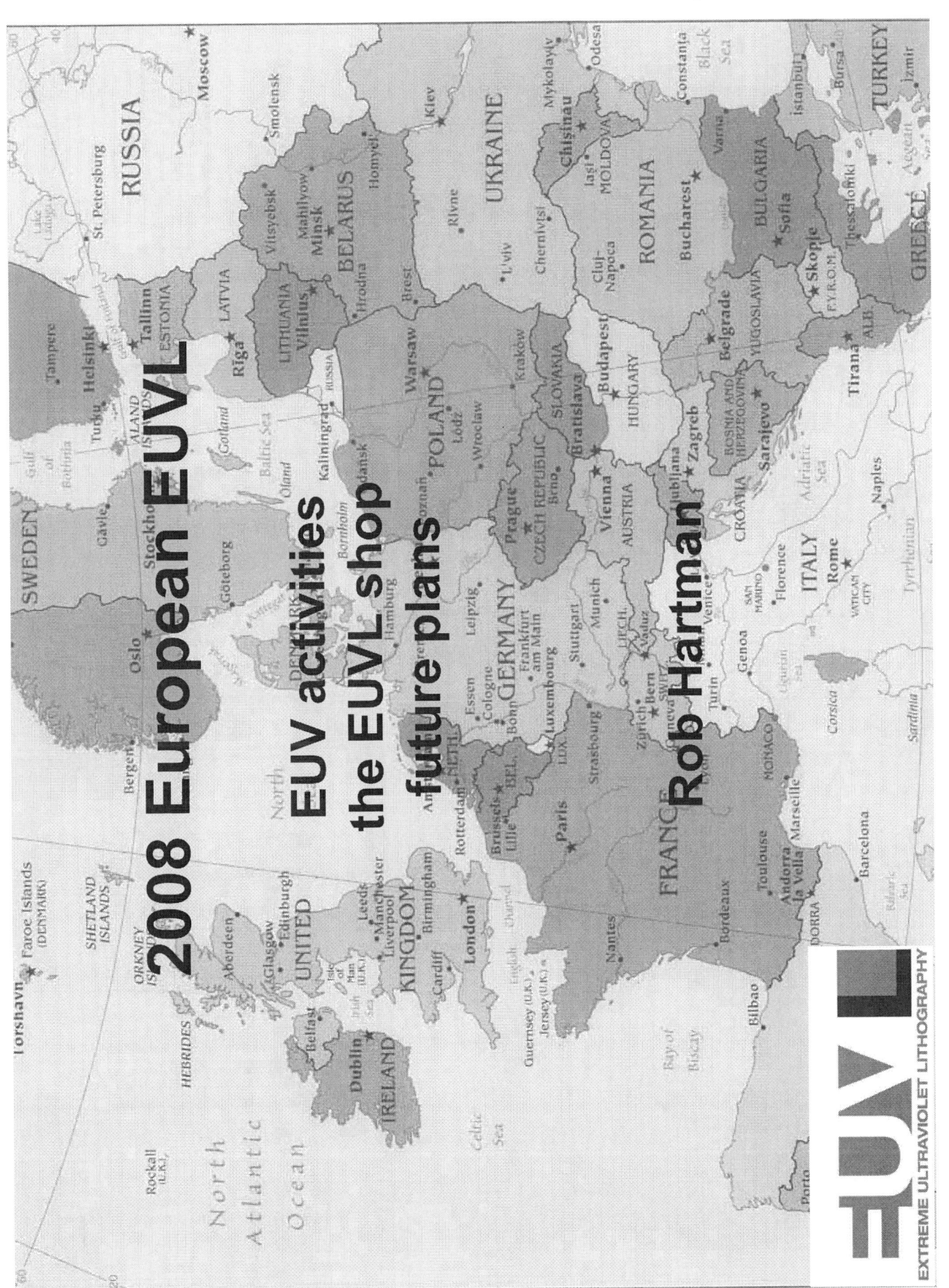

2008 European EUVL

EUV activities
the EUVL shop
future plans

Rob Hartman

EUVL
EXTREME ULTRAVIOLET LITHOGRAPHY

Co-operation in Europe

- Since 1997 started EUCLIDES: basic technology

- Followed in 2001 with MEDEA+ project cluster

- and in 2003 with a large project "more Moore" sponsored by the European Commission

- So what did we gain ???

- Come and shop !!!

EXTREME ULTRAVIOLET LITHOGRAPHY

The European EUVL SHOP

- Two Alpha Demo Tools – ASML
- HVM Collector – Media Lario Technologies
- EUV-Reflectometer – AIXUV
- Collector concept for DDP – SAGEM
- EUV quality of large and steep aspheric optics – SAGEM
- Commercial EUVL Masks – AMTC

- **and many important achievements**

EXTREME ULTRAVIOLET LITHOGRAPHY

Two full field scanning Alpha Demo tools installed

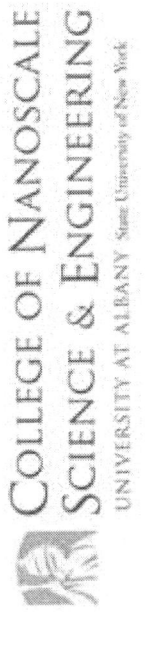

COLLEGE OF NANOSCALE
SCIENCE & ENGINEERING
UNIVERSITY AT ALBANY State University of New York

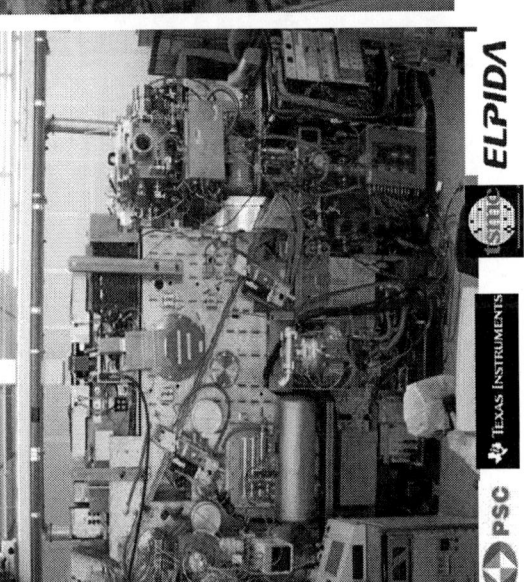

TOSHIBA Alliance

AMD Smarter Choice

- Single stage, 300mm wafers, linked to track
- Single reticle load
- Uses TWINSCAN technology (e.g. focus)
- Reflective optics
- Sn discharge source

ELPIDA

TEXAS INSTRUMENTS

PSC

SAMSUNG

- λ   13.5 nm
- NA   0.25
- Field size   26 x 33 mm^2
- Magnification 4x reduction
- Sigma   0.5

Hynix
Semiconductor

Qimonda

intel

Micron

NXP
founded by Philips

Panasonic
ideas for life

infineon

EXTREME ULTRAVIOLET LITHOGRAPHY

PHILIPS

Sn-DPP based SoCoMo enables EUV lithography

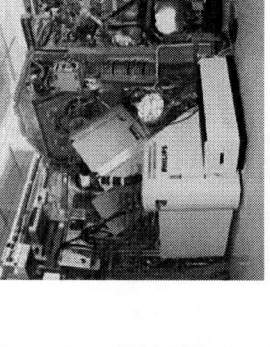

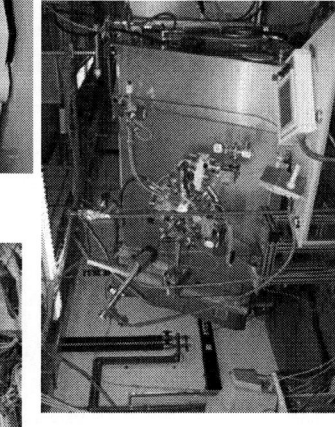

- No technical showstoppers to scale to HVM power using Philips/Xtreme Sn-DPP technology

  – Scaling with frequency and pulse energy feasible to achieve HVM requirements

- Integrated SoCoMo system based on experience from both Xtreme and Philips with complete Alpha systems

  Source + Debris Mitigation + Collector module

- Philips/Xtreme Sn-DPP SoCoMo development is already in Beta phase: Beta source is alive and being boosted to Beta power levels

EUV
EXTREME ULTRAVIOLET LITHOGRAPHY

102

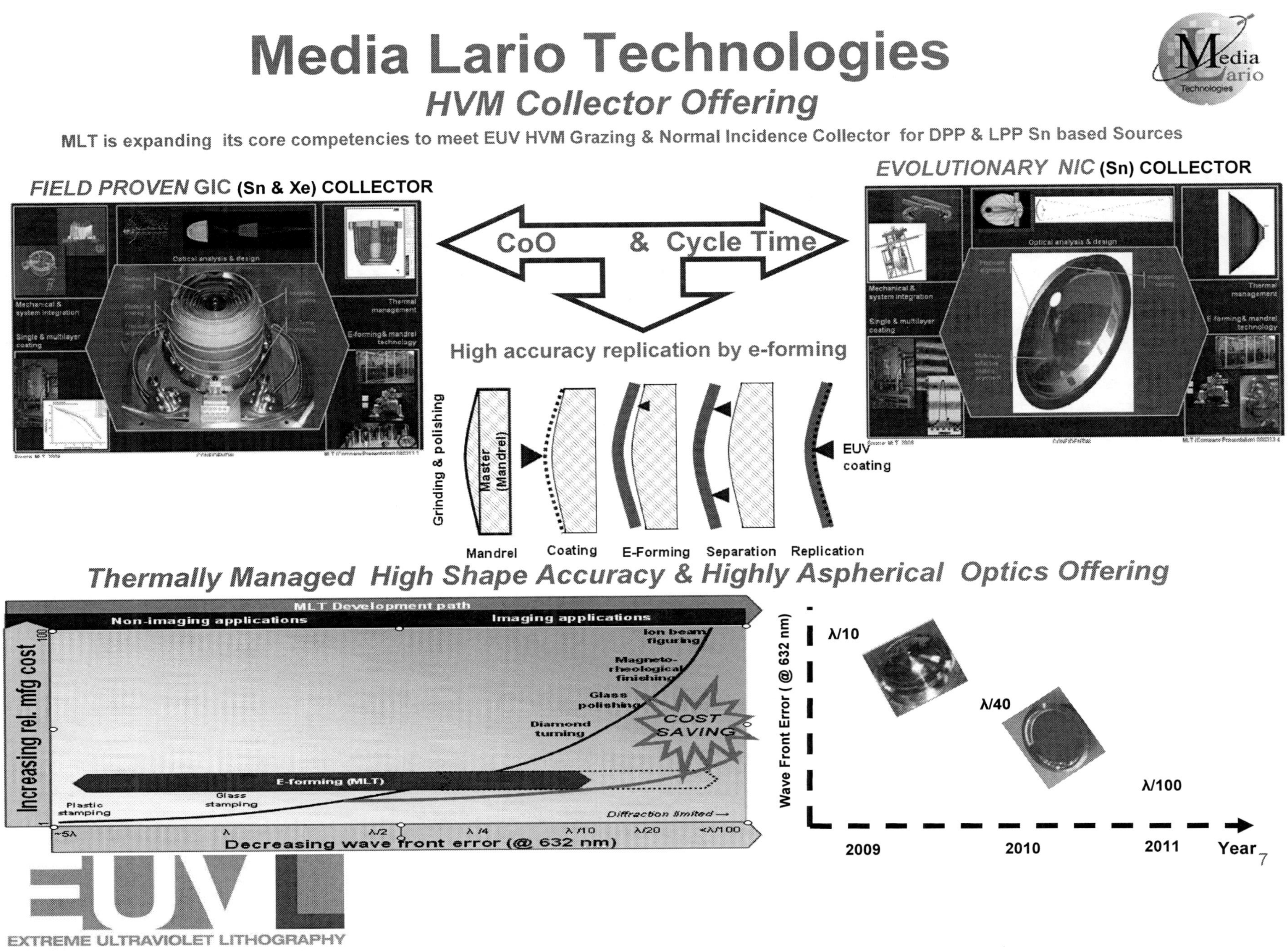

Media Lario Technologies

HVM Collector Offering

MLT is expanding its core competencies to meet EUV HVM Grazing & Normal Incidence Collector for DPP & LPP Sn based Sources

EVOLUTIONARY NIC (Sn) COLLECTOR

FIELD PROVEN GIC (Sn & Xe) COLLECTOR

CoO & Cycle Time

High accuracy replication by e-forming

Grinding & polishing — Mandrel (Master) — Coating — E-Forming — Separation — Replication

EUV coating

Thermally Managed High Shape Accuracy & Highly Aspherical Optics Offering

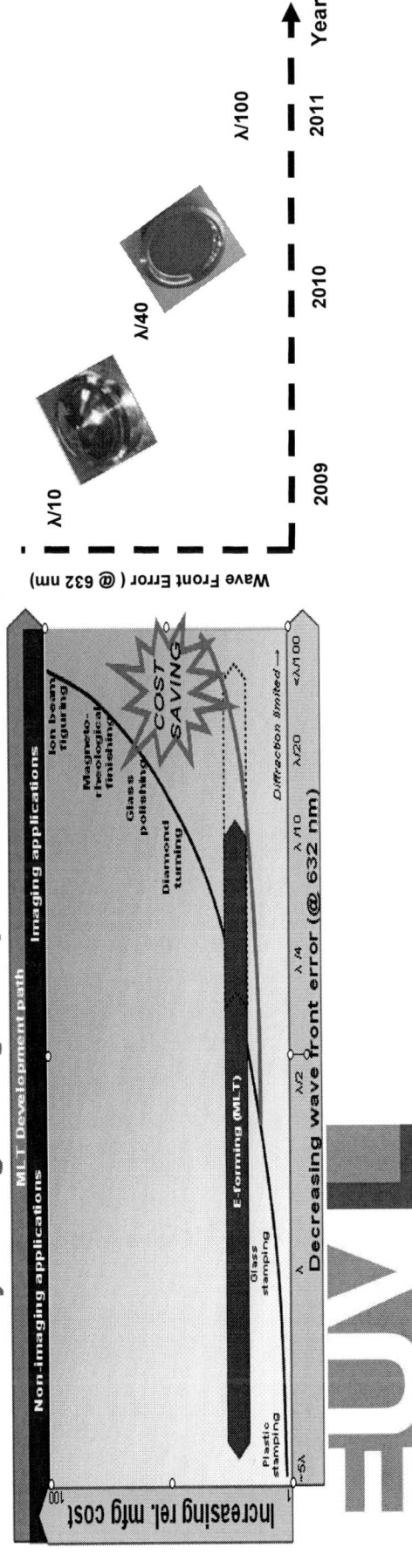

EUV EXTREME ULTRAVIOLET LITHOGRAPHY

EUV-Reflectometer for EUV Masks and Mask Blanks (EUV-MBR)

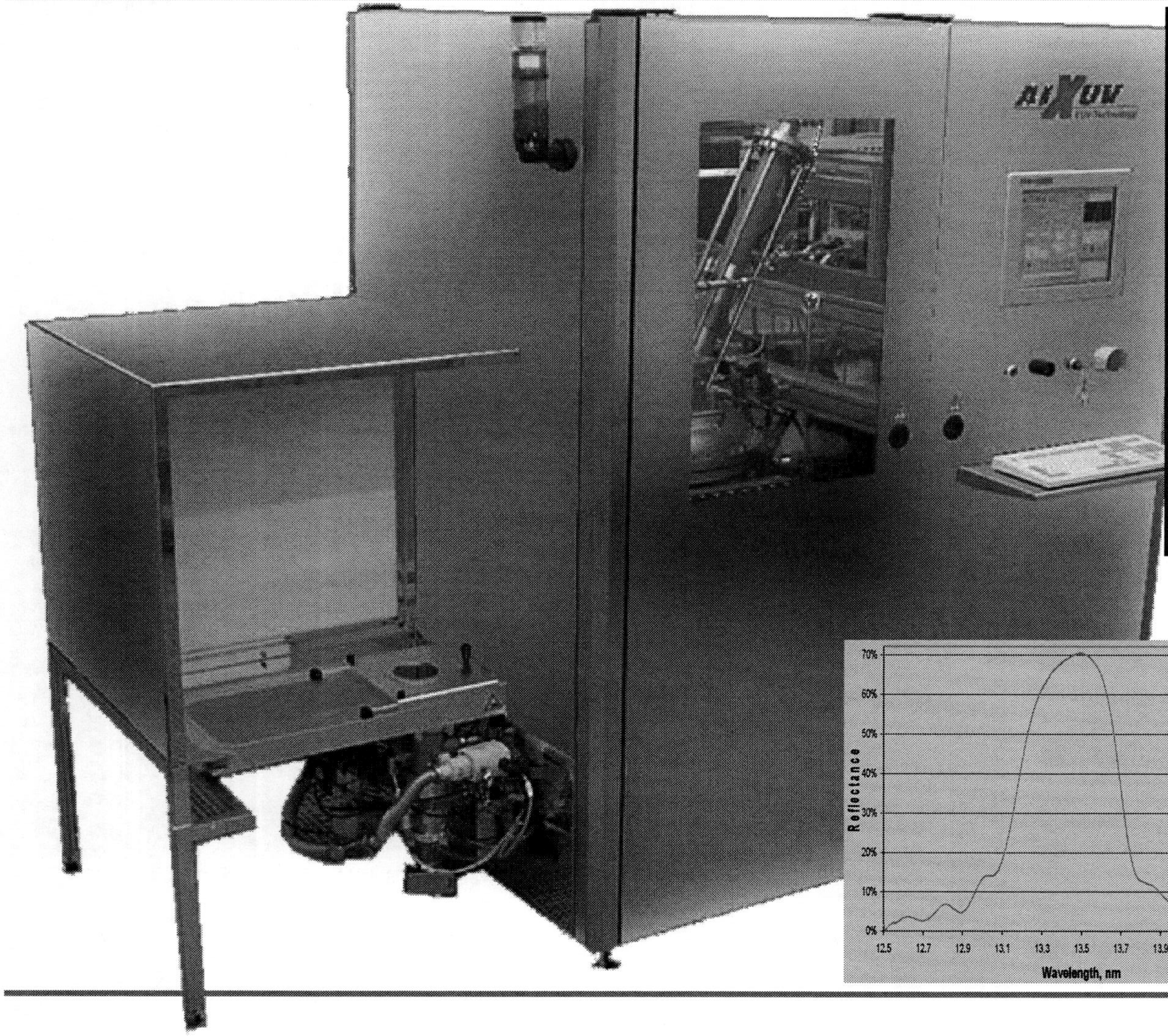

Feature		Performance
Spot Size		< 50*500 µm²
Spectral channel		1.6 pm
Measurement of 2000 channels		< 30 seconds
CWL_50		
	Precision	< 3 pm 3σ
	Accuracy	< 5 pm 3σ
Reflectivity on reflector R > 60%		
	Precision	0.3% 3σ
	Accuracy	0.5% 3σ
Reflectivity on Absorber R < 2%		
	Precision	0.05 % 3σ
	Accuracy	0.1 % 3σ

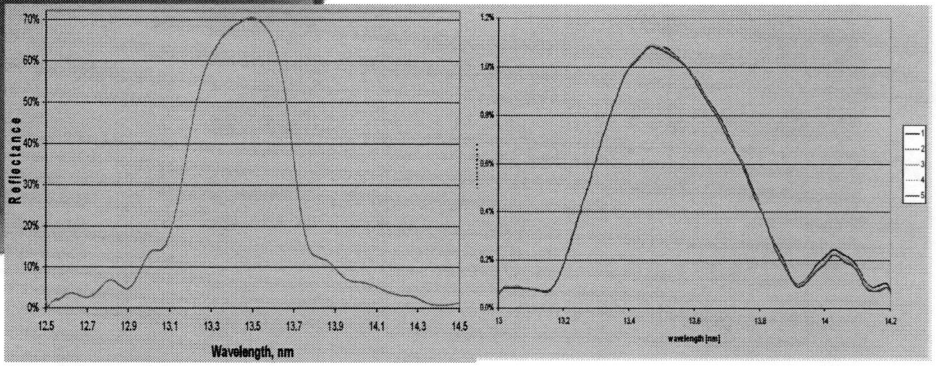

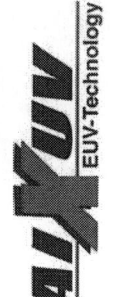

One tool is made available for routine use in Aachen

AIXUV EUV-Technology

Sensitivity to small changes
(out of center measurements)

Peak reflectance = 70.37 % ± 0.08 % (3σ)
Central wavelength = 13.4371 nm ± 0.4 pm
Wavelength of peak reflection = 13.4937 nm ± 0.8 pm
FWHM = 0.5309 nm ± 0.4 pm

ΔR = 0.06 %
Δλ = 18 pm

ΔR = 0.1 %
Δλ = 10 pm

ΔR = 0.55 %
Δλ = 2 pm

Precision and sensitivity also on absorbers

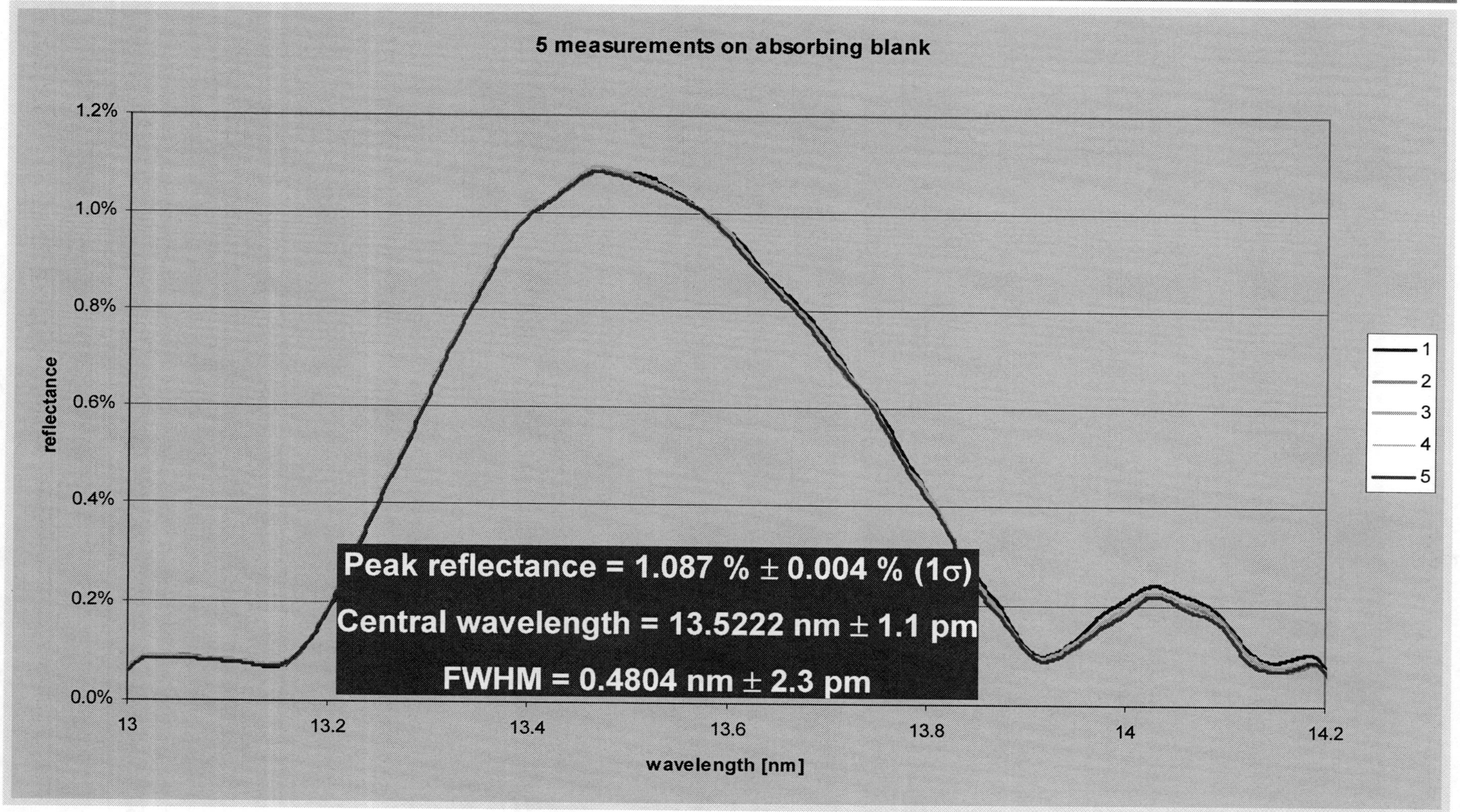

5 measurements on absorbing blank

Peak reflectance = 1.087 % ± 0.004 % (1σ)
Central wavelength = 13.5222 nm ± 1.1 pm
FWHM = 0.4804 nm ± 2.3 pm

EUV-Sources for Metrology

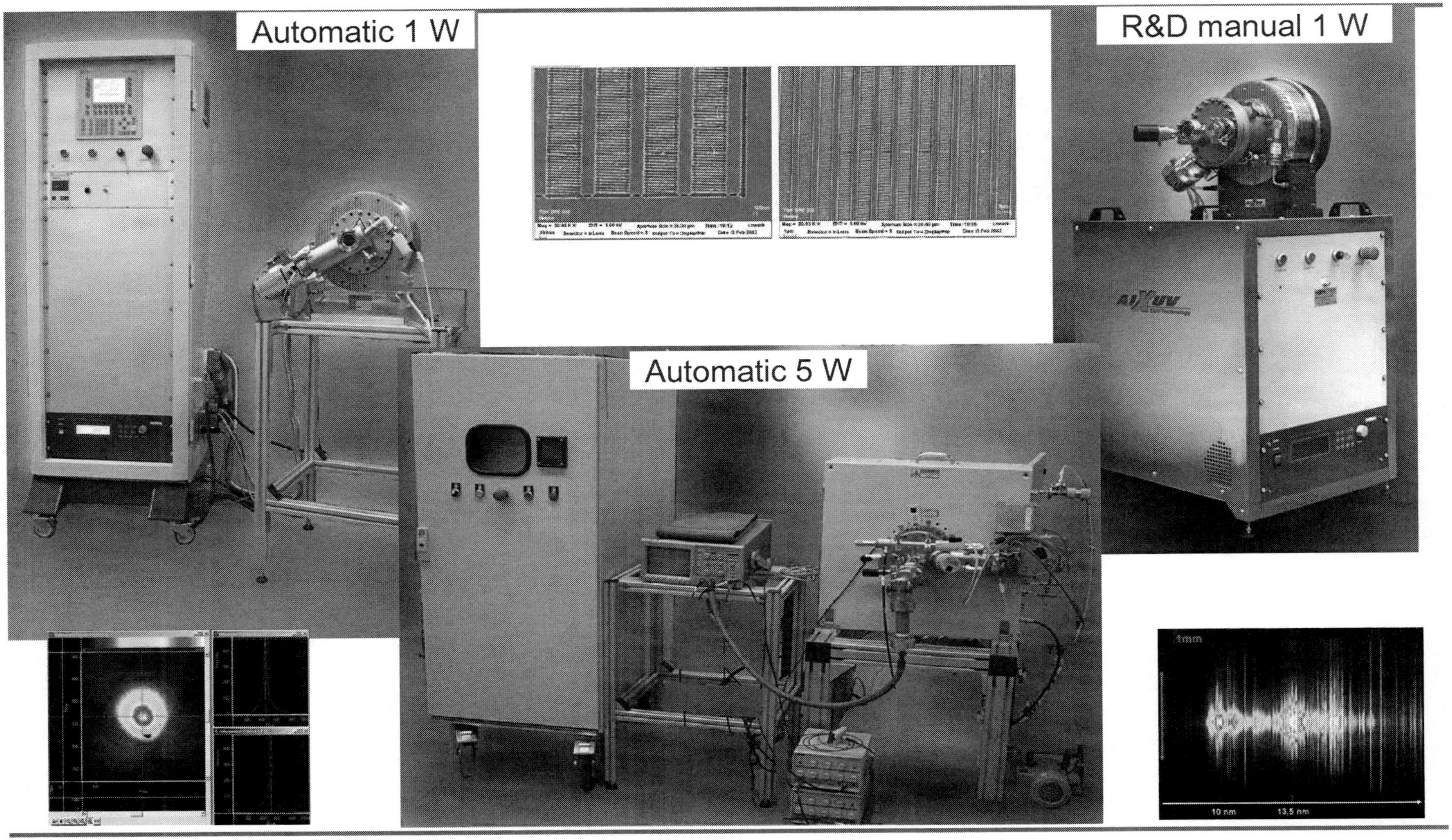

Automatic 1 W
R&D manual 1 W
Automatic 5 W
10 nm
13,5 nm
1mm

AIXUV EUV-Technology

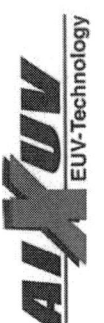

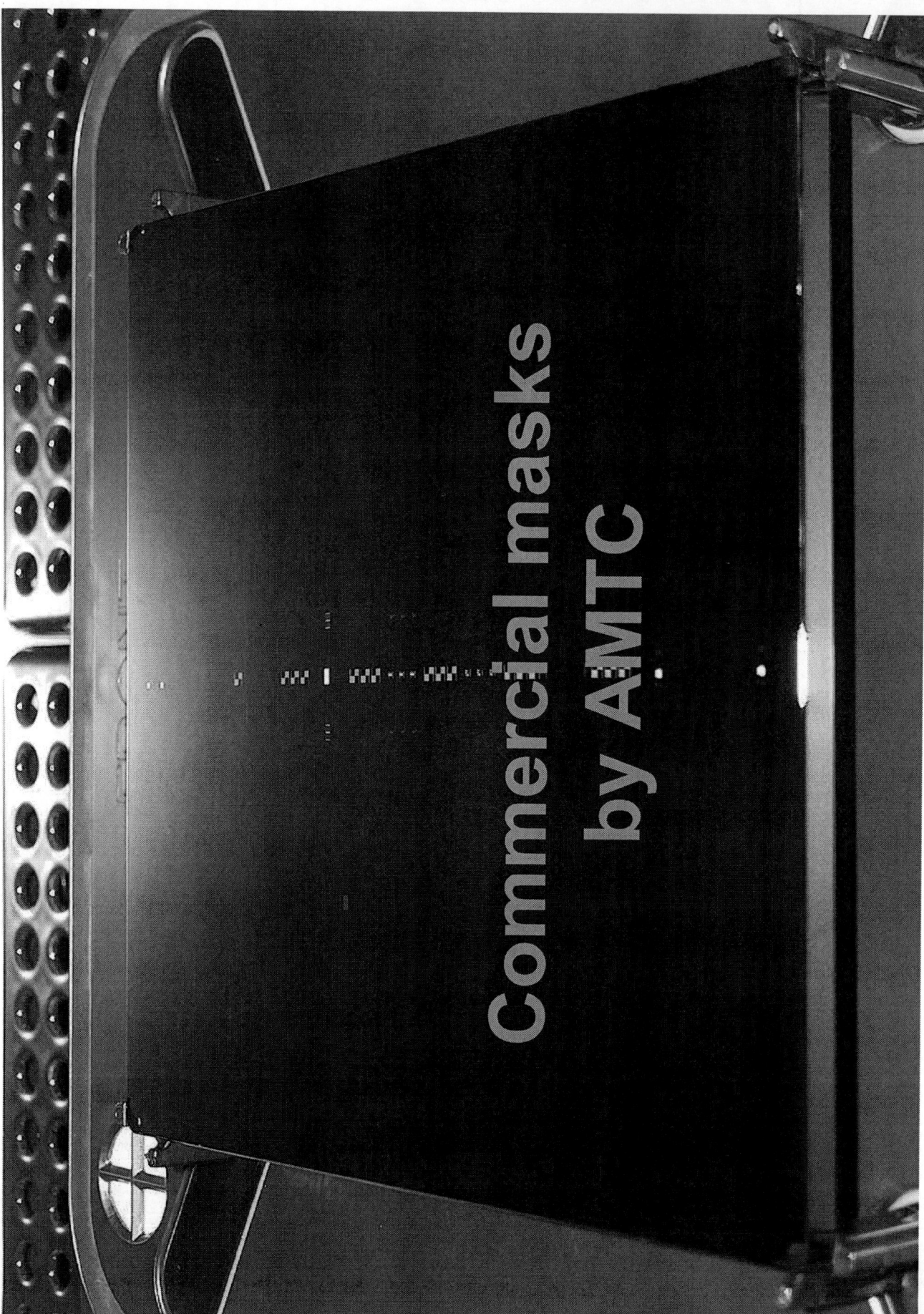

Commercial masks by AMTC

AMTC / TPI Dresden EUV Mask Status

- The Dresden campus has a fully-integrated commercial EUV mask fabrication line established and accepting orders via Toppan Photomask.

- Many deliveries already made over several years to various users of the full-field EUV exposure systems including test plates and real product demonstrations.

- All processes are established in Dresden including inspection with die-to-die capability to P72 pixel size and repair by nano-machining and e-beam.

- Defect printability studies to be published in the coming 3 weeks at the EUV Symposium and SPIE BACUS including SRAM electrical characterization and mask – wafer defect correlation on a 45nm node product design.

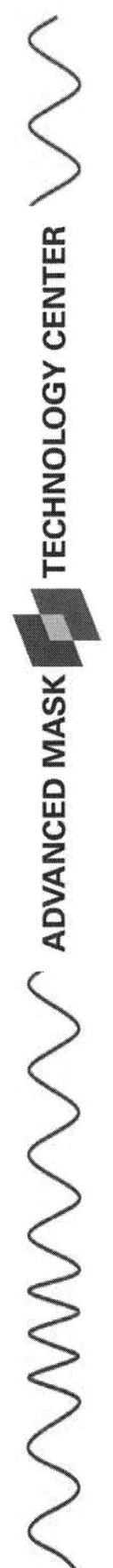

ADVANCED MASK TECHNOLOGY CENTER

Examples of EUV Mask Defects

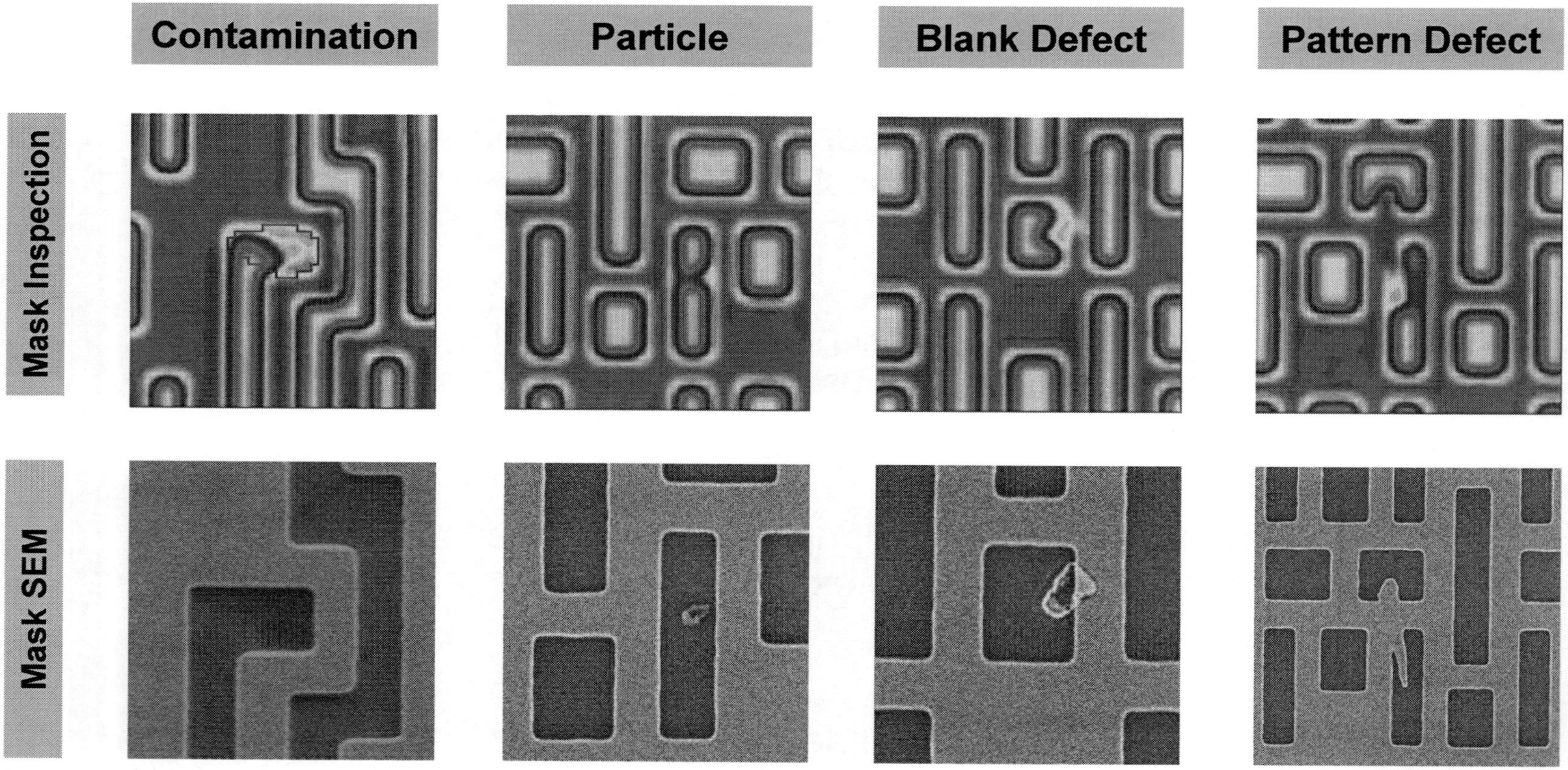

Contamination	Particle	Blank Defect	Pattern Defect

Mask Inspection

Mask SEM

- Defect types are difficult to differentiate from inspection image alone

ADVANCED MASK TECHNOLOGY CENTER

/3100 coatings in full progress @ FOM

- Data PTB, Berlin
- Local angle of incidence (s-pol)

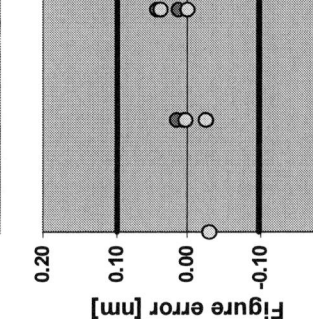

Reflectance vs Wavelength (nm)

R = 69.6%
s-polarized

Y-axis: Reflectance (0.0% – 70.0%)
X-axis: Wavelength (nm): 12.75, 13, 13.25, 13.5, 13.75, 14, 14.25

Figure error [nm] vs Radial distance [mm]
- X-direction
- Y-direction

Y-axis: Figure error [nm]: -0.20, -0.10, 0.00, 0.10, 0.20
X-axis: Radial distance [mm]: 0, 10, 20, 30, 40, 50, 60

Results /3100 optics depositions

- >69% reflectivity including improved capping layer and diffusion barriers
- Total coating stack non correctable thickness error ≤30 pm rms

Resist performance on ASML ADT
NA=0.25, σ=0.5

Dense and iso line resolution

40nm

CD DL = 38.0nm
CD ISO = 34.1nm

35nm

CD DL = 36.1nm
CD ISO = 30.4nm

32nm

Reference resist
Fuji FEVS-P1101
11.0mJ/cm2
LER=5.5nm

CD DL = 30nm
CD ISO = 21.5nm

X-sectional profiles of advanced resists

Resist A

29.5nJ/cm2
LER =3.8nm

40L80P

35L70P

32L64P

Resist B

14.2mJ/cm2
LER =4.8nm

40L80P

35L70P

32L64P

ASML ADT at IMEC

35nm 1:1

28nm 1:1

32nm SRAM device with EUV Lithography on contact level

Lithography : Exposure latitude

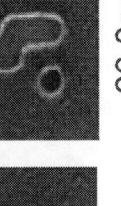
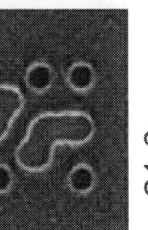
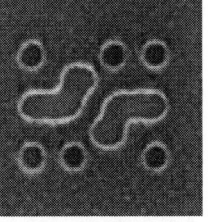

11mJ/cm^2	13 mJ/cm^2	15 mJ/cm^2	17 mJ/cm^2
50.2nm	57.4nm	61.6nm	63.0nm

After Oxide etch

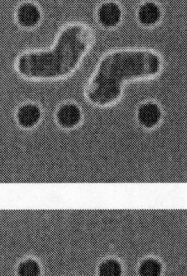
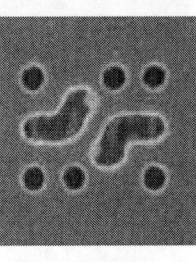
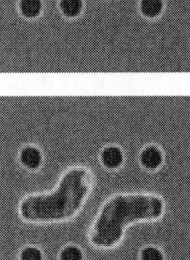
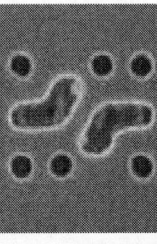
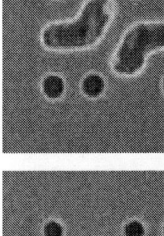

53.4nm	60.4nm	63.7nm	64.5nm

Design

Active fin
Poly fin
Contact

Reticle SEM

Simulated resist image

Electrically functional 0.186μm^2 32nm node SRAM cells demonstrated using EUVL for contact hole patterning

imec

aspire invent achieve

imec

intel

Micron

Panasonic
ideas for life

qimonda

SAMSUNG

NXP
founded by Philips

ELPIDA

hynix
Semiconductor

PSC

infineon

TEXAS INSTRUMENTS

The European EUVL plans

- new co-operative project under CATRENE

- CATRENE, an EUREKA Project, is the successor of MEDEA+

- Project is named EXEPT

EXEPT: 1350 manyears

EXEPT project is to develop technologies, tools and infrastructure components as required for high volume EUV lithography for 22nm hp structures in 2012.

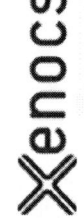

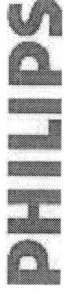

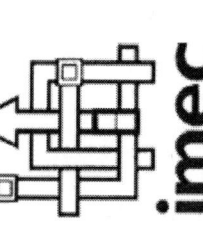

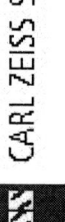

ASML EUVL Roadmap
NXE platform supports manufacturing down to 11nm

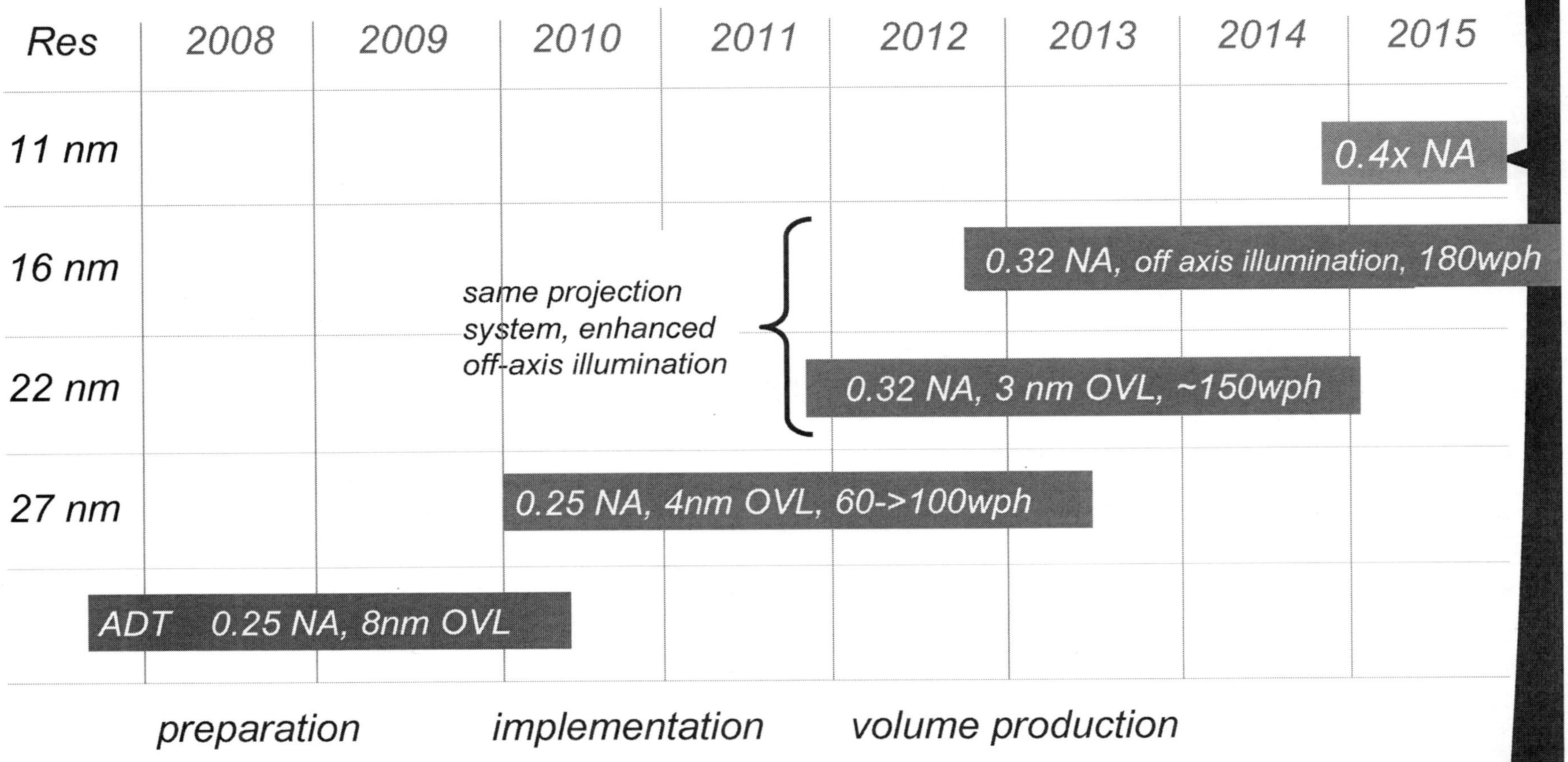

Res	2008	2009	2010	2011	2012	2013	2014	2015
11 nm								0.4x NA
16 nm					0.32 NA, off axis illumination, 180wph			
22 nm						0.32 NA, 3 nm OVL, ~150wph		
27 nm			0.25 NA, 4nm OVL, 60->100wph					
ADT	0.25 NA, 8nm OVL							

same projection system, enhanced off-axis illumination

preparation   implementation   volume production

ASML

First production system design is finished
and module manufacturing ongoing

adixen
by Alcatel Vacuum Technology

Alcatel Vacuum activities for EUV

instrumentation

- Advanced process & equipment monitoring
- Leak detection

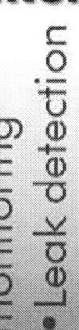

Modeling activities & basic studies

vacuum & advanced process solutions

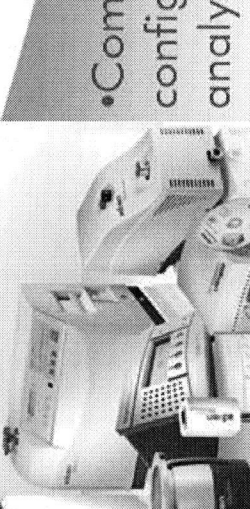

- Complete configuration & analysis

- Vacuum systems : performant , robust & ultra clean

environmental solutions

- Vacuum transport
- Contamination & AMC control in current pods (real time)
- decontaminati on system

ed © 2004, Alcatel

ALCAT

EUVL Cleaning Challenges

Drivers:

- Absence of protective pellicle
- Vulnerable surface material of sophisticated dielectric mirrors
- Fragile pattern structures (22nm/16nm)
- Exposure under vacuum

Requirements:

- 100% removal of sub-20nm particles
- Residue–free cleaning to avoid outgassing
- Elimination CD-line width and overlay shift
  - Maintain mechanical properties
  - Protect mask capping layer
  - Preserve pattern integrity

Traditional isolated cleaning methodologies are no longer effective to eliminate contamination while maximizing MTBC and the lifetime of the reticle.

Understand. Align. Innovate. Develop.

A MEMBER OF THE SINGULUS GROUP

HamaTech EUVL Progress

Understand. Align. Innovate. Develop.

- Extended proven 32/22nm cleaning technology to EUVL development

- Successfully completed EUV blank cleaning milestones at Sematech

- Achieved particle-free EUV blanks down to 30nm

- Demonstrated technical achievement of 10nm defect removal

- Developed critical specific-area cleaning techniques

- Established global network of development partnerships

A MEMBER OF THE SINGULUS GROUP

EUV Mask Cleaning Development

Understand. Align. Innovate. Develop.

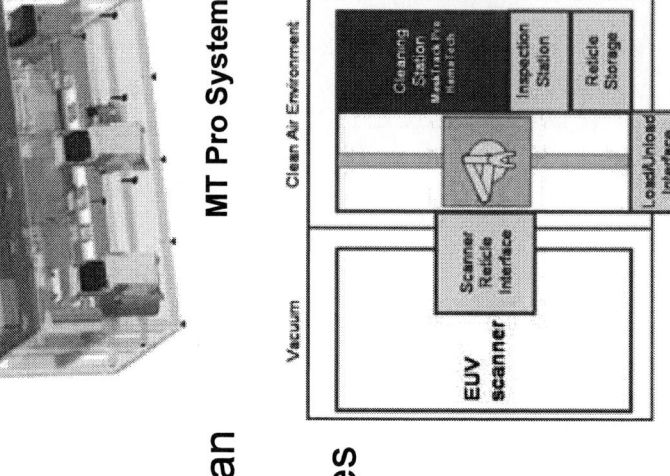

MT Pro System

In-Litho Concept

- Launch of MaskTrack *Pro*™ 1H09 (EUVL and 22nm)
- Develop innovative advanced cleaning concepts
  - New physical and chemical methods
  - Targeted Spot Cleaning (TSC)
  - Dry clean technology development
- Establish holistic mask management approach
  - Expand cleaning process to include storage, clean and metrology
  - On-demand cleaning of specific areas or particles
  - MaskTrack *Pro*™ design allows clustering
  - Aligned with development partners for
    - Handling/Storage
    - Metrology (co-operation with OEMs)
  - Standalone: in-fab concept
  - Integrated: in-litho concept
    - scanner, clean, inspection and storage

HAMATECH APE
A MEMBER OF THE SINGULUS GROUP

EUVL:

EXEPT: 1350 analysis

EXEPT project is to develop technologies, tools and infrastructure components as required for high volume EUV lithography for 22nm structures in 2012

IT IS HAPPENING in EUROPE

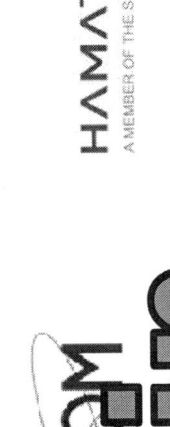

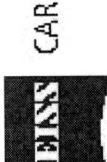

EXTREME ULTRAVIOLET LITHOGRAPHY

US Regional Update

**7th International EUVL Symposium
Lake Tahoe, CA
September 30, 2008**

Patrick Naulleau, Lawrence Berkeley National Laboratory

EUV Activity in the U.S.

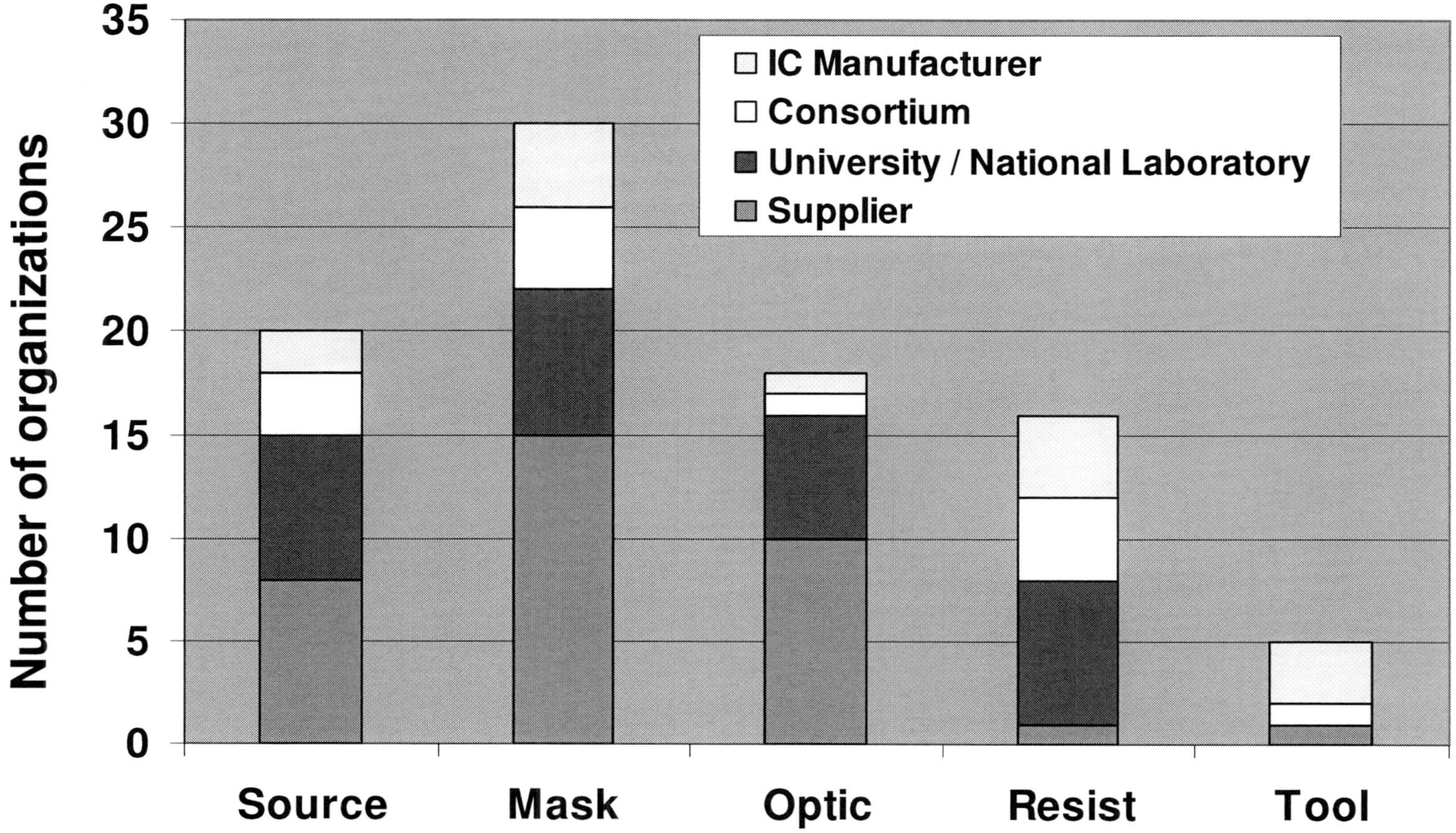

BERKELEY LAB

Number of organizations

Legend:
- □ IC Manufacturer
- □ Consortium
- ■ University / National Laboratory
- ▨ Supplier

Categories: Source, Mask, Optic, Resist, Tool

❖ **Strongest US efforts overall are in source and mask areas; IC manufacturers focus on process development – mask and resist**

LPP EUV Source Advancing from Development to Product

- Laser Produced Plasma EUV product development continues on schedule
- Multiple 5 sr collection optics in the manufacturing process
- Effective debris mitigation and stable power over 8 hrs on 320mm (1.6sr) collector

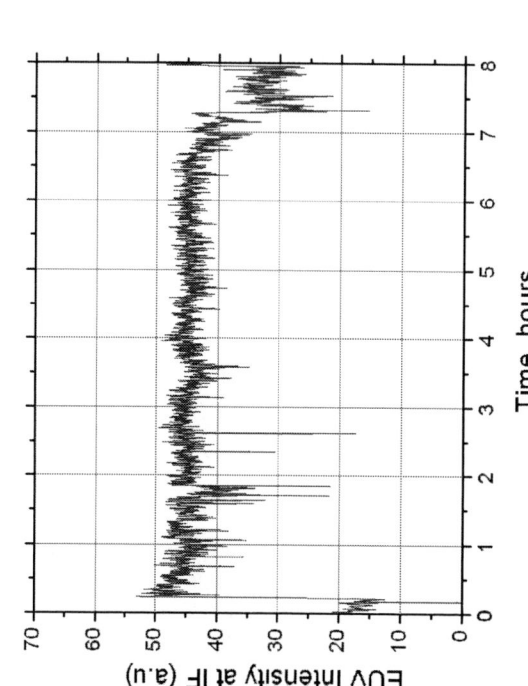

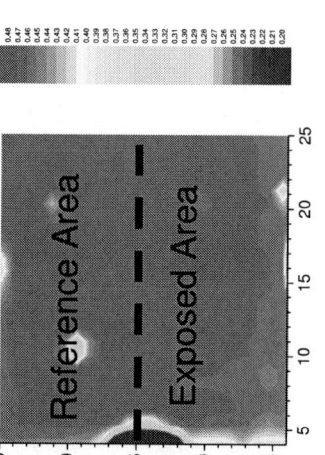

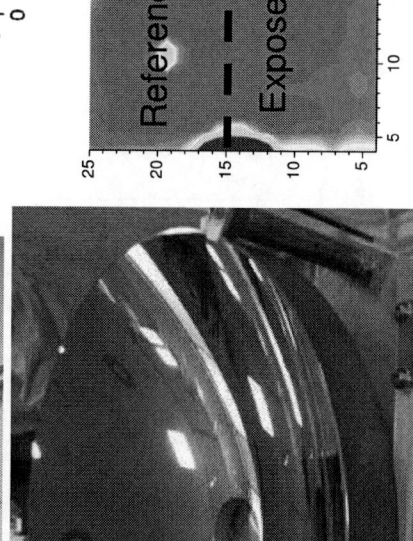

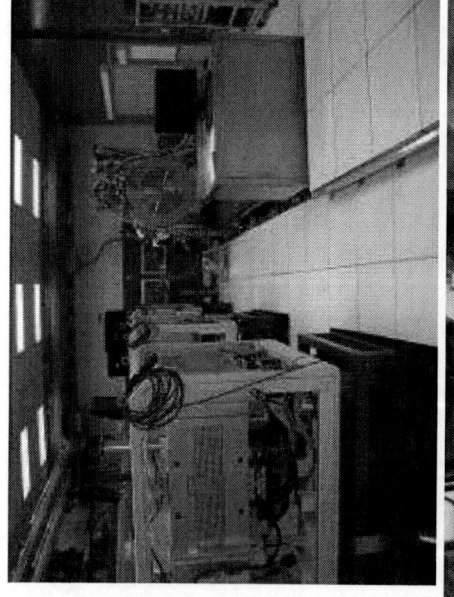

Courtesy of David Brandt, Cymer

EUV Uniformity and Collector Reflectivity over 8 Hours

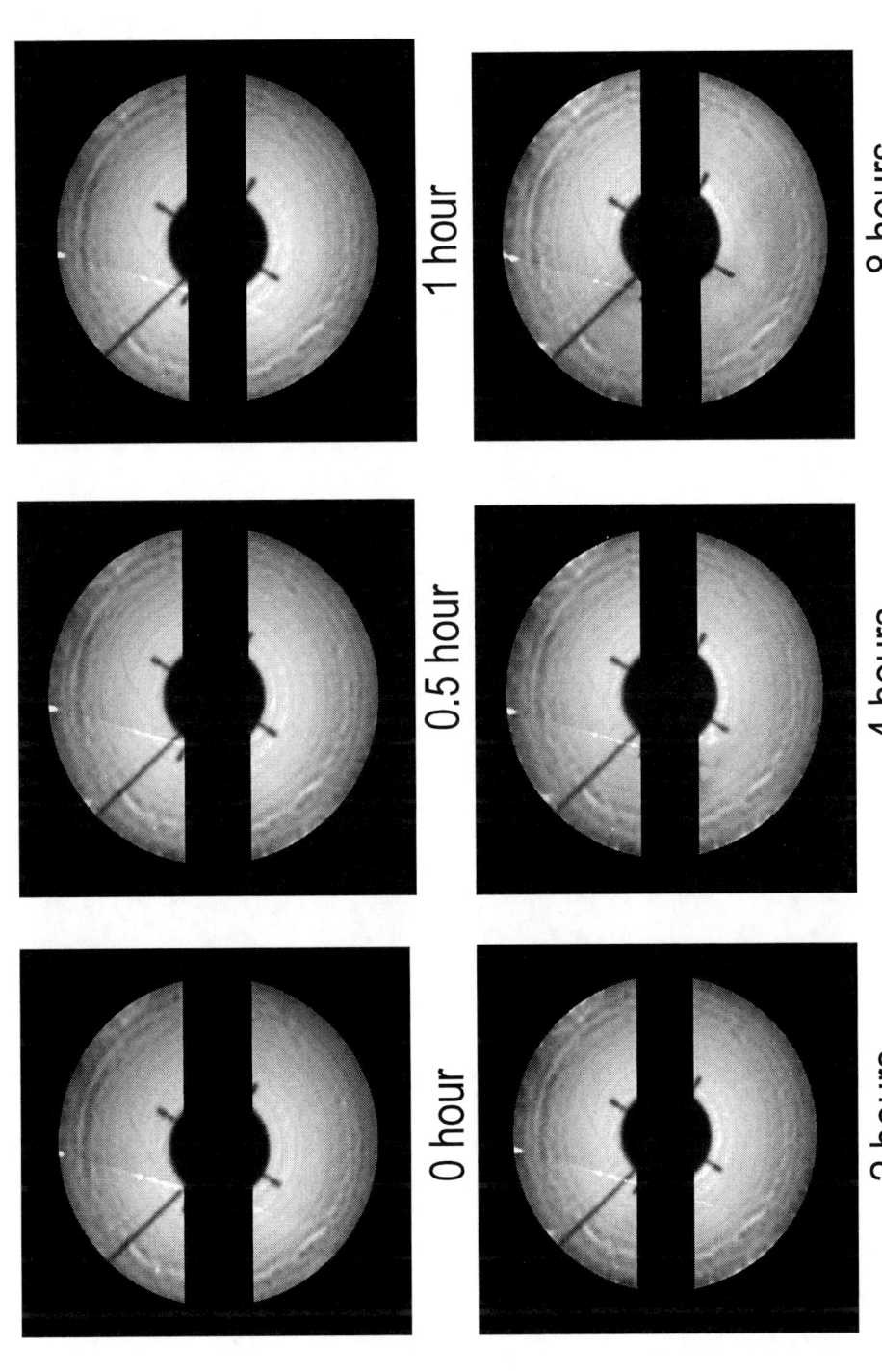

320mm Diameter (1.6sr) Collector, EUV Images taken in the Far Field with Fluorescence Converter

Courtesy of David Brandt, Cymer

Chemically amplified resists with 22-nm half pitch resolution demonstrated

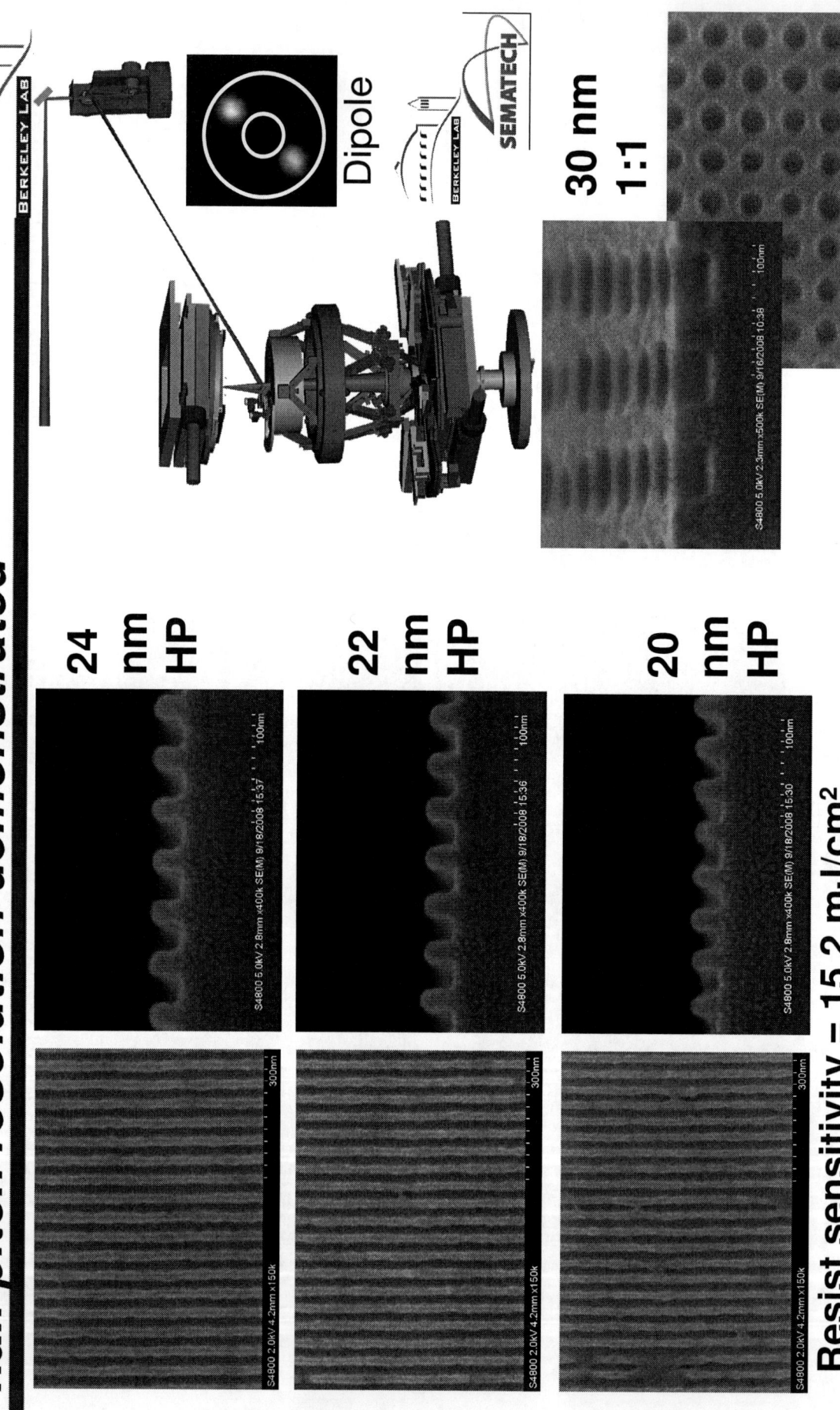

Dipole

30 nm
1:1

24 nm HP

22 nm HP

20 nm HP

Resist sensitivity = 15.2 mJ/cm²

SEM images courtesy of C. Koh, SEMATECH

Rinse Agent Improves EUVL Manufacturability

30 nm HP	1H'08 "Smooth" RLS Champs		"Fast" RLS Champion	
	Resist A	Resist B	Resist C	Resist C + Rinse Agent
MIN LWR (nm)	3.8	4.3	5.3 → 3.9 (26%↓)	
DOSE (mJ/ cm²)	14.0	11.0	8.7 → 8.7 (38%↓)	
Z-Factor[1]	2.2 E-08	2.4 E-08	3.3 E-08	1.8 E-08

New Champ!

- Pre- and/or post-processing techniques can significantly improve roughness of high resolution / fast EUV resist.

↪ Focus on continued material evolution for improved pattering and LER/ LWR reduction.

[1]Wallow, et al SPIE 69211F (2008)

(intel)

Courtesy of Gil Vandentop

Pit smoothing by cleaning

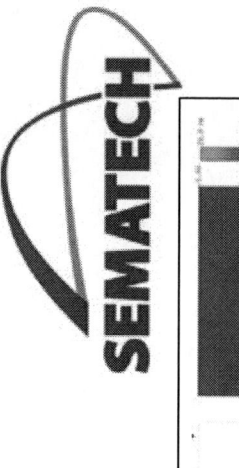

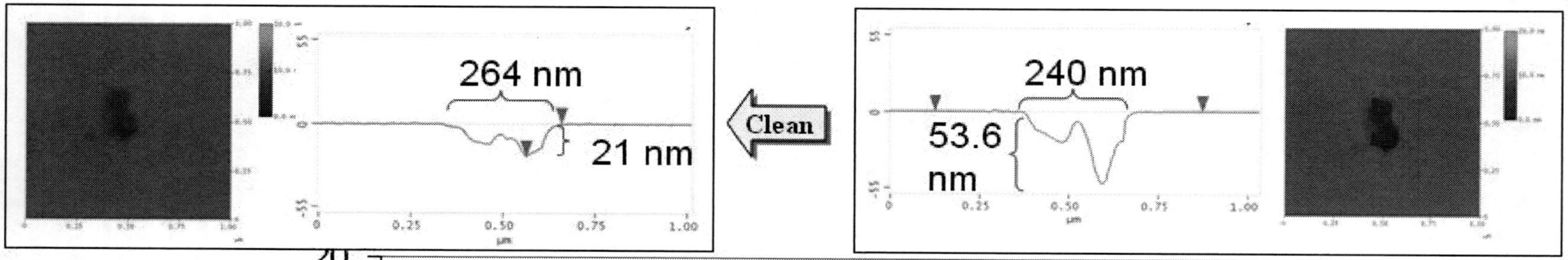

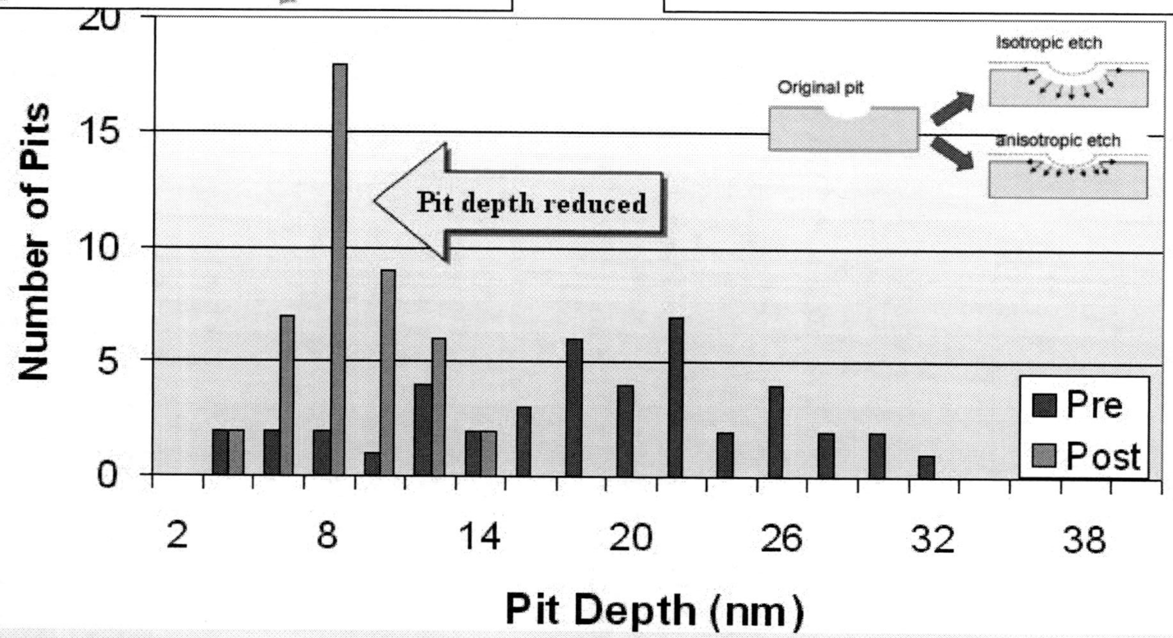

240 nm

53.6 nm

264 nm

21 nm

Clean

Original pit

Isotropic etch

Anisotropic etch

Pit depth reduced

Number of Pits

Pit Depth (nm)

Pre
Post

Courtesy of
A. Rastegar,
SEMATECH

SEMATECH has developed new cleaning processes
demonstrating reduction in pit depth distribution

25 September 2008

EUV mask patterning capability reaches 60nm (4x)

- *Enable lithography development at 22 nm node and beyond*

Mask fabrication in the pilot line

EUV Mask

EUV Blank

1) TaN based absorber
2) 2.5nm Ru cap
3) 40 pairs of Mo/Si multilayer
4) LTEM
5) Conductive layer

E-beam patterning
(250nm resist)

80nm   70nm   60nm

ICP dry etch
(87nm TaN based absorber)

80nm   70nm   60nm

Courtesy of Guojing Zhang, Intel

Integration of EUVL in semiconductor fab flow

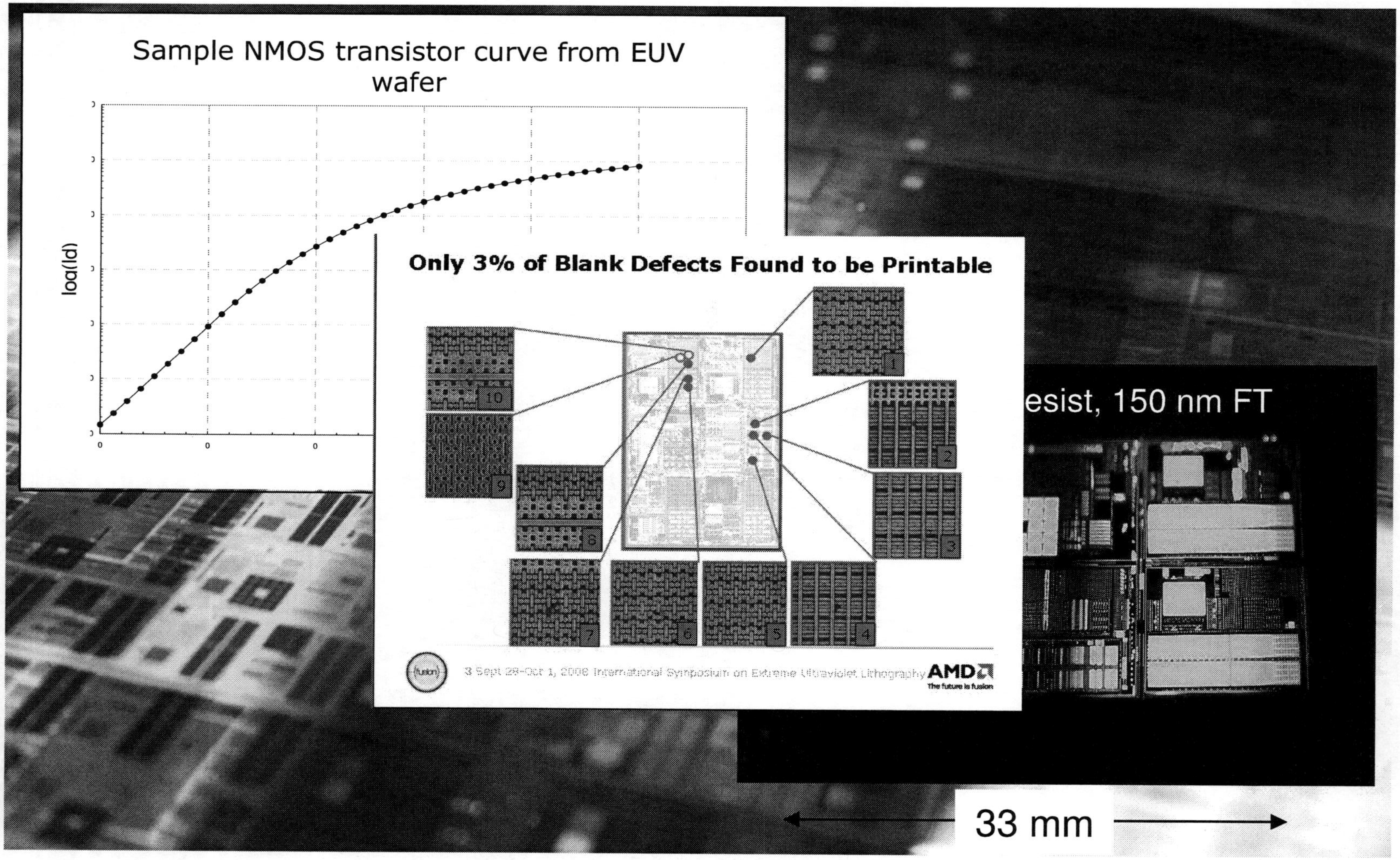

Sample NMOS transistor curve from EUV wafer

log(Id)

esist, 150 nm FT

33 mm

Only 3% of Blank Defects Found to be Printable

TOSHIBA ❖ ASML   IBM®   Courtesy of Bruno La Fontaine   AMD

Demonstration of high yield SRAM at EUV

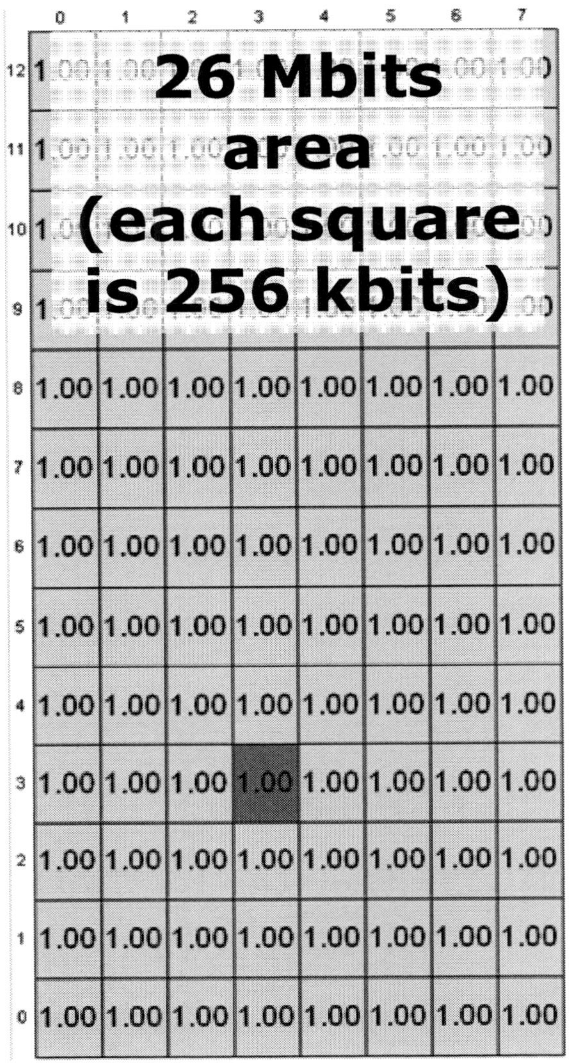

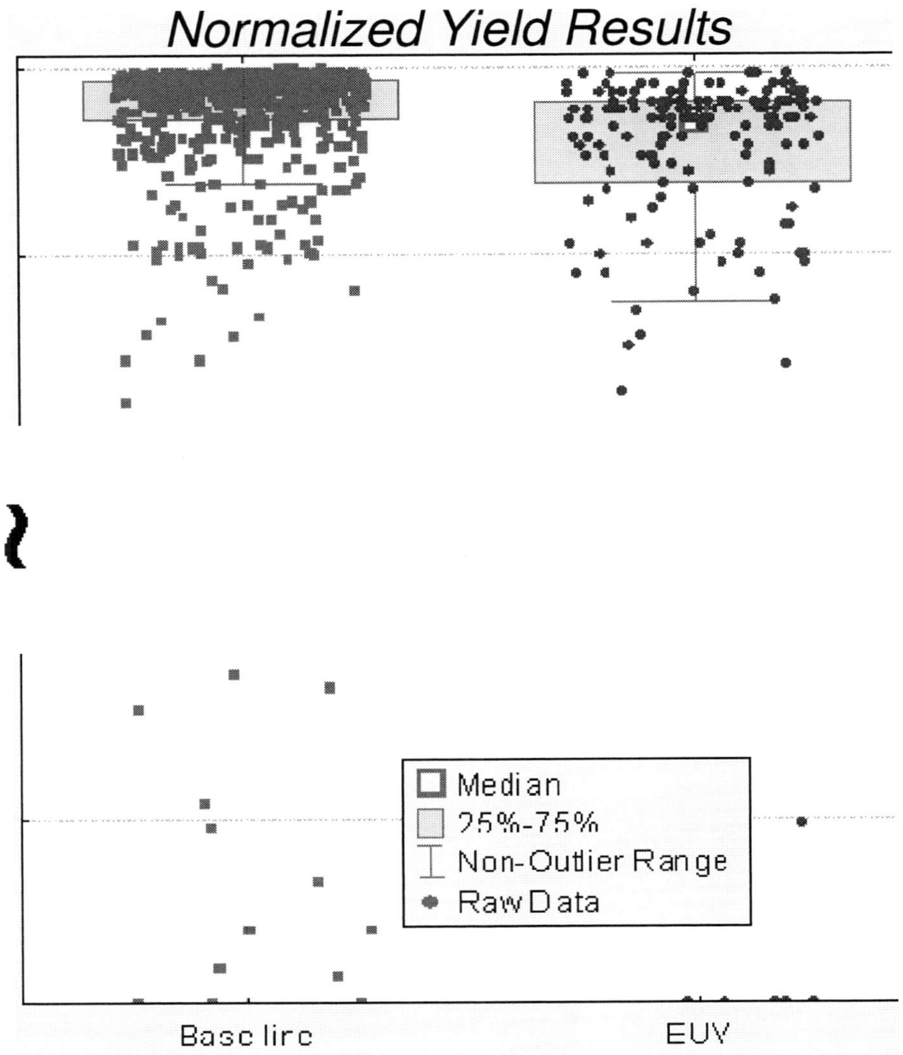

Courtesy of Anna Tchikoulaeva

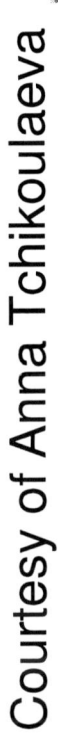

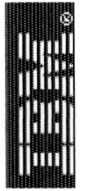

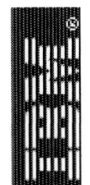

Summary

- Source/collector
  - 1MJ Energy produced at IF in 24 hour period
  - Effective debris mitigation for Sn deposition and ion erosion, long-term operation with integrated collector

- Resist
  - 22-nm HP resolution demonstrated
  - Effective LER smoothing methods demonstrated

- Mask
  - Feasibility of fast pit smoothing technique demonstrated
  - 60-nm mask CD achieved

- Integration
  - EUVL integrated into semiconductor fab flow
  - SRAM demonstrated with high yield potential

Acknowledgements

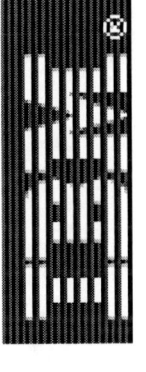

Advanced Materials Research Center, AMRC, International SEMATECH Manufacturing Initiative, and ISMI are servicemarks of SEMATECH, Inc. SEMATECH, the SEMATECH logo are registered servicemarks of SEMATECH, Inc. All other servicemarks and trademarks are the property of their respective owners.

IEUVI Update

Paolo Gargini

Chairman ITRS

Chairman Technology Strategy Committee, SIA

Director of Technology Strategy

Intel Fellow

IEEE Fellow

Outline

- Background and ITRS Lithography
- IEUVI Overview
- Optics TWG
- Resist TWG
- Mask TWG
- Source TWG
- Summary

Outline

- **Background and ITRS Lithography**
- IEUVI Overview
- Optics TWG
- Resist TWG
- Mask TWG
- Source TWG
- Summary

2008 Litho ITRS

Lithography iTWG - Meeting
July 2008

Litho Potential Solutions

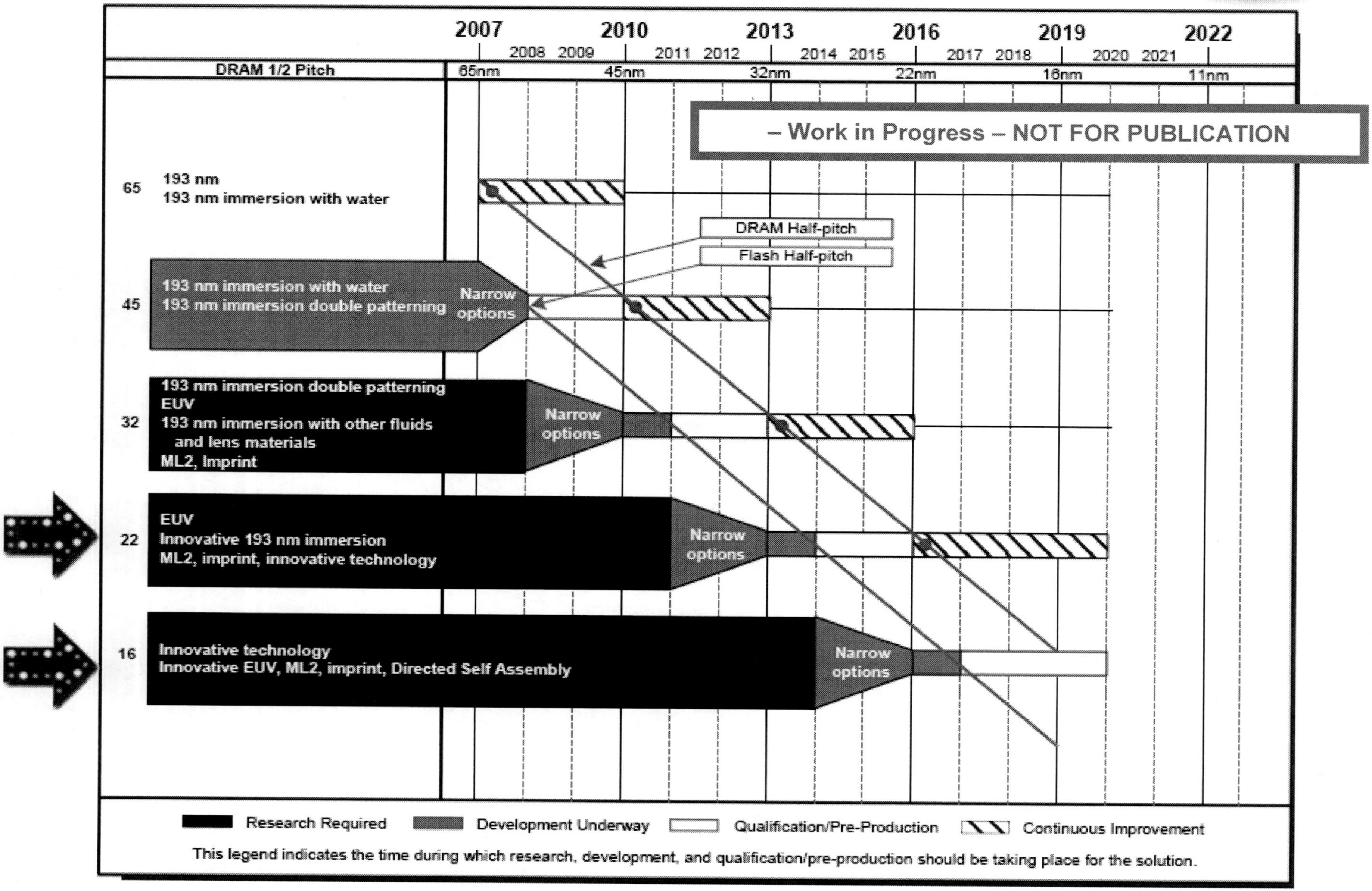

-- Work in Progress -- NOT FOR PUBLICATION

DRAM 1/2 Pitch	2007	2008	2009	2010	2011	2012	2013	2014	2015	2016	2017	2018	2019	2020	2021	2022
	65nm			45nm			32nm			22nm			16nm			11nm

65 193 nm
193 nm immersion with water

45 193 nm immersion with water
193 nm immersion double patterning

32 193 nm immersion double patterning
EUV
193 nm immersion with other fluids and lens materials
ML2, imprint

22 EUV
Innovative 193 nm immersion
ML2, imprint, innovative technology

16 Innovative technology
Innovative EUV, ML2, imprint, Directed Self Assembly

DRAM Half-pitch
Flash Half-pitch

Narrow options

Research Required
Development Underway
Qualification/Pre-Production
Continuous Improvement

This legend indicates the time during which research, development, and qualification/pre-production should be taking place for the solution.

Potential Show Stoppers (7/08)

- Resist Limitations (Shot Noise, Photo electron Blur), e.g. set upper limits on LER & Resolution.

- Cost of Ownership – all technologies

- EUV Lithography
  - Source availability at required power levels and life time.
  - Extendibility
  - Mask Defects

Conclusions

- Update Lithographic Requirements
  - Separate into DRAM, Flash, MPU
- Double patterning (3 types of approaches)
  - Double Exposure
  - DP of Uncorrelated Lines (trenches)
  - Double Patterning
- Double exposure / patterning requires a complex set of parameters when different exposures are used to define single features

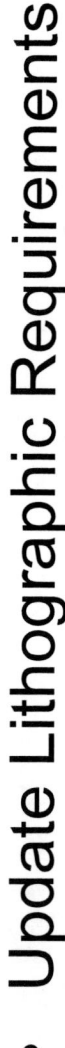

Outline

- Background and ITRS Lithography
- **IEUVI Overview**
- Optics TWG
- Resist TWG
- Mask TWG
- Source TWG
- Summary

International EUV Initiative (IEUVI)...

Asia / Pacific
- ASET
- EUVA
- SELETE

United States
- SEMATECH
- INVENT, CNSE
- SRC

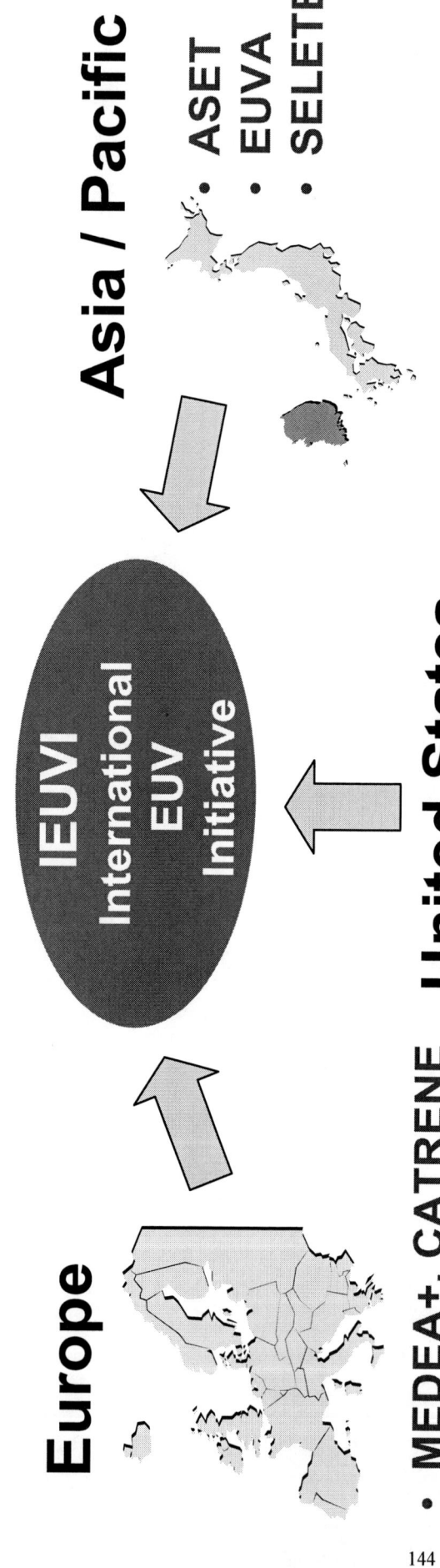

IEUVI
International
EUV
Initiative

Europe
- MEDEA+, CATRENE
- ENIAC, AENEAS
- LETI
- IMEC

IEUVI Mission: To further the coordination of collaborative efforts among leading EUVL R&D consortia. *http://www.ieuvi.org*

IEUVI – What, How, and Who

- **What:** To address infrastructural issues that need to be resolved for EUVL commercialization.

- **How:** The IEUVI collects technical inputs through its Technical Working Groups (TWGs), and identifies possible show stoppers for commercialization.

- **Who:** IEUVI TWG members include integrated device manufacturers, suppliers, national laboratories and universities.

IEUVI Organization

IEUVI Board

Chair:	Paolo Gargini	Intel
Member Organizations:	CEA / LETI	(EU)
	IMEC	(EU)
	Selete	(JP)
	EUVA	(JP)
	SEMATECH	(US)
	MEDEA+, CATRENE, ENIAC, AENEAS (EU)	

IEUVI Technical Working Groups (TWG)

Optics Contamination TWG

Chair:	Andrea Wüest	(US)	SEMATECH
Co-Chairs:	Yasuaki Fukuda	(JP)	EUVA / Canon
	Rogier Verberk	(EU)	TNO
	Tom Lucatorto	(US)	NIST

Resist TWG

Chair:	Serge Tedesco	(EU)	CEA / LETI
Co-Chair	Jacque Georger	(US)	SEMATECH / Intel
	TBD	(AP)	

Mask TWG

Chair:	Kevin Orvek	(US)	SEMATECH / Intel
Co-Chairs:	Iwao Nishiyama	(JP)	Selete / NECEL
	Jinho Ahn	(KR)	Hanyang Univ.
	Jan Hendrik Peters	(EU)	AMTC

Source TWG

Chair:	Frank Goodwin	(US)	SEMATECH
Co-Chair:	Masashi Ogawa	(JP)	EUVA
	TBD	(EU)	

Makeup of TWG Attendees

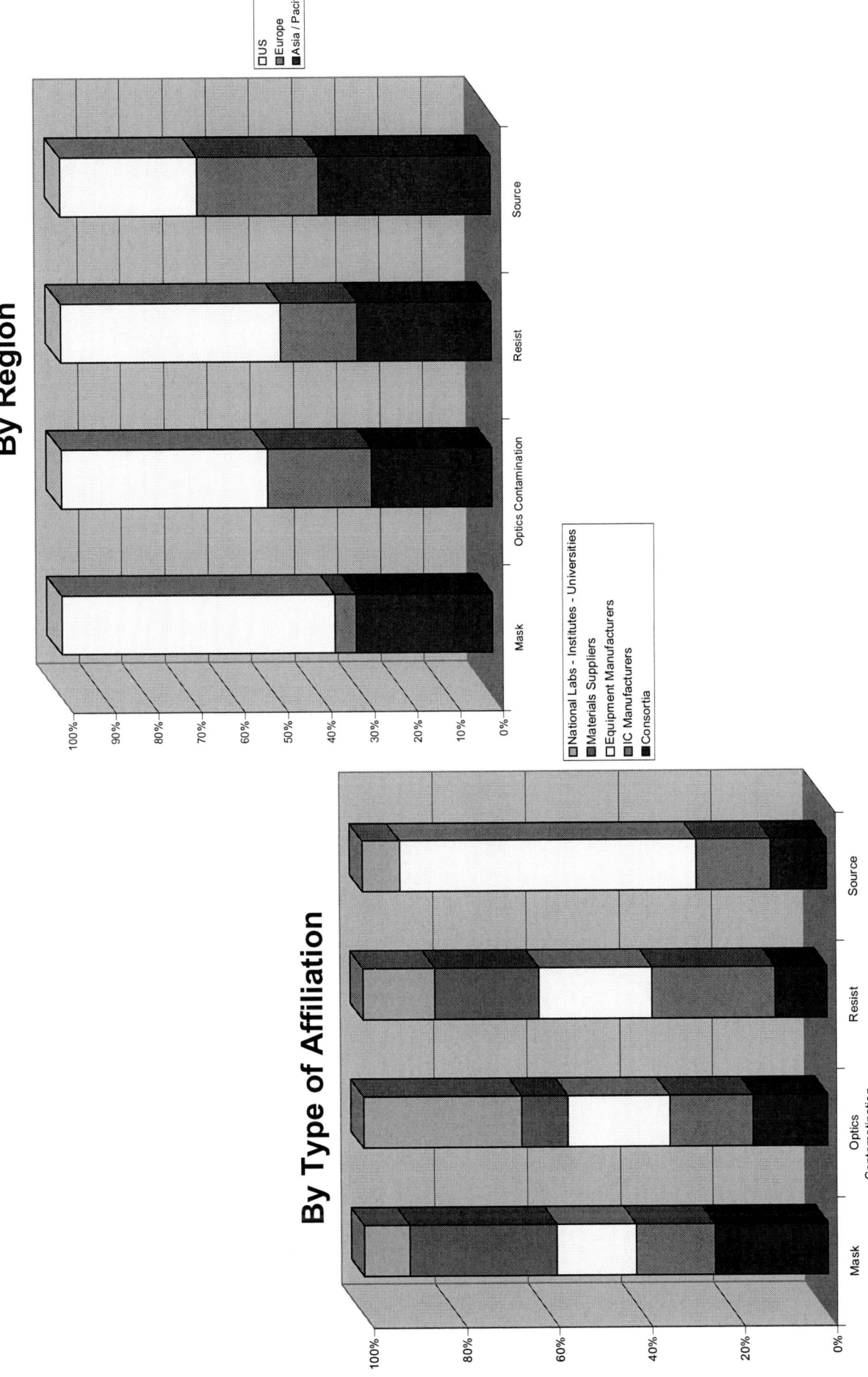

By Region

Legend: US, Europe, Asia / Pacific

Categories: Mask, Optics Contamination, Resist, Source

By Type of Affiliation

Legend: National Labs - Institutes - Universities, Materials Suppliers, Equipment Manufacturers, IC Manufacturers, Consortia

Categories: Mask, Optics Contamatination, Resist, Source

Outline

- Background and ITRS Lithography
- IEUVI Overview
- **Optics TWG**
- Resist TWG
- Mask TWG
- Source TWG
- Summary

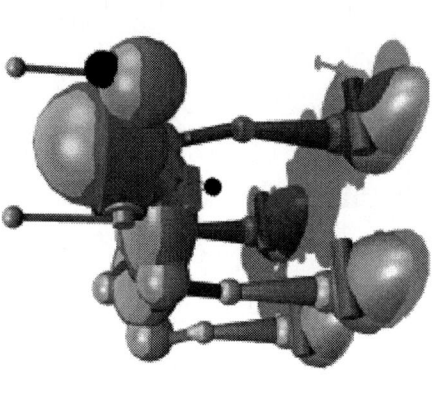

Optics TWG: Mission & Objective

- Mission:
  - Increased coordination and exchange of information among international consortia, national laboratories and commercial suppliers in the area of EUV optics and mask lifetime/contamination
  - Ensure that all critical issues are being addressed and that duplication is being avoided.

- Objective:
  - Provide forum for 1) Critical exchange of EUV optics/mask lifetime and contamination R&D in Europe, Japan, and the US and 2) Drive discussions and solutions on particular problems of high current interest.

Optics TWG: Participating Organizations

- AMD
- ASML
- Brewer Science
- Canon
- Carl Zeiss SMT
- CEA-LETI
- CNSE Albany
- Colorado State University
- Dai Nippon Printing
- Edwards Vacuum
- Energetiq
- EUVA
- EUV Technology
- FOM Rijnhuizen
- Fraunhofer IOF
- IBM

- IMEC
- Intel
- LBNL
- Nikon
- NIST
- PTB
- Rutgers University
- Samsung
- Selete
- SEMATECH
- Sumitomo
- TOK
- Toshiba
- University of Hyogo
- University of Illinois

Optics TWG: Accomplishments / Results

- Provided updates on optics lifetime / contamination results from exposure tools

- Identified and ranked critical issues

- Accelerated development of resist outgassing methodologies and protocols

- Hosted presentations on special topics from subject matter experts (e.g. photochemistry on TiO_2 by Prof. Yates, U.Virginia)

Optics TWG: Ranking of Projection Optics Critical Issues

Ranking (9/08)	Critical Issue	Gap Analysis
1	Carbon deposition (outgassing from vacuum components, wires, and other materials)	
2	Carbon deposition (resist outgassing)	
3	Optics cleaning (in situ)	
4	Oxidation (residual water)	
5	Accelerated lifetime testing/understanding scaling laws	
6	Availability of quantitative contamination model	
7	Other (refurbishing capability)	
8	Contamination due to EUV source	

Survey: 13 Responses

For HVM Implementation of EUVL

Manufacturable solutions exist, and are being optimized	
Manufacturable solutions are known but needing further development	
Manufacturable solutions are not known.	

Optics TWG: Ranking of Illumination
Optics Critical Issues

Ranking (9/08)	Critical Issue	Gap Analysis
1	Carbon deposition (Outgassing from vacuum components, wires, and other materials)	
2	Optics cleaning (in situ)	
3	Contamination due to EUV source	
4	Oxidation (residual water)	
5	Accelerated lifetime testing/understanding scaling laws	
6	Out-of-band radiation	
7	Carbon deposition (resist outgassing)	
8	Availability of quantitative contamination model	
9	Other (refurbishing capability)	

For HVM Implementation of EUVL

Survey: 13 Responses

Manufacturable solutions exist, and are being optimized

Manufacturable solutions are known but needing further development

Manufacturable solutions are not known.

Optics TWG: Development Gaps and Recommendations

1. **Carbon deposition and oxidation**
   1. Further improvement in cleaning, the application of mitigating gases and vacuum cleanliness required.

2. **Accelerated lifetime testing / Scaling laws**
   1. Effect of high EUV power levels has to be understood.
   2. Modeling of contamination can provide essential insight and enable prediction of contamination rates.

3. **Contamination due to EUV source**
   1. Reduction of debris from EUV sources required.

4. **Out-of-band radiation**
   1. Contribution of out-of-band wavelengths has to be determined.

5. **Resist outgas testing**
   1. Establish re... ...rate resist

Outline

- Background and ITRS Lithography
- IEUVI Overview
- Optics TWG
- **Resist TWG**
- Mask TWG
- Source TWG
- Summary

Resist TWG: Mission & Objective

- Mission: Increased cooperation among EUV resist development community world wide

  - Coordinate efforts to address top 3 issues

- Objective: Provide forum to share information to foster global collaboration and accelerate development of EUV resist

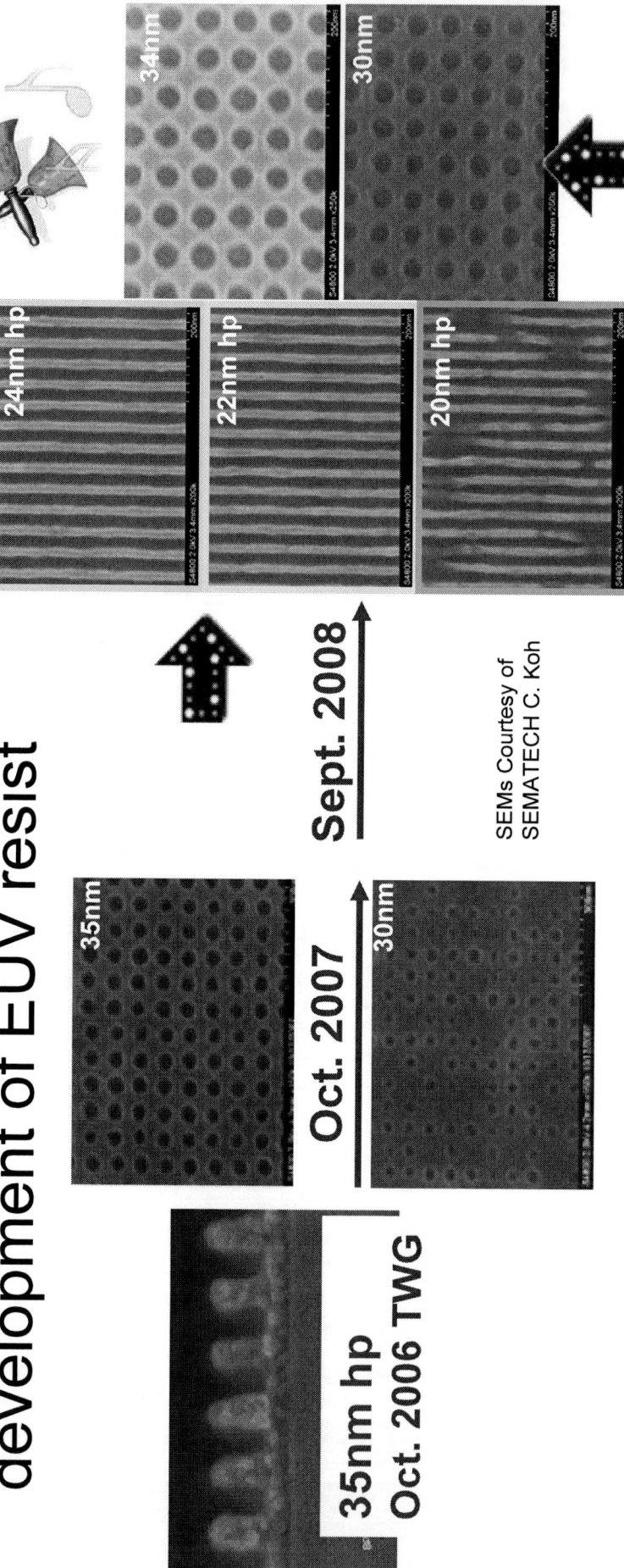

SEMs Courtesy of SEMATECH C. Koh

Resist TWG: Participating Organizations

- AIXUV
- AMD
- ASML
- ASET
- BOC Edwards
- Brewer Science
- Canon
- CEA/LETI
- Dongjin
- Energetiq
- Fujifilm
- IBM
- IMEC
- Intel
- JSR

- Nikon
- Osaka University
- Philips
- Panasonic
- Qimonda
- Rohm & Haas
- Samsung
- SELETE
- SEMATECH
- Shin-Etsu
- Sumitomo
- SUNY Albany
- Texas Instruments
- TOK
- University of Birmingham
- University of Hyogo

Resist TWG: Highlights

- Significant progress in Resolution and Sensitivity
  - 7 suppliers participant in cycle of learning evaluations
  - Chemically amplified resist is leading candidate demonstrating 22nm L/S resolution, results reported at EUV symposium
  - Evaluation of many new resist platforms including Molecular Glass, polymer bond PAG, Fullerene etc... and are showing steady progress

- 2008 TWG focus Topic: Optimizing RLS
  - 6 researchers from Univ. & Industry in Asia, US and Europe present findings/learning's on topic

- => Global collaboration is working to help reach consensus on status and assessing challenges

Resist TWG: Challenges/Gaps

Resolution: 32→22nm hp

LWR: 1.7→1.2nm

Sensitivity: 10mJ/cm²

- **Resolution, Line width Roughness, Sensitivity (RLS)**
  - Simultaneously meeting RLS requirements remains difficult
    - Line space resolution to 22nm hp at 14mJ/cm² has been demonstrated in 50nm resist thickness on BMET, but only minor improvements in LWR have been achieved to ~5nm.
    - Resist line collapse at 22nm L/S needs to be improved.
    - Dense contact resolution has improved to ~30nm, but sensitivity is poor with dose to size of 49mJ/cm².
  - Availability of tools with high resolution EUV exposure photons remains limited.

- **Outgassing**
  - MET spec of 2.6E+14 (molecules/sec*cm²) achieved by ~90% of resists tested but full field tool specs significantly more difficult and may not be practical?
    - ASML ADT spec of 5*10¹² (molecules/sec*cm²) is achieved by ~33% of resists tested
    - Nikon EUV spec of (molecules/sec*cm²) is NOT achieved by any resist tested

- **Full Field EUV resist characterization**
  - There is very little full field data available
    - Full Characterization of leading candidates is needed.
    - This includes DOF, EL, PED, PEB sensitivity, CD uniformity, defects, Etch resistance, etc…

Resist TWG: Next Steps

- Organize and facilitate TWG meeting with global experts to address keys to optimizing RLS

- Continue driving resist performance through benchmarking on MET tools
  - Evaluate full field performance of leading resists on ADT

- Line Width Roughness
  - Engage in and share fundamental research to <u>reduce resist LWR</u>
  - Encourage and help coordinate work to look at pre/post etch processes to reduce LWR

- Beyond 22nm HP, Novel system may be necessary
  - New resist platforms that are small, or dynamic in molecular size need further developing
  - Bottom up approaches need consideration

- More exposure time for resist is needed from all available EUV tools, including full field alpha tools.

Agenda

- Background and ITRS Lithography
- IEUVI Overview
- Optics TWG
- Resist TWG
- **Mask TWG**
- Source TWG
- Summary

Mask TWG: Mission & Objective

- Mission:

  Ensure EUV Mask Infrastructure Readiness for:
  - Pilot Line Production 2010 – 2012
  - High Volume Manufacturing 2013 – 2016

- Objectives:
  - Identify Required Standards
  - Coordinate industry-wide conversions, such as future mask incidence angle change.
  - Identify any gaps between current industry efforts and projected future needs
  - Highlight gaps to member organizations and IEUVI Board for action

Mask TWG: Members

- Alcatel
- AMD
- AMTC
- Asahi Glass Co.
- ASML
- Canon
- Carl Zeiss
- CSNE
- Corning Inc.
- Dai Nippon Printing
- Entegris
- Hanyang University
- Hoya Corp.
- IBM
- Intel
- IMEC
- KLA-Tencor

- Lasertec
- Lawrence Berkeley N.L.
- Nikon
- NuFlare Technology Inc.
- Ohara
- Photronics
- Qimonda
- Samsung
- SELETE
- SEMATECH
- SEMI
- Toppan Printing Co.
- Toshiba
- TOSO
- TSMC
- Veeco Instruments Inc.
- Wisconsin University

Mask TWG: 2008 Accomplishment

- Outlined format and key specifications for new version of SEMI P37 Standard to cover Substrates and Masks for pilot line timeframe.

- Member survey identified 14 areas of critical concern for Pilot Line Production 2010-2012 (see next foil).

- Identified two new issues for HVM – mask angle change and mask scanning temperature.

EUV Mask Infrastructure Readiness: Pilot Line 2010 - 2012

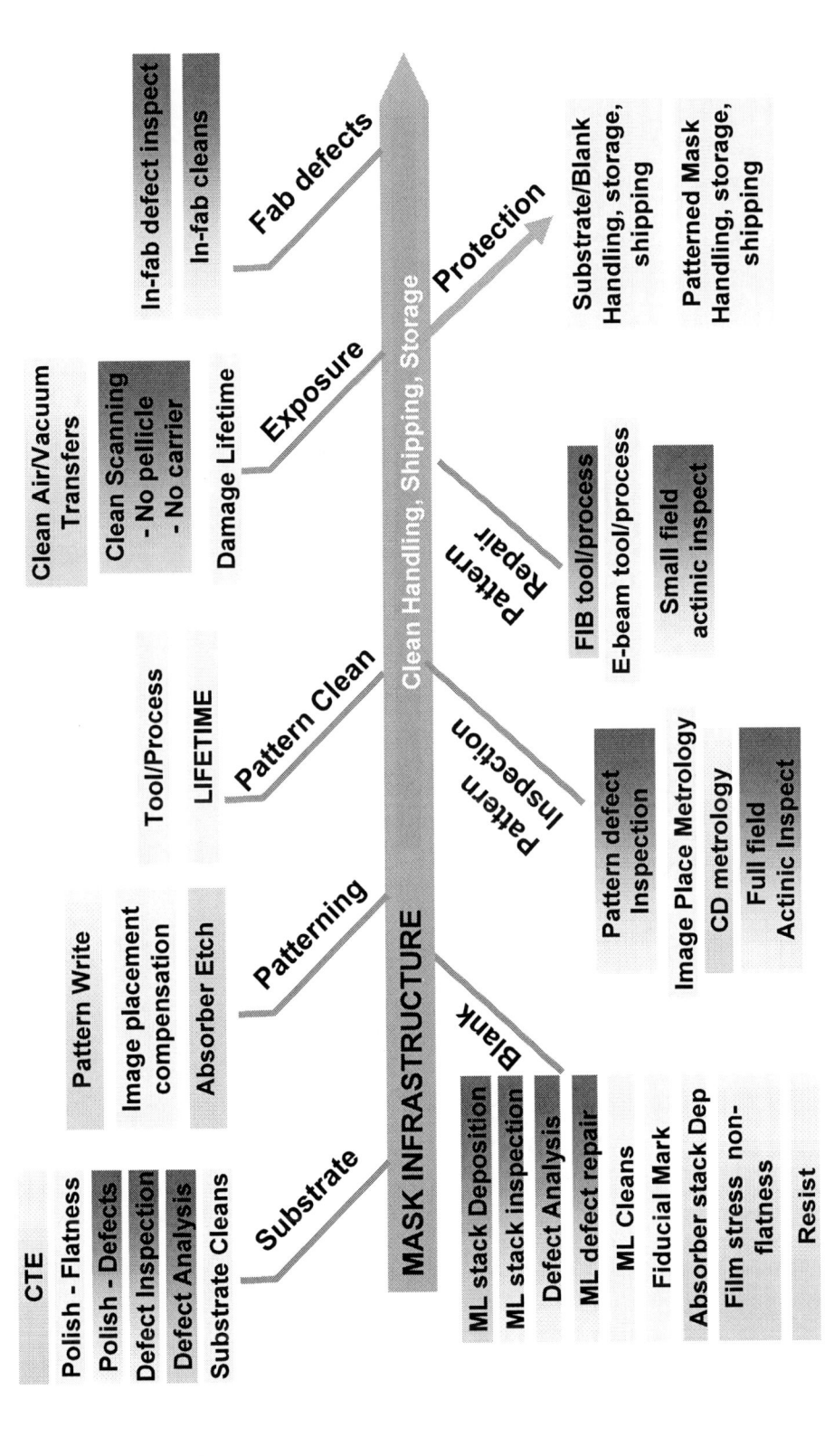

- Preliminary Results of new TWG survey
- TWG 10/2/08 will focus on items with red

Mask TWG: Development Gaps

(To be discussed at Thursday TWG)

➤ Pilot Line 2010-2012:

- Substrate: Polish defects, defect inspection & analysis
- Blanks: ML deposition, defect inspection & analysis & repair
- Pattern Inspection: Pattern defect inspection, full-field actinic inspection
- Pattern Repair: FIB Tool/process, small-field actinic inspection
- Mask EUV Scanning: Clean scanning (no pellicles)
- Fab use: In-fab inspection and cleans

➤ HVM issues (> 2013)

- When does mask incidence angle change from 6° to ≥ 8° ?
- What impact is higher mask use temperature (~ 70°C) versus mask room temperature manufacture?

Outline

- Background and ITRS Lithography
- IEUVI Overview
- Optics TWG
- Resist TWG
- Mask TWG
- **Source TWG**
- Summary

Source TWG: Goal & Objective

- Goal: Provide a platform for all stake-holders

  – Identify EUV critical issues, and develop consensus on critical technical challenges

  – Provide forum for frank technical discussions between suppliers and end-users

  – Foster collaborations in development and standardization of source metrology

- Objective: Accelerate consensus building in the industry

  – Coordinate key messages to the broader source community

  – Support source benchmarking efforts (power, out-of-band, debris)

Source TWG:
Participating Organizations

- Advanced Micro Devices
- ASML
- Canon Inc.
- Canon Inc. Optics R&D Center
- Carl Zeiss Laser Optics GmbH
- College of Nanoscale Science and Engineering
- Cymer
- EUVA
- HP
- IBM
- IMEC
- Intel
- LP Photonics
- MEDEA+
- Media Lario Technologies

- Media Lario USA, Inc.
- NEC
- Nikon Corporation
- Panasonic
- Philips Extreme UV
- Powerlase Limited
- Sagem
- Samsung
- Selete
- SEMATECH
- Texas Instruments
- University of Central Florida
- Ushio Inc.
- XTREME Technologies GmbH

Source TWG: Accomplishments

- **Drive standards development**

  - Development of specifications for collector lifetime requirements for alpha, beta and gamma level EUV sources

  - Development of standards for Intermediate Focus (IF) metrology

  - Identified interface opportunities for the industry to reduce cumulative development costs

- **Drive industry consensus on critical technical challenges for EUV source commercialization**

  - Identify and rank critical issue for EUV source technology

  - Map which of the stake holders will be addressing which EUV source critical issue and which issues are not being addressed

  - Consolidated supplier source performance data into a summary table

Source TWG:
Ranking of DPP Technical Challenges

Ranking (5/08)	Technical Challenge (previous ranking Q3/06)	Gap Analysis
1	Power at IF (3)	
2	Collector lifetime (1)	
3	Debris mitigation (1)	
4	Thermal loading of DMS and Collector	
5	Cost of ownership (4)	
6	Conversion efficiency (9)	
7	Higher efficiency collector designs (6)	
8	Spectral purity (5)	
9	Scalability	
10	Reliability and Stability	

For HVM Implementation of EUVL   Survey: 85 Responses to Source Survey

Manufacturable solutions exist, and are being optimized

Manufacturable solutions are known but needing further development

Manufacturable solutions are not known.

Source TWG:
Ranking of LPP Technical Challenges

Ranking (5/08)	Technical Challenge (previous ranking Q3/06)	Gap Analysis
1	Power at IF (3)	
2	Debris mitigation (1)	
3	Cost of ownership (4)	
4	Collector lifetime (1)	
5	Laser Power	
6	Conversion efficiency (9)	
7	Thermal loading of DMS and Collector	
8	Scalability	
9	Spectral purity	
10	No Integrated System	

For HVM Implementation of EUVL   Survey: 85 Responses to Source Survey

Manufacturable solutions exist, and are being optimized

Manufacturable solutions are known but needing further development

Manufacturable solutions are not known.

Source TWG: Sn DPP Showstoppers

Primary DPP source detractors preventing implementation of EUV as a litho solution for pilot-line and HVM

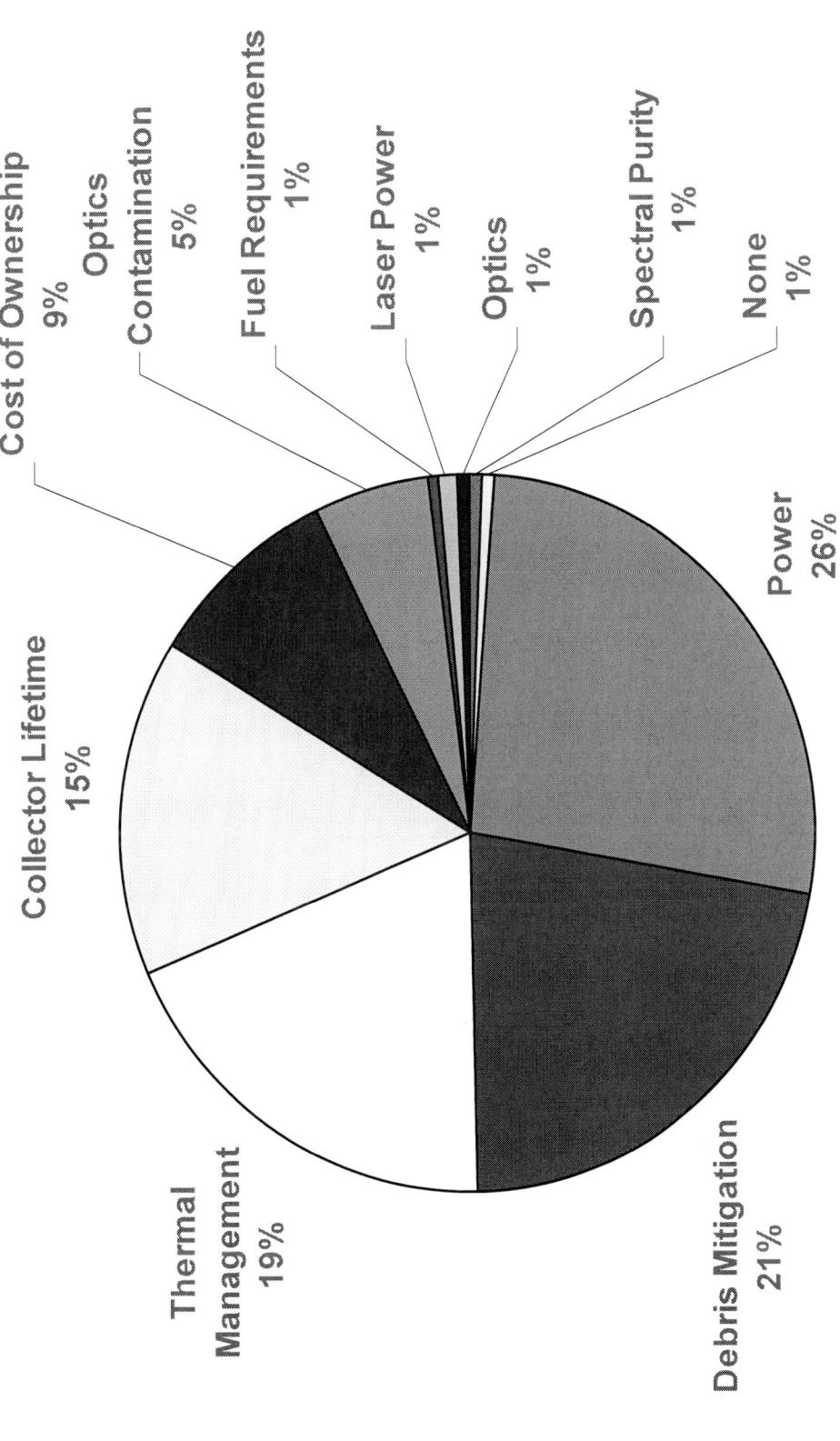

Survey: 85 Responses to Source Survey

Source TWG: Sn LPP Showstoppers

Primary LPP source detractors preventing implementation of EUV as a litho solution for pilot-line and HVM

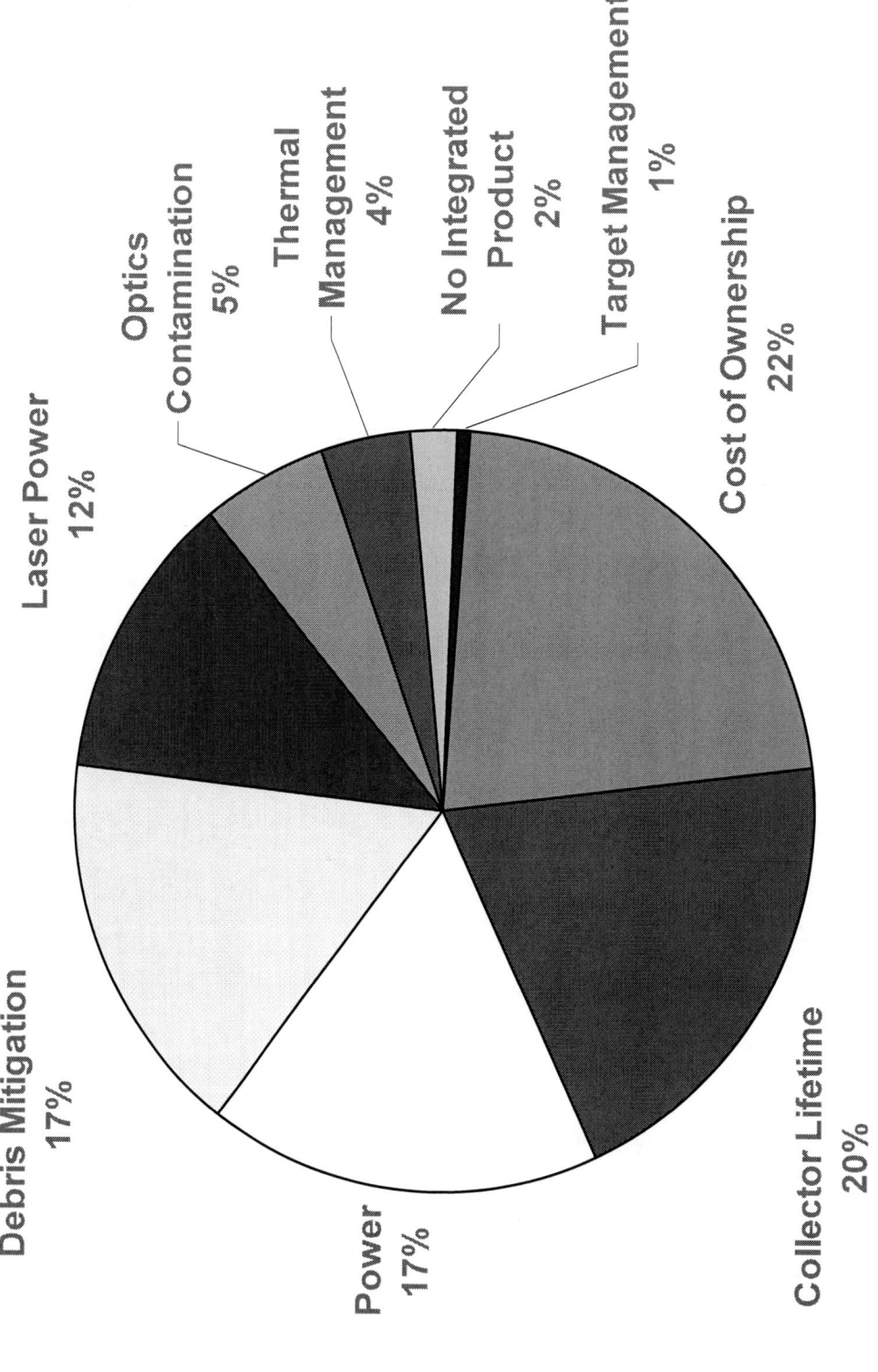

Debris Mitigation
17%

Laser Power
12%

Optics
Contamination
5%

Thermal
Management
4%

No Integrated
Product
2%

Target Management
1%

Cost of Ownership
22%

Collector Lifetime
20%

Power
17%

Survey: 85 Responses to Source Survey

Source TWG: Development Gaps

- Fundamental Understanding
  - Debris mitigation of LPP sources
  - Power scaling of sources
  - Efficiency of power transmission to IF
  - Reliability and stability
- Engineering Development
  - LPP source/collector/DMS integration
  - Improved debris mitigation and handling of fuel of DPP sources
  - Improvement of source component designs/materials/lifetimes
  - Solutions for spectral filtering, particularly near-IR
  - Design optimization of illuminator and projection optics
  - Improved cost of ownership

Source TWG: Recommendations

- Extreme ultraviolet lithography (EUVL) has missed the window to support the 32 nm half-pitch technology node, primarily due to lack of a fully integrated light source.

- For EUV to be seen as a viable lithography solution significant progress needs to be seen in addressing the critical issues facing HVM implementation.
  - Particularly in resolving
    - In-band Power at IF
    - Debris Mitigation
    - Source Component Lifetimes
    - Cost of Ownership
    - Reliable and Stable operation

- Even within the current competitive environment collaborations can accelerate source readiness
  - Collaborations can provide better use of available resources and can significantly reduce the time to market for commercial EUV sources

Summary

- It is time to step on the accelerator!

- Cooperation is still relevant(i.e., contact the IEUVI TWG Chairs if you have contributions!)

- There are not too many technology nodes left for EUV to make a difference!

http://www.ieuvi.org

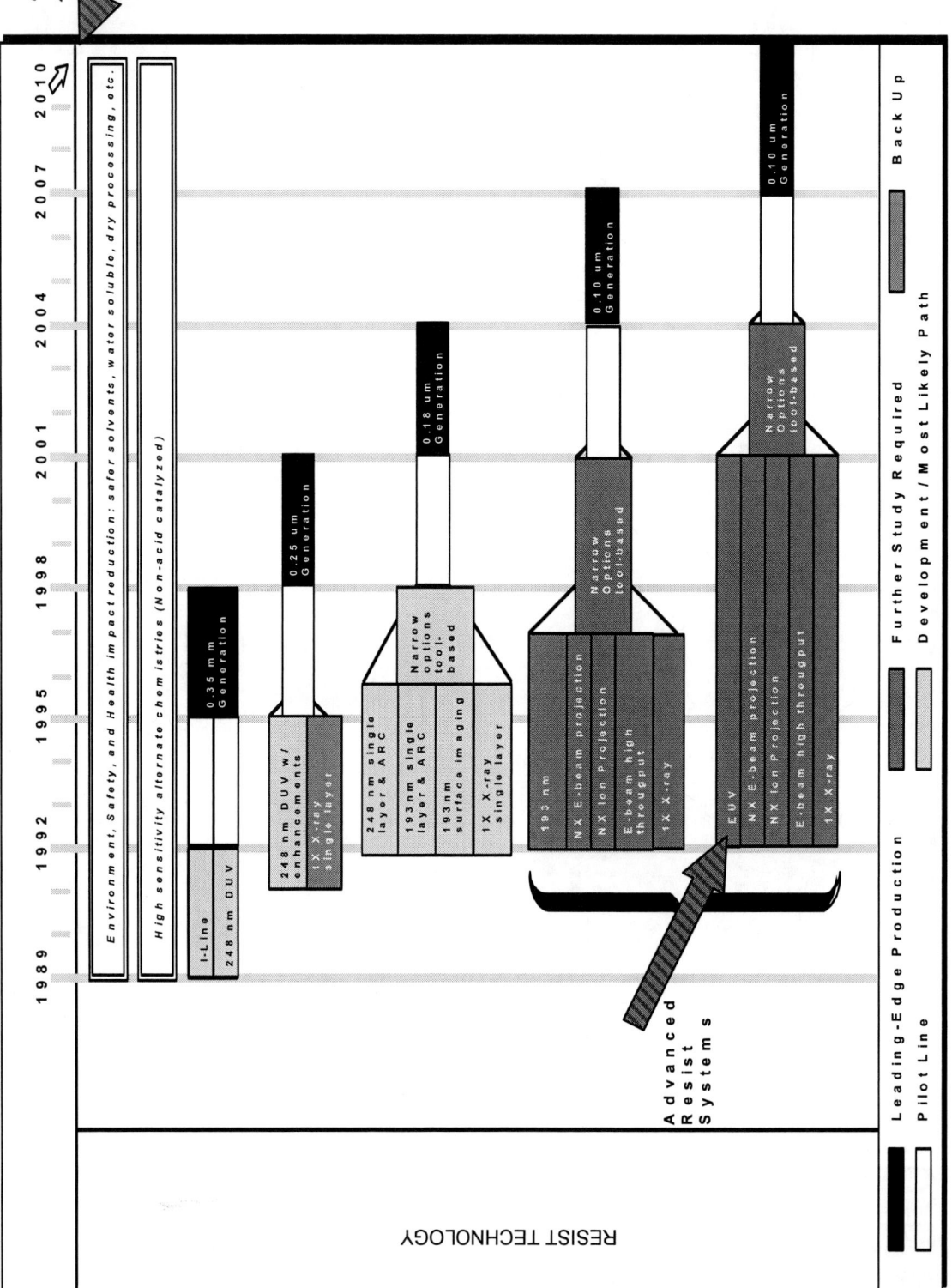

Figure 14 Critical Level Resist Technology Potential Solutions Roadmap   1994 NTRS

Semiconductor Industry Association. *The National Technology Roadmap for Semiconductors, 1994 edition.* SEMATECH:Austin, Tx, 1994.

The Fifth International Nanotechnology Conference on Communication and Cooperation (INC5)

May 18th – 21st, 2009 at UCLA

Organized by

http://www.cnsi.ucla.edu/conferences/INC5/

INC5 Preparations

- Prof. Louis Ignarro to give a keynote speech
- California Governor Arnold Schwarzenegger has been invited...
- The CNSI-UCLA to host a reception on May 18th evening
- The Getty Center, confirmed for networking
- Homepage to be launched in Q4, '08

Keynote Address
by
Prof. Louis Ignarro
UCLA, Medical School, Molecular & Medical Pharmacology
The Nobel Prize in Physiology or Medicine, 1998

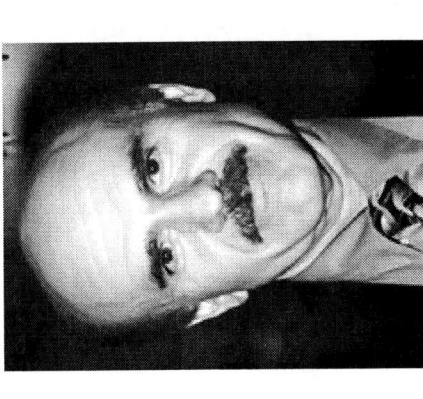

INC5, May 18th – 21st, 2009 at UCLA
http://www.cnsi.ucla.edu/conferences/INC5/

Thank you for your attention

Asia Pacific Regional Update

- Japan, Korea and Taiwan Regional Update -

2008 EUVL Symposium
Lake Tahoe, California
October 1, 2008

Ichiro Mori

Semiconductor Leading Edge Technologies, Inc

Acknowledgements

- Jinho Ahn, Hanyang University, Korea
- Anthony Yen, TSMC
- Bryan Shew,
National Synchrotron Radiation Research Center, Taiwan

Where we are ?

EUVL Development Stages

2006	2010
α-level development stage	β-level /HVM

Feasibility study

Key component development

- Accelera... Source, Resist, Mask, Optics, Contamination,...... components
- Verification of m...
- Construction of r...
- Learning things ... for next development stage
- Establishment c ... masks in HVM s, processes and

EUVA

MIRAI

Selete

EUVA R&D Project (2002 ~ 2007)

EUV Source

LPP (CO$_2$: Sn-Droplet)

IF image

2 mm

EUV Power ~60W@IF(CO$_2$:6kW)
CO2-Laser Power up to 13kW@100kHz
Mitigation with Magnetic Field

DPP (SnH$_4$-Gas)

EUV Power~60W@IF
Collector Lifetime ~5.7B Pulses
Foil-Trap & Halogen Gas Cleaning

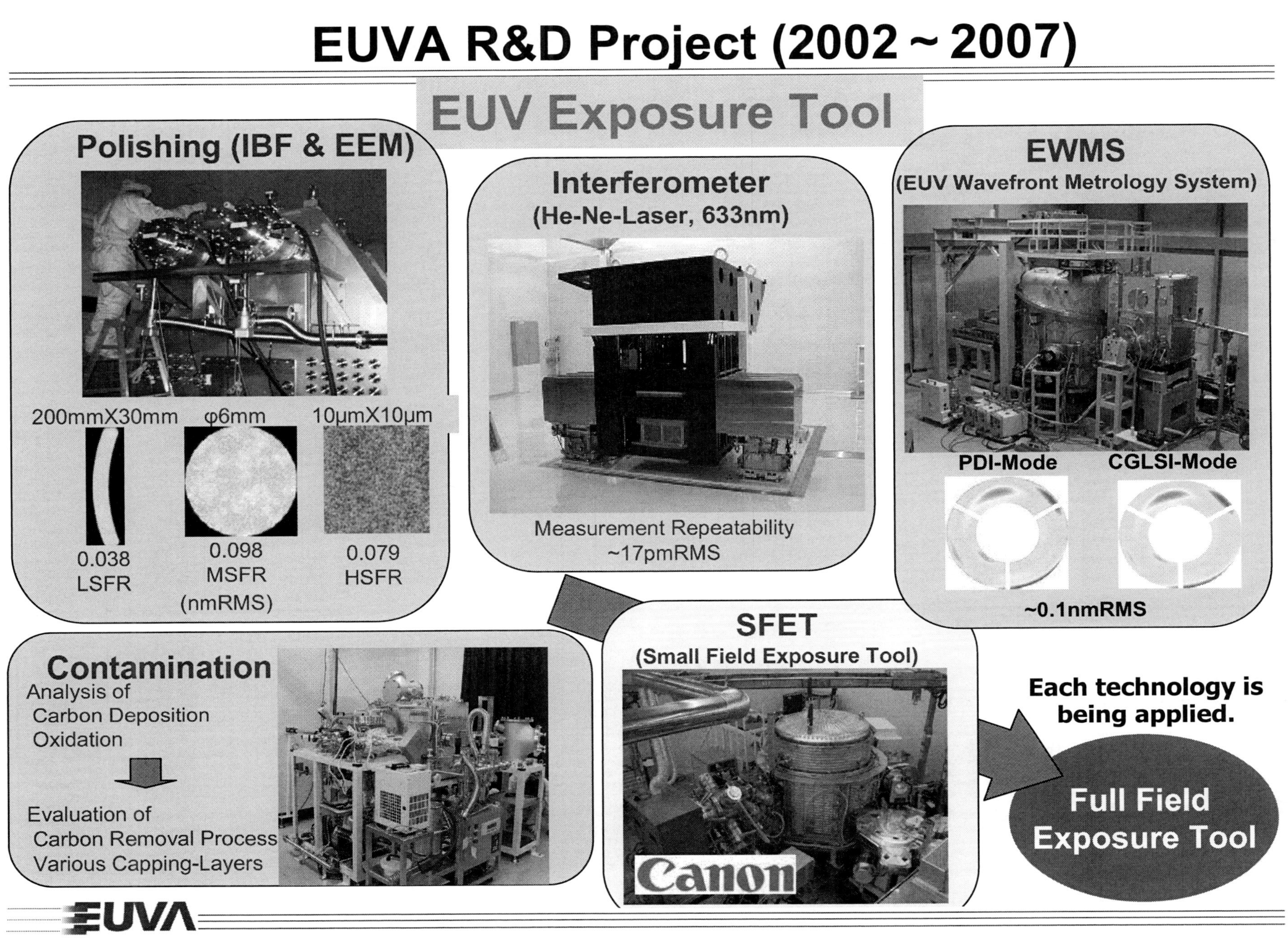

EUVA R&D Project (2002 ~ 2007)

EUV Exposure Tool

Polishing (IBF & EEM)

200mmX30mm φ6mm 10μmX10μm

0.038 LSFR   0.098 MSFR   0.079 HSFR

(nmRMS)

Interferometer
(He-Ne-Laser, 633nm)

Measurement Repeatability ~17pmRMS

EWMS
(EUV Wavefront Metrology System)

PDI-Mode   CGLSI-Mode

~0.1nmRMS

Contamination

Analysis of
Carbon Deposition
Oxidation

Evaluation of
Carbon Removal Process
Various Capping-Layers

SFET
(Small Field Exposure Tool)

Canon

Each technology is
being applied.

Full Field Exposure Tool

New EUVA Project Just Started (2008 ~ 2010)

Focus on High-Power (>100W@IF) & High-Reliability Source

Source Module（Private

Funded）
- High-Power >100W@IF
- Scalability up to 180W@IF

& Collector Module（National

Funded）
- Evaluation of Contamination Effects
- Debris Mitigation & Cleaning
- Thermal Management

DPP

Rotating Electrode

Pulsed Power Module

Fuelling Laser

Rotating Electrode, Laser for Sn Ablation
Molten Sn Bath, Pulsed Power Driver

LPP

Magnet

Plasma

IF

Sn Supply

CO2 Laser

Collector Mirror

Sn Collector

Sn Droplet
High Power Pulsed CO2 Laser
Magnetic Field Plasma Guiding

Selete's EUV program

Full-field exposure tool

Resist development

Multi-layer resist

SFET

Flare characterization & compensation

- Accelerate key component development
- Construct mask infrastructure
- Verify EUVL manufacturability

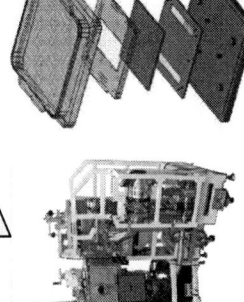

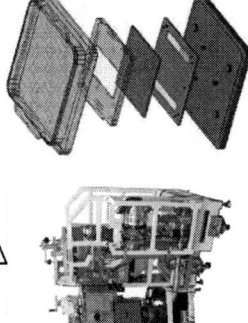

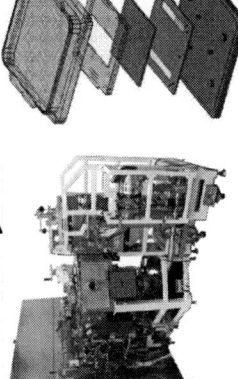

Protection of mask from particles

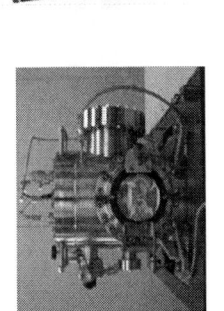

Resist out-gassing   Mask cleaning

Contamination control

EUV

Actinic blank inspection

DUV (199 nm)

Pattern defect inspection

FIB

Defect repair

Korea Update
2008 International Symposium on EUVL

Program Director: Jinho Ahn (Professor, Hanyang University)

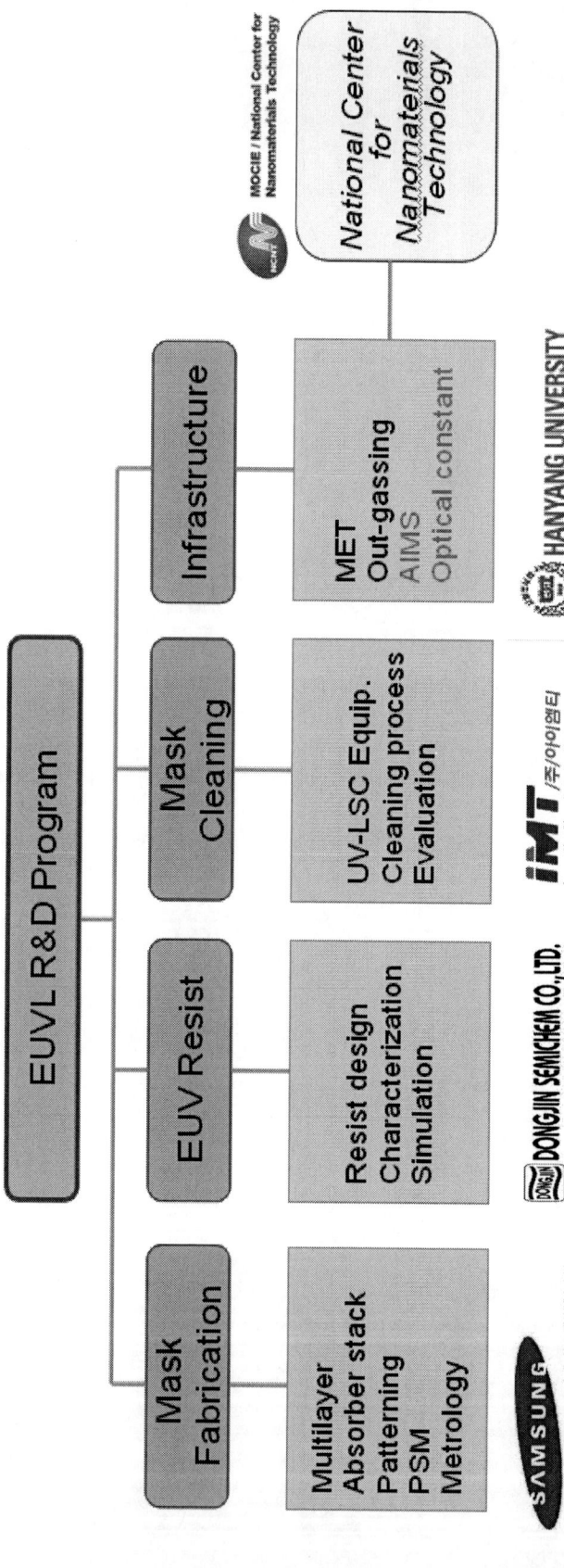

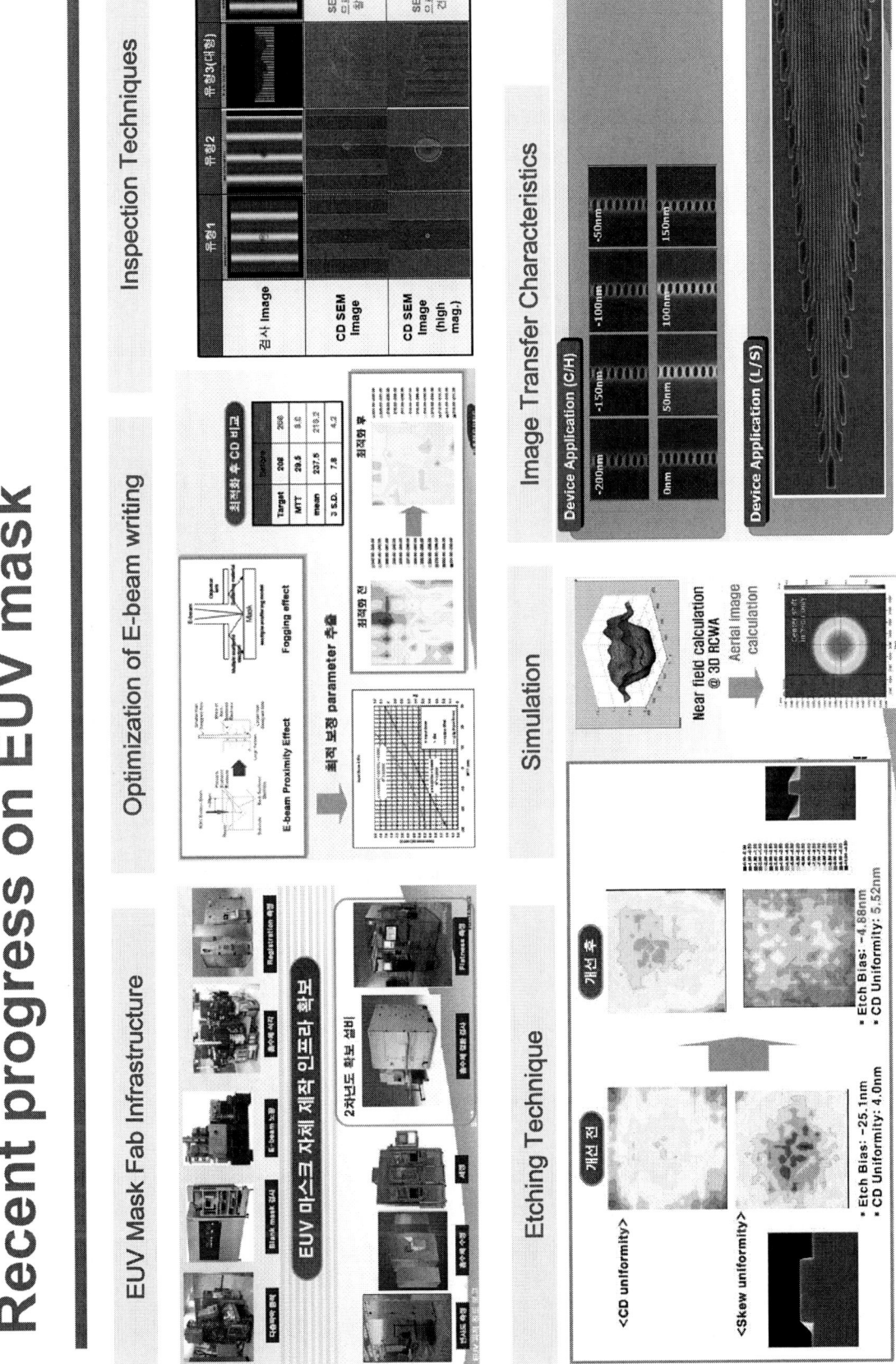

Recent Progress on EUV resist

Resist Optimization

08' 2Q

Film Thickness	Dose 40nm (mJ/cm2)	Resolution
50nm	6.2	HP 34nm

LBNL test results

Exposure : LBNL MET
Illumination : Ann (0.3/0.55)
Inspection : L/S 1:1
Film Thickness : 50nm

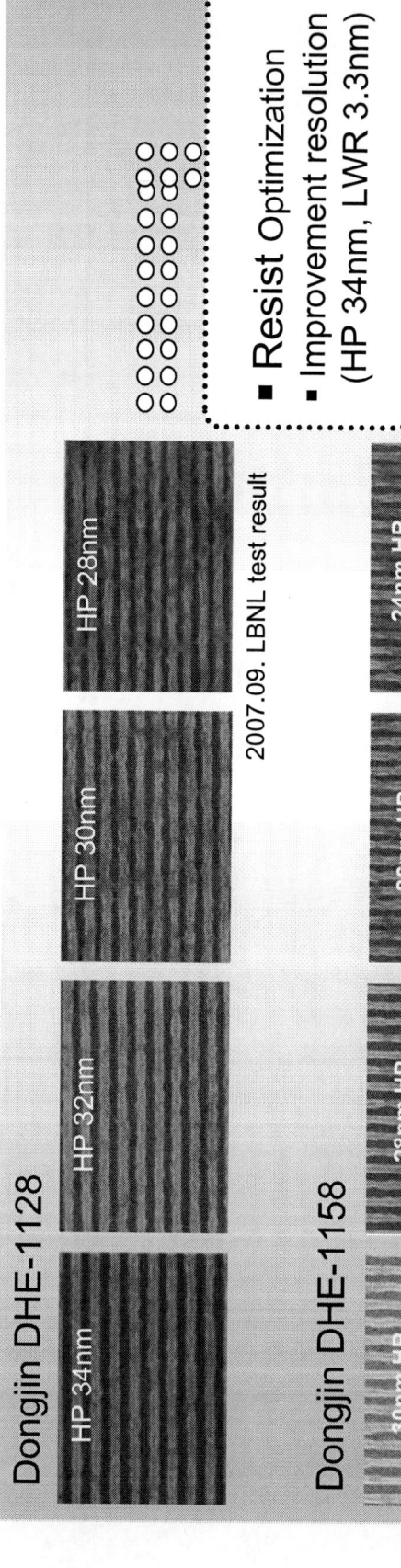

Dongjin DHE-1128

HP 34nm   HP 32nm   HP 30nm   HP 28nm

2007.09. LBNL test result

Dongjin DHE-1158

30nm HP   28nm HP   26nm HP   24nm HP

2008.05. LBNL test result

- **Resist** Optimization
- Improvement resolution (HP 34nm, LWR 3.3nm)

Recent progress on cleaning technology

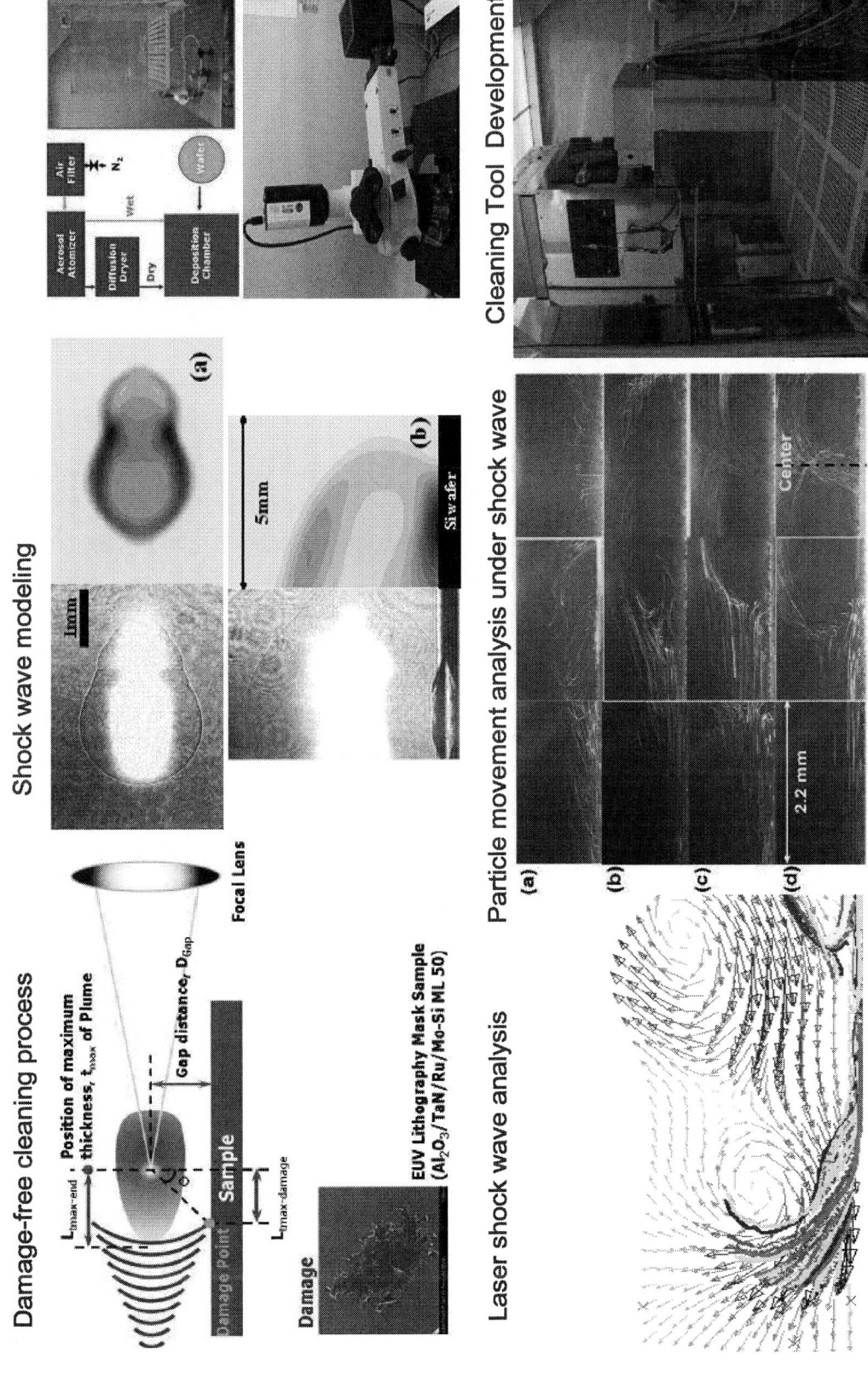

Recent progress on evaluation infrastructure

EUV Beamline and Test Bed at Pohang Accelerator Laboratory

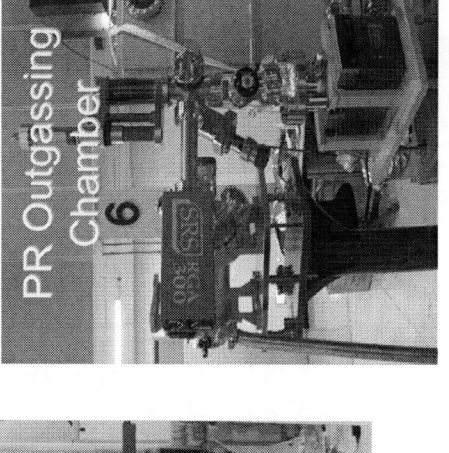

PR Outgassing Chamber

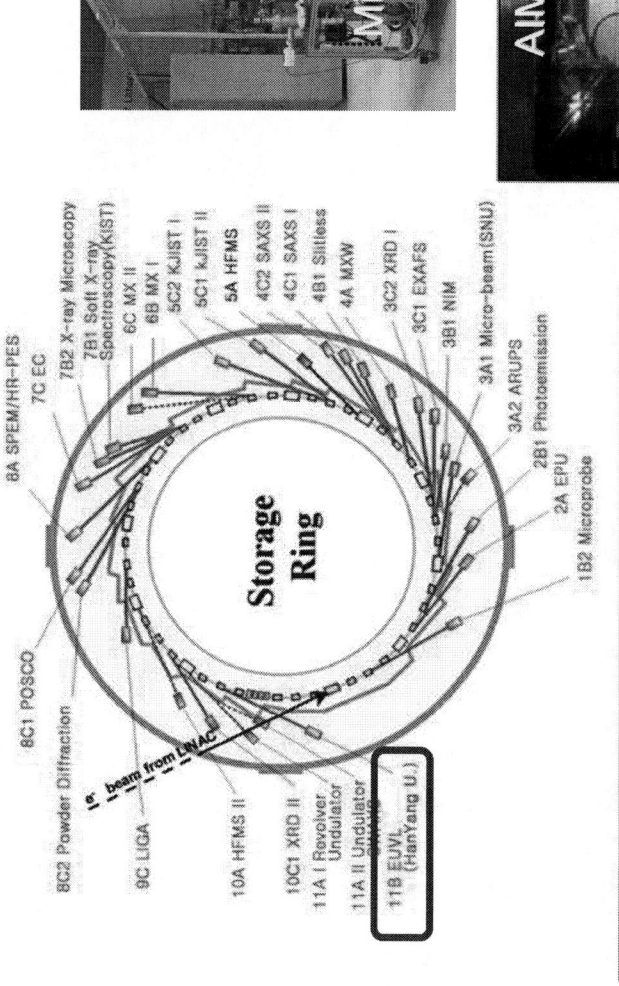

Micro Exposure Tool

Reflectometer

2008/01/10

AIMS

2008/06/14

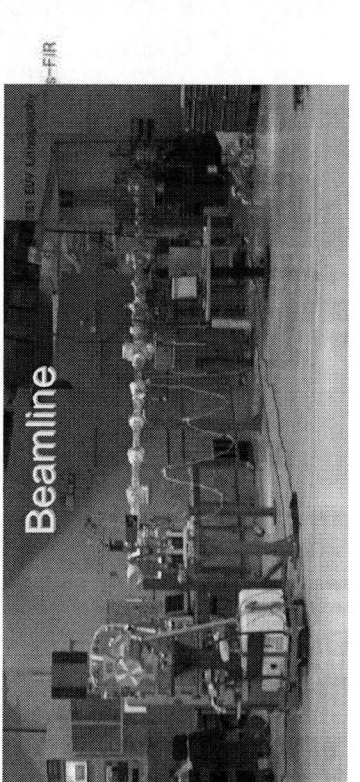

Beamline

Taiwan National Science and Technology Program for Nano-science and Nano-technology 2008~11

Investigations on Extreme UV lithography (EUVL)

- Taiwan EUVL project

Objects

- Initiate the EUVL research in Taiwan
- Establish the EUVL core facilities to meet R&D requirements both from academics and industrials.
- Develop novel EUV metrology technologies and instruments
- Integrate and accumulate the talents with various expertise for EUV-related researches

Local Consultants:

- *TSMC*
- *PSC*
- *ASML Taiwan*

EUVL Project

Group. (I) Fundamentals

- Study on **resist out-gassing** and photo chemistry
- Construct a EUV **reflectometer** for n, k, t measurement of ultra-thin resist/multilayer

Group.(II) Optics & Metrology

- Design of a high-throughput **mask inspection** system
- Construct a EUV **interferometer** for nano-lithography

Group. (III) EUVL Applications

- Study on radiation damage
- Study on anti-reflrection coating
- Mix & match lithography
- EUVL nano optical devices

SR EUV Light Source

Project Funding

NSRRC Matching

Participants:

- *Universities*
  (NCTU, NTU, NCKU, NUK)
- *National Lab.*
  (NSRRC, NDL)

Major Sponsors

- *MOEA* (Ministry of Economic Affairs)
- *NSC* (National Science Council)

Highly Integration of EUVL Researches at Hsinchu Science Park
- Universities, National Lab. and Industrial !!

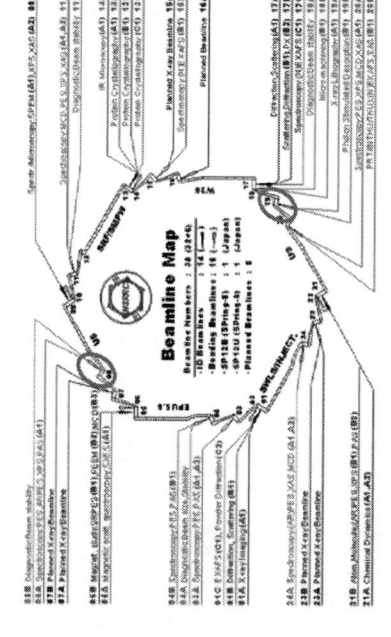

Two Beamlines for EUVL Research at NSRRC:

- **08A1 LSGM beamline**
- **19A1 Nanolithography beamline**

- **NSRRC** - National Synchrotron Radiation Research Center
- **NCTU** - National Chiao-Tung Univ.
- **NDL** - National Nano Device Lab.

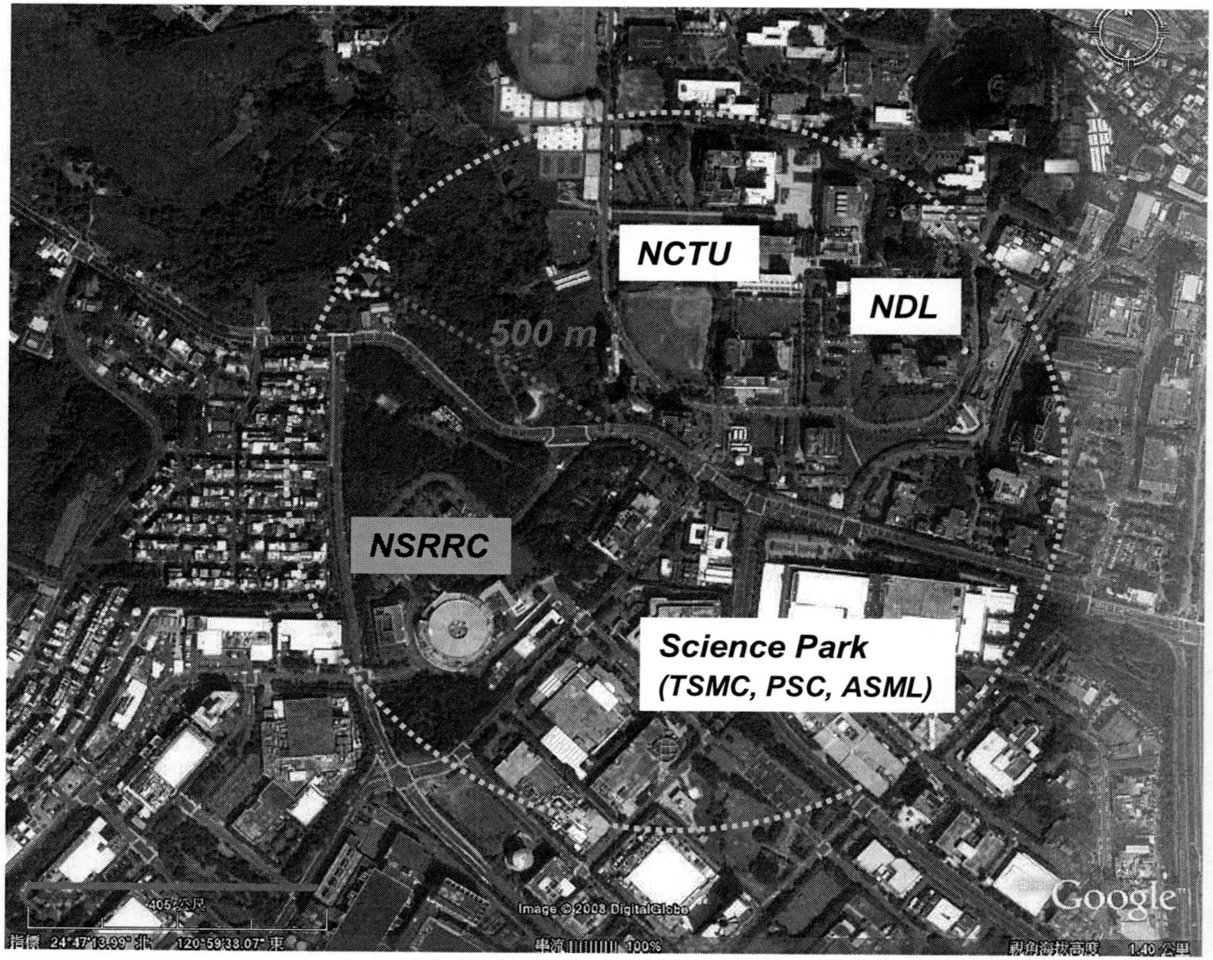

Preliminary Works
- Taiwan EUVL Project

•EUV reflectometer for n, k, t measurements

sample	n	k, μm^{-1}	t (Å)	Density (g/cm³)
PMMA (measured)	0.977	4.95	1220	1.12
Literature data	0.976	5.03	-	1.18

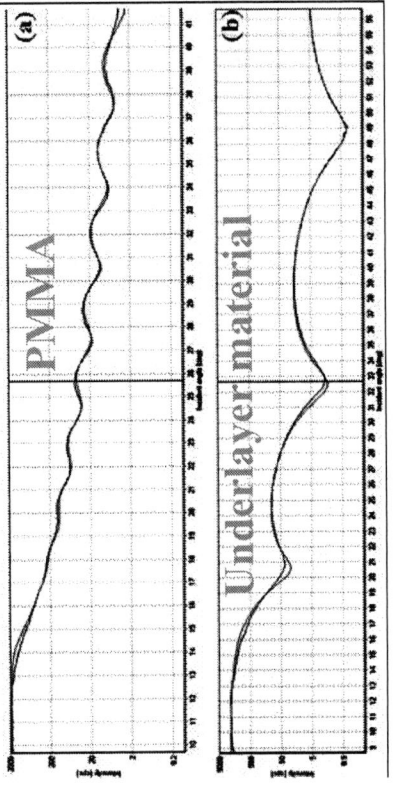

(a) PMMA

(b) Underlayer material

•Resist outgassing study by QMS

• Ionic outgassing from:

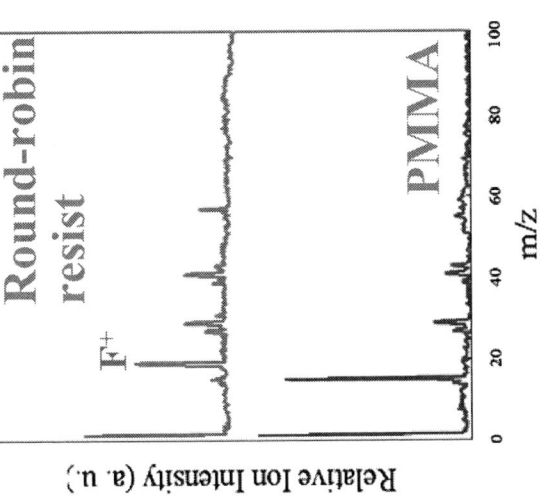

Round-robin resist

PMMA

F^+

Relative Ion Intensity (a. u.)

m/z

$C_mH_n^+$, $C_xF_y^+$ outgassing

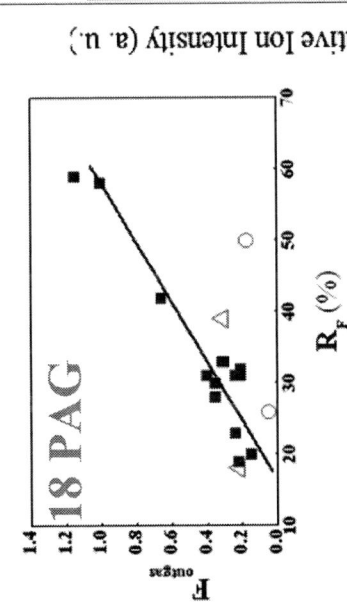

18 PAG

R_F (%)

F outgas

F^+ outgassing vs. F photoabosrption

• Results demonstrate the feasibility to evaluate the chemical and optical properties of EUV materials at NSRRC in Taiwan.

Summary

Japan update

- The new EUVA project focusing on EUV source just started in July 2008 and the universities continue to support EUV source development.
- MIRAI-Selete's EUV program covers a wide range of areas that require further efforts to get EUVL ready for volume production.

Korea update

- The Korea EUVL program, which consists of 4 sub-programs; EUVL mask, EUVL resist, EUVL mask cleaning and infrastructure, has started from 2002, and now it is just in the 3rd phase of its program.

Taiwan update

- A 3-year national project on EUVL research was just approved in Aug. 2008, which is an initiation of EUVL research program at Taiwan.

EUV Alpha Demo Tools – Stepping stones towards volume production

H. Meiling, S. Lok, B. Hultermans, E. van Setten,
B. Pierson, K. Cummings, C. Wagner, N. Harned

28 September 2008

Acknowledgements

- A. van Dijk, J. van Dijk, J. Galloway, R. Routh, J. Zimmerman, J. Mallmann (ASML).

- All other EUV teams at ASML, Zeiss, Philips, TNO, …

- A.M. Goethals, E. Hendrickx, J. Hermans, G. Lorusso, B. Baudemprez, G. Vandenberghe, K. Ronse (IMEC), A. Niroomand (Micron)

- M. Tittnich (CNSE), O. Wood (AMD), B. Lafontaine (AMD), D. Medeiros (IBM)

- Resist vendors (TOK, ShinEtsu, RHEM, FujiFilm)

Content

- ASML EUVL Roadmap

- Performance data from the AD-tools

- Stepping towards volume production

- Summary

Content

- ASML EUVL Roadmap

- Performance data from the AD-tools

- Stepping towards volume production

- Summary

ASML EUVL Roadmap
NXE platform supports manufacturing down to 11nm

Res	2008	2009	2010	2011	2012	2013	2014	2015
11 nm								0.4x NA
16 nm					0.32 NA, off axis illumination, 180wph			
22 nm					0.32 NA, 3 nm OVL, ~150wph			
27 nm			0.25 NA, 4nm OVL, 60->100wph					
ADT	0.25 NA, 8nm OVL							

same projection system, enhanced off-axis illumination

preparation implementation volume production

Content

- ASML EUVL Roadmap

- Performance data from the AD-tools

- Imaging, overlay, throughput

- Stepping towards volume production

- Summary

Two full field scanning Alpha Demo tools installed

College of Nanoscale Science & Engineering — University at Albany State University of New York

imec

λ	13.5 nm	• Single stage, 300mm wafers, linked to track
NA	0.25	• Single reticle load
Field size	$26 \times 33\ mm^2$	• Uses TWINSCAN technology (*e.g.* focus)
Magnification	4x reduction	• Reflective optics
Sigma	0.5	• Sn discharge source

SAMSUNG · PSC · TEXAS INSTRUMENTS · SEMATECH · ELPIDA · IBM · AMD Smarter Choice · TOSHIBA Alliance · ASML

Hynix Semiconductor · Qimonda · intel · Micron · NXP founded by Philips · Panasonic ideas for life · infineon

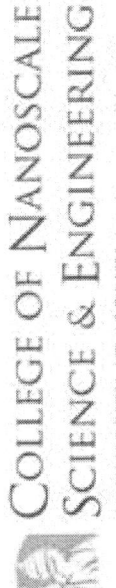

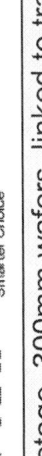

Resist advancements confirmed on full field system

AD-tool (NA = 0.25, σ = 0.5), no OPC

slit position: -12.72mm   -6.36mm   x=0mm   6.36mm   12.72mm

40 nm HP
Dose: 18.0 mJ/cm^2
April 2008

-11.2mm   -5.6mm   x=0mm   5.6mm   11.2mm

35 nm HP
Dose: 26 mJ/cm2
July 2008

32 nm HP
Dose: 29.5 mJ/cm2
Sep. 2008

source: IMEC system, Albany system

35nm HP L/S FWCDU≤1.5nm on full field EUV system

AD-tool (NA = 0.25, σ = 0.5)

Intrafield CDU TF (H lines)

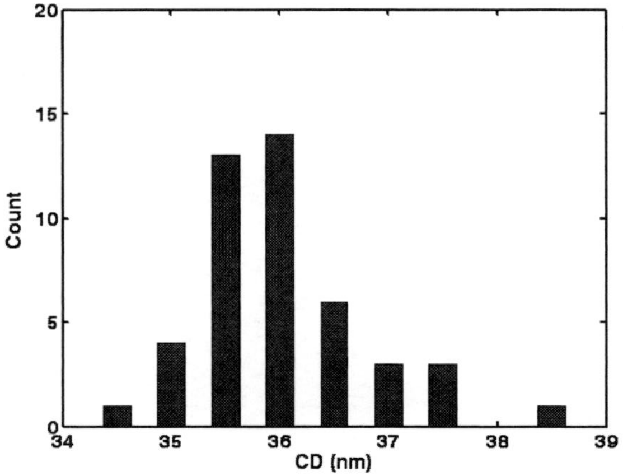

	Mean CD [nm]	CDU [nm, 3σ]
H	36.0	2.4
V	32.2	2.5

TF = 180nm focus range

Intrafield CDU (H lines)

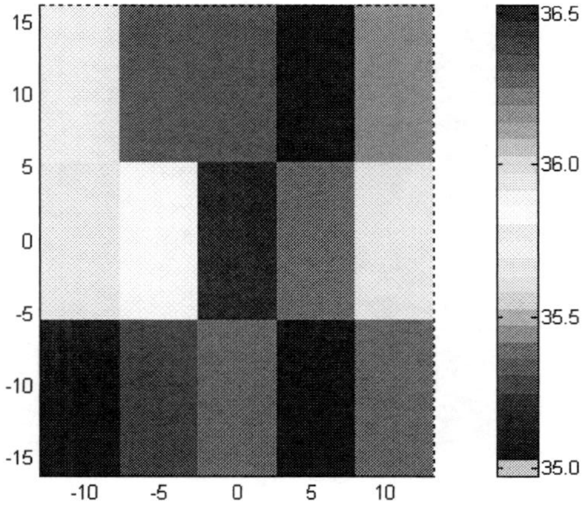

	Mean CD [nm]	CDU [nm, 3σ]
H	35.5	1.2
V	31.7	1.2

full wafer CDU (H lines)

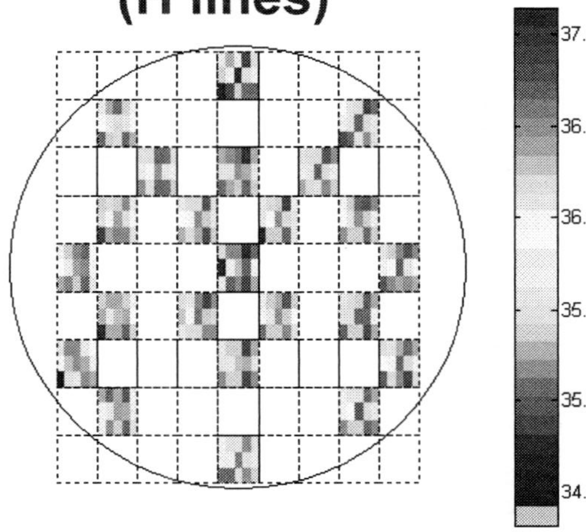

	Mean CD [nm]	CDU [nm, 3σ]
H	35.5	1.3
V	31.7	1.5

No OPC applied to correct for shadowing

source: IMEC system

30nm IL (1:6) performance on full field EUV system

AD-tool (NA = 0.25, σ = 0.5)

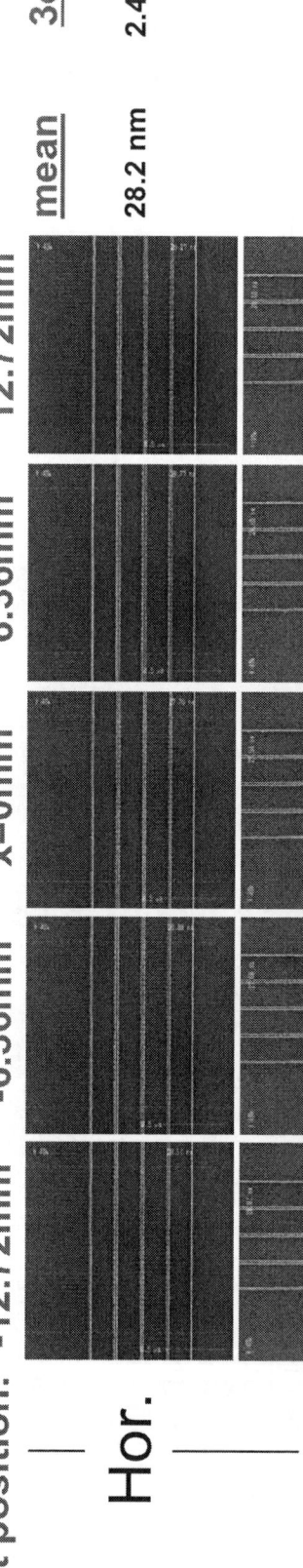

slit position:	-12.72mm	-6.36mm	x=0mm	6.36mm	12.72mm	mean	3σ
Hor.						28.2 nm	2.4 nm
Ver.						27.3 nm	2.5 nm
+45°						28.4* nm (*after SEC)	1.7 nm
-45°						30.7* nm (*after SEC)	1.3 nm

orientation

80nm SEVR40 (ShinEtsu)
Dose: 32.0 mJ/cm²

source: Albany system

COLLEGE OF NANOSCALE
SCIENCE & ENGINEERING
UNIVERSITY AT ALBANY State University of New York

ASML

30nm IL (1:6) performance on full field EUV system
AD-tool (NA = 0.25, σ = 0.5)

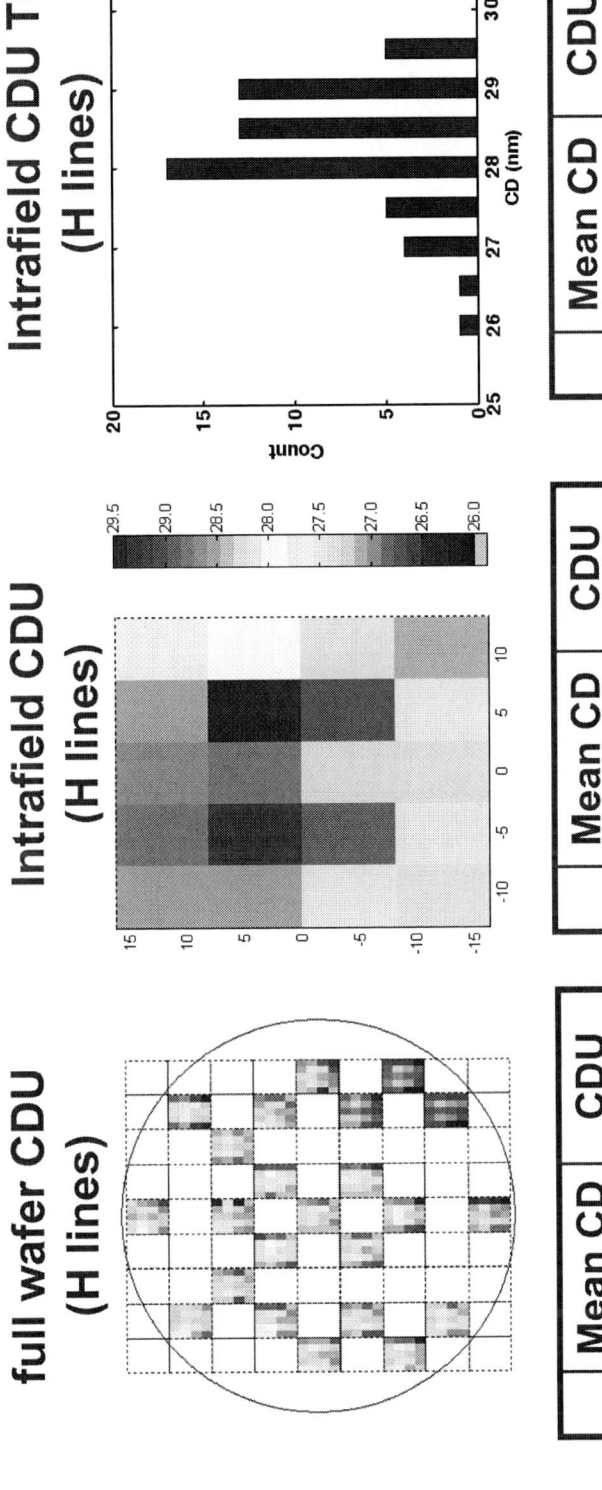

Intrafield CDU TF (H lines)

	Mean CD [nm]	CDU [nm, 3σ]
H	28.3	2.3
V	27.8	2.0

TF = 100nm focus range

Intrafield CDU (H lines)

	Mean CD [nm]	CDU [nm, 3σ]
H	28.3	2.1
V	28.0	2.1

OPC applied to correct for shadowing

full wafer CDU (H lines)

	Mean CD [nm]	CDU [nm, 3σ]
H	28.4	2.9
V	28.0	3.3

source: Albany system

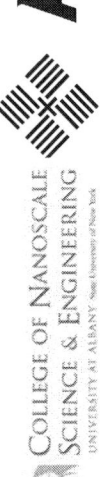

College of Nanoscale Science & Engineering
University at Albany, State University of New York

ASML

32nm HP Flash gate structure full field

AD-tool (NA = 0.25, σ = 0.5), no OPC, single exposure

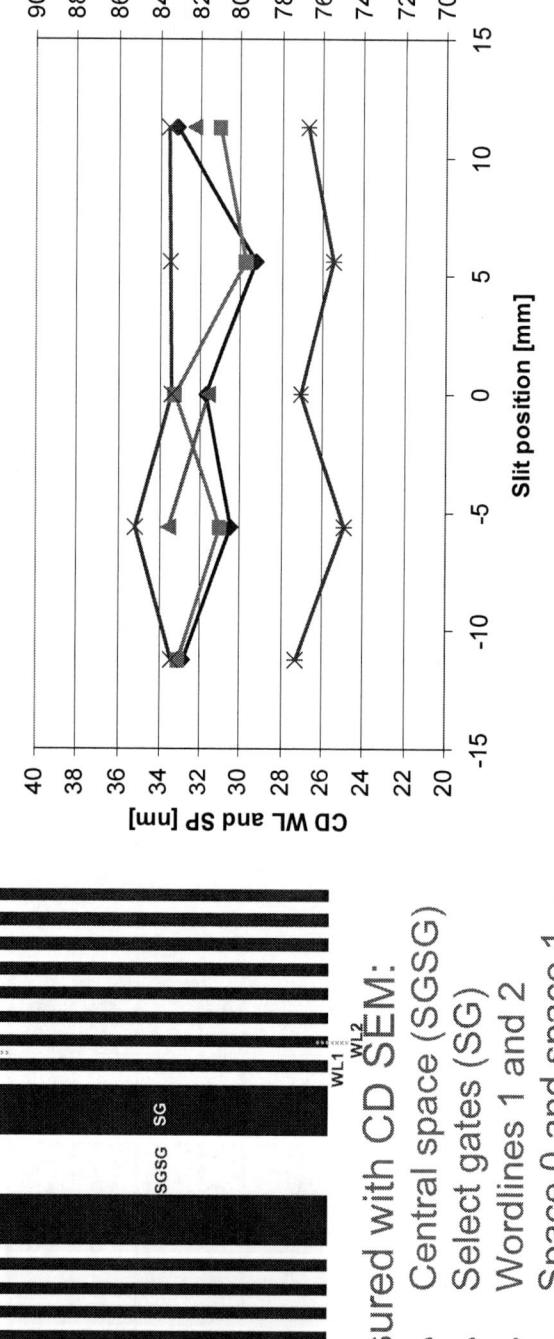

Legend: WL2, WL1, SP1, SP0, SG

CD SG [nm] (axis: 90, 88, 86, 84, 82, 80, 78, 76, 74, 72, 70)

CD WL and SP [nm] (axis: 40, 38, 36, 34, 32, 30, 28, 26, 24, 22, 20)

Slit position [mm] (axis: -15, -10, -5, 0, 5, 10, 15)

Measured with CD SEM:
- Central space (SGSG)
- Select gates (SG)
- Wordlines 1 and 2
- Space 0 and space 1

X = -11.2 mm   X = -5.6 mm   X = 0 mm   X = 5.6 mm   X = 11.2 mm

80nm SEVR-40 (ShinEtsu)
Dose: 29 mJ/cm²

COLLEGE OF NANOSCALE
SCIENCE & ENGINEERING
UNIVERSITY AT ALBANY State University of New York

ASML

source: Albany system

First scanned 28 nm HP L/S demonstrated

AD-tool (NA = 0.25, σ = 0.5), no OPC

Status Feb/2008:

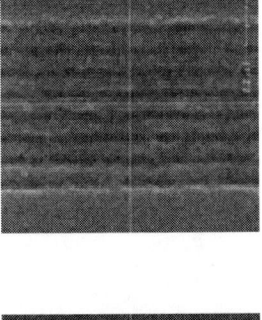

31 nm

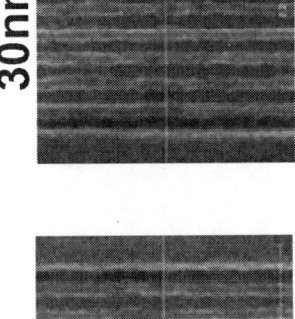

30nm

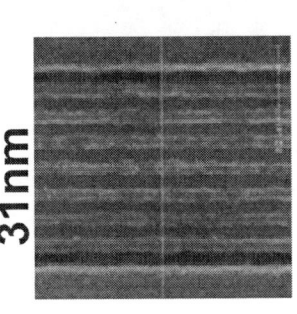

29nm

Resist A
Dose: ~25 mJ/cm^2

Status Sep/2008:

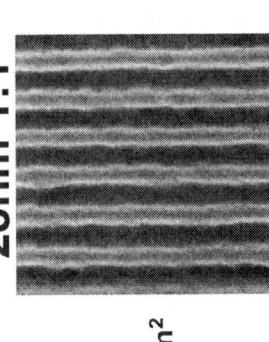

28nm 1:1

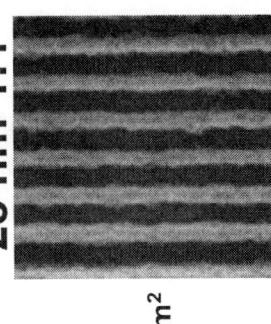

28 nm 1:1

Resist B
Dose: 29.5 mJ/cm^2

Resist C
Dose: 19.2 mJ/cm^2

Conclusion: ADT imaging capability is better explored as a results of progress in resist development. Further improvements still required.

source: Albany system, IMEC system

First scanned 28 nm HP L/S and CHs resolved

AD-tool (NA = 0.25, σ = 0.5), no OPC

Status Feb/2008:

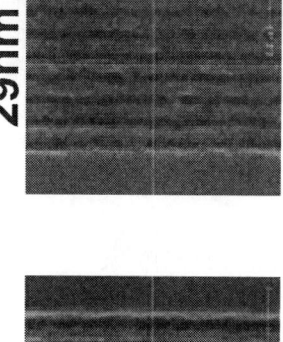

31nm

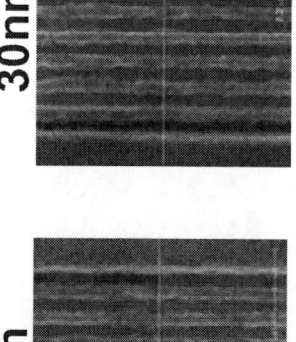

30nm

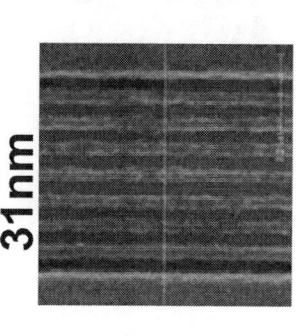

29nm

Resist A
Dose: ~25 mJ/cm²

Status Sep/2008:

28nm
aligned CH

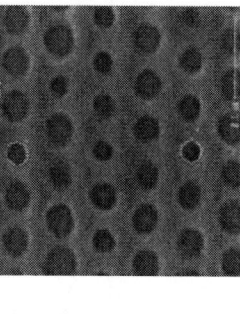

28nm
staggered CH

Holes opened

Dose: ~68 mJ/cm²

Conclusion: ADT imaging capability is better explored as a results of progress in resist development. Further improvements still required.

source: Albany system, IMEC system

Measured system throughput has improved >10×

After installation of 120 W/2π Sn DPP source upgrade

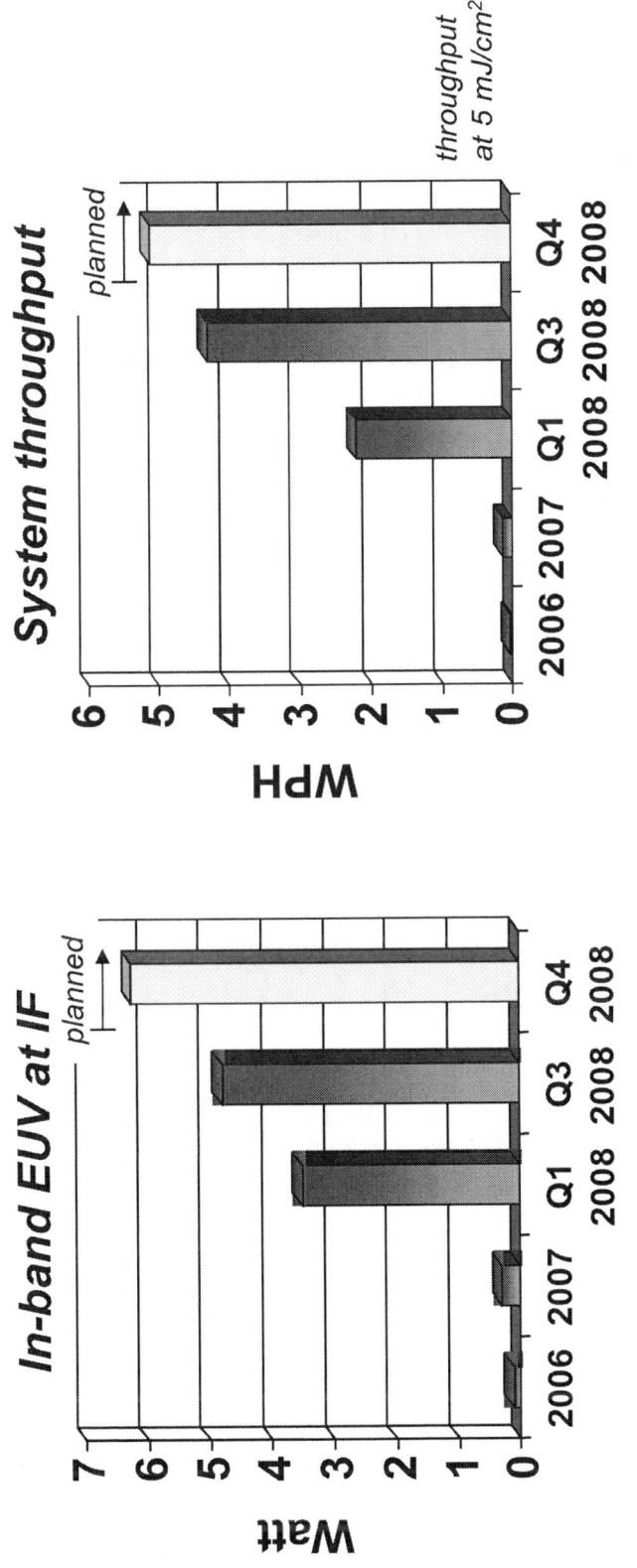

- 4Q'08: full throughput potential to be further explored with 170 W/2π source upgrade.

source: Albany system

ASML

Full field systems: >580 300mm wafers exposed

Demonstrated throughput >4 wph @ 5 mJ/cm²

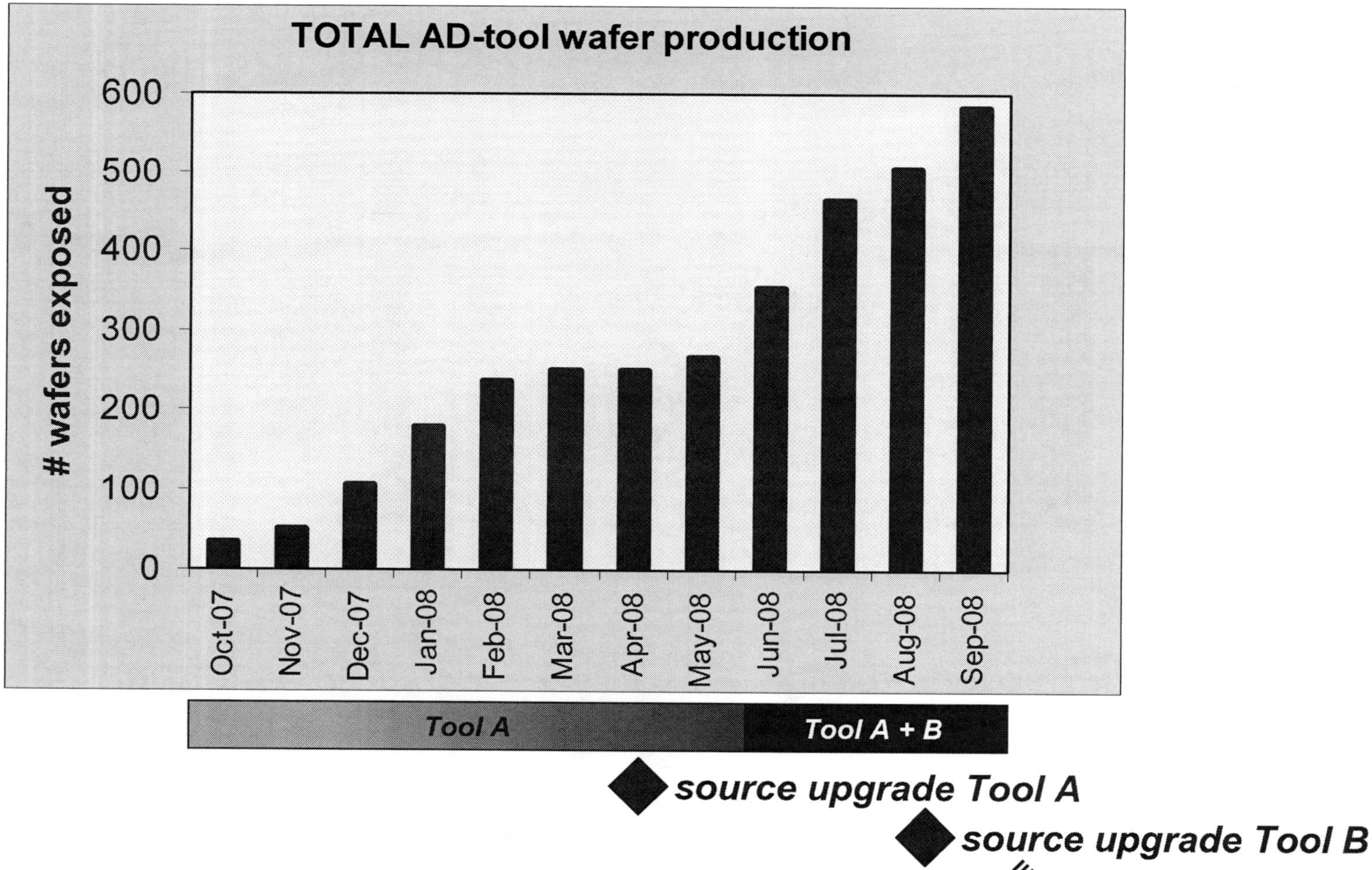

TOTAL AD-tool wafer production
wafers exposed
600
500
400
300
200
100
0
Oct-07
Nov-07
Dec-07
Jan-08
Feb-08
Mar-08
Apr-08
May-08
Jun-08
Jul-08
Aug-08
Sep-08
Tool A
Tool A + B
source upgrade Tool A
source upgrade Tool B

ASML

EUV system overlay has improved to 5.0 nm

120 W/2π source, 4.3 wph @ 5 mJ/cm²

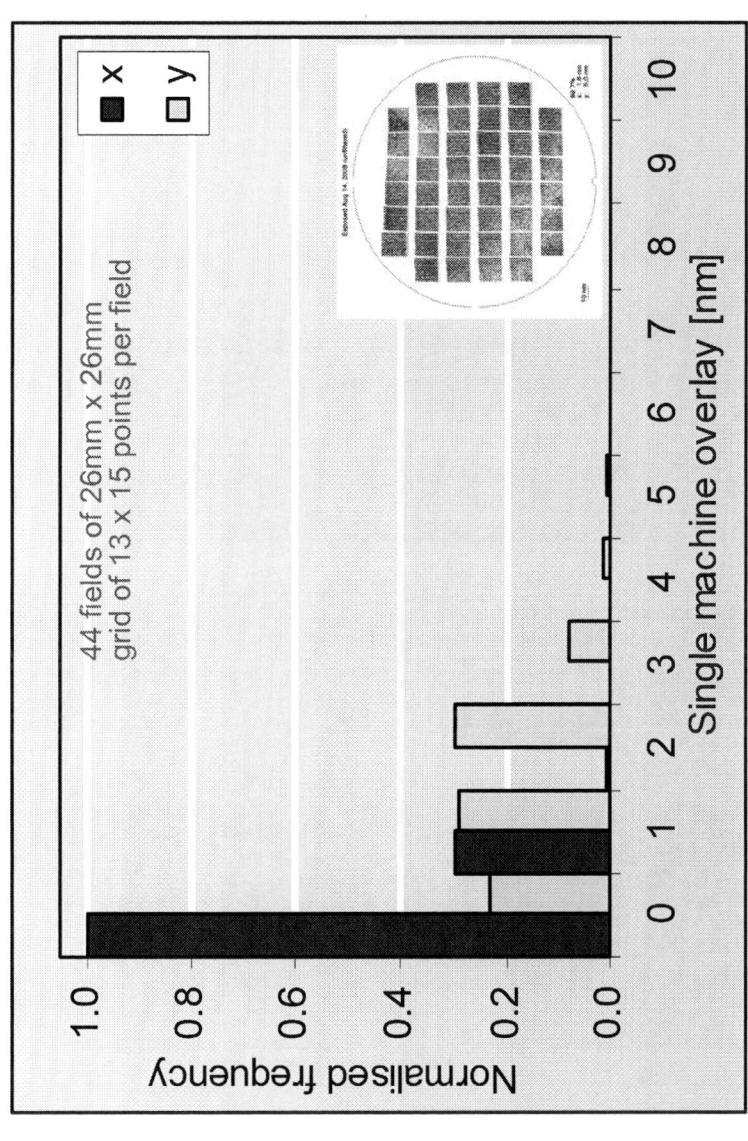

Single Machine Overlay = 5.0 nm (99.7%)

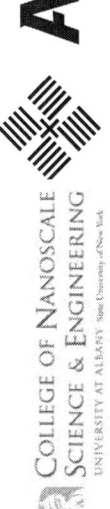

source: Albany system

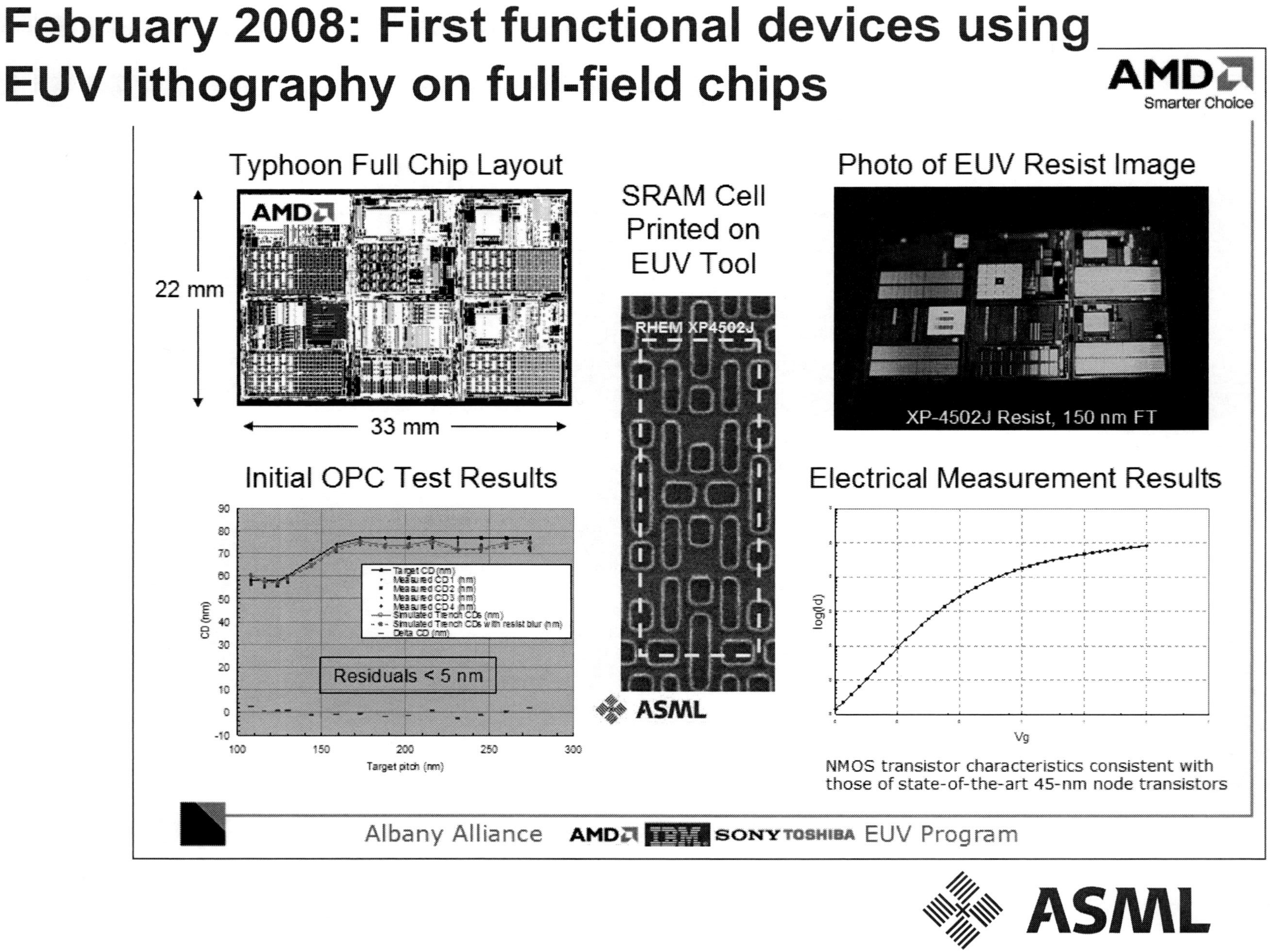
February 2008: First functional devices using EUV lithography on full-field chips
AMD Smarter Choice
Typhoon Full Chip Layout
22 mm
33 mm
SRAM Cell Printed on EUV Tool
RHEM XP4502J
ASML
Photo of EUV Resist Image
XP-4502J Resist, 150 nm FT
Initial OPC Test Results
CD (nm)
Target pitch (nm)
Target CD (nm)
Measured CD1 (nm)
Measured CD2 (nm)
Measured CD3 (nm)
Measured CD4 (nm)
Simulated Trench CDs (nm)
Simulated Trench CDs with resist blur (nm)
Delta CD (nm)
Residuals < 5 nm
Electrical Measurement Results
log(Id)
Vg
NMOS transistor characteristics consistent with those of state-of-the-art 45-nm node transistors
Albany Alliance AMD IBM SONY TOSHIBA EUV Program
ASML

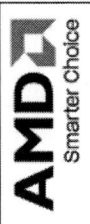

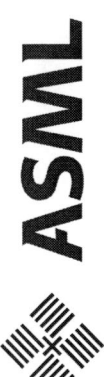

July 2008: 32 nm SRAM device circuit using EUVL demonstrated

NEWS RELEASE

IMEC reports major progress in EUV

SEMICON WEST 2008 – July 15-17, 2008 – Moscone Center – San Francisco – California – Booth #851

Leuven, Belgium – July 14, 2008 – IMEC reports functional 0.186µm² 32nm SRAM cells from which the contact layer was successfully printed using ASML's full field extreme ultraviolet (EUV) Alpha Demo Tool (ADT). IMEC also completed the integration and site acceptance test of the EUV ADT in its 300mm clean room. Stimulated by these milestones and with a concerted effort from all actors involved in EUV research, IMEC is determined to advance EUV full speed towards the 22nm node. The circuits were

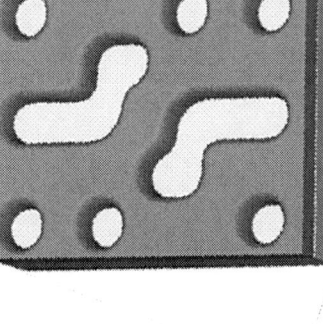 

design

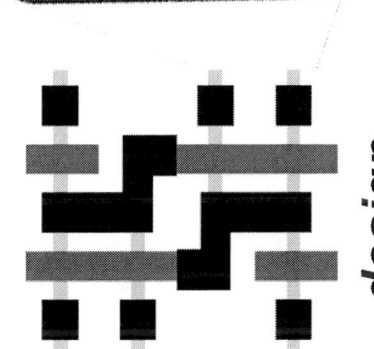

post litho

aerial image

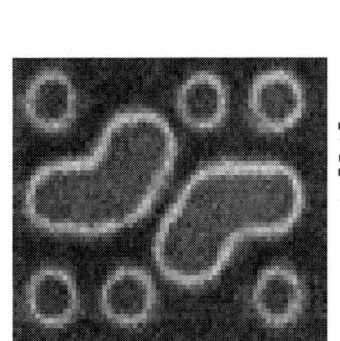

post etch

Lithography : Exposure latitude

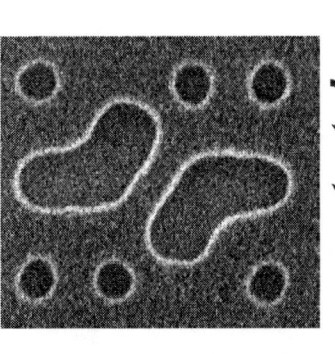 12mJ/cm² 55.3nm
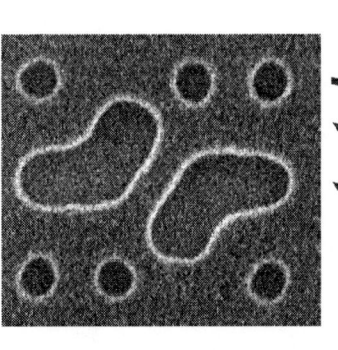 14 mJ/cm² 59.1nm
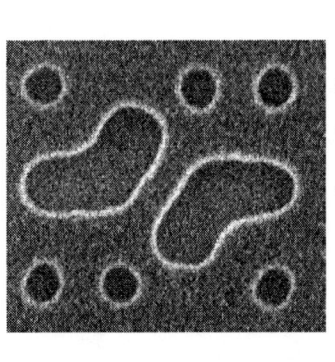 16 mJ/cm² 62.6nm
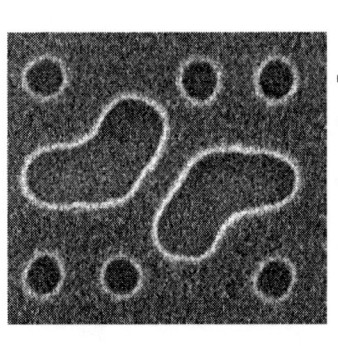 18 mJ/cm² 65.1nm

After Oxide etch

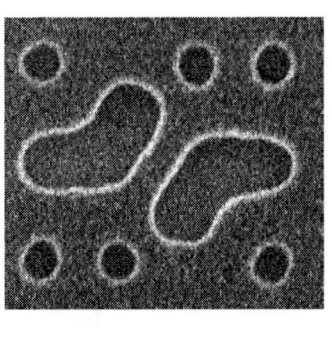 43.5nm 48.8nm 50.1nm 51.0nm

    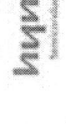 

Content

- ASML EUVL Roadmap

- Performance data from the AD-tools

- **Stepping towards volume production**

- system design, optics, source

- Summary

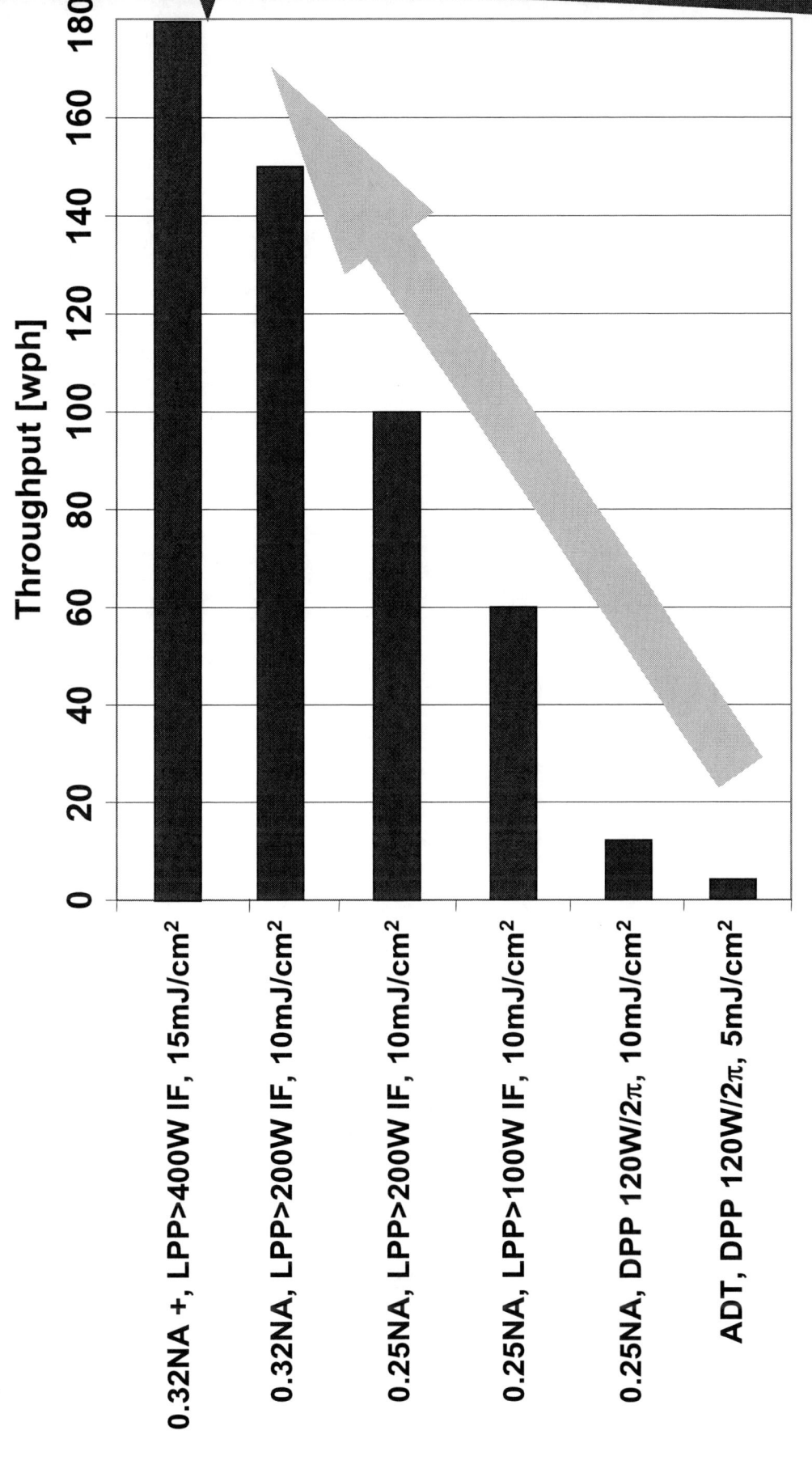

First production system design is finished
and module manufacturing ongoing

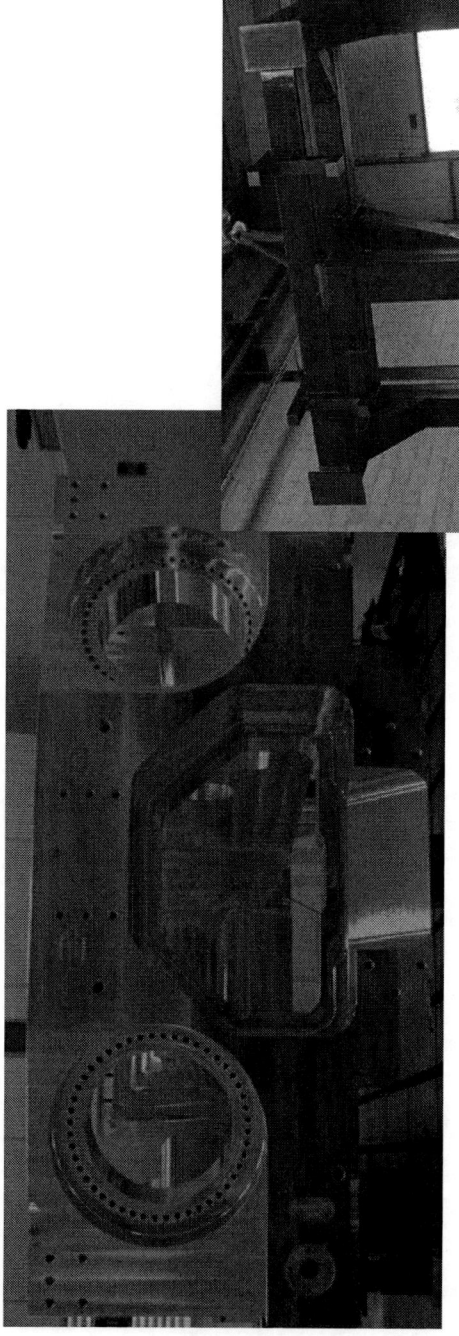

More than 10 HVM mirrors have been fabricated:
Considerable reduction of the flare has been achieved

Flare is calculated for a 2 µm line in a bright field

$$Flare \propto n_{mirrors} \cdot (MSFR/\lambda)^2$$

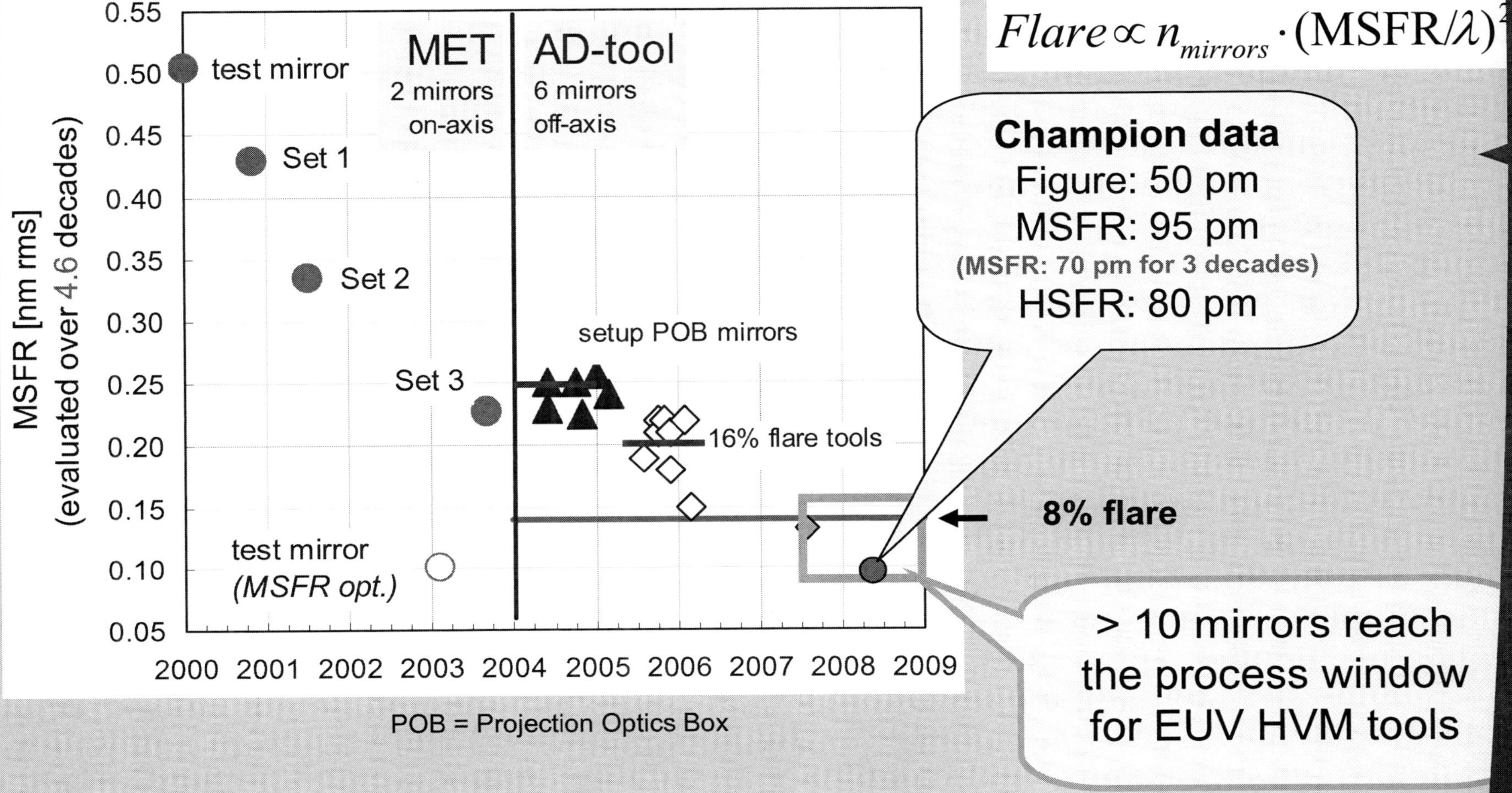

Champion data
Figure: 50 pm
MSFR: 95 pm
(MSFR: 70 pm for 3 decades)
HSFR: 80 pm

8% flare

> 10 mirrors reach
the process window
for EUV HVM tools

MSFR [nm rms]
(evaluated over 4.5 decades)

MET	AD-tool
2 mirrors	6 mirrors
on-axis	off-axis

test mirror

Set 1

Set 2

Set 3

setup POB mirrors

16% flare tools

test mirror
(MSFR opt.)

0.55 0.50 0.45 0.40 0.35 0.30 0.25 0.20 0.15 0.10 0.05

2000 2001 2002 2003 2004 2005 2006 2007 2008 2009

POB = Projection Optics Box

Enabling the Nano-Age World®

ASML

ADT has stable transmission (same solution for NXE)

17 month system data, 50-1000x improvem. by vacuum control

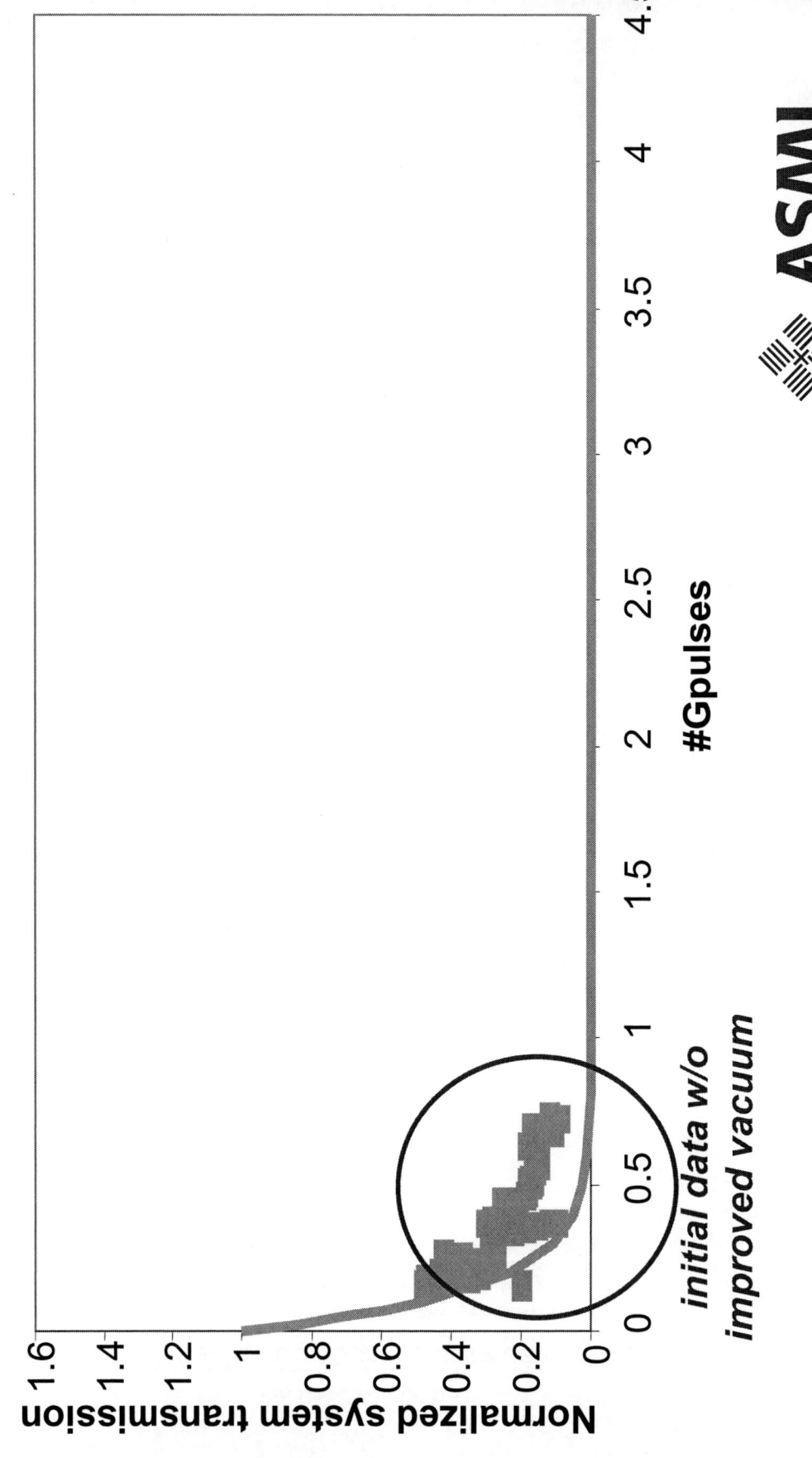

Normalized system transmission

#Gpulses

initial data w/o
improved vacuum

ASML

ADT has stable transmission (same solution for NXE)

17 month system data, 50–1000x improvem. by vacuum control

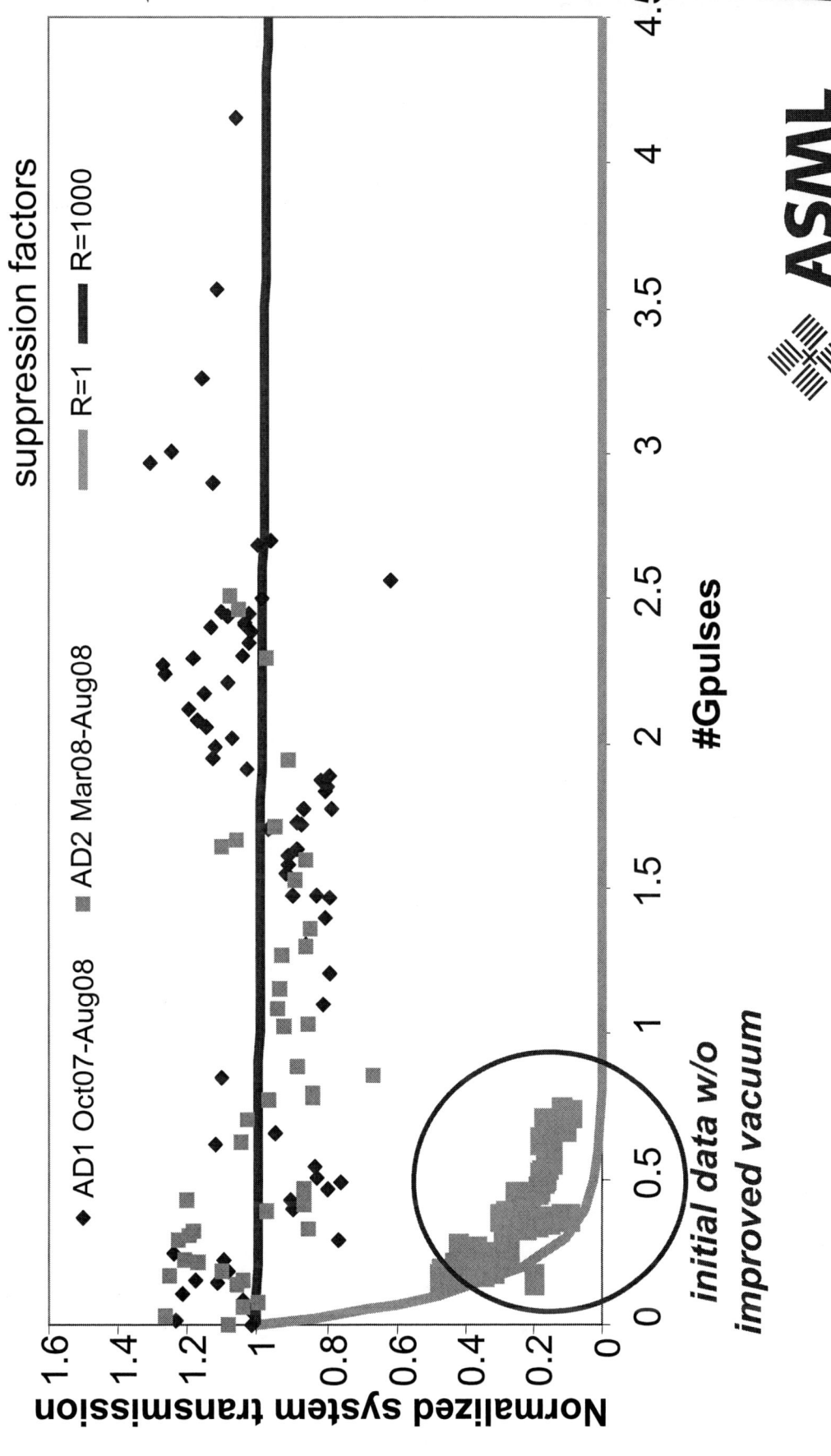

LPP Source System Production

First system to ASML 12/'08

- The first production LPP system is operational
- 100W Burst Power achieved *on schedule*
- 5sr collector ready for integration Q4 08
- Thermal capability designed for >100W at IF
- Second system being assembled

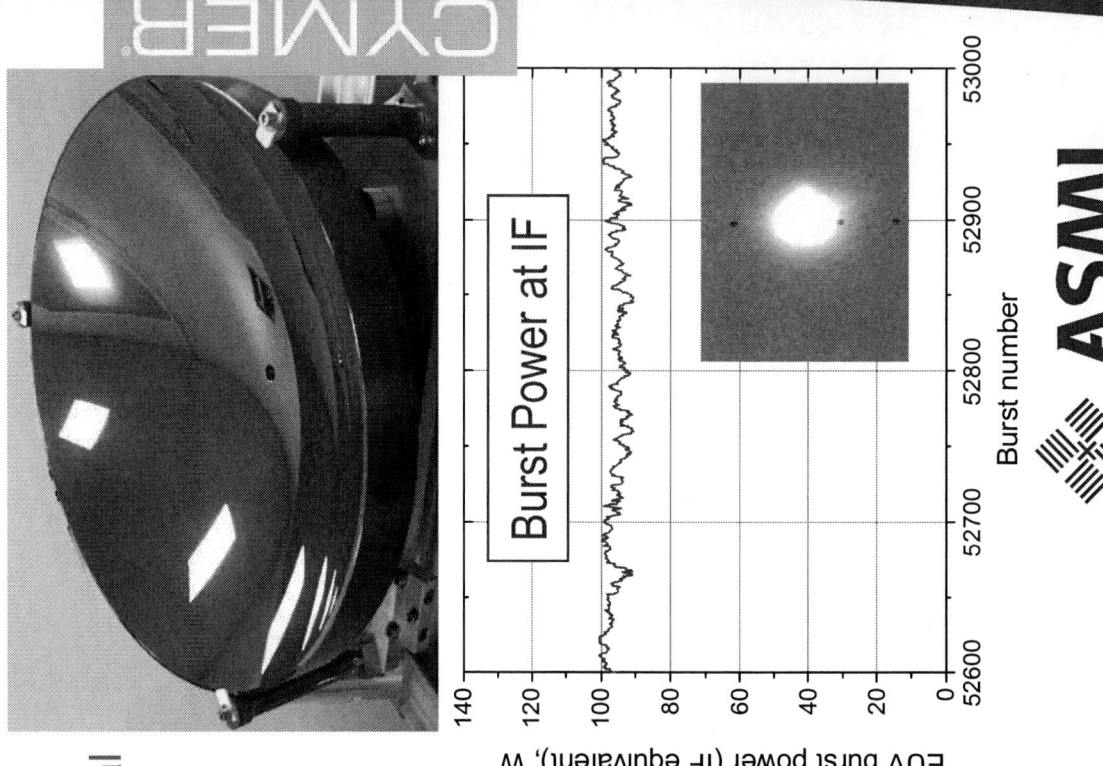

LPP Debris Mitigation Testing Results

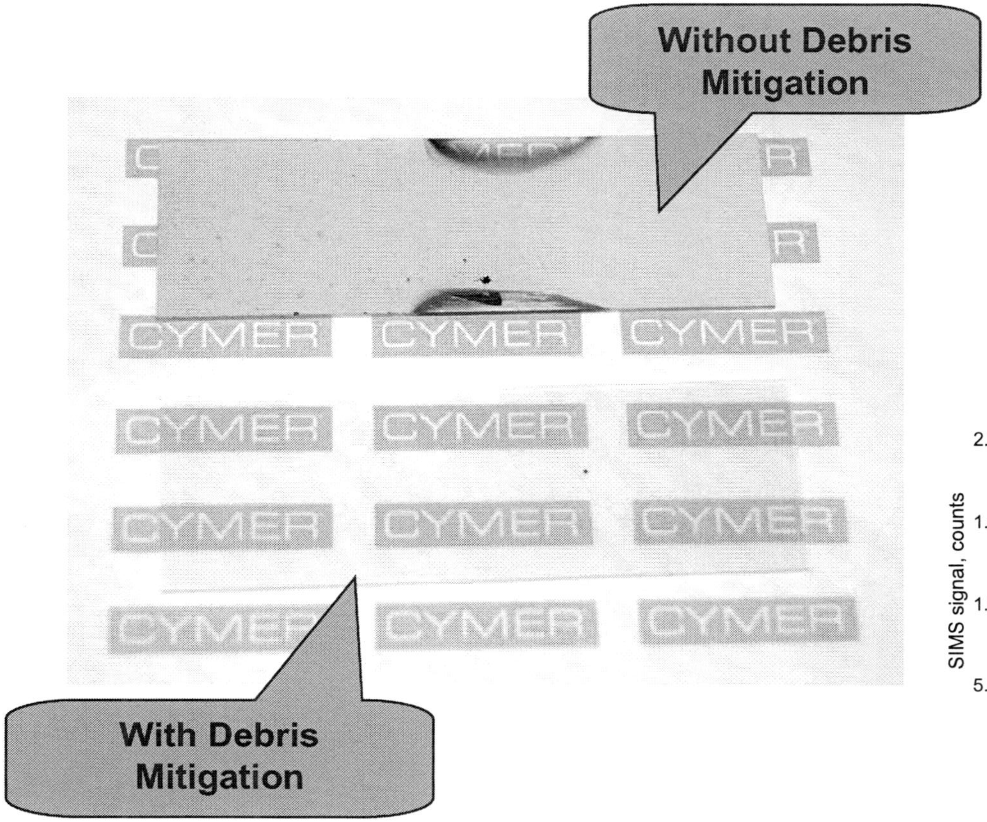

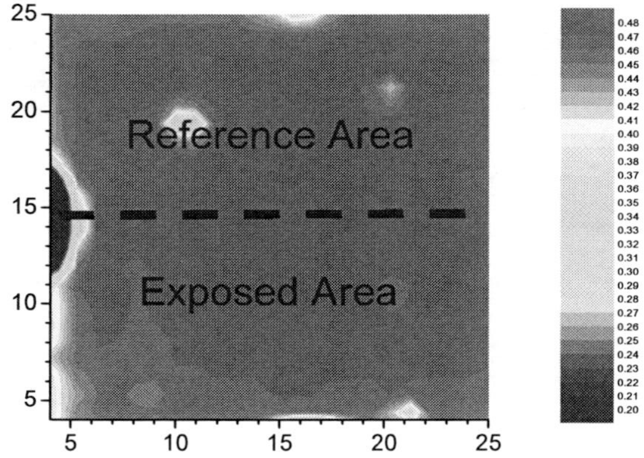

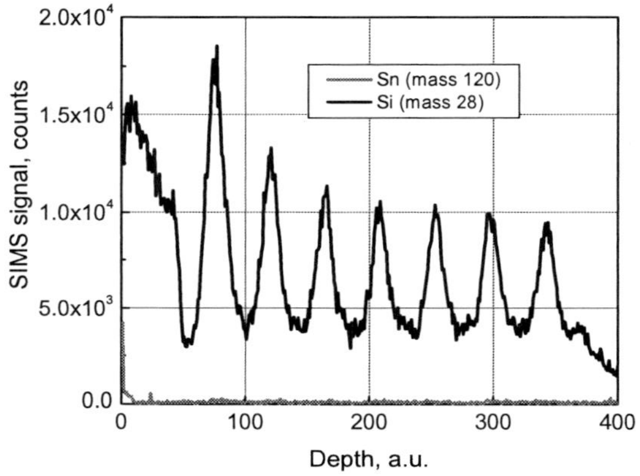

Reference Area

Exposed Area

Without Debris Mitigation

With Debris Mitigation

Sn (mass 120)
Si (mass 28)

SIMS signal, counts

Depth, a.u.

- Glass witness samples were mounted at collector position and exposed to LPP in several experiments to 30 million pulses
- SIMS analysis of 8 layer sample shows no erosion from ions
- 2D reflectivity maps shows ~1% between exposed and reference areas
- 30M pulses ~2 hours at 8% duty cycle

LPP Source output energy planned to improve 100x during 2008

DC from 5%->~100%, continuous operation from 2s to 10hrs

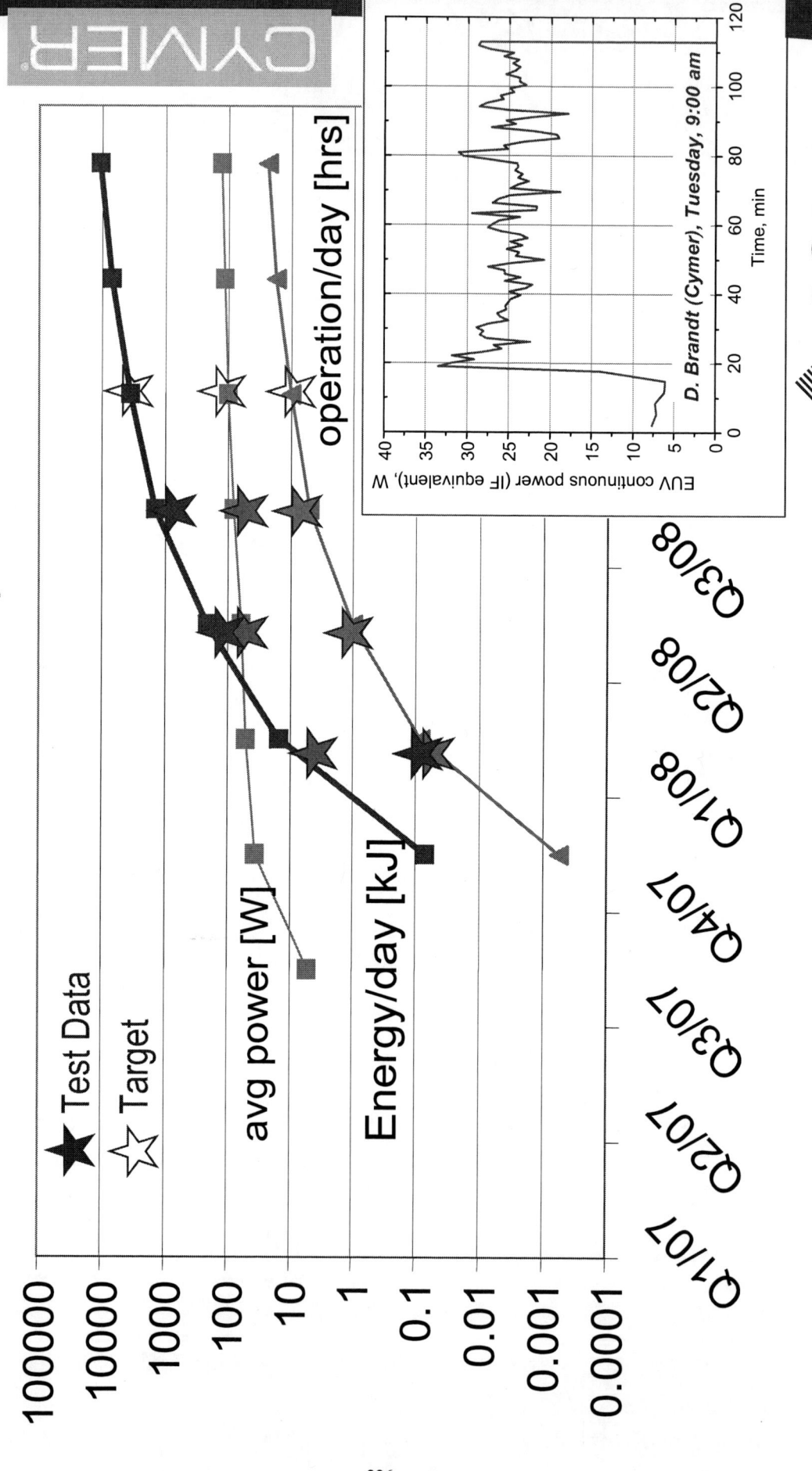

Related papers

- Monday 9:20 AM – 9:40 AM *EUV Alpha Demo Tools – Stepping Stones Towards Volume Production*, H. Meiling, S. Lok, B. Hultermans, C. Wagner, N. Harned, K. Cummings
- Monday 10:40 AM – 11:00 AM *EUV resist performance on the ASML ADT and LBNL MET*, B. Pierson, T. Wallow, A. Fumar-Pici, L. Ohara, K. van Ingen Schenau, S. Han, O. Wood, J. Mallmann, B. Kessels, R. Routh, K. Cummings
- Monday 12:20 PM – 12:30 PM *EU Regional Update*, Rob Hartman
- Tuesday 9:00 AM – 9:20 AM *LPP EUV Source Development and Productization*, D. Brandt, I. Fomenkov, A. Ershov, N. Bowering, W. Partlo, D. Myers, A. Bykanov, G. Vaschenko, O. Khodykin, J. Hoffman, C. Chrobak (Cymer, Inc.)
- Tuesday 9:40 AM – 10:00 AM *Sn DPP Source Collector Modules for Beta and HVM*, M. Corthout, M. Yoshioka, P. Zink (Phillips Extreme UV GmbH)
- Tuesday 10:40 AM – 11:00 AM *Strategy For Minimizing EUV Optics Contamination During Exposure*, R. Moors, M. van Kampen, V. Banine, J. Huijbregtse, D. Ehm, R. Verberk, E. te Sligte, A. Storm
- Tuesday 11:20 AM – 11:40 AM *Shielded Plasmas for Cleaning of EUV Mirrors*, N.B. Koster, R. Koops, K. Agovic, F. de Groote, F. Wieringa, M. Meijerink (TNO Science and Industry)
- Tuesday 12:00 PM – 12:20 PM *The Progress of EUV Reticle Dual Pod Carriers for Use in the Fab and Exposure Tools*, J. Zimmerman, L. He, R. Koops
- Wednesday 1:20 PM – 1:40 PM *Optics for EUV Lithography*, P. Kuerz, T. Boehm, U. Dinger, H. Mann, S. Muellender, M. Dahl, M. Lowisch, M. Muehlbeyer, O. Natt, S. Rennon, E. Sohmen, T. Stein, G. Wittich, W. Kaiser, W. Rupp (Carl Zeiss SMT AG)
- Wednesday 5:00 PM – 5:20 PM *Spectral Purity Filter Development for EUV HVM*, A. Yakunin, E. Kluenkov, A. Lopatin, V. Luchin, N. Salashchenko, N. Tsybin, L. Sjmaenok, M. Markosov, R. Moors, V. Banine
- …and various ADT customer papers

Content

- ASML EUVL Roadmap

- Performance data from the AD-tools

- Stepping towards volume production

- Summary

ASML

Summary

- At IMEC and Albany EUVL development programs are executed on ADT systems.

- Resist progress yielded 28 nm half pitch L/S and CH images using scanning single exposure, conventional illumination, and no OPC.

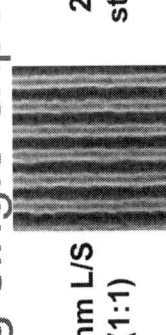

28nm L/S (1:1)

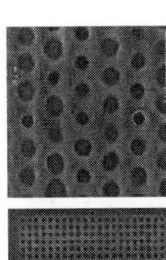

28nm dense staggered CH

  - further improvements are required to support ≤27nm node imaging.

- Hardware of first production system is coming together.

- 1st LPP source planned to be shipped to ASML end of year.

- NXE platform productivity roadmap targets 180 wph in 2012.

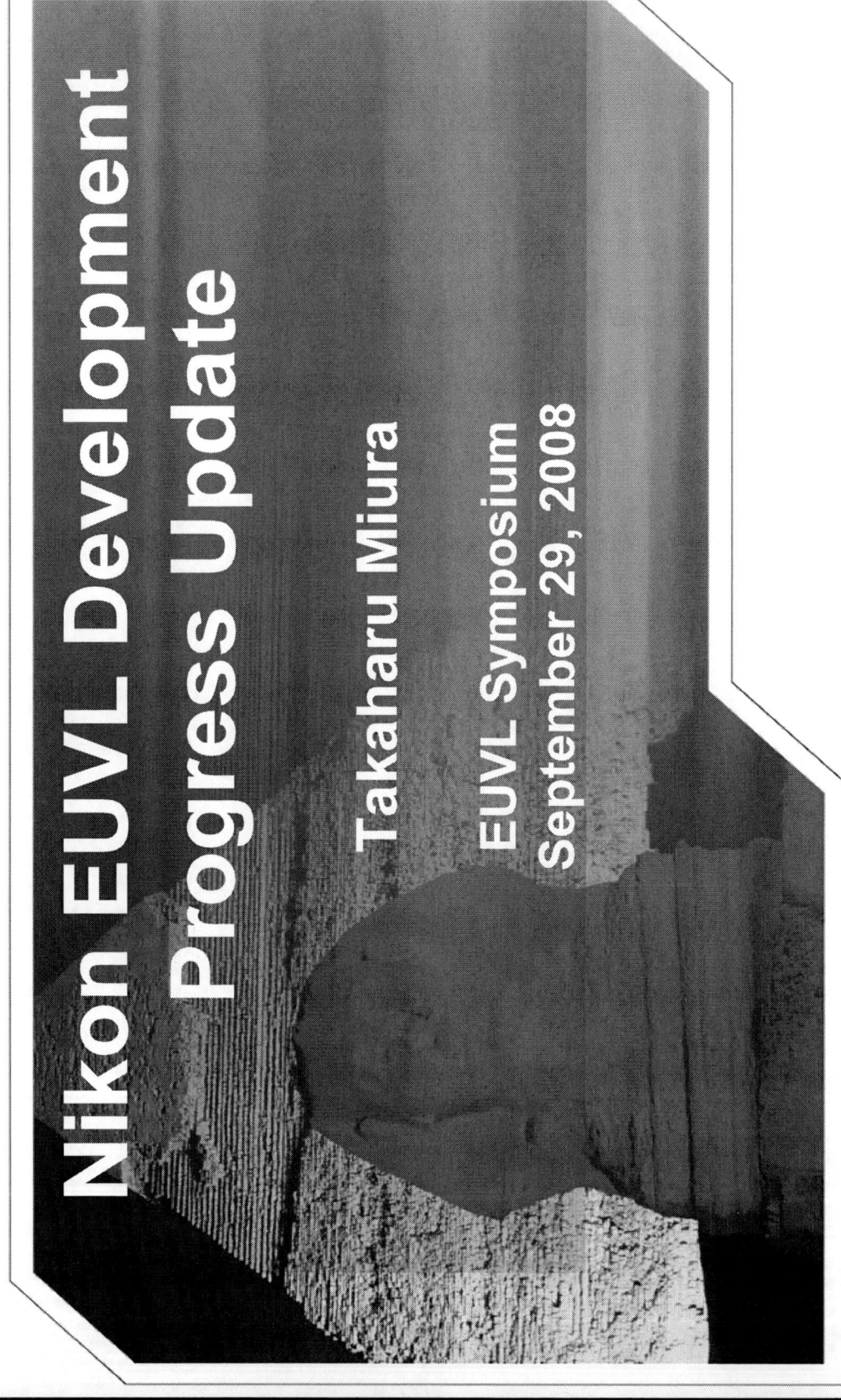

Nikon EUVL Development Progress Update

Takaharu Miura

EUVL Symposium
September 29, 2008

NIKON CORPORATION

EUVL Symposium 2008 @Lake Tahoe   T. Miura   September 29, 2008

Presentation Outline

1. Nikon EUV roadmap

2. Current status of EUV1 and latest data

3. Update progress in various areas

4. EUVL tool development challenges

5. Future projection lenses

6. Future tool realization

7. Development summary

NIKON CORPORATION

Technology Options

ITRS Year	Half Pitch	ArFr	ArFi (1.07)	ArFi (1.3)	ArFi (1.35)	HI ArFi (1.55)	HI ArFi (1.7)	EUV (0.25)	EUV (0.35)
2007	65nm	0.31							
	55nm		0.3						
2010	45nm	0.21		0.3	0.31				
	40nm			0.27	0.28				
2013	32nm	0.15		0.22	0.28	0.32		0.59	
2016	22nm	0.15		0.22	0.26	0.28		0.41	0.57
2019	16nm			0.15		0.18	0.19	0.3	0.41

k1 Factor — ArF (193 nm) / HI ArFi / EUV (13.5 nm)

Callouts:
- Path is clear to 40-45 nm
- Timing issues for high-index immersion
- EUV for late 32 nm and 22 nm beyond
- High-index DP for 22 nm?
- DP for 32 nm, 22nm?

NIKON CORPORATION

EUVL Development Roadmap

Cal. Year	2007	2008	2009	2010	2011	2012	2013
ITRS2007							
DRAM ½ p	65 nm	57 nm	50 nm	45 nm	40 nm	36 nm	32 nm
Flash ½ p	54 nm	45 nm	40 nm	36 nm	32 nm	28 nm	25 nm
MPU C. Hole	84 nm	73 nm	64 nm	56 nm	50 nm	44 nm	39 nm

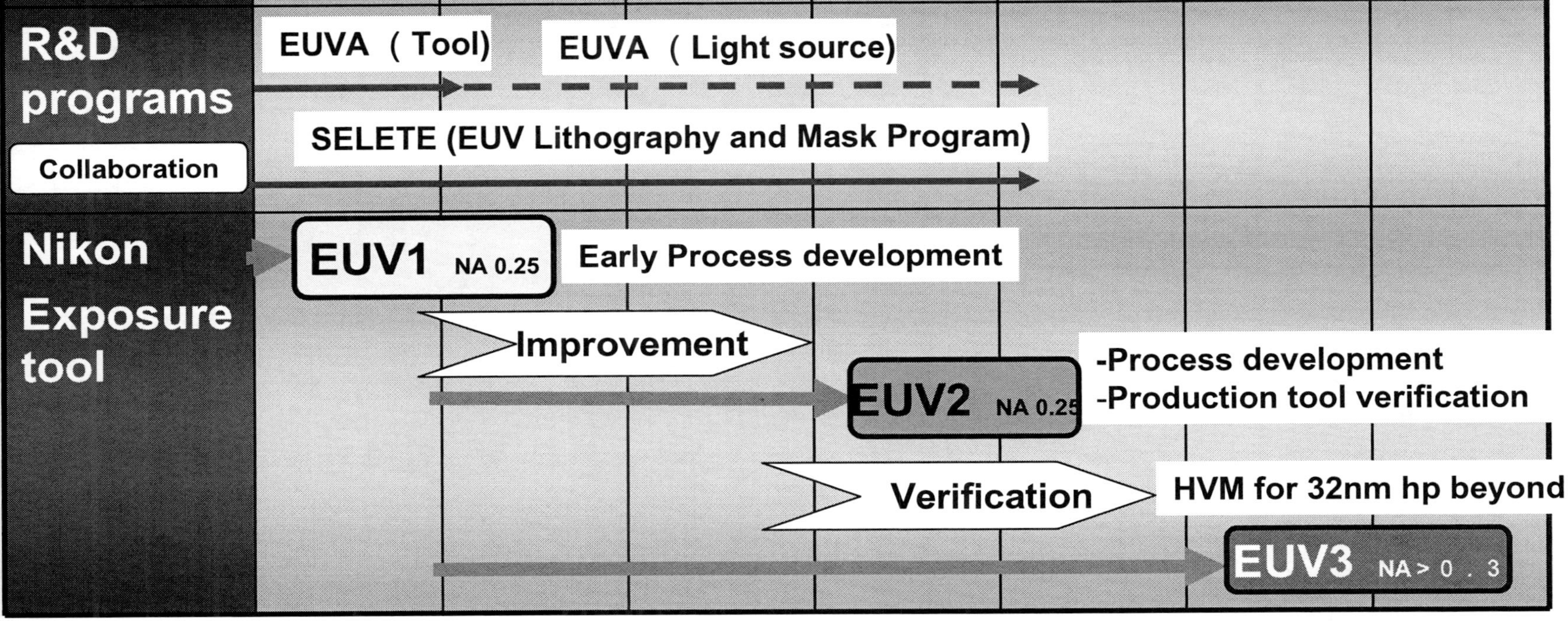

R&D programs

Collaboration

EUVA (Tool) EUVA (Light source)

SELETE (EUV Lithography and Mask Program)

Nikon Exposure tool

EUV1 NA 0.25 → Improvement → Early Process development

EUV2 NA 0.25 → Verification

-Process development
-Production tool verification

HVM for 32nm hp beyond

EUV3 NA > 0 . 3

NIKON CORPORATION

Performance Expectation

Specification Item	EUV1	EUV2 (2010)	EUV3 (2012)
Field Size	26 x 33 mm^2	26 x 33 mm^2	26 x 33 mm^2
NA and Magnification	0.25, x1/4	0.25, x1/4	>0.3, x1/4
Flare	10 %	7 %	5 %
Overlay	10 nm	5 nm	<3 nm
Throughput	5-10 wph (10W IF, 5mJ/cm^2) 76 shots	20 wph (50W IF, 10mJ/cm^2) 76 shots	100 wph (115W IF, 5mJ/cm^2 180W IF, 10mJ/cm^2) 76 shots

NIKON CORPORATION

EUV1 Tool Development Status

Module	Module/System Integration	System Operation

We are here

LDR

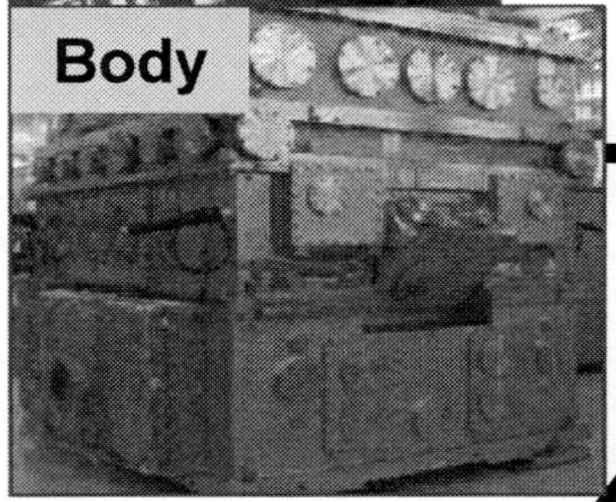

Body

Reticle Stage

Wafer Stage

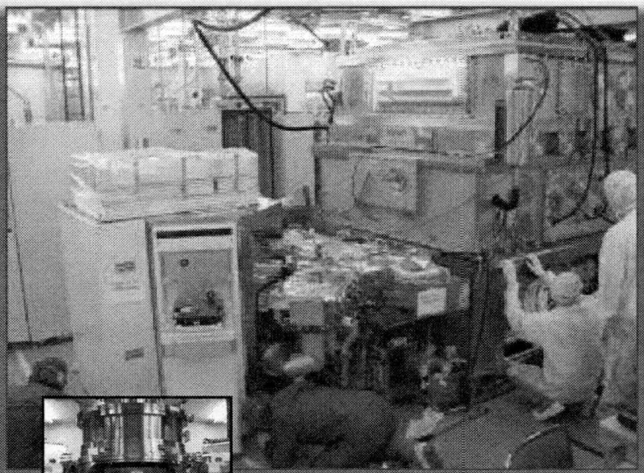

EUV Source

PO

Dynamic exposure started. Optimization underway.

Static exposures started. 28 nm L/S & CH capability.

Field Size	26 x 33 mm^2
NA and Magnification	0.25, x1/4
Sigma	0.8
Overlay	10 nm

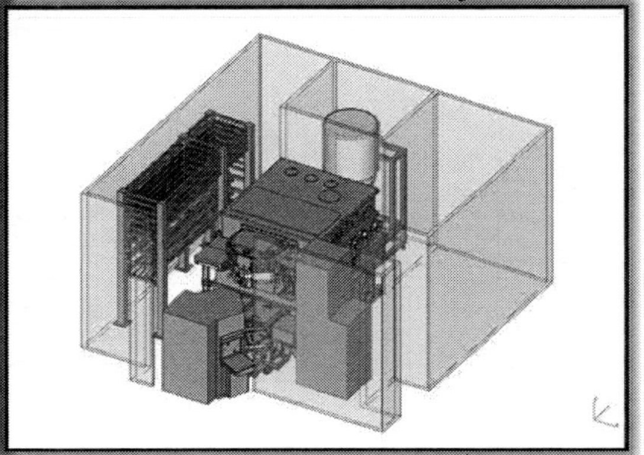

NIKON CORPORATION

EUV1 PO Wavefront Map

Extremely small WFE achieved

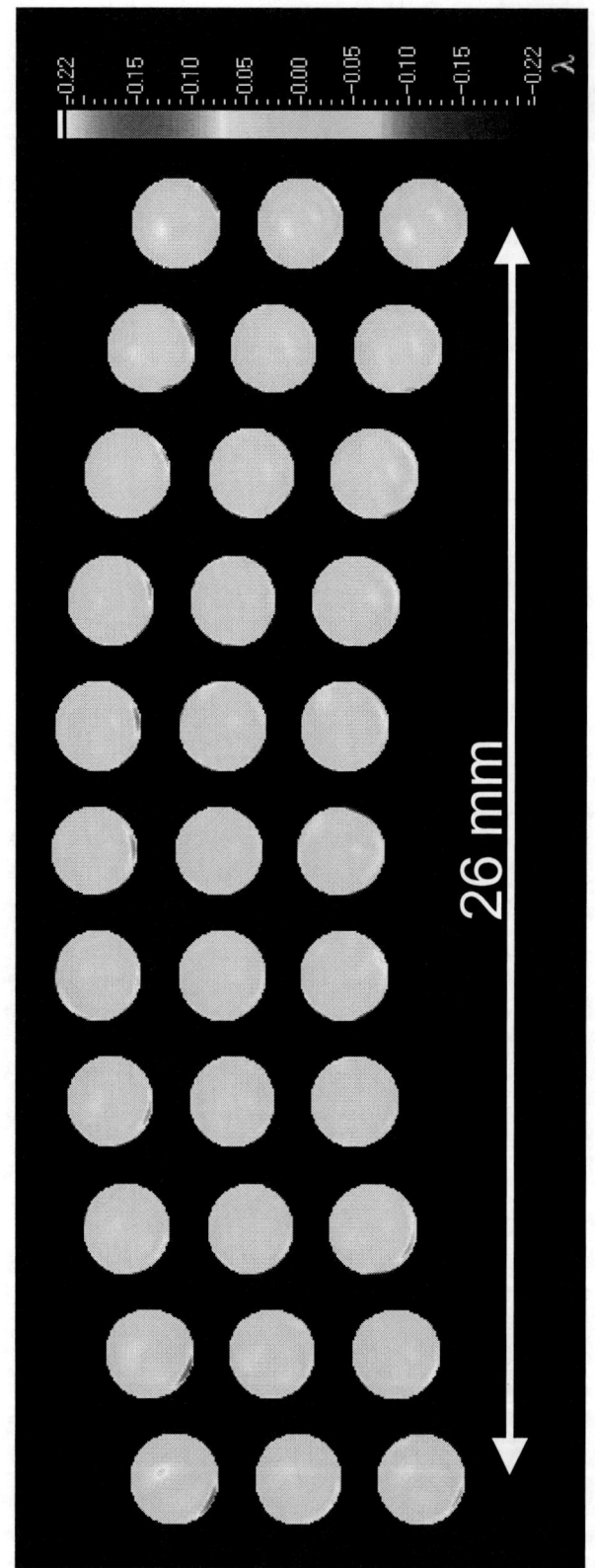

26 mm

WFE   0.4 nm RMS   (average)

Min. 0.3 nm RMS ~ Max. 0.5 nm RMS

Static Exposure Results Through Full Ring-Field

NIKON CORPORATION

By courtesy of Selete
Shown at Selete Symposium

32 nm hp Elbow pattern

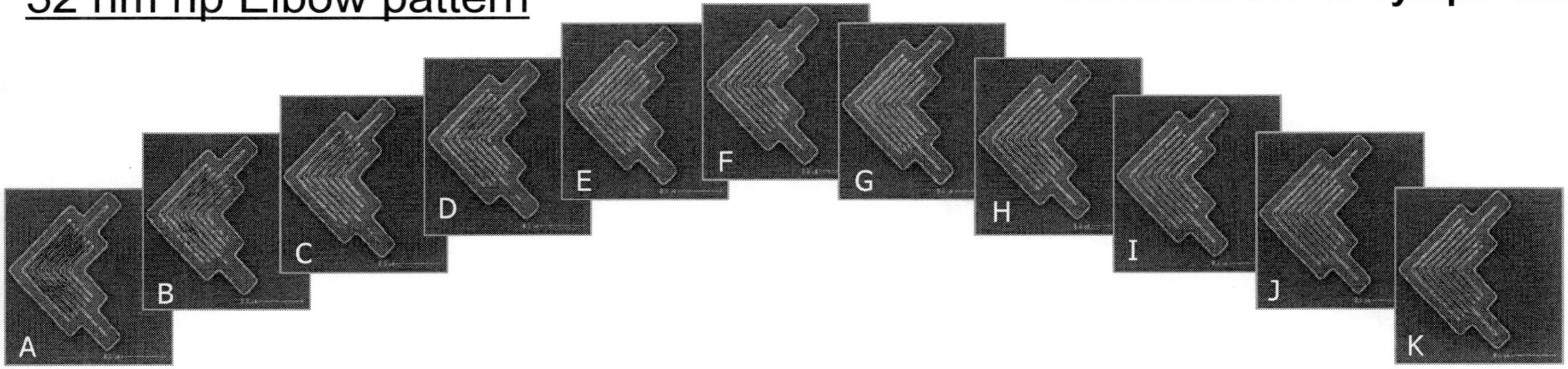

32 nm hp V-line

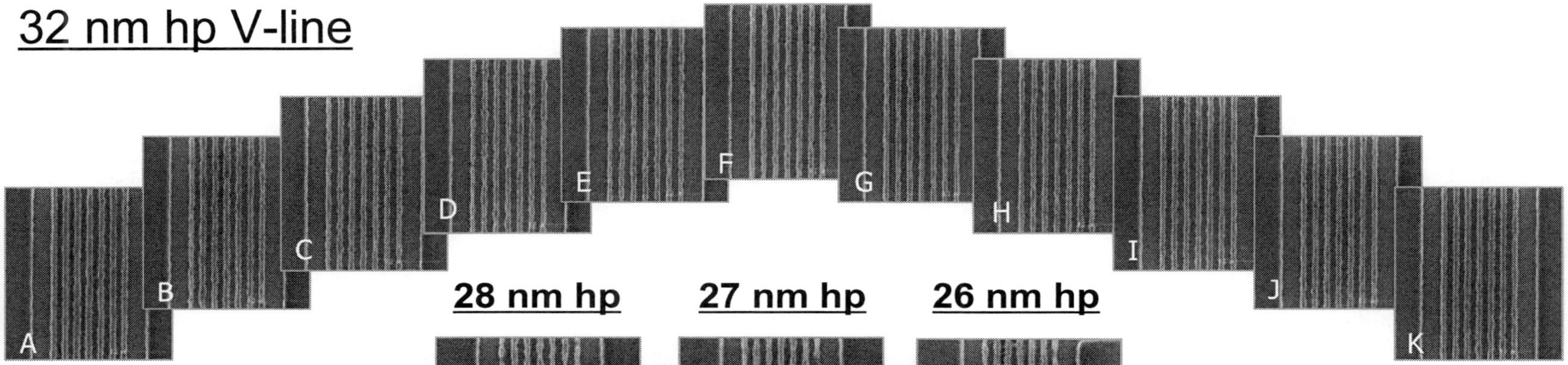

28 nm hp **27 nm hp** **26 nm hp**

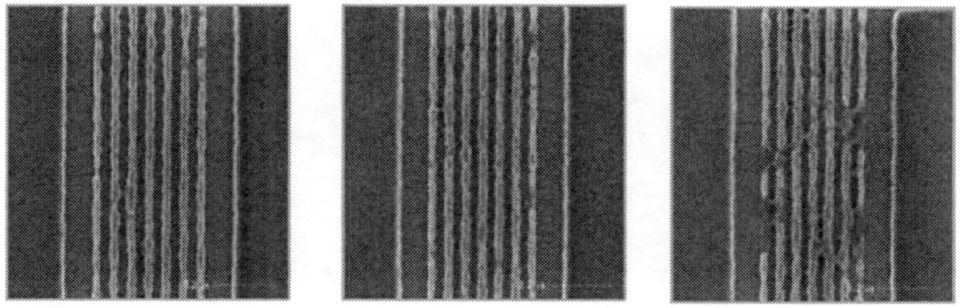

First Dynamic Exposure Result

By courtesy of Selete

26x33mm field

26x33mm field

0.6mJ/cm2

32nm L/S

10mJ/cm2

Refer to "Lithographic performance of Selete's full field EUV exposure tool" by K.Tawarayama, et.al, on October 1.

NIKON CORPORATION

Presentation Outline

1. Nikon EUV roadmap

2. Current status of EUV1 and latest data

3. **Update progress in various areas**

4. EUVL tool development challenges

5. Future projection lenses

6. Future tool expectation

7. Development summary

NIKON CORPORATION

EUV Light Source Status
Based on "Source WS", May 2008

	Cymer	Gigaphoton (EUVA)	Philips/Xtreme/Ushio
Type	LPP Sn, Droplet	LPP Sn, Droplet	DPP Sn, Rotating disc
Rep. rate	50 kHz	100 kHz	5 kHz Demonstrated feasibility of 100 kHz
Drive Laser Power	12 kW CO_2 laser	13 kW CO_2 laser	-
EUV Power*	100W @IFP Burst ~35W@IF average (>10min., Duty ~40%) ~25W @IF for 1.5 hrs operation (Duty 32%)	60 W @IFP (Sn plate)	170W@plasma 8W@IFP 100% duty Demonstrated feasibility of 3.5 kW @plasma
Plasma size	210μm (1/e^2)	(~100 um)	<1.3 mm
Collector mirror	MLM	MLM	Grazing

* Estimated IF power based on a transmissibility of a collector mirror.

Performance of integrated plasma source, collector and DMT modules must be demonstrated.

EUV Reticle Protection

- Dual Pod Concept by Canon and Nikon -

1. Reticle in Cassette (RC) in Carrier (RSP200).
2. Cassette protects the reticle in load locks.
3. Top cover stays with reticle during in-tool handling.
4. Reticle remains in RC in library to protect against vacuum accidents and contamination.

Top Cover

Reticle

Bottom Cover

RSP

Cassette (Reticle Cover)

NIKON CORPORATION

Reticle Protection with CNE Dual Pod

EUV pod installed RSP

Closed EUV pod

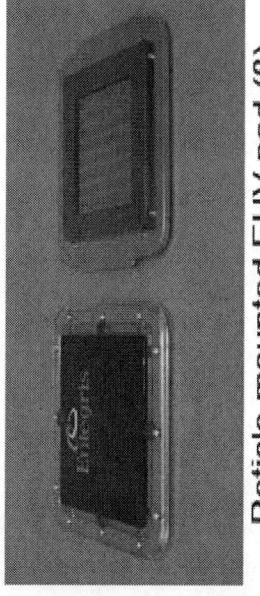

Reticle mounted EUV pod (β)

Reticle dismounted EUV pod (β)

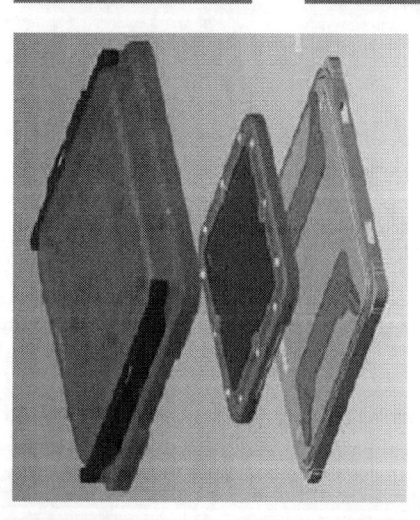

EUV mask carrier standard workshop @San Jose, Feb. 2006

1. **CNE (Canon/Nikon/Entegris) Dual Pod has been evaluated[1]:**
   · **Particle adders reported less than 0.01 particles / cycle.**

2. 1st Yellow Ballot's vote in Q1 '08 was completed unsuccessfully (rejected). Resubmit 2nd Yellow ballot for the cycle five voting (started 27th Aug.).

3. The revised yellow ballot clearly specified carrier dedication for exposure tools and general purpose carrier for all other uses.

(1) "Particle-Free Mask Handlling Techniques and a Dual-Pod Carrier", Mitsuaki Amemiya, Selete(Japan), et al. Proceedings of SPIE advanced lithography 2008, emerging lithography[6921-142]

Contamination Control Strategy

Oxidation

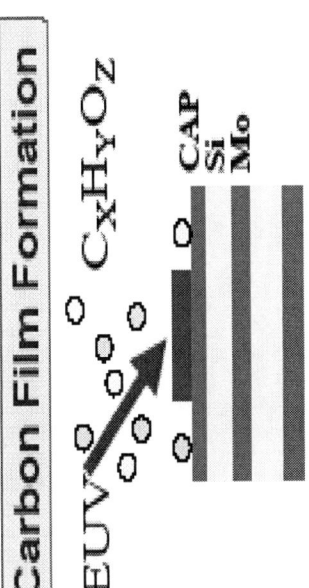

Carbon Film Formation

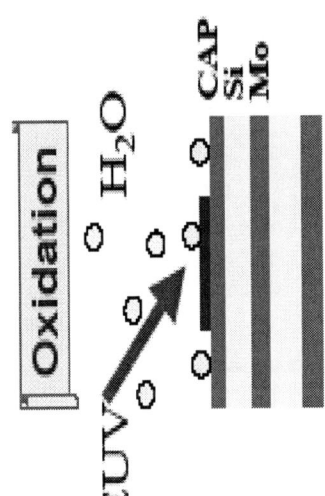

1. Long-life anti-oxidation capping layer

2. Carbon-film suppression and removal using EUV+O_2 in-situ cleaning
   - Oxygen gas introduction under EUV irradiation can suppress carbon deposition onto mirrors.

3. Experimentation facilities
   - "New SUBARU" in Himeji (Univ. of Hyogo) and new SLS facility in Kyushu now available.

NIKON CORPORATION

Long-life Anti-oxidation Capping Layer

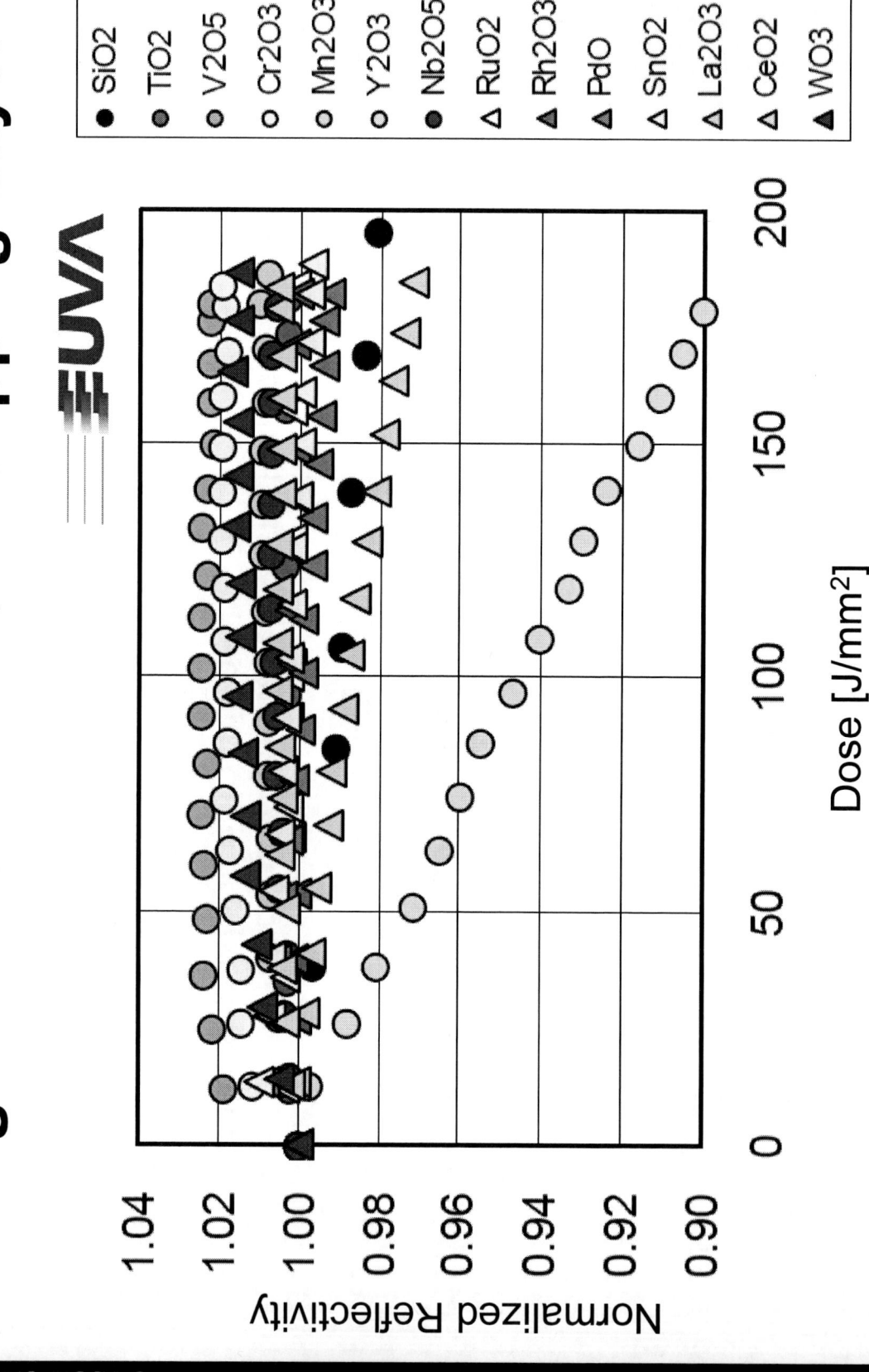

Some of our candidates show significant high anti-oxidation capabilities.

Carbon Film Suppression

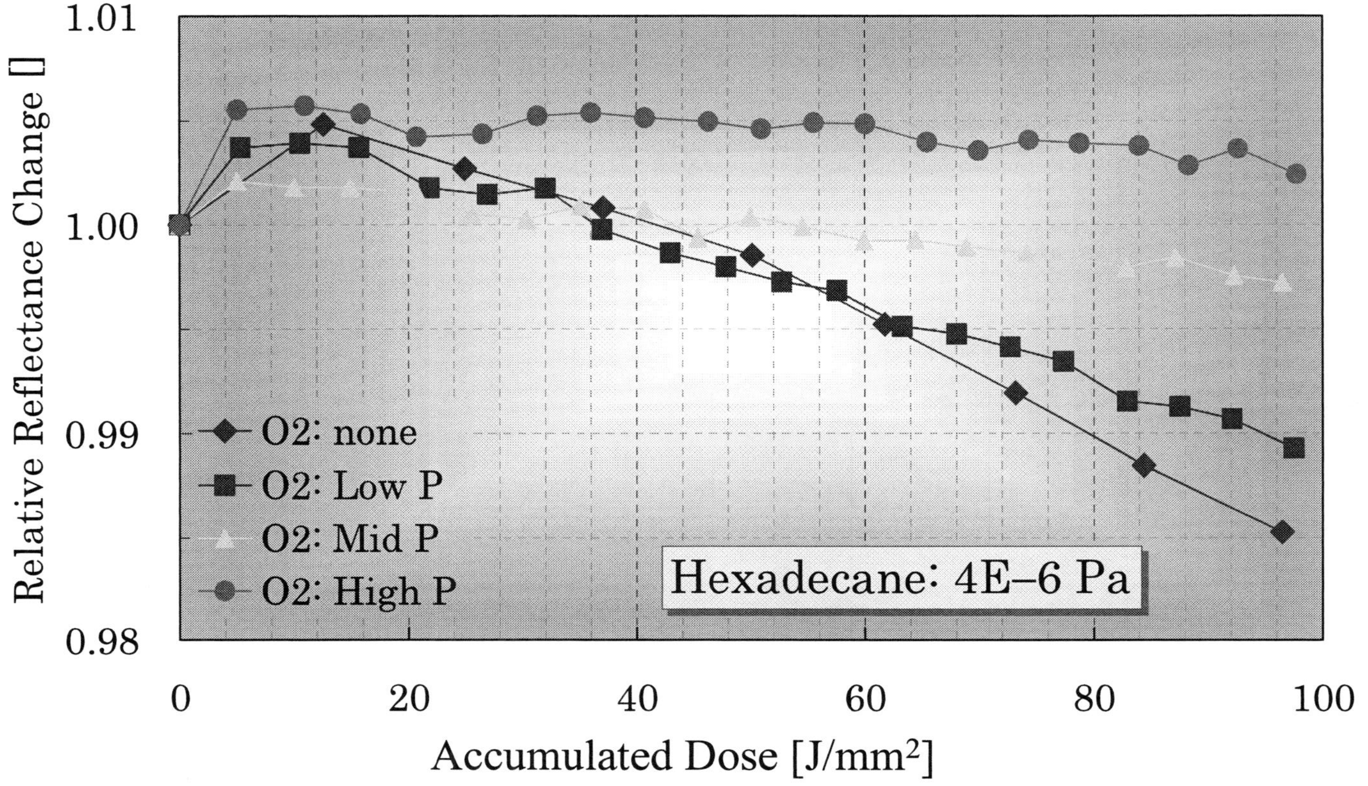

Reflectance History

EUVA

Relative Reflectance Change []

Accumulated Dose [J/mm^2]

- O2: none
- O2: Low P
- O2: Mid P
- O2: High P

Hexadecane: 4E–6 Pa

Oxygen gas introduction under EUV irradiation can suppress carbon deposition onto mirrors.

NIKON CORPORATION

New EUV Irradiation Facility

Saga Light Source

Dec. 2007:
- NTT Super ALIS operation ended.

Jan – July 2008:
- Remove experimental set up at NTT
- Installation completed at SLS

August 2008:
- Operation started at SLS
- Dedicated beam line to Nikon's contamination R&D

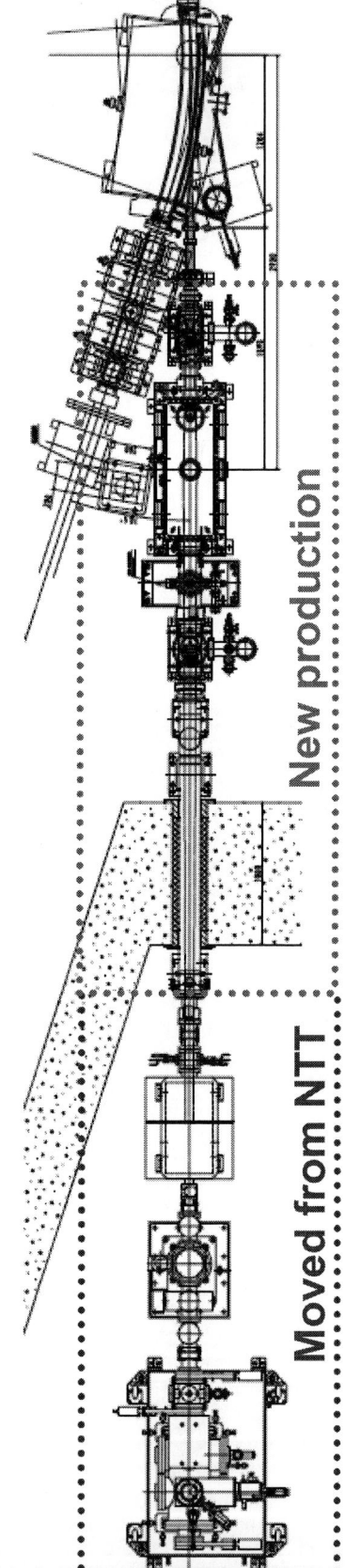

New production

Moved from NTT

NIKON CORPORATION

Presentation Outline

1. Nikon EUV roadmap

2. Current status of EUV1 and latest data

3. Update progress in various areas

4. **EUVL tool development challenges**

5. Future projection lenses

6. Future tool realization

7. Development summary

NIKON CORPORATION

EuVL Tool Development Challenges

Imaging Performance
- Low aberration, low flare optics
- RET and uniformity control

Overlay
- Thermal stability
- Thermal distortion

Reticle
- Defect-free
- Particle-free

Thermal Management
- Heat rejection of optics

Throughput
- Higher optical chain transmittance
- Higher EUV source power
- Higher resist sensitivity

CoO Improvement
- Vacuum quality and Optics lifetime
- Maintenance downtime

Mask Stage

Projection Optics

Wafer Stage

Condenser Optics

Wafer Alignment Sensor

Source

λ:13.5nm

Xe Nozzle

YAG Laser λ:1064nm

EUV Lithography

NIKON CORPORATION

Projection Optics Technology

PO technology improvement ongoing

Optics evaluation tools

Supported by NEDO

EUVA

Supported by NEDO

EUVA

Supported by NEDO

EUVA

Supported by NEDO

EUVA

Mirror fabrication tools

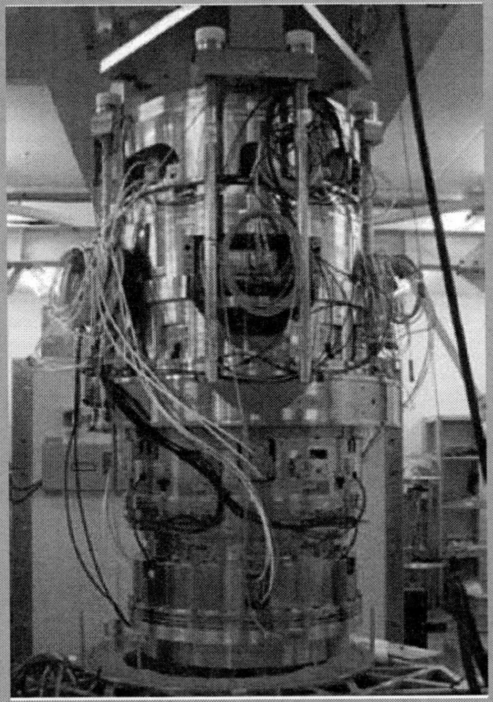

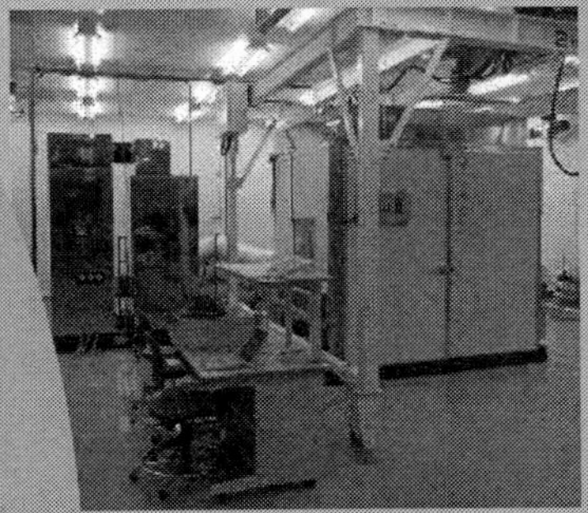

NIKON CORPORATION

Wavefront Error Improvement

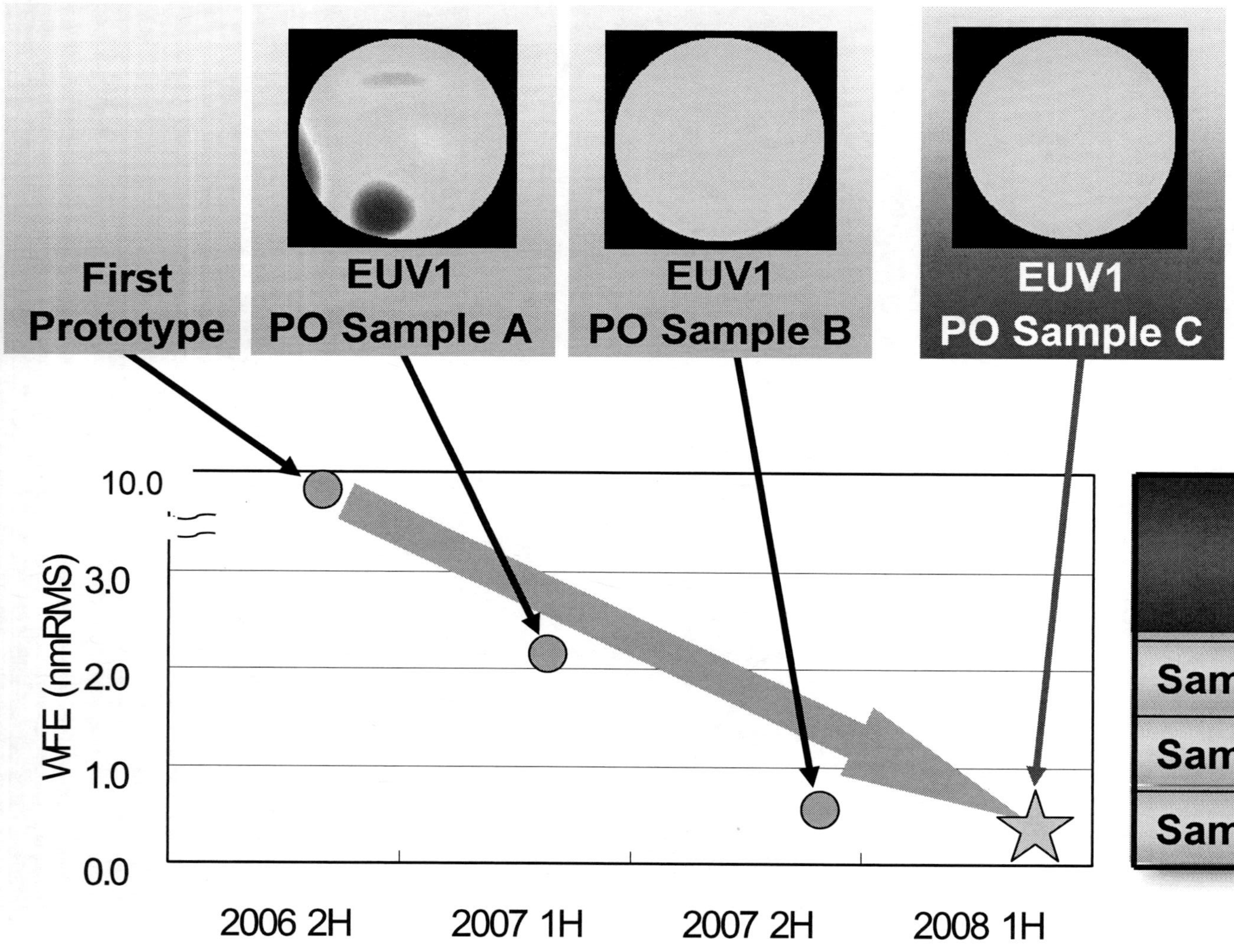

First Prototype

EUV1 PO Sample A

EUV1 PO Sample B

EUV1 PO Sample C

WFE (nm RMS) (Average)	
Sample A	2.2
Sample B	0.6
Sample C	0.4

WFE (nmRMS)

10.0
3.0
2.0
1.0
0.0

2006 2H   2007 1H   2007 2H   2008 1H

NIKON CORPORATION

Calculation of "Kirk flare"

Kirk flare ≠ Flare

Kirk flare

$$KF = \frac{I_{D\min}}{I_{B\max}}$$

$$I_D(x_d, y_d) = \int_{\text{bright field}} PSF^{SC}(x - x_d, y - y_d)\,dxdy$$

$$I_B(x_b, y_b) = (1 - TIS) + \int_{\text{bright field}} PSF^{SC}(x - x_b, y - y_b)\,dxdy$$

TIS (Total Integrated Scatter)

$$TIS = \int_0^\infty PSF^{SC0}(x, y)\,dxdy$$

Flare

$$F(x_0, y_0) = \int_{\text{bright field}} PSF^{SC}(x - x_0, y - y_0)\,dxdy$$

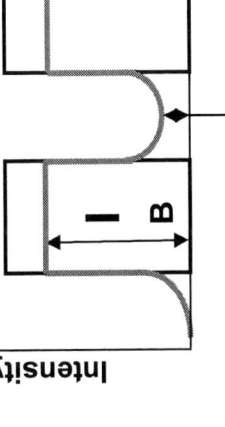

Calculated Kirk flare agreed with measurement.

	Flare	Kirk flare
Sample B PO	10%	15%
Sample C PO	6%	8%

Refer to "Improvement of Optics for EUV Exposure Tool" by K.Murakami, et.al, on October 1.

NIKON CORPORATION

Presentation Outline

1. Nikon EUV roadmap

2. Current status of EUV1 and latest data

3. Update progress in various areas

4. EUVL tool development challenges

5. Future projection lenses

6. Future tool realization

7. Development summary

NIKON CORPORATION

NA 0.25 Imaging Simulation

Process Window Vs. Illumination Condition

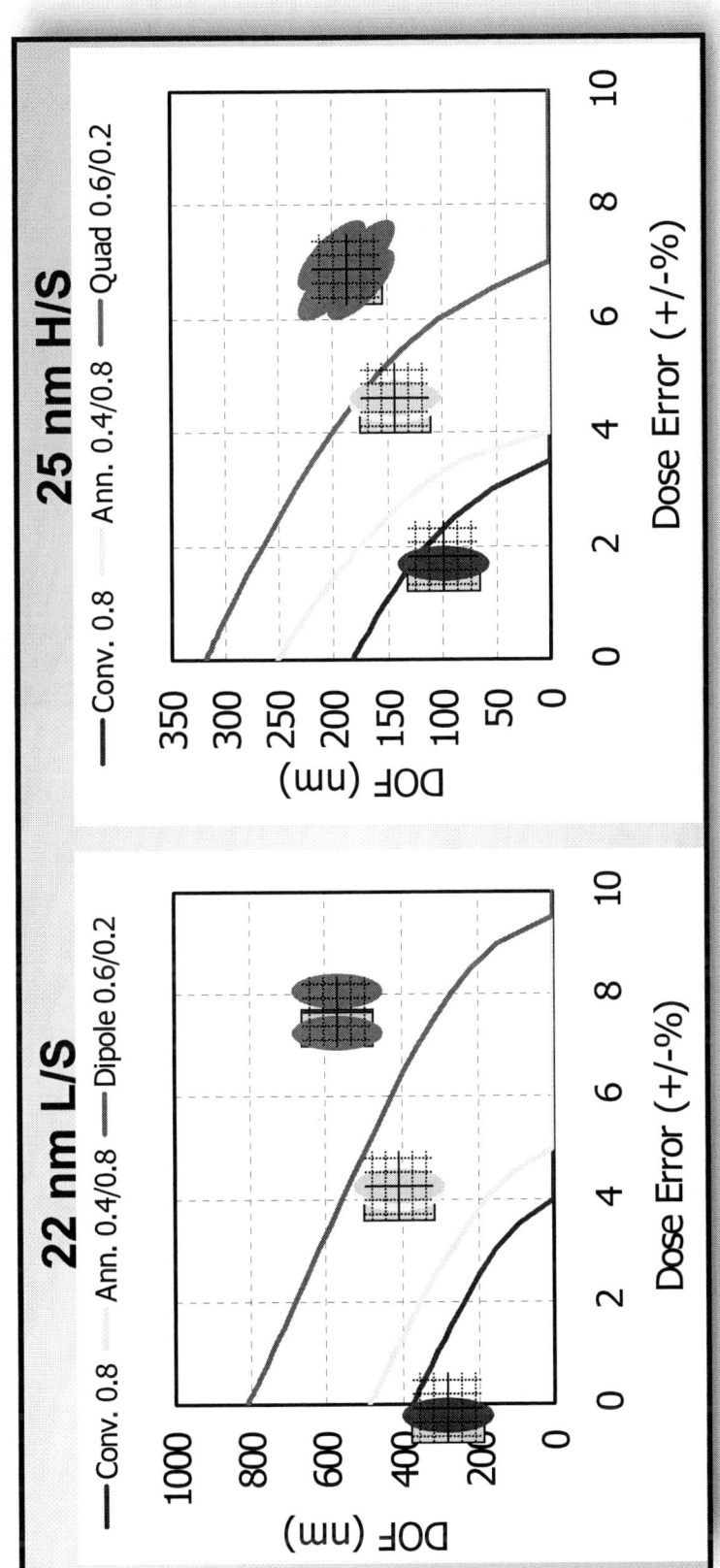

ED-Tree DOF Conditions:

Lambda: 13.5 nm, NA: 0.25,

CD error: +/-10% of CD, Mask CD error: +/-0.5 nm,

Mask contrast: 1:100, Flare: 7%*pattern density

High NA Imaging Simulation

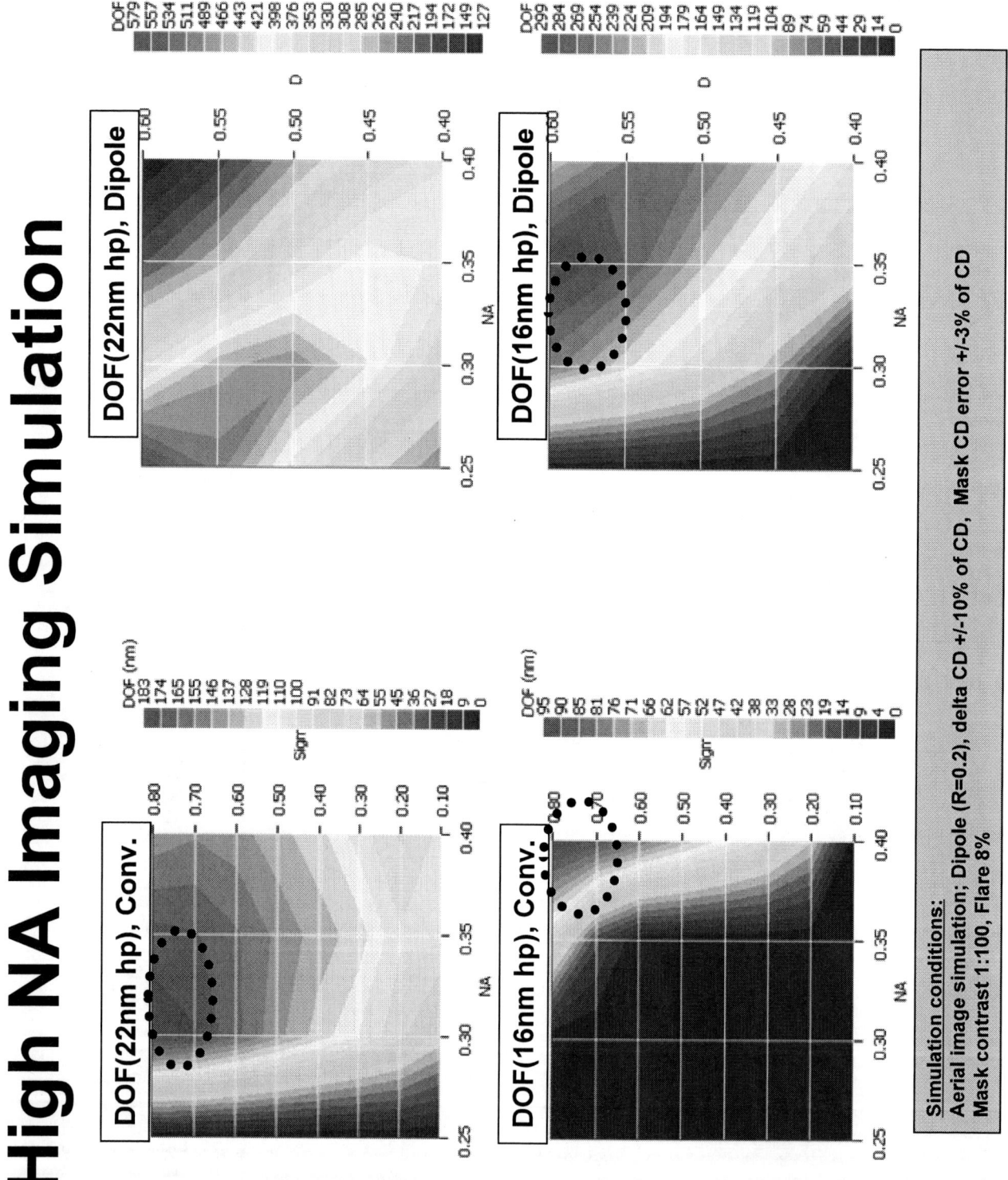

Simulation conditions:
Aerial image simulation; Dipole (R=0.2), delta CD +/-10% of CD, Mask CD error +/-3% of CD
Mask contrast 1:100, Flare 8%

High NA Imaging Simulation Summary

		DOF >200nm	Contrast > 0.5

NA		0.25	0.3	0.35	0.4	0.25	0.3	0.35	0.4
Sigma		Conv	Conv	Conv	Conv	Dipole	Dipole	Dipole	Dipole
DOF (nm) EL=+/-2%	16nm hp	0	0	49.9	95.5	0	246.5	299.8	257
	22nm hp	0	183.5	168.3	140.6	579.8	484.1	414.6	374.1
Contrast	16nm hp	0.039	0.263	0.465	0.605	0.201	0.764	0.764	0.764
	22nm hp	0.44	0.638	0.746	0.803	0.762	0.762	0.762	0.769

1. 22 nm hp
   0.25 NA + Dipole illumination or >0.3 NA + Conventional illumination

2. 16nm hp
   >0.3 NA + off-axis illumination

NIKON CORPORATION

High NA EUV Projection Optics

$$CD = K_1 \lambda / NA \qquad DOF = \lambda / NA^2$$

HP	45nm	32nm	22nm	16nm	DOF (nm)
		K1			
NA0.25	0.83	0.59	0.41	0.30	216
NA0.30		0.71	0.49	0.36	150
NA0.35		0.83	0.57	0.41	110
NA0.40			0.65	0.47	84
NA0.45			0.73	0.53	67
NA0.50			0.81	0.59	54

Design examples

6 mirror system

8 mirror system

8 mirror system (center obscuration)

NA: 0.30 0.40 0.50

General issues:

1. DOF reduction
2. Flare increase
3. Transmittance
4. Obscuration
5. Manufacture engineering

NIKON CORPORATION

Throughput Improvement

■ Current Optical Chain and source power

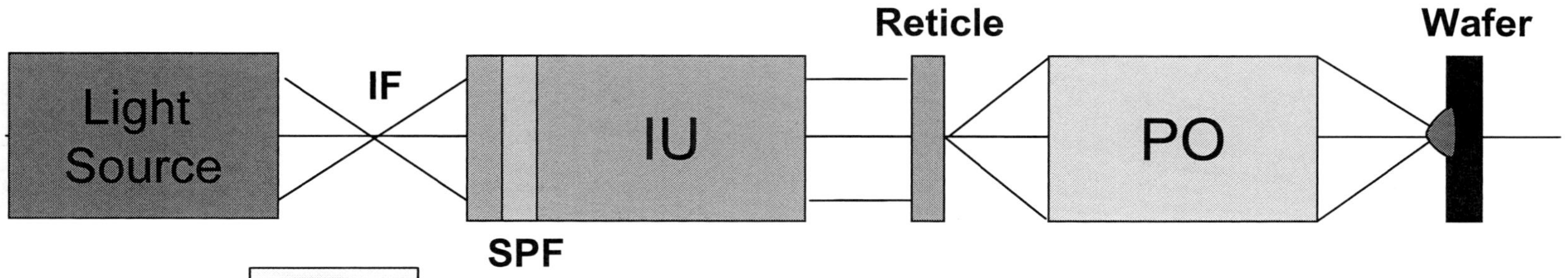

1. Source power improvement
2. optical chain transmittance improvement

■ HVM Concept

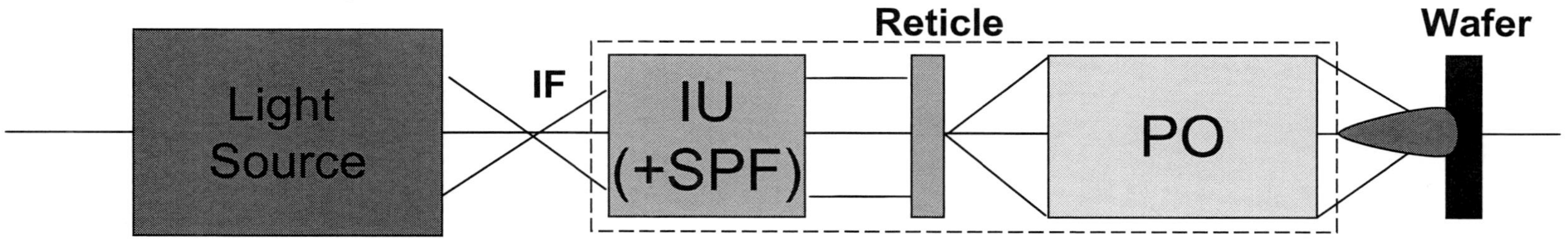

Efficient RET Illumination

1. Conventional Fly's eye mirror with pupil aperture

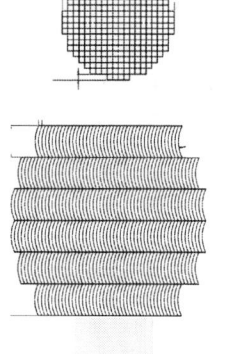

+ Easy to make
- Power loss
- Degrade uniformity on reticle

2. Efficient RET Fly's eye mirror

+ No power loss
+ No change uniformity on reticle
- Difficult to make Fly's eye mirrors

Intensity distribution in pupil

Reticle
Pupil aperture
Fly's eye 2 (Pupil plane)
Fly's eye 1
Source

Reticle
Fly's eye 2 (Pupil plane)
RET Fly's eye
Fly's eye 1
Source

Refer to "Improvement of Optics for EUV Exposure Tool" by K.Murakami, et.al, on October 1.

NIKON CORPORATION

Throughput Study

Total Number of Mirrors vs. Throughput

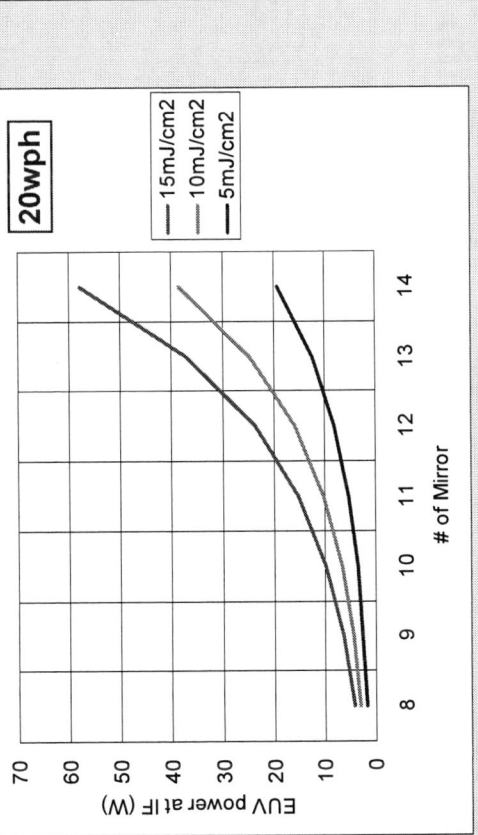

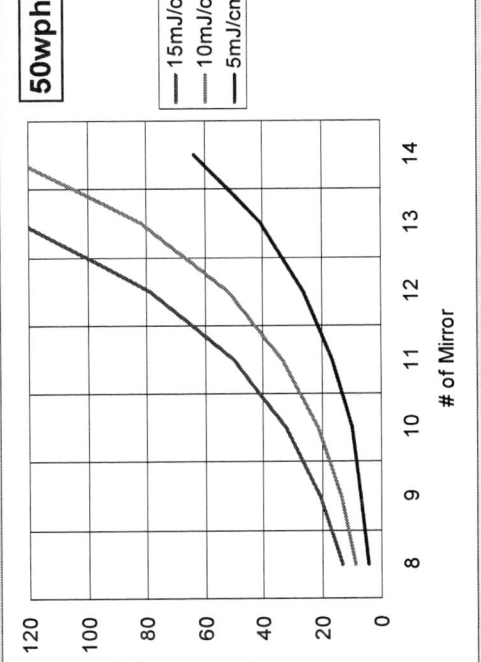

Condition;
- Mirror reflectance: 64%
- Shot number: 76 shots

Key issues
- Source power improvement
- Optical transmittance improvement

NIKON CORPORATION

EUVL HVM Tool Realization

- **Optics improvement**
  - WFE, flare, RET, distortion
- **Throughput improvement**
  - Optical chain transmittance
  - Light source IF power
  - Stepping and overhead time
- **Overlay improvement**
  - Thermal stability, heat rejection and cooling
- **EUVL facility issues**
  - New raised or recessed floor arrangement
  - Light source utility and space
- **CoO issue**
  - Consumable cost and lifetime

NIKON CORPORATION

EUVL Production Tool Realization

Consumable Cost Reduction

Tool consumables
- Optics
  - Capping layer
  - In-situ cleaning
  - Refurbishment
  - Vacuum quality

Source consumables
- Collector
  - DMT
  - Cleaning
  - Coating

DPP

LPP

EUV Source WS May, 2008

EUV Source WS May, 2008

CYMER

SIMS Depth Profile Analysis of CO_2 / Sn Exposed 8-Layer Witness Sample

Unexposed — Si 28 reference

Exposed — Si 28 exposed to LPP

- Removal rate without debris mitigation is ~0.2 layers per million pulses
- Debris mitigation suppression factor >10,000; 500 sacrificial layers
- Normal Incidence Collector lifetime projection >1E12 pulses

May 12, 2008 — Sematech EUV Source Workshop 2008 – New York

PHILIPS / XTREME

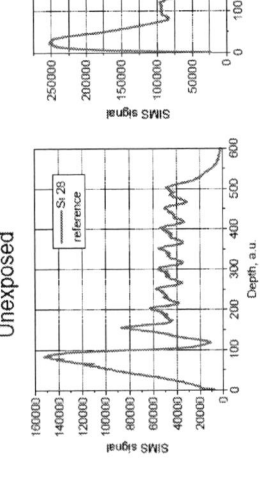

Collector lifetime

- Last year's results:
- No Sn deposition because of efficient Debris Mitigation
- New debris mitigation system shows 10 Gshot lifetime proven by accelerated lifetime tests on samples
- Enhanced reflective layers from Media Lario realize additional 6x lifetime increase

Media Lario

=> 60 Gshot or half-year lifetime already

Source Workshop, Bolton Landing, May 12, 2008

NIKON CORPORATION

Summary

- **EUVL technology targeted for 22 nm hp and beyond**

- **Current Nikon EUVL program**
  - Full field exposure tool (EUV1) integrated and starting Scan Exposure
  - Early process development for 32 nm hp node

- **World class projection optics performance**

- **EUV2 tool development**
  - Designed for both 32 nm hp and 22 nm device process development
  - Production tool verification leading to EUV3 development

- **EUV3 planned for production at 22 nm hp and beyond**

NIKON CORPORATION

Acknowledgements

1. A part of this work was conducted under *EUVA* projects. EUVA projects have been supported by New Energy and Industrial Technology Development Organization (NEDO). - Nikon gratefully acknowledges *Japan Ministry of Economy, Trade and Industry (METI) and NEDO* for their supports.

2. Nikon also participate in *Selete* program and appreciate Selete members for their useful discussion and advice.

Thank You.

NIKON CORPORATION

Development Status of Canon's Full-Field EUVL Tool

**Shigeyuki Uzawa, Tokuyuki Honda,
Hideki Morishima, Takayuki Hasegawa**

Canon Inc.

Canon

Outline

1. **Canon's Roadmap**

2. **Research for the Full-Field Tool**

   1. Issues of EUVL tool

   2. Projection Optics

   3. Extend toward sub-20nm

4. **Summary**

Outline

Canon

1. Canon's Roadmap

2. Research for the Full-Field Tool

1. Issues of EUVL tool

2. Projection Optics

3. Extend toward sub-20nm

4. Summary

EUVL Tool Development Plan

Canon

ITRS2007	2008	2009	2010	2011	2012	2013	2014	2015	2016	2017	2018	2019
2 year cycle (nm)	45		32		22		16		11		8	
DRAM (hp)	57	50	45	40	36	32	28	25	23	20	18	16
Flash (hp)	45	40	36	32	28	25	23	20	18	16	14	13

4X Generation Immersion

3x, 2x nm Generation Double Patterning

2x, 1x nm Generation EUV

HVM Platform

HVM tool (VS2) NA > 0.3

HVM tool (VS2) TP Enhancement

Prototype

SFET Installed at SELETE

6-mirror (PO1) NA0.3 In-house test

Outline

1. Canon's Roadmap

2. Research for the Full-Field Tool

1. Issues of EUVL tool

2. Projection Optics

3. Extend toward sub-20nm

4. Summary

Specification of VS2 (Preliminary) Canon

Schematic view of VS2

W x H x D: 4600mm x 3000mm x 2480mm Weight: 40t

field size	26 x 33 mm²
Magnification	4x (+/- 10 ppm)
Numerical Aperture	>0.3
Lens aberration	<0.35 nm RMS
Flare	5 %
Wafer stage	-
Reticle stage	-
Dense Lines	<25nm
Isolated lines	17nm
Dense Contacts	< 28nm
Overlay (SMO)	<3 nm
Throughput	55 wph@10mJ/cm2 100w@IF

◆ **Critical issues for EUVL tool are**

　◆ **No pellicle for EUV mask**

　　◆ Laser cleaning system

　◆ **Optics life time**

　　◆ Oxidation

　　◆ Carbon deposition and cleaning

Laser Cleaning System

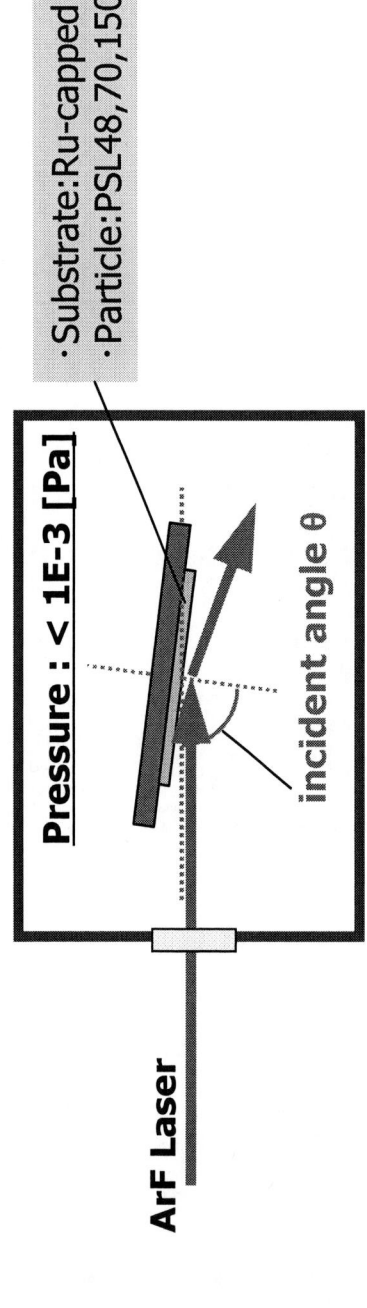

Pressure : < 1E-3 [Pa]

incident angle θ

ArF Laser

·Substrate:Ru-capped EUV mask blanks
·Particle:PSL 48,70,150nm

◆ Grazing incidence ArF laser cleaning system is developed.

◆ Removal rate of PSL particles on the Ru-capped Mo/Si ML EUV mask and damage to EUV mask are evaluated.

Laser Cleaning System

Canon

Reflectivity of Ru-capped EUV mask blanks

Cleaning results of 70 nm PSL
Measured by M3350 MAGICS

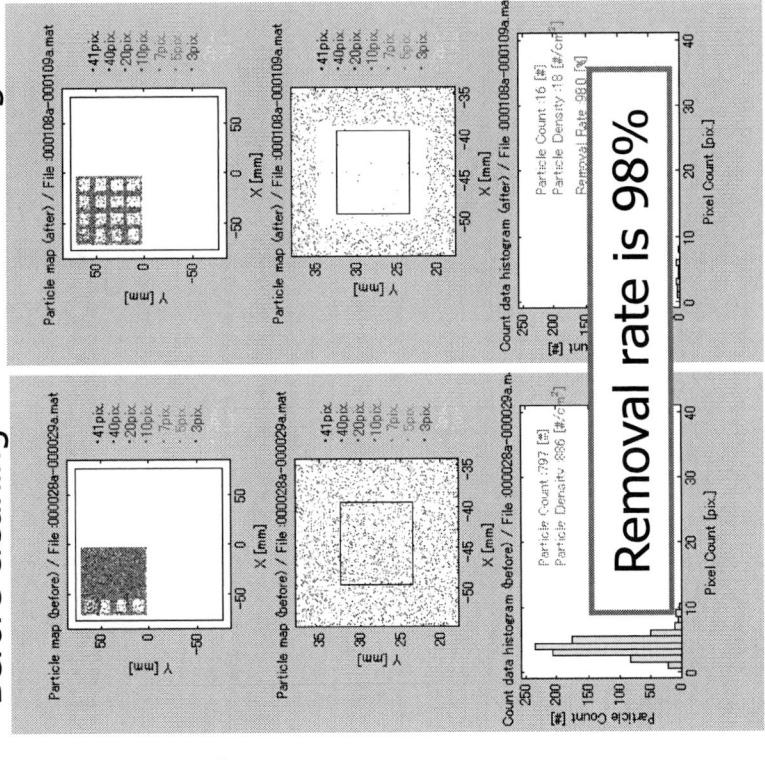

Before cleaning After cleaning

Removal rate is 98%

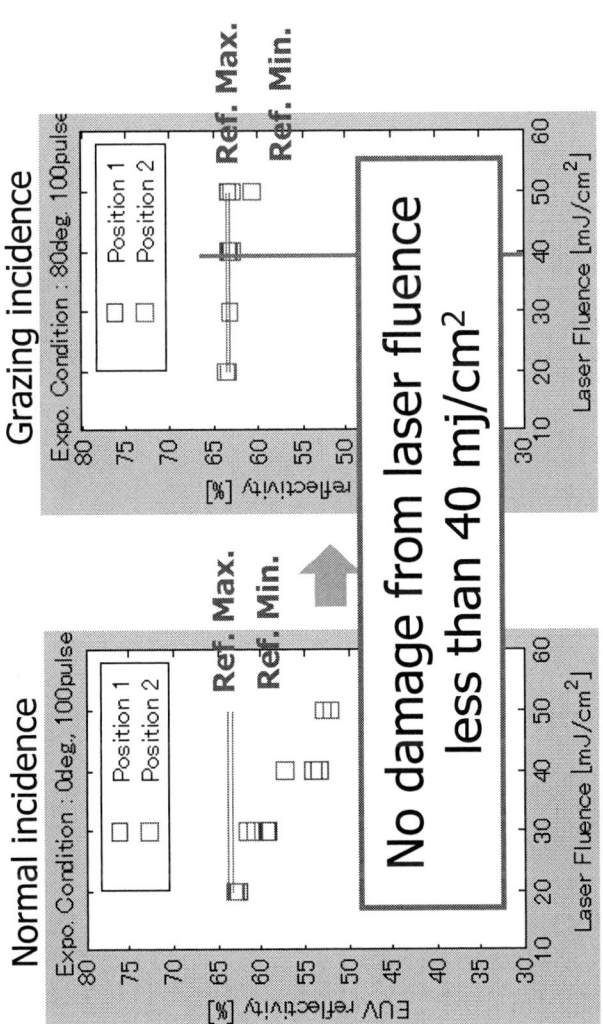

Normal incidence Grazing incidence

No damage from laser fluence less than 40 mj/cm²

◆ It is possible to remove the PSL particles on mask without damage.

Optics lifetime/Oxidation

Canon

Reflectivity degradation of Oxide-capped Mo/Si ML

Legend: SiO2, TiO2, V2O5, Cr2O3, Mn2O3, Y2O3, Nb2O5, RuO2, Rh2O3, PdO, SnO2, La2O3, CeO2, WO3

Relative reflectivity vs Dose (J/mm²)

S. Matsunari, Y. Kakutani, T. Aoki, K.Murakami, S. Kawata, T. Nakayama, S. Terashima, H. Takase, Y. Watanabe, Y.Gomei, M. Niibe, Y. Fukuda; "Anti-oxidation property of capping layer on multi-layer mirror for EUV lithography", NGL Workshop 2008 (July,2008, Tokyo)

◆ **Oxidation is unacceptable**
◆ **TiO₂ is reported as a good anti-oxidation material.**

Y. Fukuda et al., "Contamination Study at EUVA"
POSTER NUMBER: 95

Reflectivity degradation of Ru-capped Mo/Si ML

Water partial pressure
△ 1×10⁻⁶ Pa
◇ 1×10⁻⁵ Pa
□ 1×10⁻⁴ Pa
○ 9×10⁻⁴ Pa

Relative reflectivity vs Dose (J/mm²)

S.Matsunari, T.Aoki, Y.Gomei, S.Terashima, H.Takase, M.Niibe, Y.Kakutani; "Lifetime estimation and improvement of capping layer on multi-layer mirror for EUV lithography in EUVA", 4th International Workshop on Extreme Ultraviolet Lithography (Miyazaki, Japan; 2005)

Schematic view of top layers

EUV

Ru
RuSi
Si

◆ Additional SiO₂ layer formed just below Ru.

EUVA

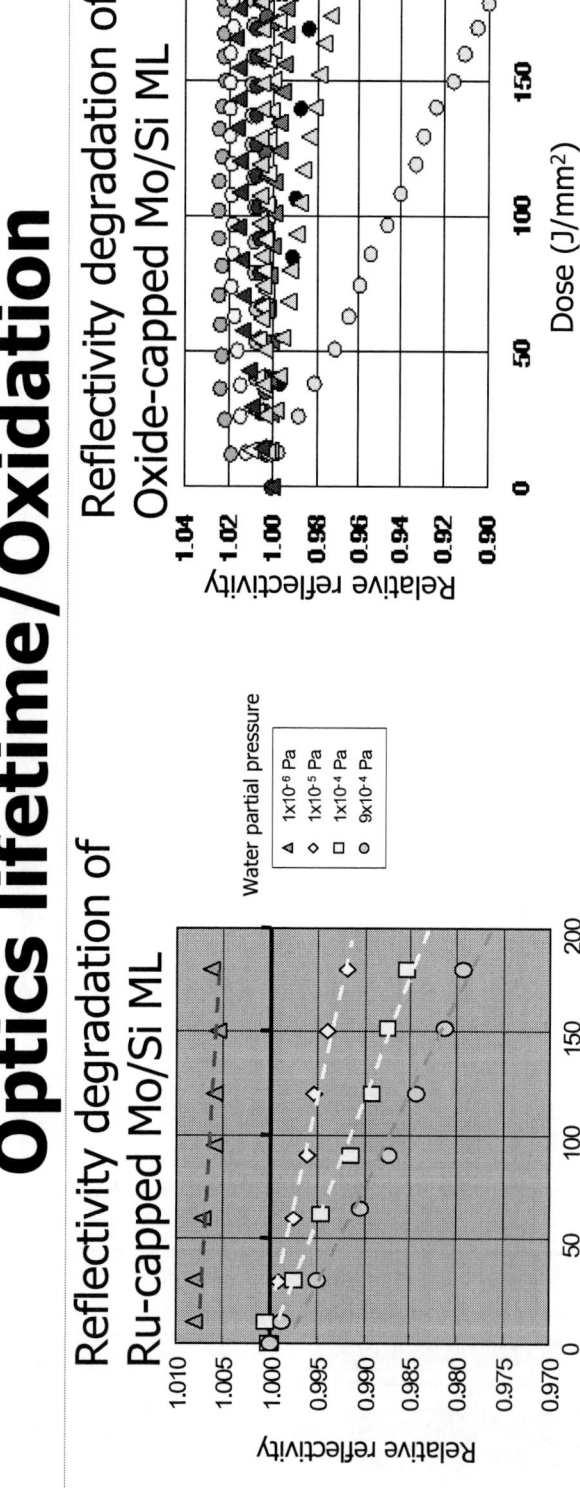

Carbon deposition rate vs. EUV intesity

Canon

Parameter: Decane pressure($C_{10}H_{22}$; MW =142)

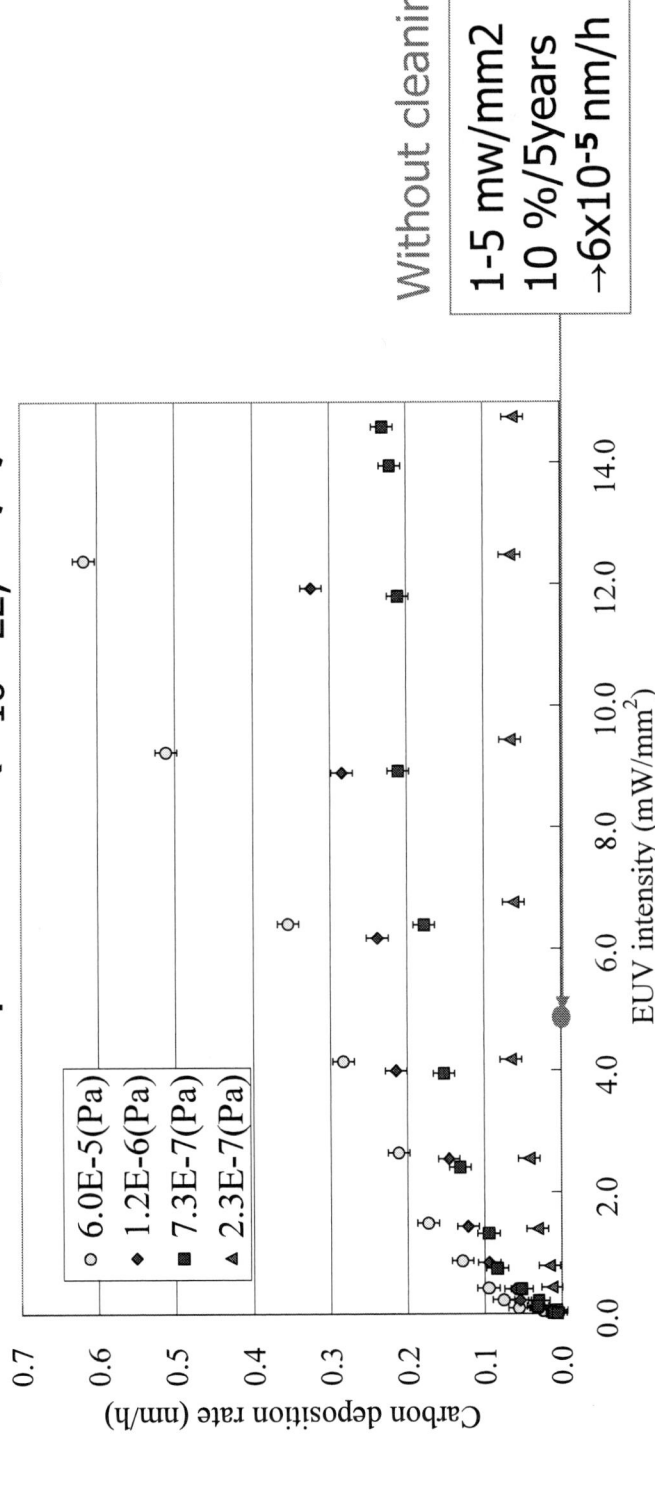

Without cleaning

1-5 mw/mm2
10 %/5years
→6x10⁻⁵ nm/h

◆ It is very difficult to avoid the carbon deposition.
◆ Periodical cleaning is needed.

Y. Fukuda et al., "Contamination Study at EUVA"
POSTER NUMBER: 95

EUVA

Canon

UV Cleaning with O₂

XPS analysis for Carbon

Sputtered carbon films cleaned with 100 Pa O_2 UV intensity of 3.9 mW/cm²

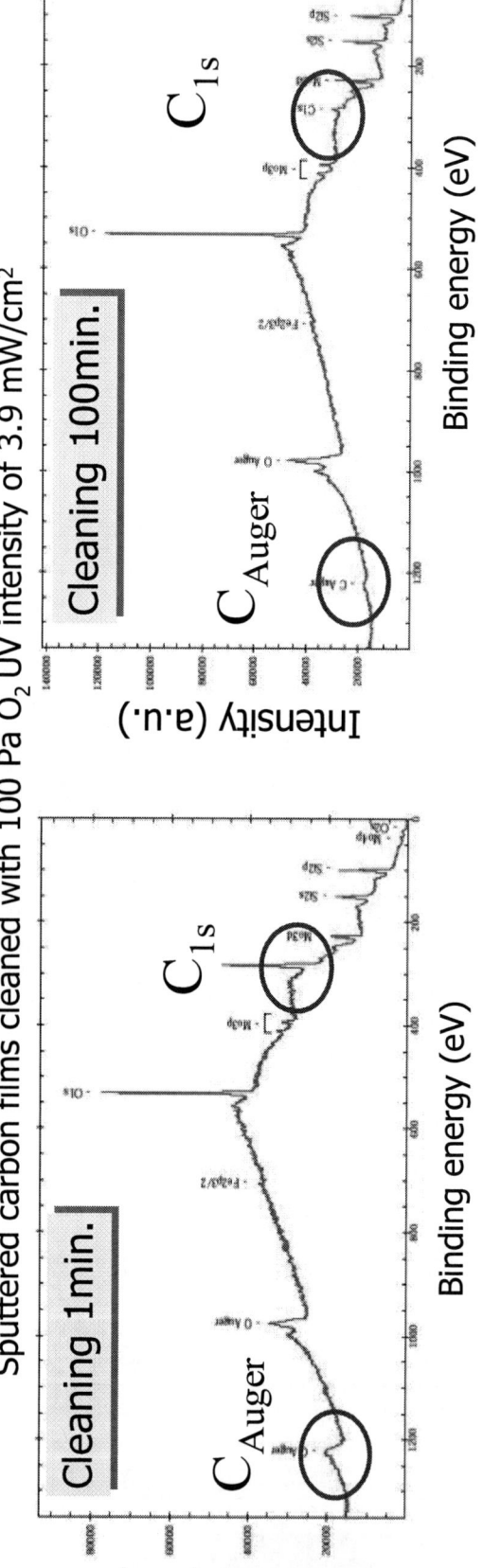

T. Aoki, H. Kondo, S. Matsunari, H. Takase, Yoshio Gomei , and S. Terashima;
"Apparatus for contamination control development in EUVA" Proceedings of
SPIE Vol.5751(SPIE, Bellingham, WA, 2005), pp. 1137–1146

◆ No significant peak for carbon is seen after 100 minutes cleaning.

◆ Cleaning rate is higher than 0.03nm/min.

Y. Fukuda et al., "Contamination Study at EUVA"
POSTER NUMBER: 95

Canon

Outline

1. Canon's Roadmap

2. Research for the Full-Field Tool

   1. Issues of EUVL tool

   2. Projection Optics

   3. Extend toward sub-20nm

4. Summary

Specification of PO1

Canon

Design of 6-mirror system

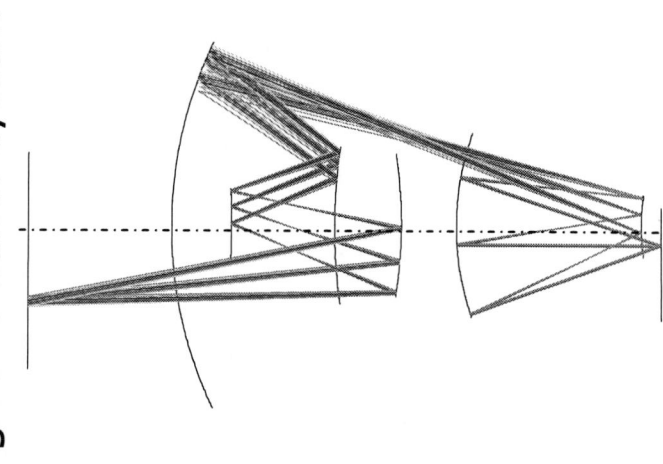

Numerical Aperture	0.3
Slit size	26 x 2 mm
Lens aberration	0.55 nm RMS
Flare	7 %

◆ PO1 is being fabricated to evaluate and improve our fabrication technology.

◆ This high NA optics is a study tool for step up to higher NA machine.

◆ Now mirrors are under polishing.

Feature of PO1 System

Schematic View of PO1 System

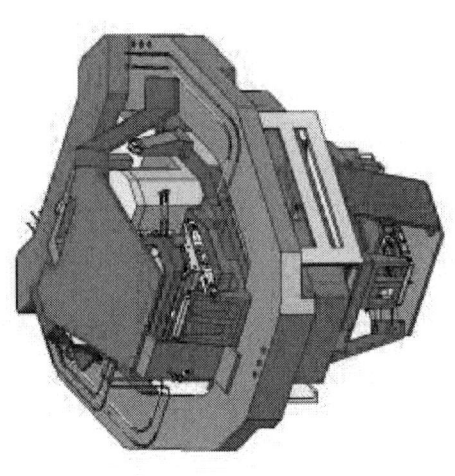

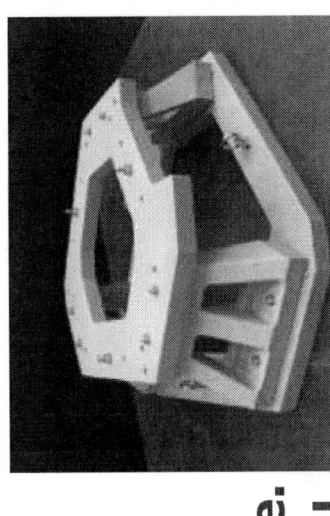

Mirror

Ceramic structure

◆ **Features**
- ◆ Low-thermal expansion ceramic structure.
- ◆ Position of 5 mirrors are controlled in real time.
- ◆ High stability of mirror position against vibration.
- ◆ Lens aberration can be compensated.

Mirror Surface Accuracy

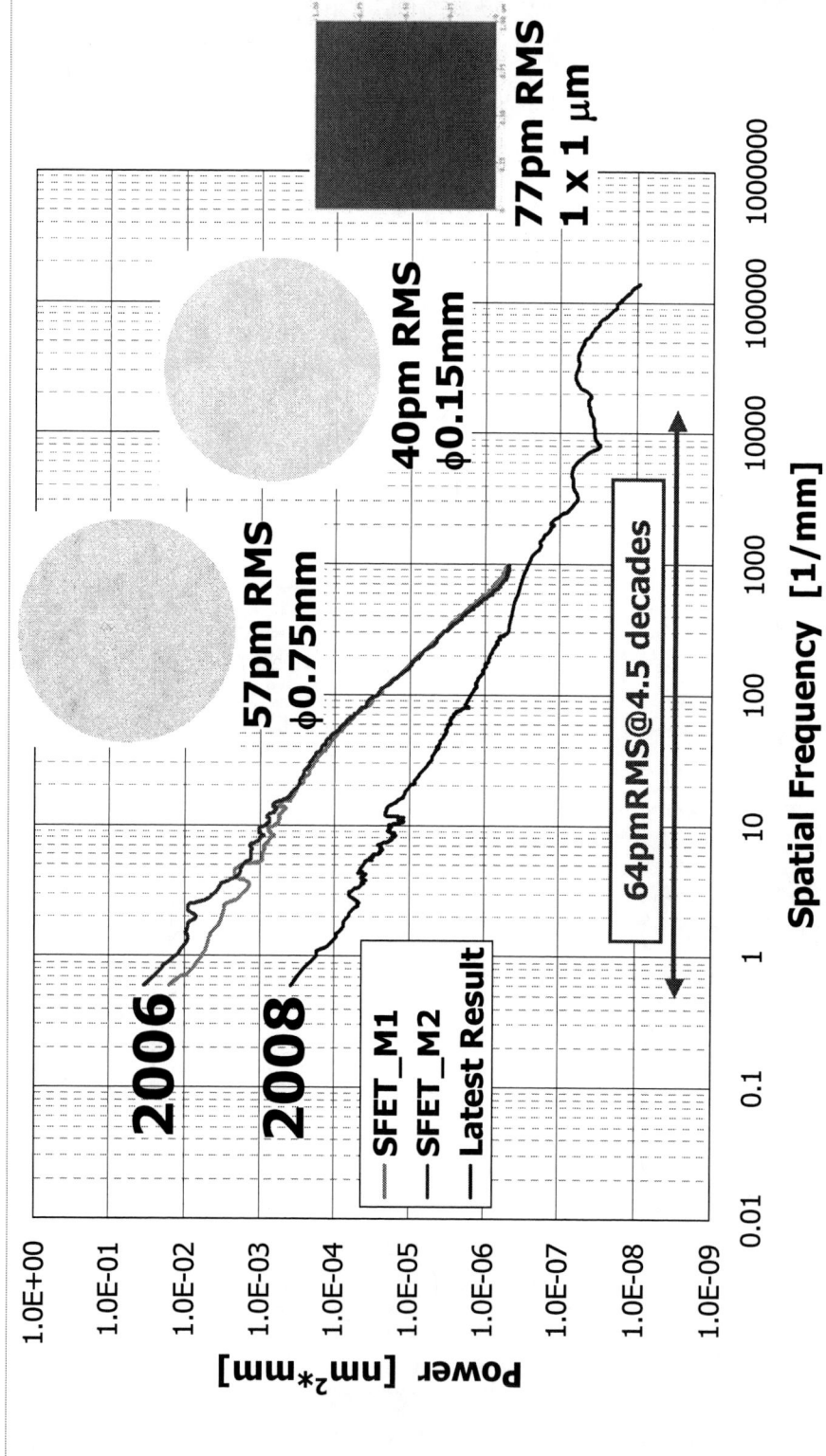

- Flare calculated from PSD curve is less than 6 %.
  - ◆ Range of spatial frequency that causes the flare is 4.5 decades.
- ◆ Current technologies satisfy the specification of HVM tools.

Outline

1. Canon's Roadmap

2. Research for the Full-Field Tool

1. Issues of EUVL tool

2. Projection Optics

3. Extend toward sub-20nm

4. Summary

Extend to sub-20nm Feature

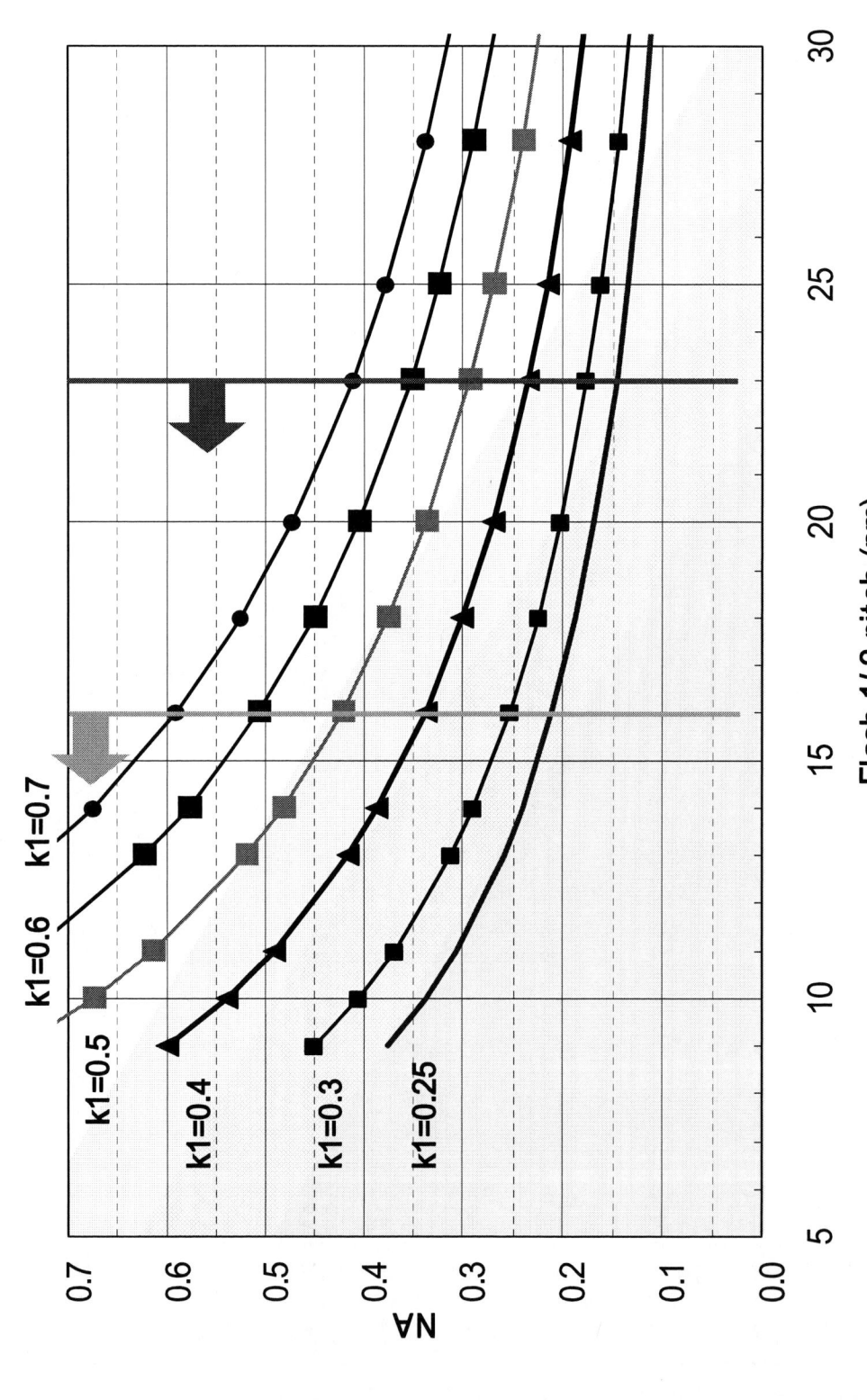

◆ **Higher NA tool is needed to resolve sub-20 nm feature.**

Examples of High NA Design

Canon

Constraining Condition of Design Canon

- Obscuration
  - Imaging results
- Apodization
  - Imaging results
- Chief Ray Angle on Object side (CRAO)
  - Imaging results
- EUV use efficiency
  - Throughput

Constraining Condition of Design Canon

- Obscuration
  - Imaging results
- Apodization
  - Imaging results
- Chief Ray Angle on Object side (CRAO)
  - Imaging results
- EUV use efficiency
  - Throughput

Obscuration

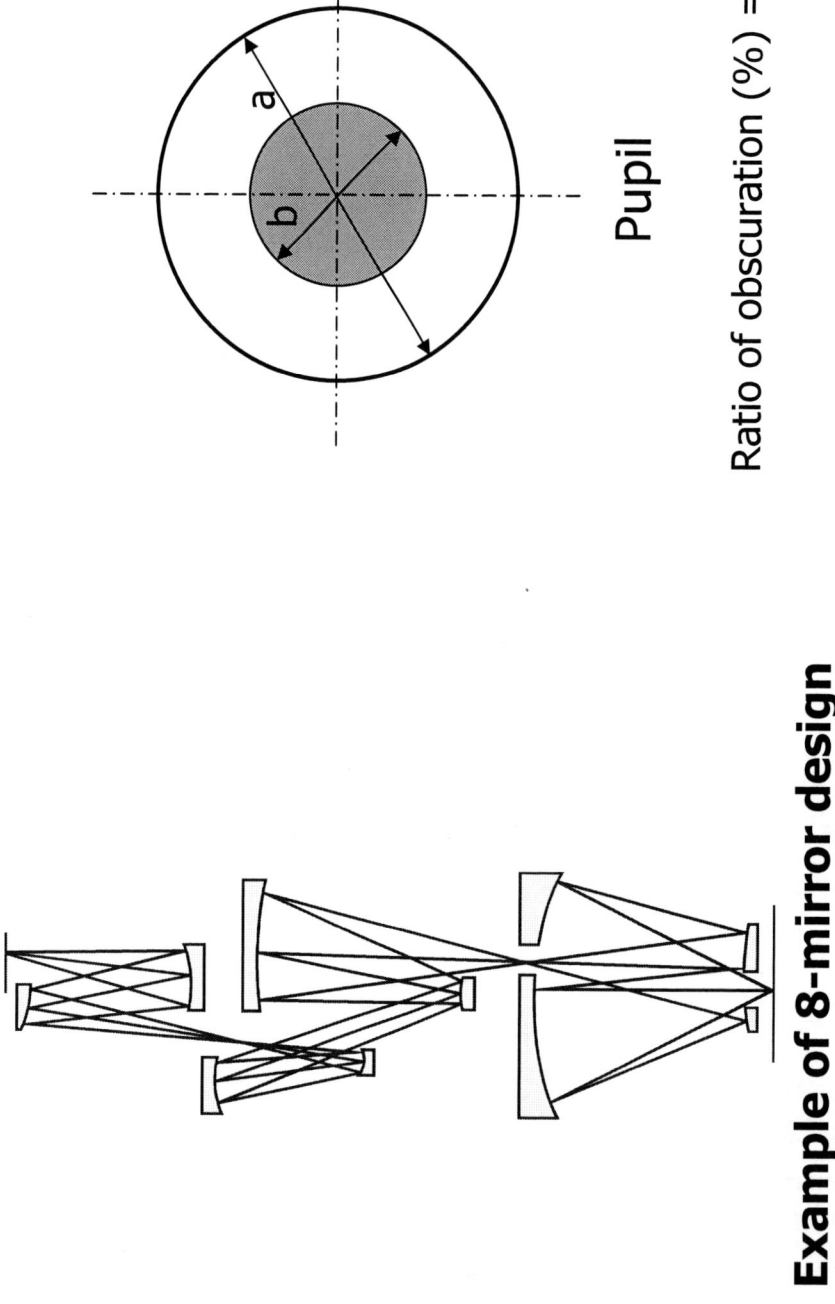

Pupil

$$\text{Ratio of obscuration (\%)} = \frac{b}{a} \times 100$$

Example of 8-mirror design

We calculate the impact of obscuration on the image.
condition of calculation : NA=0.5, σ=0.6

Calculation results of Obscuration Canon

Line & space pattern

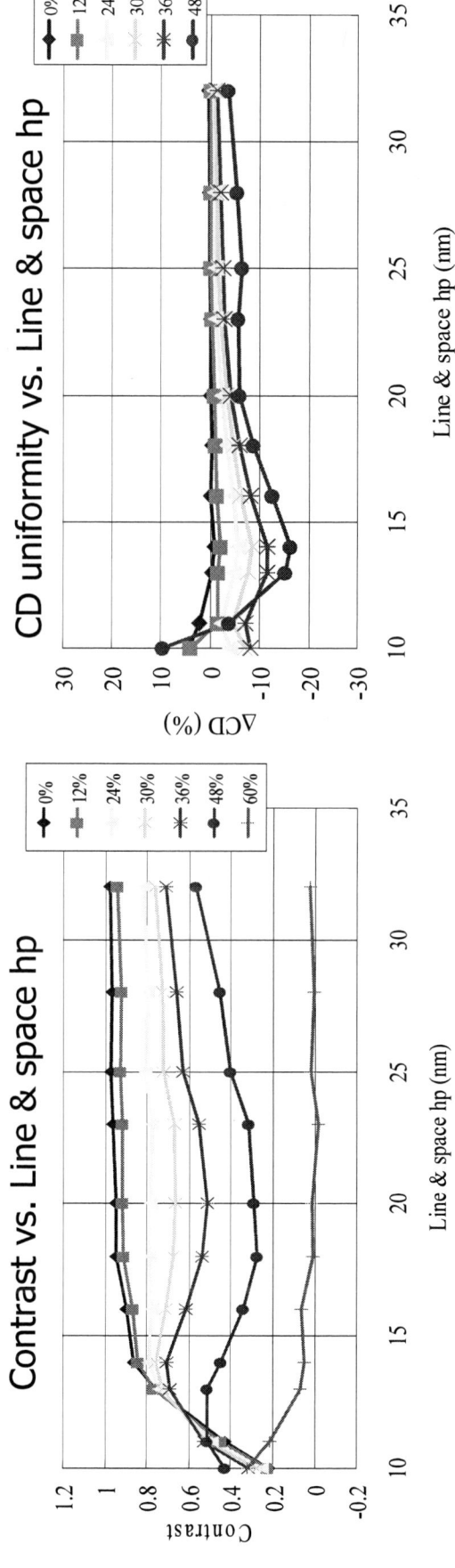

◆ Contrast and CD uniformity depend on pattern density.

Calculation results of Obscuration Canon

Elbow pattern

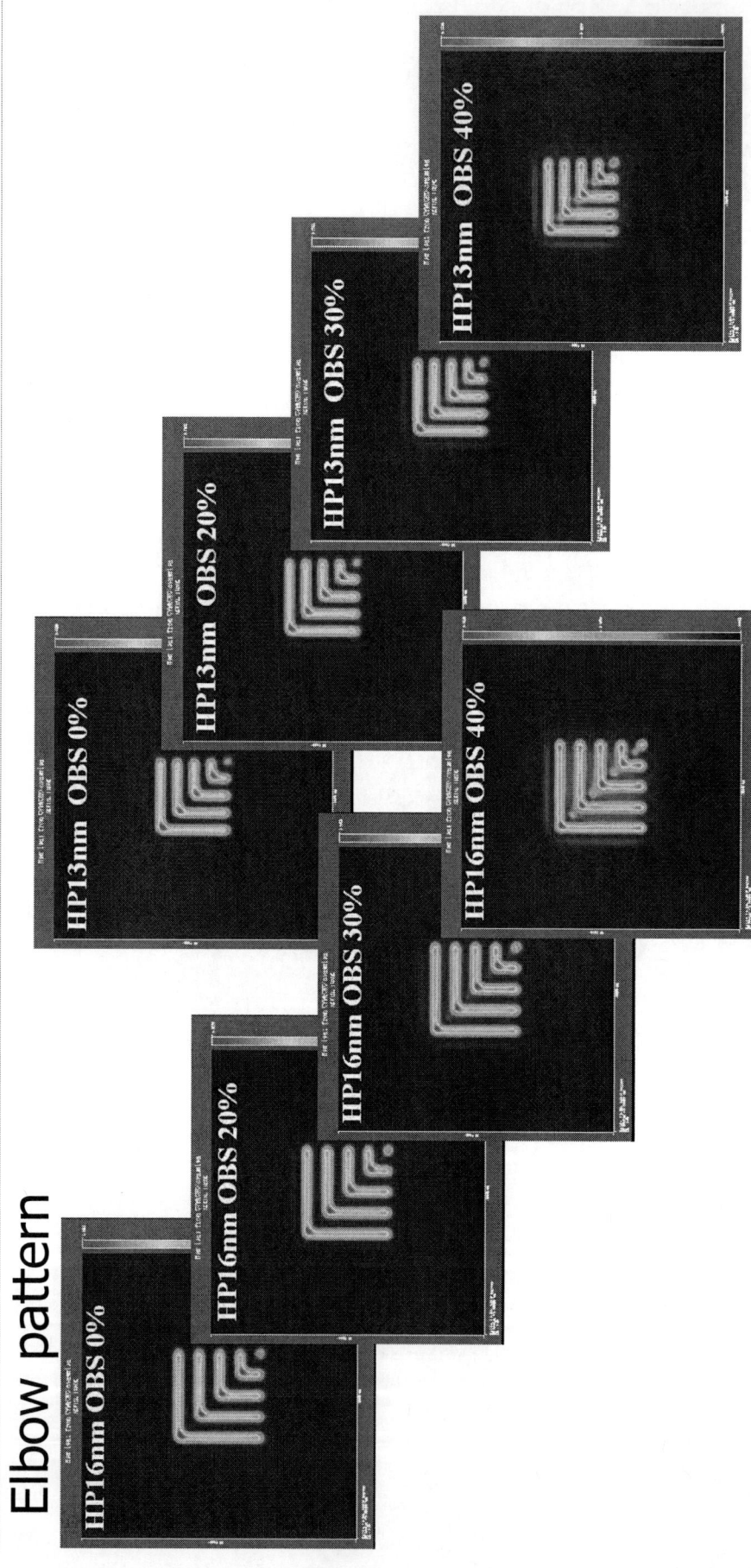

◆ **Over 30% obscuration makes a significant impact on the fidelity of the image.**

◆ **In case of higher NA optics, we will challenge the optics with obscuration less than 30%.**

Outline

1. Canon's Roadmap

2. Research for the Full-Field Tool

   1. Issues of EUVL tool

   2. Projection Optics

   3. Extend toward sub-20nm

4. Summary

EUVL Development Status

Canon

Full-field tool

◆ Laser cleaning system can remove the particles on the mask without damage.

◆ Steady progress is made in the issues of optics lifetime.

6-mirror projection optics

◆ Mirrors are under polishing.

◆ Our current technologies satisfy the specification of HVM tools.

High NA projection optics design

◆ Simulation results show:

Over 30% obscuration makes a significant impact on the fidelity of the image.

Acknowledgment

Canon

A part of this work was performed

under the management of Extreme Ultraviolet Lithography

System Development Association (EUVA)

in the Ministry of Economy Trade and Industry program

supported by

New Energy and Industrial Technology Development

Organization.

Thank you for your attention!

Canon

EUV Resist Performance on the ASML ADT and LBNL MET

Bill Pierson[1], Tom Wallow[2], Hiroyuki Mizuno[3], Anita Fumar-Pici[1], Linda Ohara[1], Karen Petrillo[4], Koen van Ingen Schenau[1], Steve Hansen[1], Sang-In Han[1], Robert Watso[1], Lior Huli[2], Obert Wood[2], Joerg Mallmann[1], Bart Kessels[1], Robert Routh[1], Kevin Cummings[1]

1 ASML; 2 AMD; 3 Toshiba America Electronic Components; 4 IBM; 5 CNSE

Acknowledgements

- Yunfei Deng, Bruno La Fontaine, (AMD)

- Eelco van Setten, Sjoerd Lok, Joep van Dijk, James Weidman, Rick Zachgo (ASML)

- Erin McLellan (IBM)

- Patrick Naulleau, Paul Denham, Brian Hoef, Gideon Jones, Jerrin Chiu, John Roller (LBNL)

- Sandy Finkey, Paula Yergeau, Scott Wright, S. M. Anwar, Dominic Ashworth, Cecilia Montgomery, Warren Montgomery, Emil Piscani, (Sematech)

- This work was performed by the Research Alliance Teams at various IBM Research and Development Facilities

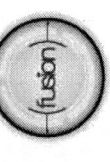

Outline

Goals of performance matching

Simulation-based assessment of metrics

Metric-based comparison of MET and ADT

- CH resist performance
- L/S resist performance

Conclusions

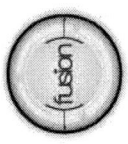

Comparison of MET and ADT

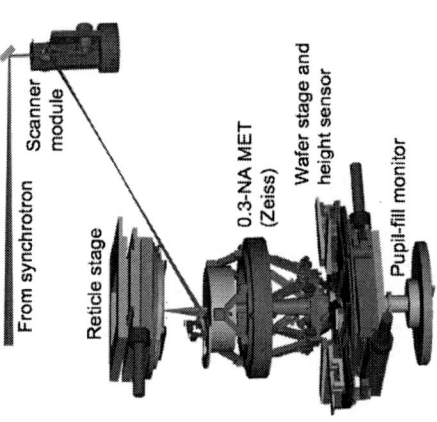

	MET	ADT
NA	0.3	0.25
Illumination	Programmable (0.35/0.55 Annular)	Conventional (0.5)
Aberration	1.4 nm (RMS)	1.1 nm (RMS)
Flare	7%	16%
Facilities	Experimental, 100mm wafers	CNSE Albany, ACT12 Track, 300mm wafers
Qualification date	2004	2007

→ **How do we effectively use both tools to enable 22nm and 16nm logic EUV demonstrations?**

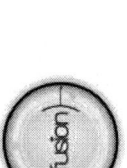

Outline

Goals of performance matching

Simulation-based assessment of metrics

Metric-based comparison of MET and ADT

- CH resist performance

- LS resist performance

Conclusions

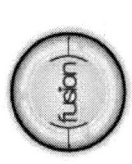

Simulation-based Assessment of Metrics

Prolith EUV resist models
- 193nm based (SIM-1, SIM-2)
- EUV resist based (SIM-3)

Process window response- ADT v MET
- defocus response- varies
- flare response- varies

Exposure latitude response tracks known performance

Make use of RELS as an exposure latitude metric

RELS metric appears to be robust

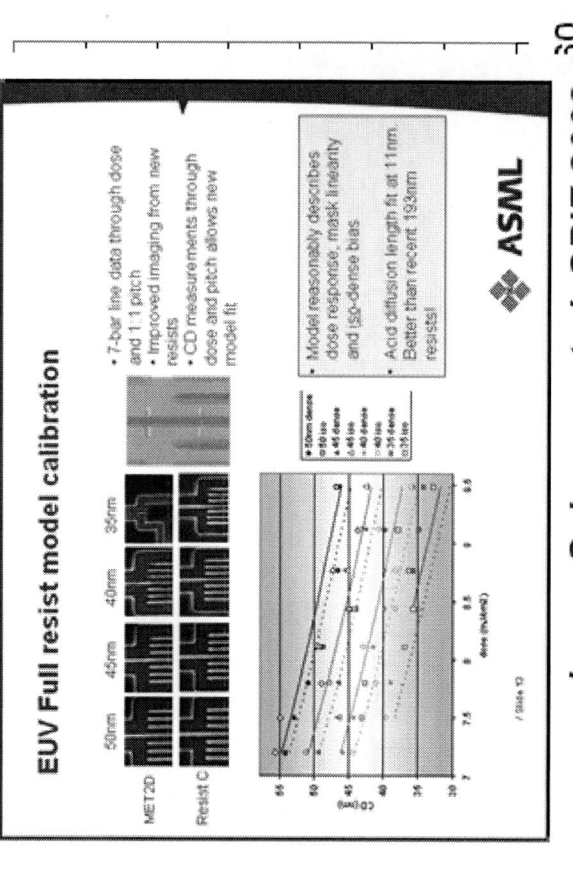

-van Ingen Schenau et al. SPIE 2008

Depth of Focus (um)

Half-pitch (nm)

Outline

Goals of performance matching

Simulation-based assessment of metrics

Metric-based comparison of MET and ADT

- CH resist performance

- LS resist performance

Conclusions

Experimental Methodology and Goals

3 'state of the art' resists (2008) qualified for use on ADT based on MET evaluations

Evaluation of MET and ADT patterning- 1:1 CH and 1:1 LS

- Assess EL metric comparison
- Optimized process conditions, 80nm FT w/BARC for all evaluations
- $E_{(size)}$ range 10-25 mJ/cm^2; LER range 3-4.5 nm 3σ

Continued simulation development

Outline

Goals of performance matching

Simulation-based assessment of methods

Metric-based comparison on MET and ADT

- CH resist performance

- L/S resist performance

Conclusions

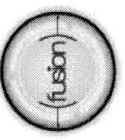

MET CH Patterning: Resist C

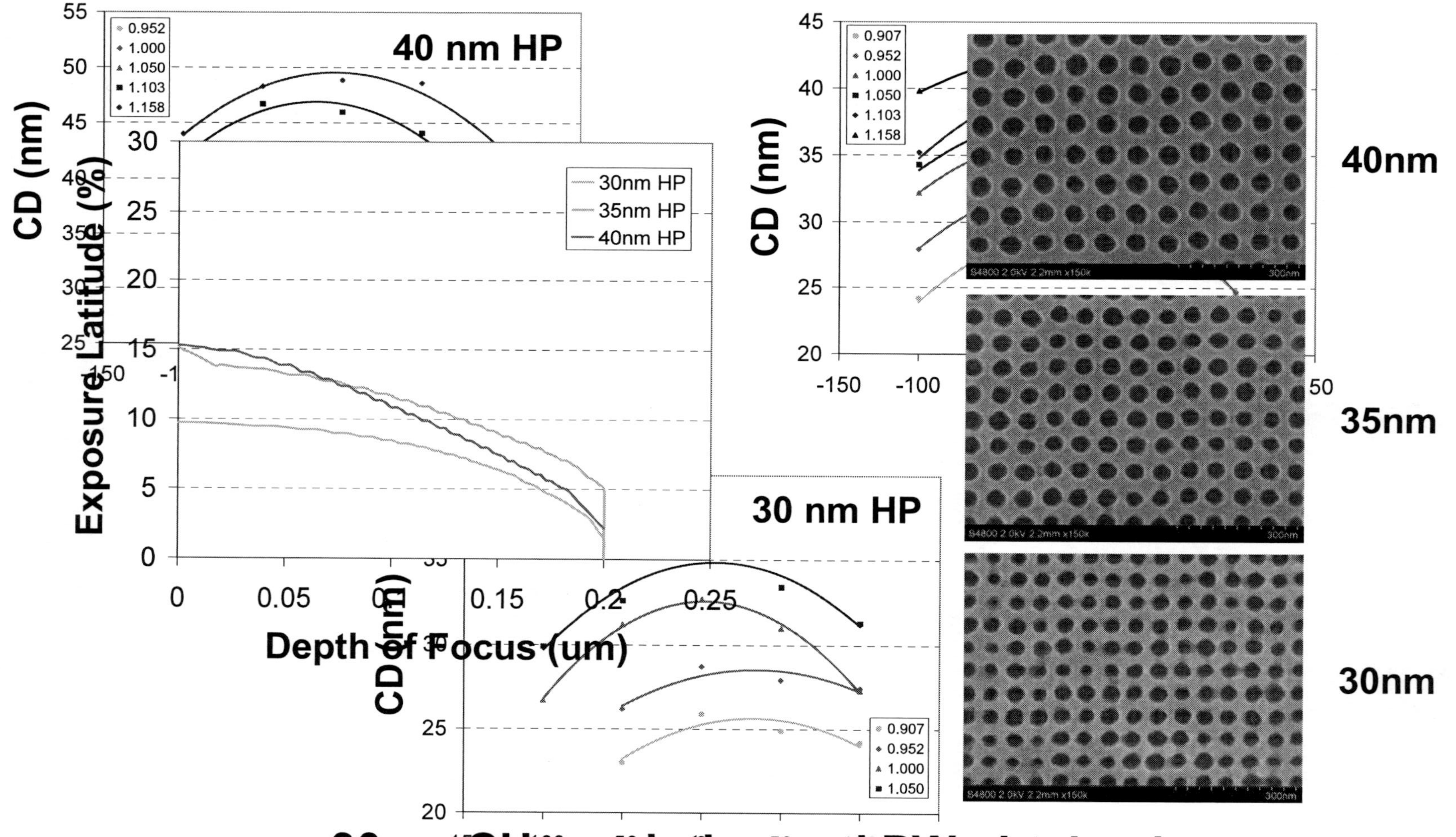

40nm

35nm

30nm

30nm CH resolution and PW obtained

ADT CH Patterning: Resist C

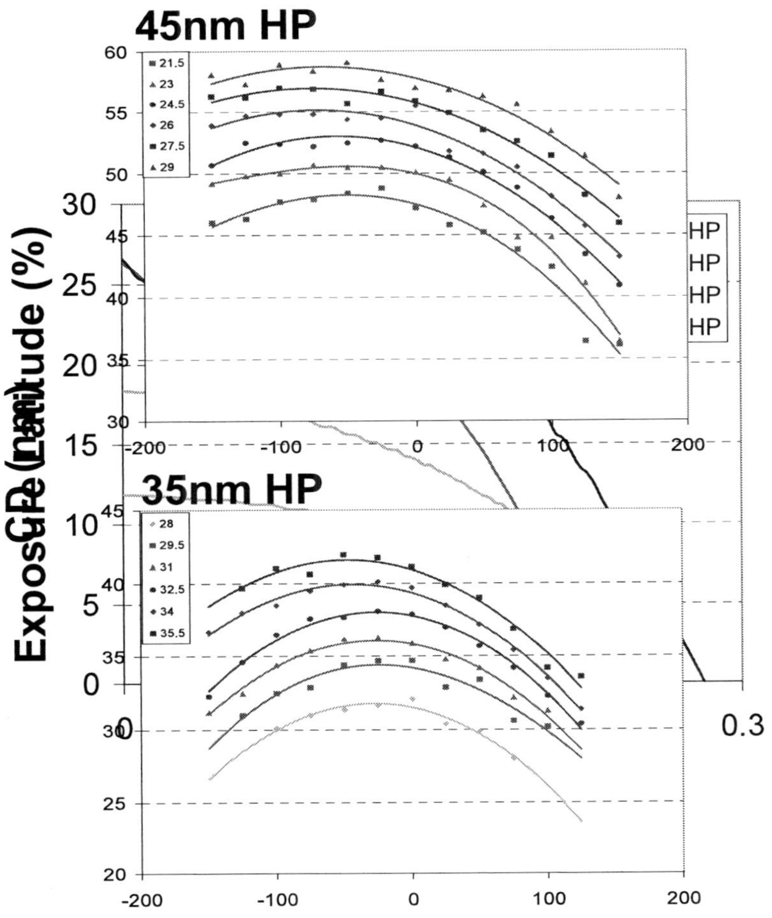

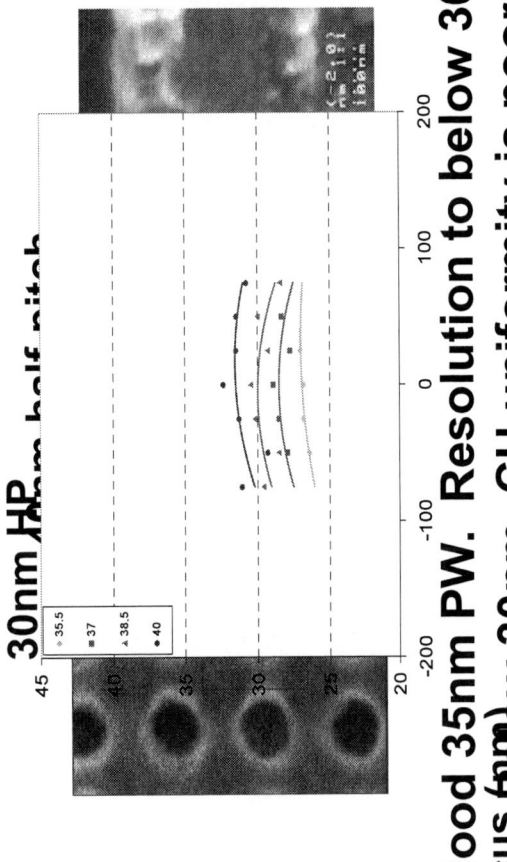

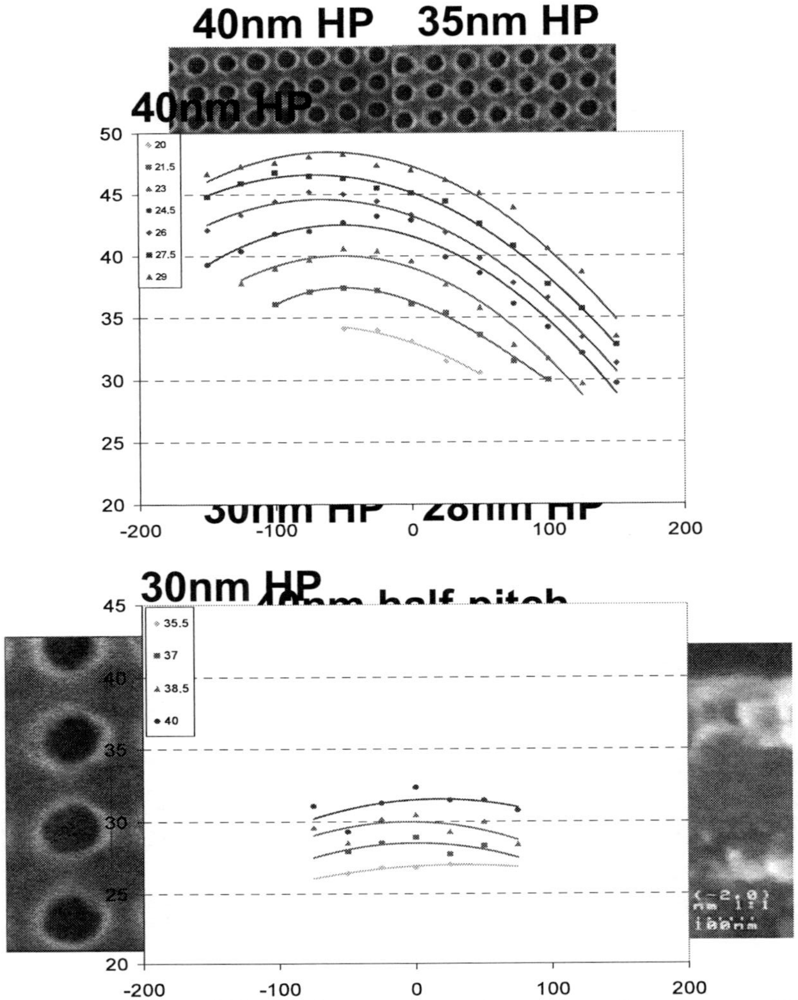

Good 35nm PW. Resolution to below 30nm.
Defocus Below 30nm, CH uniformity is poor.

ADT CH Patterning: Resist B

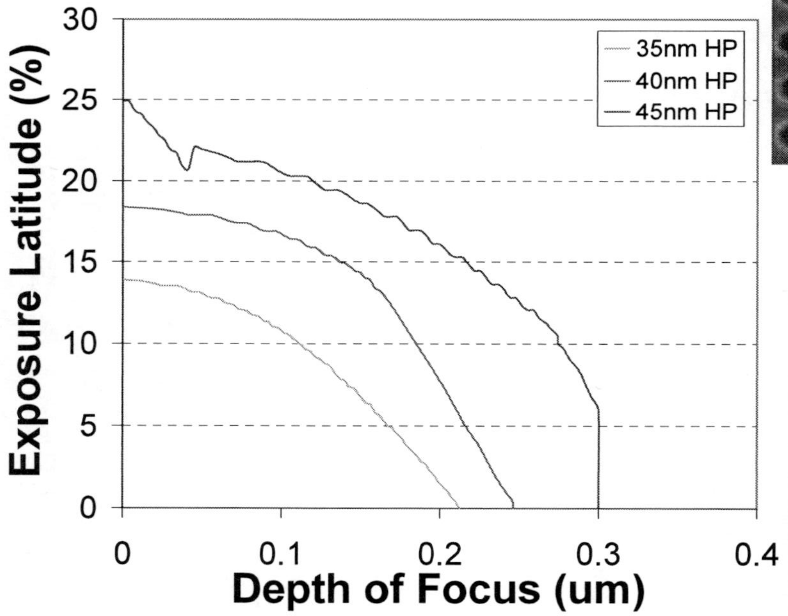

40nm HP   35nm HP   32nm HP

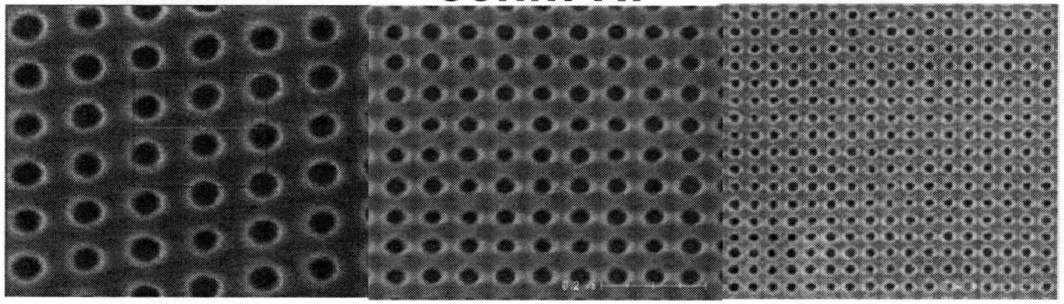

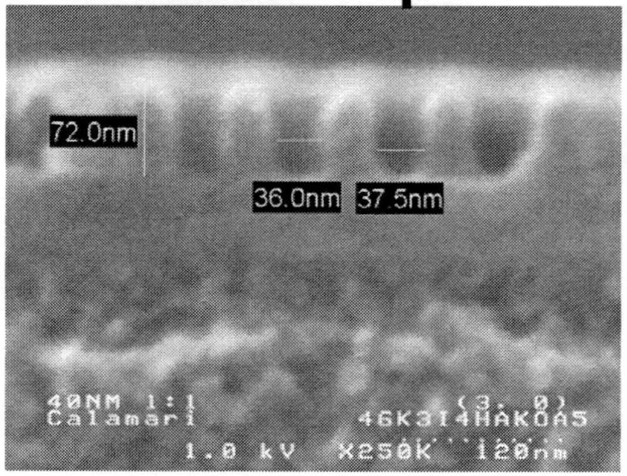

28nm HP @ 50nm FT

40nm half-pitch

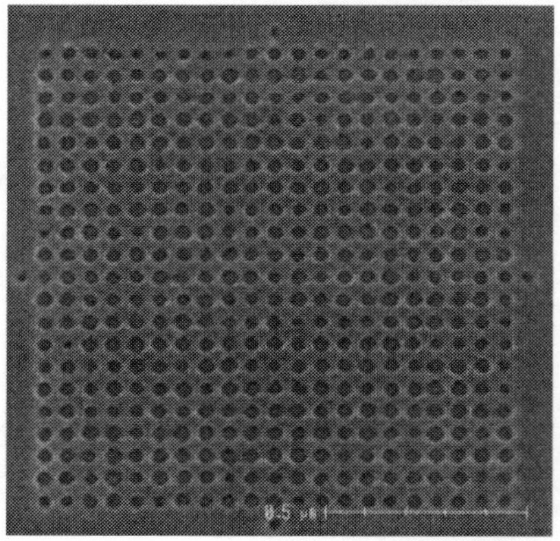

Smaller PW compared to Resist C. 32nm CH obtained.
Resolved 28nm at 50nm FT. Poor uniformity at 28nm.

ASML   AMD
The future is fusion

CH Resist Performance: RELS

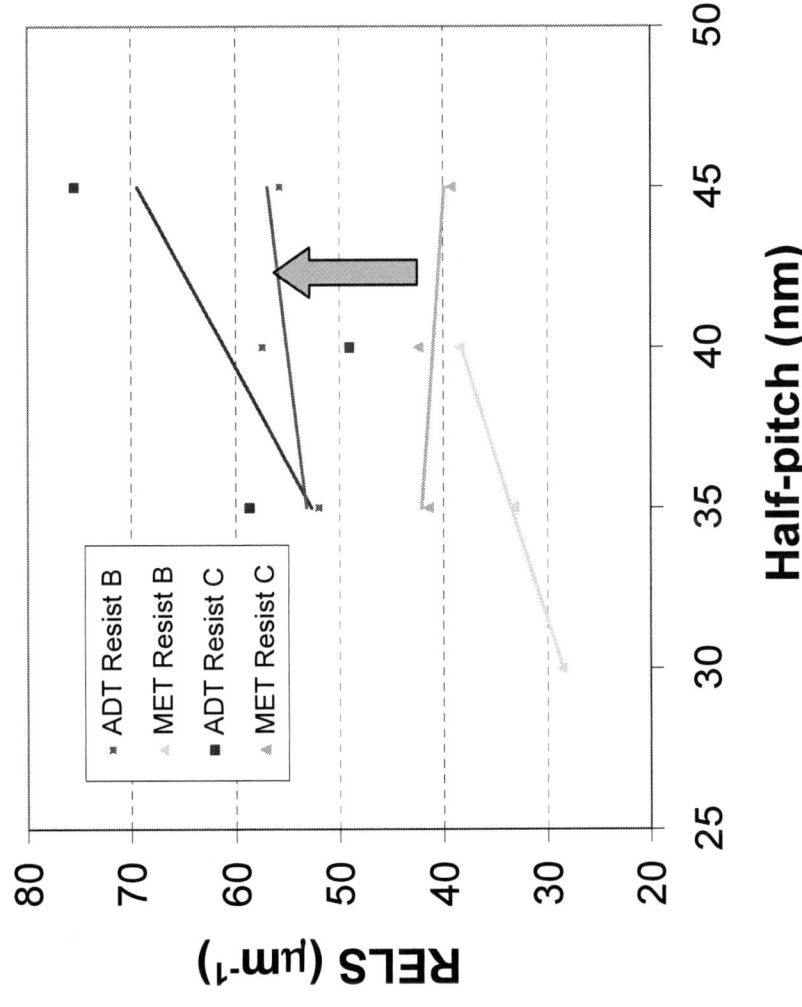

RELS, as a metric of process performance, suggests improvement

Outline

Goals of performance matching

Simulation-based assessment of metrics

Metric-based comparison of MET and ADT

- CH resist performance

- L/S resist performance

Conclusions

MET L/S Patterning: Resist A

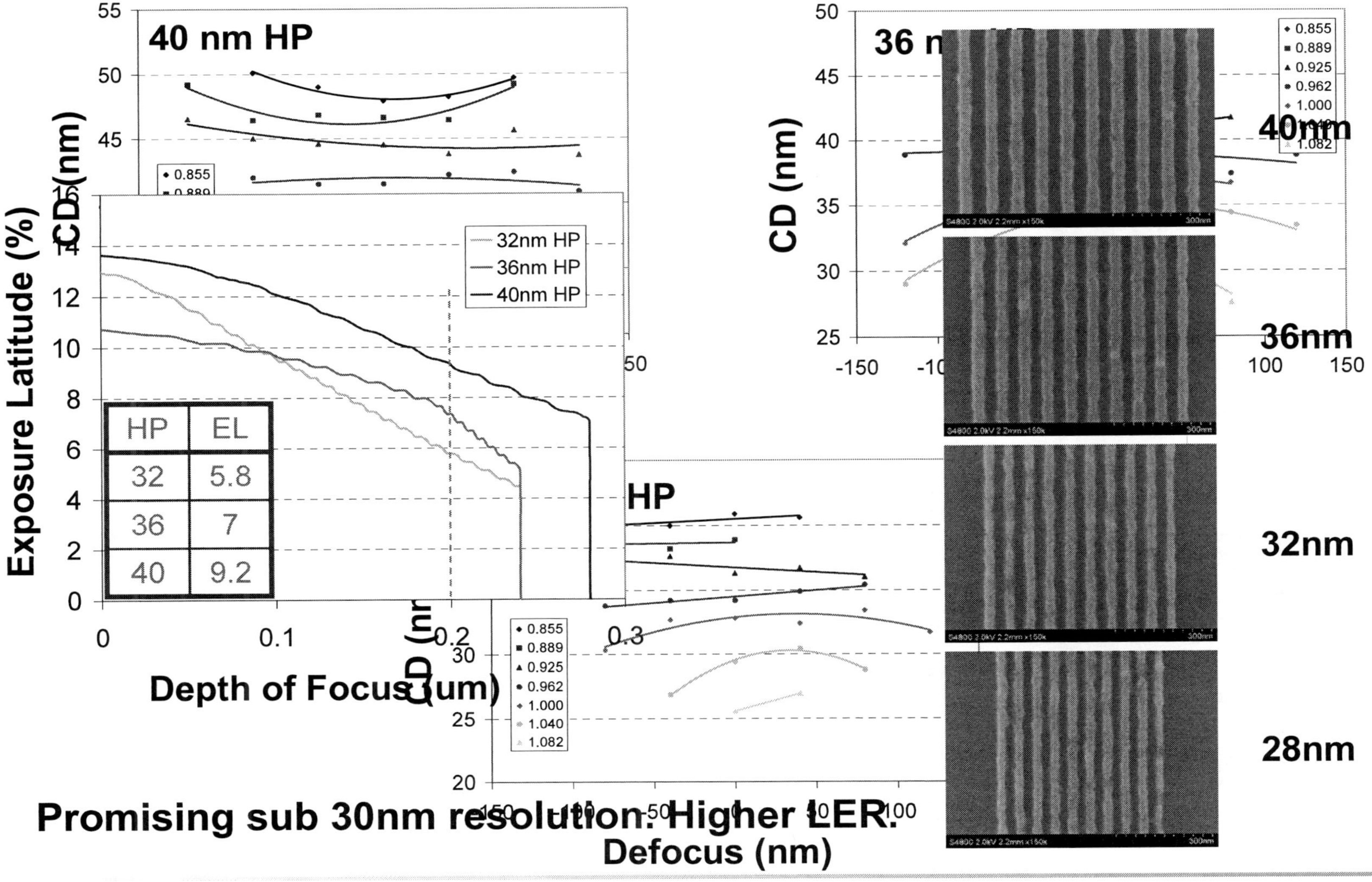

40 nm HP

40nm
36nm
32nm
28nm

36 nm

CD (nm)

HP	EL
32	5.8
36	7
40	9.2

Exposure Latitude (%)

Depth of Focus (µm)

Defocus (nm)

HP

Promising sub 30nm resolution. Higher ER. Higher resolution.

- 0.855
- 0.889
- 0.925
- 0.962
- 1.000
- 1.040
- 1.082

32nm HP
36nm HP
40nm HP

ASML

AMD
The future is fusion

fusion

ADT L/S patterning: Resist C

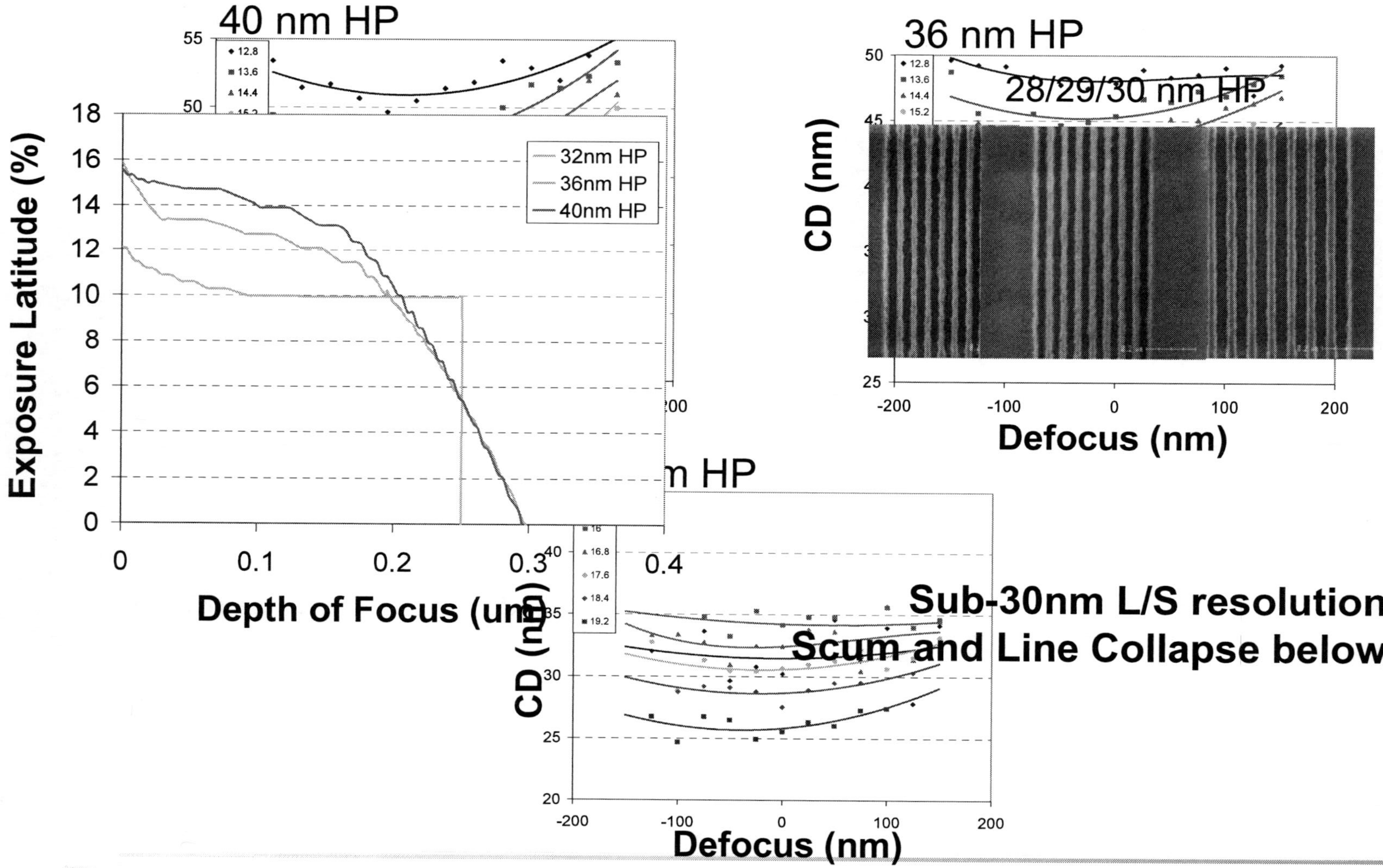

Sub-30nm L/S resolution.
Scum and Line Collapse below 27nm.

ASML

AMD
The future is fusion

ADT L/S Patterning: Resist B

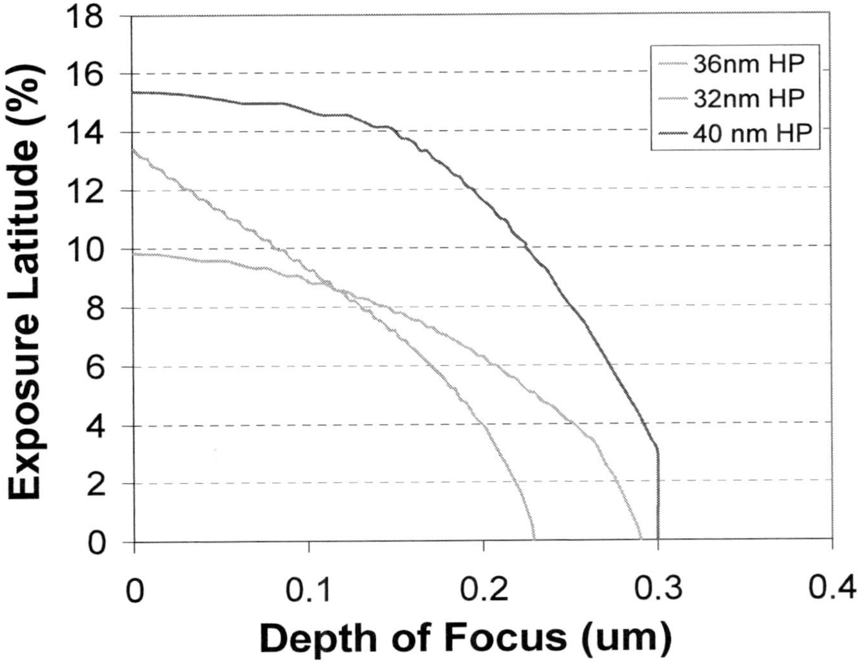

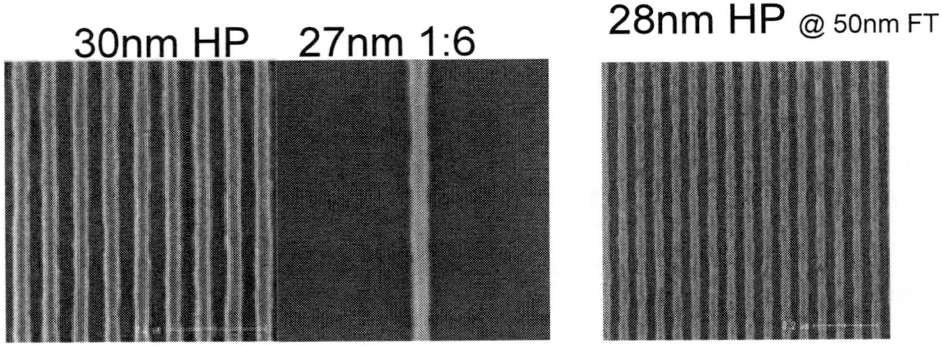

28nm HP @ 50nm FT

30nm HP   27nm 1:6

Superior LER and good DOF.
Strong Isolated line performance.
Sub30nm resolution.
Line Collapse below 27nm

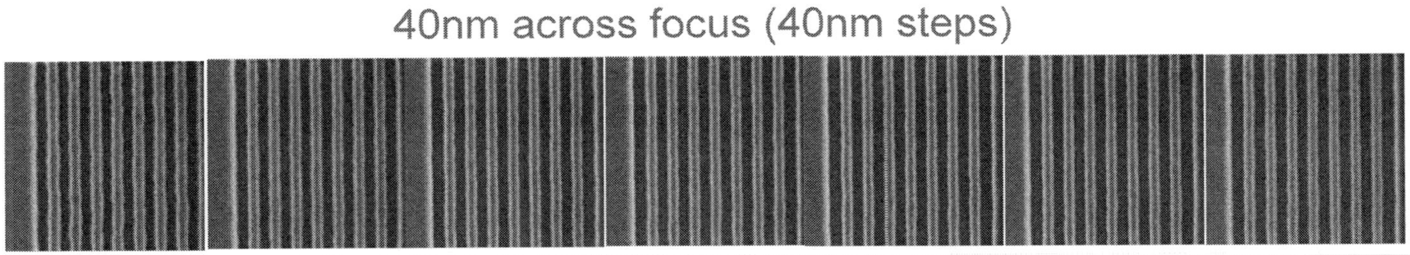

40nm across focus (40nm steps)

Exposure Latitude (%) vs Depth of Focus (um)

- 36nm HP
- 32nm HP
- 40 nm HP

LS Resist Performance: RELS

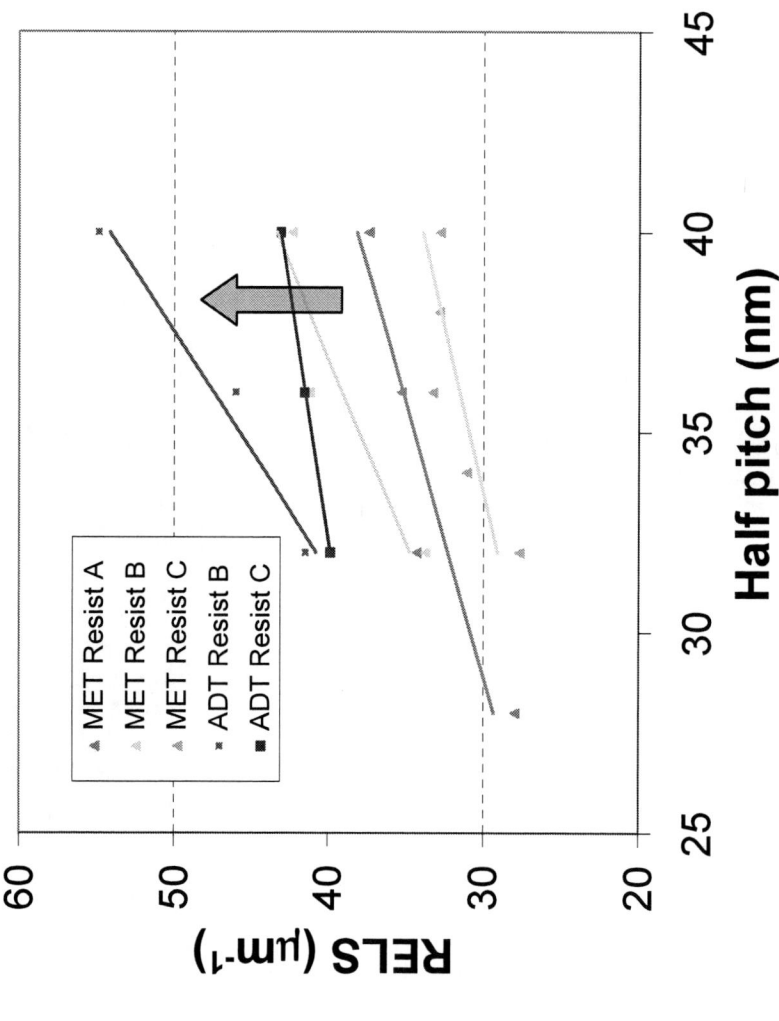

- RELS tracks well for different resists

- As metric of process performance, RELS suggests overall improvement from MET to ADT

Outline

Goals of performance matching

Simulation-based assessment of methods

Metric-based comparison on MET and ADT

- L/S resist performance

- CH resist performance

Conclusions

Summary and Conclusions

- All 3 'next generation' resists evaluated have greatly improved performance on both MET and ADT.

  - Robust process for L/S to 32nm on both tools

  - Robust process for CH to 30nm on ADT and 35nm on MET

  - Resist blur is in the 10-16nm range based on simulation matching

- Quantitative resist performance comparison is difficult due to differences in exposure tools.

- RELS metric is useful to predict relative resist performance on the MET and ADT tool.

- Under conditions studied here, RELS, when considered as a process robustness metric, indicates that process performance appears to improve on transfer from MET to ADT

Trademark Attribution

AMD, the AMD Arrow logo and combinations thereof are trademarks of Advanced Micro Devices, Inc. in the United States and/or other jurisdictions. Other names used in this presentation are for identification purposes only and may be trademarks of their respective owners.

©2008 Advanced Micro Devices, Inc. All rights reserved.

The SEMATECH Berkeley MET: learning at the 22 nm node

Patrick Naulleau

Lawrence Berkeley National Laboratory

Chris Anderson
Paul Denham
Simi George
Ken Goldberg
Brian Hoef
Gideon Jones
Dimitra Niakoula
Ryan Miyakawa
John Roller
LBNL

Chawon Koh
Warren Montgomery
Stefan Wurm
SEMATECH

Bruno La Fontaine
Tom Wallow
AMD

Andy Ma
Intel

Joo-on Park
Samsung

Outline

- Introduction
- Latest resist results
- Future plans
- Summary

SEMATECH Berkeley EUV MET: enabling learning at 22-nm half pitch and beyond

Lossless variable illumination

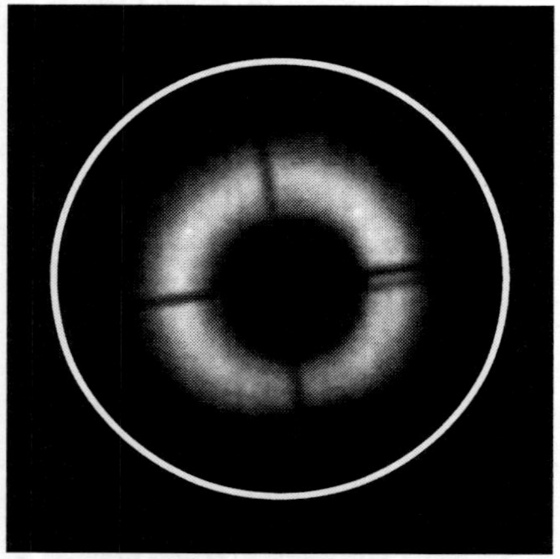

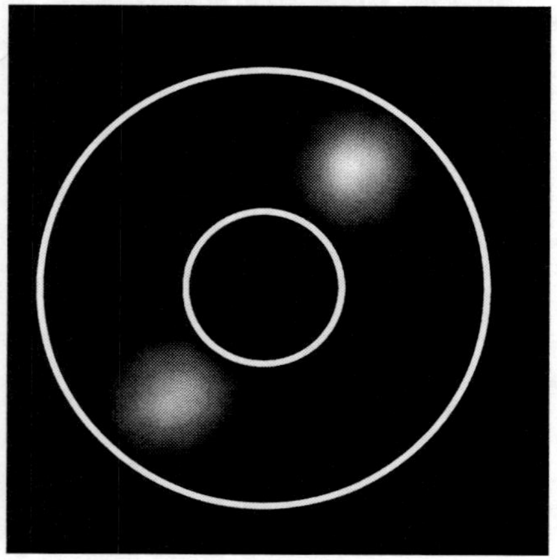

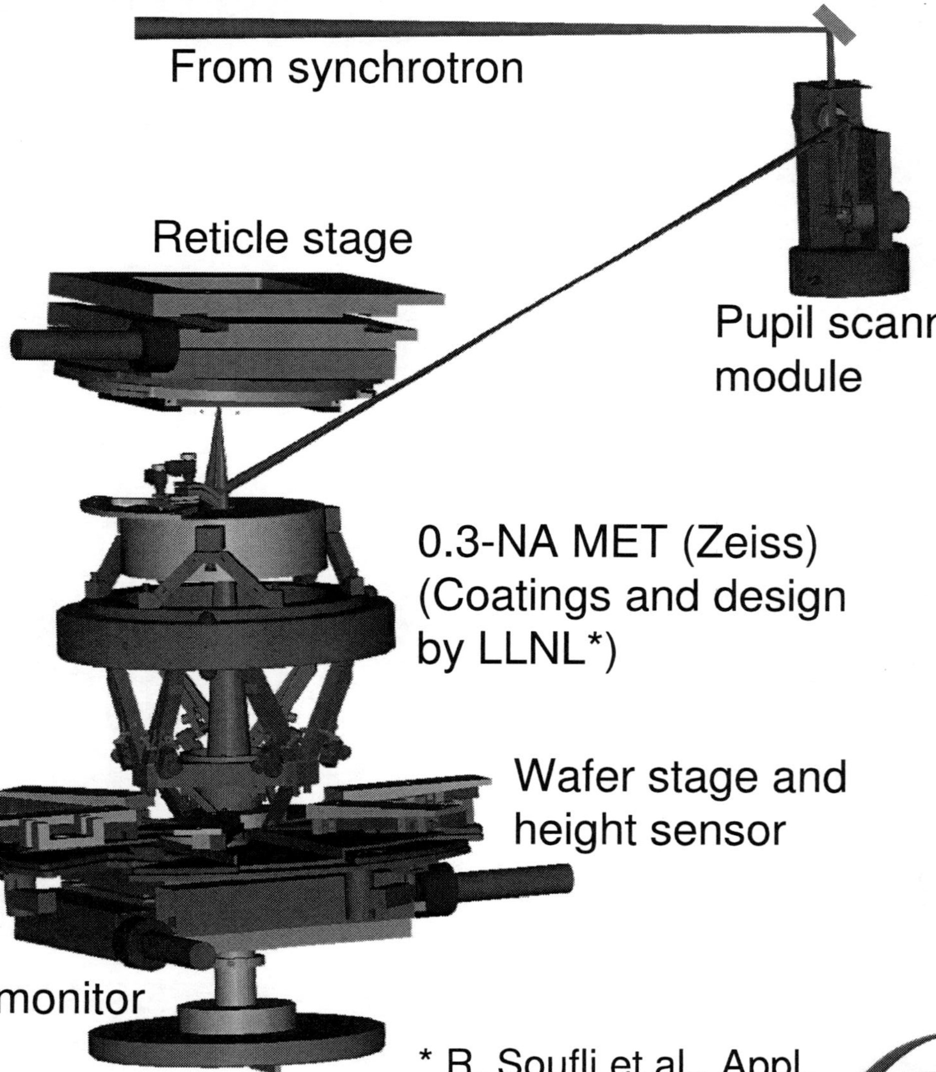

From synchrotron

Pupil scanner module

Reticle stage

0.3-NA MET (Zeiss)
(Coatings and design by LLNL*)

Wafer stage and height sensor

Pupil-fill monitor

* R. Soufli et al., Appl. Opt. 46, 3736 (2007)

Radial grating with rotated dipole illumination

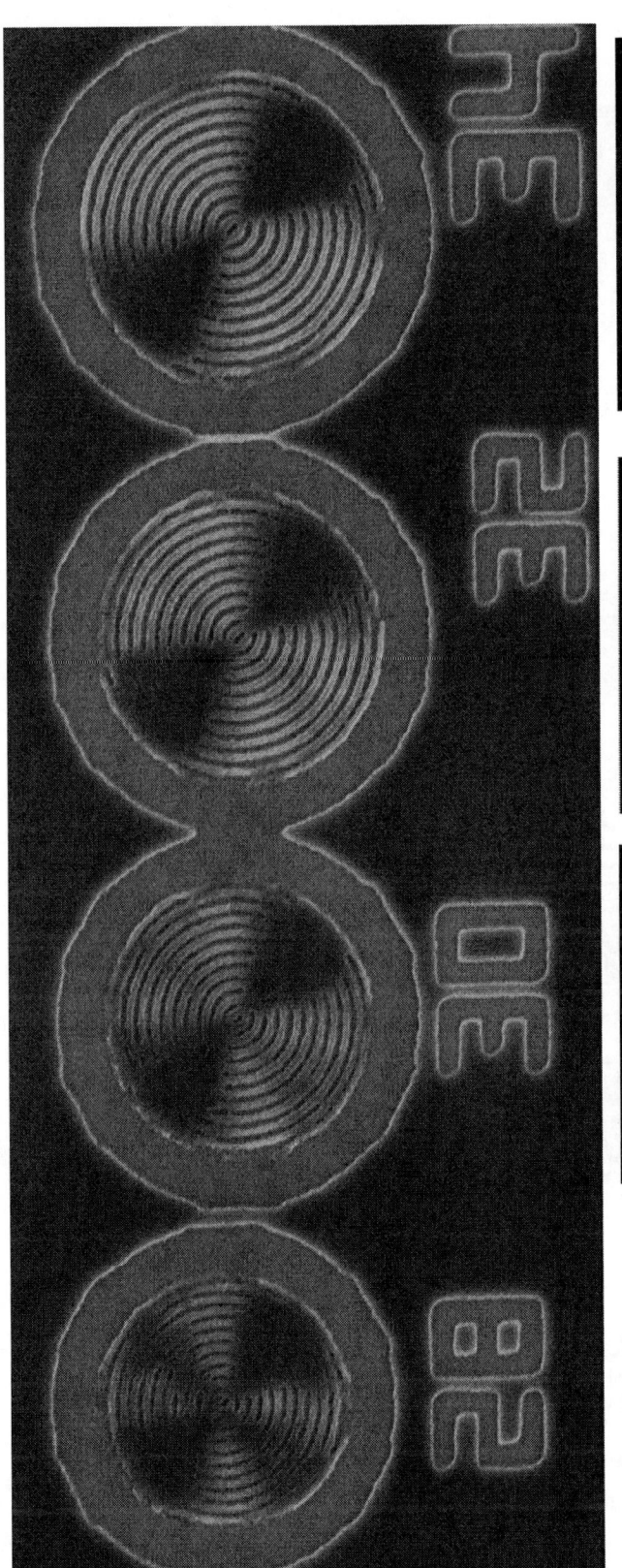

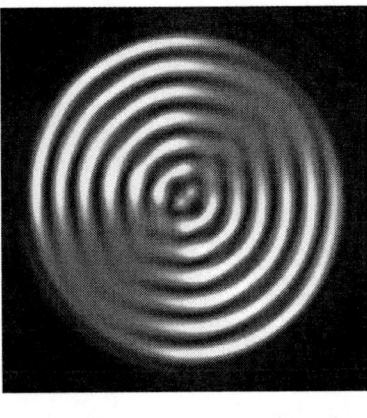

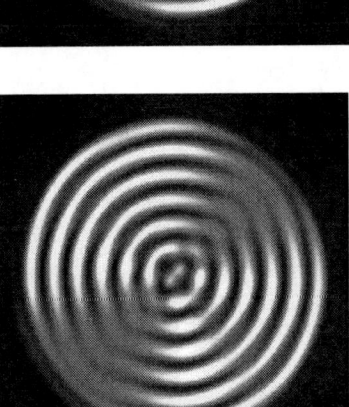

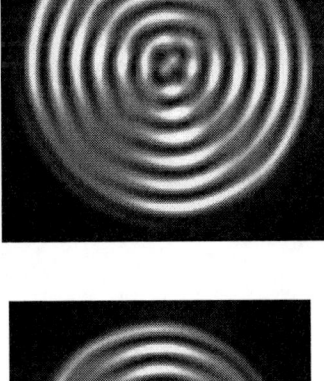

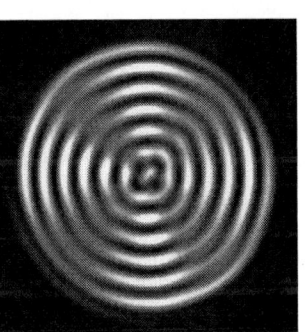

Predicted aerial images match printing results

- Introduction
- **Latest resist results**
- Future plans
- Summary

Simultaneously meeting resolution, sensitivity, and LER crucial issue for EUV resists

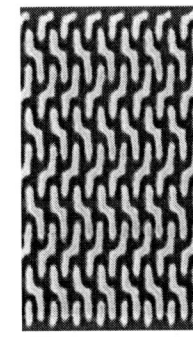

32-nm half pitch (21-nm iso) - 2013
22-nm half pitch (15-nm iso) -2016
16-nm half pitch (11-nm iso) -2019

All three requirements must be met and balanced for any technology or it will not work

Resolution

Acid diffusivity, activation energy, photon statistics, quantum efficiency, outgassing

Sensitivity

10 mJ/cm^2
10 mJ/cm2
10 mJ/cm^2

LER

1.2 nm
0.8 nm
0.6 nm

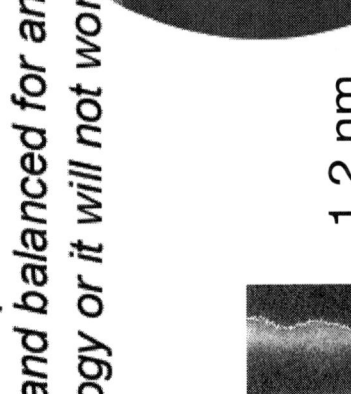

Numbers taken from 2007 ITRS

EUV resists 2x faster than previously thought

See posters tonight

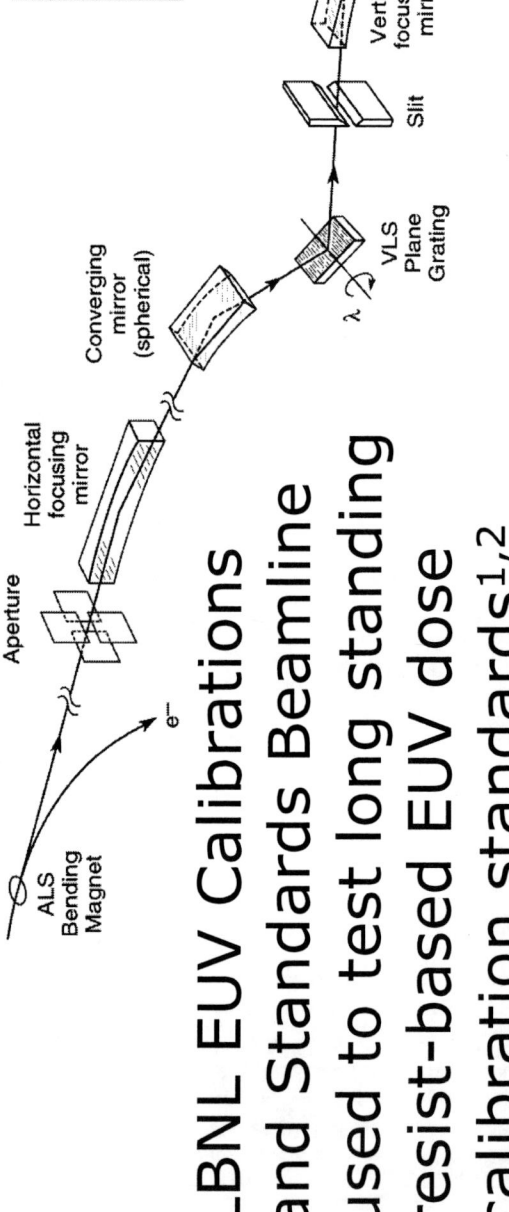

- LBNL EUV Calibrations and Standards Beamline used to test long standing resist-based EUV dose calibration standards[1,2]

- 2x error found in community-wide resist standards going back ~10 years

- EUV resists 2x faster than previously thought

- LBNL results subsequently reproduced by NIST

[1] IEUVI Resist TWG, SPIE Advanced Lithography Feb. 2008.

[2] P. Naulleau, E. Gullikson, A. Aquila, S. George, D. Niakoula, "Absolute sensitivity calibration of extreme ultraviolet photoresists," Opt. Exp. **16**, 11519-11524 (2008).

Several chemically amplified resists break the 25-nm resolution barrier

BERKELEY LAB

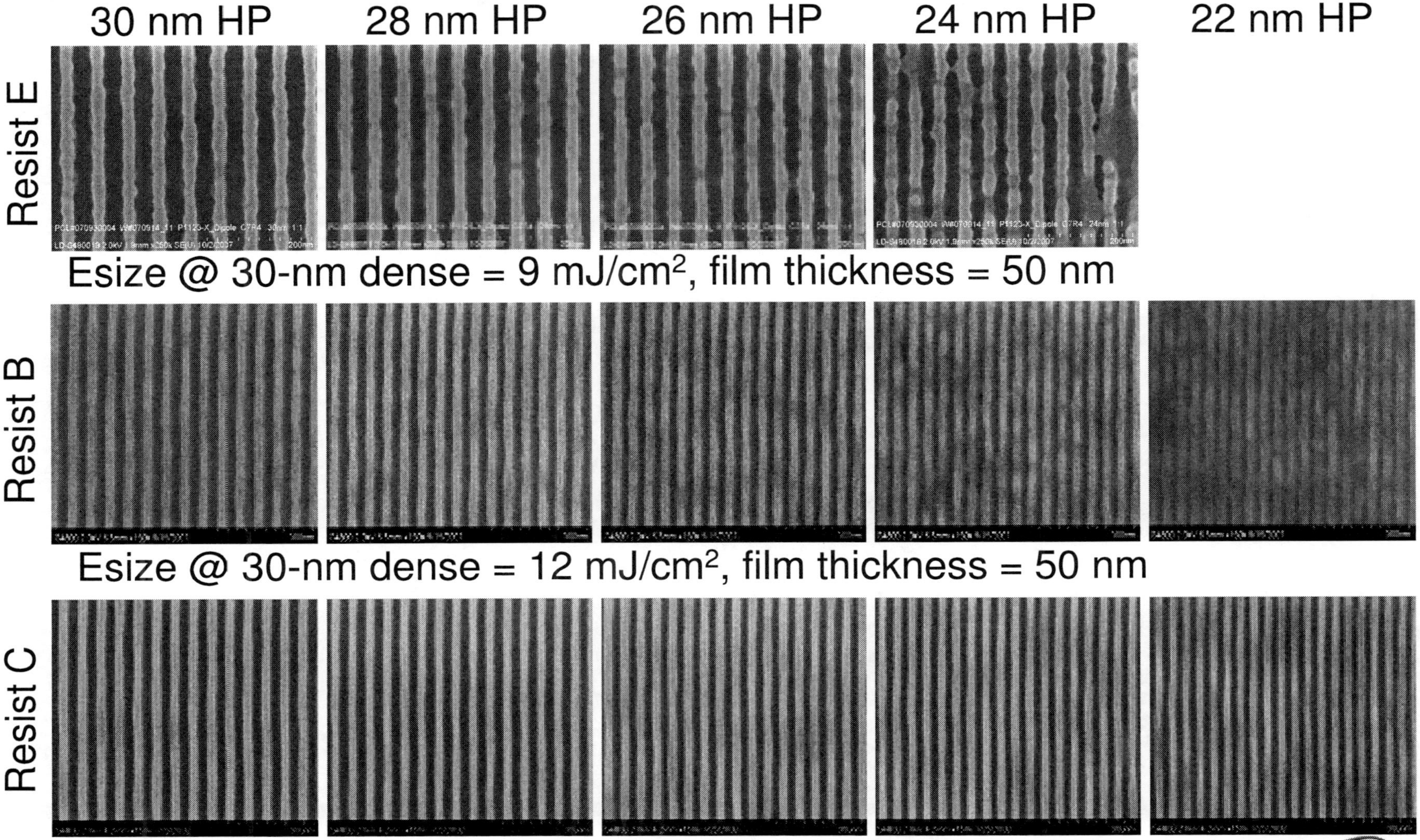

SEMATECH

Resist E — 30 nm HP | 28 nm HP | 26 nm HP | 24 nm HP

Esize @ 30-nm dense = 9 mJ/cm², film thickness = 50 nm

Resist B — 24 nm HP | 22 nm HP

Esize @ 30-nm dense = 12 mJ/cm², film thickness = 50 nm

Resist C — 24 nm HP | 22 nm HP

Esize @ 30-nm dense = 13 mJ/cm², film thickness = 50 nm

CA resists now approaching 20 nm

BERKELEY LAB

24 nm HP 22 nm HP 20 nm HP

Resist C 12.7 mJ/cm²

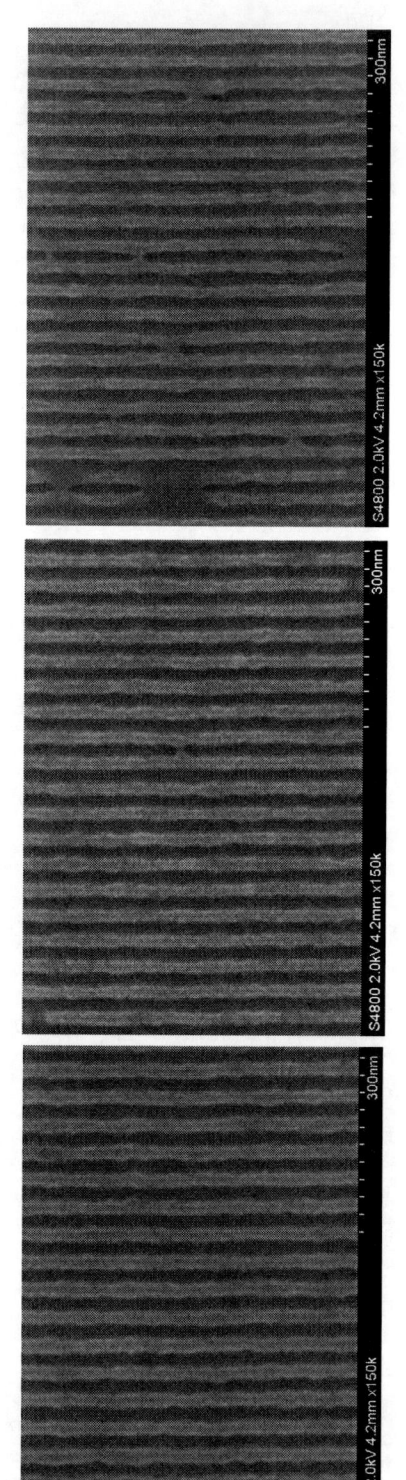

Resist D 15.2 mJ/cm²

50-nm resist thickness

SEMATECH

High pattern fidelity at small feature sizes

BERKELEY LAB

24 nm HP   22 nm HP 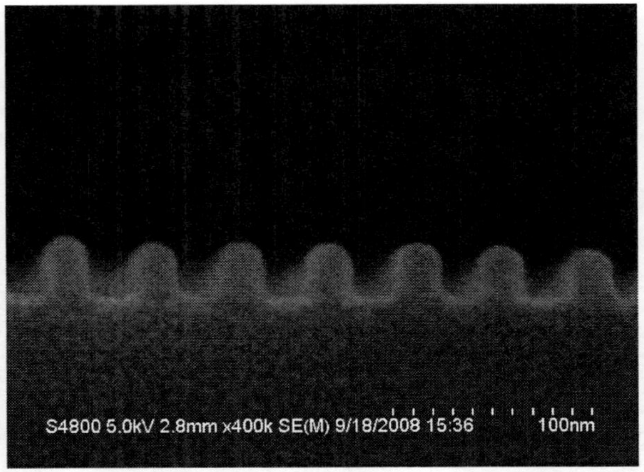   20 nm HP

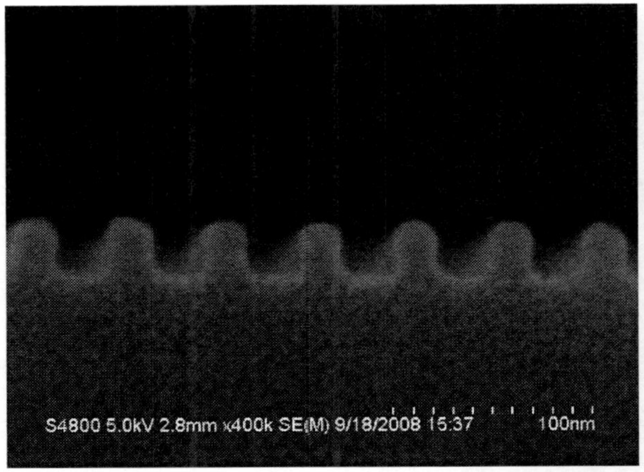

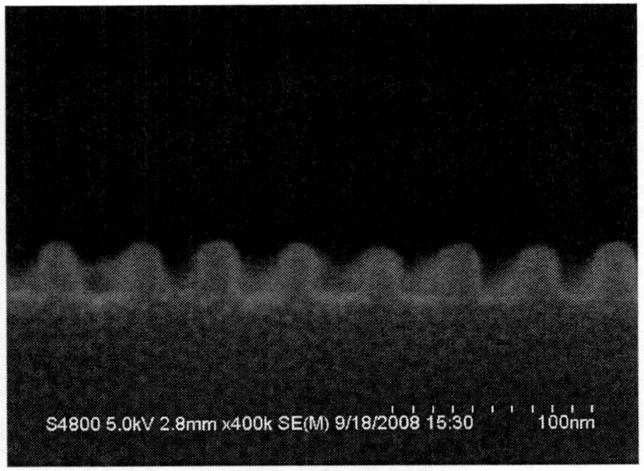

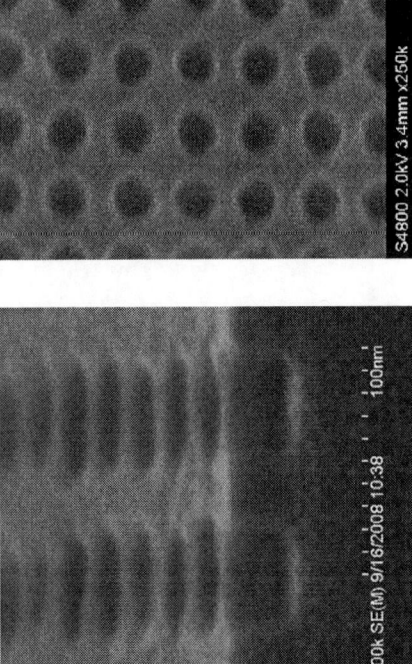

30 nm 1:1 contacts

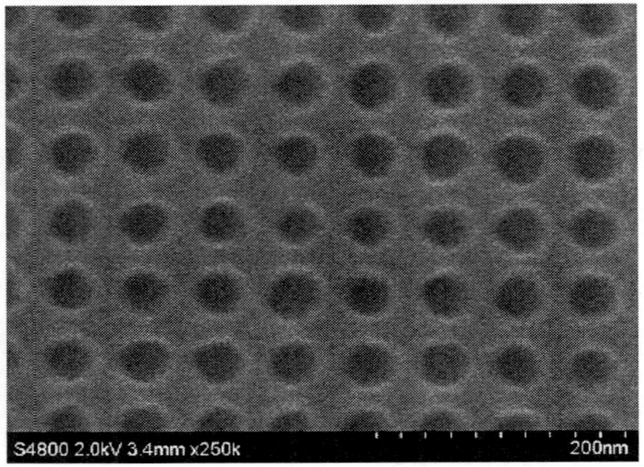

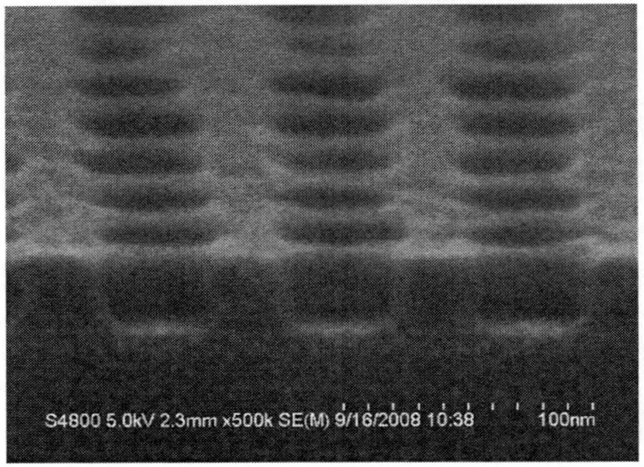

Resist D, film thickness = 50 nm

See talk Wednesday
10:30 AM: C. Koh et al.

SEMATECH

Significant improvement in resolution over past year without sacrificing LER

BERKELEY LAB

SEMATECH

24 nm HP   **22 nm HP**   **20 nm HP**

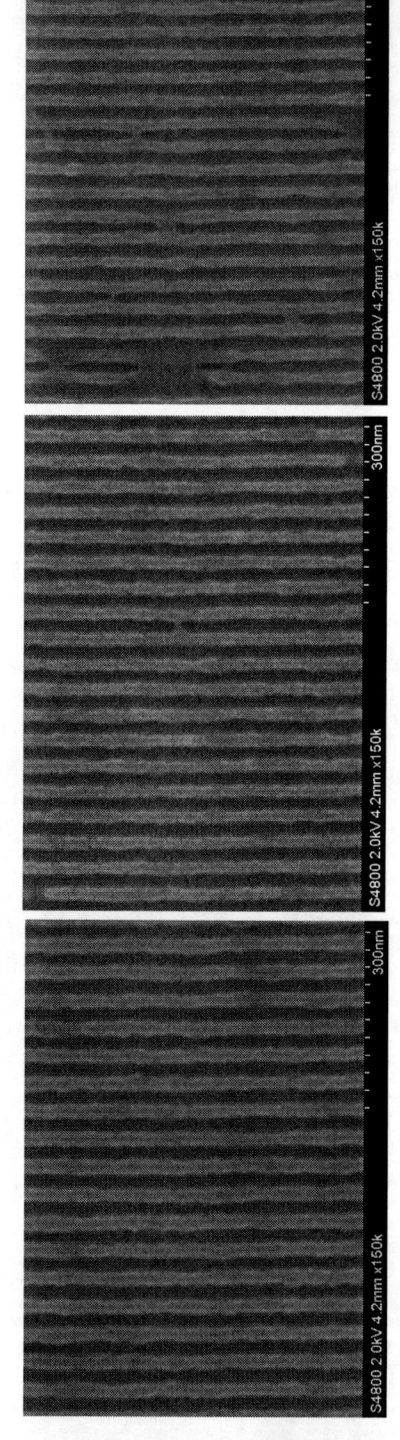

Best 10/07 10 mJ/cm²

Best 10/08 15 mJ/cm²

☐ LER ~4.0 nm

Line-space resolution: progress over time

Annular illumination

40 nm Half-pitch	36 nm Half-pitch	32 nm Half-pitch

16% EL 200nm DOF

15% EL 200nm DOF

8% EL 150nm DOF

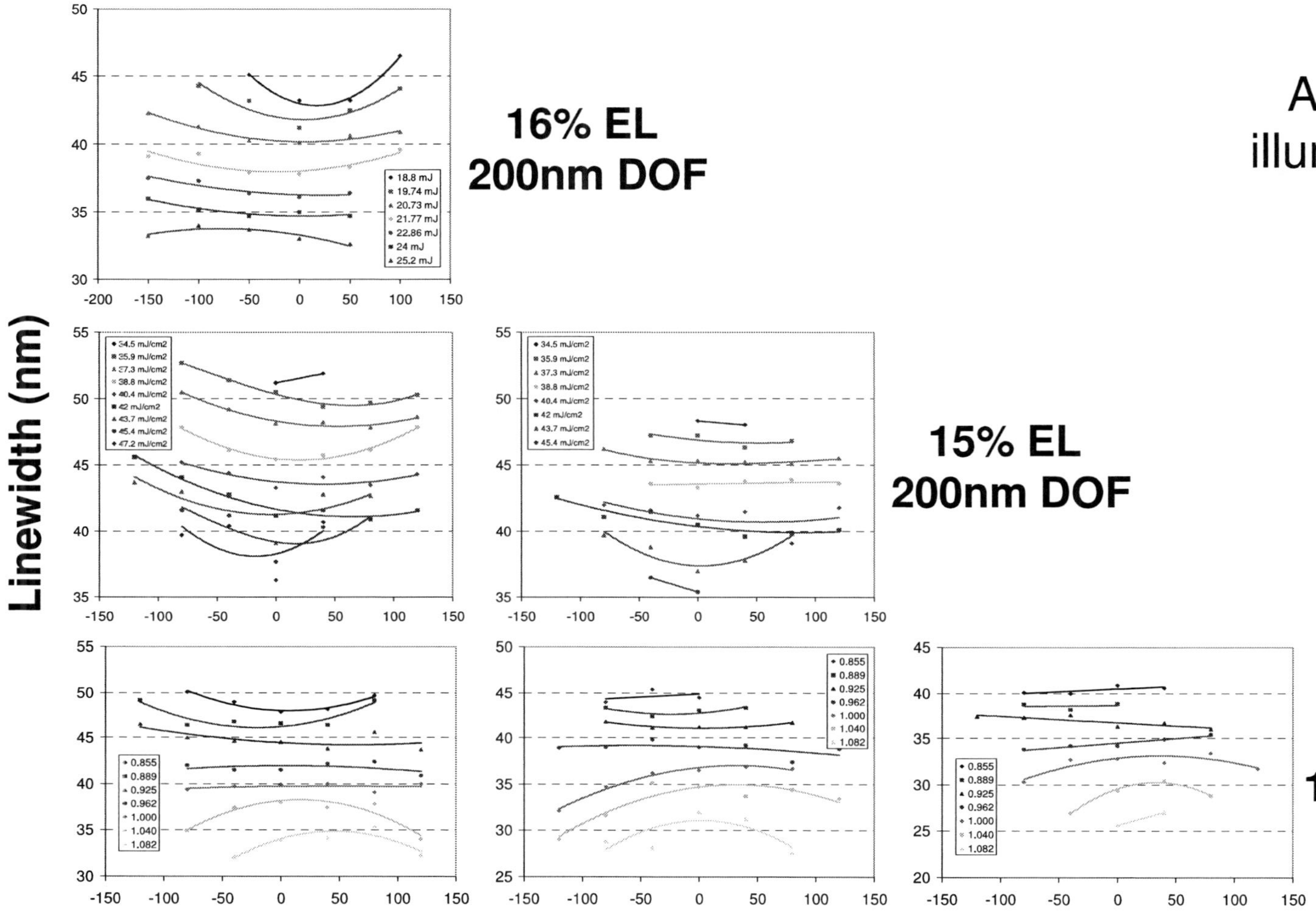

Linewidth (nm)

Defocus (nm)

2006

2007

2008

Reaching resolution limits of mask

20 nm HP

18 nm HP

17 nm HP

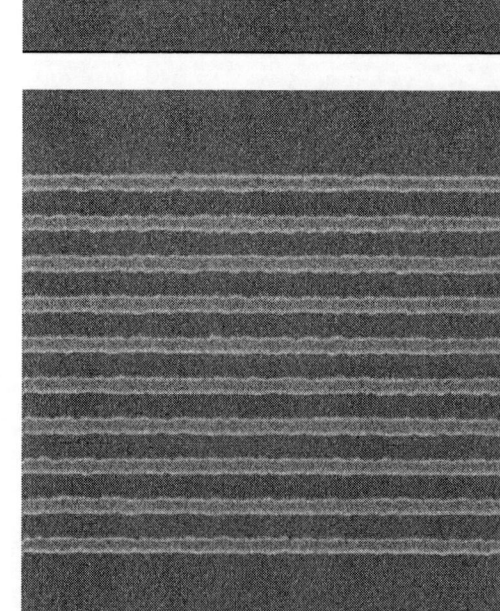
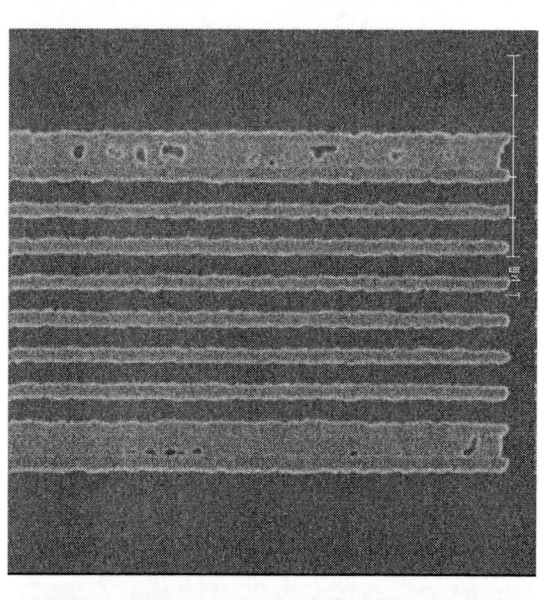
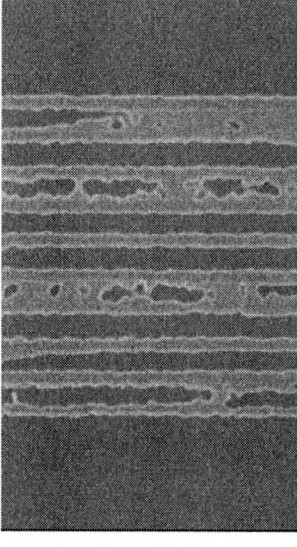
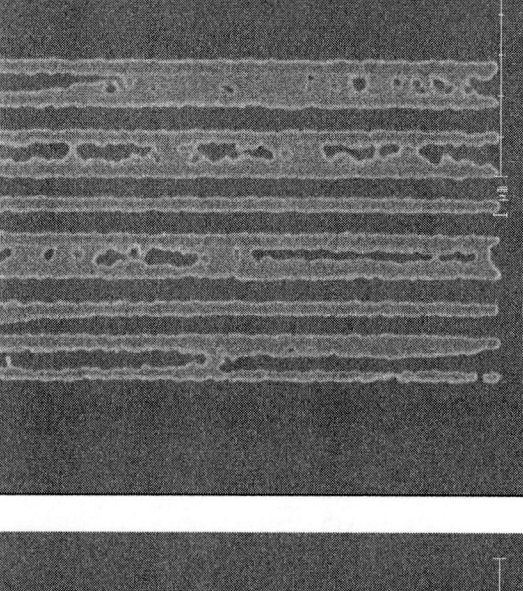

- ☐ SEM images from the SEMATECH Berkeley mask
- ☐ Fabricated Q3 2006
- ☐ Limitation due to mask patterning process used at the time
- ☐ Similar results on SU451 mask used for SEMATECH resist benchmarking

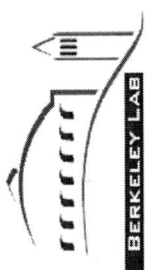

Resolution metrics show steady increase in resist performance

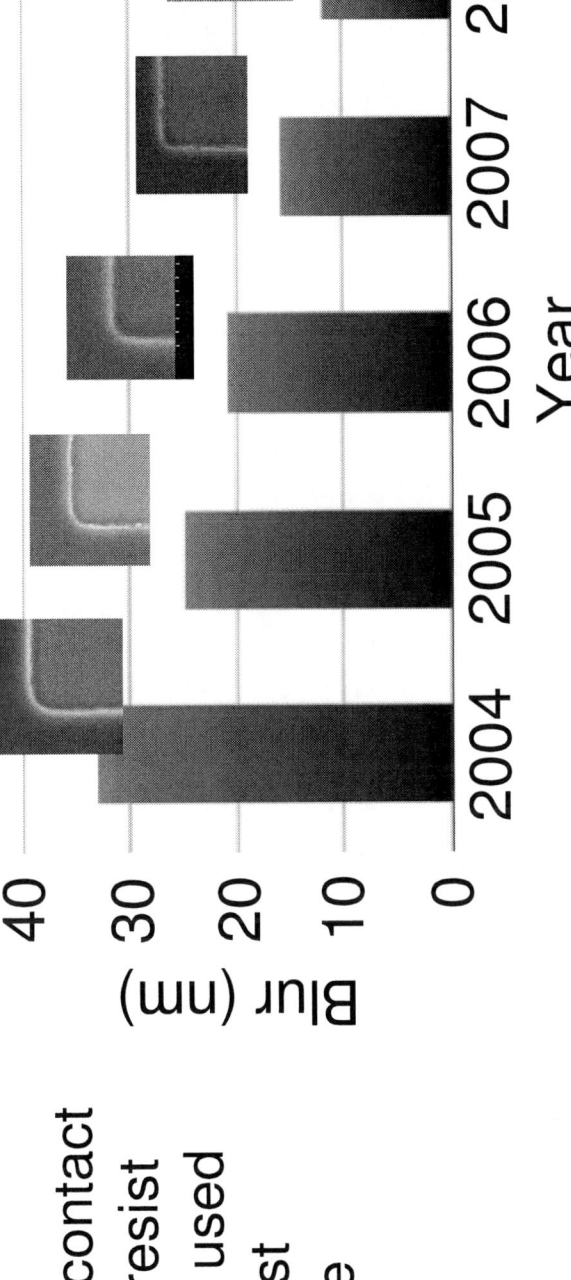

Systematic contact and corner resist blur metrics used to track resist performance over time

- Detailed analysis of corners reveals unexpected large difference between inner and outer corners
- Implications on resist fundamentals under investigation

Study of underlayers for performance optimization: Underlayer A

Film thickness = 50 nm, conventional annular illumination

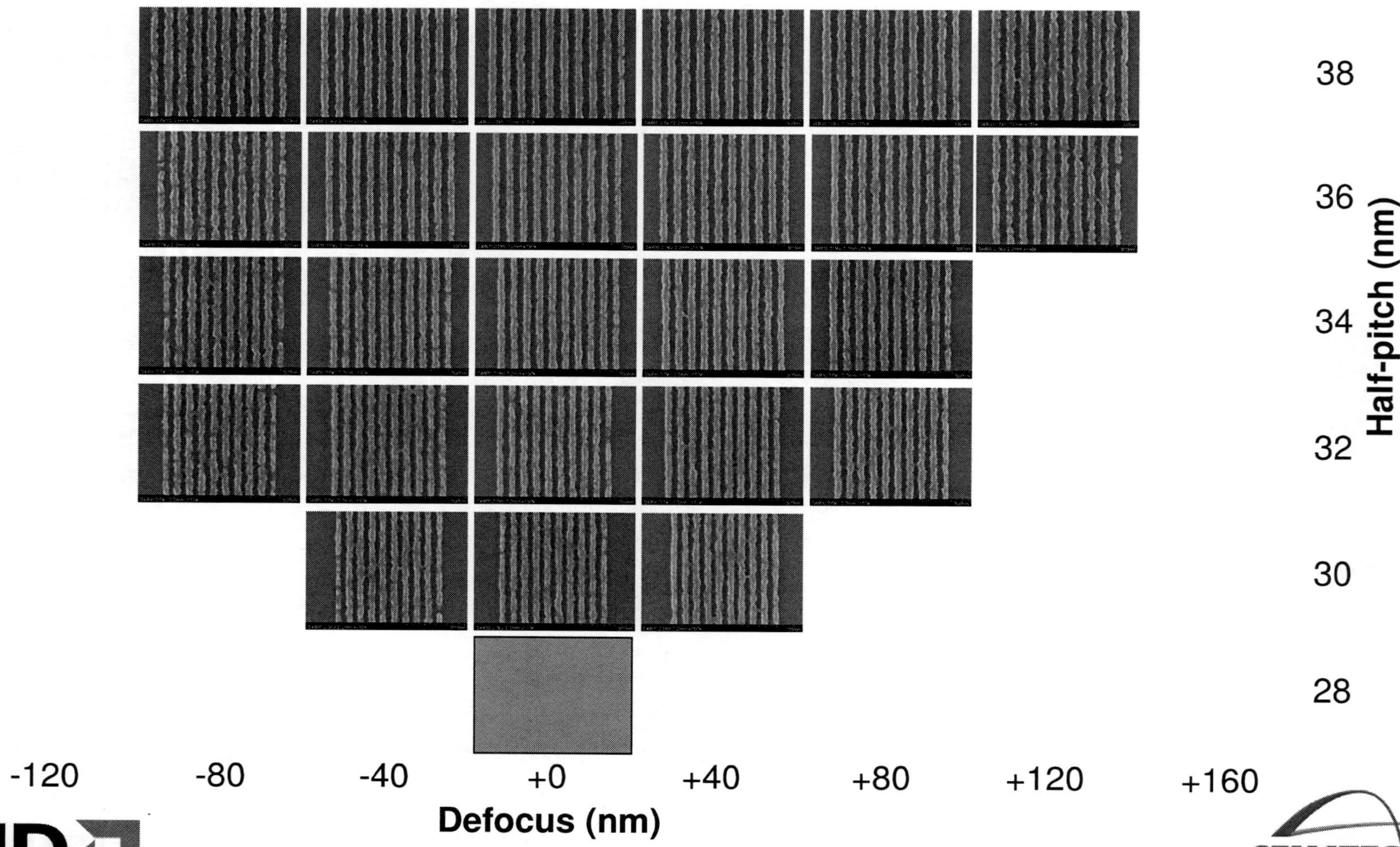

Study of underlayers for performance optimization: Underlayer B

Film thickness = 50 nm, conventional annular illumination

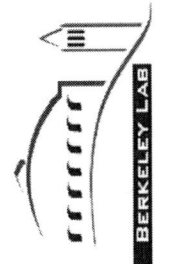

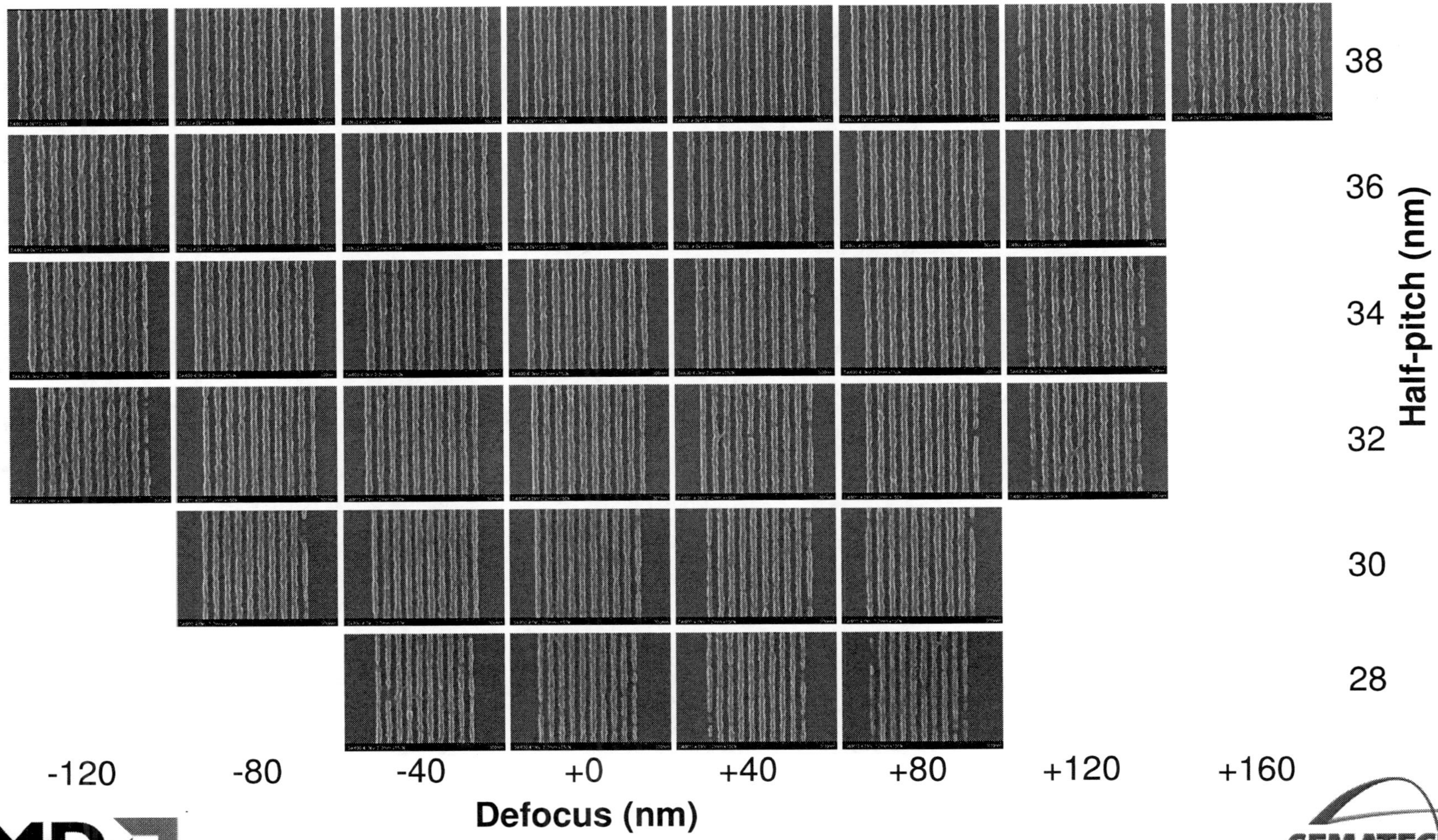

Underlayers have significant effect on LER: Underlayer A

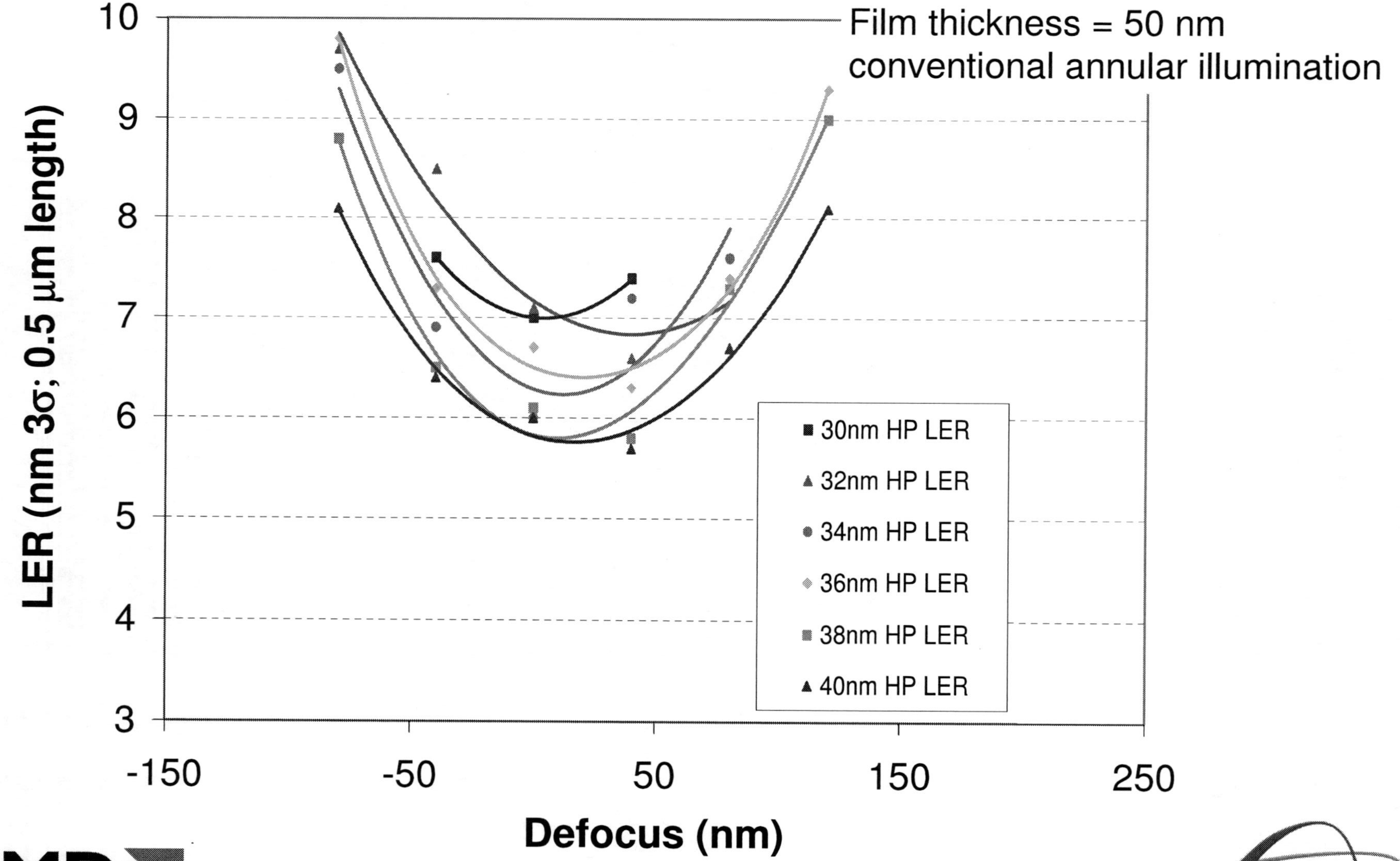

BERKELEY LAB

Film thickness = 50 nm
conventional annular illumination

Legend:
- 30nm HP LER
- 32nm HP LER
- 34nm HP LER
- 36nm HP LER
- 38nm HP LER
- 40nm HP LER

Y-axis: LER (nm 3σ; 0.5 μm length)

X-axis: Defocus (nm)

SEMATECH

AMD

Underlayers have significant effect on LER: Underlayer B

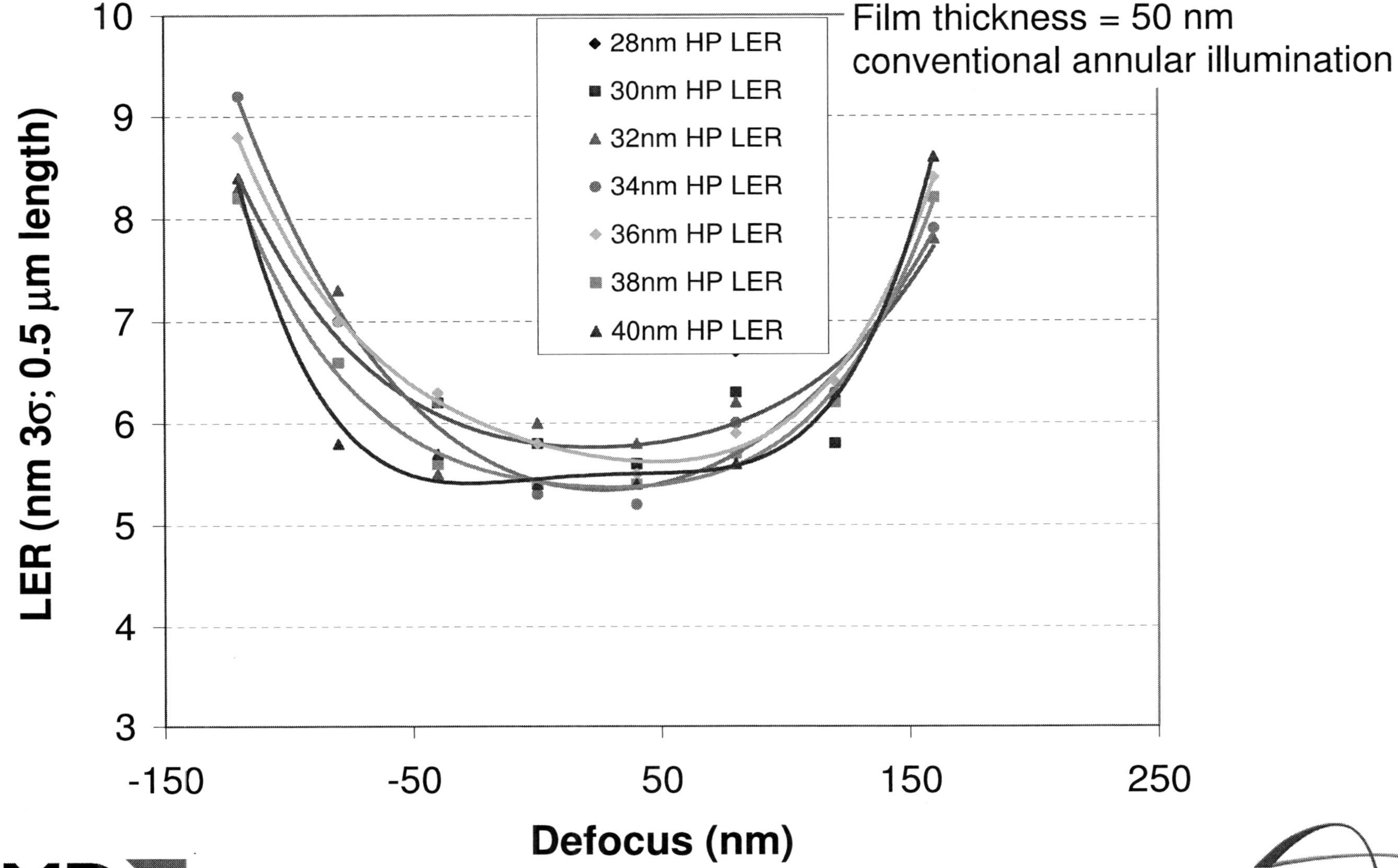

Film thickness = 50 nm
conventional annular illumination

Champion LER = 1.5 nm after correction for mask effects*

LER scatter plot after quadrature subtraction of predicted mask contribution for annular illumination*

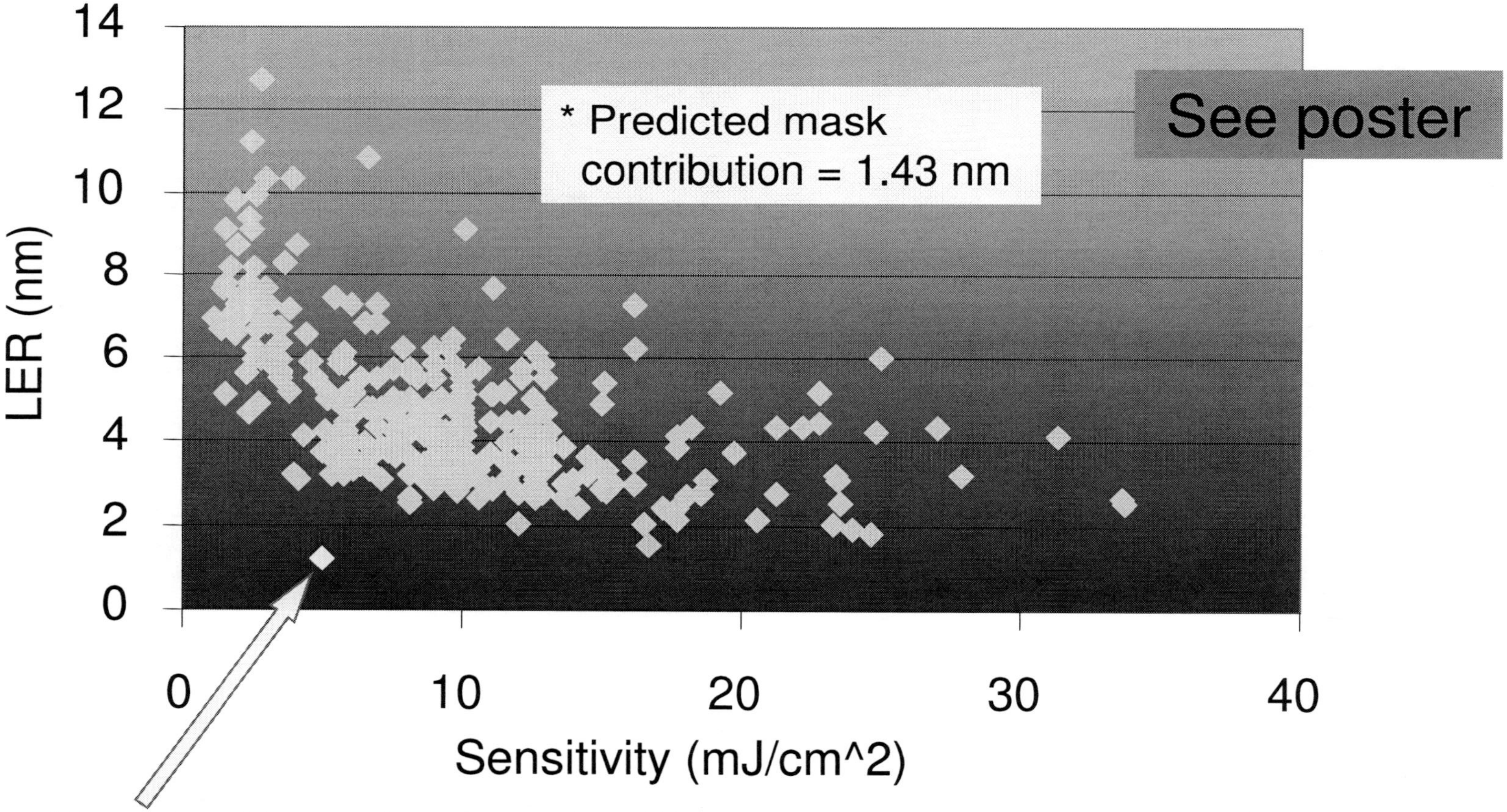

See poster

* Predicted mask contribution = 1.43 nm

Sensitivity (mJ/cm^2)

LER (nm)

Target: 32-nm dense node LER spec < 1.2 nm @ 5 mJ/cm²
Status: Best Current LER = 1.5 nm @ 17 mJ/cm²

- Introduction
- Latest resist results
- **Future plans**
- Summary

Future plans:
0.5-NA SEMATECH exposure tool in Berkeley

BERKELEY LAB

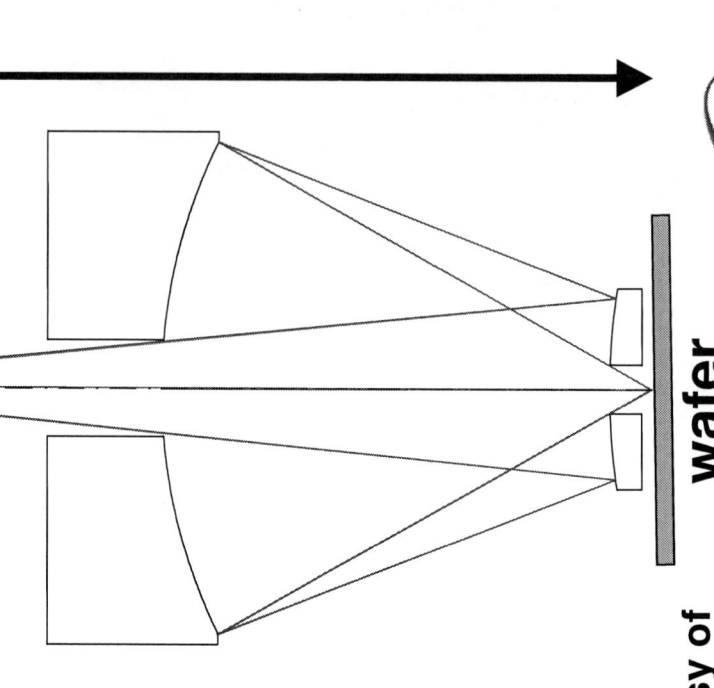

0.5 m

mask

wafer

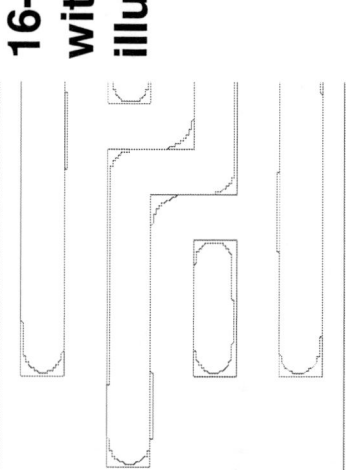

SEMATECH

- Optic design complete and optics manufactures engaged
- NA = 0.5
- Resolution = 8 nm
- Magnification = 5x
- Field of view = 200x30 μm
- Mask angle of incidence = 6°

16-nm features
with conventional
illumination

Optical model courtesy of
Russ Hudyma, Hyperion

Summary

- MET enables advanced resist and mask testing
  - MET results basis for 8 orals and 9 posters

- Significant improvements in resolution at reasonable sensitivities have been achieved over past year
  - dense line resolution improved from 28 to 22 nm
  - Contact resolution improved from 35 to 30 nm

- LER remains a significant issue
  - Mask LER contributions are of concern, but currently not limiting item

- Planning underway for a 0.5-NA microfield exposure tool supporting EUV testing at 16-nm and below

Acknowledgments

Erik Anderson
David Attwood
Kevin Bradley
Rene Delano
Jeff Gamsby
Eric Gullikson
Bob Gunion
Drew Kemp
Dimitra Niakoula
Seno Rekawa
Farhad Salmassi
Ron Tackaberry
LBNL

Shinji Tarutani
Fujifilm

Jim Thackeray
Katherine Spear
Rohm and Haas

Koki Tamura
Chris Rosenthal
Dave White
Hiroto Yukawa
TOK

Harry Levinson
Obert Wood
AMD

Bob Allen
Greg Wallraff
IBM

Ted Liang
Guojing Zhang
Intel

Seong-Sue Kim
Hwan-Seok Seo
Samsung

Supported by: **SEMATECH**

SEMATECH and the SEMATECH logo are registered servicemarks of SEMATECH, Inc.
AMD, the AMD Arrow logo and combinations thereof are trademarks of Advanced Micro Devices, Inc.

Evaluation of EUV resists at Selete

Hiroaki Oizumi, Daisuke Kawamura, Koji Kaneyama, Shinji Kobayashi, and Toshiro Itani

Selete

Semiconductor Leading Edge Technologies, Inc.

Contents

- Introduction
  - EUVL Development Program in Selete
  - Critical Issues for EUV resist & processing

- Current status of EUV litho. & resists
  - SFET status
  - Current status of EUV resists
  - New material
  - Resist out-gassing analysis

- Summary

EUVL Development Program in Selete

FY	2006	2007	2008	2009	2010
Lithography performance	SFET* *Small Field Exposure Tool	Resist materials			
		Mask optimization			
		Exposure tool evaluation - Optics quality - Flare - Optics lifetime (contamination) - Source and collector lifetime		
	EUV1** **Full Field Exposure Tool		Flare compensation		
			Resist multi-layer process		
			Lithography performance		
			Litho-verification using process/device TEG		
Reliability		Particle free mask handling			
		Contamination control technology			
Mask defectivity		Mask infra. (Inspection & repair)			

hp32 process spec.
→ β machine

We are here.

Selete's presentations in this Symposium

Selete

Please visit the following presentations

EXPOSURE TOOL EVALUATION
Lithographic Performance of Selete's Full-Field EUV Exposure Tool
K. Tawarayama, S. Magoshi, Y. Tanaka, S. Shirai, H. Tanaka ....[oral 39]

RESIST
Evaluation of EUV resists at Selete
H. Oizumi, D. Kawamura, K. Kaneyama, S. Kobayashi, T. Itani ....[oral 7]
EUV Resist Outgassing Quantification and Qualification Analysis Methods
S. Kobayashi, J. Santillan, H. Oizumi and T. Itani................[poster 51]
Development of New Negative Tone Molecular Resists Based on Phenyl Calixarene for EUVL
M. Echigo , D. Oguro , H. Oizumi, T. Itani ....[poster 31]
Non-Chemically Amplified Negative Resists for EUV Lithography
M. Shirai, H. Okamura, K. Kaneyama, T. Itani ....[poster 33]
Development of Novel Positive-Tone Resists for EUVL
T. Owada, A. Yomogita , T. Kashiwamura , H. Oizumi, T. Itani ....[poster 45]
Relation between Acid Diffusion and Resolution in Chemical Amplified EUV Resist
Y. Hirai, M. Shimizu, K. Maruyama, T. Kai, T. Shimokawa, T. Itani, D. Kawamura ....[poster 46]

SOURCES
Performance Evaluation of EUV SFET Source Collector Module 90
S. Magoshi, S. Shirai, H. Mori, K. Tawarayama, Y. Tanaka, H. Tanaka ....[poster 27]

DEVICE INTEGRATION
Evaluation of Pattern Fidelity of Half-Pitch 32-45 nm SRAM Patterns Using SFET
Y. Tanaka, H. Aoyama, S. Magoshi, K. Tawarayama, S. Shirai, H. Tanaka ....[poster 103]

TECHNOLOGY READINESS
Flare Impact to Critical Dimension Control on a Full-Field Exposure Tool
H. Aoyama, Y. Tanaka, K. Tawarayama, Y. Arisawa, T. Tanaka ....[poster 104]

Please see O. Suga's presentation for "Selete's Mask Program" in next session.

Critical Issues for EUV resist & processing *Selete*

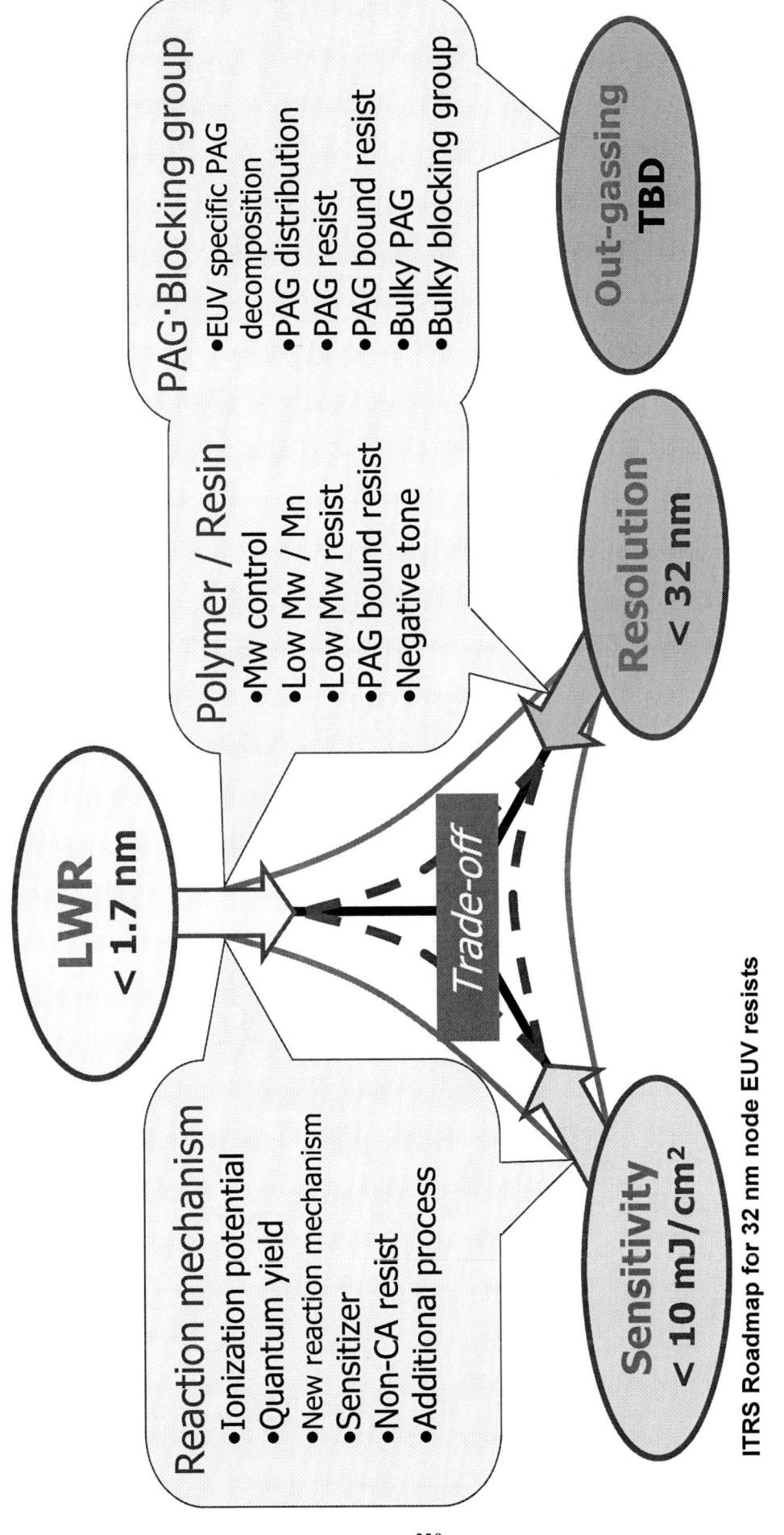

PAG·Blocking group
- EUV specific PAG decomposition
- PAG distribution
- PAG resist
- PAG bound resist
- Bulky PAG
- Bulky blocking group

Polymer / Resin
- Mw control
- Low Mw / Mn
- Low Mw resist
- PAG bound resist
- Negative tone

Reaction mechanism
- Ionization potential
- Quantum yield
- New reaction mechanism
- Sensitizer
- Non-CA resist
- Additional process

LWR
< 1.7 nm

Out-gassing
TBD

Resolution
< 32 nm

Sensitivity
< 10 mJ/cm²

Trade-off

ITRS Roadmap for 32 nm node EUV resists

Sensitivity, Resolution, LWR, are in trade-off relationships.
Resist out-gassing is specific issue for EUV.

Contents

- Introduction
  - EUVL Development Program in Selete
  - Critical Issues for EUV resist & processing

- Current status of EUV litho. & resists
  - SFET status
  - Current status of EUV resists
  - New material
  - Resist out-gassing analysis

- Summary

Small Field Exposure Tool : SFET

Mask loader

Exposure chamber

Wafer loader

Wafer track

Source

Selete corrects Resist sensitivity by a factor of _1.5_ from this September.

Purposes
- ◆ Optimize mask structure
- ◆ Develop resist materials
- ◆ Evaluate optics & source lifetime

Items	Target Specifications
NA	0.3
Field size	0.2 x 0.6 mm
Magnification	1/5
Wavefront error	<0.9 nm rms
Flare	<7% (MSFR)
Resolution	32 nm L/S
Source power	0.5W @IF
Wafer size	300 mm

Illumination condition

Annular (0.3/0.7)

Correction of Resist Sensitivity

Selete   Sept./2008

> **Selete corrects Resist sensitivity by a factor of _1.5_ from this September.**

CY	'07/3Q	'07/4Q	'08/1Q	'08/2Q	'08/3Q	'08/4Q
SFET	Critical Illumination		Koehler Illumination		Dose calibration	
Sensitivity	Resist sensitivity based (: Resist TWG (Mar. '07) recommendations)				X 1.5 — Sensor based	
EUV1			1st exposure	Dose calibration		
Berkeley MET			*Dose calibration*			

Cf. Optimal dose of MET-2D in Selete: for 45nm L&S
12 mJ/cm2 (← 18 mJ/cm2 : Old status)

See Y. Tanaka's poster 103 (DEVICE INTEGRATION).

Contents

- Introduction
  - EUVL Development Program in Selete
  - Critical Issues for EUV resist & processing

- Current status of EUV litho. & resists
  - SFET status
  - Current status of EUV resists
  - New material
  - Resist out-gassing analysis

- Summary

Selete Standard Resist 2 (SSR2)

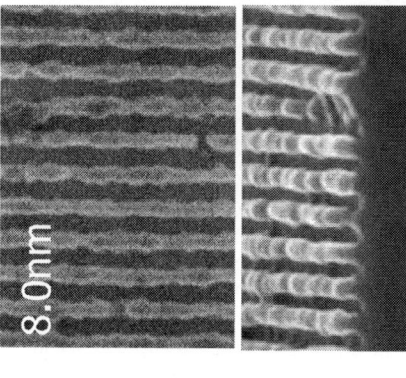
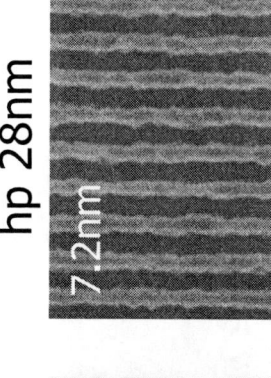
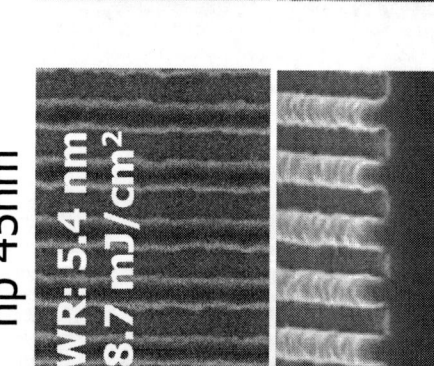
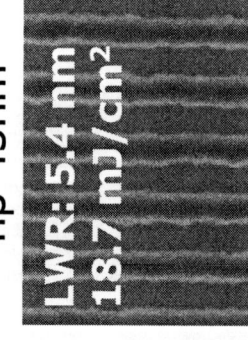

hp 45nm   hp 32nm   hp 28nm   hp 26nm

SSR2

LWR: 5.4 nm
18.7 mJ/cm^2

6.8nm   7.2nm   8.0nm

Annular (0.3/0.7)

Resist performance summary:

- Resolution limit   :   26nm hp
- E_{size} (45nm hp L/S)   :   18.7 mJ/cm^2
- Short-range LWR   :   5.4 nm
- Resist outgassing rate   :   6.4 x 10^{16} molecules/cm^2

Resist benchmarking by SFET

	hp 45nm	hp 32nm	hp 28nm	hp 26nm
SMR45	LWR: 5.6nm 13.3 mJ/cm²	5.7nm	6.5nm	9.1nm
SMR68	5.6nm 12.0 mJ/cm²	8.2nm	10.5nm	
SMR80	4.1nm 11.1 mJ/cm²	5.3nm	5.3nm	
SMR261	6.3nm 11.3 mJ/cm²	8.2nm	8.5nm	

Annular (0.3/0.7)

Resist benchmarking by SFET

More than 130 resist samples were evaluated.

Best results of resolution & sensitivity

Selete

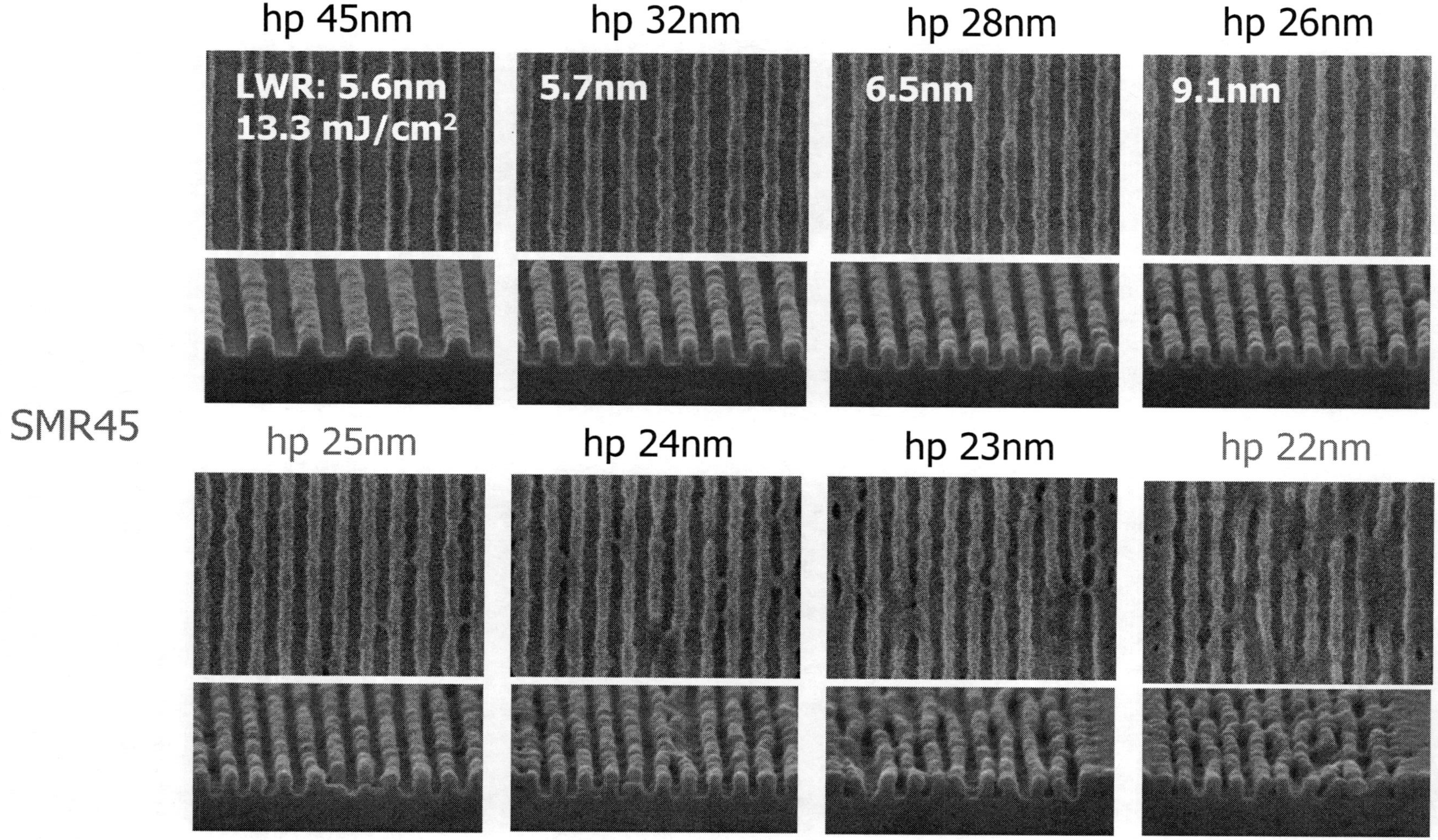

hp 26nm

9.1nm

hp 28nm

6.5nm

hp 32nm

5.7nm

hp 45nm

LWR: 5.6nm
13.3 mJ/cm²

hp 22nm

hp 23nm

hp 24nm

hp 25nm

SMR45

Annular (0.3/0.7)

hp 25nm was resolved. hp 22nm modulation achieved.

Best results of LWR & sensitivity

SMR80

hp 45nm — 11.0 mJ/cm²

hp 32nm — 5.3nm, 12.2 mJ/cm²

hp 28nm — 5.3nm

hp 26nm

Annular (0.3/0.7)

Higher sensitivity & lower LWR were obtained.

Selete Standard Resist 3 (SSR3)

SSR3

hp 45nm
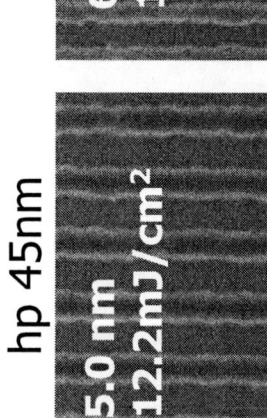
5.0 nm
12.2mJ/cm^2

hp 32nm

6.7nm
12.8mJ/cm2

hp 28nm
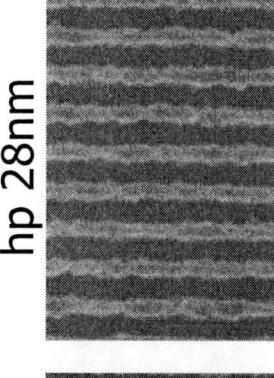

hp 26nm
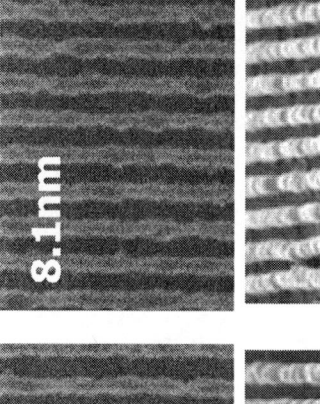
8.1nm

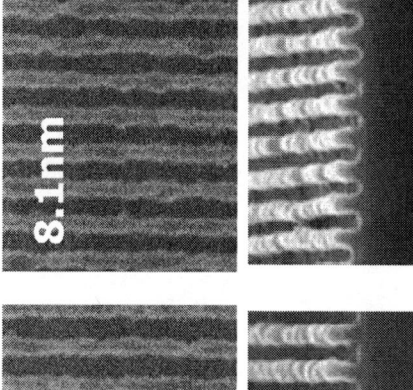
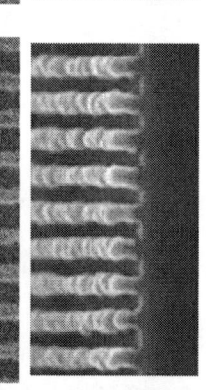

Annular (0.3/0.7)

Resist performance summary:

- Resolution limit : 25nm hp (partially)

- E_{size} (45nm hp L/S) : 12.2 mJ/cm^2

- Short-range LWR : 5.0 nm

- Resist outgassing rate : 3.4×10^{17} molecules/cm^2·s

"Sensitivity" – "Resolution" – "LWR" for various resist *Selete*

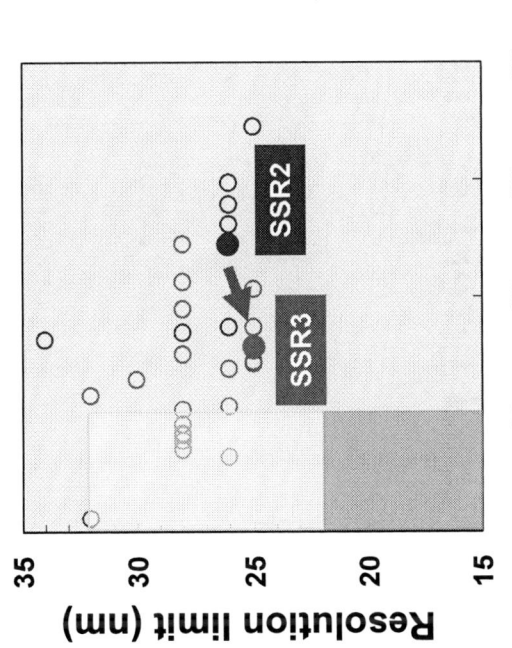

32nm | 22nm

Targets: Sensitivity (E_{size}) : 10mJ/cm²
Short-range LWR : hp x 10%

O : 2007
O : 2008

Sensitivity - Resolution

Sensitivity - LWR

Annular (0.3/0.7)

● "Resolution" of 25-26nm L&S was achieved.
● "Sensitivity" continuously is improving.
● "LWR" is still far from targets, these should be improved.

Contents

- Introduction
  - EUVL Development Program in Selete
  - Critical Issues for EUV resist & processing

- Current status of EUV litho. & resists
  - SFET status
  - Current status of EUV resists
  - New material (Low Mw resist)
  - Resist out-gassing analysis

- Summary

LER/LWR reduction by Low Mw resist

Selete

80nm L/S by EB exposure

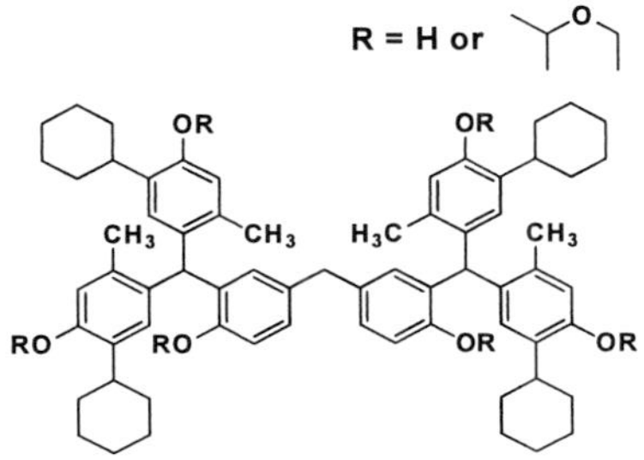

R = H or
OR
CH3
H3C
RO

3M6C-MBSA-EE

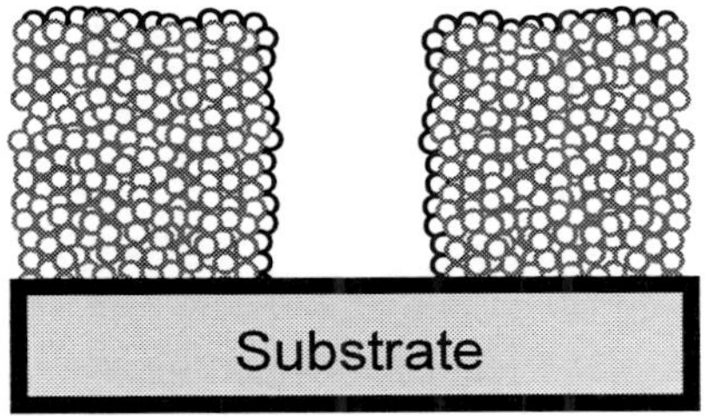

Substrate

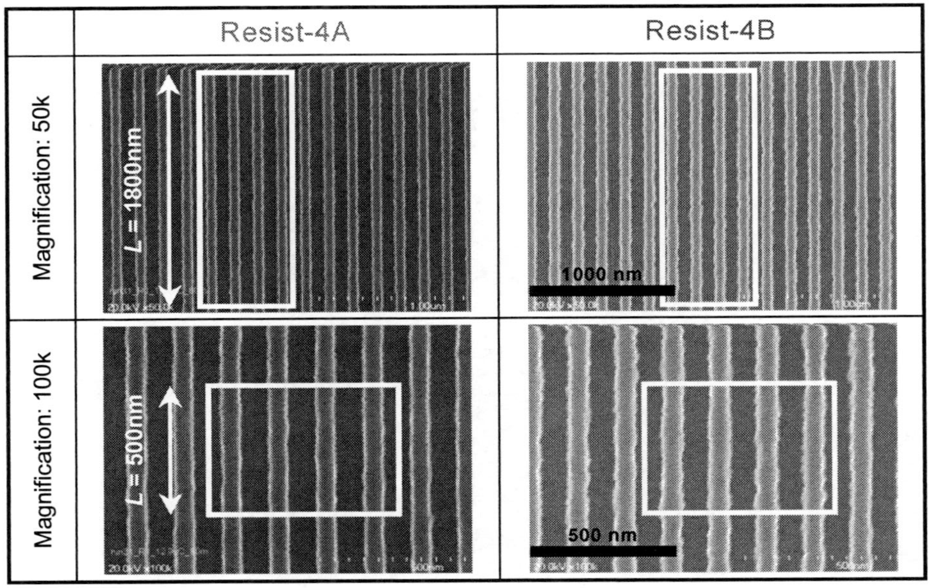

Resist-4A
Resist-4B
Magnification: 50k
Magnification: 100k
L = 1800nm
L = 500nm
1000 nm
500 nm

LER = 5.1 nm

LER = 7.4 nm

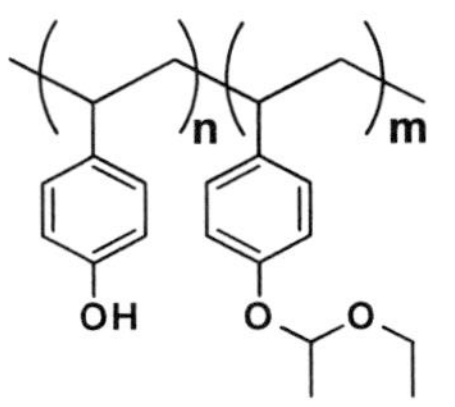

n
m
OH
O

PHS-EE

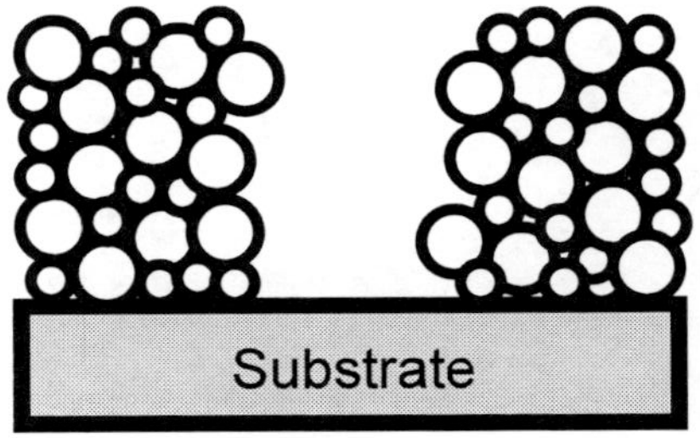

Substrate

T. Hirayama et al.: *Polym. Adv. Technol.*, **17** (2006) 116-121.

"LER" was improved in Low Mw resist.

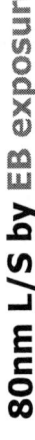

New positive-tone Low Mw resist based on calix[4]resorcinarene

hp 35nm

hp 29nm

hp 28nm

hp 40nm

hp 30nm

hp 45nm

hp 32nm

SMR172

Process Conditions
Substrate: organic layer
Thickness: 50nm
Exposure: HINA
Dev: TMAH 0.26N 30sec
SEM: S8840
Mag: 100k

Esize@29nm-hp: 12.8 mJ/cm^2

Resolution limit is good. LWR is under evaluation.

See T. Owada's poster 45 (resist).

New negative-tone Low Mw resist based on phenylcalix[4]resorcinarene

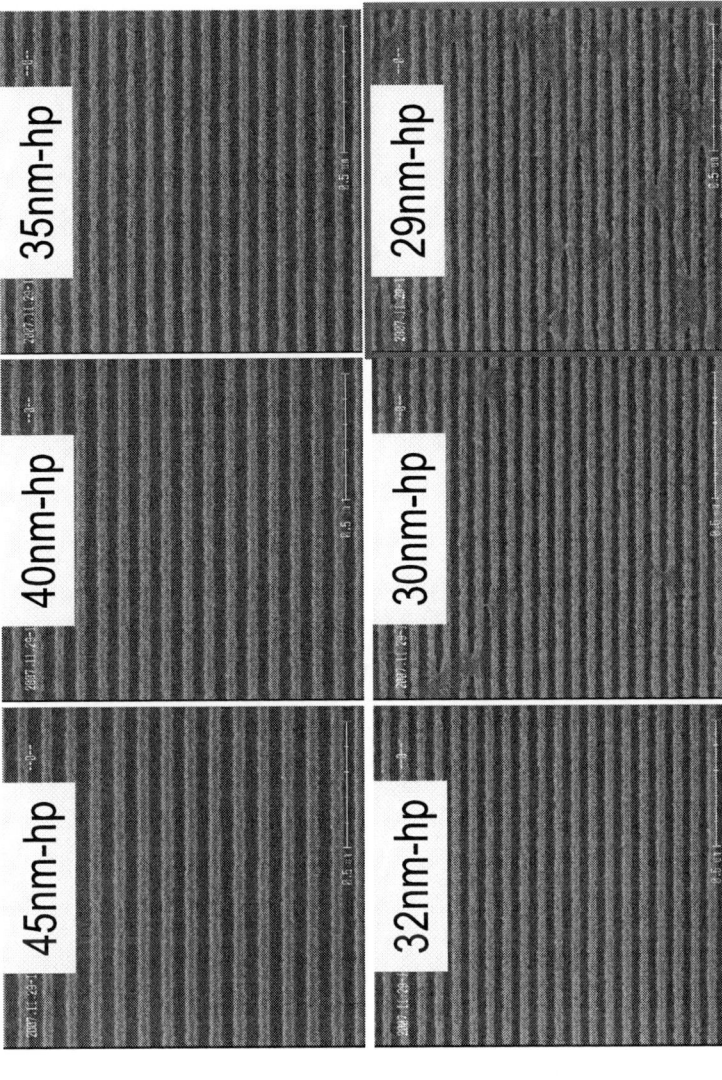

SMR177

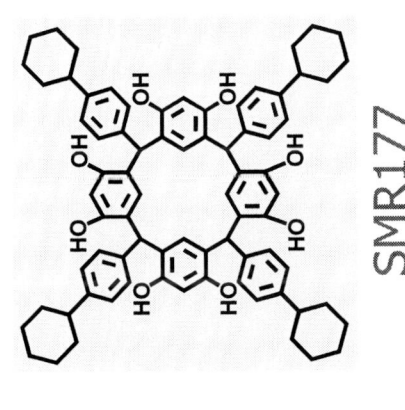

Process Conditions
Substrate: organic layer
Thickness: 60nm
Prebake: 110°C/90sec
Exposure:HINA
PEB: 110°C/90sec
Dev: TMAH 0.26N 30sec
SEM: S8840
Mag: 100k
Mask:MB24

Esize@29nm-hp: 14.8 mJ/cm^2

Resolution limit is good. LWR is under evaluation.

See M. Echigo's poster 31 (resist).

Contents

- Introduction
  - EUVL Development Program in Selete
  - Critical Issues for EUV resist & processing

- Current status of EUV litho. & resists
  - SFET status
  - Current status of EUV resists
  - New material
  - Resist out-gassing analysis

- Summary

Evaluation tools for resist outgassing

Pressure rise method | **QMS analysis**

EUV Source	: EQ-10MR (Energetiq)
Power on Wafer	: 0.03mW/cm^2
Exposure area	: 1.43 cm^2
Base pressure	: **8x10^{-7} Pa**

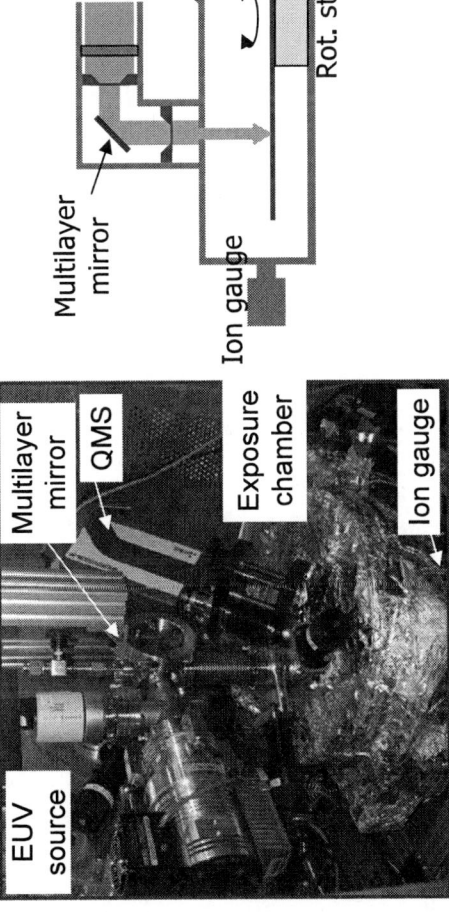

GC-MS method

EUV Source	: EQ-10MR (Energetiq)
Power on Wafer	: 0.014mW/cm^2
Exposure area	: 1.69 cm^2
Base pressure	: **1x10^{-7} Pa**

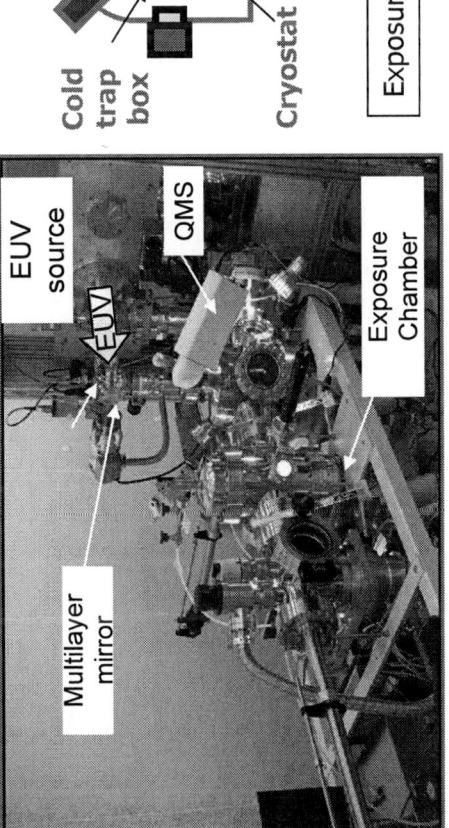

Resist outgassing evaluation method

 *Selete*

Methods	Description	Evaluation time	Selete
Pressure rise	□ Simple and quick for quantification. ■ Component identification not possible.	2 hours/sample	○
GC-MS	□ Component identification possible. ■ CO_2 cannot be detected. ■ Low throughput.	1 day/sample	○
QMS	□ In-situ qualification possible ■ Quantification not possible. ■ Qualification inaccuracy due to fragmentation effect.	2 hours/sample	○
Witness mirror	□ Contamination level directly observed. ■ Low throughput and high cost.	A few days/sample	△

□ Advantage & ■ Disadvantage

Pressure rise, GC-MS and QMS methods are applied for resist outgassing evaluations.

Resist out-gassing benchmarking   *Selete*

Pressure rise method

Resist outgassing rate

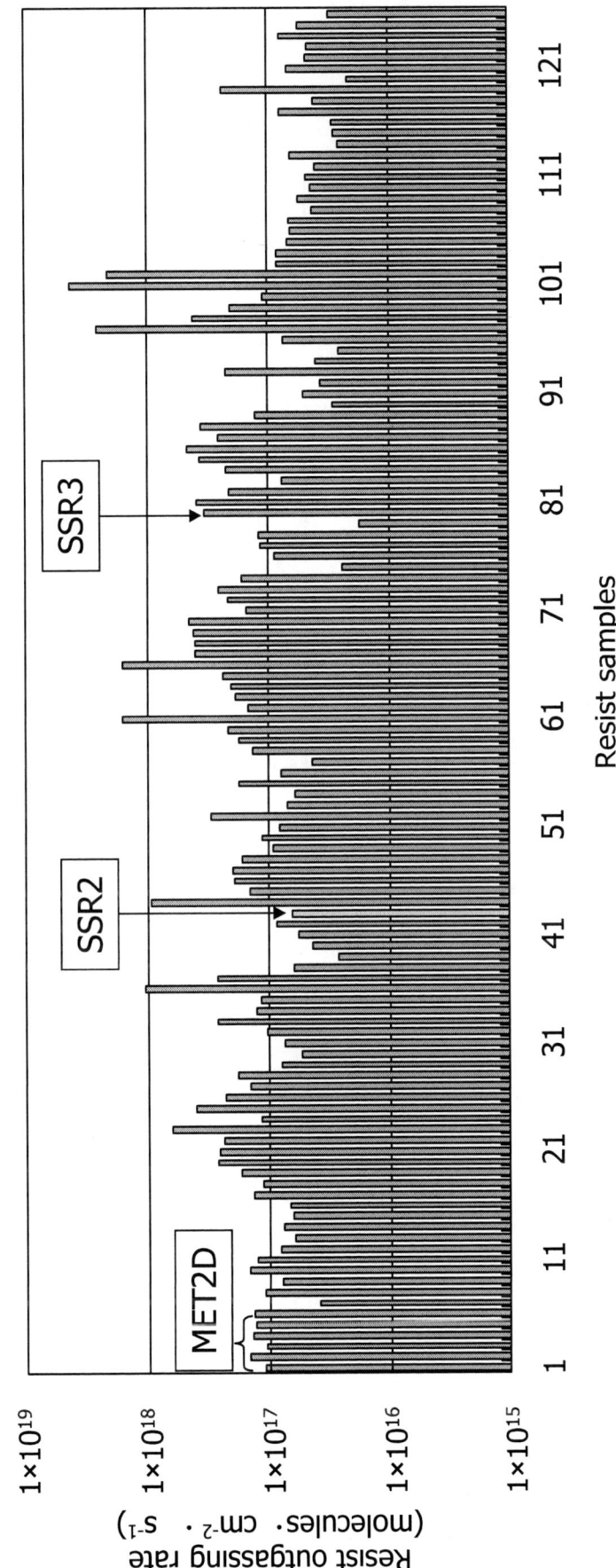

More than 120 resist samples were evaluated.

Selete resist outgassing evaluation procedure

Quantification (Quick Screening)	**Pressure rise method**	Screening of resist samples received for exposure.
Qualification (Component analysis)	**GC-MS method**	Improvement of resist samples based on new resist components.
Mechanism analysis	**QMS analysis** **GC-MS method**	Basic study using model resists to improve tools and control methods.

Resist outgassing methods are applied depending on analysis objectives.

See S. Kobayashi's poster-51 (resist).

Contents

- Introduction
  - EUVL Development Program in Selete
  - Critical Issues for EUV resist & processing

- Current status of EUV litho. & resists
  - SFET status
  - Current status of EUV resists
  - New material
  - Resist out-gassing analysis

- Summary

Summary

- **More than 130 resist samples were evaluated using SFET.**

- **Selete Standard Resist 3 (SSR3) was released.**

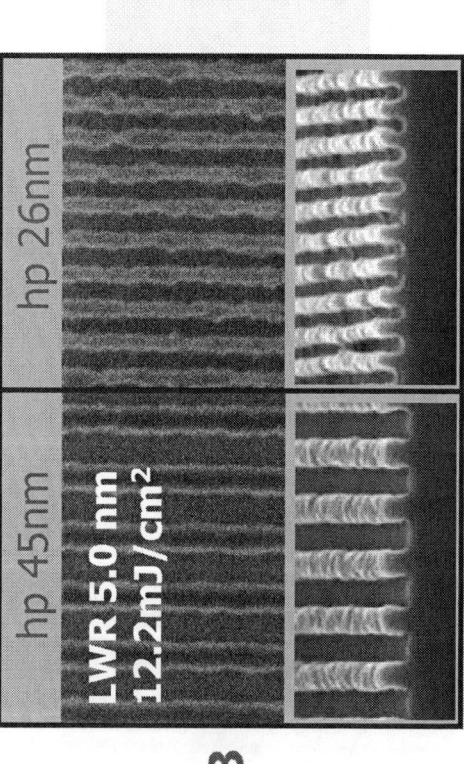

- Resolution limit : 25nm hp (partially)
- E_{size} (45nm hp L/S) : 12.2 mJ/cm^2
- Short-range LWR : 5.0 nm
- Resist outgassing rate : 3.4 x 10^{17} molecules/cm$^2 \cdot$s

- **Sensitivity improved to about 12mJ/cm^2,**
  but LWR at 4–5nm is still far from requirements.

- **"Low Mw resist" is one of the potential candidates for advanced resists.**

- **Resist outgassing methods are applied depending on analysis objectives.**

Acknowledgements

■ A part of this work is supported by New Energy and Industrial Technology Development Organization (NEDO).

■ Selete member companies (EUV Lithomask program).

■ Resist and material manufacturers.

Thank you for your kind attention !!

Advances in resist testing at PSI EUV-IL exposure tool

Vaida Auželytė, Pratap Sahoo, Menouer Saidani, Anja Weber, Harun H Solak

Laboratory for Micro and Nanotechnology, Paul Scherrer Institute Switzerland

PAUL SCHERRER INSTITUT

PSI

EUV Interference Lithography
at Swiss Light Source

- Uses specially coherent and achromatic beam at λ=13.4 nm from undulator

- Interferometer is set of diffraction gratings used to form 1D or 2D fringes

- No alignment, no focus, short exposure

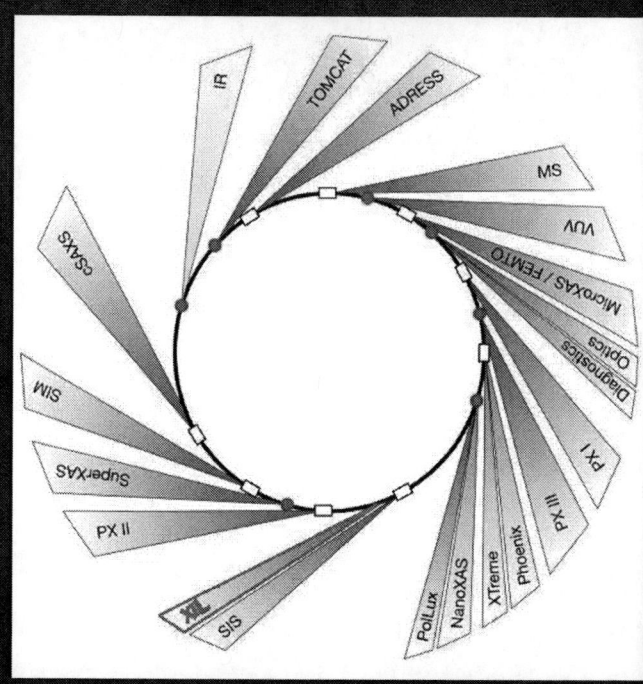

EUV Interference Lithography

2-beam interference   4-beam interference   8-beam interference

Line/space pattern   Contact hole/dot pattern

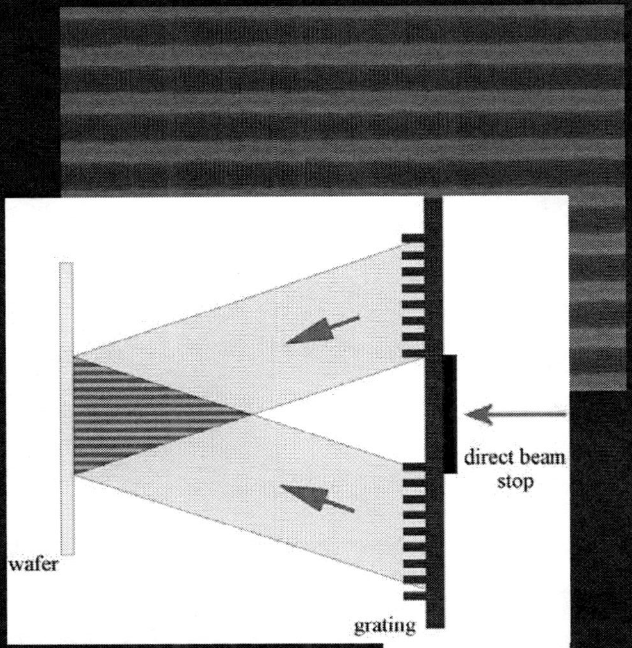

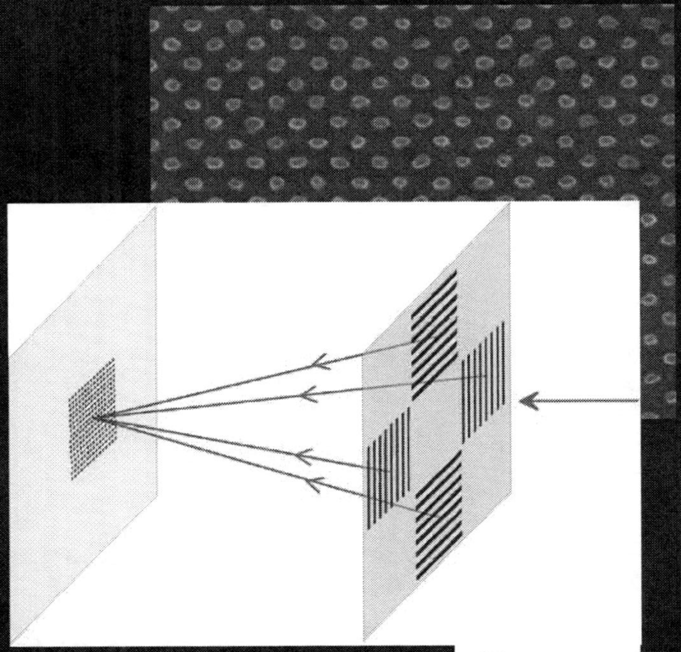

$$P_{wafer} = \frac{P_{grating}}{2}$$

$$P_{wafer} = \frac{P_{grating}}{\sqrt{2}}$$

direct beam stop

grating

wafer

P $p/2$

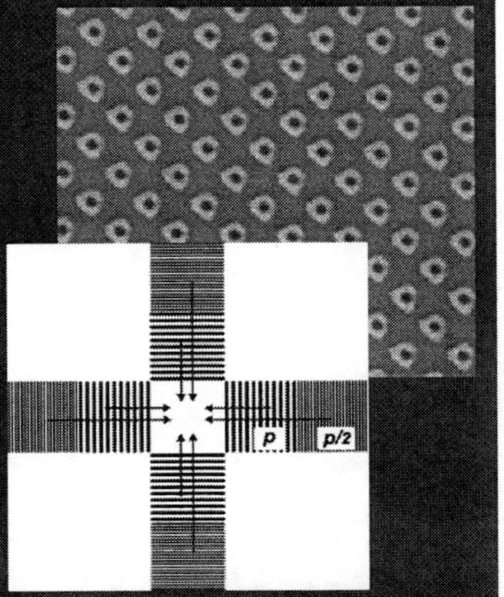

Mask fabrication

$Eff_{40nm}/Eff_{10nm} = 8$

Grating period ←→

Cr — 10-40 nm
Si₃N₄ — 100 nm (42% transmission)
Si

Electron Beam Lithography (LION-LV1) and etching processes

45 nm to 2000 nm pitches
Many pitches and patterns on one mask
Dose is dose-to-mask

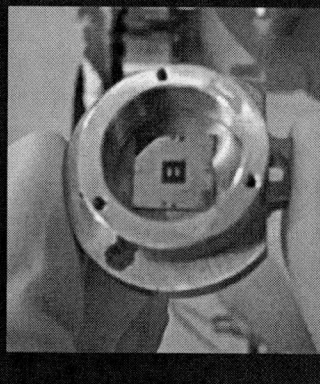

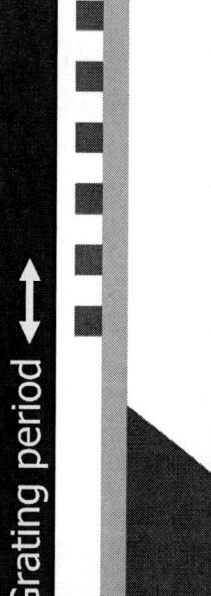

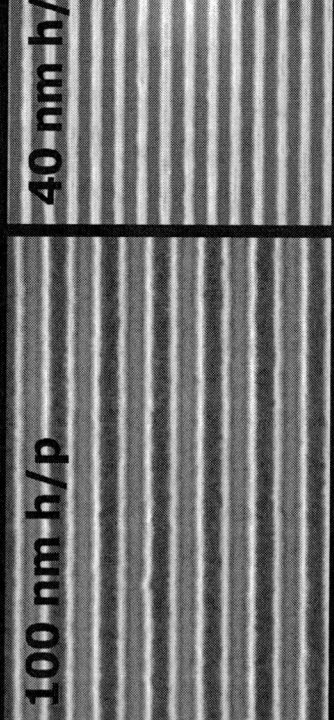

100 nm h/p

40 nm h/p

20 nm h/p

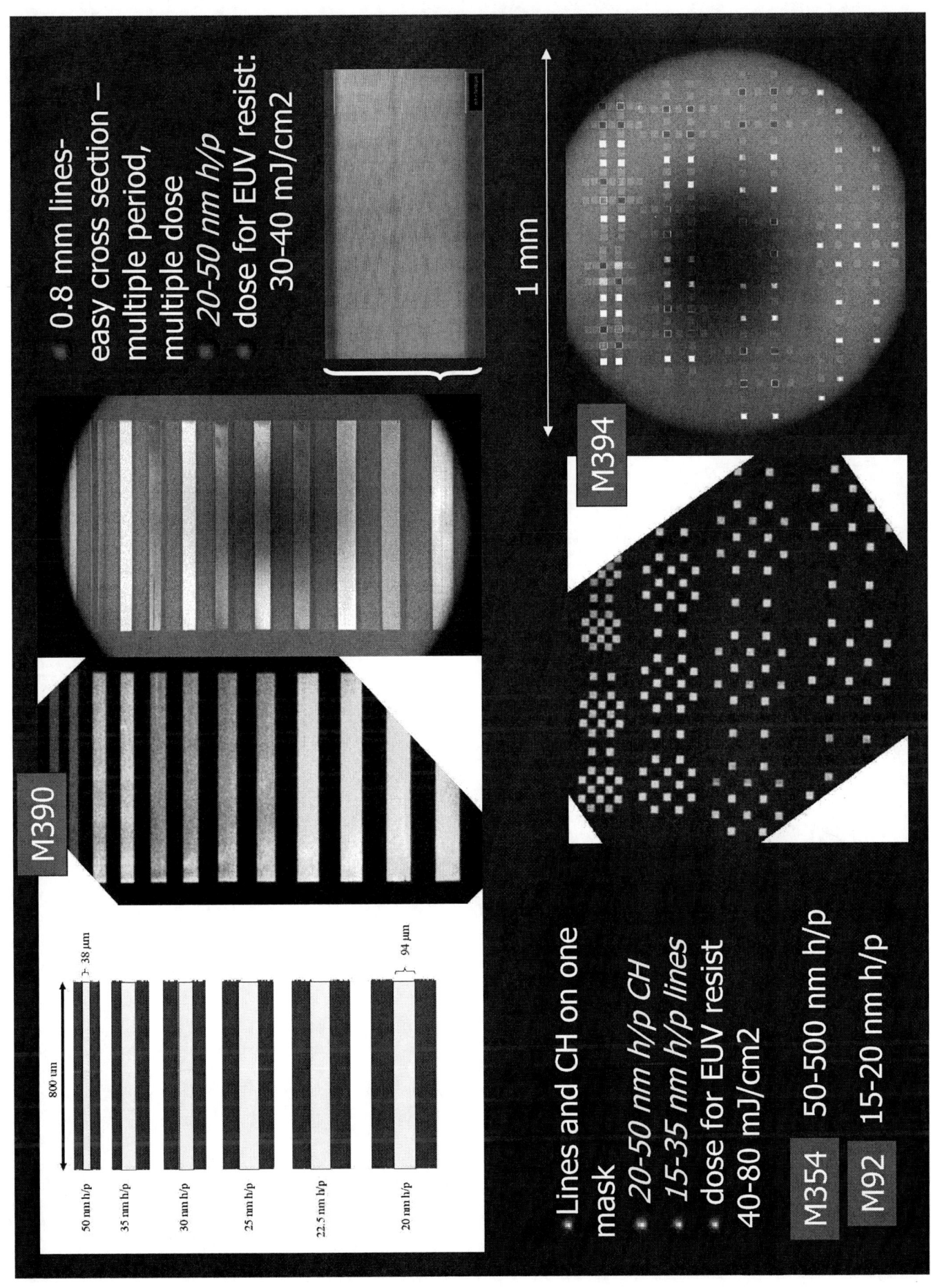

PAUL SCHERRER INSTITUT

EUV Interference Lithography

Performance

- Area: 1-4 mm² area
- Exposure time: 1-30 sec
- 12.5 nm ... 1000 nm half-pitch

calixarene

- Stabile flux, 10-50 mW/cm2
- Throughput ~ 1 wafer/hour
- Sample stage for 4", 6", 8" wafers
- Stage travel size 80×80 mm²
- Software controlled exposure

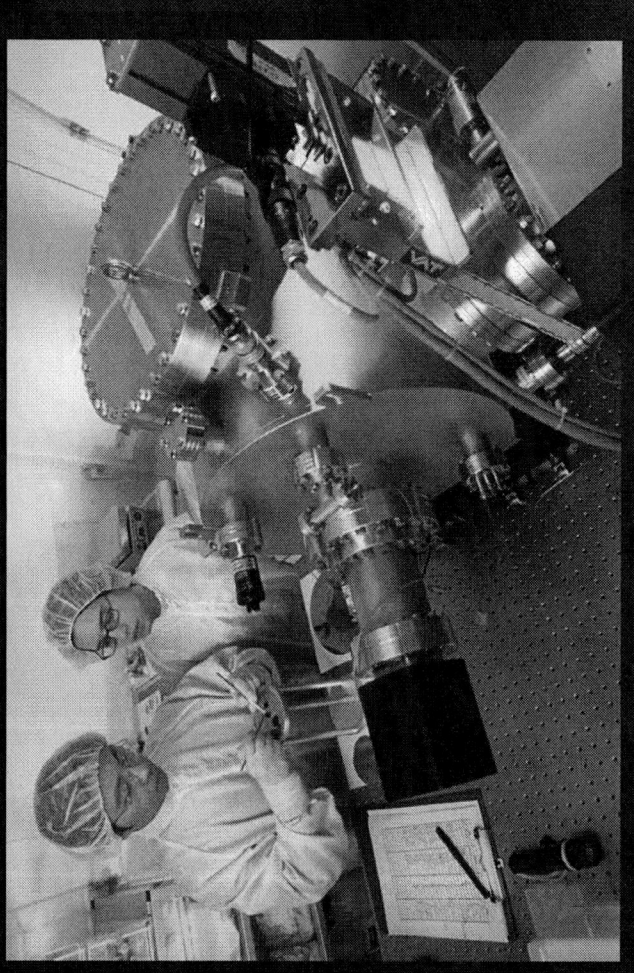

Sample processing:

- Spin coating and development in class 100
- Exposures in class 1000
- PEB right after each exposure
  – amine filtered env.

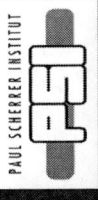

Testing resist performance

Zn-containing resist

lines

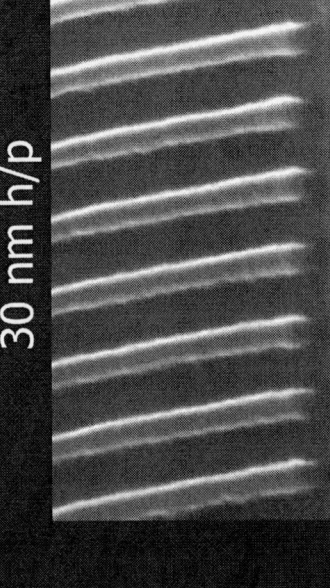

30 nm h/p

50 nm h/p

16 nm h/p

250 nm h/p

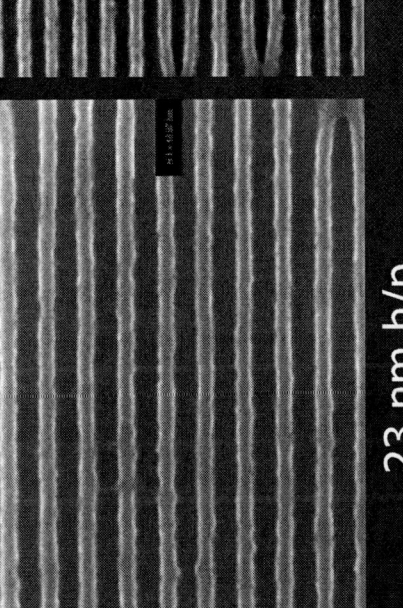

23 nm h/p

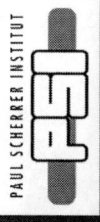

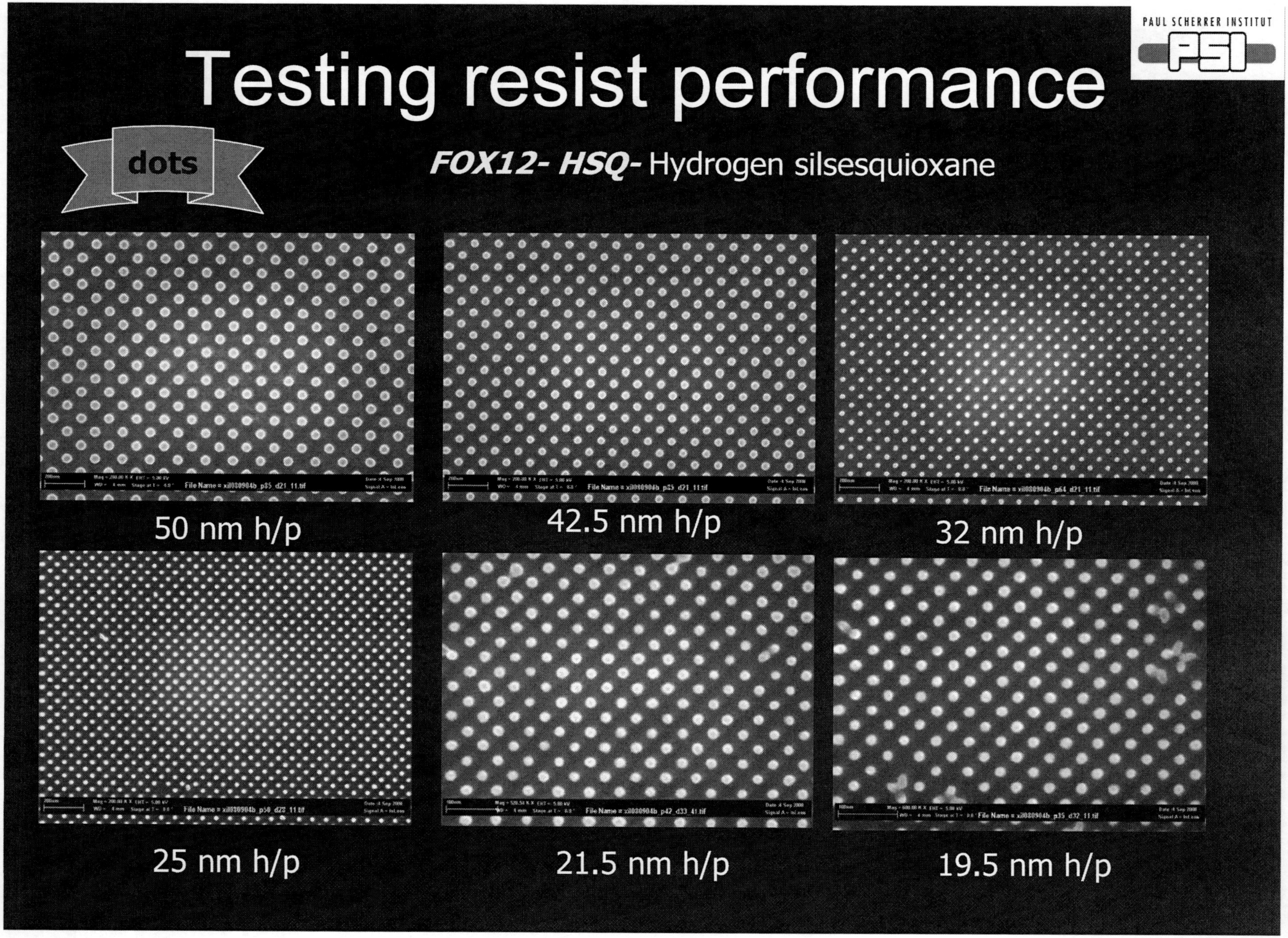
PAUL SCHERRER INSTITUT
PSI
Testing resist performance
dots
FOX12- HSQ- Hydrogen silsesquioxane
50 nm h/p
42.5 nm h/p
32 nm h/p
25 nm h/p
21.5 nm h/p
19.5 nm h/p

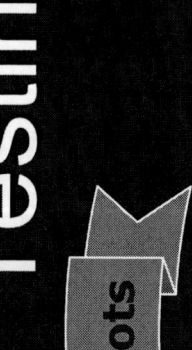

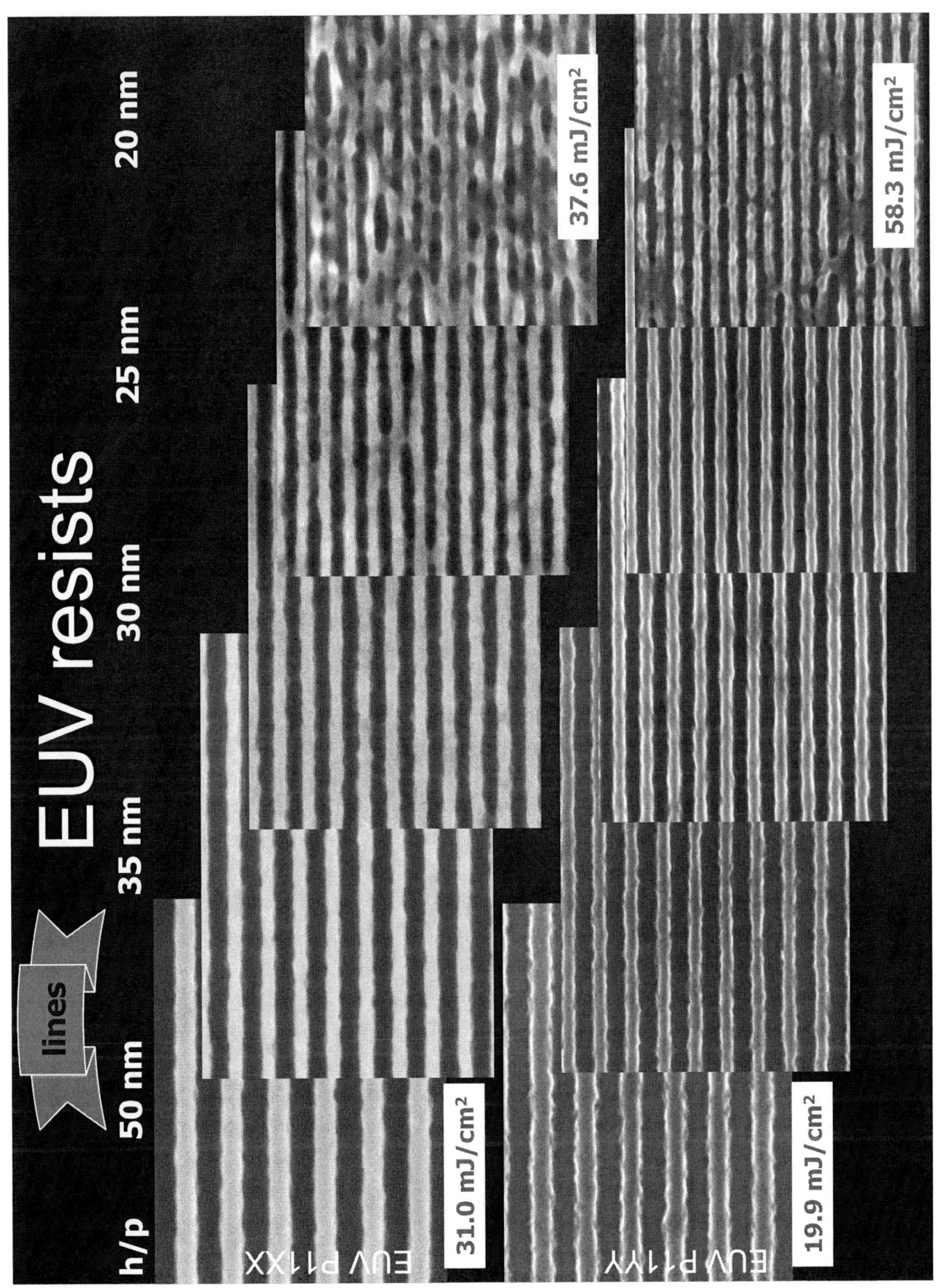

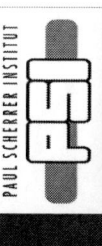

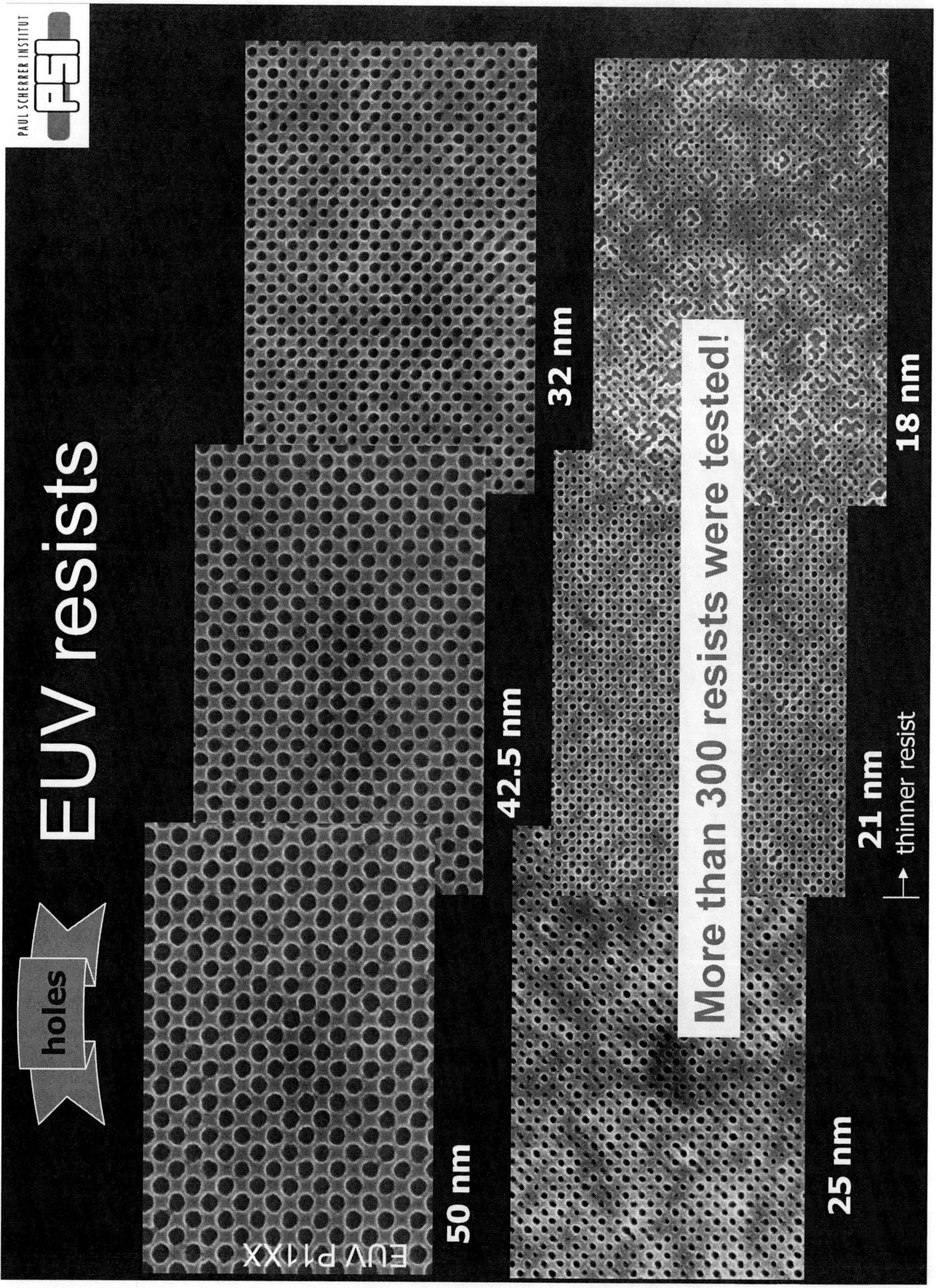
EUV resists
holes
PAUL SCHERRER INSTITUT
PSI
50 nm
EUV P11XX
42.5 nm
32 nm
25 nm
21 nm
thinner resist
18 nm
More than 300 resists were tested!

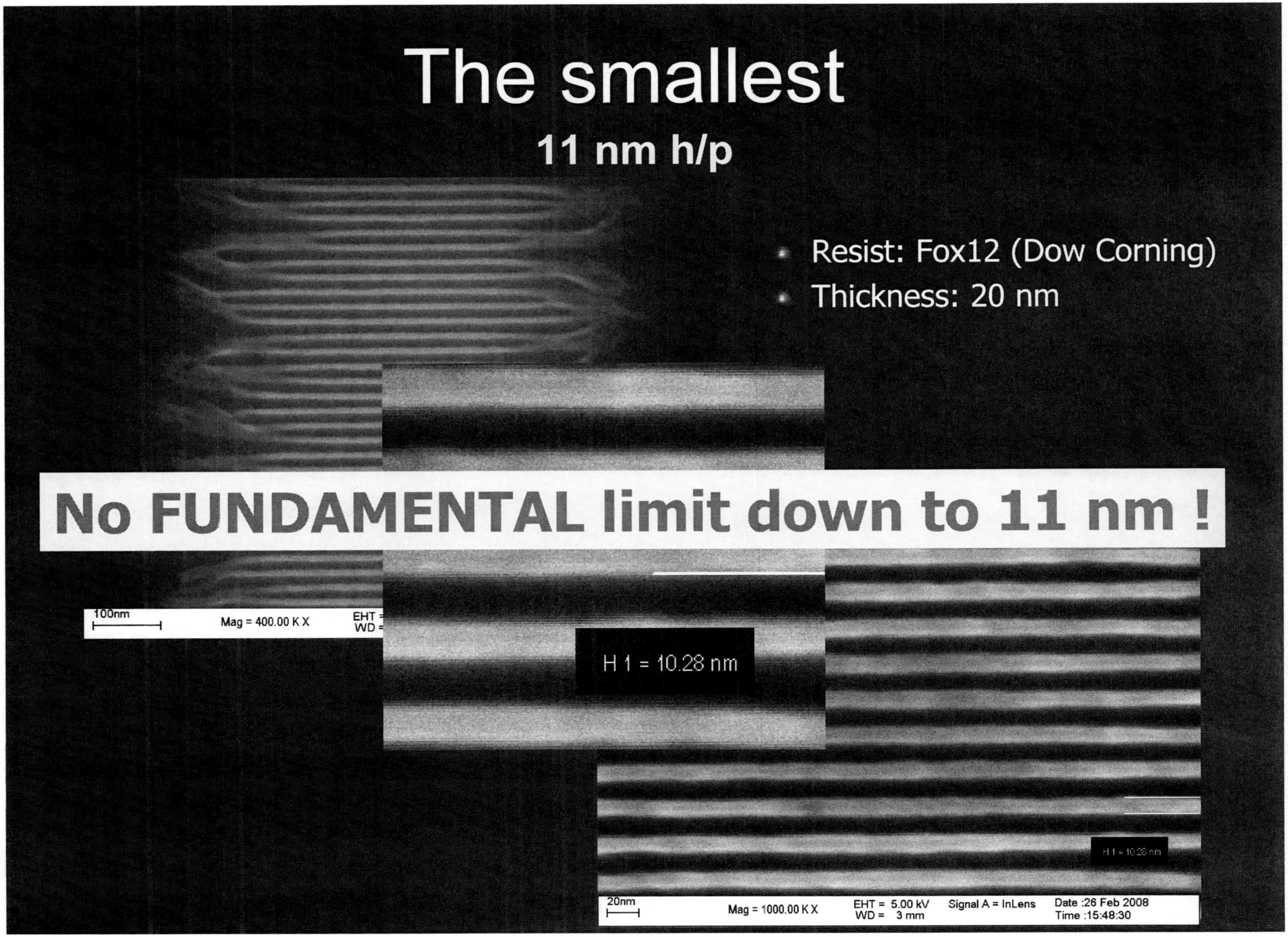

The smallest

11 nm h/p

- Resist: Fox12 (Dow Corning)
- Thickness: 20 nm

No FUNDAMENTAL limit down to 11 nm !

H 1 = 10.28 nm

H 1 = 10.28 nm

Mag = 400.00 K X   EHT =   WD =

100nm

Mag = 1000.00 K X

EHT = 5.00 kV   Signal A = InLens   Date :26 Feb 2008
WD = 3 mm   Time :15:48:30

20nm

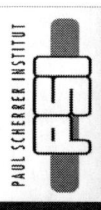

PAUL SCHERRER INSTITUT

Upgrades

2 MUSD upgrade program for *Full time facility*

➢ From 15% to 100% operation
  ➢ Dedicated undulator
➢ Sample preparation/development
  at the beamline

WHEN:
fall of 2009

AVAILABILITY:
Contact dr. Harun Solak
Harun.Solak@psi.ch

NEW electron beam writer
from VISTEC
EBPG5000plus ES
50 keV and 100keV

for mask preparation

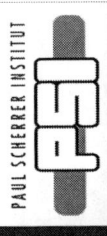

PAUL SCHERRER INSTITUT
PSI

Acknowledgements

The team:

Dr. Harun H. Solak

Anja Weber

In the audience

Dr. Pratap Sahoo

Dr. Menouer Saidani

Markus Kropf

And
many users!

Collegues at
Laboratory of Micro and Nanotechnology
And
Swiss Light Source

Positive and Negative Tone Molecular Resists for 22-nm node EUVL Patterning

Richard Lawson[1], Cheng-Tsung Lee[1], Laren M. Tolbert[2],

Clifford L. Henderson[1] *

[1] School of Chemical & Biomolecular Engineering
[2] School of Chemistry and Biochemistry
Georgia Institute of Technology

Todd R. Younkin
Intel Corporation

* Corresponding Author: cliff.henderson@chbe.gatech.edu

2008 International Symposium on Extreme Ultraviolet Lithography

RLS Tradeoff

$$\text{Resolution}^3 \times \text{LER}^2 \times \text{Sensitivity} \approx \text{material constant}$$

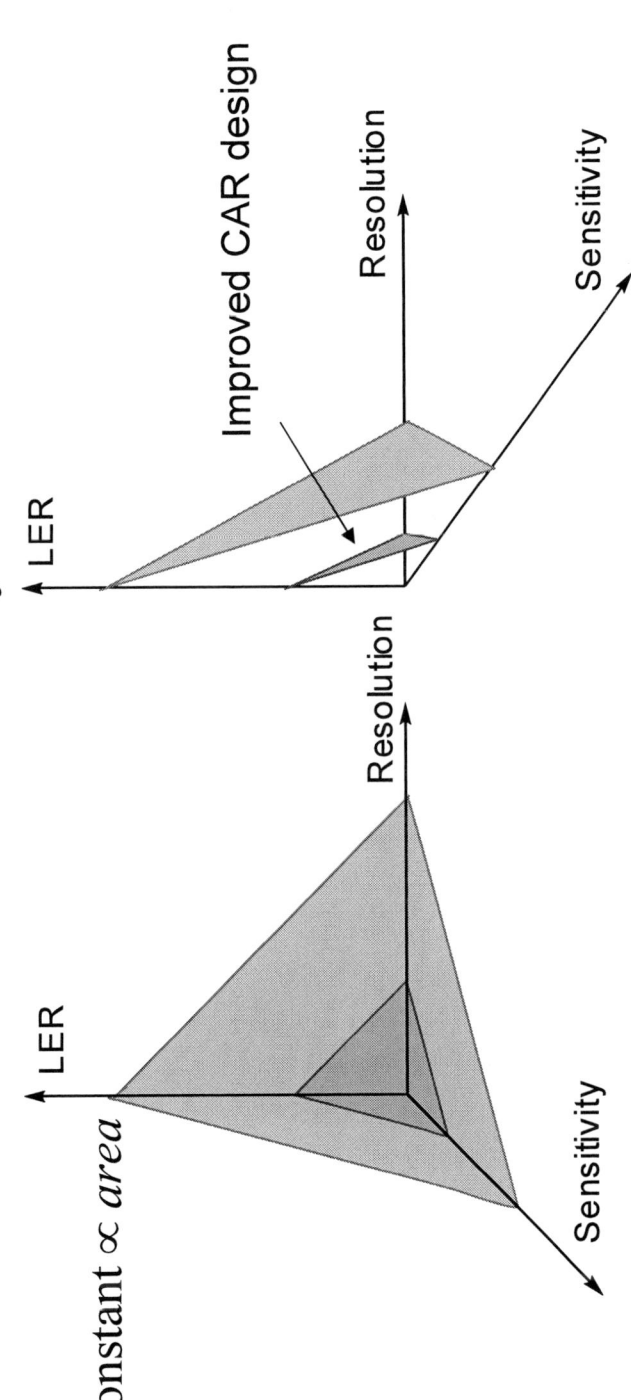

material constant $\propto$ *area*

- There exists a well known trade-off in resolution, LER, and sensitivity for chemically amplified resist materials

- RLS limitation is intrinsic to CARs → must reduce material *constant*

- Modern CAR design at minimal material *constant* - performance still does not meet the requirements.

Attempts to Solve RLS Tradeoff

- **Base Quencher** – shown to reduce LER and improve resolution, but at cost of sensitivity

- **Polymer-bound PAGs**
  - reduce photoacid diffusion length to improve resolution
  - has shown LER improvements
  - sensitivity penalty reduced by increased PAG loading
  - several years into development – only recently began 22 nm patterning, LER still greater than desired

- **Molecular Resists** – reduce pixel size to improve LER – currently no significant advantage over low MW polymers – LER performance dominated by issues other than pixel size

UNCC-GT-Intel program

Homogeneity

- Blended systems – resist/PAG/base – lead to inhomogenity and LER
- Multiple reactive sites on a molecule lead to distribution of protecting – not monodisperse = inhomogenity

Effect of PAG homogeneity

Hirayama, et al. Jap. J. App. Phys. 44, 5484, (2005)

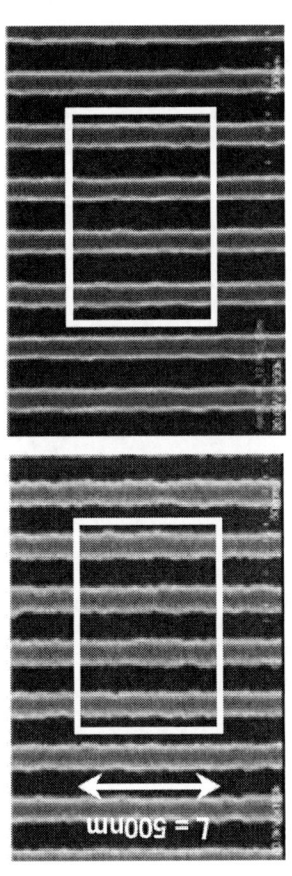

L = 500nm

Increasing homogeneity

Decreasing LER

- Single component small molecules = full synthetic control of stereochemistry and regiochemistry → monodisperse distribution of protecting groups
- No physically blended additives = no phase separation or aggregation of components
- Advantages over polymer-bound PAGs - typically complex tetramer [20] and tetrapolymers have larger polydispersity and variations in chain compositional uniformity.

Shiono, et al. SPIE 6519, 65193+ te (2007)

Positive Tone Resists

Single Component Molecular Resists

Single Component Molecular Resists

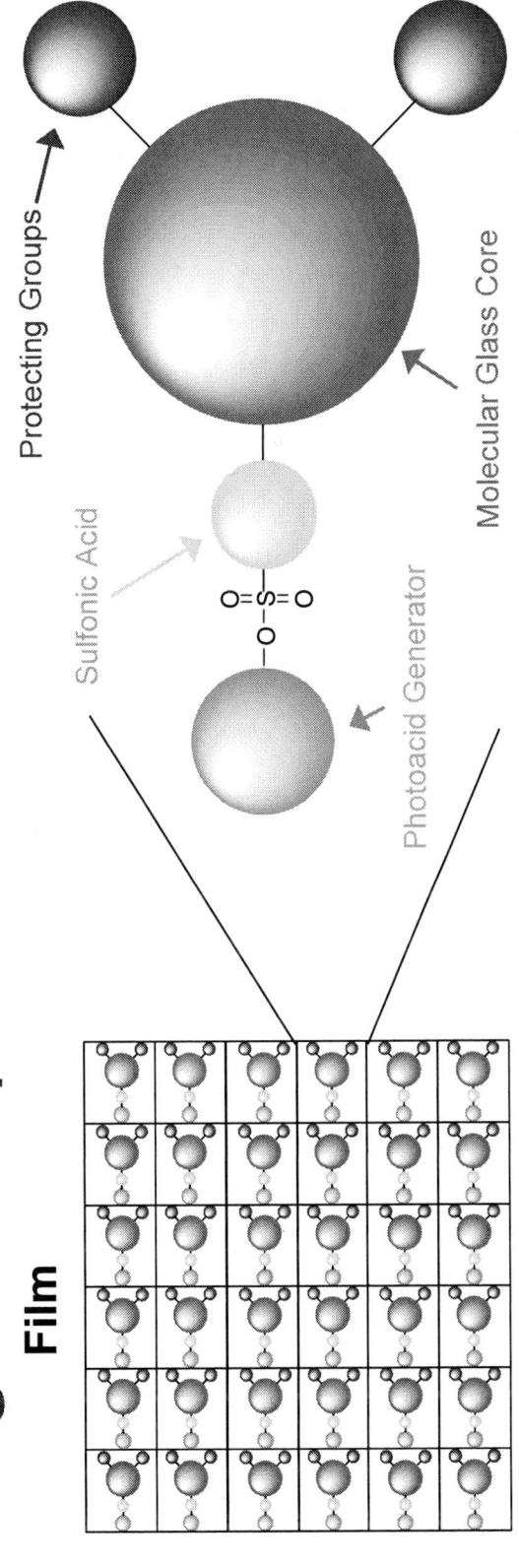

- Single component molecular resist = molecular resist that contains PAG functionality and acid labile protecting groups on a base soluble, etch resistant molecular glass core.

- Provides three distinct advantages over current CAR materials

  1. homogeneity
  2. very high PAG loading
  3. high level of diffusion control

TAS-tBoc-SbF$_6$ EUV Results

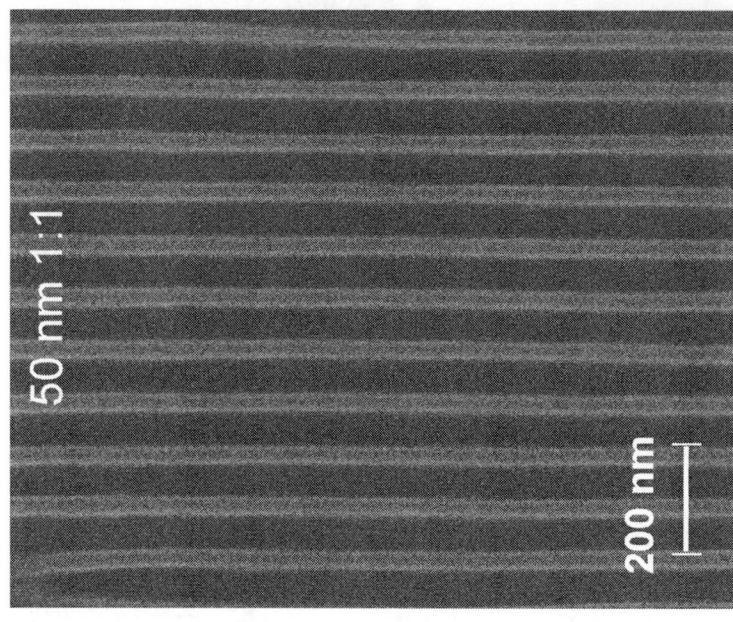

Dose-to-Mask (PSI Dec.07) 255 mJ/cm^2

50 nm 1:1

200 nm

LER (3σ) = 5.2 nm

- Passed outgassing tests at U.Wisconsin – bulky ring subs. Improves outgassing

- EUV exposures done on PSI tool in Switzerland courtesy of Intel

- 50 nm 1:1 lines resolved with low LER of 5.2 nm

NBB

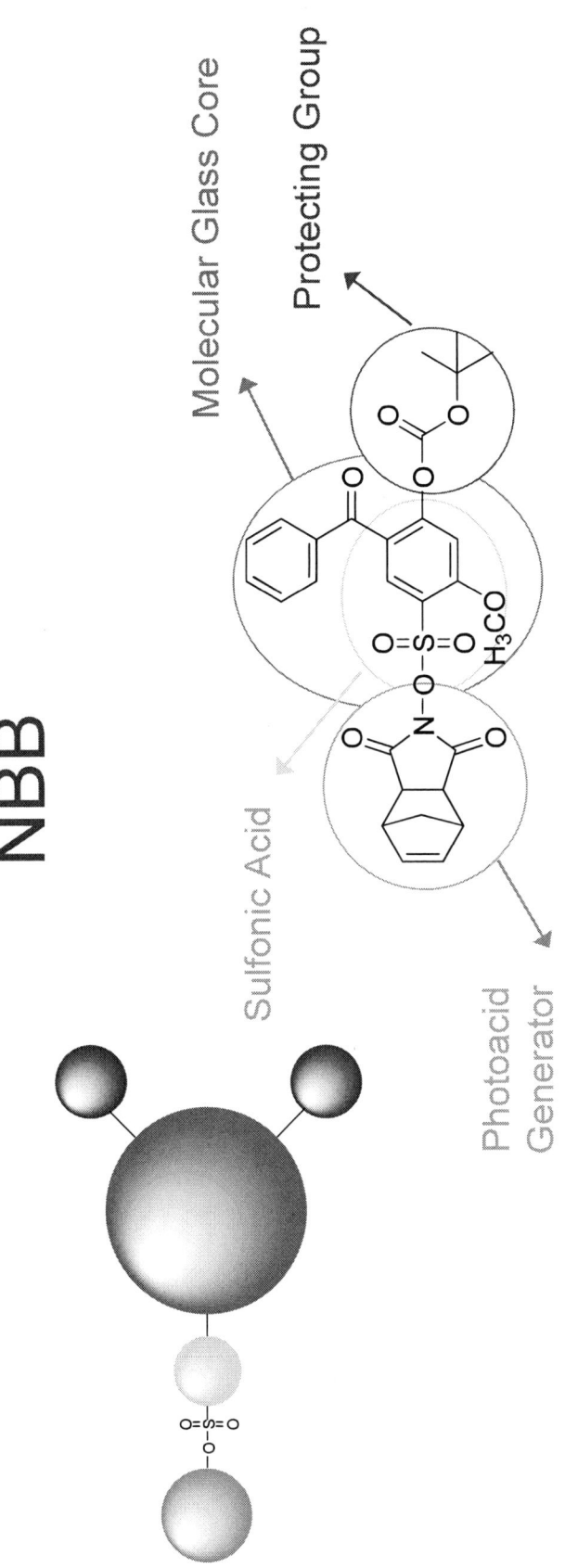

Molecular Glass Core

Protecting Group

Sulfonic Acid

Photoacid Generator

H3CO

- NBB is example of non-ionic bound sulfonic acid molecular resist
- Sulfonic acid is directly part of molecular glass core
- Norbornene dicarboximide PAG
- Superior solubility in casting solvent as compared to TAS compounds
- Has good adhesion and forms excellent films
- Zero dark loss over 30 sec. development in 0.261N TMAH

Glass Transition of NBB

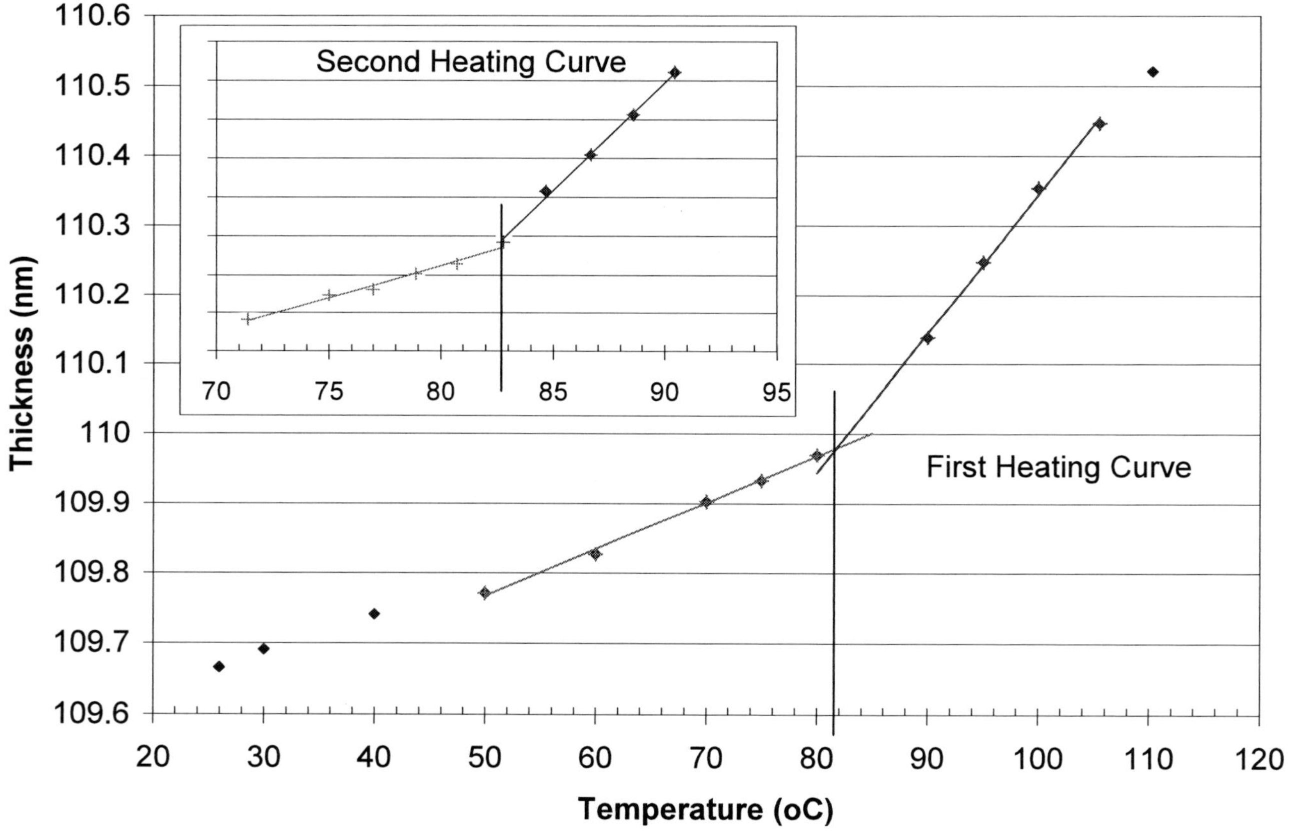

First Heating Curve

Second Heating Curve

Temperature (oC)

Thickness (nm)

- Glass transition temperature (**Tg**) determined to be **83°C** using ellipsometry with hot stage to measure volume expansion

- Tg measurement made of thin film rather than bulk sample

NBB High Resolution E-beam

- Good image quality and resolution
- 40 nm 1:3 lines/space at 1000 μC/cm^2 (100 keV), PEB = 90°C above Tg
- Suffers from pattern collapse starting at 60 nm 1:1 lines (aspect ratio > 2)
- **LER (3σ) = 3.9 nm**

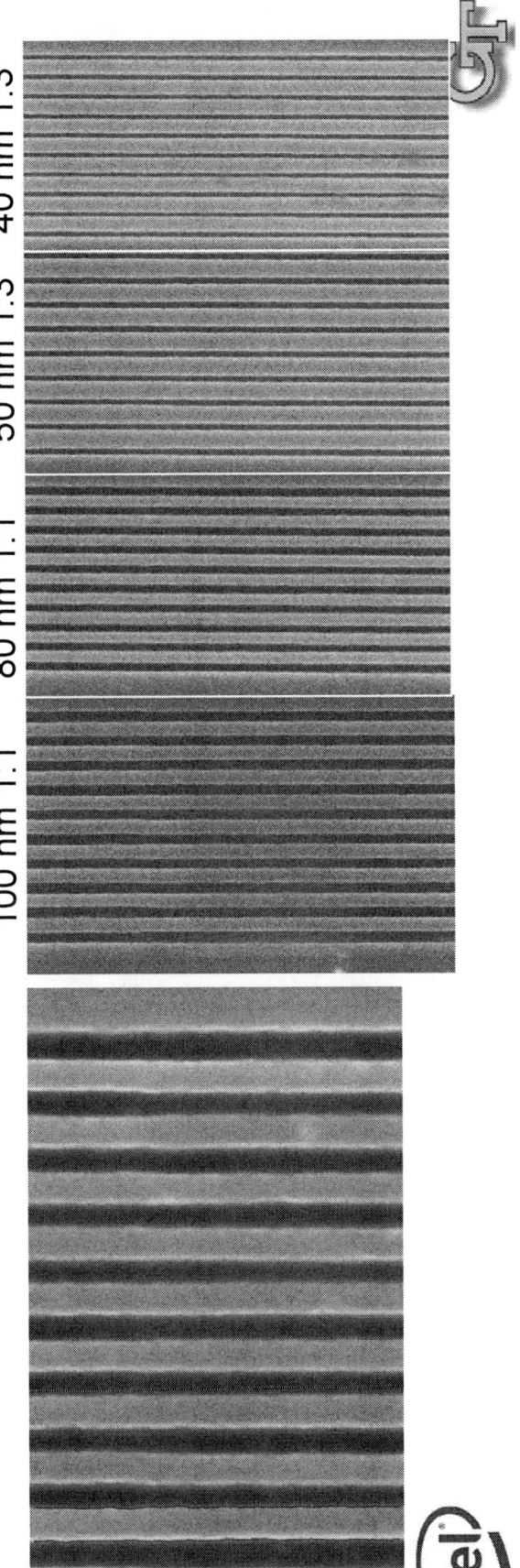

Summary – Positive Tone

- Two different types of truly single component positive tone molecular resists designed and synthesized

- Both show improved LER compared to conventional resists.

- Good resolution can still be maintained by reduction in acid diffusion

- Sensitivity is lower than expected, but can be improved.

- Binding sulfonic acid and PAG to molecular glass cores provides potential path forward in resist design required for high resolution and low LER.

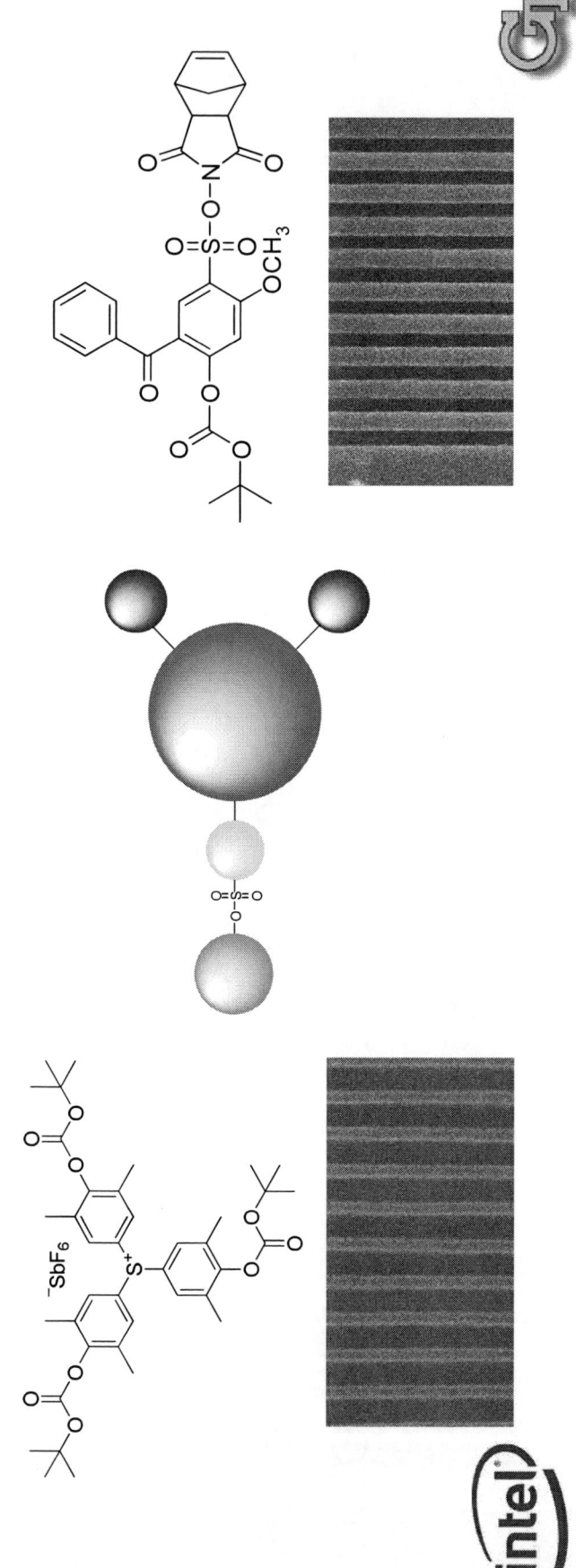

Future Designs – Positive Tone

Negative Tone Resists

Cationic Polymerization Molecular Resists

Cationic Polymerization: A New (Old) Concept

microchem.com

- Most all current high resolution CARs are based on photoacid catalyzed deprotection of protecting groups – requires active acid for reaction

- CARs based on cationic polymerization have been used for several years – primarily in MEMs fabrication (SU-8)

- Use of cationic polymerization CARs for high resolution resists offer several potential advantages over conventional deprotection CARs

  – superior mechanical strength – high MW cross-linked film versus low MW

  – superior environmental stability – acid only has to react once – highly stable cationic chain propagation controls conversion

  – intrinsic diffusion control – active cation directly attached to exponentially growing chain/network

  – no outgassing – zero mass loss process – epoxide shows no shrinkage

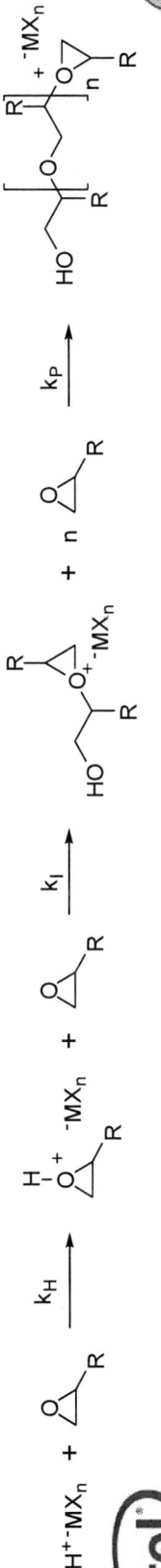

Cationic Polymerization Resists

- Four different resists investigated in this study
- Size and functionality systematically changed
- 1-Ep will not form good films – de-wets during spin-coating and PAB
  – likely strong tendency to crystallize/poor surface energy mismatch
- 3-Ep has low melting point – de-wets if PAB is above 60°C

E–Beam and EUV Contrast Curves

100 keV e-beam

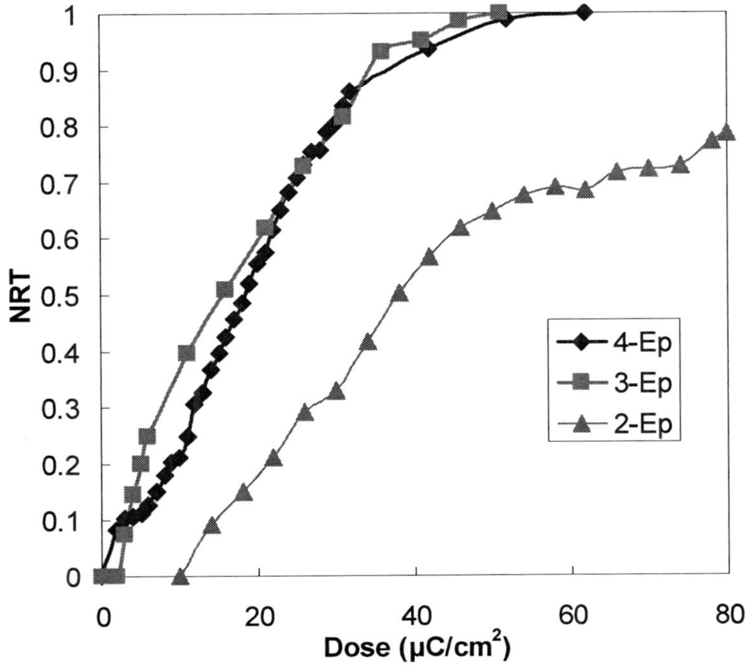

EUV

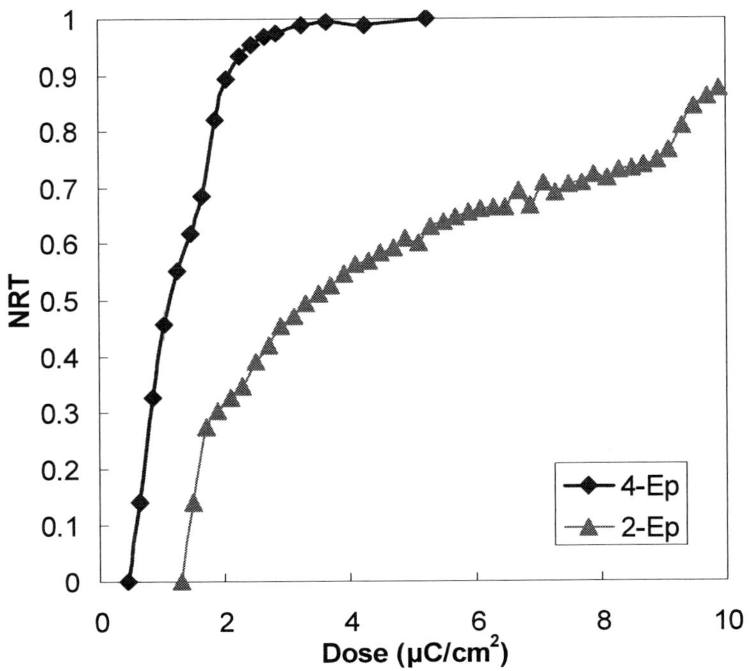

5 mol% TPS-SbF$_6$, PEB = 90°C 1 min., MIBK dev.

- E-beam: γ = 1.2 (4-Ep), 1.3 (3-Ep), 1.5 (2-Ep)
- EUV: γ = 2.3 (4-Ep) vs. 1.3 (2-Ep)
- Sensitivity for 2-Ep larger than for 4-Ep and 3-Ep

(intel)

Comparison of Epoxide MR E-beam

- Resolution increases as functionality decreases
- Maximum aspect ratio increases as functionality increases

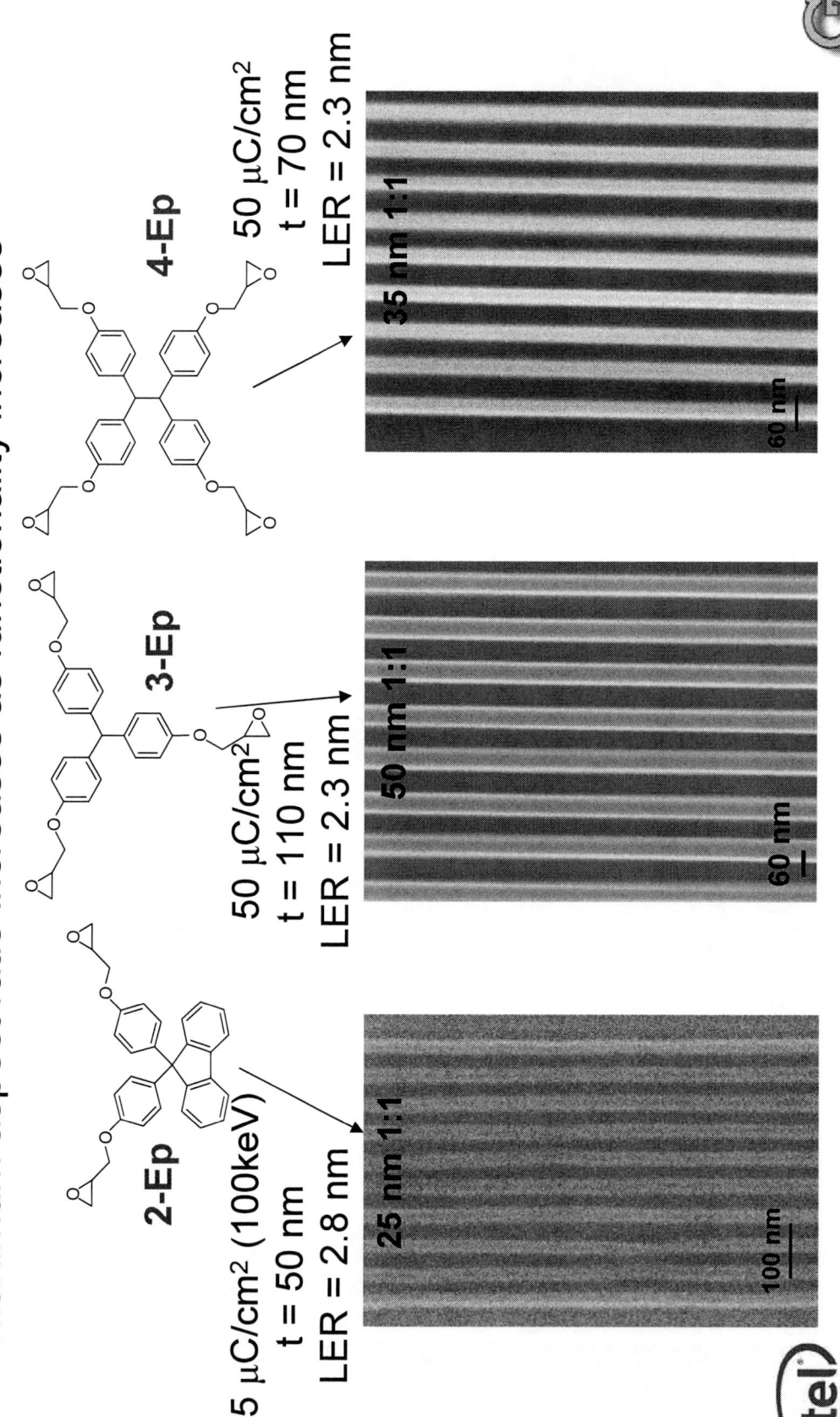

4-Ep EUV Exposures at PSI PEB 90°C

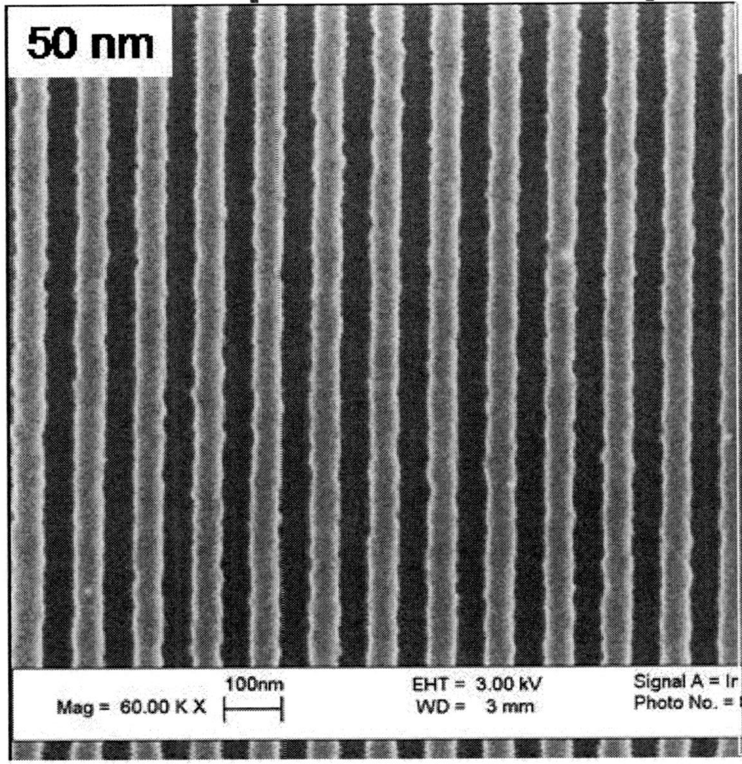

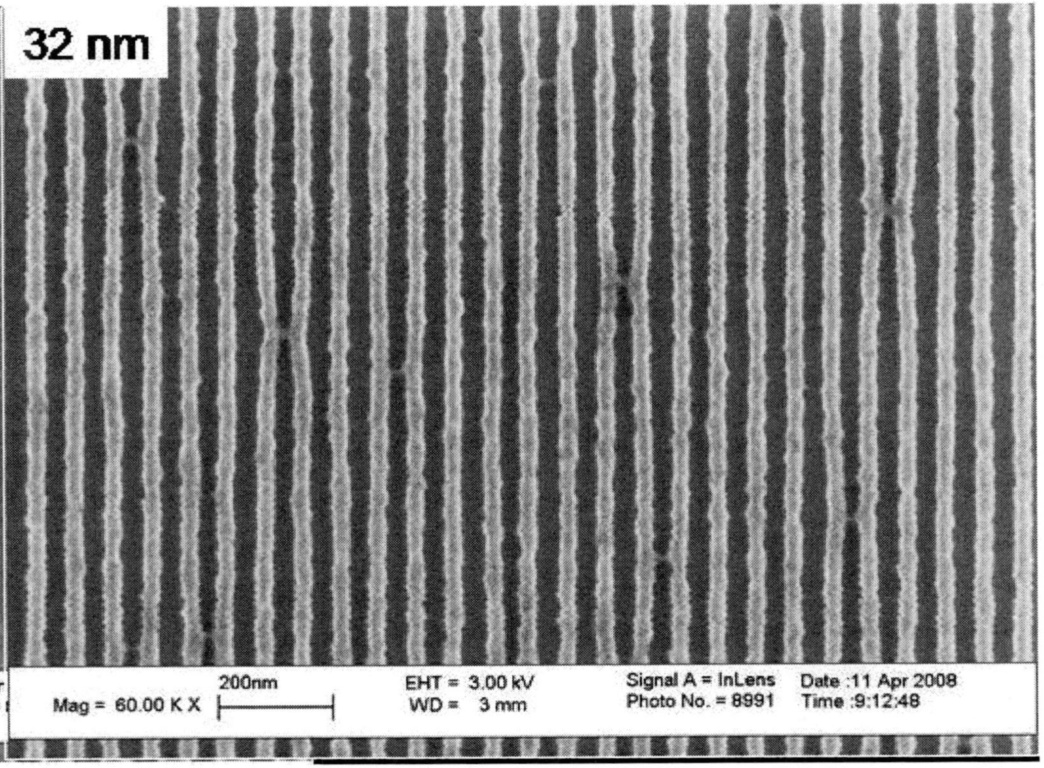

- Dose-to-Mask = 24.7 mJ/cm^2 for PEB = 90°C (March 08)
- LER (3σ) = 4.9 nm
- Resolution down to 32 nm – smaller sizes limited by pattern collapse, >2.5:1 aspect ratio

4-Ep EUV PEB 60°C – TPS-Tf$_3$C

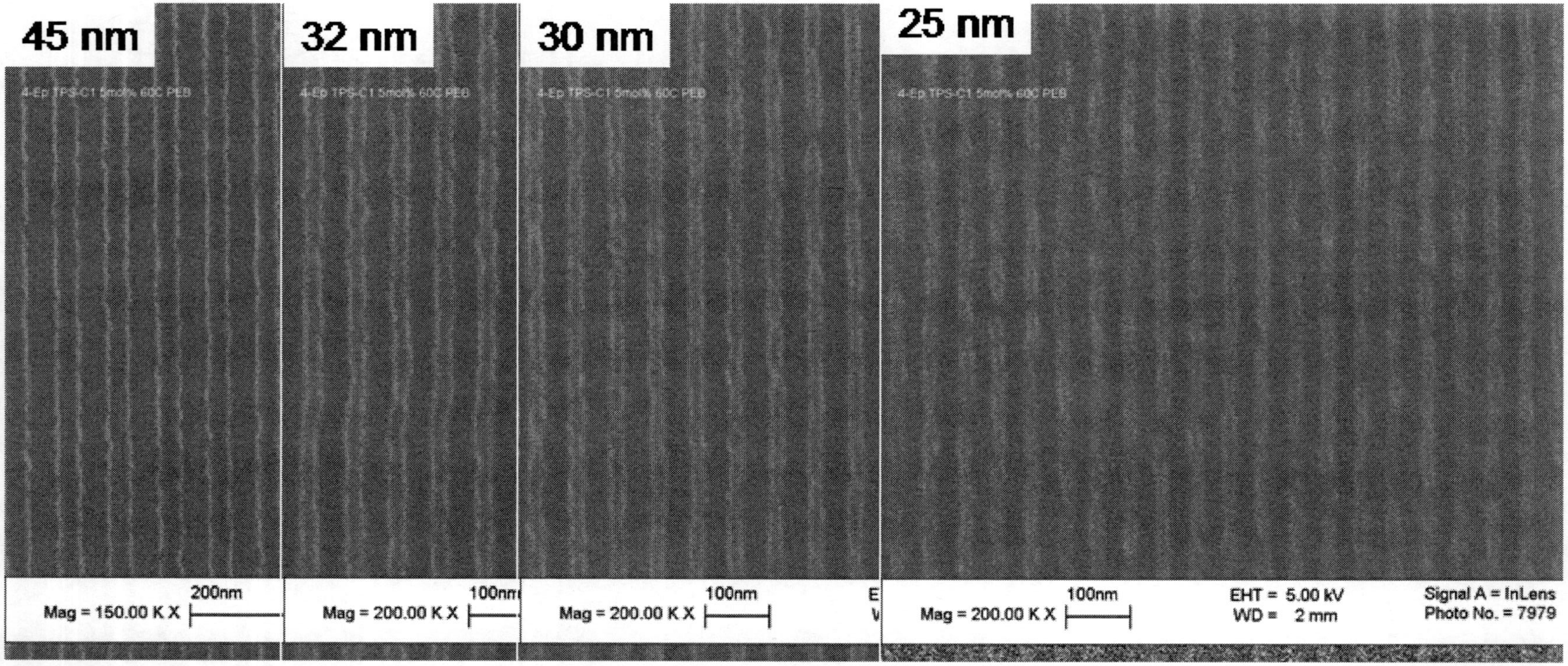

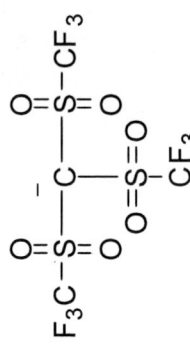

- TPS-methide instead of TPS-SbF$_6$
- Dose-to-Mask = 21.6 mJ/cm^2 (May 08)
- Near 25 nm resolution – defects in lines

4-Ep EUV PEB 60°C

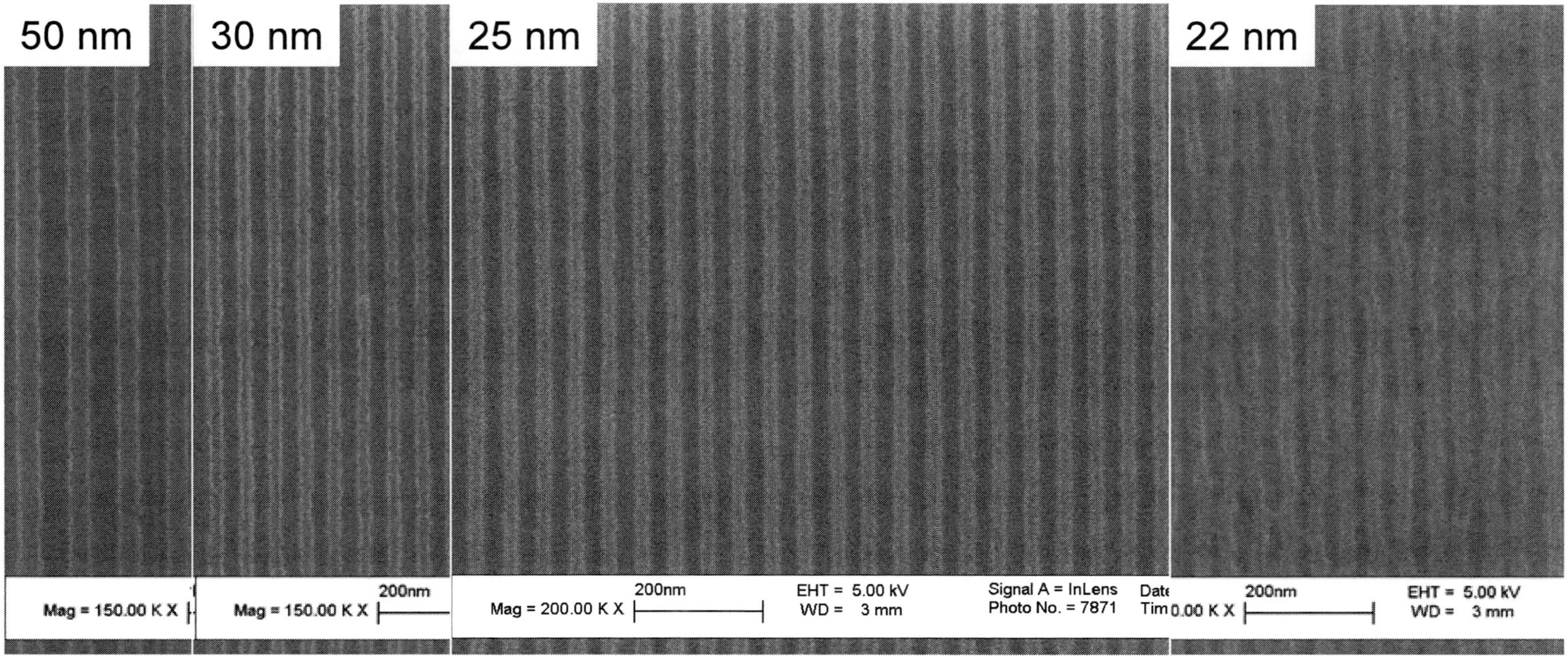

50 nm   30 nm   25 nm   22 nm

Mag = 150.00 K X   Mag = 150.00 K X   Mag = 200.00 K X

200nm   200nm   200nm

EHT = 5.00 kV   WD = 3 mm
Signal A = InLens   Photo No. = 7871
EHT = 5.00 kV   WD = 3 mm

- Modified resist formulation
- Dose-to-Mask = 45 mJ/cm^2 (July 08)
- LER (3σ) = 4.0 nm for 50 nm lines
- LER (3σ) = 4.5 nm for 25 nm lines

(intel)

Summary – Negative Tone

- Epoxide functionalized molecular resists show promise to potentially satisfy patterning requirements for 22 nm node EUV and below

- Resolution: 25 nm 1:1 and below (EUV and e-beam)

- Sensitivity: 8-15 mJ/cm^2 (EUV);   30-75 µC/cm^2 (e-beam)

- LER (3σ):   4.0-4.8 (EUV);   2.3-3.0 (e-beam)

SEMATECH
ALS
22 nm 1:1
15 mJ/cm^2
LWR: 5-6 nm

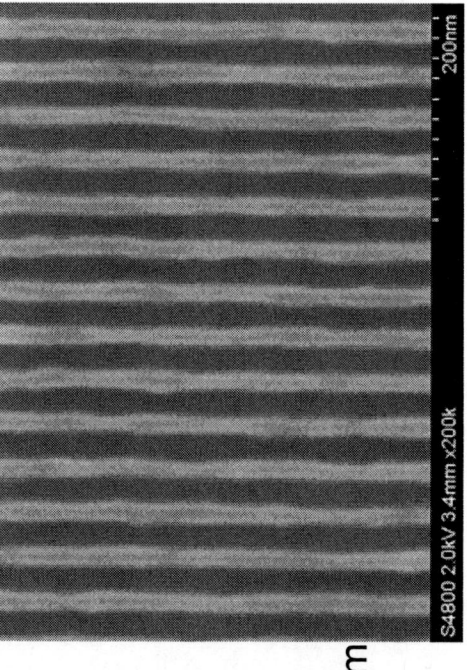

S4800 2.0kV 3.4mm x200k   200nm

4-Ep
PSI
25 nm 1:1
15 mJ/cm^2
LWR: 5-6 nm

Acknowledgements

- Steve Putna, Jeanette Roberts, and Wang Yueh at Intel for helpful discussions.

- Harun Solak, Vaida Auzelyte, and Anja Weber at PSI for EUV exposures

Accelerating the next technology revolution

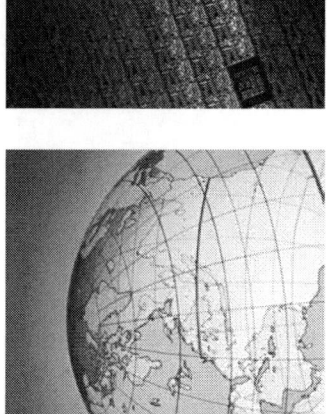
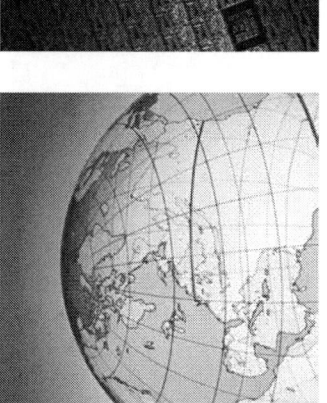

Sub-22nm Half-Pitch (HP) EUV Resist Imaging Results

Chawon Koh, Stefan Wurm (SEMATECH)
Joo-on Park (Samsung Electronics)
Andy Ma (Intel Corporation)
Patrick Naulleau (LBNL)

Outline

- Introduction
- Experimental Conditions
- Progress and Current Status of EUV Resists
- Patterning Results of 22nm HP Lines and Spaces Feature
- Patterning Results of 30nm Contact Hole Feature
- Resist Collapse and Defect of EUV Resist
- Summary
- Acknowledgement

EUV Resist Targets

Evaluate EUV resist samples with well-defined protocols and specification targets.
Focus on resolution, sensitivity, and LWR for 30nm and targeted to 22nm.

EUV Resist Specifications		32nm hp	22nm hp
Resolution (lines 1:1, nm)	½ Pitch	32	22
	MPU Gate CD	21	15
LWR (nm, 3σ)	8% of MPU Gate CD	1.7	1.2
	10% of DRAM HP	3.2	2.2
Photospeed, EUV(mJ/cm²)		10	10
Outgassing(molecules/cm²)		6.5E+14	6.5E+14

Assumptions: Sensitivity target is for 1:1 lines and spaces.
Outgassing specification is for 35-200 AMU excluding 44 AMU.

Experimental Conditions

- Resist Thickness : 50nm

- Substrate : HMDS on Bare Si (or Underlayer)

- Mask : SU451 (SEC-2 Mask)

- Illumination : LBNL MET, 45° Rot Dipole, σ_{radius} 0.1, σ_{offset} 0.57

- PAB : 90 ~ 130 °C (Dependent on Resist)

- PEB : 90 ~ 130 °C (Dependent on Resist)

- Develop : 2.38% TMAH 30~60sec + DIW rinse 30sec

RLS Evaluation Results of EUV Resists

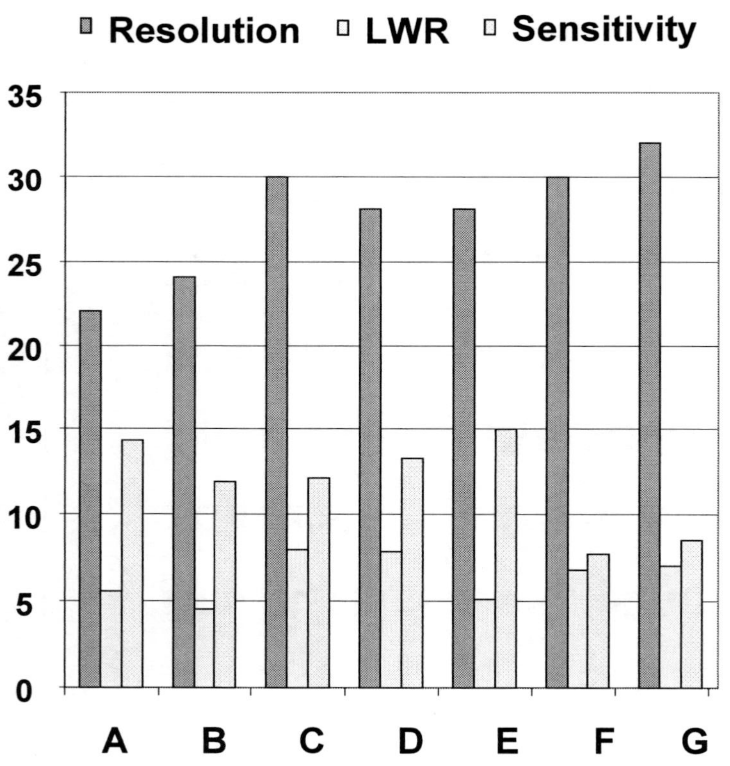

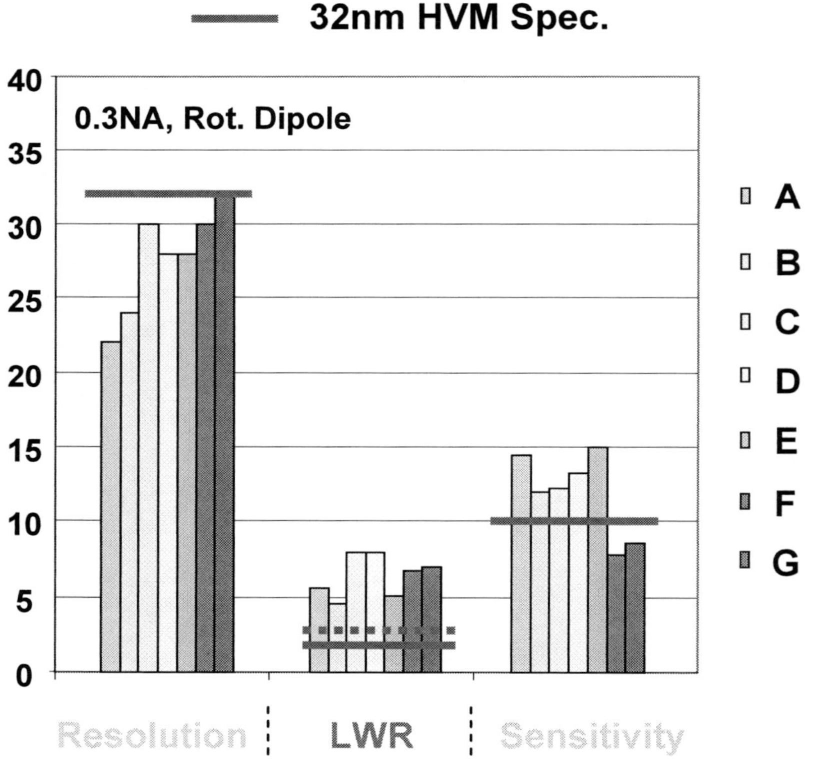

Sub 30nm Patterning Fidelity of EUV Resists

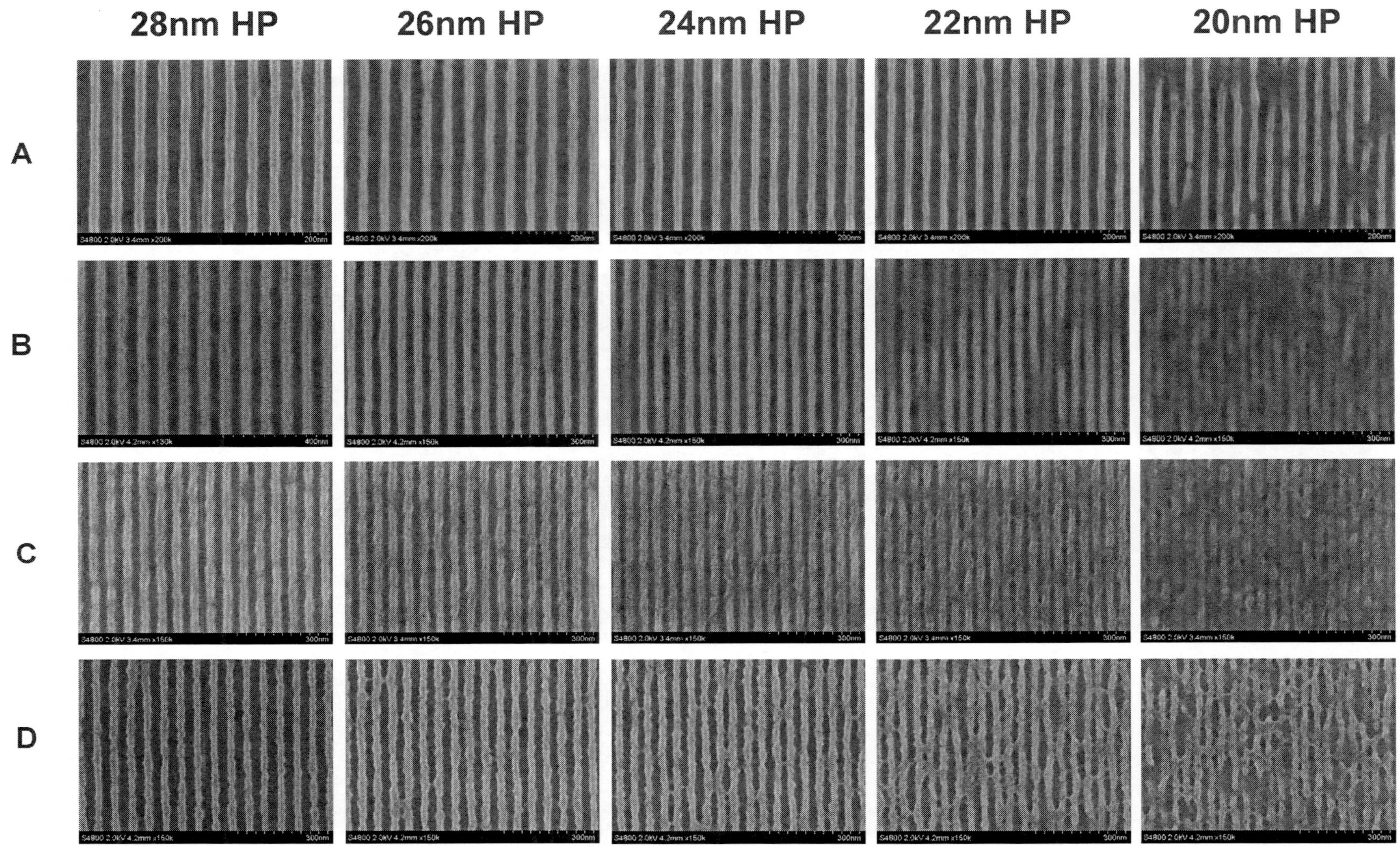

Sub 30nm Patterning Fidelity of EUV Resists

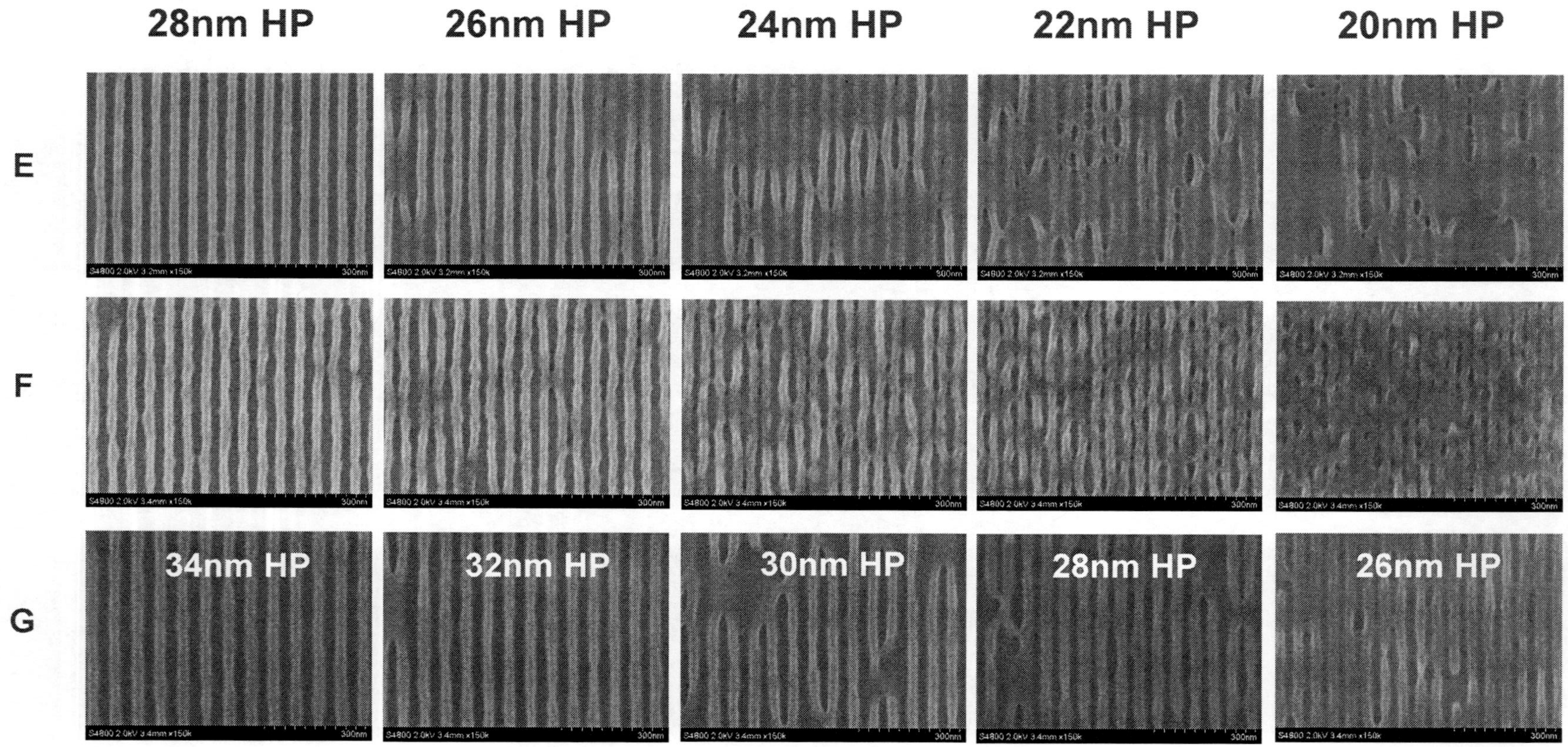

Detailed Evaluation Results of EUV Resists

	ITRS HVM Specs	[1]Resolution [nm] HP	Image Capability [nm]	LER / [2]LWR [nm]	[3]Dose [mJ/cm²]	Resist Collapse [nm] ([4]A/R)	Z-Factor
	32nm HP	32	NA	NA / 1.7 (2.56)*	10.0	2.5	
	22nm HP	22	NA	NA / 1.2 (1.76)*	10.0	2.5	
EUV Resists	A	22	20	3.8 / 5.6	14.5	26 (1.92)	2.21E-08
	B	24	22	3.2 / 4.6	12.0	28 (1.78)	1.70E-08
	C	30	28	5.3 / 8.0	12.2	30 (1.67)	9.08E-08
	D	28	26	5.2 / 7.9	13.3	30 (1.67)	8.49E-08
	E	28	26	3.5 / 5.1	15.0	30 (1.67)	4.03E-08
	F	30	28	4.8 / 6.8	7.8	34 (1.47)	4.85E-08
	G	32	30	4.5 / 7.1	8.6	36 (1.39)	5.71E-08

[1]) LBNL MET, 0.3 NA, Rotated dipole, [2]) Average of 1:1 30 nm, 32 nm, 36 nm, and 40 nm HP features
[3]) LBNL / 2.0, [4]) 50 nm resist thickness
*) The number in brackets gives LWR for DRAM hp with 8% spec.

* Z-factor : Thomas Wallow, et al, SPIE 69211F (2008)

* (LER; nm)^2 * (Sensitivity; mJ/cm2) ~ Constant [Z-factor (mJ*nm^3)]

(Resolution ; nm)^3

Outline

- Introduction

- Experimental Conditions

- Progress and Current Status of EUV Resists

- **Patterning Results of 22nm HP Lines and Spaces Feature**

- **Patterning Results of 30nm Contact Hole Feature**

- Resist Collapse and Defect of EUV Resist

- Summary

- Acknowledgement

Pattern Fidelity of EUV SMT01 and SMT02 Resist

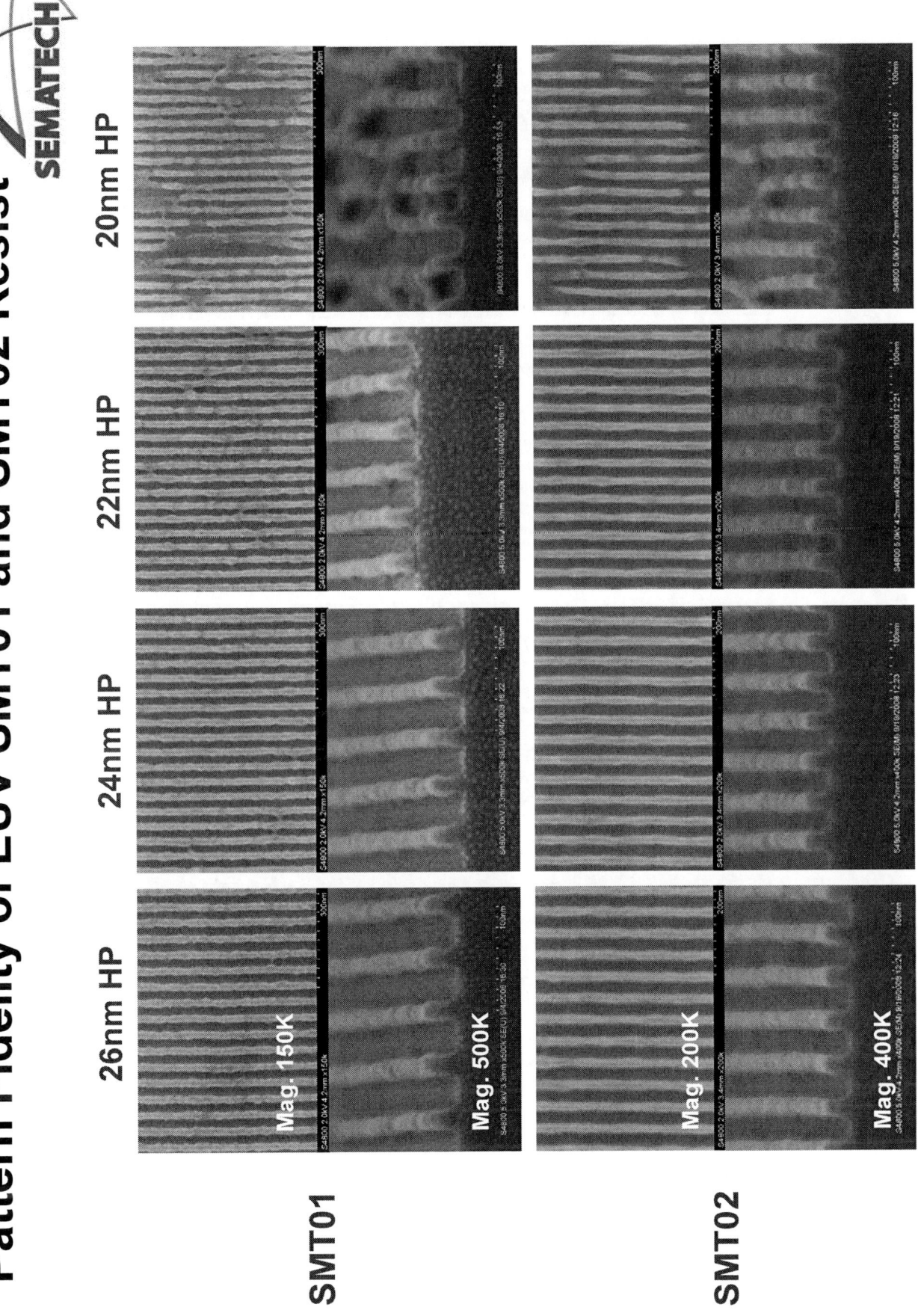

Dose Dependency of EUV SMT02 Resist

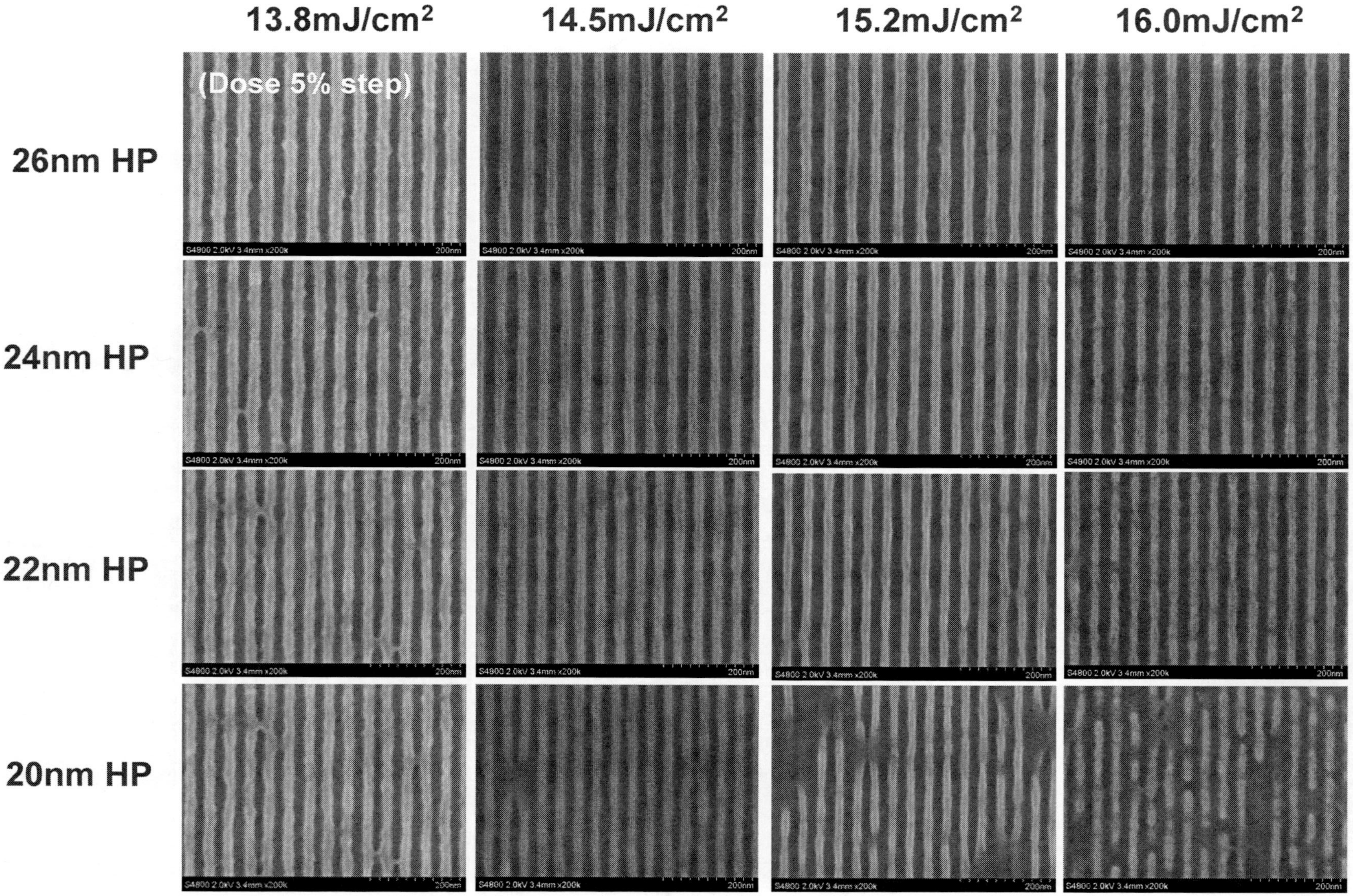

	13.8mJ/cm²	14.5mJ/cm²	15.2mJ/cm²	16.0mJ/cm²
26nm HP	(Dose 5% step)			
24nm HP				
22nm HP				
20nm HP				

Focus Dependency of EUV SMT02 Resist

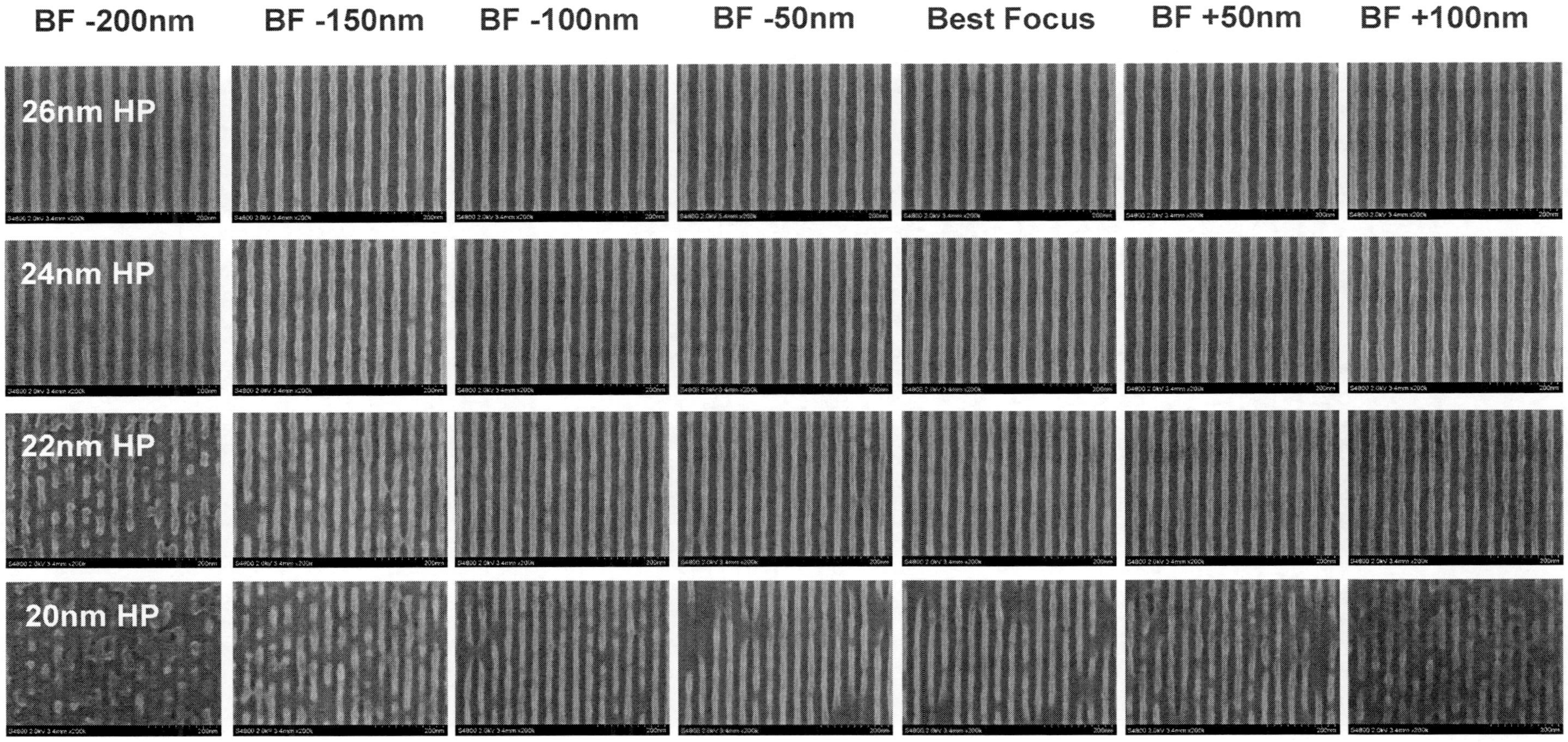

Pattern Fidelity of EUV SMT02 Resist

24nm HP

22nm HP

20nm HP

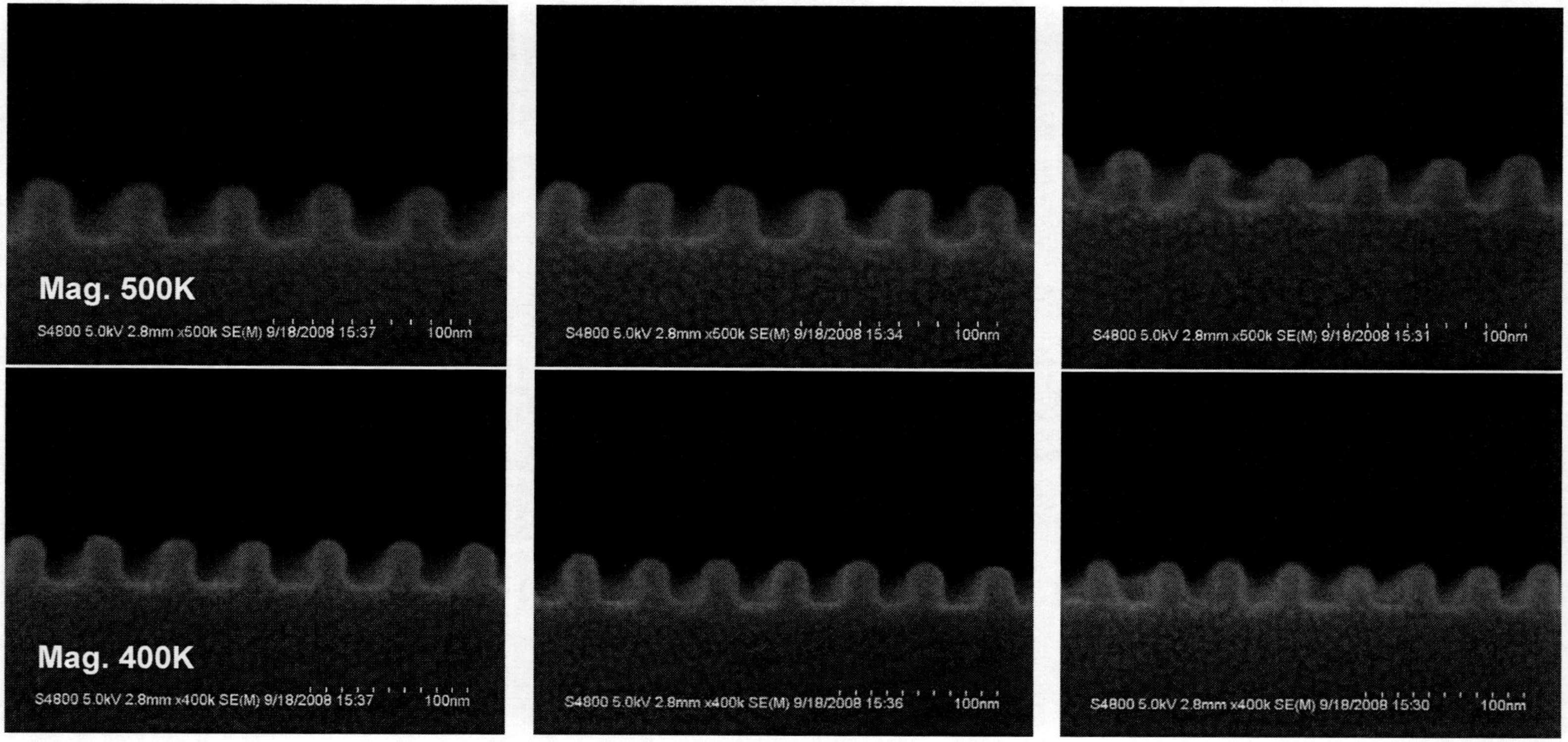

Mag. 500K

Mag. 400K

Focus dependency of EUV SMT02 Resist

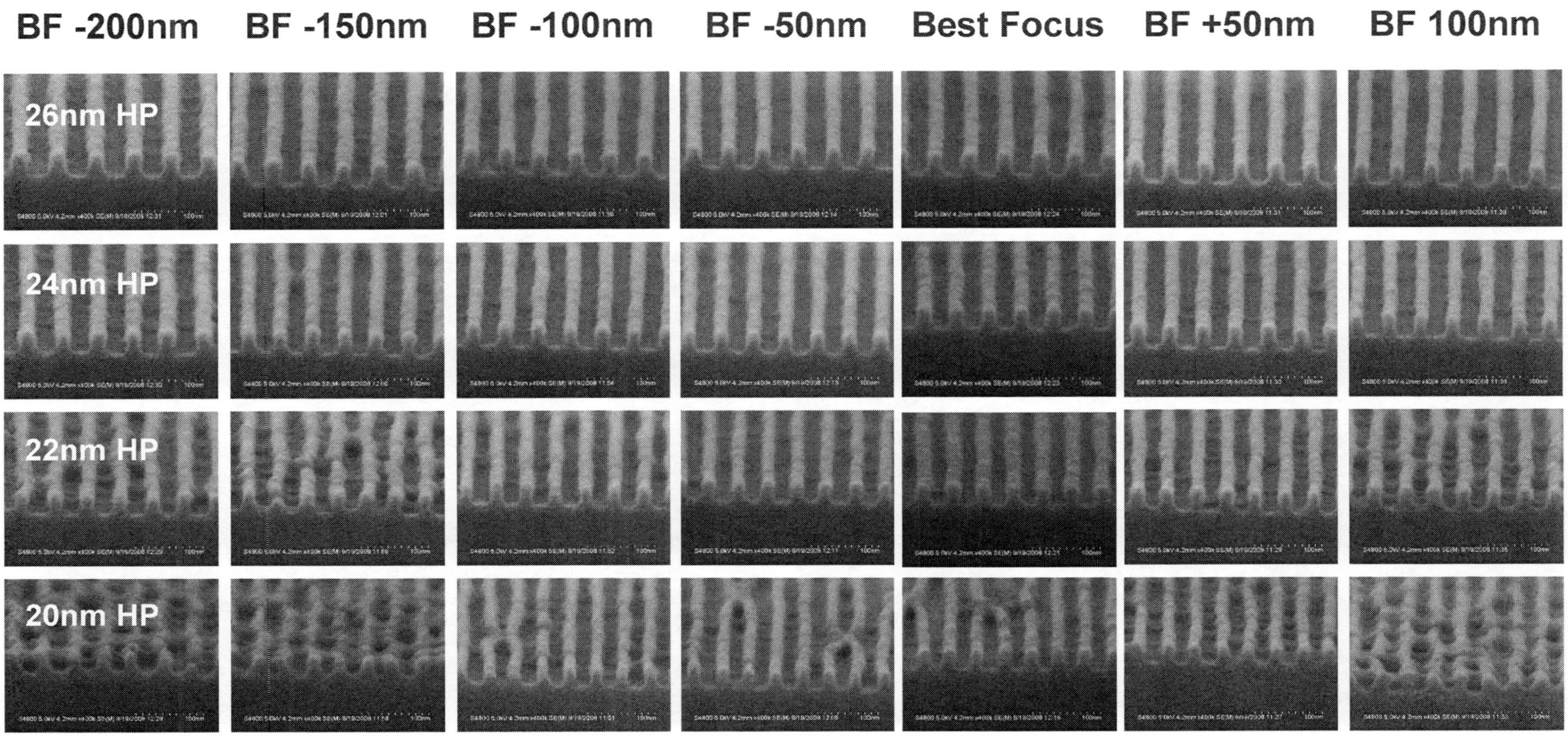

Pattern Fidelity of 30nm 1:1 Dense Contact Hole of EUV SMT02 Resist

Mag. 400K
S4800 5.0kV 2.3mm x400k SE(M) 9/16/2008 10:35   100nm

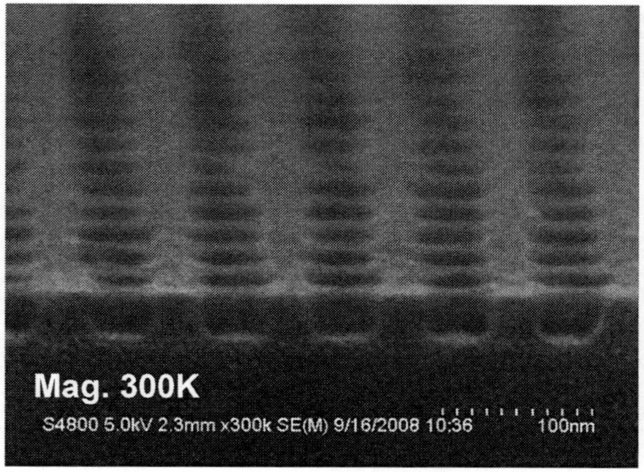

Mag. 250K
S4800 5.0kV 2.3mm x250k SE(M) 9/16/2008 10:37   200nm

Mag. 500K
S4800 5.0kV 2.3mm x500k SE(M) 9/16/2008 10:38   100nm

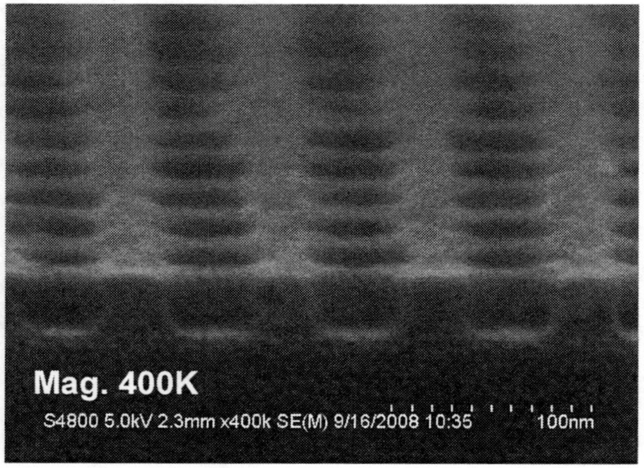

Mag. 300K
S4800 5.0kV 2.3mm x300k SE(M) 9/16/2008 10:36   100nm

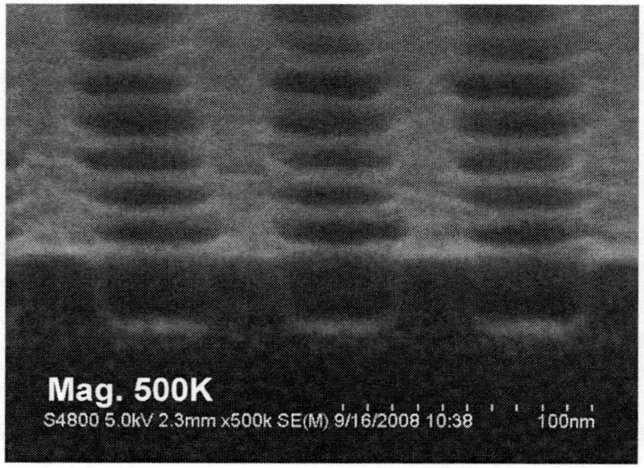

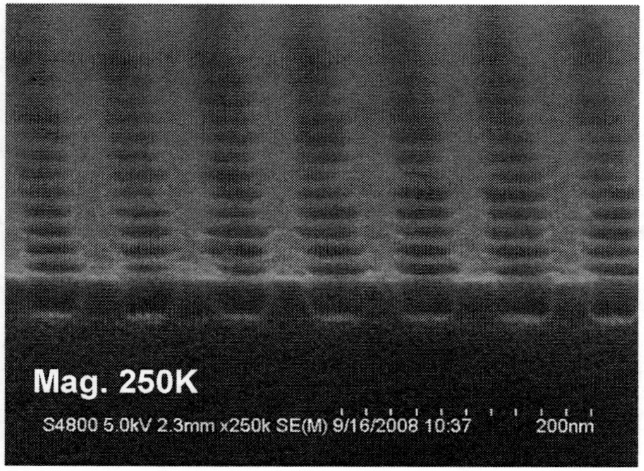

Dose Dependency of 30nm 1:1 Dense Contact Hole in SMT02 Resist

SEMATECH

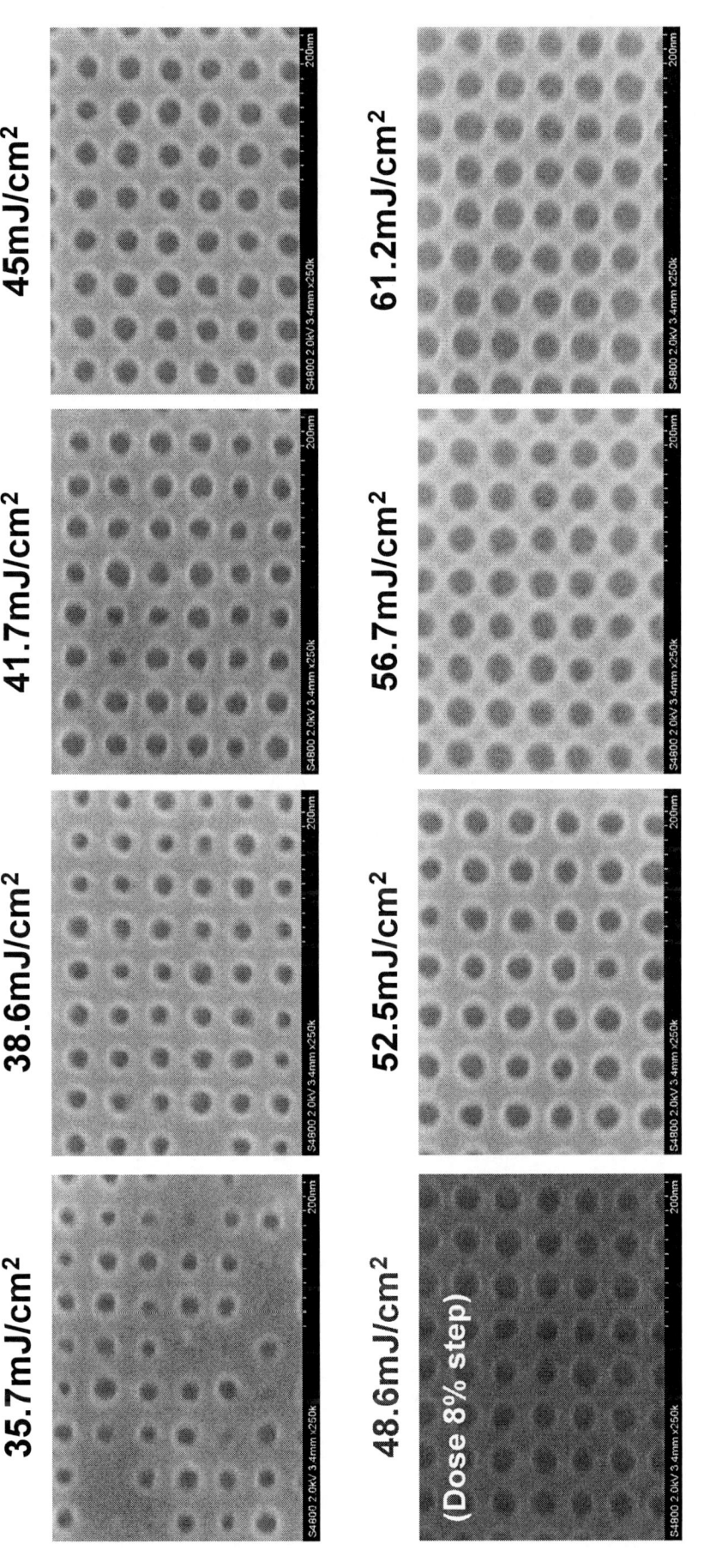

35.7mJ/cm²   38.6mJ/cm²   41.7mJ/cm²   45mJ/cm²

(Dose 8% step)   48.6mJ/cm²   52.5mJ/cm²   56.7mJ/cm²   61.2mJ/cm²

* LBNL MET, 0.3NA, Annular 0.35~0.55

% EL >15%

Focus Dependency of 30nm 1:1 Dense Contact Hole in SMT02 Resist

Dose : 48.6mJ/cm²

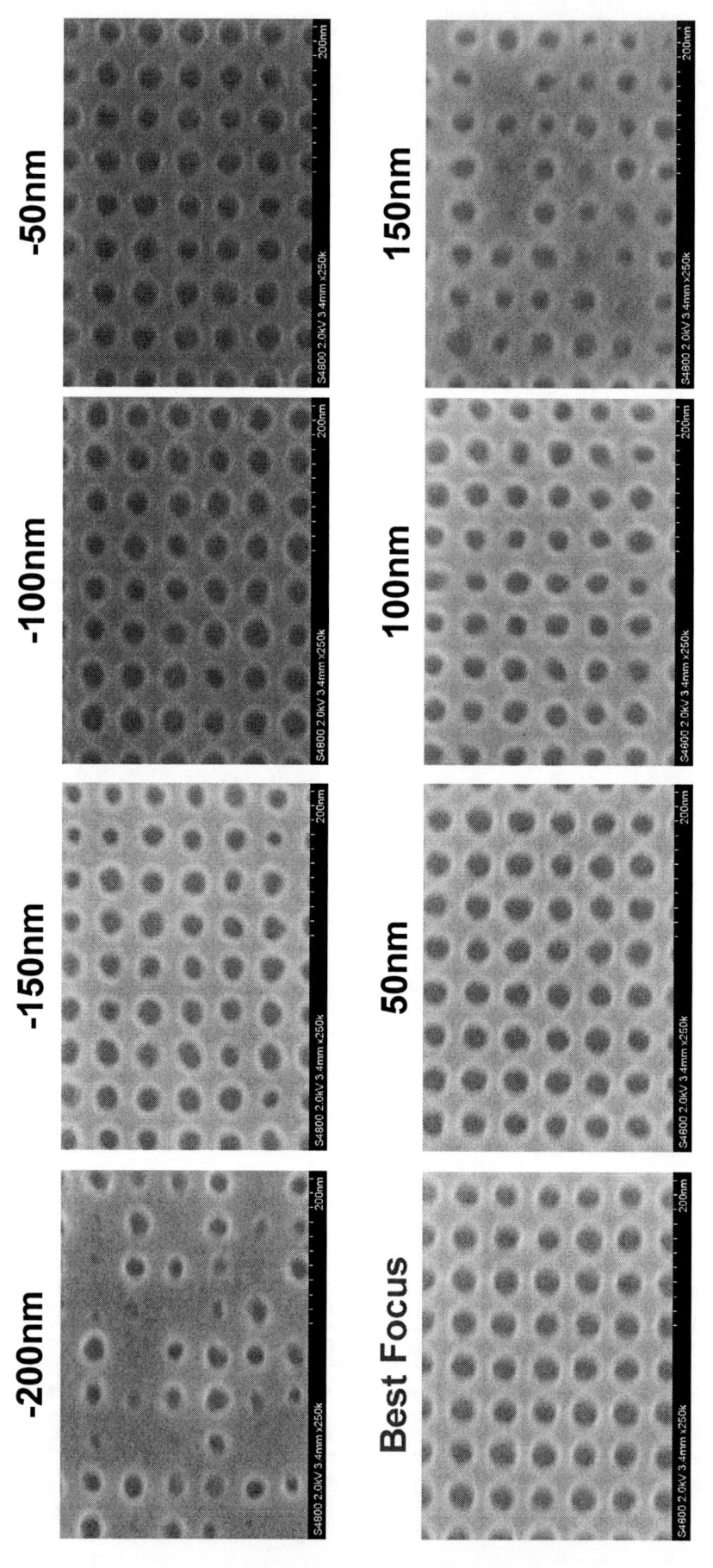

* LBNL MET, 0.3NA, Annular 0.35~0.55

DOF ~150nm

EUV Resist Development Progress

October 2007

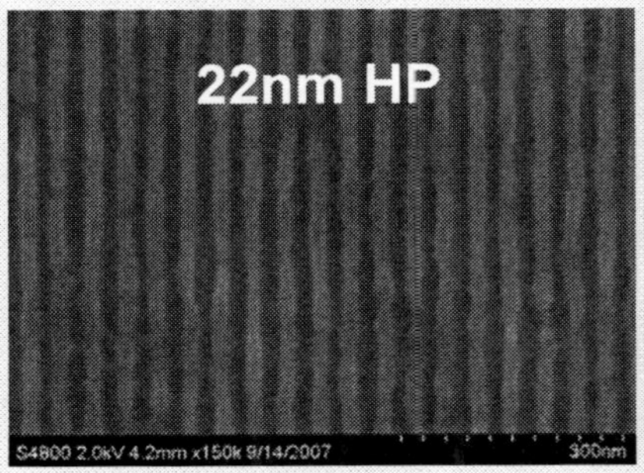

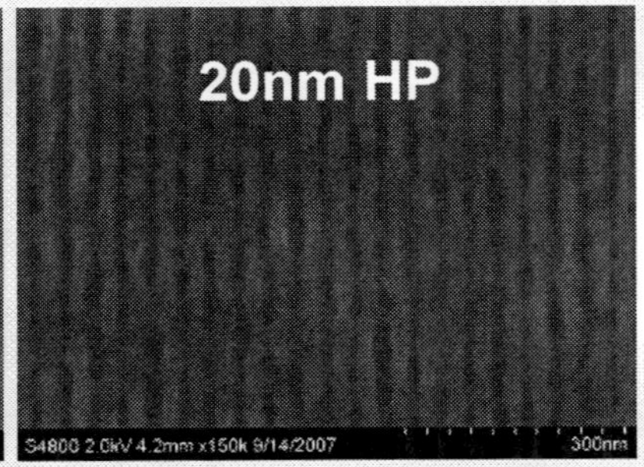

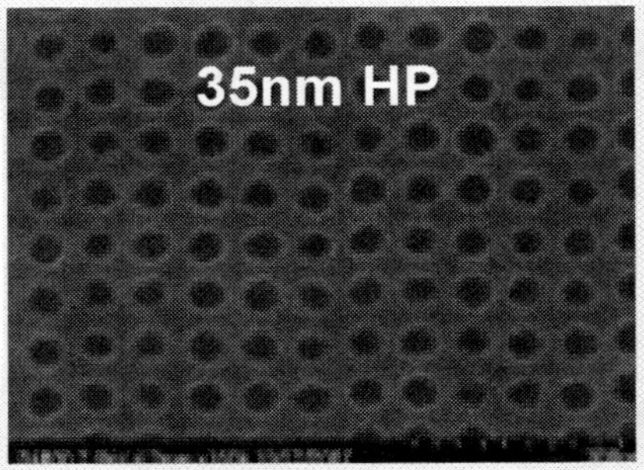

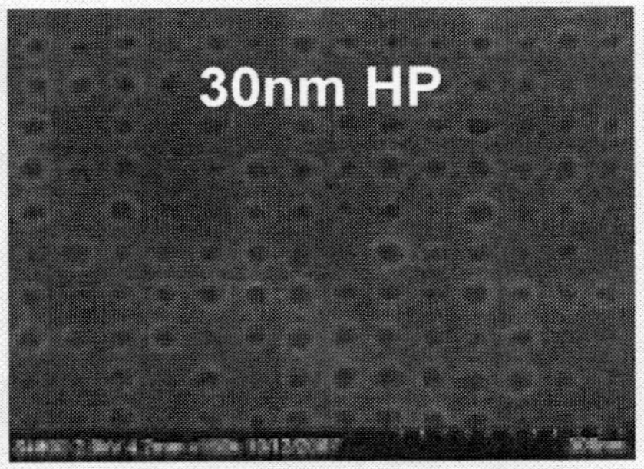

2007

30nm HP

35nm HP

20nm HP

22nm HP

July 2008

2008

30nm HP

20nm HP

22nm HP

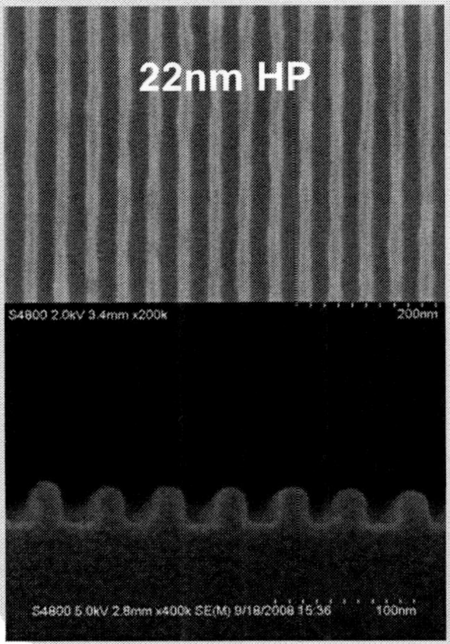

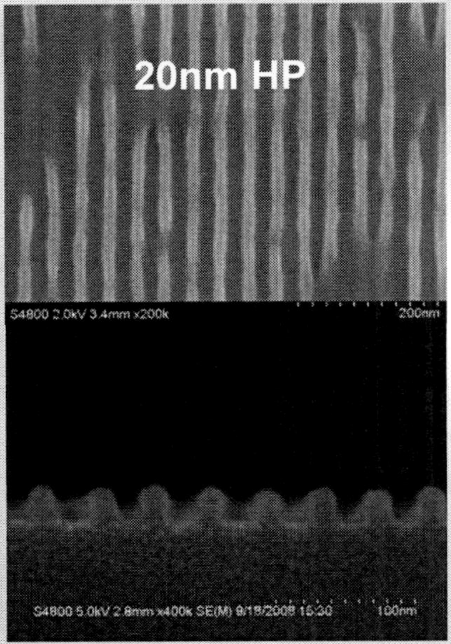

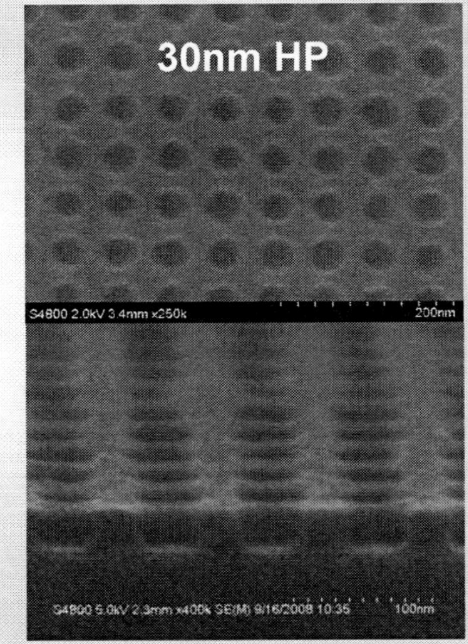

Current Performance Status of EUV Resist

· Resolution ; Looks good, but need to confirm PW
using ADT of 0.25NA and conventional illumination

· LWR ; Need 70% improvement (3.3X larger than MPU Gate spec.)
Need 42.9% improvement (1.3X larger than DRAM HP 10% spec.)

· Sensitivity ; Need 34% improvement (1.5X larger than spec.)

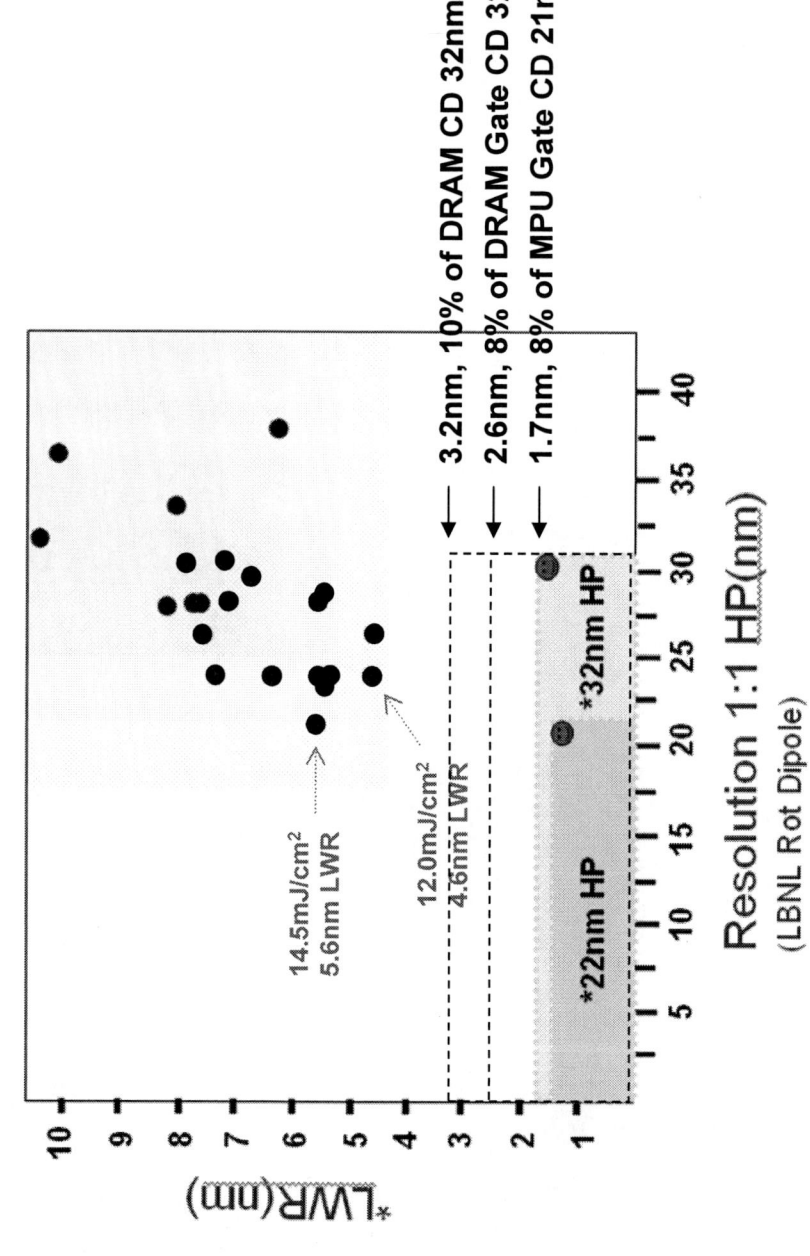

Outline

- Introduction
- Experimental Conditions
- Progress and Current Status of EUV Resists
- Patterning Results of 22nm HP Lines and Spaces Feature
- Patterning Results of 30nm Contact Hole Feature
- **Resist Collapse and Defect of EUV Resist**
- Summary
- Acknowledgement

Resist Collapse Status of EUV Resists

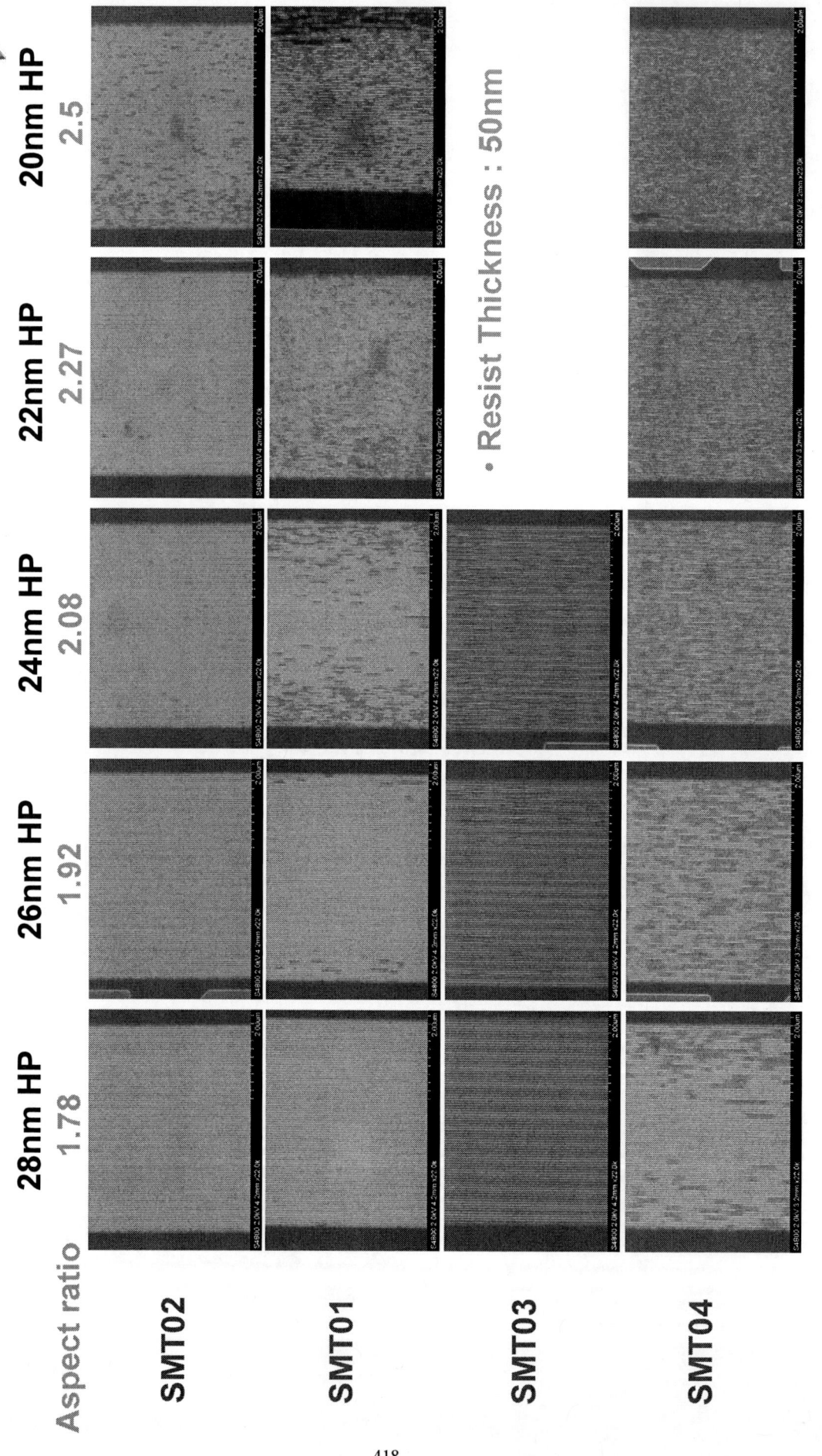

	28nm HP	26nm HP	24nm HP	22nm HP	20nm HP
Aspect ratio	1.78	1.92	2.08	2.27	2.5

- Resist Thickness : 50nm

We need to improve resist collapse to ensure fullfield patterning feasibility

Resist Collapse Status of EUV Resists

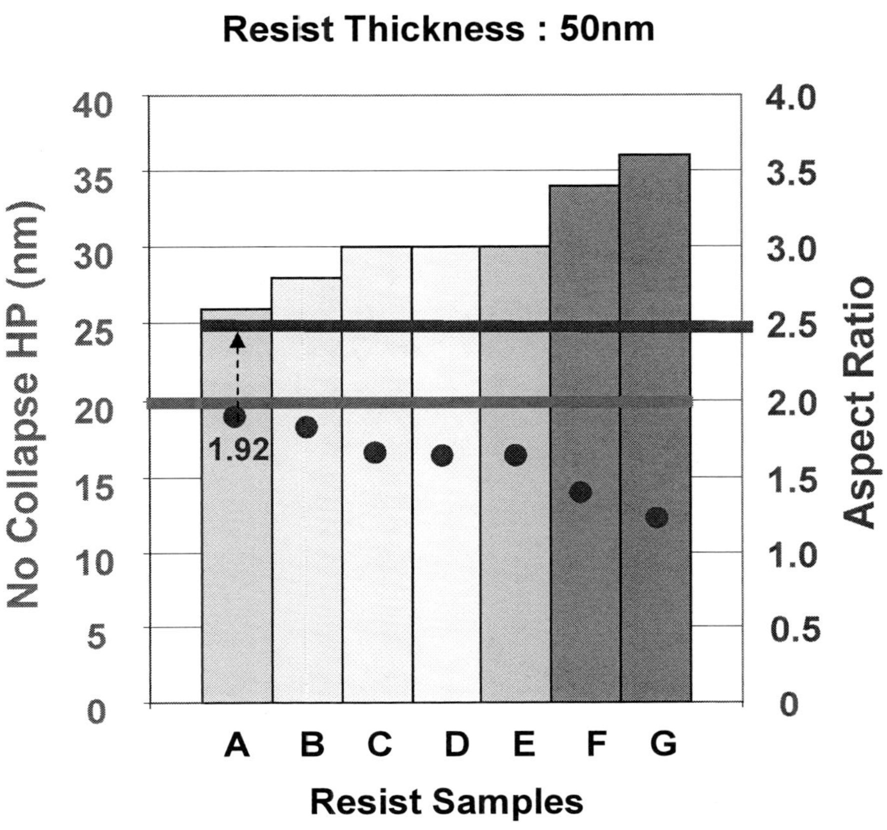

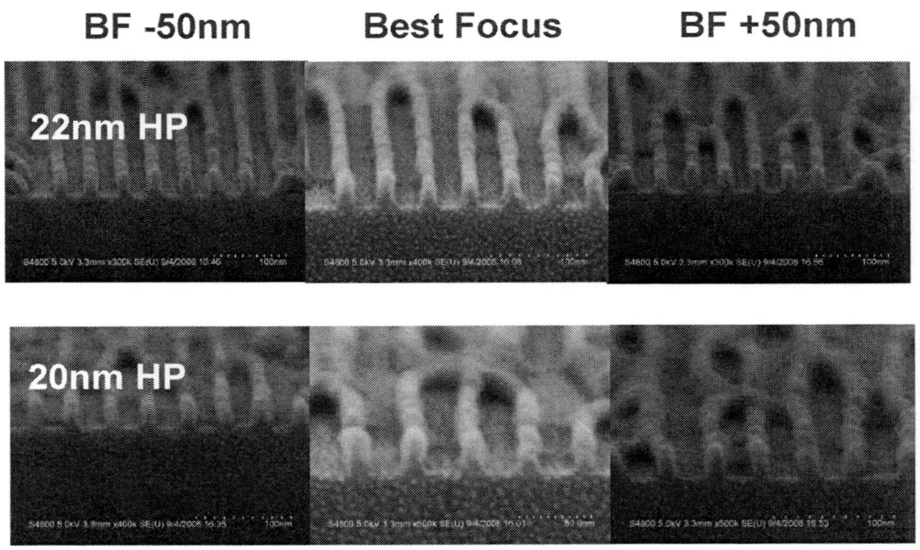

Resist Thickness : 50nm

Aspect Ratio

No Collapse HP (nm)

Resist Samples

1.92

BF -50nm   Best Focus   BF +50nm

22nm HP

20nm HP

* SMT01 Resist

Defect of EUV Resist

Microbridge defect need to be improved

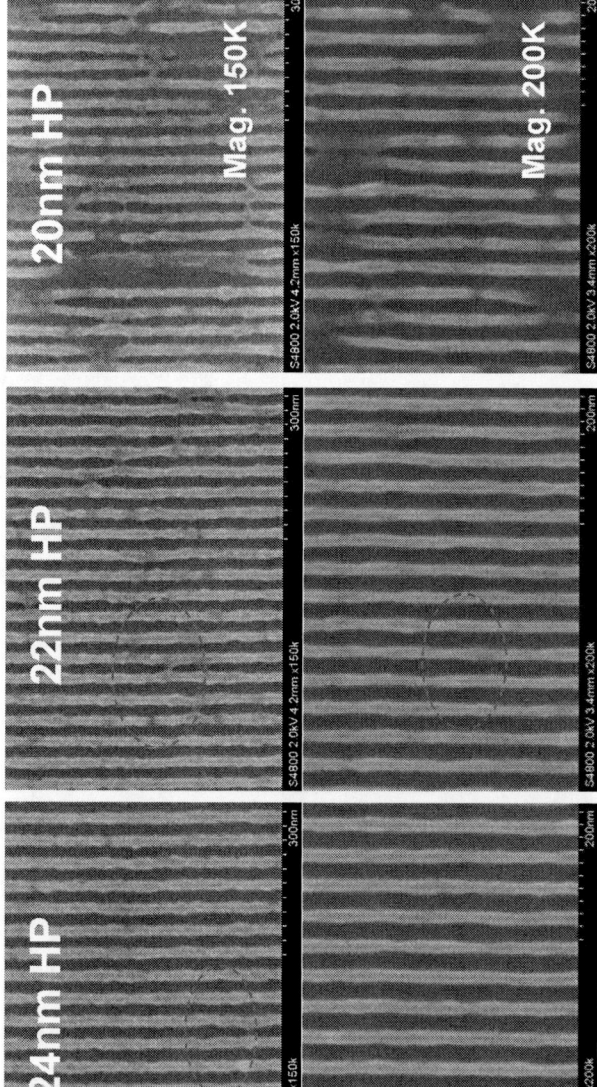

EUV Resist Performance Summary

➤ Possibility of CAR as a promising resist platform was demonstrated for sub 22nm patterning.

	Resolution (nm)	Sensitivity (mJ/cm²)	LER (nm)	LWR (nm)	Z-Factor	Relative Z-factor for MPU	Relative Z-factor for DRAM
Target (MPU)	32	10	1.414	2	6.60E-09	1.0	0.4
Target (DRAM)	32	10	2.26	3.2	1.67E-08	2.5	1.0
SMT01	22	12.1	3.4	5.4	1.49E-08	2.3	0.89
SMT02	22	14.5	3.8	5.6	2.21E-08	3.3	1.32
SMT03	24	12	3.2	4.6	1.70E-08	2.6	1.02

* Z-factor : Thomas Wallow, et al, SPIE 69211F (2008)

(Resolution ; nm)^3 * (LER; nm)^2 * (Sensitivity; mJ/cm^2) ~ Constant [Z-factor (mJ*nm^3)]

Full Characterization of EUV Resist

	Spec. for 32nm HVM	Spec. for 22nm HVM	2008 Target @ ADT 0.25NA	Current @ LBNL MET 0.3NA
*Resolution(1:1 Lines)	32nm	22nm	30nm	22nm
DOF(nm)	200nm	200nm	100nm	> 50nm (CD)
Sensitiviy, Photospeed (mJ/cm²)	10 mJ/cm²	10 mJ/cm²	<15 mJ/cm²	14.5mJ/cm²
*LWR_{Lf}(nm, 3σ)	1.7nm	1.2nm	<3.5nm	5.6nm
Outgassing(molecules/cm²)	6.5E+14	6.5E+14	6.5E+14	Pass
Pattern Collapse	AR > 2.5	AR > 2.5	AR > 2.0	1.92
*CD Uniformity(IFU/IWU/ILU)	2.6nm	1.9nm	3.5nm	TBD
*PEB Sensitivity (nm/°C)	1.0 nm/°C	1.0 nm/°C	1.5 nm/°C	TBD
PED Stability @ < 1ppb amine	>30min	>30min	>30min	TBD
Chemical Flare	<2nm	<1.5nm	< 3nm	TBD
Sidewall Profile Angle	85°	85°	85°	TBD
*Defects in spin-coated resist films	0.01/cm²	0.01/cm²	0.03/cm²	TBD
*Defects in patterned resist films	0.02/cm²	0.01/cm²	-	TBD
*Resist Thickness	50 – 90nm	35 – 65nm	40 – 60nm	50nm
*Unexposed Film Thickness Loss	< 5%	< 5%	< 5%	TBD
Etch Resistance	Similar to Novolac	Similar to Novolac	Similar to Novolac	TBD

* From ITRS 2007

Summary

- Demonstrated 22nm HP lines and space patterning, and showed imaging capability of 20nm HP lines and spaces patterning.

- Demonstrated 30nm 1:1 dense contact hole patterning (DOF 150nm, EL 20%).

- Demonstrated CAR as a promising resist platform for 22nm hp patterning.

- The study to control resist collapse and defect is important for ensuring fullfield EUV Patterning.

- Guideline of optimal resist platform among various candidates is important to early EUV implementation to 32nm HP and 22nm HP patterning considering all aspects of resist characteristics

Acknowledgement

- Dongjin ; Jung-Youl Lee, Jaehyun Kim
- Fujifilm ; Katsuhiro Yamashita, Shinji Tarutani
- JSR ; Yoshi Hishiro, Shalini Sharma
- Rohm and Hass ; Su Jin Kang, James W Thackeray
- Shinetsu ; Jun Hatakeyama, Yoshio Kawai
- Sumitomo ; Nobuo Ando, Yuko Yamashita
- TOK ; Rick Uchida, Taku Hirayama

- AMD ; Tom Wallow, Bruno La Fontaine
- CNSE ; Corbet Johnson, Martin Rodgers
- IBM ; Karen Petrillo
- Intel ; Todd R. Younkin, Ernisse Steve Putna
- Lawrence Berkeley National Laboratory ; Gideon Jones, Brian Hoef, Paul Denham, Jerrin Chiu
- Samsung Electronics ; Haisub Na, Pryool Kang
- SEMATECH ; Kevin Orvek, Jacque Georger, Warren Montgomery, Anwar Khurshid

Thank you for your attention

Corner rounding in photoresists for extreme ultraviolet lithography.

Christopher N. Anderson[1], Patrick P. Naulleau[2], Thomas Wallow[3], and Yunfei Deng[3]

[1]Applied Science & Technology program, University of California, Berkeley, California.
[2]Center for X-ray Optics, Lawrence Berkeley National Laboratory, California.
[3]Advanced Micro Devices, Sunnyvale, California.

cnanderson@berkeley.edu

EUVL Symposium, Lake Tahoe, California, 2008.10.01

A really good resist.

cnanderson@berkeley.edu

EUVL Symposium, Lake Tahoe, California, 2008.10.01

40 nm

38 nm

36 nm

34 nm

32 nm

30 nm

28 nm

26 nm

24 nm

22 nm

Rotated dipole
Sensitivity = 12.7 mJ/cm^2
LWR on 24-nm lines = 4.0 nm

cnanderson@berkeley.edu

EUVL Symposium, Lake Tahoe, California, 2008.10.01

P. Naulleau, et. al. EIPBN 2008

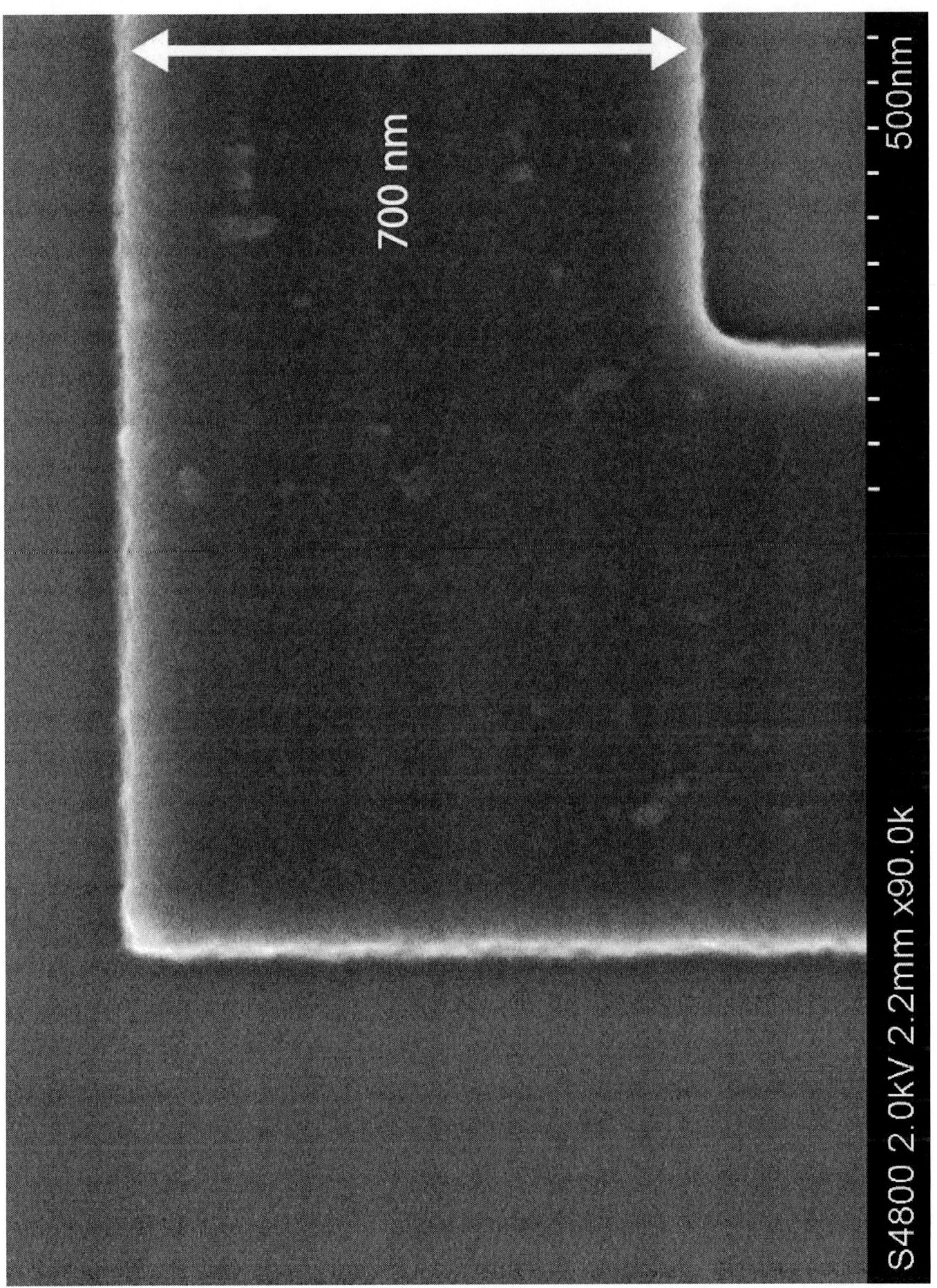

700 nm

S4800 2.0kV 2.2mm x90.0k   500nm

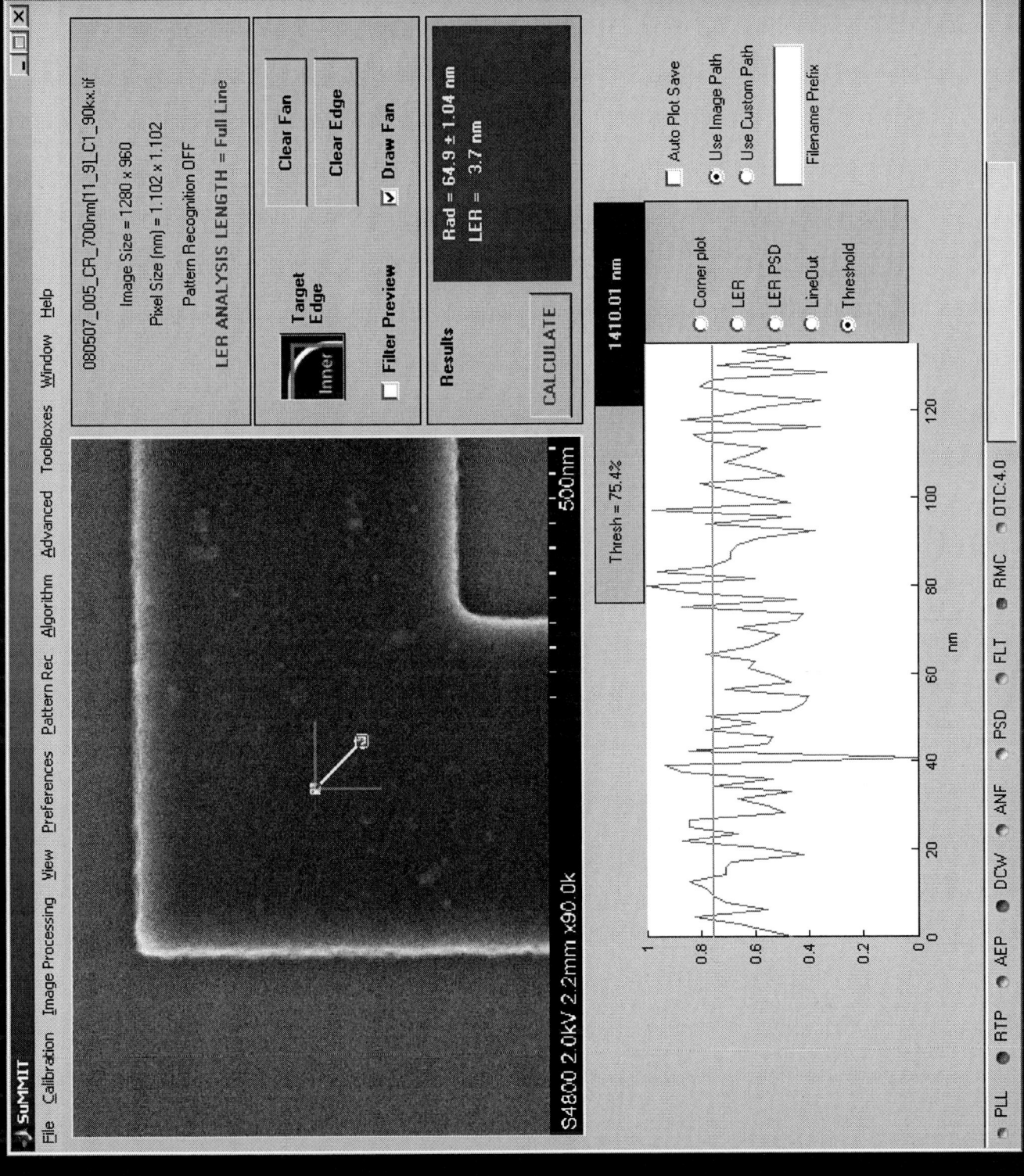

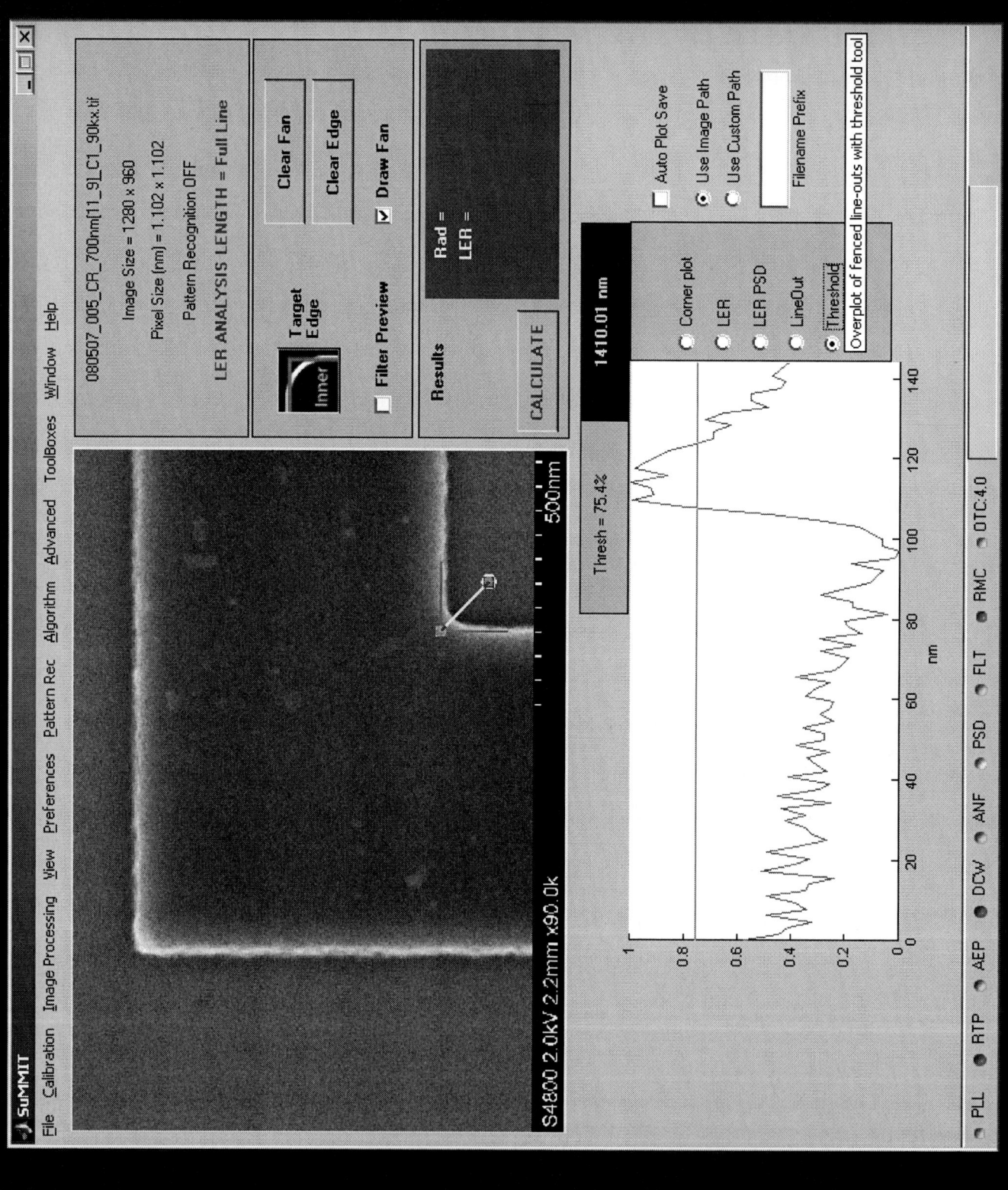

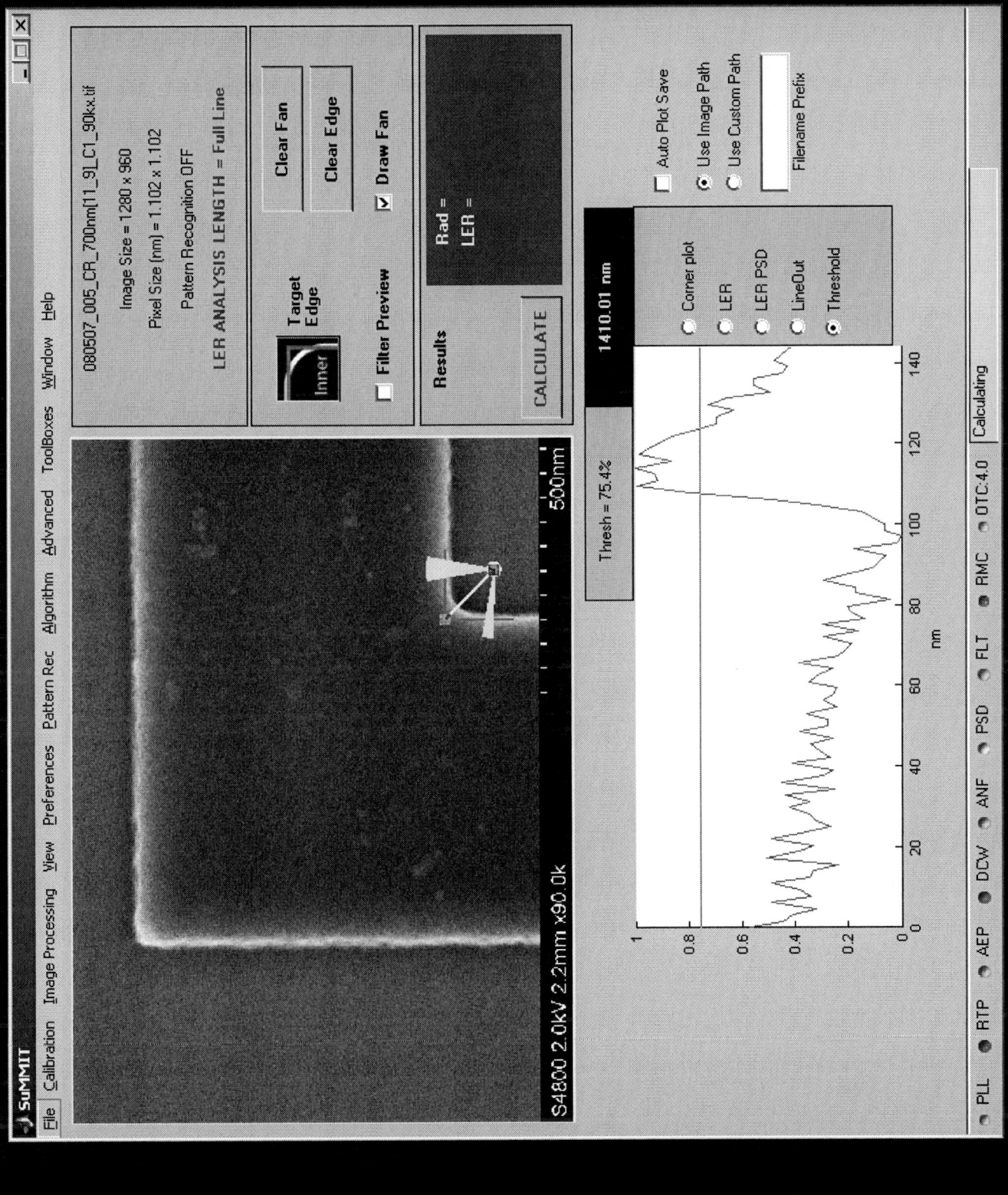

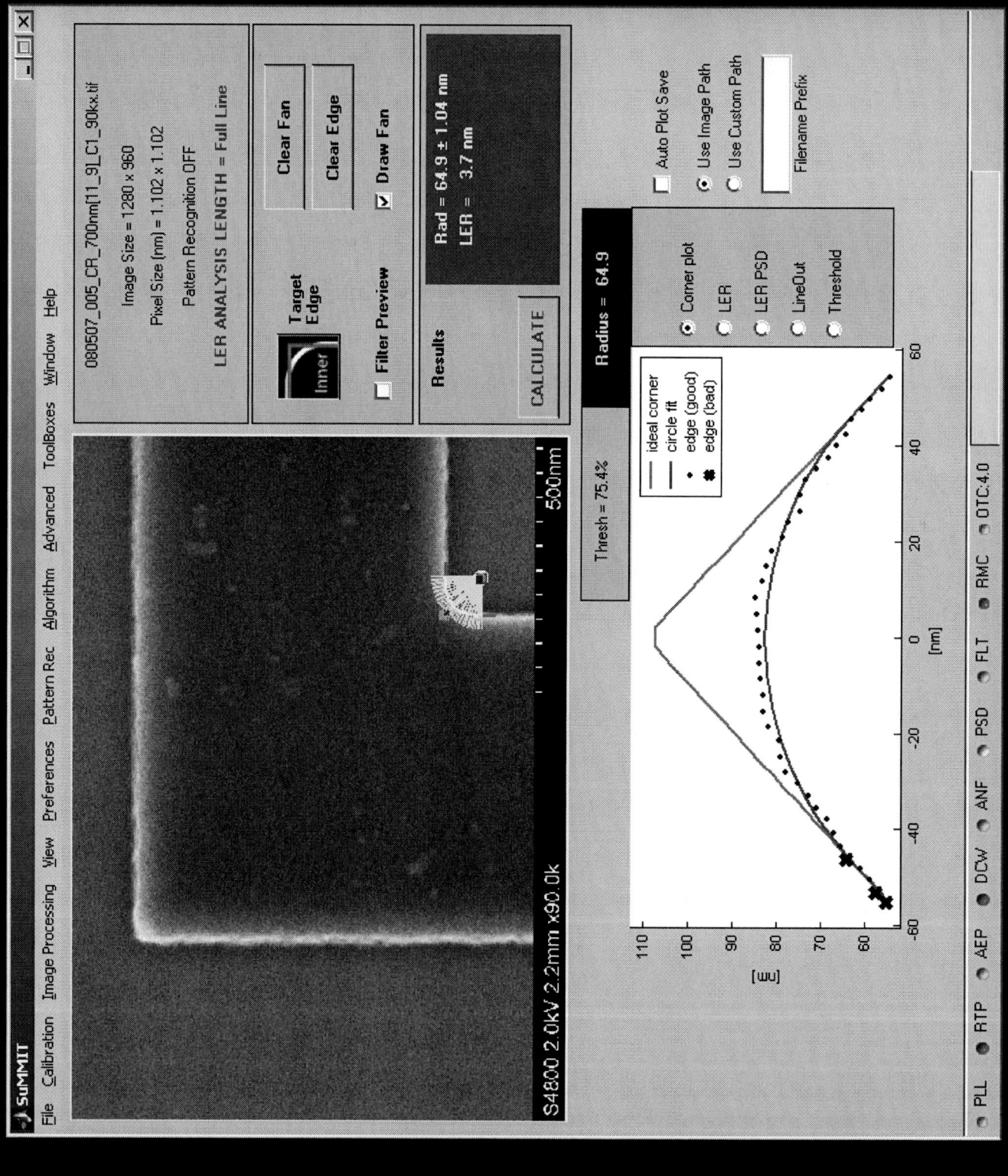

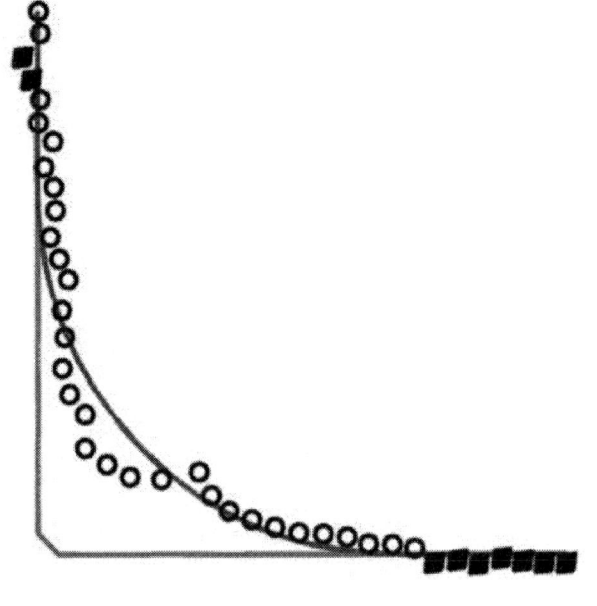

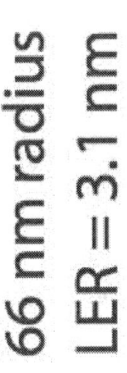

66 nm radius
LER = 3.1 nm

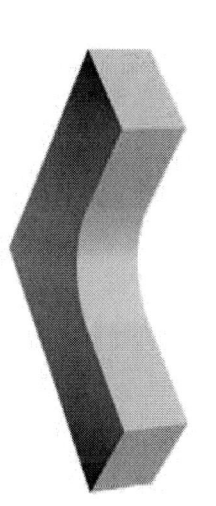

42 nm radius
LER = 6.0 nm

cnanderson@berkeley.edu

EUVL Symposium, Lake Tahoe, California, 2008.10.01

What about smaller features?

cnanderson@berkeley.edu

EUVL Symposium, Lake Tahoe, California, 2008.10.01

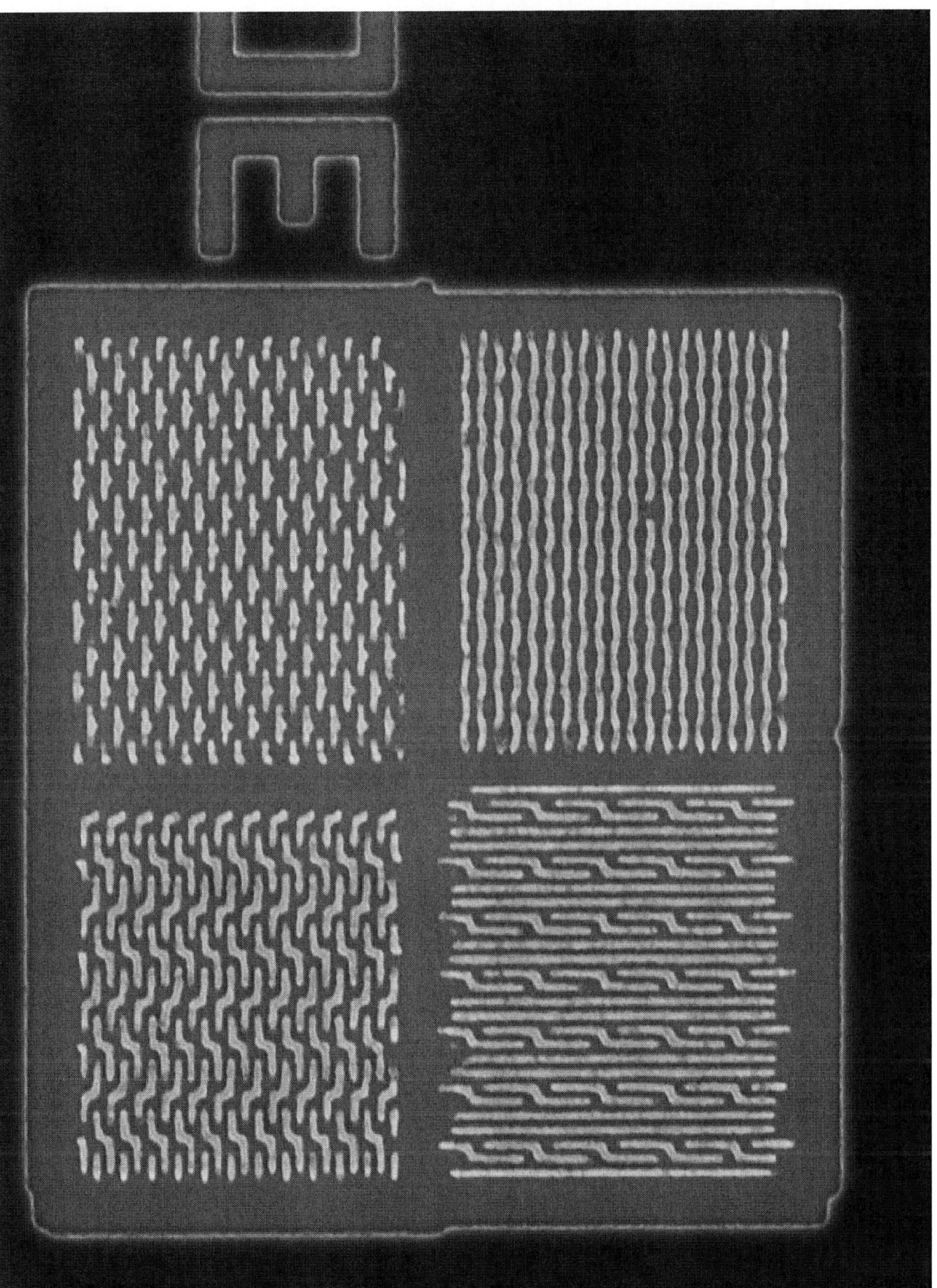

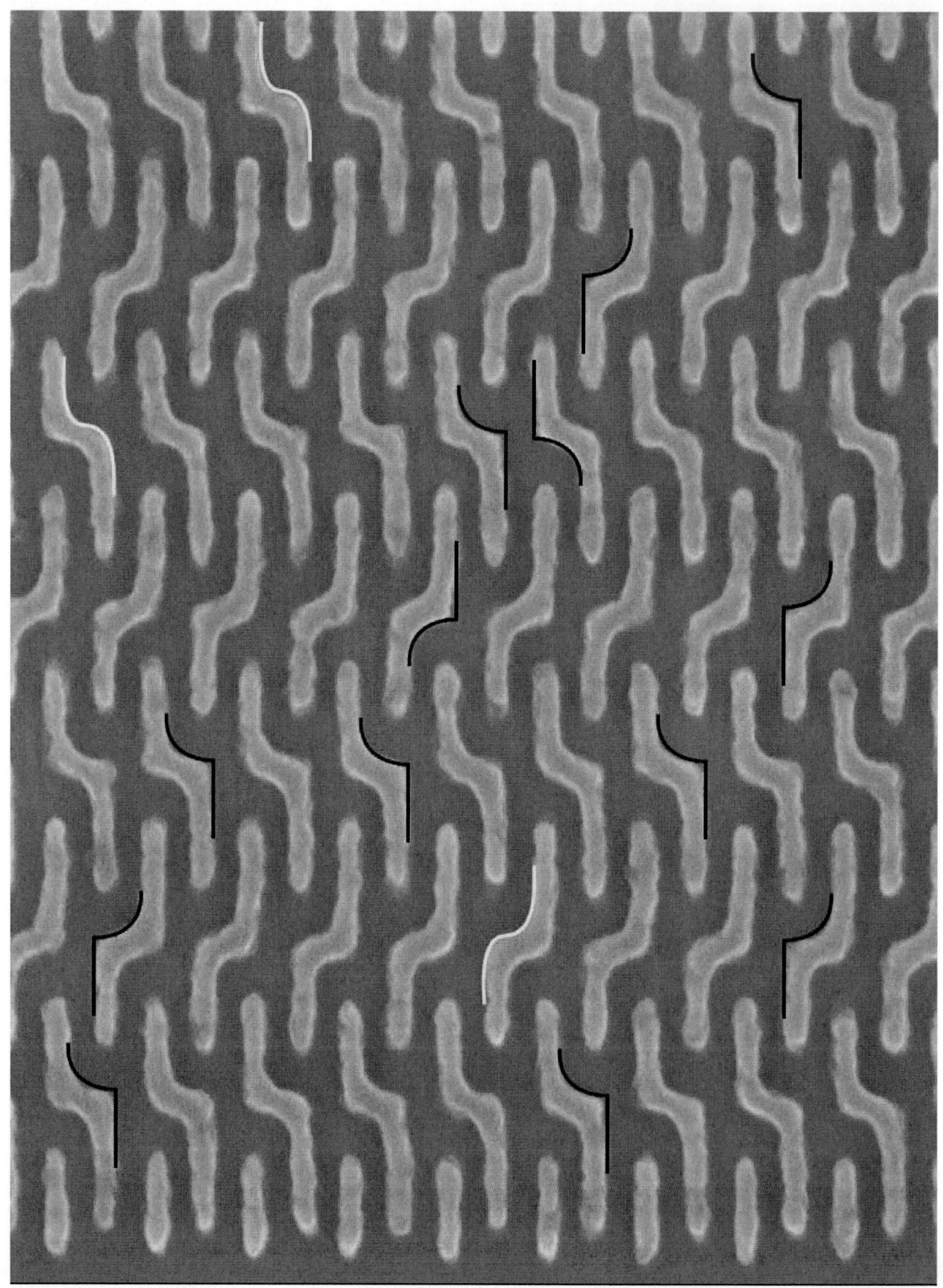

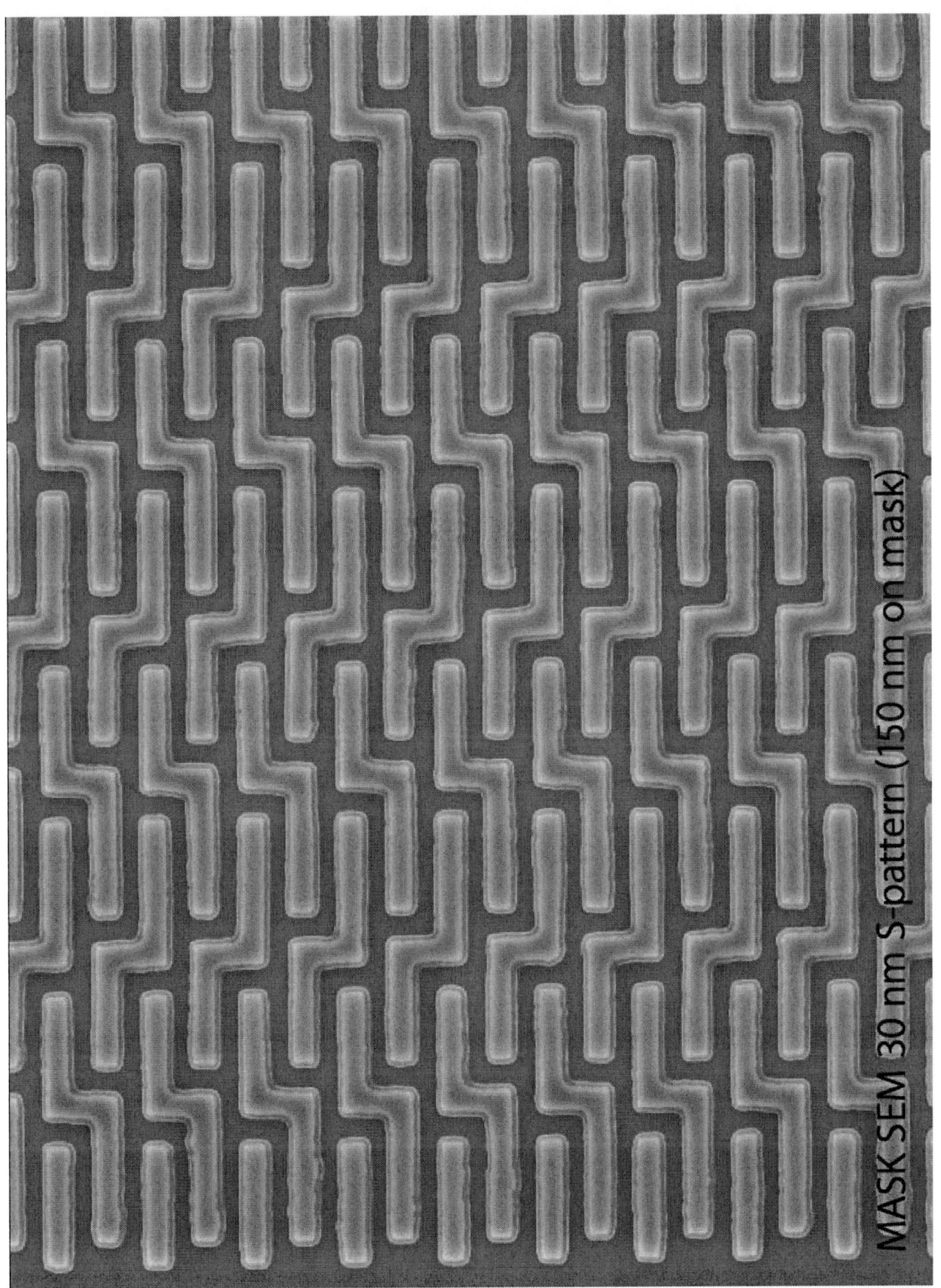

MASK SEM 30 nm S-pattern (150 nm on mask)

Another really good resist.

cnanderson@berkeley.edu

EUVL Symposium, Lake Tahoe, California, 2008.10.01

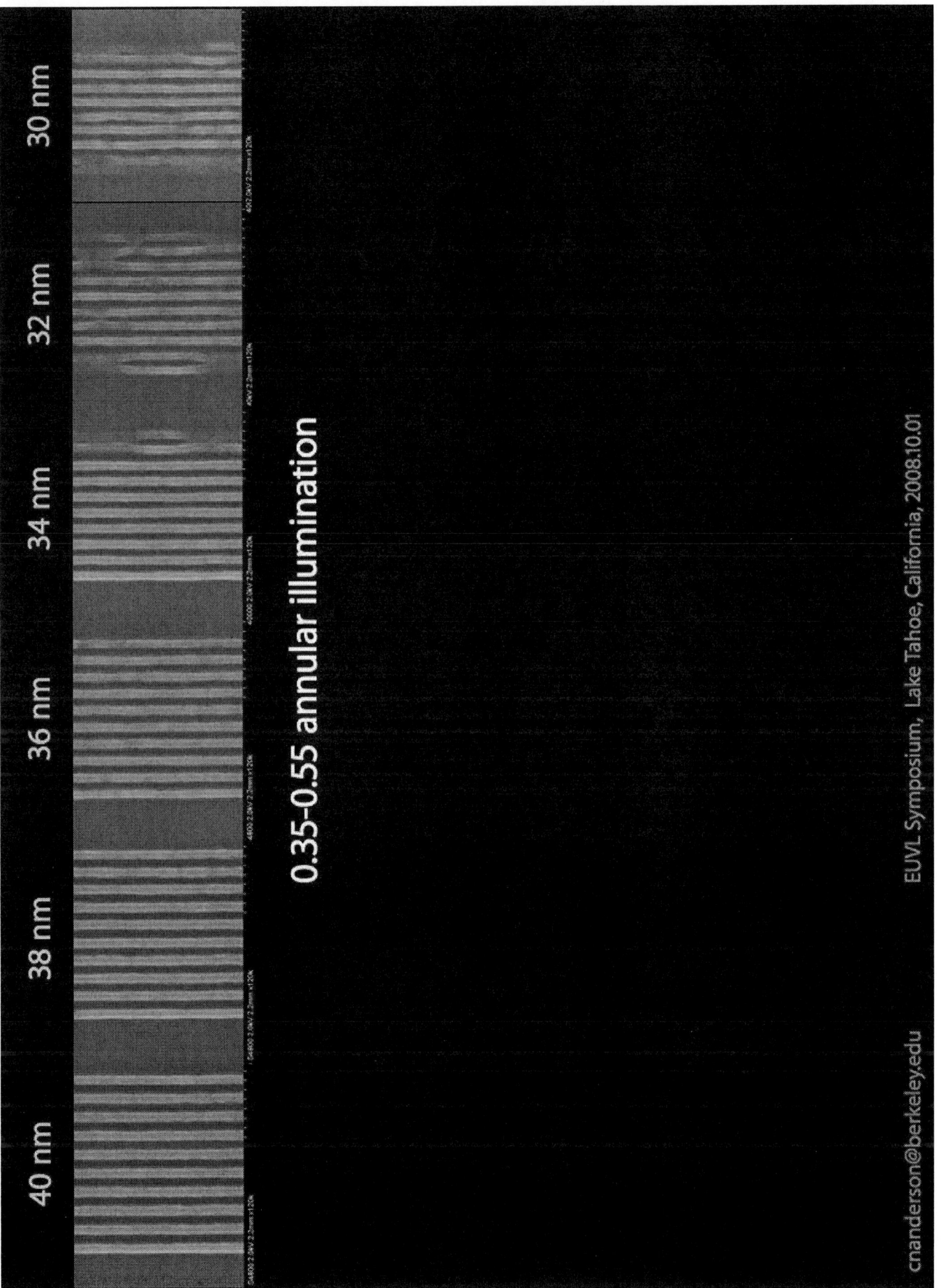

0.35-0.55 annular illumination

40 nm   38 nm   36 nm   34 nm   32 nm   30 nm

cnanderson@berkeley.edu   EUVL Symposium, Lake Tahoe, California, 2008.10.01

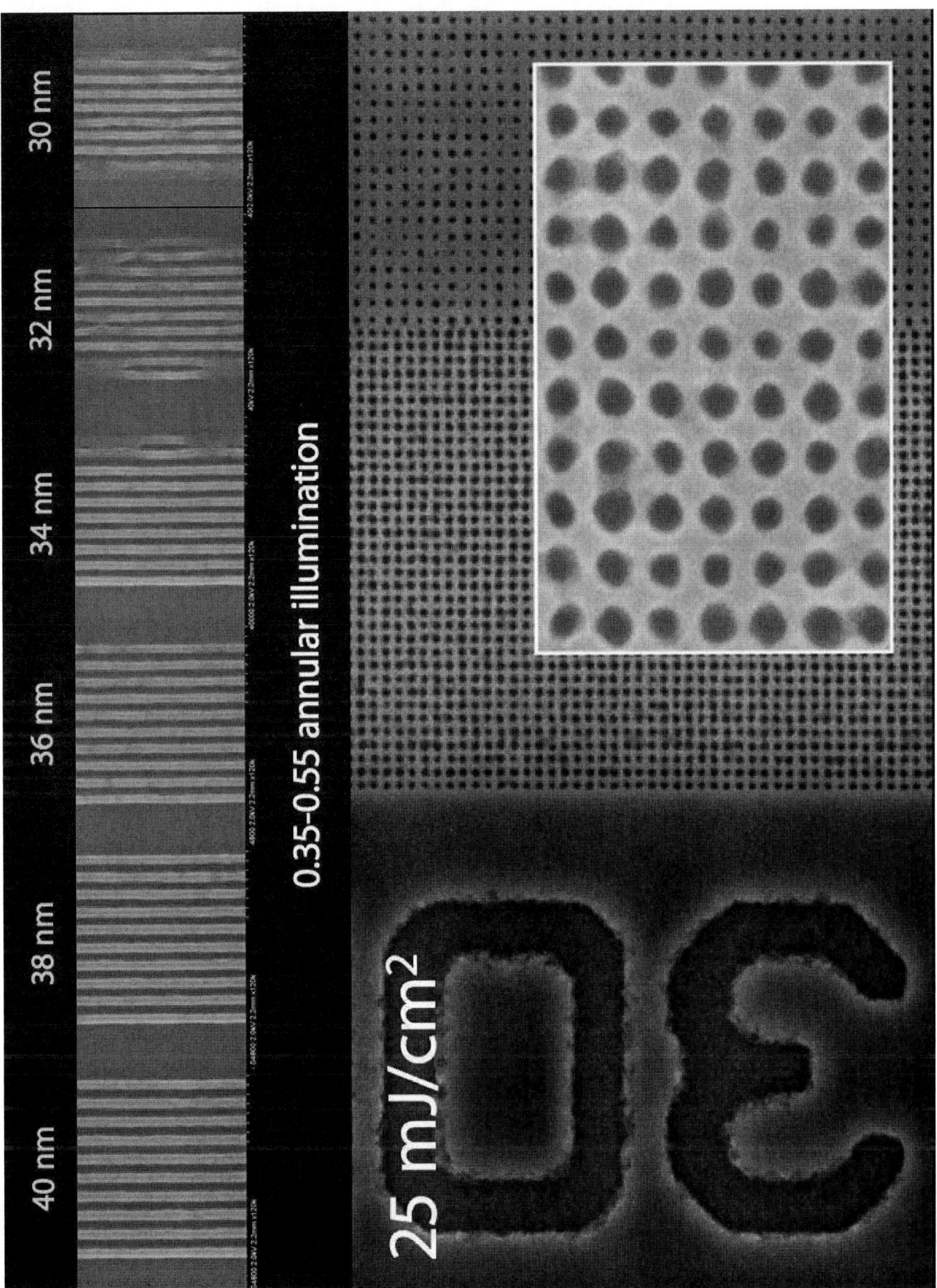

What about corners?

cnanderson@berkeley.edu

EUVL Symposium, Lake Tahoe, California, 2008.10.01

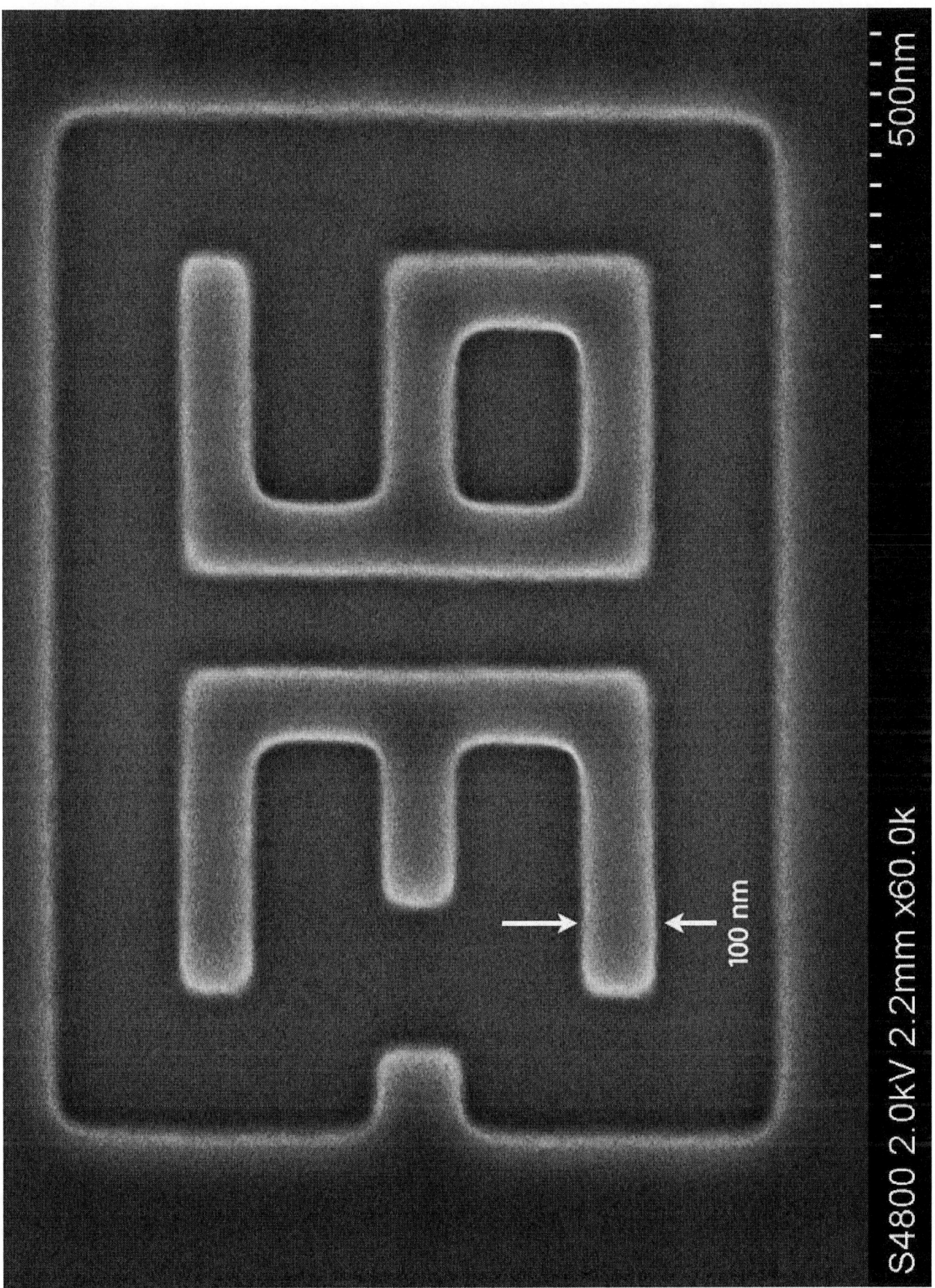

S4800 2.0kV 2.2mm x60.0k   100 nm   500nm

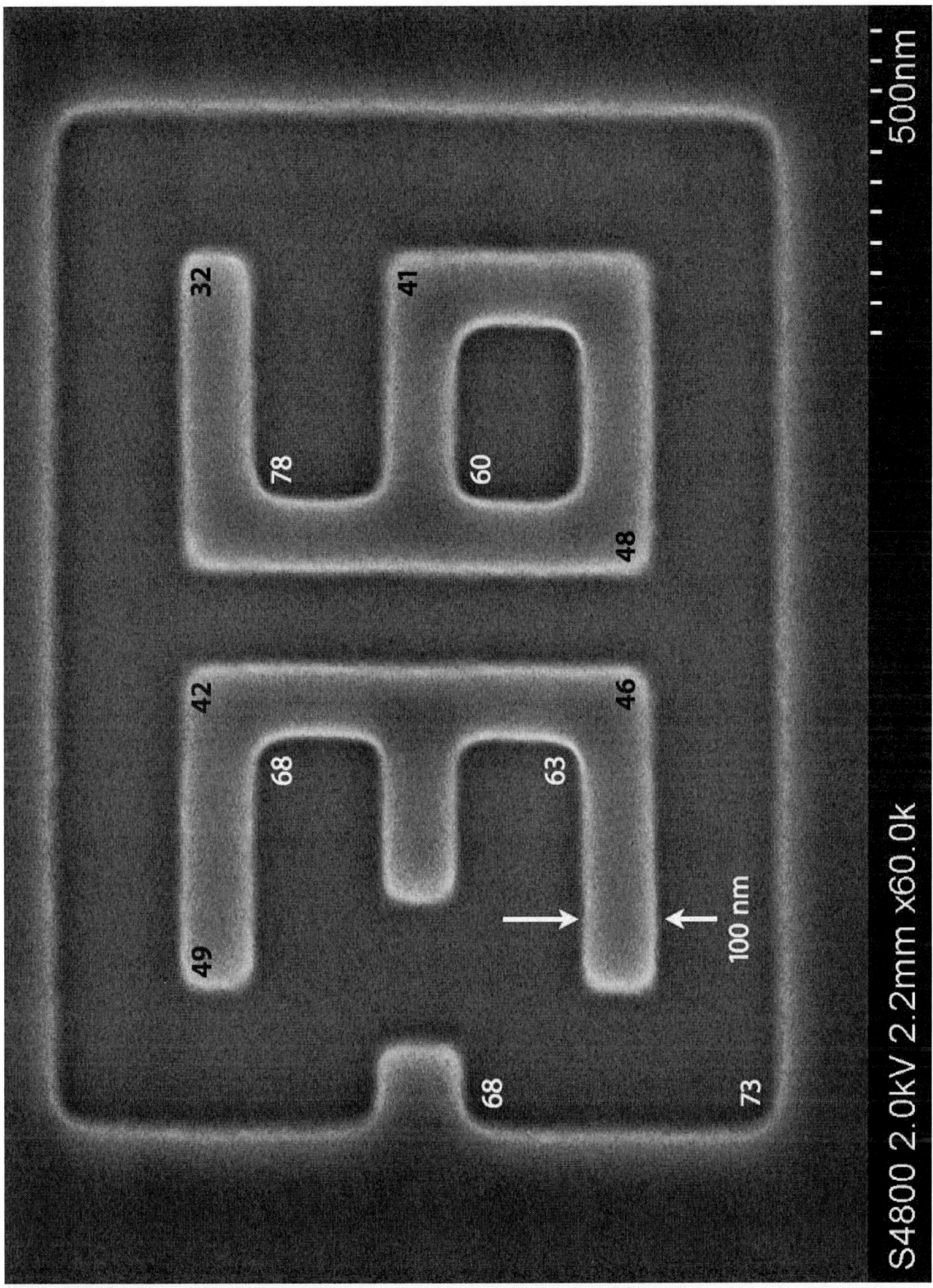

Learning

90 degrees of resist

270 degrees of resist

- What is causing the corner bias?
- What can we learn from it?
- Can we learn to mitigate it?

Q1: Does corner bias show up in aerial image?

cnanderson@berkeley.edu

EUVL Symposium, Lake Tahoe, California, 2008.10.01

Thin mask modeling.

EUVL Symposium, Lake Tahoe, California, 2008.10.01

cnanderson@berkeley.edu

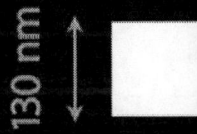

130 nm

cnanderson@berkeley.edu

EUVL Symposium, Lake Tahoe, California, 2008.10.01

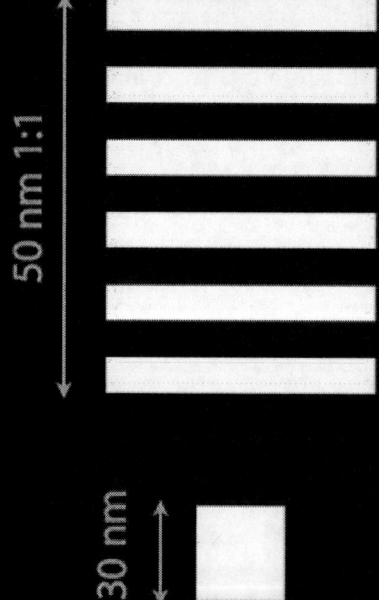

50 nm 1:1

130 nm

cnanderson@berkeley.edu

EUVL Symposium, Lake Tahoe, California, 2008.10.01

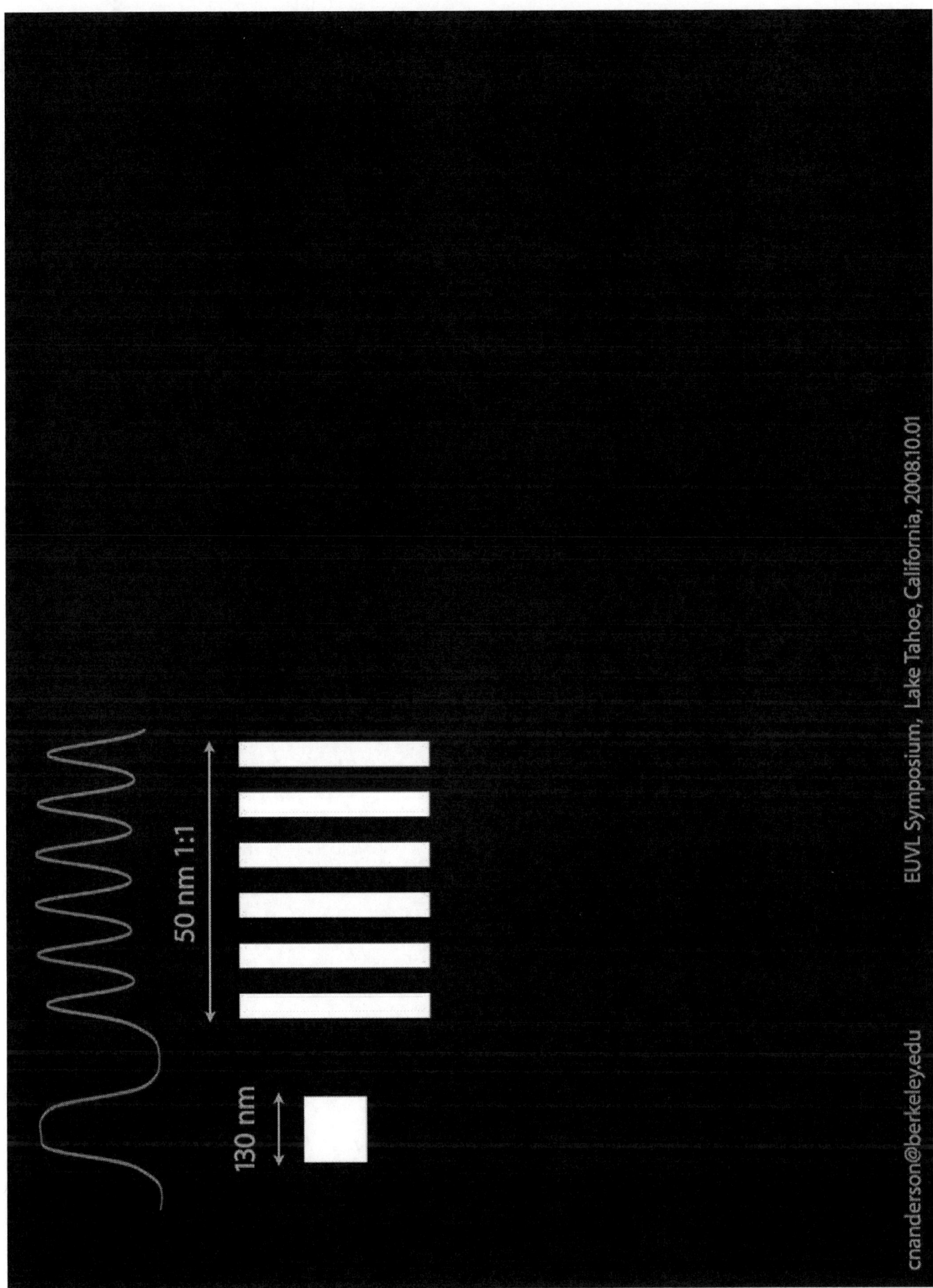

cnanderson@berkeley.edu

EUVL Symposium, Lake Tahoe, California, 2008.10.01

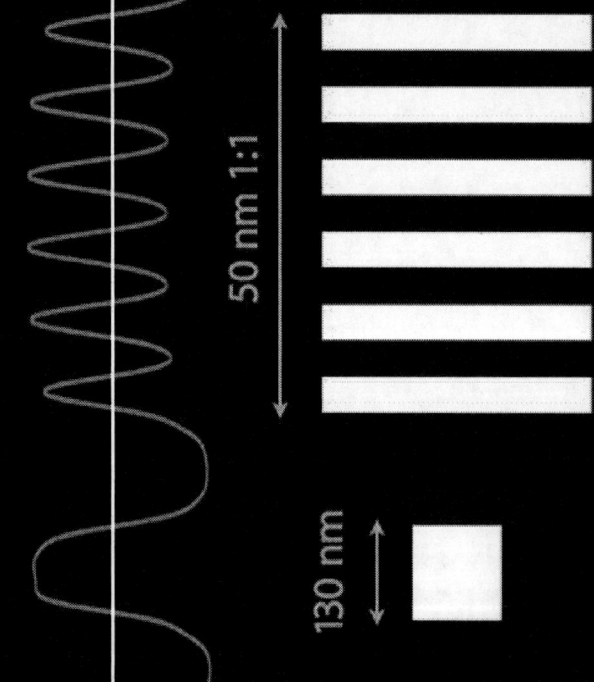

50 nm 1:1

130 nm

cnanderson@berkeley.edu

EUVL Symposium, Lake Tahoe, California, 2008.10.01

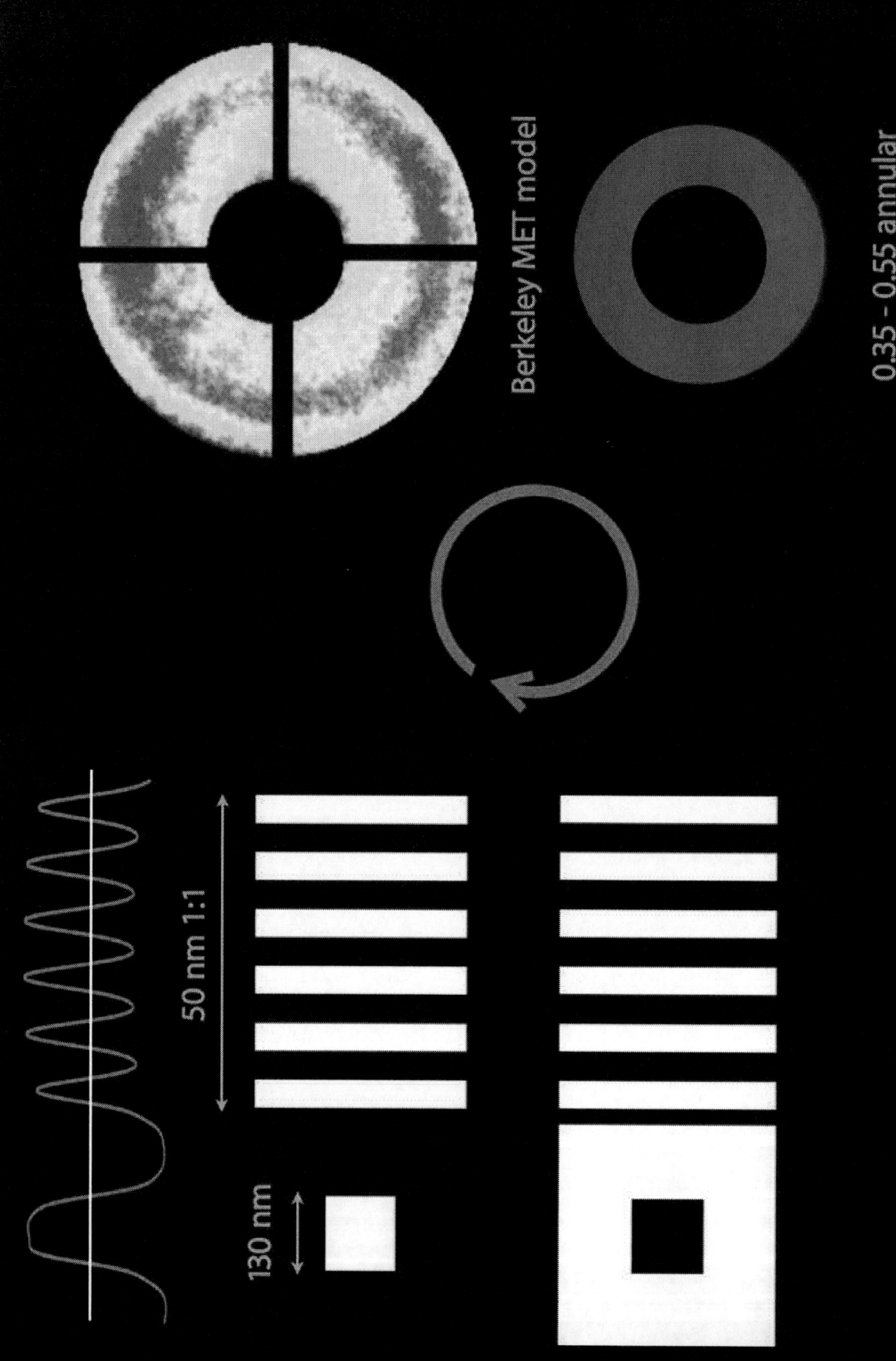

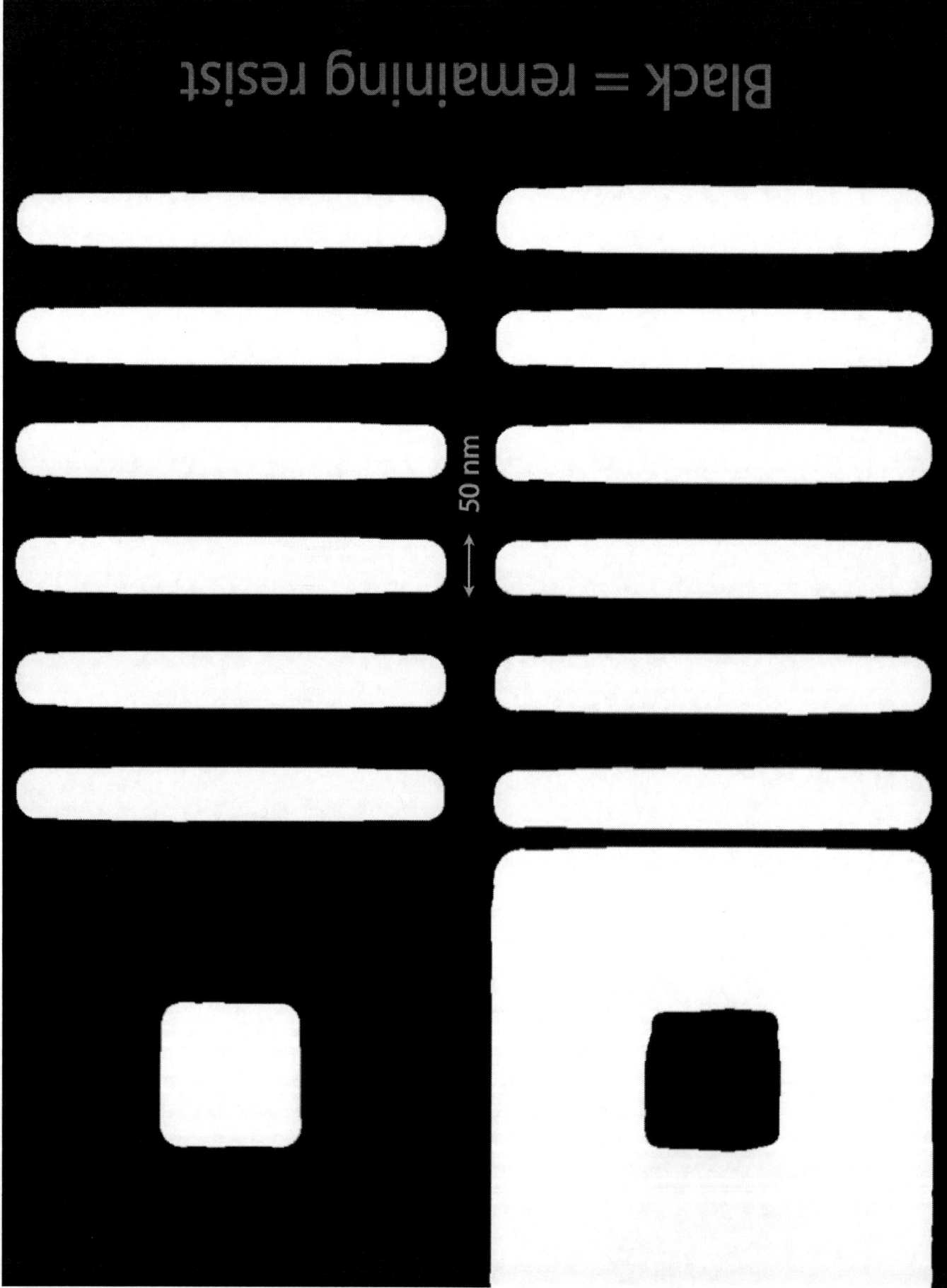

130 nm

cnanderson@berkeley.edu

EUVL Symposium, Lake Tahoe, California, 2008.10.01

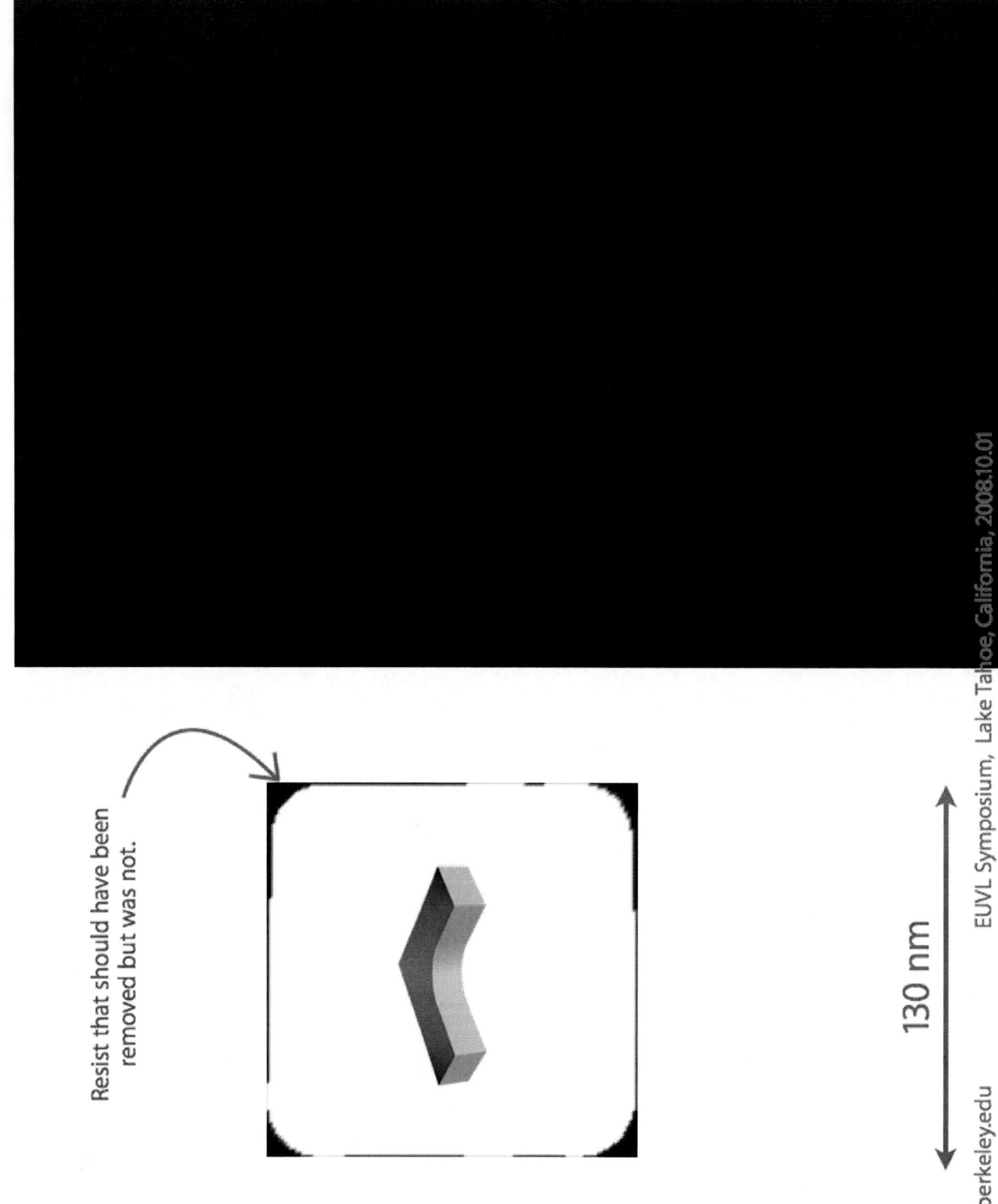

Resist that should have been removed but was not.

130 nm

cnanderson@berkeley.edu

EUVL Symposium, Lake Tahoe, California, 2008.10.01

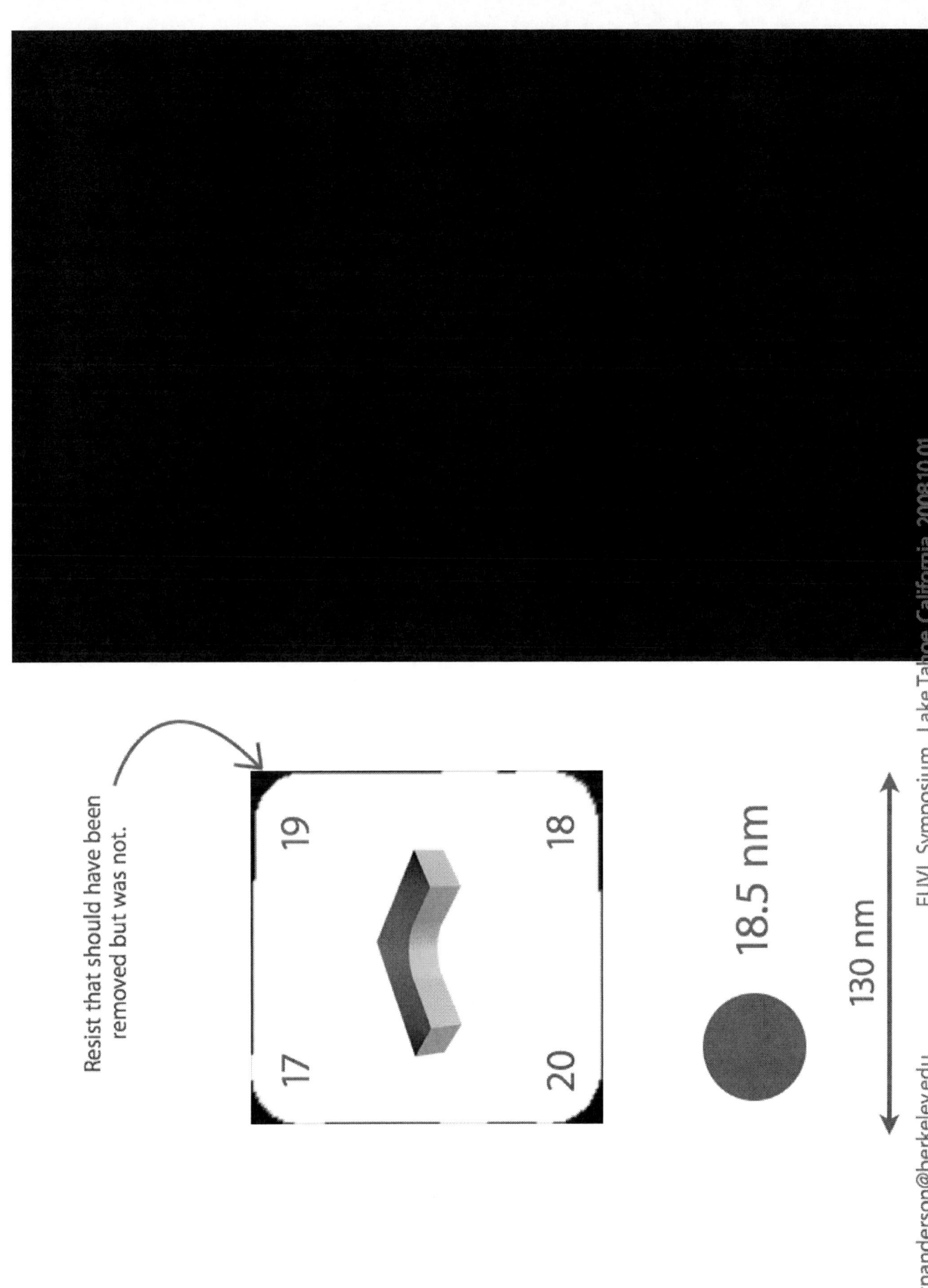

Resist that should have been removed but was not.

18.5 nm

130 nm

cnanderson@berkeley.edu   EUVL Symposium, Lake Tahoe, California, 2008.10.01

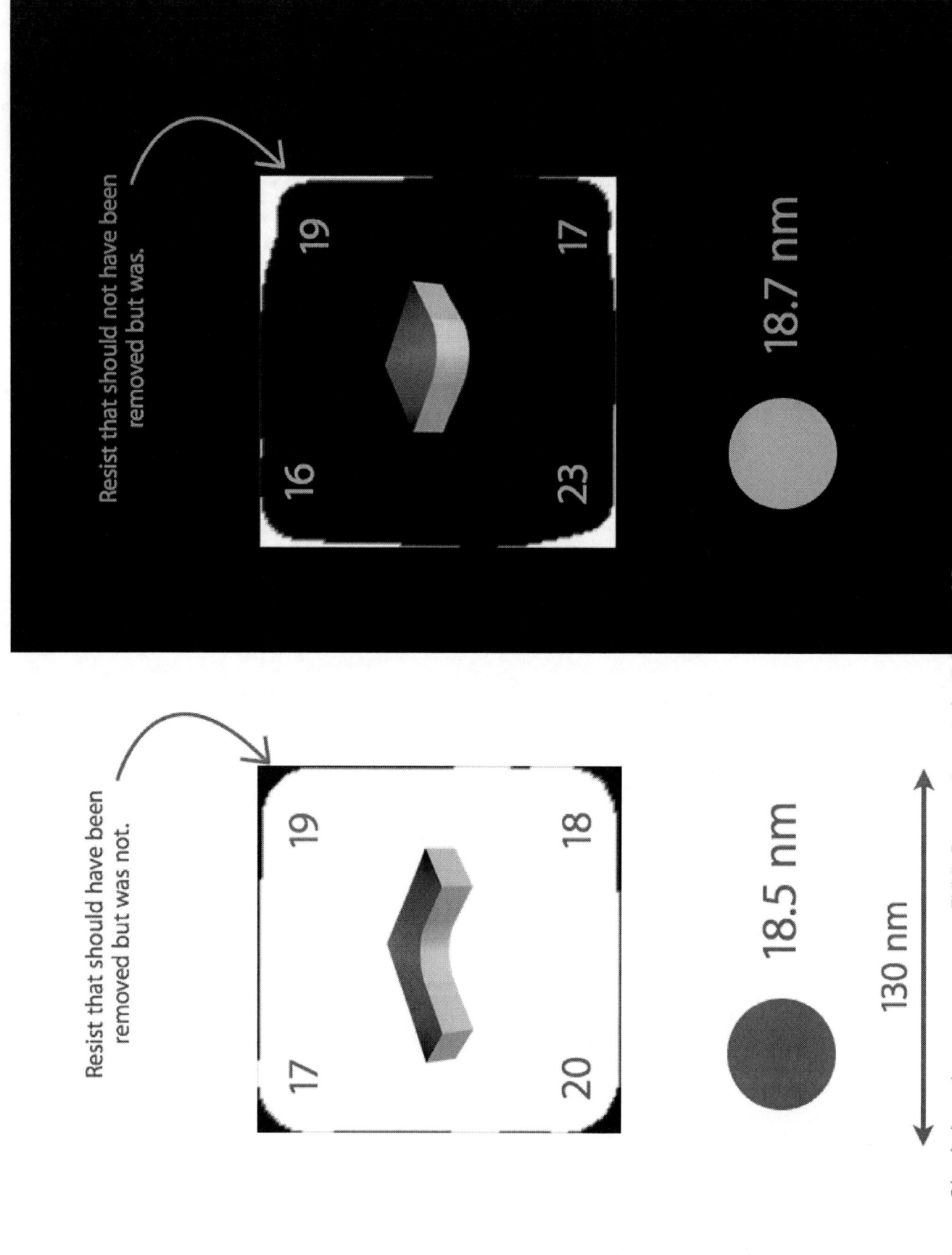

Thick mask modeling.

Yunfei Deng and Tom Wallow

- Cr 70 nm / SiO$_2$ 20 nm
- 40 bilayer of Mo/Si multilayer
- 4° angle of incidence
- Berkeley MET model

cnanderson@berkeley.edu

EUVL Symposium, Lake Tahoe, California, 2008.10.01

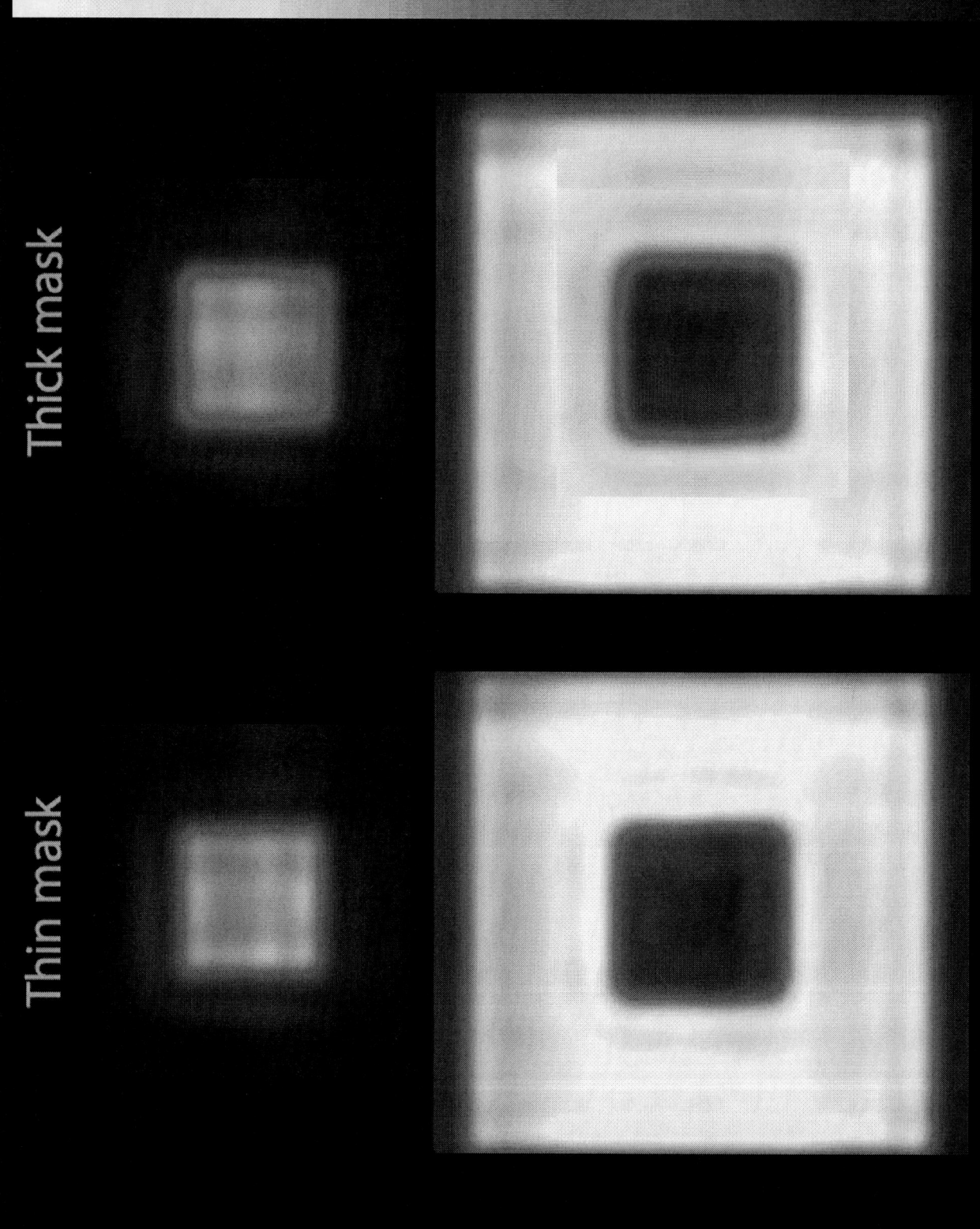

Q2: Do resist models predict corner bias?

cnanderson@berkeley.edu

EUVL Symposium, Lake Tahoe, California, 2008.10.01

PROLITH modeling
RHEM EUV2D

cnanderson@berkeley.edu

EUVL Symposium, Lake Tahoe, California, 2008.10.01

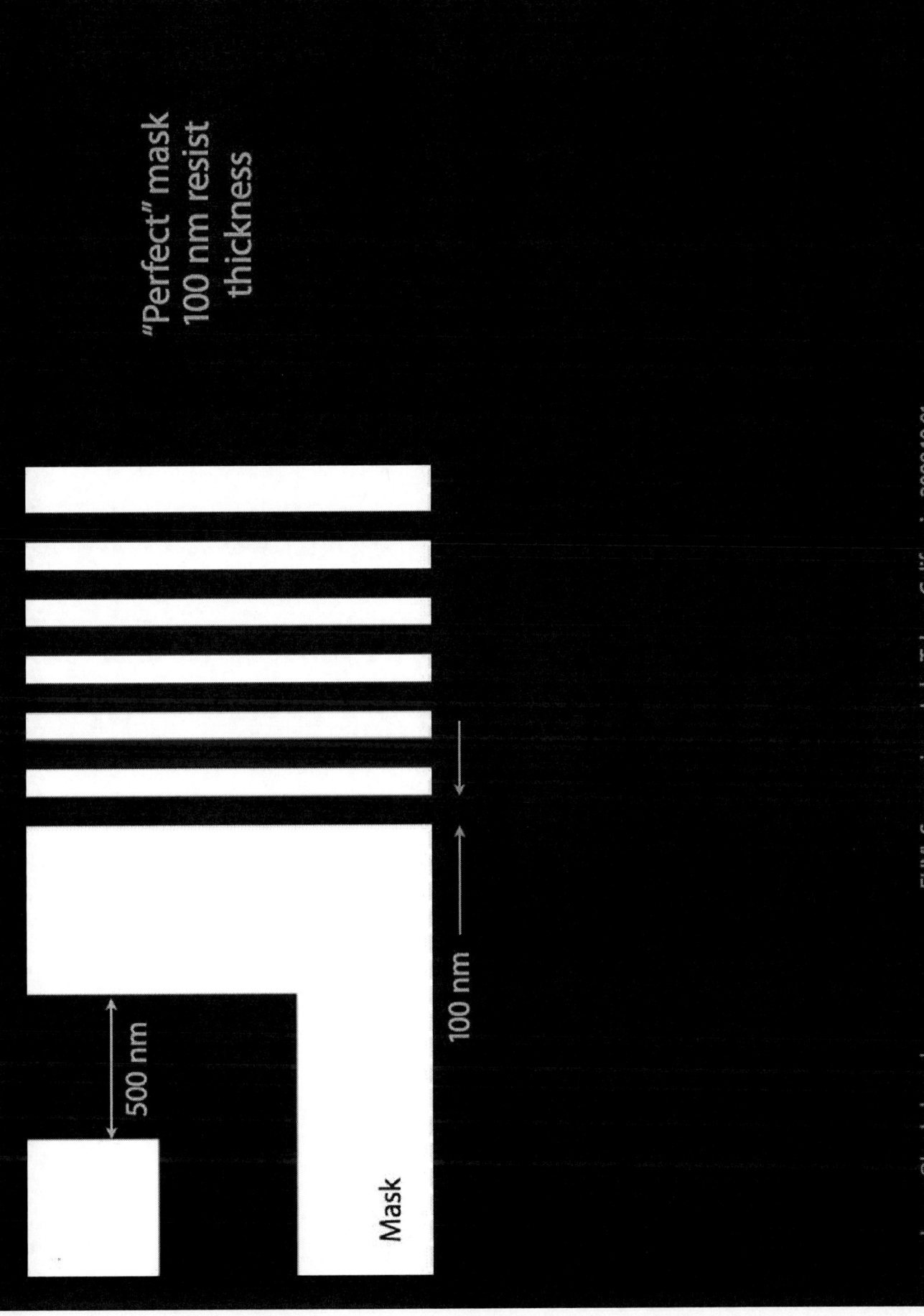

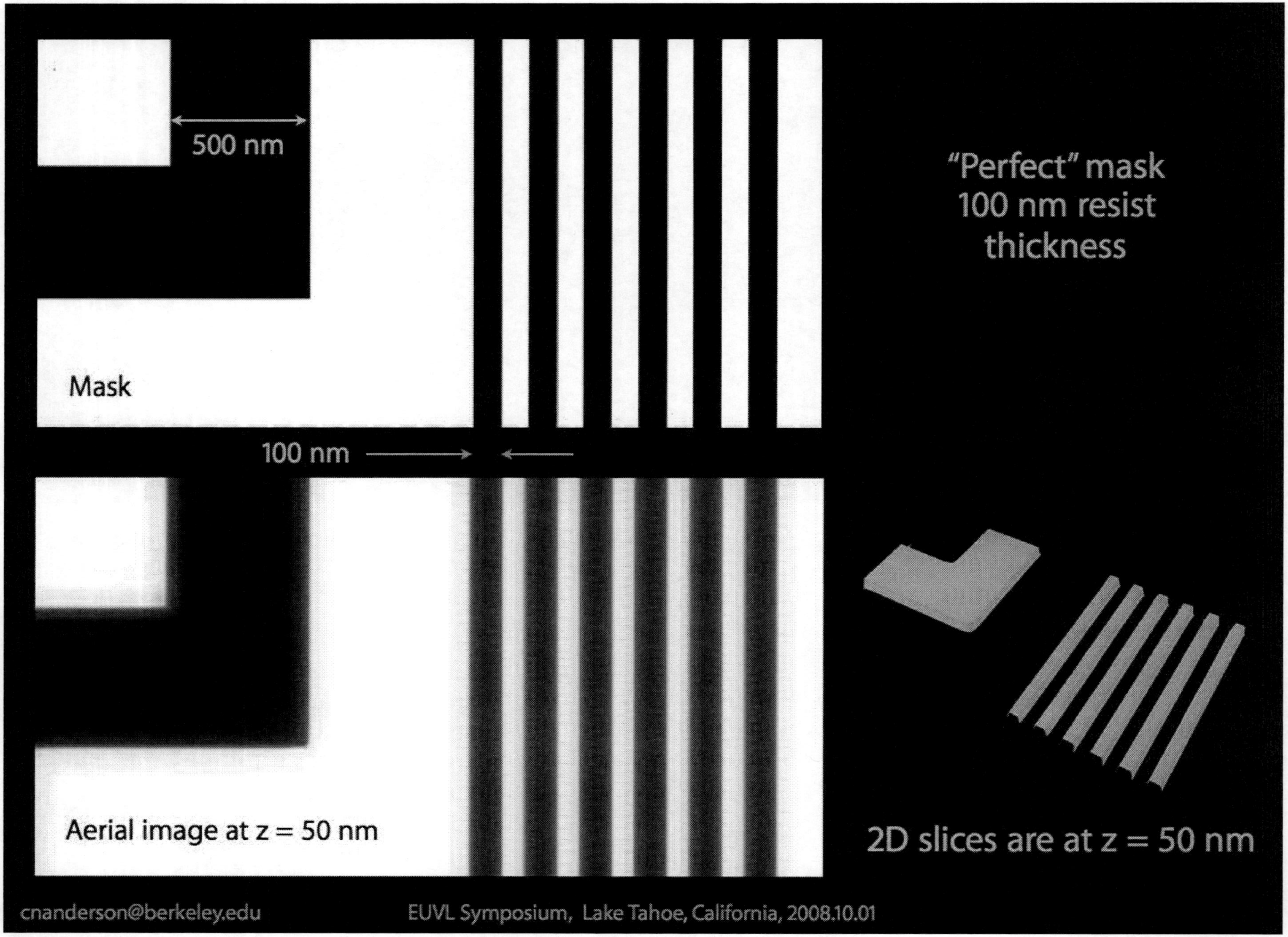

"Perfect" mask
100 nm resist thickness

2D slices are at z = 50 nm

500 nm

Mask

100 nm

Aerial image at z = 50 nm

cnanderson@berkeley.edu

EUVL Symposium, Lake Tahoe, California, 2008.10.01

Before PEB

EUVL Symposium, Lake Tahoe, California, 2008.10.01

cnanderson@berkeley.edu

Before PEB

21   20

cnanderson@berkeley.edu   EUVL Symposium, Lake Tahoe, California, 2008.10.01

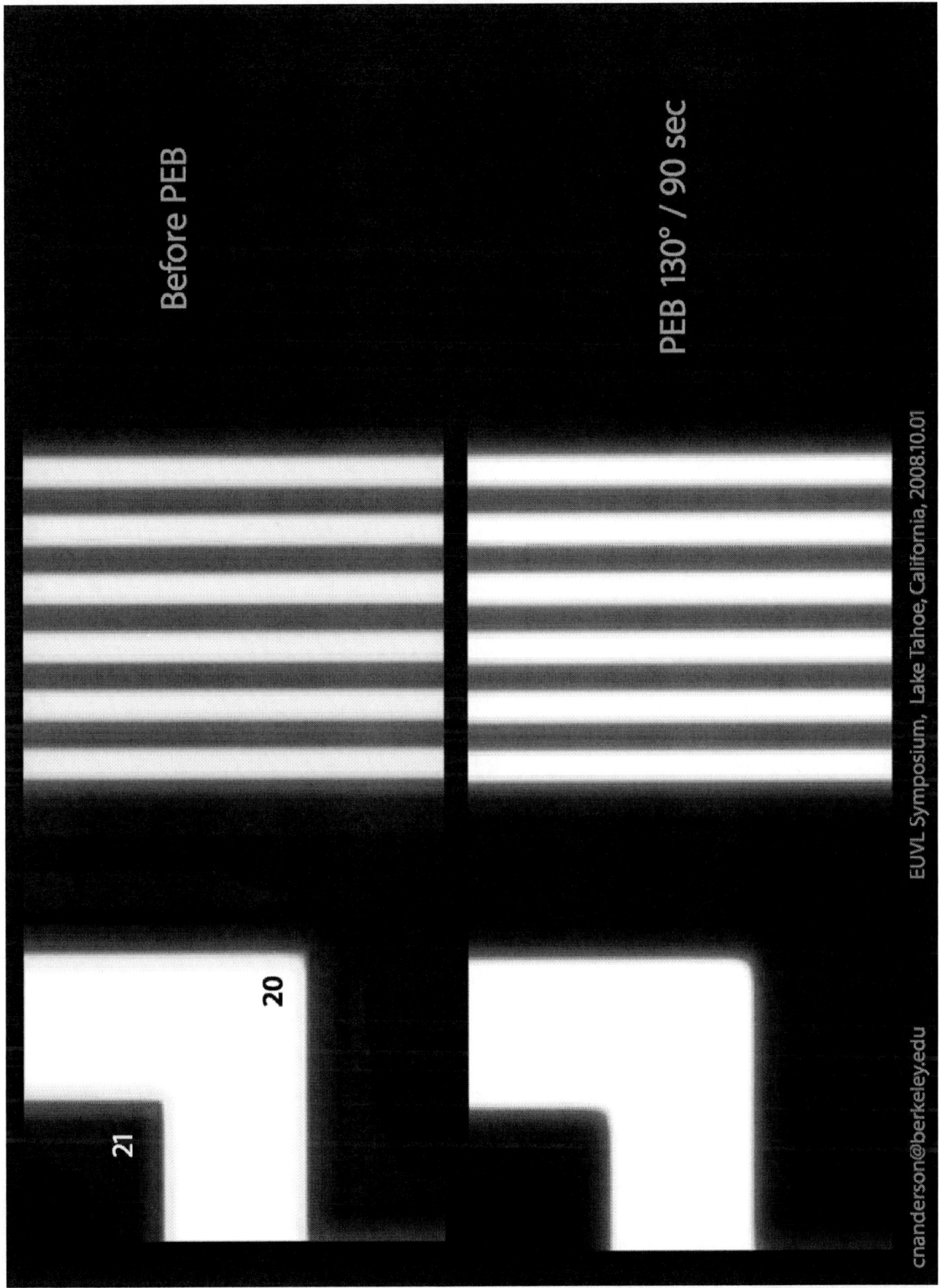

Before PEB

PEB 130° / 90 sec

20

21

cnanderson@berkeley.edu

EUVL Symposium, Lake Tahoe, California, 2008.10.01

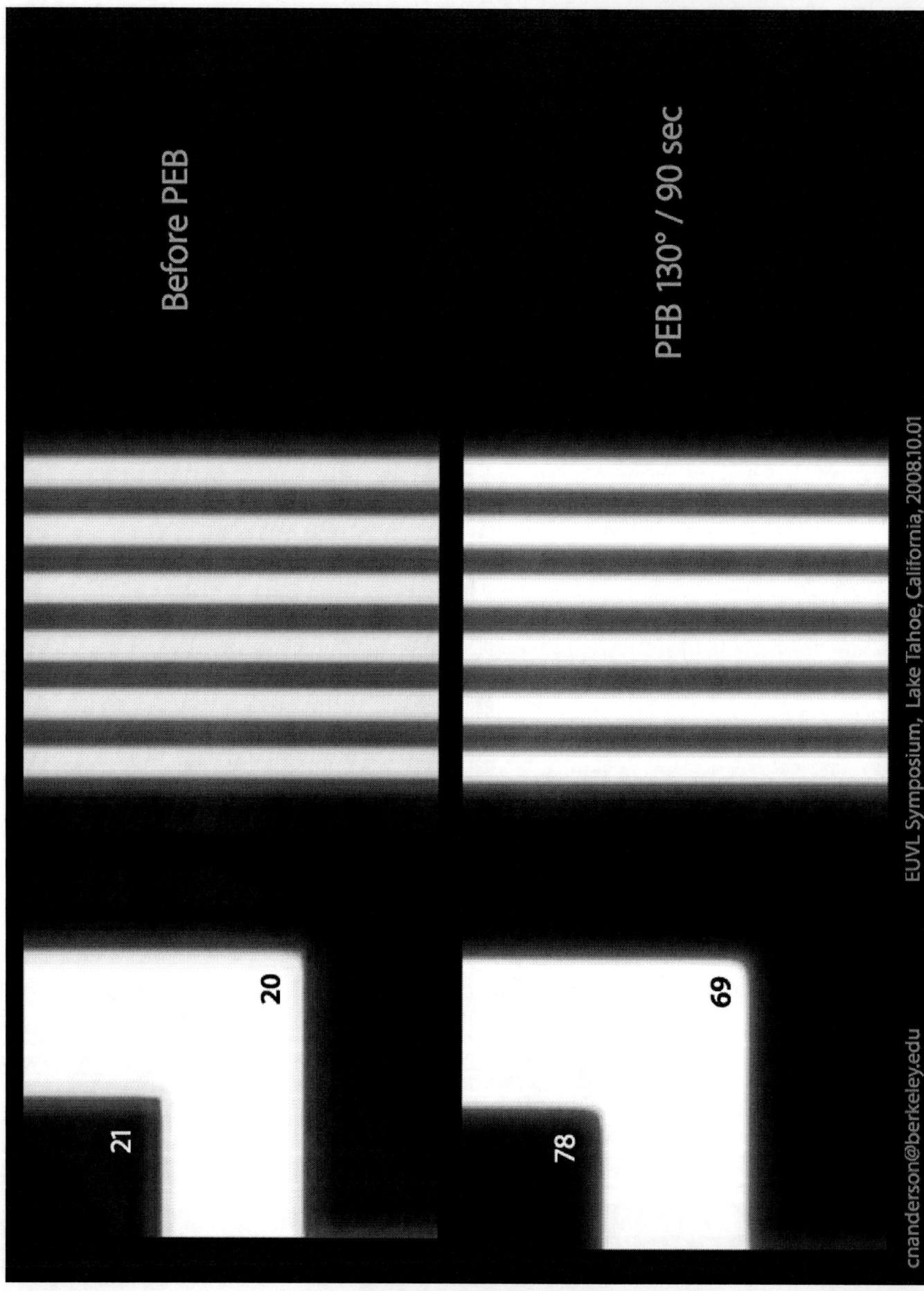

Develop 45 sec
white = resist

500 nm

cnanderson@berkeley.edu

EUVL Symposium, Lake Tahoe, California, 2008.10.01

Develop 45 sec
white = resist

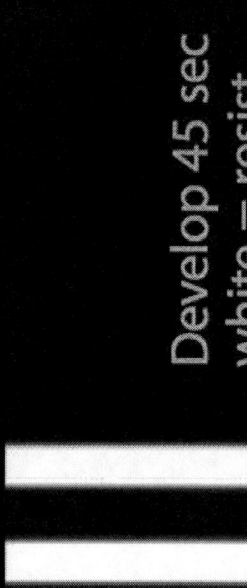

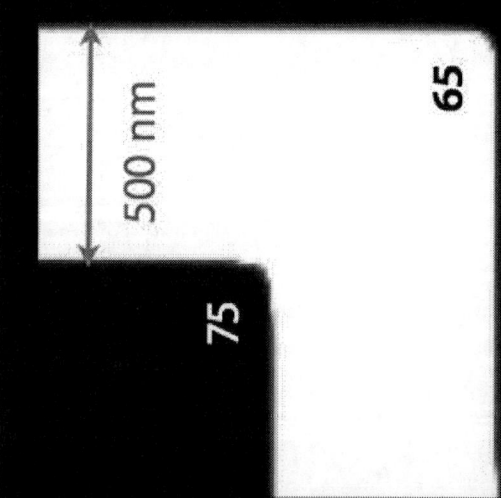

500 nm

75

65

cnanderson@berkeley.edu

EUVL Symposium, Lake Tahoe, California, 2008.10.01

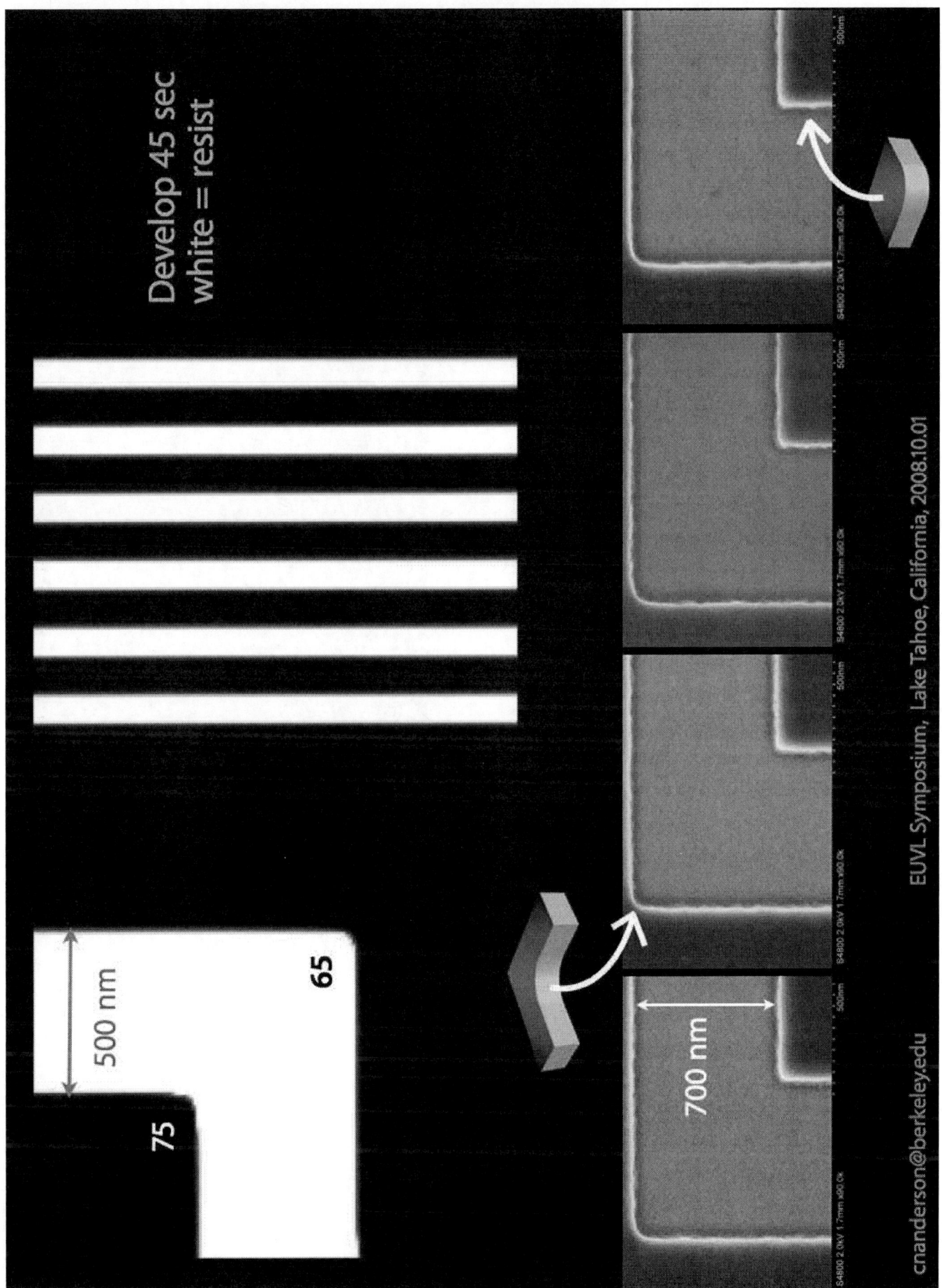

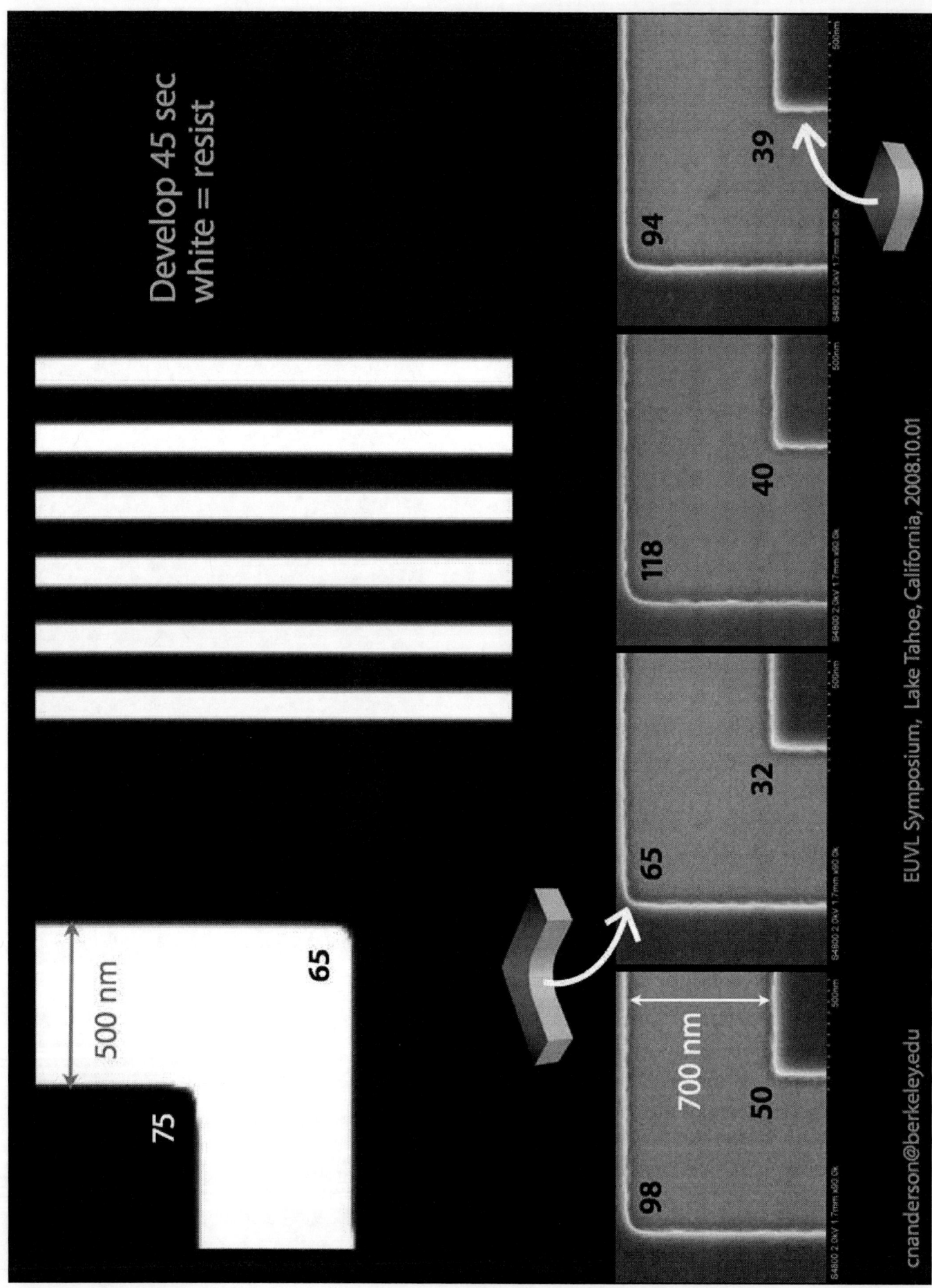

Other models that have not shown bias.

Single blur (HOST) model (C. Ahn et. al.)
Dual blur (HOST) model (Tanaka et. al.)

Aerial image

*

Final acid image

cnanderson@berkeley.edu   EUVL Symposium, Lake Tahoe, California, 2008.10.01

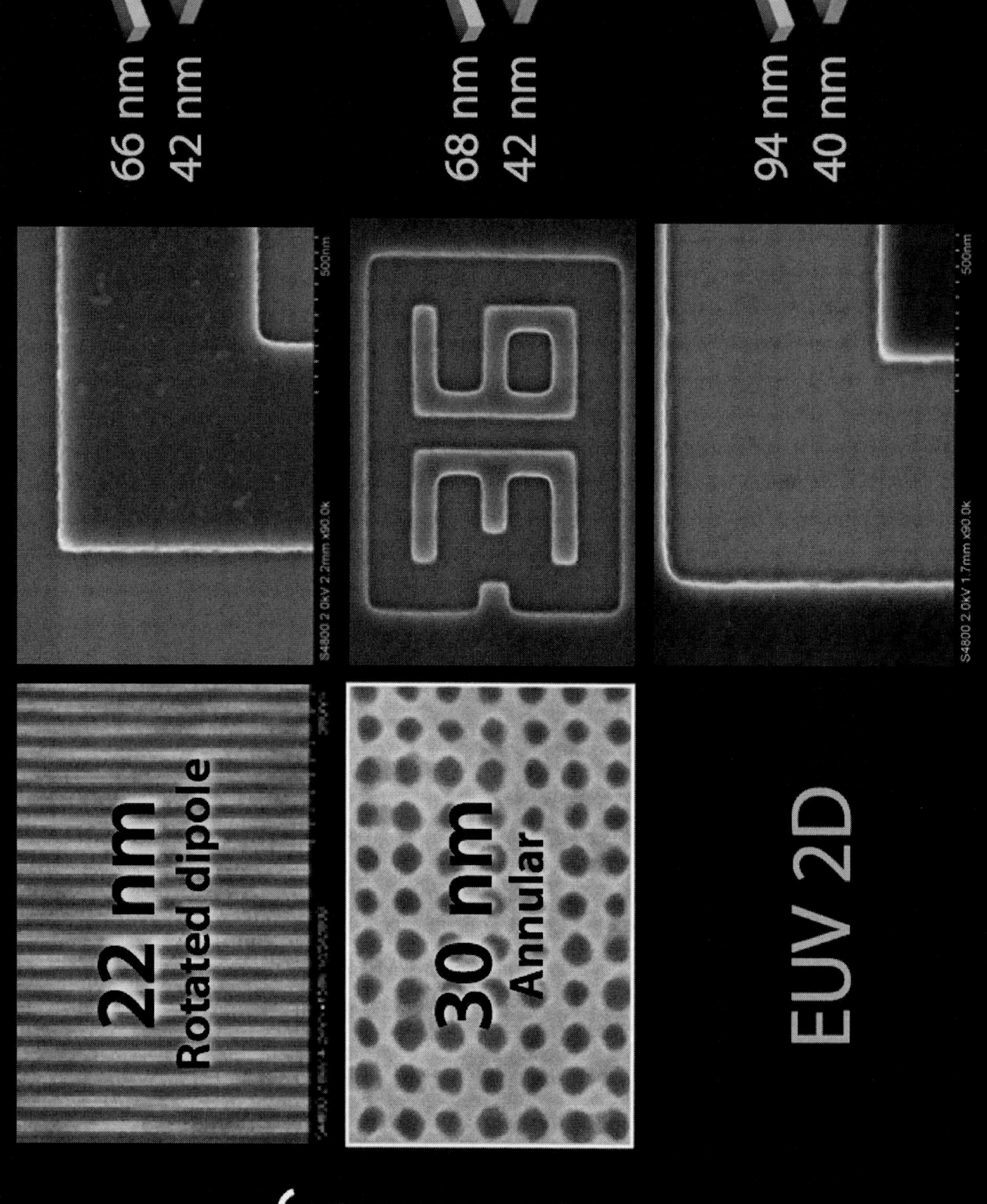

What is causing the bias?

- Aerial image? No.
- Resist blur/diffusion? Possible. But... current resist models don't capture it.

Things to think about.

- New EUV resist models.
- Resist material/structural properties.
- OPC strategies.

Closing remarks

Acknowledgements

CXRO staff
Paul Denham
Brian Hoeff
Gideon Jones
Jerrin Chu
Ken Goldberg

Others
Kim Dean

Advanced Materials Research Center, AMRC, International SEMATECH Manufacturing Initiative, and ISMI are servicemarks of SEMATECH, Inc. SEMATECH, the SEMATECH logo are registered servicemarks of SEMATECH, Inc. All other servicemarks and trademarks are the property of their respective owners.

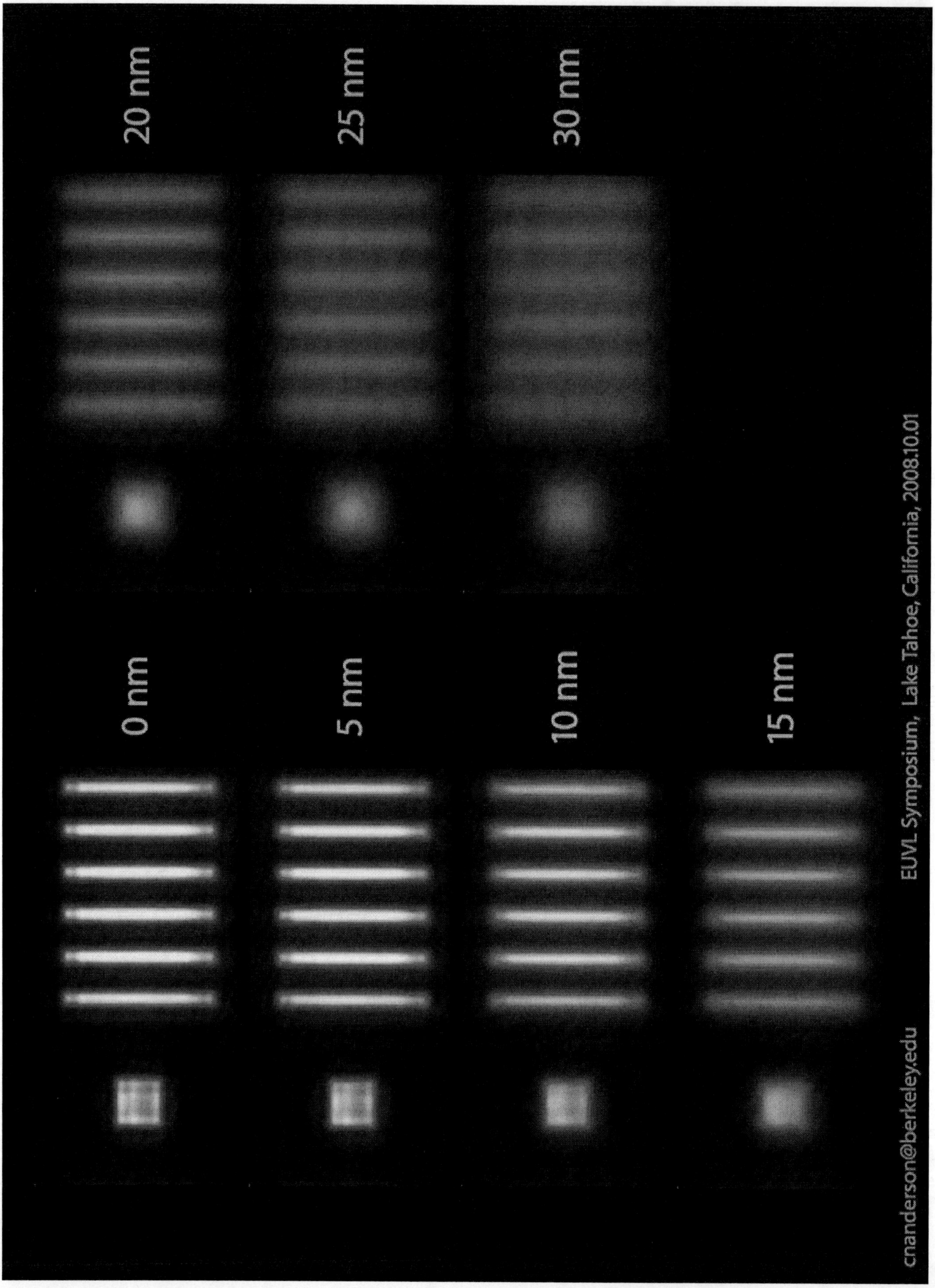

20 nm

25 nm

30 nm

0 nm

5 nm

10 nm

15 nm

cnanderson@berkeley.edu

EUVL Symposium, Lake Tahoe, California, 2008.10.01

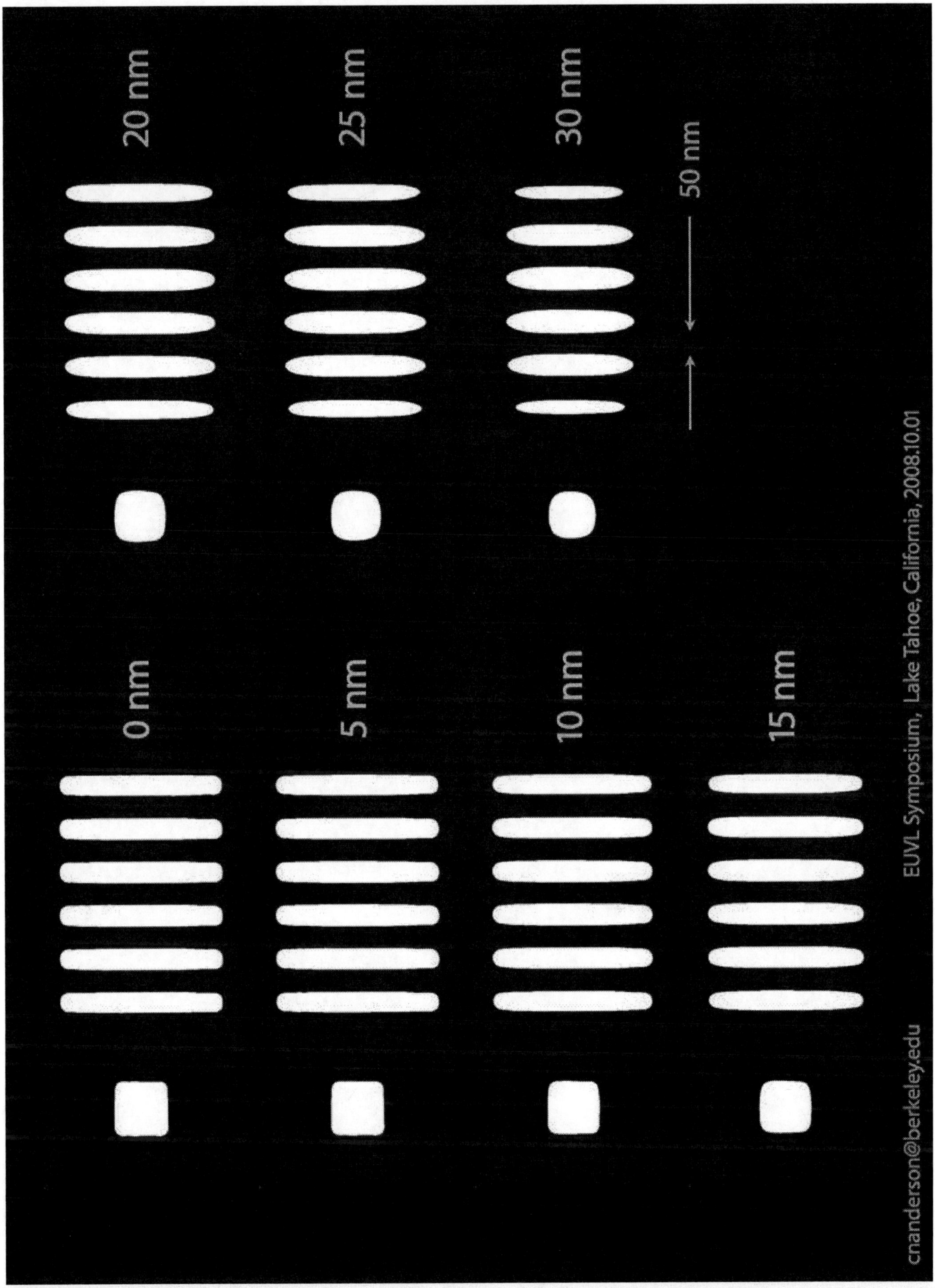

20 nm

25 nm

30 nm

50 nm

0 nm

5 nm

10 nm

15 nm

EUVL Symposium, Lake Tahoe, California, 2008.10.01

cnanderson@berkeley.edu

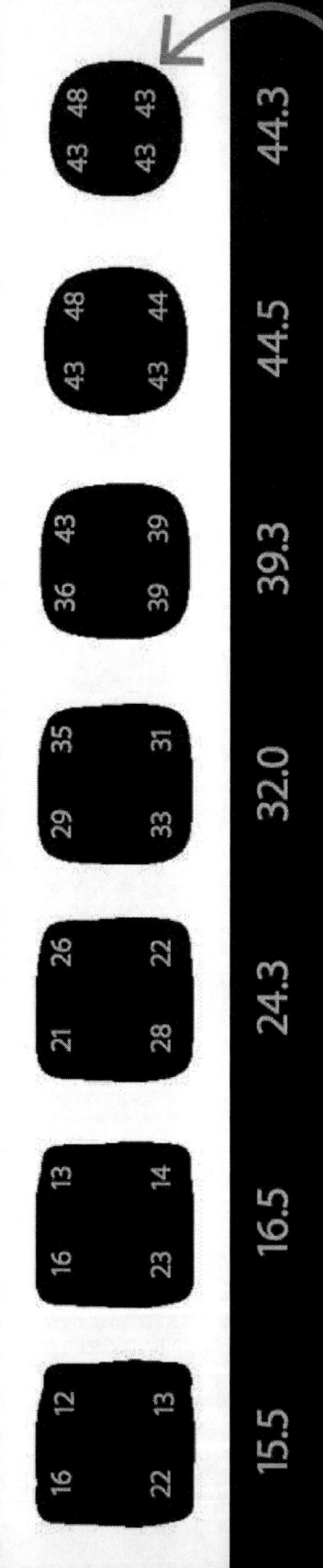

cnanderson@berkeley.edu

EUVL Symposium, Lake Tahoe, California, 2008.10.01

Reconciling Resist Resolution Metrics

Gregg M. Gallatin, Applied Math Solutions, LLC
ggallatin@charter.net

Patrick P. Naulleau, Lawrence Berkeley Lab
pnaulleau@lbl.gov

Christopher N. Anderson, Lawrence Berkeley Lab
cnanderson@lbl.gov

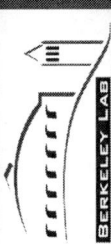

Applied Math Solutions, LLC

Applied Math Solutions, LLC

Outline

- RLS Tradeoff

- Deprotection distributions in 1D, 2D, and 3D

  - Two parameters

- Resist Resolution Metrics

  - LER based: "Blur" fit to PSD
  - CD based: MTF, Corner Rounding, Contact Hole, ...

- Data Comparison of PSD "Blur" and Contact Hole Metric

- Model Comparison of PSD and Contact Hole Metric

- Conclusion

Applied Math Solutions, LLC

"RLS" Tradeoff

Line Edge Roughness →   LER

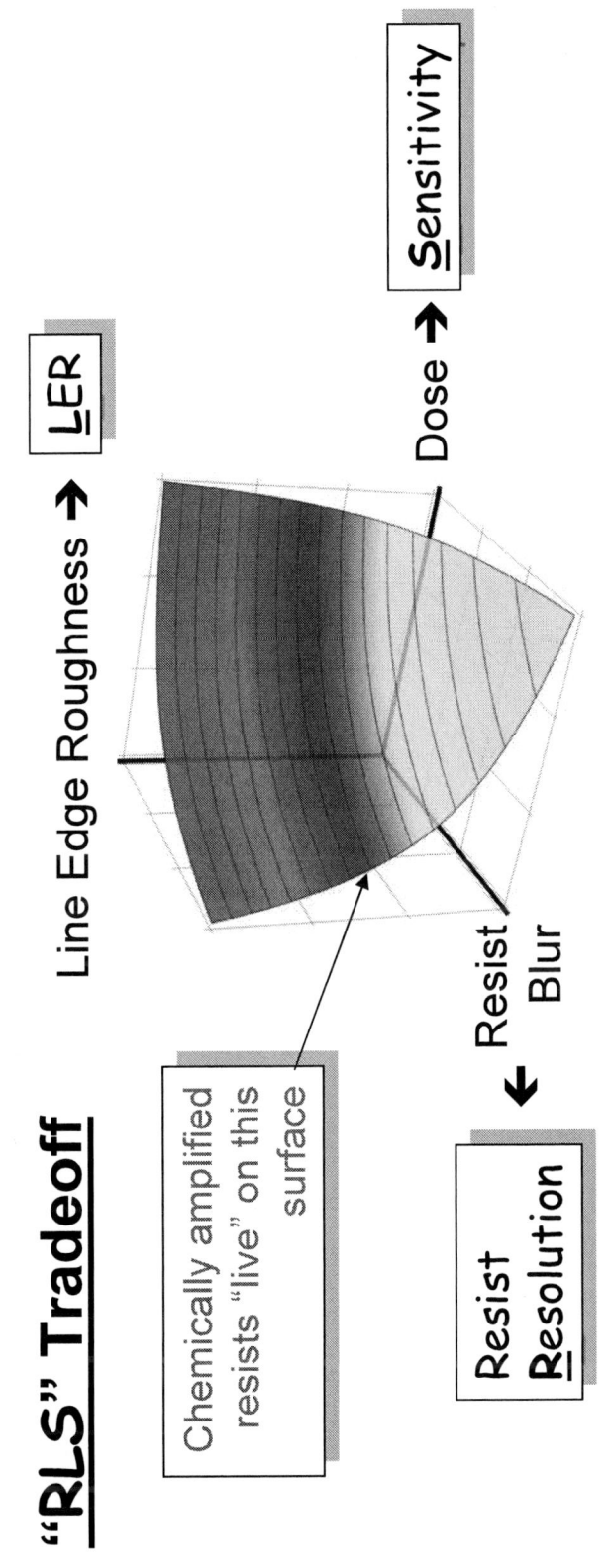

Chemically amplified resists "live" on this surface

Resist
Resolution

Resist
Blur

Dose →   **S**ensitivity

Standard chemically amplified resists → Very difficult to get Resist Blur, LER and Dose all arbitrarily small at the same time.

Also called other things....
  "Lithographic Uncertainty Principle"
  "Triangle of Death"

Lammers, et al., SPIE 2007, Bristol, et al., SPIE 2007, Brainard, et al., SPIE 2004, Gallatin SPIE 2005, and many others...

Applied Math Solutions, LLC

PEB Reaction-Diffusion Equations

$$\rho_D(\vec{x}, t) = 1 - \exp\left[-k \int_0^t dt\, e^{Dt\vec{\nabla}^2} \delta(\vec{x})\right]$$

Deprotection
Density

Deprotection
Rate $\dfrac{nm^d}{\sec}$
Units =

$d = \#$ Dimensions

Diffusion Range
= Resist "Blur"

$$\boxed{R = \sqrt{Dt}}$$

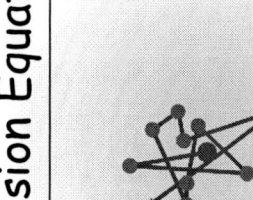

Acid
Diffusion

$\Gamma(a,b) =$ Incomplete Gamma Function

3D: $\rho_D(\vec{r}, t) = 1 - \exp\left[-\dfrac{kt}{4\pi R^2 r}\left(1 - erf\left(\dfrac{r}{2R}\right)\right)\right]$

2D: $\rho_D(\vec{r}, t) = 1 - \exp\left[-\dfrac{kt}{4\pi R^2}\Gamma\left(0, \dfrac{r}{2R}\right)\right]$

1D: $\rho_D(\vec{r}, t) = 1 - \exp\left[-kt\left(\dfrac{1}{\sqrt{\pi}R}e^{-r^2/2R^2} - \dfrac{r}{2R^2}\left(1 - erf\left(\dfrac{r}{2R}\right)\right)\right)\right]$

Applied Math Solutions, LLC

Analytic form of the deprotection "blur"

- Matches full numerical simulation Hinsberg, et. al, SPIE 03

- And experimental shape Hoffnagle, Opt. Letts. 02

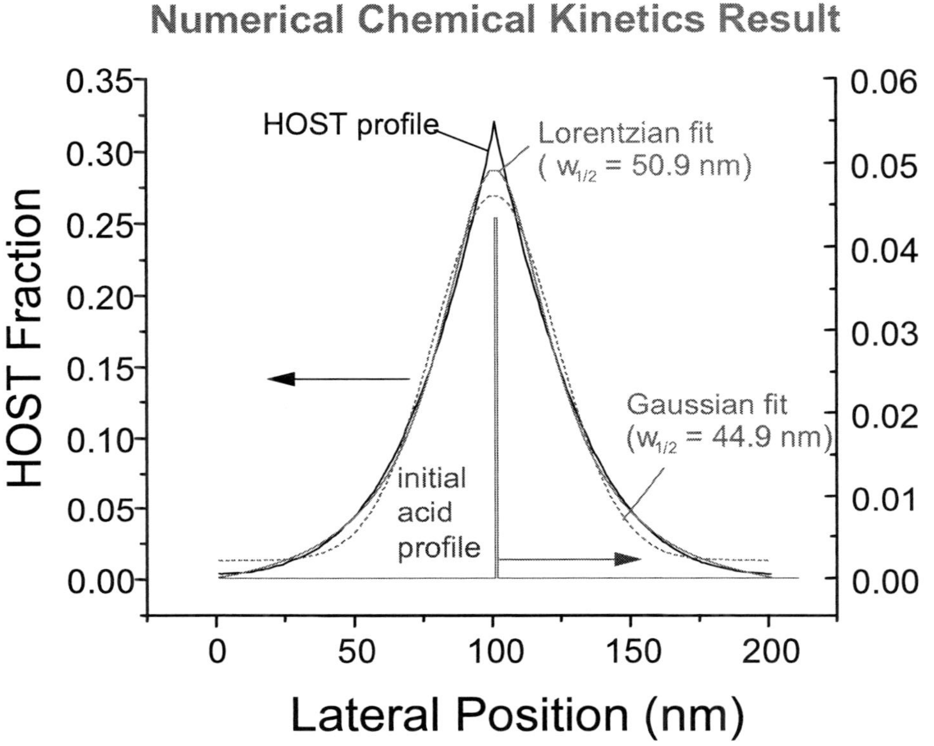

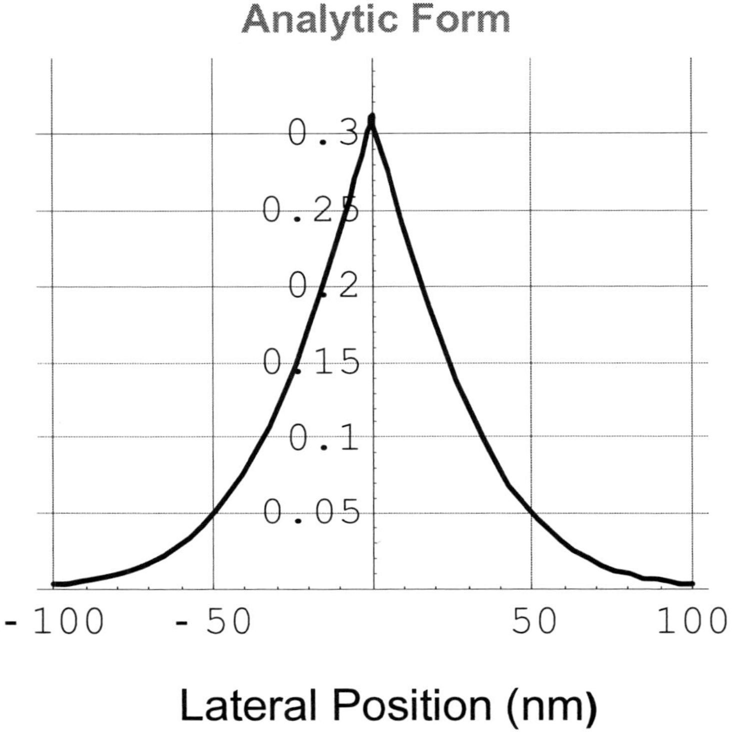

Analytic Form

Numerical Chemical Kinetics Result

Applied Math Solutions, LLC

Graphs of Deprotection Density

For all 3 Graphs

$R = 15nm$

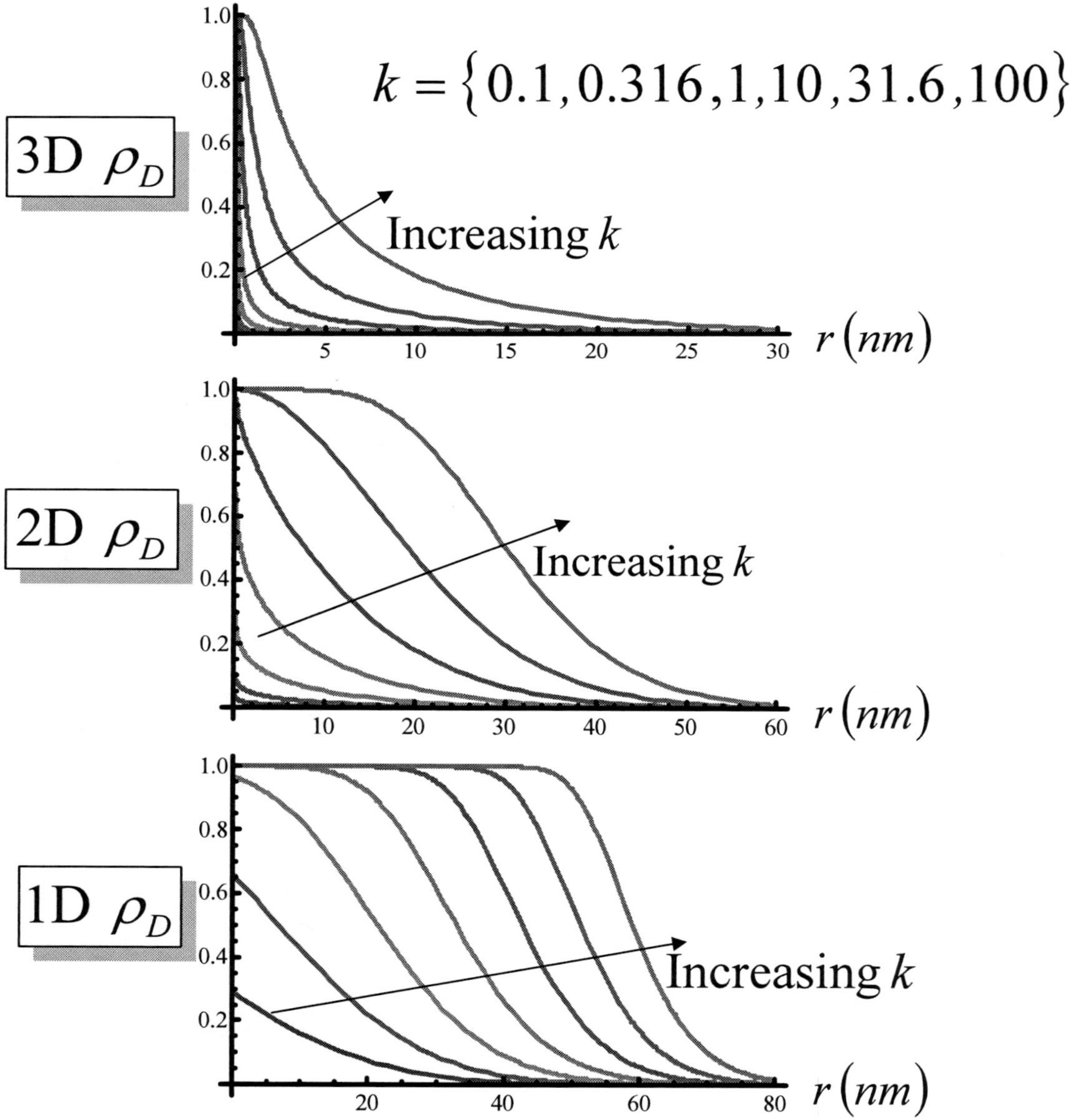

$k = \{0.1, 0.316, 1, 10, 31.6, 100\}$

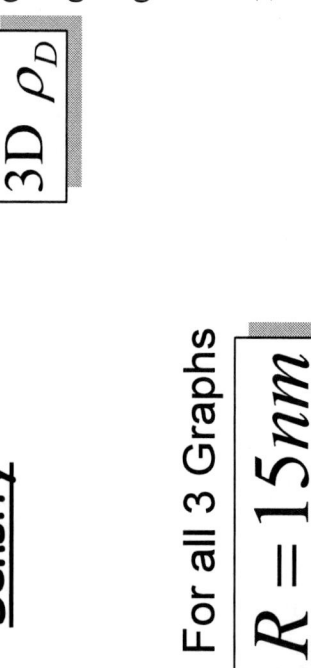

3D ρ_D — Increasing k — $r\,(nm)$

2D ρ_D — Increasing k — $r\,(nm)$

1D ρ_D — Increasing k — $r\,(nm)$

Kang, et. al., SPIE 6519 (2007)

Applied Math Solutions, LLC

Graphs of Deprotection Density

For all 3 Graphs

$$R = 15nm$$

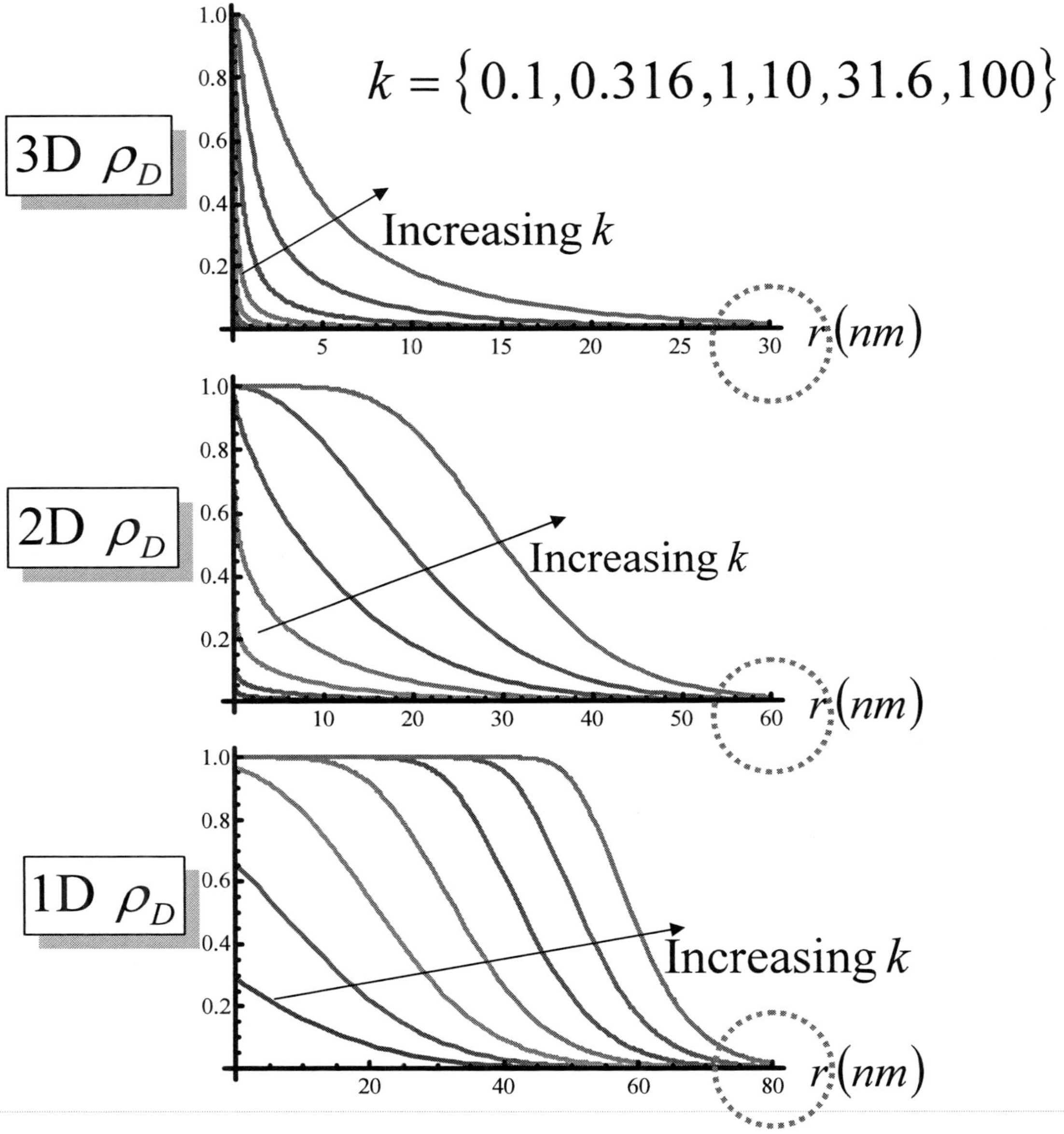

$$k = \{0.1, 0.316, 1, 10, 31.6, 100\}$$

3D ρ_D

2D ρ_D

1D ρ_D

Increasing k

Increasing k

Increasing k

$r(nm)$

$r(nm)$

$r(nm)$

Kang, et. al., SPIE 6519 (2007)

Applied Math Solutions, LLC

Graphs of Deprotection Density

$$k = \{0.1, 0.316, 1, 10, 31.6, 100\}$$

3D ρ_D — Increasing k — $r(nm)$

2D ρ_D — Increasing k — $r(nm)$

1D ρ_D — Increasing k — $r(nm)$

For all 3 Graphs

$$R = 15nm$$

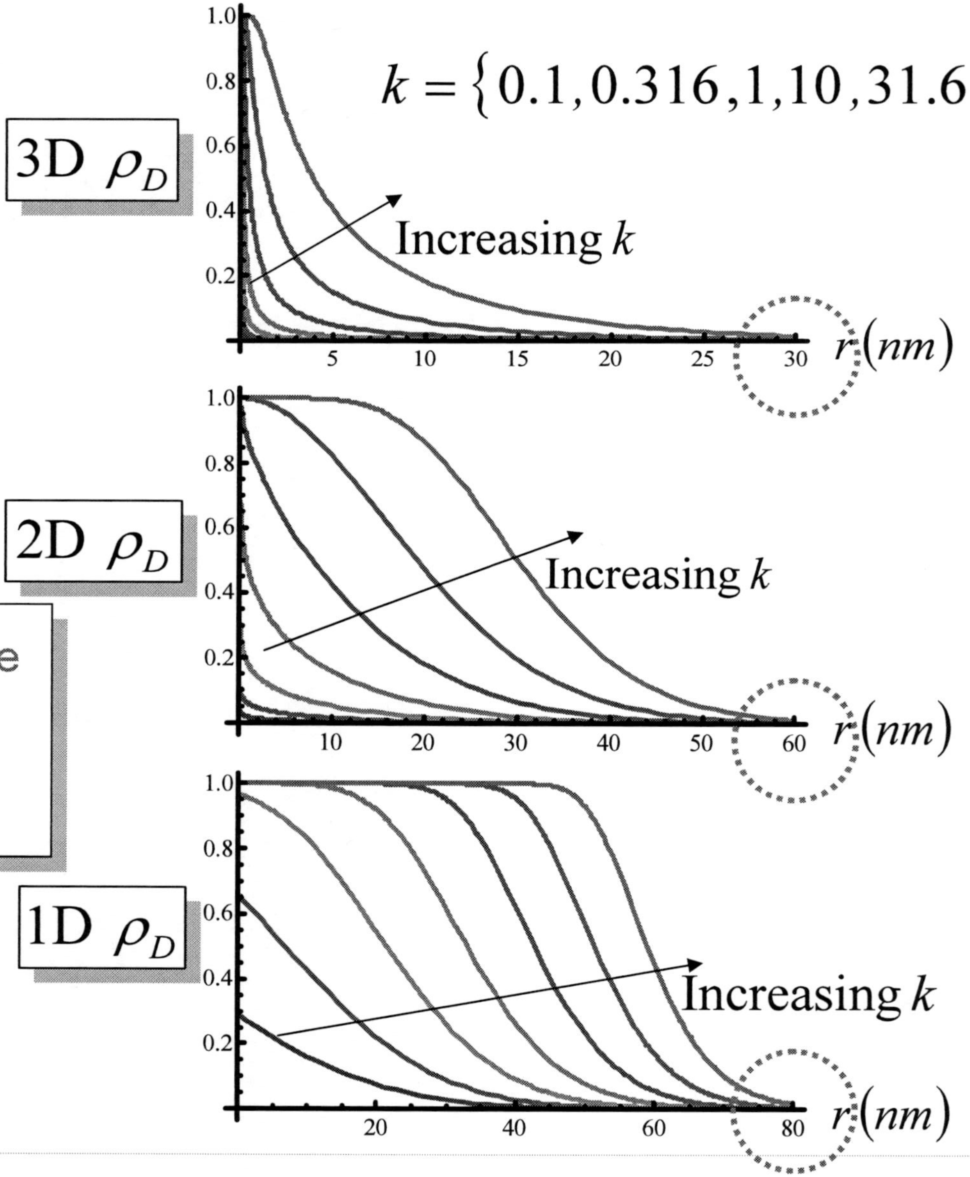

The value of k and the space dimensionality both have a very strong effect on the effective resist blur.

Kang, et. al., SPIE 6519 (2007)

Applied Math Solutions, LLC

LER Metrology

Apply edge finding algorithm and determine best fit straight line for each edge

1.

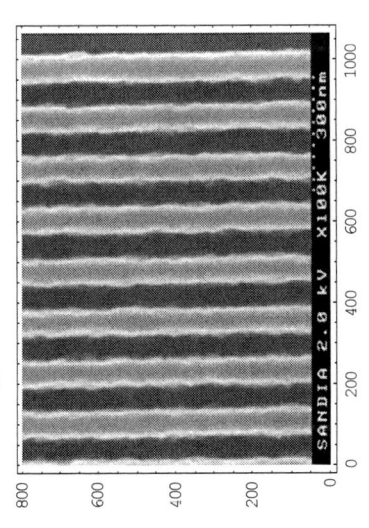

2. Subtract straight line fit to obtain roughness residual

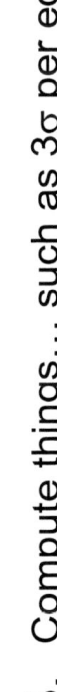

3. Compute things... such as 3σ per edge

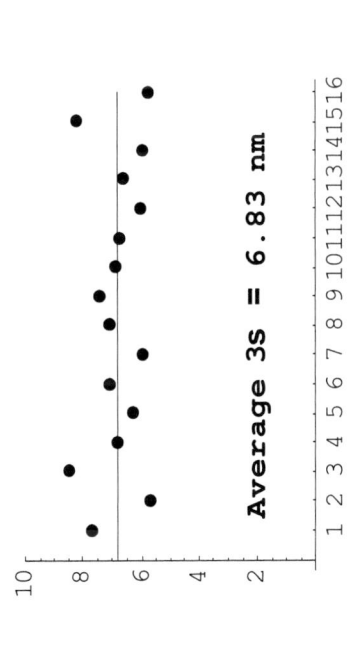

Average 3s = 6.83 nm

...FFT and square to obtain frequency content = "PSD"

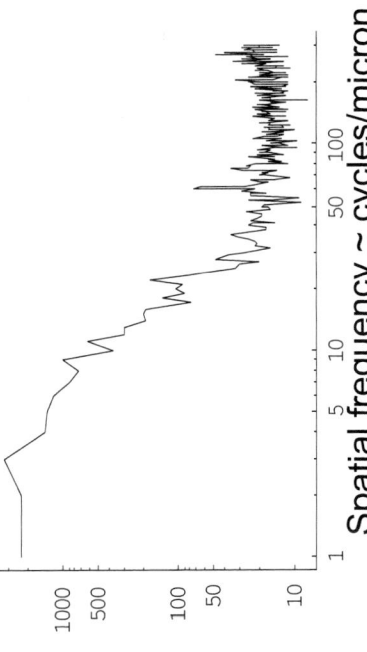

Spatial frequency ~ cycles/micron

Applied Math Solutions, LLC

Analytic PSD

$$PSD(\beta) = N \times \frac{1}{(R\beta)^3} \left[\begin{array}{l} 2(R\beta)e^{-2(R\beta)^2}\left(\sqrt{2\pi} - 2\sqrt{\pi}e^{(R\beta)^2}\right) \\ + 2\pi\left(1 - 2(R\beta)^2\right)erf(R\beta) \\ + \pi\left(4(R\beta)^2 - 1\right)erf(\sqrt{2}R\beta) \end{array} \right]$$

Valid for small k

Normalization factor ~ σ_{LER}^2

PSD "shape": Depends <u>only</u> on $R\beta = R2\pi\nu$

ν = Spatial Frequency (cycles/micron)

Can determine resist parameters from roughness data

- rms roughness → σ_{LER} → N

- Intrinsic resist "blur" R is determined by fitting the analytic PSD "shape" to $|FFT(data)|^2 / N$

Applied Math Solutions, LLC

Effect of the k parameter on the PSD

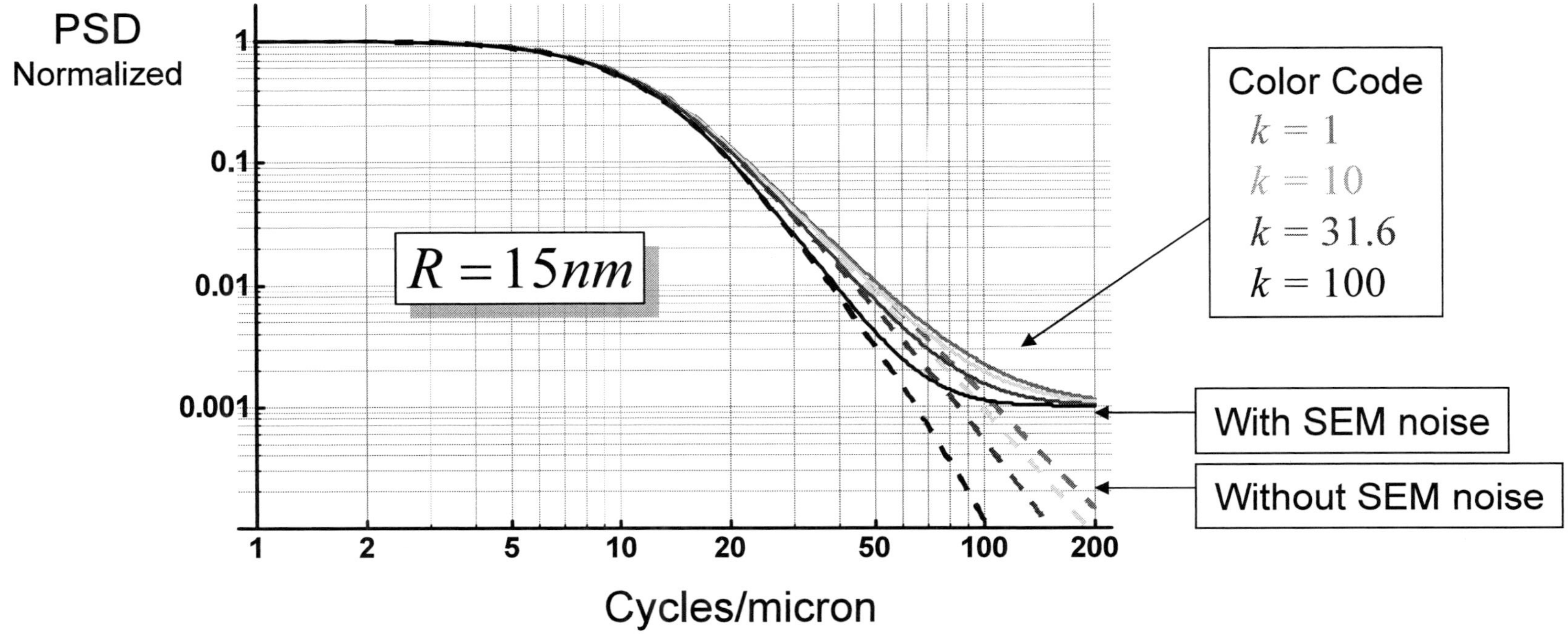

PSD Normalized

Cycles/micron

$R = 15nm$

Color Code
$k = 1$
$k = 10$
$k = 31.6$
$k = 100$

With SEM noise

Without SEM noise

Very little difference in PSD shape for $k \lesssim 30$

➡ Analytic PSD is a good approximation in this range
➡ Difficult to use it to determine k

Applied Math Solutions, LLC

Contact Resolution Metric: CD vs. Dose → Resist "Blur" = Resist Resolution

Through-Dose Contact Printing as Resolution Metric

- 50-nm contacts with different levels of resist blur

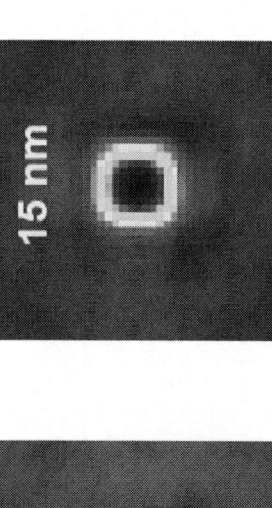

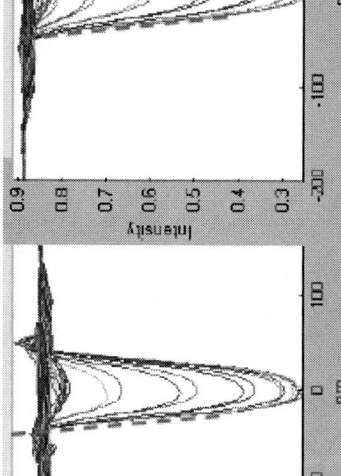

Patrick Naulleau
Lawrence Berkeley National Laboratory
University at Albany
Chris Anderson
University of California, Berkeley
Ryoung-han Kim, Bruno La Fontaine, Tom Wallow
AMD

Applied Math Solutions, LLC

Results from the Sematech Sponsored RLS Project

Red dots = "5435" (EUV2D)
Green dots = "5271" (MET2D)
Blue dots = "5496"
Black dots = "EH27"
Red Circle = MET1K (old data)

Green Circle = EUV2D (old data)
Blue Circle = "A"
Black Circle = "B"
Red Squares = "C" (PEB temps)
Green Square = "D"

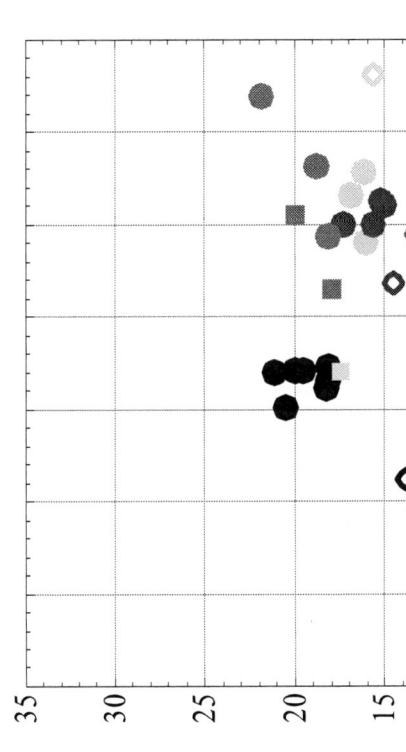

R (nm)
Determined
by fitting
analytical PSD
to data PSD

Applied Math Solutions, LLC

Modeling the Contact Resolution Metric

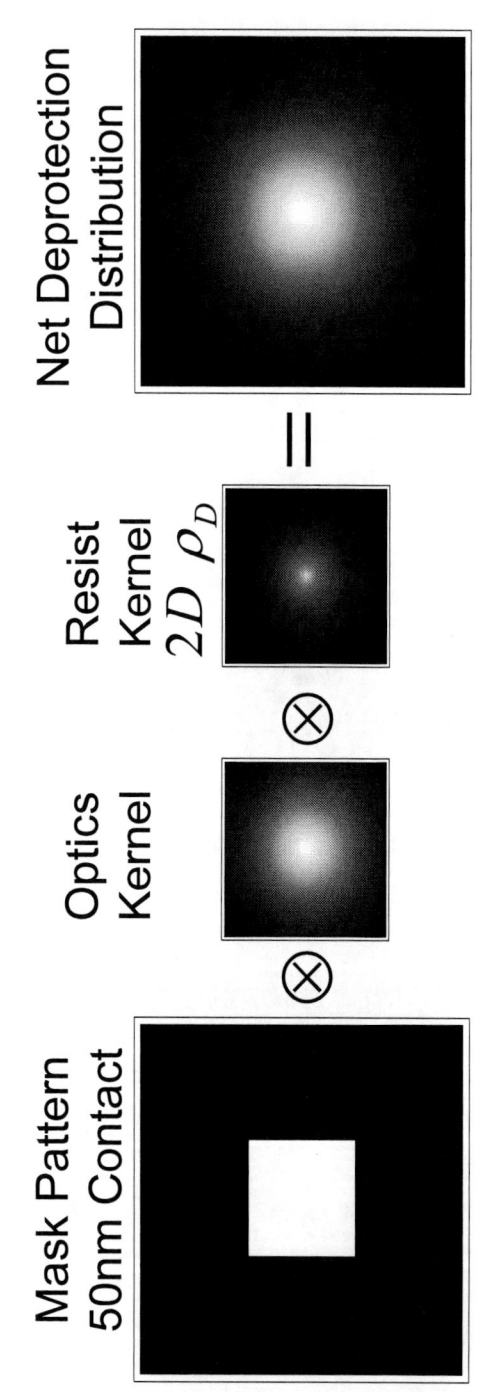

Net Deprotection Distribution

Resist Kernel $2D$ ρ_D

$=$

Optics Kernel

$\otimes$

Mask Pattern 50nm Contact

$\otimes$

1. Compute the Net Deprotection Distribution for particular R and k

2. Pick a different k and do a least squares fit to find the R which produces the best match of the new Net Deprotection Distribution to the one computed in Step 1.

Applied Math Solutions, LLC

How does Least Squares Fit *R* value depend on *k*?

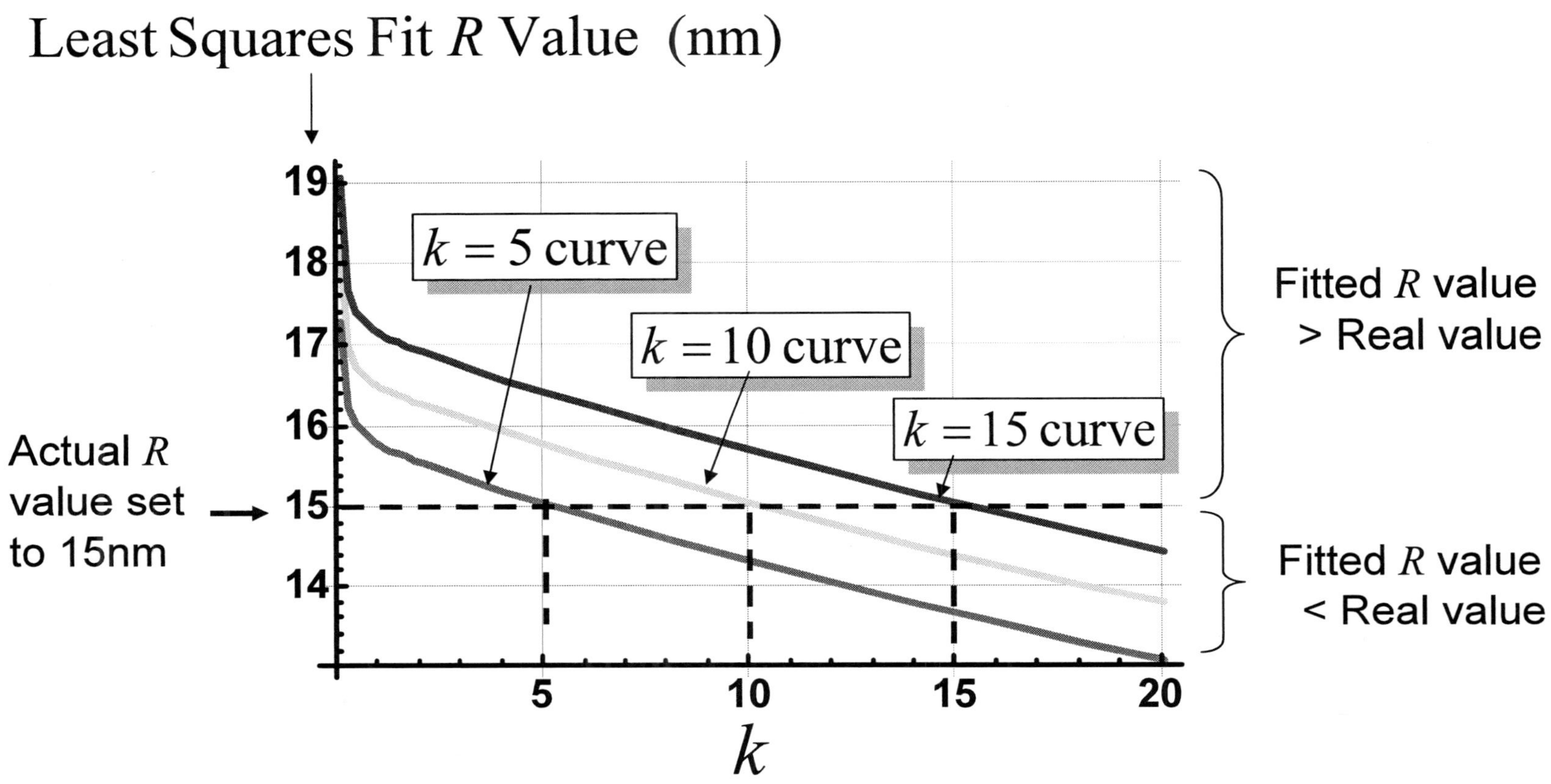

Least Squares Fit *R* Value (nm)

Fitted *R* value
> Real value

Fitted *R* value
< Real value

k = 5 curve

k = 10 curve

k = 15 curve

Actual *R*
value set
to 15nm

19
18
17
16
15
14

5   10   15   20

k

Applied Math Solutions, LLC

How does Least Squares Fit R value depend on k?

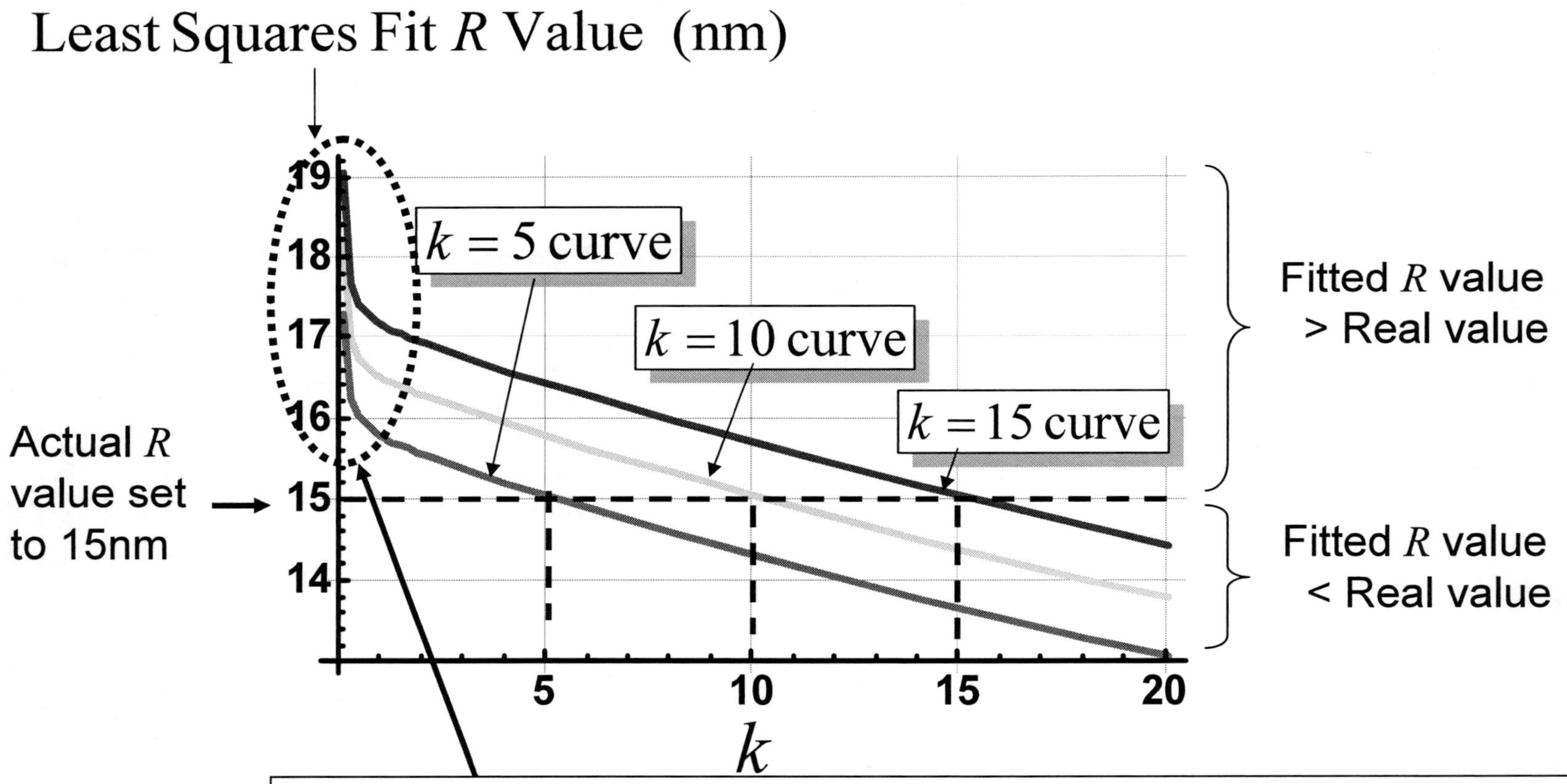

Least Squares Fit R Value (nm)

Fitted R value > Real value

Fitted R value < Real value

$k = 5$ curve

$k = 10$ curve

$k = 15$ curve

k

Actual R value set to 15nm

If you fit the data using a k value much smaller than the real value then the computed "Blur" is larger than the PSD determined R value

Results from the Sematech Sponsored RLS Project

Red dots = "5435" (EUV2D)
Green dots = "5271" (MET2D)
Blue dots = "5496"
Black dots = "EH27"
Red Circle = MET1K (old data)

Green Circle = EUV2D (old data)
Blue Circle = "A"
Black Circle = "B"
Red Squares = "C" (PEB temps)
Green Square = "D"

Cluster of generically larger "blur" values can be accounted for by the effect of k on the fitted "blur" value.

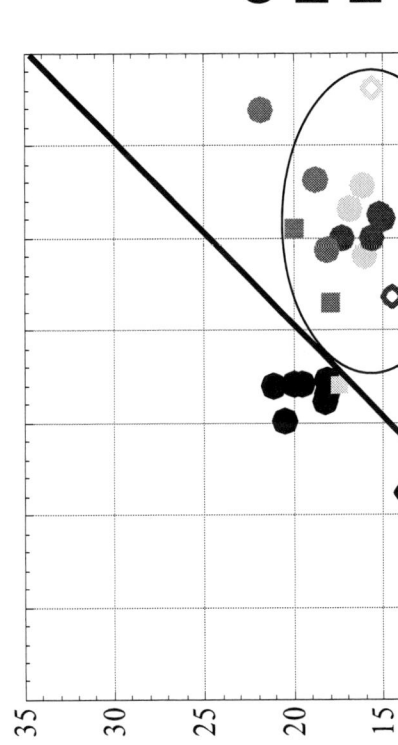

R (nm)
Determined by fitting analytical PSD to data PSD

Contact hole resolution measurement (nm)

Applied Math Solutions, LLC

Conclusions

- "Size/Shape" of the deprotection distribution depends strongly on the deprotection rate constant k.

    - Small k and Large R ~ Small R and Large k

- This dependence can account for at least part of the disconnect between the PSD determined resist blur and contact hole method.

- Difficult to "see" the effect of k when looking at just the PSD

- Refit contact hole data allowing for the value of k

- Getting the correct deprotection shape is very important for getting the CD's correct when modeling EUV printing capabilities.

- Thank you for your attention!

Feasibility Study on High-Sensitivity Chemically Amplified Resist by Polymer Absorption Enhancement in Extreme Ultraviolet Lithography

T. Kozawa[1,2], K. Okamoto[1,2], J. Nakamura[1,2], and S. Tagawa[1,2]

[1]Osaka University, [2]JST-CREST

Trade-off relationships between resolution, sensitivity, and LER

Trade-off relationship

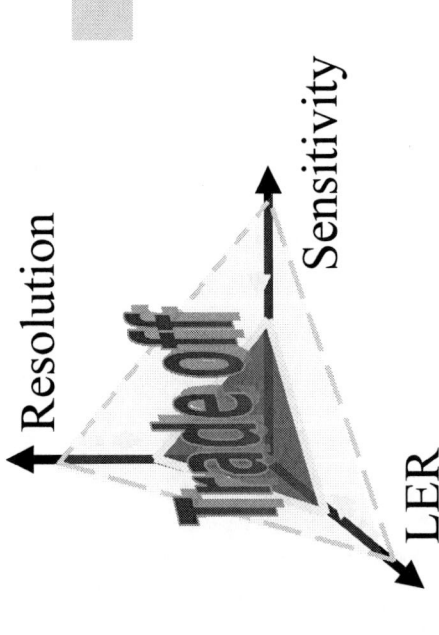

Solution

Increase in pattern formation efficiency

Polymer absorption enhancement

Acid generation efficiency enhancement

    ex. increasing AG conc.

Deprotection efficiency enhancement

    ex. lowering E_a

Acid diffusion confinement

    ex. control of acid diffusion by matrix

Feasibility of high absorption process

Current status

Sensitivity : 10~20 mJ cm^{-}

Abs. coefficient : 3.8 μm^{-1} (PHS)

Thickness : 50 nm (35-65 nm)

Transparency : 83%

Increase in efficiency can be used for the improvement of sensitivity, resolution, and LER at will.

X 1/4 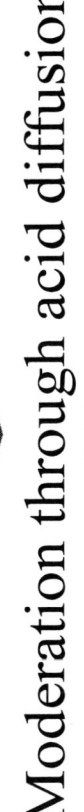 Sensitivity : <5 mJ cm^{-2}

Abs. coefficient : 16 μm^{-}

Thickness : 20 nm

Transparency : 73%

Polymer absorption enhancement ← Side wall degradation

The gradient of acid concentration is determined by absorption coefficient.

Moderation through acid diffusion

Restriction in acid diffusion

+

Effect of secondary electrons

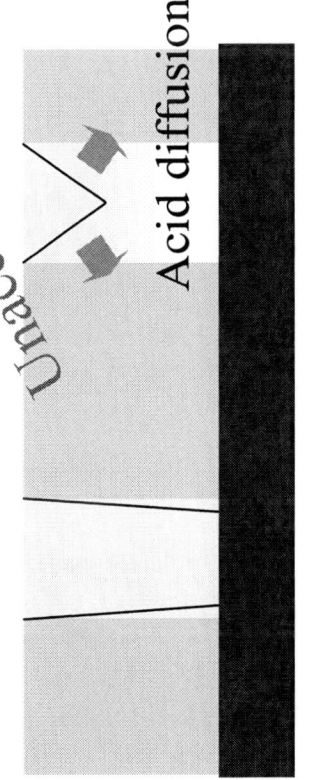

Difference between DUV and EUV resists

Resolution blur

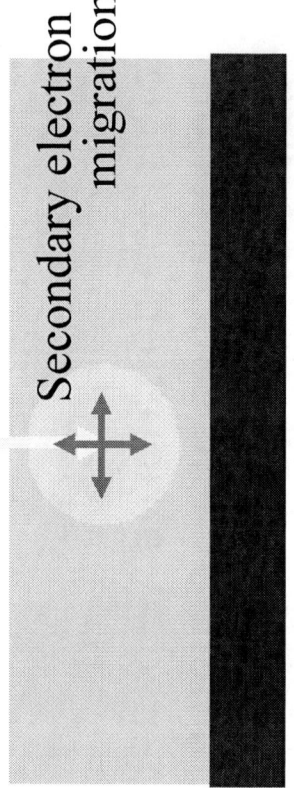

EUV

Secondary electron migration

Blur in vertical direction

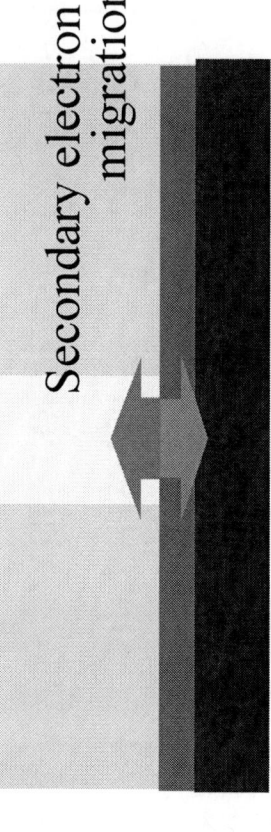

EUV

Secondary electron migration

DUV: Reflection from substrate

DUV

BARC

EUV: No reflection

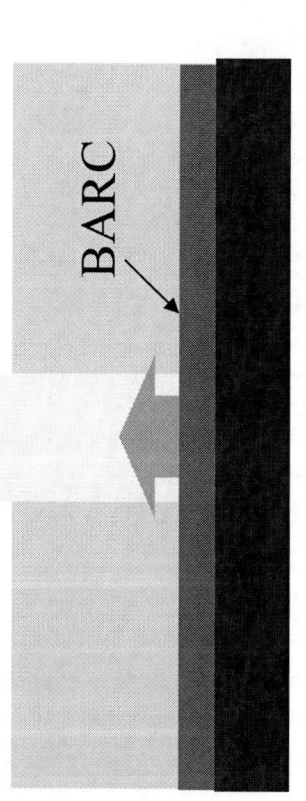

EUV

Secondary electron migration

Electron emission from surface

EUV photon

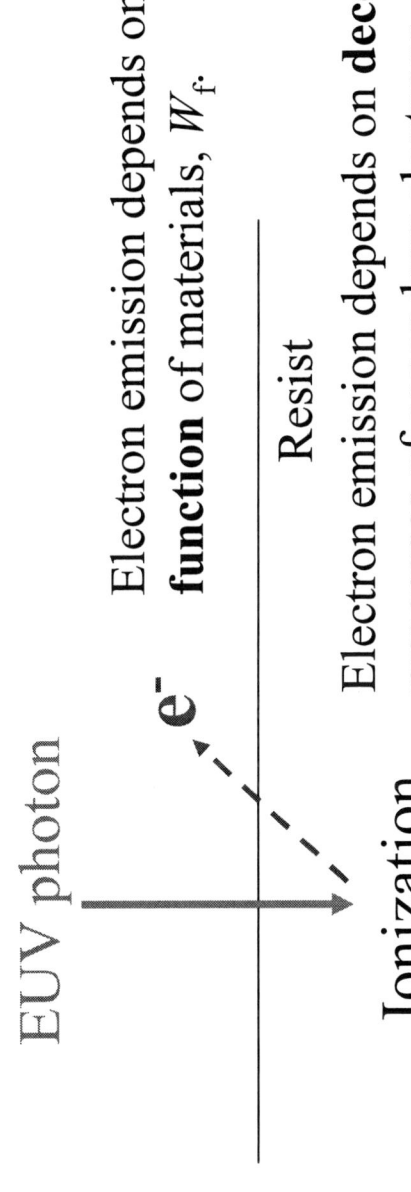

Electron emission depends on **work function** of materials, W_f

Resist

Electron emission depends on **deceleration processes** of secondary electrons.

Total-reflection model : $W_f = \infty$ Total-transmission model : $W_f = 0$

Initial acid distribution

$$C_{acid}(x,y,z) = \alpha \int_{z_s}^{ft+r_e} \int_{-\infty}^{\infty} \int_{-\infty}^{\infty} PSF\left(\sqrt{(x-x')^2 + (y-y')^2 + (z-z')^2}\right) I(x',y',z')\,dx'dy'dz'$$

$$+ f_{excitation} C_{AG} E_{EUV} \alpha I(x,y,z)$$

Total-reflection model : $z_s = -r_e$

Total-transmission model : $z_s = 0$

Parameters used in simulation

r_o (nm)	4
fp (nm)	0.69
N_+ (photon^{-1})	4.2
S	2.8
R_{AG} (nm) → TPS-tf	2.4
C_{AG} (wt%)	20
ε	4
$\phi_{polymer+}$	1
$p_{1/2}$ (nm)	22
L (nm)	12-22
ft (nm)	20-75
r_e (nm)	20
A (mJ cm^{-2})	5, 20
α (µm^{-1})	**3.8, 16**
$f_{excitation}$ (dm^3 mol^{-1} eV^{-1})	1.4×10^{-2}
D_{acid} (nm^2 s^{-1})	1
$D_{quencher}$ (nm^2 s^{-1})	1

Line & Space Pattern (22 nm half pitch)

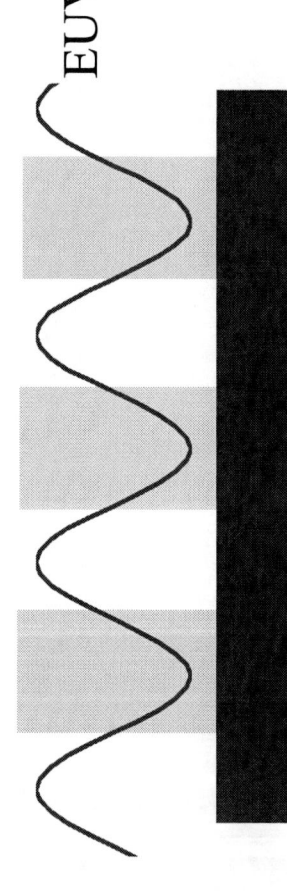

EUV

Optical image $I(x,y,z) =$

$$
\begin{cases}
\dfrac{A}{2}\left[\cos\left(\dfrac{\pi(x-2np_{1/2})}{L}\right)+1\right]\exp(-\alpha|z|) & \text{if } |x-2np_{1/2}| \leq L \\[2ex]
0 & \text{if } |x-2np_{1/2}| > L
\end{cases}
$$

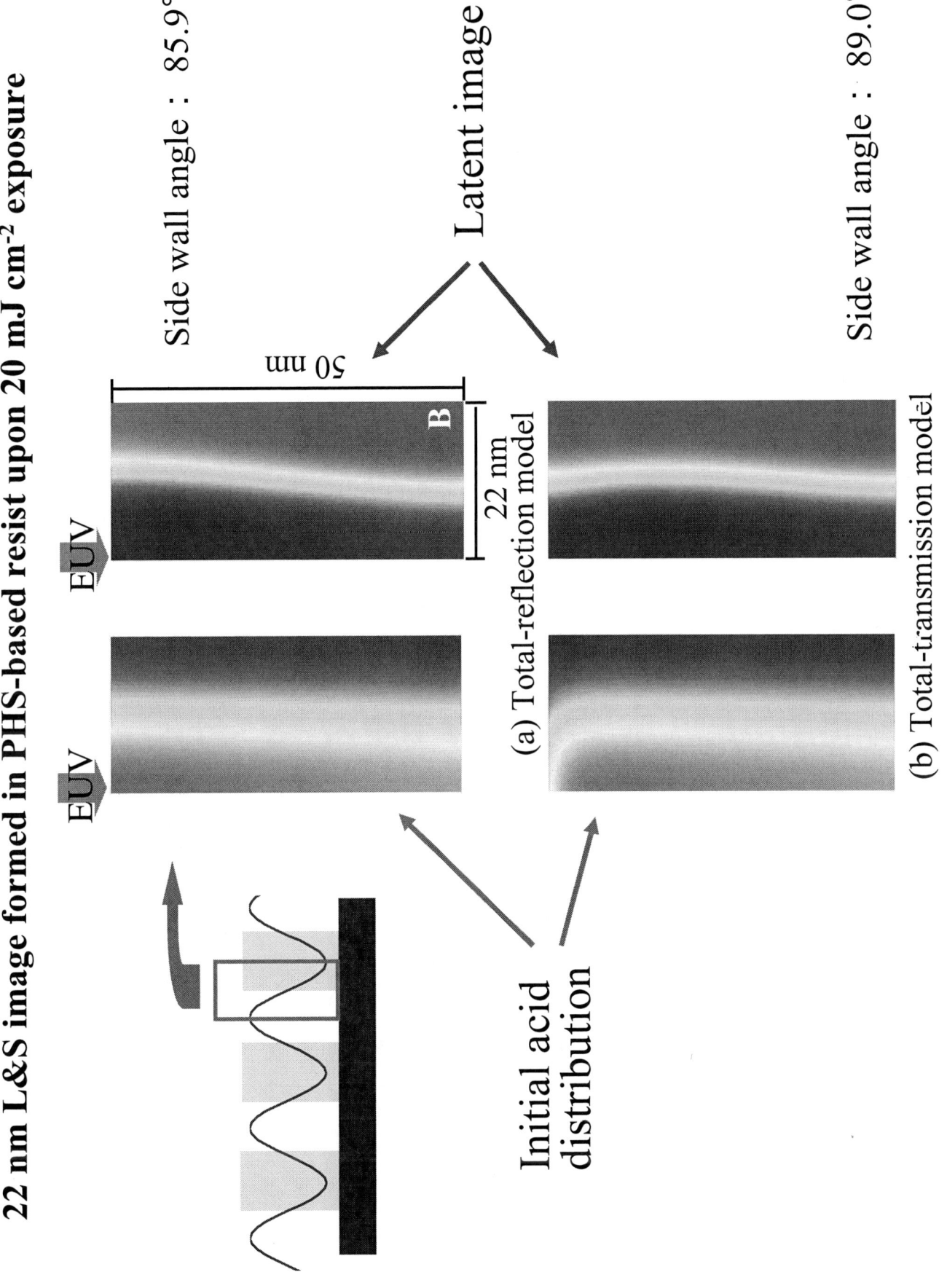

22 nm L&S image formed in PHS-based resist upon 20 mJ cm⁻² exposure

Acid diffusion (22 nm L&S, high absorption, 5 mJ cm^{-2})

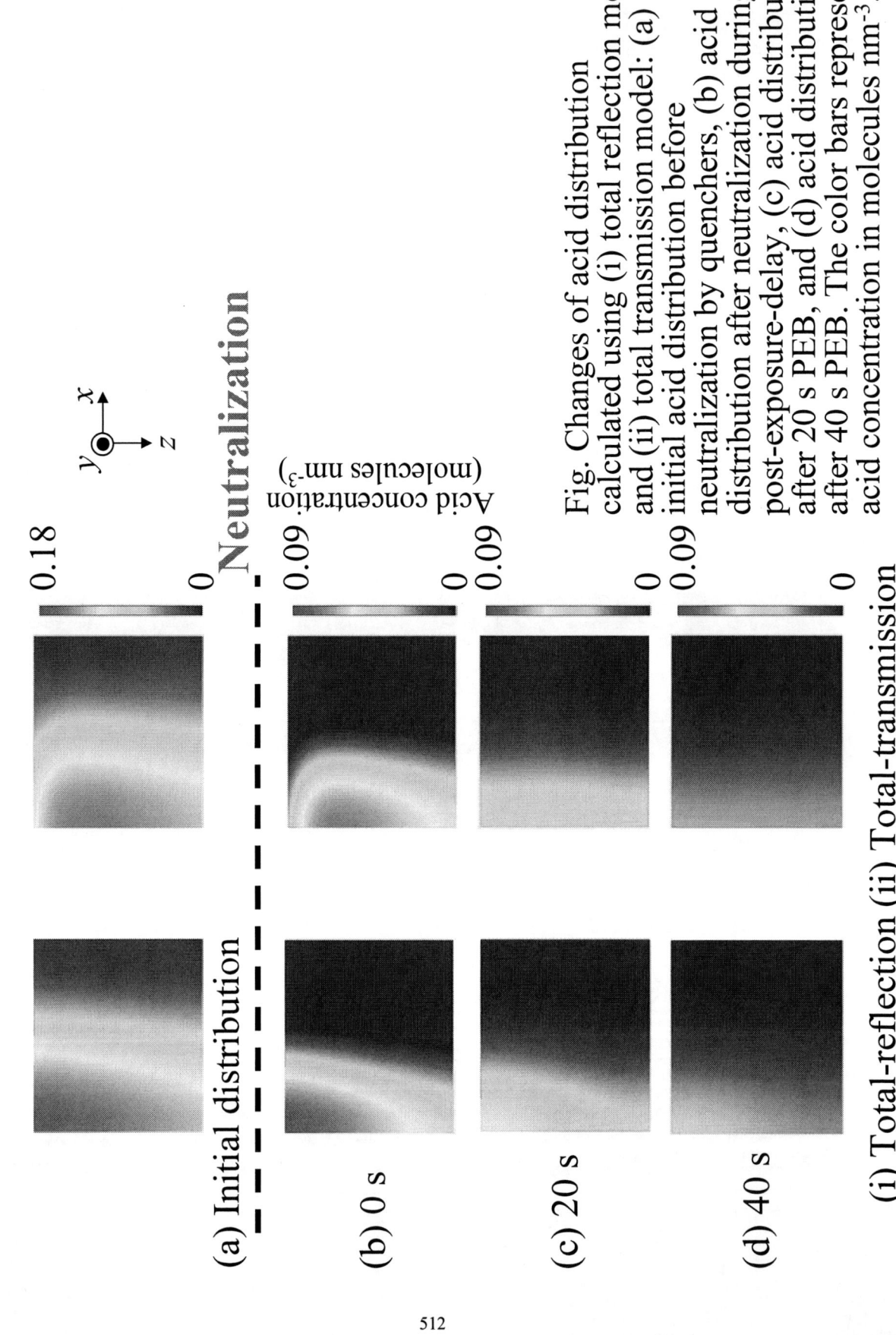

Fig. Changes of acid distribution calculated using (i) total reflection model and (ii) total transmission model: (a) initial acid distribution before neutralization by quenchers, (b) acid distribution after neutralization during post-exposure-delay, (c) acid distribution after 20 s PEB, and (d) acid distribution after 40 s PEB. The color bars represent acid concentration in molecules nm^{-3}.

Change in latent image (22 nm L&S, high absorption, 5 mJ cm^{-2})

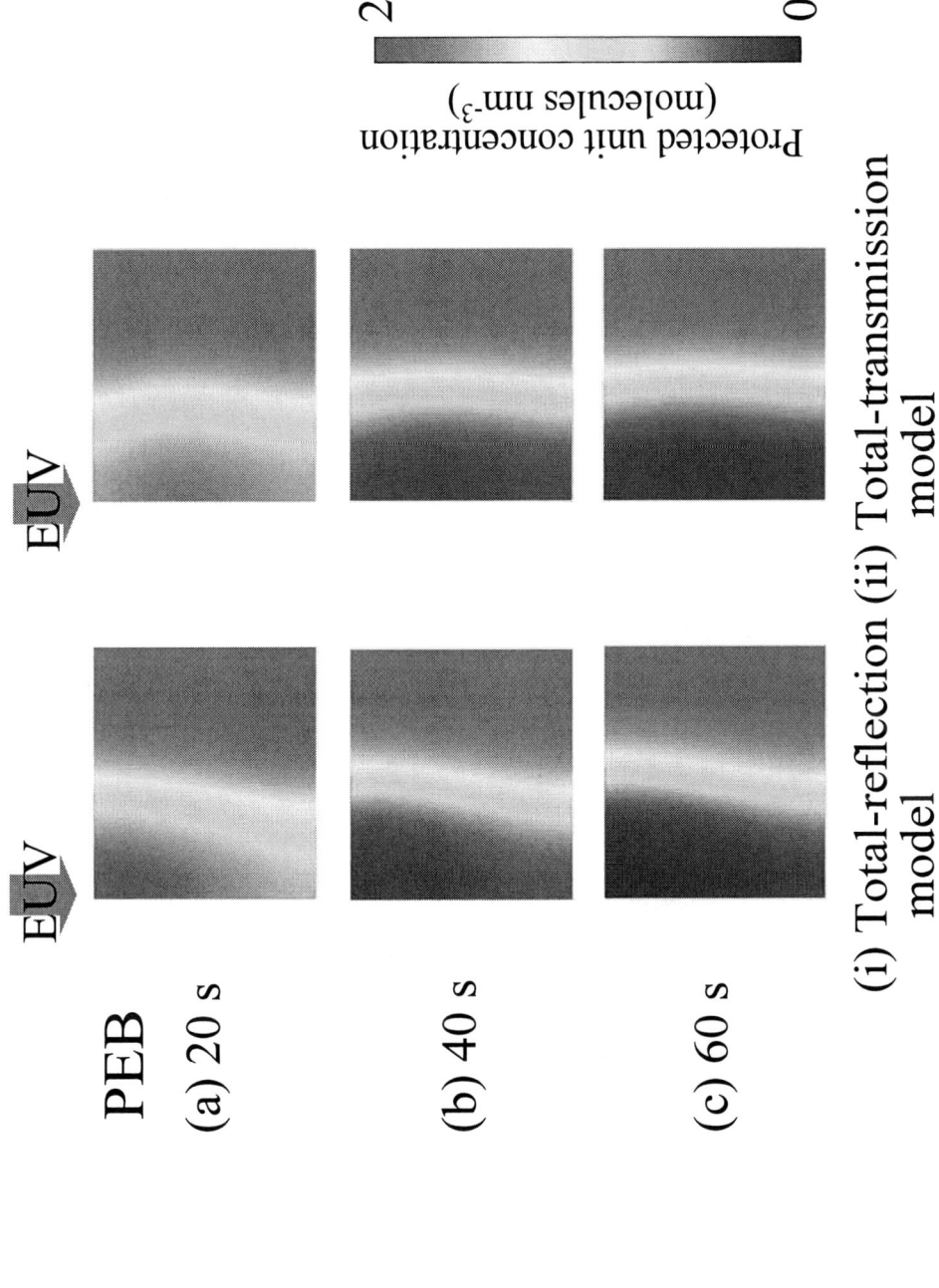

(i) Total-reflection (ii) Total-transmission model model

Fig. Latent images at (a) 20, (b) 40, and (c) 60 s PEB, calculated using (i) total reflection model and (ii) total transmission model. The color bars represent the concentration of protected groups in molecules nm^{-3}.

Effect of optical image quality

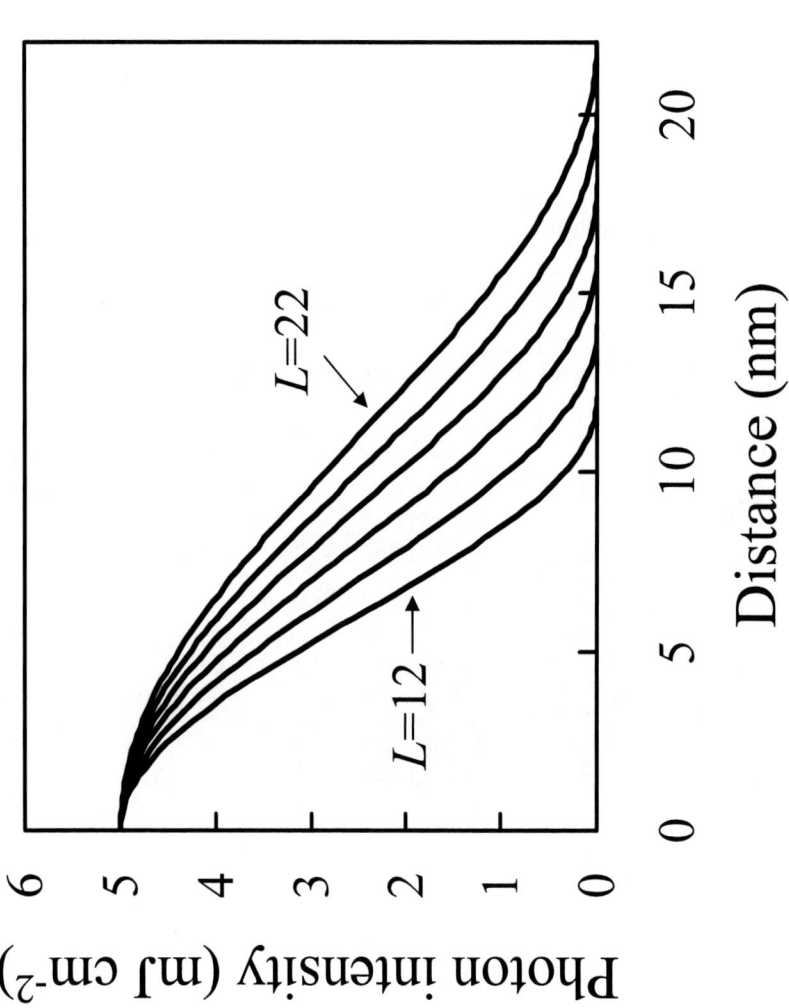

Fig. Aerial images of incident EUV photons at the resist surface. L was varied from 22 to 12 nm with decrement of 2 nm.

$$I(x,y,z) = \begin{cases} \dfrac{A}{2}\left[\cos\left(\dfrac{\pi(x-2np_{1/2})}{L}\right)+1\right]\exp(-\alpha|z|) & \text{if} \quad |x-2np_{1/2}| \leq L \\[2ex] 0 & \text{if} \quad |x-2np_{1/2}| > L \end{cases}$$

Relationship between EUV optical image and initial acid image

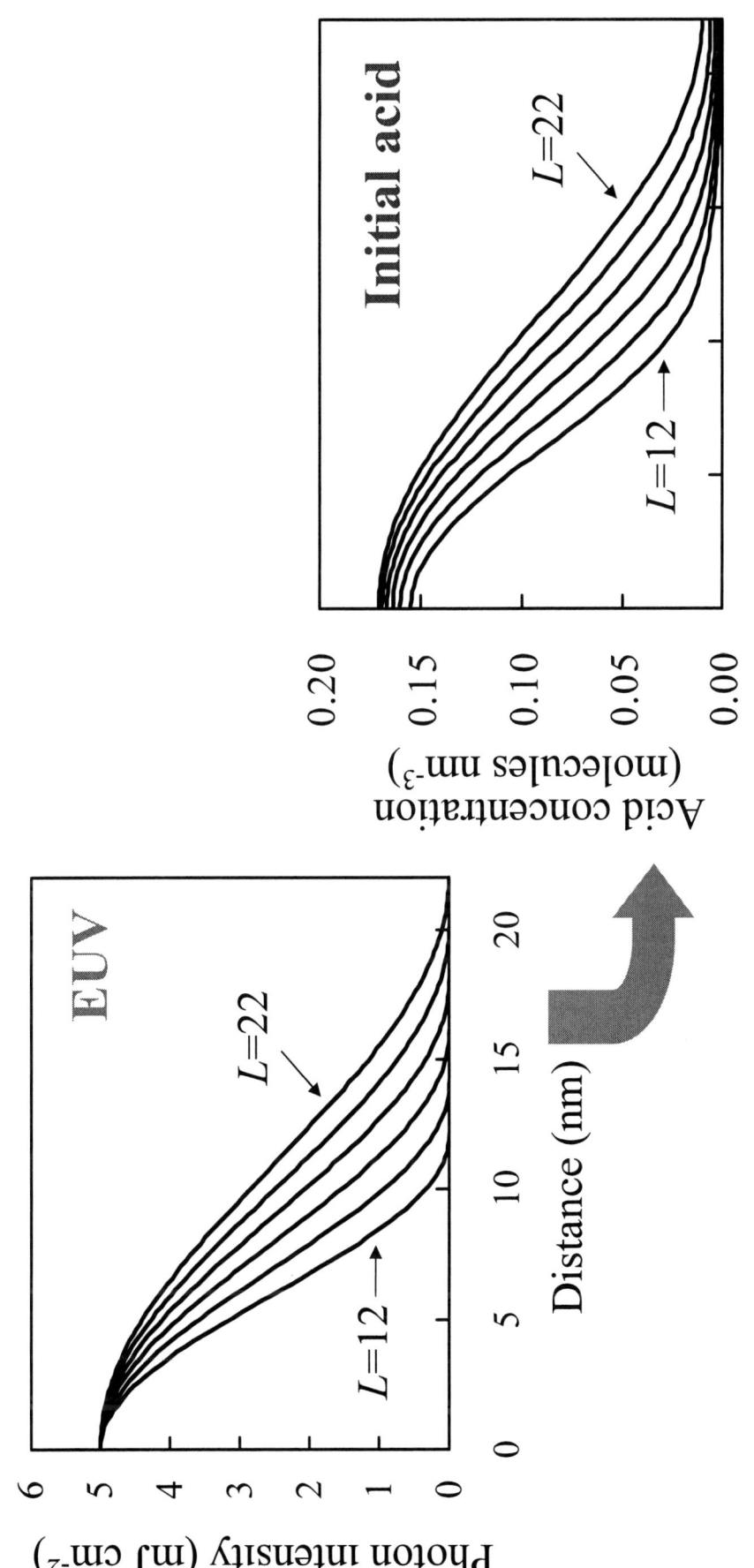

Fig. Initial acid distribution at the resist surface calculated using the total reflection model and the aerial images shown in Fig. 1. Each line corresponds to the aerial images of $L = 12$, 14, 16, 18, 20, and 22 nm from the left to the right.

Relationship between EUV optical image and latent image

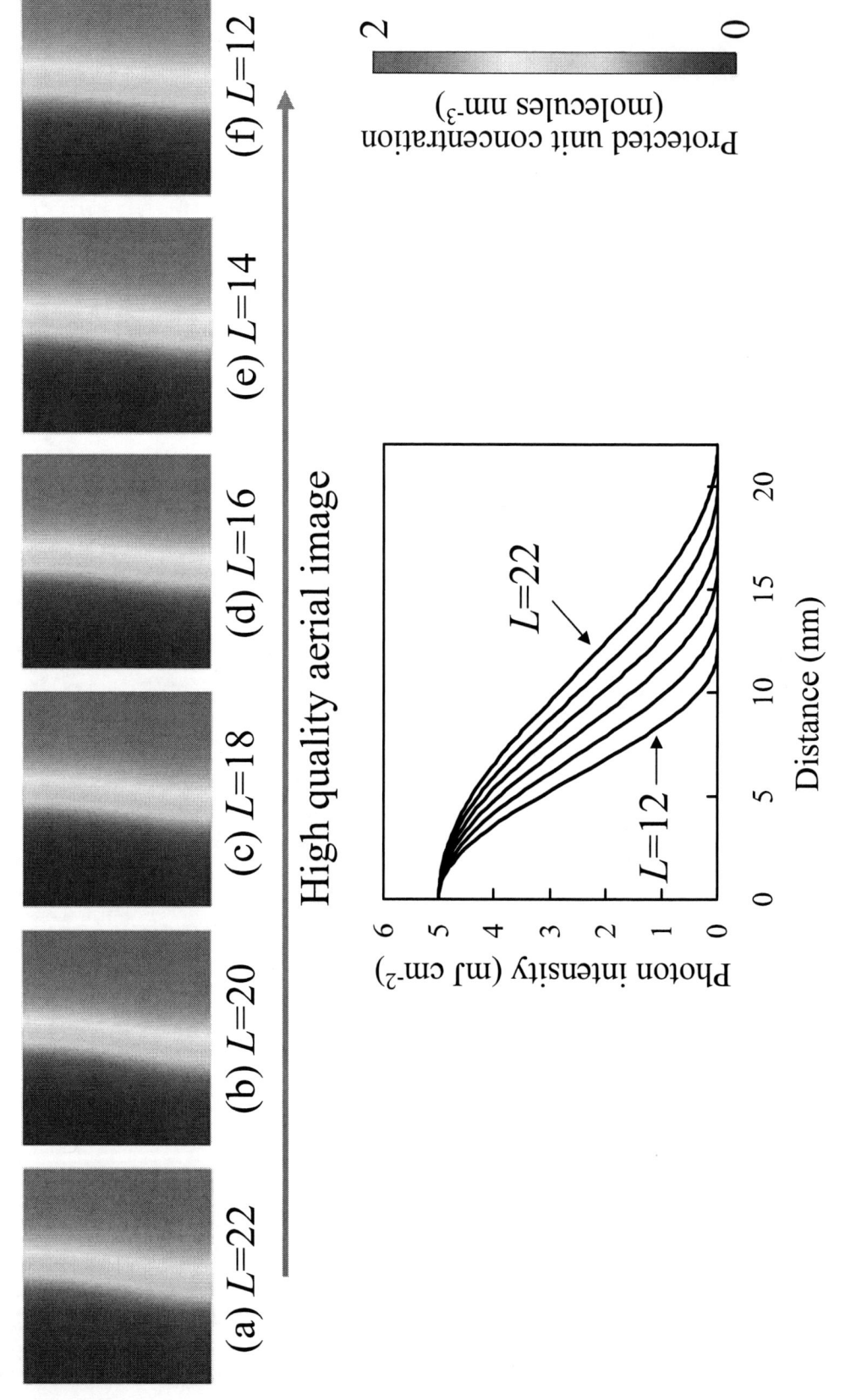

Fig. Latent images after the complete quenching of acids. The images calculated using the aerial images with L = (a) 22, (b) 20, (c) 18, (d) 16, (e) 14, and (f) 12 nm. The color bars represent the concentration of protected groups in molecules nm^{-3}.

Relationship between optical image quality and side wall angle

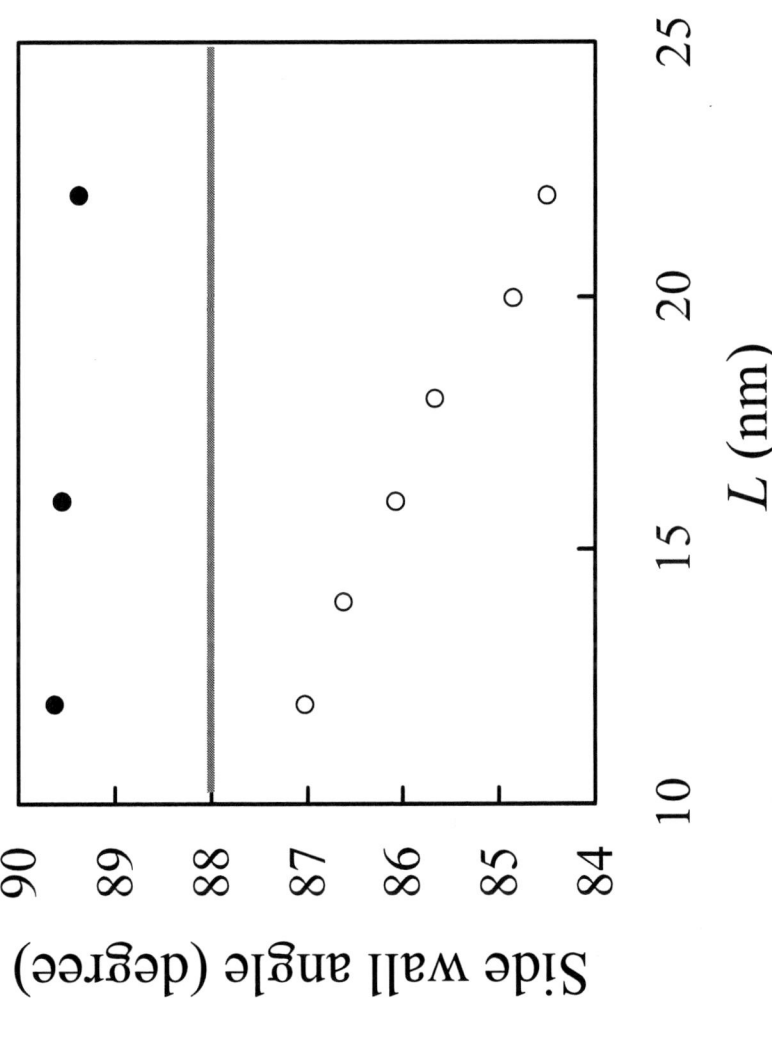

Fig. Relationship between side wall angles of latent images and aerial image qualities, L. The open and solid circles represent the angles of latent images calculated using the total reflection and total transmission models, respectively. The side wall angles were estimated using the contour line at half of the initial concentration of protected groups.

Comparison between high absorption and PHS-based resists

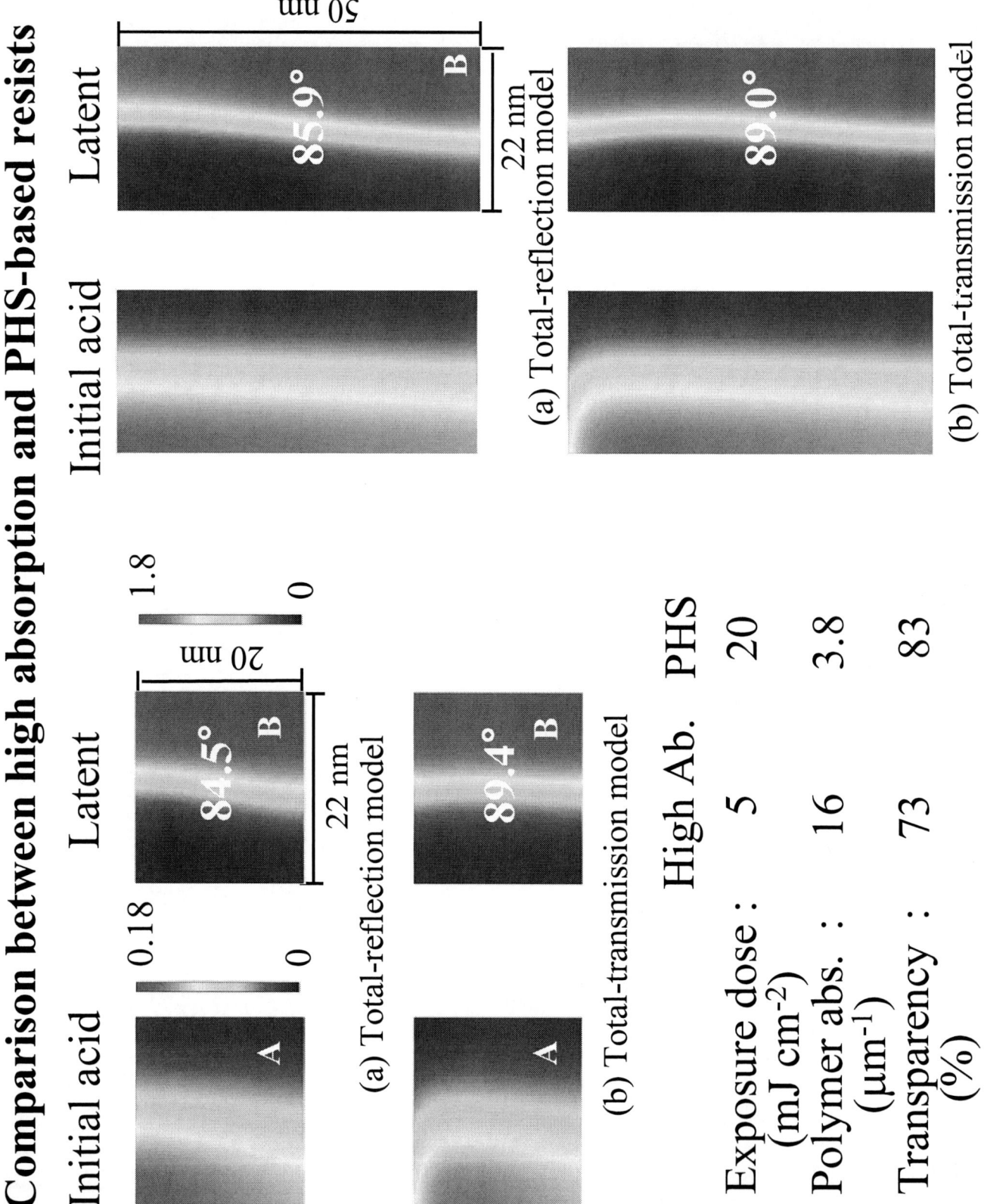

	High Ab.	PHS
Exposure dose : (mJ cm^{-2})	5	20
Polymer abs. : (μm^{-1})	16	3.8
Transparency : (%)	73	83

Summary

The feasibility of a high-absorption resist process was investigated by a simulation based on EUV sensitization mechanisms. Compared with PHS-based resists, the fourfold enhancement of polymer absorption is feasible without side wall degradation partly due to the long migration range of secondary electrons, although it is necessary to reduce the resist thickness to 20 nm.

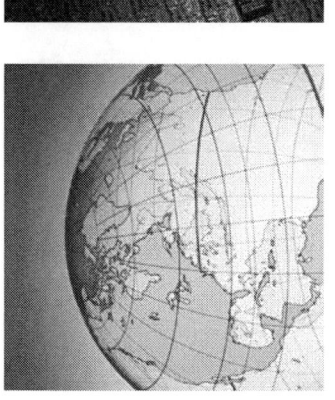

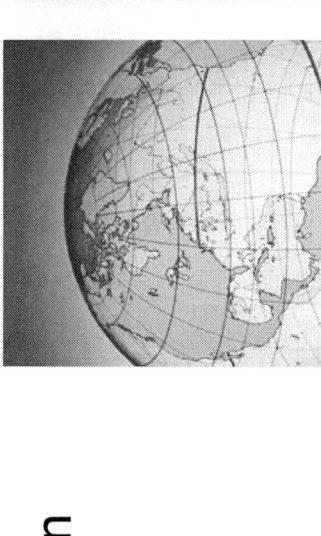

Accelerating the next technology revolution

Cost Implications of EUV Lithography Technology Decisions

Andrea F. Wüest, SEMATECH
Andrew J. Hazelton, Nikon Corporation
Greg Hughes, SEMATECH
Lloyd C. Litt, SEMATECH
Frank Goodwin, SEMATECH

Overview

- Introduction / Motivation
- Calculation Procedure
- Results
- EUVL Considerations
- Conclusions

Motivation

- Leading edge litho cost will increase dramatically for the 32 nm half-pitch (hp) node

- Miniaturization of devices continuing at the same pace

  → Identify parameters for cost-effective lithography at 32 nm and 22 nm hp

Overview

- Introduction / Motivation
- Calculation Procedure
- Results
- EUVL Considerations
- Conclusions

Candidate Technologies

- 32 nm hp
  - ArFi DPL: LELE, Freeze, Spacer
  - High-index ArFi SE
  - EUVL

- 22 nm hp
  - ArFi DPL: LELE, Freeze, Spacer
  - High-index ArFi DPL: LELE
  - EUVL
  - Nanoimprint

SE: Single Exposure, DPL: Double Patterning

LELE: Litho-Etch-Litho-Etch

Process Flows (schematic)

Deposit hardmask(s) — **Coat, expose, develop**

ArFi SE

LELE (Line) — **Etch hardmask, Strip resist** — **Coat, expose, develop**

Freeze — **Freeze resist** — **Coat, expose, develop**

Spacer — **Etch hardmask** — **Deposit spacer, Etch back spacer** — **Remove hardmask lines** — **Coat, expose, develop (cut mask)**

EUV

Imprint — **Imprint**

DPL

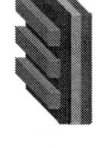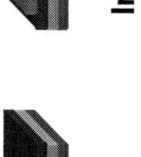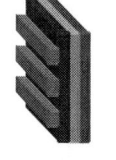

For all flows at end:
Etch hardmask, Strip resist,
Etch pattern, Strip hardmask

Assumptions

- All technologies are **equally reliable** and **support equal yield.**
- All technologies **meet manufacturing requirements.**

- Double Patterning
  - Stepper overlay
  - Mask registration
  - Mask yield (with 30 hour write time)
  - Yield even though with more processing steps

- EUVL
  - Defect-free masks
  - Source power, tool transmission, and resist sensitivity enable throughput
  - Tool reliability supports uptime
  - Mask and optics meet lifetime requirements

- Imprint
  - 1× defect-free masks
  - Tool meets throughput (15× improvement over today)
  - Imprint defects
  - Mask lifetime
  - Mask 1× inspection

Calculation Procedure

General

$$COO = \underbrace{\frac{\overbrace{(C_{fixed} + C_{recur})}^{\$\,/\,yr}}{\underbrace{T}_{wafer\,/\,h} \cdot U \cdot Y \cdot \underbrace{24 \cdot 365}_{h\,/\,yr}}}_{\$\,/\,wafer} + \underbrace{C_{materials}}_{\$\,/\,wafer} + \underbrace{C_{other}}_{\$\,/\,wafer}$$

C_{fixed} Depreciation, Floor space ($/yr)

C_{recur} Utilities, Consumables, Labor ($/yr)

T Throughput (wafer / h)

U Utilization (%)

Y Yield (%)

$C_{materials}$ Resist, etc. ($/wafer)

C_{other} Other ($/wafer)

Calculation Procedure

Linear relationship

Lithography

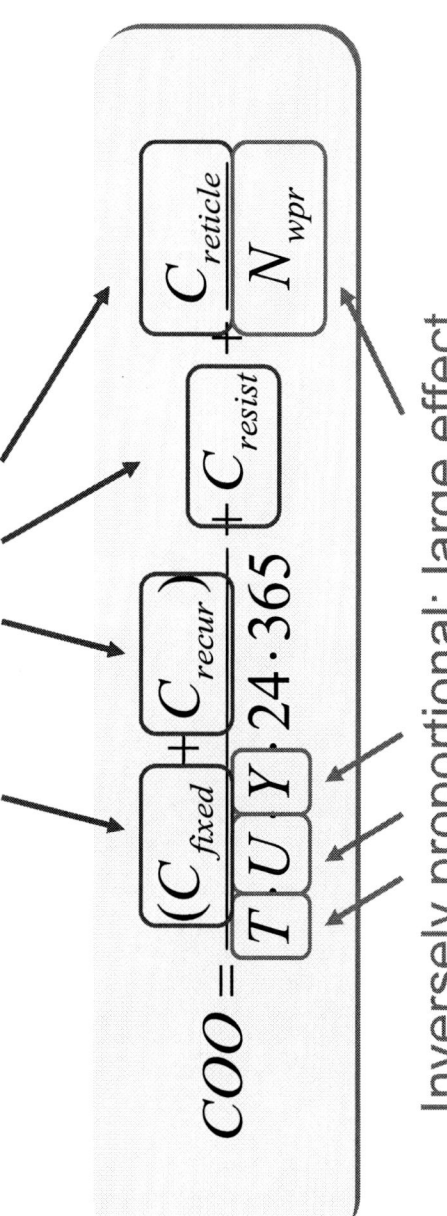

$$COO = \frac{(C_{fixed} + C_{recur})}{T \cdot U \cdot Y \cdot 24 \cdot 365} + C_{resist} + \frac{C_{reticle}}{N_{wpr}}$$

Inversely proportional: large effect

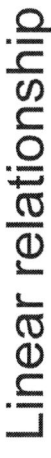

C_{fixed}	Depreciation, Floor space ($/yr)
C_{recur}	Utilities, Consumables, Labor ($/yr)
T	Throughput (wafer / h)
U	Utilization (%)
Y	Yield (%)

C_{resist}	Resist, etc. ($/wafer)
$C_{reticle}$	Reticle ($)
N_{wpr}	Wafer / reticle

Calculation Procedure

- Calculate COO for each process step
  - Litho
  - Deposition
  - Etch
  - Metrology
  - Clean

- Process flow → # of different process steps

- Total COO = Σ (Process Cost × #Process Steps)

- Only **one critical layer** calculated

- Normalized to 45 nm ArFi SE

Model Parameters

32 nm hp

	45 nm hp ArFi SE	ArFi DPL			HI ArFi	EUVL
		LELE	Freeze	Spacer		
Tool Cost	$40M	$49M	$49M	$49M	$50M	$54M
Throughput / wph	125	180	180	180	120	50
Tool Cost / TPT (M$/wph)	0.3	0.3	0.3	0.3	0.4	1.1
Reticle Cost	$200k	$584k	$584k	$466K	$396k	$178K

22 nm hp

	45 nm hp ArFi SE	ArFi DPL			HI ArFi DPL LELE	EUVL	NIL
		LELE	Freeze	Spacer			
Tool Cost	$40M	$52M	$52M	$52M	$53M	$89M	
Throughput / wph	125	200	200	200	135	100	
Tool Cost / TPT (M$/wph)	0.3	0.3	0.3	0.3	0.4	0.9	0.3
Reticle Cost	$200k	$1176K	$1176K	$752K	$1176K	$252K	$622k

- Tool cost based on historical extrapolation, scales with throughput and resolution

Model Parameters

- Reticle cost (SEMATECH model, G. Hughes)
  - Mask Cost = [(Capital Cost term × Write Time) + Material Cost] / Yield
  - 2.5× data growth per node for optical
  - 2× data growth for EUV and Imprint
  - Mask yield based on ITRS difficulty
    - 45 nm: 70%
    - EUVL: 77% (Yield targets looser)
    - DPL: 63% (Mask registration tighter)
    - Imprint: 54% (Defect specs much tighter, minimum feature 1×).

- Utilization fixed at 83%, yield at 98% for all technologies.
- Non-litho process costs based on SEMATECH data
- ~ 250 total parameters (~ 25 per litho technology)

Overview

- Introduction / Motivation

- Calculation Procedure

- Results

- EUVL Considerations

- Conclusions

Overview – 20,000 Wafers/Mask

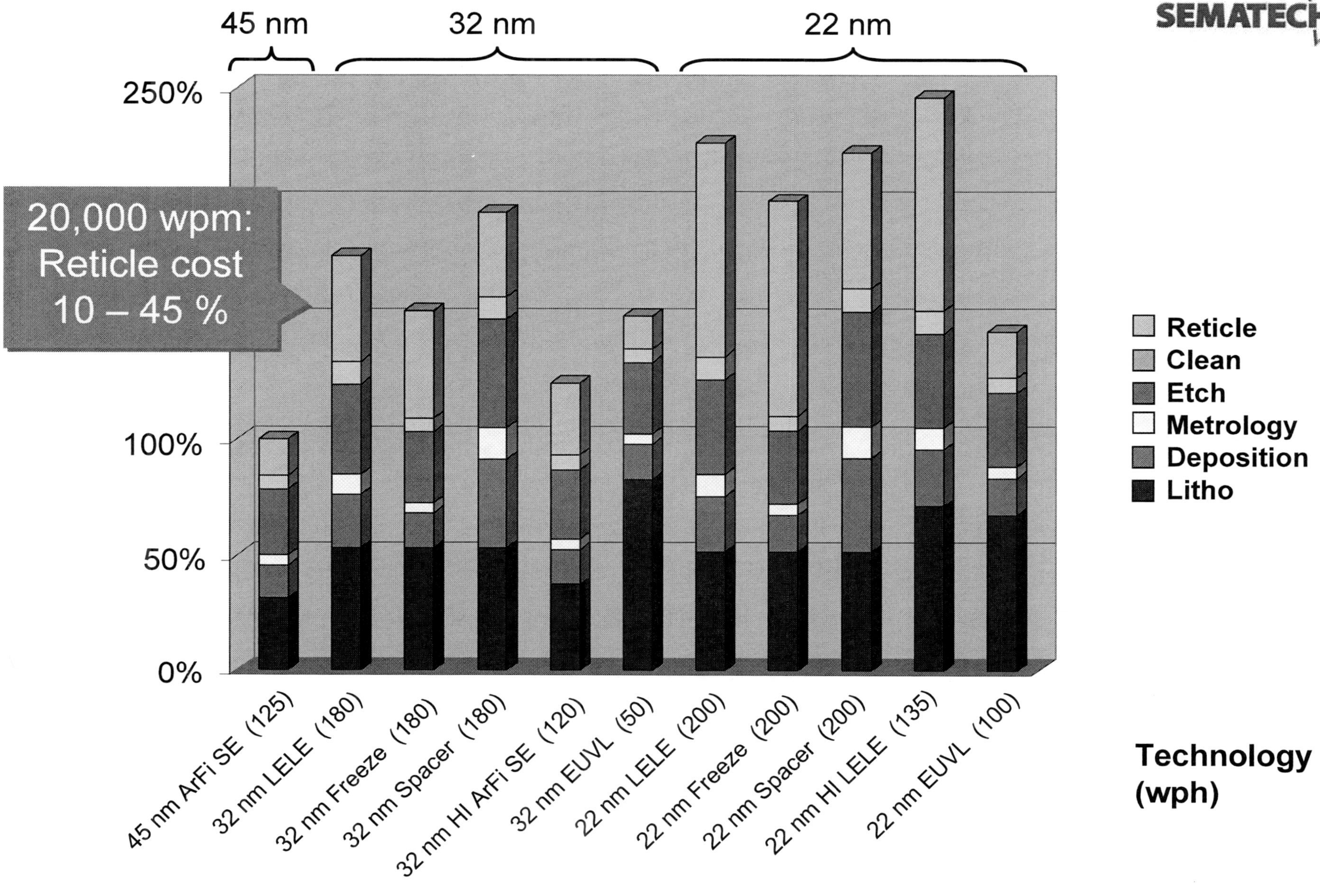

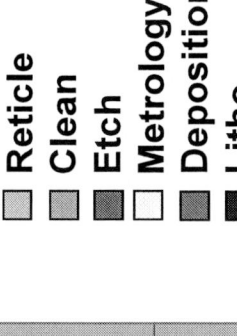

Legend:
- Reticle
- Clean
- Etch
- Metrology
- Deposition
- Litho

Technology (wph)

20,000 wpm: Reticle cost 10 – 45 %

45 nm:
- 45 nm ArF! SE (125)

32 nm:
- 32 nm LELE (180)
- 32 nm Freeze (180)
- 32 nm Spacer (180)
- 32 nm HI ArF! SE (120)
- 32 nm EUVL (50)

22 nm:
- 22 nm LELE (200)
- 22 nm Freeze (200)
- 22 nm Spacer (200)
- 22 nm HI LELE (135)
- 22 nm EUVL (100)

Y-axis: 0%, 50%, 100%, 250%

Overview – 1,000 Wafers/Mask

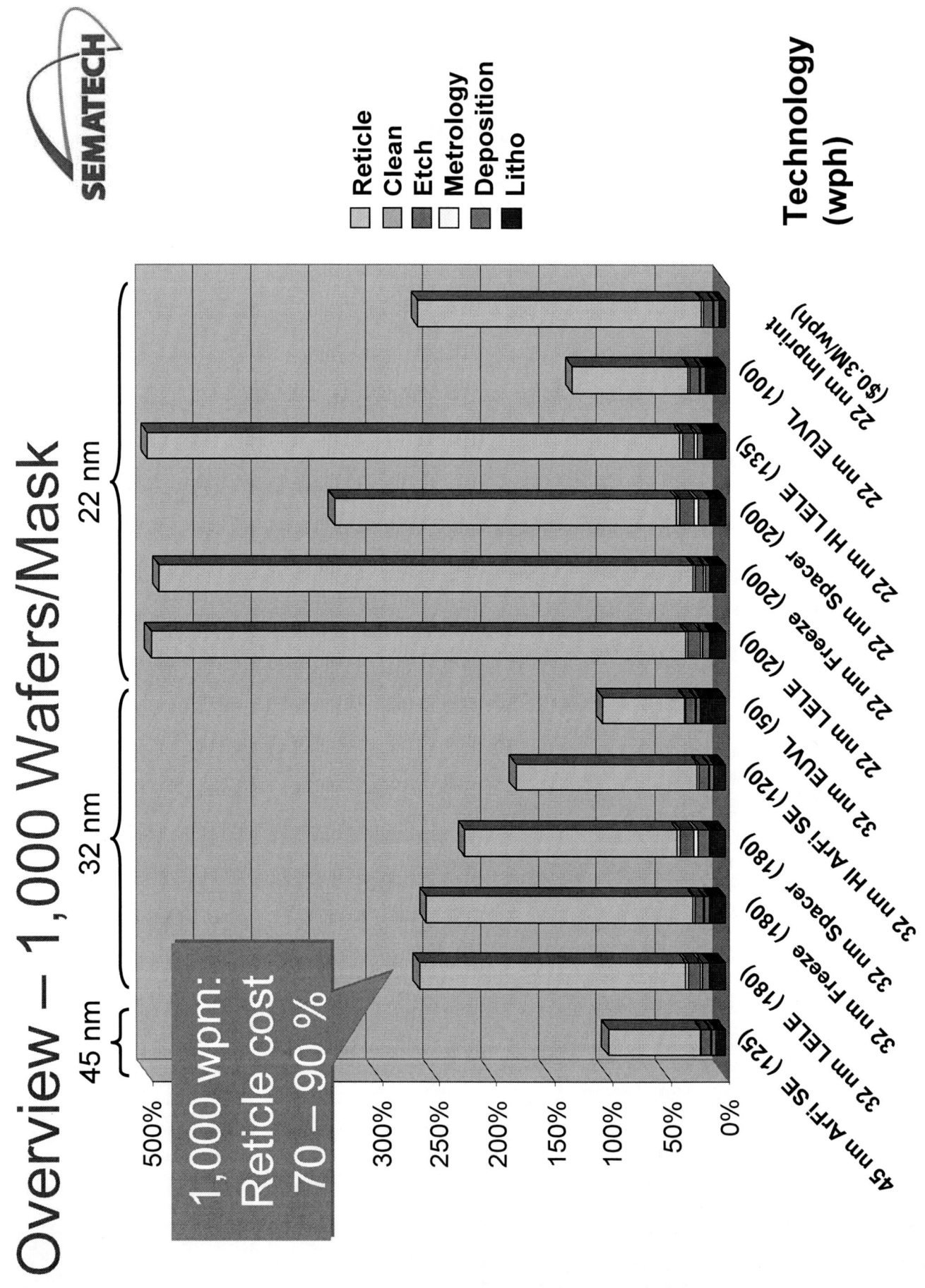

Error Estimation (20,000 Wafers/Mask)

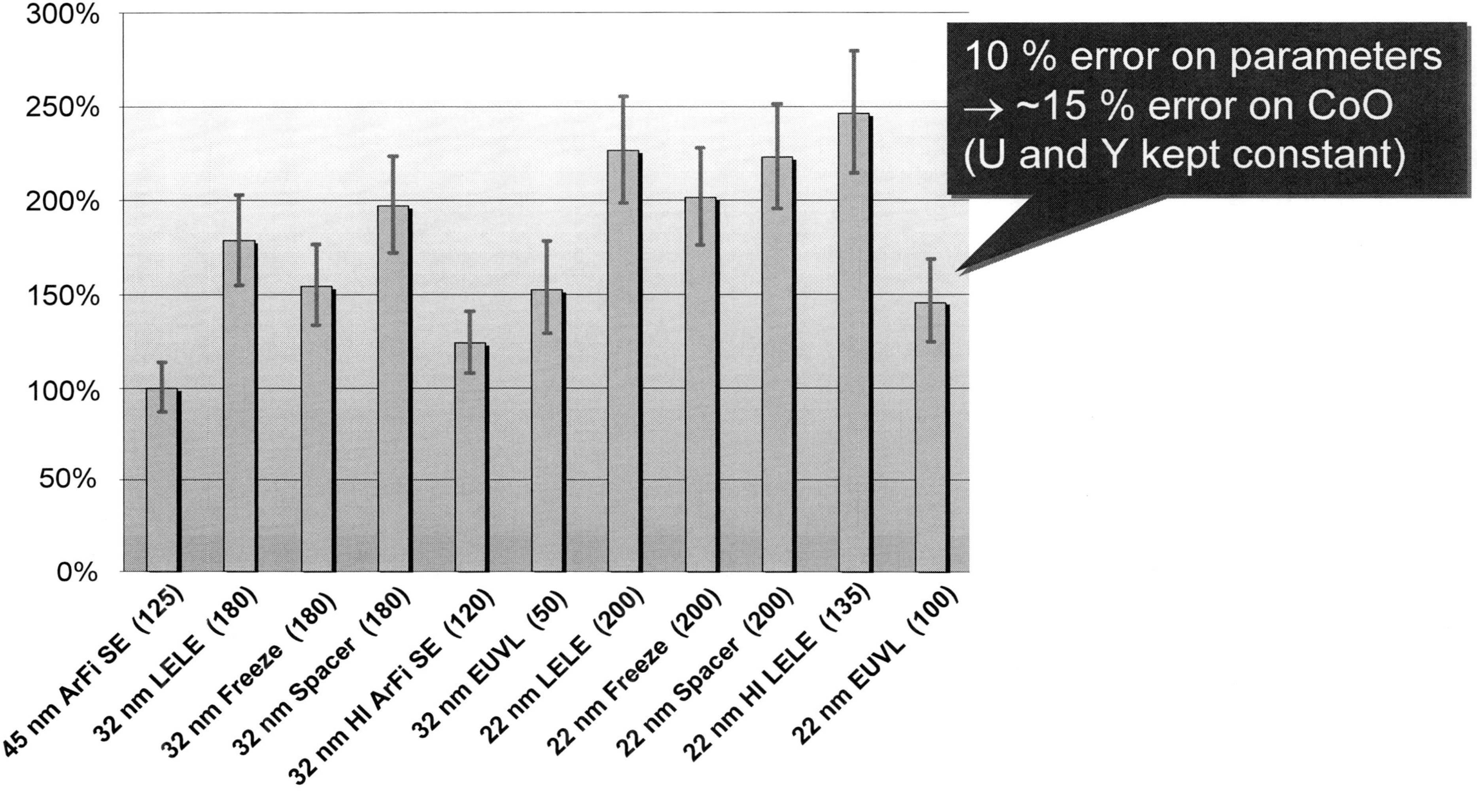

10 % error on parameters → ~15 % error on CoO (U and Y kept constant)

45 nm ArFi SE (125)
32 nm LELE (180)
32 nm Freeze (180)
32 nm Spacer (180)
32 nm HI ArFi SE (120)
32 nm EUVL (50)
22 nm LELE (200)
22 nm Freeze (200)
22 nm Spacer (200)
22 nm HI LELE (135)
22 nm EUVL (100)

300%
250%
200%
150%
100%
50%
0%

Results - 32 nm hp (20,000 w/m)

SEMATECH

Legend:
- Reticle
- Clean
- Etch
- Metrology
- Deposition
- Litho

High EUV litho cost because of capital cost

DPL most expensive because of reticle, deposition and etch costs

45 nm

32 nm

Technology (wph)	ArFi SE (125)	ArFi DPL LELE (180)	ArFi DPL Freeze (180)	ArFi DPL Spacer (180)	HI ArFi SE (120)	EUVL (50)

Y-axis: 0%, 20%, 40%, 60%, 80%, 100%, 120%, 200%

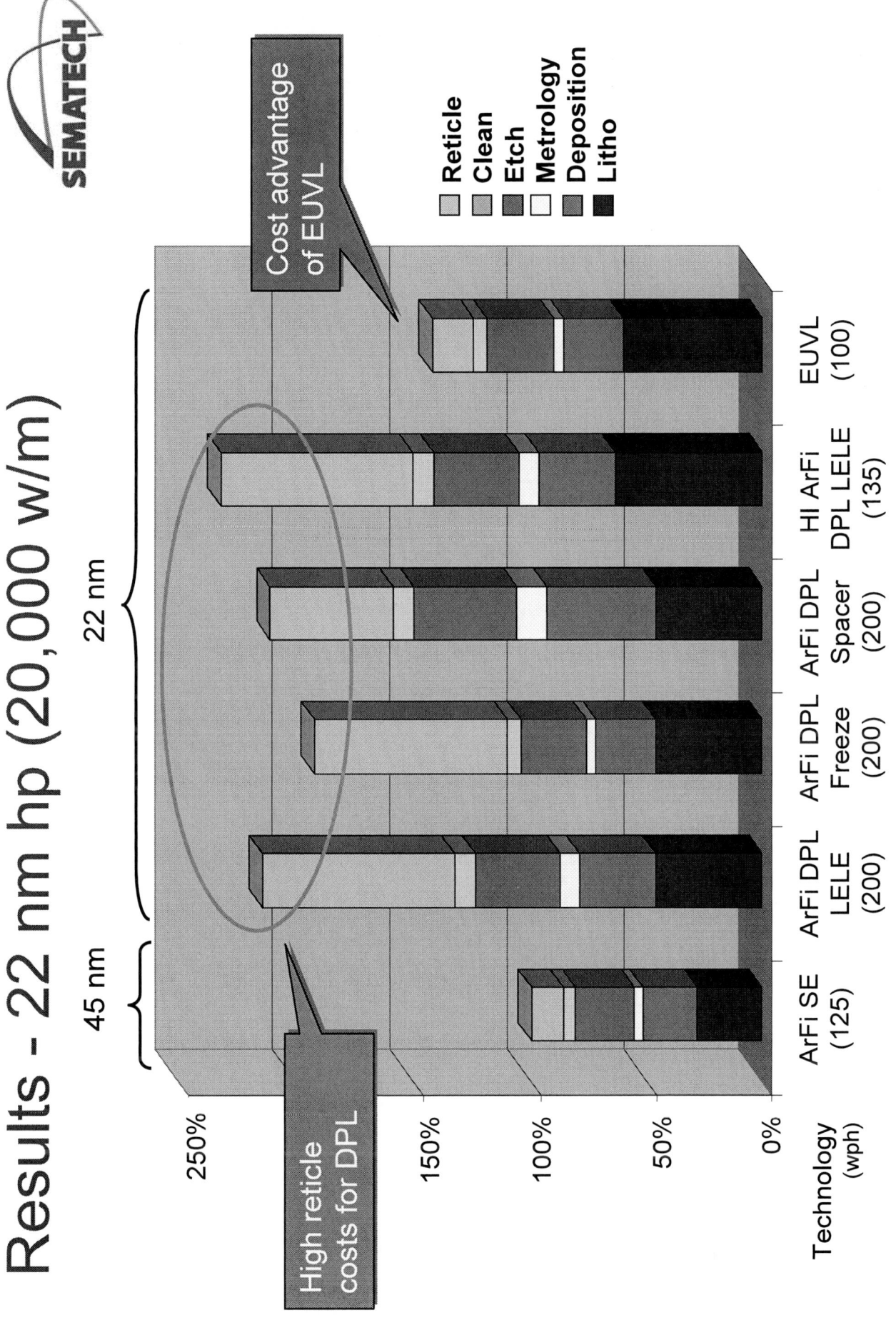

Overview

- Introduction / Motivation
- Calculation Procedure
- Results
- EUVL Considerations
- Conclusions

EUVL Cost Parameters

- Contribution of individual parameters to COO?

- Sensitivity Analysis:
  - What are the target values of cost parameters such that EUVL remains as cost-effective as LELE?

EUVL Cost Parameters

- Contribution of individual parameters to COO?

- Sensitivity Analysis:
  - What are the target values of cost parameters such that EUVL remains as cost-effective as LELE?

EUVL: Equipment Costs

SEMATECH

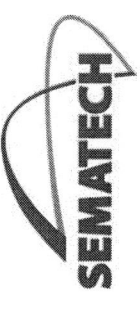

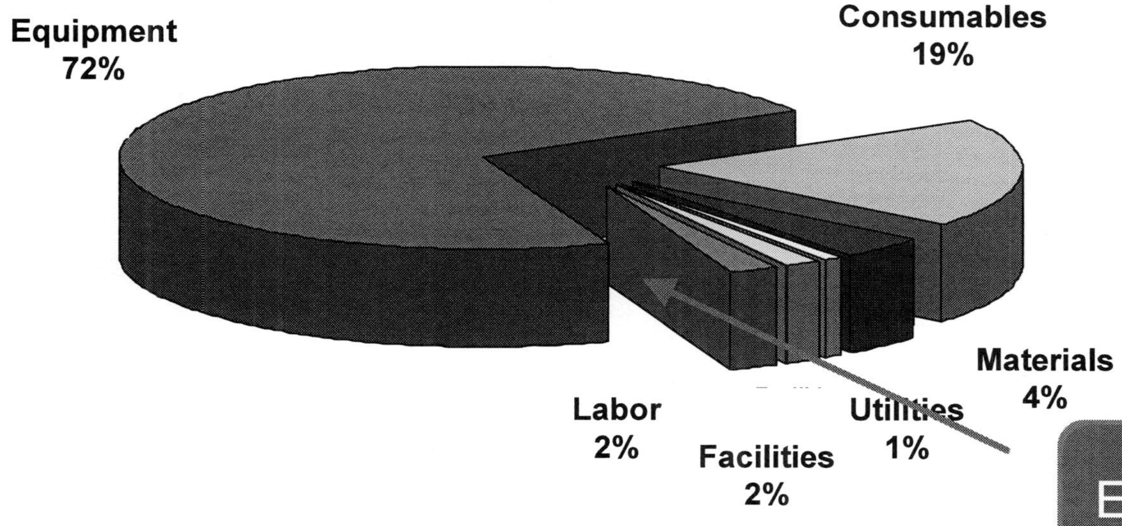

Equipment
72%
Consumables
19%
Materials
4%
Utilities
1%
Facilities
2%
Labor
2%

32 nm (50 wph)
Litho cost excluding reticle

Equipment main cost factor

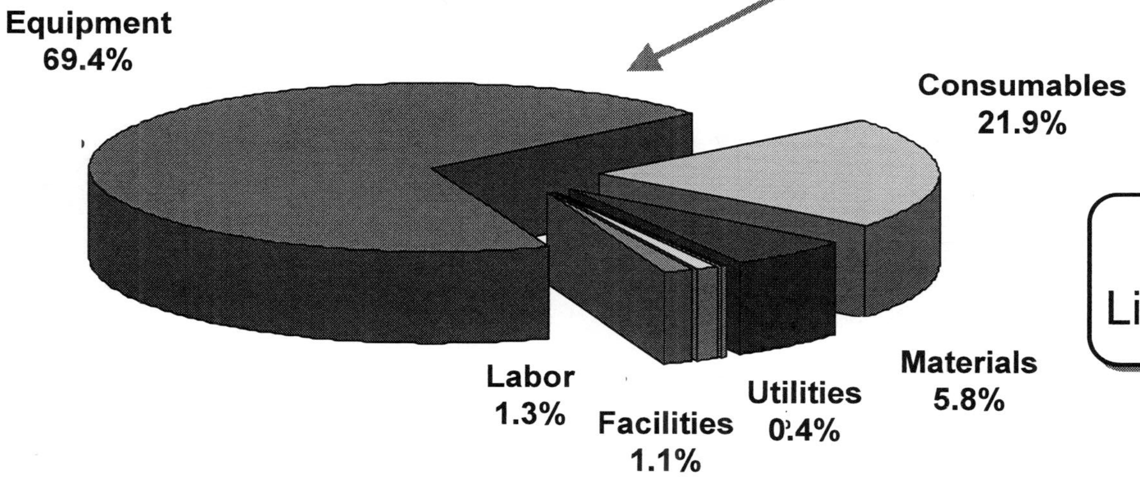

Equipment
69.4%
Consumables
21.9%
Materials
5.8%
Utilities
0.4%
Facilities
1.1%
Labor
1.3%

22 nm (100 wph)
Litho cost excluding reticle

EUVL: Consumable Costs

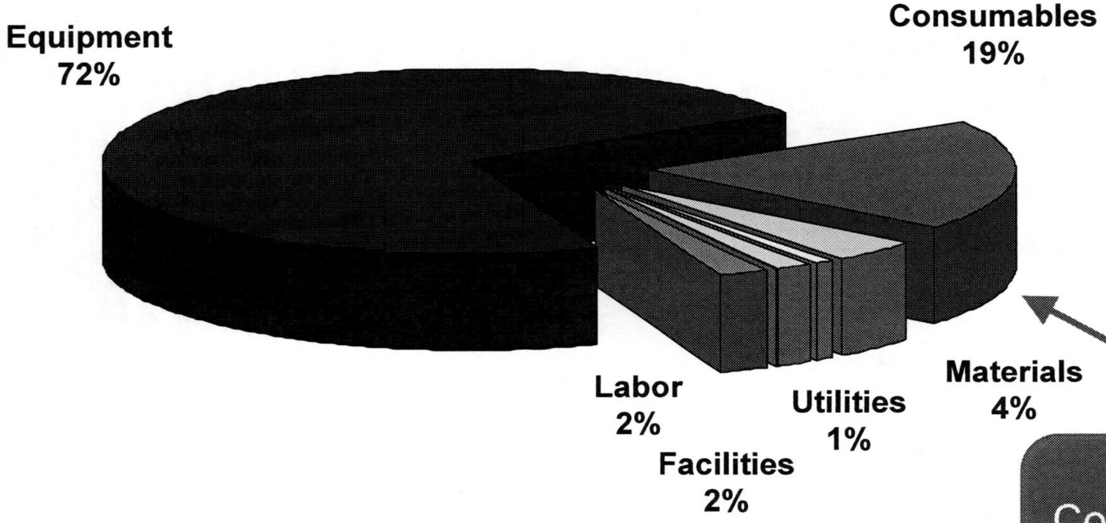

32 nm (50 wph)
Litho cost excluding reticle

Consumables and materials (resist) are significant cost factor

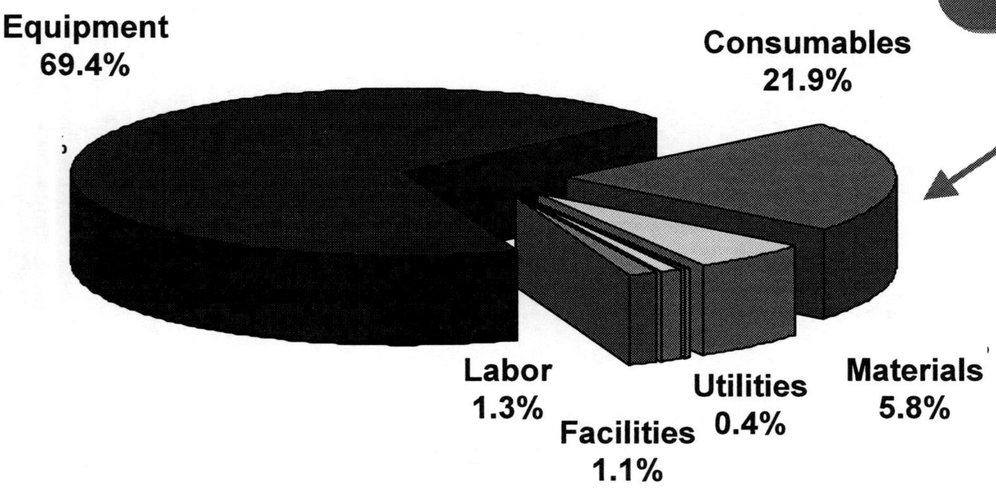

22 nm (100 wph)
Litho cost excluding reticle

Consumables 19%

Materials 4%

Utilities 1%

Facilities 2%

Labor 2%

Equipment 72%

Consumables 21.9%

Materials 5.8%

Utilities 0.4%

Facilities 1.1%

Labor 1.3%

Equipment 69.4%

EUVL: Labor, Facilities, Utilities Costs

32 nm (50 wph)
Litho cost excluding reticle

Consumables
19%

Materials
4%

Utilities
1%

Facilities
2%

Labor
2%

Equipment
72%

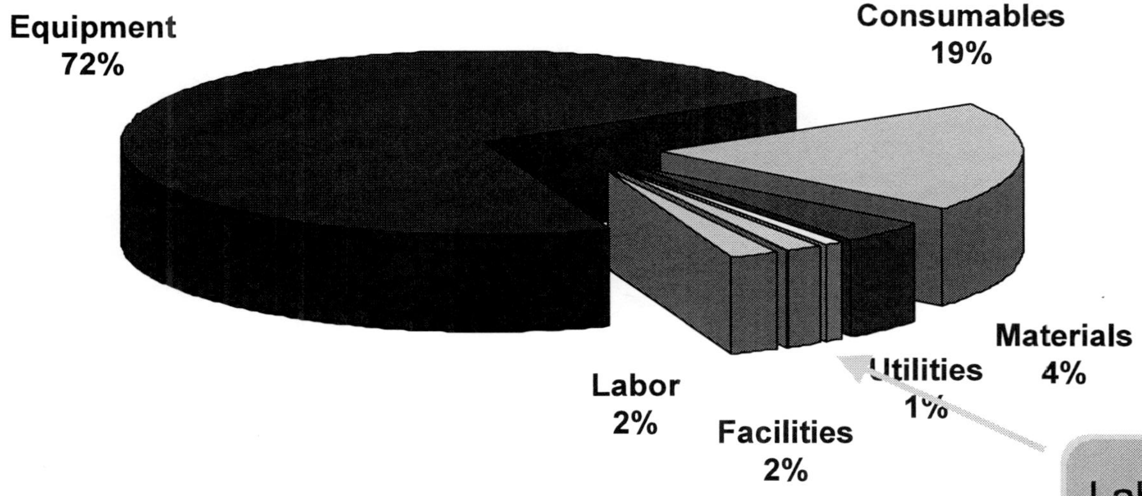

Labor, facilities and utilities
very small contribution

22 nm (100 wph)
Litho cost excluding reticle

Consumables
21.9%

Materials
5.8%

Utilities
0.4%

Facilities
1.1%

Labor
1.3%

Equipment
69.4%

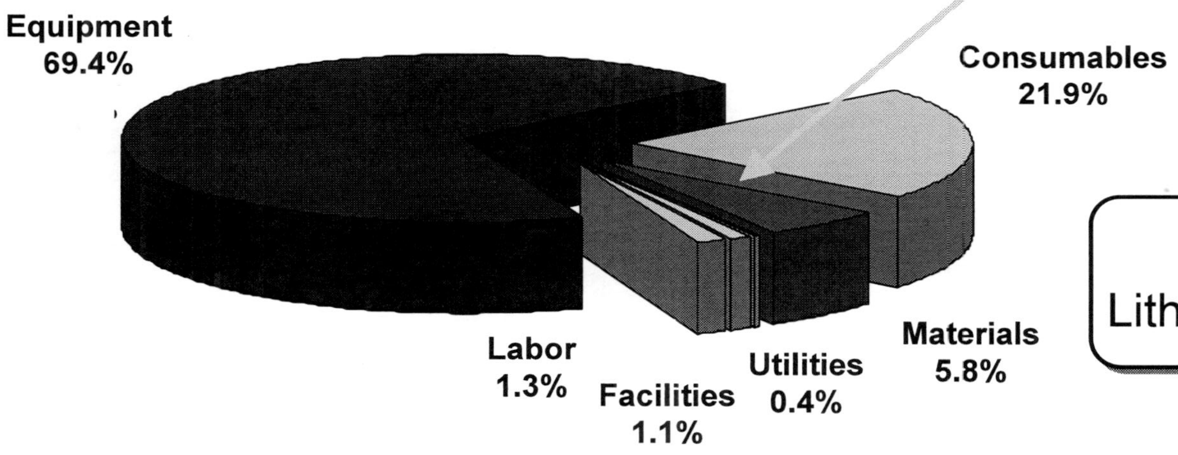

EUVL: Power Consumption Costs

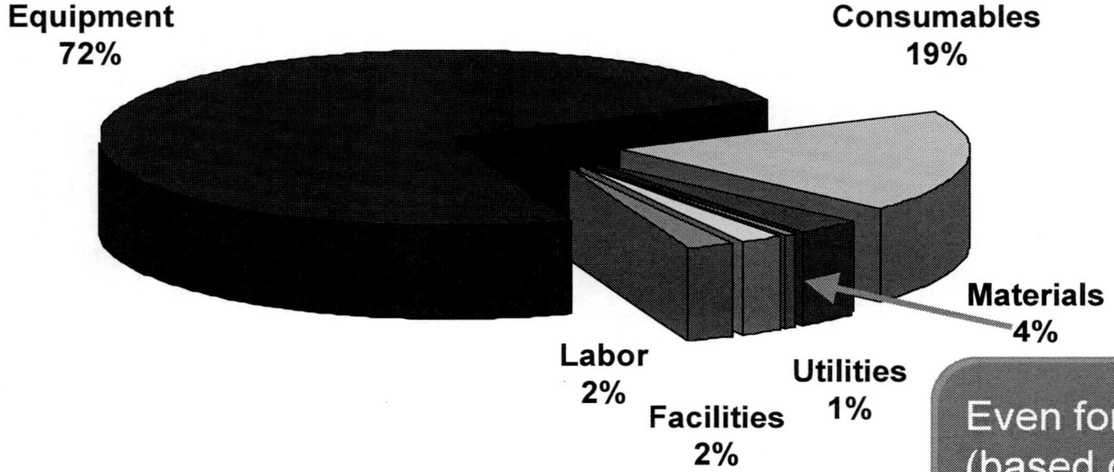

32 nm (50 wph)
Litho cost excluding reticle

Even for power consumption of 200 kW (based on 20 kW CO_2 laser for LPP) facilities cost is negligible: 1-2 % of total litho cost excluding reticle.

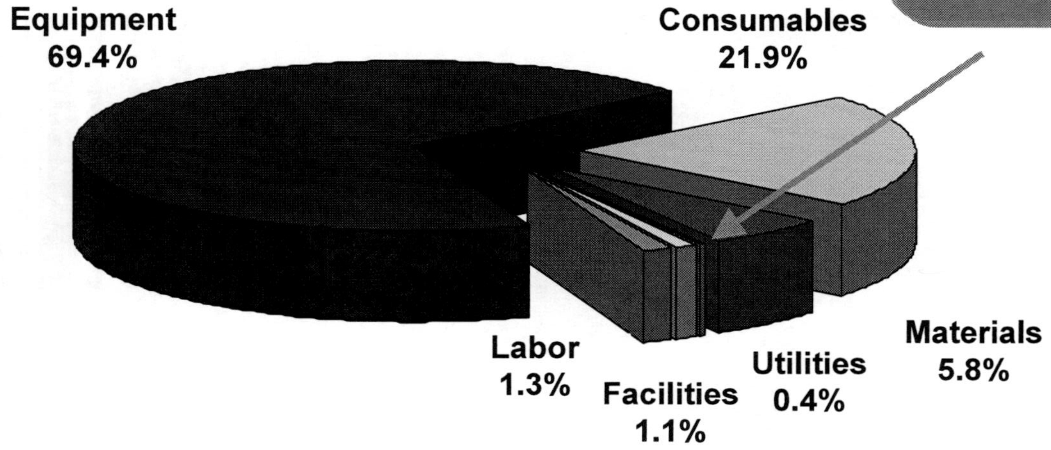

22 nm (100 wph)
Litho cost excluding reticle

Consumables 19%

Materials 4%

Utilities 1%

Facilities 2%

Labor 2%

Equipment 72%

Consumables 21.9%

Materials 5.8%

Utilities 0.4%

Facilities 1.1%

Labor 1.3%

Equipment 69.4%

EUVL Cost Parameters

- Contribution of individual parameters to CoO?

- Sensitivity Analysis:
  - What are the target values of cost parameters such that EUVL remains as cost-effective as LELE?

EUVL Throughput Sensitivity Analysis

20,000 wpm

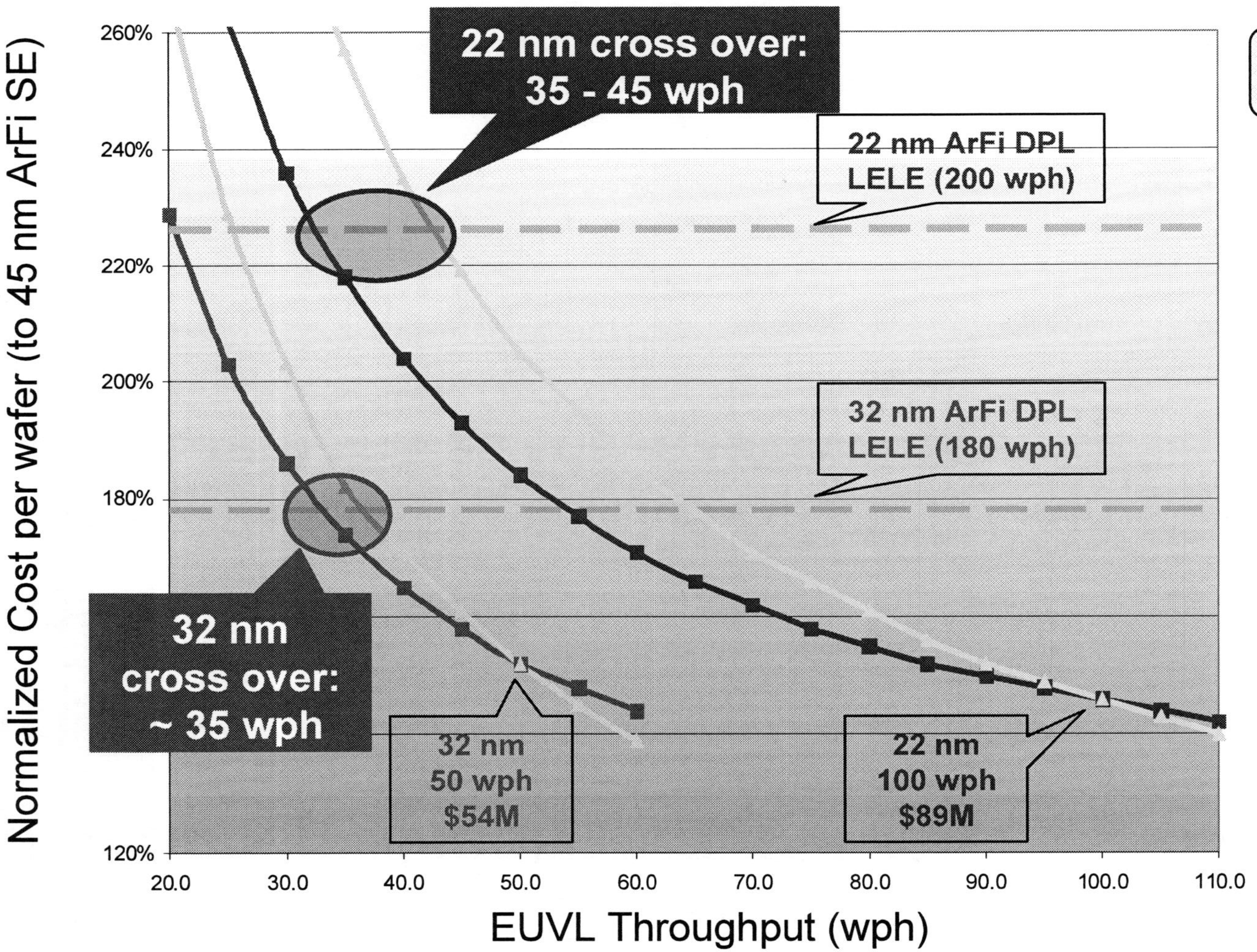

22 nm cross over: 35 - 45 wph

22 nm ArFi DPL LELE (200 wph)

32 nm ArFi DPL LELE (180 wph)

32 nm cross over: ~ 35 wph

22 nm 100 wph $89M

32 nm 50 wph $54M

Normalized Cost per wafer (to 45 nm ArFi SE)

EUVL Throughput (wph)

- Constant and variable tool price as function of throughput assumed

EUVL Sensitivity Analysis

- Repeat analysis for different wafer/mask values
- Perform same analysis for EUVL utilization

Comparing to

LELE 180 wph	LELE 200 wph	LELE 83 %	LELE 83 %

Wafers / mask	EUVL Throughput (wph)		EUVL Utilization	
	32 nm	22 nm	32 nm	22 nm
1,000	<10	<5	<10 %	<10 %
20,000	35 / 40	35 / 45	60%	35%
50,000	45 / 45	50 / 60	75%	50%
100,000	50 / 50	65 / 70	83%	60%

Throughput > 30 wph needed (U kept fixed)

Utilization can be lower than LELE (TPT kept fixed)

The numbers on the left/right correspond to variable/constant tool price.

Overview

- Introduction / Motivation
- Calculation Procedure
- Results
- EUVL Considerations
- Conclusions

Conclusions

- At 22 nm hp, EUVL has cost advantages over DPL because of fewer process steps and lower reticles costs

- EUVL equipment cost contributes strongly, facilities negligible

- EUVL throughput > 30 wph required to be as cost-effective as LELE (under assumptions made)

- Except for EUVL, reticles are significant cost component (even at 20,000 wpm)

Acknowledgements

- Dennis Fandel, Jackie Ferrelll, Jacque Georger, Chawon Koh, Bob Rulliffson, Phil Seidel, Larry Smith, Robert Wright (SEMATECH)

- Will Conley (Freescale), Rob Crowell (TEL),Hiroyuki Mizuno (Toshiba), Nick Stacey (Molecular Imprints), Obert Wood (AMD)

- Céline Lapeyre (CEA-LETI Minatec), Gary Zhang (Rohm and Haas)

AUTHOR INDEX

Agovic, K. 1119
Ahn, B.-S. 646, 721
Allred, B. 1479
Amano, T. 1565, 1593
Amemiya, M. 1564, 1631
Anazawa, T. 1611
Anderson, C. N. 312, 426, 487, 1389, 1390, 1500
Antohe, A. 602, 1634, 1650
Aoyama, H. 1174, 1593
Aquila, A. 1560
Aramaki, F. 1565
Arisawa, Y. 1593
Auželyte, V. 363
Ayers, T. 1461
Bacuita, T. 1433
Banine, V. 1018, 1055, 1481
Bear, W. L. 1418, 1449, 1668
Bender, M. 602
Bergeson, S. 1479
Bernath, R. 956
Bianucci, G. 1041
Bollanti, S. 1391
Borisov, V. M. 996
Borisova, G. N. 996
Brainard, R. 1529
Brandt, D. C. 890
Brunner, T. 1227
Burkhardt, M. 1227
Caudillo, R. 1433
Cha, B. 663
Chiba, H. 1275
Cho, H. K. 646, 721, 848
Cho, W. 721
Choi, S. 1201
Cloutier, P. 1529
Conte, A. 1666
Corthout, M. 935
Cummings, K. 199, 291
Cunado, J. 956
De Groote, F. P. J. 1119
Denbeaux, G. 602, 1634, 1650
Deng, Y. 426, 1227
Denham, P. 312, 1560
Di Lazzaro, P. 1391
Echigo, M. 1499
Ehm, D. 1055
Eichenlaub, S. 663
Ellwi, S. 956
Endo, A. 916
Engelstad, R. 689
Fan, Y. 602, 1634
Flora, F. 1391
Fulford, B. 956
Fumar-Pici, A. 291, 602
Gallatin, G. M. 487

Garg, R. 602, 1634, 1650
Gargini, P. 136
George, S. A. 312, 956, 1530, 1560
Gielissen, K. 1481
Glushkov, D. 1481
Goethals, A. M. 742, 1242
Goldberg, K. 312, 602, 721, 867, 1563, 1633
Goldstein, M. 1321, 1416, 1480
Goo, D. 1201
Goodwin, F. 520, 602, 1417, 1480
Grove, D. 1461
Gullikson, E. 583, 623, 1560
Gustafson, D. 1461
Hada, H. 1501, 1513
Hagiwara, R. 1565
Han, H. 646, 721
Han, S. I. 291
Han, W. 71, 1201
Hansen, S. 291
Harned, N. 199, 1055
Hartley, J. 689, 1417
Hartman, R. 1, 97
Hasegawa, T. 264
Hay, N. 956
Hayes, A. 709
Hazelton, A. J. 520
He, L. 1155
Henderson, C. L. 376
Henderson, I. 956
Hirai, Y. 1542
Hirayama, T. 1501, 1513
Hoef, B. 312
Holfeld, C. 602
Honda, T. 264
Hotta, K. 1462
House, M. 663
Hudyma, R. 1321, 1341, 1416
Hughes, G. 520
Huh, S. 602, 646, 867
Huijbregtse, J. 1055
Huli, L. 291
Hultermans, B. 199
Ikuta, Y. 563
Irie, M. 1513
Ishihara, S. 1583
Itani, T. 335, 1499, 1531, 1542, 1561
Ivanov, A. S. 996
Iwai, T. 1513
Iwamoto, F. 742
Iwasaki, T. 805
Iwashita, J. 1513
Jak, M. 1018
Jonckheere, R. 742
Jones, G. 312
Kadaksham, A. J. 663

AUTHOR INDEX

Kai, T. 1542
Kamo, T. 1174
Kamptaprasad, R. 956
Kaneyama, K. 335
Kang, I.-Y. 721
Kashiwamura, T. 1531
Kawamura, D. 335, 1542
Kearney, P. 709
Kemp, D. 1530, 1560
Kemp, J. 1479
Kempen, A. 1055
Kessels, B. 291
Khristoforov, O. B. 996
Kim, D. 646, 721, 848
Kim, G. 721
Kim, H. 646, 721
Kim, I. 1201
Kim, S.-S. 646, 721, 848
Kinoshita, H. 848, 1501, 1562
Kirukhin, Y. B. 996
Kishimoto, J. 848
Kluenkov, E. 1018
Knight, L. 1479
Koay, C.-S. 1227
Kobayashi, S. 335, 1561
Koh, C. 312, 398, 1417
Koida, K. 1648
Komori, H. 1482
Kondo, H. 1275
Kools, J. 1041
Koops, R. 1119
Koster, N. B. 1119
Kozawa, T. 505
Kuerz, P. 1300
La Fontaine, B. 312, 602, 867, 1227
Landoni, C. 1666
Langer, E. 602
Laubis, C. 1667
Lawson, R. 376
Lee, C.-T. 376
Lee, D. 646, 721, 848
Lee, J. 1201
Lee, S. 721, 1583
Levinson, H. J. 21
Liang, T. 583, 829
Lin, C. C. 709
Litt, L. C. 520
Loginova, E. 1529
Lok, S. 199
Lopatin, A. 1018
Lopez, J. 1479
Luchin, V. 1018
Lyons, A. 689
Lytle, W. M. 1077
Ma, A. 312, 398, 829

Madey, T. E. 1141, 1529
Magoshi, S. 1174, 1483
Mallmann, J. 291
Manini, P. 1666
Maruyama, K. 1542
Mbanaso, C. 1634, 1650
McIntyre, G. 1227
Meijerink, M. G. H. 1119
Meiling, H. 199
Mezi, L. 1391
Mimura, T. 1513
Mirkarimi, P. 709
Mishchenko, V. A. 996
Miura, T. 230
Miyakawa, R. 312, 1632, 1633
Mizokoshi, H. 1462
Mizuno, H. 291, 1227
Mochi, I. 867, 1563
Montgomery, W. 312
Moors, R. 1055
Mori, H. 1483
Mori, I. 1, 183, 551
Morishima, H. 264
Murakami, K. 1275
Murra, D. 1391
Nakamura, J. 505
Nanver, L. K. 1667
Naulleau, P. 10, 124, 312, 398, 426, 487, 602, 867, 1321, 1367, 1389, 1390, 1416, 1500, 1530, 1560, 1563, 1632, 1633
Neumann, M. J. 1077
Niakoula, D. 312, 1560
Nihtianov, S. N. 1667
Niibe, M. 1648
Nishihara, K. 975
Nishimura, N. 1564, 1583
Nishiyama, I. 551, 1583, 1611
Nishiyama, Y. 1565, 1611
Nomura, K. 1275
Oguro, D. 1499
Ohara, L. 291
Oizumi, H. 335, 1499, 1531, 1561, 1611
Okamoto, K. 505
Okoroanyanwu, U. 602
Omata, K. 829
Onodera, J. 1501
Orvek, K. 689, 1634, 1650
Oshino, T. 1275
Ota, K. 1564, 1583, 1631
Owada, T. 1531
Park, C. 1201
Park, J. 312, 398, 1201, 1201
Park, S.-J. 829
Partlow, M. 1461
Peters, J. H. 602

AUTHOR INDEX

Petrillo, K. 291
Pierson, B. 199
Pierson, W. 291
Prokofiev, A. V. 996
Rabellino, L. 1666
Raghunathan, S. 1417
Raju, R. 1077
Randive, R. 709
Rastegar, A. 663
Reiss, I. 709
Rekawa, S. 1530, 1560
Ren, L. 1417
Richardson, M. 956
Riddle, S. 1666
Roberts, J. 1433
Roller, J. 312
Ronse, K. 742
Rossinger, M. 602
Routh, R. 291
Ruzic, D. N. 1077
Sahoo, P. 363
Saidani, M. 363
Salamassi, F. 623
Salashchenko, N. 1018
Salmaso, G. 1041
Sanche, L. 1529
Santillan, J. J. 1561
Sarubbi, F. 1667
Sasaki, T. 1562
Sato, H. 1462
Schietinger, C. 1461
Scholze, F. 1667
Schwarz, M. 1341
Seo, H.-S. 646, 721, 848
Shevelko, A. 1479
Shigemura, H. 551, 1565, 1593
Shiina, K. 1565
Shimizu, M. 1542
Shimokawa, T. 1542
Shin, H. 1077
Shiono, D. 1501, 1513
Shirai, S. 1174, 1483
Shroff, Y. A. 623
Sidelnikov, Y. 1481
Sjmaenok, L. 1018
Sligte, E. te 1055
Soer, W. 1018, 1481
Sohn, J. 689
Solak, H. H. 363
Spiller, E. 709
Sporre, J. 1077
Stepanenko, N. 742
Storm, A. 1055
Suga, O. 551, 805, 1564, 1565, 1583, 1593, 1611, 1631

Sugiyama, T. 709
Szilagyi, J. 956
Tagawa, S. 505
Taguchi, T. 551, 1564, 1583, 1631
Takenoshita, K. 956
Tamura, T. 829
Tanaka, H. 1174, 1483
Tanaka, T. 805, 1593
Tanaka, Y. 1174, 1483
Tawarayama, K. 1174, 1483
Tchikoulaeva, A. 774
Teramoto, Y. 1462
Terasawa, T. 551, 805, 1565
Thomas, M. 1341
Tolbert, L. M. 376
Tomie, T. 805
Torre, A. 1391
Trogisch, S. 602
Tsybin, N. 1018
Uzawa, S. 264
Van Der Mullen, J. 1481
Van Herpen, M. 1481
Van Ingen-Schenau, K. 291
Van Kampen, M. 1055
Van Setten, E. 199
Vandentop, G. 1433
Vanneer, R. 1055
Verberk, R. 1055
Vinokhodov, A. Y. 996
Wagner, C. 199
Wallow, T. 291, 312, 426, 602
Warisawa, S. 1583
Watanabe, T. 848, 1501, 1562
Watso, R. 291
Weber, A. 363
Wieringa, F. P. 1119
Wood, O. 10, 291, 602, 1227, 1367
Wüest, A. 520, 602, 1529, 1634
Wurm, S. 1, 312, 398, 1367, 1417, 1480
Xiong, X. 1418, 1449, 1668
Yakshinskiy, B. 1141, 1529
Yakunin, A. 1018
Yamamoto, T. 1583
Yamane, T. 805
Yamatani, D. 1462
Yan, P.-Y. 623
Yankulin, L. 602, 1634
Yasaka, A. 1565
Yeo, J.-H. 1201
Yokokoji, O. 1562
Yokoyama, T. 1462
Yomogita, A. 1531
Yoshioka, M. 935
Yun, H. 663, 709
Zakhor, A. 1632

AUTHOR INDEX

Zalkind, S. 1141
Zhang, G. 829, 1563
Zimmerman, J. 1155
Zocchi, F. E. 1041

SEMATECH
2706 Montopolis Drive
Austin, Texas 78741

ISBN 978-1-61567-661-3

International Symposium on Extreme Ultraviolet Lithography 2008

(2008 EUVL Symposium)

Lake Tahoe, California, USA
28 September – 1 October 2008

Volume 2 of 3

International Symposium on Extreme Ultraviolet Lithography 2008

(2008 EUVL Symposium)

Lake Tahoe, California, USA
28 September – 1 October 2008

Volume 2 of 3

ISBN: 978-1-61567-661-3

Printed from e-media with permission by:

Curran Associates, Inc.
57 Morehouse Lane
Red Hook, NY 12571

Some format issues inherent in the e-media version may also appear in this print version.

Copyright© (2008) by SEMATECH
All rights reserved.

Printed by Curran Associates, Inc. (2009)

For permission requests, please contact SEMATECH
at the address below.

SEMATECH
2706 Montopolis Drive
Austin, Texas 78741

Phone: (512) 356-3500
Fax: (512) 356-7848

www.sematech.org

Additional copies of this publication are available from:

Curran Associates, Inc.
57 Morehouse Lane
Red Hook, NY 12571 USA
Phone: 845-758-0400
Fax: 845-758-2634
Email: curran@proceedings.com
Web: www.proceedings.com

TABLE OF CONTENTS

Volume 1

OPENING PRESENTATIONS

Welcome to the 2008 EUVL Symposium! 1
Stefan Wurm, Ichiro Mori, Rob Hartman

Program Logistics & Overview 10
Obert Wood, Patrick Naulleau

EUV Lithography's Future 21
Harry J. Levinson

Samsung Lithography Strategy 71
Woosung Han

EUV Activities the EUVL Shop Future Plans 97
Rob Hartman

US Regional Update 124
Patrick Naulleau

IEUVI Update 136
Paolo Gargini

Asia Pacific Regional Update - Japan, Korea and Taiwan Regional Update 183
Ichiro Mori

TECHNICAL SESSION: EXPOSURE TOOL

EUV Alpha Demo Tools – Stepping Stones Towards Volume Production 199
H. Meiling, S. Lok, B. Hultermans, E. van Setten, B. Pierson, K. Cummings, C. Wagner, N. Harned

Nikon EUVL Development Progress Update 230
Takaharu Miura

Development Status of Canon's Full-Field EUVL Tool 264
Shigeyuki Uzawa, Tokuyuki Honda, Hideki Morishima, Takayuki Hasegawa

TECHNICAL SESSION: RESIST I

EUV Resist Performance on the ASML ADT and LBNL MET 291
Bill Pierson, Tom Wallow, Hiroyuki Mizuno, Anita Fumar-Pici, Linda Ohara, Karen Petrillo, Koen van Ingen-Schenau, Steve Hansen, Sang-In Han, Robert Watso, Lior Huli, Obert Wood, Joerg Mallmann, Bart Kessels, Robert Routh, Kevin Cummings

The SEMATECH Berkeley MET: Learning at the 22 nm Node 312
Patrick Naulleau, Chris Anderson, Paul Denham, Simi George, Ken Goldberg, Brian Hoef, Gideon Jones, Dimitra Niakoula, Ryan Miyakawa, John Roller, Chawon Koh, Warren Montgomery, Stefan Wurm, Bruno La Fontaine, Tom Wallow, Andy Ma, Joo-on Park

Evaluation of EUV Resists at Selete 335
Hiroaki Oizumi, Daisuke Kawamura, Koji Kaneyama, Shinji Kobayashi, Toshiro Itani

Advances in Resist Testing at PSI EUV-IL Exposure Tool 363
Vaida Auželyte, Pratap Sahoo, Menouer Saidani, Anja Weber, Harun H. Solak

Positive and Negative Tone Molecular Resists for 22-nm Node EUVL Patterning 376
Richard Lawson, Cheng-Tsung Lee, Laren M. Tolbert, Clifford L. Henderson

TECHNICAL SESSION: RESIST II

Sub-22nm Half-Pitch (HP) EUV Resist Imaging Results 398
Chawon Koh, Stefan Wurm, Joo-on Park, Andy Ma, Patrick Naulleau

Corner Rounding in Photoresists for Extreme Ultraviolet Lithography 426
Christopher N. Anderson, Patrick P. Naulleau, Thomas Wallow, Yunfei Deng

Reconciling Resist Resolution Metrics 487
Gregg M. Gallatin, Patrick P. Naulleau, Christopher N. Anderson

Feasibility Study on High-Sensitivity Chemically Amplified Resist by Polymer Absorption Enhancement in Extreme Ultraviolet Lithography 505
T. Kozawa, K. Okamoto, J. Nakamura, S. Tagawa

TECHNICAL SESSION: COST OF OWNERSHIP

Cost Implications of EUV Lithography Technology Decisions 520
Andrea F. Wüest, Andrew J. Hazelton, Greg Hughes, Lloyd C. Litt, Frank Goodwin

Volume 2

TECHNICAL SESSION: MASK I

An Overview of a Development Program for EUVL Mask Technologies in Selete 551
Osamu Suga, Tsuneo Terasawa, Hiroyuki Shigemura, Takao Taguchi, Iwao Nishiyama, Ichiro Mori

Development Status of EUVL Mask Blank and Substrate 563
Yoshiaki Ikuta

Multilayer Defect Compensation to Enable Quality Masks for EUVL Production 583
Ted Liang, Eric Gullikson

EUV Reticle Contamination and Cleaning 602
U. Okoroanyanwu, E. Langer, A. Fumar-Pici, T. Wallow, O. Wood, B. La Fontaine, C. Holfeld, J. H. Peters, M. Bender, M. Rossinger, S. Trogisch, F. Goodwin, A. Wüest, S. Huh, G. Denbeaux, Y. Fan, A. Antohe, L. Yankulin, R. Garg, K. Goldberg, P. Naulleau

High Transmission EUVL Pellicle Development 623
Yashesh A. Shroff, Pei-Yang Yan, Farhad Salamassi, Eric Gullikson

Applying Thinner Absorber to the EUVL Mask: EUV Printability and Integration Issues 646
Hwan-Seok Seo, Dong Gun Lee, Hoon Kim, Sungmin Huh, Byung-Sup Ahn, Hakseung Han, Dongwan Kim, Seong-Sue Kim, Han Ku Cho

TECHNICAL SESSION: MASK II

Study of Pit Defect Formation on EUV Blank Substrates 663
Abbas Rastegar, Sean Eichenlaub, Arun John Kadaksham, Matt House, Brian Cha, Henry Yun

Investigation of a Compensation Method for Pattern Placement Shifts of Chucked EUVL Masks 689
J. Sohn, K. Orvek, R. Engelstad, A. Lyons, J. Hartley

High Throughput Defect Mitigation 709
Patrick Kearney, C. C. Lin, T. Sugiyama, H. Yun, R. Randive, I. Reiss, P. Mirkarimi, E. Spiller, A. Hayes

Characterization of EUV Mask Defects: Printability and Repair Process 721
Hakseung Han, Donggun Lee, Hwan-Seok Seo, Kenneth A. Goldberg, Hoon Kim, Byung-Sub Ahn, In-Yong Kang, Wonil Cho, Sanghyeon Lee, Suyoung Lee, Geunbae Kim, Dongwan Kim, Seong-Sue Kim, HanKu Cho

Mask Defect Printability in Full Field EUV Lithography – Part 2 742
R. Jonckheere, F. Iwamoto, N. Stepanenko, A. M. Goethals, K. Ronse

TECHNICAL SESSION: METROLOGY/INSPECTION

A Practical Approach to EUV Reticle Inspection 774
Anna Tchikoulaeva, et.al.

Development of Actinic Mask Blank Inspection Technology at Selete 805
Tsuneo Terasawa, Takeshi Yamane, Teruo Iwasaki, Toshihiko Tanaka, Osamu Suga, Toshihisa Tomie

EUVL Blank Defect Inspection Capability at Intel 829
Andy Ma, Ted Liang, Seh-Jin Park, Guojing Zhang, Tomoya Tamura, Kazunori Omata

Analysis of Sub-22-nm Aerial Image Using Coherent Scattering Microscopy 848
Dong Gun Lee, Junki Kishimoto, Takeo Watanabe, Hiroo Kinoshita, Hwan-Seok Seo, Dongwan Kim, Seong-Sue Kim, HanKu Cho

Aerial Image Linewidth Measurement Capabilities of the Actinic Inspection Tool 867
Kenneth A. Goldberg, Iacopo Mochi, Patrick Naulleau, Bruno LaFontaine, Sungmin Huh

TECHNICAL SESSION: SOURCE I

Laser Produced Plasma Source System Development 890
    David C. Brandt, et.al.
CO_2 Laser-produced Sn Plasma Source for EUV Lithography 916
    Akira Endo, et.al.
EUV Sources based on DPP 935
    Marc Corthout, Masaki Yoshioka
Time-Multiplexed Solid-State Laser-driven EUV Sources for Beta-Tools and HVM 956
    K. Takenoshita, R. Bernath, R. Kamptaprasad, J. Szilagyi, S. A. George, J. Cunado, M. Richardson, B. Fulford, I.
    Henderson, N. Hay, S. Ellwi

TECHNICAL SESSION: SOURCE II

Guidelines and Promising Approach for LPP-EUV Light Source for HVM 975
    K. Nishihara, et.al.
Technological Aspects of Sn RDE Source Development for HVM Lithography 996
    V. M. Borisov, G. N. Borisova, A. S. Ivanov, Yu. B. Kirukhin, O. B. Khristoforov, V. A. Mishchenko, A. V.
    Prokofiev, A. Yu. Vinokhodov
Spectral Purity Filter Development for EUV HVM 1018
    A. Yakunin, V. Banine, N. Salashchenko, E. Kluenkov, A. Lopatin, V. Luchin, N. Tsybin, L. Sjmaenok, W. Soer, M.
    Jak
Design and Fabrication Considerations of EUVL Collectors for HVM 1041
    G. Bianucci, J. Kools, G. Salmaso, F. E. Zocchi

TECHNICAL SESSION: CONTAMINATION AND PARTICLES

Strategy for Minimizing EUV Optics Contamination During Exposure 1055
    N. Harned, R. Moors, M. van Kampen, V. Banine, J. Huijbregtse, R. Vanneer, A. Kempen, D. Ehm, R. Verberk, E.
    te Sligte, A. Storm
Effective Debris Detection, Mitigation and Cleaning Methods for Source - Collector Optics 1077
    David N. Ruzic, Ramasamy Raju, J. Sporre, H. Shin, W. M. Lytle, M. J. Neumann

Volume 3

TECHNICAL SESSION: CONTAMINATION AND PARTICLES (cont,)

Shielded Plasma's for Cleaning EUV Mirrors 1119
    N. B. Koster, R. Koops, K. Agovic, F. P. J. de Groote, F. P. Wieringa, M. G. H. Meijerink
Carbon Accumulation on Model MLM Cap Layer: Interaction of Benzene Vapor with $TiO2$ Surface 1141
    Boris Yakshinskiy, Shimon Zalkind, Theodore E. Madey
Progress on EUV Reticle Dual Pod Carriers for use in the Fab and Exposure Tools 1155
    John Zimmerman, Long He

TECHNICAL SESSION: EXPOSURE TOOL EVALUATION

Lithographic Performance of Selete's Full Field EUV Exposure Tool 1174
    Kazuo Tawarayama, Hajime Aoyama, Takashi Kamo, Shunko Magoshi, Yuusuke Tanaka, Seiichiro Shirai,
    Hiroyuki Tanaka
Full-field Patterning Test with ADT for 30-nm Node Device Application 1201
    Doohoon Goo, Insung Kim, Joo-On Park, Jeonghoon Lee, Changmin Park, Jinhong Park, Jeong-Ho Yeo,
    Sungwoon Choi, Woosung Han
Flare Evaluation of an ASML Alpha Demo Tool 1227
    Hiroyuki Mizuno, Martin Burkhardt, Chiew-seng Koay, Greg McIntyre, Tim Brunner, Bruno La Fontaine, Yunfei
    Deng, Obert Wood
Implementing Full field EUV Lithography using the ADT 1242
    Anne-Marie Goethals, et.al.

TECHNICAL SESSION: OPTICS

Improvement of Optics for EUV Exposure Tools 1275
Katsuhiko Murakami, Tetsuya Oshino, Hiroyuki Kondo, Hiroshi Chiba, Kazushi Nomura

Optics for EUV Lithography 1300
Peter Kuerz, et.al.

Projection Optics for a 0.5-NA Microstepper Upgrade 1321
Michael Goldstein, Russ Hudyma, Patrick Naulleau

Projection Architectures for High NA EUVL 1341
Russ Hudyma, Mike Thomas, Mark Schwarz

CLOSING ADDRESS

2008 EUVL Symposium - Closing Address 1367
Stefan Wurm, Obert Wood, Patrick Naulleau

POSTERS: EXPOSURE TOOL

Ultra-High Resolution Extreme Ultraviolet Lithography by Incoherent to Coherent Conversion 1389
Christopher N. Anderson, Patrick P. Naulleau

MOSAIC - A New Way to Measure Optical Aberrations 1390
Christopher N. Anderson, Patrick P. Naulleau

First Italian EUV Micro Exposure Tool at 14.4 nm based on Kr DMS 1391
S. Bollanti, P. Di Lazzaro, F. Flora, L. Mezi, D. Murra, A. Torre

Lithographic Modeling of a 0.5-NA Microstepper Optic 1416
Patrick Naulleau, Michael Goldstein, Russ Hudyma

**Impact of Flare and Aberrations on Patterning Performance - Simulation with the EUV Full Field
Alpha Tool Conditions** 1417
Liping Ren, Frank Goodwin, Stefan Wurm, Chawon Koh, Sudharshanan Raghunathan, John Hartley

Multiple Catadioptric Simplified Extreme Ultraviolet Whole Lithography Machine and System 1418
Wynn L. Bear, Xiangwen Xiong

Reliability and Productivity Improvements on the Intel MET 1433
Roman Caudillo, Jeanette Roberts, Terence Bacuita, Gilroy Vandentop

**Composite Double Reflection Simplified Extreme Ultraviolet Whole Lithography Machine and
System** 1449
Wynn L. Bear, Xiangwen Xiong

POSTERS: SOURCE

Spectral Purity Filter Life-Time Testing on EQ-10 Source 1461
Chuck Schietinger, Dave Grove, Travis Ayers, Matthew Partlow, Debbie Gustafson

**Dependence of Laser Parameter on Conversion Efficiency in High-Repetition-Rate Laser-Ablation-
Discharge EUV Source** 1462
Takuma Yokoyama, Hiroshi Mizokoshi, Yusuke Teramoto, Daiki Yamatani, Hiroto Sato, Kazuaki Hotta

EUV Spectrometers for Source Development, Characterization and Optimization 1479
Scott Bergeson, Bryce Allred, Jershon Lopez, Jeffrey Kemp, Larry Knight, Alexander Shevelko

High Power from Low Etendue EUV Light Sources 1480
Michael Goldstein, Stefan Wurm, Frank Goodwin

The Characterization of the Ion Beam from a Sn-based DPP with Respect to the Ignition Parameters 1481
K. Gielissen, Y. Sidelnikov, W. Soer, M. van Herpen, D. Glushkov, V. Banine, J. van der Mullen

Present Status of Laser-produced Plasma EUV Light Source 1482
Hiroshi Komori, et.al.

Performance Evaluation of EUV SFET Source Collector Module 1483
Shunko Magoshi, Seiichiro Shirai, Hideto Mori, Kazuo Tawarayama, Yuusuke Tanaka, Hiroyuki Tanaka

POSTERS: RESISTS

Development of New Negative-tone Molecular Resists based on Alkylphenyl Callxarene for EUVL 1499
Masatoshi Echigo, Dai Oguro, Hiroaki Oizumi, Toshiro Itani

Survey and Comparison of Deprotection Blur Metrics for Extreme Ultraviolet Photoresists 1500
Christopher N. Anderson, Patrick P. Naulleau

Investigation of CA Resist Decomposition by EUV and EB Exposure 1501
Daiju Shiono, Taku Hirayama, Hideo Hada, Junichi Onodera, Takeo Watanabe, Hiroo Kinoshita

Novel Polyphenol Base Molecular Resist Having High Thermal Resistance 1513
Taku Hirayama, Takeyoshi Mimura, Jun Iwashita, Makiko Irie, Daiju Shiono, Hideo Hada, Takeshi Iwai

Electrons in EUV Resist Activation 1529
Theodore E. Madey, B. V. Yakshinskiy, E. Loginova, L. Sanche, P. Cloutier, R. Brainard, A. Wuest

DUV Source Integration into the 0.3 NA Berkeley SEMATECH MET for OOB Exposure Studies 1530
Simi A. George, Patrick P. Naulleau, Senajith Rekawa, Drew Kemp

Development of Novel Positive-tone Photoresists for EUVL 1531
Takanori Owada, Akinori Yomogita, Takashi Kashiwamura, Hiroaki Oizumi, Toshiro Itani

Relation between Acid Diffusion and Resolution in Chemically Amplified EUV Resists 1542
Y.uuki Hirai, Makoto Shimizu, Ken Maruyama, Toshiyuki Kai, Tsutomu Shimokawa, Toshiro Itani, Daisuke Kawamura

EUV Photoresists Twice as Fast as Previously Thought 1560
Patrick Naulleau, Eric Gullikson, Andrew Aquila, Paul Denham, Simi George, Drew Kemp, Dimitra Niakoula, Seno Rekawa

EUV Resist Outgassing Quantification and Qualification Analysis Methods 1561
Shinji Kobayashi, Julius Joseph Santillan, Hiroaki Oizumi, Toshiro Itani

Development of Partially Fluorinated EUV Resist Polymers for Sensitivity Improvement 1562
Takashi Sasaki, Osamu Yokokoji, Takeo Watanabe, Hiroo Kinoshita

POSTERS: MASK

Mask Effects on Line-Edge Roughness (LER) 1563
Patrick P. Naulleau, Kenneth A. Goldberg, Iacopo Mochi, Guojing Zhang

Experimental Study on Flatness of Electrostatically Chucked Reticle 1564
K. Ota, T. Taguchi, M. Amemiya, N. Nishimura, O. Suga

FIB Mask Repair Technology for EUV Lithography 1565
Tsuyoshi Amano, Yasushi Nishiyama, Hiroyuki Shigemura,Tsuneo Terasawa, Osamu Suga, Kensuke Shiina, Fumio Aramaki, Ryoji Hagiwara, Anto Yasaka

Analysis of Entrapped Object Size Effects on Out-of-Plane Distortion of the EUVL Mask in Electrostatic Chucking 1583
S. Lee, T. Yamamoto, K. Ota, N. Nishimura, T. Taguchi, I. Nishiyama, O. Suga, S. Warisawa, S. Ishihara

POSTERS: DEFECT INSPECTION

A Study of Optical Inspection on EUVL Mask for 32 nm Half Pitch Node Device and Beyond 1593
Yukiyasu Arisawa, Hiroyuki Shigemura, Tsuyoshi Amano, Hajime Aoyama, Toshihiko Tanaka, Osamu Suga

POSTERS: RETICLE CONTAMINATION

Characterization of EUV-Deposited Carboneous Contamination 1611
Toshihisa Anazawa, Yasushi Nishiyama, Hiroaki Oizumi, Osamu Suga, Iwao Nishiyama

Experimental Study of Particle-free Mask Handling Techniques using the MPE Tool 1631
Mitsuaki Amemiya, Kazuya Ota, Takao Taguchi, Osamu Suga

POSTERS: OPTICS AND ML COATINGS

Iterative Procedure for in situ Optical Testing using an Incoherent Source 1632
Ryan Miyakawa, Patrick P. Naulleau, Avideh Zakhor

Lateral Shearing Interferometry for EUV Optical Testing 1633
Ryan Miyakawa, Patrick P. Naulleau, Ken Goldberg

POSTERS: OPTICS CONTAMINATION

Resist Outgassing Measurements and Calibrations for High Volume Manufacturing.. 1634
Greg Denbeaux, Alin Antohe, Rashi Garg, Chimaobi Mbanaso, LeonidYankulin, Yu-Jen Fan, Kevin Orvek, Andrea Wüest

Experiment of Contamination Generation by EUV Irradiation with the Use of High-mass Hydrocarbon Gas 1648
Masahito Niibe, Keigo Koida

EUVL Optics Contamination from Resist Outgassing: Status Overview 1650
Kevin Orvek, Greg Denbeaux, Alin Antohe, Rashi Garg, Chimaobi Mbanaso

Moisture and Hydrocarbon Management for EUVL Tools: Ultra High Vacuum and Purge Gas Purification Solutions 1666
Andrea Conte, Cristian Landoni, Paolo Manini, Larry Rabellino, Sarah Riddle

POSTERS: DEVICE INTEGRATION

Characterization of New EUV Stable Silicon Photodiodes... 1667
F. Scholze, C. Laubis, F. Sarubbi, Lis K. Nanver, S. N. Nihtianov

POSTERS: COST OF OWNERSHIP

The Comprehensive Cost for the Mainstream NGL and Simplified Extreme Ultraviolet Lithography Method.. 1668
Wynn L. Bear, Xiangwen Xiong

Author Index

MIRAI **Selete**

An Overview of a Development Program for EUVL Mask Technologies in Selete

Osamu Suga, Tsuneo Terasawa, Hiroyuki Shigemura,
Takao Taguchi, Iwao Nishiyama
and Ichiro Mori

*Millennium Research for Advanced Information Technology (MIRAI)
- Semiconductor Leading Edge Technologies, Inc. (Selete)*

MIRAI *Outline* **Selete**

I. *EUVL Mask Program of Selete*

II. *Current Status And Progress*
   Hard Defect Free Mask

   — *Phase Defect/ Mask Blank Inspection*

   — *Pattern Defect/ Pattern Inspection*

   — *Pattern Defect/ Pattern Repair*

III. *Summary*

EUVL Mask Program

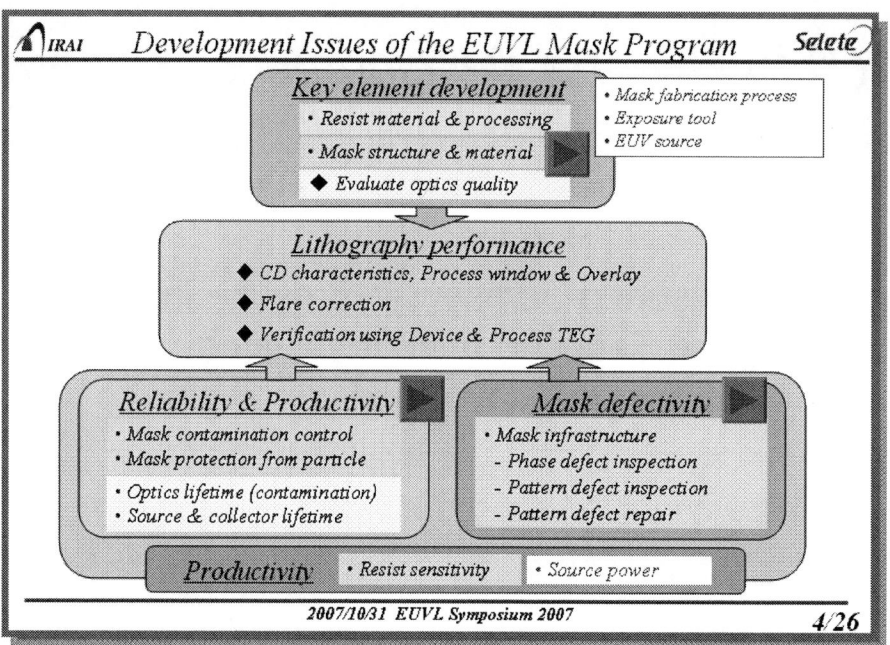

EUVL Mask Program

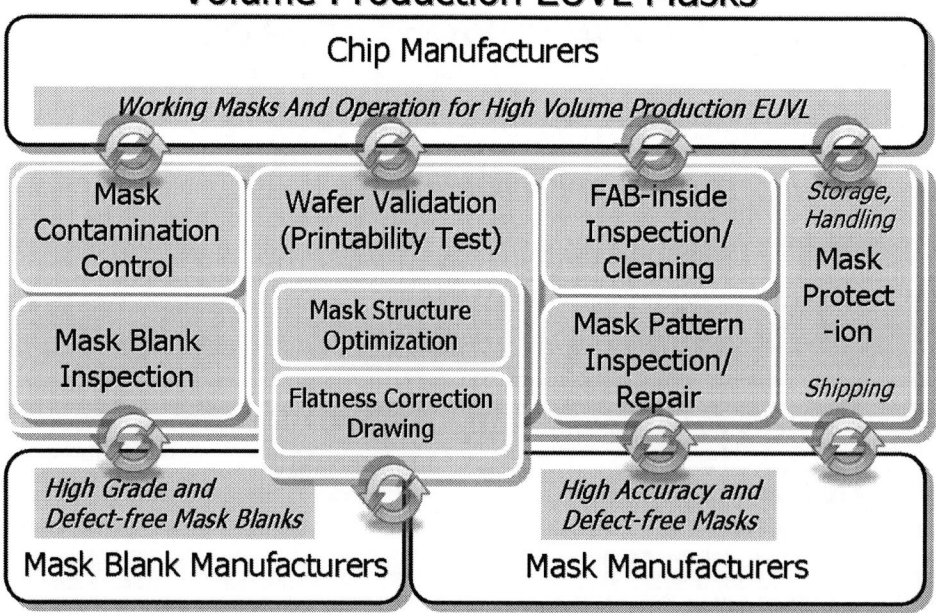

 IRAI ## *Hard and Soft Defect of EUVL Mask*

- *Characterization*
- *Cleaning Technology*

- *Pattern Inspection Technology*
- *Pattern Repair Technology*

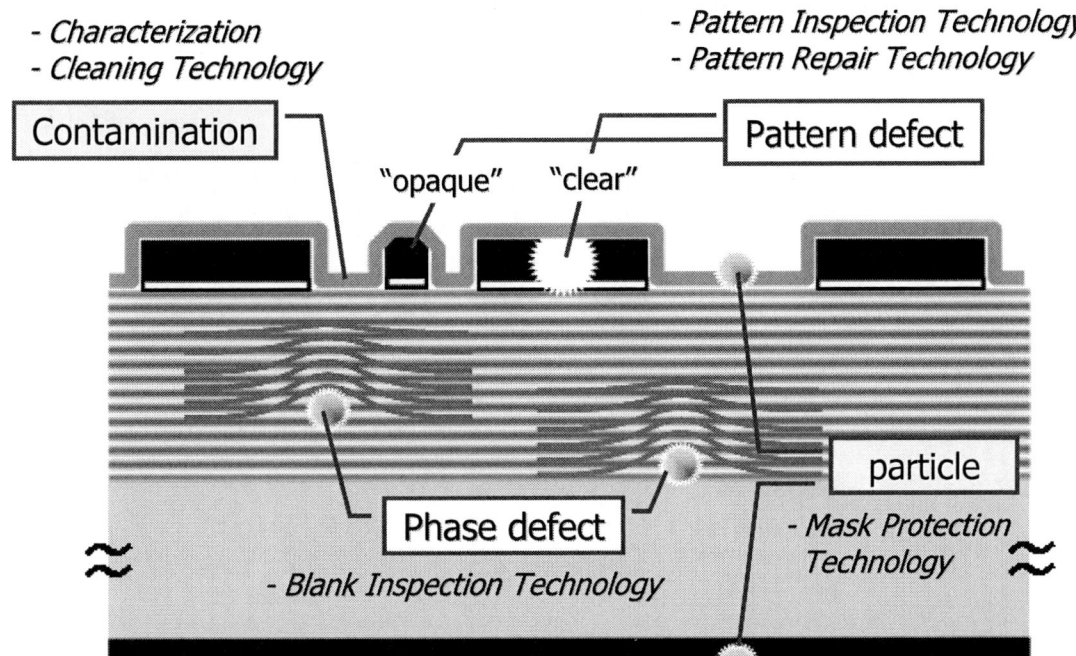

Contamination

"opaque"   "clear"

Pattern defect

Phase defect

- *Blank Inspection Technology*

particle

- *Mask Protection Technology*

 IRAI ## *EUVL Mask Blank Inspection Technology*

Development Methods

■ Full-field mask blank inspection technology is being developed from the MIRAI Project (MIRAI-2) concerning an Actinic Inspection Proof-of-Concept (PoC) tool.

"PoC" Tool (MIRAI-2)

- *actinic inspection: EUV LPP Source*
- *static dark-field image*
- *measurement area: 2x30 mm*
- *smallest detective defect: 1.2nm(height) x60nm(Width) @natural defect*

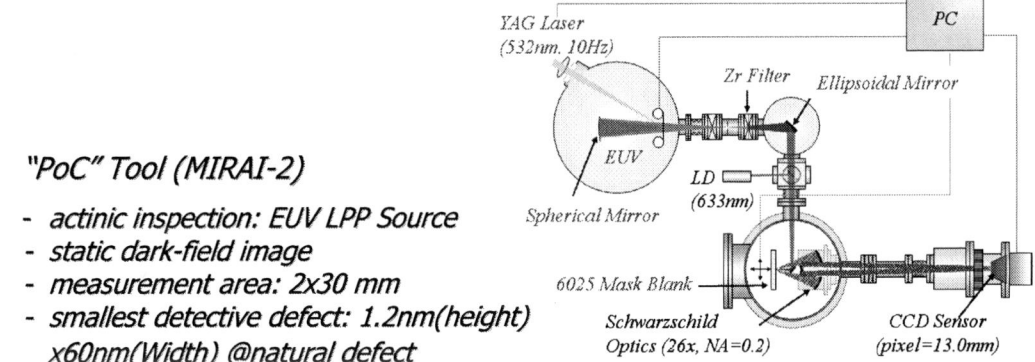

EUVL Mask Blank Inspection Technology

Images of the 1st Results retrieved from the Actinic, Full-field Mask Blank Inspection Tool

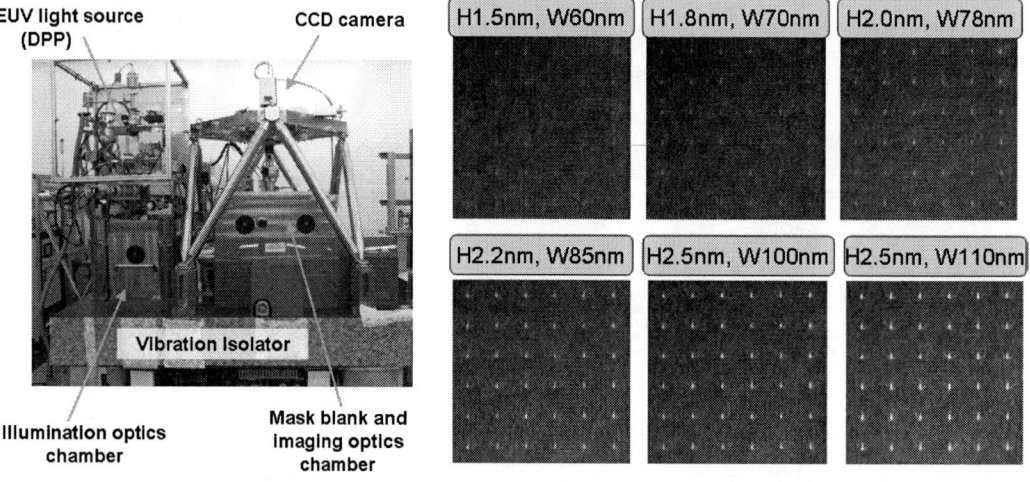

- **EUVL Symposium 2008 [oral 20]**
  T. Terasawa, T. Yamane, T.Iwasaki, T.Tanaka, O. Suga, T. Tomie, "Development of actinic mask blank inspection technology at Selete"

EUVL Mask Inspection Technology

Development Methods

- Selete believed that further developing current Photomask inspection technology would be the fastest and most efficient way to actualize EUVL mask inspection technology.

- Selete started this project by detecting and attempting to address the key technical issues which exist with the most sophisticated and current DUV light inspection system, which is manufactured by Nuflare Technology Inc.

"NPI-5000PLUS"
(Nuflare Technology Inc.)

- *the shortest wavelength light: 198.5nm*
- *the smallest pixel size: 50nm*
- *simultaneous inspection using both of transmissive mode and reflective mode*

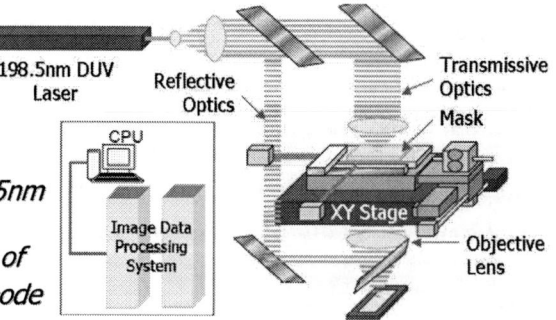

 *IRAI*　　　*EUVL Mask Inspection Technology*　　　*Selete*

The Key limitations of EUVL Mask

- ■ Only One Inspection Option (Single Reflective Mode)
  - Resulting in low Mask Pattern Resolution and Defect Sensitivity.

Proposed Approach to Solution

- ☐ Simulations to Find the Most Effective Solution
- ☐ Development of Inspection Tool Resolution Enhancement Technology
- ☐ Optimization of Mask Structure for Inspection

IRAI　　　*Simulation of Mask Pattern Resolution*　　　*Selete*

Simulated by "EM-Suite" @Panoramic Technology Inc.

The Effects of AR-Layer reflectivity and Tool Illumination Mode on Pattern Resolution

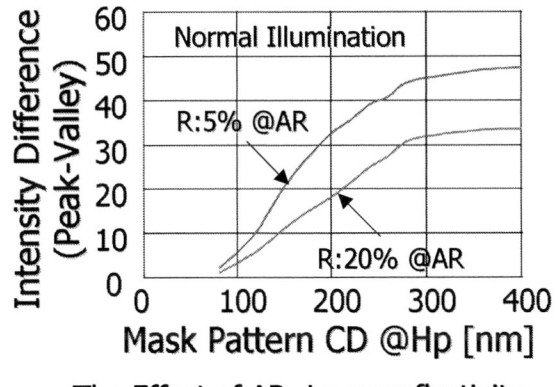

The Effect of AR- Layer reflectivity

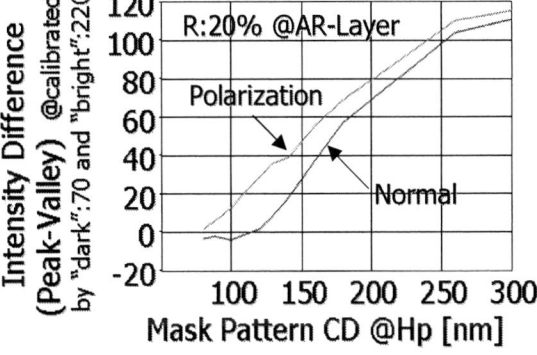

The Effect of illumination mode

– **EUVL Symposium 2008 [poster 72]**
　Y. Arisawa, H. Shigemura, T. Amano, H. Aoyama, T. Tanaka, O. Suga,
　"A study of optical inspection on EUVL mask for 32nm half pitch node device and beyond"

"AR (Anti Reflective)-Layer"

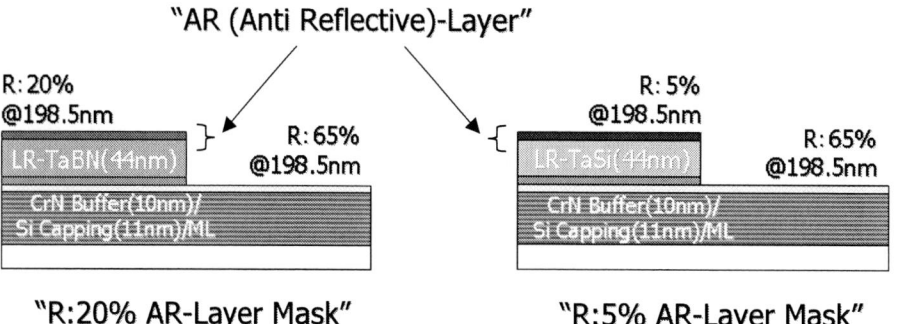

"R:20% AR-Layer Mask" "R:5% AR-Layer Mask"

Simulation of Mask Defect Sensitivity

Simulated by "EM-Suite" @Panoramic Technology Inc.

The Effects of AR-Layer reflectivity and
Tool Illumination Mode on Defect Sensitivity

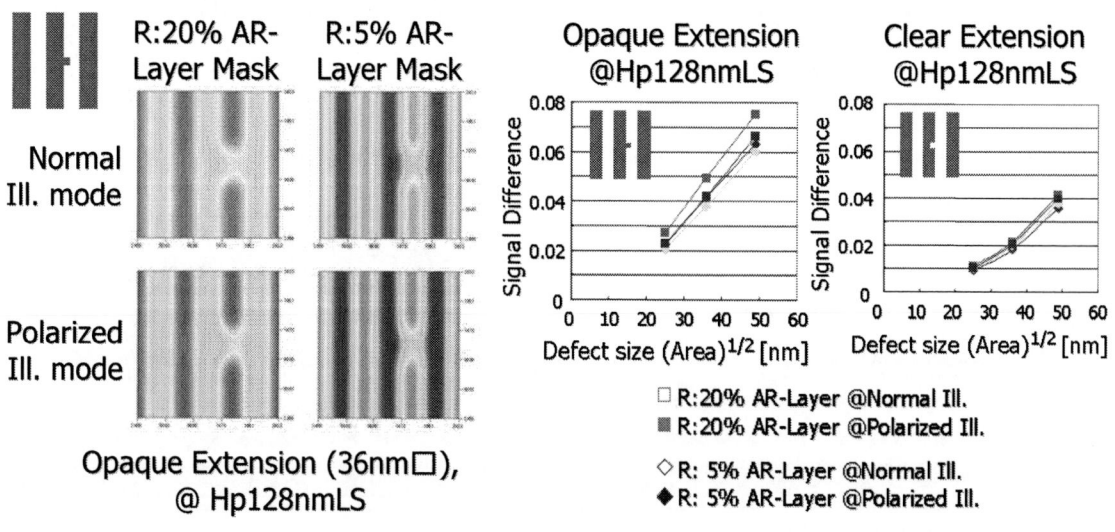

A New EUVL Mask Inspection Tool

■ Resolution Enhancement Technology (Polarized Illumination) and New Inspection Algorithm for Reflective Inspection

■ Low Noise Mechanical and Optical Systems

 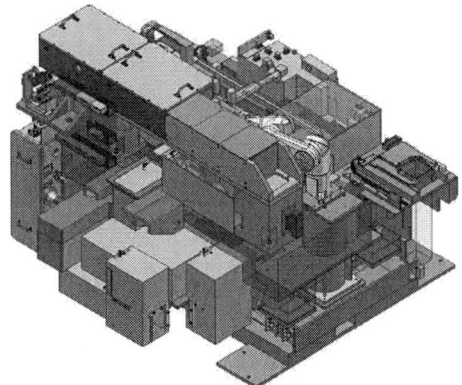

A New EUVL Mask Inspection Tool with 198.5nm DUV light
(NuFlare Technology Inc./ Advanced Mask Inspection Technology Inc.)

Impact on Mask Pattern Resolution

The Effect of Illumination Mode @R:20% AR-Layer

The Effect of Reflectivity AR-Layer @Normal Illumination

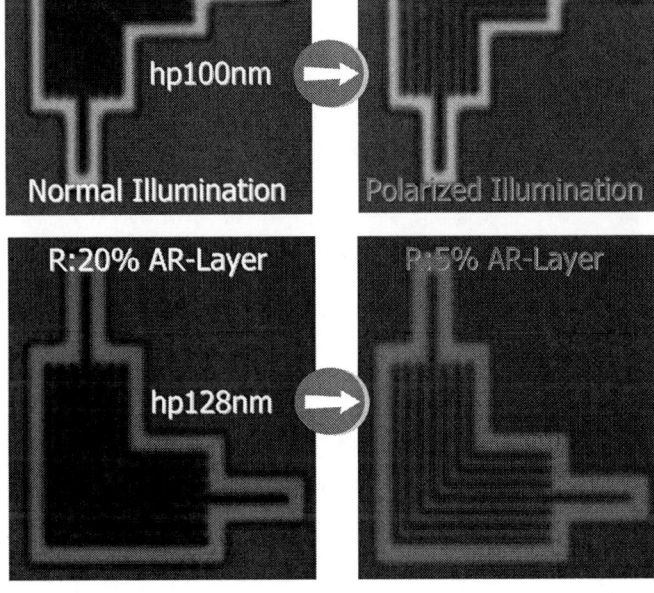

 *IRAI* ## EUVL Mask Repair Technology *Selete*

Development Method

- Selete believed that further developing current Photomask repair technology would be the fastest and most efficient way to actualize EUVL mask repair technology.
- Selete started this project by detecting and attempting to address the key technical issues which exist with the most sophisticated and current FIB repair system, which is manufactured by SII-NanoTechnology Inc.

"SIR-7" (SII-NanoTechnology Inc.)

- *high repair accuracy: $3\sigma \leqq 4nm$*
- *low accelerated voltage: 10~15kV*
- *Gas Assisted Etching (GAE) by FIB*

IRAI ## EUVL Mask Repair Technology *Selete*

The Key Limitations of FIB Repair

- Damage such as Absorber Side-etching, Mixing Layer and Reflectivity Loss, etc.
- Restricted Accuracy

Proposed Solution Approach

- ☐ Development of a High Accuracy Repair Tool
- ☐ Mask Damage Analysis and Development of Mask Damage Mitigation Repair Processes
- ☐ Optimization of Mask Structure for Repair

 IRAI

A New EUVL Mask Repair Tool

Selete

- New FIB/SEM Double-beam System for High Visibility and Three Dimensional FIB Repair with SEM-Imaging and a Rotated Mask Stage
- High Selectivity Etching Control

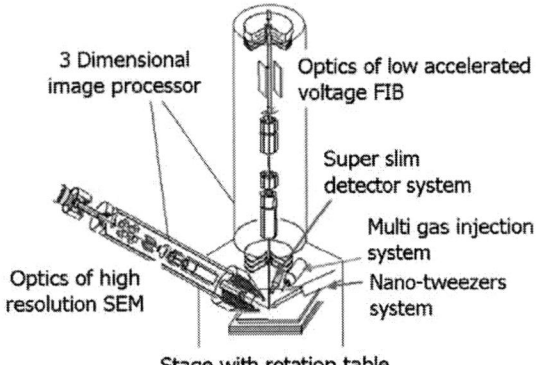

A new EUVL Mask Repair Tool with the FIB/SEM Double-beam System (SII-NanoTechnology Inc.)

 IRAI

Side-etching Analysis of TaBN Absorber

Selete

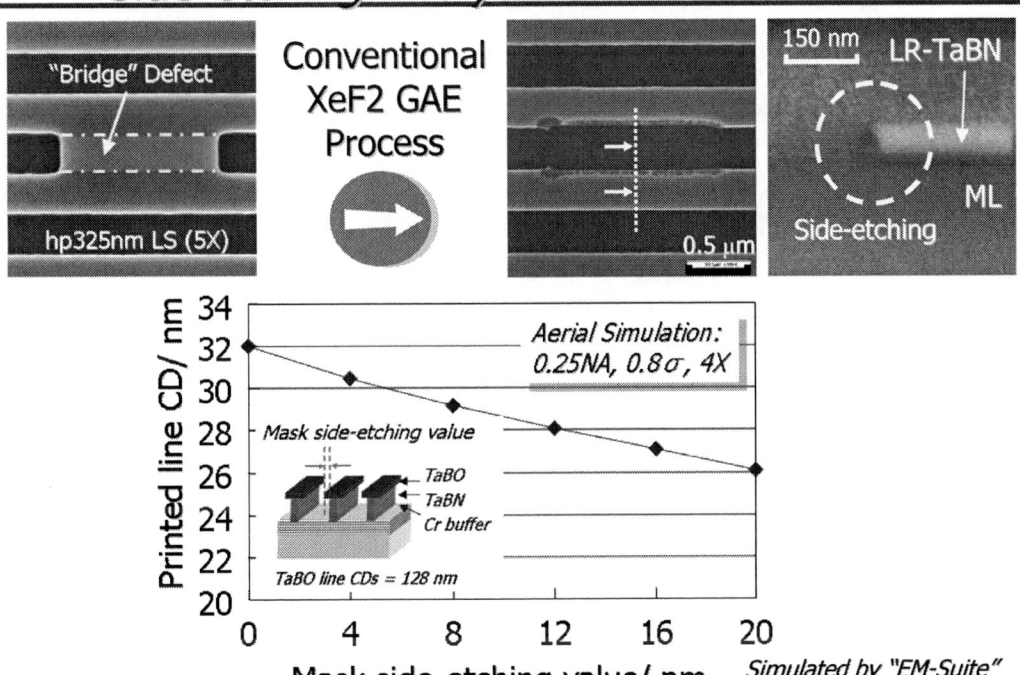

Side-etching Analysis of TaBN Absorber

IRAI   Selete

Conventional Process: XeF2

Conventional GAE Process: XeF2

Ga+   XeF₂

TaBO
TaBN

Absorber extension defect

New GAE Process: XeF2+H2O

H₂O

150 nm   LR-TaBN

ML

Side-etching

0.5 µm

New Process with XeF2+H2O

150 nm

No Side-etching

0.5 µm

Top View   Cross Section

– EUVL Symposium 2008 [poster 66]
T. Amano, Y. Nishiyama, H. Shigemura, T. Terasawa, O. Suga, K. Shiina*, F. Aramaki*,
T. Kozakai*, A. Yasaka*, "FIB mask repair technology for EUV lithography"

Damage Analysis of Mixing Layer

IRAI   Selete

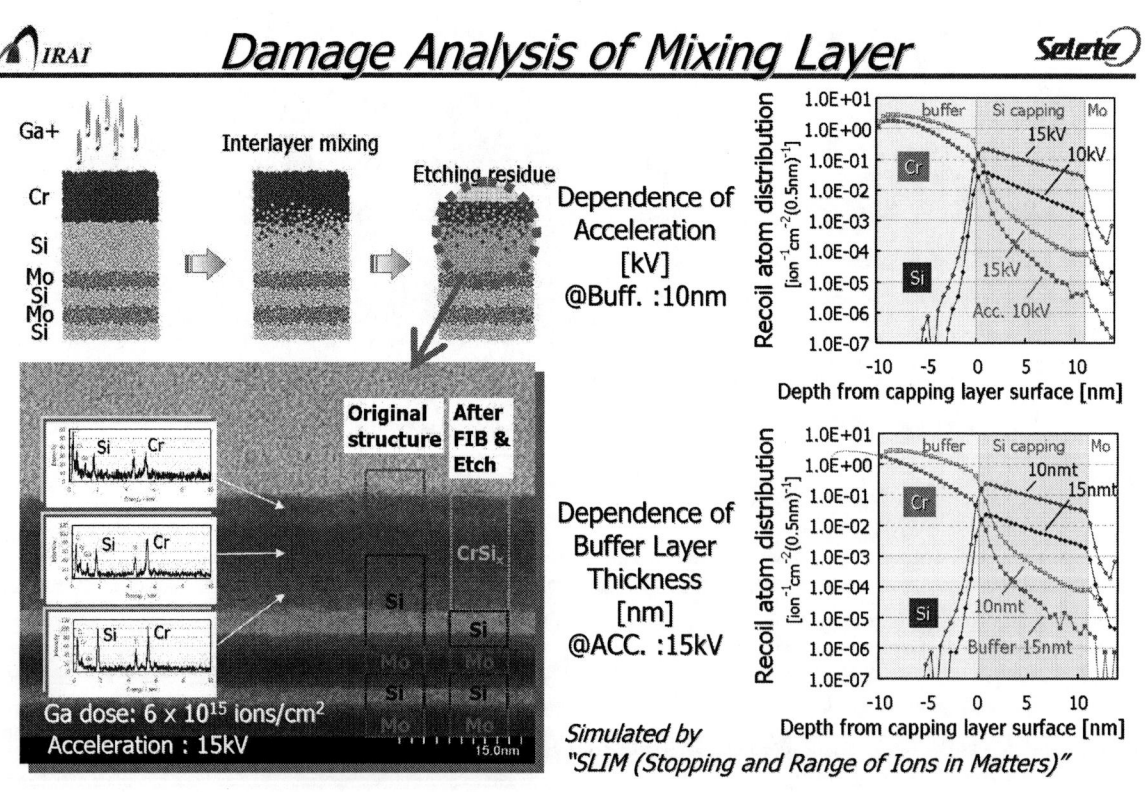

Ga+

Interlayer mixing

Etching residue

Cr
Si
Mo
Si
Mo
Si

Dependence of Acceleration [kV] @Buff. :10nm

Original structure | After FIB & Etch

CrSiₓ

Si | Si

Mo | Mo
Si | Si
Mo | Mo

Ga dose: 6 × 10¹⁵ ions/cm²
Acceleration : 15kV

15.0nm

Dependence of Buffer Layer Thickness [nm] @ACC. :15kV

Simulated by
"SLIM (Stopping and Range of Ions in Matters)"

560

 IRAI **Printability Verification Using SFET**

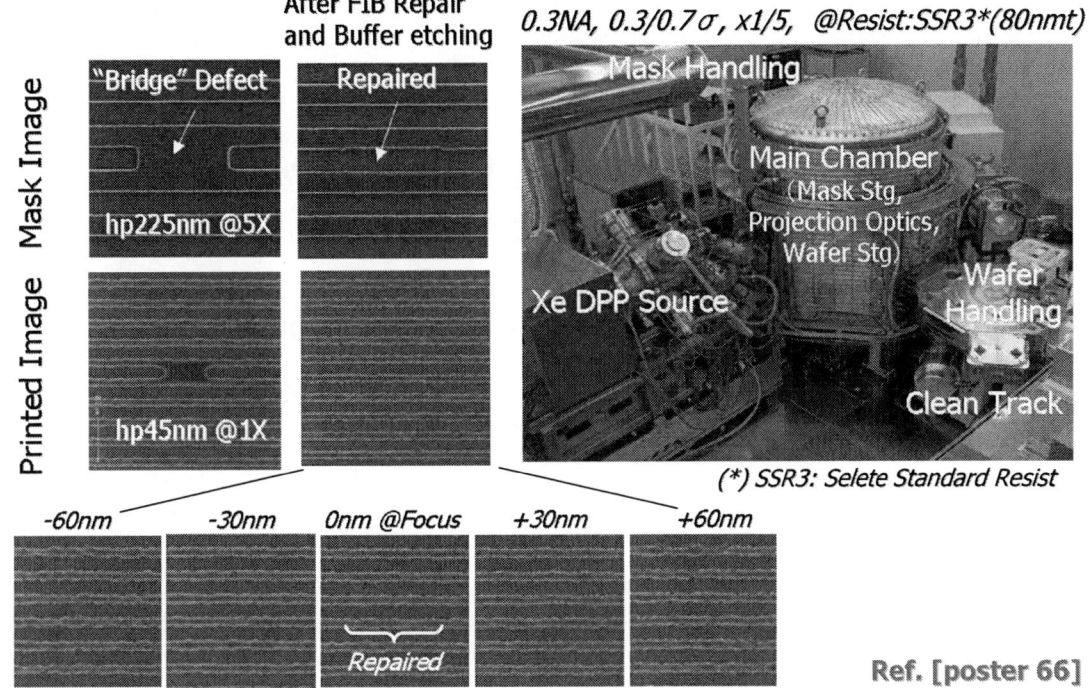

(*) SSR3: Selete Standard Resist

Ref. [poster 66]

 IRAI *Summary*

1. The Purpose of Selete's "EUVL Mask Program" is to develop mask technologies and infrastructures for high volume production EUVL masks for the hp32nm generation and beyond.

2. Selete is mainly focusing on developing EUVL masks without soft and/or hard defects ("Defect-free Masks").

3. In order to reduce effects of hard defects, mask blank inspection technology and mask pattern inspection/repair technologies are being developed.

Acknowledgement

This work was supported by
New Energy and Industrial Technology
Development Organization (NEDO).

We would like to thank Nuflare technology Inc.,
Advance Mask Inspection Technology Inc., SII-
NanoTechnology Inc. and finally all of the
people working on this program at Selete Inc.

Thank you for your attention !

Development status of EUVL mask blank and substrate

Asahi Glass Company
Yoshiaki Ikuta

AGC *Grow Beyond*

2008.Sep.28 EUVL symposium, Lake Tahoe, CA

Outline

1. Standard structure of AGC blank

2. Polished LTEM substrate

3. Reflective and capping films-coated substrate (EUV substrate)

4. Absorber

5. Resist

6. Summary

AGC *Grow Beyond*

1. Standard structure of AGC blank

Resist (Posi or Nega, 200~300nm)

Anti-reflective layer
Absorber } (Ta-based, 84nm)

Capping layer (Ru, 2.5nm)

Reflective layer
(Mo/Si multilayer)

Polished substrate (QZ or LTEM, 6025)

Conductive film (CrN, 70nm)

EUV substrate

AGC *Grow Beyond*

1. Standard structure of AGC blank

2. Polished LTEM substrate

3. Reflective and capping films-coated substrate
   (EUV substrate)

4. Absorber

5. Resist

6. Summary

AGC *Grow Beyond*

2. Polished LTEM substrate ~ Material

■ AGC's low thermal expansion material (LTEM) "AZ" met SEMI P37 class A requirements.

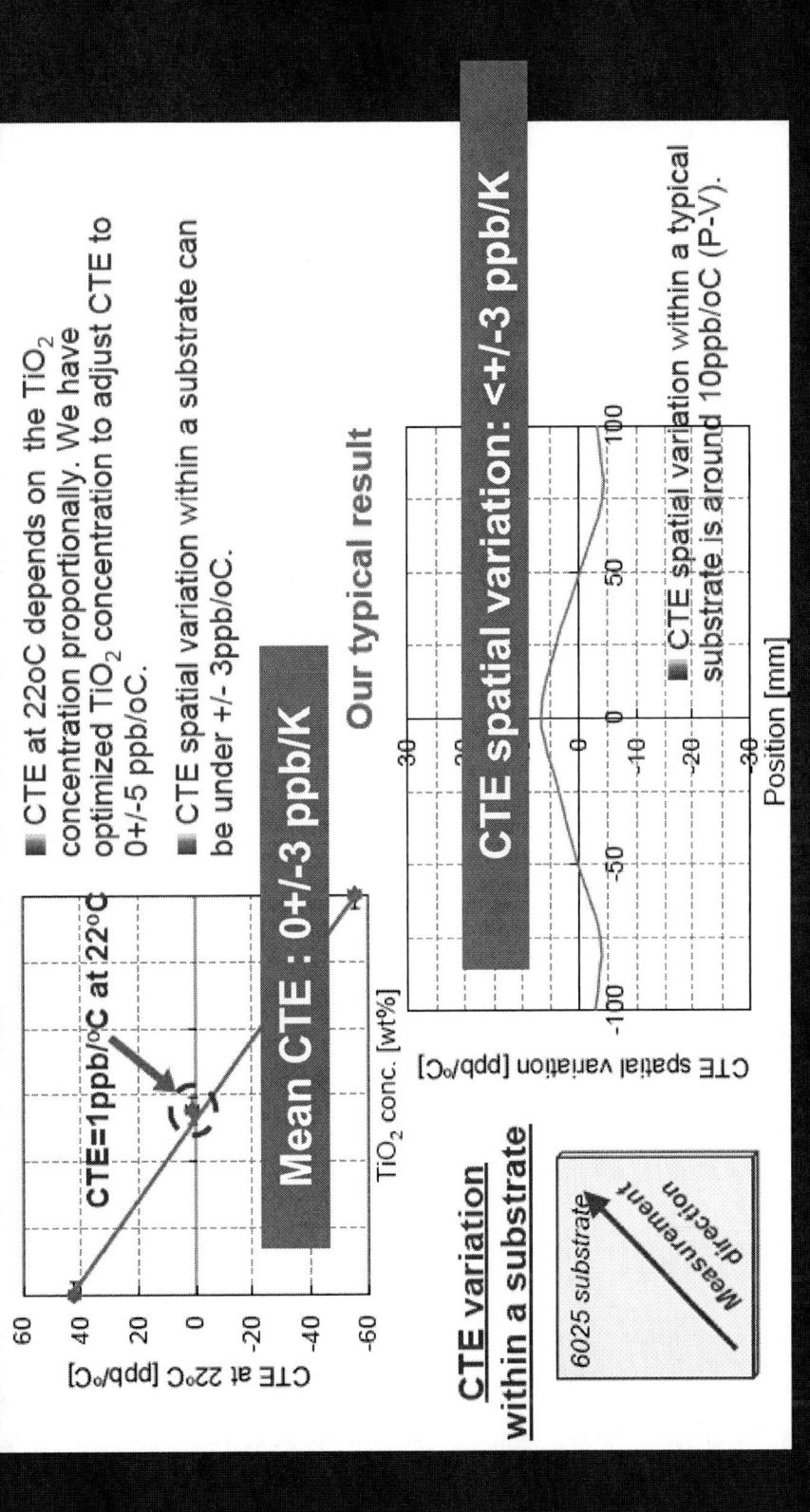

■ CTE at 22oC depends on the TiO_2 concentration proportionally. We have optimized TiO_2 concentration to adjust CTE to 0+/-5 ppb/oC.

■ CTE spatial variation within a substrate can be under +/- 3ppb/oC.

Mean CTE : 0+/-3 ppb/K

Our typical result

CTE spatial variation: <+/-3 ppb/K

■ CTE spatial variation within a typical substrate is around 10ppb/oC (P-V).

CTE variation within a substrate

Kenji Okamura, 2007 International EUVL Symposium in Sapporo

AGC *Grow Beyond*

2. Polished LTEM substrate ~ Material

■ Our main focus on LTEM shifted to improve the productivity of the current LTEM "AZ" and to develop the modified LTEM with lower CTE dependence on temperature.

Supply capability of current LTEM   Modified LTEM

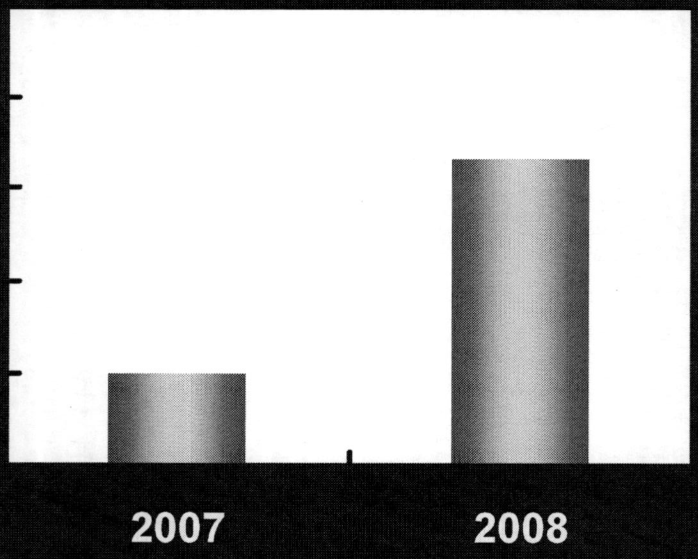

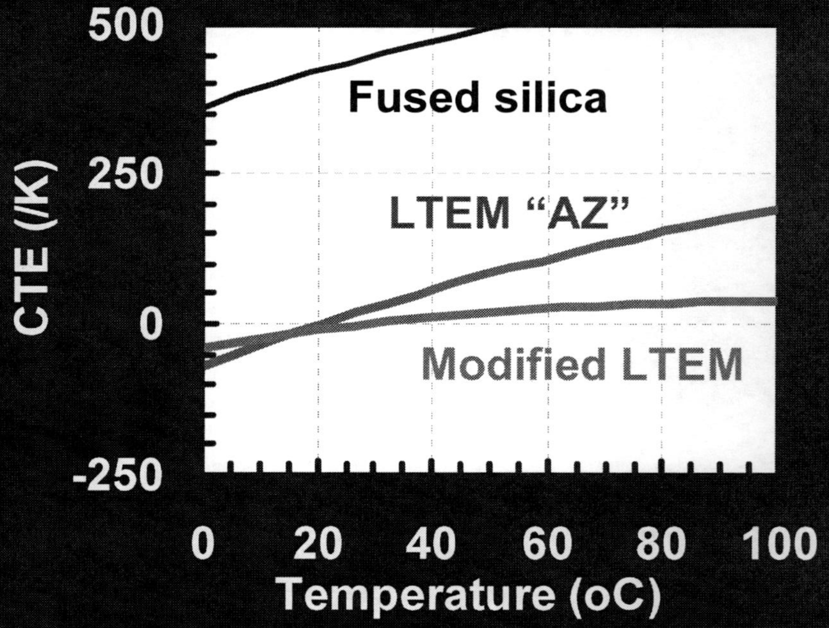

Fused silica

LTEM "AZ"

Modified LTEM

CTE (/K) — Temperature (oC)

500 · 250 · 0 · -250

0 · 20 · 40 · 60 · 80 · 100

2007 · 2008

AGC *Grow Beyond*

2. Polished LTEM substrate ~ Flatness

- The continuous development of the local polishing technique realized ~30 nm substrate flatness on both front and back surfaces, which almost met SEMI P37 class D (<30nm).

Flatness 3D map

	2007	2008
Front surface	38 nm	32 nm
Back surface	48 nm	31 nm

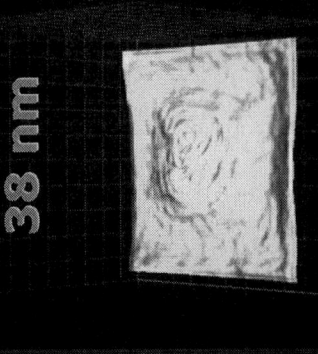

Q.A.=142x142mm

AGC *Grow Beyond*

2. Polished LTEM substrate ~ Defect

■ Substrate defect has been continuously decreased by improving the polishing and wet cleaning processes. At the substrate inspection, no pits can be observed and the un-removal particle is the dominant defect.

Average defect count of 10 LTEM substrates

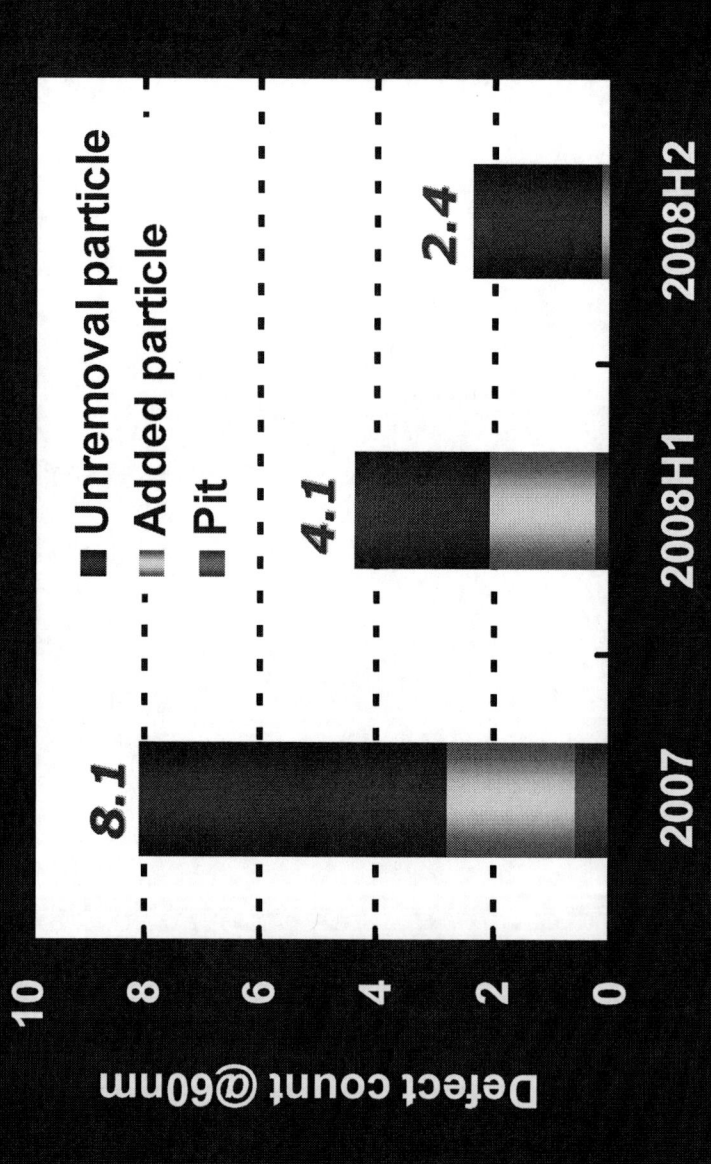

Q.A.=142x142mm

AGC *Grow Beyond*

1. Standard structure of AGC blank

2. Polished LTEM substrate

3. Reflective and capping films-coated substrate
   (EUV substrate)

4. Absorber

5. Resist

6. Summary

AGC *Grow Beyond*

3. EUV substrate ~ Defect

■ AGC also continued the development of the multilayer (ML) and capping film deposition process. Its cleanliness was verified by the ML over ML technique, and the defect analysis was carried out with SEM/FIB/EDX.

Adder defect map of ML over ML

Adder defect analysis

Pixel Histogram

5adders
@73nm

■ 27 ···	3
■ 26 ···	0
■ 12 ···	0
■ 11 ···	0
■ 10 ···	2
■ 7 ···	0
■ 6 ···	0
■ 3 ···	
■ 2 ···	
■ 1 ···	
(Pixels)	

Type D

Type A

Type C

Type B

Q.A.=142x142mm

50ML/Ru

substrate

inspection-1

50ML/Ru

50ML/Ru

substrate

inspection-2

AGC *Grow Beyond*

3. EUV substrate ~ Integrated performance

- **Structure**
  CrN/LTEM/50ML/Ru

- **LTEM CTE properties**
  Mean   -4 ppb/K
  Spatial variation +/-4 ppb/K

- **Defect**

- **Substrate flatness**

Front 82 nm

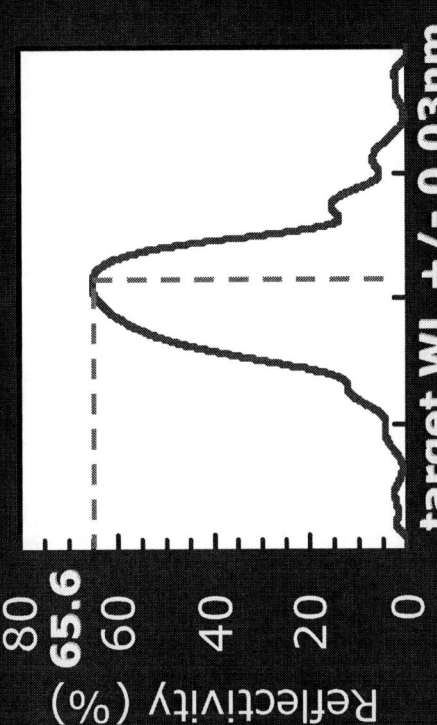

Back 75 nm

- **Optical properties**

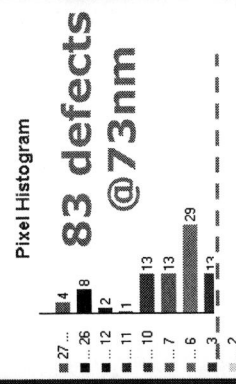

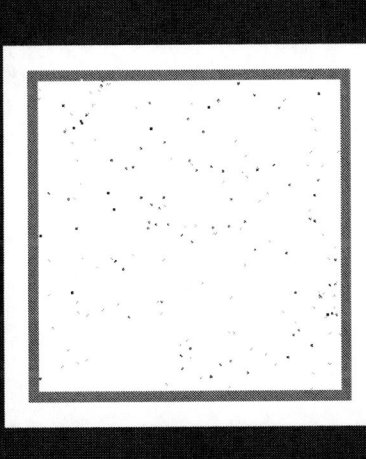

Pixel Histogram

83 defects
@73nm

AGC Grow Beyond

3. EUV substrate ~ Bow

- Blank bow can be controlled <1um.

Front surface flatness change upon front and backside films coating

Polished LTEM substrate

CrN/LTEM/50ML/Ru

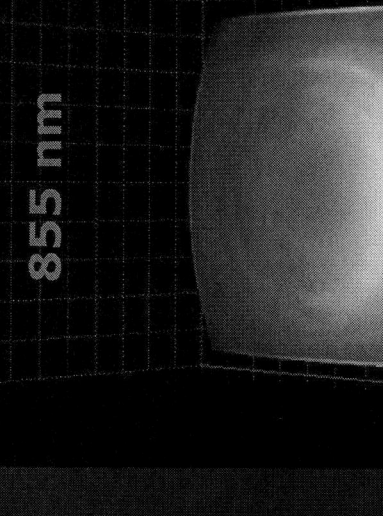

96 nm

855 nm

Q.A.=142x142mm

AGC *Grow Beyond*

1. Standard structure of AGC blank

2. Polished LTEM substrate

3. Reflective and capping films-coated substrate (EUV substrate)

4. Absorber

5. Resist

6. Summary

AGC *Grow Beyond*

4. Absorber ~ Film properties

■ AGC has developed Ta-based absorber and anti-reflective films, whose properties almost met SEMI P38 class A.

	Semi P38 Class A	AGC Ta-based absorber
Absorber film thickness (nm)	-	77
Anti-reflective film thickness (nm)	-	7
Thickness uniformity (range, %)	-	<3
Reflectivity at EUV (%)	<0.5	<0.5 @ 13.5nm
Film structure	-	amorphous
Surface roughness (RMS, nm)	-	0.26
Film stress (MPa)	-200 to 200	-56
Reflectivity at 257 nm (%)	<5	8.3

AGC *Grow Beyond*

4. Absorber ~ Defect

■ Here are two examples of adder defects upon absorber/antireflective films' deposition.

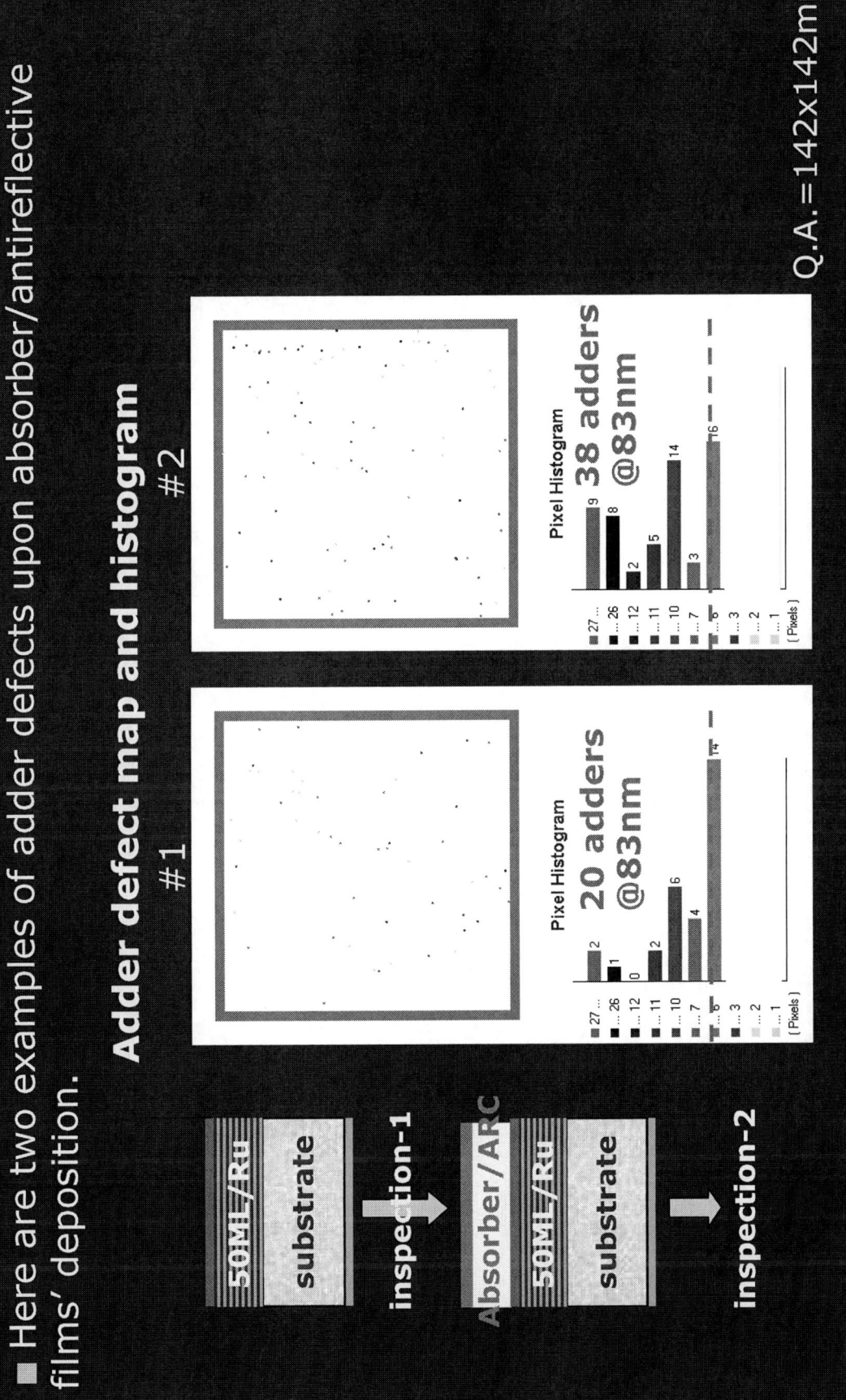

AGC *Grow Beyond*

4. Absorber ~ Etch result

■ Here are customers' patterning results of AGC blank.

100nm Line&Space

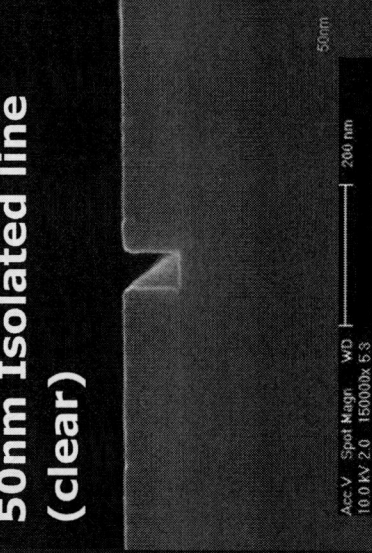

50nm Isolated line (clear)

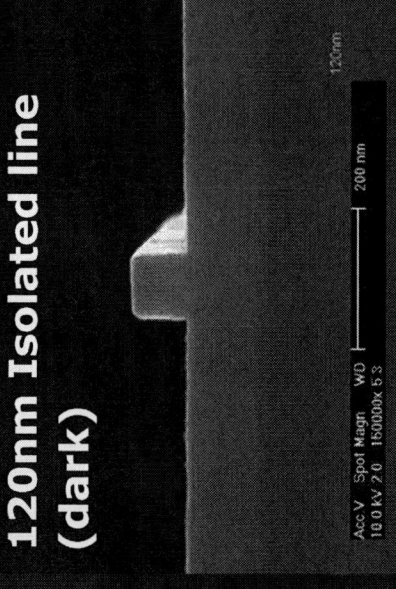

150nm Contact hole

120nm Isolated line (dark)

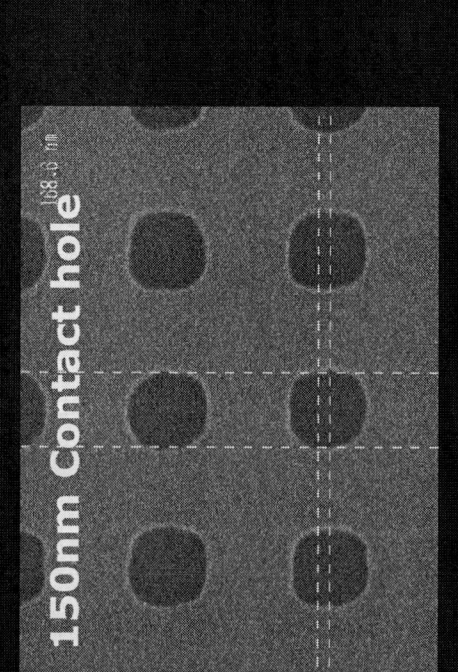

AGC *Grow Beyond*

4. Absorber ~ Pattern inspection

■ We are also developing the anti-reflective film suitable for the next generation pattern inspection tool whose light source will be ~200nm DUV light.

DUV reflectivity spectrum

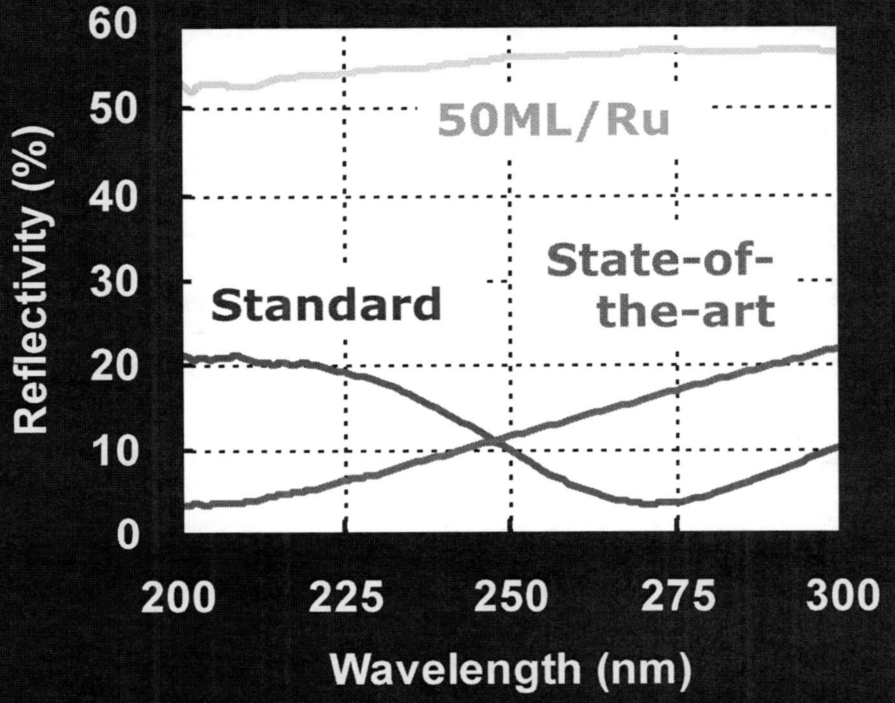

Contrast to 50ML/Ru

	257nm	199nm
Standard	79	62
State-of-the-art	44	88

AGC *Grow Beyond*

1. Standard structure of AGC blank

2. Polished LTEM substrate

3. Reflective and capping films-coated substrate (EUV substrate)

4. Absorber

5. Resist

6. Summary

AGC *Grow Beyond*

5. Resist

■ Both positive and negative resists are available. The thickness can be ~200nm.

Thickness uniformity of e-beam resist

2007

Avg. 303nm, PV 2.0nm

2008

Avg. 202nm, PV 3.1nm

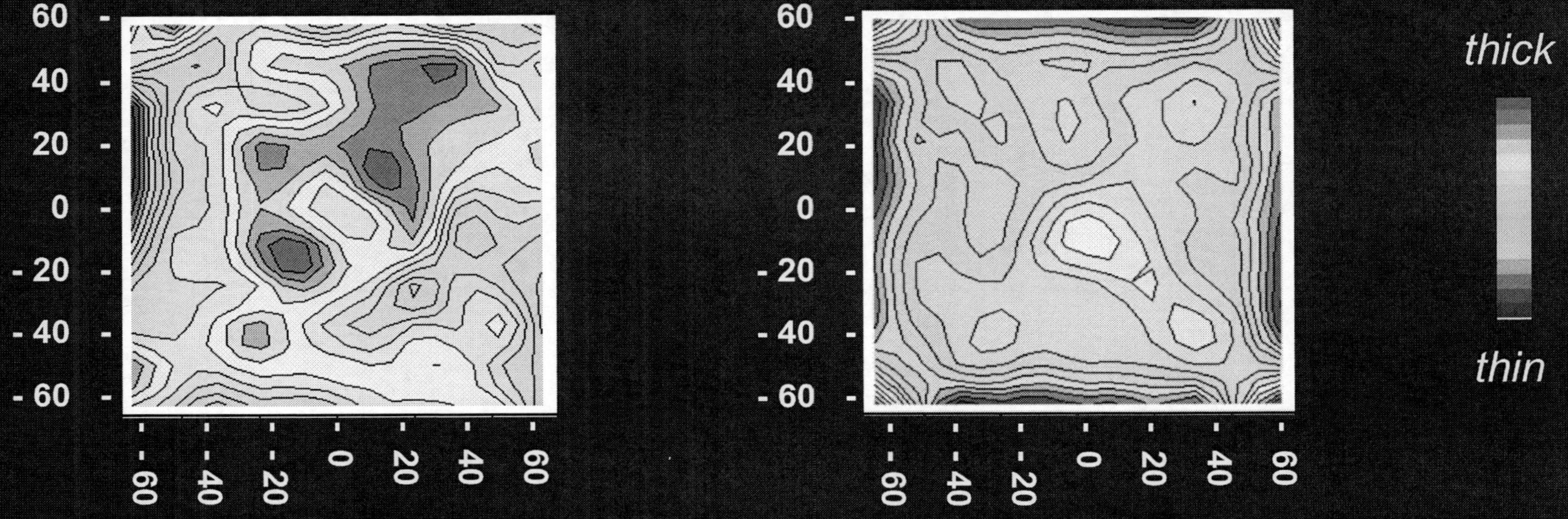

thick

thin

Q.A.=132x132mm

AGC *Grow Beyond*

6. Summary

■ AGC has developed the EUVL mask blanks suitable for EUV lithography process development and has delivered full stack blanks to customers.

　1. EUV substrate (ML/Ru-coated LTEM substrate) met the CTE, flatness, and optical properties requirements of SEMI P37/P38.

　2. The absorber and anti-reflective coatings also met the optical properties and film properties requirements of SEMI P38.

　3. Positive and negative type resists are available.

　4. Defect is the only remaining issue to be solved.

■ AGC continues the following activities so that AGC will be able to realize EUVL mask blanks suitable for HVM.

　1. Defect reduction through continuous developments of polishing, cleaning and film deposition processes,

　2. Developments of thinner absorber and resist,

　3. Productivity improvement of EUVL mask blank processes.

Multilayer Defect Compensation to Enable Quality Masks for EUVL Production

Ted Liang

Components Research
Intel Corporation

Eric Gullikson

CXRO

Lawrence Berkeley National Laboratory

2008 EUVL Symposium, Lake Tahoe, CA

Motivation

- **Availability of defect-free mask is a well-known risk for EUVL**

- **Feasibility of producing defect-free ML blanks has not been demonstrated yet**
  - Current trend of progress not sufficient
  - Is the current strategy (tool and process) sufficient?

- **Can we produce <u>quality masks</u> from 'defective' ML blanks?**

- **This presentation**
  - Discusses the challenges and focus areas in ML coating
  - Describes mitigation strategy to use defective blank

Outline

- **Introduction: ML defects**
  - Requirement vs. trend
- **Quality mask using 'defective' blank**
  - Focus areas for ML deposition
  - Mitigation strategy
- **Defect compensation**
  - Models
  - Examples
- **Summary**

Four Major Categories of EUV Mask Defects

I: Substrate defects: 'propagate' to ML surface This talk:
II: ML defects: substrate and deposition process Embedded ML defects
III: Absorber pattern defects: mask fabrication process
IV: 'Soft' defects: contaminations from handling and use

No Printable defects >30nm

Absorber (TaN, 80nm)

ML cap (Ru, 2.5nm)

ML (Mo-Si, 280nm) (~3nm Mo/4nm Si)

Substrate (LTEM, ¼")

Conductive layer

At issue: Defects on ML Blank Are Complex and Plenty

- Two general modes of generation
  - Growth from seeded substrate imperfections
  - During deposition

- Dependent on deposition conditions
  - Off-normal: 'enlarging' defects – decoration effect
  - Near-normal: 'flattening' defects
  - With etch back: 'smoothing' defects

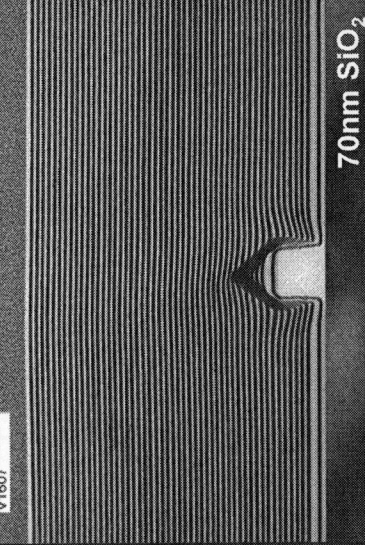

Smoothing

70nm SiO$_2$

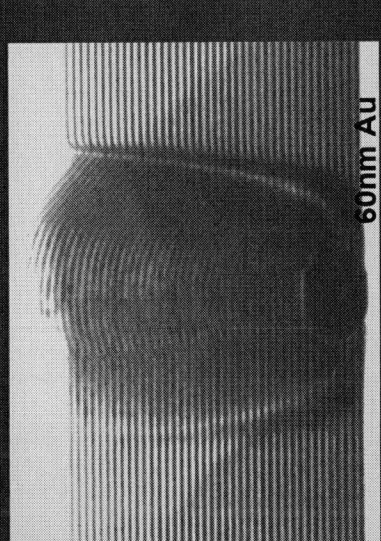

60nm Au

IBD near-normal

60nm Au

IBD off-normal

Defect Status: A Reality Check

- Desired: Zero defect, defect-free, 0.003/cm^2, ... @30nm

Today's Status
(not champion data)

M1350: ~50 - 100 @70nm

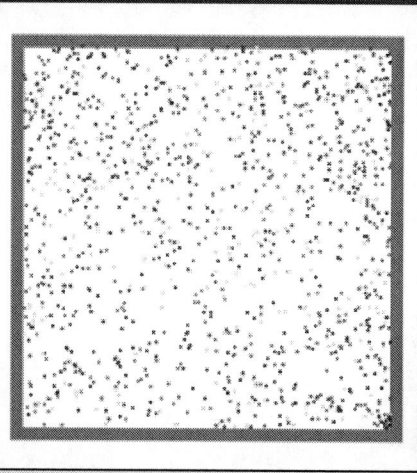

M7360: ~1000 @50nm

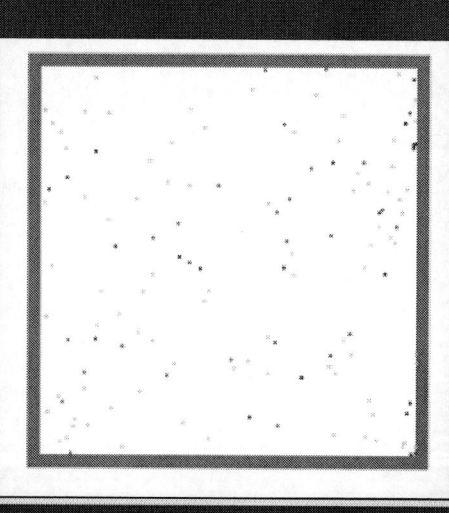

- >1000x reduction is needed
- Current progress unlikely to be sufficient
  - Good progress in substrate polishing
  - Only incremental progress in ML coating
- Is current strategy (tool, process) sufficient?
  - Unlikely: IBD technology >10yrs old; Need innovations

Accelerate ML Defect Reduction

- Substrate defect rendering – rendering 'invisible', residual

Two paths

▲ **Standard ML coating:**
'Invisible' defects on substrate can not be eliminated, but **become printable**

▲ **Rendering processes:**
Make 'invisible' defects smaller, become non-printable

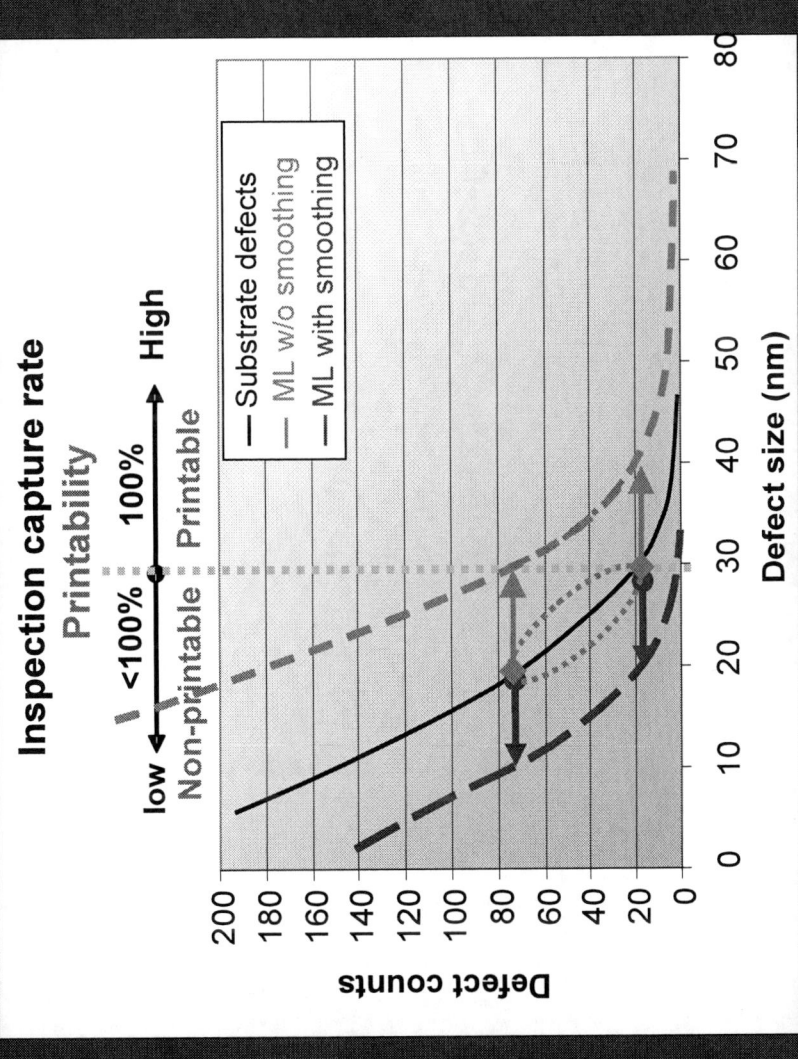

- Substrate defects include adders from loading
- Possible to avoid adders during coating

Example: Substrate Defect Rendering by 'Smoothing'

- Effective for sub-threshold defects on substrate (always the goal)
  - Must not be used for large, detectable defects (for example >40nm)
- Intel had funded and collaborated with LLNL for many years developing ion beam processes – simultaneously smooth particles and pits
- Benefits demonstrated on quartz substrates

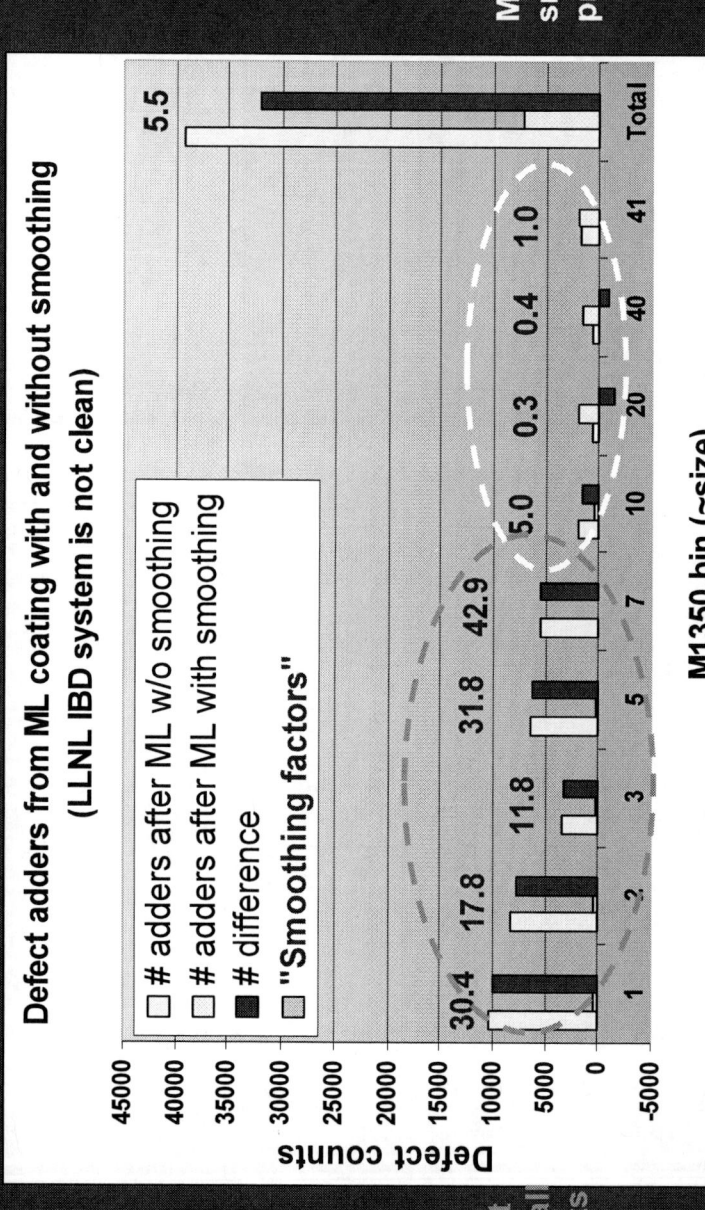

Most adders from smoothing process are large

Smoothing most effective for small substrate defects

Mitigation Strategy for Defects on ML Blank

- **What if the yield for defect-free blanks is low?**
  - EUVL CoO too high
  - Use defective blanks – allow 'few' defects

- **Must develop mitigation strategy for allowing the use of defective blanks**

- **Assess two potential mitigation options**
  - Pattern placement to 'avoid' and cover defects – before fabrication
  - Pattern edit to compensate defects – after fabrication

OPC-like Compensation during Beam Write

- **Conceptually, blank defects could be compensated by adjusting the absorber patterns during E-beam writing, similar to OPC procedures**

- **Necessary input parameters**
  - Accurate defect location on blank (x, y, $\Delta r \sim 30nm$)
  - Accurate optical strength of isolated defect (Aerial image)
  - Rule-based model to determine amount of correction

- **In practice, this is extremely difficult to implement**
  - ML defect optical signal may be below AIM tool detection
  - Logistics: timing of blank data and EB data difficult to manage

Covering Defects by Absorber Patterns

- ML defect effect on 30nm 1:1 L:S – evaluate 3 defect sizes

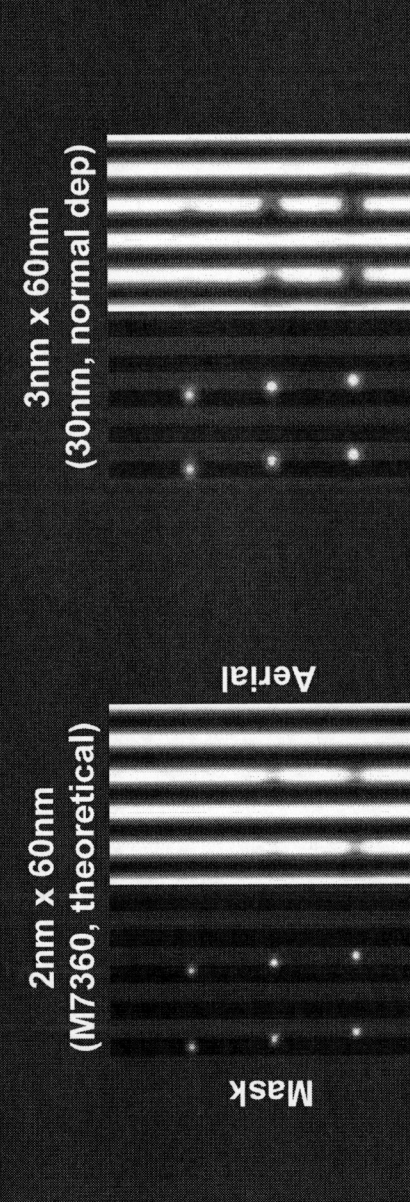

Mask

Aerial

2nm x 60nm
(M7360, theoretical)

3nm x 60nm
(30nm, normal dep)

3nm x 90nm
(M7360, actual)

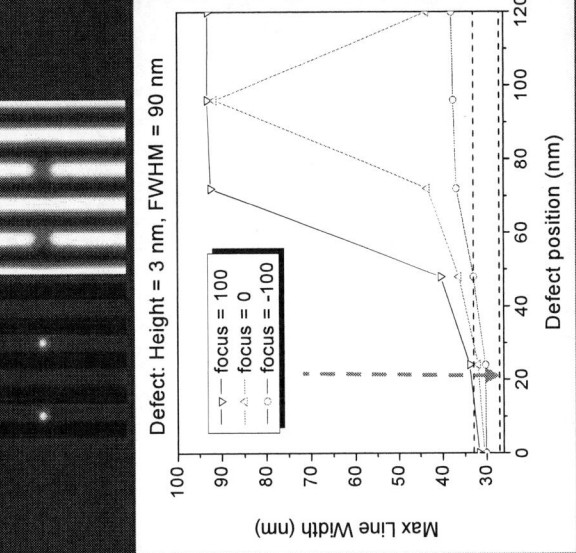

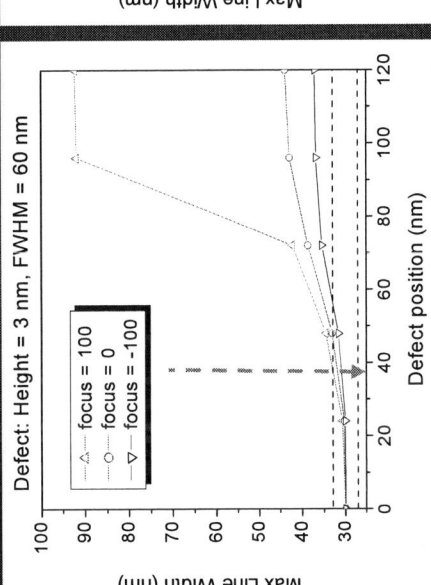

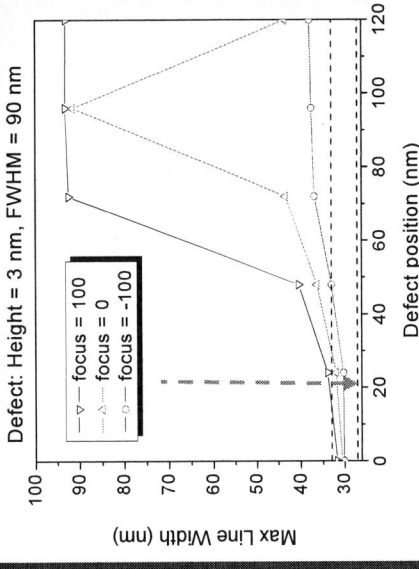

- Printability highly sensitive to defect proximity (20 - 30nm tolerance)
- Aligning pattern to cover >1 defect unlikely (no rotation)

intel

CD Change Sensitivity to Defect Proximity

- Confirmed with resist printing on MET: 50nm 1:1 L:S

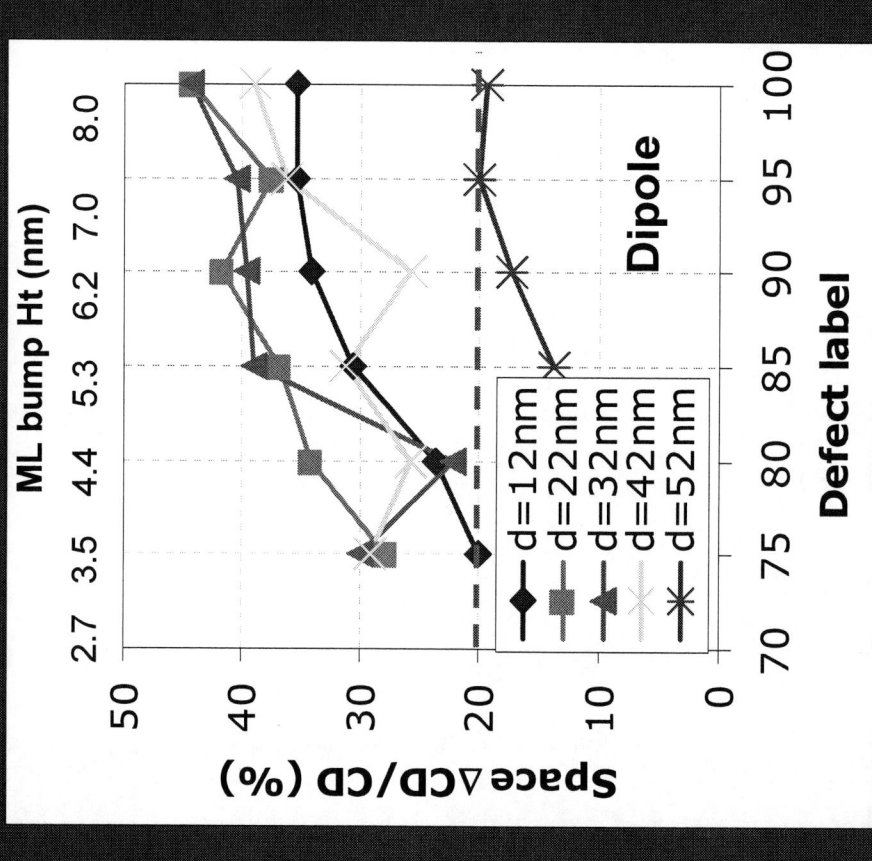

Defect near center of space

Defect under center of line

JVST 2007

Defect Compensation on Finished Mask

- Edit the patterns after the mask is made
  - Similar to rule-based OPC

Nested line 1:1

x ↕

Absorber CD cut — Defect distance (x)

line #1
line #2

Cut out
Absorber

Iso-line

x ↕

Absorber CD cut — Defect distance (x)

Defect #1
Defect #2

Absorber line

Buried ML defect

Simulations show the possibility for such compensation

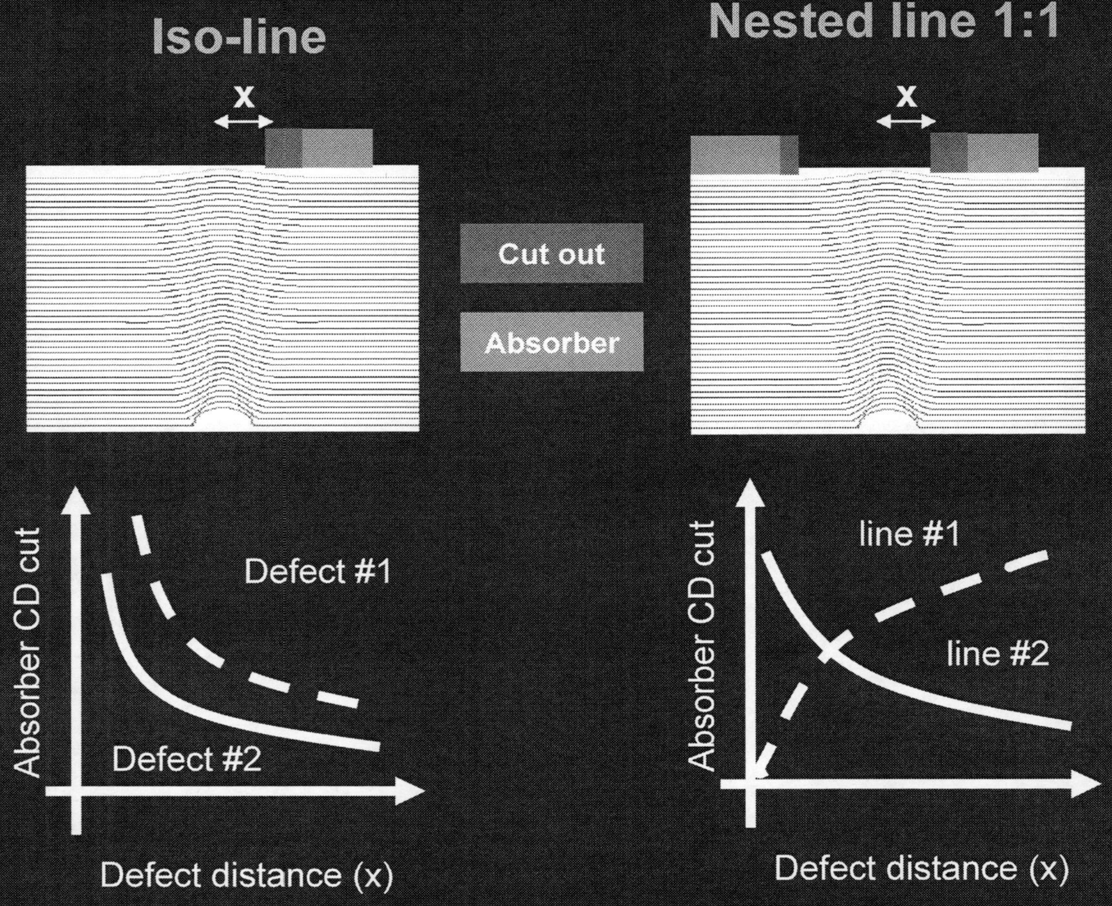

Mask Process Flow and Required Infrastructure

- **Process flow**

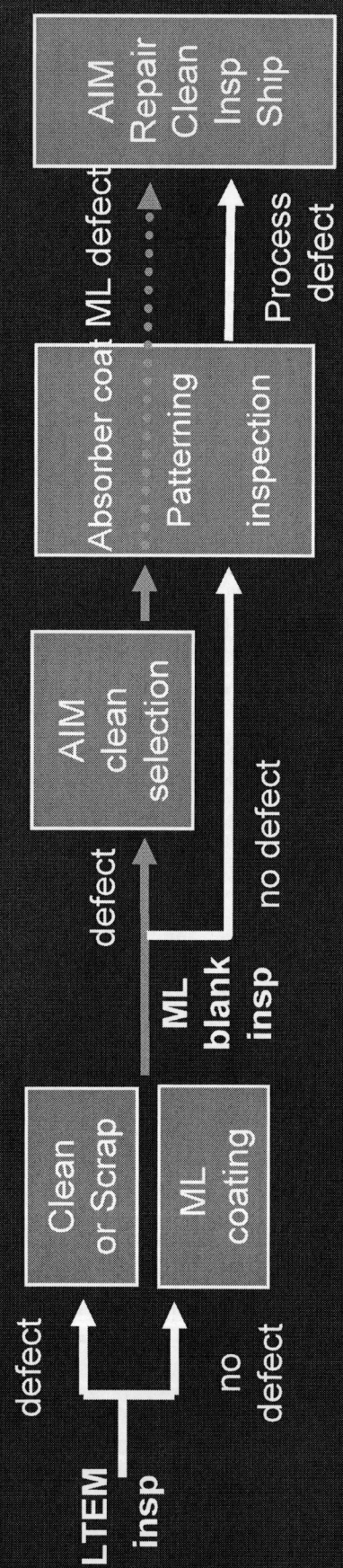

- **Necessary infrastructure/capability to ensure implementation**
  - Substrate inspection: 30nm - 40nm sensitivity (100% capture)
  - Blank inspection: 2nm x 60nm ML sensitivity
  - Aerial image tool: with high res mode
  - Damage-free repair tool: allows rework – FEBIE/D

Examples to Demonstrate Feasibility

- Edit a <u>50nm line</u> to compensate a <u>6nm x 90nm ML phase</u> bump at 65nm proximity

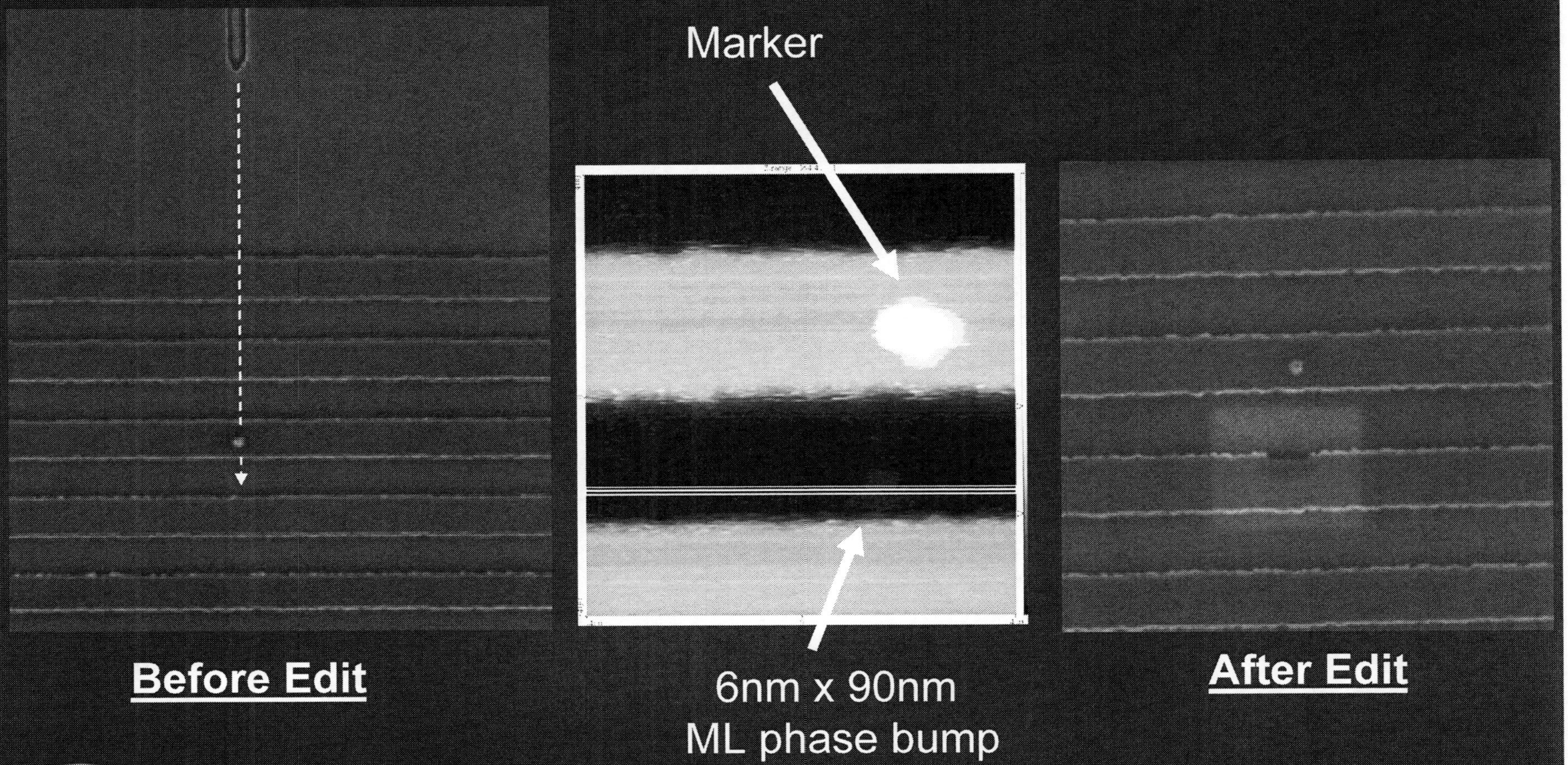

Marker

6nm x 90nm
ML phase bump

Before Edit

After Edit

Examples to Demonstrate Feasibility (con't)

- Print on MET (5X, 0.35NA, Annular)

Resist image (50nm 1:1)

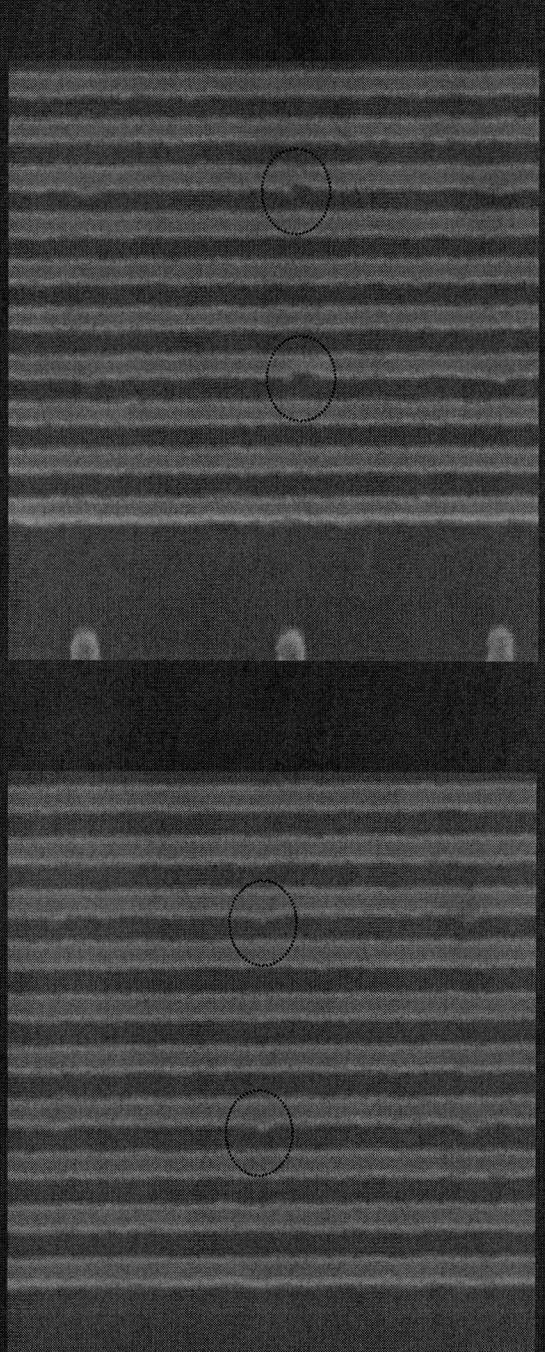

Before edit

After edit

- **It works, but need to improve accuracy**
  - Not sufficient by relying on AFM measurements
- **Underscores the need for AIM to ensure accurate editing !!!**
- **E-beam repair: damage-free, allows repetitive editing**

Summary

- Discussed difficult challenges in current strategy and trend of progress in the production of defect-free ML blanks
  - ML deposition must planarize defects seeded by substrate imperfections

- Described the need for mitigation techniques, such as smoothing, to render substrate defects that are 'invisible' (<30nm) or <100% capture (<40nm) during inspection

- Discussed limitations of mitigating ML defects by absorber pattern OPC and covering during pattern generation

- Presented a viable process for producing quality masks using 'defective' ML blanks via defect compensation on a finished mask
  - Benefits: Enable quality mask; Extending current blank inspection tool may meet needs
  - Aerial imaging tool is a 'must have' in order to implement defect compensation

- Smoothing + compensation is the path to quality 'defect-free' EUV mask

Acknowledgements

— Guojing Zhang, Pei-Yang Yan, Seh-Jin Park, Andy Ma, Armando Cobarrubia, Erdem Ultanir (ex-Intel), Gilroy Vandentop

— Erik Anderson, Patrick Naulleau, Ken Goldberg at LBNL

— Paul Mirkarimi, Sherry Baker, Eberhard Spiller* at LLNL, SO*

— Thorsten Hofmann, Klaus Edinger at Nawotec/Zeiss

Backup: Pit smoothing

Pit smoothing

100 nm

EUV Reticle Contamination and Cleaning

Uzodinma Okoroanyanwu, E. Langer, A. Fumar-Pici*, T. Wallow, O. Wood, and B. La Fontaine
AMD
*ASML
C. Holfeld, J. H. Peters
AMTC
M. Bender, M. Rossinger, S. Trogisch
Qimonda
F. Goodwin, A. Wüest, S. Huh
SEMATECH
G. Denbeaux, Y. Fan, A. Antohe, L. Yankulin, R. Garg
CNSE
K. Goldberg, P. Naulleau
LBNL

Outline

- Contamination of reticle

- Wafer printing at SEMATECH-Berkeley MET

- Aerial image measurements at SEMATECH-Berkeley AIT

- Cleaning results

- Conclusion

Reticle description

— *MET1* mask

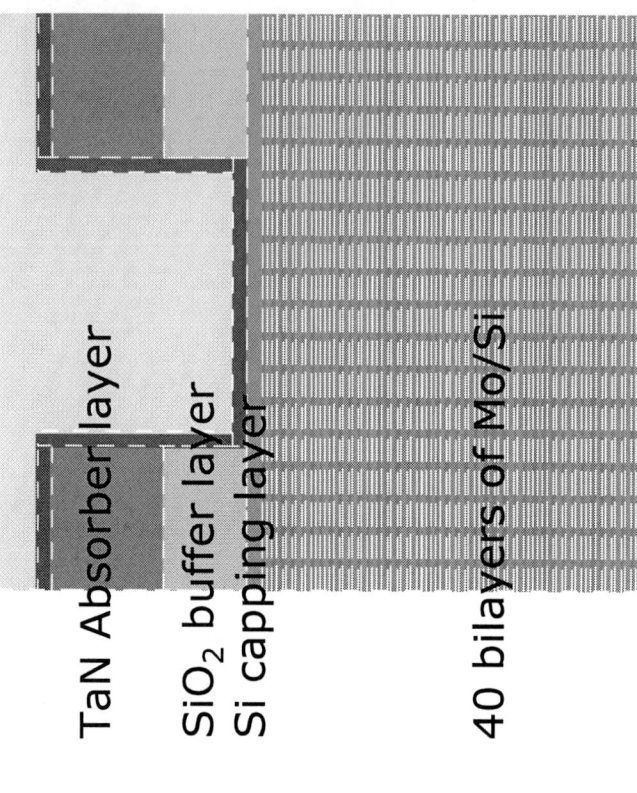

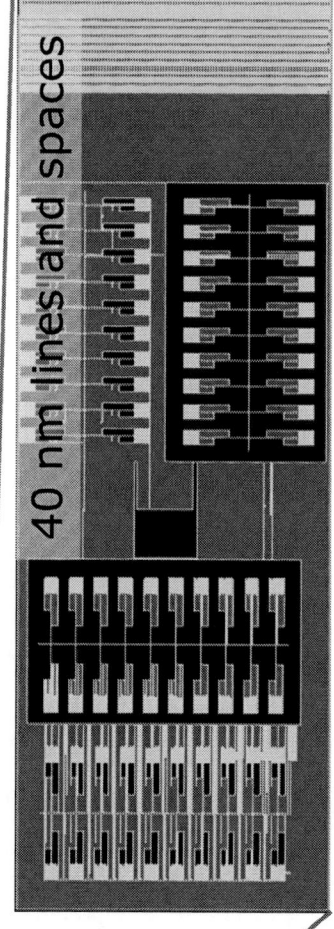

TaN Absorber layer

SiO$_2$ buffer layer
Si capping layer

40 bilayers of Mo/Si

40 nm lines and spaces

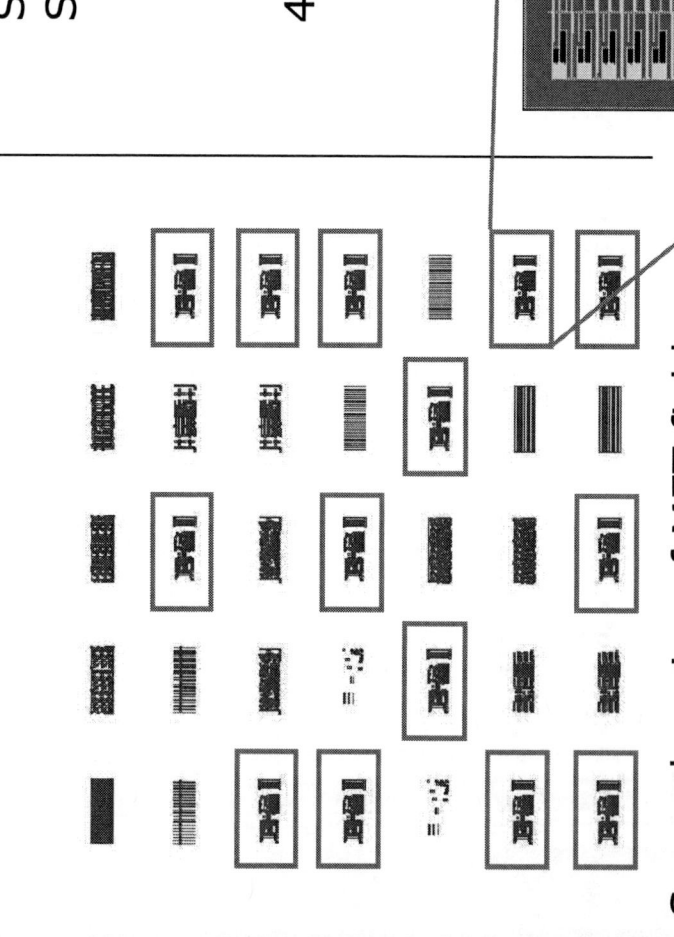

Several copies of MET fields dedicated to contamination studies

Engineered contamination process

- Microscope for <u>m</u>ask <u>i</u>maging and <u>c</u>ontamination <u>s</u>tudies (MIMICS tool) at the College of Nanoscale Science and Engineering

- The *MET1* test mask was contaminated with hydrocarbons and exposed to EUV radiation for a duration of 10h, 20h, and 36h

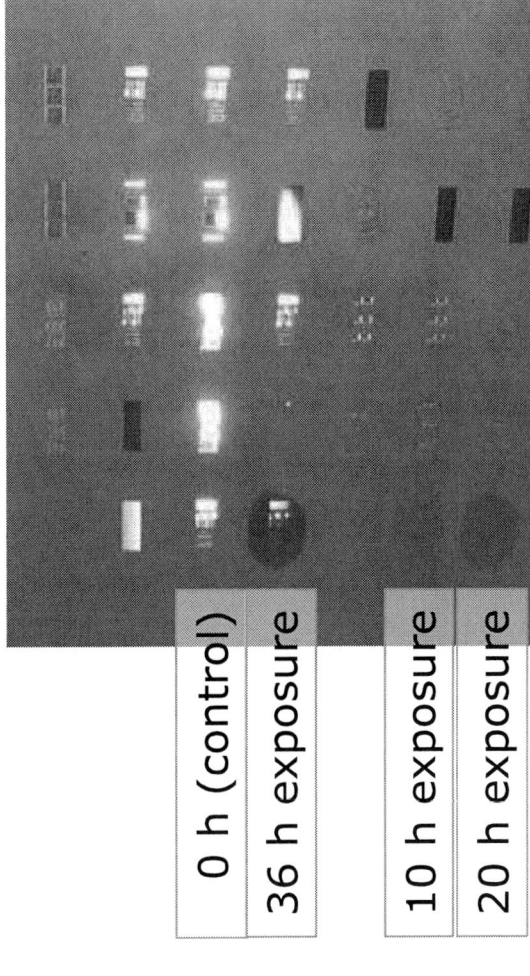

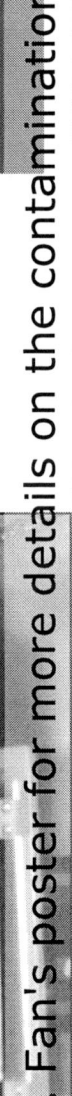

0 h (control)

36 h exposure

10 h exposure

20 h exposure

See Y. Fan's poster for more details on the contamination process

Contamination analysis of MET1 mask
Reflected FTIR spectra of contaminated sub-fields

— Carboxylic and amide groups

— Significant increase in the intensity of the absorption bands with contamination time

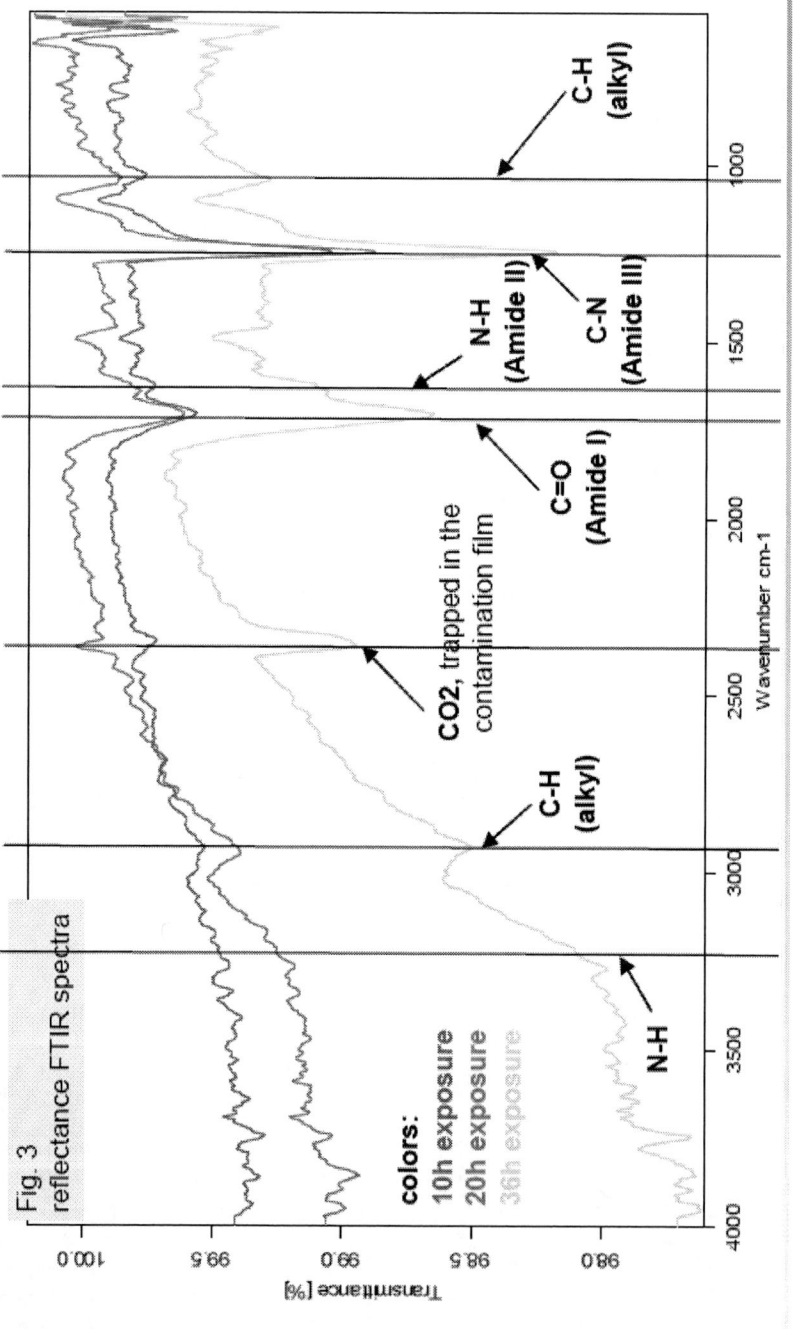

Fig. 3
reflectance FTIR spectra

colors:
10h exposure
20h exposure
36h exposure

Comparison to MET mask under normal use
Auger Emission Spectra

— Analysis of normally occurring contamination shows clear signals of carbon and oxygen on a Berkeley MET mask

— Blue curve: control field (Cr and N come from the absorber)

— Red curve: contaminated field

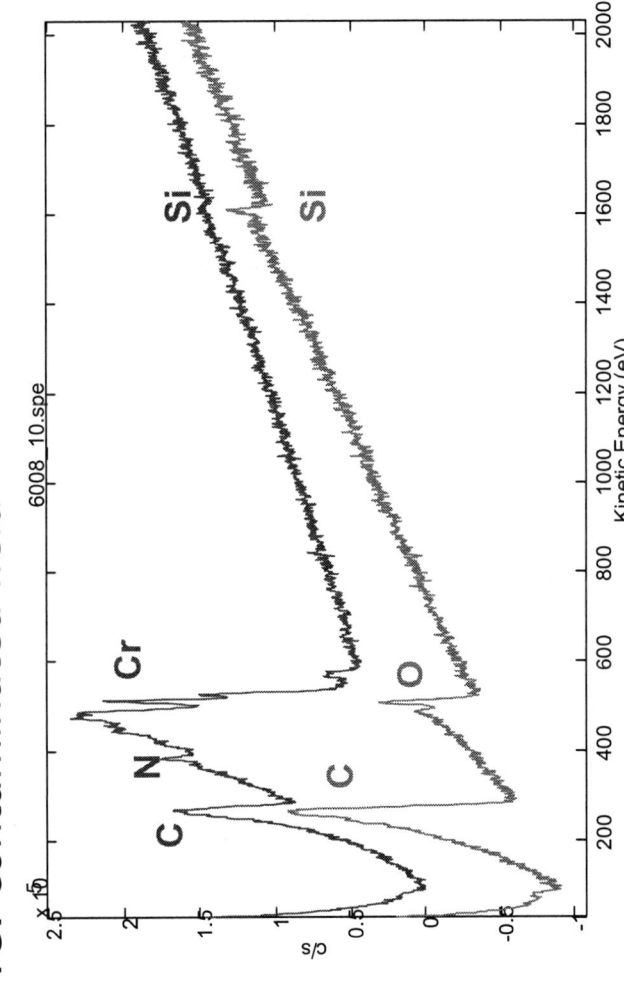

Wafer printing experiments
Imaging at the Berkeley MET

– Sample images of the 40 nm lines and spaces in the control sub-field

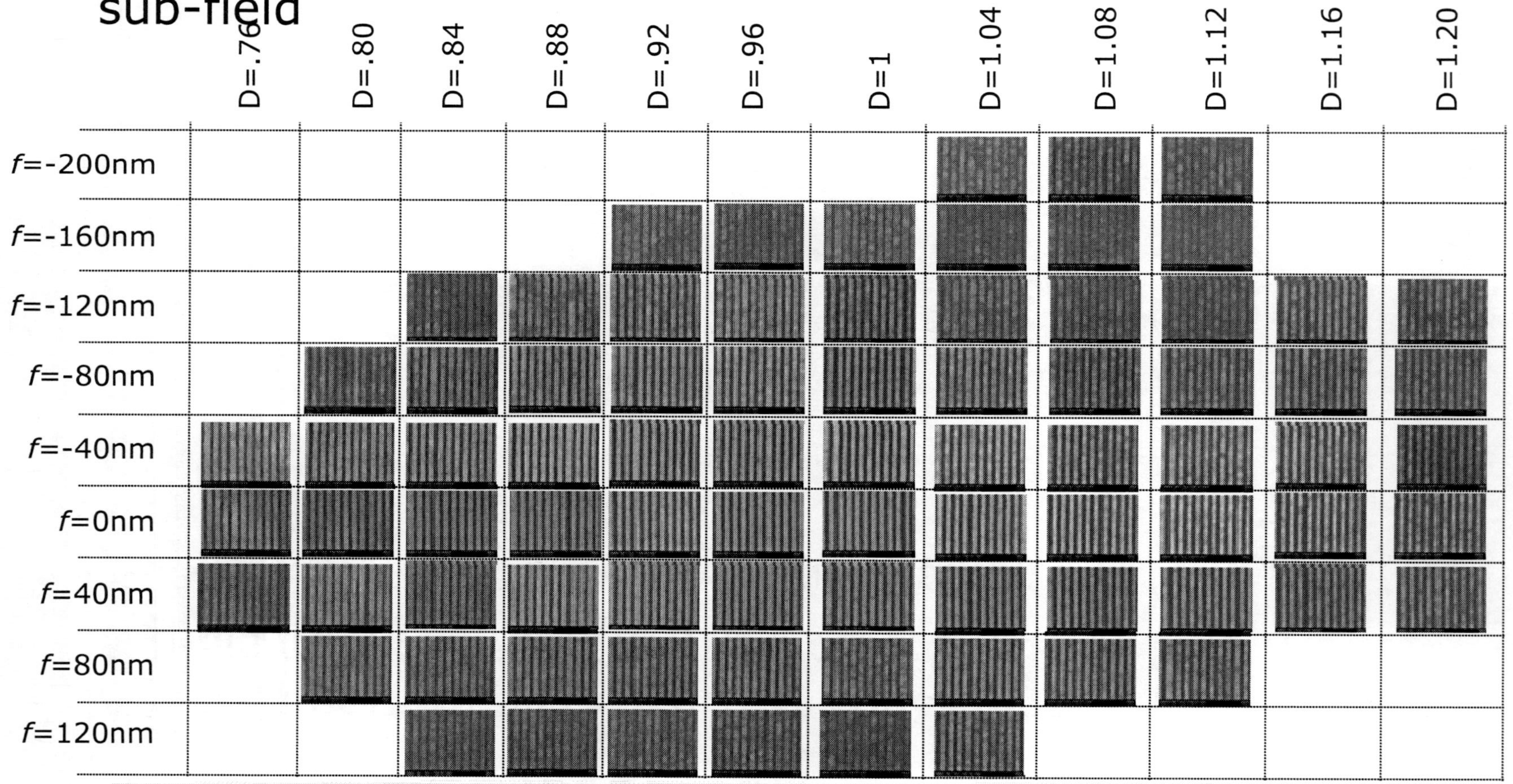

Analysis of wafer prints

- Bossung plots from the control and most contaminated sub-fields

- Resist: TOK P1123

- Qualitative differences between prints: Bossung curves are more curved on contaminated features

→ It is possible that a different portion of the aerial image is sampled once the mask is contaminated

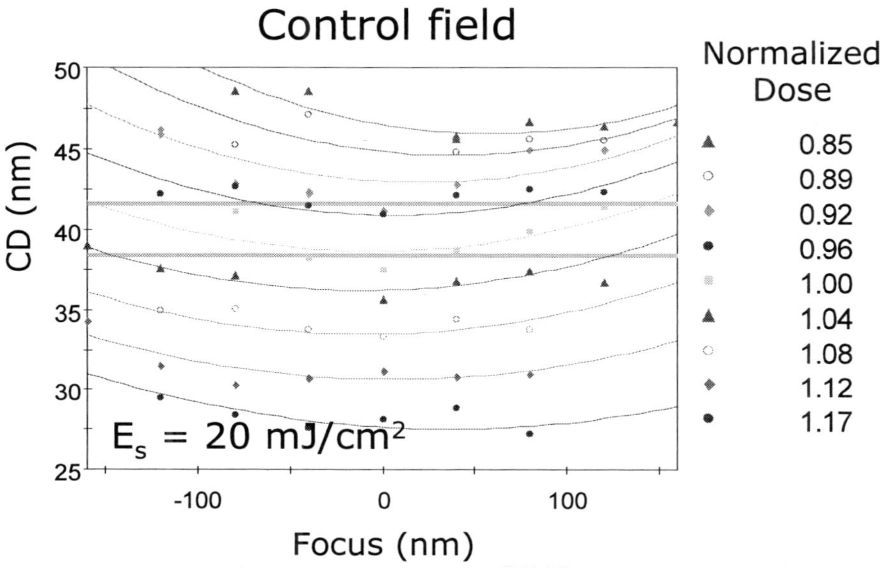

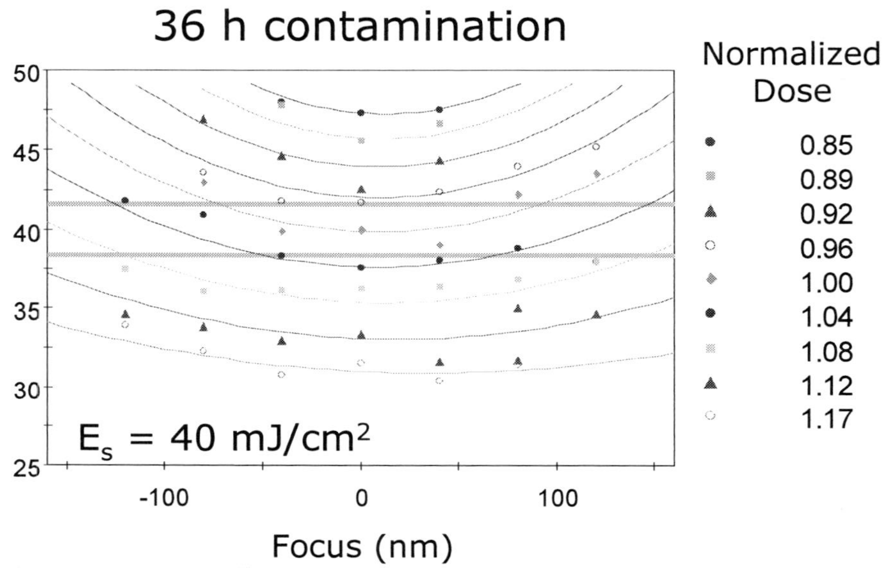

36 h contamination

Normalized Dose
0.85
0.89
0.92
0.96
1.00
1.04
1.08
1.12
1.17

$E_s = 40 \ mJ/cm^2$

Focus (nm)

Control field

Normalized Dose
0.85
0.89
0.92
0.96
1.00
1.04
1.08
1.12
1.17

$E_s = 20 \ mJ/cm^2$

Focus (nm)

CD (nm)

The future is fusion

Analysis of wafer prints

— Process windows

— Apparent loss of depth of focus for most contaminated sub-field

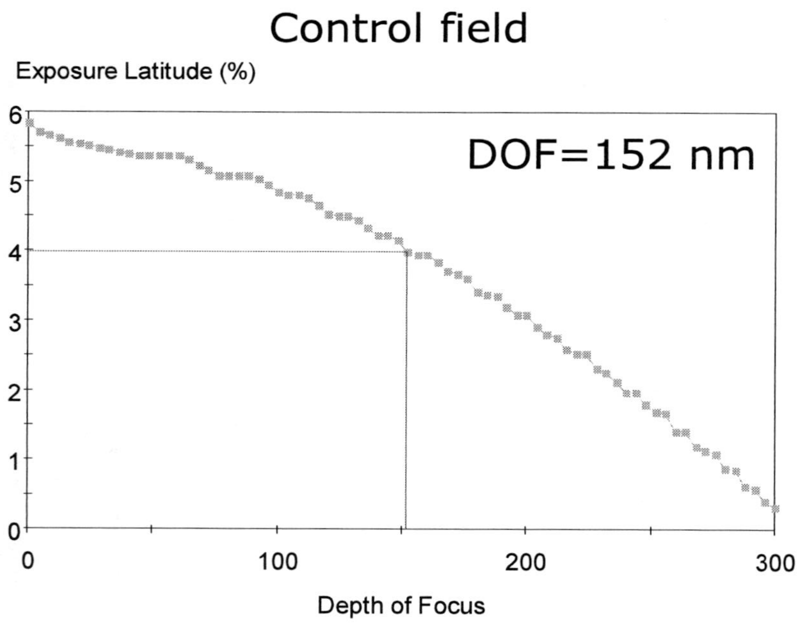

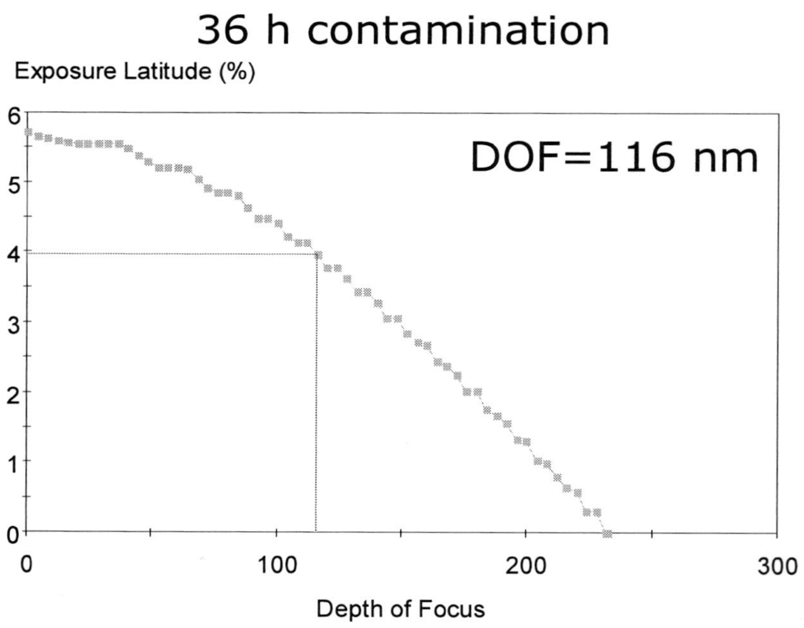

Control field

DOF=152 nm

Exposure Latitude (%)

Depth of Focus

36 h contamination

DOF=116 nm

Exposure Latitude (%)

Depth of Focus

ΔDOF=-36 nm

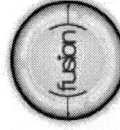

Summary of wafer print results

– Main effect is an increase of the dose to size: 2×

– We also observe a degradation of the depth of focus (~14%)

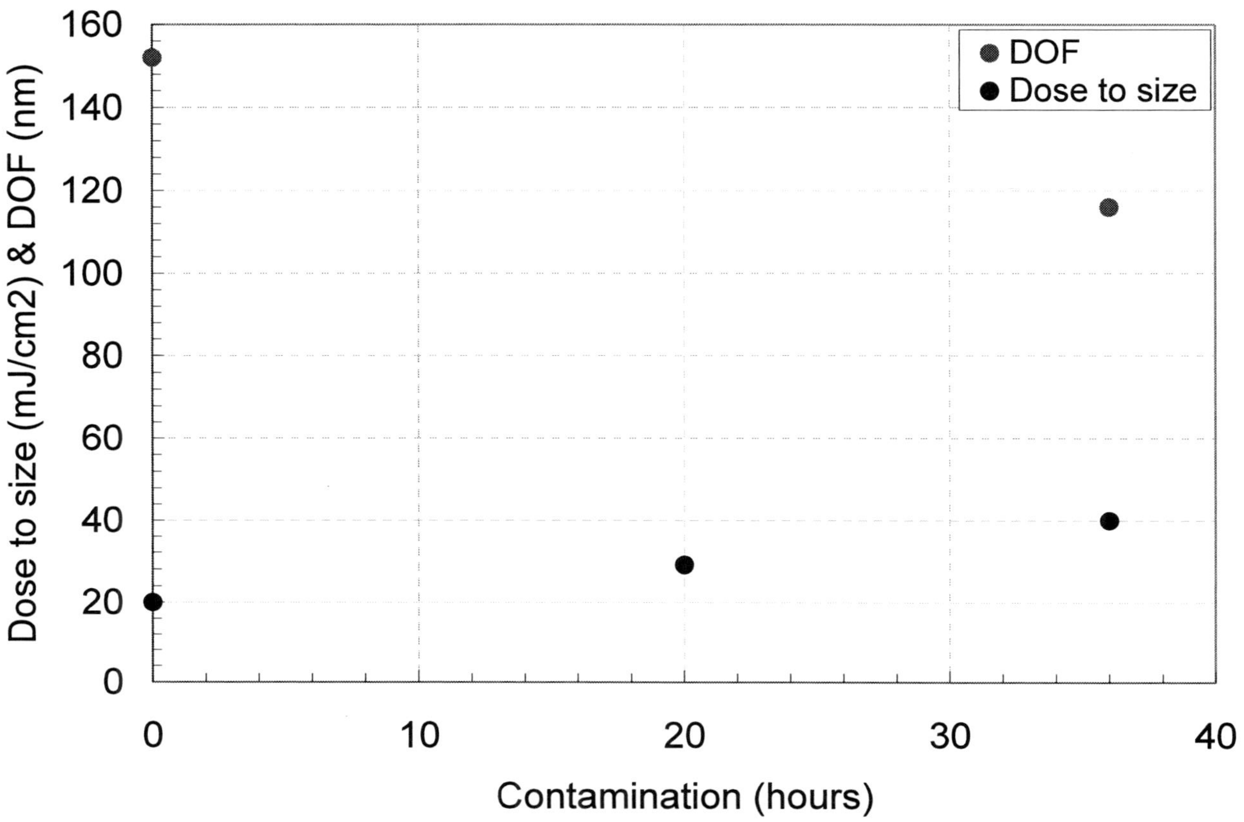

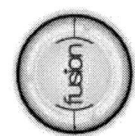

Predicting observed trends

— Simulations using Panoramic TEMPESTpr2

— Most direct consequence of contamination is an increase in the dose to size, as observed experimentally

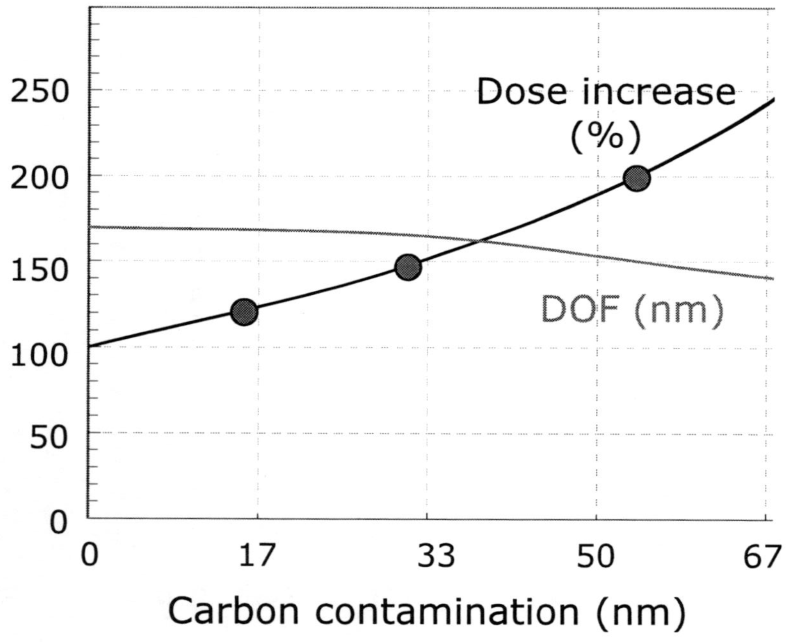

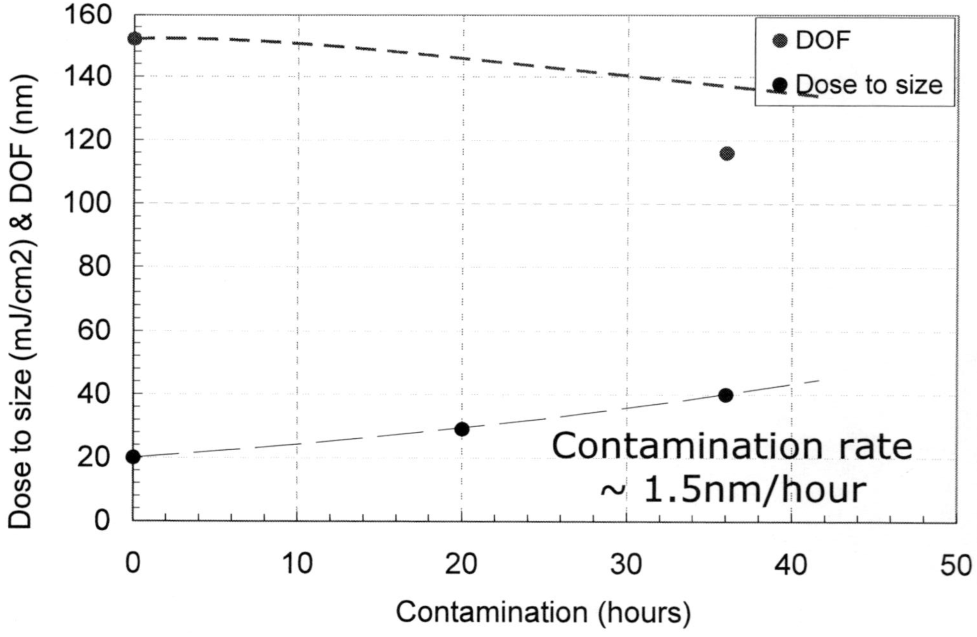

See Y. Fan's poster for more details on modeling

Aerial image measurements
Mask review using the Berkeley AIT

– Images of 36 nm and 40 nm lines and spaces (180 nm & 200 nm on the mask)

– Significant improvements to the AIT occurred between the initial analysis of the contaminated mask and that of the cleaned mask

1st series: December 2007

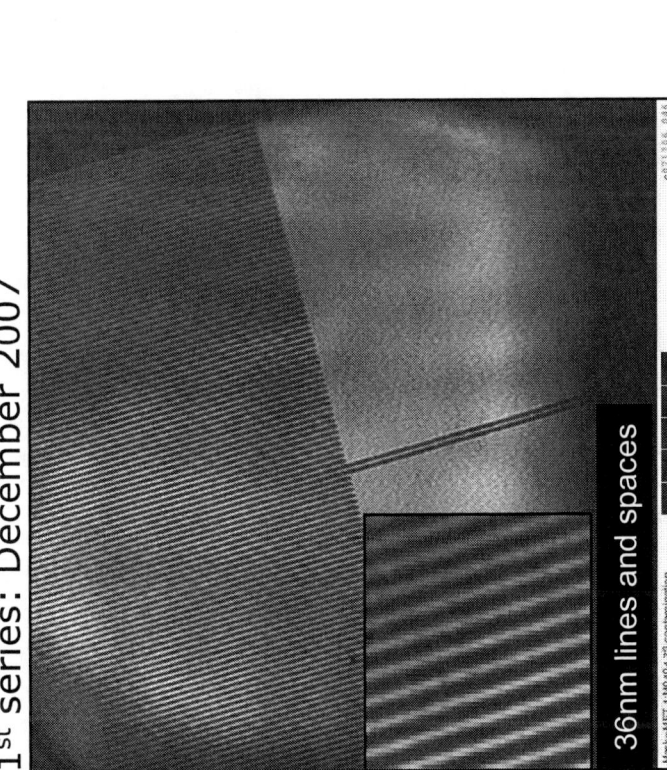

2nd series: August 2008

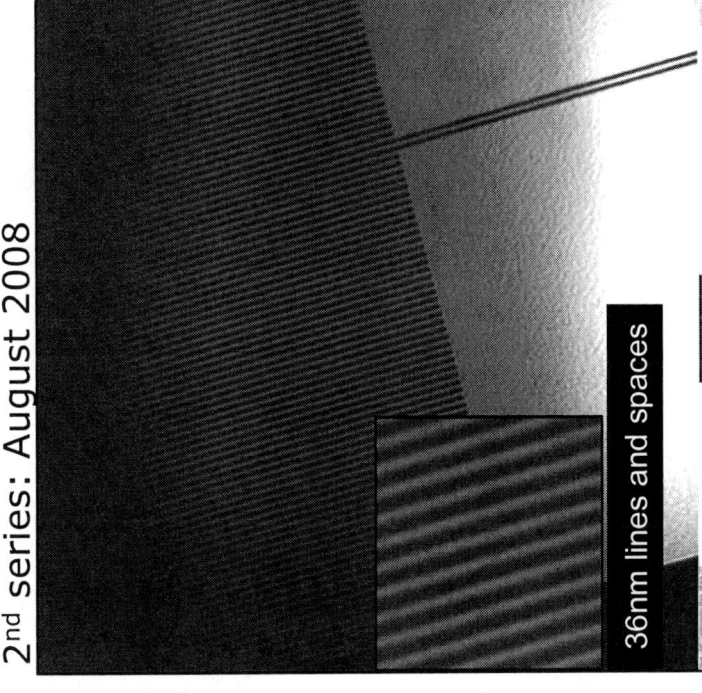

Analysis of AIT images
Best focus images

– Images of control and contaminated sub-fields

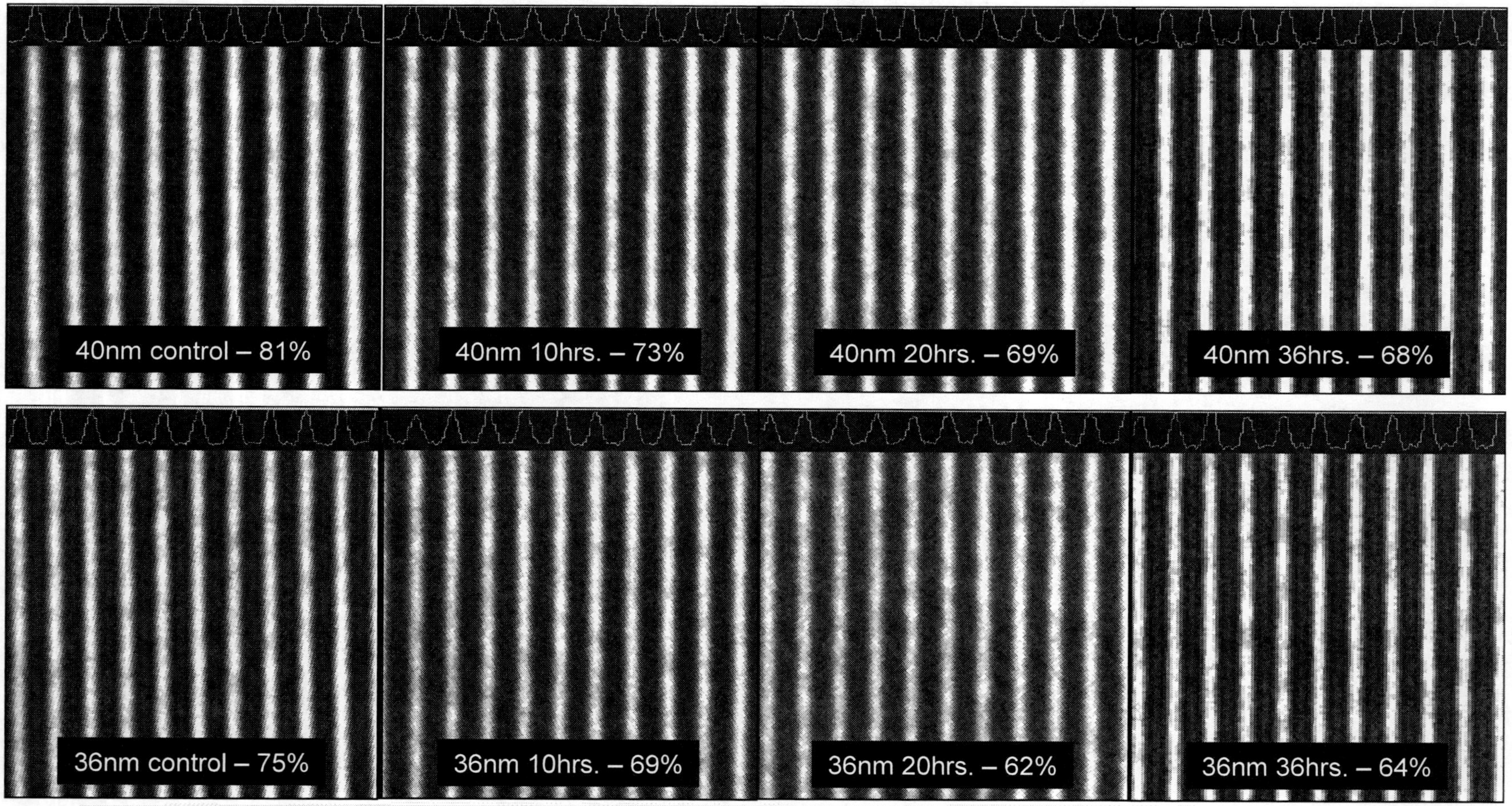

40nm 36hrs. – 68%

40nm 20hrs. – 69%

40nm 10hrs. – 73%

40nm control – 81%

36nm 36hrs. – 64%

36nm 20hrs. – 62%

36nm 10hrs. – 69%

36nm control – 75%

AMD
The future is fusion

Analysis of AIT images

– Loss of contrast observed in contaminated sub-fields

– Similar behavior observed for 36 nm and 40 nm lines and spaces

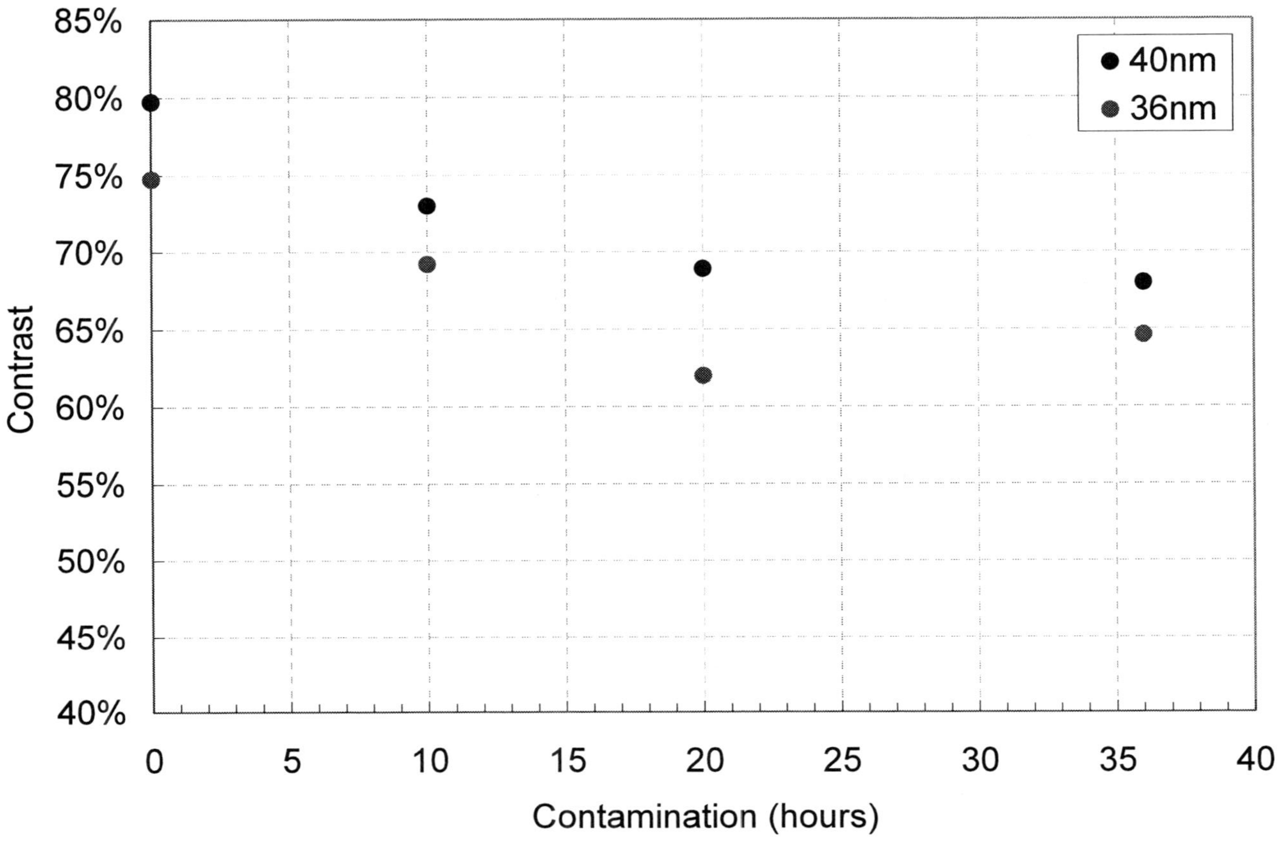

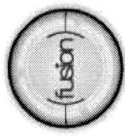

EUV Mask cleaning

– Optical images before and after cleaning

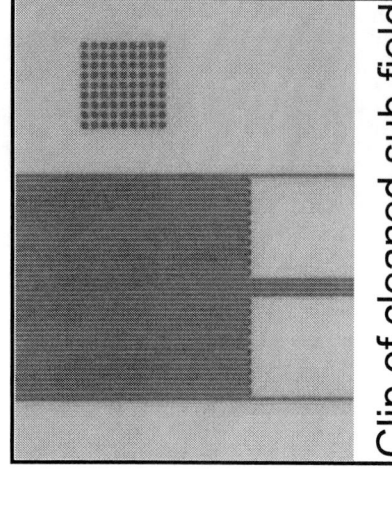

Clip of cleaned sub-field

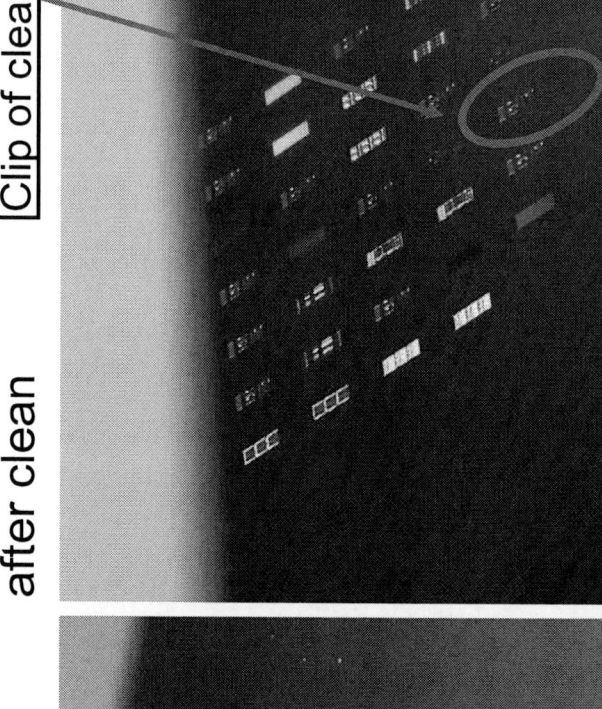

after clean

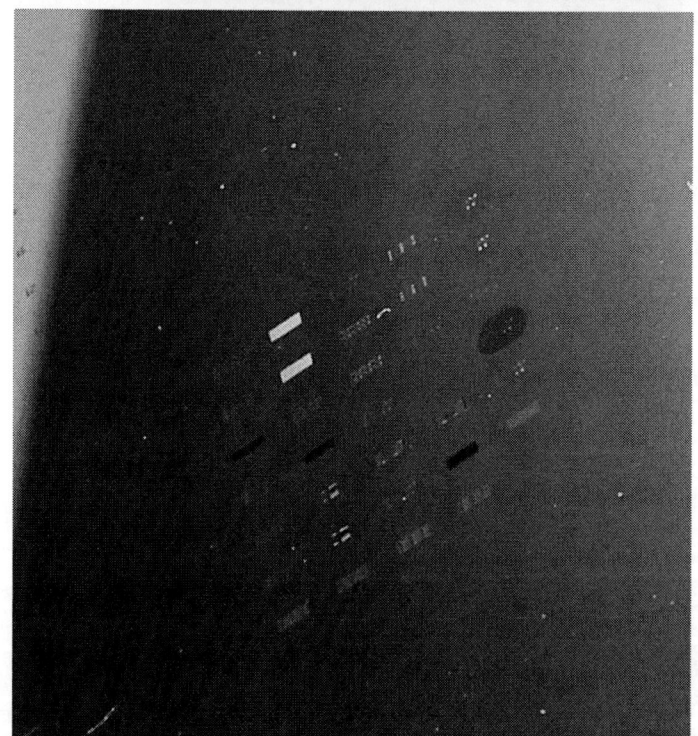

before clean

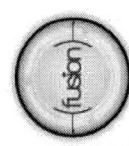

Analysis of wafer prints after mask clean

— After clean, the dose to size for control and contaminated sub-fields are the same (within 4%)

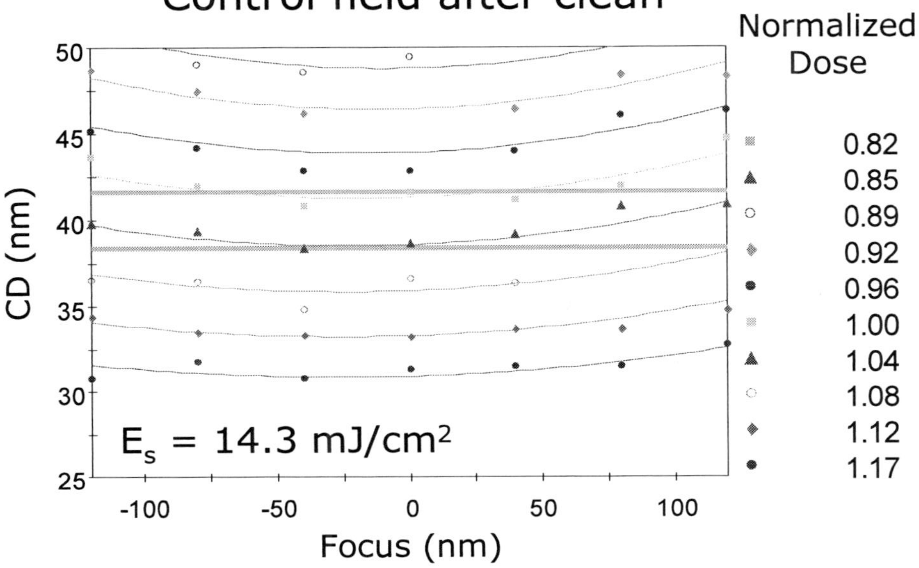

36 h contamination after clean

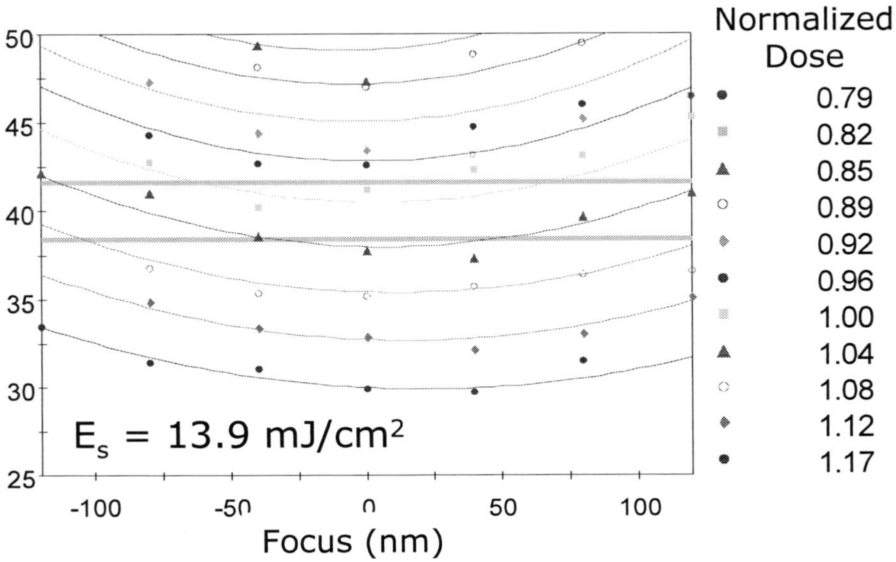

Normalized Dose
0.79
0.82
0.85
0.89
0.92
0.96
1.00
1.04
1.08
1.12
1.17

$E_s = 13.9\ mJ/cm^2$

Control field after clean

Normalized Dose
0.82
0.85
0.89
0.92
0.96
1.00
1.04
1.08
1.12
1.17

$E_s = 14.3\ mJ/cm^2$

CD (nm) vs Focus (nm)

Analysis of wafer prints after mask clean

– Slight reduction in the observed depth of focus (15 nm) after mask clean

   – *Probably not significant*

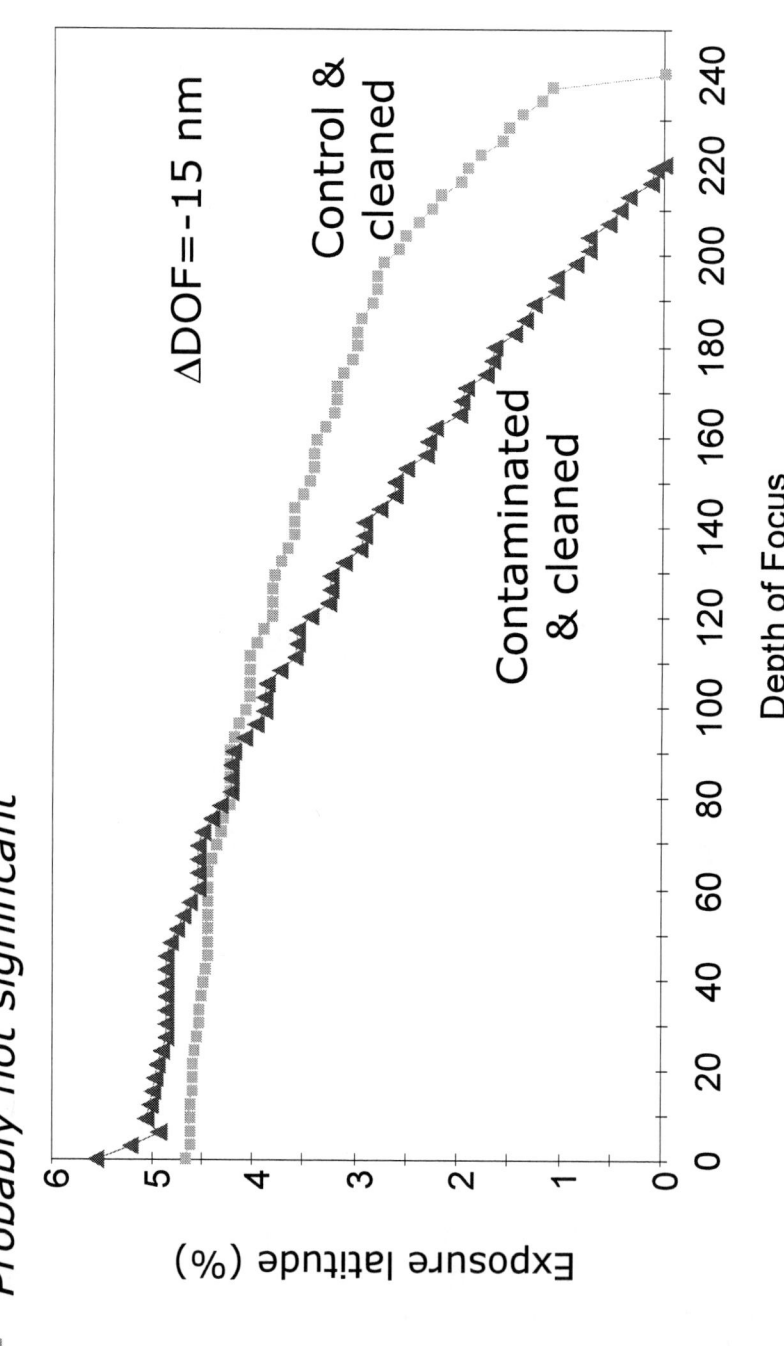

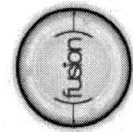

Analysis of AIT images after mask clean

– Control and highly contaminated fields, after mask clean

– No significant loss of contrast

Cleaned Contaminated area

Cleaned control

40nm high contam. – 80%

36nm high contam. – 79%

40nm control – 81%

36nm control – 80%

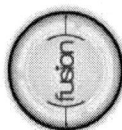

Before and after clean comparisons
AIT aerial images

– Full recovery of contrast after the mask was cleaned

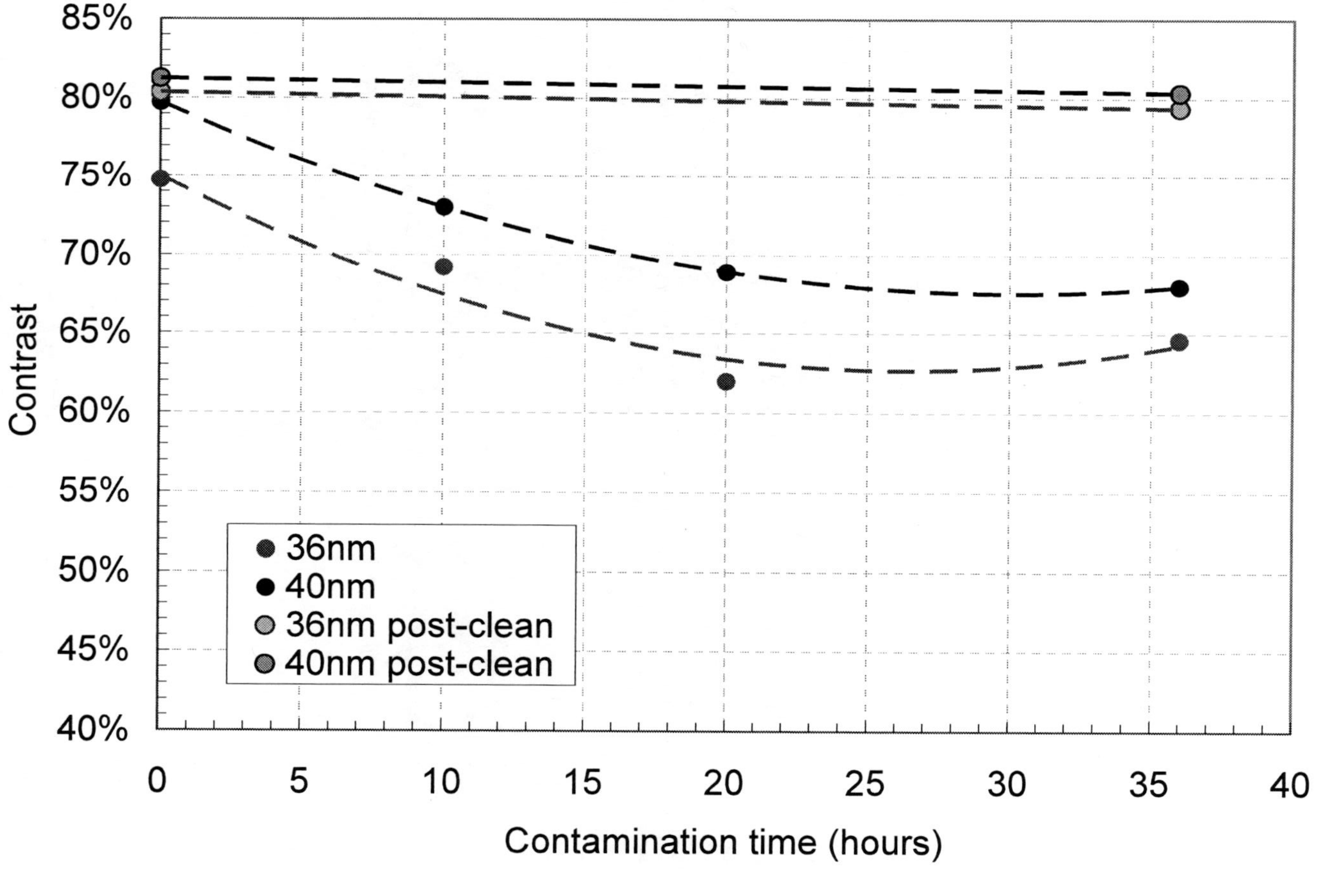

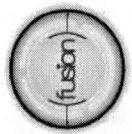

Conclusions

- We have assessed the impact of contamination on the printing of EUV masks
  - The main effect is an increase of dose to size with increasing contamination
  - Other, more subtle effects such as a loss of contrast and depth of focus have been observed

- Fair agreement between print-based results, aerial image results, and simulations
  - Experimental trends are predicted by simulations

- Increase in dose to size can be used as a metric to trigger a mask clean

- Further studies with additional contamination/clean cycles are required to better understand reticle lifetime (e.g. how many times a reticle can be cleaned?)

Trademark Attribution

AMD, the AMD Arrow logo and combinations thereof are trademarks of Advanced Micro Devices, Inc. in the United States and/or other jurisdictions. Other names used in this presentation are for identification purposes only and may be trademarks of their respective owners.

©2008 Advanced Micro Devices, Inc. All rights reserved.

High transmission EUVL pellicle development

Yashesh A. Shroff, Pei-Yang Yan,
Farhad Salamassi*, Eric Gullikson*

Intel Corporation

*Lawrence Berkeley Lab

2008 EUVL Symposium, Lake Tahoe

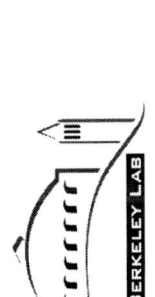

Overview

- Motivation & specs
- Thermal modeling
- High transmission prototype development
  - Illumination non-uniformity
- Full-scale pellicle HVM mechanical test
- DUV/AIMS study
- Summary & future work

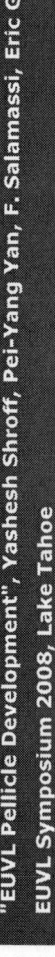

"EUVL Pellicle Development", Yashesh Shroff, Pei-Yang Yan, F. Salamassi, Eric Gullickson
EUVL Symposium 2008, Lake Tahoe

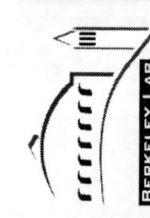

Concept

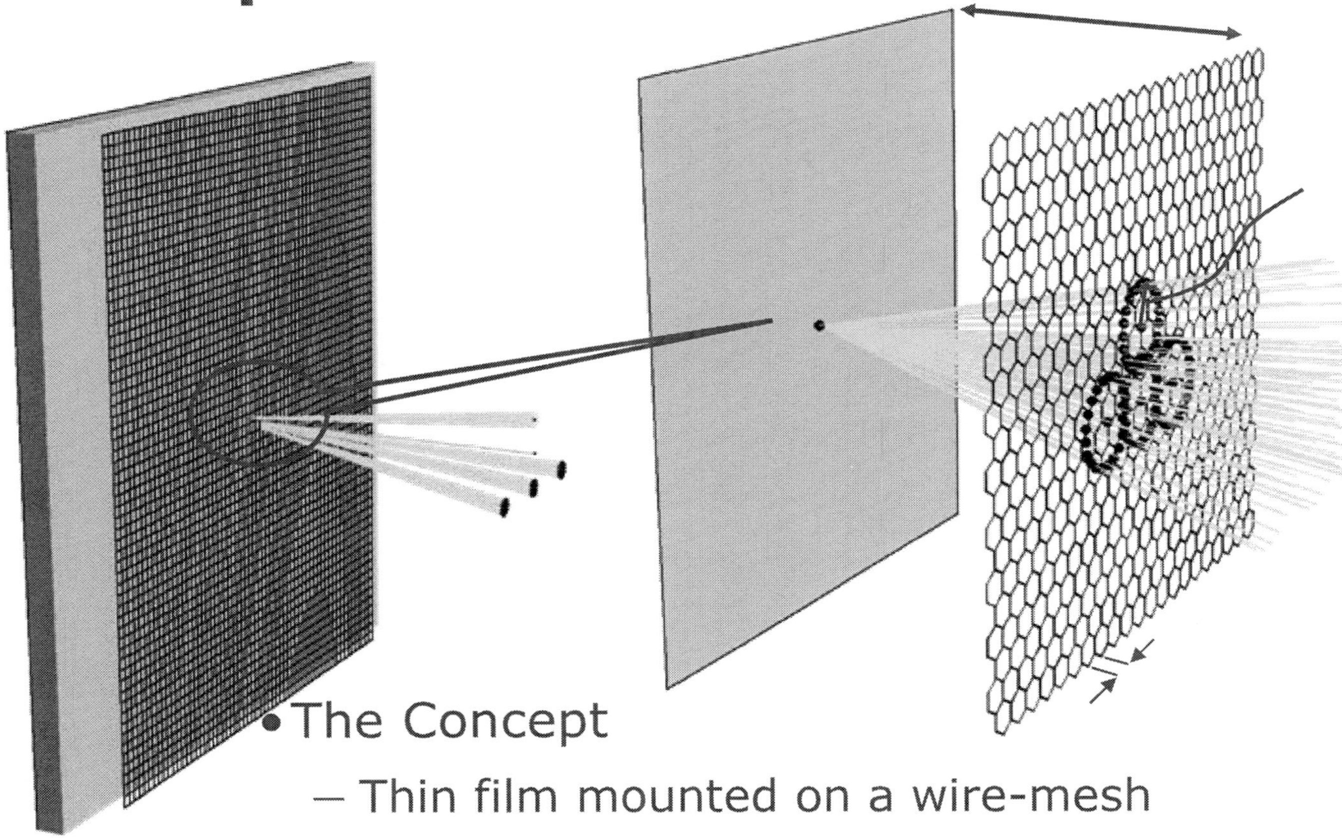

- Mesh apodizes the illuminator exit pupil and PO entrance pupil.
- Mesh film absorbs with both passes.

- The Concept
  - Thin film mounted on a wire-mesh
  - Mesh located "far" from reticle plane to defocus defects
  - Transmission requires a high percentage open area
  - Illumination uniformity requires partial coherence

"EUVL Pellicle Development", Yashesh Shroff, Pei-Yang Yan, F. Salamassi, Eric Gullickson
EUVL Symposium 2008, Lake Tahoe

Motivation & Goal

- The aim of this research is to create an EUV pellicle as a backup to pellicle-less operation
  - Operating within litho tool without added particles
- Mesh based pellicle's target performance:
  - EUV transmission: double-pass loss < 30%
  - Robustness: HVM capable operation
  - Imaging: minimal contrast and uniformity loss

EUV Pellicle Specifications	
Frame length/width	149/122mm
Height	5 mm
Transmission	>70%
Film thickness	50nm ± 2nm
Illumination non-uniformity	± 0.6%
Film stress	Tensile (<50MPa)
#wafers/pellicle	> 2000
Stiffness	193nm frame

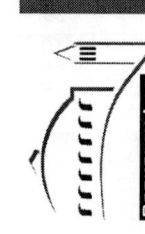

"EUVL Pellicle Development", Yashesh Shroff, Pei-Yang Yan, F. Salamassi, Eric Gullickson
EUVL Symposium 2008, Lake Tahoe

Overview

- Motivation & specs

- **Thermal modeling**

- High transmission prototype development
  - Illumination non-uniformity

- Full-scale pellicle HVM mechanical test

- DUV/AIMS study

- Summary & future work

"EUVL Pellicle Development", Yashesh Shroff, Pei-Yang Yan, F. Salamassi, Eric Gullickson
EUVL Symposium 2008, Lake Tahoe

Thermal model

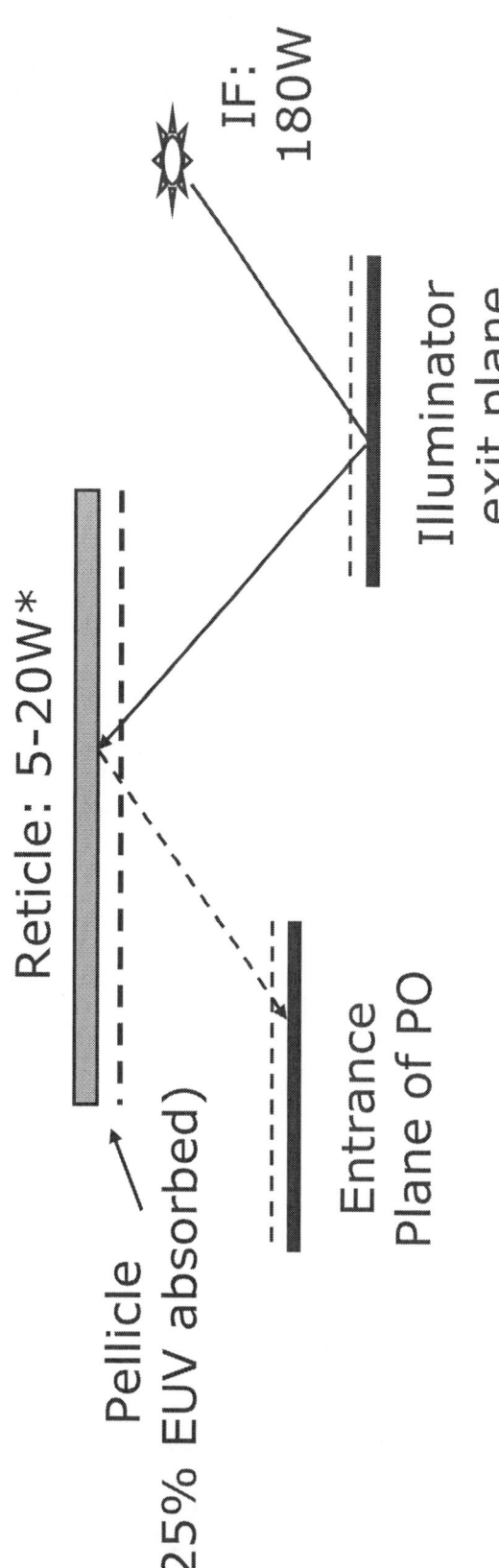

Reticle: 5-20W*

Pellicle
(25% EUV absorbed)

IF:
180W

Illuminator
exit plane

Entrance
Plane of PO

- Instantaneous power at slit can result in high thermal load.
- Transient model is developed to calculate peak rise in pellicle temperature

* OoB power & IU transmission of EUV + OoB spectrum dependent

"EUVL Pellicle Development", Yashesh Shroff, Pei-Yang Yan, F. Salamassi, Eric Gullickson
EUVL Symposium 2008, Lake Tahoe

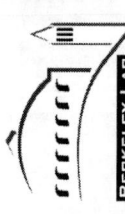

Cooling time

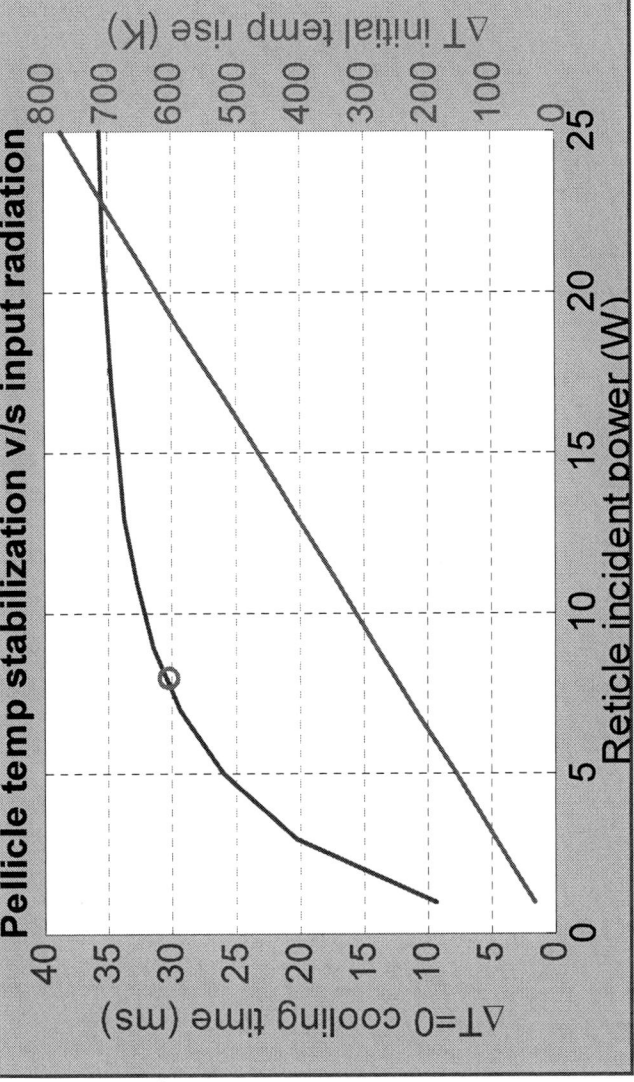

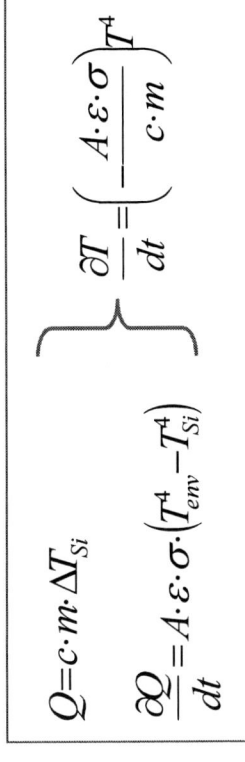

Membrane Parameters:

- Pellicle thickness, d = 50nm
- Exposure slit area, A = 10cm^2
- Exposure slit time, t = 10ms

Constants

- Si specific heat, c = 0.7J/g-k
- Emissivity, ε = 0.5
- Radiative loss, σ = 5.67e-12
- Si density, ρ = 2.33g/cm3

$$Q = c \cdot m \cdot \Delta T_{Si}$$

$$\left.\frac{\partial Q}{\partial t} = A \cdot \varepsilon \cdot \sigma \cdot (T_{env}^4 - T_{Si}^4)\right\} \quad \frac{\partial T}{\partial t} = \left(-\frac{A \cdot \varepsilon \cdot \sigma}{c \cdot m}\right) T^4$$

Assumes 25% reticle incident power absorption by pellicle

"EUVL Pellicle Development", Yashesh Shroff, Pei-Yang Yan, F. Salamassi, Eric Gullickson
EUVL Symposium 2008, Lake Tahoe

Overview

- Motivation & specs
- Thermal modeling
- **High transmission prototype development**
  - Illumination non-uniformity
- Full-scale pellicle HVM mechanical test
- DUV/AIMS study
- Summary & future work

"EUVL Pellicle Development", Yashesh Shroff, Pei-Yang Yan, F. Salamassi, Eric Gullickson
EUVL Symposium 2008, Lake Tahoe

Requirements for reaching pellicle transmission limits

- GOAL: total 2-pass transmission: 70%
  - Mesh transmission: 97%
  - Membrane transmission: 93%

- Improving the mesh
  - Reduce linewidth: 1-3um
  - Increase pitch: 70-100um

- Improving the membrane
  - Reliable membrane thickness of 50nm

→ Lithographically define the line
  - Commercially available mesh min LW only ~10um
  - Improve robustness by increasing aspect ratio of mesh wires
  - Use Si: better thermal stability with membrane

"EUVL Pellicle Development", Yashesh Shroff, Pei-Yang Yan, F. Salamassi, Eric Gullickson
EUVL Symposium 2008, Lake Tahoe

Pellicle Process at LBNL

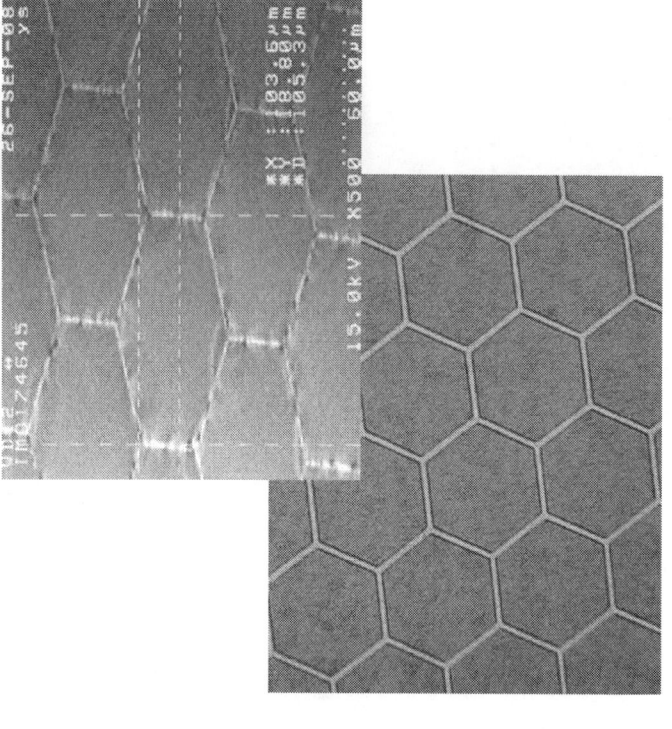

1. Spin release layer on Si wafer.

2. Deposit Si + Ru capping layer using magnetron sputtering.

3. Spin on photoresist.

4. Pattern mesh using contact lithography.

5. Electroplate Ni mesh.

6. Remove resist.

7. Attach to frame and release from substrate.

Pellicle 50 nm Si with 100 micron pitch and 3 micron linewidth mesh. Average transmission at 13.5 nm is 75.7%

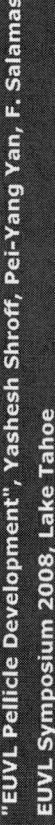

"EUVL Pellicle Development", Yashesh Shroff, Pei-Yang Yan, F. Salamassi, Eric Gullickson
EUVL Symposium 2008, Lake Tahoe

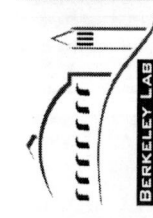

Transmission results

Linewidth	Pitch	T_{mesh} expected	T_{mesh} measured	$T_{pellicle}$ measured	Double-pass txm
1 μm	35 μm	94%	94.5%	79.5%	63.2%
1 μm	100 μm	98%	97.5%	81.9%	67.1%

Membrane: 50nm Si + 2nm Ru cap with transmission = 84% @ 13.5 nm

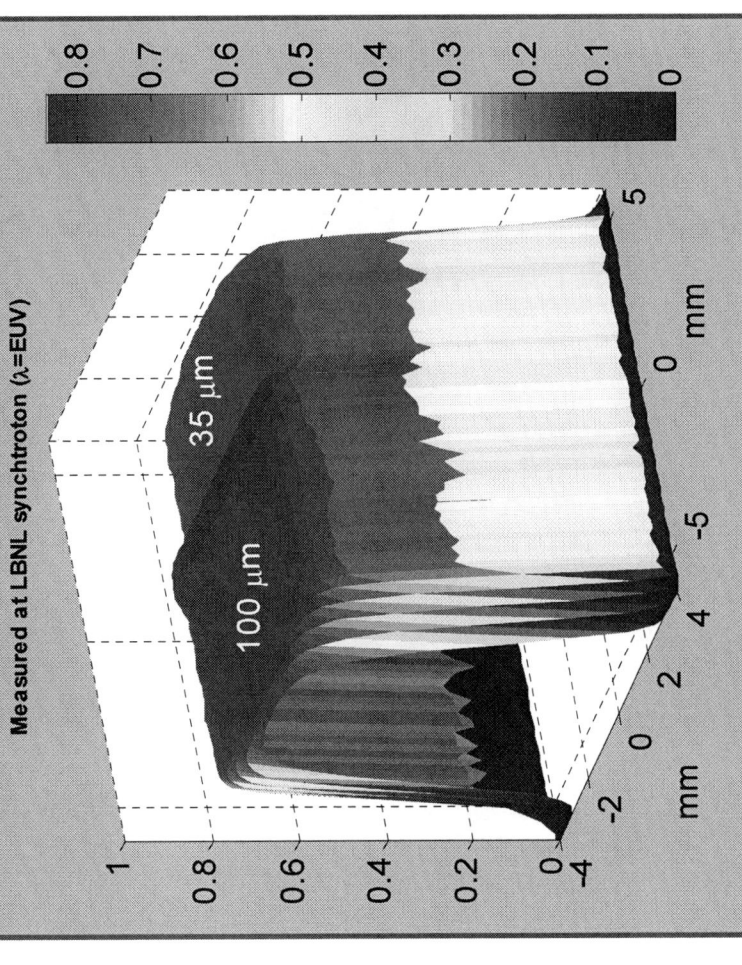

Measured at LBNL synchtroton (λ=EUV)

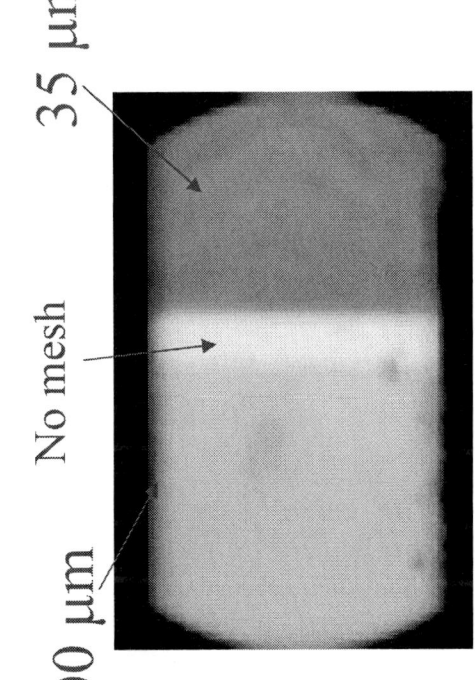

"EUVL Pellicle Development", Yashesh Shroff, Pei-Yang Yan, F. Salamassi, Eric Gullickson

EUVL Symposium 2008, Lake Tahoe

Impact on reticle plane non-uniformity

- Non-uniformity scales with mesh linewidth. Modeled results:

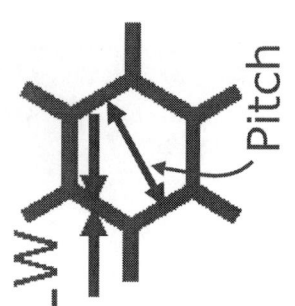

LW / Pitch

		Mesh1	Mesh2	Mesh3
Linewidth (LW)		10u	3u	1u
Pitch		250u	100u	35u
Non-uniformity	5mm	1.4%	0.25%	0.08%
	2mm	7.4%	0.50	0.14%
Mesh only txm		94%	94%	98%

- Decreasing stand-off distance dramatically impacts uniformity due to the severely increased apodization of the pupil by the mesh

- We recommend minimum stand-off distance of 5mm, consistent with current DUV technologies.

"EUVL Pellicle Development", Yashesh Shroff, Pei-Yang Yan, F. Salamassi, Eric Gullickson
EUVL Symposium 2008, Lake Tahoe

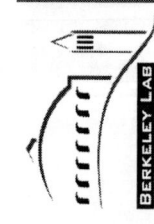

Pellicle transmission improvement

80nm Si membrane on 10/250um mesh

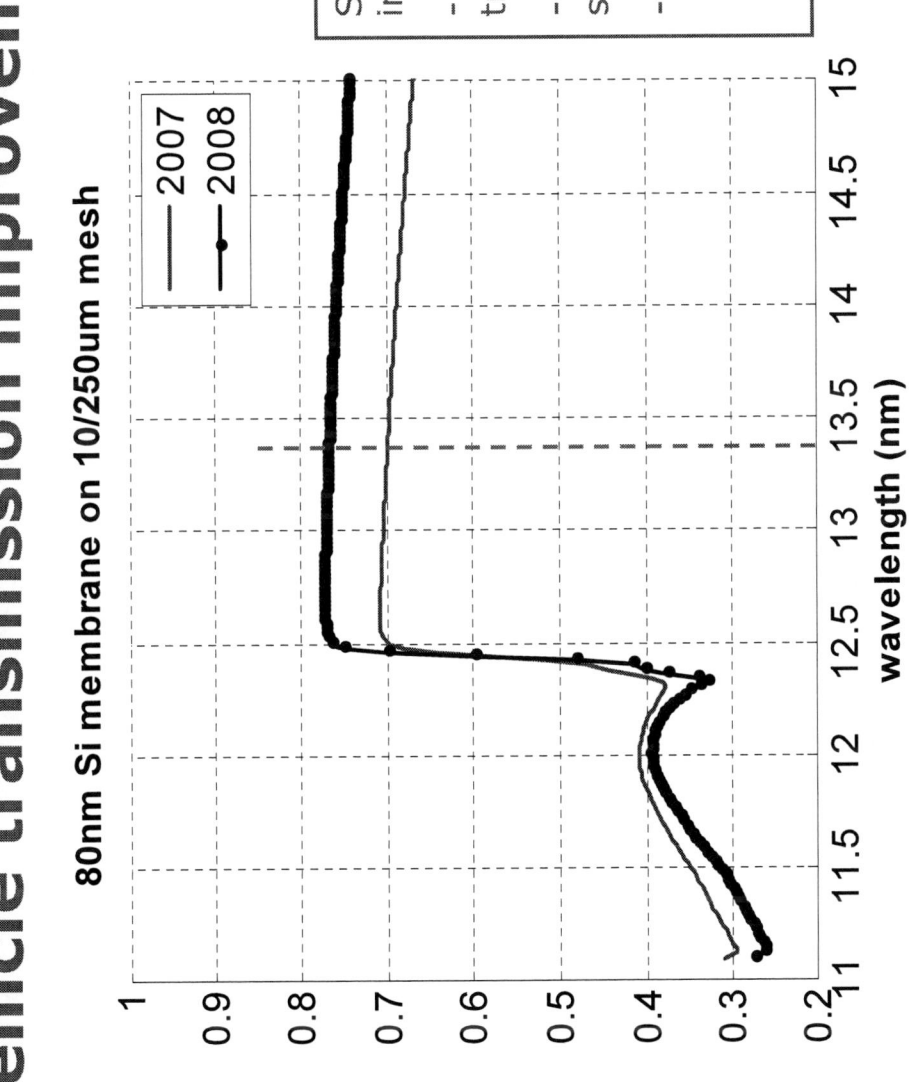

Silicon deposition process improvement:

- 6% increase in 1-pass transmission

- Measurement at LBNL synchrotron

- Mesh linewidth/pitch: 10μm / 250μm

"EUVL Pellicle Development", Yashesh Shroff, Pei-Yang Yan, F. Salamassi, Eric Gullickson
EUVL Symposium 2008, Lake Tahoe

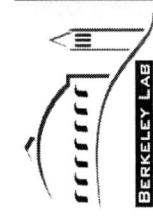

Overview

- Motivation & specs
- Thermal modeling
- High transmission prototype development
  - Illumination non-uniformity
- **Full-scale pellicle HVM mechanical test**
- DUV/AIMS study
- Summary & future work

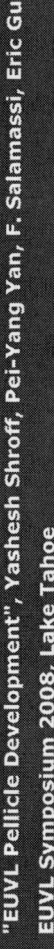

"EUVL Pellicle Development", Yashesh Shroff, Pei-Yang Yan, F. Salamassi, Eric Gullickson
EUVL Symposium 2008, Lake Tahoe

Full-size pellicle development

- Lowest Si membrane thickness achieved: 70nm
  - Corresponds to 85% EUV txm

- Mesh properties:
  - Linewidth: 10um
  - Pitch: 250um
  - Transmission: 91%

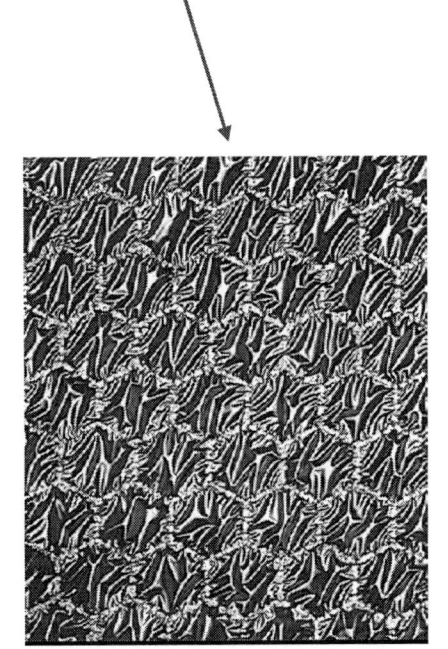

"EUVL Pellicle Development", Yashesh Shroff, Pei-Yang Yan, F. Salamassi, Eric Gullickson
EUVL Symposium 2008, Lake Tahoe

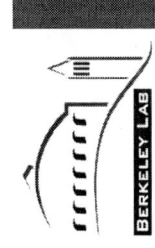

Full-size pellicle scan test

Mechanical stability test to verify robustness.

Repeated 6g acceleration, 2m/s scan velocity expected in 180WPH HVM system.

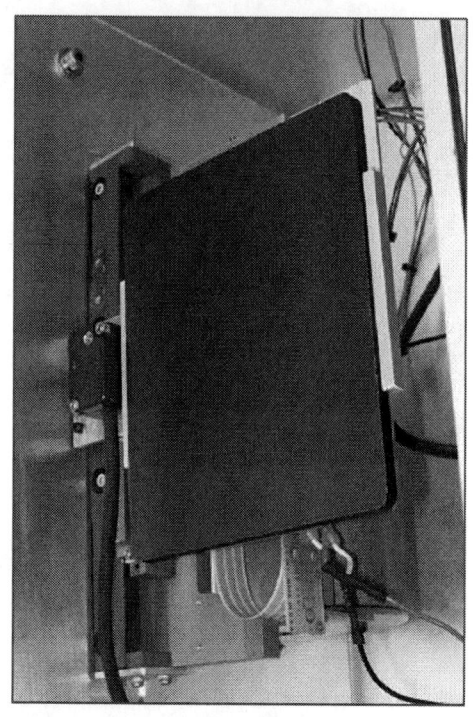

- Vacuum environment chamber
- Track length 75mm
- 50min test, ramping from 1G→6G
  - Continuous scan for 10' at 6G, limited by motor reliability concerns
- Full-size pellicles tested:
  - 70nm membrane
  - 100nm membrane

"EUVL Pellicle Development", Yashesh Shroff, Pei-Yang Yan, F. Salamassi, Eric Gullickson
EUVL Symposium 2008, Lake Tahoe

Scan profile

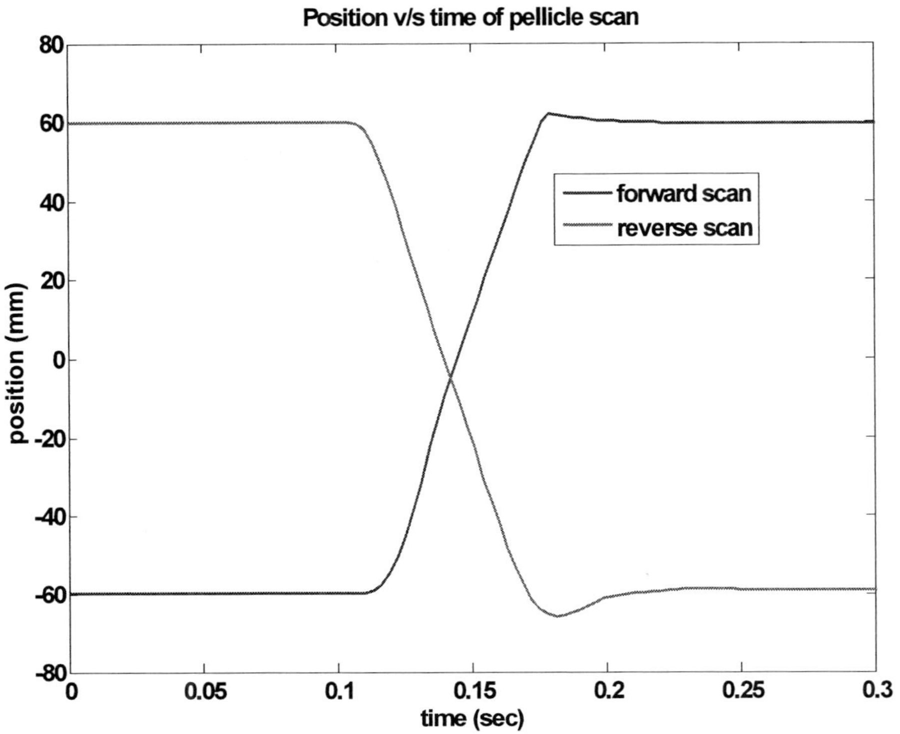

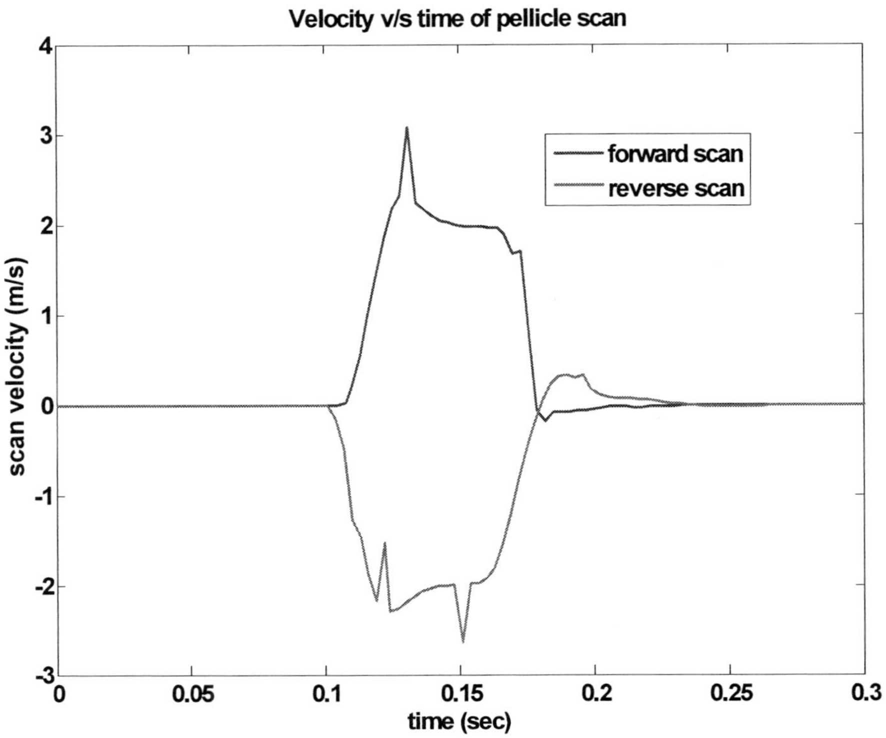

Position v/s time of pellicle scan

Velocity v/s time of pellicle scan

"EUVL Pellicle Development", Yashesh Shroff, Pei-Yang Yan, F. Salamassi, Eric Gullickson
EUVL Symposium 2008, Lake Tahoe

Scan test: 6G, continuous (100nm membrane on 10um linewidth / 250μm pitch mesh)

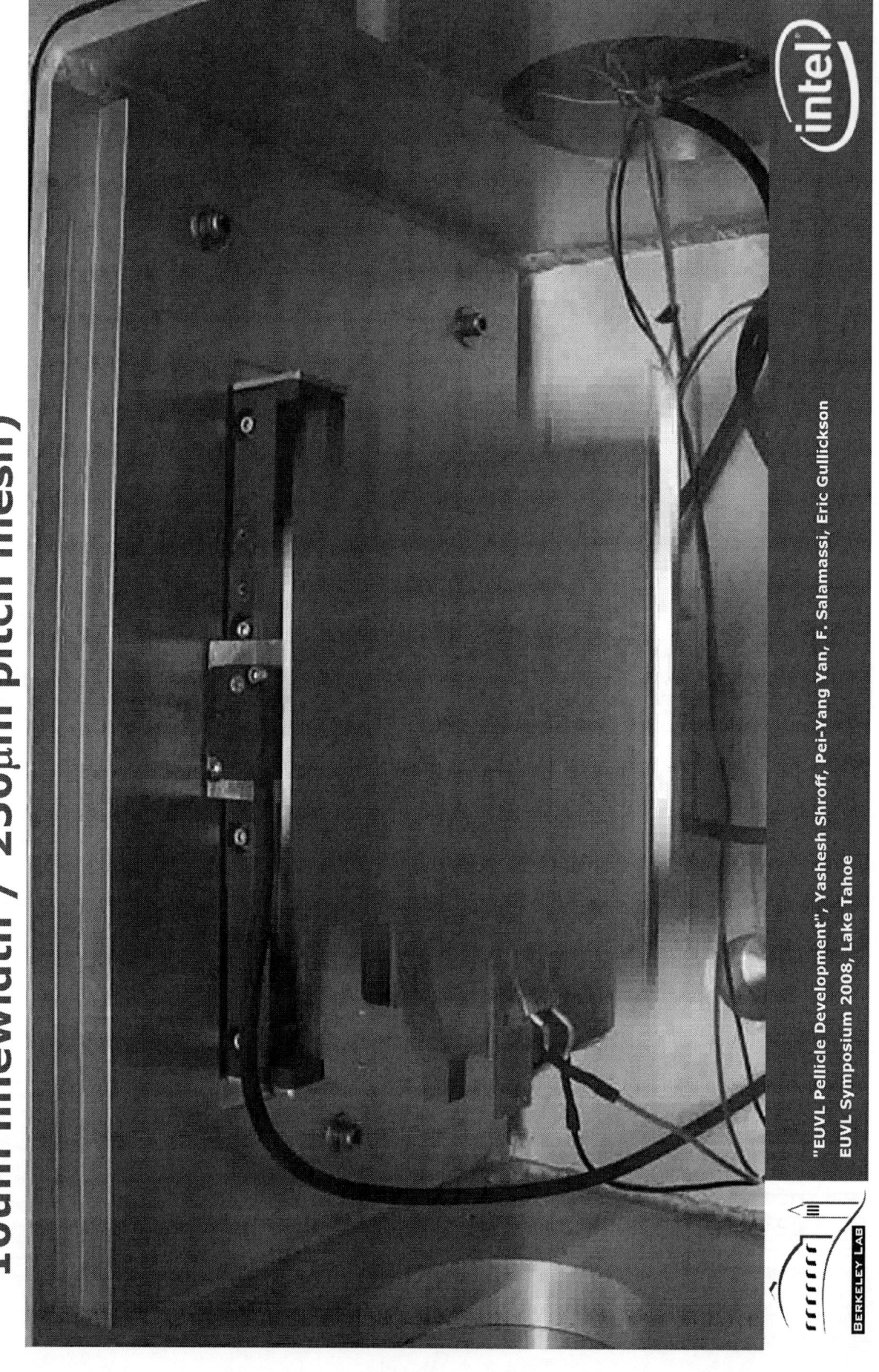

"EUVL Pellicle Development", Yashesh Shroff, Pei-Yang Yan, F. Salamassi, Eric Gullickson
EUVL Symposium 2008, Lake Tahoe

Scan test: 6G with 500ms pause every cycle

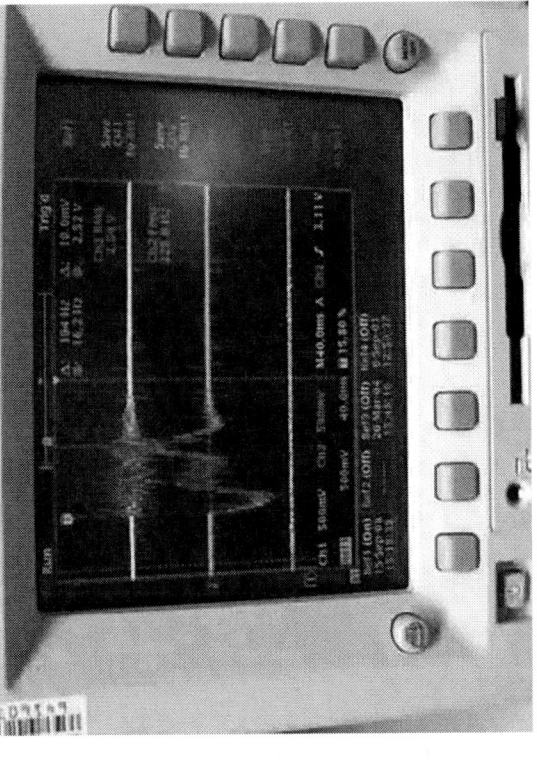

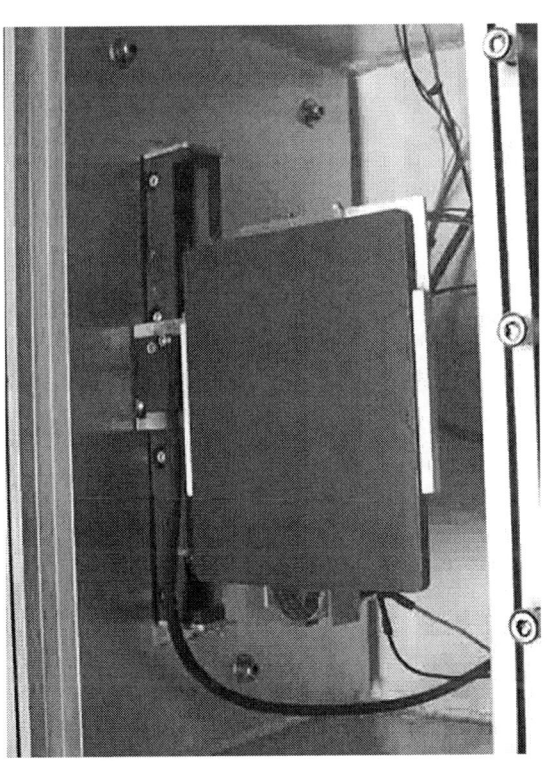

Accelerometer

- Two pellicles are tested: 70nm and 100nm membranes with 10/250um (LW/pitch) supporting mesh.

- Vacuum environmental chamber was kept at 70mT pressure

- Reticle plate was not used (load concerns)

"EUVL Pellicle Development", Yashesh Shroff, Pei-Yang Yan, F. Salamassi, Eric Gullickson

EUVL Symposium 2008, Lake Tahoe

Overview

- Motivation & specs
- Thermal modeling
- High transmission prototype development
- Illumination non-uniformity
- Full-scale pellicle HVM mechanical test
- **DUV/AIMS study**
- Summary & future work

"EUVL Pellicle Development", Yashesh Shroff, Pei-Yang Yan, F. Salamassi, Eric Gullickson
EUVL Symposium 2008, Lake Tahoe

DUV AIMS test

- 150nm 1:1 features exposed with 248nm wavelength AIMS tool.

- Mesh placed 2mm above the reticle plane

- CD variation is within measurement error of tool
  - Impact of pellicle on CD needs to be ascertained w/ a higher resolution tool.

- Illumination conditions: $\sigma = 0.6$; NA = 0.68

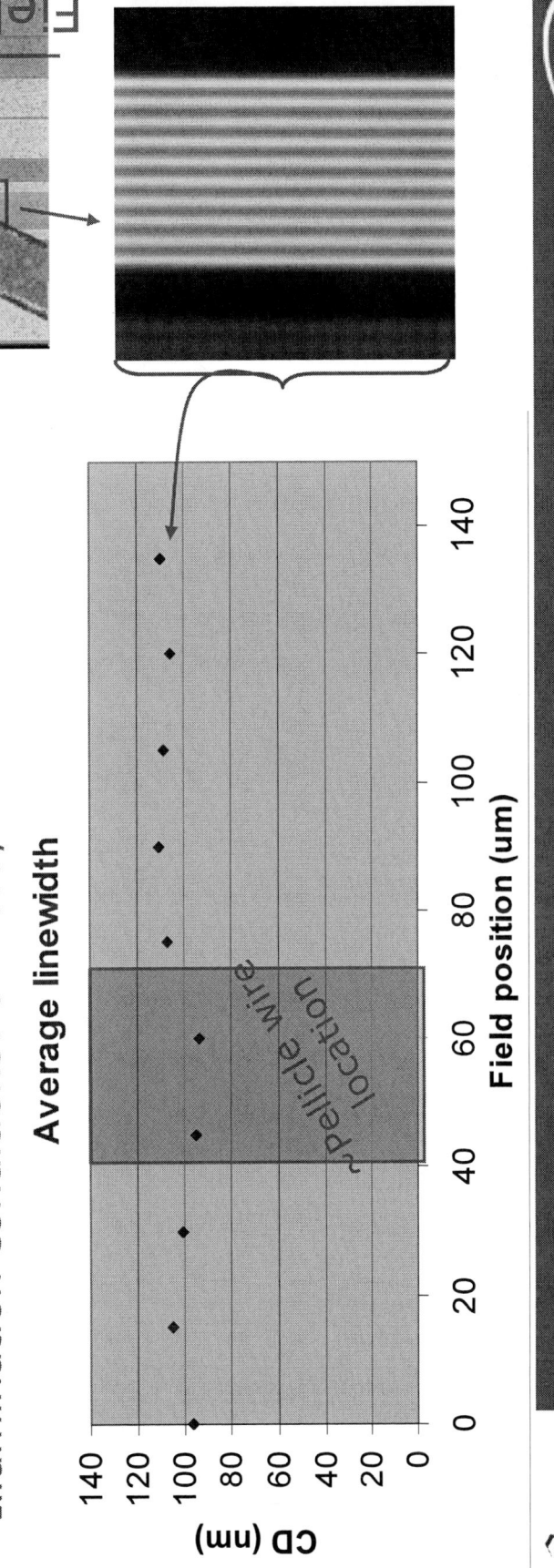

Average linewidth

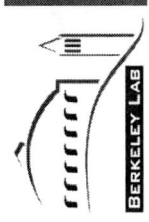

"EUVL Pellicle Development", Yashesh Shroff, Pei-Yang Yan, F. Salamassi, Eric Gullickson
EUVL Symposium 2008, Lake Tahoe

Summary

- Full-size pellicles fabricated with membrane thickness of 70%. Expected transmission is 61% double pass

- New process developed to achieve near-best transmission of 67% on small-size samples with tensile membrane stress.

- Thermal model completed & shows that CTE compatible mesh + Si membrane is necessary to avoid differential thermal expansion related issues.

- Scan test with in-house scanner shows pellicle robustness at repeated 6G acceleration with peak velocities of 2m/s.

Future work

- Vibration test to satisfy zero-particle adder criteria

- Full-scale pellicle for 70% double-pass txm

"EUVL Pellicle Development", Yashesh Shroff, Pei-Yang Yan, F. Salamassi, Eric Gullickson
EUVL Symposium 2008, Lake Tahoe

Acknowledgements

- Bob Gunion and Paul Denham (LBNL) for writing very challenging device drivers for the scan test motor and flexible UI and device setup

- Armando Cobarrubia (Intel), for SEM support

- Guojing Zhang and Kangmin Hsia (Intel), for IMO support

- Luxel Corp (OR) for fabricating full-size mesh

- Committed support from senior management at Intel for funding and providing resources for the project.

"EUVL Pellicle Development", Yashesh Shroff, Pei-Yang Yan, F. Salamassi, Eric Gullickson
EUVL Symposium 2008, Lake Tahoe

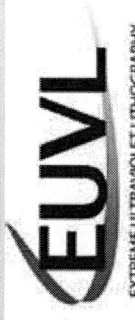

2008 International Symposium on Extreme Ultraviolet Lithography

Applying Thinner Absorber to the EUVL Mask: EUV printability and integration issues

Hwan-Seok Seo*, Dong Gun Lee, Hoon Kim, Sungmin Huh, Byung-Sup Ahn, Hakseung Han, Dongwan Kim, Seong-Sue Kim, and Han Ku Cho

Memory R&D Center
Samsung Electronics Co., Ltd.

*hwanseok.seo@samsung.com

Outline

- Introduction
  - Mask shadowing effect in EUVL
  - Phase-shifting mask in EUVL
- Simulation
  - n, k measurements at EUV wavelength
  - Reflectivity, H-V bias, phase, contrast, and NILS
- Evaluation of mask performances
  - CSM analyses
  - MET analyses
  - Issues in EUVL masks applying thinner absorber
- Summary & Conclusions

Motivation: Mask shadowing effect in EUVL

- Mask shadowing effect arises from the non-telecentric off-axis illumination on the mask combined with the 3-D mask topography.

  Geometrically in vertical L/S patterns,

  Space CD (printed) = CD (designed) − 2d·tanθ·M

  Line CD (printed) = CD (designed) + 2d·tanθ·M

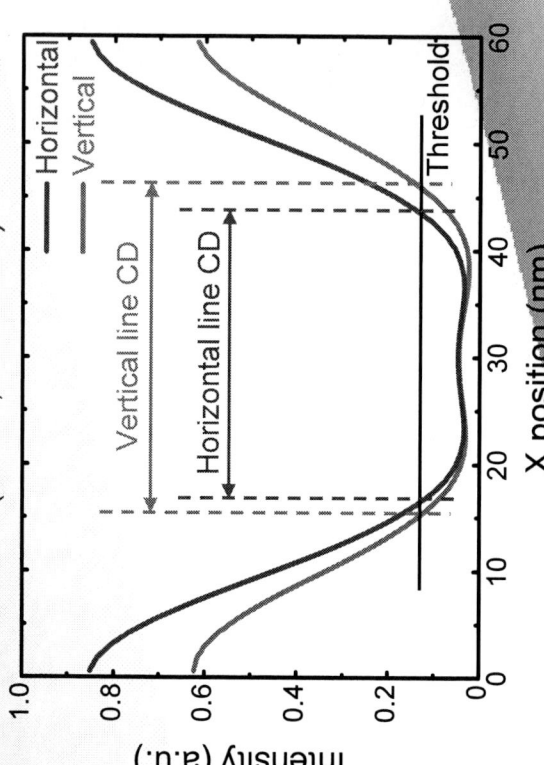

Aerial images from a L/S pattern with d = 80 nm at CD = 30 nm (θ = 8°, NA =0.35)

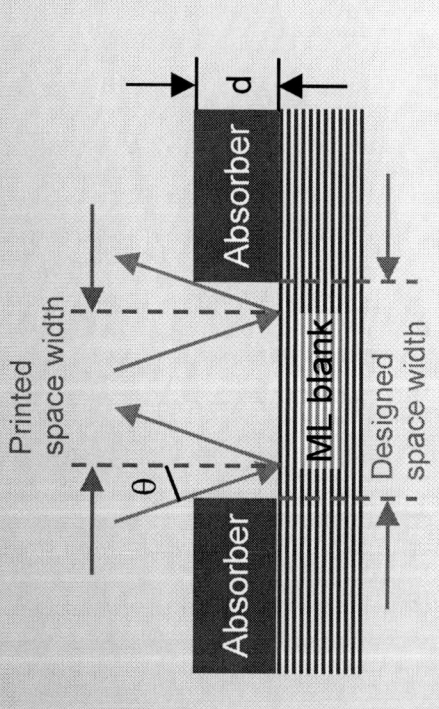

Printed space width

Absorber   d   Absorber

ML blank

Designed space width

θ

Horizontal (non-shadowing)   Vertical (shadowing)

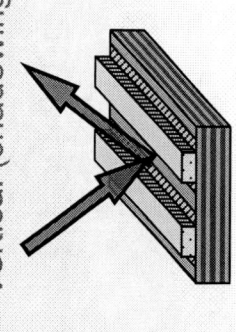

- Shadowing effect (H-V bias) increases with d ↑, θ ↑, CD ↓.

- H-V bias deteriorates lithography process margin by reducing overlapping PW in both directions.

- OPC to correct H-V bias becomes a bigger burden with increasing mask shadowing.

Phase-shifting mask in EUVL

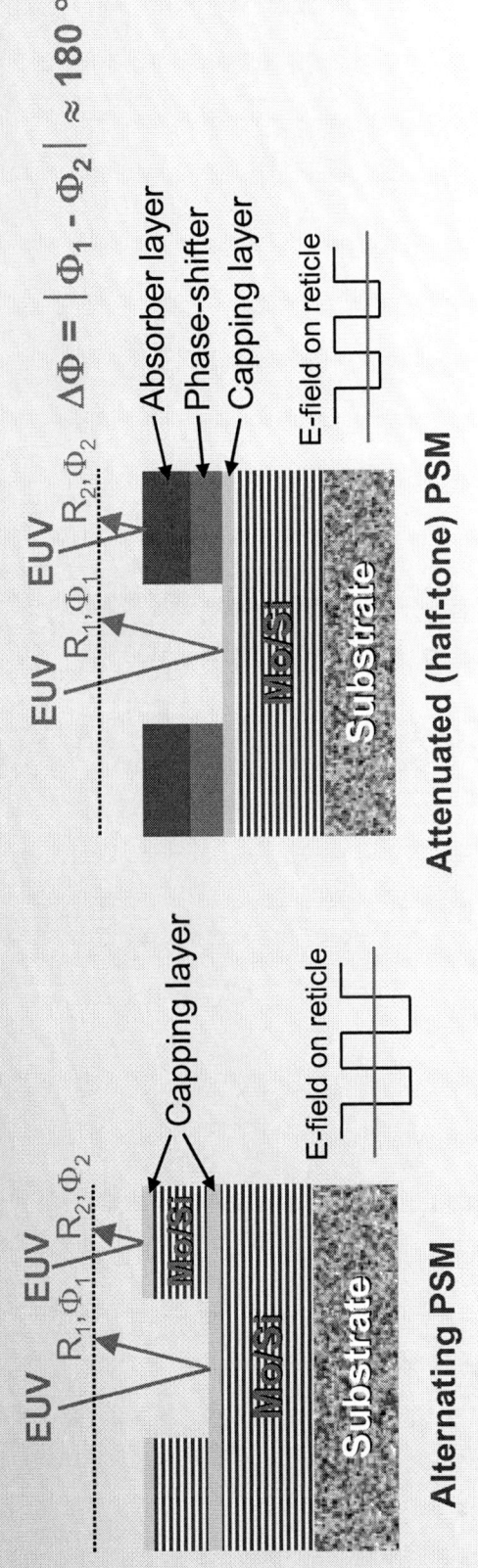

Alternating PSM

Attenuated (half-tone) PSM

$$\Delta\Phi = |\Phi_1 - \Phi_2| \approx 180°$$

- Applying PSM, EUVL could be more easily extended to 22 nm node and below.
- However, it is hard to make a strong phase-shifting structure in EUVL masks, due to
  - Small variations in refractive indices among materials at EUV wavelengths.
  - Complexity in process (e.g. ML etch in alt. PSM) compared to EUV BIM.
  - Difficulty in obtaining optimum phase-attenuation relation in att. PSM structure.
  - High k_1 factor in EUVL which restricts PSM effect on the resolution enhancement.
- In addition, application of PSM using thinner absorber stack can be a solution to correct shadowing problems.

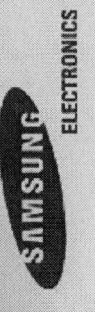

Experimental procedure

1. In-house mask blank fabrication

IBD sputter at Samsung

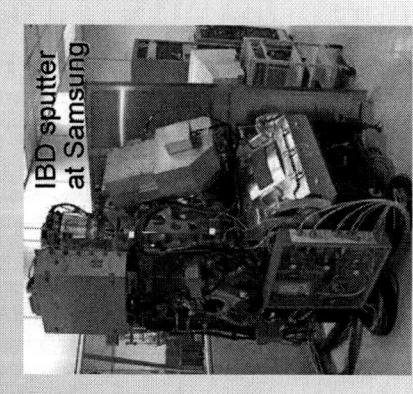

2. EUV n & k analyses at ALS BL 6.3.2

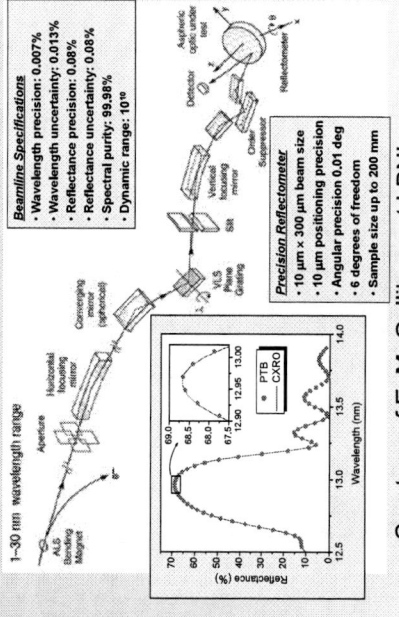

Courtesy of E. M. Gullikson at LBNL

3. Simulation & Mask design

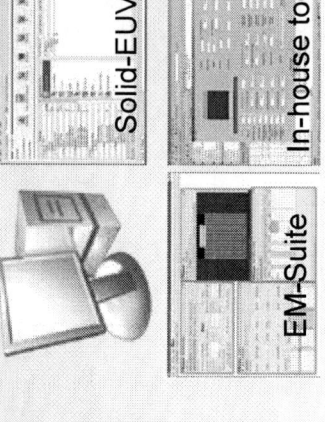

Solid-EUV

In-house tool

EM-Suite

4. Mask patterning at Samsung Mask shop

Incident electron beam

Objective Lens

resist

mask

5. Evaluation with EUV CSM at NewSUBARU

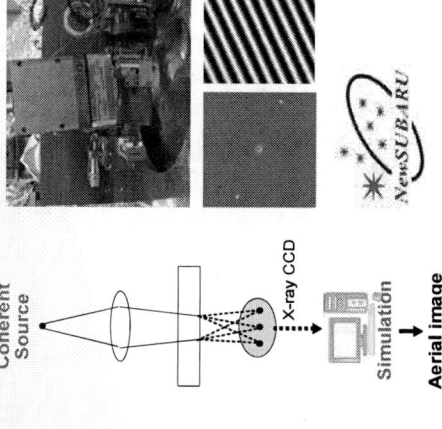

Coherent Source

X-ray CCD

Simulation

Aerial image

NewSUBARU

6. Evaluation with EUV MET at LBNL

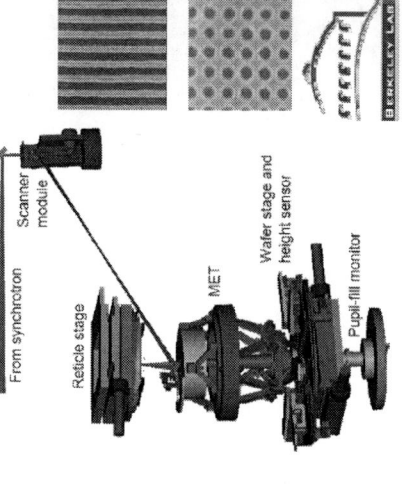

From synchrotron

Scanner module

Reticle stage

MET

Wafer stage and height sensor

Pupil-fill monitor

BERKELEY LAB

Courtesy of P. Naulleau at LBNL

SAMSUNG ELECTRONICS

Analysis of EUV refractive index

- n and k values of absorber and capping materials in actual EUV wavelengths are analyzed to design reliable thin absorber structure through rigorous lithography simulation.

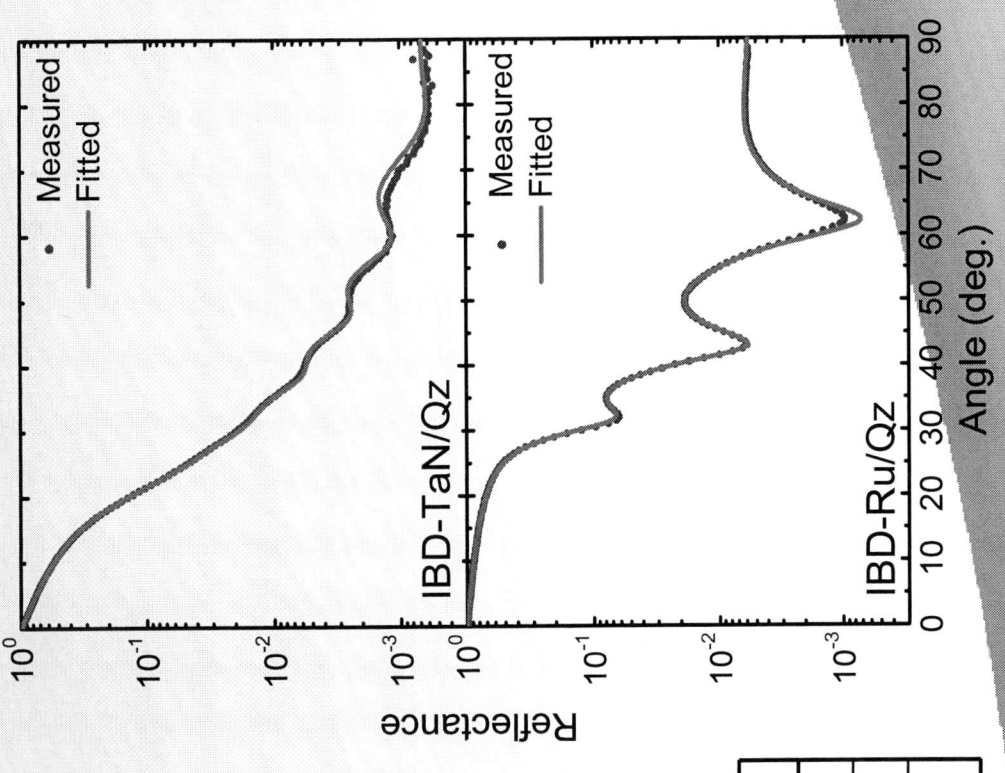

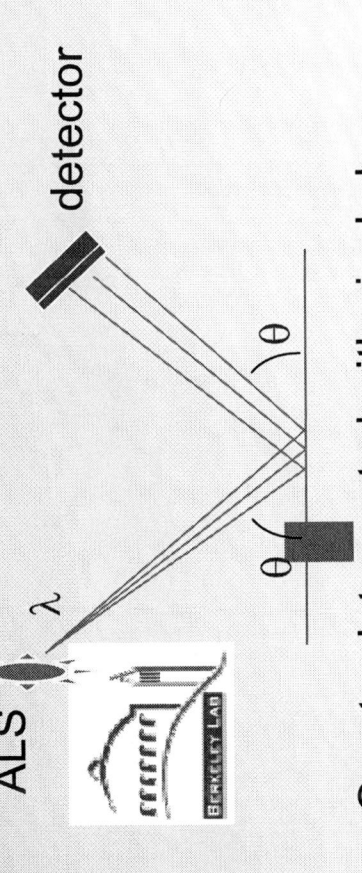

ALS

detector

Quartz plate coated with single layer

Fresnel reflectivity

$$R_s = \left| \frac{\sin\theta - \sqrt{\tilde{n}^2 - \cos^2\theta}}{\sin\theta + \sqrt{\tilde{n}^2 - \cos^2\theta}} \right|^2 \quad , \quad \tilde{n} = 1 - \delta + i\beta$$

	n (= 1 - δ)	k (= β)	Remarks
IBD-TaN*	0.9474	0.0313	measured
IBD-Ru*	0.8890	0.0178	measured
Bulk-TaN	0.9260	0.0436	http://www.cxro.lbl.gov/optical_constants

* Thanks to E. M. Gullikson at LBNL for the analysis

EUV reflectivity: simulations vs. measurements

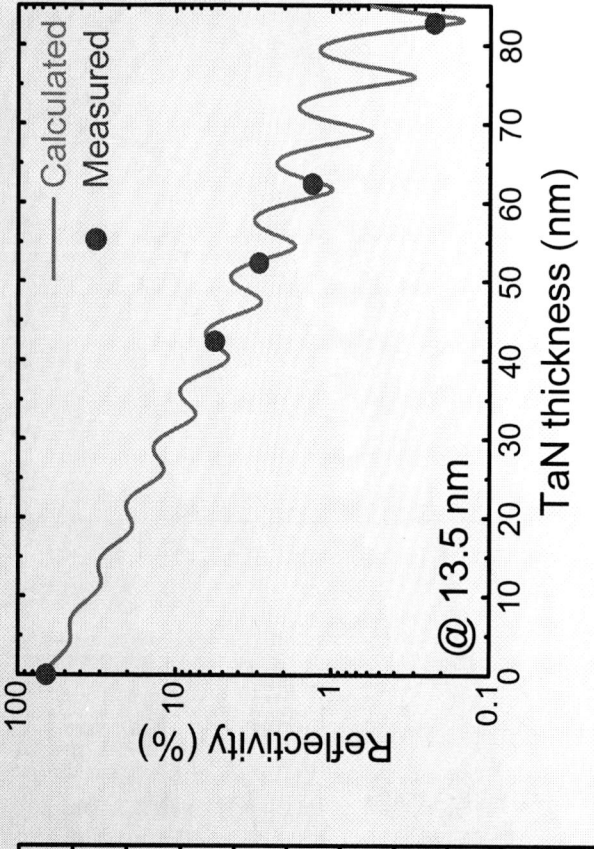

Verification of measured n, k values

- Both simulation and measurement exhibit very similar reflectivity values and spectrum shapes in each thickness.

- Simulation results using experimental refractive indices of TaN and Ru match well with measured data in all ranges.

H-V bias simulation

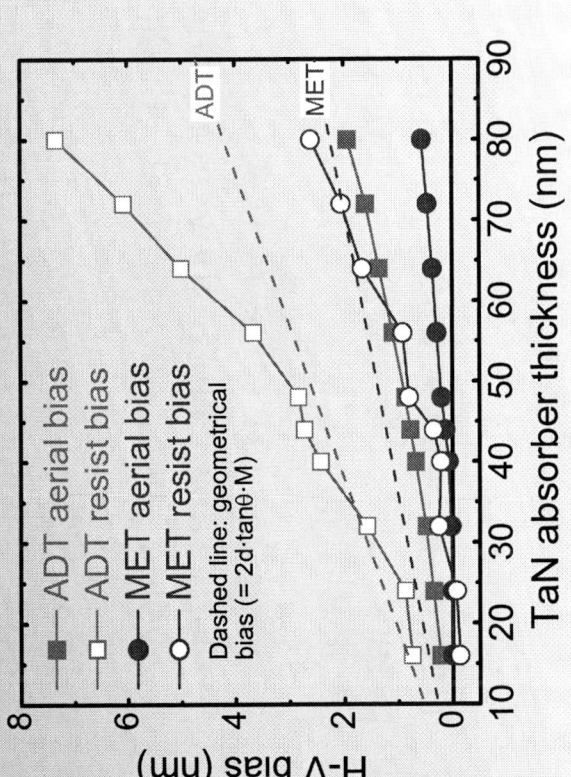

Simulation Conditions

- Simulator: waveguide model with Solid-EUV
- 32 nm HP 1:1 L/S
- Parameters:

	ADT	MET
Incident angle (θ)	6 deg.	4 deg.
Reduction factor (M)	1/4	1/5
Illumination (σ)	0.5	0.55/0.35

- Resist model: MET-2D

- As TaN absorber thickness becomes thinner, H-V bias on the printed image is further reduced.

- H-V bias reduction rate (in 40–80 nm thickness range)
  - ~0.6 nm per 10-nm-reduction in TaN thickness for MET resist bias.
  - ~1.3 nm per 10-nm-reduction in TaN thickness for ADT resist bias.

- The degree of H-V bias reduction using thinner absorber is greater than that estimated from geometry.

- **Application of thin absorber** to the EUVL mask is one of the best solutions to correct shadowing problems.

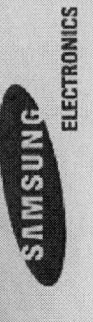

Simulations: phase-shifting effects

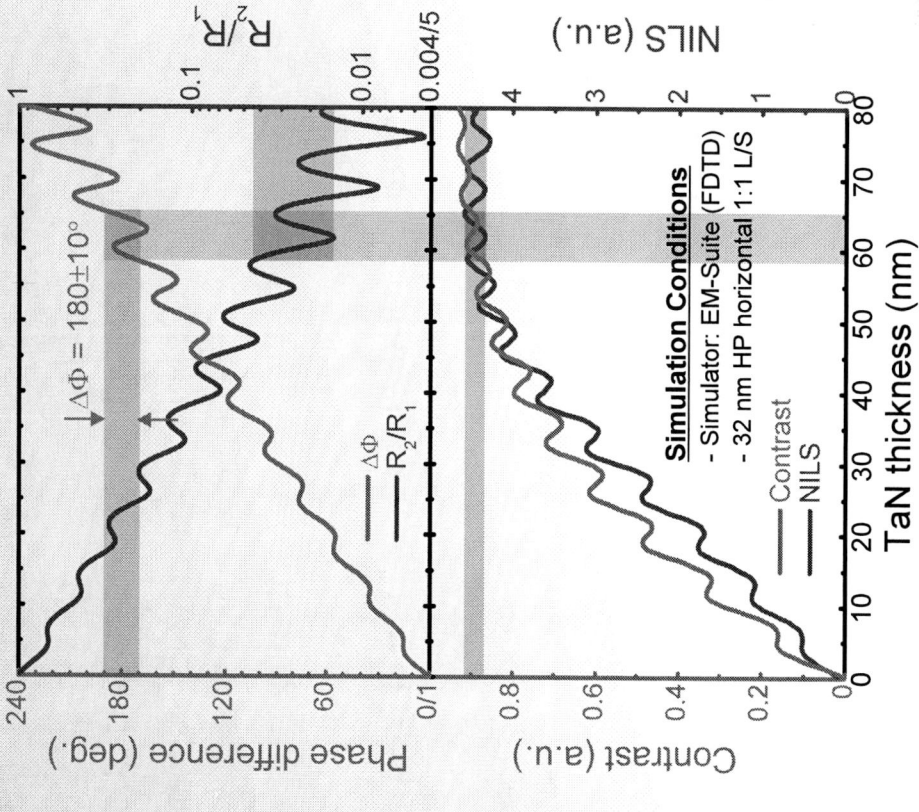

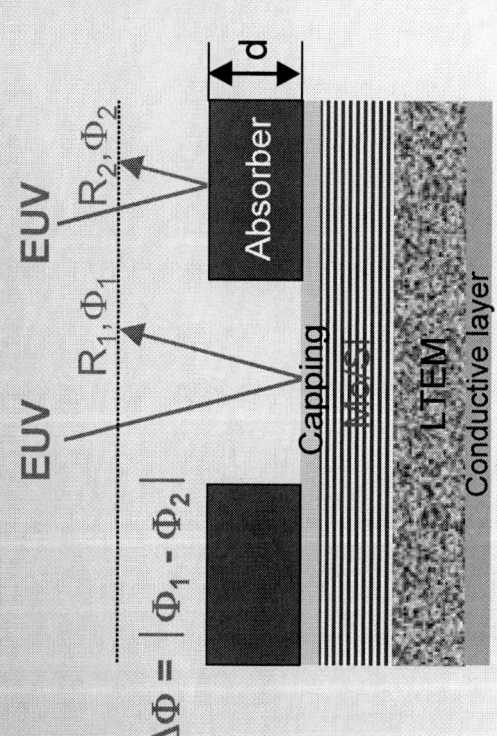

$$\Delta\Phi = |\Phi_1 - \Phi_2|$$

- Thickness between **58 and 65 nm** meets out-of-phase ($\Delta\Phi = 180\pm10°$) for IBD-TaN.
  - $R_2/R_1 = 1.5 \sim 4.5\ \%$
  - Contrast & NILS show maximum values over out-of-phase condition.

- Finding out-of-phase condition ($\Delta\Phi = 180°$) is a way to reduce absorber thickness without loss of image contrast and process margin.

EUV coherent scattering microscope (CSM)

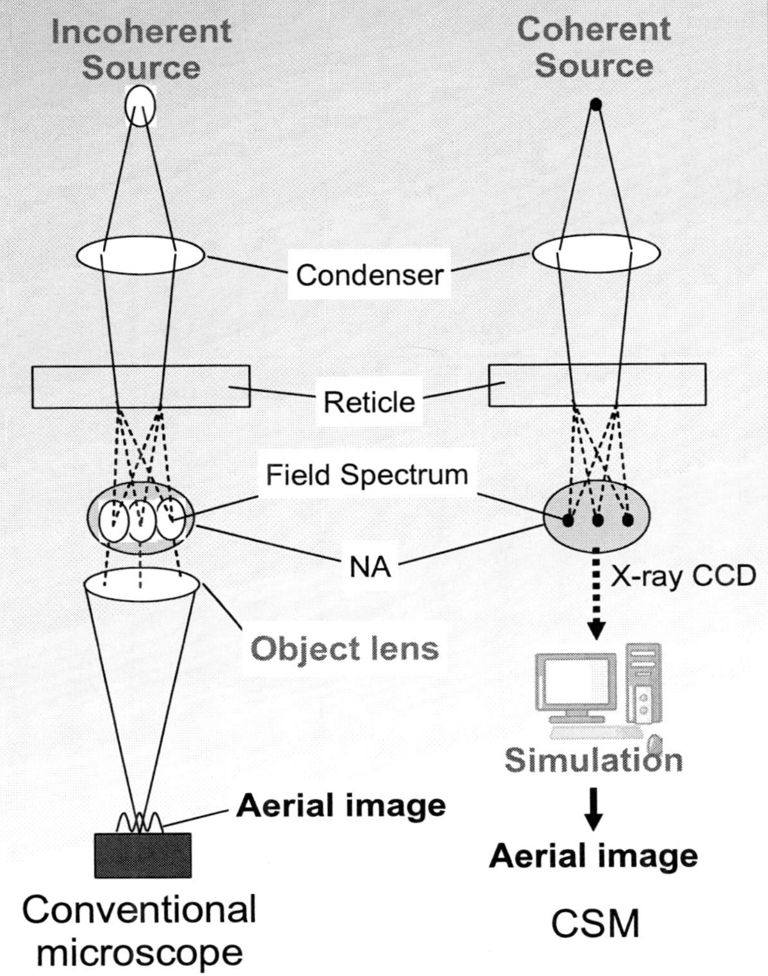

Incoherent Source
Coherent Source
Condenser
Reticle
Field Spectrum
NA
X-ray CCD
Object lens
Simulation
Aerial image
Aerial image
Conventional microscope
CSM

Refer to D. Lee et al, "Analysis of Sub-22-nm Aerial Image Using Coherent Scattering Microscopy", Metrology session on Monday.

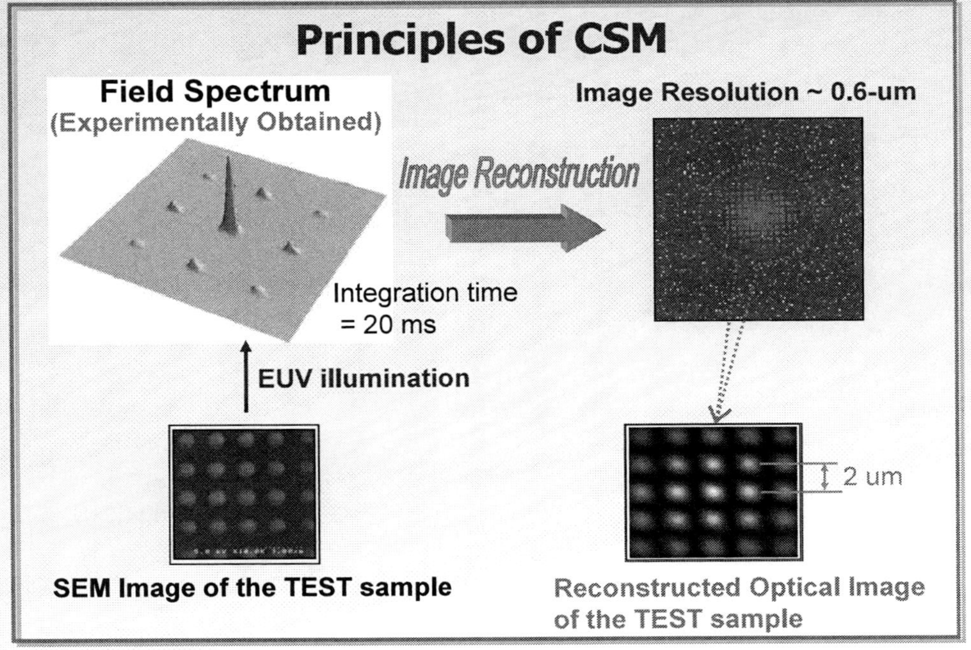

Principles of CSM
Field Spectrum
(Experimentally Obtained)
Image Resolution ~ 0.6-um
Image Reconstruction
Integration time = 20 ms
EUV illumination
2 um
SEM Image of the TEST sample
Reconstructed Optical Image of the TEST sample

- The CSM reconstructs aerial images by phase-retrieval algorithm from the field spectra produced by coherent diffractions of EUV beam.
- Intrinsic properties of EUVL mask (e.g. phase, actinic CD, NILS, DOF) can be evaluated by the CSM without aberration coming from lens and resist.

SAMSUNG ELECTRONICS

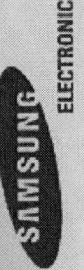

Performances of EUVL mask applying phase-shifting thin absorber

Wafer images @ 28 nm HP 1:1 L/S by MET

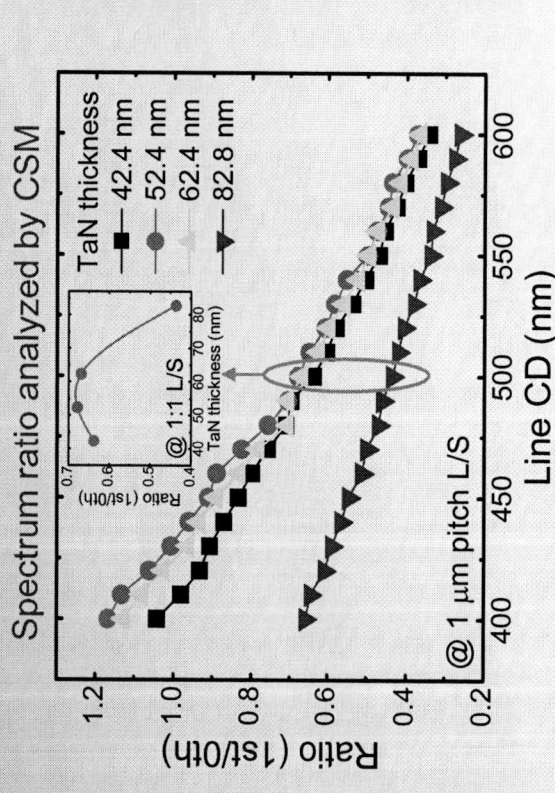

TaN 42.4 nm

TaN 52.4 nm

TaN 62.4 nm

TaN 82.8 nm

Spectrum ratio analyzed by CSM

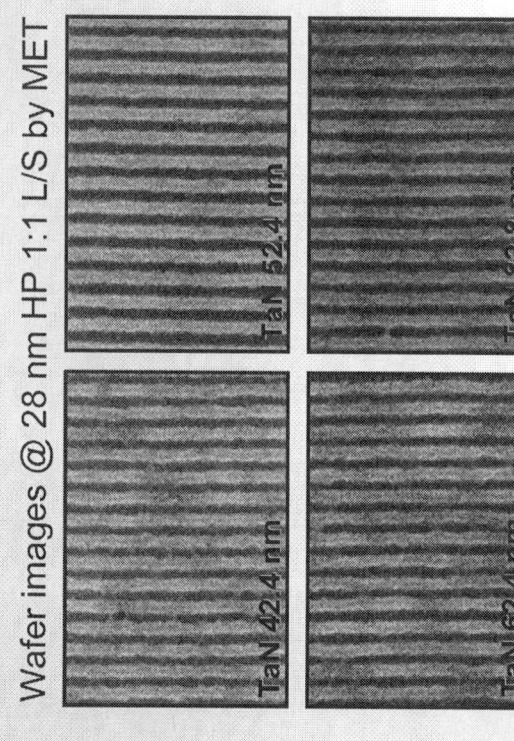

- **52.4-nm-thick** TaN shows the highest $1^{st}/0^{th}$ spectrum ratio. ⇒ Close to thickness of **out-of-phase** ($\Delta\Phi = 180°$).

- MET results also show the best image quality at the 52.4-nm-thick TaN.

- H-V bias reduced by 50-60 %, when we apply 52.4-nm-thick TaN instead of 82.8-nm-thick TaN.

H-V bias results from CSM & MET

SAMSUNG ELECTRONICS

Dose variation @best focus, 40nm HP L/S

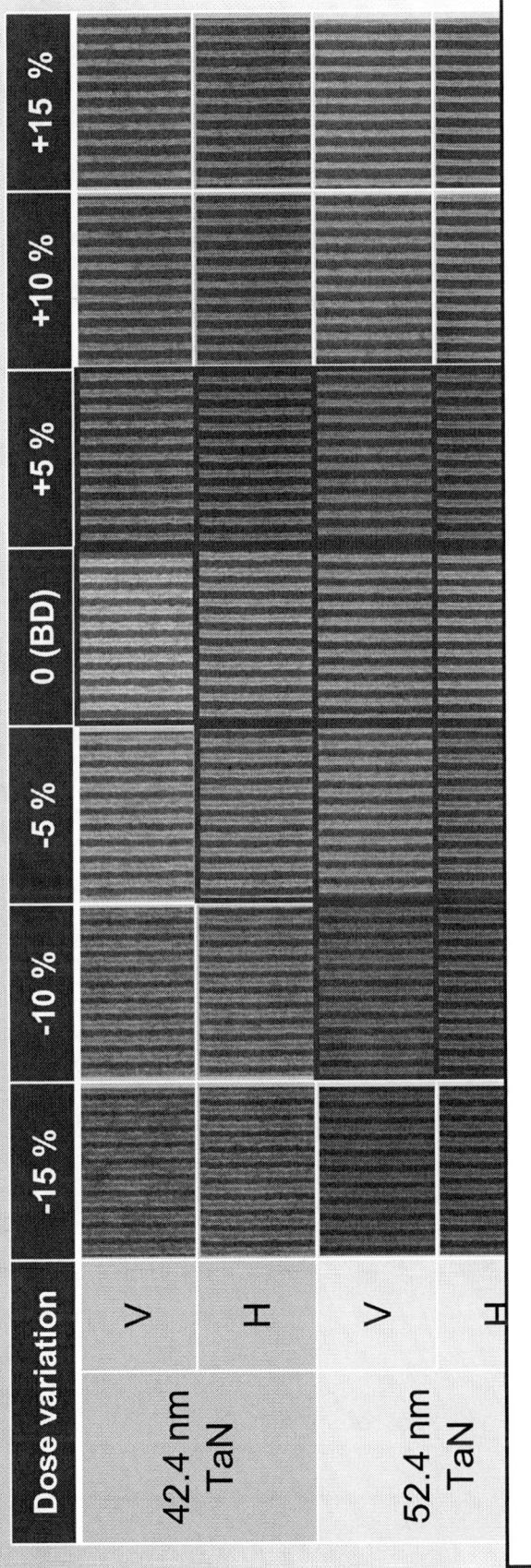

Dose variation	-15 %	-10 %	-5 %	0 (BD)	+5 %	+10 %	+15 %
42.4 nm TaN — V							
42.4 nm TaN — H							
52.4 nm TaN — V							
52.4 nm TaN — H							

EL (Exposure Latitude; %dose/nm)

- EL is acceptable over 52.4 nm (out-of-phase) in thickness.
- For vertical L/S (shadowing), EL decreases in thick absorber.

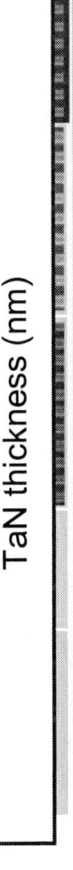

Blue area means allowable dose variation which meets ±10 % tolerance from the target CD.

SAMSUNG ELECTRONICS

DOF & LWR of phase-shifting EUVL mask

Defocus	-150 nm	-100 nm	-50 nm	0 (BF)	+50 nm	+100 nm	+150 nm
TaN 42.4nm							
TaN 52.4nm							
TaN 62.4nm							
TaN 82.8nm							

*SEM images from Horizontal (non-shadowing) L/S patterns

- Out-of-phase condition (52.4-nm-thick TaN) also shows
  - The largest DOF margin and
  - The lowest LWR & LER properties on the printed resist images.

SAMSUNG
ELECTRONICS

An issue of EUVL phase-shifting mask: low OD

Fig. Resist pattern printability on the wafer when we apply additional exposures on the absorber area

Add. Exp.	As it is	1X	2X	3X
TaN 42.4nm				N/A
TaN 52.4nm				
TaN 62.4nm				N/A
TaN 82.8nm				

- SEM images from 40nm HP L/S patterns

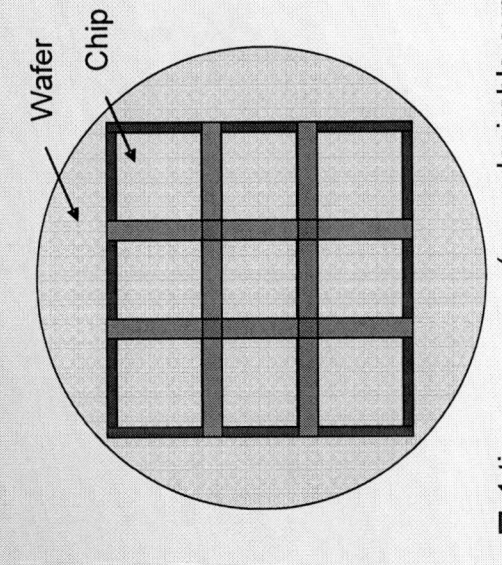

Wafer
Chip

■ 1 time exposure (no undesirable exp.)
■ 2 times exposures (1X undesirable exp.)
■ 4 times exposures (3X undesirable exp.)

- When multiple chips are printed on the wafer using a full-field scanner, EUV light reflected from the absorber area of adjacent shots could affect printability of patterns which are located in the border region.

- Due to the lower OD (optical density), deteriorations in pattern printability as well as CD reduction get worse with decreasing absorber thickness.

- Thin absorber structure needs an appropriate light-shielding layer to increase OD in border area.

SAMSUNG
ELECTRONICS

Issues & future action items

- Collaboration with commercial blank suppliers.
  - Finding out-of-phase condition of commercial absorber materials.
  - Applying an appropriate ARC layer with thin absorber for DUV inspection.

- Solving low OD problems in thinner absorber.
  - Applying a light-shielding layer in border region, etc.

- Evaluation of thin absorber mask with a full-field EUV scanner (ADT).
  - Further evaluation and confirmation on the properties of thin absorber EUVL mask (resolution, H-V bias, CD, PW).

- Developing PSM-oriented new material and structure compatible with sub-22-nm node EUVL.

Summary & conclusions

- Phase-shifting EUVL masks applying thinner absorber are investigated to minimize shadowing problems.

- Simulation results using practical refractive indices show that absorber thickness can be reduced to that of out-of-phase ($\Delta\Phi = 180°$) ranges without degradation of image contrast and process margin.

- EUV CSM and MET results indicate that 52.4-nm-thick TaN is close to $\Delta\Phi = 180°$ since it shows the highest $1^{st}/0^{th}$ field spectrum intensity ratio as well as the best resolution.

- Thinner absorber mask with out-of-phase condition also shows low H-V bias (50 % reduction) and improved PW & LER/LWR properties.

- Combined with low OD, EUV light leakage from the adjacent shots deteriorates pattern printability in thinner absorber masks. Thus, an appropriate light-shielding layer should be introduced.

Acknowledgements

The authors would like to thank...

- Mask blank development center (MBDC) at SEMATECH for providing ML blanks used in these experiments,

- Dr. Eric M. Gullikson at LBNL for analysis of refractive indices of absorber and capping materials using ALS beamline,

- Laboratory of Advanced Science and Technology for Industry (Prof. H. Kinoshita and Dr. T. Watanabe) at University of Hyogo for their helps to use beamline in the NewSUBARU synchrotron facility,

- Dr. Patrick Naulleau, Paul Denham, Brian Hoef, and Gideon Jones for their helps on the wafer printing test using EUV MET at ALS in LBNL.

Accelerating the next technology revolution

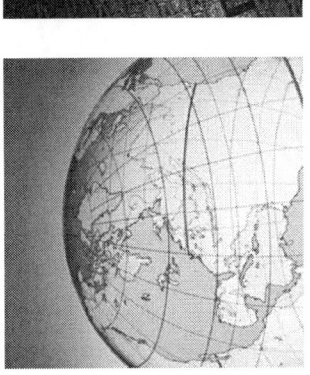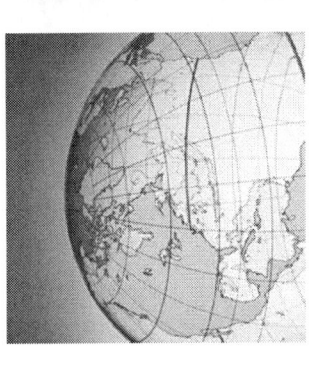

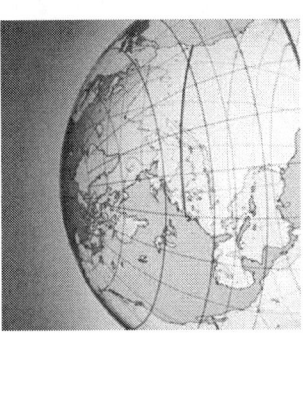

Study of Pit Defect Formation on EUV Blank Substrates

Abbas Rastegar
Sean Eichenlaub, Arun John Kadaksham, Matt House, Brian Cha, Henry Yun

SEMATECH-Albany
EUVL Symposium, Lake Tahoe, October 2008

Outline

- **Current status of EUV substrate defectivity**

- **Pit defects on EUV substrates**
  - Distribution of native pits

- **Origin of the pits**
  - EUV substrate manufacturing
  - CMP and polishing
  - Surface layers on LTEM and Quartz

- **Pit smoothing by cleaning process**

EUV substrate defect trend

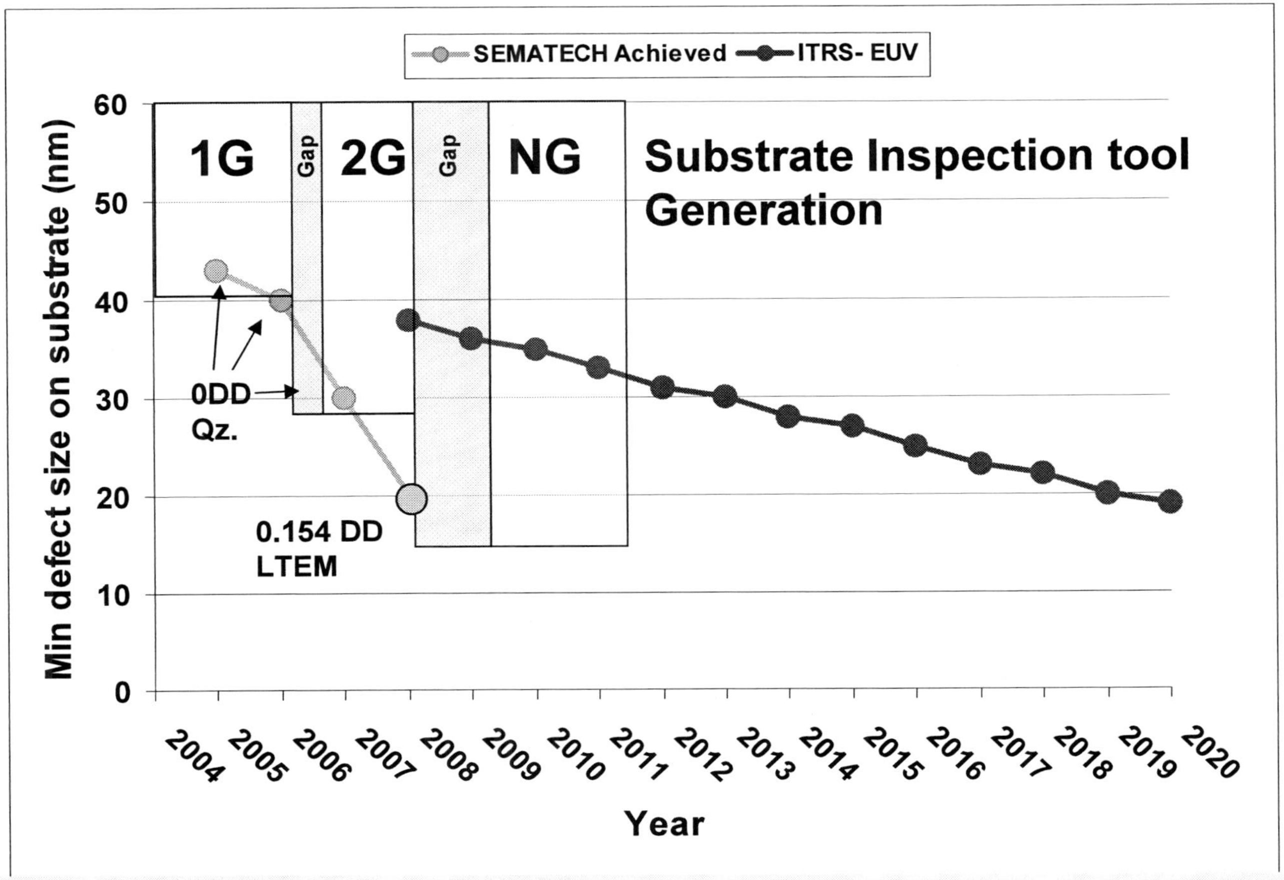

Substrate defect reduction progress is faster than ITRS requirement

Lasertec M7360 detection of Native defects

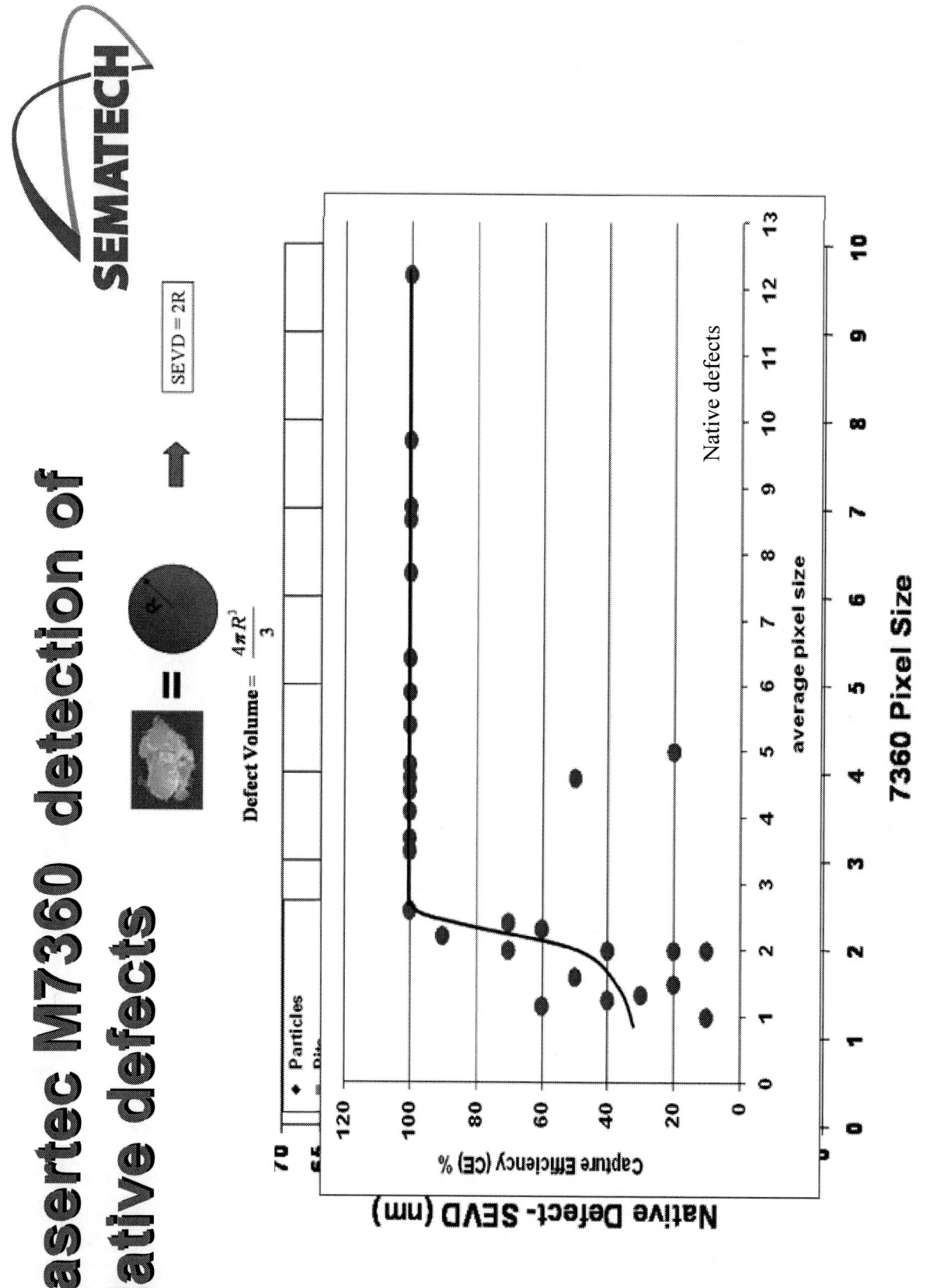

$$\text{Defect Volume} = \frac{4\pi R^3}{3}$$

SEVD = 2R

Defectivity of EUVL LTEM substrates@ 20 nm SEVD

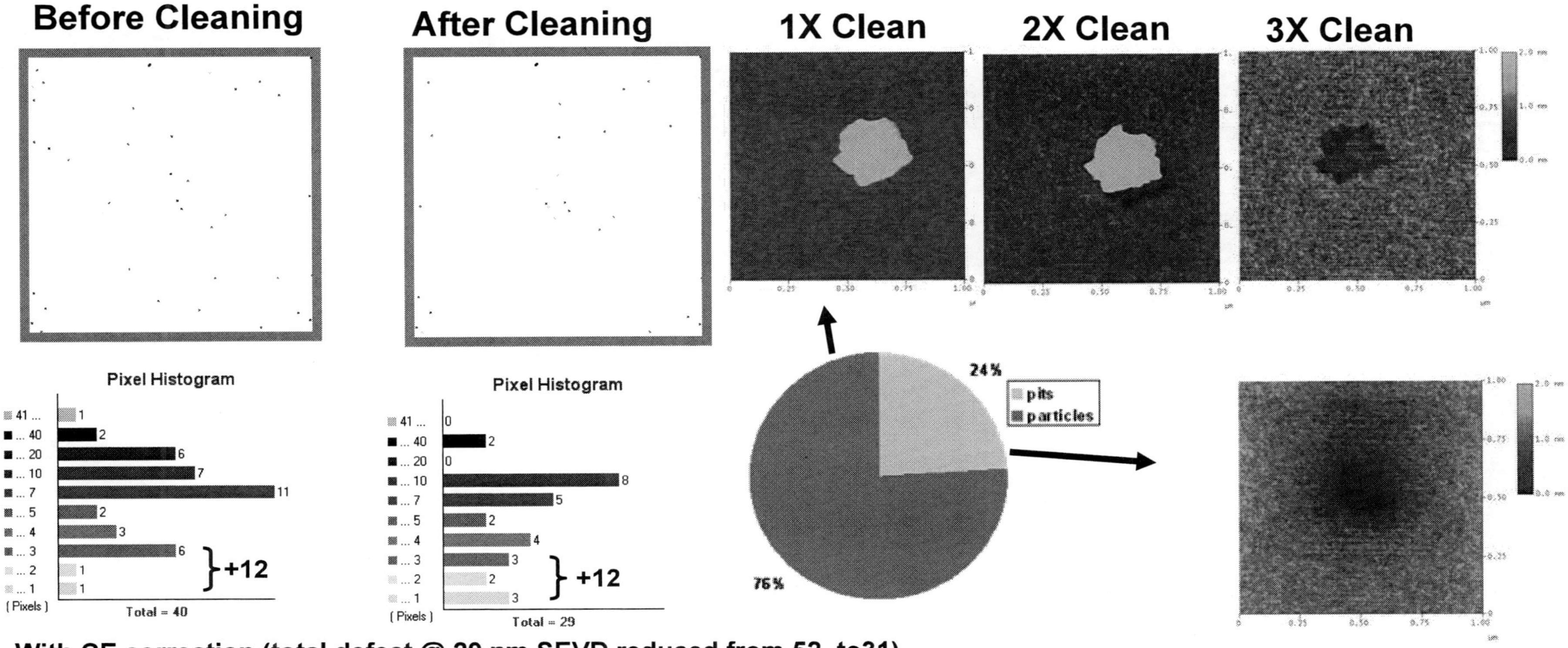

Before Cleaning

After Cleaning

1X Clean **2X Clean** **3X Clean**

Pixel Histogram

Pixel Histogram

With CE correction (total defect @ 20 nm SEVD reduced from 52 to31)
→ Worse case Defect density estimate @ 20 nm SEVD= 0.154 defect/cm² on LTEM

pits
particles

24%

76%

Embedded particles and pits are dominant defects

Native pits on EUVL substrates

(M7360 inspections)

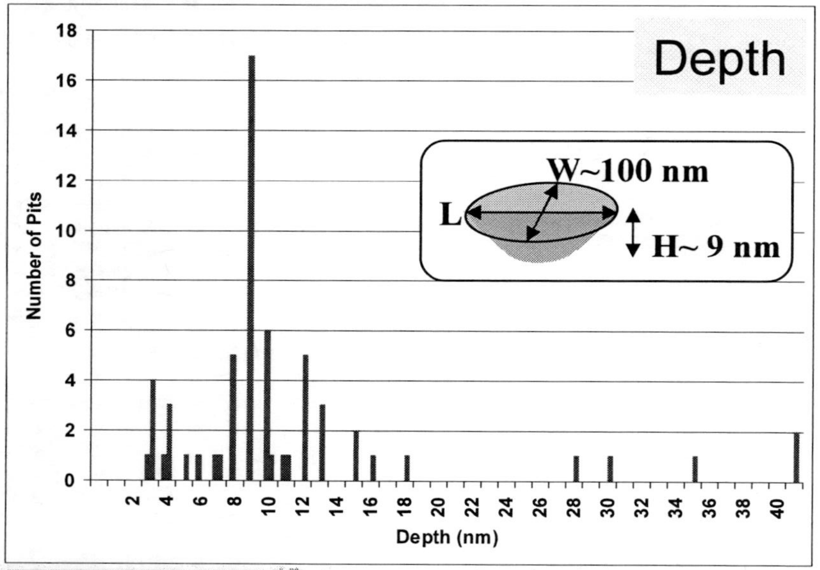

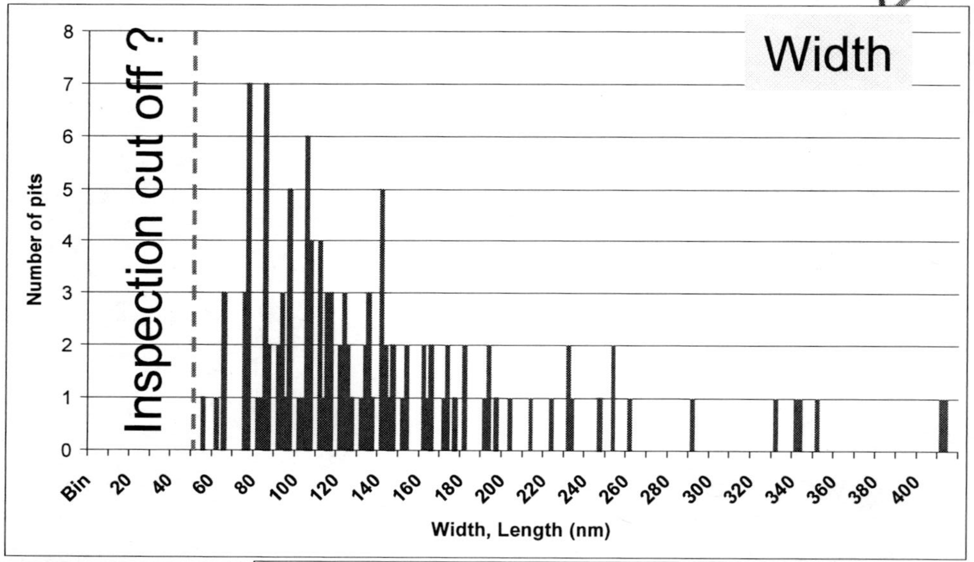

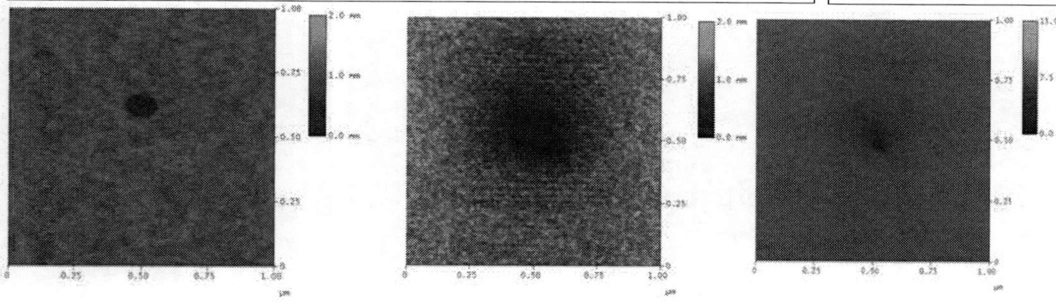

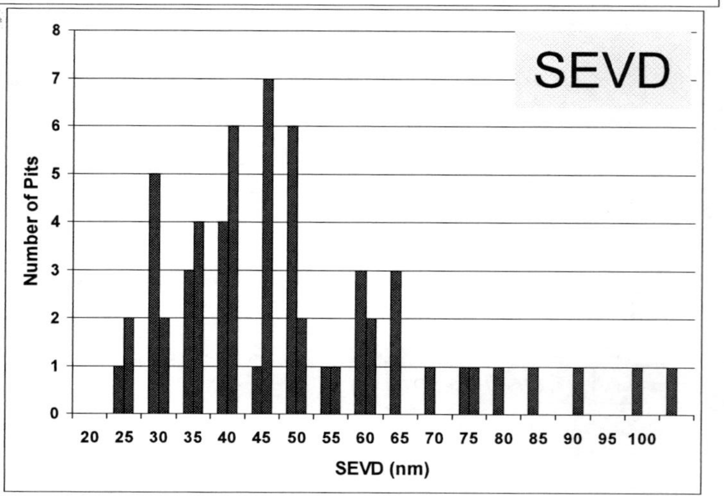

SEMATECH

Width

Inspection cut off ?

Width, Length (nm)

Number of pits

Bin

Depth

W~100 nm

H~9 nm

L

Depth (nm)

Number of Pits

SEVD

SEVD (nm)

Number of Pits

Detected pits are shallow and wide (Aspect ratio H/W = ~0.1)

Outline

- Current status of EUV substrate defectivity

- Pit defects on EUV substrates
  - Distribution of native pits

- **Origin of the pits**
  - **EUV substrate manufacturing**
  - **CMP and polishing**
  - **Surface layers on LTEM and Quartz**

- Pit smoothing by cleaning process

EUV substrate manufacturing

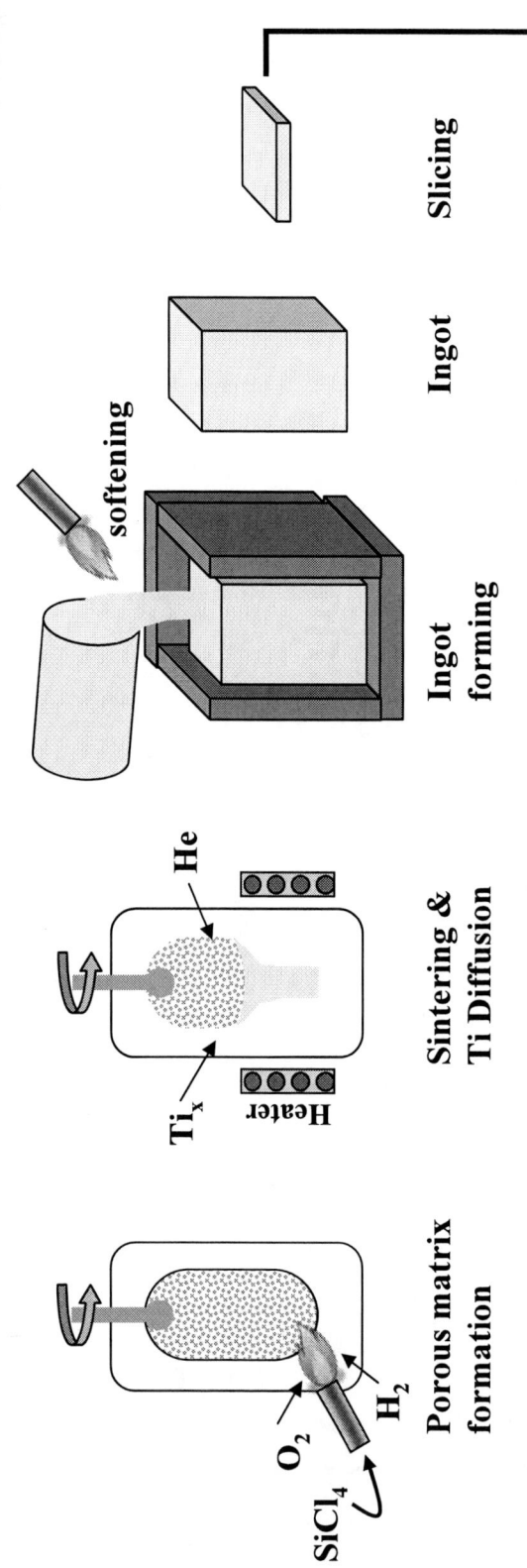

Porous matrix formation

$SiCl_4$ O_2 H_2

Ti_x He Heater

Sintering & Ti Diffusion

softening

Ingot forming

Ingot

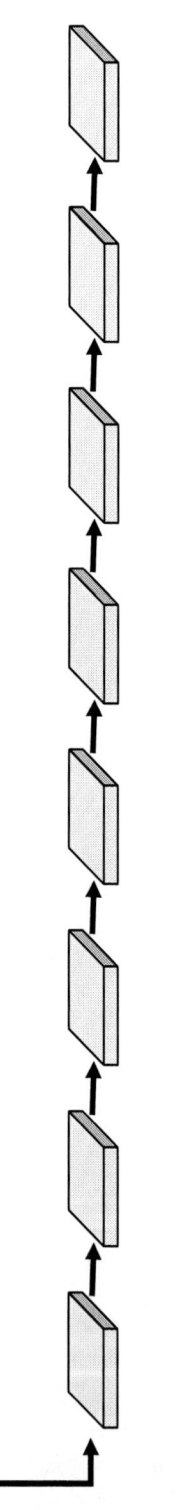

Slicing

Grinding   Cleaning   CMP   Cleaning   IBF   Cleaning   Back CrN   Final clean

To understand pit formation, one should know how substrates are made

Sources of pit in substrate

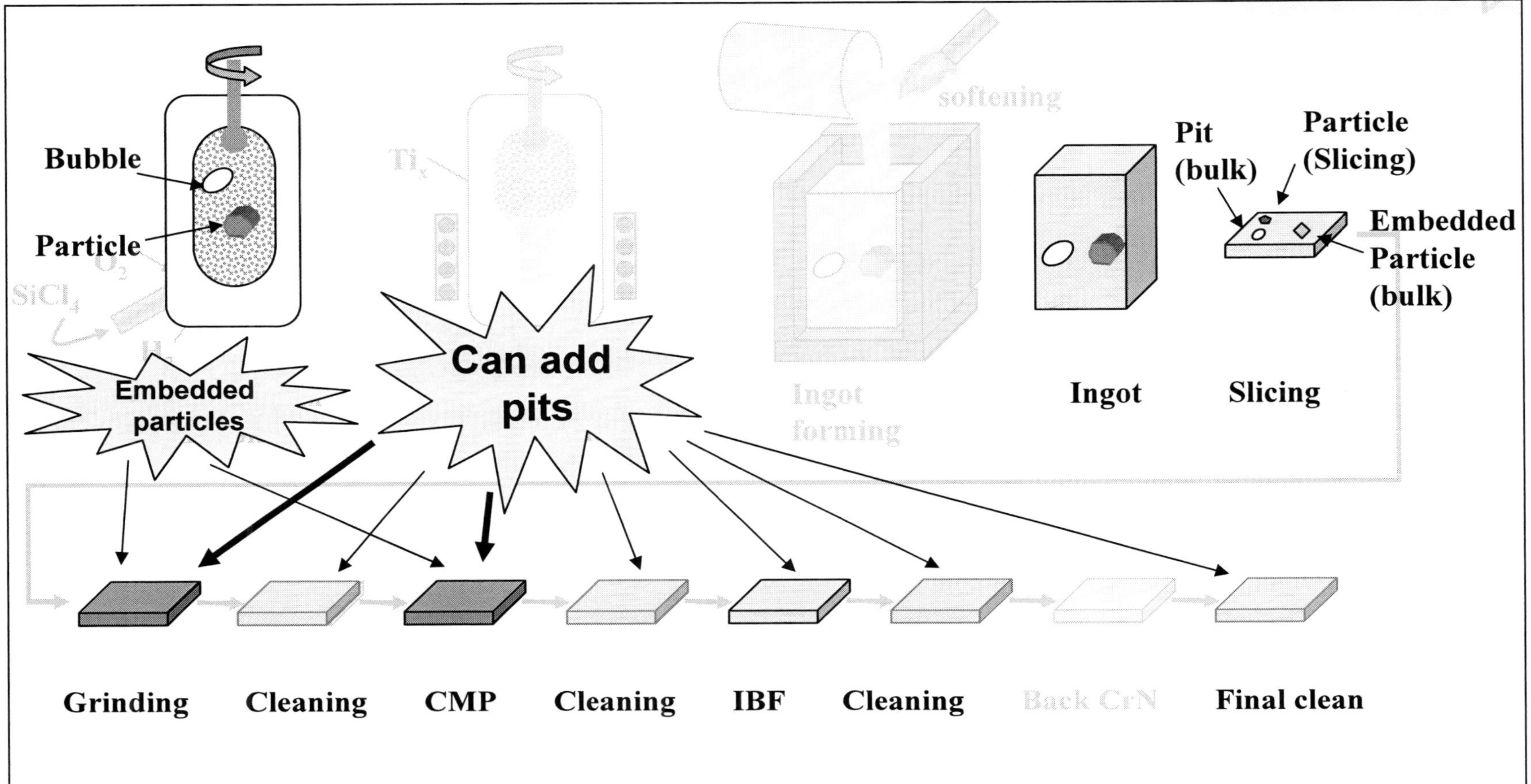

Pit (bulk)

Particle (Slicing)

Embedded Particle (bulk)

softening

Ingot forming

Bubble

Particle

SiCl₄

Embedded particles

Can add pits

Ingot   Slicing

Grinding   Cleaning   CMP   Cleaning   IBF   Cleaning   Back CrN   Final clean

CMP is the main step for creating pits and embedded particles

Glass CMP

Surface Water/Contamination layer

Redeposition/stress layer

bulk

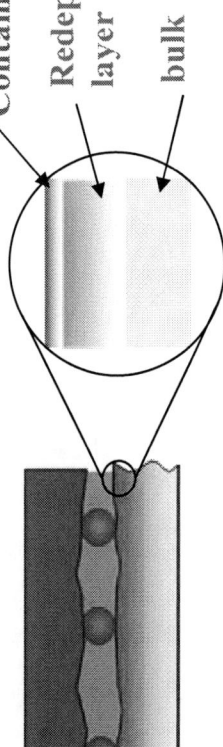

Polishing lap

Slurry

Glass substrate

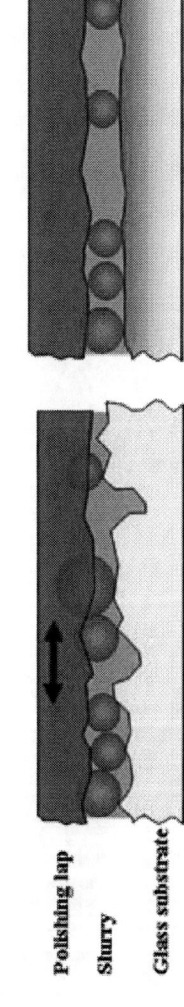

- **Polishing mechanism in glass**
  - **Mechanical**: tangential plastic flow of fine glass particles underneath the hard abrasive particles ((Beilby 1903)
  - **Mechano-Chemical**: Chemical reactions govern polishing under mechanical forces (Cook 1990)

$$(\equiv Si-O-Si \equiv)+(H_2O) \leftrightarrows 2(\equiv Si-OH)$$

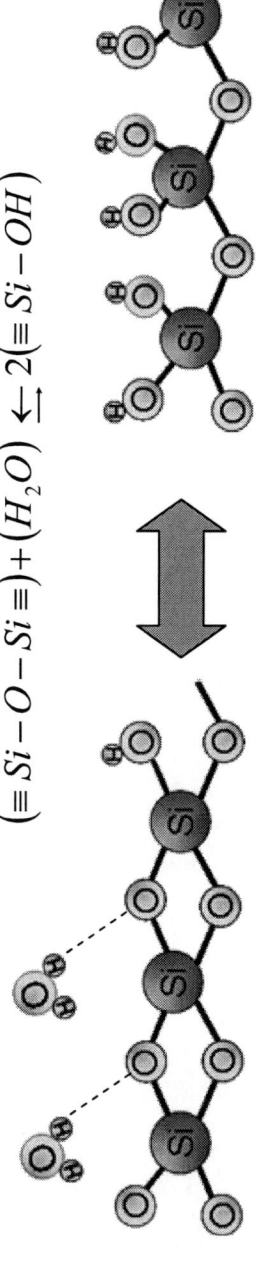

CMP is the most critical step for pit creation and removal

Surface layers in Quartz

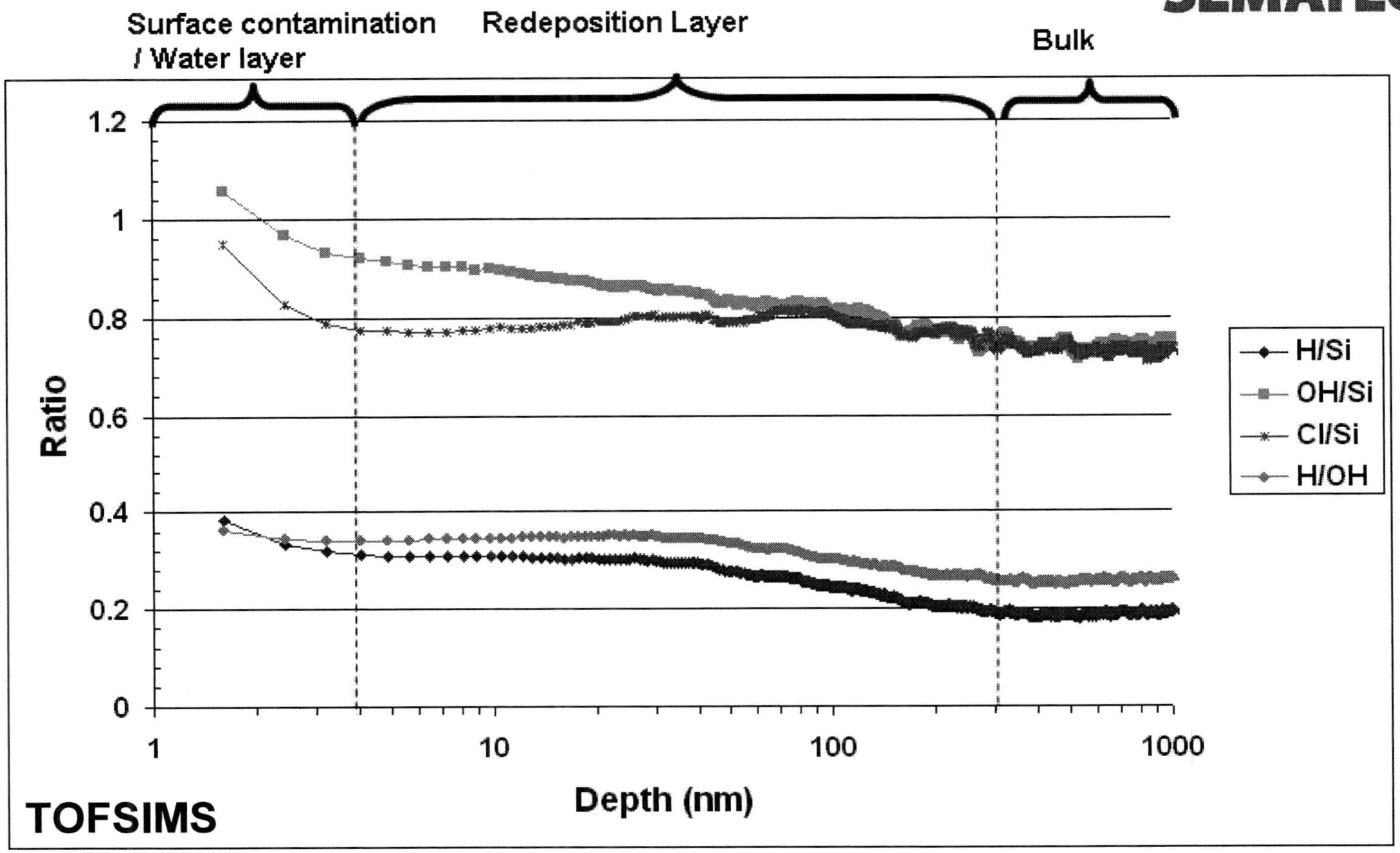

Surface contamination   Redeposition Layer   Bulk
/ Water layer

TOFSIMS

There are 2 distinctive layers on the surface of quartz, as predicted.

Surface layers in LTEM

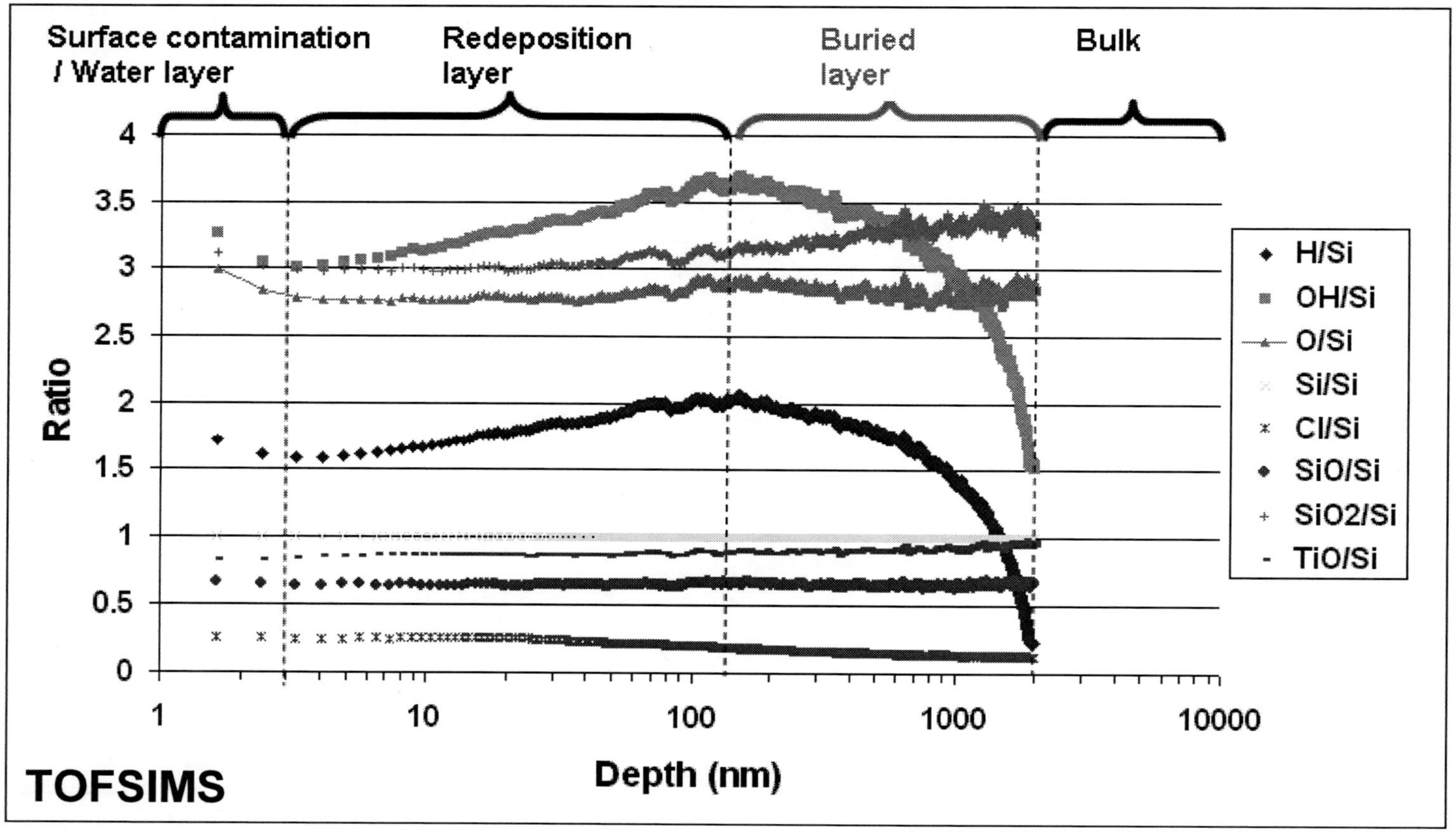

There are 3 distinctive layers on the surface of LTEM.

Surface layers ; Quartz vs. LTEM

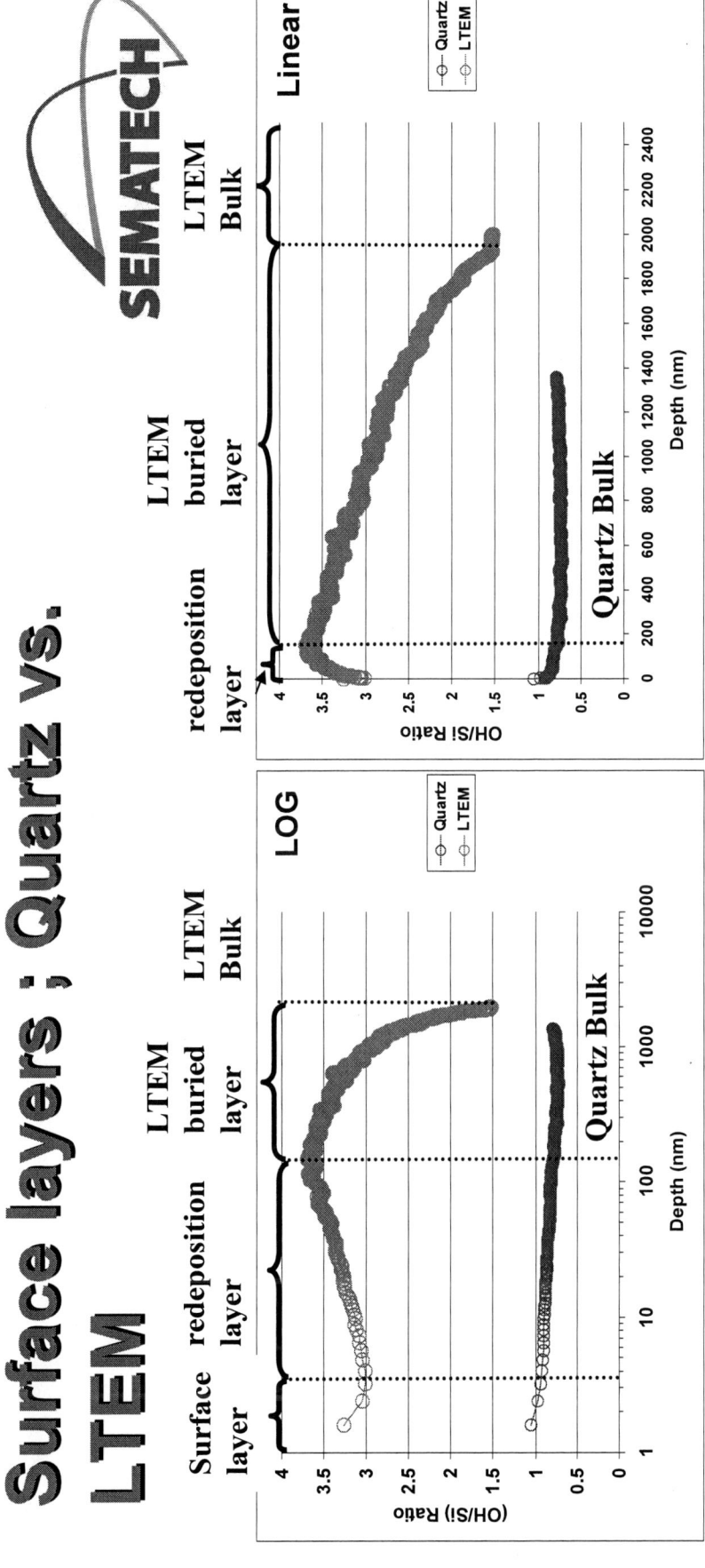

- Water/contamination layer; (3-4 nm) OH decreases by depth in both LTEM and quartz

- Redeposition (a.k.a Thermoplastic) layer; (~ 100 nm in both quartz and LTEM) OH vs. depth increases in LTEM and decreases in quartz

- LTEM buried layer; (1 -2 μm) in LTEM only and does not appear in quartz ;OH concentration reduced by more than half by reaching bulk

LTEM has a unique buried layer, which does not appear on quartz.

Water diffusion in LTEM

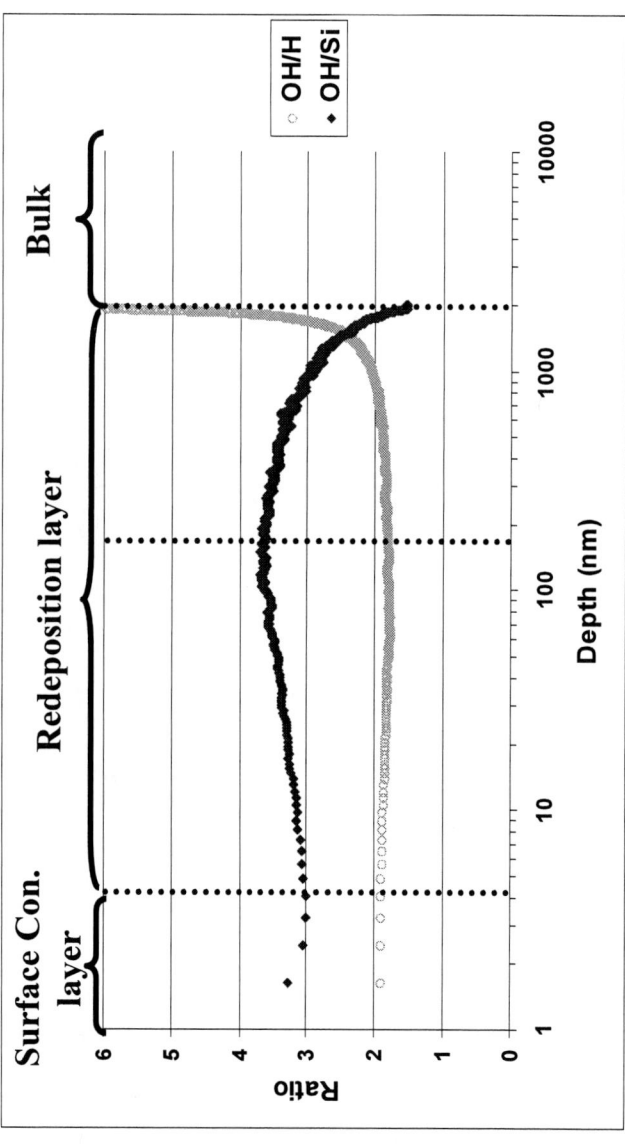

- The top few layers of the surface is hydrolyzed $(\equiv Si - O - Si \equiv) + (H_2O) \leftrightarrows 2(\equiv Si - OH)$

- The ratio of ([OH]/[H])=2 remains constant in all these layers indicating diffusion of molecular water in these layers { H_2O + $(OH)_{si} \rightarrow 2[OH]$ + [H]$\rightarrow$ ([OH]/[H])=2 }

Water diffuses in LTEM in depth of few microns in all surface layers

Mechanism of water diffusion in LTEM

$$(\equiv Si - O - Si \equiv) + (H_2O) \underset{\longleftarrow}{\longrightarrow} 2(\equiv Si - OH)$$

Hydrogen bonding of water molecules

Breaking of Si-O-Si bonds

Water diffusion through rings

- It is known that water diffuses in Silica glass by two mechanisms (L. M. Cook)
  - Slow (D=~10^{-18}cm^2/s) by breaking Si-O-Si links and forming SiOH
  - Fast (D=~10^{-15}cm^2/s) by diffusion of molecular water

- Diffusion coefficient (D) depends on static pressure and applied uniaxial stress
  - Tensile stress increase→ Exponential increase of D
  - Compressive stress& Hydrostatic pressure increase → Exponential decrease of D

Polishing/CMP impact on LTEM

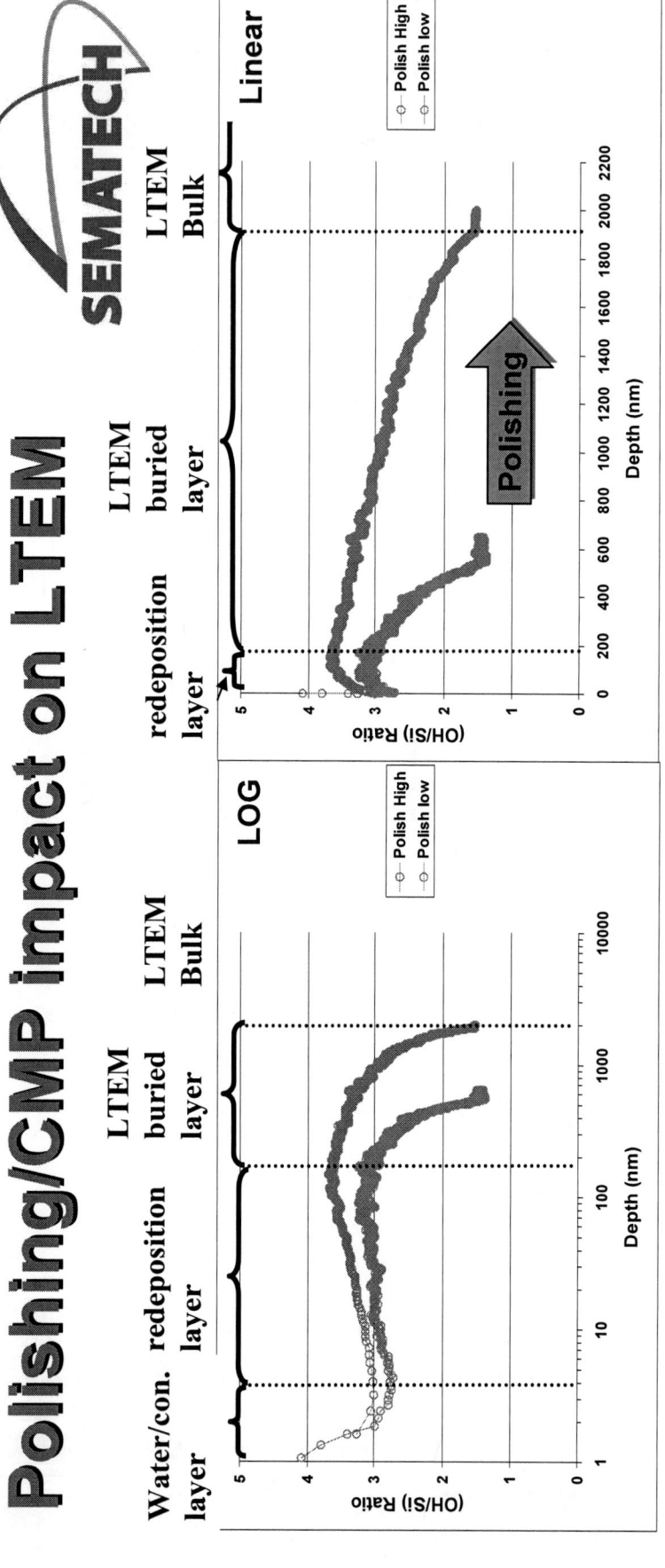

- Water/contamination and Redeposition layers do not change under different polishing conditions.
- The thickness of LTEM buried layer increases by increasing polishing.
- Water diffusion depth increases by polishing

Polishing has deepened the LTEM buried layer but does not change other surface layers.

Outline

- Current status of EUV substrate defectivity

- Pit defects on EUV substrates
  - Distribution of native pits

- Origin of the pits
  - EUV substrate manufacturing
  - CMP and polishing
  - Surface layers on LTEM and Quartz

- **Pit smoothing by cleaning process**

Pit smoothing by cleaning; Concept

Original pit

Isotropic etch

Pit enlarges

anisotropic etch

**Pit widen
with reduced depth**

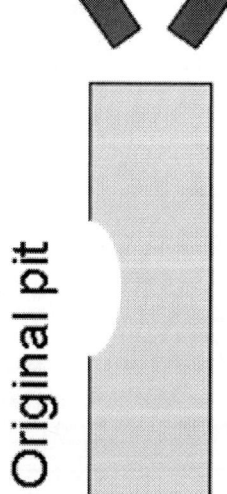

240 nm

53.6 nm

Before smoothing

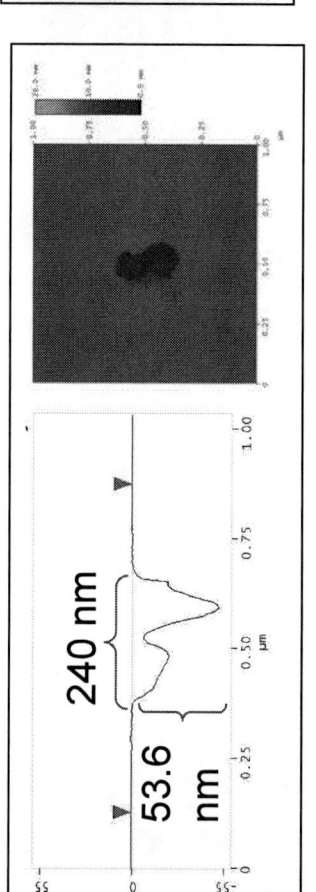

264 nm

21 nm

After smoothing

Pit smoothing by Cleaning;
Pit depth distribution

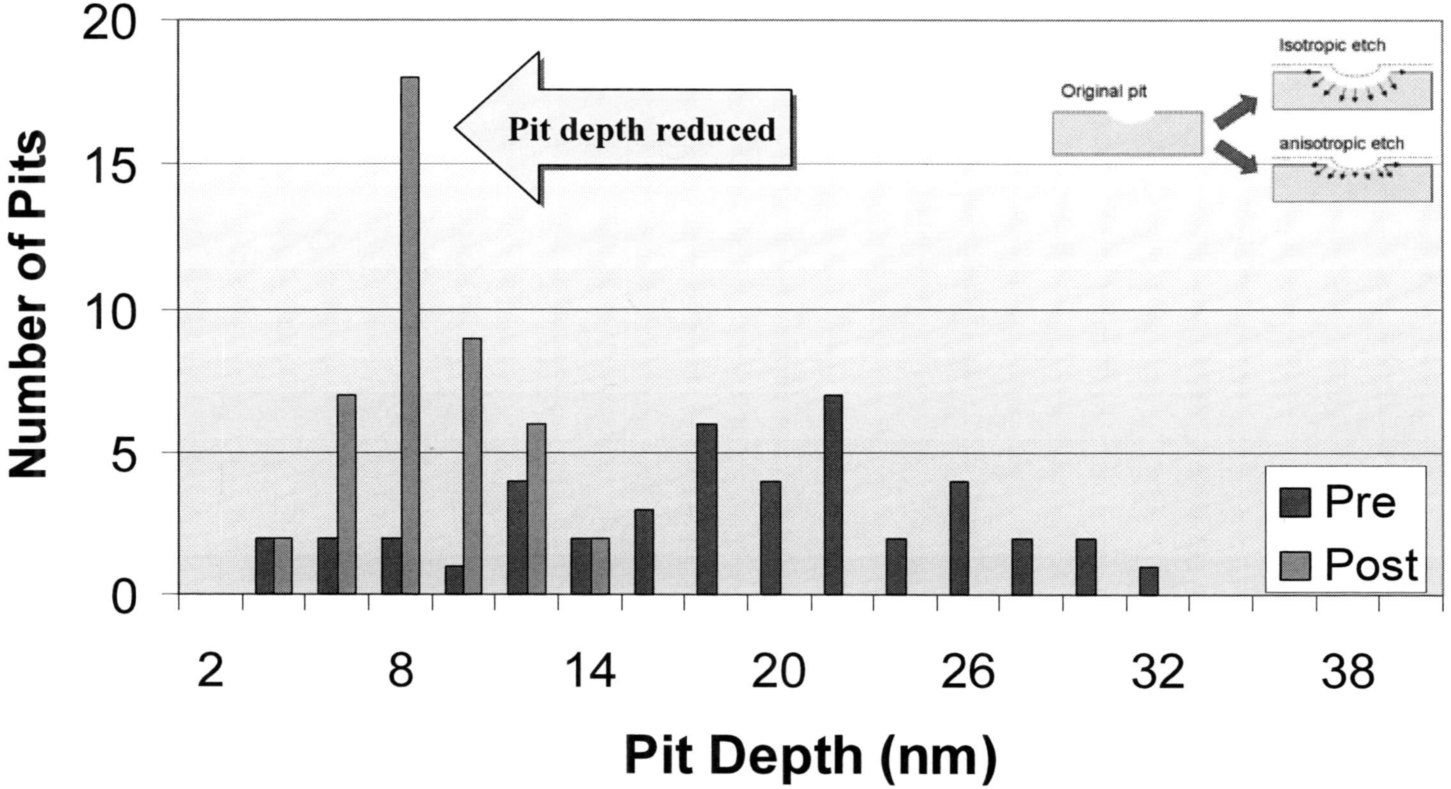

SEMATECH has developed new cleaning processes demonstrating reduction in pit depth distribution

Pit smoothing by Cleaning; Defect density

After Clean/Smooth

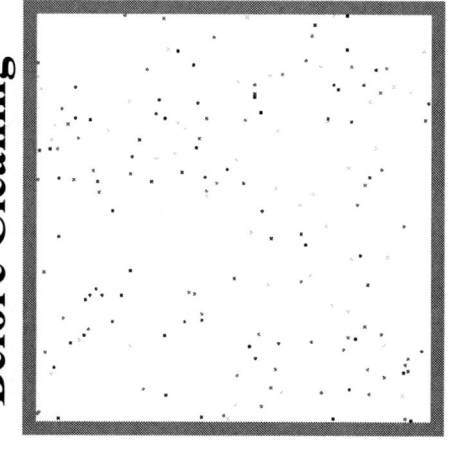

Before Cleaning

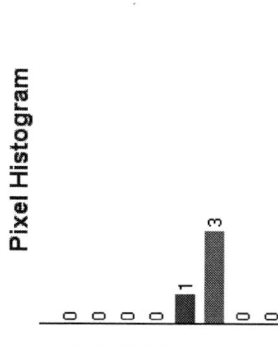

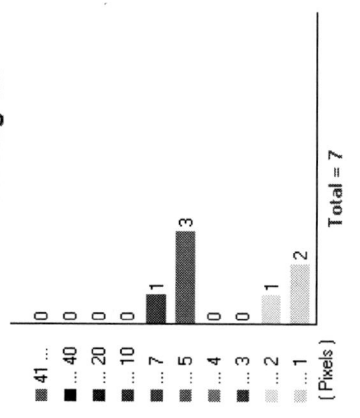

Pixel Histogram

Total = 7

Pixel Histogram

Total = 181

Pit smoothing by cleaning is clean!

Summary and conclusions

- Defect density of LTEM substrates @ 20 nm SEVD was shown to be 0.154 defects/cm^2 (worse case). The remaining defects are embedded particles or pits

- SEMATECH performed a detailed fundamental study of surface of LTEM and quartz to enable glass/blank suppliers to improve their CMP processes for achieving lower defectivity

- For the first time we report that LTEM has a unique buried layer under the redeposition layer that is absent in quartz. All LTEM substrates from different suppliers show similar behavior.

- Our data indicates that water diffuses few microns deep in LTEM

- Polishing/CMP increases the depth of LTEM buried layer as well as water diffusion depth

- SEMATECH demonstrated pit smoothing by cleaning based on anisotropic etching of surface layers

Acknowledgement

- **AFM Imaging**
  - Timothy Owen
  - Patrick Tracy
  - Jon Underwood
  - Mark Maloney

- **Defect Inspection, Classification, and Marking**
  - Nancy Lethbridge

- **SEM/EDS/FIB Characterization**
  - John Underwood
  - Timothy Owen

- **Cleaning Support**
  - Nancy Lethbridge
  - Nelson Lethbridge
  - Lenny Gwenden

- **TOFSIMS Support**
  - Dr. Albert Schnieders

- **References**
  - Lee. M. Cook – J.non-Cryst. Solids 120(1990)152-171
  - A. Kaller, Int. J. Glass Sci.& Tech. Vol64 (1991)

Backup

TOFSIMS analysis

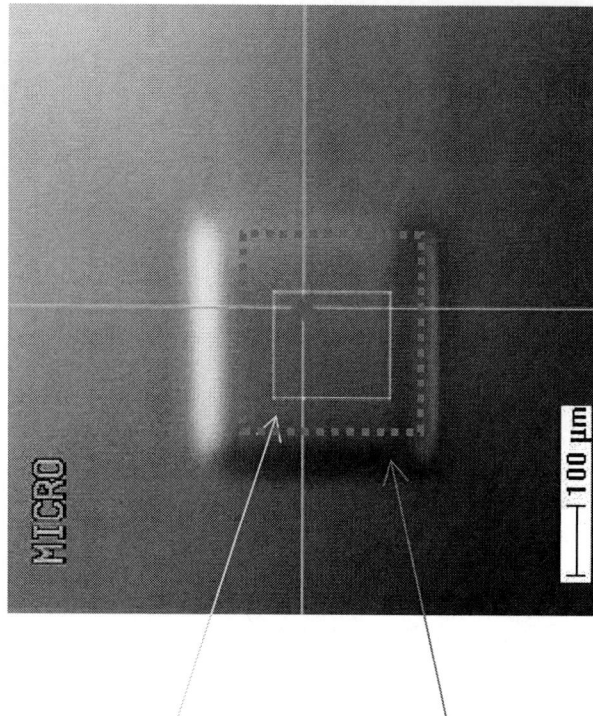

Field of view: ≈ 800 x 750 μm²

analyzed area

sputter crater

A05804.bmp

- Analysis of Crater Center
  - ion/energy: Bi_1^+, 25 keV
  - target current: 0.80 pA
  - analyzed area: 150 x 150 μm²
  - polarity: negative

- IONTOF TOF.SIMS 5-300 instrument

- Sputtering
  - ion/energy: Cs^+, 2 keV
  - target current: 130 nA
  - sputtered area: 300 x 300 μm²

Sources of stress in substrate

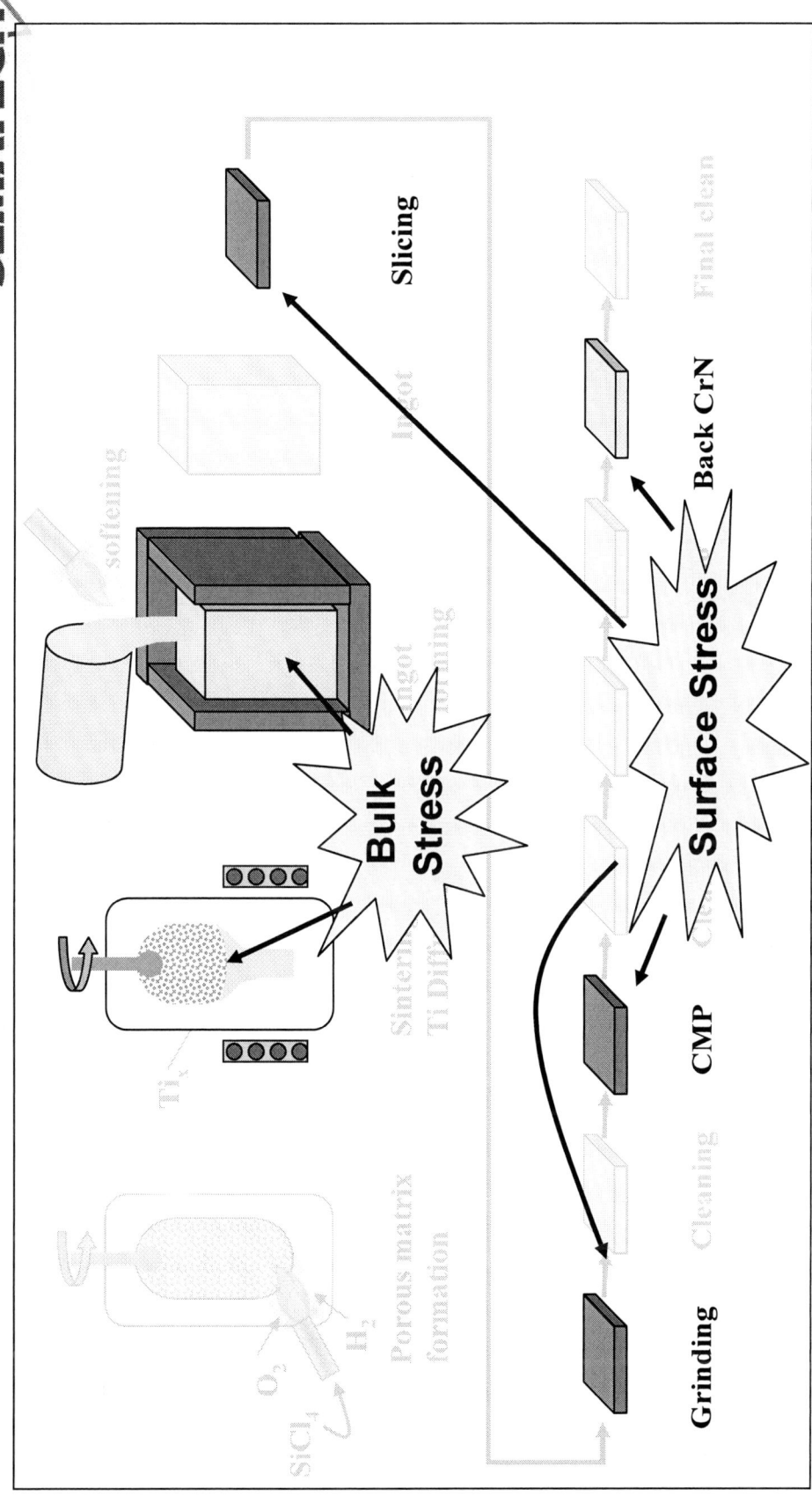

EUV substrate defectivity

SEMATECH champion blank 2005

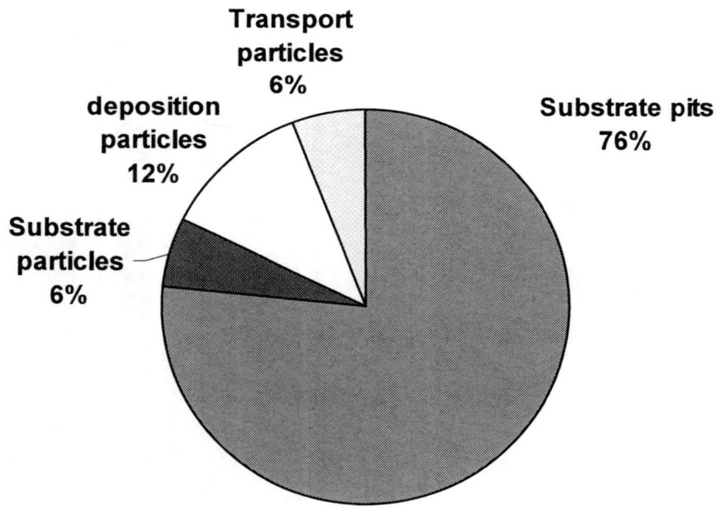

Substrate pits
76%

Transport particles
6%

deposition particles
12%

Substrate particles
6%

18 Defects @ 73 nm

SEMATECH champion blank 2008

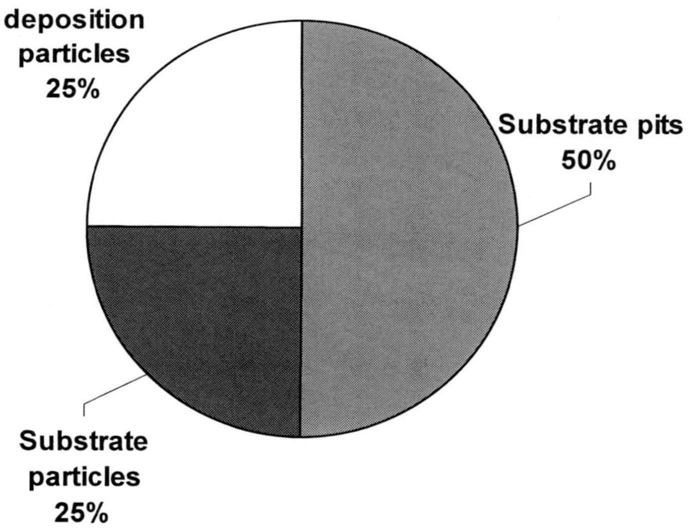

Substrate pits
50%

deposition particles
25%

Substrate particles
25%

8 Defects @ 53 nm

Substrate Defects = Substrate Particles + Substrate Pits

Pit remains dominant defect type

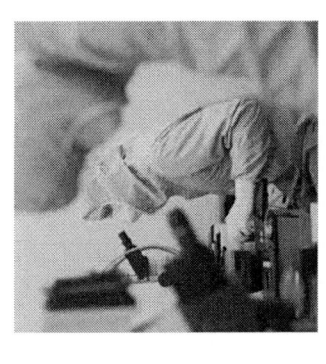
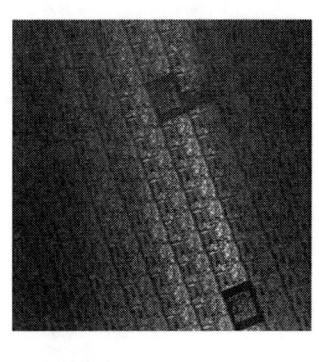
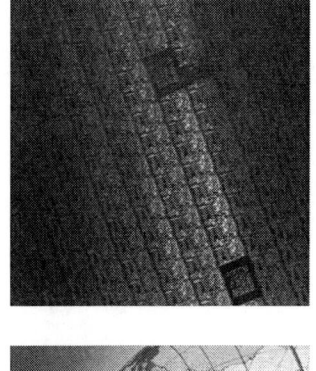

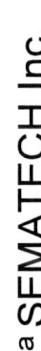

Accelerating the next technology revolution

Investigation of a Compensation Method for Pattern Placement Shifts of Chucked EUVL Masks

J. Sohn[a], K. Orvek[a], R. Engelstad[b], A. Lyons[c], and J. Hartley[c]

[a] SEMATECH Inc.

[b] CMC, University of Wisconsin - Madison

[c] CNSE, State University of New York at Albany

Outline

- **Introduction**

- **EUVL Mask Flatness Compensation Methodologies**

- **SEMATECH Integrated EUVL Mask Test Plan**
  - EUVL Mask Flatness Compensation Methodology for Patterning
  - Experimental Data for EUVL Mask Flatness Measurements
  - EUVL Mask Layout

- **Preparing SEMI Standards for non-flatness compensation**

- **Summary and Conclusion**

- **Acknowledgments**

Introduction

- ## ITRS Lithography

Table LITH5c EUVL Mask Requirements – Near-term years

Year of Production	2008	2009	2010	2011 PILOT LINE	2012	2013	2014 HVM intro	2015
Mask substrate flatness (nm, peak-valley)[O]	65	57	51	46	41	36	32	29

Table LITH5d EUVL Mask Requirements – Long-term years

Year of Production	2016	2017	2018	2019	2020	2021	2022
Mask substrate flatness (nm, peak-valley)[O]	26	23	20	18	16	14	13

Manufacturable solutions exist, and are being optimized

Manufacturable solutions are known

Manufacturable solutions are NOT known

Introduction

➤ In EUVL, chucked mask flatness variations contribute image placement (IP) errors to overlay budgets.

➤ To minimize the overlay contribution severe restrictions are placed on non-flatness of the mask and chuck surfaces.

  ▪ Sub-50nm mask flatness requirements for Pilot Line are very tight for volume shipments

    • Extra polishing adds cost and defects

  ▪ Sub-30nm flatness requirements for HVM are very high risk.

➤ Industry has proposed compensation methods to reduce the level of flatness required by inputting IP corrections during mask write.

➤ SEMATECH has begun a major program to validate the non-flatness compensation strategy and to develop the methodology.

  ▪ Without mask non-flatness compensation methods EUVL could lose COO advantage over alternative lithography techniques.

Published Compensation Techniques

- NuFlare Technology Inc.
  - Analytical calculations based on surface non-flatness maps.
  - Demonstrated technique did not add undesired IP errors to a written test mask.
  - Examined reproducibility using vacuum chuck on IP metrology tool.

- University of Wisconsin/ Intel Corporation/ SEMATECH
  - Finite Element Model based on surface non-flatness maps.
  - Chucking model validated to first order against experimental data for Out-of-Plane distortions.

SEMATECH Project

- Write masks with various compensation techniques, print EUV wafers, validate techniques with IP and overlay data.
  - Full-mask non-flatness metrology performed with University at Albany
  - Work with University of Wisconsin, NuFlare, and SEMATECH member companies to include various compensation techniques
  - Work with Mask Shops to implement compensation during e-beam writing
  - Work with ASML to expose wafers on ADT and perform overlay metrology
- Key assumption of compensation strategy is that scanner e-chucks can be made flat enough to neglect their surface contribution.
  - This is the place to concentrate on super polishing, not each substrate!

EUVL Mask Flatness Compensation Methodology for Patterning

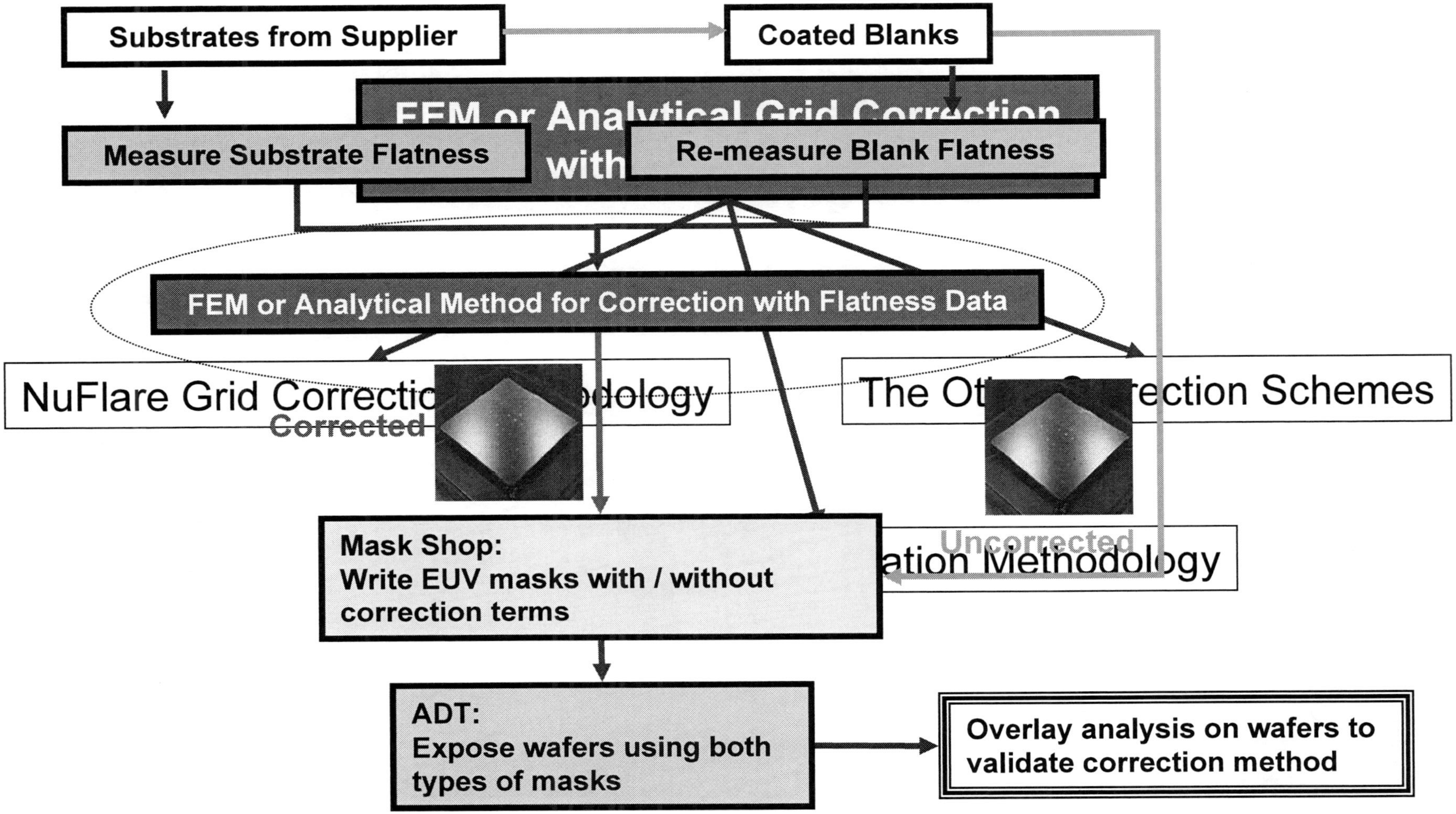

Substrates from Supplier

Coated Blanks

FEM or Analytical Grid Correction with

Measure Substrate Flatness

Re-measure Blank Flatness

FEM or Analytical Method for Correction with Flatness Data

The Other Correction Schemes

Uncorrected
...ation Methodology

...odology

NuFlare Grid Correction Methodology
Corrected

Mask Shop:
Write EUV masks with / without correction terms

ADT:
Expose wafers using both types of masks

Overlay analysis on wafers to validate correction method

Albany Experimental Set-up

12 inch Diameter
Beam Expander

Vertical Holder for Substrate/Blank Flatness Measurements

Example Substrate Flatness Data
Quality Area in 142 x 142 mm²

SEMATECH

Front	Back	Thickness Variation

Ultra-polished LTEM Substrate

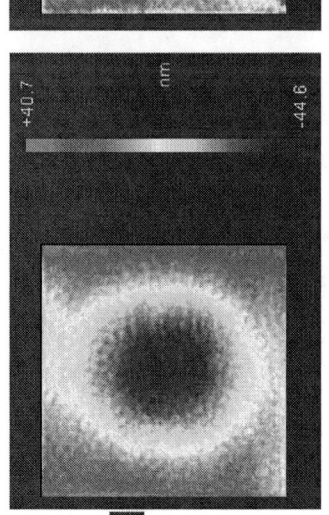

P-V = 85 nm P-V = 63 nm P-V = 125 nm

Standard LTEM Substrate

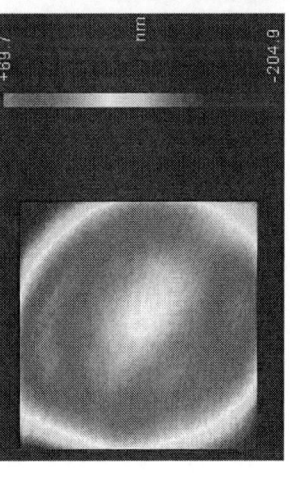

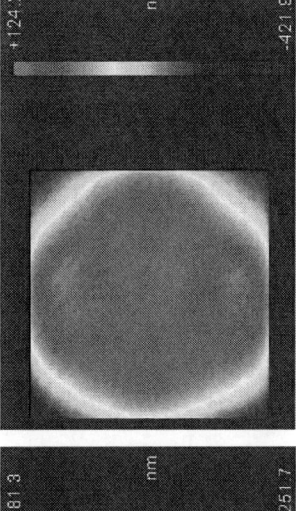

P-V = 275 nm P-V = 333 nm P-V = 546 nm

- Anticipate small uncompensated IP errors on ultra-polished substrate.
- Anticipate large uncompensated IP errors (>6nm) on standard substrate

Mask Layout

- Designed for 5x MET and 4x ADT use
- 7 x 18 XPA groups for IP and overlay testing of compensation techniques
- Each group has uncompensated mark and up to 3 compensated marks
- New CD resolution features for resist development.
- Features for mask contamination studies.

MET Field

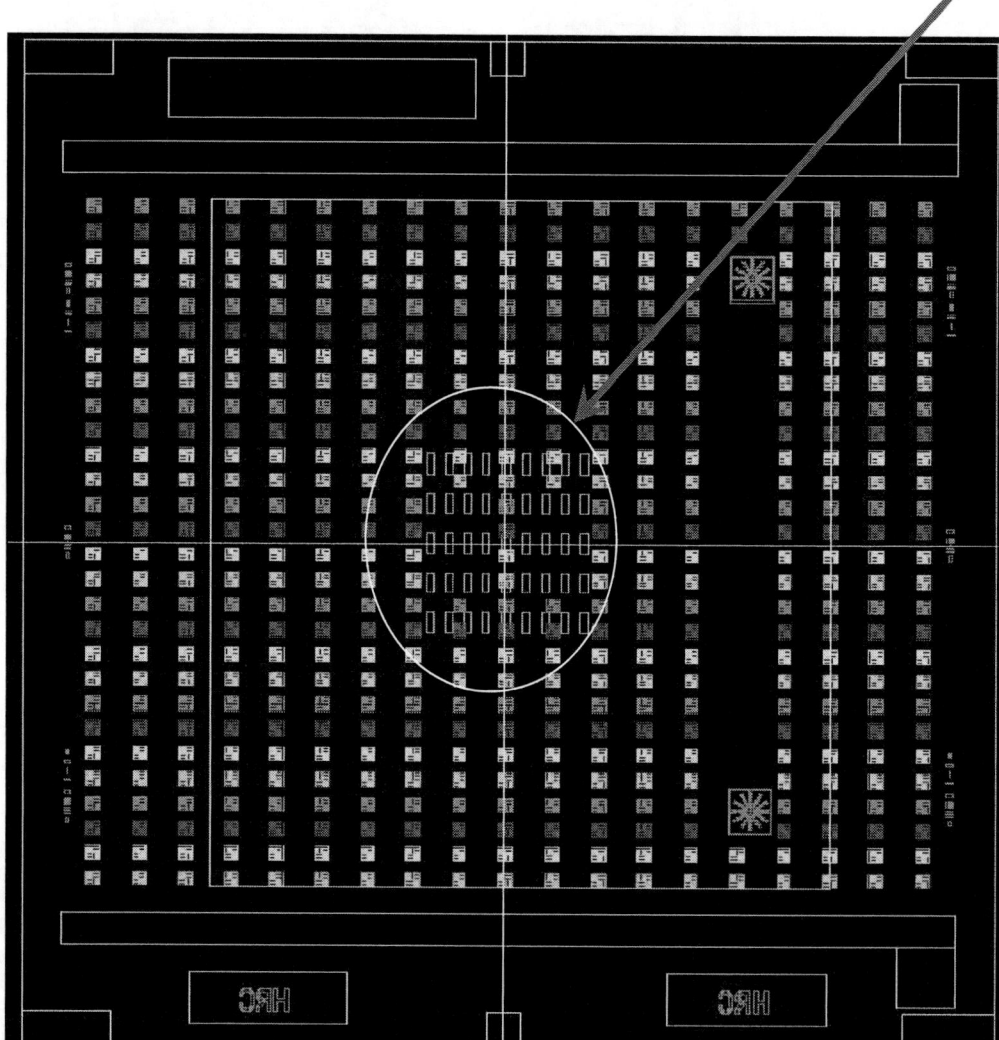

XPA Group with 5x Berkeley CD Subfield

Unit Pattern repeats in a 7x18 array across mask.

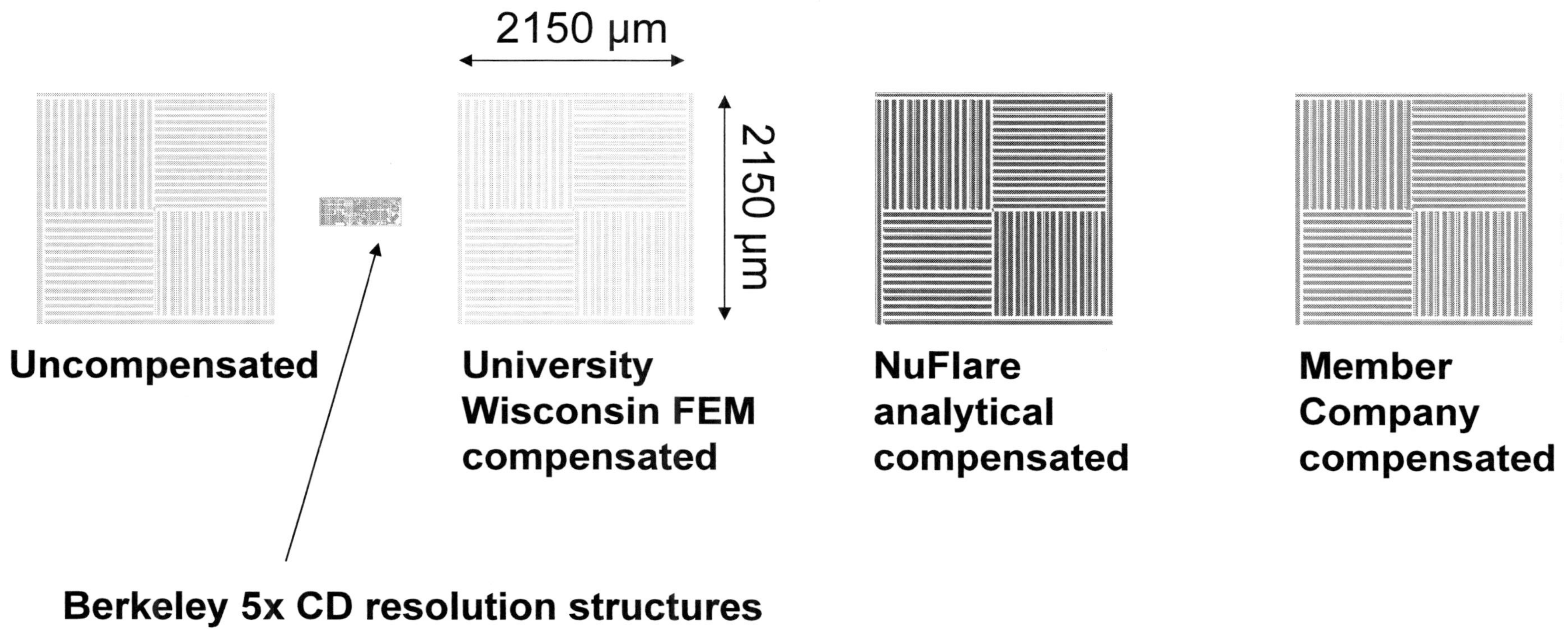

2150 µm

2150 µm

Member Company compensated

NuFlare analytical compensated

University Wisconsin FEM compensated

Uncompensated

Berkeley 5x CD resolution structures

Fitting Experimental Data to E-beam Grid

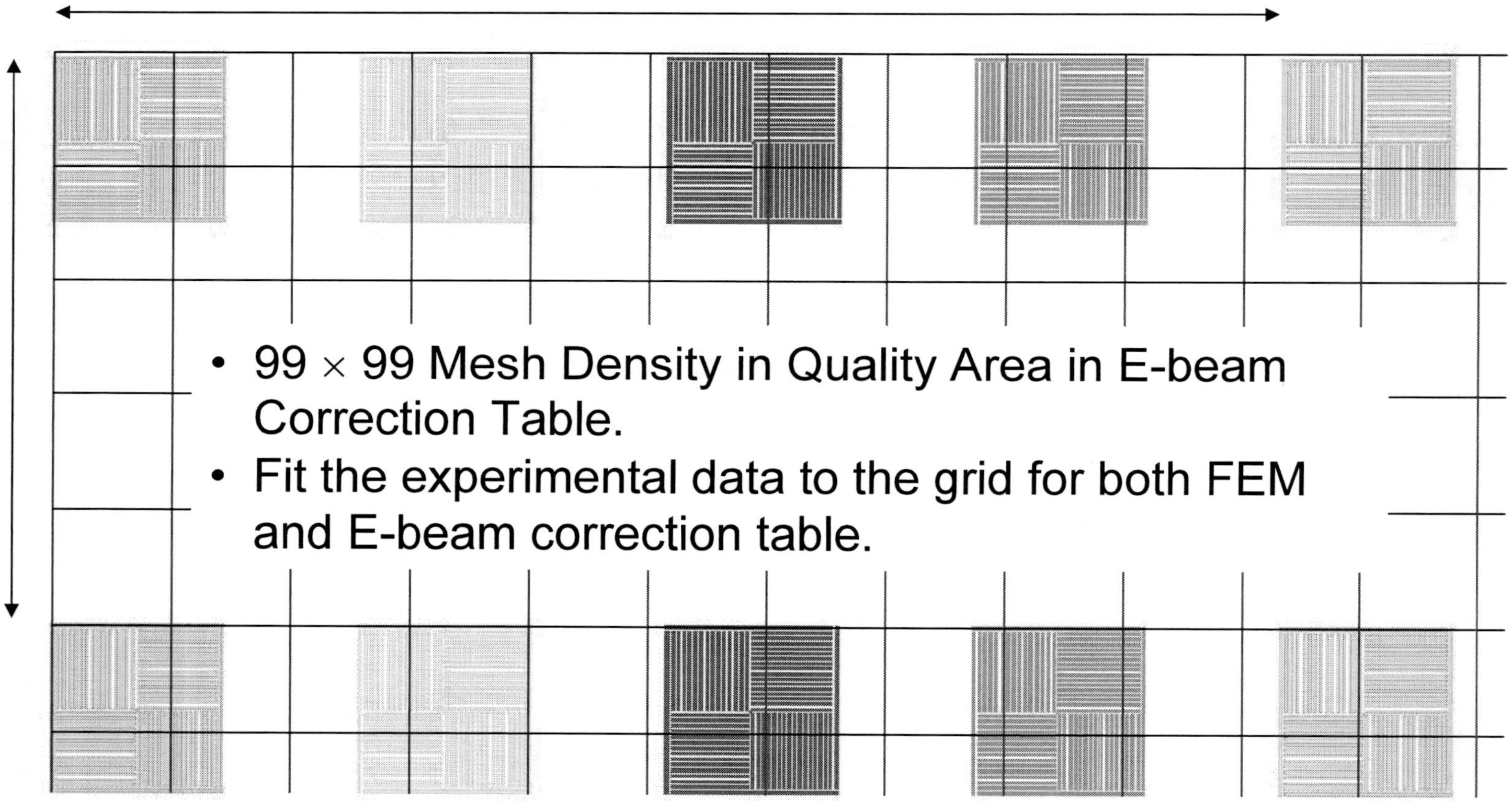

15 mm

7.2 mm

- 99 × 99 Mesh Density in Quality Area in E-beam Correction Table.
- Fit the experimental data to the grid for both FEM and E-beam correction table.

1 ~ 4 data points in each XPA

Experimental Grid Density

- Anticipated chucked frontside non-flatness, for one LTEM substrate

- ~ 300 nm Center to upper corner flatness difference in printed field.

- We anticipate that >100 nm flatness difference can be detected with image placement measurement.

- The experimental grid density easily captures the low order flatness variations.

Preparing SEMI Standards for Non-flatness Compensation Strategy

- P37 (EUV Substrates) and P38 (EUV Blanks, Masks) are due for update.

  - Incorporate option of using non-flatness compensation to allow for relaxed substrate/blank non-flatness specifications.

- P40 (e-chuck) is also due for update

  - Incorporate tighter flatness specification upon the e-chuck surface.

    - Required to be able to neglect chuck-to-chuck variations and model chuck as flat surface

- SEMI EUV MASK Task Force is working on revisions, plan is to go to Yellow Ballot in November.

OLD SEMI Standard P37-1102:
Specification for Extreme Ultraviolet Lithography Mask Substrates

- ## Flatness Specification

Flatness Error in Flatness Quality Area

Class	Frontside Flatness, within Flatness Quality Area	Backside Flatness, within Flatness Quality Area	Low Order Thickness Variation (LOTV), within Flatness Quality Area (See Note2.) $\lambda_{spatial}>$ (edge length)	Units
A	100 peak-to-valley	100 peak-to-valley	100	nm
B	75 peak-to-valley	75 peak-to-valley	75	nm
C	50 peak-to-valley	50 peak-to-valley	50	nm
D	30 peak-to-valley	30 peak-to-valley	30	nm

NOTE 1: $\lambda_{spatial}$ is the spatial period of the flatness error.

NOTE 2: Evaluated after removing wedge angle.

SEMATECH

Recommendation for Revising SEMI Standards
Combining P37 and P38:
Specification for Extreme Ultraviolet Lithography Substrates and Pilot-
Line Masks

➤ Specify flatness only for Pilot line masks required to achieve yield.
➤ Type A for 2010-2012 device requirements, Type B for advanced.

☐ Substrate Flatness Specifications for Type-A EUV Masks

Compensation Options	Frontside Flatness, within Flatness Quality Area	Backside Flatness, within Flatness Quality Area	Units
No Mask Pattern Compensation for nonflatness	30 peak-to-valley	30 peak-to-valley	nm
With use of Mask Pattern Compensation for nonflatness	250 peak-to-valley	250 peak-to-valley	nm

☐ Substrate Flatness Specifications for Type-B EUV Masks

Compensation Options	Frontside Flatness, within Flatness Quality Area	Backside Flatness, within Flatness Quality Area	Units
No Mask Pattern Compensation for nonflatness	23 peak-to-valley	23 peak-to-valley	nm
With use of Mask Pattern Compensation for nonflatness	250 peak-to-valley	250 peak-to-valley	nm

OLD SEMI Standard P40-1103:

Specification for Mounting Requirements and Alignment Reference Locations for Extreme Ultraviolet Lithography Masks

- The mounting surface flatness

Flatness of the Mounting Surface

Any Square Region with Specified Edge Length (millimeters)	Peak-to-Valley Flatenss (nanometers)
150	48
75	24
40	12
25	8
20	6
10	3

Recommendation for Revising SEMI Standards

SEMI Standard P40:

Specification for Mounting Requirements and Alignment Reference Locations for Extreme Ultraviolet Lithography Masks

- ## The mounting surface flatness

Flatness of the Mounting Surface

Any Square Region with Specified Edge Length (millimeters)	Peak-to-Valley Flatenss (nanometers)
150	30

Summary and Conclusion

- IP compensation strategy has been proposed to relax the required mask nonflatness specifications.

- SEMATECH is running a Mask Integrated Test to validate the compensation Strategy with EUV wafer overlay data.

- Mask nonflatness compensation in e-beam writing has to work in order to reduce mask COO and minimize defects from polishing procedure.

- We will require extremely flat chucks in order for the modeling to work.

- Validity of the correction methodology will be reported next year with overlay analysis from SEMATECH.

Acknowledgments

- ASML: John Zimmerman and Kevin Cummings

- SEMATECH: Sean Eichenlaub, Abbas Rastegar, Sean Huh, and Chawon Koh

- CNSE: Sudharshanan Raghunathan

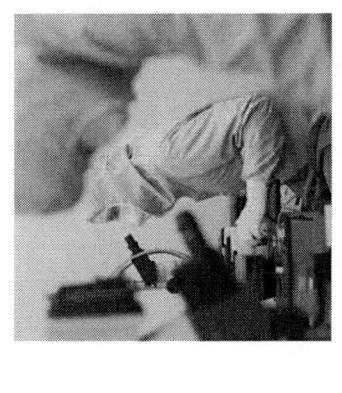
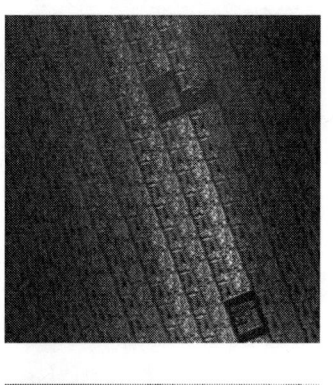
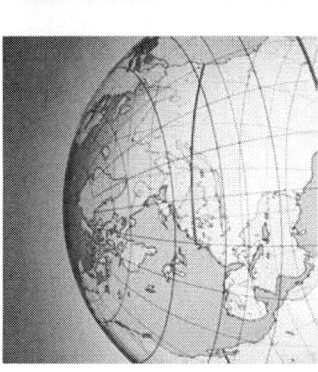

Accelerating the next technology revolution

High throughput defect mitigation

Patrick Kearney, C.C. Lin, T. Sugiyama, H. Yun, R. Randive, I. Reiss, P. Mirkarimi, E. Spiller, A. Hayes

SEMATECH

EUVL mask blank defects are a challenge

Bumps or pits on the substrate or particles added during ML deposition can cause a defect.

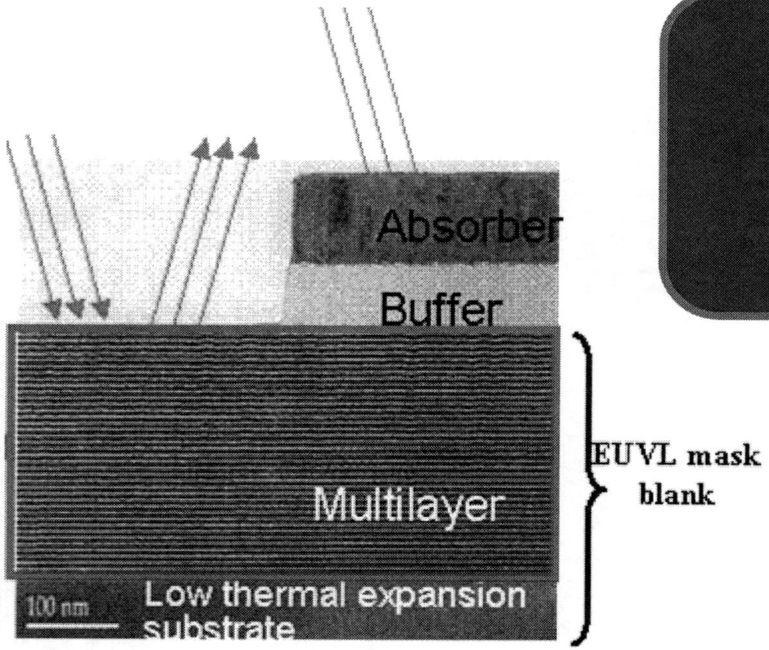

What can go wrong?

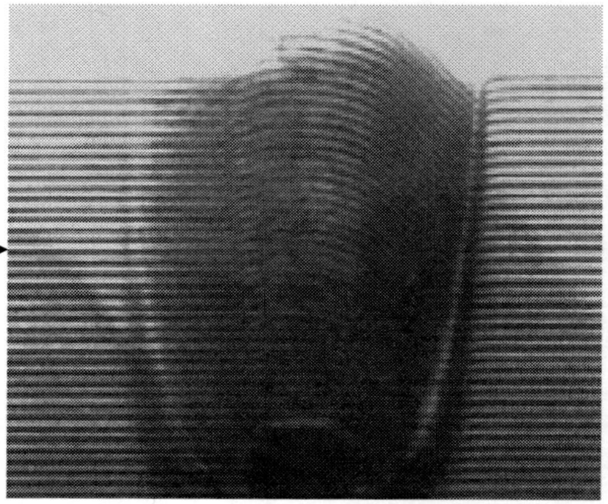

Particle on substrate

Pit in substrate

Particle in ML

EUV mask construction

Absorber

Buffer

Multilayer

Low thermal expansion substrate

EUVL mask blank

100 nm

SEMATECH EUV Mask Blank Development Center (MBDC)

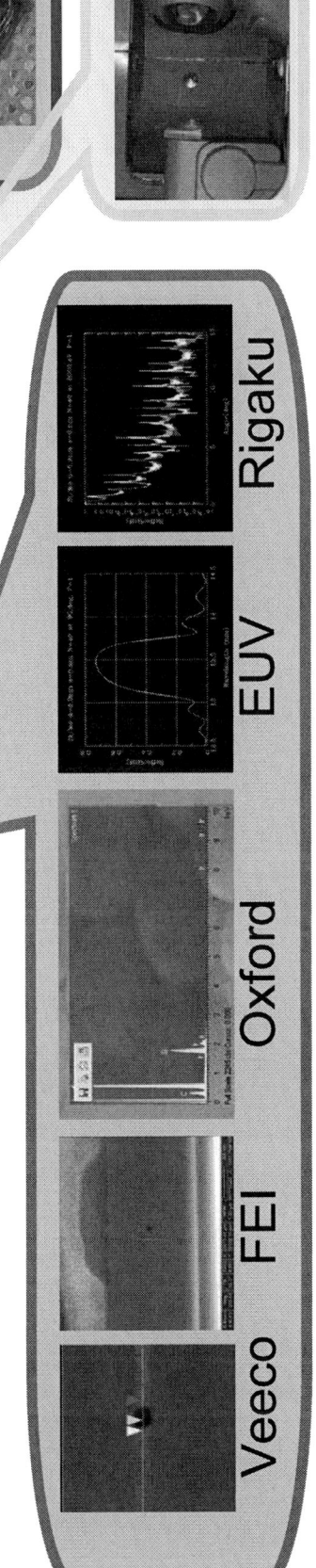

- Advanced Cleaning-Hamatech
- Advanced Inspection-Lasertec
- Advanced Handling-Entegris
- Advanced ML Deposition-Veeco
- ML Blank Metrology

Veeco   FEI   Oxford   EUV   Rigaku

The MBDC has focused on reducing total defects on blanks

- We buy the highest quality QZ and LTEM substrates.

- We use our advanced cleaning to remove any particles.

- We work with Veeco to deposit low defect MLs.

- Our goal is reducing total defects.

Unfortunately incoming pits and scratches were exceedingly high and started to dominate our Pareto.

This year we focused on smoothing to make the process tolerant to substrate defects

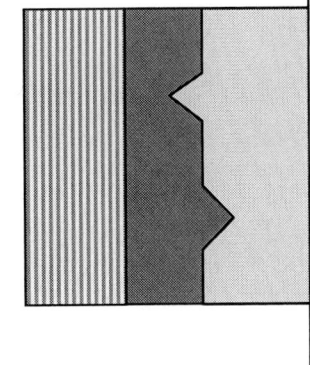
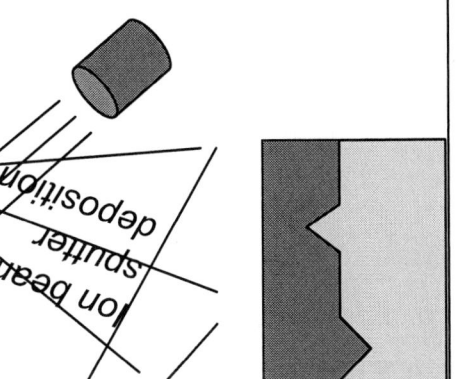
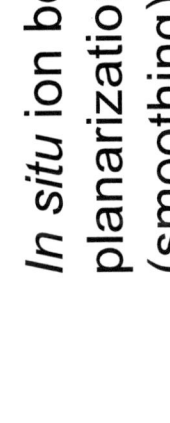
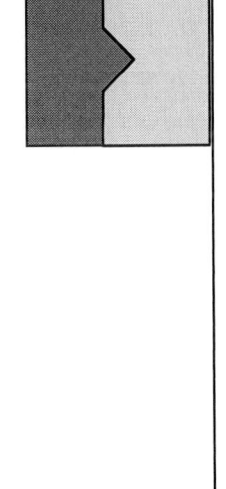

Ion beam sputter deposition

Ion etch

Substrate with pits and bumps

In situ ion beam planarization (smoothing)

Deposit ML on smoothed substrate

How we smooth defects

Veeco Nexus tool

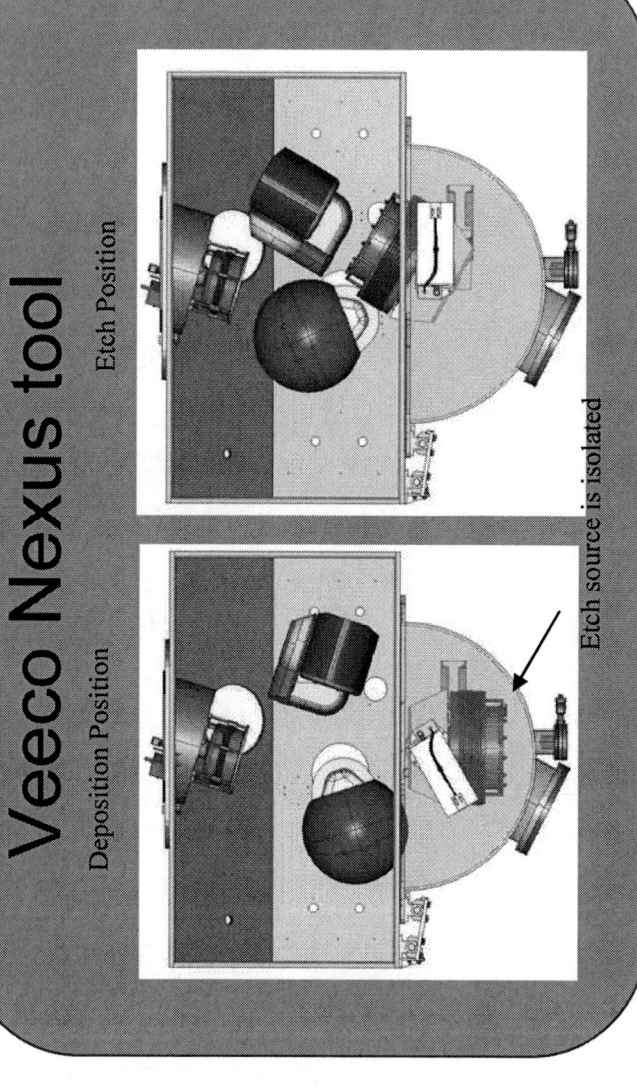

Deposition Position

Etch Position

Etch source is isolated

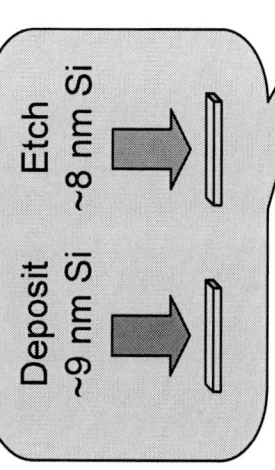

Deposit ~9 nm Si   Etch ~8 nm Si

Deposit ~9 nm Si   Etch ~8 nm Si

Repeat 69X

Repeat 16X

Smoothing works, but is too slow

- It can do the job.
  - Demonstrated smoothing of 20 nm pits and 40 nm bumps to a height small enough to prevent printing.
  - Resulting surface is smooth enough to grow a high quality ML on.

- It's too slow.
  - Original process takes 12 hours.
  - Blank manufacturers want a 2-hour process.
  - Slow smoothing is too expensive.

Methods to speed up smoothing

- Higher etch and deposition rates
  - Current parameters were chosen not to overstress the ion sources.

- Speed up motion
  - There is a lot of complicated motion required to do 85 smoothing cycles without the target turret, substrate fixture, and PAM module colliding.
  - Initially the concern was defects, so motion was conservative; can probably go much faster, but will defects allow it?

- Can we do fewer longer cycles?
  - We know that doing all the deposition and then all the etch does not have the required smoothing properties, but could we do half as many cycles of twice as much deposition/etch and get satisfactory smoothing?

- Descope process from 20 nm to 10 nm pits.
  - Smoothing time is driven by the depth of the largest pit on the substrate.
  - In the last two years, the pits we see have gotten smaller, and we anticipate this trend will continue.
  - Most plates have only sub-10 nm pits now.

Smoothing speedup progress

	Process of Record	High Power Process	Fewer Thicker Cycles	Fast Motion	10 nm Process
Deposition Time	2.2	1.3	1.3	1.3	0.8
Etch Time	4.1	1.4	1.4	1.4	0.8
Motion Time	5.6	5.6	2.8	0.7	0.4
Total Time	11.9 hours	8.3 hours	5.5 hours	3.4 hours	2.05 hours

Fast smoothing defect levels

- No big change in smoothing defect levels resulted from the speedup work.
  - Still at ~1.25 defects per square cm larger than 74 nm after smoothing.

- Fast process successfully smoothes 10 nm pits to <1 nm high.

- Were able to get surprisingly good ML-added defect levels after fast smoothing.
  - After dozens of smoothing samples, we still got a median defect level of 9 defects.
  - Best ever median defect level was 5.5 defects at 67 nm.
  - Worry was all the stressed Si we were depositing would cause premature flaking.

Pin chuck results

- We tested a pin type e-chuck to reduce backside contamination by reducing the chuck contact area.

- Backside defect levels were reduced by ~20X.

- One worry with the pin chuck is that the reduced contact area with the chuck would allow the mask substrate to heat up too much.

  – Mask substrate remained <140°C even during the high power smoothing process.

- For more details, see the poster by A. Hayes at Bacus.

Conclusions

- Smoothing speed increased ~6X.
  - 12 hours → 2.05 hours
  - Much more manufacturable process.
- Smoothing speedup did not influence smoothing defect levels greatly.
  - No improvement or worsening of defect level seen.
  - There is still a defectivity gap between the current levels and manufacturing requirements.
- Demonstrated ML deposition still clean after smoothing.
  - Worry was that smoothing would contaminate the chamber and make MLs dirty.
  - Median defect levels after a great deal of smoothing within a factor of ~2 of best value ever.
- Pin chuck reduces backside contamination by ~20X.
  - Temperature of smoothed plate remains within specification even with the pin chuck.

2008 International Symposium on Extreme Ultraviolet Lithography

Characterization of EUV Mask Defects: Printability and Repair Process

Hakseung Han*, Donggun Lee, Hwan-Seok Seo, Kenneth A. Goldberg**, Hoon Kim, Byung-Sub Ahn, In-Yong Kang, Wonil Cho, Sanghyeon Lee, Suyoung Lee, Geunbae Kim, Dongwan Kim, Seong-Sue Kim, and HanKu Cho

Memory R&D Center Samsung Electronics Co., Ltd.
**CXRO, Lawrence Berkeley National Lab.

*hakseung.han@samsung.com

OUTLINE

Introduction

Defect printability – ADT printing test

- L&S, Contact
- 45nm, 32nm HP
- Extrusion, Intrusion defects

Defect printability – Simulation study

- Comparison with ADT result

Repair process assessment using AIT, MET test

- Nano-machining, Ebeam Repair
- L&S(37.5nm HP), Contact(50nm HP)

Summary

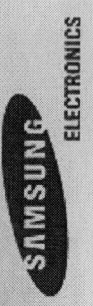

Introduction

Tighter Defect spec.
- Mask Defects have always been challenges to EUV Lithographers.
- EUV Lithography era requires much tighter mask defect specification.

Complex mask stack
- Complex mask stacks require deeper study on mask defect printability.
- Repair process development is needed for different Absorber.

So,
- We studied printability using ADT & Simulation
- And also assessed repair process using available infra.

SAMSUNG
ELECTRONICS

Absorber defect printability

Fabrication of programmed defect Mask

- Silicon Capping on 40 bilayers on LTEM Plate
- 32nm/45nm HP L&S, 50nm Contact with intrusion & extrusion type defects

Printability study using ADT(@IMEC)

- Conventional illumination
- No Shadowing illumination
- Measured at best focus and dose

Comparison with Simulation

- Sentaurus lithography(EUV) – Ver. B- 2008.06.
- Threshold CD analysis
- No corner rounding considered

ADT Printability (45nm HP) - Intrusion

- **Intrusion type defects were formed ranging from ~14nm(4X).**
- **Intrusions larger than ~40nm on mask are printable.**

Measured defect size (on mask) [nm]

Mask SEM Image

| 14.1nm | 25.4nm | 29.0nm | 41.0nm | 60.6nm | 78.4nm |

Wafer SEM Image

Printable

SAMSUNG ELECTRONICS

ADT Printability (32nm HP) - Intrusion

- **Intrusions larger than ~40nm on mask are printable.**
- **Difficult to analyze due to large resist LER.**

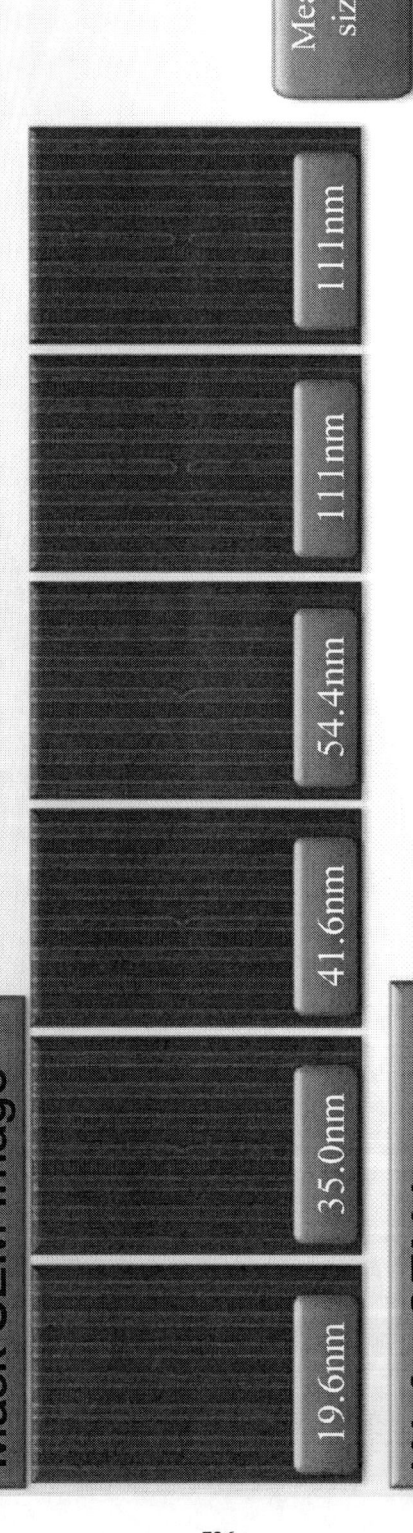

Measured defect size (on mask) [nm]

| 19.6nm | 35.0nm | 41.6nm | 54.4nm | 111nm | 111nm |

Mask SEM Image

Wafer SEM Image

SAMSUNG ELECTRONICS

ADT Printability (45nm HP) - Extrusion

- **Extrusions larger than ~40nm on mask are printable.**
- **Failed to get printability about extrusions on 32nm Line .**

Mask SEM Image

24.4nm 26.5nm 47.8nm 57.7nm 71.6nm 86.5nm 104.2n

Wafer SEM Image

SAMSUNG ELECTRONICS

ADT Printability (45nm HP) - Contact

- Defect size defined from virtual edge
- Defects larger than ~35nm are printable.

Defect size Definition

Mask SEM Image

24.2nm | 27.4nm | 33.0nm | 39.1nm | 41.9nm | 47nm | 53.2nm | 57.0nm

Wafer SEM Image

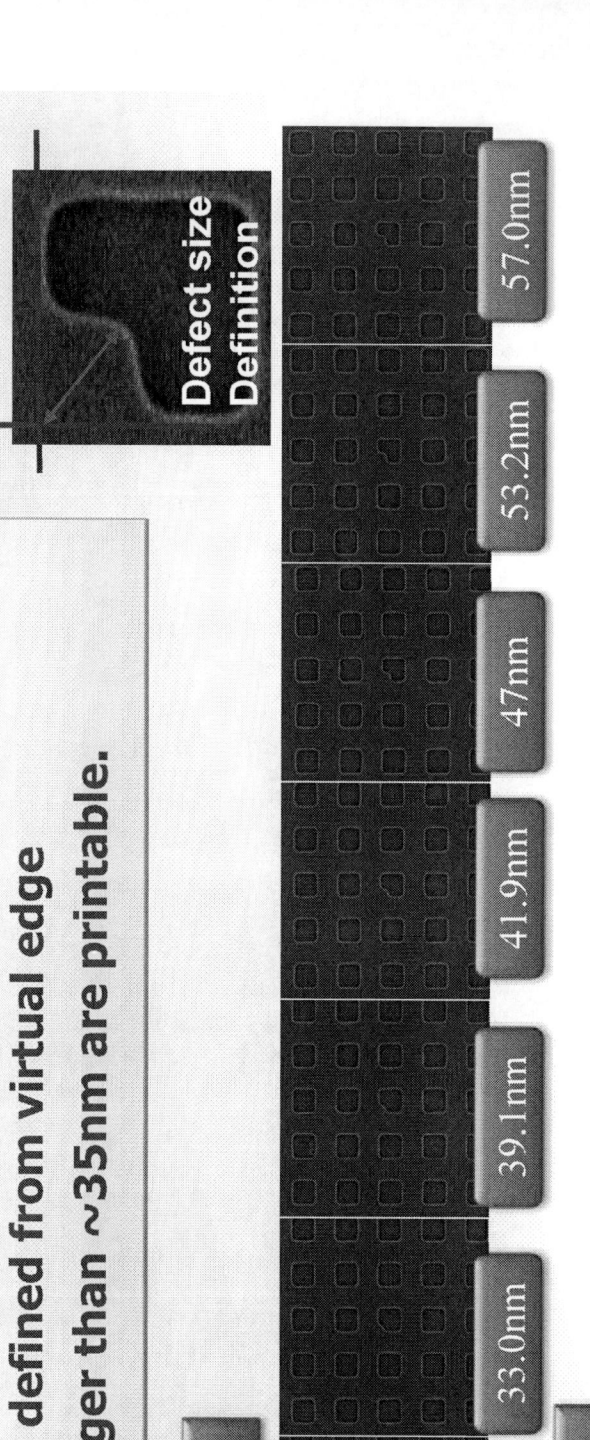

SAMSUNG ELECTRONICS

Printability Simulation(45nmHP) - Intrusion

- Th'shold CD was obtained with varying defect width without shadowing effect.
- 15nm(60nm on mask) square makes CD change more than 5%.
- Less printable than ADT test → LER & Process dominant.

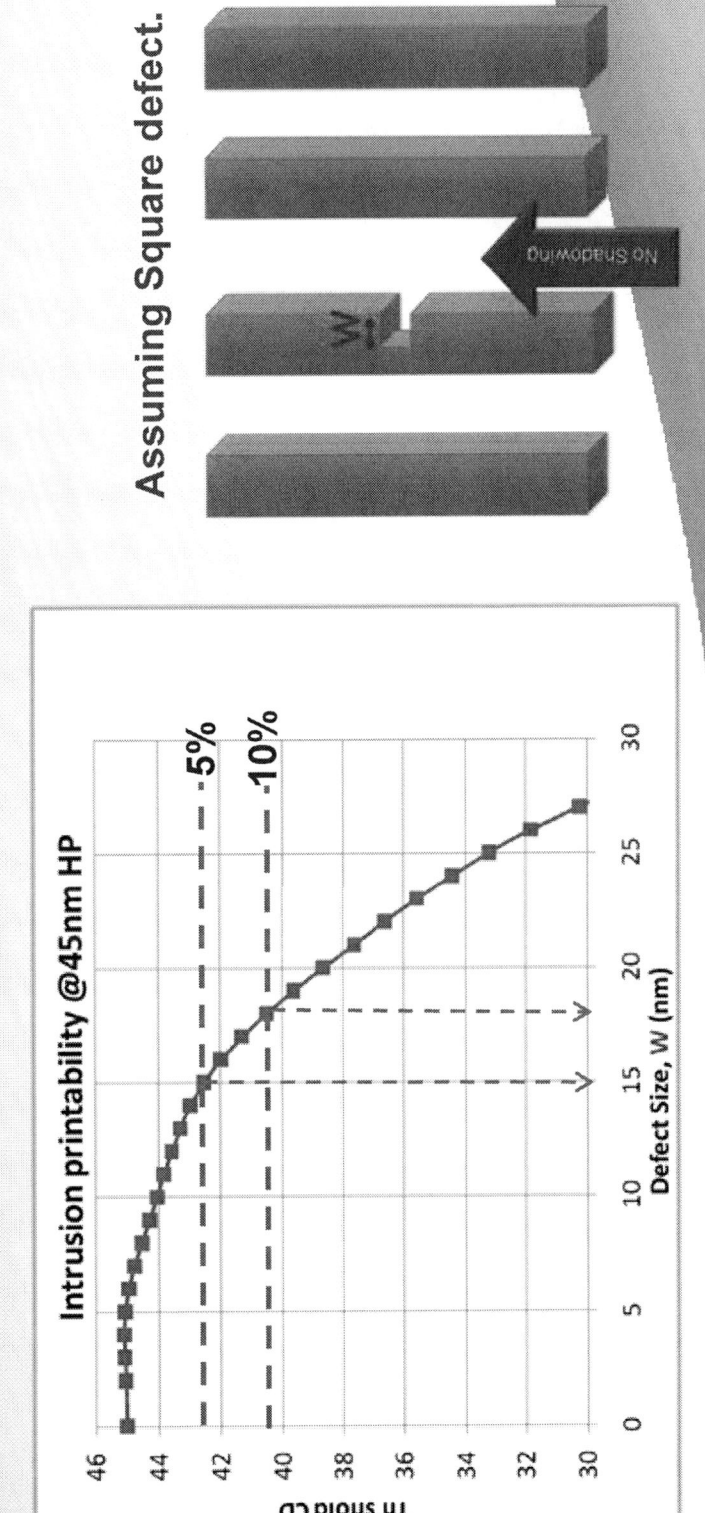

Assuming Square defect.

Printability Simulation(32nmHP) - Intrusion

- 12nm(48nm on mask) square makes CD change more than 5%.

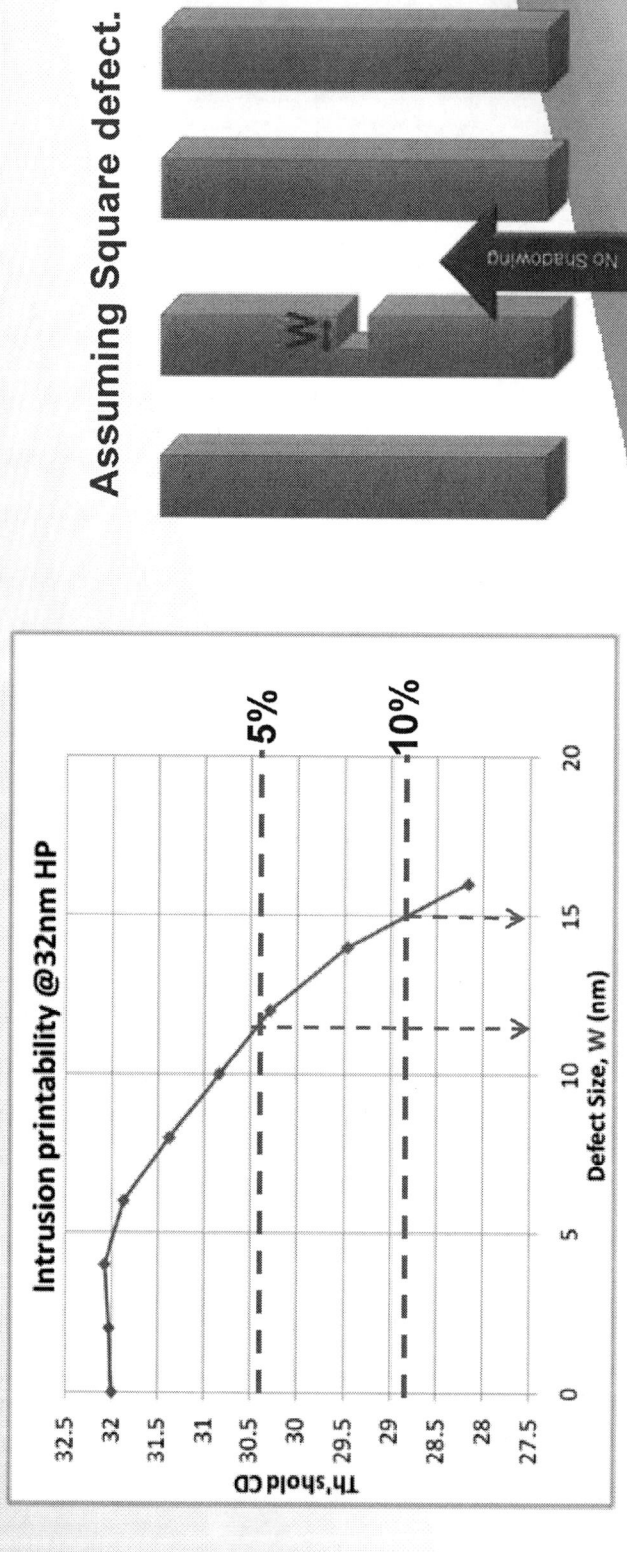

Assuming Square defect.

Printability Simulation(45nmHP) - Extrusion

- 8nm(32nm on mask) square makes CD change more than 5%.
- Printable defect size from simulation roughly matches ADT printability data between 5% and 10%.
- Extrusion type shows much tighter spec. than intrusion.

Assuming Square defect.

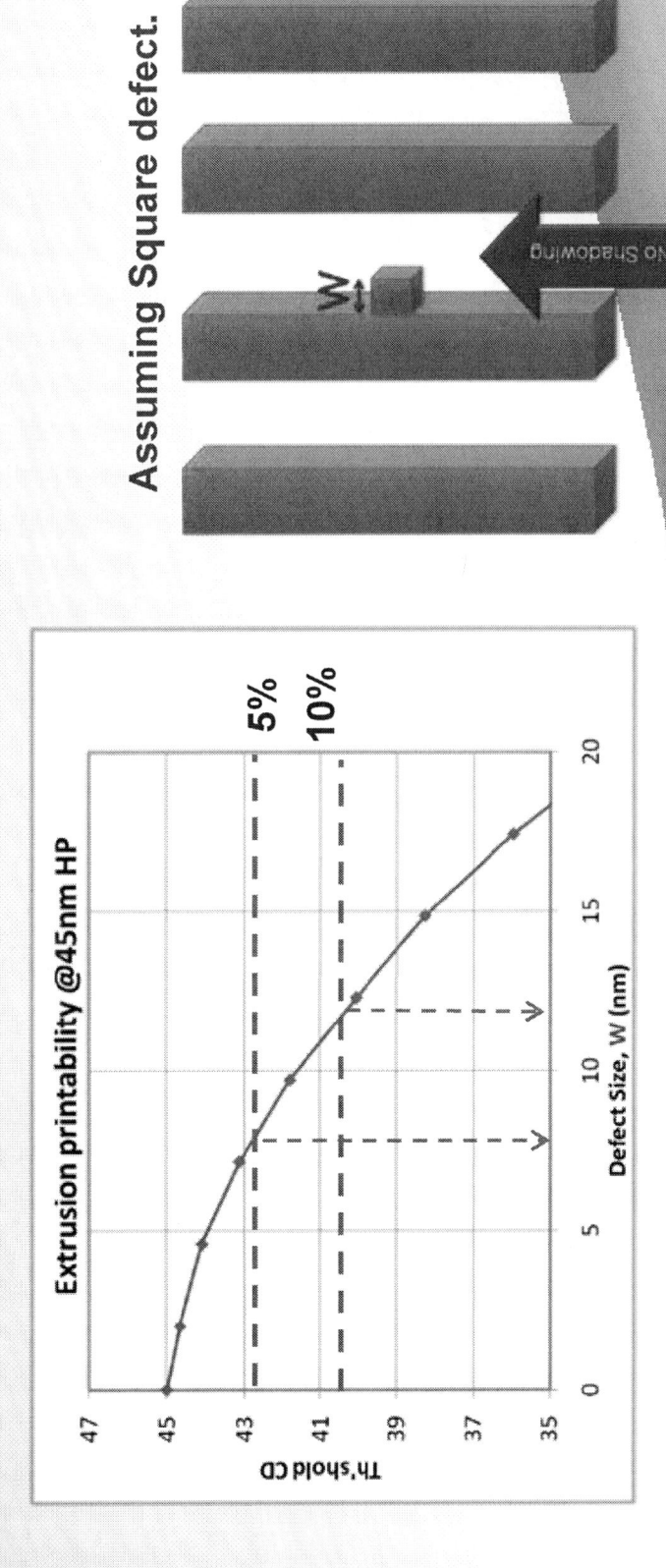

SAMSUNG
ELECTRONICS

Printability Simulation(32nmHP) - Extrusion

- 6nm(24nm on mask) square makes CD change more than 5%.

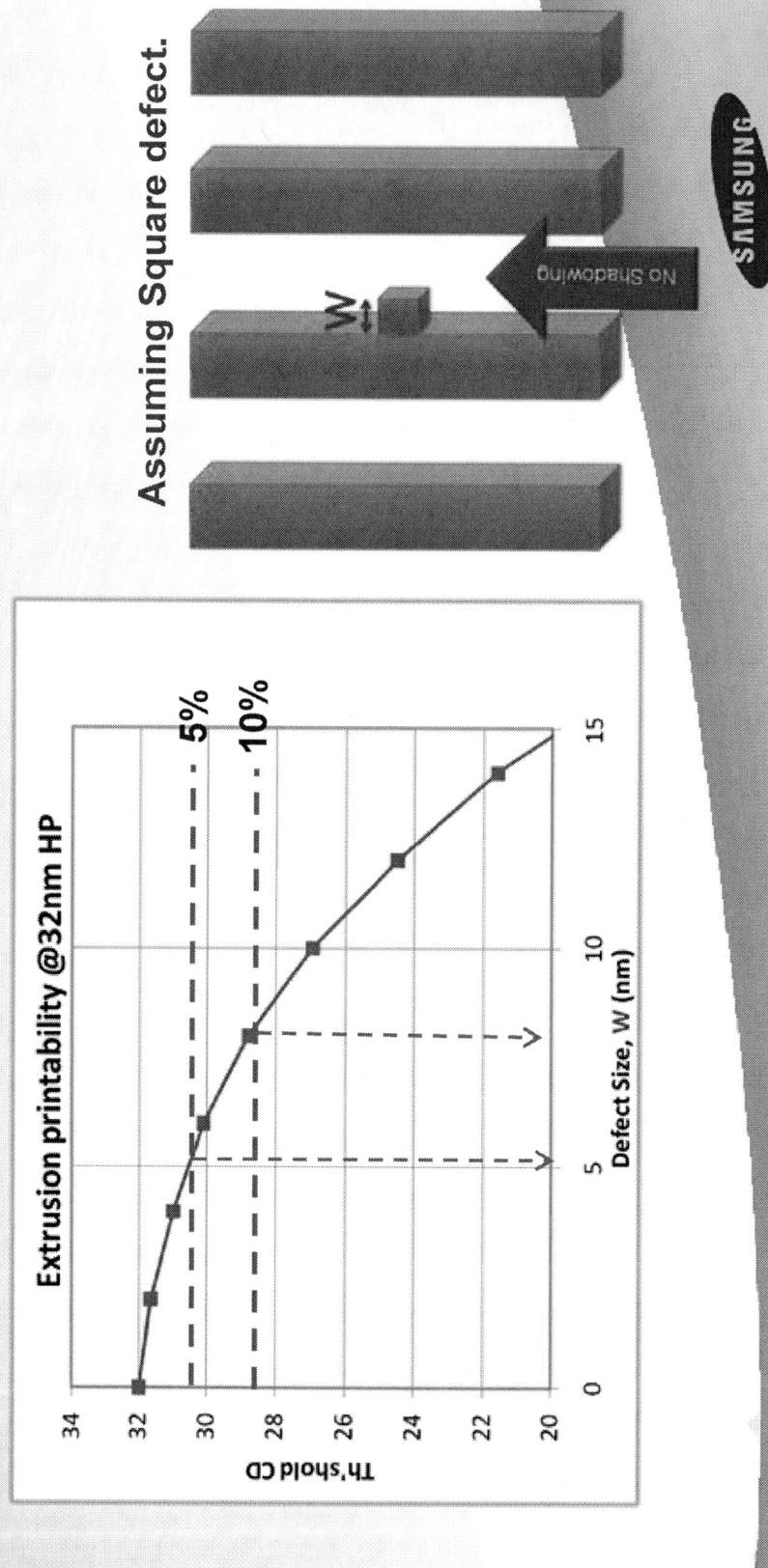

Assuming Square defect.

Simulation Comparison with ADT result

- **Simulation underestimates printability for intrusion type.**
- **It is likely test result could be misunderstood due to process effect.**

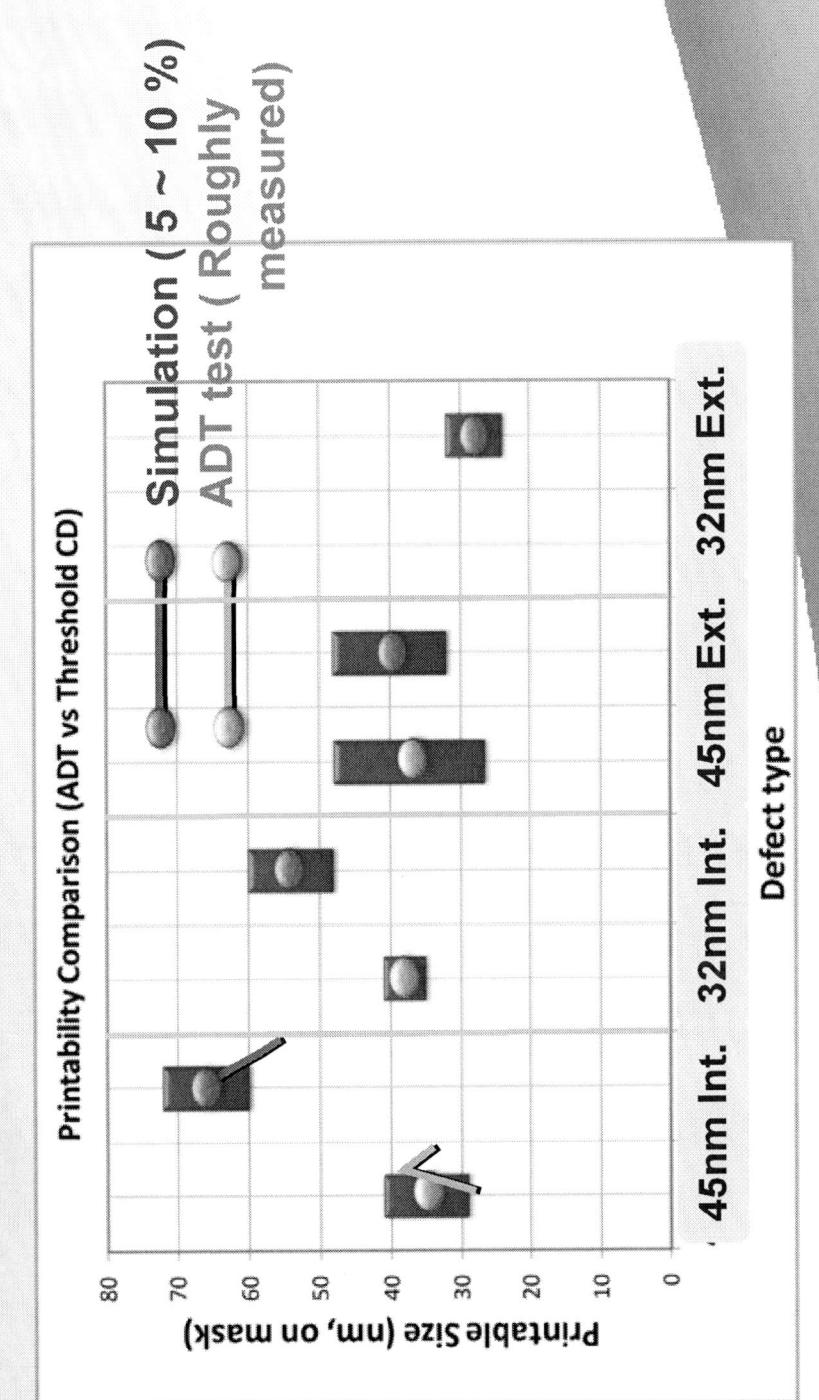

Repair Process Assessment

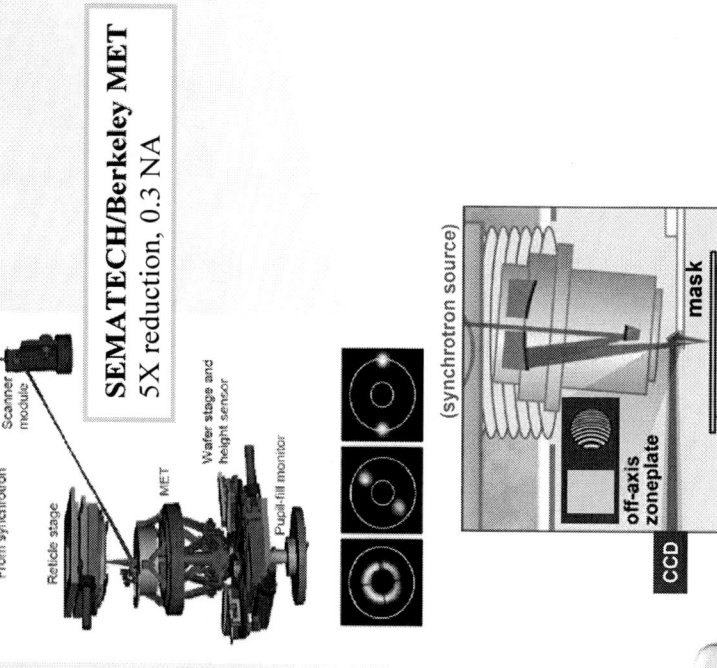

SEMATECH/Berkeley MET
5X reduction, 0.3 NA

SEMATECH/Berkeley AIT
~0.35 NA, 0.2σ

Mask preparation

- Ru Capping Mask for NMrepair & Si Capping Mask for Ebeam repair
- 37.5nm HP L&S, 50nm missing contact hole (on wafer)

Nanomachining Repair process

- 45nm Node Tool
- Small extrusion on 37.5nm HP and 50nm missing Contact

Ebeam Repair process

- 45nm Node Tool
- 3xHP Bridge on 37.5nm HP

Assessment using AIT and MET

- Obtained Aerial image and printed resist image

SAMSUNG ELECTRONICS

Repair Process Development- Ebeam Repair

- **450nm Bridge Defects were repaired and evaluated using the AIT.**
- **Process tuning needed due to Buffer layer(B) → Ru Cap. Preferred.**
- **After fine tuning, repair processes (C,D) worked well.**

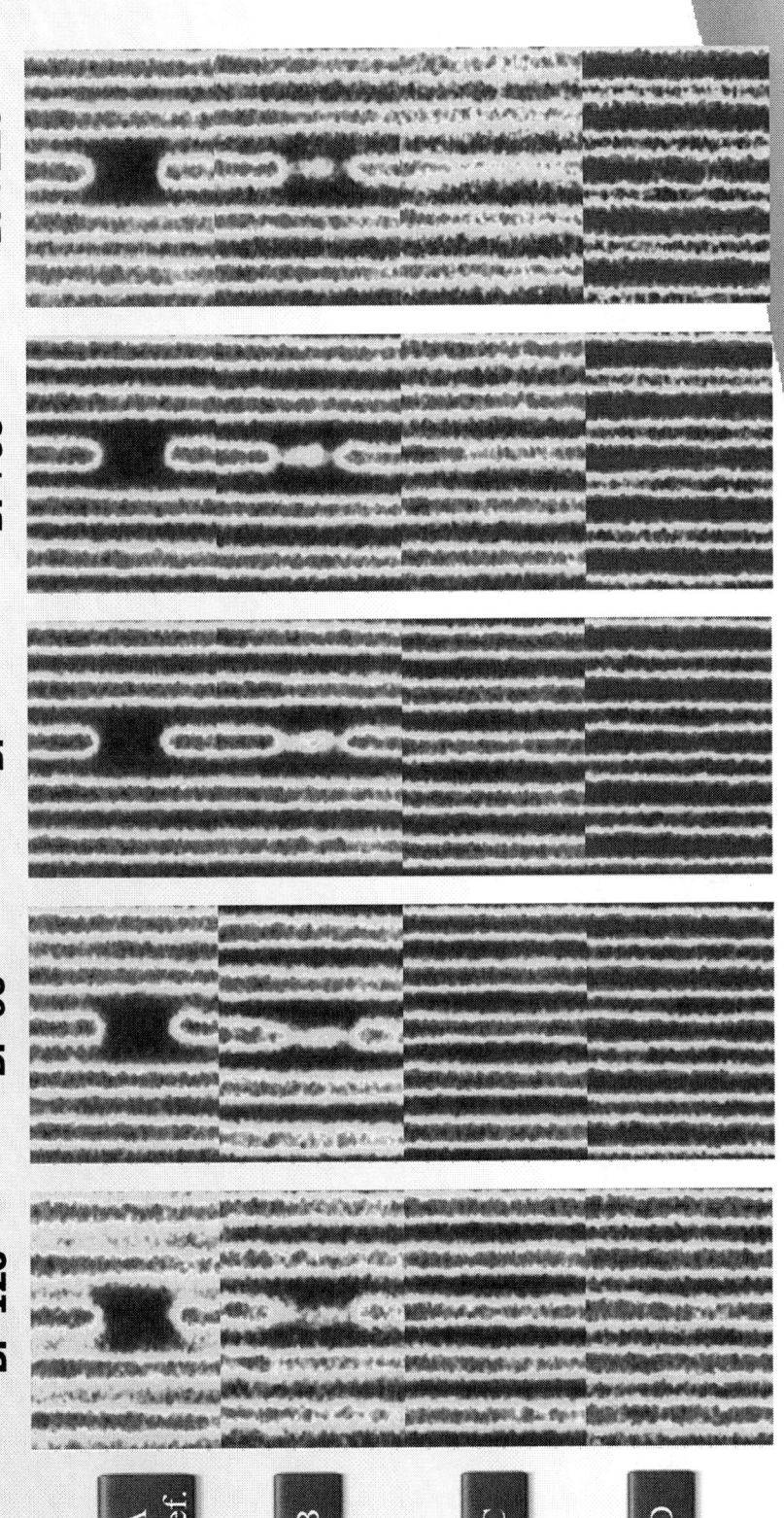

Ebeam Repair Process

- **Measured Th'hold CD through focus.**
- **Case D showed enough process margin about 200nm.**

C
D
Delta CD
Delta CD
Spec(10%)
8
7
6
5
4
3
2
1
0
-150
-100
-50
0
50
100
150

SAMSUNG ELECTRONICS

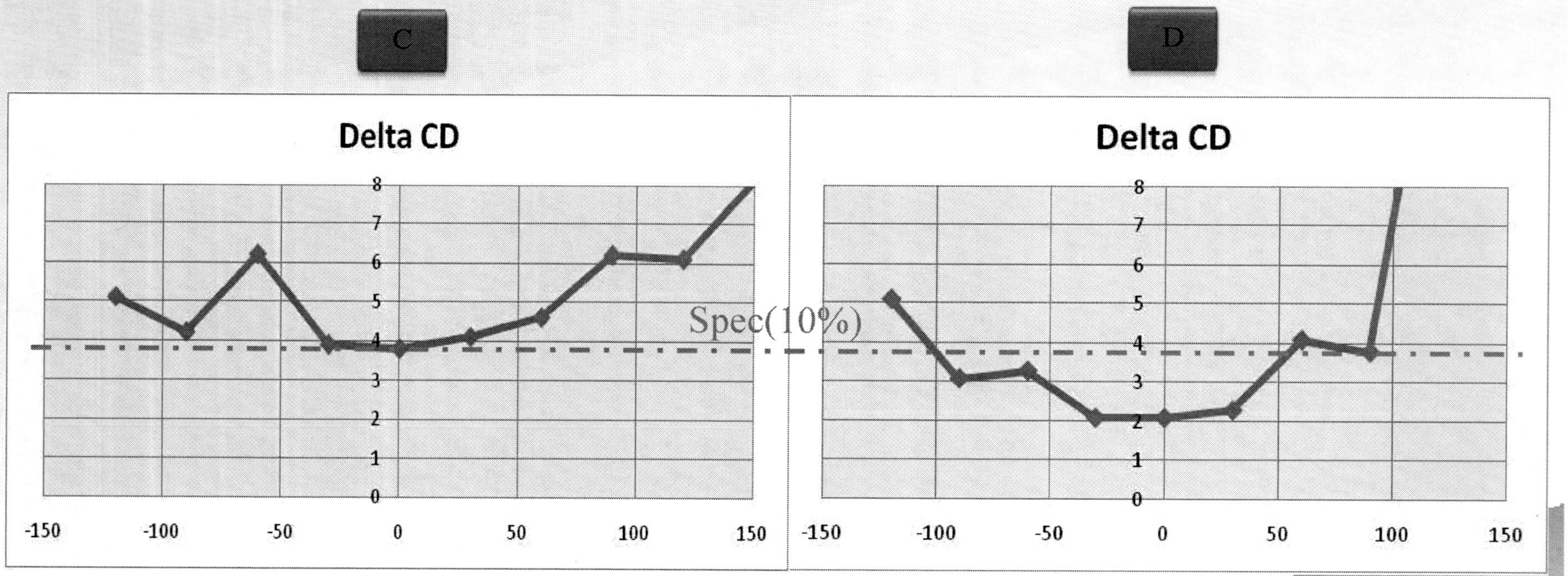

Nano-machining Repair Process

- **Extrusion(75nm x 150nm) on 150nm Line was repaired and evaluated using Nano-machinning repair.**
- **Showed good feasibility.**

Refer to paper by Dr. Suyoung Lee at BACUS 08 for more details

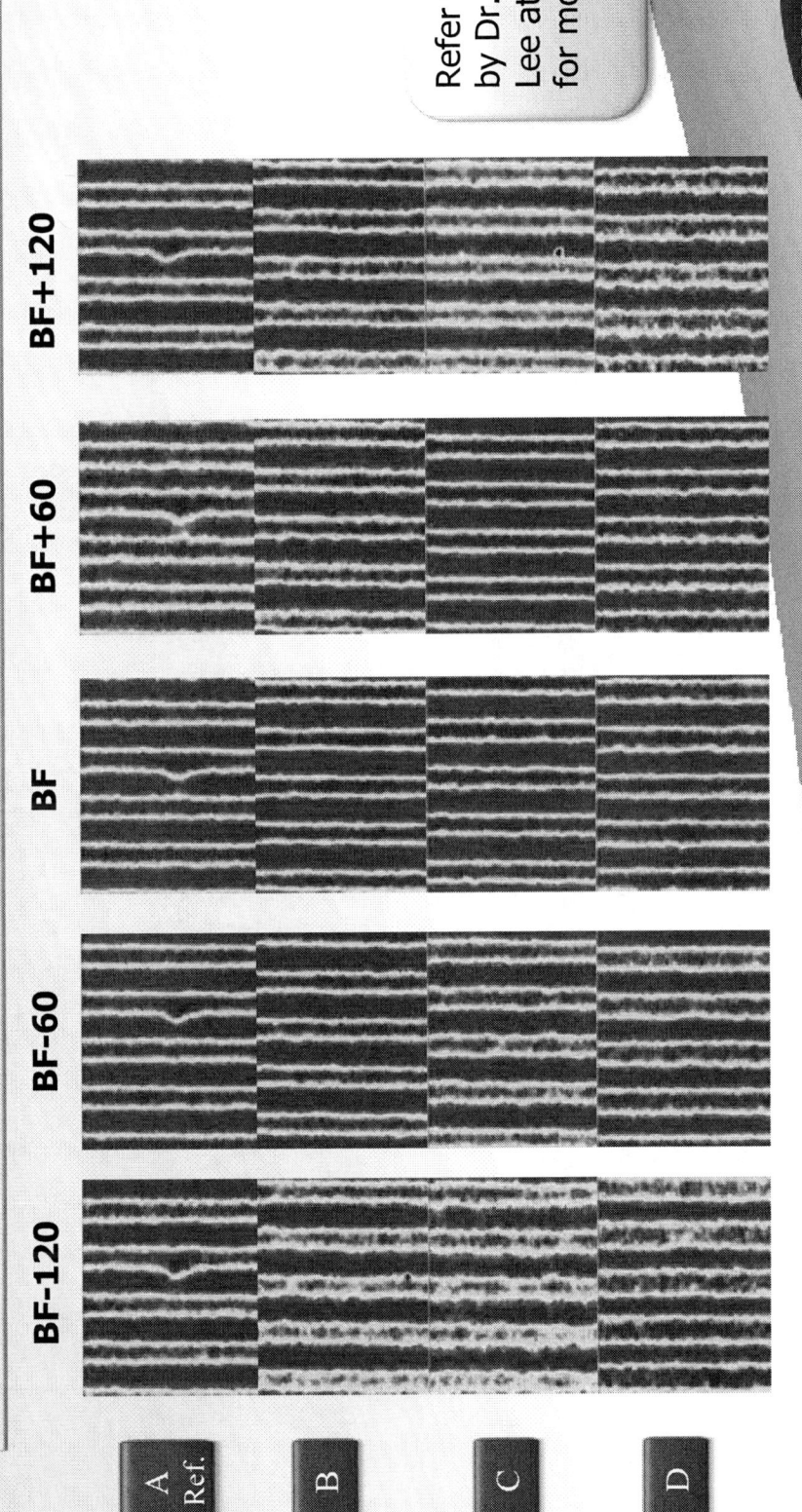

Nano-machining Repair Process

- Confirmed again using MET.
- 5x demagnification exposure and 30nm HP L&S images.
- Appears process margins are enough.

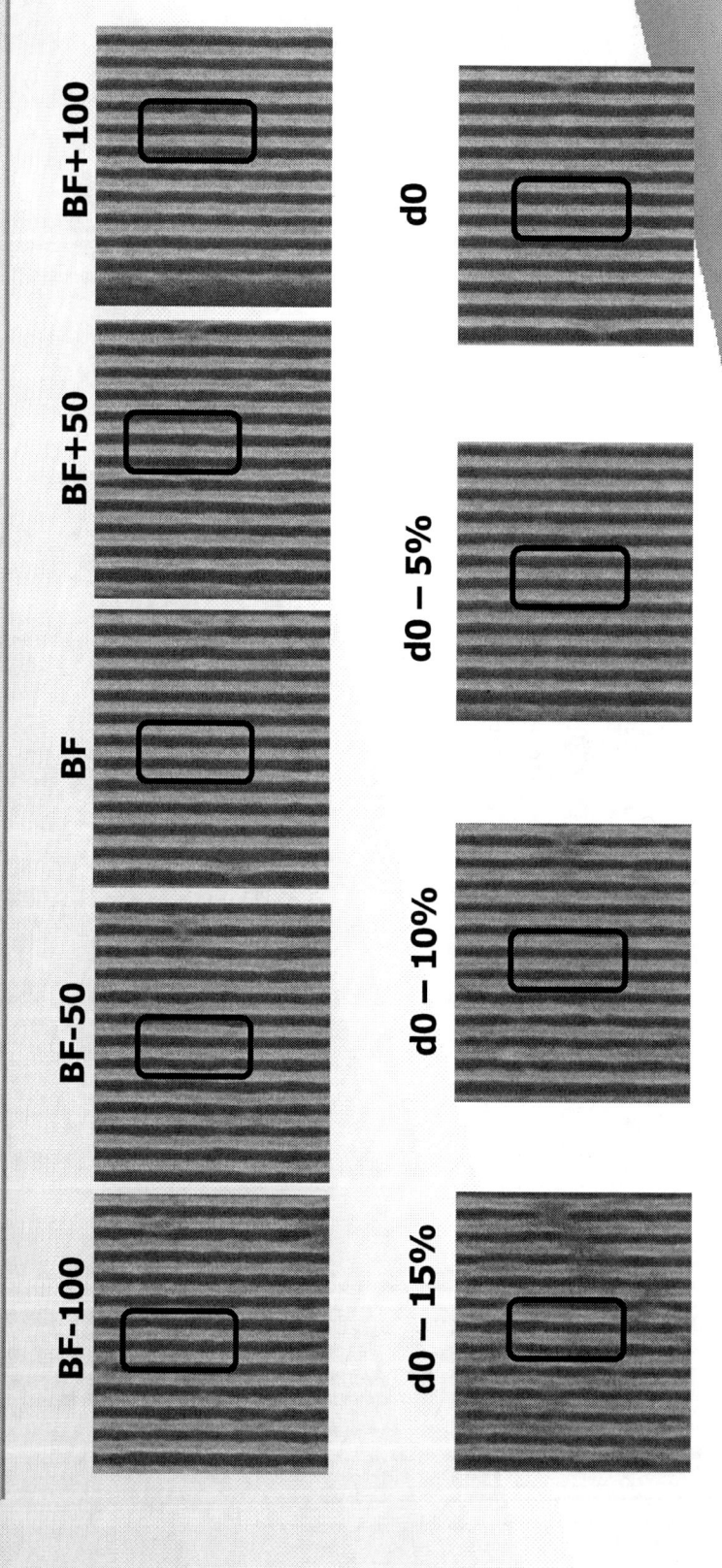

Nano-machining Repair Process

- **Missing contact hole repaired and evaluated using AIT and showed good feasibility.**
- **Aerial imaging is very useful for defect printability-related analysis, much more convenient than using wafer exposure.**

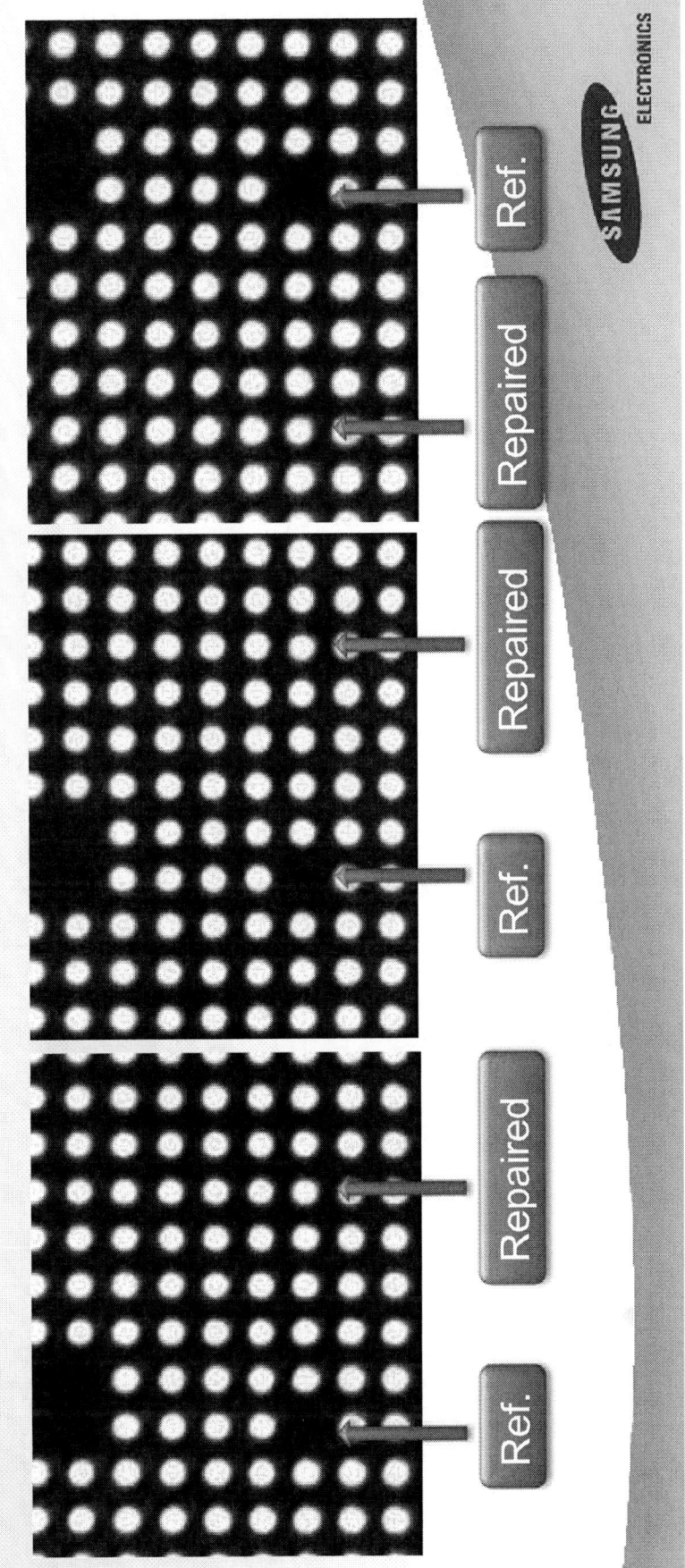

Summary

Defect printability study using ADT

- For intrusion and extrusion, ~40nm Absorber defects are printable at 32 and 45 nm HP.
- For contact hole, 35nm absorber defect on an edge is printable.
- Large resist LER make it difficult to accurately analyze printability.

Printability Simulation and comparison with ADT result

- Quite difference between Threshold simulation and ADT result.
- Simulation underestimates for intrusion.
- Extrusion type showed much tighter spec. than intrusion.
- Printability of small defects highly depends on process condition.

Repair process development study

- Ebeam Repair showed good result at 37.5nm HP bridge repair.
- Nano-machining worked well for 37.5nm HP small extrusion and 50nm missing contact.
- Standalone EUV AIMS is key tool for repair and defect related EUV work.

Acknowledgement

The authors would like to thank...

- Jinhong Park and Junghoon Lee at Process Development Team for measuring wafer printed at ADT and other ADT related engineers in IMEC and ASML,
- Iacopo Mochi for measuring aerial image from the AIT,
- Dr. Patrick Naulleau, Paul Denham, Brian Hoef, and Gideon Jones for their helps for the wafer printing test using EUV MET at ALS in LBNL.

SAMSUNG ELECTRONICS

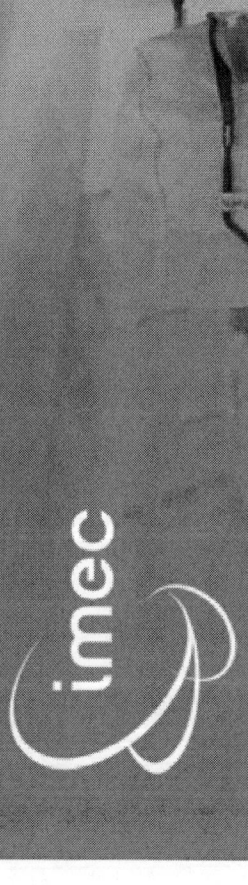

Mask defect printability in Full Field EUV Lithography – Part 2

R. Jonckheere, F. Iwamoto *, N. Stepanenko °,
A.M. Goethals, K. Ronse

*on assignment from Panasonic Corporation
° on assignment from Qimonda

EUVL Symposium, 1 October 2008, Lake Tahoe

4 types of mask defects

Absorber defect

- opaque type
- clear type

Particle

Opaque absorber like defect with varying material and thickness

Part 1 was presented at least year's EUVS (Sapporo):
Detailed simulation study covering all defect types

Local R%-loss

Local carbon deposition

ML defect

- in the ML
- on the substrate

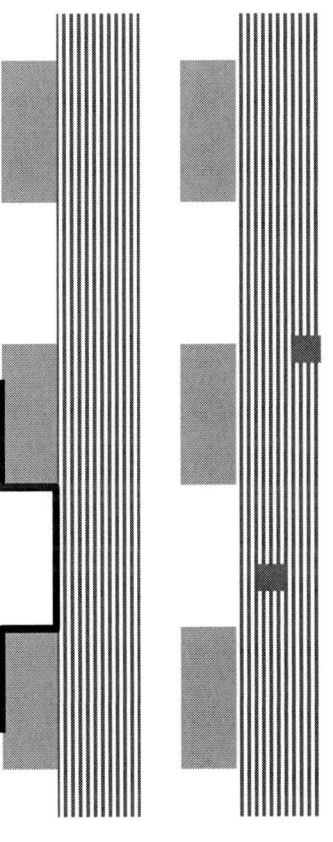

imec

This presentation

Absorber defect

- opaque type
- clear type

Simulation update +
Experimental validation on ADT

THIS PRESENTATION

Simulation update +
First experimental evidence

ML defect

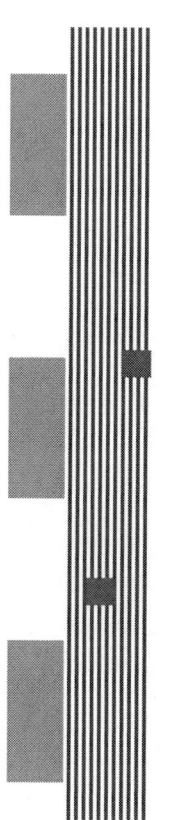

- in the ML
- on the substrate

imec

Outline

- Introduction

- ## Further simulation study

  - Absorber defects
    - Simulation update: role of defect location
  - ML defects
    - Simulation update:
      - What is worse: Pits or bumps ?
      - Impact of cross-sectional shape

- ## Experimental printability study on EUV-ADT

  - Analysis of absorber defect printability and correlation to simulation (programmed approach)

  - Possible evidence of ML defects (non-programmed) and analysis by simulation

Note: Defect sizes given at wafer level

imec

Absorber defects *(shown for opaque)*

Center

Absorber   Absorber

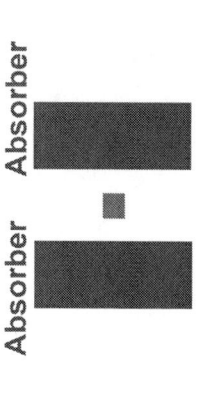

Quasi-edge

Absorber   Absorber

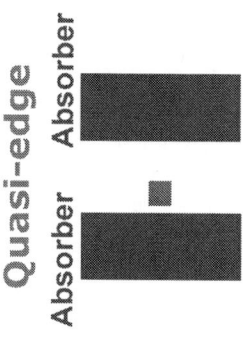

- Most printing location: center or quasi-edge ?

- Analysis for 4 cases
  - Shadowing
  - Resist image w/ and w/o flare
  - Broadband exposure
  - Variation of pitch

- Outcome: matter of "contrast"

imec

Variation of pitch
(absorber opaque defect)

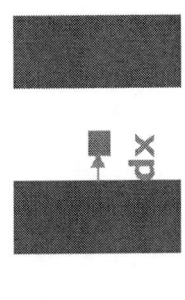

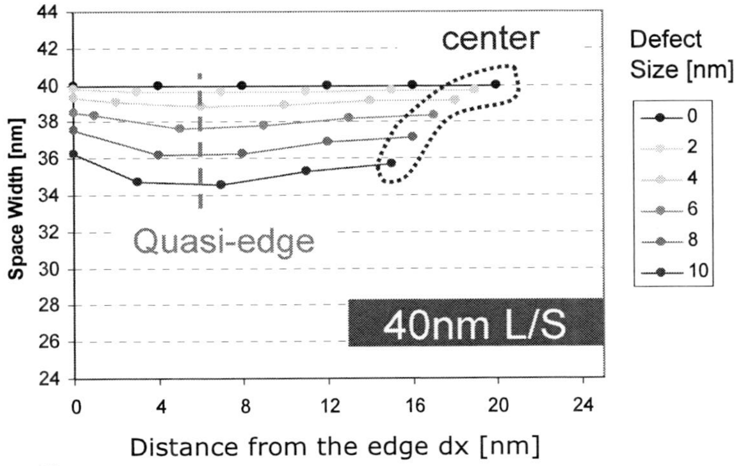

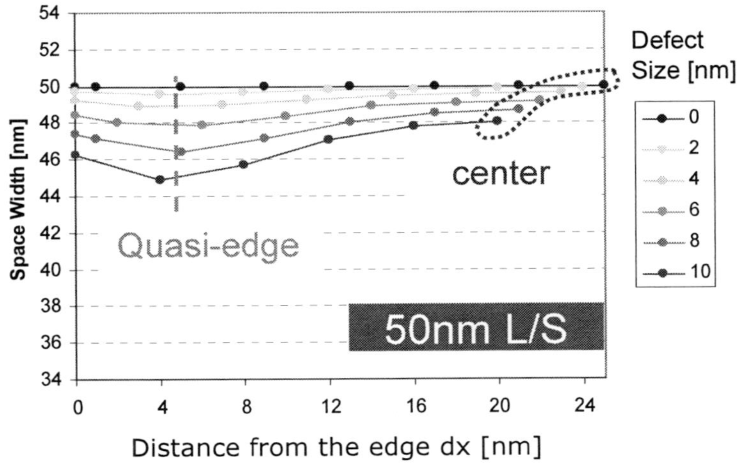

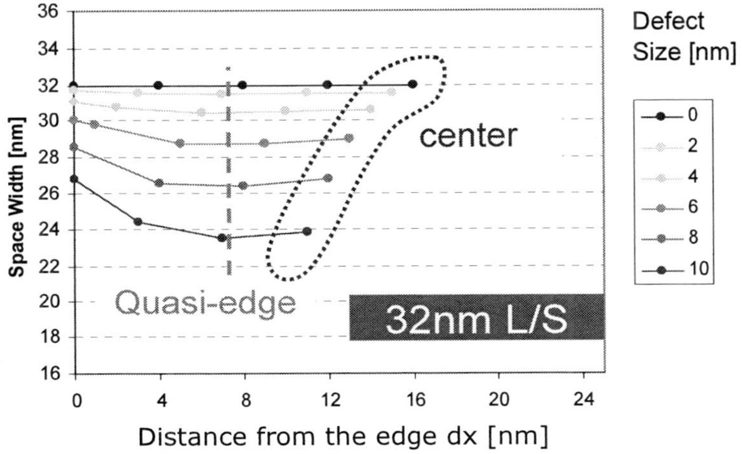

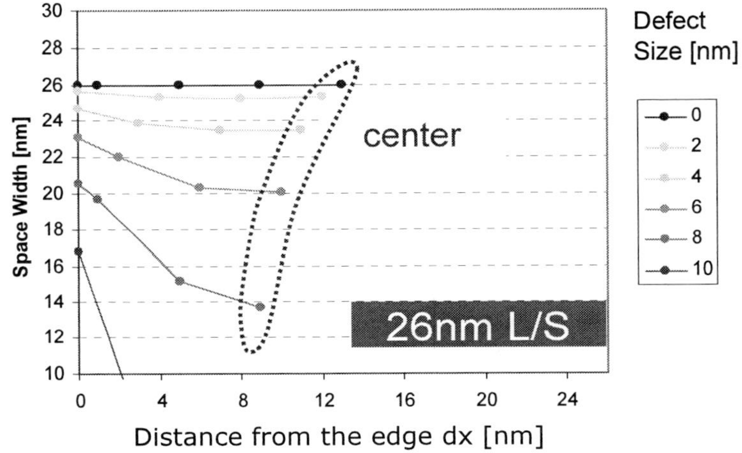

50nm L/S

Defect Size [nm]: 0, 2, 4, 6, 8, 10

center / Quasi-edge

Distance from the edge dx [nm]

Space Width [nm]

26nm L/S

Defect Size [nm]: 0, 2, 4, 6, 8, 10

center

Distance from the edge dx [nm]

Space Width [nm]

40nm L/S

Defect Size [nm]: 0, 2, 4, 6, 8, 10

center / Quasi-edge

Distance from the edge dx [nm]

Space Width [nm]

32nm L/S

Defect Size [nm]: 0, 2, 4, 6, 8, 10

center / Quasi-edge

Distance from the edge dx [nm]

Space Width [nm]

- Center position becomes the worst case when the half pitch is smaller than 32nm.
- The most critical location is dependent on the pitch.

imec

Variation of pitch
(absorber opaque defect)

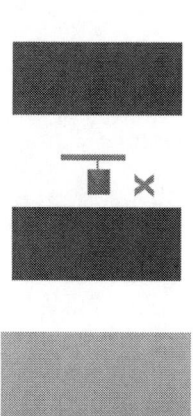

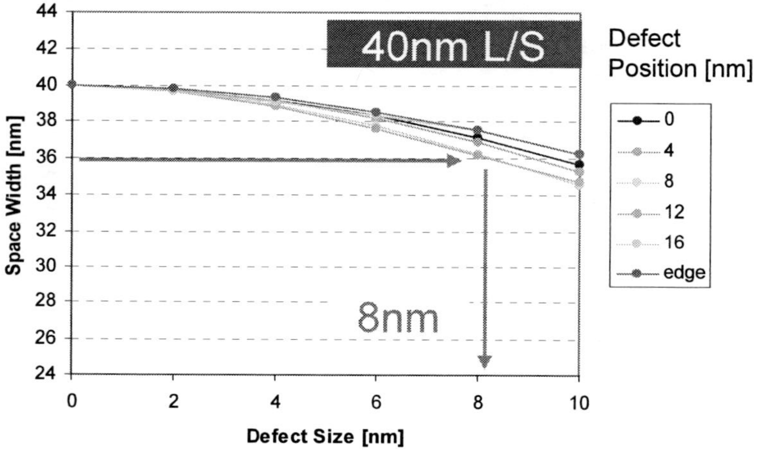

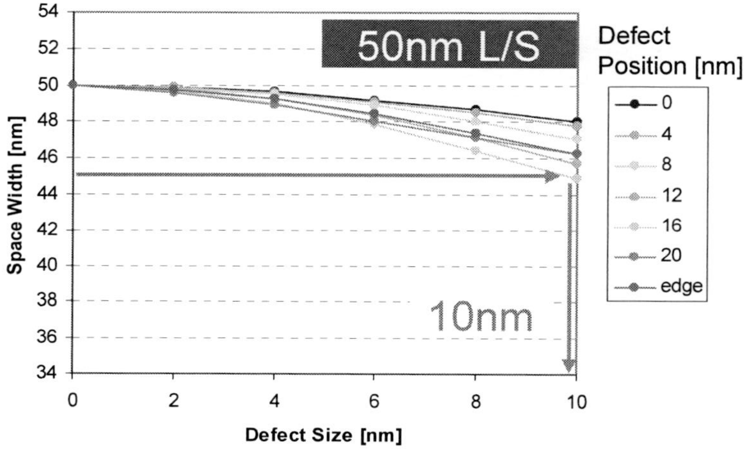

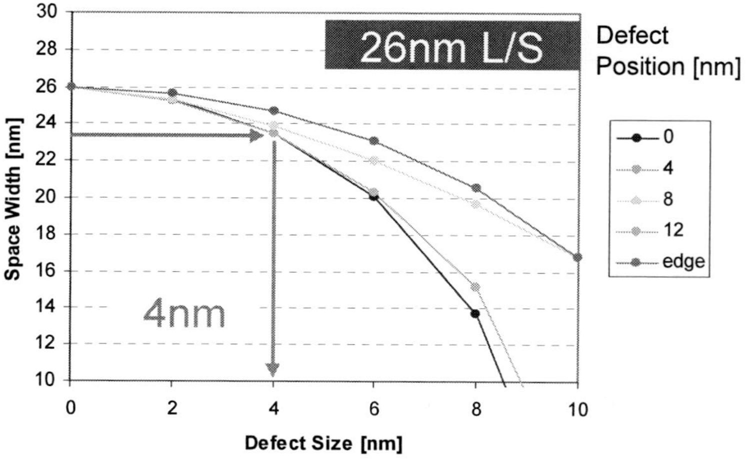

50nm L/S

Defect Position [nm]: 0, 4, 8, 12, 16, 20, edge

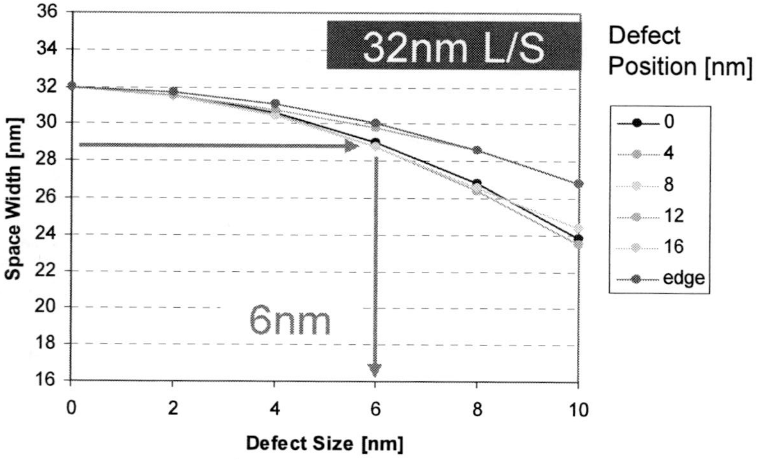

10nm

40nm L/S

Defect Position [nm]: 0, 4, 8, 12, 16, edge

8nm

26nm L/S

Defect Position [nm]: 0, 4, 8, 12, edge

4nm

32nm L/S

Defect Position [nm]: 0, 4, 8, 12, 16, edge

6nm

- Critical defect size becomes smaller with shrink of the pitch.
- Rule of Thumb (critical defect size at mask = CD at wafer) slightly underestimates printability.

imec

748

Outline

- Introduction

- **Further simulation study**

  – Absorber defects

    • Simulation update: role of defect location

  – ML defects

    • Simulation update:

      – What is worse: Pits or bumps ?

      – Impact of cross-sectional shape

    Note: Defect sizes
    X,Y given at wafer level
    Height (Z) at mask level (4X)

- **Experimental printability study on EUV-ADT**

  – Analysis of absorber defect printability and correlation to simulation (programmed approach)

  – Possible evidence of ML defects (non-programmed) and analysis by simulation

imec

Reminder from Part 1
Only for bump

- Main effect from distorted height near the top of the ML
- Material dependence only when the defect is located near the surface
- For simulation most easy to define defects as substrate defects and with defect height = top height
  - Swing nature with (top) height

Substrate defect with Gaussian shape, for defect width = top width = 5nm FWHM

Simultaneous variation of defect height and top height

2D simulation

Space Width

0.04
0.038
0.036
0.034
0.032
0.03
0.028

Top Height (on mask)

Space Width

Smoothing

Defect Height (on mask)

imec

ML defect : role of cross-sectional shape
(repeat from Part 1) **Only for bump and with fixed height**

Substrate defect, with Defect Height = Top Height=5nm (at Mask)

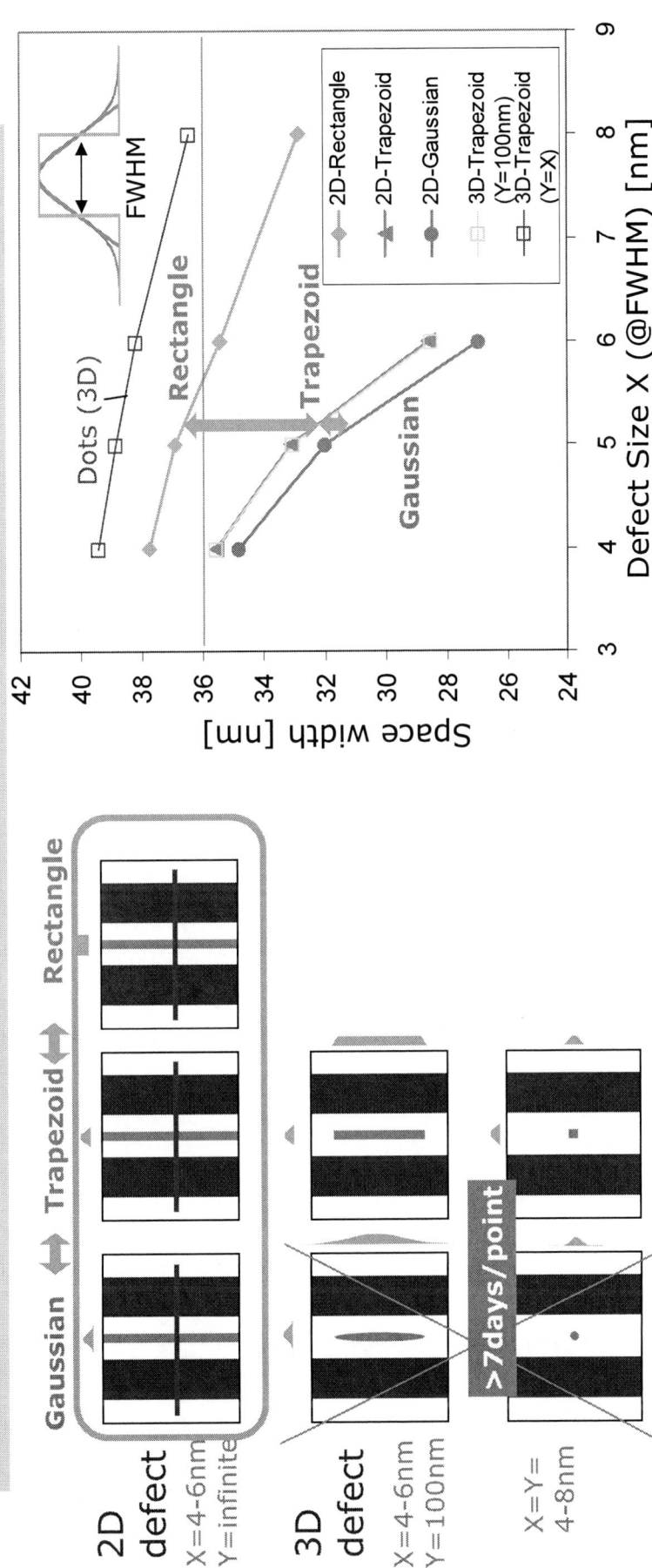

- As a dot defect requires long calculation time, we focused on line defects first (2D).
- The result of trapezoid defect follows the trend of Gaussian defect, and rectangular shape defect prints less.
- The result of elongated 3D trapezoid defect agrees completely with 2D trapezoid defect.

- Critical Size for such dot defect (trapezoidal) is around 8nm (vs. 10nm for an absorber opaque defect).

imec

Critical defect size for ML defect (XZ-2D)

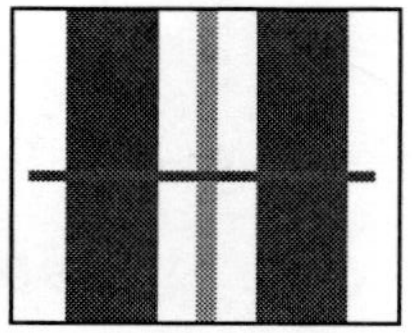

Note:
Sizes X,Y wafer level
Height z mask level

X=0-8nm
Y=infinite

Z=1-8nm(at mask)

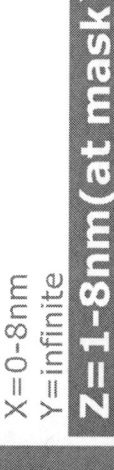

Rectangle

~3.5nm

defect height
1
2
3nm
4
5
6
7
8

Space Width [nm]
Defect Size [nm]

~3.5nm

defect height
1
2
3nm
4
5
6
7
8

Space Width [nm]
Defect Size [nm]

Trapezoid

~3nm

defect height
1
2
3
4nm
5
6
7
8

Space Width [nm]
Defect Size [nm]

~3nm

defect height
1
2
3
4nm
5
6
7
8

Space Width [nm]
Defect Size [nm]

Bump

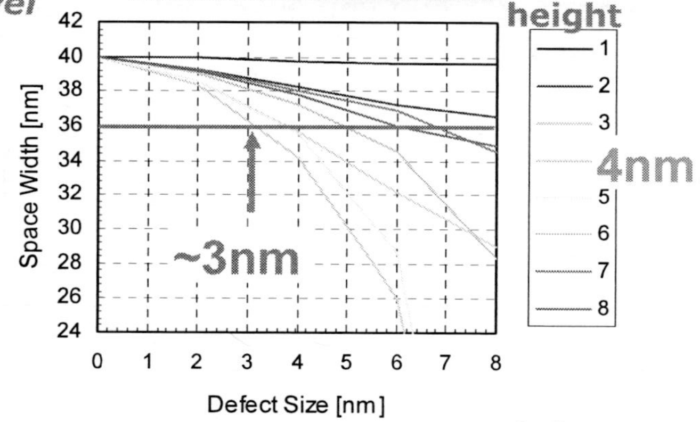

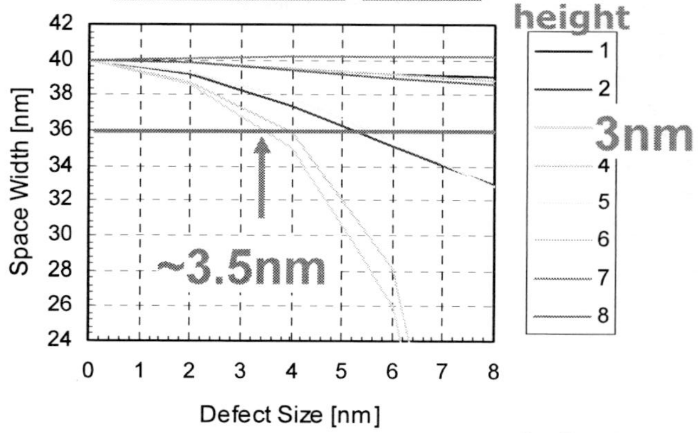

Pit

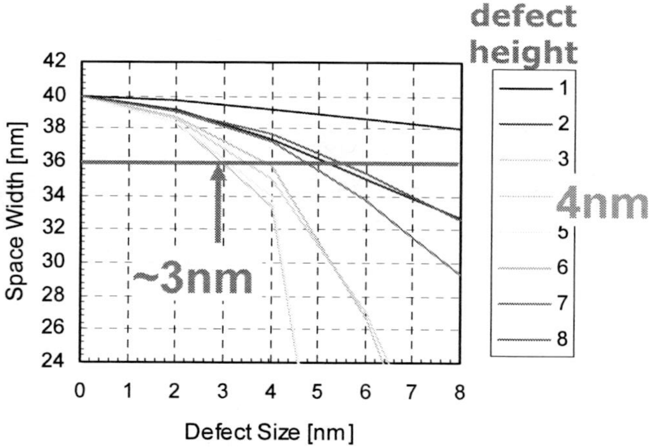

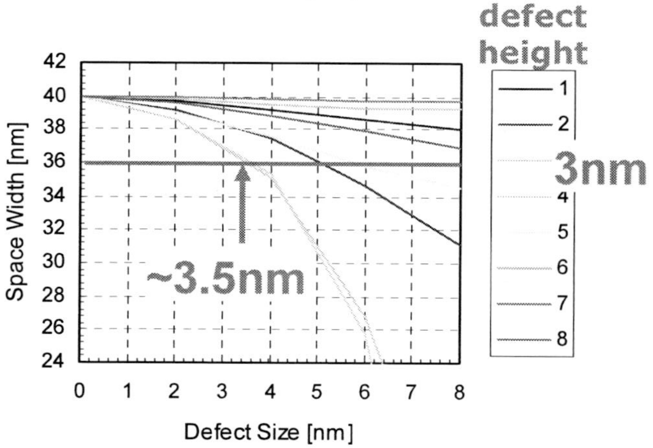

- At worst case of defect height for each type, the critical sizes are about the same; 3~3.5nm for all defects

imec

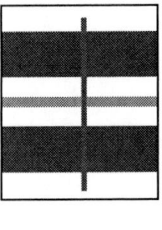

Critical defect size vs. defect height

Note:
Sizes X,Y wafer level
Height Z mask level

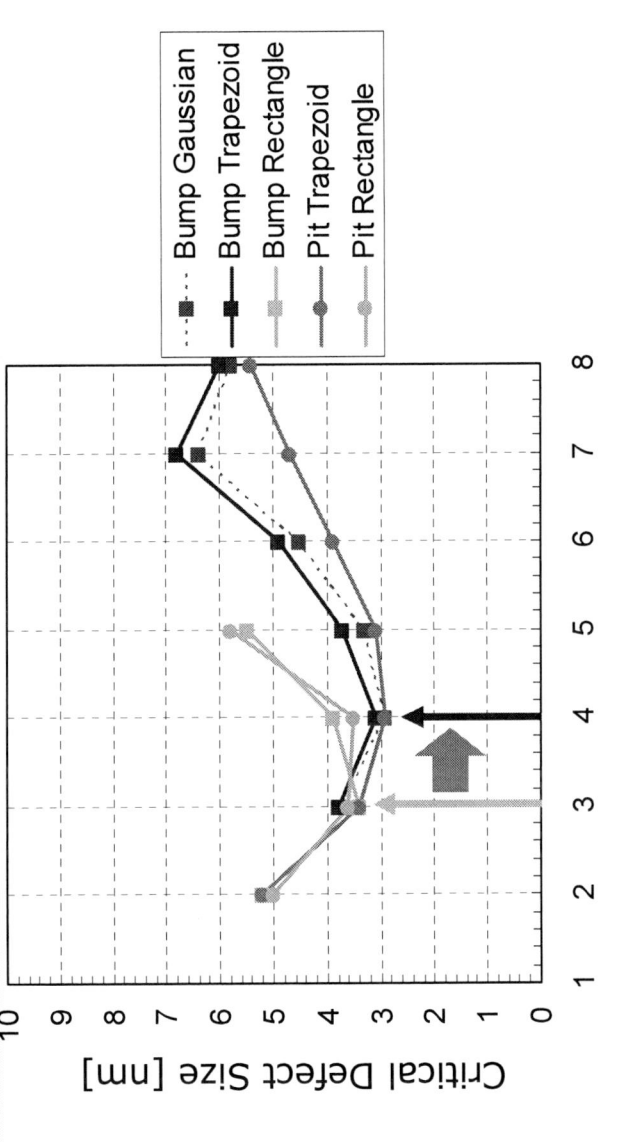

Legend:
- Bump Gaussian
- Bump Trapezoid
- Bump Rectangle
- Pit Trapezoid
- Pit Rectangle

Y-axis: Critical Defect Size [nm]
X-axis: Top Height (= Defect Height) [nm]

- Critical defect size is determined by the top height (= defect height)
- Defect shape also determines the printability
  - Critical defect sizes at the worst case of defect height are almost same ; 3~3.5nm but the tolerance for the height variation is different
  - Trapezoid pits are slightly more printable than trapezoid bumps taking into account of the height variation
- The worst position of the top height depends on the cross section shape

imec

Outline

- Introduction

- Further simulation study
  - Absorber defects
    - Simulation update: role of defect location
  - ML defects
    - Simulation update:
      - What is worse: Pits or bumps ?
      - Impact of cross-sectional shape

- **Experimental printability study on EUV-ADT**
  - Analysis of absorber defect printability and correlation to simulation (programmed approach)
  - Possible evidence of ML defects (non-programmed) and analysis by simulation

Programmed absorber defect on Reticle

on Reticle (4X)

Design

Designed Defect Size (wafer level)

absorber

Defect size =
8 to 32 (2nm step),
36 and 40nm

Pinspots
(absorber opaque, at center)

16nm   18nm   20nm   22nm

Pinholes
(absorber clear, at center)

22nm   24nm   26nm   28nm

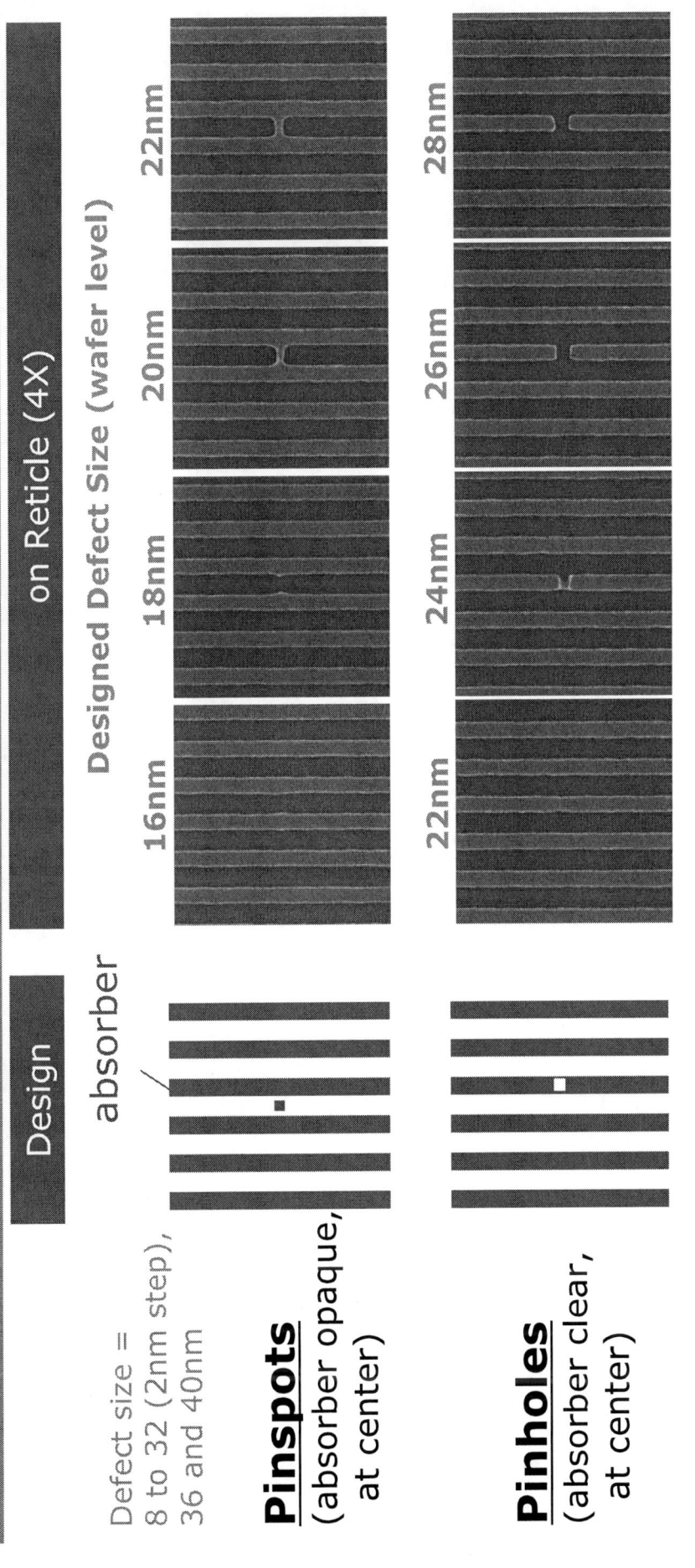

- Programmed pinspots showed-up on the reticle as extentions or even bridging: 16nm programmed size does not show-up yet.
- Same for pinholes : 22nm programmed size does not show-up yet.

imec

Programmed absorber defect on Reticle

Design | **on Reticle (4X)**

Designed Defect Size (wafer level)

Defect size =
8 to 32 (2nm step),
36 and 40nm

Extensions
(absorber opaque, at edge)

10nm 18nm 20nm 22nm

Ratbites
(absorber clear, at edge)

10nm 24nm 26nm 28nm

- Edge defects have better fidelity on the reticle.
- Note that simulated critical size is around 10nm, resp. 17nm.
- Further analysis of experimental printing therefore focused on edge defects.

imec

Measurement of defect size on Reticle

Square shape defect for simulation

Designed shape also square

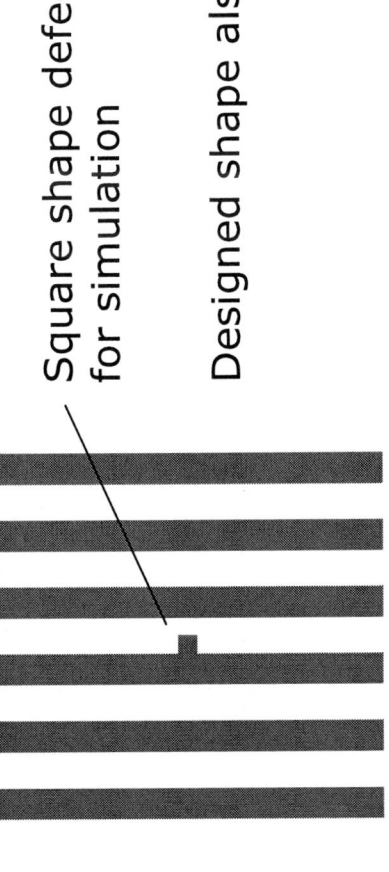

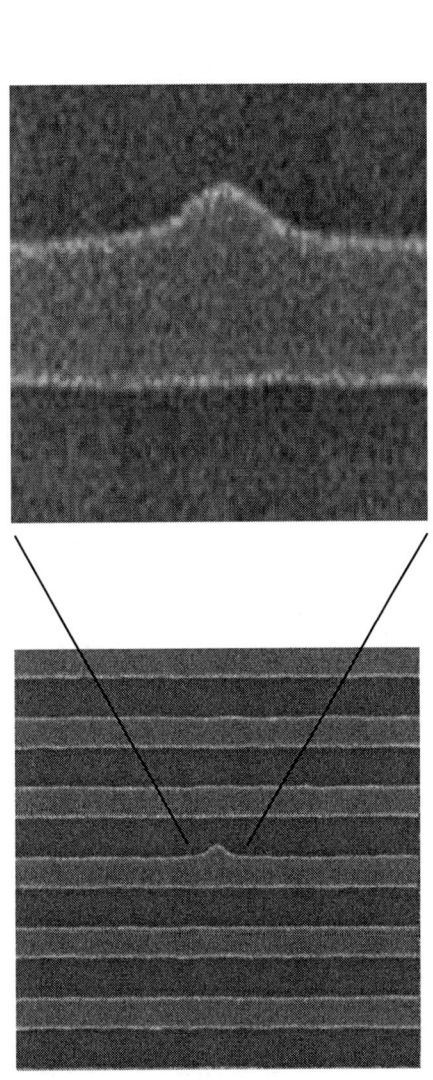

$X = height$
$Y = FWHM$

$$Defect\ Size = \sqrt{X \times Y}$$

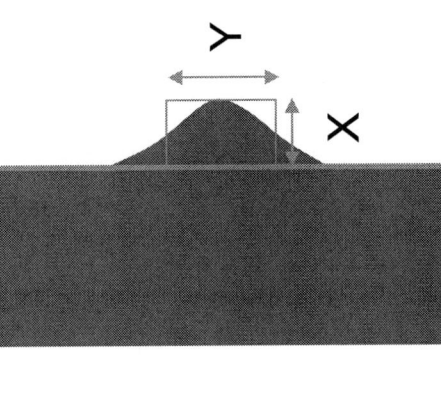

- Note: Manually measured on the SEM pictures provided by mask shop

imec

Actual defect size on Reticle

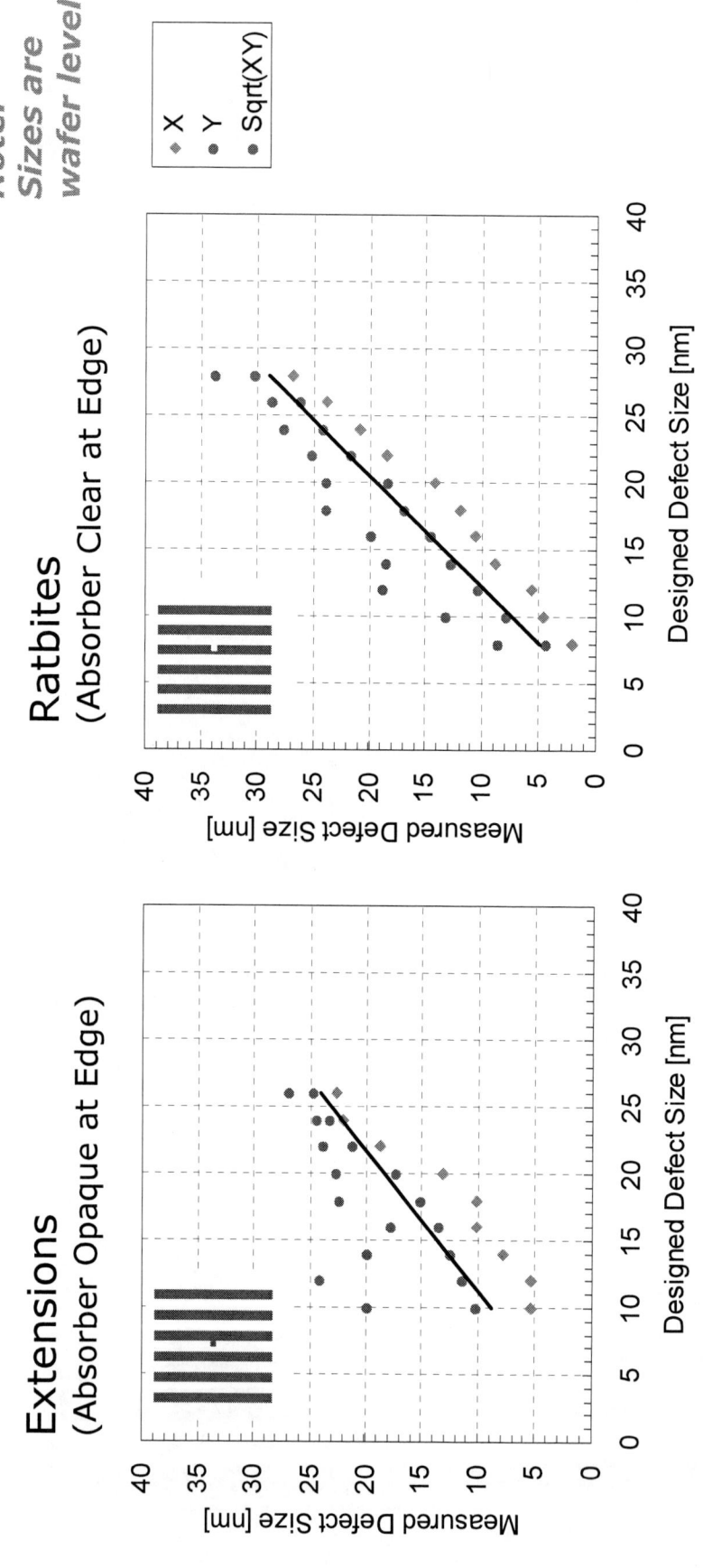

Note:
Sizes are wafer level

Extensions
(Absorber Opaque at Edge)

Ratbites
(Absorber Clear at Edge)

- Measured defect size in X was typically smaller than Y.
- Sqrt(X.Y) follows the design size reasonably well.

imec

Programmed absorber defect on Reticle and **on Wafer**

Extensions
(absorber opaque, at edge)

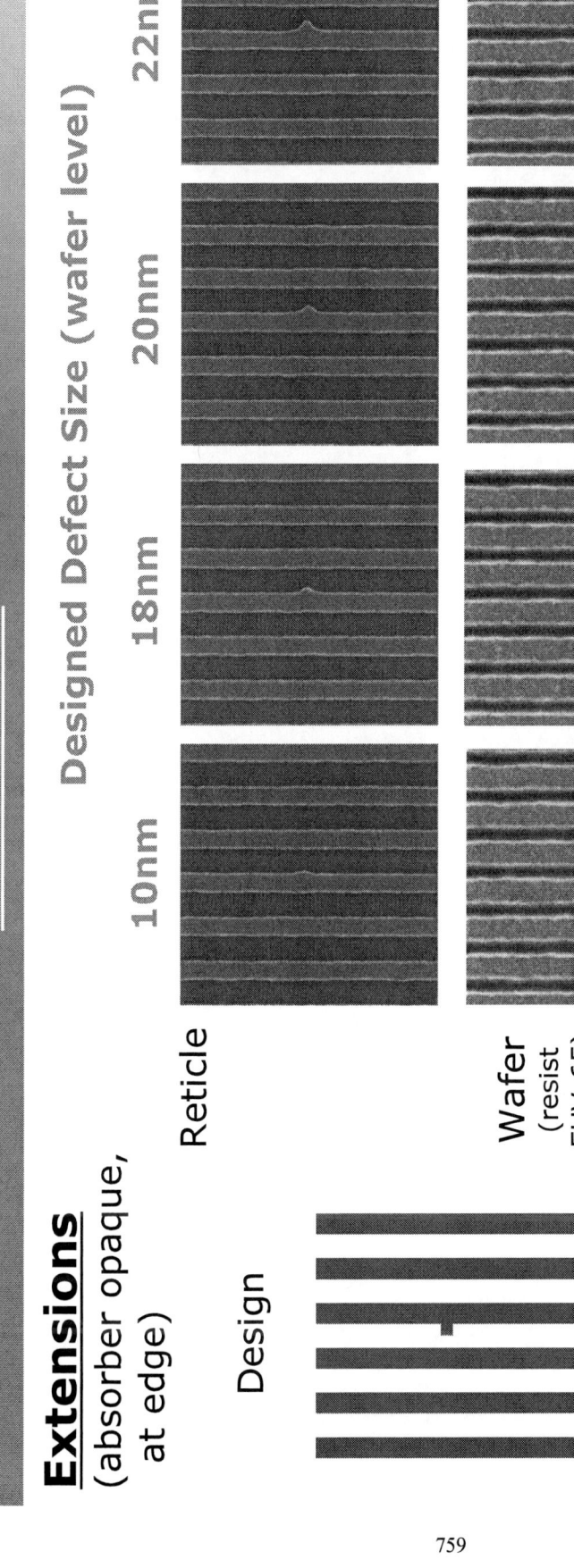

Designed Defect Size (wafer level)

22nm 20nm 18nm 10nm

Reticle

Wafer
(resist EUV-65)

Design

Defect size =
8 to 32 (2nm step),
36 and 40nm

imec

Experimental results absorber defects

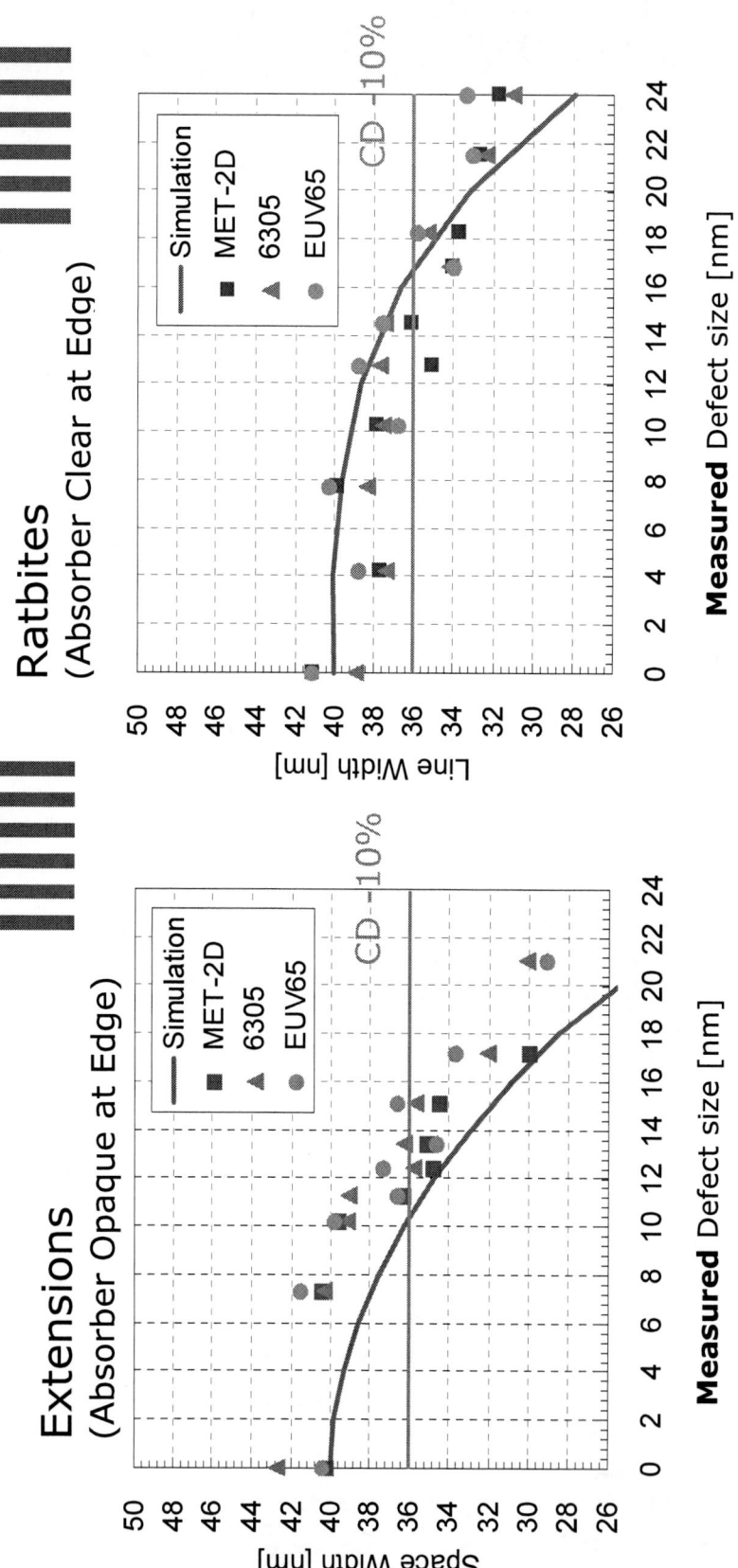

Ratbites
(Absorber Clear at Edge)

Extensions
(Absorber Opaque at Edge)

- Extensions: simulation overestimates printability
  - Critical defect size ~14nm experimentally vs. ~10nm by simulation
- Ratbites: simulation slightly underestimates for small defects, overestimates for large ones
  - Critical defect size ~15nm experimentally vs. ~17nm by simulation

imec

Outline

- Introduction

- Further simulation study
  - Absorber defects
    - Simulation update: role of defect location
  - ML defects
    - Simulation update:
      - What is worse: Pits or bumps ?
      - Impact of cross-sectional shape

- Experimental printability study on EUV-ADT
  - Analysis of absorber defect printability and correlation to simulation (programmed approach)
  - Possible evidence of ML defects (non-programmed) and analysis by simulation

imec

Detection of mask defects using wafer inspection

- Subdie (A) consists of 40nm L/S, with 3 variations:
  - A: without programmed defects
  - A': with opaque absorber defects
  - A*: intended for substrate defects
- Evaluation of A' on wafer by KLA2800
  - Broadband DUV (260-380nm)
  - 0.28µm pixel size
  - Detection by cell-to-cell comparison
  - Very sensitive setting possible
    - It gives many detections (LER related)
  - Analysis was confined to repeating defects only (= reticle defects)

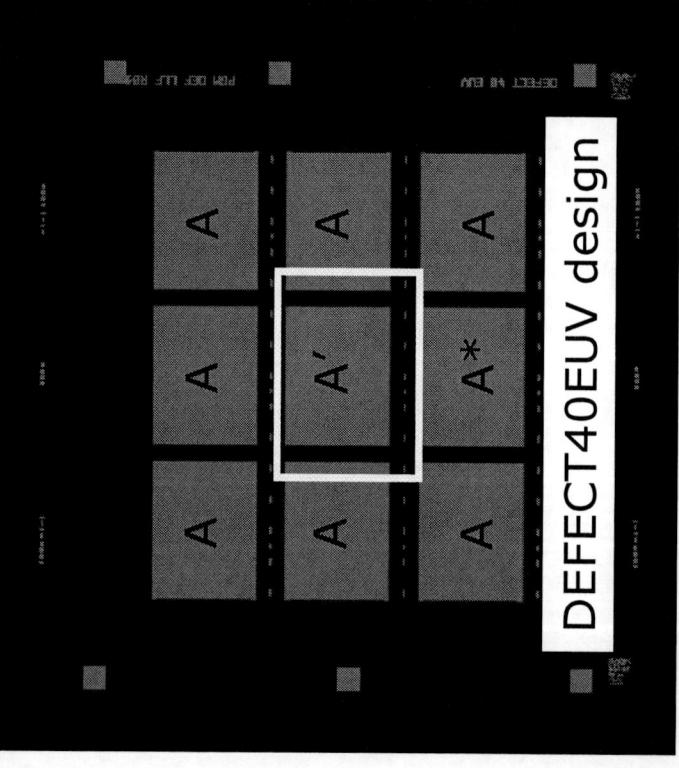

DEFECT40EUV design

9 blocks of programmed defects in A'

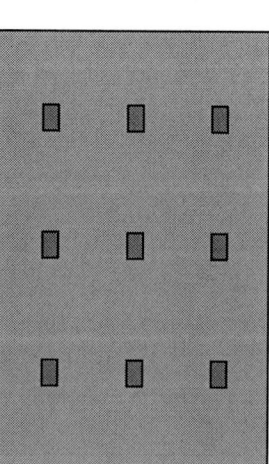

Detail of A'

imec

Detail of detected result (part of 1 block)

(design sizes at wafer level, nm)

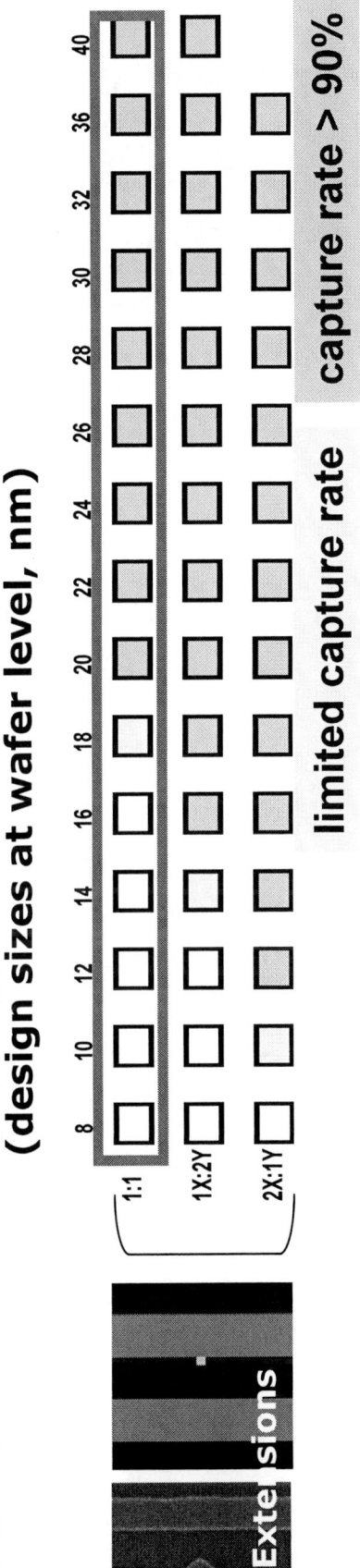

limited capture rate | capture rate > 90%

Extensions

Comparison to printing analysis by SEM

Causing >10% CD deviation | Causing visible deformation or bridging

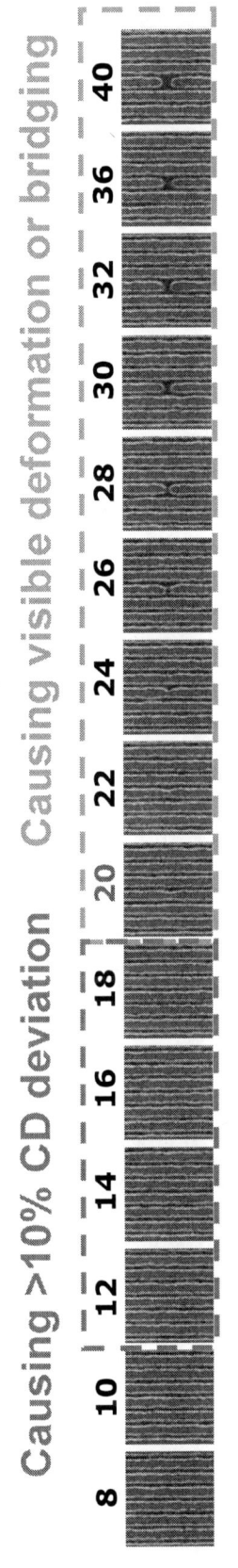

This illustrates the sensitivity of the wafer inspection technique for defects >~ 60nm on mask *(designed 18nm 1:1 found as ~15nm)*.

imec

Wafer inspection result (1 die)

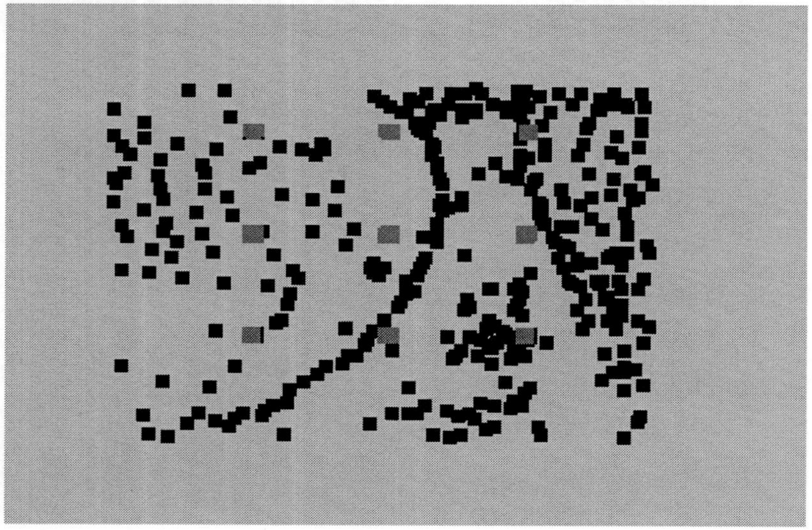

- **Programmed** defects detected
- Detection of **other** repeaters, considered as natural defects, consisting of
  - Particles
  - ML defects

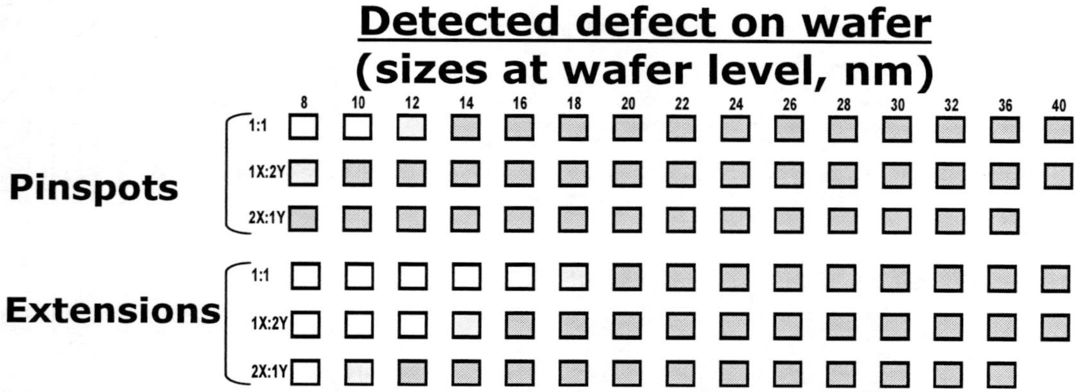

?

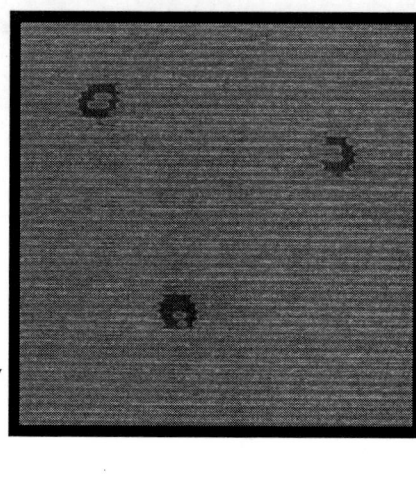

Detected defect on wafer (sizes at wafer level, nm)

	8	10	12	14	16	18	20	22	24	26	28	30	32	36	40
Pinspots 1:1															
Pinspots 1X:2Y															
Pinspots 2X:1Y															
Extensions 1:1															
Extensions 1X:2Y															
Extensions 2X:1Y															

imec

Isolated ML defect w/o absorber (= in a clear area)
Rectangular Bump Type

X-Z 2D situation ;
infinite line ML defect

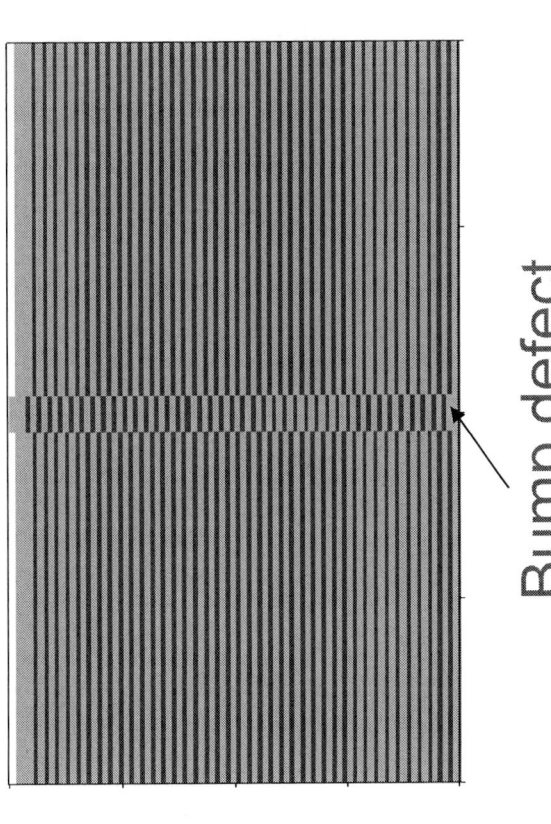

Bump defect

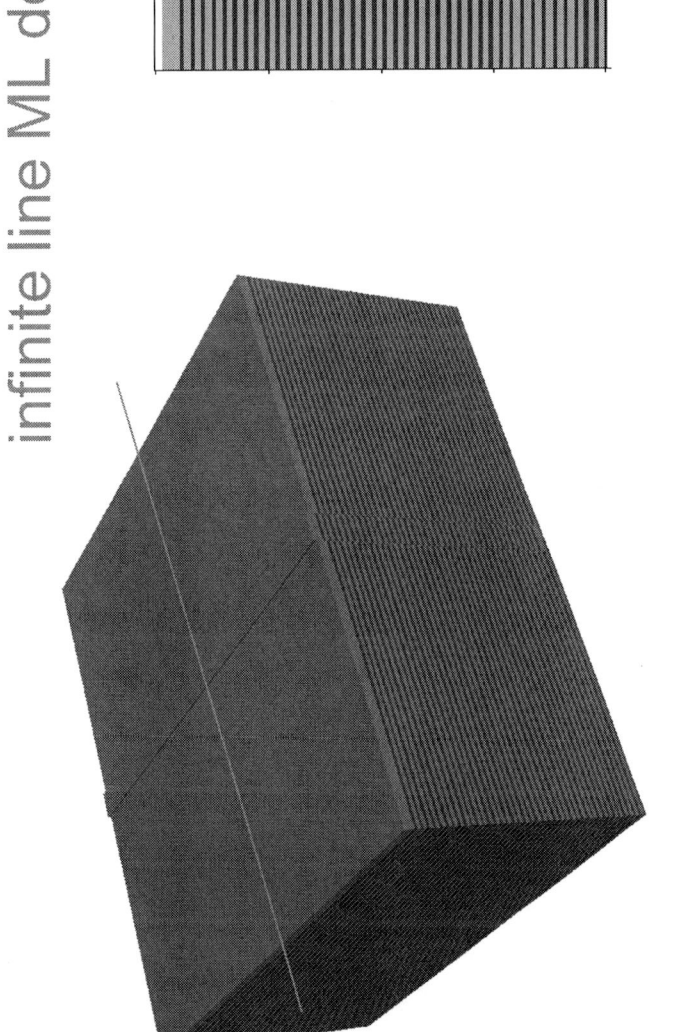

- When do isolated ML defects print as standalone features ?

imec

2D simulation

Isolated ML defect in a clear area
Rectangular Bump Type

Width (FWHM) = 4, 8, 16, 32, 64nm (at wafer)

Height = 3.36nm (at mask) ~ λ/4

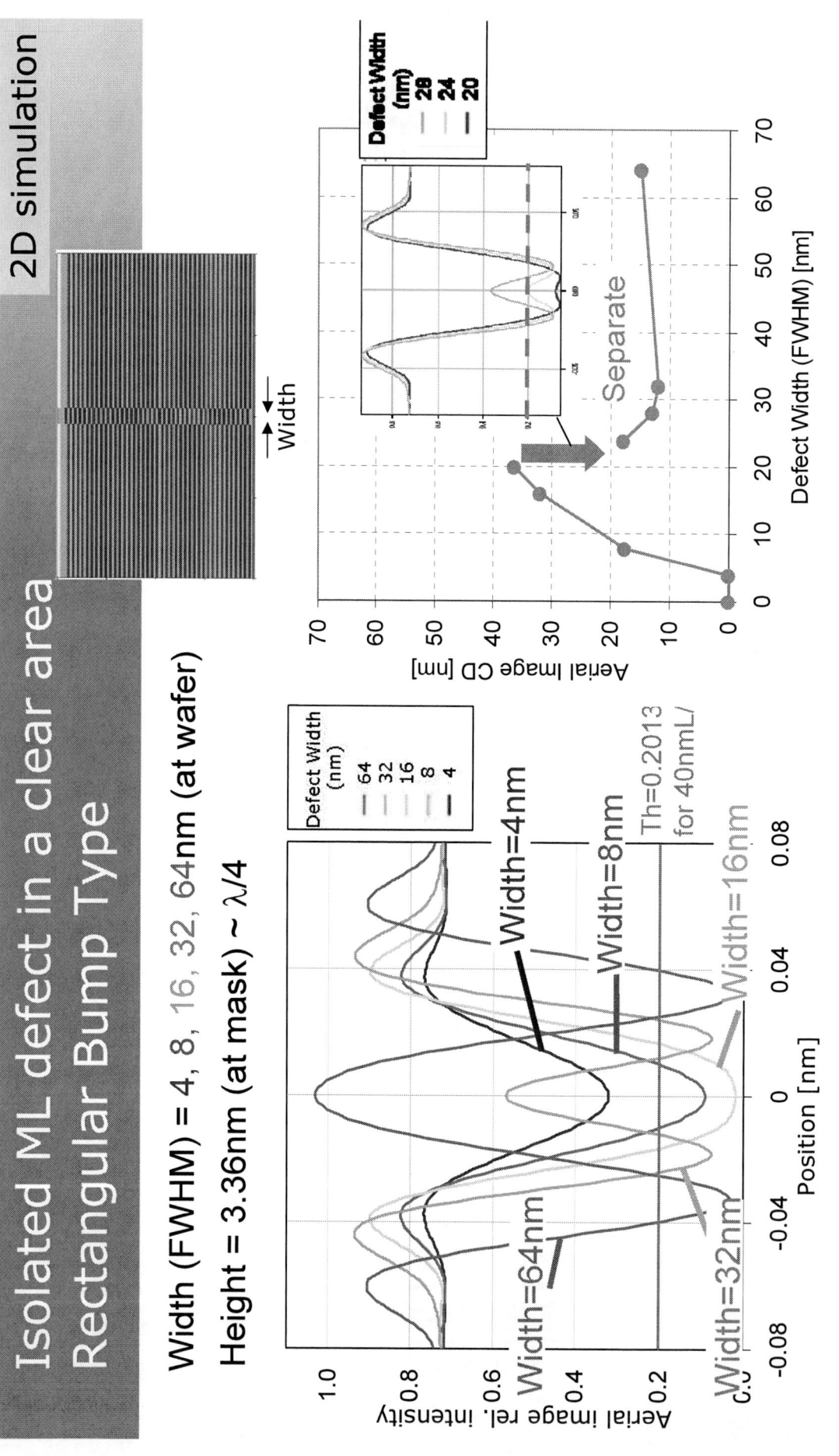

- Phase edge prints as 2 separate lines from a defect size ~24nm.
- This reminds us of CPL in 193nm lithography.

imec

ML defect (3D situation, large defect)

bump defect with rectangular cross-section
(on substrate and no smoothing)
height= 4nm (at mask) and size X=240nm, Y=160nm

Aerial Image

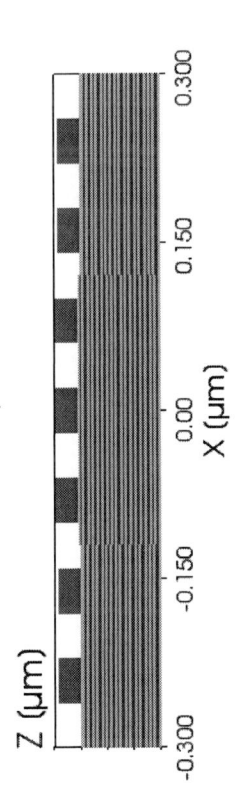

- Simulation predicts that large ML defects cause a ring-shaped defect.
- Such natural defects were found during the experimental verification.

imec

Conclusions

- ## Absorber type defect
  - The dependence of the critical defect size and its location as a function of CD was shown.
  - Experimental defect printability study done on the ADT for 40nm L/S, with 3 different resists, largely confirms the simulation results.

- ## ML defect
  - Simulation summary:
    - Height (at top of ML) determines printability the most (phase nature $\lambda/4$).
    - Bumps and pits print about the same; cross-sectional shape is important
    - Critical defect size
      - for line-ML bumps or pits is 3~3.5nm at the worst case of the defect height (at top of ML).
      - Critical defect size for dot-type ML bumps of this height is about the same as an opaque absorber type defect (~8nm).
  - ML defect may print as a hollow feature from a certain defect size onwards.
  - Wafer inspection is a powerful technique to find natural defects on reticle.
  - Likely evidence for ML defect printability found on the ADT printed wafer.

Acknowledgements

- Jan Hermans, Bart Baudemprez, Gian Lorusso, E. Hendrickx (EUV-ADT tool team)

- Rudi De Ruyter (Mask design and tape-out)

- Gino Marcuccilli (KLA) and Dieter Van Den Heuvel (Wafer inspection on KLA2800)

- Synopsys for support on Solid-EUV

- On-site Hitachi team for support in the CD-SEM analysis work

- ASML, especially the on-site team

- Our partners in the Advanced Lithography Program

imec

aspire invent achieve

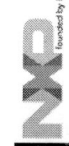

Back-up slides

aspire invent achieve

imec

Simulation conditions

- **Solid-EUV**
  - 3D calculations, except where shown
  - First used the FDTD algorithm, changed to Waveguide (confirmed well-matching)
- **Aerial image only**
  - Printed size based on optimized threshold for correct CD in defect-less case
- **40nm L/S**
- **Mimic the ASML Alpha Demo Tool**
  - NA 0.25, 0.5 sigma, 6° AOI, except
    - λ = 13.6 nm, monochromatic
    - Flare = 0
- **Schott-Lithotec mask stack as published** (as example)
- **Situation without shadowing** (= vertical lines, center of slit)

SCHOTT-Lithotec

(source: U. Dersch @EUV Symposium, SanDiego 2005)

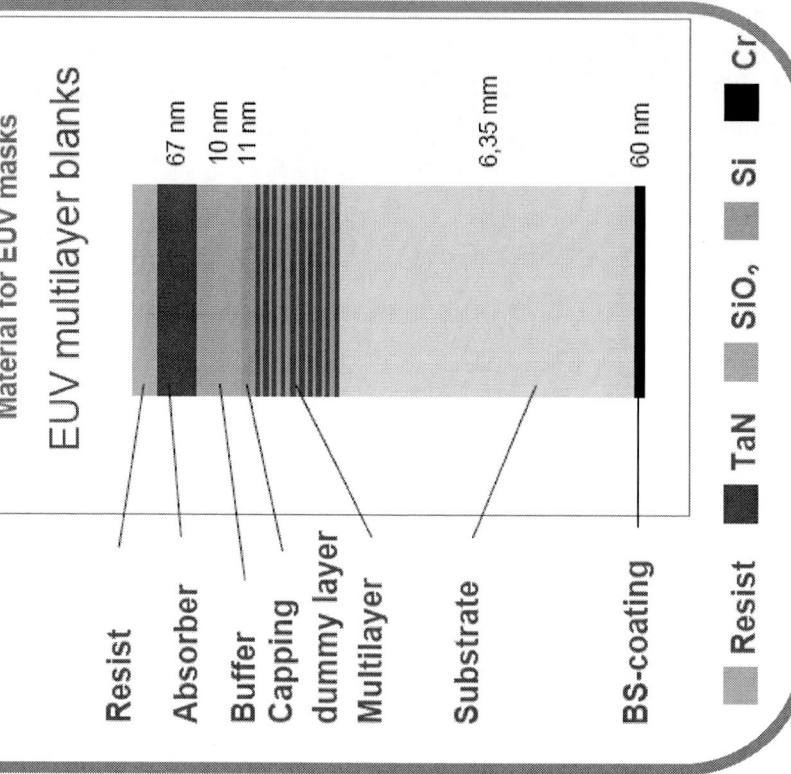

Future work

- Printability analysis for 30nm isolated line

- Experimental verification for ML defect
  - Further analysis of natural defects found
  - Reticles with programmed ML defects in progress

imec

A Practical Approach to EUV Reticle Inspection

Anna Tchikoulaeva, Matthias Zinke, Martin Schmidt,
Yunfei Deng, Tom Wallow, Uzodinma Okoroanyanwu,
Obert Wood, and Bruno La Fontaine
AMD
Hiro Mizuno
Toshiba
Chiew-Seng Koay and Karen Petrillo
IBM
Christian Holfeld and Jan Hendrik Peters
AMTC
Thomas Laursen, Youri van Dommelen, Brian Lee, Bill Pierson,
Robert Routh, and Kevin Cummings
ASML
Sumanth Kini
KLA-Tencor

EUV Reticle Inspection

- Direct reticle inspection
  - Only Die-to-Die capability available at this time
  - No reliable detection of multilayer defects
- Inspection of wafer prints
  - Detection of printable defects only
  - Recipe optimization required to achieve sensitivity comparable to that of reticle inspection
- Electrical detection: yield maps
  - Detection of electrically relevant defects only
  - Ultimately this is what matters for manufacturing
  - However this is a poor test of defectivity for process control: interesting only for the purpose of such studies

Experiment

- 45-nm technology test chip: 1st metal level mask with EUVL, 5 Wafers á 17 fields

- Processing done at AMD's Fab36 except for M1 level exposed using ASML Alpha Demo Tool

- Mask fabricated at AMTC using Schott-Lithotec (SL) blank. Typical defectivity for SL blanks from that period: 30 to 60 defects/cm² (80 nm PSL equivalent)

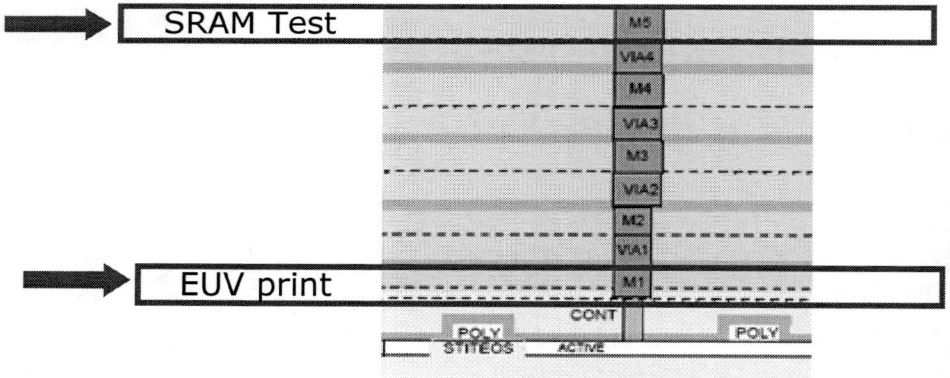

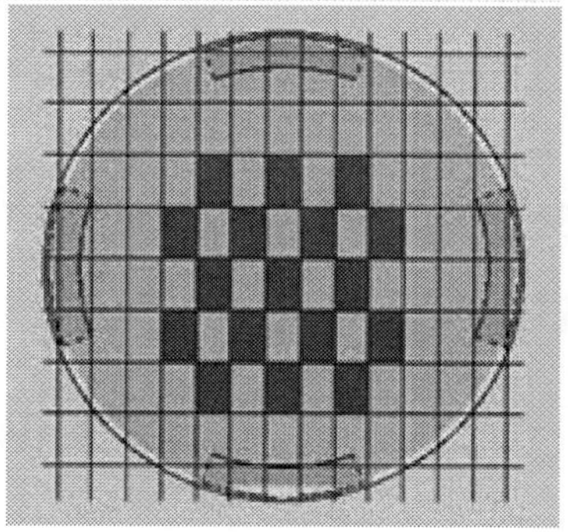

EUV Mask

SRAM Test

EUV print

EUV Print Layout

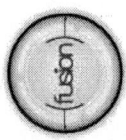

Reticle Inspection Areas

R1
(area 2)

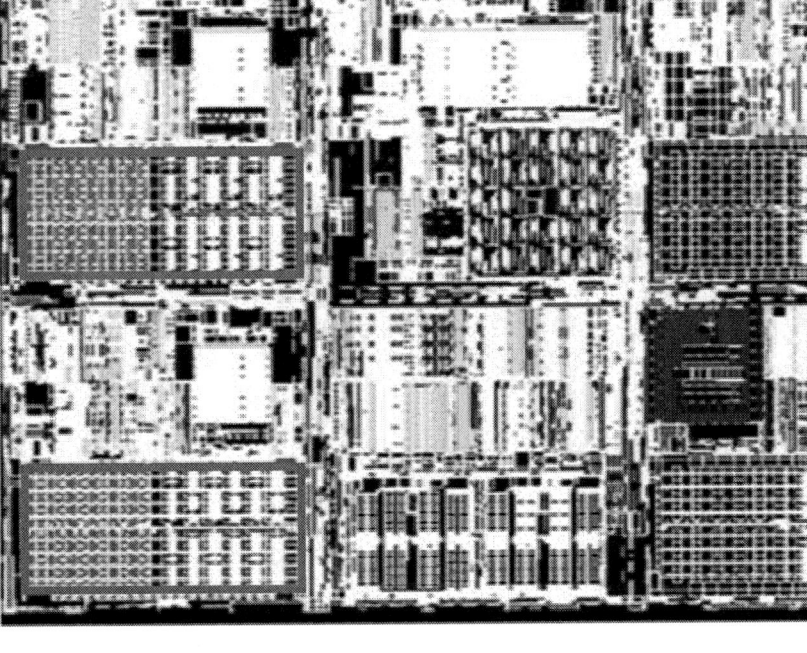

R1
(area 1)

KLA-Tencor 587

Die-to-Die Reflected Inspection

72nm Inspection Pixel Size

Total area inspected 13.5 cm^2

Reticle Inspection Results

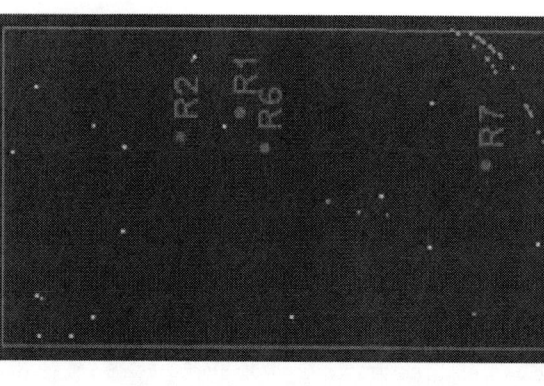

R1 (part 1)

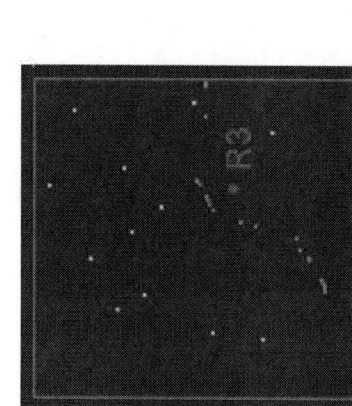

R1 (part 2)

Characteristic traces of contamination with unknown origin

No false defects during inspection

Small amount of other defects if contamination is neglected

AMD
The future is fusion

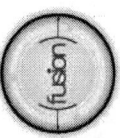

Reticle Inspection: Defect Distribution

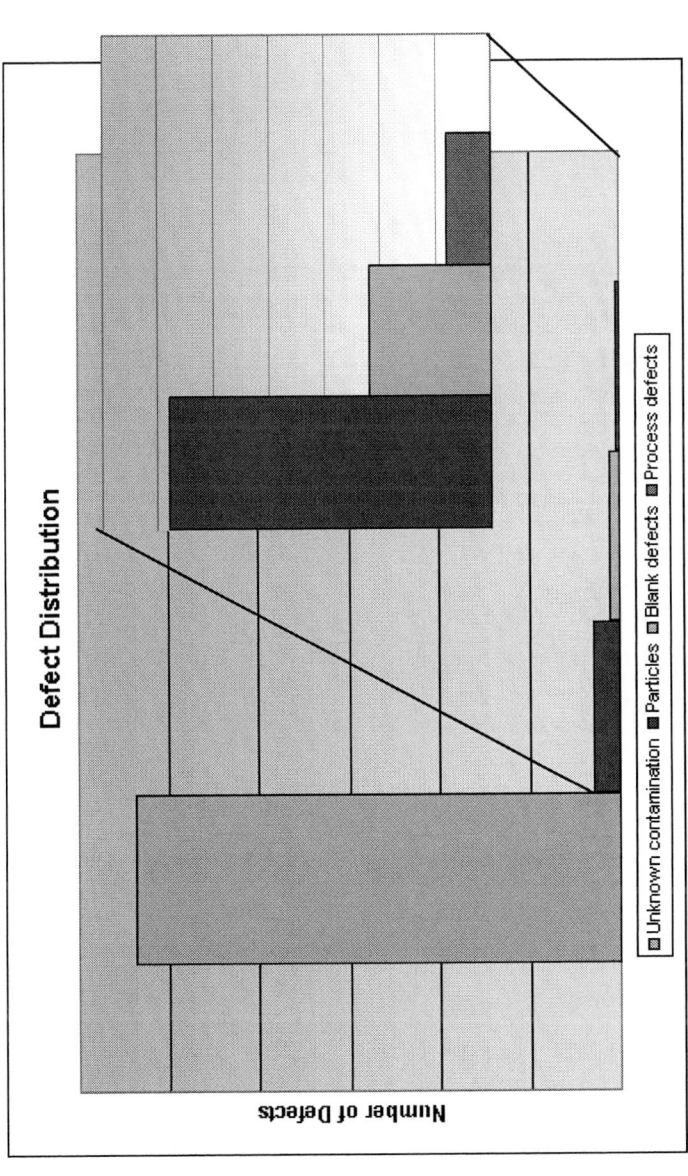

Mask defectivity is dominated by a characteristic contamination of unknown origin

All other defects account for 3.7 defects/cm^2 (mask blank: 30-60 defects/cm^2)

Noticeable number of blank defects: 11 (e.g. embedded particles)

Few pattern defects only (similar to standard masks)

The future is fusion

Wafer Inspection Areas

Inspected:
- Device wafers
- Bare silicon wafers printed at different dose and focus

R1 (area 2)

R2 (upper)

R2 (lower)

KLA-Tencor 2800

Cell-to-Cell Inspection

90nm Inspection Pixel Size

Total inspected area on wafer: 1.42 cm^2

R1 (area 1)

R3

Stacked Repeater Map
(Wafer inspection results)

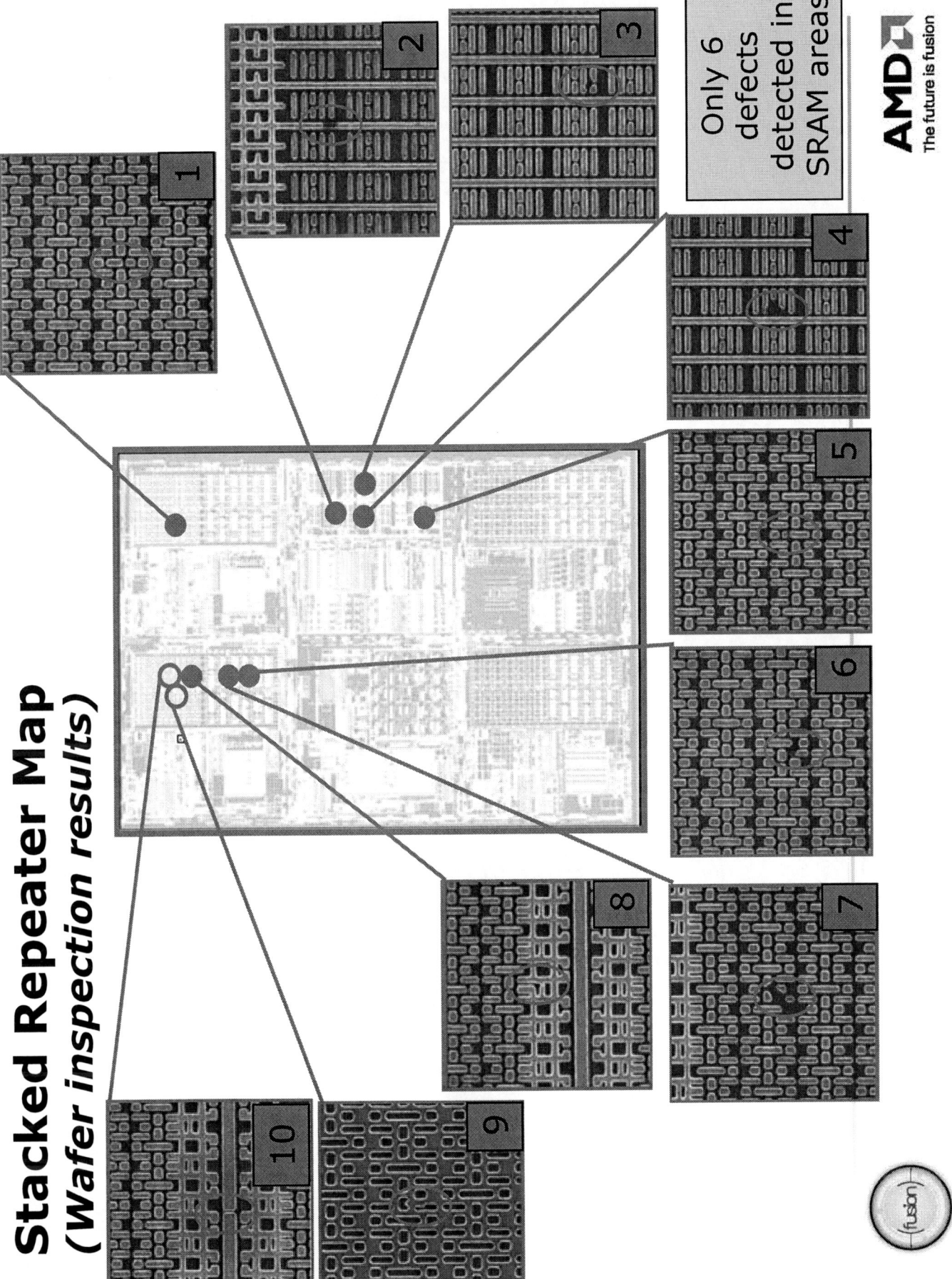

Only 6 defects detected in SRAM areas

AMD
The future is fusion

Comparing Wafer and Reticle Inspection
Defects found using both methods – SRAM Area 2

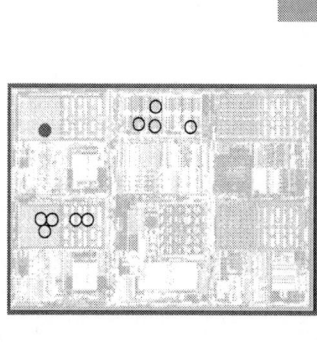

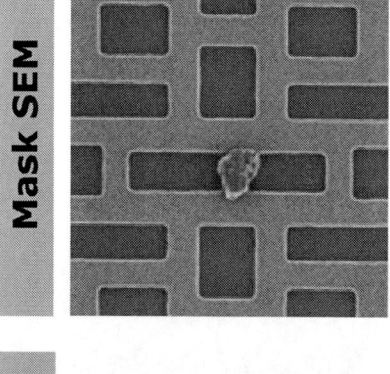

Mask SEM

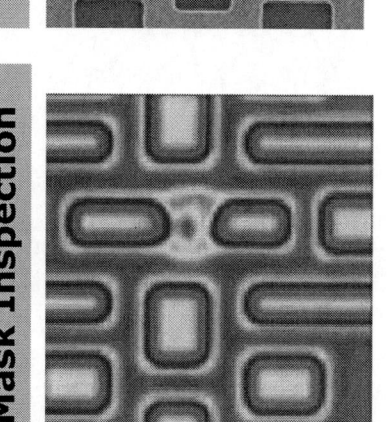

Mask Inspection

Wafer SEM

Repeater #1

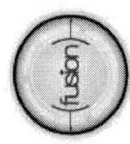

Comparing Wafer and Reticle Inspection
Defects found using both methods – SRAM Area 1

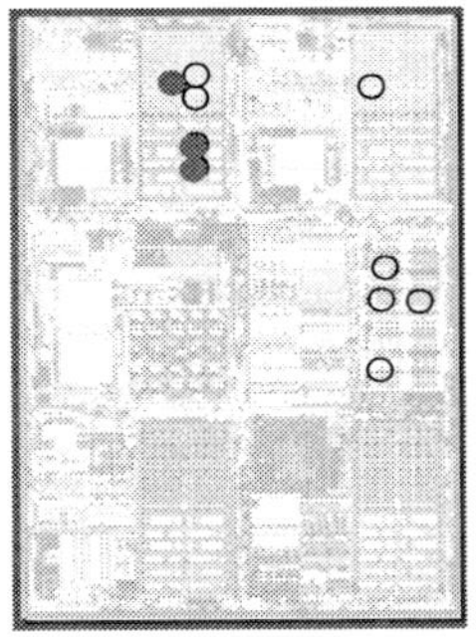

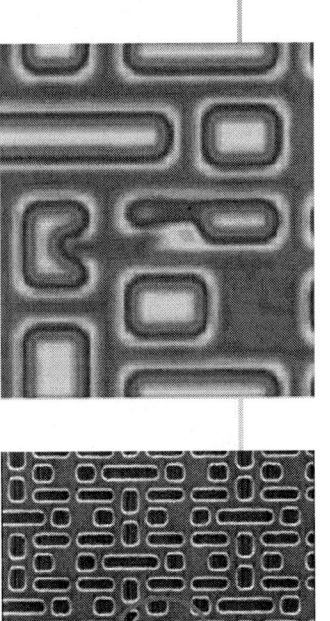

Wafer SEM | Mask Inspection | Mask SEM

Repeater #6

Repeater #7

Repeater #9

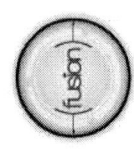

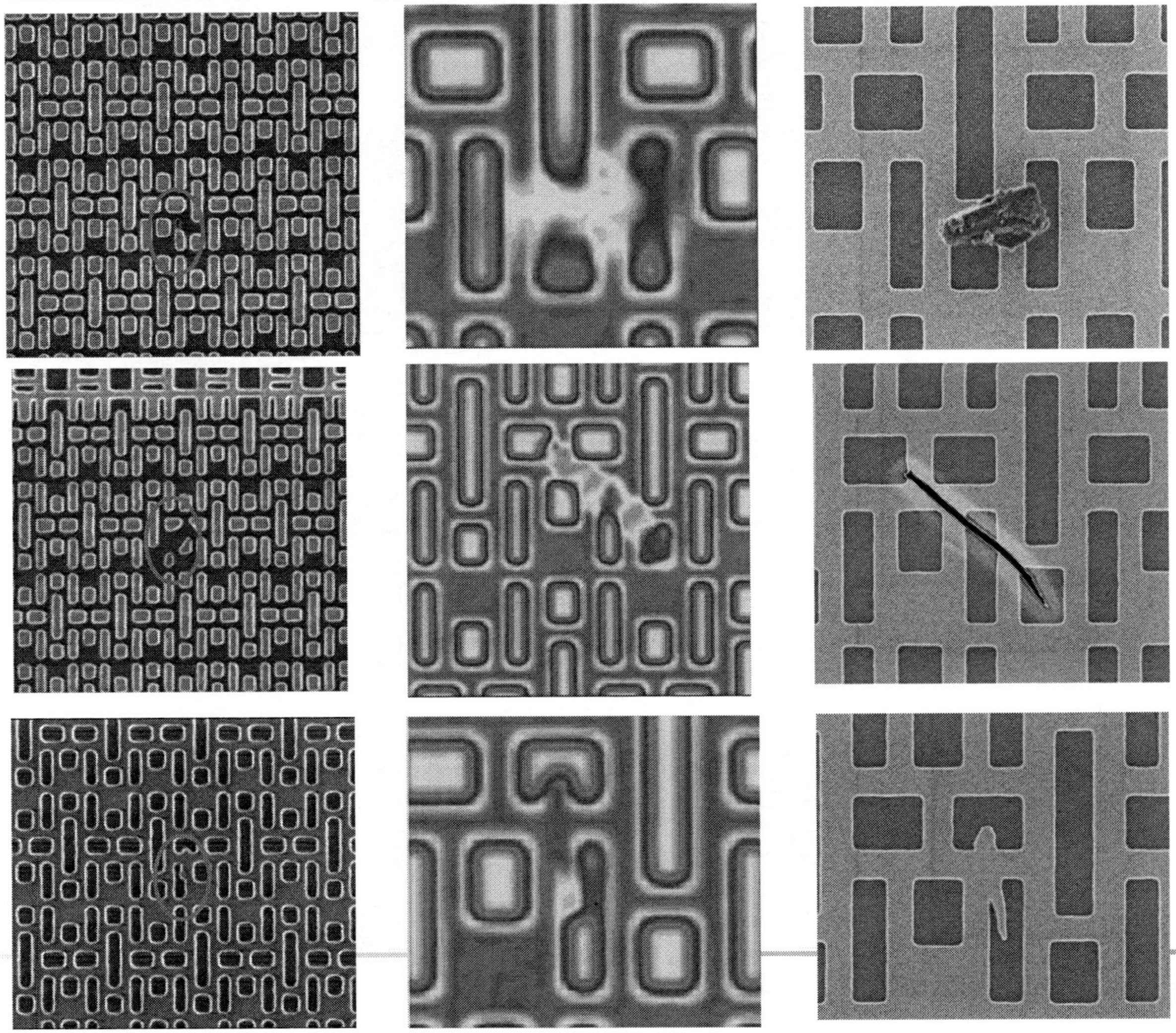

AMD
The future is fusion

fusion

Comparing Wafer and Reticle Inspection
Defects found using both methods – SRAM Area 1

Mask SEM

Mask Inspection

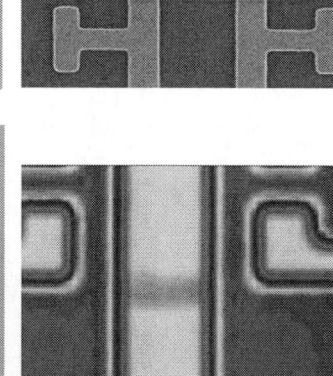

Wafer SEM

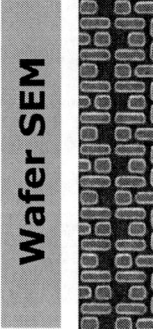

Repeater #10

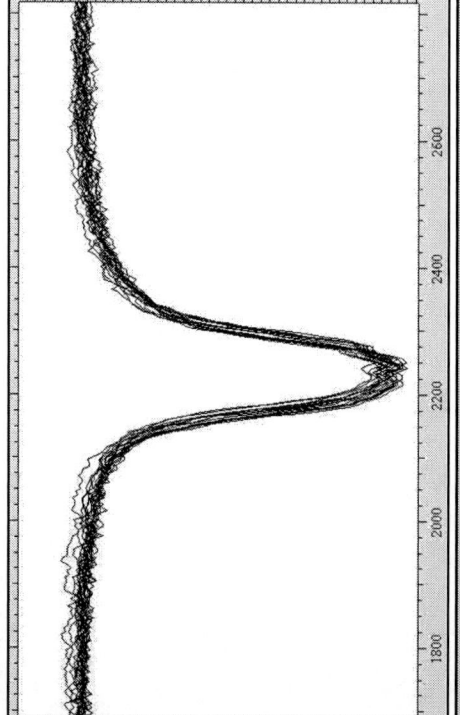

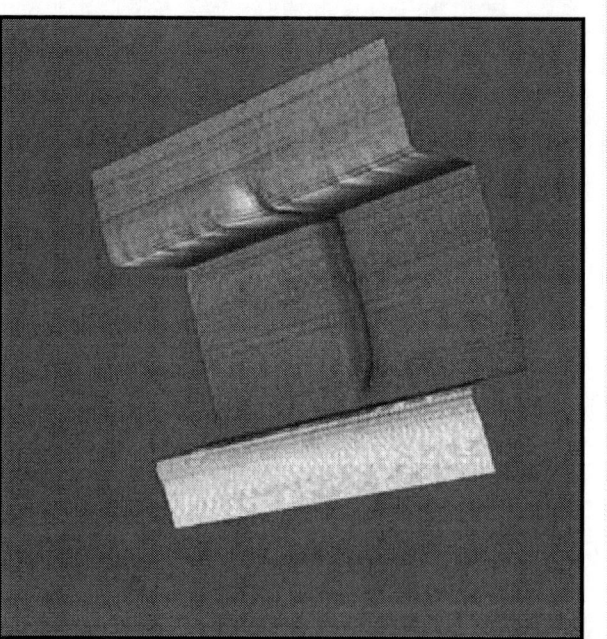

AFM scan of the defect area indicates a depth of ~20nm

Comparing Wafer and Reticle Inspection
Defect found only with wafer inspection – SRAM Area 1

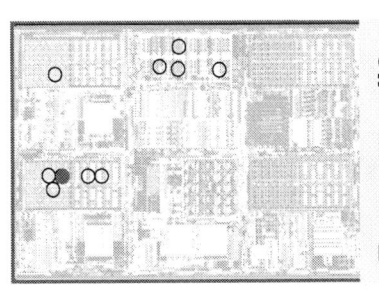

Repeater #8

Wafer SEM

Mask Inspection

not found

Mask SEM

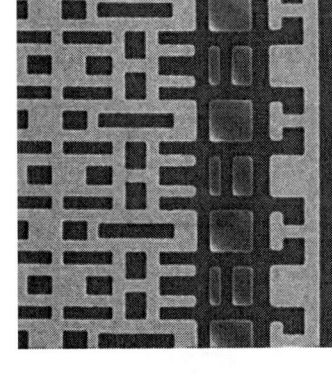

Comparing Wafer and Reticle Inspection
Defects in areas only inspected on wafers

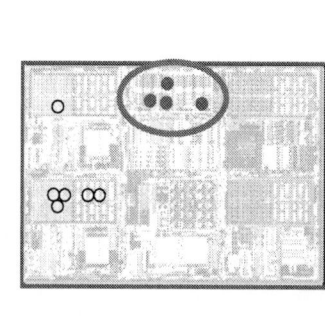

Repeater #2   Repeater #3   Repeater #4   Repeater #5

Wafer SEM

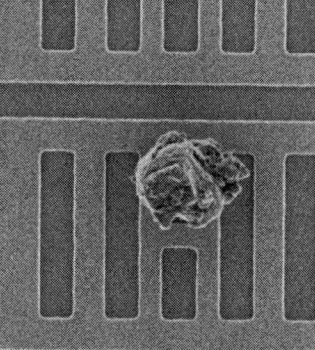

Mask SEM

AMD
The future is fusion

Comparing Reticle Defects to Wafer Prints

SRAM areas inspected on the mask

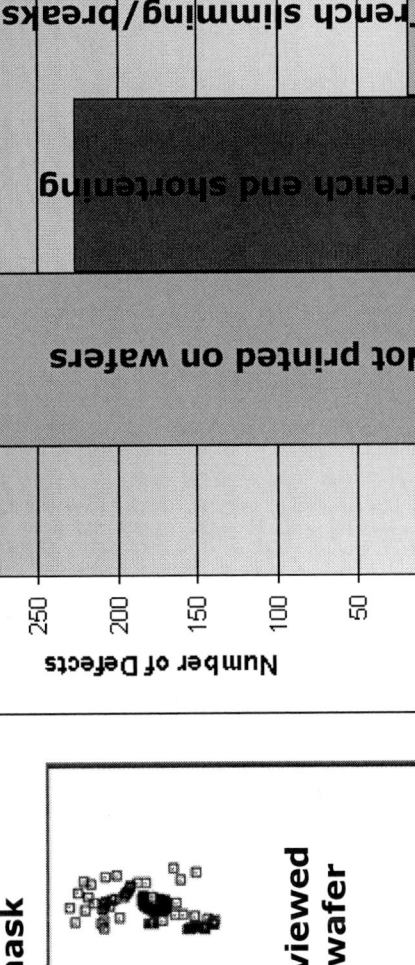

Wafer
Defect Distribution

Number of Defects

Sites reviewed on the wafer

☐ Not printable ■ Line end shortening ☐ Line slimming/break ■ Dark extention

Only 45% of reticle defects printable

Comparing Reticle Defects to Wafer Prints
Defects of Unknown Origin → Line End Shortening

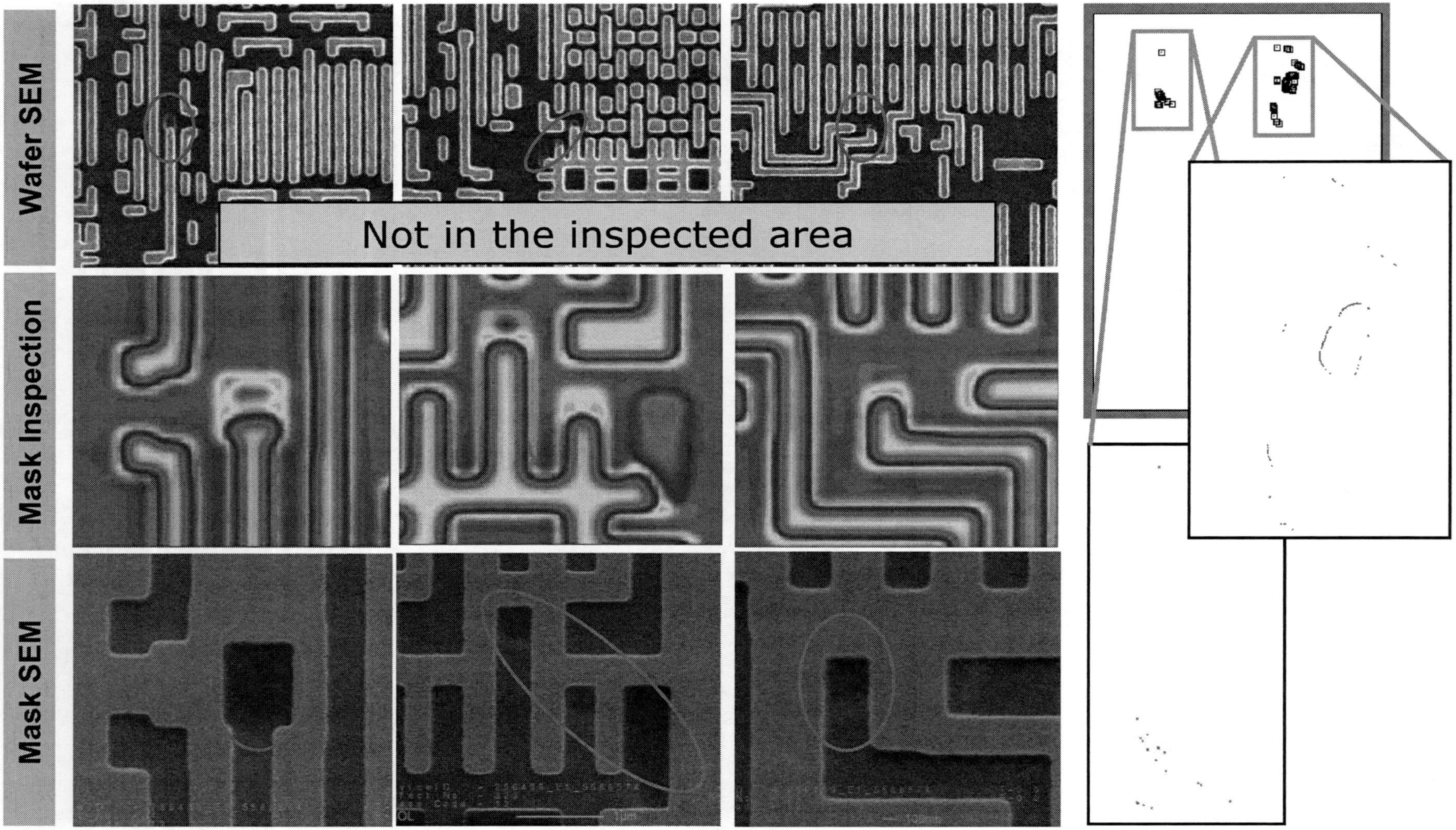

Not in the inspected area

Wafer SEM

Mask Inspection

Mask SEM

Comparing Reticle Defects to Wafer Prints
Dark Extension Defects

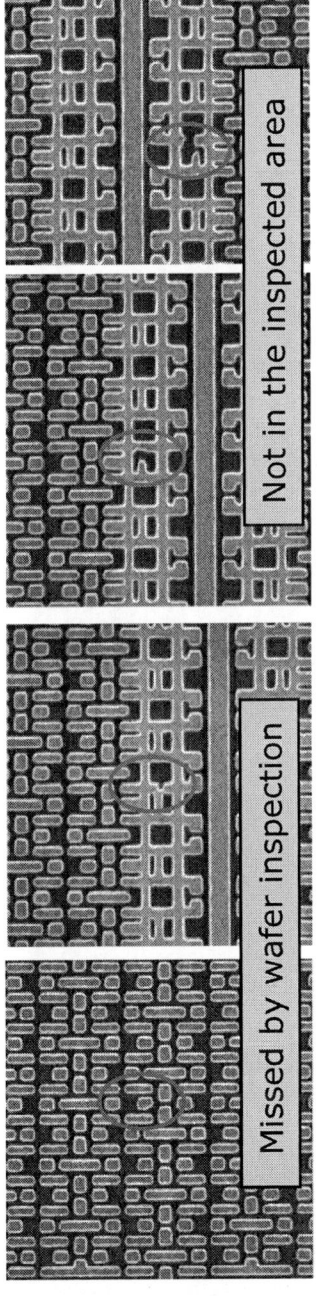

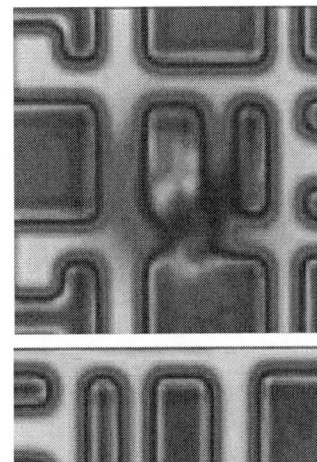

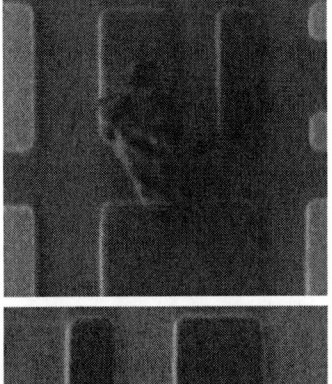

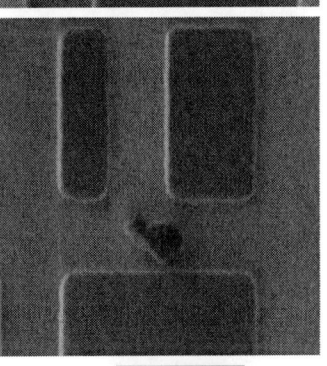

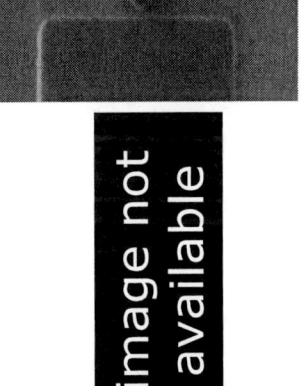

Missed by wafer inspection

Not in the inspected area

image not available

Wafer SEM

Mask Inspection

Mask SEM

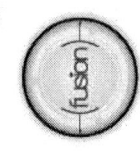

Electrical Inspection
SRAM Test Areas

R1
(area 1)

R1
(area 2)

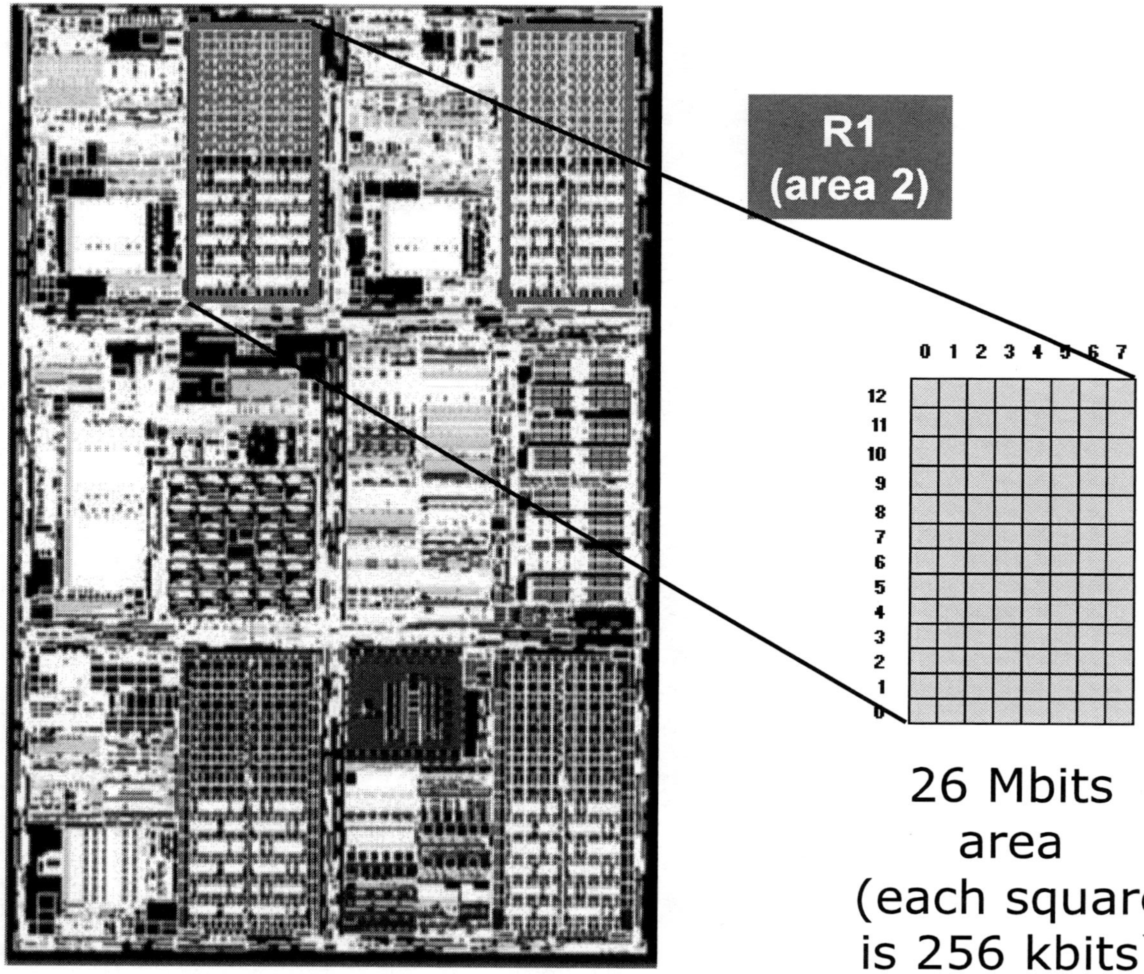

0 1 2 3 4 5 6 7

12
11
10
9
8
7
6
5
4
3
2
1
0

26 Mbits
area
(each square
is 256 kbits)

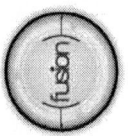

Normalized Yield Results

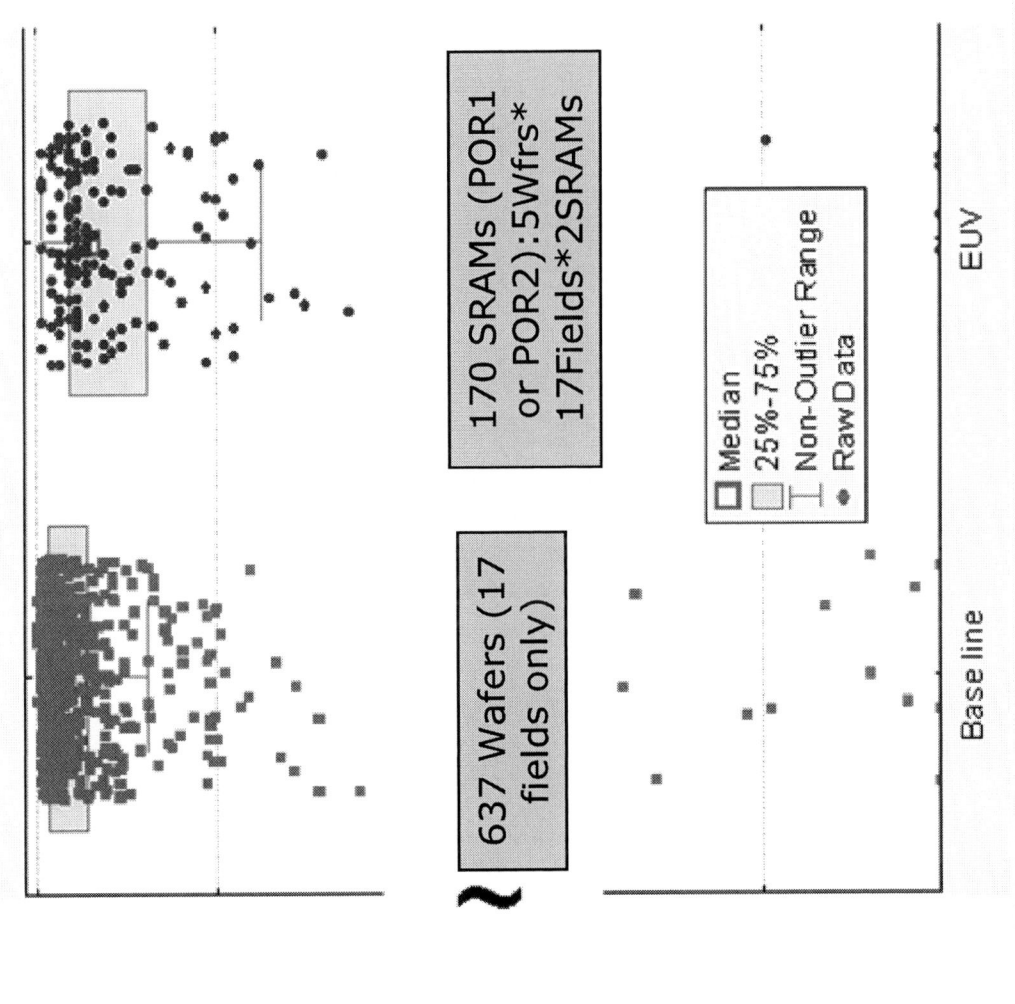

170 SRAMs (POR1 or POR2):5Wfrs* 17Fields*2SRAMs

637 Wafers (17 fields only)

Median
25%-75%
Non-Outlier Range
Raw Data

Base line

EUV

Repeater Mapping in the SRAM Area

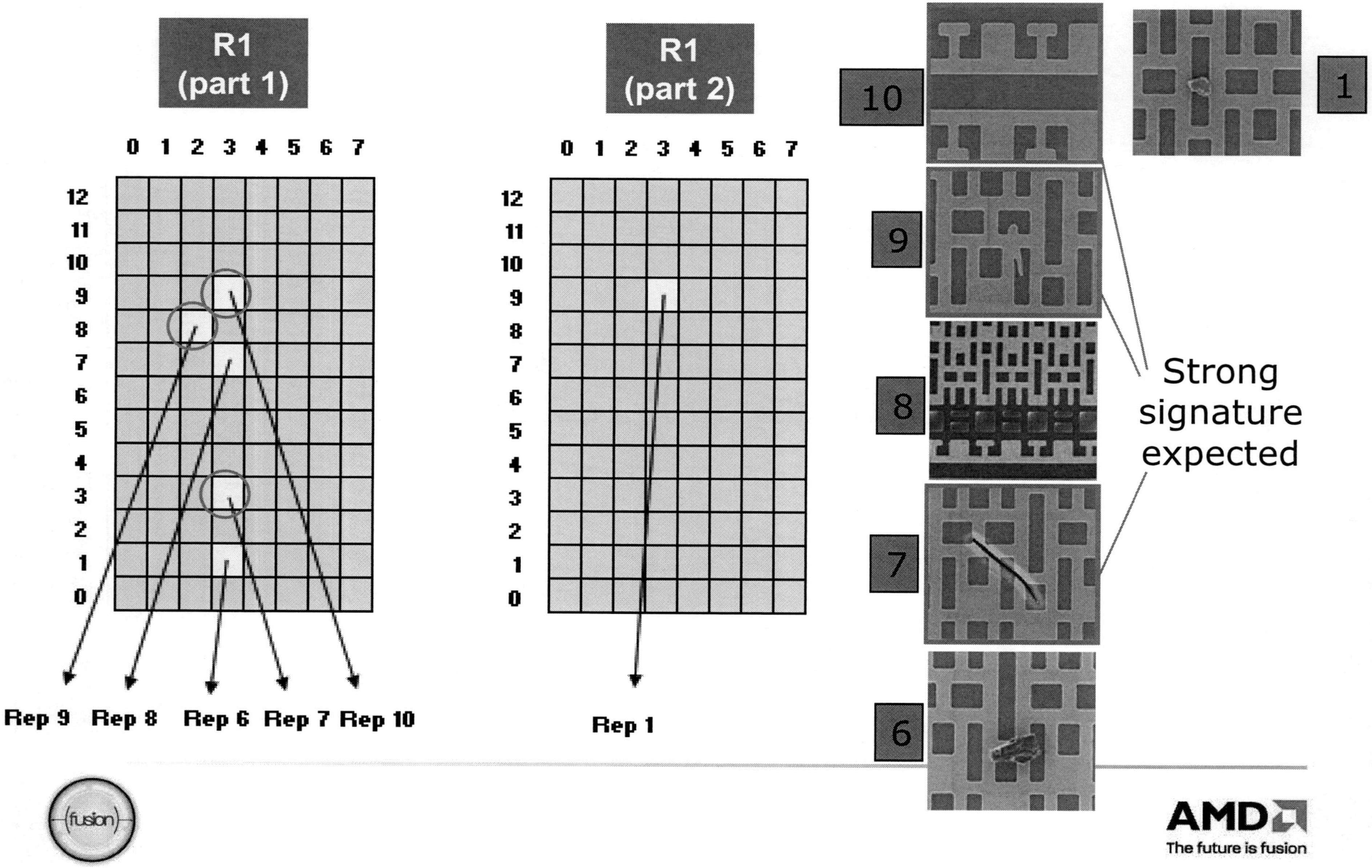

R1 (part 2)

R1 (part 1)

Strong signature expected

Rep 1

Rep 9   Rep 8   Rep 6   Rep 7   Rep 10

AMD
The future is fusion

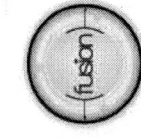

Repeater Mapping in the SRAM Area

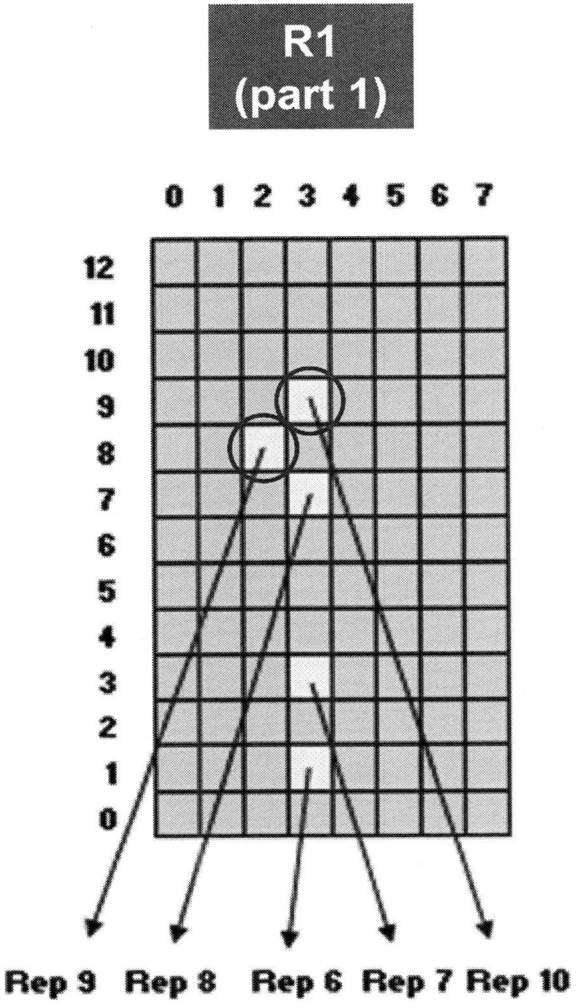

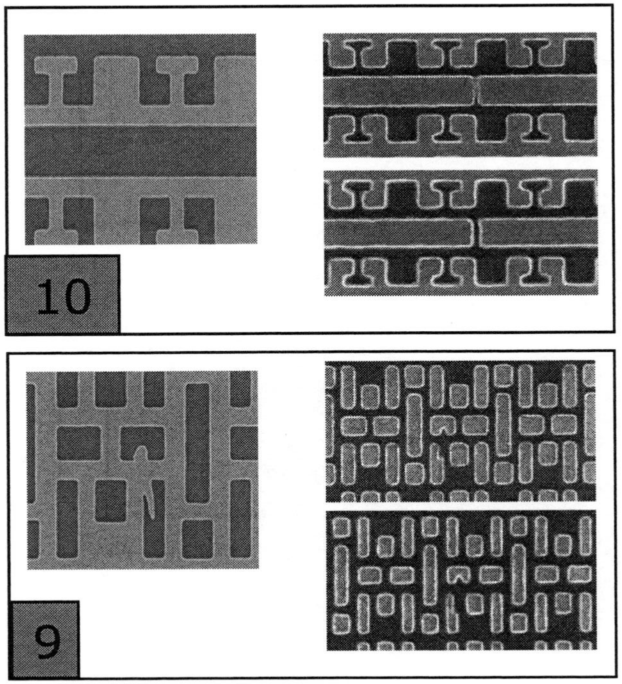

R1 (part 2)

R1 (part 1)

Rep 1

Rep 9   Rep 8   Rep 6   Rep 7   Rep 10

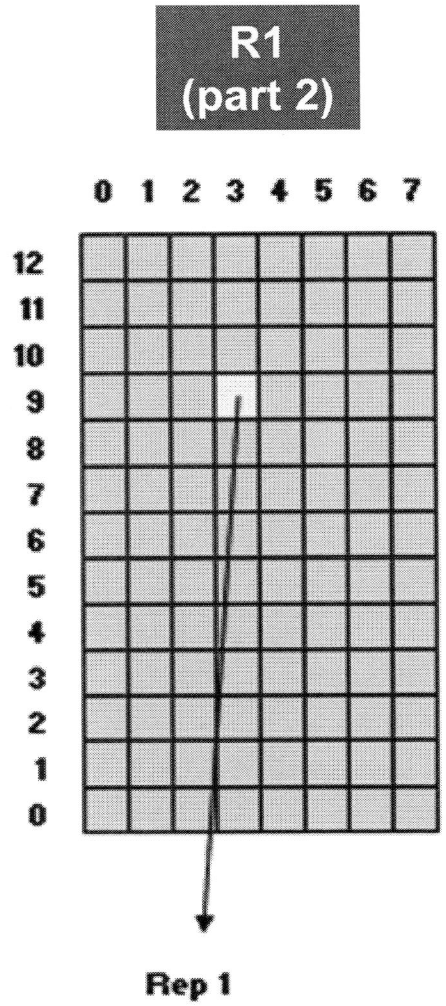

10

9

Depending on exposure and focus conditions defect printability might change

Best yielding SRAM R1 Area 1

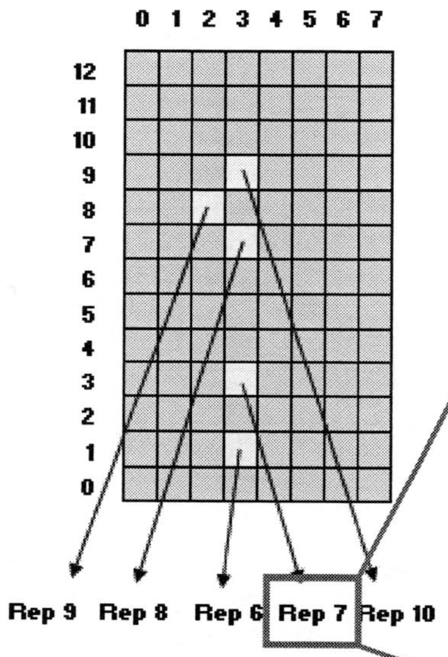

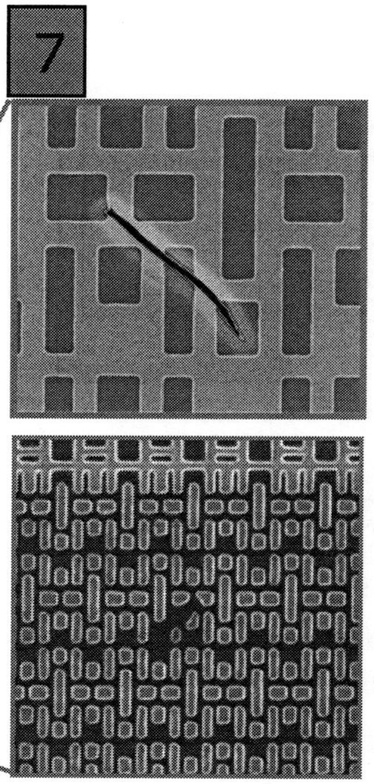

99.9992% yield
or
2 bits out of 256 kbits are failing

Best yielding SRAM R1 Area 2

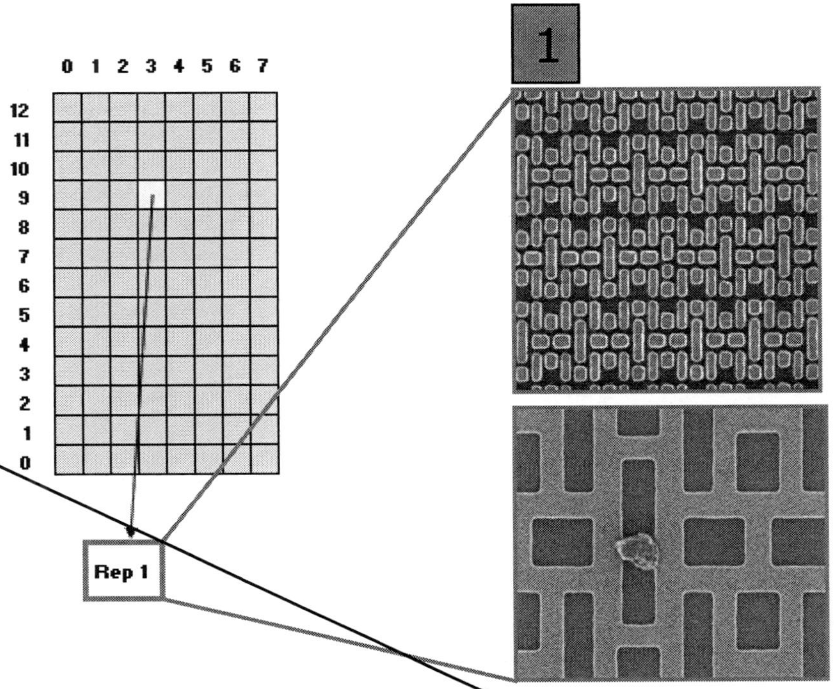

99.9992% yield
or
2 bits out of 256 kbits are failing

Best yielding SRAM R1 Area 2

98.8289% yield
or
3 bits out of 256 kbits are failing

50.0854% yield
or
128 kbits out of 256 kbits are failing

	0	1	2	3	4	5	6	7
12	1.00	1.00	1.00	1.00	1.00	1.00	1.00	1.00
11	1.00	1.00	1.00	1.00	1.00	1.00	1.00	1.00
10	1.00	1.00	1.00	1.00	1.00	1.00	1.00	1.00
9	1.00	1.00	1.00	1.00	1.00	1.00	1.00	1.00
8	1.00	1.00	1.00	1.00	1.00	1.00	1.00	1.00
7	1.00	1.00	1.00	1.00	1.00	1.00	1.00	1.00
6	1.00	1.00	1.00	0.99	0.50	1.00	1.00	1.00
5	1.00	1.00	1.00	1.00	1.00	1.00	1.00	1.00
4	1.00	1.00	1.00	1.00	1.00	1.00	1.00	1.00
3	1.00	1.00	1.00	1.00	1.00	1.00	1.00	1.00
2	1.00	1.00	1.00	1.00	1.00	1.00	1.00	1.00
1	1.00	1.00	1.00	1.00	1.00	1.00	1.00	1.00
0	1.00	1.00	1.00	1.00	1.00	1.00	1.00	1.00

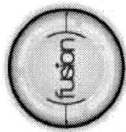

Best yielding SRAM R1 Area 2

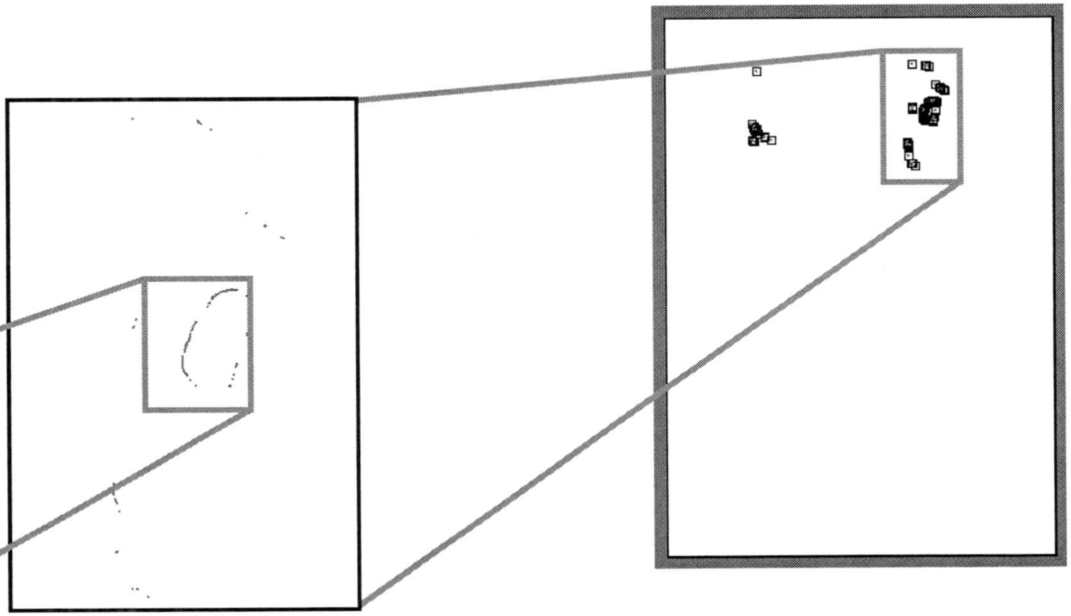

Related to the defects
of unknown origin

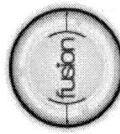

Summary

- A 45 nm test chip reticle was used to assess the printability of mask defects in EUV lithography

- *Only approximately 3% of mask blank defects appear printable*

- The sensitivity of state-of-the-art wafer and reticle inspection tools was compared

  - Each tool detected a different sub-set of defects, with some common defects

- Reticle inspection yielded no false defects

- Wafer inspection requires some optimization to allow a better comparison to reticle inspection

- **SRAM devices printed using EUVL show yield potential in line with the 193 nm baseline**

  - SRAM failure maps correlate well with repeater defect maps

Acknowledgments

From AMD F36:
Johannes Steinmetz, Wolfram Grundke, Rolf Seltman, Kai Frohberg, Holger Schuehrer, Ines Becker, Guenther Ewald, Axel Pawlowitsch, Thomas Merbeth, Volker Heinig

Part of this work was performed by the Research Alliance Teams at various IBM Research and Development Facilities

Trademark Attribution

AMD, the AMD Arrow logo and combinations thereof are trademarks of Advanced Micro Devices, Inc. in the United States and/or other jurisdictions. Other names used in this presentation are for identification purposes only and may be trademarks of their respective owners.

© 2008 Advanced Micro Devices, Inc. All rights reserved.

Reticle Inspection Observations

- A few sources of the mask defects could be identified:

  - contamination   unknown origin

  - particles   handling, shipping

  - blank defects   blank manufacturing

  - pattern defects   mask manufacturing

- No false defects were encountered during inspection!

- No mechanical pattern damage from manufacturing, handling or printing were observed.

- Defect source is difficult to determine without SEM review.

Comparing Reticle Defects to Wafer Prints
Line Slimming/Break Defects: Unknown Origin

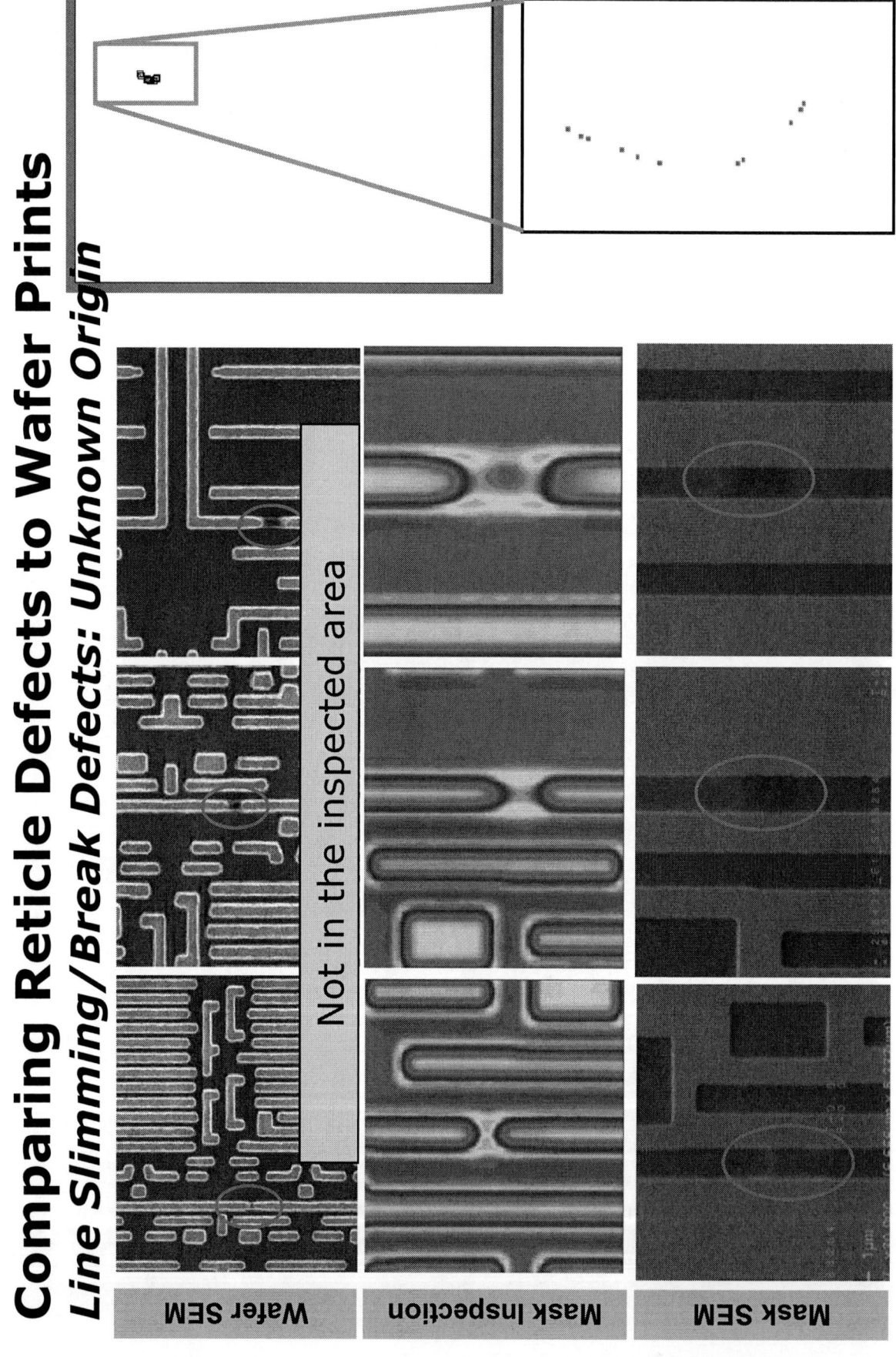

Exposure Setup 1, Bare Si Wafer

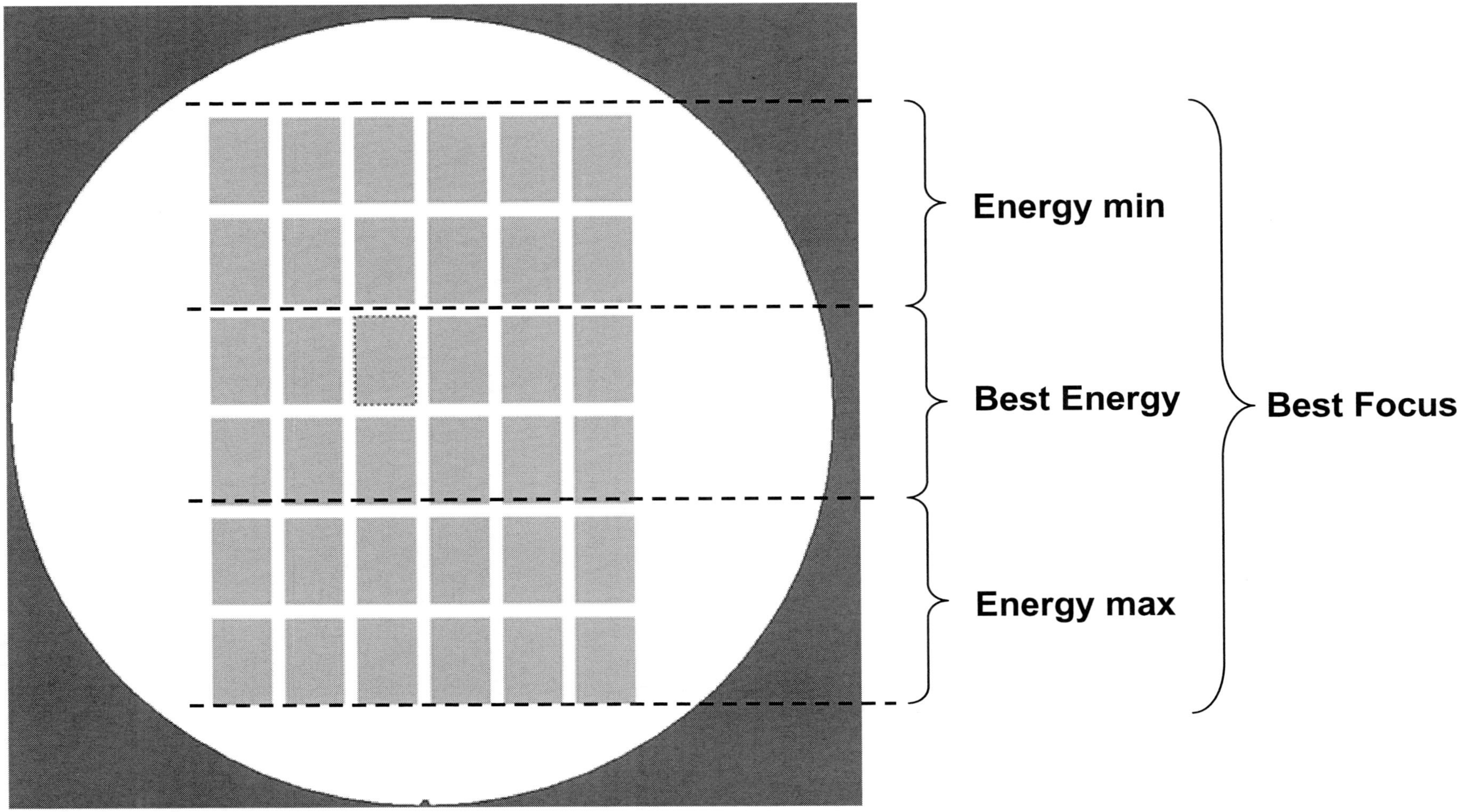

Energy min

Best Energy

Energy max

Best Focus

Exposure Setup 2 , Bare Si Wafer

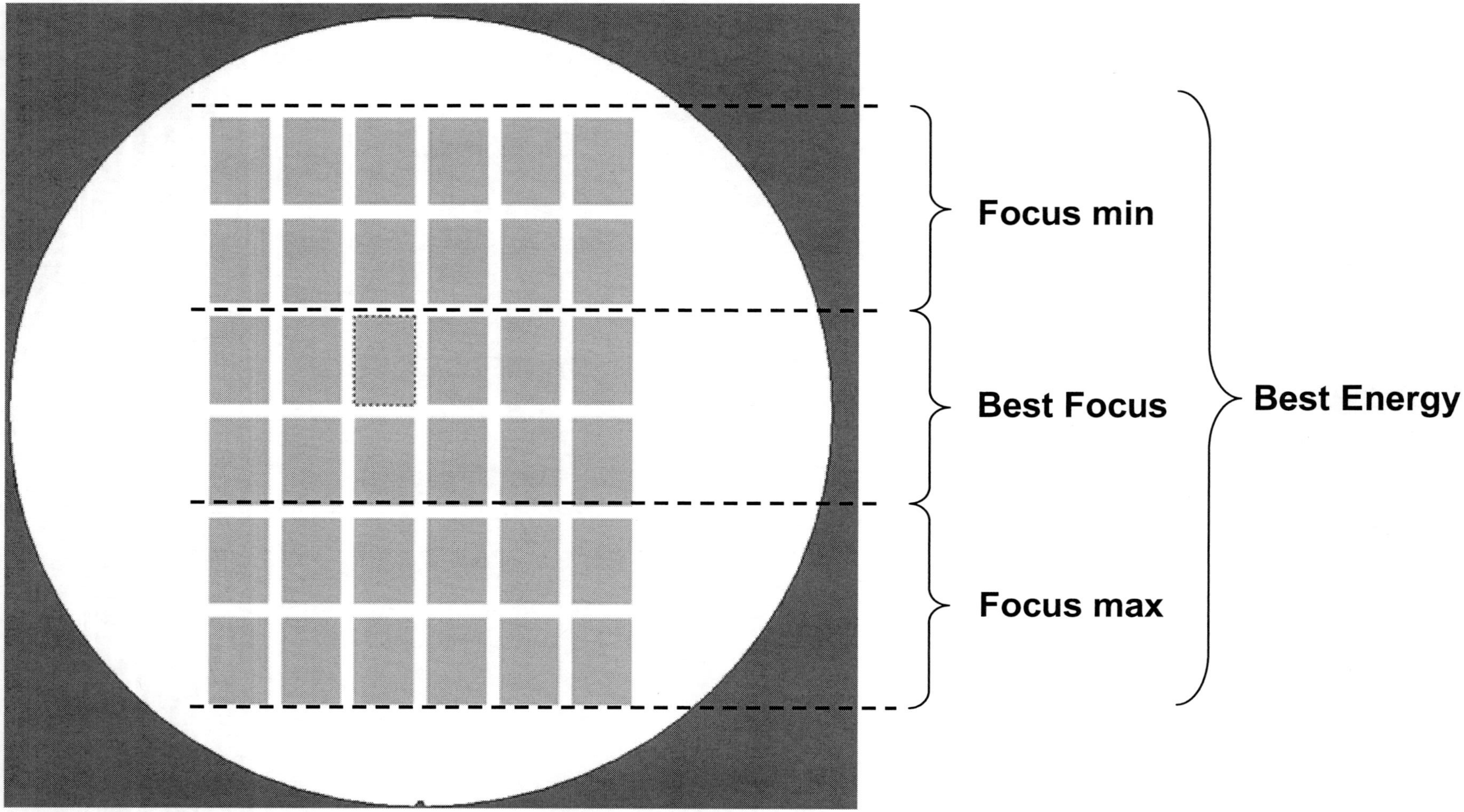

Best Energy

Focus min

Best Focus

Focus max

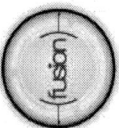

Development of actinic mask blank inspection technology at Selete

Tsuneo Terasawa, Takeshi Yamane, Teruo Iwasaki, Toshihiko Tanaka, Osamu Suga,

MIRAI- Semiconductor Leading Edge Technologies, Inc. (Selete)

Toshihisa Tomie

National Institute of Advanced Industrial Science and Technology (AIST)

Semiconductor Leading Edge Technologies, Inc.

Sept. 29th, 2008 EUVL Symposium

Outline

1. Introduction
   - Impact of phase defect
   - Positioning of Selete activity

2. Actinic mask blank inspection system design

3. Key elements of inspection system
   - Stage system
   - Inspection optics
   - EUV light source

4. Experimental of phase defect detection

5. Summary

Outline

1. **Introduction**
   - Impact of phase defect
   - Positioning of Selete activity

2. Actinic mask blank inspection system design

3. Key elements of inspection system
   - Stage system
   - Inspection optics
   - EUV light source

4. Experimental of phase defect detection

5. Summary

Impact of phase defect on CD

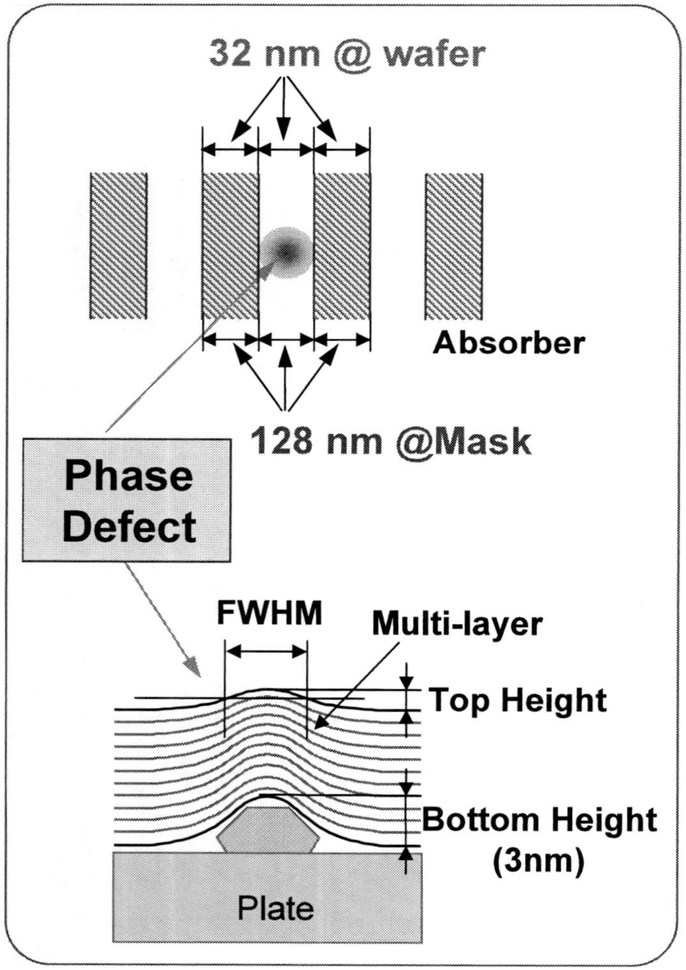

Critical phase defect at 32-nm L/S

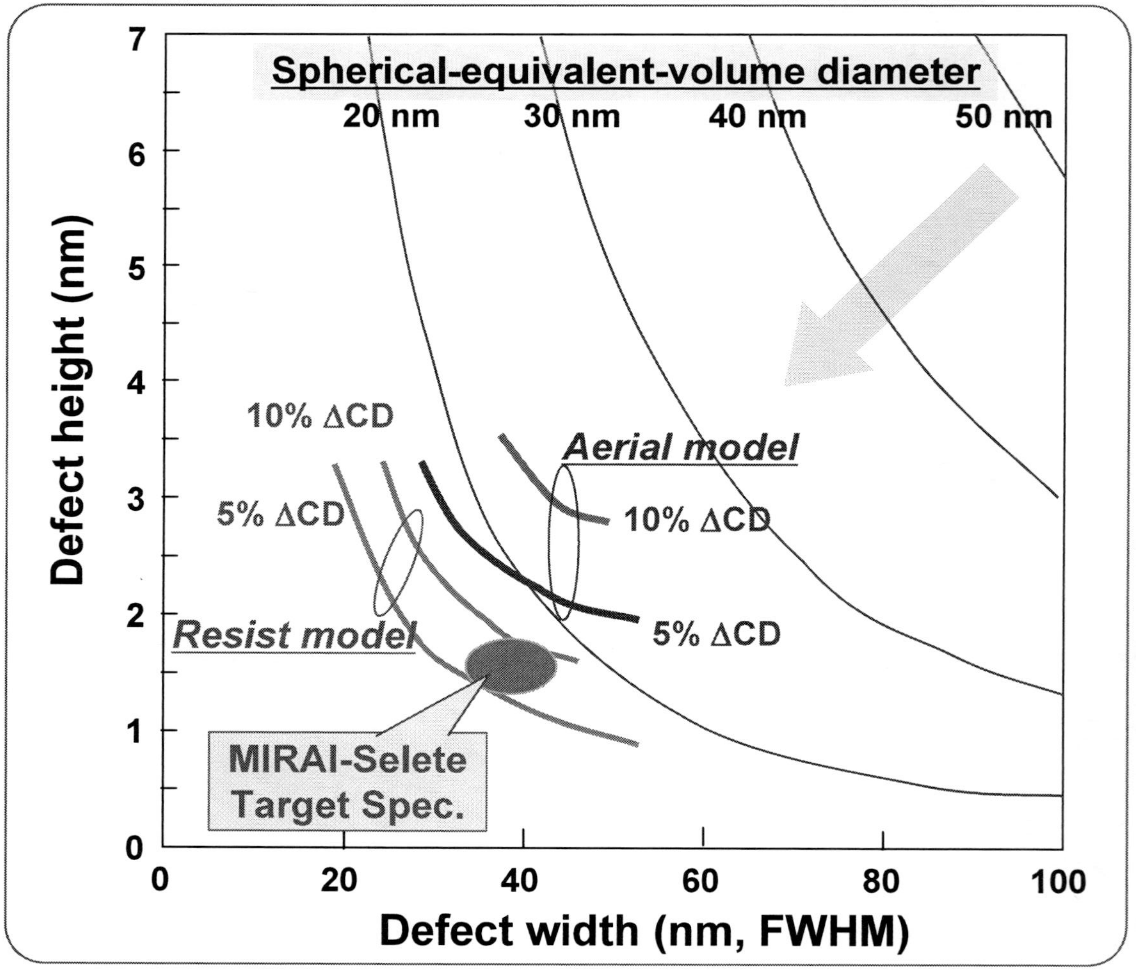

Positioning of Selete activity

MIRAI I and II (~F.Y. 2005)

-Proof of concept of actinic mask blank inspection system with demonstration of minimum critical defect detection.

-Sensitivity limiting factor analysis and design completion of manufacturing - ready prototype with accumulated experimental data

MIRAI -Selete

- Implementation of full-field mask blank inspection with a reasonable inspection speed.

Development of actinic full-field mask blank inspection tool for:

- Multilayer process development support
- Defect free mask blank fabrication support
- Outgoing / incoming qualification of premium mask blanks

Outline

1. Introduction
   - Impact of phase defect
   - Positioning of Selete activity

2. Actinic mask blank inspection system design

3. Key elements of inspection system
   - Stage system
   - Inspection optics
   - EUV light source

4. Experimental of phase defect detection

5. Summary

Concept of actinic inspection tool

1. Plasma EUV light source.
2. Dark field detection with Schwarzschild optics.
3. Two-dimensional CCD camera.
4. Chief ray angle of 0 degree.

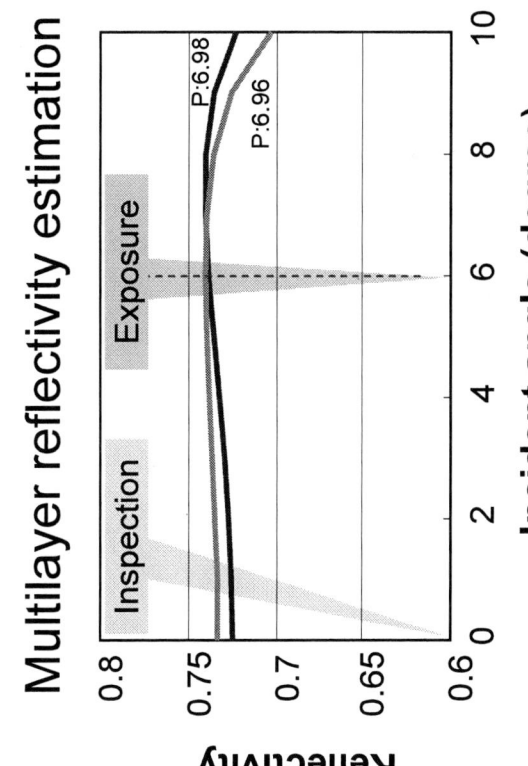

Multilayer reflectivity estimation

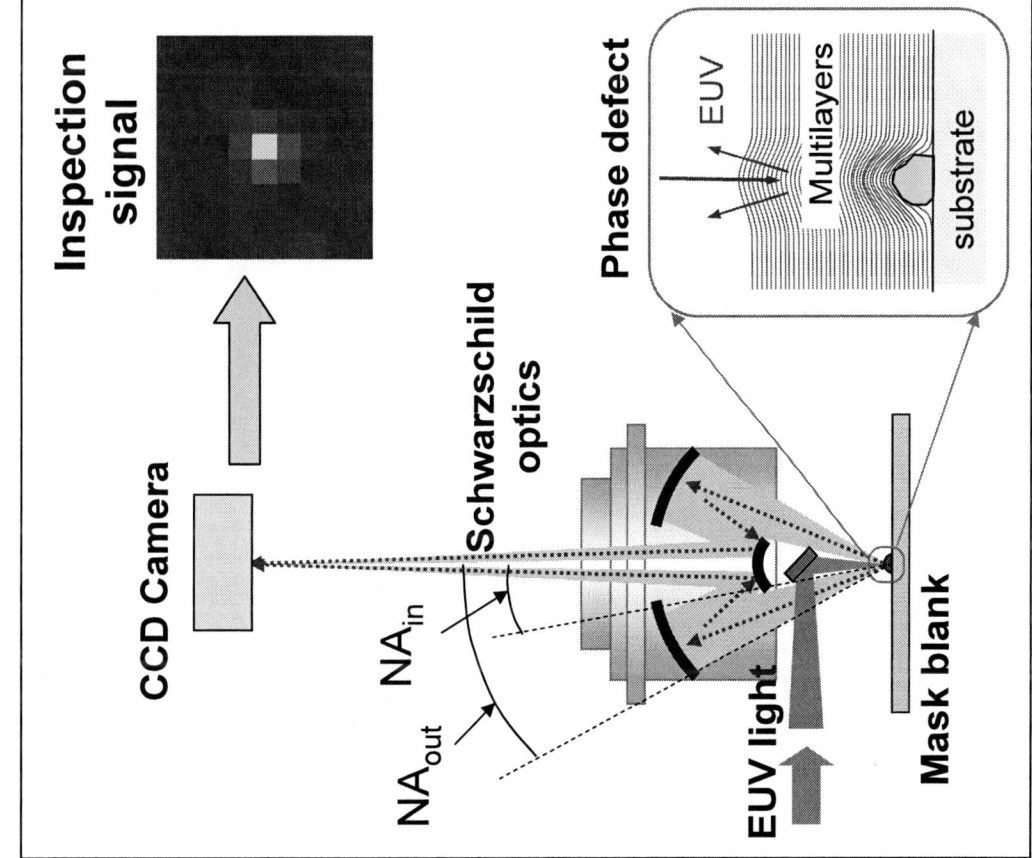

Pixel size

- Pixel size of 500 nm is chosen together with 26X imaging optics.
- Inspection throughput of 2 hours per mask blank is possible with a data transfer rate of 10 Mega pixels/sec.
- Large depth of focus is obtained.

Pixel intensity vs. Defocus

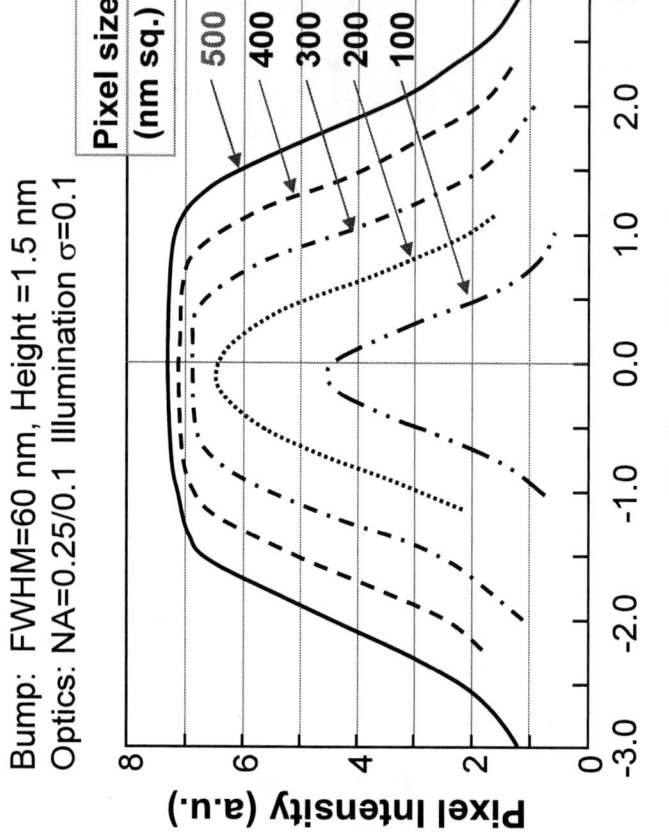

Bump: FWHM=60 nm, Height =1.5 nm
Optics: NA=0.25/0.1 Illumination σ=0.1

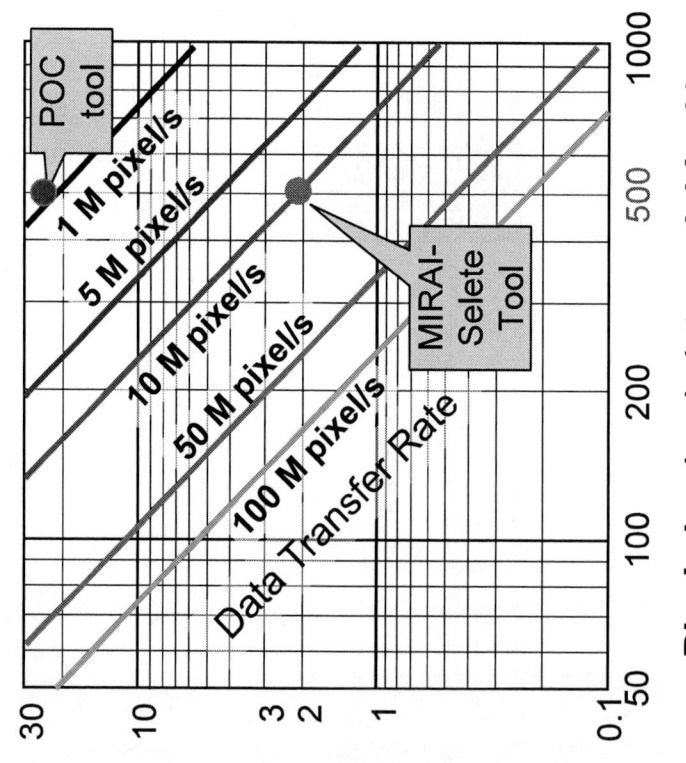

Configuration

	POC tool	Full-field inspection tool
Light source	LPP, 10 Hz (Source size:0.1 mm)	DPP, 1.9 kHz, 10 W /2πsr. (Source size:0.4 mm)
Illumination optics	Spherical mirror + Ellipsoidal mirror + Plane mirror (Illu. Area : 0.5 mm sq.)	Ellipsoidal mirror + Plane mirror (Illu. Area : 0.5 mm sq.)
Imaging optics	•26X Schwarzschild optics • Inner NA=0.1, •Outer NA=0.2	•26X Schwarzschild optics • Inner NA=0.1 •Outer NA=0.2~0.27 (Variable)
Sensor	Back illuminated CCD	•Back illuminated CCD •Imaging operation •TDI operation
Mask stage	•Stroke X,Y: 10mm,2mm •Manually controlled •Focus position alignment	•Stroke X,Y: 169 mm,169 mm •Continuous move, •Laser interferometer feed back •Focus position alignment

Outline

1. Introduction
   - Impact of phase defect
   - Positioning of Selete activity

2. Actinic mask blank inspection system design

3. **Key elements of inspection system**
   - Stage system
   - Inspection optics
   - EUV light source

4. Experimental of phase defect detection

5. Summary

IRAI

Mask stage

Selete

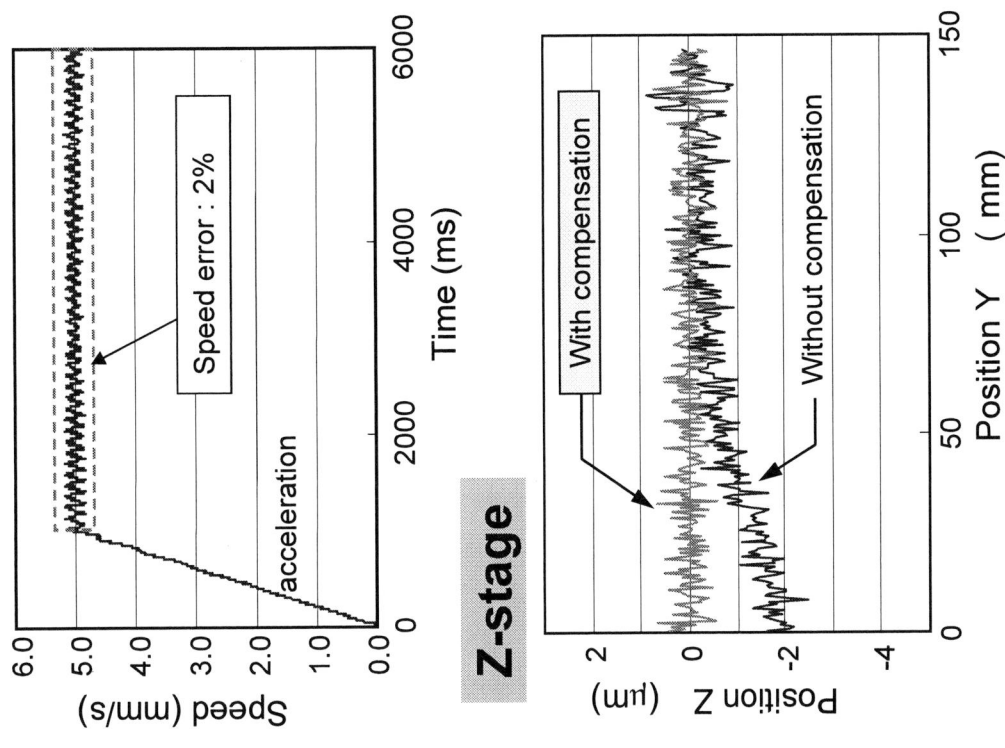

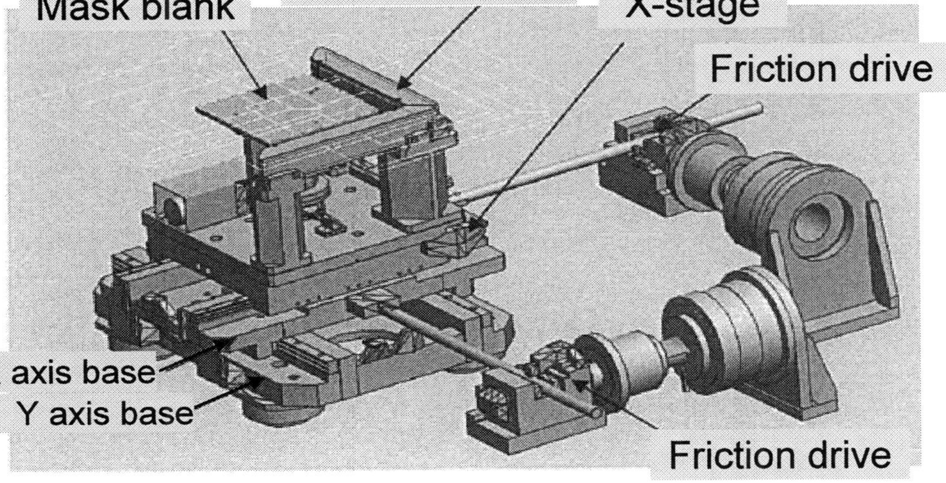

Mask blank
Mirror for Laser interferometer
X-stage
Friction drive
X axis base
Y axis base
Friction drive

◆Stroke	X,Y : 169 mm Z fine stage : 60 μm
◆Drive system	X,Y : Friction drive Z fine stage: piezo actuator

X,Y-stage

6.0
5.0
4.0
3.0
2.0
1.0
0.0
Speed (mm/s)
Speed error : 2%
acceleration
0
2000
4000
6000
Time (ms)

Z-stage

2
0
-2
-4
Position Z (μm)
With compensation
Without compensation
0
50
100
150
Position Y (mm)

Inspection optics

	Illumination optics	Imaging optics
Configuration	Ellipsoidal mirror + Plane mirror (Illu. Area : 0.5 mm sq.) 	Schwarzschild optics (Concave and convex mirrors) 
Magnification	2.0	1/26
NA at mask side	•Illumination NA=0.06	•Inner NA=0.1 (fixed) •Outer NA=0.2~0.27 (Variable)
Total reflectivity	23 %	42.7 %

EUV light source

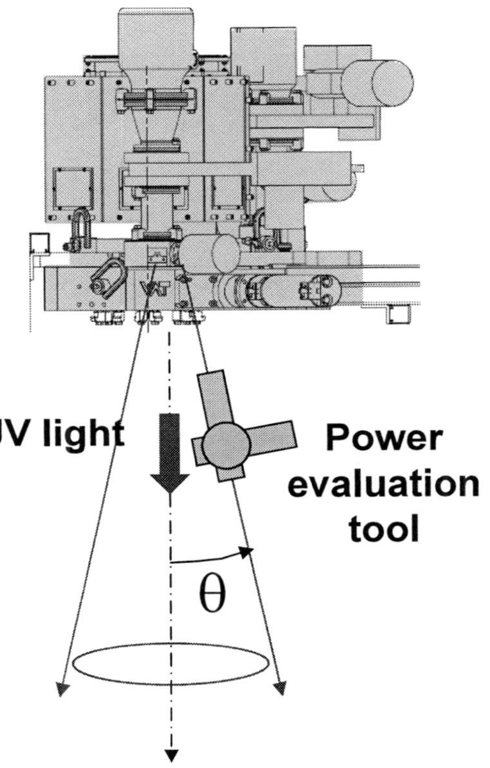

Light source

EUV light

Power evaluation tool

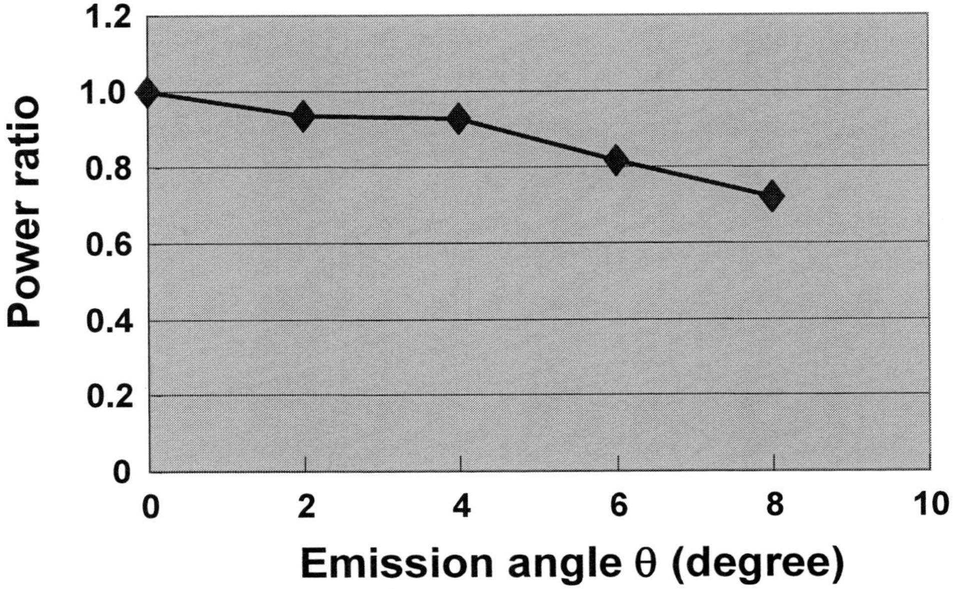

θ

Power ratio vs Emission angle θ (degree)

- ◆ Emission angle enlarged to 8 degree.
- ◆ Total power of 53 mW (in-band) was obtained.

Outline

1. Introduction
   - Impact of phase defect
   - Positioning of Selete activity

2. Actinic mask blank inspection system design

3. Key elements of inspection system
   - Stage system
   - Inspection optics
   - EUV light source

4. Experimental of phase defect detection

5. Summary

Actinic EUVL mask blank inspection system *Selete*

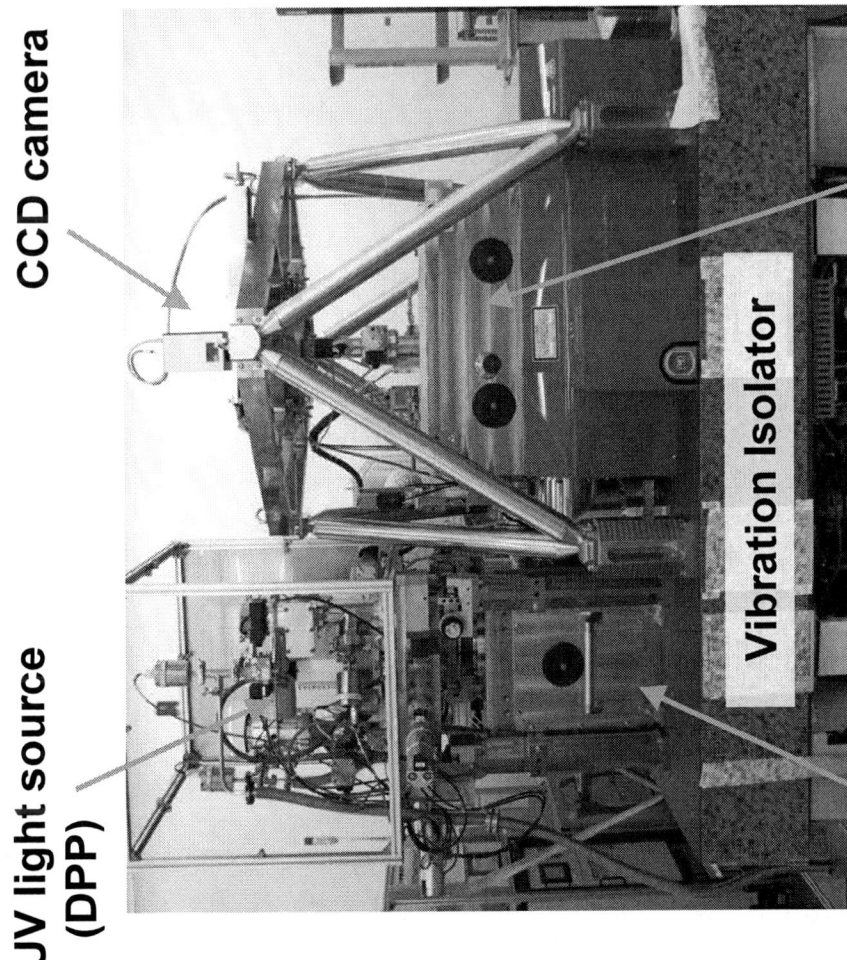

- Installed in MIRAI-Selete clean room.
- Static inspection images obtained.
- TDI operation not yet.

Programmed phased defect array

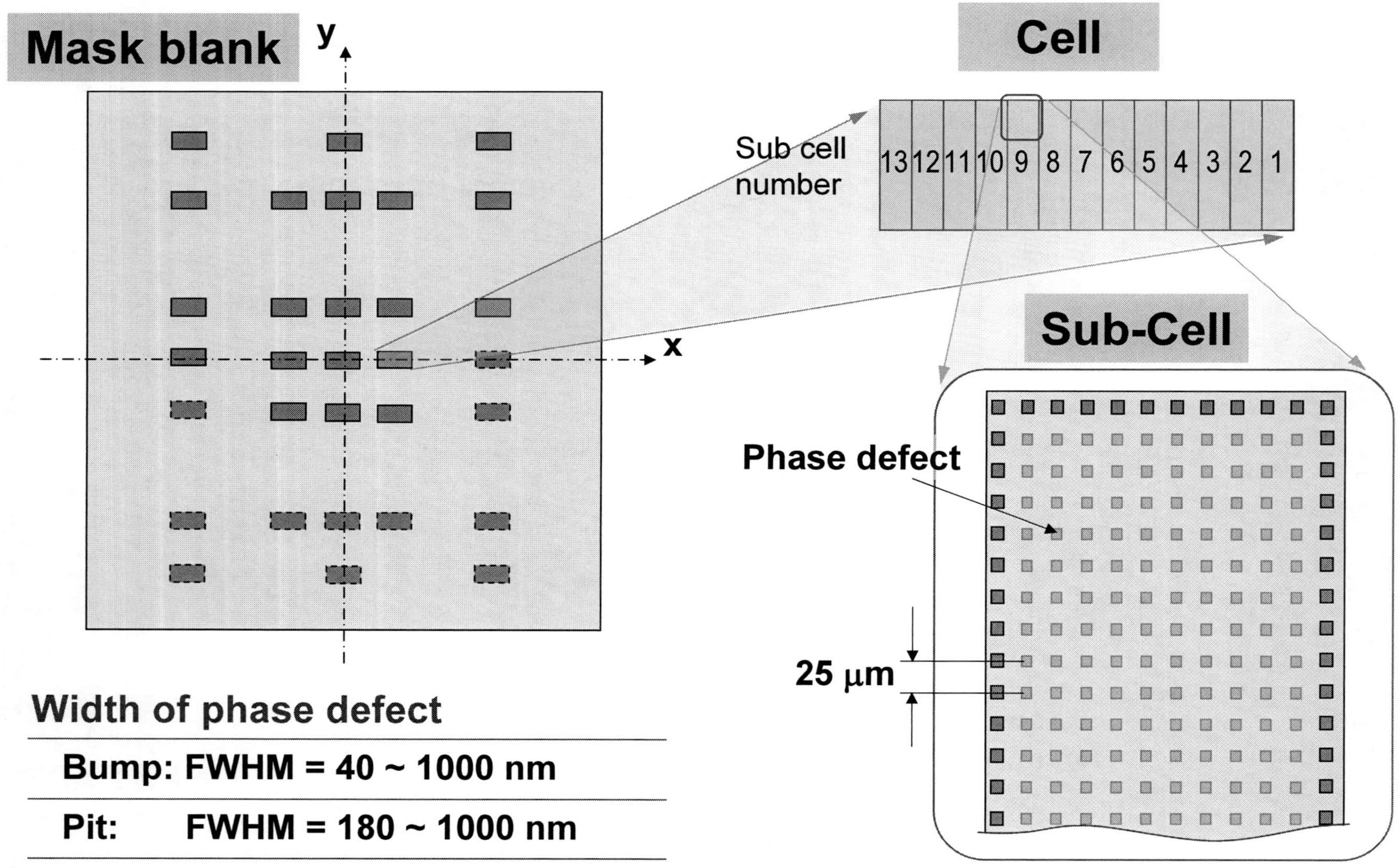

Mask blank

Cell

Sub cell number

13 12 11 10 9 8 7 6 5 4 3 2 1

Sub-Cell

Phase defect

25 μm

Width of phase defect

Bump: FWHM = 40 ~ 1000 nm

Pit: FWHM = 180 ~ 1000 nm

Programmed phase defect sizes

Bump defect cells

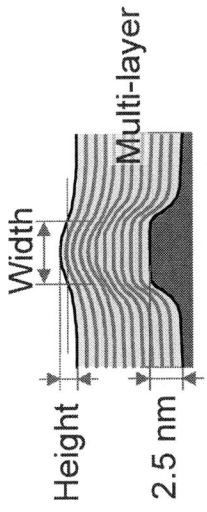

Sub cell Number	1	2	3	4	5	6	7	8	9	10	11	12	13
Width (nm)	60	70	78	85	100	110	120	130	140	160	190	210	250
Height (nm)	1.5	1.8	2.0	2.2	2.5	2.5	2.5	2.5	2.5	2.5	2.5	2.5	2.5

Pit defect cells

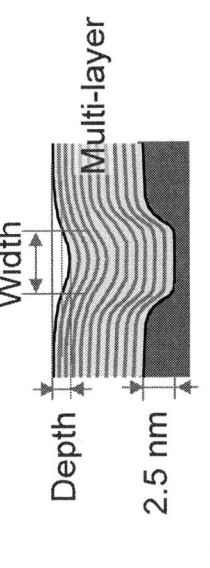

Sub cell Number	1	2	3	4	5	6	7	8	9	10	11	12	13
Width (nm)	120	190	230	250	270	300	320	360	420	460	520	760	1000
Depth (nm)	2.3	2.5	2.5	2.2	2.5	2.5	2.5	2.5	2.5	2.5	2.5	2.5	2.5

6 inch mask blank

Phase defect detection signal - Bump

(Step and repeat imaging mode: Stitch images)

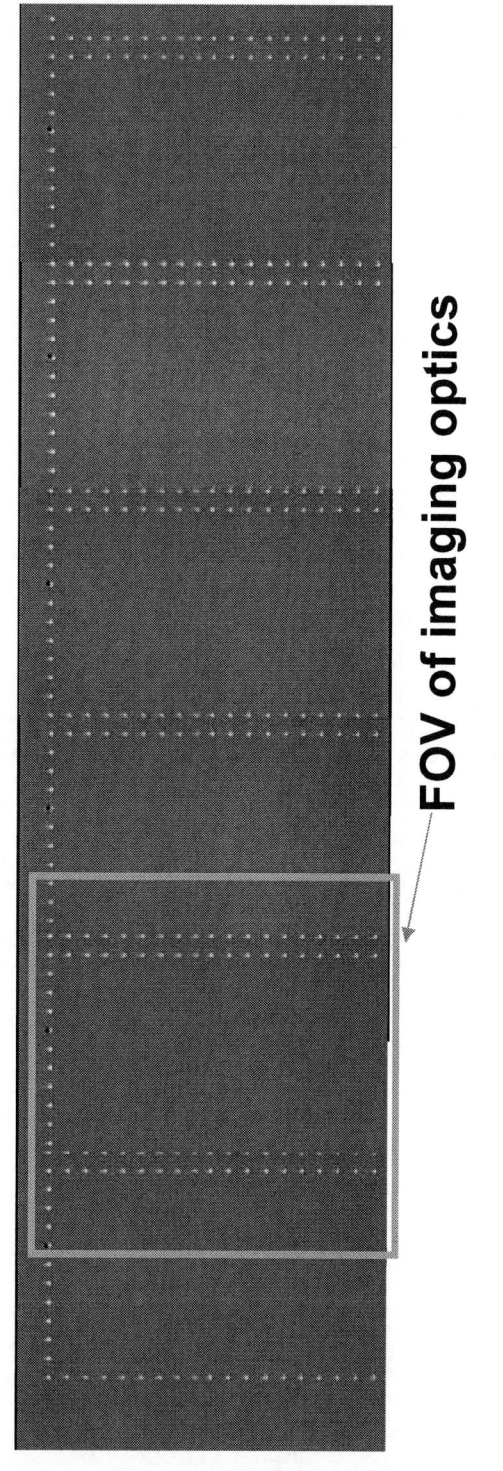

FOV of imaging optics

Inspection signals of small bump defects

(Step and repeat imaging mode)

H:1.5 nm, W:60 nm

H:1.8 nm, W:70 nm

H:2.0 nm, W:78 nm

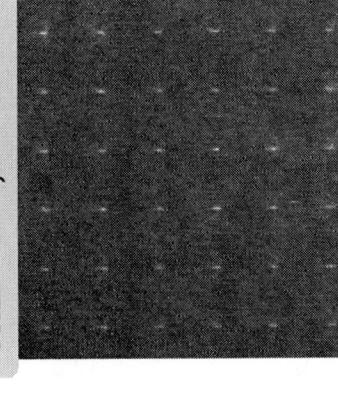

H:2.2 nm, W:85 nm

H:2.5 nm, W:100 nm

H:2.5 nm, W:110 nm

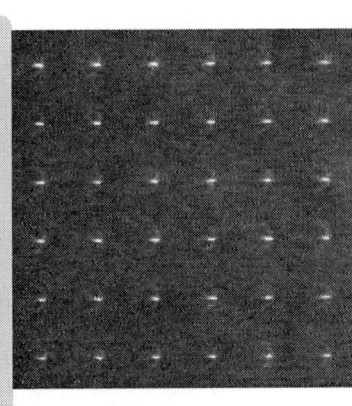

All of programmed defects sucessfully identified as bright spots.

Phase defect detection signal - Pit

(Step and repeat imaging mode: Stitch images)

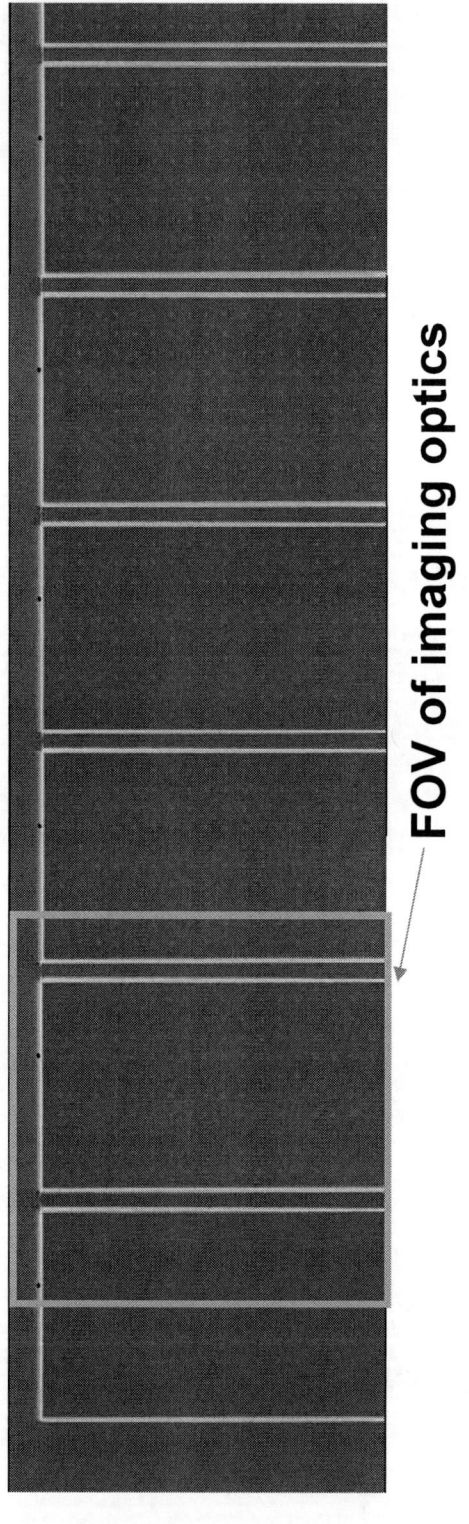

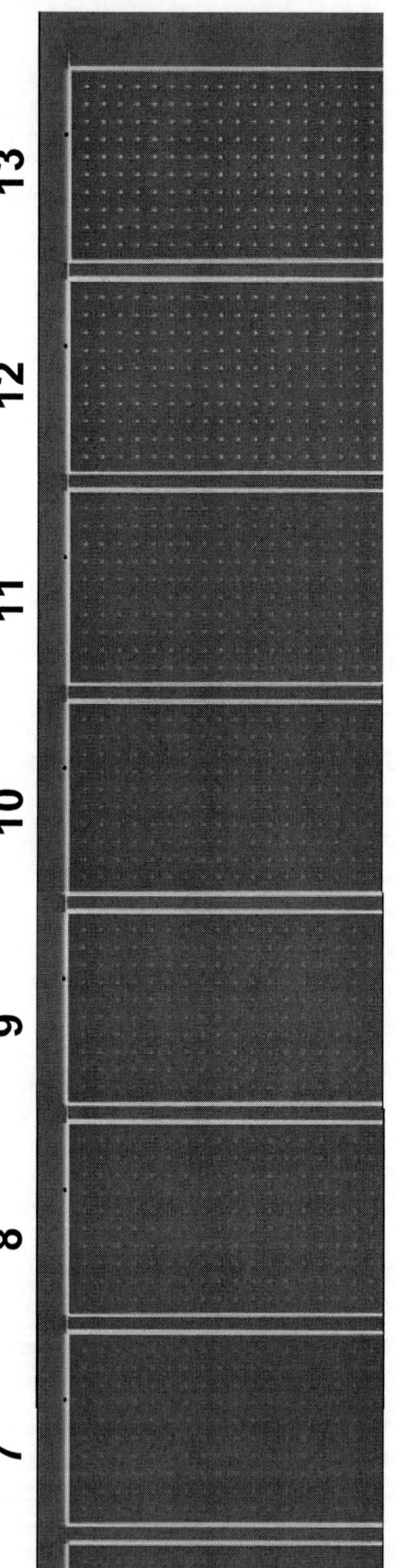

FOV of imaging optics

 IRAI

Effect of outer NA of inspection optics

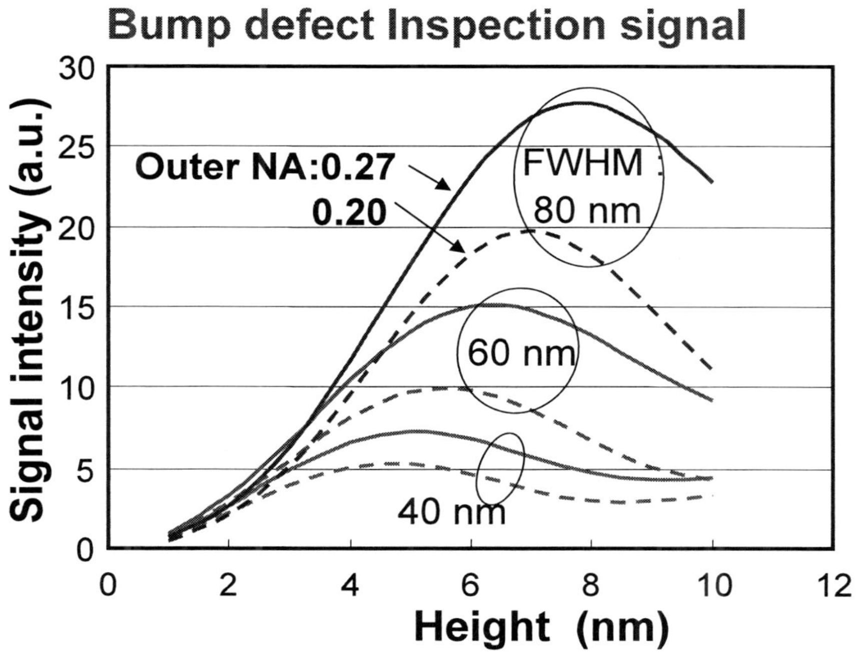

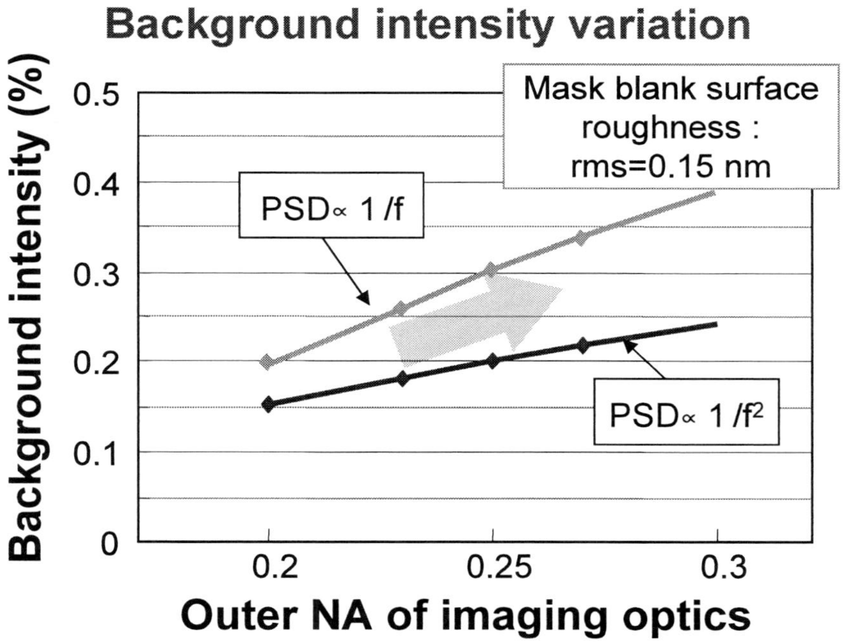

Selete

Bump defect inspection signal

Signal intensity (a.u.) vs Height (nm)

Outer NA:0.27, 0.20

FWHM 80 nm, 60 nm, 40 nm

Background intensity variation

Mask blank surface roughness : rms=0.15 nm

$PSD \propto 1/f$

$PSD \propto 1/f^2$

Background intensity (%) vs Outer NA of imaging optics

Simulation results suggest:

By extending outer NA of dark-field inspection optics from 0.2 to 0.27,

- inspection intensity of tall phase defects will increase,
- background intensity will increase by 40%~80% .

 T. Yamane, et al., SPIE Photomask Technology 28th Annual Symposium (6-10 October 2008)

Outline

1. Introduction
   - Impact of phase defect
   - Positioning of Selete activity

2. Actinic mask blank inspection system design

3. Key elements of inspection system
   - Stage system
   - Inspection optics
   - EUV light source

4. Experimental of phase defect detection

5. Summary

Summary

Activities completed / in progress:

1. Actinic full-field mask blank inspection tool using dark field imaging was installed in MIRAI-Selete clean room.

2. The performance of each subsystem was also checked and verified to meet the set requirements in the tool specification.

3. The smallest defect of 1.5 nm in top-height and 60 nm in width on a test mask was visually identified with static imaging mode.

4. An impact of increased NA on detection signal was analyzed.

Activities planned:

1. Detect phase defects with TDI operation mode.

2. Adjust optics more precisely to improve an inspection resolution.

3. Collect / compile inspection data of natural multilayer defects.

Acknowledgments

We would like to acknowledge:

- H. Kashima, N. Arasawa, and H. Toyama of Hitachi Ltd.
  (Total inspection system)

- K. Omori, T. Namikawa, and Y. Sugiyama of Nikon Corporation.
  (Inspection optics)

- T. Shoki, K. Yamashiro, Y. Usui, and O. Nagarekawa of HOYA Corporation.
  (Programmed defect mask blank)

- Y. Minami and M. Kadoi of Litho Tech Japan Corporation.
  (EUV light source)

This work was supported by New Energy and Industrial Technology Development Organization (NEDO).

EUVL Blank Defect Inspection Capability at Intel

Andy Ma, Ted Liang, Seh-Jin Park, Guojing Zhang, Tomoya Tamura*, Kazunori Omata*

Components Research, Intel Corporation

* Lasertec Corporation

2008 EUVL Symposium, 09/29/2008

Outline

- **Introduction**
- **Current status**
  - Defect inspection performance
  - Damage assessment
- **Roadmap vs Inspection capability**
- **Future plan**
- **Summary**

Introduction

- **Defect-free EUV mask blank remains one of the greatest commercialization challenges for EUVL**

- **Defect inspection capability is necessary for EUVL defect-free mask blank development**

  – ~25 nm @ 2011

- **Blank inspection tool is critical for**

  – EUV mask pilot line incoming blank inspection
  – EUV mask blank cleaning process development
  – EUV mask handling development
  – EUV mask blank defect reduction improvement at suppliers

High Sensitivity EUV Mask Inspection System (M7360) installed at Intel

Tool Performance

- Inspection sensitivity (> 98% C.E.)

  QZ: 45nm, ML: 50nm

- Detection sensitivity

  QZ: 31.5nm, ML: 39nm

  - > ~ 10X sensitivity improvement from 1G inspection tool (M1350)

  - Tool uptime: > 95% (July to current)

2G EUV Blank Defect Insp. Tool

	M1350 (1G)	M7360 (2G)
Laser Wavelength	488 nm	266nm
Inspection Power	200 mW on Qz 100 mW on ML	550mW on Qz 200mW on ML
Detection system	Normal scan mode & dense scan mode with one detector	Optical dense scan mode with two detectors

(intel) Lasertec

Defect inspection Performance

- **Programmed defects**
  - QZ substrate
  - ML blank

- **Particles**
  - SiO_2 spheres on QZ
  - SiO_2 spheres on ML

- **Natural defects**
  - Quartz substrate
  - ML blank

M7360 has higher sensitivity than M1350 on QZ PDM

#7 (99.8%)

M7360
(266nm)

M1350
(488nm)

Capture Rate: 99.8%
SiO2 sphere Equiv.: 44.7 nm

#	13	12	11	10	9	8	7	6	5	4	3	2	1
Defect width (nm FWHM)	90	110	130	150	200	220	250	270	325	430	580	810	1050
Defect height (nm)							2.8						

(intel) Lasertec

M7360 has higher sensitivity than M1350 on ML PDM

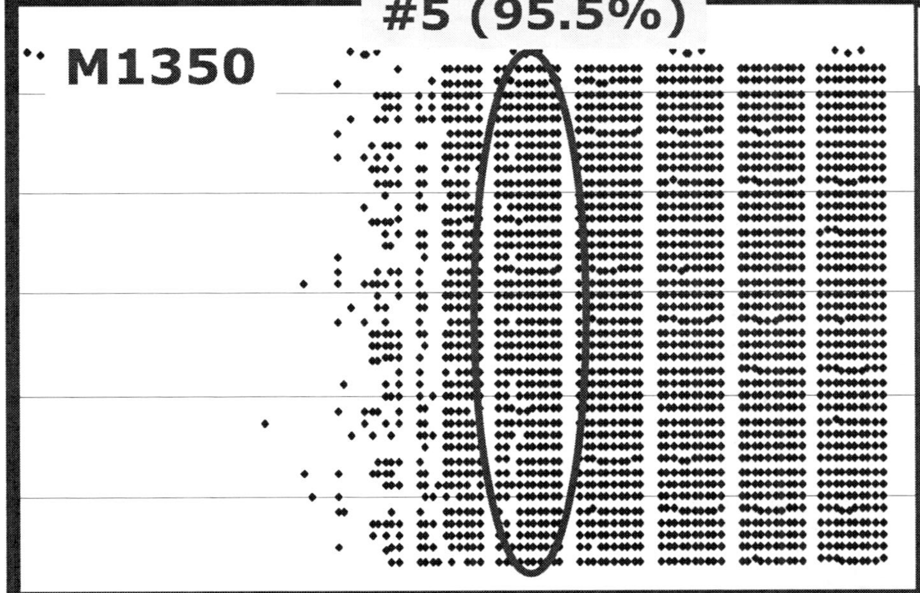

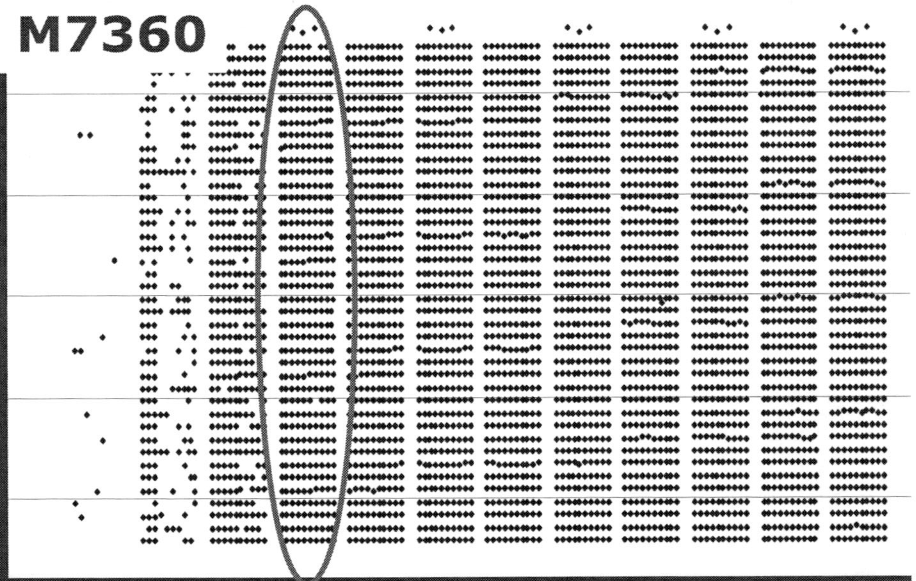

#9 (98%)

M7360

#5 (95.5%)

M1350

Capture Rate: 98%
SiO$_2$ sphere Equiv.: 49.2 nm

#	13	12	11	10	9	8	7	6	5	4	3	2	1
Defect width (nm FWHM)	70	75	85	90	95	100	120	140	145	155	170	195	300
Defect height (nm)	1.5	2.3	2.7	2.9	3.0	3.1	3.2	3.4	3.5	3.5	3.5	3.5	3.5

Lasertec

45nm particle inspection capability on Quartz

M7360

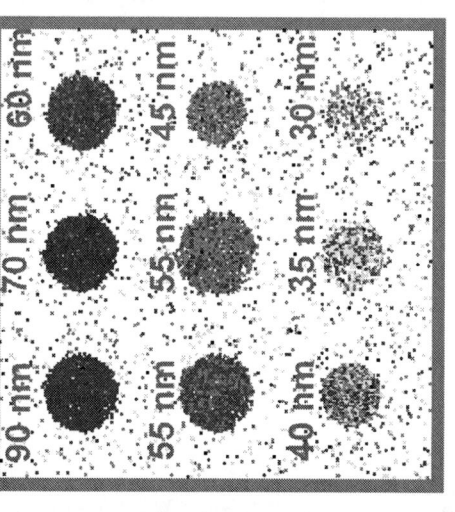

SiO2 Particles Size (nm)	# of Particles Detected	Capture Rate (%)	Ave. M7360 Pixel Size
90	2032	100.0	13.15
70	2205	100.0	11.26
60	2064	100.0	9.19
55	2129	99.9	7.91
50	1967	99.8	5.88
45	2178	96.7	4.26
40	1733	82.1	3.13
35	714	47.1	2.37
30	365	39.0	1.76

M1350

SiO2 Particles Size (nm)	# of Particles detected	Capture rate	Average M1350 pixel size
90nm	1825	1.00	8.36
70nm	1935	1.00	4.72
60nm	1886	0.61	1.40
55nm	955	0.18	0.39

: Current Capability

: Process Improvement

50 nm particle inspection capability on ML Blank

M7360

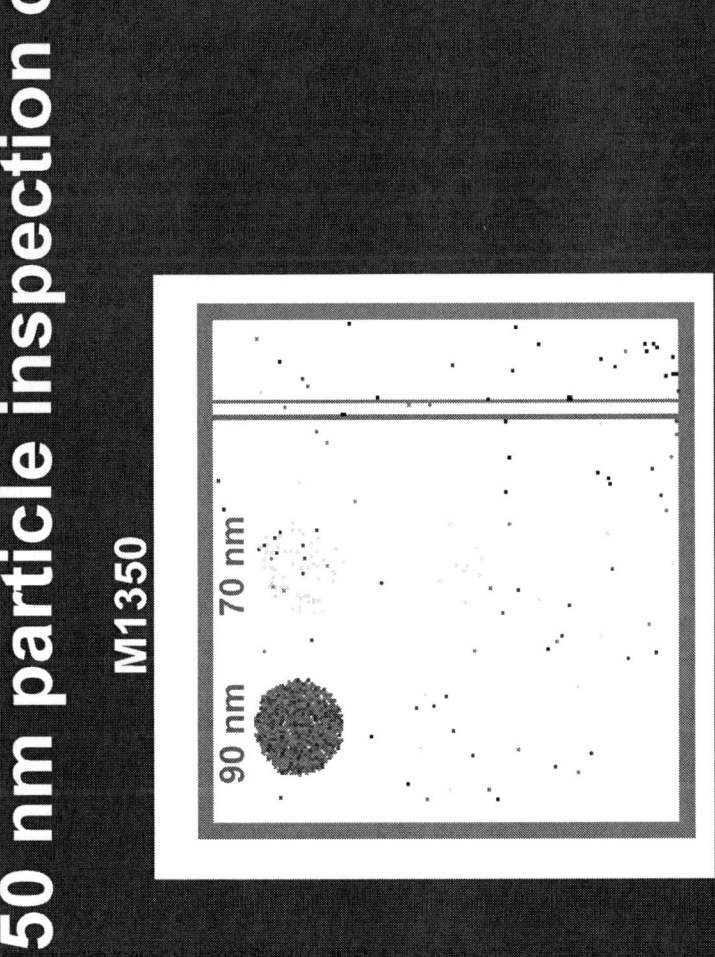

SiO2 Particles Size (nm)	# of Particles Detected	Capture Rate (%)	Ave. M7360 Pixel Size
90	2590	100.0	15.27
70	1917	100.0	11.21
60	2218	100.0	8.63
55	1935	100.0	7.28
50	2043	100.0	5.73
45	1975	94.0	3.43
40	1442	34.0	0.78

M1350

SiO2 Particles Size (nm)	# of Particles Detected	Capture Rate (%)	Ave. M1350 Pixel Size
90	1764	100.0	6.53
70	802	16.0	0.34

: Current Capability

: Process Improvement

Natural defect capture rate analysis on Qz (10X inspections)

SiO2 Equiv.	Pixel	Detected Defect Average	Capture rate (%)	Detected 1X (%)
31.5	1	12	57.0	0.02
34.3	2	14	72.0	0.01
37.2	3	10	81.0	0
40.5	4	5	92.0	0
44.0	5	2	96.0	0
47.8	6	2	100.0	0
52.0	7	1	100.0	0
56.5	8	2	100.0	0
61.4	9	1	100.0	0
66.7	11	1	100.0	0
72.5	19	1	100.0	0
78.8	20	1	100.0	0

Inspection sensitivity @ 44nm
Detection sensitivity @ 31.5nm

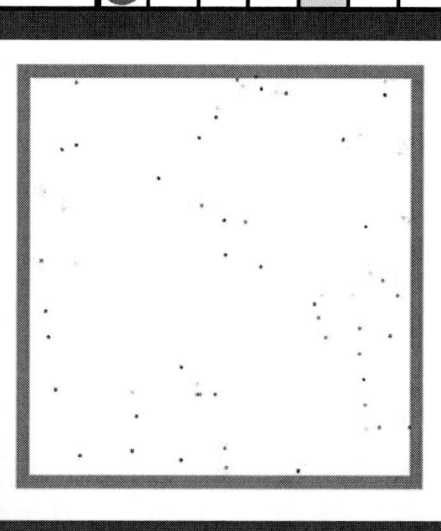

Natural defect capture rate analysis on RuML Blank (4X inspections)

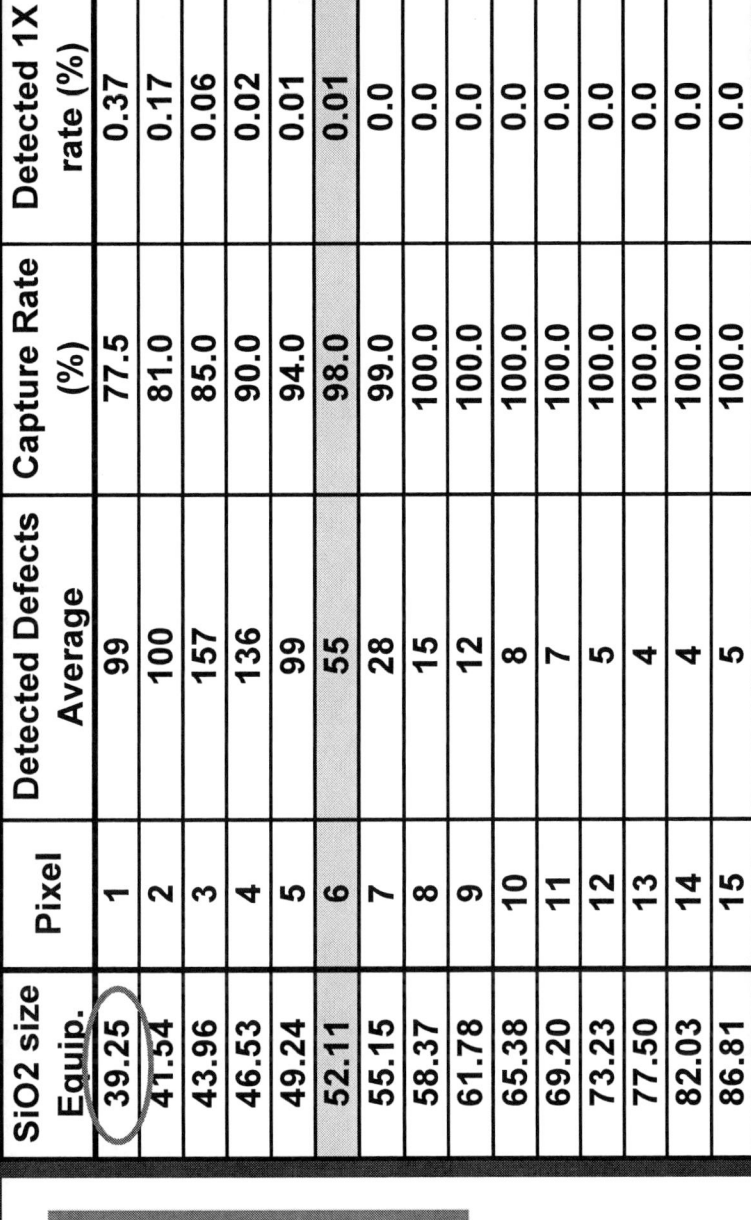

SiO2 size Equip.	Pixel	Detected Defects Average	Capture Rate (%)	Detected 1X rate (%)
39.25	1	99	77.5	0.37
41.54	2	100	81.0	0.17
43.96	3	157	85.0	0.06
46.53	4	136	90.0	0.02
49.24	5	99	94.0	0.01
52.11	6	55	98.0	0.01
55.15	7	28	99.0	0.0
58.37	8	15	100.0	0.0
61.78	9	12	100.0	0.0
65.38	10	8	100.0	0.0
69.20	11	7	100.0	0.0
73.23	12	5	100.0	0.0
77.50	13	4	100.0	0.0
82.03	14	4	100.0	0.0
86.81	15	5	100.0	0.0

Inspection sensitivity @ 52 nm
Detection sensitivity @ 39 nm

EUV reflectivity control in multiple inspections on M7360

- **EUV reflectivity sensitive to inspection laser power**
  - Higher inspection laser power will degrade EUV reflectivity

Example TEM of surface damage area (550mW)

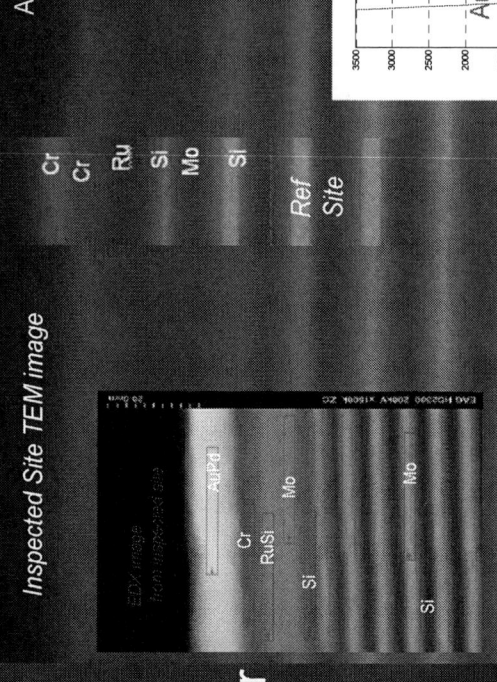

- **Possible root cause**
  - Silicide formation due to heat
  - Oxidation at Si layer right under Ru-cap layer

- Cr and AuPd are coated after inspection to protect the plate during TEM sample preparation
- Damage creates pit on the surface; sample shrinks

Tests: 300mW to 550 mW (Si-cap ML and Ru-Cap ML)
200mW to 325 mW (Ru-cap ML)

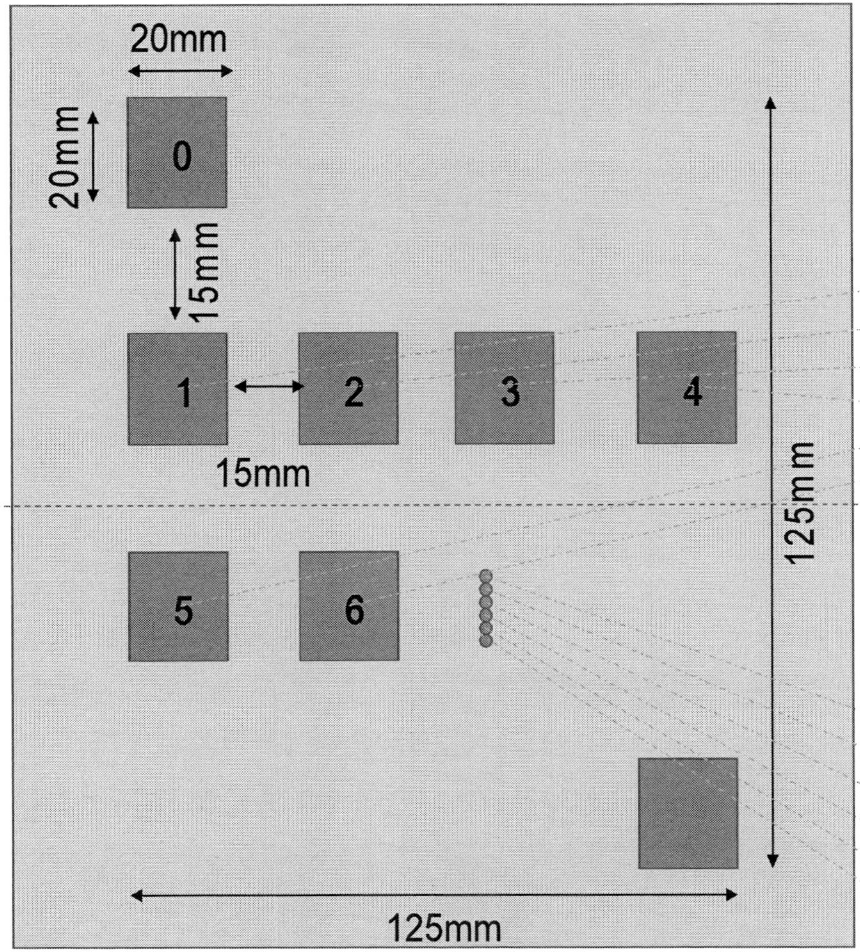

Area #0 is the position for focus adjustment.

Inspection tests
#1-6 each area is inspected 10 times at different power.

Center

#1: 300mW (-52.5, 17.5)
#2: 350mW (-17.5, 17.5)
#3: 400mW (17.5, 17.5)
#4: 450mW (52.5, 17.5)
#5: 500mW (-52.5, -17.5)
#6: 550mW (-17.5, -17.5)

1 beam review tests
Each position is reviewed 10 times at different power.

(10, -10.0): 300mW
(10, -12.5): 350mW
(10, -15.0): 400mW
(10, -17.5): 450mW
(10, -20.0): 500mW
(10, -22.5): 550mW

125mm

125mm

20mm

15mm

15mm

20mm

0

1

2

3

4

5

6

Inspection laser power has been defined

Ru-Cap: ≤ 300 mW; Si-Cap: 550 mW

Criteria: ML reflectivity non-uniformity < 0.5 % and
ML Centroid wavelength non-uniformity < 0.024 nm

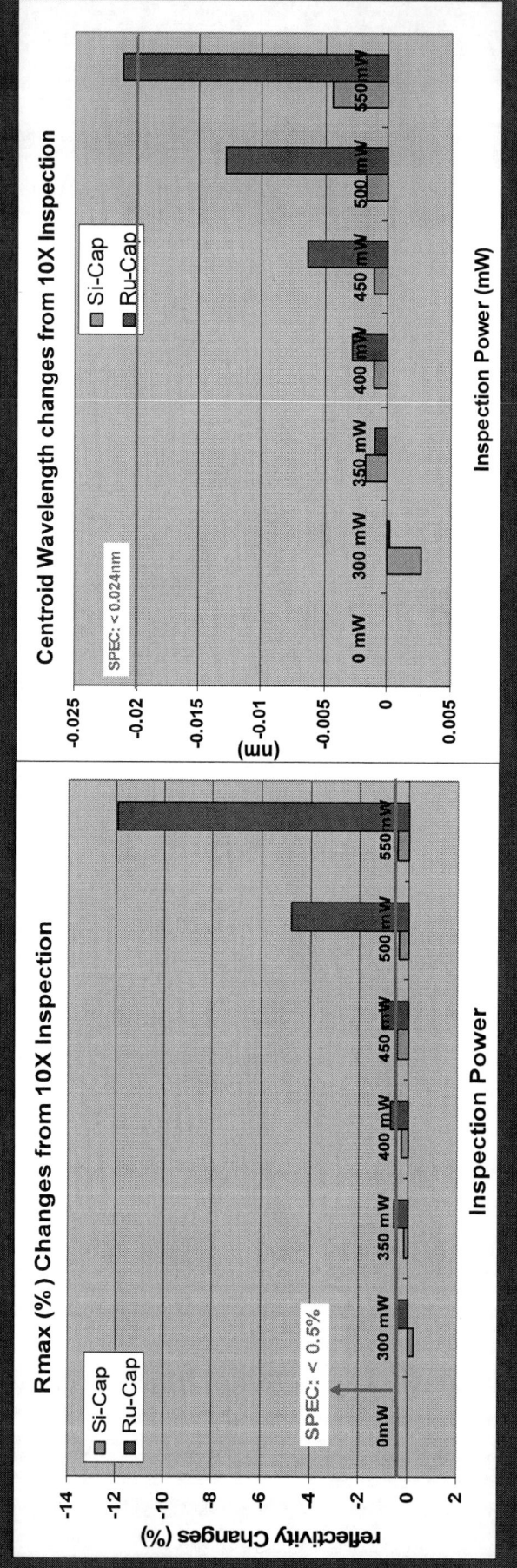

- Ru-Cap ML: ≤ 300 mW Laser inspection power to prevent damage
- Si-Cap ML blank can sustain up to 550mW of Laser inspection power

(intel) Lasertec

Effect of Inspection Laser Power to Ru-ML blank

Criteria: ML reflectivity non-uniformity < 0.5 %

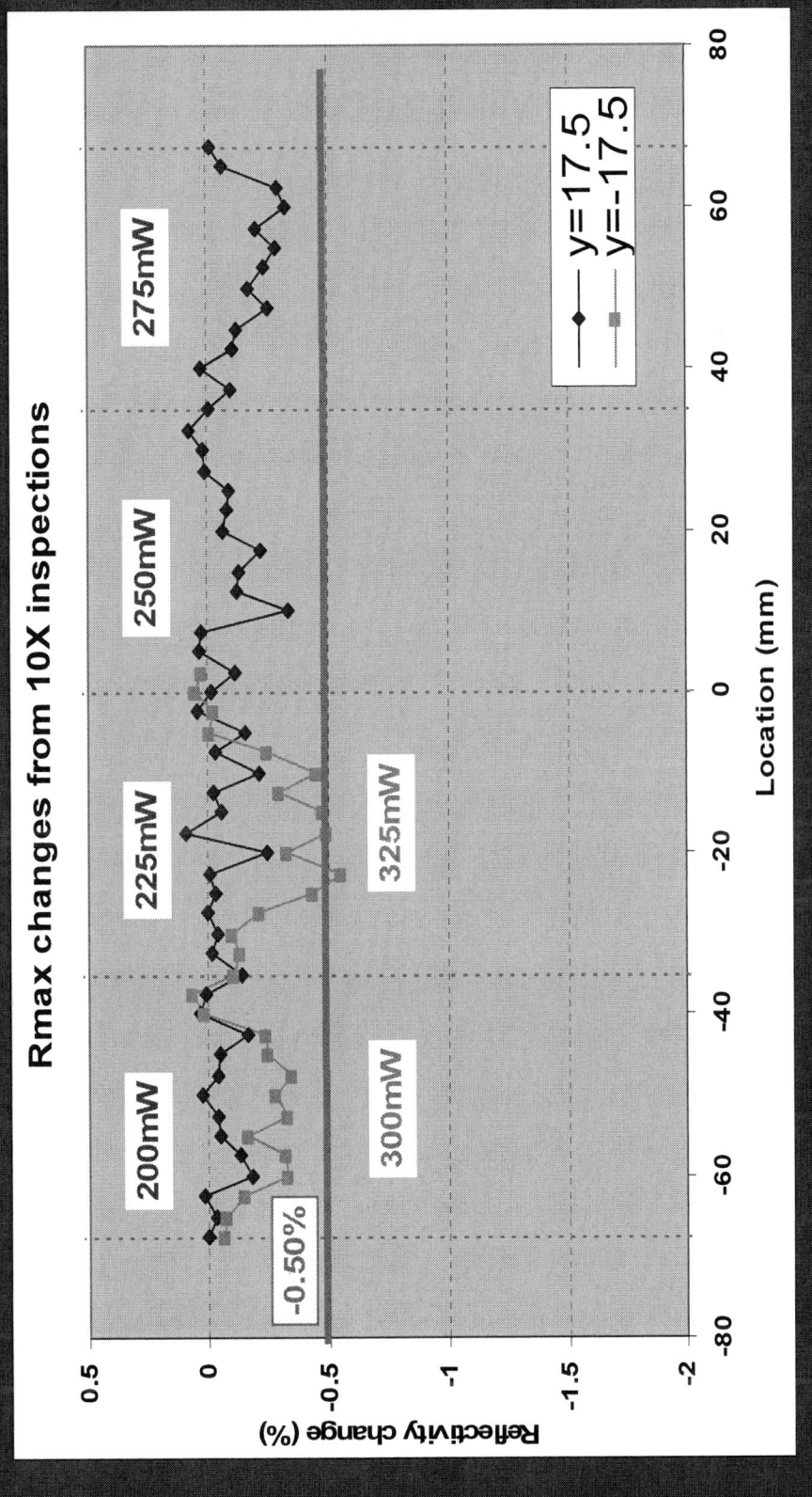

Recommendation: Ru-Cap ML: ≤ 300 mW

(intel) Lasertec

Defect Inspection Requirements & Risk Assessment

Specified item	Spec	2008	2009	2010	2011
Mask Substrate Defect cut-off size	25 nm	45	40	35	25
Defect Density	0 (< 0.005)	0.09	0.02	0.01	0.005
M7360Defect Inspection capability		45 40	35	35	3G tool

Specified item	Spec	2008	2009	2010	2011
Multiliers Defect cut-off size	≤ 30	50	45	35	25
Defect Density	0.0005	0.04	0.009	0.002	0.0005
M7360 Defect Inspection capability		50 45	40	3G tool	3G tool

Legend:
- : Current Capability
- : Process Improvement
- : Tool Upgrade

- 3G tool : Medium Risk
- 3G tool : High Risk

- 3G inspection tool needs to meet 25 nm defect inspection sensitivity
- Inspection capability gap is being addressed to allow sufficient time for tool development

intel

Lasertec

Future Plan

- **Upgrade with Advanced Spatial Filter (ASF) and compensation of shading waveform using the "high-speed tilt-mirror"**
  - upgrade is scheduled for Dec. 2008

- **Improve inspection capability**
  - Qz: from 45nm to 35 nm
  - ML: from 50nm to 40 nm

- **Working with SEMATECH to complete the gap analysis for 3G inspection tool development**
  - Started in July 2008

Summary

- New inspection capability established at Intel
- Options defined to improve M7360 sensitivity

	M7360		
Available	Q3/ 2008	Q4/ 2008	Q1/2009
QZ Insp.	45nm	40nm	35
ML Insp.	50nm	45nm	40
QZ Sens.	31.5 nm		
ML Sens.	39 nm		

: Current Capability
: Process Improvement
: Tool Upgrade

- The need of 3rd generation blank inspection tool being discussed
  - Development time is long

Acknowledgements

- Hoya Corporation: Qz and ML programmed defect masks

- David Pui and Jing Wang, University of Minnesota: SiO_2 sphere particles deposition

2008 International Symposium on Extreme Ultraviolet Lithography

Analysis of Sub-22-nm Aerial Image Using Coherent Scattering Microscopy

Dong Gun Lee*[1], Junki Kishimoto[2], Takeo Watanabe[2], Hiroo Kinoshita[2], Hwan-Seok Seo[1], Dongwan Kim[1], Seong-Sue Kim[1], and HanKu Cho[1]

1. Memory R&D Center, Samsung Electronics Co., LTD.
2. LASTI, University of Hyogo

*gun2.lee@samsung.com

Outline

- Introduction to conventional EUV microscopy and coherent scattering microscopy

- Sub-22-nm aerial image analysis using coherent scattering microscopy
  - Patterning of EUVL mask down to 12.5-nm half pitch (1x)
  - Shadowing effect measurements
  - NILS & through-focus aerial image analyses

- Summary & conclusions

Introduction of Microscopy

- Microscopy as an aerial image measurement tool
  ; Microscopy simulates the scanner with NA&σ emulation

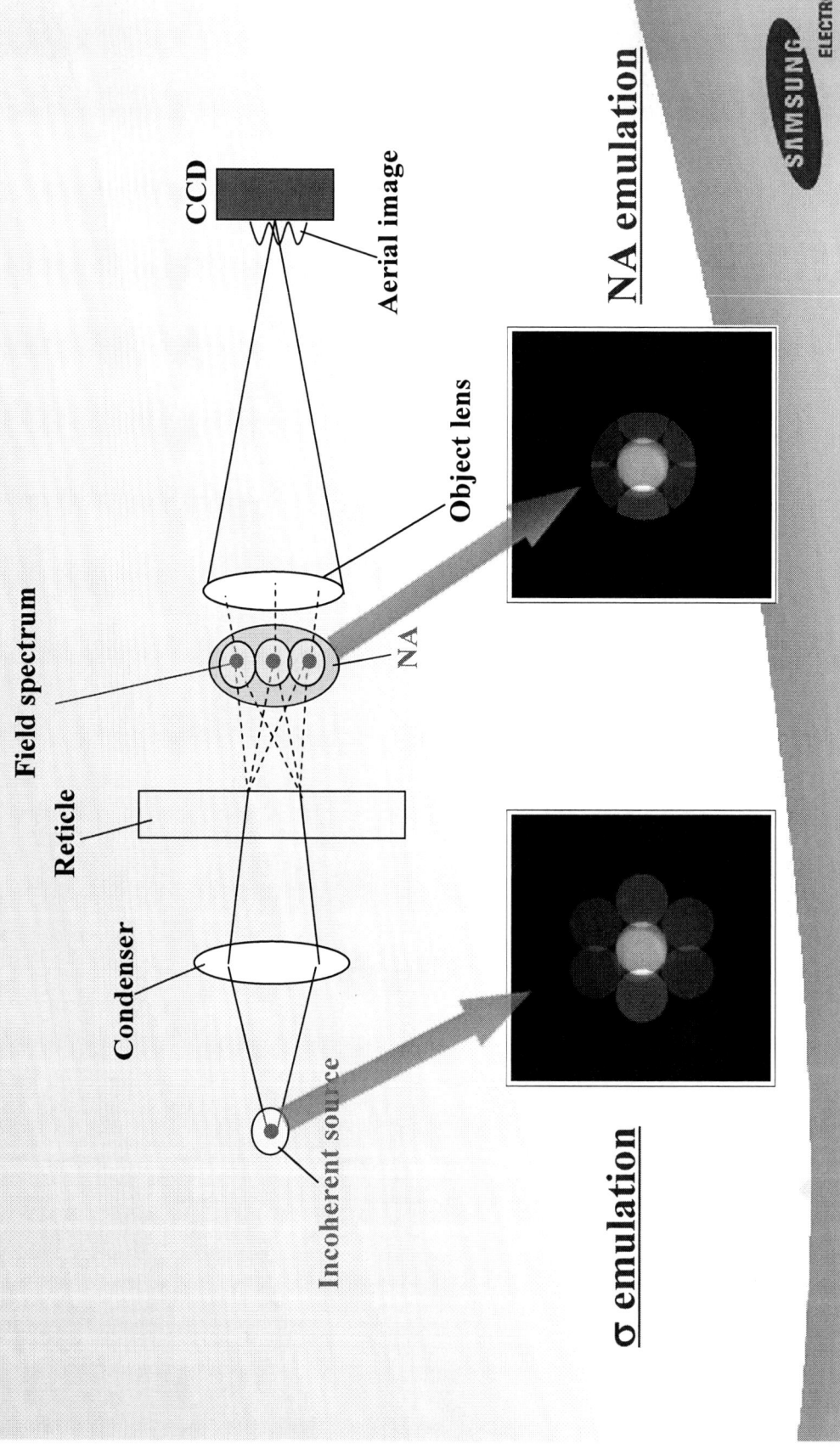

- **Microscopy spectrum**

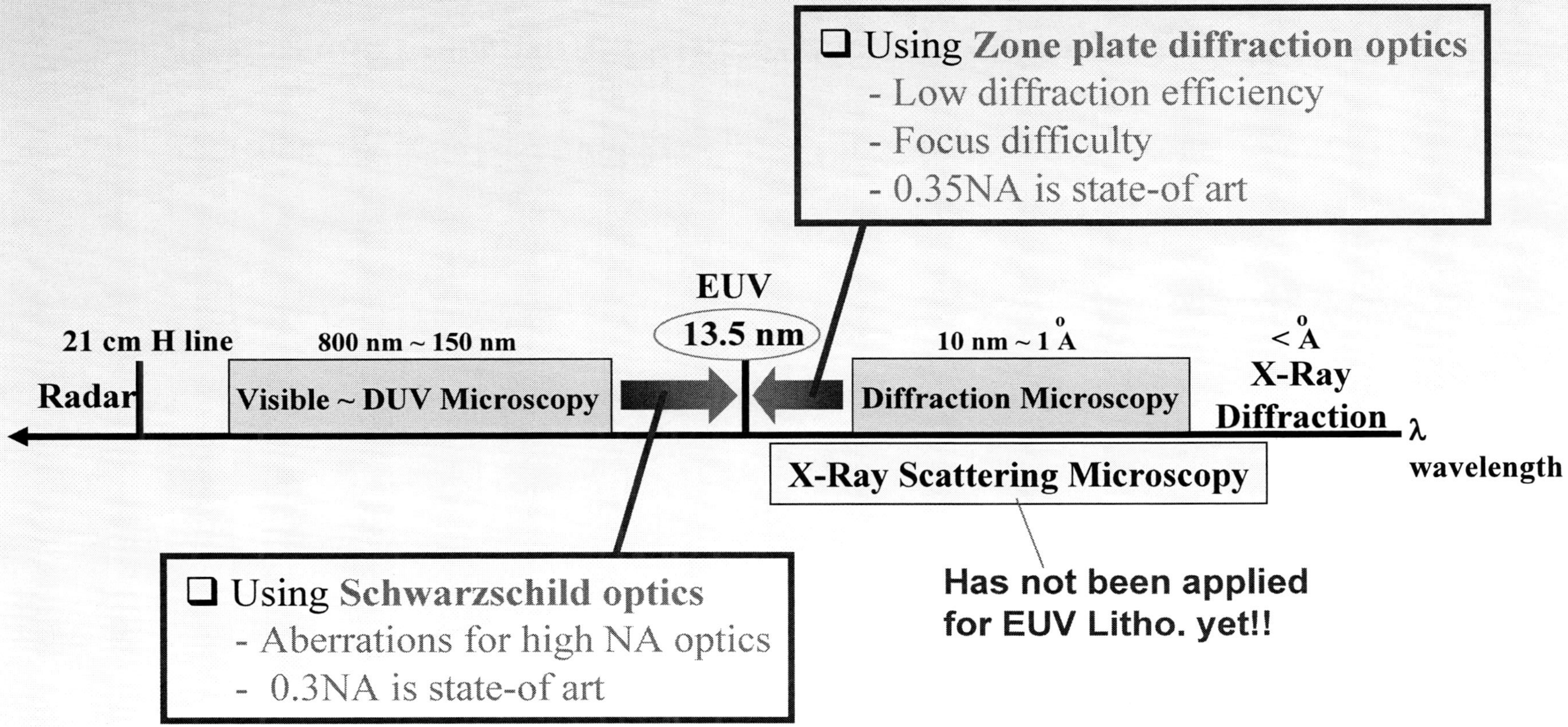

□ Using Zone plate diffraction optics
- Low diffraction efficiency
- Focus difficulty
- 0.35NA is state-of art

□ Using Schwarzschild optics
- Aberrations for high NA optics
- 0.3NA is state-of art

Radar | 21 cm H line | 800 nm ~ 150 nm | Visible ~ DUV Microscopy | EUV 13.5 nm | 10 nm ~ 1 Å | Diffraction Microscopy | X-Ray Scattering Microscopy | < Å | X-Ray Diffraction | wavelength λ

Has not been applied for EUV Litho. yet!!

■ To overcome disadvantages of the conventional microscopy, an alternative microscopy, like x-ray scattering microscopy, could be preferable for the application of EUV litho.

SAMSUNG ELECTRONICS

EUV scattering microscopy as an aerial image measurement tool: The shift of paradigm

ArF (193 nm)

EUV (13.5 nm)

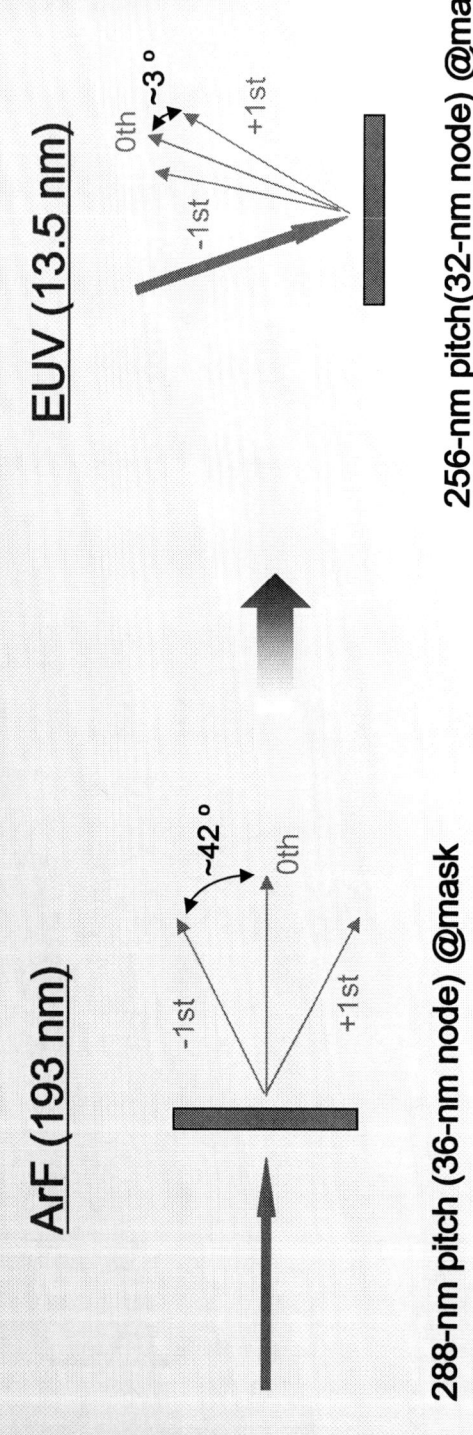

288-nm pitch (36-nm node) @mask

256-nm pitch(32-nm node) @mask

- Pattern shrinks by about 1/2.5, but the wavelength shrinks by only 1/14

➢ Possible to measure 2-D scattering spectrum at once !!

SAMSUNG ELECTRONICS

Coherent Scattering Microscopy

- **High NA (>0.59 for 4X) Aerial Image Analysis Tool without aberrations**
- **Test of EUVL Extendibility down to 12-nm node**

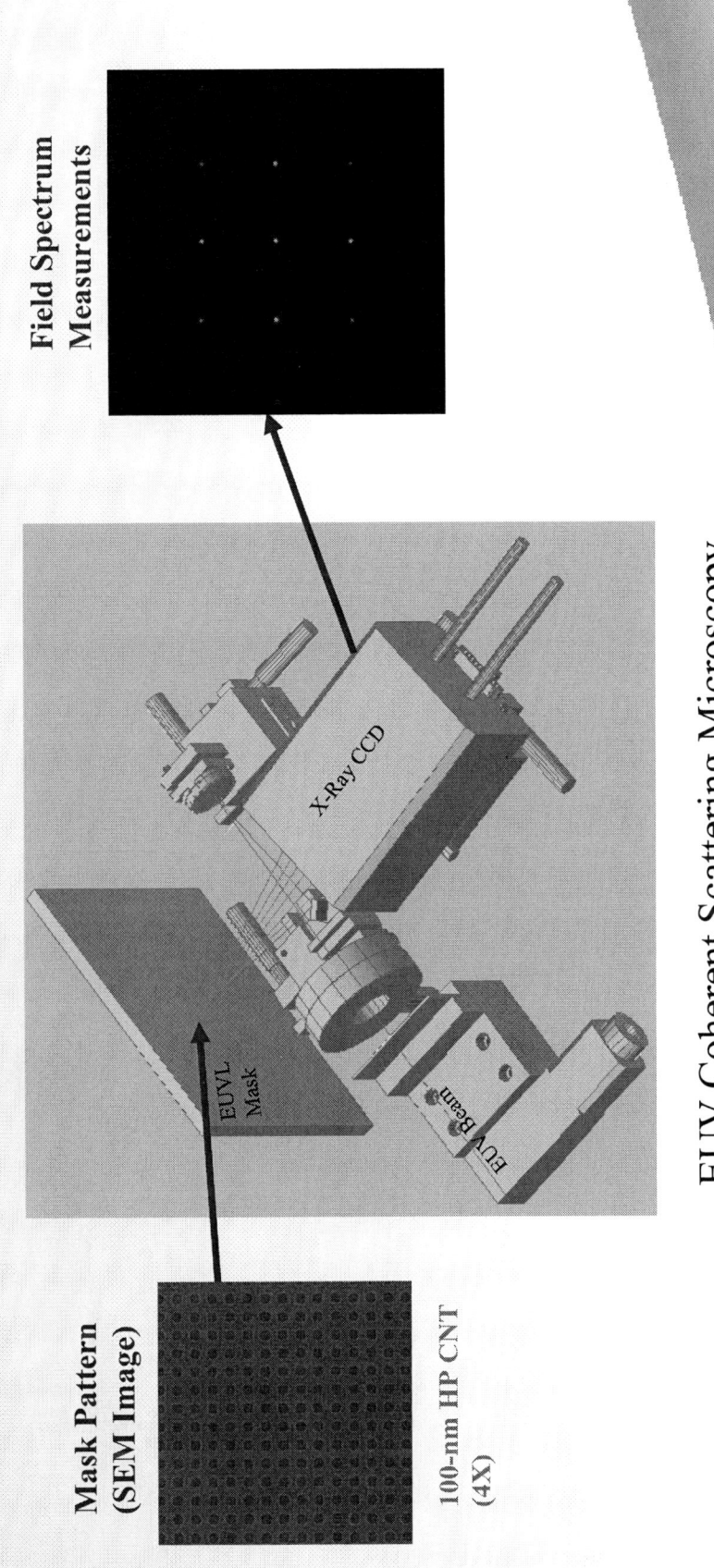

Field Spectrum Measurements

Mask Pattern (SEM Image)

100-nm HP CNT (4X)

EUV Coherent Scattering Microscopy

SAMSUNG ELECTRONICS

From Field Spectrum To Aerial Image

- Reconstruction of mask image from the field spectrum
- Calculation of aerial image with various NA & σ conditions from the mask image

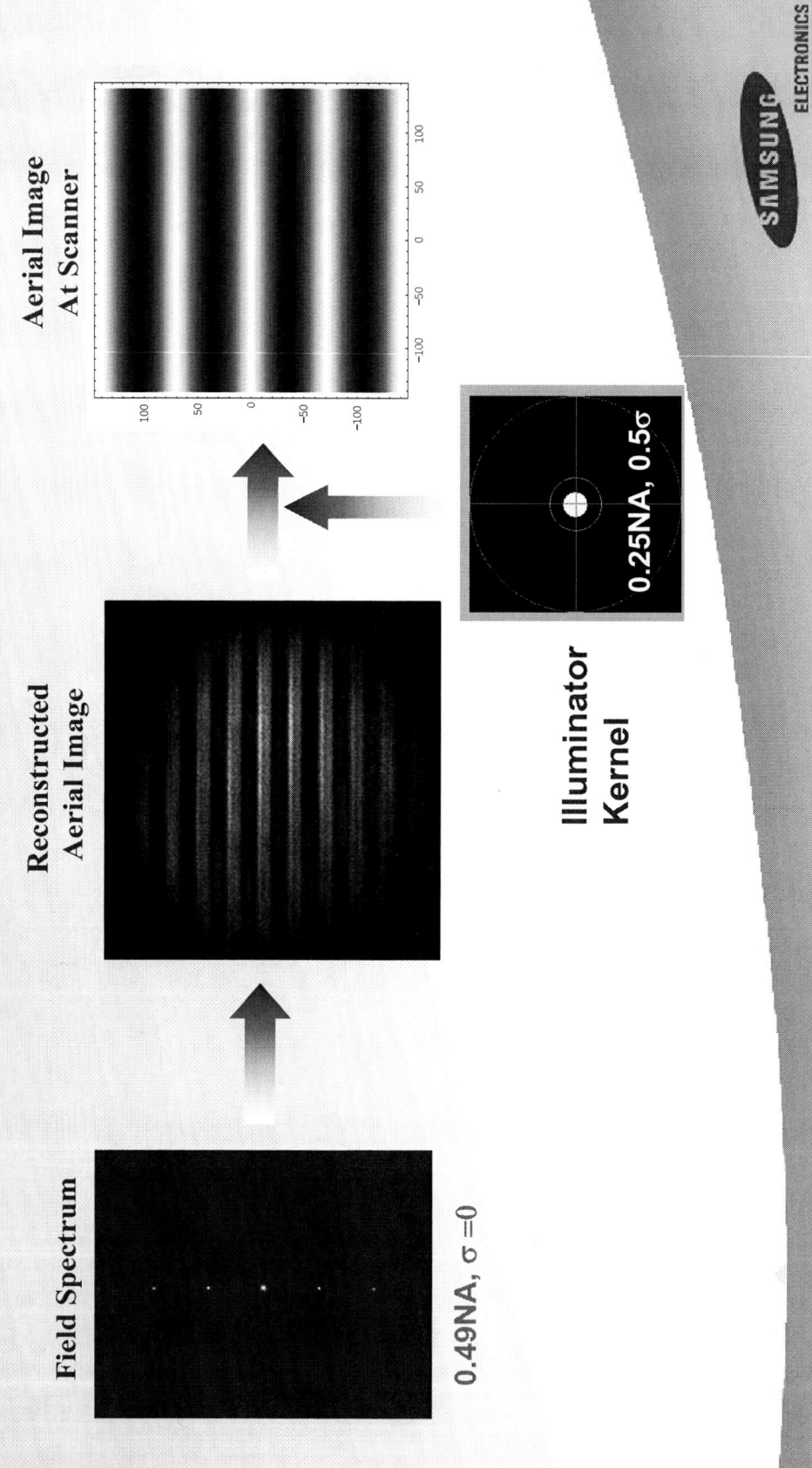

Aerial Image
At Scanner

Reconstructed
Aerial Image

Field Spectrum

0.49NA, σ =0

Illuminator
Kernel

0.25NA, 0.5σ

Sub-22 nm Aerial Image Analysis

● EUV mask patterning down to 50-nm half pitch at mask

SEM Image of L/S Pattern on Multilayer

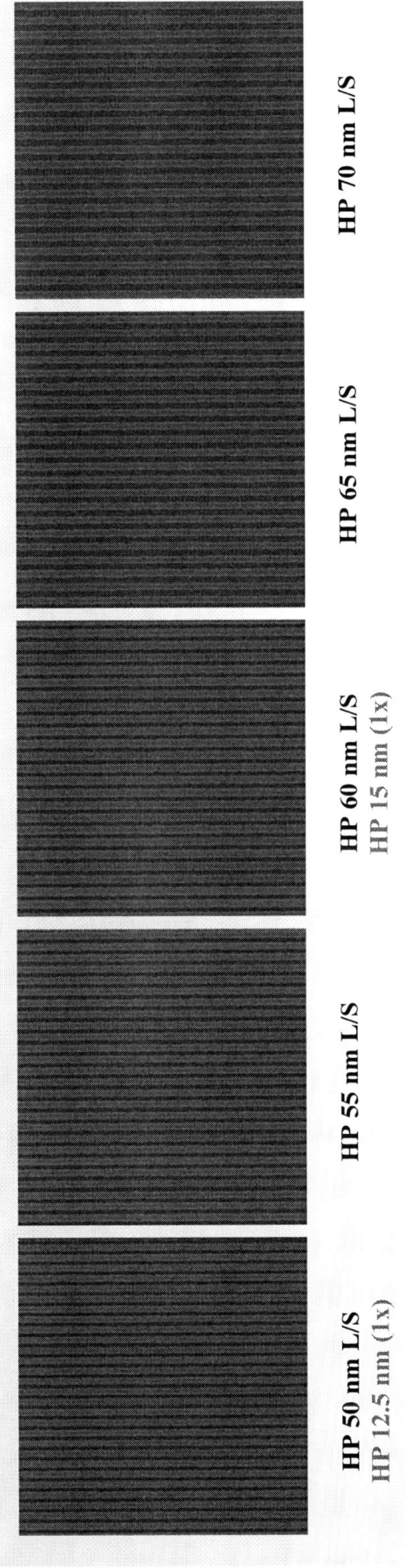

HP 50 nm L/S
HP 12.5 nm (1x)

HP 55 nm L/S

HP 60 nm L/S
HP 15 nm (1x)

HP 65 nm L/S

HP 70 nm L/S

CSM field spectrum measurement results

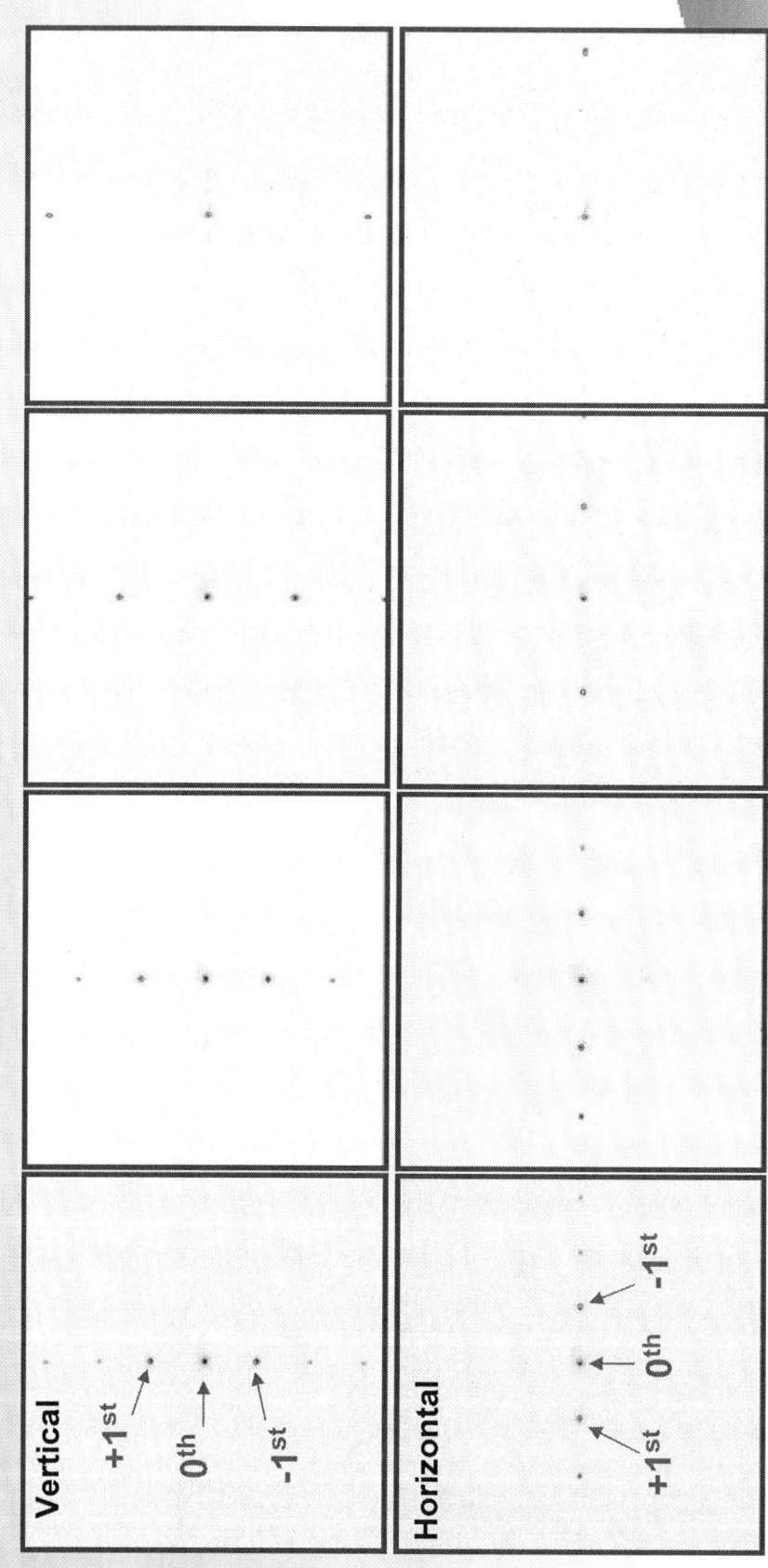

CSM field spectrum measurement results

For horizontal L/S pattern (with shadowing effect)

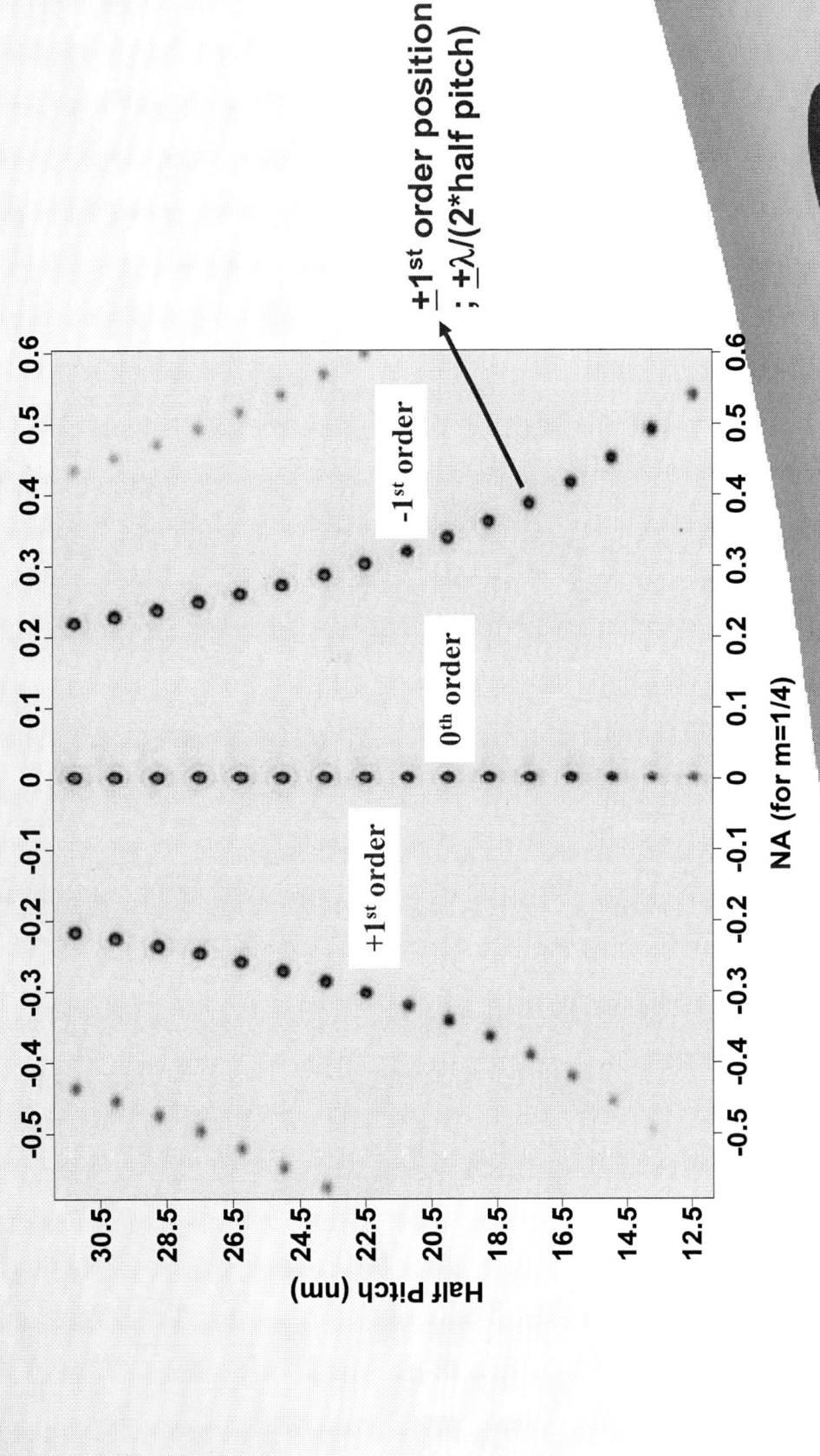

CSM measurements vs. 3D simulation (S. lithography EUV)

For 26 nm L/S Pattern with 50 HP (4X)

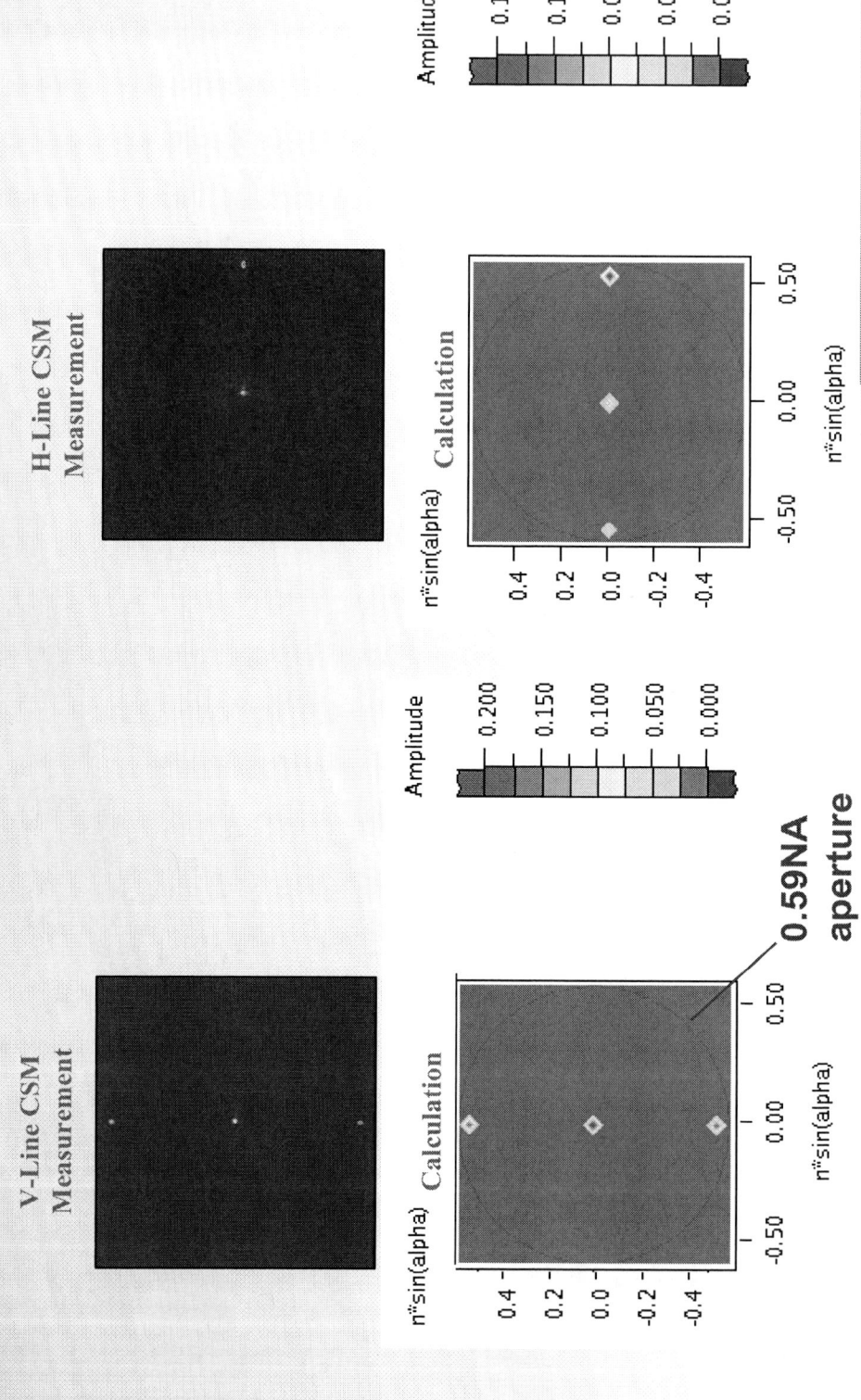

CSM measurements vs. 3D simulation (S. Lithography EUV)

- Experiment & Simulation Conditions

NA/ σ	Mask CD (CL, 4X)	Half Pitch (1X)	Target CD (1X)
0.45/ 0.9	26 nm	12.5 nm	12.5 nm

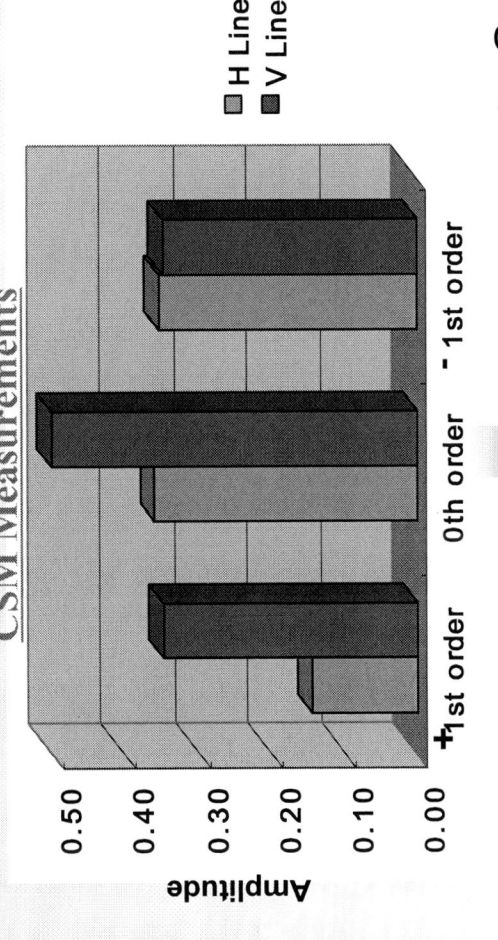

CSM Measurements

- Good agreement !!

	H-Line Contrast	V-Line Contrast	H-Line NILS	V-Line NILS
CSM Measurements	0.62	0.59	1.95	1.87
3-D Calculation	0.68	0.58	1.91	1.87

Measured and Calculated HV Bias

Shadowing Effects For L/S Pattern

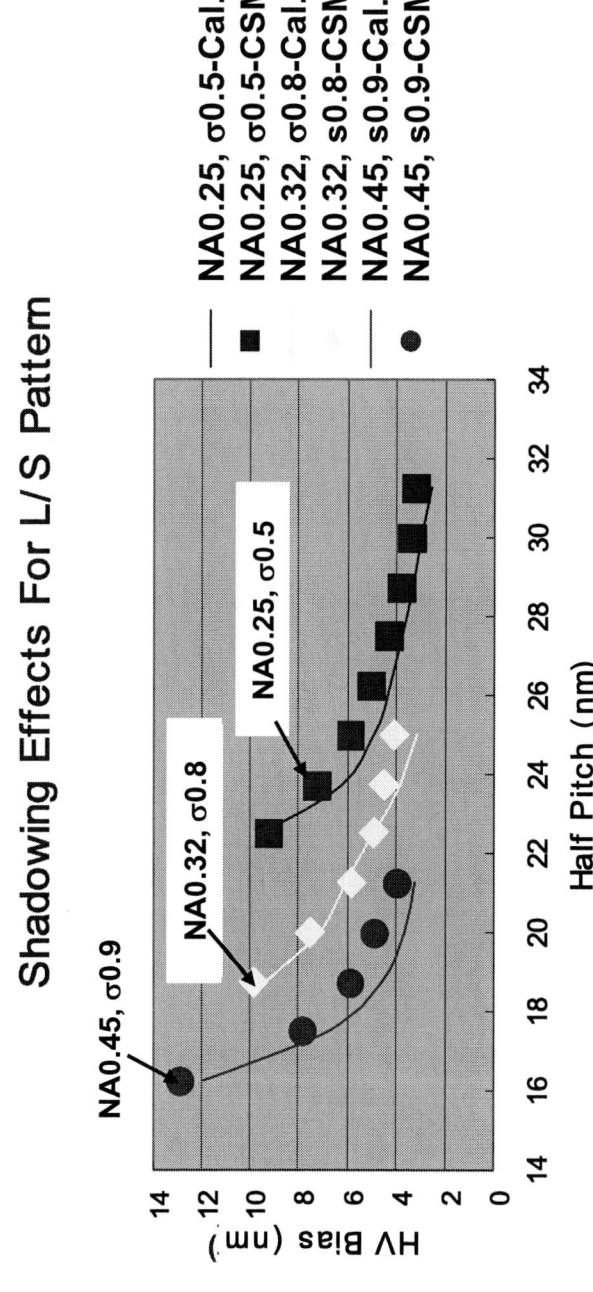

NA0.45, σ0.9

NA0.32, σ0.8

NA0.25, σ0.5

— NA0.25, σ0.5-Cal.
■ NA0.25, σ0.5-CSM
— NA0.32, σ0.8-Cal.
◆ NA0.32, s0.8-CSM
— NA0.45, s0.9-Cal.
● NA0.45, s0.9-CSM

For small k value & large MEEF

- HV Bias due to the shadowing effect also increases with reducing the pattern pitch.
- Higher NA reduces the HV bias.

NILS depending on pattern pitch and NA

NILS (Normalized Image Log Slpoe)

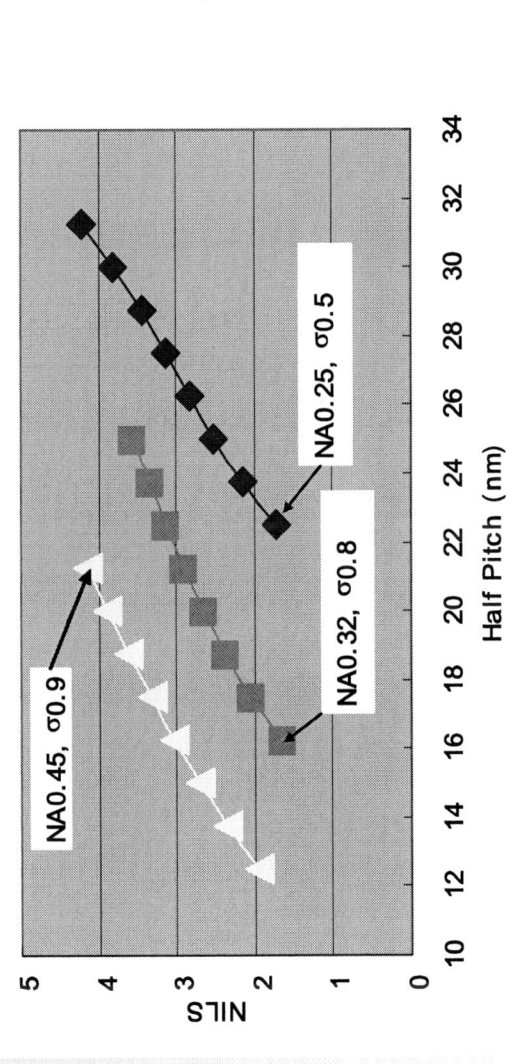

- NA0.25, σ0.5
- NA0.32, σ0.8
- NA0.45, σ0.9

- Mask pattern was biased by ~ -6 nm (1X)
- In exposure conditions with NA = 0.25 and σ = 0.5, NILS are larger than 2.0 for the patterns with half pitch above 24 nm

• EUVL mask CNT patterning down to 50-nm half pitch at mask

SEM Image of CNT Pattern

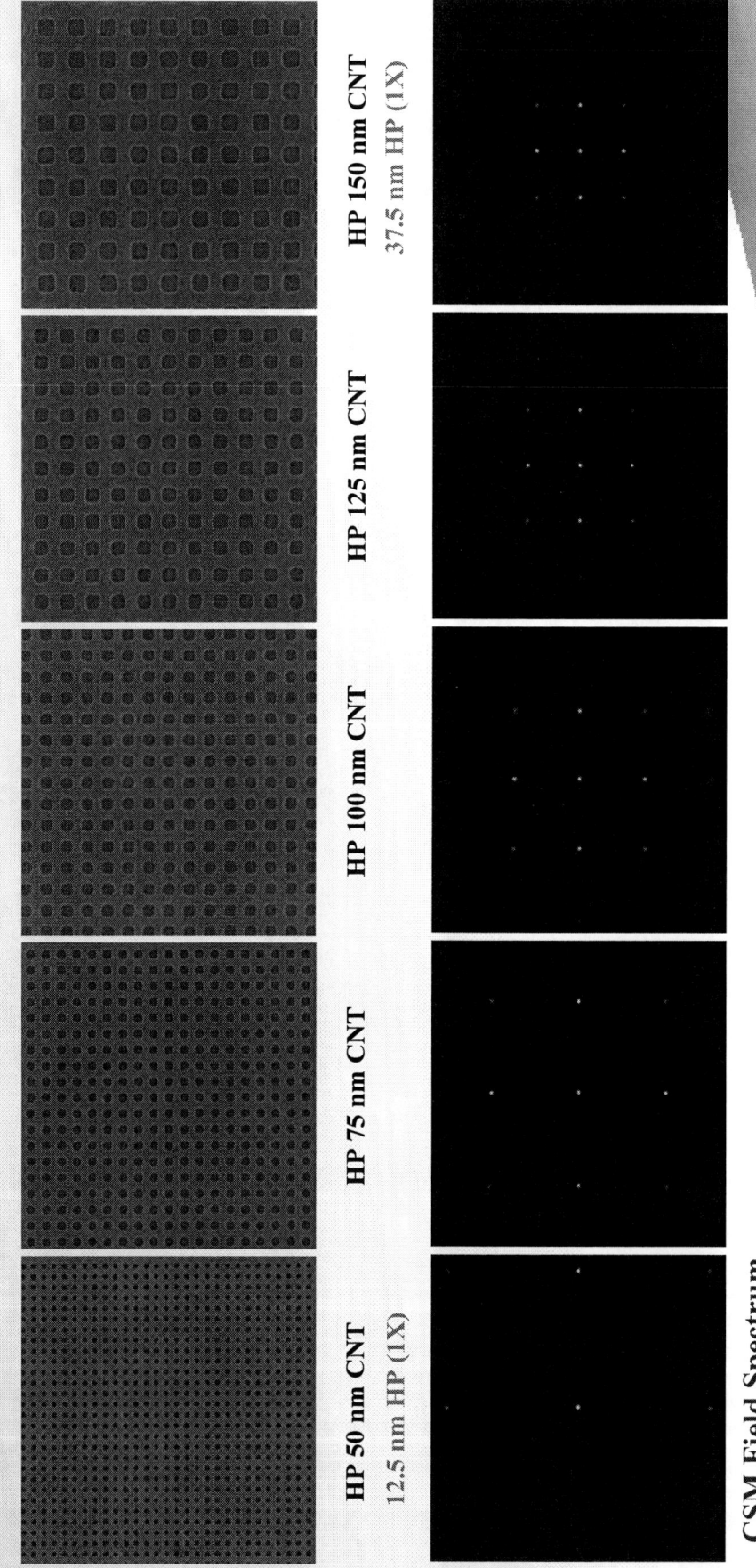

HP 50 nm CNT	HP 75 nm CNT	HP 100 nm CNT	HP 125 nm CNT	HP 150 nm CNT
12.5 nm HP (1X)				37.5 nm HP (1X)

CSM Field Spectrum

HV bias measurement results for contact hole

Shadowing Effects For Contact

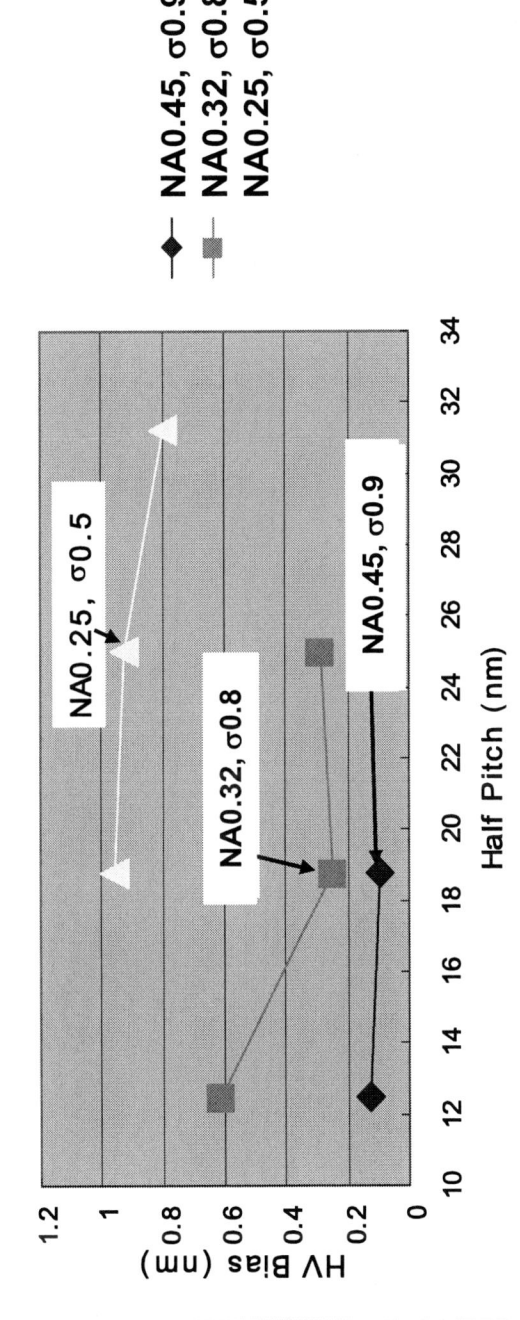

- NA0.45, σ0.9
- NA0.32, σ0.8
- NA0.25, σ0.5

- CSM shows that the HV Bias of the contact is less than 1 nm in the analysis region

Through-focus aerial image measurements

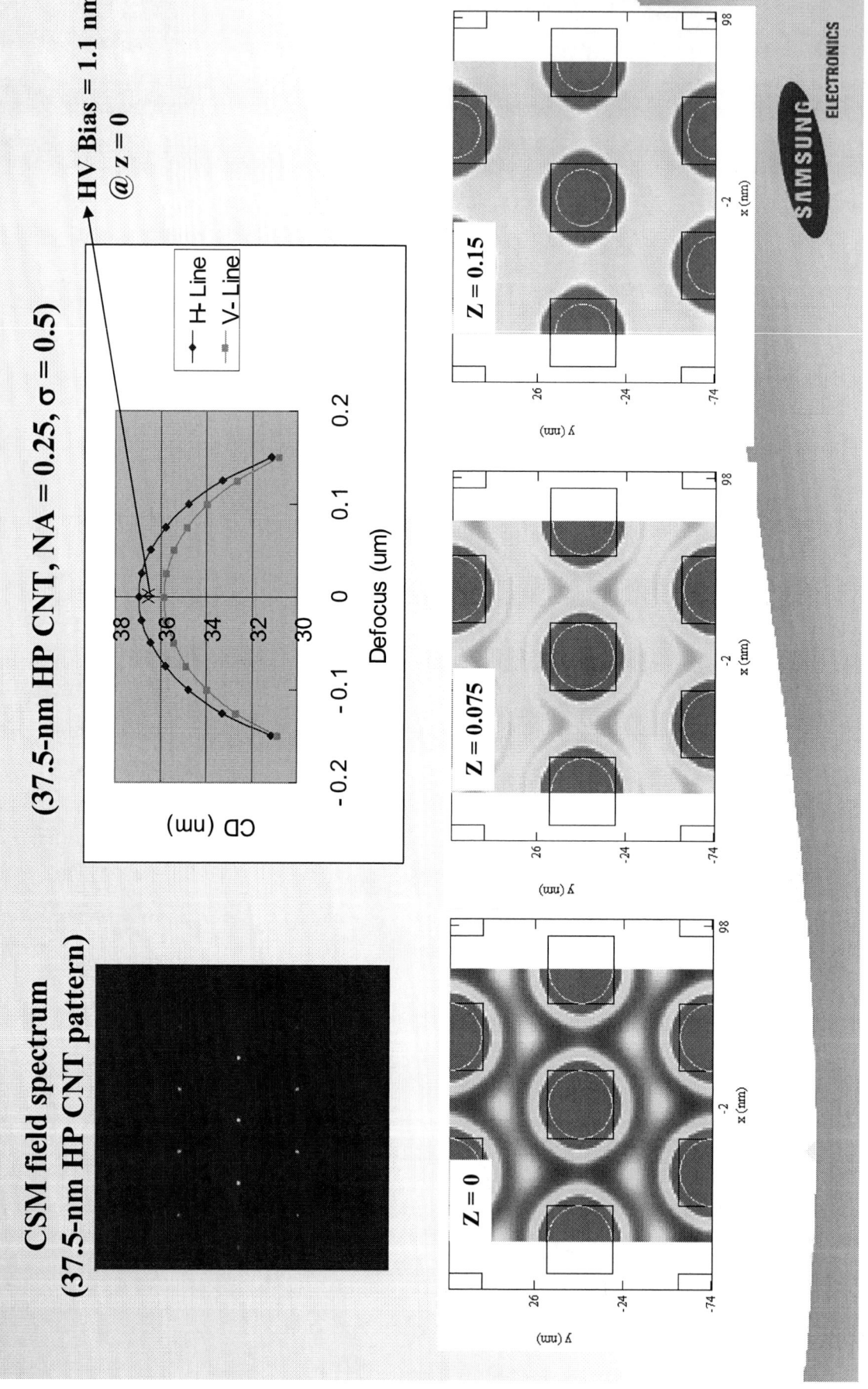

Summary & conclusions

- We have developed a CSM tool for sub-22-nm node EUV lithography.
  - High NA of 0.59 which enables us to measure aerial images down to 12-nm node without aberrations
  - Emulating exposure tool with the variable NA & σ
  - Reliable actinic CD measurements without focus alignment (for 32-nm node pattern, CD repeatability 0.14 nm in 3σ)

- Aerial image analysis results:
  - HV bias for L/S pattern increases with decreasing pattern size.
  - But, HV bias for contact hole below 32-nm half pitch is less than 1 nm.

- Thin absorber is preferable to reduce shadowing effects.

Acknowledgements

The authors would like to thank....

- Laboratory of Advanced Science and Technology for Industry (Prof. H. Kinoshita and Dr. T. Watanabe) at University of Hyogo for their helps to use beamline in the NewSUBARU synchrotron facility.

Aerial Image Linewidth Measurement Capabilities of the Actinic Inspection Tool

The SEMATECH Berkeley Actinic Inspection Tool

AIT

An EUV-wavelength mask inspection microscope

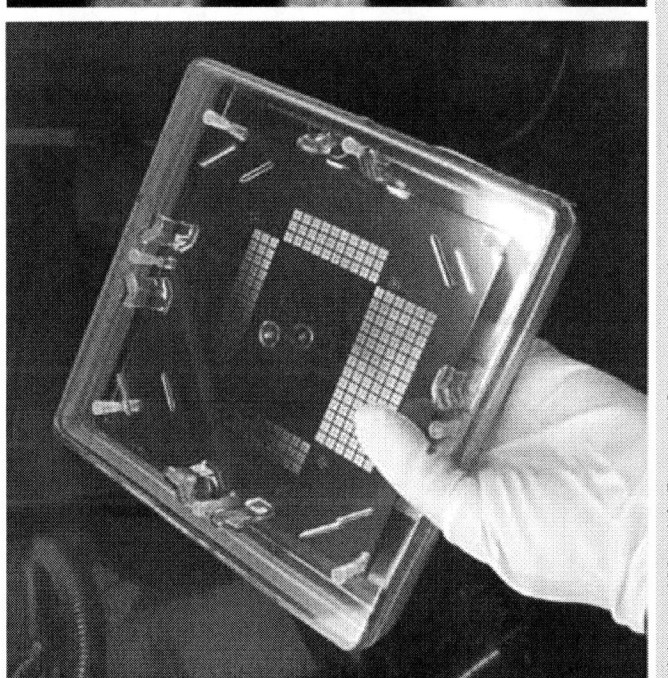

Advanced Materials Research Center, AMRC, International SEMATECH Manufacturing Initiative, and ISMI are servicemarks of SEMATECH, Inc. SEMATECH, the SEMATECH logo, Advanced Technology Development Facility, ATDF, and the ATDF logo are registered servicemarks of SEMATECH, Inc. All other servicemarks and trademarks are the property of their respective owners.

Aerial Image Linewidth Measurement Capabilities of the Actinic Inspection Tool

LBNL: Kenneth A. Goldberg,
Iacopo Mochi, Patrick Naulleau

AMD: Bruno LaFontaine

SEMATECH: Sungmin Huh

AIT
The SEMATECH Berkeley
Actinic Inspection Tool

Upgrades Resolution (Uniformity)

Aberrations Modeling, Measurement, Reduction

Linewidth Measurement, Repeatability

AIT: An EUV Zoneplate Microscope

$\lambda = 13.4 \pm 0.01$ nm, *tunable*

{0.25, 0.30, 0.35} **NA** (4x)

907x–1000x mag

25–35 sec/exposure

250 images / 8h

CCD

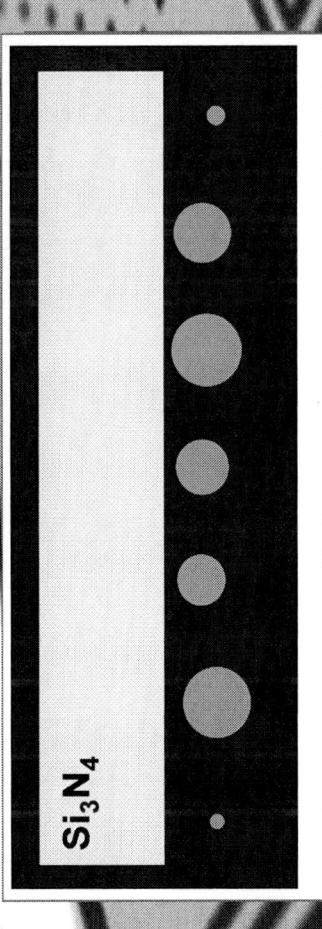

Si$_3$N$_4$

5 lenses—different mag and NA

SEMATECH Berkeley Actinic Inspection Tool

New CCD + higher mag = higher contrast.
Higher-NA lens → Improved resolution.

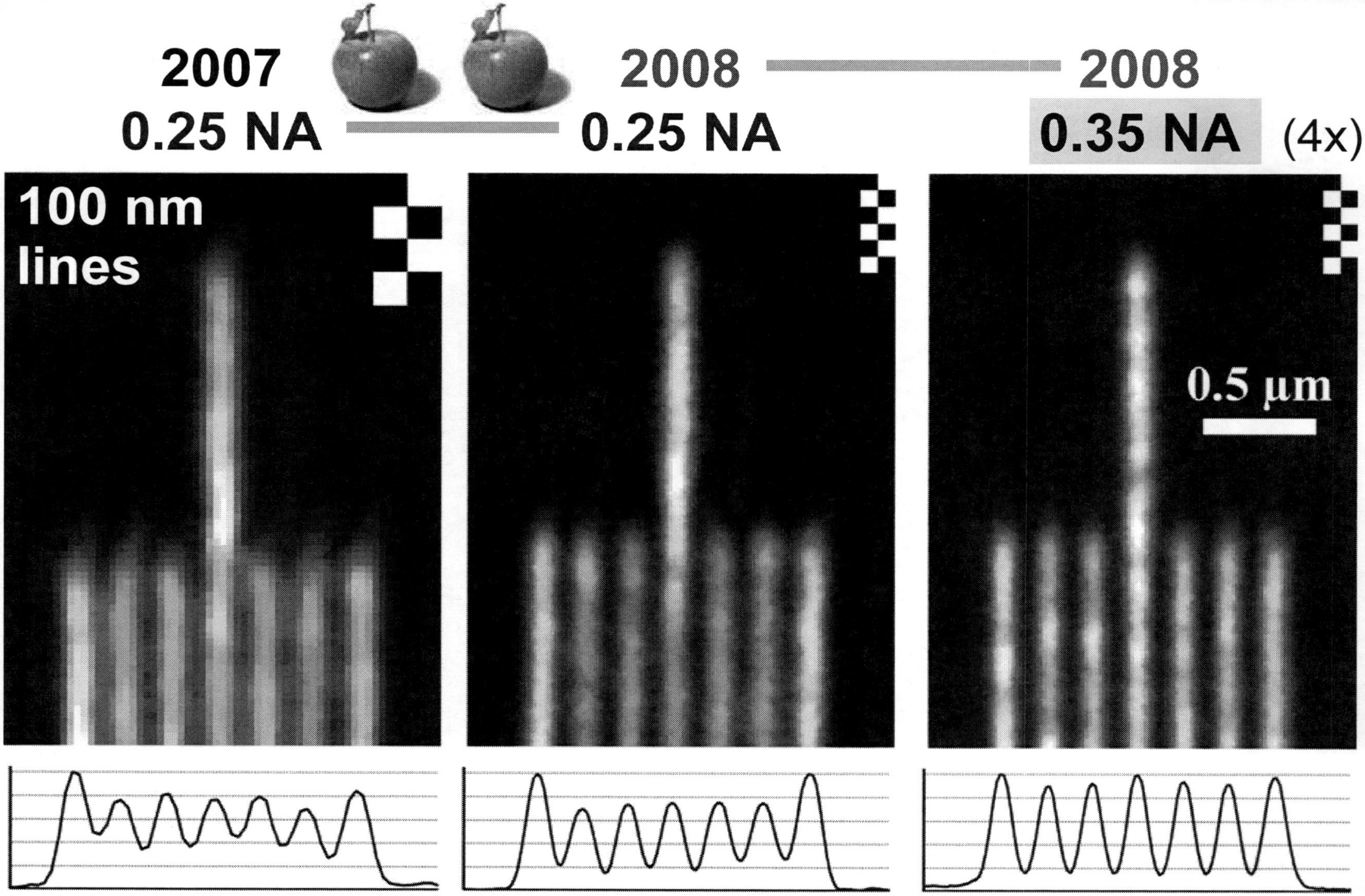

EIPBN 2008

2008 ——— 0.35 NA (4x)

0.5 μm

2008 ——— 0.25 NA

2007 —— 0.25 NA

100 nm lines

SEMATECH Berkeley Actinic Inspection Tool

Contrast Transfer Function (CTF)
We now achieve much higher contrast below 225 nm hp

2007 and 2008 darkfield CTF

150 nm lines

2007

2008

2008 ↑NA

Legend:
- 2007: 0.25 NA
- 2008: 0.25 NA
- 2008: 0.35 NA
- 2008: 0.35 NA

contrast

Mask [nm] 0

4x [nm] 0

SEMATECH Berkeley Actinic Inspection Tool

Improving performance through alignment

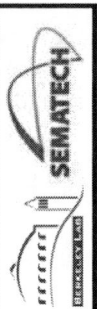

The off-axis zoneplates have a **small field of view**, 5–8 μm, so alignment is critical.

imaging performance

alignment (aberration minimization)

detailed feedback

through-focus image analysis

Accurate measurement of **contrast, defects, line-width, etc.** all rely on low aberrations.

SEMATECH Berkeley Actinic Inspection Tool

175 nm Contacts
32-µm-wide area

5 µm

SEMATECH Berkeley Actinic Inspection Tool

175 nm Contacts
16-µm-wide area

2 µm

SEMATECH Berkeley Actinic Inspection Tool

175 nm Contacts
8-μm-wide area

1 μm

SEMATECH Berkeley Actinic Inspection Tool

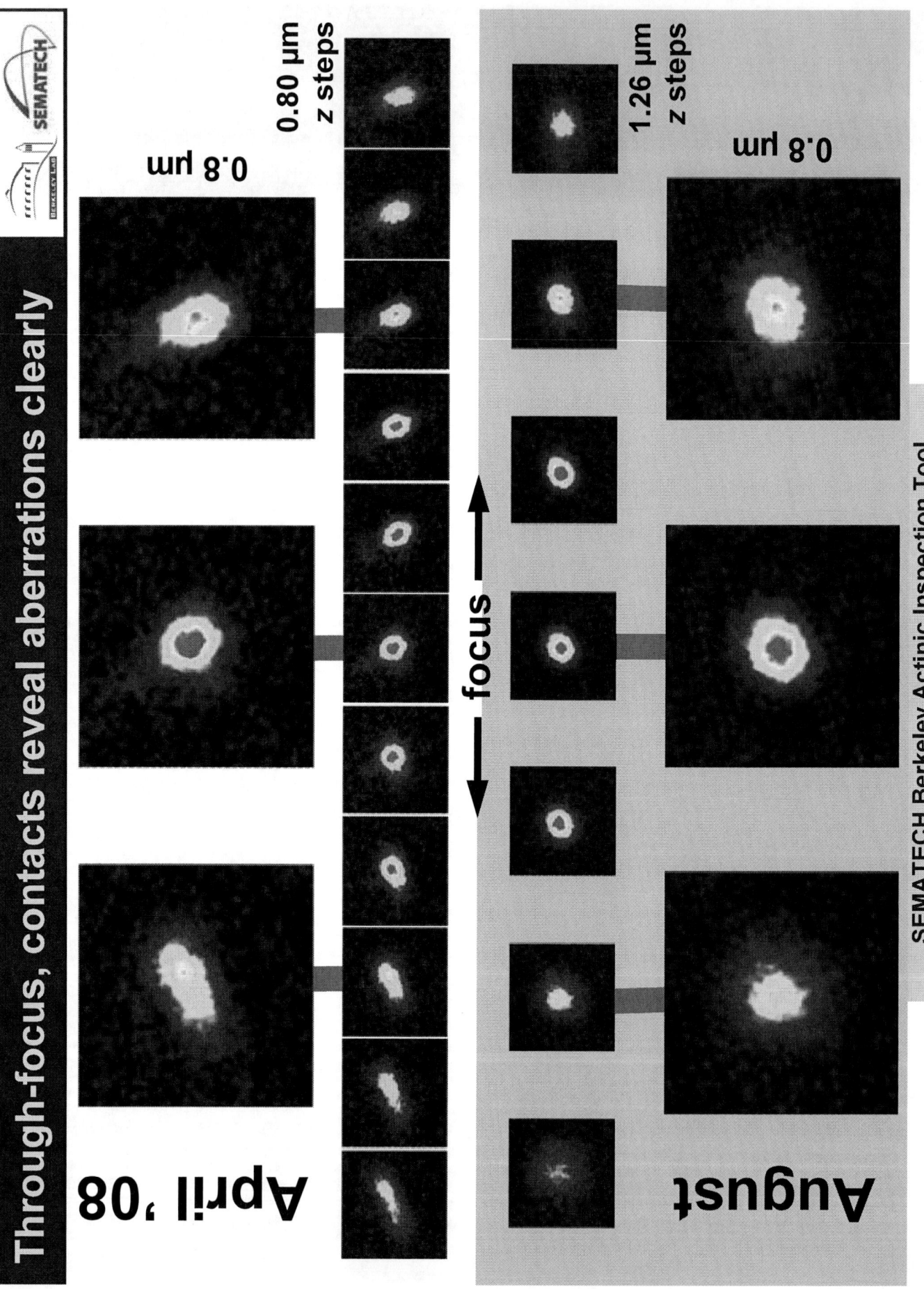

Through-focus, contacts reveal aberrations clearly

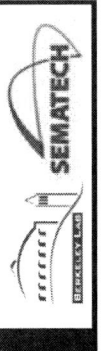

April '08

0.8 μm

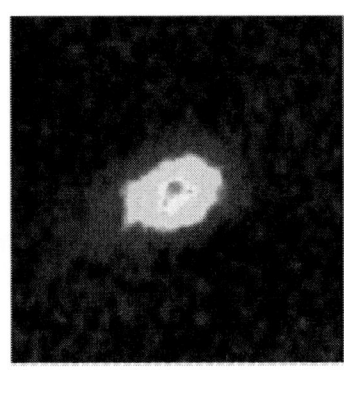

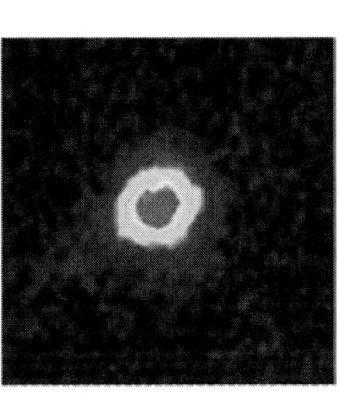

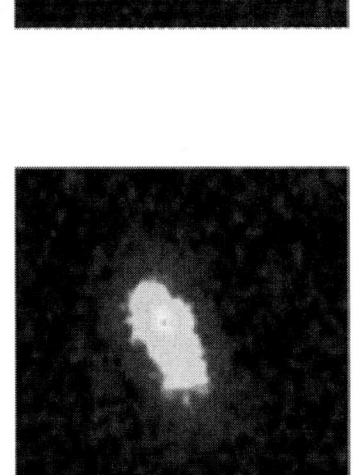

Astigmatism

Astigmatic Displacement (AD) ≥ 2.0 μm (125 nm *wafer*)

RMS Wavefront Error ≥ **0.23 Waves RMS**

Astigmatism

Astigmatic Displacement (AD) ≥ 0.3 μm (19 nm *wafer*)

RMS Wavefront Error ≥ **0.08 Waves RMS**

August

0.8 μm

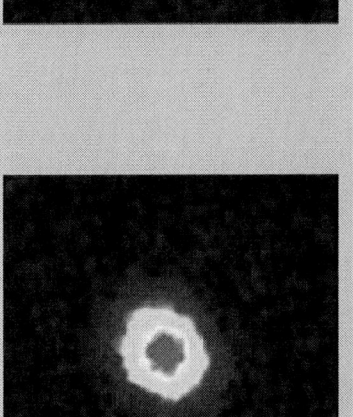

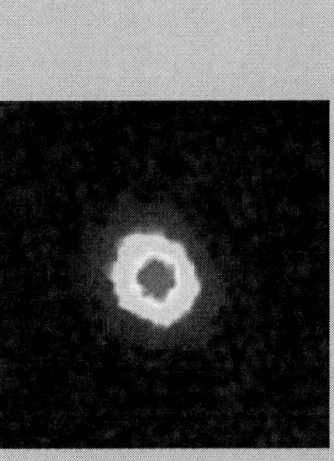

SEMATECH Berkeley Actinic Inspection Tool

Models help assess our position within the field

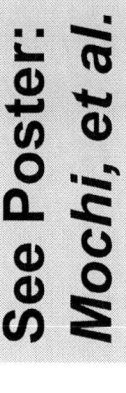

See Poster: *Mochi, et al.*

Measured Astigmatism

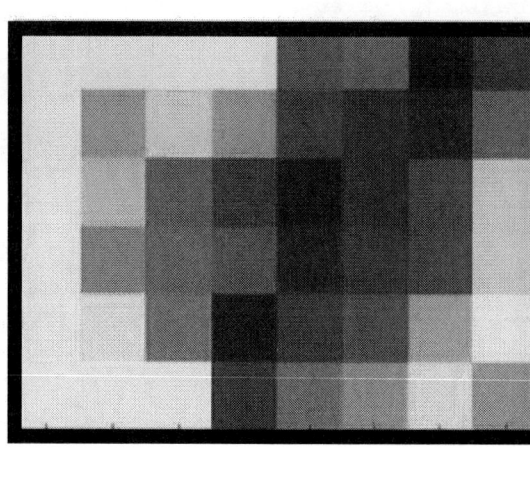

22 µm
one image

Zoneplate field of view
Field-dependent Astigmatism Model

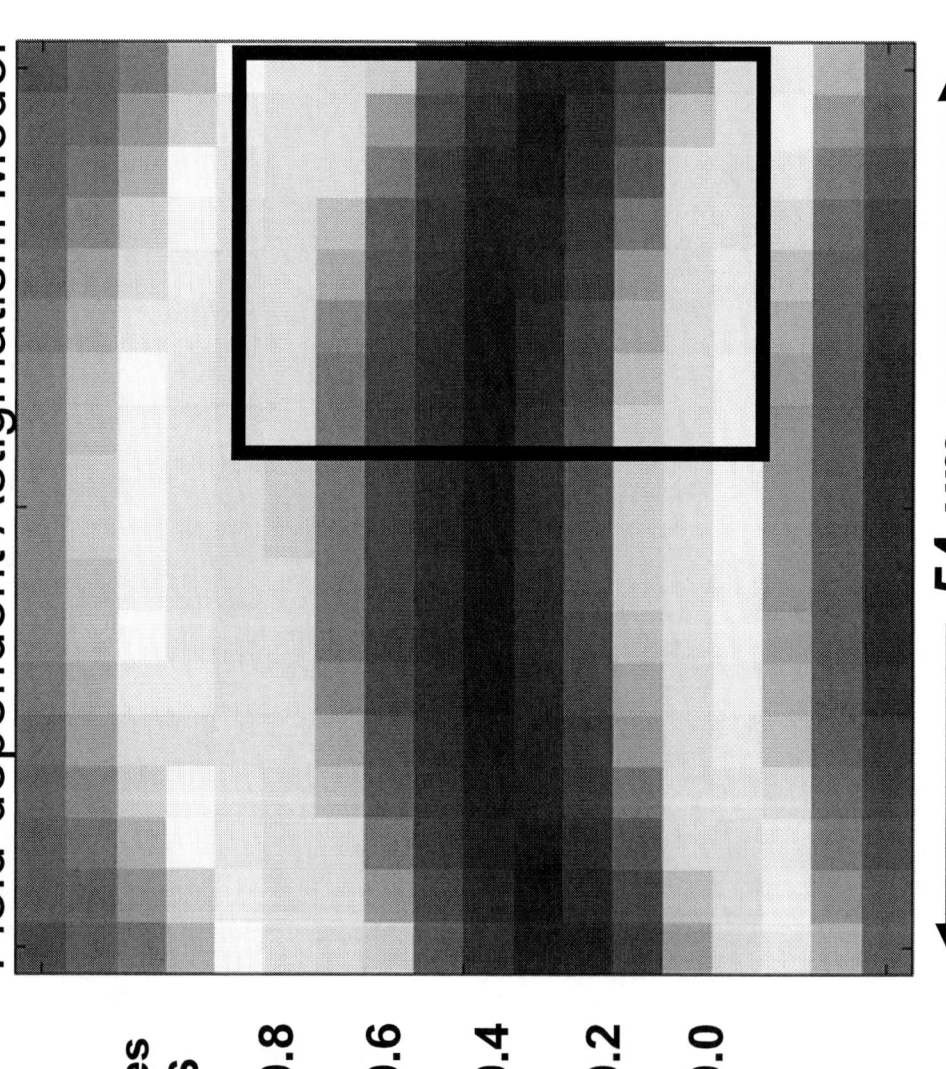

54 µm

Waves RMS

0.8 0.6 0.4 0.2 0.0

SEMATECH Berkeley Actinic Inspection Tool

Testing line-width measurement capabilities

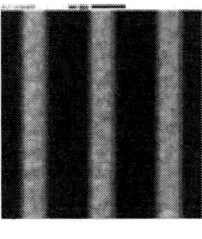
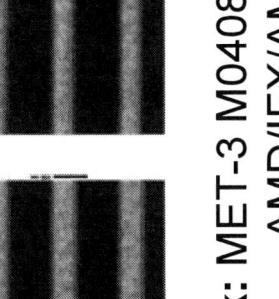

Mask: MET-3 M0408-01
AMD/IFX/AMTC

1. Measuring Biased lines

- 1:1 lines = intensity-threshold reference.
- 10 regions per image *through-focus*.
- LW based on best-focus images.

Mask: PDM-AIT[1]
Samsung/SEMATECH

2. Repeatability testing

- *through-focus* 10x per location.
- 10 regions per image.
- best focus = highest contrast.
- Assume 1:1 lines, calculate a global "best threshold" value.
- → *Statistics.*

SEMATECH Berkeley Actinic Inspection Tool

Measuring CD from biased lines — 1000 nm pitch

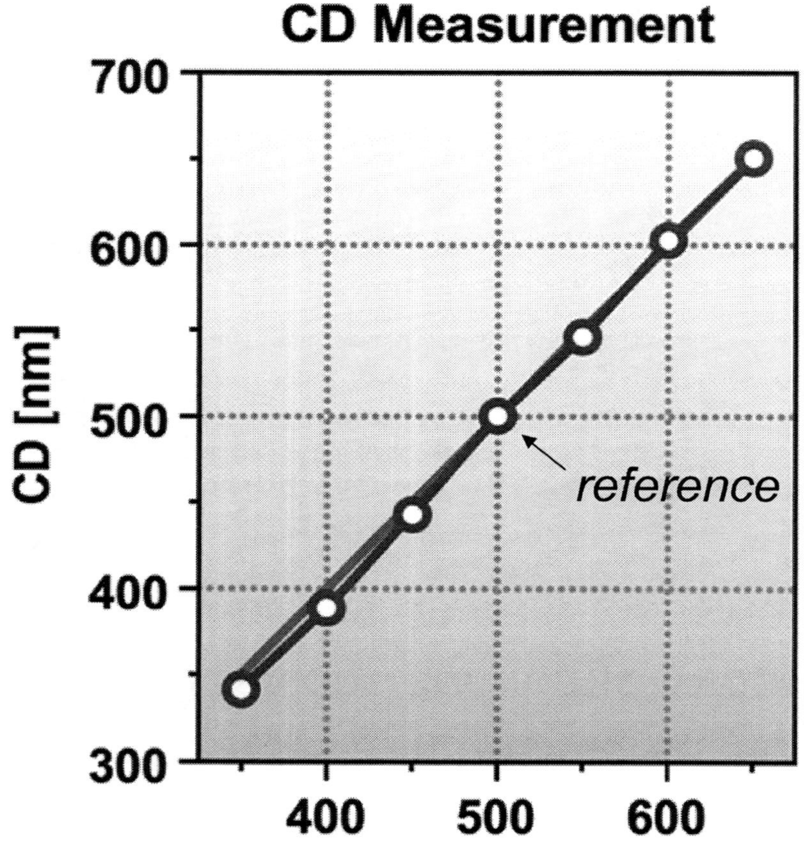

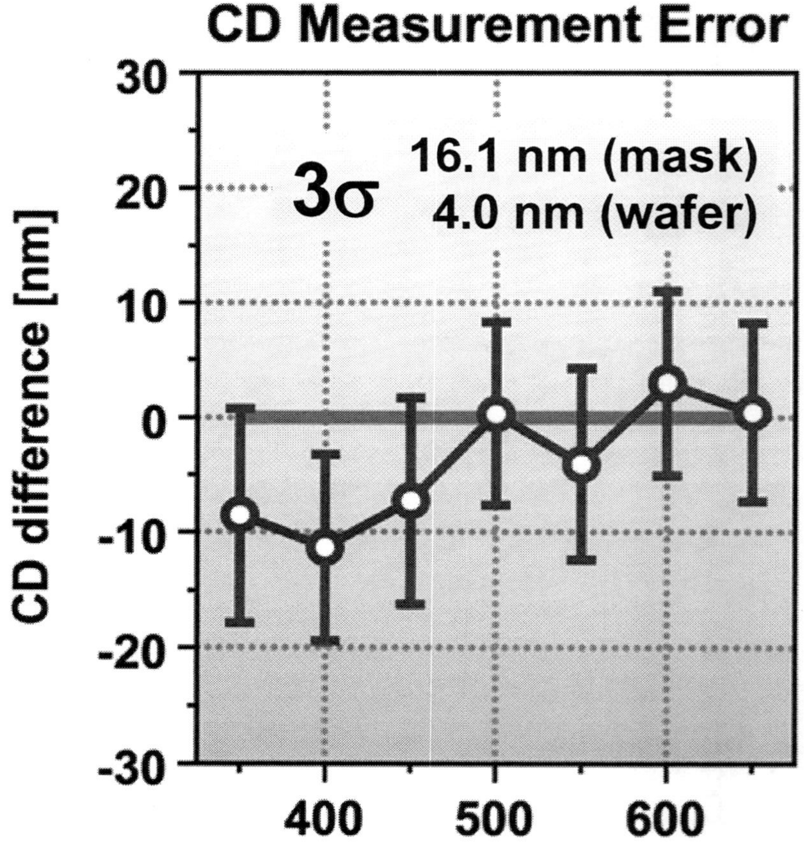

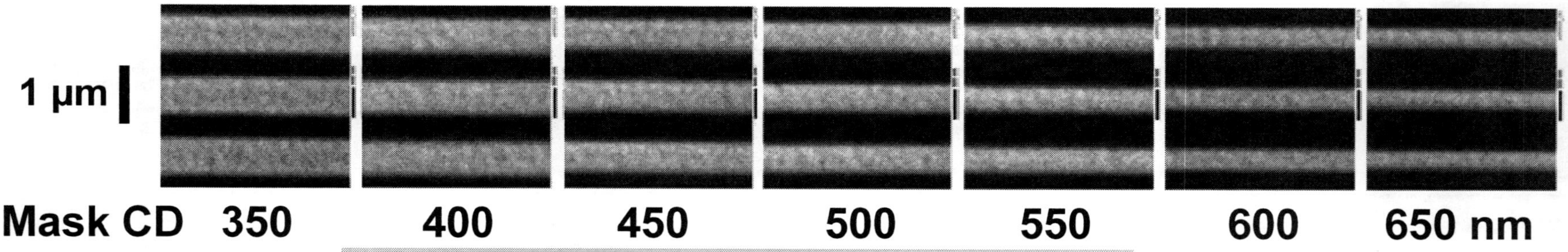

CD Measurement Error

3σ 16.1 nm (mask)
       4.0 nm (wafer)

CD difference [nm]

CD Measurement

reference

CD [nm]

SEMATECH Berkeley Actinic Inspection Tool

1 µm

Mask CD   350   400   450   500   550   600   650 nm

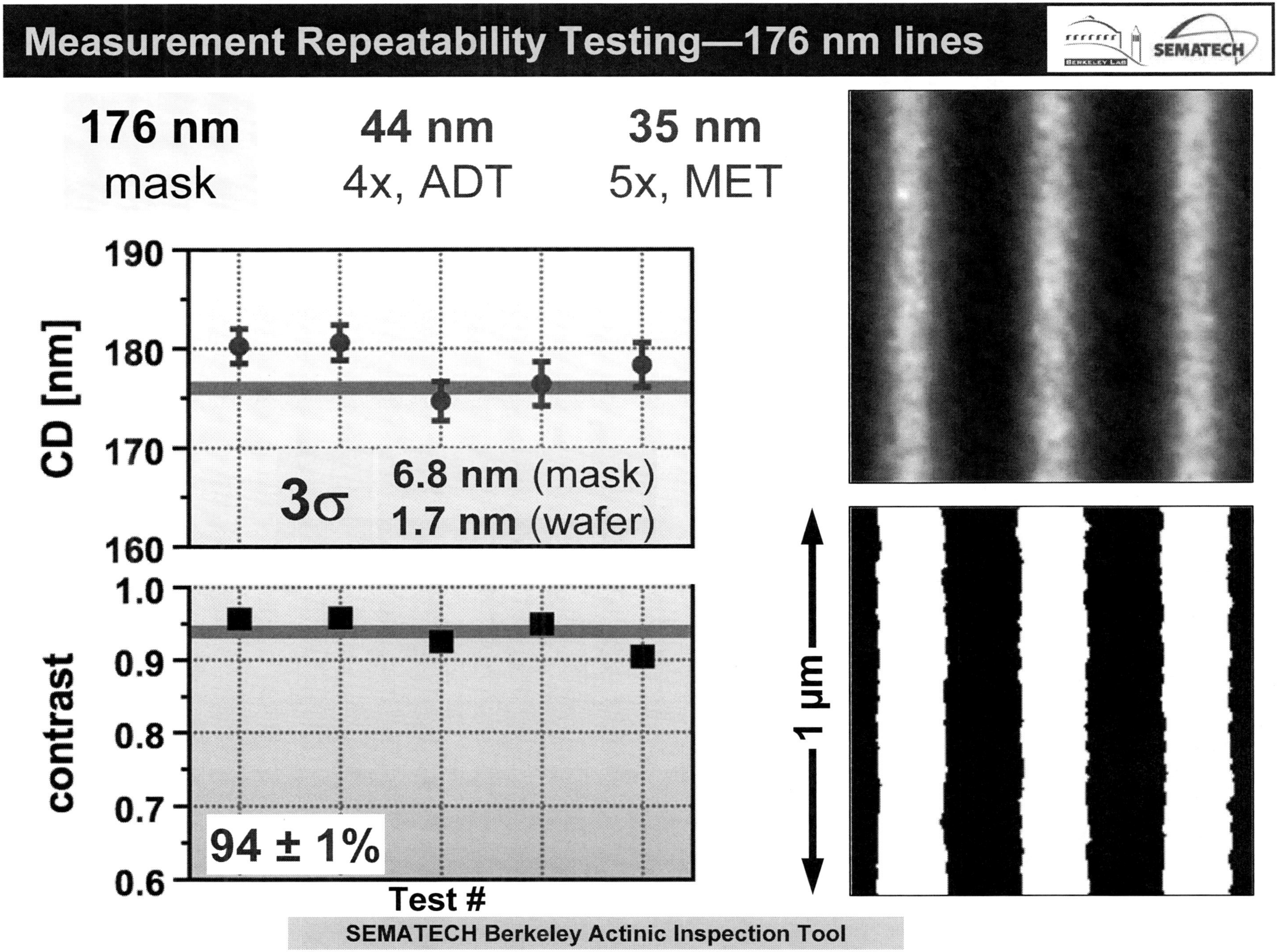

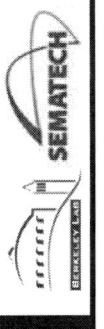

Measurement Repeatability Testing—176 nm lines

176 nm
mask

44 nm
4x, ADT

35 nm
5x, MET

3σ   6.8 nm (mask)
     1.7 nm (wafer)

94 ± 1%

CD [nm]

contrast

Test #

1 μm

SEMATECH Berkeley Actinic Inspection Tool

Measurement Repeatability Testing—132 nm lines

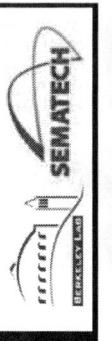

132 nm
mask

33 nm
4x, ADT

26 nm
5x, MET

3σ : 6.0 nm (mask)
1.5 nm (wafer)

CD [nm]

contrast

Test #

88 ± 1%

1 µm

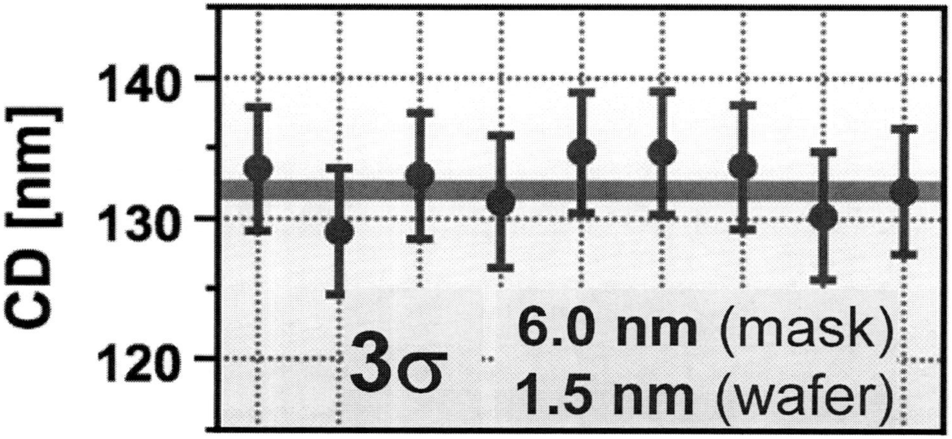

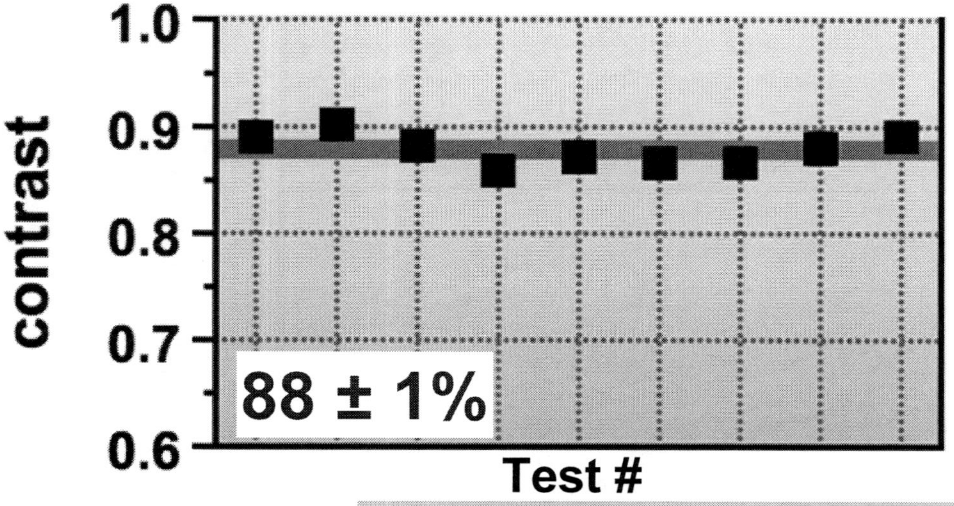

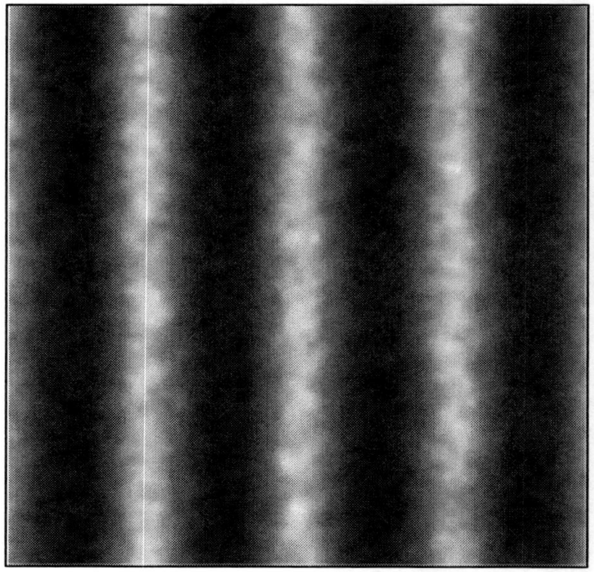

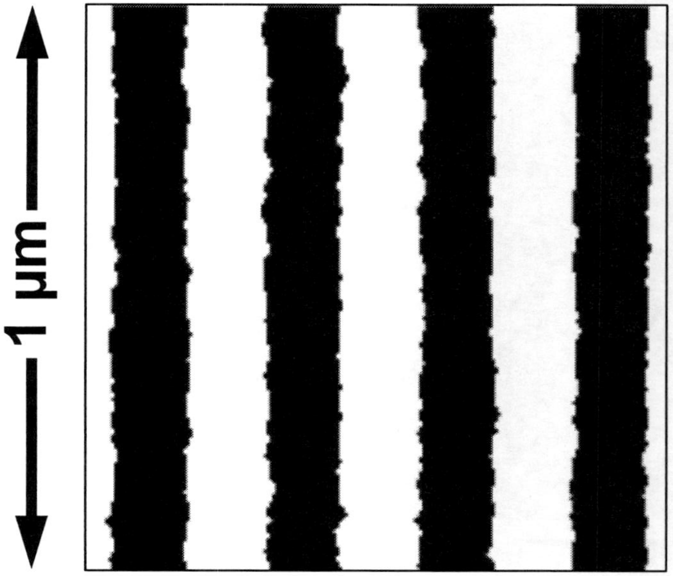

SEMATECH Berkeley Actinic Inspection Tool

Measurement Repeatability Testing—110 nm lines

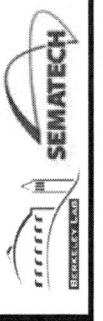

110 nm mask 28 nm 4x, ADT 22 nm 5x, MET

1 µm

CD [nm]

3σ ... 3.7 nm (mask)
... 0.9 nm (wafer)

contrast

74 ± 4%

Test #

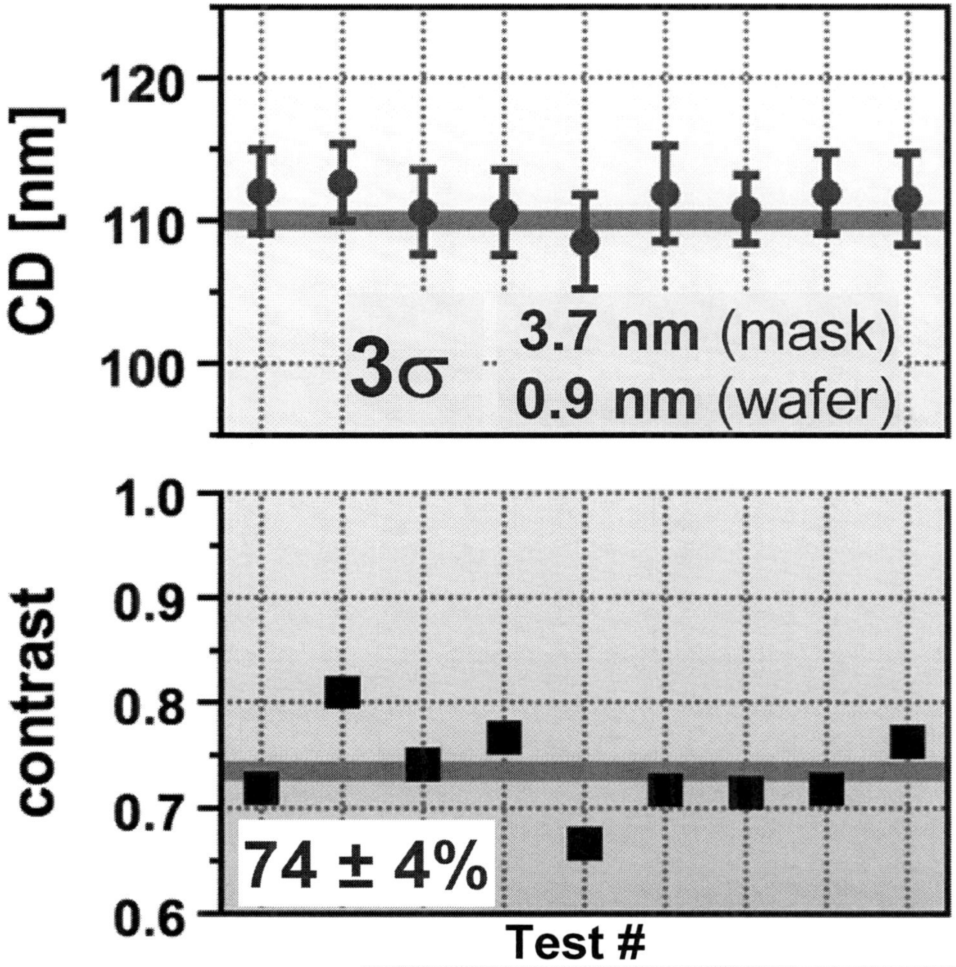

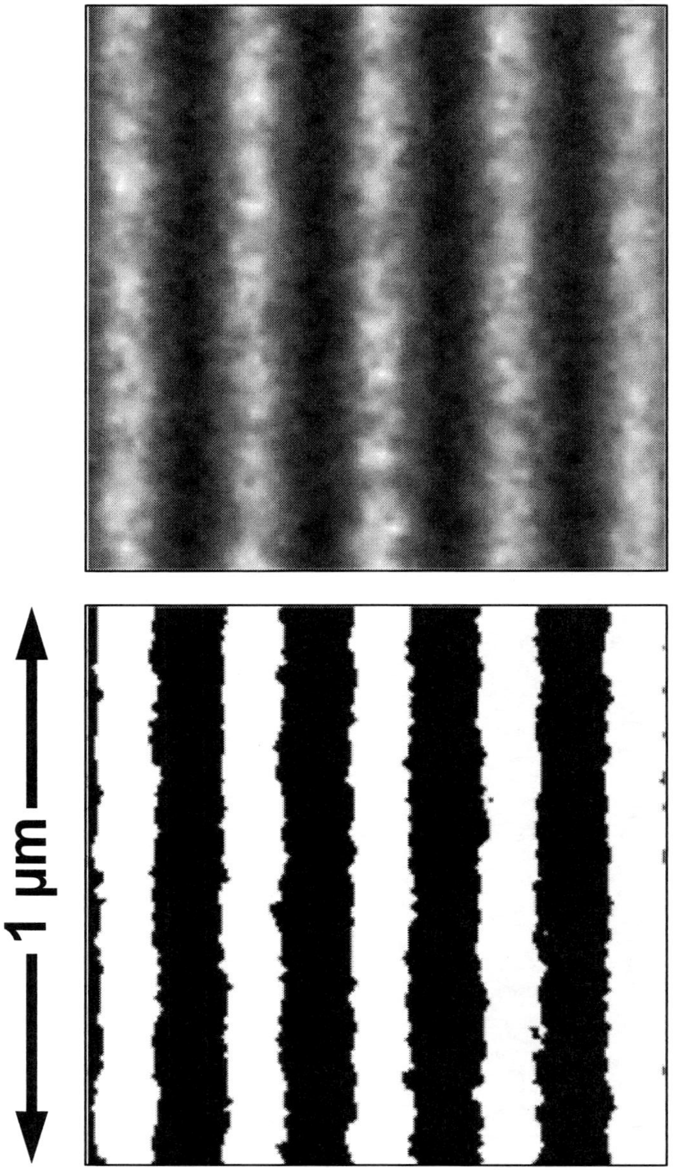

SEMATECH Berkeley Actinic Inspection Tool

Measurement Repeatability Testing—88 nm lines

88 nm
mask

22 nm
4x, ADT

18 nm
5x, MET

3σ

8.0 nm (mask)
2.0 nm (wafer)

CD [nm]

70 ± 6%

contrast

Test #

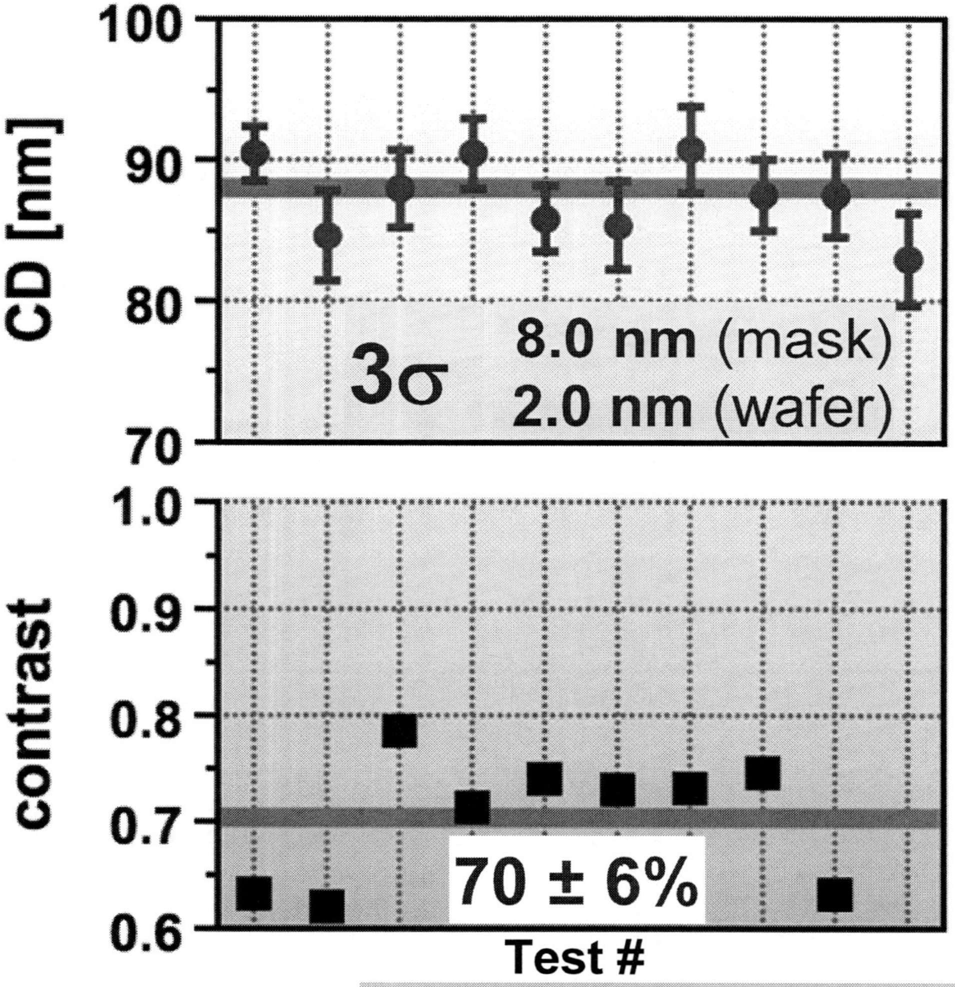

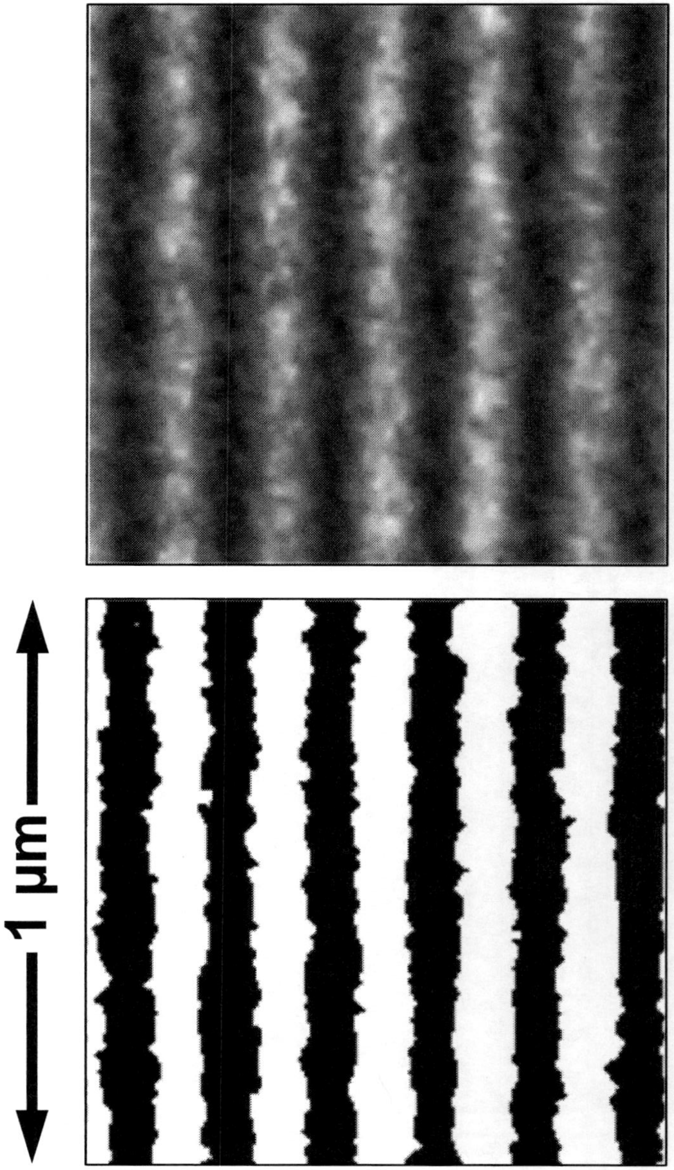

SEMATECH Berkeley Actinic Inspection Tool

1 µm

Measurement Repeatability Testing Summary

Contrast Transfer Function

CD Measurement 3σ

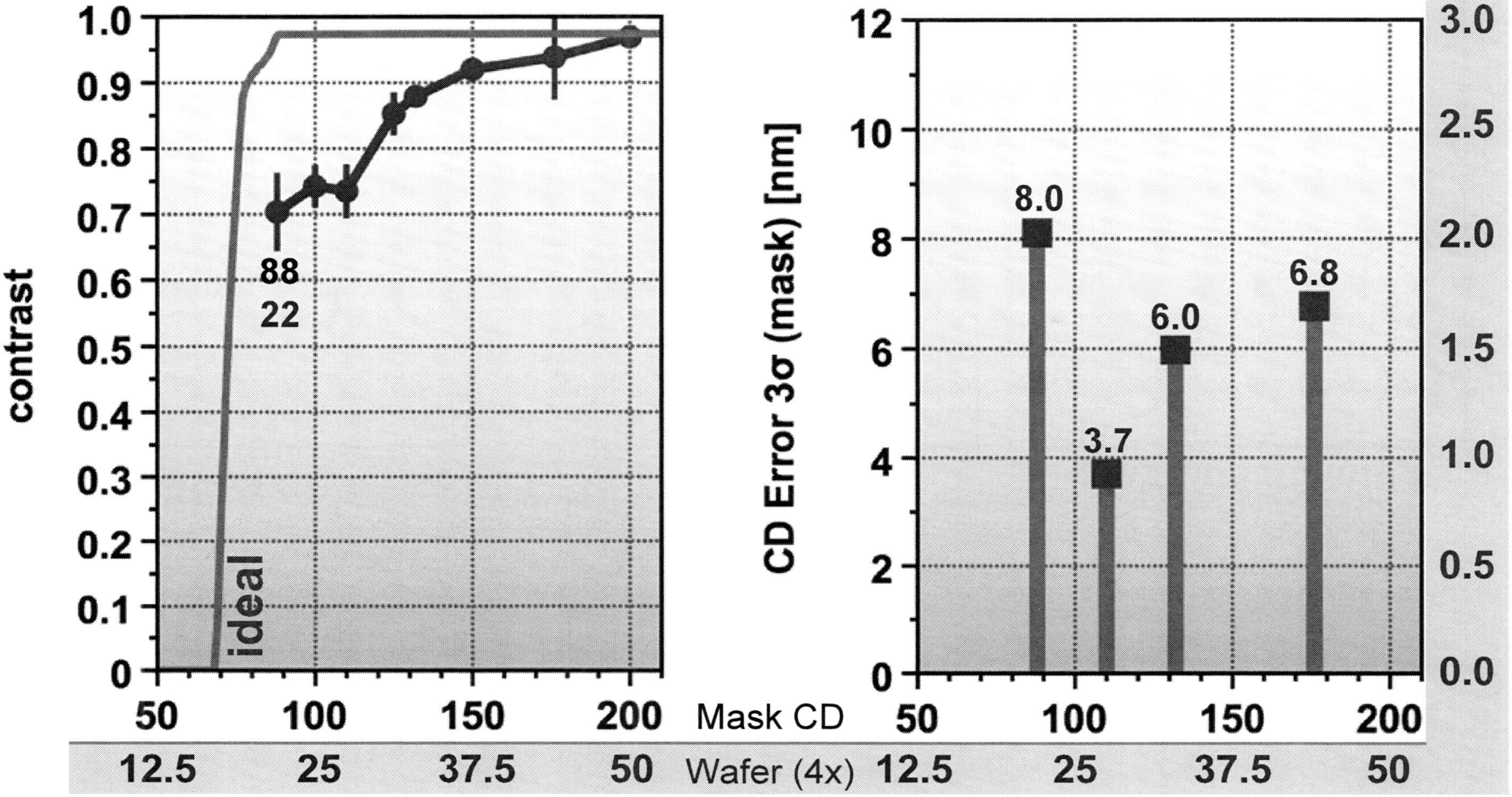

SEMATECH Berkeley Actinic Inspection Tool

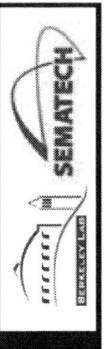

Through-focus analysis of 180 nm CD (mask) lines

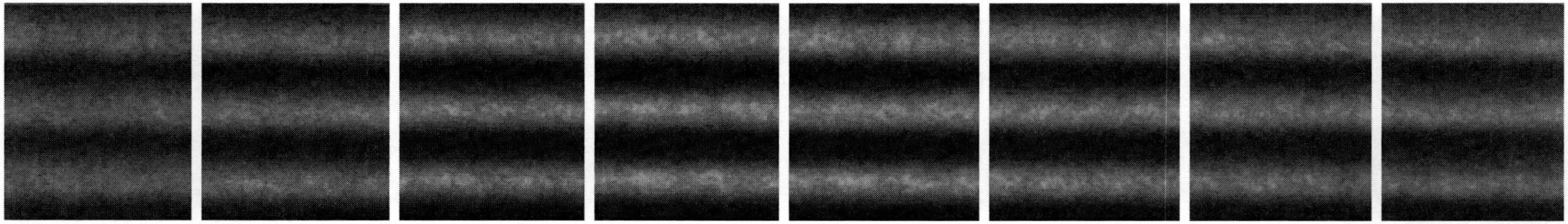

Defocus Z [μm]

best

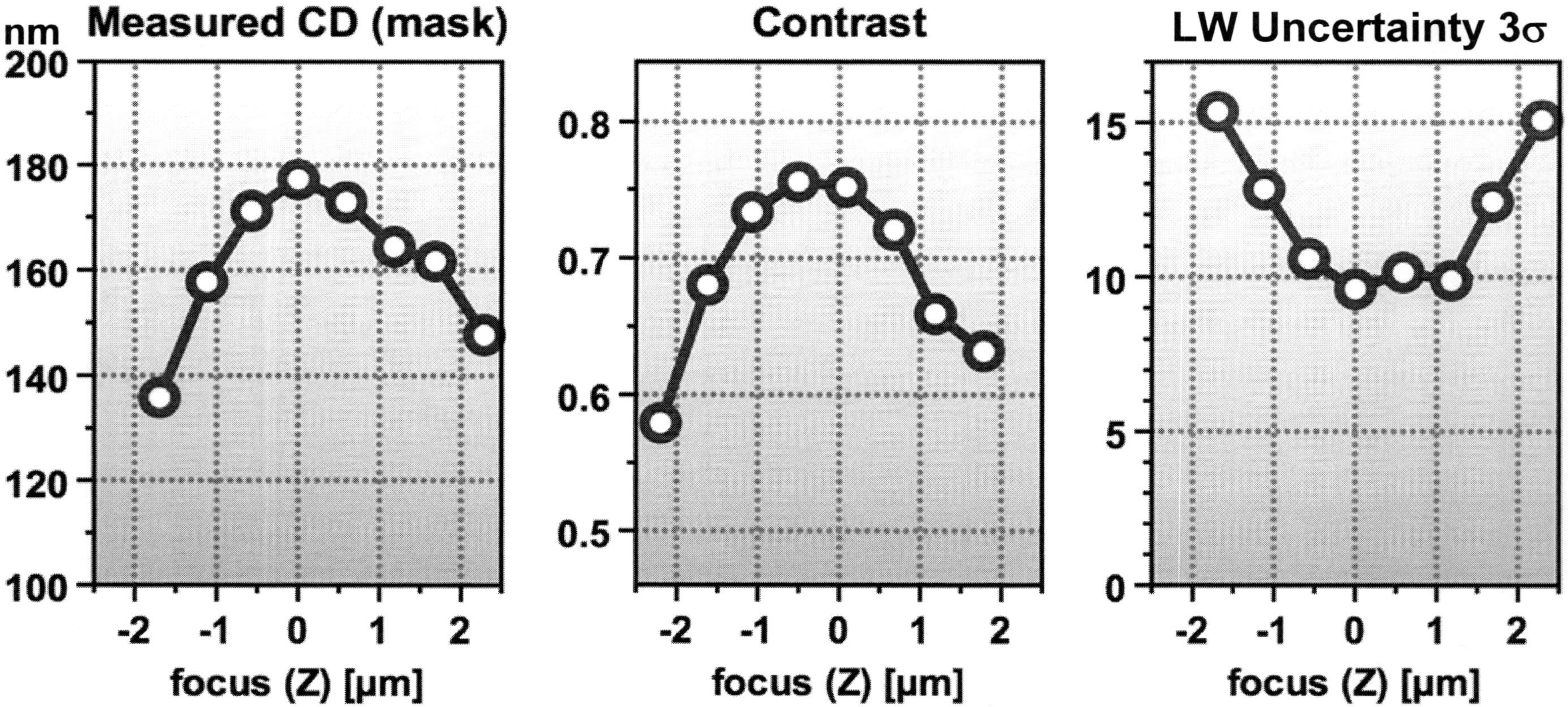

Measured CD (mask)

Contrast

LW Uncertainty 3σ

focus (Z) [μm]

SEMATECH Berkeley Actinic Inspection Tool

Linewidth: Main Issues and solutions

- **Signal-to-noise ratio**

  Measurement dependence on exposure time

- **Illumination uniformity**

  X-Scanning illuminator; soon XY scanning

- **System stability**

  *Mask lateral shift during **z** motion*

  affects illumination and aberrations

  Improved Z actuator, λ-scan method

- **Zoneplate and illumination alignment**

  How stable and repeatable can we make it?

  What sorts of feedback are available to correct it?

  Characterizing aberrations, refining alignment

SEMATECH Berkeley Actinic Inspection Tool

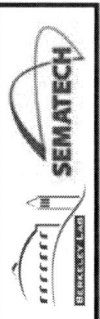

Conclusion

- **System upgrades:** improved resolution, contrast, 0.35 NA (4x)

- **Aberrations:** reduced by **alignment & analysis**
  Astigmatism is down, from 0.23 λ to 0.08 λ RMS

- **Contrast: 71% for 88-nm lines** (22 nm wafer)

- **Linewidth:** Intensity calibration for accurate measurement is challenging

- **Linewidth 3σ:** Repeatability measurements show
  ~5 nm 3σ (1.25 nm on wafer)

- **More improvements** will come from a combination of **hardware & software solutions.**

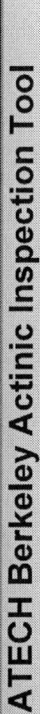

SEMATECH Berkeley Actinic Inspection Tool

Special Thanks!

CXRO
David Attwood
Seno Rekawa
Drew Kemp
Nathan Smith
Paul Denham
Ron Tackaberry
Eric Gullikson
Erik Anderson

Samsung
Hakseung Han

For More Information: `http://Goldberg.LBL.Gov`

SEMATECH Berkeley Actinic Inspection Tool

Laser Produced Plasma Source System Development

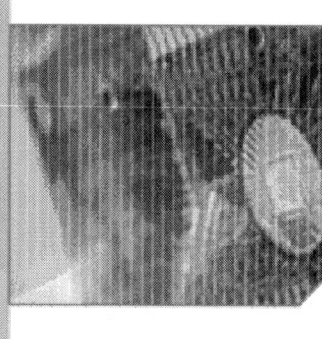

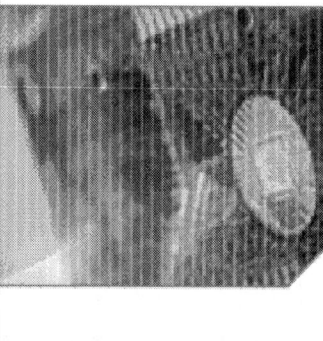

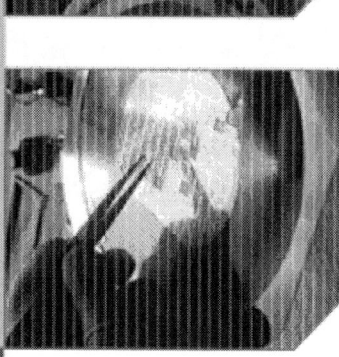

Sematech EUVL Symposium 2008

David C. Brandt*, Igor V. Fomenkov, Alex I. Ershov, William N. Partlo, David W. Myers, Georgiy O. Vaschenko
Oleh V. Khodykin, Alexander N. Bykanov, Jerzy R. Hoffman, Christopher P.Chrobak, Norbert R. Böwering

Shailendra Srivastsava, David Vidusek, Silvia De Dea, Richard Hou

Contents

- Overview of Progress
- Power Status
- Debris Mitigation Status
- Technology
- Roadmap and Summary

Laser Produced Plasma EUV Source Development Continues on Schedule

Source System Sub-Technologies

- High Power CO_2 Laser
- High Reflectivity MLM Collector
- Liquid Sn Droplet Generation
- Debris Mitigation / Collector Lifetime
- Vacuum Technology
- Beam Transport and Focusing
- Droplet Targeting Control
- Intermediate Focus Protection
- Plasma and Intermediate Focus Metrology
- System Control and Scanner Interface

Manufacturing Bay #1

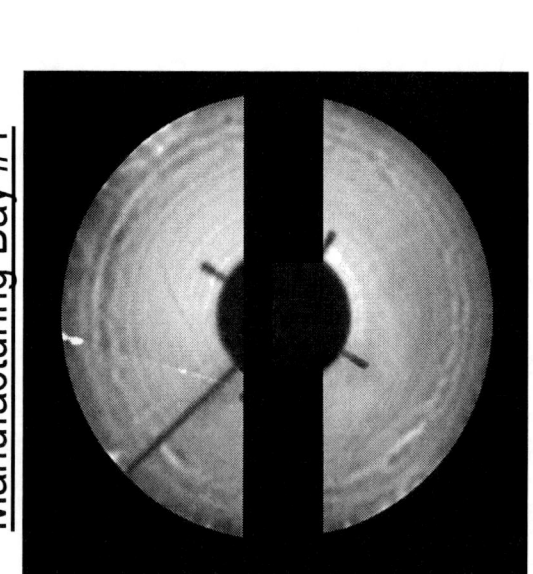

EUV Far Field Image after 8 hrs

Manufacturing of First Generation EUV Laser-Produced-Plasma (LPP) Source Systems

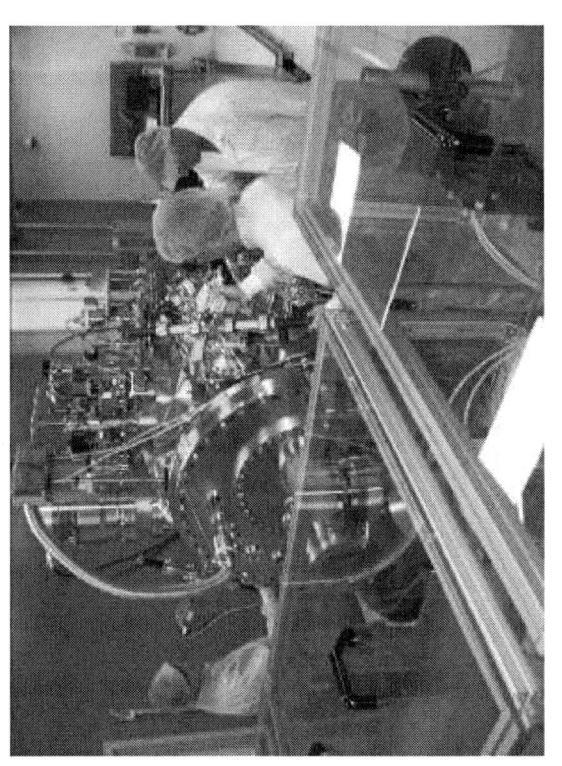

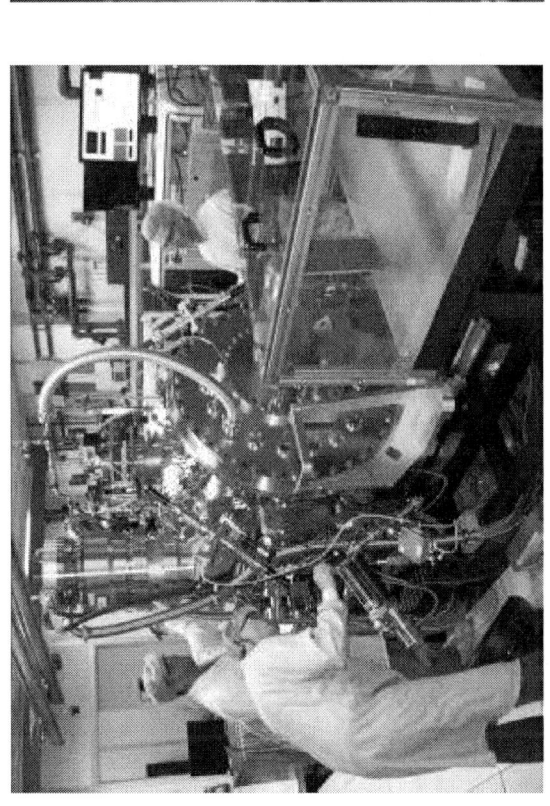

- Manufacturing Bays in class 10,000 clean-room
- Multiple systems to be built, qualified, and shipped in 2009
- First system planned to be complete by end of 2008

Multiple 5 sr Collection Optics in Final Stages of the Manufacturing Process

- Manufacturing processes
  - Blank machining
  - Shaping the figure
  - Coarse polishing
  - Super polishing
  - MLM coating
  - Reflectivity measurement

- Coated Collector expected to be integrated into first LPP system in Q4

- Good High Spatial Frequency Roughness (HSFR) is required for high reflectivity

AFM measurements
1.8 μm x 1.9 μm
0.452 nm RMS

50μm Droplet Stability Demonstrated over 500 hours

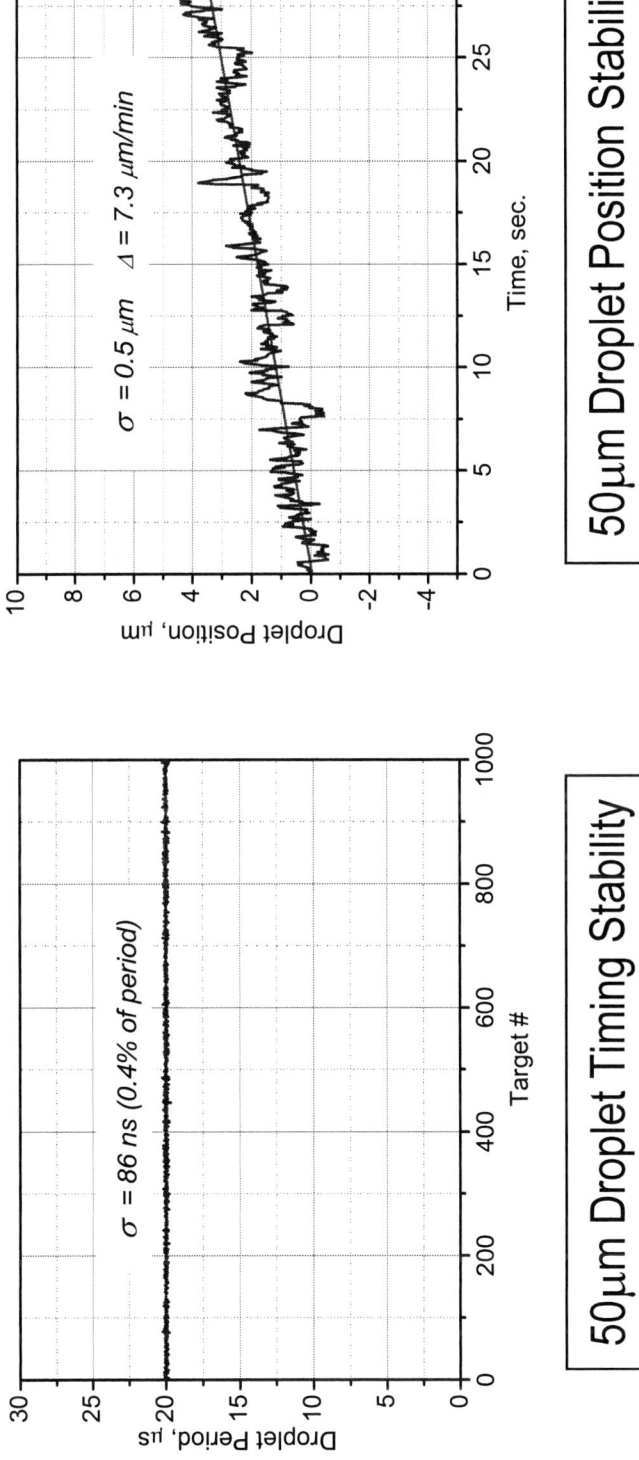

50μm Droplet Timing Stability

50μm Droplet Position Stability

- 50μm diameter droplet stability tested over 500 hours showed no degradation of timing or short term position stability

- Droplets produced as small as 30μm diameter are meeting stability and timing requirements

- The Sn droplet diameter will be reduced to 10 microns in time for second generation EUV sources

Contents

- Overview of Progress
- Power Status
- Debris Mitigation Status
- Technology
- Roadmap and Summary

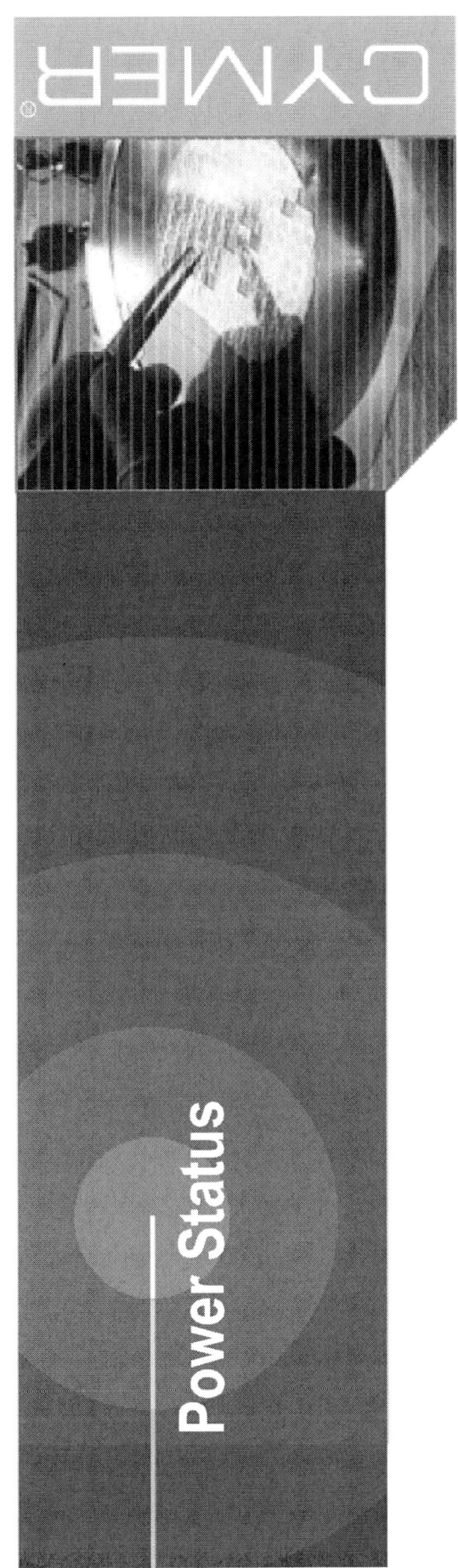

CYMER

Power Status

100W Burst Power Demonstrated in Q4 2007

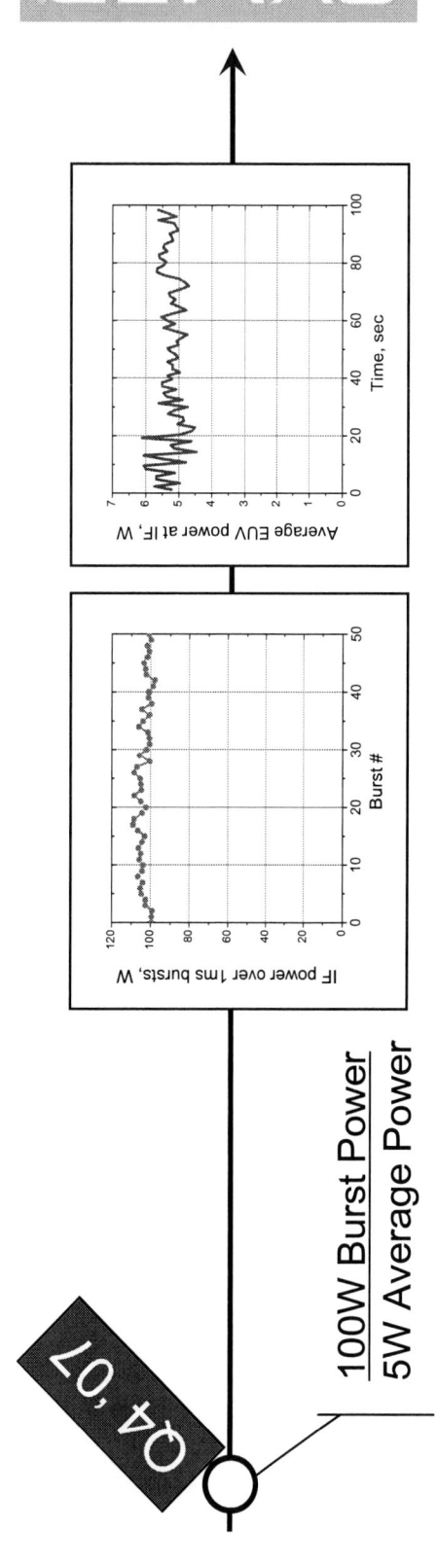

100W Burst Power
5W Average Power

Q4 '07

At the end of 2007, the power achievement was:

- 100W burst power
  - Confirmed the capability to operate at 3% conversion efficiency using droplet targets

- 5W average power
  - Increasing average power required solutions for thermal limitations in the laser, beam delivery and chamber components

35W Average Power Demonstrated in Q1 2008

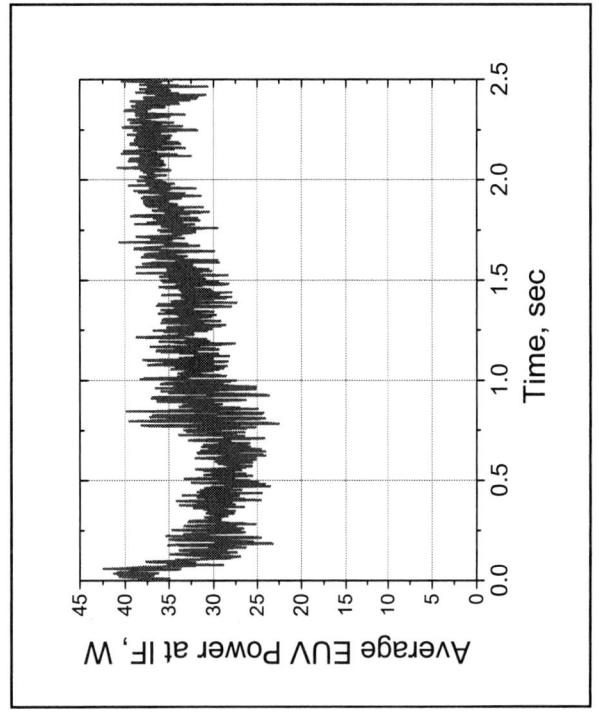

Q4 '07

100W Burst Power
5W Average Power

Q1 '08

35W Average Power
for 2.5 sec

In Q1 08 the average power was increased to 35W

- Duration was limited to 2.5 seconds

- No collector was used, power measured at plasma

1.5 Hours Run Time at 25W Demonstrated in Q2 2008

Q4, 07
100W Burst Power
5W Average Power

Q1, 08
35W Average Power
for 2.5 sec

Q2, 08
25W Average Power
for 1.5 hours

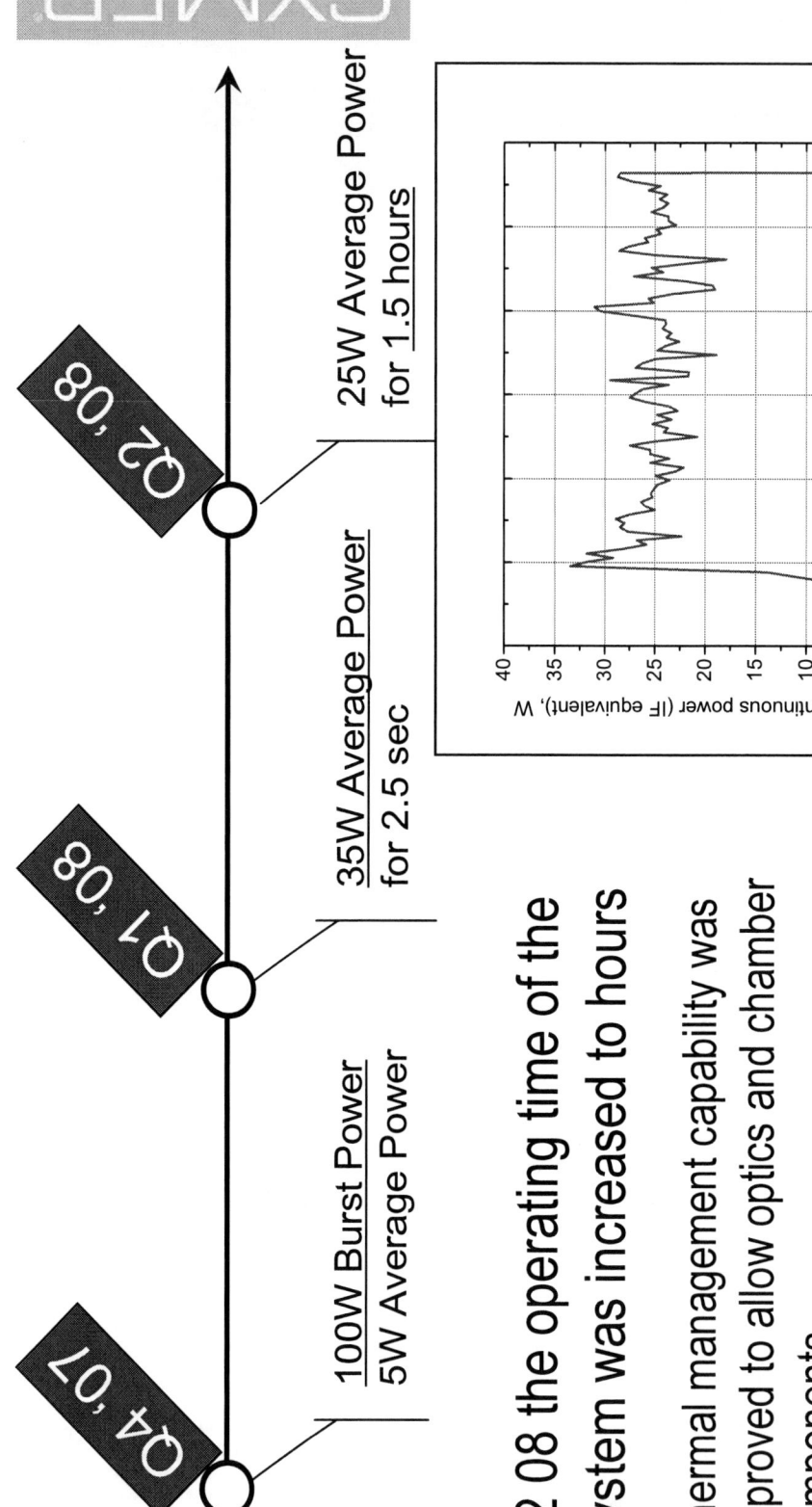

In Q2 08 the operating time of the system was increased to hours

- Thermal management capability was improved to allow optics and chamber components

- No collector was used, power measured at plasma

Multiple 8 Hour Runs with 320mm MLM Collector

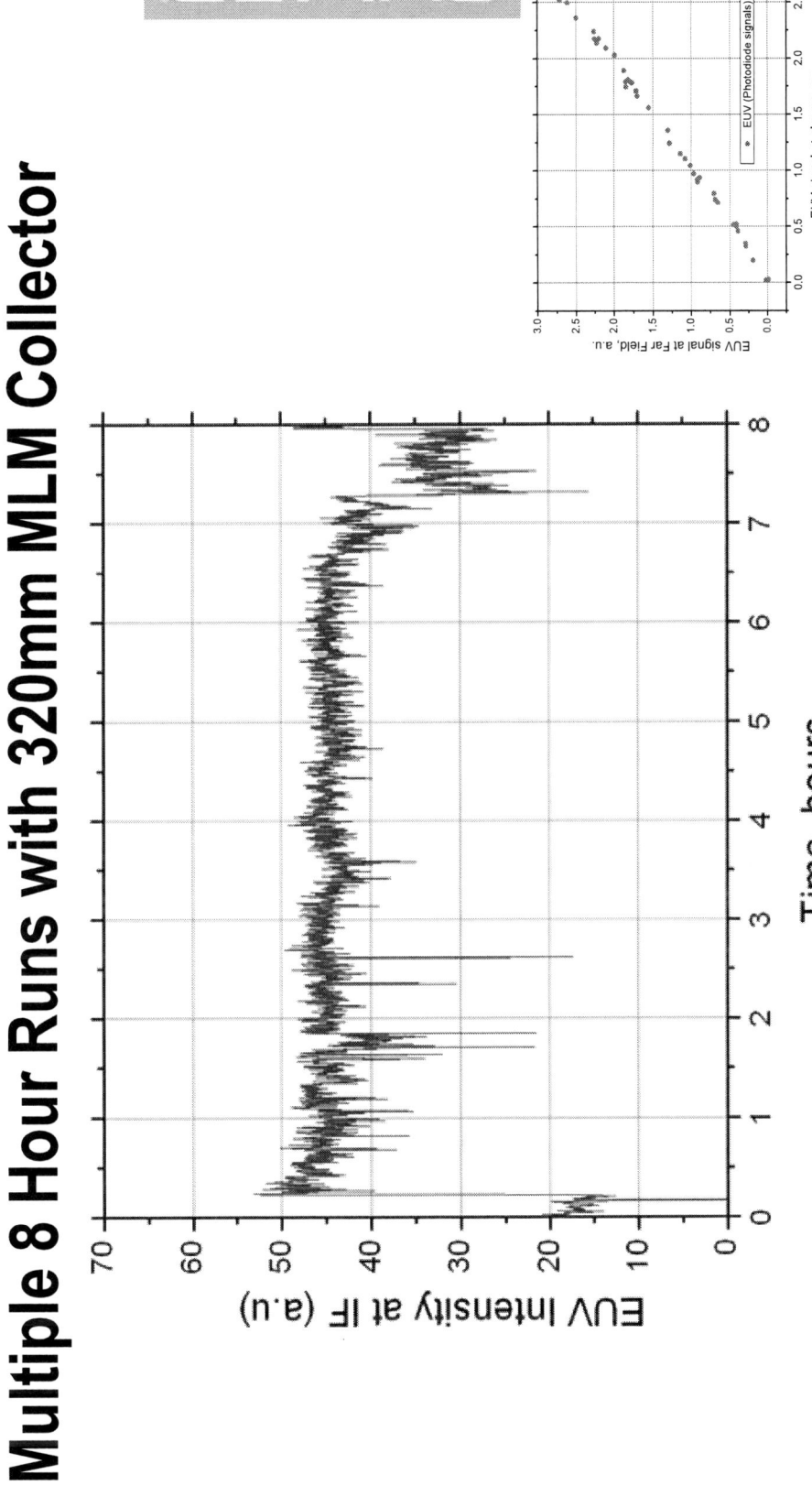

- Eight (8) hours exposure on 320mm (1.6sr) collector measured in the far field with a fluorescence converter and CCD camera
- Photo diode with 1μm thick Zr foil used to calibrate far field intensity to EUV photo diode looking at the plasma

2008 LPP Source Output Energy Plan

DC from 5%->~100%, continuous operation from 2s to 10hrs

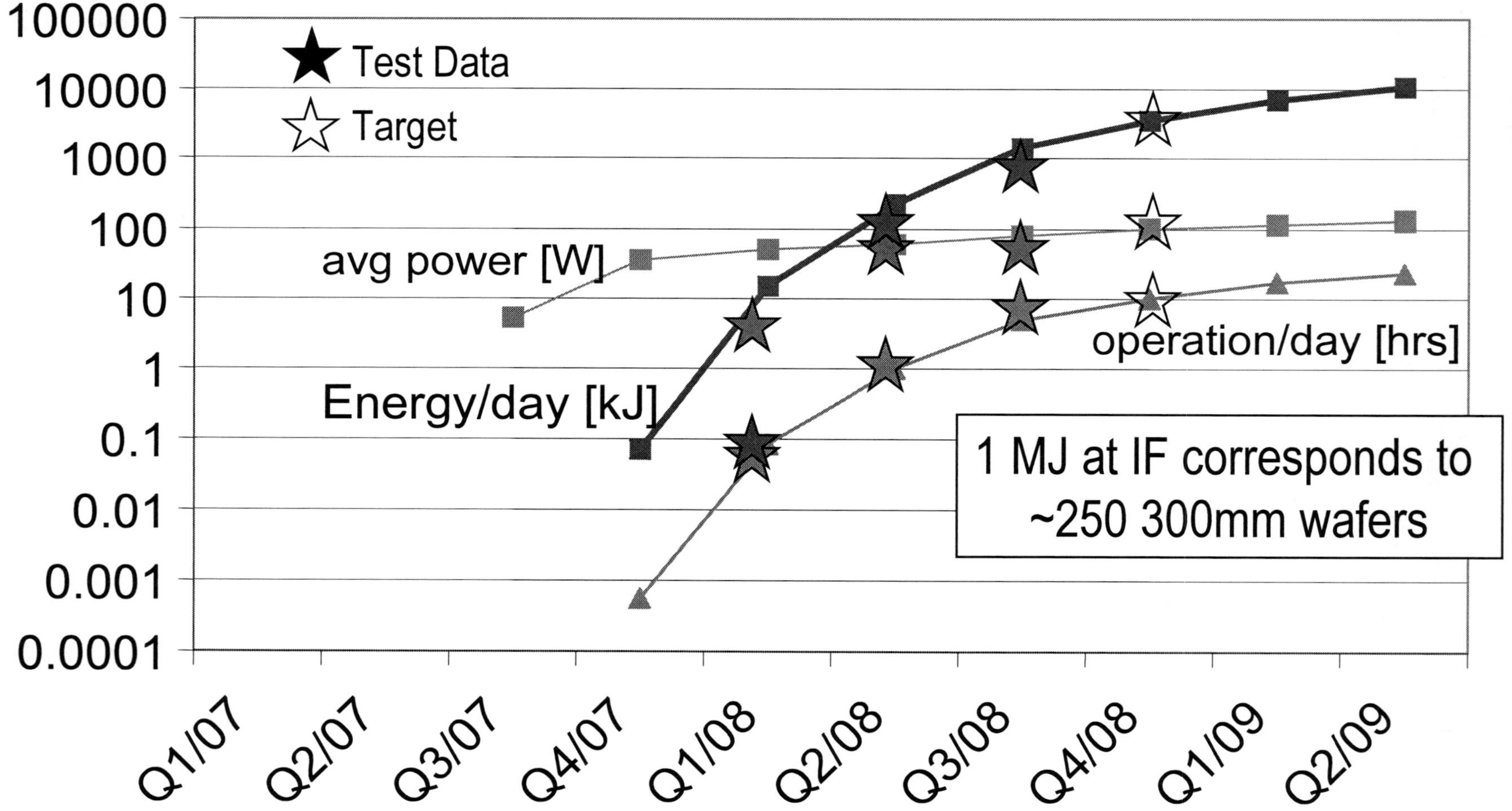

CYMER

Debris Mitigation

Debris Mitigation Stops Erosion from Ions

Ion Measurements

Sn Ions are stopped before they reach the collector surface

— with Debris Mitigation

Faraday Current, a.u. / Sn Ion Energy, eV

Previously showed elimination of erosion over 3 Mpulses with a small amount of Sn Deposition

Si 28
Sn 120

Depth, a.u.

Laser Pulse

tin ions

electrons

All measurements taken at collector surface

Exposure of Witness Samples Shows No Degradation of MLM Coating

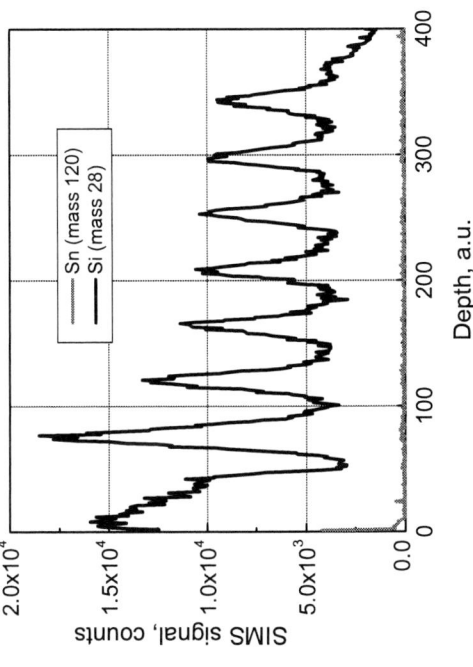

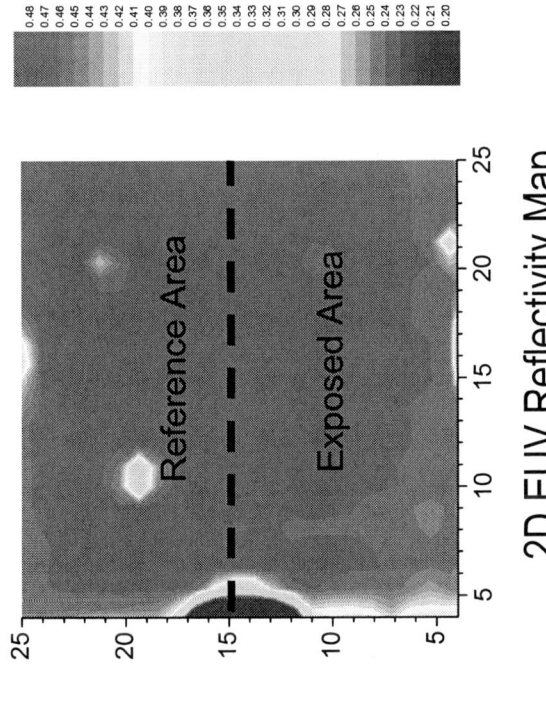

2D EUV Reflectivity Map

8 Layer MLM Sample Post Exposure

- 2D reflectivity maps shows <1% between exposed and reference areas

- SIMS analysis of 8 layer sample shows no erosion from ions

- Exposure parameters
  - 2 hours exposure
  - 60W / 10% duty cycle
  - Reflectivity measurement from NIST

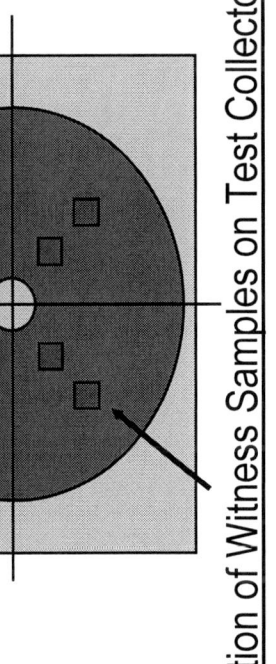

Position of Witness Samples on Test Collector

EUV Uniformity and Collector Reflectivity over 8 Hours

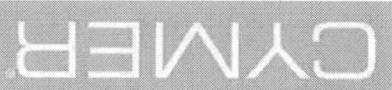

CYMER®

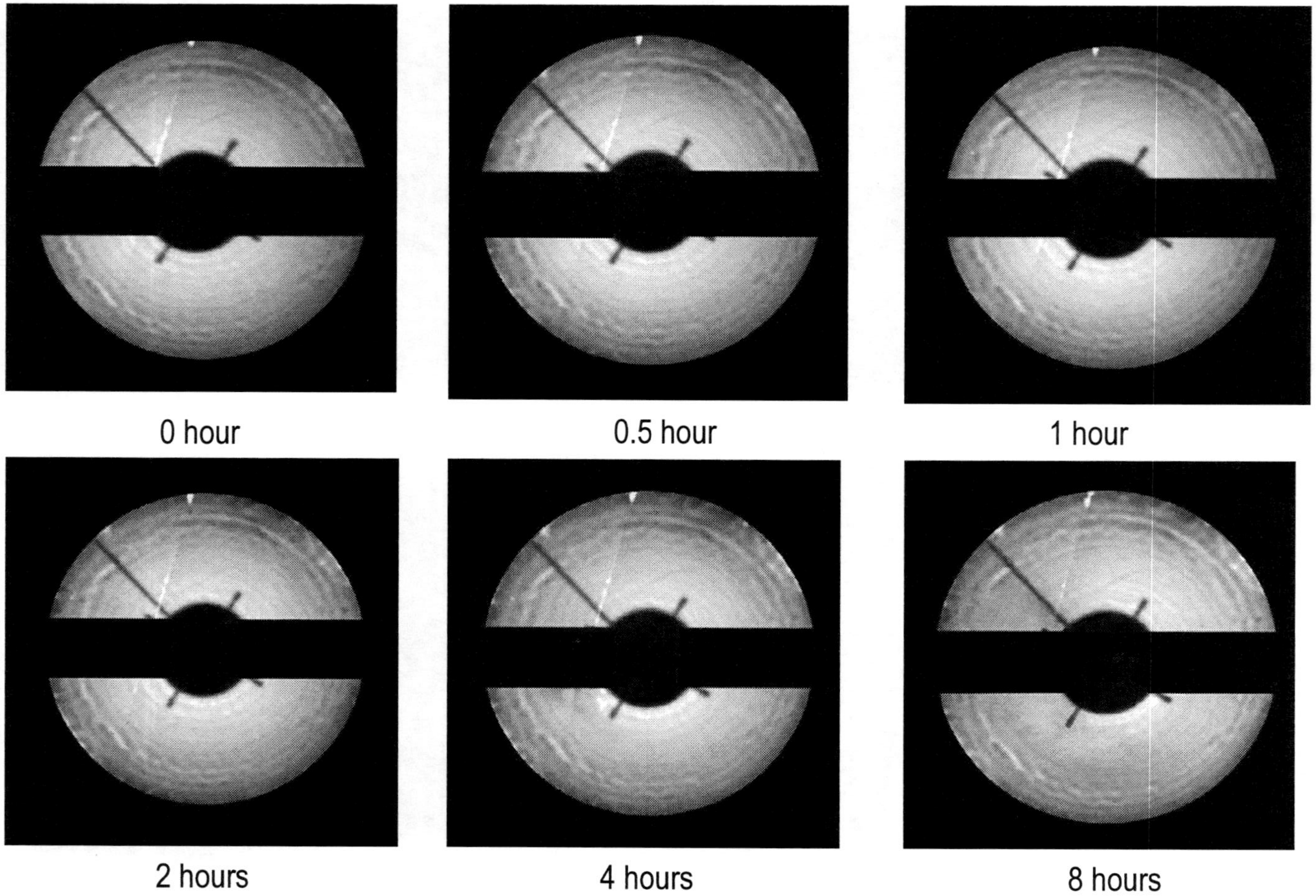
0 hour
0.5 hour
1 hour
2 hours
4 hours
8 hours

320mm Diameter (1.6sr) Collector, EUV Images taken in the Far Field with Fluorescence Converter

CYMER

Technology

Size of EUV Emitting Region, 100W Bursts, 50kHz

EUV source size from in-band pinhole camera measurements viewing the plasma source at 90° angle

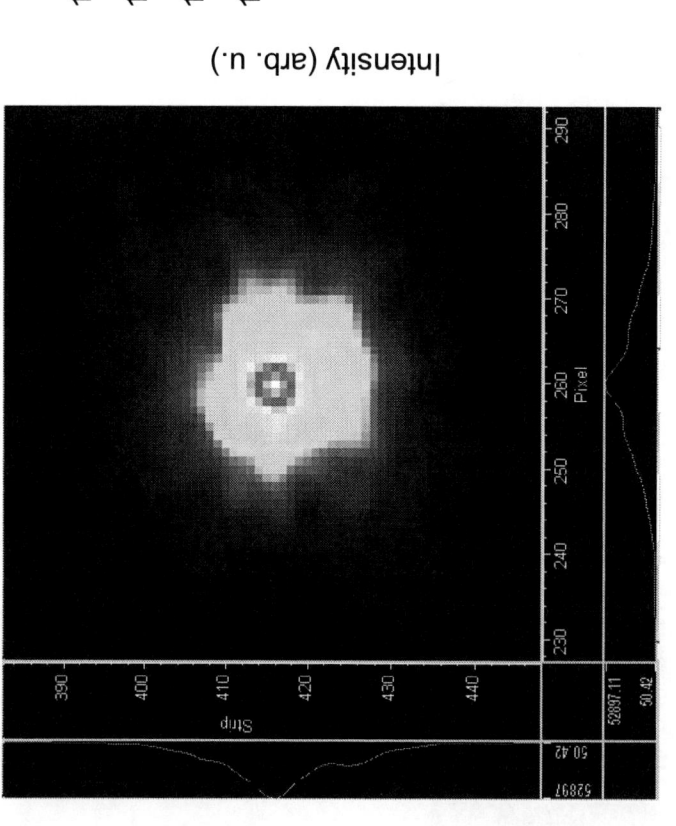

Intensity Profile

90 μm (FWHM)
210 μm (1/e^2)

Diffusion Barrier MLM Coating EUV Reflectivity

Peak Reflectance

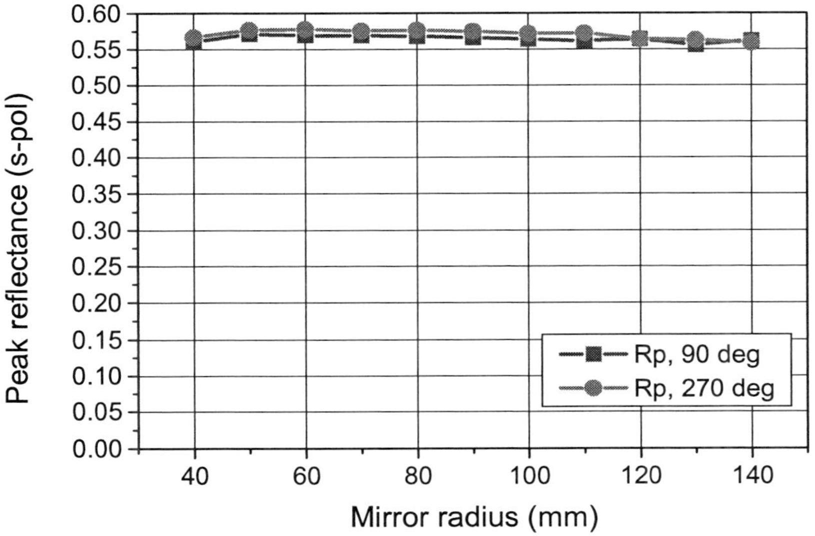

Peak reflectance (s-pol)
0.60
0.55
0.50
0.45
0.40
0.35
0.30
0.25
0.20
0.15
0.10
0.05
0.00
40
60
80
100
120
140
Mirror radius (mm)
Rp, 90 deg
Rp, 270 deg

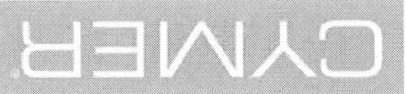

Reflectivity Curves

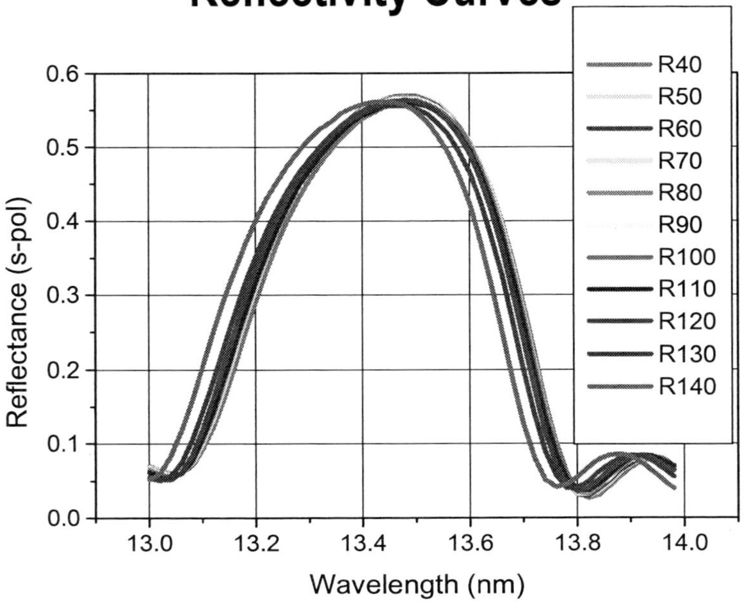

Reflectance (s-pol)
0.6
0.5
0.4
0.3
0.2
0.1
0.0
13.0
13.2
13.4
13.6
13.8
14.0
Wavelength (nm)
R40
R50
R60
R70
R80
R90
R100
R110
R120
R130
R140

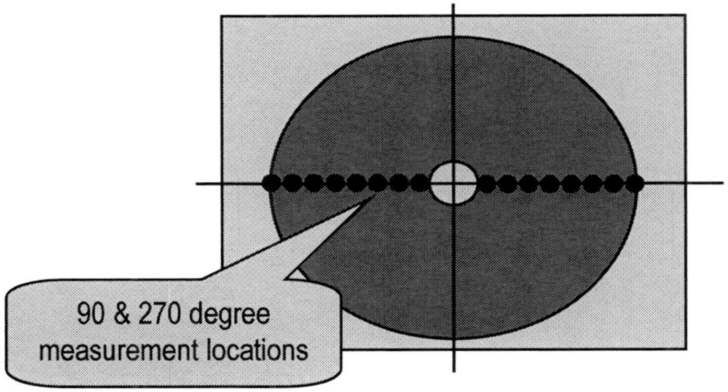

90 & 270 degree measurement locations

- 57% EUV peak reflectivity measured
- Graded coating with diffusion barrier layers for high temperature stability
- EUV measurements made at PTB

Source-Scanner Mechanical Interface Provides Cleanliness from Contamination

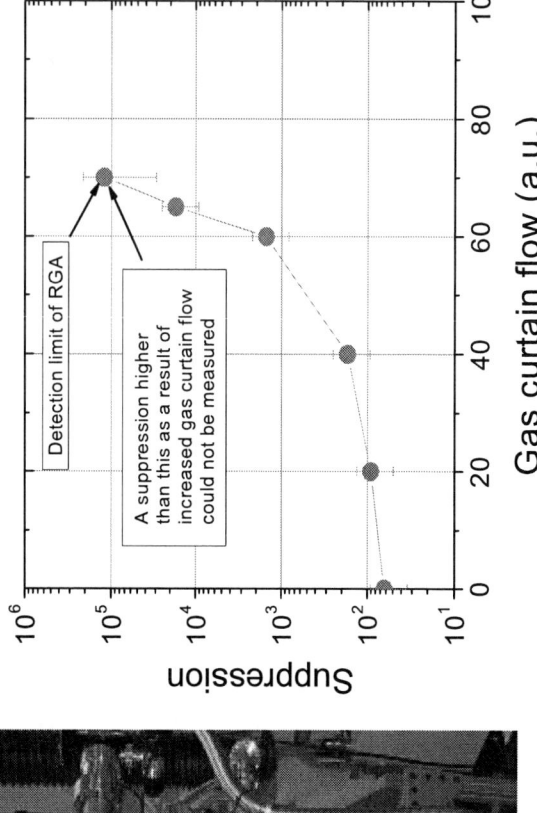

- Intermediate Focus Protection (IFP) is essential to keep contamination from passing through to the illumination optics
- IF Protection module experiments have demonstrated cleanliness at the source-scanner interface to the limit of detection

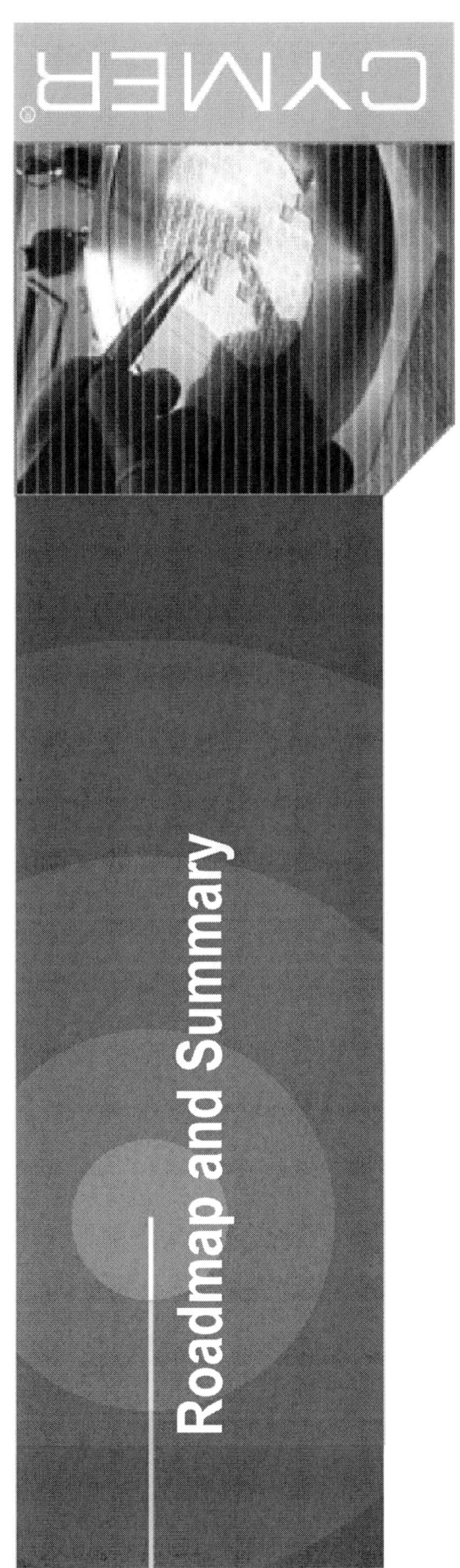

CYMER

Roadmap and Summary

LPP EUV Source Roadmap

CYMER

EUV Source Power Roadmap

	Pilot	HVM I	HVM II
Drive laser power (kW)	11	19	>20
In-band CE (%)	3.0	3.5	4.0
Collection Efficiency (sr)	5	5.2	5.5
Collector Reflectivity (%)	>60	>60	>60
Optical Transmission (%)	80	85	90
Total EUV power at IF (W)	**>100**	**>200**	**>400**

2007 2008 2009 2010 2011 2012 2013 2014

Laser Produced Plasma R&D

HVM – EUV Light Source Generations

Pilot

HVM I

HVM II

HVM III

Summary

- Cymer continues to meet it's EUV source development schedule

- Manufacturing of first pilot systems is in progress

- Run time up to eight (8) hours with stable performance demonstrated

- Debris Mitigation effectivity demonstrated to protect the collector from erosion due to ions and Sn deposition

- Integrated system testing with 320mm (1.6sr) collector has shown stable transmission of EUV power to the far field and good distribution of EUV energy

- Cymer is committed to commercializing an HVM EUV light source for the sub-32nm node

CYMER

Please see our poster
"LPP EUV Source Development
for HVM Lithography"

Acknowledgements

- ASML
  Erik Loopstra, Vadim Banine, Hans Meiling, Noreen Harned

- University of Illinois at Urbana Champaign
  David N. Ruzic, Martin J. Neumann, Ben Masters

- University of Central Florida
  Martin Richardson, Simi George

- University of California at San Diego
  Mark Tillack, Yezheng Tao

- Fraunhofer Institut f. Angewandte Optik und Feinmechanik
  Torsten Feigl, Sergiy Yulin, Nicolas Benoit, Norbert Kaiser

- Lawrence Berkeley National Laboratory
  Eric Gullikson, Franklin Dollar

- National Institute of Standards and Technology
  Steven Grantham, Charles Tarrio

CO_2 laser-produced Sn plasma source for EUV lithography

Akira Endo[1], Y. Ueno[2], K. Nowak[2], T. Yabu[2], T. Yanagida[2], T. Suganuma[2], T. Asayama[2], H. Someya[2], H. Hoshino[2], M. Nakano[2], M. Moriya[2], T. Nishisaka[2], T. Abe[2], H. Komori[2], A. Sumitani[2], H. Nagano[1], Y. Sasaki[1], S. Nagai[1], Y. Watanabe[1], G. Soumagne[1], B. T. Ishihara[1], O. Wakabayashi[1], K. Kakizaki[1], H. Mizoguchi[3]

1 EUVA / Gigaphoton Inc., 2 EUVA / Komatsu Ltd., 3 Gigaphoton Inc.

International EUVL Symposium
September 30, 2008
Lake Tahoe, California

Acknowledgments
This work was partly supported by the New Energy and Industrial Technology Development Organization -NEDO- Japan.

Outline

▲ Introduction
  - LPP clean source concept and roadmap

▲ Update of CO_2 laser produced Sn plasma source
  - Laser output power and operation duty
  - Sn plasma guiding by magnetic field
  - Sn droplet irradiation optimization

▲ LPP/EUV direction to HVM

▲ Summary

LPP Source Roadmap

	1st Mid term 2004/9	2nd Mid term 2006/3	EUVA Final 2008/3	HVM source-1 2010 planning
EUV Power (IF)	5.7W [1]	10W [1]	50W [2]	110W [2] /140W [3]
Stability	---	$\sigma < \pm 10\%$	$\sigma < \pm 5\%$	$3\sigma < \pm 0.3\%$
Laser	YAG:1.5kW	CO_2:2.6kW	CO_2: 7.5kW	CO_2: 10kW
Laser freq.	10kHz	100kHz	100kHz	100kHz
CE (source)	0.9%	0.9%	2.5%	4%
Target	Xe-Jet	SnO_2 choroid	Sn-Droplet	Sn-Droplet

EUVA project

New EUVA project

Gigaphoton

Technology for <10W
Nd:YAG Laser, Liquid Xe jet

Technology for 115-200W
CO2 Laser, Sn droplet target
Magnetic field mitigation

Note)
Primary source to IF EUV transfer efficiency :
1) 43%
2) 28% with SPF
3) 36% without SPF

Clean Light Source Concept

Requirement for EUV source for HVM

- High EUV power >115 W
- EUV Stability
- Collector mirror lifetime
- Low CoG / CoO

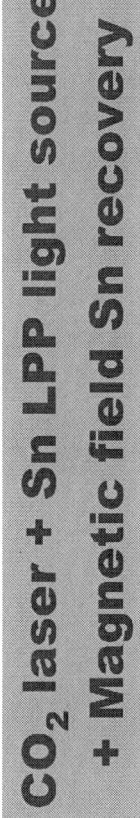

CO$_2$ laser + Sn LPP light source + Magnetic field Sn recovery

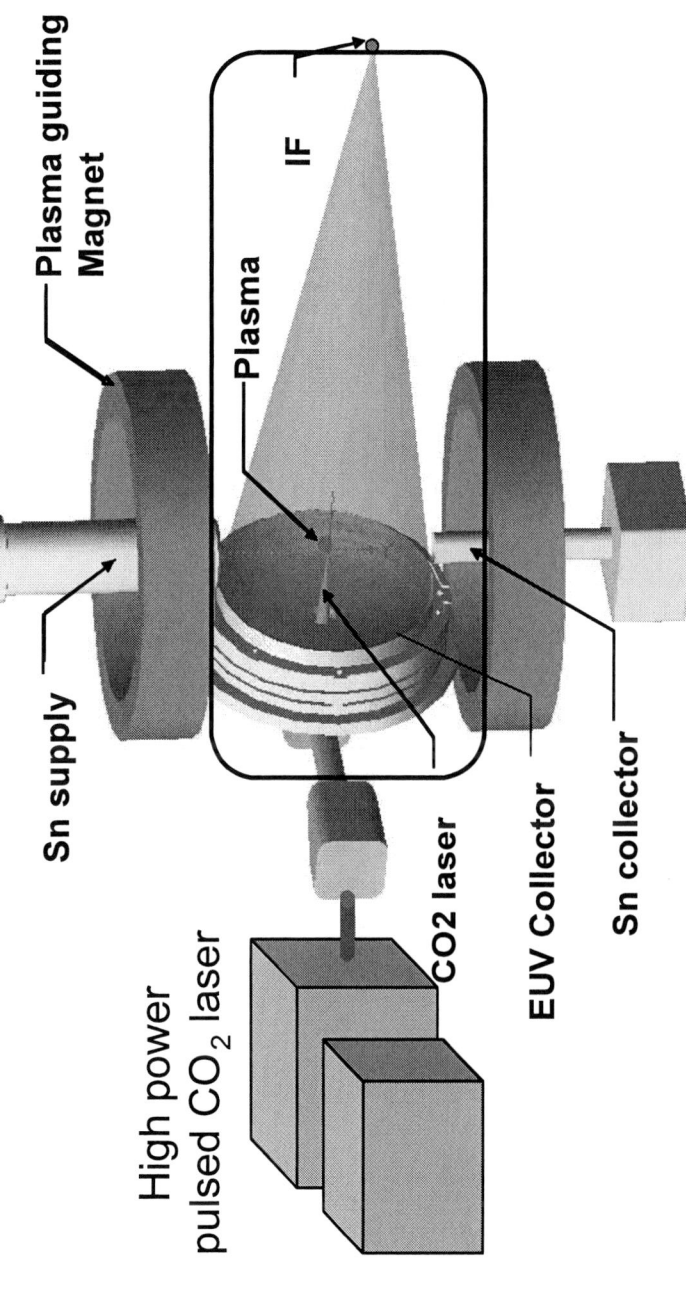

Plasma guiding Magnet

IF

Plasma

Sn supply

High power pulsed CO$_2$ laser

CO2 laser

EUV Collector

Sn collector

High power CO2 laser MOPA system

Laser Power : 13 kW
Pulse Width : 20 ns
Repetition Rate : 100 kHz
Pulse energy stability : 2% (3s, 500 pulses)

■ Laser System

60w

3 kW

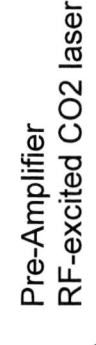

13 kW

100 W at I/F equivalent

Oscillator
Wave length: 10.6um
Rep. rate :100kHz
Pulse width :20 ns (FWHM)

Pre-Amplifier
RF-excited CO2 laser

Main-Amplifier
RF-excited CO2 laser

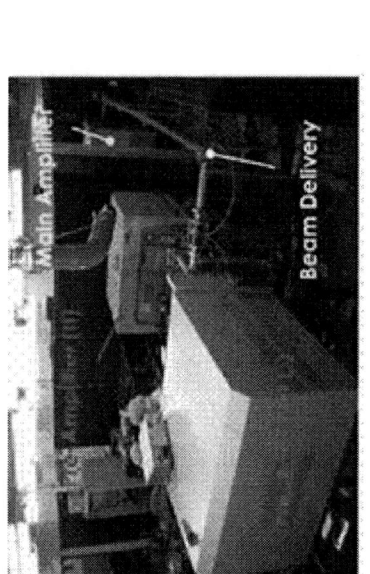
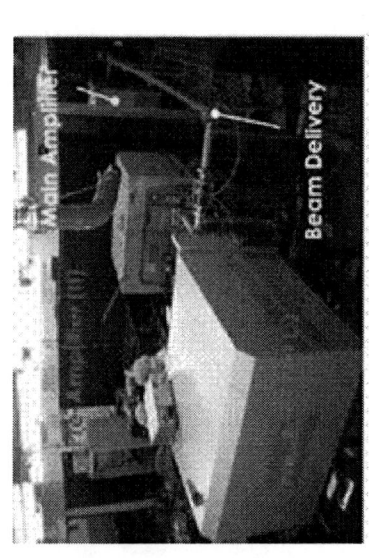

Laser beam profile

Long time operation

1hr non-stop operation of 100kHz CO_2 laser, 13kW, 10% duty cycle

On: 1.6 ms
Off: 14.4 ms

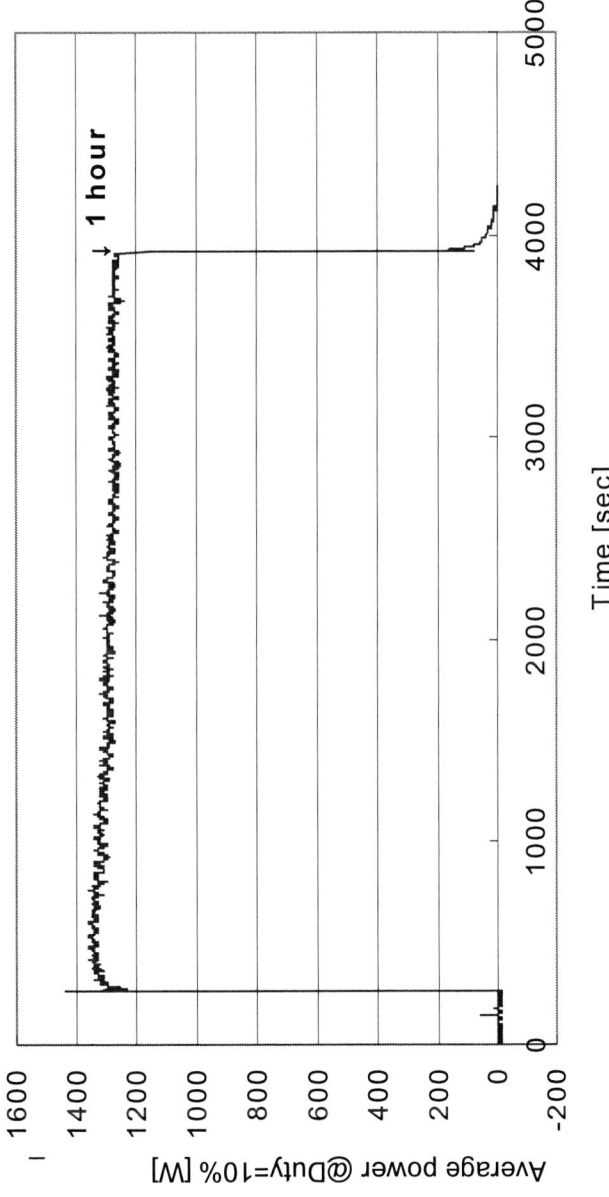

- 100% duty cycle / Pre-Amplifier operation demonstrated with 2.7 kW output.
- 100% duty cycle / Full MOPA system will be demonstrated by improved optics until year end.

Magnetic field plasma guiding

Superconducting magnet was installed for:
1) Investigation of the Tin ion flux in "Real" large space.
2) Optimization of Tin fuel evacuation.

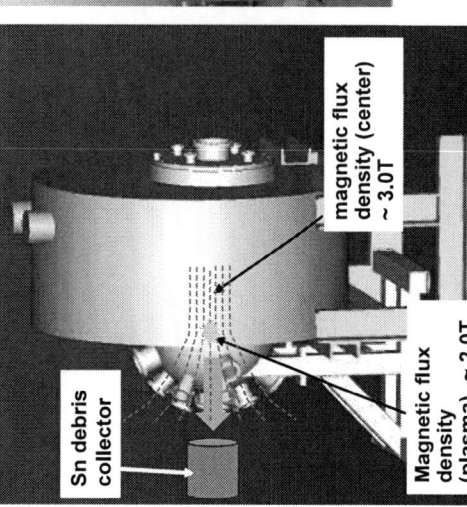

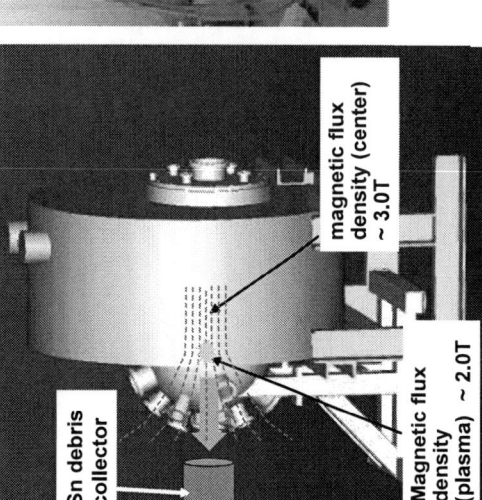

Sn debris collector

magnetic flux density (center) ~ 3.0T

Magnetic flux density (plasma) ~ 2.0T

Visible image of Sn ion flow in magnetic field

Laser : CO2 laser, Target : Sn plate

Magnetic field : 2T

Without magnetic field

Magnetic field plasma guiding

Experimental set up

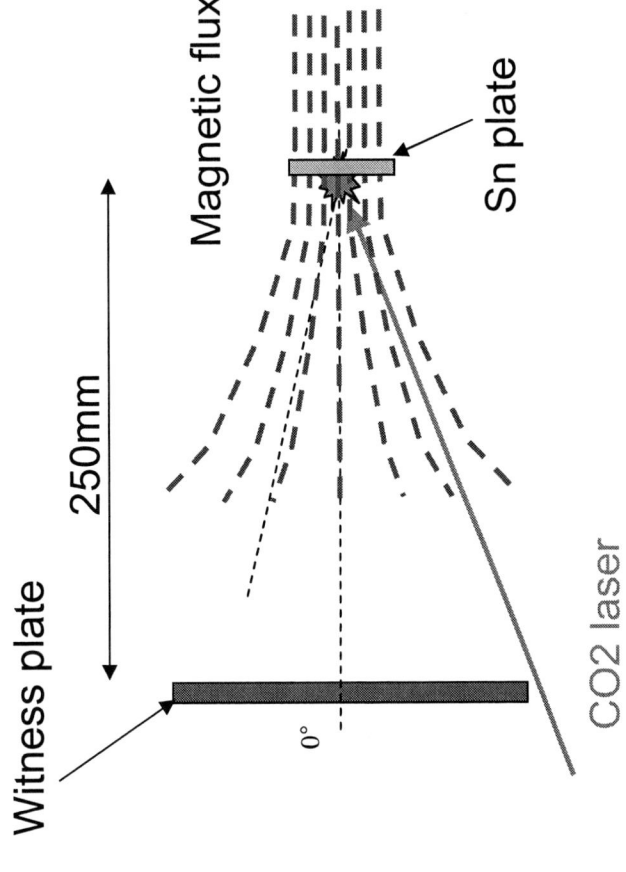

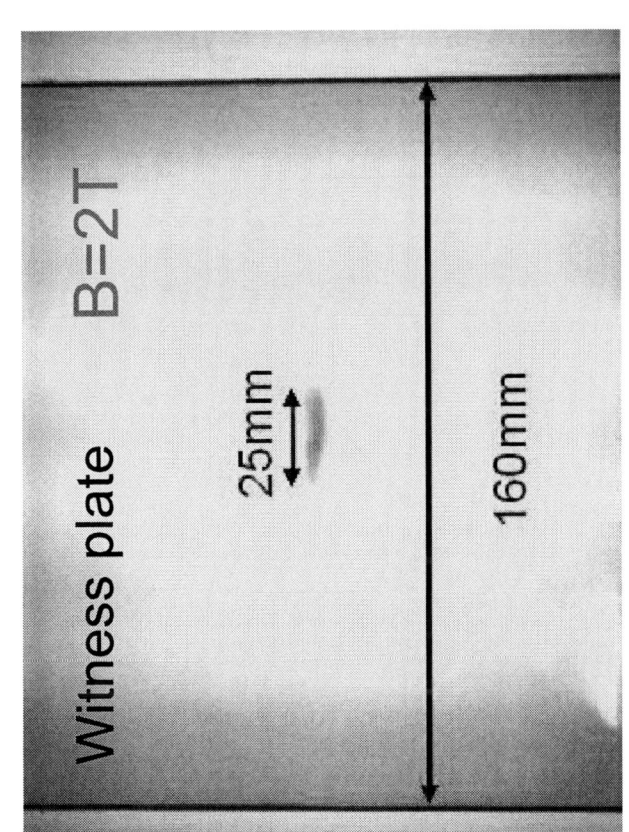

Tin ions are effectively confined and guided by the magnetic field.

Magnetic field plasma guiding

Mo/Si sample mirror test

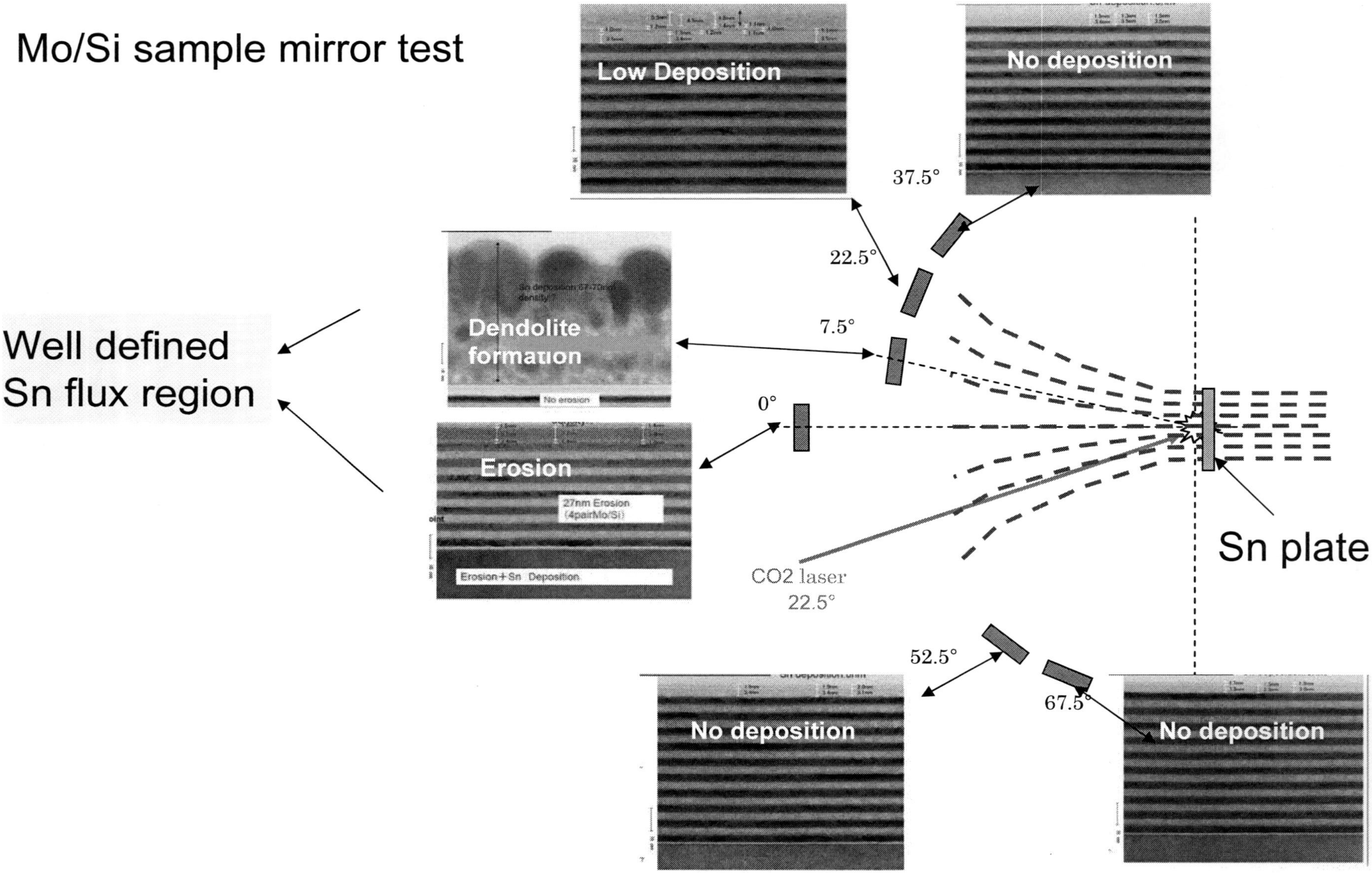

EUVA GIGAPHOTON

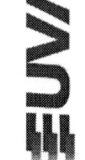

EUV generation with high power CO_2 system

EUV measured at primary source

Droplet:
267kHz, D=100um
60m/s

Laser beam

Laser:
70mJ, 89kHz (=267/3)

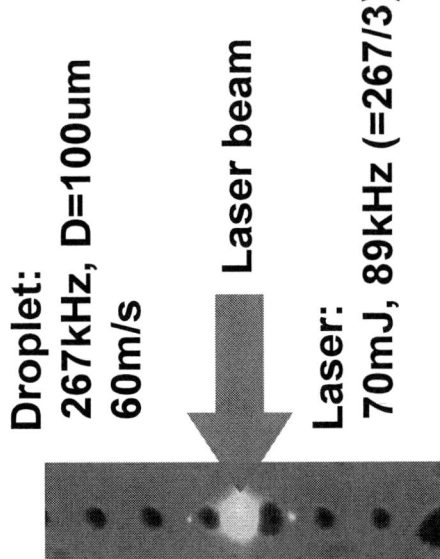

Droplet Image
2 μs after CO_2 irradiation

#1 burst

1σ=12%

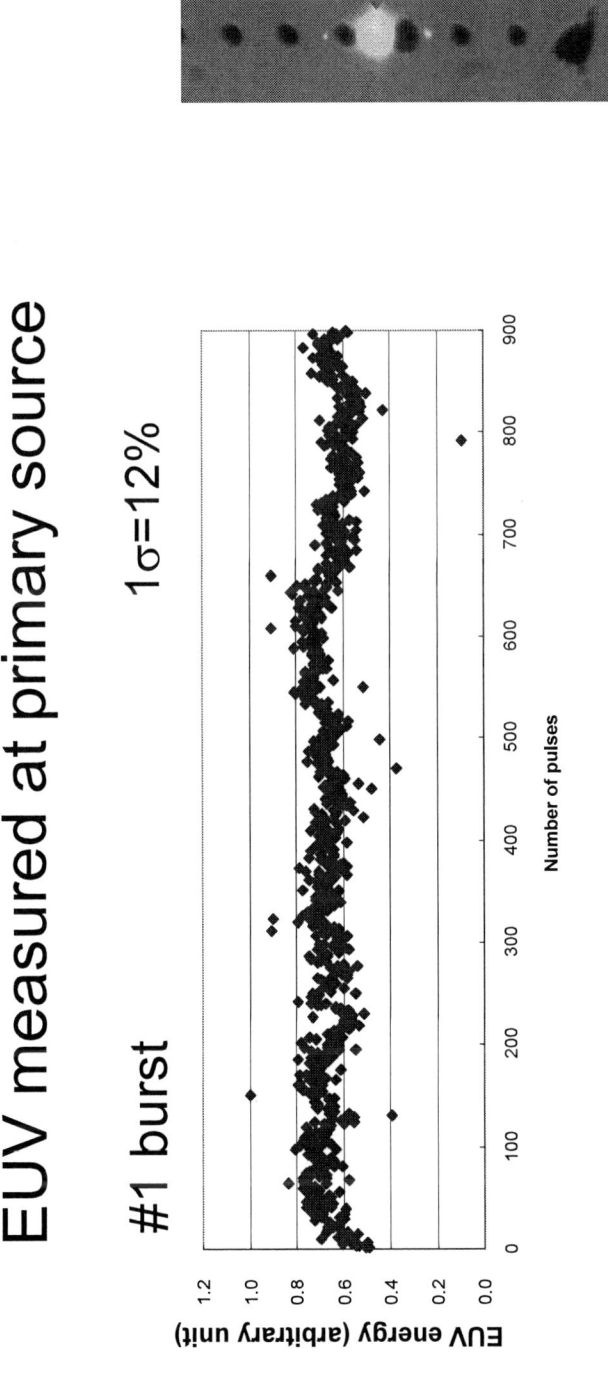

#2 burst

1σ=14%

GIGAPHOTON

EUV from Xe droplet with pre-pulsed CO_2 laser

Pre-pulse: 5mJ Nd:YAG
Main Pulse: 60mJ CO2
(Pulse width 25 ns)

Optimum delay: 200ns
Max Conversion Efficiency: 0.4%

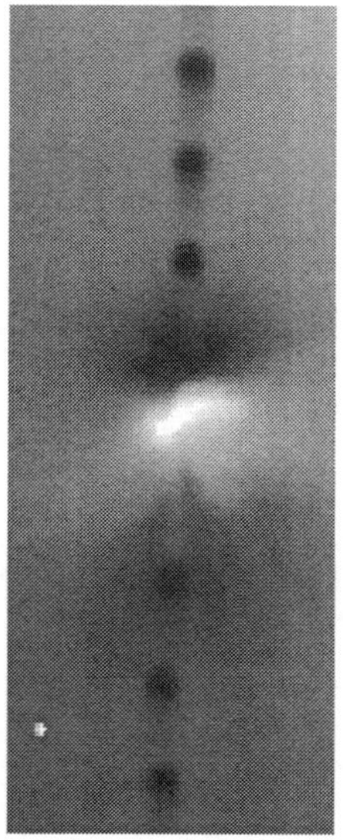

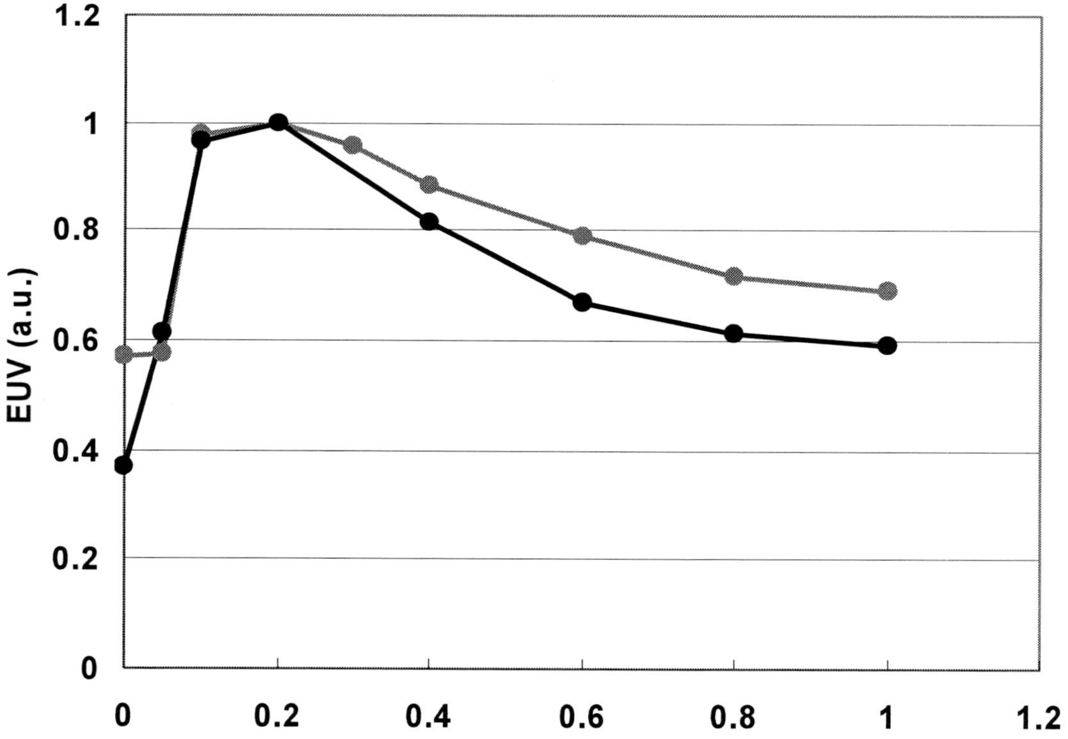

Droplet : ϕ40μm
Jet : ϕ30μm

Legend:
- Droplet
- JET

Chart axes:
- Y-axis: EUV (a.u.), from 0 to 1.2
- X-axis: Delay Time (μs) between pre-pulse and CO_2 laser, from 0 to 1.2

Ref: Appl. Phys. B 83, 213 (2006)

Xe ion energy dependence on delay time

Ion Energy measurment:

✓ Time of Flight Method (TOF)

Laser Energy
✓ Pre-Pulse (Nd:YAG Laser): 5mJ
✓ Main-Pulse(CO_2 Laser): 65mJ

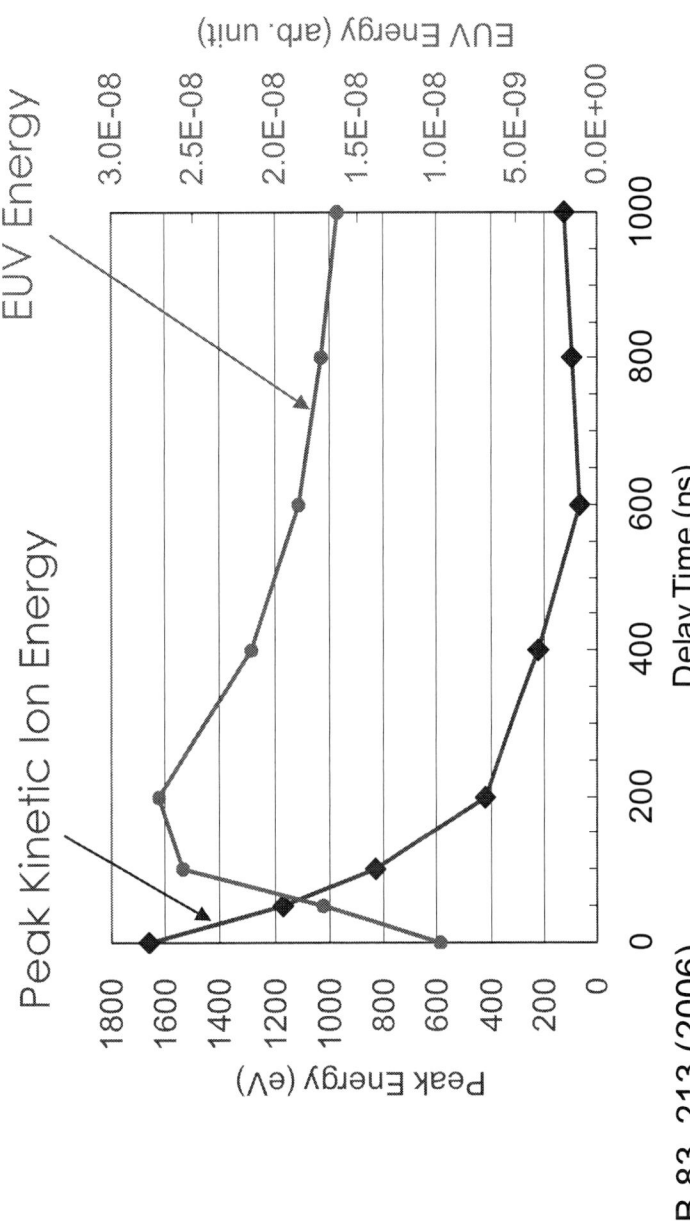

Ref: Appl. Phys. B 83, 213 (2006)

The peak ion kinetic energy decreases with delay time for the CO2 laser irradiation.
0.5 T magnetic field can fully control tin ions.

Sn droplet target

Laser irradiation onto a Sn droplet (Image 4 µs after irradiation)

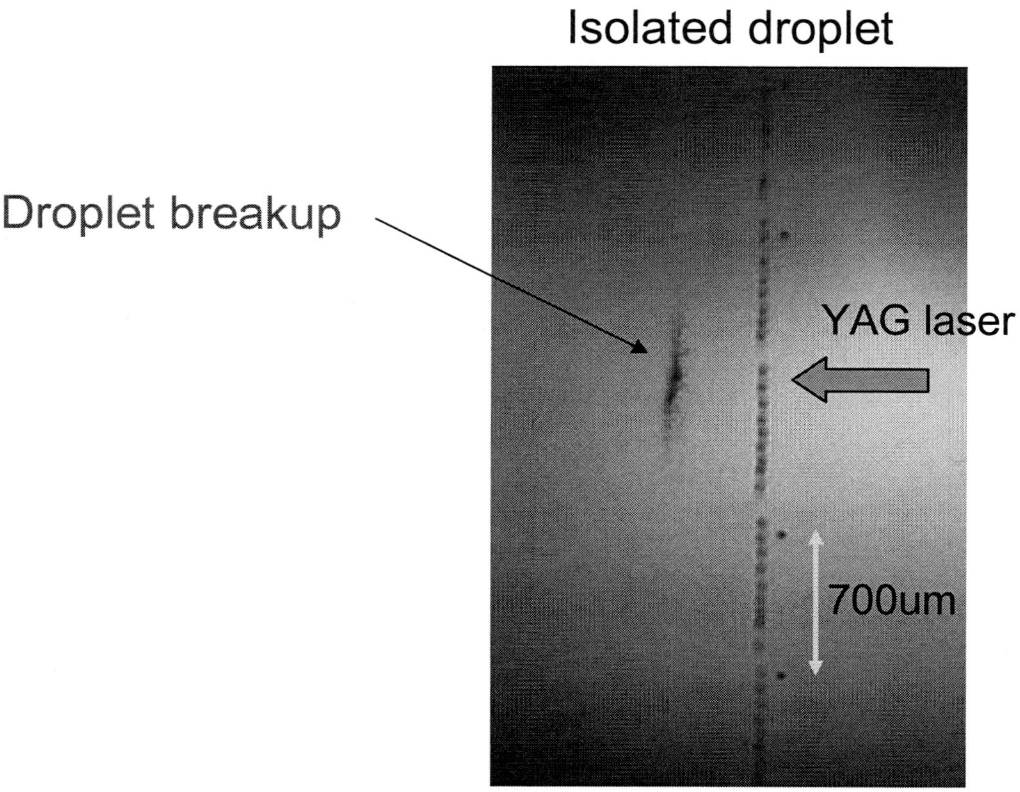

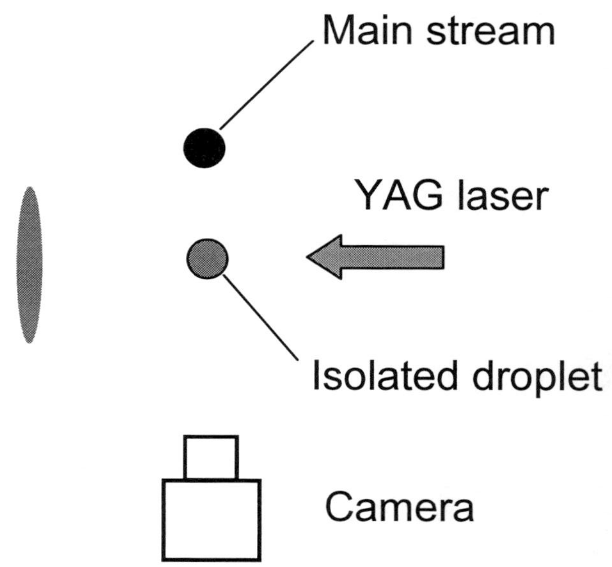

Top view

Main stream

YAG laser

Isolated droplet

Camera

Isolated droplet

YAG laser

700um

Droplet breakup

Ref: *Proc. SPIE* **6921** (2008) 69210T

Double pulse laser irradiation onto Sn droplet

Experimental setup

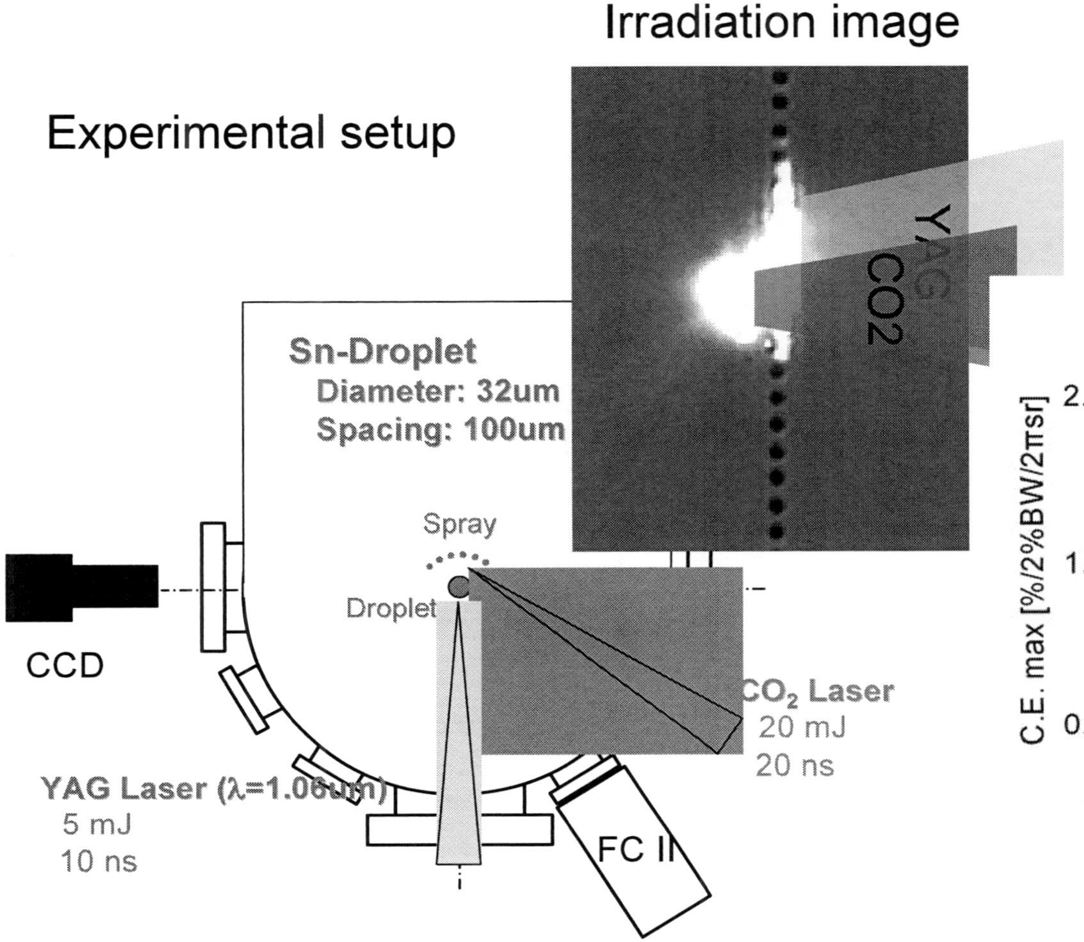

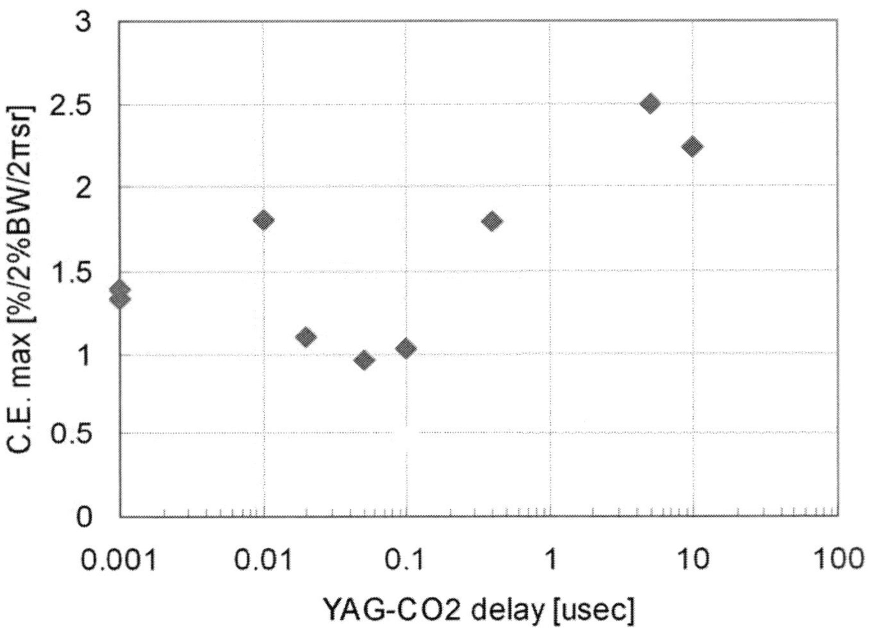

Irradiation image

Sn-Droplet
Diameter: 32um
Spacing: 100um

CO₂ Laser
20 mJ
20 ns

Spray

Droplet

FC II

YAG Laser (λ=1.06um)
5 mJ
10 ns

CCD

The maximum conversion efficiency of 2.5 % is obtained at a YAG-CO2 delay time of about 5ms.

GIGAPHOTON

UVA

Target expansion dependence on pre-pulse laser intensity

KYUSHU UNIVERSITY LASER LABORATORY

Nd:YAG Laser Beam(Δt = 10 ns, Spot size: 40 μm)　Sn solid pellet target

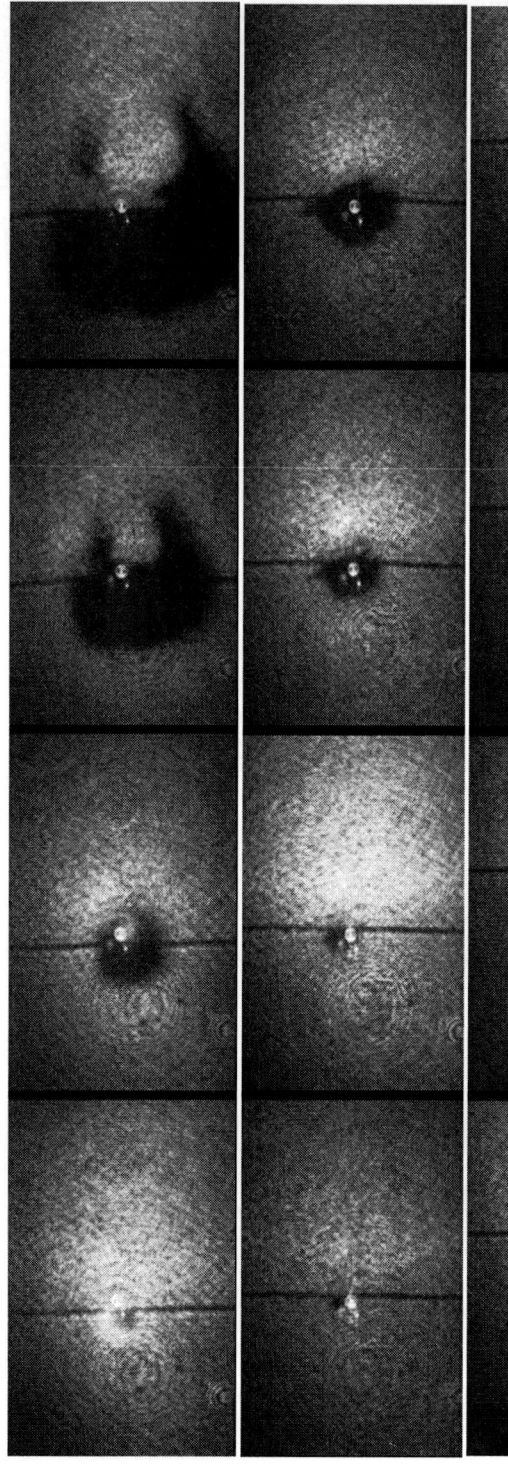

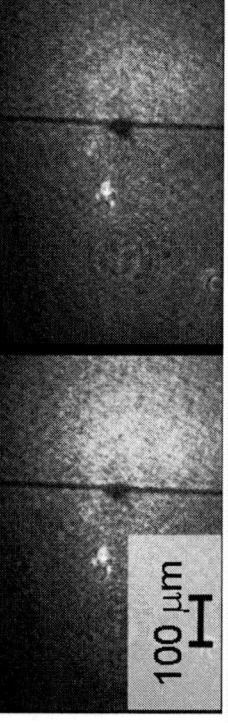

100 μm

4.8×10^{11} W/cm^2

1.6×10^{11} W/cm^2

3.4×10^{9} W/cm^2

5.4×10^{8} W/cm^2

- The expanding speed increases with increased the intensity of the pre-pulse.
- The target hardly expanded at lower intensity of 5.4×10^8 W/cm^2

Optimization of target irradiation

EUV critical issues:
Reliable high power source and long life collector module.

Realization of high conversion efficiency (CE) and clean plasma.

1. Fast ion management.
2. Real mass limited target.

EUV LPP light source roadmap

			ETS (Internal use only)	SD (1st Gen.) (proto/ integration possible)	HVM(2nd Gen.) (product)
Timing			2009/1Q	2009/4Q	2011/1Q
Power (Source to IF:34%) (R=0.6, 4sr(0.64), T=0.9)			100W	140W	280W
Drive laser			10kW	10kW	20kW
CE			3.5%	4.0%	4.0%
Target			Tin droplet	Tin droplet	←
Mitigation			Single magnet & ionization	magnet & ionization	←
C1 Mirror	Spec.		4sr 60 Bi-layer R>60%	TBD Heat Protected	TBD
	Life		200Bpls	TBD	TBD
Tool interface (I/F)			No	Yes	Yes
Duty			>75%	TBD	TBD

GIGAPHOTON

EUVA

Power roadmap

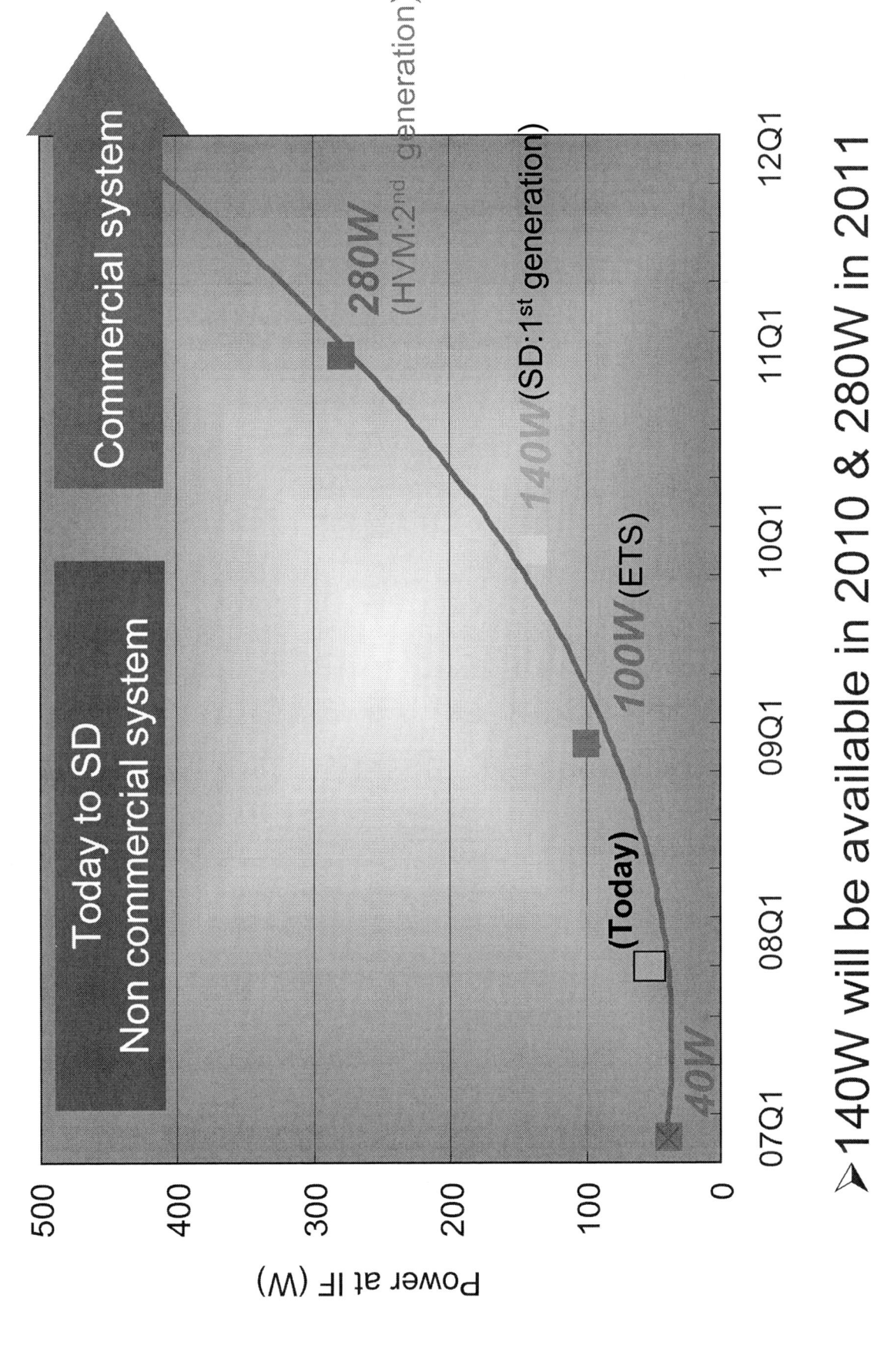

➤ 140W will be available in 2010 & 280W in 2011

Summary

- **LPP source at EUVA (non-integrated setup)**
  - Further advance of component technology is reported
    - 13 kW drive laser output power; 100 W in-band EUV at I/F equivalent.
    - Improving operating duty of drive laser . 100% duty cycle Pre-Amplifier operation demonstrated with 2.7 kW output power.
    - Magnetic field plasma guiding of CO_2 laser produced Sn plasma.
      - Sn plasma is guided by magnetic field.
        - → Basic technology for Sn evacuation is established.
      - Sn droplet irradiation optimization is to be completed.
    - Clean LPP source fully possible.

- **Next step (integrated setup)**
  - Integrated system demonstration with advanced component technology and mirror lifetime evaluation.

PHILIPS

EUV sources based on DPP

EUVL Symposium
Lake Tahoe, CA, September 30, 2008

Marc Corthout, Masaki Yoshioka, et al.

PHILIPS

Overview

- Alpha source experience in the field

- Collector lifetime as key integration topic

- Power scaling fundamentals

- Beta source productization

PHILIPS

- Alpha source experience in the field

  - Collector lifetime as key integration topic

  - Power scaling fundamentals

  - Beta source productization

PHILIPS

DPP sources in the field

- 24 DPP sources in use for wafer exposures and EUV R&D

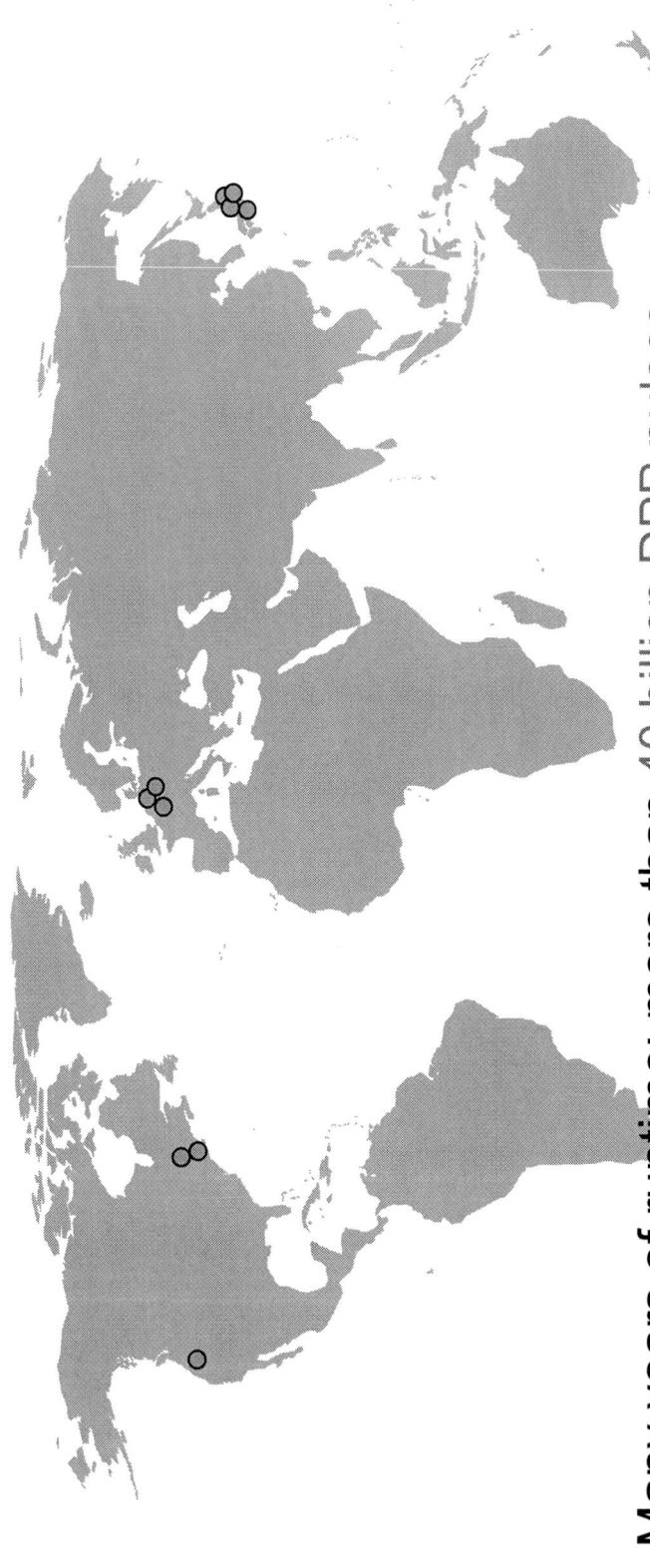

- Many years of runtime: more than 40 billion DPP pulses used for exposures and source testing
- Up to 4-5W IF power continuously, scanning > 4 wafers per hour (5 mJ/cm2 resist)
- Continuous Improvement Process ongoing to support customers

PHILIPS

Xe DPP: XTS 13-150 IF Source upgrade

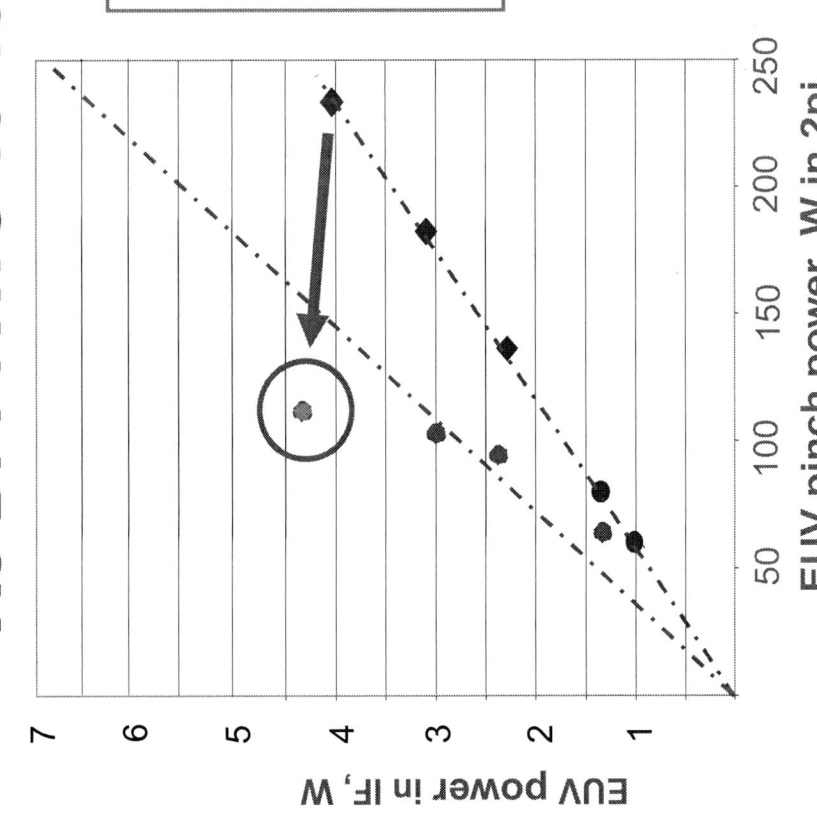

> 4 W IF power with more efficient collection to IF →
smaller thermal load →
longer Xe electrode lifetime

Debris mitigation improvement →
collector lifetime extension

- \> 4W IF power directly measured in 100% duty cycle operation

PHILIPS

Sn DPP Alpha reliability improvement

- Sn present in SoCoMo in 3 phases: liquid, gas & solid state

- Lots of testing and engineering required to avoid unwanted Sn deposition at source, sensors, and collector after many days of operation

With dose control

Well within 0.5%
Dose stability

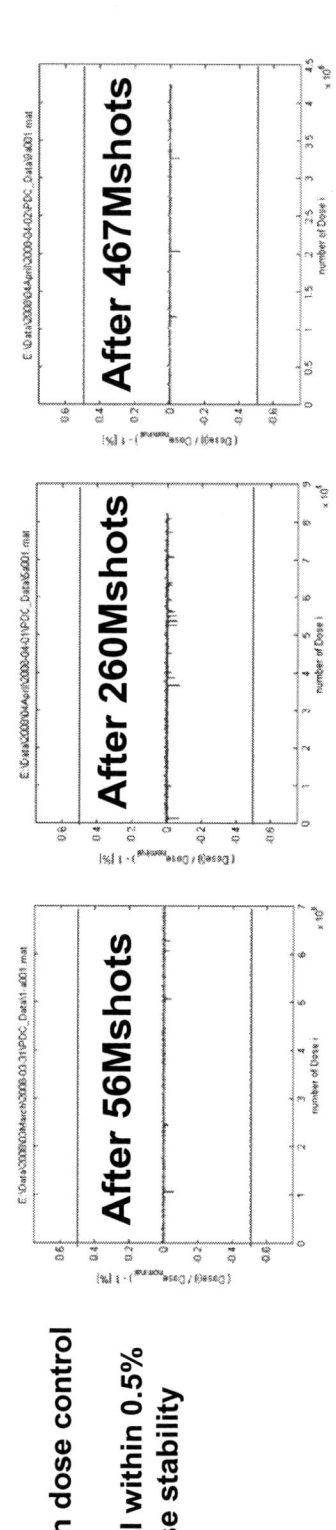

After 56Mshots

After 260Mshots

After 467Mshots

→ Debris mitigation as key technology

PHILIPS

- Alpha source experience in the field

- Collector lifetime as key integration topic

- Power scaling fundamentals

- Beta source productization

PHILIPS

Sn Debris Mitigation Generations:
running for many years parallel to source development

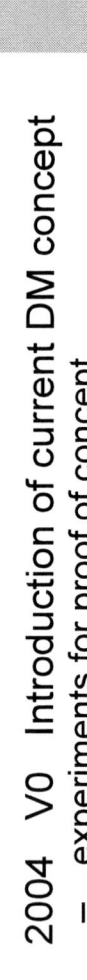

- **2003** Early experiments on glass plates

- **2004 V0** Introduction of current DM concept
  - experiments for proof of concept

- **2005 V1** DM system for research tool
  - experiments for debris mitigation with samples

- **2006 V2** DM system for Sn-SoCoMo
  - experiments for debris mitigation with collector shells

- **2007 V3** DM system for full Alpha collection angle
  - advanced system for high power (>170 W source)
  - long life and efficient water-cooling solution

- **2008 V4** DM system for Beta collection angle
  - further improvement of mitigation efficiency
  - beta source power levels

PHILIPS

Collector Lifetime

Collector exposed for weeks has been cut in samples and analyzed

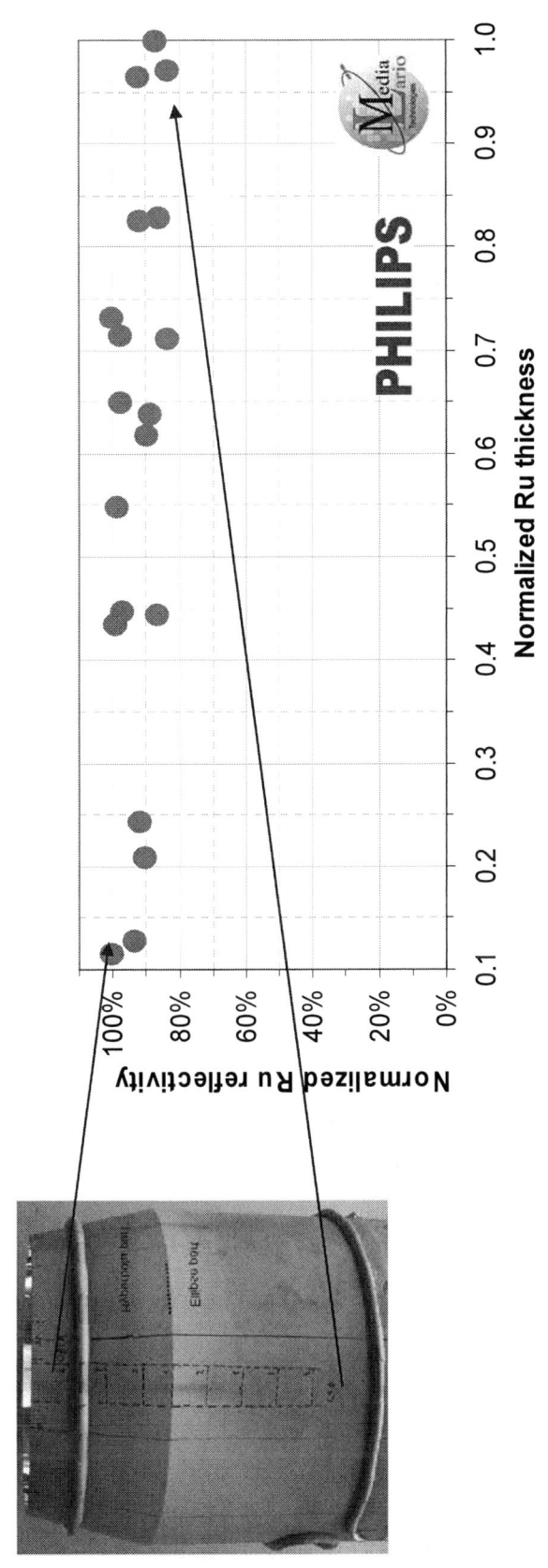

Reflectivity was not reduced:

--> Sn deposition was negligible due to very efficient mitigation

--> Sputtering of reflective Ru layer was observed:

Ru coating of Grazing Incidence collector works as sacrificial layer keeping full reflectivity despite significant material removal

PHILIPS

Collector Lifetime

Results for Alpha type hardware: 20 Gshots = many months of operation

Improvements for Beta and HVM phase:

- Thicker Ru layer:
  factor 6 already realized

- More space between pinch and collector
  factor 5 reduced sputter rates

- Improved source operating :
  > factor 5 less fast ions

Enabling collector lifetime significantly above 1 year under full time HVM use

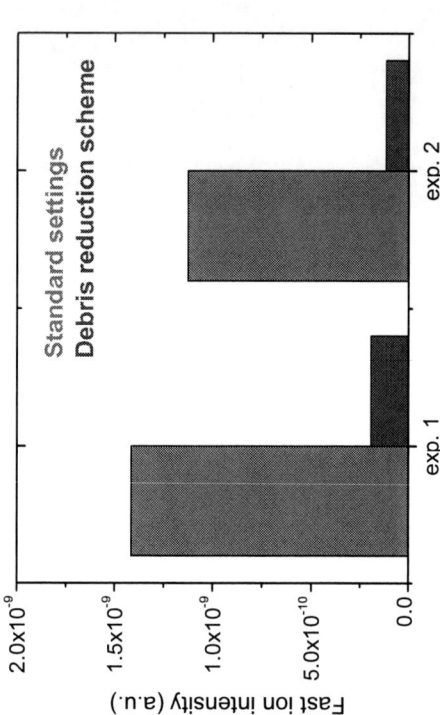

PHILIPS

- Alpha source experience in the field

- Collector lifetime as key integration topic

- **Power scaling fundamentals**

- Beta source productization

PHILIPS

Power scaling of Sn-DPP enables high EUV power

- Higher EUV powers up to HVM ranges can be achieved by

  – Increased pulse energy (E)

  – Increased pulse frequency (f)

  – Improved conversion efficiency (CE) from electrical to EUV

  (electrical wall plug eff. is 90%)

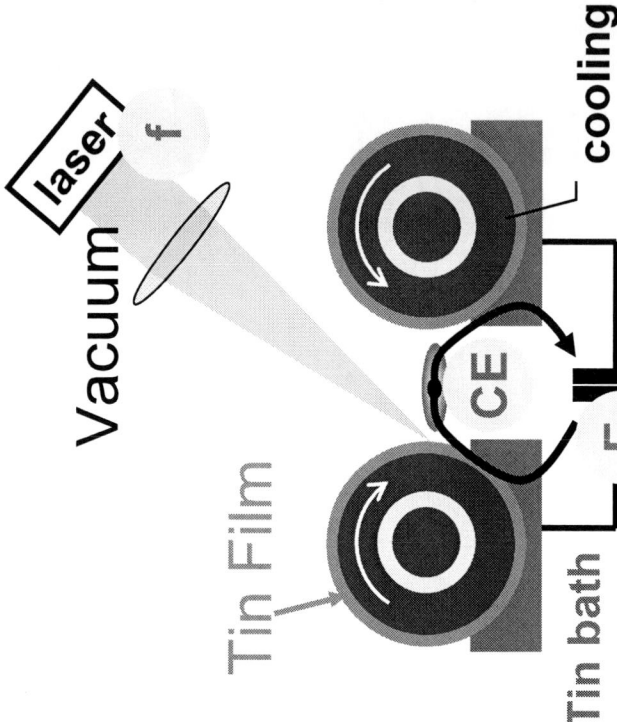

Power scaling parameters

PHILIPS

Energy scaling to above 80 mJ per pulse

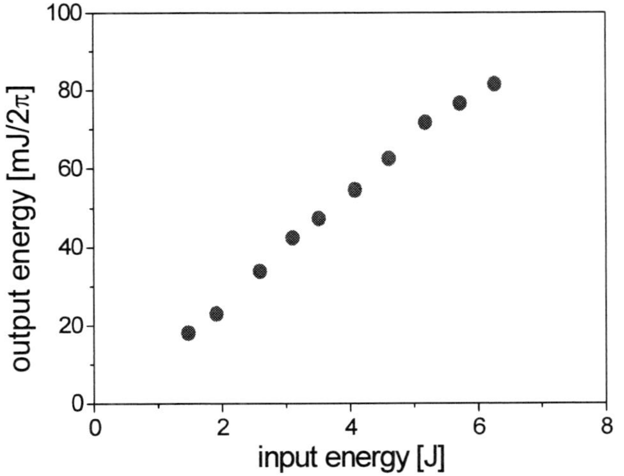

Frequency scaling to 100kHz

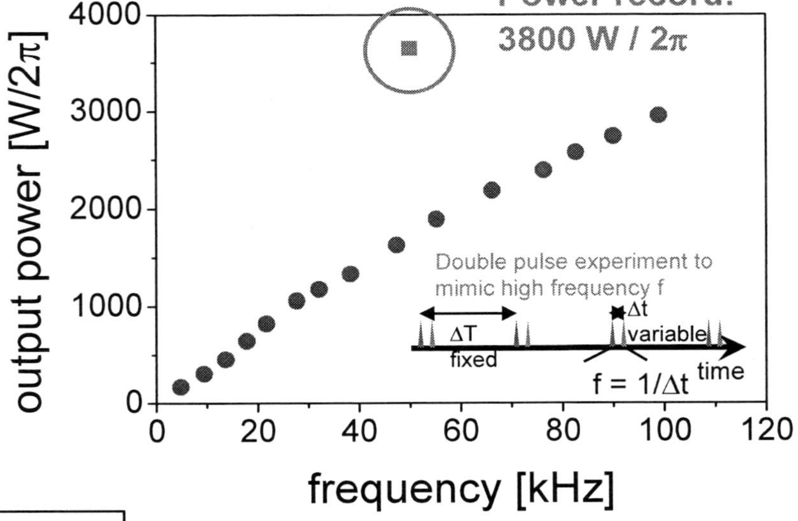

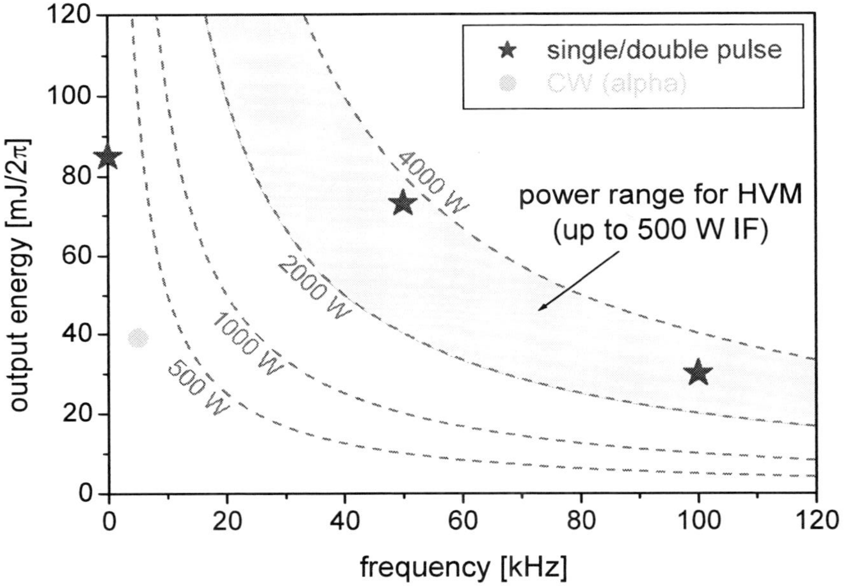

Power record:

3800 W / 2π

Double pulse experiment to mimic high frequency f

f = 1/Δt

Energy and frequency scaling without losing efficiency

HVM requirements are in scalable Sn-DPP range

power range for HVM (up to 500 W IF)

- ★ single/double pulse
- ● CW (alpha)

4000 W
2000 W
1000 W
500 W

PHILIPS

Fast plasma decay time
enables continuous operation

5500 pulses at **40kHz** continuous operation without power loss

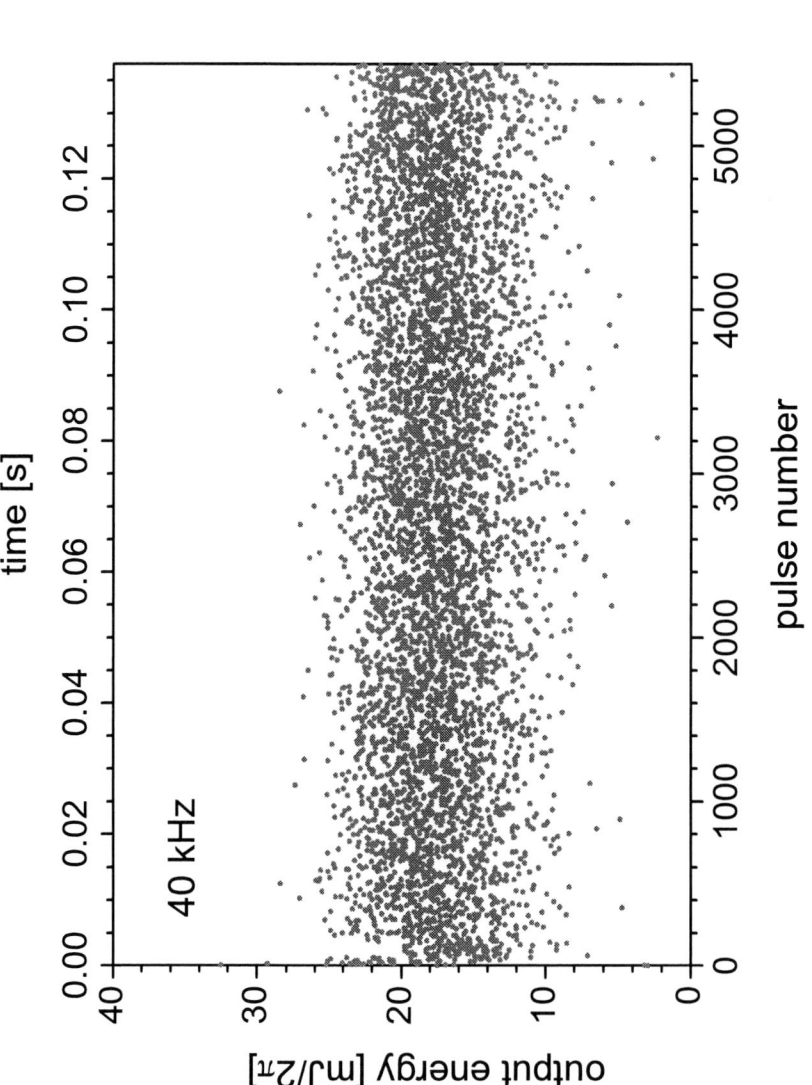

PHILIPS

- Alpha source experience in the field

- Collector lifetime as key integration topic

- Power scaling fundamentals

- Beta source productization

PHILIPS

SoCoMo designs ready

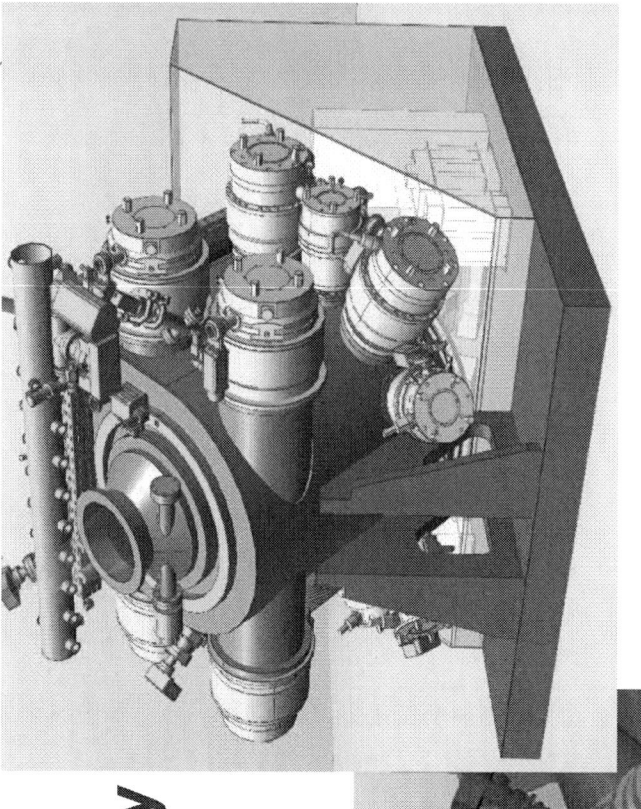

Commercial SoCoMo's contain:

+ Source
+ Debris mitigation
+ Grazing incidence collector
+ Vacuum chamber and pumps
+ Alignment tooling and metrology

PHILIPS

Releasing Beta components

- Lamp head
- Capacitor bank
- Power stages
- Laser system
- Periphery and controls

- Tin Handling Box for Sn supply and electrode cooling
  - Electrodes absorb > 50% of dissipated power
  - Improved cooling via tin pumping

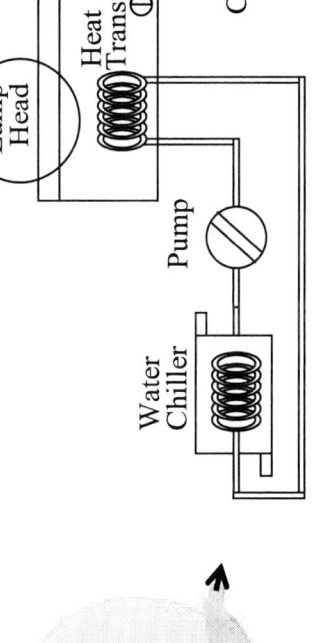

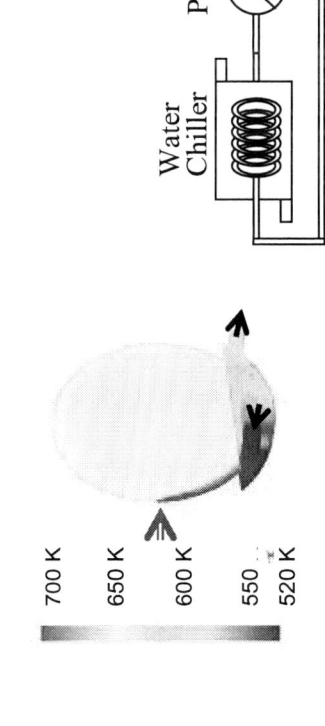

700 K
650 K
600 K
550 K
520 K

PHILIPS

Beta power results

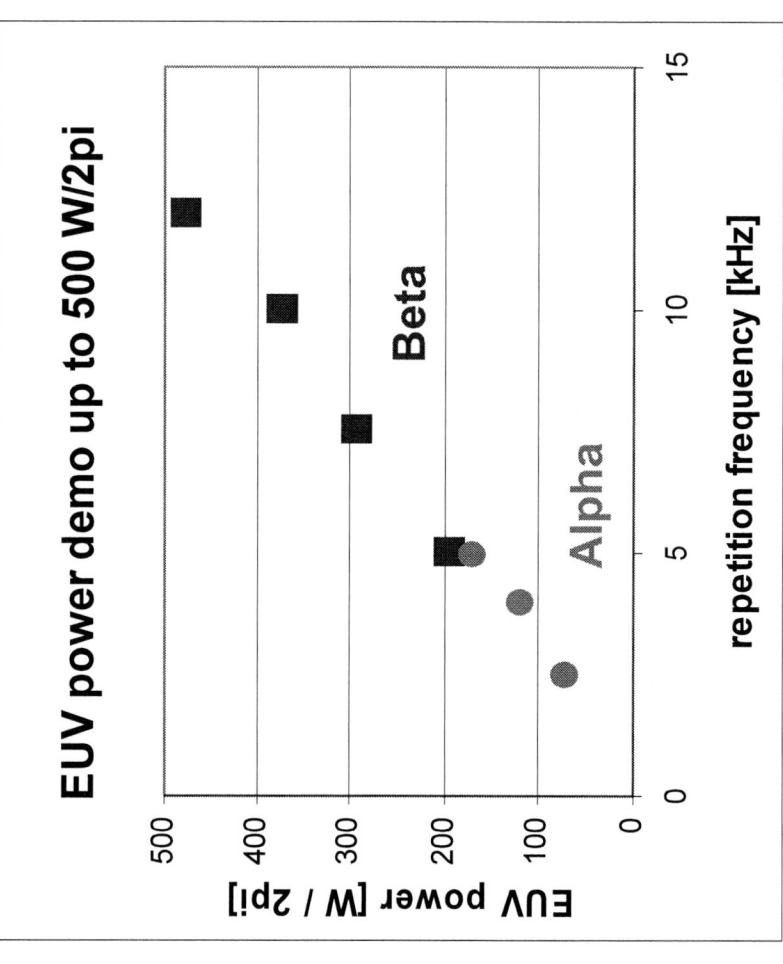

EUV power demo up to 500 W/2pi

Corresponding to 50W@IF

PHILIPS

Conclusions

- In the field: Xe and Sn DPP sources enable the EUV progress reported at this conference
  - Accumulated 40 billion pulses made, about 600 wafers scanned
  - Collector lifetime proven in practice

- In the lab: Scalability of Sn DPP to HVM levels proven
  - 500W @ IF (3800W in 2 pi)
  - Collector lifetime > 1 year

- In productization: Beta SoCoMo modules designed and tested
  - Beta source up and running at 500W in 2 pi

- Intense cooperation between XTREME GmbH and Philips EUV has enabled DPP to combine all lessons learnt into an industrial platform to enable EUV production start-up in 2010.

PHILIPS

Acknowledgements

The PHILIPS Extreme UV and XTREME technologies teams

ASML

Nikon

Media Lario

Zeiss

The financial support of the German Research Ministry under contract numbers 13N8865 and 13N8866 is gratefully acknowledged.

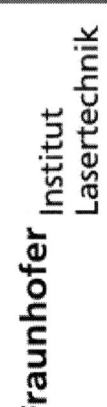

2008 International Symposium on EUVL

Time-multiplexed solid-state laser-driven EUV sources for Beta-tools and HVM

K. Takenoshita, R. Bernath, R. Kamptaprasad, J. Szilagyi, S. A. George, J. Cunado, M. Richardson,

Townes Laser Institute, College of Optics & Photonics, UCF

B. Fulford, I. Henderson, N. Hay, S. Ellwi

Powerlase Ltd. Crawley, UK

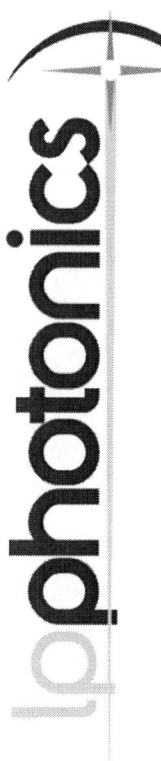

LP Photonics

Company spun off from long-term UCF stake in EUV development at one of the biggest optics schools in the nation, with university support and State investment .

Large inventory of patents and knowhow (> 40 patents) in EUV source and related technology

Building sources for alpha and beta steppers and related equipment – extendable to HVM

Modular, low-risk strategy and source architecture drawing on commercial optics and laser technology

Forming specific relationships with strategic partners

Minimized-risk SoCoMo development path

Multiplexed Commercial Solid-State Lasers

leveraging multi-$B developments in DP 1µm SSL's

Minimal-tin doped liquid droplet target

CE's 2.0-2.5% at high power
1% tin inventory of other sources

Mass-limited laser plasma regime

Room-temp mixed target is fully ionized
Negligible particulate debris
Two mitigation schemes for stopping ions

Low-risk multiplexed SS laser architecture

Time-synchronized separate 1μm commercial lasers

Powerlase Starlase
Industrial DPSSL's
Reliable/stable/long life
Hot-swappable

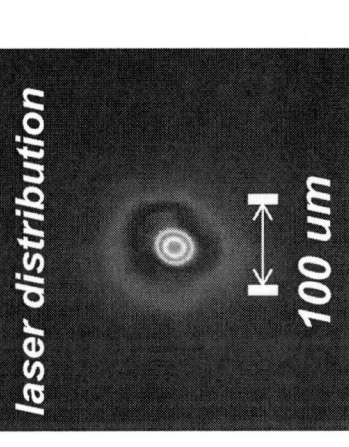

laser distribution

100 um

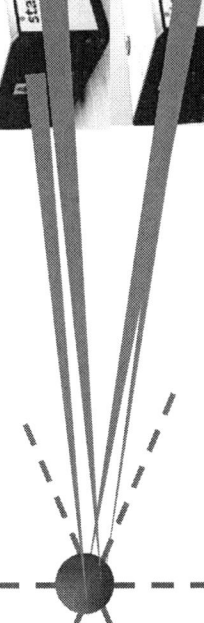

High CE

EUV

OOB

Low OOB

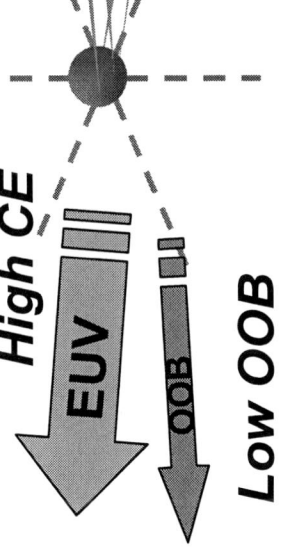

Modular architecture – low-risk fab-line operation
4 x 1.6 kW @ 28 kHz @ 100% duty cycle for Beta tool

Scattered 1μm laser light not transmitted by EUV optics
Compact footprint – competitive CoO

lp photonics

powerlase

Collector Mirror

Conservative implementation strategy

Collection angle	5 sr
Solid angle at IF	0.2 sr
Source –IF distance	1.0 m
Collector diameter	547 mm
Overall reflectivity	> 50%

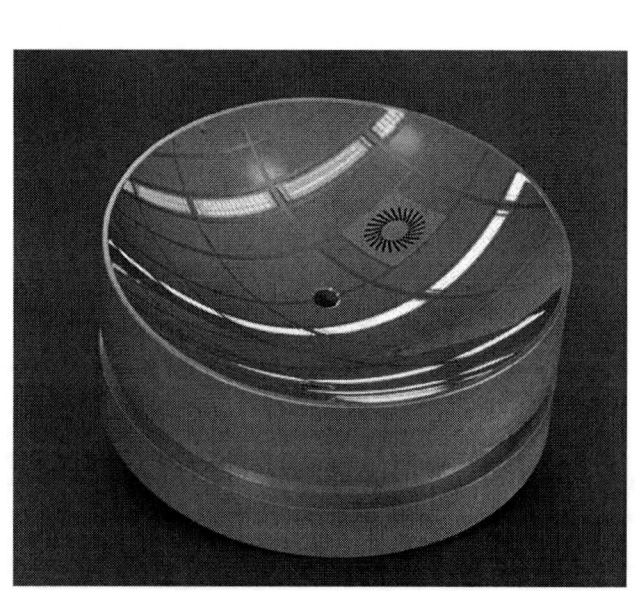

~ 40 % transmission.... less ~ 10% for mitigation

powerlase

Minimal tin-doped liquid target

TARGET → meniscus plasma → **PLASMA** → non-uniform ion and particle emission → **DEBRIS**

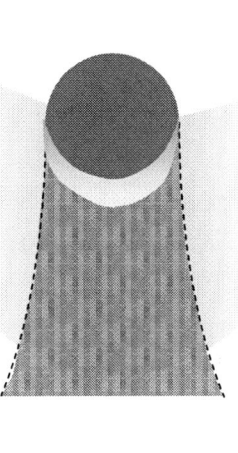

100 μm dia solid target
10^{16} tin atoms/target

solid tin

Particle Debris

tin-doped liquid → spherical plasma → isotropic ion emission

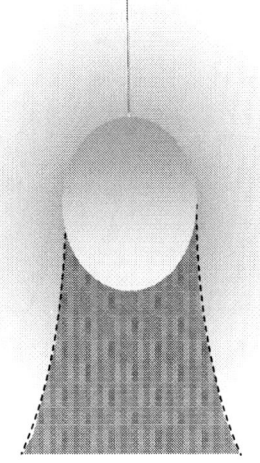

30 μm dia doped target
10^{13} Tin atoms/target

100 - 1000 times less tin/target – same EUV emission
No particulate debris – primarily ions

powerlase

lp photonics

Minimal debris over hours of operation

Many hours of continuous operation demonstrated.
Minimal accumulation of tin particulates under optimum CE conditions
3 hrs @ 6.6 kHz
7 x 10⁷ shots

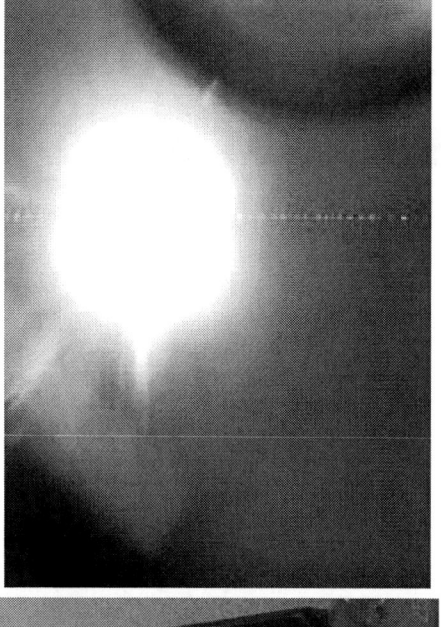

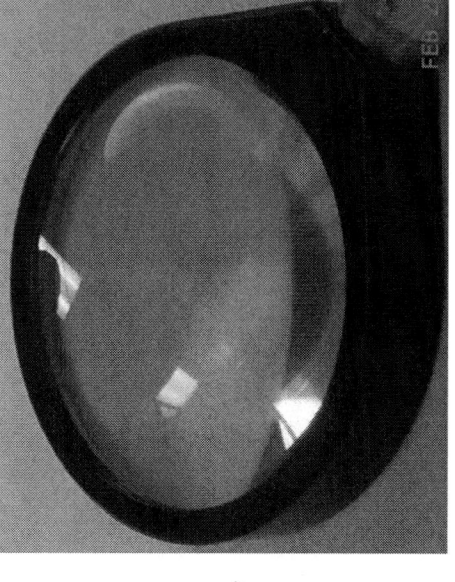

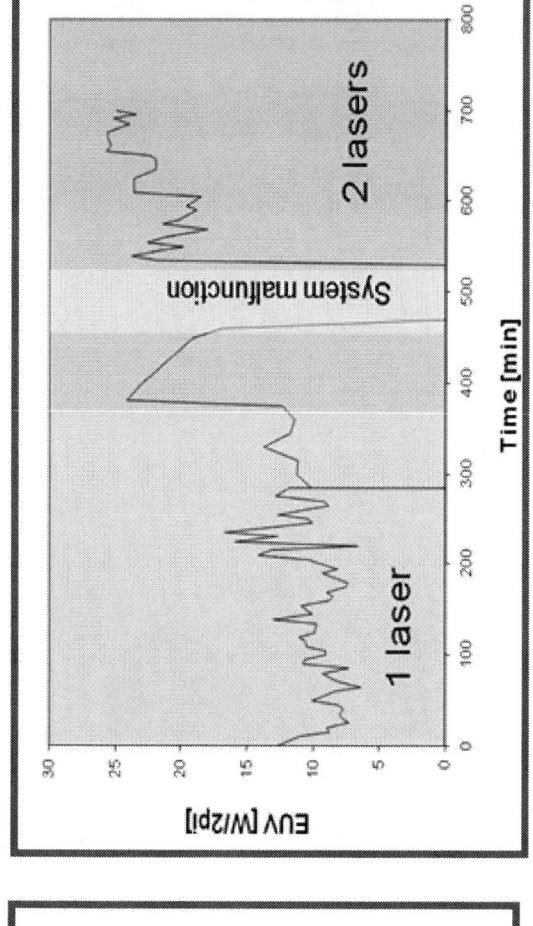

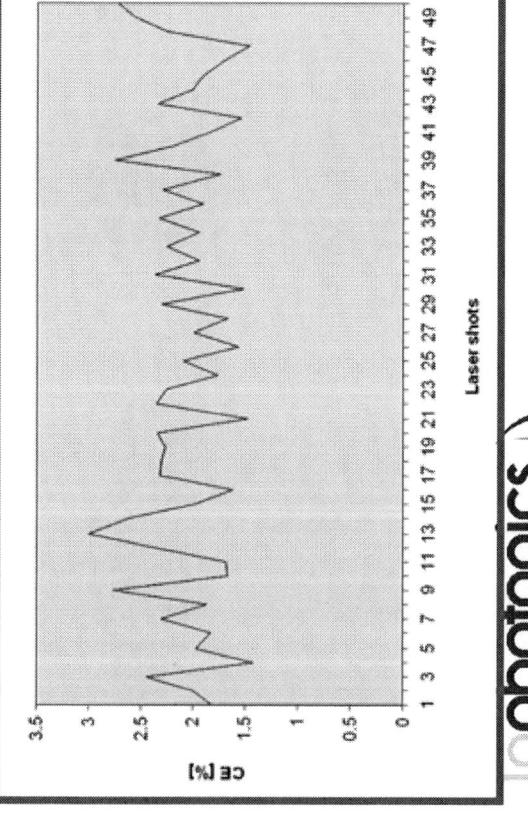

Minimal Mirror Erosion

Ion energy distributions

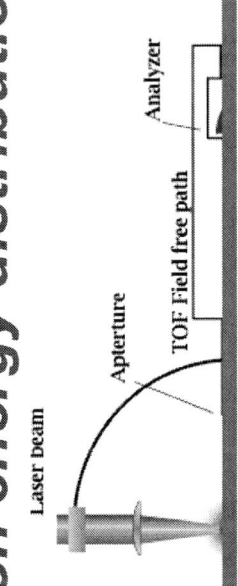

Sn$^+$
Sn^{2+}
Sn^{3+}
Sn^{4+}
Sn^{5+}

1.0e+08

1.0e+07

1.0e+06

$\frac{shot}{dE}$

Laser beam

Apterture

Analyzer

TOF Field free path

Sputtering rate

Si : 1.4 x 10^8 shots/nm

Mo: 1.3 x 10^8 shots/nm

Without mitigation
at 7 kHz

After 1.4 x 10^{10} shots
40 bi-layers reduced to 25
Reflectivity 73% to 66%

~ 600 hours
operation for mirror
failure

powerlase

lp photonics

Two mitigation methods

Dual Mitigation Layers

Repeller Field Inhibitor: Charged particles
low energy ions
Reduce mirror deposition

Magnetic Foil Trap: Highly charged Ions/neutrals
Reduce mirror erosion

High NA, high transmission design
No buffer gas used
No moving mechanics used

IF

RFI

MFT

Source

EUV

Collector mirror

powerlase

lp photonics

Ion Mitigation Effects

Magnetic Foil Trap Mitigation installed

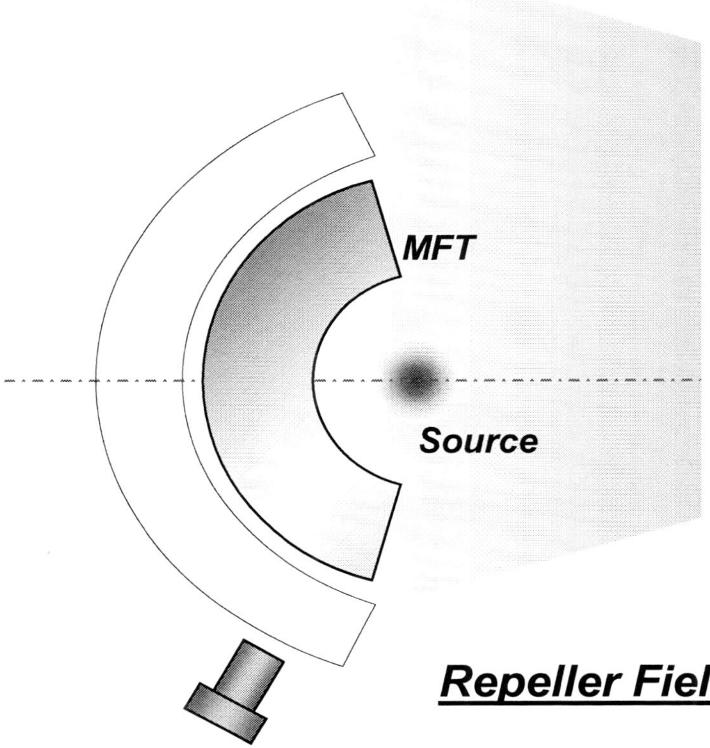

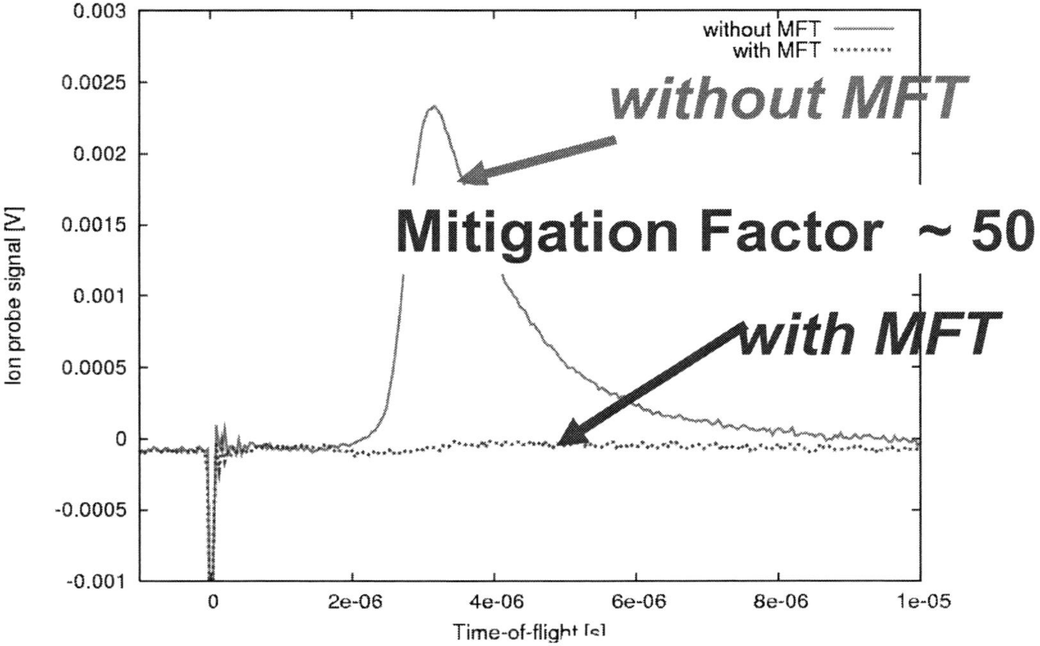

Mitigation Factor ~ 50

without MFT

with MFT

Repeller Field Mitigation Factor ~ 20   Total MF ~ 1000

Predicted mirror lifetime - 30,000 hr roadmap Currently operates several hours at 100% duty cycle

Out of band radiation

Complete out-of-band spectral characterization

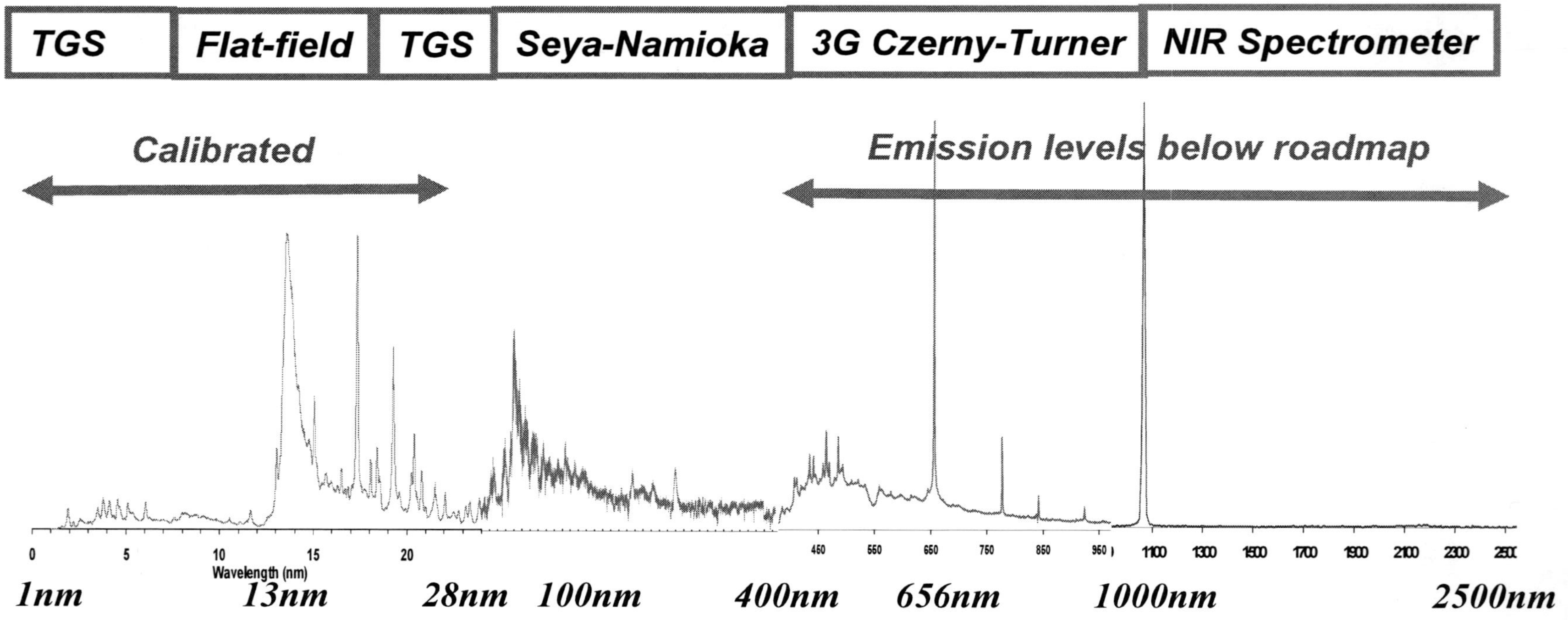

TGS	Flat-field	TGS	Seya-Namioka	3G Czerny-Turner	NIR Spectrometer

No apparent strong out-of-band sources in the VUV..

EUV optics transmission of laser light

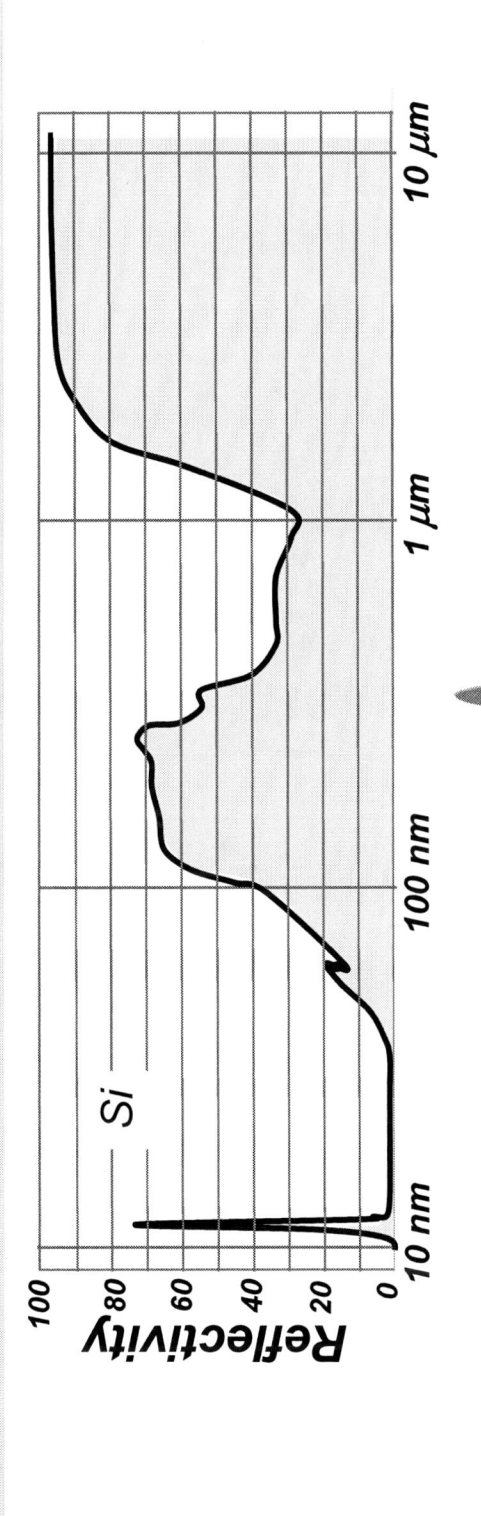

Reflectivity (100, 80, 60, 40, 20, 0)

Si

10 nm   100 nm   1 µm   10 µm

Nd:YAG
1.064 nm

CO₂
10.6 µm

Scattered solid state laser
light attenuated by 600,000

*8 mirror
system*

65% transmission of
scattered 10 µm laser light
transmitted to the wafer

*Scattered SS laser light does not
pose a threat to resist heating?*

powerlase

photonics

2007 Results Two-beam multiplexing

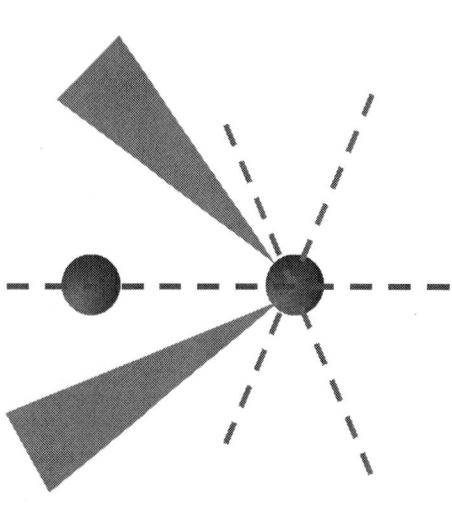

Temporal Multiplexing

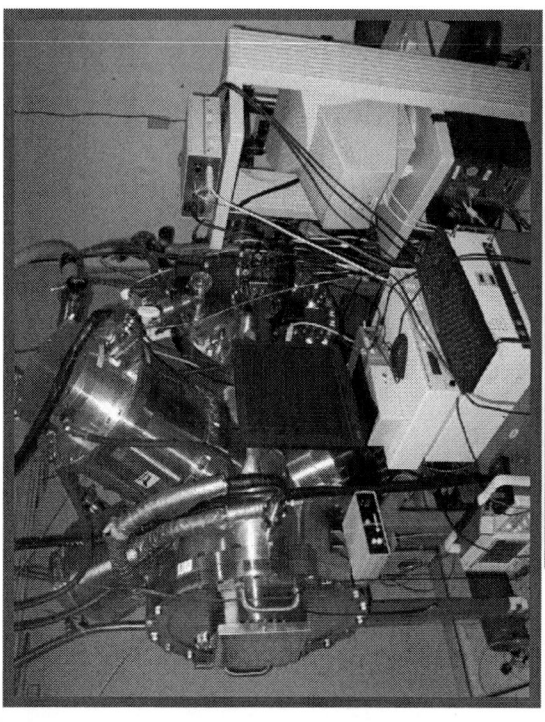

Two 500 W lasers @ 3.0 kHz
30 kHz target source
CE ~ 2.4%
100% duty cycle
EUV power 24 W

powerlase—

lp photonics

Introduction of new 1.6 kW SS laser

CE = 2.0 %
EUV Power = 27 W
100% duty cycle
Hours of operation

Activation of new 1.6 kW laser
Size: 24 x 42 x 150 cm

Rep. rate: 6 – 9 kHz
Pulse energy: 150 – 200 mJ
Pulse duration: 24 – 30 ns
Focus diameter: 80 – 100 μm

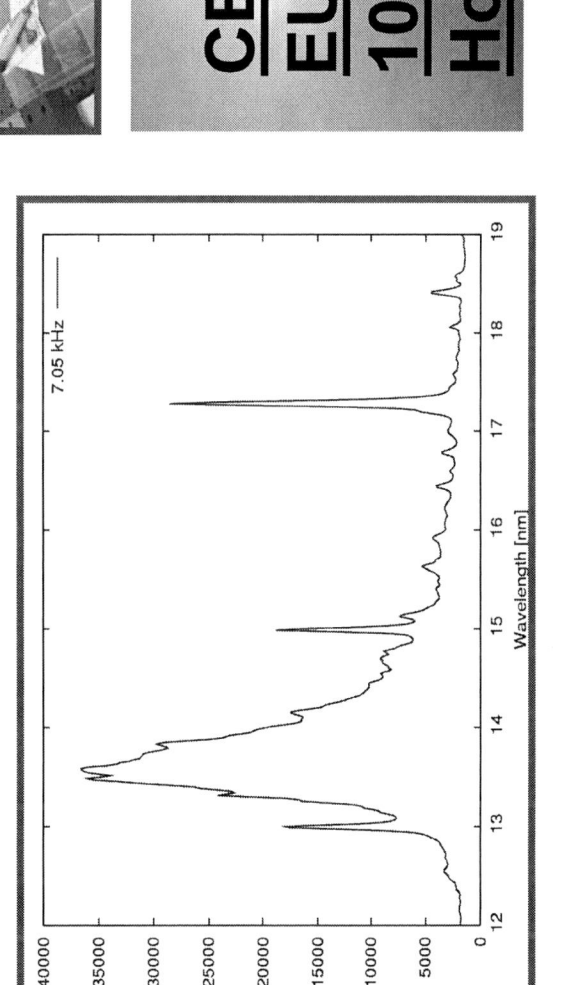

lp photonics

powerlase

EUV source with two multiplexed 1.6 kW lasers

LASER 1
1.4 kW @ 7 kHz

LASER 2
1.4 kW @ 7 kHz

14 kHz EUV source
CE = 2.0%
100% duty cycle

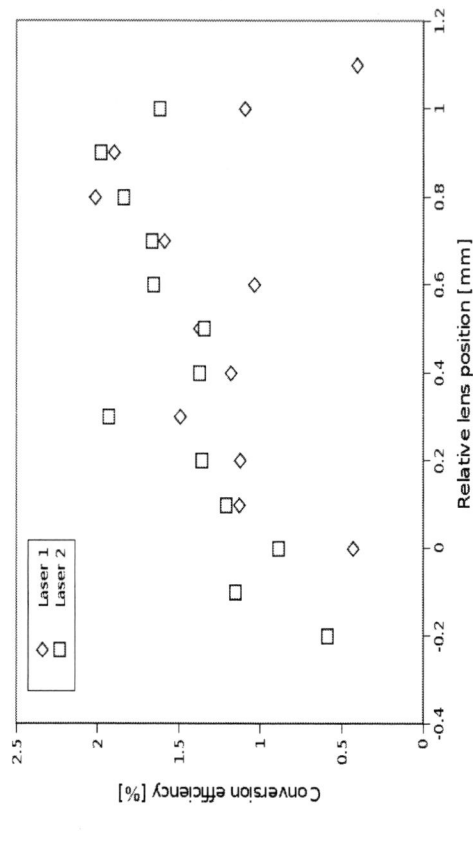

20 W source at IF with 100% duty cycle
....4 lasers will produce 40 W at IF

powerlase

lp photonics

Direct path to Beta Tool SoCoMo

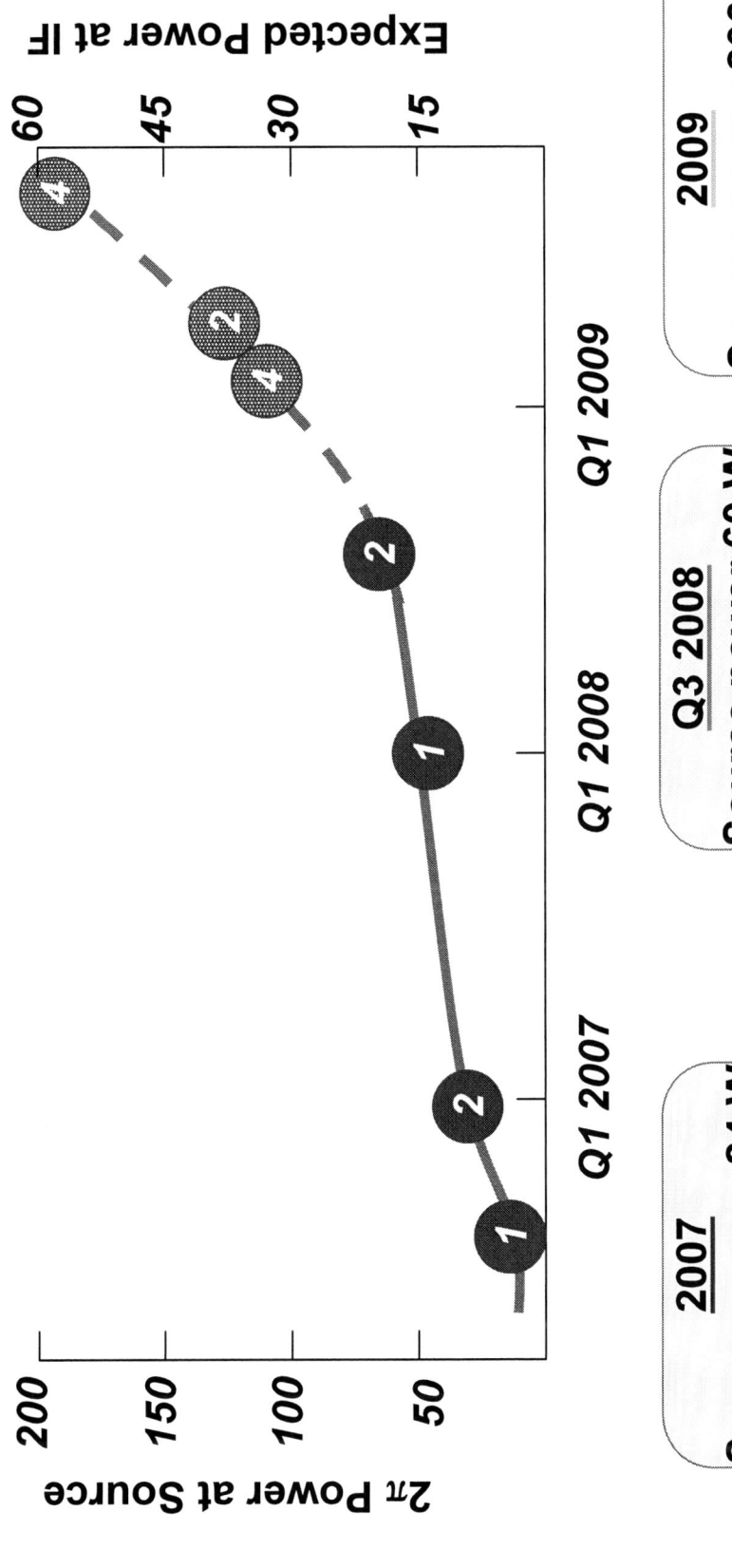

2007
Source power 24 W
Lasers (2 x 800W)
Duty cycle 100%
Debris ~ 0.1 Bshots
Power @ IF ~ 8W

Q3 2008
Source power 60 W
Lasers (2 x 1600 W)
Duty cycle 100%
Debris ~ 0.1B shots
Power @ IF ~ 20W

2009
Source power 200 W
Lasers (4 x 2500W)
Duty cycle 100%
Debris >10 B shots
Power @ IF > 60W

Extendibility to HVM requirements

2009 - 2010

Improved CE - 2.4 % measured previously at high power.

Extend total collection towards 2π

Higher power SS Lasers will be available

EUV source power at IF of 120 W possible with 4 lasers

Summary

LP Photonics is building a compact, 4-laser multiplexed SS laser Beta Tool SoCoMo for 2009

Advantages of high CE, low debris, high RR, low tin inventory source

Compact, commercial laser drivers at 1 um with no out-of-band and laser light transport issues

Multiplexed laser architecture ensure high stepper and fabline up-time.

Footprint and cost scenarios within roadmap expectations

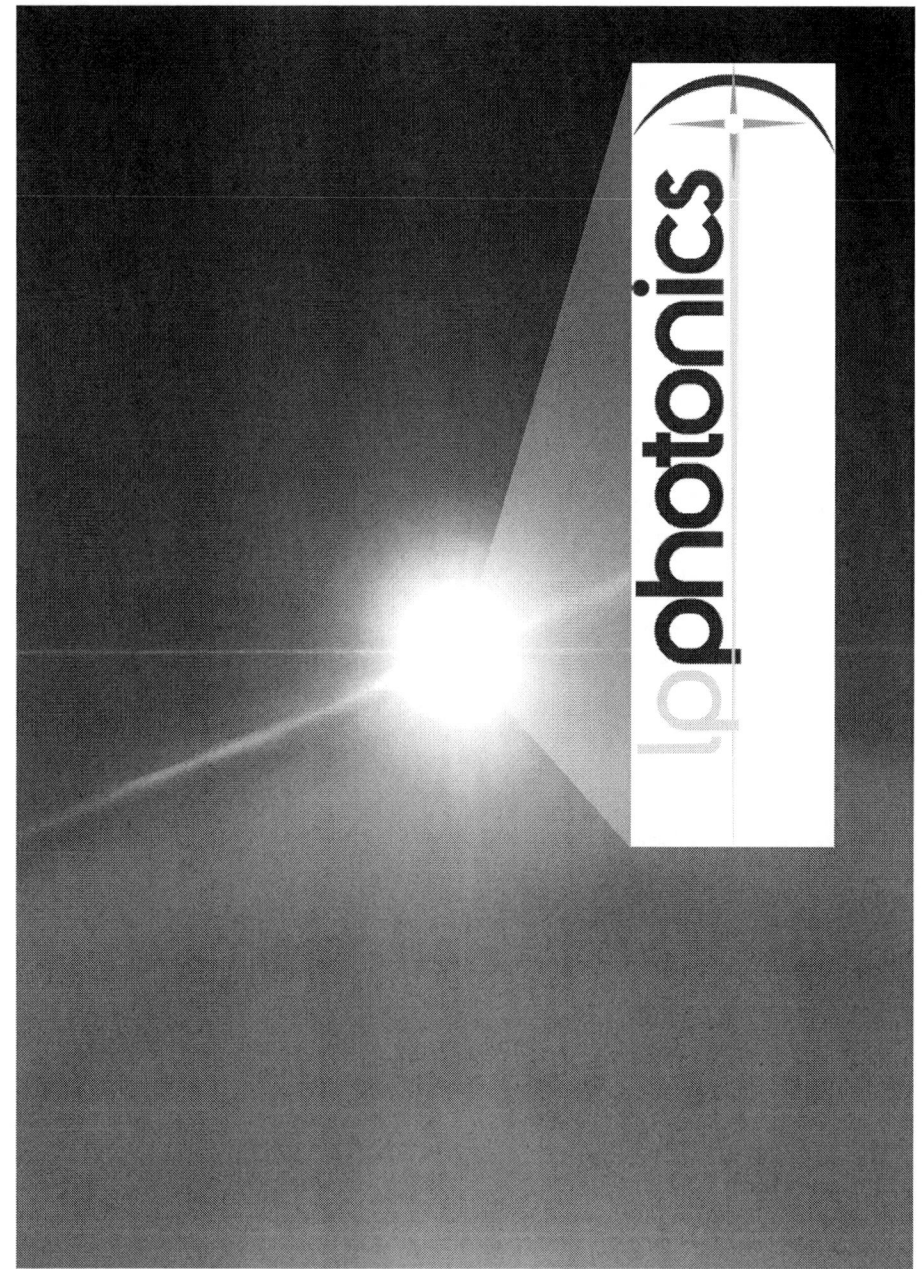

www.lpphotonics.com

Guidelines and Promising Approach for LPP-EUV Light Source for HVM

K. Nishihara, A. Sunahara[1], S. Fujioka, Y. Shimada[1], A. Sasaki[2], M. Nunami, T. Aota, H. Tanuma[3], K. Fujima[4], M. Murakami, V. Zhakhovskii, K. Nagai, H. Nishimura, M. Miyanaga, Y. Izawa, K. Mima, T. Okada[5], M. Kubodera[6], Y. Yuba[7], Y. Ueno[8], G. Sumagne[8], H. Komori[8], A. Endo[8], and A. Sumitani[8]

Institute of Laser Engineering, Osaka University, [1] Institute for Laser Technology, [2] Japan Atomic Energy Agency, [3] Tokyo Metropolitan University, [4] Yamanashi University, [5] Kiyushu University, [6] Miyazaki University, [7] Osaka University, [8] EUVA

outline of talk

· double pulse scheme for high CE

· EUV power of 500-1000 W with laser power of 13 – 20 kW

· debris mitigation

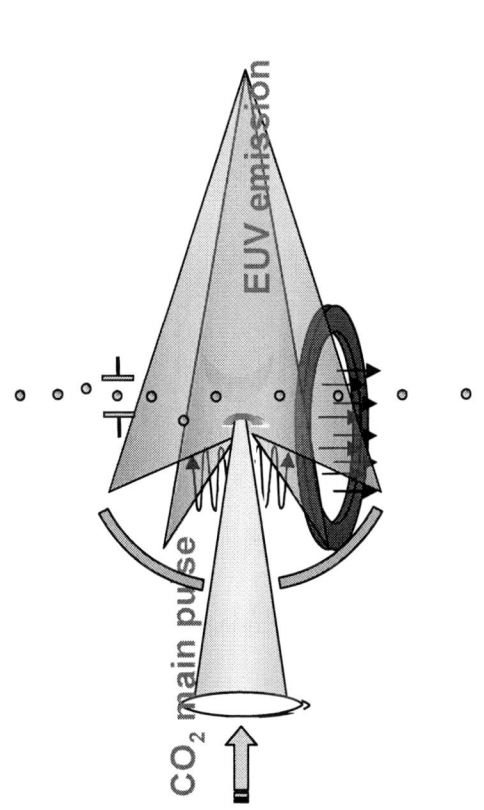

double pulse 1

We proposed double laser pulse irradiation scheme with tin droplets in EUVL06 & EUVL07 symposiums.

concept of double pulse irradiation scheme

(EUVA)

tin droplet shield

isolated 40-μm Sn droplets

(UVA Osaka) emission

YAG prepulse laser

CO_2 main pulse

Sn^{n+}

4mm

1mm

(EUVA)

original Sn droplets position

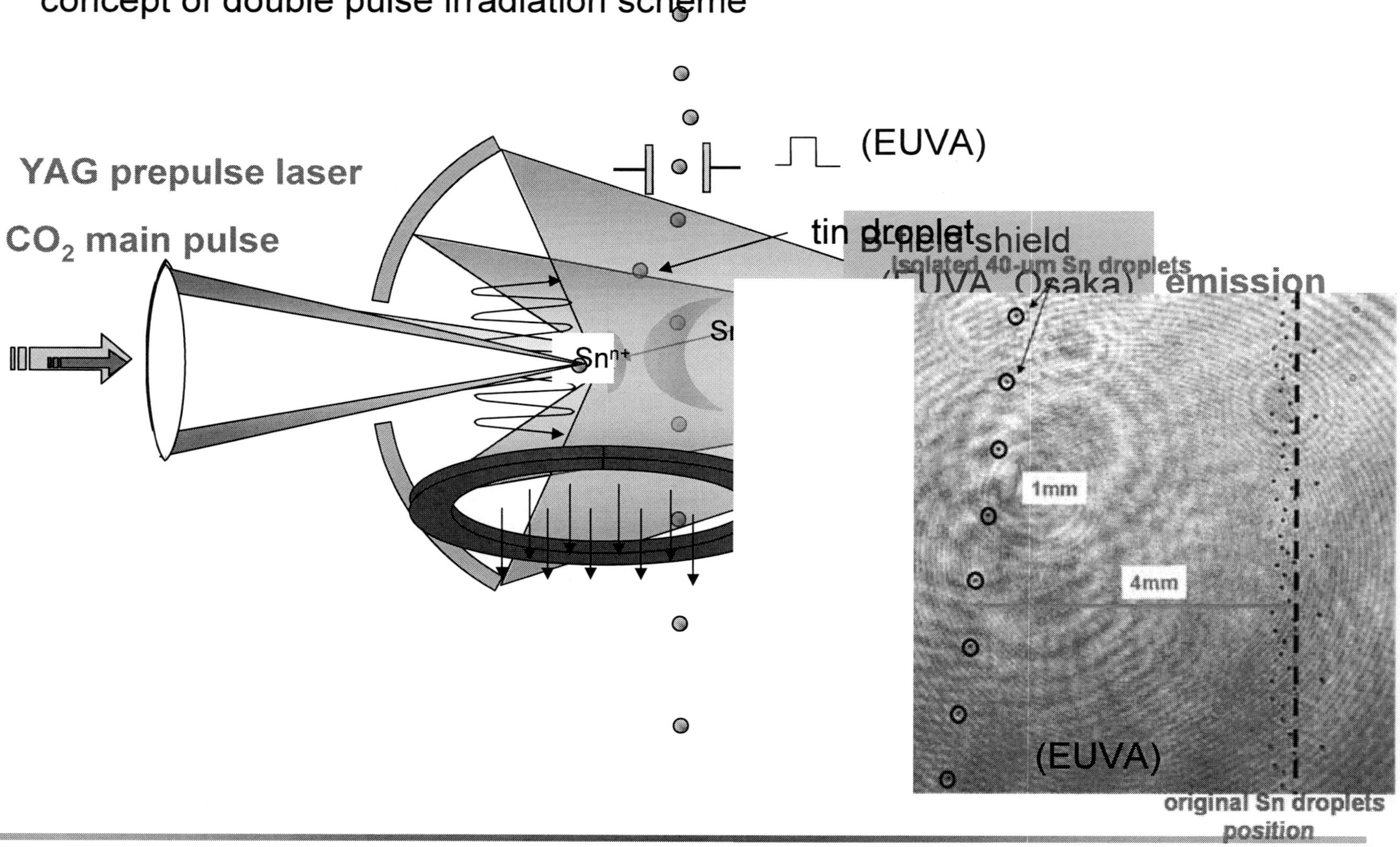

Nishihara et al SPIE08

double pulse 2

Double pulse irradiation scheme with a droplet target results in high conversion efficiency up to 5 – 6 %

benchmarked radiation hydro code, STAR simulation

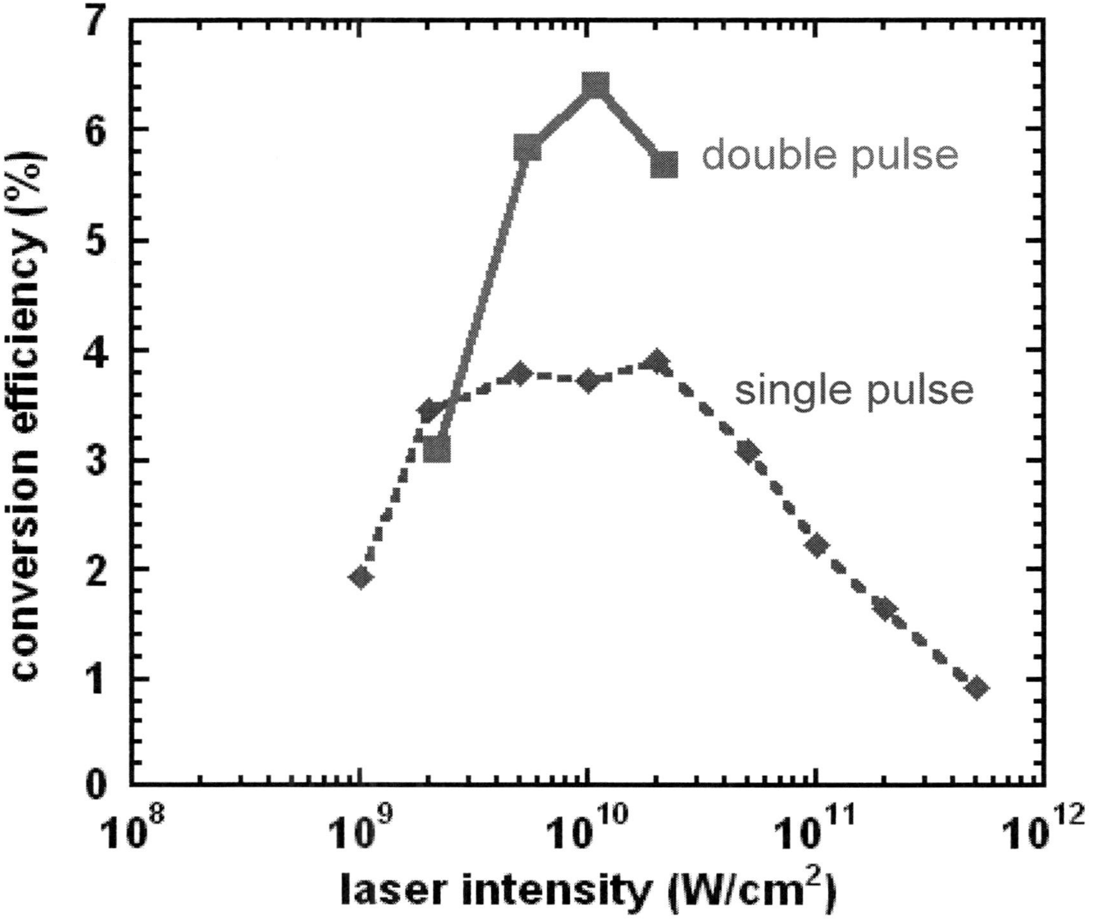

Nishihara et al EUVL07, SO-O2

executive summary

summary

1. EUV source power of 500 – 1000 W / 2π can be obtained in double pulse irradiation scheme with a tin droplet target and 13 – 20 kW CO_2 main pulse.

2. Conversion efficiency of > 4% has been experimentally achieved from a Sn droplet with 36 μm in diameter with the two-color laser, double pulse irradiation scheme.

3. We have also discussed advantages of the scheme, such as reduction of high energy ions, ion mitigation by B-field, and Sn cleaning by hydrogen radicals.

CO$_2$ laser 1

CO$_2$ laser with shorter duration results in higher conversion efficiency.

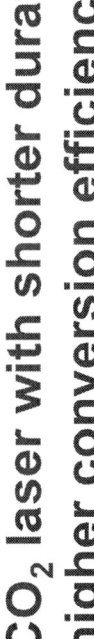

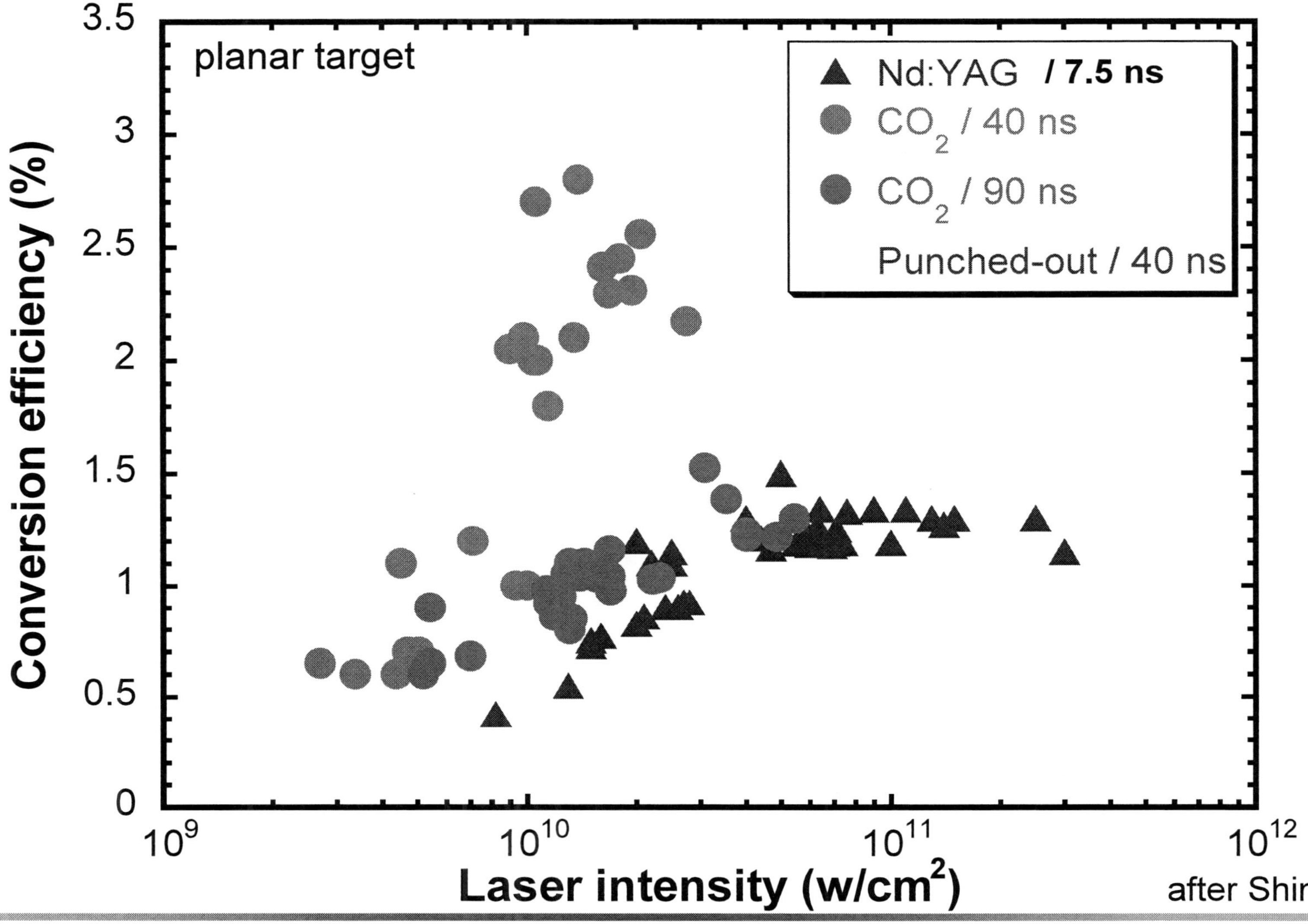

after Shimada

CO$_2$ laser 2 | **High conversion efficiency is obtained even for relatively low intensity, by providing low density plasma in punchout target**

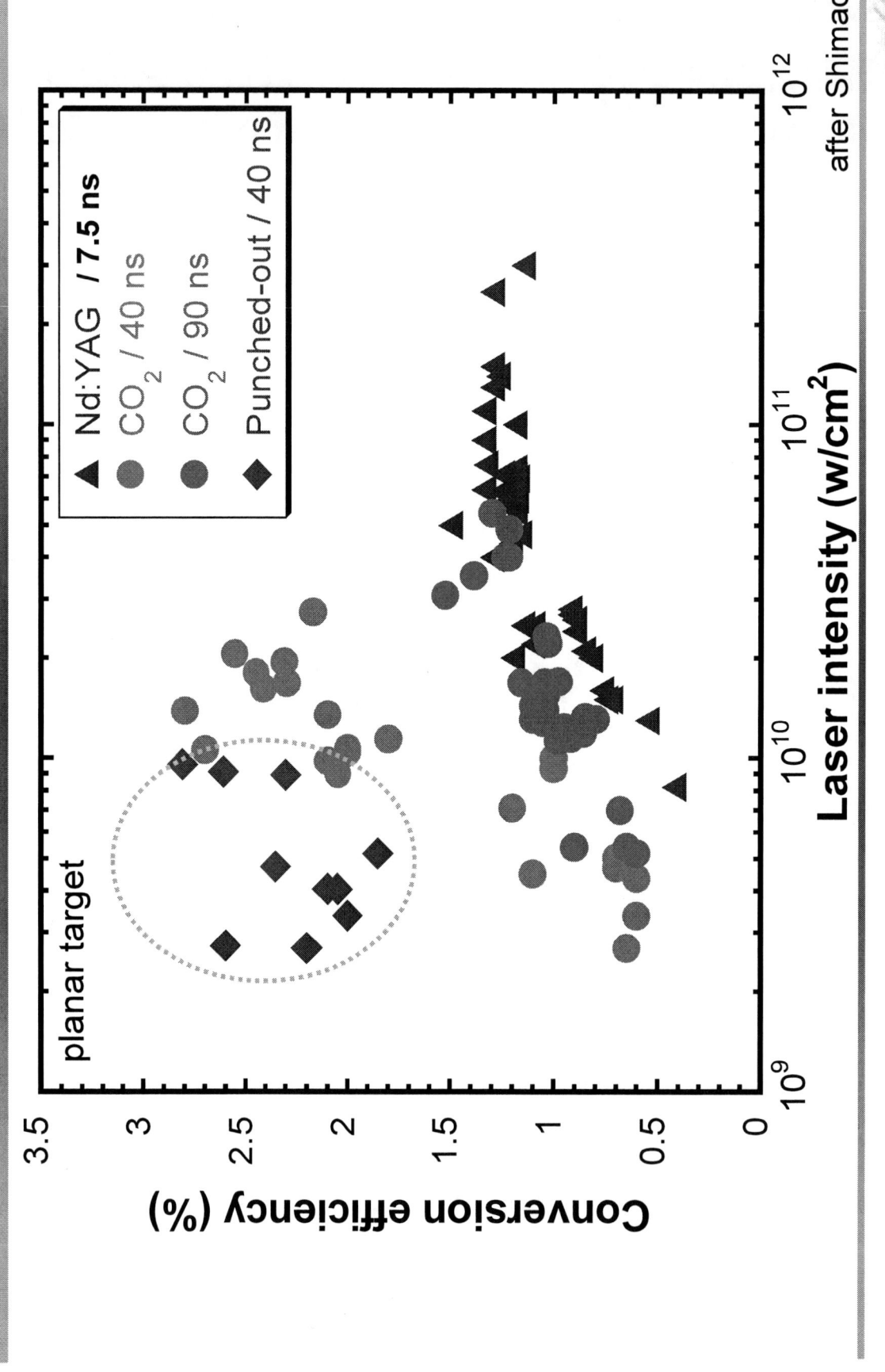

CO₂ laser 2

High efficiency in CO_2 laser is mainly due to spectrum narrowing peaked at 13.5 nm.

dependence of EUV spectra
on laser wavelength and
pulse duration

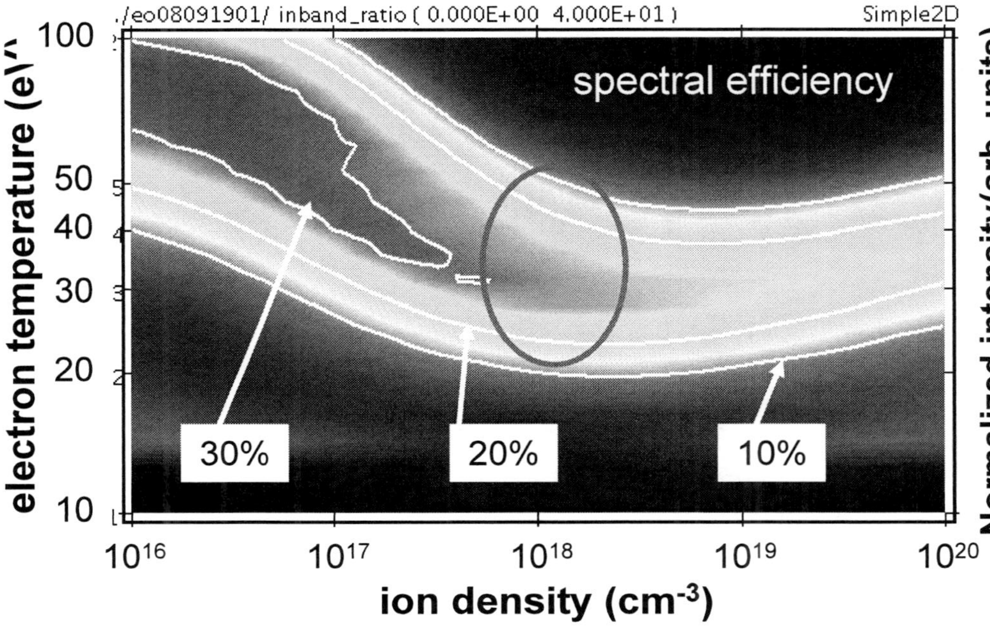

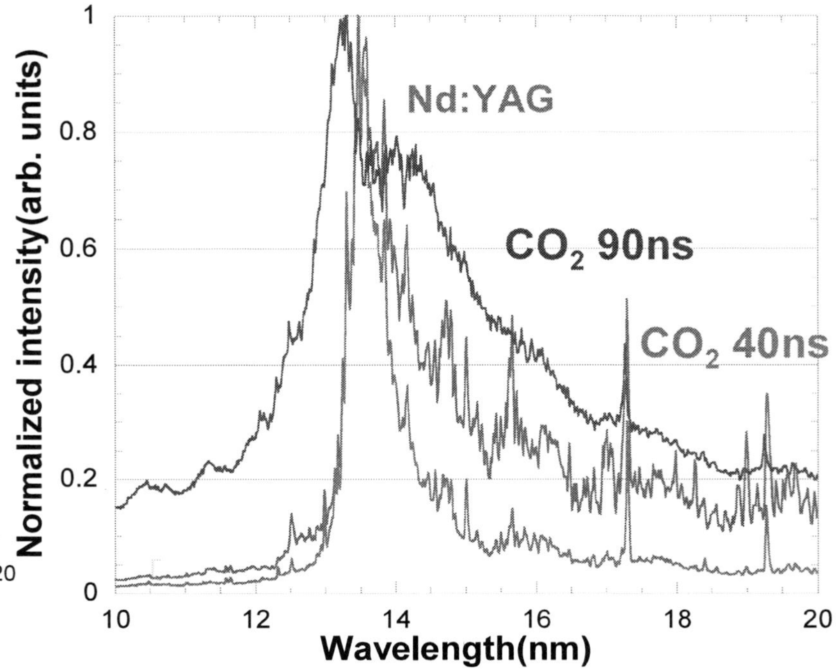

spectral efficiency (%)
(13.5 nm with 2 % band width
to total radiation)

Nd:YAG : 4×10^{10} W/cm², 7.5 ns
CO_2 : 2×10^{10} W/cm², 40 ns or 90 ns

after Shimada

Punchout target provides low density plasma before main pulse as double pulse scheme.

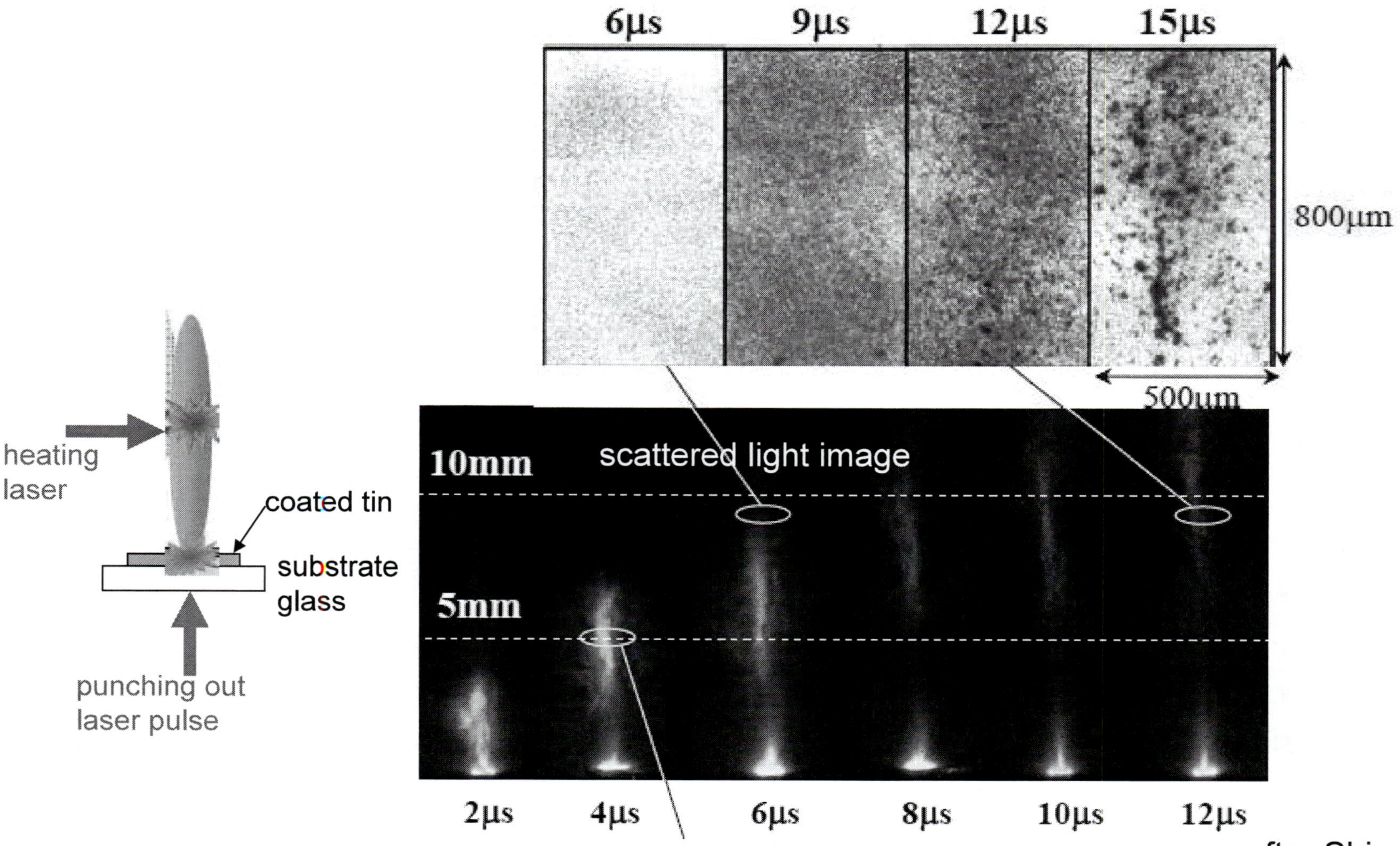

punchout

800μm

6μs   9μs   12μs   15μs

500μm

scattered light image

10mm

5mm

2μs   4μs   6μs   8μs   10μs   12μs

coated tin

substrate glass

punching out laser pulse

heating laser

after Shimada

double pulse 3

Double pulse scheme with two color lasers is performed for tin droplet target, which provides low density plasma

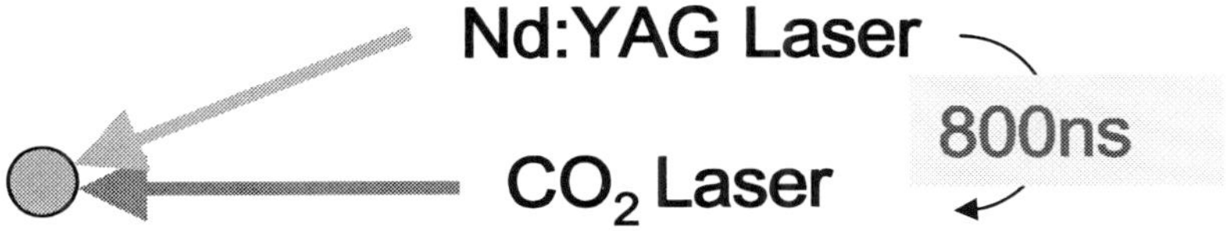

Nd:YAG Laser

CO₂ Laser

800ns

Target size:30um
Nd;YAG : $2.6×10^{11}W/cm^2$
CO_2: : $1.6×10^9W/cm^2$

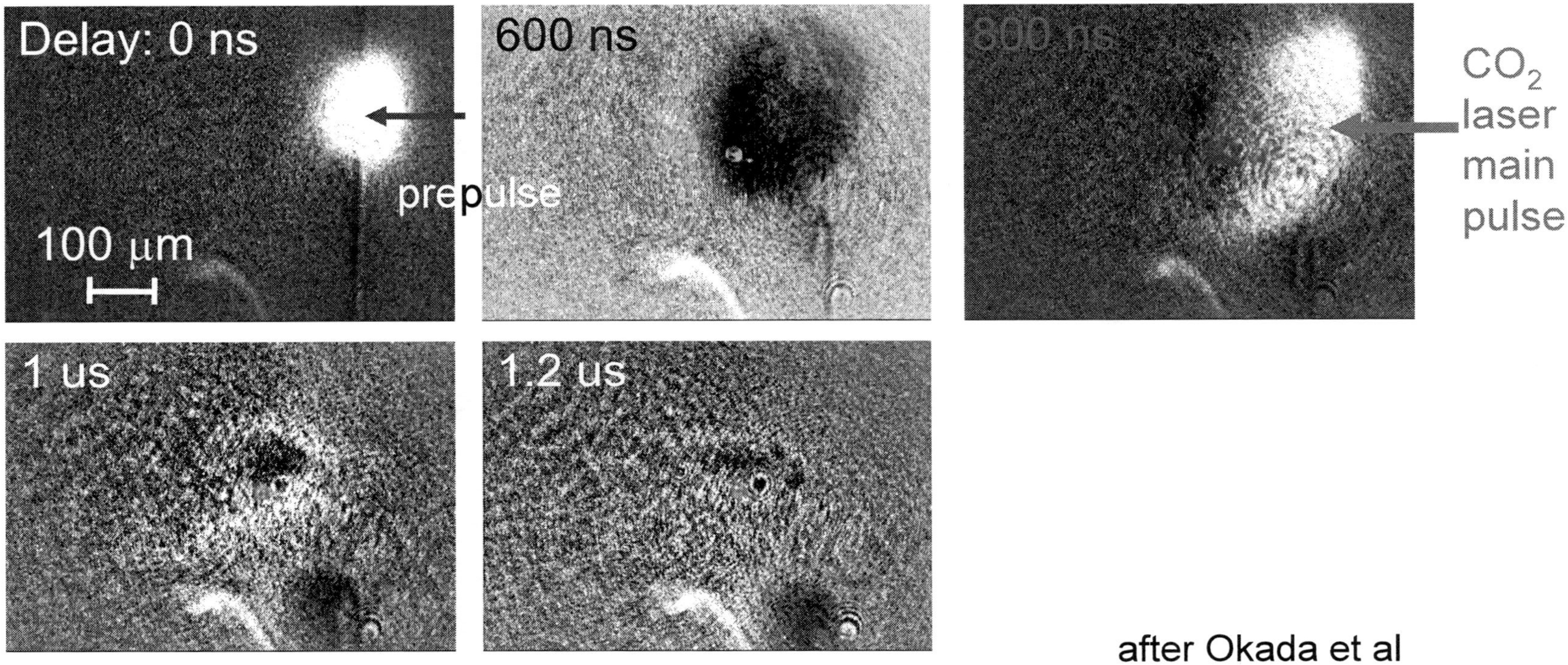

CO_2 laser main pulse

Delay: 0 ns

prepulse

100 µm

600 ns

800 ns

1 us

1.2 us

after Okada et al

After main pulse, no clusters remain

double pulse 4

CE of > 4% is obtained from 36 μmφ pure Sn droplet in two-color lasers, double pulse irradiation scheme.

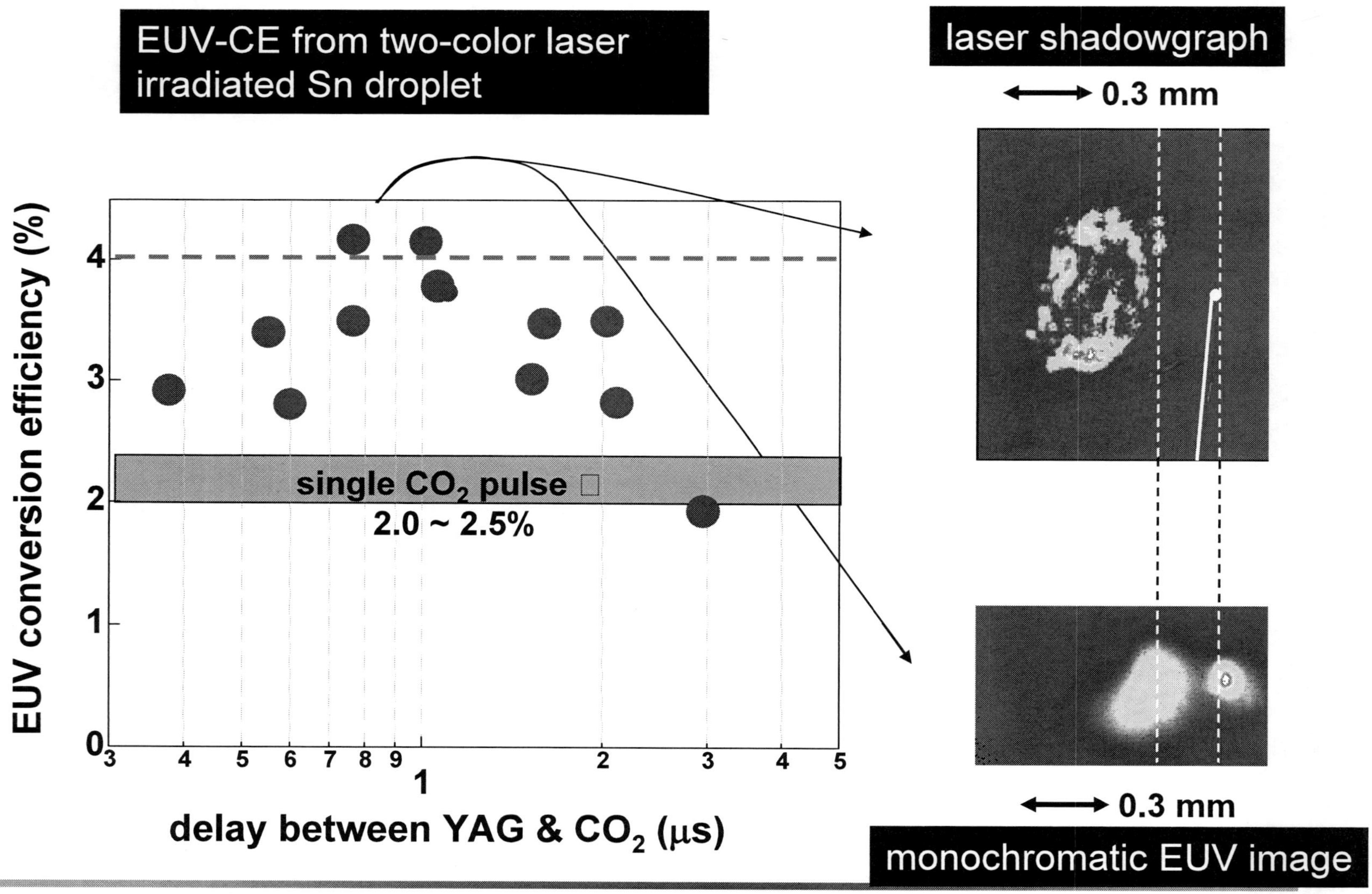

laser shadowgraph

→ 0.3 mm

EUV-CE from two-color laser irradiated Sn droplet

monochromatic EUV image

→ 0.3 mm

single CO_2 pulse
2.0 ~ 2.5%

EUV conversion efficiency (%)

delay between YAG & CO_2 (μs)

Fujioka et al APL (08)

optimization 1

Atomic model is further improved by adjusting energy levels for individual ions to experimental data.

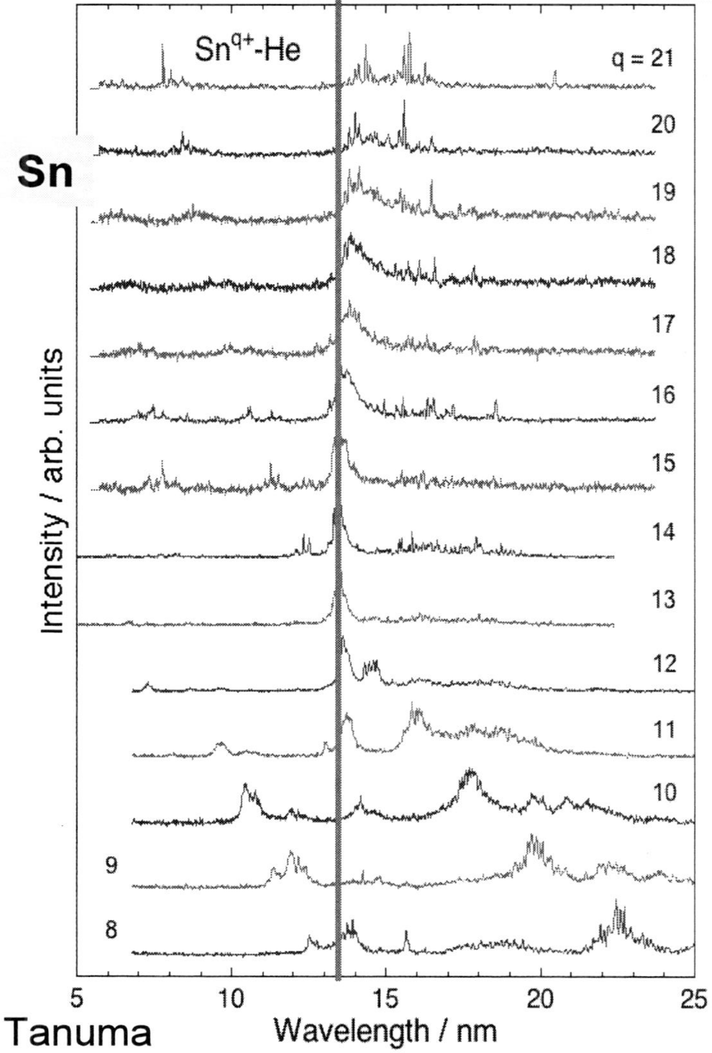

n_i: 10^{17} cm^{-3}
T_e: 52 eV

n_i: 10^{19} cm^{-3}
T_e: 52 eV

emissivity (W/cm³/Å)

5 10^9

5 10^{11}

0

0

wavelength (nm)

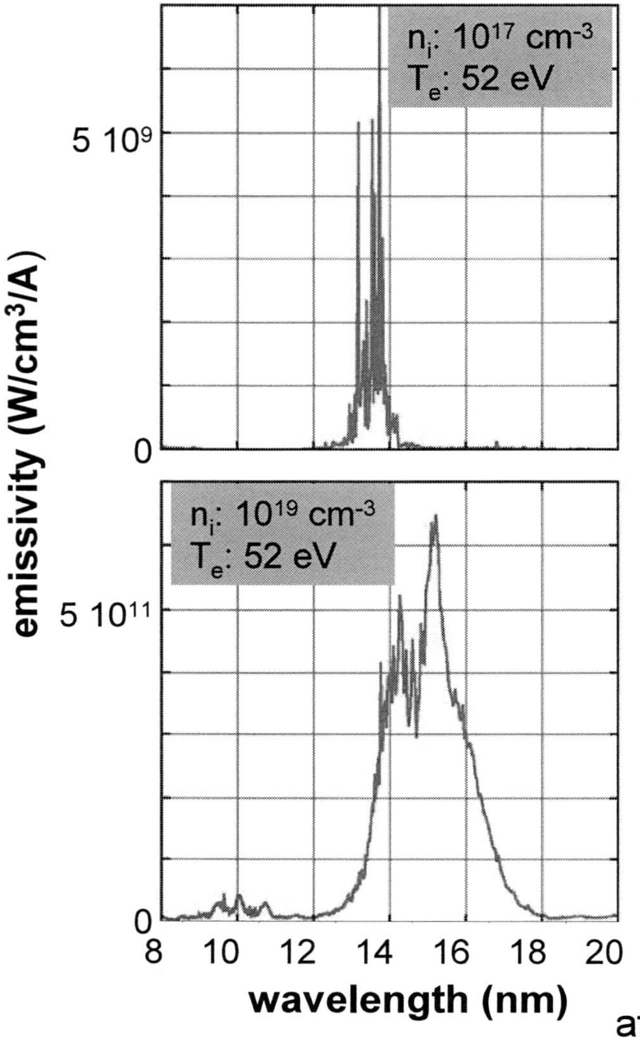

after Sasaki

after Tanuma

Sn^{q+}-He

q = 21
20
19
18
17
16
15
14
13
12
11
10
9
8

Sn

Intensity / arb. units

Wavelength / nm

CES: Sn^{+q} + Xe $\rightarrow$ Sn^{+q-1} (n,l) $\rightarrow$ Sn^{+q-1} (n',l') + hν

optimization 2

EUV power of 500 – 1000 W can be obtained with CO_2 laser < 15 kW at optimized conditions (power balance model*)

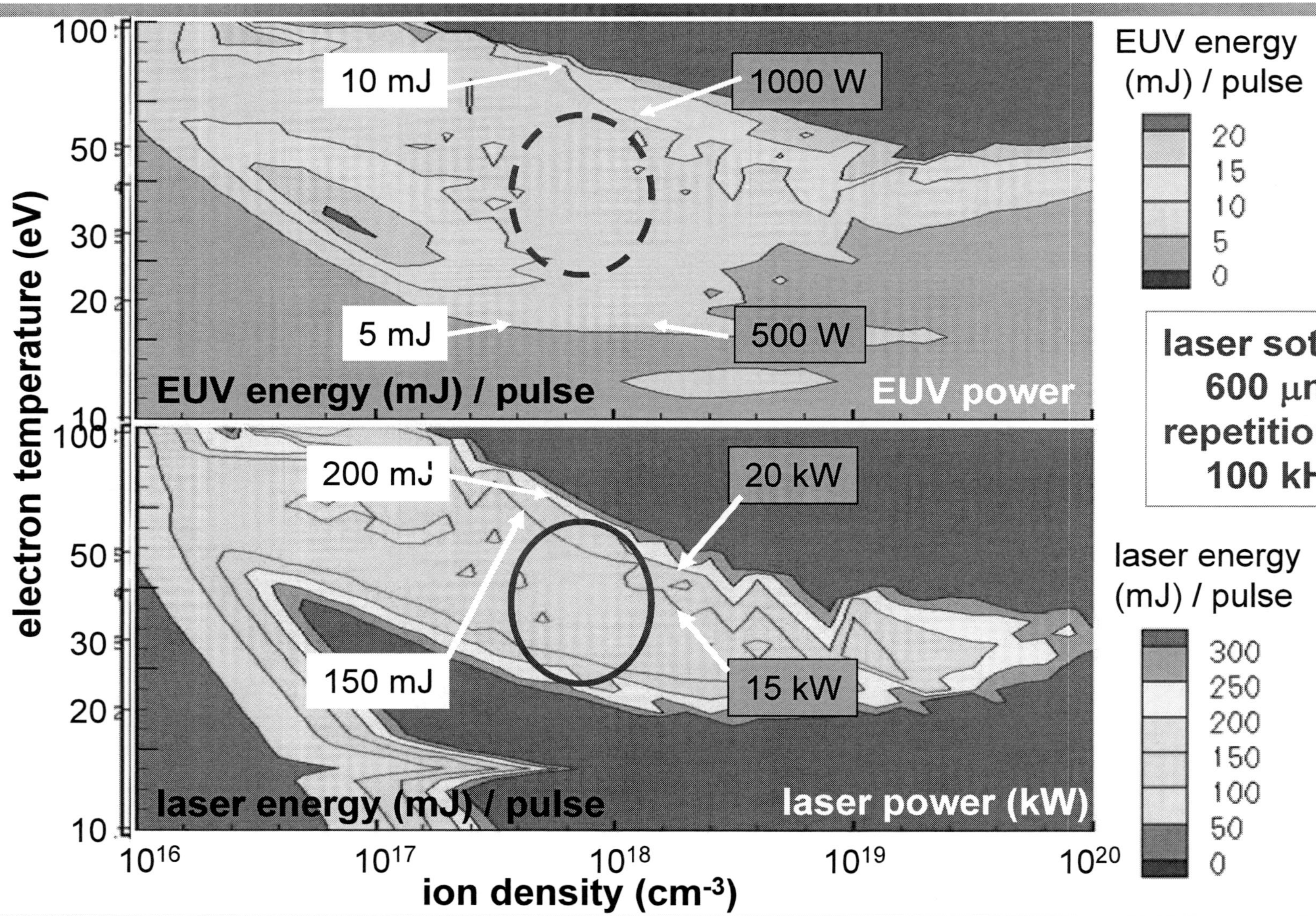

EUV energy (mJ) / pulse

20
15
10
5
0

laser sot size 600 μm, repetition rate 100 kHz

laser energy (mJ) / pulse

300
250
200
150
100
50
0

EUV power

1000 W

500 W

10 mJ

5 mJ

EUV energy (mJ) / pulse

laser power (kW)

20 kW

15 kW

200 mJ

150 mJ

laser energy (mJ) / pulse

electron temperature (eV)

100

50

30

20

10

ion density (cm^{-3})

10^{16} 10^{17} 10^{18} 10^{19} 10^{20}

*Nishihara et al EUVL04, PoP (08)

optimization 3

EUV power of 500 – 1000 W can be obtained with CO laser 13 - 20 kW even for 1.5 times longer pulse than the optimum

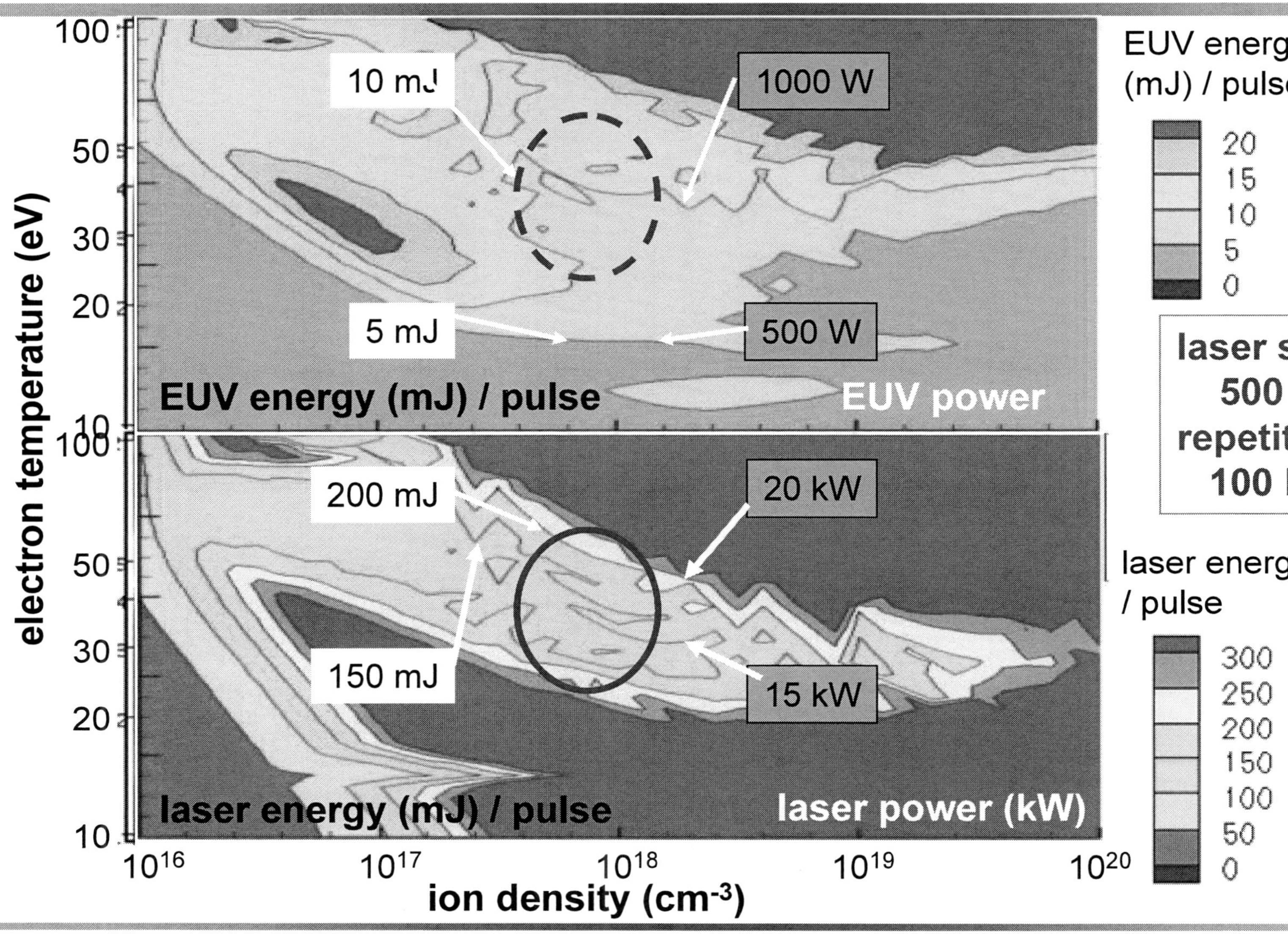

EUV energy (mJ) / pulse

20
15
10
5
0

laser sot size
500 μm,
repetition rate
100 kHz

laser energy (mJ) / pulse

300
250
200
150
100
50
0

1000 W

500 W

10 mJ

5 mJ

EUV power

EUV energy (mJ) / pulse

20 kW

15 kW

200 mJ

150 mJ

laser power (kW)

laser energy (mJ) / pulse

electron temperature (eV)

100
50
30
20
10

100
50
30
20
10

10^{16} 10^{17} 10^{18} 10^{19} 10^{20}

ion density (cm^{-3})

optimization 4

Conversion efficiency of 5 – 6 % can be achieved by CO_2 laser with pulse duration of 10 – 20 ns.

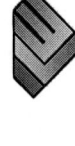

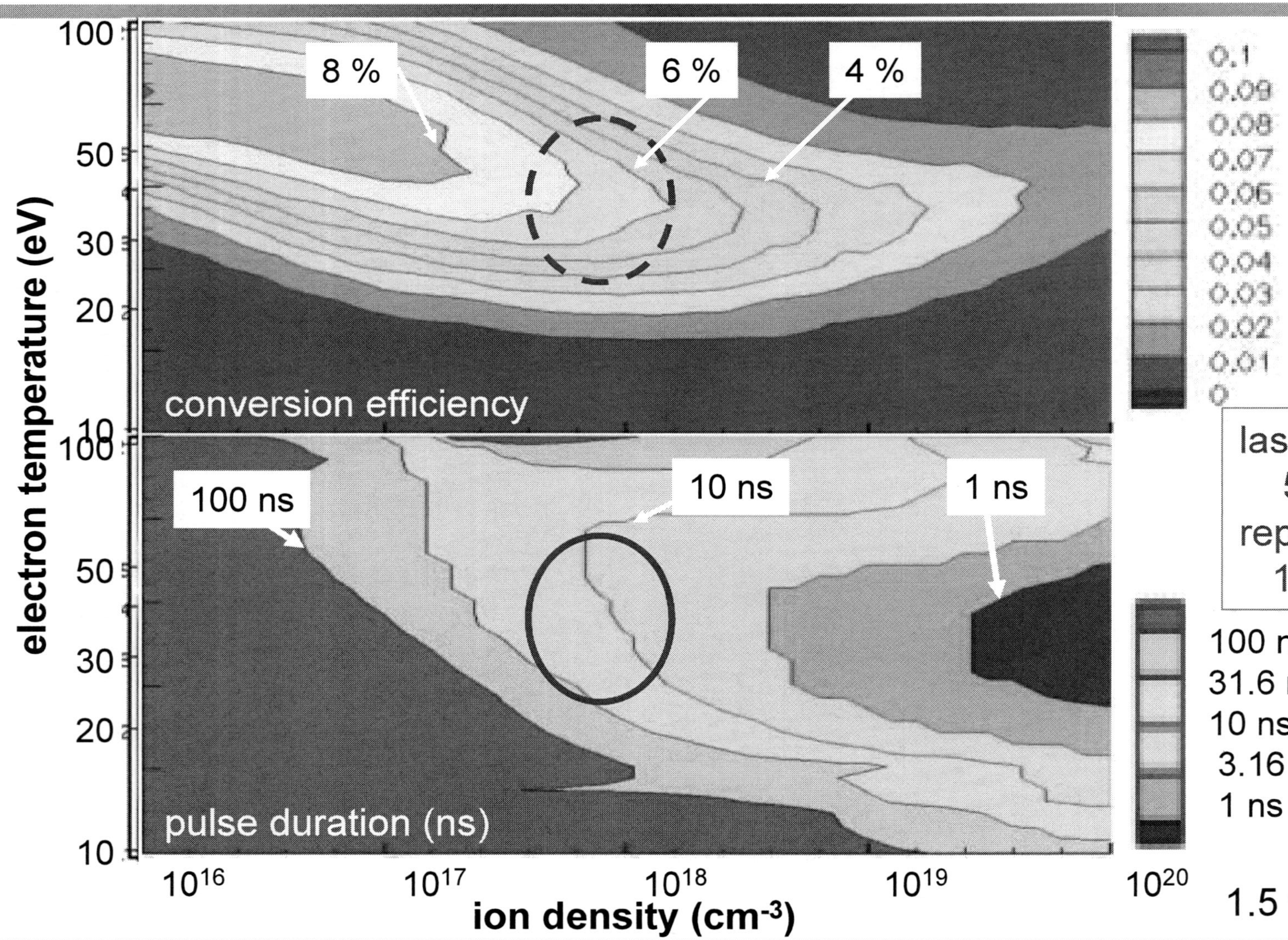

8 %
6 %
4 %
conversion efficiency
0.1
0.09
0.08
0.07
0.06
0.05
0.04
0.03
0.02
0.01
0
100 ns
10 ns
1 ns
pulse duration (ns)
100 ns
31.6 ns
10 ns
3.16 ns
1 ns
electron temperature (eV)
100
50
30
20
10
10^{16}
10^{17}
10^{18}
10^{19}
10^{20}
ion density (cm^{-3})

laser sot size 500 μm, repetition rate 100 kHz

1.5 times longer

Reduction of ion energy is experimentally observed in punch-out target and double pulse laser irradiation.

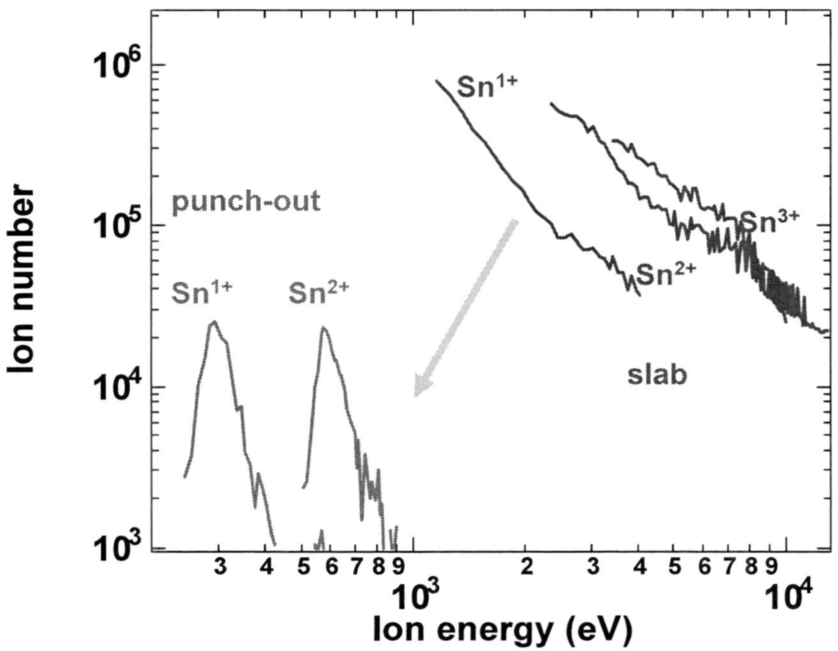

debris 1

□ Single pulse
○ Dual pulses ($\Delta\tau = 100$ ns)

Fast ions

Ion current (A/cm²) — $\cos\theta$, $\cos^6\theta$

double pulse — Kinetic energy (keV) — ○ Sn^+ □ Sn^{2+}

single pulse — ESA signal (arb. units) — Kinetic energy (keV) — ○ O^+ □ O^{2+}

Ion number vs Ion energy (eV): Sn^{1+}, Sn^{2+}, Sn^{3+} (slab); punch-out Sn^{1+}, Sn^{2+}

Maximum ion energy decreased by the use of low density target.

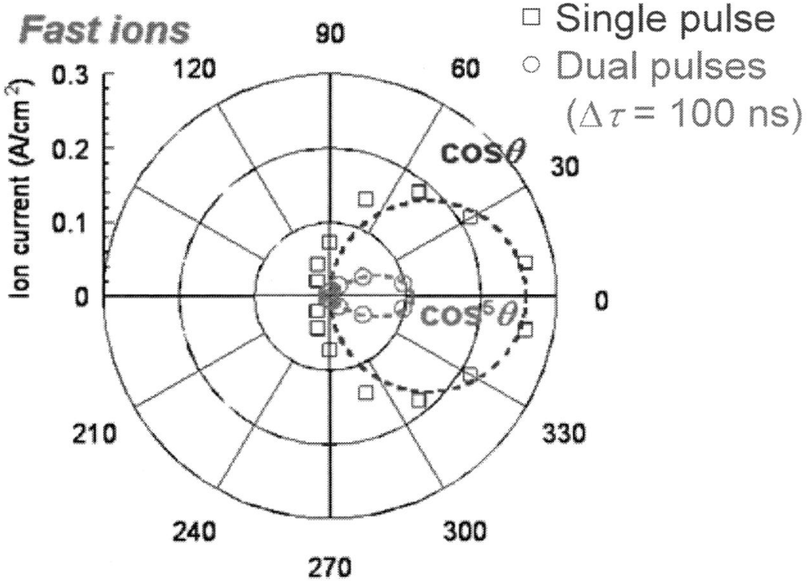

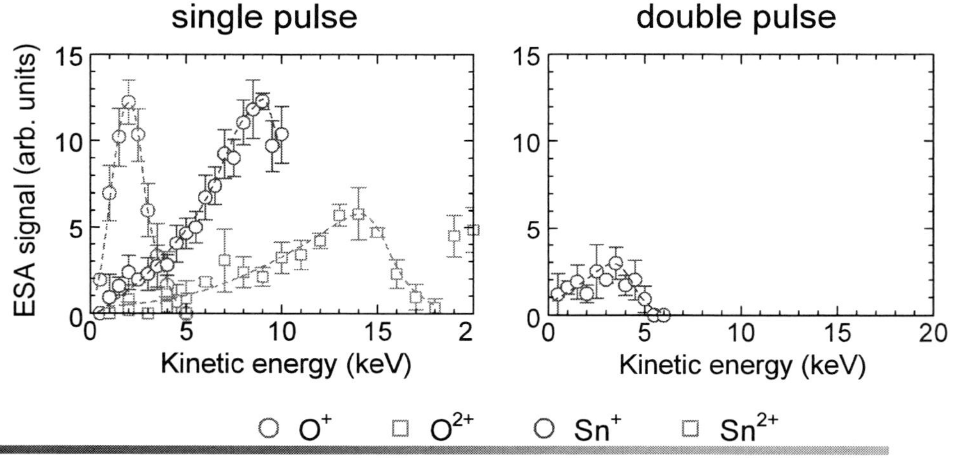

T. Higashiguchi et al., Appl. Phys. Lett. **91**, 151503 (2007) (double-pulse)
S. Fujioka et al., J. Plasma Fusion Res. **82**, 609 (2006) (punch-out)

B-field 1

High energy ions are rapidly exhausted along B-field using a single coil.

Color corresponds to ion energy:

3D PIC simulations

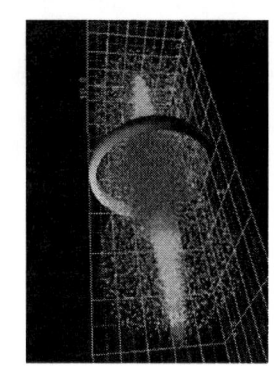

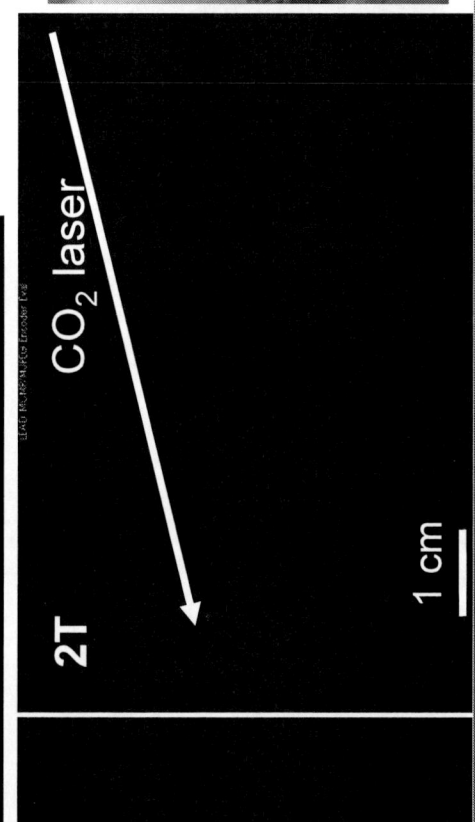

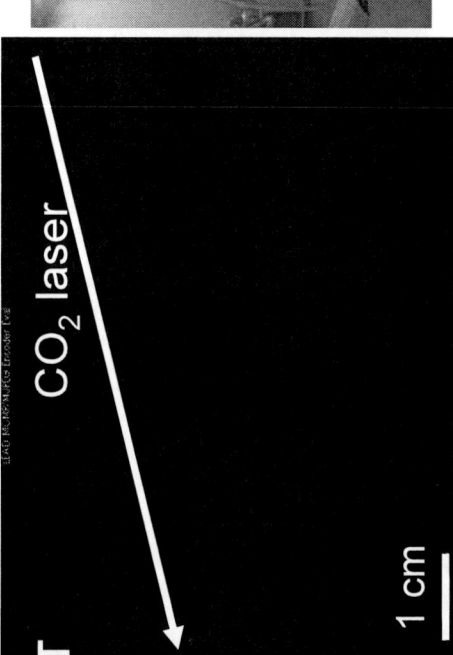

CO_2 laser

2T

0T

1 cm

1 cm

EUVA

B-field 2

Ions are rapidly exhausted along B-field even in oblique incident of laser light

Dots are ions, color indicates energy

B-field

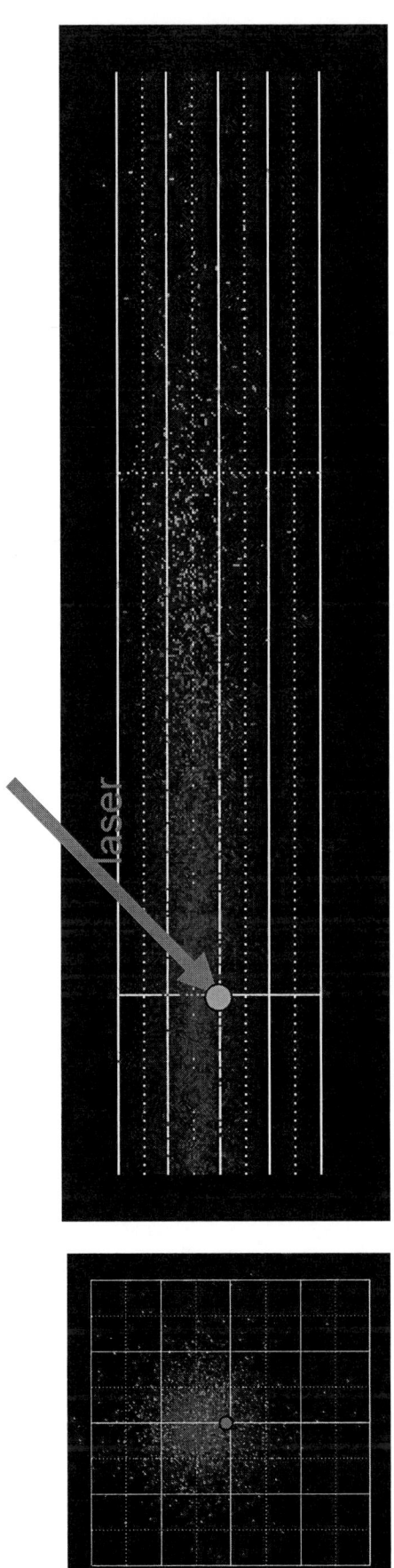

laser

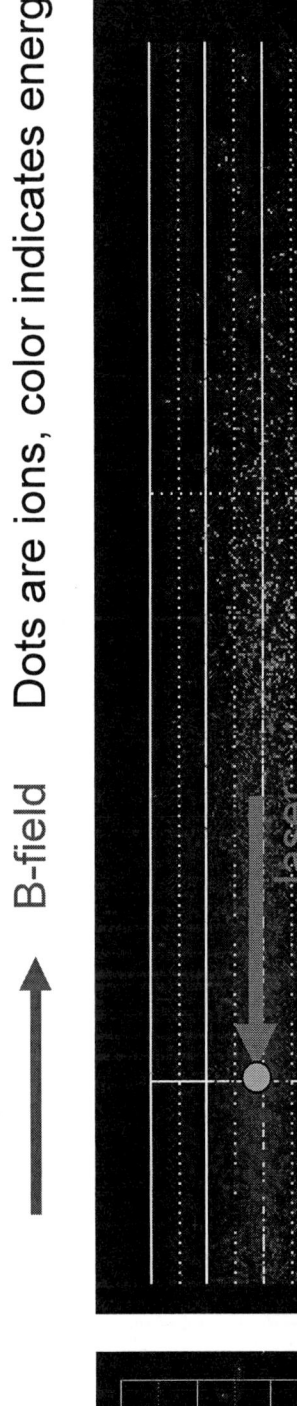

laser

minimum mass

An atom emits a few 10's EUV photons during ablation. Minimum mass for 5 mJ EUV is about 1,7 ng, 8 μm$^{\phi}$.

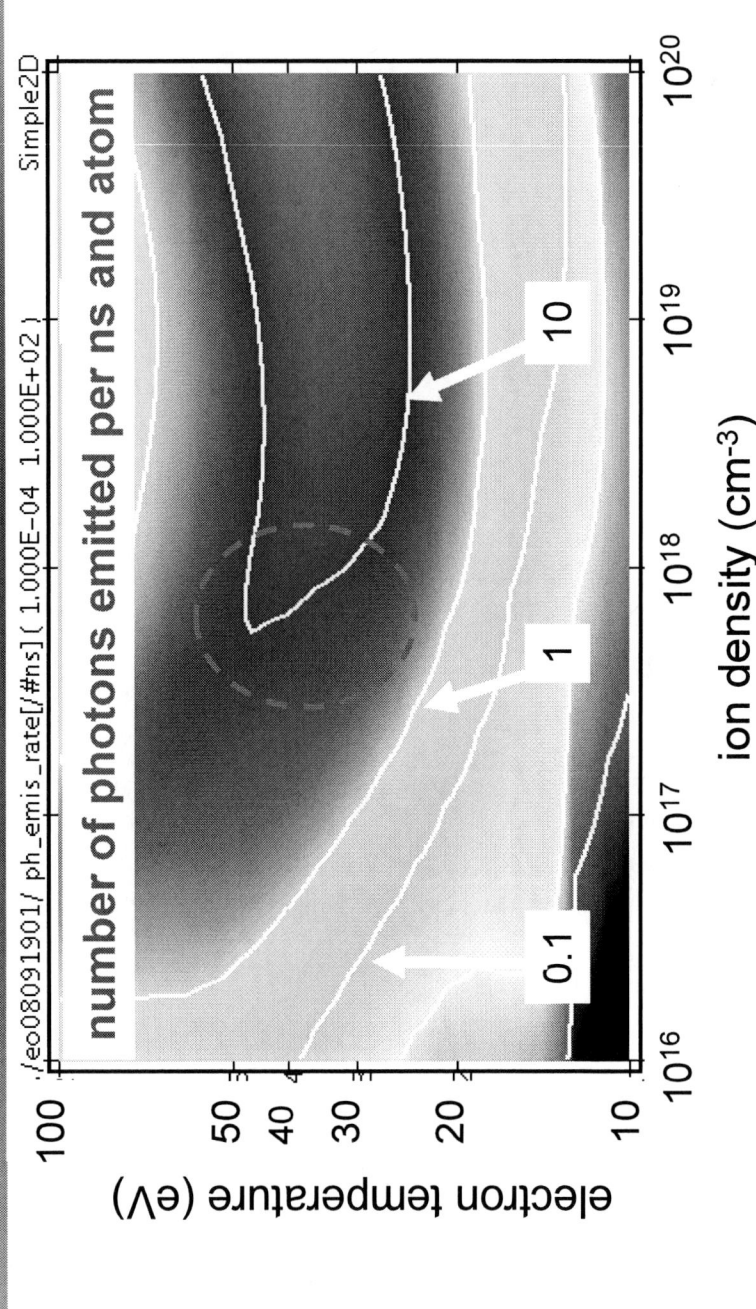

number of photons emitted per ns and atom

./eo08091901/ ph_emis_rate[/#ns] (1.000E-04 1.000E+02) Simple2D

electron temperature (eV)

ion density (cm^{-3})

EUV

Minimum mass target, which emits EUV photons corresponding to 5 mJ (3.4×10^{14}), is about 1.7 ng, and 8 μm in diameter

minimum number of atoms : 3.4×10^{14} / (10/ns x 4ns) = 4.25×10^{12} atoms

emission time of each atom: $\tau_e = l / c_s = 100(\mu m) / 2.3 \times 10^6 \, (cm / sec) = 4.4\,(ns)$

T: 50eV, Z:12

debris 3

Atoms, clusters and solid states are coexisting and expands after prepulse (MD simulation)

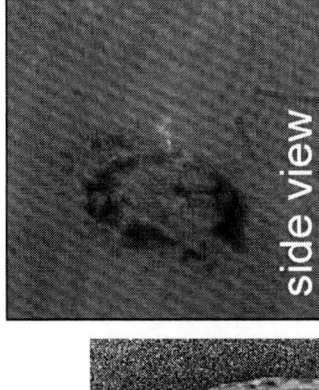

side view

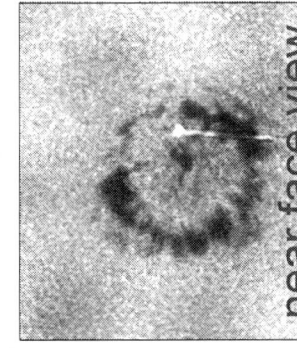

near face view

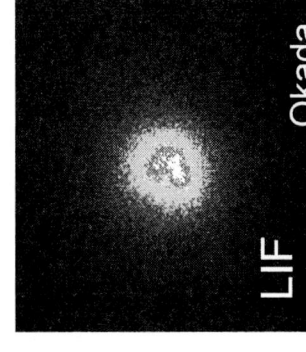

LIF

Okada

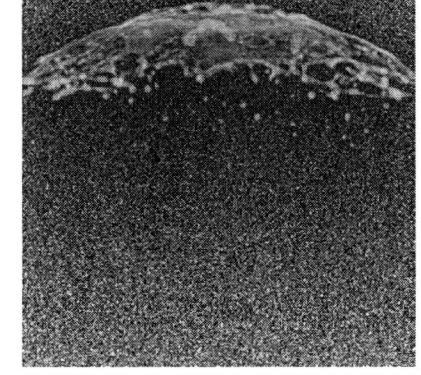

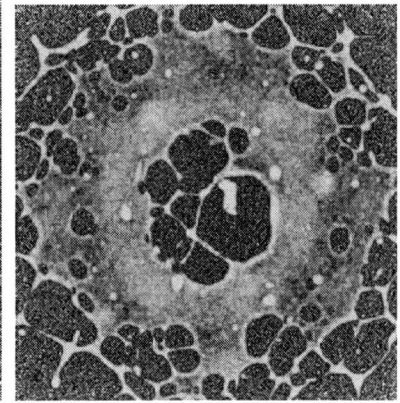

Aluminum sphere R_0 = 57.4 nm, F= 0.4 J/cm2, t = 0.4 ps t =178 ps, $V_{shell} \sim$ 560 m/s

after Zhakhovskii

cleaning

Tin cleaning by hydrogen radical is a promising scheme

etching rate : 60 nm/min for initial thickness of 300 nm

vacuum evaporated Sn film

	before (50nm)	2min after	10min after
condition A			residual
condition B		Clean	clean

750 n m

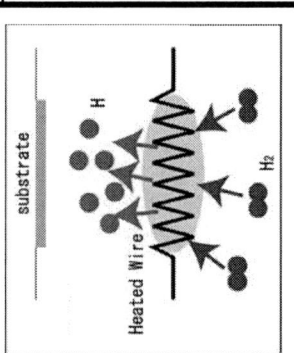

After Yuba et al (Osaka Univ. Eng. Sci.)

summary

1. EUV source power of $500 - 1000$ W $/2\pi$ can be obtained in double pulse irradiation scheme with a tin droplet target and $13 - 20$ kW CO_2 main pulse.

2. Conversion efficiency of $> 4\%$ has been experimentally achieved from a Sn droplet with 36 μm in diameter with the two-color laser, double pulse irradiation scheme.

3. We have also discussed advantages of the scheme, such as reduction of high energy ions, ion mitigation by B-field, and Sn cleaning by hydrogen radicals.

The State Research Center of the Russian Federation - Troitsk Institute for Innovation and Fusion Research (TRINITI)

Technological aspects of Sn RDE source development for HVM lithography

V.M. Borisov,
G.N. Borisova, A.S. Ivanov, Yu.B. Kirukhin, O.B. Khristoforov,
V.A. Mishchenko, A.V. Prokofiev, A.Yu. Vinokhodov

TRINITI

2008 Int.Symposium on EUV Lithography Sept.28 –Oct.1, 2008, Lake Tahoe, CA

Outline

- **Sn RDE source concept and design features**
- **Power scaling parameters:**
  - Input energy
  - Conversion and collection efficiencies
  - Pulse repetition frequency
- **EUV power up to HVM ranges**
- **Conclusion**

Sn RDE source design features

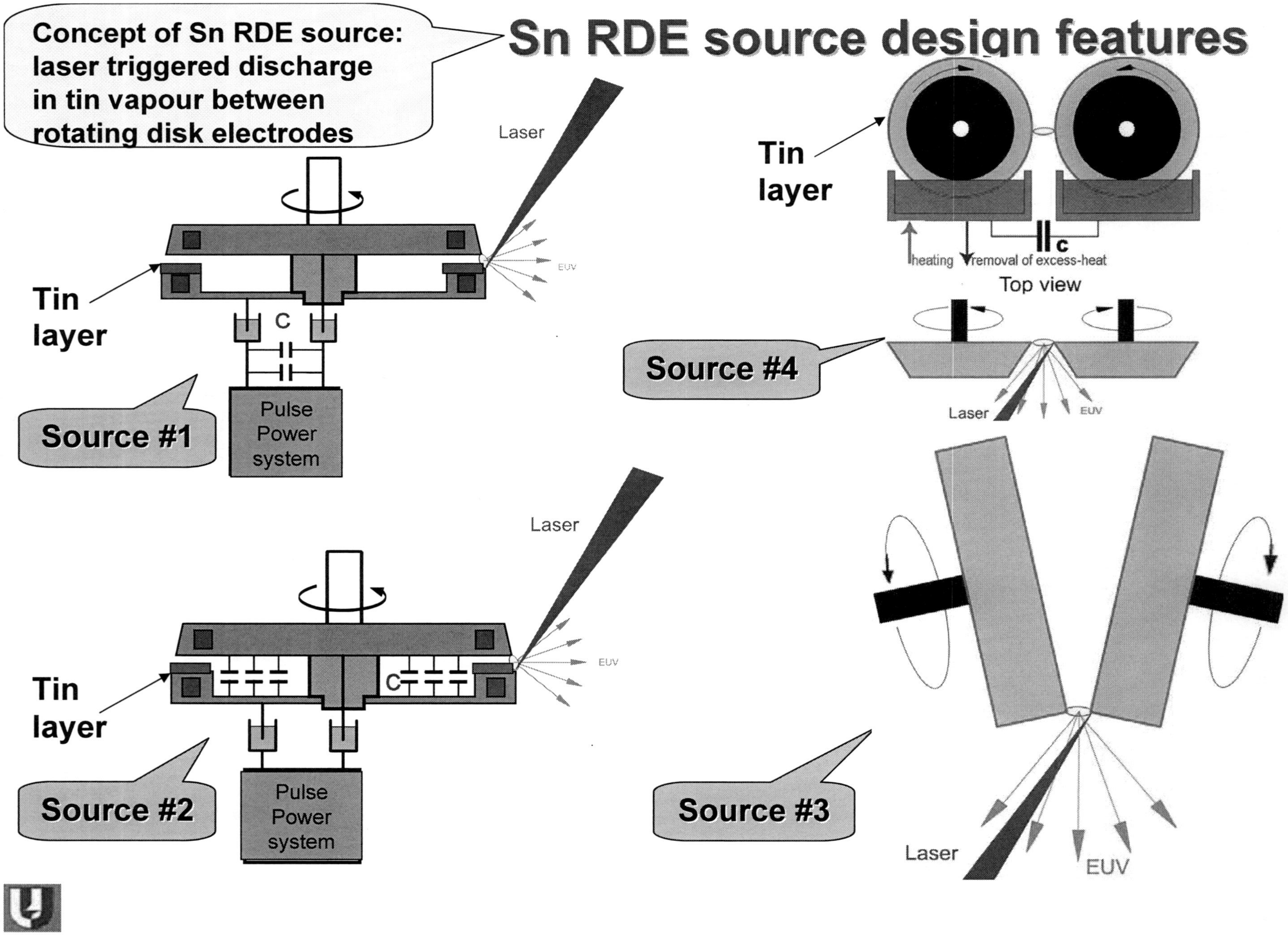

Concept of Sn RDE source: laser triggered discharge in tin vapour between rotating disk electrodes

Laser

Tin layer

Source #1

Pulse Power system

C

EUV

Laser

Tin layer

Source #2

Pulse Power system

C

EUV

Tin layer

heating

removal of excess-heat

C

Top view

Laser

EUV

Source #4

Laser

EUV

Source #3

source #2

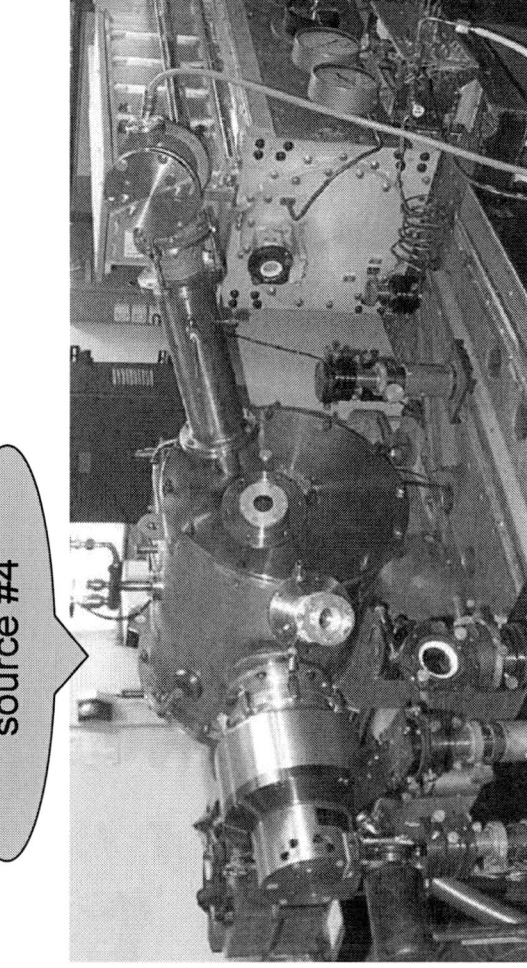

source #4

source #1

source #3

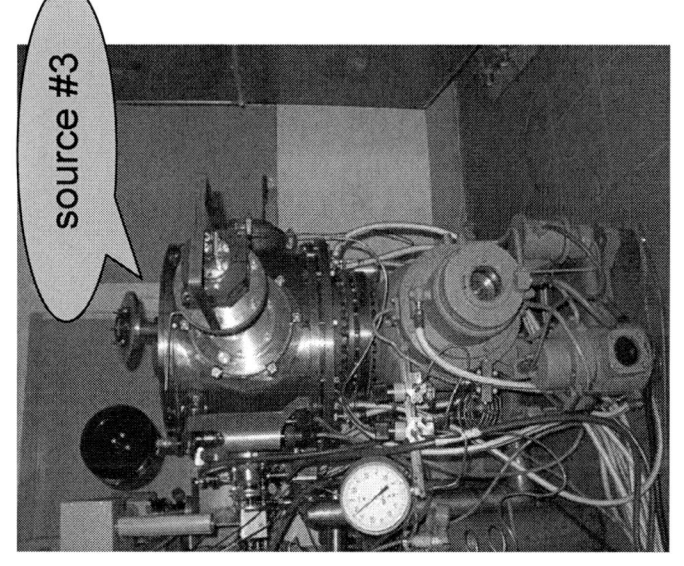

Experimental Sn RDE sources developed at TRINITI

coaxial source

has minimal electrical circuit inductance (L=4.4 nH)

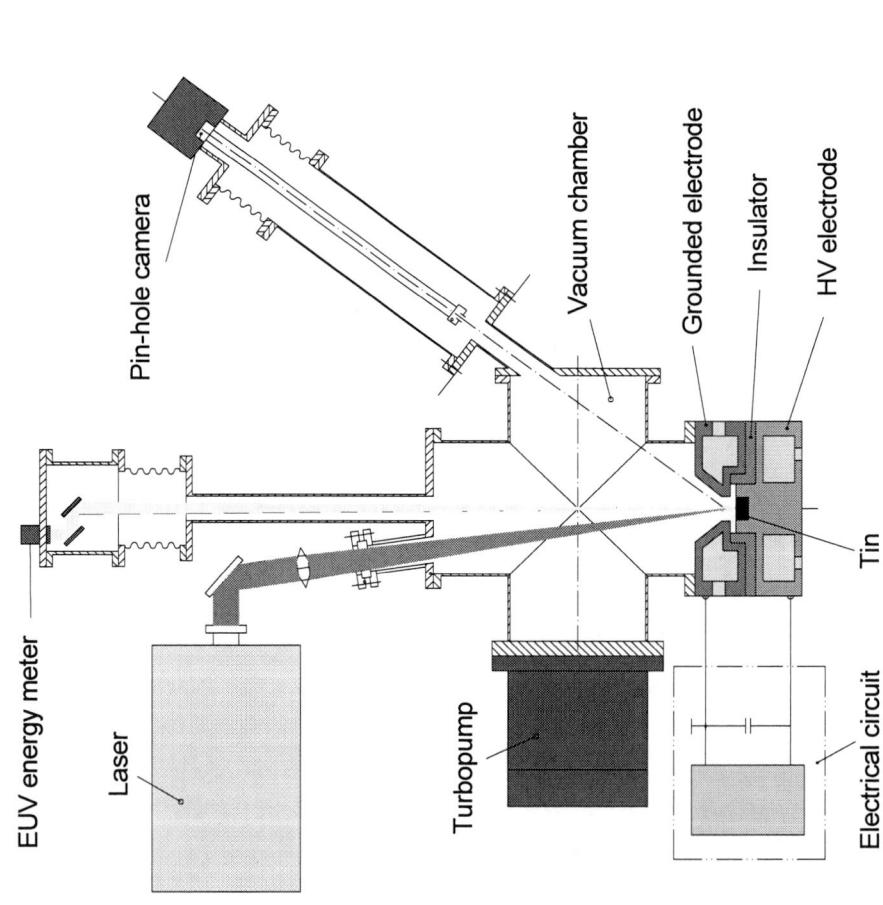

Tin delivery and tin surface regeneration systems

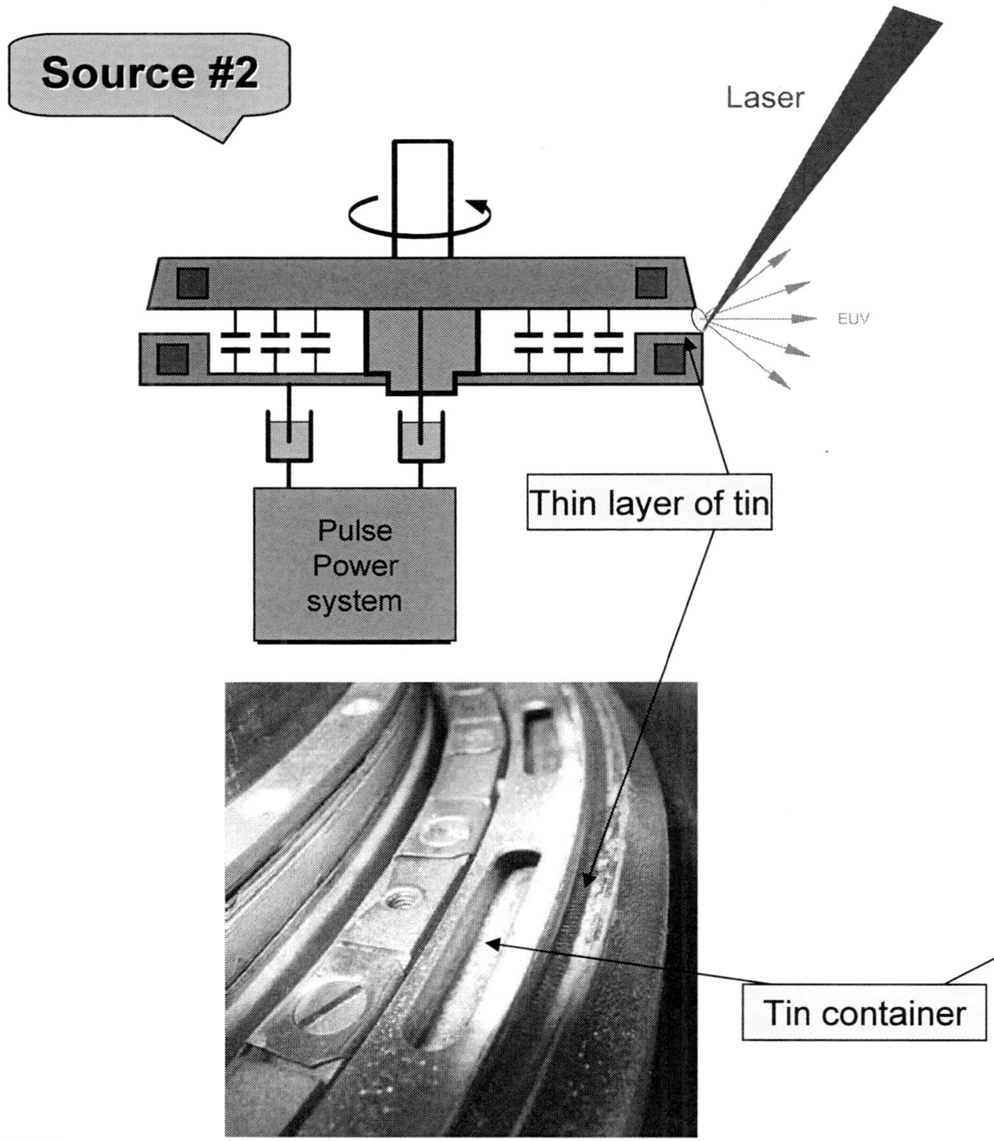

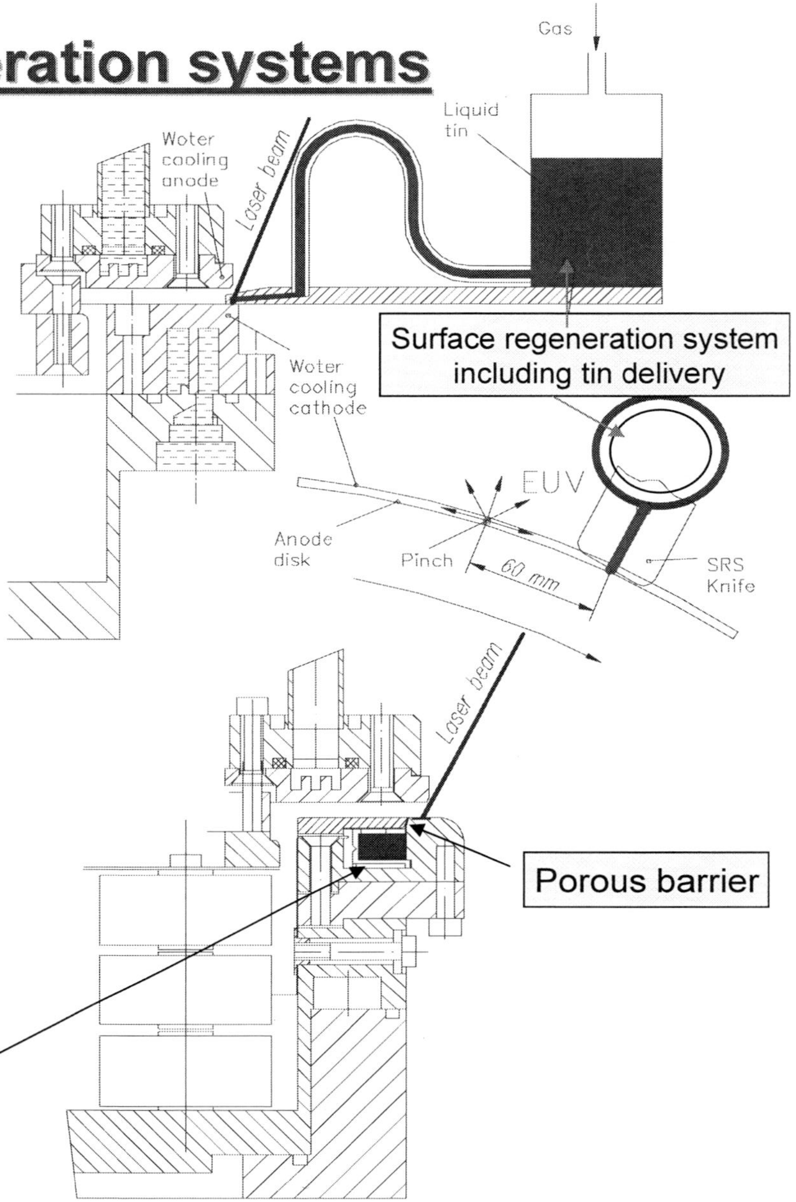

Source #2

Gas

Liquid tin

Surface regeneration system including tin delivery

SRS Knife

EUV

60 mm

Pinch

Anode disk

Laser beam

Water cooling cathode

Water cooling anode

Porous barrier

Tin container

Laser

EUV

Thin layer of tin

Pulse Power system

Tin delivery and sprinkling

Source #3, #4

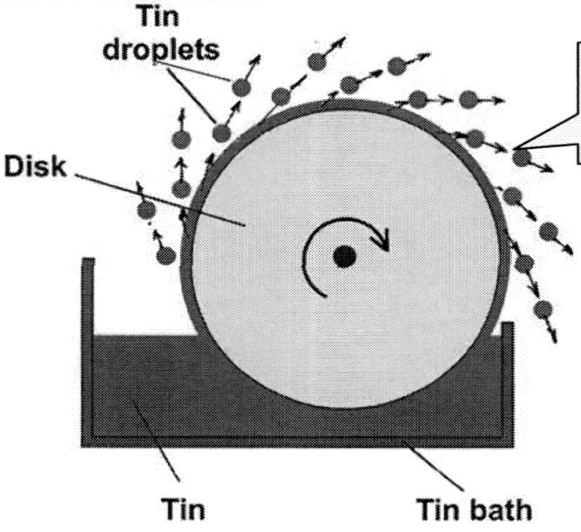

∅disk=120mm
V>2m/s → tin sprinkling

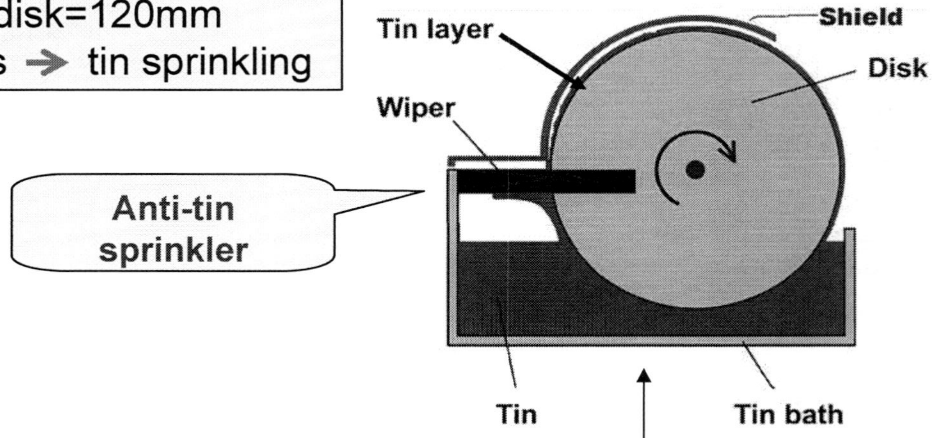

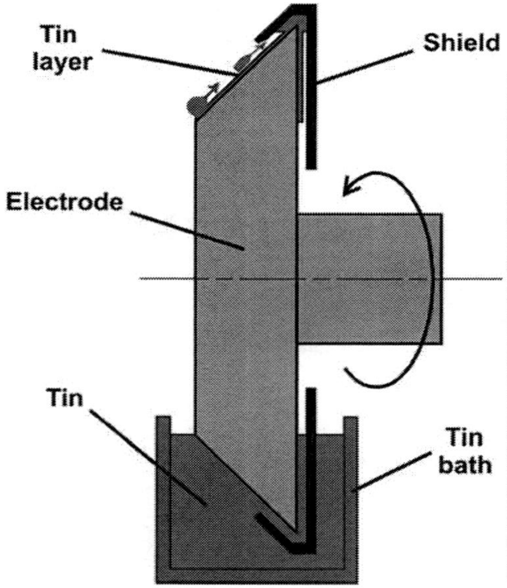

Anti-tin sprinkler

Tin droplets

Disk

Tin bath

Tin

Shield

Disk

Tin layer

Wiper

Tin bath

Tin

Examples of anti-tin sprinkling systems

V=20 m/c without tin sprinkling

Shield

Tin layer

Electrode

Tin bath

Tin

Surface of rotating disk electrode covered by tin layer has a conical form.
The excess of tin converted in droplets does not left a conic surface but moves to a cone base and then tin droplets are accumulated by the electrode shield and come back to the bath.

EUV energy vs. voltage and capacitance

The EUV energy can linearly depend on voltage and capacitance with coaxial source

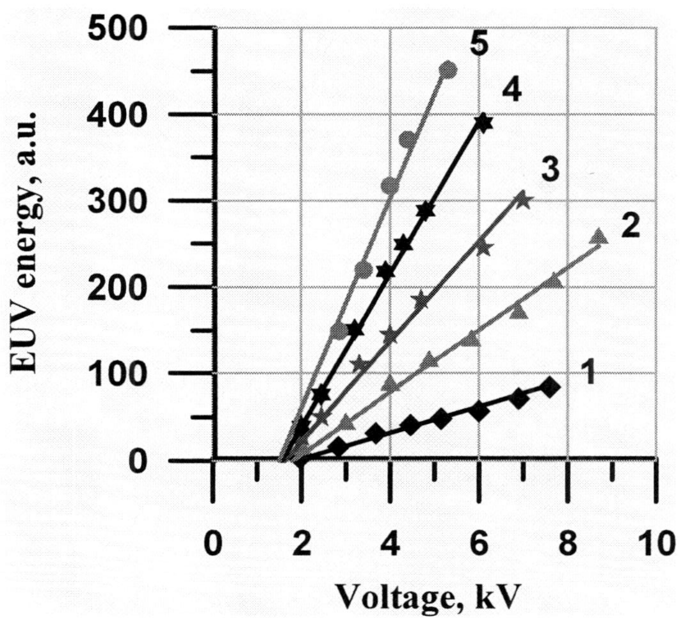

EUV energy, a.u. vs **Voltage, kV**

1 – 0.25 µF   2 – 0.5 µF
3 – 1.0 µF   4 – 1.5 µF
5 – 1.8 µF

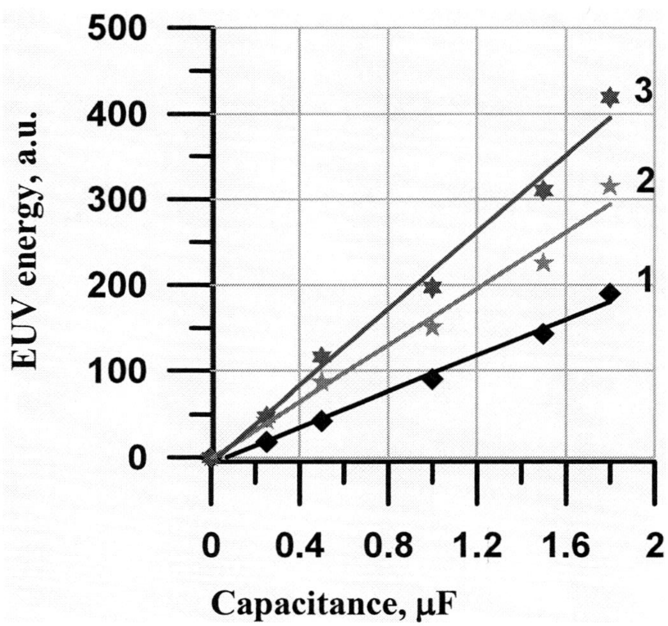

EUV energy, a.u. vs **Capacitance, µF**

1 – 3 kV
2 – 4 kV
3 – 5 kV

Collection efficiency

Only EUV light emitted by small plasma volume (⌀~1.3 mm) can be collected. Collection efficiency **K**% was calculated on the measured distributions of EUV radiation with the CCD-camera as the relation of EUV energy into a circle with diameter 1.3 mm to EUV energy on all field of the CCD-camera (background emission).

Source #2 $E_{in}=9.9$ J

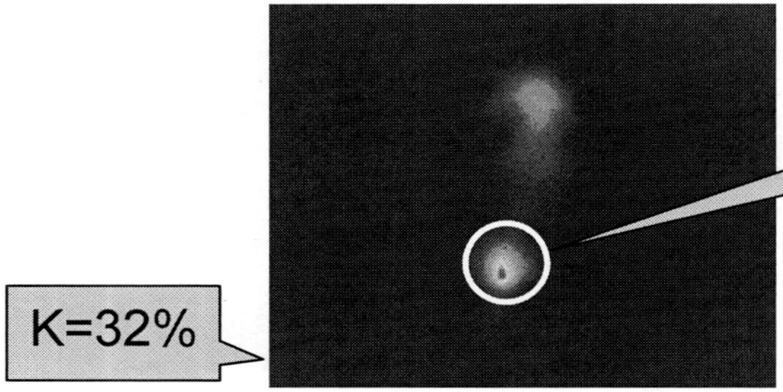

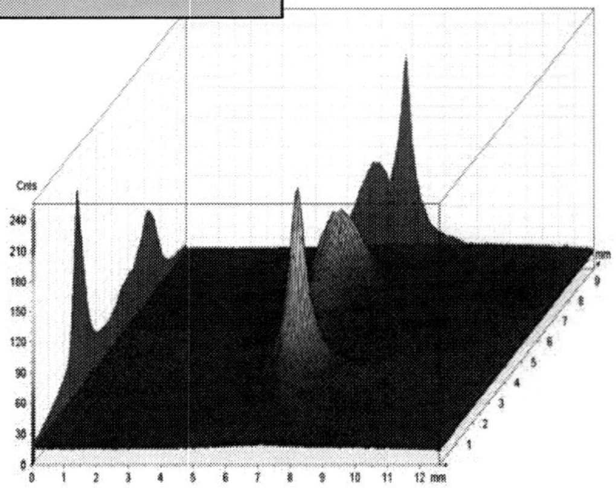

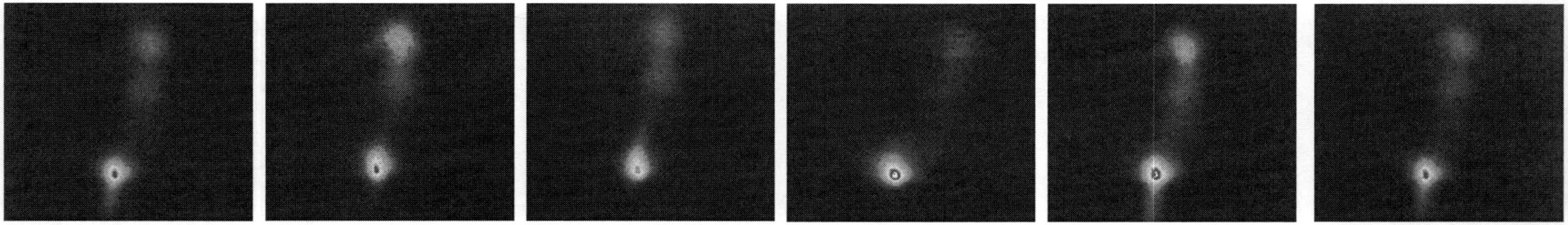

⌀ 1.3 mm

K=32%

Continuous operation (20 minutes) of source #2 with input power 19.8 kW (9.9J x 2kHz) at interelectrode gap *d*=4mm and pulse to pulse distance △x= 2 mm.

EUV power was measured as 270 W in 2π, but collected EUV power was estimated as 90 W in 2π only (**K**≈30%) with conventional discharge condition

Collection efficiency obtained with coaxial source

_K_ > 80% is achieved at input energy 12 J / pulse with optimized discharge condition

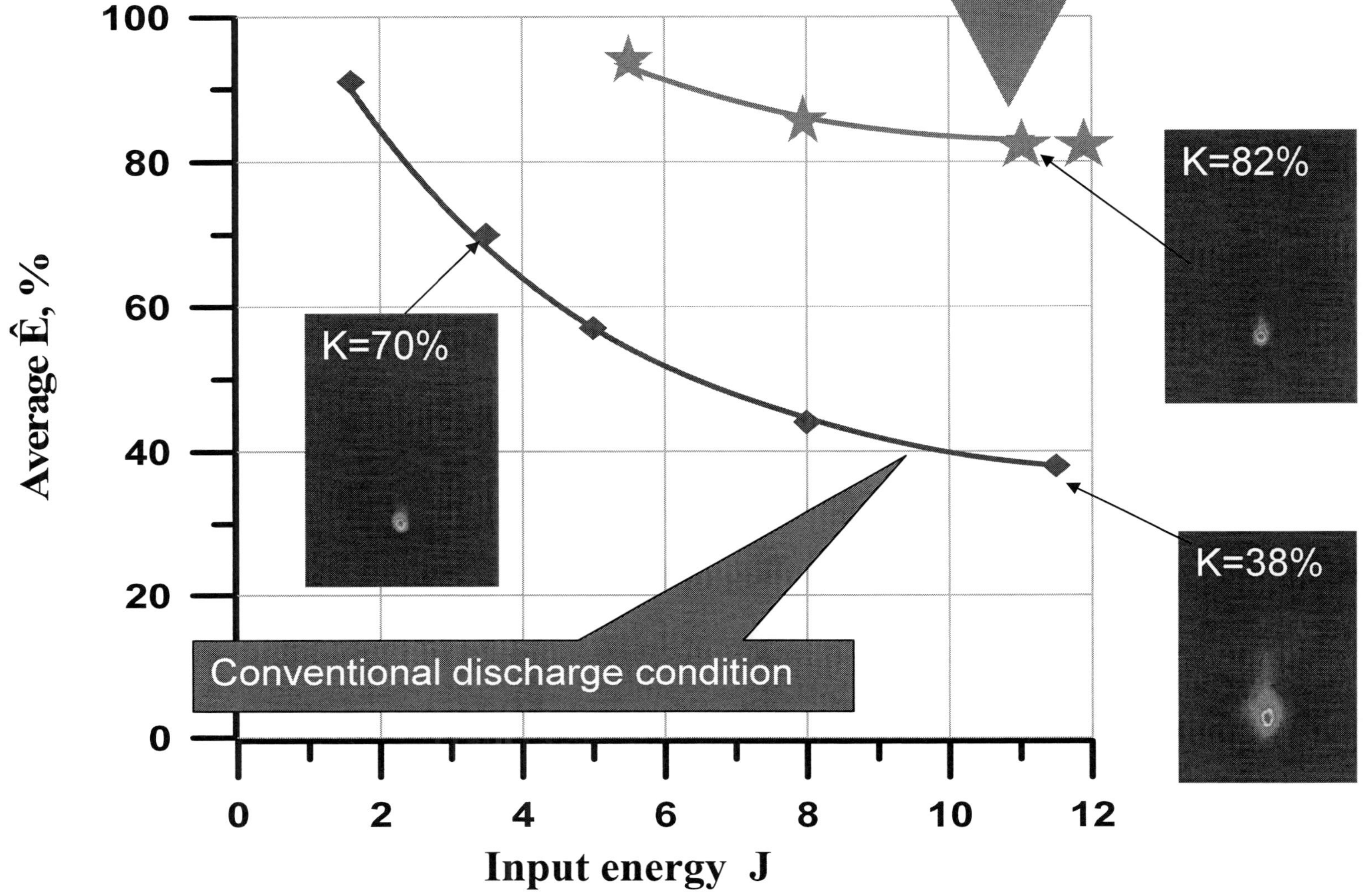

K=82%

K=38%

K=70%

Conventional discharge condition

Average Ẽ, %

Input energy J

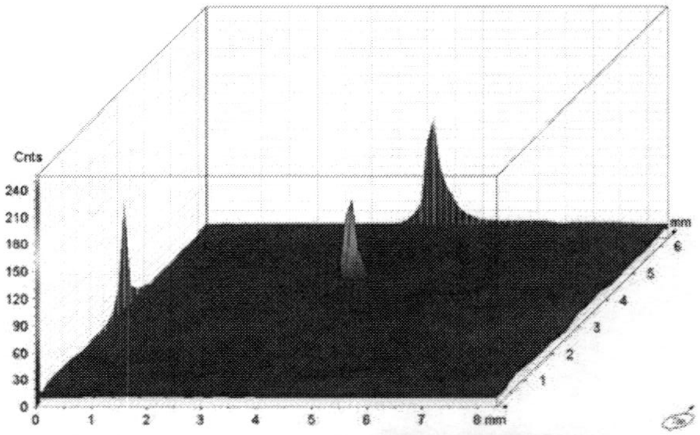

coaxial source

A

K

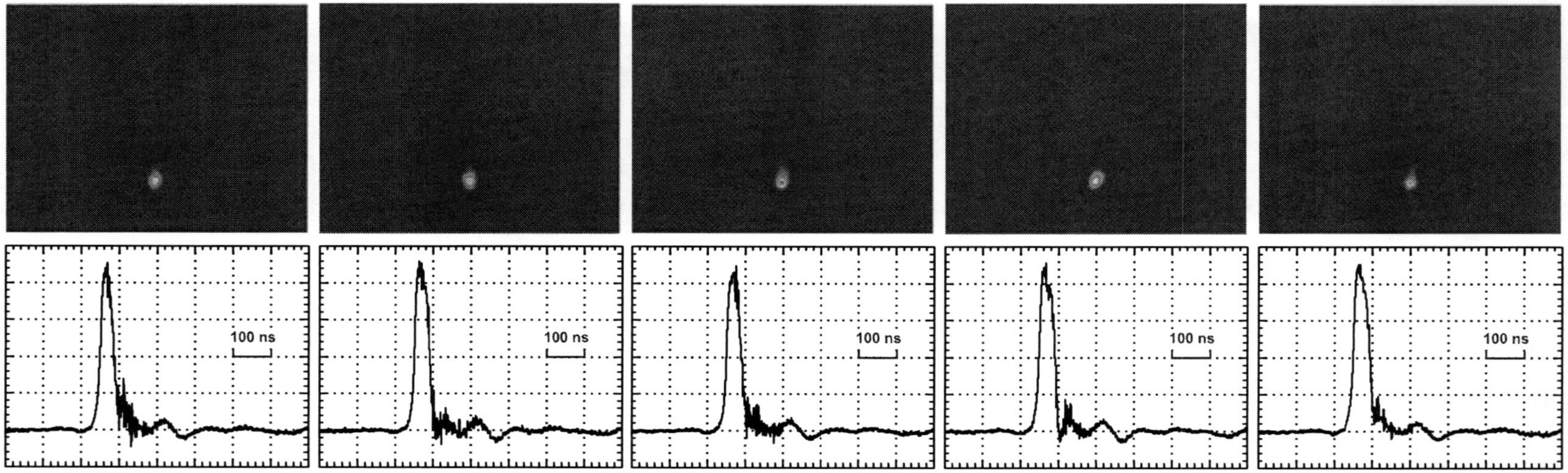

E_{in} = 12J, K = 82%

Good pulse-to-pulse stability with optimized discharge condition

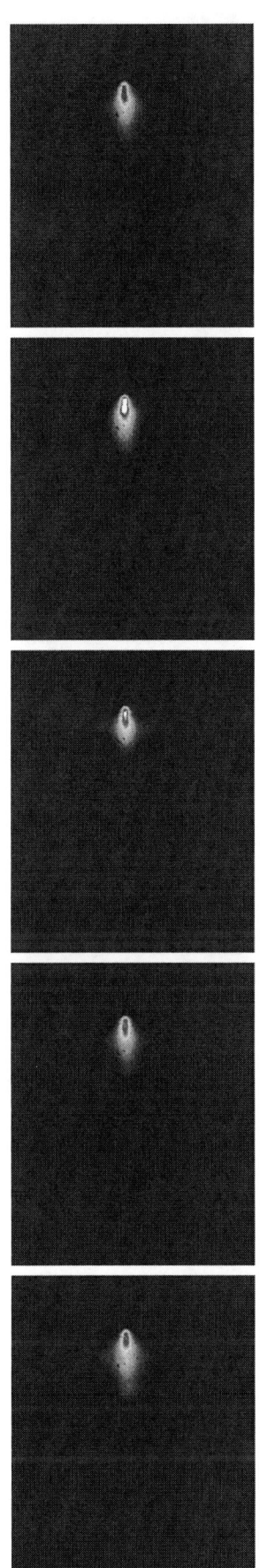

Source #3

$E_{in} = 7.8$ J, K= 76%

Conversion efficiency

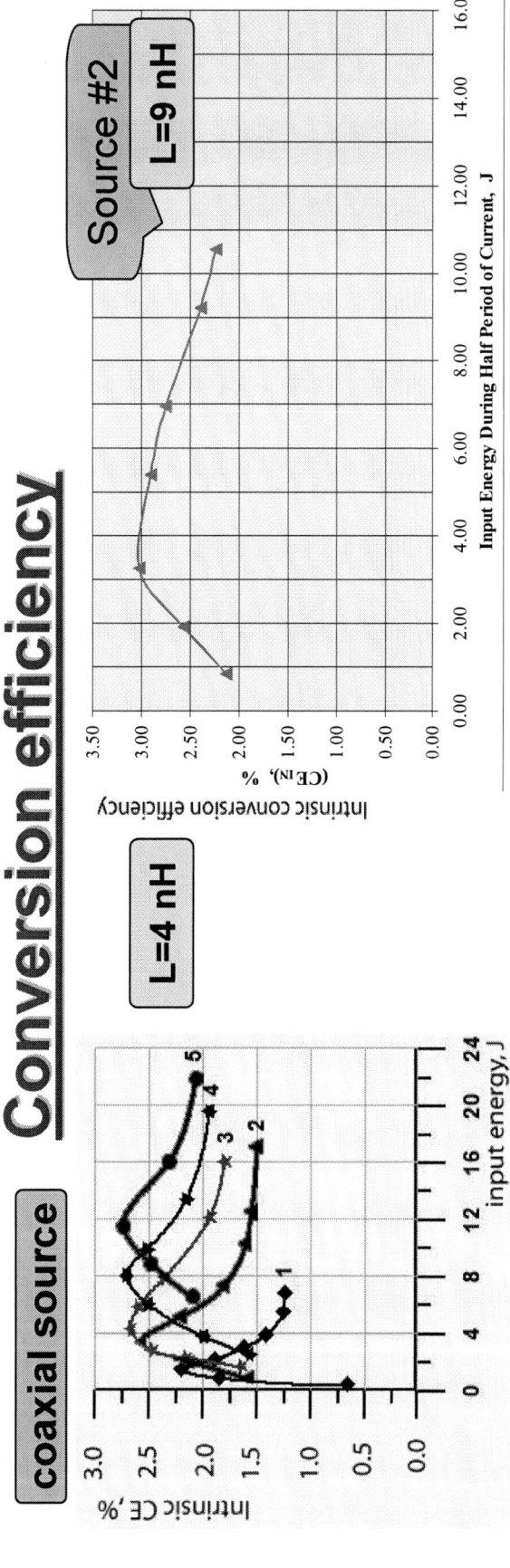

coaxial source

L=4 nH

Source #2 L=9 nH

Intrinsic conversion efficiency as a function of energy put into the plasma during first half period of the current

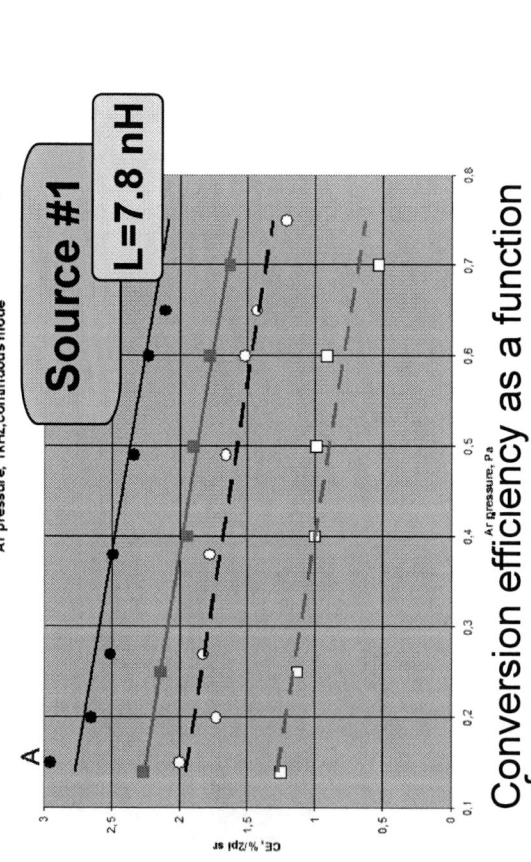

Conversion efficiency as a function of the stored energy into the last capacitor battery

Source #1 L=7.8 nH

Conversion efficiency as a function of gas pressure

Continuous source operation

Source #4 with XeF laser
V=0.85m/s, Δx=0.85mm

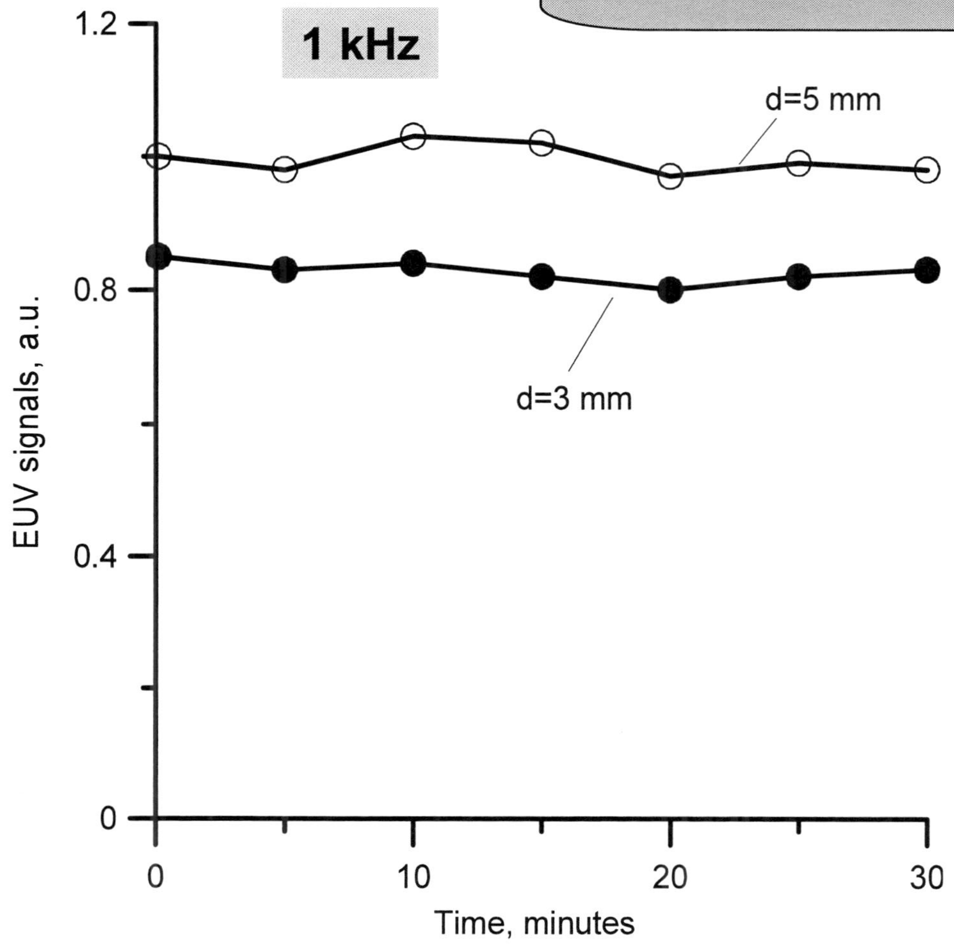
1 kHz
EUV signals, a.u.
1.2
0.8
0.4
0
d=5 mm
d=3 mm
0
10
20
30
Time, minutes

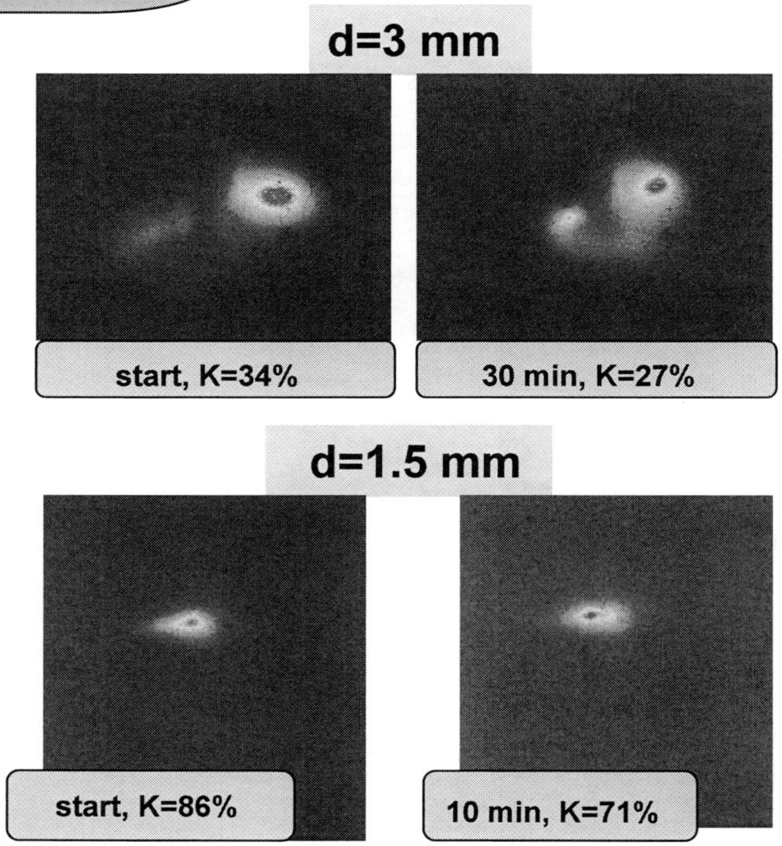
d=3 mm
start, K=34%
30 min, K=27%
d=1.5 mm
start, K=86%
10 min, K=71%

Degradation of collection efficiency during a source operation time

Source #4 with XeF laser

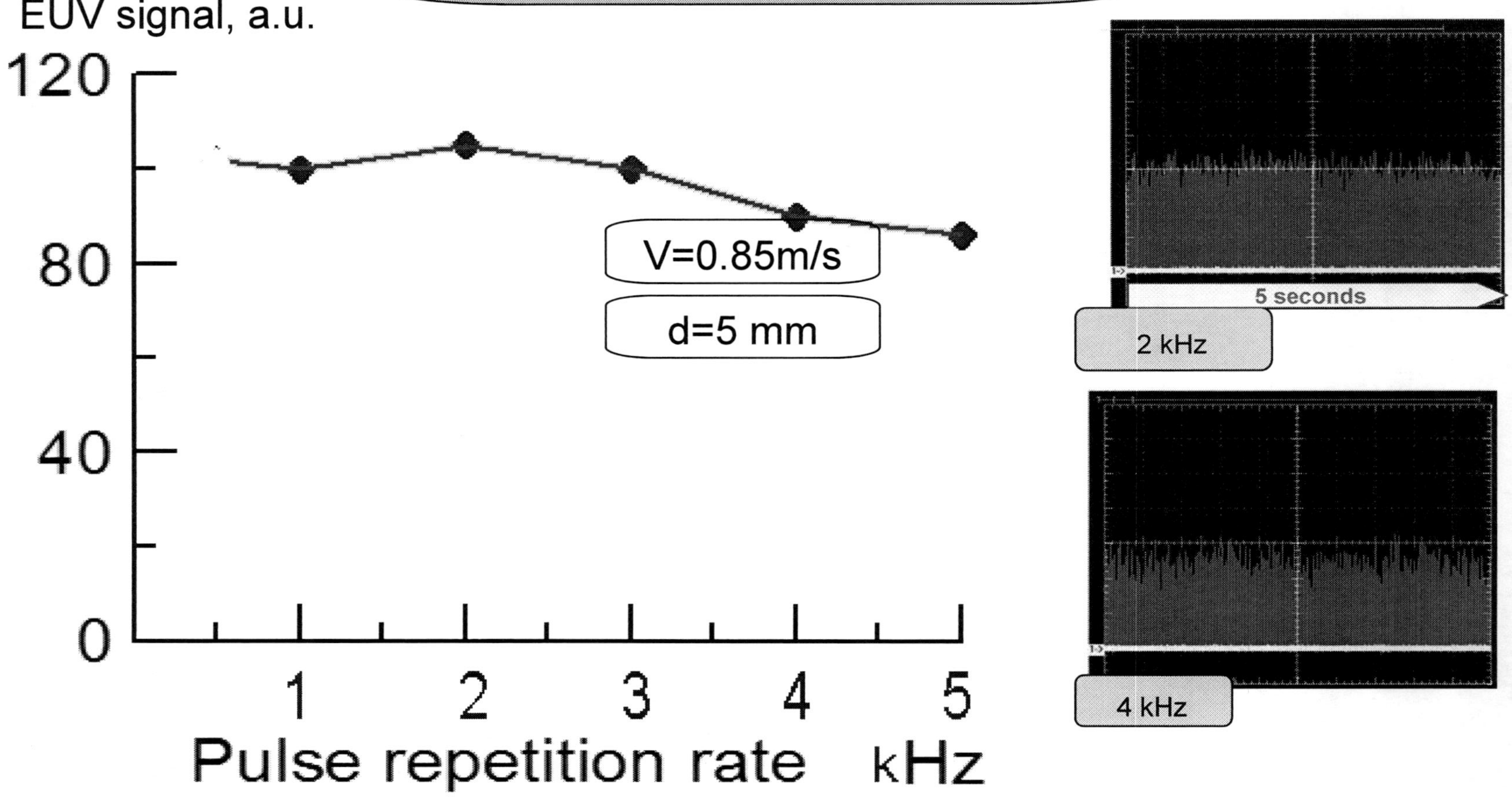

EUV signal, a.u.
120
80
40
0
1
2
3
4
5
Pulse repetition rate kHz
V=0.85m/s
d=5 mm
5 seconds
2 kHz
4 kHz

EUV signals, measured during ~5 s,
after 10 s as a source started to operate

Source #4 with XeF laser

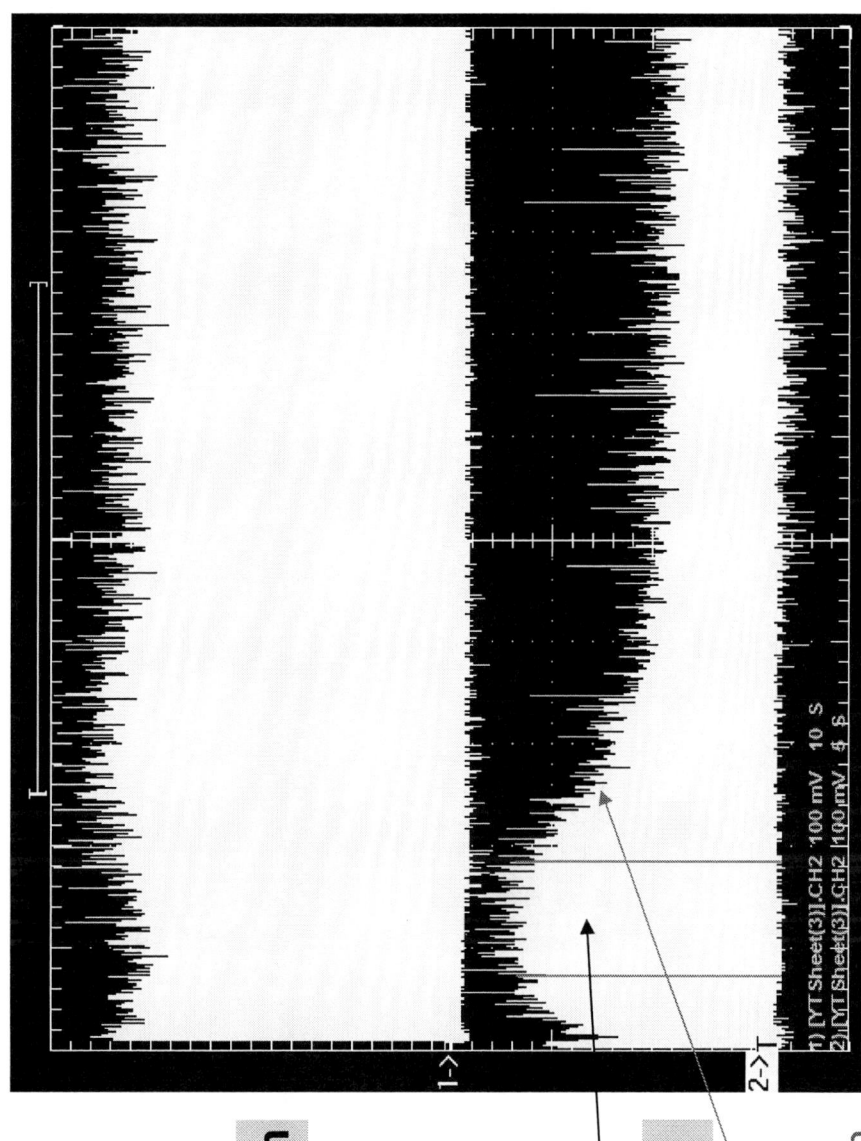

EUV signals as a function of source operation time

50 sec

1000 Hz , Δx=0.85mm

Region of measurement in the previous figure

2000 Hz, Δx=0.425mm

Drop of EUV signals during source operation time if repetition frequency >1000Hz, pulse to pulse distance Δx<0.85mm and XeF laser was used

Source #4 with XeF laser

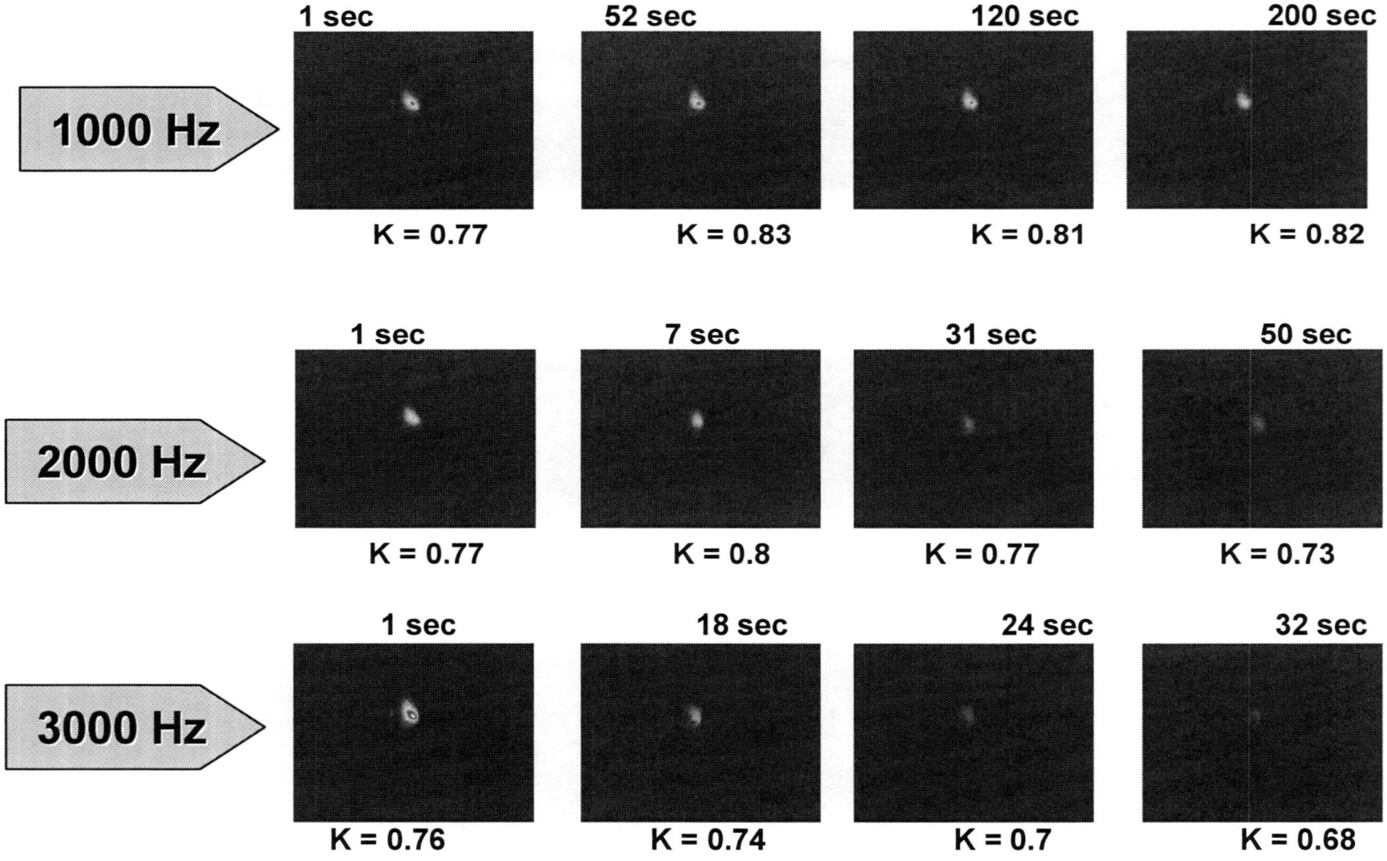

1000 Hz

1 sec	52 sec	120 sec	200 sec
K = 0.77	K = 0.83	K = 0.81	K = 0.82

2000 Hz

1 sec	7 sec	31 sec	50 sec
K = 0.77	K = 0.8	K = 0.77	K = 0.73

3000 Hz

1 sec	18 sec	24 sec	32 sec
K = 0.76	K = 0.74	K = 0.7	K = 0.68

Images of pinch for different source operation time

Illustrate EUV light degradation during source operation time if XeF laser was used

Source #4 with Nd:YAG laser

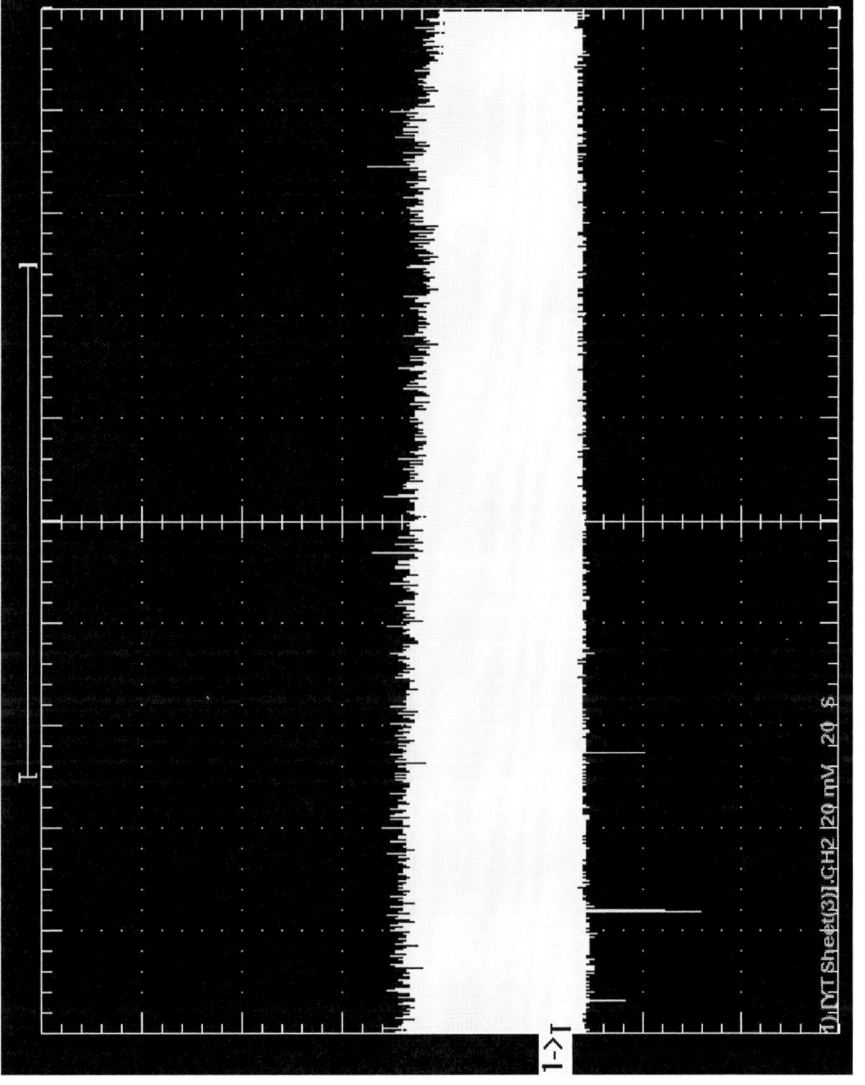

all scale 200 s

There are not drop of EUV dignals during source operation time
if Nd:YAG laser was used (f=4 kHz, $\Delta x \approx 0.2$ mm, ε in=3.14 J, d=1.8 mm)

Source #4 with Nd:YAG laser

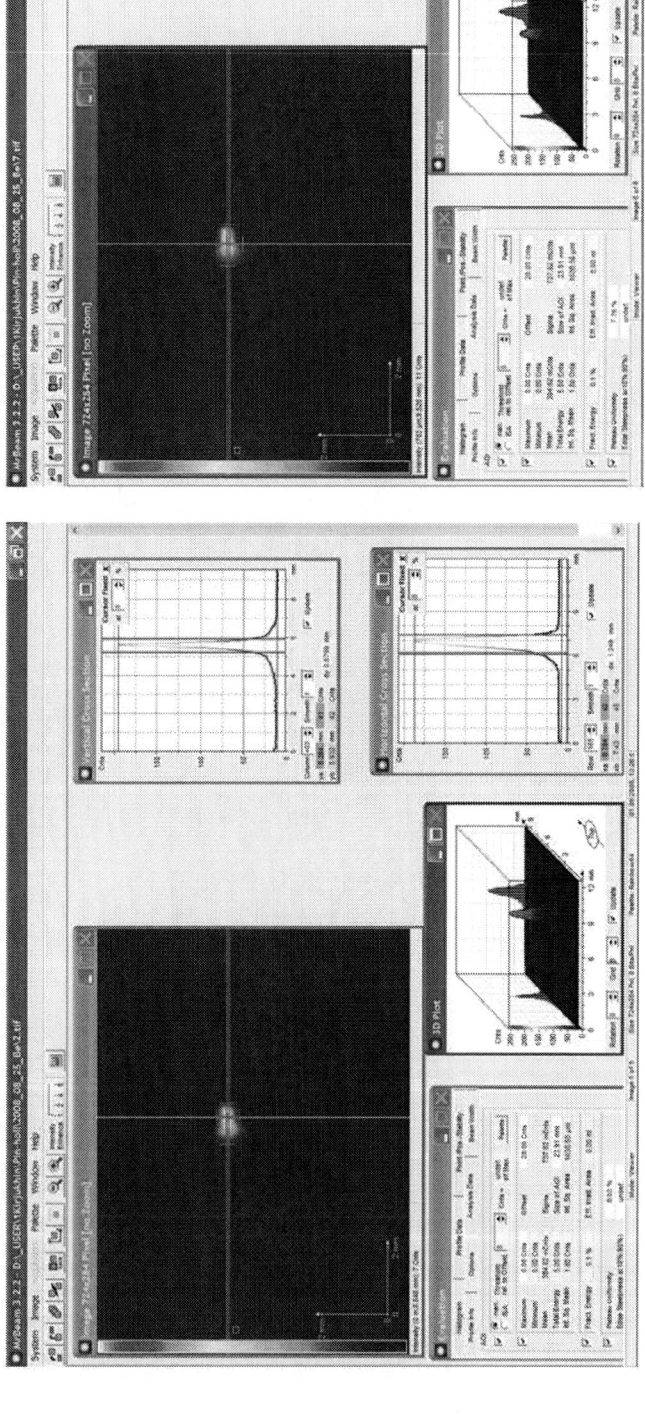

EUV image of discharge at once after source start to operate

EUV image of discharge after 200 seconds of source operation

There are no degradation of EUV light and collect efficiency during source operation time if Nd:YAG laser was used ($\varepsilon = 3.14$ J/pulse, $f = 4$ kHz)

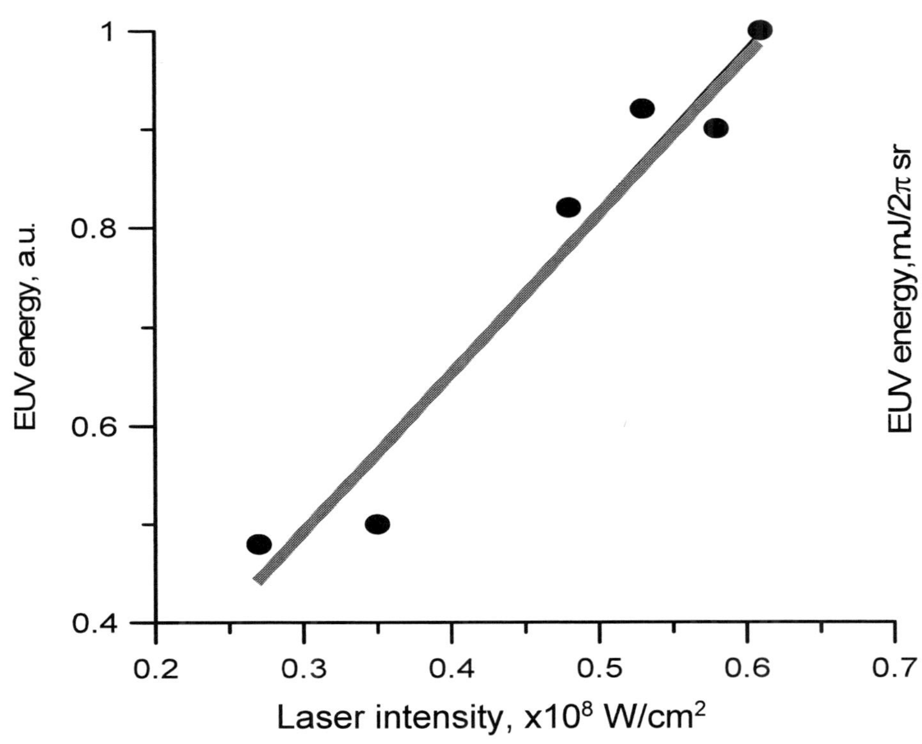

EUV energy as a function of Nd:YAG intensity obtained for source #1 at 2 Hz with Nd:YAG (AO8) at XTREME techn

Laser intensity, x10^8 W/cm^2

EUV energy, mJ/2π sr

EUV energy as a function of Nd:YAG intensity for continuous operation of source #4 at 4 kHz, ε=3.14J

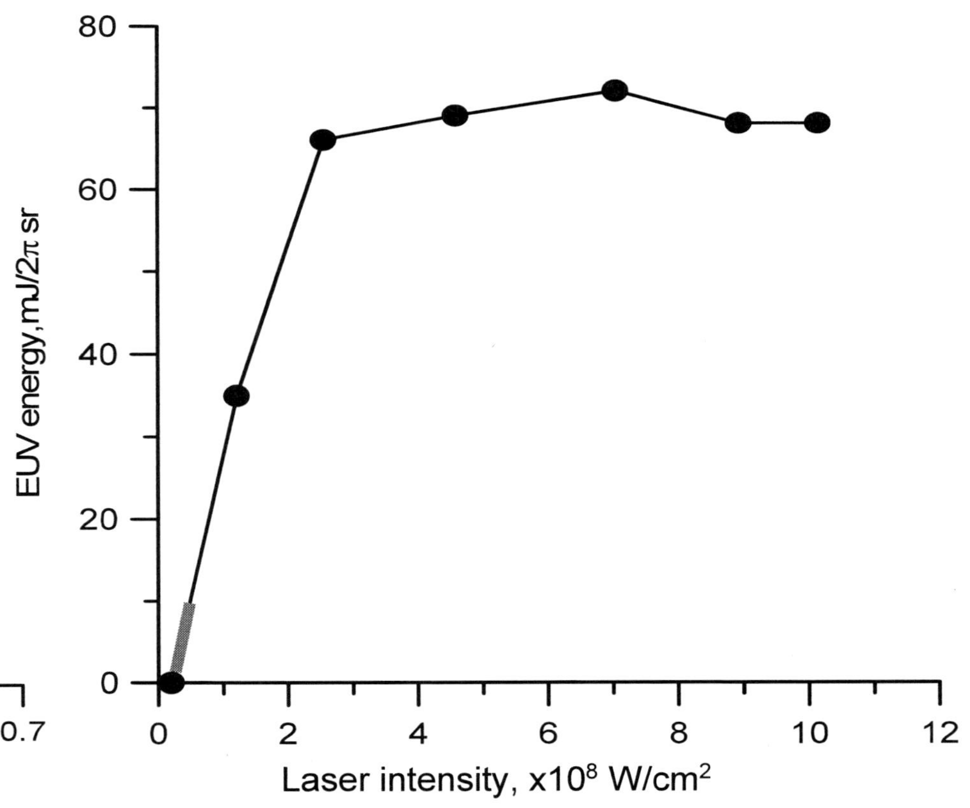

Laser intensity, x10^8 W/cm^2

EUV energy, a.u.

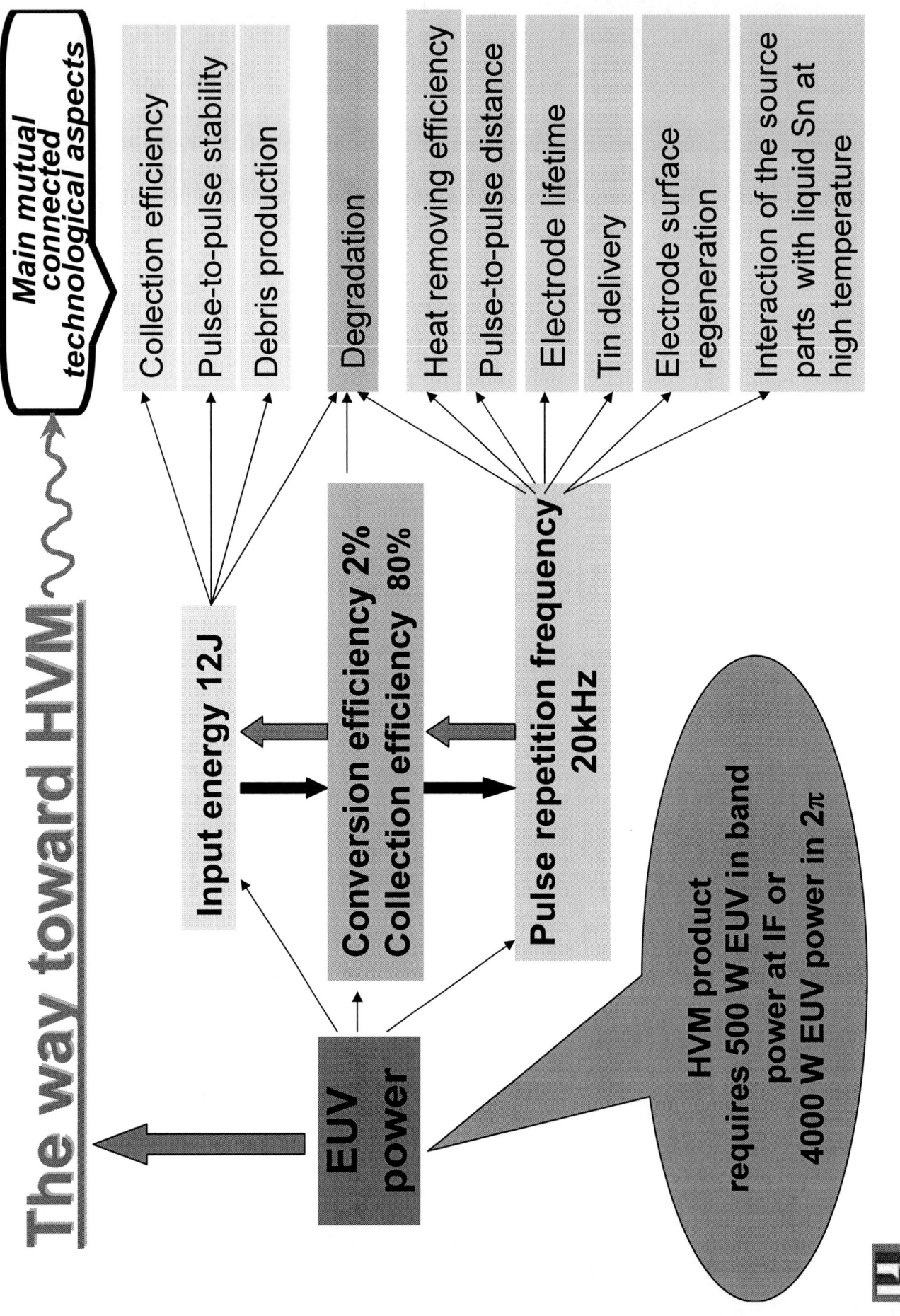

Summary

□ Main technological aspects of Sn RDE source have been studied with four different designs. The optimal variant of Sn RDE source design for HVM is determined.

□ Intrinsic conversion efficiency in the range of 2% and collection efficiency in the range of 80% up to 12 J input pulse energy is demonstrated.

□ Our experience with different sources and the obtained results indicate that Sn RDE source can be scalable to HVM power level ($\approx$4000W in 2π) with input pulse energy about 12J , pulse repetition frequency about 20kHz, using Nd:YAG laser.

Acknowledgements

We gratefully thank XTREME technologies GmbH and USHIO Inc. for support and cooperation, especially,

Mr.Tatsushi Igarashi, Mr.Masaki Yoshioka, Dr.Jürgen Kleinschmidt, Dr.Kazuaki Hotta, Dr.Yusuke Teramoto

ASML

Spectral Purity Filter Development for EUV HVM

A. Yakunin, V. Banine, ASML

N. Salashchenko, E. Kluenkov, A. Lopatin, V. Luchin, N. Tsybin, IPM

L. Sjmaenok, Phystex

W. Soer, M. Jak, Philips

Outline

- Why SPF?
- Achieved up to now
- New challenges and testing
- How to meet new challenges
- Summary and acknowledgements

Philips *PhysTeX* IPM ASML

Outline

- Why SPF?
- Achieved up to now
- New challenges and testing
- How to meet new challenges
- Summary and acknowledgements

Philips *PhysTeX* IPM ASML

Spectral purity filters for discharge and laser produced plasmas must filter different wavelengths

- Radiation behind Intermediate Focus (IF):
  - DPP:
    - Broad-band EUV
    - DUV-visible-IR
  - LPP:
    - Narrow-band EUV
    - DUV-visible-IR
    - Scattered drive laser radiation (10.6 μm)
- Purpose of the SPF
  - DPP: filtering of DUV
  - LPP: filtering of DUV and scattered drive laser radiation

DUV and drive laser radiation must be filtered for HVM to maintain imaging and overlay specs

- Origin of parasitic out of band (OoB):
  - 10.6 µm CO_2 radiation – partially reflected from plasma
  - 130-400 nm DUV radiation is emitted together with EUV
- Associated challenge from OoB:
  - IR heating of the wafer stage and POB mirrors
  - DUV induced image deterioration

First HVM tool (for 100+ W at IF)

Challenge	At wafer	Exceeding requirement by
Wafer stage heating	Overlay	5-10x
POB mirror heating	Overlay	2-5x
Resist sensitivity to DUV	Imaging	10-30x

Philips PhysTeX   IPM   ASML

Outline

- Why SPF?
- Achieved up to now
- New challenges and testing
- How to meet new challenges
- Summary and acknowledgements

Philips PhysTeX IPM ASML

Filtering solution for DUV (without IR) has been previously presented

Coating on a mirror:
DUV suppression 3-5 x
with 7% EUV loss

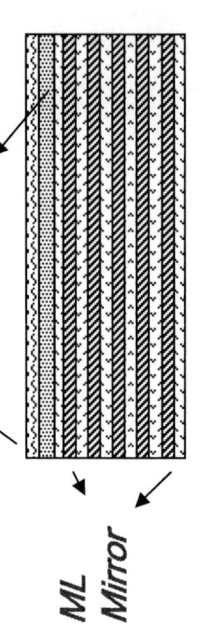

Coating
Normal cap layer

ML
Mirror

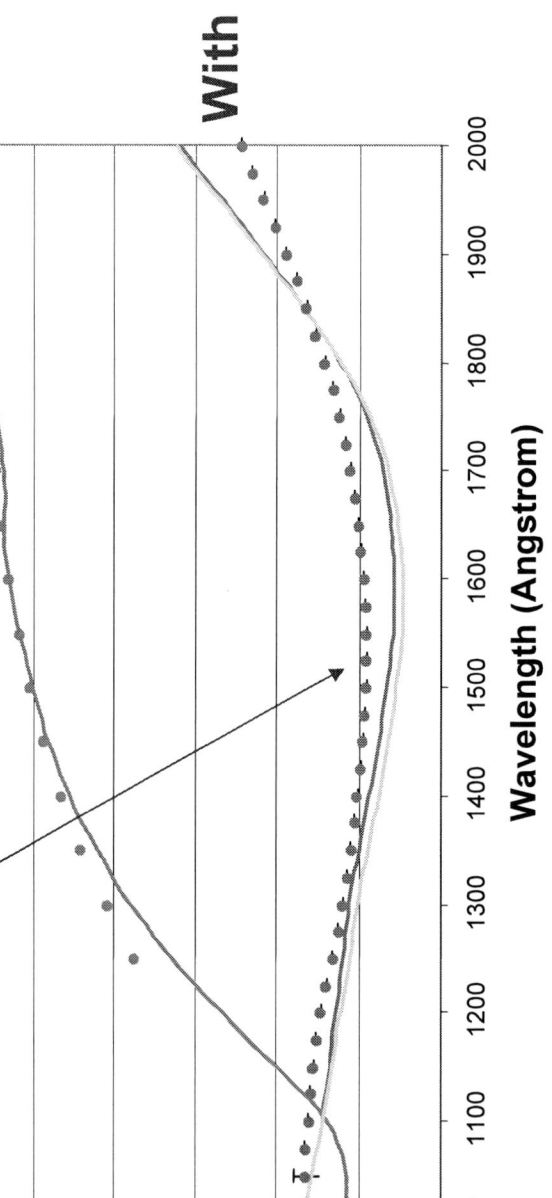

Without coating

With coating

For higher suppression in DUV multiple mirrors can be coated

EUVL symposium 2006-2007

Filtering solution for IR and DUV with a Si/Zr SPF has been previously presented

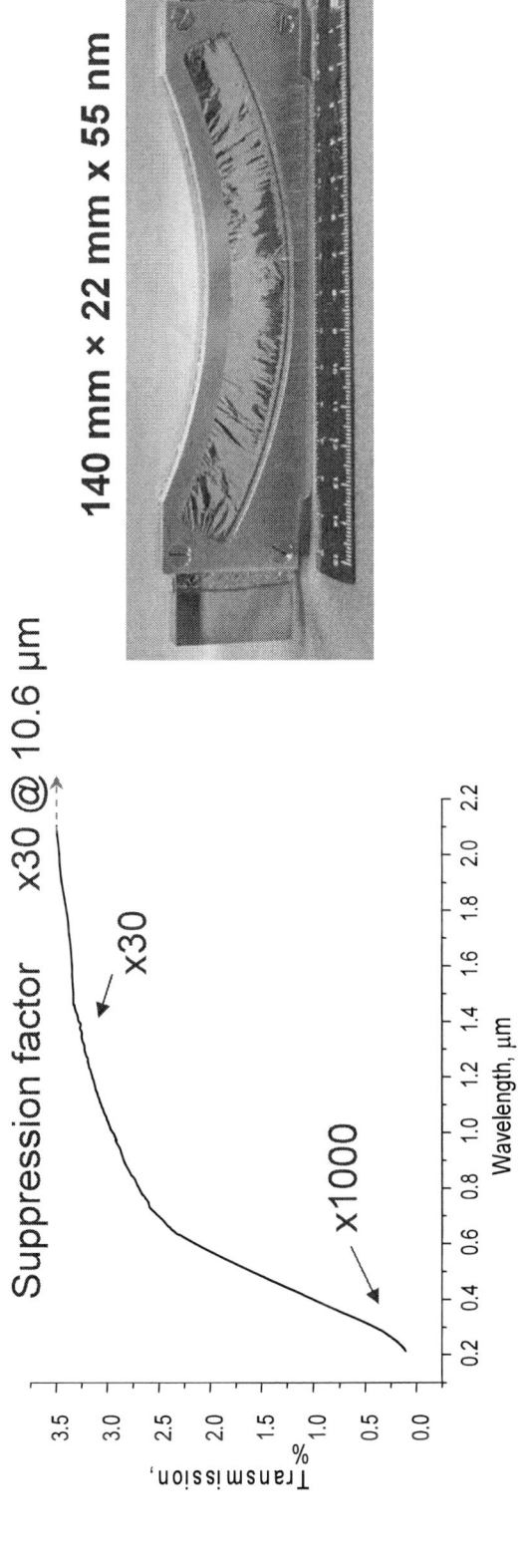

140 mm × 22 mm × 55 nm

- Multilayer structure $N = 25$, $h_{Zr} = 1.6$ nm, $h_{Si} = 0.6$ nm ↔ strength (where N-number of layers and h corresponding thicknesses)

- Si ↔ DUV suppression x1000

- Zr ↔ High up to 80% EUV transmission

EUVL symposium 2006-2007

Philips *PhysTeX*  **ASML**

Si/Zr SPF under high power exposures in vacuum maintain transmission requirement for ~120 days has been previously demonstrated

• Testing Si/Zr film sample

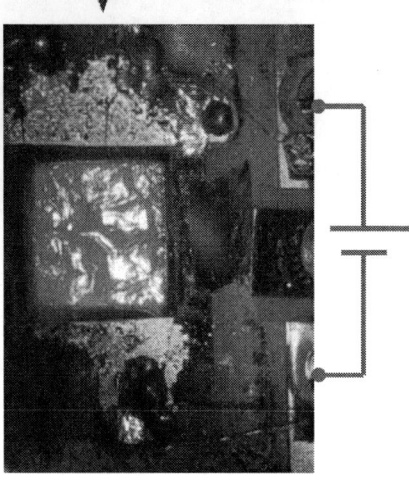

55 nmSi/Zr film

Incident power	W/cm²	0.7	1.1	2.3
Duration	days	25	60	37
Transmission @ 13.5 nm (before)	%	76.53	75.48	74.46
Transmission @ 13.5 nm (after)	%	76.04	72.25	72.13
Relative transmission loss	%	0.6	4.3	3.1
Relative transmission loss	%/day	0.026	0.071	0.085
Extrapolated 10% transmission loss	days	390	140	118

Extrapolated 10% EUV transmission loss ~120 days with corresponding filter temperature of 600 C

Philips *PhysTeX* **ASML**

Outline

- Why SPF?
- Achieved up to now
- **New challenges and testing**
- How to meet new challenges
- Summary and acknowledgements

Philips *PhysTeX* IPM ASML

Power load on SPF for LPP is significantly higher than previously expected

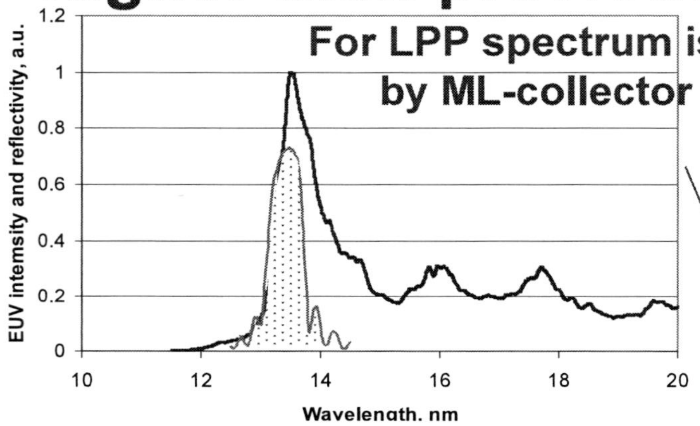

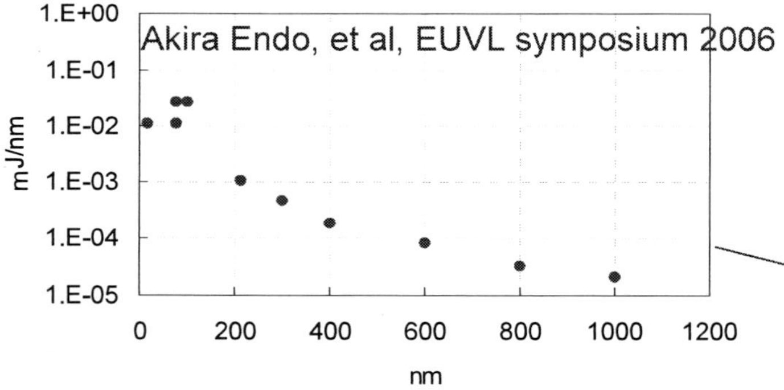

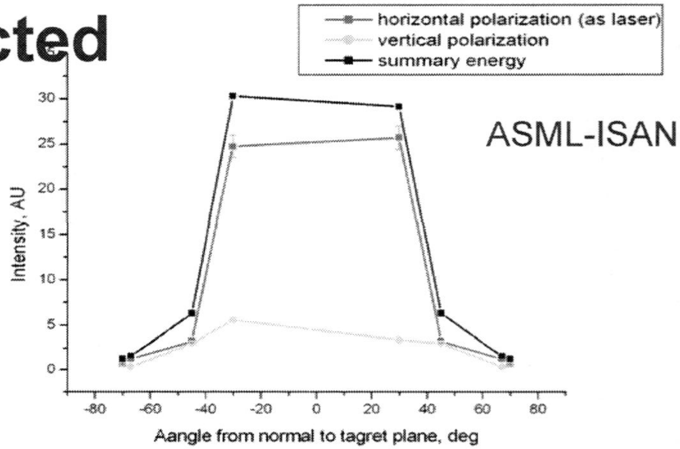

For LPP spectrum is filtered by ML-collector

ASML-ISAN

Significant amount of laser light is scattered and collected

Wavelength name	Power relative to in-band
in-band EUV	1
out-of-band EUV	1-2
DUV-IR	0.3-0.7
IR-scatter	3-5

Akira Endo, et al, EUVL symposium 2006

- CO2 is dominant component in HVM LPP spectrum

- With about 150 cm2 of filter surface the requirement for energy flux is: 3-6 W/cm² with uniform filling for 100+ W at IF

- New SPF power load target value including scaling and non-uniformity: <u>20+ W/cm² vs 2+ /cm² before</u>

Philips PhysTeX IPM ASML

Test rig has been built for SPF characterization

The purpose:

- Validate SPF performance at higher heat
  Loads than demonstrated previously
- Verify pulsed heat load endurance

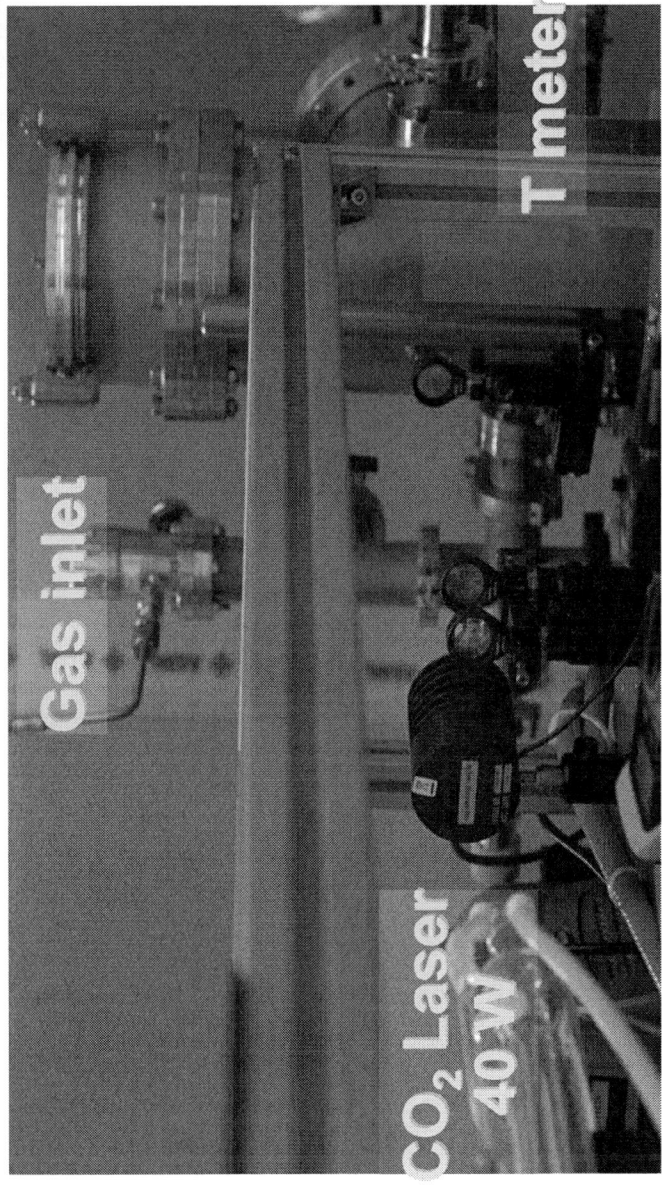

Philips *PhysTeX* **ASML**

Preliminary results from pulsed power load testing does not show additional problems compared to DC testing

Pulse power mode heating   **DC versus AC (50 kHz, 100 ns pulse)**

Gas	Pressure AU	Power: average / peak W/cm²		
		0.5	1.2	1.8
		100	240	360
Background	0.01	318	458	546
Gas	1	290	441	520
Gas	3	255	420	495
Gas	6	220	390	470
Gas	10		355	
Gas	100		190	
Gas	500		155	

- 1.2 W/cm2, DC
- 1.8 W/cm2, DC
- 1.2 W/cm2, pulse
- 1.8 W/cm2, pulse

Temperature, C — Gas pressure, AU

Peak power during 100 ns pulse P=360 W/cm²

Philips /PhysTeX IPM ASML

Existing Si/Zr SPF solution needs to be improved

High power CO_2 exposure

20 W/cm2 CO_2

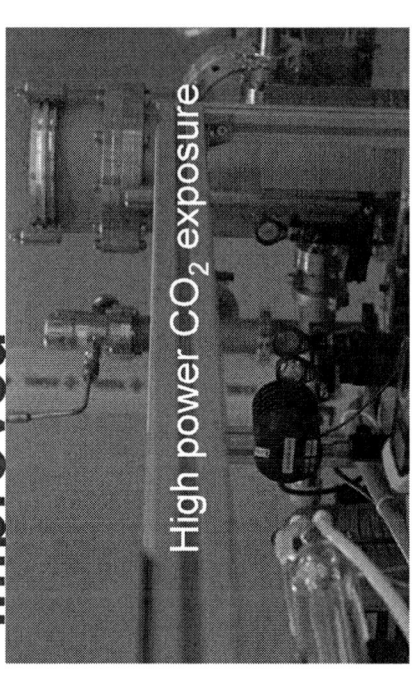

After exposure: back illuminated

- Critical parameter: Filter temperature should be <u>T<600 °C</u>

- Inter-diffusion and silicide formation-> transparency for DUV and IR

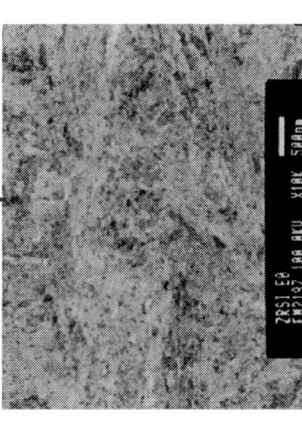

Before exposure

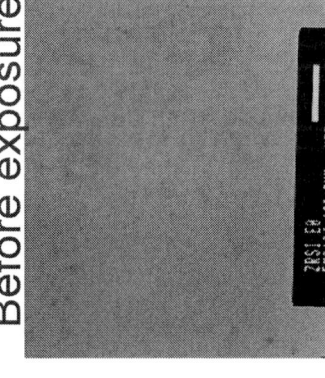

After exposure

The filter remained mechanically intact (!) but became transparent for IR

Philips *PhysTeX* IPM ASML

Outline

- Why SPF?
- Achieved up to now
- New challenges and testing
- **How to meet new challenges**
- Summary and acknowledgements

Philips *PhysTeX* IPM ASML

Several items have been identified for improving the current SPF for the higher power load

- Add effective gas cooling

- Make material improvements:

  - Increase reflectivity of light from the front surface (current reflection is about 50%)

  - Increase emissivity (currently about 30%) for cooling

  - Use chemically stable materials

Effectiveness of gas cooling has been tested

Current measurement point

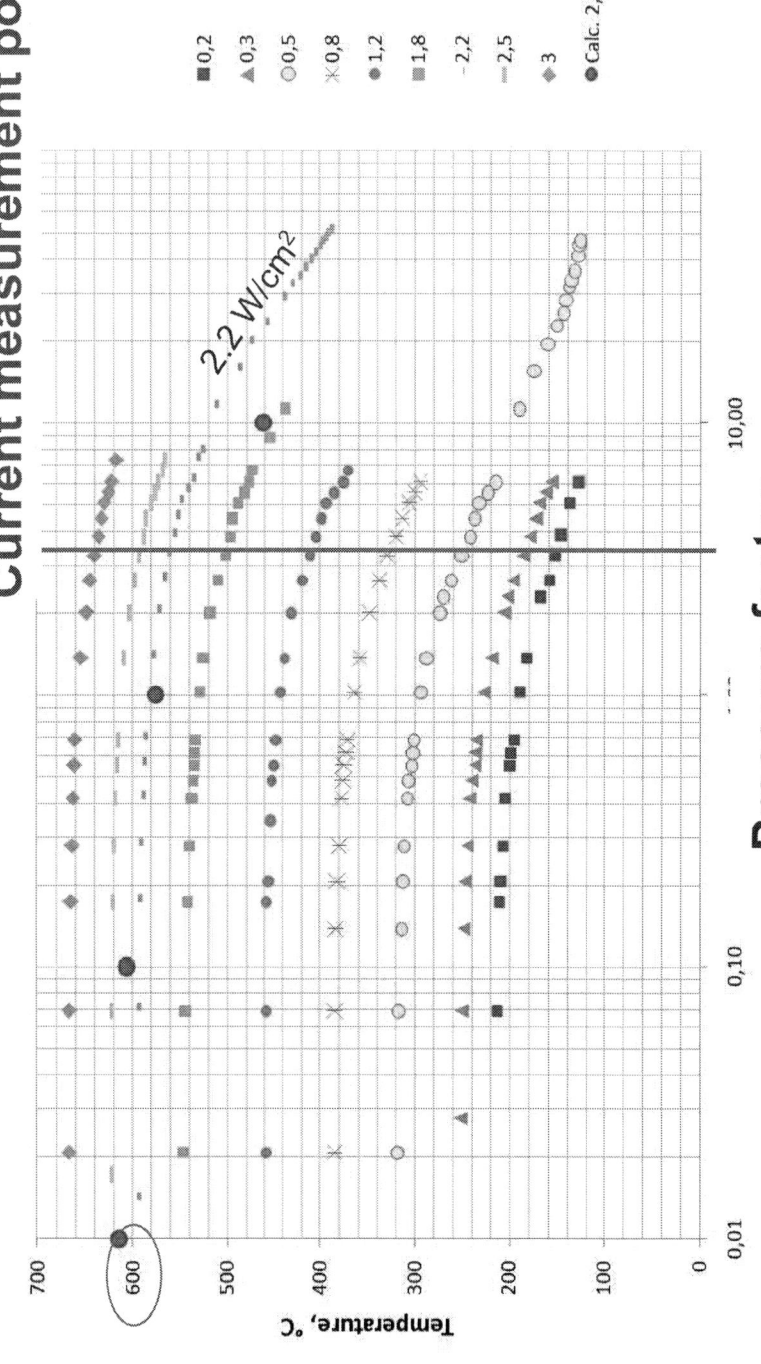

With a pressure factor of 100 the acceptable power (<600 °C) can be 12 W/cm², thus 5x scaling with respect to the current performance

Additional filter optimization is possible

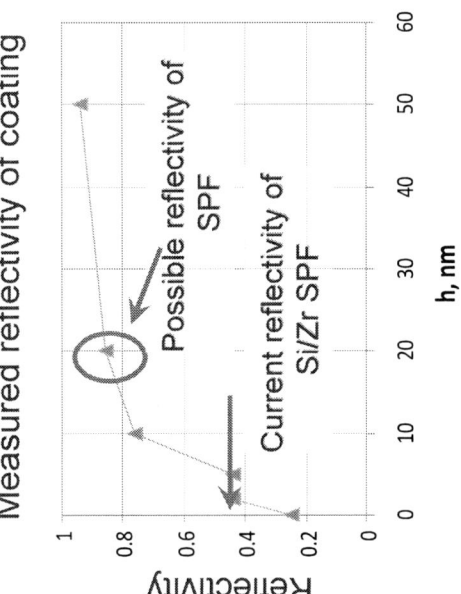

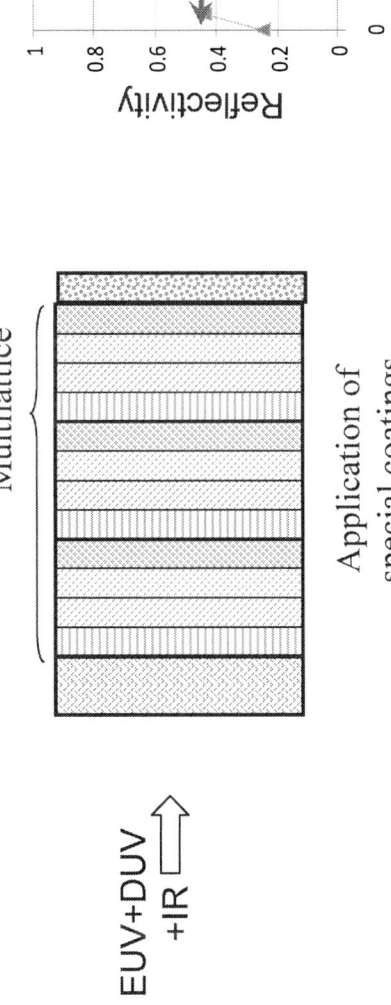

- Decrease of absorption -> 2x
- Increase surface emissivity -> 3x

} 6x*

- *Temperature can be reduced (1000 °C -> 530 °C)

Philips *PhysTeX* **ASML**

Total potential improvement using all options is possible without significant impact to chemical structure of the Si/Zr SPF

- Utilizing all possible options - cooling, improving absorption and emmissivtity - transmissive filter incident powers of 20+ W/cm^2 seem to be possible without significant change to chemical structure of the filter

- Additional use of non-chemically active pairs can bring the tolerable temperatures up to 1000-1200 °C and thus to tolerable fluxes ~2x than mentioned above

Philips *PhysTeX* IPM **ASML**

An alternative grid filter for shielding the CO_2 radiation has been investigated

Simulation of grid performance

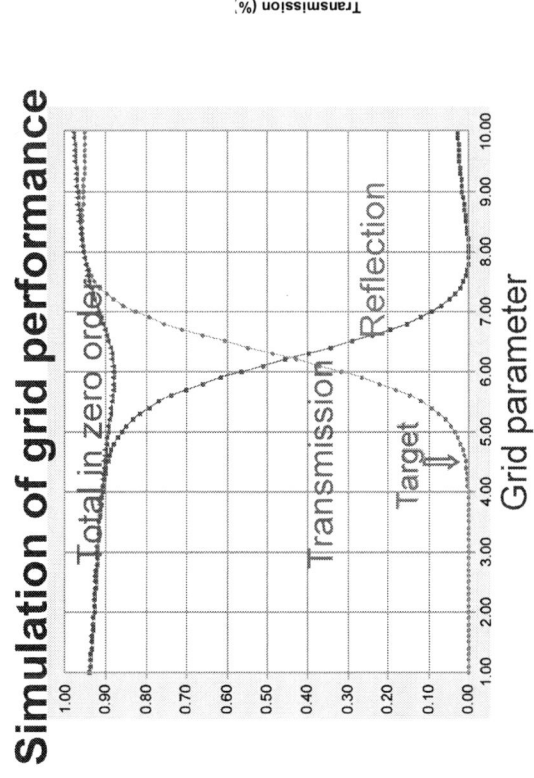

Measurement

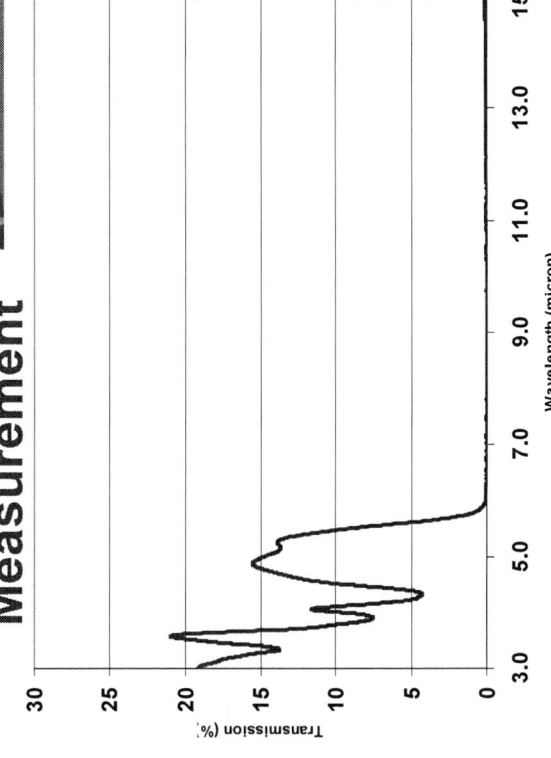

- Suppression of 10.6µm is as large as 1000x
- A sample of mesh with 80% of geometrical transmission has been made and manufacturability of a large grid is being investigated.
- Due to the chemical stability and reflection of IR, heat balance on the filter is satisfactory

Philips *PhysTeX* IPM ASML

Outline

- Why SPF?
- Achieved up to now
- New challenges and testing
- How to meet new challenges
- **Summary and acknowledgements**

Philips *PhysTeX* IPM ASML

Summary

- Solutions for SPF both for LPP and DPP have been investigated

- Solutions for DPP are identified:

  - transmissive filter (1000x DUV suppression vs 80% EUV transmission) and

  - special ML coatings (nx3 DUV suppression with 4-5% EUV loss)

- Solutions for LPP are identified and further development is under way:

  - transmissive filter power management and composition improvement is feasible for the required 20+ W/cm^2 load

  - grid solution satisfies the transmission for IR requirement (0.1%) and the manufacturability on the large scale is being investigated

Philips *PhysTeX* IPM ASML

Acknowledgements

- Authors are grateful to

- Konstantin Koshelev and ISAN team

- Georg Soumagne (GPI)

- Akira Endo (EUVA)

- Fred Bijkerk and FOM team

- Noreen Harned and Hans Meiling (ASML)

DESIGN AND FABRICATION CONSIDERATIONS
OF EUVL COLLECTORS FOR HVM

G. Bianucci *, J. Kools, G. Salmaso, F. E. Zocchi

Media Lario Technologies – Località Pascolo, 23842 Bosisio Parini (LC) – Italy

* e-mail: giovanni.bianucci@media-lario.com

2008 International Symposium on Extreme Ultraviolet Lithography, September 28 – October 1, 2008, Lake Tahoe, Ca

Outline

- Grazing incidence collector and HVM requirements

- Optical design

- Lifetime

- Thermal management

- Roadmap

- Evolution of electroforming towards normal incidence mirrors

- Conclusions

Gracing Incidence Collector and HVM requirements

HVM requirements

- Throughput → High collection efficiency
  - >25% (Point source in 2π)
- Throughput → High thermal load
  - 6 kW increasing to 12 kW in 2nd generation HVM
- Lifetime → Large source-collector distance
- Lifetime → Sacrificial layer
- Performance → Image quality
- Performance → optical stability

Integrated cooling

Wolter optics

Ru reflective layer

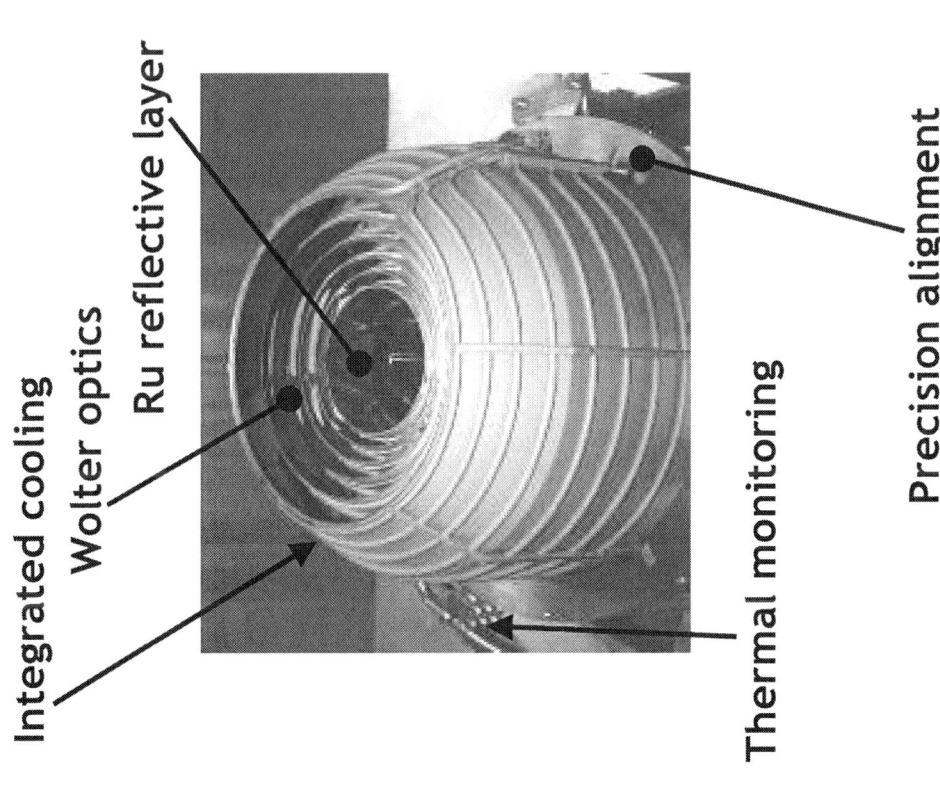

Thermal monitoring

Precision alignment

Designed-in optical efficiency & lifetime

HVM optical design doubles the photons delivered to IF and enables lifetime increase

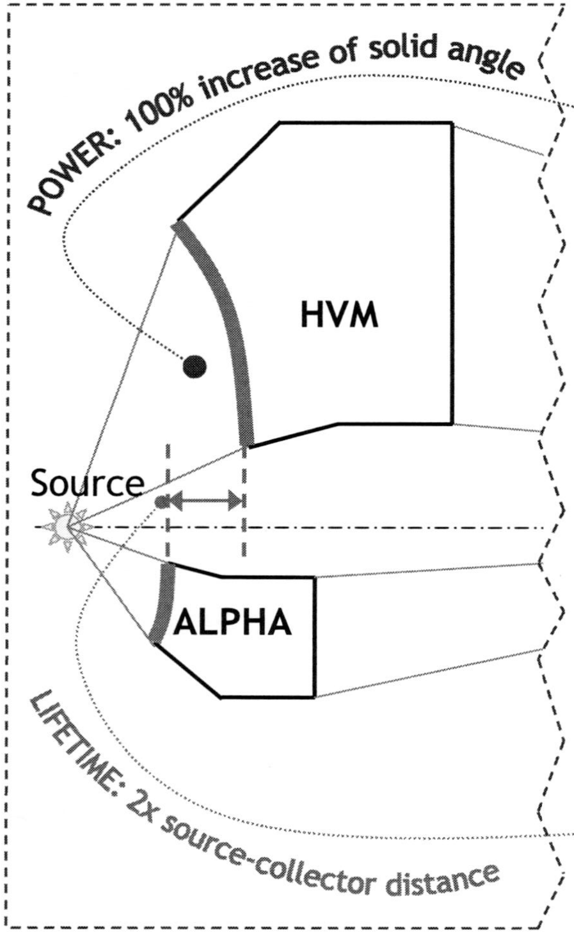

2x increase in collection solid angle allows HVM optical designs with 2x collection efficiency improvement factor

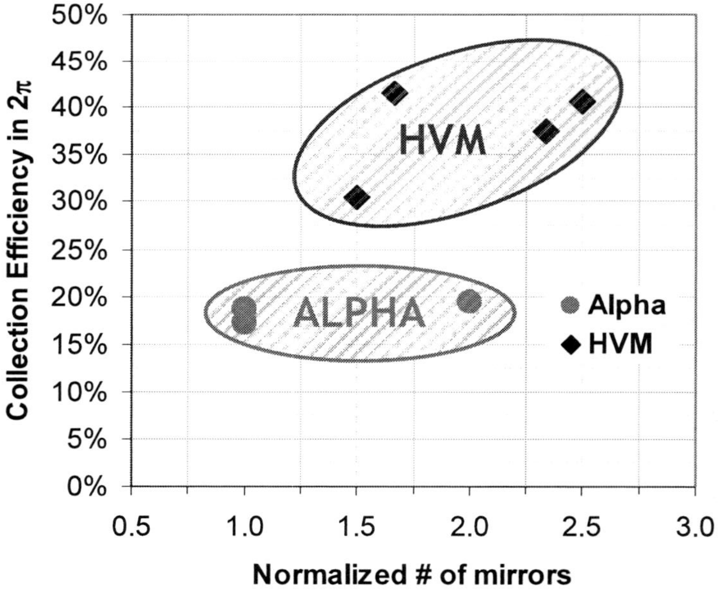

2x source-collector distance allows more effective debris mitigation designs (5x improvement)

Collection Efficiency in 2π

- Alpha
- ◆ HVM

HVM

ALPHA

Normalized # of mirrors

POWER: 100% increase of solid angle

HVM

ALPHA

Source

LIFETIME: 2x source-collector distance

>1-year lifetime with sacrificial reflective layer PHILIPS

- Philips and MLT have proven in joint experiments that Ru reflective layer works as sacrificial layer, maintaining reflectivity in erosion regime

- Ru thickness, effective debris mitigation, and reduced debris emission lead to >1-year lifetime projection for HVM collector

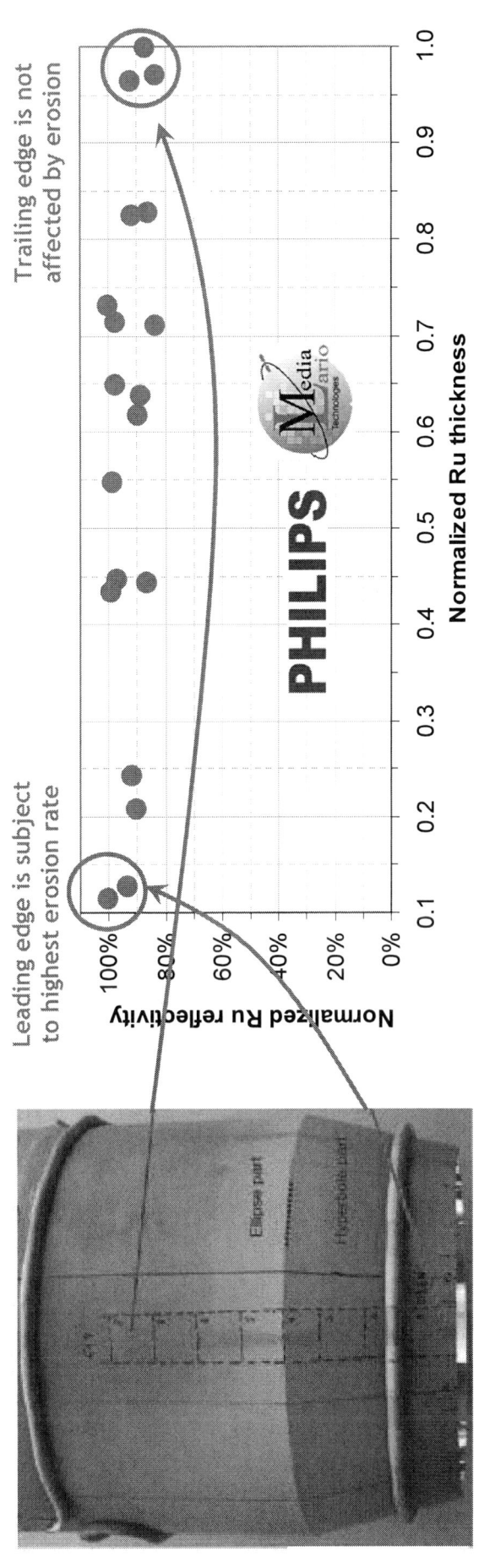

Reflectivity remains constant in all collector's areas subject to different erosion rates, thus ensuring constant IF power and far-field optical quality during the entire collector's lifetime

Process development for thick Ru deposition

- PVD films of refractory metals tend to become rougher as thickness increases

- High HFSR results in diffused scattering

- Reflectivity, as measured with a finite collection angle detector, will decrease

- MLT is developing deposition techniques to increase reflectivity by increasing atomic surface mobility during growth

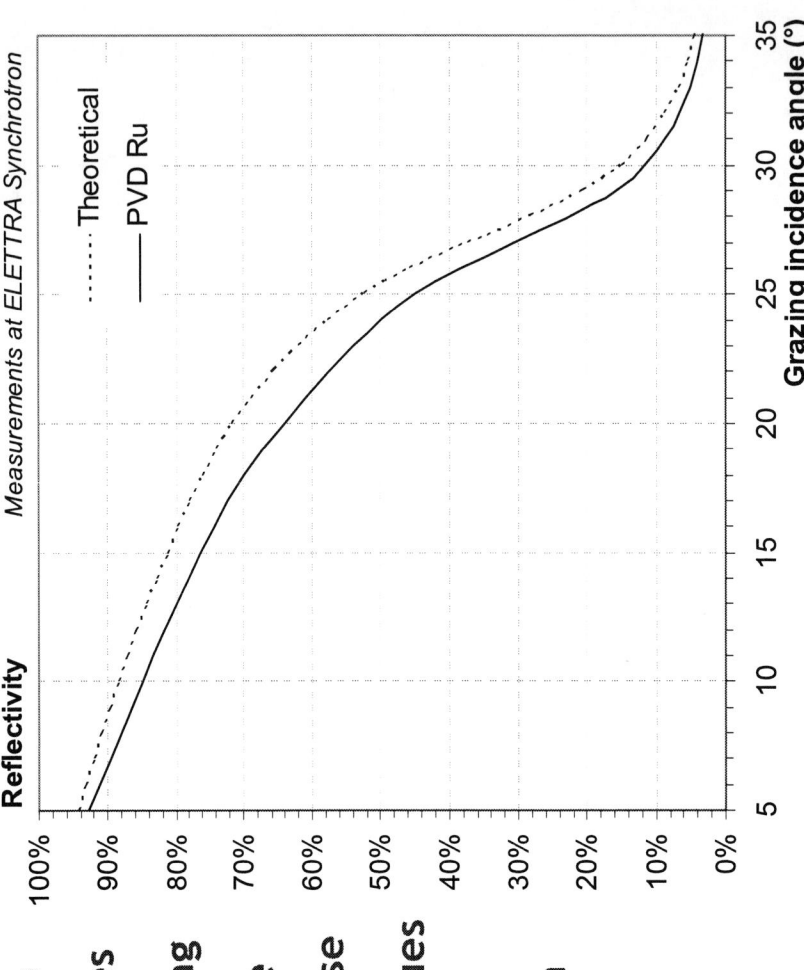

Measurements at ELETTRA Synchrotron

Scalable cooling technology to HVM thermal load

- Cooling technology developed and proven at Alpha level on Sn and Xe DPP sources
- Thermo-optical design validated in Alpha experiments *(2006 EUVL Symposium, 2008 SPIE Advanced lithography)*

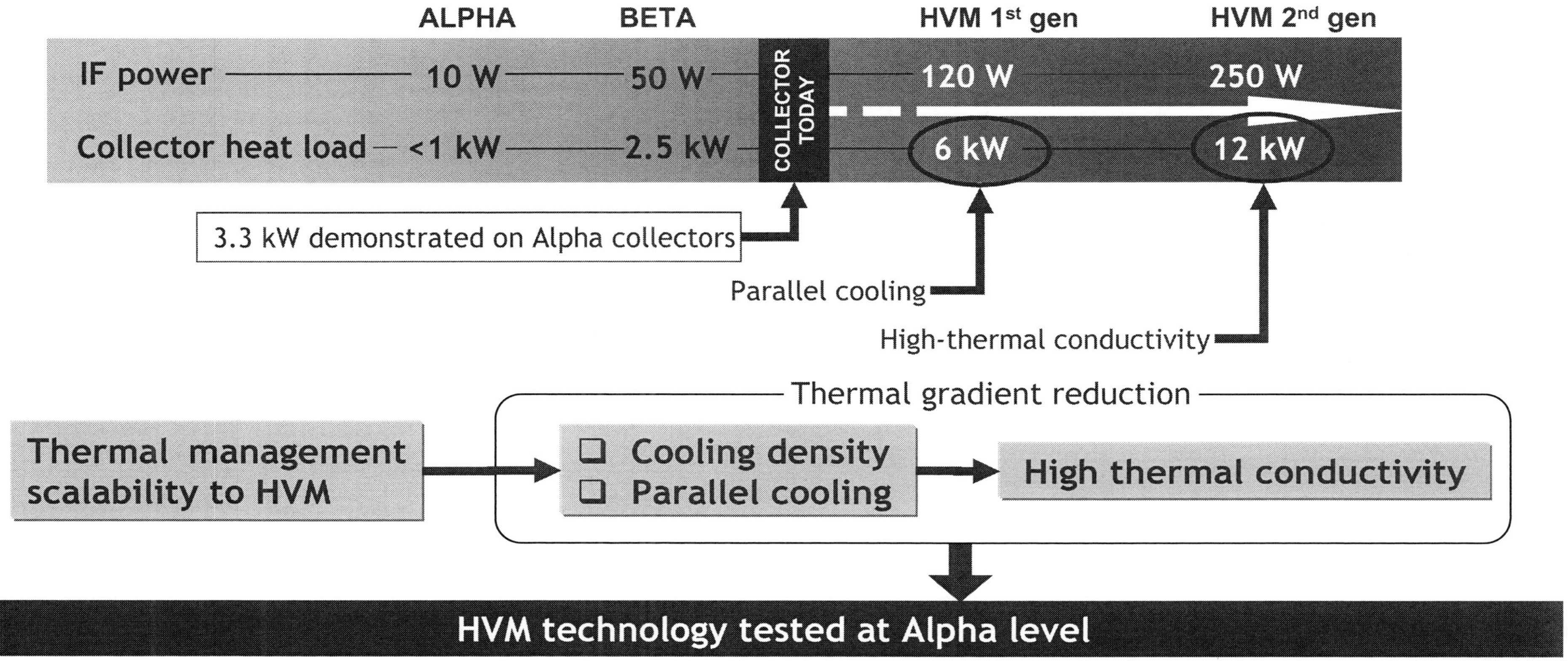

	ALPHA	BETA	HVM 1st gen	HVM 2nd gen
IF power	10 W	50 W	120 W	250 W
Collector heat load	<1 kW	2.5 kW	6 kW	12 kW

COLLECTOR TODAY

3.3 kW demonstrated on Alpha collectors

Parallel cooling

High-thermal conductivity

Thermal management scalability to HVM
- ☐ Cooling density
- ☐ Parallel cooling

Thermal gradient reduction

High thermal conductivity

HVM technology tested at Alpha level

High density and parallel cooling implemented in Alpha

1st generation cooling

3rd generation cooling

- 1st **gen cooling introduced in 2005**
  - *Single-circuit cooling*

- 2nd **gen cooling introduced in 2006**
  - *Dual-circuit cooling*

- 3rd **gen cooling introduced in 2007**
  - *High density cooling*
  - *2 parallel circuits*
  - *New fabrication & integration method*

50% reduction of slope error with new cooling

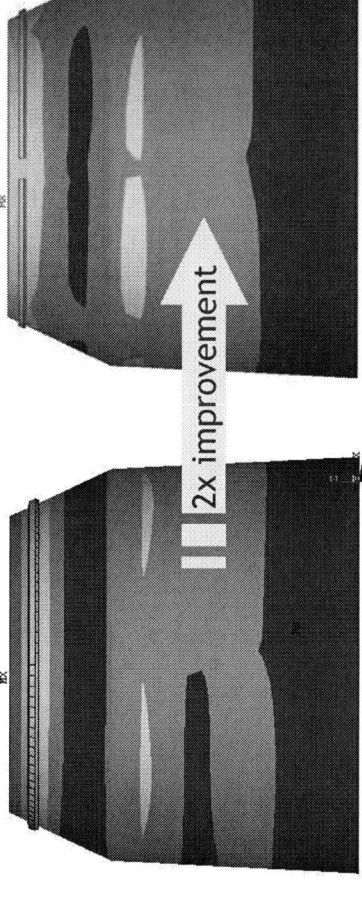

2x improvement

HVM high thermal conductivity material tested at Alpha level

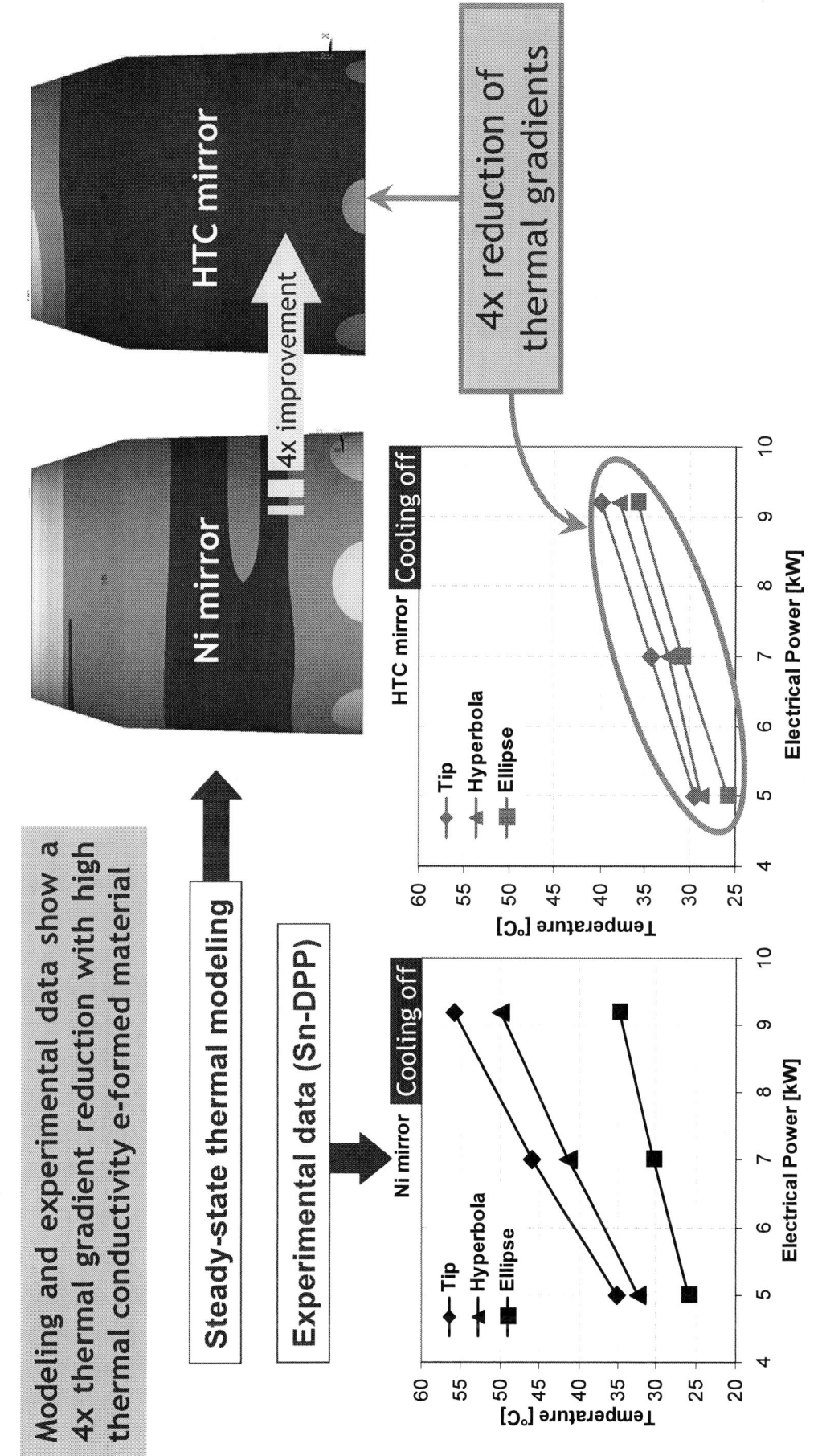

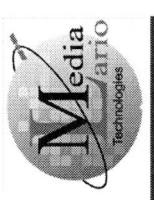

Grazing Incidence Collector product roadmap

2005	2006	2007	2008	2009	2010	2011

ALPHA

16% CE
Single circuit cooling
3-mo lifetime

ALPHA

16% CE
Dual-circuit
3-mo lifetime

ALPHA+

12% CE
High-density, 2-parallel circuits
9-mo lifetime

HVM 1st gen

25→30% CE
High-density multi-parallel cooling
1-yr lifetime

HVM 2nd gen

35% CE
High thermal conductivity
>1-yr lifetime

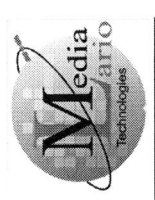

Evolution of electroforming towards normal incidence mirrors

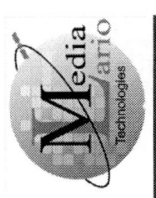

Evolution from grazing to normal incidence mirrors

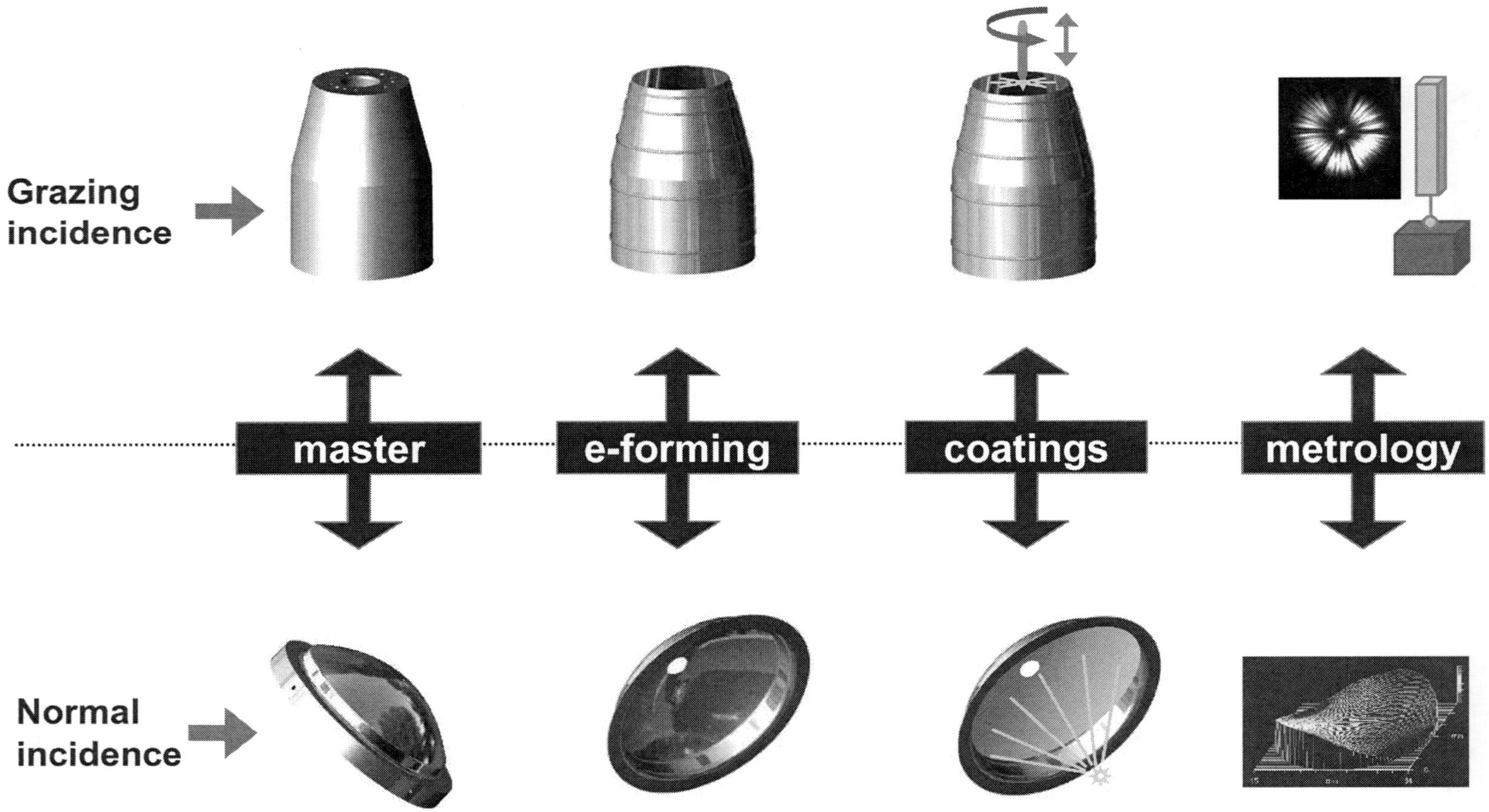

Grazing incidence

Normal incidence

master

e-forming

coatings

metrology

Evolution of e-forming to Normal Incidence Mirror applications

MLT is expanding its core competencies to meet Normal Incidence Mirror application requirements in size, performance, and CoO

Mo/Si multilayer deposition demonstrated on electroformed substrates

- Peak reflectivity: 61.4 %
- Peak wavelength shift: -0.01 nm
- FWHM: 0.52 nm
- HFSR = 1.3 Å rms

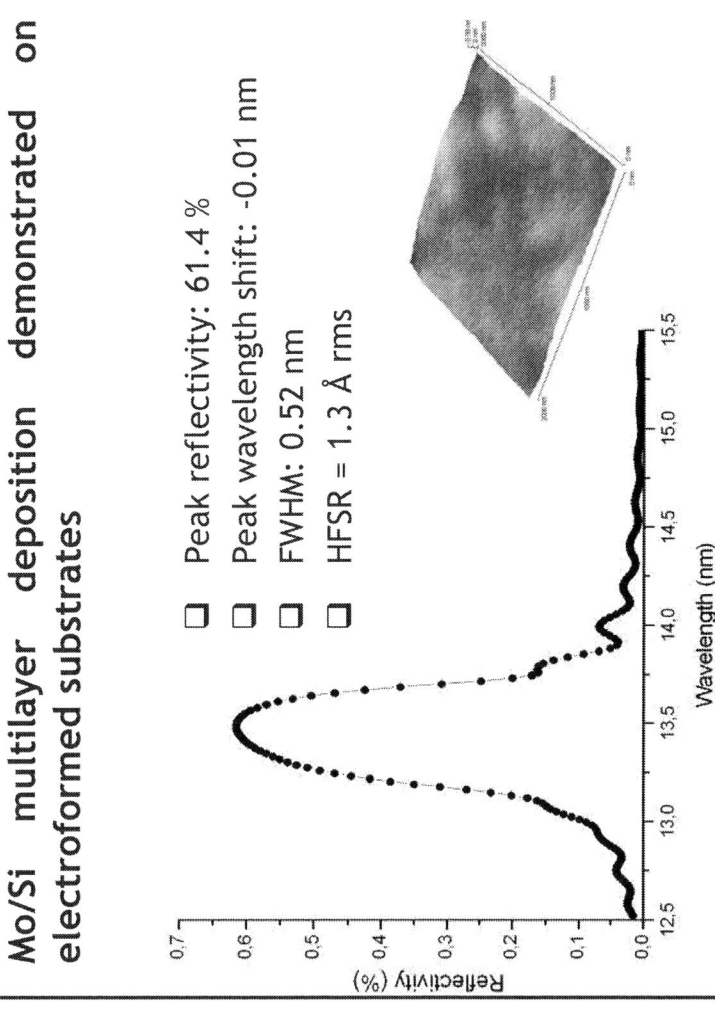

Extendibility of electroforming and master technologies to normal incidence mirror demonstrated

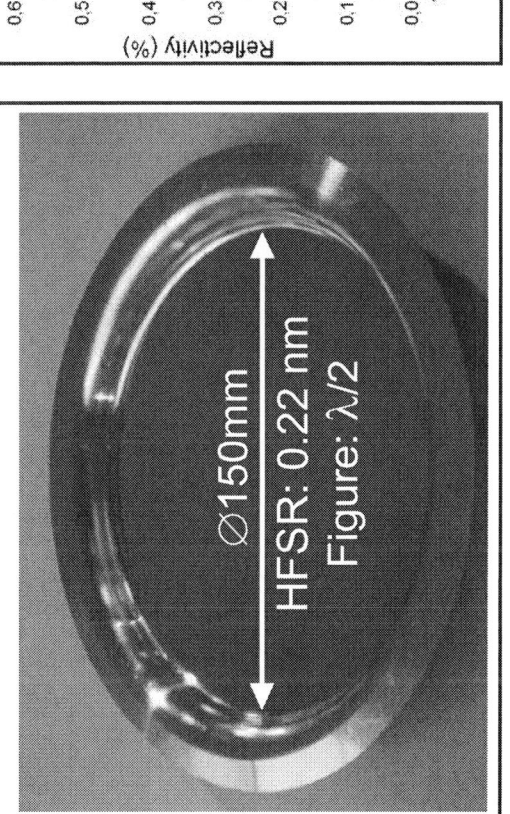

Ø150mm
HFSR: 0.22 nm
Figure: λ/2

Conclusions

- HVM Grazing Incidence Collectors design and manufacturing leverages off scalable technology solutions tested at Alpha level

- >25% collection efficiency (2x with respect to Alpha collectors)

- >1-yr collector lifetime projection

- Cooling technology scalable to HVM thermal loads

- Roadmap: HVM collector available in 2009

- MLT expanding its core competencies to Normal Incidence Mirrors

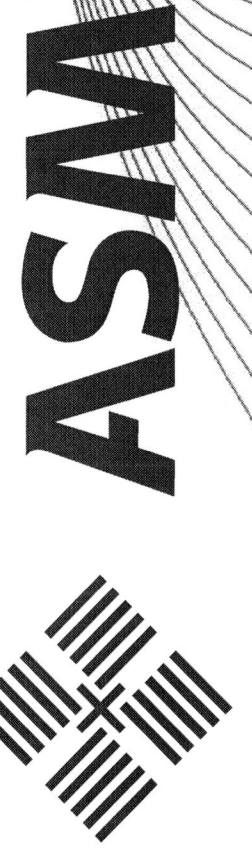

Strategy for Minimizing EUV Optics Contamination During Exposure

ASML: N.Harned, R. Moors, M. van Kampen, V. Banine,
J.Huijbregtse, R. Vanneer, A. Kempen,
CZO: D. Ehm
TNO: R. Verberk, E. te Sligte, A. Storm

Outline

- Carbon contamination of EUV mirrors
- Optics lifetime strategy EUV tools for High Volume Manufacturing
  - Clean vacuum
  - Local vacuum improvement
  - In situ cleaning
- Validation of this strategy: experimental test results
- Summary

During early Alpha Demo Tool integration it was noted that a rapid change in transmission occurred

Normalized system transmission vs #Gpulses

initial data w/o improved vacuum

Carbon contamination of EUV mirrors was confirmed as the cause of transmission loss

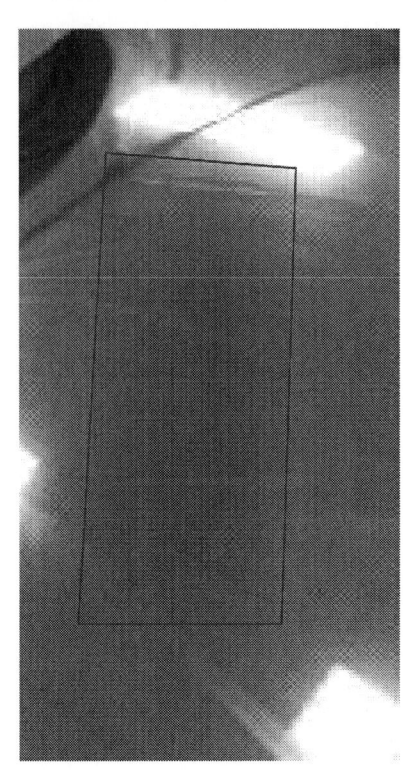

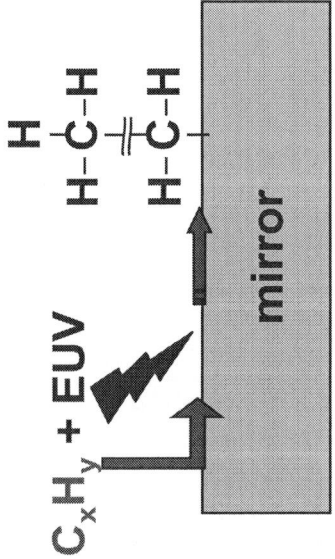

1% relative reflection loss per nm carbon

Experiments using pulsed EUV have been on-going for years

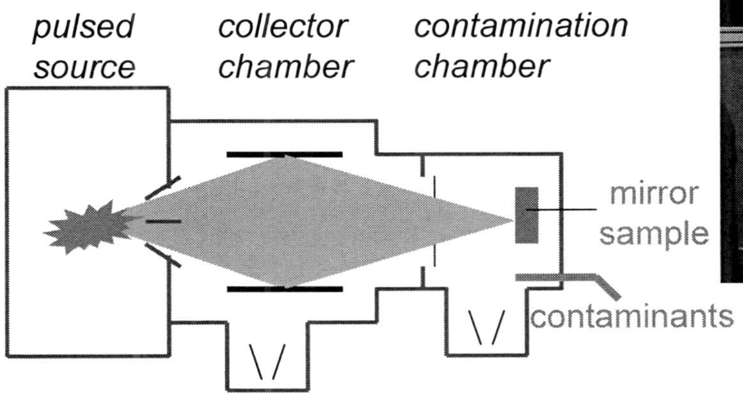

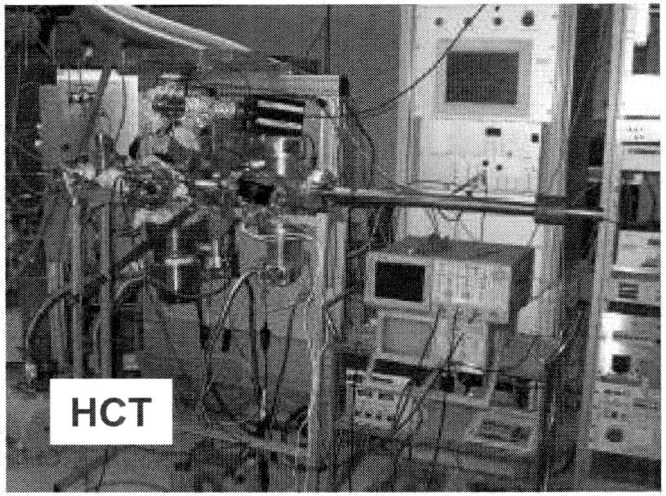

Test chamber

Source

Courtesy of TNO Science and Industry

in-situ XPS

Transfer chamber

ASML

pulsed source

collector chamber

contamination chamber

mirror sample

contaminants

HCT

Effect of environment on contamination rate was one of the early findings

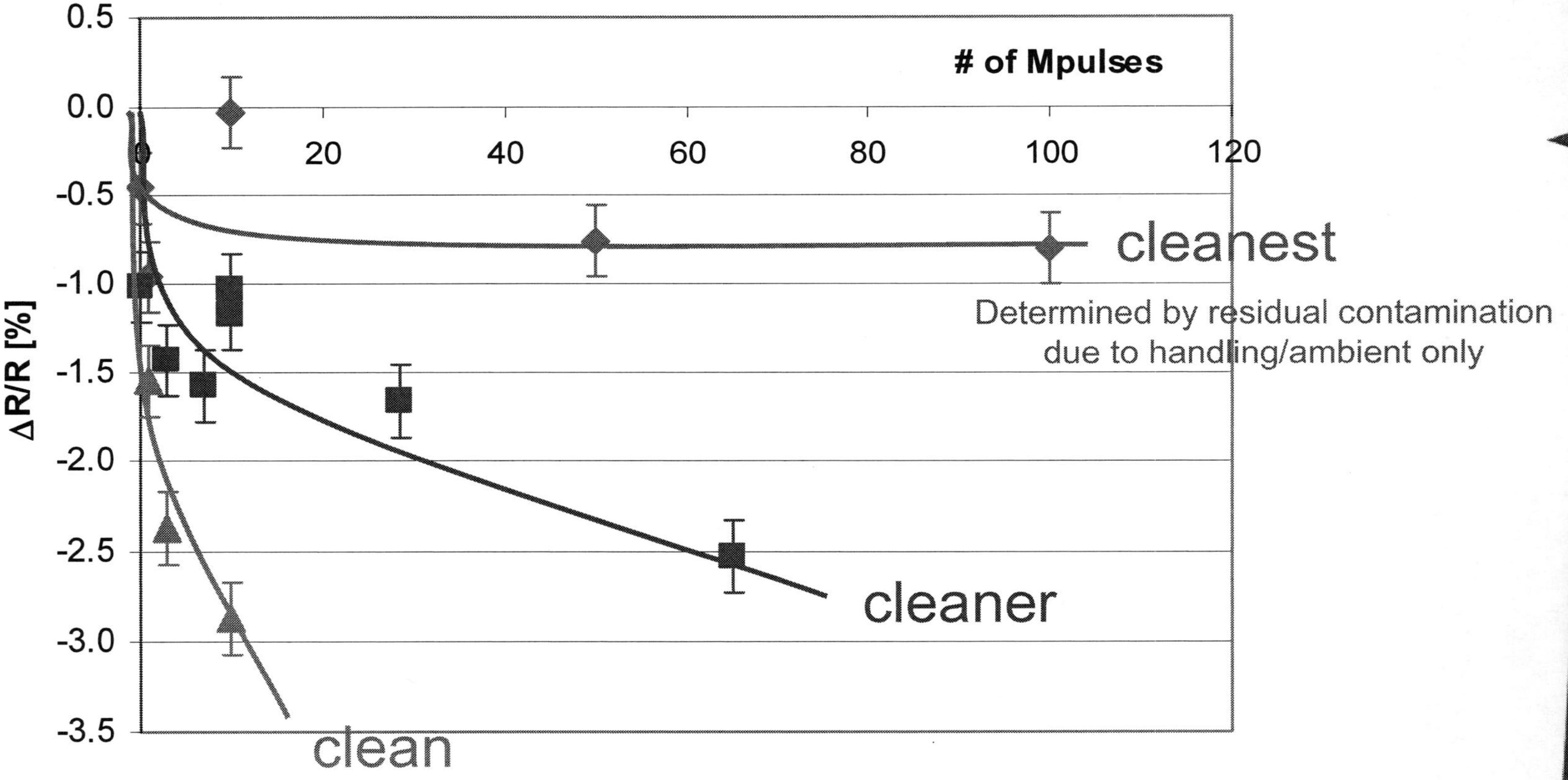

0.5
0.0
-0.5
-1.0
-1.5
-2.0
-2.5
-3.0
-3.5
ΔR/R [%]
0
20
40
60
80
100
120
# of Mpulses
cleanest
Determined by residual contamination due to handling/ambient only
cleaner
clean

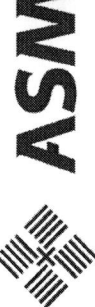

→ Large effect of vacuum cleanliness on carbon contamination

ASML

Optics reflectivity is not influenced by H_2O partial pressure, but is influenced by C-growth

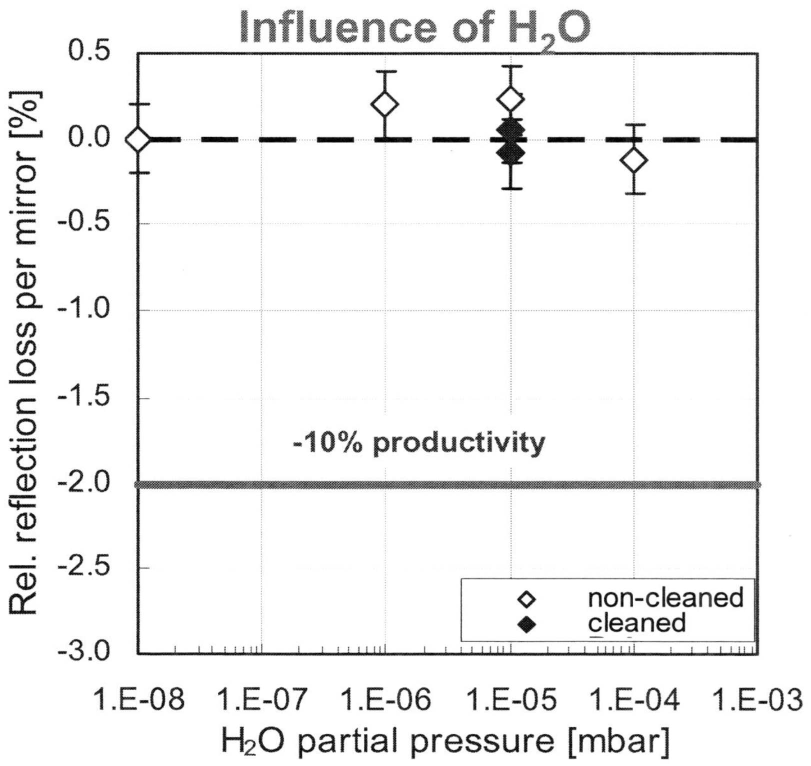

Influence of C-growth

Rel. reflection loss per mirror [%]

-10% productivity

◇ non-cleaned
◆ cleaned

Exposure time [Mpulse]

Influence of H_2O

Rel. reflection loss per mirror [%]

-10% productivity

◇ non-cleaned
◆ cleaned

H_2O partial pressure [mbar]

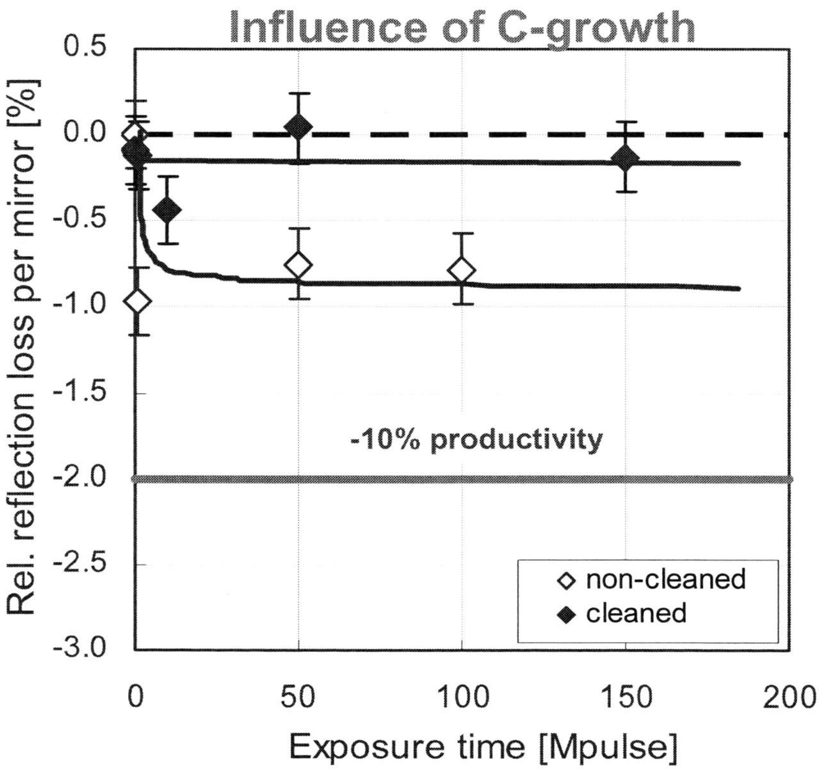

 ASML

Conclusion: H_2O at 1×10^{-6} is ok; $CxHy$ at 1×10^{-9} is critical

Optics lifetime strategy for EUV HVM is to balance cleaning with the rate of contamination

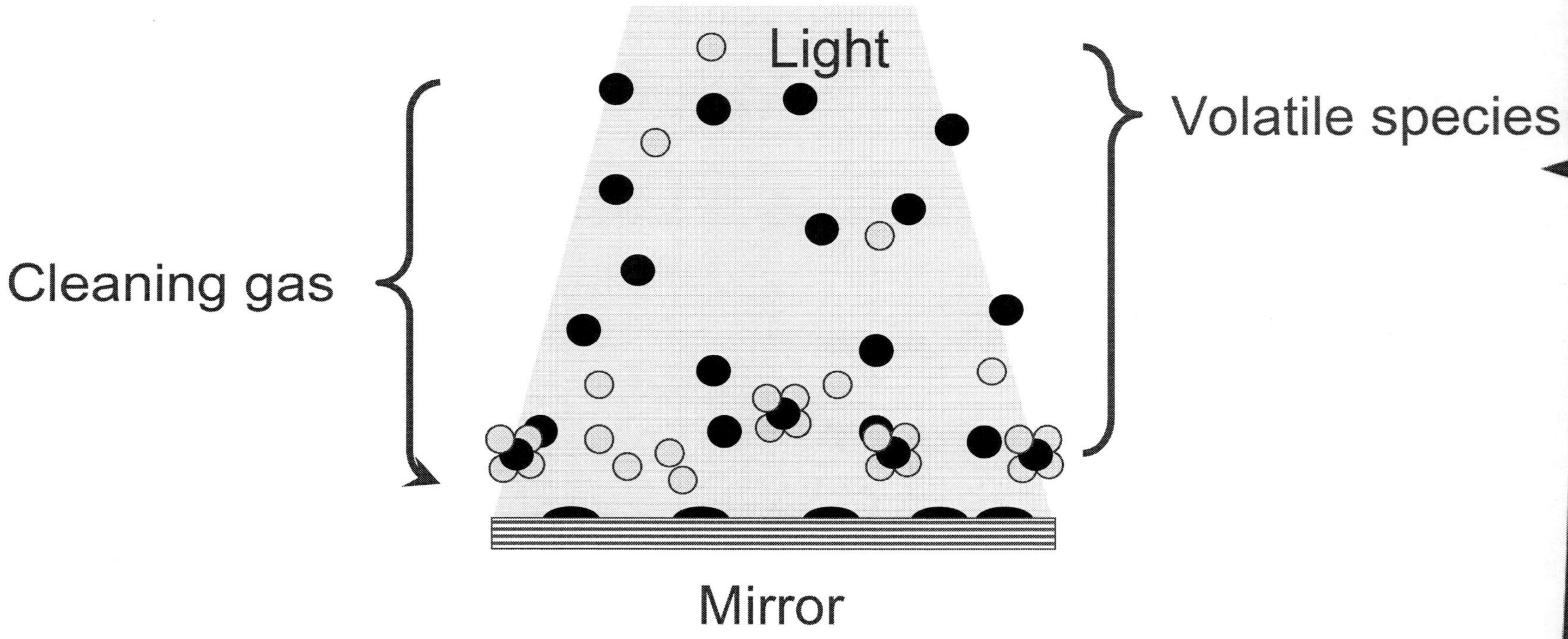

Volatile species

Light

Cleaning gas

Mirror

1. Diffusion causes transport to mirror, gas adsorbs to mirror

2. EUV causes carbon layer formation from adsorbed gas

3. Cleaning gas transforms carbon layer into volatile species

Three main actions are needed to achieve the target of 30,000 hours optics lifetime

1. Minimal carbon growth rate via good vacuum quality & materials selection

2. Local improved vacuum conditions

3. In-situ Cleaning

Achieving an Ultra Clean Vacuum (UCV) is highly dependent on material selection and procedures

- Outgassing budget for all modules have been determined

- Every material is tested for outgassing compliance

- UCV cleaning of all materials and modules is integrated as a part of Supply Chain and internal assembly

- UCV procedures and training for handling and manufacturing, in-house and at suppliers, is done

- Every module undergoes qualification for outgassing

High vacuum test chamber + mass spec for qualification

Bake-out of electronics shows compliance to specification

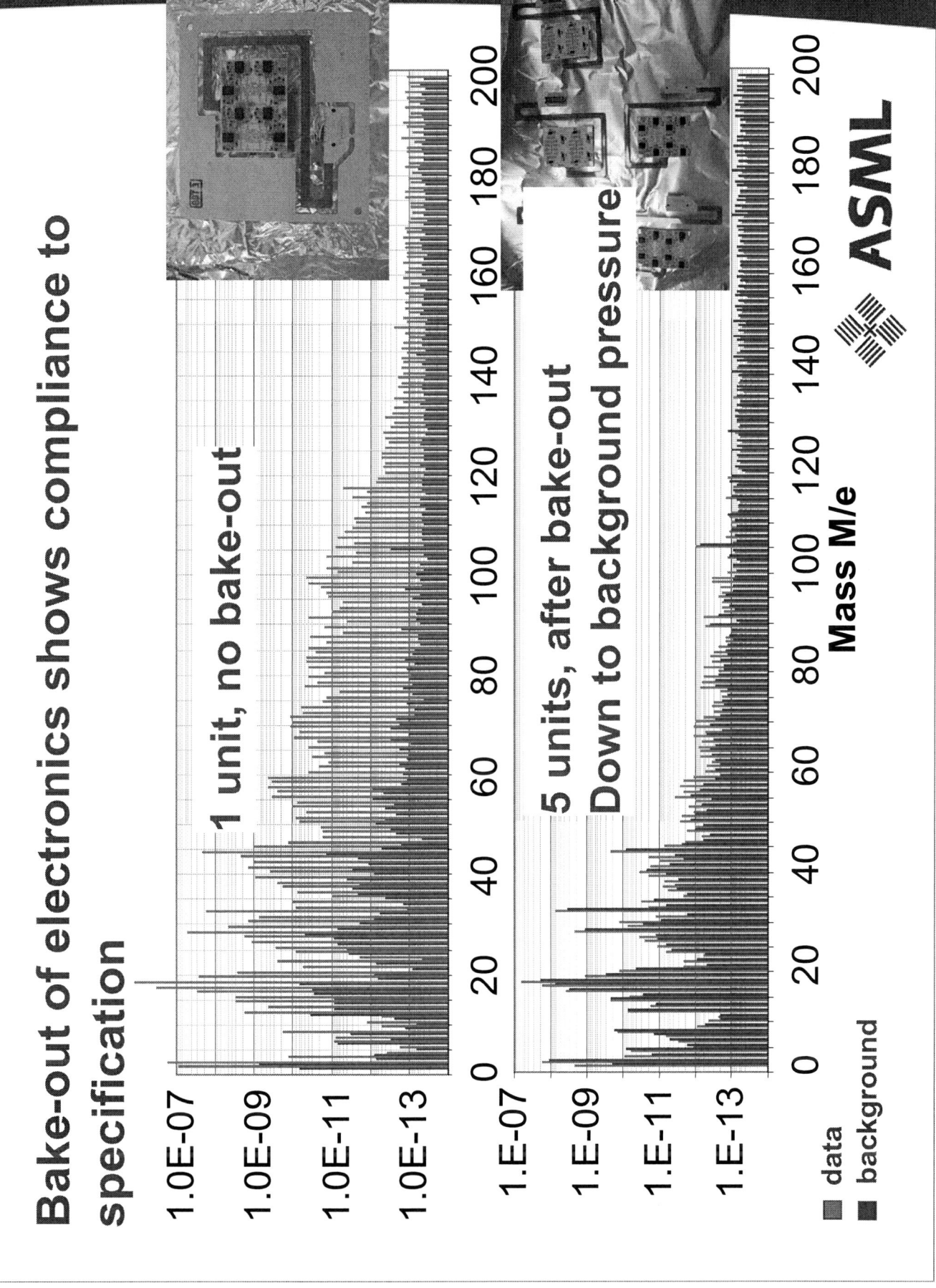

Multiple vacuum chamber concept for EUV platform is necessary

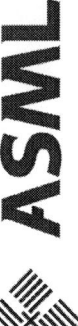

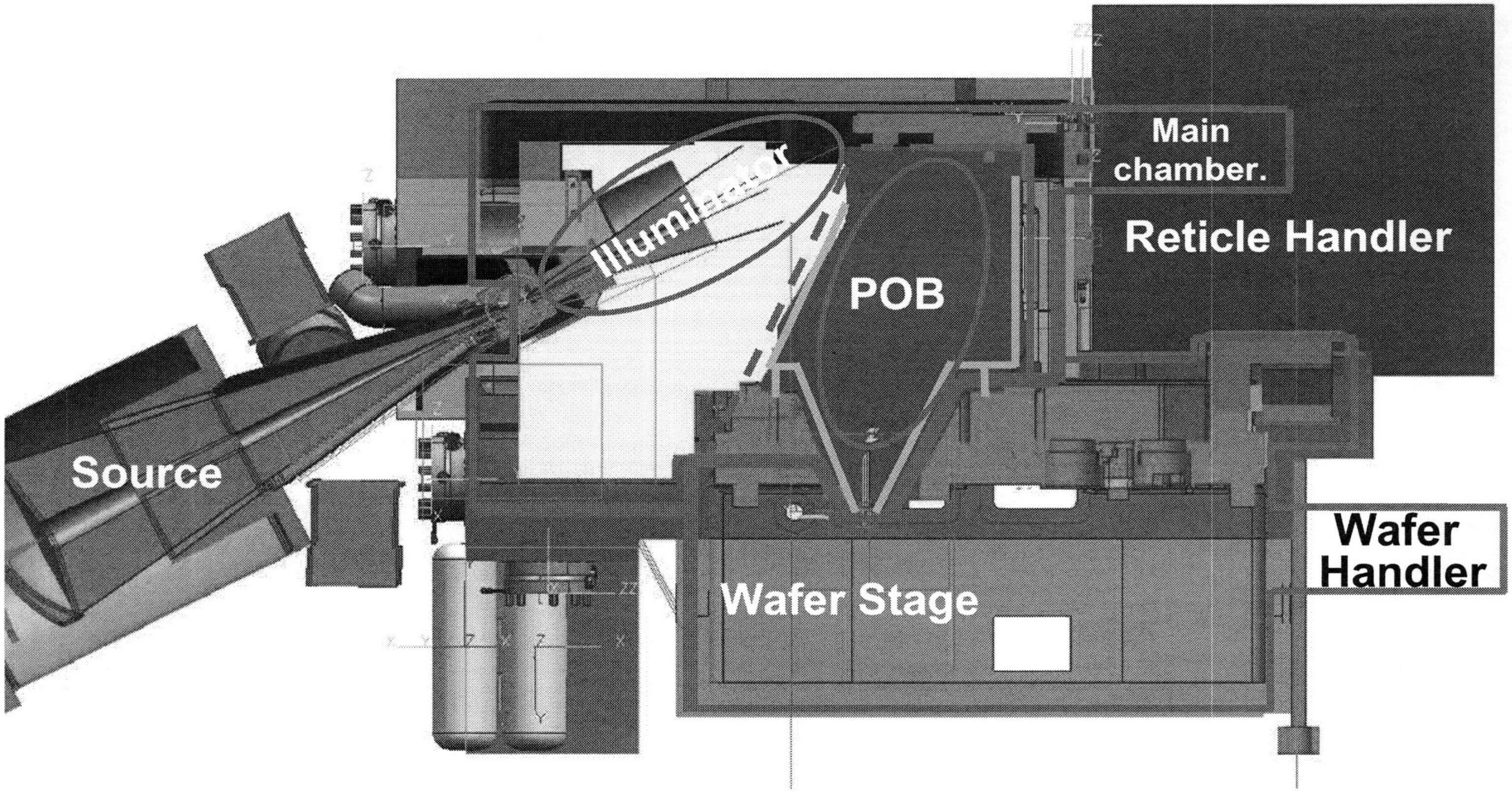
Main
chamber.
Reticle Handler
Illuminator
POB
Source
Wafer Stage
Wafer
Handler

Schematic example

ASML

AD1 main chamber RGA Spectra meets requirements (Spec increased to 200 amu)

Date: 30-1-2007

Total pressure	1.4E-7	mbar
Water	1.0E-7	mbar
Sum 45-100	9.8E-11	mbar
Sum 100-200	4.8E-11	mbar
Detection limit (per amu)	4.2E-13	mbar

Background AD1 30-1-2007 (p=1.4e-7)

Partial pressure [mbar]

m/z

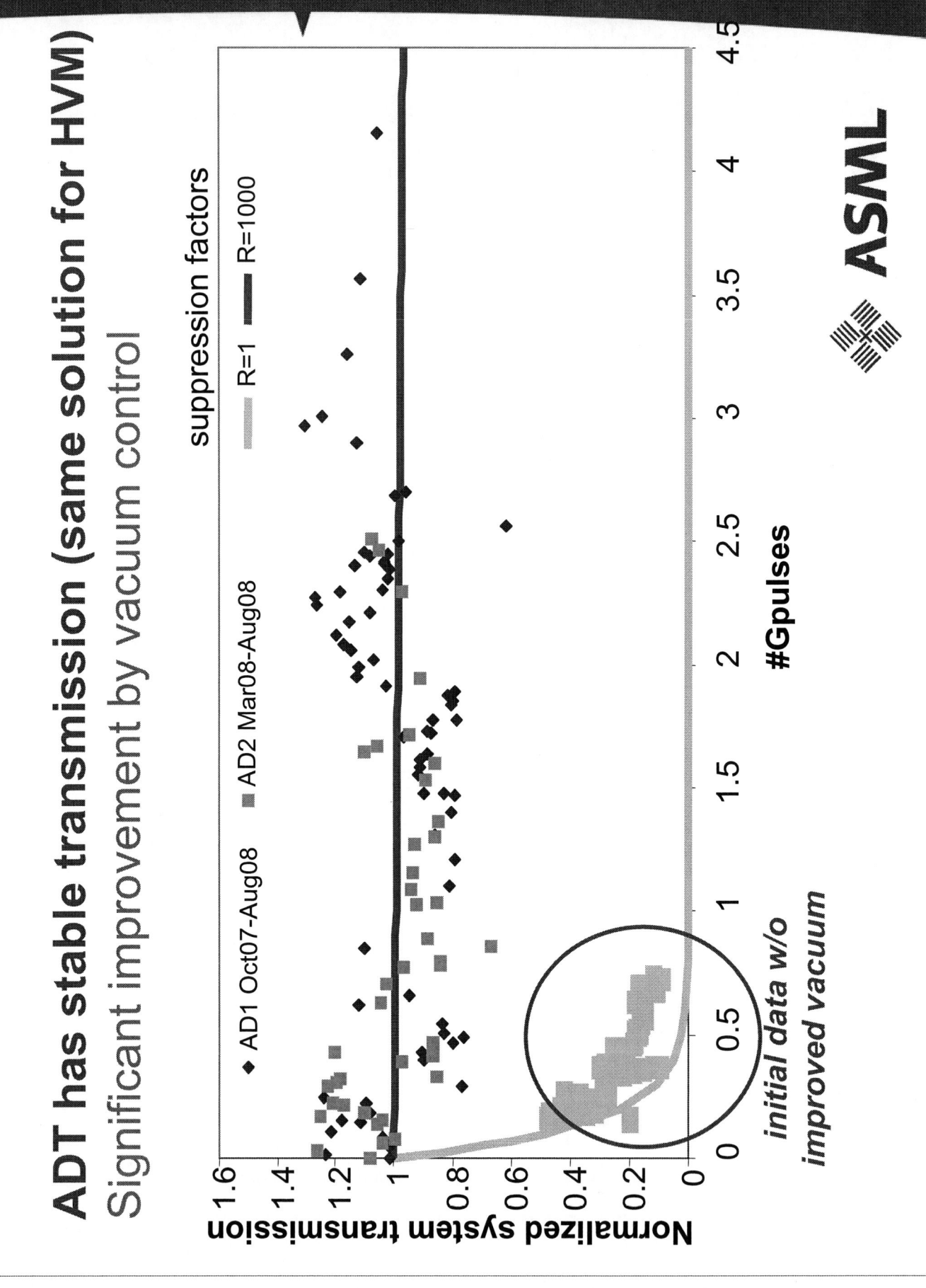

Improved vacuum is effective for even the highest energy density on mirrors

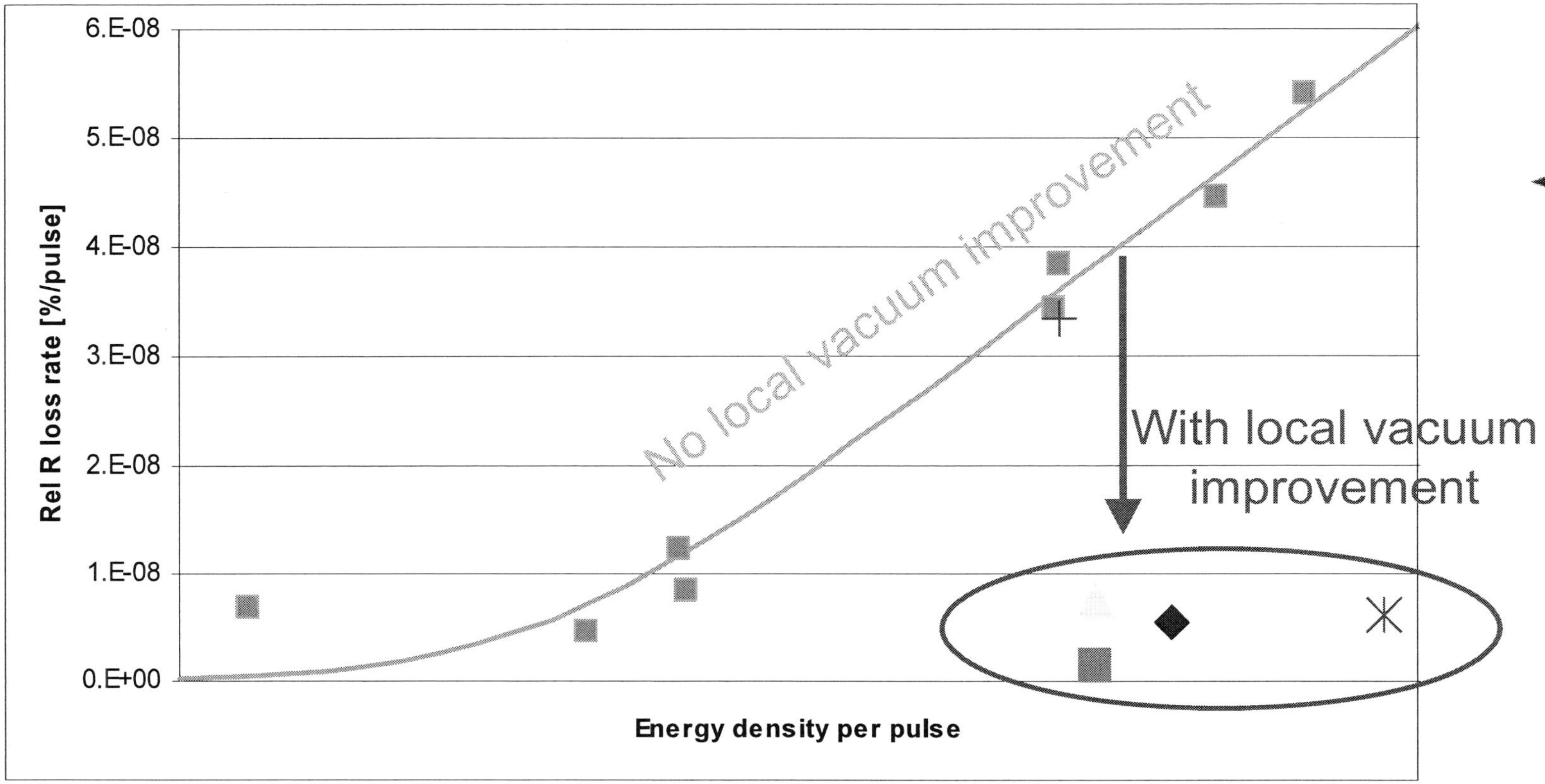

Rel R loss rate [%/pulse]

6.E-08
5.E-08
4.E-08
3.E-08
2.E-08
1.E-08
0.E+00

No local vacuum improvement

With local vacuum improvement

Energy density per pulse

→ Reflectivity loss greatly reduced after vacuum improvement

ASML

In-situ Cleaning System must be implemented

- Cleaning System design is optimized with respect to:
  - cleaning rate
  - limited heat load towards optical surfaces
  - cleanliness, low outgassing of materials
  - manufacturability and reproducibility
  - optics lifetime
  - reliability

- In-situ cleaning must be implemented for all optical surfaces (mirrors, sensors, fiducials)

Cleaning of carbon from mirrors has been demonstrated

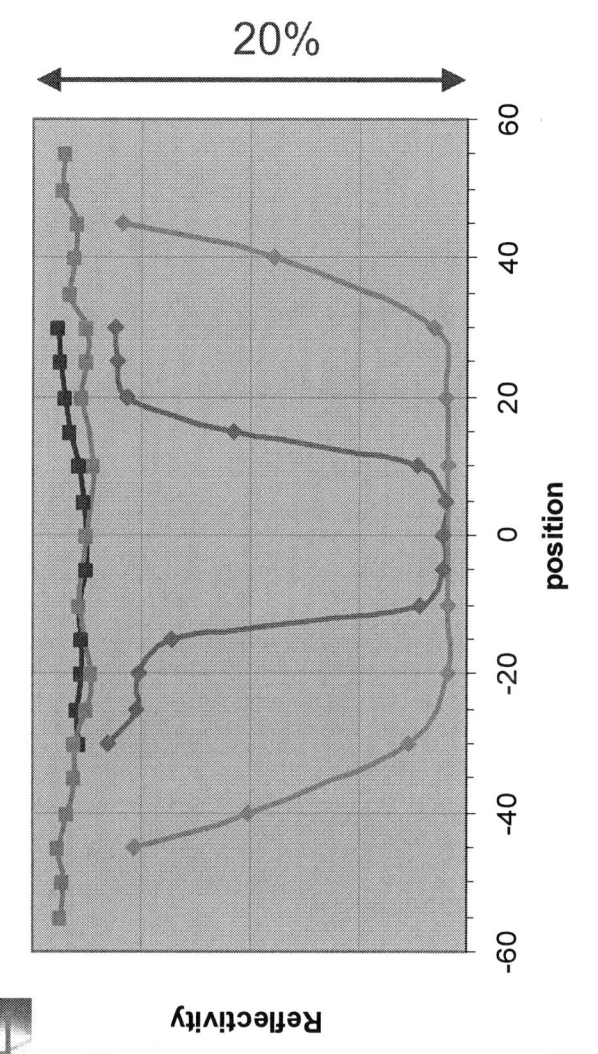

→ Reflectivity fully restored after cleaning

Recovery of reflectivity of accelerated oxidation (high p_{O2}) has been demonstrated

MLM oxidised by exposure to elevated Oxygen partial pressure.

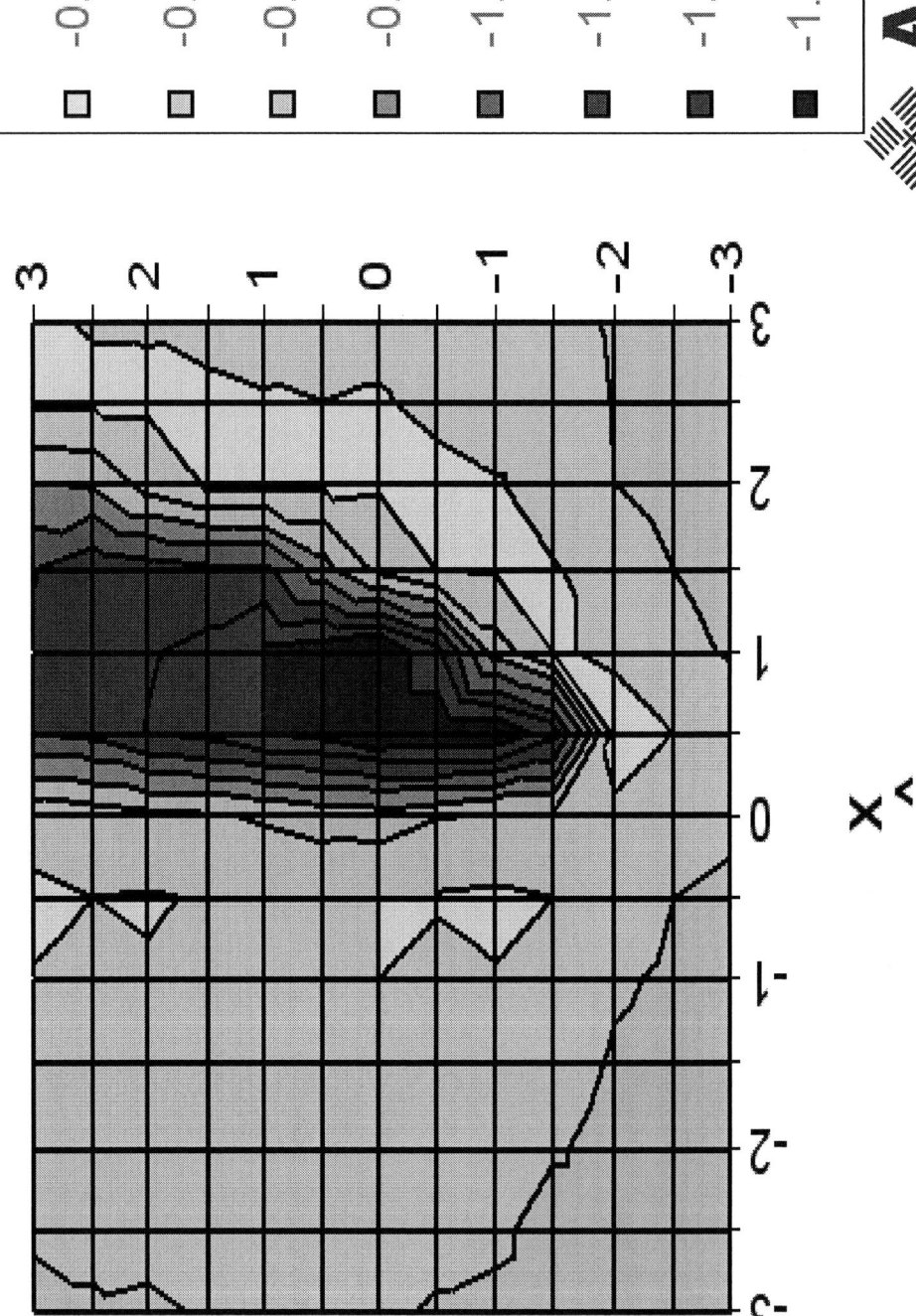

Cleaning tests have validated cleaning efficiency and control of potential cleaning side effects

Set-up for accelerated cleaning experiments

220 nm equivalent carbon contamination cleaned without degradation over the entire mirror surface

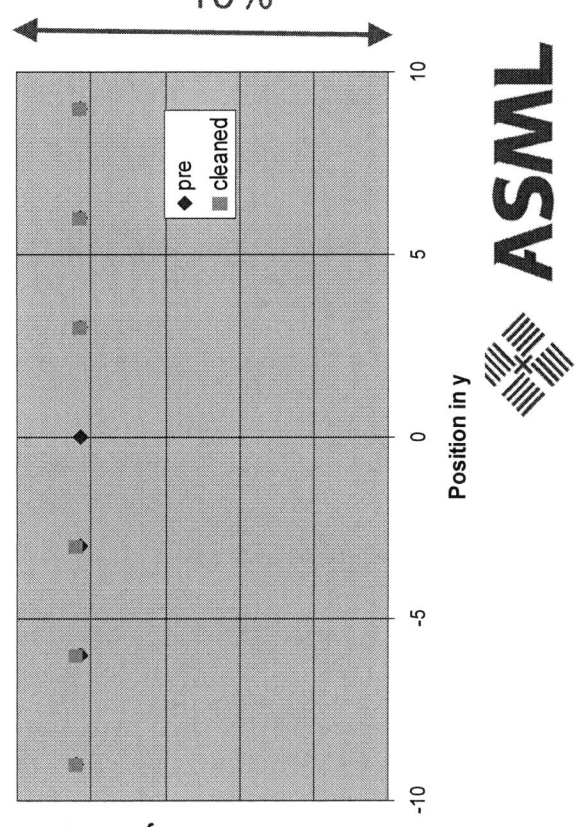

Preventing carbon contamination of optics needs good vacuum / local protection / in-situ cleaning

Residual carbon growth
- <0.1nm / day / mirror

In-situ cleaning of residual contamination
- All mirrors, sensors, etc
- <10min / day

Vacuum quality
- Pre treatment / cleaning
- Machine construction
- Handling
- Resist outgassing

Local protection near optics
- Suppression of contaminates
- Gas lock for wafer stage / resist

Summary

- Without appropriate measures, EUV induced carbon contamination will limit optics lifetime in HVM tools

- Appropriate measures are:

  - Strict material selection / pretreatment criteria

  - Local optimized vacuum conditions around the optics

  - Implementation of in situ cleaning for all optics (mirrors, sensors, fiducials)

- Both ADT data and data from experimental set-ups supports the planned strategy for controlling optics contamination for HVM

Acknowledgement

ASML:
Monika Lubomska, Antoine Kempen, Luc Stevens, John de Kuster
Zeiss:
Stefan Schmidt, Moritz Becker, Thomas Stein
TNO:
Rik Jansen, Herman Bekman, Michel van Putten, Jetske Stortelder, Emiel van Kimmenade
FOM:
Eric Louis, Robbert vd. Kruijs, Tim Tsarfati, Fred Bijkerk
Philips:
Peer Zalm
PTB:
Roman Klein, F. Scholze, Christian Laubis
MSU/ISAN:
Vladimir Ivanov
ALS:
Eric Gullikson

Effective Debris Detection, Mitigation and Cleaning Methods for Source - Collector Optics

Prof. David N. Ruzic

Dr. Ramasamy Raju, J. Sporre, H. Shin, W. M. Lytle, Dr. M. J. Neumann

2008 International Symposium on Extreme Ultraviolet Lithography, Lake Tahoe,

Sep 28-Oct 1, 2008

Center for Plasma Material Interactions
http://starfire.ne.uiuc.edu

Outline

- Introduction

- Source Contamination
  - Damage Mechanism
  - Ion and Neutral Debris measurement
  - Erosion and life time prediction
  - Debris detection at Intermediate Focus (IF)

- Collector Optics Contamination
  - Cleaning of Sn by RIE

- Particle Contamination of Mask and Wafer
  - PACMAN cleaning of nano particles

- Line Edge Roughness

- Conclusions

- Acknowledgements

Center for Plasma Material Interactions
http://starfire.ne.uiuc.edu

Introduction

- Development of a reliable source collector module (SoCoMo) to channel clean EUV photons to a EUV scanner is highly crucial for HVM

But, there are some hurdles to achieve this goal:

- Generation of energetic ions and neutrals
- Deposition of condensable fuel materials
- Carbon and oxygen contamination on optics surfaces
- Nano-particle contamination on mask and wafer
- Line Edge Roughness resulting from too little source power

So, effective debris detection, mitigation and reliable cleaning techniques are needed to achieve HVM

Center for Plasma Material Interactions
http://starfire.ne.uiuc.edu

Outline

- Introduction
- **Source Contamination**
  - Damage Mechanism
  - Ion and Neutral Debris measurement
  - Erosion and life time prediction
  - Debris detection at Intermediate Focus (IF)
- Collector Optics Contamination
  - Cleaning of Sn by RIE
- Particle Contamination of Mask and Wafer
  - PACMAN cleaning of nano particles
- Line Edge Roughness
- Conclusion
- Acknowledgements

Center for Plasma Material Interactions
http://starfire.ne.uiuc.edu

Damage mechanism

Debris from the plasma source is the critical issue:

EUV Pinch

- EUV Photons
- Fast Ions
- Electrode Debris
- Condensables
- Chemical attack

Collector Optics

Intermediate Focus

Degradation of the collector optic can be measured in terms of minutes for some HVM conditions.

Fast ions generated in the pinch can lead to significant collector erosion.

Electrode materials generated during the plasma pinch from surrounding surfaces.

Condensable metal vapor from advanced fuels (*Sn*).

Chemical interactions between collector mirror and debris (*Sn* and *C and O contamination*).

Degradation at elevated temperatures of because of the enhanced thermal interdiffusion of the high and low index materials within the mirror structure.

Lopez, E.V., et al., *"Origins of debris and mitigation through a secondary RF plasma system for discharge-produced EUV sources,"* Microelectronic Engineering, 2005. 77: p. 95-102.

Center for Plasma Material Interactions
http://starfire.ne.uiuc.edu

ILLINOIS
UNIVERSITY OF ILLINOIS AT URBANA-CHAMPAIGN

Ion debris measurement

XTREME Commercial EUV Emission Device (XCEED)

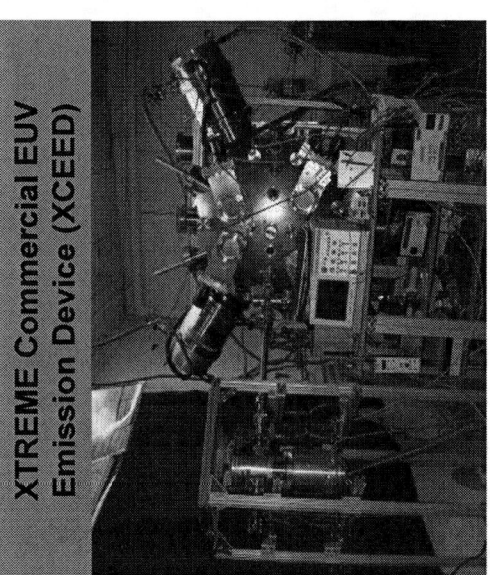

Electrostatic Spherical Sector Ion Energy Analyzer (ESA)

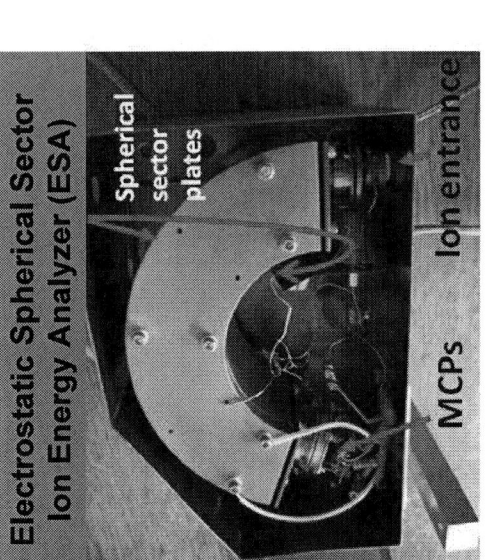

Spherical sector plates

MCPs   Ion entrance

Experimental Set-up for ESA Calibration

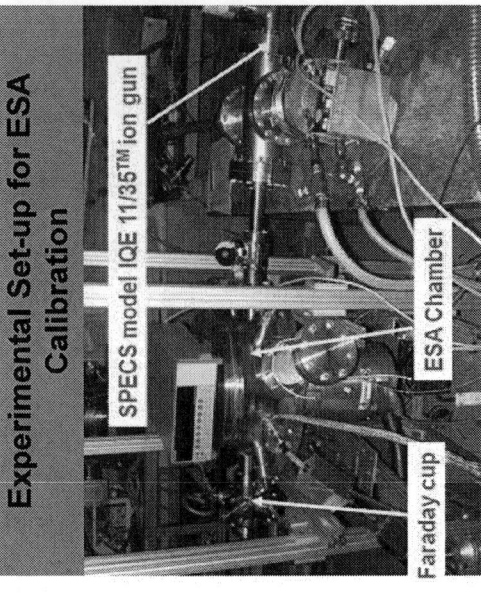

SPECS model IQE 11/35™ ion gun

Faraday cup

ESA Chamber

$$\frac{E}{q} = -\frac{\Delta V}{(r_1/r_2 - r_2/r_1)}$$

Ions are discriminated based on energy-to-charge ratio

The spherical sectors are charged to equal voltages of opposite signs

The electric field created inside the device, turns ions by 160°, where they impact the micro-channel plates

ESA is calibrated using an Ion Gun and a Faraday Cup

Multiple Charge State Resolution

E/q=2.667 keV

Distance to source = 150 cm

$t = \frac{d \cdot \sqrt{m}}{\sqrt{2E}}$

Keith C. Thompson, E. L. Antonsen, M. R. Hendricks, B. E. Jurczyk, M. Williams, D. N. Ruzic, "Experimental test chamber design for optics exposure testing and debris characterization of a xenon discharge produced plasma source for extreme ultraviolet lithography", *Microelectronic Engineering*, 83 (2006) 476.

Center for Plasma Material Interactions
http://starfire.ne.uiuc.edu

ESA capabilities

Charge resolved measurement

Isotope separation

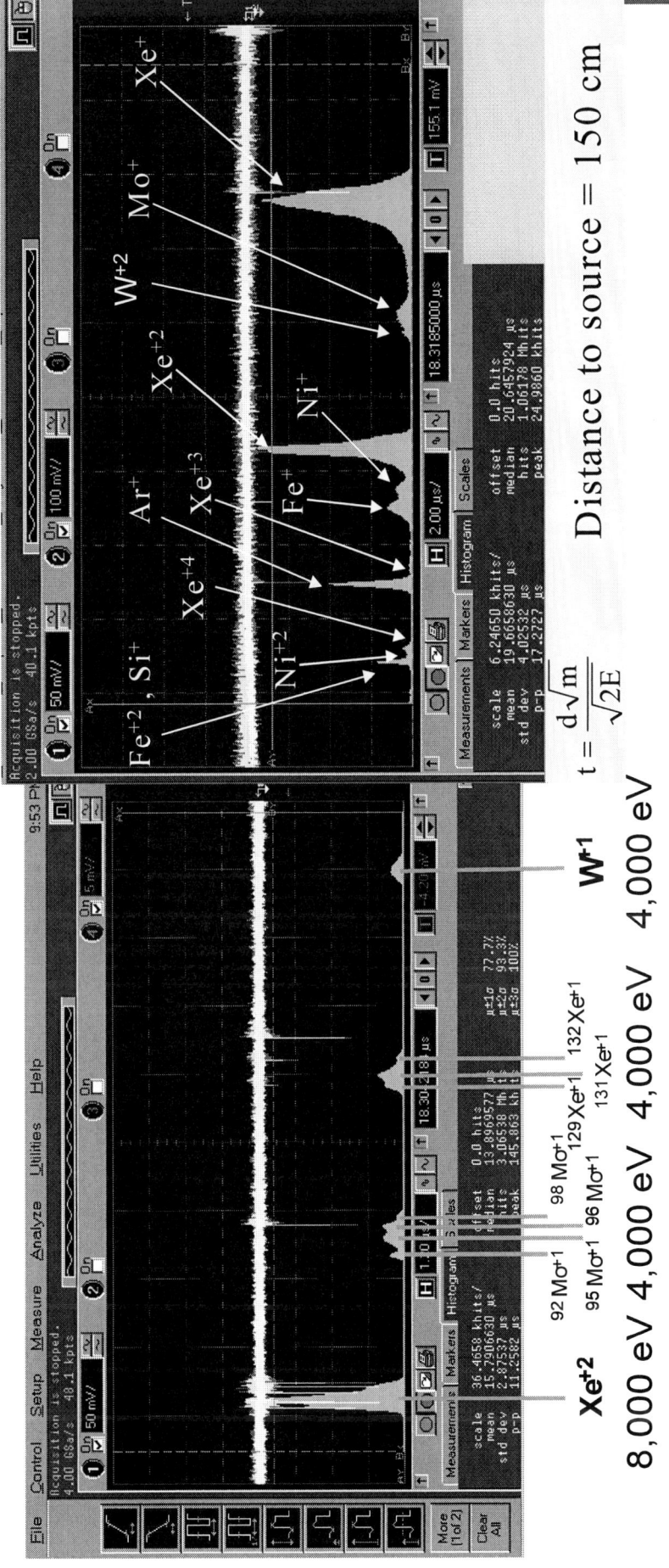

$$t = \frac{d\sqrt{m}}{\sqrt{2E}}$$

Distance to source = 150 cm

8,000 eV   4,000 eV   4,000 eV   4,000 eV   4,000 eV

TOF-ESA sensitivity is high enough to resolve isotopic abundance of Mo and Xe present in the debris field.

TOF-ESA shows higher charge states and can show the impact of mitigation schemes on the debris field for collector lifetime estimation.

Center for Plasma Material Interactions

http://starfire.ne.uiuc.edu

ILLINOIS
UNIVERSITY OF ILLINOIS AT URBANA-CHAMPAIGN

Absolute Ion Spectra

A single full energy analysis of the Xe fueled DPP EUV source reveals the various energetic ion species present in the plasma debris.

K. C. Thompson, S. N. Srivastava, J. Sporre and D. N. Ruzic, *International EUVL symposium, Sapporo, Japan* (2007)

Remember this shape!

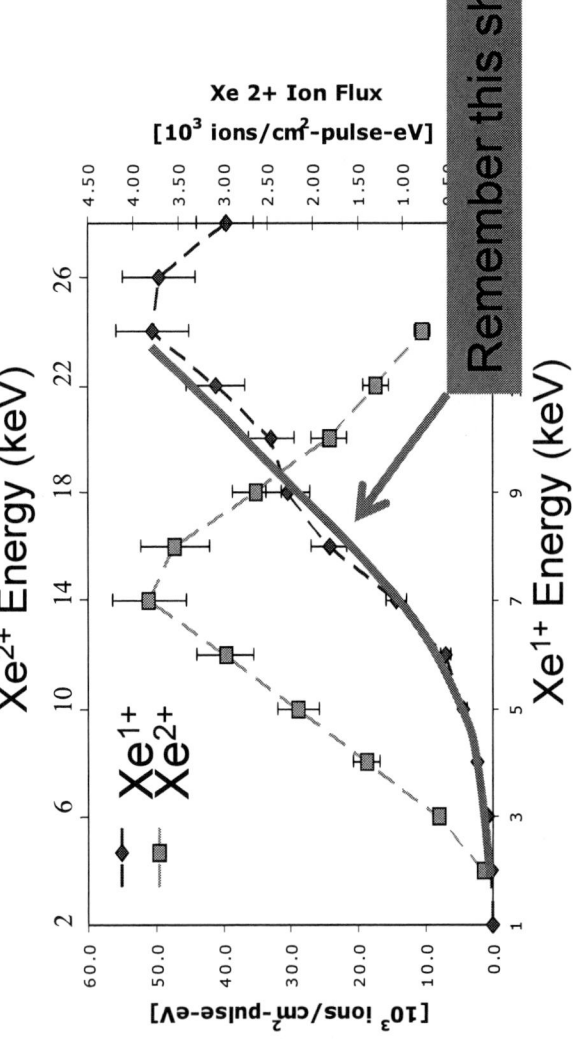

Different ion energies are measured by changing the diverting plate potential. This allows for an energy spectral analysis of ion debris.

Center for Plasma Material Interactions
http://starfire.ne.uiuc.edu

Do we have only ionic debris?

- No. Many debris mitigation systems may just neutralize the ions and not alter their energy.
- Also, low-energy neutrals, all the way down to thermal, can cause deposition and therefore also degrade reflectivity. In addition there seem to be other effects too......

Xe^{8+} to Xe^{10+} radiates

Recombination (and acceleration) occurs during expansion

We see Xe^{4+}, Xe^{3+}, Xe^{2+}, Xe^+

Shouldn't Xe^0 be there too?

Specie	Predicted Time (us)	Actual Time (us)
Fe2+, Si+	11.04, 11.07	11.15
Ni2+	11.32	11.31
Xe4+	11.97	11.87
Xe3+	13.82	13.77
Fe+	15.61	15.33
Ni+	16.00	15.80
Xe2+	16.93	16.93
Xe+	23.94	23.59

$E/q = 2.667$ keV

Distance to source = 150 cm

$$t = \frac{d\sqrt{m}}{\sqrt{2E}}$$

Center for Plasma Material Interactions

http://starfire.ne.uiuc.edu

Neutral Particle Detection

Using micro-channel plates similar to those used in the ESA tool, detect neutral particles by diverting ions from the beam before allowing the remaining particles to impact. ESA spherical plates are used to divert ions

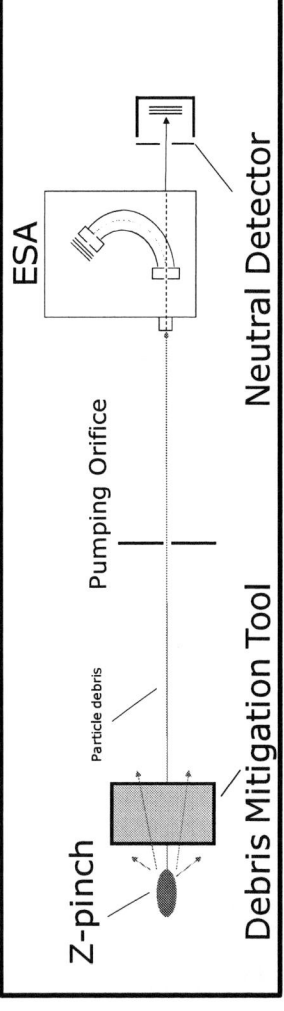

Z-pinch

Particle debris

Pumping Orifice

ESA

Debris Mitigation Tool

Neutral Detector

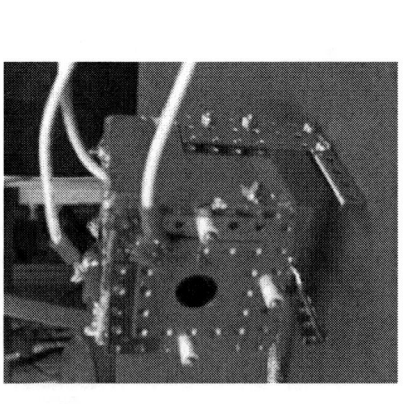

The Neutral Detector (ND) MCPs can be calibrated in a way similar to that for the calibration of the ICE machines

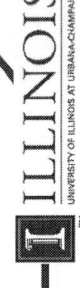

Center for Plasma Material Interactions

http://starfire.ne.uiuc.edu

Energy Spectra

Calibrated, measured, total, ions and neutral debris with no buffer gas

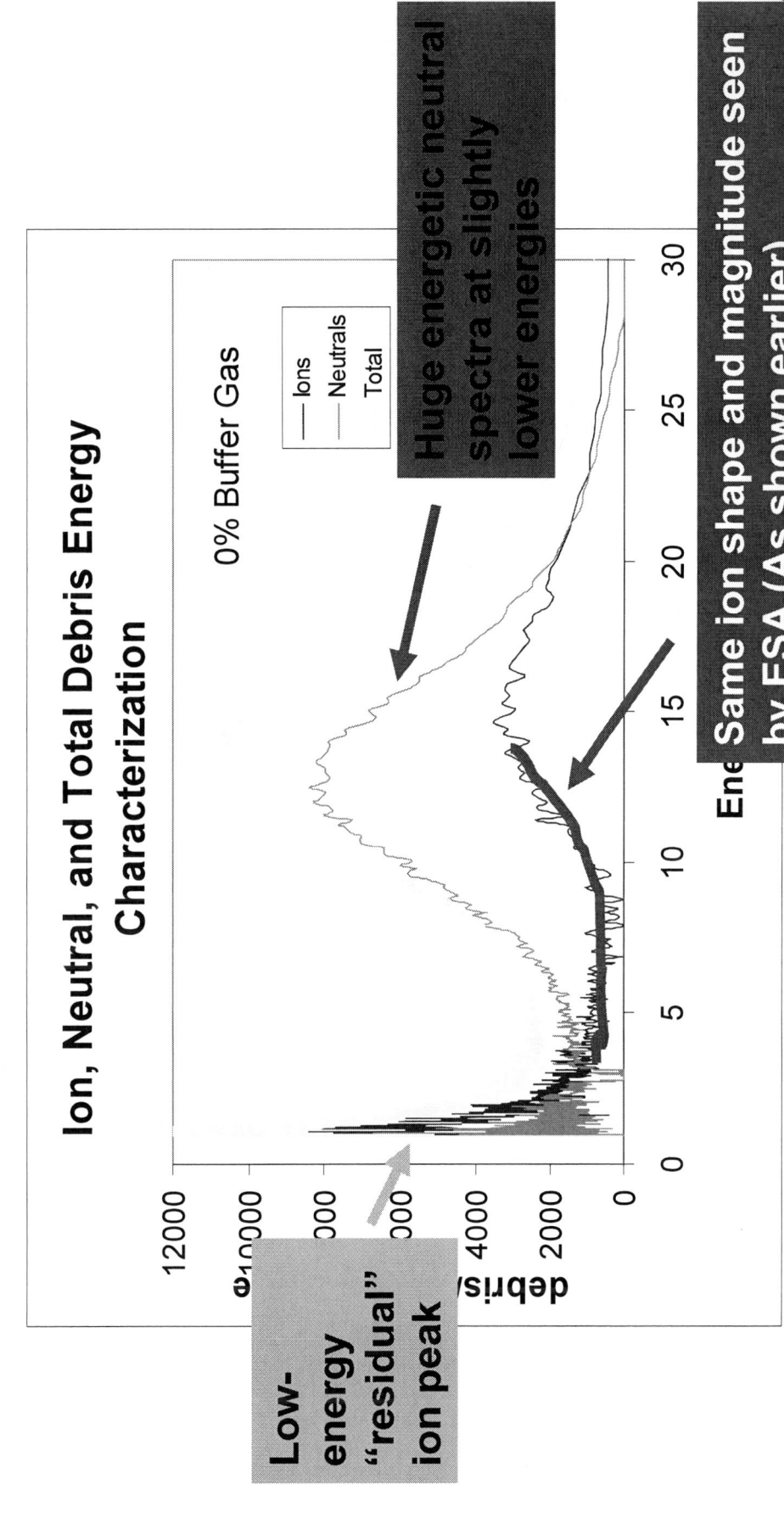

Ion, Neutral, and Total Debris Energy Characterization

0% Buffer Gas

Legend: Ions — Neutrals — Total

Huge energetic neutral spectra at slightly lower energies

Low-energy "residual" ion peak

Same ion shape and magnitude seen by ESA (As shown earlier)

Center for Plasma Material Interactions

http://starfire.ne.uiuc.edu

ILLINOIS
UNIVERSITY OF ILLINOIS AT URBANA-CHAMPAIGN

Debris mitigation – Buffer gas

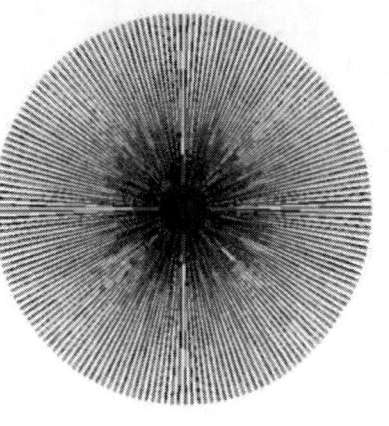

- **Collimated Debris Tool/Buffer Gas**
  - Designed by XTREME Technologies
  - Similar to collimator used in CVD/PVD processes

Debris Mitigation comes at the cost of decreased light intensity

Debris Mitigation comes at the cost of decreased light intensity

Buffer gas flow dramatically reduces energetic ion flux when the highest gas pressures are inside a foil trap so that the gas scattering deflects the ions into the foils.

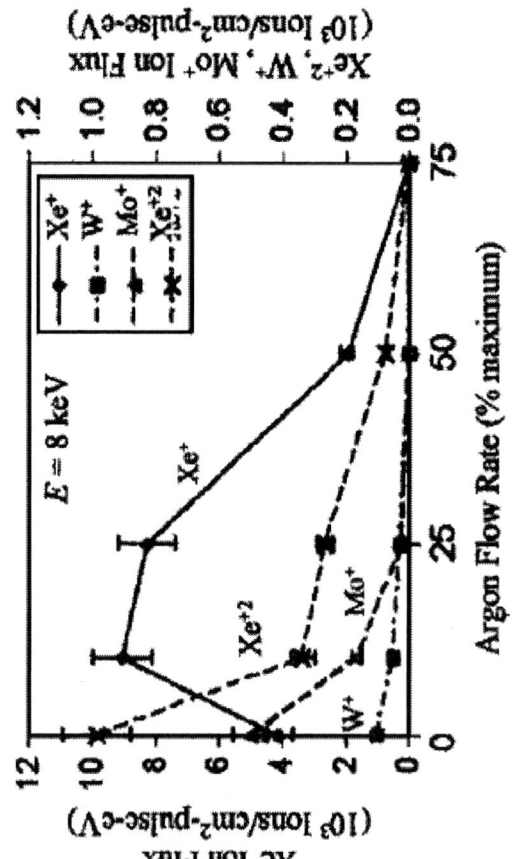

Erik L. Antonsen, K. C. Thompson, M. R. Hendricks, D. A. Alman, B. E. Jurczyk, D. N. Ruzic, "Ion debris characterization from z-pinch extreme ultraviolet light source", *Journal of Applied Physics*, 99 (2006) 063301.

D. N. Ruzic, "Chapter 36. Origin of Debris in EUV Sources and its Mitigation" in *EUV Sources for Lithography*, ed. Vivek Bakshi, SPIE Press, Bellingham, Washington, 2006.

Center for Plasma Material Interactions
http://starfire.ne.uiuc.edu

ILLINOIS
UNIVERSITY OF ILLINOIS AT URBANA-CHAMPAIGN

Electric field ion mitigation

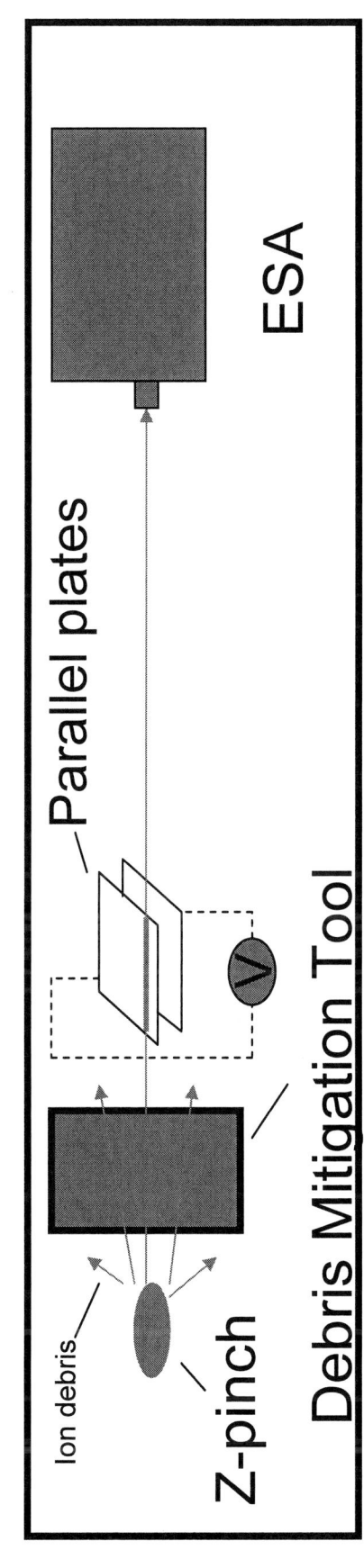

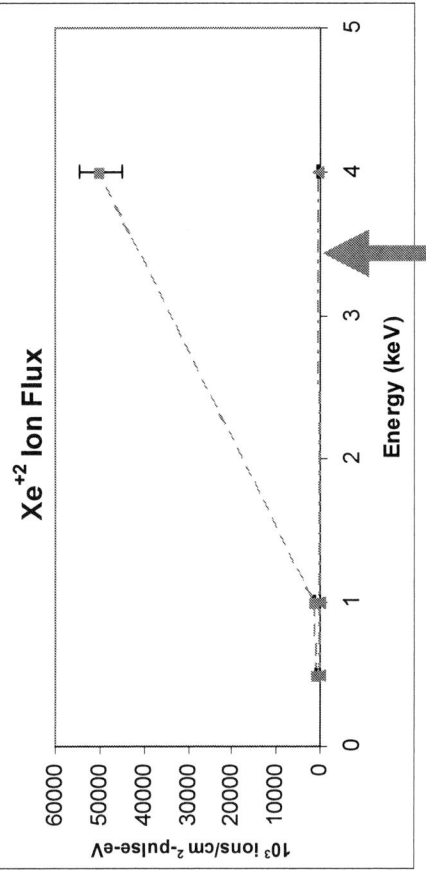

Decrease by a factor of 88 for 4 keV Xe^{2+} ions using electric field post-foil-trap mitigation technique

Decrease by a factor of 3.75 for 4 keV Xe^{+} ions using electric field post-foil-trap mitigation technique

Center for Plasma Material Interactions
http://starfire.ne.uiuc.edu

I ILLINOIS
UNIVERSITY OF ILLINOIS AT URBANA-CHAMPAIGN

Ion energy reduction with H$_2$ (INERT)

By adding a small amount of lower-mass fuel (like hydrogen) the pinch expansion parameters are changes such that the heavy ions are not accelerated to the same energies. This works independent of collector type.

Adding 5% H$_2$ gas – same EUV signal

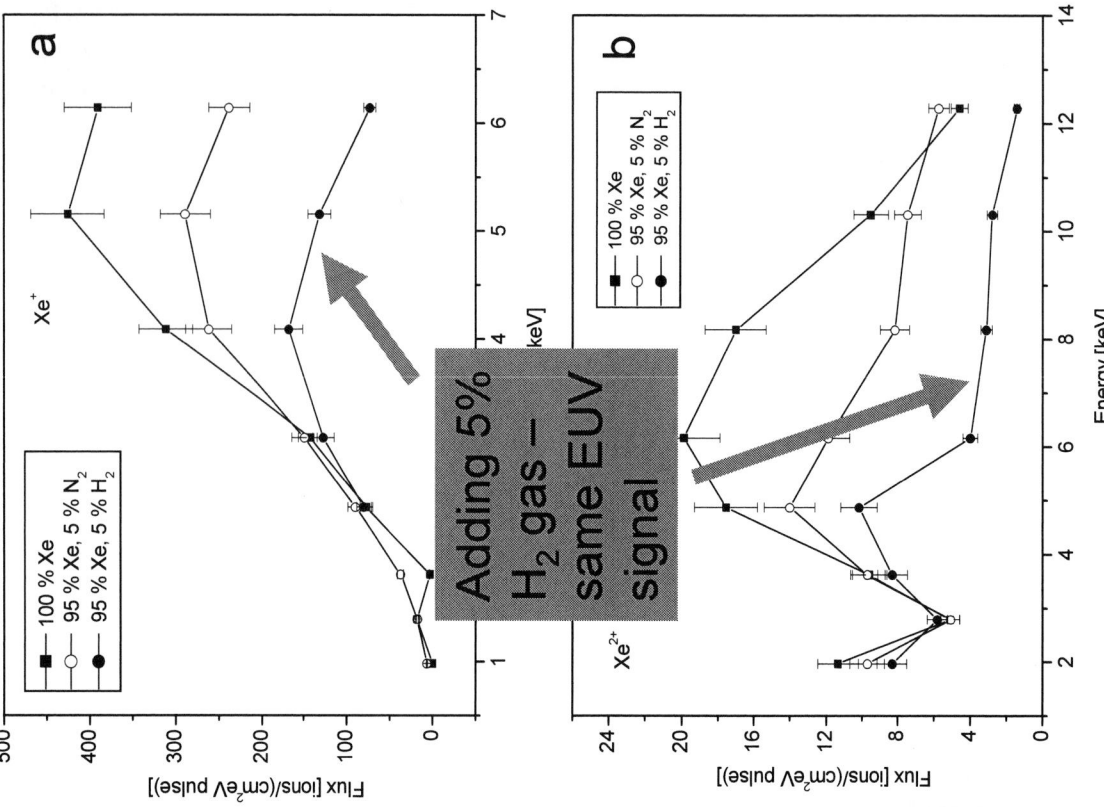

XTS Commercial EUV Source

UIUC Custom Test Chamber

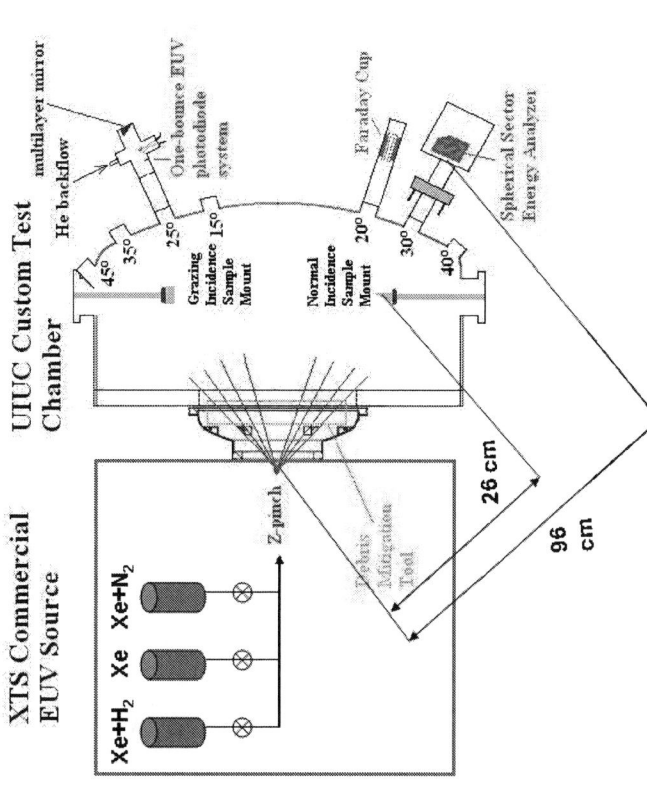

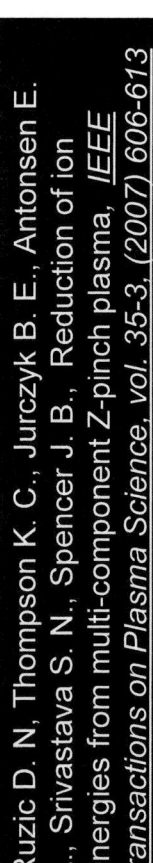

Ruzic D. N, Thompson K. C., Jurczyk B. E., Antonsen E. L., Srivastava S. N., Spencer J. B., Reduction of ion energies from multi-component Z-pinch plasma, *IEEE transactions on Plasma Science, vol. 35-3, (2007) 606-613*

Center for Plasma Material Interactions
http://starfire.ne.uiuc.edu

What problems will these energetic ions and neutrals cause?

Reflectivity degradation from collector optics is an effect of debris interaction with collector system

- Sputtering will occur, destroying multilayers, or roughening grazing incident collectors

- Energy flux to the surfaces can promote carbon, oxygen contamination

Center for Plasma Material Interactions
http://starfire.ne.uiuc.edu

Xe-Exposure

(Measured and Predicted Erosion)

Measured

SEM was performed to measure
Erosion for eight samples

Erosion ranges from 10-50 nm
for different materials

Predicted

$$Y_{sputt.} = \frac{atoms}{ion} \longrightarrow \textbf{SRIM Calculation}$$

$$N = \text{total atoms eroded} = Y_{sputt.} \cdot Ion_{flux}\left(\frac{1}{cm^2 \cdot shots}\right) \cdot shots$$

$$\Delta T(cm) = Erosion = N \cdot \frac{m}{\rho}\left(\frac{1}{cm^2 shots} \cdot \frac{g}{g/cm^3}\right) \cdot shots$$

Agreement is generally good. Means "accelerated lifetime testing" is accomplished simply by making a flux measurement.

Missing predicted erosion flux is from energetic neutral attoms

ESA measured

Center for Plasma Material Interactions
http://starfire.ne.uiuc.edu

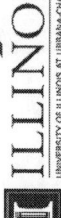

Calculated results for Si/Mo

Sputtering yield calculation (SRIM) using SRIM from Ru, Si and Mo

MLM Material	Xe (6keV @10°)	Sn (6keV @10°)
Ruthenium	4.38	4.65
Silicon	2.87	2.99
Molybdenum	1.94	1.93

S. N. Srivastava, K.C. Thompson, E. L. Antonsen, H. Qiu, J.B Spencer, D. Papke, D. N. Ruzic, "Lifetime measurements on collector optics from Xe and Sn extreme ultraviolet sources" J. of Applied Physics, 102, 023301, (2007)

Xe and Sn fluxes needed to erode one layer of Ru, Si and Mo

MLM Material	Xe (6keV @10°)	Sn (6keV @10°)
Ruthenium	2.34×10^{15}	2.20×10^{15}
Silicon	6.17×10^{15}	5.93×10^{15}
Molybdenum	1.06×10^{16}	1.07×10^{16}

Thus the fluence required to remove one bi-layer of Si/Mo is 6.17×10^{15} $+1.06 \times 10^{16}$

$$= 1.677 \times 10^{16} \text{ particles/cm}^2$$

Center for Plasma Material Interactions
http://starfire.ne.uiuc.edu

Lifetime prediction

9 million shots erode one bi-layer with no mitigation

Therefore reflectivity is reduced by 10% in 6 minutes (at 10kHz)

Clearly, mitigation schemes are needed, and are present. Source suppliers show that they have sufficient mitigation to achieve reasonable lifetimes.

However, what goes into the scanner other than EUV photons as a result of these mitigation systems?

Center for Plasma Material Interactions
http://starfire.ne.uiuc.edu

Debris measurement at IF

Measurement at out of line of sight from the pinch:

Ion and neutral debris: Farady cup, ESA and Neutral detector

Carbon and oxygen contamination: Si witness plate exposure

Presence of plasma: Triple Langmuir Probe

Collector Mock-up

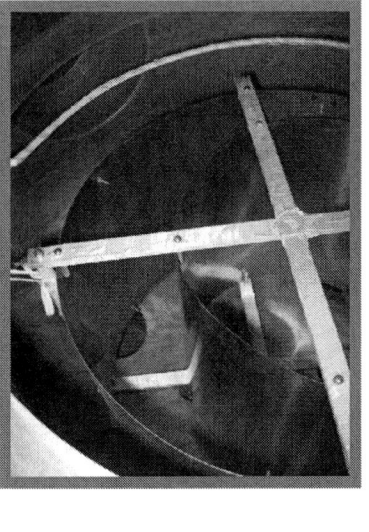

XTS 13-35 device

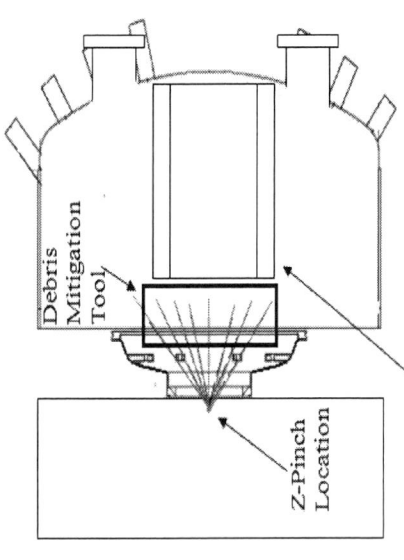

XTS 13-35
EUV Source

XCEED
Test Chamber

Debris Mitigation Tool

Z-Pinch Location

2-Shell Mock Collector Optic

- The XCEED test chamber has four 6"CF gate valves located equidistant from the chamber centerline.
- A mock collector optic is installed and completely obscures the line of sight between the pinch and these gate valves.
- The Z-pinch was run with 200 sccm Ar, 85 sccm Xe at 350 Hz for each experiment

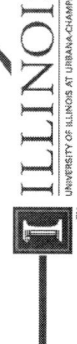

Center for Plasma Material Interactions
http://starfire.ne.uiuc.edu

Charged particle detection
Faraday cup

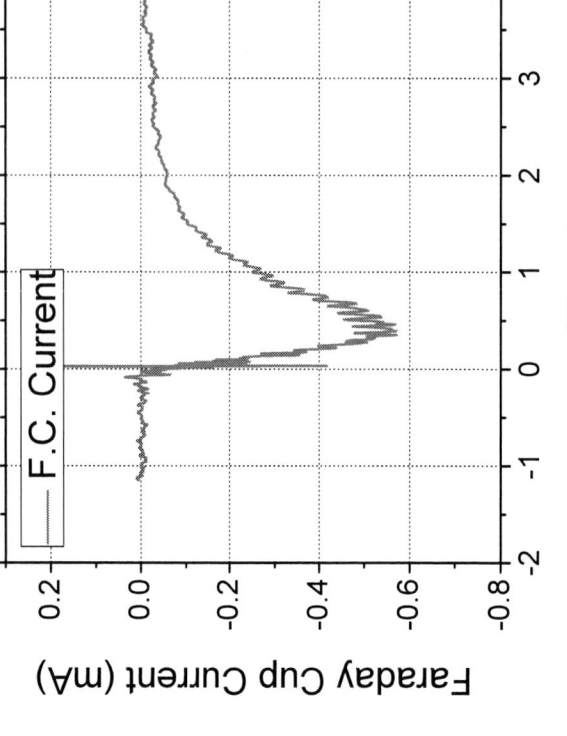

Covered

Un covered

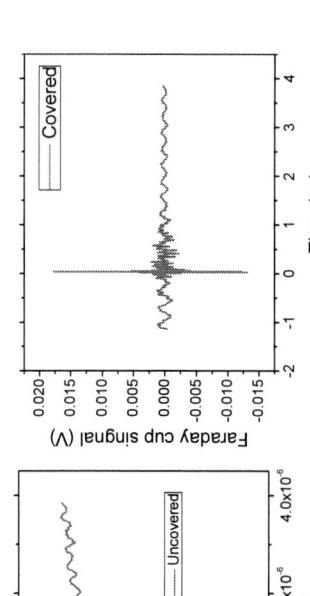

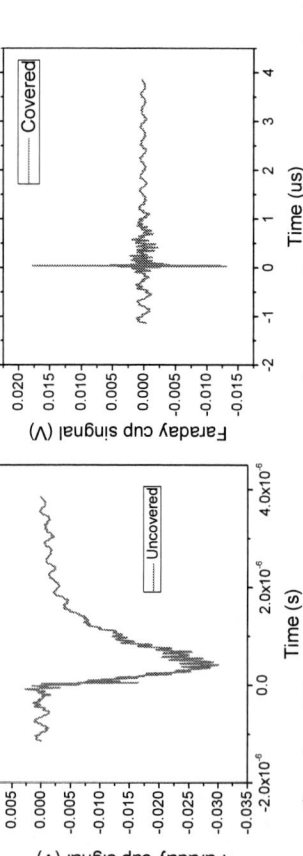

- A Faraday cup placed in one of the off-axis gate valve flanges revealed a strong electron presence following the ejection of photons (signified by the positive peak at ~ t = 0). To test for E-M noise contribution, the Faraday cup was fitted with a non conducting, opaque plastic. The signal is not noise.

- Despite being obscured from the pinch, there is still a strong electron presence outside of the collector shells. ~3E-8 Coulumb were accumulated in less than 4μs after the pinch.

Center for Plasma Material Interactions
http://starfire.ne.uiuc.edu

Neutral Particle Detection

J.Sporre, R.Raju and D. N. Ruzic, Poster Number: 106

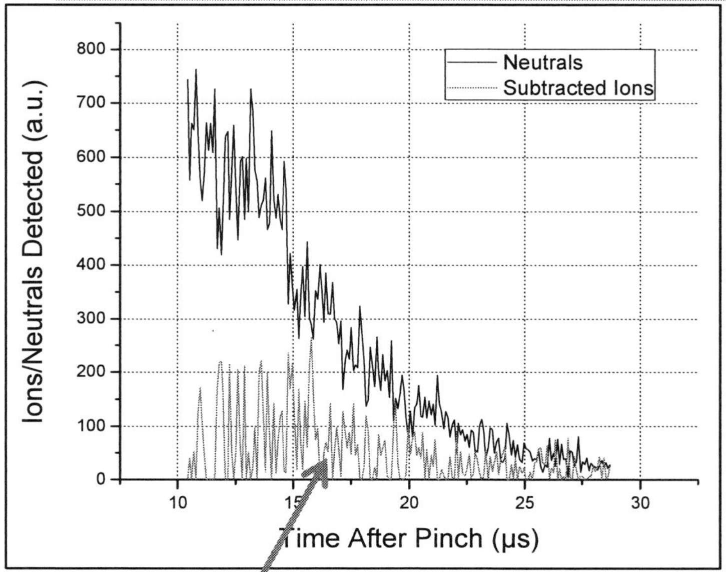

No ions were observed using ESA

Without Collector Mock-up   With Collector Mock-up

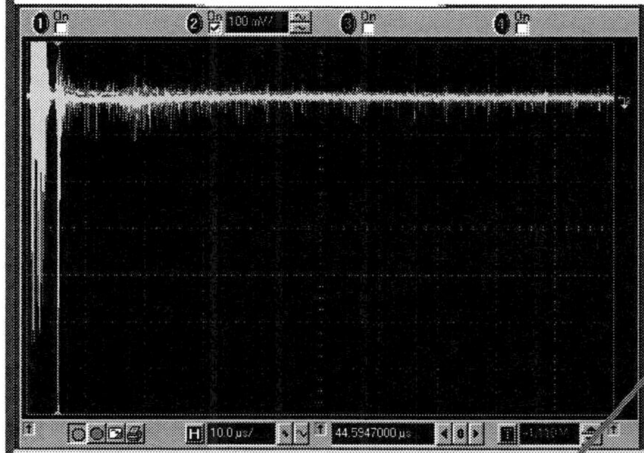

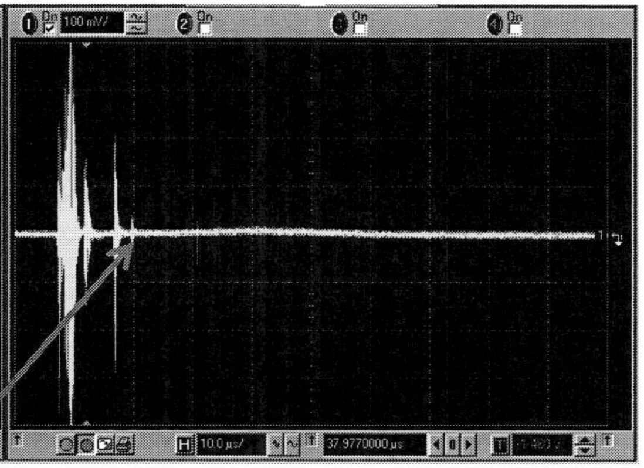

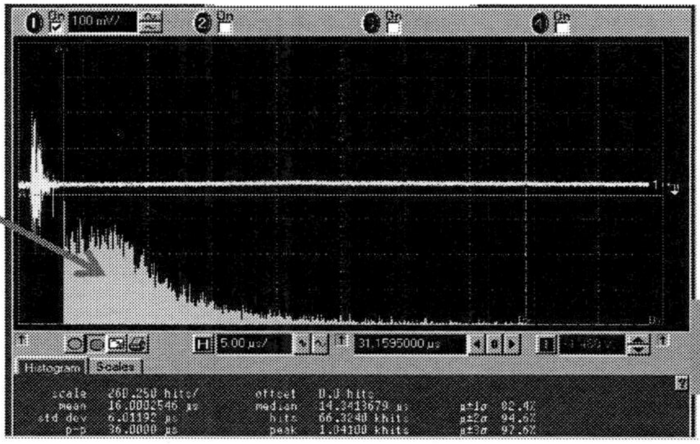

Energetic Neutrals were observed

- A set of microchannel plates, placed behind a set of spherical sector energy isolation plates allows for the segregation of neutral particle flux from the total measured particle flux.
- A large amount of energetic neutral atoms are present !

Center for Plasma Material Interactions
http://starfire.ne.uiuc.edu

ILLINOIS
UNIVERSITY OF ILLINOIS AT URBANA-CHAMPAIGN

Contamination

- Si witness plate samples were exposed for 3 different shot durations. Each sample location contained 4 samples, two that were partially masked and exposed to the open chamber, and two that were covered to prevent direct energetic particle interaction. These are not in line-of-sight of the pinch.

Cover Flap

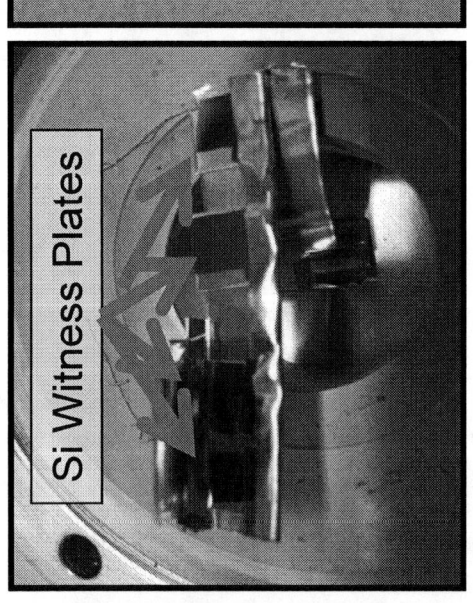

Si Witness Plates

XPS Analysis

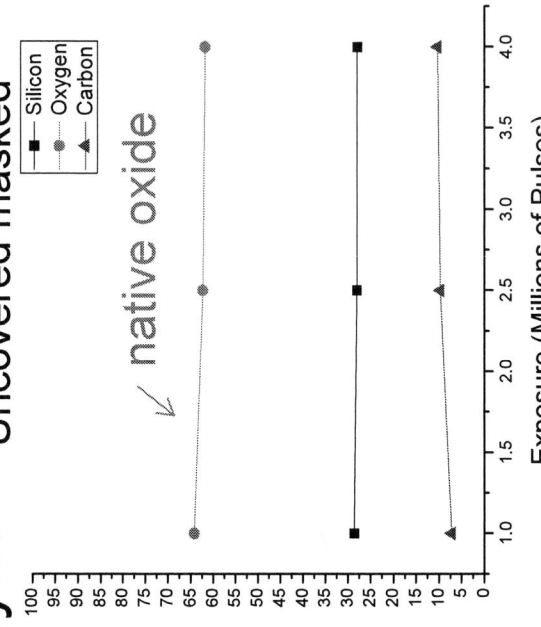

Uncovered masked

native oxide

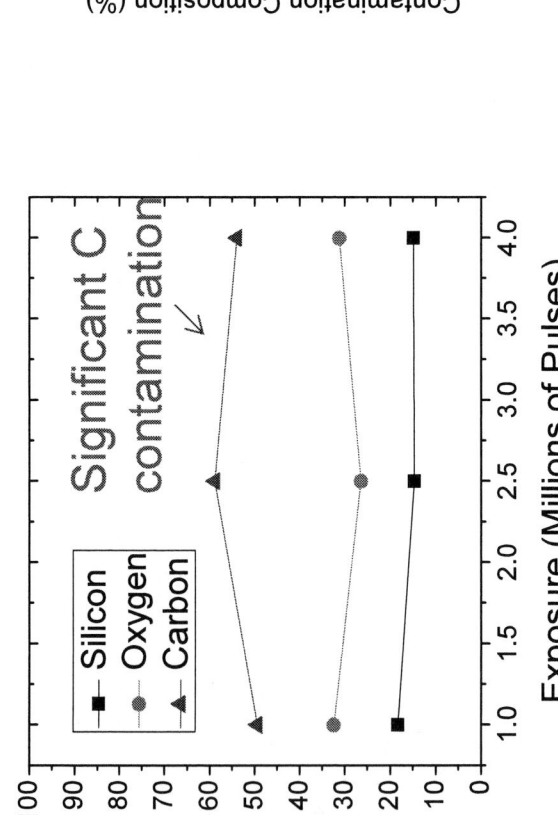

Uncovered un-masked

Significant C contamination

Center for Plasma Material Interactions
http://starfire.ne.uiuc.edu

XPS analysis

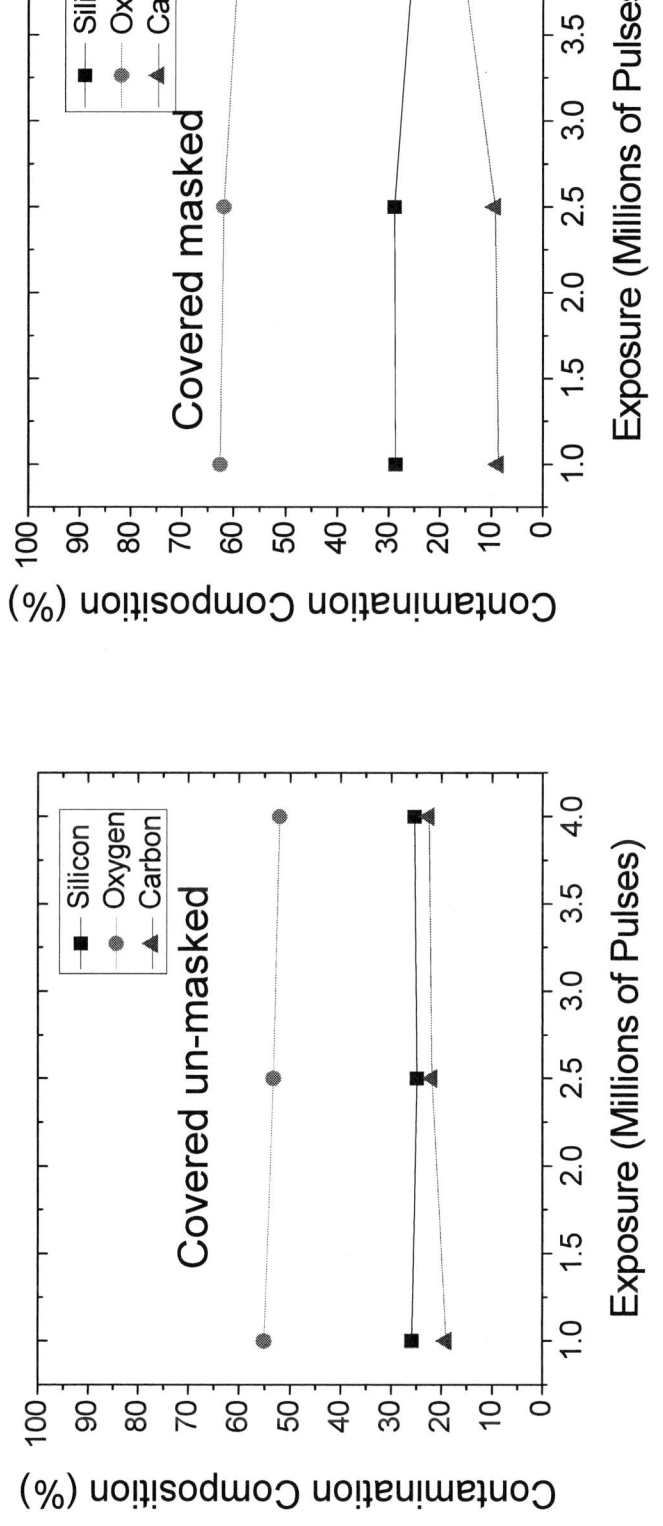

- The exposed samples revealed an increase of carbon composition from ~10% up to ~60%, while the covered witness plates increased only from ~10% to 20%.
- This shows that both UV-cracked or energetic reflected and thermal carbon-containing molecules are present and deposit.
- The contamination ratios were independent of shot exposure time, but thickness probably increased.

Center for Plasma Material Interactions
http://starfire.ne.uiuc.edu

SEM Analysis

Carbon film build-up

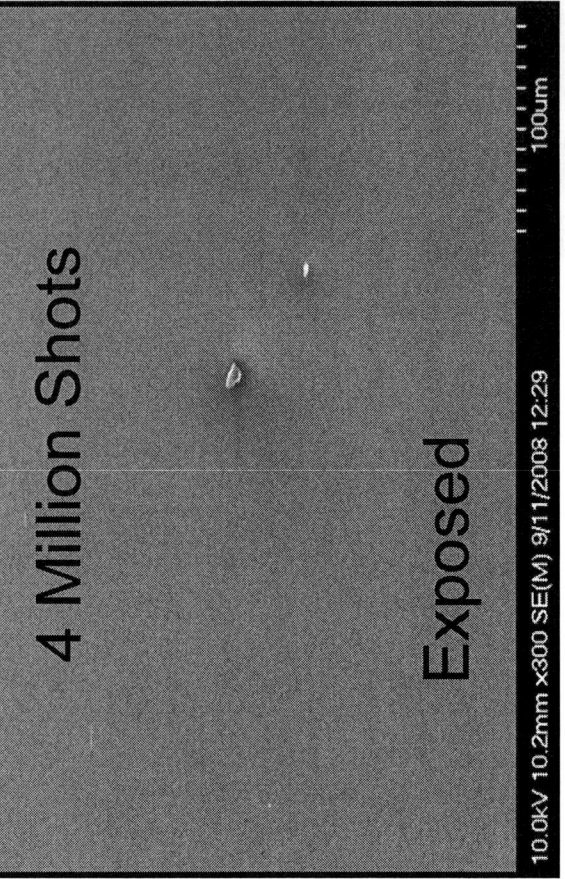

4 Million Shots

Exposed

10.0kV 10.2mm x300 SE(M) 9/11/2008 12:29   100um

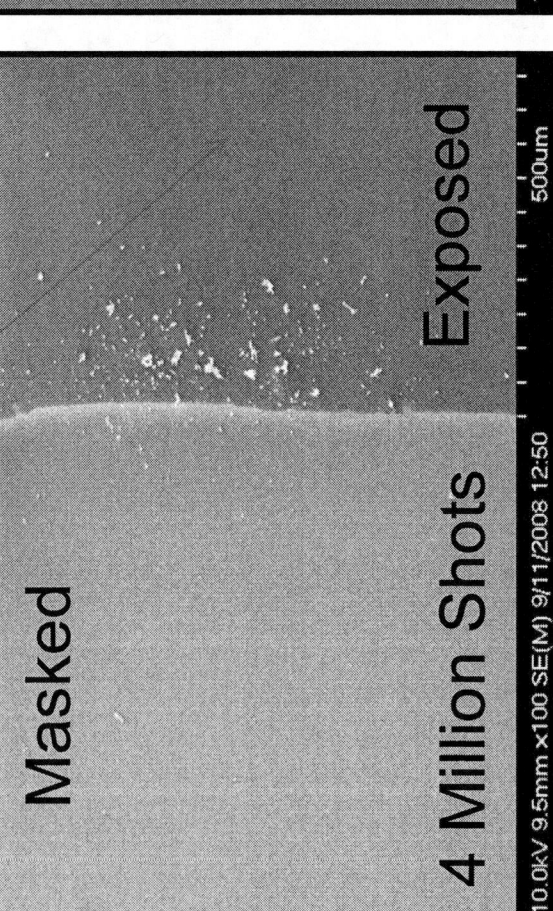

Masked

Exposed

4 Million Shots

10.0kV 9.5mm x100 SE(M) 9/11/2008 12:50   500um

- Contamination present in each of the samples was uniformly deposited along the exposed surfaces. A distinct line between the masked and exposed part of the witness plate is evident in the SEM analysis.

Center for Plasma Material Interactions
http://starfire.ne.uiuc.edu

Triple probe diagnostics

A triple probe was placed immediately behind the mock collector optic to detect a plasma. Experiments are still in progress.

Ion saturation current

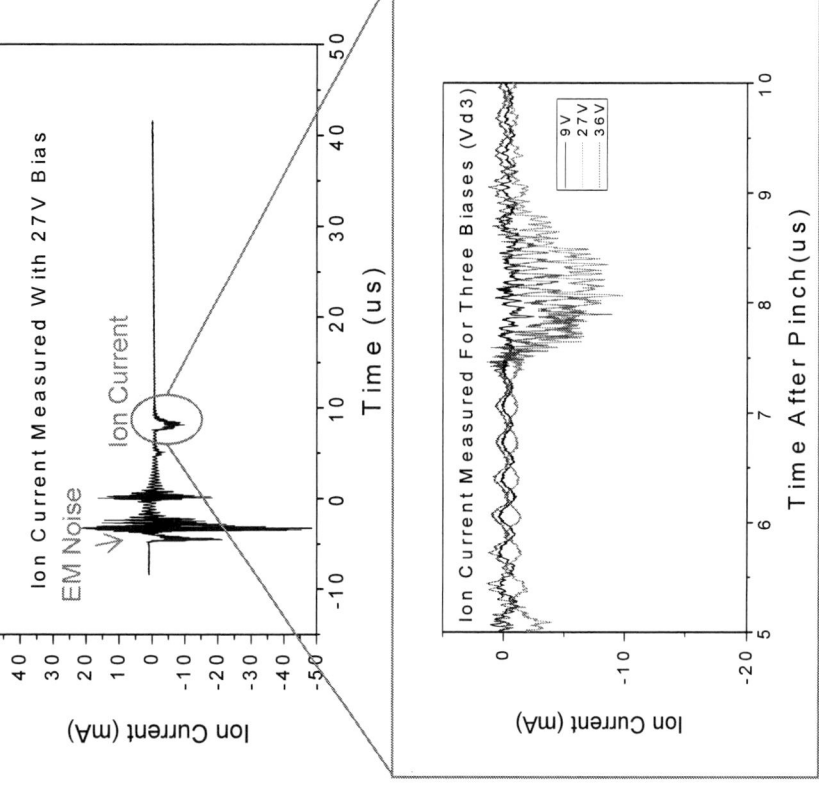

Triple probe set-up

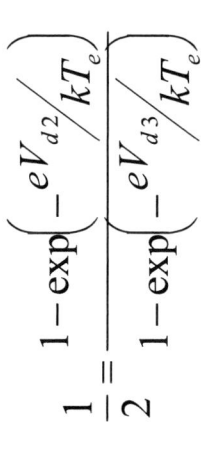

$$\frac{1}{2} = \frac{1-\exp\left(-\dfrac{eV_{d2}}{kT_e}\right)}{1-\exp\left(-\dfrac{eV_{d3}}{kT_e}\right)}$$

Center for Plasma Material Interactions
http://starfire.ne.uiuc.edu

IF detection system

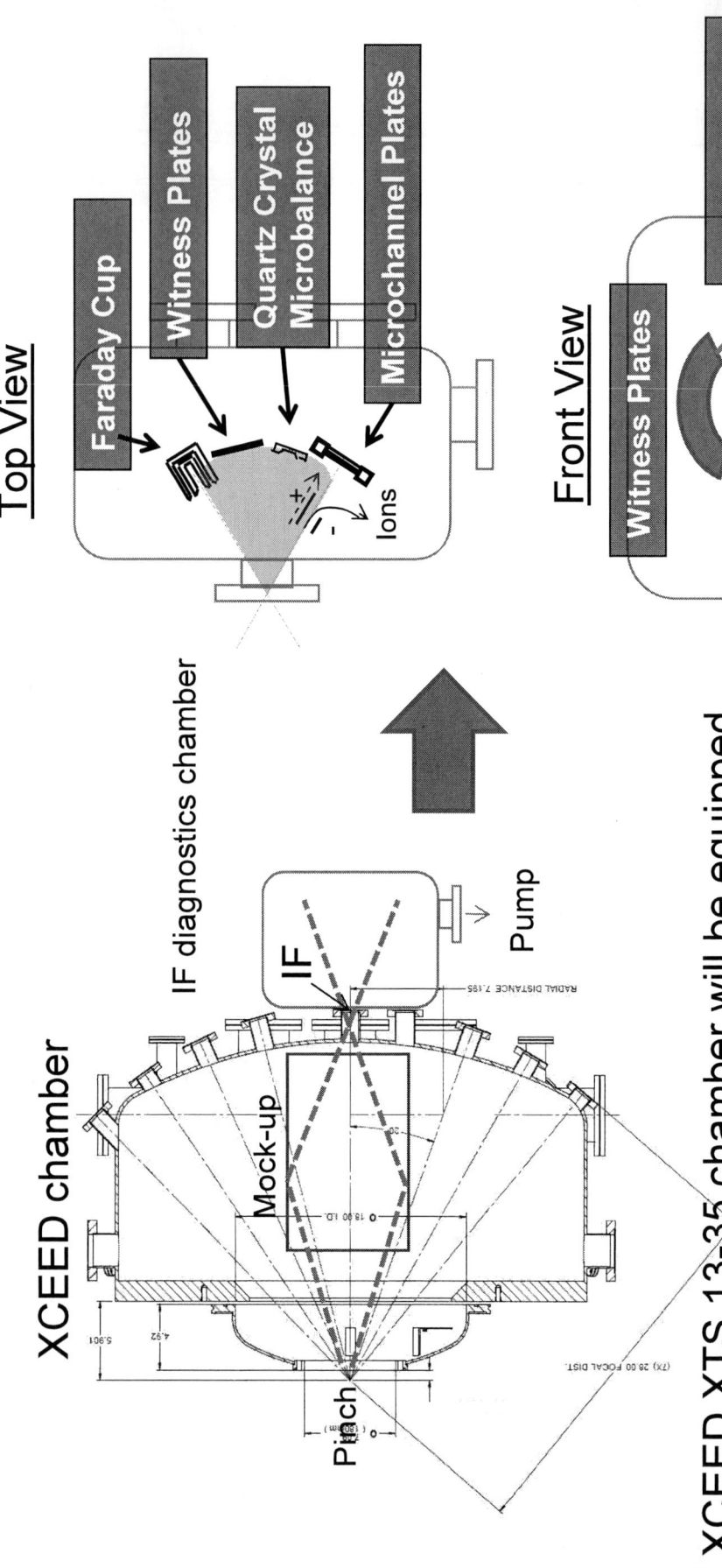

XCEED XTS 13-35 chamber will be equipped with diagnostics tools to measure debris at IF. More diagnostics may be added such as a triple probe or RGA.

Center for Plasma Material Interactions
http://starfire.ne.uiuc.edu

Outline

- Introduction
- Source Contamination
  - Damage Mechanism
  - Ion and Neutral Debris measurement
  - Erosion and life time prediction
  - Debris detection at Intermediate Focus (IF)
- **Collector Optics Contamination**
  - Cleaning of Sn by RIE
- Particle Contamination of Mask and Wafer
  - PACMAN cleaning of nano particles
- Line Edge Roughness
- Conclusions
- Acknowledgements

Center for Plasma Material Interactions
http://starfire.ne.uiuc.edu

Contamination of Collector Optics

- Carbon contamination can happen anywhere
- Sn-DPP or LPP sources are welcomed because of their high conversion efficiency
- But Sn is a condensable fuel, which degrades the optics and reduces the mirror lifetime
  - Deposition takes place
  - Surface becomes rougher because erosion takes place simultaneously.
  - Off-normal events (mistakes) may occur

To improve the collector optics lifetime, Sn cleaning of the mirror surface is essential, especially if that process can correct roughness variations and mistakes too.

Center for Plasma Material Interactions
http://starfire.ne.uiuc.edu

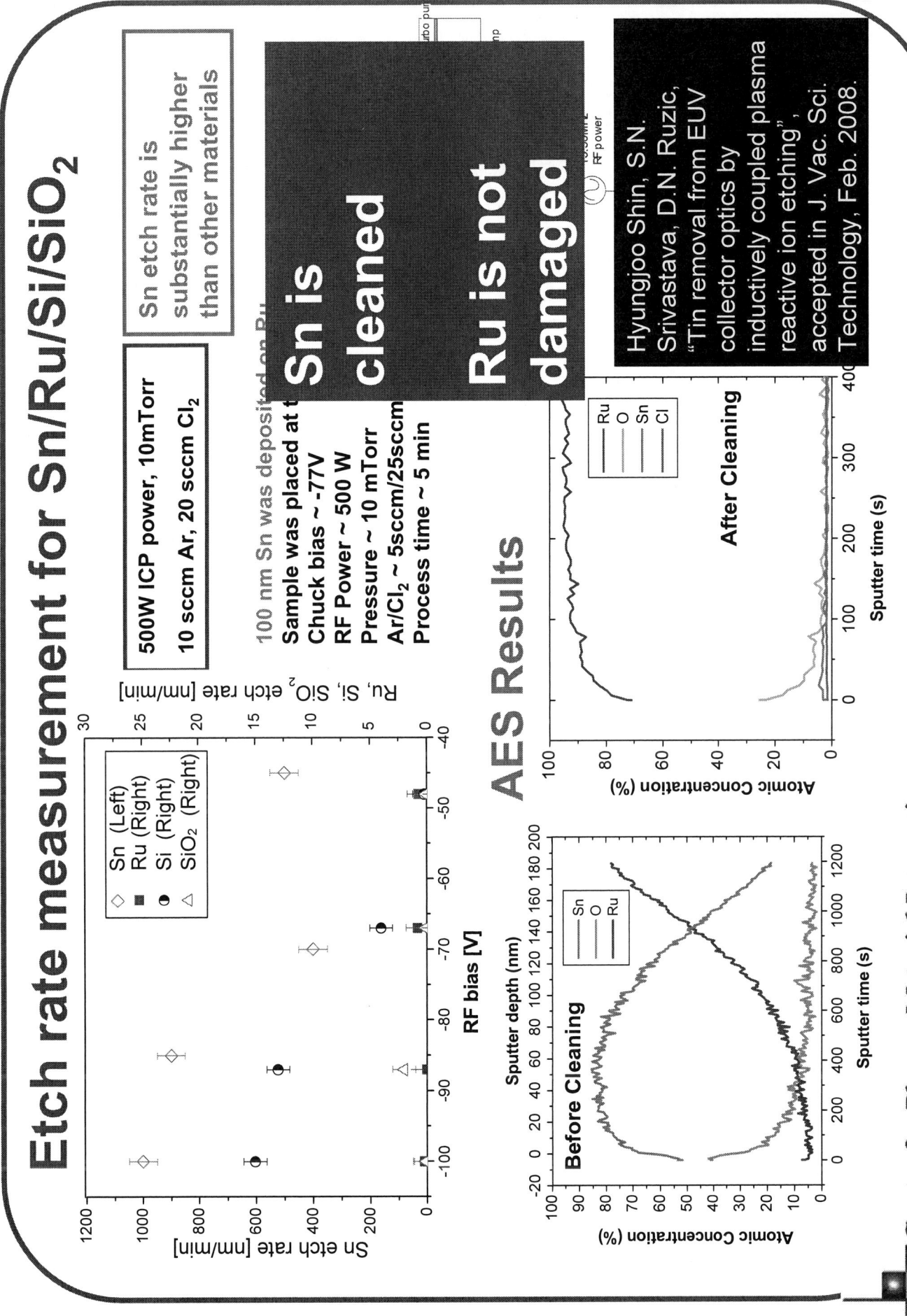

Etch rate measurement for Sn/Ru/Si/SiO$_2$

Sn etch rate is substantially higher than other materials

500W ICP power, 10mTorr

10 sccm Ar, 20 sccm Cl$_2$

100 nm Sn was deposited on Ru

Sample was placed at t

Chuck bias ~ -77V

RF Power ~ 500 W

Pressure ~ 10 mTorr

Ar/Cl$_2$ ~ 5sccm/25sccm

Process time ~ 5 min

Sn is cleaned

Ru is not damaged

Hyungjoo Shin, S.N. Srivastava, D.N. Ruzic, "Tin removal from EUV collector optics by inductively coupled plasma reactive ion etching", accepted in J. Vac. Sci. Technology, Feb. 2008.

AES Results

Center for Plasma Material Interactions

http://starfire.ne.uiuc.edu

ILLINOIS
UNIVERSITY OF ILLINOIS AT URBANA-CHAMPAIGN

Full blown Integrated system In XTS EUV Source

RF coil power: 400 W
Chlorine gas pressure:20mTorr and 5mTorr,
Bias to mock-up:-70V DC
Cleaning time:8 min for 2000 nm Sn samples

2000 nm Sn samples : Effect of gap between the cells

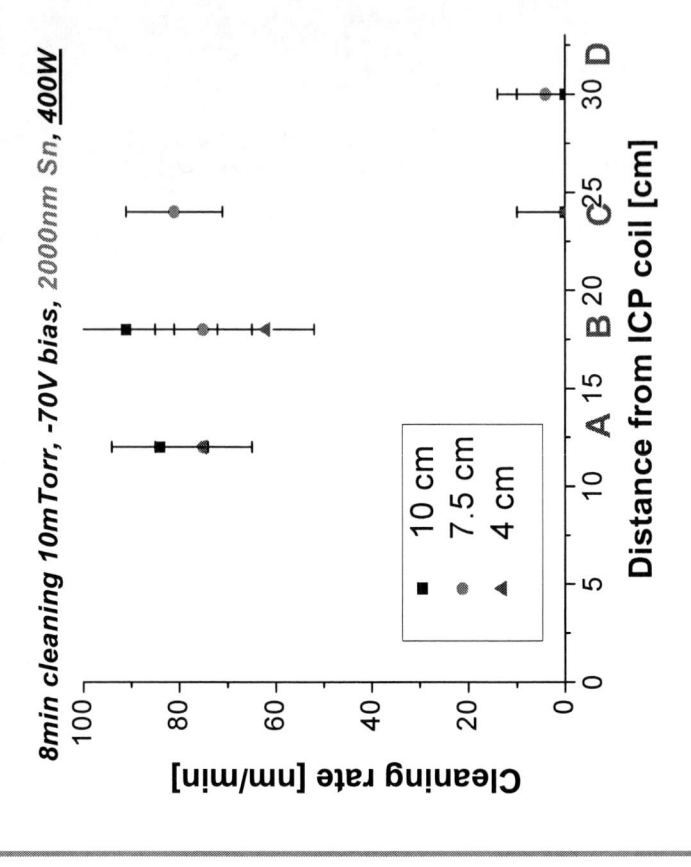

8min cleaning 10mTorr, -70V bias, 2000nm Sn, 400W

Legend:
- 10 cm
- 7.5 cm
- 4 cm

Y-axis: Cleaning rate [nm/min]
X-axis: Distance from ICP coil [cm]

Chlorine gas

Collector mock-up

Z-pinch electrode

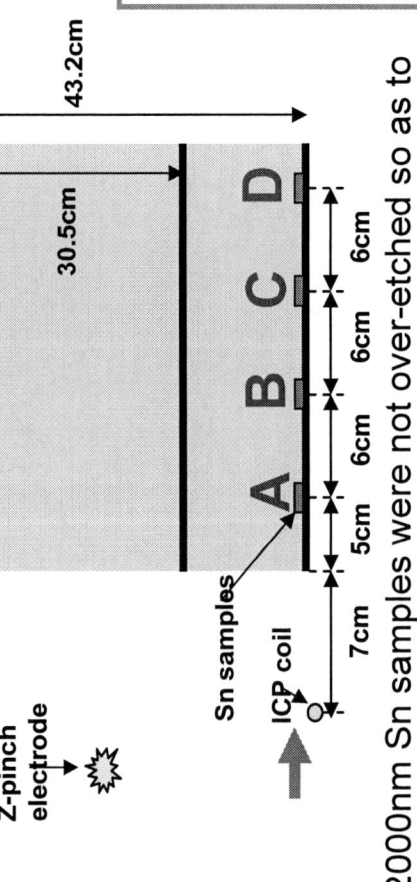

43.2cm

30.5cm

Sn samples

ICP coil

A B C D

7cm 5cm 6cm 6cm 6cm 6cm

- 2000nm Sn samples were not over-etched so as to provide etch rates with -70 V DC bias and 400W ICP power.
- Cleaning is observed on the samples located with 4 cm gap width.
- Samples with 10cm gap width show less removal at location C (24 cm from the ICP coil) compared to samples with 7.5 cm gap width; samples with 7.5 cm gap width have more residence time of etchant because they are located at opposite side of pumping port.

Center for Plasma Material Interactions
http://starfire.ne.uiuc.edu

Effect of parameters

RF coil power: 400 W
Chlorine gas pressure:20mTorr and 5mTorr,
Bias to mock-up:-70V DC
Cleaning time:10 min for 100 nm Sn samples

Effect of cleaning time

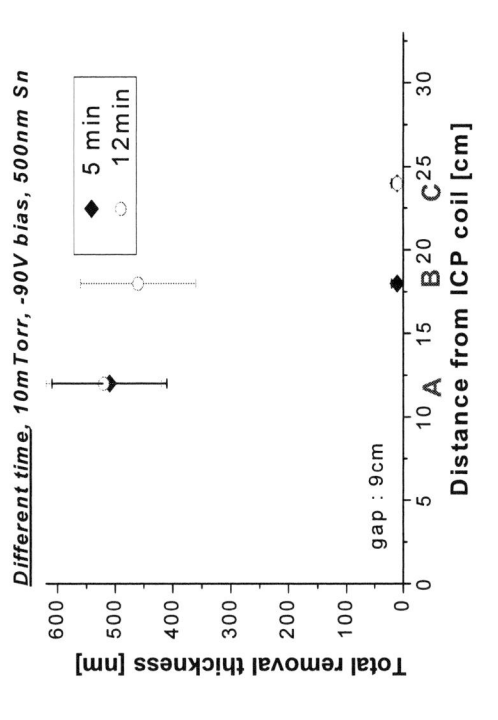

Different time, 10mTorr, -90V bias, 500nm Sn

Different cleaning time

•Even in 5 minutes, 500nm Sn was completely removed at location A. After 12-minutes cleaning, sample at location B also showed almost complete removal. Samples at location C, however, showed no measurable removal.

Effect of DC bias

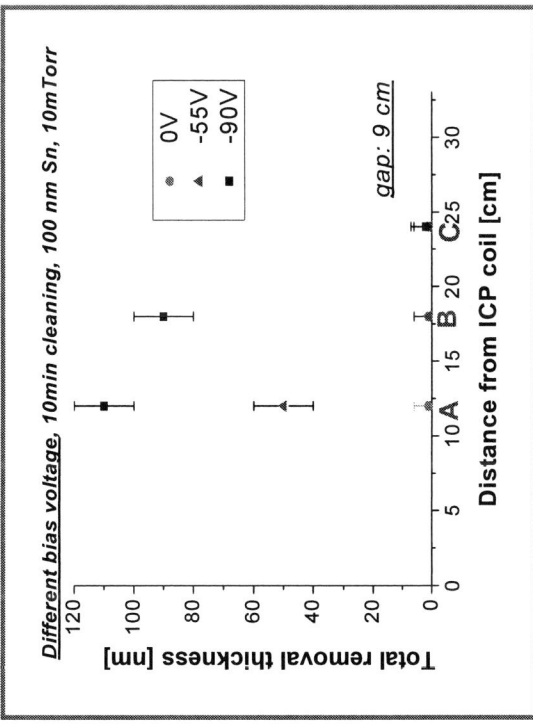

Different bias voltage, 10min cleaning, 100 nm Sn, 10mTorr

Bias effect

• Only bias voltage was changed keeping other conditions same (only Cl_2, 100nm Sn samples, 10 mTorr, 200W ICP power, 10 min. cleaning time)
• Without any bias, Sn removal does not occur; With -90 V DC, Sn was completely removed. Second sample showed some residue left; With -55 V DC bias, the total removal was smaller compared to -90 V DC.

H.Shin, R.Raju and D. N. Ruzic,
Poster Number: 101

Center for Plasma Material Interactions
http://starfire.ne.uiuc.edu

ILLINOIS
UNIVERSITY OF ILLINOIS AT URBANA-CHAMPAIGN

Outline

- Introduction
- Source Contamination
  - Damage Mechanism
  - Ion and Neutral Debris measurement
  - Erosion and life time prediction
  - Debris detection at intermediate focus (IF)
- Collector Optics Contamination
  - Cleaning of Sn by RIE
- **Particle Contamination of Mask and Wafer**
  - PACE cleaning of nano particles
- Line Edge Roughness
- Conclusions
- Acknowledgements

Center for Plasma Material Interactions
http://starfire.ne.uiuc.edu

Contamination of wafer and mask

- EUV Radiation induced carbon, oxygen contamination
  - Out gassing of photoresist
  - Residual hydrocarbon in chamber
     Reflectivity drop of ML mirrors
    Reduce life time

- Particle contamination
     Affects the printing pattern

Center for Plasma Material Interactions
http://starfire.ne.uiuc.edu

Plasma Assisted Cleaning by He Metastable Atom Neutralization (PACMAN)

Diagram and setup of the plasma cleaning chamber

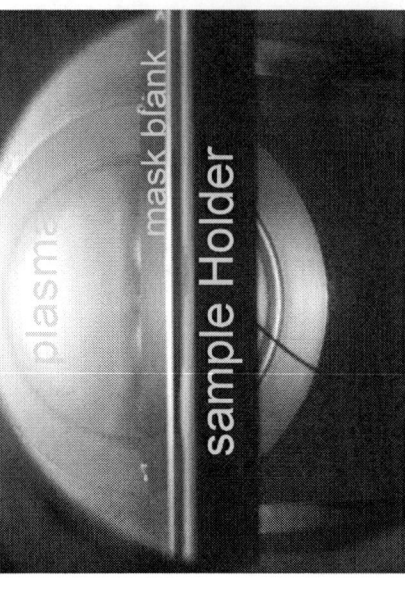

Plasma Source
Vacuum Chamber
Sample Transfer Arm

Sample Introduction

Main Chamber Turbo

- The plasma cleaning technique developed within the Center for Plasma Material Interactions can clean an entire mask/wafer at once
  - No need to focus a cleaning tool onto a specific particle in order to remove it
- Damage has not been observed for the plasma cleaning technique
  - Does not cause the thermo stress such as those present in laser shockwave cleaning or the pattern damage associated with megasonics/cavitation
- A 100% effective cleaning technique will be needed in order for high volume manufacturing and integration of EUV technology

Center for Plasma Material Interactions
http://starfire.ne.uiuc.edu

Mechanism of particle removal

- Plasma creates electrons, helium ions, and helium metastables
- He ions directed to the surface by plasma sheath electric field
- Metastables(non ionized) arrives to the surface/particle
- He ions and metastables impart energy to atoms of the particle
- Particle bonds break
- Volatilization of particle occurs
- Particle is removed similar to etching

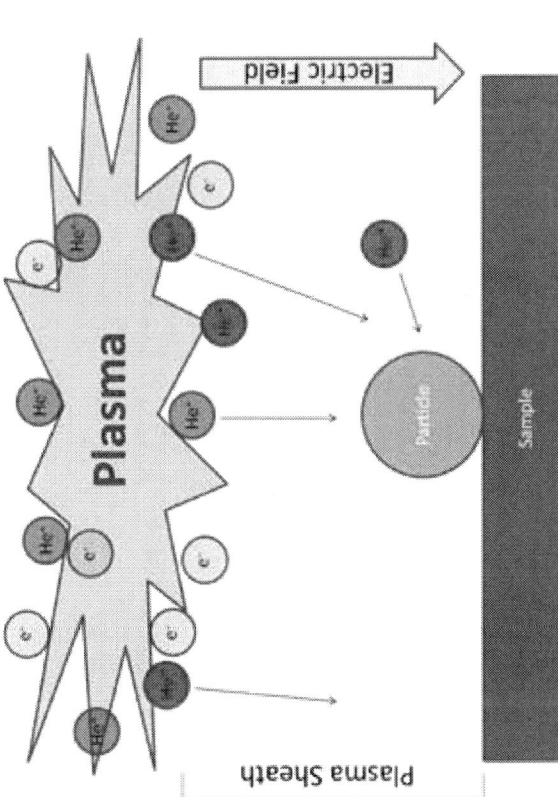

Shrinkage of the PSL particle in PACMAN process

Particles etched
Surface is not damaged

Particles "shrinking" during PACMAN processing confirms an etching like process

Center for Plasma Material Interactions

http://starfire.ne.uiuc.edu

Cleaning results of PSL particles

- 30 nm polystyrene latex test particles can be fully removed from silicon wafers
- The developed plasma cleaning technique is a quick non-contact cleaning process

Large particle is from the physical marking of the wafer

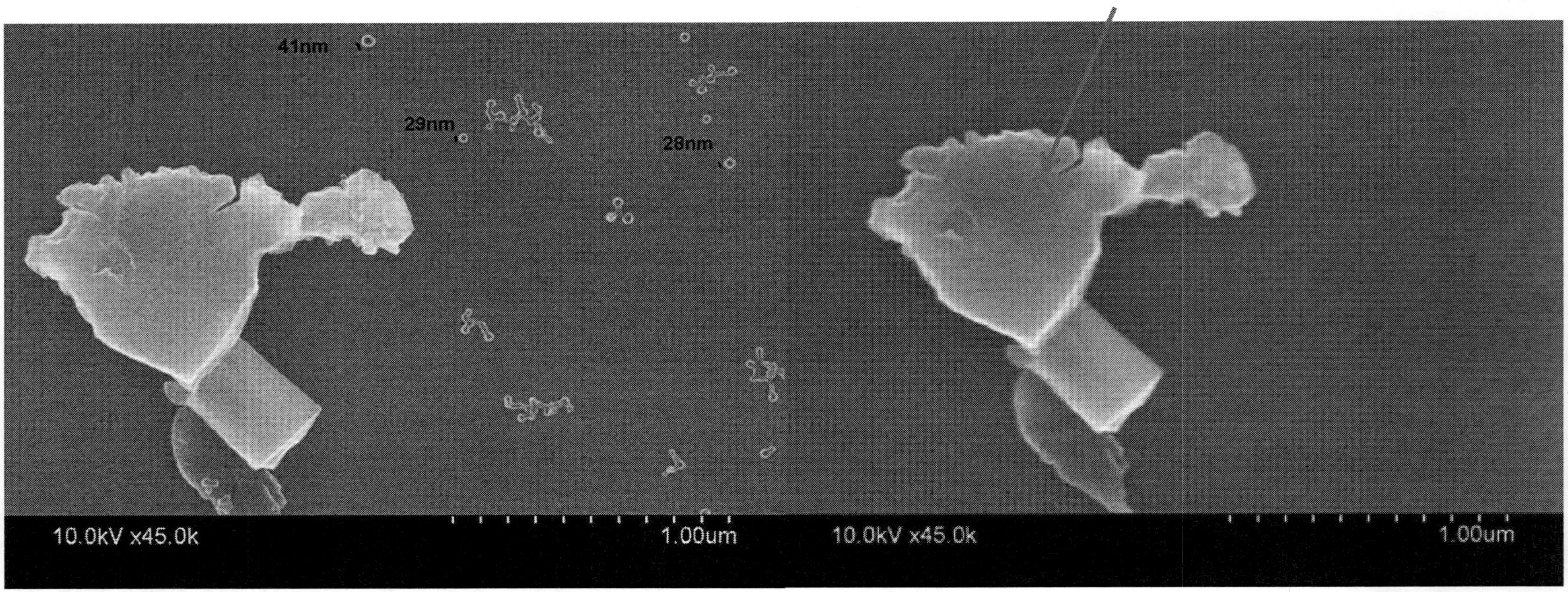

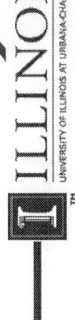

10.0KV x45.0k   1.00um   10.0KV x45.0k   1.00um

41nm   29nm   28nm

Reference Location Before   **Reference Location After**

W. Lytle, C. Das, R. Raju, M. Neumann, D. Ruzic, Poster Number: 83

100 % removal of 30-200 nm PSL particles

Center for Plasma Material Interactions
http://starfire.ne.uiuc.edu

ILLINOIS
UNIVERSITY OF ILLINOIS AT URBANA-CHAMPAIGN

Plasma Bombardment AND He Metastables are needed

- Wafer held in He metastable flux but not in plasma

Before

After

No etching observed !

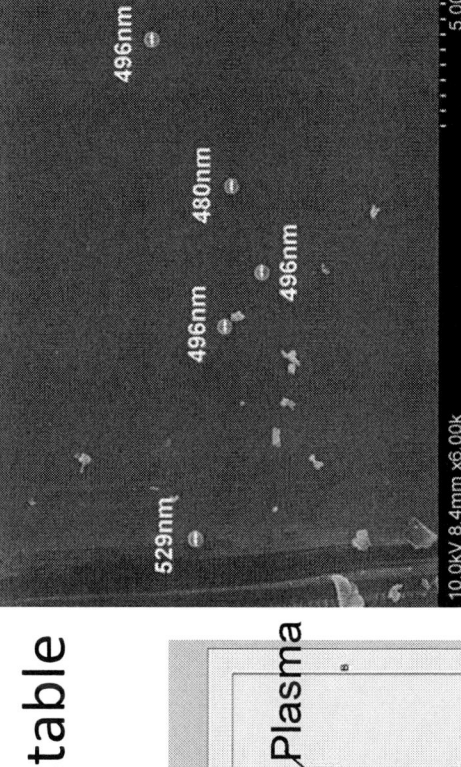

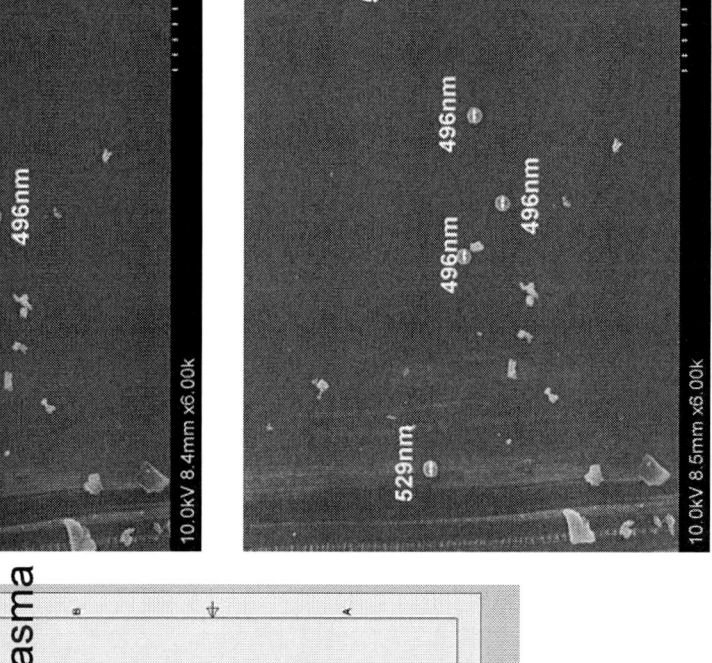

Center for Plasma Material Interactions
http://starfire.ne.uiuc.edu

Outline

- Introduction
- Source Contamination
  - Damage Mechanism
  - Ion and Neutral Debris measurement
  - Erosion and life time prediction
  - Debris detection at intermediate focus (IF)
- Collector Optics Contamination
  - Cleaning of Sn by RIE
- Particle Contamination of Mask and Wafer
  - PACMAN cleaning of nano particles
- **Line Edge Roughness**
- Conclusions
- Acknowledgements

Center for Plasma Material Interactions
http://starfire.ne.uiuc.edu

Line Edge Roughness

With feature sizes shrinking to the nanometer scale, deformities in the trenches become more noticeable. This roughness is known as Line Edge Roughness (LER) or Line Width Roughness (LWR).

LER is caused by inhomogeneous photocatalysts in photoresist and varying light intensity. LER worsens as intensity of light is decreased.

LER causes in homogeneity in etching and depositing steps that occur during microchip manufacturing, limiting chip size and reliability.

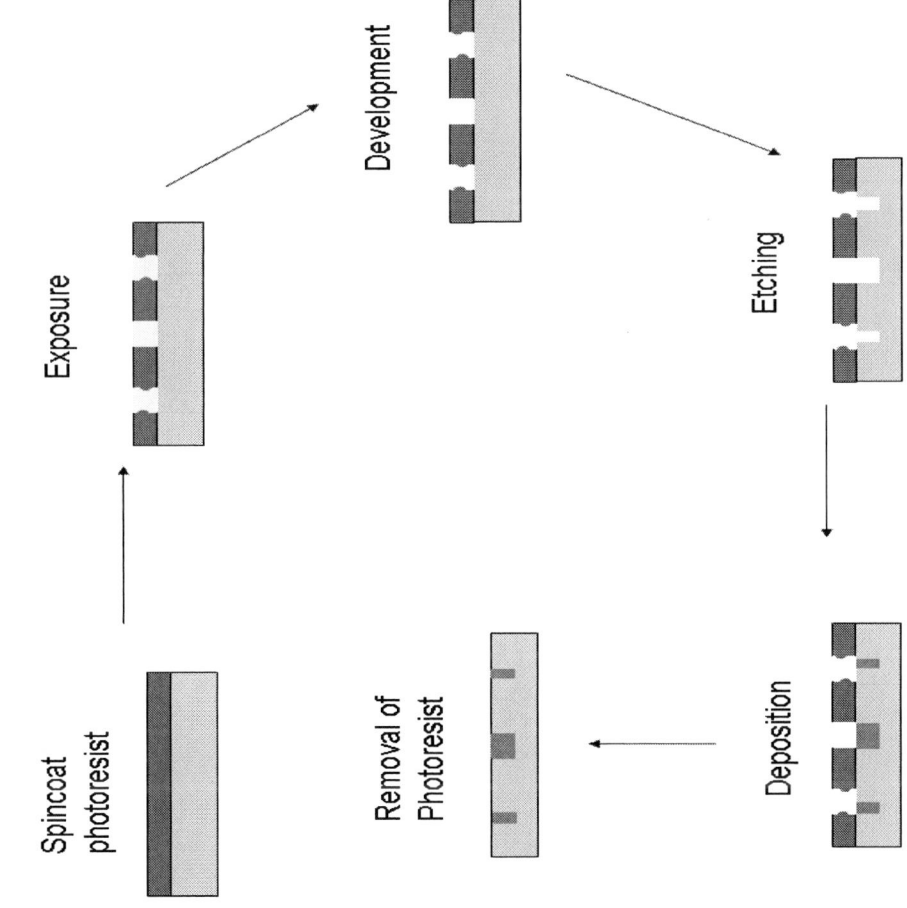

Spincoat photoresist

Exposure

Development

Etching

Deposition

Removal of Photoresist

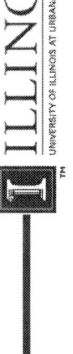

Center for Plasma Material Interactions
http://starfire.ne.uiuc.edu

LER results

Our technique reduces LER by one-third
(patent being applied for with INTEL)

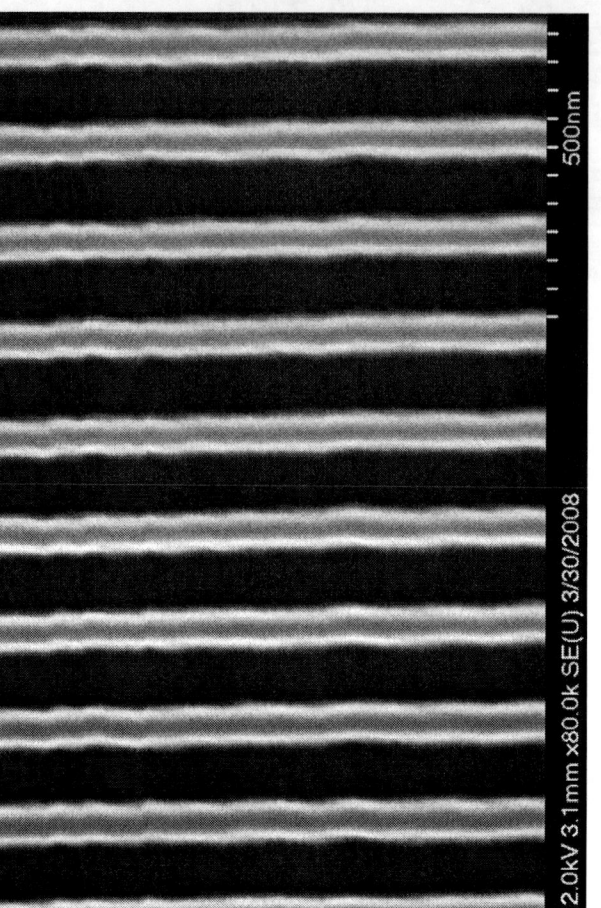

Processed sample 3σ LER of 8.5nm

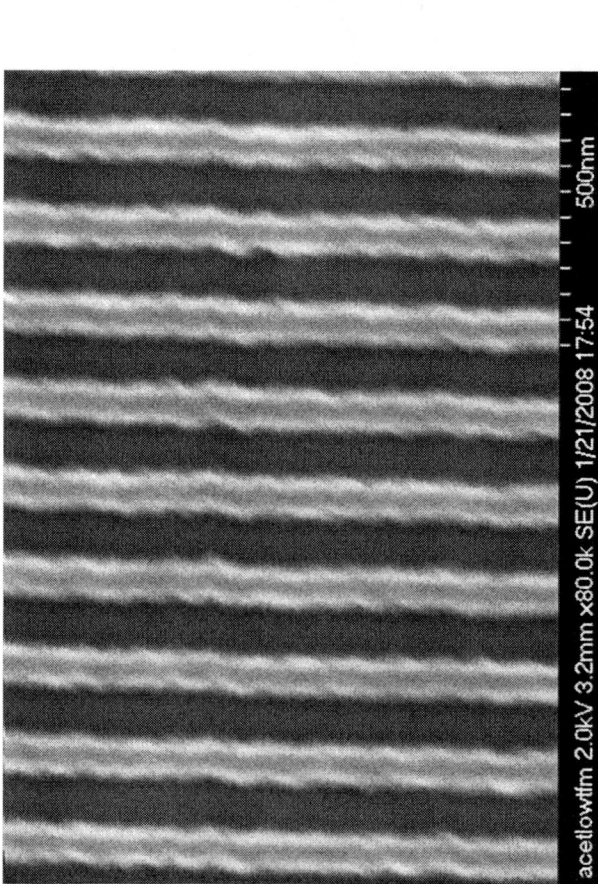

Unprocessed sample 3σ LER of 12.3nm

Center for Plasma Material Interactions
http://starfire.ne.uiuc.edu

Conclusions

HVM Requirements

- Need more sensitive resist and /or more source power

- Need to develop an effective detector to measure debris at Intermediate Focus – new Sematech project

- Need reliable cleaning technique for collector optics

- Potential mitigation techniques to reduce the reflectivity loss due to carbon and oxidation on projection optics and mask

- Need reliable technique to clean mask

Center for Plasma Material Interactions
http://starfire.ne.uiuc.edu

Acknowledgements

- SEMATECH (Frank Goodwin)
- INTEL (Robert Bristol and Ted Liang)
- USHIO / EUVA
- XTREME
- ASML
- CYMER
- NIST
- MRL UIUC

Thank you for your attention

Center for Plasma Material Interactions
http://starfire.ne.uiuc.edu

AUTHOR INDEX

Agovic, K. 1119
Ahn, B.-S. 646, 721
Allred, B. 1479
Amano, T. 1565, 1593
Amemiya, M. 1564, 1631
Anazawa, T. 1611
Anderson, C. N. 312, 426, 487, 1389, 1390, 1500
Antohe, A. 602, 1634, 1650
Aoyama, H. 1174, 1593
Aquila, A. 1560
Aramaki, F. 1565
Arisawa, Y. 1593
Auželyte, V. 363
Ayers, T. 1461
Bacuita, T. 1433
Banine, V. 1018, 1055, 1481
Bear, W. L. 1418, 1449, 1668
Bender, M. 602
Bergeson, S. 1479
Bernath, R. 956
Bianucci, G. 1041
Bollanti, S. 1391
Borisov, V. M. 996
Borisova, G. N. 996
Brainard, R. 1529
Brandt, D. C. 890
Brunner, T. 1227
Burkhardt, M. 1227
Caudillo, R. 1433
Cha, B. 663
Chiba, H. 1275
Cho, H. K. 646, 721, 848
Cho, W. 721
Choi, S. 1201
Cloutier, P. 1529
Conte, A. 1666
Corthout, M. 935
Cummings, K. 199, 291
Cunado, J. 956
De Groote, F. P. J. 1119
Denbeaux, G. 602, 1634, 1650
Deng, Y. 426, 1227
Denham, P. 312, 1560
Di Lazzaro, P. 1391
Echigo, M. 1499
Ehm, D. 1055
Eichenlaub, S. 663
Ellwi, S. 956
Endo, A. 916
Engelstad, R. 689
Fan, Y. 602, 1634
Flora, F. 1391
Fulford, B. 956
Fumar-Pici, A. 291, 602
Gallatin, G. M. 487

Garg, R. 602, 1634, 1650
Gargini, P. 136
George, S. A. 312, 956, 1530, 1560
Gielissen, K. 1481
Glushkov, D. 1481
Goethals, A. M. 742, 1242
Goldberg, K. 312, 602, 721, 867, 1563, 1633
Goldstein, M. 1321, 1416, 1480
Goo, D. 1201
Goodwin, F. 520, 602, 1417, 1480
Grove, D. 1461
Gullikson, E. 583, 623, 1560
Gustafson, D. 1461
Hada, H. 1501, 1513
Hagiwara, R. 1565
Han, H. 646, 721
Han, S. I. 291
Han, W. 71, 1201
Hansen, S. 291
Harned, N. 199, 1055
Hartley, J. 689, 1417
Hartman, R. 1, 97
Hasegawa, T. 264
Hay, N. 956
Hayes, A. 709
Hazelton, A. J. 520
He, L. 1155
Henderson, C. L. 376
Henderson, I. 956
Hirai, Y. 1542
Hirayama, T. 1501, 1513
Hoef, B. 312
Holfeld, C. 602
Honda, T. 264
Hotta, K. 1462
House, M. 663
Hudyma, R. 1321, 1341, 1416
Hughes, G. 520
Huh, S. 602, 646, 867
Huijbregtse, J. 1055
Huli, L. 291
Hultermans, B. 199
Ikuta, Y. 563
Irie, M. 1513
Ishihara, S. 1583
Itani, T. 335, 1499, 1531, 1542, 1561
Ivanov, A. S. 996
Iwai, T. 1513
Iwamoto, F. 742
Iwasaki, T. 805
Iwashita, J. 1513
Jak, M. 1018
Jonckheere, R. 742
Jones, G. 312
Kadaksham, A. J. 663

AUTHOR INDEX

Kai, T. 1542
Kamo, T. 1174
Kamptaprasad, R. 956
Kaneyama, K. 335
Kang, I.-Y. 721
Kashiwamura, T. 1531
Kawamura, D. 335, 1542
Kearney, P. 709
Kemp, D. 1530, 1560
Kemp, J. 1479
Kempen, A. 1055
Kessels, B. 291
Khristoforov, O. B. 996
Kim, D. 646, 721, 848
Kim, G. 721
Kim, H. 646, 721
Kim, I. 1201
Kim, S.-S. 646, 721, 848
Kinoshita, H. 848, 1501, 1562
Kirukhin, Y. B. 996
Kishimoto, J. 848
Kluenkov, E. 1018
Knight, L. 1479
Koay, C.-S. 1227
Kobayashi, S. 335, 1561
Koh, C. 312, 398, 1417
Koida, K. 1648
Komori, H. 1482
Kondo, H. 1275
Kools, J. 1041
Koops, R. 1119
Koster, N. B. 1119
Kozawa, T. 505
Kuerz, P. 1300
La Fontaine, B. 312, 602, 867, 1227
Landoni, C. 1666
Langer, E. 602
Laubis, C. 1667
Lawson, R. 376
Lee, C.-T. 376
Lee, D. 646, 721, 848
Lee, J. 1201
Lee, S. 721, 1583
Levinson, H. J. 21
Liang, T. 583, 829
Lin, C. C. 709
Litt, L. C. 520
Loginova, E. 1529
Lok, S. 199
Lopatin, A. 1018
Lopez, J. 1479
Luchin, V. 1018
Lyons, A. 689
Lytle, W. M. 1077
Ma, A. 312, 398, 829

Madey, T. E. 1141, 1529
Magoshi, S. 1174, 1483
Mallmann, J. 291
Manini, P. 1666
Maruyama, K. 1542
Mbanaso, C. 1634, 1650
McIntyre, G. 1227
Meijerink, M. G. H. 1119
Meiling, H. 199
Mezi, L. 1391
Mimura, T. 1513
Mirkarimi, P. 709
Mishchenko, V. A. 996
Miura, T. 230
Miyakawa, R. 312, 1632, 1633
Mizokoshi, H. 1462
Mizuno, H. 291, 1227
Mochi, I. 867, 1563
Montgomery, W. 312
Moors, R. 1055
Mori, H. 1483
Mori, I. 1, 183, 551
Morishima, H. 264
Murakami, K. 1275
Murra, D. 1391
Nakamura, J. 505
Nanver, L. K. 1667
Naulleau, P. 10, 124, 312, 398, 426, 487, 602,
867, 1321, 1367, 1389, 1390, 1416, 1500,
1530, 1560, 1563, 1632, 1633
Neumann, M. J. 1077
Niakoula, D. 312, 1560
Nihtianov, S. N. 1667
Niibe, M. 1648
Nishihara, K. 975
Nishimura, N. 1564, 1583
Nishiyama, I. 551, 1583, 1611
Nishiyama, Y. 1565, 1611
Nomura, K. 1275
Oguro, D. 1499
Ohara, L. 291
Oizumi, H. 335, 1499, 1531, 1561, 1611
Okamoto, K. 505
Okoroanyanwu, U. 602
Omata, K. 829
Onodera, J. 1501
Orvek, K. 689, 1634, 1650
Oshino, T. 1275
Ota, K. 1564, 1583, 1631
Owada, T. 1531
Park, C. 1201
Park, J. 312, 398, 1201, 1201
Park, S.-J. 829
Partlow, M. 1461
Peters, J. H. 602

AUTHOR INDEX

Petrillo, K. 291
Pierson, B. 199
Pierson, W. 291
Prokofiev, A. V. 996
Rabellino, L. 1666
Raghunathan, S. 1417
Raju, R. 1077
Randive, R. 709
Rastegar, A. 663
Reiss, I. 709
Rekawa, S. 1530, 1560
Ren, L. 1417
Richardson, M. 956
Riddle, S. 1666
Roberts, J. 1433
Roller, J. 312
Ronse, K. 742
Rossinger, M. 602
Routh, R. 291
Ruzic, D. N. 1077
Sahoo, P. 363
Saidani, M. 363
Salamassi, F. 623
Salashchenko, N. 1018
Salmaso, G. 1041
Sanche, L. 1529
Santillan, J. J. 1561
Sarubbi, F. 1667
Sasaki, T. 1562
Sato, H. 1462
Schietinger, C. 1461
Scholze, F. 1667
Schwarz, M. 1341
Seo, H.-S. 646, 721, 848
Shevelko, A. 1479
Shigemura, H. 551, 1565, 1593
Shiina, K. 1565
Shimizu, M. 1542
Shimokawa, T. 1542
Shin, H. 1077
Shiono, D. 1501, 1513
Shirai, S. 1174, 1483
Shroff, Y. A. 623
Sidelnikov, Y. 1481
Sjmaenok, L. 1018
Sligte, E. te 1055
Soer, W. 1018, 1481
Sohn, J. 689
Solak, H. H. 363
Spiller, E. 709
Sporre, J. 1077
Stepanenko, N. 742
Storm, A. 1055
Suga, O. 551, 805, 1564, 1565, 1583, 1593, 1611, 1631

Sugiyama, T. 709
Szilagyi, J. 956
Tagawa, S. 505
Taguchi, T. 551, 1564, 1583, 1631
Takenoshita, K. 956
Tamura, T. 829
Tanaka, H. 1174, 1483
Tanaka, T. 805, 1593
Tanaka, Y. 1174, 1483
Tawarayama, K. 1174, 1483
Tchikoulaeva, A. 774
Teramoto, Y. 1462
Terasawa, T. 551, 805, 1565
Thomas, M. 1341
Tolbert, L. M. 376
Tomie, T. 805
Torre, A. 1391
Trogisch, S. 602
Tsybin, N. 1018
Uzawa, S. 264
Van Der Mullen, J. 1481
Van Herpen, M. 1481
Van Ingen-Schenau, K. 291
Van Kampen, M. 1055
Van Setten, E. 199
Vandentop, G. 1433
Vanneer, R. 1055
Verberk, R. 1055
Vinokhodov, A. Y. 996
Wagner, C. 199
Wallow, T. 291, 312, 426, 602
Warisawa, S. 1583
Watanabe, T. 848, 1501, 1562
Watso, R. 291
Weber, A. 363
Wieringa, F. P. 1119
Wood, O. 10, 291, 602, 1227, 1367
Wüest, A. 520, 602, 1529, 1634
Wurm, S. 1, 312, 398, 1367, 1417, 1480
Xiong, X. 1418, 1449, 1668
Yakshinskiy, B. 1141, 1529
Yakunin, A. 1018
Yamamoto, T. 1583
Yamane, T. 805
Yamatani, D. 1462
Yan, P.-Y. 623
Yankulin, L. 602, 1634
Yasaka, A. 1565
Yeo, J.-H. 1201
Yokokoji, O. 1562
Yokoyama, T. 1462
Yomogita, A. 1531
Yoshioka, M. 935
Yun, H. 663, 709
Zakhor, A. 1632

AUTHOR INDEX

Zalkind, S. 1141
Zhang, G. 829, 1563
Zimmerman, J. 1155
Zocchi, F. E. 1041

SEMATECH
2706 Montopolis Drive
Austin, Texas 78741

ISBN 978-1-61567-661-3

International Symposium on Extreme Ultraviolet Lithography 2008

(2008 EUVL Symposium)

Lake Tahoe, California, USA
28 September – 1 October 2008

Volume 3 of 3

International Symposium on Extreme Ultraviolet Lithography 2008

(2008 EUVL Symposium)

Lake Tahoe, California, USA
28 September – 1 October 2008

Volume 3 of 3

ISBN: 978-1-61567-661-3

Printed from e-media with permission by:

Curran Associates, Inc.
57 Morehouse Lane
Red Hook, NY 12571

Some format issues inherent in the e-media version may also appear in this print version.

Copyright© (2008) by SEMATECH
All rights reserved.

Printed by Curran Associates, Inc. (2009)

For permission requests, please contact SEMATECH
at the address below.

SEMATECH
2706 Montopolis Drive
Austin, Texas 78741

Phone: (512) 356-3500
Fax: (512) 356-7848

www.sematech.org

Additional copies of this publication are available from:

Curran Associates, Inc.
57 Morehouse Lane
Red Hook, NY 12571 USA
Phone: 845-758-0400
Fax: 845-758-2634
Email: curran@proceedings.com
Web: www.proceedings.com

TABLE OF CONTENTS

Volume 1

OPENING PRESENTATIONS

Welcome to the 2008 EUVL Symposium! ... 1
Stefan Wurm, Ichiro Mori, Rob Hartman

Program Logistics & Overview ... 10
Obert Wood, Patrick Naulleau

EUV Lithography's Future ... 21
Harry J. Levinson

Samsung Lithography Strategy ... 71
Woosung Han

EUV Activities the EUVL Shop Future Plans ... 97
Rob Hartman

US Regional Update ... 124
Patrick Naulleau

IEUVI Update ... 136
Paolo Gargini

Asia Pacific Regional Update - Japan, Korea and Taiwan Regional Update ... 183
Ichiro Mori

TECHNICAL SESSION: EXPOSURE TOOL

EUV Alpha Demo Tools – Stepping Stones Towards Volume Production ... 199
H. Meiling, S. Lok, B. Hultermans, E. van Setten, B. Pierson, K. Cummings, C. Wagner, N. Harned

Nikon EUVL Development Progress Update ... 230
Takaharu Miura

Development Status of Canon's Full-Field EUVL Tool ... 264
Shigeyuki Uzawa, Tokuyuki Honda, Hideki Morishima, Takayuki Hasegawa

TECHNICAL SESSION: RESIST I

EUV Resist Performance on the ASML ADT and LBNL MET ... 291
Bill Pierson, Tom Wallow, Hiroyuki Mizuno, Anita Fumar-Pici, Linda Ohara, Karen Petrillo, Koen van Ingen-Schenau, Steve Hansen, Sang-In Han, Robert Watso, Lior Huli, Obert Wood, Joerg Mallmann, Bart Kessels, Robert Routh, Kevin Cummings

The SEMATECH Berkeley MET: Learning at the 22 nm Node ... 312
Patrick Naulleau, Chris Anderson, Paul Denham, Simi George, Ken Goldberg, Brian Hoef, Gideon Jones, Dimitra Niakoula, Ryan Miyakawa, John Roller, Chawon Koh, Warren Montgomery, Stefan Wurm, Bruno La Fontaine, Tom Wallow, Andy Ma, Joo-on Park

Evaluation of EUV Resists at Selete ... 335
Hiroaki Oizumi, Daisuke Kawamura, Koji Kaneyama, Shinji Kobayashi, Toshiro Itani

Advances in Resist Testing at PSI EUV-IL Exposure Tool ... 363
Vaida Auželyte, Pratap Sahoo, Menouer Saidani, Anja Weber, Harun H. Solak

Positive and Negative Tone Molecular Resists for 22-nm Node EUVL Patterning ... 376
Richard Lawson, Cheng-Tsung Lee, Laren M. Tolbert, Clifford L. Henderson

TECHNICAL SESSION: RESIST II

Sub-22nm Half-Pitch (HP) EUV Resist Imaging Results ... 398
Chawon Koh, Stefan Wurm, Joo-on Park, Andy Ma, Patrick Naulleau

Corner Rounding in Photoresists for Extreme Ultraviolet Lithography ... 426
Christopher N. Anderson, Patrick P. Naulleau, Thomas Wallow, Yunfei Deng

Reconciling Resist Resolution Metrics 487
  Gregg M. Gallatin, Patrick P. Naulleau, Christopher N. Anderson

**Feasibility Study on High-Sensitivity Chemically Amplified Resist by Polymer Absorption
Enhancement in Extreme Ultraviolet Lithography** 505
  T. Kozawa, K. Okamoto, J. Nakamura, S. Tagawa

TECHNICAL SESSION: COST OF OWNERSHIP

Cost Implications of EUV Lithography Technology Decisions 520
  Andrea F. Wüest, Andrew J. Hazelton, Greg Hughes, Lloyd C. Litt, Frank Goodwin

Volume 2

TECHNICAL SESSION: MASK I

An Overview of a Development Program for EUVL Mask Technologies in Selete 551
  Osamu Suga, Tsuneo Terasawa, Hiroyuki Shigemura, Takao Taguchi, Iwao Nishiyama, Ichiro Mori

Development Status of EUVL Mask Blank and Substrate 563
  Yoshiaki Ikuta

Multilayer Defect Compensation to Enable Quality Masks for EUVL Production 583
  Ted Liang, Eric Gullikson

EUV Reticle Contamination and Cleaning 602
  *U. Okoroanyanwu, E. Langer, A. Fumar-Pici, T. Wallow, O. Wood, B. La Fontaine, C. Holfeld, J. H. Peters, M.
  Bender, M. Rossinger, S. Trogisch, F. Goodwin, A. Wüest, S. Huh, G. Denbeaux, Y. Fan, A. Antohe, L. Yankulin, R.
  Garg, K. Goldberg, P. Naulleau*

High Transmission EUVL Pellicle Development 623
  Yashesh A. Shroff, Pei-Yang Yan, Farhad Salamassi, Eric Gullikson

Applying Thinner Absorber to the EUVL Mask: EUV Printability and Integration Issues 646
  *Hwan-Seok Seo, Dong Gun Lee, Hoon Kim, Sungmin Huh, Byung-Sup Ahn, Hakseung Han, Dongwan Kim, Seong-
  Sue Kim, Han Ku Cho*

TECHNICAL SESSION: MASK II

Study of Pit Defect Formation on EUV Blank Substrates 663
  Abbas Rastegar, Sean Eichenlaub, Arun John Kadaksham, Matt House, Brian Cha, Henry Yun

Investigation of a Compensation Method for Pattern Placement Shifts of Chucked EUVL Masks 689
  J. Sohn, K. Orvek, R. Engelstad, A. Lyons, J. Hartley

High Throughput Defect Mitigation 709
  Patrick Kearney, C. C. Lin, T. Sugiyama, H. Yun, R. Randive, I. Reiss, P. Mirkarimi, E. Spiller, A. Hayes

Characterization of EUV Mask Defects: Printability and Repair Process 721
  *Hakseung Han, Donggun Lee, Hwan-Seok Seo, Kenneth A. Goldberg, Hoon Kim, Byung-Sub Ahn, In-Yong Kang,
  Wonil Cho, Sanghyeon Lee, Suyoung Lee, Geunbae Kim, Dongwan Kim, Seong-Sue Kim, HanKu Cho*

Mask Defect Printability in Full Field EUV Lithography – Part 2 742
  R. Jonckheere, F. Iwamoto, N. Stepanenko, A. M. Goethals, K. Ronse

TECHNICAL SESSION: METROLOGY/INSPECTION

A Practical Approach to EUV Reticle Inspection 774
  Anna Tchikoulaeva, et.al.

Development of Actinic Mask Blank Inspection Technology at Selete 805
  Tsuneo Terasawa, Takeshi Yamane, Teruo Iwasaki, Toshihiko Tanaka, Osamu Suga, Toshihisa Tomie

EUVL Blank Defect Inspection Capability at Intel 829
  Andy Ma, Ted Liang, Seh-Jin Park, Guojing Zhang, Tomoya Tamura, Kazunori Omata

Analysis of Sub-22-nm Aerial Image Using Coherent Scattering Microscopy 848
  *Dong Gun Lee, Junki Kishimoto, Takeo Watanabe, Hiroo Kinoshita, Hwan-Seok Seo, Dongwan Kim, Seong-Sue
  Kim, HanKu Cho*

Aerial Image Linewidth Measurement Capabilities of the Actinic Inspection Tool 867
  Kenneth A. Goldberg, Iacopo Mochi, Patrick Naulleau, Bruno LaFontaine, Sungmin Huh

TECHNICAL SESSION: SOURCE I

Laser Produced Plasma Source System Development 890
David C. Brandt, et.al.

CO₂ Laser-produced Sn Plasma Source for EUV Lithography 916
Akira Endo, et.al.

EUV Sources based on DPP 935
Marc Corthout, Masaki Yoshioka

Time-Multiplexed Solid-State Laser-driven EUV Sources for Beta-Tools and HVM 956
K. Takenoshita, R. Bernath, R. Kamptaprasad, J. Szilagyi, S. A. George, J. Cunado, M. Richardson, B. Fulford, I. Henderson, N. Hay, S. Ellwi

TECHNICAL SESSION: SOURCE II

Guidelines and Promising Approach for LPP-EUV Light Source for HVM 975
K. Nishihara, et.al.

Technological Aspects of Sn RDE Source Development for HVM Lithography 996
V. M. Borisov, G. N. Borisova, A. S. Ivanov, Yu. B. Kirukhin, O. B. Khristoforov, V. A. Mishchenko, A. V. Prokofiev, A. Yu. Vinokhodov

Spectral Purity Filter Development for EUV HVM 1018
A. Yakunin, V. Banine, N. Salashchenko, E. Kluenkov, A. Lopatin, V. Luchin, N. Tsybin, L. Sjmaenok, W. Soer, M. Jak

Design and Fabrication Considerations of EUVL Collectors for HVM 1041
G. Bianucci, J. Kools, G. Salmaso, F. E. Zocchi

TECHNICAL SESSION: CONTAMINATION AND PARTICLES

Strategy for Minimizing EUV Optics Contamination During Exposure 1055
N. Harned, R. Moors, M. van Kampen, V. Banine, J. Huijbregtse, R. Vanneer, A. Kempen, D. Ehm, R. Verberk, E. te Sligte, A. Storm

Effective Debris Detection, Mitigation and Cleaning Methods for Source - Collector Optics 1077
David N. Ruzic, Ramasamy Raju, J. Sporre, H. Shin, W. M. Lytle, M. J. Neumann

Volume 3

TECHNICAL SESSION: CONTAMINATION AND PARTICLES (cont,)

Shielded Plasma's for Cleaning EUV Mirrors 1119
N. B. Koster, R. Koops, K. Agovic, F. P. J. de Groote, F. P. Wieringa, M. G. H. Meijerink

Carbon Accumulation on Model MLM Cap Layer: Interaction of Benzene Vapor with TiO2 Surface ... 1141
Boris Yakshinskiy, Shimon Zalkind, Theodore E. Madey

Progress on EUV Reticle Dual Pod Carriers for use in the Fab and Exposure Tools 1155
John Zimmerman, Long He

TECHNICAL SESSION: EXPOSURE TOOL EVALUATION

Lithographic Performance of Selete's Full Field EUV Exposure Tool 1174
Kazuo Tawarayama, Hajime Aoyama, Takashi Kamo, Shunko Magoshi, Yuusuke Tanaka, Seiichiro Shirai, Hiroyuki Tanaka

Full-field Patterning Test with ADT for 30-nm Node Device Application 1201
Doohoon Goo, Insung Kim, Joo-On Park, Jeonghoon Lee, Changmin Park, Jinhong Park, Jeong-Ho Yeo, Sungwoon Choi, Woosung Han

Flare Evaluation of an ASML Alpha Demo Tool 1227
Hiroyuki Mizuno, Martin Burkhardt, Chiew-seng Koay, Greg McIntyre, Tim Brunner, Bruno La Fontaine, Yunfei Deng, Obert Wood

Implementing Full field EUV Lithography using the ADT 1242
Anne-Marie Goethals, et.al.

TECHNICAL SESSION: OPTICS

Improvement of Optics for EUV Exposure Tools 1275
Katsuhiko Murakami, Tetsuya Oshino, Hiroyuki Kondo, Hiroshi Chiba, Kazushi Nomura

Optics for EUV Lithography 1300
Peter Kuerz, et.al.

Projection Optics for a 0.5-NA Microstepper Upgrade 1321
Michael Goldstein, Russ Hudyma, Patrick Naulleau

Projection Architectures for High NA EUVL 1341
Russ Hudyma, Mike Thomas, Mark Schwarz

CLOSING ADDRESS

2008 EUVL Symposium - Closing Address 1367
Stefan Wurm, Obert Wood, Patrick Naulleau

POSTERS: EXPOSURE TOOL

Ultra-High Resolution Extreme Ultraviolet Lithography by Incoherent to Coherent Conversion 1389
Christopher N. Anderson, Patrick P. Naulleau

MOSAIC - A New Way to Measure Optical Aberrations 1390
Christopher N. Anderson, Patrick P. Naulleau

First Italian EUV Micro Exposure Tool at 14.4 nm based on Kr DMS 1391
S. Bollanti, P. Di Lazzaro, F. Flora, L. Mezi, D. Murra, A. Torre

Lithographic Modeling of a 0.5-NA Microstepper Optic 1416
Patrick Naulleau, Michael Goldstein, Russ Hudyma

Impact of Flare and Aberrations on Patterning Performance - Simulation with the EUV Full Field Alpha Tool Conditions 1417
Liping Ren, Frank Goodwin, Stefan Wurm, Chawon Koh, Sudharshanan Raghunathan, John Hartley

Multiple Catadioptric Simplified Extreme Ultraviolet Whole Lithography Machine and System 1418
Wynn L. Bear, Xiangwen Xiong

Reliability and Productivity Improvements on the Intel MET 1433
Roman Caudillo, Jeanette Roberts, Terence Bacuita, Gilroy Vandentop

Composite Double Reflection Simplified Extreme Ultraviolet Whole Lithography Machine and System 1449
Wynn L. Bear, Xiangwen Xiong

POSTERS: SOURCE

Spectral Purity Filter Life-Time Testing on EQ-10 Source 1461
Chuck Schietinger, Dave Grove, Travis Ayers, Matthew Partlow, Debbie Gustafson

Dependence of Laser Parameter on Conversion Efficiency in High-Repetition-Rate Laser-Ablation-Discharge EUV Source 1462
Takuma Yokoyama, Hiroshi Mizokoshi, Yusuke Teramoto, Daiki Yamatani, Hiroto Sato, Kazuaki Hotta

EUV Spectrometers for Source Development, Characterization and Optimization 1479
Scott Bergeson, Bryce Allred, Jershon Lopez, Jeffrey Kemp, Larry Knight, Alexander Shevelko

High Power from Low Etendue EUV Light Sources 1480
Michael Goldstein, Stefan Wurm, Frank Goodwin

The Characterization of the Ion Beam from a Sn-based DPP with Respect to the Ignition Parameters 1481
K. Gielissen, Y. Sidelnikov, W. Soer, M. van Herpen, D. Glushkov, V. Banine, J. van der Mullen

Present Status of Laser-produced Plasma EUV Light Source 1482
Hiroshi Komori, et.al.

Performance Evaluation of EUV SFET Source Collector Module 1483
Shunko Magoshi, Seiichiro Shirai, Hideto Mori, Kazuo Tawarayama, Yuusuke Tanaka, Hiroyuki Tanaka

POSTERS: RESISTS

Development of New Negative-tone Molecular Resists based on Alkylphenyl Callxarene for EUVL 1499
Masatoshi Echigo, Dai Oguro, Hiroaki Oizumi, Toshiro Itani

Survey and Comparison of Deprotection Blur Metrics for Extreme Ultraviolet Photoresists 1500
Christopher N. Anderson, Patrick P. Naulleau

Investigation of CA Resist Decomposition by EUV and EB Exposure 1501
Daiju Shiono, Taku Hirayama, Hideo Hada, Junichi Onodera, Takeo Watanabe, Hiroo Kinoshita

Novel Polyphenol Base Molecular Resist Having High Thermal Resistance 1513
Taku Hirayama, Takeyoshi Mimura, Jun Iwashita, Makiko Irie, Daiju Shiono, Hideo Hada, Takeshi Iwai

Electrons in EUV Resist Activation 1529
Theodore E. Madey, B. V. Yakshinskiy, E. Loginova, L. Sanche, P. Cloutier, R. Brainard, A. Wuest

DUV Source Integration into the 0.3 NA Berkeley SEMATECH MET for OOB Exposure Studies 1530
Simi A. George, Patrick P. Naulleau, Senajith Rekawa, Drew Kemp

Development of Novel Positive-tone Photoresists for EUVL 1531
Takanori Owada, Akinori Yomogita, Takashi Kashiwamura, Hiroaki Oizumi, Toshiro Itani

Relation between Acid Diffusion and Resolution in Chemically Amplified EUV Resists 1542
Y.uuki Hirai, Makoto Shimizu, Ken Maruyama, Toshiyuki Kai, Tsutomu Shimokawa, Toshiro Itani, Daisuke Kawamura

EUV Photoresists Twice as Fast as Previously Thought 1560
Patrick Naulleau, Eric Gullikson, Andrew Aquila, Paul Denham, Simi George, Drew Kemp, Dimitra Niakoula, Seno Rekawa

EUV Resist Outgassing Quantification and Qualification Analysis Methods 1561
Shinji Kobayashi, Julius Joseph Santillan, Hiroaki Oizumi, Toshiro Itani

Development of Partially Fluorinated EUV Resist Polymers for Sensitivity Improvement 1562
Takashi Sasaki, Osamu Yokokoji, Takeo Watanabe, Hiroo Kinoshita

POSTERS: MASK

Mask Effects on Line-Edge Roughness (LER) 1563
Patrick P. Naulleau, Kenneth A. Goldberg, Iacopo Mochi, Guojing Zhang

Experimental Study on Flatness of Electrostatically Chucked Reticle 1564
K. Ota, T. Taguchi, M. Amemiya, N. Nishimura, O. Suga

FIB Mask Repair Technology for EUV Lithography 1565
Tsuyoshi Amano, Yasushi Nishiyama, Hiroyuki Shigemura, Tsuneo Terasawa, Osamu Suga, Kensuke Shiina, Fumio Aramaki, Ryoji Hagiwara, Anto Yasaka

Analysis of Entrapped Object Size Effects on Out-of-Plane Distortion of the EUVL Mask in Electrostatic Chucking 1583
S. Lee, T. Yamamoto, K. Ota, N. Nishimura, T. Taguchi, I. Nishiyama, O. Suga, S. Warisawa, S. Ishihara

POSTERS: DEFECT INSPECTION

A Study of Optical Inspection on EUVL Mask for 32 nm Half Pitch Node Device and Beyond 1593
Yukiyasu Arisawa, Hiroyuki Shigemura, Tsuyoshi Amano, Hajime Aoyama, Toshihiko Tanaka, Osamu Suga

POSTERS: RETICLE CONTAMINATION

Characterization of EUV-Deposited Carboneous Contamination 1611
Toshihisa Anazawa, Yasushi Nishiyama, Hiroaki Oizumi, Osamu Suga, Iwao Nishiyama

Experimental Study of Particle-free Mask Handling Techniques using the MPE Tool 1631
Mitsuaki Amemiya, Kazuya Ota, Takao Taguchi, Osamu Suga

POSTERS: OPTICS AND ML COATINGS

Iterative Procedure for in situ Optical Testing using an Incoherent Source 1632
Ryan Miyakawa, Patrick P. Naulleau, Avideh Zakhor

Lateral Shearing Interferometry for EUV Optical Testing 1633
Ryan Miyakawa, Patrick P. Naulleau, Ken Goldberg

POSTERS: OPTICS CONTAMINATION

Resist Outgassing Measurements and Calibrations for High Volume Manufacturing............................ 1634
Greg Denbeaux, Alin Antohe, Rashi Garg, Chimaobi Mbanaso, LeonidYankulin, Yu-Jen Fan, Kevin Orvek, Andrea Wüest

Experiment of Contamination Generation by EUV Irradiation with the Use of High-mass Hydrocarbon Gas 1648
Masahito Niibe, Keigo Koida

EUVL Optics Contamination from Resist Outgassing: Status Overview 1650
Kevin Orvek, Greg Denbeaux, Alin Antohe, Rashi Garg, Chimaobi Mbanaso

Moisture and Hydrocarbon Management for EUVL Tools: Ultra High Vacuum and Purge Gas Purification Solutions 1666
Andrea Conte, Cristian Landoni, Paolo Manini, Larry Rabellino, Sarah Riddle

POSTERS: DEVICE INTEGRATION

Characterization of New EUV Stable Silicon Photodiodes.. 1667
F. Scholze, C. Laubis, F. Sarubbi, Lis K. Nanver, S. N. Nihtianov

POSTERS: COST OF OWNERSHIP

The Comprehensive Cost for the Mainstream NGL and Simplified Extreme Ultraviolet Lithography Method.. 1668
Wynn L. Bear, Xiangwen Xiong

Author Index

Shielded plasma's for cleaning EUV mirrors

TNO | Knowledge for business

Sematech EUVL symposium

September 30, 2008

Lake Tahoe

N.B. Koster, R. Koops, K. Agovic, F.P.J. de Groote, F.P. Wieringa, M.G.H. Meijerink

Contents

- Shielded Microwave Induced Remote Plasma (SMIRP)
- Cleaning rate Hydrogen Plasma
- Cleaning rate Helium Plasma
- EUV mirror damage study
- Future work
- Conclusions
- Acknowledgements

Shielded Microwave Induced Remote Plasma (SMIRP)

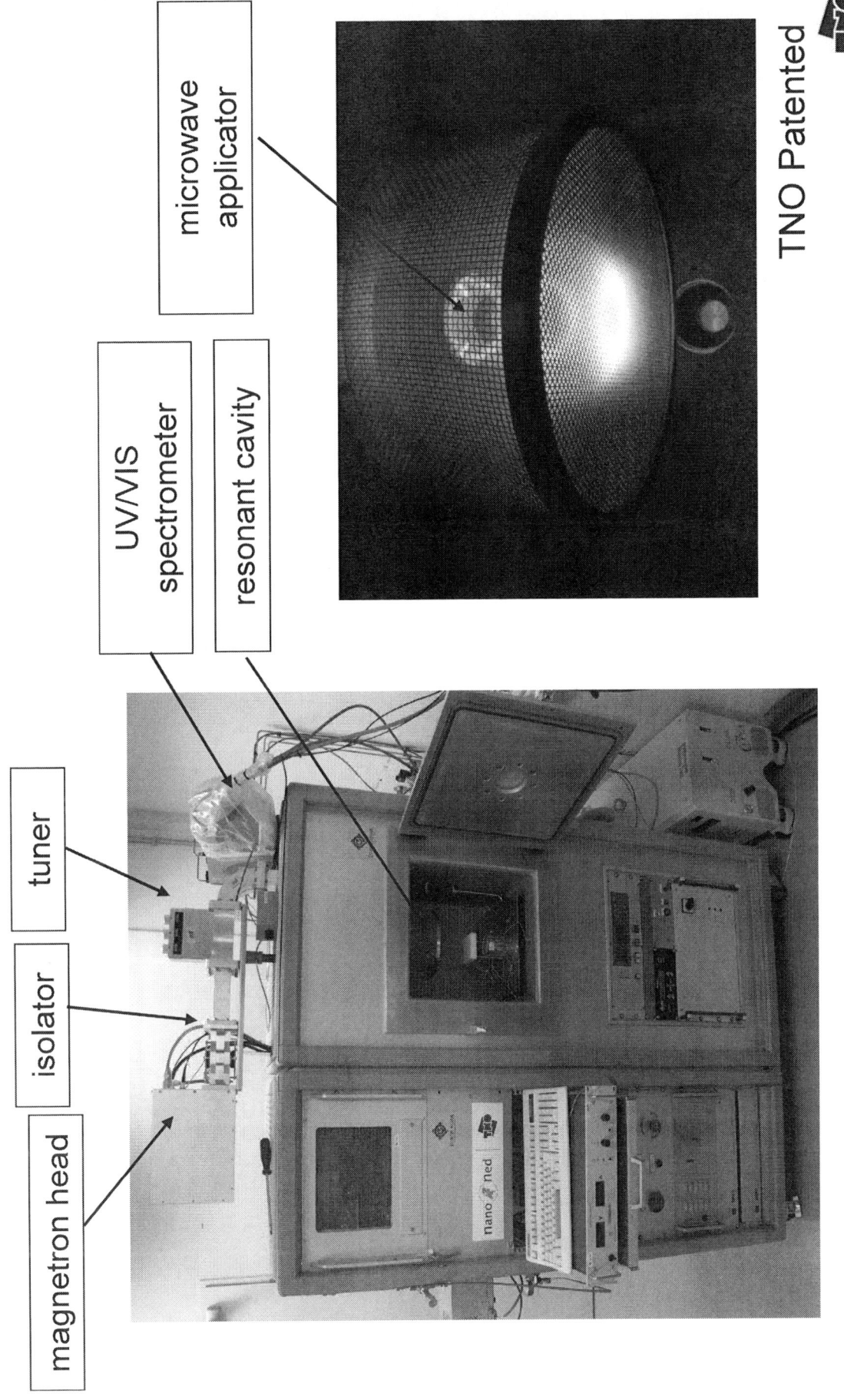

microwave applicator

UV/VIS spectrometer

resonant cavity

tuner

isolator

magnetron head

TNO Patented

Shielded Microwave Induced Remote Plasma (SMIRP)

Advantages

- Resonant cavity for high efficiency power coupling

- No EM fields outside plasma formation region

- Constant load for power supply

- Very short pulses possible (μs to s)

- Several gasses possible: H_2, O_2 (He, N_2, CO and CO_2 under investigation)

- Power easy adjustable (peak power, duty cycle or pulsewidth)

- Wide working pressure range (0.01 - 10 mbar)

- Low thermal load

Shielded Microwave Induced Remote Plasma (SMIRP)

Field-induced arcing and redeposition

Check for metal deposition

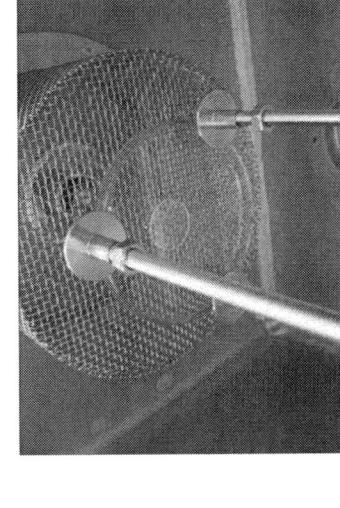

XPS-analysis: 100% clean!

Shielded antenna

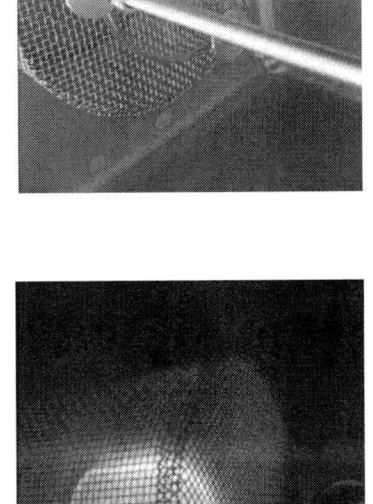

No arcing

Non-shielded antenna

Arcing !

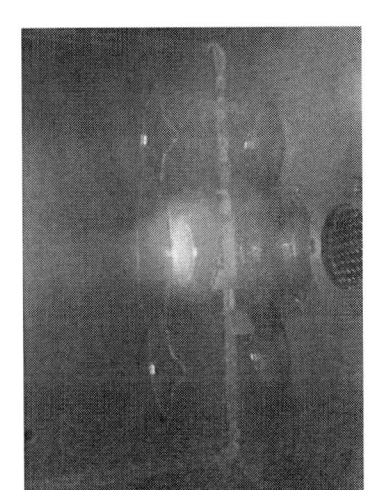

Contents

- Shielded Microwave Induced Remote Plasma (SMIRP)
- Cleaning rate Hydrogen Plasma
- Cleaning rate Helium Plasma
- EUV mirror damage study
- Future work
- Conclusions
- Acknowledgements

Cleaning rate hydrogen plasma

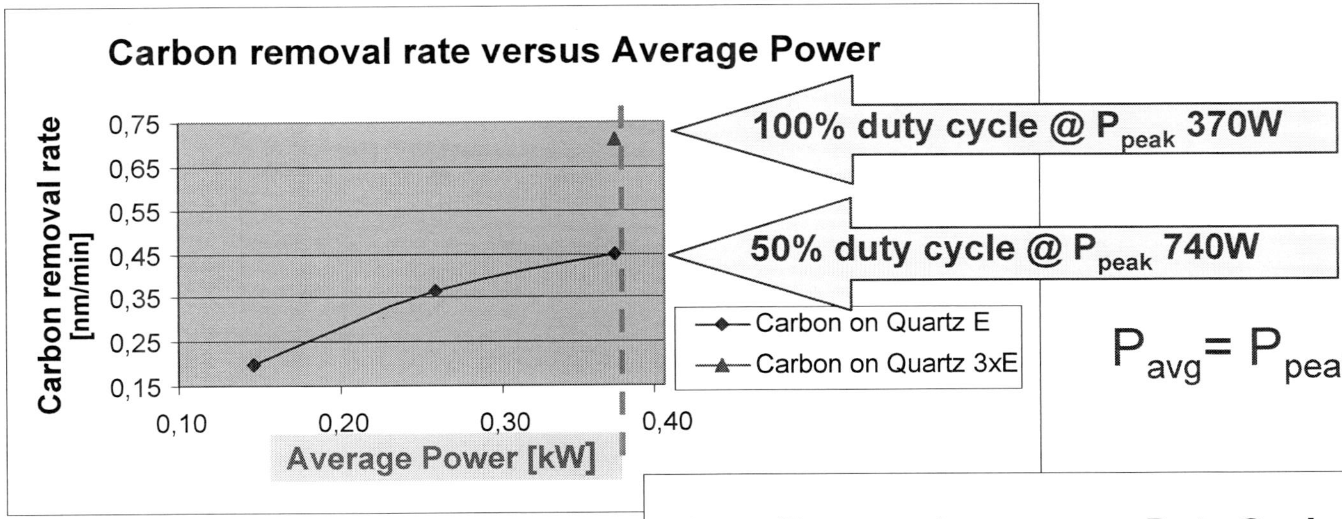

Carbon removal rate versus Average Power

100% duty cycle @ P_{peak} 370W

50% duty cycle @ P_{peak} 740W

- ◆ Carbon on Quartz E
- ▲ Carbon on Quartz 3xE

$$P_{avg} = P_{peak} * \text{Duty Cycle}$$

Carbon removal rate [nm/min] vs Average Power [kW]

Temperature versus Duty Cycle

- ◆ Carbon on Quartz E
- ■ Metal on Quartz E
- ▲ Carbon on Quartz 3xE

Temperature [Celsius] vs Duty Cycle [%]

Pulse: 130, 250, 500 µs and CW

Pressure: 100 Pa

Flow: 100 sccm

Distance: 1 mm

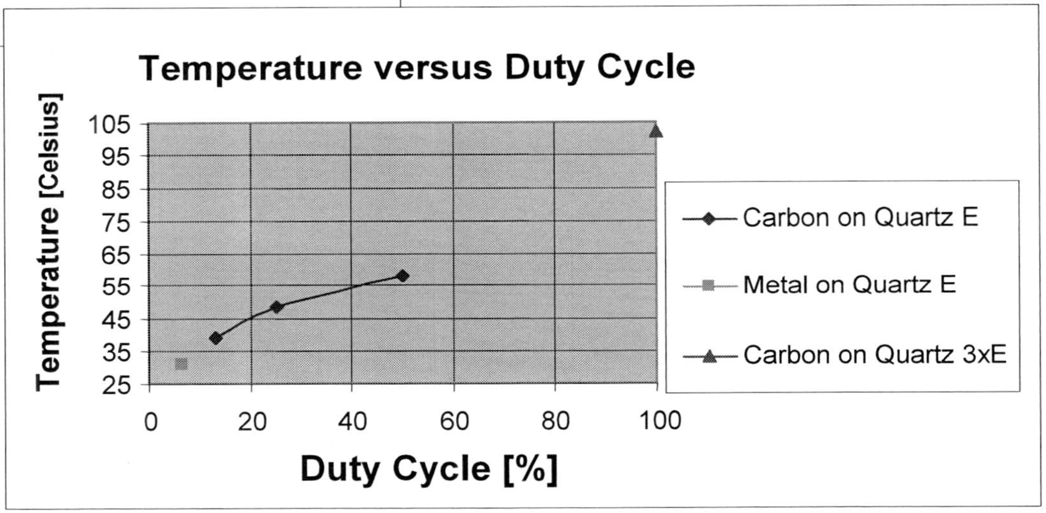

Cleaning rate hydrogen plasma

Carbon on quartz GO flats exposed to shielded H_2 plasma with different pulse widths and constant duty cycle.

Peak power: 500 W

Pressure: 100 Pa

Duration: 30 min

Distance: 30 mm

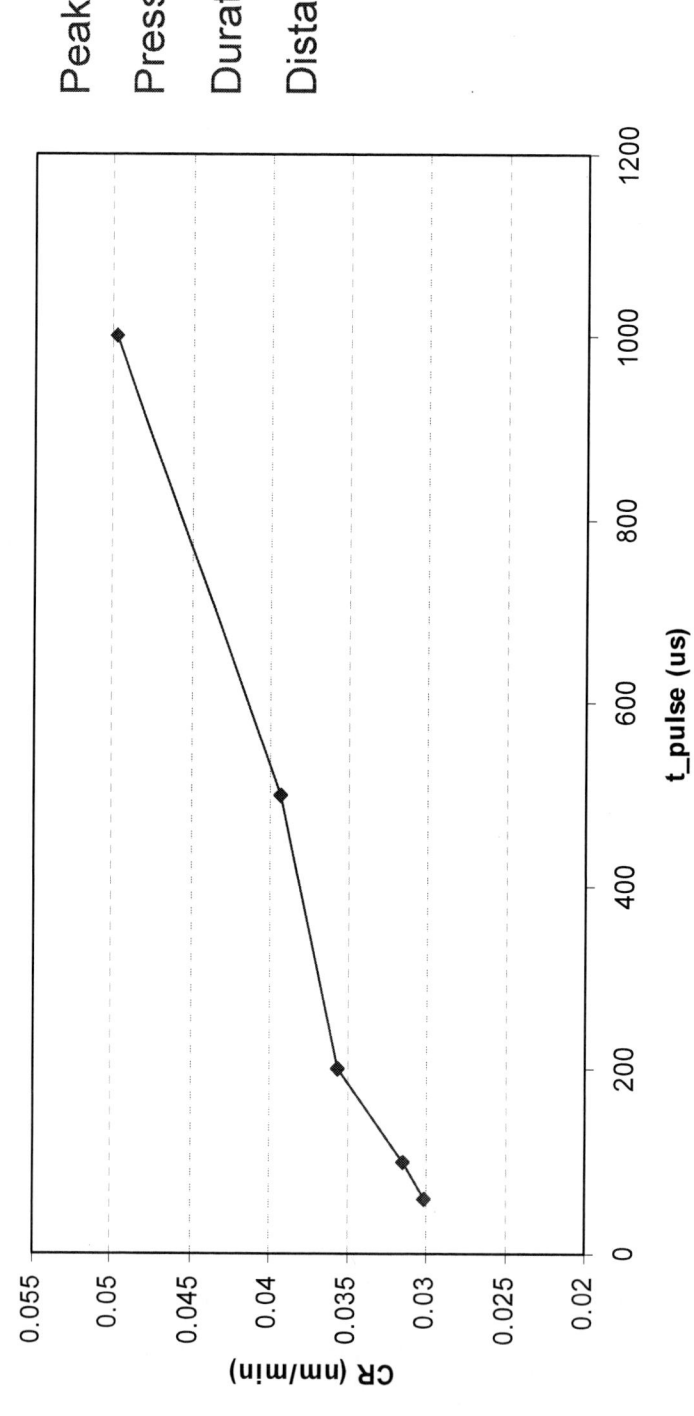

Cleaning Rate @ constant duty cycle (10%)

Contents

- Shielded Microwave Induced Remote Plasma (SMIRP)
- Cleaning rate Hydrogen Plasma
- Cleaning rate Helium Plasma
- EUV mirror damage study
- Future work
- Conclusions
- Acknowledgements

Cleaning rate helium plasma

Carbon on quartz GO flats exposed to shielded He plasma with different pulse periods

Peak power: 500 W

Pressure: 200 Pa

Duration: 30 min

Distance: 30 mm

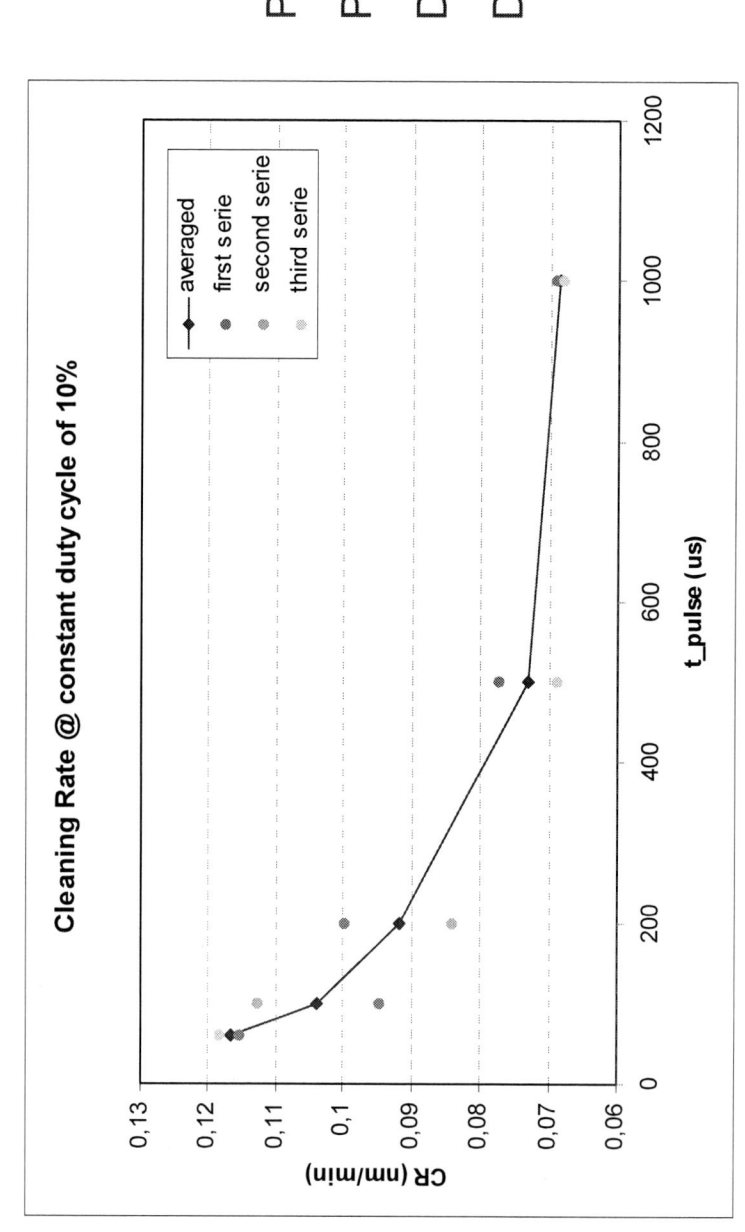

Helium plasma cleaning: radicals or ions?

Applying an increasing repellant electric field initially showed a distinct floor level in the C-removal rate, which might be associated with He*-cleaning without ion-etching.

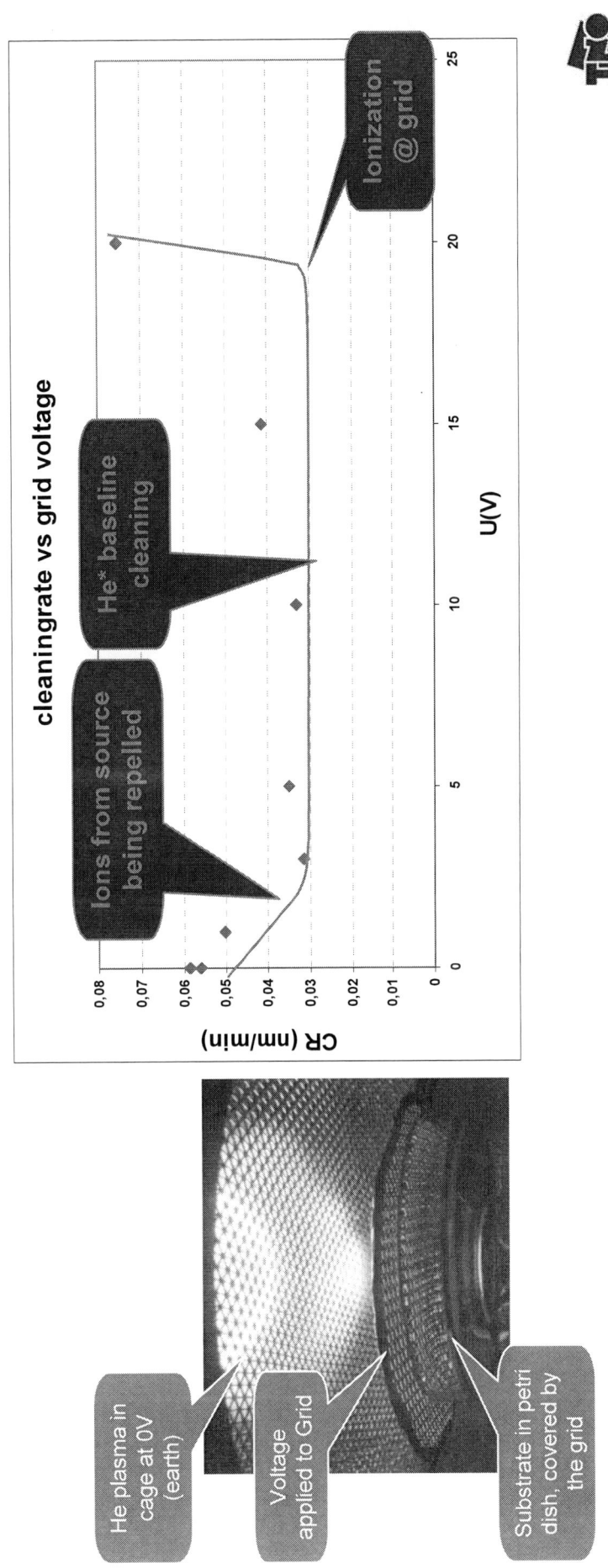

cleaningrate vs grid voltage

Ions from source being repelled

He* baseline cleaning

Ionization @ grid

CR (nm/min)

U(V)

He plasma in cage at 0V (earth)

Voltage applied to Grid

Substrate in petri dish, covered by the grid

Helium and hydrogen plasma compared

- Carbon succesfully removed from Quartz GO flats
- Carbon removal rate with helium is 3 to 4 times higher than with H_2 plasma
- Sputtering by He-ions seems to occur
- He-radical chemistry still unclear

Reactive species	He^+ & He^*	He^* only (estimate[1])	H^+ & H^*
CR (nm/hr)	5.0	~2.0	1.5
T_{max} (°C)	45	<45	53

(1) Based on results from ion-exclusion tests on earlier sheet

Contents

- Shielded Microwave Induced Remote Plasma (SMIRP)
- Cleaning rate Hydrogen Plasma
- Cleaning rate Helium Plasma
- EUV mirror damage study
- Future work
- Conclusions
- Acknowledgements

EUV mirror damage study

Hydrogen plasma

1 uncontaminated EUV MLM sample mirror hydrogen plasma exposed at settings for 6 nm C-removal for damage study

P_{peak} = 1000 W
duty cycle = 6.5% (50 W average power)
exposure time = 30 min
P = 100 Pa

Temperature after exposure 31 ^{0}C

EUV-mirror below shielding

Pulsed H_2 plasma exposure

EUV mirror damage study

Hydrogen plasma

30 min Hydrogen Plasma cleaned

	Pre	Post	Difference
Date	9-11-06	5-05-07	25.3 weeks
R (%)	67.22	67.17	-0.07 % ΔR/R
CTW96 (nm)	13.550	13.553	0.003 nm
CTW50 (nm)	13.507	13.503	-0.003 nm
FWHM (nm)	0.549	0.546	-0.003 nm
SE peak (nm)	13.43	13.43	0.003 nm

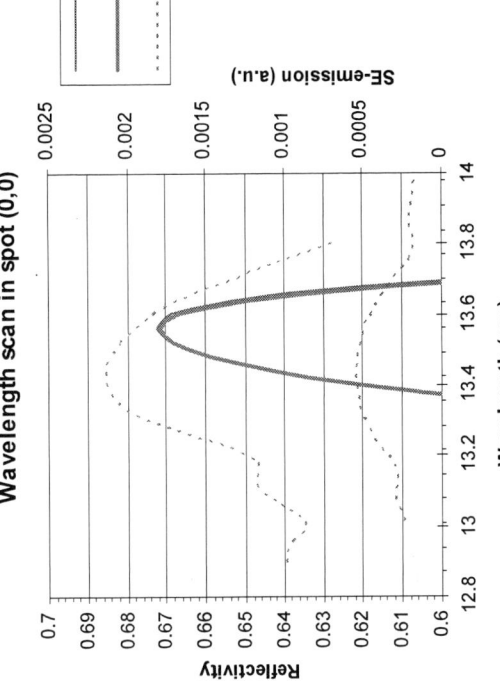

Lab witness

	Pre	Post	Difference
Date	18-11-06	5-05-07	24.0 weeks
R (%)	66.75	66.71	-0.06 % ΔR/R
CTW96 (nm)	13.550	13.553	0.004 nm
CTW50 (nm)	13.505	13.503	-0.003 nm
FWHM (nm)	0.548	0.546	-0.002 nm
SE peak (nm)	13.44	13.47	0.023 nm

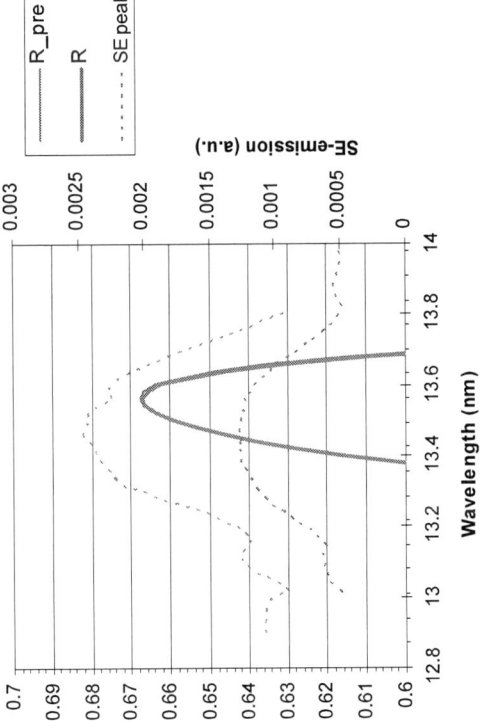

EUV mirror damage study

Hydrogen plasma

Hydrogen cleaned EUV mirror reflectometry results

- 30 min plasma exposed: DR/R = - 0.07%
- Lab witness: DR/R = - 0.06%
- Reflectivity is OK, although no reflectivity gain due to cleaning
- Lab witness showed small STOCC buildup after 24 weeks (but pre measurement already showed low reflectivity)
- CTW50 shifts are < 10 pm, so temperature increase cannot have exceeded 40 degrees.

- Conclusion: No damage to MLM found!

EUV mirror damage study

Helium plasma

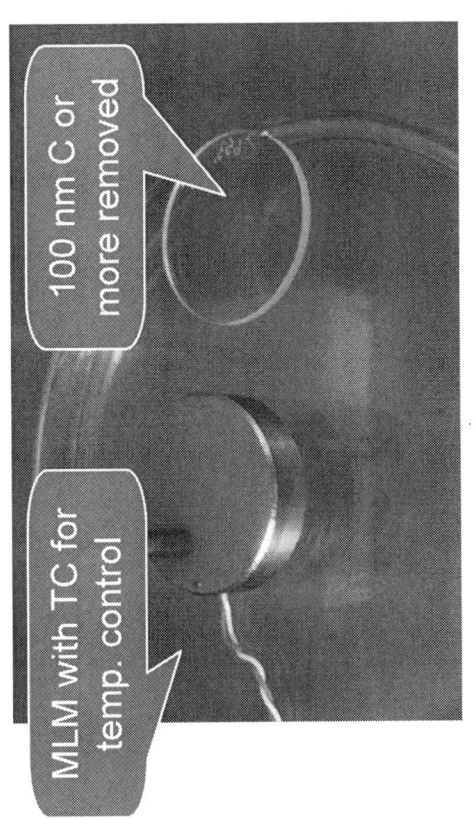

MLM with TC for temp. control

100 nm C or more removed

property	gain
Reflectivity R	-4.39%
CTW96	0.016nm
CTW50	0.002nm
FWHM	-0.002nm

A blank (not contaminated!) EUV MLM has been exposed to a He-plasma dose equivalent to ~100nm C-removal without any ion deflection. Reflectrometry measurements revealed:

The shifts in the reflection peak (CTW@ 50% and 96% peak height) indicate sputtering of ~1.3 nm capping-layer).
This is confirmed by the decrease in FWHM.
The heavy reflection loss is probably due to oxidation of the exposed Molybdenum (Mo).

Contents

- Shielded Microwave Induced Remote Plasma (SMIRP)
- Cleaning rate Hydrogen Plasma
- Cleaning rate Helium Plasma
- EUV mirror damage study
- Future work
- Conclusions
- Acknowledgements

Future work

- Investigate He* chemistry
- Optimize pulse length for maximum radical formation

New research tool for plasma cleaning with lots of in-situ analytical instruments at Delft Nanolab facility (Van Leeuwenhoek Laboratory)

- Ellipsometer
- UV/VIS spectrometer
- Interferometer
- Ion analyzer
- Langmuir probe
- Mass Spectrometer

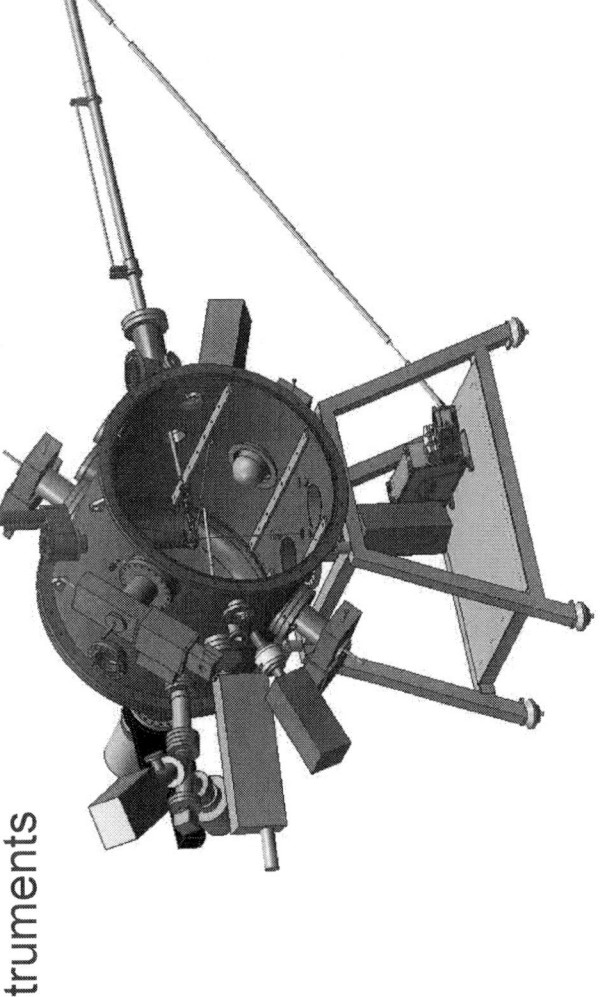

Contents

- Shielded Microwave Induced Remote Plasma (SMIRP)
- Cleaning rate Hydrogen Plasma
- Cleaning rate Helium Plasma
- EUV mirror damage study
- Future work
- Conclusions
- Acknowledgements

Conclusions

H_2 shielded plasma looks very promising for EUV mirror cleaning:

- cleaning rate easy adjustable
- surface temperature below d-spacing shift threshold
- limited heat load
- no damage to MLM

He plasma cleaning looks promising:

- higher cleaning rates than with H_2
- lower heat load
- shorter pulses gives higher cleaning rate
- But: damage to MLM
- chemistry not understood

Acknowledgements

We wish to thank the following parties for the assistance with this research:

Fabrication of Multilayers:
FOM Institute for Plasma Physics Rijnhuizen

Reflectometry:
Center for X-Ray optics and Berkeley Lab
Physikalisch-Technischen Bundesanstalt

This work was performed as part of a TNO internal research program

Carbon accumulation on model MLM cap layer: Interaction of benzene vapor with TiO_2 surface

Boris Yakshinskiy, Shimon Zalkind, and Theodore E. Madey

Dept. of Physics and Astronomy, Rutgers, The State University of New Jersey, Piscataway NJ 08854 USA

email: yaksh@physics.rutgers.edu

Outline

1. Characterize thermal-induced interaction benzene-TiO_2

2. Characterize electron-induced defect formation on $TiO_2(011)$

3. Characterize carbon deposition upon irradiation in benzene vapor; compare with C growth from MMA

4. Characterize mitigation effects in O_2 + benzene

5. Estimate cross-section for electron-induced reaction

6. Characterize the role of secondary electrons

Methods

TPD: Temperature Programmed Desorption

*clean surface
*dose gas (benzene, MMA)
*heat at linear rate
*determine energies, lifetimes

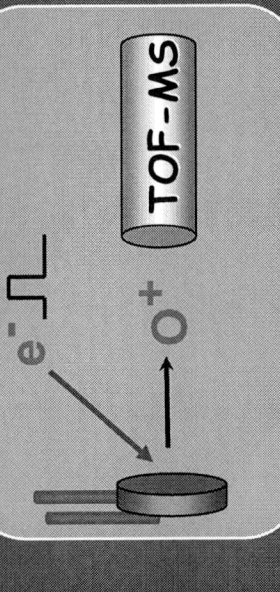

ESD: Electron stimulated desorption and reaction

XPS: X-ray photoelectron spectroscopy

LEIS: Low energy ion scattering

Secondary electron yield (synchrotron radiation)

Temperature Programmed Desorption (TPD) of C_6H_6 from clean and C-covered $TiO_2(011)$

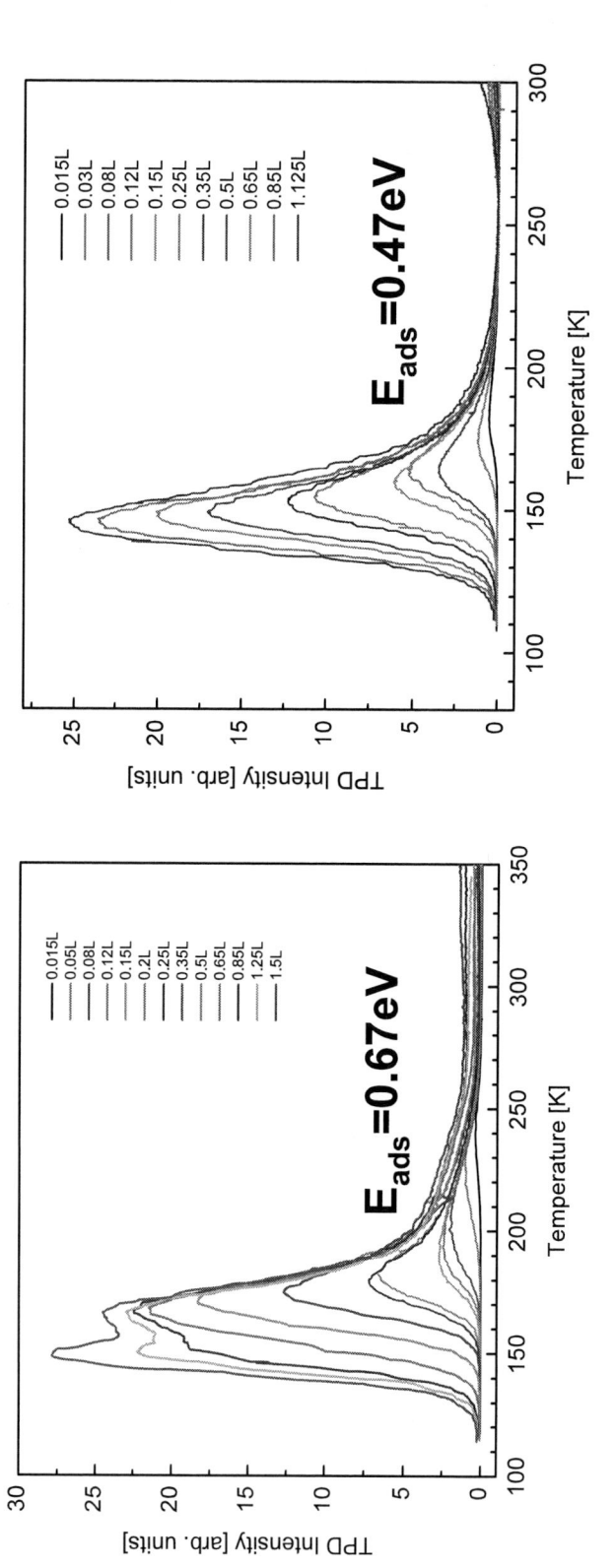

- Adsorption energy decreases with increasing coverage
- No carbon accumulation

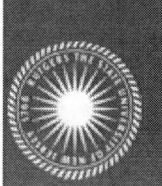

Steady-state benzene and MMA coverage

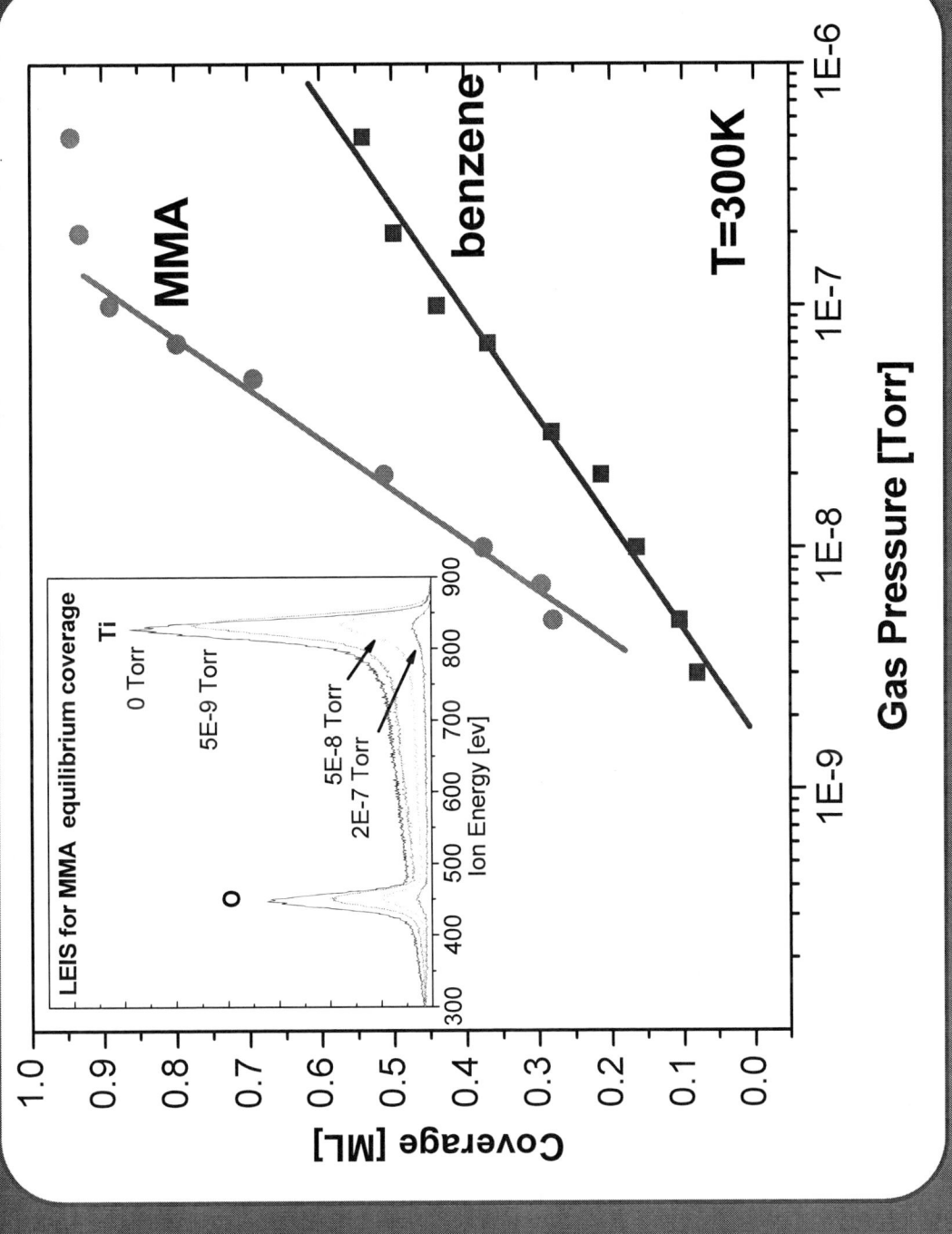

LEIS control. Coverage depends on Log p.

Temkin isotherm

Electron irradiation induces defects on $TiO_2(011)$

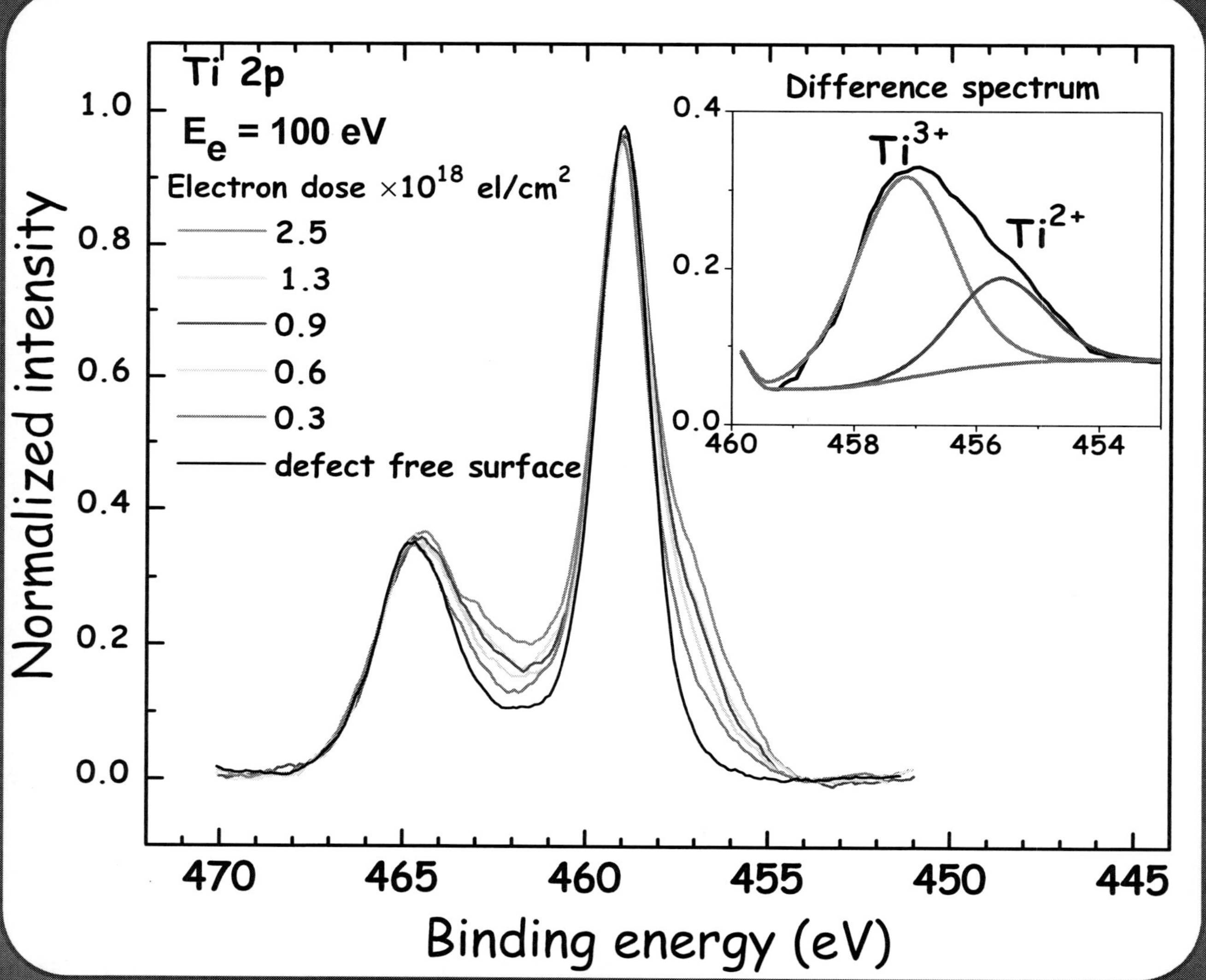

Ti 2p

$E_e = 100$ eV

Electron dose $\times 10^{18}$ el/cm^2

- 2.5
- 1.3
- 0.9
- 0.6
- 0.3
- defect free surface

Difference spectrum

Ti_i^{3+}

Ti_i^{2+}

Normalized intensity

Binding energy (eV)

Anion vacancy defects are produced by electron stimulated desorption of surface oxygen atoms.

Electron irradiation induces defects on TiO₂(011)

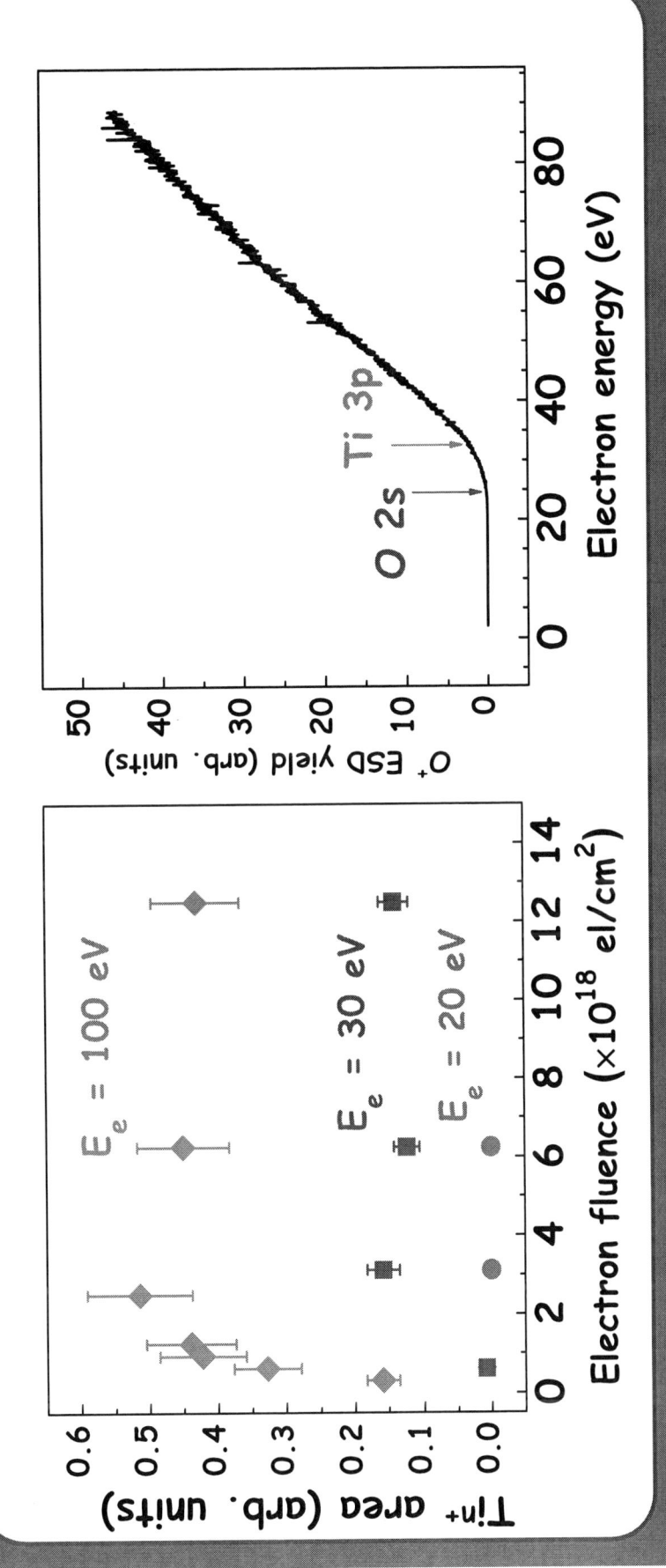

Electron energy > 25 eV leads to O vacancy formation on TiO₂

C growth under electron irradiation

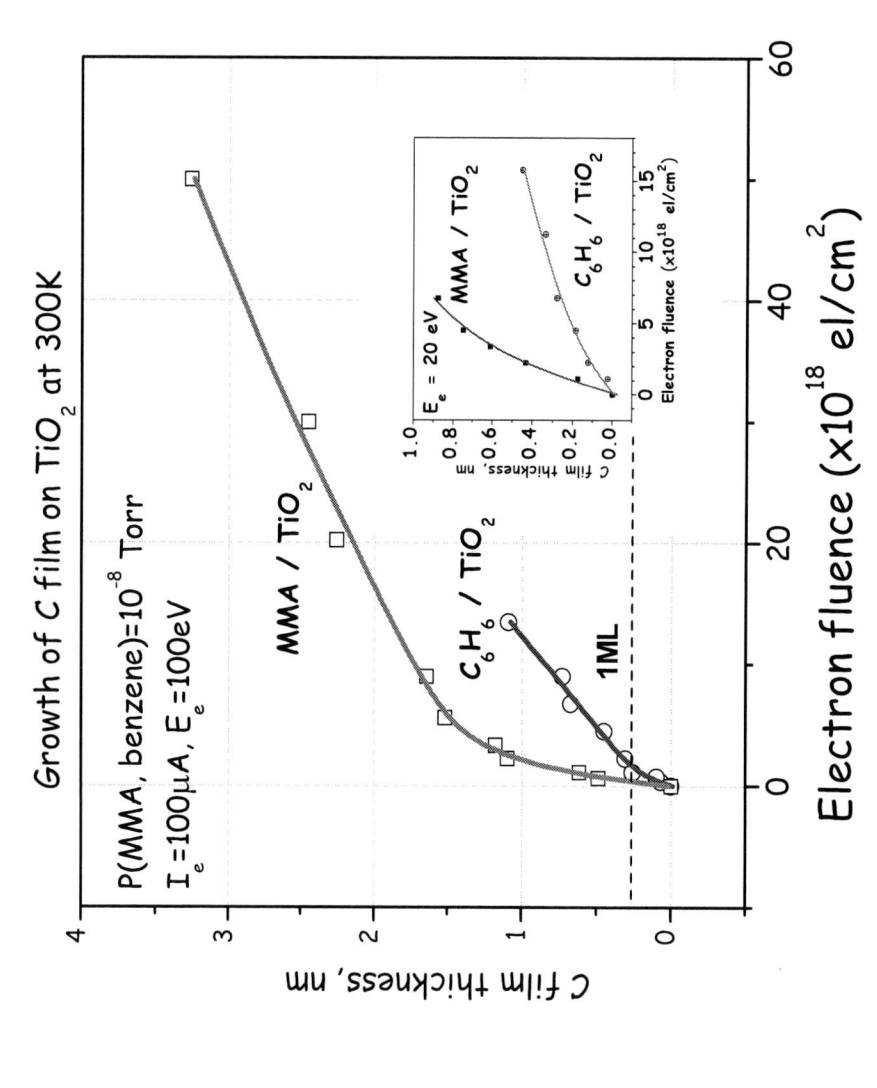

Growth of C film on TiO₂ at 300K

$P(MMA, benzene) = 10^{-8}$ Torr
$I_e = 100\mu A$, $E_e = 100 eV$

MMA / TiO₂

C_6H_6 / TiO₂

1ML

Electron fluence ($\times 10^{18}$ el/cm²)

C film thickness, nm

Inset:

$E_e = 20$ eV

MMA / TiO₂

C_6H_6 / TiO₂

Electron fluence ($\times 10^{18}$ el/cm²)

C film thickness, nm

Initial C growth rate from MMA >> than from benzene

C growth on TiO₂ under electron irradiation in benzene vapor: effects of substrate temperature

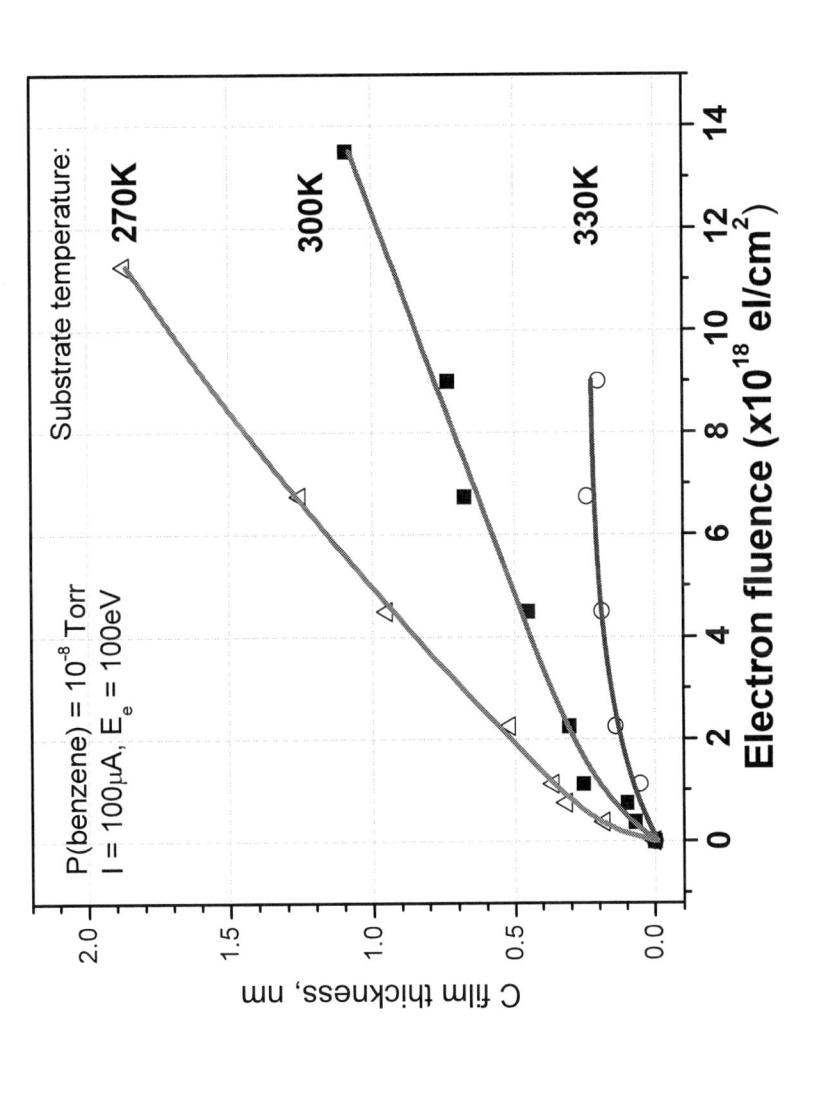

Surface concentration, lifetime

C growth on TiO$_2$ under electron irradiation: effects of benzene pressure; mitigation in O$_2$

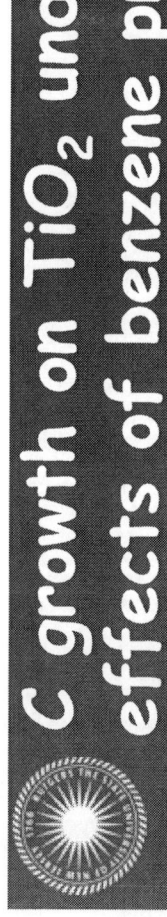

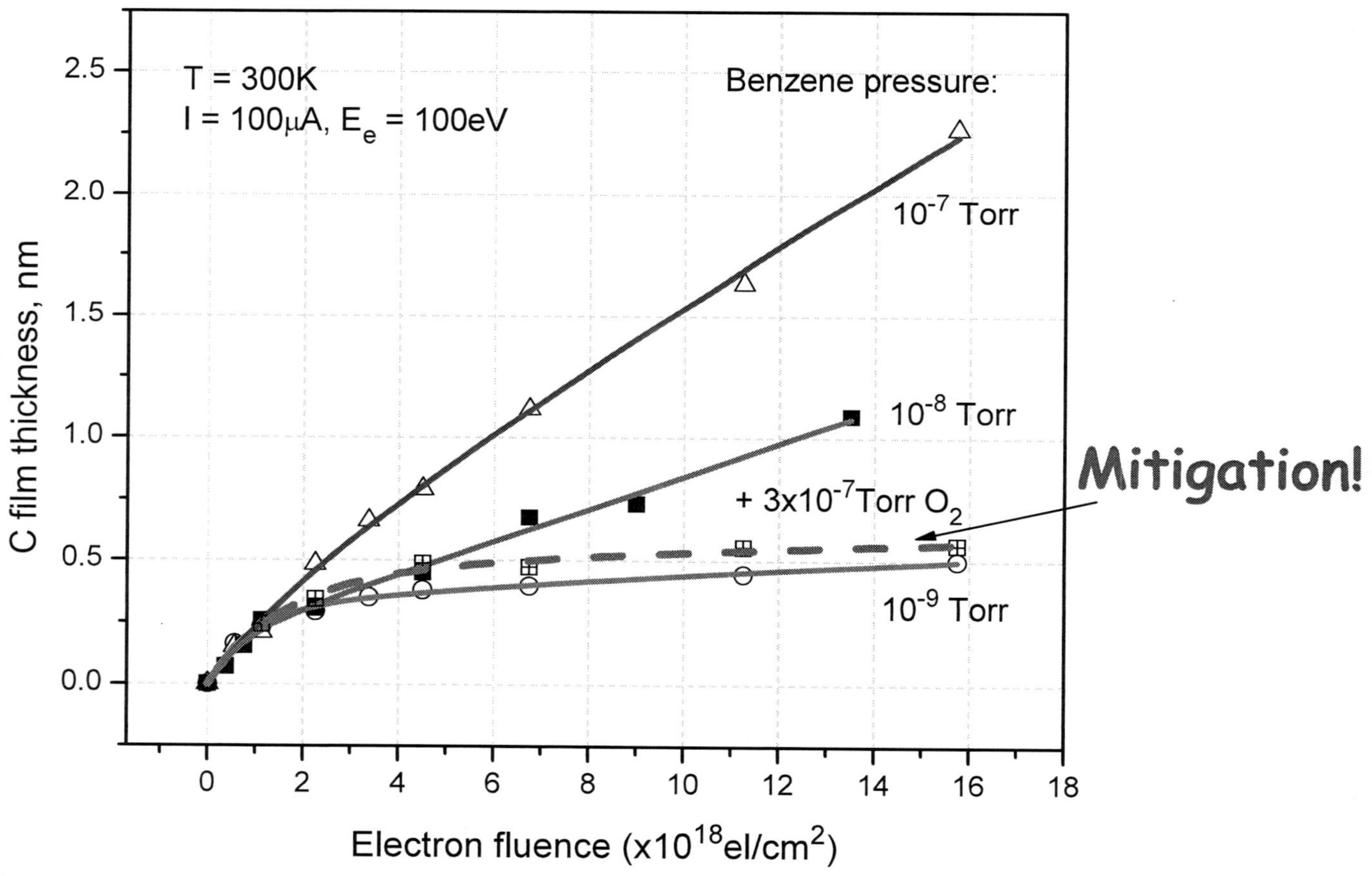

Benzene pressure:

10^{-7} Torr

10^{-8} Torr

+ 3x10^{-7} Torr O$_2$

10^{-9} Torr

Mitigation!

T = 300K
I = 100μA, E$_e$ = 100eV

C film thickness, nm

Electron fluence (x10^{18}el/cm^2)

Surface concentration

Carbon film removal from TiO_2: electron bombardment + O_2

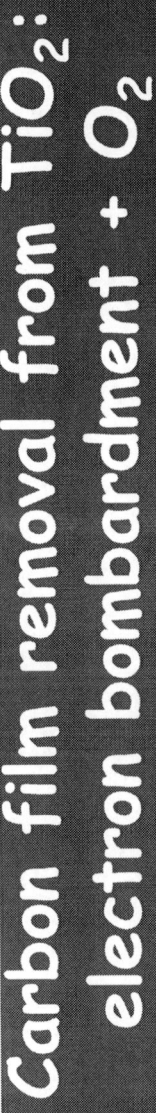

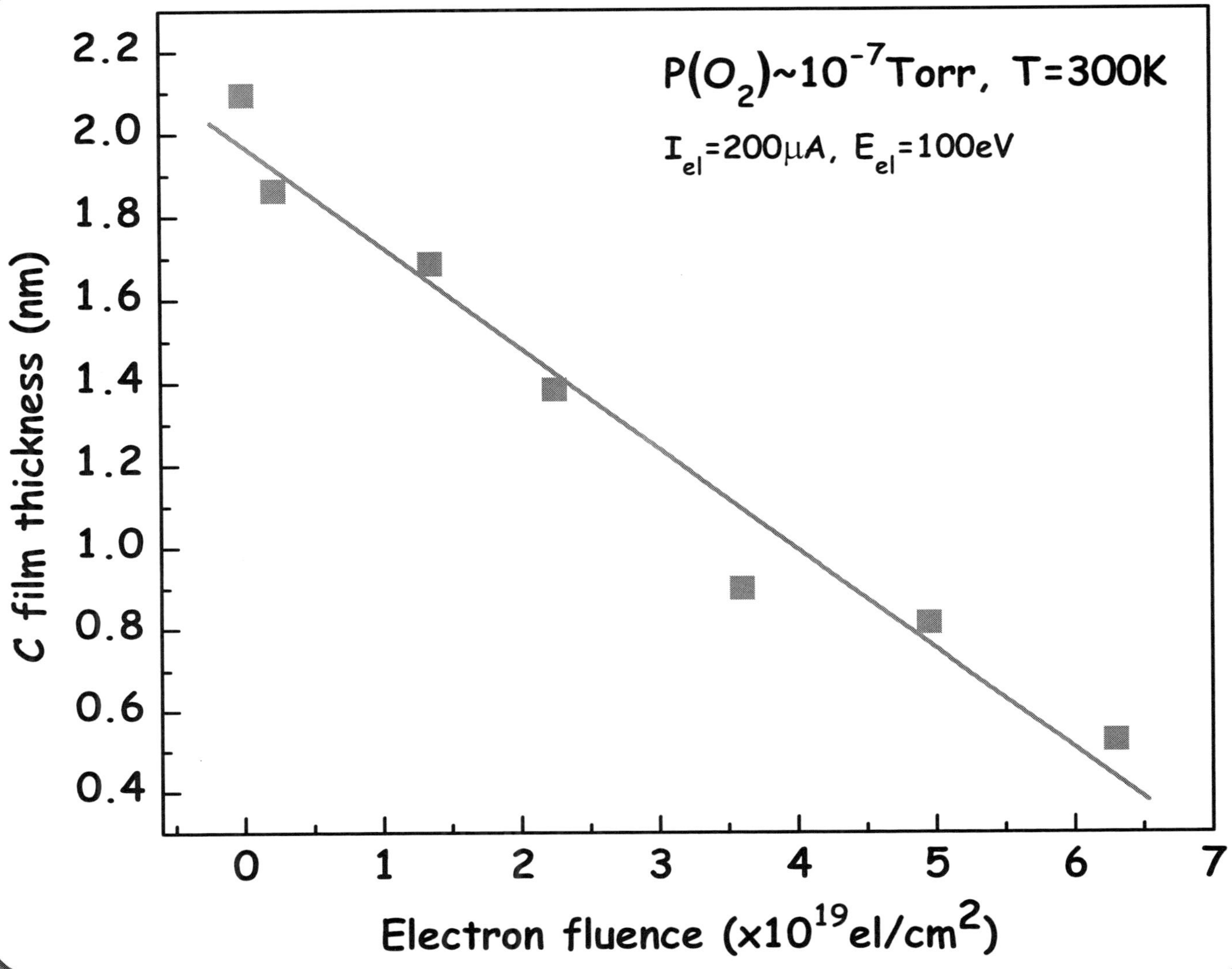

Cross section for e-stimulated reaction of C_6H_6 on clean and C-covered $TiO_2(011)$

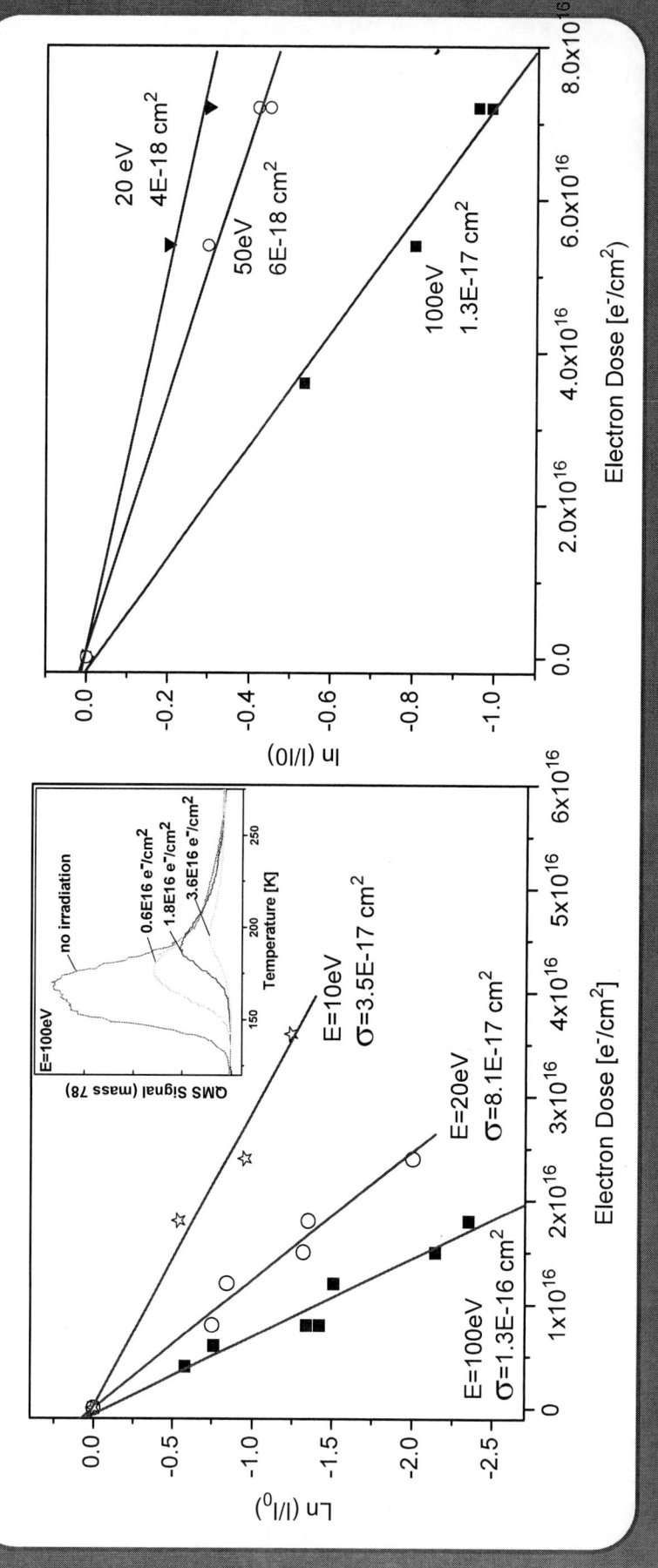

• Importance of low energy electrons

See JVST B, 26 (6) in press

SEY from C-covered MLMs at normal incidence

October run 2007, U3C beamline

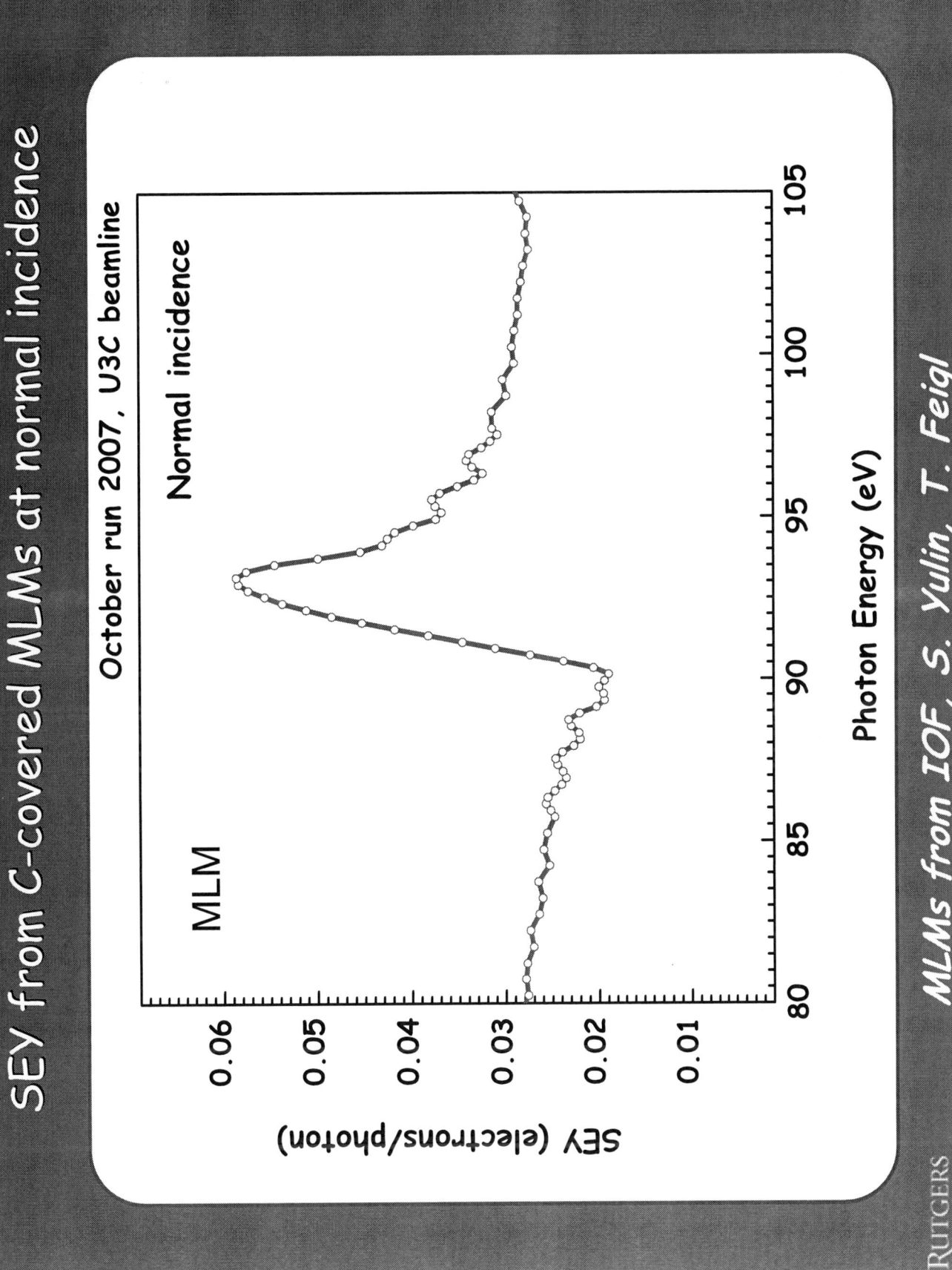

MLMs from IOF, S. Yulin, T. Feigl

Summary

o Coverage-dependent adsorption energy, Temkin isotherm

o Defects (O-vacancies) caused by electron energies > 25 eV

o Radiation-induced carbon accumulation on TiO_2 depends on benzene pressure and substrate temperature

o Electron energies ~20eV cause C – growth on TiO_2, so secondaries important

o Mitigation by irradiation of TiO_2 in benzene + O_2

o Cross section for e–induced reaction

o Resonances in SEY from MLMs

The work is supported by Intel

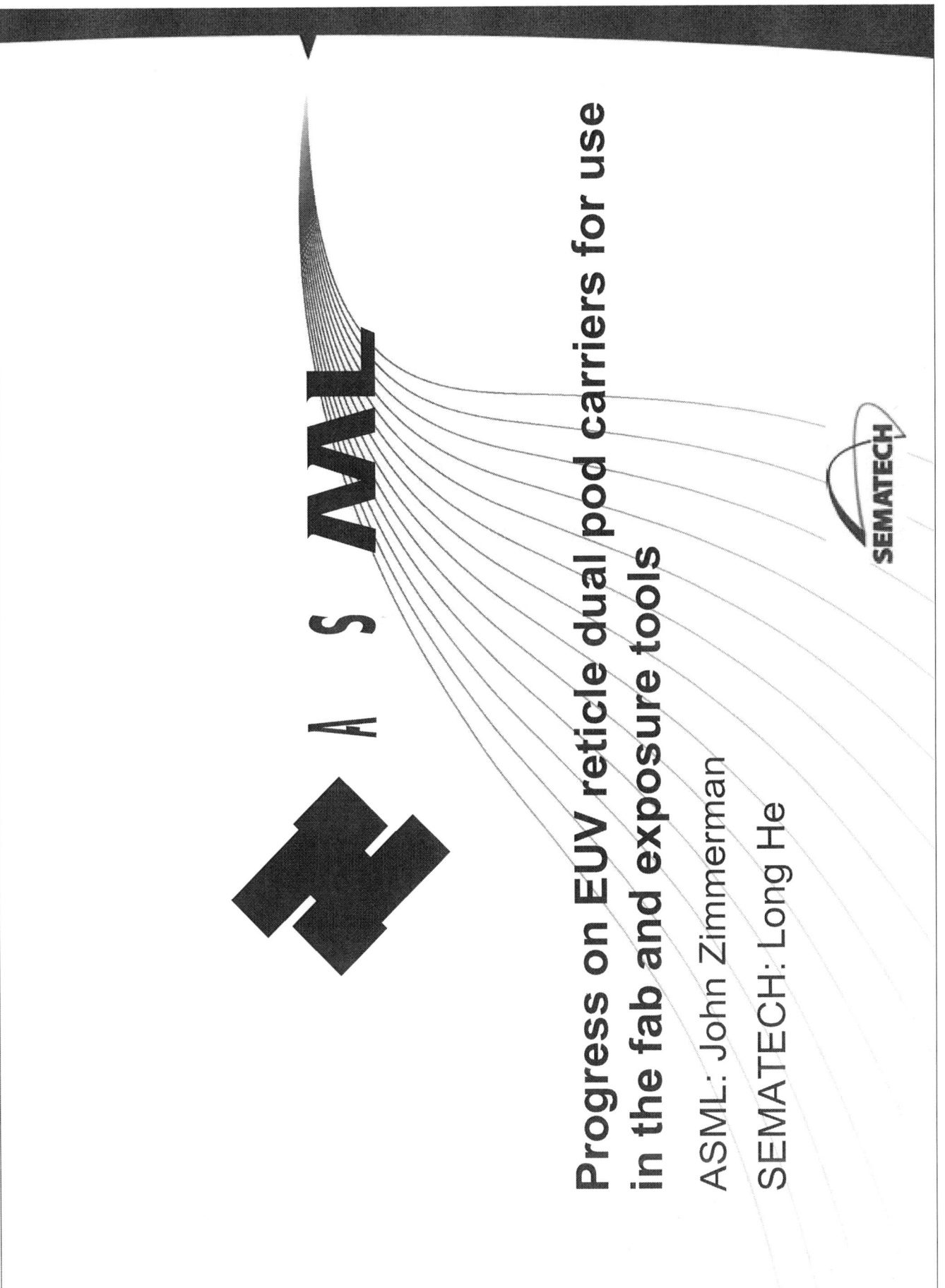

Progress on EUV reticle dual pod carriers for use in the fab and exposure tools

ASML: John Zimmerman

SEMATECH: Long He

Outline

- Introduction
- Outgas results
- Particle results
- Summary

SEMATECH ASML

Outline

- Introduction
  - Description
  - Use within exposure tool
- Outgas results
- Particle results
- Summary

EUV-pods have been been made to SEMI draft 4466 which can be used in fab equipment

- EUV-pod = Carrier with reticle in inner pod
  - In fab transport
  - Shipping

Entegris

SEMATECH

ASML

The inner pod is used to move the reticle in an exposure tool

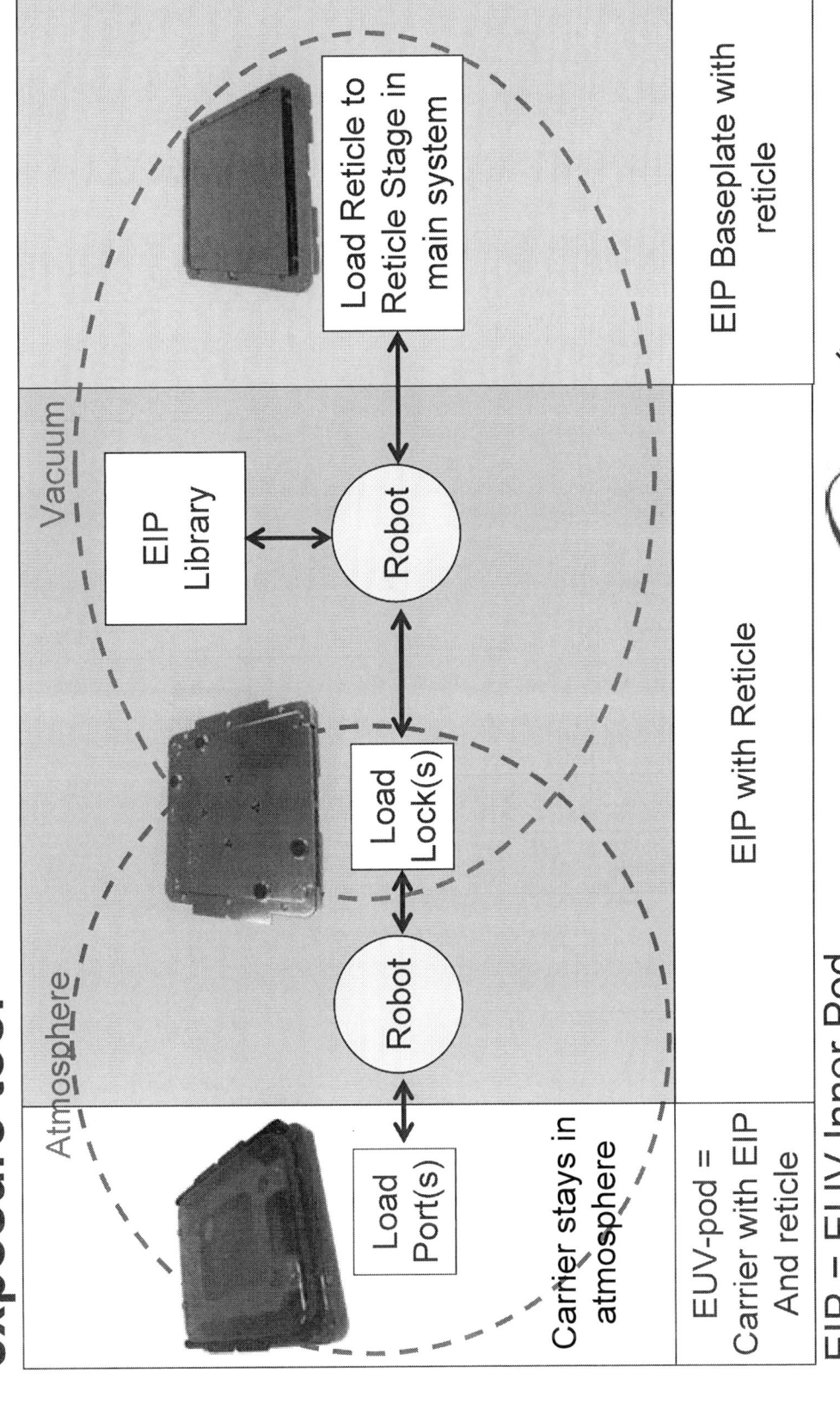

EIP = EUV Inner Pod

2 Different SEMI (draft) compliant implementations have been / are being tested at SEMATECH

A-pod

Type A without windows: implementation of SEMI Draft 4466

S-pod

Differences

Size:
Large → small
Centrality:
±0.25 → ±0.55 mm
Pockets:
0 → 5

Type B: Largest implementation of SEMI Draft 4466

Previously reported

This presentation

Outline

- Introduction
- Outgas results
  - Test set up
  - Compare to SEMI 4466 draft guidelines
  - Compare to ASML requirements
- Particle results
- Summary

The outgas test rig is used to vacuum test parts to ensure NXE compliance

- Test rig consists of
  - Chamber
  - Roughing pump
  - Turbo pump
  - Residual Gas Analyzer
    - 6 decades

A-pod EIP Base plate in test chamber

Materials and manufacturing considerations are needed to ensure compliance to ASML NXE vacuum requirements

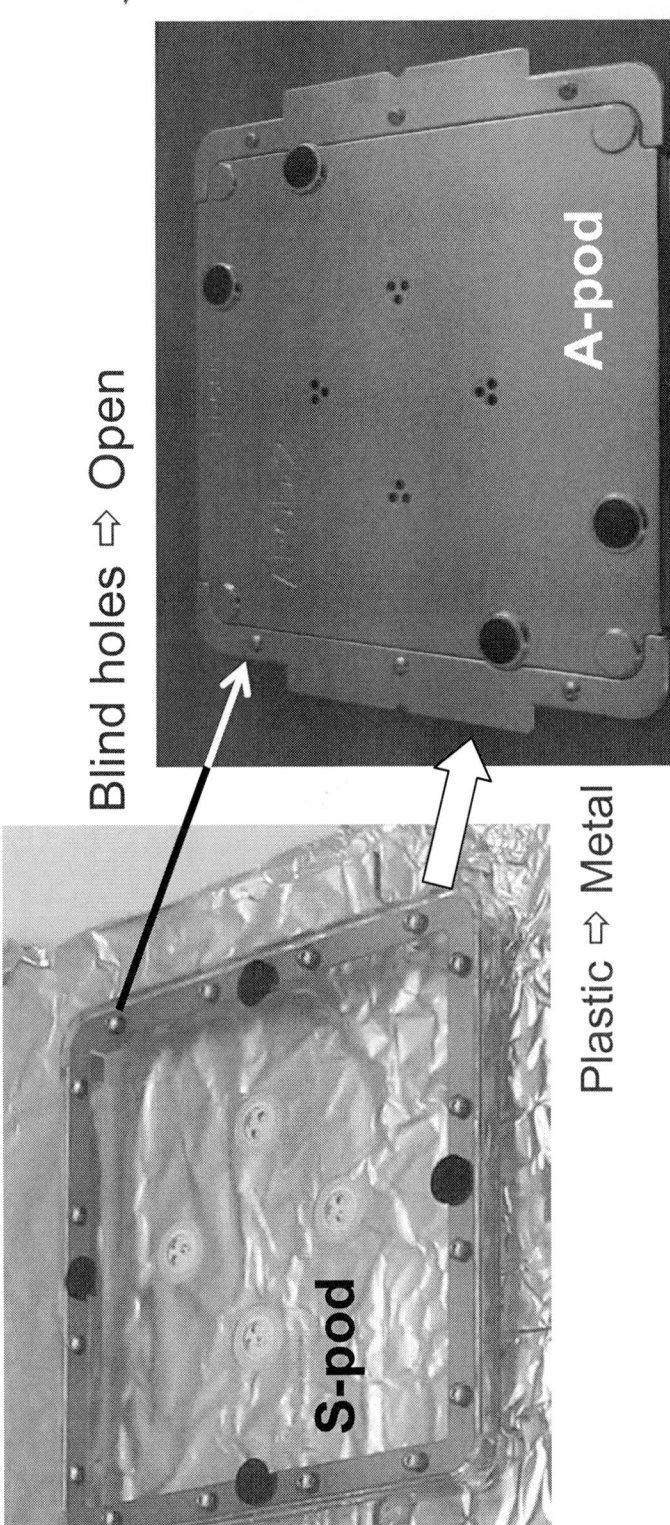

Blind holes ⇨ Open

A-pod

S-pod

Plastic ⇨ Metal

Virtual leaks removed
No Oil based cutting fluids

Both covers meet draft SEMI 4466

SEMATECH

ASML

Outgassing rates for the A-pod EIP compare well to the draft 4466 SEMI standard guidelines for outgassing

Complete EIP		Measured	Guideline	
	Q [total]	6.18E-05	6.50E-04	*
	Q [CxHy]	1.65E-07	6.50E-06	*

Baseplate				
	Q [total]	1.89E-05	1.20E-05	**
	Q [CxHy]	< LDL	1.20E-07	**

* After 10 minutes   ** After 1hr

All values are mbar•l/s

Remaining compliance can be met by changing remaining plastic parts with metal

All ASML hydrocarbon outgassing requirements are easily met for the A-pod

Hydrocarbon Outgas Rate (Arb units)

<= Top | Base Plate =>

Specification

Top (complete)   Top w/o filters or elastomeric parts   Elastomeric Parts   Filters (Quantity 4)   Base plate (complete)   Base plate w/o Plastic   Platsic pieces

SEMATECH

ASML

In order to meet ASML H$_2$O requirements, the use of plastic parts are being minimized

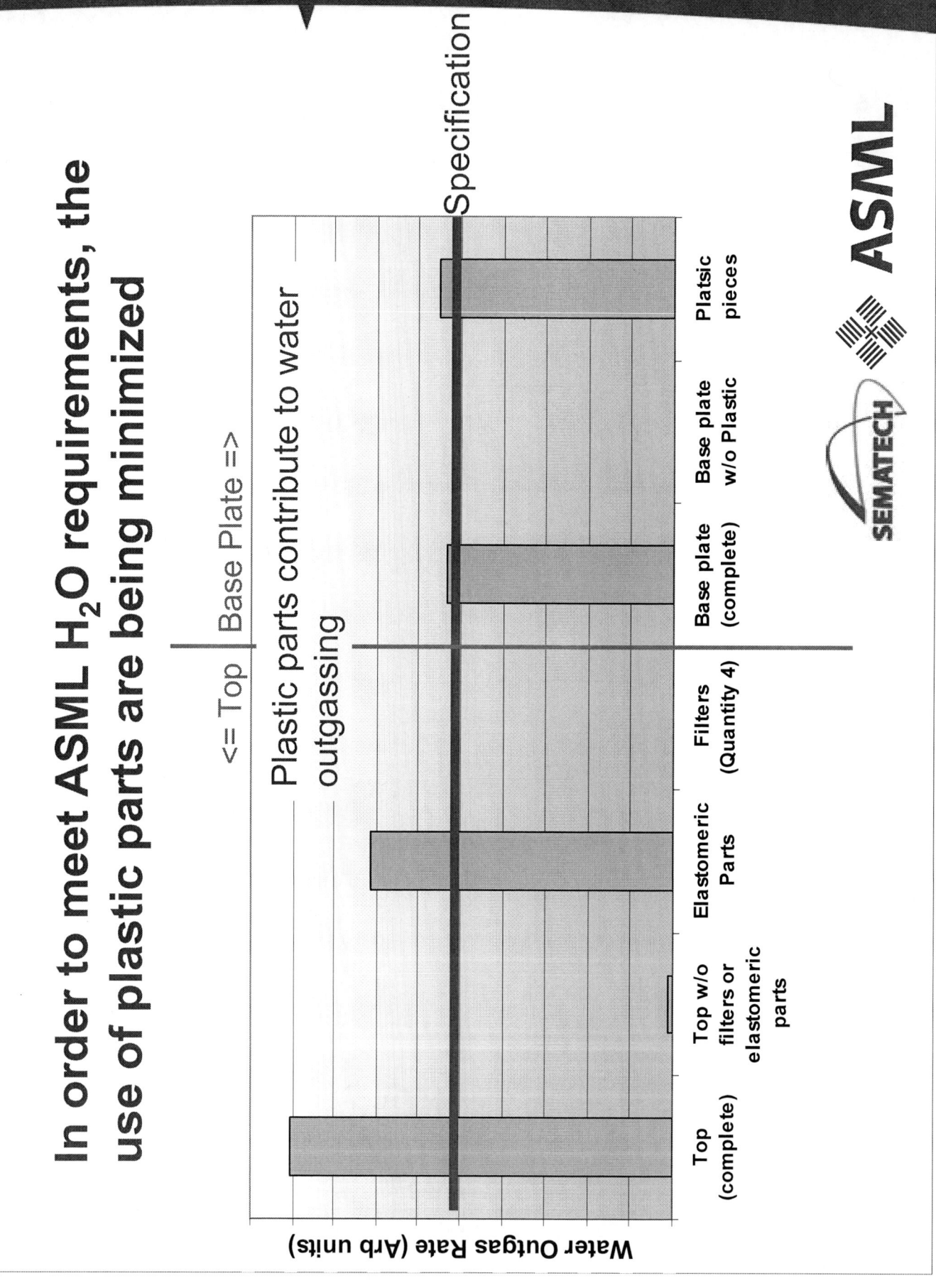

Outline

- Introduction
- Outgas results
- Particle results
  - Shipping
  - Storage
  - Atmospheric Handling
  - Vacuum Handling
- Summary

Shipping results of the A-pod compare well to earlier EUV-pod testing

- ≤ 0.9 particles/roundtrip shipment, ≥ 53 nm inspection capability
- Shipped by commercial shippers

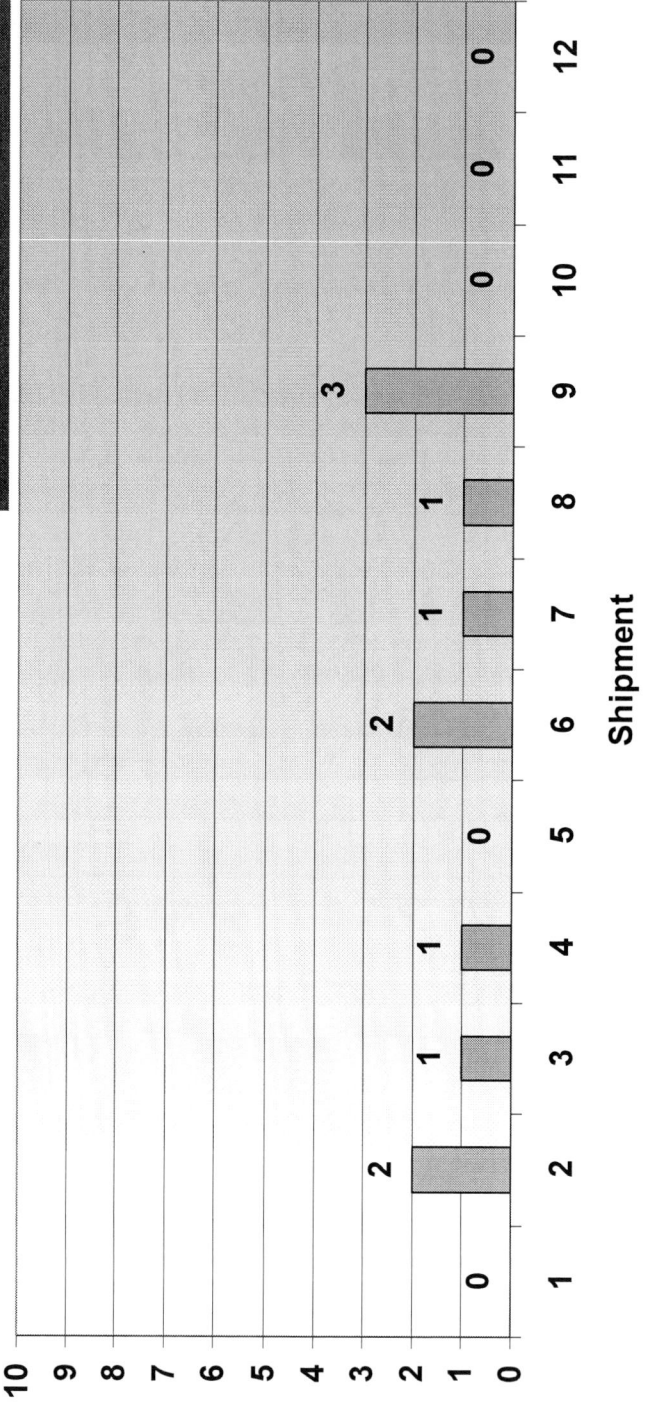

Integrated shipping and storage of A-pod data shows good capability

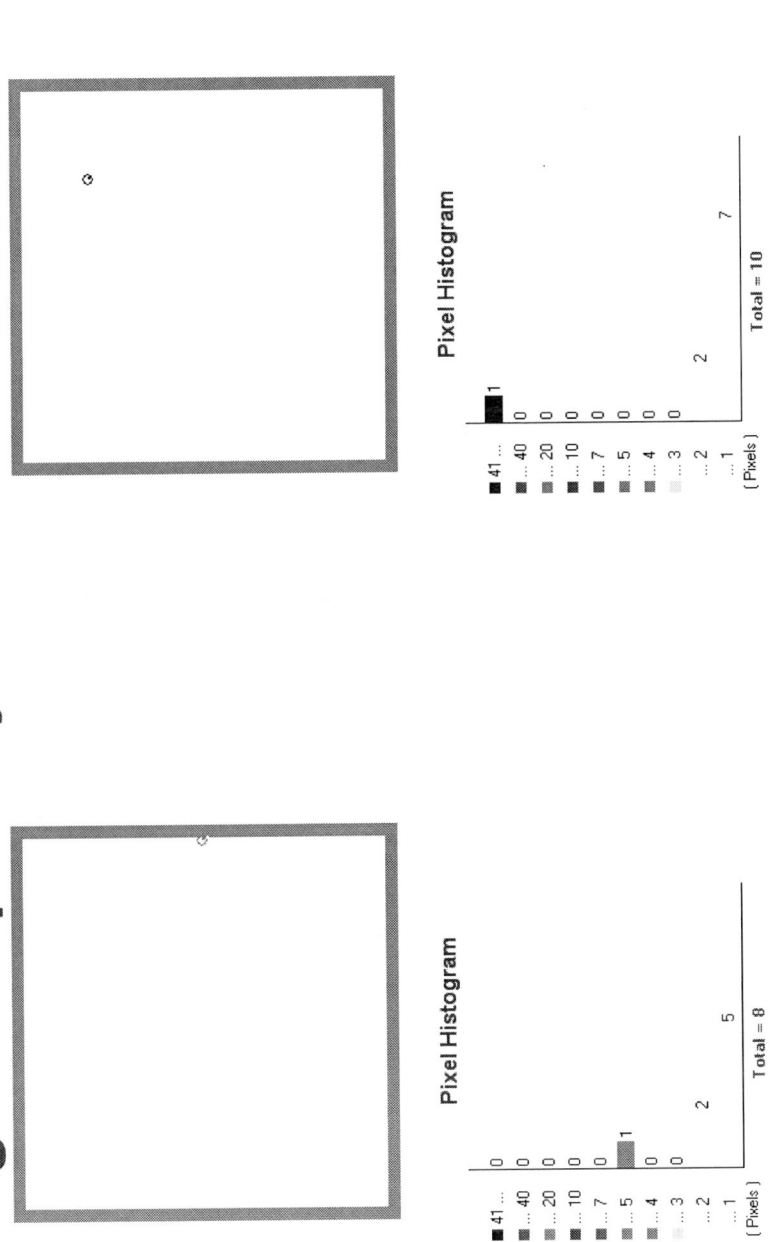

- 1 adder in a 4 week storage test after shipping
- Stored on a clean room rack

Atmospheric handling results of the A-pod demonstrate robustness of mechanical interface

- A substrate is transferred between the A-pod and a RSP-200 using robots at SEMATECH

- The substrate was inspected every 30 transfers over a 142mm x 142mm area @ 53nm PSL

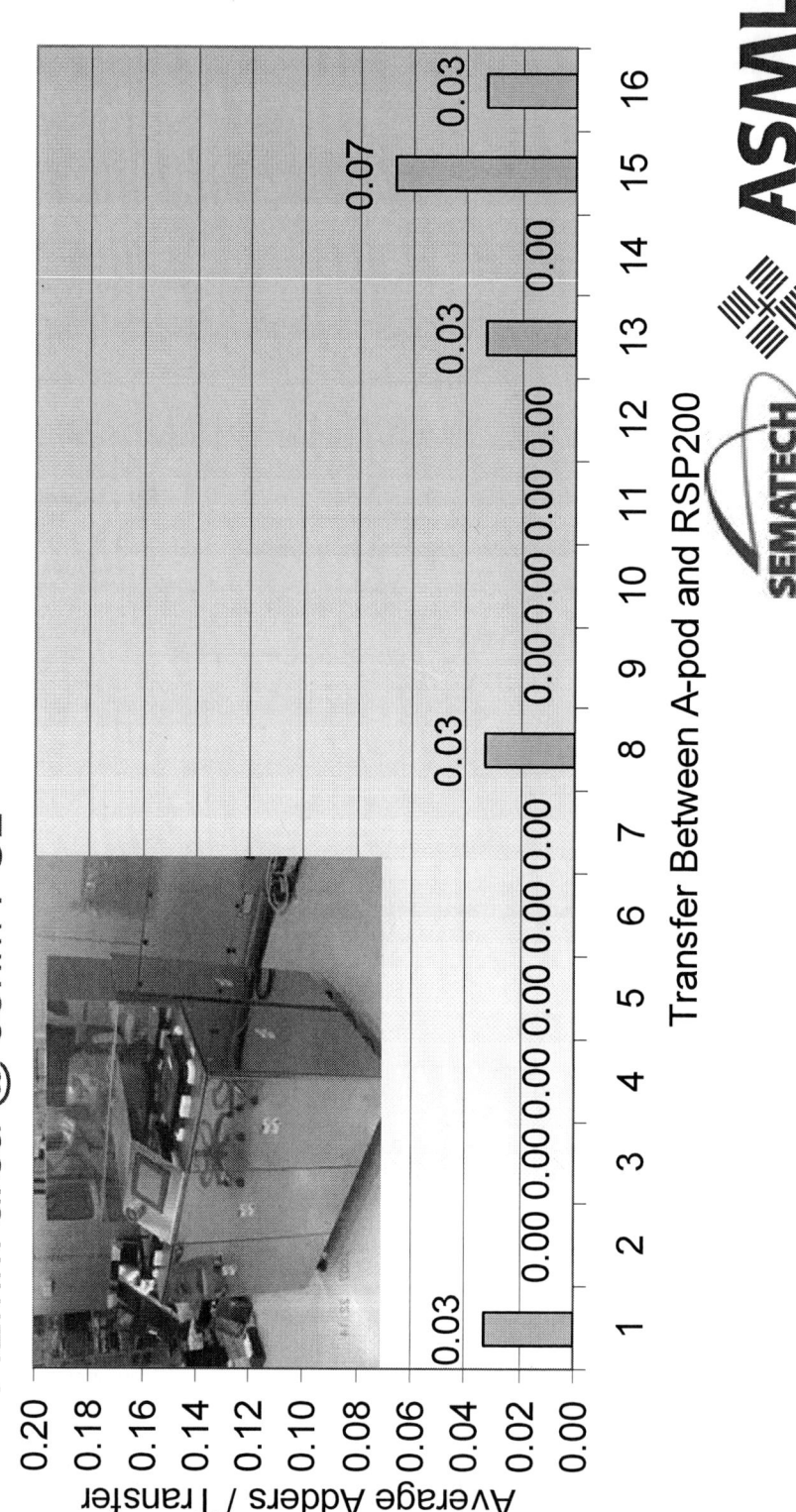

Vacuum handing results are excellent and exceed requirements for NXE

- Requirements
  - ≤ 0.001 particle per RH cycle @ 50nm PSL over pattern area
  - ≤ 0.01 particle per RH cycle @ 50nm PSL over rest of reticle
- Three vacuum mask handling tests were done
  - Complete reticle handling cycle from carrier to cover removal
  - Pumping and venting of the inner pod in the loadlock
  - Inner pod cover removal and placement cycling in vacuum

Particle Adders >= 53nm in 142x142 mm^2

	Cycles	Adders per cycle
Complete Cycle	370	0.00
Pump and Vent	60	0.0
Open/Close	1100	0.000

Outline

- Introduction
- Outgas results
- Particle results
- Summary

SEMATECH ASML

Summary

- The EUV Inner Pod can meet outgas requirements for NXE with minor material changes
- Shipping results are quite good
  - More data is required to increase sample size and confirm the low particle counts
- Particle results of the A-pod exceed requirements for use in NXE and within the fab
- Similar particle results are expected for the final base plate configuration of A-Pod

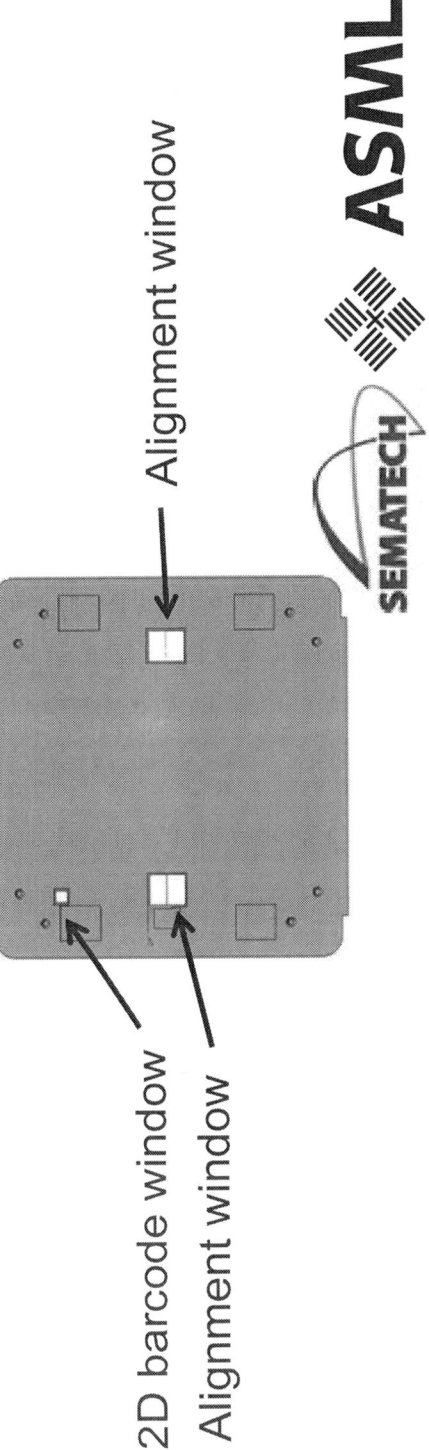

Alignment window

2D barcode window
Alignment window

Lithographic performance of Selete's full field EUV exposure tool

Kazuo Tawarayama*, Hajime Aoyama,
Takashi Kamo, Shunko Magoshi,
Yuusuke Tanaka, Seiichiro Shirai,
and Hiroyuki Tanaka

Selete

Semiconductor Leading Edge Technologies, Inc.

'08 EUVL Symposium

K.Tawarayama

Outline

- Introduction
- Tool performance
  - Static exposure
  - Dynamic exposure
- Future work
- Summary

EUVL Development Program in Selete

Selete

FY	2006	2007	2008	2009	2010
Lithography performance	SFET*	EUV1**			
		Resist materials			
		Mask optimization			
		Exposure tool evaluation - Optics quality - Flare - Optics lifetime (contamination) - Source and collector lifetime			
		Flare compensation			
		Resist multi-layer process			
		Lithography performance			
		Litho-verification using process/device TEG			
Reliability		Particle free mask handling			
		Contamination control technology			
Mask defectivity		Mask infra. (Inspection & repair)			

*Small Field Exposure Tool

** Full Field Exposure Tool

hp32 process spec. → β machine

EUVL Development Program in Selete

Selete

FY	2006	2007	2008	2009	2010
Lithography performance	SFET*	EUV1**			
Reliability					
Mask defectivity					

*Small Field Exposure Tool

** Full Field Exposure Tool

Resist materials

Mask optimization

Exposure tool evaluation
- Optics quality - Flare
- Optics lifetime (contamination)
- Source and collector lifetime

Flare compensation

Resist multi-layer process

Lithography performance

Litho-verification using process/device TEG

Particle free mask handling

Contamination control technology

Mask infra. (Inspection & repair)

hp32 process spec.

β machine

Full field EUV tool at Selete

EUV1 Tool Specifications

EUV1: For 45nm hp node process development & 32nm hp node R&D

Specification Item	EUV1
Field Size	26 x 33 mm^2
NA and Magnification	0.25, x1/4
Resolution	Dense line: 45 nm @hp Isolated line: 25 nm (Target 32 nm @hp)
Flare	10 %
Overlay	target 10 nm (3s)
Wafer Size	300 mm
Throughput (10W & 5mJ/cm^2)	5-10 WPH

NIKON CORPORATION

Nikon

EUV1 tool status in Selete

Tool is ready for exposure

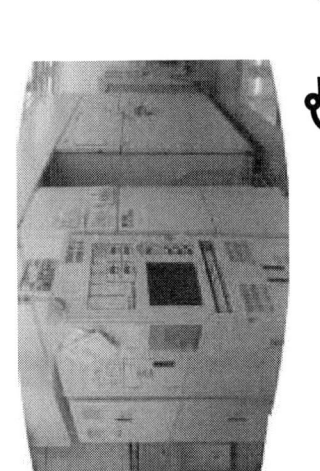

Outline

- Introduction
- **Tool performance**
  - Static exposure
  - Dynamic exposure
- Future work
- Summary

Various pattern for static exposure

1st static exposure results : Dense & isolated

35 nm hp

32 nm hp

30 nm hp

40 nm hp

1st static exposure results : Elbow patterns

45 nm hp

35 nm hp

1st static exposure results : Contact holes

32 nm C/H

30 nm C/H

28 nm C/H

45 nm C/H

40 nm C/H

35 nm C/H

NA/sigma=0.25/0.8

resist: SSR2 0.6umt

PO adjustment NOT finalized

Static pattern across the slit

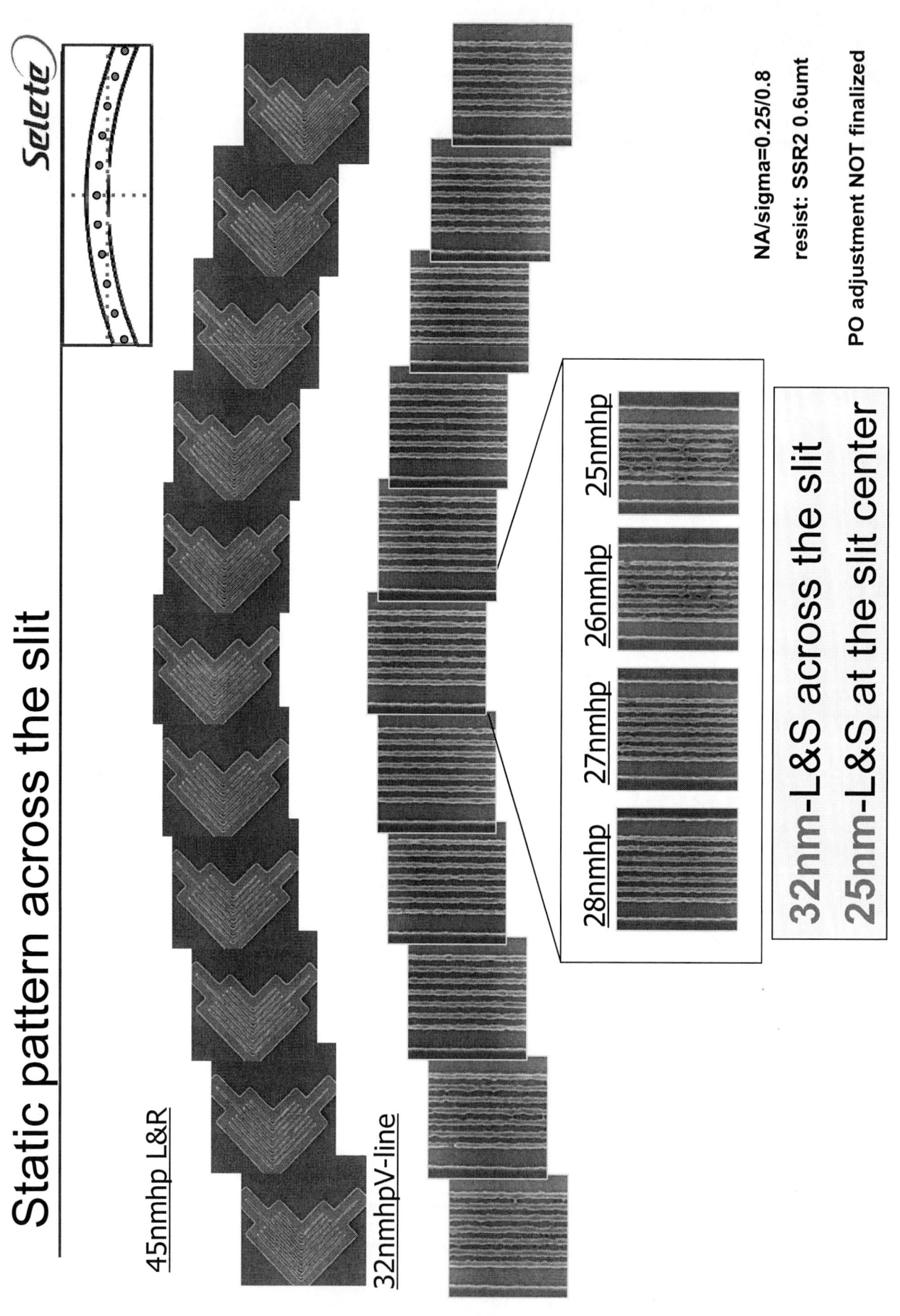

45nmhp L&R

32nmhpV-line

28nmhp   27nmhp   26nmhp   25nmhp

32nm-L&S across the slit
25nm-L&S at the slit center

NA/sigma=0.25/0.8

resist: SSR2 0.6umt

PO adjustment NOT finalized

Static pattern across the slit

45nmhp L&R

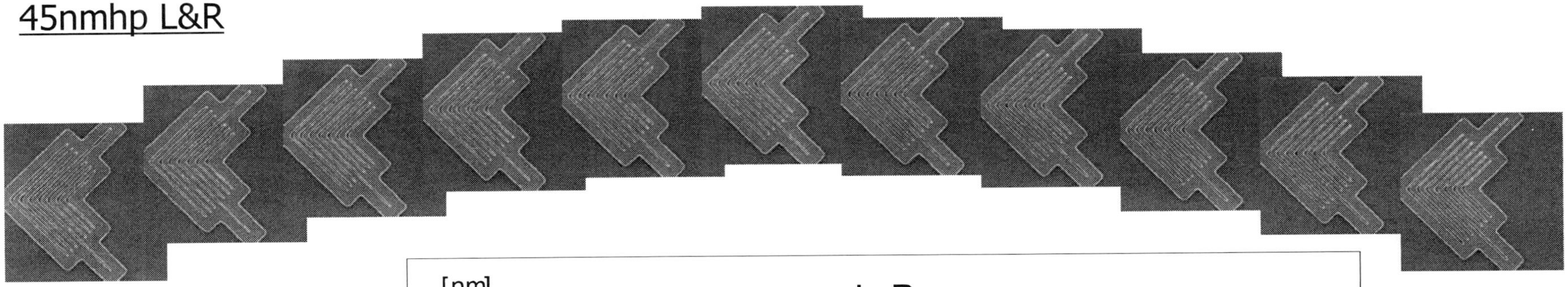

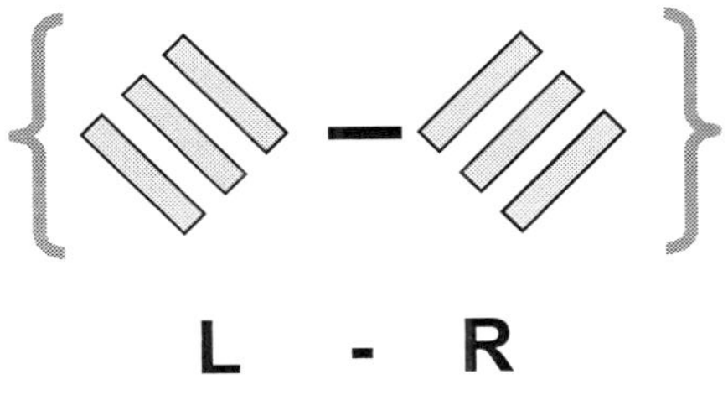

L - R

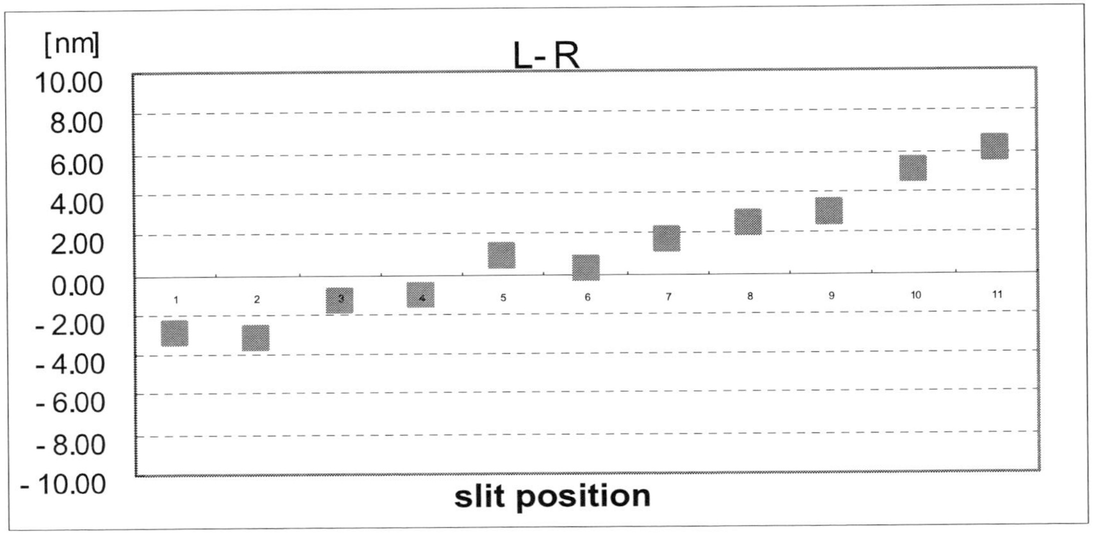

slit position

Experimental verification of shadowing effect

NA/sigma=0.25/0.8
resist: SSR2 0.6umt
PO adjustment NOT finalized

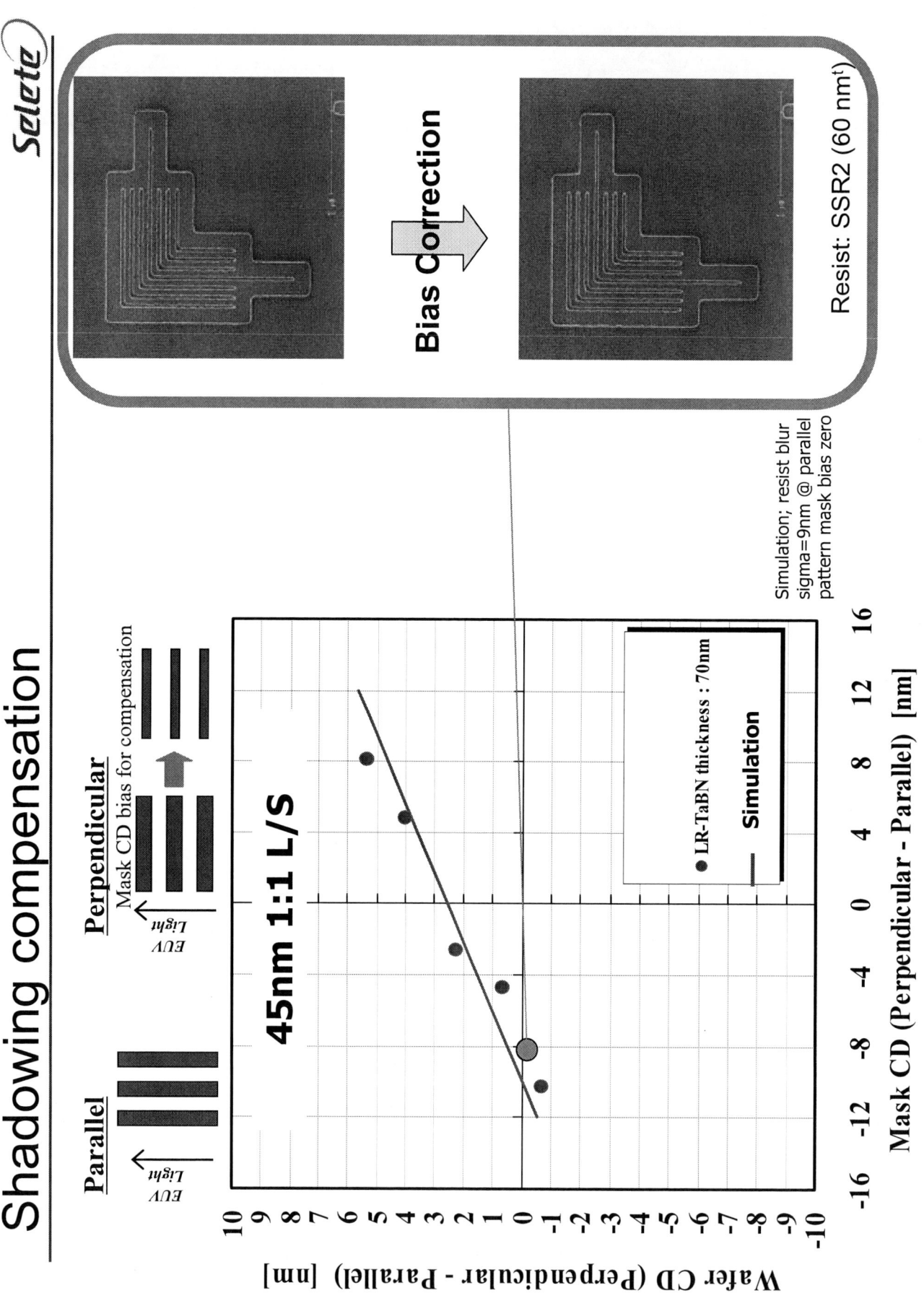

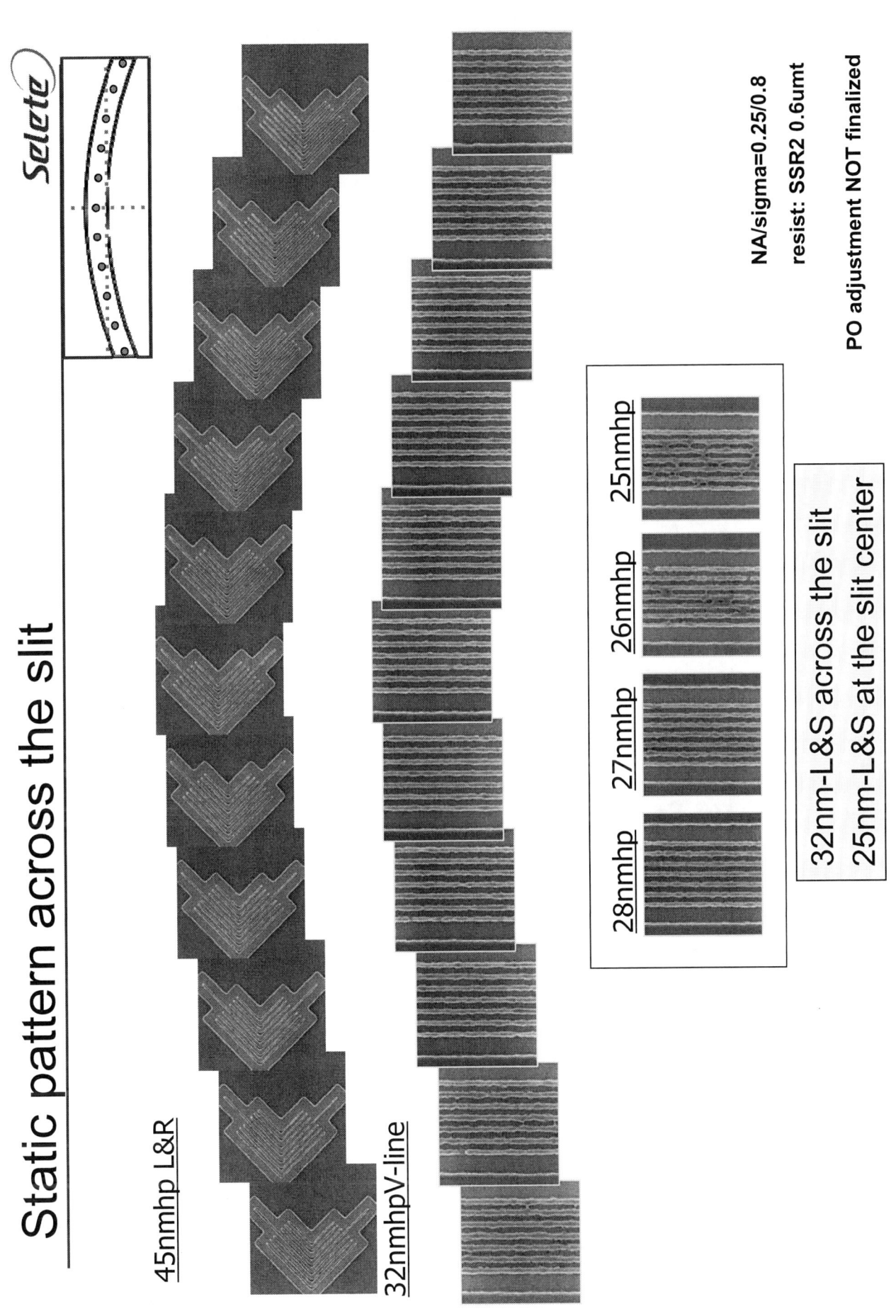

Static pattern across the slit

Selete

45nmhp L&R

32nmhpV-line

| 28nmhp | 27nmhp | 26nmhp | 25nmhp |

32nm-L&S across the slit
25nm-L&S at the slit center

NA/sigma=0.25/0.8

resist: SSR2 0.6umt

PO adjustment NOT finalized

Static pattern CDU

Selete

45nmhp L&R

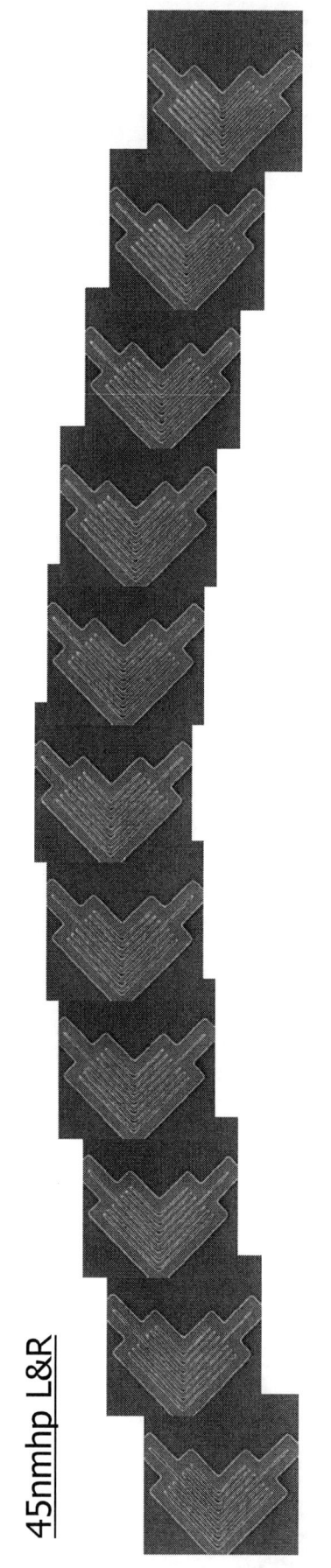

Static CDU 45nm L&S

4direction V/H/L/R

11x3 sites

NA/sigma=0.25/0.8

resist: SSR2 0.6umt

PO adjustment NOT finalized

Ave: 42.92

3sigma: 8.4 [nm]

Selete

Static CDU -45nm-

Grating mask

Resist:SSR2 60nmt

Dose:12mJ

Ave. 49.9 [nm]
3sigma 5.3 [nm]

459points
Contains Mask err. & Illmn. unif

60.00
59.00
58.00
57.00
56.00
55.00
54.00
53.00
52.00
51.00
50.00
49.00
48.00
47.00
46.00
45.00
44.00
43.00
42.00
41.00
40.00

Flare effect -dependency on surrounding area- *Selete*

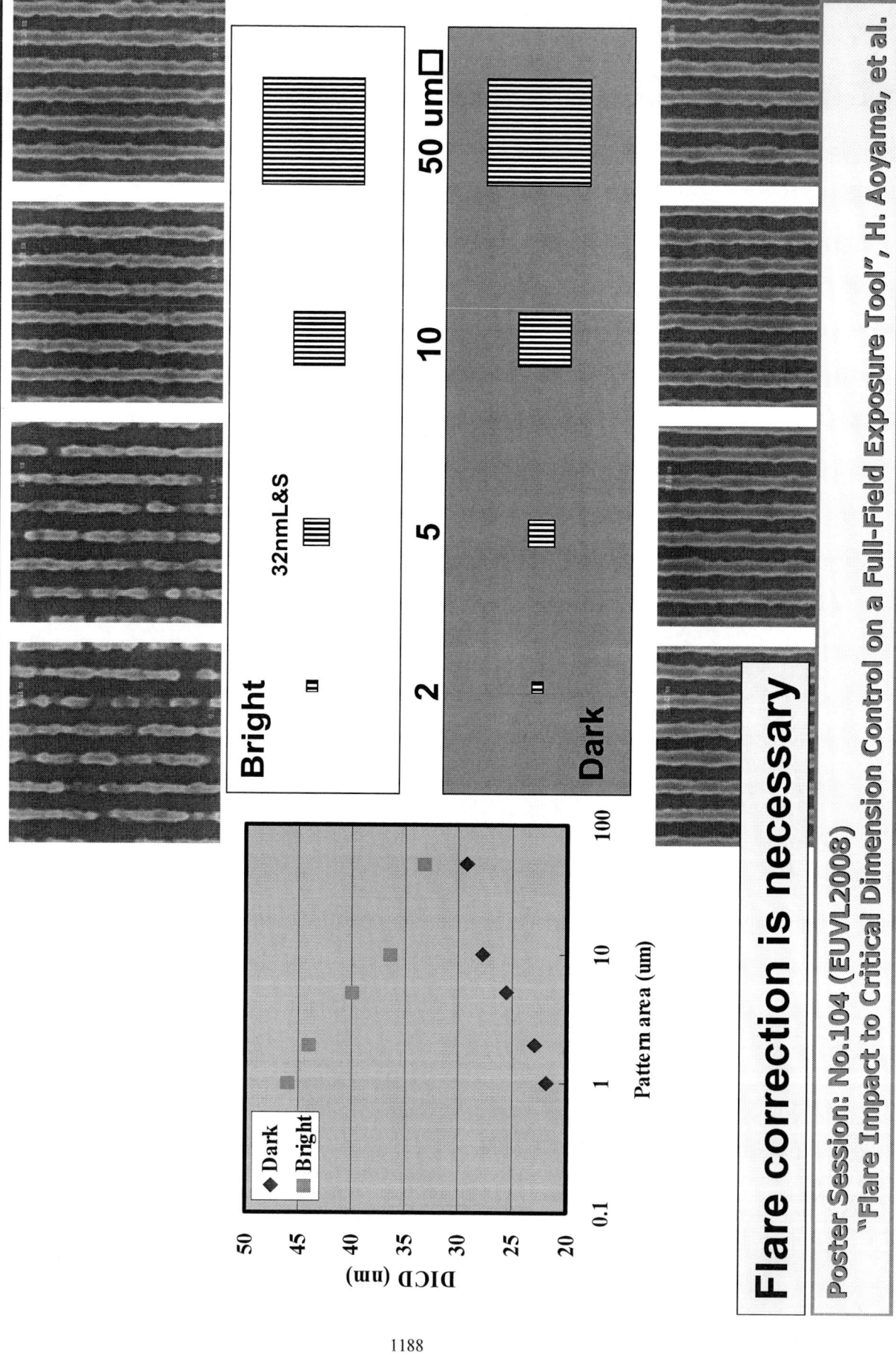

Bright

32nmL&S

Dark

50 um □

Flare correction is necessary

Poster Session: No.104 (EUVL2008)
"Flare Impact to Critical Dimension Control on a Full-Field Exposure Tool", H. Aoyama, et al.

Outline

- Introduction
- **Tool performance**
  - Static exposure
  - Dynamic exposure
- Future work
- Summary

Mask for full field – Mask CD uniformity – *Selete*

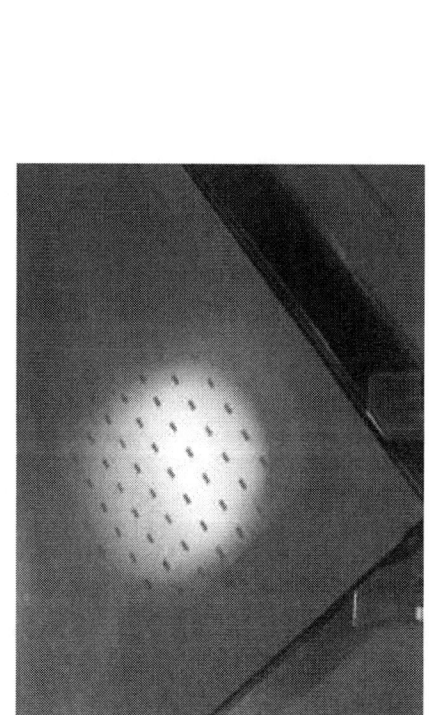

Mask CDU

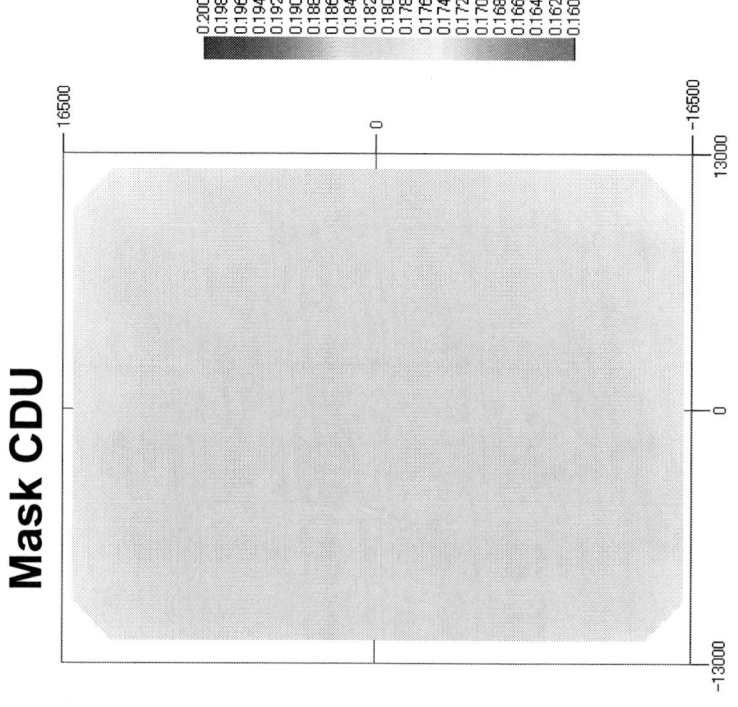

Mean: 184.1 nm
3σ: 4.2 nm

45nm on Wafer

Full field mask is ready for exposure

Full-field Scanning exposure

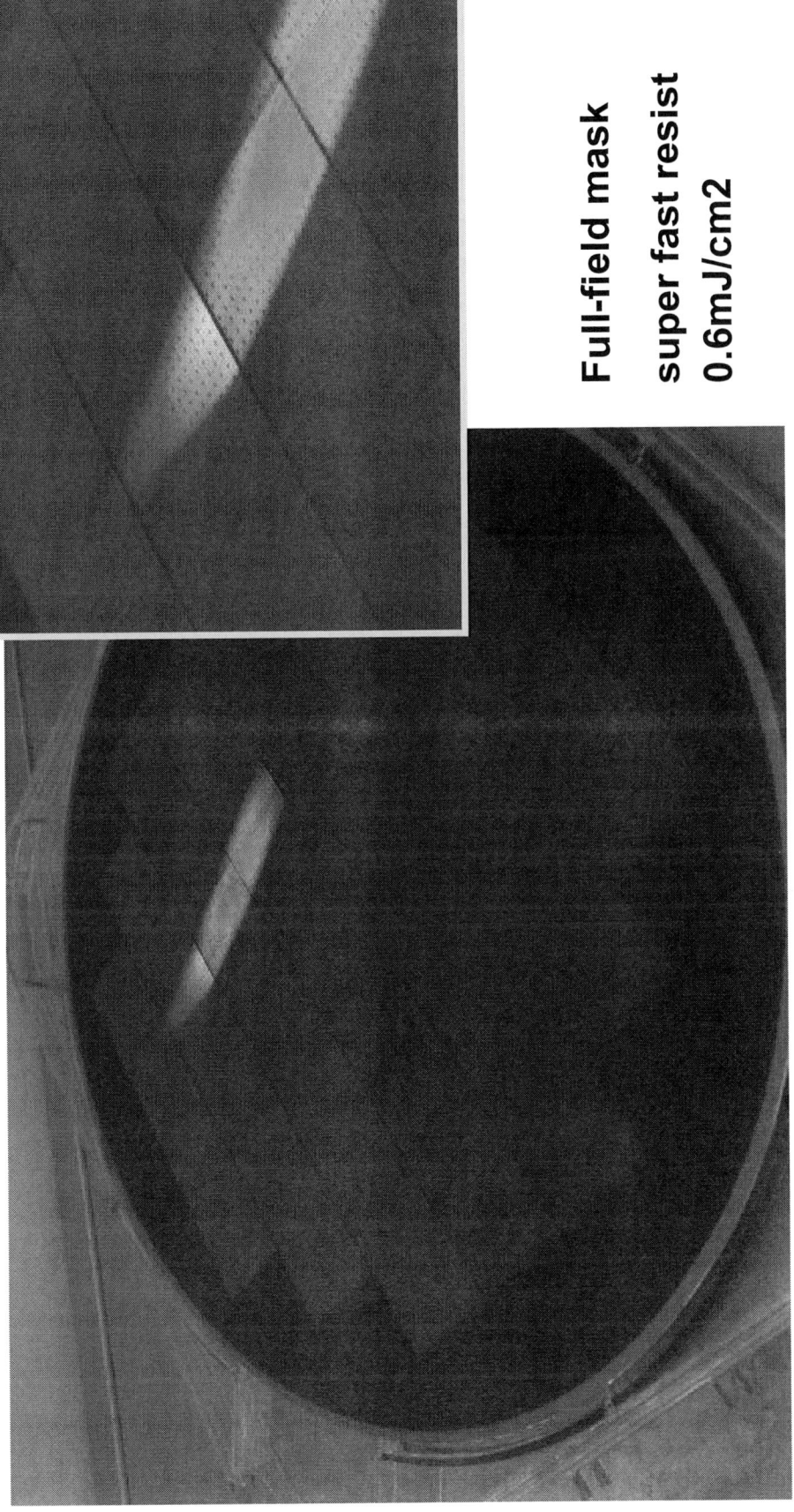

Full-field mask

super fast resist

0.6mJ/cm2

1st full field exposure across the wafer was obtained

Dynamic Scan image 45nm L&S

Selete

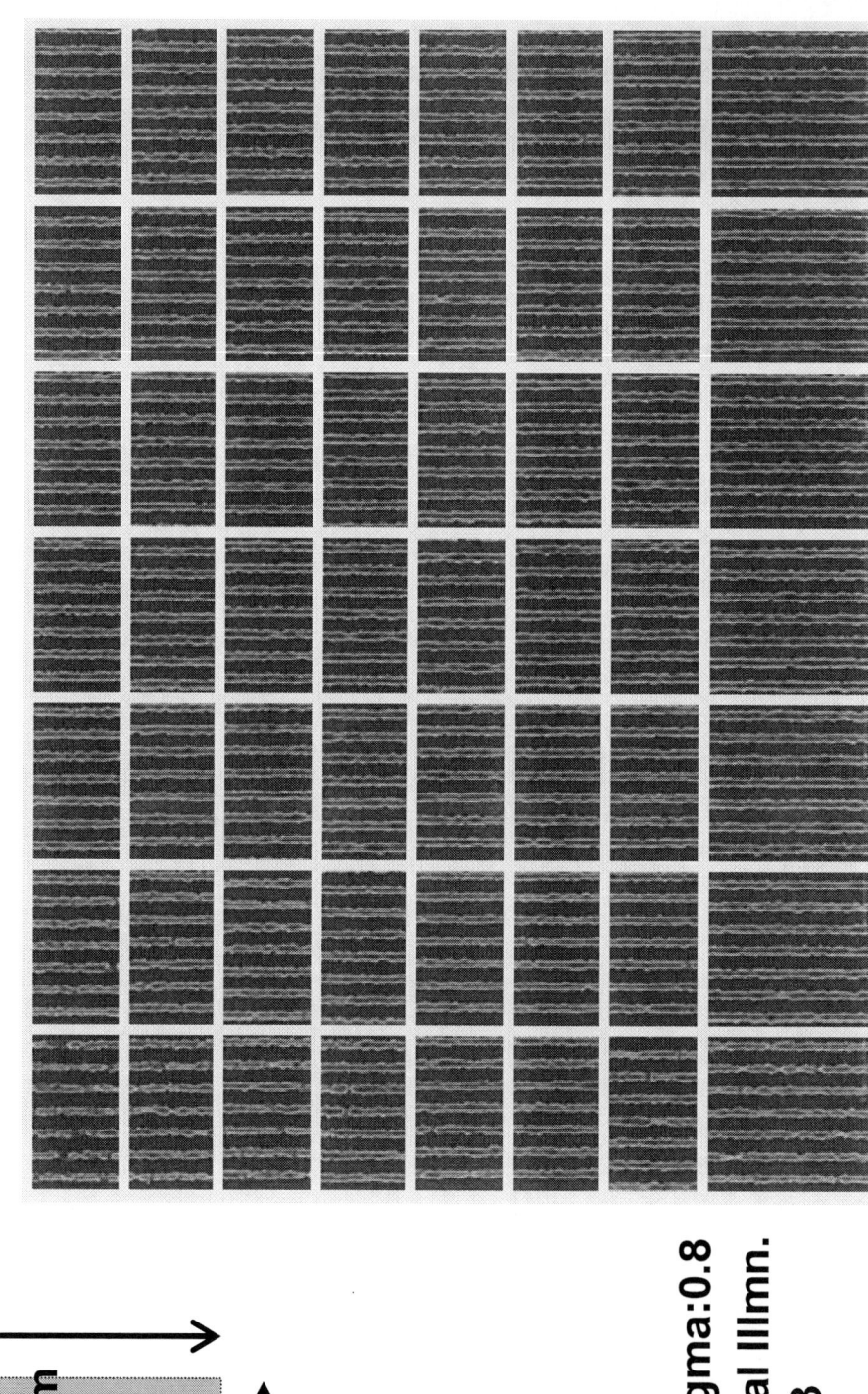

33mm
26mm

NA:0.25 , sigma:0.8
Conventional Illmn.
resist: SSR3
10 mJ/cm2

PO adjustment NOT finalized
Scan tuning not optimized

45nm L&S scanning image across the chip was obtained

Dynamic Scan CDU - 45nm L&S -

45nm V-line CDU

3sigma 7.6nm

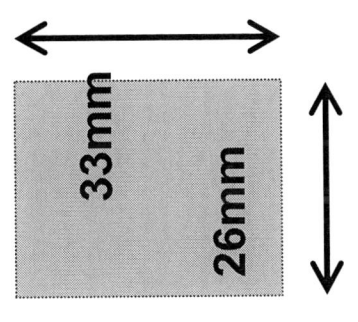

33mm

26mm

NA:0.25 , sigma:0.8
Conventional Illmn.
resist: SSR3
10 mJ/cm2

Good CDU was obtained

Dynamic Scan image 32nm L&S

Selete

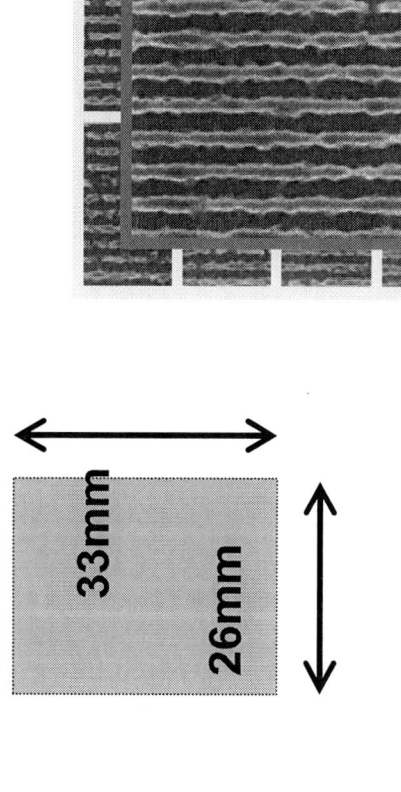

33mm
26mm

NA:0.25 , sigma:0.8
Conventional Illmn.
resist: SSR3
10 mJ/cm2

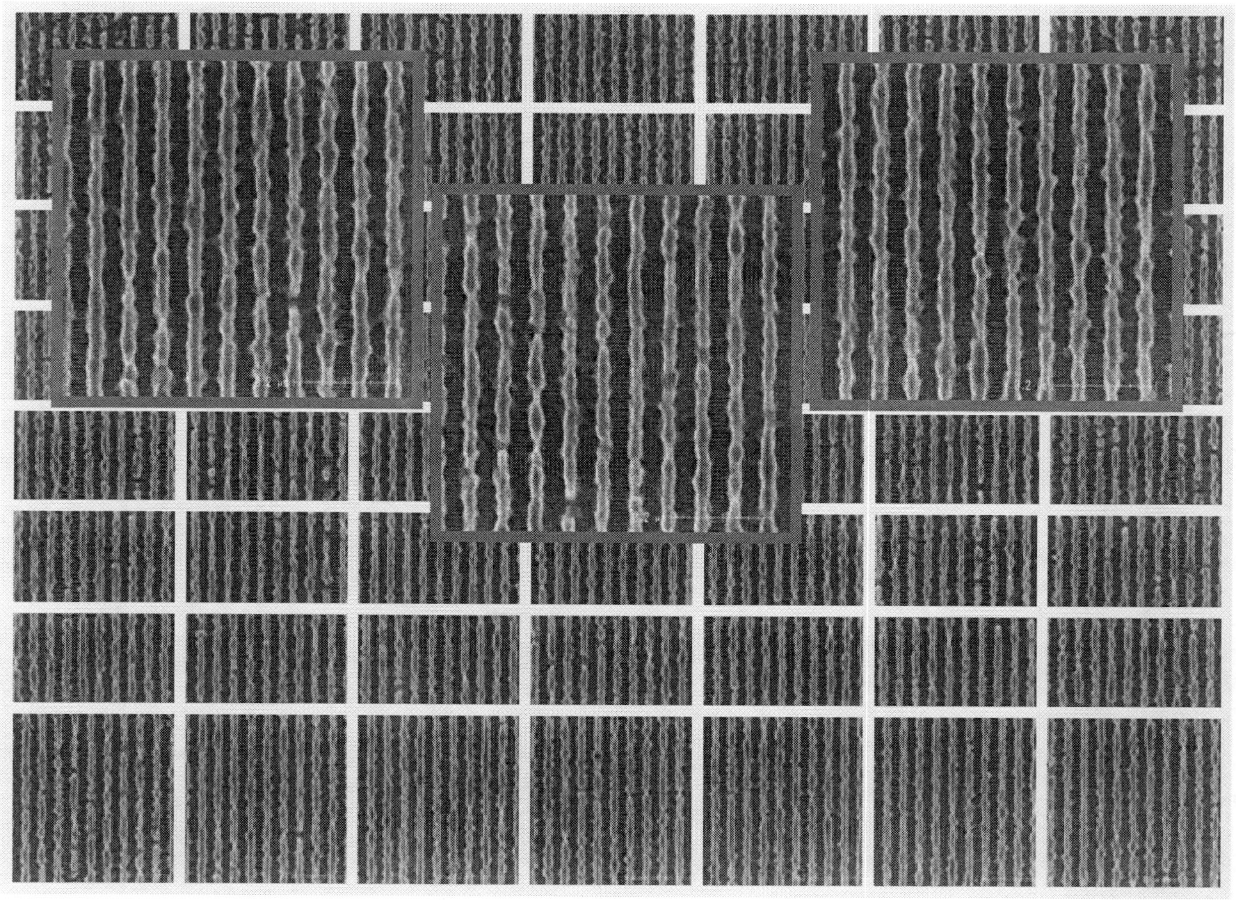

PO adjustment NOT finalized
Scan tuning not optimized

32nm L&S scanning image across the chip was obtained

Outline

- Introduction
- Tool performance
  - Static exposure
  - Dynamic exposure
- **Future work**
- Summary

EUVL Development Program in Selete

Selete

FY	2006	2007	2008	2009	2010

SFET*
*Small Field Exposure Tool

EUV1**
** Full Field Exposure Tool

Lithography performance
- Resist materials
- Mask optimization
- Exposure tool evaluation
  - Optics quality - Flare
  - Optics lifetime (contamination)
  - Source and collector lifetime
- Flare compensation
- Resist multi-layer process
- Lithography performance
- Litho-verification using process/device TEG

Reliability
- Particle free mask handling
- Contamination control technology

Mask defectivity
- Mask infra. (Inspection & repair)

hp32 process spec. → β machine

EUVL Development Program in Selete

Selete

FY	2006	2007	2008	2009	2010

SFET*

*Small Field Exposure Tool

EUV1**

** Full Field Exposure Tool

Lithography performance

- Resist materials
- Mask optimization.
- Exposure tool evaluation
  - Optics quality - Flare
  - Optics lifetime (contamination)
  - Source and collector lifetime
- Flare compensation
- Resist multi-layer process
- Lithography performance
- Litho-verification using process/device TEG

Reliability

- Particle free mask handling
- Contamination control technology

Mask defectivity

- Mask infra. (Inspection & repair)

hp32 proces. spec.

β machine

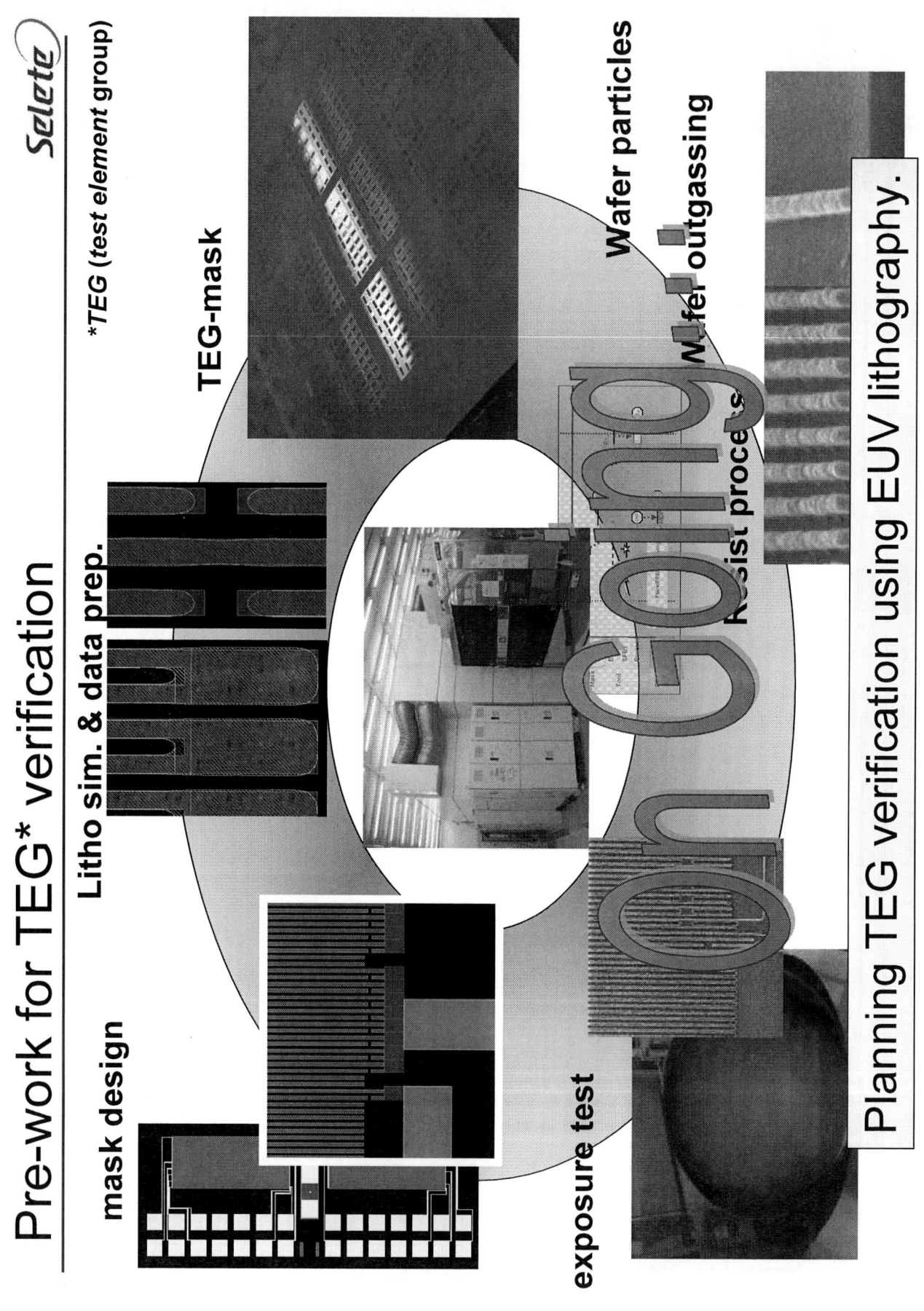

Summary

- High resolution images is demonstrated with Nikon full field exposure tool.

  – 32nm L&S pattern across the slit with static exposure

  – 25nm L&S pattern resolution at the slit center with static exposure

  – 32nm L&S pattern across the chip with dynamic exposure

- Full field scan is getting ready for exposure verification.

  – Yield test module (Process Liability test) plan is ongoing.

Acknowledgements

Special thanks to …

Nikon Tool Setup Team

Toshiba: PL-Project team

Selete: All colleagues

also Tsukuba-san shrine

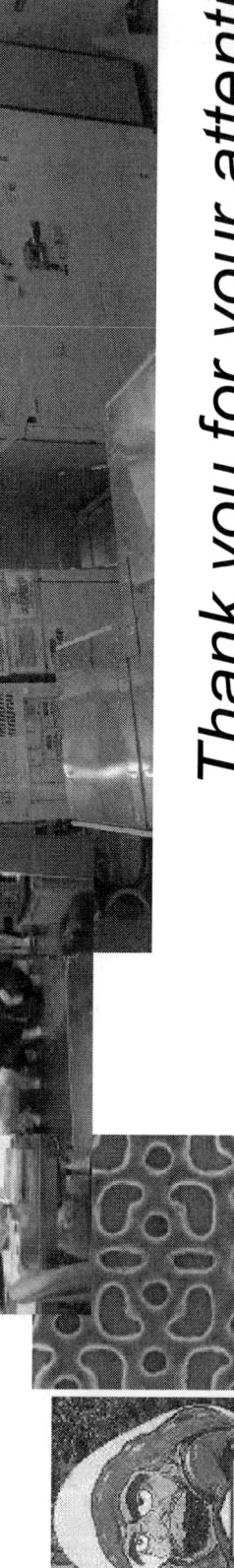

Thank you for your attention

EUVL symposium 2008

Full-field Patterning Test with ADT for 30-nm node Device Application

Doohoon Goo*, Insung Kim, Joo-On Park, Jeonghoon Lee, Changmin Park, Jinhong Park, Jeong-Ho Yeo, Sungwoon Choi, Woosung Han

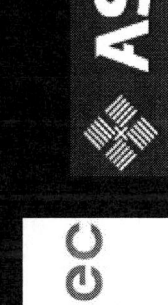

Outline

- **Aims of ADT characterization**

- **Characterization results of ADT @IMEC**
  - Overlay (Mix&Match Overlay)
  - CD Uniformity
  - Resolution
  - Flare
  - Dose latitude and MEEF

- **Summary**

- **Acknowledgement**

Litho tech. development trend

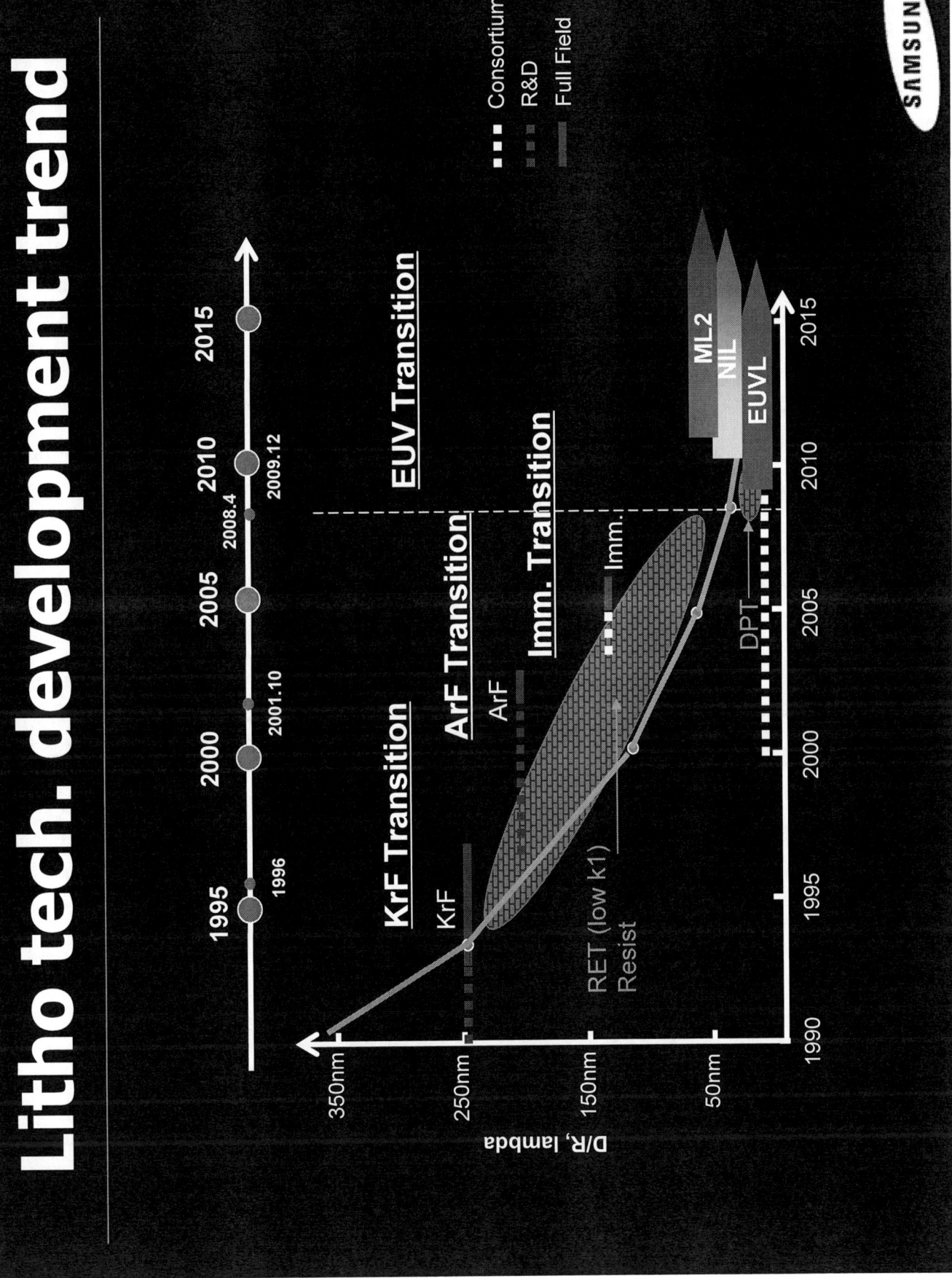

EUV Strategy Roadmap

| 2005 | 2006 | 2007 | 2008 | 2009 | 2010 | 2011 | 2012 | 2013 |

Now → 2008

Tools

ASET

Consortium

SEMATECH

IMEC

Design Rule

ADT

PPT

HVM

D4x

D3x

D2x

D1x ?

F/Field char.
@EUVL sympo.

Integration
@SPIE

D3x
@EUVL sympo.

D2x
@EUVL sympo.

SAMSUNG

Experiment

Mask
- IMEC mask (LTEM)
  - JIIRA2 mask
- SAMSUNG mask (LTEM)
  - C/H , MMO, Flare
  - L/S, CDU mask

Resist
- P1195, P-1101 (Fuji)
- P-1123 (TOK)
- XP-6305, MET-1K (R&H)

Tools
- ADT @ IMEC
- MET @ LBNL for comparison and screening

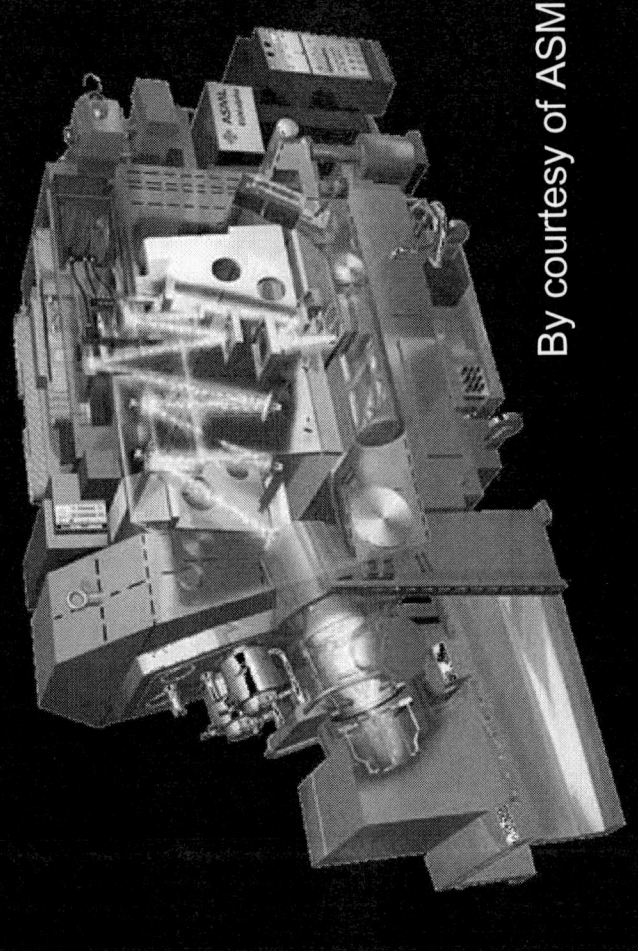

By courtesy of ASML

SAMSUNG

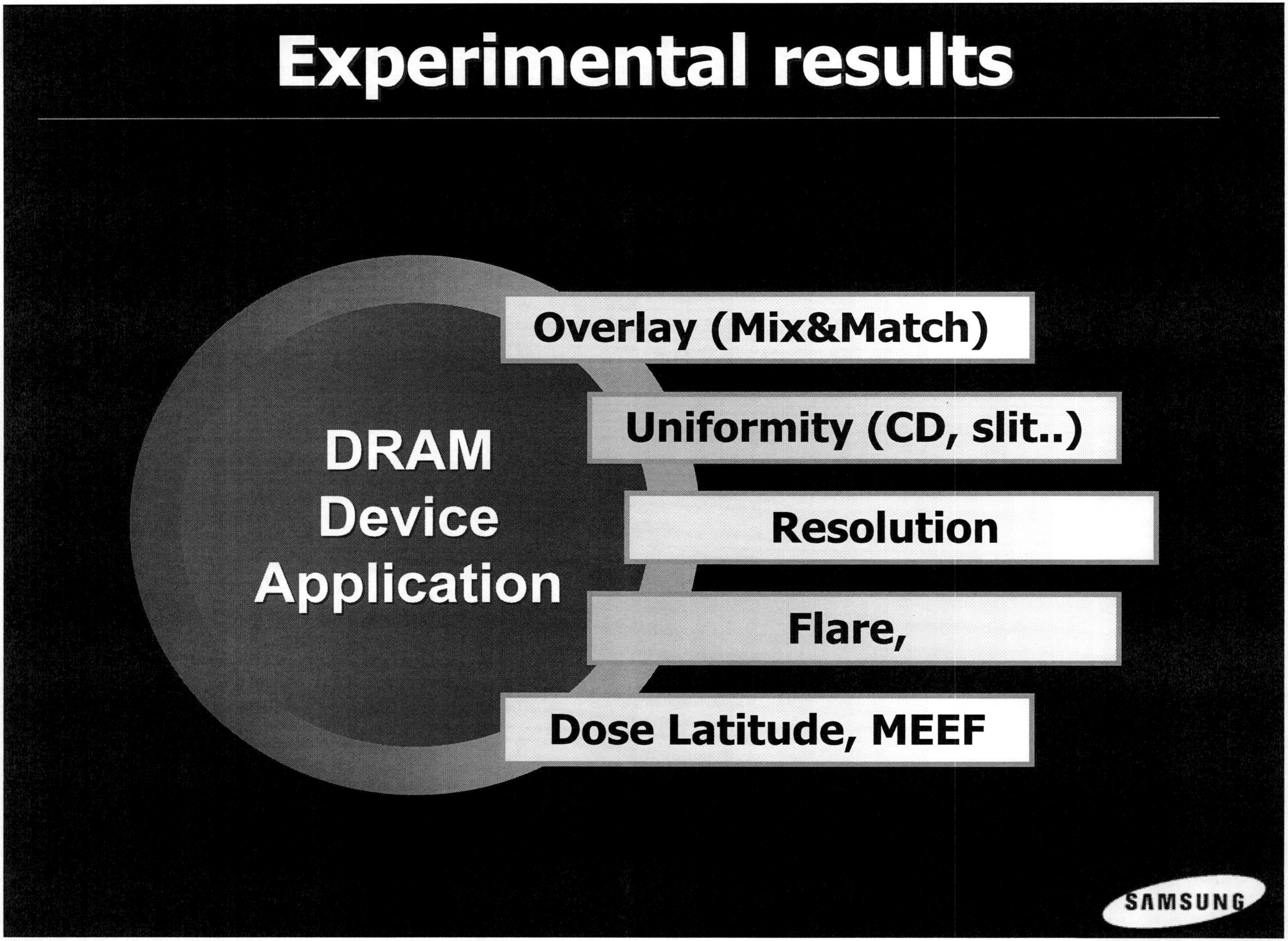

Experimental results

DRAM Device Application

- Overlay (Mix&Match)
- Uniformity (CD, slit...)
- Resolution
- Flare,
- Dose Latitude, MEEF

SAMSUNG

Experimental results

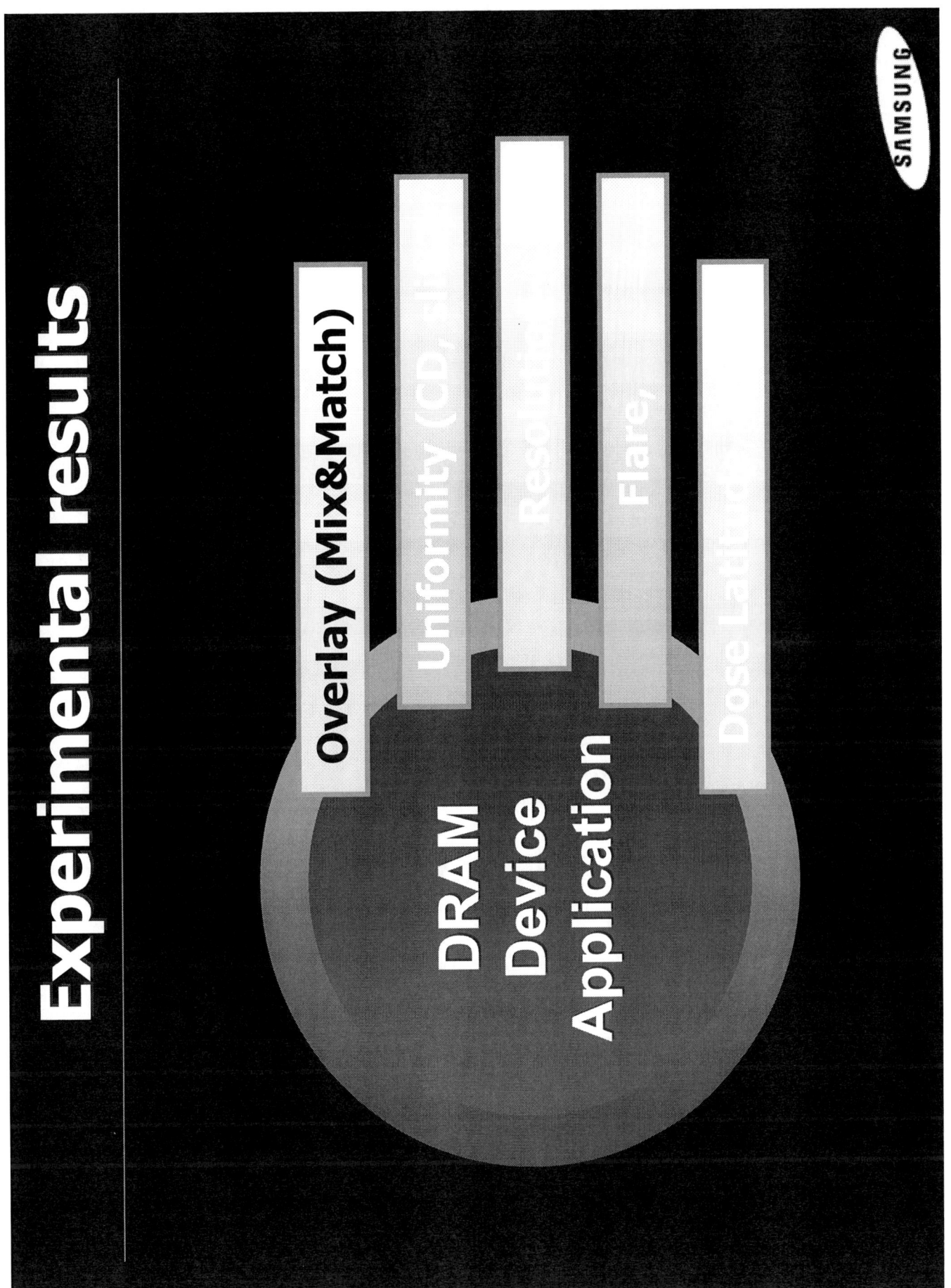

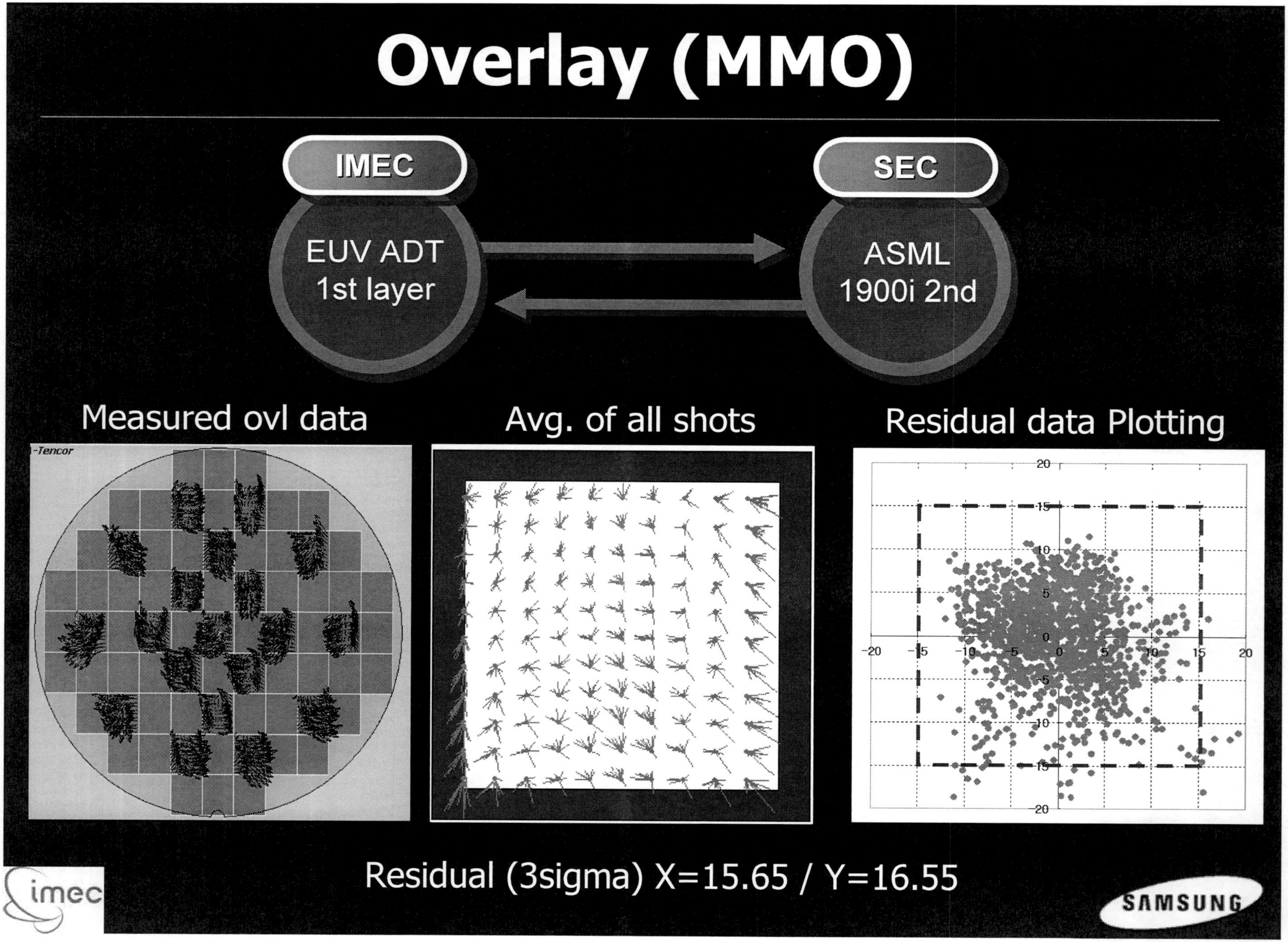
Overlay (MMO)
IMEC
EUV ADT
1st layer
SEC
ASML
1900i 2nd
Measured ovl data
-Tencor
Avg. of all shots
Residual data Plotting
20
15
10
5
0
-5
-10
-15
-20
Residual (3sigma) X=15.65 / Y=16.55
imec
SAMSUNG

Overlay : scan direction matching

- **Same scan direction**
- **Opposite scan direction**

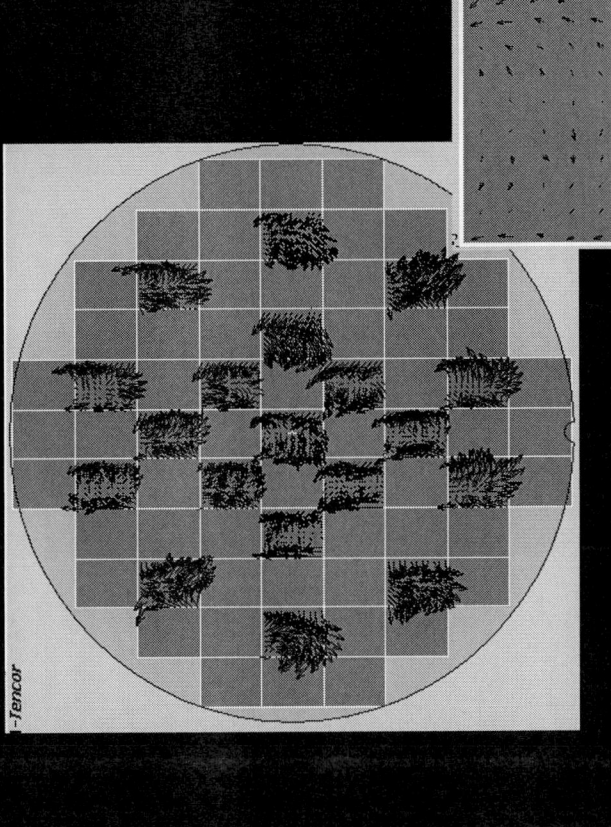

Residual (3s) X: 20.9nm, Y: 17.5nm   Residual (3s) X: 15.0nm, Y: 17.3nm

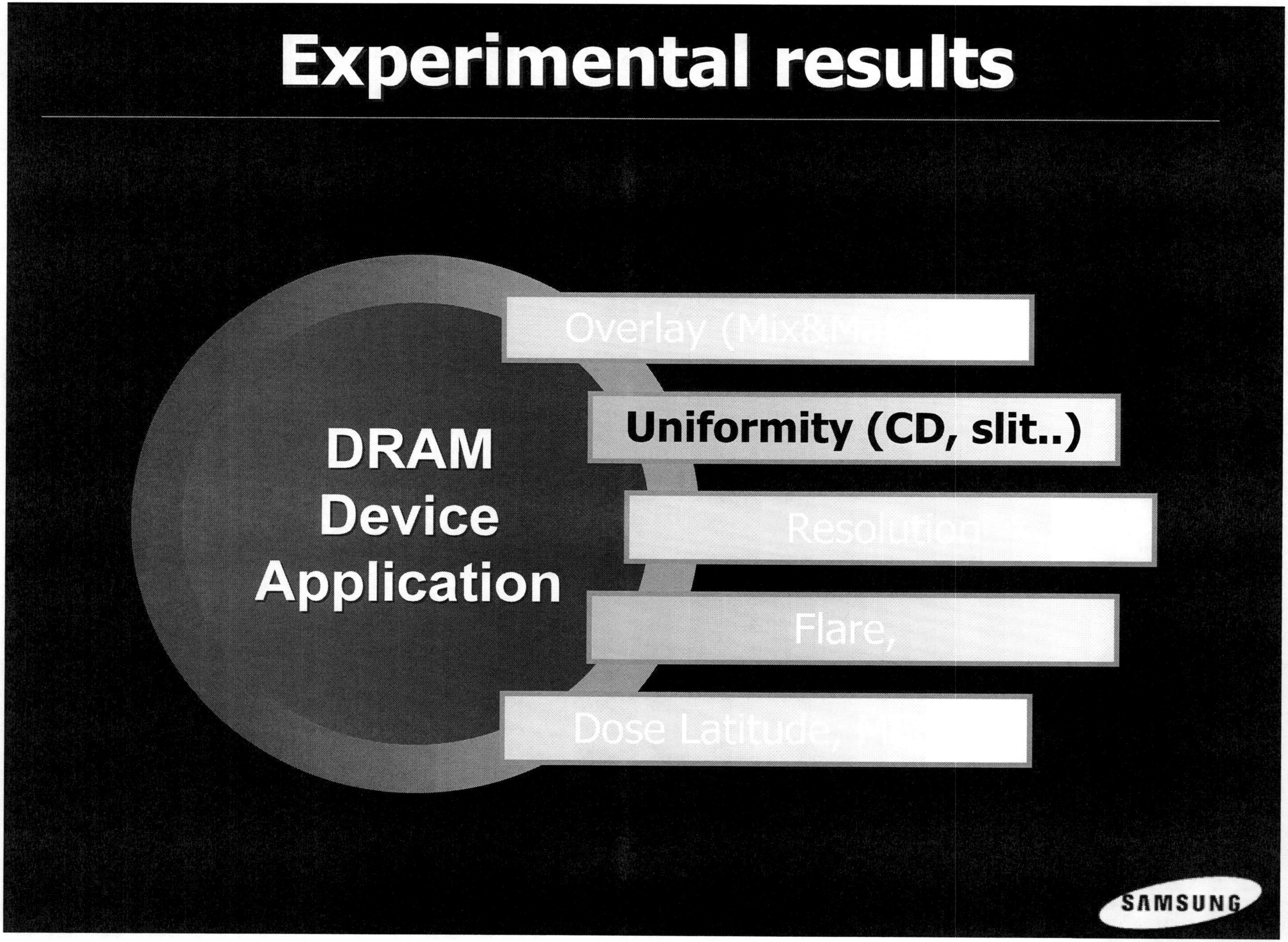

Experimental results

Overlay (Mixing)

Uniformity (CD, slit..)

Resolution

Flare,

Dose Latitude,

DRAM Device Application

SAMSUNG

Slit Uniformity

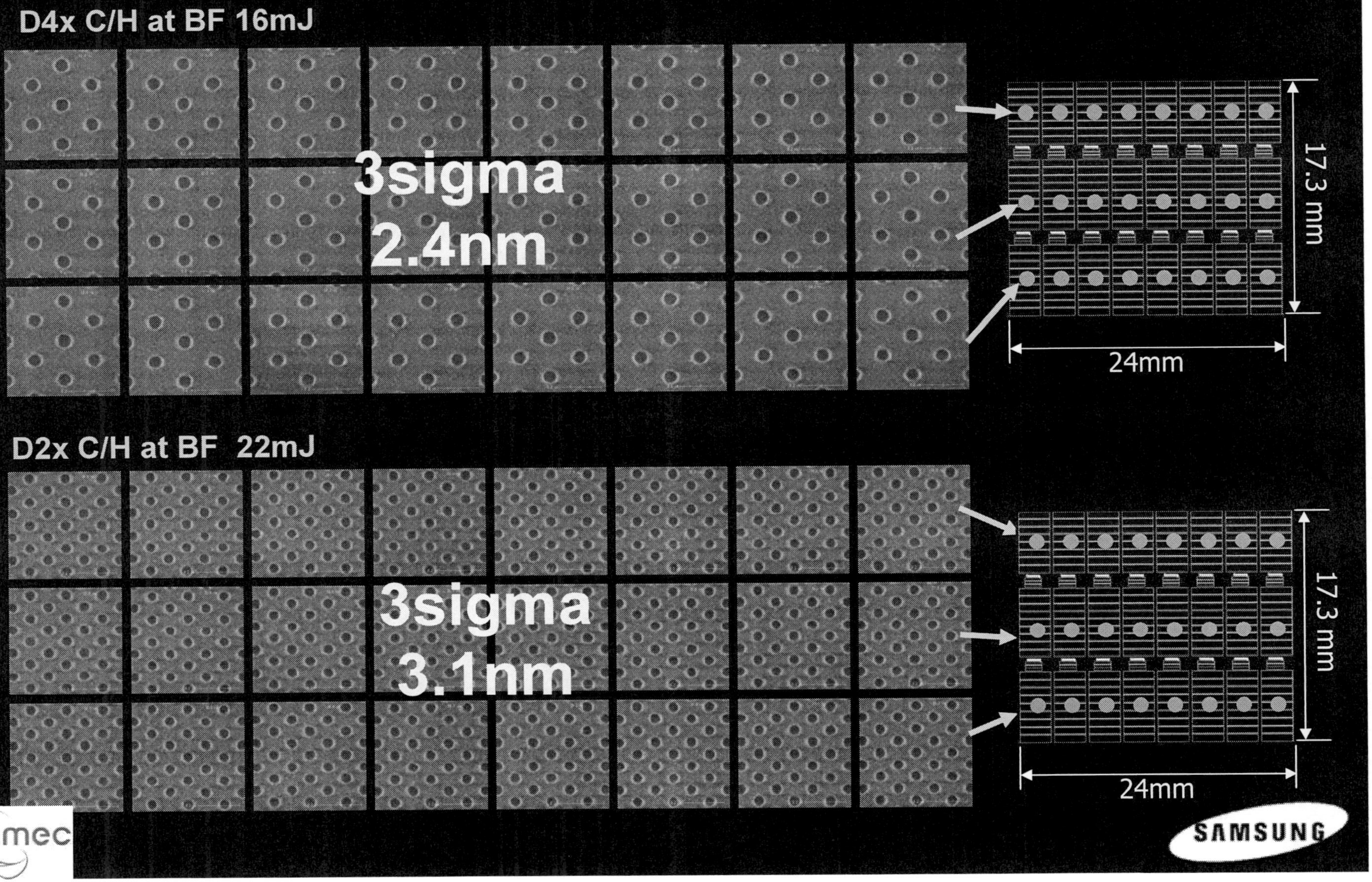

17.3 mm

24mm

17.3 mm

24mm

SAMSUNG

D4x C/H at BF 16mJ

3sigma
2.4nm

D2x C/H at BF 22mJ

3sigma
3.1nm

imec

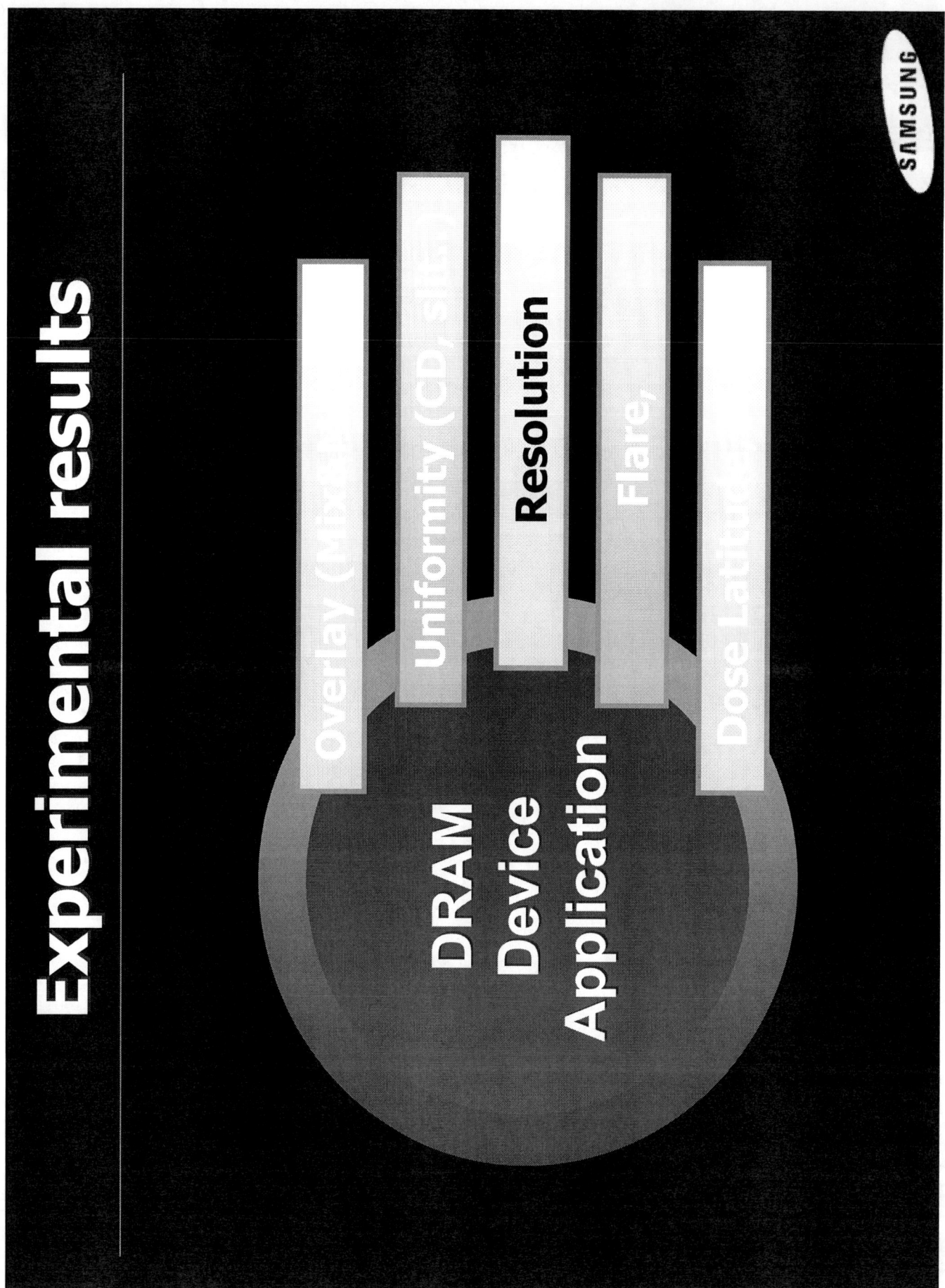

C/H pattern comparison

Block edge

w/o OPC

w/ OPC

EUVL

D4x   D3x   D2x

Well-defined

Immersion (1.35NA)

Not-defined

Additional process needed

1:1 Line&Space pattern w/ ADT

	28nm	30nm	32nm	34nm	35nm	36nm

Resist A
Dose=14mJ/cm2

Resist B
Dose= 11mJ/cm2

Resist C
Dose=11mJ/cm2

Hp 30nm Resist C @LBNL

DRAM cell pattern

Driving Circuit

D3x   D2x

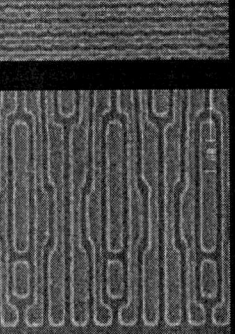

EUVL

EUVL

D3x   D2x

Well-defined

Immersion

Tap Not-defined   Not-defined

EUVL shows superior patterning performance to immersion ArF Lithography

SAMSUNG

Etching results of C/H pattern

D4x edge

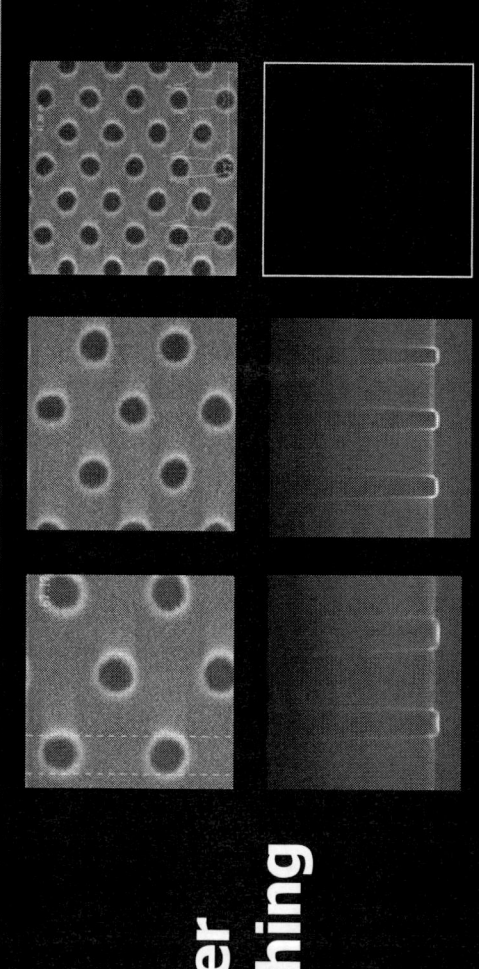

D4x D3x D2x

Before Etching

Well-defined

After Etching

Etching condition is not optimized yet

Experimental results

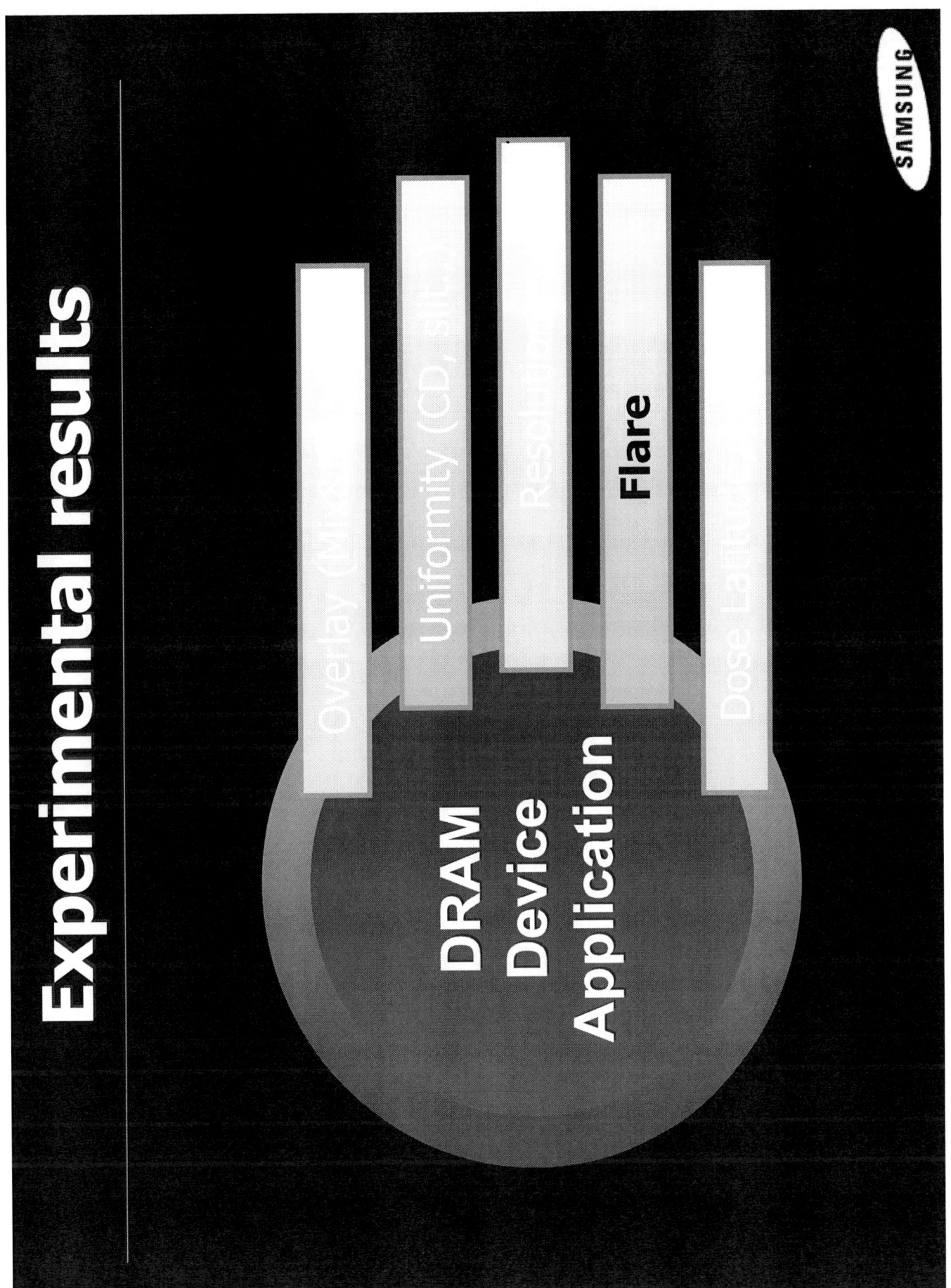

Flare evaluation

Mask information for the JIIRA2 mask of IMEC

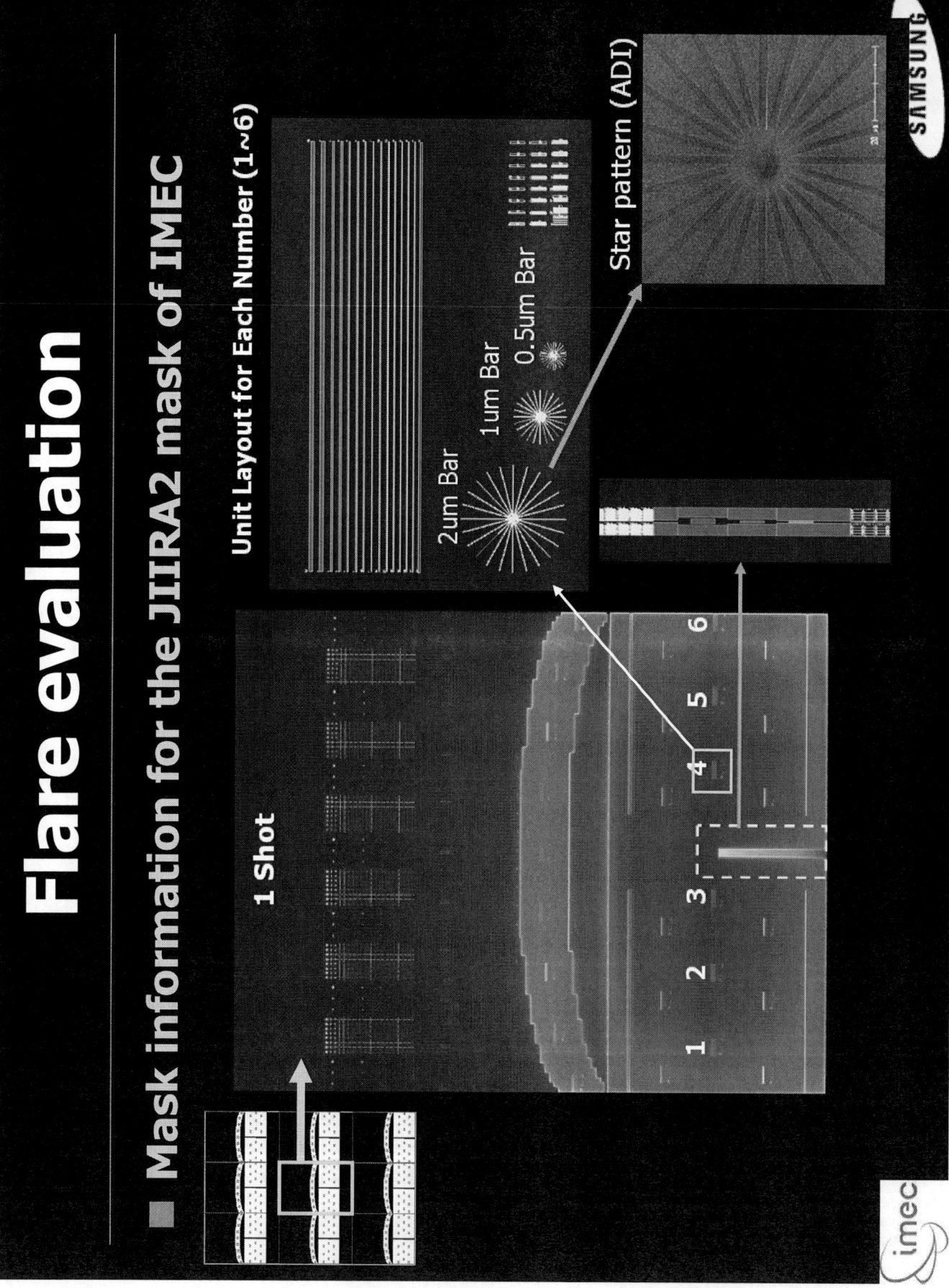

Unit Layout for Each Number (1~6)

2um Bar
1um Bar
0.5um Bar

Star pattern (ADI)

1 Shot

Flare calculation @2um Line

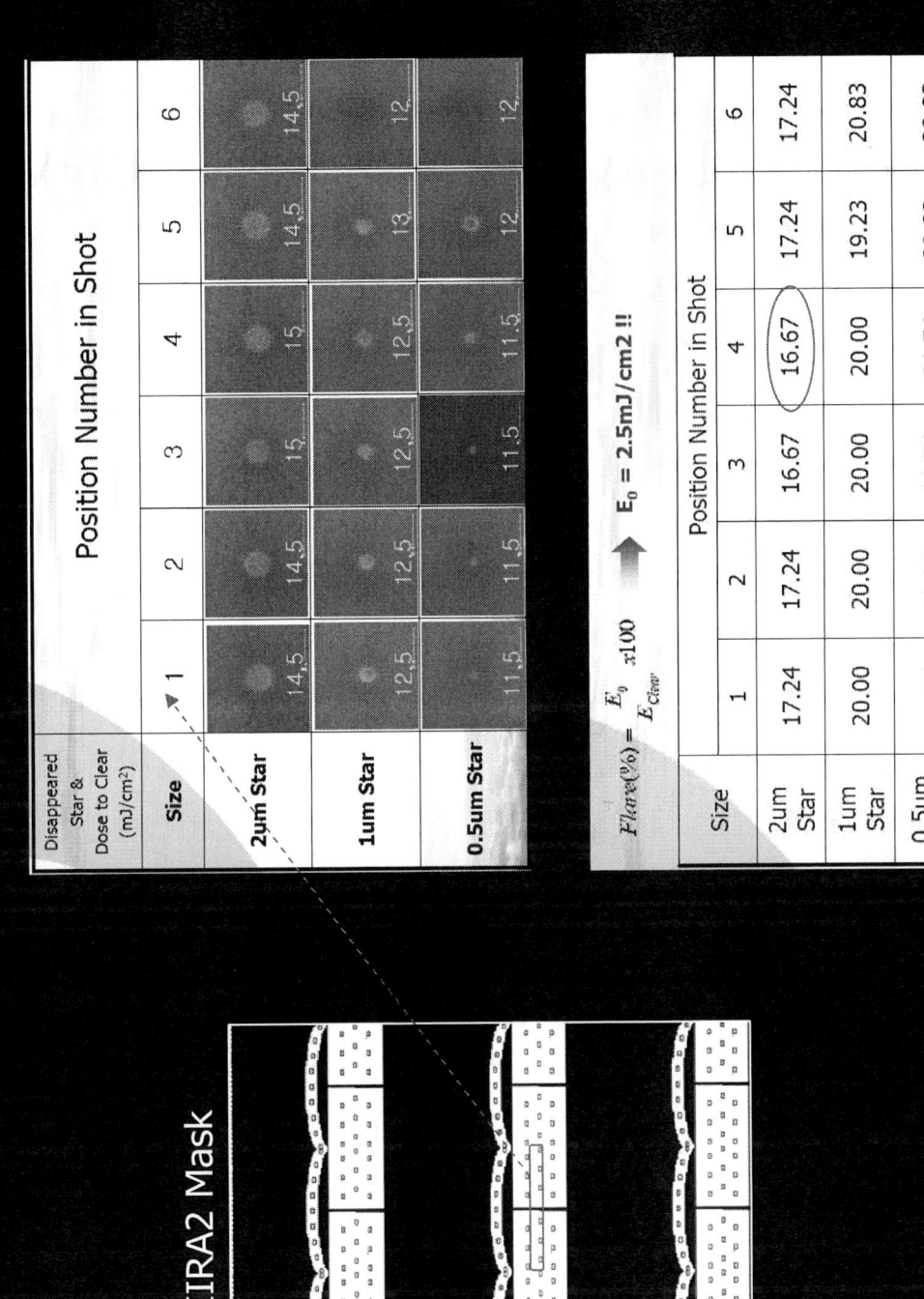

JIIRA2 Mask

Disappeared Star & Dose to Clear (mJ/cm²)	Position Number in Shot					
Size	1	2	3	4	5	6
2um Star	14.5	14.5	15	15	14.5	14.5
1um Star	12.5	12.5	12.5	12.5	13	12
0.5um Star	11.5	11.5	11.5	11.5	12	12

$$Flare(\%) = \frac{E_0}{E_{Clear}} \; x100 \qquad E_0 = 2.5mJ/cm2\ !!$$

Size	Position Number in Shot					
	1	2	3	4	5	6
2um Star	17.24	17.24	16.67	16.67	17.24	17.24
1um Star	20.00	20.00	20.00	20.00	19.23	20.83
0.5um Star	21.74	21.74	21.74	21.74	20.83	20.83

SAMSUNG

Method of Flare Compensation

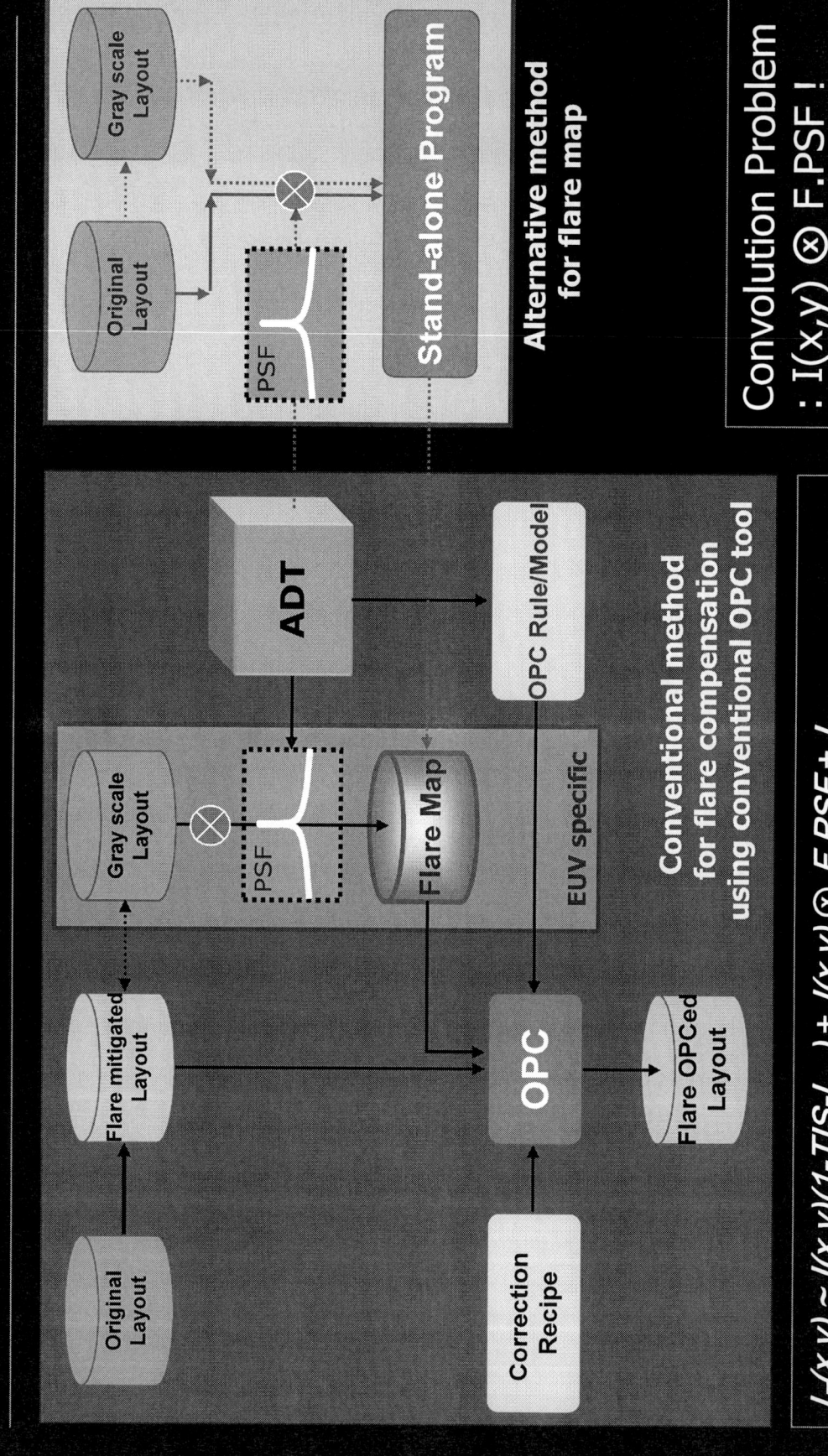

Alternative method for flare map

Convolution Problem
: $I(x,y) \otimes F.PSF$!

Conventional method for flare compensation using conventional OPC tool

$I_F(x,y) \sim I(x,y)(1\text{-}TIS\text{-}I_{DC}) + I(x,y) \otimes F.PSF + I_{DC}$

TIS : Total Integrated Scattering

I_{DC} : DC offset

SAMSUNG

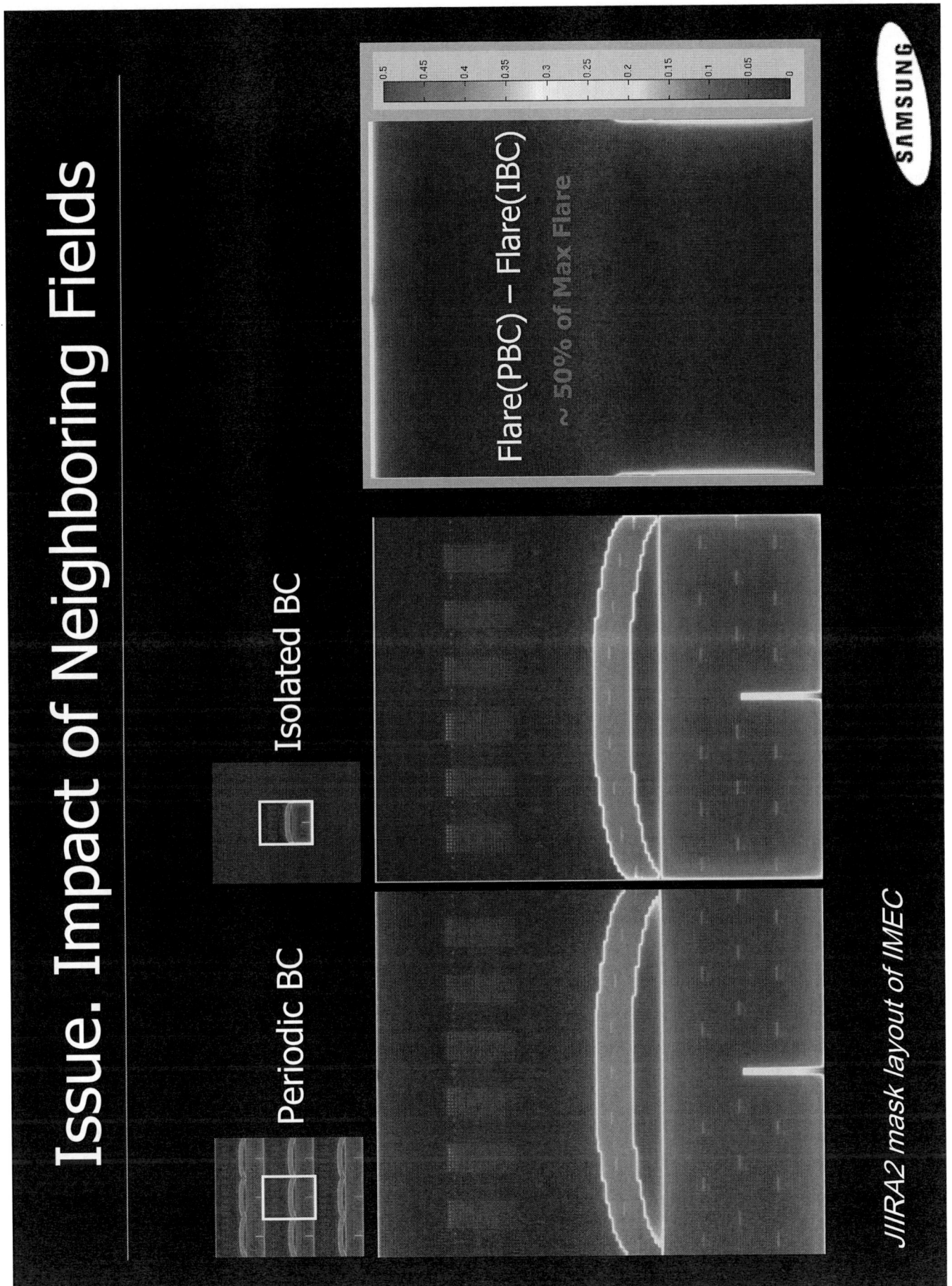

Experimental results

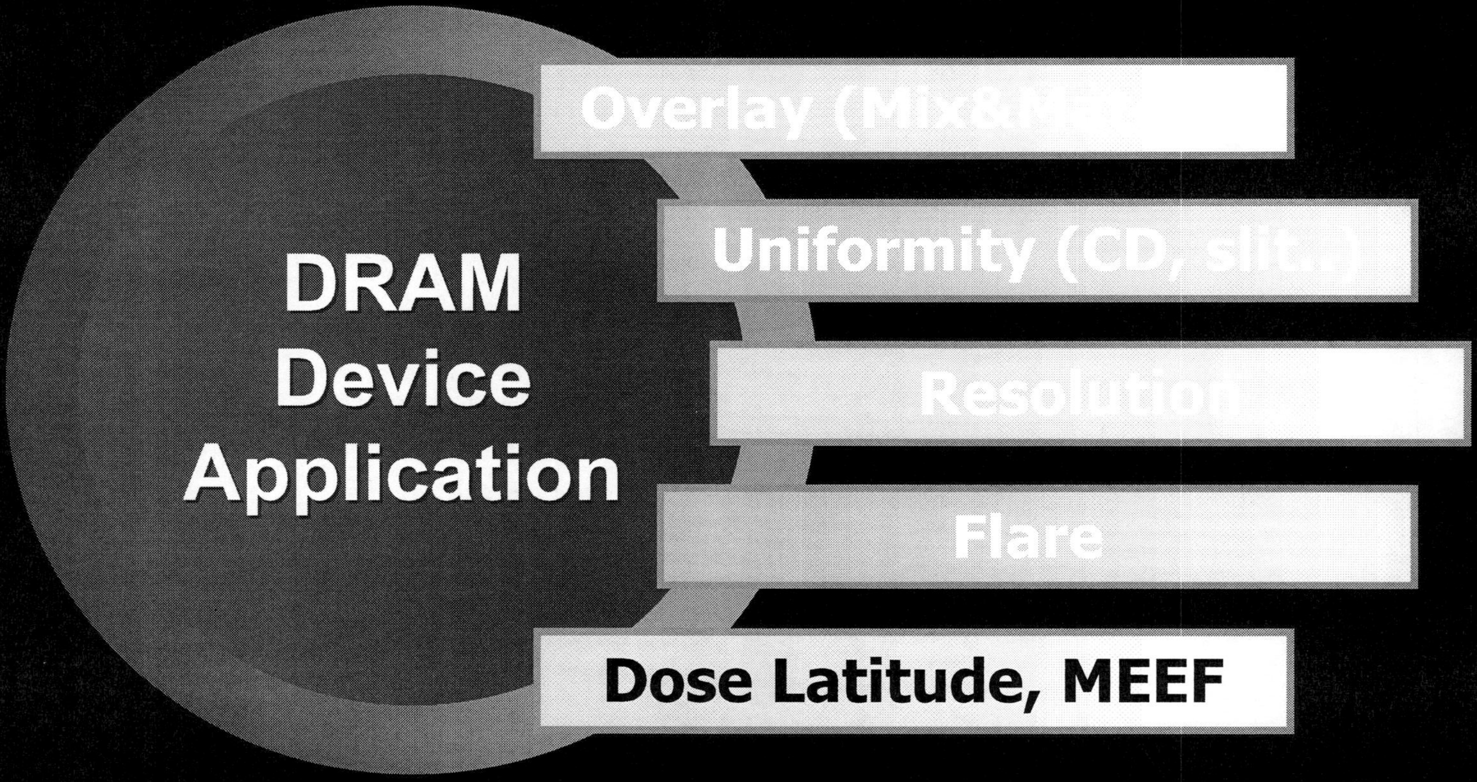

SAMSUNG

- Overlay (Mix&Mat)
- Uniformity (CD, slit...)
- Resolution
- Flare
- Dose Latitude, MEEF

DRAM Device Application

Dose latitude & MEEF of C/H

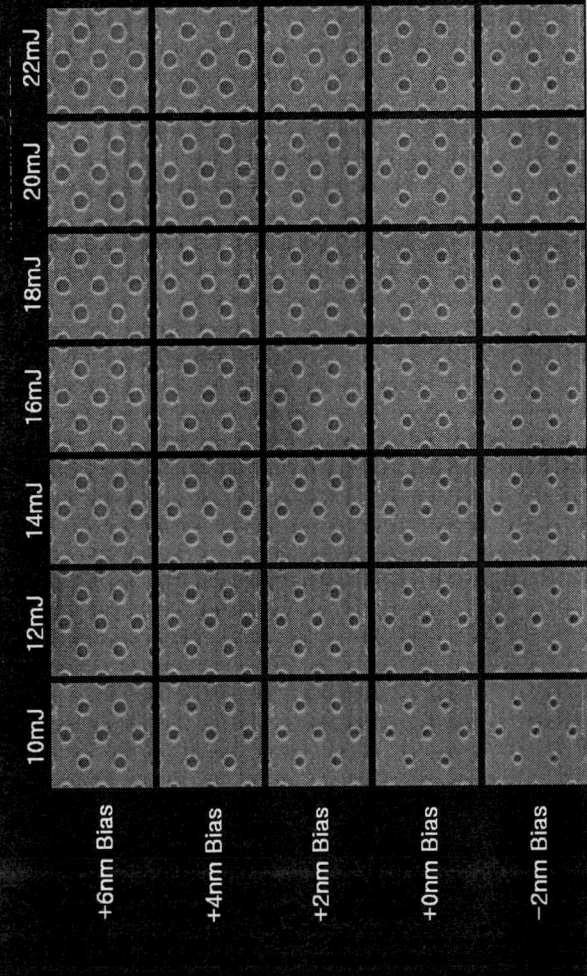

- **Dose latitude at BF**
  - < 0.1nm/%dose

We are in the high-k1 technology era 20 years ago !

- **MEEF at 16mJ**
  - 1.2

Summary

- **We characterized ADT at IMEC with**
  - MMO, scan direction behavior
    - X, Y ~ 15nm
  - CD uniformity
    - Full-field slit uniformity < 2.4 nm (3sigma) for D4x
  - Resolution
    - Line&space pattern ~ hp 32nm, C/H ~hp 35nm resolved, MEEF ~1.2
  - Flare
    - Setup the protocol and compensation solution addressed

- **We found out that**
  - D4x C/H layer printed with ADT can be applied to the real D3x DRAM device.
  - EUVL is a promising solution for small C/H patterns comparing with immersion lithography.

- **EUV-integrated DRAM will be presented in SPIE 2009**

Acknowledgement

SAMSUNG

- **IMEC**
  - Jan Hermans, Kurt Ronse

- **ASML**
  - IMEC support team

- **SEC**
  - EUV team

Thank You !

SAMSUNG

Flare evaluation of an ASML Alpha Demo Tool

Hiroyuki Mizuno,

Martin Burkhardt[1], Chiew-Seng Koay[1], Greg McIntyre[1], Tim Brunner[1], Bruno La Fontaine[2], Yunfei Deng[2], and Obert Wood[2]

Toshiba America Electronic Components, Inc.

[1]IBM Advanced Lithography Research

[2]Advanced Micro Devices, Inc.

Outline

- Introduction
- Flare evaluation methods and early results
  - Donut pattern method
    - What are donuts ?
    - Point Spread Function data from 2 resists
  - LS method
    - How to determine flare distance ?
    - Results from 3 resists
    - How to determine flare magnitude, η ?
    - Results from 3 resists
- Summary

Introduction

- **The cause of EUV flare is mirror surface roughness.**

- **Flare specification for the ADT is 16%; the actual value is ~12%.**

- **We were successful in our 45nm device demonstration using rule based-OPC derived from mirror power spectral density (PSD) data.**
  - **Residuals thru pitch were less than 5 nm.**

- **Do the following patterns measure "actual" flare ?**
  - **Donut patterns**
    - to determine flare distance
  - **LS pattern**
    - traditional e-beam proximity effect correction

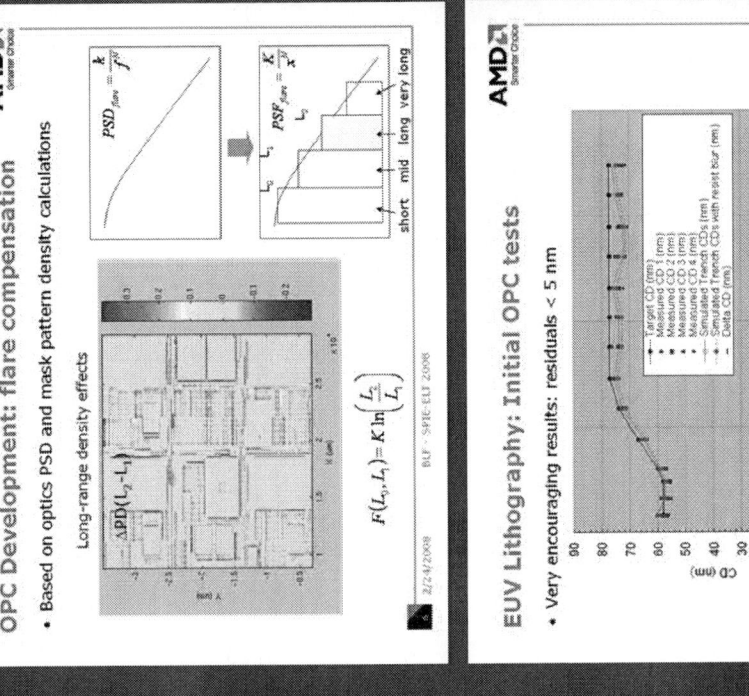

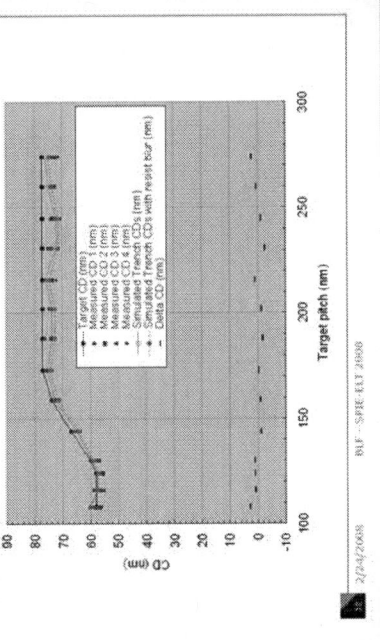

Today, I will show early results of Donut and LS evaluation method.

Outline

- Introduction
- **Flare evaluation methods and early results**
  - Donut pattern method:
    - What are donuts ?
    - Point Spread Function data from 2 resists
  - LS method:
    - How to determine flare distance ?
    - Results from 3 resists
    - How to determine flare magnitude, η ?
    - Results from 3 resists
- Summary

What are donuts ?

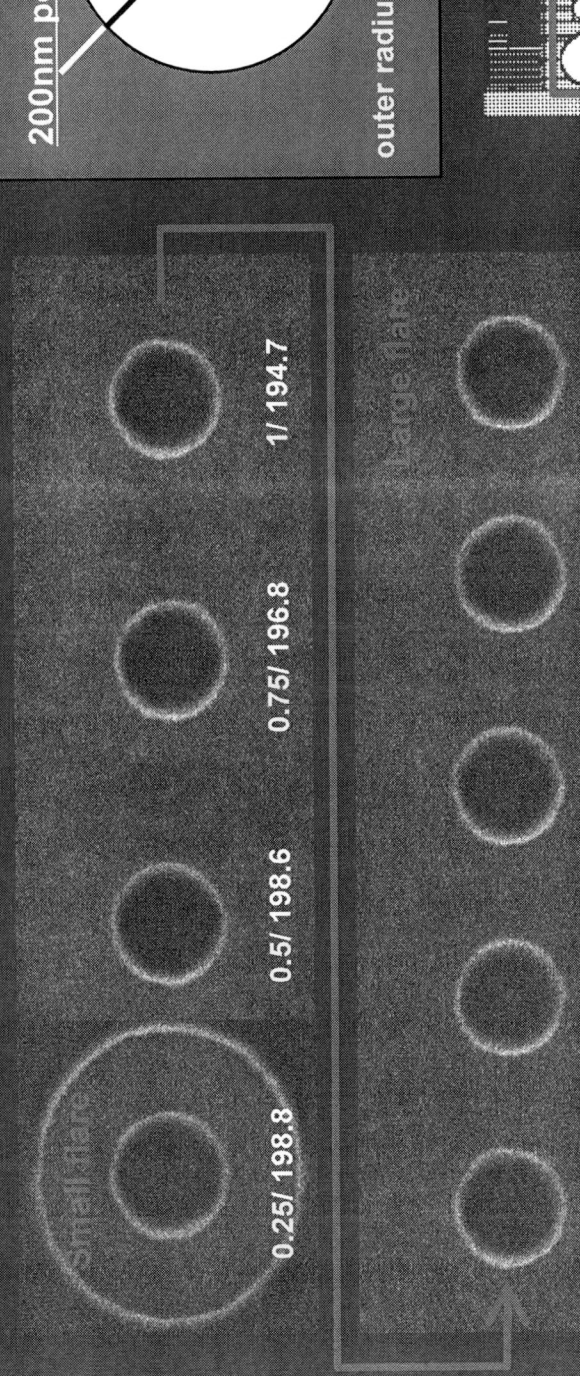

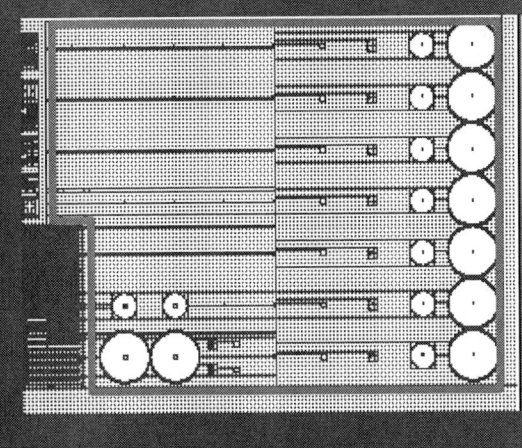

200nm post

outer radius (μm)/ inner CD (nm)

Small flare

0.25/ 198.8   0.5/ 198.6   0.75/ 196.8   1/ 194.7

large flare

2.5/ 190.2   5/ 191.2   10/ 190.3   25/ 187.5   50/ 183.4

- CD of inner post is affected by outer radius of clear annulus.
- CD measurements of resist images of donut patterns yield a value for flare distance.

How to determine Flare PSF from Donuts data

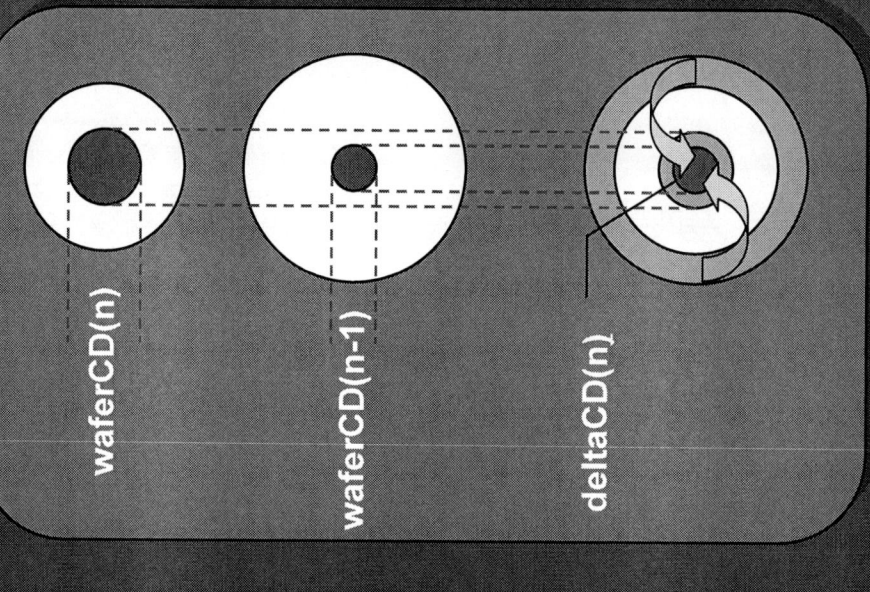

waferCD(n)

waferCD(n-1)

deltaCD(n)

- Process:
  - Determine CD sensitivity to dose
  - Going from smallest to largest donut outer rings:
    - Determine change in CD due to each additional ring

      deltaCD(n) = waferCD(n) – waferCD(n-1)

    - Determine dose required to cause this amount of CD change

      deltaDose(n) = deltaCD(n) / DoseSlope

    - Assuming spillover intensity to center is constant for any given ring, determine the magnitude of the spillover to cause the required change in dose at the center

      spillover(n) = deltaDose(n) / $(2\pi(r_{outer} - r_{inner}))$

    - Plot this spillover intensity vs. donut radius. This should be a crude approximation to the Flare Intensity PSF

Flare evaluation by donut patterns

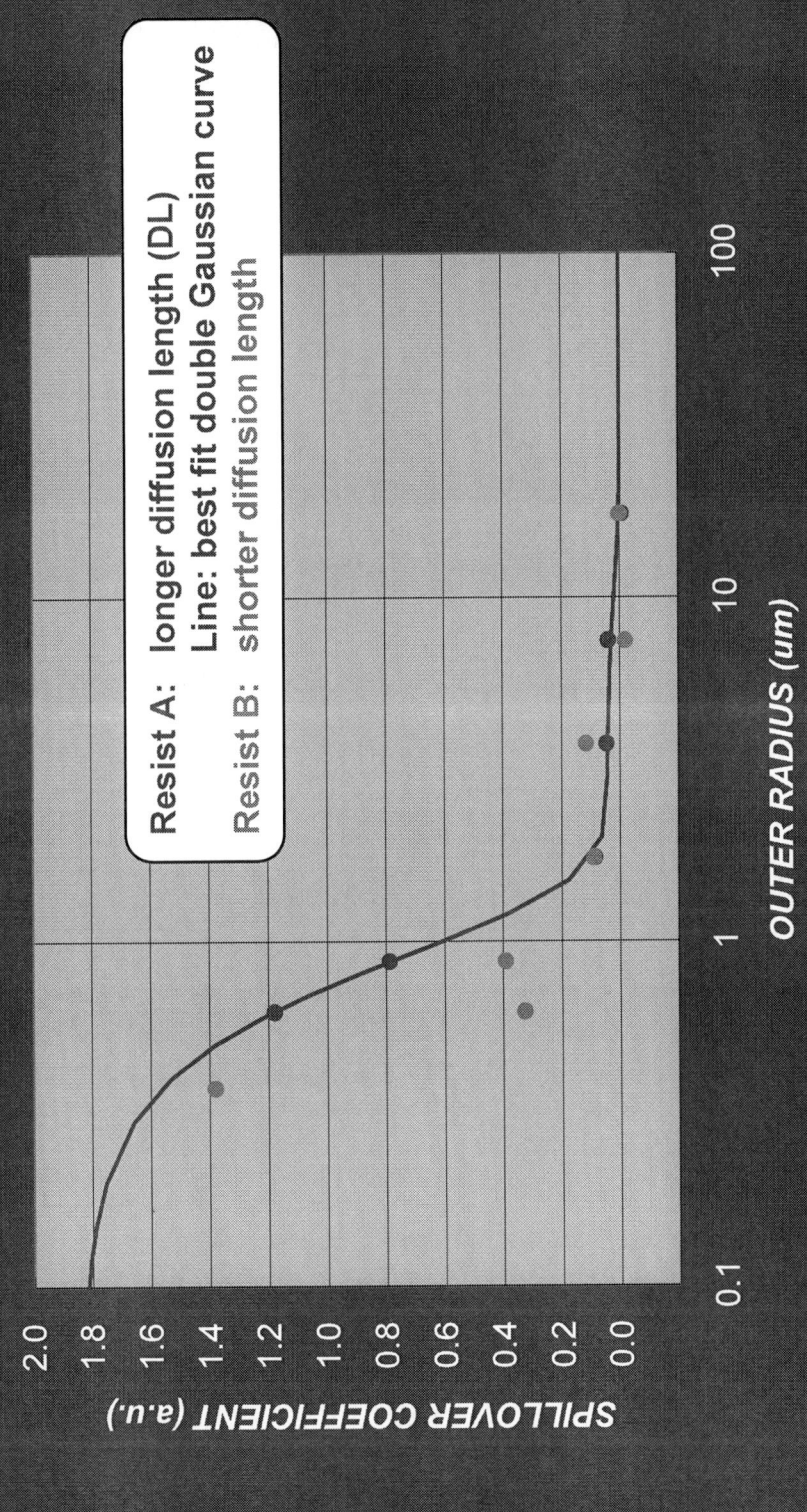

Resist A: longer diffusion length (DL)
Line: best fit double Gaussian curve
Resist B: shorter diffusion length

There is a huge spillover area, especially at < 2 µm outer radius. Both Resist A and B show the same trends even though they have different diffusion lengths.

Outline

- Introduction
- **Flare evaluation methods and early results**
  - Donut pattern method:
    - What are donuts ?
    - PSF data from 2 resists
  - LS method:
    - How to determine flare distance ?
    - Results from 3 resists
    - How to determine flare magnitude, η ?
    - Results from 3 resists
- Summary

How to determine flare distance ? (50nm LS pattern)

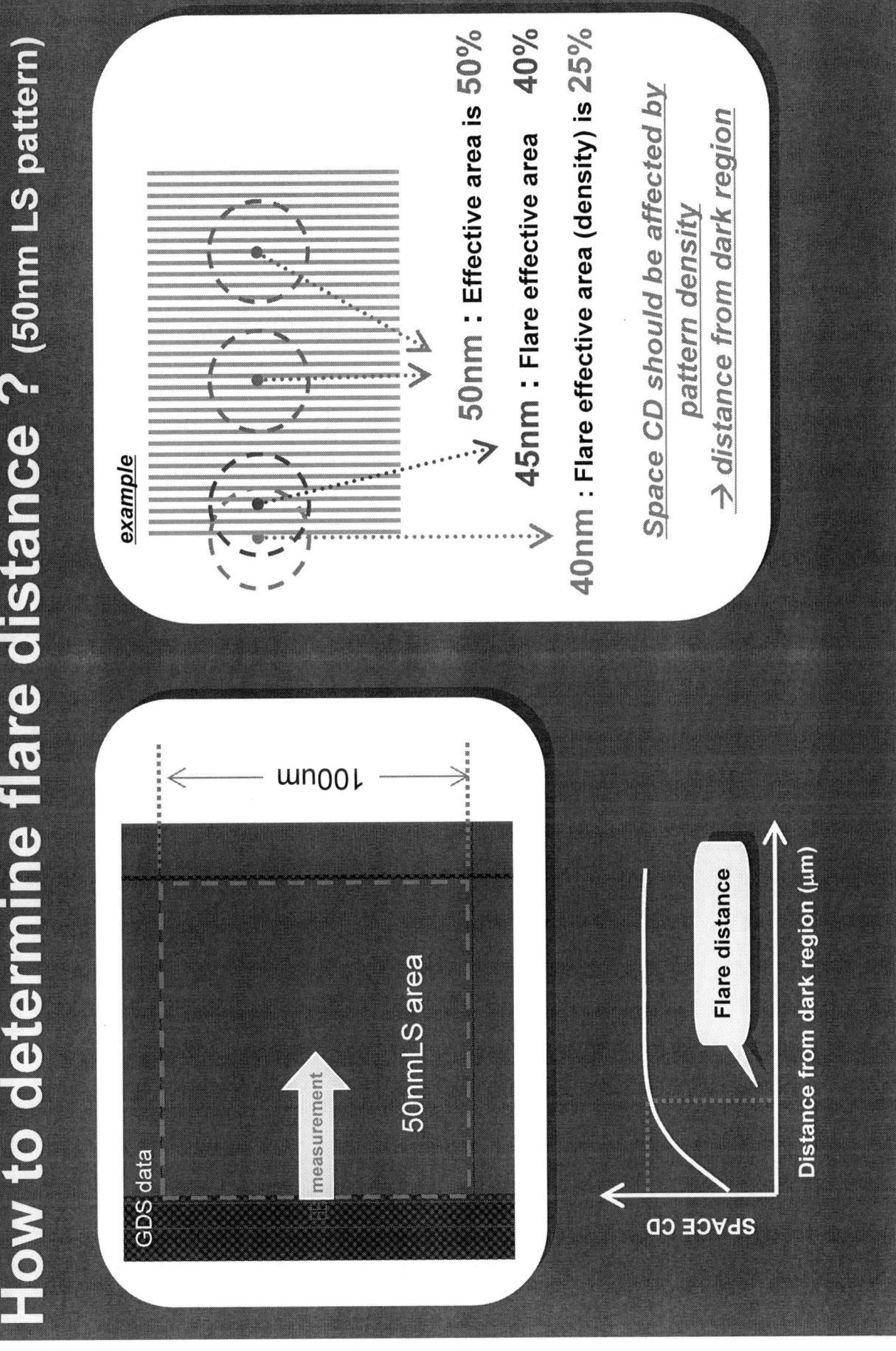

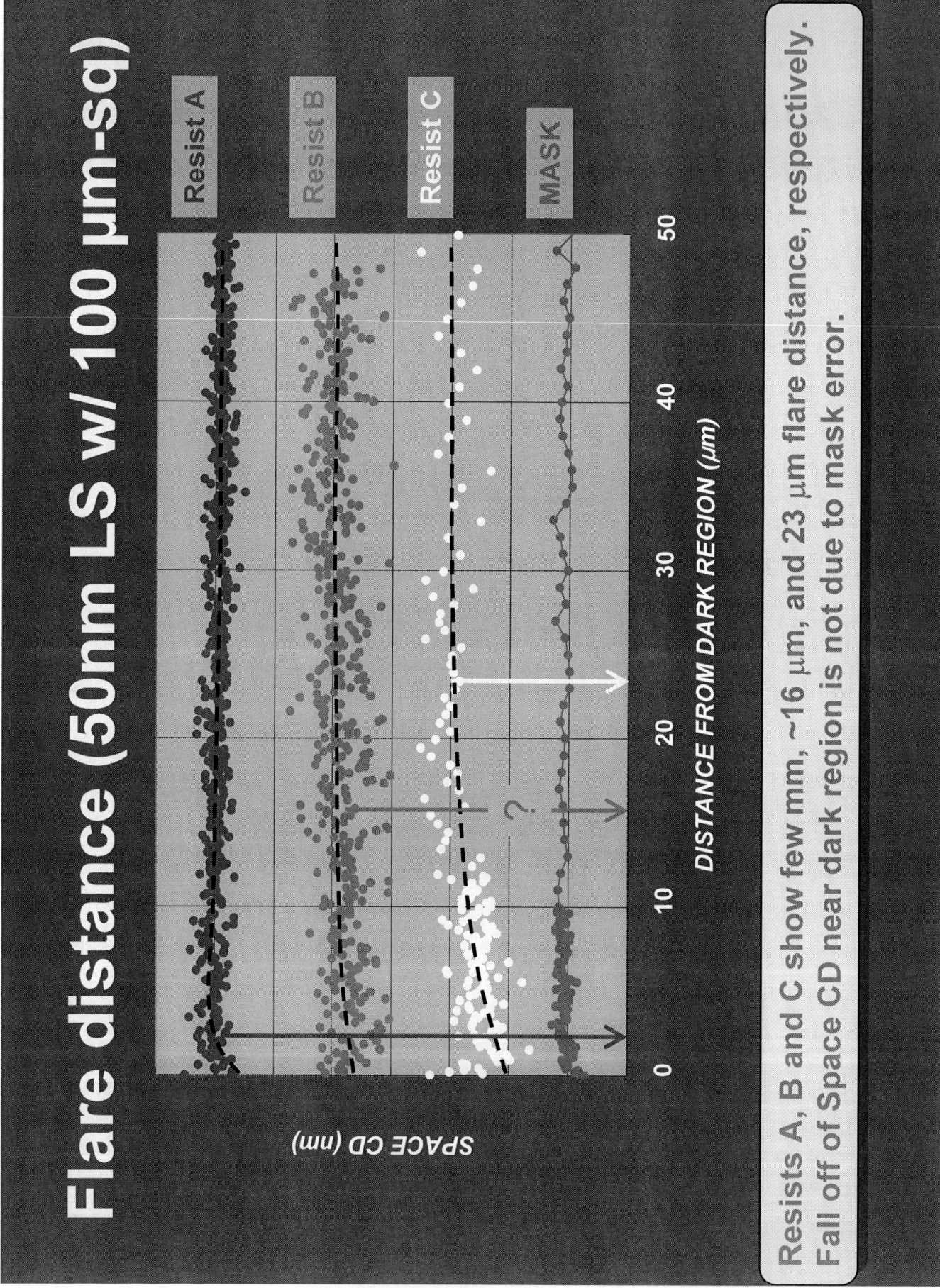

Flare distance (50nm LS w/ 100 μm-sq)

Resists A, B and C show few mm, ~16 μm, and 23 μm flare distance, respectively. Fall off of Space CD near dark region is not due to mask error.

How to determine flare magnitude, η ?

- Find optimum dose at the center of a large area that contains LS patterns at various pattern densities:

- Calculate η by fitting the dose data to the following equation:

$$\frac{Dx}{D_{50}} = \frac{1+\eta}{1+2\eta x}$$

$\left\{\begin{array}{l} \text{Dx: Optimized dose for various density} \\ \text{D}_{50}\text{: Optimized dose at 50\% density} \\ \text{x : pattern density} \end{array}\right.$

Line width (nm)
50 40 32

Pattern density (%): 1, 10, 20, 30, 40, 50, 60, 70, 80, 90, 99

NOVACD pattern for evaluating η
(each square is 100 µm-sq area)

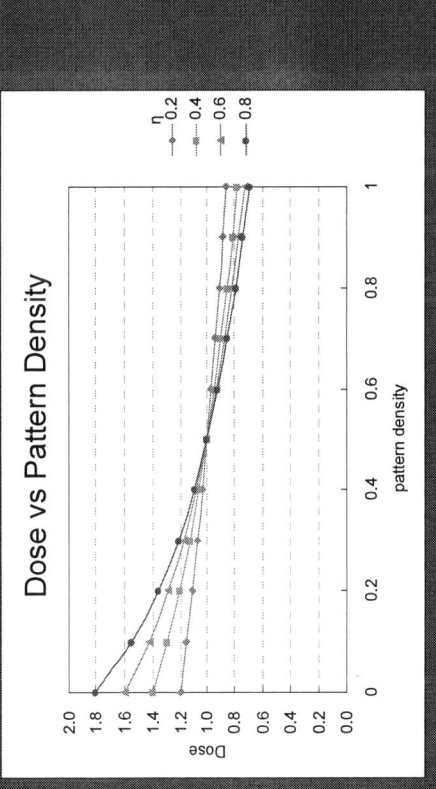

Dose vs Pattern Density

η: 0.2, 0.4, 0.6, 0.8

How was $\dfrac{D_x}{D_{50}} = \dfrac{1+\eta}{1+2\eta X}$ derived ?

- **Assumption (1): pattern area > flare distance > LS pitch**

$$\underline{E_x = (0.5 + \eta X)\ D_x}$$

Ex: Optimized energy
X : pattern density
Dx: Optimized dose
(We can think of η as ratio of flare in BF.)

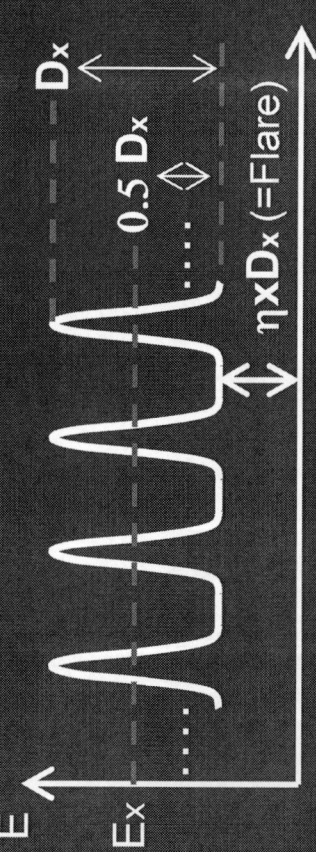

- **Assumption (2): optimum energy for target CDs should be at the same Ex, even if its pattern densities are different. (in this case, base-line density is 50%.)**

$$\underline{E_x = (0.5 + \eta X)\ D_x = (0.5 + \eta\ 0.5)\ D_{50} = E_{50}}$$

$$\boxed{D_x = \dfrac{1+\eta}{1+2\eta X}\ D_{50}}$$

Parameter η (flare magnitude)

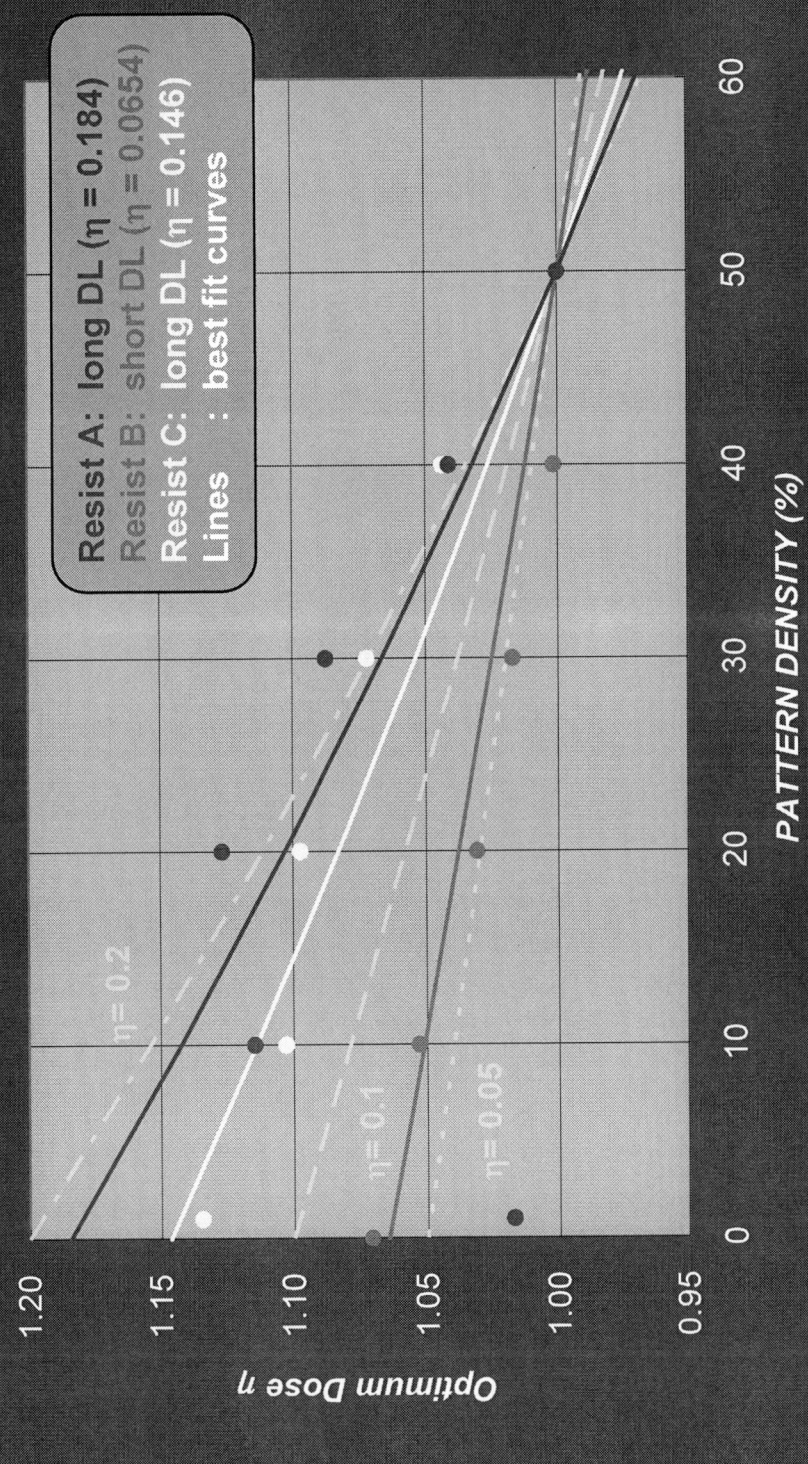

Optimum η values are 18.4, 6.45, & 14.6% for 3 Resists A, B & C, respectively. Long DL resists show higher η values. η strongly depends on resist material.

Summary

- Early results of 2 unique flare evaluation methods are shown.

  – Donut patterns: huge spillover area at < 2 μm outer radius for Resist A and B. (No data yet for Resist C.)

  – LS pattern: resists with long diffusion lengths show higher η values. η strongly depends on resist material.

		Resist A long DL	Resist B short DL	Resist C long DL
Magnitude	η	18.4 %	6.45 %	14.6%
	PSF	Same trend		No data
Distance	LS edge	Few μm	~16 μm	23 μm

- These results will be utilized for our next device demo.

Acknowledgements

This work was performed by the Research Alliance Teams at various IBM Research and Development Facilities

- **ASML**
  - Kevin Cummings
  - Bill Pierson
  - Sang-In Han
  - Rick Zachgo

- **College of Nanoscale Science and Engineering**
  - Sudhar Raghunathan

Implementing full field EUV lithography using the ADT

Anne-Marie Goethals[1], E. Hendrickx[1], R. Jonckheere[1], G. F. Lorusso[1], B.Baudemprez[1], J. Hermans[1], D. Laidler[1], A. Niroomand[2], F. Van Roey[1], A. Van Dijk[3], L. Romijn[3], N. Stepanenko[4], V. Timoshkov[4], F. Iwamoto[5], A. Myers[6], Y. Hyun[7], C. Lim[7], I. Pollentier[1], M. Leeson[5], J-F. de Marneffe[1], S. Demuynck[1] and K. Ronse[1]

1IMEC, Leuven, Belgium
2On assignment from Micron
3ASML
4On assignment from Qimonda
5On assignment from Panasonic Corporation
6On assignment from Intel
7On assignment from Hynix

International EUVL Symposium, 29 Sept -1 Oct 2008, Lake Tahoe

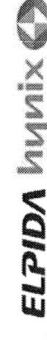

Introduction
IMEC EUV tool set and program

Outgassing Tool
(EUV Technology)

First Light Apr 2007

35nm vertical lines & spaces

40nm horizontal lines & spaces

Images, September 2007

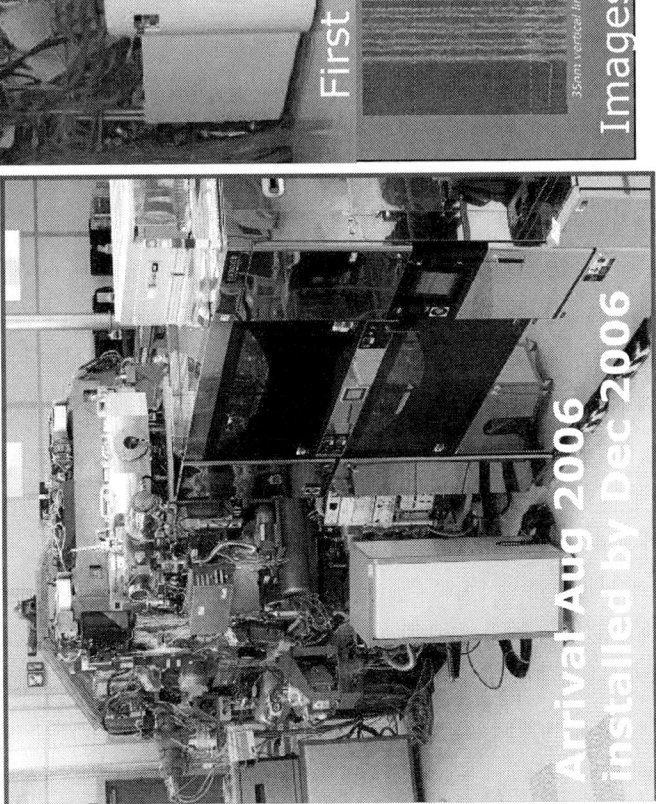

Arrival Aug 2006
Installed by Dec 2006

FAT on Sept 19th, 2008

ASML ADT SAT on June 4th, 2008

IMEC EUV program

EUV Resists and process
EUV Reticles*
ASML ADT assessment

* session mask-2 : Mask defect printability in full field EUV lithography-Part2, R. Jonckheere

Outline

- Introduction

- **Resist and process performance**
  - Interference lithography exposures
  - ADT screening/benchmarking results

- ASML EUV ADT performance

- Process implementation

- Conclusion

- Acknowledgements

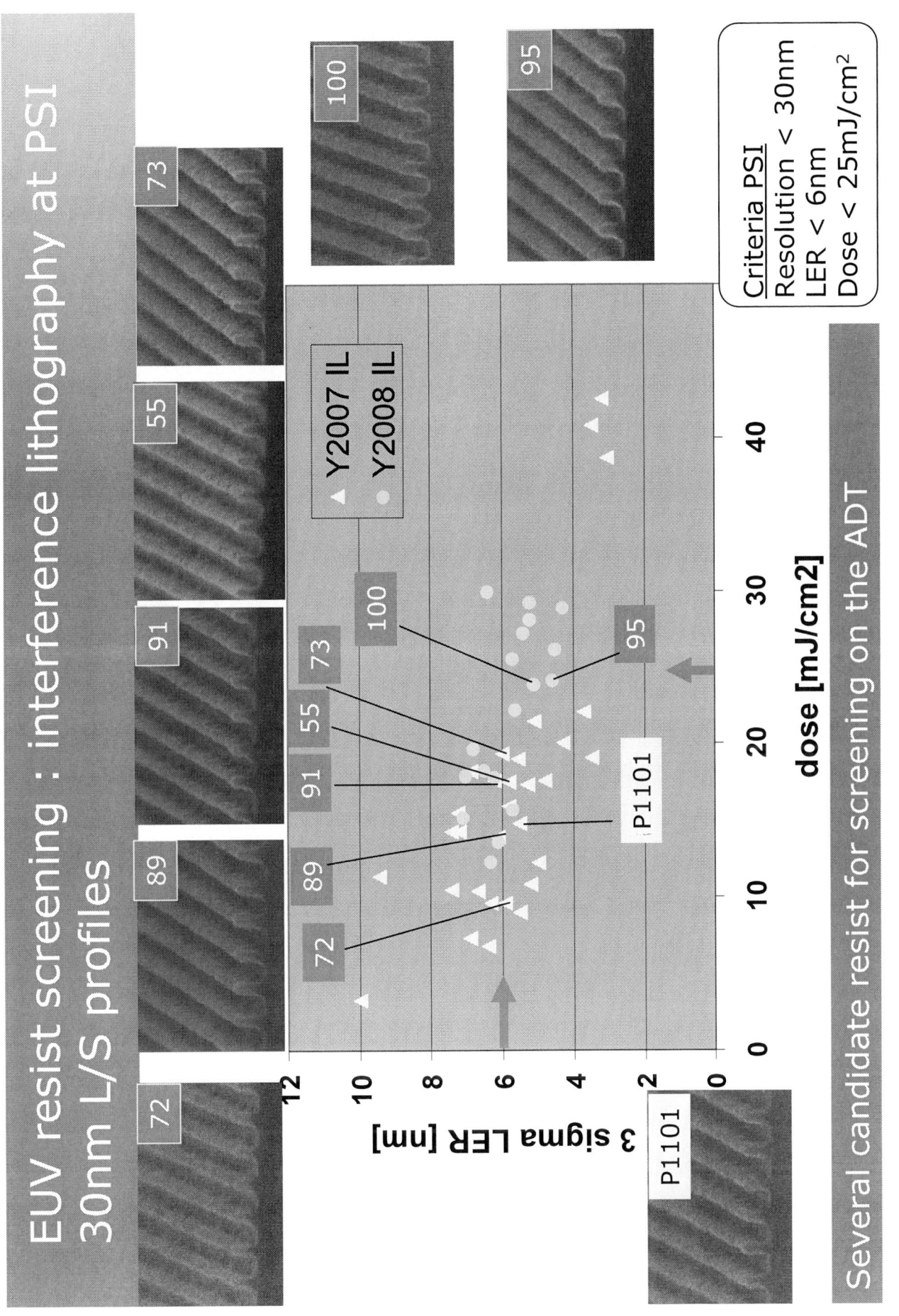

EUV resist screening with interference litho and on the ASML EUV ADT (NA=0.25, σ=0.5)

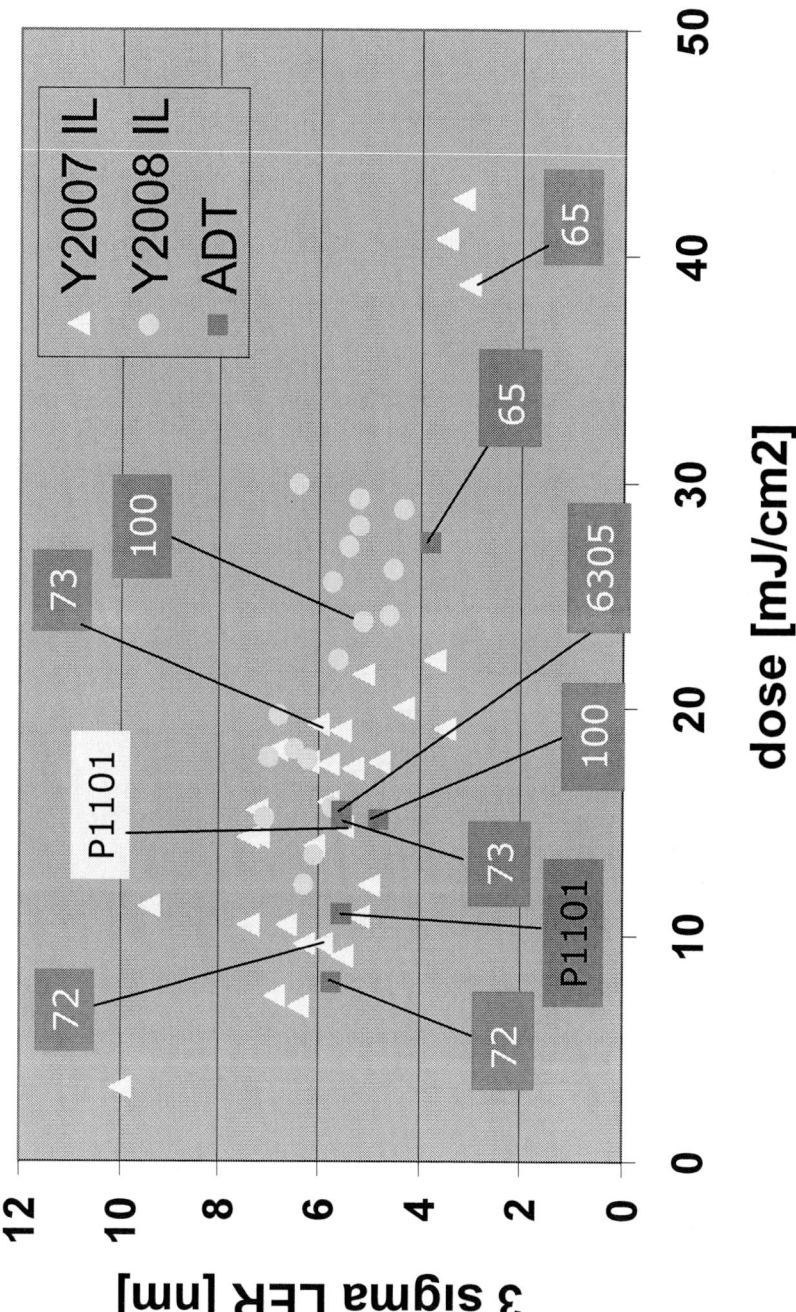

Similar LER values but lower dose to size on the ADT for same resists as compared to interference litho

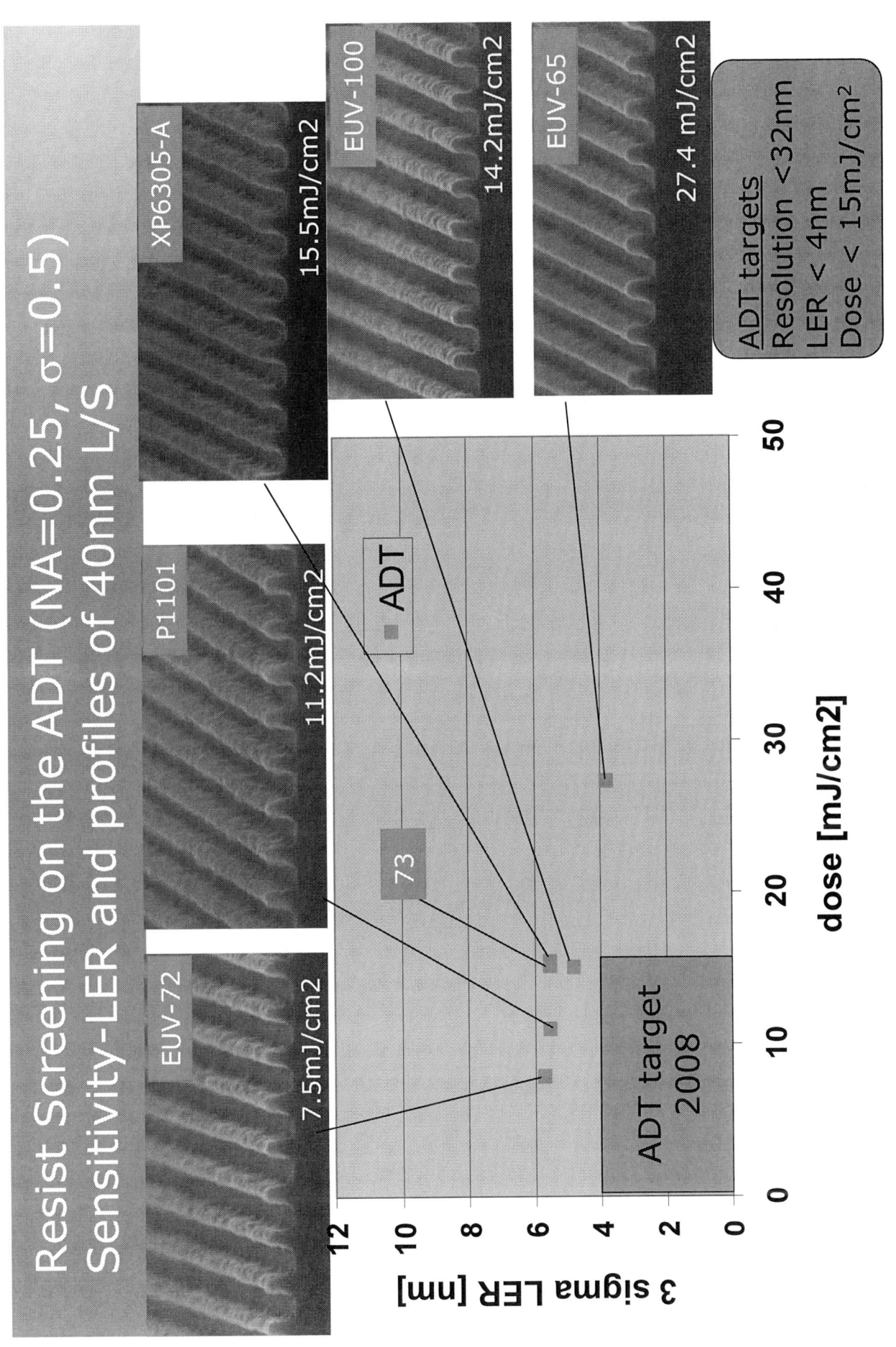

Resist benchmarking on ADT
MEEF for 40nm L/S (Horizontal)

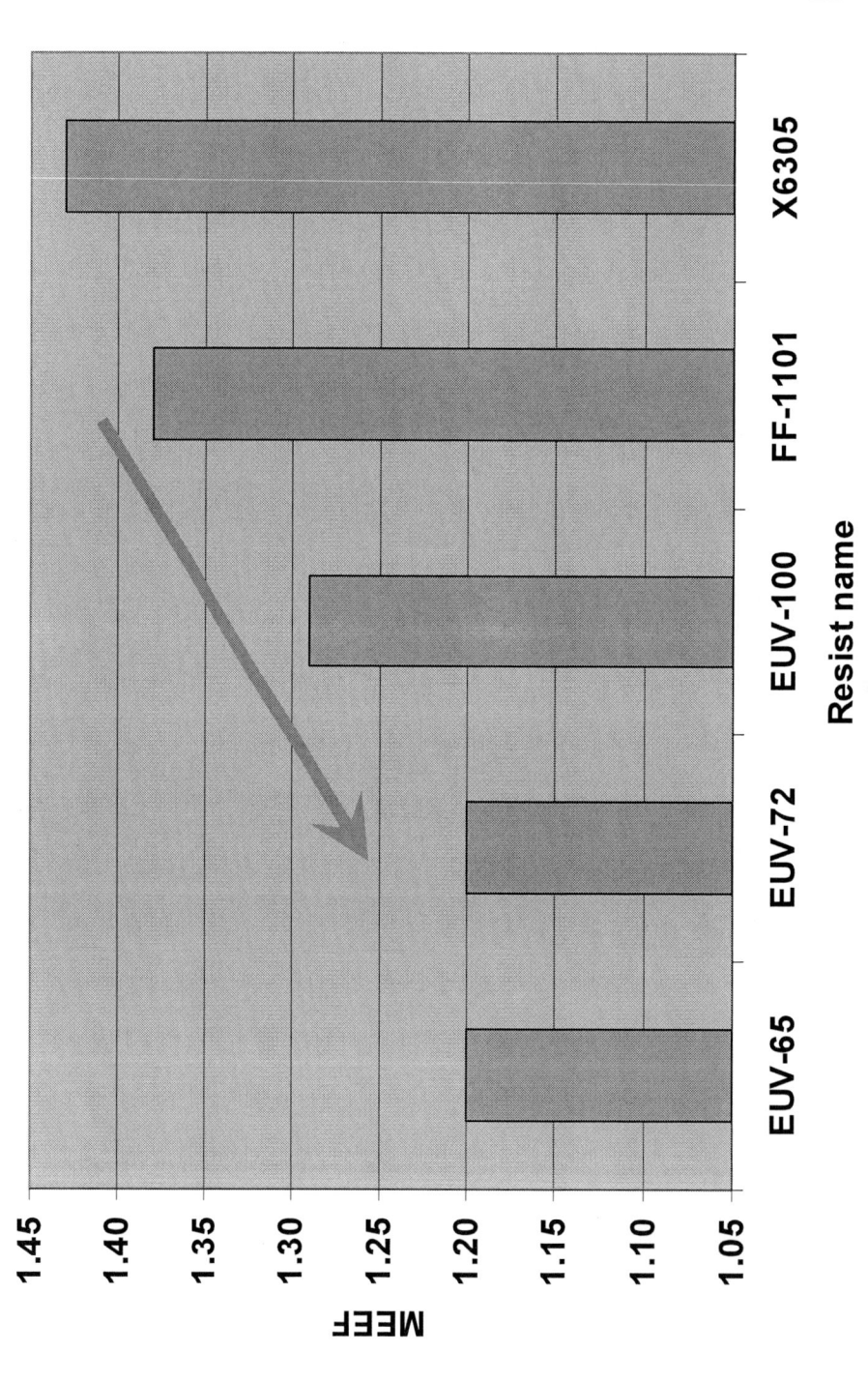

NA=0.25
σ=0.5

Newer resist materials have lower MEEF

Resist benchmarking
Processing latitudes for different features

Focus latitude

Exposure latitude

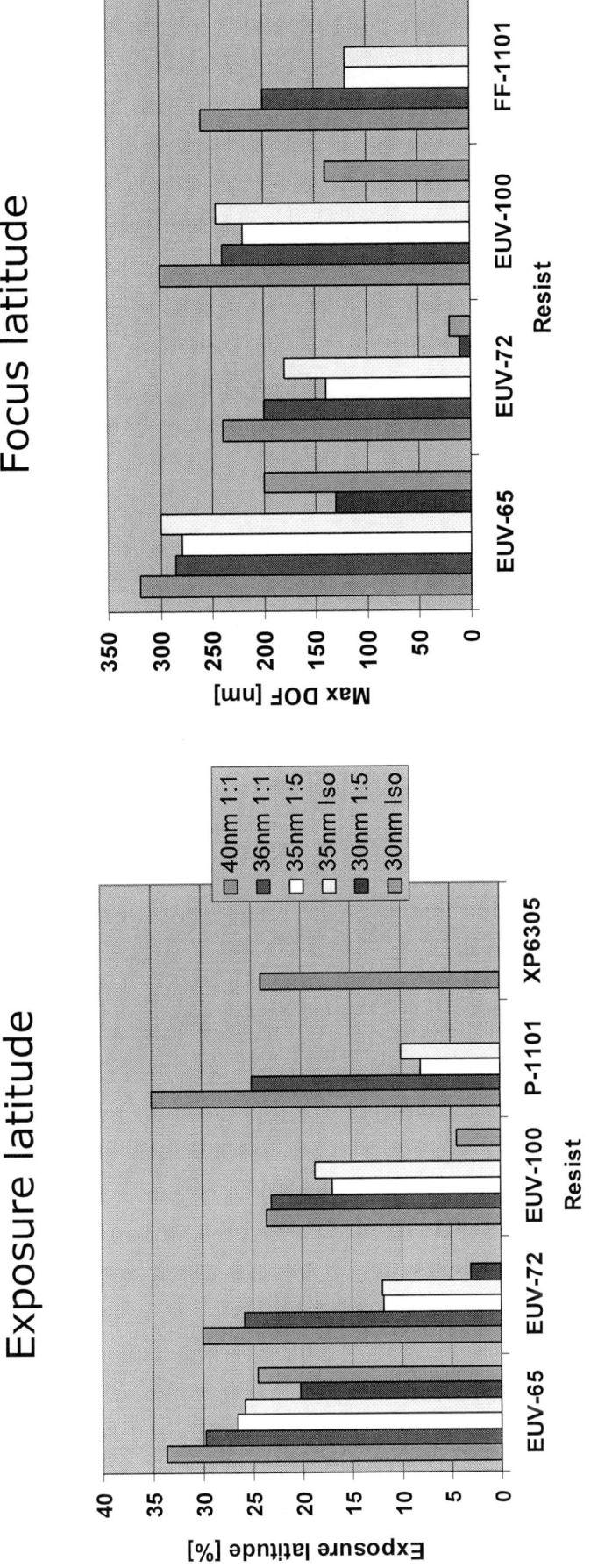

Large process window for dense feature
Limited process window for 30nm 1:5 and 30nm iso lines

Resist benchmarking on ASML ADT 40nm to 32nm HP resolution

	P1101	EUV-100	EUV-72	XP6305	EUV-65
40nm	E=11.0 mJ LER5.5nm	E=14.3 mJ LER=4.8nm	E=7.5 mJ LER=5.7nm	E=15.5 mJ LER=5.5nm	E=27.4 mJ LER=3.8nm
34nm	11.0 mJ	14.3 mJ	7.5 mJ	15.5 mJ	27.4 mJ
32nm	11.0 mJ	14.3 mJ	7.5 mJ	15.5 mJ	27.4 mJ

Linear resolution down to 32nm HP for most resists

Resist benchmarking on ASML ADT
30nm to 26nm HP resolution

	P1101	EUV-100	EUV-72	XP6305	EUV-65
30nm	11.6 mJ	14.3 mJ	Not available / Not resolved		28.6 mJ
28nm	11.6 mJ	14.3 mJ			28.6 mJ
26nm	11.6 mJ	14.3 mJ			28.6 mJ

Resolution on ASML ADT : X-sections of EUV-100
35nm to 30nm HP, dose 14.2 mJ/cm²

35L70P

34L68P

32L64P

30L60P

Resist profiles under different flare conditions
Resist EUV-65, dose=29.8mJ/cm²

Light field (50%)

Flare 6-7%

Dark field

Flare 2-3%

40L80P

35L70P

32L64P

30L60P

Resist performing well also under higher flare conditions

Resist benchmarking on ASML ADT
Isolated line performance (X-sections)

Resist EUV-100
Dose=10.6mJ/cm2

35nm iso line

Resist EUV-65
dose= 27.4mJ/cm2

30nm iso line

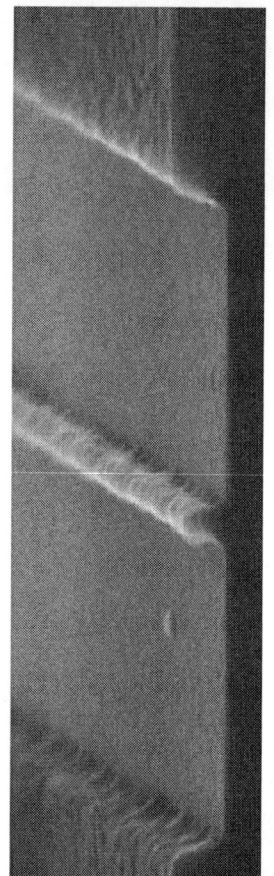

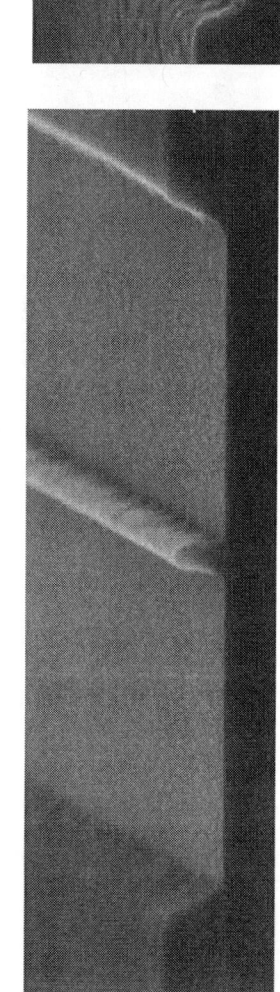

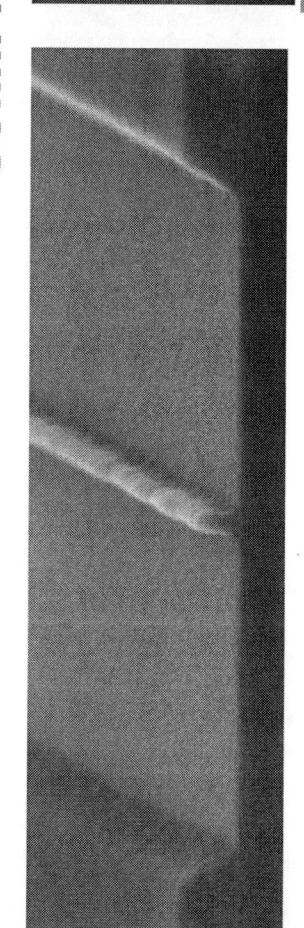

30L180P

80nm resist thickness

Contact hole resolution on the ADT
Resolution down to 34nm

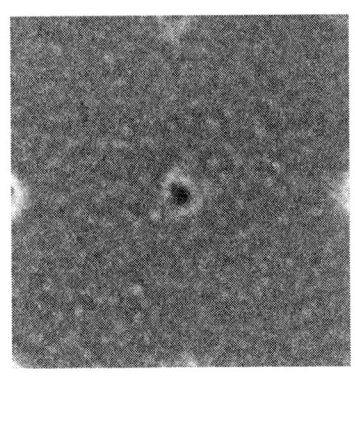

38nm 1:1 — E = 19 mJ/cm²

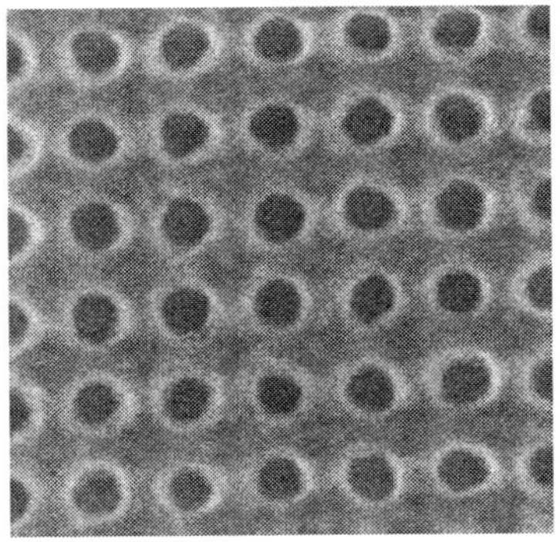

E = 19 mJ/cm²

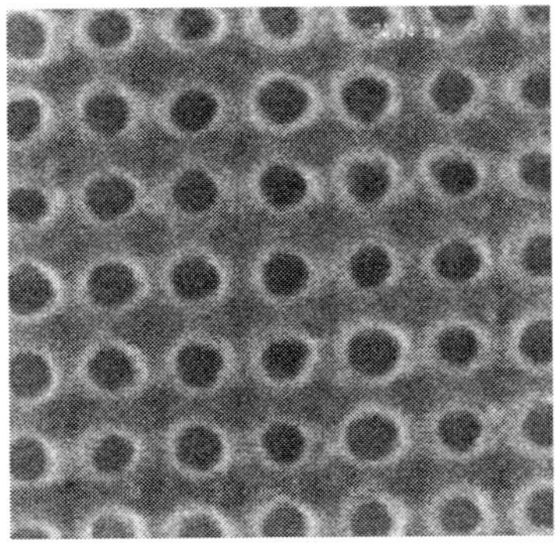

36 nm 1:1 — E = 22 mJ/cm²

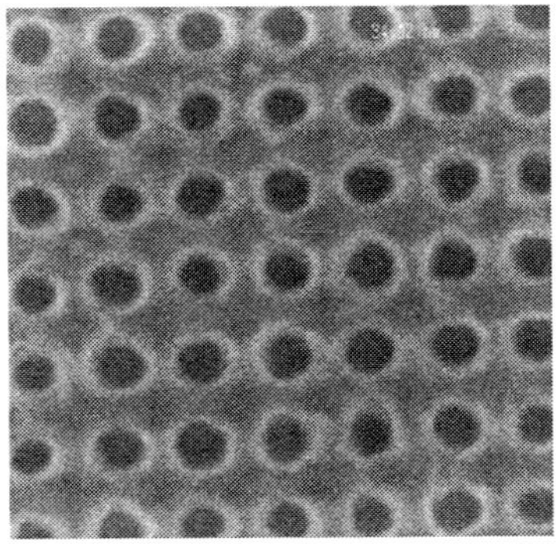

E = 21 mJ/cm²

35nm 1:1 — E = 22 mJ/cm²

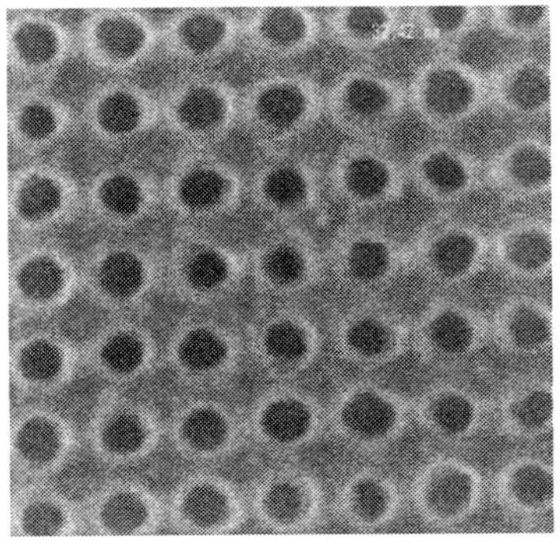

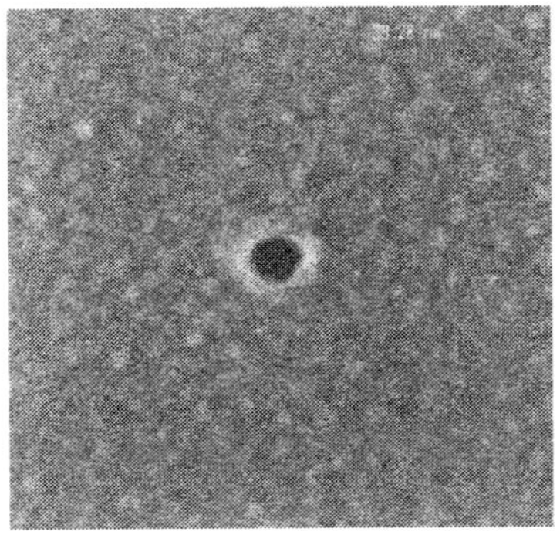

E = 24 mJ/cm²

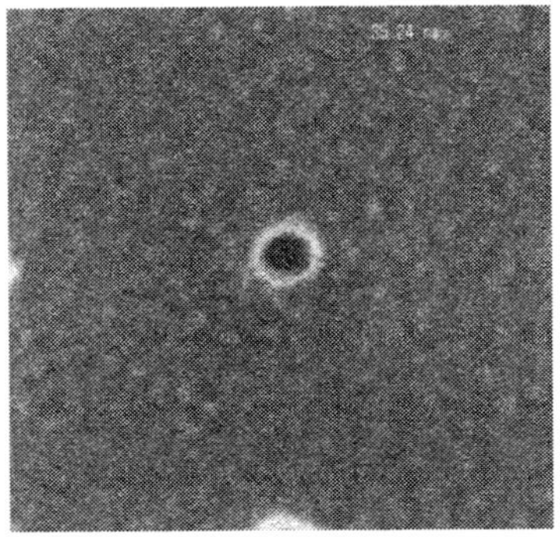

34nm 1:1 — E = 23 mJ/cm²

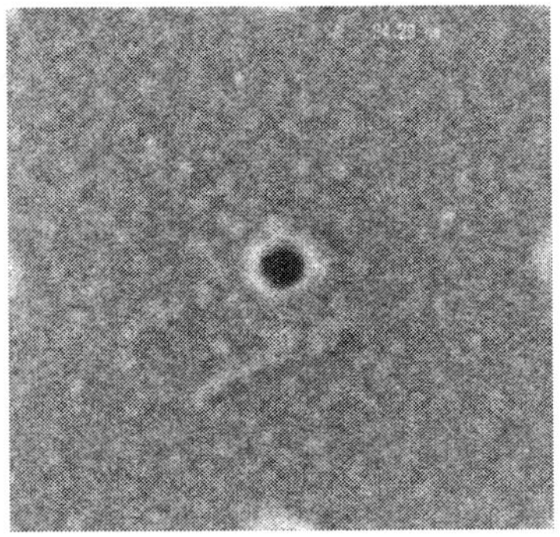

E = 25mJ/cm²

Resist: Fuji FEVS-P1101
Thickness: 80nm

Resolution down to 34 nm Contact holes

Outline

- Introduction

- Resist and process performance

- **ASML EUV ADT performance**
  - Imaging performance : 40nm and 35nm CDU
  - monitoring

- Process implementation

- Conclusions

- Acknowledgments

ASML EUV ADT SAT (NA=0.25, σ=0.5) 40nm V and H LS full wafer CDU

- CD of 40nm H and V lines on mask
- Sampling plan
  - 13 slit positions
  - 4 scan positions
  - 23 dies in best focus
- Result (nm):

	V	H
<CD>	37.7	40.9
3σ	3.1	3.2

- Well within specification without any correction applied
- EUV ADT accepted June 4th, 2008 with 16W/2π DPP source

Resist RHEM XP6305-A
80nm thick, Dose=14.4 mJ/cm²

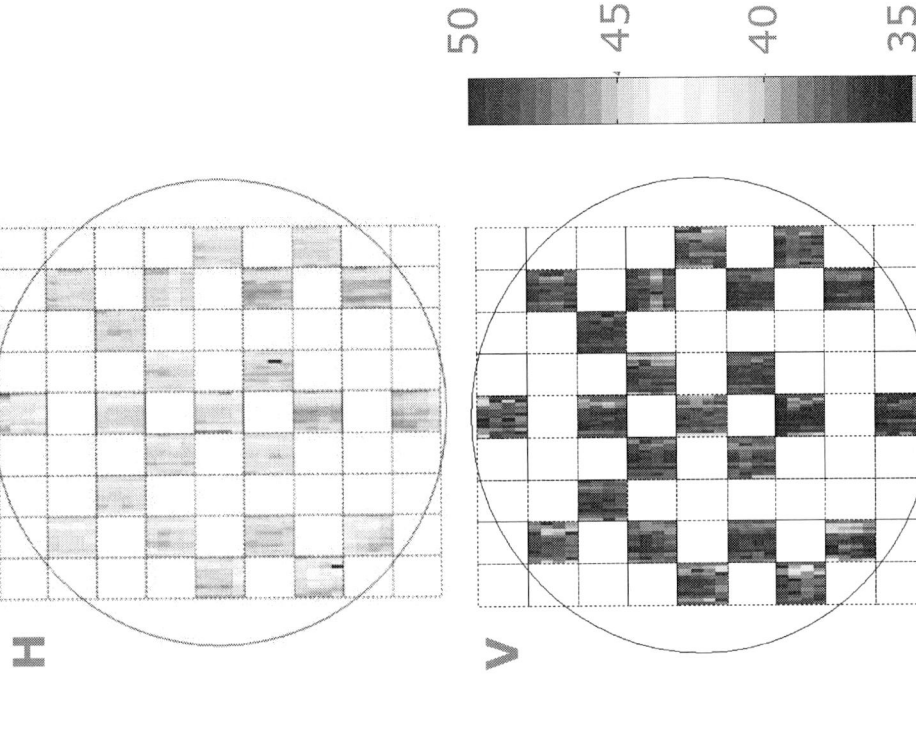

EUV ADT SAT : 40nm V and H LS CD fingerprint

- Slit uniformity was identified as a strong contribution to slit CD fingerprint (at 16W/2π DPP source)

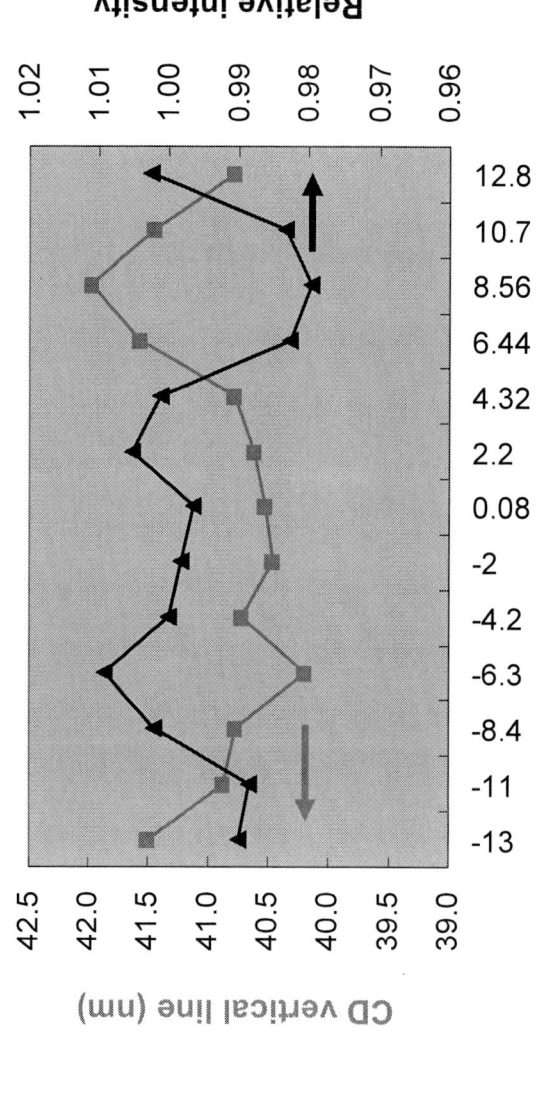

- Other contributions to slit CD signature
  - Image plane deviation and astigmatism (<1nm CD through slit)
  - Reticle: MEEF is 1.3 for 40nm LS, low influence (max 1.2nm CD at wafer)
  - Shadowing: seen mainly as the H/V CD offset

EUV ADT Monitor
40nm V and H LS CD and CDU

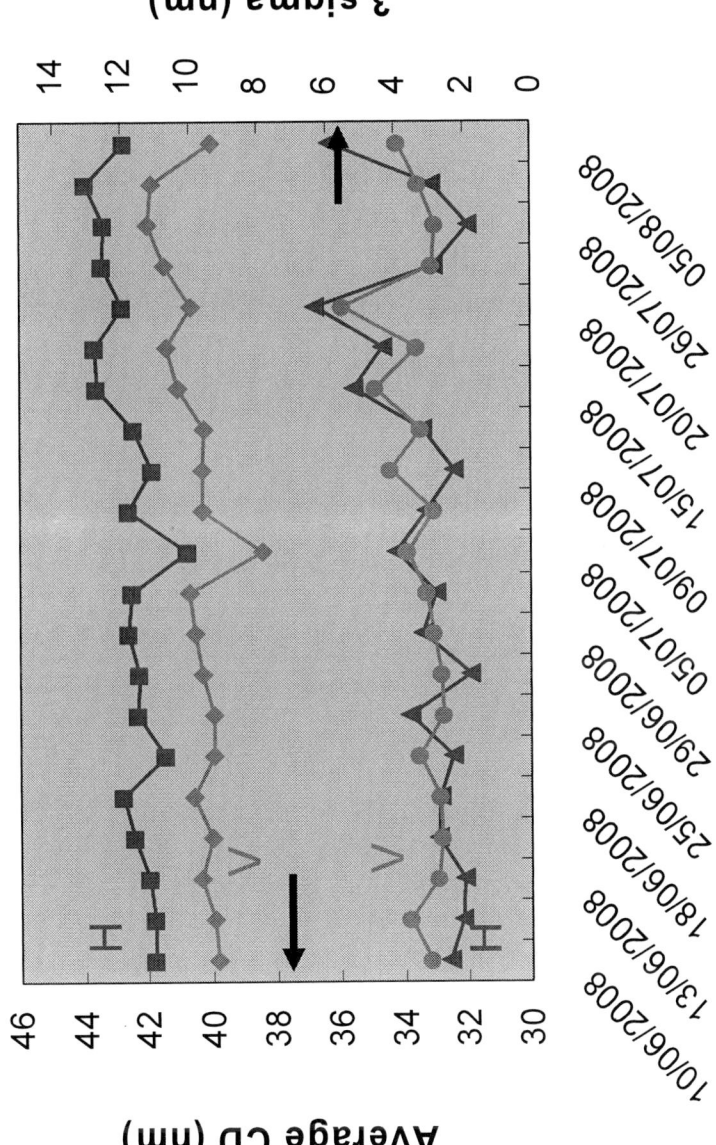

EUV ADT intrafield CD monitoring

Average CD and intrafield CD variation is fairly stable over time

ADT Monitor: Optical column transmission

- Contamination can affect the reflectivity of the mirrors, which could reduce power and affect imaging

- EUV power was measured at 2 positions over time
  - Intermediate focus
  - At wafer level

- The ratio of both measures the transmission of the column

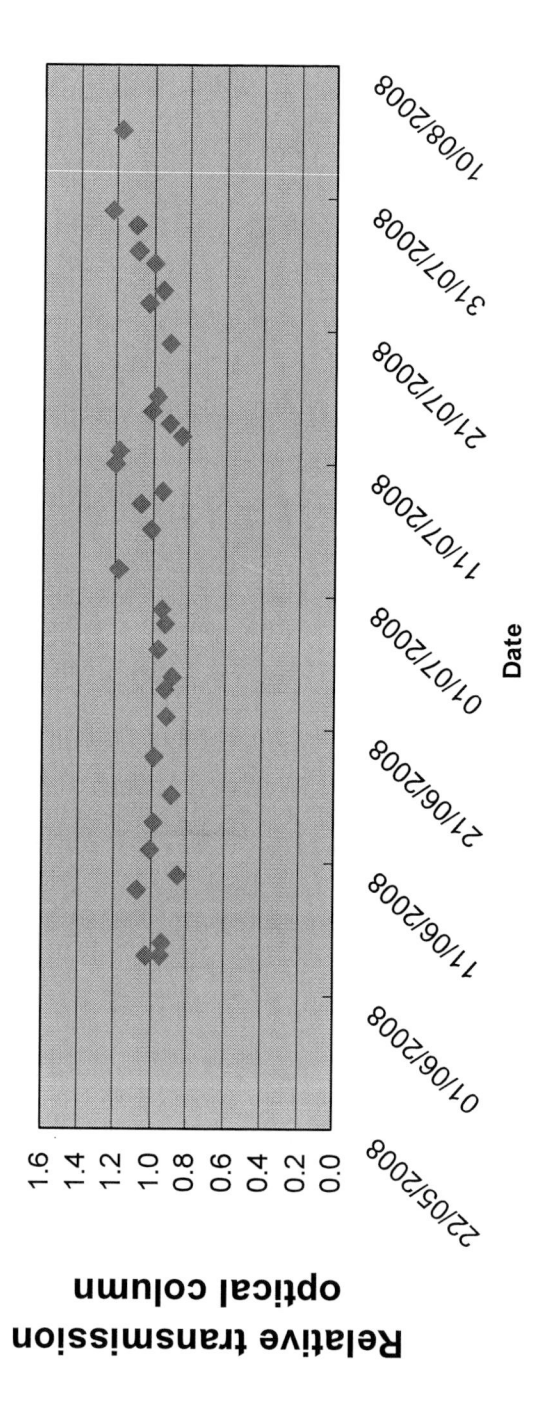

Within the error margin, no change in optical column transmission over time

ADT Current Status
Installation of 120W/2π source

- As of Sept 5, EUV ADT was equipped with 120W/2π source
- Significant improvement in slit uniformity within field and from field to field -> better CDU expected

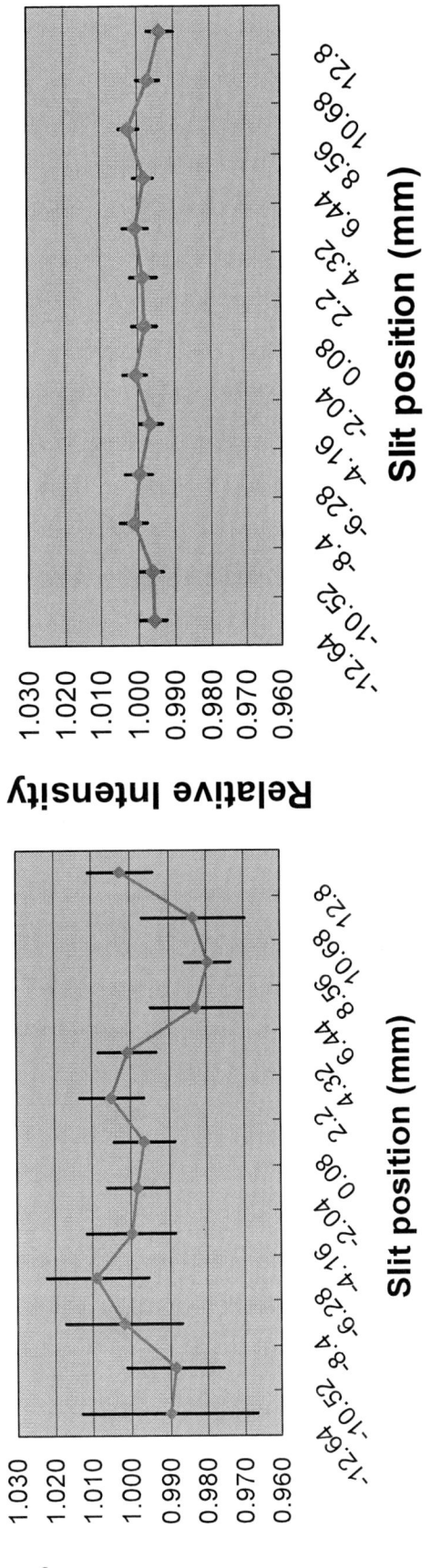

120W/2π source

16W/2π source

EUV ASML ADT (NA=0.25, σ=0.5) champion data
35nm V and H LS full wafer CDU

- CD of 35nm H and V lines
- Sampling plan
  - 5 slit positions
  - 3 scan positions
  - 23 dies in best focus

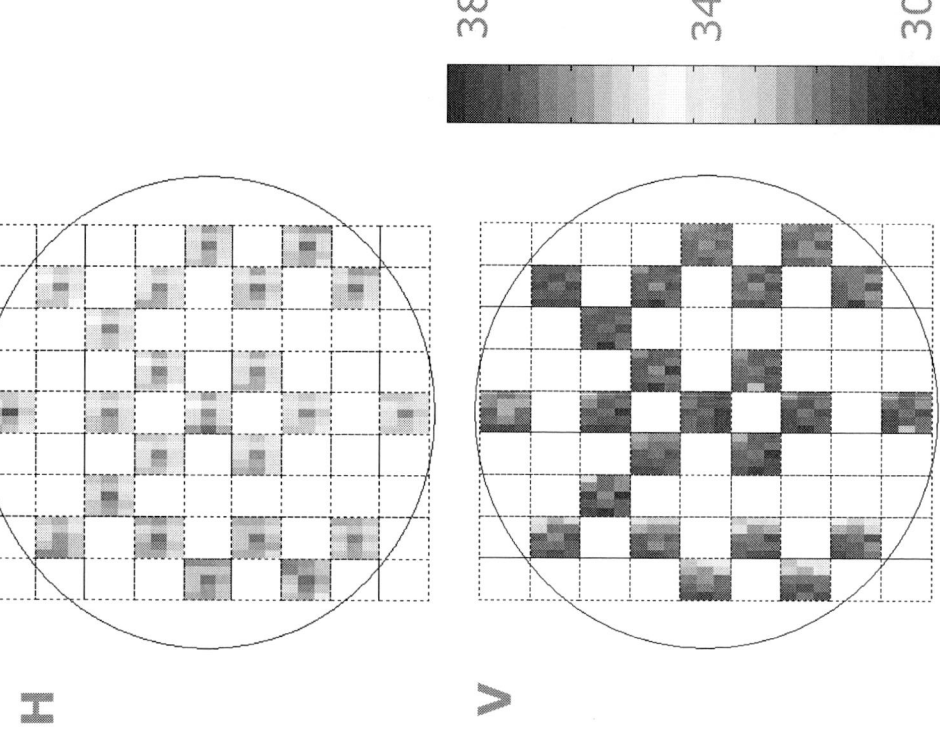

- Result (nm):

	V	H
<CD>	31.7	35.5
3σ	1.5	1.3

- *No correction and no OPC applied*
- *Use of better performing resist*

Resist EUV-65
80nm thick
Dose= 31 mJ/cm²

EUV ASML ADT (NA=0.25, σ=0.5) 35nm V and H LS intrafield CD fingerprint

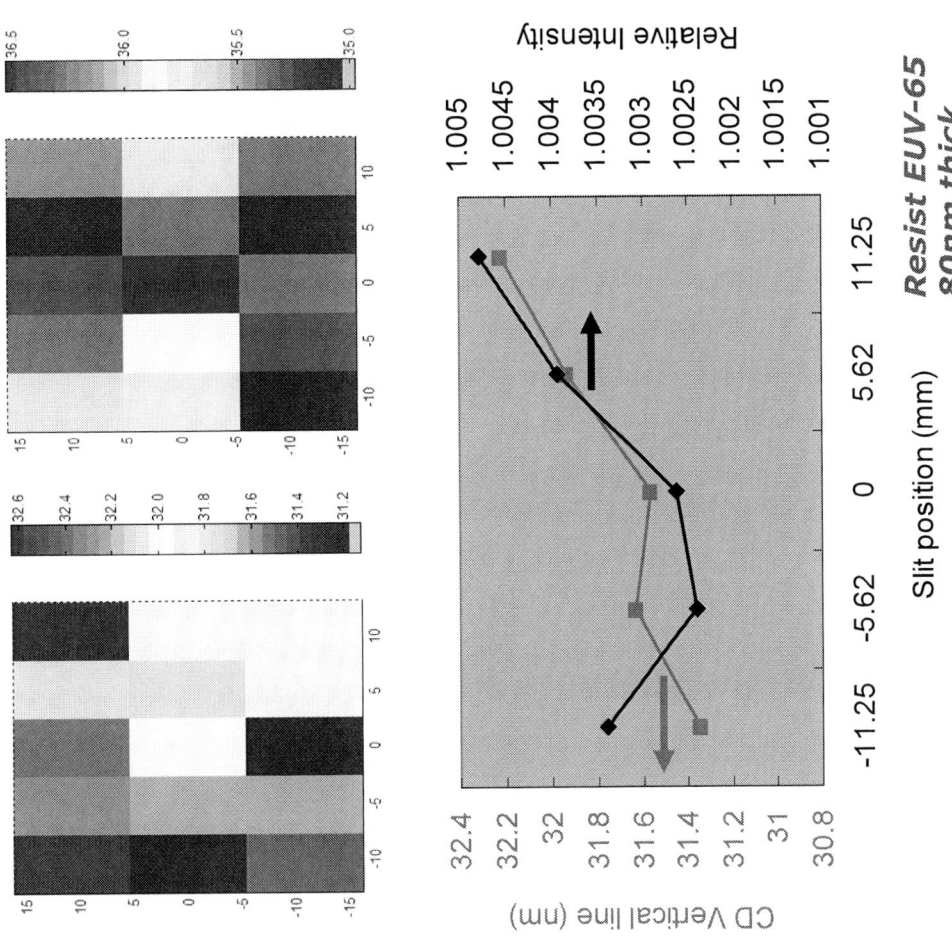

- Intrafield CDU

	V	H
<CD>	31.7	35.5
Field range	1.5	1.5
Slit range	0.9	0.6

- *No OPC applied for shadowing correction*

- Influence from slit uniformity on CD through slit is much less apparent with new source

Resist EUV-65 80nm thick

Improvement in intrafield CDU by use of a better resist and by improvement in slit uniformity

35nm and 32nm FLASH pattern through slit
ADT (NA=0.25, σ=0.5), no OPC, single exposure

- 35nm FLASH (70nm pitch) through slit

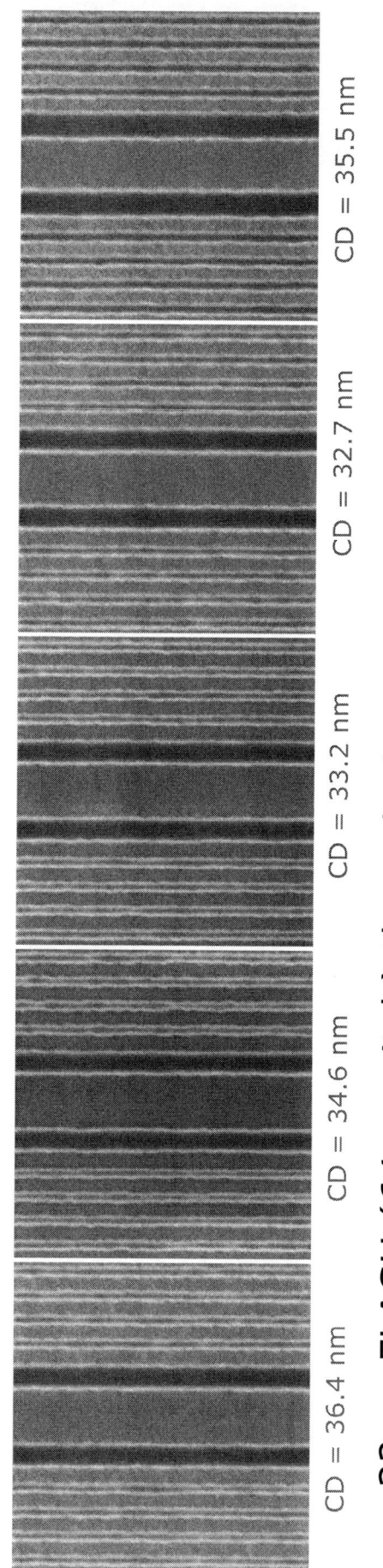

CD = 36.4 nm CD = 34.6 nm CD = 33.2 nm CD = 32.7 nm CD = 35.5 nm

- 32nm FLASH (64nm pitch) through slit

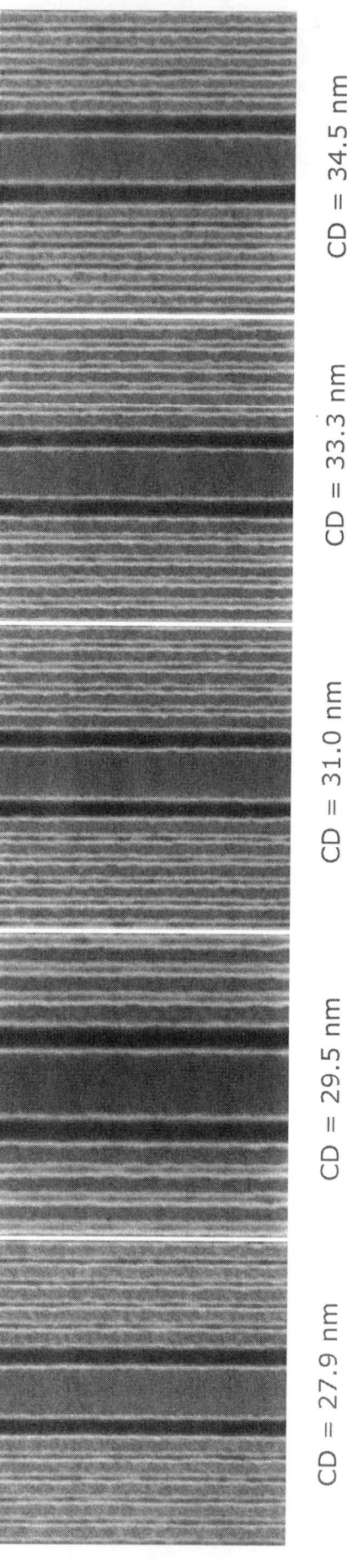

CD = 27.9 nm CD = 29.5 nm CD = 31.0 nm CD = 33.3 nm CD = 34.5 nm

28nm FLASH pattern resolved!
ADT (NA=0.25, σ=0.5), no OPC, single exposure

- FLASH patterns from 70nm to 50nm pitch in slit center

- 56nm pitch FLASH pattern still is resolved

- 50nm pitch is below resolution

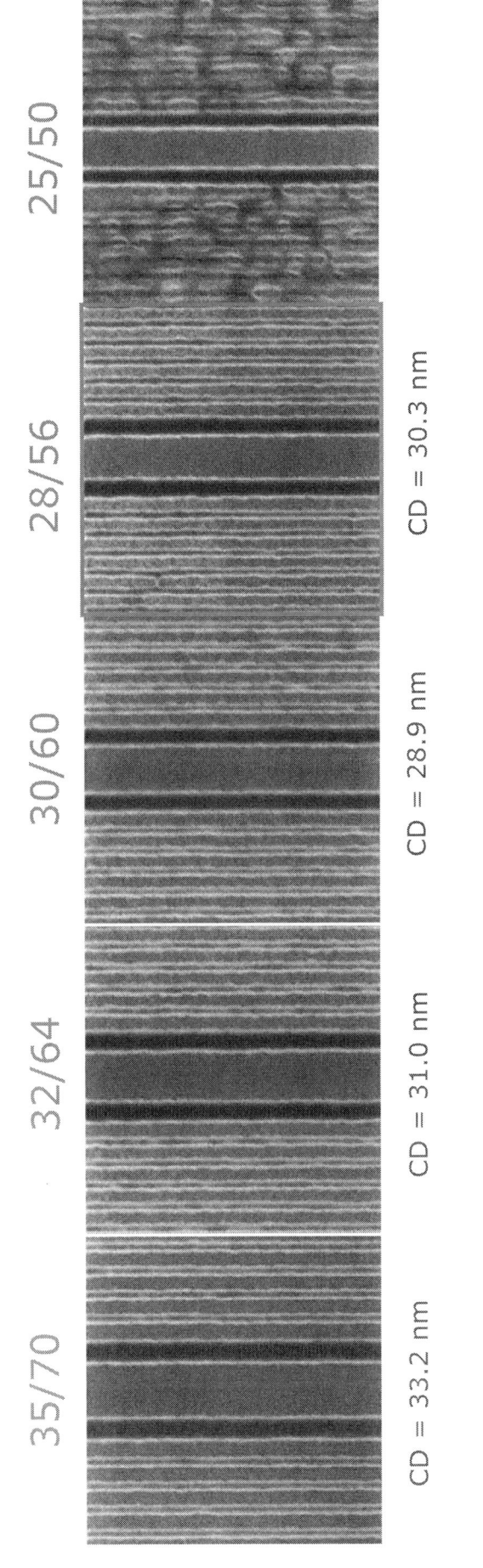

25/50 28/56 30/60 32/64 35/70

CD = 30.3 nm   CD = 28.9 nm   CD = 31.0 nm   CD = 33.2 nm

Resist EUV-65
65nm thick

Outline

- Introduction

- Resist and process performance

- ASML EUV ADT performance

- **Process implementation**
  - 32nm SRAM contact

- Conclusion

- Acknowledgements

Process implementation : 32nm node SRAM
Contact layer with EUV lithography, active/poly with immersion litho

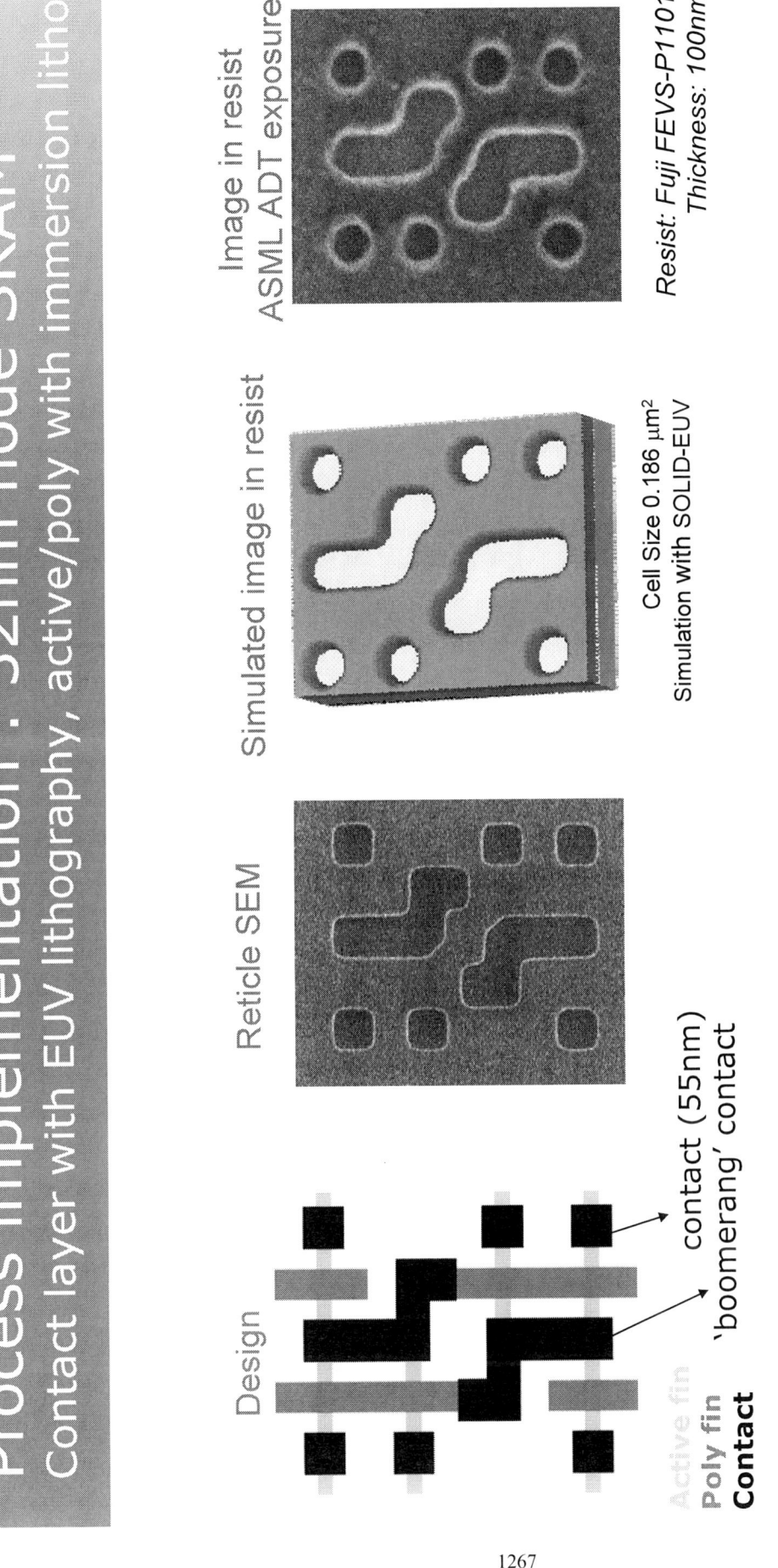

Image in resist
ASML ADT exposure

Resist: Fuji FEVS-P1101
Thickness: 100nm

Simulated image in resist

Cell Size 0.186 μm²
Simulation with SOLID-EUV

Reticle SEM

Design

Active fin
Poly fin
Contact

contact (55nm)
'boomerang' contact

Patterning requirements
- 55nm contact holes and boomerang contact hole imaging on 124nm minimum pitch
- Overlay requirement of 15nm (3 sigma)

- Contact printed on ASML EUV ADT
- Active/poly level printed with immersion litho on ASML /1700i

Process implementation
Contact hole etching

Pattern transfer using metal hard mask

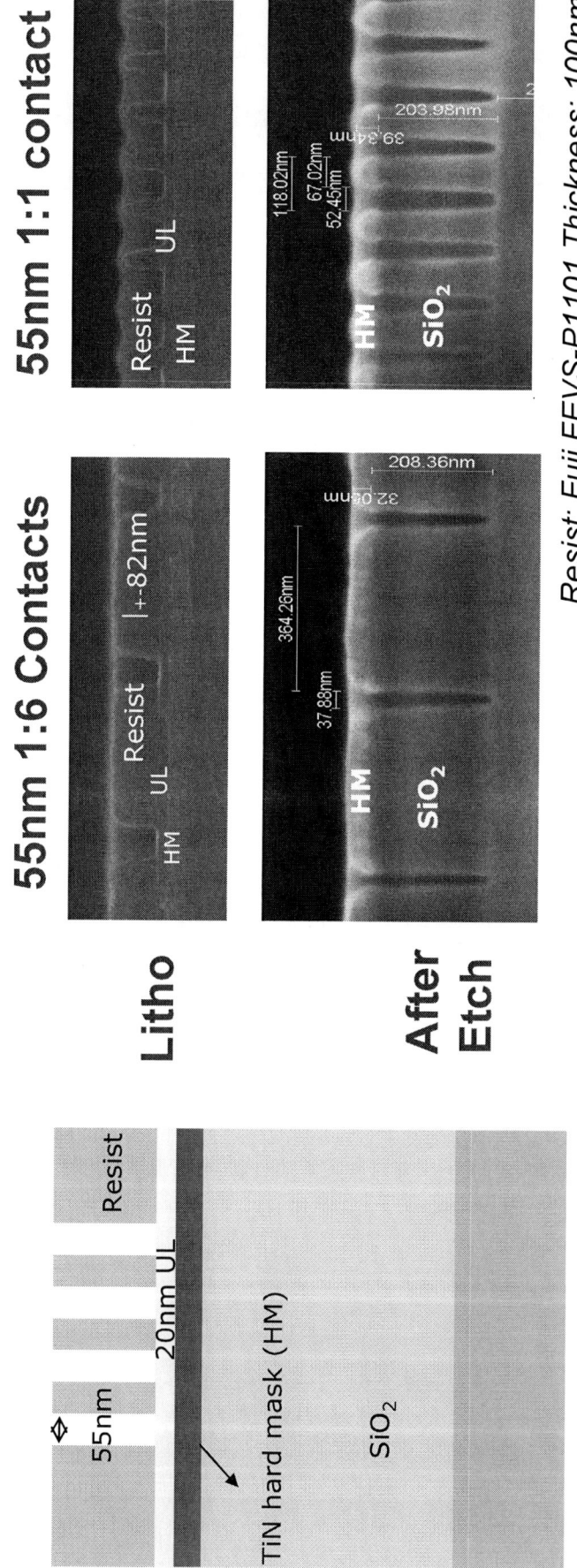

55nm 1:6 Contacts

55nm 1:1 contact

Litho

After Etch

Resist: Fuji FEVS-P1101 Thickness: 100nm
Use of an organic underlayer for substrate compatibility

80nm resist thickness sufficient for etching 30nm Hard Mask

Overlay performance
Measured Overlay Error |m|+3σ per Wafer

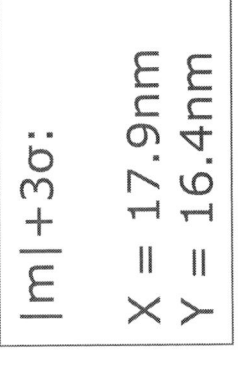

|m|+3σ:

X = 17.9nm
Y = 16.4nm

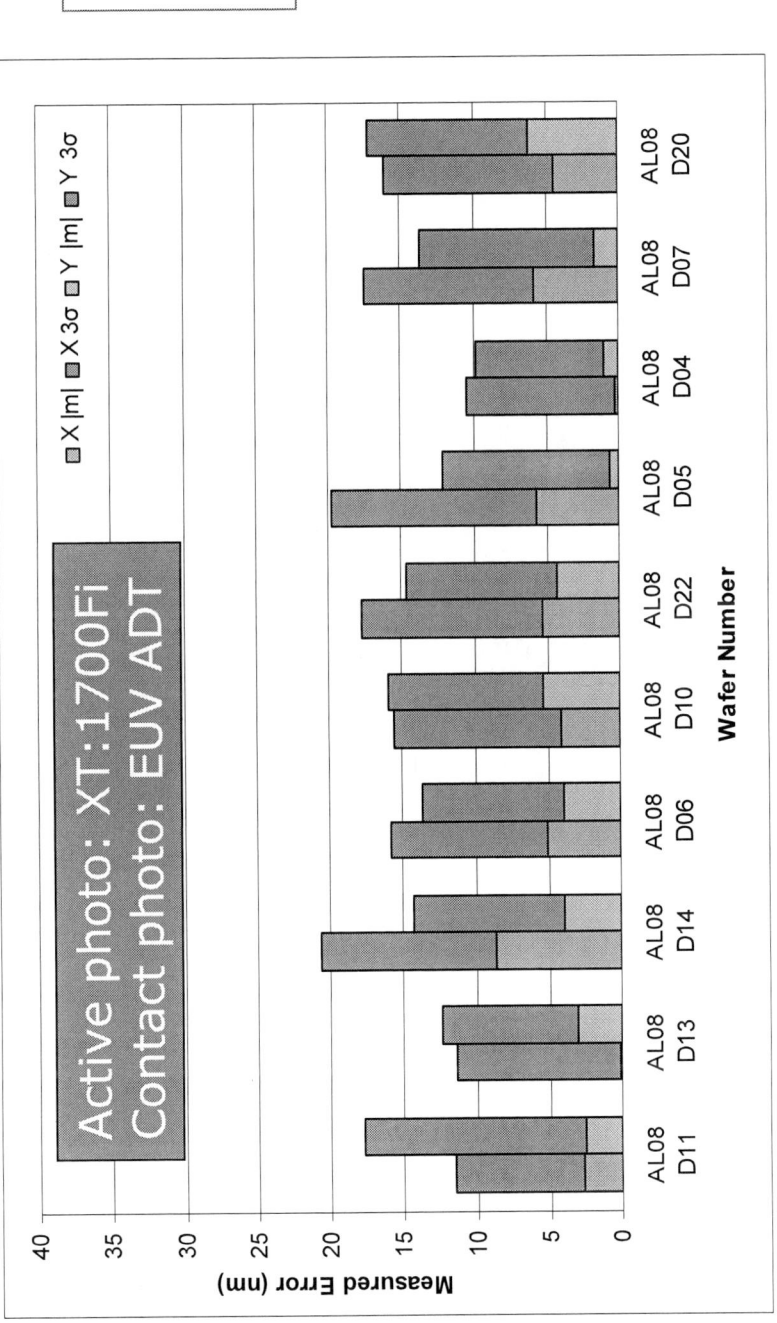

Overlay numbers are consistent with expected overlay performance between a wet and a dry tool. The EUV ADT has shown good overlay stability.

Process Implementation
32nm SRAM device after litho and after Hard Mask etch

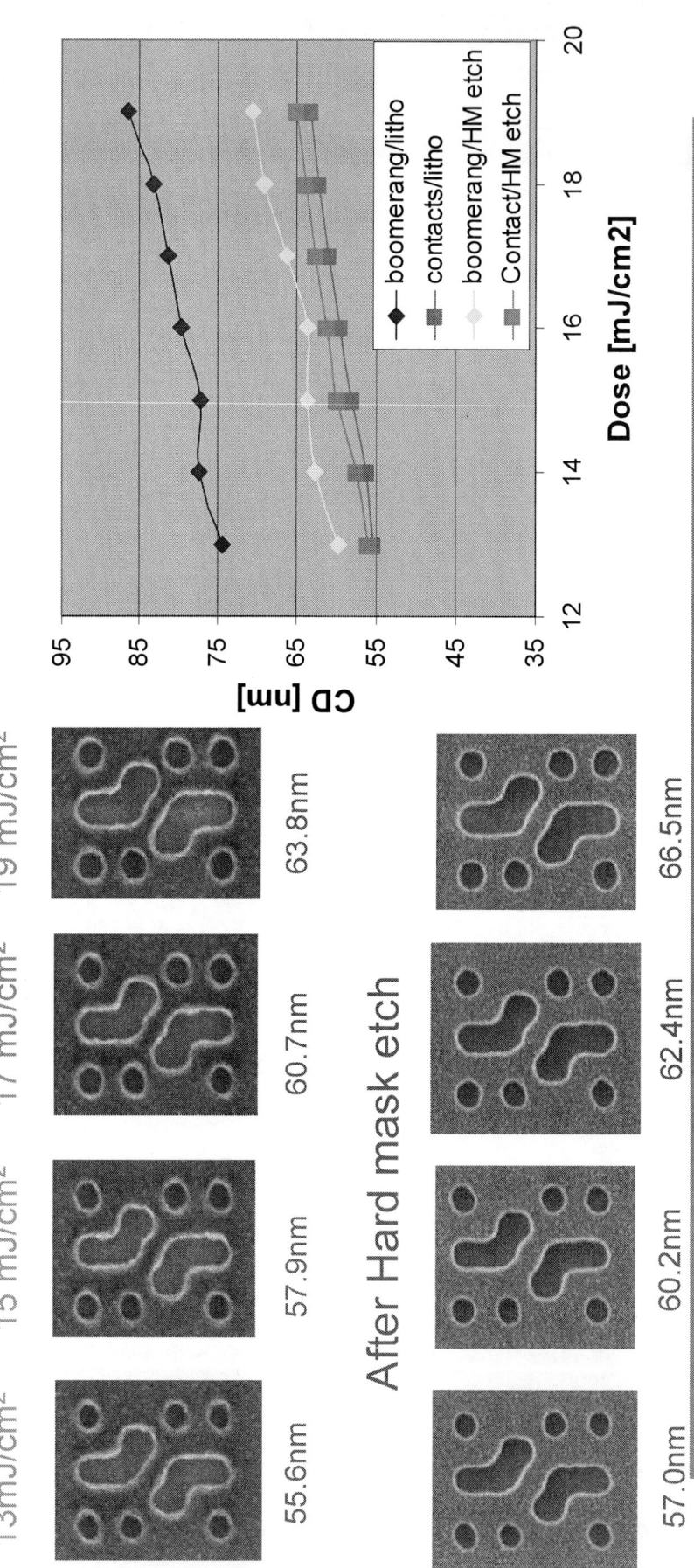

Lithography : Exposure latitude

13mJ/cm²   15 mJ/cm²   17 mJ/cm²   19 mJ/cm²

55.6nm   57.9nm   60.7nm   63.8nm

After Hard mask etch

57.0nm   60.2nm   62.4nm   66.5nm

Boomerangs after HM etch become on average 14nm smaller
CH become 1.5nm larger after HM etch

Process Implementation
32nm SRAM device after litho and after etch

Lithography : Exposure latitude

11 mJ/cm²	12mJ/cm²	13 mJ/cm²	14mJ/cm²	15 mJ/cm²	16 mJ/cm²	17 mJ/cm²
50.2nm	56.2nm	57.4nm	60.8nm	61.6nm	63.3nm	63.0nm

After Oxide etch

53.4nm	58.7nm	60.4nm	63.3nm	63.7nm	64.5nm	64.5nm

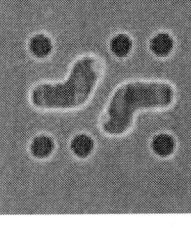
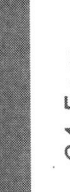
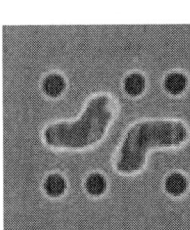

Electrically functional 0.186 μm^2 32nm SRAM cells demonstrated using EUV for contact hole patterning

Summary and conclusions

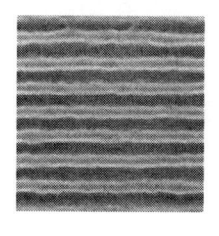

28nm L/S

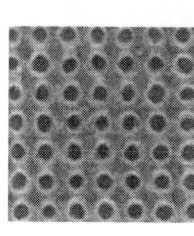

34nm CH

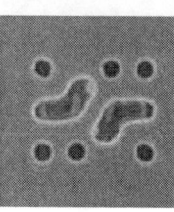

SRAM cell

- ASML EUV ADT has been recently accepted resulting in significant acceleration in EUVL learning

- Clear progress in resist performance over the last 6 months (on ADT)
  - Linear resolution for several resists down to 32nm HP
  - Ultimate resolution of 28nm HP resolution achieved
  - Improved LER observed

- ADT shows good imaging performance for 35nm HP in terms of CDU and capability for 32nm and below.

- Electrically functional 0.186um^2 32nm node SRAM cells have been demonstrated with EUV lithography on the contact hole level

Acknowledgements

Many thanks to:

- **ASML**
  - Noreen Harned, Hans Meiling, John Zimmerman, Bas Hultermans
  - ASML EUV team at IMEC
  - Sjoerd Lok, Joep Van Dijk

- **IMEC**
  - Geert Vandenberghe, Staf Verhaeghen, Tom Vandeweyer , Anabela Veloso, Alain Vandervorst, Danny Goossens

- **PSI**
  - Harun Solak, Anja Weber , Vaida Auzelyte

- **Resist and BARC suppliers**

- **Mask shops**

- **IMEC IIAP partners**

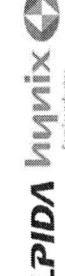

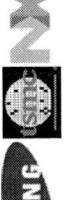

aspire invent achieve

imec

Improvement of optics for EUV exposure tools

Katsuhiko Murakami, Tetsuya Oshino, Hiroyuki Kondo,
Hiroshi Chiba and Kazushi Nomura

Nikon Corporation

October 1st, 2008, EUVL Symposium 2008

NIKON CORPORATION

Outline

EUV1 projection optics

Metrology, polishing and coating

Wavefront error and exposure result

MSFR and flare

On-body PO control system

EUV1 illumination optics

Evaluation of pupil fill

Future improvement of optics for high throughput

High-efficiency RET illumination

High-efficiency SPF

Summary

NIKON CORPORATION

Outline

EUV1 projection optics

Metrology, polishing and coating

Wavefront error and exposure result

MSFR and flare

On-body PO control system

EUV1 illumination optics

Evaluation of pupil fill

Future improvement of optics for high throughput

High-efficiency RET illumination

High-efficiency SPF

Summary

NIKON CORPORATION

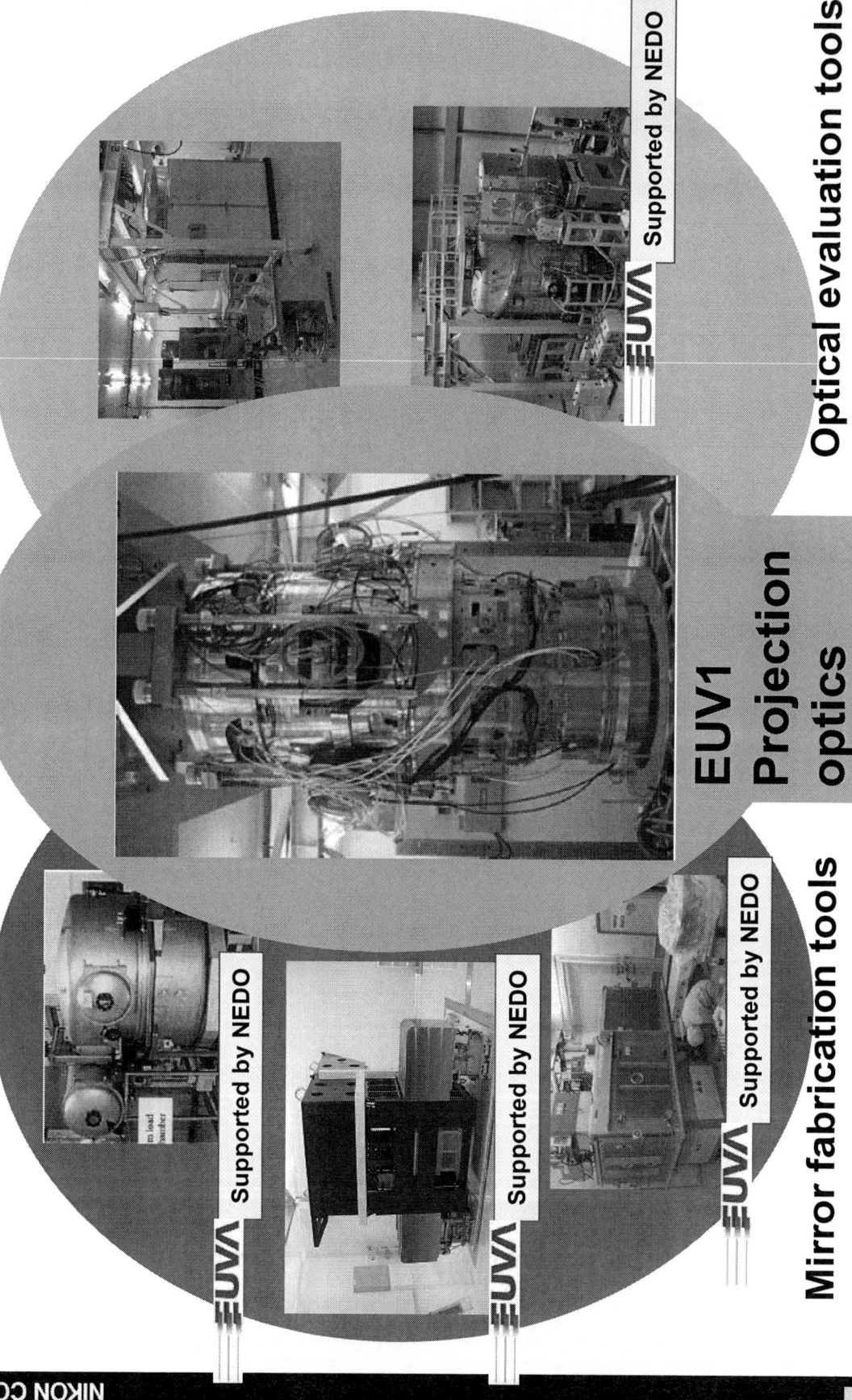

Improvement of aspheric figure metrology

Supported by NEDO

Repeatability

6 pm RMS

Reproducibility

re-mount

17 pm RMS

High repeatability
interferometer

High repeatability interferometer has sufficient repeatability for metrology of aspheric mirrors for EUV PO.

The state of the art for mirror polishing
Current best data

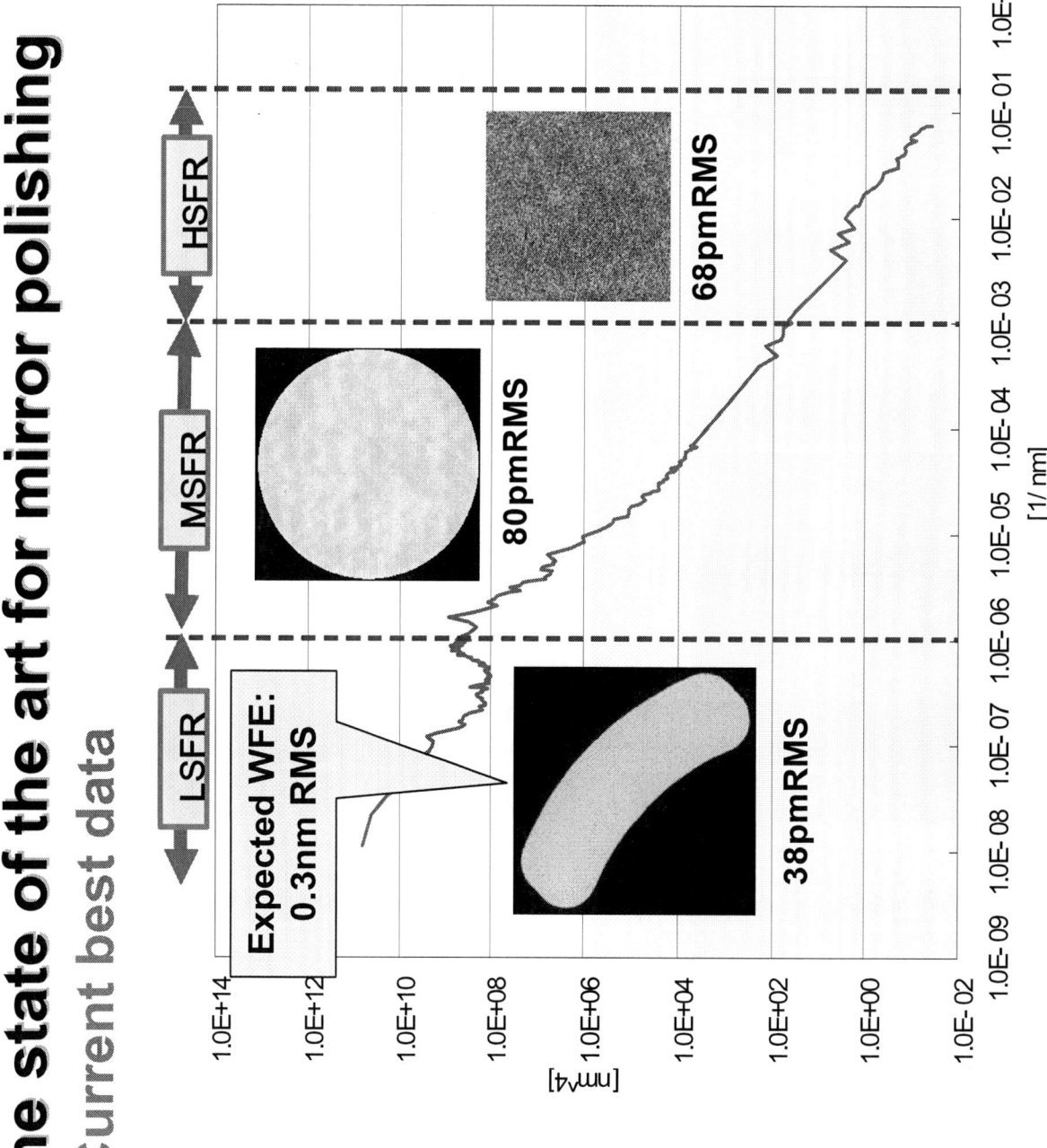

Performance of PO mirror coatings

Mo/Si multilayer coating with Ru capping layer

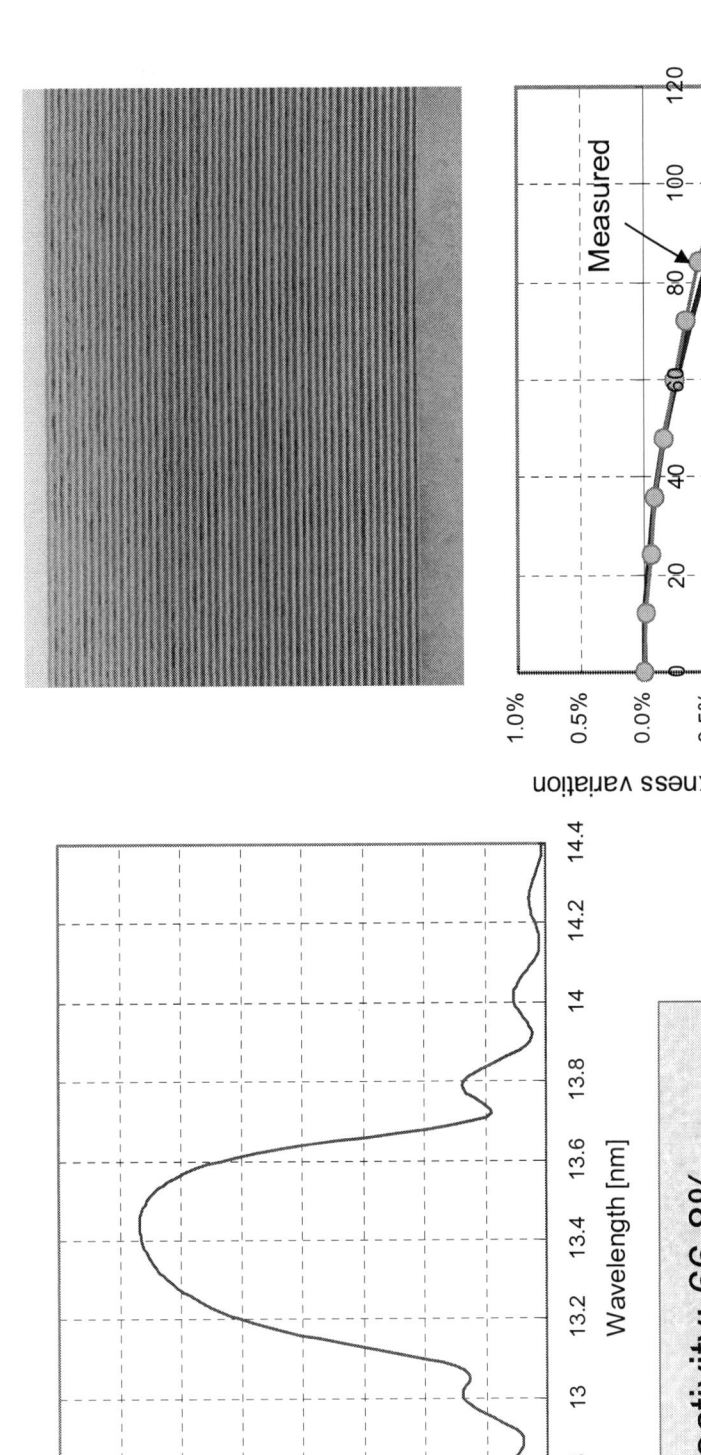

Reflectivity: 66.8%

Internal stress: -173MPa

Thickness variation: 0.1%

Wave front error improvement

WFE (nmRMS) (average)	
2nd Prototype	2.2
EUV1 #1PO	0.6
EUV1 #2PO	0.4

EUV1 #2PO wavefront map

- Extremely small WFE achieved -

WFE   0.4 nm RMS   (average)

Min. 0.3nm RMS ~ Max. 0.5nm RMS

NIKON CORPORATION

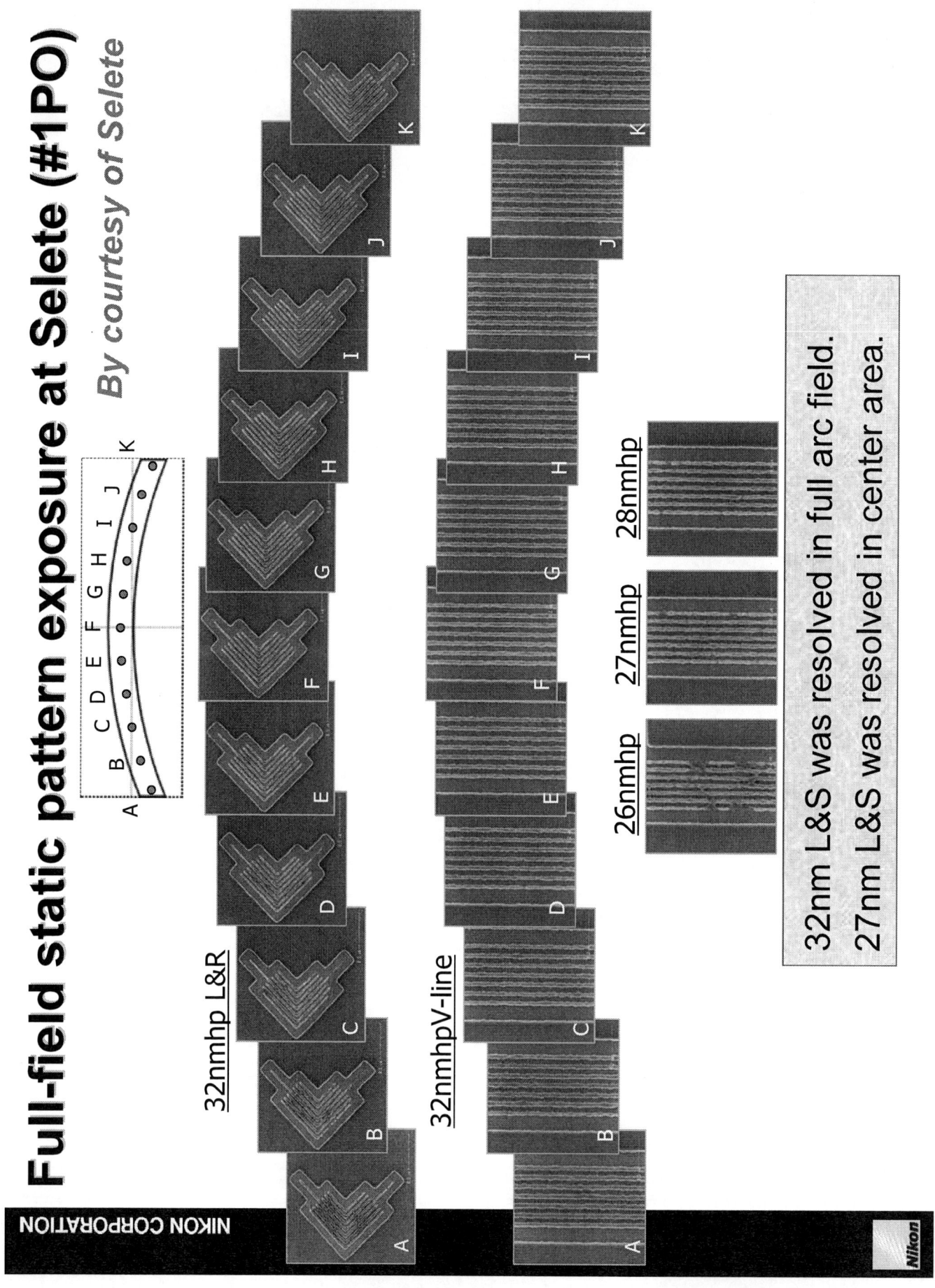

Improvement of MSFR

Extremely small MSFR was achieved on all 6 mirrors in EUV1.

NIKON CORPORATION

Kirk test of #1 PO at Selete

By courtesy of Selete

PSD (power spectral density) of each mirror

$$PSD_1(f_1), PSD_2(f_2), \cdots, PSD_6(f_6)$$

Accumulated PSD on pupil

$$SystemPSD(f) = \sum_{i=1}^{6} \alpha_i^2 PSD_i(\alpha_i f)$$

PSF (point spread function) by scattering

$$PSF^{SC}(r) = \left(\frac{4\pi}{\lambda^2 z}\right)^2 SystemPSD\left(\frac{r}{\lambda z}\right)$$

Flare $\quad Flare = \int 2\pi r \cdot PSF^{SC}(r)dr$

r in bright area

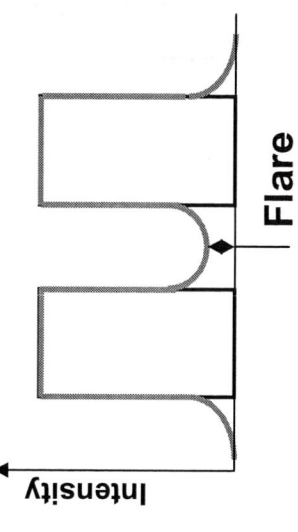

Measured flare did not agreed with calculation using PSD of each mirror. We have investigated the reason.

26x2mm bright field

Measured flare with Kirk test

Calculated flare

Flare [%]

radius (flare range) [um]

NIKON CORPORATION

Calculation of "Kirk flare"

By courtesy of Selete

Kirk flare ≠ Flare

Kirk flare $KF = \dfrac{I_{D\min}}{I_{B\max}}$

$$I_D(x_d, y_d) = \int_{\text{bright field}} PSF^{SC}(x - x_d, y - y_d)\,dxdy$$

$$I_B(x_b, y_b) = (1 - TIS) + \int_{\text{bright field}} PSF^{SC}(x - x_b, y - y_b)\,dxdy$$

TIS (Total Integrated Scatter)

$$TIS = \int_{\infty} PSF^{SC0}(x, y)\,dxdy$$

Flare

$$F(x_0, y_0) = \int_{\text{bright field}} PSF^{SC}(x - x_0, y - y_0)\,dxdy$$

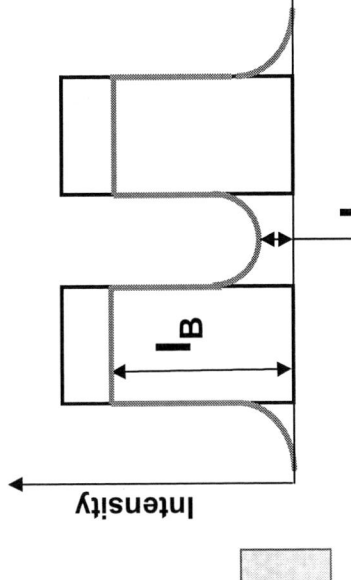

26x2mm bright field

Measured flare with Kirk test

Calculated Kirk flare

Flare [%] — radius (flare range) [um]

	Flare	Kirk flare
EUV1 #1PO	10%	15%
EUV1 #2PO	6%	8%

2μm Kirk pattern in bright field

Calculated Kirk flare agreed with measurement.

NIKON CORPORATION

Impact of flare on imaging

Uniform intensity loss Image blur

$$I(x,y) = (1 - TIS)\,I_0(x,y) + PSF^{SC}(x,y) \otimes I_0(x,y)$$

$$= \{(1 - TIS)\,\delta(x,y) + PSF^{SC}(x,y)\} \otimes I_0(x,y)$$

$$TIS = \int_\infty PSF^{SC0}(x,y)\,dxdy$$

Impact on flare on imaging can be calculated using TIS and PSFsc.

TIS: Uniform intensity loss due to scattering

PSFsc: Image blur due to flare

On body PO control system in EUV1
Fine adjustment unit of mirror position

By courtesy of Selete

Before adjustment
45-nm elbow pattern

Astigmatism was remained

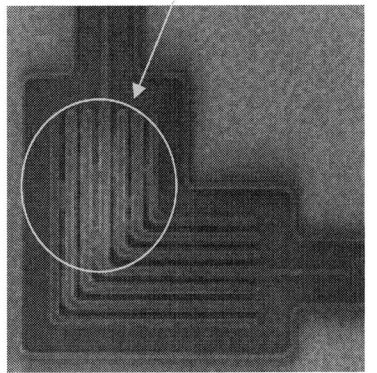

After adjustment

Astigmatism was removed.

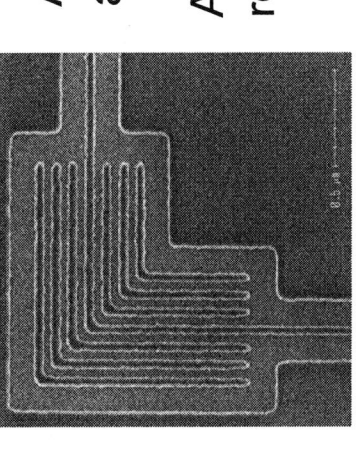

Mirror

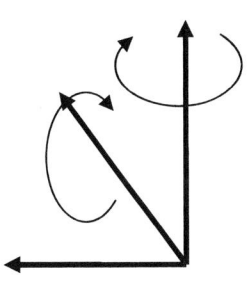

Fine adjustment unit can control position of each mirror in 6 DOF in vacuum.

Fine adjustment system of EUV1 PO is effective to control its optical performance.

NIKON CORPORATION

Outline

EUV1 projection optics
Metrology, polishing and coating
Wavefront error and exposure result
MSFR and flare
On-body PO control system

EUV1 illumination optics
Evaluation of pupil fill

Future improvement of optics for high throughput
High-efficiency RET illumination
High-efficiency SPF

Summary

NIKON CORPORATION

EUV1 IU (Illumination Unit) test stand

EUV light measurement

Visible light measurement

Vacuum chamber
(Contains IU mirrors)

EUV light source

Side view

Visible light source

Back view

Performance of EUV1 IU was evaluated with IU test stand.

NIKON CORPORATION

Pupil image in arc field

Coherence factor = 0.8

Uniformity: +/- 0.5%

Uniform pupil fill in arc field was confirmed.

Left hand: Measured with visible light on IU test stand

Right hand: Measured with pinhole reticle exposure on EUV1 body

NIKON CORPORATION

Outline

EUV1 projection optics

Metrology, polishing and coating

Wavefront error and exposure result

MSFR and flare

On-body PO control system

EUV1 illumination optics

Evaluation of pupil fill

Future improvement of optics for high throughput

High-efficiency RET illumination

High-efficiency SPF

Summary

NIKON CORPORATION

Future improvement of optics for high throughput

Fly's eye mirrors for RET

Specially designed fly's eye mirrors for RET illumination (annular, dipole, etc.) will minimize transmission loss of EUV power.

Reflection-type SPF

Membrane-type SPF looses about 50% of EUV power.

Reflection-type SPF can eliminate OoB radiation with much higher efficiency.

High-efficiency RET illumination

Conventional fly's eye mirror with pupil aperture (EUV1)

+ Easy
- Power loss
- Degrade uniformity on reticle

Pupil aperture

Reticle

Fly's eye 2 (Pupil plane)

Fly's eye 1

Source

Intensity distribution in pupil

RET fly's eye mirror

+ No power loss
+ No uniformity change on reticle
- Difficult to make

Reticle

Fly's eye 2 (Pupil plane)

RET Fly's eye

Fly's eye 1

Source

Specially designed RET fly's eye mirrors can minimize photon loss.

High-efficiency SPF (spectral purity filter)

Membrane-type SPF

+ OoB light is almost eliminated.

− EUV light is absorbed by about 50%.

− Difficult to cool – Thermal damage

− Fragile

EUV transmittable thin membrane (~100 nm thick)

Reflection-type SPF

+ Small additional reflectivity loss for EUV light.

+ Easy and effective cooling – Applicable to HVM.

+ Durable

+ Tunable

− Eliminate only a specific λ in OoB.

Multilayer + SPF coating

Mirror substrate

Reflection-type SPF is more efficient and robust than membrane-type.

NIKON CORPORATION

Multilayer design for reflection-type SPF

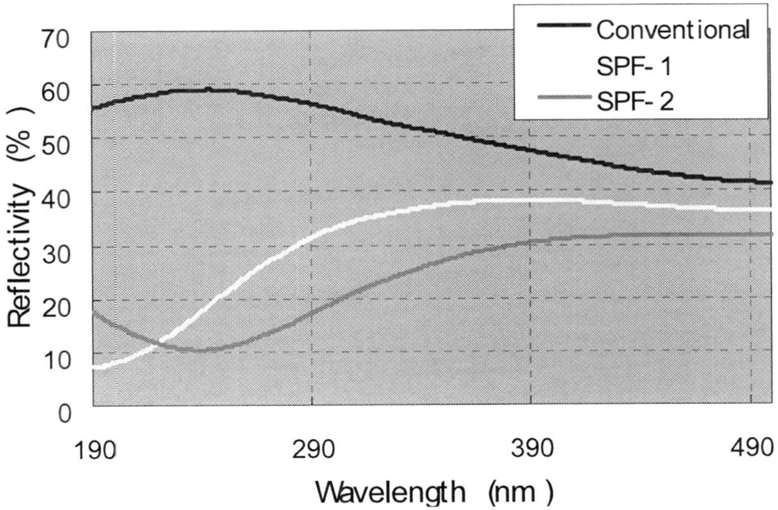

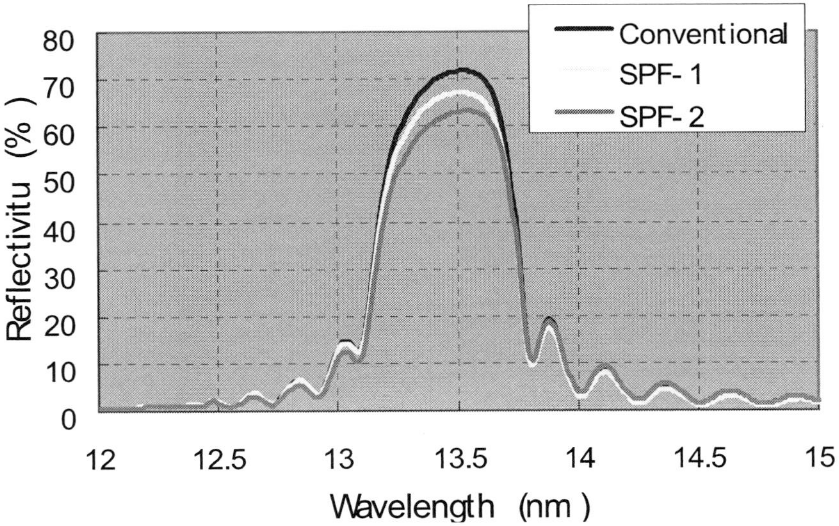

UV reflectivity of multilayer coatings can be reduced with a small loss of EUV reflectivity.

NIKON CORPORATION

Summary

EUV1 projection optics

Metrology, polishing and coating technologies for EUV projection optics were improved.

Wavefront error of 0.4nmRMS in average was achieved.

32nm L&S was resolved in full arc field.

Calculated Kirk flare was agreed with measurement. Impact on flare on imaging can be calculated using TIS and PSF^{SC}.

Fine adjustment system of EUV1 PO is effective to control its optical performance.

EUV1 illumination optics

Uniform pupil fill in arc field was confirmed with IU test stand and on body.

Future improvement of optics for high throughput

Specially designed RET fly's eye mirrors can minimize photon loss.

Reflection-type SPF is more efficient and robust than membrane-type SPF.

NIKON CORPORATION

Acknowledgements

A part of this work was conducted under EUVA project. EUVA project has been supported by NEDO.

Nikon gratefully acknowledges METI and NEDO for their support.

Nikon also participate in Selete program and appreciate Selete members for their useful discussion and advice.

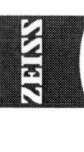

EUV Symposium: Oct. 2008, Lake Tahoe

Optics for EUV lithography

Peter Kuerz, Thure Boehm, Udo Dinger, Hans-Juergen Mann, Stephan Muellender, Manfred Dahl, Martin Lowisch, Michael Muehlbeyer, Oliver Natt, Siegfried Rennon, Erik Sohmen, Thomas Stein, Gero Wittich, Bernd Thüring, Winfried Kaiser, Wolfgang Rupp

Eric Louis, Fred Bijkerk

Alpha Demo Tools: The functionality of full-field EUV optics has been successfully demonstrated

28 nm
29 mJ/cm^2

28 nm
19 mJ/cm^2

28 nm
68 mJ/cm^2

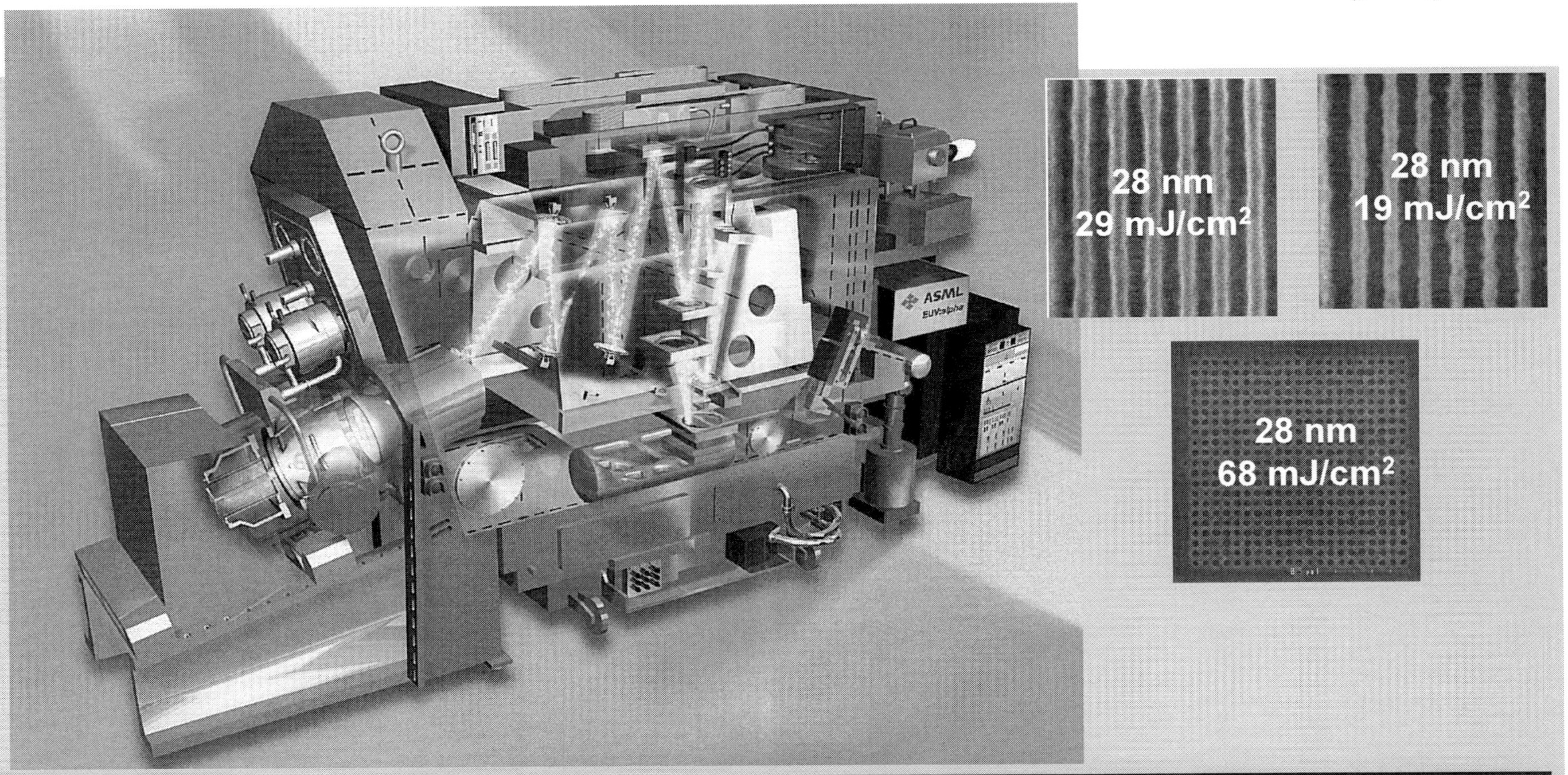

1st full field EUV scanner able to print 2x nm dense features in a single exposure

See also: Hans Meiling et al.: this conference

Enabling the Nano-Age World®

Overview

From the Alpha Demo Tool to High Volume Tools:
EUV enters the preproduction phase

» Setup of an EUV infrastructure

» Progress in key technology areas
- Optics metrology and fabrication
- Coatings
- System metrology

» Production of first systems is in full progress

The future: high NA EUV tools enable resolutions down to 11 nm

Enabling the Nano-Age World®

EUV Production Tools – Introduction

- λ   13.5 nm
- NA   0.25
- Field   26 x 33 mm²
- Magnification 4x

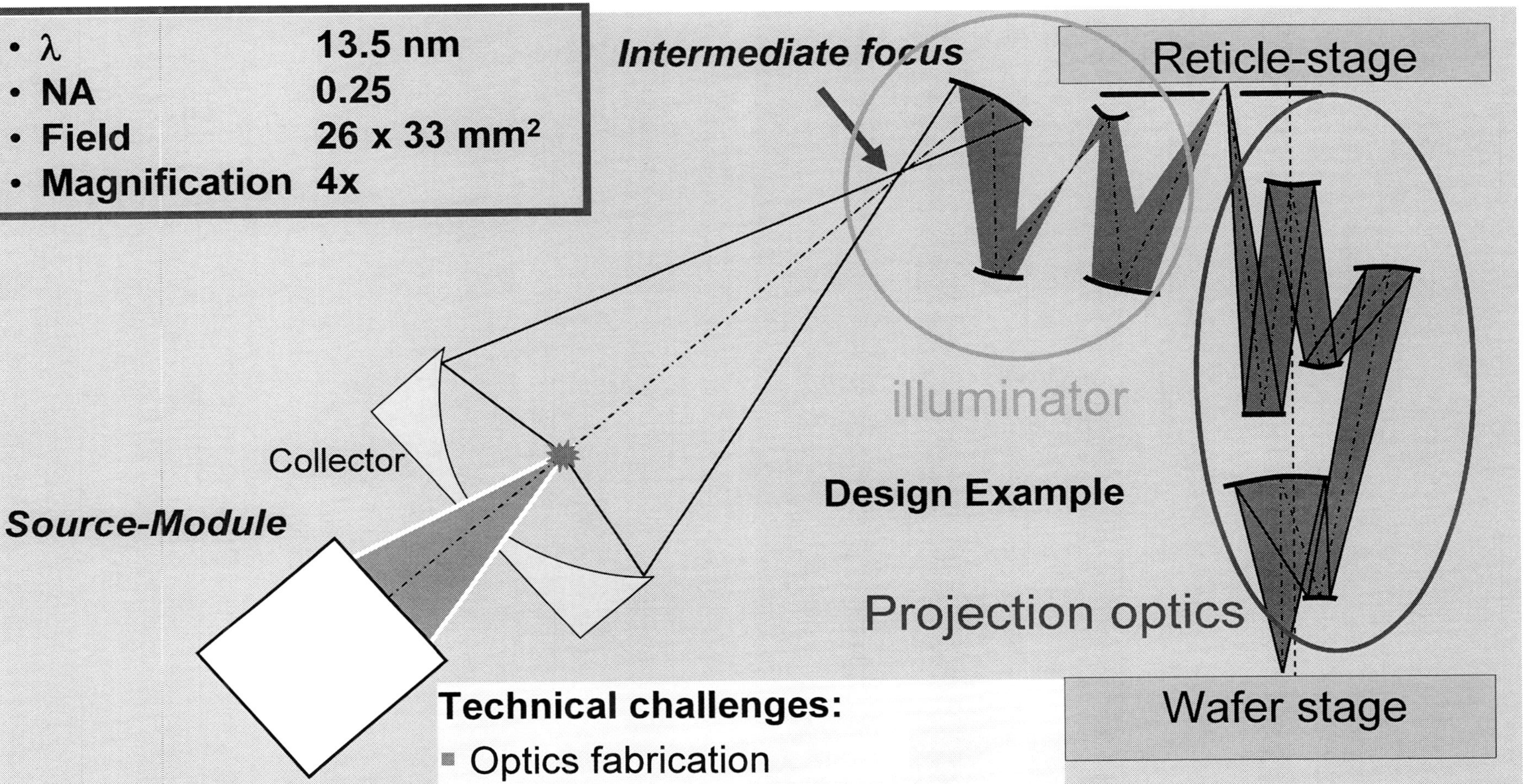

Intermediate focus

Reticle-stage

illuminator

Design Example

Projection optics

Wafer stage

Collector

Source-Module

Technical challenges:
- Optics fabrication
- Coating of EUV mirrors
- Alignment and qualification

Fabrication of EUV mirrors – The technical challenge

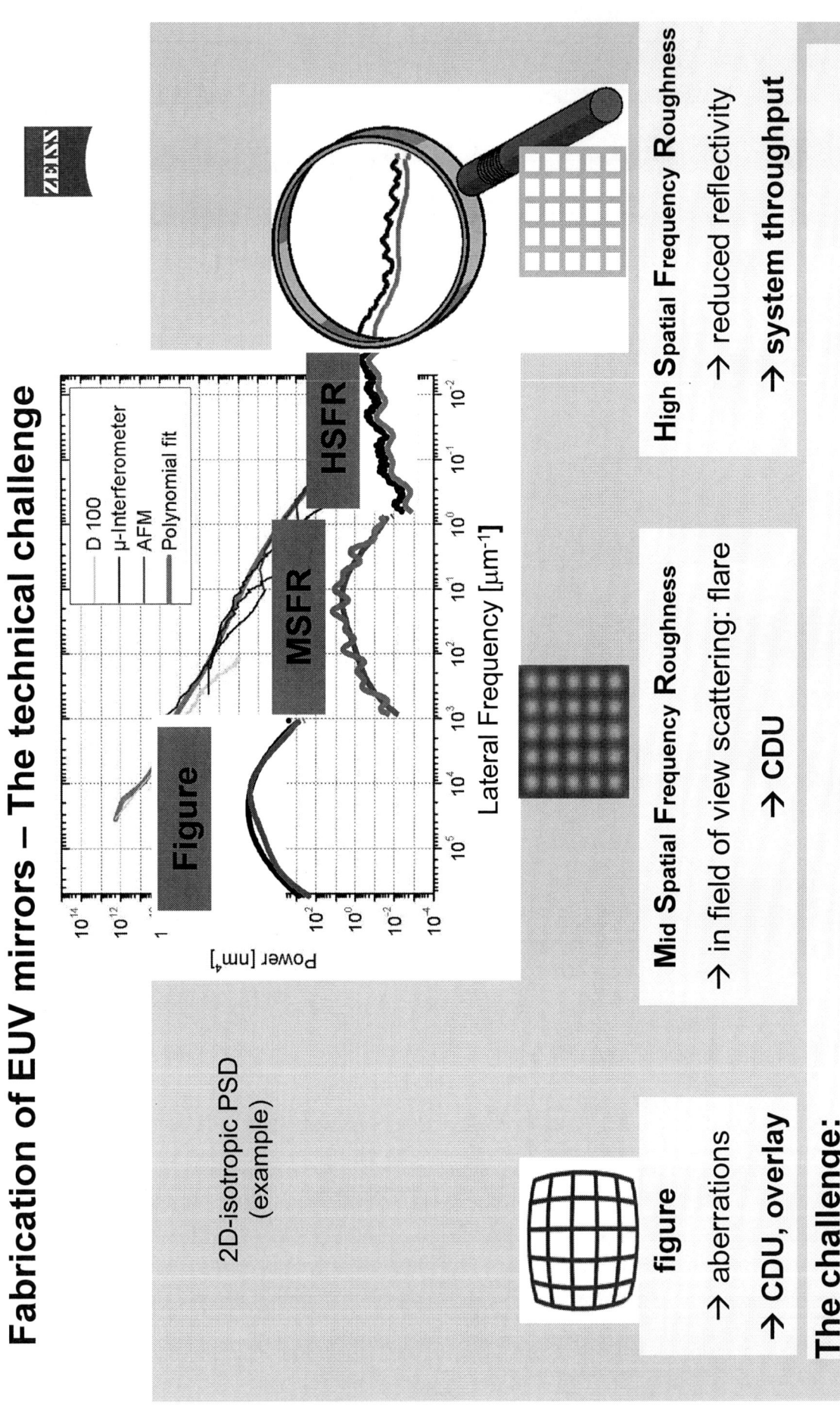

An EUV optics fabrication line for HVM tools is fully operational

optics fabrication infrastructure

Computer Controlled Polishing for Deterministic Processes

Ion Beam Figuring: Atomic Level Figure Control

Fast Magneto Rheological Figuring

metrology infrastructure

Measurement of surface figure:

- statistical errors (repeatability): ~ 10 pm rms

- statistical + adjustment errors (reproducibility): ~ 20 pm RMS

Enabling the Nano-Age World®

More than 10 HVM mirrors have been fabricated:
Considerable reduction of the flare has been achieved

Flare is calculated for a 2 µm line in a bright field

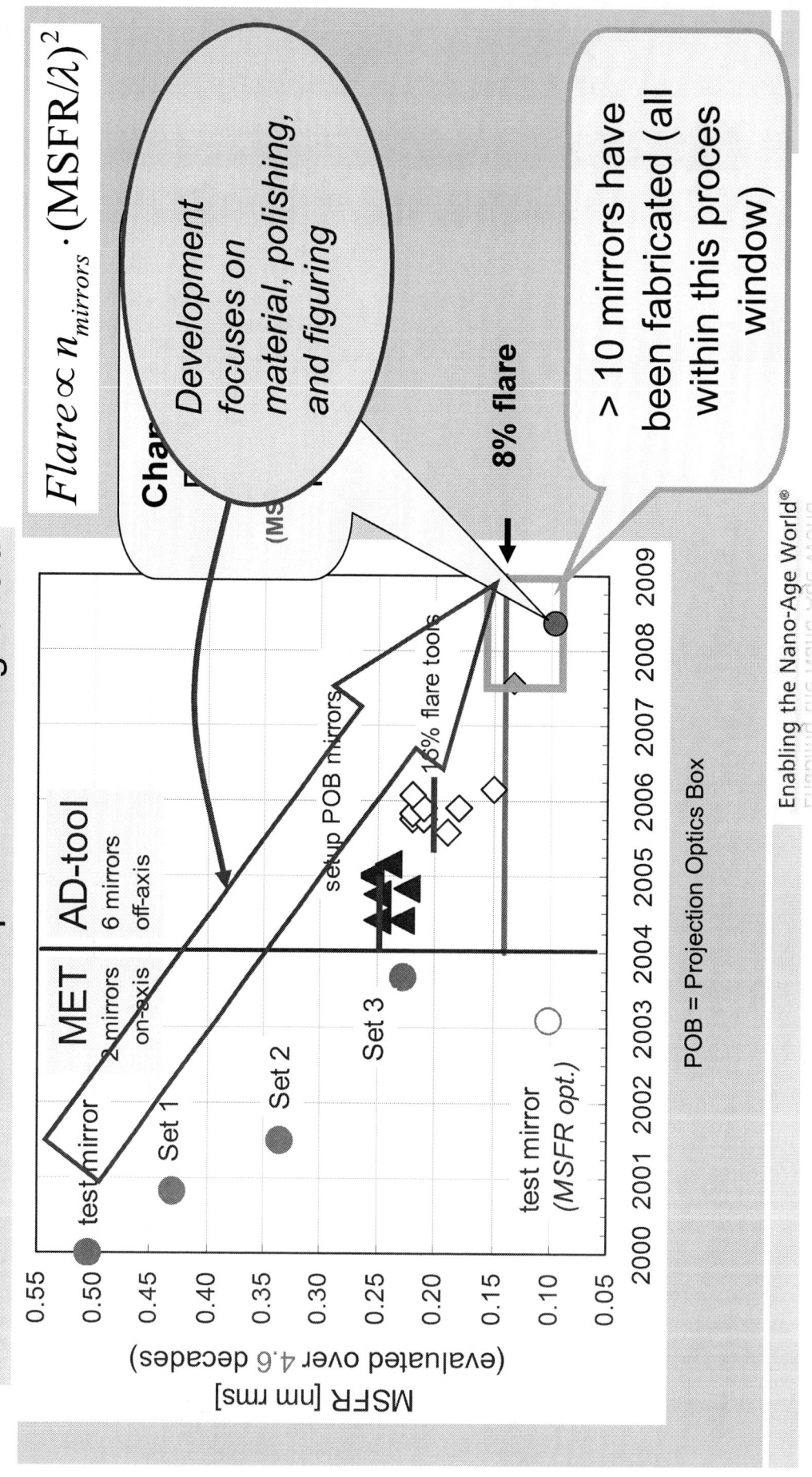

$$Flare \propto n_{mirrors} \cdot (MSFR/\lambda)^2$$

Development focuses on material, polishing, and figuring

8% flare

> 10 mirrors have been fabricated (all within this proces window)

POB = Projection Optics Box

Enabling the Nano-Age World®

Coating of EUV mirrors – The technical challenge

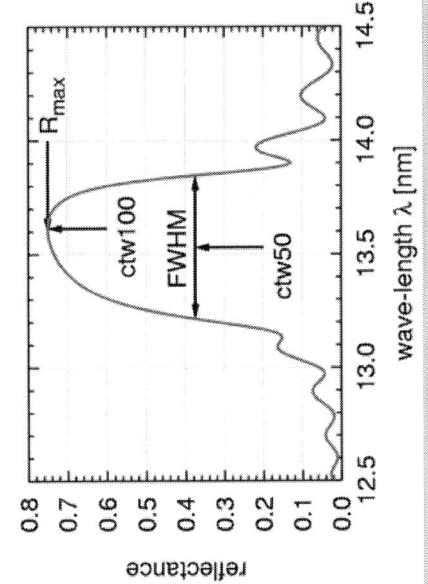

R_max, ctw100, FWHM, ctw50

reflectance vs wave-length λ [nm]

State of the art profile control on an EUV illuminator mirror

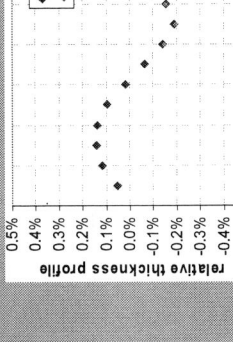

480 mm

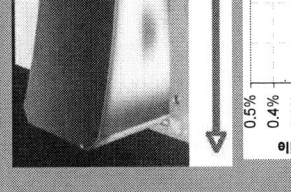

relative thickness profile vs s [mm]

y=0
x=0

Challenges

high peak reflectance and large FWHM

wave-length matching requires a few ‰

control of absolute thickness

lateral uniformity of a few ‰ for d

film stress less than ~ 50 MPa

thermal stability of several 100°C

EUV coatings: Mo/Si
Bragg reflectors

Si
Mo
d
θ
γ
substrate

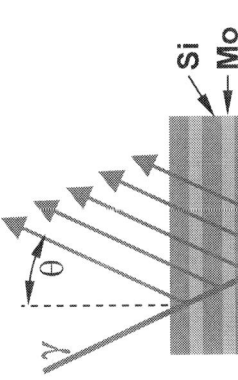

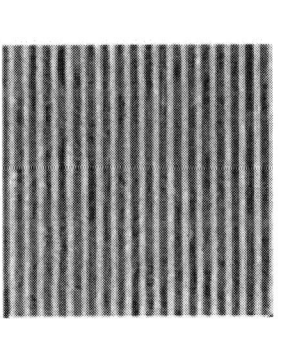

> 50 bilayers

Enabling the Nano-Age World®

A coating infrastrucure for HVM tools has been been set up and first mirrors have been successully coated

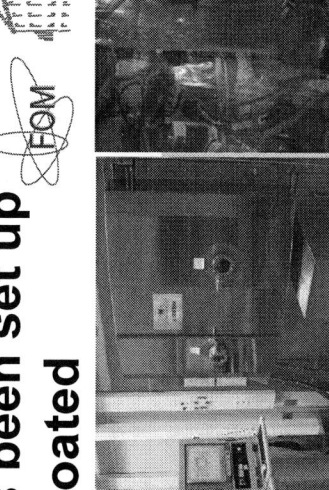

Coater at Zeiss

Coater at FOM

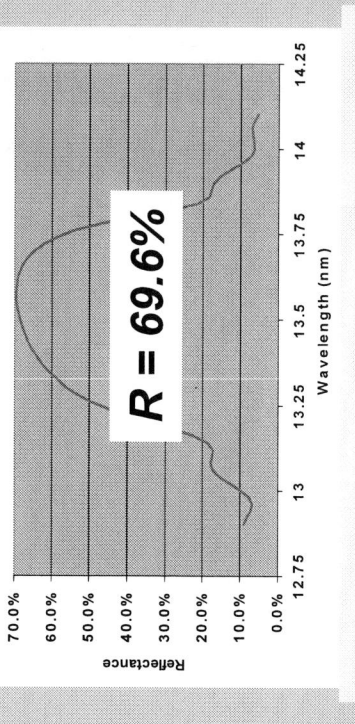

R = 69.6%

- HVM Mirror coated at FOM, measured at PTB
- Local angle of incidence (s-pol)

Total coating stack non correctable thickness error ≤30 pm rms

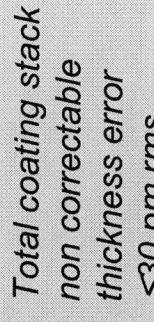

Progress in technology development...
Peak reflectivities of first HVM mirrors coated at FOM/Rijnhuizen and Zeiss are considerably higher than for the Alpha Demo Tool mirrors

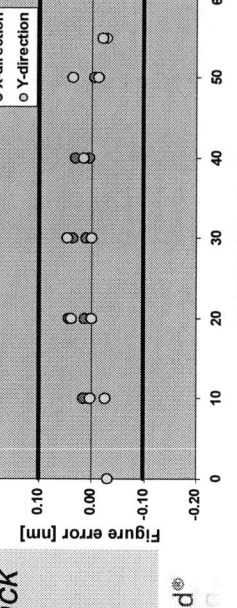

... results in ~50% higher transmission of the POBox

Enabling the Nano-Age World®

EUV metrology enables at wavelength qualification with high accuracy: a test stand demonstrates the capabilities

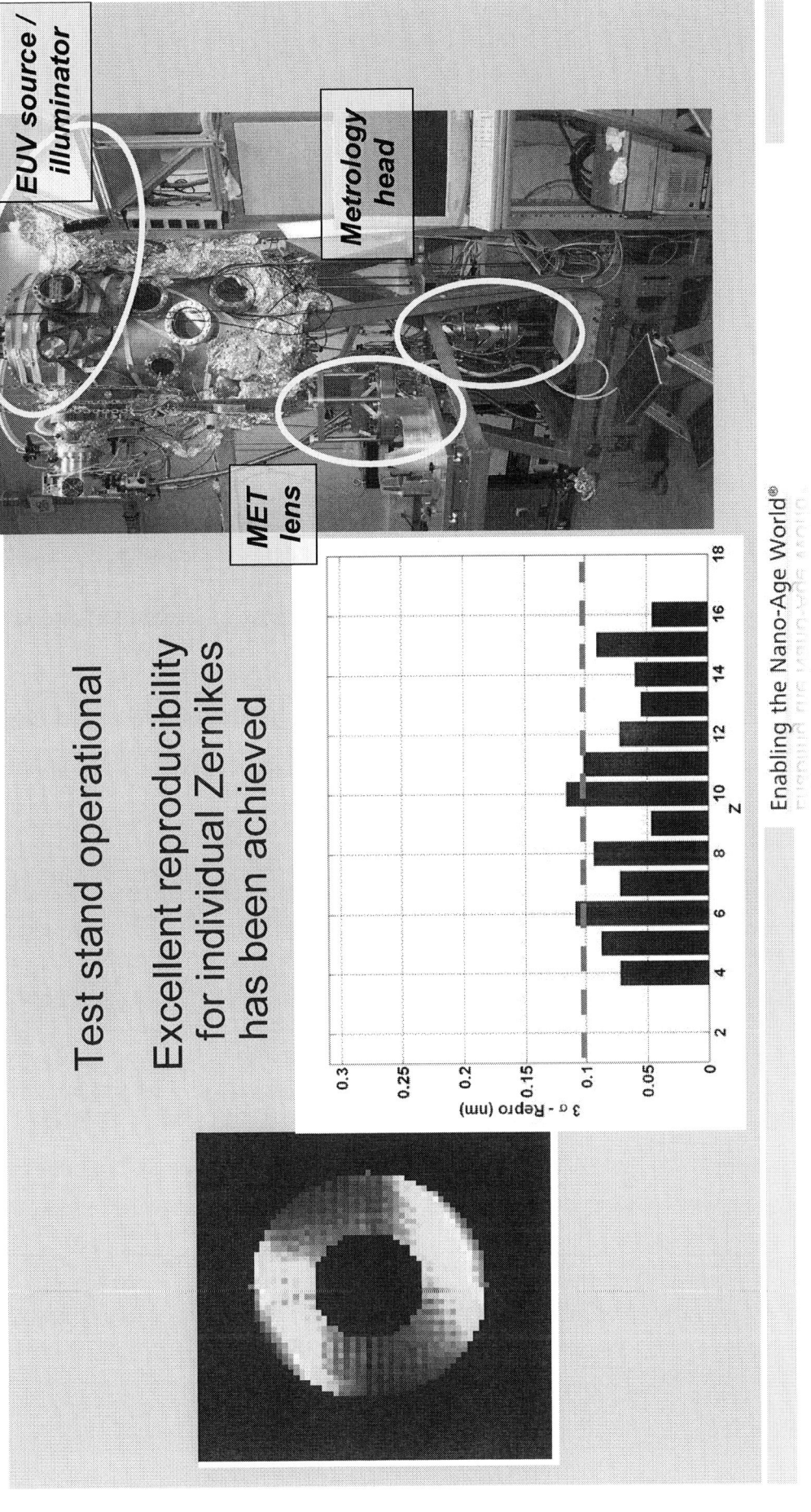

EUV source / illuminator

Metrology head

MET lens

Test stand operational

Excellent reproducibility for individual Zernikes has been achieved

Enabling the Nano-Age World®

System Metrology for the HVM production line

ZEISS

On site EUV metrolgy in Oberkochen...

248 nm metrology...

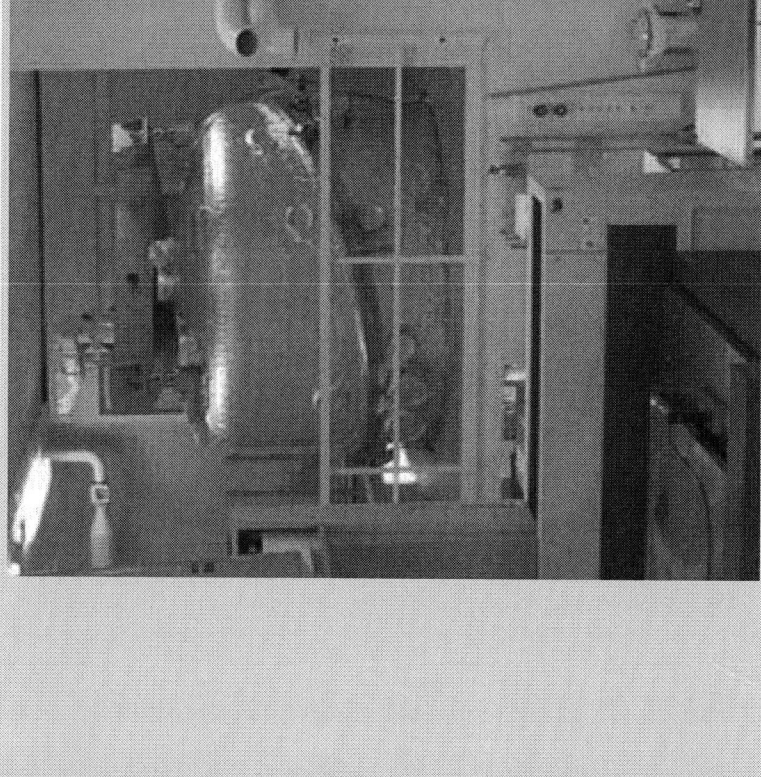

... enables a final alignment and
@ wavelength qualification of the optics

... enables an efficient
assembly and alignment process

Enabling the Nano-Age World®

Overview

From the Alpha Demo Tool to High Volume Tools:
EUV enters the preproduction phase

» Setup of an EUV infrastructure

» Progress in key technology areas
  - Optics metrology and fabrication
  - Coatings
  - System metrology

» Production of first systems is in full progress

The future: high NA EUV tools enable resolutions down to 11 nm

Enabling the Nano-Age World®

EUV Optics: The future

EUV is introduced as a high k1 technology...

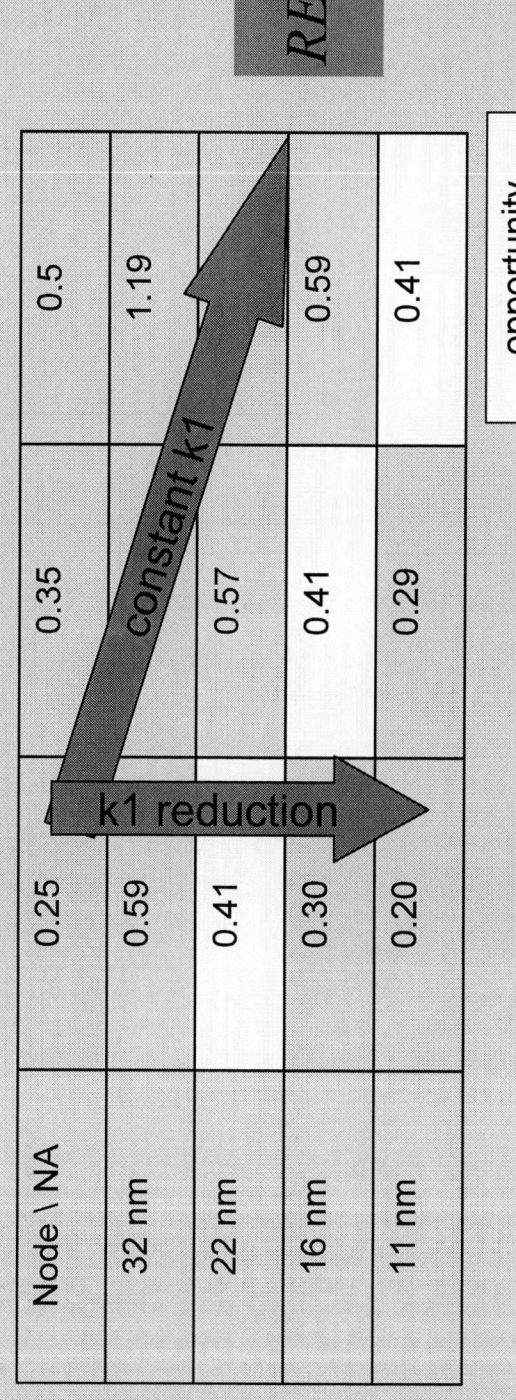

$$RES = \frac{k_1 \lambda}{NA}$$

Node \ NA	0.25	0.35	0.5
32 nm	0.59		1.19
22 nm	0.41	0.57	
16 nm	0.30	0.41	0.59
11 nm	0.20	0.29	0.41

... and will at higher NA and lower k factors enable resolutions down to 11 nm.

The path to NA > 0.3

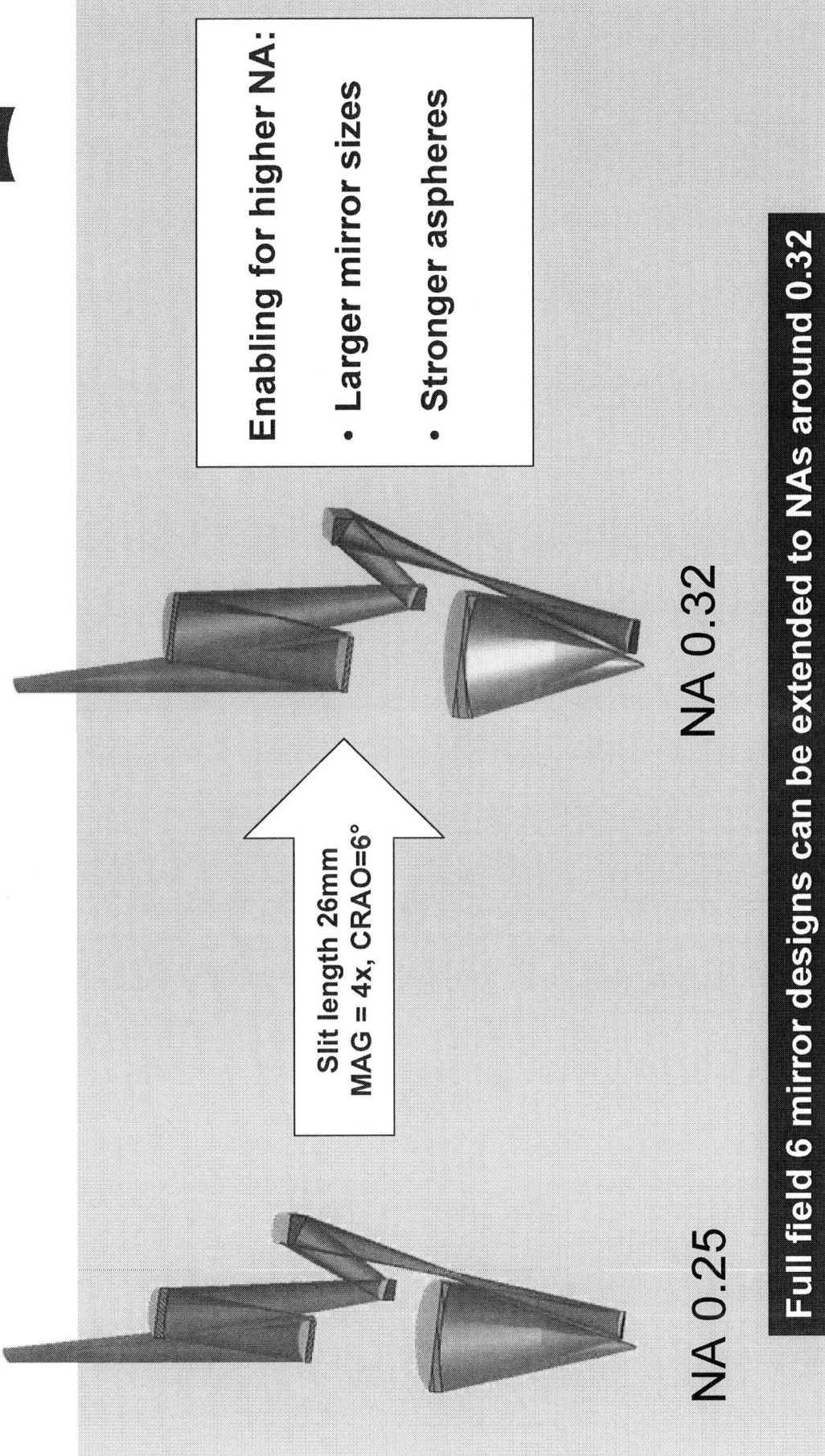

Enabling for higher NA:

- Larger mirror sizes
- Stronger aspheres

Slit length 26mm
MAG = 4x, CRAO=6°

NA 0.25

NA 0.32

Full field 6 mirror designs can be extended to NAs around 0.32

Enabling the Nano-Age World®

Apodization is limiting 6M designs

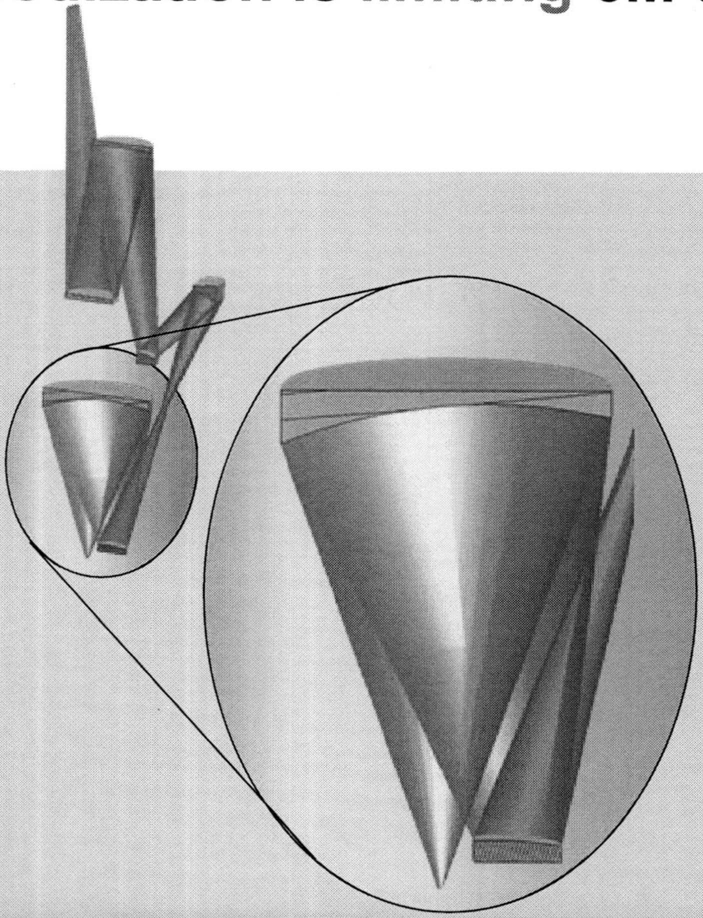

center of field (x=0) pupil

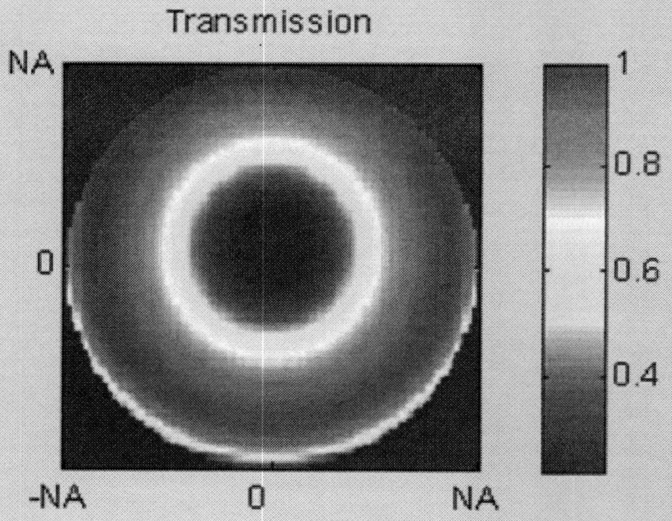

Transmission

The larger NA introduces a high angular load on surfaces which cause significant apodisation effects

Balancing of optical and coating design is needed to achieve a (quasi) rotational symmetric apodisation uniform over the field

Enabling the Nano-Age World®

NA > 0.4 : 2 solutions

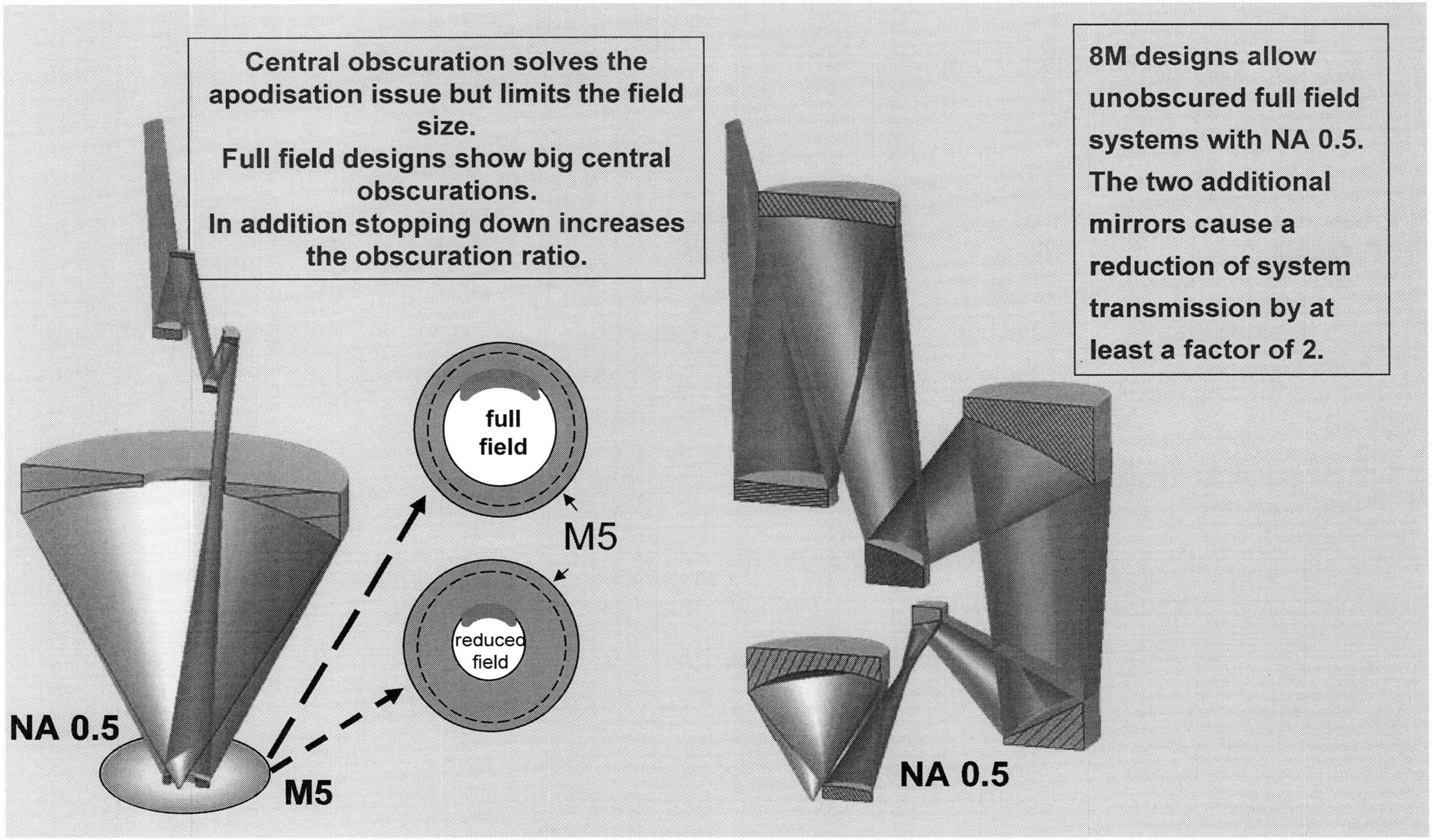

8M designs allow unobscured full field systems with NA 0.5. The two additional mirrors cause a reduction of system transmission by at least a factor of 2.

NA 0.5

Central obscuration solves the apodisation issue but limits the field size.
Full field designs show big central obscurations.
In addition stopping down increases the obscuration ratio.

M5

full field

reduced field

M5

NA 0.5

Enabling the Nano-Age World®

Impact of CRAO change

- **C**hief **R**ay **A**ngle on **O**bject side (α)
- EUV optics has to be non-telecentric on reticle side.
- Actual standard is 6°, this limits the NA to < 0.4 (Mag 4x).
- For larger NAs the CRAO has to be increased accordingly.

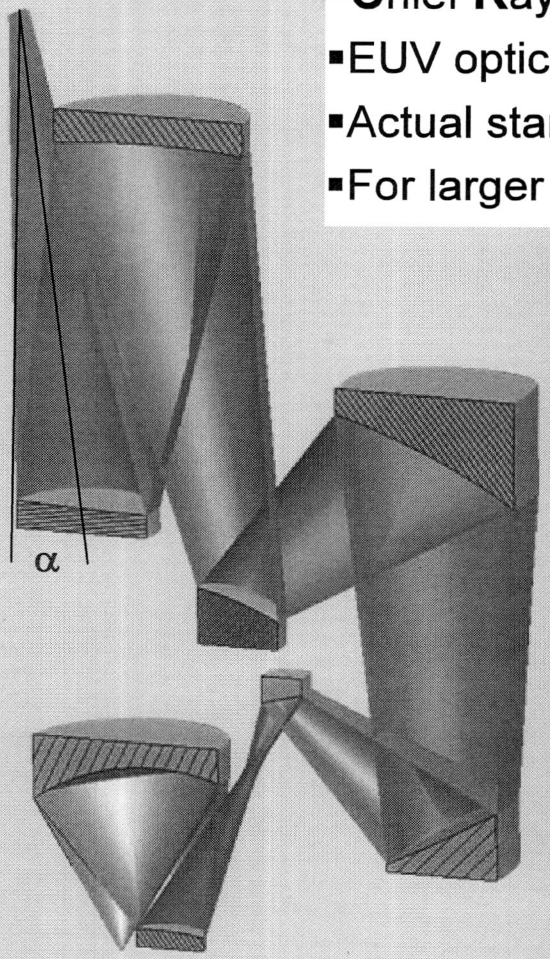

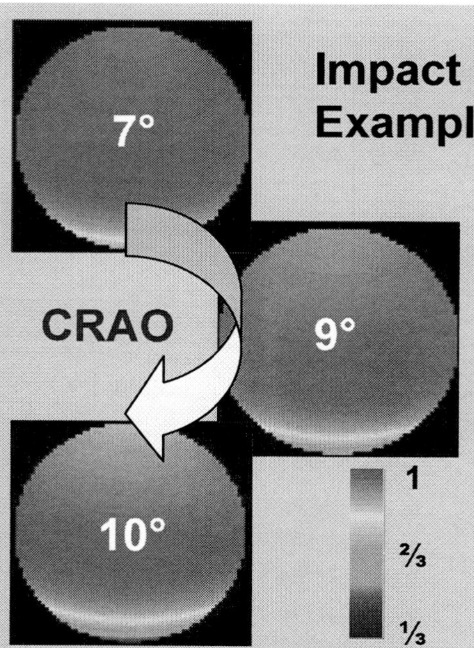

Impact of CRAO change on contrast
Example: NA=0.45

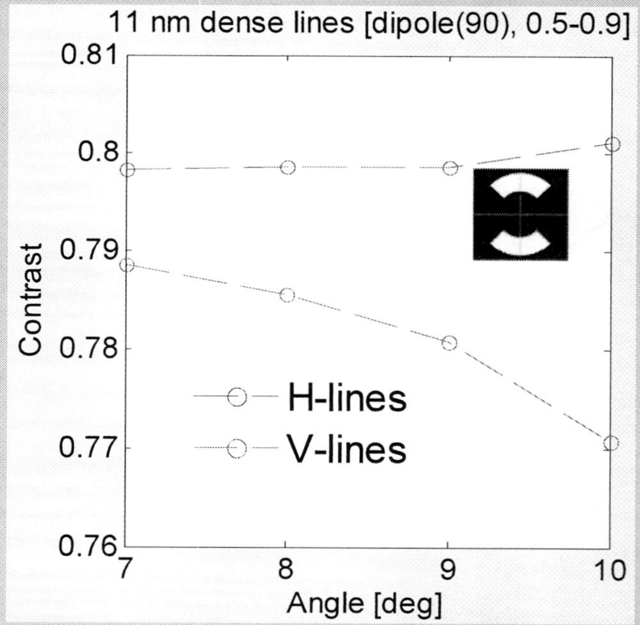

11 nm dense lines [dipole(90), 0.5-0.9]

Pupil apodization in dependence of different CRAO for NA=0.45 system

Enabling the Nano-Age World®

ASML/Zeiss EUV Roadmap
NXE platform supports manufacturing down to 11nm

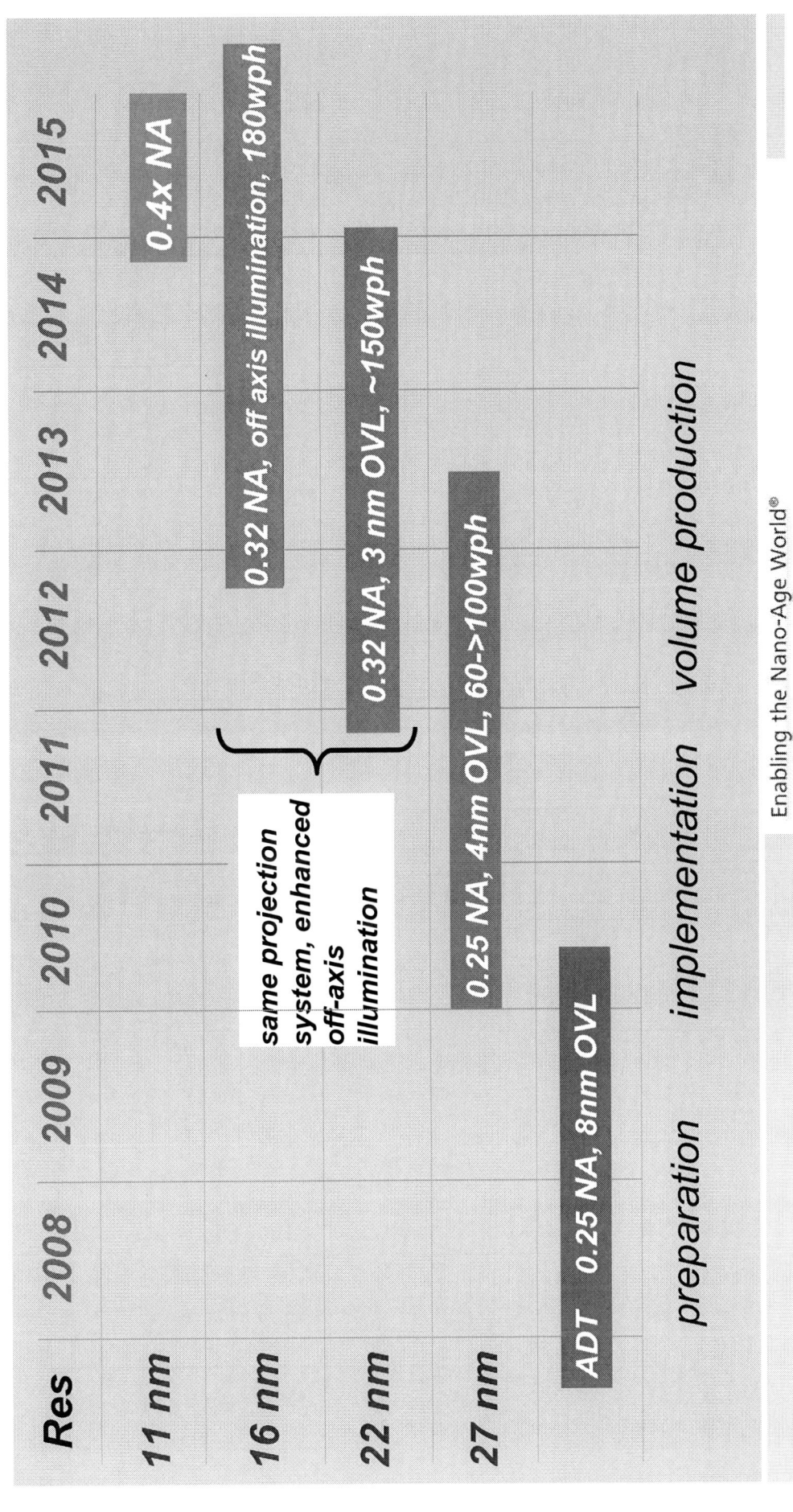

Summary

28 nm imaging has been demonstrated with Alpha Demo Tool optical systems

HVM optics fabrication is in full progress

- » A fabrication infrastructure has been set up
- » Key specs secured by technology advancements
- » > 10 mirrors have been fabricated, assembly of the first systems is on its way

Future development:

- » There are solutions visible for high NA design with > 0.4 NA.
  The challenge will be to find full field designs with optimum transmission to enable high productivity.
- » Off axis illumination will allow reduction of k1 to ~0.4.
- » These together will enable the printing of 11nm dense features in single exposure mode and even beyond. Improvements in polishing and coating technologies are expected to support this progress.

Enabling the Nano-Age World®

Extending NA to 0.7 for EUVL

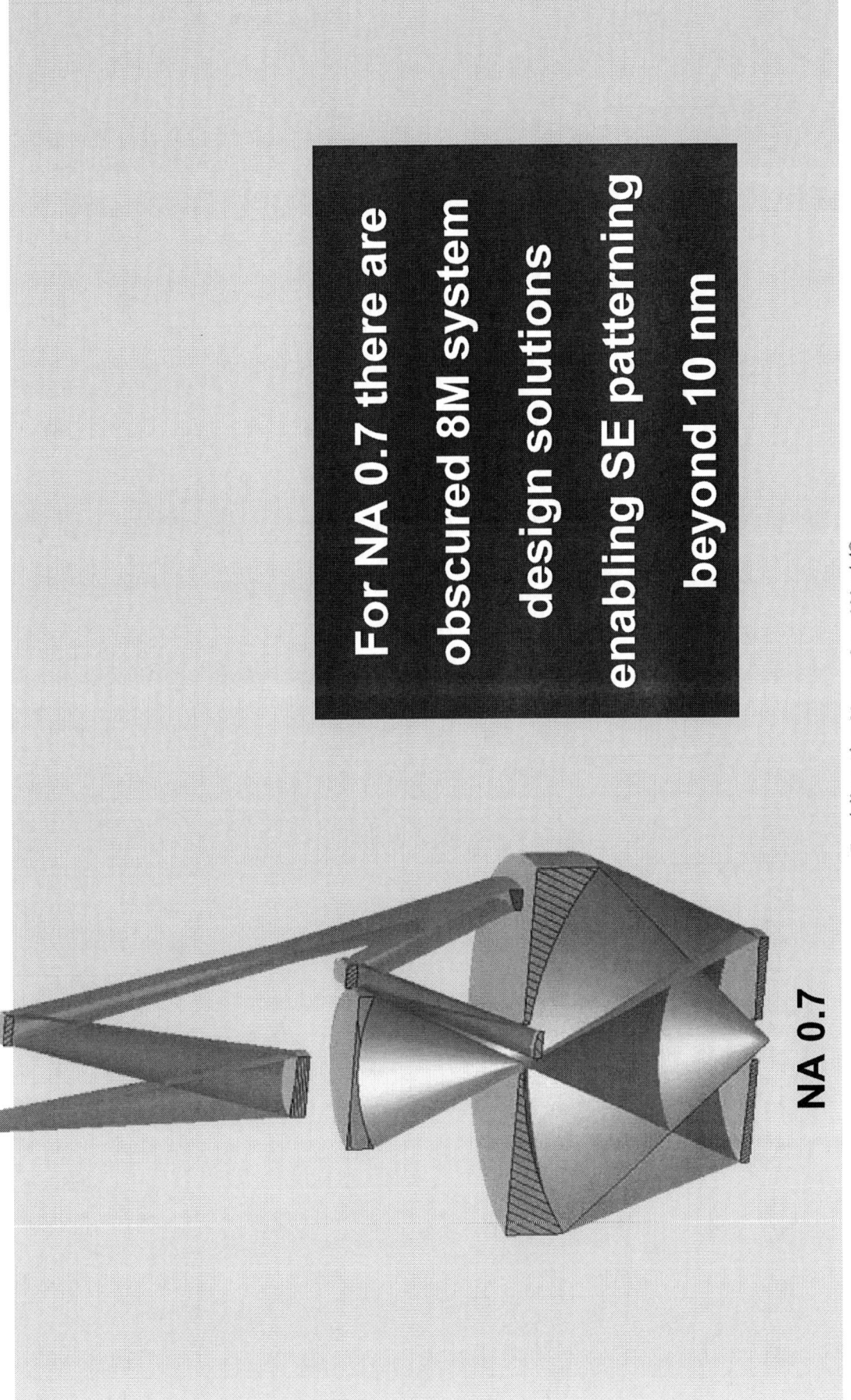

For NA 0.7 there are obscured 8M system design solutions enabling SE patterning beyond 10 nm

NA 0.7

Enabling the Nano-Age World®

ACKNOWLEDGEMENT

Thanks

to the EUV teams at ASML and Carl Zeiss SMT and our partners

The activities received funding
by the European Commission in the project "More Moore"
and by various national European governments
including the German Federal Ministry of Education and Research
in the program MEDEA+.

Enabling the Nano-Age World®

Accelerating the next technology revolution

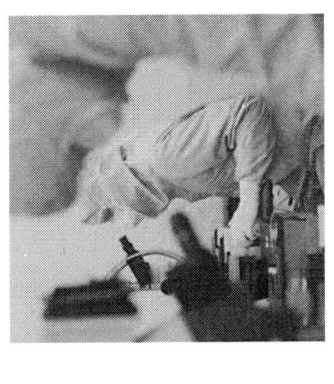
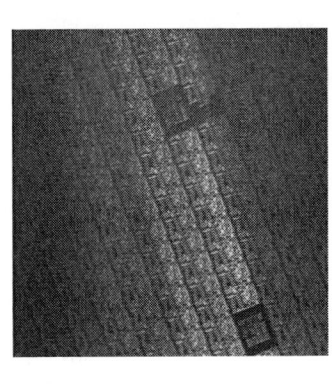

SEMATECH

Projection Optics for a 0.5-NA Microstepper Upgrade

Michael Goldstein[1], Russ Hudyma[2], Patrick Naulleau[3]

[1] Sr. Principal Physicist, Intel assignee at SEMATECH
[2] Managing Partner, Hyperion Development LLC
[3] Lead Scientist EUVL Program, Lawrence Berkeley National Lab

Acknowledgements: **Stefan Wurm[4]**

[4] Associate Director, SEMATECH Lithography Division

Lithography Roadmap

☐ The compelling argument for EUVL is an intrinsic capability to scale resolution for future technology nodes.

☐ However, very early tool access is needed to develop mask and resist solutions.

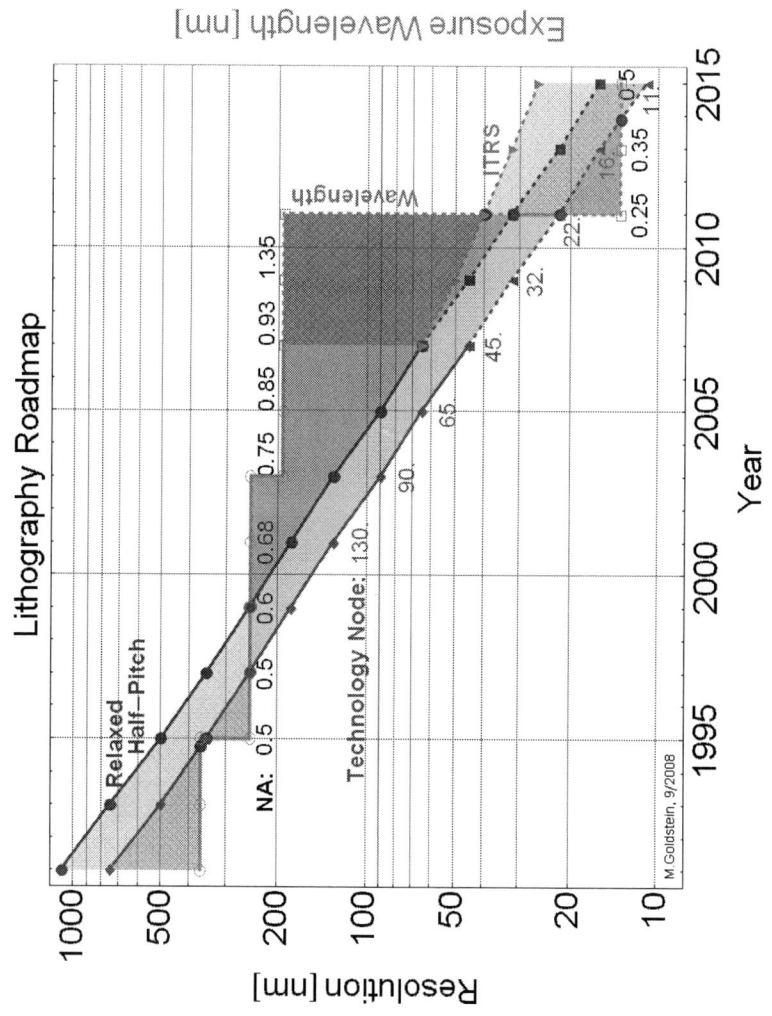

Lithography Roadmap

- The compelling argument for EUVL is an intrinsic capability to scale resolution for future technology nodes.

- However, very early tool access is needed to develop mask and resist solutions.

Rayleigh k1 Factor [P/2 = k1 λ / NA]			Half Pitch [nm]			
Lithography			32	22	16	11
Ex	NA	λ				
SE	1.35	193	0.22	0.15	0.11	0.08
DP	1.35	193	0.45	0.31	0.22	0.15
SE	1.50	193	0.25	0.17	0.12	0.09
DP	1.50	193	0.50	0.34	0.25	0.17
SE	1.65	193	0.27	0.19	0.14	0.09
DP	1.65	193	0.55	0.38	0.27	0.19
SE	0.25	13.5	0.59	0.41	0.30	0.20
SE	0.35	13.5	0.83	0.57	0.41	0.29
SE	0.50	13.5	1.19	0.81	0.59	0.41

SE ≡ Single Exposure
DP ≡ Double Patterning

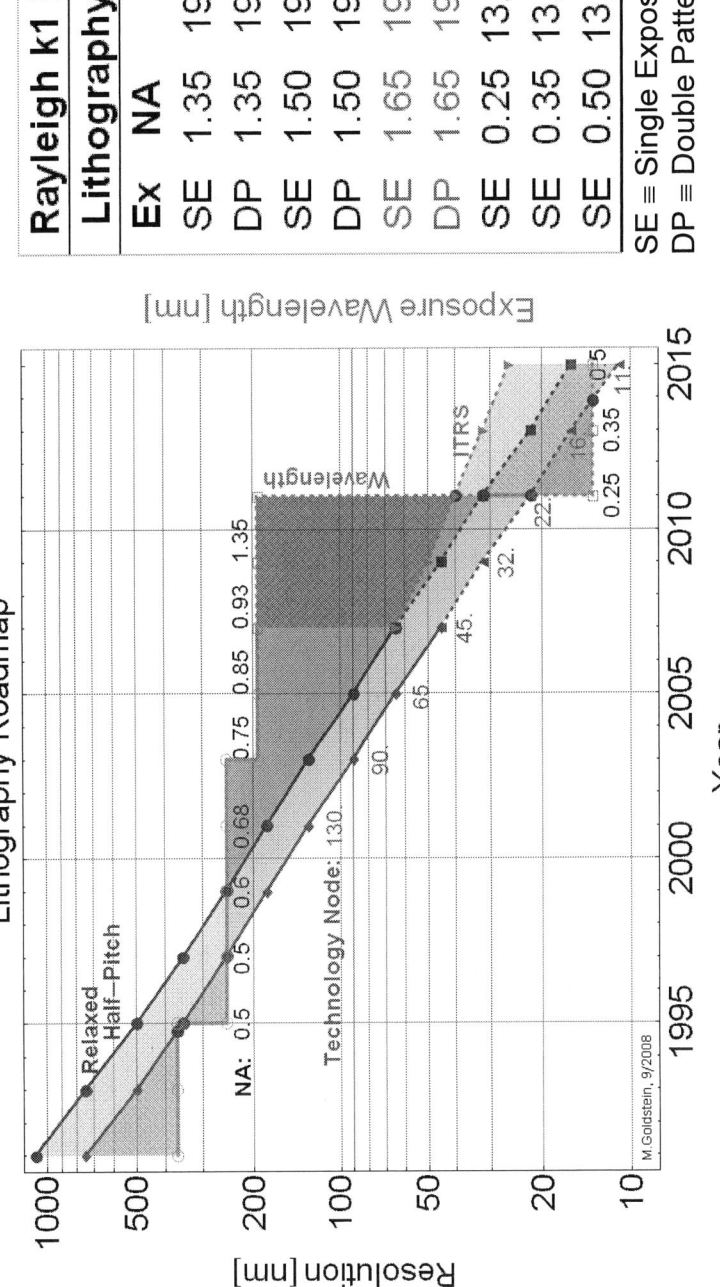

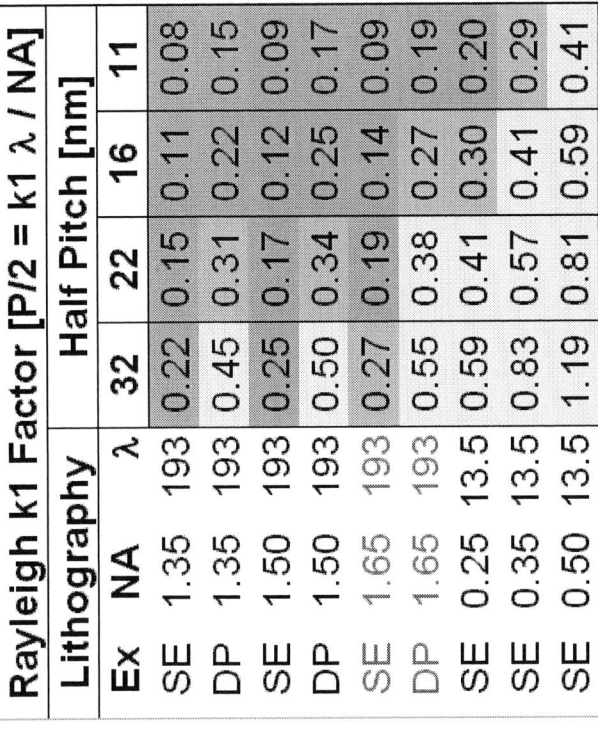

Research Levels

Half Pitch:
Research:
Development:
HVM year:

22 nm	16 nm	11 nm
2007	2009	2011
2009	2011	2013
2011 (Node) (2013 Relaxed Pitch)	2013 (Node) (2015 Relaxed Pitch)	2015

SEMATECH

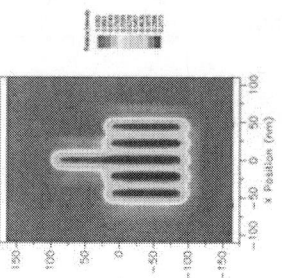

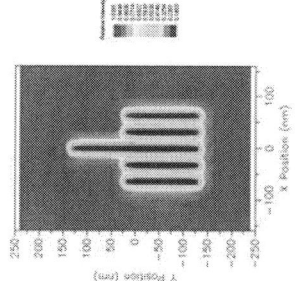

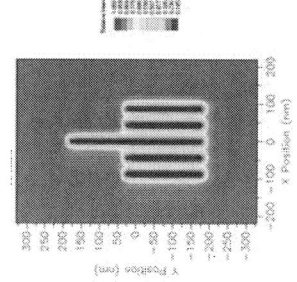

--- 0.50NA ---
1.0nm WFE, 4%Flare
0.7/0.35 Annular Sigma

--- 0.35NA ---
1nm WFE, 8% Flare
0.7 Sigma

--- 0.25NA ---
1nm WFE, 8% Flare
0.7 Sigma

MET Research at the 16 nm Node

☐ First full-field lithography tools at the 16 nm node (τ_o = 2013) will need highly advanced resist/mask solutions.

☐ Research typically starts ~4 years in advance of HVM. ($\tau \approx \tau_o$ – 4 years ≈ 2009).

☐ The SEMATECH High NA MET (MET2) project started in Q4'2007.

☐ Planned first light is in the 2010–2011 timeframe
  ▲ 2 to 3 years before HVM node compared to a ~4 years POR.
  ▲ 4 to 5 years before relaxed pitch introduction.

☐ To ensure a long useful life we designed the MET2 to be capable of early 11nm node development.

The Existing MET (the MET1)

☐ SEMATECH's flagship mask and resist testing is;

▲ performed using a 0.3 NA EUV Micro-Exposure Tool (MET1),

▲ located at Lawrence Berkeley National Laboratory's Advanced Light Source synchrotron radiation facility (the ALS).

From synchrotron

Scanner module

Reticle stage

0.3-NA MET (Zeiss)
(Coatings and design by LLNL*)

Wafer stage and height sensor

Pupil-fill monitor

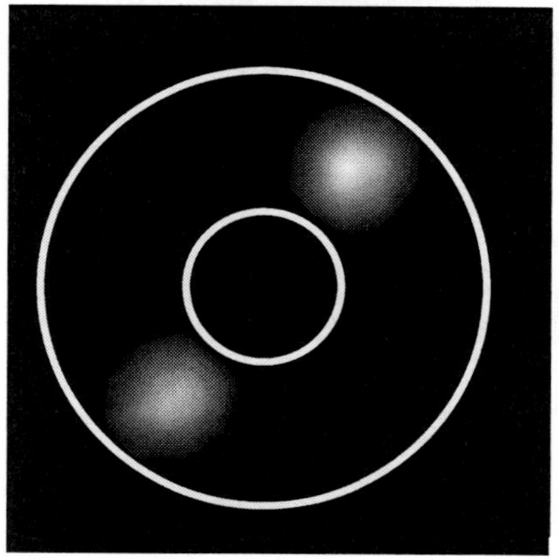

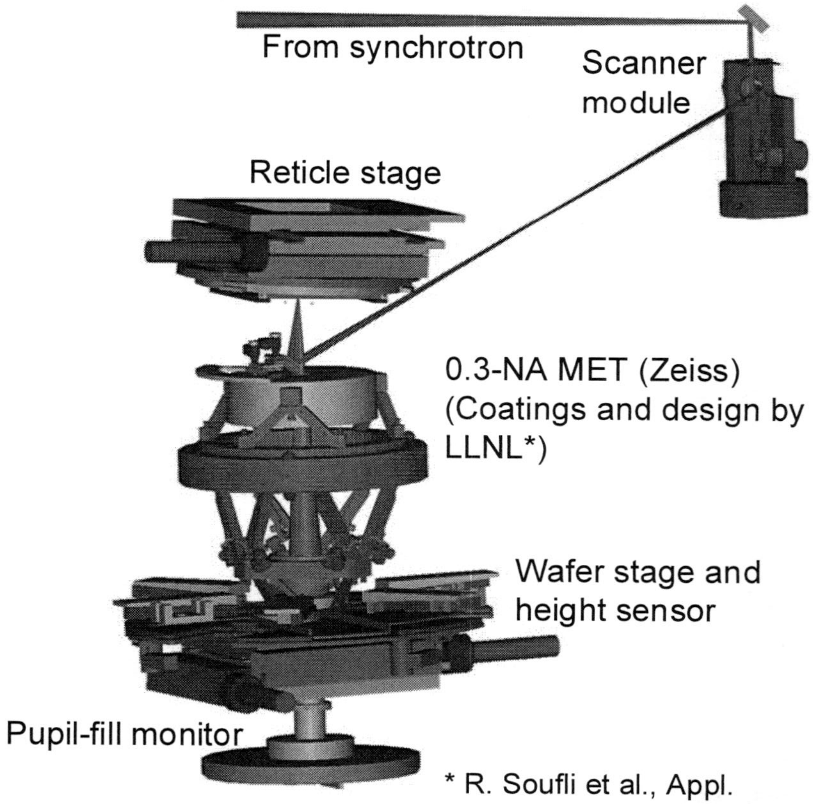

* R. Soufli et al., Appl. Opt. 46, 3736 (2007)

<u>Annular</u>
32 nm L/S

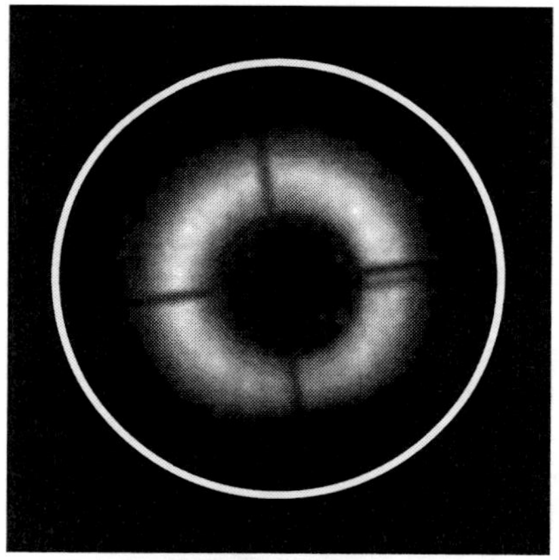

<u>Dipole</u>
22 nm L/S

Current Capability
☐ MET1

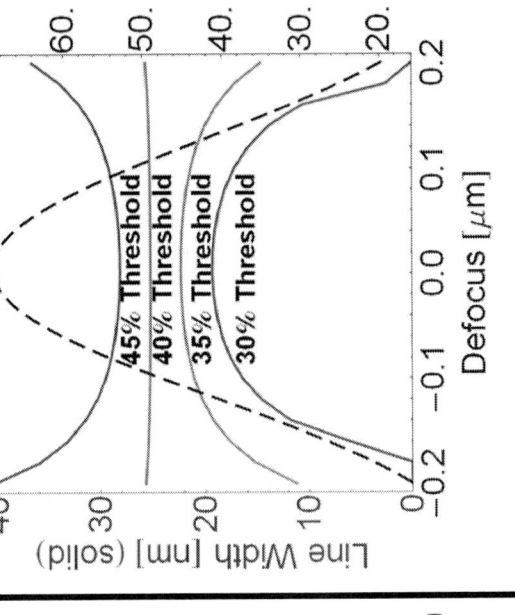

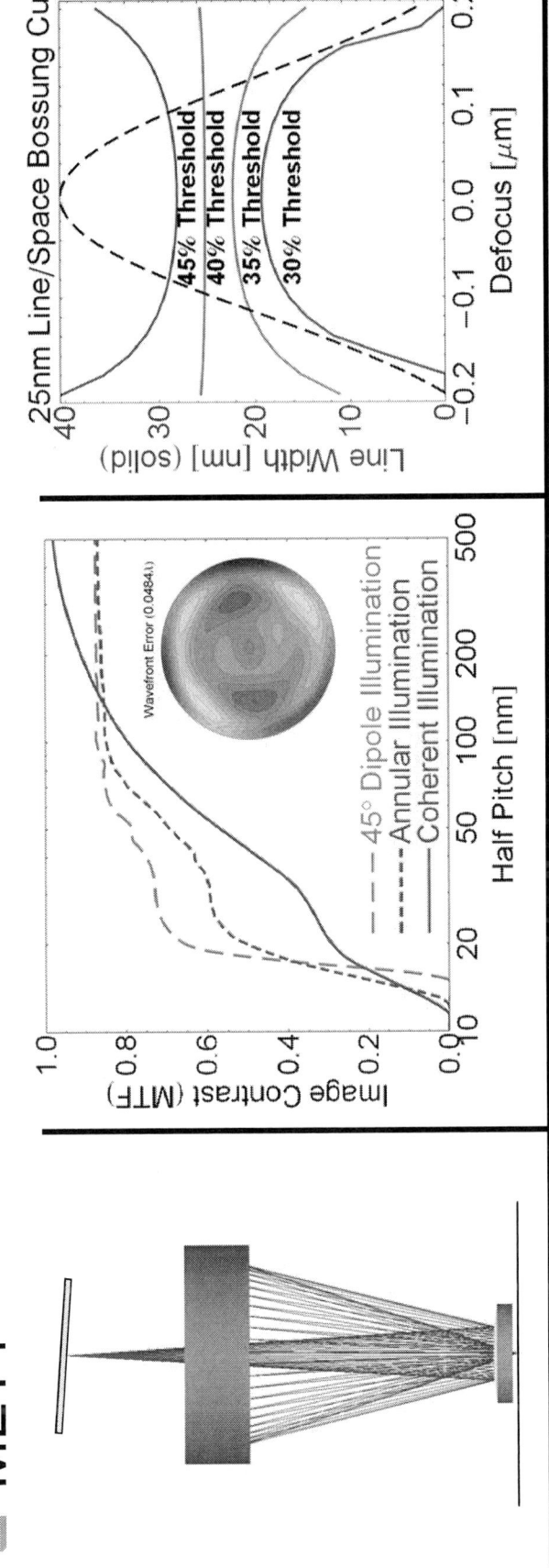

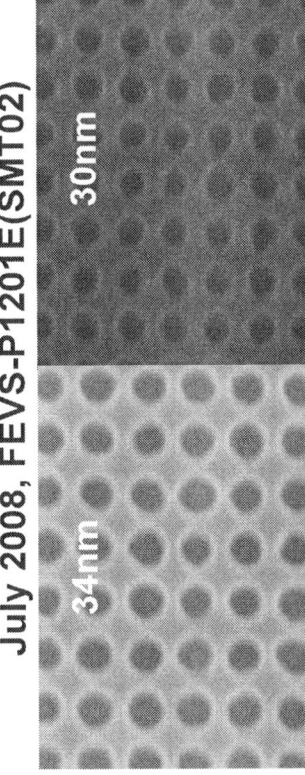

July 2008, FEVS-P1201E(SMT02)

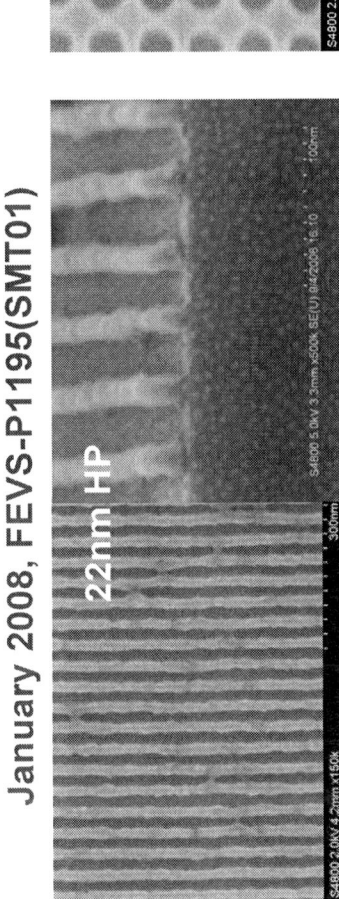

January 2008, FEVS-P1195(SMT01)

**SEM images courtesy of Chawon Koh
(Samsung assignee at SEMATECH)**

MET2 Design Options

- ☐ Three optical designs analyzed.
  - ▲ Near-equal radii: Larger field size, but with larger secondary mirror
  - ▲ Non-equal radii: Field reduced ~20%, fits packaging constraints
  - ▲ Four bounce: Issues: primary aspect ratio, secondary size, obscuration, and wavefront errors

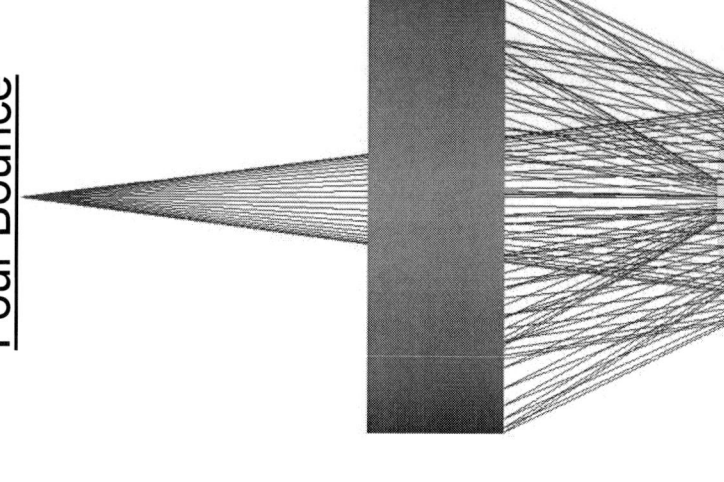

Four Bounce

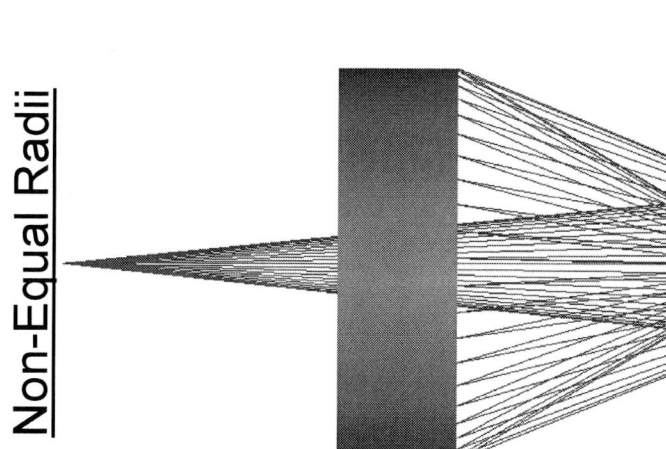

Non-Equal Radii

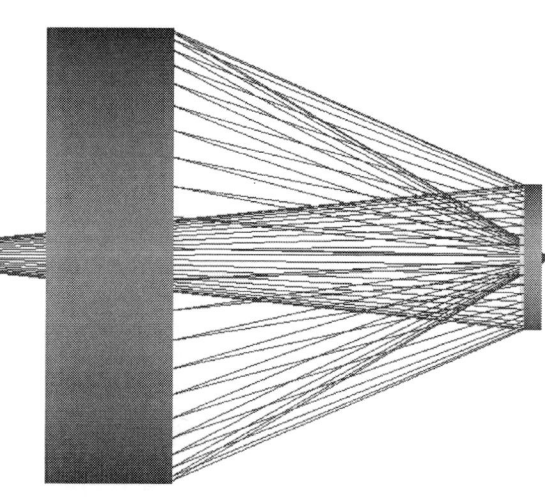

Near-Equal Radii

MET2 Design Options

Near-equal radii design:

- Scale up of existing MET1
- Minimizes Petzval Sum, reduced field curvature increases field size
- Large secondary complicates platform mechanics
- Small secondary – mask spacing limits maximum pupil fill

Near-Equal Radii

Metric	Spec	Units	Comment
Aperture	0.50	#	
Demagnification	5X	#	
Track Length / Back FL	474.164 / 5.0	mm	
Central Obscuration	7.2	%	
Centered WFE	37.5	mλ	Minor additional optimization may be possible
Centered Field Size	400	µm	
Primary Diameter	91.75	mm	
Secondary Diameter	287.3	mm	Not Compliant
Aspheric Departure	M1: 41.6 M2: 60.9	µm	
Aspheric Slope	M1: -6.1 M2: -2.5	µm/ mm	

MET2 Design Options

☐ Four bounce design:

▲ Large secondary complicates platform mechanics

▲ Very high aspect ratio primary (not manufacturable)

▲ Mid-spatial frequency wavefront errors

▲ Increased central obscuration

Four Bounce

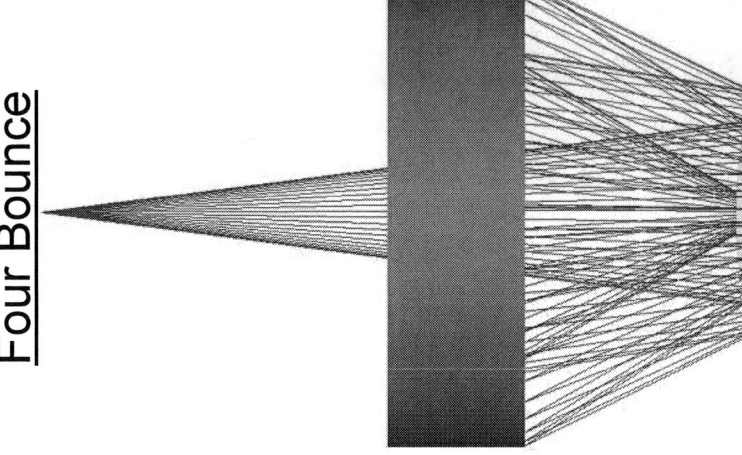

Metric	Spec	Units	Comment
Aperture	0.45	#	Not Compliant
Demagnification	4X	#	
Track Length / Back FL	474.164 / 5.0	mm	
Central Obscuration	12	%	Not Compliant
Centered WFE	28	mλ	Mid freq. comp.
Centered Field Size	300	μm	
Primary Diameter	144.8	mm	
Secondary Diameter	278.3	mm	Not Compliant
Aspheric Departure	Not computed	μm	
Aspheric Slope	Not computed	μm/ mm	

MET2 Design Options

Non-equal radii design:

- Slight (~20%) field size reduction from equal radii design
- Best compromise of performance and mechanical retrofit with existing METs
- This design was selected with a 6° chief ray angle at the mask

Non-Equal Radii

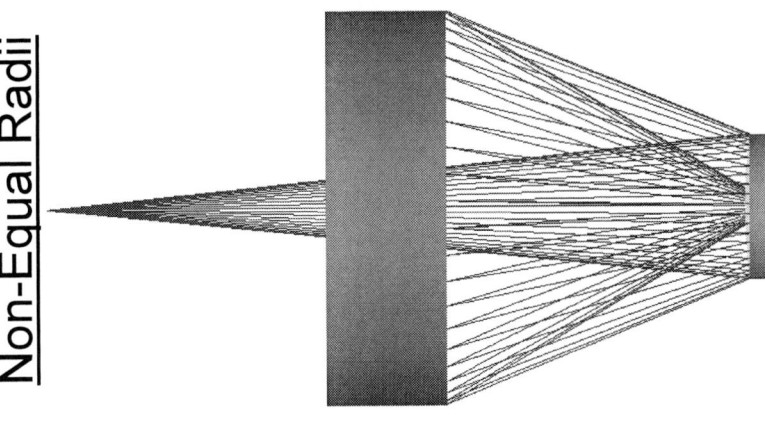

Metric	Spec	Units	Comment
Aperture	0.50	#	
Demagnification	5X	#	
Track Length / Back FL	474.165 / 5.0	mm	
Central Obscuration	7.2	%	
Centered WFE	47.2	mλ	Minor additional optimization may be possible
Centered Field Size	300	μm	
Primary Diameter	91.09	mm	
Secondary Diameter	248.31	mm	
Aspheric Departure	M1: 38.9 M2: 59.8	μm	
Aspheric Slope	M1: -5.9 M2: -2.9	μm/ mm	

Aspheric Departure

☐ Aspheric departure calculated from best RMS spherical fit.

▲ Primary Mirror: Maximum and Range = 31 and 44 μm

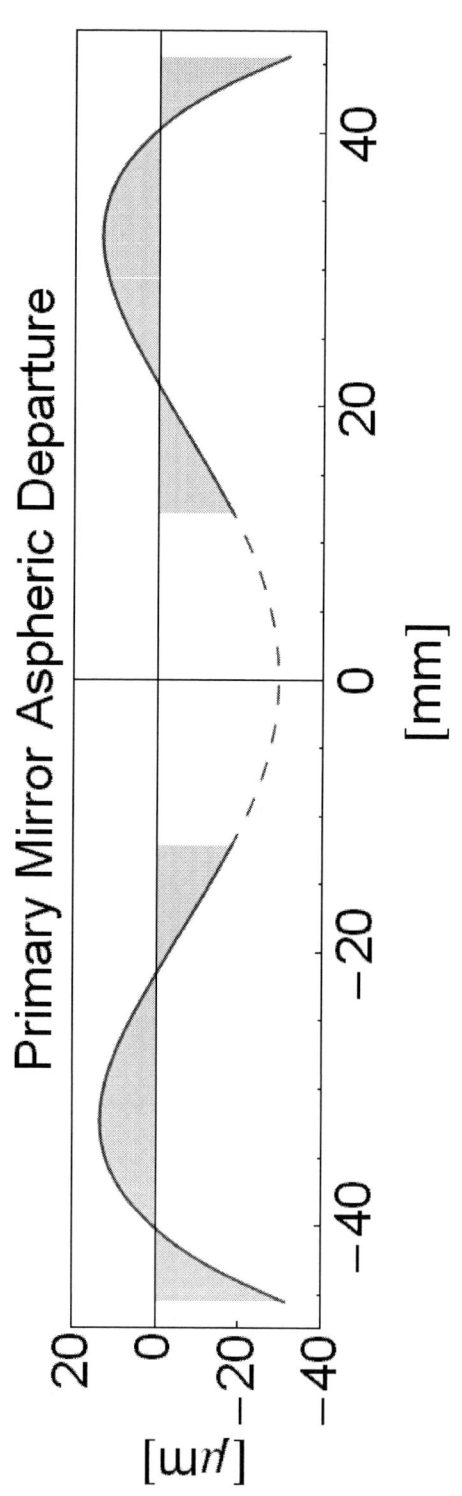

Primary Mirror Aspheric Departure

▲ Secondary Mirror: Maximum and Range = 39 and 56 μm

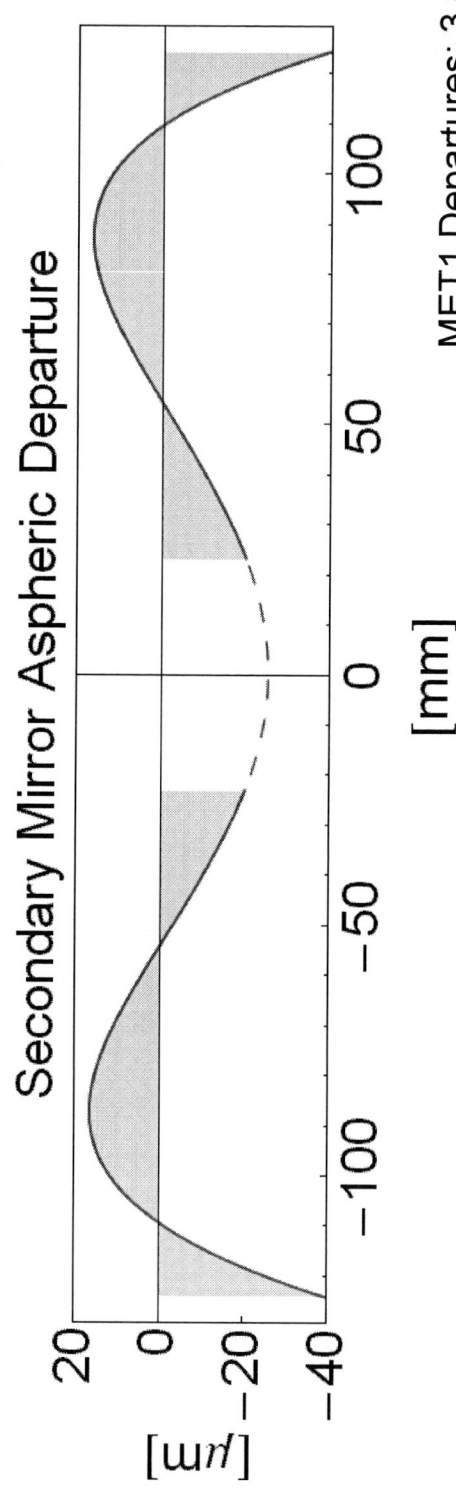

Secondary Mirror Aspheric Departure

MET1 Departures: 3.8 & 5.6 μm

Ray Trace Analysis

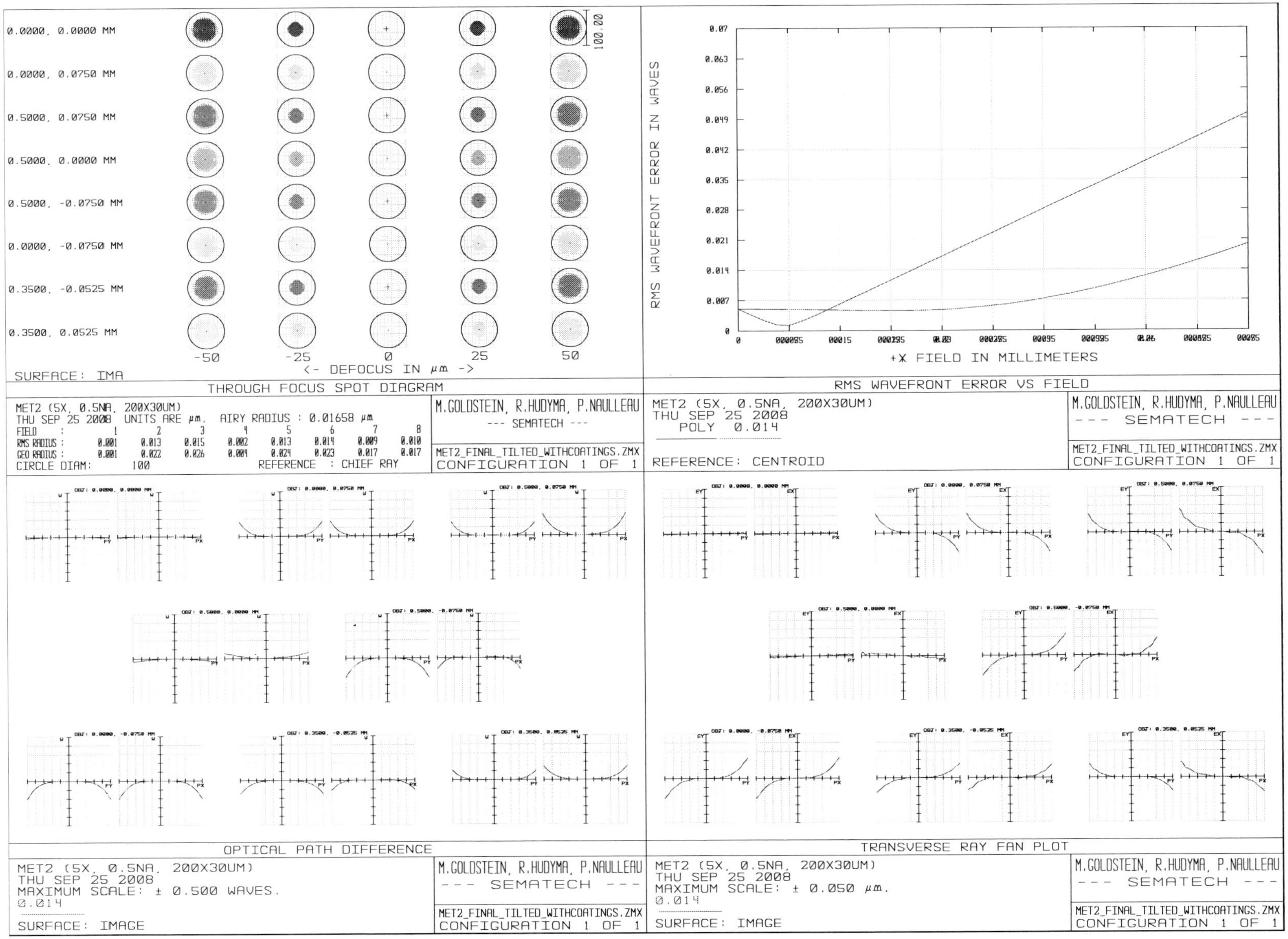

Field Wavefront Performance

☐ A 6° mask tilt (1.2° wafer tilt) reduced field size ~30×200μm

▲ Lateral offsets primarily induce astigmatism

▲ Axial offsets primarily induce spherical aberration

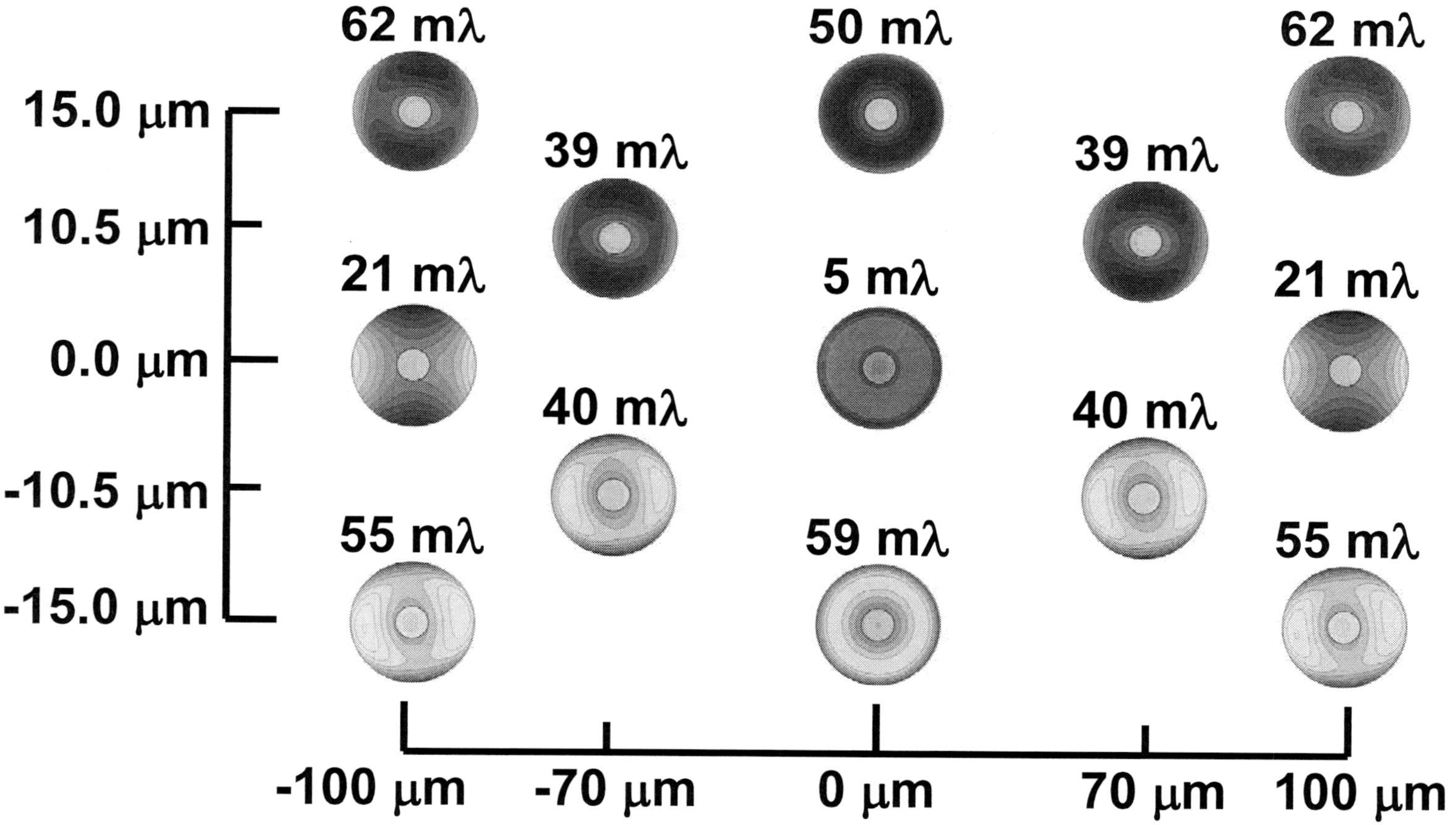

Interference Coatings

☐ Uniform coatings were designed for optical and actinic interferometry correlation.

▲ Molybdenum / Silicon multilayer stacks, oxide capped (preliminary)

▲ Thickness specified relative to local surface normal direction

▲ Transmission = 44%

▲ Pupil apodization (P-V)/(P+V) = 4.7%

▲ Polarized illumination induces a slight on-axis astigmatism (~6mλ)

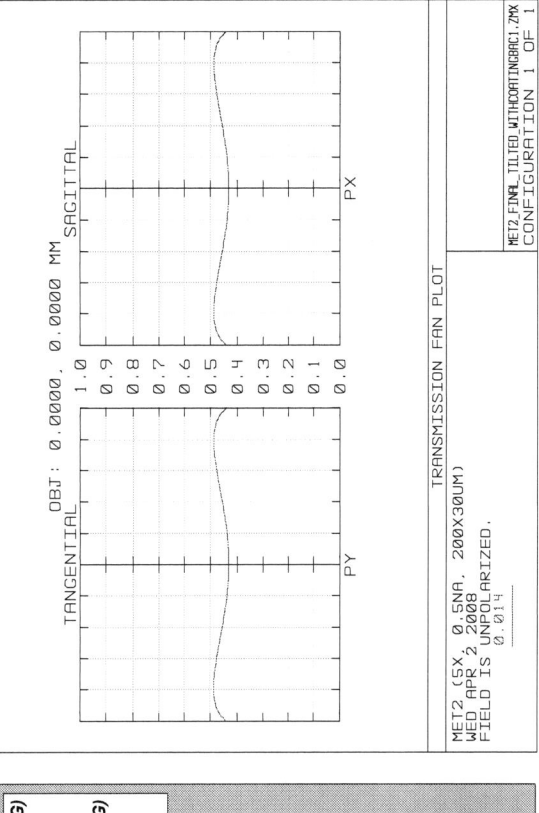

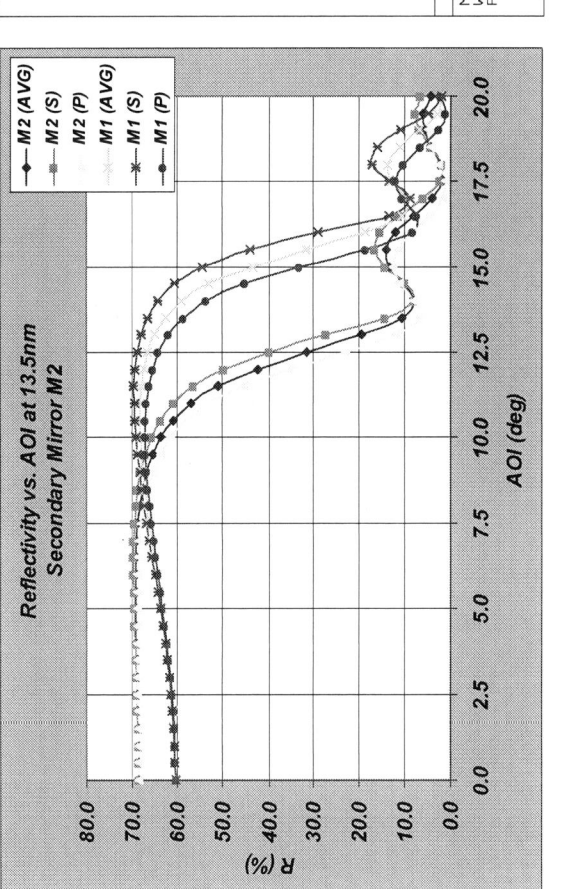

Edge of Field Performance

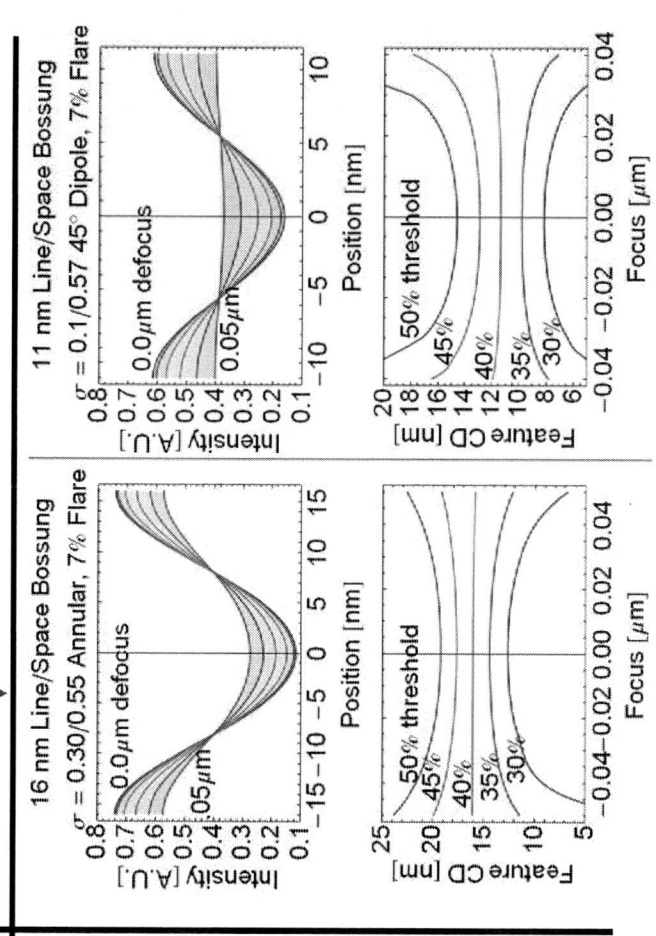

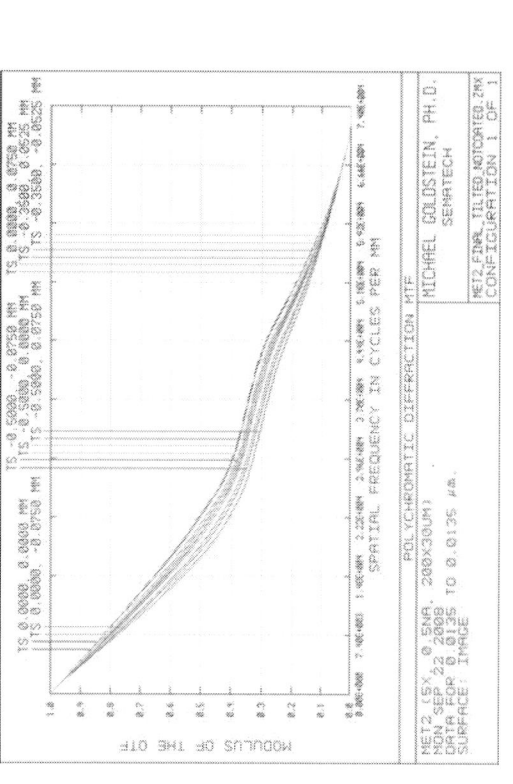

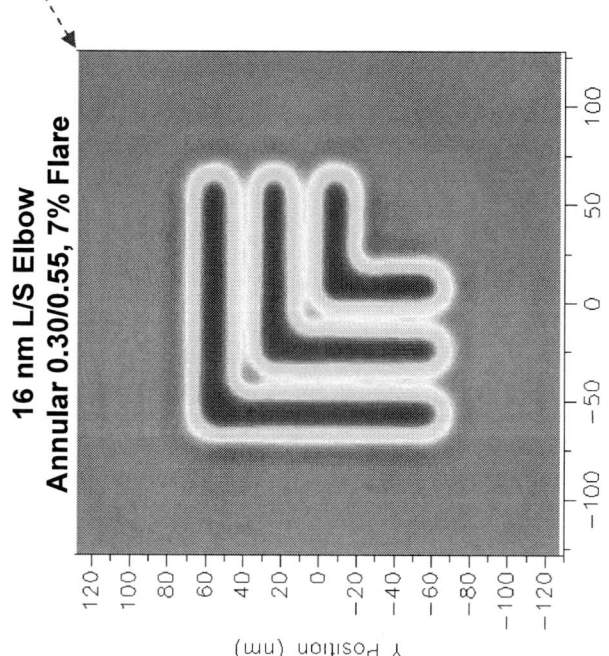

16 nm L/S Elbow
Annular 0.30/0.55, 7% Flare

Partial Coherence Setting

- Grouped and isolated features exhibit iso-coherence points near 12 and 27 nm, respectively

- Larger partial incoherence (σ) improves grouped line contrast for resolution below the lower iso-coherence point.

- Smaller partial incoherence (σ) improves isolated line contrast for resolution below the iso-coherence point.

Grouped Line Contrast

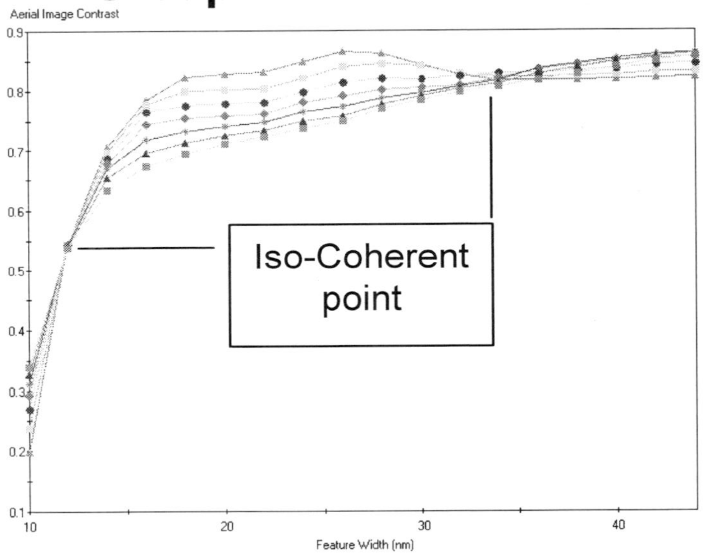

Isolated Line Contrast

Annular

$\sigma_{inner}=0.30$

σ_{Outer}
- 0.5
- 0.55
- 0.6
- 0.65
- 0.7
- 0.75
- 0.8

Iso-Coherent point

Iso-Coherent point

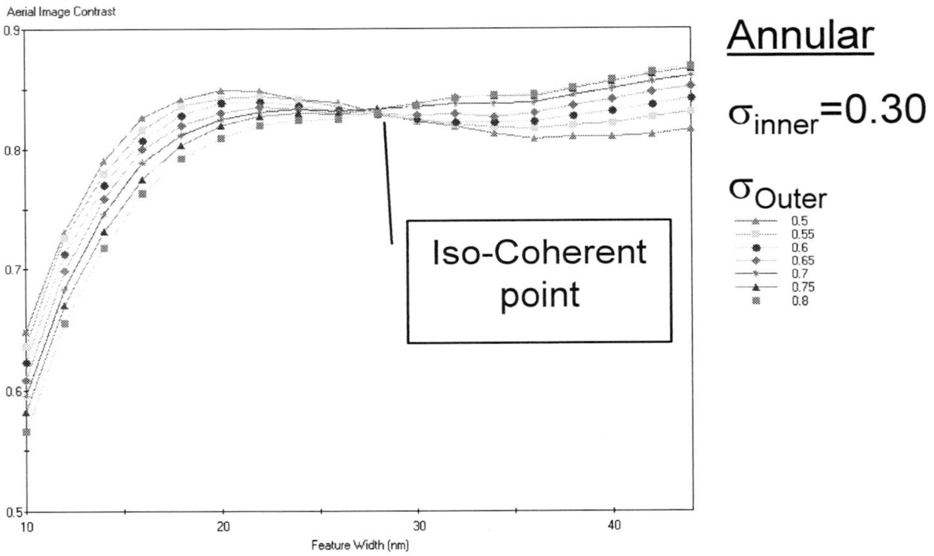

Process Window: equal Line/Space Features at 16nm SEMATECH

Pupil fill: annular 0.35-0.55
Central field point, no flare
16-nm half pitch lines

120 nm DOF with 10% EL

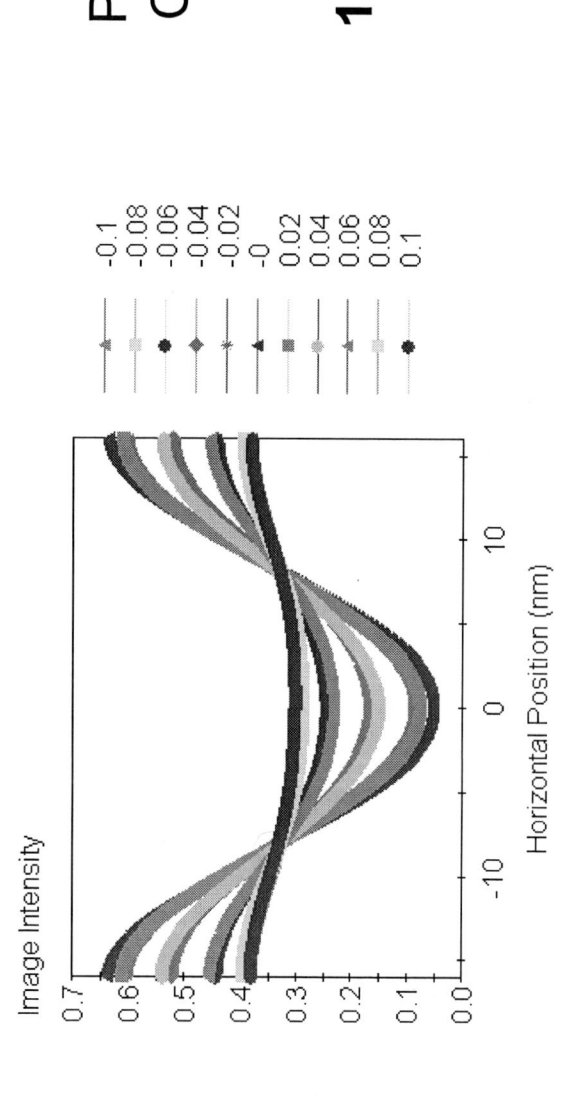

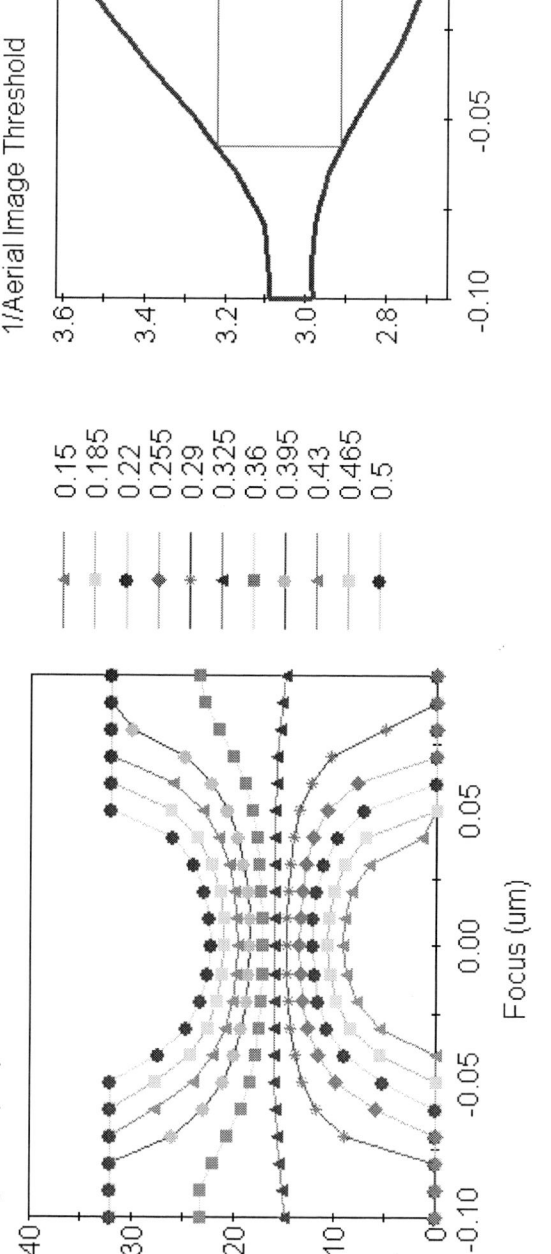

Process Window: equal Line/Space Features at 12nm ⟋ SEMATECH

Pupil fill: annular 0.35-0.55
Central field point, no flare
12-nm half pitch lines

70 nm DOF with 10% EL

Visit our poster for additional
modeling results on the MET2.

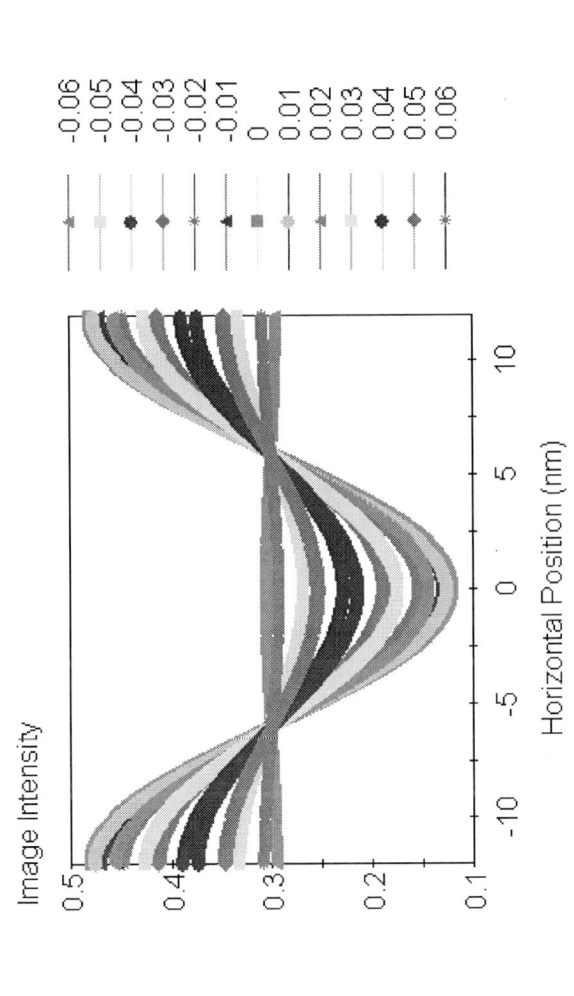

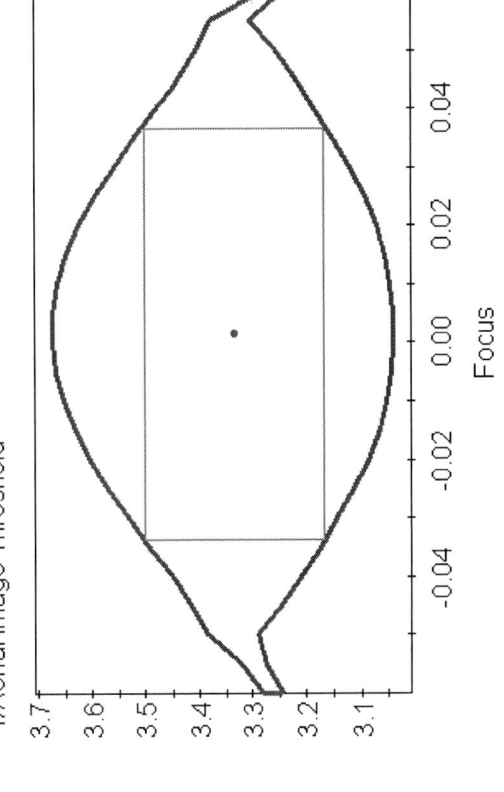

Conclusion

- SEMATECH, Berkeley, and Hyperion started a high NA EUV MET project in Q4 2007.

- The aperture is 0.5, demagnification is 5×, chief ray angle at the mask is 6°, and the diffraction limited field size is ~30×200 microns.

- Lithographic simulations predict a useful process window for 16 nm and 11 nm node research.

- Supplier engagement is starting, and first light is anticipated at the ALS in the 2010 – 2011 time frame.

Overview

1) Review obscured high numerical aperture concept presented at 2006 Litho Forum

2) Pupil dimming – the effect that complicates development of EUVL projection systems at NA > 0.50

3) Discuss 2 strategies for reduction in incidence angles at NA's > 0.60

   a) Reduction of ring field radius

   b) Use of free-form optical surfaces (e.g., surfaces that lack rotational symmetry)

Motivation for studying EUV Extensibility

- Intel supported the foundation of this study back in 2004-2005

- Out of simple intellectual curiosity, Hyperion has extended the work a bit

- Motivating questions:

  1) What are the ultimate resolution limits of EUV?

  2) What projection architectures are required to achieve these limits?

  3) Is the resulting numerical aperture/field size consistent with high volume manufacture?

  4) What new risk factors arise?

Review of Basic Ideas

HYPERION
DEVELOPMENT

"Standard" system enables patterning 0.35<NA< 0.40

Mask

M1
M2
M4
M5
M6
M7
M3
M8
Wafer

1630.0 mm

Reduction = 4x
NA = 0.35
RFW = 2.0mm
Dist. < 0.20nm
Comp. RMS < 25mλ

Mean ray incidence angle
Mask: 8.3° M1: 10.7° M2: 9.1° M3: 5.1° M4: 15.4° M5:
19.4° M6: 2.6° M7: 11.9° M8: 2.9°

Parameter	Values	Notes
Design	8-mirror	-
Obscuration	No	-
Resolution	22.0 nm	$R = k_1\lambda/NA$ (k_1= 0.57)
Numerical aperture	0.35	Maximum 0.42 to 0.45
Field size	22 mm x 2.0 mm	Scanning ring field
Wavelength	13.5 mm	MoSi
RMS Wavefront error	< 25.0 mλ	Field composite
Distortion	< 0.15 nm	Lower possible
Field curvature	< 1.0 nm	No astigmatism or FC
Chief ray angle at mask	8.3°	-

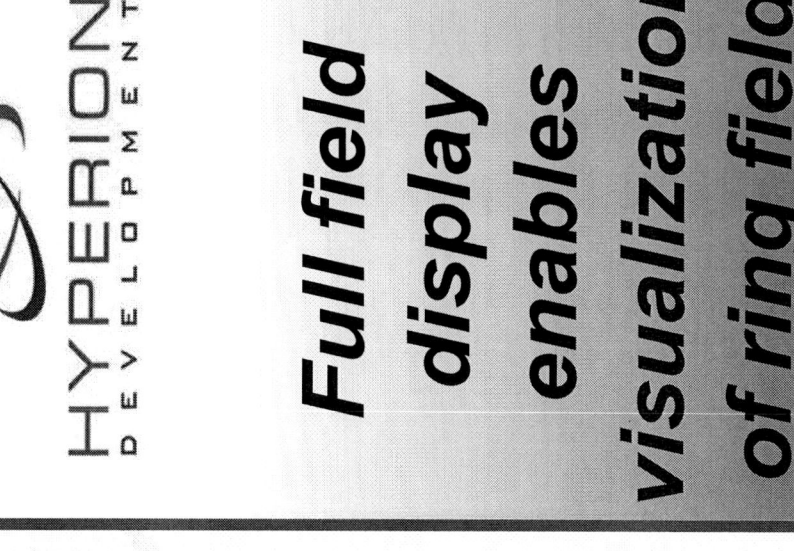

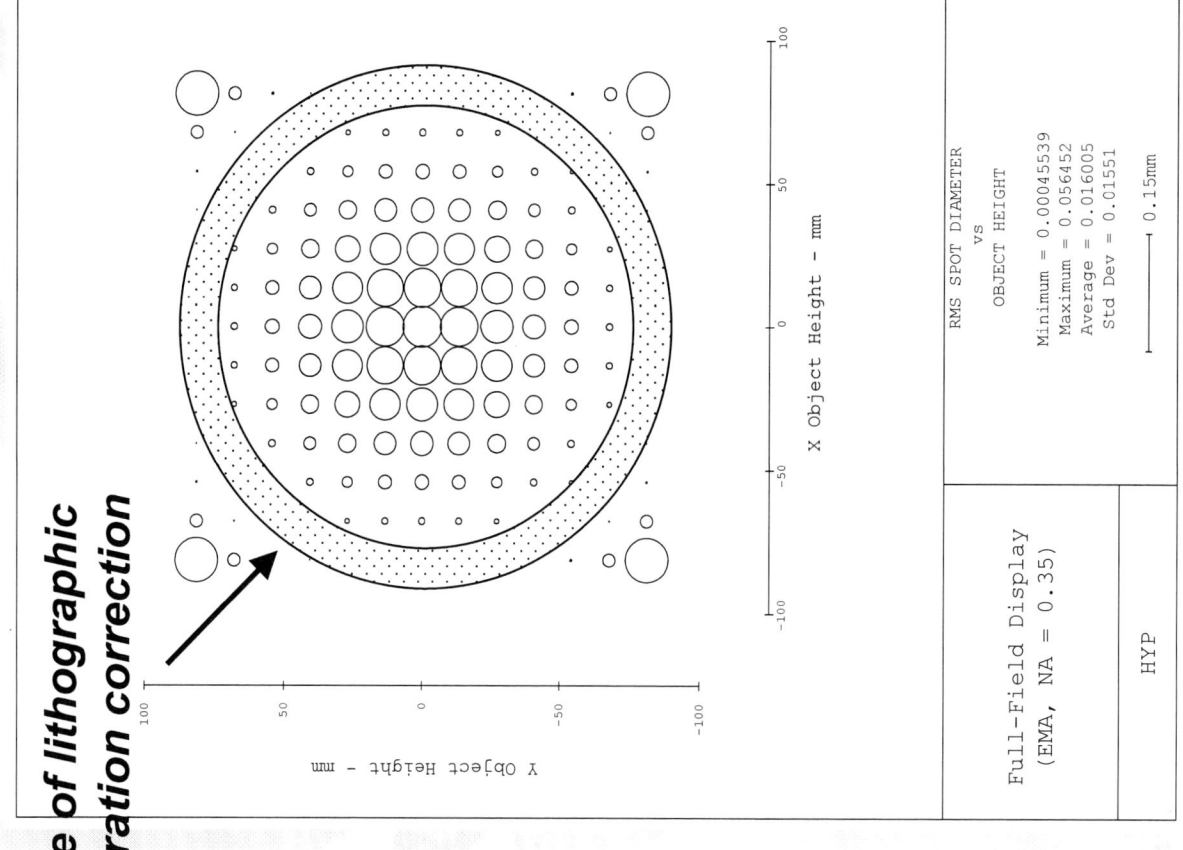

Conceptual 0.53NA 8-mirror system drives resolution

HYPERION
DEVELOPMENT

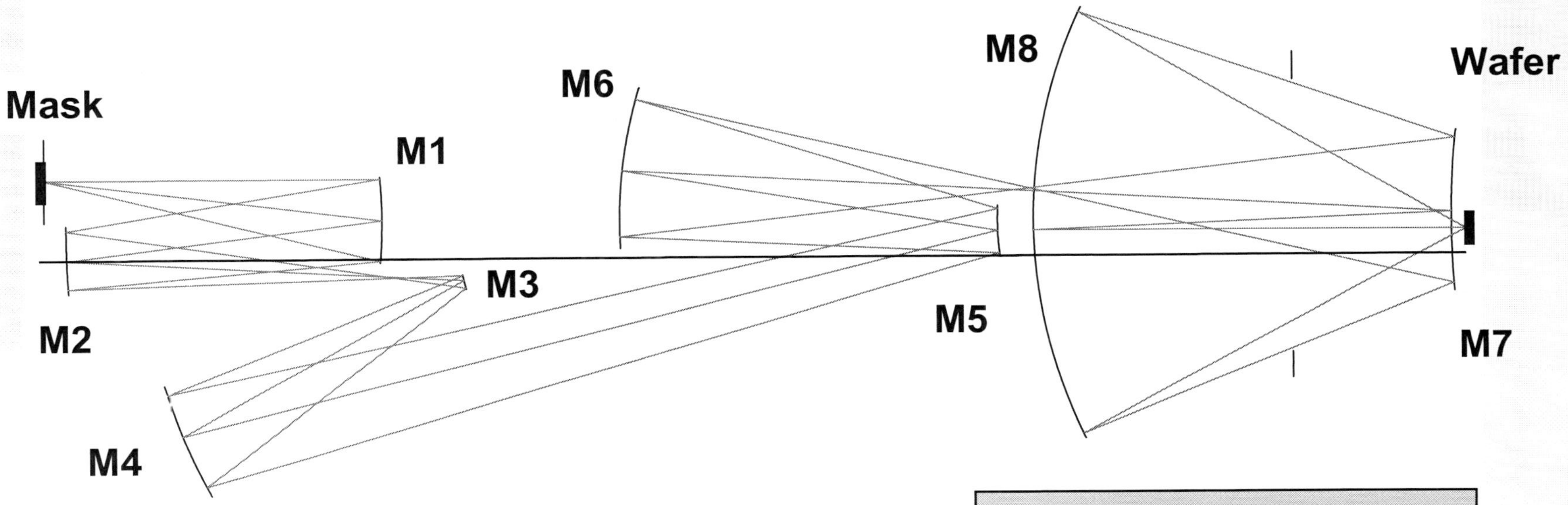

Wafer

M7

M8

M5

M6

M1

M3

M2

M4

Mask

Reduction = 4x
NA = 0.53
RFW = 1.50mm
Dist. < 0.15nm
Comp. RMS < 22mλ

Mean ray incidence angle
Mask: 7.1° M1: 7.2° M2: 5.2° M3: 16.2° M4: 7.5°
M5: 11.9° M6: 3.2° M7: 2.8° M8: 1.2°

1975.0 mm

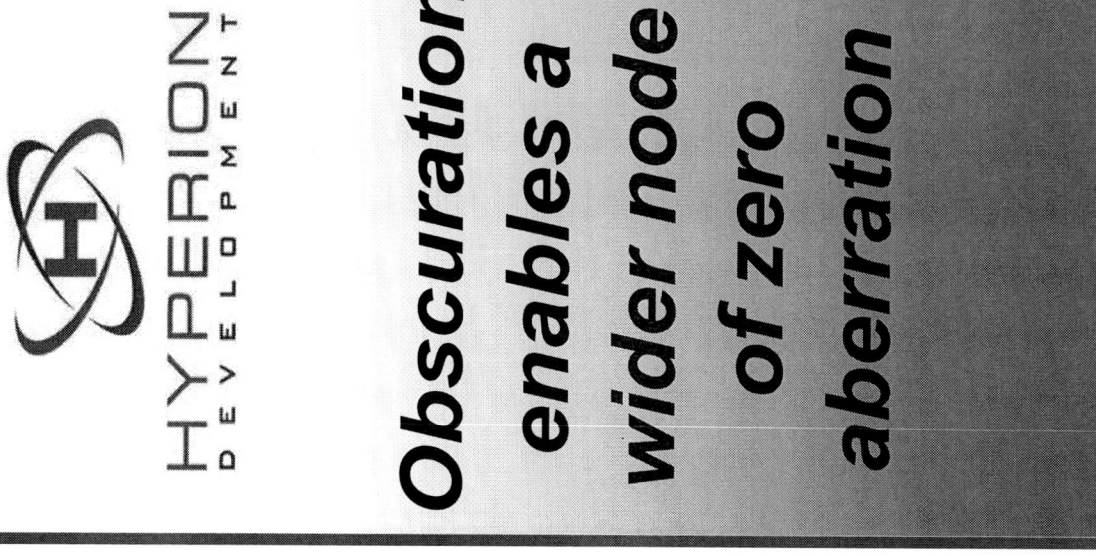

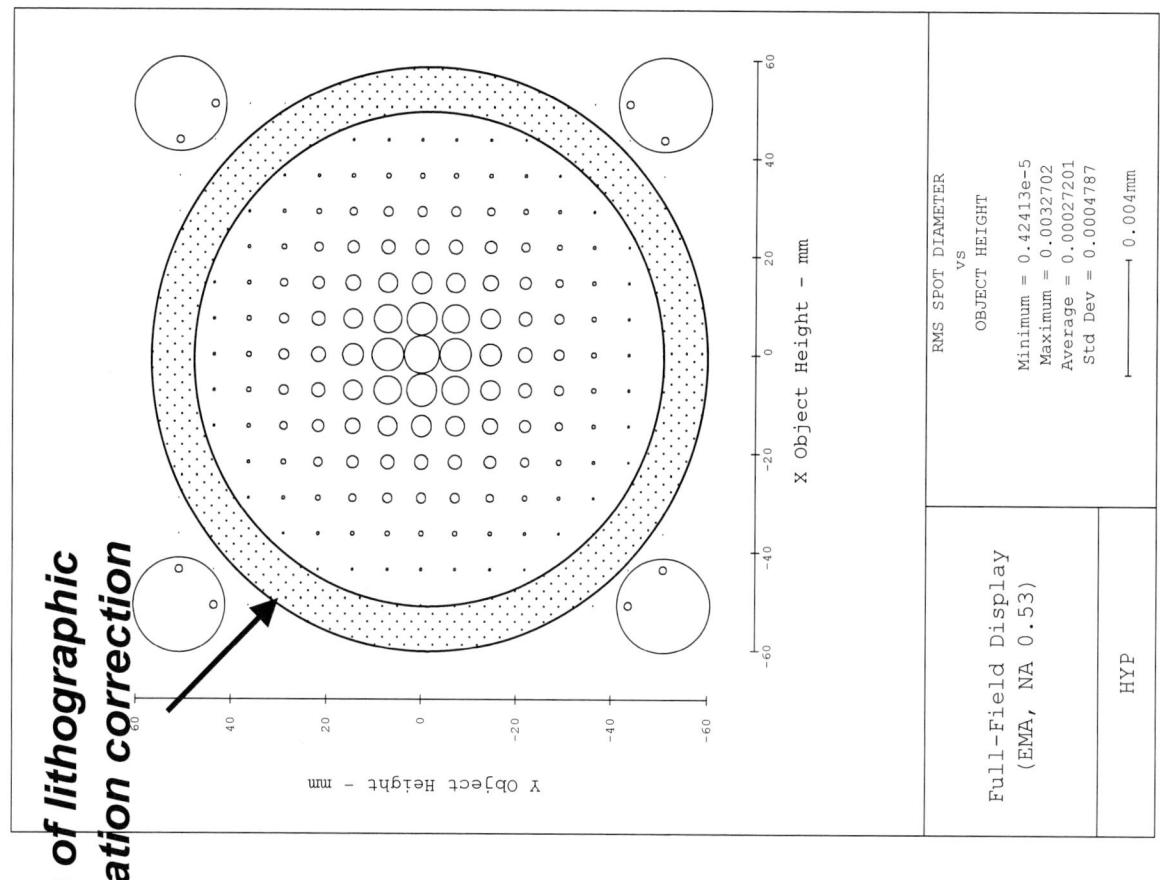

Design minimizes pupil obscuration function

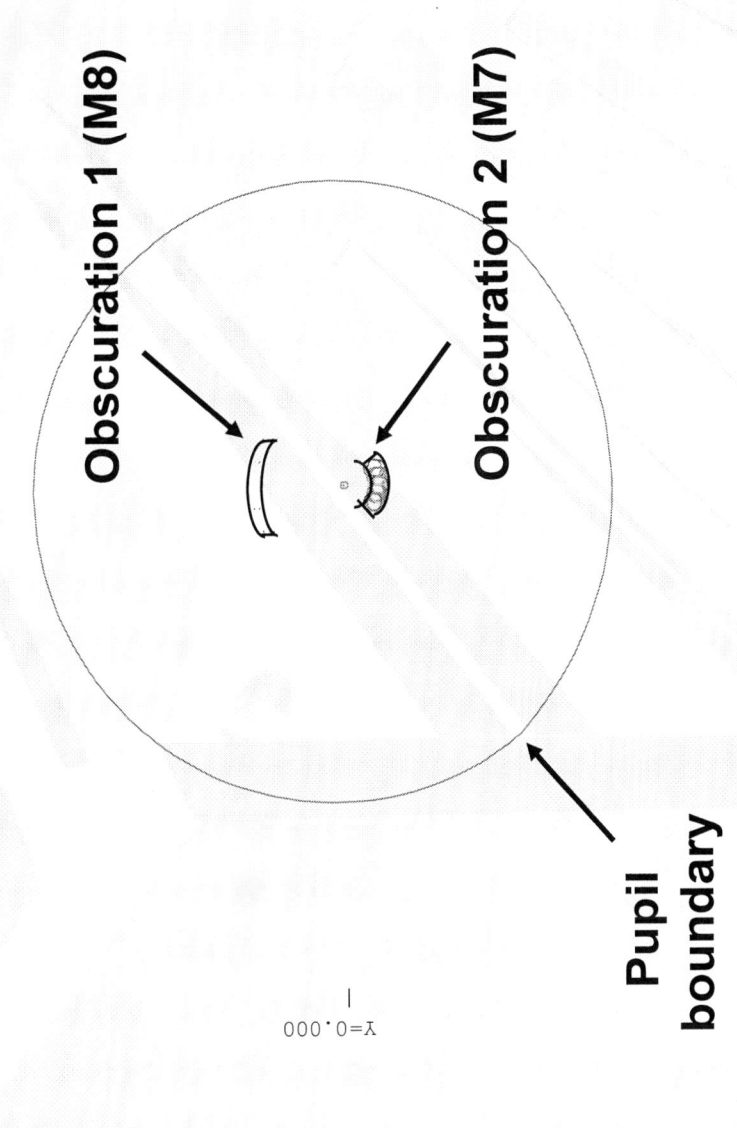

"Pupil Dimming" & The Challenge of NA Scaling

HYPERION
DEVELOPMENT

Mean incidence angle scaling poses design challenges

MoSi Reflectivity Properties

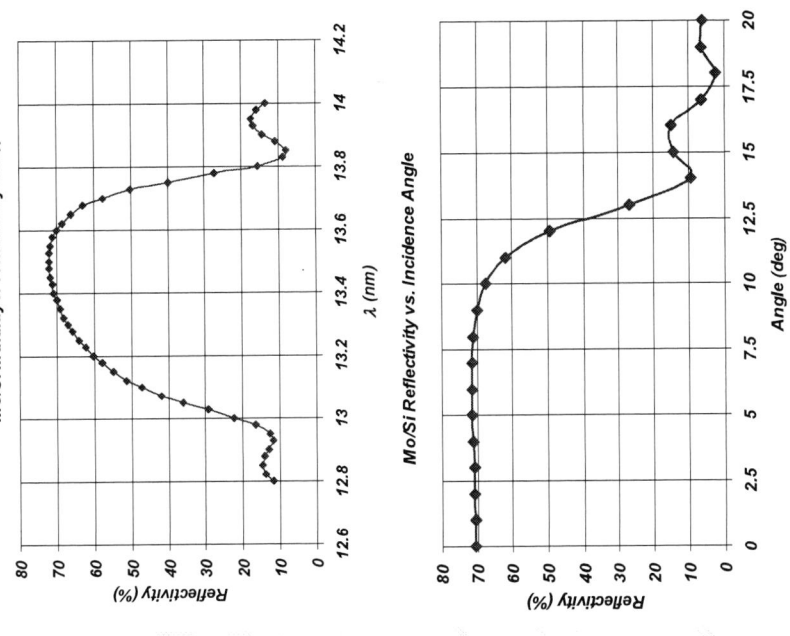

Mean incidence angle scales with NA

Mirror	0.35NA	0.45NA
M1	10.7°	12.0°
M2	9.1°	10.3°
M3	5.1°	5.7°
M4	15.4°	17.4°
M5	19.4°	22.6°
M6	2.6°	3.1°
M7	11.9°	15.3°
M8	2.9°	3.2°

Apodization functions at 0.35NA

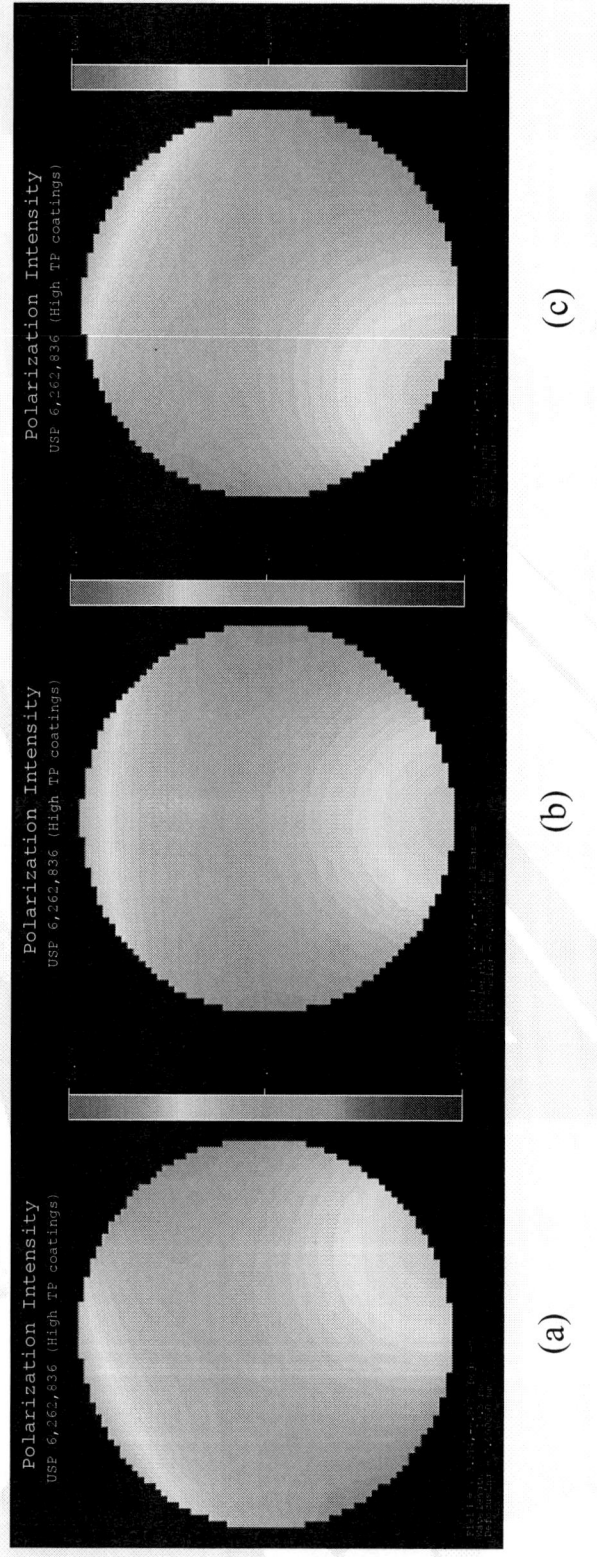

(a)

(b)

(c)

Figure 9. The exit pupil apodization functions of projection system described NA 0.35 show insignificant amplitude errors. The projection system uses a fairly sophisticated multilayer coating design set that effectively maximizes throughput and is process capable using current coating technology. The data is presented across the "smile" starting with leftmost (a), center (b), and rightmost (c) field points, respectively, along the central radius of the arcuate ring field.

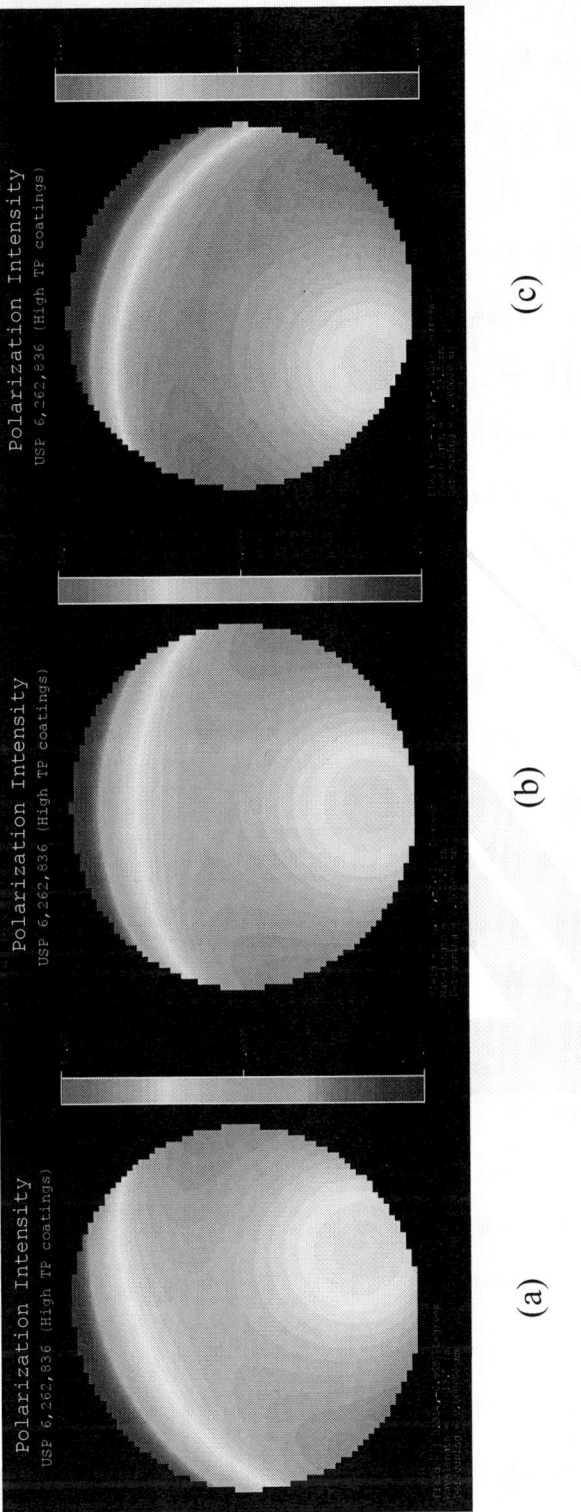

Apodization functions at 0.45NA illustration pupil dimming

Polarization Intensity
USP 6,262,836 (High TP coatings)

Polarization Intensity
USP 6,262,836 (High TP coatings)

Polarization Intensity
USP 6,262,836 (High TP coatings)

(a)

(b)

(c)

Figure 10. Pupil dimming (blue) becomes apparent in as the projection system is scaled from NA 0.35 to NA0.45. Again the projection system uses a fairly sophisticated multilayer coating design set that has been reoptimized, but the result fails to overcome this effect. This egg-shaped pupil causes a loss of effective numerical aperture and an orientation dependent imaging bias. These apodization maps across the exit pupil are again from the leftmost (a), center (b), and rightmost (c) field points along the central ring field radius.

Learning

- High incidence can be accommodated
  - Graded coatings (i.e., Δt vs. R)
  - "Broadband coatings" (chirped multilayer despace)
- Generally trade uniformity vs. transmission
- Rule of thumb is to keep AOI on (m−1) mirrors to less 20 degrees and use the multilayer tricks on the mth mirror

New Thinking

Alternate Thinking

HYPERION
DEVELOPMENT

- Let assume the obscured architecture with all it's problems

- Two options

  - Decrease the ring field radius (RFR) to ~15mm to lower angles of incidence

  - Trade manufacturing complexity to completely eliminate ring field architecture in an attempt to move to a slot field

Conceptual 8-mirror at 0.65NA drives resolution with low AOI's

HYPERION DEVELOPMENT

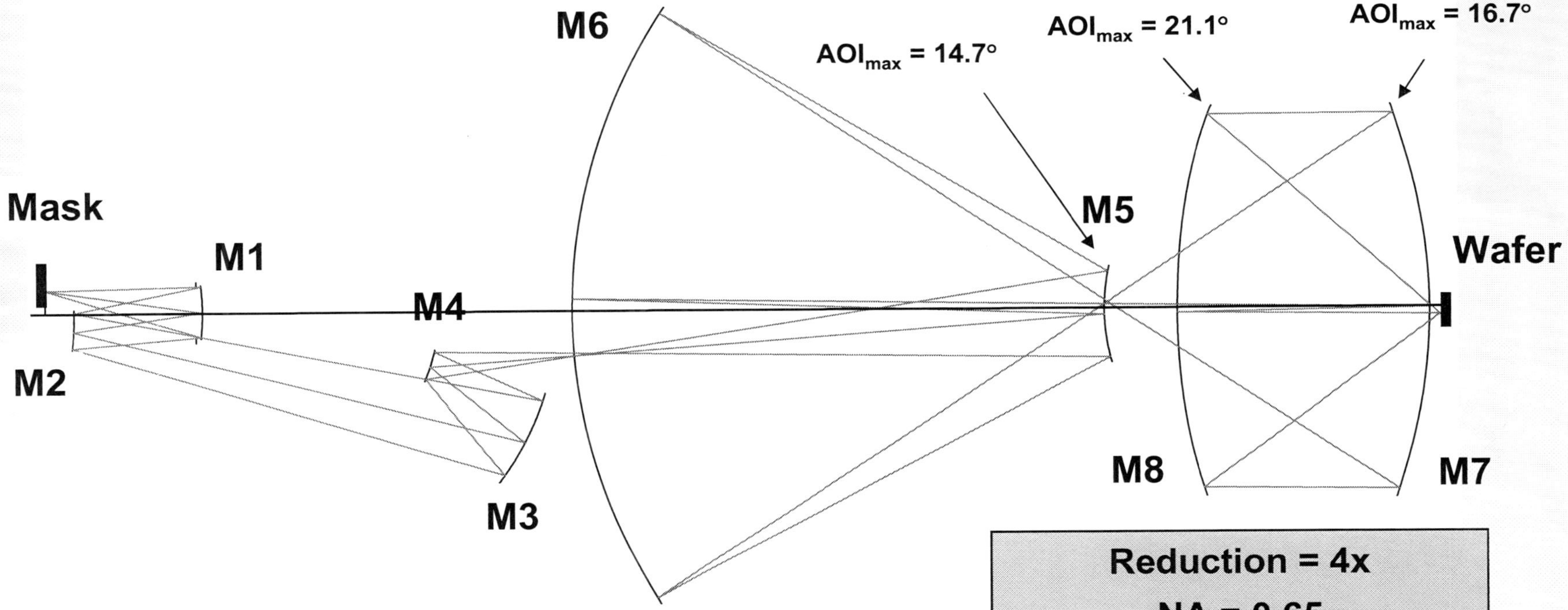

Mask

M1

M4

M2

M3

M6

M5

M8

M7

Wafer

AOI_{max} = 16.7°

AOI_{max} = 21.1°

AOI_{max} = 14.7°

Reduction = 4x
NA = 0.65
FW = 15.0 x 1.50mm
Comp. RMS < 50mλ
R = 9nm HP (k1 = 0.43)

Mean ray angle of incidence (AOI)
Mask: 8.0° M1: 8.3° M2: 11.5° M3: 12.5° M4: 22.5° M5: 3.0° M6: 0.5° M7: 0.9° M8: 0.8°

1812.5 mm

Can we broaden the aberration node to enable a slot field?

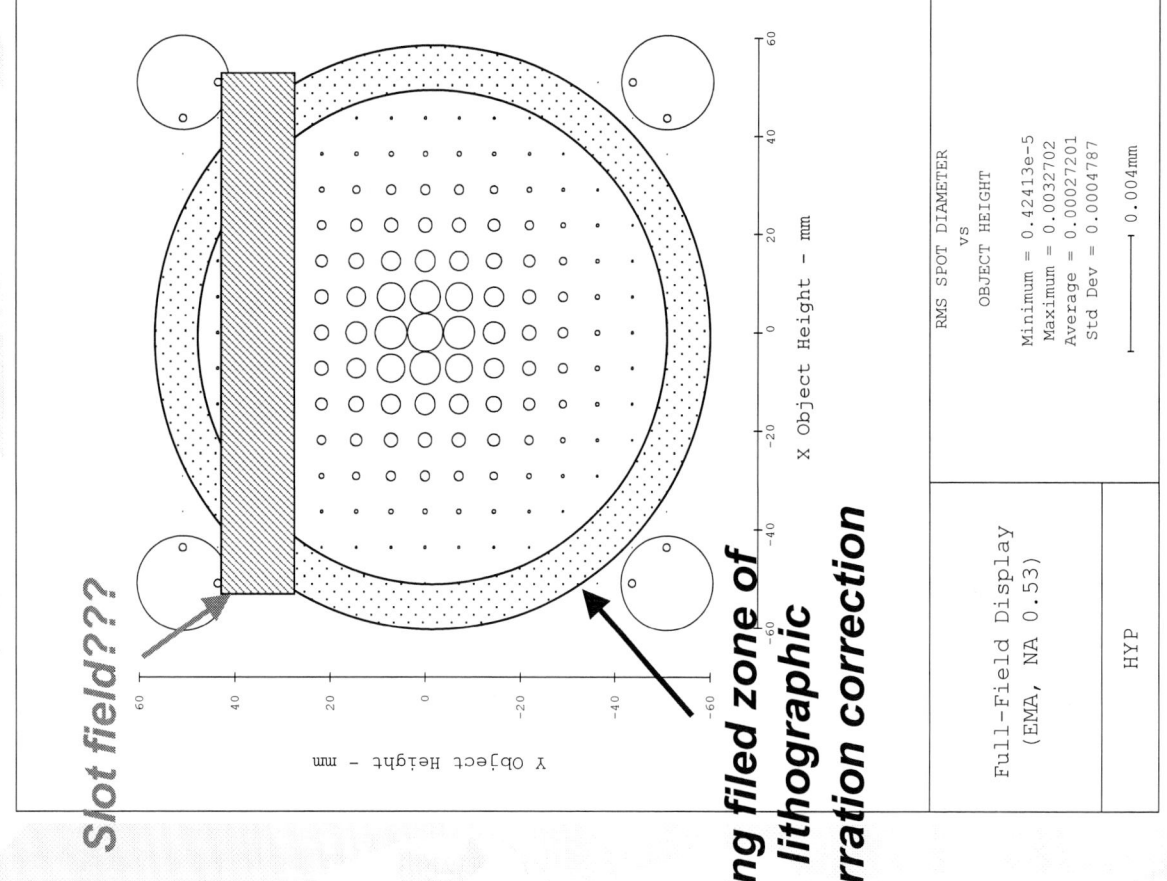

Basis function option 1: "Cosine Surface"

$$f(x,y,z) = z - \sum_{i_y=0}^{n_y-1}\sum_{i_x=0}^{n_x-1} a_{i_x,i_y} \cos\left(i_x\pi\left(\frac{x-x_0}{x_1-x_0}\right)\right)\cos\left(i_y\pi\left(\frac{y-y_0}{y_1-y_0}\right)\right)$$

$$\frac{df(x,y,z)}{dx} = \frac{\pi}{x_1-x_0}\sum_{i_y=0}^{n_y-1}\sum_{i_x=0}^{n_x-1} a_{i_x,i_y}\, i_x \sin\left(i_x\pi\left(\frac{x-x_0}{x_1-x_0}\right)\right)\cos\left(i_y\pi\left(\frac{y-y_0}{y_1-y_0}\right)\right)$$

$$\frac{df(x,y,z)}{dy} = \frac{\pi}{y_1-y_0}\sum_{i_y=0}^{n_y-1}\sum_{i_x=0}^{n_x-1} a_{i_x,i_y}\, i_x \cos\left(i_x\pi\left(\frac{x-x_0}{x_1-x_0}\right)\right)\sin\left(i_y\pi\left(\frac{y-y_0}{y_1-y_0}\right)\right)$$

$$\frac{df(x,y,z)}{dz} = 1$$

Basis function option 2:
"Legendre Surface"

$$f(x, y) = \sum_{n=0}^{k} \sum_{m=0}^{j} A_{nm} P_n(x) P_m(y)$$

where, P_n and P_m are Legendre polynomials

Notes are basis functions

- Based on work in rear projection television

- Cosine surface provides a very smooth surface formulation with no discontinuities

- With there are a number of polynomials orthogonal on real axis, note so many are orthogonal in 2D

- Taking Legendre polynomials as a product $P(x)P(y)$ results in an orthogonal set over square or rectangle

Using FFS, we can morph the ring field

Broad node of zero aberration to enable slot field

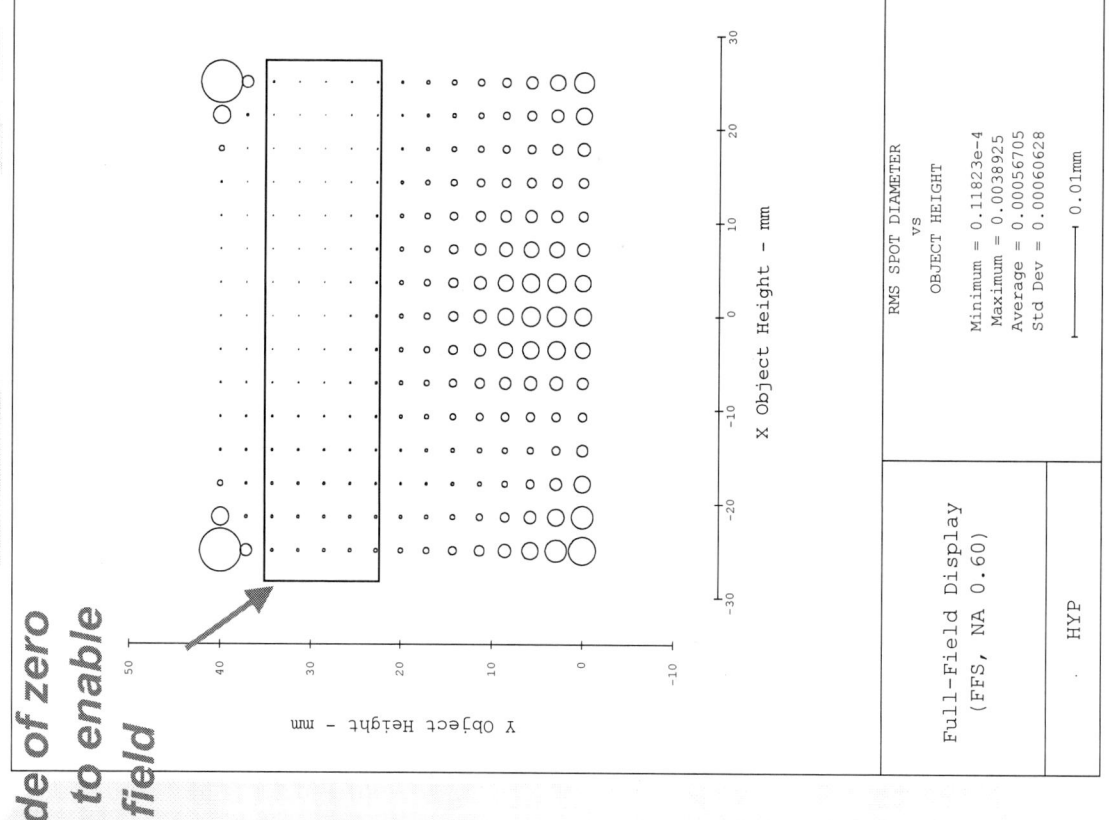

Distortion needs some fine tuning, but close

HYPERION DEVELOPMENT

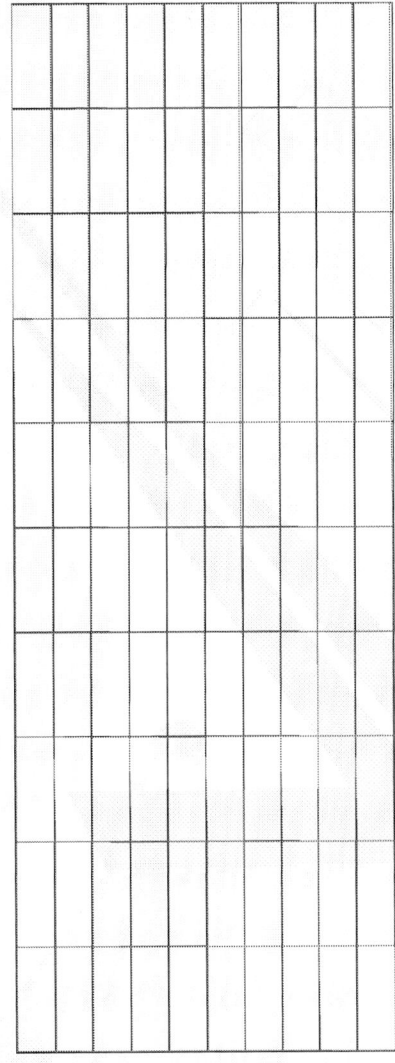

HORIZONTAL FOV

VERTICAL FOV

———— Reference

———— Actual FOV

Conceptual 8-mirror at 0.60NA using free-form surfaces

FSS = Free Form Surface

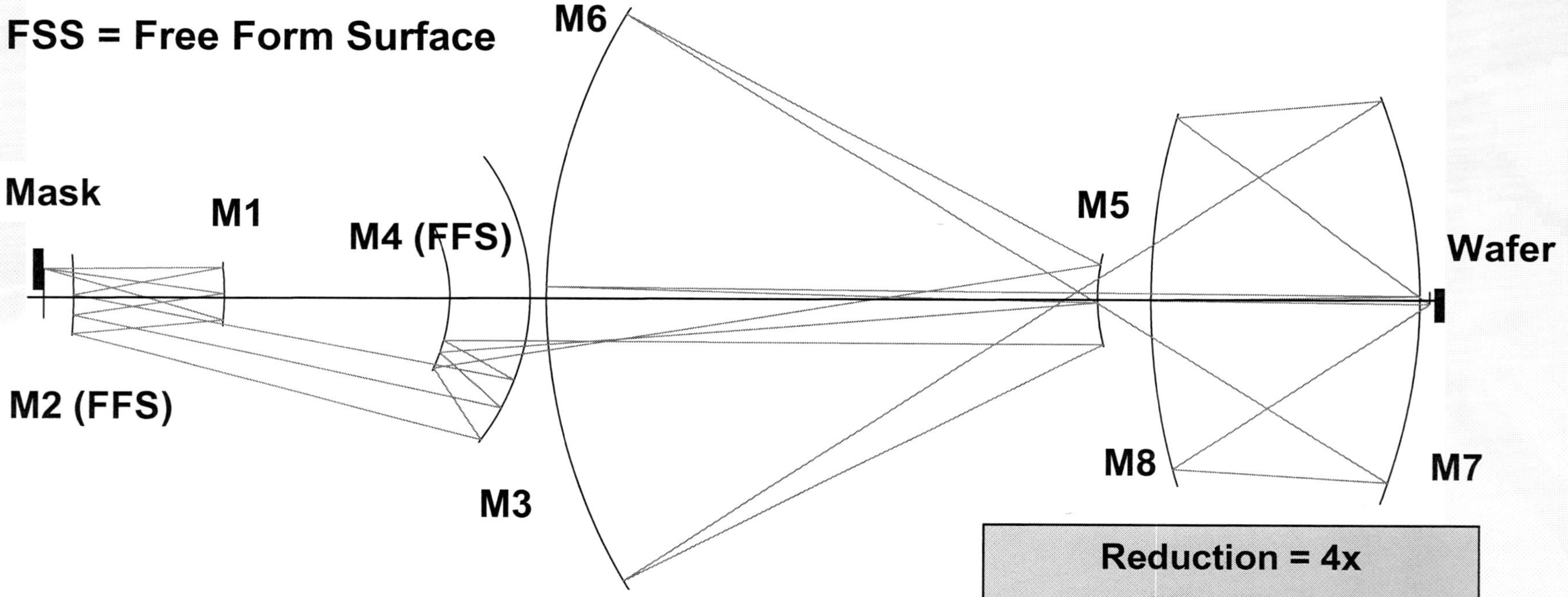

Reduction = 4x
NA = 0.60
FW = 15.0 x 1.50mm
Comp. RMS < 60mλ
Dist = TBD

Mean ray angle of incidence (AOI)
Mask: 8.0° M1: 8.3° M2: 11.5° M3: 12.5° M4: 22.5° M5: 3.0° M6: 0.5° M7: 0.9° M8: 0.8°

1812.5 mm

Mask
M1
M2 (FFS)
M3
M4 (FFS)
M5
M6
M7
M8
Wafer

Conclusions

- EUV projection solutions exist for 22 nm, 16 nm, 11 nm, and 9 nm half-pitch

- DOF becomes stressed as half pitch is decreased

- Unobscured systems can be considered for numerical apertures up to 0.40 with <u>reasonable</u> incidence angles

- Obscured system can be considered for numerical apertures in excess of 0.50 with reasonable incidence angle

- Using of free form surfaces for high NA EUVL projection systems could be a realistic option is high accuracy CGH testing can be implemented

Acknowledgements

- Hyperion would like to acknowledge the funding of Intel Corp. for portions of this work and in particular the valuable contributions of Manish Chandhok and Melissa Shell

- I would also like to acknowledge my co-authors for many fruitful discussions and assistance in preparation of this presentation

Accelerating the next technology revolution

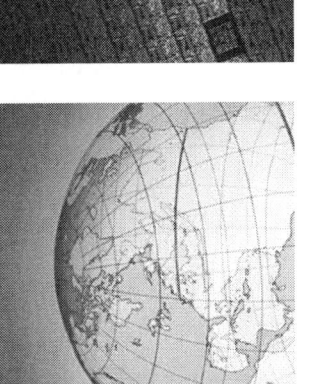
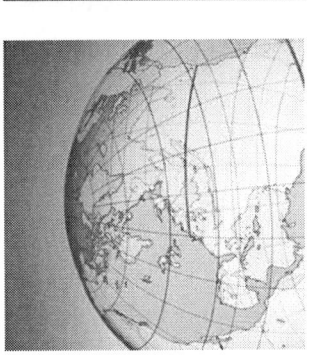

2008 EUVL Symposium
- Closing Address
Lake Tahoe, October 1, 2008

Stefan Wurm

Obert Wood

Patrick Naulleau

SEMATECH

Attendance by Geographic Region

SEMATECH

**United States
129 (41%)**

**Asia / Pacific
117 (37%)**

**Europe
69 (22%)**

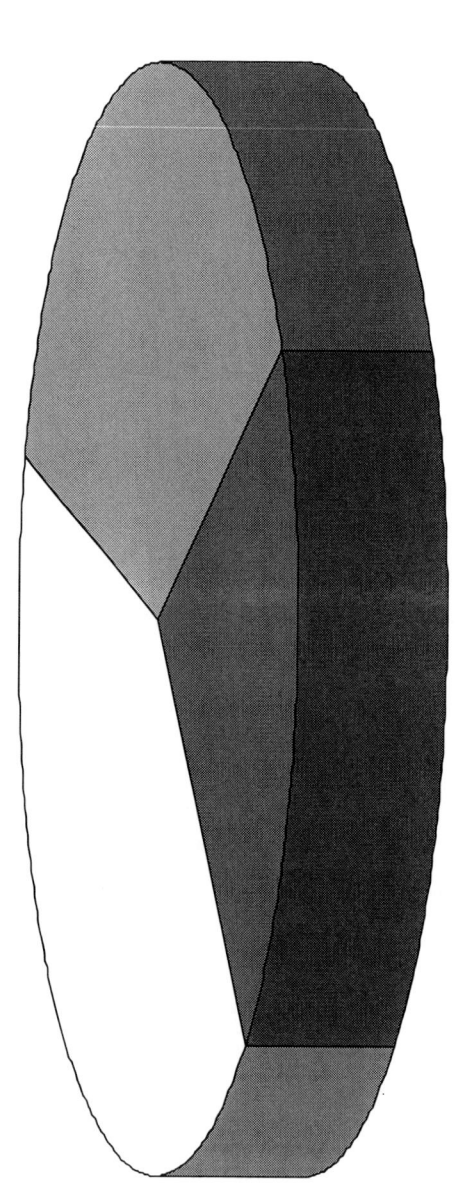

- **315 registered as of September 28**
- 325 registered as of October 1

7th International Program Steering Committee Meeting
Lake Tahoe, October 1, 2008

2008 EUVL Symposium Highlights

Integration / Tool

- Operational Alpha tools support device development

- All major scanner suppliers announced tool roadmaps supporting extendibility of EUV to HVM

- Functional devices made using EUV lithography on full-field chips. EUV demonstrating SRAM yield (45 nm node)

- EUVL used to fabricate electrically functional 32-nm SRAM cells

- EUVL contact hole printing verified for 3x hp nodes to be very cost effective (no OPC)

Source

- Demonstration of fully integrated LPP source collector module with effective mitigation of Sn deposition and ion erosion

- LPP integrated source systems run for 8 hours with 3-4 W power at IF

- DPP collector lifetime of several months proven in alpha systems

- 3X improvement in generated EUV power to 500 W in 2π continuous mode shown in DPP source

Mask

- Electrical device data show much fewer critical defects than expected from mask blank defectivity data (45 nm node data)

- Proof of concept for 67% dual pass transmission pellicle demonstrated for EUVL wavelength

- Fast defect removing cleans process developed and defect smoothing time reduced by 6X to 2.0 hours

- Commercially available mask blanks with 5 defects at 73 nm size

Resist Highlights

- ## Small field tool:
  - 20-nm hp resolution as confirmed with cross sections
  - 30-nm 1:1 contact holes images and cross section SEM confirming resolution

- ## Full field tool:
  - 28-nm half pitch LS resolution and 28-nm 1:1 contact hole resolution (conventional illumination, no OPC)

- ## Resolution / LER / Sensitivity (RLS) Challenge
  - Supplier are addressing the RLS challenge in a systematic way. Good progress towards achieving targets for resolution, LER, and sensitivity.

7th International Program Steering Committee Meeting
Lake Tahoe, October 1, 2008

2008 EUVL Focus Areas

EUV Focus Areas 2003-2007

2003	2004	2005	2006	2007
1. Source power and lifetime including condenser optics lifetime	1. Availability of defect free mask	1. Resist resolution, sensitivity & LER met simultaneously	1. Reliable high power source & collector module	1. Reliable high power source & collector module
2. Availability of defect free mask	2. Lifetime of source components & collector optics	2. Collector lifetime	2. Resist resolution, sensitivity & LER met simultaneously	2. Resist resolution, sensitivity & LER met simultaneously
3. Reticle protection during storage, handling and use	3. Resist resolution, sensitivity & LER met simultaneously	3. Availability of defect free mask	3. Availability of defect free mask	3. Availability of defect free mask
4. Projection and illuminator optics lifetime	▪ Reticle protection during storage, handling and use	4. Source power	4. Reticle protection during storage, handling and use	4. Reticle protection during storage, handling and use
5. Resist resolution, sensitivity and LER	▪ Source power	▪ Reticle protection during storage, handling and use	5. Projection and illuminator optics quality & lifetime	5. Projection and illuminator optics quality & lifetime
6. Optics quality for 32-nm half-pitch node	▪ Projection and illuminator optics lifetime	▪ Projection and illuminator optics quality & lifetime		

2008 EUV Focus Areas

Key Focus Areas	Rank*
Long-term source operation with 100 W at IF and 5MJ/day	1.2
Defect free masks through lifecycle & inspection/review infrastructure	2.2
Resist resolution, sensitivity & LER met simultaneously	2.6
Reticle protection during storage, handling and use#	4.1
Projection / illuminator optics and mask lifetime#	4.7

*) Average of 23 steering committee member votes

#) Still significant concerns

EUV Focus Areas 2003-2008

2003	2004	2005	2006	2007	2008
1. Source power and lifetime including condenser optics lifetime	1. Availability of defect free mask	1. Resist resolution, sensitivity & LER met simultaneously	1. Reliable high power source & collector module	1. Reliable high power source & collector module	1. Long-term source operation with 100 W at IF and 5MJ/day
2. Availability of defect free mask	2. Lifetime of source components & collector optics	2. Collector lifetime	2. Resist resolution, sensitivity & LER met simultaneously	2. Resist resolution, sensitivity & LER met simultaneously	2. Defect free masks through lifecycle & inspection/review infrastructure
3. Reticle protection during, storage, handling and use	3. Resist resolution, sensitivity & LER met simultaneously	3. Availability of defect free mask	3. Availability of defect free mask	3. Availability of defect free mask	3. Resist resolution, sensitivity & LER met simultaneously
4. Projection and illuminator optics lifetime	▪ Reticle protection during storage, handling and use	4. Source power	4. Reticle protection during storage, handling and use	4. Reticle protection during storage, handling and use	• Reticle protection during storage, handling and use
5. Resist resolution, sensitivity and LER	▪ Source power	▪ Reticle protection during storage, handling and use	5. Projection and illuminator optics quality & lifetime	5. Projection and illuminator optics quality & lifetime	• Projection / illuminator optics and mask lifetime
6. Optics quality for 32-nm half-pitch node	▪ Projection and illuminator optics lifetime	▪ Projection and illuminator optics quality & lifetime			

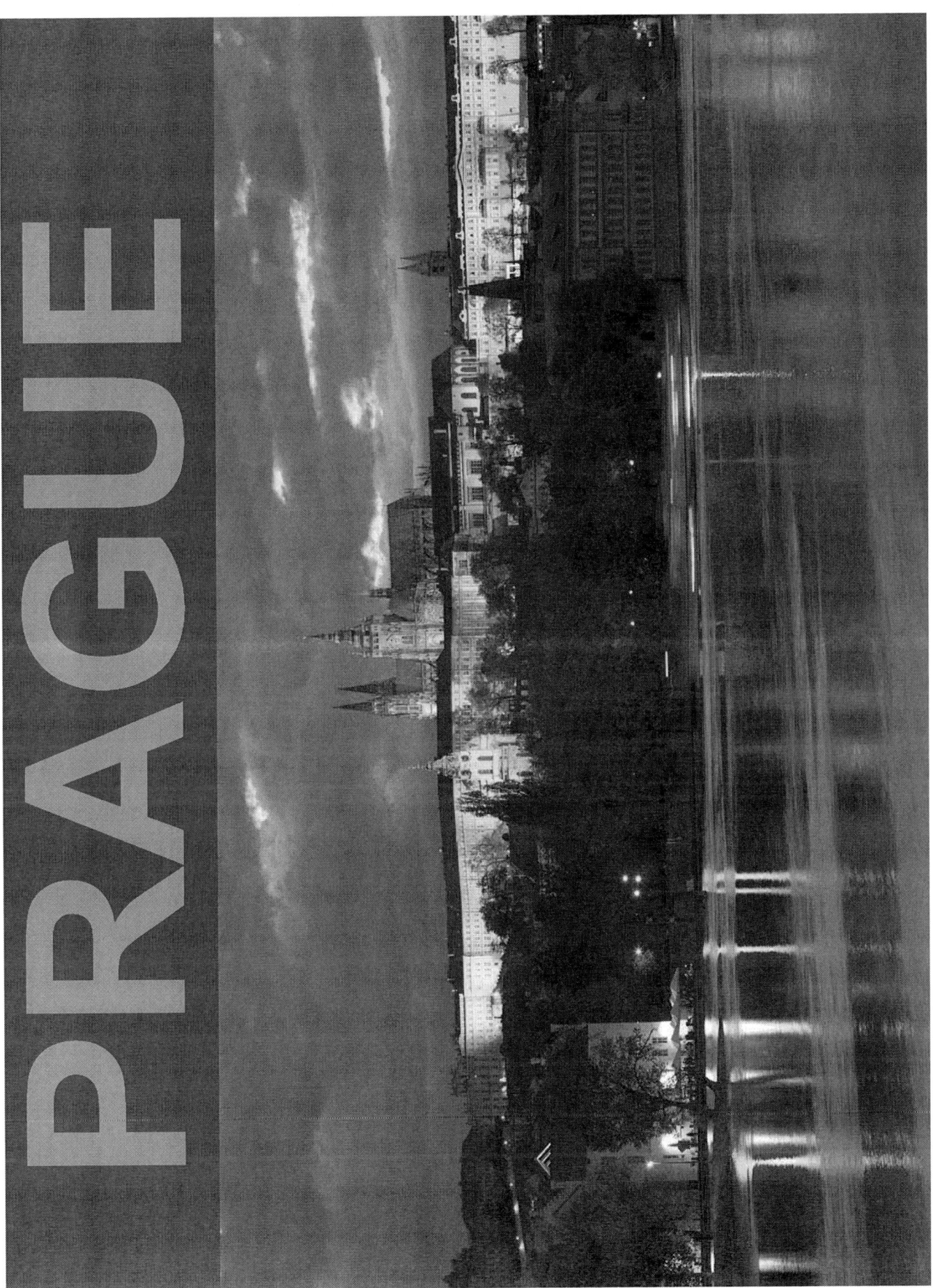

PRAGUE

EUVL
EXTREME ULTRAVIOLET LITHOGRAPHY

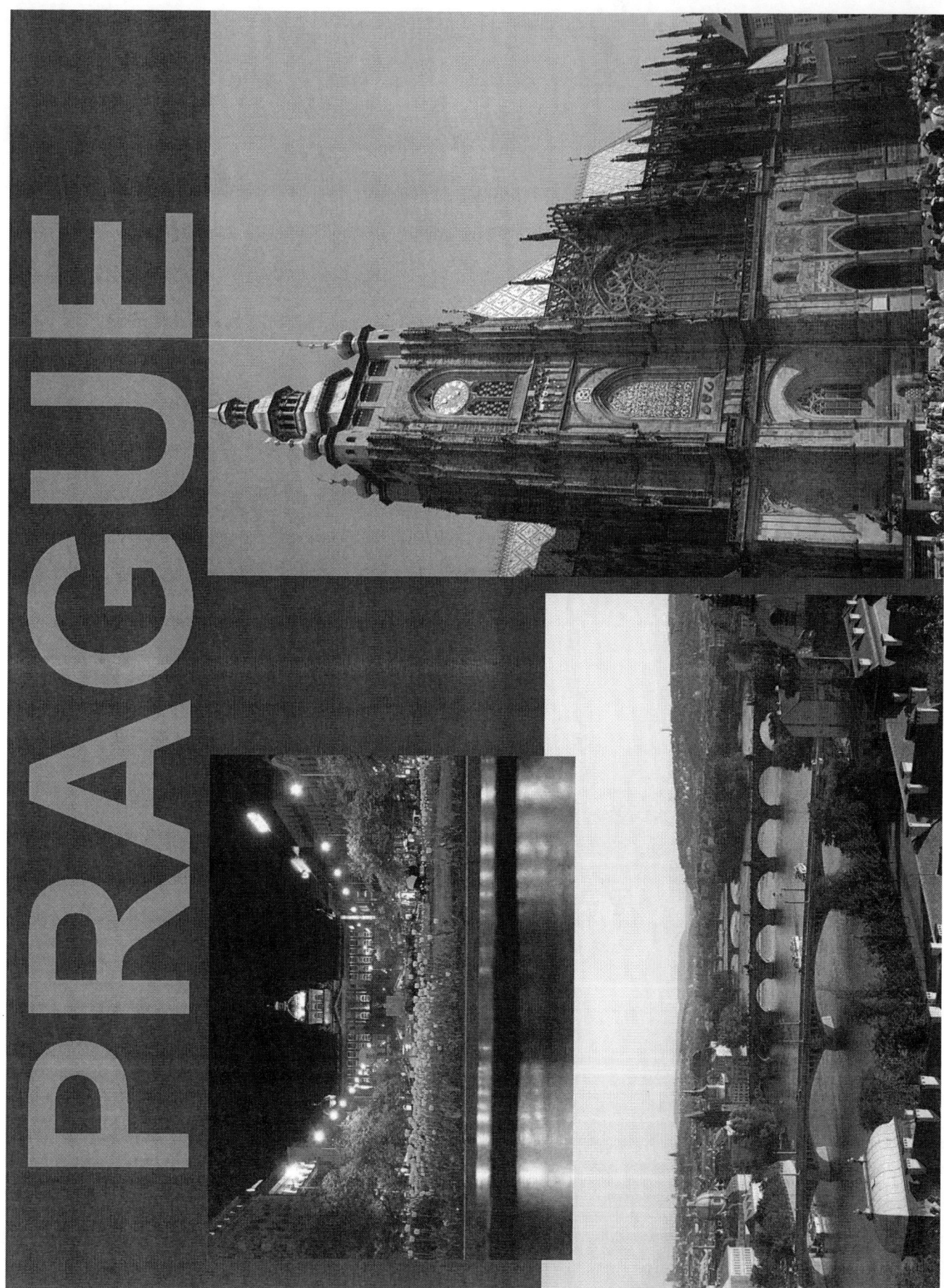

Clarion Congress Hotel Prague****

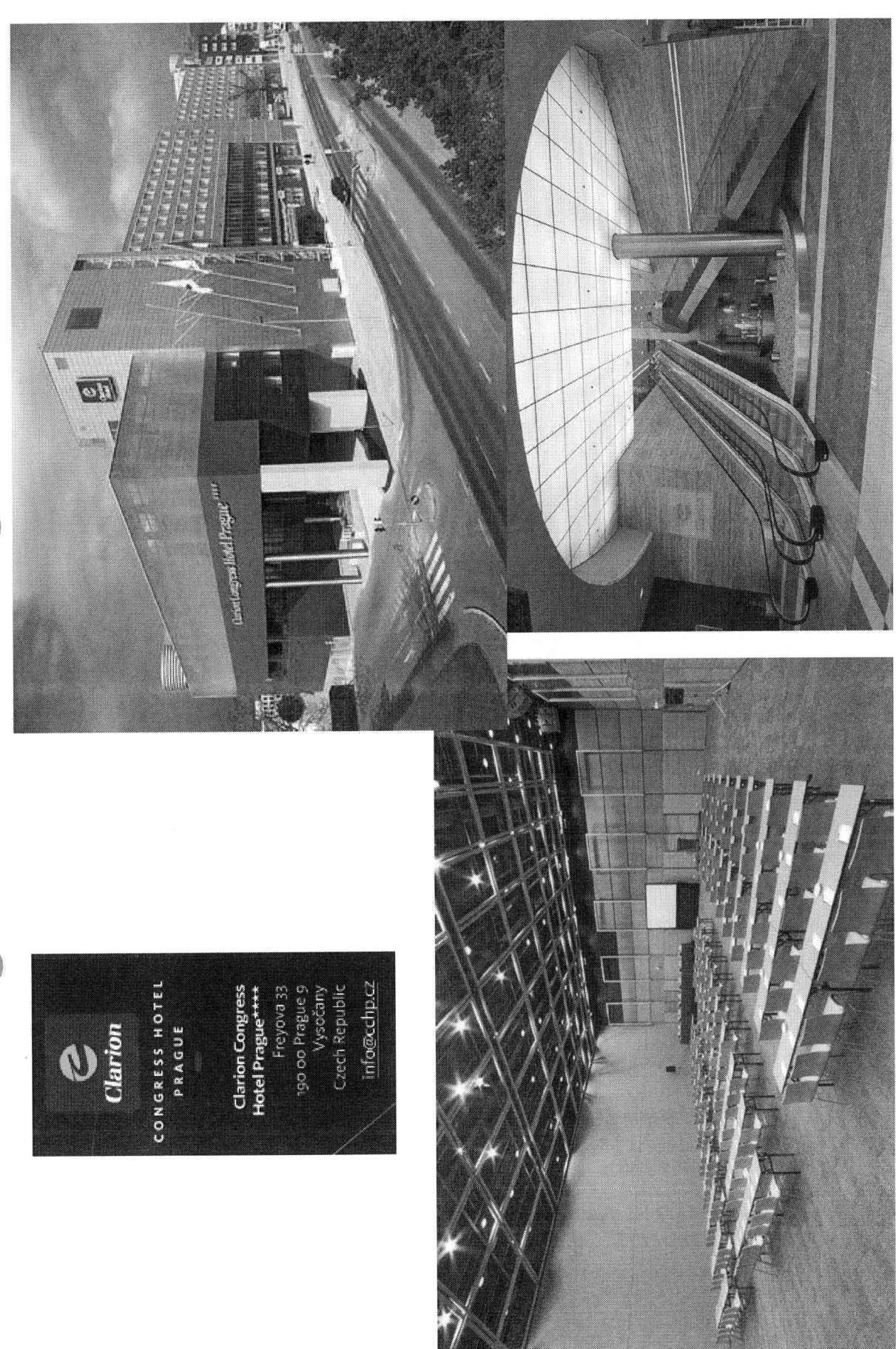

Clarion Congress Hotel Prague**
Freyova 33
190 00 Prague 9
Vysočany
Czech Republic
info@ccchp.cz

Clarion Congress Hotel Prague****

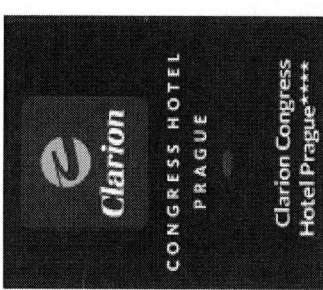

2009 EUVL Symposium

Sunday 18 October
-
Friday 23 October 2009

Thank You!!!

- Program Committee (Obert, Patrick, Nishiyama-san, Winfried)
- All paper/poster presenters
- Steering Committee & Session Chairs
- Audio Visual (Mario, Randall)
- Logistic Support (Beverly, Jill, Jessica, Marcia, Marcy, Lisa, Kris, Kristy,….)

Thank You

Stefan Wurm (SEMATECH / AMD) - Symposium Chair
Ichiro Mori (Selete / Toshiba) – Symposium Co-Chair
Rob Hartman (ASML) - Symposium Co-Chair

Ultra-high resolution extreme ultraviolet lithography by incoherent to coherent conversion

Christopher N. Anderson[1*] and Patrick P. Naulleau[2]

Why EUV IL?

Photoresist is still a major challenge for EUV.

Resist development is hampered by the lack of availability of high-resolution exposure tools.

Interference lithography (IL) is a potential pathway to the development of low-cost exposure tools for resist testing.

If synchrotrons were easy to come by...

If powerful spatially coherent EUV sources were readily available, EUV IL tools would already exist in very major resist development lab around the world.

When you have spatial coherence, IL is straightforward [1]. Amplitude division coherent IL tools routinely demonstrate sub-50-nm patterning [2]. Unfortunately, stand-alone coherent EUV sources are not mature enough to support the rapid development of coherent IL tools.

Incoherence ≠ game over

White light interference has been done for years. A common configuration for generating structured patterns is the achromatic two-grating interferometer [3].

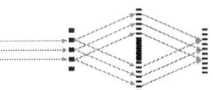

Light is split by a grating and recombined by a second grating with twice the spatial frequency of the first.

Structured patterning with two different low-coherence approaches

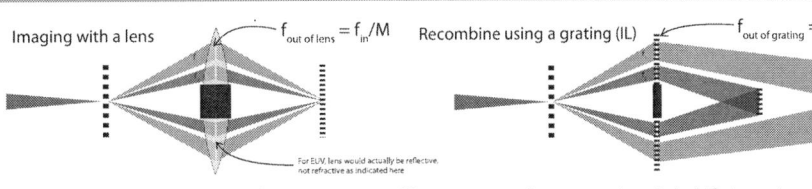

Illumination:
Kohler illuminator- monochromatic with a full cone of angles defined by the illumination NA.

Object:
Stacked sinusoidal transmission gratings with spatial frequencies f_1 and f_2 ($f_1 > f_2$). Zero order is not shown because it is blocked at the recombining lens/grating. The diffracted rays from grating f_1 (f_2) are labeled f_1 (f_2) and are colored pink (purple).

Imaging with a lens $f_{out\,of\,lens} = f_{in}/M$

The local curvature of the lens **scales** incoming spatial frequencies. Diffracted rays from different object spatial frequencies converge to one point in one z plane. The perfect lens images all object spatial frequencies.

Recombine using a grating (IL) $f_{out\,of\,grating} = f_{in} + mf_{grating}$

Constant grating pitch **shifts** incoming spatial frequencies. Diffracted rays from different object spatial frequencies do not converge to one z plane. The grating images only one object spatial frequency to any particular z plane.

IL: robust against grating noise

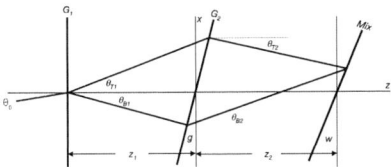

Zero LER aerial image (almost)

For IL, can view incoherent aerial image as a linear superposition of coherent interference patterns from each spatial carrier within the NA of the illumination cone.

Each illumination angle interacts with a different area of the grating - no two input angles produce the same coherent interference pattern.

Through illumination incoherence, line imperfections in gratings are completely homogenized in the final

The second grating

The second grating in the incoherent IL configuration functions like an achromatic cylindrical lens with a focal length that changes with object spatial frequency.

Serving this function, the second grating only images one object spatial frequency to the wafer; all other spatial frequencies are defocused at the wafer (their fringe contrast vanishes) and they contribute to a DC offset or flare.

This is how zero LER aerial images are produced.

Grating tilt, depth of focus, and alignment specs

In practice it will be difficult to maintain parallelism between the gratings and the wafer. Is fringe contrast sensitive to small parallelism errors in the direction of grating shear?

The depth of focus (DOF) of the parallel incoherent two-grating interferometer is inversely proportional to the patterned spatial frequency and the NA of the illumination [3].

$$DOF = \cos^2\theta_0 * (NA * f_{patterned})^{-1}$$

Rescent work [4] has shown that for on-axis transmission implementations, the tilt angle g between the gratings must satisfy:

$$g < (f_{patterned} * z_1 * NA^2)^{-1}$$

For off-axis (at angle θ_0) reflection-based implementations, the spec is given by:

$$g << \cos^2\theta_0 * (4NA * f_{patterned} * z_1\tan\theta_0)^{-1}$$

An EUV implementation with z_1 = 20 mm, NA = 4°, and $1/f_{patterned}$ = 30 nm gives:

g < 30 µrad   (on-axis transmisson-based)
g < 500 nrad   (θ_0 = 30° reflection-based)

Off-axis implementations should be avoided in practice.

Grating rotation

Rotation between gratings introduces a shear in the direction perpendicular to the grating lines that changes with illumination angle and color.

Shear must be less than the illumination lateral coherence length. With NA and z specs above, β < 7 µrad.

Multipitch patterning

Imaging optics clearly support multipitch patterning; the pitch range is set by the NA of the optic, the central obstruction and the pupil fill.

Incoherent IL also supports multipitch patterning; gratings can be stacked in the direction perpendicular to grating shear.

Recombining grating must also be stacked with twice the spatial frequency of the first (object)grating.

Summary

Incoherent IL naturally spatially filters / homogenizes all unwanted features in object; aerial image has almost zero LER. This is **not** true for coherent IL.

Perfect gratings are not required for incoherent IL. In fact, incoherent IL works quite well with high LER gratings.

On-axis transmission-based incoherent IL tools are feasible but present several significant fabrication challenges.

DOF similar to microfield exposure tools can be used for contrast studies without double exposure.

References

1] M. Wei, D.T. Attwood and I.K. Gasichon, JVSTB 12, 3661 (1994)
2] H. Solak et al. Microelectronic Engineering, Volume 67-68 , 56 - 62 (2003)
3] E.N. Leith and B.J. Chang Applied Optics Vol. 12 1957-1963 (1973)
4] Y.S. Cheng, Applied Optics Vol. 23 3057 - 3059 (1984)
5] C. Anderson and P. Naulleau, Applied Optics Vol. 47 No. 9 1327 - 1335 (2008)

Affiliation
1] Applied Science & Technology program, UC Berkeley
2] Center for X-ray Optics, LBNL, Berkeley, California

Contact* / Visit
cnanderson@berkeley.edu
www.ocf.berkeley.edu/~cnanders (electronic version available)

MOSAIC— a new way to measure optical aberrations.

Christopher N. Anderson[1*] and Patrick P. Naulleau[2]

A need

Accurate characterization of the aberrations in high numerical aperture (NA) EUV projection optics is critical for EUVL.

Existing at-wavelength metrologies (PSPDI [1], LSI [2]) will face significant engineering challenges in migrating to NA > 0.3.

Non-EUV metrologies are easier to implement at higher NA, however they cannot probe phase effects in EUV multilayers.

Existing metrologies can not characterize EUV projection optics in their final lithographic printing configuration.

What is MOSAIC?

MOSAIC: Metrology of optical system aberrations by incoherent curvature sensing.

Why MOSAIC?
It does not require high spatial coherence.
No reference wave.
The optic is characterized in its working environment.
At-wavelength; captures phase effects in EUV multilayers.
Measures wavefront, not point-spread function.
Easy to implement in EUVL projection tools.
Scalable to any NA.

A question.

Assume that we can control the location where an optic is probed during image formation, i.e., by illuminating the object with a low-NA source at a prescribed incidence angle.

Does measuring the z position of best focus as a function of pupil probe location tell us anything about the aberrations in the optic?

Example probe locations

The answer turns out to be yes.

How?

Any two dimensional function W(x,y) confined to the unit circle or a subset of the unit circle is uniquely determined[1] by adequate[2] sampling of the second derivatives of the function in two orthogonal directions, i.e., x and y.

1. Up to all Zernike polynomials with non-vanishing second derivatives: $Z = \rho\cos(\theta)$ [x], $Z = \rho\sin(\theta)$ [y], and $Z = \rho^2\sin(2\theta)$ [xy].

2.) There must be enough samples in (x, y) or (ρ, θ) to capture all spatial frequency content of the highest order Zernike polynomial present in the function. For max radial order n, this requires ceil(n/2) radial samples and 2(n-2) + 1 azimuthal samples. I.E., including all terms up to Z36 (n = 7) requires 4 radial samples and 11 azimuthal samples.

Curvature + least squares ⟶ function recovery

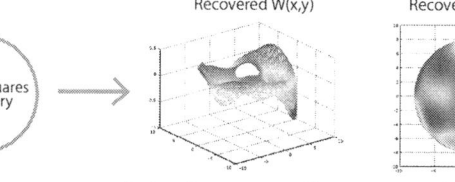

Once the second derivatives of a function W(x,y) are adequately sampled in two orthogonal directions, a least-squares Zernike decomposition [3] can be used to recover the original function up to all Zernike terms with non-vanishing second derivatives.

The least-squares recovery method produces a system of linear equations that can be solved using standard techniques. Assuming that the samples of the second derivatives are accurate, the recovered function perfectly resembles the original one[1].

Speedy Gram Schmidt

When the polynomial functions, i.e., Zernike polynomials, used in the wavefront decomposition are not orthogonal over the discrete sample space (and they won't be) a modified Gram Schmidt Orghogonalization [3] can be used to efficiently solve the system of linear equations.

Efficient generation of the Zernike polynomials and their second derivatives [4] enables rapid construction of an orthonormal basis out to any order Zernike polynomial, ensuring that the least-squares decomposition does not alias contributions from higher-order terms in the function under test.

Obtaining the second derivatives

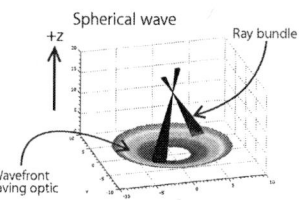

Geometrical rays, by definition, propagate perpendicular to the wavefront or surface of constant phase. The longitudinal (z) distance required for all rays within a small ray bundle to converge (focus) is related to the local slope and curvature (second derivative) of the wavefront. If the local slope can be reasonably approximated, and the focus distance accurately measured, the second derivative can be obtained.

Two directions

By imaging a pair of diffraction gratings with orthogonal grating lines and determining the plane of best focus for each grating, the second derivatives in orthogonal directions can be obtained. If an additional pair of gratings rotated by 45° are used, the 45° astigmatism term can be obtained while acquiring a second independent measurement of all other Zernike coefficients.

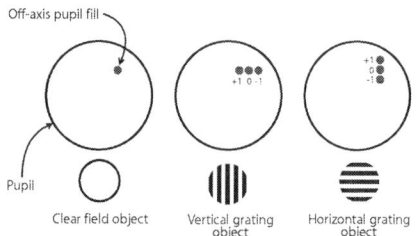

Grating spatial frequency determines the spread of the diffracted orders at the pupil; it needs to be small enough that there is minimal overlap between neighboring probe zones.

Get it all in one shot

Holographic masks enable neighboring features of the object to be imaged by different localized sections of the optic.

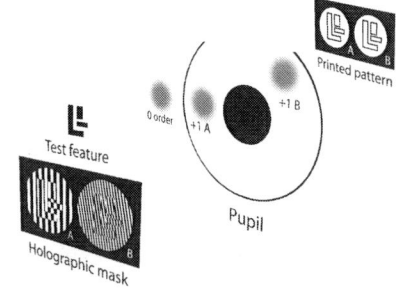

The holographic mask contains a test feature encoded on two different spatial carriers A and B. Off-axis illumination steers the zero order (undiffracted) light outside the pupil. The + 1 order diffracted light from carriers A and B (which contain the test feature information in phase) probe the imaging optic at different locations and are imaged to the detector (wafer). An outline of the test feature is observed at the detector.

MOSAIC is almost ready

A visible-wavelength side-by-side comparison between MOSAIC and LSI [2] at 0.3 NA is in the pipeline for the next couple of months.

Our goal is to implement MOSAIC at the SEMATECH Berkeley MET before SPIE 09. We are currently exploring how illumination NA and object spatial frequency affect the determination of best focus in lithographically patterned features.

Since I am sure some of you are wondering...

For accurate determination of the second derivatives, the maximum tolerable focus resolution is about two orders of magnitude larger than the RMS wavefront deviation from a sphere. For EUV projection optics with 0.3 nm RMS wavefront tolerances, focus steps of 30 nm are required.

Thanks for reading. We hope to present experimental data on MOSAIC at SPIE 09.

2008 Int. Symposium on Extreme Ultraviolet Lithography, Lake Tahoe 29 Sept. – 1 Oct. 2008

First Italian EUV micro exposure tool at 14.4 nm based on Kr DMS

S. Bollanti, P. Di Lazzaro, F. Flora, L. Mezi, D. Murra, A. Torre

ENEA CR Frascati, Via E. Fermi 45

00044 Frascati, Italy

Schematic of the MET developed at ENEA (FIRB RBNE01ABPB national project)

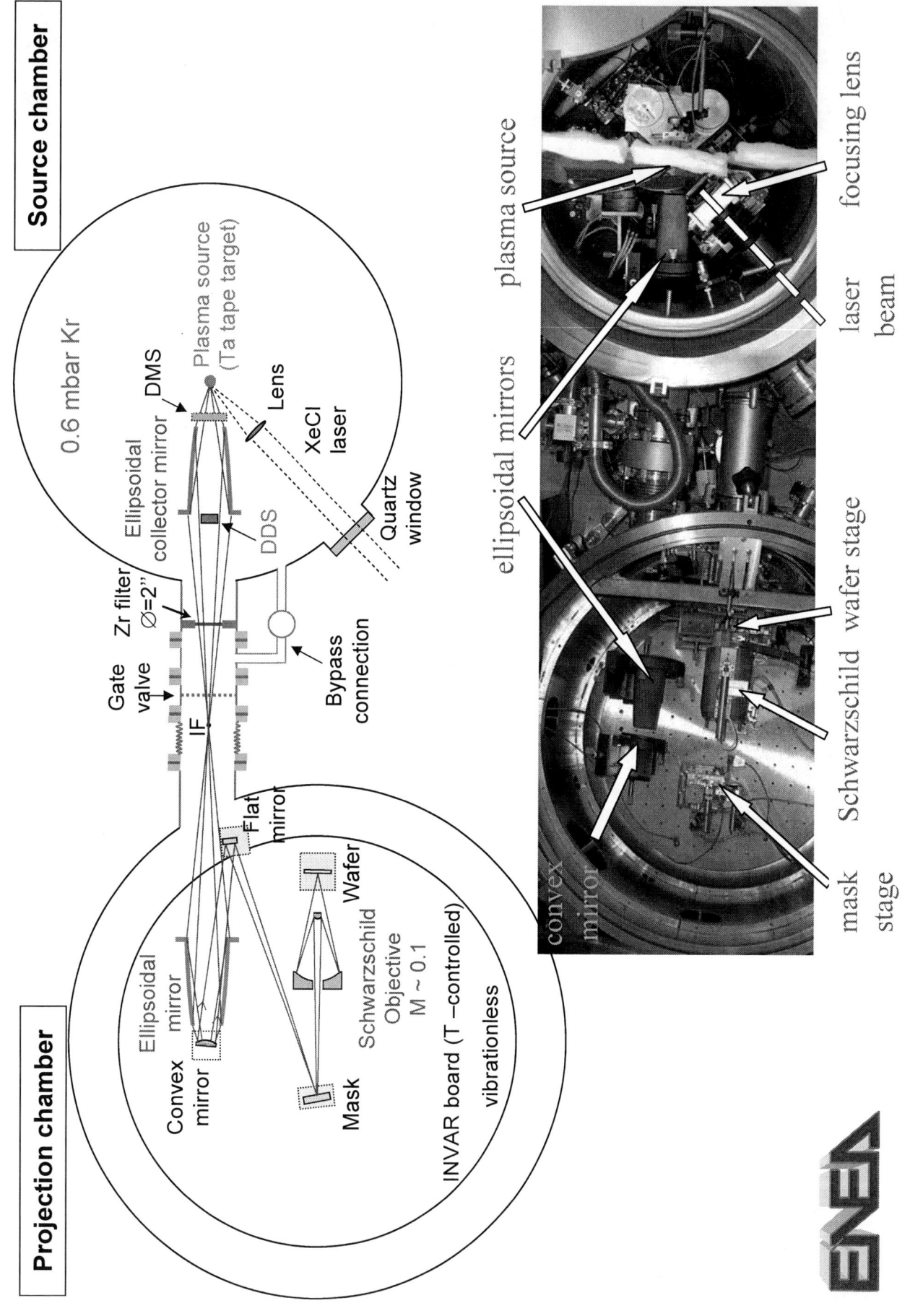

Schematic of the laboratory plant

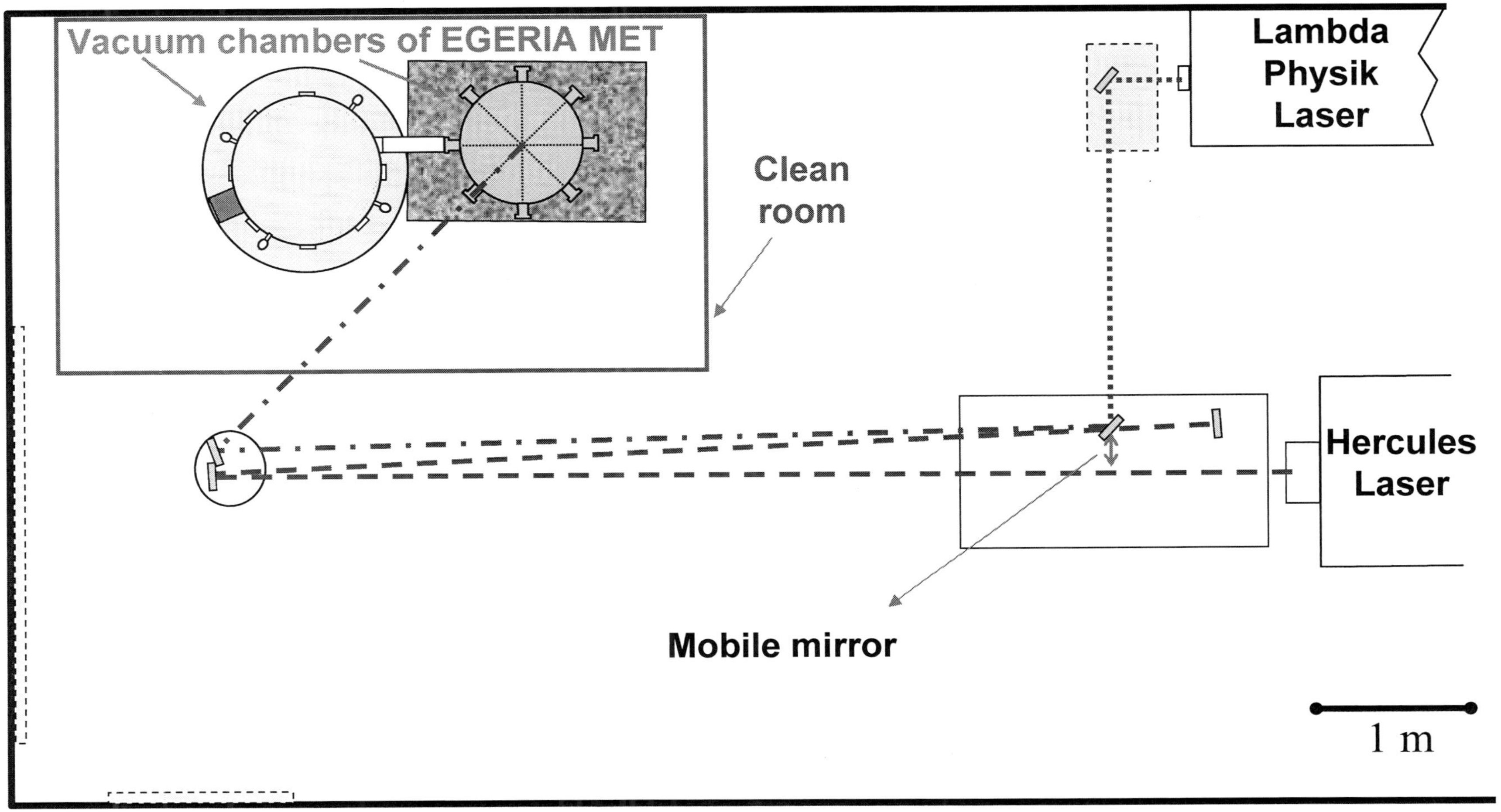

Lambda Physik Laser

Hercules Laser

Mobile mirror

Clean room

Vacuum chambers of EGERIA MET

1 m

The EGERIA-MET inside the clean-room

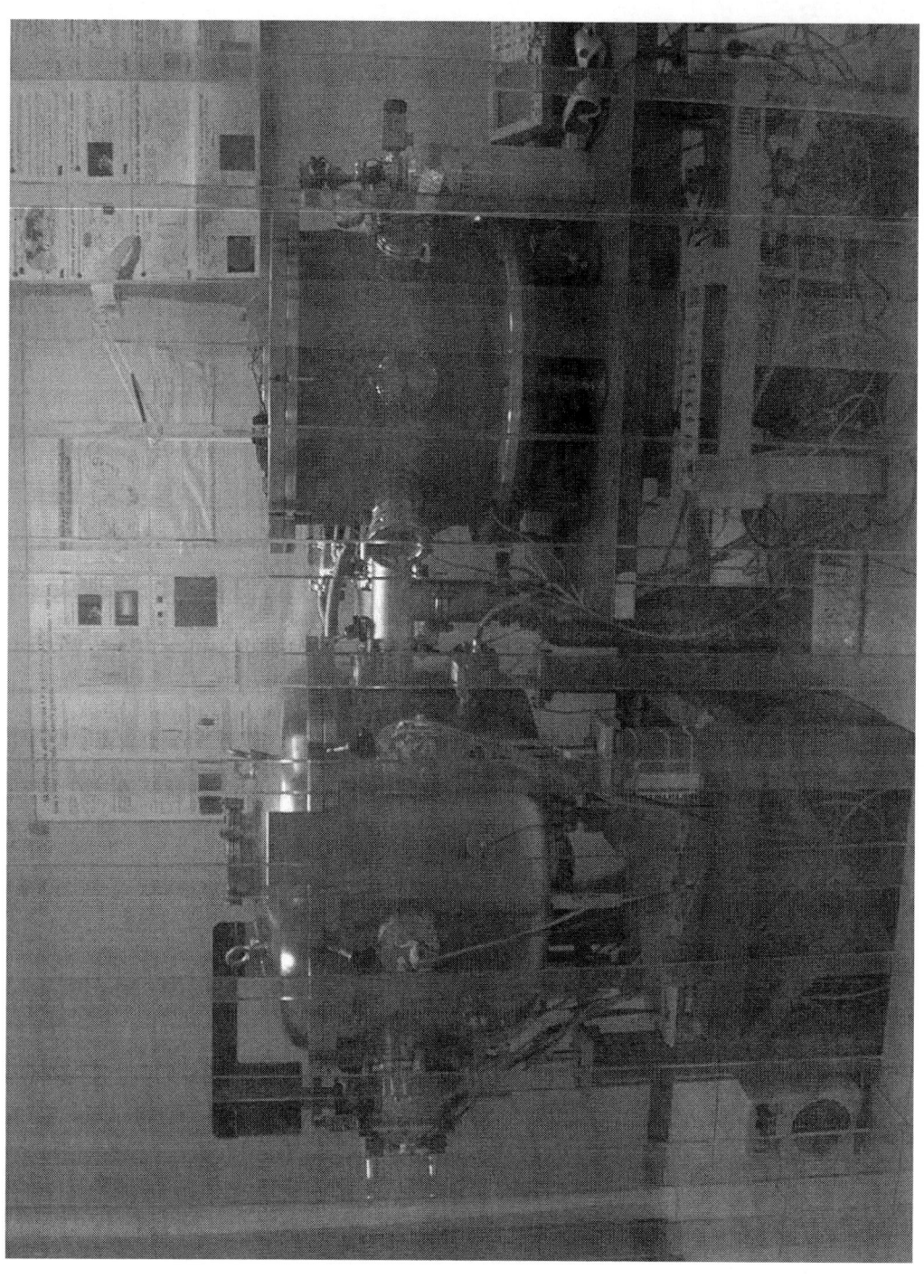

The excimer laser laser driving systems

High-energy-per-pulse Hercules laser realised in ENEA

Active medium: XeCl
Wavelength: 308 nm
Optical resonator: PBUR M=7
Energy per pulse: 6 J
Pulse duration (FWHM): 120 ns
Repetition rate (max): 5 Hz
Preionising X-ray diode modulator (resonant charging with transformer)
Main discharge modulator (LC-inversion)

Lambda Physik LPX305 laser

Active medium: XeCl - Wavelength: 308 nm
Optical resonator: PBUR
Energy per pulse: 0.5 J - Pulse duration (FWHM): 30 ns
Repetition rate (max): 50 Hz

Plasma debris emission

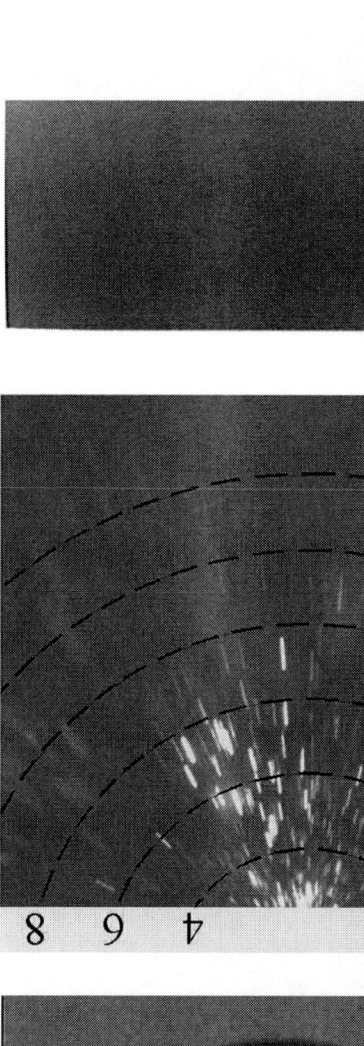

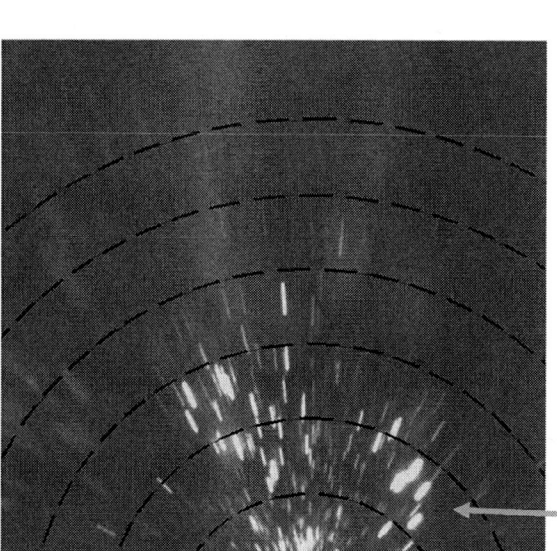

screened

Glass plate put at 5-cm distance from plasma in vacuum, during 10⁴ pulses

Particulate debris (picture captured 0.5 ms after laser pulse)

Visible light emission from atomic debris

ENEA

The ENEA Debris Mitigation System (DMS)

developed within the frame of the European
"More Moore" Integrated Project

- The source chamber is filled with ~ 1mbar Kr which can stop in 10 cm atomic debris and also particle debris having a size-speed product $\varnothing \cdot V \leq (1\ \mu m) \cdot (70\ m/s)$.

- larger debris, slowed-down by the gas, can be stopped by the high speed fan.

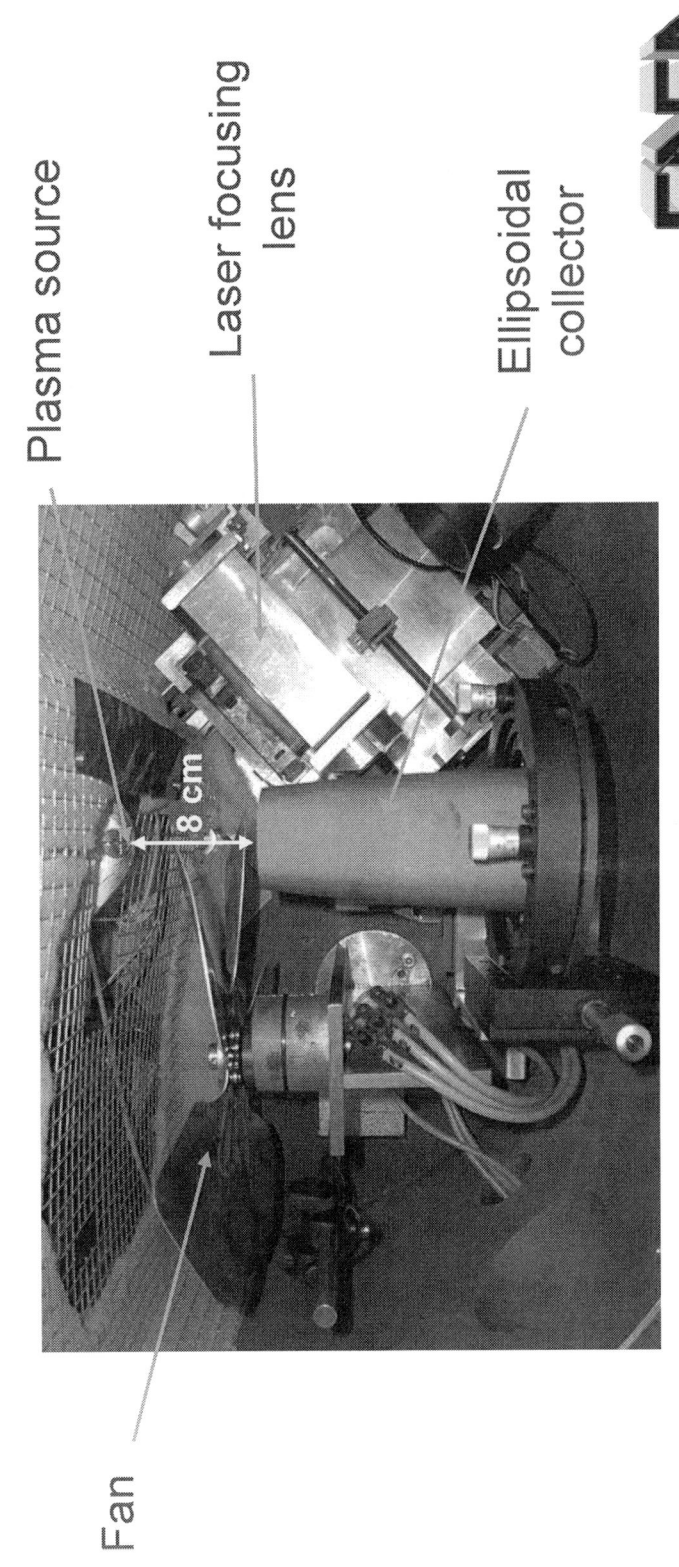

Plasma source

Laser focusing lens

Ellipsoidal collector

Fan

8 cm

Transmission spectrum of He Ar and Kr gases in the EUV

and debris range of flight (RoF) of atomic and droplet debris in such gases

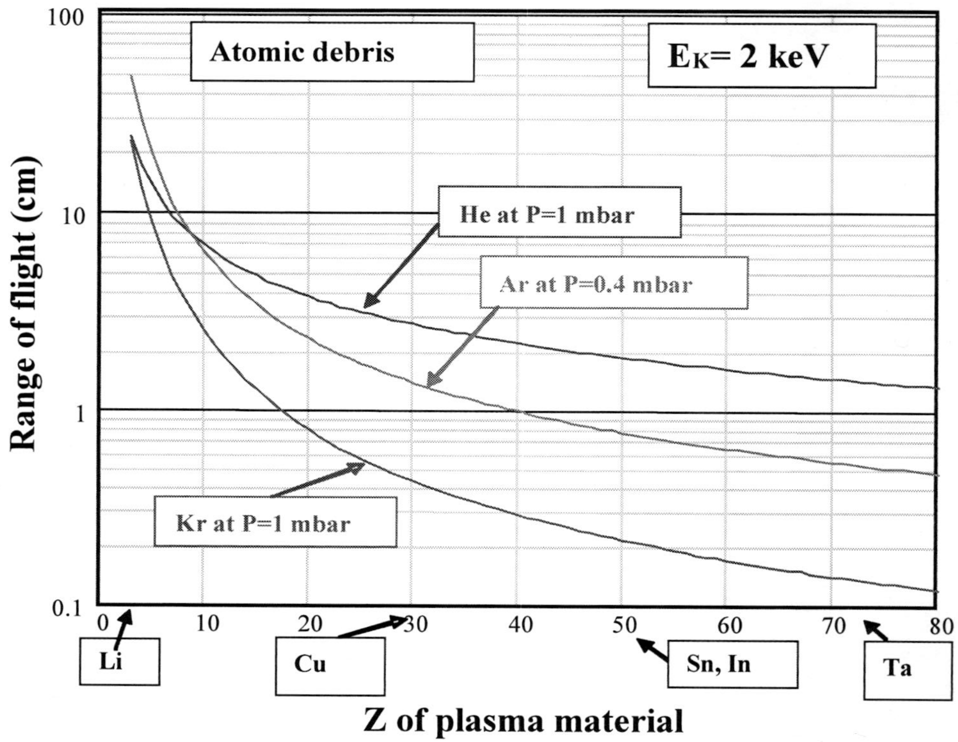

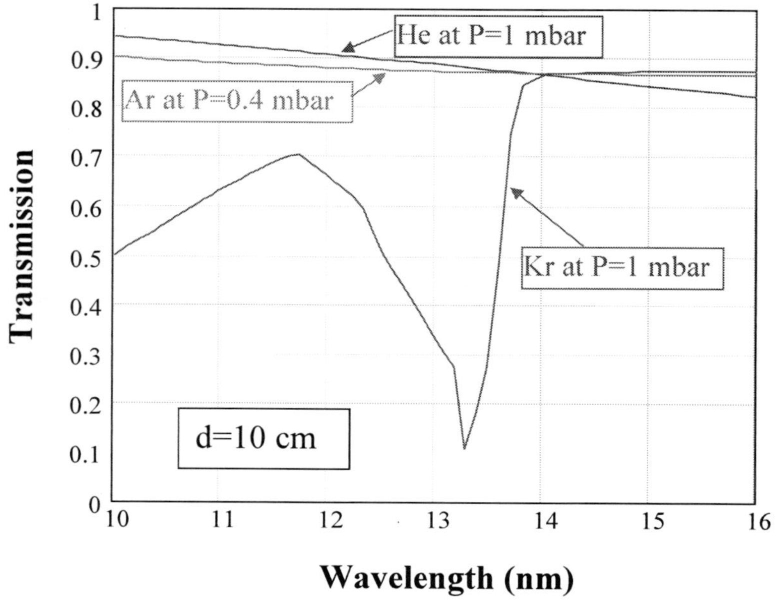

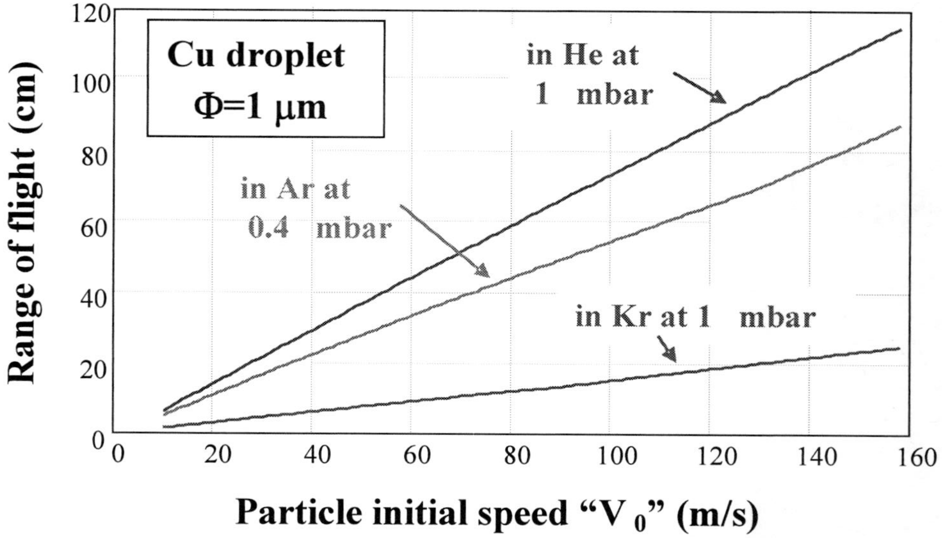

Experimental Cu debris maximum emission speed vs their diameter for LPX305 laser (blue line) and theoretical maximum debris speed allowed for a debris Range-of-Flight RoF ≤ 10 cm (violet lines) in 1 mbar Kr (solid) or in 0.4 mbar Ar (dashed) gases.

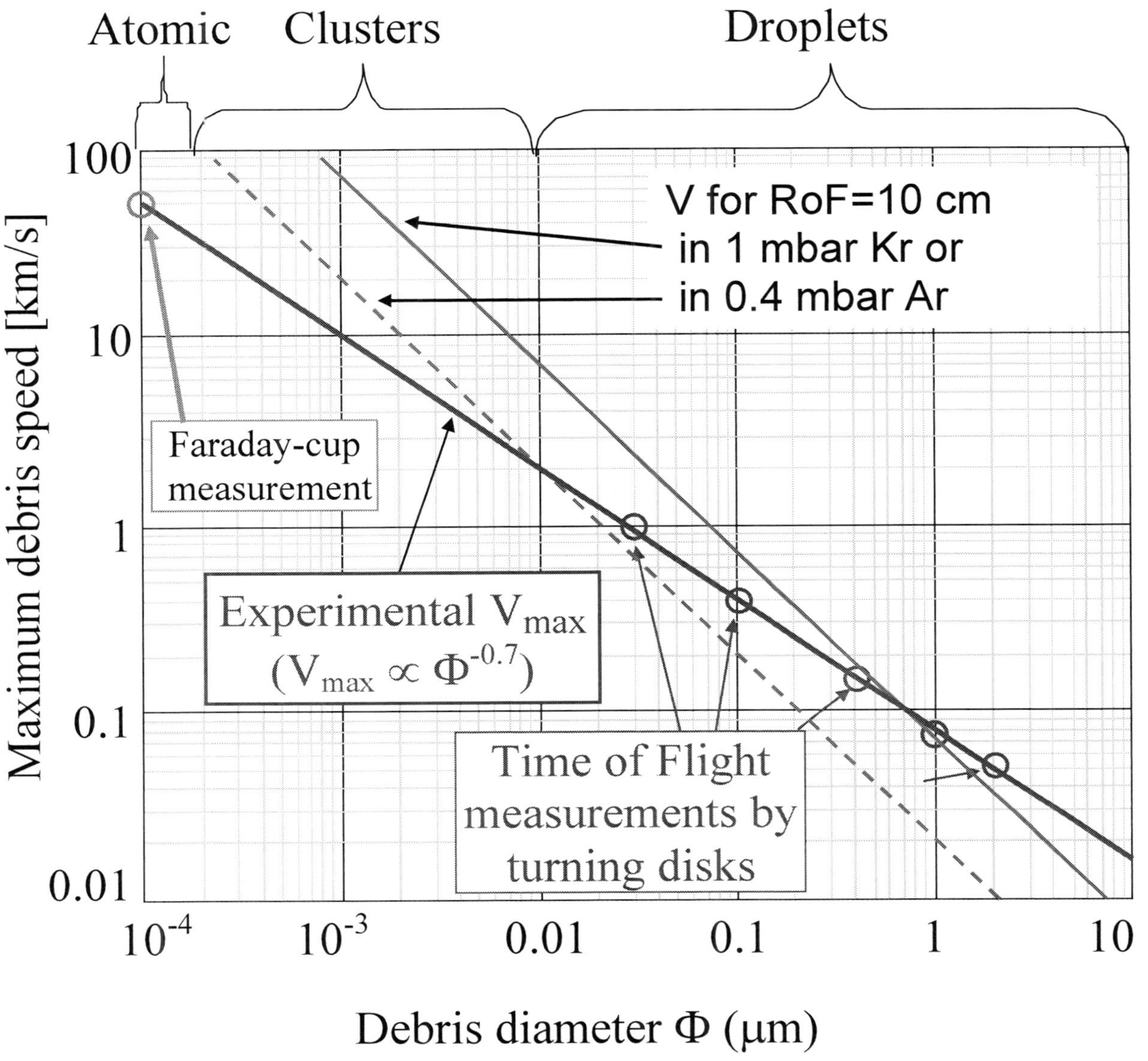

EUV transmiss.= 86% at 14 nm in 10 cm for both gases

The crossing between blue and violet lines gives the maximum speed of residual debris to be stopped by the fan.

In case of Ar gas, a 20x faster fan is needed to stop the residual debris!!

Atomic   Clusters   Droplets

V for RoF=10 cm in 1 mbar Kr or in 0.4 mbar Ar

Faraday-cup measurement

Experimental V_{max} ($V_{max} \propto \Phi^{-0.7}$)

Time of Flight measurements by turning disks

Maximum debris speed [km/s]

Debris diameter Φ (µm)

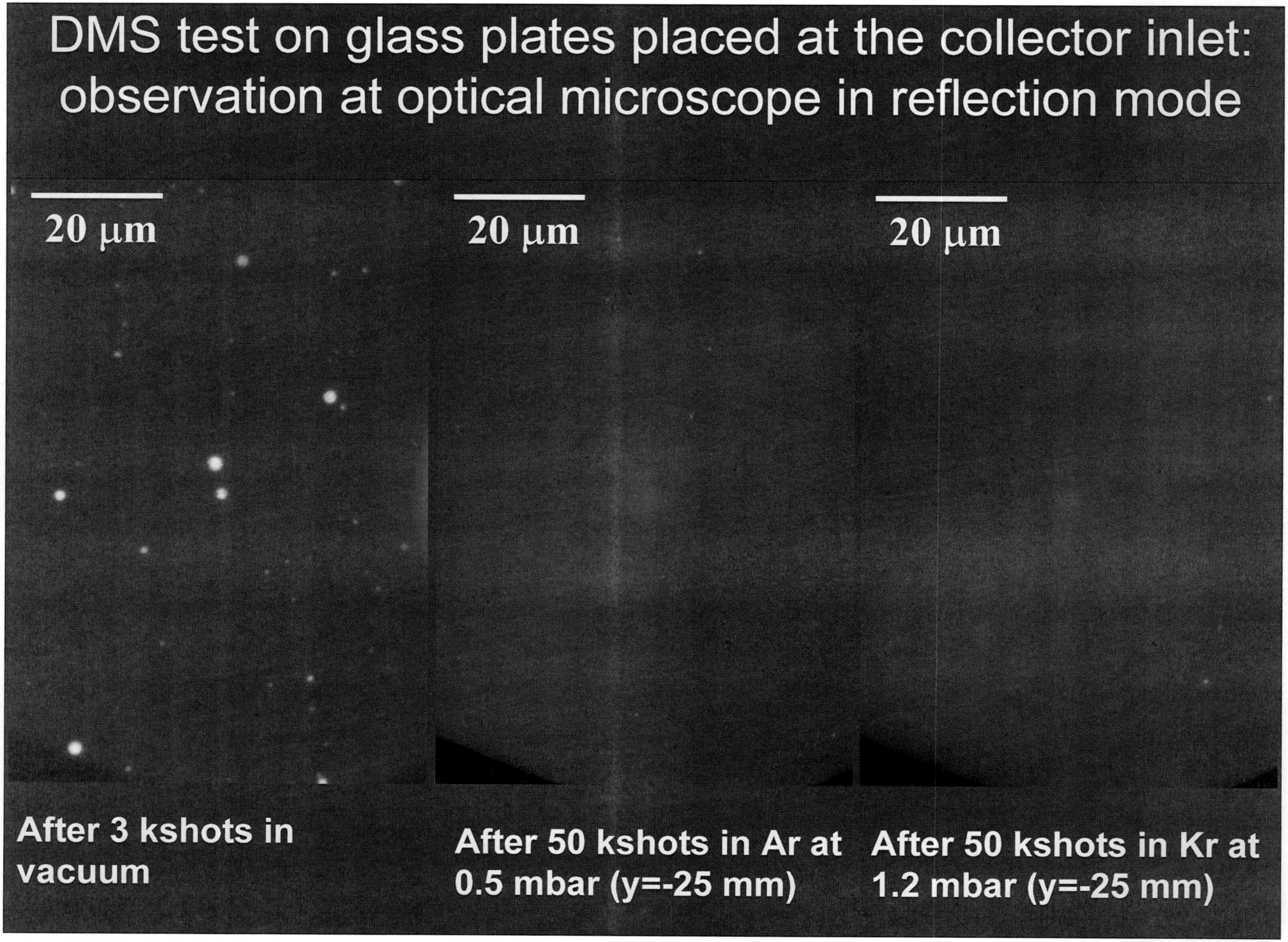

DMS test on glass plates placed at the collector inlet: observation at optical microscope in reflection mode

20 μm

20 μm

20 μm

After 3 kshots in vacuum

After 50 kshots in Ar at 0.5 mbar (y=-25 mm)

After 50 kshots in Kr at 1.2 mbar (y=-25 mm)

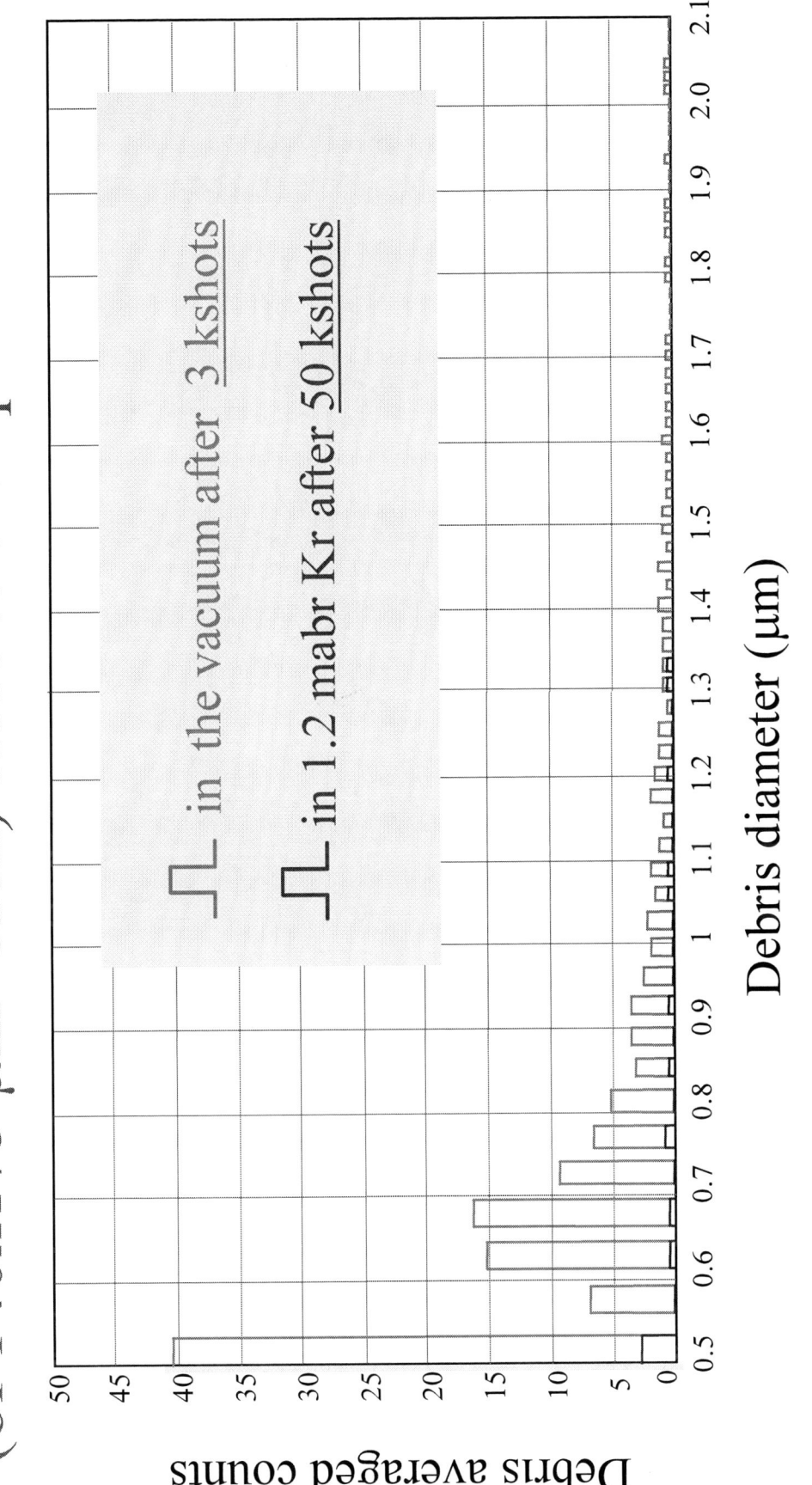

The Cu droplet counts, averaged over 11 images (of 140x175 μm² each) close to the optical axis

Debris mitigation factor (DMf) at the optical axis (target normal):
{ for Ar case: DMf ~ 400
{ for Kr case: DMf ~ 1000

The atomic and cluster debris

Much higher DMf values would be expected for smaller debris ($\Phi <$ 0.5 µm) like atomic and cluster debris. As a matter of fact, the DMf measurement for such debris ($\sim$ 400 as obtained from optical density comparison among the plates) is limited by the huge clusters formation in the proximity of the source due to plasma-ions / gas interaction as observed at SEM .

SEM image of glass plate after 3 kshots in the vacuum or 50 kshots in 1.2 mbar Kr

Count of small size (200<Φ<400 nm) debris, deposited on the gas plate after 50 kshots in 1.2 mbar Kr or 0.5 mbar Ar, for different positions along the vertical axis "Y" (counts in single images of 140x175 μm² each)

Lower regions of the plate, corresponding to the lowest border of the collector mirror, are reached by the fan shovel after a shorter delay (Tcut) from the laser shot, so that only very fast debris (V>800 m/s) can reach these regions.

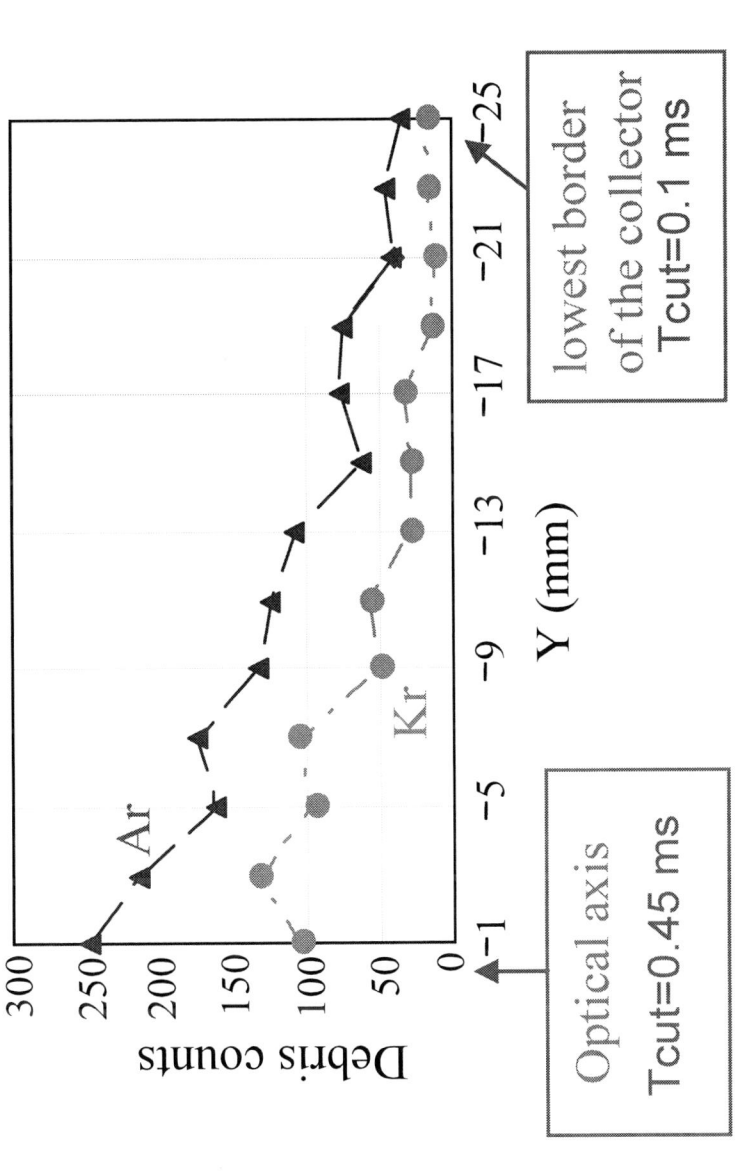

The possibility to reduce cluster formation by generating a gas depression region around to the EUV LPP source is under consideration.

The ellipsoidal mirrors

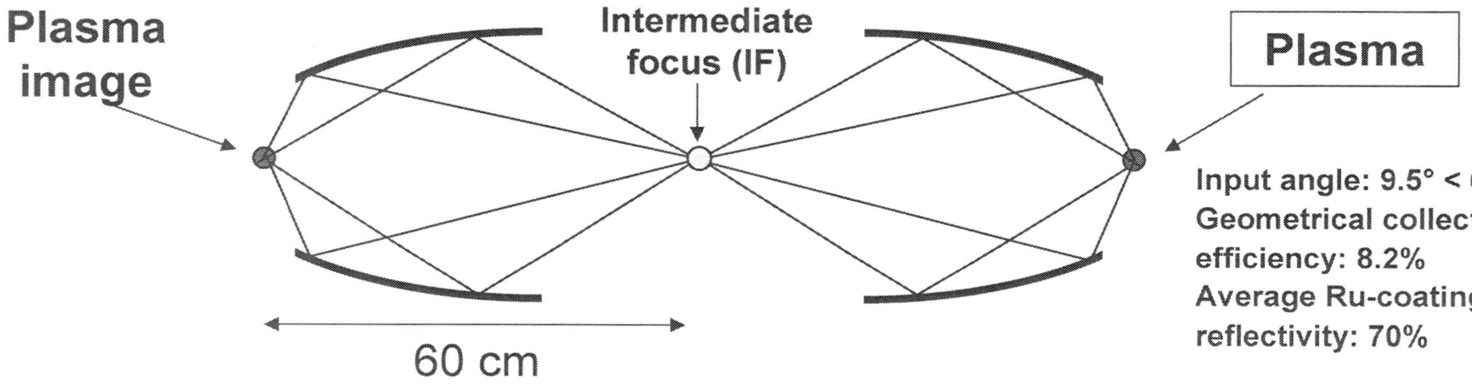

Plasma

Input angle: 9.5° < θ < 18°
Geometrical collection efficiency: 8.2%
Average Ru-coating reflectivity: 70%

Plasma image

Intermediate focus (IF)

60 cm

Media Lario Techologies Ru-coated ellipsoidal mirror

Reflection surface

Typical EUV radiation transverse distribution at IF after 0.8-μm-thickness Al filter on HD-810 Gafchromic dosimetric film

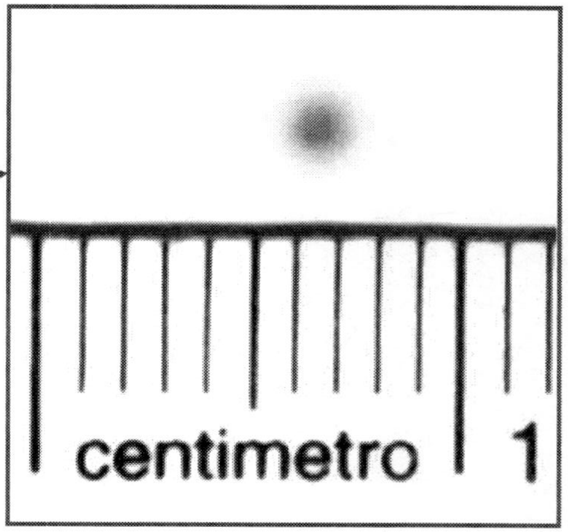

centimetro 1

The spherical aberration introduced by the high numerical aperture of the convex ML mirror can be partially balanced taking the ellipsoidal mirrors about 1.2 mm away from each other.

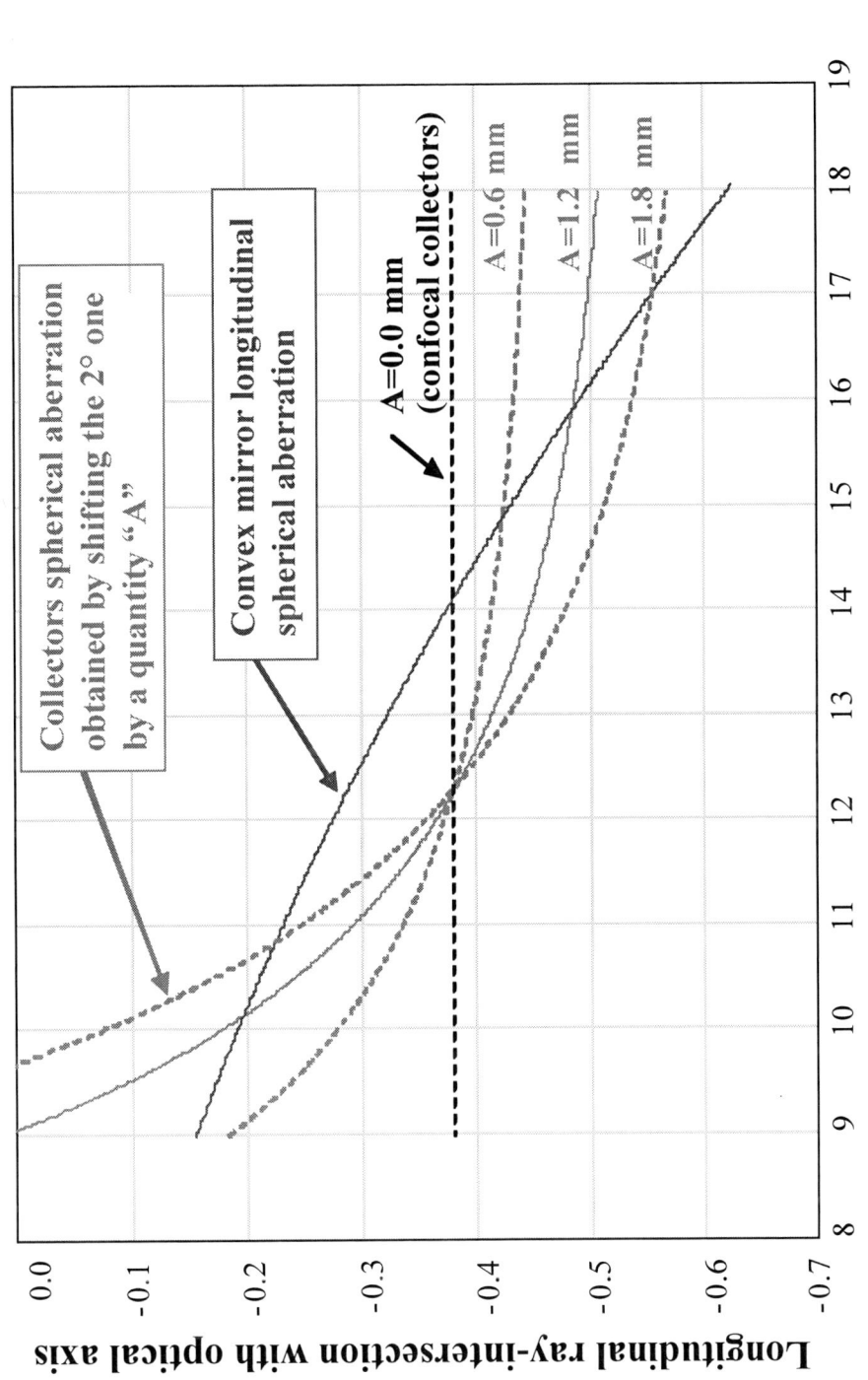

The estimated EUV spectrum at IF (from Ta target) and the multilayer (ML) mirrors developed at INFN of Legnaro (Padova, Italy):

Spectral transmission of 0.15-μm thick Zr filter (dashed line) Kr gas at 1 mbar·42 cm (dotted line), and their product giving EUV radiation spectral shape at the IF (solid line)

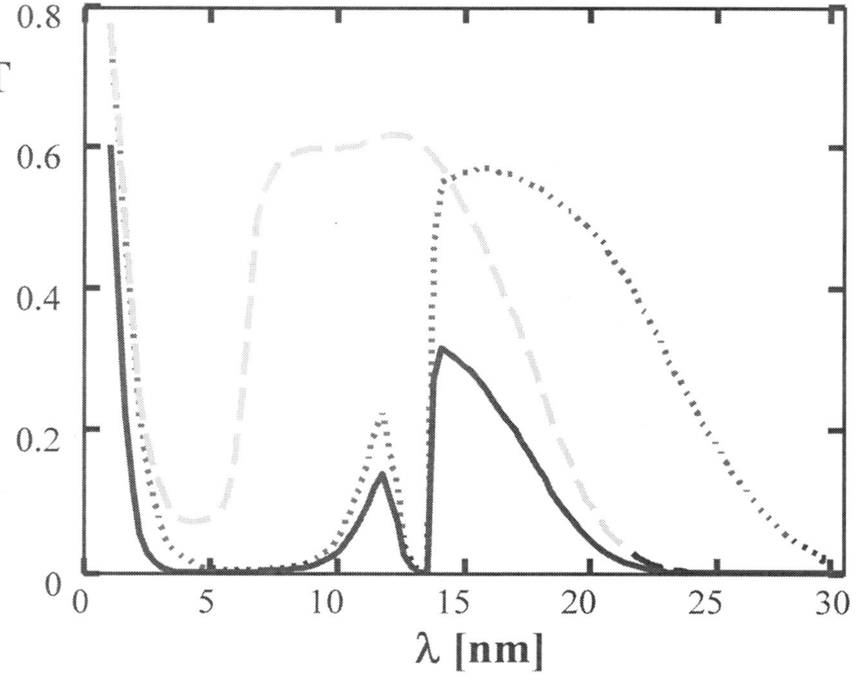

Narrowband wavelength selection at 14.4 nm is obtained with Mo/Si ML mirrors

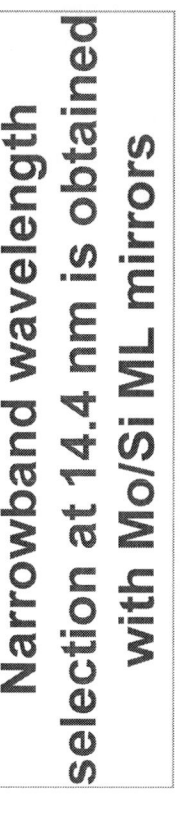

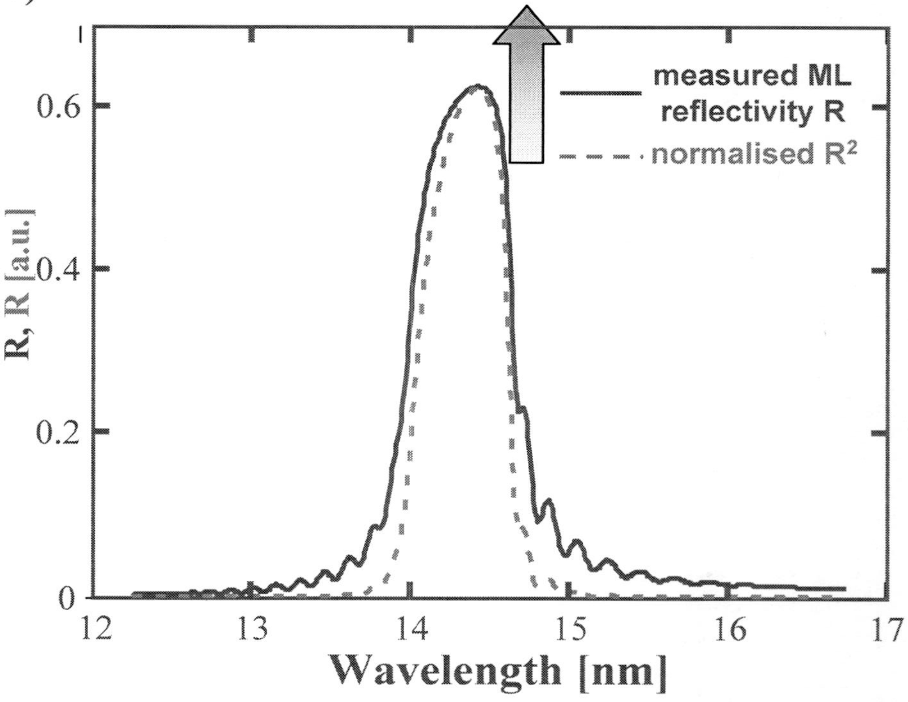

The mask

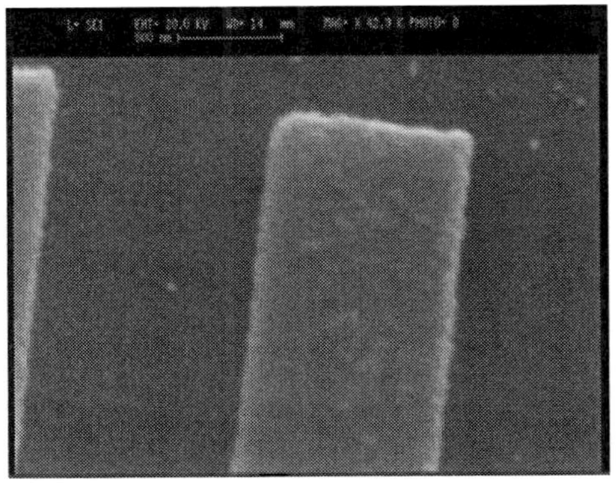

Edge structure SEM image of a 1 μm line

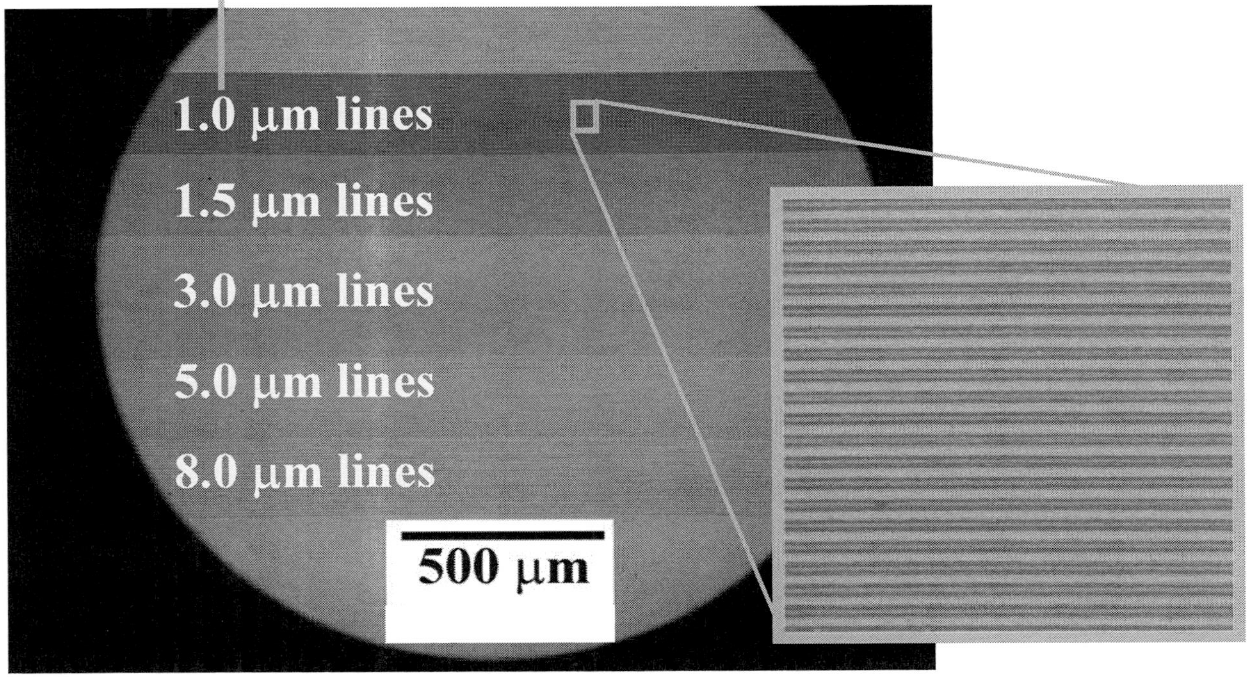

1.0 μm lines
1.5 μm lines
3.0 μm lines
5.0 μm lines
8.0 μm lines

500 μm

Optical microscope picture of different half-pitch line-space patterns of the mask

The projection optics: a Schwarzschild objective (SO) in a modified configuration (MSO), i.e. with shifted object plane while keeping concentric mirrors.

R_1= 144.23 mm R_2= 45.06 mm

Z_i= 36.26 mm Z_o= 340.22 mm

Φ_1= 74 mm Φ_2= 12.7 mm

M=1/9.5

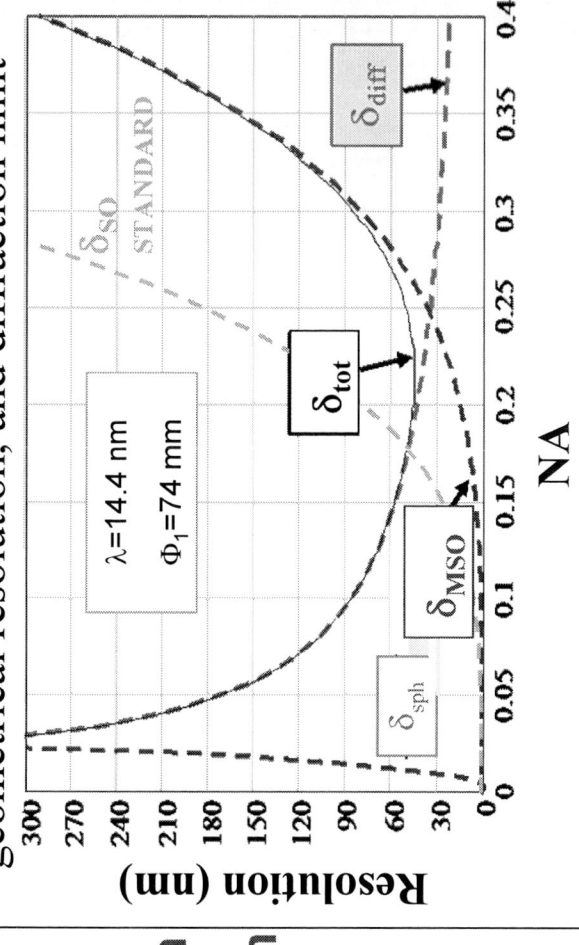

Single mirror (δ_{sph}); standard SO and MSO geometrical resolution, and diffraction limit

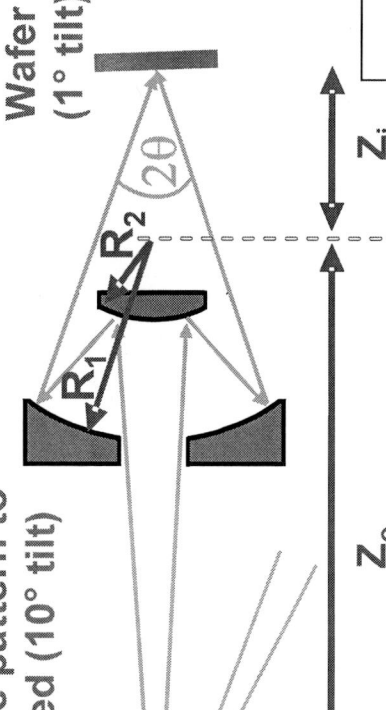

Mask with the pattern to be reproduced (10° tilt)

Wafer (1° tilt)

$$NA=0.23 \begin{cases} \lambda=14.4 \text{ nm} \Rightarrow \delta_{diff}=38 \text{ nm} \\ \text{MSO} \Rightarrow \delta_{geo}=27 \text{ nm} \end{cases}$$

$$\delta_{tot}= \sqrt{(\delta_{diff}^2+ \delta_{geo}^2)} = \underline{47 \text{ nm}}$$

Sensitivity to SO misalignment

The 47-nm resolution limit is true for ideal spherical optics and perfect alignment

The overlapping of the two curvature centres is very critical:

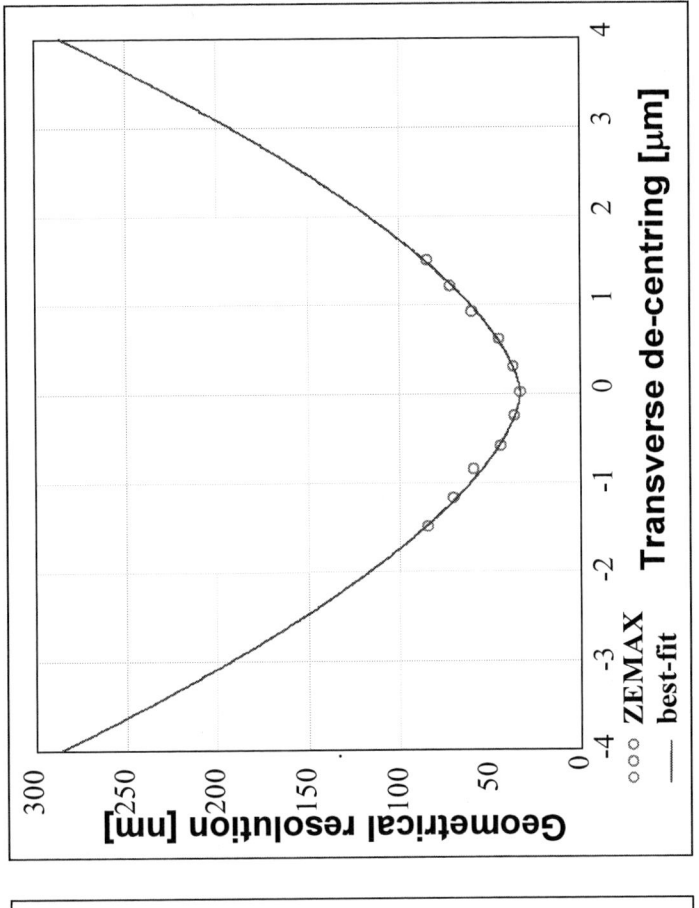

SO alignment by modified Foucault technique at $\lambda=308$ nm

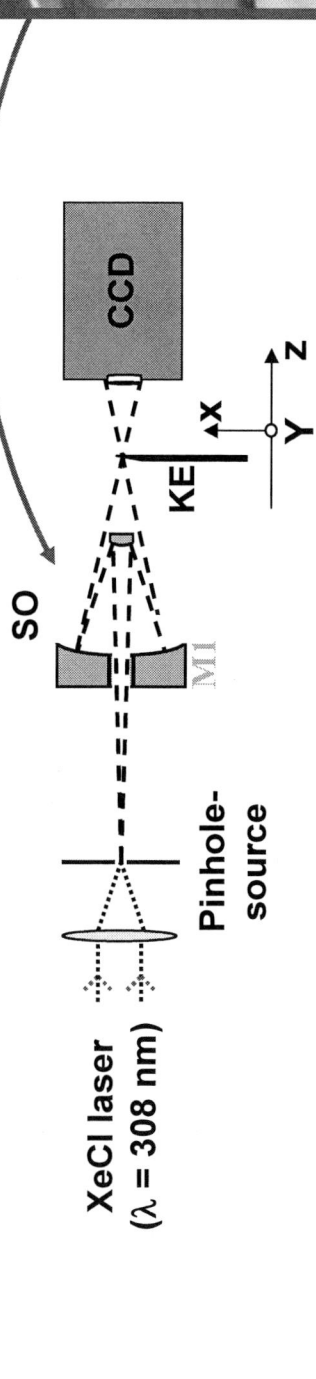

XeCl laser ($\lambda = 308$ nm)

Pinhole-source

SO

MI

KE

CCD

Z

X

Y

→ Experimental

→ 2-D ray-tracing for ZM2=0.8 mm of longitudinal misa-lignment (spherical aberration)

→ Ray-tracing with simulated diffraction effect

Sequence of Foucault-grams vs Knife-edge Z position (X=0):

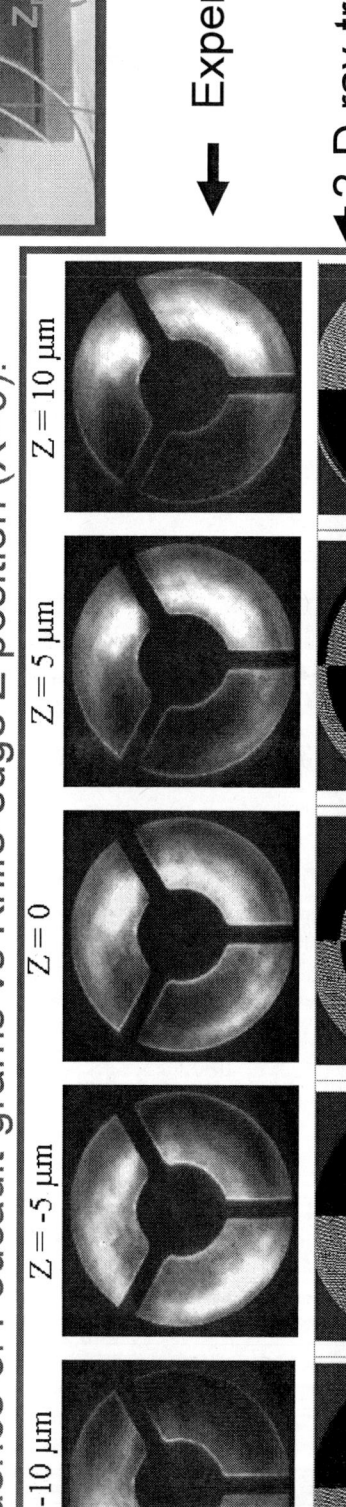

Z = 10 µm Z = 5 µm Z = 0 Z = -5 µm Z = -10 µm

The optimization of each of the three alignment parameters (Z_{M2}, θx_{M1} and θy_{M1}) is reached by interpolation among widely misaligned conditions (where the corresponding aberration is much larger than the diffraction limit at 308 nm)

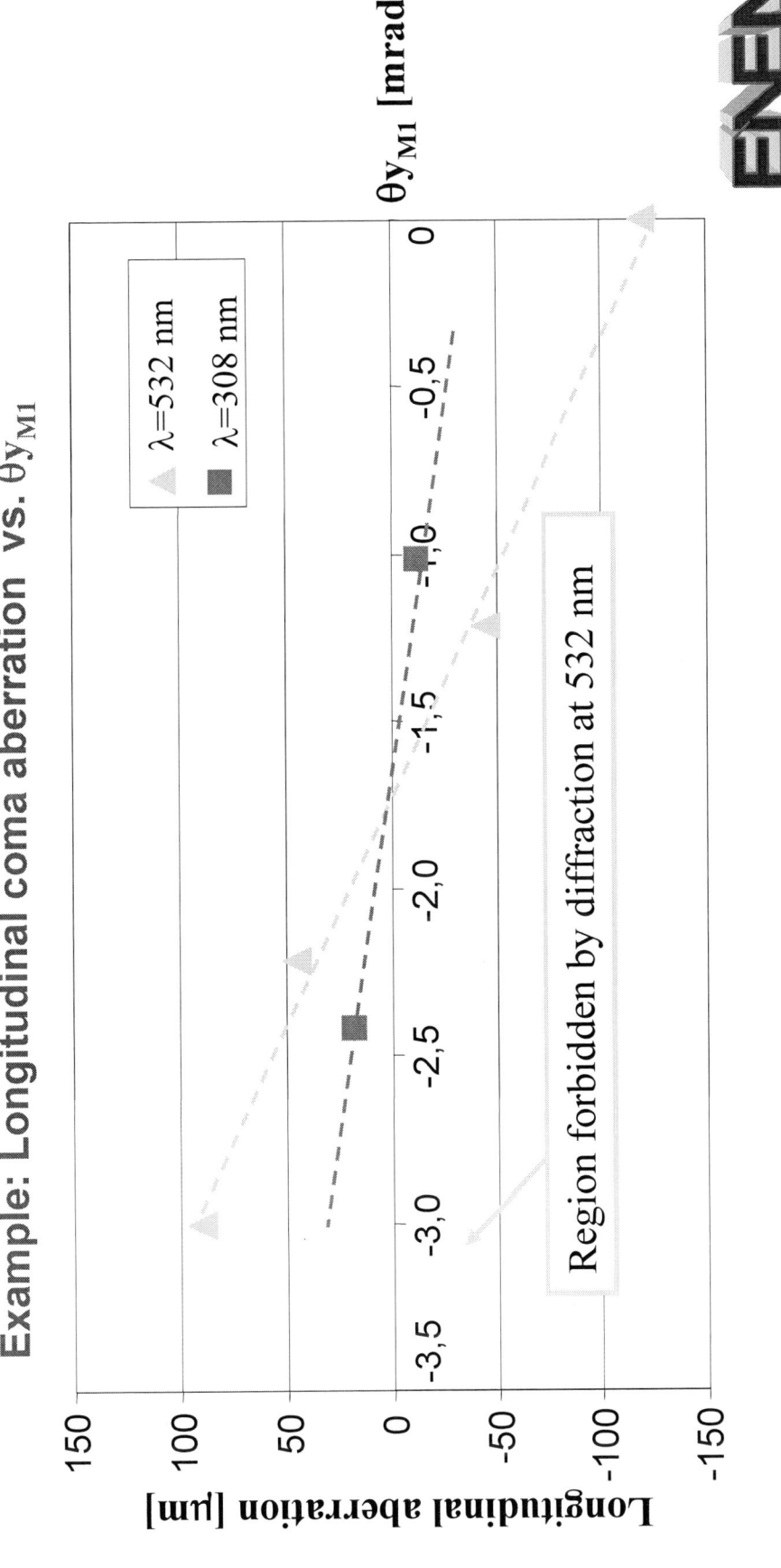

Example: Longitudinal coma aberration vs. θy_{M1}

PMMA resist exposure results: wide field

100 nm thick PMMA; molecular weight: 996.000. EUV radiation dose: ~ 10 mJ/cm²

10 µm

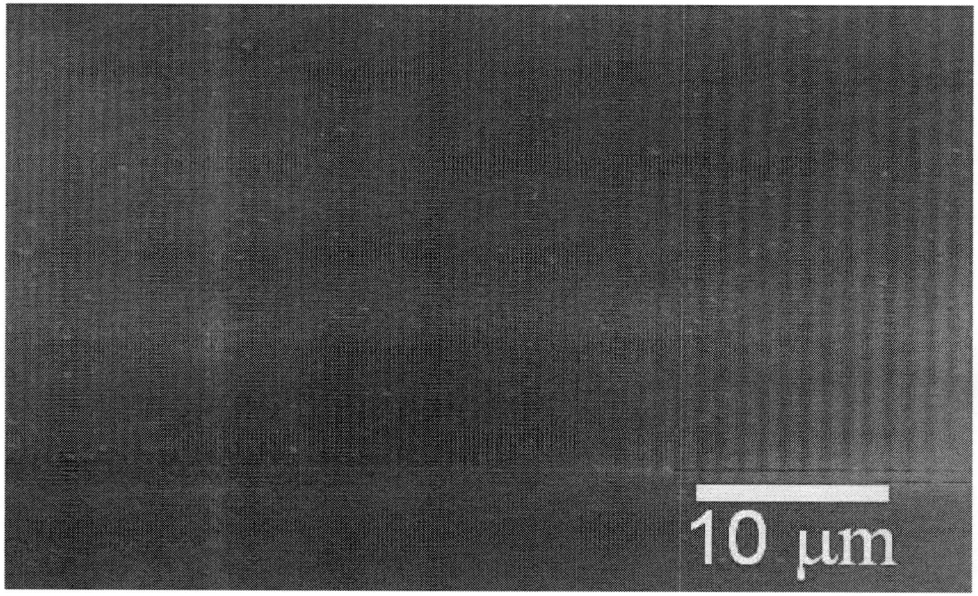

540 nm lines

1 µm

320 nm lines

1 µm

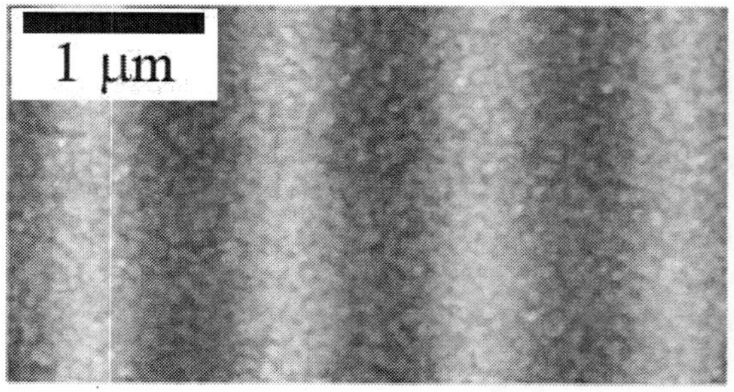

AFM images. Same modulation amplitude (20 nm) obtained for both the 540 nm and 320 nm lines!

PMMA resist exposure results: high resolution patterns.

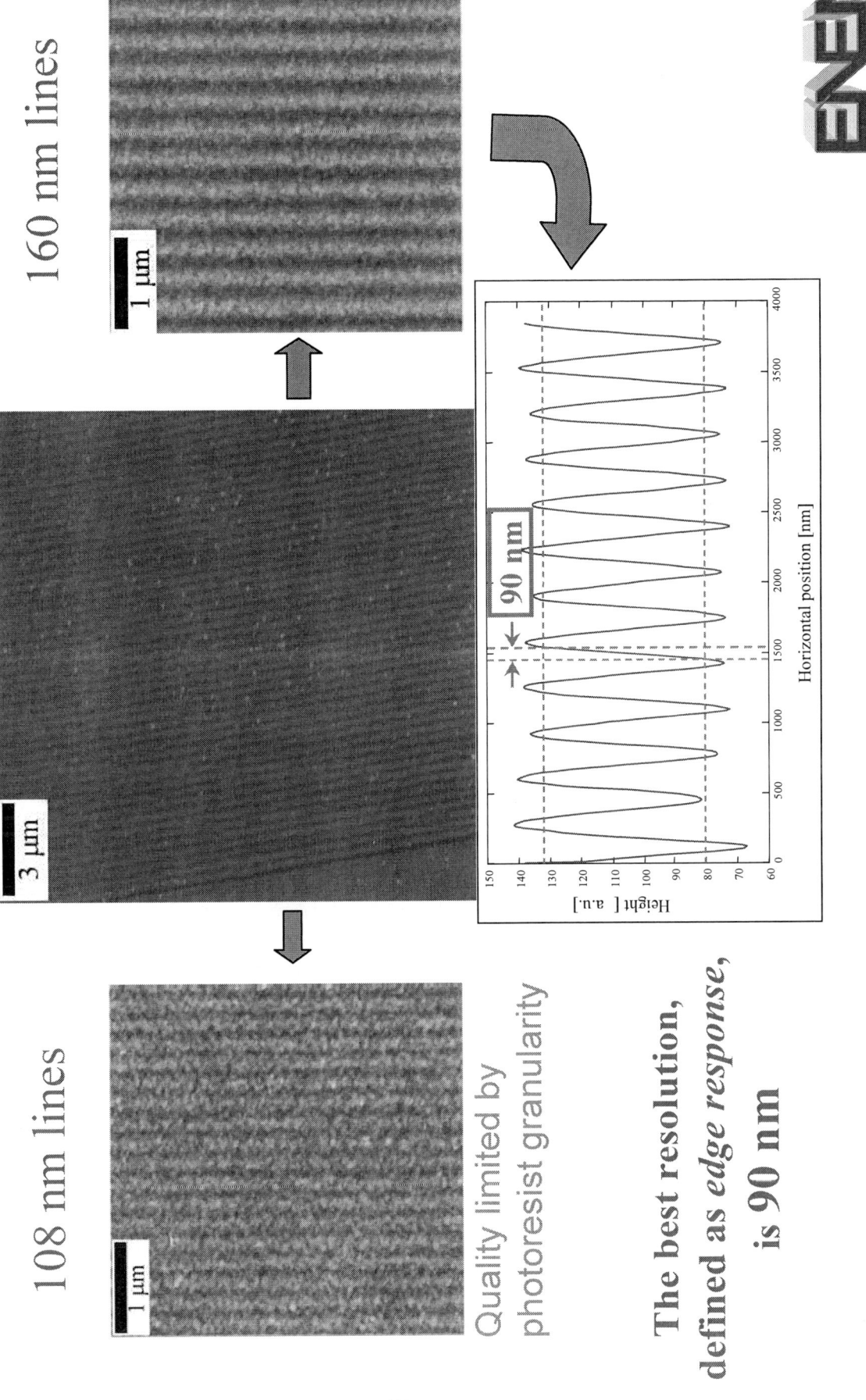

160 nm lines

108 nm lines

Quality limited by photoresist granularity

The best resolution, defined as *edge response*, is 90 nm

Estimation of optical and lithographic resolution on the base of the Modulation Transfer Function (MTF)

The modulation amplitude for the 160 nm lines is 80% of that of the 320 nm or larger lines:

$f_{80} = 1/320$ periods/nm

Optical resolution from MTF (see http://www.normankoren.com/Tutorials/MTF.htm):

$$\delta_{opt.} = 1/f_9 = 0.23/f_{80} \sim 75 \text{ nm}$$

Lithographic resolution:

$$\delta_{litho.} = \delta_{opt.} \cdot K_1/0.61 \sim 50 \text{ nm}$$

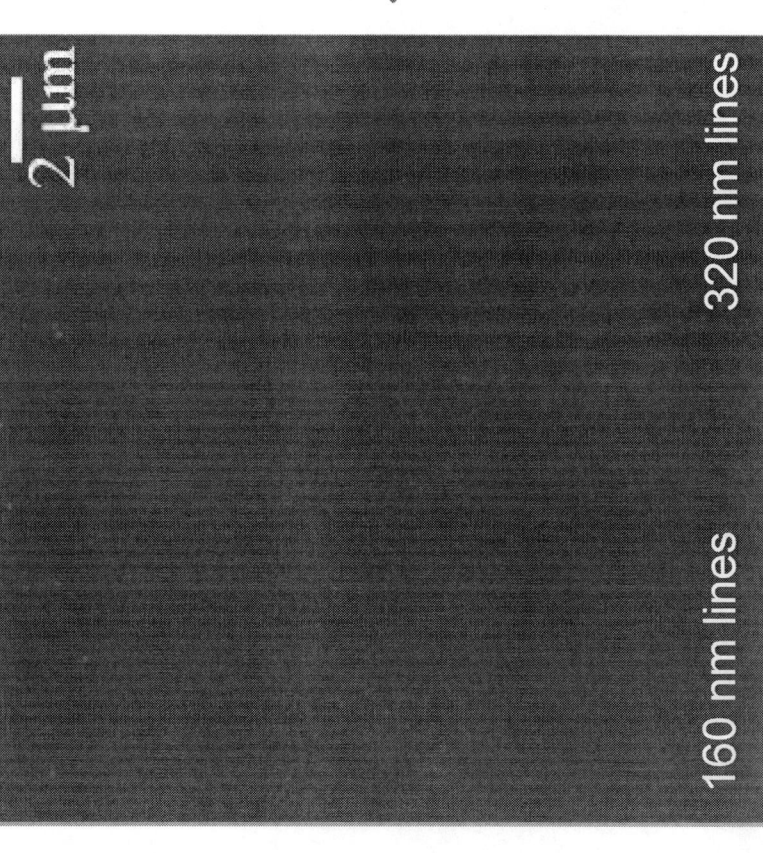

2 μm

160 nm lines

320 nm lines

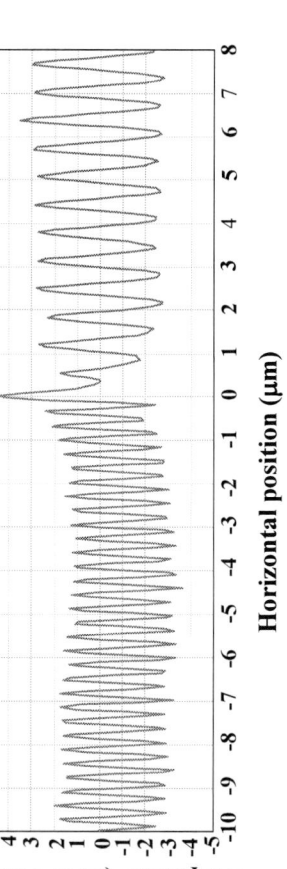

CONCLUSIONS

- The first (ever) MET working at 14.4 nm has been developed and tested. This wavelength is compatible with DMS based on Kr gas.

- The DMS demonstrated a much higher efficiency with Kr rather than Ar gas. More than 3 orders of magnitude of droplet debris mitigation factor has been achieved with Kr. Further research is in progress.

- An innovative method (Foucault variant) for UV alignment of the projection optics has been developed.

- An optical resolution of 75-90 nm for MET is obtained.

- The MET-EGERIA is a scientific tool available for collaboration with other R&D institutes or industries.

Lithographic modeling of a 0.5-NA microstepper optic

Patrick Naulleau,[1] Michael Goldstein,[2] Russ Hudyma[3]
[1]Center for X-ray Optics, Lawrence Berkeley National Laboratory; [2]SEMATECH, Albany, NY; [3]Hyperion Development LLC, San Ramon, CA;

MOTIVATION/INTRODUCTION

- High volume insertion for EUV expected at 22-nm node
- Current generation 0.3-NA MET and 0.25-NA alpha tools lack resolution to support 22-nm node mask and resist development
- SEMATECH has launched 0.5-NA MET program to address the 22-nm node development need
- Optics design completed
- Here we present lithographic modeling results based on the design

OPTICAL DESIGN

- Compatible with 0.3-NA MET
- Non-equal radii design
- NA = 0.5
- Resolution = 8 nm
- Magnification = 5x
- Field of view = 200x30 µm
- Mask illumination angle = 6°
- obscuration by area = 7.1%

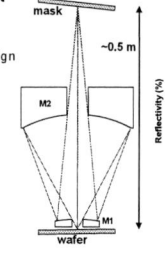

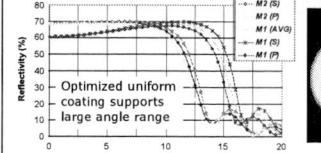

Reflectivity vs angle of incidence on M1 & M2

- Optimized uniform coating supports large angle range

- Angle range on M1: 3.4° - 13.6°
- Angle range on M2: 1.2° - 4.1°

Pupil apodization with multilayers

- Apodization map essentially constant across field of view

CONCLUSIONS/SUMMARY

- METs are key enablers of EUV infrastructure progress: Resist development, mask architecture development, mask defect printability studies, ...
- 0.5-NA MET needed for sub-22-nm development
- Design of a two-mirror 0.5-NA EUV optic with a 200x30 µm field of view completed
- Integration into an exposure tool at Berkeley's Advanced Light Source synchrotron is planned
  - The system will support variable illumination
- The designed 0.5-NA optic supports resolutions of:
  - 12-nm 1:1 lines (conventional illumination)
  - 8-nm isolated lines (conventional illumination)
  - 14-nm 1:1 contacts (conventional illumination)
  - 8-nm dense lines (dipole illumination)
  - 11-nm 1:1 contacts (quadrupole illumination)

Astigmatism and spherical dominate

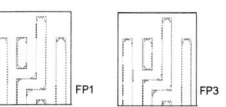

Design aberrations across the field

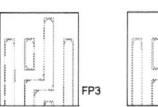
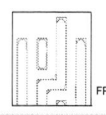

AERIAL IMAGE MODELING RESULTS: See aberration map for field point definitions

- Through focus contrast transfer functions (CTFs) at FP3
- Annular illumination 0.35-0.55
- 10-nm focus steps

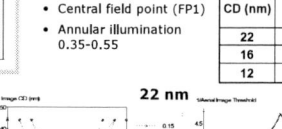

16- nm features at various fields points

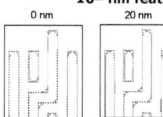

16- nm features through focus

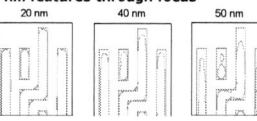
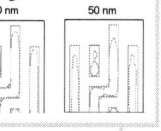

Dense Line Process Windows

- Central field point (FP1)
- Annular illumination 0.35-0.55

CD (nm)	DOF (nm) @ 10% EL
22	80
16	120
12	70

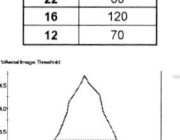

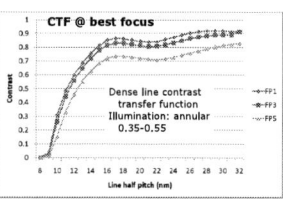

Dense Contact Process Windows

- Plots for central field point (FP1)
- Annular illumination 0.35-0.55
- DOF at 10% exposure latitude

CD (nm)	32	22	16
FP1 DOF (nm) @ 10% EL	120	80	60
FP3 DOF (nm) @ 10% EL	100	60	50
FP5 DOF (nm) @ 10% EL	90	60	60

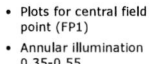
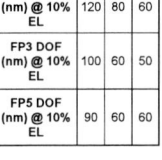

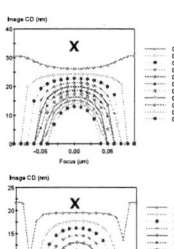
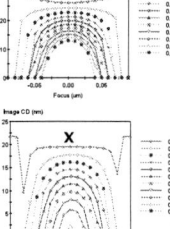
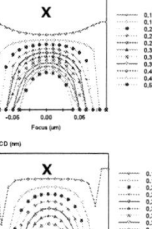
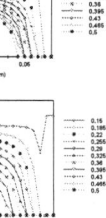
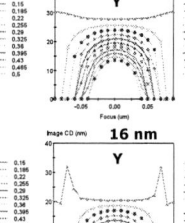
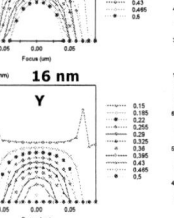
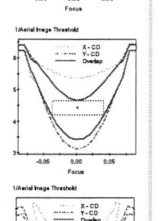
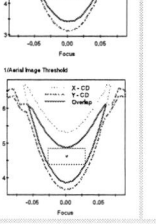

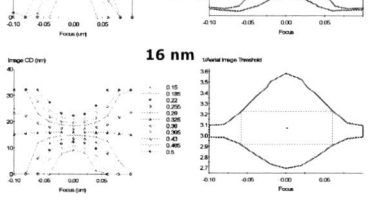
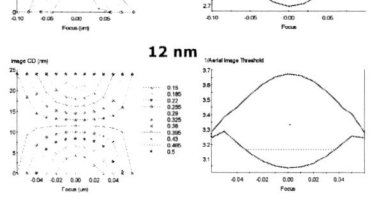

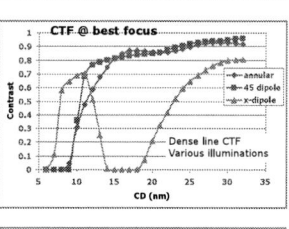
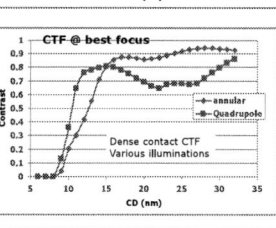

16-nm overlapping process windows

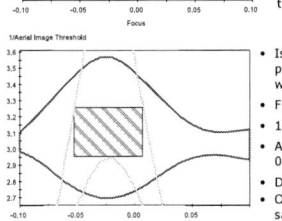

- H/V overlapping process window
- Field point 5
- 16-nm dense lines
- Annular illumination 0.35-0.55
- DOF = 110 nm
- Effect of astigmatism evident

- Iso/Dense overlapping process window (PW)
- Field point 5
- 16-nm lines
- Annular illumination 0.35-0.55
- DOF = 60 nm
- Overlapping PW same size as Iso PW

Isolated Line Process Windows

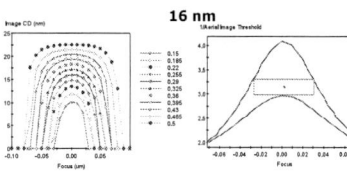
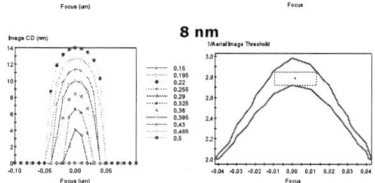

- Plots for central field point (FP1)
- Annular illumination 0.35-0.55

CD (nm)	FP1 DOF (nm) @ 10% EL	FP3 DOF (nm) @ 10% EL	FP5 DOF (nm) @ 10% EL
22	80	80	80
16	60	60	60
12	40	40	40
8	20 @ 20% EL	20 @ 20% EL	20 @ 20% EL

Funding provided in part by

Impact of Flare and Aberrations on Patterning Performance Simulation with the EUV Full Field Alpha Tool Conditions

Liping Ren,[1] Frank Goodwin,[1] Stefan Wurm,[1] Chawon Koh,[1] Sudharshanan Raghunathan,[2] and John Hartley [2]

[1] SEMATECH Inc., [2] College of Nanoscale Science and Engineering, University at Albany State University of New York, 255 Fuller Road, Albany, NY 12203

1. Introduction:

The great promise of extreme ultraviolet lithography (EUVL) comes from its tremendous reduction in wavelength which enables large operational K1 factor. But short wavelength also comes with various challenges as well, such as high flare and aberrations. In comparison with 1% flare and 5mλ lens aberrations of traditional DUV or 193nm scanners, typically, flare and aberrations in the early EUV full field alpha tools are in the range of 12% to 16% and 50 to 100mλ RMS, respectively. In EUVL, flare and aberration induced patterning performance issues significantly gain in importance.

In this poster, the impact of flare and aberrations on patterning performance is evaluated by Panoramic and Prolith lithographic simulation with illumination, flare and aberration conditions of the Albany ASML EUV full field alpha demo tool (ADT).

2. Kirchhoff or Rigorous Electromagnetic (EM) Simulation for EUV Lithography:

In order to achieve good image contrast, the absorber thickness of EUV mask is relative large compared to the wavelength. As a consequence, for EUV lithography the simple thin mask approximation (Kirchhoff) may need to be replaced with rigorous electromagnetic (EM) simulation. However, the refractive index of most materials in the EUV region is near unity, so it is expected that EUV will have only small interactions with absorber and multilayer. The interactions with EM wave at 193nm and 13.5nm are simulated by Panoramic as shown below: the wavefront is strongly disturbed by the attenuated PSM materials. In contrast, at EUV wavelength, wavefront is hardly disturbed by absorber. From this comparison, it can be expected that relatively simple modification in Kirchhoff simulation can mimic the shadowing effect with a simple mask bias, which had been verified by other researchers [1-4]

Interaction of EM field with 193 Att-PSM mask (MoSi)

Interaction of EM field with EUV mask (TaN)

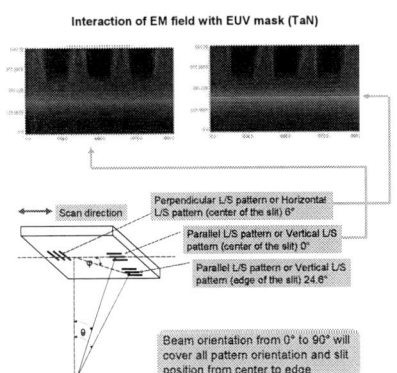

3. Impact of the Flare and Aberrations on Aerial Image Contrast and Process Windows:

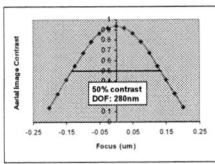

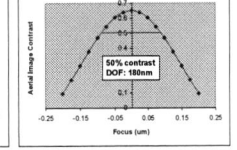

Aerial image and DOF for 32nm hp 1:1 Line/Space feature with perfect lens

Aerial image and DOF for 32nm hp 1:1 Line/Space feature with 16% flare

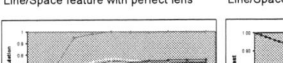

Effects of flare and aberration on aerial image contrast vs. feature hp

Aerial image contrast vs. flare level for 105nm hp 1:1 dense L/S

Simulation conditions:
32nm 1:1 L/S; flare: 0-20%; NA=0.25 Conventional, Sigma: 0.5
Kirchhoff diffraction, DOF was defined at 50% areal image contrast

4. Impact of Different Types of Aberrations on Process Windows:

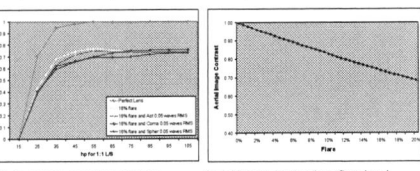

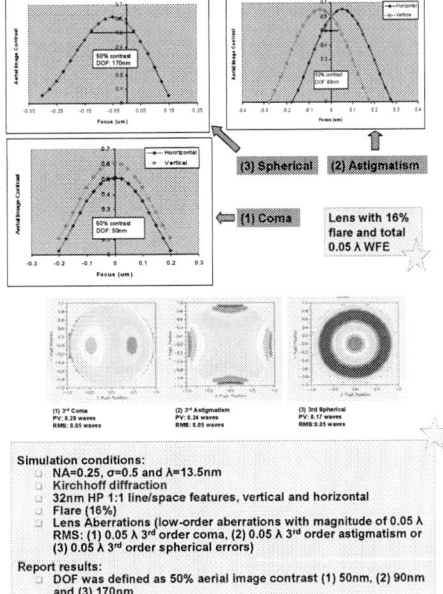

(3) Spherical   (2) Astigmatism

(1) Coma

Lens with 16% flare and total 0.05 λ WFE

Simulation conditions:
- NA=0.25, σ=0.5 and λ=13.5nm
- Kirchhoff diffraction
- 32nm HP 1:1 line/space features, vertical and horizontal
- Flare (16%)
- Lens Aberrations (low-order aberrations with magnitude of 0.05 λ RMS: (1) 0.05 λ 3rd order coma, (2) 0.05 λ 3rd order astigmatism or (3) 0.05 λ 3rd order spherical errors)

Report results:
- DOF was defined as 50% aerial image contrast (1) 50nm, (2) 90nm and (3) 170nm

5. Aberration Induced Best Focus, CD and Image Placement Errors (IPEs):

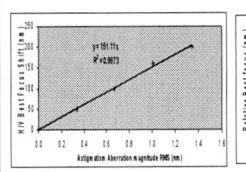

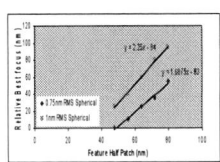

Best focus difference for Horizontal and Vertical 1:1 32nm L/S features with lens aberration of 3rd order astigmatism

Relative best focus as function of the feature half patch with 3rd order spherical aberrations

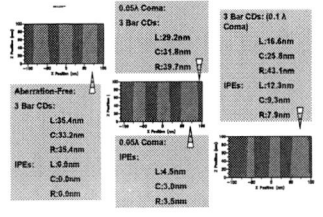

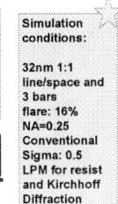

Simulation conditions:

32nm 1:1 line/space and 3 bars
flare: 16%
NA=0.25
Conventional
Sigma: 0.5
LPM for resist and Kirchhoff Diffraction

32nm 3-bar feature CD difference and image placement errors with 0.05 λ and 0.1 λ 3rd order coma aberrations

6. Conclusions:

As compared to the flare and aberration of the traditional 193 or DUV scanners, the impact of flare and aberrations on patterning performance of the EUV full field alpha tool should not be ignored: 16% flare could result in 30% reduction in areal image contrast for 32nm hp equal line/space features shown by simulation under current EUV full field alpha tool illumination conditions (ASML ADT at Albany, Conventional, NA=0.25 and Sigma=0.5).

In addition, the impact of different types of lens aberrations on depth of focus (DOF) for combination of 32nm horizontal and vertical line/space feature was simulated under same conditions with assumption of 16% flare: 0.05λ WFE of spherical gives 170nm process windows for the specific patterns, while 0.05λ WFE of astigmatism and coma only allow 90nm DOF and 50nm DOF, respectively. It is also predicted by simulation that 1nm rms 3rd order astigmatism induces 151nm best focus shift between horizontal and vertical line/space with 32nm feature half patch, 1nm rms 3rd order spherical aberration results in 2.25 of the relative best focus to the feature hp ratio, and 0.7 to 1.4nm rms 3rd order coma causes 10nm-27nm CD difference and 3 to 12nm IPEs for 32nm 3-bar features, these simulation results provide the convenient way for lithographic characterization of low-order aberrations in ADT.

7. References and Acknowledgement:

1. H. Kang, et al., Proc. Of SPIE, vol. 6921, 692131-4 (2008);
2. P. Naulleau et al., SEMATECH technology transfer #06104785A-TR;
3. F. Goodwin et al., 5th EUVL symposium, Barcelona, 2006;
4. M. Sugawara et al., 5th EUVL Symposium, Barcelona, 2006.

Acknowledgement: Authors would like to thank P. Naulleau, LBNL, J. Biafore, KLA-Tencor and J. Thackeray, R&H for their helpful discussions on lens aberration, resist and lithographic simulation.

EXTREME ULTRAVIOLET LITHOGRAPHY

Multiple Catadioptric Simplified Extreme Ultraviolet Whole Lithography Machine and System

Wynn.L.Bear,
Xiangwen,Xiong * .

ABSTRACTS

In this paper, first we proposed the theoretical basis of the principle of a whole formation image with optics mutilation lens, and then make use of such the theory and principle to deduce a mathematic model and formulas of judgment. A basic principle that has no disturbances for two cross beams in their propagation is introduced. Based on such the principle and methods, we put forward a mixed light principle by using the laser light and EUV. Based on the mixed light principle we invented a design and manufacturing method for such the mixed light. According to above principle, a multiple Catadioptric multiband lithography methods, and composite double reflection multilane methods had been form. In this paper, a comprehensive research and systematic analysis for the multiple Catadioptric lithography been described, including the principle and structure of the systems of system, chemical-resist, flare, resolution, line-edge roughness and sensitivity, as well as the research of the EUV mask and EUV integrated system, etc. This is a pleasant surprise to the conclusion is that they are almost all compatible for laser resist and EUV. By the RLS trade-off and the size of the power, we get the quality of qualified, extremely cost-effective and yield for the IC Manufacturing of 45, 32, 22 nm node and less, and be use for the factory immediately. Such the simplified EUVL system also facilitates people to upgrade to a standard EUVL machine at anytime.

We consider that the current lithography problem is like the CPU with difficulty from single-core to multi-core. Our simplified EUVL system, in essence, using the core of the carrier and functions of the platform to develop into the leading edge Sci & Tech integrate the cutting edge application system as a whole to solve the technology problem for current lithography, and to match the trend and progress as the CPU of computer from the single-core age enter multi-core. Its performance has been significant improvements like the advanced multi-core CPU also.

INTRODUCTION

2008 International Symposium on Extreme Ultraviolet Lithography, SEMATECH.
September 28 - October 1, 2008
Lake Tahoe, California, USA.
Wynn.L.bear, Wynn Bear International Cutting-edge High-Tech Institute, CA, USA
Xiangwen,Xiong * , Wynn Bear International advanced Business Machines Company, China.
Tel: (028) 38902-0677, Fax: (028) 38902-0677, Email: wynnbear@gmail.com, wynnbear.inc@gmail.com.
(Wynn.bear is the English name, Xiangwen, xiong is the pinying name).

We have proposed new methods that called simplified Extreme Ultraviolet Lithography to let the EUVL to use to the production line immediately. Our goal is that to let the EUVL technology should be a place in the 32nm node, to become mainstream lithography technology in the 22nm node, and in 16 nm and less are an absolute dominance. At the same time we also proposed the cost analysis for our technology with other two kinds of competition technology (NIL and MEBDW, etc), the cost comparison shows that the simplified EUVL is the only next-generation lithography technology for its tremendous cost advantages and the economy. We proposed the basic principles and implementation examples for the catadioptric multiband simplified Extreme Ultraviolet whole Lithography Machine and system in this paper. We will be the first to make the simplified Extreme Ultraviolet Lithography to first use in the actual production process, at the same time to let the standard Extreme Ultraviolet Lithography to win more time and funding to enable it to further improve and enhance to fully capture the market and co-exist with the simplified EUVL for the respective cost advantages of owner. The use of our technology, semiconductor manufacturers will more quickly be advanced technology and be cost-effective, the semiconductor industry could soon see more beautiful and better prospects for hope.

Like an F-16 fighter in comparison to an F-15 or an F-35 fighter in comparison to an F-22, a simplified EUVL is likely to become the mainstream of NGL for silicon-based semiconductor manufacturing. The Simplified EUVL includes the Catadioptric multiband simplified and Bireflectance multiband simplified types, etc. The Catadioptric multiband simplified EUVL is the use of the whole imaging principle of a mutilation lenses and reflection imaging to integrate as a whole, and a basic principle that has no disturbances for two cross beams in their propagation. Such a method and system let the high density lines and low density on the chip been created by different band light, respectively. It will immediately be applied to production lines.

THE BASE PRINCIPLE OF THE CATADIOPTRIC MULTIBAND SIMPLIFIED EUVL

The Catadioptric multiband simplified type EUVL may be the best strategy given the current state of EUVL technology and be immediately applied to the actual factory production process. It is the use of the light from a 193 nm ArF source and an EUV source. We cut out a small hole in the lens of a 193 nm ArF lithography system and then let the EUV light beam pass through the small hole in the lens. The EUV light and the 193 nm ArF laser then are mixed together in this clever and unique way. The EUV light creates the high density lines on the chip, and the 193nm ArF laser is for general lines. This machine will be used immediately in the factory production line. About cut out a small hole in the lens of a 193 nm ArF lithography system, the hole that cut out by us will damage the lines graph on the chip or not? Do not worry about this problem. This is a whole imaging principle of mutilation convex lens. A convex lens that been any mutilation will not affect the integrity of graphs and imaging, it only will be stronger or weaker. Such the convex lens and system will be cut into any shape that we need, so that we get any space shape to set the EUV light and EUVL device & parts to let the 193 nm ArF laser and Extreme Ultraviolet beam been mixed together in such the way. The EUV light creates the high density lines on the chip, and the 193nm ArF laser is for general lines. This machine will be used immediately in the factory production line.

1419

The standard "EUVL is now well into the alpha tool demonstration phase, in which critical elements of the technology and infrastructure are being evaluated at the system level. However, the industry still faces a significant number of challenges to prepare for the arrival of EUV beta and pilot production tools in the next few years. These challenges include progress in key critical technology issues such as Demonstration of reliable high-power source, collector modules, simultaneously meeting to resist resolution, sensitivity and line edge roughness specifications and fabricating defect-free masks, etc" [1]. It needs a lot of new idea, developments and solutions for the advancement of extreme ultraviolet lithography into cost-effective, manufacturable solutions. Our simplified Extreme Ultraviolet whole Lithography fully meet these needs. The Lithography is so important to advanced IC manufacturing, so we must be careful to treat it.

The base principle of the whole lens and the mutilation lens

The imaging of the convex lens is a very ancient topic that used the principle of light refraction and the convex lens on the convergence of light. The figure – 1 is shown the imaging principle of a standard convex lens, AB is the object, A′ B′ is the image. Now, we set up a barrier or obstacle between the lens and the objects, what would the results be? The result is the exactly same, only the imaging is weaker. It is shown in figure – 2.

Undoubtedly, the size of the obstacles is less than Lens.

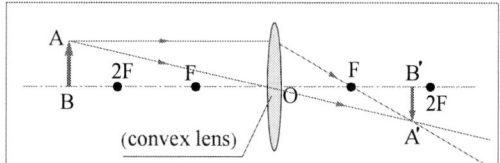   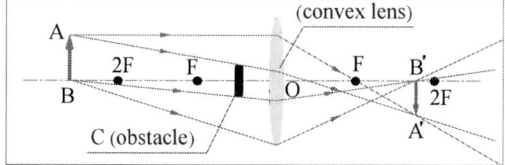

Figure–1   **Figure–2**

The image definition (real image principle) and image intensity are two important factors for optical lithography, so that a detailed description of the above-mentioned 2 factors for mathematical models and formulas are necessary.

The concrete expression for the whole lens and the mutilation lens

The Formula -1 is the concrete expression for the Figure–1.

$$F(x_1) = |T_1| = \begin{cases} 1/F = 1/U + 1/V \\ \text{(U is object distance, V is image distance, F is focal length)} \\ II = K \cdot \dfrac{AB}{A'B'} \cdot f \\ \text{(II is the image intensity, K is light intensity, f is the strength coefficient)} \end{cases}$$

Formula -1

1420

The Formula -2 is the concrete expression for the Figure–2.

$$F(x_2) = |T_2| = \begin{cases} 1/F = 1/U + 1/V \\ \text{(U is object distance, V is image distance, F is focal length)} \\[2mm] II = K \cdot \dfrac{AB}{A'B'} \cdot \dfrac{S-S'}{S} \cdot f \\ \text{(II is the image intensity, K is light intensity, f is the strength coefficient,} \\ \text{S is the surface projected area of the lens, S' is the surface projected area of the obstacle)} \end{cases}$$

Formula -2

The Base Principle of the Dual-System Combination

The principle of the mutilation lens is the base for our catadioptric multiband simplified Extreme Ultraviolet Lithography technology. Our system will be successful when we use another system to replace the barriers in the figure-2. Now, we give more graphics and text to explain them.

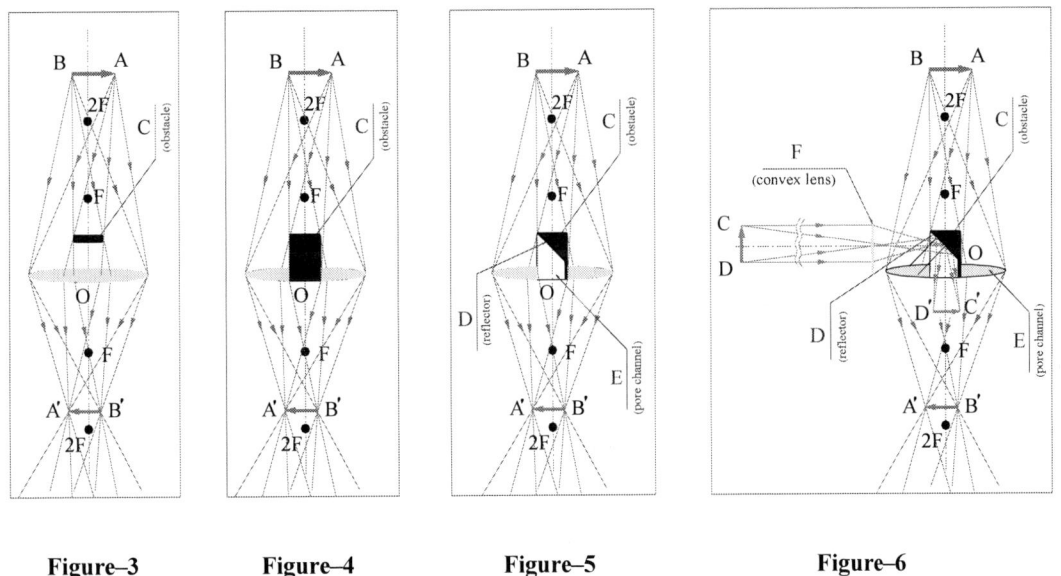

| Figure–3 | Figure–4 | Figure–5 | Figure–6 |

The more detailed for the figure – 1 is shown in figure – 3. Figure – 4 is the more detailed for the figure – 2. Figure – 5 is the concrete structure for obstacle and projection system. In the Figure – 5, a pore channel and reflector had been created; a new light beam be passing through this structure.

Figure – 6 is shown a new projection system of imaging pass by the Figure – 5, and integrated as a whole with it. The figure – 6 is only shown a basic principle for the integration of the multi-beam. Actually, we cut out any shape that we need, so that we can set a lot of beams and device into it. Base on such the basic principle that have no disturbances for two cross beams in their propagation, such the beams will imaging independently. It realizes the Catadioptric multiband simplified Extreme Ultraviolet whole Lithography method and system when we set the EUV beam and reflective illuminators on such a place.

The Types of Mutilation Lenses

There are many types of mutilation lenses be proposed by us. Although the convex lens that been any mutilation will not affect the integrity of graphs and only is the imaging stronger or weaker, we continue to provide a wide range of choices for the complexity of lithography. They had been shown in figure – 7.

More details will be provided in the future research and papers.

Figure–7

THE CATADIOPTRIC MULTIBAND SIMPLIFIED EXTREME ULTRAVIOLET WHOLE LITHOGRAPHY MACHINE AND SYSTEM

Base on the basic principle that has no disturbances for two cross beams in their propagation and the imaging principle of the mutilation lens, we let the 193 nm ArF laser lithography system and the low-power EUVL system to integrate as a whole.

An Example for Catadioptric Multiband EUVL System

Such the Catadioptric multiband simplified Lithography Machine and system is shown in figure-8. The 193 nm laser Lithography system is used the mutilation lenses, and the EUV Lithography system is used the low power EUV source. The laser Lithography system creates the low density lines, and the EUV system do the high density.

The 193nm excimer laser is a well-market test of mature Lithography technology. The system of the important factors includes the optical system design, optics materials, excimer laser source, resists, etc. Its performance be greatly improved by using to polarized illumination, higher-index materials, solid mask immersion, and double exposure/patterning, etc [2].

Extreme Ultraviolet Lithography (EUVL) is a leading candidate for next generation lithography technology, with wavelengths of 13.0nm, and is expected to support multiple technology generation from 32 to 22nm node and less.

Figure 8 is a typical example for the Catadioptric multiband simplified Lithography Machine and system.

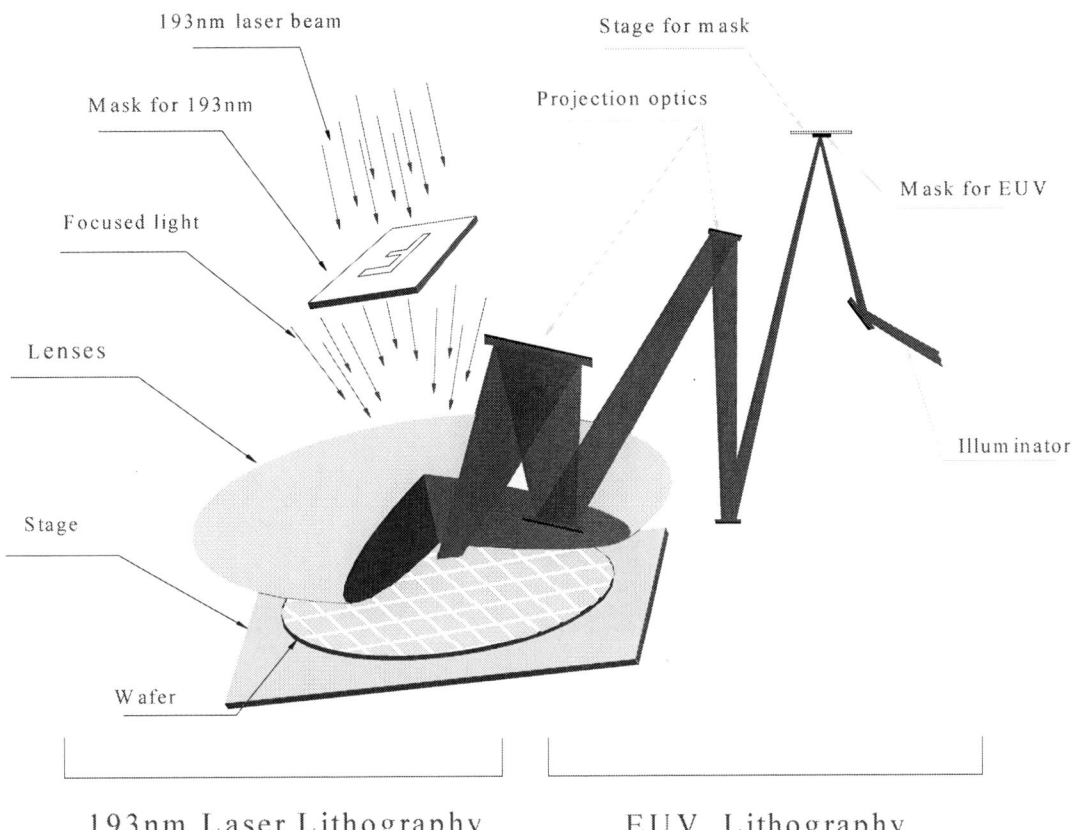

193nm laser beam

Mask for 193nm

Focused light

Lenses

Stage

Wafer

Stage for mask

Projection optics

Mask for EUV

Illuminator

193nm Laser Lithography   EUV Lithography

Figure–8

Immersion Technology for Catadioptric Multiband EUV Lithography

Because the EUV Lithography generally exists in a vacuum environment, so that the solid immersion lens is the best option for the Catadioptric multiband simplified Lithography systems. Figure – 9 is shown such a solid immersion lenses system for it. But, if we give it a special vacuum channel for EUV beam, then the immersion Lithography also be considered, or be given a special sealing for immersion fluid, it is shown if figure-10.

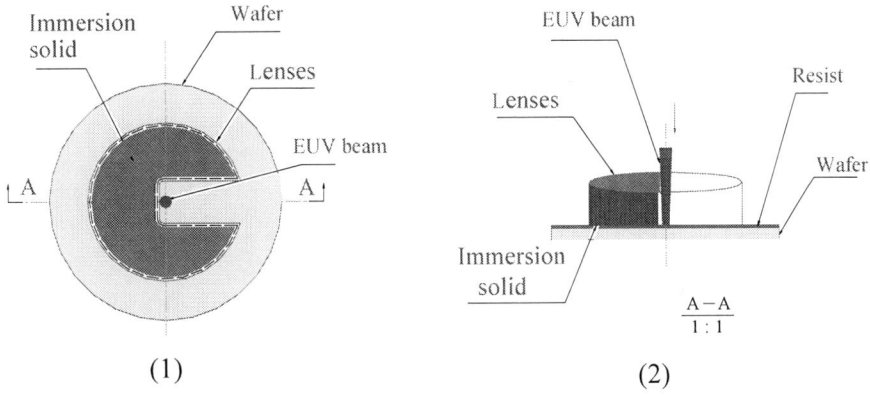

Immersion solid

Wafer

Lenses

EUV beam

A   A

(1)

EUV beam

Resist

Lenses

Wafer

Immersion solid

$\dfrac{A-A}{1:1}$

(2)

1423

Figure–9

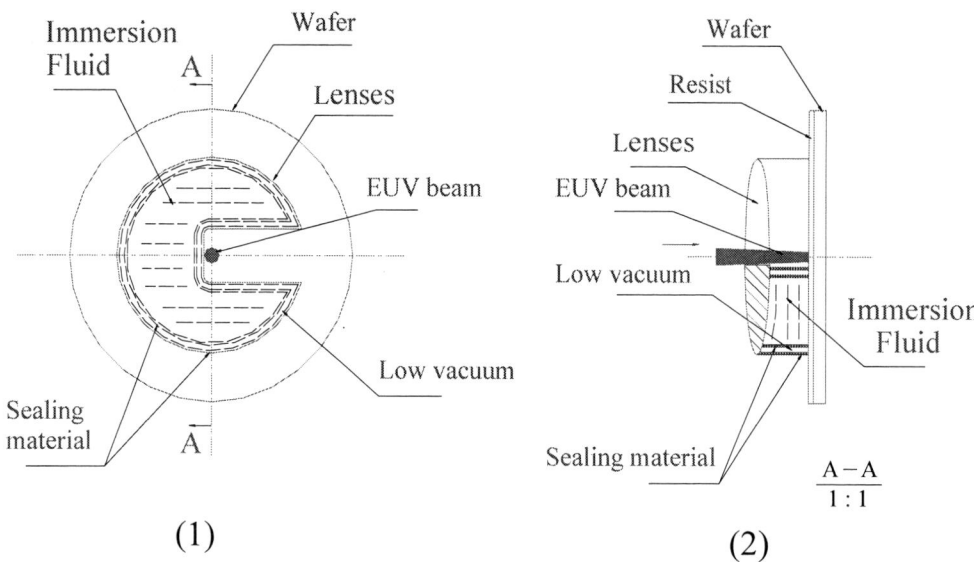

(1)   (2)

Figure–10

The Impact for Flare on Catadioptric Multiband EUVL System

With the imaging of the mutilation lens being adopted, the impact of flare on performances for the Catadioptric Multiband EUV lithography System is crucial. To analyze and control the flare in such the method and system is more beneficial. We smear a layer of absorption on the surface of the hole in the mutilation lens, such the absorption layer can absorb the flare, at the same time the heat dissipation on the absorption layer is easy when we set a passageway for it. Such the passageway need an especially device for it is in vacuum environment. Truthfully, we take the implementation of cooling in the interval time that the replacement of wafer.

Curved Lenses to Sustain Hyper NA for the Catadioptric Multiband EUVL System

Despite the surface of the lens curve down a bright "star" will forms, the center range of a reflective light field will be form "dark hole", the current Curved lens has already been entered the stages to actual use. Such the Curved lens more improve the efficiency of the imaging.

Figure–11 is shown the Catadioptric Multiband EUVL System by using such the curved lens. Whether "preserving the incident angle in resist to improve DOF, or Preserving the physical angle in coupling medium to improve resolution" [2], the Curved lens Lithography system be used to further meet the actual needs.

Figure–12 is shown a Catadioptric Multiband EUVL System that using a quartering shot EUV beam. The different shapes in lenses are shown in Figure–7.

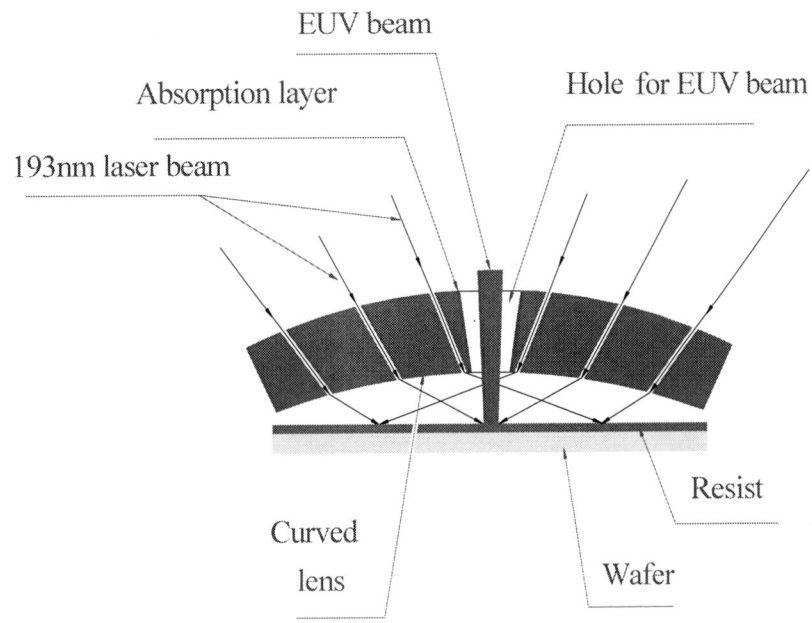

Figure–11

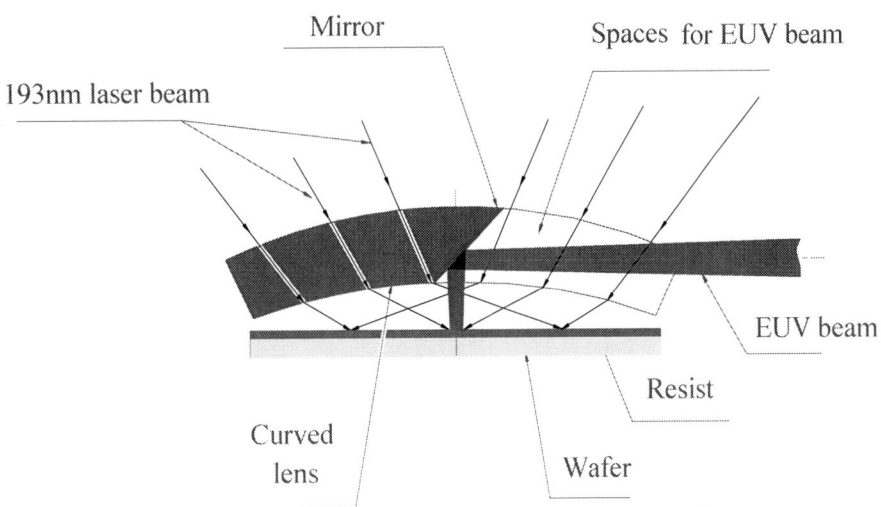

Figure–12

The eccentric structure and forms still be accepted and used, here shown a typical example in Figure–13 and Figure–14. The Eccentric structure will help us to realize the extreme accuracy overlay control and dynamic measurement, and in the back of our paper will be detailed.

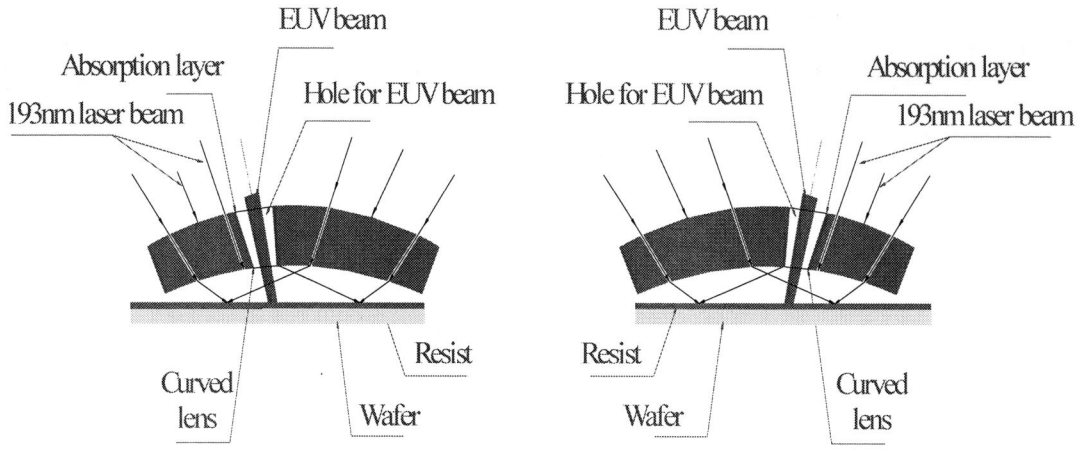

Figure–13   **Figure–14**

Extreme accuracy dynamic overlay control

When we used the eccentric structure and forms like the Figure–13 and Figure–14, the patterning create by laser and EUV will not be coincided. We are enabling to use the dynamic measure method by laser interferometer to realize the extreme accuracy overlay control. The concrete execution measure and process are shown in Figure–15.

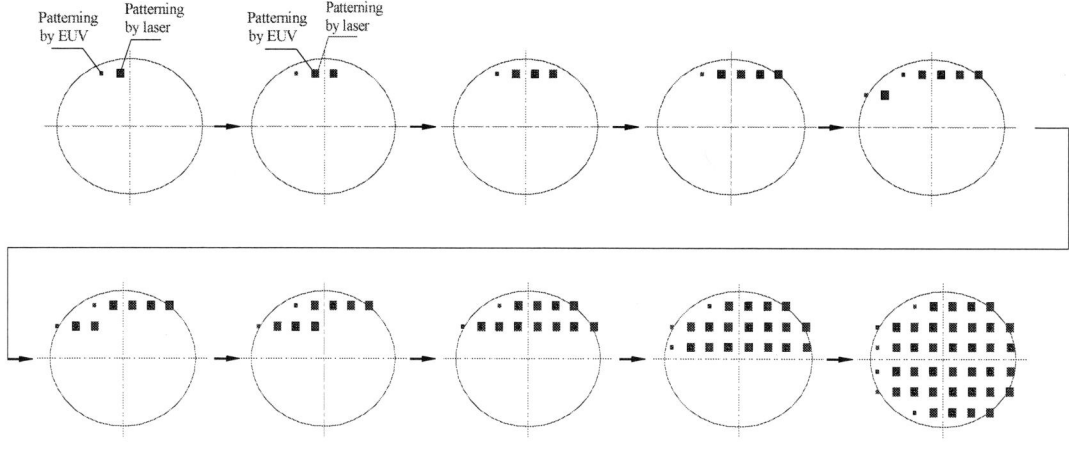

Figure–15

THE TECHNICAL CONDITION & IMPACT TO THE CATADIOPTRIC MULTIBAND EUV LITHOGRAPHY

Doubtlessly, our simplified Extreme Ultraviolet Lithography first use in the actual production process than NIL even is the MEBDW. We let the standard Extreme Ultraviolet Lithography to win more time and funding to enable it to further improve and enhance to fully capture the market and co-exist with the simplified EUVL for the respective cost advantages of owner. Nevertheless, our technology involved a mixed light source from a 193 nm ArF and an EUV, so that all the issues related to 193 nm ArF lithography and EUV lithography must be a comprehensive analysis.

Incomplete lens and interference of cross beams

Based on such the whole imaging principle of mutilation lenses and the basic principle that has no disturbances for two cross beams in their propagation, above issues will be no problem.

Photoresists

Almost all researchers that focused on EUV photoresists have found that to deal with the problems encountered at 193 nm laser, which are becoming very similar to those for EUV. Although we also are experimenting with the process, but we boldly assume that such effective materials are both sensitive to the 193 and EUV is existent [11], [6].

The Impact of Flare

This is not a problem, even better with our program. It has been described in page 6.

Resolution, line-edge roughness and sensitivity

The College of Nanoscale Science and Engineering (CNSE) at the University at Albany in New York, and other university and research institution were done a lot of successful study in Resolution, line-edge roughness and sensitivity, "new series of resists (MET-2A, 2B. 2C, 2D) also show great promise for good resolution, LER and sensitivity."[9]. If you would like to ask our technology is really met the core of progress, and then on this issue, we will provide a great advantage. As you know, our simplified Extreme Ultraviolet Lithography is using the mixed source by the EUV light and the 193 nm ArF laser, it means that the majority of the energy of this lithography system be provided by the laser; In other words, we reduce the exposure area of the EUV to guarantee the provision of adequate energy.

When it was sufficient energy, the chip would have been greatly improving the Resolution, line-edge roughness of the lines or not? Yes! At the College of Nanoscale Science and Engineering (CNSE), they even put forward a new theory named the RLS trade-off [12], [11]. Robert Brainard, an associate professor at the CNSE said that "the EUV resist challenges be summed up succinctly. They are the three basic properties — resolution, line-edge roughness and sensitivity. My colleagues and I call it the RLS trade-off. You need all three properties for EUV [and EUV resists] to be successful, and when you improve one, you make another one worse."

About the photoresists and line edge roughness (LER), Dr. Wurm, S said that "All three lithography technologies currently competing for insertion at the 32 nm hp node pose serious challenges in the resist area that must be resolved. Specifically, meeting line edge roughness (LER) specifications will be difficult for all technologies but more so for EUV than for double exposure lithography (DEL) and 193+." [13]

"We focus on cross-comparisons in which selected EUV and 193 nm resist are evaluated using both EUV and 193 nm lithography. Simulation methods linking 193 nm and EUV performance will be described as well. Results from simulation indicate that image blur in current generation 193 nm photoresists is comparable to that of many EUV resists..." [17].

Obviously, we would rather sacrifice part sensitivity to gain enough resolution and line-edge roughness; the key is that we can.

Yield

Our output is one of the major advantages. We use the mixed light source that has sufficient energy easily to let yield to reach 160 – 190 wph.
.

Mask

Our technology may do not need the Optical Proximity Correction (OPC) and the Phase-Shifting Mask (PSM). EUV mask even though has still some problems, but you can use high-quality and reliable source that is low-power for our lithography systems. Obviously, it will be good to solve such the problem.

Cost advantage

Our cost is much lower than NIL and MEBDW, even still much lower than double patterning lithography. We will describe it in a specialized paper.

A CHANGE CONCEPT FOR IC DESIGNER TO MEET CHALLENGES IN THE NEW SITUATION

IC designers should have changed the concept. In the past, they are in all important units to use the same lines. Now, they should distinguish the most important element, sub important and common, etc, and then in the most important area to use the smallest lines, and in sub important by second-class lines, and others.

In a word, different lines are different cost for the chip. We use the smallest lines to create the unit that bear most arduous tasks; in the unit that only is ordinary tasks we use the general lines. Take the

Highway as an example; it includes interstate highways, trunk roads, and the local highway, etc.

CONCLUSIONS AND DIRECTIONS FOR FUTURE RESEARCH

Our simplified EUVL system using the multi-system instead of the single system to solve the technology problem of current lithography, and let the lithography technology from the single system age enter the multi-system to match the trend as the CPU of computer from the single-core age enter multi-core. Its performance has been significant improvements like the advanced multi-core CPU also. However, from the view of the correlation and order relationship, in the development of the multi-core (multi-system) Lithography technology should be prior to CPU multi-core technology, rather than falling behind in it. As a result, the time to manufacturing technology researchers like us is very tight.

At present, the Progress and development in the Simplified Extreme Ultraviolet Lithography, includes the Catadioptric multiband simplified EUVL and Bireflectance multiband simplified is being made on a number of fronts in our company. The whole imaging of the mutilation lenses have been developed to improve machine operation and better adaptability are being implemented. Except Modeling and simulation tools to visualize the possibility of the simplified EUVL machine and other characteristics, we have already begun to create such the lens and related devices.

So that, we do not intend to independent research and development any devices & equipment to relate the EUVL, our wish is to find some companies or organizations that have such the ability to carry out cooperative study and research to speed up the progress of the study. You may also remember a case that is about the engine, for more low-cost and be the high-power, it uses a wide range of metal materials prices, and multi-cylinder structure, but no one material prices and one cylinder. Another example is that a fighter general use one engine, but the commercial airliner is two, and four with the transport, and Stratofortress achieved eight.

The Simplified Extreme Ultraviolet Lithography of our company efforts to the interactions with industry and other researchers indicate the following as significant works for future to fully realize:

1. We realize the whole imaging by using the mutilation lenses.

2. Such the mutilation lenses, many types can be manufactured by us.

3. We mainly use the 193nm dry lithography machine for our Simplified EUVL method and system.

4. The Curved Lenses is strongly recommended by us.

5. A type of the photoresists both to meet the requirements of the EUV and 193 nm laser is needed to be further verified.

6. A type of the photoresists is that higher resolution & line-edge roughness, and lower

sensitivity will be needed.

7. A EUV source that is high-quality & reliable and low-power are more conducive to EUV mask and projection optics will be strongly needed. In other words, it is no longer just focused on the size of the power but to pay attention to overall balance to meet the needs of the EUV mask and the higher resolution & line-edge roughness, etc.

8. Our Simplified Extreme Ultraviolet Lithography will be given a special design to achieve the lowest cost. If the design to be improved to adapt this Manufacturing technology, we will further reduce the cost.

9, Based on above methods and research, to be found the best structures, installations and system by contrast.

10, This lithography system not only the Catadioptric multiband simplified EUVL machine, but also be used for the independent 193nm laser lithography machine, and EUV lithography machine. Such the function will be very useful to the factory.

11. When we use the standard lens to replace the last several lenses that have the holes in the exposed system, we resume the standard 193nm laser lithography machine.

12, Whether the multiple Catadioptric multiband lithography methods, or composite double reflection multilane methods, in essence, this is from a single system develop into the multi-system; It is the same trend like CPU of computer from the single-core age enter a age of the multi-core.

13, Our Simplified EUVL system facilitate people to upgrade to a standard EUVL machine at anytime.

14, Our Simplified EUVL method is the use of the principle of a whole formation image with optics mutilation lens, and the basic principle that has no disturbances for two cross beams in their propagation to integrated a core system like the advanced multi-core CPU, instead of Simplified joint.

ACKNOWLEDGMENT

It is a hopeless understatement to say that I am deeply grateful that except many professors and other critical thinkers, including this symposium's organizers, etc, I would like to especially appreciate the Board of Reviewing Editors of the *Science* magazine, and Ph.D Phillip D. Szuromi, the Supervisory Senior Editor; Ph.D Gilbert J. Chin, the professional Senior Editor. It is they true support the EUVL is the most possible IC manufacturing technology to meet the Moore's Law, and to be the next generation lithography.

REFERENCES

[1], Semiconductor Manufacturing Technology, www.sematech.org.

[2], Burn J. Lin, Optical lithography—present and future challenges, C. R. Physique 7 (2006) 858–874

[3], Kurt Ronse, Optical lithography—a historical perspective, C. R. Physique 7 (2006) 844–857.

[4], Runcai Miao, Zongli Yang, Visualization of low-frequency liquid surface acoustic waves by means of optical diffraction CJ3. Applied Phys. Lett. 2002，80(17)：3033~3035

[5], YANG Zong—li, ji Shu—li, MIAO Run-cai, ZHANG Zong—quan, Effects of cascade nonlinear medium spatial configuration on optical field characteristics. journal of shanxi university of science & technology, Apr. 2004, Nol, 22

[6], Brainard, Robert L.; Barclay, George G.; Anderson, Erik H.; Ocola, Leonidas E. Resists for next generation lithography. Microelectronic Engineering (2002), 61-62 707-715.

[7], Brainard, Robert L.; Cobb, Jonathan; Cutler, Charlotte A. Current status of EUV photoresists. Journal of Photopolymer Science and Technology (2003), 16(3), 401-410.

[8], Brainard, Robert L.; Trefonas, Peter; Lammers, Jeroen H.; Cutler, Charlotte A.; Mackevich, Joseph F.; Trefonas, Alexander; Robertson, Stewart A. Shot noise, LER, and quantum efficiency of EUV photoresists. Proceedings of SPIE (2004), 5374 (Pt. 1), 74-85.

[9], Thomas Köhler, Robert L. Brainard, Patrick P. Naulleau, David Van Steenwinckel, Jeroen H. Lammers, Kenneth A. Goldberg, Joseph F. Mackevich, and Peter Trefonas Performance of EUV Photoresists on the ALS Micro Exposure Tool, Proceedings of SPIE 2005, 5753, pp 754-764.

[10], Koehler, Thomas; Brainard, Robert L.; Naulleau, Patrick P.; Lammers, Jeroen H.; Steenwinckel, David Van; Goldberg, Kenneth A.;Mackevich, Joseph F.; Trefonas, Peter; Performance of EUV Photoresists on the ALS Micro Exposure Tool, 3rd International EUVL Symposium 2004.

[11], http://www.semiconductor.net/article/CA6529237.html.

[12], http://www.cnse.albany.edu/.

[13], Wurm, S. Byers, J. Zimmerman, P. Wallow, T. Dean, K. Photoresist challenges and potential solutions for the 32 nm half-pitch node and beyond. Microprocesses and Nanotechnology, 2007 Digest of papers, 2007, 428-429.

[14], Mccall, Monnikue M ; Han, Hakseung ; Cho, Wonil ; Goldberg, Kenneth ; Gullikson, Eric ; Jeon, Chan-Uk ; Wurm, Stefan. Determining the Critcial Size of EUV Mask Substrate Defects. Journal: International Society for Optical Engineering (SPIE); Journal Volume: 6921.

[15], Stefan Wurm, Hakseung Han, Kenneth A. Goldberg, Anton Barty, Eric M. Gullikson,Yoshiaki Ikuta, Toshiyuki Uno, Obert R. Wood. EUV MET printing and actinic imaging analysis on the effects of phase defects on wafer CDs. Proceedings of SPIE, 2007.

[16], Stefan Wurm, Hakseung Han, Patrick Kearney, Wonil Cho, and Chan-Uk Jeon,Eric Gullikson. EUV mask blank defect inspection strategies for 32-nm half-pitch and beyond. Proc. SPIE, Vol. 6607, 66073A (2007); DOI:10.1117/12.729029.

[17], Jones, Juanita ; Pathak, Piyush ; Wallow, Thomas ; LaFontaine, Bruno ; Deng, Yunfei ; Kim, Ryoung-han ; Kye, Jongwook ; Levinson, Harry ; Naulleau, Patrick ; Anderson, Chris. A Comparison of Photoresist Resolution Metrics using 193 nm and EUV Lithography. SPIE Advanced Lithography, San Jose, CA, Feb. 24-29, 2008.

[18], Jones, Juanita ; Anderson, Christopher ; Naulleau, Patrick ; Niakoula, Demitra ; Hassanein, Elsayed ; Brainard, Robert ; Gallatin, Gregg ; Dean, Kim. Influence of base and PAG on deprotection blur in EUV photoresists and some thoughts on shot noise. 2008 Internation coference on electron,

ion, and photon beam technology and nanofabrication, Portland, OR, May 27-30, 2008.

[19], Yumi Nakajima, Takashi Sato, Ryoichi Inanami, Tetsuro Nakasugi, and Tatsuhiko Higashiki. Aberration budget in extreme ultraviolet lithography. Emerging Lithographic Technologies XII, SPIE, 2008.

[20], S. Bollanti, P. D Lazzaro, F. Flora, L. Mezi1, D. Murra and A. Torre. Conventional and modified Schwarzschild objective for EUV lithography: design relations. Applied Physics B: Lasers and Optics. 603-610, 2006.

[21], M Chandhok, SH Lee, J Roberts, BJ Rice, HB Cao. Lithographic flare measurements of Intel's microexposure tool optics. Journal of Vacuum Science & Technology B: Microelectronics and Nanometer Structures -- January 2006 -- Volume 24, Issue 1, pp. 274-278.

[22], J. G. Hartley, S. Raghunathan, and A. Govindaraju. Electrical characterization of multilayer masks for extreme ultraviolet lithography. Journal of Vacuum Science & Technology B: Microelectronics and Nanometer Structures -- November 2005 -- Volume 23, Issue 6, pp. 2891-2895.

[23], Heidi B. Cao, Wang Yueh, Bryan J. Rice, Jeanette Roberts, Terence Bacuita, and Manish Chandhok. Sources of line-width roughness for EUV resists. Advances in Resist Technology and Processing XXI, SPIE, 2004.

[24], Manish Chandhok, Sang H. Lee, and Terence Bacuita. Effects of flare in extreme ultraviolet lithography: Learning from the engineering test stand. Journal of Vacuum Science & Technology B: Microelectronics and Nanometer Structures -- November 2004 -- Volume 22, Issue 6, pp. 2966-2969.

[25], O. R. Wood II, D. Back, R. Brainard, G. Denbeaux, D. Goldfarb, F. Goodwin, J. Hartley, K. Kimmel and C. Koay, etc. Initial experience establishing an EUV baseline lithography process for manufacturability assessment. Proc. SPIE, Vol. 6517, 65170U (2007); Emerging Lithographic Technologies XI.

[26], Gregory Denbeaux, Rashi Garg, Chimaobi Mbanaso, Justin Waterman, Leonid Yankulin, Alin Antohe, Yu-Jen Fan, and Warren Montgomery, etc. Extreme ultraviolet resist outgassing and its effect on nearby optics. Proc. SPIE, Vol. 6921, 69211G (2008).

Reliability and productivity improvements on the Intel MET

Roman Caudillo, Jeanette Roberts, Terence Bacuita, Gilroy Vandentop

Intel Corporation

September 29, 2008

International EUVL Symposium 2008, Lake Tahoe, CA USA

Background

- **The Intel MET, the world's first commercial EUV exposure tool, was installed in Intel's development fab in 2004**
  - 0.3 NA, 2 mirror PO (5x demag), 600um x 600um static field, annular illumination (0.55 sigma outer/0.36 sigma inner), Xe-fueled z-pinch plasma, λ = 13.5 nm

- **Since its installment, the Intel MET has demonstrated significant improvements in reliability and productivity:**
  - Yearly uptime has steadily increased from 34% to a current value of 87%
  - Productivity has increased from less than 1 J/cm^2 of dose delivered per day to nearly 7 J/cm^2 per day

- **MET productivity is enabling critical resist screening and development**

Schematic of MET

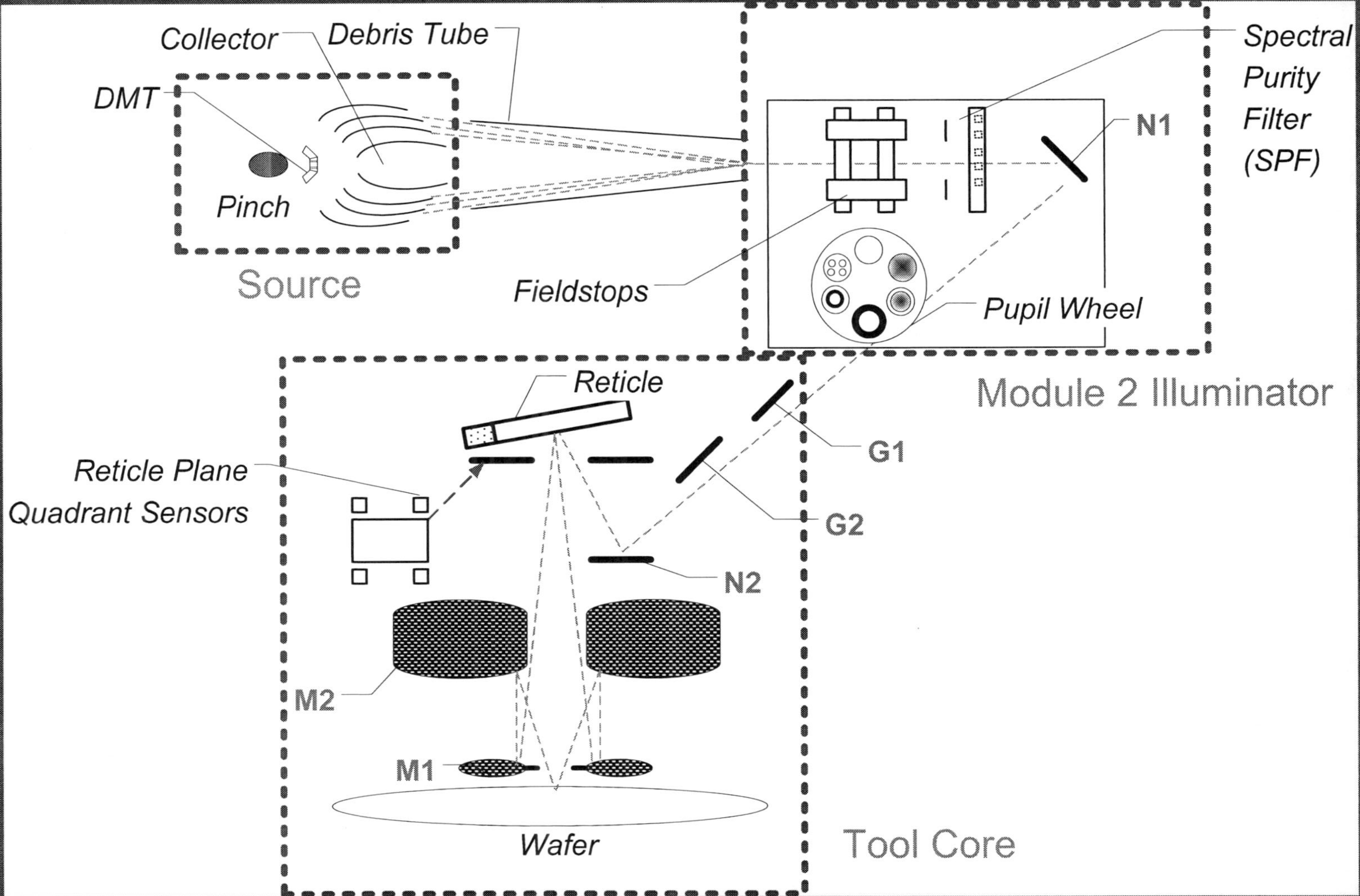

Outline

- **Review MET reliability and productivity trends**
  - Yearly Uptime %
  - Total Dose Delivered

- **Productivity is enabled by increased wafer plane power**
  - New illuminator optics installed in Dec. 2007
  - Appropriately selected field stop position may help extend optics lifetime

- **Original Projection Optics**
  - Tool flare has increased at a rate of ~1% per 2000 J/cm^2 of dose delivered but is still low compared to other EUV tools

- **High productivity has enabled critical resist development and resulted in consistent imaging of 28 nm HP lines with low LWR (~4 nm)**

Reliability as measured by uptime

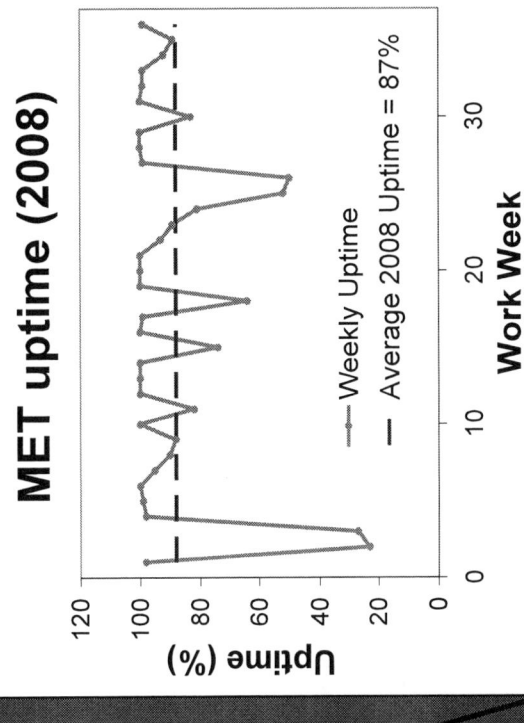

MET uptime (2008)

- Weekly Uptime
- Average 2008 Uptime = 87%

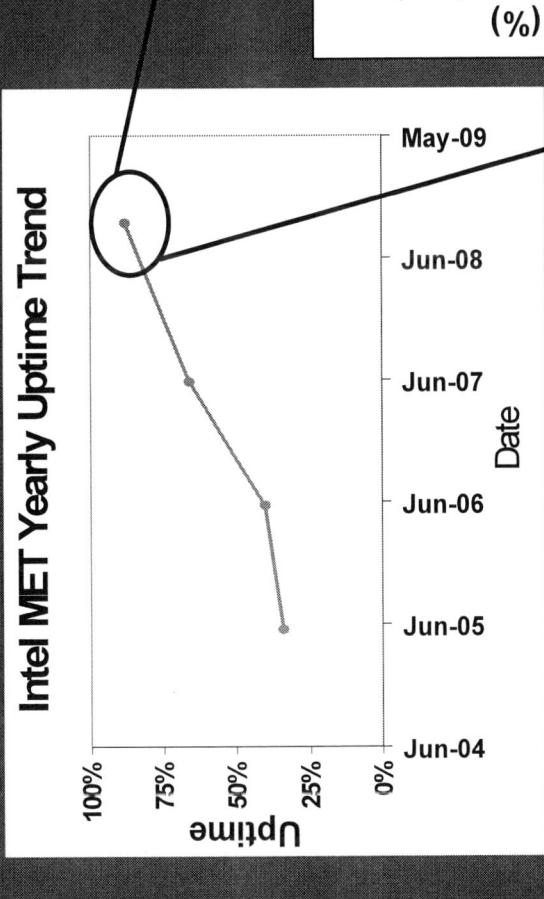

Intel MET Yearly Uptime Trend

- MET operates on a 12 hour x 7 day/week schedule

- 2008 Average uptime = 87%

Productivity as measured by cumulative dose delivered

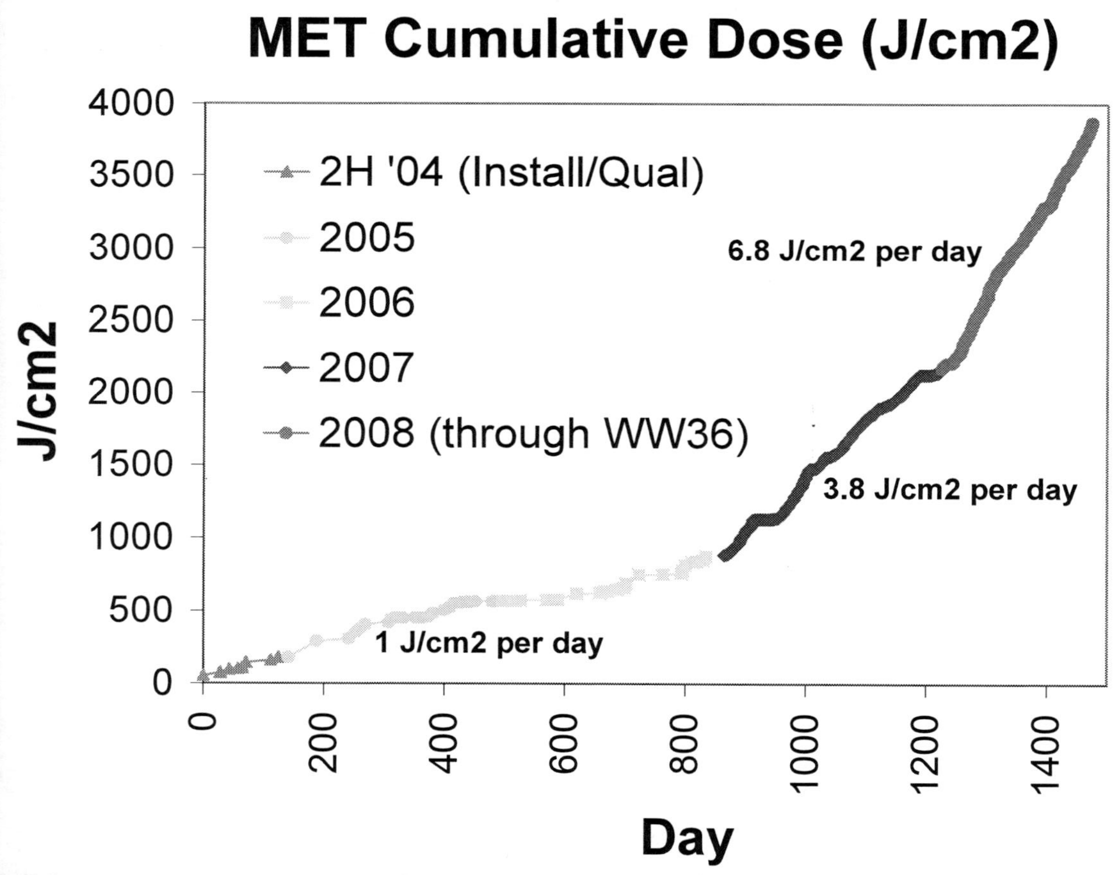

MET Cumulative Dose (J/cm2)

Note: Dose calculated using updated dose calibration (based on NIST calibrated photodiode)

Legend:
- 2H '04 (Install/Qual)
- 2005
- 2006
- 2007
- 2008 (through WW36)

6.8 J/cm2 per day

3.8 J/cm2 per day

1 J/cm2 per day

X-axis: Day (0, 200, 400, 600, 800, 1000, 1200, 1400)

Y-axis: J/cm2 (0, 500, 1000, 1500, 2000, 2500, 3000, 3500, 4000)

Continuous improvement in dose delivered

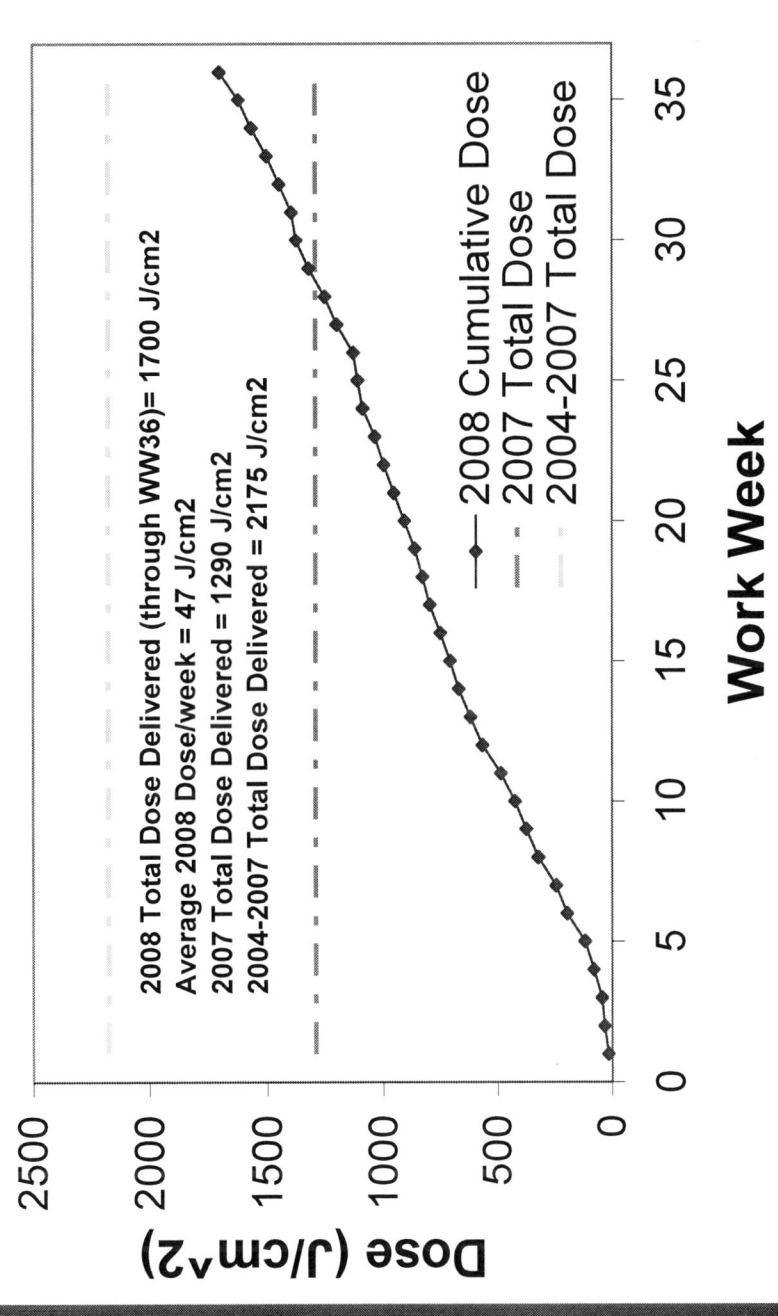

MET dose delivered (2008 vs previous)

2008 Total Dose Delivered (through WW36)= 1700 J/cm2
Average 2008 Dose/week = 47 J/cm2
2007 Total Dose Delivered = 1290 J/cm2
2004-2007 Total Dose Delivered = 2175 J/cm2

— 2008 Cumulative Dose
-- 2007 Total Dose
-- 2004-2007 Total Dose

Work Week

Dose (J/cm^2)

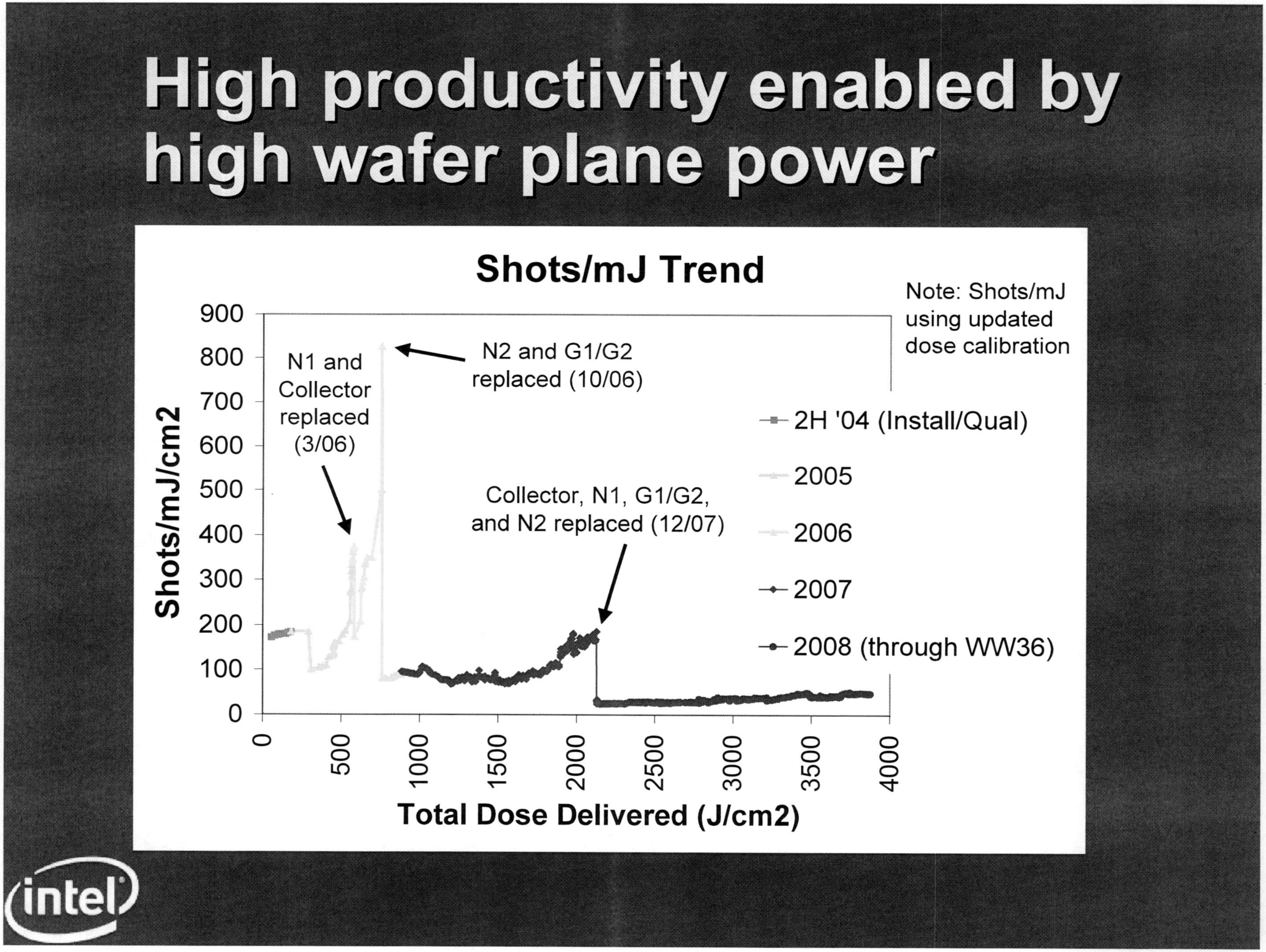

High productivity enabled by high wafer plane power

Shots/mJ Trend

Note: Shots/mJ using updated dose calibration

- 2H '04 (Install/Qual)
- 2005
- 2006
- 2007
- 2008 (through WW36)

N2 and G1/G2 replaced (10/06)

Collector, N1, G1/G2, and N2 replaced (12/07)

N1 and Collector replaced (3/06)

Total Dose Delivered (J/cm2)

Shots/mJ/cm2

2008 Wafer plane power trend

- Illuminator optics (Collector, N1, G1/G2, and N2) were replaced in Dec. 2007

- Appropriately selected field stop position may help extend high wafer plane power by reducing EUV light on optics

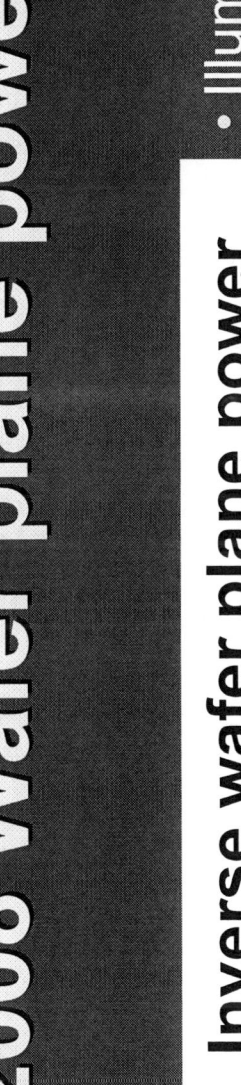

Inverse wafer plane power

Avg. rate of increase 2008:
1.4 shots/mJ/cm2 per 100 J/cm2

• 2008 (through WW36)

Shots/mJ/cm2

Dose Delivered (J/cm2)

Measuring flare on MET

MEASURED FLARE – MASK FLARE = TOOL FLARE

$$\frac{\text{Absorber Reflectivity}}{\text{ML Reflectivity}}$$

$$\frac{E0}{E0_{\text{island feature}}}$$

KEY POINTS:

- Flare measurements made using resist clearing method (*Kirk's Method*)

- Projection Optics have never been replaced on Intel MET

- Contamination may increase MSFR and HSFR on PO, thus increasing tool flare over time

Tool flare trend over time

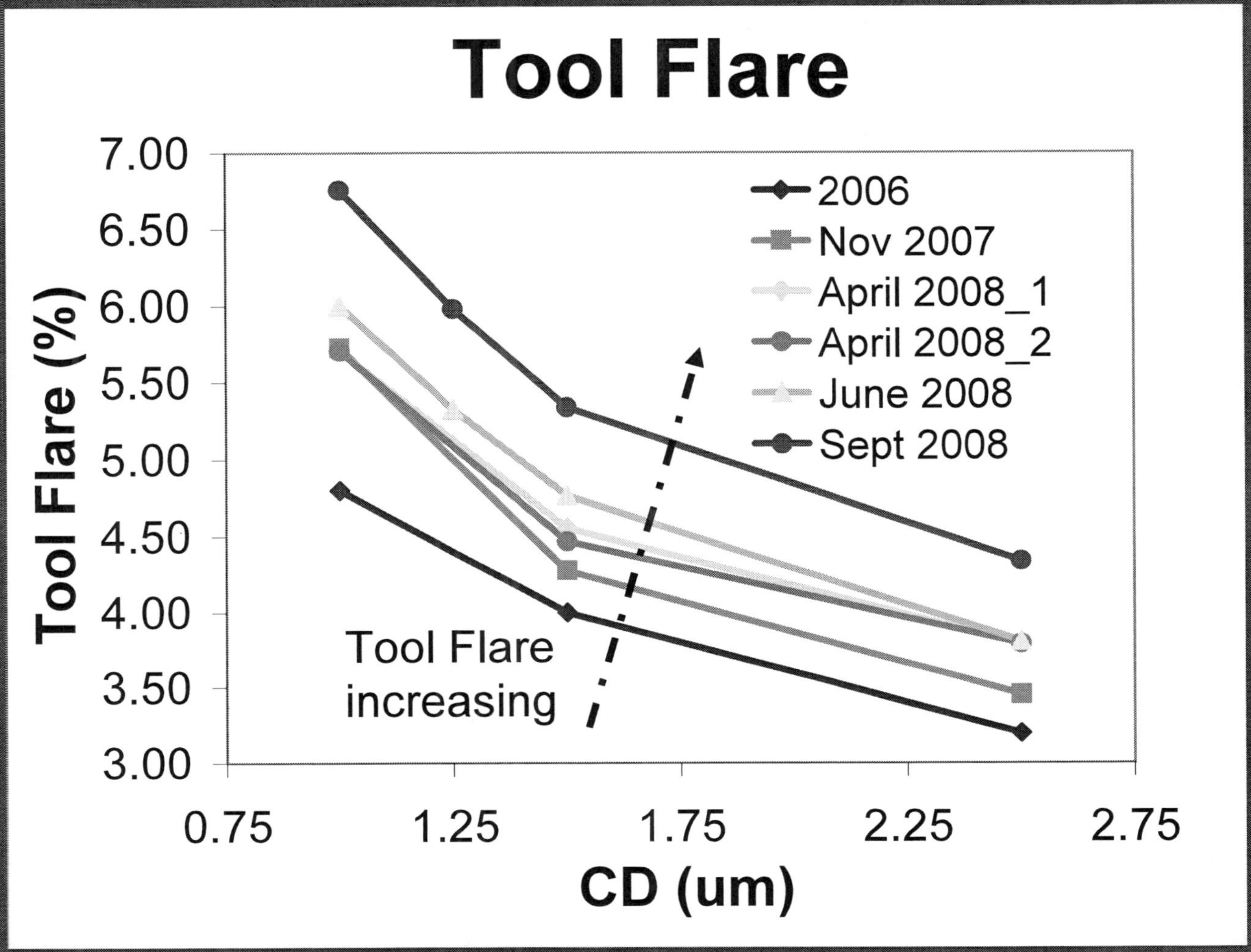

Tool Flare

Rate of tool flare increase as a function of dose delivered

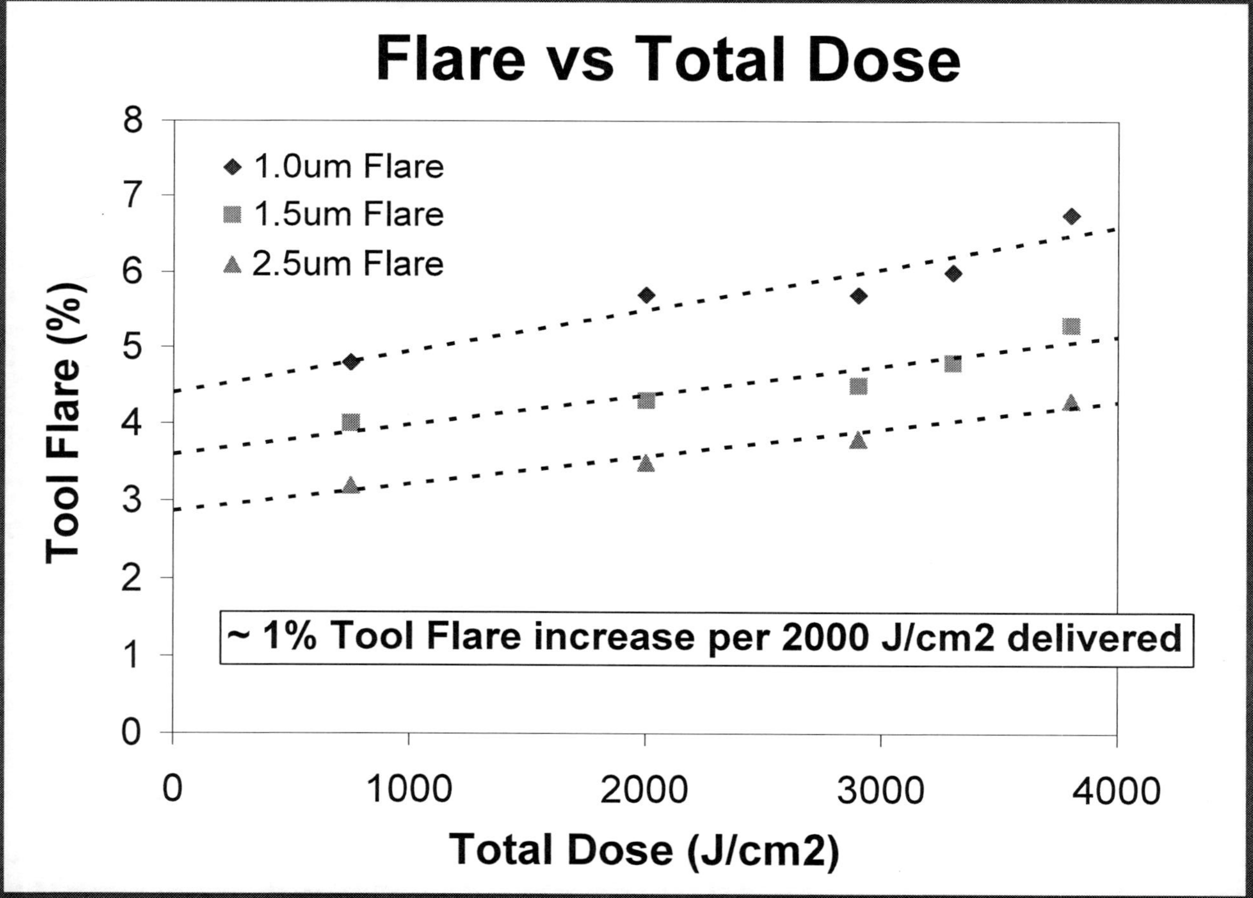

Flare vs Total Dose

- ♦ 1.0um Flare
- ■ 1.5um Flare
- ▲ 2.5um Flare

Tool Flare (%) (y-axis: 0 to 8)

Total Dose (J/cm2) (x-axis: 0 to 4000)

~ 1% Tool Flare increase per 2000 J/cm2 delivered

MET productivity enabling resist screening/development

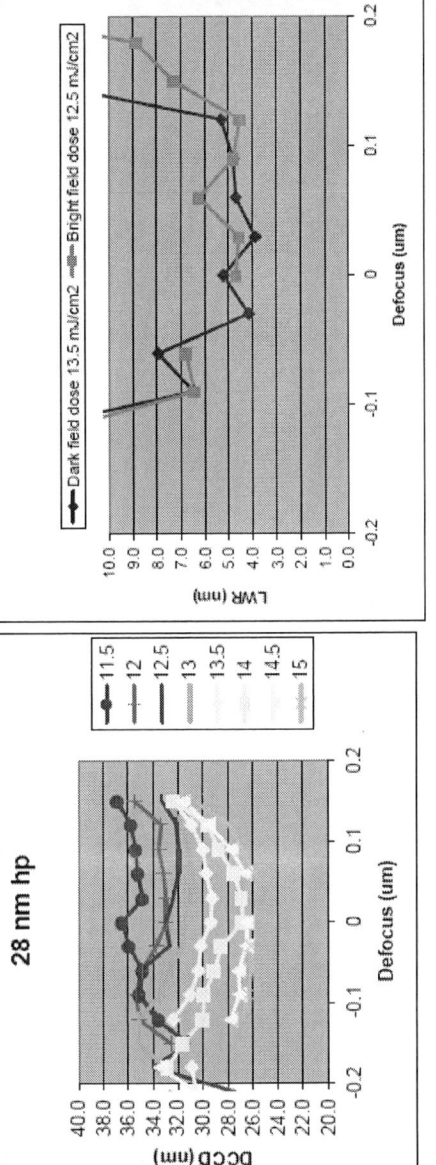

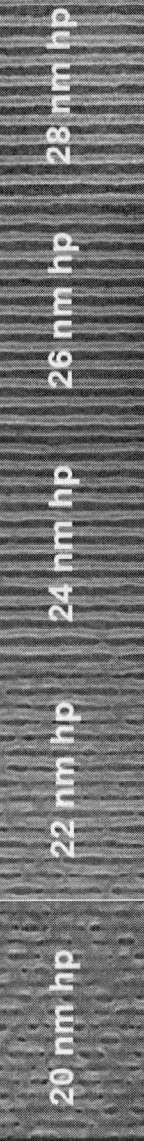

- 122 resists screened in 2008

- Consistent imaging of 28nm HP lines
  - LWR < 4nm, Dose = 14 mJ/cm2

- Best resolution down to 26nm HP

Future Plans for improvement on the Intel MET

- **MET Contamination Mitigation Experiments will test** contamination mitigation and cleaning strategies in the real tool environment of the MET

- **Heating of N1 mirror** is expected to increase desorption rate of contaminant species and help mitigate further contamination
  - Mitigation/cleaning effect can be monitored by tracking shots/mJ/cm^2 in real time

- Mitigation experiments using **controlled directional dosing of O$_2$ near N1** can be used to test known O$_2$ mitigation/cleaning effects in a real tool environment
  - Mitigation/cleaning effect can be monitored by measuring shots/mJ/cm^2 in real time, thus providing real time feedback

Conclusions

- The steady increase in uptime, coupled with a significant increase in wafer plane power after the latest illuminator optics replacement, has resulted in a non-linear improvement in dose delivered per day

- The increased reliability and productivity of the Intel MET has enabled significant resist screening and development and resulted in consistent imaging of 28nm HP lines with low LWR (~4 nm).

- The continually improving uptime and rate of dose delivery on the Intel MET serve as encouraging data points for the viability of future EUVL tools and EUV technology in general

Acknowledgements

- EUV team members at Intel: Jim Clarke, Todd Younkin, Steve Putna, Manish Chandhok, Kent Frasure, Jim Ryan, and Ed Johnson

- Tetsunori Murachi for 2006 flare data collection

- Erik Sohmen at Carl Zeiss SMT AG

EXTREME ULTRAVIOLET LITHOGRAPHY

Composite Double Reflection Simplified Extreme Ultraviolet Whole Lithography Machine and System

Wynn.L.Bear,
Xiangwen,Xiong * .

ABSTRACTS

We have developed a general mechanism which employs the theoretical basis of the principle of a whole formation image with optics mutilation lens, and a principle and phenomenon that has no disturbances for two cross beams in their propagation is introduced to model a mixed light principle by using the laser light and EUV, with the objectives of creating a design and manufacturing method for such the mixed light by multiple Catadioptric multiband lithography methods, and composite double reflection multilane methods. We have proposed multiple Catadioptric multiband lithography in a past paper, a comprehensive research and systematic analysis for the composite double reflection lithography methods been described in this paper, including the principle and structure of the system of systems, chemical-resist, flare, resolution, line-edge roughness and sensitivity, as well as the research of the EUV mask and EUV integrated system, etc. Such the lithography methods use double EUV source or the mixed light by a 193 nm ArF source and an EUV. This is a pleasant surprise to the conclusion is that they are almost all compatible for laser resist and EUV when the mixed light source is used. Our current work afforded a new strategy of the RLS trade-off and the size of the power to get the quality of qualified, extremely cost-effective and yield for the IC Manufacturing of 45, 32, 22 nm node and less, and be use for the factory immediately also. At the same time, it still facilitates people to upgrade to a standard EUVL machine at anytime.

We are using the core of the carrier and functions of the platform to develop into the leading edge technology integrate the cutting edge application system as a whole to solve the technology problem for current lithography, at the same time matching the trend and progress as the CPU from the single-core age enter advanced multi-core, and significant improve the performance.

INTRODUCTION

2008 International Symposium on Extreme Ultraviolet Lithography, SEMATECH.
September 28 - October 1, 2008
Lake Tahoe, California, USA.
Wynn.L.bear, Wynn Bear International Cutting-edge High-Tech Institute, CA, USA
Xiangwen,Xiong * , Wynn Bear International advanced Business Machines Company, China.
Tel: (028) 38902-0677, Fax: (028) 38902-0677, Email: wynnbear@gmail.com, wynnbear.inc@gmail.com.
(Wynn.bear is the English name, Xiangwen, xiong is the pinying name).

We have proposed new lithography methods and systems to let the EUVL technology be a place in the 32 nm node, to become mainstream lithography technology in the 22 nm node, and in 16 nm and less be an absolute dominance. Based on the cost analysis for our technology with other two kinds of competition technology (NIL and MEBDW, etc), it be shown that the EUVL and simplified EUVL is the only next-generation lithography technology for its tremendous cost advantages and the economy. We are continuing proposed the basic principles and implementation examples for the composite double reflection multilane simplified Extreme Ultraviolet whole Lithography Machine and system in this paper. Our goal is still being the first to make the simplified Extreme Ultraviolet Lithography to first use in the actual production process, at the same time to let the standard Extreme Ultraviolet Lithography to win more time and funding to enable it to further improve and enhance to fully capture the market and co-exist with the simplified EUVL for the respective cost advantages of owner for the use of our technology, the semiconductor manufacturers more quickly be advanced technology and be cost-effective, the semiconductor industry soon see more beautiful and better prospects for hope.

All the simplified Extreme Ultraviolet lithography methods are like an F-16 fighter in comparison to an F-15 or an F-35 fighter in comparison to an F-22, the simplified EUVL is likely to become the mainstream of NGL for silicon-based semiconductor manufacturing. The Composite double reflection multilane simplified EUVL is the use of the whole imaging principle of a mutilation lenses and reflection imaging to integrate as a whole, and a basic principle that has no disturbances for two cross beams in their propagation. Such a method and system let the high density lines and low density on the chip been created by different band light, respectively. It will immediately be applied to production lines.

THE BASE PRINCIPLE AND APPLICATION OF THE COMPOSITE DOUBLE REFLECTION MULTILANE SIMPLIFIED EUV LITHOGRAPHY

The Composite double reflection multilane simplified type EUVL still be the best strategy given the current state of EUVL technology and be immediately applied to the actual factory production process. It is the use of the double EUV source or the mixed light by a 193 nm ArF source and an EUV. The power size of the EUV source is no longer a question, and even the power of everyone source is lower. The few sources of low power EUV light create the high density lines on the chip, and the remaining sources of EUV light are used for general lines, or the 193 nm ArF laser source. This machine will be still used immediately in the factory production line. Our invention and methods will immediately eliminate the pressure on competitors. In the future, we still further improve such the lithography machine.

The standard "EUVL is now well into the alpha tool demonstration phase, it still needs a lot of new idea, developments and solutions for the advancement of extreme ultraviolet lithography into cost-effective, manufacturable solutions. The Lithography is so important to advanced IC manufacturing, so we must be careful to treat it.

The base principle of the composite double reflection multilane Lithography

About the principle and phenomenon that has no disturbances for two cross beams in their propagation, there is evidence to show that it is absolutely correct and reliable long time ago. These are theory foundation and the implementation basis for our lithography method, at the same time we use the double EUV source or the mixed light by a 193 nm ArF source and an EUV to solve the "progress in key critical technology issues such as Demonstration of reliable high-power source, collector modules, simultaneously meeting to resist resolution, sensitivity and line edge roughness specifications and fabricating defect-free masks, etc" [1].

The Simplified Lithography System by using the mixed light source of laser and EUV

Based on such the principle and phenomenon that has no disturbances for two cross beams in their propagation, we developed a 193 nm ArF laser (or other lasers) beam and an EUV the mixed light source to share power problems. Figure-1 is shown the principle of a standard composite double reflection multilane lithography method by using a 193 nm ArF laser (or other lasers) beam and an EUV. Now, the power of the lithography system is common supplied by the laser and EUV. Based on the current level of human technology, the laser power is already big, so that the problem of EUV power will be eliminated.

A typical example is shown in the figure-1.

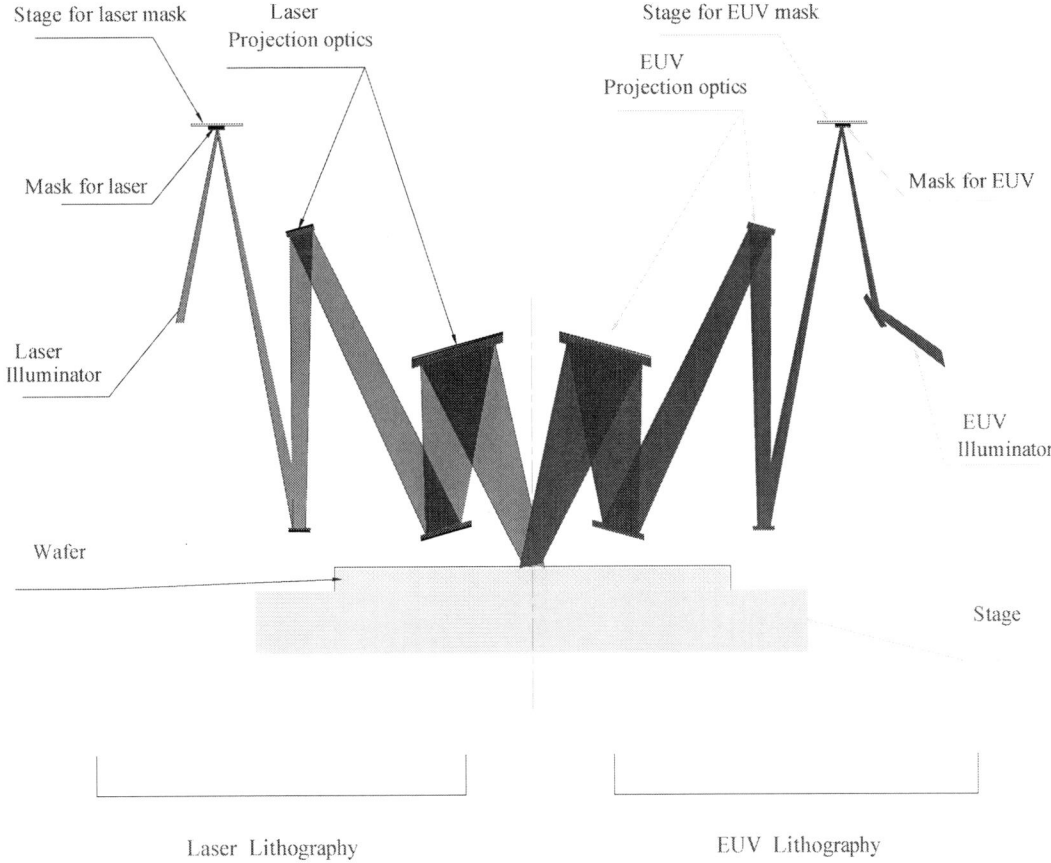

Figure–1

In the figure-1, laser lithography system and EUV system are both effect to a same area of a die or a field on the chip. Now, the high density lines and low density on the chip been created by the laser or EUV, respectively. Base on the independent ray propagation theory of light, they have no disturbances each other. We use many types of laser to better match it, such the laser the wavelength including 193, 157, 120, 100, 80 nm and others. Generally speaking, laser create the high density lines on the chip, and low density been created by EUV.

The Simplified Lithography System by using the double EUV source

Based on the independent ray propagation theory of light, we developed a double EUV mixed light source to joint share power problems. Figure-2 is shown the principle of a standard composite double reflection multilane lithography method by using a double EUV mixed light. Now, the power of the lithography system is common supplied by the two EUV source. Based on such the system and method, the problem of EUV power will be reduced.

A typical example is shown in the figure-2.

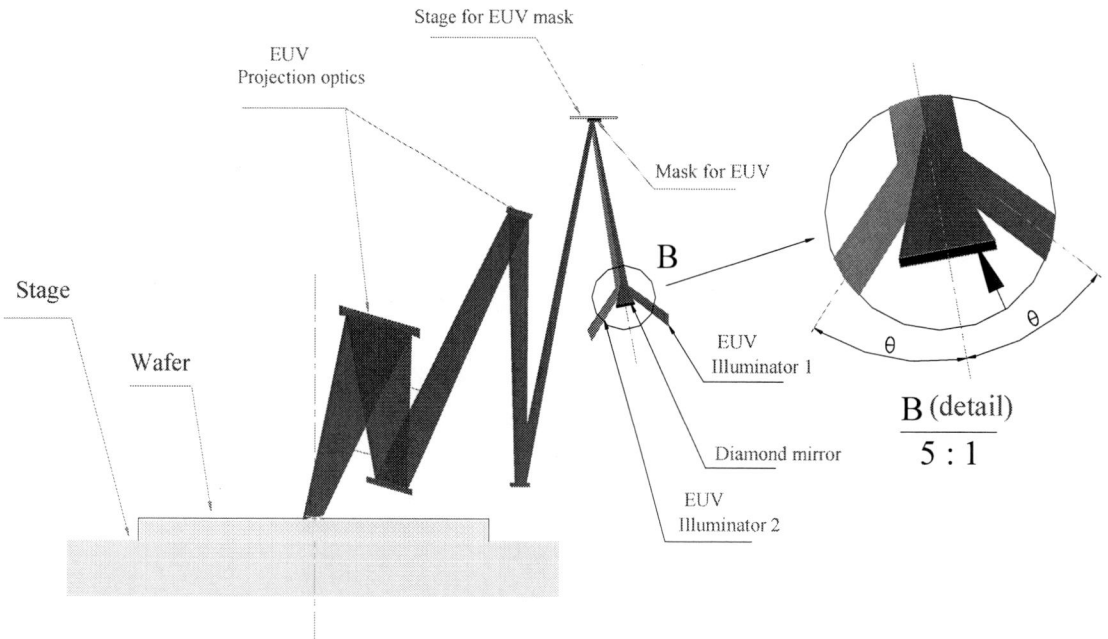

Figure–2

In the figure-2, two beams of the double EUV mixed source are both effect to a same area of a die or a field on the chip. Now, the high density lines and low density on the chip been created by the two EUV beam, respectively. Base on the independent ray propagation theory of light, they have no disturbances each other. Diamond mirror in the figure-2 is the reflector that has isosceles triangle's section, which will be the same intersection angle of light reflection to the same direction.

Generally speaking, a high quality EUV beam create the high density lines on the chip, and low density been created by generally quality EUV.

The Simplified Lithography System by using the quadruple EUV source

Based on the theoretic and method that two beams of the double EUV mixed source are both effect to a same area of a die or a field on the chip, the lithography system have received the double energy. However, the value of energy does not mean that no longer need to be increased for the high yield, so that we design the quadruple EUV source system. Figure-3 is shown the principle of a standard composite reflection multilane lithography method by using a quadruple EUV source. Now, the power of the lithography system is common supplied by the four EUV source. Based on such the system and method, the problem of EUV power will almost not be existed.

A typical example is shown in the figure-3.

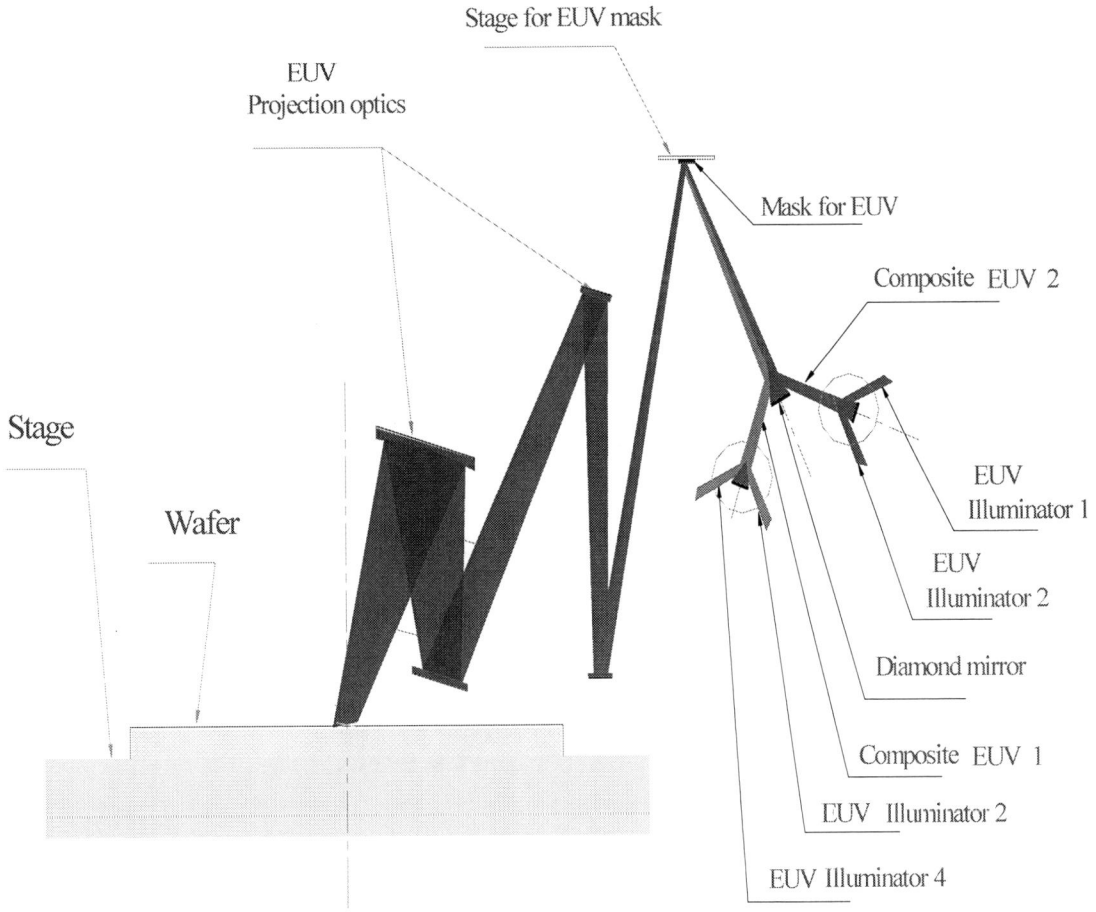

Figure–3

In the figure-3, four beams of the double EUV mixed source are all effects to a same area of a die or a field on the chip. Now, the high density lines and low density on the chip been created by the four EUV beam, respectively. Base on the independent ray propagation theory of light, they have no disturbances each other. Diamond mirror in the figure-3 still is the reflector that has isosceles triangle's section, which will be the same intersection angle of light reflection to the same direction.

Also, a high quality EUV beam creates the high density lines on the chip, and low density been created by generally quality EUV.

The Simplified Lithography System by using the double Projection optics

Based on such the principle and phenomenon that has no disturbances for two cross beams in their propagation, we developed a EUV double Projection optics Lithography System to solve power problems. Figure-4 is shown the principle of a composite double Projection optics lithography method and system by using double EUV Projection optics. Now, the power of the lithography system is common supplied by the double reflective system, it will be good for eliminating the problem of EUV power.

A typical example is shown in the figure-4.

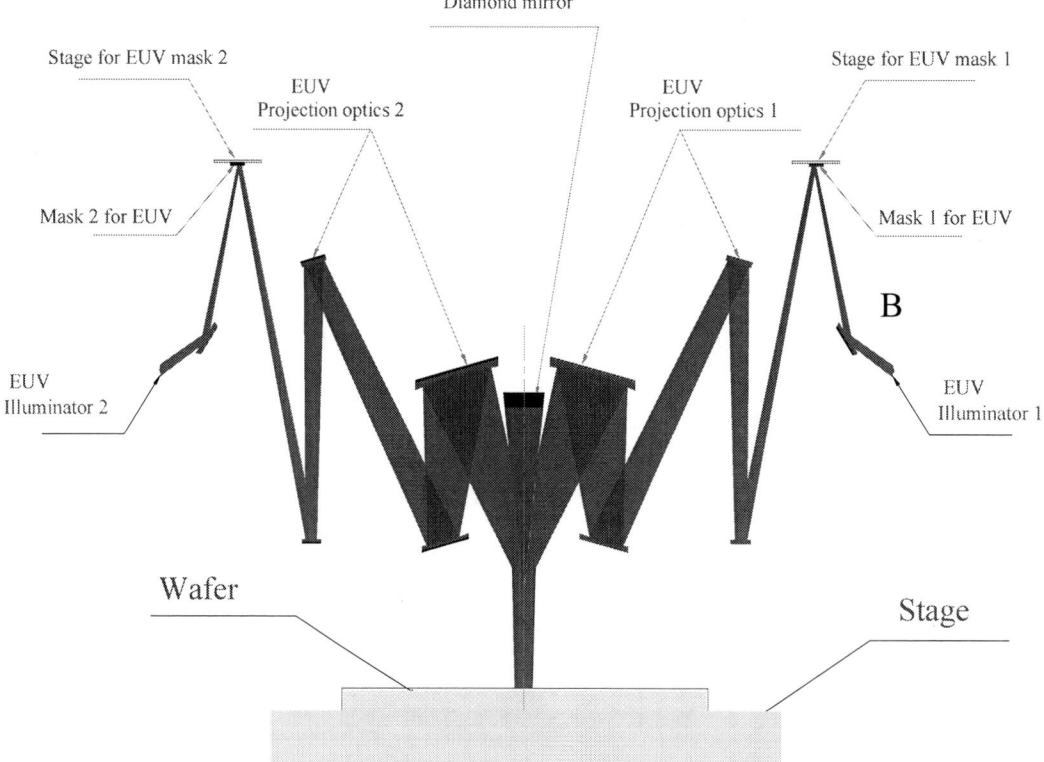

Figure–4

In the figure-4, the double Projection optics Lithography System are both effect to a same area of a die or a field on the chip. Now, the high density lines and low density on the chip had been created by the different Projection optics Lithography System, respectively. Base on the independent ray propagation theory of light, they have no disturbances each other. Diamond mirror in the figure-4 still is the reflector that has isosceles triangle's section, which will be the same intersection angle of light reflection to the same direction.

Based on such the principle and method, the Simplified Lithography System by using the quadruple Projection optics is manifest.

THE TECHNICAL CONDITION & IMPACT TO THE COMPOSITE DOUBLE REFLECTION MULTILANEEUV LITHOGRAPHY

Doubtlessly, our simplified Extreme Ultraviolet Lithography has an advantage first to use in the actual production process than NIL even is the MEBDW. Also, we let the standard Extreme Ultraviolet Lithography to win more time and funding to enable it to further improve and enhance to fully capture the market and co-exist with the simplified EUVL for the respective cost advantages of owner. Nevertheless, our technology involved a mixed system by a 193 nm ArF and an EUV, so that all the issues related to 193 nm ArF lithography and EUV lithography must be a comprehensive analysis.

The interference of cross beams and double light

Based on the principle and phenomenon that has no disturbances for two cross beams in their propagation, the interference of cross beams and double light is not a problem.

Photoresists

Almost all researchers that focused on EUV photoresists have found that to deal with the problems encountered at 193 nm laser, which are becoming very similar to those for EUV. Although we also are experimenting with the process, but we boldly assume that such effective materials are both sensitive to the 193 (and other types, such as 157, 120, 100, 80 nm and less) and EUV is existent [11], [6].

The Impact of Flare

Such a problem should be probably the same with the existing EUV Lithography system.

Resolution, line-edge roughness and sensitivity

We still concerned about the College of Nanoscale Science and Engineering (CNSE) at the

University at Albany in New York, and other university and research institution were done a lot of successful study in Resolution, line-edge roughness and sensitivity, "new series of resists (MET-2A, 2B. 2C, 2D) also show great promise for good resolution, LER and sensitivity."[9]. We will provide a great advantage for Resolution, line-edge roughness by immolating some sensitivity. Our composite simplified Extreme Ultraviolet Lithography is the use of the double EUV source and quadruple, or the double Projection optics and quadruple, etc, it means that all the energy of this lithography system be provided by the multiple-sources, and the power is not a problem again. When it was sufficient energy, the chip would have been greatly improving the Resolution, line-edge roughness of the lines or not? Yes! At the College of Nanoscale Science and Engineering (CNSE), they even put forward a new theory named the RLS trade-off [12], [11]. Robert Brainard, an associate professor at the CNSE said that "the EUV resist challenges be summed up succinctly. They are the three basic properties — resolution, line-edge roughness and sensitivity. My colleagues and I call it the RLS trade-off. You need all three properties for EUV [and EUV resists] to be successful, and when you improve one, you make another one worse.

About the photoresists and line edge roughness (LER), Dr. Wurm, S said that "All three lithography technologies currently competing for insertion at the 32 nm hp node pose serious challenges in the resist area that must be resolved. Specifically, meeting line edge roughness (LER) specifications will be difficult for all technologies but more so for EUV than for double exposure lithography (DEL) and 193+." [13]

"We focus on cross-comparisons in which selected EUV and 193 nm resist are evaluated using both EUV and 193 nm lithography. Simulation methods linking 193 nm and EUV performance will be described as well. Results from simulation indicate that image blur in current generation 193 nm photoresists is comparable to that of many EUV resists…" [17].

Obviously, we would rather sacrifice part sensitivity to gain enough resolution and line-edge roughness.

Yield

Our output is one of the major advantages. We use the multiple-sources and system that has sufficient energy easily to let yield to reach 160 – 190 wph.
.

Mask

Our technology may do not need the Optical Proximity Correction (OPC) and the Phase-Shifting Mask (PSM). EUV mask even though has still some problems, but you can use high-quality and reliable source that is low-power for our lithography systems. Obviously, it will be good to solve such the problem.

Cost advantage

Our cost is much lower than NIL and MEBDW, even still much lower than double patterning lithography. We will describe it in a specialized paper.

A CHANGE CONCEPT FOR IC DESIGNER TO MEET CHALLENGES IN THE NEW SITUATION

IC designers should have changed the concept. In the past, they are in all important units to use the same lines. Now, they should distinguish the most important element, sub important and common, etc, and then in the most important area to use the smallest lines, and in sub important by second-class lines, and others.

In a word, different lines are different cost for the chip. We use the smallest lines to create the unit that bear most arduous tasks; in the unit that only is ordinary tasks we use the general lines. Take the Highway as an example; it includes interstate highways, trunk roads, and the local highway, etc.

CONCLUSIONS AND DIRECTIONS FOR FUTURE RESEARCH

For catch up the advanced multi-system lithography technology like the multi-core CPU, the time to us is very tight. At present, the Progress and development in the Composite double reflection multilane simplified EUVL is being made on a number of fronts in our company. Such the method and system have been developed to improve machine operation and better adaptability is being implemented. Except Modeling and simulation tools to visualize the possibility of the simplified EUVL machine and other characteristics, we have already begun to create such the diamond mirror and related devices.

At present, our wish is to find some companies or organizations that have the ability to carry out cooperative study and research with EUVL to speed up the progress of the study. You may also remember a case that is about the engine, for more low-cost and be the high-power, it uses a wide range of metal materials prices, and multi-cylinder structure, but no one material prices and one cylinder. In addition, the aircraft of different use is also used in a number of engines.

The Composite double reflection multilane simplified EUVL of our company efforts to the interactions with industry and other researchers indicate the following as significant works for future to fully realize:

1. We realized the phenomenon that has no disturbances for two cross beams in their propagation.

2. We realized the diamond mirror; Furthermore, such the reflector, many types can be manufactured by us.

3. We mainly use the 193nm dry lithography machine for our Simplified EUVL method and system.

4. A type of the photoresists both to meet the requirements of the EUV and 193 nm laser is

needed to be further verified.

5. A type of the photoresists is that higher resolution & line-edge roughness, and lower sensitivity is consumingly being needed.

6. A EUV source that is high-quality & reliable and low-power or high are more conducive to EUV mask and projection optics will be strongly needed. In other words, it is no longer just focused on the size of the power but to pay attention to overall balance to meet the needs of the EUV mask and the higher resolution & line-edge roughness, etc.

7. Our Simplified Extreme Ultraviolet Lithography will be given a special design to achieve the lowest cost. If the design to be improved to adapt this Manufacturing technology, we will further reduce the cost.

8, Based on above methods and research, to be found the best structures, installations and system by contrast.

9, This lithography system not only the Composite double (or quadruple) reflection multilane simplified EUVL machine, but also be upgraded to the standard EUV lithography machine. Such the function will be very useful to the factory.

10, our composite double reflection multilane methods, from a single system develop into the multi-system; it meets the need for CPU of computer from the single-core enter the multi-core.

11, Our Simplified EUVL system facilitates people to upgrade to a standard EUVL machine at anytime.

12, Our Simplified EUVL method is the use of the basic principle that has no disturbances for two cross beams in their propagation to be integrated a core system like the advanced multi-core CPU, instead of Simplified joint.

ACKNOWLEDGMENT

It is a hopeless understatement to say that I am deeply grateful that except many professors and other critical thinkers, including this symposium's organizers, etc, I would like to especially appreciate the Board of Reviewing Editors of the Science magazine, and Ph.D Phillip D. Szuromi, the Supervisory Senior Editor; Ph.D Gilbert J. Chin, the professional Senior Editor. It is they true support the EUVL is the most possible IC manufacturing technology to meet the Moore's Law, and to be the next generation lithography.

REFERENCES

[1], Semiconductor Manufacturing Technology, www.sematech.org.

[2], Burn J. Lin, Optical lithography—present and future challenges, C. R. Physique 7 (2006) 858–874

[3], Kurt Ronse, Optical lithography—a historical perspective, C. R. Physique 7 (2006) 844–857.

[4], Gregory Denbeaux, Rashi Garg, Chimaobi Mbanaso, Justin Waterman, Leonid Yankulin, Alin Antohe, Yu-Jen Fan, and Warren Montgomery, etc. Extreme ultraviolet resist outgassing and its effect on nearby optics. Proc. SPIE, Vol. 6921, 69211G (2008).

[5], YANG Zong—li, ji Shu—li, MIAO Run-cai, ZHANG Zong—quan, Effects of cascade nonlinear medium spatial configuration on optical field characteristics. journal of shanxi university of science & technology, Apr. 2004, Nol, 22

[6], Brainard, Robert L.; Barclay, George G.; Anderson, Erik H.; Ocola, Leonidas E. Resists for next generation lithography. Microelectronic Engineering (2002), 61-62 707-715.

[7], Brainard, Robert L.; Cobb, Jonathan; Cutler, Charlotte A. Current status of EUV photoresists. Journal of Photopolymer Science and Technology (2003), 16(3), 401-410.

[8], Brainard, Robert L.; Trefonas, Peter; Lammers, Jeroen H.; Cutler, Charlotte A.; Mackevich, Joseph F.; Trefonas, Alexander; Robertson, Stewart A. Shot noise, LER, and quantum efficiency of EUV photoresists. Proceedings of SPIE (2004), 5374 (Pt. 1), 74-85.

[9], Thomas Köhler, Robert L. Brainard, Patrick P. Naulleau, David Van Steenwinckel, Jeroen H. Lammers, Kenneth A. Goldberg, Joseph F. Mackevich, and Peter Trefonas Performance of EUV Photoresists on the ALS Micro Exposure Tool, Proceedings of SPIE 2005, 5753, pp 754-764.

[10], Koehler, Thomas; Brainard, Robert L.; Naulleau, Patrick P.; Lammers, Jeroen H.; Steenwinckel, David Van; Goldberg, Kenneth A.;Mackevich, Joseph F.; Trefonas, Peter; Performance of EUV Photoresists on the ALS Micro Exposure Tool, 3rd International EUVL Symposium 2004.

[11], http://www.semiconductor.net/article/CA6529237.html.

[12], http://www.cnse.albany.edu/.

[13], Wurm, S.   Byers, J.   Zimmerman, P.   Wallow, T.   Dean, K.   Photoresist challenges and potential solutions for the 32 nm half-pitch node and beyond.   Microprocesses and Nanotechnology, 2007 Digest of papers, 2007, 428-429.

[14], Mccall, Monnikue M ; Han, Hakseung ; Cho, Wonil ; Goldberg, Kenneth ; Gullikson, Eric ; Jeon, Chan-Uk ; Wurm, Stefan. Determining the Critcial Size of EUV Mask Substrate Defects. Journal: International Society for Optical Engineering (SPIE); Journal Volume: 6921.

[15], Stefan Wurm, Hakseung Han, Kenneth A. Goldberg, Anton Barty, Eric M. Gullikson,Yoshiaki Ikuta, Toshiyuki Uno, Obert R. Wood. EUV MET printing and actinic imaging analysis on the effects of phase defects on wafer CDs. Proceedings of SPIE, 2007.

[16], Stefan Wurm, Hakseung Han, Patrick Kearney, Wonil Cho, and Chan-Uk Jeon,Eric Gullikson. EUV mask blank defect inspection strategies for 32-nm half-pitch and beyond. Proc. SPIE, Vol. 6607, 66073A (2007); DOI:10.1117/12.729029.

[17], Jones, Juanita ; Pathak, Piyush ; Wallow, Thomas ; LaFontaine, Bruno ; Deng, Yunfei ; Kim, Ryoung-han ; Kye, Jongwook ; Levinson, Harry ; Naulleau, Patrick ; Anderson, Chris. A Comparison of Photoresist Resolution Metrics using 193 nm and EUV Lithography.   SPIE Advanced Lithography, San Jose, CA, Feb. 24-29, 2008.

[18], Jones, Juanita ; Anderson, Christopher ; Naulleau, Patrick ; Niakoula, Demitra ; Hassanein, Elsayed ; Brainard, Robert ; Gallatin, Gregg ; Dean, Kim. Influence of base and PAG on deprotection blur in EUV photoresists and some thoughts on shot noise.   2008 Internation coference on electron, ion, and photon beam technology and nanofabrication, Portland, OR, May 27-30, 2008.

[19], Yumi Nakajima, Takashi Sato, Ryoichi Inanami, Tetsuro Nakasugi, and Tatsuhiko Higashiki. Aberration budget in extreme ultraviolet lithography. Emerging Lithographic Technologies XII, SPIE, 2008.

[20], S. Bollanti, P. D Lazzaro, F. Flora, L. Mezi1, D. Murra and A. Torre. Conventional and modified Schwarzschild objective for EUV lithography: design relations. Applied Physics B: Lasers and Optics. 603-610, 2006.

[21], M Chandhok, SH Lee, J Roberts, BJ Rice, HB Cao. Lithographic flare measurements of Intel's microexposure tool optics. Journal of Vacuum Science & Technology B: Microelectronics and Nanometer Structures -- January 2006 -- Volume 24, Issue 1, pp. 274-278.

[22], J. G. Hartley, S. Raghunathan, and A. Govindaraju. Electrical characterization of multilayer masks for extreme ultraviolet lithography. Journal of Vacuum Science & Technology B: Microelectronics and Nanometer Structures -- November 2005 -- Volume 23, Issue 6, pp. 2891-2895.

[23], Heidi B. Cao, Wang Yueh, Bryan J. Rice, Jeanette Roberts, Terence Bacuita, and Manish Chandhok. Sources of line-width roughness for EUV resists. Advances in Resist Technology and Processing XXI, SPIE, 2004.

[24], Manish Chandhok, Sang H. Lee, and Terence Bacuita. Effects of flare in extreme ultraviolet lithography: Learning from the engineering test stand. Journal of Vacuum Science & Technology B: Microelectronics and Nanometer Structures -- November 2004 -- Volume 22, Issue 6, pp. 2966-2969.

[25], O. R. Wood II, D. Back, R. Brainard, G. Denbeaux, D. Goldfarb, F. Goodwin, J. Hartley, K. Kimmel and C. Koay, etc. Initial experience establishing an EUV baseline lithography process for manufacturability assessment. Proc. SPIE, Vol. 6517, 65170U (2007); Emerging Lithographic Technologies XI.

Spectral Purity Filter Life-Time Testing On EQ-10 Source

Chuck Schietinger, Dave Grove, Travis Ayers, Luxel, Friday Harbor, Washington
Matthew Partlow, Debbie Gustafson, Energetiq Technology, Woburn, Massachusetts

Background

In the fabrication of semiconductor devices EUV lithography will likely be the technology of choice for the 22nm node and beyond. A free-standing thin film will continue to be used as a spectral purity filter (SPF) between the plasma source and the optics. The SPF serves the following functions: 1.) blocking DUV where the photo resist maybe sensitive, 2.) blocking IR, critical for reducing thermal loading of sensitive optics, 3.) debris mitigation, and 4.) allowing control of pressure differentials and gas flows.[1] The SPF life-time can impact both cost-of-ownership and available light.

Energetiq and Luxel have been working jointly in understanding and improving the life-time of SPF's in the EQ-10 Electrodeless Z-pinch™ EUV source. The EQ-10 EUV source is designed for stability in metrology and research applications.[2,3,4]

Test Filters

The number of variables was intentionally limited in this test set. Filter films were 200nm to 250nm thick Zirconium and meshed with a square 70 line per inch nickel mesh. The five thin film spectral purity filters tested in order of life-time, best filter was #1 and the shortest life-time filter was #5.

Filter #	Filter Description
1	Zirconium Baseline 236nm
2	Zirconium 202 nm
3	Zirconium 250nm
4	Zirconium 202 nm less wrinkled
5	Zirconium 250 nm **Extremely Flat Film**

Test Set-Up

All test filters were mounted 35cm from the Energetiq Xenon Gas Pinch Plasma. There was no mitigation between the source and the filter. The filters were each exposed to between 15 and 100 million pluses of the EUV source. Standard operating conditions of the source are 280V, 1900Hz.

Results

Leak Rate is used as the primary measure of filter damage. New filters have initial leak rates of less than ~ 5 1D4 mBarL/sec. As the filter develops small holes from either particle impact or thermal expansion induced movement the leak rate increases. The SPF leak rate is determined using a rate of rise approach.[1] First the SPF is inserted into the beamline while the source is operating. At fixed intervals of time, a section of the beam line behind the SPF is isolated from its pump with a gate-valve. The background rate of rise in the small section is determined using a Baratron gauge. The volume of the small section is estimated from known dimensions. By multiplying the net rate of rise by the volume we have the effective leak rate through the SPF. The rate of rise for the five test filters is presented in the center graph.

References

1. Forbes R, Powell, Terry A. Johnson. Filter Windows for EUV Lithography, SPIE, Vol. 4343 (2001).

2. P.A. Blackborow, M. J. Partlow, S. F. Horne, M. M. Besen, D. K. Smith, and D. S. Gustafson, "EUV Source Development at Energetiq," in Emerging Lithographic Technologies XII, Edited by Michael J. Lercel, Proceedings of the SPIE, Volume 6921(1), pp. 69211J (2008).

Flat
Roughness <14nm

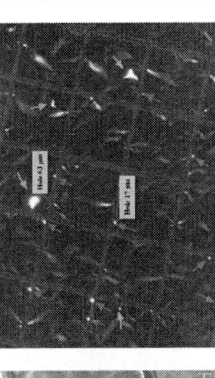

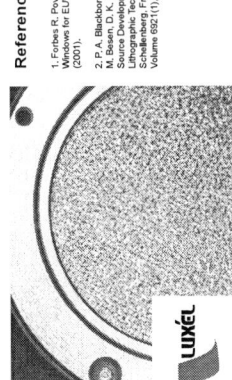

Filter #1
Filter #2
Filter #3
Filter #4
Filter #5

Leak Rate [T/min] (norm to 5 mT dome pressure)

Million Pulses

Wrinkled
Roughness >3µm

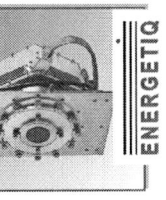

Hole Generation

After exposure to million of source pulses the filters develop holes ranging in size from less than 1µm to over 100µm. It is believed that there are two basic mechanisms which produce the holes in the films, fast moving particles and thermal-mechanical induced holes. The extremely fast moving particles however small generally penetrate the thin film. However if the film is low in stress and not stretched tightly, then the hole will be limited in size. If the film is under tension (flat) the hole can propagate all the way to the support mesh which can result in tears larger than 100µm. The highly wrinkled filters had the best performance.

The thin films have very little thermal mass. Therefore the film can heat at rates exceeding the pulse rate of the light source. The film heats at a rate different than the frame or the nickel mesh. The film and mesh both expand at different rates because of the difference in C/TE and their temperatures. This thermal expansion movement can create holes in the film or cause the particle generated holes to enlarge as the film and mesh move during every heating and cooling cycle. The holes next to the mesh bars are likely not the result of particles as seen in the lower right photo, the red arrows. The green arrows show probable particle generated holes.

There is another important thermal effect related to changes in surface oxide of the filters over time. Thicker oxides generally cause the filter to darken increasing the visible and IR emissivity. This will increase the filter temperature, causing more rapid and larger thermal cycling. A thicker oxide will also decrease EUV transmission. Auger Electron Spectroscopy was used to profile the filters before and after exposure. There was little or no deposition coming from the plasma source. There was thermal oxide growth on some of the filters after long exposure. This oxide was thicker in the center of the filters indicating higher temperatures.

Topography

The filters with the longest and shortest lifetime were mapped with a con-focal microscope to generate a 3D topographical map and to measure the waviness. The filter with the best lifetime was Filter #1 with a local waviness (roughness) of greater than 3µm within a mesh cell. The filter with the shortest lifetime had virtually no waviness and a roughness of less than 14nm (resolution of measurement) within a mesh cell.

Conclusion

The life-time of a thin film SPF is determined predominantly by a combination of particle impact holes and thermal induced mechanical damage. The filter life-time can be improved by fabrication of highly wrinkled films which are not under tension. Flat films tear more easily and the holes tend to be larger.

Additional SPF testing including other key variables is ongoing.

References

3. Blackborow, Paul A., Gustafson, Deborah S., Smith, Donald K., Besen, Matthew M.; Horne, Stephen F.; D'Agostino, Robert J.; Minami, Youichi; Denbeaux, Gregory; "Application of the Energetiq EQ-10 electrodeless Z-Pinch EUV light source in outgassing and exposure of EUV photoresist" in Emerging Lithographic Technologies XII, Edited by Lercel, Michael J., Proceedings of the SPIE, Volume 6517, pp. 65171W (2007).

4. S. F. Horne, M. M. Besen, D. K. Smith, P. A. Blackborow, and R. D'Agostino, "Application of a high-brightness electrodeless-Z-pinch EUV source for metrology, inspection, and resist development", in Emerging Lithographic Technologies X, Edited by Lercel, Michael J, Proceedings of the SPIE, Volume 6151, pp. 201–210 (2006).

5. M. Partlow, S. Horne, M. Besen, D. Smith, P. Blackborow, and D. Gustafson, "Baseline Design for the Energetiq EQ-10 EUV Source," in International Symposium on Extreme Ultraviolet Lithography, Sapporo, Japan, Sematech. 2007, Proceedings available from SEMATECH, Austin, TX.

Dependence of Laser Parameter on Conversion Efficiency in High-Repetition-Rate Laser-Ablation-Discharge EUV Source

Takuma Yokoyama, Hiroshi Mizokoshi, Yusuke Teramoto, Daiki Yamatani, Hiroto Sato and Kazuaki Hotta

Extreme Ultraviolet Lithography System Development Association (EUVA)
Gotemba Research and Development Center
1-90 Komakado, Gotemba, Shizuoka 412-0038, Japan
Phone: +81-(0)550-87-3000, Fax: +81-(0)550-87-3200

Background

- Laser Ablation (Assisted) Discharge is one of the most expected EUV sources for next generation lithography

- It is necessary to improve conversion efficiency

- Detail phenomena of laser ablation and discharge, and relation between them have not been discussed yet

- Observation of laser ablation and discharge plasma
- Discussion on relation between CE and laser parameter

Laser ablation discharge system

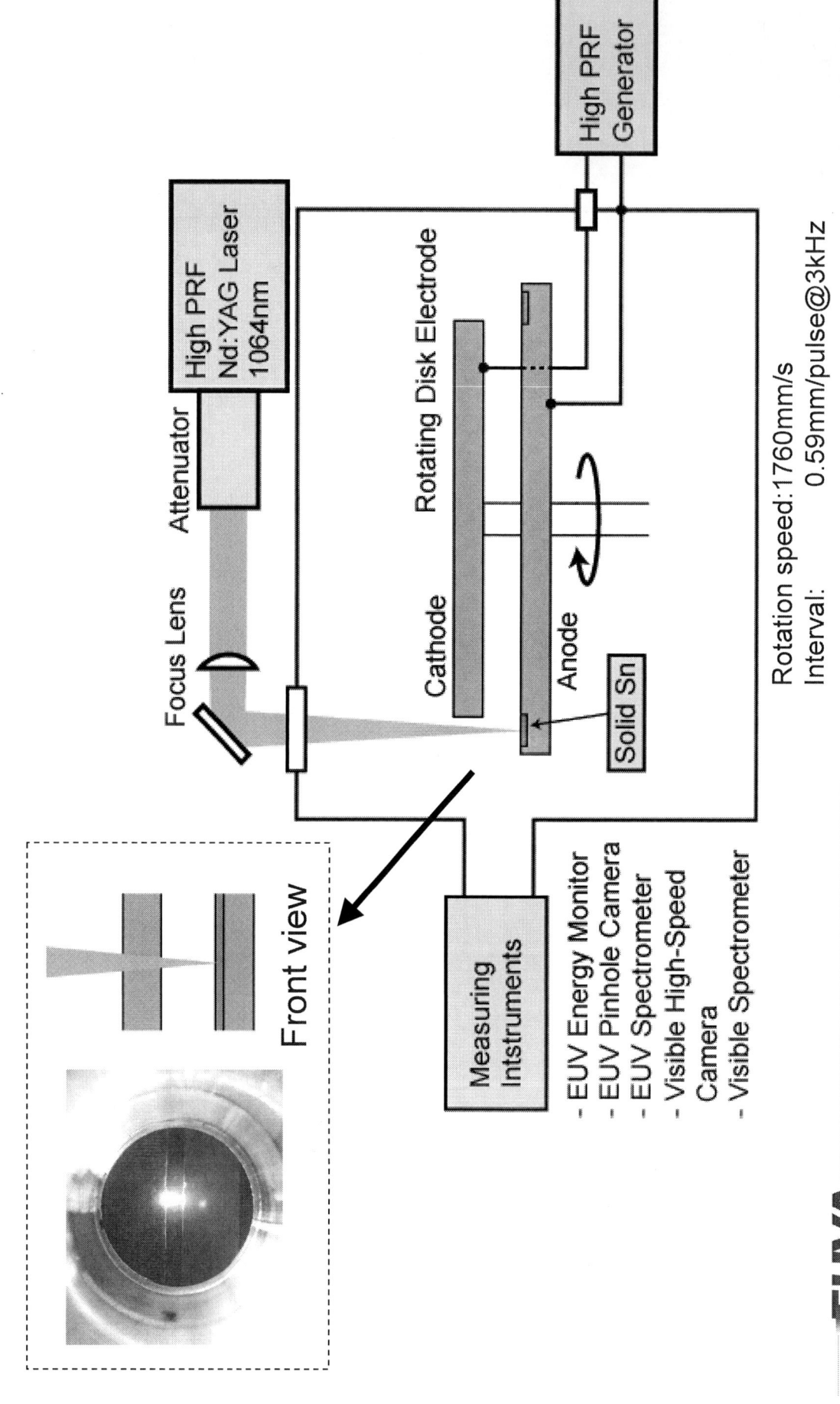

Source power and CE dependence on pulse repetition frequency

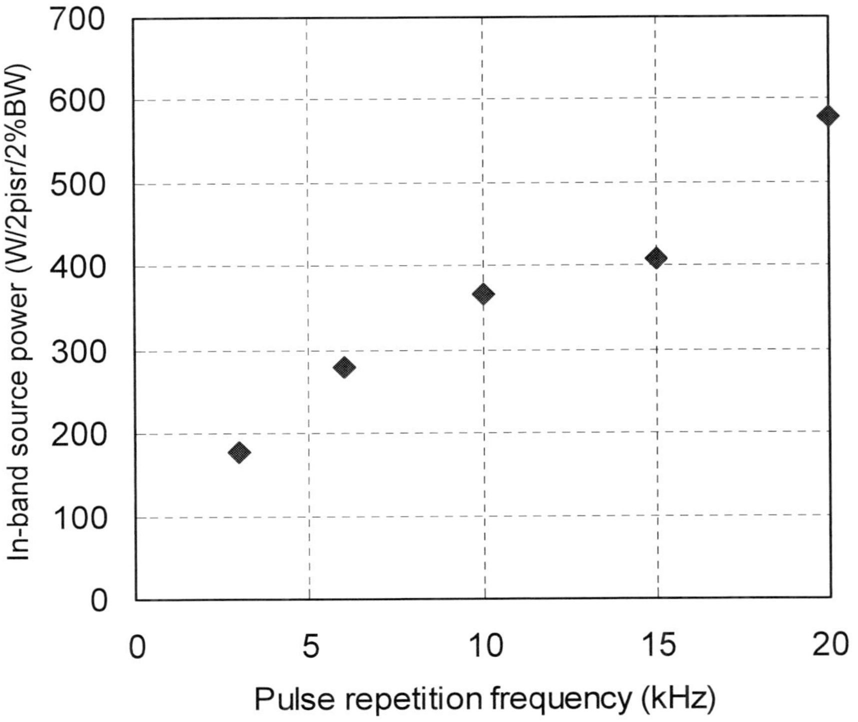

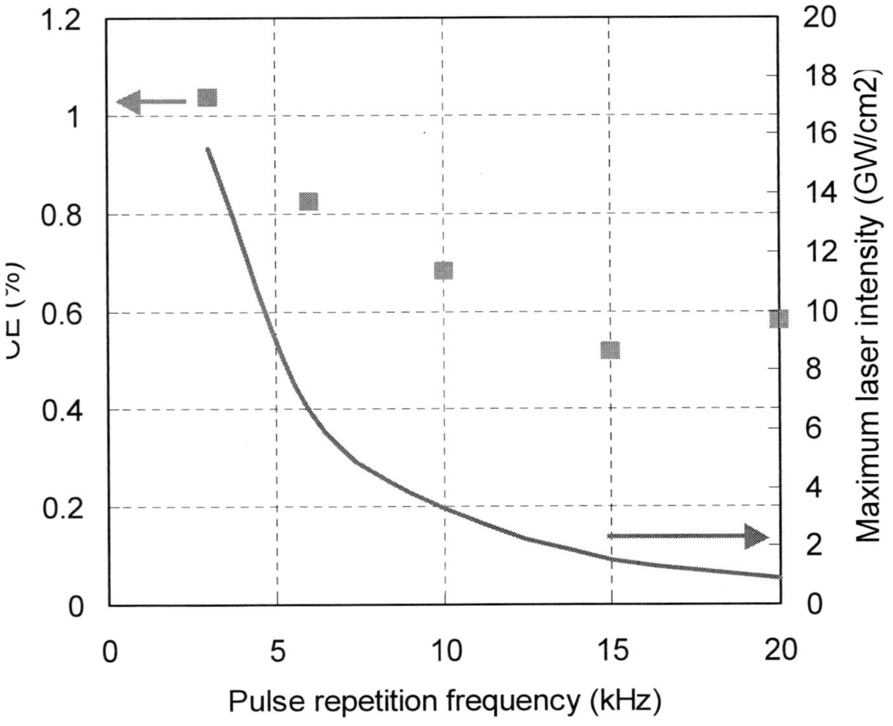

Maximum In-band source power : 579W (20kHz)
Maximum conversion efficiency : 1.04% (3kHz)

Decrease of CE in higher frequency was caused by fall of laser intensity
It is possible to improve CE by using higher intensity laser

CE dependence on laser intensity

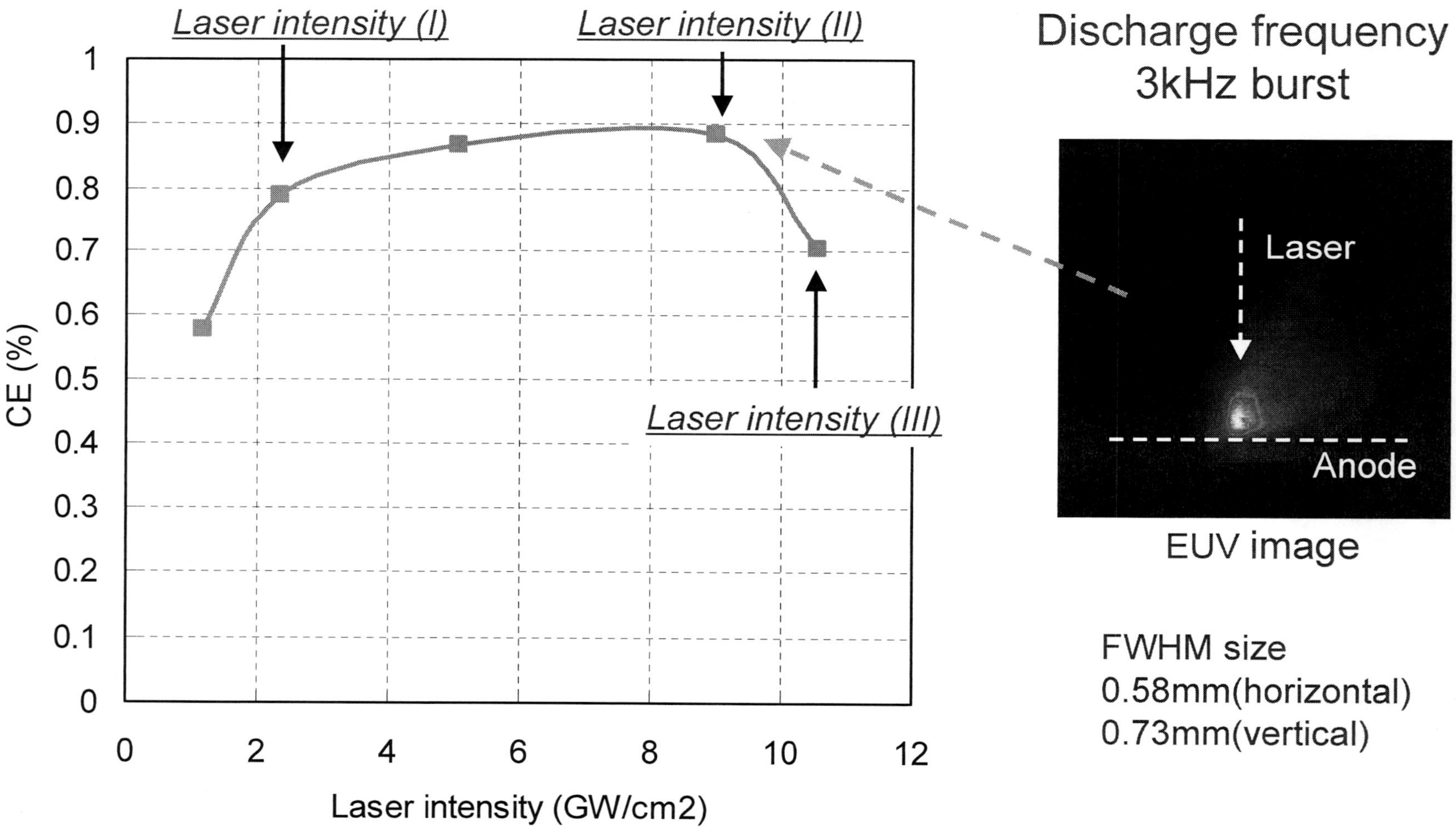

Discharge frequency : 3kHz burst

Laser

Anode

EUV image

FWHM size
0.58mm(horizontal)
0.73mm(vertical)

Laser intensity (I)

Laser intensity (II)

Laser intensity (III)

CE (%)

Laser intensity (GW/cm2)

Visible images of laser ablation plasma (Non-discharge)

Visible images were taken with high-speed camera

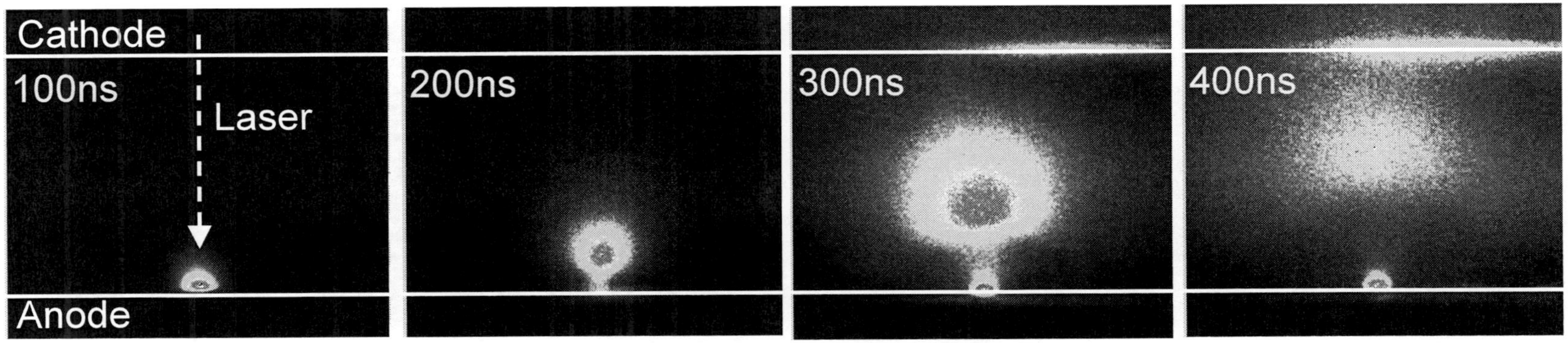

Cathode			
Laser →			
100ns	200ns	300ns	400ns
Anode			

Laser intensity (I)

Cathode			
Laser →			
100ns	200ns	300ns	400ns
Anode			

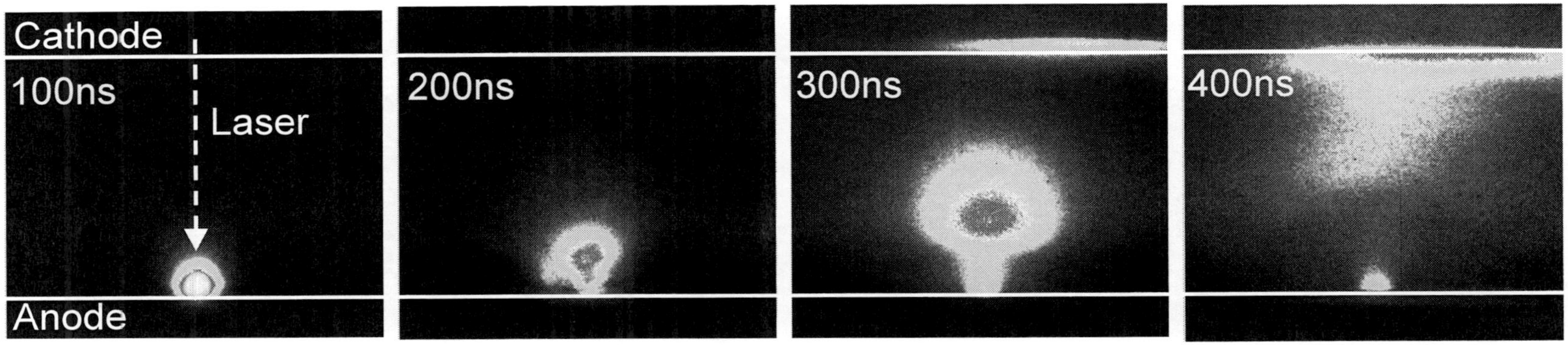

Laser intensity (II)

Diffusion veolucity of laser ablation plasma (dense ablation plume) seems almost similar in all laser conditions

Structure of laser ablation plasma (Non-discharge)

Band-pass filters were used in condition of *Laser intensity (II)*

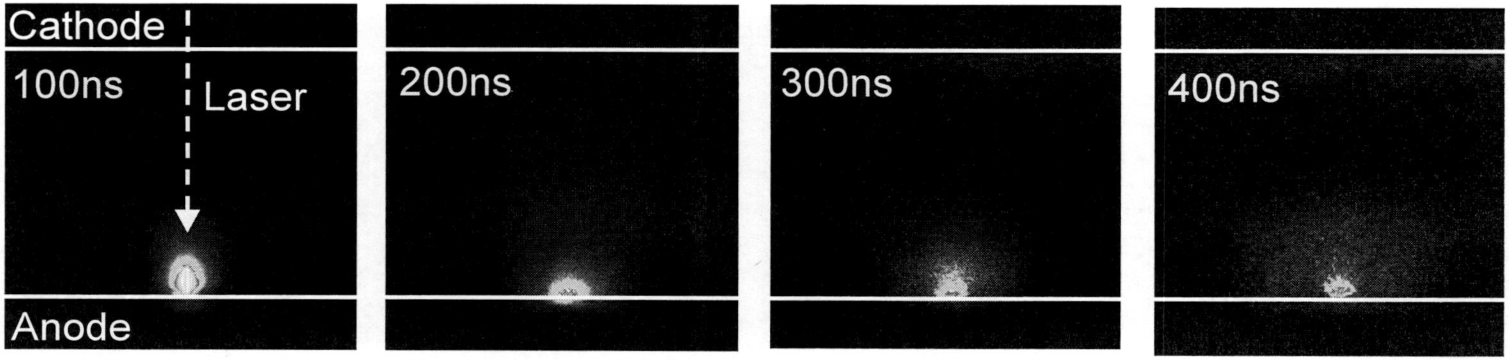

Atomic line (Sn I 452.5nm)

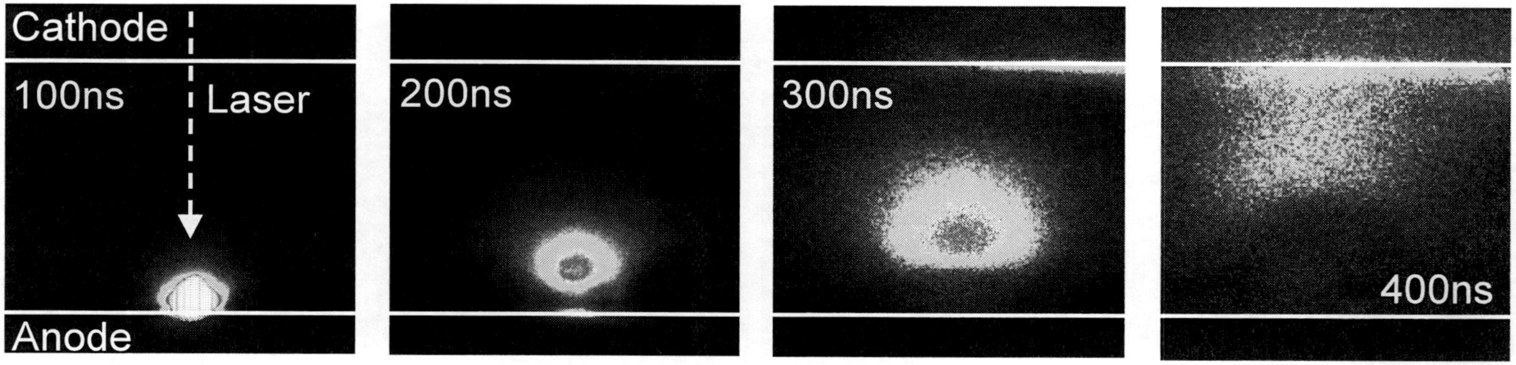

Ion line (Sn II 579.9nm)

Neutral vapor was continuously supplied from irradiation spot
Ionic gas was flying with expansion toward cathode

Front of laser ablation plasma (Non-discharge)

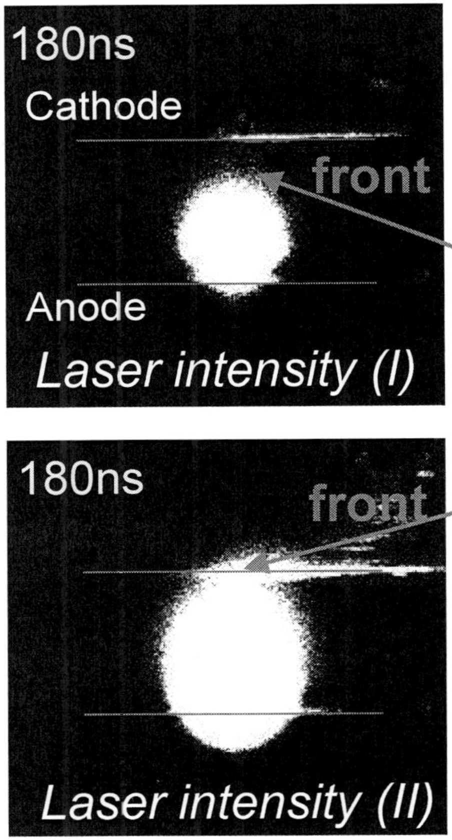

Weak emission (front of laser ablation plasma) was observed by increasing gain of high-speed camera

Velocity of
front of laser ablation plasma
was higher with stronger laser intensity

⬄

Diffusion velocity of
dense part of laser ablation plasma (in page 6)
was almost similar with various laser intensity

180ns

Cathode

front

Anode

Laser intensity (I)

180ns

front

Laser intensity (II)

Time-resolved picture
taken at 180ns

Electron number density of laser ablation plasma (Non-discharge)

Electron number density was estimated from stark width of visible line spectrum

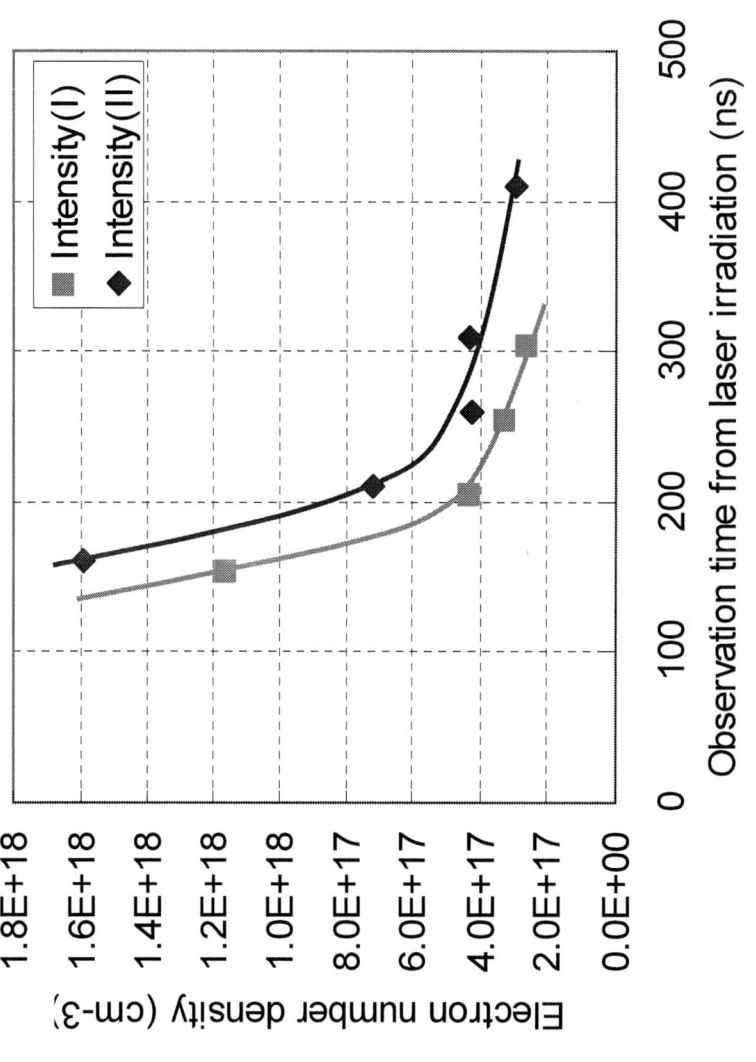

Electron number density increased with higher laser intensity

Transition of Z-pinch (Discharge)

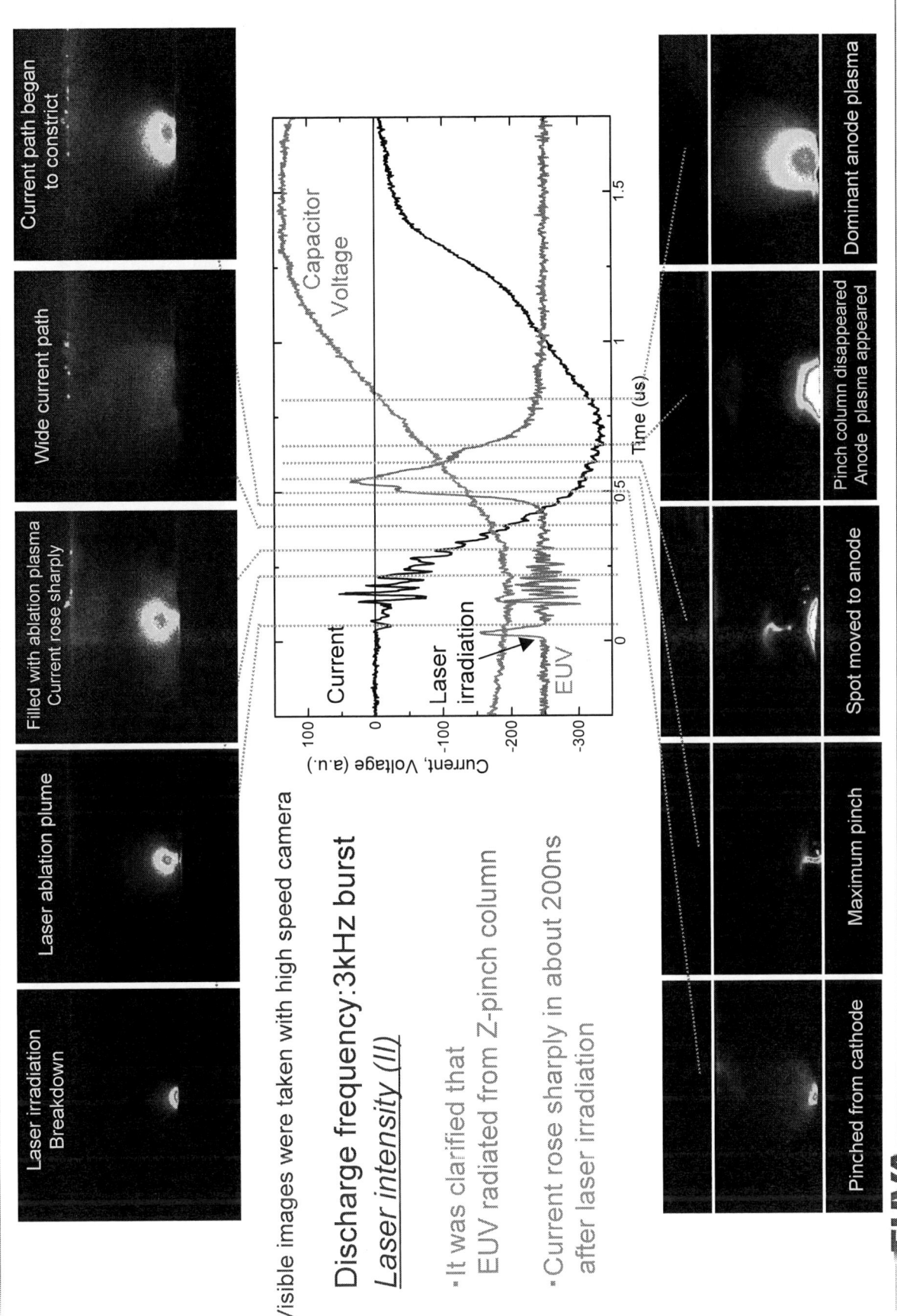

Visible images were taken with high speed camera

Discharge frequency:3kHz burst
Laser intensity (II)

- It was clarified that EUV radiated from Z-pinch column
- Current rose sharply in about 200ns after laser irradiation

EUV Spectrum (Discharge)

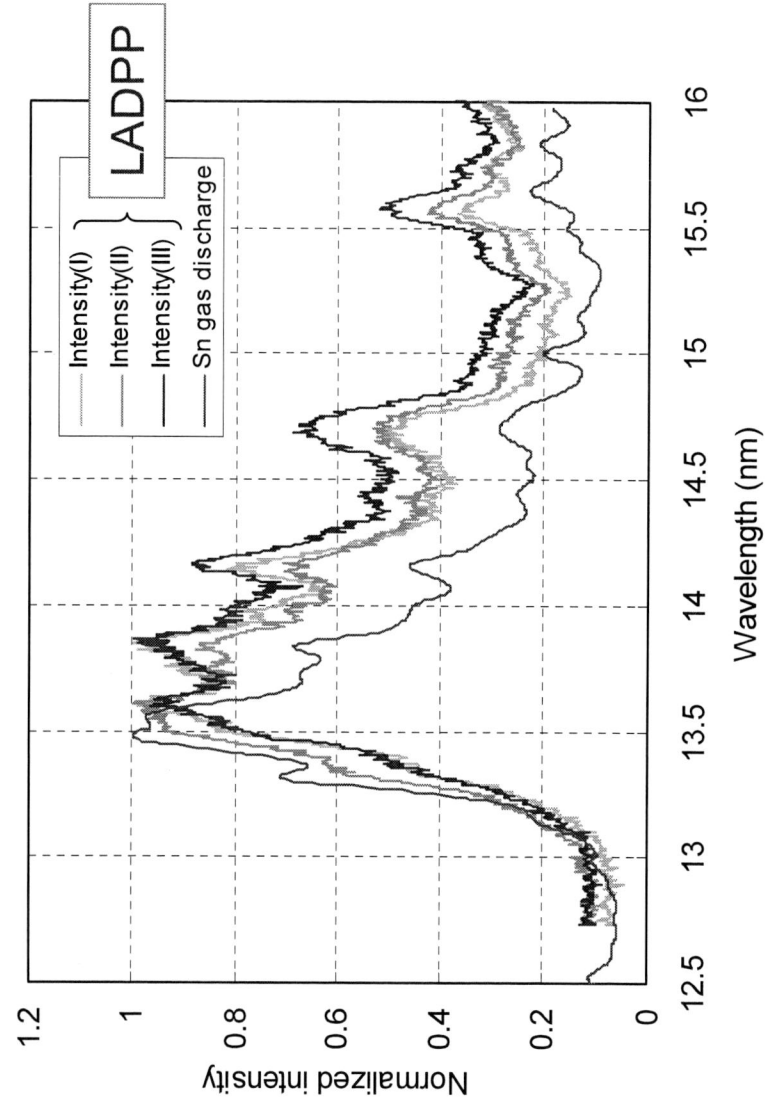

Discharge with laser intensity(II) was most efficiently heated in confinement
(except Sn gas discharge),
which corresponded to trend of CE (intensity(II) > intensity(I,III))

Current-rise-delay from breakdown

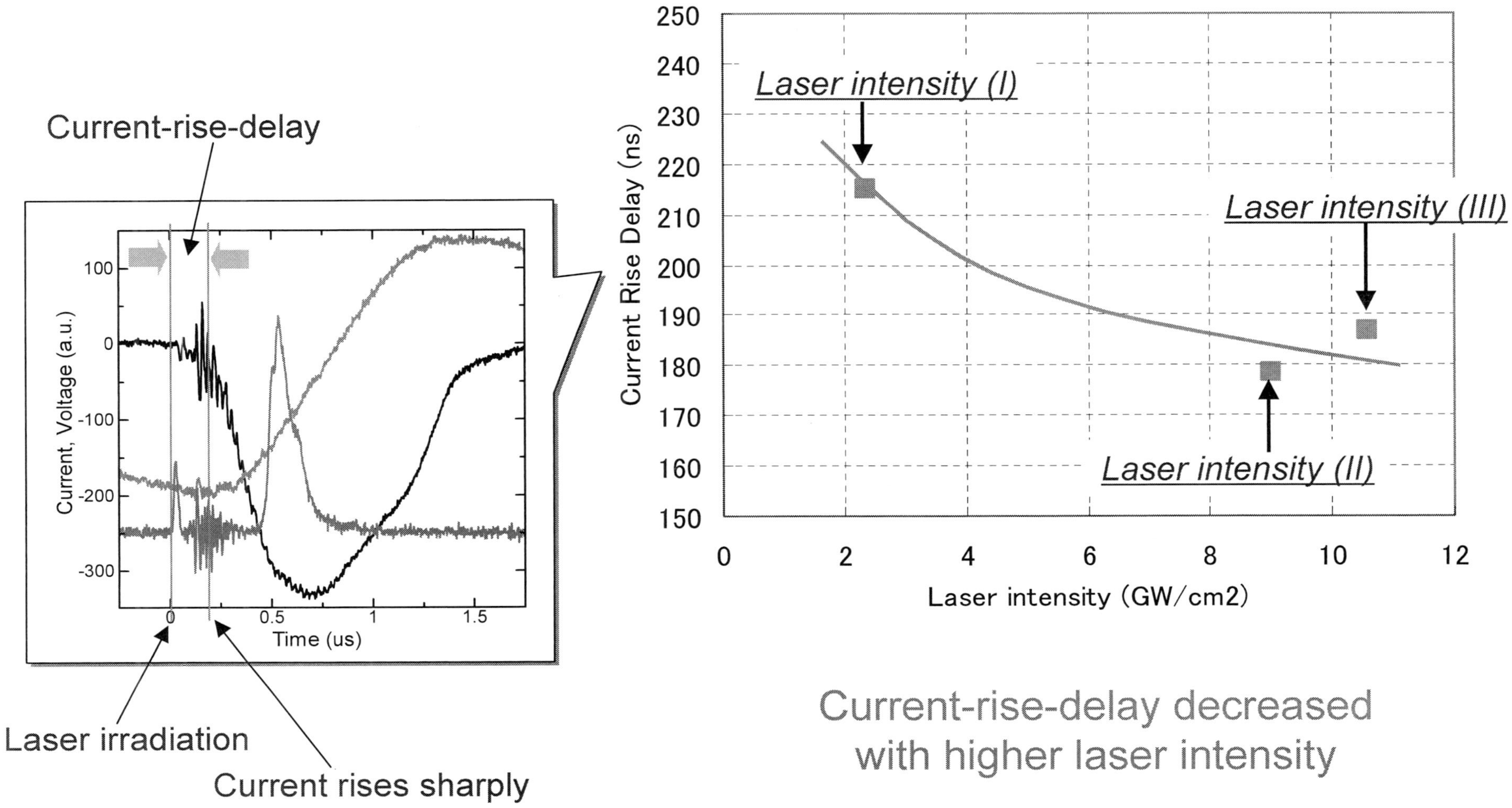

Current Rise Delay (ns)

Laser intensity (I)

Laser intensity (III)

Laser intensity (II)

Laser intensity (GW/cm2)

Current-rise-delay decreased
with higher laser intensity

Current-rise-delay

Current, Voltage (a.u.)

Time (us)

Laser irradiation

Current rises sharply

Current-rise-delay depends on front of laser ablation plasma

Front of laser ablation plasma

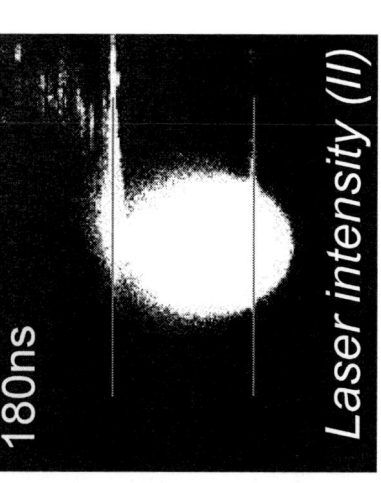

180ns

Laser intensity (II)

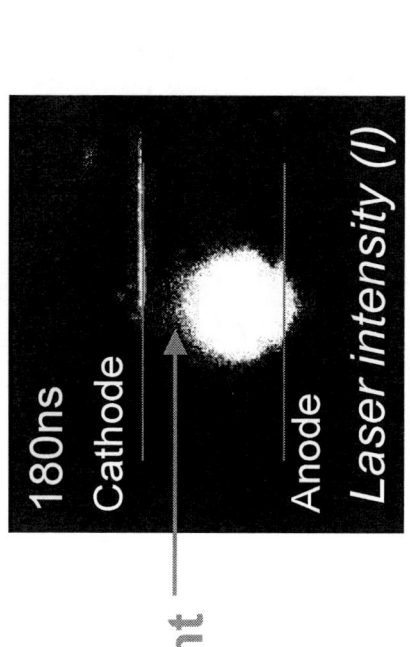

180ns

Cathode

front

Anode

Laser intensity (I)

- When front of laser ablation plasma arrived at cathode, current began to rise sharply
- Velocity of front of laser ablation plasma was higher with stronger laser intensity

Current-rise-delay decreased with higher laser intensity

Initial density distribution at current rise time

Initial density distribution at current rise time determines following development of Z-pinch column and efficiency of EUV radiation

- If laser intensity is too high, initial density increases too much

- If laser intensity is too low, current-rise-delay increases
  - { Expansion of initial density distribution
    Decrease of density }

Images of laser ablation plasma

Cathode | Laser
100ns | 200ns | 300ns | 400ns
Anode

Diffusion speed of laser ablation plasma is constant with various laser intensity
Initial density distribution is determined by current-rise-delay

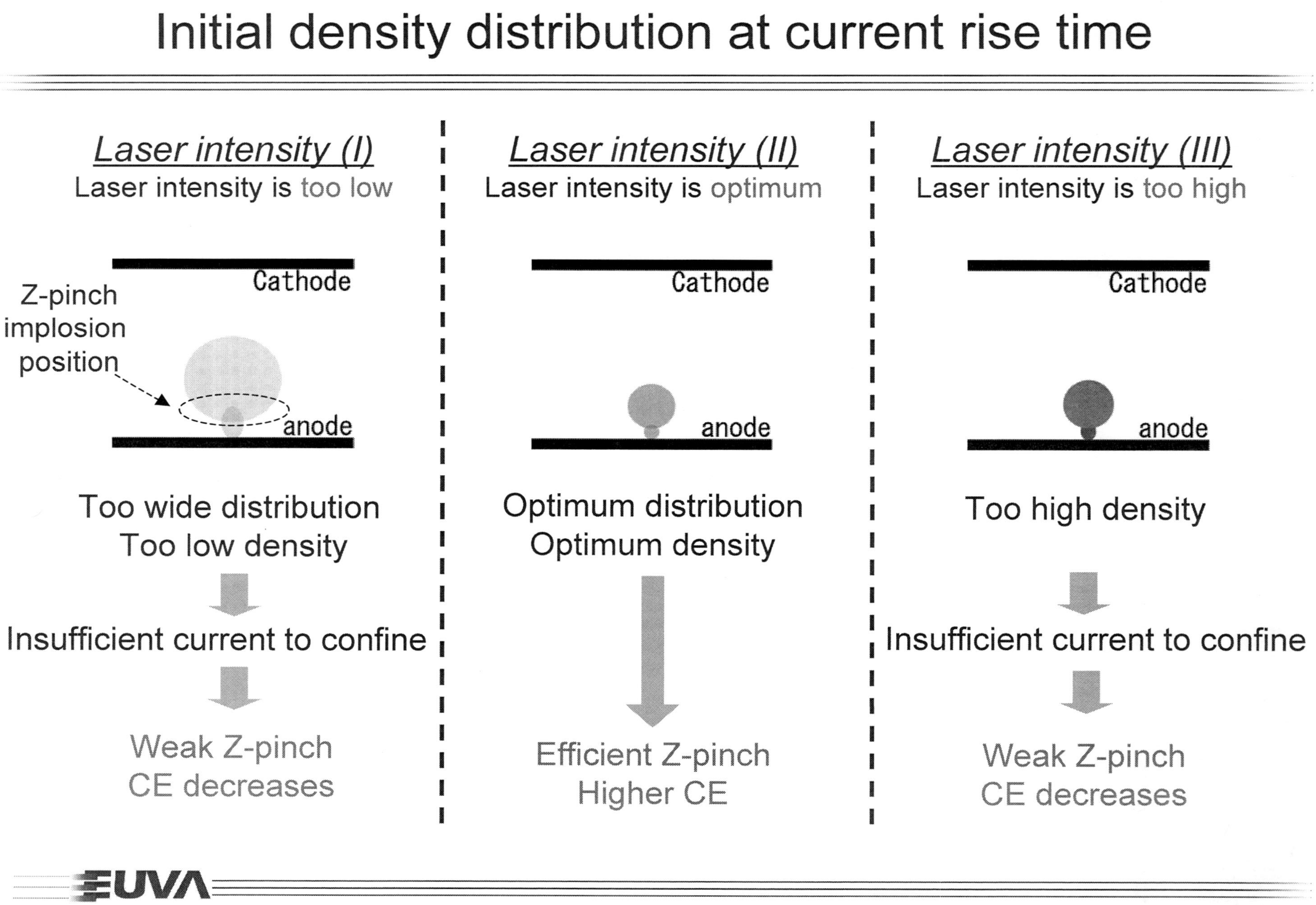

Initial density distribution at current rise time

Laser intensity (I)
Laser intensity is too low

Cathode

anode

Z-pinch implosion position

Too wide distribution
Too low density

Insufficient current to confine

Weak Z-pinch
CE decreases

Laser intensity (II)
Laser intensity is optimum

Cathode

anode

Optimum distribution
Optimum density

Efficient Z-pinch
Higher CE

Laser intensity (III)
Laser intensity is too high

Cathode

anode

Too high density

Insufficient current to confine

Weak Z-pinch
CE decreases

Summary

- In-band source power :579W
  Conversion efficiency :1.04%

- It was clarified that EUV radiated from Z-pinch discharge in laser-ablation-discharge system

- Initial density distribution is important to form best Z-pinch discharge for effective EUV radiation, which is determined by laser intensity.

- Current-rise-delay is one of the most significant factor to determine initial density distribution.

Acknowledgments

This work has been supported by

New Energy and Industrial Technology Development Organization (NEDO).

EUV Spectrometers for Source Development, Characterization and Optimization

Scott Bergeson[1], Bryce Allred[1], Jershon Lopez[1], Jeffrey Kemp[1], Larry Knight[1], and Alexander Shevelko[1,2]

[1]Brigham Young University, Provo, Utah 84602,   [2]P.N.Lebedev Physical Institute, Moscow, Russia, 119991

Introduction: We have developed a family of spectrometers to characterize plasma sources in the x-ray and EUV spectral regions. The von Hamos spectrometer covers 0.04 to 1.6 nm. The grazing incidence spectrometer covers 2 to 80 nm. The wide range spectrometer covers 2 to 100 nm and 5 to 250 nm, depending on the grating.

Von Hamos

The von Hamos spectrometer is a crystal spectrometer. We use a curved mica crystal to simultaneously diffract and focus the x-rays onto a linear CCD detector. This instrument typically has a resolving power greater than 1000.

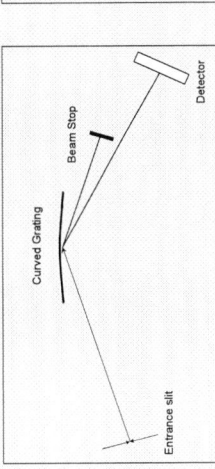

Grazing Incidence Spectrometer

The grazing incidence spectrometer uses a spherical grating at grazing incidence. Depending on the grating, it covers 2 to 80 nm with a resolving power of 100 to 250.

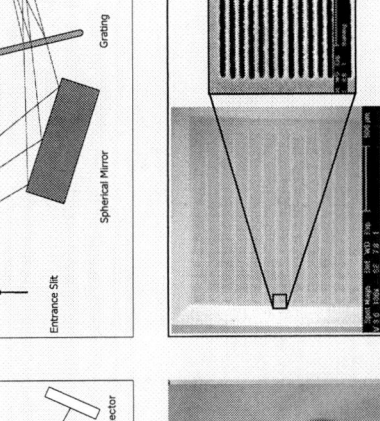

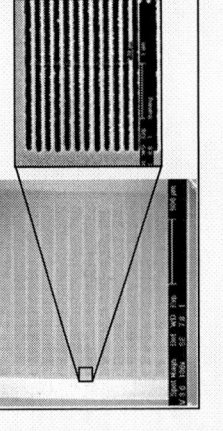

Wide Range Spectrometer: 2 to 250 nm

Transmission grating spectrometers can cover an extremely wide spectral range because transmission properties are independent of wavelength. Higher orders can be suppressed by the grating design.

The grating consists of wires 200 nm deep of Si_3N_4, with 25 nm Au deposited on each side (50nm total). The transmission constants for these materials and the index of refraction δ are found in [1], and also online at the Lawrence Berkeley Lab Center for X-ray Optics (CXRO) online database [2].

These instruments can be calibrated using a known emission source, such as the NIST synchrotron. The ability to measure the actual photon flux density is a tremendous advantage in characterizing EUV emission sources.

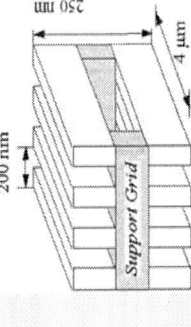

Experimental setup

X-ray source: A plasma was created using a frequency doubled Nd:YAG laser focused on a cylindrical target that rotates at the same frequency as the laser on a vertical micro screw. In this way, the target moves vertically and rotationally, exposing fresh target material after each shot.

Alignment: Initially the optical components in the spectrometer must be aligned so that the beam path passes through the center of the grating with normal incidence and then focuses on the detector. Translational stages are used in the chamber to optimize X-ray transmission through the grating in order to maximize the spectrum's zeroth order.

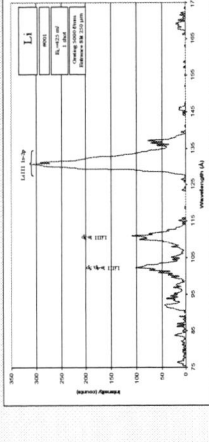

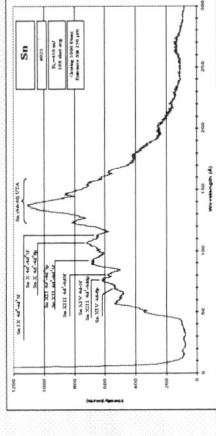

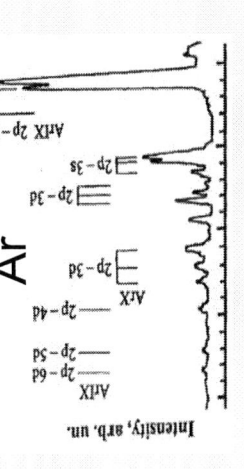

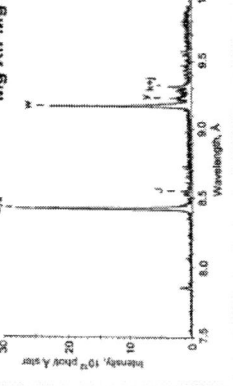

1. B. L. Henke, E.M. Gullikson, and J. C. Davis, Atomic Data and Nuclear Data Tables 54, 181 (1993).
2. http://henke.lbl.gov/optical_constants/, Lawrence Berkeley Laboratory Center for X-ray Optics.

High Power From Low Etendue EUV Light Sources

Michael Goldstein[1], Stefan Wurm[2], Frank Goodwin[2]

[1]Sr. Principal Physicist, Intel assignee at SEMATECH, [2]SEMATECH

Abstract:

Power and reliability specifications for EUV light sources are preeminent barriers impeding the adoption of EUV lithography in manufacturing. The benefit gained in recent years from scaling pulse energy and repetition rate has been strongly moderated by losses incurred from increased mirror debris protection. Here, we present an alternative regime where low etendue sources are multiplexed to higher power for greater optical efficiency and system reliability. Two commercially available plasma sources will be modeled with high efficiency optical designs suited to the individual source properties. The first is a low etendue and low power plasma source with an intrinsic collector multiplexed using transformed Kirkpatrick-Baez relay mirrors. The second is an electrodeless EUV plasma source with a middle range etendue and power comprising tilted sets of nested hybrid collectors. A comparison of the methods, limitations, and practical expectations will be reported.

Multiplexing Benefits:

Multiplexing Improves Reliability

- Reliability and serviceability increase significantly.
- System Reliability $\approx 1.0 - (1.0 - \text{Source Reliability})^N$

Source Debris (fully ionized plasma w/ fixed T & ρ)

- Electric discharge sources are not fuel/mass rate limited reactions. Debris scales with input power and roughly by source volume, ~R^3.
- EUV power scales by source surface area, ~R^2.
- Multiplexing N source w/ fixed etendue reduces debris on the order of ~$N^{3/2}$. Example: Debris drops ~27× per source when multiplexing 9 sources.

Multiplexing Very Low Etendue Sources:

- Source: R = 1.0 mm, $\theta_{full\ angle}$ = 0.5°
- Modified 1× Kirkpatrick-Baez optical design

- Multiplexing of 42 source is shown
  Etendue≈1.0 mm² sr, NA_{IL}=0.3, Transmission ≥ 75%

Half Angle = 17.7°
Solid Angle = 0.3 sr

100 Watts total will require ~3.2 Watts from each source.

M2 Footprint [R$_s$=1 mm]

M1 Footprint [R$_s$=1 mm]

Encircled Energy

Iso-Etendue Curves

Etendue = 5 mm² Str
Etendue = 3 mm² Str
Etendue = 1 mm² Str

A small source example (by NanoUV):

Power = 8.8 mWatts[***]
Spot Radius = 2.5 mm
Divergence = 0.18°
Rep. Rate = 2000 Hz
[***] λ=13.5 nm, Δλ/λ≈4%

Reference
www.nanouv.com
Peter Choi
EUV Source Workshop
Bolton Landing, NY
May 12, 2008

Multiplexing Mid-Etendue Sources:

- Nine mid-etendue sources can be multiplexed in a hexagonal + center source arrangement.

Iso-Etendue Curves

Etendue = 6 mm² Str
Etendue = 3 mm² Str
Etendue = 1 mm² Str

R ~ 2.4 mm

Example source:
- EnergetiQ
  R ~ 0.4 mm
  P ~ 10 Watt[***]
  Divergence ~ 2π sr.
  Rep. Rate > 1000 Hz
  [***] λ=13.5 nm, Δλ/λ≈2%

Reference
www.metastech.com
S.F.Horne et al.,
"Application of a high-brightness electrodeless Z-pinch EUV source for metrology, inspection, and resist development," Proc. of SPIE (2006).

Etendue (mm² sr.)	
• Total System	= 5.03
• Center Collector	= 0.99
• Edge Collectors	= 0.41
• Fill Fraction	= 84%

Power (Watts)	
• 10 or 25 Watts / 2π sr / Source	
• Total Power = ~43 or ~107 Watts	

Temporal Multiplexing:

- Increased reliability & frequency w/ multiple lasers
- Reduced debris w/ smaller plasmas & higher freq.
- Increased collection possible w/ smaller plasmas

Conclusion:

- Multiplexing increases reliability, and collector lifetime from smaller sources produce less debris
- Kirkpatrick-Baez relays are suited for very low etendue source multiplexing
- Hybrid Wolter and elliptical collectors are suited to multiplex mid-range etendue sources
- A low/mid-etendue source can achieve high power with improved reliability and collector lifetime

Advanced Materials Research Center, AMRC, International SEMATECH Manufacturing Initiative, and ISMI are servicemarks of SEMATECH, Inc. SEMATECH, the SEMATECH logo are registered servicemarks of SEMATECH, Inc. All other servicemarks and trademarks are the property of their respective owners.

The Characterization of the Ion Beam from a Sn-based DPP with Respect to the Ignition Parameters

K. Gielissen[1], Y. Sidelnikov[2], W. Soer[3], M. van Herpen[3], D. Glushkov[4], V. Banine[4], J. van der Mullen[1]

[1]Eindhoven University of Technology, The Netherlands, [2]ISAN, Troitsk, Russia, [3]Philips Research, Eindhoven, The Netherlands, [4]ASML, Veldhoven, The Netherlands

Abstract

Based on several years of development work, multiple source vendors and research institutions dealing with EUV lithography are demonstrating a strong rise in performance of EUV sources. One of the EUV source types, namely Discharge Produced Plasma (DPP) source, powers the installed EUV ASML scanners. One of the critical aspects of the integrated source performance proved to be lifetime of the collector optics in the source-collector assembly. The factor which determines this life time is ion sputtering of the material at the surface of the collector. The ions are in turn being generated by the DPP itself. This research focuses on the characterization of the fast ionic debris in the range up to 100 keV with extreme sputtering efficiencies. The ion beam emitted from the DPP is analyzed using different time-of-flight techniques while the discharge was monitored by a current probe and gated MCP imaging. By varying the ignition parameters the maximum kinetic energy of the emitted fast ions could be reduced and thus the lifetime of the collector can be increased with a factor of 5-10x.

The EUV-source

- DPP (discharge produced plasma)
- Rotating liquid Sn-electrodes
- Laser triggered (Nd-YAG pulse)
- Vacuum operation (p = 10^{-5} mbar)

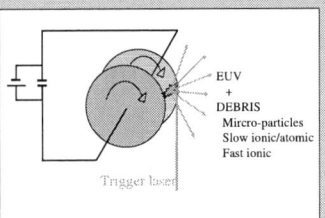

Figure 1: Drawing of the EUV-source. A voltage is applied to the rotating electrodes through a capacitor bank. The trigger laser pulse is used to initiate the discharge, which creates the EUV radiation.

Origin of fast ionic debris

- Gated MCP images of the discharge and the fast ionic debris.

- Single pulse image
  200 ns pulse width
  0.25 mm aperture

- Oscilloscope pictures show the time-of-flight of the captured ions together with a simultaneously measured Faraday cup signal.

- MCP images show origin of fast ionic debris is close to the cathode surface.

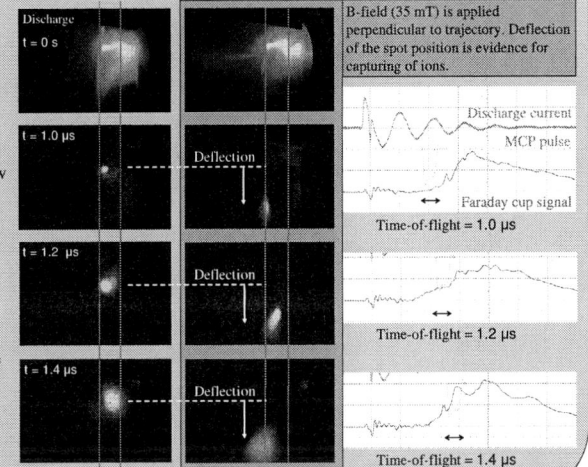

EUV emission

- Measurement of in-band EUV emission simultaneously with Faraday cup measurement of fast ion emission.
- The number of ions n_i emitted by DPP per pulse in 2π with kinetic energy $E_{kin} > 10$ keV was calculated from Faraday cup signal.

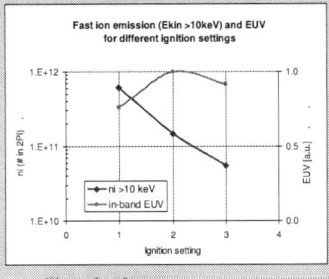

Conclusions

- Max of ion emission does not correspond to max EUV emission
- Suppression of ion beam possible with only minor EUV loss

Ion beam characterization

- Experiments under vacuum conditions
- Enhanced faraday cup configuration for noise reduction and signal amplification
- Electrostatic spectrometer used for measuring the ion spectrum for different E/Z values.
- Measurements for different ignition settings

The Faraday cup signal was recorded simultaneously with the ion spectrum. Each curve is the average of > 100 pulses.

Conclusions

- Highly charged Sn ions are present in fast ionic debris signal from the Faraday cup.
  E_{kin} up to 70 keV (Sn^{+14})

- The average charge state of the Sn ions does not change with the ignition settings.

- By changing the ignition settings of the DPP source, the maximum kinetic energy of the fast ions can be reduced with a factor of 10.

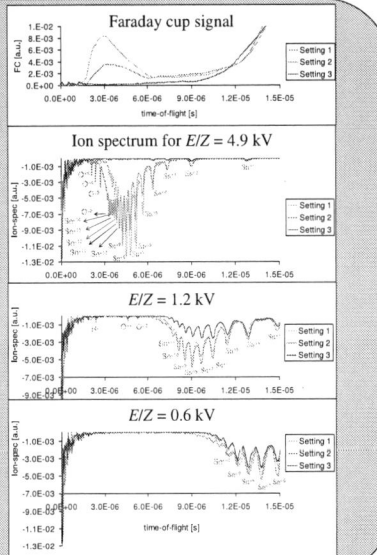

Acknowledgements: Luc Stevens, John de Kuster, The measurements presented in this poster are based on the experimental approach developed at ISAN by the research group of K. Koshelev.

International EUVL symposium, September 28 – October 1, 2008, Lake Tahoe

Present status of laser-produced plasma EUV light source

Hiroshi Komori2, Y. Ueno2, K. Nowak2, T. Yabu2, T. Yanagida2, T. Suganuma2, T. Asayama2, H. Someya2, H. Hoshino2,
M. Nakano2, M. Moriya2, T. Nishisaka2, T. Abe2, A. Sumitani2, H. Nagano1, Y. Sasaki1, S. Nagai1, Y. Watanabe1,
G. Soumagne1, T. Ishihara1, O. Wakabayashi1, K. Kakizaki1, A. Endo1, H. Mizoguchi3
1 EUVA / Gigaphoton Inc., 2 EUVA / Komatsu Ltd., 3 Gigaphoton Inc.

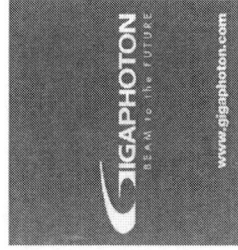

www.gigaphoton.com

Abstract

The laser produced plasma EUV light source development for high-volume manufacturing is gaining further momentum. Key light source components have been developed during the last years and component integration into a light source system is now ongoing. This poster describes the key technologies of our CO2 laser produced Sn light source, i.e., the laser, the droplet and the magnetic mitigation, as well as outlines the development of a test light source and the high-volume EUVL light source roadmap.

§ Light Source Concept & Concept History

Gigaphoton LPP Light Source
- Sn Droplet
- High power pulsed CO2 laser
- Magnetic-field Plasma Guiding

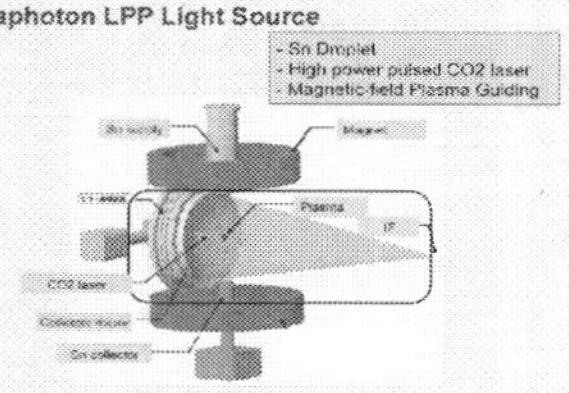

2001: Concept of CO2 laser based LPP source. (Patent applied in 2001)
2001: Concept of MOPA CO2 laser based LPP source.
(Patent applied in 2001)
2002 /09: EUVA light source project start (with Gigaphoton, USHIO and Komatsu)
2033: Concept of Magnetic field ion mitigation (Patent applied in 2004)
2004 /09: EUV 5.7 W/IF was demonstrated (Nd:YAG and Xe jet)
2006 /03: EUV 10 W/IF was demonstrated (CO2 and SnO2 choroid liquid jet)
2007 /02: EUV 40 W/IF was demonstrated (CO2 and Sn target)
2007 /10: EUV 60 W/IF was demonstrated (CO2 and Sn target)

§ High-power, short-pulse CO2 laser

Laser Power: 13 kW
Pulse Width: 20 ns
Repetition Rate: 100 kHz
Pulse energy stability : 2% (3σ, 500 pulses)

13 kW → 100 W at I/F equivalent

20 kW (250nJ at 100kHz)

AMP1 : RF-excited CO2 laser
AMP2 : RF-excited CO2 laser
AMP3 : RF-excited CO2 laser

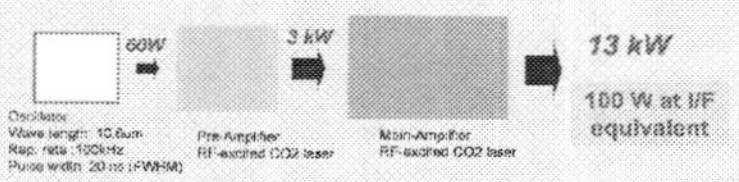

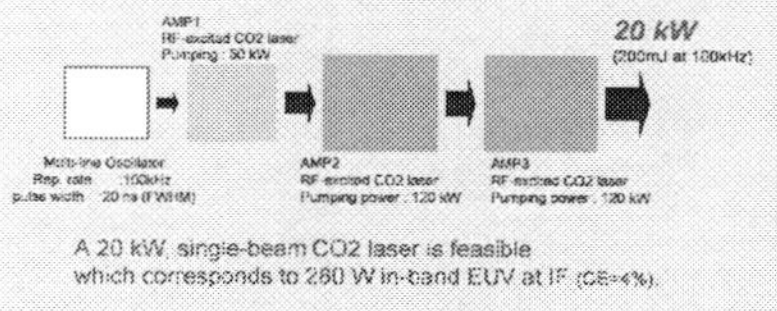

A 20 kW, single-beam CO2 laser is feasible which corresponds to 280 W in-band EUV at I/F (CE=4%).

§ EUV Generation

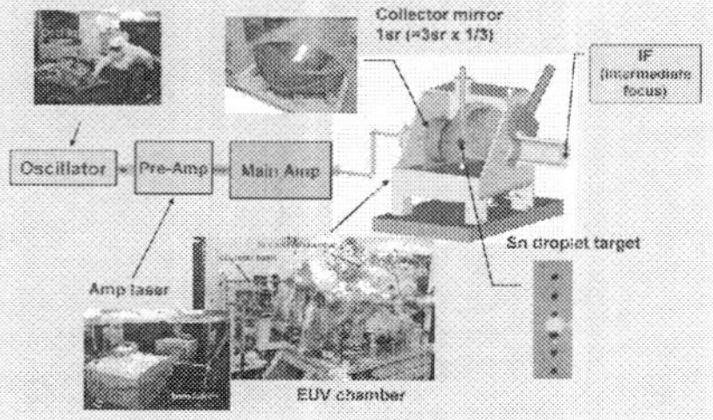

Collector mirror
1sr i=3sr x 1/3)

EUV measured at primary source

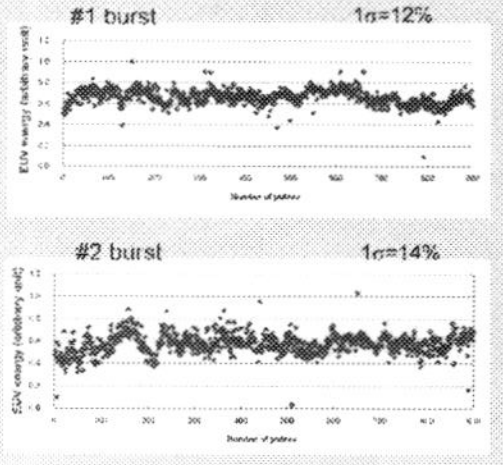

#1 burst   1σ=12%
#2 burst   1σ=14%

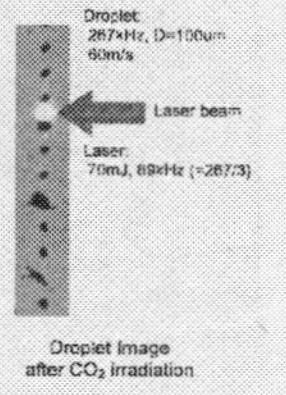

§ Sn Droplet Target & Magnetic Mitigation

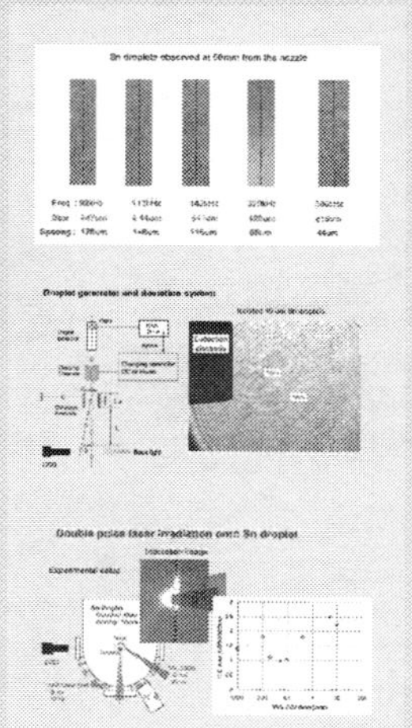

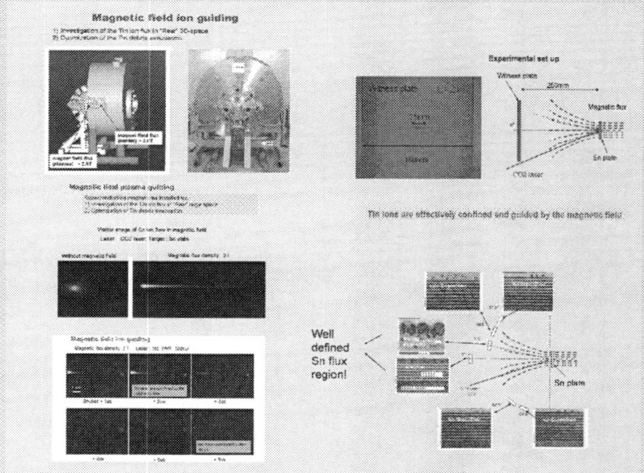

Magnetic field ion mitigation

Experimental set-up

Well defined Sn flux region!

The ions are effectively confined and guided by the magnetic field.

§ Light Source Roadmap

EUV LPP light source roadmap

Power roadmap

Today to SD — Non commercial system

Output system

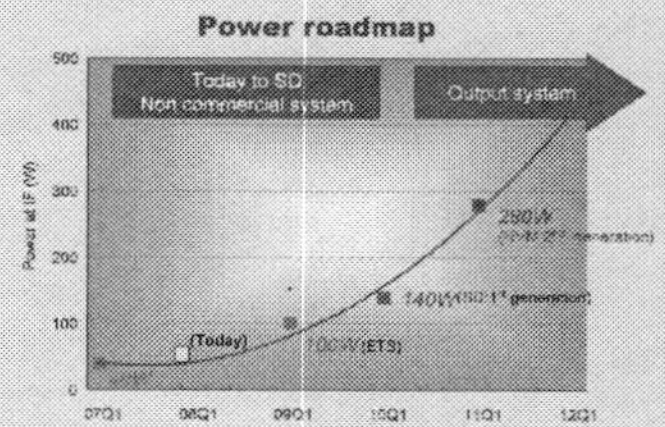

§ Summary

- key light source components and technologies have been developed
- light source system development ongoing with the following roadmap
  - 2008/E 100W, internal test source
  - 2009/E 140W, 1st generation
  - 2011/B 280W, 2nd generation
- due to scalability, high conversion efficiency and efficient mitigation LPP source is most promising source candidate for HVM EUVL

This work was partly supported by
The New Energy and Industrial Technology Development Organization (NEDO), Japan

Performance Evaluation of EUV SFET Source Collector Module

Shunko Magoshi, Seiichiro Shirai, Hideto Mori,

Kazuo Tawarayama, Yuusuke Tanaka, and

Hiroyuki Tanaka

Semiconductor Leading Edge Technologies, Inc.

EUVL Symposium 2008 at Lake Tahoe

Introduction

High-power, stable and long-life EUV SoCoMo is anxiously awaited for HVM EUV scanner. In order to accelerate SoCoMo development, it is necessary to evaluate its lithographic performance and to give feedback to SoCoMo suppliers.

Through using SFET for pattern exposure, Selete is studying what lithographic issues are there in its SoCoMo and how to evaluate them. However, the SFET SoCoMo is one of the earliest commercial DPP sources and had a simple configuration with no measurement tool to evaluate its own performance.

Therefore, we had introduced the monitors and applied the evaluation method to it as follows;

■Collector position monitor for Collector's optical alignment

■IF spot position monitor for evaluation of plasma fluctuation

■Pupilgram method for evaluation of far-field pattern behind IF

This poster presents the real-life data of the SFET SoCoMo evaluated with these tools, and also presents "the source requirements from *user's* viewpoint".

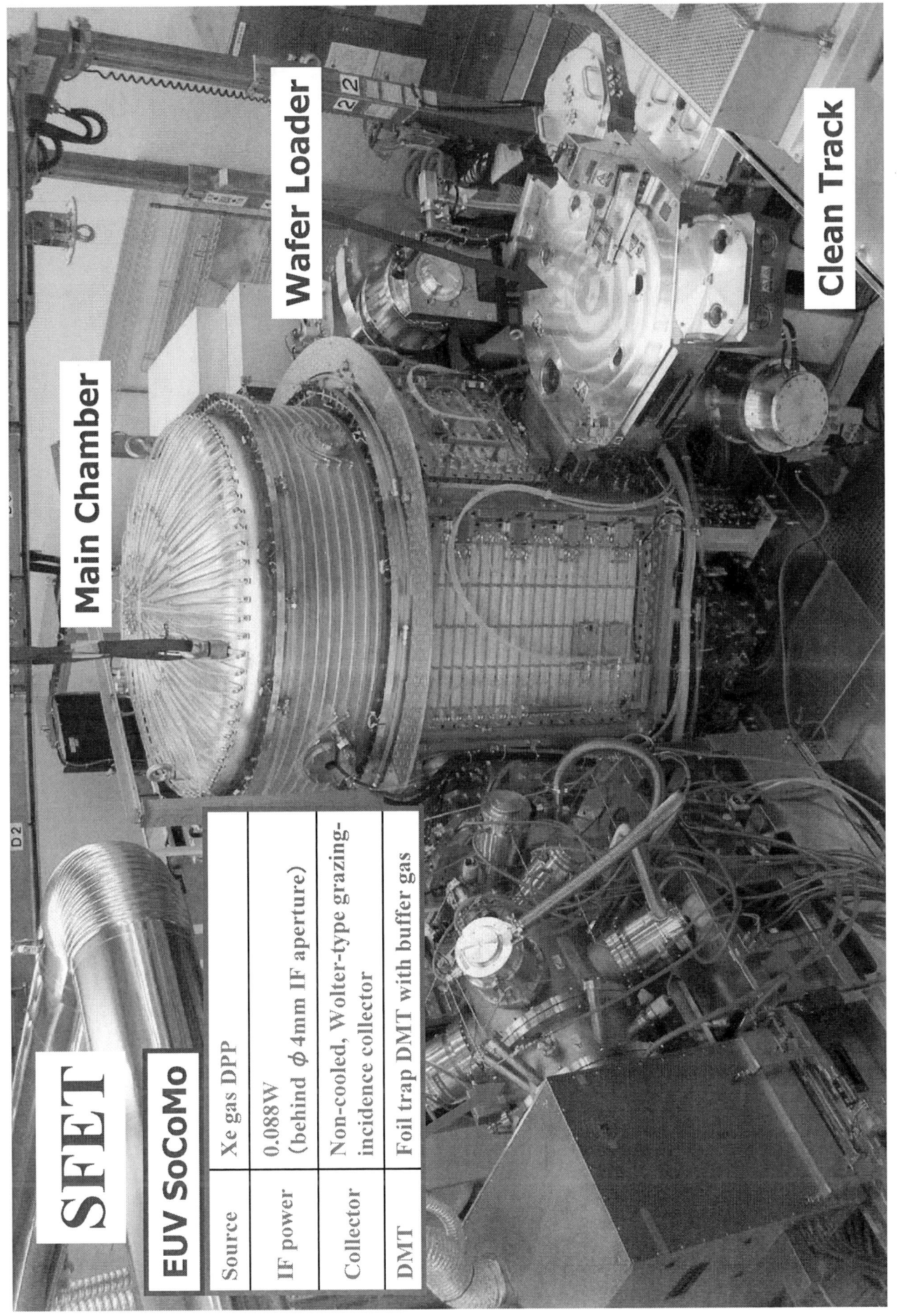

SFET Source Activity

Trend of SFET EUV Source Operation (2007/02~2008/09)

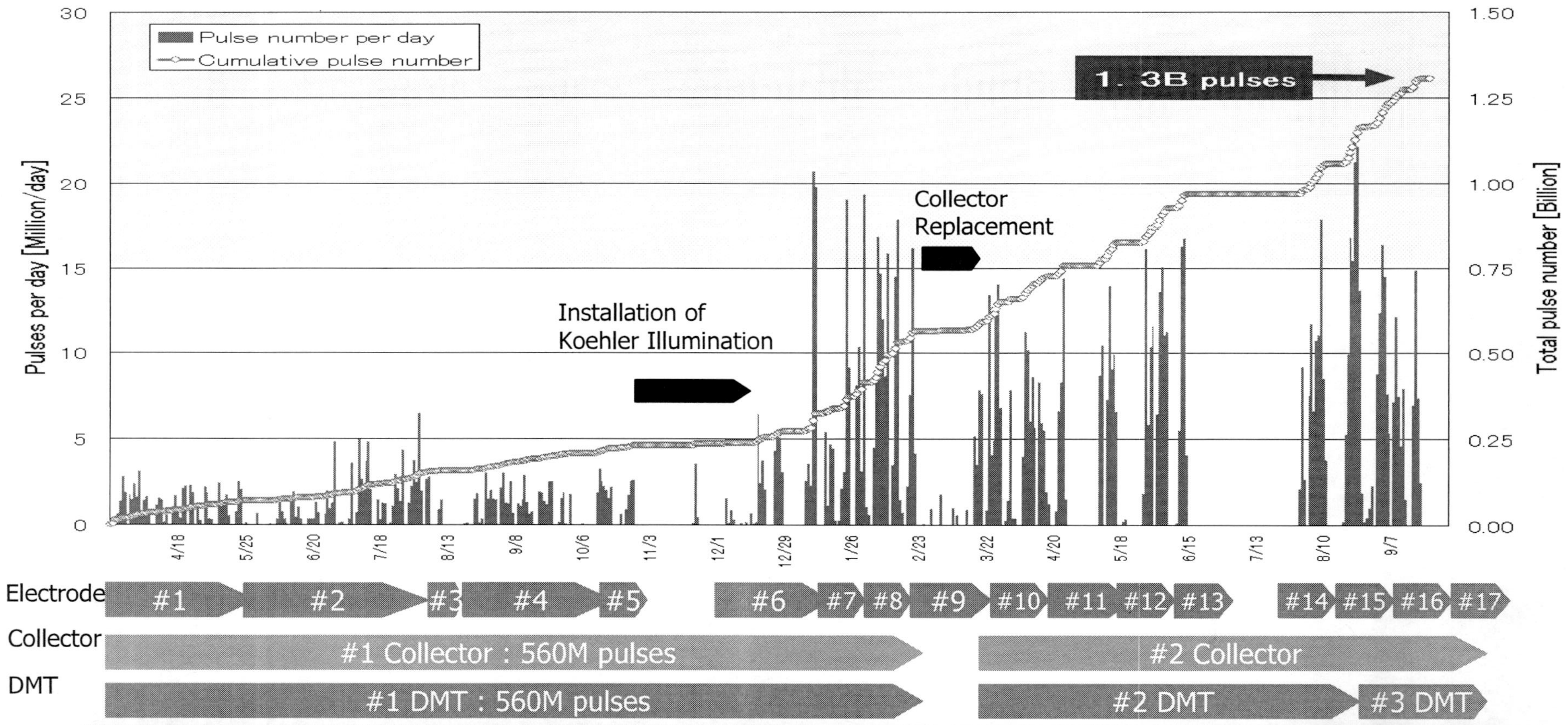

- Total pulse number had reached 1.3 billion pulses.
- 17 electrodes, 2 Collectors, 3 DMTs were evaluated.

Change in EUV light intensity at wafer plane   *Selete*

To evaluate SoCoMo performance, SFET's light intensity sensor has to be used because SFET source has no energy monitor.

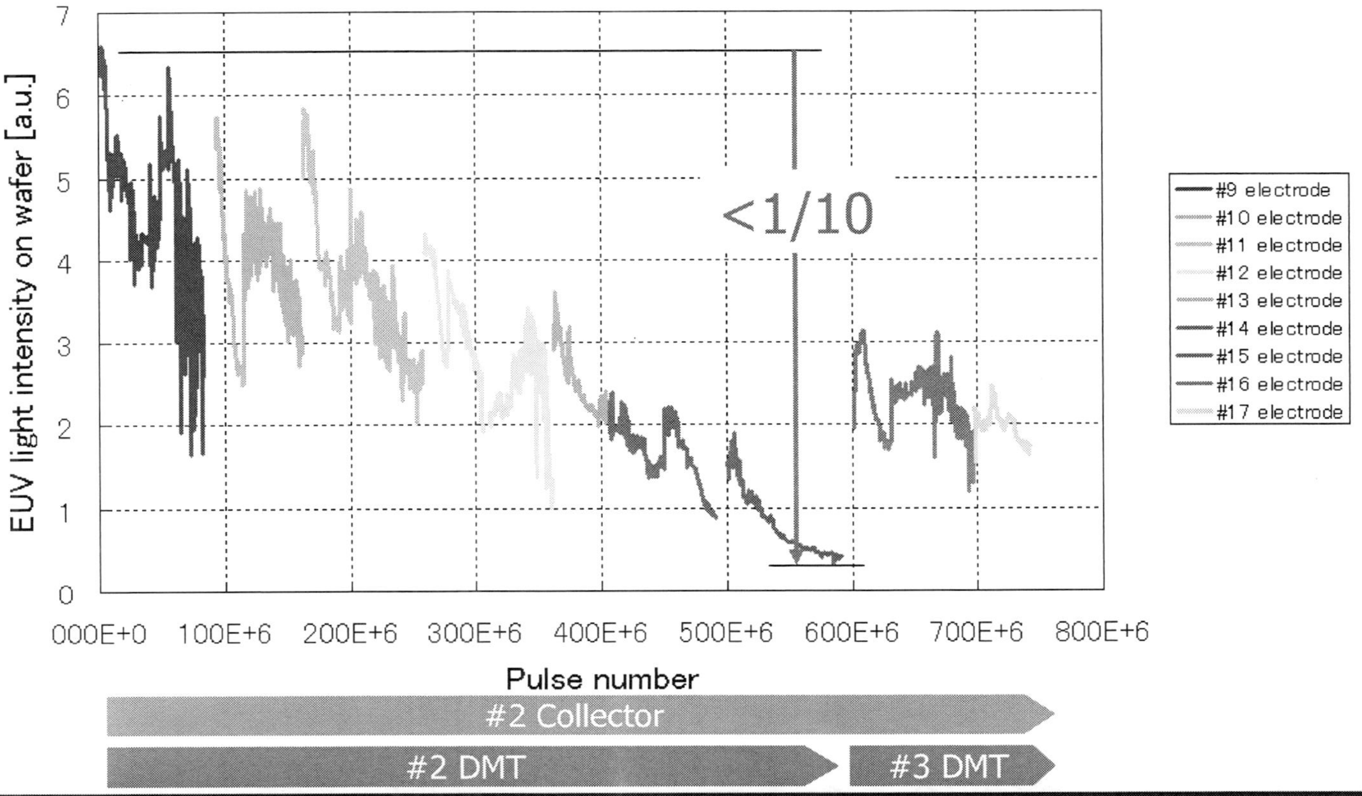

Legend:
- #9 electrode
- #10 electrode
- #11 electrode
- #12 electrode
- #13 electrode
- #14 electrode
- #15 electrode
- #16 electrode
- #17 electrode

<1/10

EUV light intensity on wafer [a.u.]

Pulse number

#2 DMT   #2 Collector   #3 DMT

EUV light intensity was dropped to less than tithe of initial value after 600M-pulse radiation. What happened?

Collector positioning mechanism

To align optical axis from plasma to IF aperture, Collector position monitors were added to existing adjusting screws.

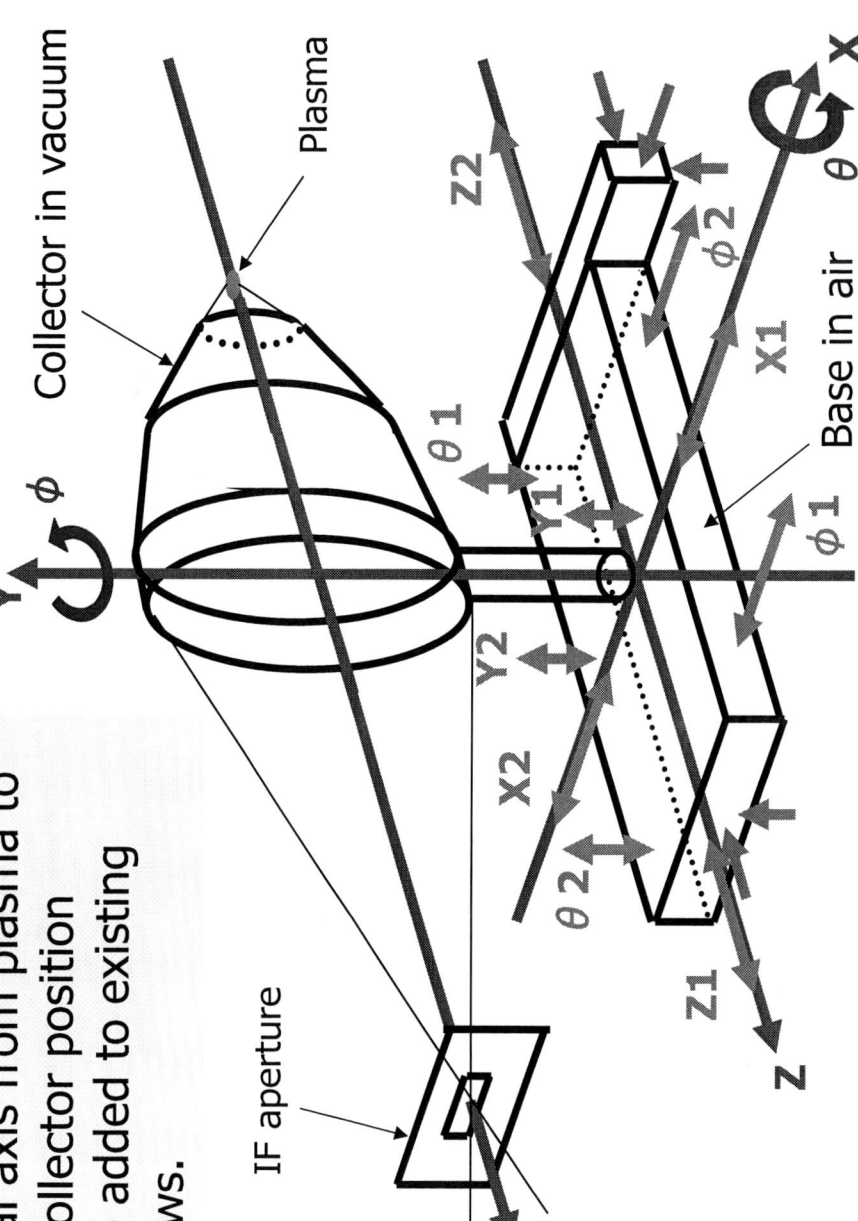

Plasma

Collector in vacuum

IF aperture

Base in air

Position monitors
Z, Combined with multi-axes

Adjusting screws
X1,X2,Y1,Y2,Z1,Z2,
$\theta 1, \theta 2, \phi 1, \phi 2$

IF spot position monitor

To evaluate EUV light spot position at IF, IF spot monitor was introduced, in which movable mirror was loaded between collector and IF, and 2nd IF spot on screen was observed using CCD.

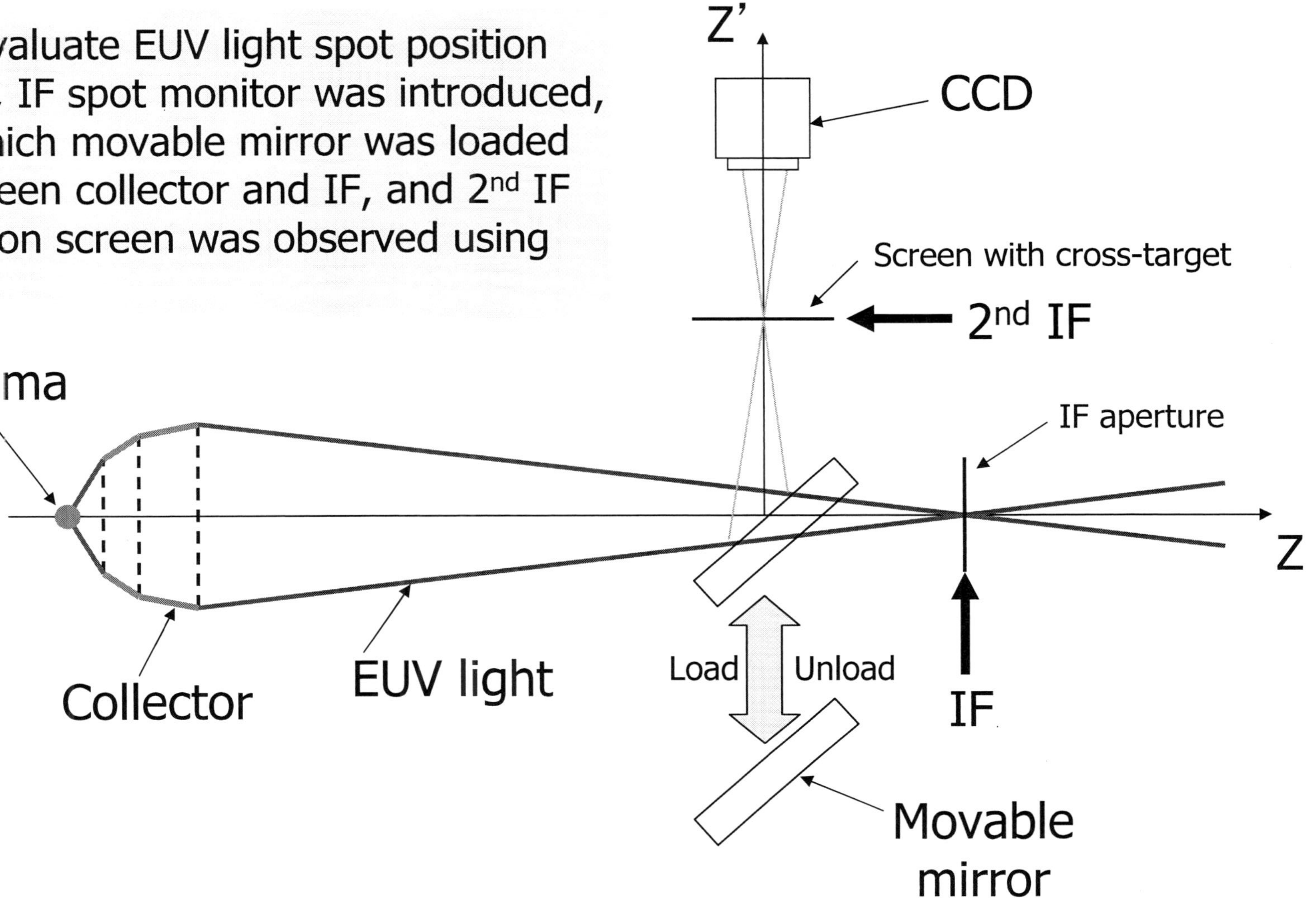

Collector adjustment using IF spot monitor

After 20-million-pulse radiation, shift and defocus of IF spot were observed on IF spot monitor. IF spot was adjusted to and focused on target by Collector moving, so that EUV light intensity at wafer plane increased by 30%.

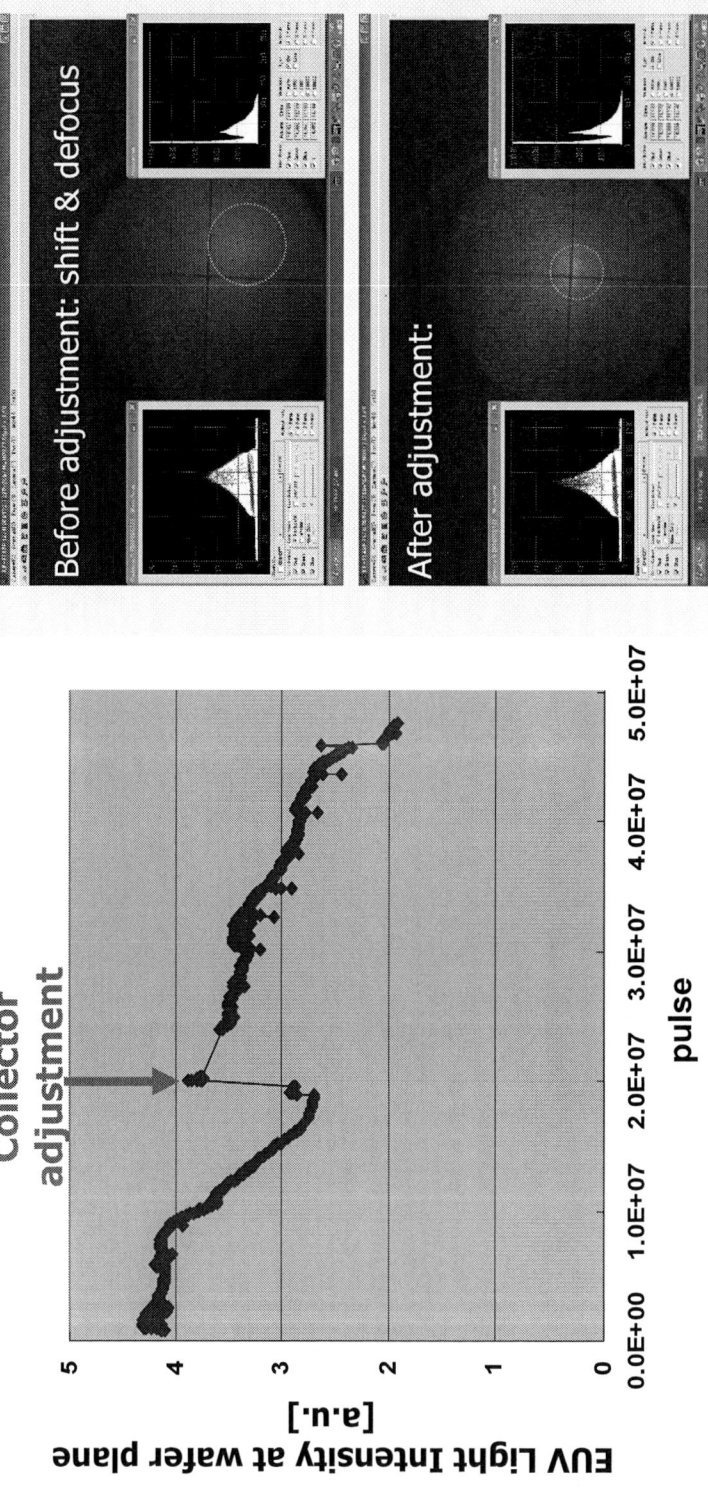

Before adjustment: shift & defocus

After adjustment:

Screenshots of IF spot monitor

Adjusting Collector with IF spot monitor can maintain higher EUV power at wafer plane. By the way, why IF spot shifts and defocuses?

Plasma fluctuation model caused by cathode erosion

Cathode erosion	Plasma variation
Conical	• expansion • intensity reduction • moving away from collector
Asymmetric	• wander off axis • non-axisymmetrical deformation

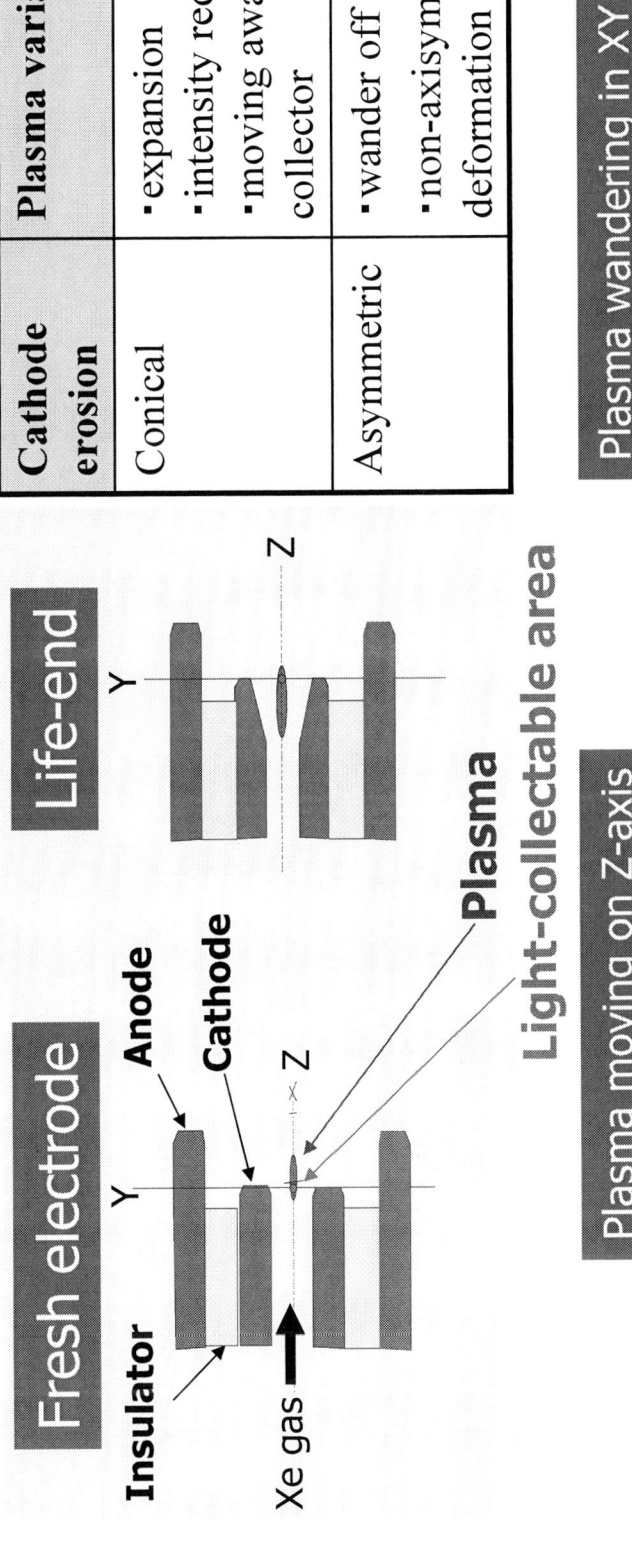

Fresh electrode

Life-end

Anode

Cathode

Insulator

Xe gas

Plasma

Light-collectable area

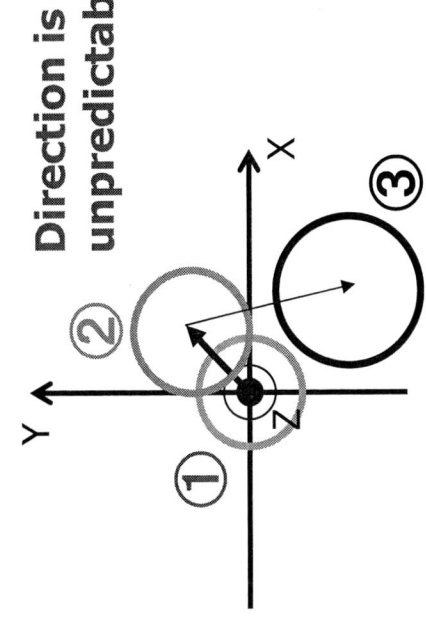

Plasma wandering in XY plane

Direction is unpredictable

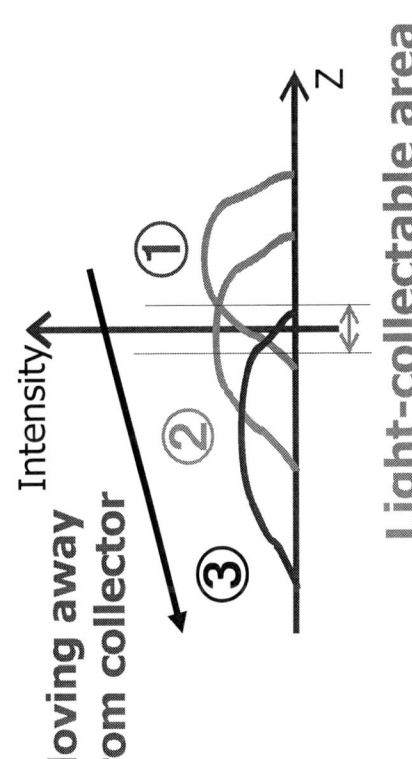

Plasma moving on Z-axis

Moving away from collector

Light-collectable area

Plasma fluctuation through electrode lifetime

To measure plasma's intensity distribution, Collector was moved to Z-axis direction and EUV light intensity at wafer plane was measured.

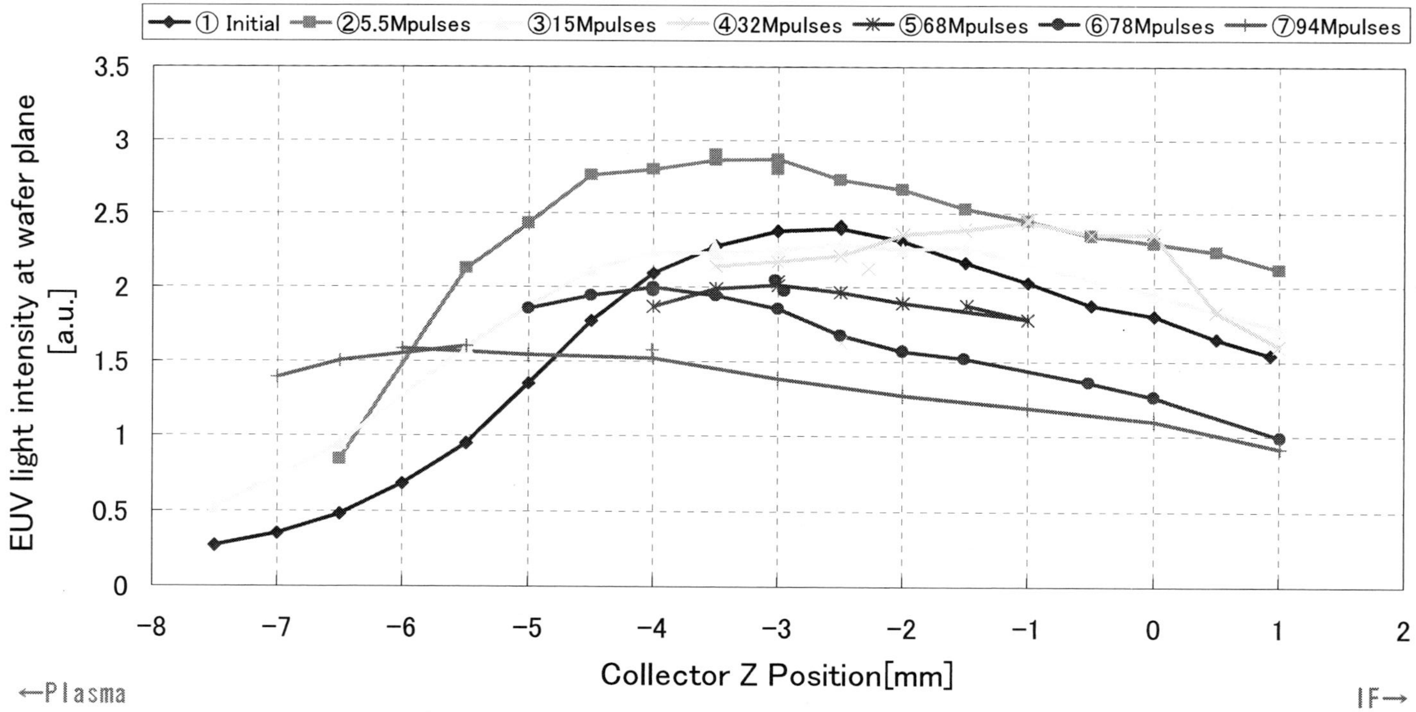

Plasma fluctuation throughout an electrode lifetime was confirmed. However, it is too difficult to know exact plasma profile by this indirect method. Plasma should be observed directly.

Contribution of each component to light reduction (delete)

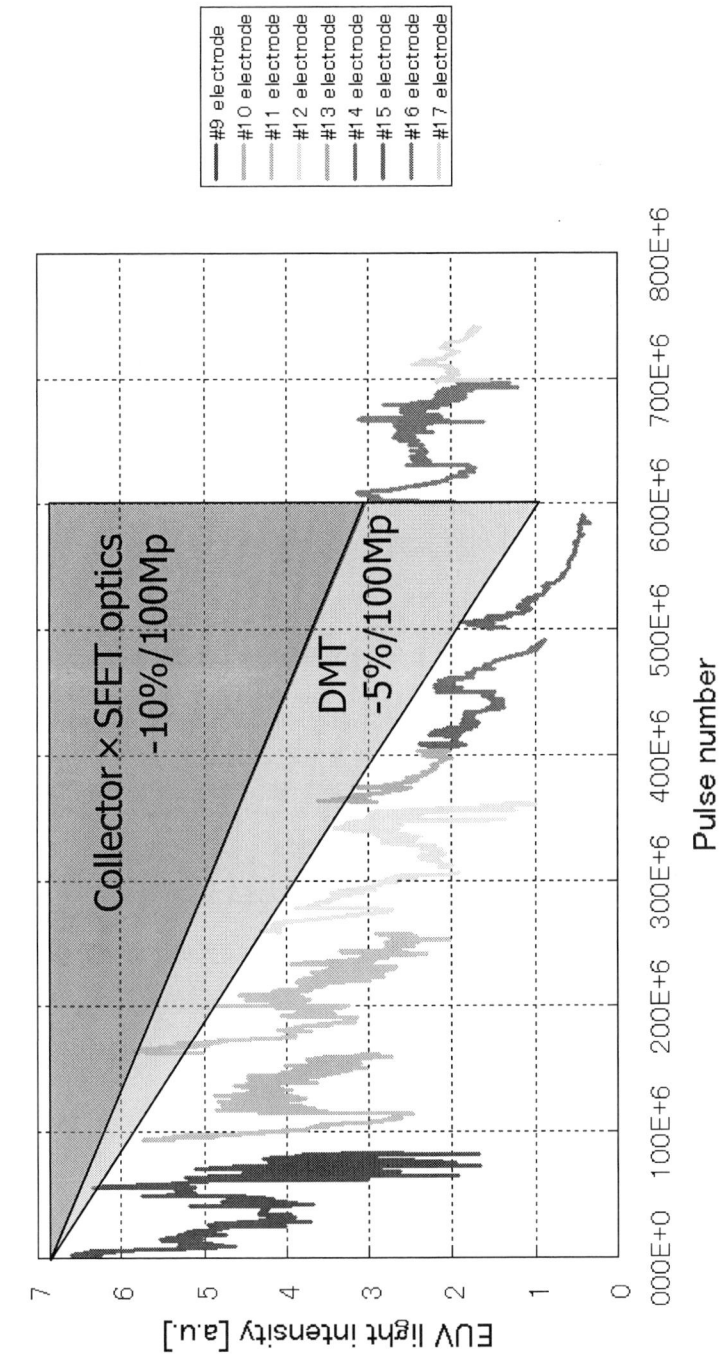

EUV light intensity fluctuation mainly causes plasma variation by electrode erosion. What happened at DMT? (We cannot isolate Collector damage now.)

Pupilgram method

1. 750x750um pinhole window on the mask
   * SFET field-size: 1000x3000um
2. Use high sensitivity resist
   3.5 times faster E_0 than MET-2D
3. Expose wafer with HUGE DEFOCUS and huge dose.

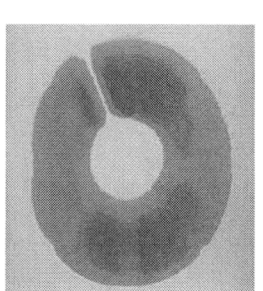

Contour map

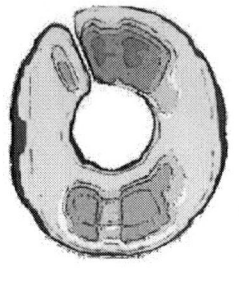

Overlay all pictures

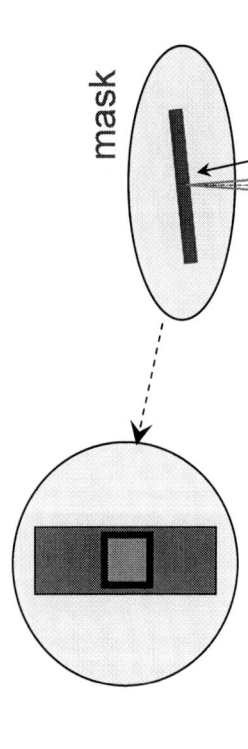

mask

Light from illumination unit

wafer

Center shift 4.3um
Distortion 2.3%
(minor / major axis) =1.89/1.934 [mm]

+4.5mm

wafer

+z

* *Toshihiko Tsuji (Canon)*

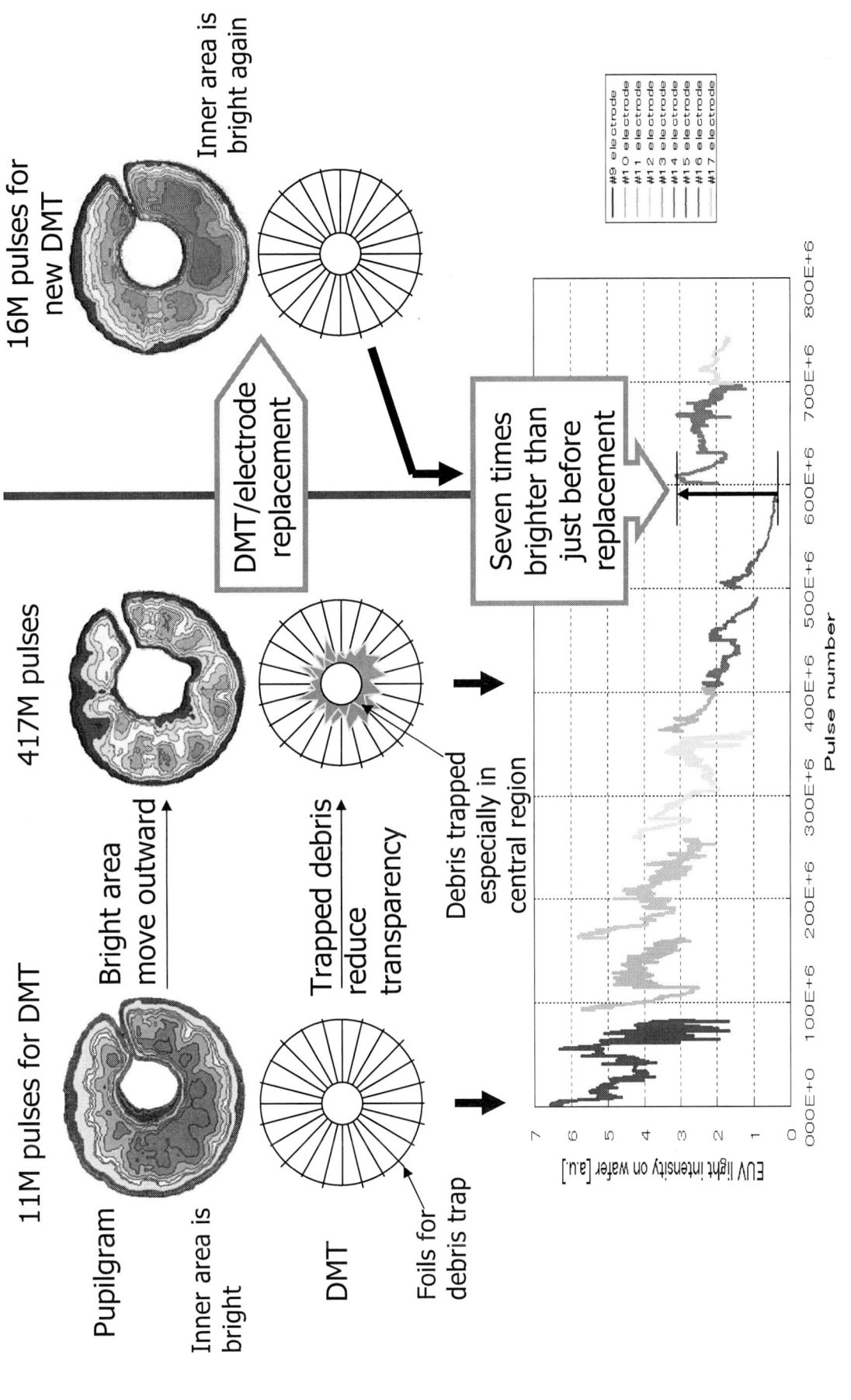

Summary

■ Causes and effects of SFET SoCoMo variation

	Time scale	Variation mode	Possible causes
Short term	Shortly after starting radiation (in seconds)	◆ Unstable intensity	◆ Gas flow rate
	Patterning wafers (in hours)	◆ Intensity drift	◆ Thermal distortion of Collector ◆ Temperature of capacitor
Long term	Electrode lifetime (in weeks)	◆ Intensity reduction ◆ Far field pattern (FFP) modification	◆ Plasma fluctuation by electrode erosion
	Collector/DMT lifetime (in months)		◆ Reproducibility of electrode replacement ◆ Optical misalignment of collector/DMT ◆ Collector/DMT degradation

IRAI Source Requirements from *User's* Viewpoint   *Selete*

■ *"Don't move, plasma!"*

■ SoCoMo should monitor its own performance.

➢ Plasma position, profile, and power should be measured *directly*, especially during pattern exposure.

➢ Optical alignment from plasma to IF should be evaluated and adjusted accurately.

➢ IF power and FFP behind IF should be measured *daily* in SoCoMo keeping connected to scanner, not separated.

➢ Damage level of electrode, DMT and Collector should be monitored *individually*.

Acknowledgements

The authors would like to thank SFET development teams in USHIO/XTREME and Canon for their continuous efforts to improve SFET performance.

This work was partly supported by New Energy Industrial Technology Development Organization (NEDO).

Development of new negative-tone molecular resists based on alkylphenyl calixarene for EUVL

Masatoshi Echigo[1], Dai Oguro[1], Hiroaki Oizumi[2] and Toshiro Itani[2]

[1]Mitsubishi Gas Chemical Company, Inc.; Hiratsuka Research Lab. , [2]Semiconductor Leading Edge Technologies, INC.

E-Mail : echigo@mgc.co.jp

INTRODUCTION

Approach from Resist Base Materials for LER Reduction

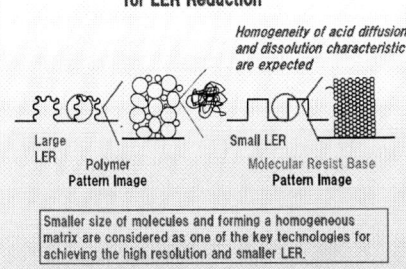

Homogeneity of acid diffusion and dissolution characteristic are expected

Smaller size of molecules and forming a homogeneous matrix are considered as one of the key technologies for achieving the high resolution and smaller LER.

MGC's proposal for EUV Molecular Resist Base

Molecular Design Concepts for Resist Base Materials

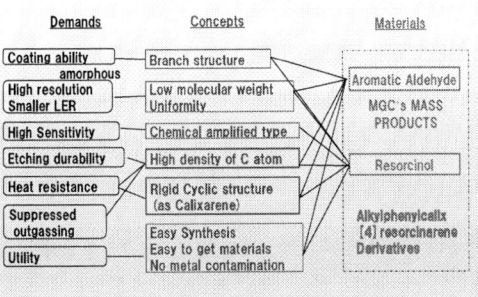

Motive Force for Negative-tone Molecular Resist

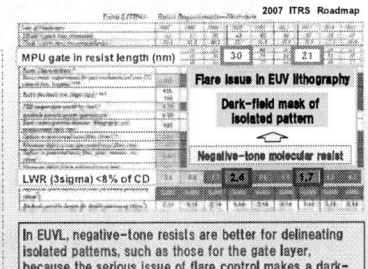

In EUVL, negative-tone resists are better for delineating isolated patterns, such as those for the gate layer, because the serious issue of flare control makes a dark-field mask pattern preferable.

Synthesis and Solubility Evaluation of Phenylcalix [4] resorcinarene derivatives

R1	R2	PGME	0.26N TMAHaq
methyl	H	Insoluble	N/A
methyl	methyl	Insoluble	N/A
n-propyl	H	Barely Soluble	Soluble
i-butyl	H	Barely Soluble	Soluble
cyclohexyl	H	Soluble	Soluble
4-n-propylcyclohexyl	H	Soluble	Insoluble

MGC's Development of Phenylcalix [4] resorcinarene for EUV Molecular Resist Base

Advanced Negative-tone molecular resist base

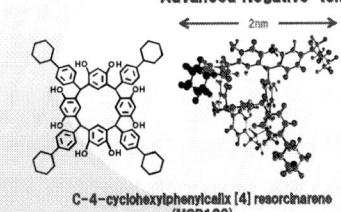

C-4-cyclohexylphenylcalix [4] resorcinarene (MGR108)

Fundamental Chemistry
· Amorphous
· High Heat Resistance (High Tg >300°C)
· Good Solubility for PGME and TMAHO.26N%aq.
· High Etching Durability (Etching rate is same as PHS)

High-NA Small-Field Exposure Tool (HINA)

(1) Wavelength: 13.5nm (SR: Super-ALIS, SBL-1)
(2) Two-aspherical-mirror projection optics
 Magnification: 1/5, NA: 0.3, field size: 0.5mm x 0.3mm
(3) Two illumination systems
 Koehler illumination with σ ~ 0.6 (Normal, Annular)
 coherent illumination with σ ~ 0
(4) Mask size: 6025 reticle; Wafer size: 6"φ

Resist Formulation

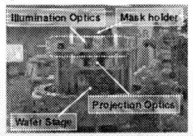

| Resin | MGR108 | Crosslinker | |
| PAG | | Quencher | |

EUV Patterning Evaluation

EUV Patterning Result

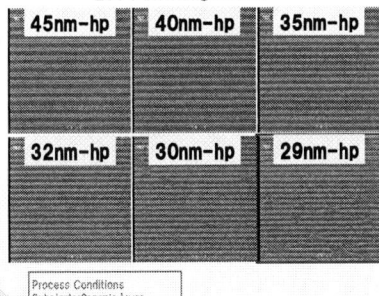

45nm-hp 40nm-hp 35nm-hp
32nm-hp 30nm-hp 29nm-hp

Process Conditions
Substrate:Organic layer
Thickness: 60nm
PB&PEB: 110°C/90s
Dev.: TMAH 0.26N 30s
SEM: S8840

29nm-hp@14.8mJ/cm2 pattern was resolved.

Crosslinking Mechanism / FT-IR

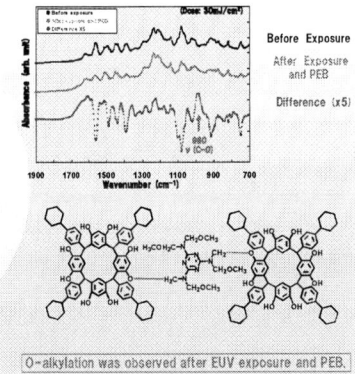

Before Exposure
After Exposure and PEB
Difference (x5)

O-alkylation was observed after EUV exposure and PEB

EUV Outgassing Evaluation

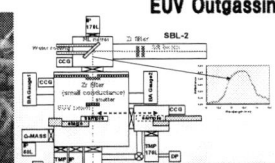

Outgassing Evaluation Apparatus

(1) Exposure wavelength: 13.5nm (SR: Super-ALIS, SBL-2)
(2) Illumination optics:
 A pair of grazing toroidal mirror, planar folding 40-pairs Mo/Si mirror, Zr filters
(3) EUV exposure area : 10 mm ×20 mm in square
(4) EUV intensity: 0.01~2 mW/cm2
(5) Back pressure: <1 x 10-7 Pa (resist sample in main chamber)

Outgassing Evaluation Method Pressure Rising Method

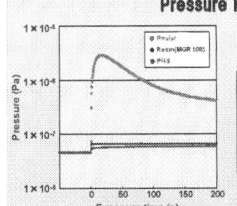

Sample	Outgassing rate at 400mW/cm2 (molecules/cm2/s)
Resist	3.8×10^{16}
Resin (MGR108)	3.3×10^{14}
Resin (PHS)	2.9×10^{14}

The outgassing rate of the MGR108 was on the same order as that of a conventional resin, PHS.

Outgassing Analysis

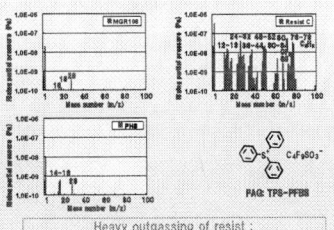

PAG: TPS-PFBS

Heavy outgassing of resist : Decomposition of PAG by EUV exposure

Summary

We investigated EUV pattering performance of new Negative-tone molecular resist based on C-4-cyclohexylphenylcalix [4] resorcinarene (MGR108) and their EUV outgassing.

An advanced Negative-tone molecular resist based on C-4-cyclohexylphenylcalix [4] resorcinarene (MGR108) provides a resolution of 29-nm hp at exposure dose of 14.8mJ/cm2

The outgassing rate of MGR108 was on the same order as that of conventional resin, PHS. On the other hand, the main cause of its resist outgassing is the EUV-induced decomposition of the PAG.

We'll develop the Negative-tone molecular resist based on phenylcalix [4] resorcinarene for high sensitivity (≦10mJ/cm2), high resolution (≦32nm), low LER (≦1.3nm) and low outgassing.

Now, we are also developing the new Positive-tone molecular resist.

ACKNOWLEDGMENT

This work was partially supported by the New Energy and Industrial Technology Development Organization (NEDO).

Survey and comparison of deprotection blur metrics for extreme ultraviolet photoresists

Christopher N. Anderson[1*] and Patrick P. Naulleau[2]

Motivation

Small perturbations of resist and process parameters almost always change patterning performance in EUV photoresists; explanations of these effects are often speculative at best.

The interplay between resolution, LER, and other factors is not well understood.

To better understand resists, and how to improve them, we wish to develop resist metrics that can deconvolve the effects of resolution, LER, and other factors that contribute to patterning performance.

Resolution vs. deprotection blur

In a practical sense, resist resolution is often defined as the smallest sized 1:1 lines that pattern with an exposure threshold greater than some level.

This definition folds in effects like pattern collapse and top loss; observed "resolution" may be larger than size limits determined solely by fundamental resist properties.

What would pattern if we could illuminate resist with a delta function of photons? One would expect that the pattern size should be determined by the diffusion of photo generated acids.

Point-spread function resist model[1]

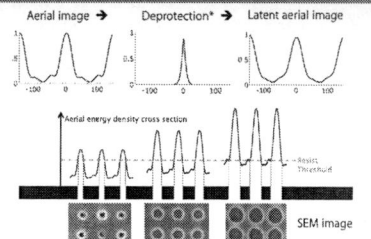

Reduced blur -> reduced contrast -> reduced exposure latitude.

MET calibration resists

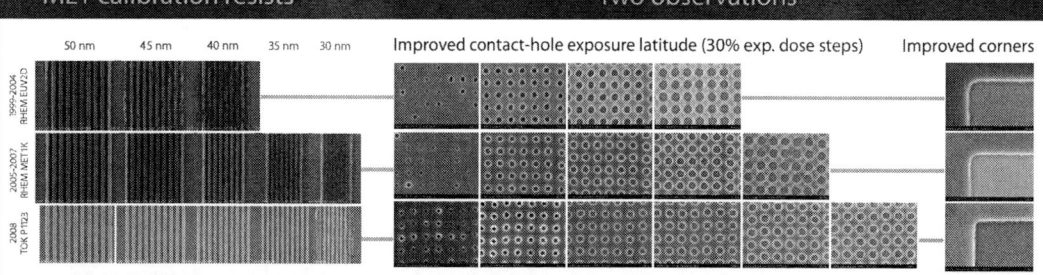

Two observations

Improved contact-hole exposure latitude (30% exp. dose steps)

Improved corners

Two blur metrics

The contact-hole and corner rounding blur metrics have been described in detail in the literature [2-3].

Here we use them to monitor the deprotection blur of several resist platforms as the wt. % of base and photo acid generator (PAG) are varied [4], and also as the post-exposure bake (PEB) temperature is varied [5].

Varying base wt. %

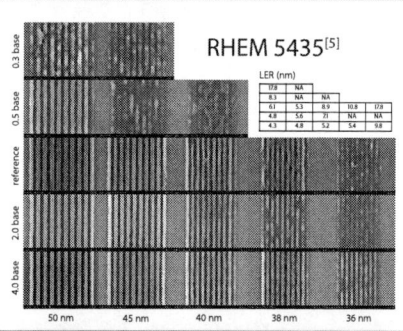

RHEM 5435[5]

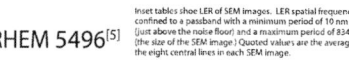

RHEM 5496[5]

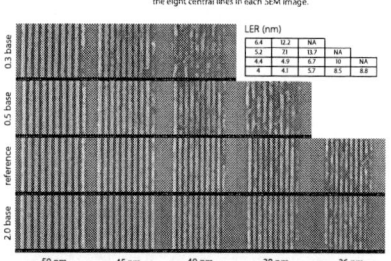

EH27[4,6]

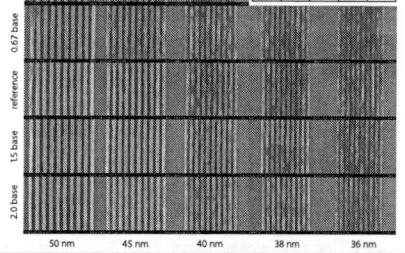

Varying PEB temperature

TOK EUVR P1123[5]

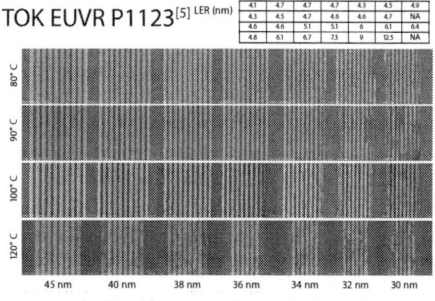

Varying PAG wt. %

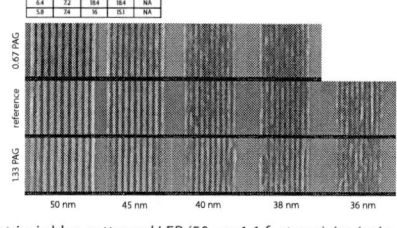

Intrinsic blur, patterned LER (50 nm 1:1 features), intrinsic LER (100 nm 1:1 features), dose-to-size, patterning limit, and LER coherence length are summarized to the right.

Results summary

Summary and Conclusions

Reduced deprotection blur is not the dominant mechanism behind improved patterning ability and improved intrinsic LER with increased base wt. % in EUV resists.

Blur numbers from each metric do not agree in an absolute sense in all platforms, however the same relative trends are observed through process parameters

Modest improvements in patterned LER can be achieved with reduced PEB temperatures. In the TOK resist we tested, intrinsic LER did not change with PEB temperature.

Increased PAG wt. % improves patterning and LER in EH27 resist while reducing dose-to-size. This interesting result is currently being investigated in more detail.

Acknowledgements

Jim Thackeray and Katherine Spear, Rohm and Haas; Koki Tamura, Chris Rosenthal, and David White, TOK; Robert Brainard, University at Albany; Tom Wallow, AMD; Paul Denham, Brian Hoef, Gideon Jones, Jerrin Chu, and Ken Goldberg, CXRO at LBNL. This work was supported by SEMATECH and the EUV ERC.

Investigation of CA resist decomposition by EUV and EB exposure

Daiju Shiono [1,2], Taku Hirayama [1], Hideo Hada [2], Junichi Onodera, Takeo Watanabe [2], Hiroo Kinoshita [2]

[1] Tokyo Ohka Kogyo Co., Ltd.,
[2] LASTI, University of Hyogo

29 September 2008, EUVL symposium

Contents

- Motivation of this work

- Advantages of Prot-Mad-2 as a resist material for decomposition analysis

- Analysis method and experimental conditions

- Analysis result of EB exposure and PEB
  - ◆ Contrast curve
  - ◆ Analysis of decomposition material by EB exposure
  - ◆ Comparison of decomposition behavior and surface roughness

- Summary and next work

Motivation of this work 1

➤ The goal of this work is to achieve CD control demand by controlling the decomposition reaction of resist material.

➤ By optimizing the under-layer and baking conditions, LER/ LWR can be improved. (A. Ma et. al., Proc. SPIE, 6921 (2008) 69213O-1)

➤ However further LER/ LWR improvement looks needed
for 22 and 16 nmhp.

Achieved 3.6 nm of LWR @ 26 nm resolution with PAB/PEB optimization and under-layer material

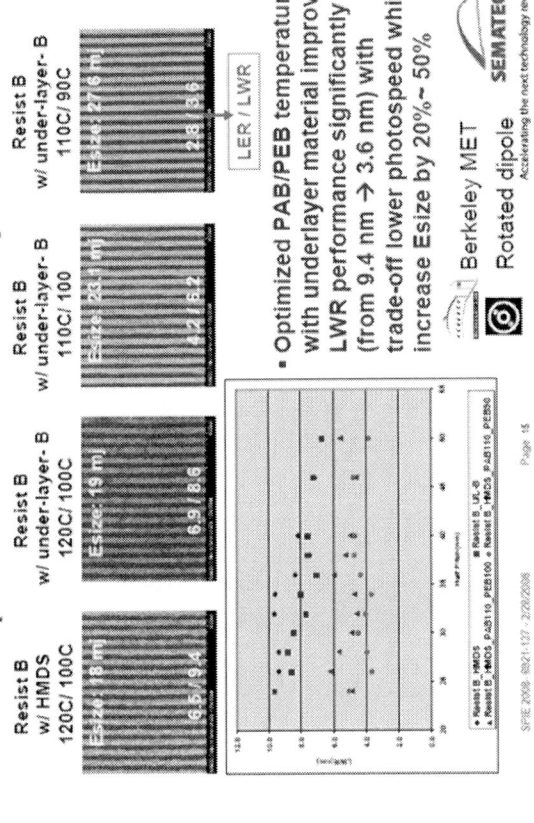

- Optimized PAB/PEB temperature with underlayer material improved LWR performance significantly (from 9.4 nm → 3.6 nm) with trade-off lower photospeed which increase Esize by 20%~ 50%

Berkeley MET
Rotated dipole

A. Ma et. al., Proc. SPIE, 6921 (2008) 69213O-1.

Motivation of this work 2

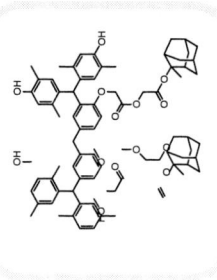

Prot-Mad-2

➤ We try to analyze decomposition behavior of the resist material to know what is happened in the resist film during exposure and PEB process.

➤ In this presentation, we will report decomposition analysis of the resist film after exposure and PEB process using our molecular glass material shown in the right because this has very simple structure attached with just two protecting groups so that it would become much more easier to analyze and assay the products generated during the process.

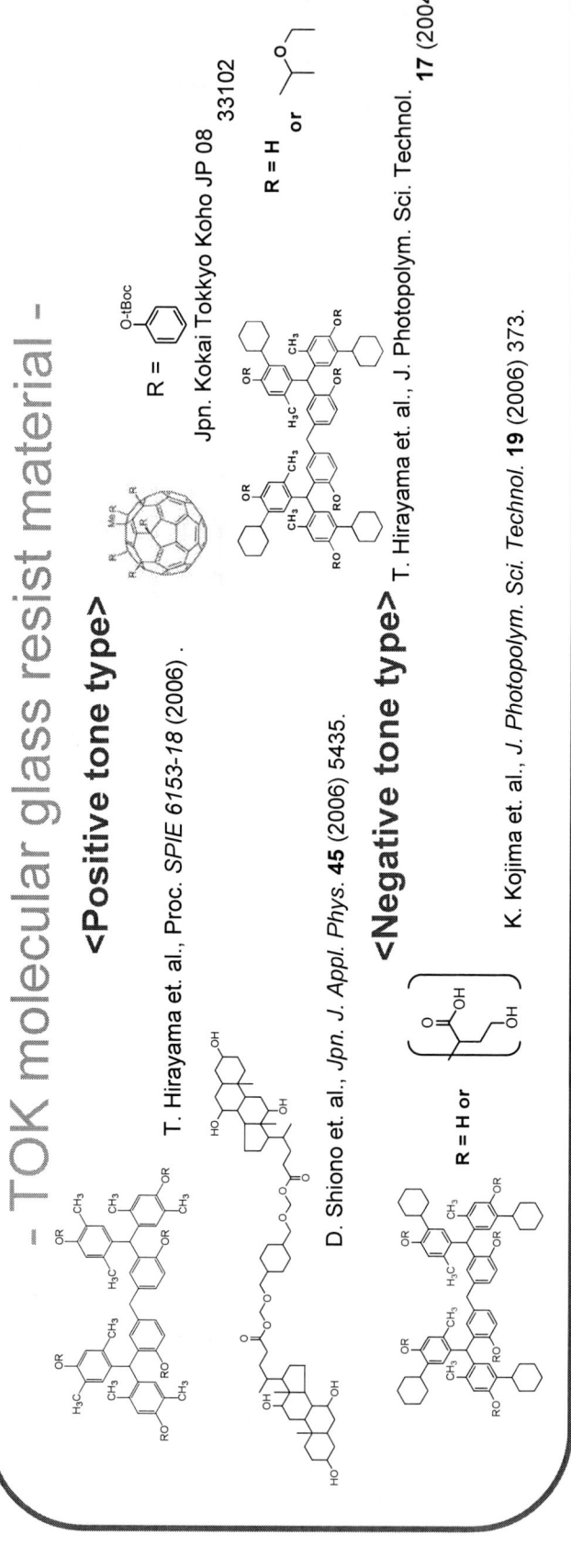

– TOK molecular glass resist material –

<Positive tone type>

T. Hirayama et. al., Proc. *SPIE 6153-18* (2006) .

D. Shiono et. al., *Jpn. J. Appl. Phys.* **45** (2006) 5435.

<Negative tone type>

R = H or

K. Kojima et. al., *J. Photopolym. Sci. Technol.* **19** (2006) 373.

R =

Jpn. Kokai Tokkyo Koho JP 08 33102

R = H   or

T. Hirayama et. al., J. Photopolym. Sci. Technol. **17** (2004) 435.

Advantages of Prot-Mad-2 as a resist material for decomposition analysis

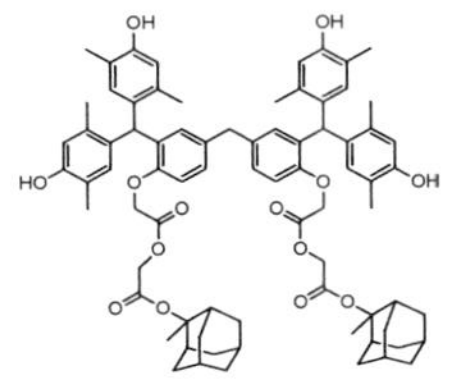

< an example of HPLC data >

Prot-Mad-2

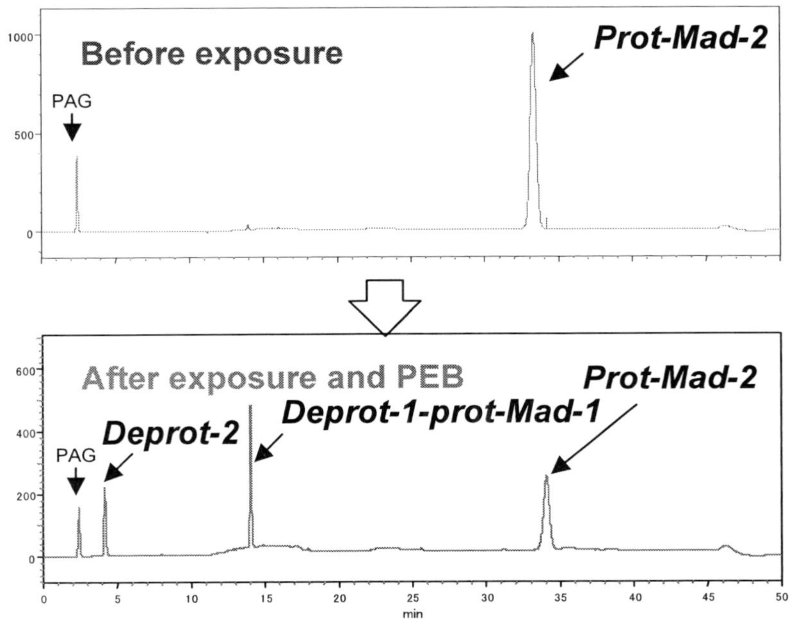

GPC chart of conventional polymer

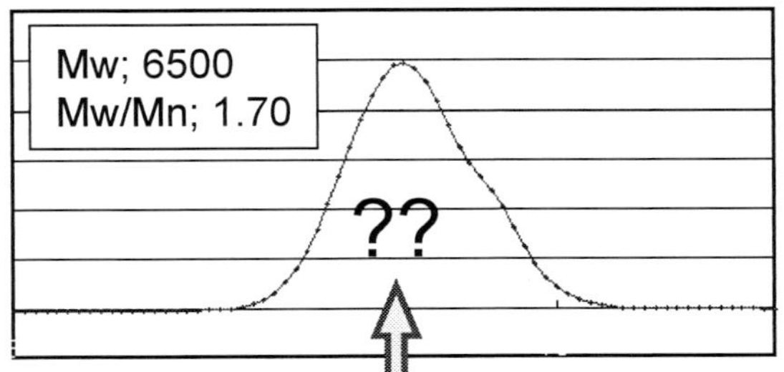

Mw; 6500
Mw/Mn; 1.70

??

Complicated !!

Conventional polymer consists of many component mixture. It is difficult to analyze what happens to each base polymer molecule after and before exposure.

Before exposure

Prot-Mad-2

PAG

After exposure and PEB

Prot-Mad-2

Deprot-1-prot-Mad-1

Deprot-2

PAG

From HPLC analyses, we can obtain the information of qualitative and quantitative analysis at exposure decomposition process.

< Advantages of Prot-Mad-2 for decomposition analysis >
➤ High purity (> 99%)
➤ Just only 2-part protection

Resist sample formulation and conditions at EB exposure

No.	Resist sample	Base material	PAG	quencher
1	Resist-A	Prot-Mad-2	$C_4F_9SO_3^{\ominus}$	none
2	Resist-B	Prot-Mad-2	$C_4F_9SO_3^{\ominus}$	trioctylamine

Exposure conditions

PAB;   100°C for 90s
Exposure; HL-800D (70kV), Hitachi Ltd.
PEB;   100°C for 90s

HPLC analysis

Measurement tool; SCL-10AVP (SHIMADZU corporation)
Column; SUPERIOREX ODS (SHISEIDO fine chemical)

Sample preparation for HPLC analysis

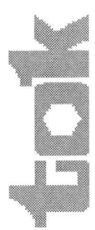

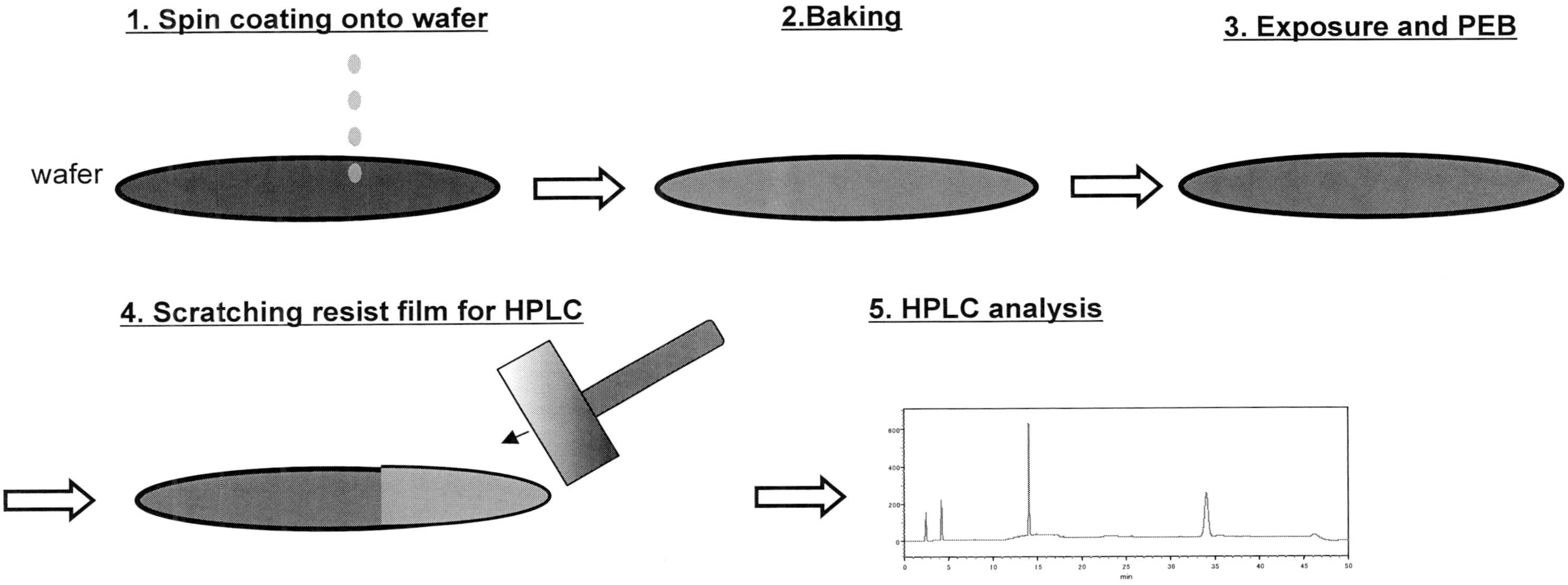
1. Spin coating onto wafer
2.Baking
3. Exposure and PEB
wafer
4. Scratching resist film for HPLC
5. HPLC analysis
800
400
200
0
0
5
10
15
20
25
30
35
40
45
50
min

Condition of HPLC analysis
Measurement tool; SCL-10AVP (SHIMADZU corporation)
Column; SUPERIOREX ODS (SHISEIDO fine chemical)

HPLC; High performance liquid chromatography

EB exposure result of Resist-A (without quencher)

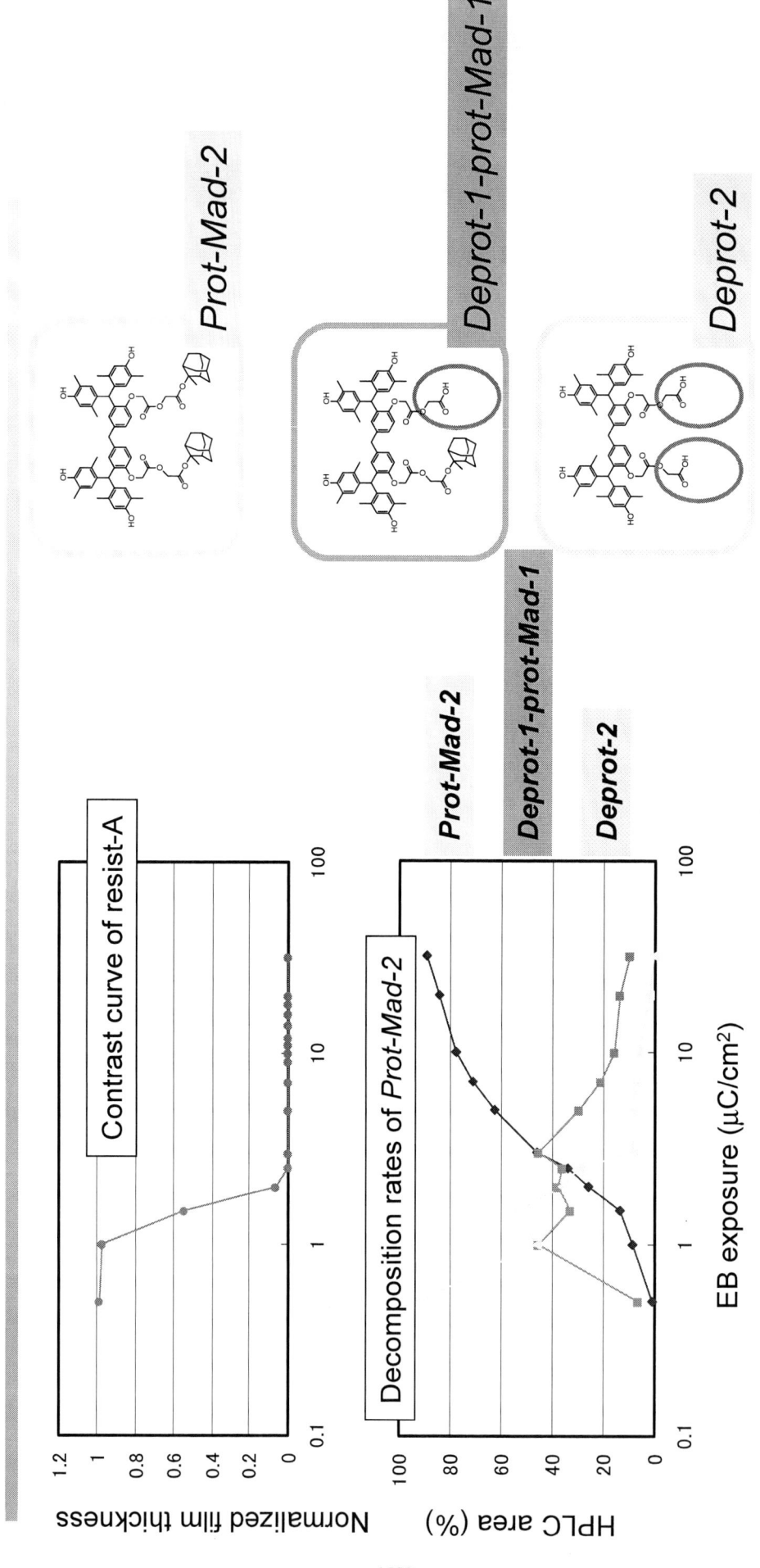

➢ Resist film thickness decreases gradually following EB exposure increase.

➢ Decomposition of Prot-Mad-2 occurs gradually following EB exposure increase.

EB exposure result of Resist-B (with quencher)

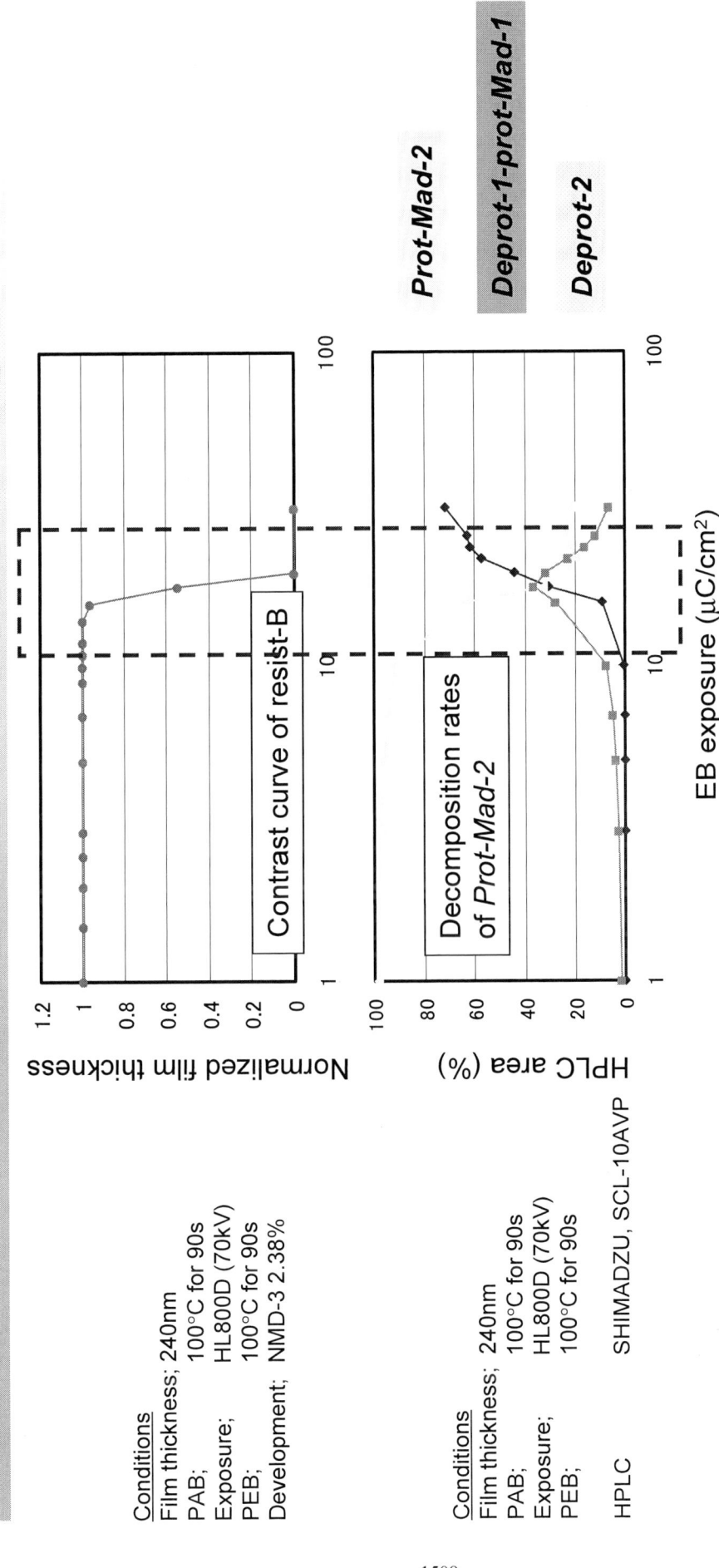

➢ Resist film thickness decreases rapidly around 10μC/cm² as decomposition products of *Deprot-1-prot-Mad-1* and *Deprot-2* increase.

➢ Decomposition reaction of Prot-Mad-2 scarcely occurs at the area of below 10μC/cm².

Surface roughness of resist-B at each decomposition rates

Prot-Mad-2

Deprot-1-prot-Mad-1

Deprot-2

Decomposition rates of *Prot-Mad-2*

HPLC area (%)

Surface roughness of resist-B

Surface roughness RMS (nm)

EB exposure (μC/cm²)

Conditions
Film thickness; 240nm
PAB;   100°C for 90s
Exposure;   HL800D (70kV)
PEB;   100°C for 90s

HPLC   SHIMADZU, SCL-10AVP

Conditions
Film thickness; 150nm
PAB;   100°C for 90s
Exposure;   HL800D (70kV)
PEB;   100°C for 90s
Development; NMD-3 2.38%

Surface roughness
measurement tool; Atomic force
                  micro scope (AFM)
measurement area; 1μm* 1μm
Measurement mode; tapping mode
Probe; Si probe

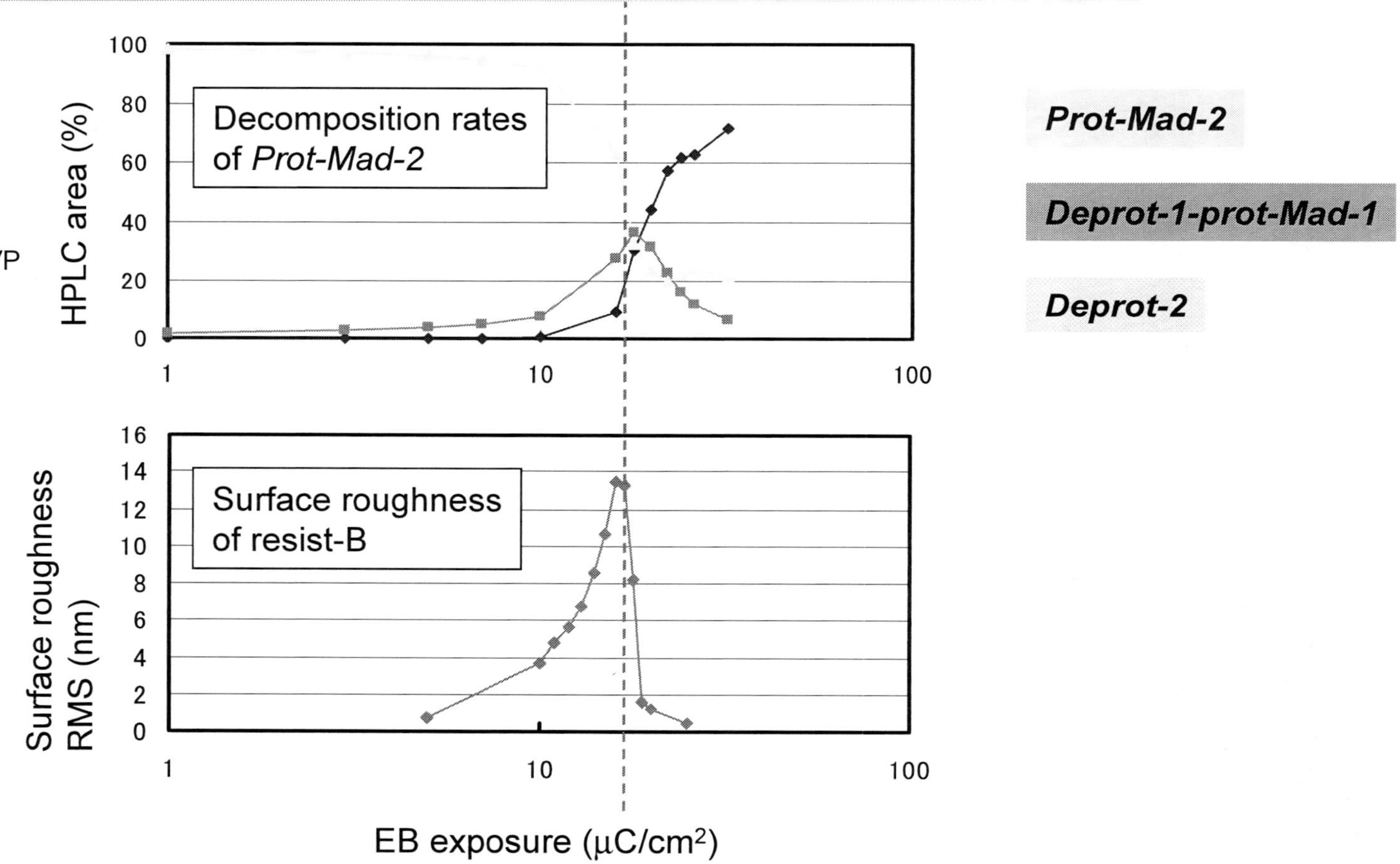

➢ Surface roughness increases as the dose increases and achieves maximum surface roughness at a dose of around 26μC/cm² where area percentage of *Deprot-1-prot-Mad-1* is maximum in the HPLC spectrum.

Hypothesis of acid decomposition on Prot-Mad-2

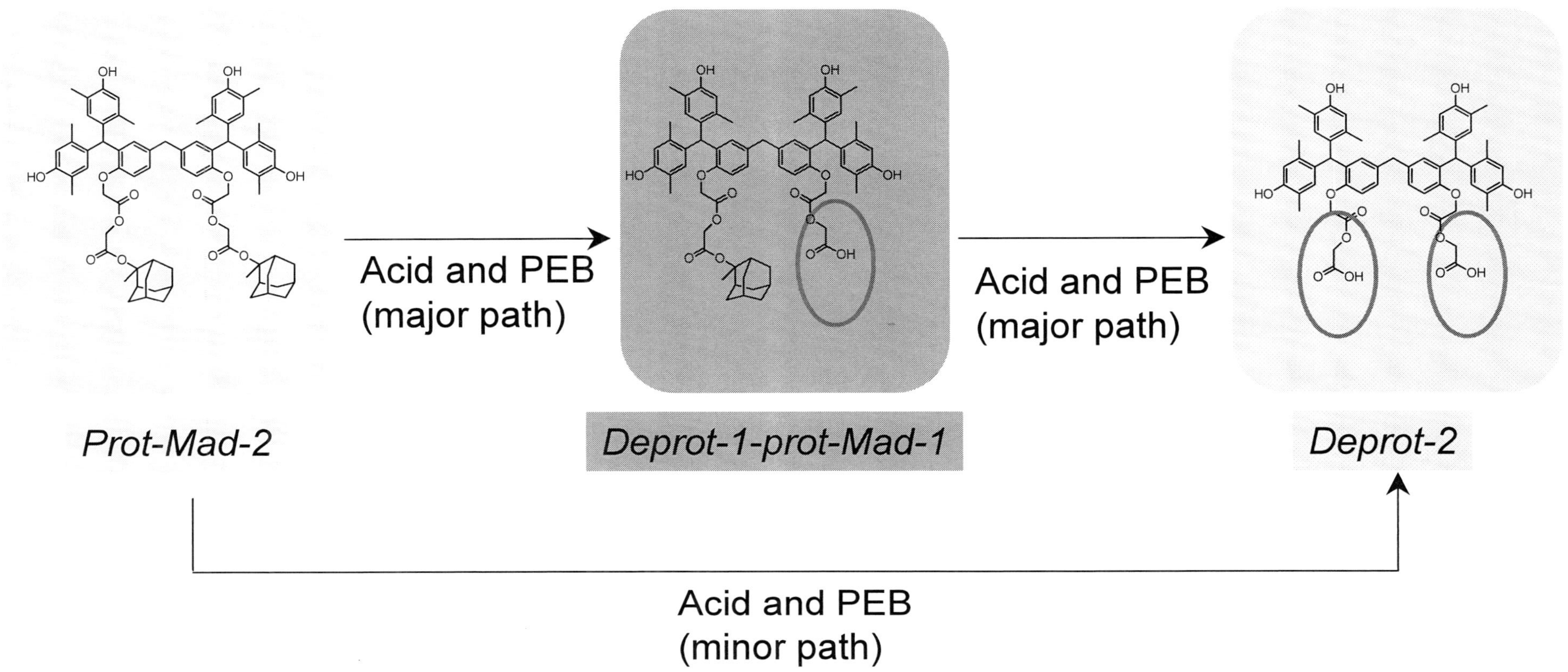

Prot-Mad-2

Acid and PEB (major path)

Deprot-1-prot-Mad-1

Acid and PEB (major path)

Deprot-2

Acid and PEB (minor path)

Summary and next work

- We analyzed decomposition behavior of *Prot-Mad-2* at EB exposure and PEB by HPLC.

  - Major decomposition products are *Deprot-1-prot-Mad-1* and *Deprot-2*.

  - Without quencher, decomposition reaction of Prot-Mad-2 occurs gradually following EB exposure increase.

  - With quencher, Resist film thickness decreases rapidly around $10\mu C/cm^2$ as decomposition products of *Deprot-1-prot-Mad-1* and *Deprot-2* increase.

  - From the exposure tests, dissolution contrasts of resist-A and-B are attributed to decomposition to *Deprot-1-prot-Mad-1* and *Deprot-2*.

  - Decomposition reaction of *Prot-Mad-2* seems to be stepwise.

- We measured surface roughness of EB exposure and PEB by AFM.

  - Decomposition reaction of Resist-B occurs drastically at around the dose of $26\mu C/cm^2$ which is a dose of maximum surface roughness.

- Next work

  - Measuring dissolution rates of *Prot-Mad-2*, *Deprot-1-prot-Mad-1* and *Deprot-2*, respectively.

  - Checking the relation between dissolution rates of each material and roughness property.

  - Measuring thermal property of each material.

  - EUV exposure and its analysis

Novel Polyphenol Base Molecular Resist Having High Thermal Resistance

°Taku Hirayama, Takeyoshi Mimura, Jun Iwashita, Makiko Irie,

Daiju Shiono, Hideo Hada and Takeshi Iwai

TOKYO OHKA KOGYO CO., LTD.

2008 International Symposium, 1st of October 2008, Lake Tahoe, California   T.Hirayama, TOK

Contents

- Introduction
- Concept
- Evaluation results of new molecular glasses
  - ◆ Thermal property
  - ◆ Dissolution curve
  - ◆ Resolution on EB exposure
- Summary

Key lithography-related characteristics and resist requirements on the ITRS 2007

Table Key lithography-related characteristics and resist requirements on the ITRS 2007.

Year of production	2007	2010	2013	2016
DRAM half pitch (nm)	65	45	32	22
MPU Gate in resist (nm)	42	30	21	15
MPU Physical Gate Length (nm)	25	**18**	**13**	**9**
Gate CD control (3 sigma) (nm)	2.6	1.9	1.3	0.9
Low frequency line width roughness (3 sigma, <8% of CD) (nm)	3.4	2.4	1.7	1.2

193 nm immersion with water
193 nm immersion double patterning

193 nm immersion double patterning
EUV
193 nm immersion with other fluids and lens materials
ML2, Imprint

EUV
Innovative 193 nm immersion
ML2, Imprint, Innovative technoligy

Yellow character : Manufacturable solutions are known
Pink character : Interim solutions are known
Red character : Manufacturable solutions are NOT known

ML2: maskless lithography
EUV: extreme ultraviolet

Molecular size comparison

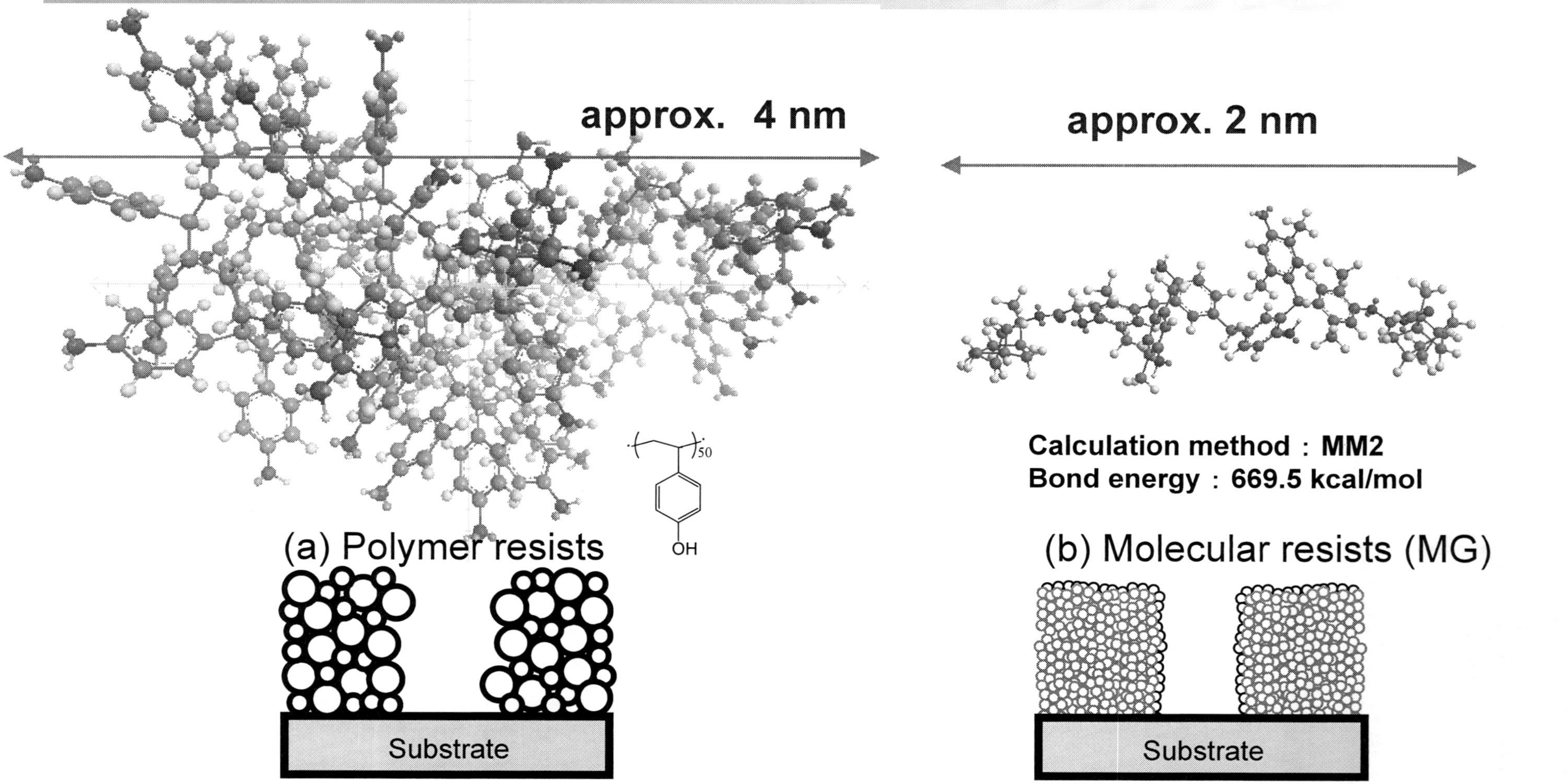

approx. 4 nm

approx. 2 nm

Calculation method : **MM2**
Bond energy : **669.5 kcal/mol**

(a) Polymer resists

(b) Molecular resists (MG)

Substrate

Substrate

Figure Schematic illustrations of cross sectional resist pattern based on (a) the conventional polymeric material and (b) the low molecular base matrix.

*T. Kadota, et.al., Proc. SPIE, **4345** (2001) 891.*

Molecular glass (MG) candidates on our work so far

\<Positive tone type resist\>

R =

Jpn. Kokai Tokkyo Koho JP 08 33102

R = H or

T. Hirayama et. al., J. Photopolym. Sci. Technol. **17** (2004) 435.

T . Hirayama et. al., *SPIE 6153-18* (2006) .

D. Shiono et. al., *Jpn. J. Appl. Phys.* **45** (2006) 5435.

\<Negative tone type resist\>

R = H or

K. Kojima et. al., *J. Photopolym. Sci. Technol.* **19** (2006) 373.

Distribution of protecting group number

Partially protected 25X-MBSA (25X-MBSA-P)

where R is H or protecting group

Protecting ratio is determined to be approx. 32 mol% by using 13C-NMR

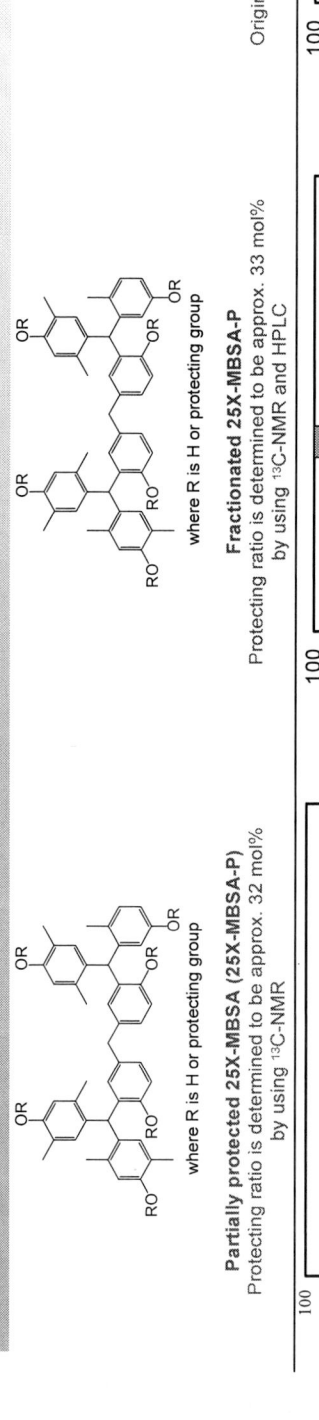

Fractionated 25X-MBSA-P

where R is H or protecting group

Protecting ratio is determined to be approx. 33 mol% by using 13C-NMR and HPLC

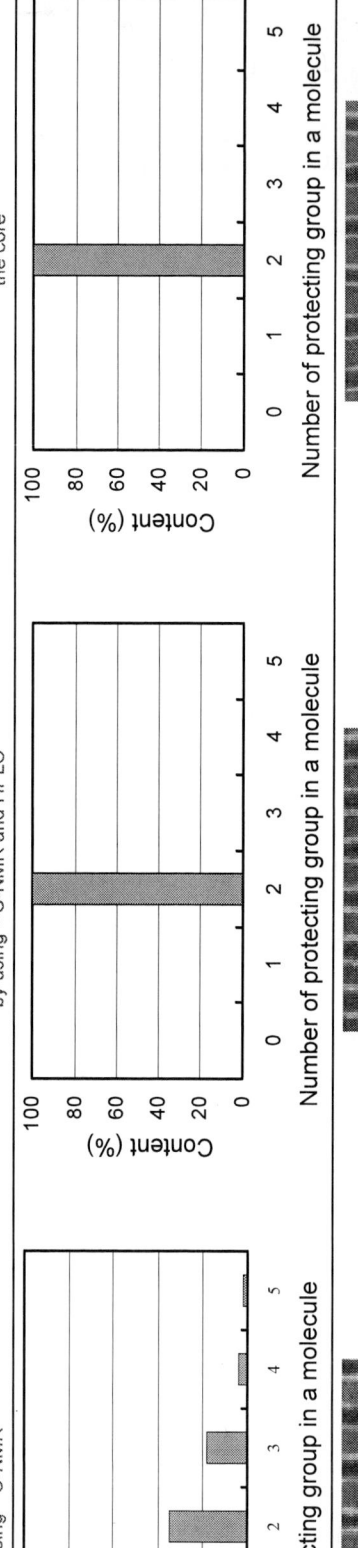

Prot-Mad-2

Originally designed to have two protecting groups onto the core

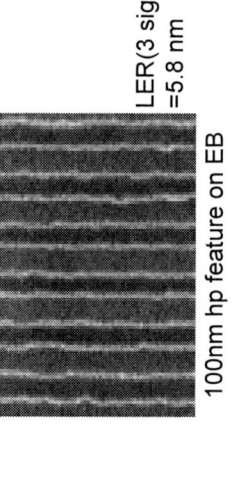

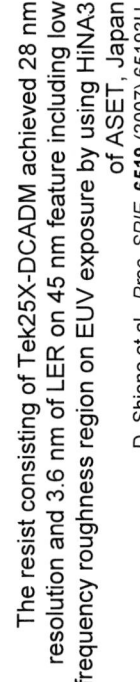

100nm hp feature on EB

LER(3 sigma) =9.9 nm

100nm hp feature on EB

LER(3 sigma) =5.8 nm

100nm hp feature on EB

LER(3 sigma) =5.6 nm

30-nm hp

28-nm hp

Thickness : 58 nm
Dose : 12.2 mJ/cm²

The resist consisting of Tek25X-DCADM achieved 28 nm resolution and 3.6 nm of LER on 45 nm feature including low frequency roughness region on EUV exposure by using HiNA3 of ASET, Japan

D. Shiono et al., *Proc. SPIE*, **6519** (2007) 65193U.

Film thickness;100nm
Exposure;HL800D (70kV)
Development;NMD-3 2.38% of 0.26 N TMAH for 60s puddle

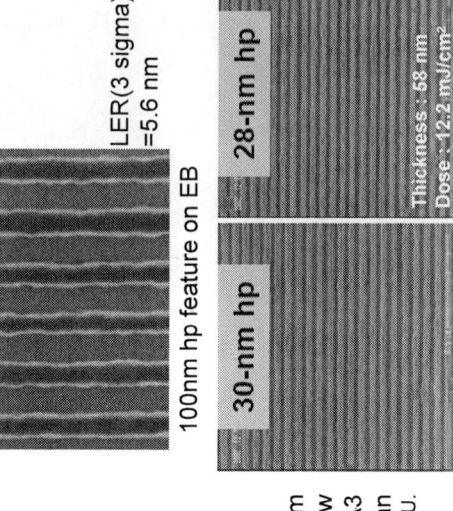

EUV Process Technology

Concept

- To achieve better resolution using MG material…

  - Number of protecting group increases, and position of protecting groups should be defined in the structure in order to increase dissolution contrast to be comparable to common polymeric materials

  - Higher thermal resistance should be realized, also to be similar to polymeric materials

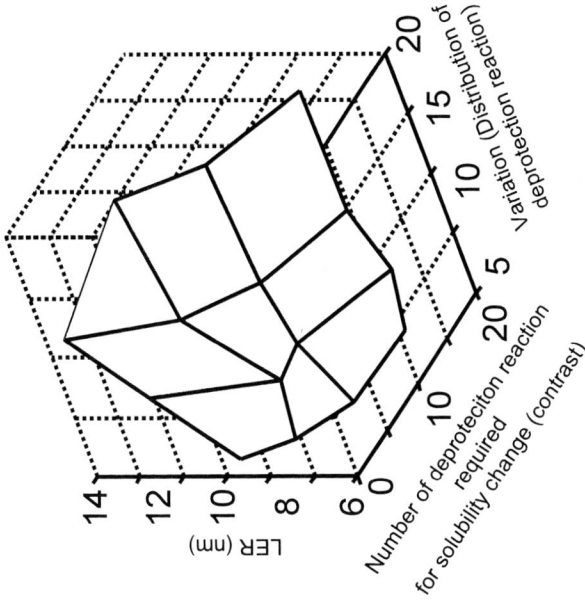

LER strongly depends on the number of de-protection reaction for solubility change and its distribution.

H. Fukuda: *Jpn. J. Appl. Phys.*, **42** (2003) 3748.

MGs based on polyphenol

- In many of papers concerning MG material and also our previous reports, some of "un-protected" MG show enough high glass transition temperature, Tg, but after protection reaction, the Tg of "(partially or fully) protected" MGs was usually getting lower than that of unprotected due to , maybe, lack of hydrogen bonding and intermolecular interaction

A. D. Silval et al., *Proc. SPIE*, **6923** (2008) 69231L-1.
T. Hirayama et. al., *J. Photopolym. Sci. Technol.*, **17** (2004) 435.

- According to our concept for MGs which have polyphenol cores and protecting groups of precise number at the specific position in the structure, we prepared several candidates to increase number of protecting group and thermal resistance by increasing number of protecting groups and molecular weight

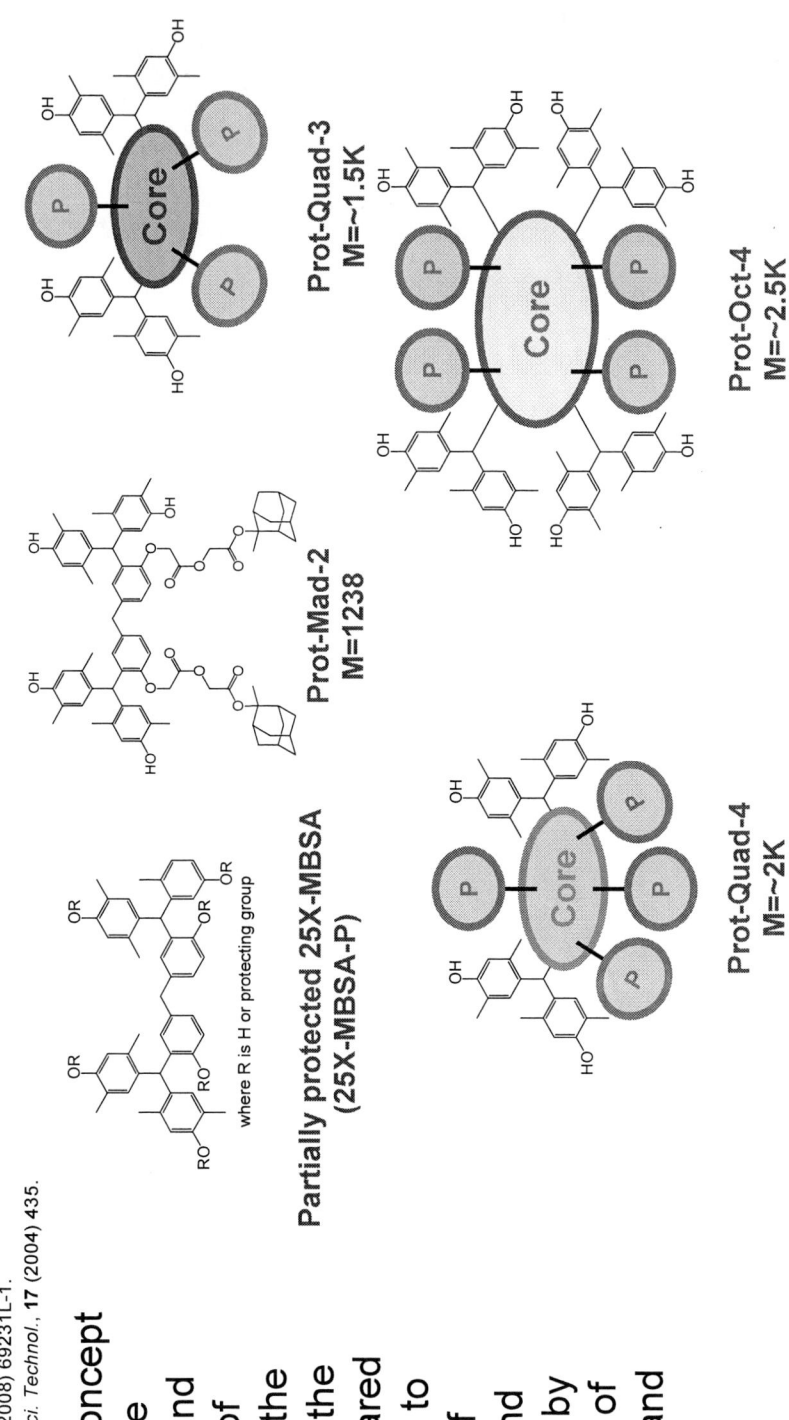

How to define the thermal resistance of "protected" base matrix

- The *Tg* is a good indicator for thermal resistance property of the resist on realistic lithography process and is one of the important factor for not only polymeric resist but MG resist

- In some case, MG material of which alkaline soluble group is fully or partially protected doesn't show apparent glass transition behavior on thermal analysis such as DSC so that we considered how we could define the thermal property and decided to measure "thermal flow starting temperature" on contact hole feature by the following procedure;

Development →

The resist film is exposed by 248nm light and baked as PEB treatment
Note; the resist solution used is formulated with "protected base resin", PAG, amine and solvent, meaning that those are model RESIST solutions, not "Un-protected base resin solution"

The exposed film is developed and 170 nm of isolated hole feature is obtained

Post bake →

Post bake is applied at various temperature and the CD is measured.
The post bake temperature where 10% of CD shrink occurs, is defined as "thermal flow starting temperature"

CD Shrink is observed by CD-SEM

Post bake temperature	No post bake	150 °C	155 °C	160 °C
Common polymeric resist (PHS/Acryl hybrid)				

10% of CD shrink is observed at 159 °C from the fitting curve of post bake temperature-CD plots

Relationship between Tg of unprotected MGs and thermal flow temperature

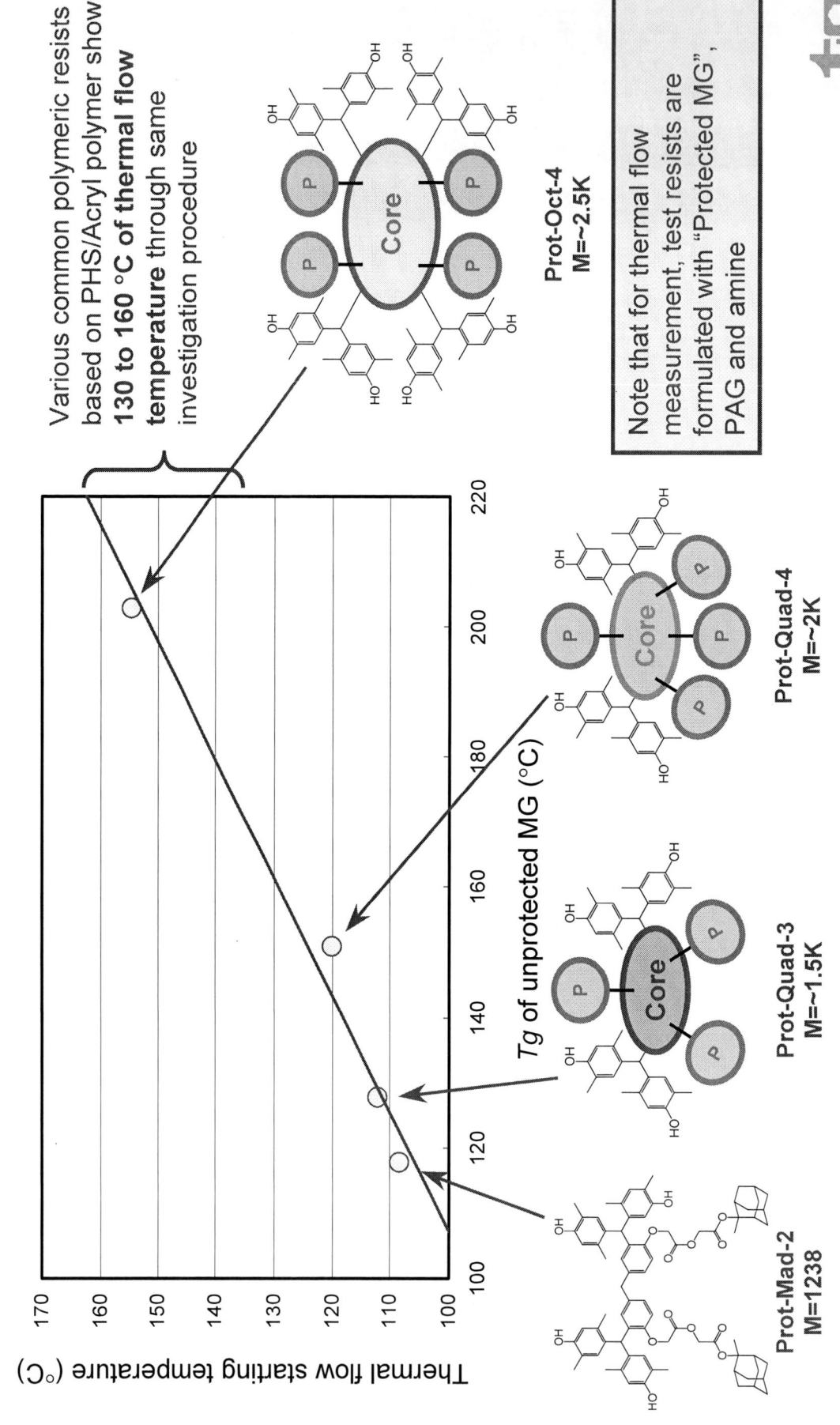

Tg dependency on molecular weight

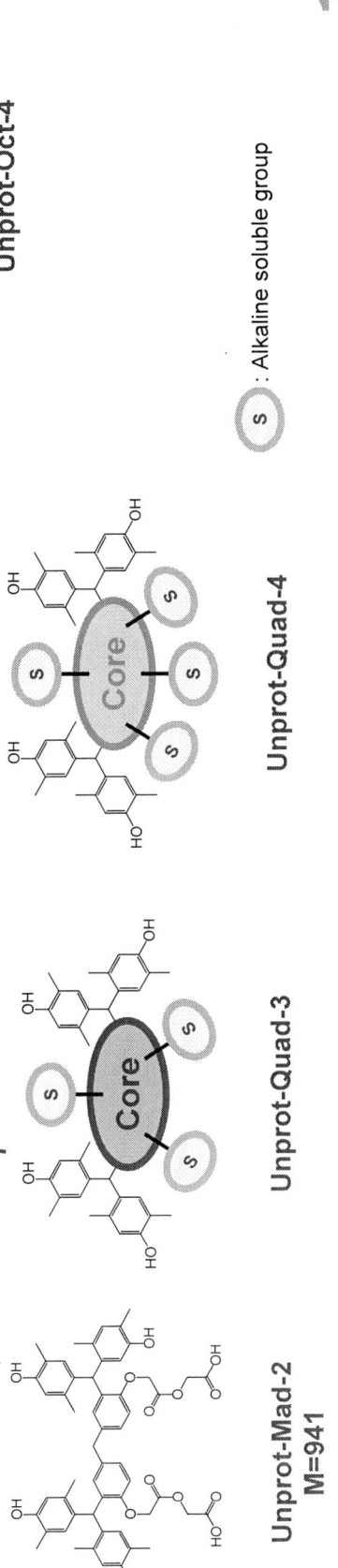

Contrast curve

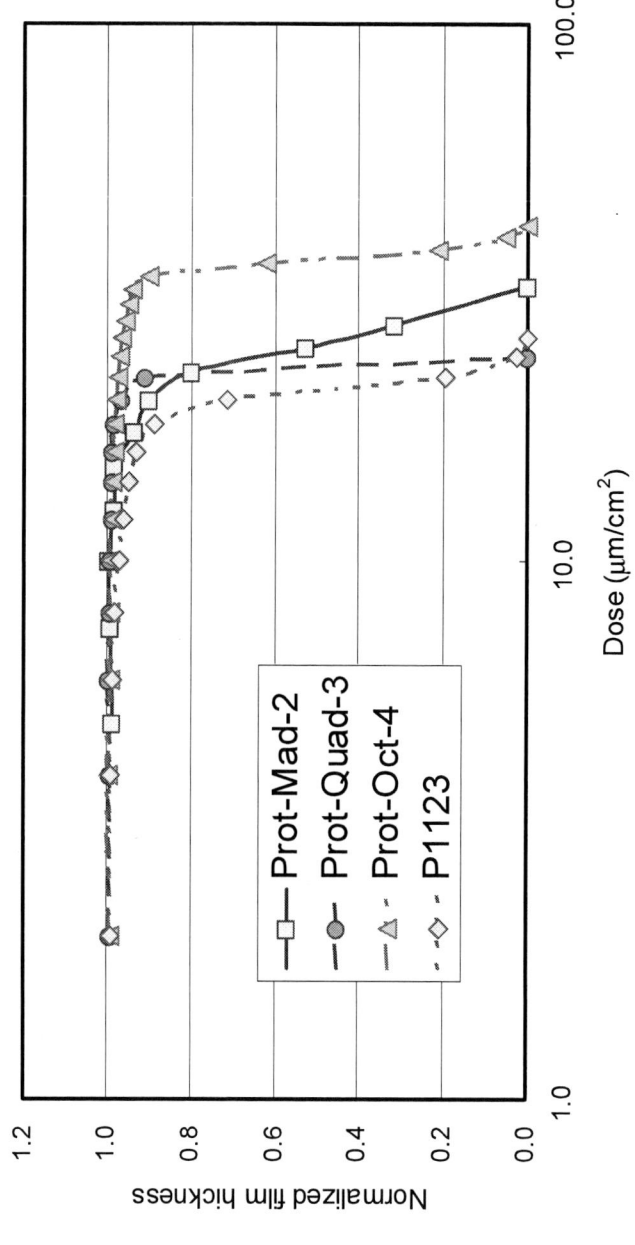

The resist consisting of Prot-Oct-4 shows good dissolution contrast as well as that with Prot-Quad-3

Substrate:Si HMDS treated 90°C–36s,
Thickness: :100nm
Tool: HL800D 70keV
PAG and amine formulation is the same among all the test samples including Prot-Mad-2, Prot-Quad-3 and Prot-Oct-4. EUVR-P1123 is TOK's EUV resist of which formulation has been optimized for the process

Exposure results on EB tool

Sample	EUVR-P1123	Test resist using Prot-Oct-4
LWR	11.9 nm	11.3 nm
EL	8.0%	7.3%
Sensitivity	48.0 µC/cm^2	30.0 µC/cm^2

Beam size (nm) 100
90
80
70
60
50

EUVR-P1123 is TOK's EUV resist of which formulation has been optimized for the process

Substrate: Bare-Sili n (HMDS treatment, 90 °C-36 sec)

Resist Film Thickness: 60 nm

EB Writer: Hitachi HL-800D (VSB, 70 kV, 7 A/cm^2)

Target pattern: 100 nm LS (duty ratio, 1:1)

Development: NMD-3, 2.38 wt% TMAHaq, 60 sec, LD-nozzle

Rinse: Distilled water, 15 sec

Summary

- We have synthesized polyohenol base molecular glass (MG) candidates based on the concept to increase number of protecting groups and molecular weight and to attach the protecting groups in the specific positions of the MG structure

- In order to define thermal resistance of base materials, measurement of thermal flow starting temperature using contact hole feature is utilized

- Prot-Oct-4 which has four protecting group showed enough high thermal flow starting temperature ~154 °C, comparable to that of common polymeric material

- Test resist formulated with Prot-Oct-4 partially achieved 50 nm resolution on EB exposure tool (70 kV, VSB) and it seems to be comparable to the result of EUVR-P1123.
  - ◆ EUV exposure test is needed to confirm the performance of this material

Appendix; EUV exposure result of EUVR-P1123 at PSI

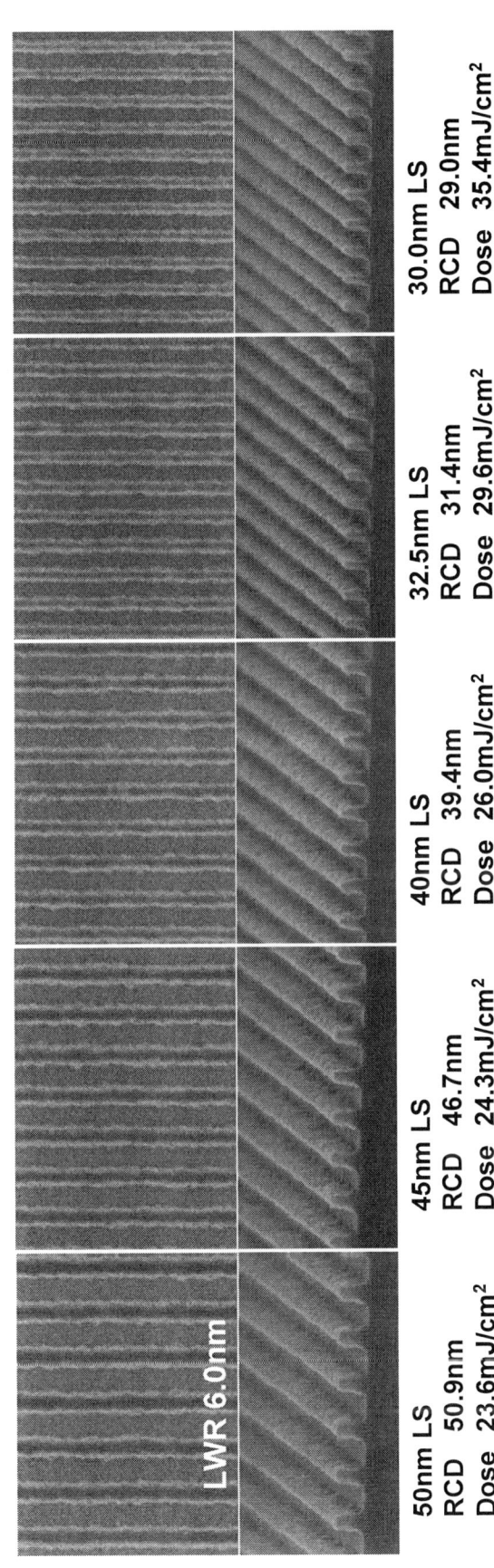

LWR 6.0nm

50nm LS
RCD 50.9nm
Dose 23.6mJ/cm^2

45nm LS
RCD 46.7nm
Dose 24.3mJ/cm^2

40nm LS
RCD 39.4nm
Dose 26.0mJ/cm^2

32.5nm LS
RCD 31.4nm
Dose 29.6mJ/cm^2

30.0nm LS
RCD 29.0nm
Dose 35.4mJ/cm^2

25nm LS
RCD 28.5nm
Dose 41.0mJ/cm^2

Process conditions
- Substrate Si
- Resist   EUVR-P1123, 80nm FT
- PAB   120 °C for 60 sec
- EUV tool   Interferometer@PSI
- PEB   100 °C for 60 sec

Data courtesy of IMEC

Acknowledgements

The authors would like to thank Mr.Akira Yoshitomo and Mr.Tatsuya Iwai of Honshu Chemical Industry co., Ltd. for their experimental supports.

RUTGERS

Electrons in EUV Resist Activation

Theodore E. Madey[1], B.V. Yakshinskiy[1], E. Loginova[1], L. Sanche[2], P. Cloutier[2], R. Brainard[3], A. Wuest[4]

[1]Dept. of Physics and Astronomy, Rutgers, The State University of New Jersey, Piscataway NJ 08854 USA;
[2]University of Sherbrooke, Sherbrooke, Quebec, Canada; [3]College of Nanoscience and Engineering, University at Albany, Albany NY 12203; [4]SEMATECH, Albany NY 12203

Abstract

We address the mechanisms by which EUV photons break chemical bonds and activate EUV resists. Secondary electrons are created when the primary EUV-generated photoelectrons excite and release other electrons. Low-energy secondary electrons (0 -15 eV) can effectively induce dissociation of organic molecules due to the high dissociation cross sections, break bonds in resist molecules and can be a primary source of resist activation. The main process is termed Dissociative Electron Attachment (DEA), and is especially effective for molecules containing fluorine atoms, including Photoacid Generators (PAG). In the present experiments resists having 0% PAG, 7.5% PAG and 15% PAG were studied. The electron energy distributions under 13.5 nm irradiation and secondary electron yields of the resist-coated samples were measured at the National Synchrotron Light Source (NSLS).

Goals

- Address the extent to which secondary electrons play a role in EUV resist activation. Model resist layers are deposited on Si substrate.

- Measure threshold energies and cross sections for electron-stimulated dissociation of resist layer.

- Measure secondary electron yield (SEY) and secondary electron kinetic energy distributions in wavelength range around 13.5 nm.

- Infer implications for future EUV resist design.

Parameters of the resists

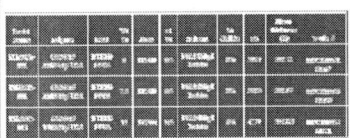

The resists are synthesized and spin-coated onto Si wafers to ~30 nm thickness (by Prof. Robert Brainard).

Electron energy distribution for the resist with 0% PAG

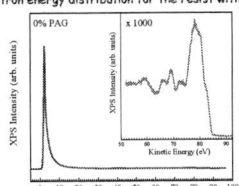

Photon energy 92 eV (13.5 nm); measured on beamline U5UA of NSLS

Electron energy distribution for the resist with 7.5% PAG

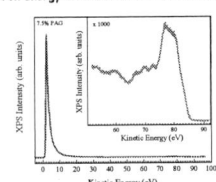

Photon energy 92 eV (13.5 nm); measured on beamline U5UA of NSLS; the inset-zoomed Valence Band

Electron energy distribution for the resist with 15% PAG

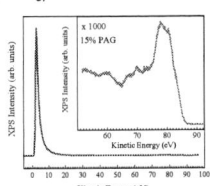

Photon energy 92 eV (13.5 nm); measured on Beamline U5UA of NSLS; the inset-zoomed Valence Band

Evolution of the Valence Band under 92 eV photon irradiation

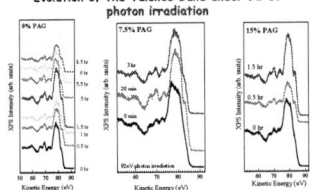

Photon flux density ~1.5×10^13 photons/cm²sec at 92eV, which corresponds to an accumulated dose of ~800 mJ/cm² for 1 hour

Secondary Electron Yield for the resists with different PAG concentration vs. photon energy

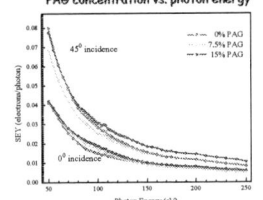

Measured on Beamline U3C of NSLS for 2 incident angles

Secondary Electron Yield for the resists with different PAG concentration vs. photon energy

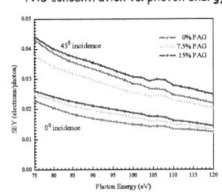

More photons are absorbed in the upper surface layers at 45° incidence in the range of 2-3 nm

Secondary Electron Yield for a Si wafer (covered by native oxide film)

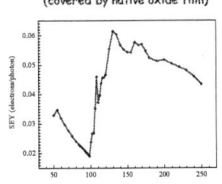

Structure above 100 ev is due to Si core level excitations, it is suppressed when 30 nm resist films cover Si substrates

Electron transmission for resist with 0% PAG

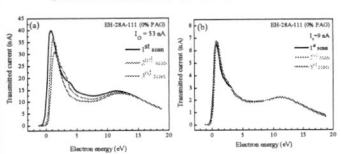

The resists charge easily at 53nA current. The onset shifts to higher values because incoming electrons must overcome the potential created by the trapped charges.

Mass spectra of negative ions desorbing from resist surface under 20eV electron bombardment

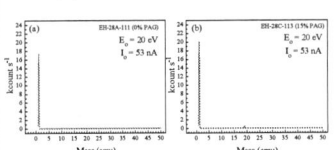

0% PAG resist exhibits only H⁻ ions
7.5% and 15% PAG resists exhibit H⁻ and F⁻ ion signals

Electron-stimulated desorption of H⁻ and F⁻ anions vs. incident electron energy for resist with 0% PAG

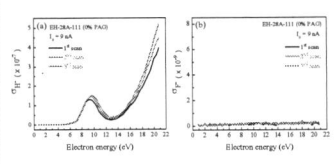

Only H⁻ anions desorb from the resist surface with 0% PAG

Electron-stimulated desorption of H⁻ and F⁻ anions vs. incident electron energy for resist with 7.5% PAG

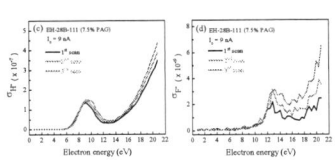

Both H⁻ and F⁻ anions desorb from the resist surfaces with 7.5% and 15% PAG

Electron-stimulated desorption of H⁻ and F⁻ anions vs. incident electron energy for resist with 15% PAG

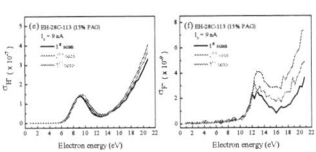

Both H⁻ and F⁻ anions desorb from the resist surfaces with 7.5% and 15% PAG

Mechanism of the anion desorption

- The threshold energy for the H⁻ anion yield is around 6 eV. The peak at 9 eV is due to dissociative electron attachment (DEA):

$$e^- + R \rightarrow R^- \rightarrow [R\text{-}H] + H^-$$

- At energies > 14 eV, H⁻ occurs via dipolar dissociation (DD):

$$e^- + R \rightarrow R^+ + e^- \rightarrow [R\text{-}H]^+ + H^- + e^-$$

- The threshold energy for the F⁻ anion yield lies at ~10 eV and the DEA peak is found around 12.7 eV. Desorption of F⁻ ion via DD process occurs above 17 eV.

Mechanism of the bond breaking

- Most initial excitations by 92 eV photons lead to production of ~80 eV primary electrons inside the film. These electrons can create excited atoms and molecules, radicals, ions, and secondary electrons with a most probable energy below ~10 eV. LEEs have thermalization distances of the order of 1-10 nm.

- In all cases low energy electrons are breaking C-H and C-F bonds in the resist polymer. Threshold for H⁻ desorption (breaking of C-H bonds) is lower than threshold for F⁻ desorption (breaking of C-F) bonds.

- Other bonds may be broken at energies < 6eV (e.g., C-C bonds) that do not lead to desorption of charged fragments.

- The valence band features change with irradiation time, as expected for changes in electronic structure for an "exposed" resist.

Summary

- Secondary electrons emitted from resists have very low energies <10 eV; electron-stimulated desorption of anions by such LEEs can provide quantitative data concerning bond-breaking in EUV resists.

- Similar bond-breaking events are generated extensively in the bulk of the resists by the LE secondaries released by primary EUV photons in nanolithographic processes.

- Next step could be SEY and ESD experiments on "neat" polymers and PAGs.

- Future experiments could involve SEY measurements from resist films (a few nm to ~10 nm) of controlled thickness on Si or Si-adsorbate-covered substrates.

The work is supported by SEMATECH

2008 Int. Symp. on EUVL, Sept. 28-Oct. 2, Lake Tahoe, CA

DUV source integration into the 0.3 NA Berkeley SEMATECH MET for OOB exposure studies

Simi A. George, Patrick P. Naulleau, Senajith Rekawa, and Drew Kemp

Why worry about Out of Band (OOB)?

1. EUV sources emit OOB radiation

Measured emission from a plasma source

Measured OOB in the 160 nm – 715 nm region at 0.15 mJ/sr is 10 % of the EUV in-band energy for this particular source

2. Multilayer mirror reflectivity

- Reflectivity of a Mo/Si Optic is comparable to EUV for these wavelengths at ~ 63%
- Filtering solely by the Projection Optics is not adequate

3. Resists

Typical EUV resist DUV sensitivity values

Wavelength (nm)	Relative Sensitivity (mJ/cm²)
13.5	1
157	0.17
193	2.1
248	0.04

- EUV resists with Polystyrene platforms are absorptive in the DUV
- EUV resists shown to be sensitive to 193 nm and 248 nm, but 5x as sensitive to 157 nm
- These wavelengths will expose the resist with a low-resolution image of mask pattern

Mask aerial image inspection of the OOB radiation effects

So what happens if you send OOB radiation through to the multilayer mirror (MLM) mask in an EUV Lithography system?

We utilize the 2D mask modeling for aerial images

E-M suite generated reflection coefficients are used to define material properties at each wavelength.

Aerial images are generated for pure EUV illumination as well as for four OOB wavelengths of interest, 157 nm, 193 nm, 248 nm, and 365 nm

Complex Index of Refraction @ 92 eV

Material		δ (1+δ)	β (1-k)	n(1-1.5+βi)
Si	MLM	8.6020E-4	1.8252E-3	0.99914
Mo	MLM	7.5800E-2	6.3800E-3	0.92420
Ru	Capping layer	0.11305	0.016829	0.88700
*TaBN	Absorber	0.04820	0.02860	0.95180
*TaON	ARC	0.04480	0.02504	0.95520

*RBS measured values provided by supplier for specified thickness

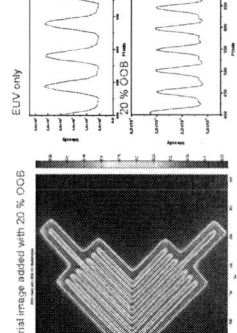

Current MET mask parameters input into E-M Suite, MAST

40 bilayer Mo/Si with Ru capping layer

MET pupil map, no flare or spider

At 20% OOB, we see a 33% drop in contrast

At 10% OOB, 12 % drop in contrast , not a linear dependence

2D elbow mask, 50 nm CD

Pure EUV Illumination

OOB Illumination: Effect of four wavelengths

EUV aerial image added with 20 % OOB

SEMATECH Berkeley MET could be upgraded with integrated OOB source

Possible Illumination design for integrating OOB source to MET

ALS Undulator BL12.0.1.3

Scanner module

Mask stage

0.3-NA 5x projection optic

Wafer stage and height sensor

Pupil-fill monitor

OOB Source: Energetiq LDLS 1000

- OOB path coincident with EUV path
- Passes through scanning illuminator
- Source etendue small enough to support MET
- UV exposure would be added through second exposure step

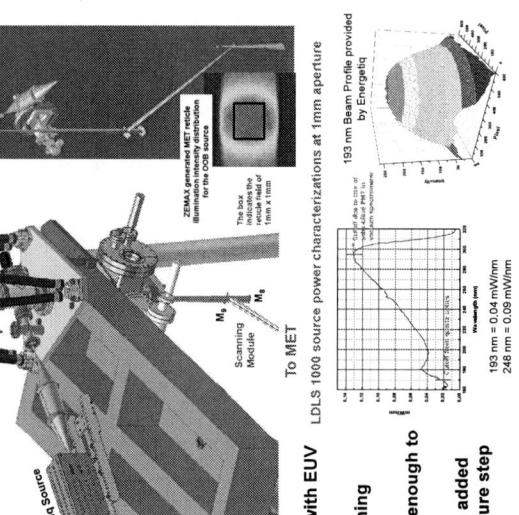

Integrated beam path to the MET projection optic

ZEMAX generated MET reticle illumination intensity distribution for the OOB source

The box indicates the reticle field of 1mm x 1mm

193 nm Beam Profile provided by Energetiq

LDLS 1000 source power characterizations at 1mm aperture

193 nm = 0.04 mW/mm
248 nm = 0.09 mW/mm

Calculated Exposures

Estimation of exposure times assuming 2 mJ/cm² dose, for the available wavelengths of interest

	Transmission filters		UV Mirrors (Opto-Sigma)			Critical wavelengths from the resist experiments		
λ (nm)	193	248	193_p	193_s	248_p	248_s		
Total throughput to wafer (μW)	0.37	1.2	0.372	1.04	2.51	4.6		
Exposure Time (s)	5.6	6.5	2.3	0.96	0.62			

Critical wavelengths from the resist experiments

157nm, $E_0(EUV) = 5*E_0$ (157 nm)
193 nm, $E_0(EUV) = 0.5*E_0$ (193 nm)
248 nm, $E_0(EUV) = E_0$ (248 nm)

- Dose value determined from reported dose to clear data for a 10 mJ/cm² EUV resist
- Estimated dose times accounts for 6 MLM bounces, UV optics, and filter profiles
- Does not account for mask absorber effects

Summary

Completed

- OOB is an important concern for EUV systems
- Preliminary modeling on 2d mask patter shows drop in contrast with OOB, detailed study to be completed.
- Berkeley MET currently has extremely high spectral purity
- MET could be upgraded for direct simulation of OOB effects on imaging
- Optical design for broadband beam path completed with ZEMAX
- Current source does not support critical 157-nm wavelength
- Explicitly include OOB resist sensitivity in the calculations
- Study impact of absorber scattering on relative spectral filtering

Acknowledgements: Dr. Eric Gullikson , Farhad Salmassi, and Energetiq team

Contact: sageorge@lbl.gov

Center for X-Ray Optics, Lawrence Berkeley National Laboratory, Berkeley, CA 94720

Development of Novel Positive-tone Photoresists for EUVL

a)Idemitsu Kosan Co.,Ltd.

b)Semiconductor Leading Edge Technologies, Inc. (Selete)

a)Takanori Owada, Akinori Yomogita, Takashi Kashiwamura
b)Hiroaki Oizumi, Toshiro Itani

E-mail :takanori.owada@si.idemitsu.co.jp

Our concept for EUV resist

Low molecular weight amorphous materials have the advantage of having definite structures and low molecular weights that would solve the current serious problem of line-edge roughness (line width roughness).

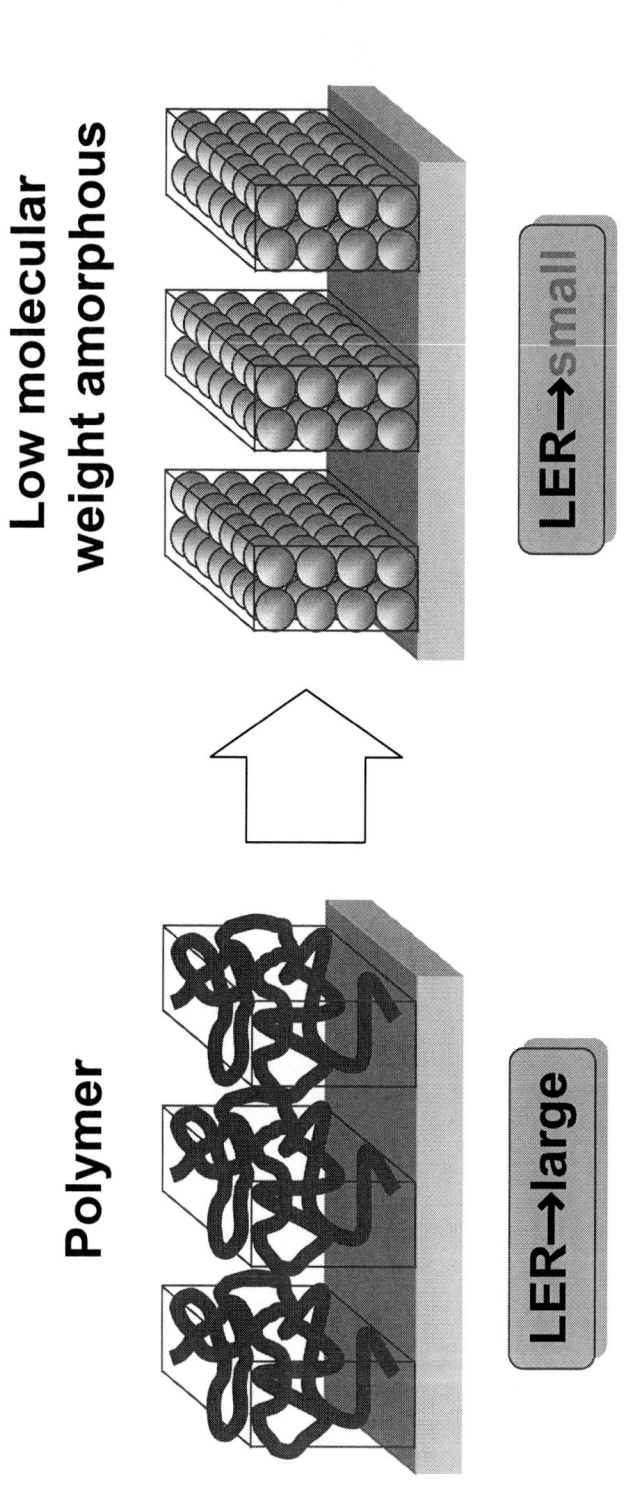

Polymer

Low molecular weight amorphous

LER→large

LER→small

IDEMITSU

Previous work (and in this study)

We had developed tetramethylcalix[4]-resorcinarene (C4-RA)

*H.Ishii et al.:J.Photopolym.Sci.Technol.,**16**(2003)685

calixarenes have unique characteristics such as …
→small molecular size
→excellent thermal stability
→many reactive groups such as hydroxyl groups and so on.

We found it shows good characters for resist and also it shows
bad characters (etching durability and so on.)

C4-RA : Molecular weight=544

We focus on the NEW Cyclic low molecular Resists
in order to improve the performance of C4-RA.

Synthesis of Cyclic Low Molecular Resists(CLM-Resists)

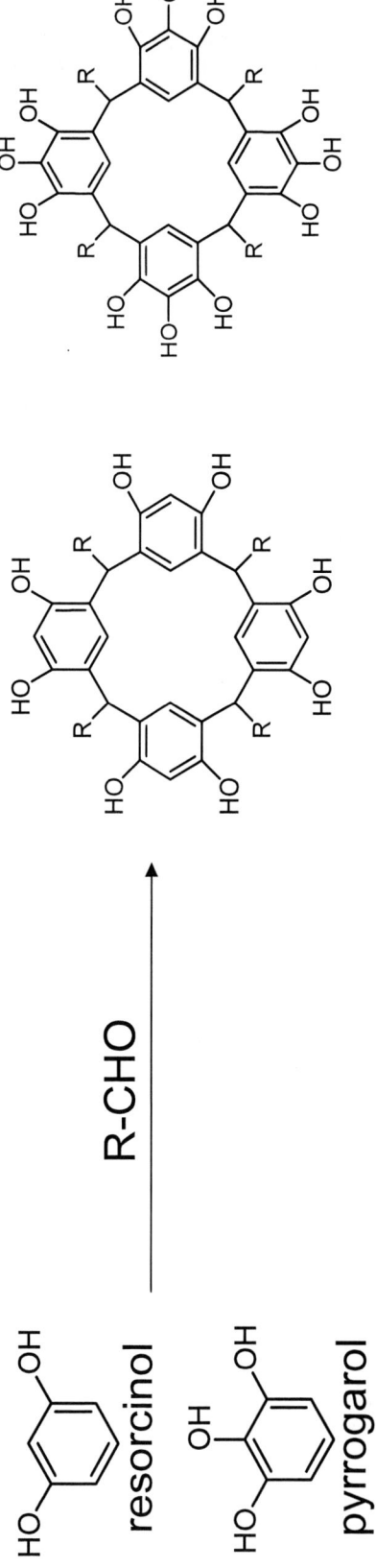

resorcinol

pyrrogarol

R-CHO

Protection Group

Cyclic Low Molecular Resists

Easy preparation of molecular resists (2 Steps)

We developed new platform, CLM-D resist.

Thermal Properties of CLM-Resist

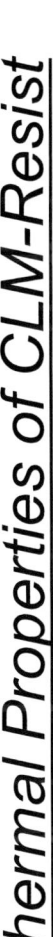

179.6Cel
-4.9%

390.0Cel
-27.9%

TG %

Evaluated conditions
N2 (200ml/min) flow
10°C/min

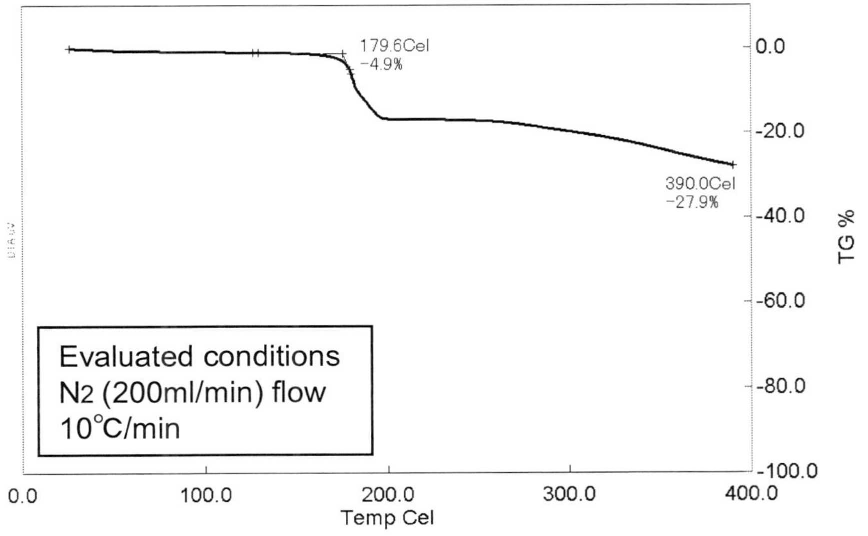

TG thermogram of CLM-D resist

Evaluated conditions
<1st heating>
N2 (200ml/min) flow
15°C/min

Endo

Temp. (°C)

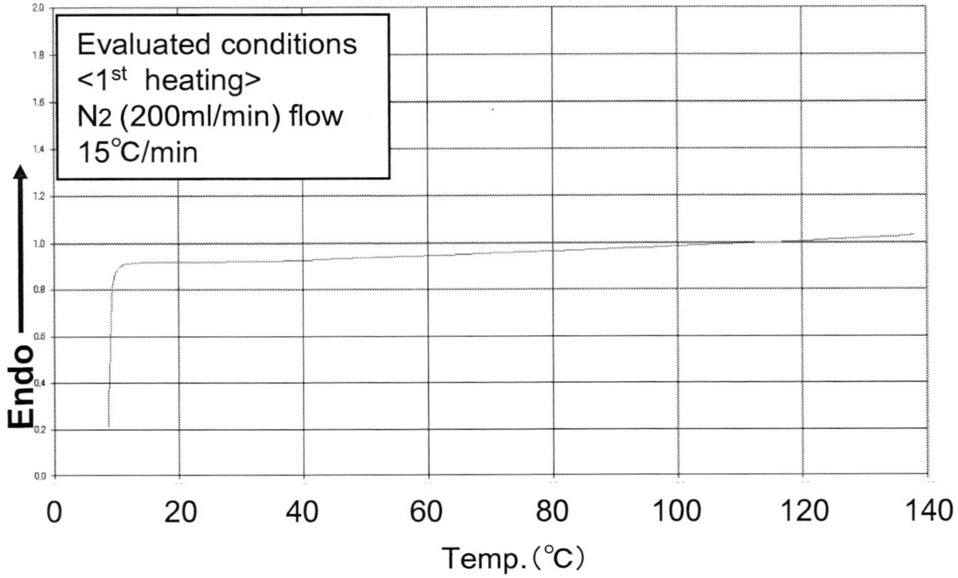

DSC thermogram of CLM-D resist

High decomposition temperature(above 170°C) and No Tg.
⇒This shows High Heat Resistance.

IDEMITSU

Selete

Evaluations of CLM-Resist

<u>High-NA(NA=0.3) Small Field Exposure Tool (HiNA)</u>

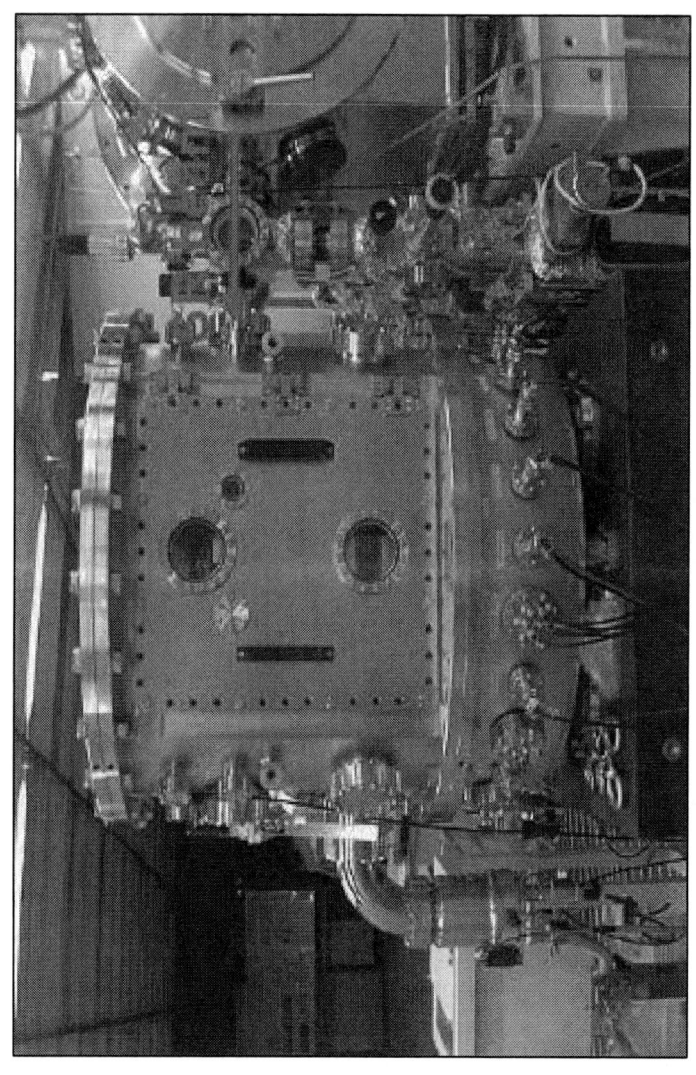

EUV imaging experiments were performed using the high-numerical-aperture (NA=3), small field EUV exposure tool (HiNA) .

EUV exposure result （CLM-D resist)

35nm-hp	40nm-hp	45nm-hp
29nm-hp	30nm-hp	32nm-hp
28nm-hp		

Process Conditions
Substrate: Si
Thickness: 50nm
Exposure:HINA
Dev: TMAH 0.26N 30sec
SEM: S8840
Mag: 100k
Mask:MB24

45nm-hp@13.3mJ/cm²

28nm L/S was resolved.

EUV exposure result (CLM-D resist with BARC)

Process Conditions
Substrate: BARC
Thickness: 50nm
Exposure:HINA
Dev: TMAH 0.26N 30sec
SEM: S8840
Mag: 100k
Mask:MB24

45nm-hp@12.8mJ/cm^2

45nm-hp	40nm-hp	35nm-hp
32nm-hp	30nm-hp	29nm-hp
		28nm-hp

 IDEMITSU

 Selete

Evaluation of Outgassing of CLM-D Resist by EUV Lithography

resist	CLM-D+ PAG+ quencher	CLM-D
film thickness (nm)	79.0	84.0
out gas rate (molecule/s/cm^2@0.4W)	4.4E+17	9.0E+15

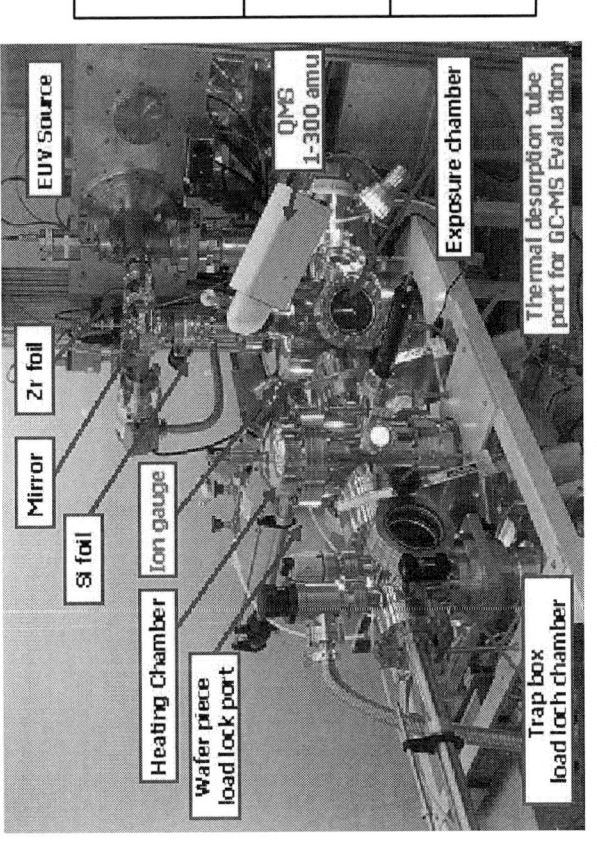

Resist outgassing analysis tool at Selete

To improve outgassing performance, we need to optimize the structure of protection group, PAG and quencher.

IDEMITSU

Summary

- We have been investigating new CLM-Resist (CLM-D) for EUV.

- CLM-D resist was resolved sub-30nm L/S.

Future Work

- Optimization of Protecting group, PAG and quencher in order to achieve the low outgassing and the other required performance.

- LER&LWR: Making sure of the affects due to the molecular grass resists.

- We have been developing the new concept resists.

Acknowledgements

This work was partially supported by the New Energy and Industrial Technology Development Organization (NEDO).

Relation between Acid Diffusion and Resolution in Chemically Amplified EUV Resists

Yuuki Hirai[1], Makoto Shimizu[1], Ken Maruyama[1], Toshiyuki Kai[1], Tsutomu Shimokawa[1], Toshiro Itani[2], and Daisuke Kawamura[2]

[1]JSR Corporation, [2]Selete

JSR Corp.-Selete/ 2008 International Symposium on EUVL / Sep. 29, 2008

Contents

∧ **Background**

∧ **Objectives**

∧ **Experimental Method**

∧ **Results and Discussion**

1. Acid Diffusion and Lithographic Performance

2. Characterization of Resist Platform

∧ **Summary**

Background

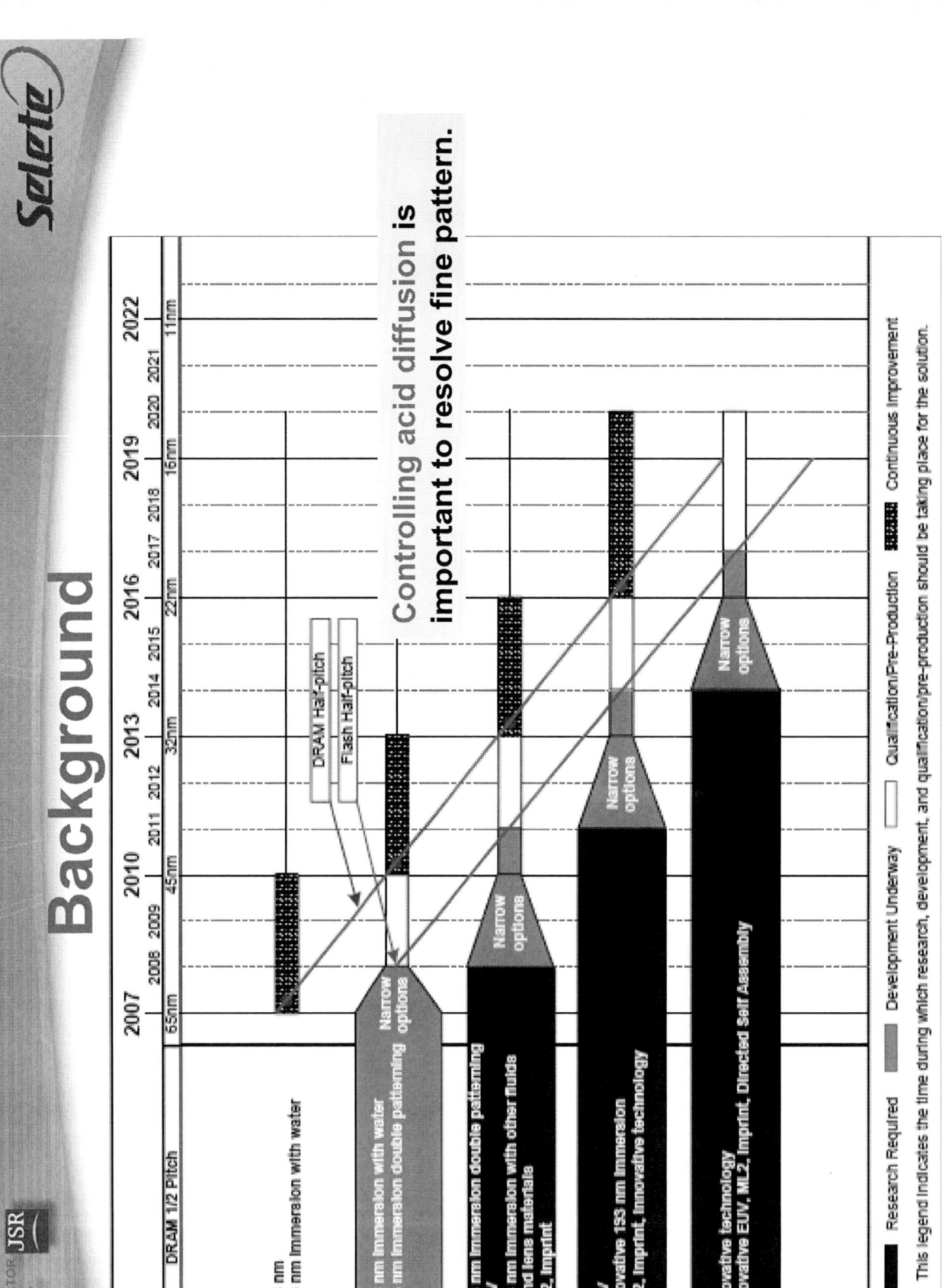

Controlling acid diffusion is important to resolve fine pattern.

2007 ITRS, Fig. LITH5 Lithography Exposure Tool Potential Solutions

Objectives

■ To measure the acid diffusion coefficient (D_{dev}) for various materials.

■ To understand the relationship between the acid diffusion coefficient (D_{dev}) and the lithographic performance of resist.

Materials

■ Resist Matrices

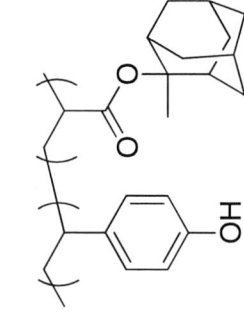

Mw: ca. 10000

HS/MAdA

Mw: ca. 2000
R: H or protecting group

Noria[1]

1) Nishikubo et al., Angew. Chem. Int. Ed. 2006, **45**, pp. 7948-7952.

■ Characteristics of PAGs Used in This Study

PAG	TPST	TPSN	TPSO
PAG Structure	$CF_3SO_3^-$	$C_4F_9SO_3^-$	$C_8F_{17}SO_3^-$
van der Waals volume of acid anion (Å^3)	84	165	272

JSR SEMICONDUCTOR MATERIALS LABORATORY / Selete

Comparison of Experimental Methods

■ Exposed Resist Powder Test [1]

A
B
Si wafer

Layer A: resin, PAG (exposed)
Layer B: resin, PAG, quencher (not exposed)

Advantage
1. All types of resins can be measured and evaluated.
2. Measurement system is similar to photo resist. (No special materials for measurement)

1) *Proc. SPIE*, **5376**, pp.790-800(2004).

■ Other Known Methods

Method	Advantage	Disadvantage	Ref.
Alcoholic top-coat	High productivity	Not applicable to PHS resist (dissolve)	*Proc. SPIE*, **6519**, 65193K-1(2007).
FT-IR "sandwich"	Measures acid transit	Use of special acid feeder and detector layer	*Proc. SPIE*, **3999**, pp.665-674(2000).

Exposed resist powder test method is more suitable for present study than other methods.

Experimental Method[1]

Exposure dose : E

KrF

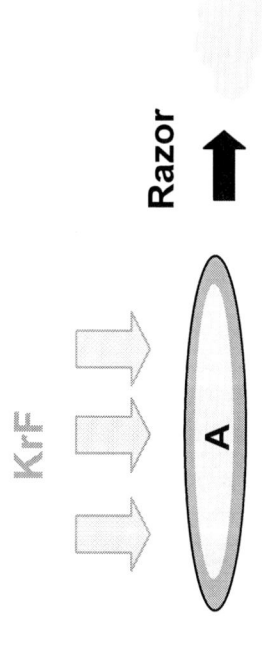

Layer A: resin, PAG

Razor → Press

Layer B: resin, PAG, quencher

Acid diffusion

Hot plate

PEB

Air flush

Dev.

Measuring thickness loss ΔL

The film thickness loss ΔL was measured for the each exposure dose E. ΔL and E were fitted to Fick's equation to calculate acid diffusion constant D.

1) *Proc. SPIE*, **5376**, pp.790-800(2004).

Curve Fitting with Fick's equation[1] (1)

•Fick's diffusion equation

$$\partial C / \partial t = D \cdot \partial^2 C / \partial x^2 \quad (1)$$

$C(x, t)$: Concentration of acid, D: Diffusion coefficient

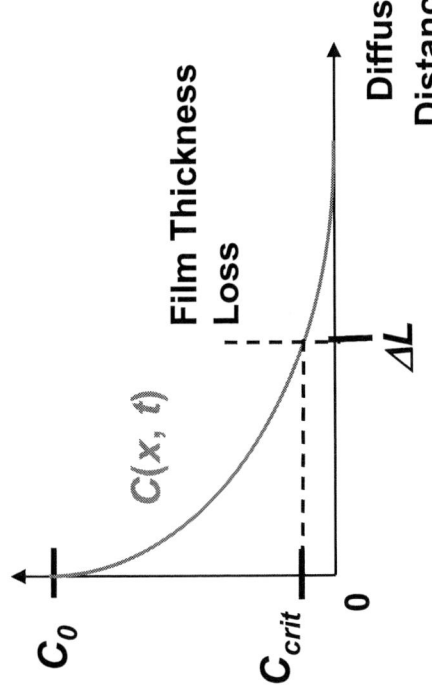

Acid Concentration

C_0

$C(x, t)$

Film Thickness Loss

C_{crit}

ΔL

0

Diffusion Distance x

•**1st assumption** : acid concentration of layer A, C_0, is constant during PEB.

Initial condition: $C(x>0, 0) = 0$, $C(0, 0) = C_0$

Boundary condition: $C(0, t>0) = C_0$

$$C(x, t) = C_0 \cdot erfc[x/\{2(D \cdot t)^{1/2}\}] \quad (2)$$

erfc: the error function complement

•**2nd assumption:** C_0 is proportional to dose given to the acid source, E.

$$C(x, t) = k \cdot E \cdot erfc[x/\{2(D \cdot t)^{1/2}\}] \quad (3)$$

k : Constant for PAG decomposition by dose, E: dose given to the acid source

•**3rd assumption:** ΔL was determined by critical acid concentration C_{crit}.

$$C(\Delta L, t_{PEB}) = C_{crit} = k \cdot E \cdot erfc[\Delta L/\{2(D \cdot t_{PEB})^{1/2}\}] \quad (4)$$

C_{crit} : concentration at which matrix resin turned from soluble to insoluble in developer, t_{PEB} : PEB time

1) *Proc. SPIE*, **5376**, pp.790-800(2004).

Curve Fitting with Fick's equation (2)

• E_{crit} : thickness loss was observed for the first time at dose E_{crit}.

$$C(0, t_{PEB}) = C_{crit} = k \cdot E_{crit} \quad (5)$$

From (4) and (5), the next formula is deduced.

$$\Delta L = 2(D \cdot t_{PEB})^{1/2} erfc^{-1}(E_{crit}/E) \quad (6)$$

ΔL: thickness loss, E_{crit}: the dose
at which thickness loss was observed for the first time

D can be calculated from the film thickness loss observed at normalized dose, E / E_{crit}.

✓ **In this paper, D is calculated from the film thickness loss instead of the acid diffusion length, so the acid diffusion constant is written as D_{dev}.**

EUV Experimental Conditions

–Pattern Exposure

- Exposure Tool : CANON SFET (NA 0.30, Mag. 1/5,)
  - Annular (σ 0.7/0.3)

–Resist Process

Resin	PAG	Sub.	ADH	BARC	Thick.	PAB	PEB	DEV.	Post Bake
HS/MAdA	TPST	Si	90C 20s		50nm^t	130C 60s	110C60s	2.38% TMAH 60s	90C 30s
	TPSO								
	TPSN						100C60s		
							110C60s		
							120C60s		

✓ Exposure dose is calibrated according to the announcement (or guidelines) from Selete.

(2008 International Symposium on EUVL, Resist-1, H. Oizumi, D. Kawamura, K. Kaneyama, S. Kobayashi and T. Itani)

I. Acid Diffusion and Lithographic Performance (1)

■ Formulation of Photoresists

Resin	HS/MAdA		
PAG	TPST	TPSN	TPSO
PAG content	24.2mmol / resin 100g		

■ Acid Diffusion Measurement Condition

{ Layer A: resin, PAG (exposed)
  Layer B: resin, PAG, quencher

Thickness of Layer B:300nm,
PAB:130C60s, PEB:variable60s,
Dev.:60s

A
B
Si wafer

✓ Formulation of layer B was same as EUV resist.

Effects of PEB temperature and PAG types were examined.

1. Acid Diffusion and Lithographic Performance (2)

Effect of PEB temp. change on D_{dev} and lithographic performance

	100C60s	110C60s	120C60s
Resin	HS/MADA		
PAG	TPSN		
PEB	100C60s	110C60s	120C60s
D_{dev} (nm^2/sec)	5.5	43	210
hp 45nm			
E_{size} [mJ/cm^2] EL [%]	14.9 / 25.0%	11.8 / 17.8%	10.2 / 12.2%

D_{dev} vs. E_{size}

$E_{size}@45nm\ hp$ (mJ/cm^2)

Ddev (nm^2/sec)

D_{dev} vs. EL

$EL@45nm\ hp$ (%)

Ddev (nm^2/sec)

Evaluation Condition
Exposure Tool: SFET at Selete, on-Bare-Si, FT: 50 nm,
PAB: 130C-60s, PEB: various, Dev:2.38% TMAH 60s.

D_{dev} showed good correlation with E_{size} and EL.

1. Acid Diffusion and Lithographic Performance (3)

Effect of PAG change on D_{dev} and lithographic performance

Resin	HS/MADA		
PAG	TPST	TPSN	TPSO
PEB	110C60s		
D_{dev} (nm²/sec)	40	43	31
hp 45nm	10.9 15.4%	11.8 17.8%	12.0 18.8%
E_{size} [mJ/cm²] EL [%]			
hp 32nm	11.9 7.5%	-- --	-- --
E_{size} [mJ/cm²] EL [%]			
Resolution	hp 30nm	hp 35nm	hp 40nm

Evaluation Condition
Exposure Tool: SFET at Selete,
on-Bare-Si, FT: 50 nm,
PAB: 130C-60s, PEB: 110C-60S,
Dev:2.38% TMAH 60s.

Top-loss of the pattern (deleted was) increased with the bulkiness of the PAG acid. As a result, the relationship between D_{dev} and resolution limit was not clear.

1. Acid Diffusion and Lithographic Performance (4)

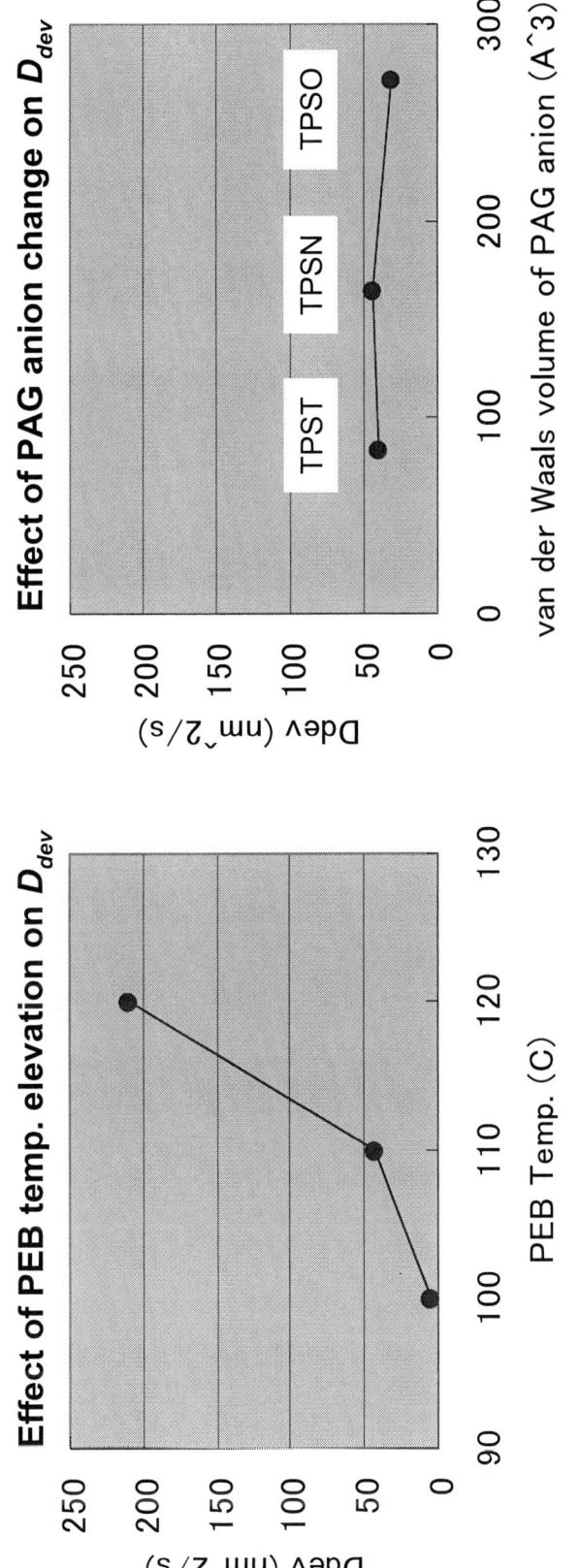

Effect of PAG anion change on D_{dev}

TPST TPSN TPSO

van der Waals volume of PAG anion (A^3)

D_{dev} (nm^2/s)

Effect of PEB temp. elevation on D_{dev}

PEB Temp. (C)

D_{dev} (nm^2/s)

Compared with PEB temp. elevation, change of PAG anion size did not affect the D_{dev} obviously.

2. Characterization of Resist Platform (1)

■ Formulation of Photoresists

Resin	HS/MAdA	Noria	
	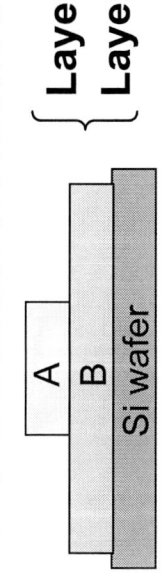		
PAG	TPST	TPSN	TPSO
PAG content	12.1mmol / resin 100g		

■ Acid Diffusion Measurement Condition

Thickness of Layer B:300nm,
PAB:130C60s, PEB:110C60s,
Dev.:60s

Layer A: resin, PAG (exposed)
Layer B: resin

A
B
Si wafer

√ Layer B contained NO quencher to eliminate the effect of localization of quencher.

D_{dev} of HS/MAdA and Noria were measured under the same condition.

2. Characterization of Resist Platform (2)

Acid diffusion constant for HS/MAdA and Noria matrices (PEB: 110C60s)

**Attention :
Measurement
condition is different
from that in pp. 11- 14.**

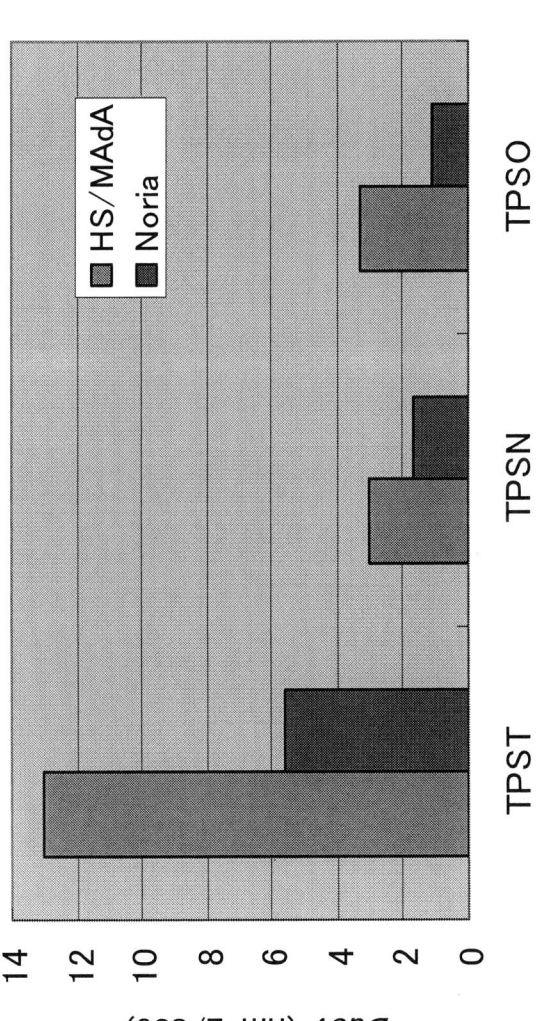

Noria showed smaller D_{dev} than HS/MAdA in all PAGs.

 **Noria-based resist was optimized and exposed by EUV.**

2. Characterization of Resist Platform (3)

Linearity and resolution of Noria with PEB 130C60s was measured.

SEM images :

45nm hp	40nm hp	35nm hp	32nm hp	30nm hp	28nm hp

Linearity :

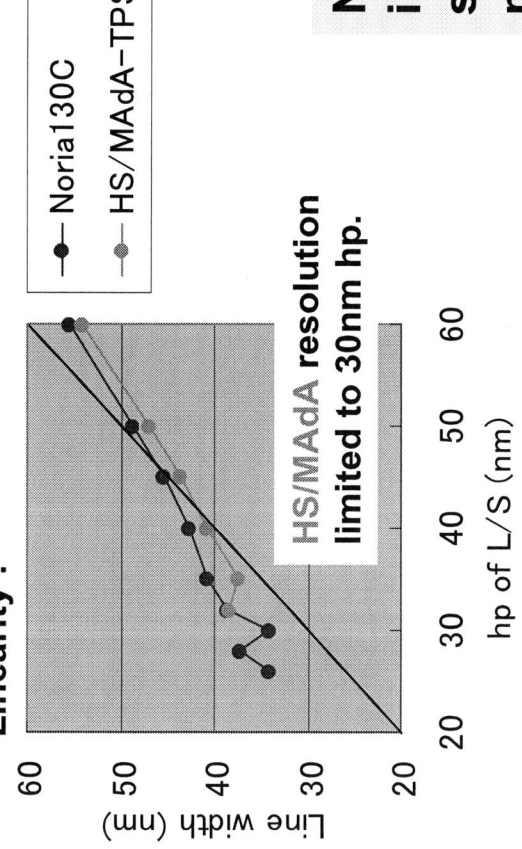

Evaluation Condition
Dose: 9.6mJ/cm^2
Exposure Tool: SFET at Selete,
on-BARC, FT: 50 nm,
PAB: 130C-60s, PEB: 130C-60S,
Dev:2.38% TMAH 60s.

Noria resolved 28nm hp by annular illumination system, suggesting the short acid diffusion length in Noria matrix.

Summary

➢ Acid diffusion constant (D_{dev}) determined using the exposed powder test method showed good correlation with resist sensitivity and EL.

➢ Resist top-loss of the pattern was increased with the bulkiness of the anion of the PAG present in PHS based resist. It limits the understanding of the relationship between D_{dev} and resolution.

➢ D_{dev} of EUV resist increased significantly with increasing PEB temperature but it remained constant by changing the size of the PAG anion.

➢ D_{dev} determined for Noria matrix was smaller than that determined for PHS matrix. L/S pattern printed with Noria matrix showed superior linearity than PHS based matrix. Smaller D_{dev} and better linearity demonstrate promising potential to resolve sub-30 nm hp pattern.

EUV photoresists twice as fast as previously thought

Patrick Naulleau, Eric Gullikson, Andrew Aquila, Paul Denham, Simi George, Drew Kemp, Dimitra Niakoula, Seno Rekawa
Center for X-ray Optics, Lawrence Berkeley National Laboratory

THE PROBLEM

- Resist sensitivity is a crucial issue for EUV resists
- EUV tools have been universally calibrated against a >10-year resist dose standard (EUV-2D) established on the now decommissioned Sandia 10x exposure tool
- From EUV-2D the resist standards have been transferred to a variety of baseline resists for tools around the world.
  - Transfers not always performed in well-controlled manner or adequately verified
- The EUV-2D dose standard was transferred to the Berkeley MET tool using E_{size} of 6.8 mJ/cm^2
- Aerial-image characteristics differences between the 10x and MET put into question the transfer of E_{size} from one tool to the other (required detailed wavefront and flare characteristics of the Sandia 10x not available)

THE SOLUTION

- Use Berkeley Calibrations and Standards EUV beamline to determine E0 by exposure and measurement of contrast curves in RHEM MET-1K and TOK-P1123
- Transfer the E0 calibration to the Berkeley MET
- Measure E0 to E_{size} ratios on Berkeley MET for RHEM MET-1K and TOK-P1123 to determine absolute E_{size} values for these two baseline resist

Resist Calibration Procedure

1. Measure beam profile at wafer plane
   - Scanning pinhole test
2. Remove pinhole and install wafer on sample stage
3. Move sample out of beam path and measure total power at wafer plane using calibrated photodiode
   - See "Calibrating the Detector" section
4. Generate exposure matrix on wafer while tracking relative beam current with in-situ detector
5. Measure total beam power after matrix exposure for self consistency test

RESIST CALIBRATION RESULTS

- Two independent measurements of RHEM XP-4502D (MET-1K) and TOK EUVR-P1123
  - Beam power changed by 10x between two measurements 1 month apart
  - Processing conditions
    - MET-1K- PAB: 130°/60s, PEB: 120°/90s, Dev: 45s
    - P1123- PAB: 120°/90s, PEB: 100°/90s, Dev: 60s
    - Contact bake plates used
    - Puddle develop
  - E_{size}/E_0 ratio measured using two independent techniques
    - Separate clear field and line-space field exposures
    - Single brightfield resolution cell exposure
      - Avoids errors due to potential reflectivity differences between different fields on reticle

THE BOTTOM LINE

All resist sensitivities reported prior to June 2008 were 1.9 times slower than actual numbers

Absolute E$_0$ measurement results in MET baseline resists

	9/07	10/07	Average
Synchrotron ring current (mA)	20	250	NA
MET-1K E_0 (mJ/cm^2)	7.6	7.3	7.45
EUVR-P1123 (mJ/cm^2)	5.6	6.0	5.8

E_{size}/E_0 ratio measurement using MET

	Separate fields	Single field*	Average
MET-1K E_{size}/E_0	1.84	1.80	1.82
EUVR-P1123 E_{size}/E_0	1.82	1.70	1.76

* E_0 determined from large open area in brightfield cell

Comparison of E_{size} for MET baseline resists for old versus new calibration

	New calibration	Old calibration	Ratio
MET-1K E_{size} (mJ/cm^2)	13.6	25	1.84
EUVR-P1123 E_{size} (mJ/cm^2)	10.2	20	1.96

Funding provided in part by **SEMATECH**

CALIBRATION & STANDARDS BEAMLINE (BL6.3.2)

Reflectometer Specs
- λ precision: 0.007%
- λ accuracy: 0.013%
- Reflect. precision: 0.08%
- Reflect. accuracy: 0.08%
- Spectral purity: 99.98%
- Dynamic range: 10^{10}

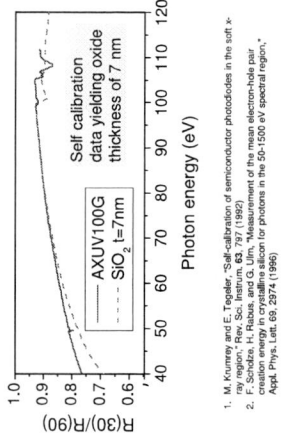

Prolith modeling with RHEM-provided EUV-2D model[*]

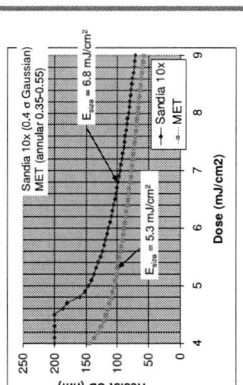

Modeled E_{size} differences between an assumed ideal 10x system and MET, demonstrates the potential problem with the E_{size} transfer used. The conditions assumed here would give rise to a 30% dose calibration transfer error.

*S. Robertson et al., Proc. SPIE 5037, 900-905 (2003)

CALIBRATING THE DETECTOR

Self Calibration[1]

- Measuring generated photocurrent as function of photon energy at two different incident angles
- Diode front oxide thickness determined from ratio of energy response
- From oxide thickness diode is responsivity can be determined[2]

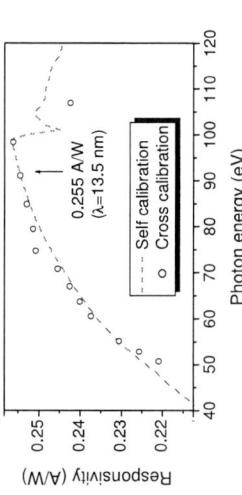

1. M. Krumrey and E. Tegeler, "Self-calibration of semiconductor photodiodes in the soft x-ray range," Rev. Sci. Instrum. 63, 797 (1992)
2. F. Scholze, H. Rabus, and G. Ulm, "Measurement of the mean electron-hole pair creation energy in crystalline silicon for photons in the 50-1500 eV spectral region," Appl. Phys. Lett. 69, 2974 (1996)

Cross Calibration

- Self calibration procedure verified with cross calibration to NIST calibrated SXUV-100 diode

Modeling verifies observed E_{size}/E_0 ratio

- EUV-2D resist modeled verifies observed E_{size}/E_0 ratio
- Model predicts 1.83, measurement results = 1.88

ACKNOWLEDGEMENTS

- Gideon Jones, Brian Hoef, and Jerrin Chui, LBNL
- Jim Thackeray, RHEM
- Dave White and Koki Tamura, TOK
- Warren Montgomery and Stefan Wurm, SEMATECH
- ALS synchrotron supported by DOE, Office of Science, Basic Energy Sciences

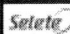

EUV resist outgassing quantification and qualification analysis methods
Shinji Kobayashi, Julius Joseph Santillan, Hiroaki Oizumi and Toshiro Itani (MIRAI-Selete)

Background & Objectives

- One of the most significant factor in EUV resist outgassing evaluation is the variation of analysis methods applied.
- Pressure rise method is used for quick quantification prior to exposure.
  GC-MS method is used for detailed quantification / quantitative analysis and QMS analysis is applied for in-situ mechanism analysis.
- It is practical to maximize the advantages of each of these methods based on the analysis objective.

1. Introduction

Major issues in EUV resist development

Resist outgassing evaluation method

Methods	Description	Evaluation time	Selete
Pressure rise	□ Simple and quick for quantitative analysis. ■ Component identification not possible.	2 hours/sample	○
GC-MS	□ Component identification possible. ■ CO_2 cannot be detected. ■ Low throughput.	1 day/sample	○
QMS	□ In-situ qualitative analysis possible ■ Quantitative analysis not possible. ■ Qualitative analysis inaccuracy due to fragmentation effect.	2 hours/sample	○
Witness mirror	□ Contamination level directly observed. ■ Low throughput and high cost.	A few days/sample	△

Resist outgassing is one of the main concerns for resist development.
Reduction of resist outgassing to keep a clean exposure machine is important.

Pressure rise, GC-MS and QMS methods are applied for resist outgassing evaluations.

2. Evaluation tools

Pressure rise method **QMS analysis**

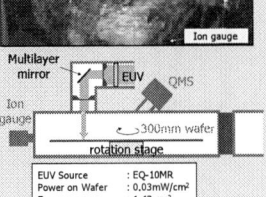

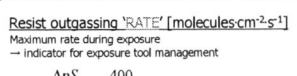

Resist outgassing 'RATE' [molecules·cm⁻²·s⁻¹]
Maximum rate during exposure
→ indicator for exposure tool management

$$J_{400} = \frac{\Delta p S_e}{RTA} N_A \frac{400}{I} \quad (400mW \cdot cm^{-2} \text{ assumed})$$

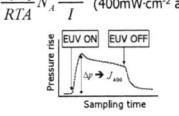

Resist outgassing 'AMOUNT' [molecules·cm⁻²]
AMOUNT dependence on exposure dose
→ indicator for resist improvement

$$N_D = \sum_{i=0}^{t} \frac{\Delta P_i S_e}{RTA}(t_{i+1} - t_i) N_A$$

- D_p : pressure rise
- S_e : eff. pumping speed
- R : Gas constant
- T : temperature
- A : area of exposure
- NA : avogadro's no.
- I : EUV intensity

subscript
- i : time
- D : established dose

EUV Source : EQ-10MR
Power on Wafer : 0.03mW/cm²
Exposure area : 1.43 cm²
Base pressure : 8x10⁻⁷ Pa

GC-MS method

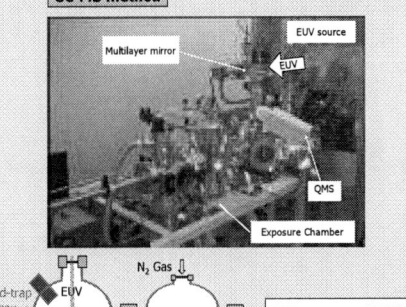

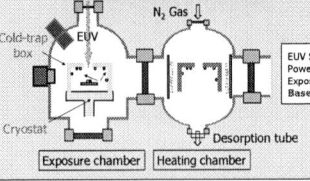

EUV Source : EQ-10MR
Power on Wafer : 0.014mW/cm²
Exposure area : 1.69 cm²
Base pressure : 1x10⁻⁷ Pa

3. Results

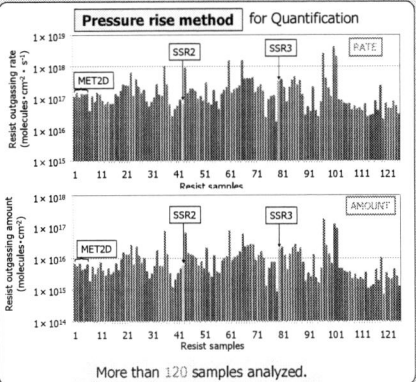

More than 120 samples analyzed.

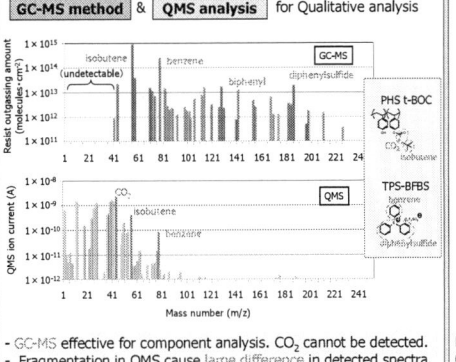

- GC-MS effective for component analysis. CO_2 cannot be detected.
- Fragmentation in QMS cause large difference in detected spectra.

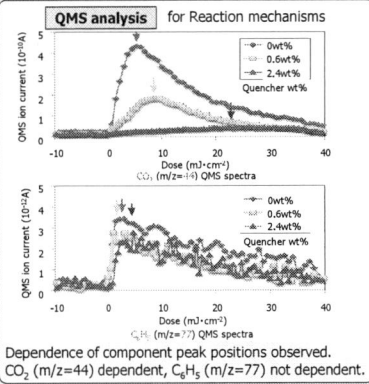

Dependence of component peak positions observed.
CO_2 (m/z=44) dependent, C_6H_5 (m/z=77) not dependent.

4. Selete resist outgassing evaluation procedure

Quantification (Quick Screening)	Pressure rise method Screening of resist samples received prior to exposure.
Component analysis	GC-MS method Improvement of resist samples based on new resist components.
Mechanism analysis	QMS analysis GC-MS method Basic study to improve tools and control methods.

5. Summary

- Quantification (Quick Screening)
  Resist outgassing rate and amount evaluations were performed for more than 120 samples using the pressure rise method, prior to exposure.
- Component analysis
  GC-MS effective and accurate in the analysis of resist outgassing components (CO_2 cannot be detected).
- Mechanism analysis
  QMS is highly recommended for component reaction mechanism analysis during exposure.
  GC-MS method is also applied to provide more accurate component identification for mechanism analysis.

Selete applies resist outgassing methods depending on the analysis objectives.

Acknowledgements

- A part of this work is supported by New Energy and Industrial Technology Development Organization (NEDO).
- Selete member companies (EUV Lithomask program).
- We would also like to thank Dr. Eiichi Nomura for many stimulating discussions.

Errata: Pressure rise method (equation)

Resist outgassing 'amount' [unit : molecules·cm⁻²]

pre-corrected equation
$$N_D = \sum_{i=0}^{t} \frac{\Delta P_i S_e}{RTA}(t_{i+1} - t_i) N_A$$

corrected equation
$$N_D = \sum_{i=0}^{t} \frac{\Delta P_i S_e}{RTA}(t_{i+1} - t_i) N_A$$

29th Sep. 2008/ 2008 EUVL Symposium

Development of Partially Fluorinated EUV Resist Polymers for Sensitivity Improvement

Takashi SASAKI[1], Osamu YOKOKOJI[1], Takeo WATANABE[2] and Hiroo KINOSHITA[2]

1. Research Center, ASAHI GLASS CO., LTD.
2. Laboratory of Advanced Science and Technology for Industry, University of Hyogo

1. Introduction

Demands for EUVL resist

	ITRS 2006 Update
Sensitivity	5-15 mJ/cm²
Resolution	32 nm 1:1 L/S
LWR (3σ)	1.7 nm
Outgas	6.5x10¹⁴ molecules/cm²

▷ Demands for EUVL Resist are severe.

LSR trade-off relationship

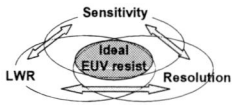

▷ Trade-off problem is one of the most critical issue in EUVL Resist.

For 32nm L/S and beyond, Sensitivity, Resolution and LWR are critical issues.

2. Approach

Acid Generation Mechanism on EUV exposure

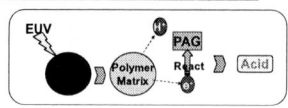

▷ Secondary electrons which were generated from a resist polymer matrix would be triggered a PAG's acid generation.

Introduction of F atoms

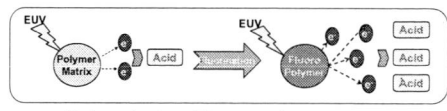

▷ By introduction of F atoms, absorption coefficient of resist and acid generation yield are increased.

Can we enhance a resist sensitivity by an introduction of F atoms?

3. In this study

Characteristics of two-type fluoropolymers are investigated, under EB and EUV exposure.

1. Side Chain Fluorinated unit — TFSt
2. Main Chain Fluorinated unit — FIT

4. Exposure Tools

▷ EUV exposure

Resist Evaluation System of BL3 in New SUBARU

▷ EB exposure

Exposure tool: ELS-7500
Acceletaion voltage: 50kV

---

5. Resist Sensitivity under EUV and EB Exposure with Side Chain Fluorinated Resist

Samples

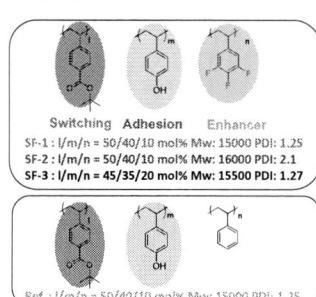

Switching Adhesion Enhancer

SF-1 : l/m/n = 50/40/10 mol% Mw: 15000 PDI: 1.25
SF-2 : l/m/n = 50/40/10 mol% Mw: 16000 PDI: 2.1
SF-3 : l/m/n = 45/35/20 mol% Mw: 15500 PDI: 1.27

Ref. : l/m/n = 50/40/10 mol% Mw: 15000 PDI: 1.25

Sensitivity curves

> EB exposure   >EUV exposure

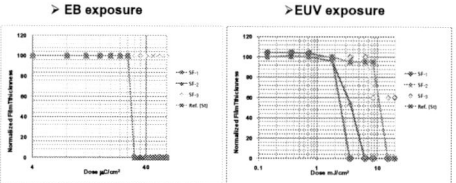

• Samples: Polymers with 5wt% TPS –Nf in PGMEA
• PB: 100°C 90sec, PEB: 120°C 90sec
• Development: NMD-3 60sec

@ EUV exposure, SF-1 showed a good sensitivity & contrast compared to non-fluorinated styrene polymer.

Etching Durability

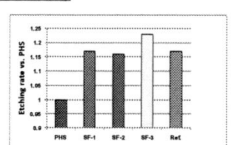

SF-1 showed almost same etching durability relative to non-fluorinated styrene polymer.

Presumed Scheme

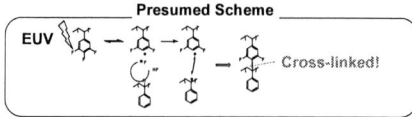

In the case of SF-3, Cross-linking reaction may be occurred.

---

6. Resist Sensitivity and Outgassing under EUV Exposure with Main Chain Fluorinated Resist

Samples

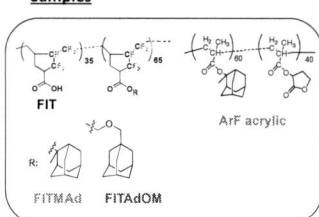

FIT   ArF acrylic

R: FITMAd FITAdOM

Sensitivity curves   ### Outgassing

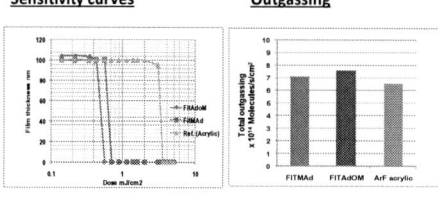

• Samples: Polymers with 4.5wt% TPS –Nf in PGMEA
• PB: 100°C 90sec, PEB: 120°C 90sec
• Development: NMD-3 60sec

FIT showed a good sensitivity & contrast.
Outgas level of FIT was almost same with non-fluorinated samples.
Small amount of HF was detected in outgas.

Molecular Weight Change

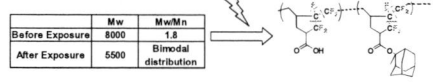

	Mw	Mw/Mn
Before Exposure	8000	1.8
After Exposure	5500	Bimodal distribution

MW was converted to lower after EUV exposure. Polymer main chain of FIT may be broken by irradiating with EUV light.

Resolution in EB exposure

FITMAd 80uC/cm² 100nm L/S

---

7. Summary

• We have been investigating fluorinated resist polymers for EUV lithography.
• Side chain fluorinated styrene polymers showed a high sensitivity relative to non-fluorinated styrene polymer.
• Main chain fluorinated type polymer FIT showed a good sensitivity.
• FIT's Polymer main chain may be broken by irradiating with EUV light.
• These results indicate that the resist sensitivity can be enhanced by introducing fluorinated unit.

8. Future Plan

• Patterning properties will be evaluated by an optimizing polymer structures and resist formulations.
• Acid generation efficiency of fluoropolymer will be estimated.
• The effect of HF gas will be evaluated.

AGC

2008 International Symposium on EUVL

Mask effects on Line-edge roughness (LER)

Patrick P. Naulleau,[1] Kenneth A. Goldberg,[1] Iacopo Mochi,[1] Guojing Zhang[2]
[1]Center for X-ray Optics, Lawrence Berkeley National Laboratory;   [2]Intel Corporation, Santa Clara, CA

MOTIVATION and METHODOLOGY

- Line-edge roughness (LER) remains the most significant challenge facing EUV resists
- Observed LER can arise from various non-resist sources including the mask
- Can recently observed lower bounds on LER performance be attributed to the mask?
- We use aerial-image modeling to study importance of mask roughness (pattern and multilayer) and projection optics roughness on LER
  - Scalar image modeling
  - Thin mask approximation

Mask multilayer (ML) roughness

- Model mask roughness using single-surface approximation[1]
  - Only need topographic profile of multilayer top surface

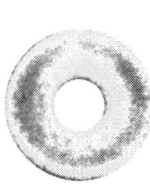

AFM image of 2×2mm area MET mask multilayer surface

Synthesized mask with phase roughness

- Consider mask with slope error meeting target specs (0.7 mrad)
- Effect of ML roughness significantly reduced relative to pattern roughness

MODELING PARAMETERS

Mask pattern roughness (LER)

- SEM image converted to modeling mask
- Modeling mask uses grayscale to define the edge transition region
  - Grayscale edge transition enables sub-pixel edge positioning for high fidelity LER reproduction at reasonable sampling levels
  - Modeling mask final size is 1020×1020 pixels with 0.84-nm size in mask coordinates and 1.8 nm in wafer coordinates.
  - Mask LER is 8.9 nm in mask coordinates and 1.8 nm in wafer coordinates.

SEM image of 50-nm coded features on MET mask

Generated modeling mask

Projection optics roughness (flare)

- Berkeley MET wavefront
- 7% flare in full bright field
- Stochastic flare model
  - Roughness added to pupil, not simply DC offset
- Wavefront quality = λ/15
- Pupil map also generated without flare for direct comparison

MET pupil map with synthesized roughness derived from AFM measurements of optics

- Isotropic PSD measured from AFM
  - RMS roughness = 0.24 nm
  - RMS slope error = 1.3 mrad
- PSD used to synthesize new rough phase surface with size & pixel density matching pattern roughness model
- Final modeling mask generated by multiplying synthesized phase and generated pattern roughness mask

1. E. Gullikson, C. Cerjan, D. Stearns, P. Mirkarimi, D. Sweeney, "Practical approach for modeling extreme ultraviolet lithography mask defects," J. Vac. Sci. Technol. B 20, 81-86 (2002)

LER MODELING RESULTS

Mask and coherence effects

- Simulation results for three different modeling masks
- Pattern LER dominates at best focus and low coherence
- ML roughness dominates at low σ and/or large defocus
- Combined LER is rss of individual terms

Dipole (σ=0.2, r=0.51)

Annual (0.35-0.55)

- Direct comparison of various illumination settings
- Less coherence is better

Full rough mask

Implications on LER measured in resist

After quadrature subtraction of predicted mask effects (1.43-nm for annular illumination)

Measured LER scatter plot versus resist sensitivity

Projection optics roughness

Annular (0.35-0.55)

Dipole (σ=0.2, r=0.51)

- Direct comparison of the flare and no-flare cases with full rough mask. As evidenced by the modeling results
- Effect of flare (projection optics scatter) is negligible

VISUALIZING THE IMAGE SPECKLE

- Multilayer phase roughness leads to image speckle
- Speckle can be directly visualized in clear field image
- Speckle contrast too low for direct resist imaging
- Observe with modeling and aerial image microscope

Modeling

- Assume ideal 0.0875-NA system (0.35 4× equivalent)
- Assume MET mask roughness characteristics
- Model 4.3×4.3-μm area (mask coordinates)
- As expected from LER results, speckle increases with defocus and coherence

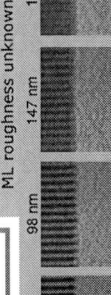

Best Focus
σ=0.1, γ=1.2%   σ=0.3, γ=1.1%   σ=0.5, γ=0.9%

50-nm eq. wafer defocus
σ=0.1, γ=11%   σ=0.3, γ=9%   σ=0.5, γ=6%

γ = rms speckle contrast

- Through focus speckle contrast with σ = 0.2
- Speckle increase saturates

Aerial image μscope (AIT)*

- 0.0875-NA zoneplate used (0.35 4× equivalent)
- σ<0.2 (anisotropic)
- Open area through focus images (wafer equivalent defocus, Z)
- NOT MET mask, actual mask ML roughness unknown

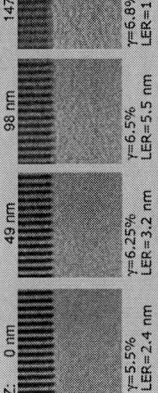

Z: 0 nm	49 nm	98 nm	147 nm	169 nm
γ=5.5% LER=2.4 nm	γ=6.25% LER=3.2 nm	γ=6.5% LER=5.5 nm	γ=6.8% LER=13.2 nm	γ=6.9%

- Expected speckle observed with AIT
- Increases with defocus, but quickly saturates as predicted
- Weak focus dependence of speckle contrast (γ) relative to modeling likely due to residual aberrations

*see poster Mochi et al.

CONCLUSIONS

- Projection optics flare insignificant contributor to LER
- Both mask pattern and ML roughness important in observed LER
  - Pattern roughness dominates at best focus
  - ML roughness dominates with defocus: limits process window
  - Multilayer effects become more important at low-k₁ (low σ) illumination settings are used
- Numerical compensation for mask effects is required to achieve accurate resist LER characterization below levels of ~3.5 nm
  - Current MET mask has multilayer roughness 2× 32-nm node spec and pattern roughness 7.5× 32-nm spec
- Mask effects not dominant cause for the observed LER lower limit
- Significant mask improvement needed for 22-nm node LER target

Funding in part by SEMATECH

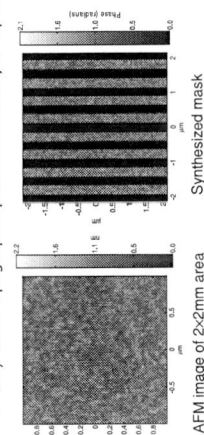

Experimental Study on Flatness of Electrostatically Chucked Reticle

K. Ota, T. Taguchi, M. Amemiya, N. Nishimura and O. Suga

MIRAI-Semiconductor Leading Edge Technologies, Inc., Japan

Abstract

The flatness of the reticle front surface has to be less than about 50nm@p-v, because the illumination light of an EUV scanner is not telecentric and the non-flatness of the reticle surface generates some overlay errors on the wafer. We have to manage not only the front surface of the reticle but also the backside of the reticle and the chuck surface. If the reticle is completely chucked without any gaps between the chuck and the reticle, we can expect that the flatness of the front surface of the chucked reticle is predicted by the thickness variation of the reticle and the flatness of the chuck. However, if the reticle has high spatial frequency non-flatness, the chuck cannot perfectly compensate it because the chucking pressure is not infinitely large. In such case, the flatness of the front surface of the reticle depends on the chucking pressure. In addition, the friction between the chuck and the reticle may disturb the reticle slipping on the chuck; it means the chucking repeatability is a concern. MIRAI-Selete has been investigating this flatness issue and carrying out many experiments using various electrostatic chucks and substrates. We measured the surface flatness of substrates with various parameter sets; chucking pressure, chucking period, timing of voltage applying to two electrode sets arranged at the inner and the outer part horizontally.

1. Two Electrostatic Chucks Used

- Selete experimental setup
  - We used two electrostatic chucks, which were installed in Selete MPE Tool.

- ESC1 (by Hitachi Chemical Co. Ltd.)
  - Chucking pressure >15kPa@70V, SiC, Flat-type

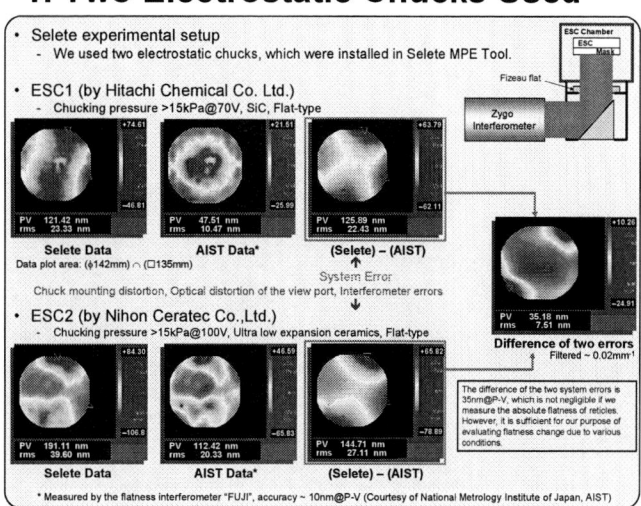

* Measured by the flatness interferometer "FUJI", accuracy ~ 10nm@P-V (Courtesy of National Metrology Institute of Japan, AIST)

2. Chucking Repeatability

- Two sets of dipolar electrodes can be controlled separately.
- Flatness repeatability of the chucked substrate evaluated.
  - The substrate was chucked and de-chucked each time.
  - Six condition was tested with ESC1.
    (1) Inner & outer simultaneously on @80V, (2) Inner 1st, outer 2nd @80V,
    (3) Outer 1st, inner 2nd @80V, (4) Inner & outer simultaneously on @640V,
    (5) Inner 1st, outer 2nd @640V, (6) Outer 1st, inner 2nd @640V
- How to calculate the repeatability.
  - The substrate was chucked and de-chucked four to six times at each condition.
  - A pair of data set was used for obtaining a repeatability result: pixel to pixel subtracted.
  - Total repeatability was calculated by averaging the results.

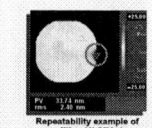

Chucking repeatability vs. chucking conditions

	80V ~ 19kPa			640V ~ 1.25MPa		
	Sim.	In→Out	Out→In	Sim.	In→Out	Out→In
P-V (nm)	28.8	33.9	9.8	9.0	8.1	8.0
Rms (nm)	2.6	3.5	1.0	1.5	1.2	1.3

- Repeatability at 640V was 1.2-1.5nm@rms, which is comparable to the measurement repeatability.
- A primary factor of degrading repeatability at 80V was chucking incompleteness at the right side (see the upper right figure) ; there is a gap between the electrodes.
- We may avoid such an issue by using higher chucking pressure. Generally, a flat-type electrostatic chuck generates higher chucking pressure than a pin-type chuck.

3. Surface Non-flatness Transfer

- When a reticle is chucked in an EUV scanner, non-flatness of the chuck surface and the reticle back surface transfers to the reticle front surface and it causes OPD (out-of-plane distortion) and IPD (in-plane distortion).
  - The surface flatness of ESC2 and the front surface flatness of the chucked reticle were compared; the front surface flatness was measured at four voltage conditions.
  - The thickness variation of the substrate was measured separately and removed from the front surface data.

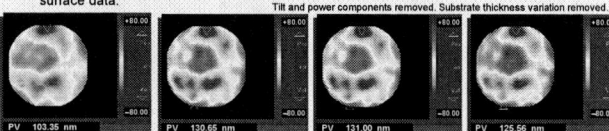

- The point is that higher chucking pressure may bring about larger OPD/IPD at higher spatial frequency region, e.g. >0.1mm⁻¹. However, we did not observe any increase of OPD at the frequency region.

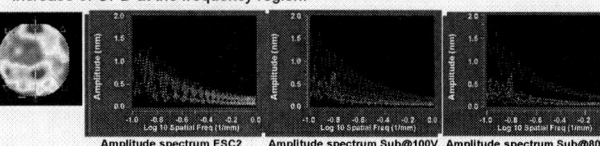

4. Particle Crushing

- We have observed an indirect evidence that a particle between the chuck and the substrate was crushing. We estimate the particle was larger than several microns..

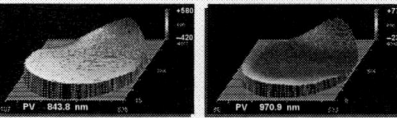

Interference fringes of a chucked substrate
Captured every 4 seconds. ESC2 was used and the chucking pressure was 15kPa@100V.

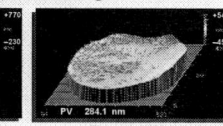

- The risk that a particle on a flat-type chuck degrades the flatness of a reticle will be reduced.

Conclusions

- We obtained good chucking repeatability [1.2-1.5nm@rms] by using higher chucking pressure [1.25MPa].
- When the chucking pressure was increased from 15kPa to 960kPa, we did not observe any increase of OPD at higher spatial frequency region > 0.1 mm⁻¹.
- A several microns particle between the chuck and the substrate was crushed and no OPD was observed.

→ *There is every possibility that a flat-type chuck is better for EUV reticles than a pin-type chuck.*

This work has been supported by NEDO.

FIB mask repair technology for EUV lithography

Tsuyoshi Amano, Yasushi Nishiyama, Hiroyuki Shigemura, Tsuneo Terasawa, Osamu Suga

MIRAI-Semiconductor Leading Edge Technologies, Inc. (Japan)

Kensuke Shiina, Fumio Aramaki, Ryoji Hagiwara, Anto Yasaka
SII NanoTechnology, Inc. (Japan)

Semiconductor Leading Edge Technologies, Inc.

2008 International Symposium on EUVL

Outline

- Introduction

- Experimental condition

- Results and discussions

    - FIB-GAE* repair

    - Defect Printability

    - Repair accuracy

    - Applications

- Summary

*FIB: Focused Ion Beam

GAE: Gas Assisted Etching

Introduction

Technical issues of EUVL mask pattern defect repair are:

➢ Damage-free repair on reflective multi layer

Refer to BACUS 2008 presentation

"Ga implantation and interlayer mixing during FIB repair of
EUV mask defects"
Y. Nishiyama et al. [7122-91]

➢ Side-etching of TaBN layer ⎫
➢ Defect printability   ⎬ This study
➢ Repair accuracy   ⎭

We evaluated repair process using FIB-GAE technique.

 **IRAI**

Conventional defect repair process

 **Selete**

Presented at PMJ2008   SEM images of repaired region

Top view

Cross sectional view

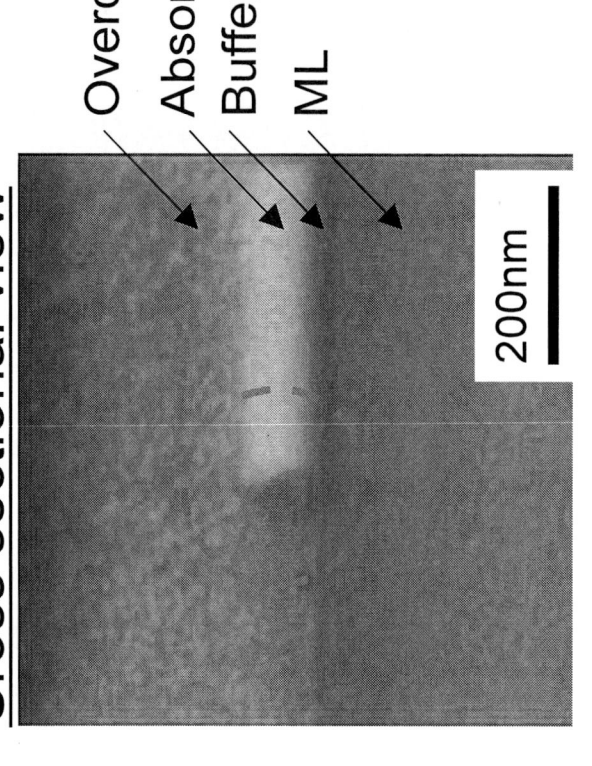

- Overcoat
- Absorber
- Buffer
- ML

200nm

hp325nm

500nm

✓ XeF_2 is an excellent etching assist gas during FIB repair.

✓ The chemical reactivity with Ta-based absorber material is hard to be controlled

because XeF_2 itself reacts with underlying Ta-nitride without FIB irradiation.

Influence of TaBN side-etching on EUVL

IRAI　　　　　　　　　　　　　　　　　　　　　　**Selete**

Presented at PMJ2008

CD degradation for patterns vs absorber side-etching values

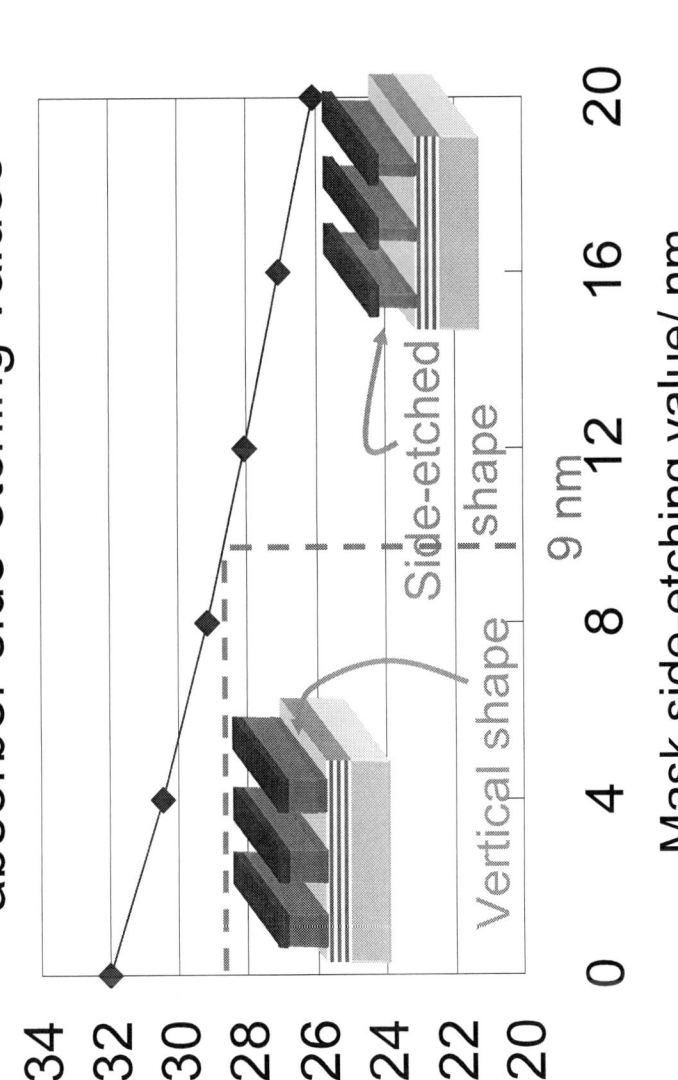

Mask side-etching value/ nm

Mask side-etching value

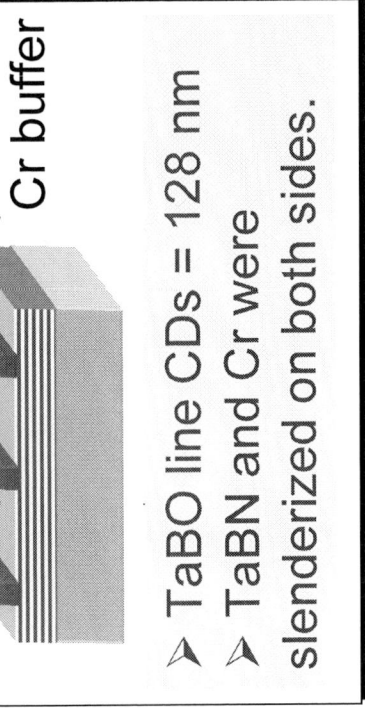

- TaBO
- TaBN
- Cr buffer

➢ TaBO line CDs = 128 nm
➢ TaBN and Cr were slenderized on both sides.

✓ Side-etched absorber layer affects printed line CD
✓ The 9 nm of the side-etching will cause 10 % of CD variation.

 We evaluated repair process focused on pattern topography

Experimental Condition

● EUVL mask structure (Hoya blanks)

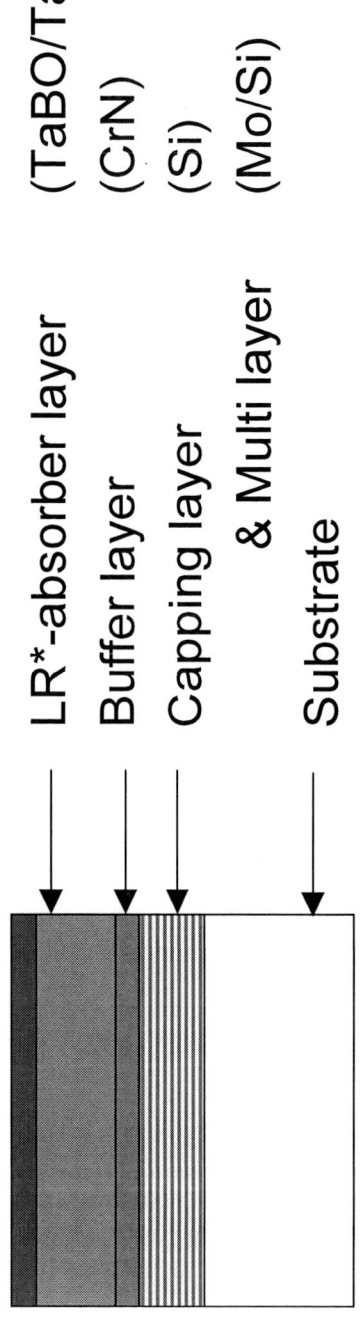

LR*-absorber layer   (TaBO/TaBN)

Buffer layer   (CrN)

Capping layer   (Si)

   & Multi layer   (Mo/Si)

Substrate

* Low reflectivity

● Repair tool and condition

FIB system:	Prototype system for EUVL mask
Acceleration Voltage:	15 kV
Probe current:	2 pA
Etching assist gas:	XeF_2
Etching control gas:	H_2O

Experimental Condition

- SFET* exposure condition

  Exposure condition: NA: 0.3 (central obscuration 30%)

  Sigma (inner/outer): 0.3/ 0.7

  Magnification: 1/ 5

  Incident angle: 6 deg.

  Resist: SSR3**

  * SFET: Small Field Exposure Tool

  ** Selete Standard Resist 3

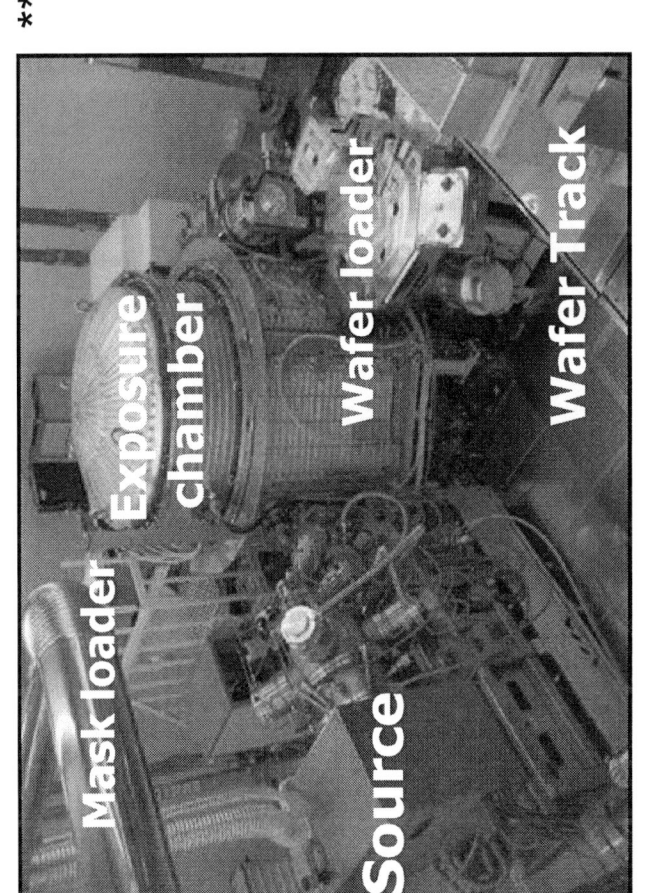

Experimental Condition

● Defect repair test pattern

Base pattern: hp225 nm (45 nm @1/5) L/S

Defect type: 4 x 1 Bridge

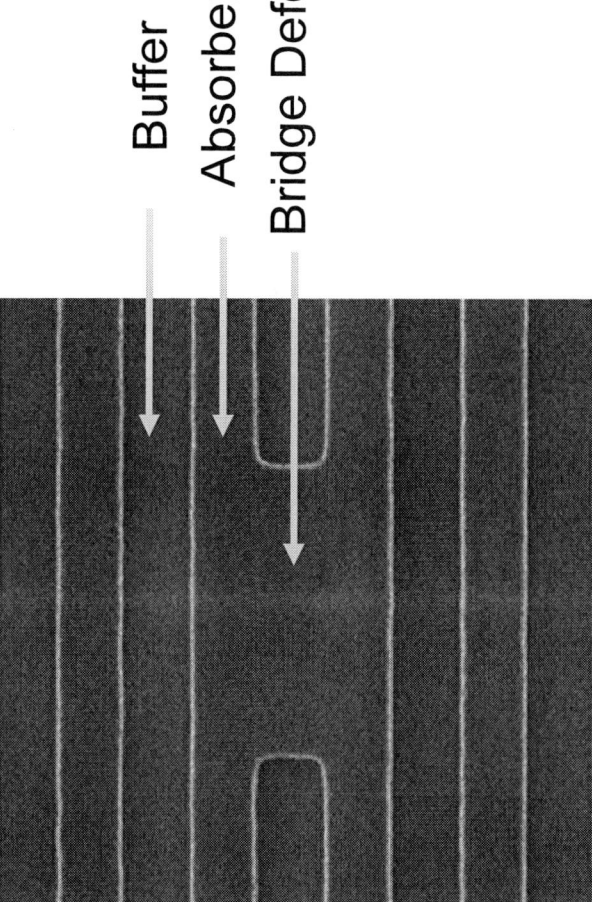

Buffer

Absorber

Bridge Defect

Mechanism of TaBN side-etching

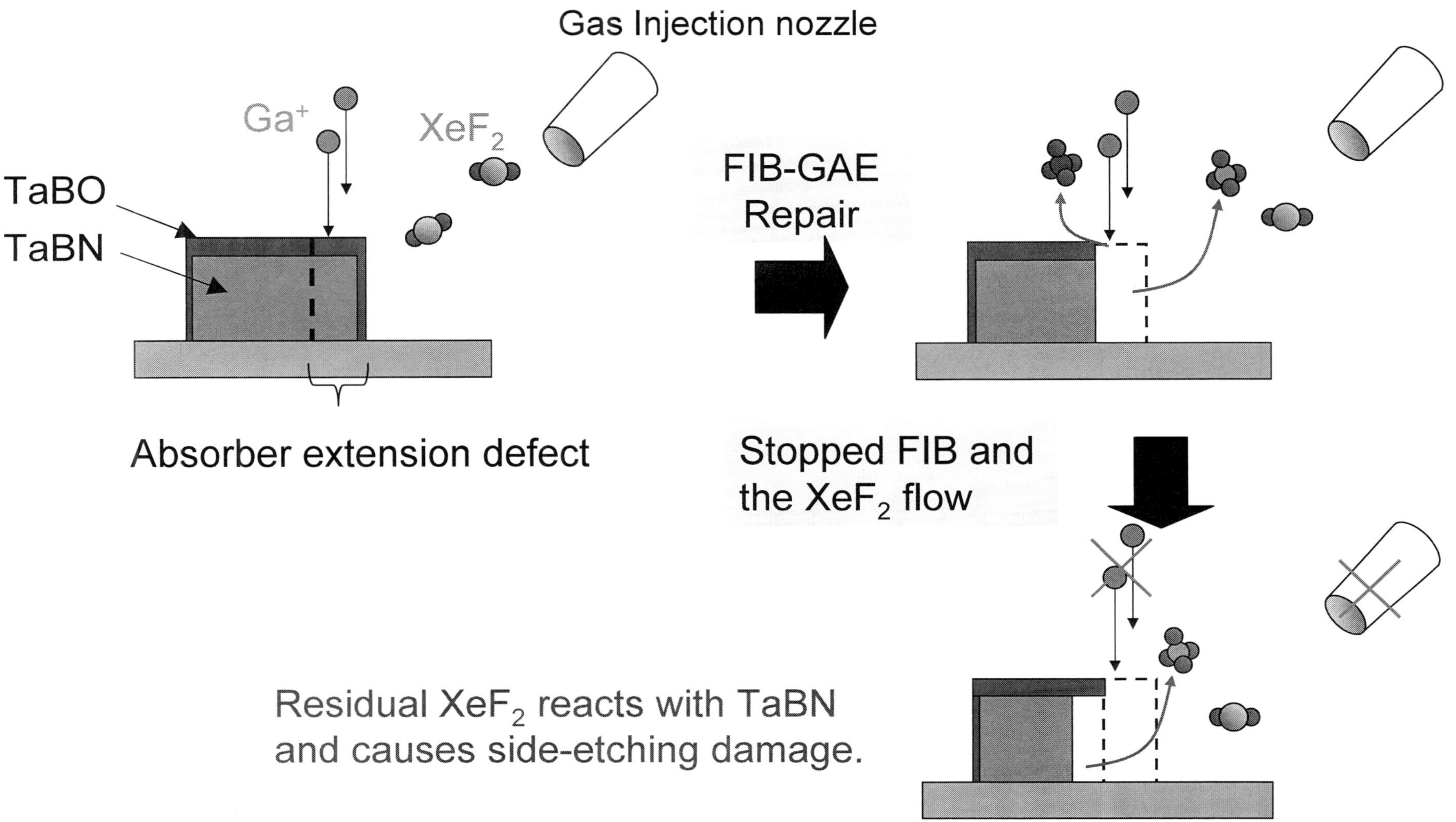

Gas Injection nozzle

XeF_2

Ga^+

TaBO

TaBN

Absorber extension defect

FIB-GAE Repair

Stopped FIB and the XeF_2 flow

Residual XeF_2 reacts with TaBN and causes side-etching damage.

Concept of new process

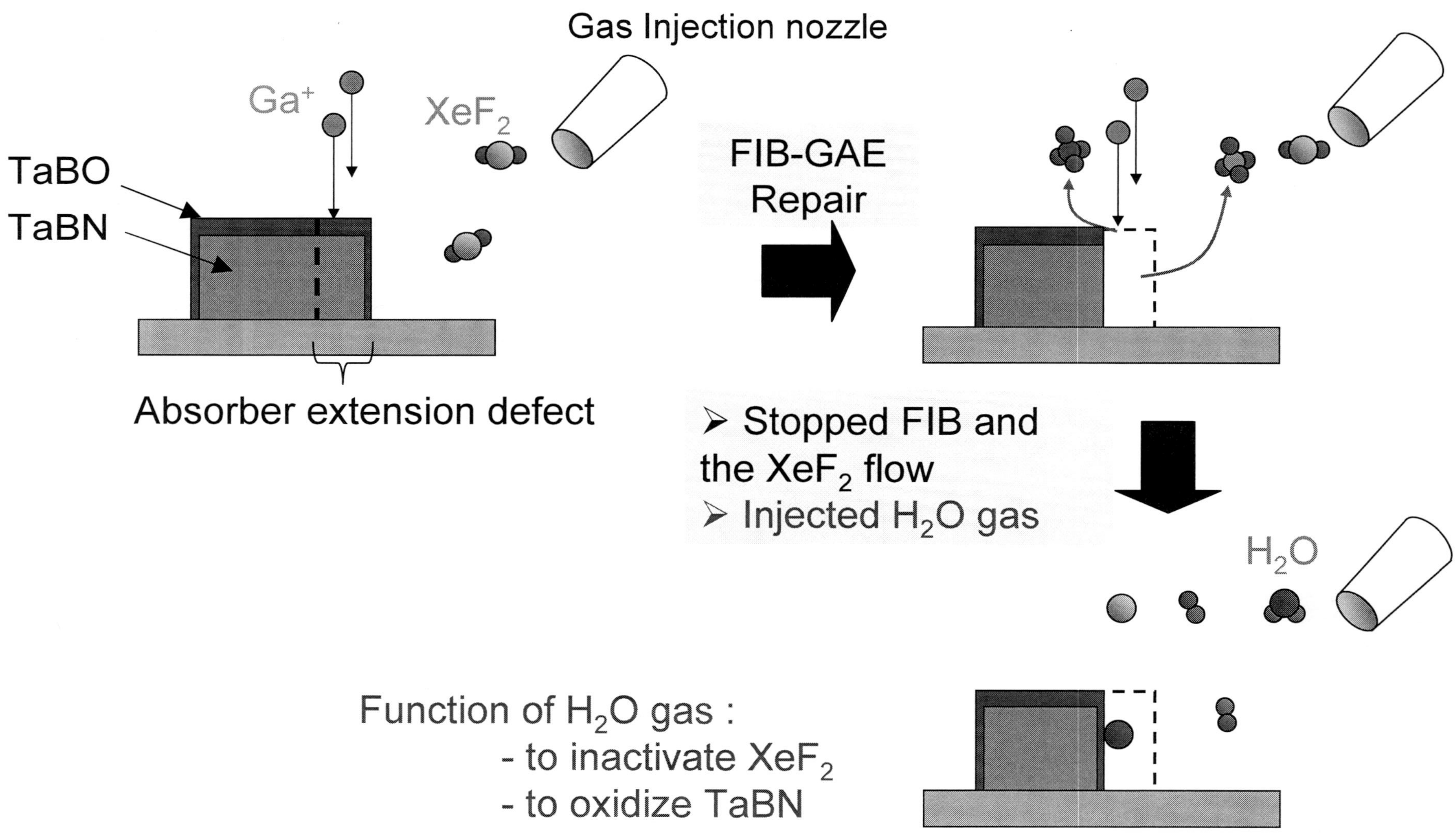

Gas Injection nozzle

FIB-GAE Repair

XeF$_2$

Ga$^+$

Absorber extension defect

TaBO
TaBN

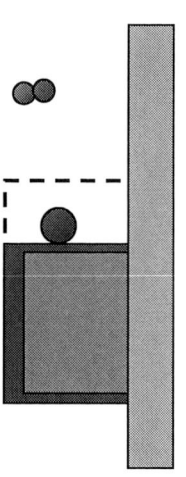

H$_2$O

- Stopped FIB and the XeF$_2$ flow
- Injected H$_2$O gas

Function of H$_2$O gas :
- to inactivate XeF$_2$
- to oxidize TaBN

MIRAI

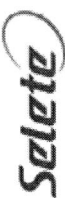

SEM images of repaired region

Selete

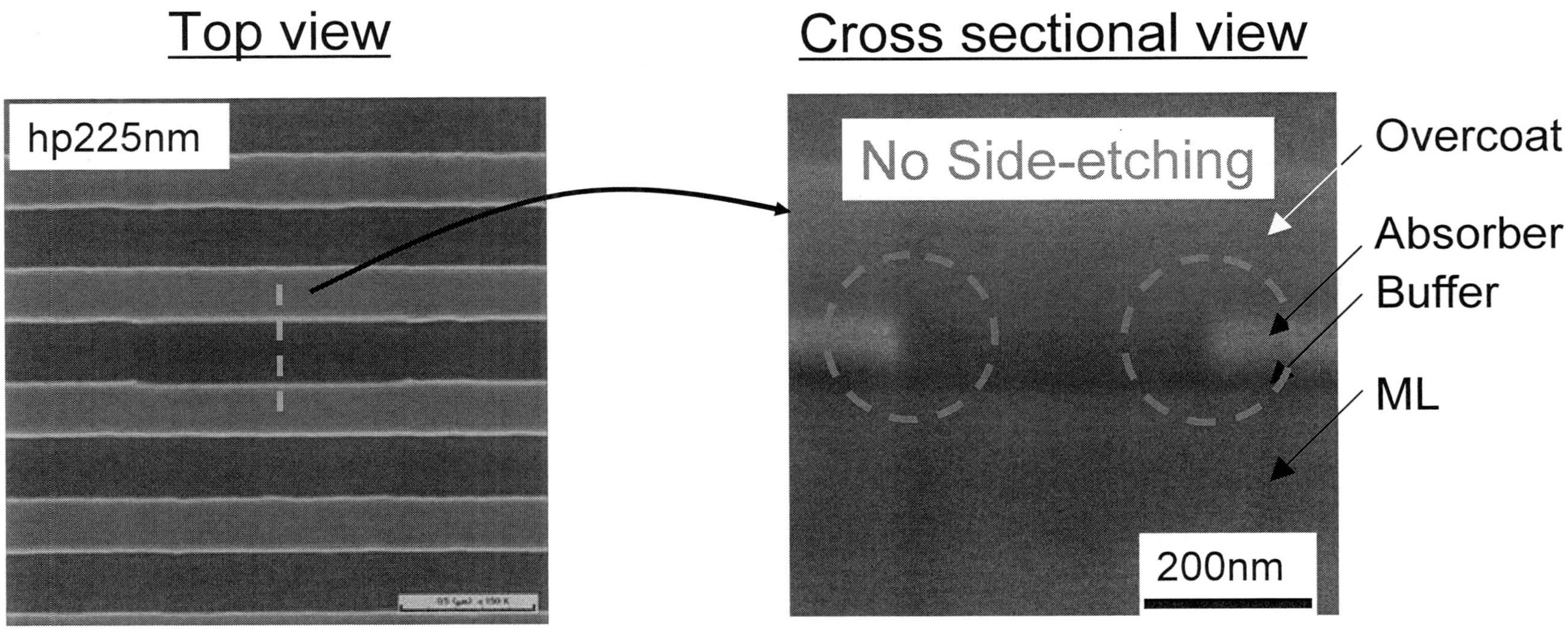
Top view
Cross sectional view
hp225nm
No Side-etching
Overcoat
Absorber
Buffer
ML
200nm

✓ Newly developed repair process could make pattern side-wall vertical shape.

Printing result for defect repaired region *Selete*

Mask image@5x

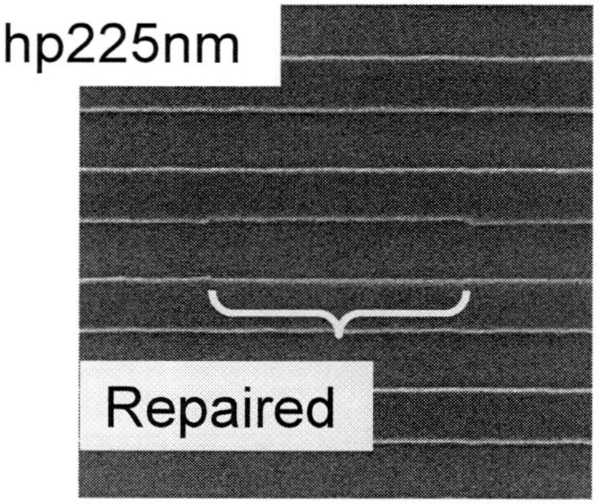

hp225nm

Repaired

Printed image@1x

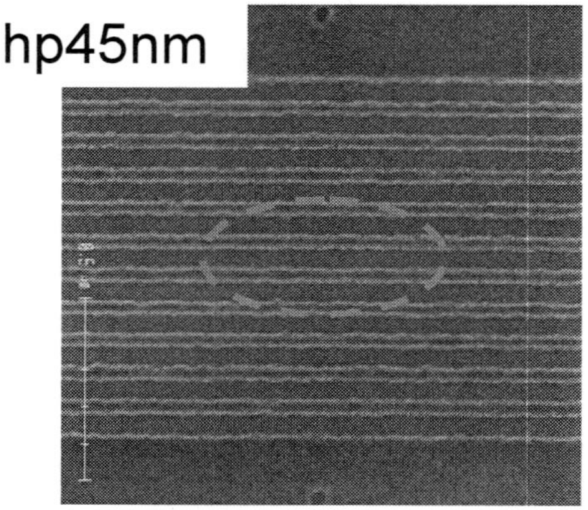

hp45nm

Focus:
-60nm@wafer   -30nm   0nm   +30nm   +60nm

Repaired

200nm

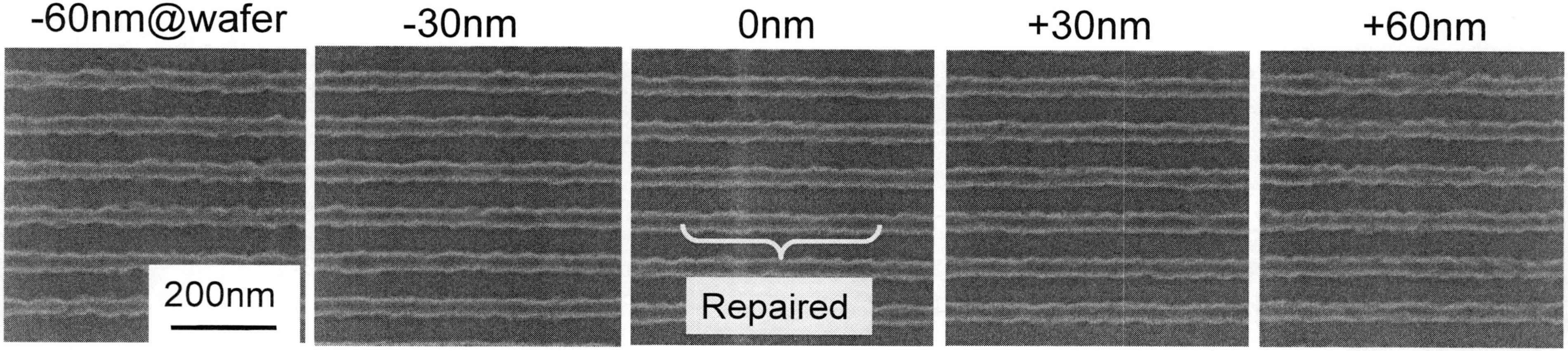

√ The bridge defect was completely repaired.
√ The defocus characteristics of defect repaired region were excellent.

FIB scan damage evaluation

Mask image@5x

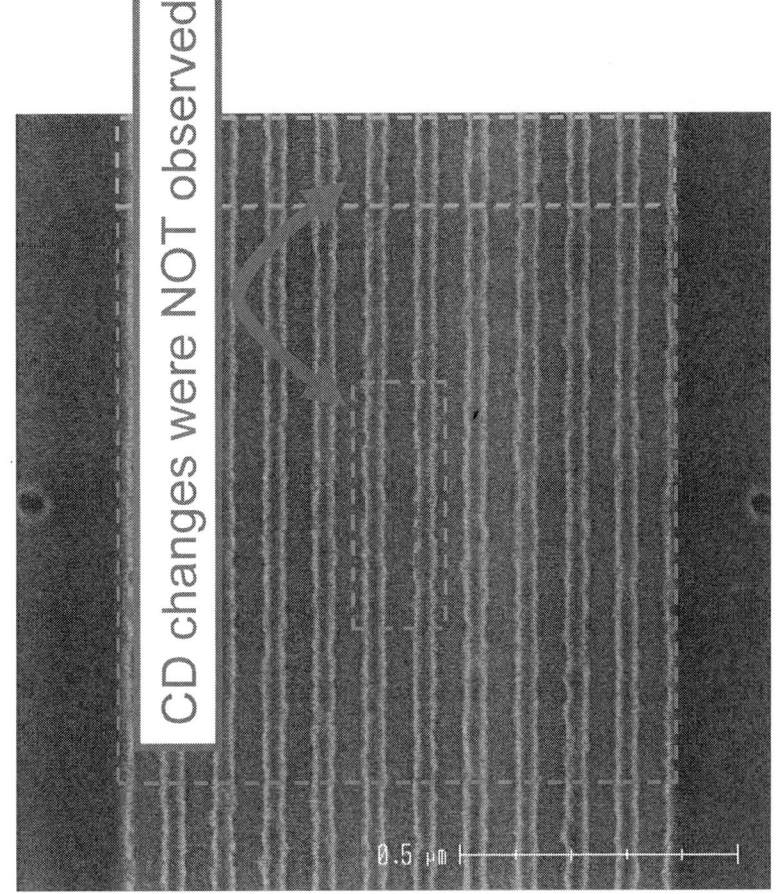

1. Rough whole scan

2. Fine partial scan

3. Repair scan

No FIB scan

Printed image@1x

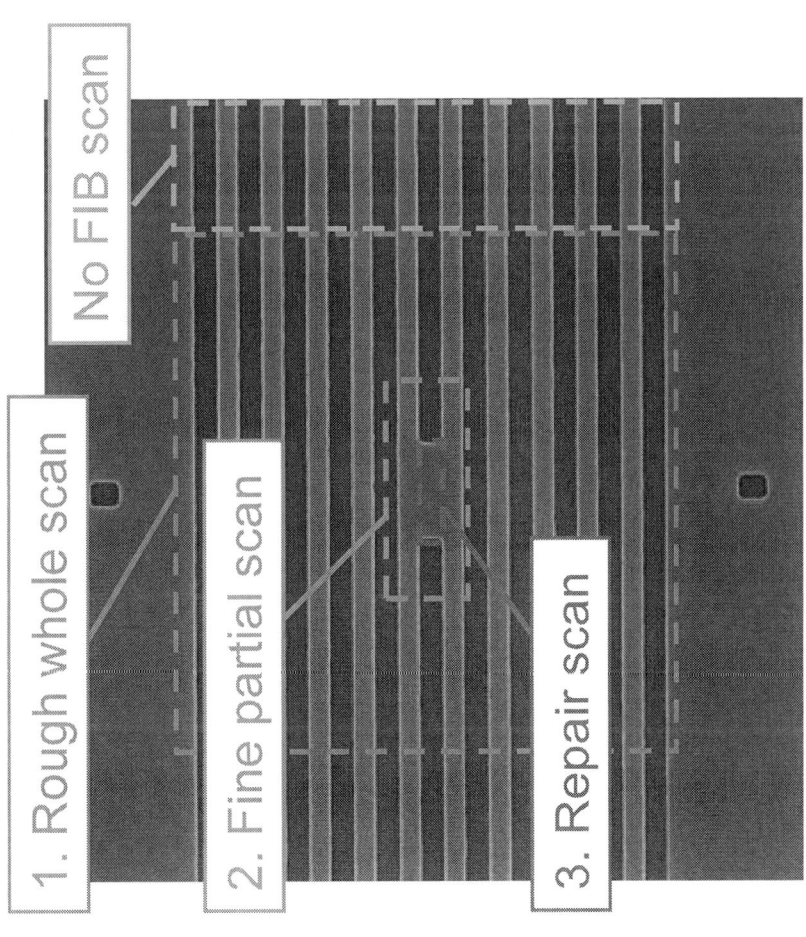

CD changes were NOT observed

0.5 μm

√ The SFET printed image verified that there was no scan damage in repaired region.

Repair performance of bridge defects

IRAI · **Selete**

Placement accuracy

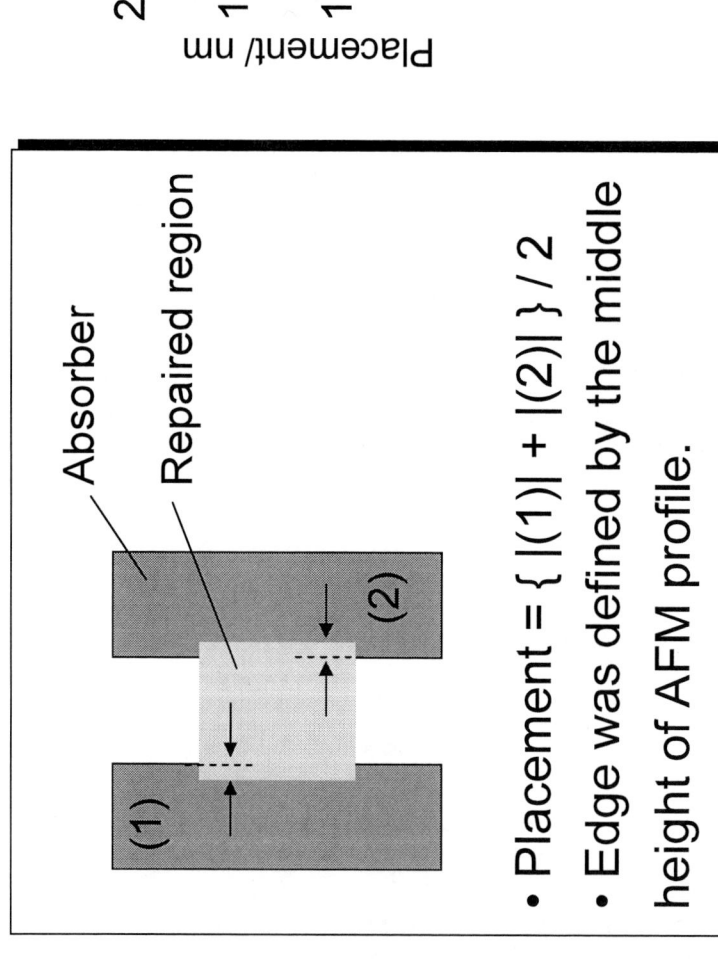

Range= 10.7 nm
3 sigma= 8.7 nm

Placement/ nm (y-axis: 0, 5, 10, 15, 20)
Number of repair/ - (x-axis: 1, 2, 3, 4, 5)

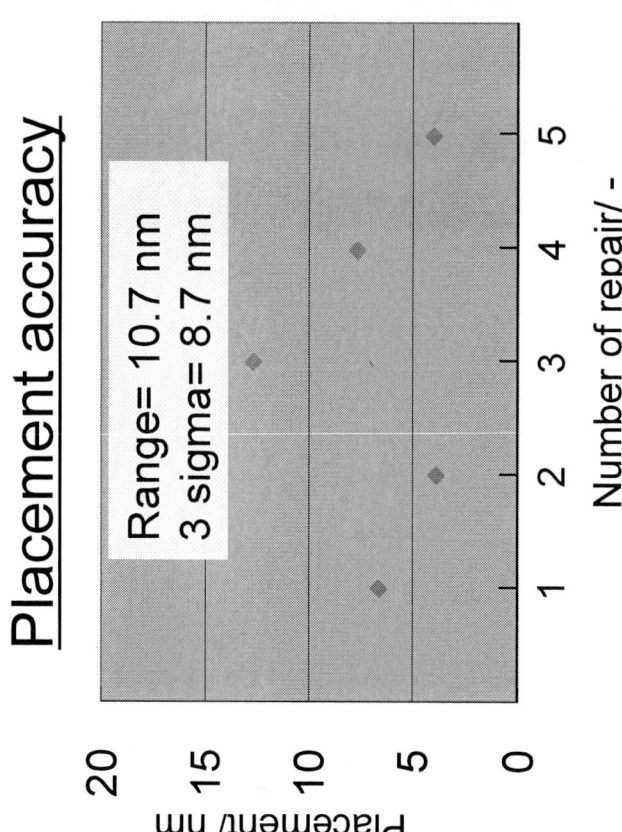

Absorber

Repaired region

(1)

(2)

- Placement = { |(1)| + |(2)| } / 2
- Edge was defined by the middle height of AFM profile.

✓ The placement accuracy was good for prototype FIB system.
✓ Analysis of variability factor and improvement of repair accuracy are future work.

 IRAI

Repair performance of bridge defects Selete

3D profile of repaired region

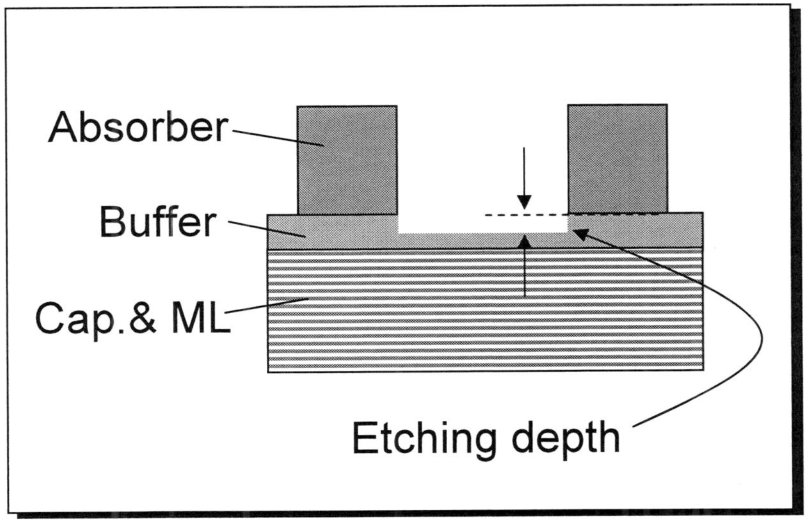

- Absorber
- Buffer
- Cap.& ML
- Etching depth

Depth controllability

Range= 1.1 nm
3 sigma= 1.3 nm

Depth/ nm (0, 1, 2, 3, 4, 5)

Number of repair / - (1, 2, 3, 4, 5)

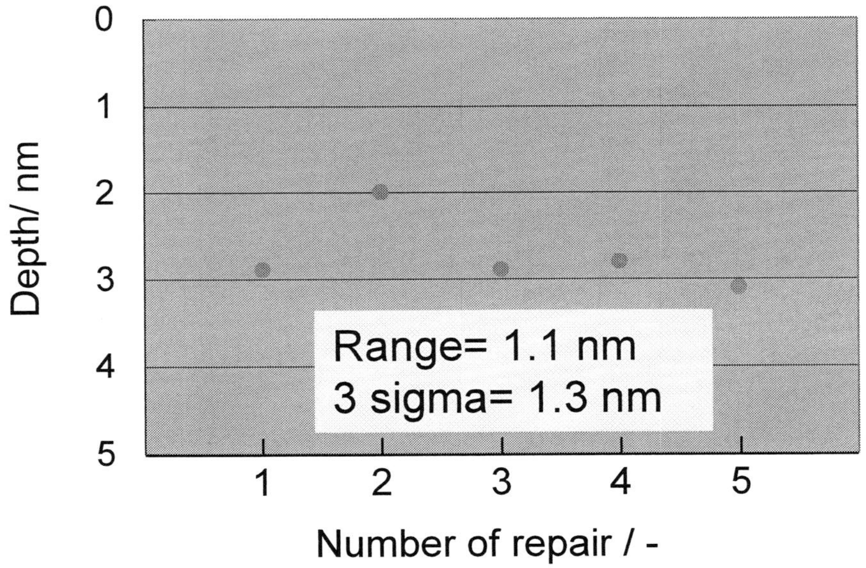

✓ An excellent depth controllability was confirmed.
✓ The etching selectivity of absorber layer against buffer layer was insured to be more than 10:1.

Application of FIB-GAE to narrow pitched pattern *Selete*

Base pattern: hp100nm L/S
Defect type: 2 bridge defects

Before repair

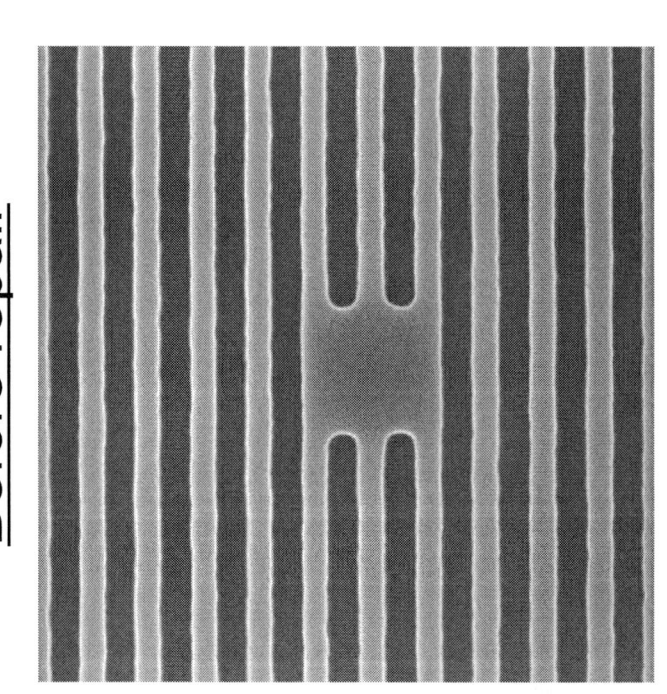

After repair

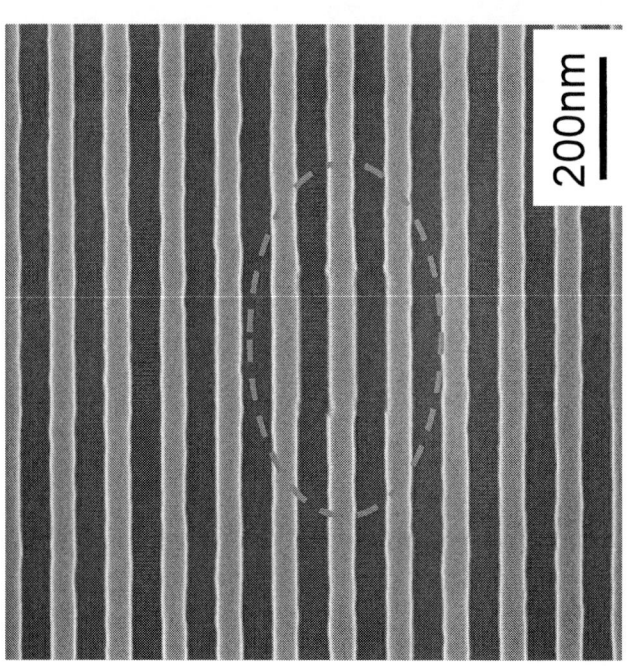

200nm

√ The defects were repaired without side-etching.
√ The FIB-GAE technique respond to repair large defects in narrow pitched mask pattern.

Summary

- The newly developed defect repair FIB process could make pattern side-wall in vertical shape.

- The SFET printed images verified that the bridge defect was completely repaired and no FIB scan damage was observed.

- The placement accuracy and etching depth controllability were excellent.

- We confirmed that FIB-GAE technique has enough potential for defect repair in hp100 nm L/S pattern.

Acknowledgement

We would like to thank to
- Tsukasa Abe of Dai Nippon Printing Co., Ltd for mask fabrication support.
- Tomokazu Kozakai of SII NanoTechnology Inc. for technical support.

This work was supported by New Energy and Industrial Technology Development Organization (NEDO).

2008 International Symposium on Extreme Ultraviolet Lithography

September 28 - October 1, 2008   Lake Tahoe, California

Analysis of entrapped object size effects on Out-of-Plane Distortion of the EUVL mask in electrostatic chucking

[1]S. Lee*, [1]T. Yamamoto, [2]K. Ota, [2]N. Nishimura, [2]T. Taguchi, [2]I. Nishiyama, [2]O. Suga, [1]S. Warisawa, [1]S. Ishihara

[1]The University of Tokyo, Tokyo, Japan

[2]Selete, Tsukuba, Japan

EUVL Symposium 2008

THE UNIVERSITY OF TOKYO

Introduction

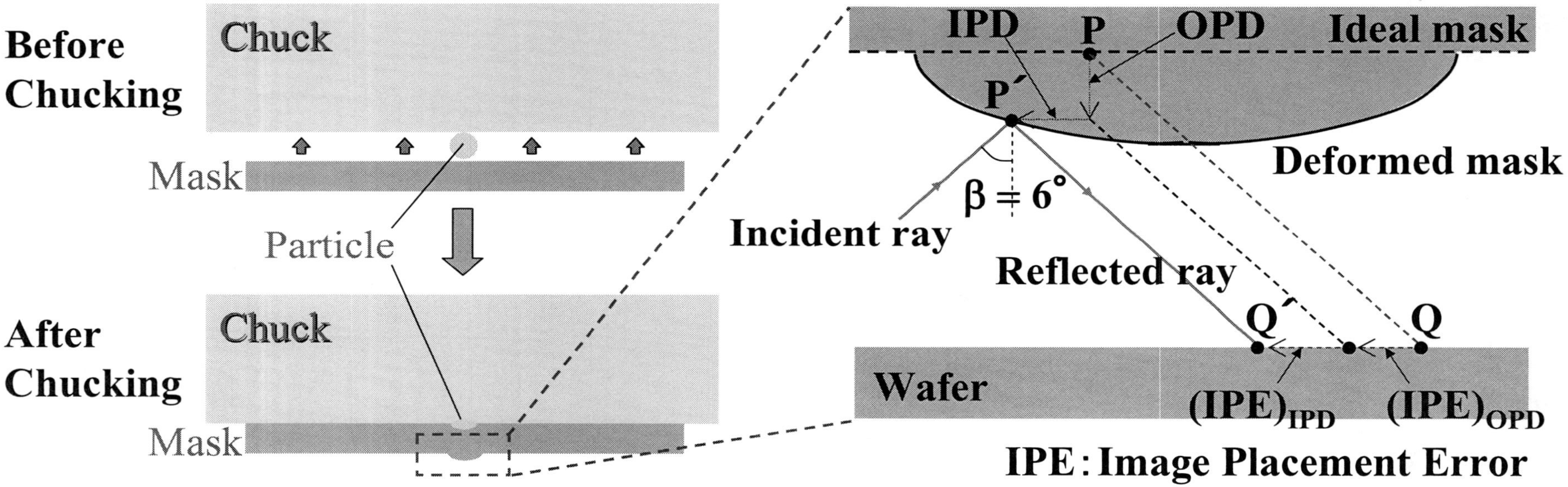

OPD: Out-of-Plane Distortion < 20 nm
IPD: In-Plane Distortion < 2 nm

- This paper investigates an entrapped particle size effect on the OPD by the deformation analysis of the mask, and entrapped particle in EUVL chucking.

- The analysis utilizes finite element (FE) models.

IPE: Image Placement Error

Proc. of SPIE Vol. 6349 634938-11

THE UNIVERSITY OF TOKYO

Approach of Modeling

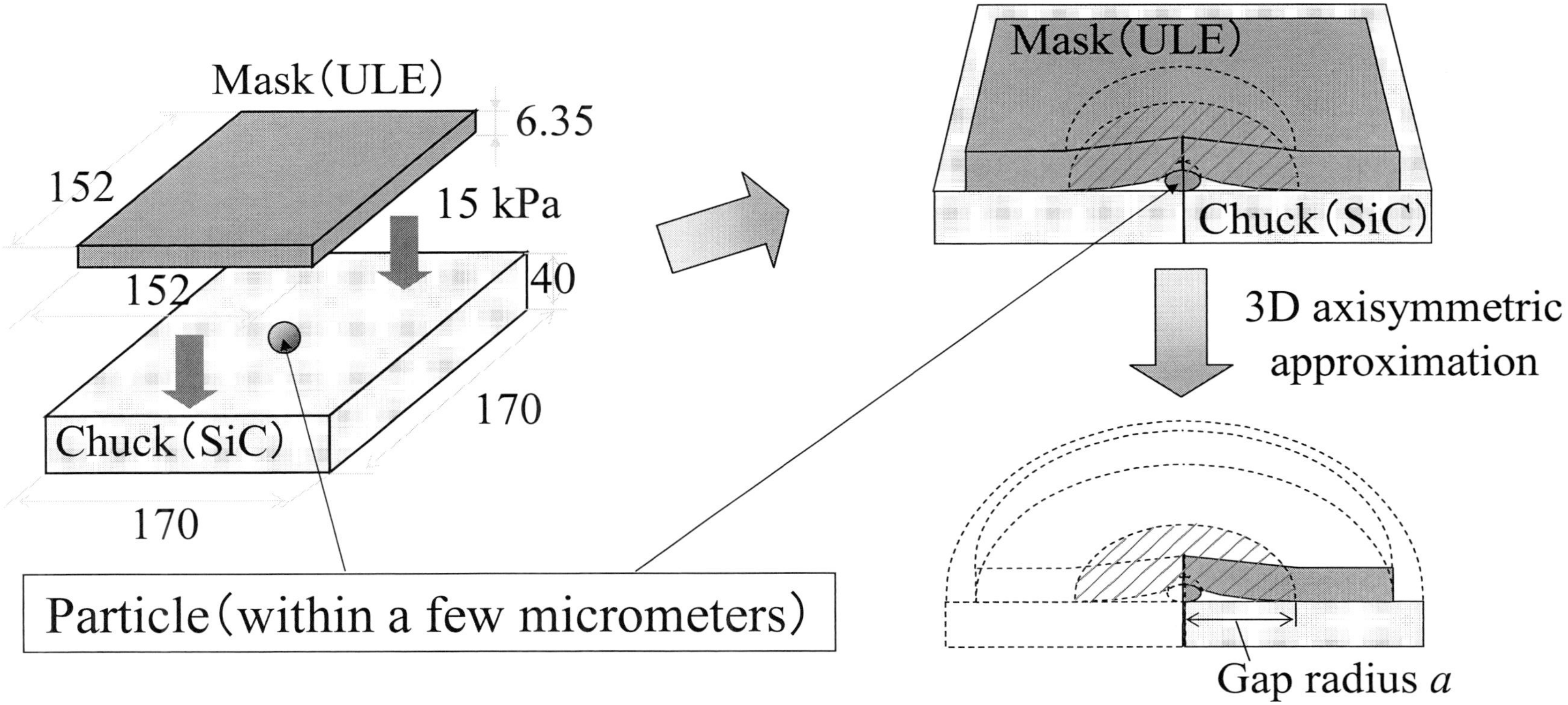

Mask (ULE)

6.35

15 kPa

152

152

40

170

170

Chuck (SiC)

Particle (within a few micrometers)

Mask (ULE)

Chuck (SiC)

3D axisymmetric approximation

Gap radius a

- The area outside the gap radius a induced by the entrapped particle which has an influence on OPD is not considered.

- The analysis utilizes 3D axisymmetric finite element (FE) models.

THE UNIVERSITY OF TOKYO

Simulation Model

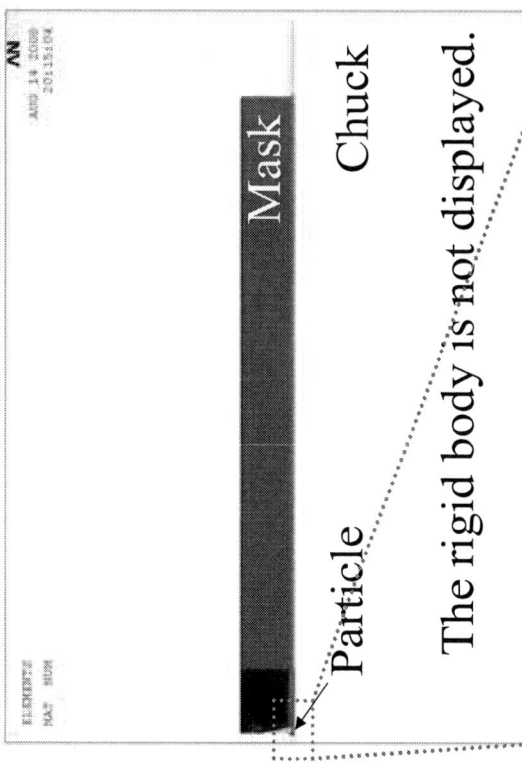

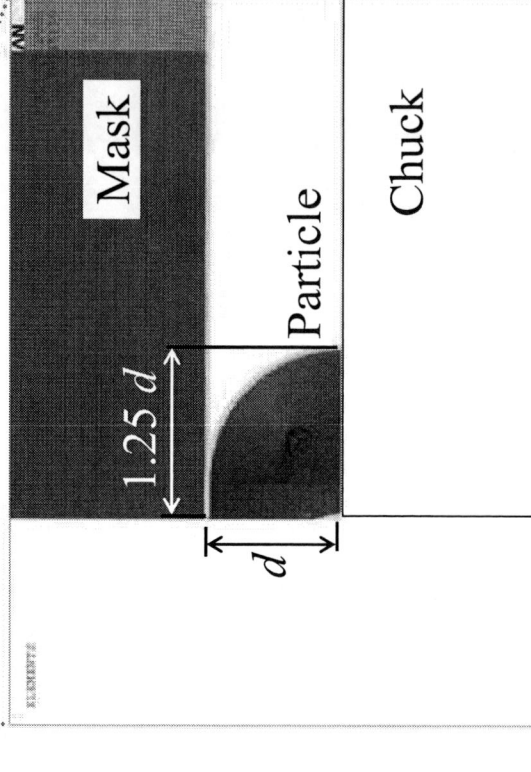

The rigid body is not displayed.

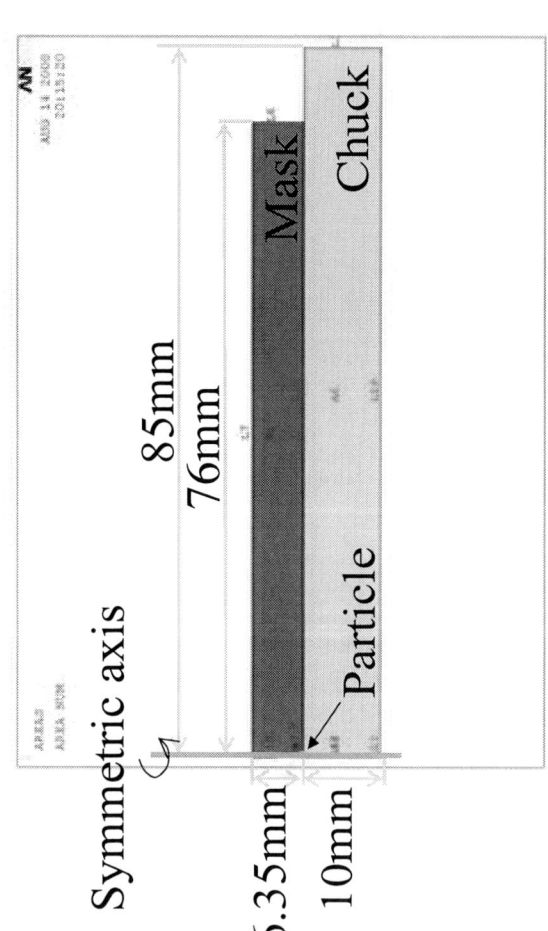

- Mask / Particle : ULE®
  (Elastic body, E [GPa] : 73, ν : 0.17)

- Chuck : Rigid body

- Clamping pressure [kPa] : 15

- Particle size d [μm] : 2, 3, 4, 5, 8, 10, 20

Relation between OPD and Particle Size

Deformation of vertical direction

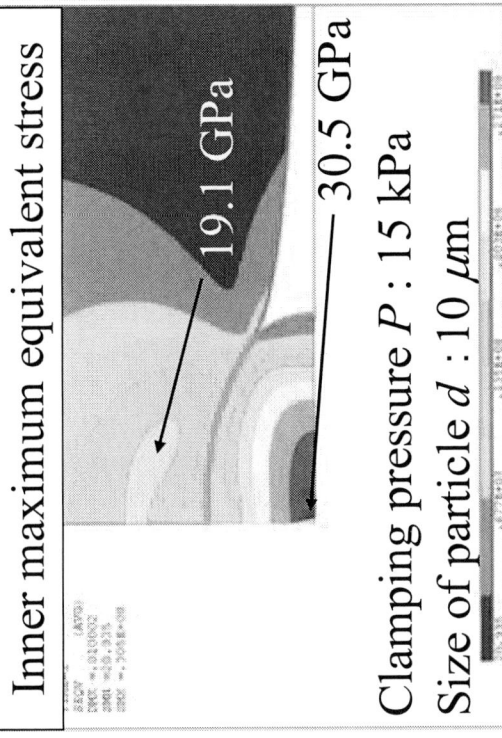

OPD : 37.5 nm

Clamping pressure P : 15 kPa
Size of particle d : 10 μm

Inner maximum equivalent stress

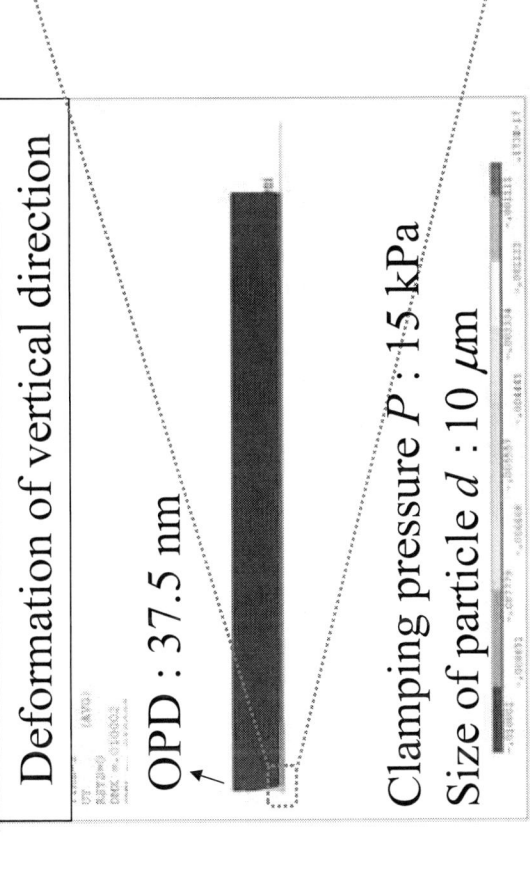

19.1 GPa

30.5 GPa

Clamping pressure P : 15 kPa
Size of particle d : 10 μm

- When the size of the particle decreases half, OPD approximately decreases to one tenth.

- If the size of the particle is equal to or less than 8 μm, OPD becomes lower than the permissible limit (20 nm).

OPD on the surface of the mask above particle

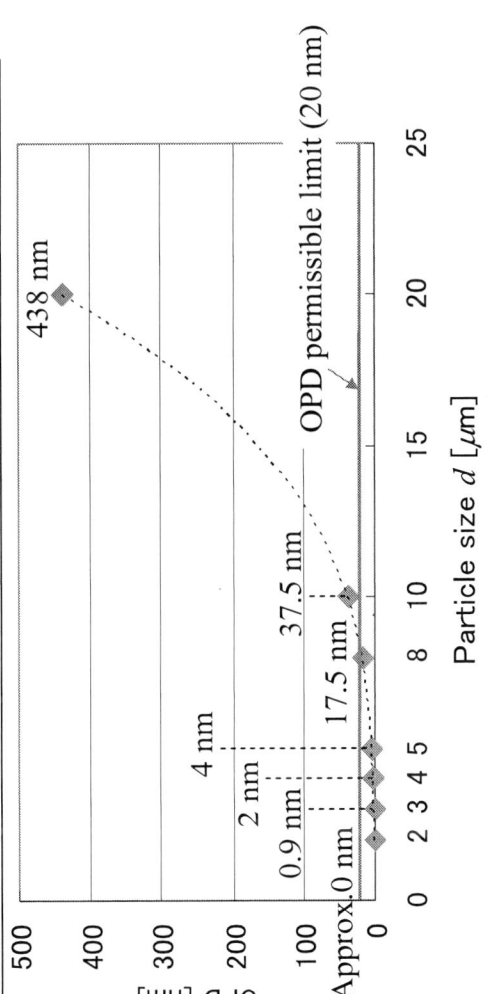

THE UNIVERSITY OF TOKYO

Relation between Gap Radius and Particle Size

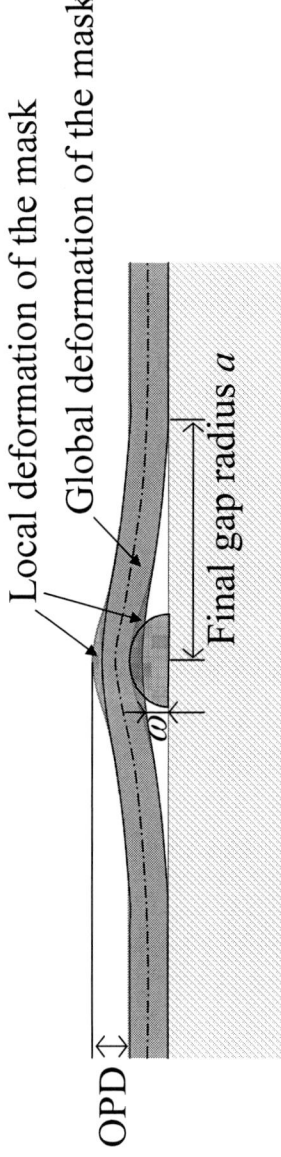

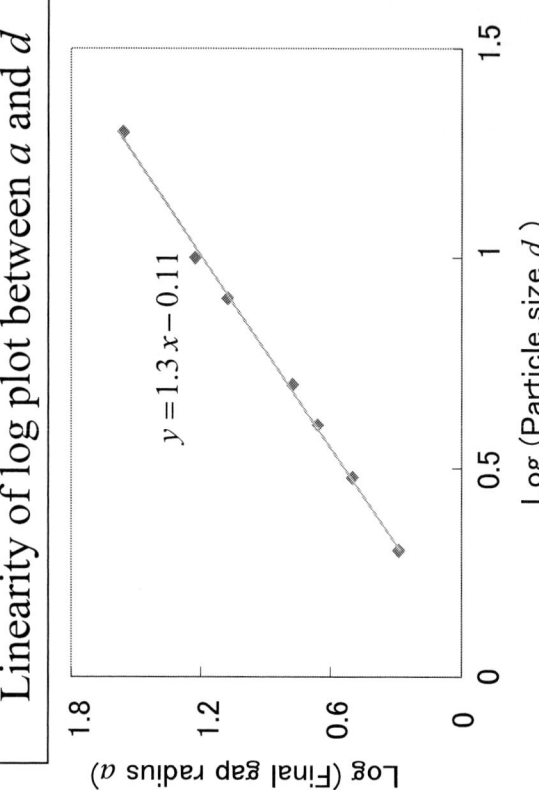

Linearity of log plot between a and d

$y = 1.3x - 0.11$

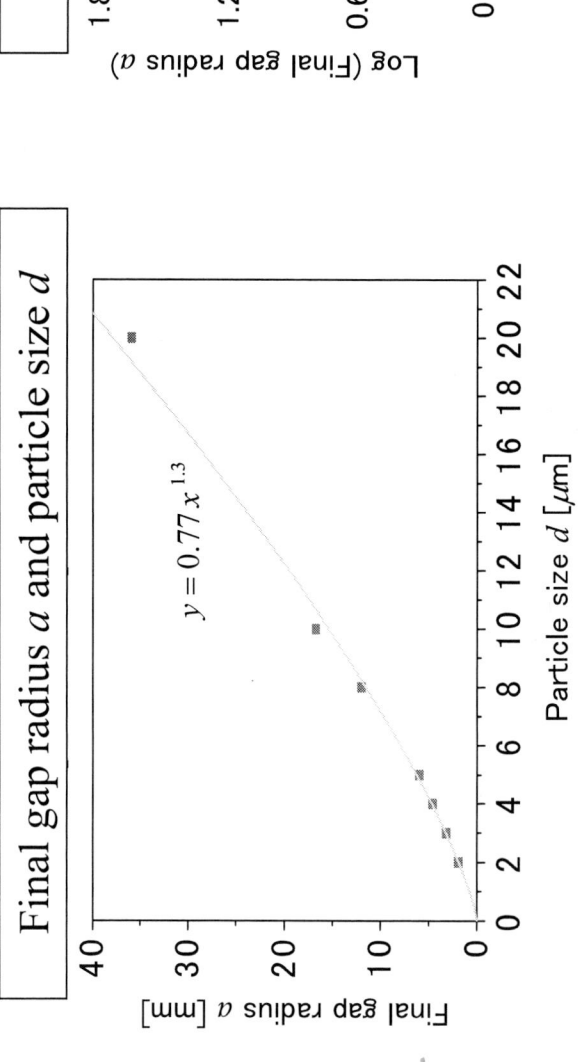

Final gap radius a and particle size d

$y = 0.77x^{1.3}$

- The relation between the final gap radius and the particle size is represented by the equation, $a = 0.77\,d^{1.3}$.

 THE UNIVERSITY OF TOKYO

Discussion on the Relation between Gap Radius and Particle size

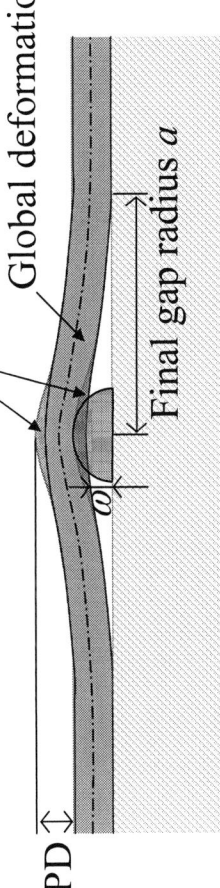

Local deformation of the mask

Global deformation of the mask

Final gap radius a

OPD

ω

Contact area of the particle

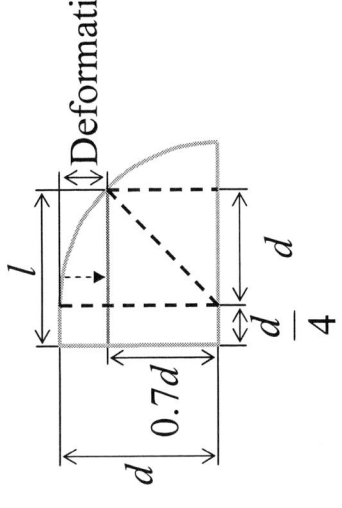

Deformation of the particle

Deformation of the particle

l

d

d

$\dfrac{d}{4}$

$0.7d$

d

$$A \approx \pi l^2 = \pi (0.71)^2 d^2 \propto d^2$$

$$\sigma = E\varepsilon = \text{Const.}$$

$$F \approx P\pi a^2 = \sigma A \propto d^2$$

$$\therefore a \propto d \approx d^{1.3}$$

Strain of particles

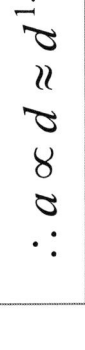

0.3

Strain of particles ε

Particle size d [μm]

σ = Stress applying on a particle

E = Young's modulus (73 GPa)

A = Contact area of particle

l = Radius of the contact area of the particle

P = Clamping pressure (15 kPa)

F = Force applying on a particle

THE UNIVERSITY OF TOKYO

Relation between OPD and Particle Size

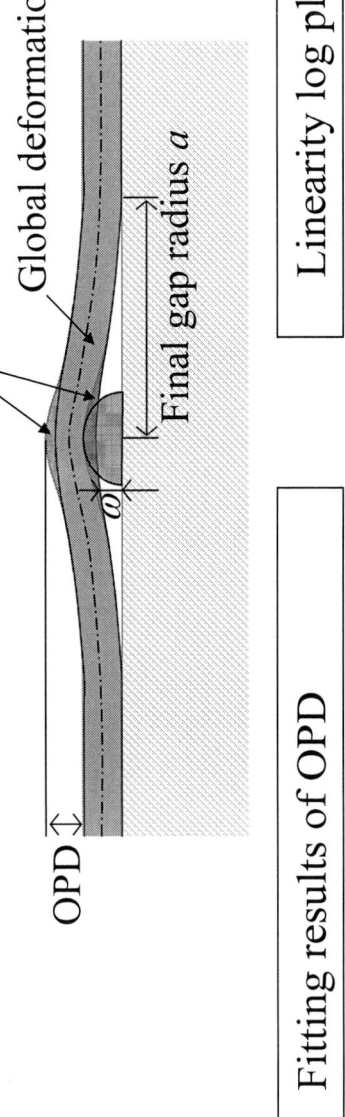

Local deformation of the mask

Global deformation of the mask

Final gap radius a

OPD

ω

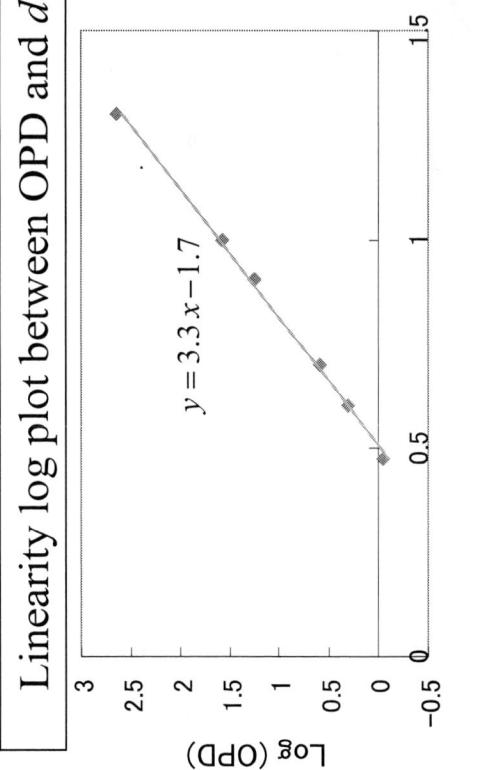

Linearity log plot between OPD and d

$y = 3.3\,x - 1.7$

Log (OPD)

Log (Particle size d)

Fitting results of OPD

$y = 0.022\ x^{3.3}$

OPD [nm]

Particle size d [μm]

- The relation between the OPD and the particle size is represented by the equation, **OPD = 0.022 $d^{3.3}$** .

- It is necessary to discuss this relation from the mechanical dynamics view point.

THE UNIVERSITY OF TOKYO

Discussion on the Relation between OPD and Particle Size

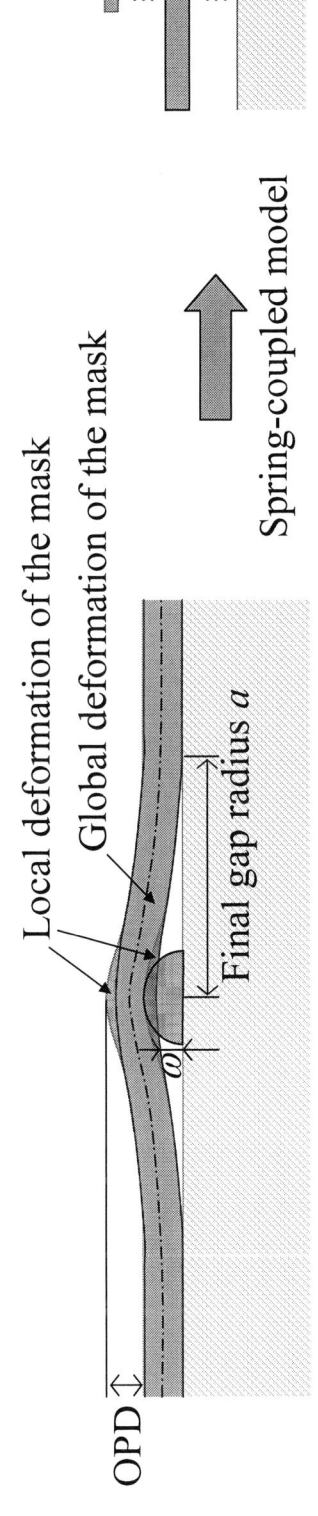

OPD

Local deformation of the mask

Global deformation of the mask

Final gap radius a

ω

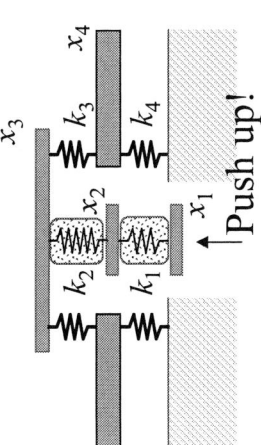

Spring-coupled model

When a particle is pushed up by particle size d from the bottom, the OPD is calculated as a following equation.

$x_1 = d$ (Particle size)

$x_1 - x_2$ = Particle deformation

$x_2 - x_3$ = Local deformation of the mask

x_3 = OPD

$x_4 = \omega$ (Deflection of the mask)

X_i = Initial length

σ = Stress applying on the particle

E = Young's modulus (73 GPa)

A = Contact area of the particle

k_i = Spring constant

$$x_3 = \dfrac{\frac{1}{k_3} + \frac{1}{k_4}}{\frac{1}{k_1} + \frac{1}{k_2} + \frac{1}{k_3} + \frac{1}{k_4}} x_1$$

$$\sigma = E\varepsilon \quad F = kx = \sigma A = E\varepsilon A = EA x/X \quad \Rightarrow \quad k = E\,{A}/{X}$$

If A is extremely small ($k_1, k_2 << k_3, k_4$) ;

$$x_3 = \dfrac{\left(\frac{1}{k_3} + \frac{1}{k_4}\right)}{\frac{X_1}{E} + \frac{X_2}{E}} A x_1 \propto x_1^3 \quad (\because A \propto x_1^2)$$

$$\therefore \text{OPD} \propto d^3 \approx d^{3.3}$$

 THE UNIVERSITY OF TOKYO

Summary and Conclusions

- This paper analyzed an entrapped particle size effect on the deformation of the EUVL mask by using FE models.

- After chucking, it is found that the stress inside the particle and strain of the particle are constant.

- The final gap radius a is approximately represented by a function of the particle size as $a = 0.77\, d^{1.3}$.

- If the particle size is equal to or less than 8 μm, the OPD becomes lower than the permissible limit (20 nm).

- The OPD is approximately represented by a function of the particle size as $OPD = 0.02\, d^{3.27}$.

- These results are discussed from the mechanical dynamics view point.

THE UNIVERSITY OF TOKYO

A study of optical inspection on EUVL mask for 32 nm half pitch node device and beyond

Yukiyasu Arisawa, Hiroyuki Shigemura, Tsuyoshi Amano, Hajime Aoyama, Toshihiko Tanaka, Osamu Suga

MIRAI-Semiconductor Leading Edge Technologies, Inc. (Japan)

Semiconductor Leading Edge Technologies, Inc.

2008 International Symposium on EUVL

MIRAI

Contents

Selete

- Introduction
- Simulation conditions
  - Mask structures
  - Parameters
- Result and discussion
  - Defect detectivity
- Summary

Introduction

We are developing pattern inspection technologies for EUVL masks using DUV light (199nm). In order to verify the potential of DUV inspection, we simulated inspection images captured by magnifying optics for various parameters.

In this work, we focus on investigating the relation between image contrast of base pattern and defect detectivity.

Item	Parameter	
Absorber stuck	Material	
	Thickness	
Optics	Degree of polarization	
	Illumination type	
	NA	
	Sigma	
Base pattern	Pitch	
	Type (L/S, CH ···)	
	Size	
Defect	Type (opaque extension, clear extension)	
	Size	

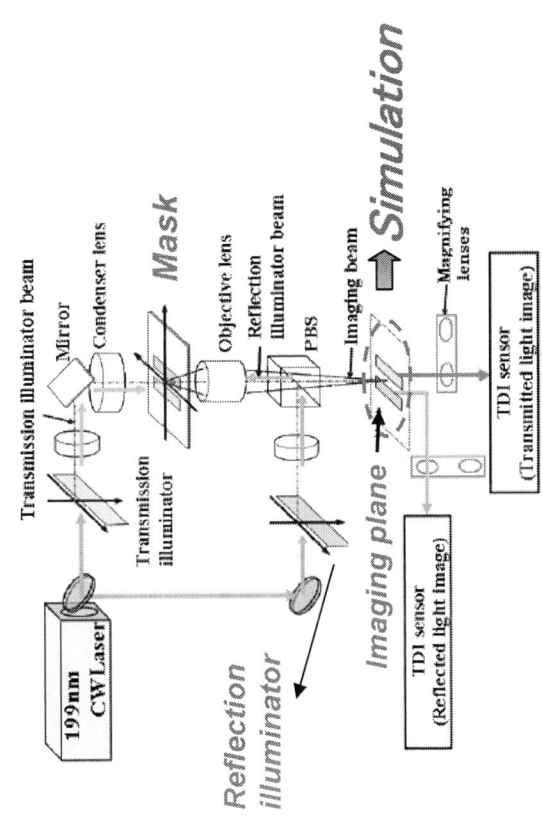

Schematic view of 199 nm inspection optics.

Mask structures for simulation

- We prepared two types of masks.
  - R5-Mask: Reflectivity of absorber stack ≈ 5%@199nm
  - R19-Mask: Reflectivity of absorber stack ≈ 19%@199nm

- Absorber stack materials were chosen so that the contrast of R5-Mask is higher than that of R19-Mask.

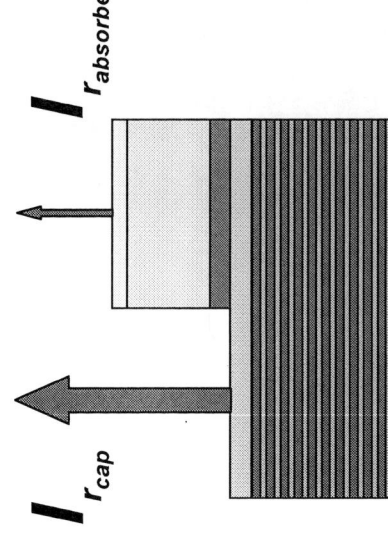

$$Contrast = \frac{I_{r_{cap}} - I_{r_{absorber}}}{I_{r_{cap}} + I_{r_{absorber}}}$$

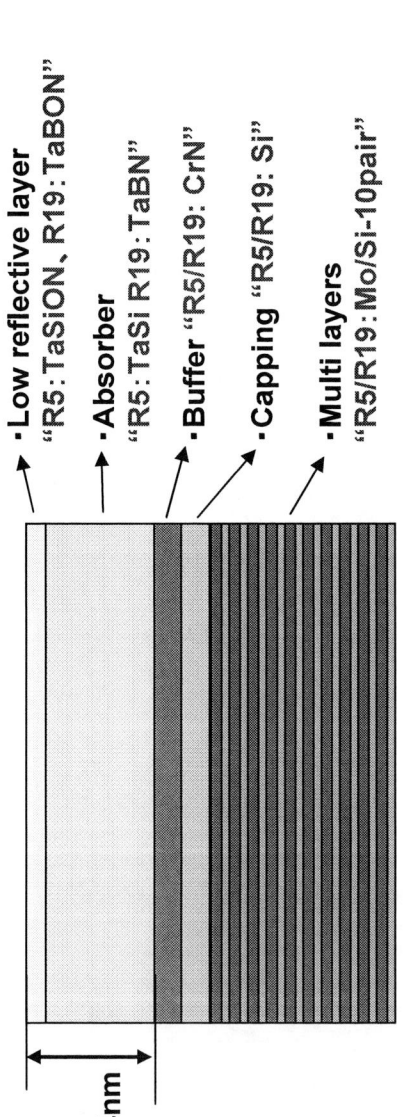

- Low reflective layer
  "R5: TaSiON, R19: TaBON"
- Absorber
  "R5: TaSi R19: TaBN"
- Buffer "R5/R19: CrN"
- Capping "R5/R19: Si"
- Multi layers
  "R5/R19: Mo/Si-10pair"

44nm

Simulation conditions

■ Simulator & optics conditions

• EM-Suite (Panoramic Technology Inc.)

• 3D rigorous calculation by Finite Difference Time Domain (FDTD) method

• λ =199 nm, NA=0.75, σ =1.0

■ Input to simulator

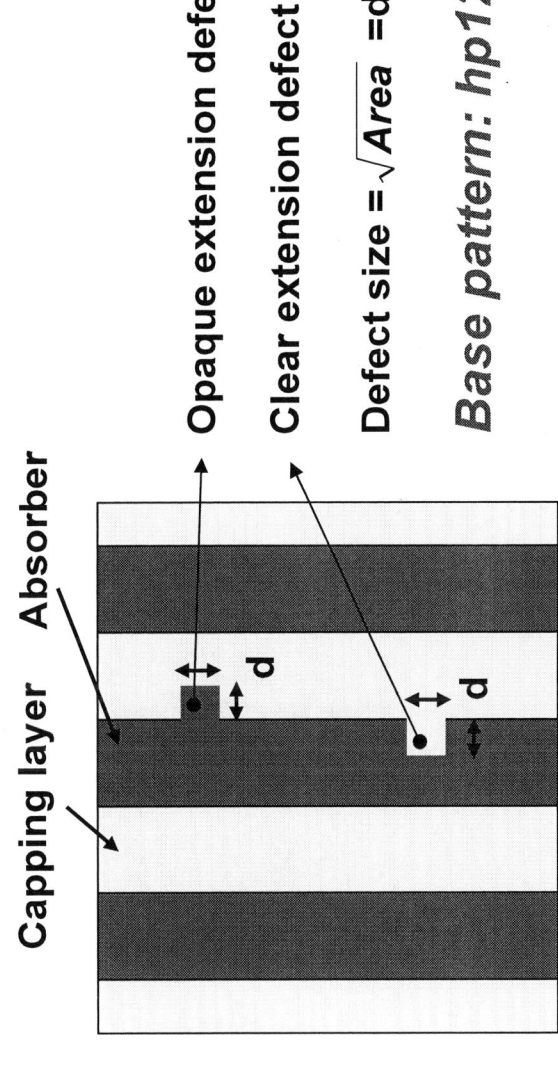

Capping layer   Absorber

Opaque extension defect (square)

Clear extension defect (square)

Defect size = $\sqrt{Area}$ =d

Base pattern: hp128 L/S (1:1)

Experimental result #1

128nm(on mask) L/S pattern with opaque extension defect

R19-Mask

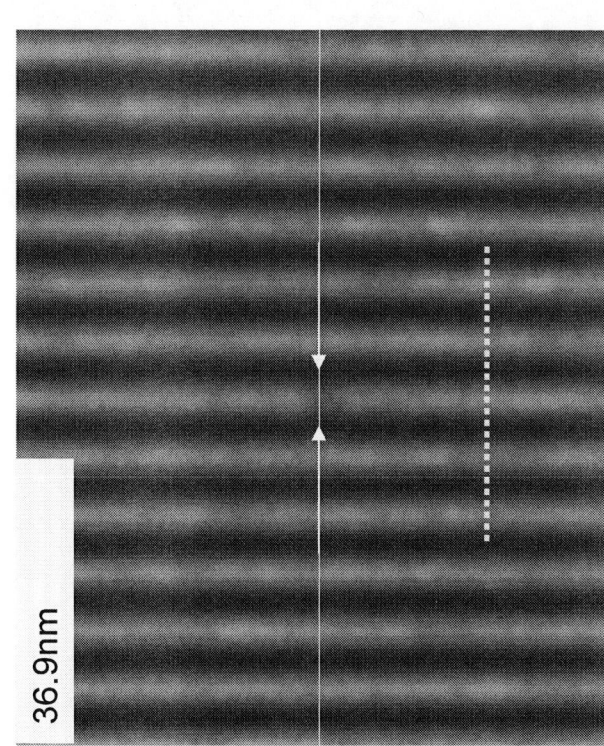

36.9nm

Inspection image

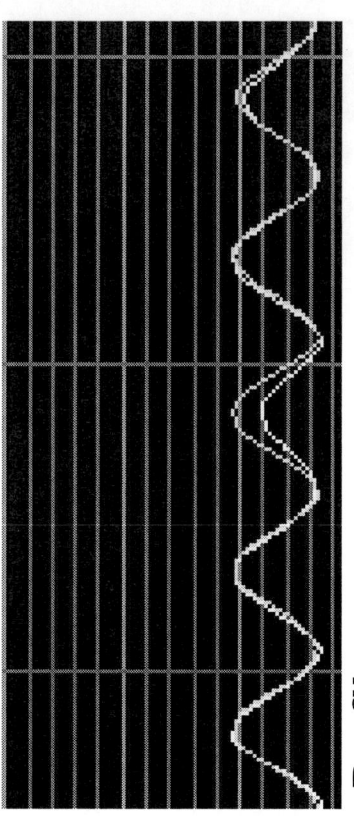

Profile

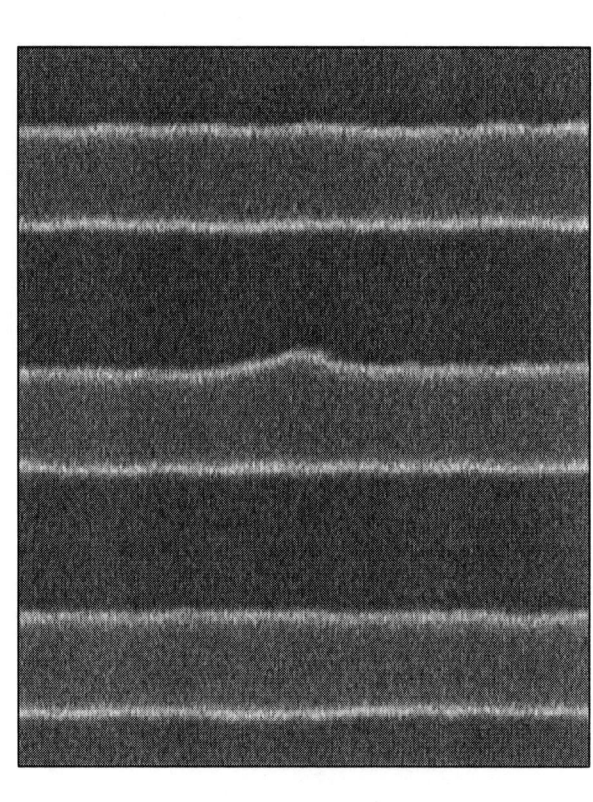

SEM image

Defect size: 36.9 nm
(Square root of defect area)

Experimental result #2

128nm(on mask) L/S pattern with clear extension defect

R19-Mask

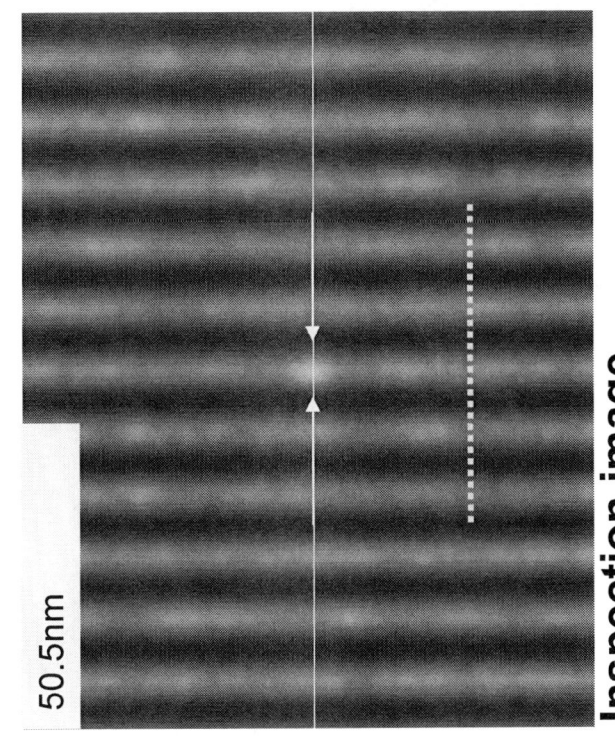

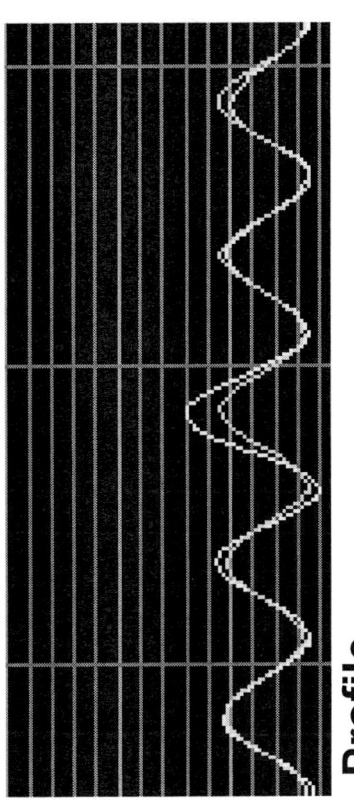

50.5nm

Inspection image

Profile

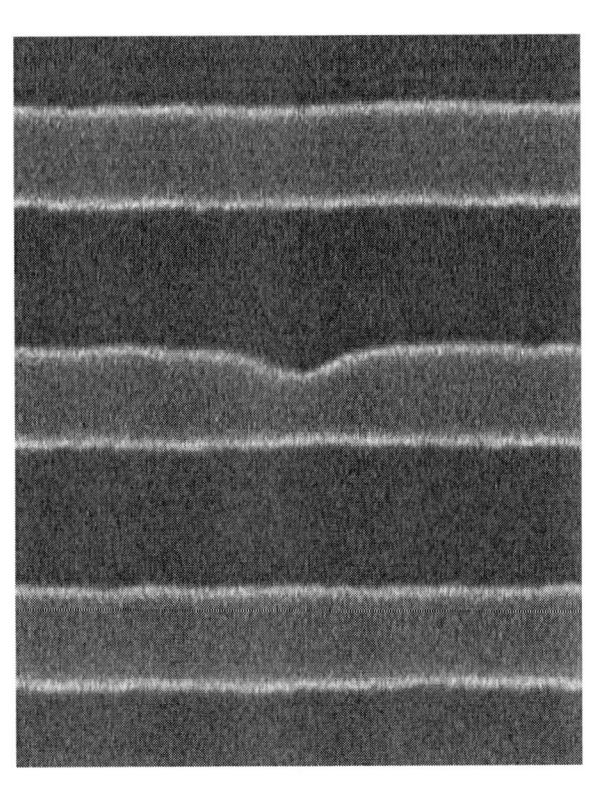

SEM image

Defect size: 50.5 nm
(Square root of defect area)

 IRAI Simulation result on a **square-shaped** clear extension defect Selete

R19-Mask

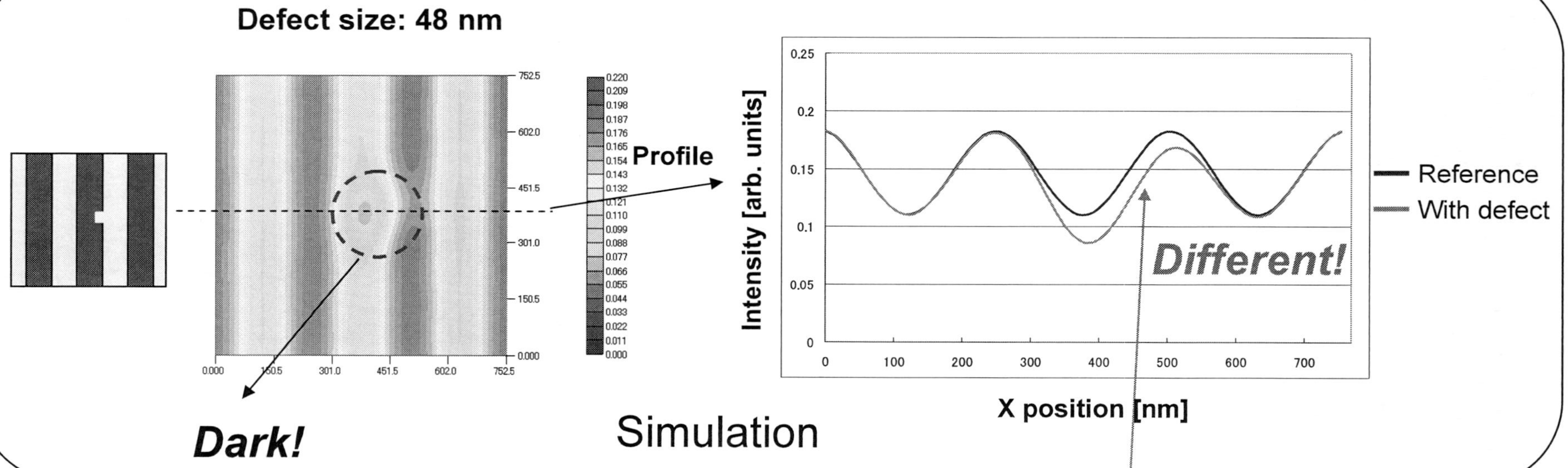

Defect size: 48 nm

Profile

Dark!

Simulation

Reference
With defect

Different!

X position [nm]

Intensity [arb. units]

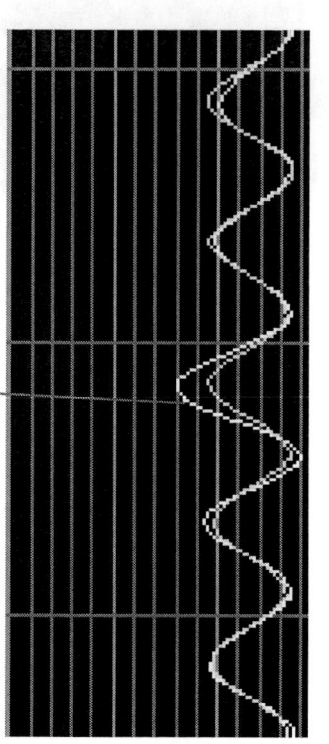

Experimental profile

✓ On the assumption that a defect is square in shape, there is the difference between the simulation and the experiment. The simulation result seems to be that of a opaque extension defect.

Simulation for a triangle-shaped defect

Triangle clear extension defect

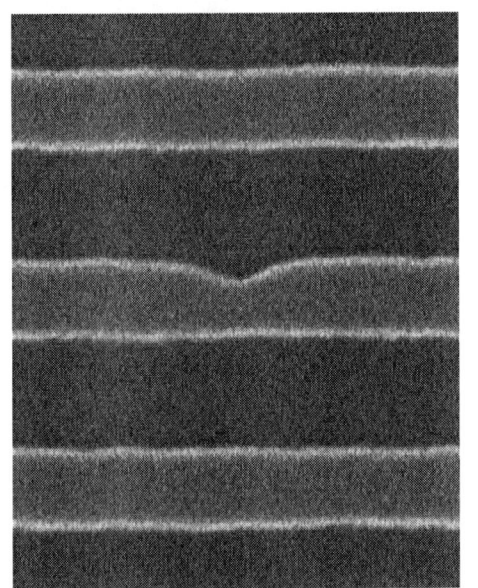

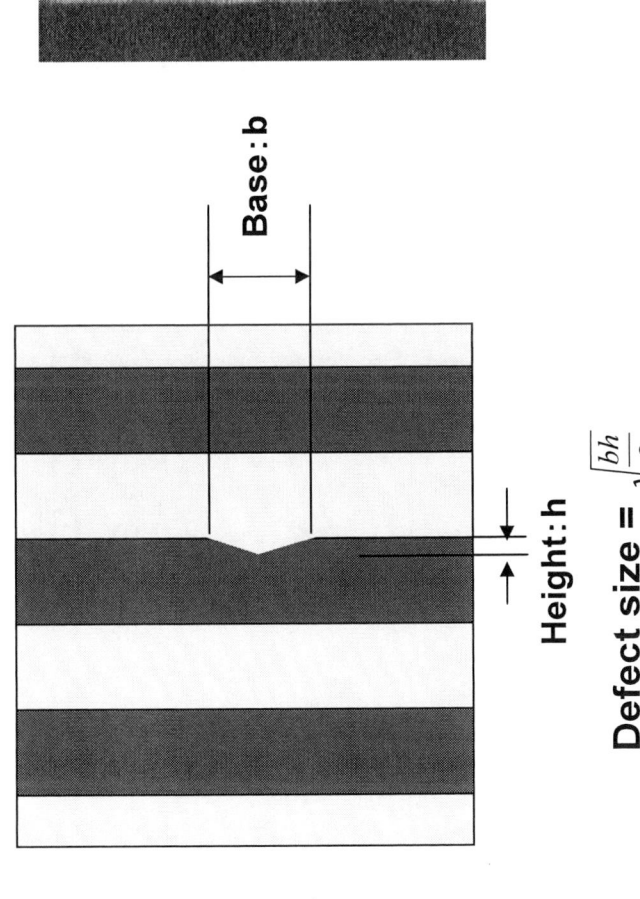

Base : b

Height : h

$$\text{Defect size} = \sqrt{\frac{bh}{2}}$$

√ Thinking that the difference between the simulation and the experiment originates in the input shape of defect, we simulated on triangle defects.

IRAI Simulation result on a *triangle-shaped* clear extension defect **Selete**

R19-Mask

Defect size: 48 nm

Bright!

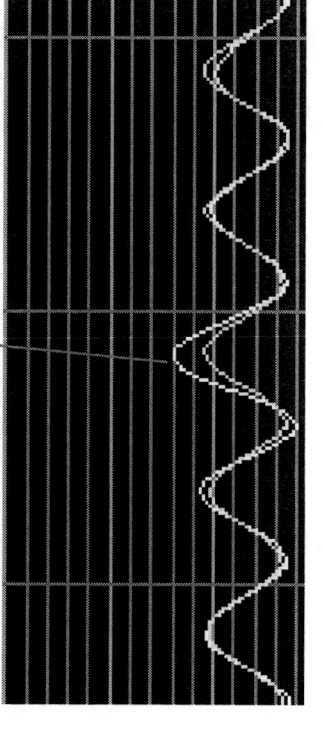

— Reference
— With defect

Intensity [arb. units]

X position [nm]

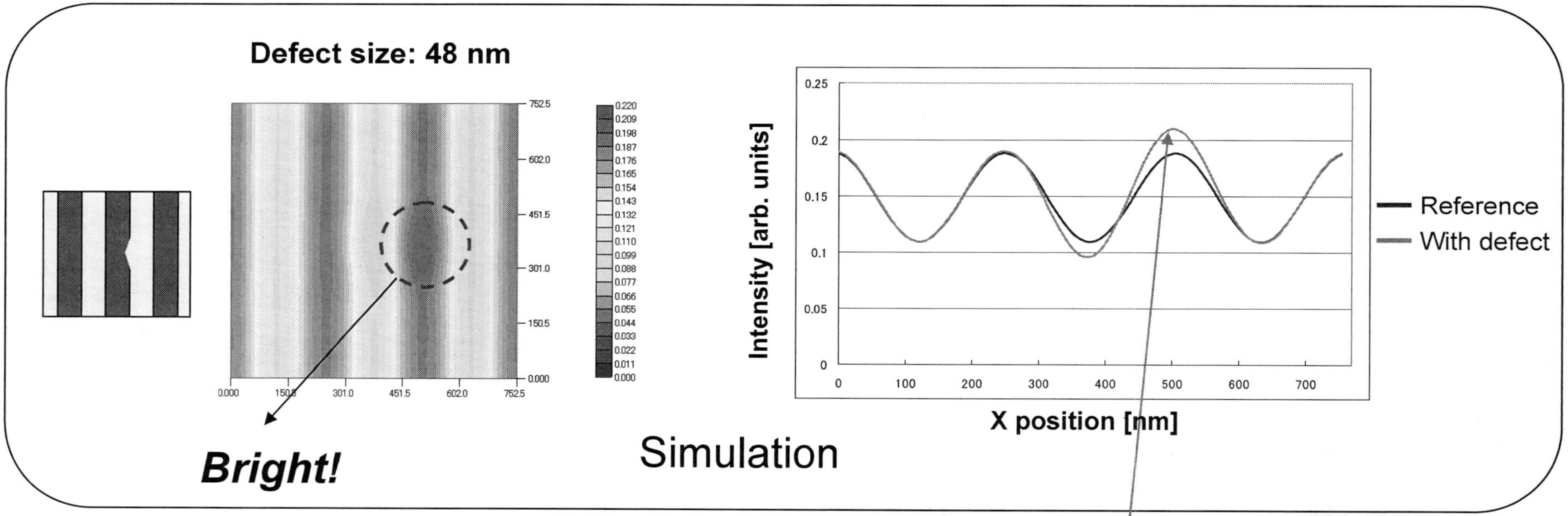

Simulation

Experimental profile

Calculated image intensity profiles
are similar to experimental results.
Defect shape is very important for
simulation.

*We describe the results on
triangle-shaped defects after this.*

Simulation for clear extension defects

Calculated images

R19-Mask

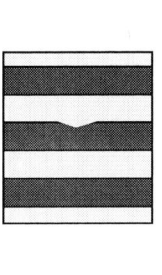

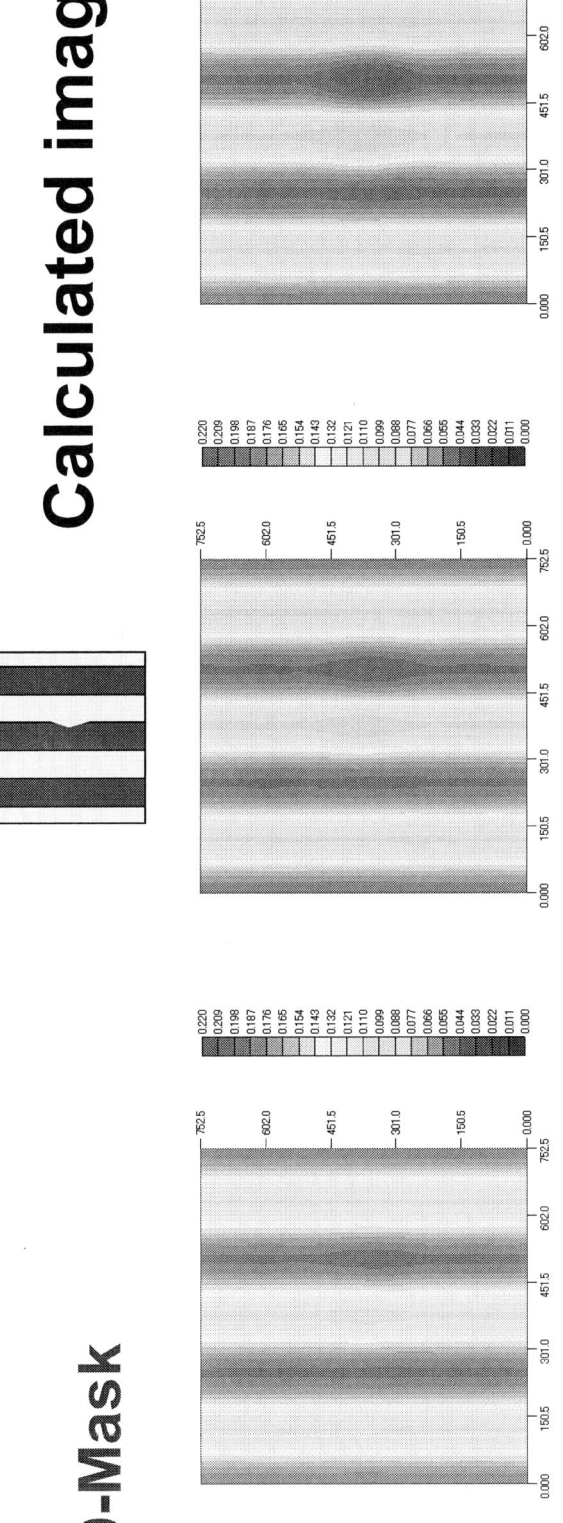

Defect size 〉 ... 24nm .... 36nm 48nm

R5-Mask

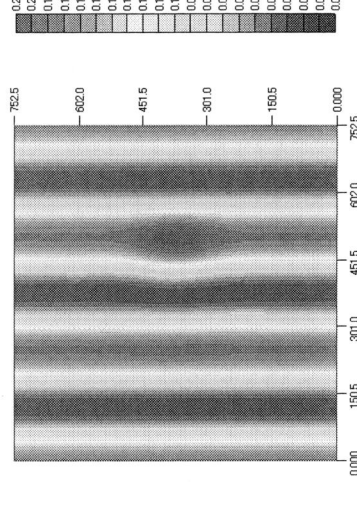

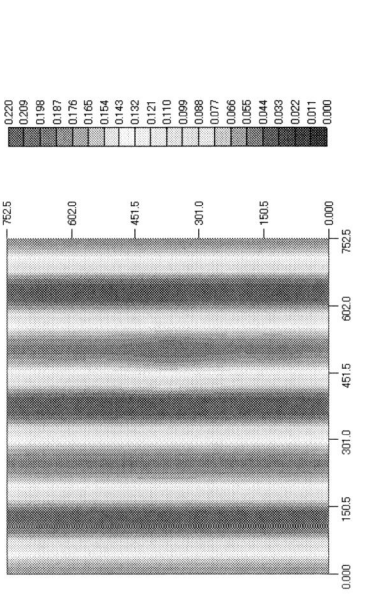

Simulation for clear extension defects

Intensity profiles

— Reference
— With defect

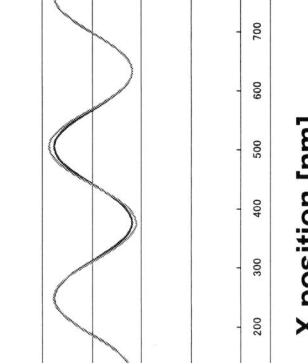

R19-Mask

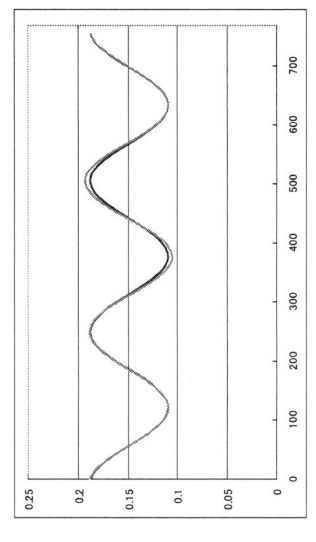

Defect size > ···· 24nm ···· 36nm ···· 48nm ····

R5-Mask

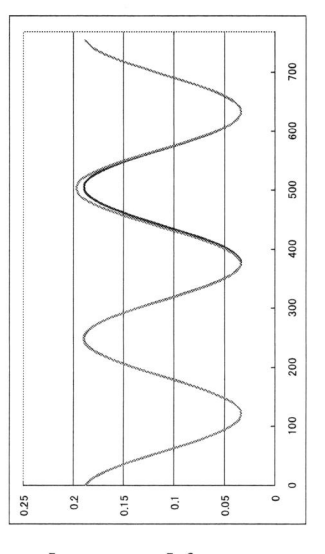

Simulation for opaque extension defects

Calculated images

R19-Mask

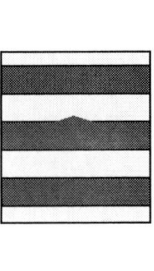

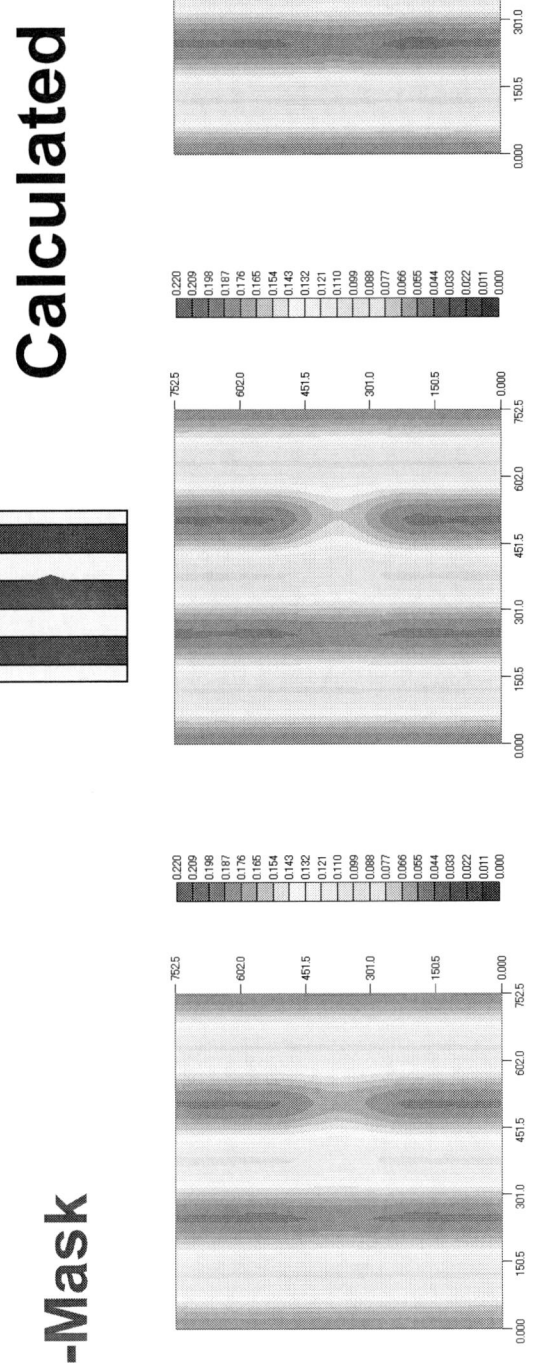

Defect size 〉 ··· 24nm ············ 36nm ············ 48nm

R5-Mask

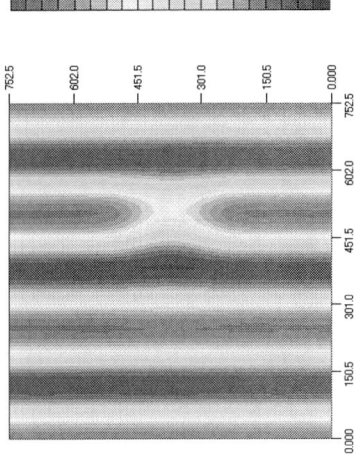

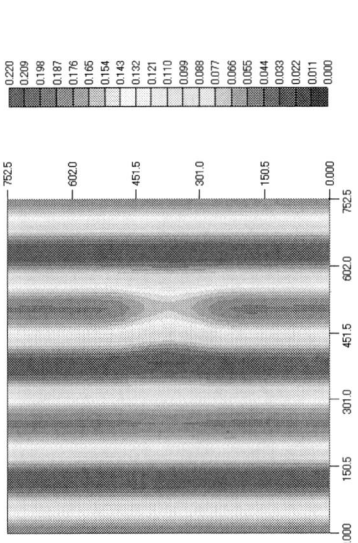

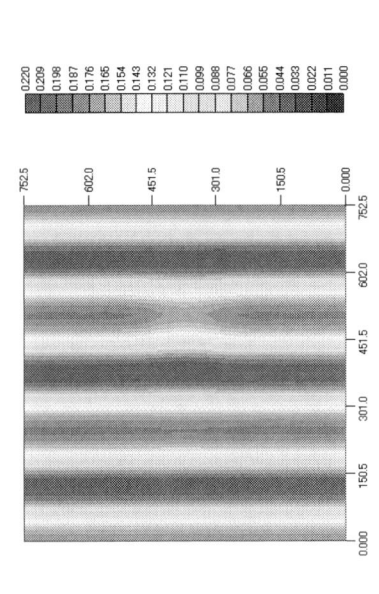

Simulation for opaque extension defects

Selete

Intensity profiles

— Reference
— With defect

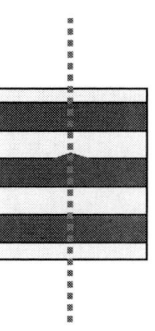

R19-Mask

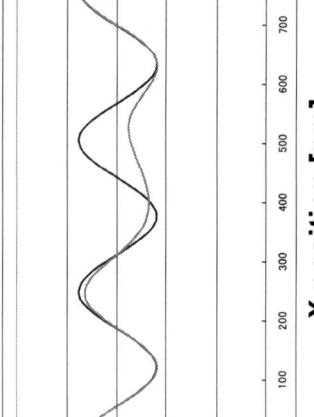

Defect size > ··· 24nm ··········· 36nm ·········· 48nm ···········

R5-Mask

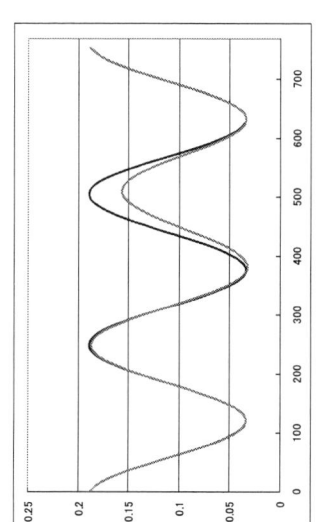

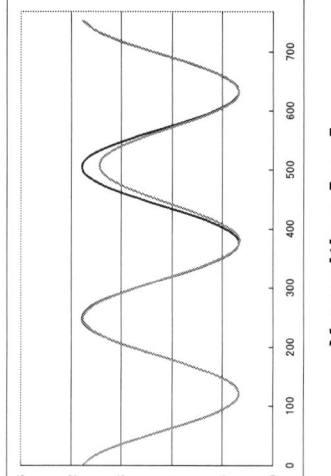

Mask structure vs. defect detectivity (1)

Clear extension defects

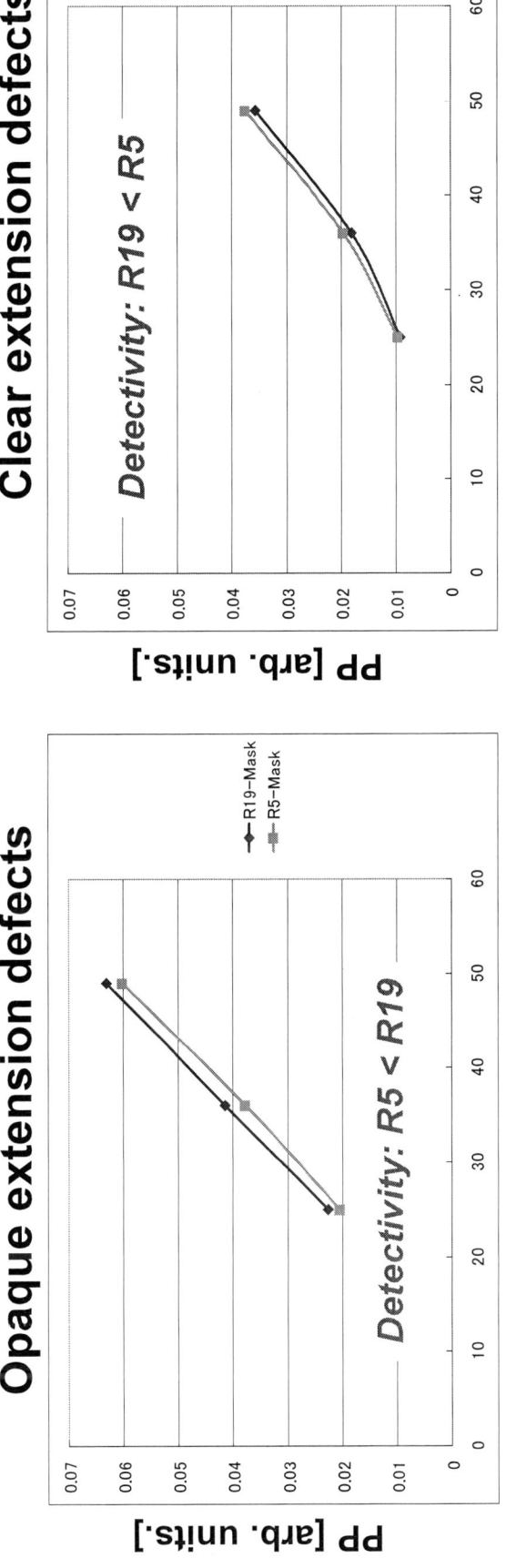

Detectivity: R19 < R5

Opaque extension defects

Detectivity: R5 < R19

Opaque defect detectivity of R19-Mask is higher than that of R5-Mask. On the other hand, for clear defects, R5-Mask has slightly higher detectivity than R19-Mask.

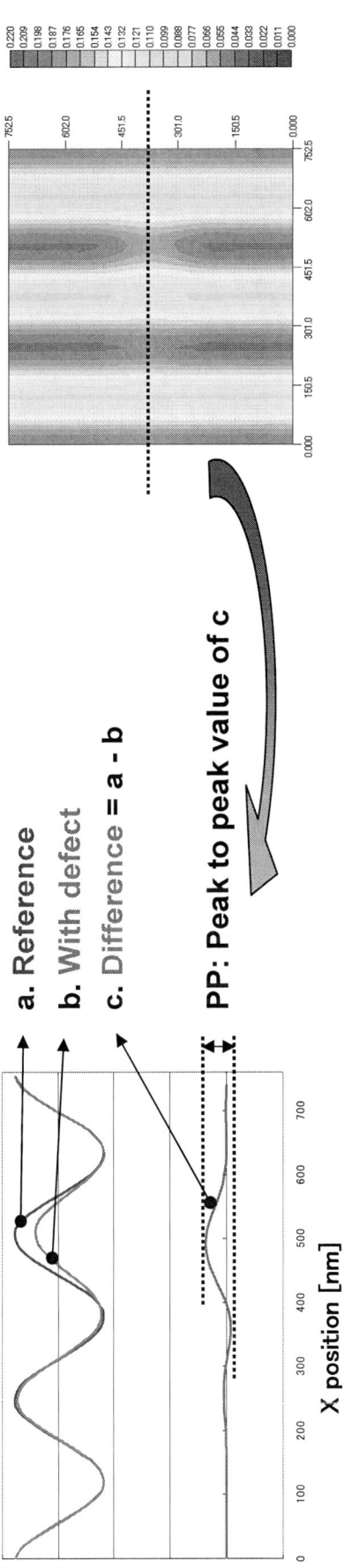

a. Reference
b. With defect
c. Difference = a - b

PP: Peak to peak value of c

Mask structure vs. defect detectivity (2)

Clear extension defects

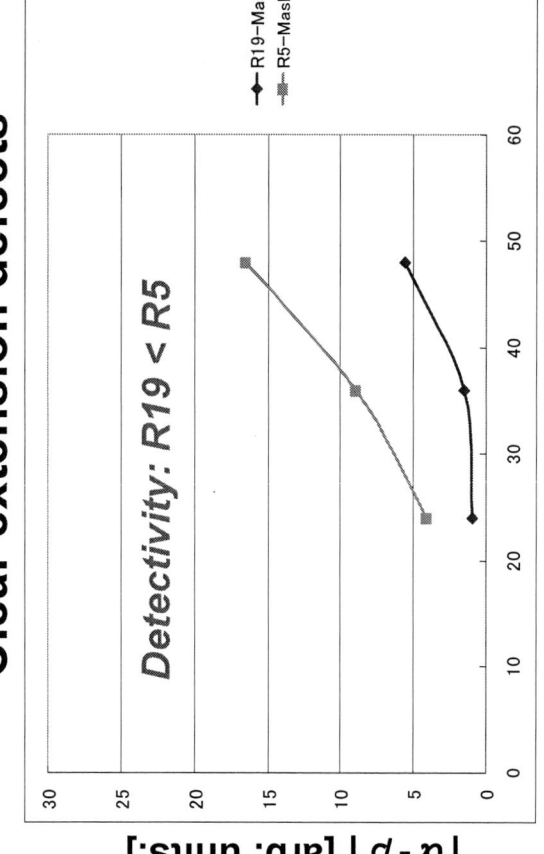

Detectivity: R19 < R5

Opaque extension defects

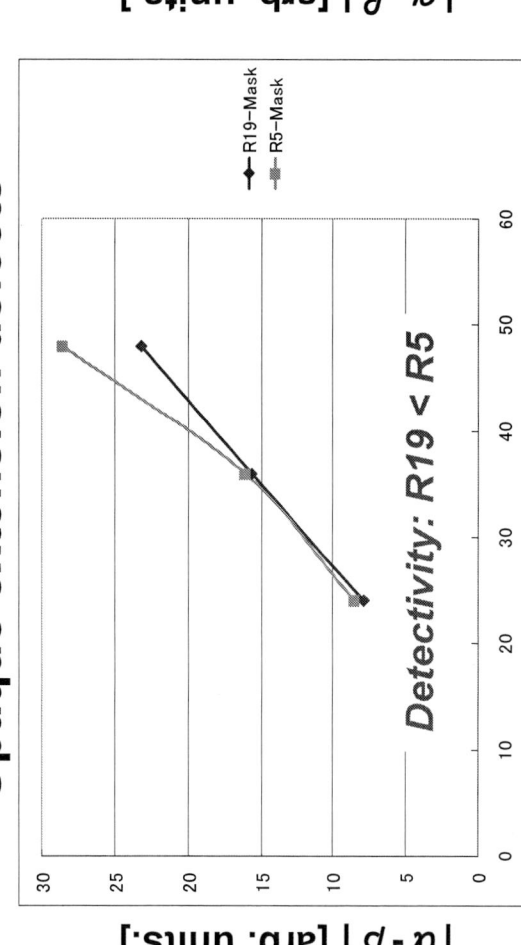

Detectivity: R19 < R5

To predict pixel intensity, we integrated the intensity of calculated image.

- Detectivity of R5-Mask is higher than that of R19-Mask both for opaque and clear defects.

- For clear defects, the difference between them is marked.

$$\alpha = \iint I_0(x,y)\,dx\,dy \qquad \beta = \iint I_{defect}(x,y)\,dx\,dy$$

Summary

- We rigorously simulated inspection images using a wavelength of 199 nm.

- The different in image contrast of hp128nm L/S between the two mask structures were clearly demonstrated.

- Calculation of peak to peak value of intensity profile shows

  · Opaque defect detectivity: R5-Mask < R19-Mask

  · Clear defect detectivity: R19-Mask < R5-Mask

- Integration of image Intensity shows

  · Opaque defect detectivity: R19-Mask < R5-Mask

  · Clear defect detectivity: R19-Mask < R5-Mask

Defect detectivity depends on image signal processing.

Future work

- Simulations based on parameters not taken up in this work

- Experiments (accumulating data)
  → Comparison of simulations and experiments

Acknowledgement

This work was supported by New Energy and Industrial Technology Development Organization (NEDO).

Characterization of
EUV–Deposited Carboneous Contamination

Toshihisa Anazawa, Yasushi Nishiyama, Hiroaki Oizumi, Osamu Suga, Iwao Nishiyama
MIRAI-Semiconductor Leading Edge Technologies, Inc.

Semiconductor Leading Edge Technologies, Inc.

Background: Carboneous Contamination

Selete

➤ Contamination of EUV Mask

● Carbon Deposition ← Hydrocarbon, Organic Molecules

competition process

● Oxidation ← Oxygen, Water Vapor

Carbon Deposition occurs
in Usual Vacuum Condition

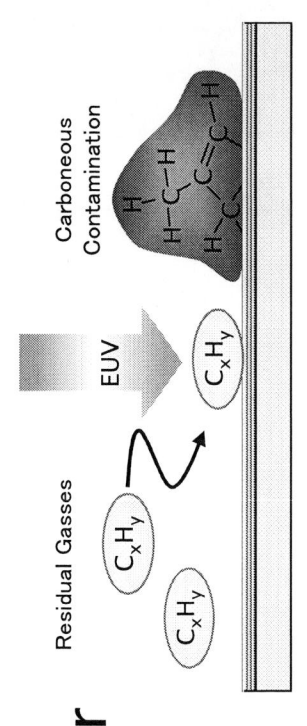

Residual Gasses

C_xH_y

C_xH_y

C_xH_y

EUV

Carboneous
Contamination

➤ Effects of Contamination on EUV Mask

● Decrease of Throughput

● CD variation

● Deterioration of Exposure Latitude
(In Case of Inhomogeneous Contamination)

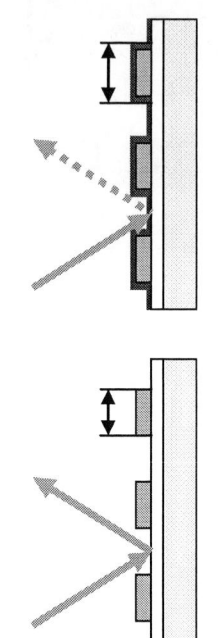

Background: Previous Studies

Densities of carboneous contamination

- 1.82 ± 0.12 g/cm³ [EB, RBS/SEM]
  R. Kurt et al., Proc. SPIE, 4688, 0277, (2002).

- ~1.125 g/cm³
  J. Hollensheada et al., J. Vac. Sci. Technol. B 24, 64 (2006).

- 0.8 ~ 1.2 g/cm³ [SR, GIXR]
  Y. Nishiyama et al., Proc. of SPIE, 6921, 692116, (2008).

- 1.9 g/cm³ [SR, RBS/TEM (similar to this study)]
  T. Anazawa et al., unpublished.

Varies data to data.
Which is actual value ?

EB: Electron Beam
RBS: Rutherford Backscattering Spectrometry
SEM: Scanning Electron Microscopy
SR: Synchrotron Radiation
GIXR: Grazing-Incidence X-ray Reflectivity

Chemical composition of carboneous contamination

- Intel MET
  N1 mirror: C : O : Si ~70 % : 20 % : 10 %
  G1, G2 mirror: C : O : Si ~ 85 % : 10 % : 5 %
  Manish Chandhok,
  IEUVI Optics Contamination / Lifetime TWG (1st Mar. 2007)

- Albany MET
  G2: C : O : Si : P : N = 74 : 20 : 2 : 2 : 1
  Andrea Wüest et al.,
  IEUVI Optics Contamination / Lifetime TWG (1st Nov. 2007)

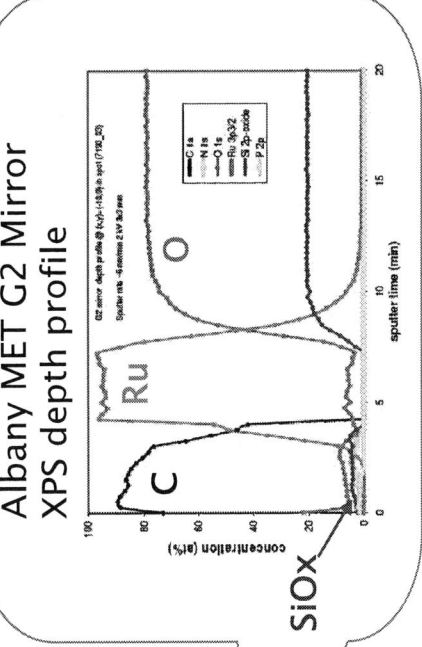

Albany MET G2 Mirror
XPS depth profile

Objective of This Study

- **Evaluation Thickness and Density of Contamination Layer**
  - for Exposure Simulation → Lifetime Prediction

- **Characterization of Carboneous Contamination**
  - for Modeling of the Deposition Process → Lifetime Prediction
    (e.g. Reaction Process and Product, Numbers of Photoelectrons, Adsorptionability of Gasses, etc.)
  - for Development of Cleaning Process
    (e.g. Relevance of Simulated Contamination, Reaction Design, etc.)

Some candidates of carboneous contamination

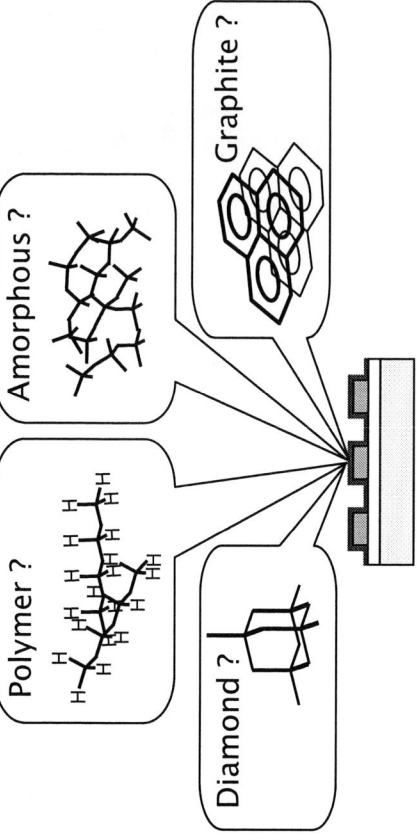

An example of reflectivity simulation

Y. Nishiyama et al., Proc. SPIE 6921, 692116 (2008)

IRAI Experimental: Sample

Synchrotron-Deposited Carboneous Contamination

Deposited at NTT Synchrotron Facilities Super-ALIS SBL-1

Wavelength: 1~20 nm (Grazing incidence mirror + Zr-filter)
Dose: ~30,000 A·s (Correspond to 500~1,000 J/cm^2)
Vacuum: $1.2 \times 10^{-4} \sim 4.0 \times 10^{-6}$ Pa (Partial pressures of hydrocarbon < 1%)

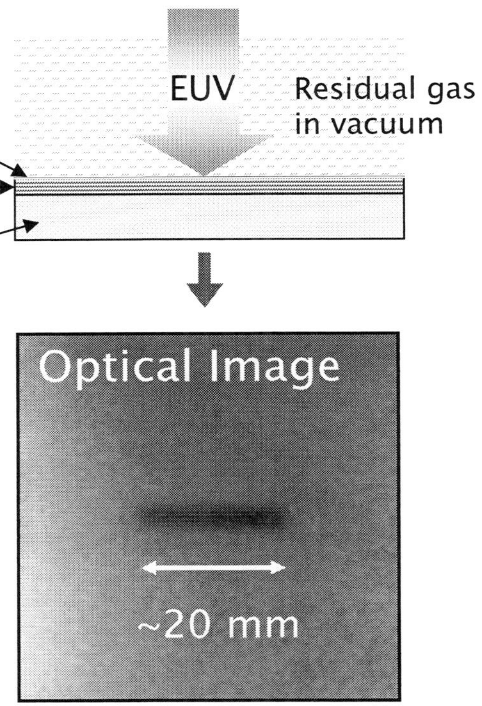

EUV Reflectivity

Reflectivity vs Wavelength (nm): 13.0, 13.5, 14.0
Reflectivity axis: 0.0, 0.2, 0.4, 0.6
Initial
Contaminated
4.7 % Down

Residual gas in vacuum
EUV
Si Capping Layer
{Mo/Si} Multilayer
Quartz

Optical Image
~20 mm

Layer Thickness: few nm

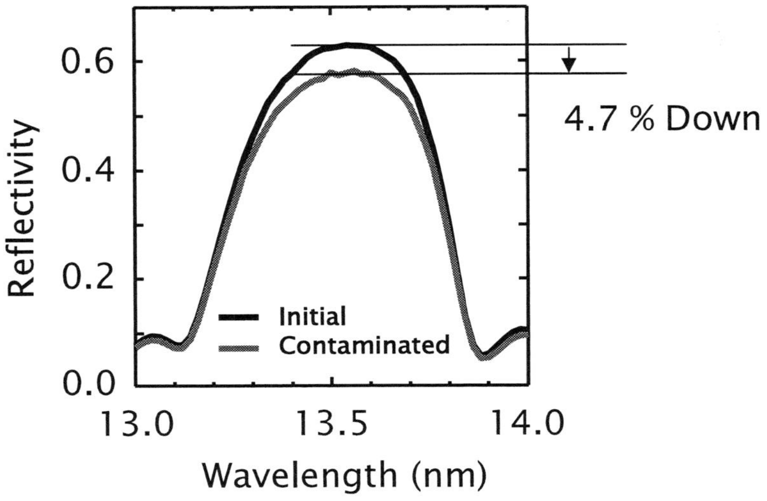

Experimental: Analysis Flow

Multiple analysis was applied to the same contamination sample.

Non-destructive ➝ Destructive

Sensitive ⬌ Insensitive

■ Reflection measurement
· EUV reflectivity

Non-Destructive   Area: mm

■ Spectroscopic Ellipsometory
· film thickness distribution

Non-Destructive   Area: mm

■ GIXR (Grazing-Incidence X-ray Reflectivity)
· film thickness, Weight density

Non-Destructive   Area: mm

■ REELS (Reflective Electron Energy Loss Spectroscopy)
· chemical bondings of carbon atoms

Weak-Contamination   Area: µm

■ XPS (X-ray Photoemission Spectroscopy)
· elements, chemical states

Weak-Contamination   Area: Sub-mm

■ HR-RBS/HR-ERDA (High-Resolution RBS/ERDA*)
· composition, area density / hydrogen amount

* ERDA: Recoil Detection Analysis

Semi-Destructive   Area: mm

■ TEM (Transmission Electron Microscopy)
· film thickness

Destructive   Area: nm

Weight Density

Result: Spectroscopic Ellipsometry

Single Point Measurement / Fitting Result

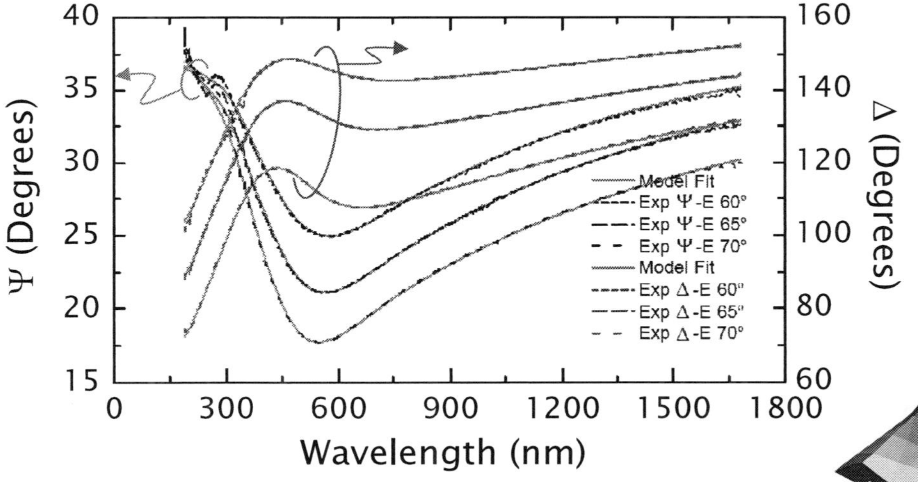

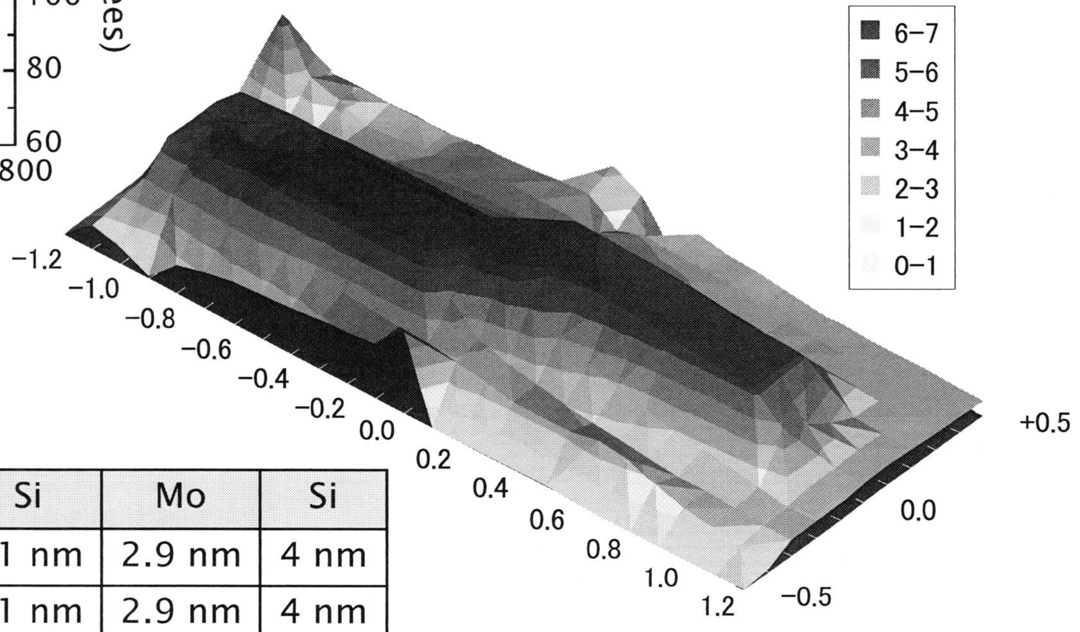

Δ (Degrees)

Ψ (Degrees)

Wavelength (nm)

Model Fit
Exp Ψ-E 60°
Exp Ψ-E 65°
Exp Ψ-E 70°
Model Fit
Exp Δ-E 60°
Exp Δ-E 65°
Exp Δ-E 70°

Legend: 6–7, 5–6, 4–5, 3–4, 2–3, 1–2, 0–1

Layer	Carboneous	SiO_2	Si	Mo	Si
Reference	–	0.6 nm	11 nm	2.9 nm	4 nm
Contamination	6.7 nm (Max)	0.6 nm	11 nm	2.9 nm	4 nm

● Carboneous Layer Thickness: 6.7 nm (at Maximum point)

Result: GIXR

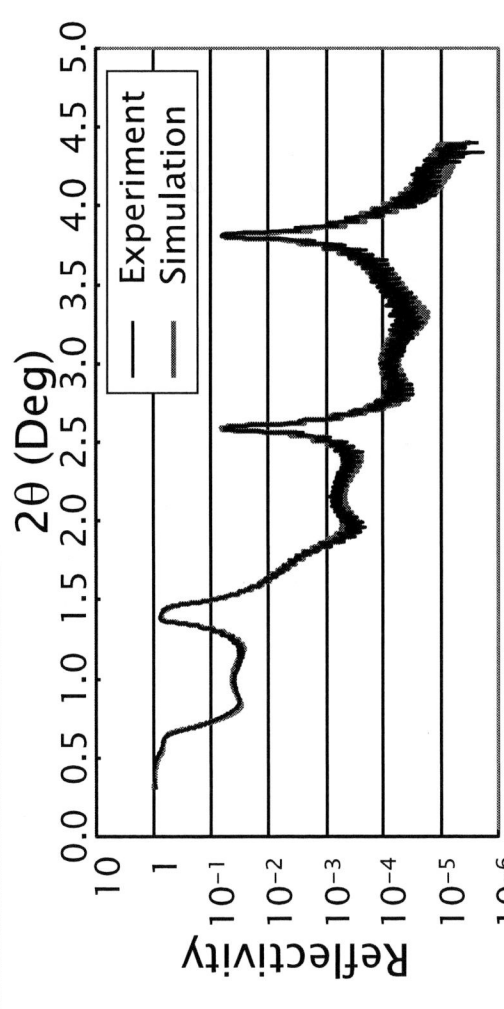

[Measurement Conditions]
Cu – Kα_1: 40kV–20mA
Incidence Slit: 0.05mmW × 1.0mmH
Sampling: 0.002° /step
Accumulation Time:
  10 sec /step (Low Angle Region)
  30 sec /step (High Angle Region)

Fitting results

Structure		Density (g/cm³)	Calculated Thickness (nm)	Nominal Thickness (nm)	Roughness (nm)	R
Carboneous Layer		1.49	9.01	–	0.57	
Si		2.34	11.51	11.0	0.56	
[Mo/Si] 40 Pairs	Mo	9.89	3.03	2.9	0.24	0.02
	Interface (Si-Mo)	4.56	0.94	– (6.9)	0.59	
	Si	2.36	3.09	4.0	0.00	
SiO₂ Substrate		2.20	–	–	0.30	

(Calculated Thickness 7.06 for the [Mo/Si] pair; Nominal Thickness 6.9)

The Density of the Carboneous Film and the Mo/Si Interface Layers are calculated using the Ratio of C:H=7:3 (RBS) and Mo:Si=3:7 Respectively.

- ● Carboneous Layer Thickness: 9.0 nm
- ● Carboneous Layer Density: 1.5 g/cm³

IRAI Result: TEM

►Thickness Measurement by TEM (at <u>most thick region</u>)

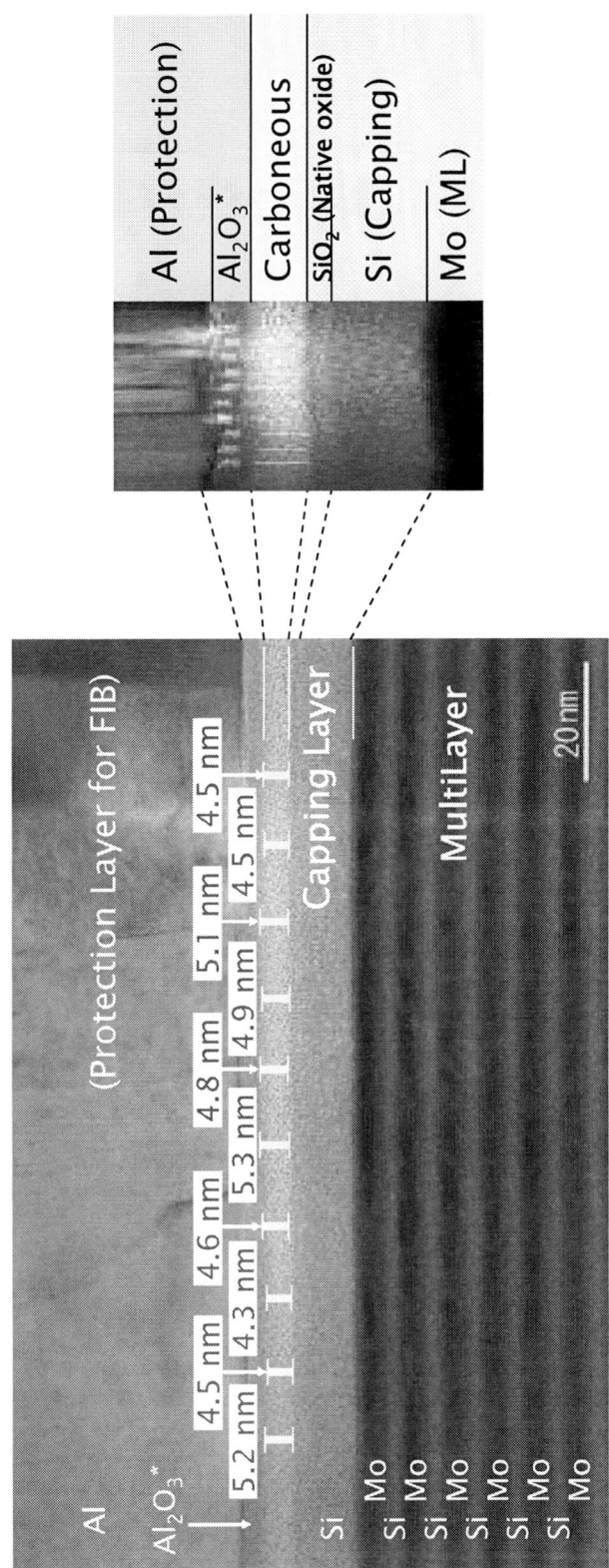

Al (Protection)

Al_2O_3*

Carboneous

SiO_2 (Native oxide)

Si (Capping)

Mo (ML)

*To Avoid Temperature Rising, the Initial Rate of Evaporation Deposition of the Protection Layer Aluminium was slowed down. So, They Reacted with Residual Gases (like water) in the Evaporation Chamber those hit at the Deposited Film to Become Alumina-like Substances.

● Carboneous Layer Thickness: 4.8±0.3 nm (average 10 points)

 *Selete*

Result: HR–RBS / TEM

◆Weight Density of Contamination Layer

HR–RBS

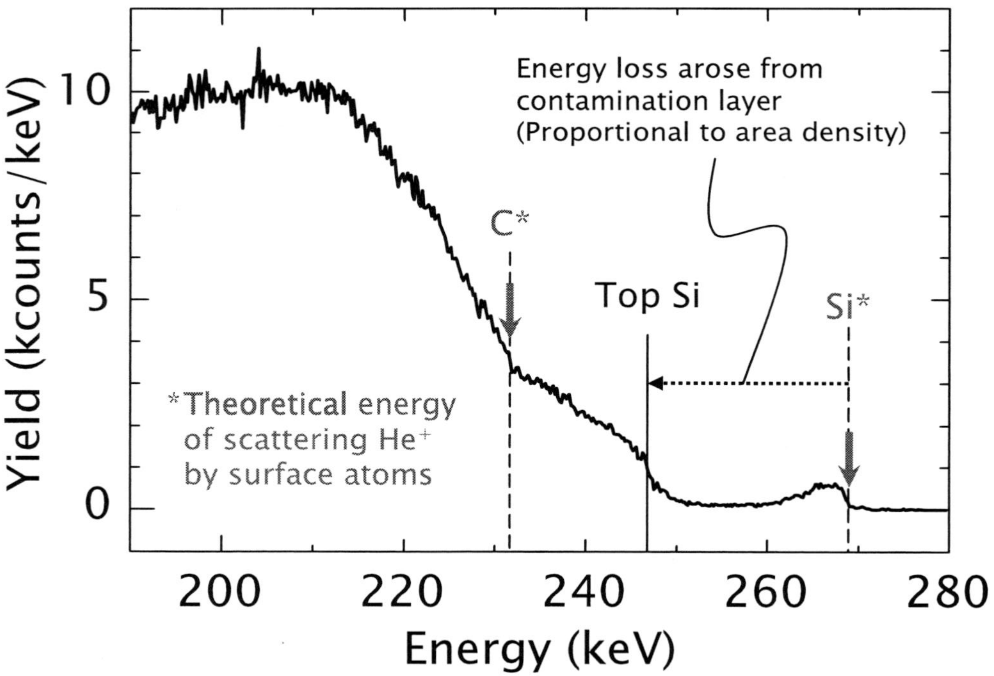

He⁺

Energy loss

Si (11.0 nm)

Mo(2.9 nm)
/Si(4.0 nm)
40 Pairs

Qz (6.3 mm)

Energy loss arose from
contamination layer
(Proportional to area density)

Si*

Top Si

C*

*Theoretical energy
of scattering He⁺
by surface atoms

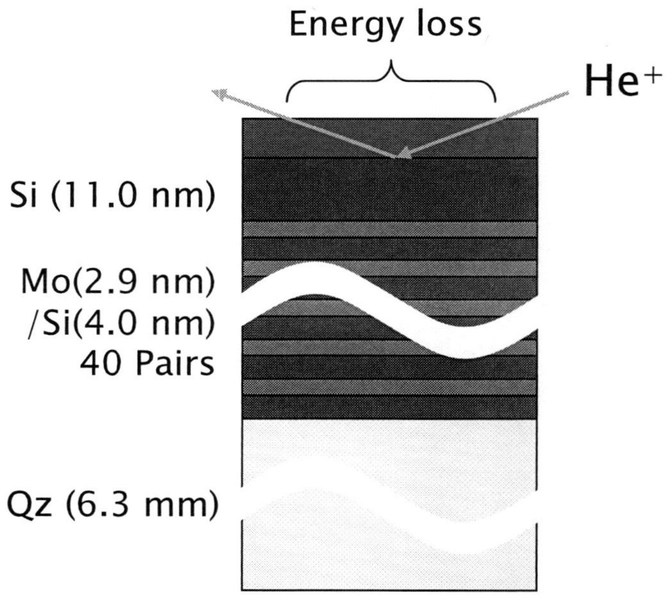

Energy (keV)

Yield (kcounts/keV)

● The average density of the contamination is 2.2 g / cm³ .
  (Combination with thickness from TEM = 4.8 nm)

IRAI Result: HR-RBS / HR-ERDA

Selete

Element Analysis by Scattering Simulation Fitting

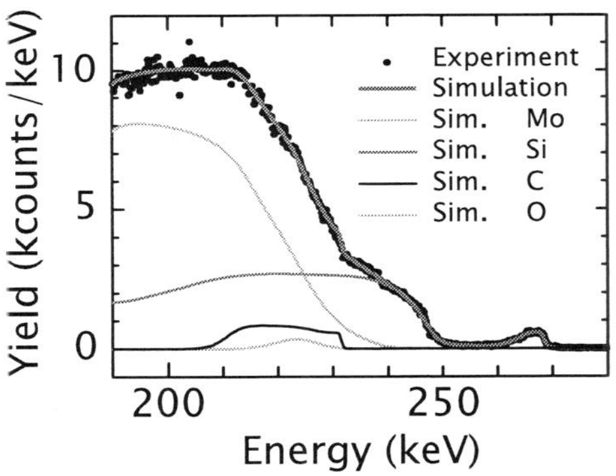

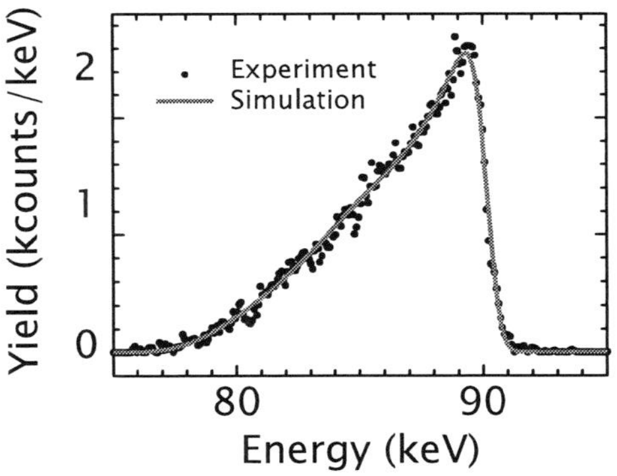

● Si is detected within and top of contamination.

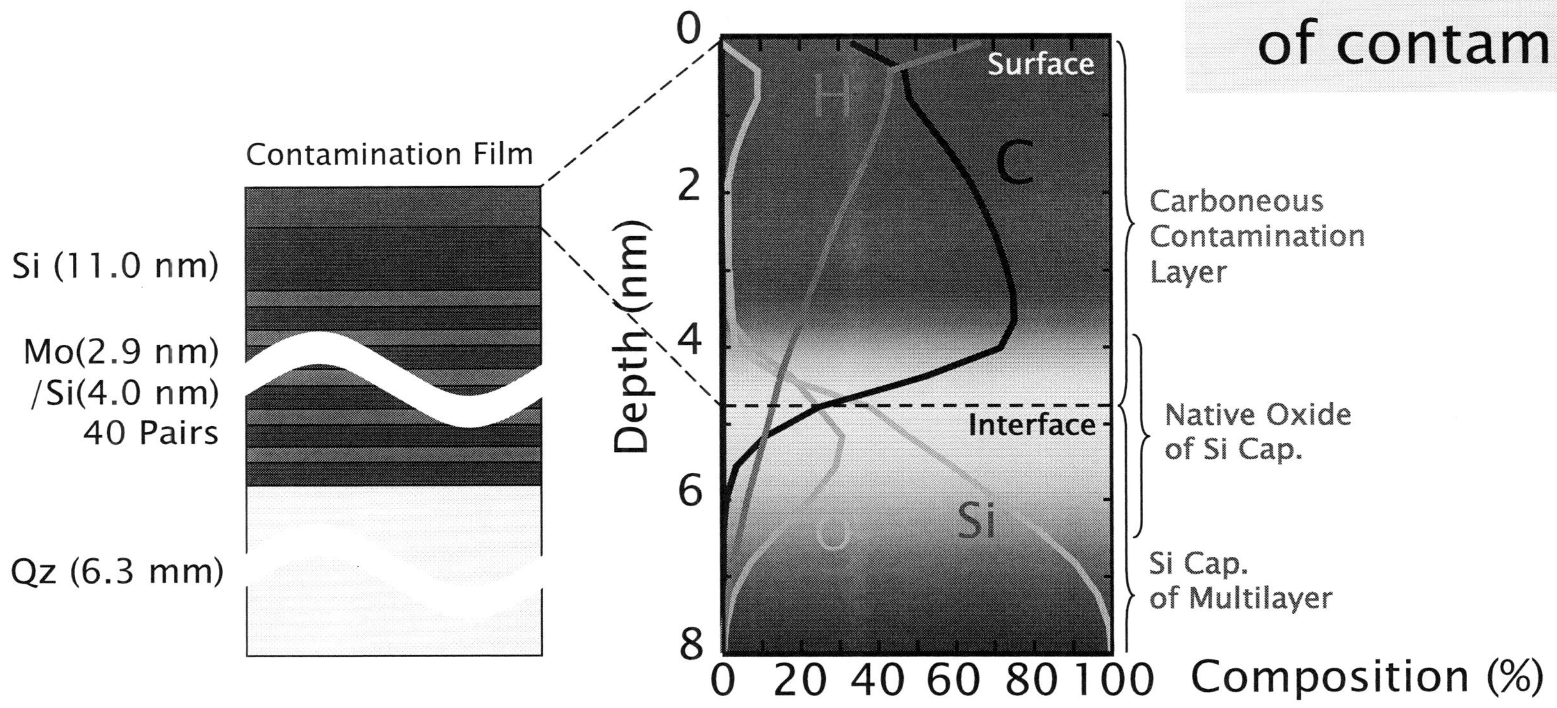

HR-RBS (left)

HR-ERDA (right)

Result: HR–RBS / HR–ERDA

→ Integrated Amount of Atoms within Contamination Layer

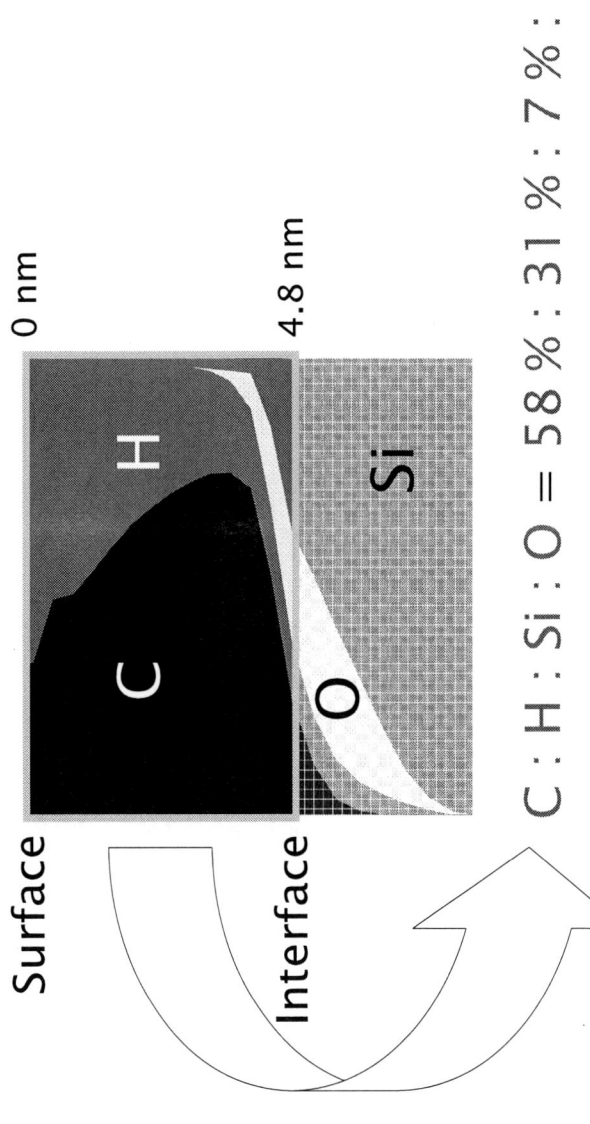

0 nm

4.8 nm

Surface

Interface

C : H : Si : O = 58 % : 31 % : 7 % : 4 %

● Averaged amount of hydrogen atom > 30 %.
● The fraction of hydrogen decreases at deeper region that may be the result of prolonged irradiation.

● There's little oxygen detected in contamination layer.

Result: XPS (Element Analysis)

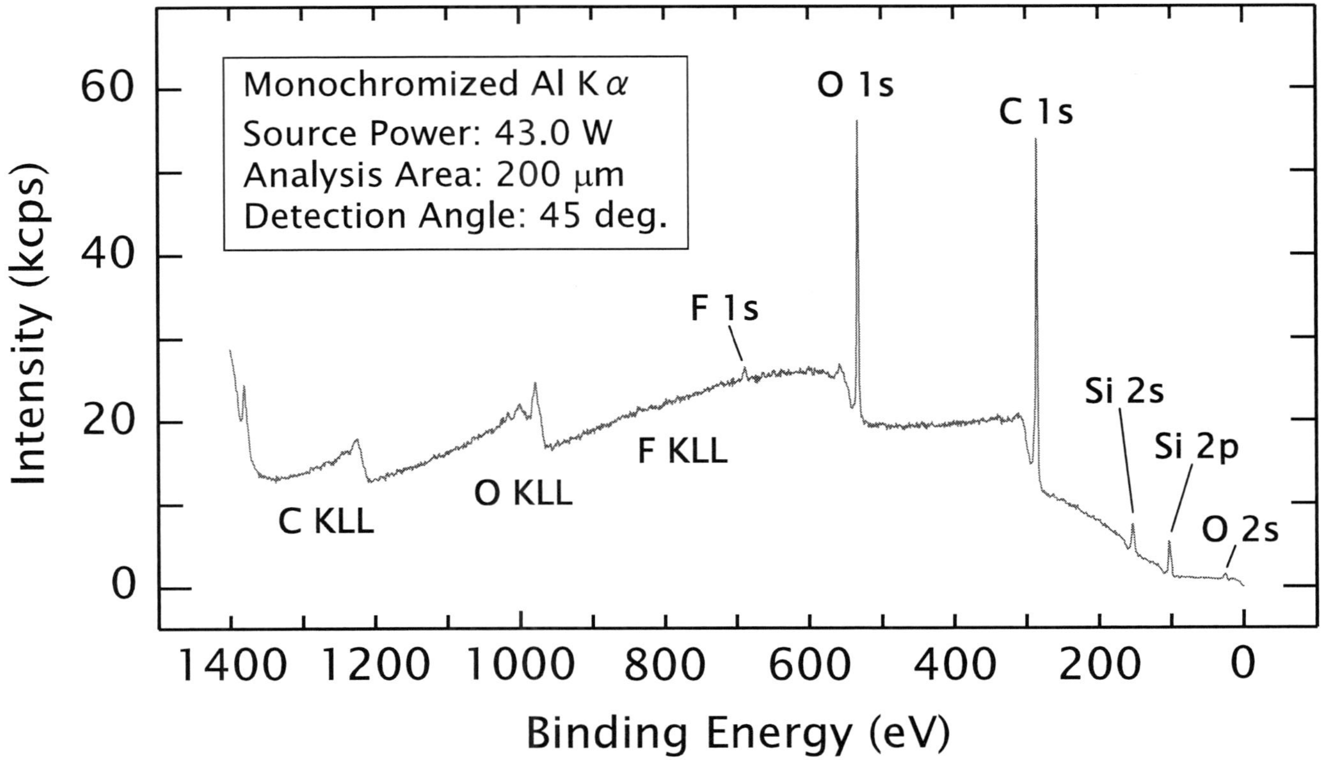

Monochromized Al Kα
Source Power: 43.0 W
Analysis Area: 200 μm
Detection Angle: 45 deg.

● Si and F were detected.

Element	C	O	F	Si
Ratio (%)	69.2	23.4	0.7	6.7

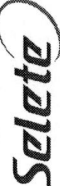

MIRAI Result: XPS (Chemical Shift)

Selete

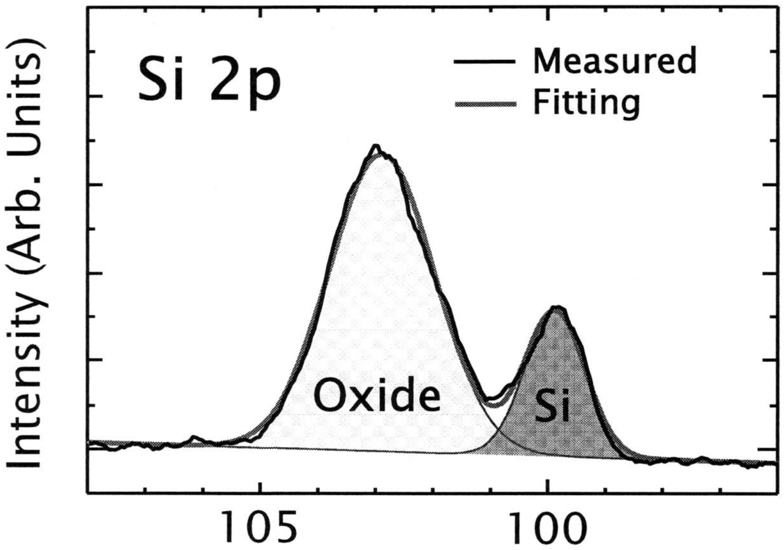
Intensity (Arb. Units)
Si 2p
Measured
Fitting
Oxide
Si
105
100
Binding Energy (eV)

	Si	Oxide
B.E. (eV)	99.9	102.9
Ratio (%)	21.8	78.2

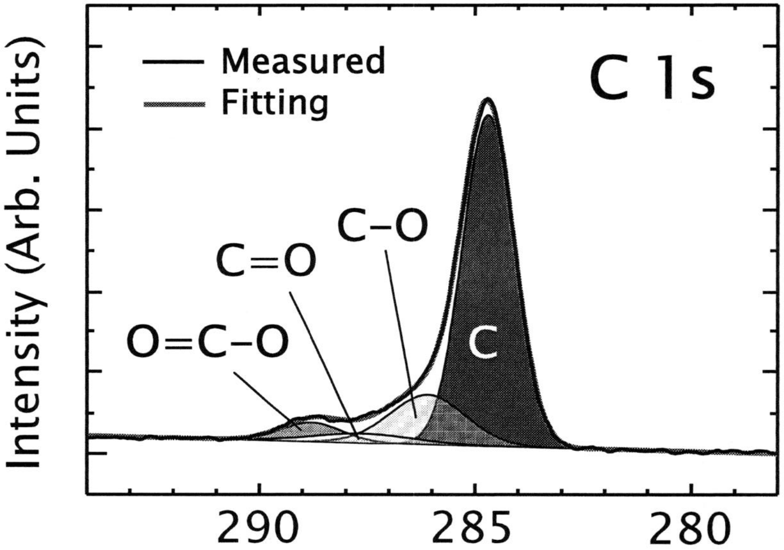
Intensity (Arb. Units)
Measured
Fitting
C 1s
C–O
C=O
O=C–O
C
290
285
280
Binding Energy (eV)

	C	C–O	C=O	O=C–O
B.E. (eV)	284.7	286.1	287.6	288.8
Ratio (%)	74.7	16.5	3.8	5.0

- Considerable oxygen atoms exist in native oxide (at interface) and as carbonyl- or carboxyl- spices (may be at the surface).
- This result agrees with RBS (little oxygen in contamination).

Result: EELS (ELNES)

ELNES

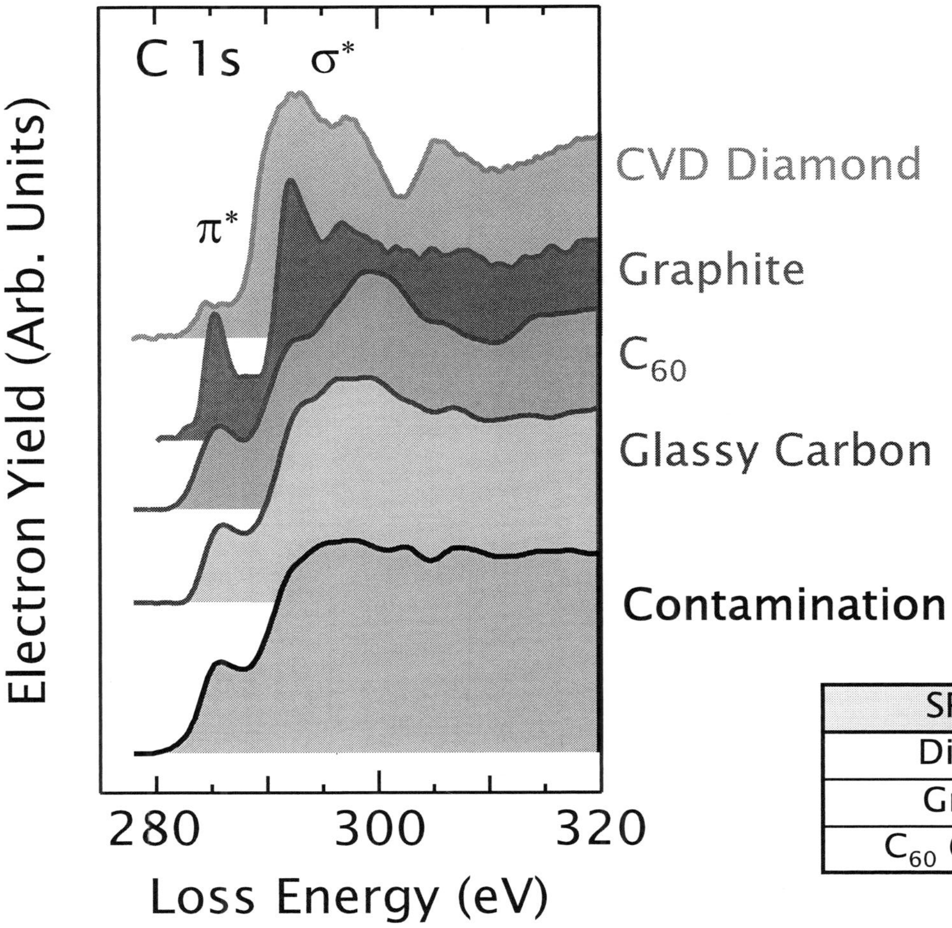

Electron Yield (Arb. Units)

C 1s σ* π*

CVD Diamond
Graphite
C_{60}
Glassy Carbon
Contamination

Loss Energy (eV)
280   300   320

ELNES Fitting

Electron Yield (a. u.)

C K–Edge

—— Experiment
······ Fitting

Loss Energy (eV)
280   290   300   310

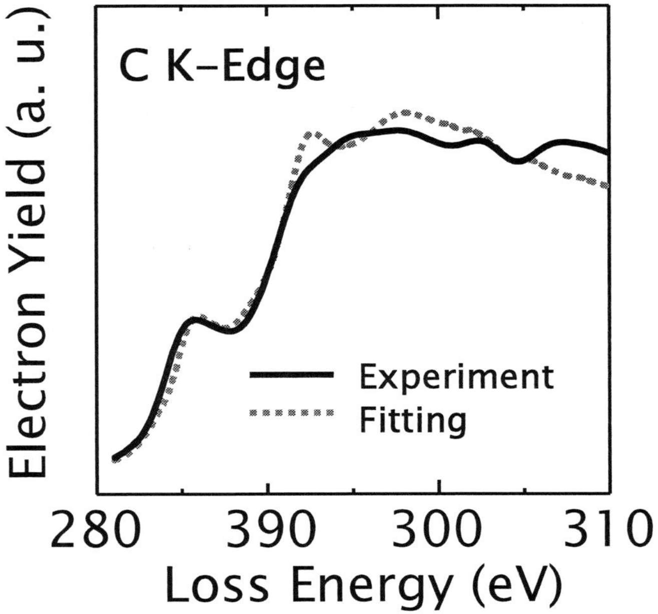

SR (R=5.56%)	Ratio (%)	Shift (eV)
Diamond (sp^3)	14.8	-2.2
Graphite (sp^2)	30.1	0.5
C_{60} (Distorted sp^2)	55.1	-0.1

- Contamination film may mainly consists of distorted sp^2.
  (But this result would need further studies i.e. measurement damage.)

 # Discussion: Thickness and Density

- Thicknesses and densities much differ by measurement methods.
- Thickness from GIXR seems to be too large. (cf. TEM, XPS escape depth)
  Hydrogen inclusion gradient in depth
  → density gradient may affect accuracy..?
- Density from RBS/TEM could be too large in this study. (cf. Graphite)
  In general, the most reliable thickness value is obtained from TEM. However, the density calculated by (area density)/(TEM thickness) seems too large.

If the carboneous contamination film is so weak in heating, TEM sample preparation process (Al-evaporation in this study) could cause a shrinkage of film..?

More accurate thickness measurement is necessary and we're testing now.

Summary of results in this study

	Thickness	Density
Ellipso.	6.7 nm	NA
GIXR	9.0 nm	1.5 g/cm^3
TEM	4.8 nm	2.2 g/cm^3
RBS	NA	

Density of carboneous materials

Sample	Density	Sample	Density
Diamond	3.6 g/cm^3	DLC*	1.2~3.5 g/cm^3
Graphite	2.2 g/cm^3	Glassy carbon	1.5 g/cm^3
C$_{60}$	1.7 g/cm^3		

Previous Studies (represented from p.3 above)

Ref.	Analysis	Depo.	Density
1	RBS/SEM	EB	1.82±0.12 g/cm^3
2			~1.125 g/cm^3
3	GIXR	SR	0.8 ~ 1.2 g/cm^3
4	RBS/TEM	SR	1.9 g/cm^3

1 R. Kurt et al., Proc. SPIE, **4688**, 0277, (2002).
2 J. Hollensheada et al., J. Vac. Sci. Technol. **B 24**, 64 (2006).
3 Y. Nishiyama et al., Proc. of SPIE, **6921**, 692116, (2008).
4 T. Anazawa et al., unpublished.

Discussion: Density and Photochemistry

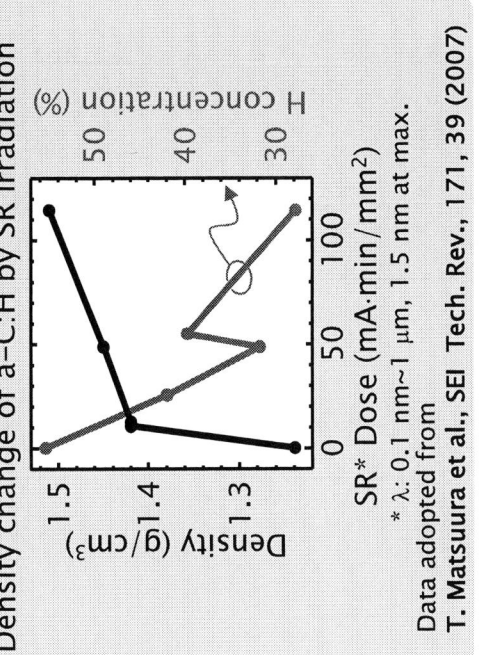

Density change of a–C:H by SR irradiation

SR* Dose (mA·min/mm²)

* λ: 0.1 nm~1 μm, 1.5 nm at max.

Data adopted from
T. Matsuura et al., SEI Tech. Rev., 171, 39 (2007)

- Irradiation of SR on a–C:H (hydrogenated amorphous carbon, a kind of so-called DLC) causes decreases of hydrogen amount and increases of their densities. **The gradient of hydrogen amount detected by ERDA must show that a similar phenomena occurs on carboneous contamination deposition process by EUV.** The contamination at deeper region is irradiated longer time and reaction will proceed more.

- **This suggests, the density of carboneous contamination strongly depends on the conditions of deposition, i.e. vacuum, flux and dose of photon, etc.** For example, if the pressure of residual gas is low enough and the optics was irradiated by strong light for long time, the contamination will well 'burned' and the density of the contamination should become high.

- **The existence of much amount of hydrogen suggests that contamination investigated here is not a lump of carbon atom but it may be like a highly carbonized polymer.**

- *If the source of Si is not residual gases, reaction with capping layer Si or SiO$_2$ must be considered.*

Discussion: Reflectivity

Comparison with Reflectivity Simulation

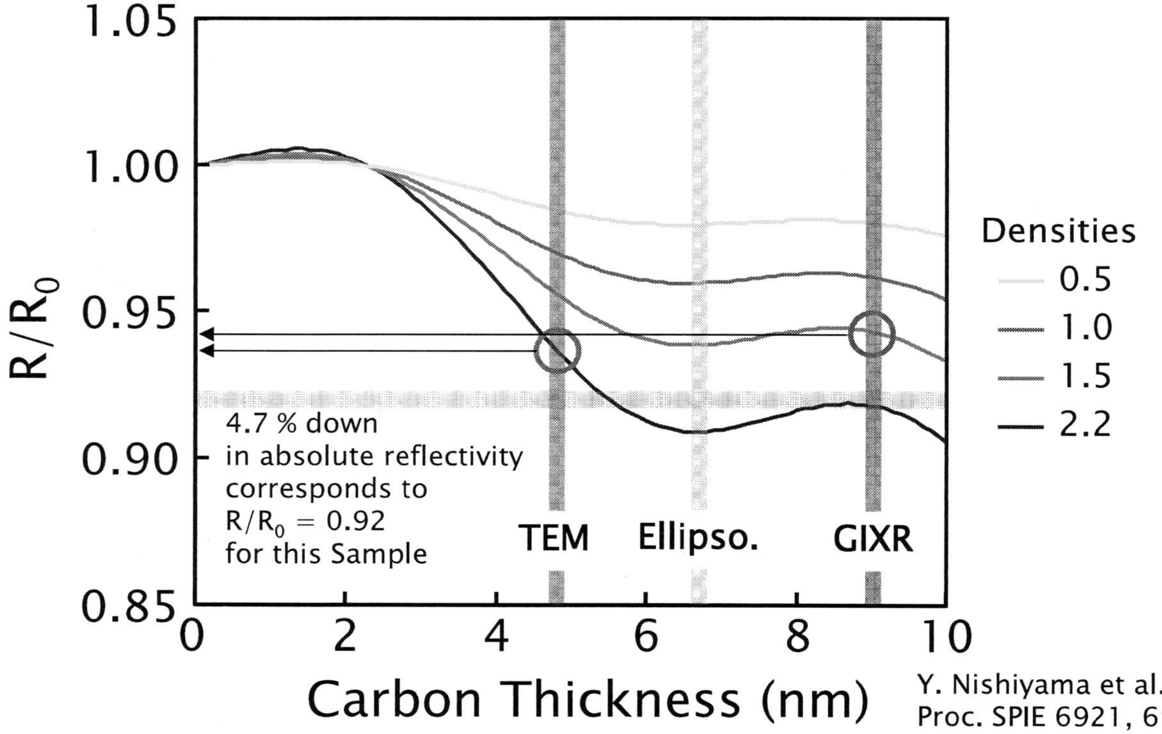

AFM image of contamination film

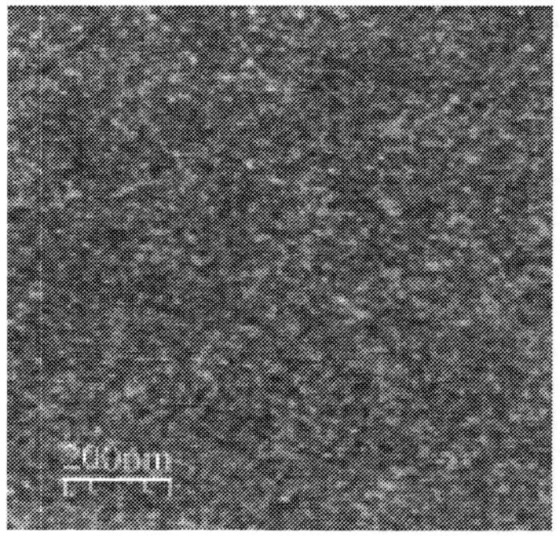

Peak-to-valley: 3.9 nm
RMS roughness: 0.37 nm

cf. Si capping layer: 0.1~ 0.2 nm RMS

Densities
— 0.5
— 1.0
— 1.5
— 2.2

R/R_0

4.7 % down
in absolute reflectivity
corresponds to
$R/R_0 = 0.92$
for this Sample

TEM   Ellipso.   GIXR

Carbon Thickness (nm)

Y. Nishiyama et al.,
Proc. SPIE 6921, 692116 (2008)

- R/R_0 calculated by using the thickness and density from both of TEM/RBS and GIXR didn't agree with measured one.

- Reflectivity down seems somewhat larger than simulation. The roughness of surface or interface deteriorates reflectivity?

Current Conclusion

◎ Related to exposure simulation...

√ The impact of contamination depends on deposition conditions. Cautions would be needed in accelerated deposition, especially.

√ In reflectivity or exposure simulations, consideration is needed about how the value of the thickness and density are evaluated.

◎ Related to modeling of deposition process...

√ The photo-reaction continues to proceed after contamination deposition. This proceeding may affect precise deposition rate calculations.

√ The source of Si should be identified as soon as possible.

◎ Related to development of cleaning process...

√ Carboneous contamination may preserve the character of polymer. The cleaning rate with hydrogen radical* (for example) could be higher than graphite or FIB-deposited carbon (=simulated contamination).

* Nishiyama et al., CC–02, EUVL Symposium 2007; Anazawa et al., 5B–3, EIPBN 2008.

Summary

➤ The identical carboneous contamination has been charecterized with various analytical techniques:

● Evaluated thickness and weight density values are varies with measurement techniques.

● Contamination contains mainly C, H, Si, O and trace of F.

● Contamination contains much hydrogen.

● The fraction of hydrogen decreases with layer depth, suggesting the proceeding of photo-reactions.

➤ Future work:

● Charcterization of the contamination by exposure tool.

This Work was Supported by NEDO*
(*New Energy and Industrial Technology Development Organization)

Experimental Study of Particle-free mask handling techniques using the MPE tool

Mitsuaki Amemiya, Kazuya Ota, Takao Taguchi and Osamu Suga

MIRAI-Semiconductor Leading Edge Technologies, Inc., Japan

Abstract

One of the critical issues for EUVL masks is clean and particle-free mask handling. A dual pod carrier has been proposed for one of the solutions of this issue. We reported that the number of particle adders on the front side of a mask in the dual pod during the process from a load port to putting on the Electrostatic chuck (ESC) in vacuum could be made less than 0.01 particle/cycle (>= 46 nm). Through another experiments we found that a lot of particles have added on the backside of a mask after chucking on the ESC. It is necessary to decrease the number of particles added on the backside by chucking. In addition it is a serious issue whether the particle adders on the backside would travel to the front side. We have done some experiments about the travel of the particles. At first we measured particle adders on the front side using a substrate after chucking and the PSL (Polystyrene Latex)-substrate that was dispersed on the backside, and then, we examined whether particles would released or not from the substrate during the transfer-process including vacuuming and venting. This result showed that there was no released particle for 100nm- and 2um-PSL and the displacement of the PSL particles was within 0.3um(s). From these experiments we consider that there is very little probability that particles on the backside of a mask will travel to the front side.

1. Backside Contamination by Chucking

2. Transfer-Experiment using CrN-substrate

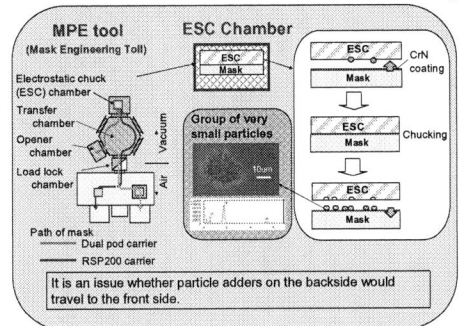

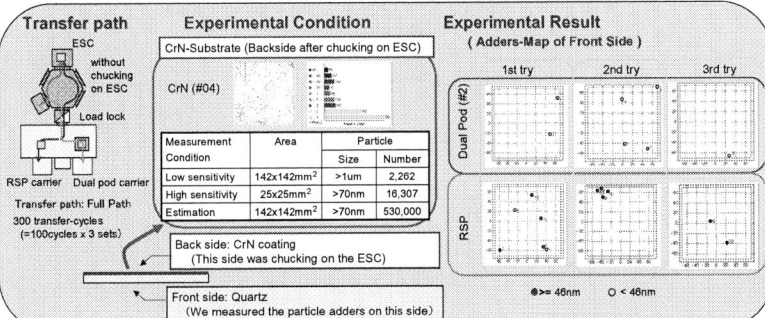

3. Transfer-Experiment using PSL-substrate

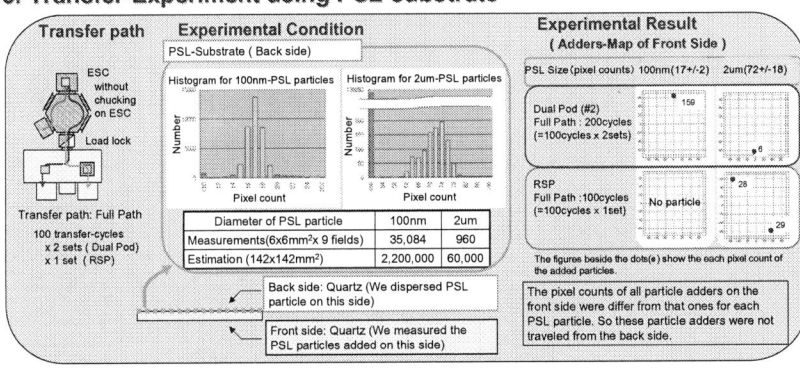

4. Transfer-Experimental Result

Substrate	CrN	100nm-PSL		2um-PSL	
Measured particles (ref)	P1	P2	P3	P2	P3
Dual Pod (#2)	0 (300)	1 (200)	0 (200)	1 (200)	0 (200)
RSP	10 (300)	0 (100)	0 (100)	2 (100)	0 (100)

This table shows the number of particle adders during the mask-transfer "Full Path". The numbers in ( ) show handling cycles. Inspection area is 142mmx142mm.
(ref) Measured particles   P1:Real particles (>=46nm)
P2:Real particles (>=50nm)
P3:PSL particles(100nm or 2um)

1. The transfer experiments with CrN-substrate
We observed some particle adders for the RSP carrier but no particle adders (>=46nm) for the dual pod carrier.
2. The transfer experiments with PSL-substrate
No PSL particle was added on the each front side for the RSP and the dual pod carrier.

5. Measurement of displacement of PSL particles

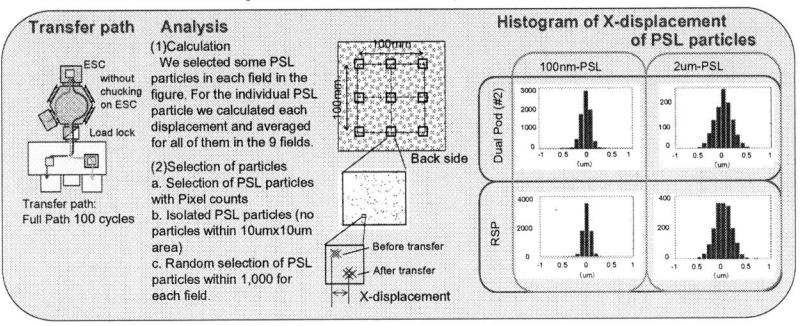

6. Displacement of PSL particles

		100nm-PSL	2um-PSL
Dual Pod Carrier Full Path: 100cycles	# of selected PSL particles	7294	1001
	# of released PSL particles	0	0
	X-Displacement (um)	0.10 (s)	0.18 (s)
RSP Carrier Full Path: 100cycles	# of selected PSL particles	7324	1674
	# of released PSL particles	0	0
	X-Displacement (um)	0.08 (s)	0.24 (s)
Measurement error	X-Displacement (um)	0.06 (s)	0.12 (s)

We observed no released PSL particle from the substrate during this transfer experiments. The standard deviations of the displacements of 100nm- and 2um-PSL particles were less than 0.1um and 0.3um respectively. These values are comparable to or slightly higher than the measurement error.

Summary

We consider that there is very little probability that particles on the backside of a mask will travel to the front side from the following experimental results.
1. Through the transfer-experiments with CrN-substrate, we observed no particle adders (>=46nm) for dual pod carrier.
2. Through the transfer-experiments using the substrate with 2 millions PSL particle(100nm dia.) on its backside, we observed no PSL particle added on its front-side.
3. We measured the number of released PSL particles and its displacement during the mask-transferring-process. For the case of 1,674 PSL particles of 2um diameter we observed no released particle and its displacement was less than 0.3 um. For the case of 7,324 PSL particles of 100nm diameter we observed no released particle and its displacement was less than 0.1 um.

Acknowledgements

1. This work is supported by NEDO.
2. We appreciate Entegris staff for providing, designing and manufacturing dual pods.
3. We appreciate the effort of Rorze staff in designing and manufacturing the MPE Tool.
4. We would like to thank Masami Yonekawa of Canon for dispersing PSL particles.

2008 International Symposium on Extreme Ultraviolet Lithography

Iterative procedure for in-situ optical testing using an incoherent source

Ryan Miyakawa[1,2], Patrick P. Naulleau[2], and Avideh Zakhor[3]

[1] Applied Science and Technology Group, UC Berkeley, [2] Center for X-ray Optics, Lawrence Berkeley National Laboratory, [3] Dept of Electrical Engineering, UC Berkeley

Contact: Rhmiyakawa@lbl.gov

OPTICAL TESTING WITH AN INCOHERENT SOURCE

Interferometry has long been the standard method for characterizing the aberrations in optical systems. However, since most exposure tools use light from incoherent sources, interferometry cannot be used in-situ. We propose an in-situ iterative procedure that solves for the aberrations in an incoherent imaging system.

Modeling an imaging system with an incoherent source requires the summation over four dimensions, which can be computationally demanding. Implementing the SOCS algorithm outlined below can help to speed up computation times.

HOPKINS INTEGRAL FOR INCOHERENT PROPAGATION OF INTENSITY

$$I = \int\int\int\int [TCCs] \times [object]$$

Transmission cross coefficients function:
$T(f_1, f_2 ; g_1, g_2)$

(Contains all the information about the illumination, coherence, and optics)

SOCS ALGORITHM

The SOCS (Sum Of Coherent Systems) algorithm decomposes the transmission cross-coefficients of the 4-D Hopkins integral using singular value decomposition. The resulting spectral components are grouped by their relative eigenvalue magnitudes, and the smallest terms are thrown out. Performing the SOCS decomposition dramatically increases computation times of the incoherent propagation which makes the iterative reconstruction feasible.

SPECTRAL DECOMPOSITION OF THE TRANSMISSION CROSS COEFFICIENTS

$T(f_1, f_2 ; g_1, g_2) = T_1 + T_2 + T_3 + T_4 + \ldots$ (Spectral decomposition)

$\ldots$ where $T_i = \lambda_i v_i v_i^+$

$$I = \int\int\int\int [TCCs] \times [object] \longrightarrow \Sigma_i |[T_i] * [object]|^2$$

EXPERIMENT SETUP

Incoherent source with finite extent

Partially coherent illumination

Test pattern

$P(x,y) = ???$

Wafer exposure through focus

Development

Binary intensity data

ITERATIVE PROCEDURE FOR RECONSTRUCTION

Propagate to lens

Modulate with pupil trial function

Propagate to various planes

Try new pupil function

Convolve aerial image with resist blur model

Binarize result

Match result to experimental data to generate error function

TRUNCATION OF INTENSITY SUM

The contribution from each of the "coherent systems" is weighted by the eigenvalue of the respective spectral component. Since most of the weight is concentrated in a fraction of the terms, the summed intensity is well approximated by a truncated sum:

$$\sum_i^{N_{max}} |[T_i] * [obj]|^2$$

$$\sum_i^{P < N_{max}} |[T_i] * [obj]|^2$$

Relative eigenvalue weights

**Most of the weight is concentrated in the first few coefficients

DESIGNING THE TEST PATTERN TO CAPITALIZE ON CHARACTERISTIC BEHAVIOR OF ABERRATIONS

Some aberrations have a characteristic signature that can be seen more easily with specific features. Astigmatism for example will print sharp lines for a horizontal pattern at a different location than a vertical pattern. Coma, as seen below, will broaden a lines in a specific direction. Designing a mask that takes advantage of these characteristic signatures will provide more information to the reconstruction algorithm.

Coma turns broadens lines in one direction

ADVANTAGES OF PRINT-BASED ITERATIVE PROCEDURE

- Does not place requirement on source coherence
- Can be performed in-situ without swapping in additional optical elements
- Does not require spatial filters which need large photon flux
- Scales well to lower wavelengths or higher numerical apertures

COMPUTATIONAL CONSIDERATIONS FOR ITERATIVE INCOHERENT IMAGING

Assume that the source has M x M pixels and the object has N x N pixels

Summation over source:

Computations	Num of operations
Coherent image for one source point x (MxM) sources	$(2 N \ln N + N^2) \times (MxM)$
Total:	$N^2M^2 + 2M^2 N \ln N$

Hopkins with SOCS decomposition:

Computations	Num of operations
Generate TCC matrix	$N^2 M \ln M$
Perform SVD Sum	N^2
Sum spectral component transforms	$P N \ln N$
Total:	$N^2 (M+1) \ln M + P N \ln N$

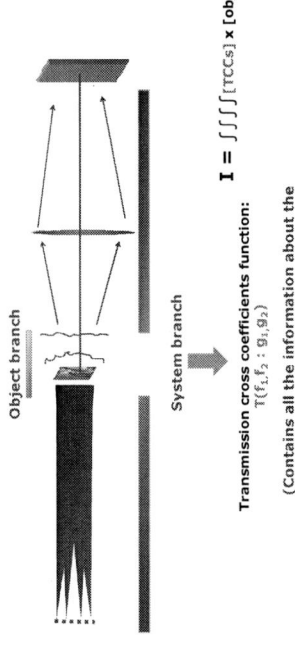
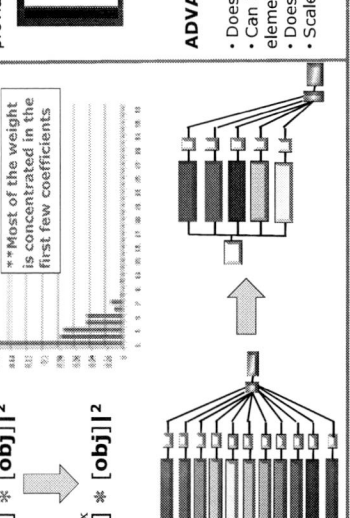

Lateral shearing interferometry for EUV optical testing

Ryan Miyakawa[1,2], Patrick P. Naulleau[2], and Ken Goldberg[2]

[1] Applied Science and Technology Group, UC Berkeley, [2] Lawrence Berkeley National Laboratory

Contact: Rhmiyakawa@lbl.gov

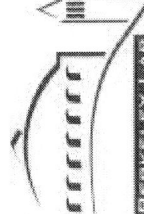

WHY LSI?

As demands increase for high quality optics in the Extreme Ultraviolet (EUV) and Soft X-ray (SXR) regions for applications in high resolution microscopy and lithography, it is becoming increasingly important to have a reliable means for measuring these optics. In Lateral Shearing Interferometry (LSI), the test wavefront is interfered with a shifted (sheared) copy of itself. The shear at EUV wavelengths is created by a low frequency diffraction grating. LSI is an attractive alternative to other interferometric techniques because of its simple setup and loose requirements on coherence and photon flux.

LSI SETUP FOR EUV AND SXR WAVELENGTHS

- Object beam diffracts off of diffraction grating
- Multiple orders propagate at slightly different angles and interfere on the CCD.
- Optical pathlength difference is the function $f(x) = W(x+s) - W(x)$, which approximates the discrete derivative of the wavefront $W(x)$
- Relaxed coherence requirements
- Can operate without large amounts of photon flux
- Robust against mechanical vibrations

GRATING DISTORTION TO THE SPHERICAL WAVE CARRIER

Spatial frequencies incident upon a periodic grating will diffract based on the grating equation:

$$f_{out} = f_{in} + m f_g$$

Since spatial frequencies are related to the sin of the angle, the angle of the outgoing wave will experience a different shift for each incoming angle. A complex waveform that is composed of many spatial frequencies will thus not preserve its shape when it is diffracted into a given order.

- Plane wave illumination diffracts into plane waves
- Grating distorts spherical wave illumination in diffracted orders

+1 order diffracted spherical wave decomposition:

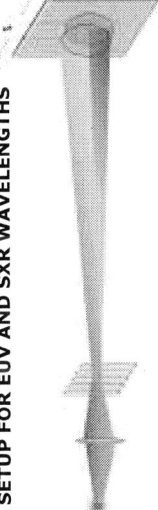

astig   coma

The astigmatism and coma aberrations due to the grating distortion means that the sheared beams are not identical. These errors need to be corrected for when performing the LSI analysis by subtracting out the null interferogram aberrations.

SYSTEM ERROR DUE TO GRATING AND DETECTOR TILT

- Improper alignment of the grating or CCD can exacerbate systematic errors. Astigmatism in particular is very sensitive to grating tilt

- Errors can be determined analytically by treating the grating as a hologram and performing a coordinate transform to the grating coordinates

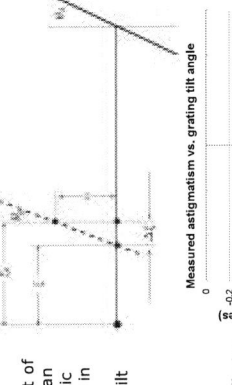

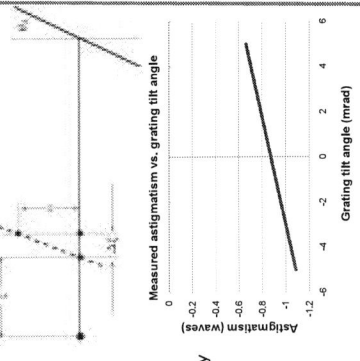

Measured astigmatism vs. grating tilt angle

SYSTEMATIC METHOD TO ALIGN THE LSI SETUP

Although LSI is sensitive to grating and detector tilt angles, it can **leverage** this sensitivity to ensure its own proper alignment. The alignment process is briefly outlined:

- Align CCD to be perpendicular to the optical axis by finding the CCD angle that has the highest density of fringes.

- Collect interferograms for many (~30) different grating tilt angles and measure the resulting tilt and astigmatism in the fringes. Repeat for at least 3 different grating translations.

- Align grating to the tilt of its translation stage axis by noting that aberrations will be independent of grating displacement at this tilt angle

Tilt vs. grating tilt at various β

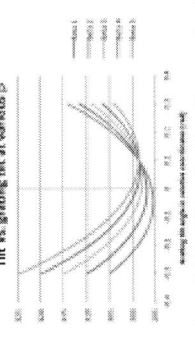

- Perform a polynomial fit of tilt and astigmatism as a function of grating tilt angle using the translation stage tilt as the origin:

$$T(\theta) = \Sigma c_i x^n, \quad A(\theta) = \Sigma d_i x^n$$

Derivatives of C^{th}s

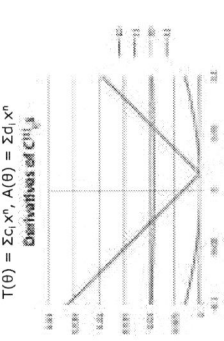

- Either match the resulting linear term c_1 or d_1 of the fit to its analytic model depending on the slope of the model around recovered angle

MONITORING ALIGNMENT USING A LINEAR SYSTEMS ANALYSIS

LSI can be thought of as a linear system D that maps a wavefront of N Zernikes to its X and Y derivatives. Since the dimension of the codomain W is twice that of the domain V, the solution wavefront is overspecified.

$$D: V \to W \qquad D\{v\} = \begin{vmatrix} \partial_x v \\ \partial_y v \end{vmatrix}$$

Where $Dim\{V\} = N$, and $Dim\{W\} = 2N$

Since **Rank{D} ≤ N**, there exists a set $F = W \setminus D\{V\}$, where $F = D\{V\}^\perp$, such that vectors $f_i \in F$ are not mapped by **D** and thus represent forbidden solutions (solutions that cannot come from a real wavefront).

We can then define an error function to determine the validity of the data **w**:

$$E = \Sigma_i |\langle f_i, w \rangle|^2$$

If E is non-zero, then the data vector **w** has a nonzero projection onto **F**, which means that the system has drifted from proper alignment. This technique allows for a **real-time** method to asses the health of the LSI system.

PRECISION OF THE LSI ALIGNMENT

Since LSI is so drastically affected by grating alignment it is important to know how effective the alignment procedure will be in the presence of mechanical uncertainties. The following plots show the relationship between these mechanical uncertainties and the eventual aberrations in the fringes from the resulting calibration errors. For clarity we show only astigmatism, which is the largest systematic error.

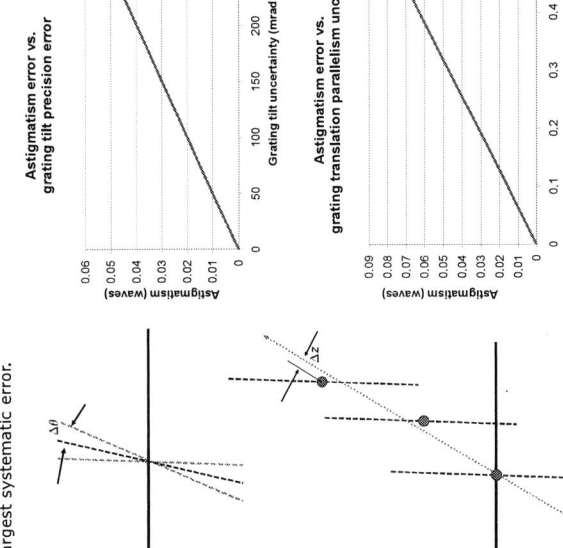

Translation axis

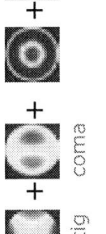
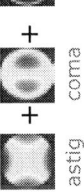

Astigmatism error vs. grating tilt precision error

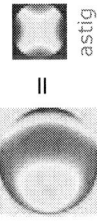

Astigmatism error vs. grating translation parallelism uncertainty

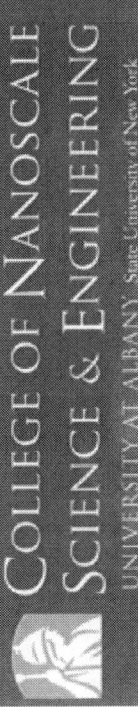

Resist Outgassing Measurements and Calibrations for High Volume Manufacturing

Greg Denbeaux, Alin Antohe, Rashi Garg, Chimaobi Mbanaso, Leonid Yankulin, Yu-Jen Fan

College of Nanoscale Science and Engineering
University at Albany

Kevin Orvek, Andrea Wüest
SEMATECH

COLLEGE OF NANOSCALE
SCIENCE & ENGINEERING
UNIVERSITY AT ALBANY State University of New York

gdenbeaux@uamail.albany.edu

cnse.albany.edu

College of Nanoscale Science & Engineering
NanoFab Complex at the University at Albany

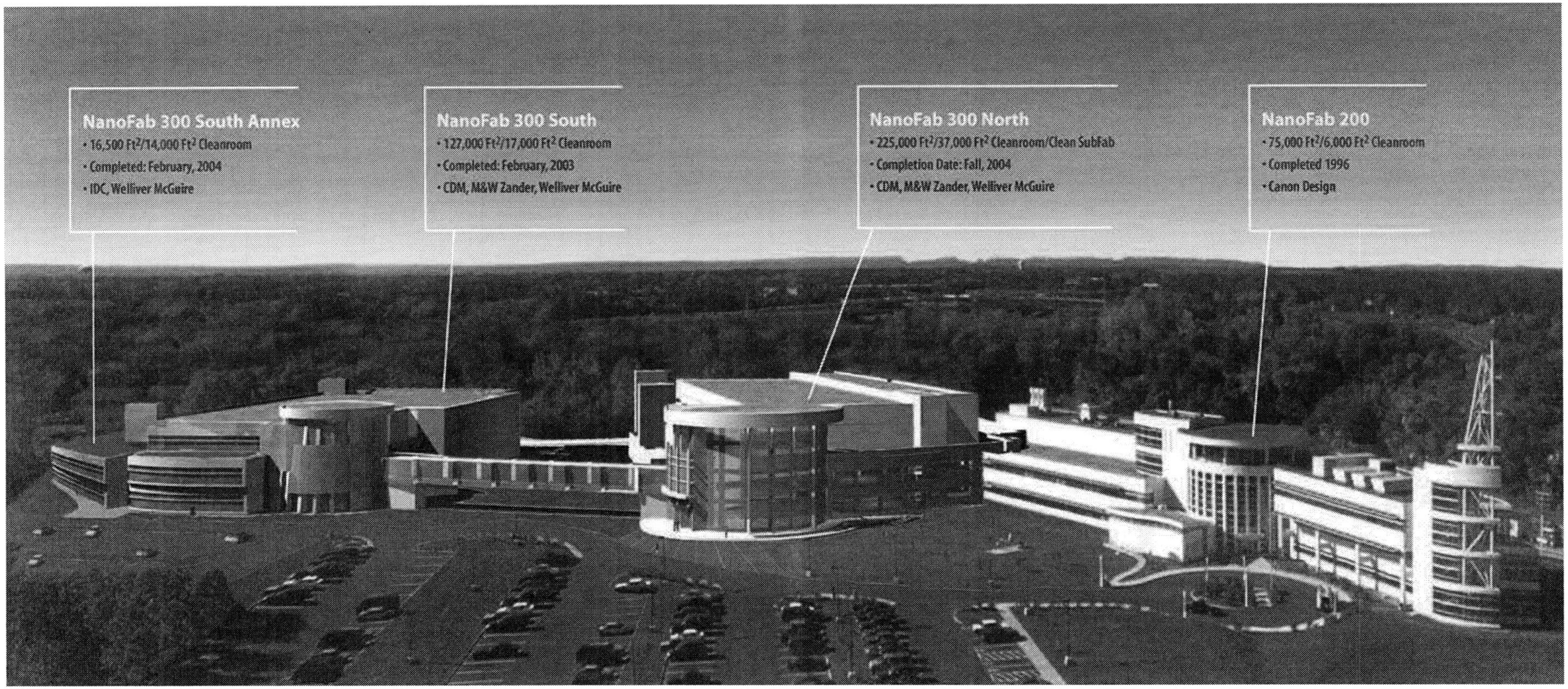

COLLEGE OF NANOSCALE
SCIENCE & ENGINEERING
UNIVERSITY AT ALBANY State University of New York

NanoFab 300 South Annex
- 16,500 Ft²/14,090 Ft² Cleanroom
- Completed: February, 2004
- IDC, Welliver McGuire

NanoFab 300 South
- 127,000 Ft²/17,000 Ft² Cleanroom
- Completed: February, 2003
- CDM, M&W Zander, Welliver McGuire

NanoFab 300 North
- 225,000 Ft²/37,000 Ft² Cleanroom/Clean SubFab
- Completion Date: Fall, 2004
- CDM, M&W Zander, Welliver McGuire

NanoFab 200
- 75,000 Ft²/6,000 Ft² Cleanroom
- Completed 1996
- Canon Design

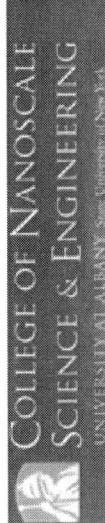

COLLEGE OF NANOSCALE
SCIENCE & ENGINEERING
UNIVERSITY AT ALBANY State University of New York

Diagrams of old and new vacuum chambers

New Chamber

- 300 mm wafer loading
- Collector for higher power
- Spinning rotor gauge and capacitance manometer for accurate pressure calibrations

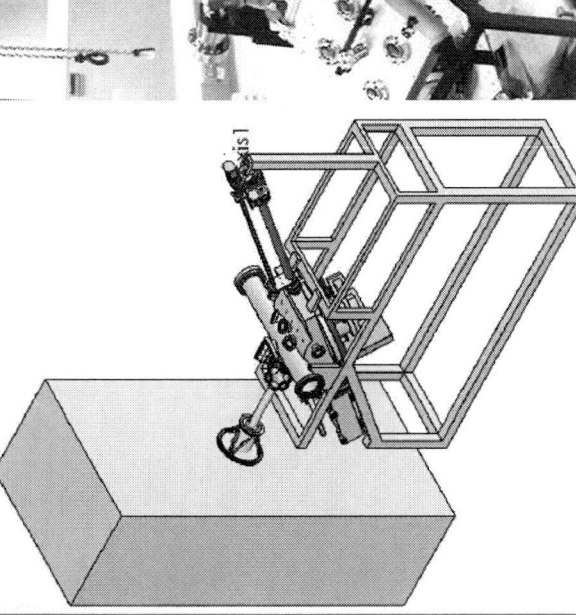

Loadlock

resist

EUV Source
Mass Spectrometer

Old Chamber

- Only an ion gauge for pressure measurements and mass spectrometer calibrations
- Limited to 1" wide pieces of wafers for outgassing

SEMATECH

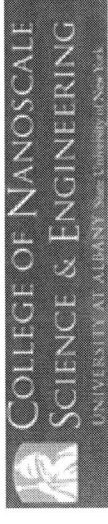

Exposure conditions for Upcoming Witness Plate Testing

- Energetiq EUV source
  - ~10 W into 2π in 2% BW
  - ~50 W into 2π within 11-15 nm
  - Typically 1250 Hz
- Exposed through filter for spectral purity and vacuum separation
  - Zr filters ~11-17 nm bandwidth
  - Zr/Si filters ~ 11-15 nm bandwidth
- Direct beam
  - Few mW/cm² within a few cm² area (limited by filter size)
- Focused from collector
  - Few hundred mW/cm² within a few mm² area
- For outgassing tests, misalign collector so only use direct beam (few mW/cm²)
- For witness plate tests
  - Focused from collector onto mirror (few hundred mW/cm²)
  - Direct beam onto 300 mm wafer (few mW/cm²)
    - Exposure time 1-2 hours per 300 mm wafer

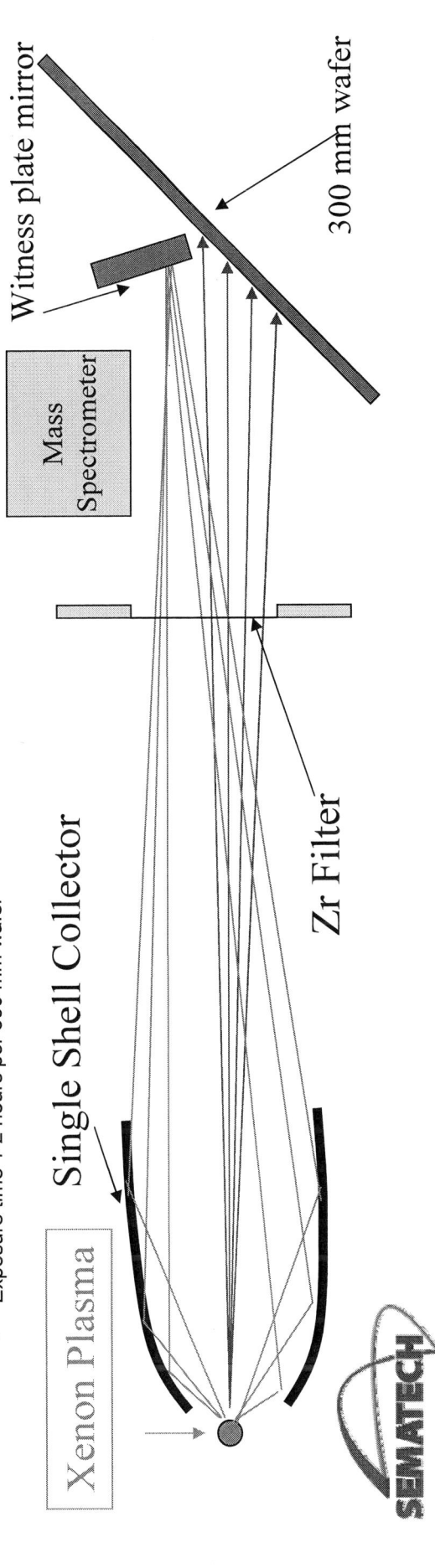

Collector alignment and power measurements

Preliminary measurements based on photoresist exposures show 100-200 mW/cm² intensity in focus

Direct EUV Beam (low intensity)

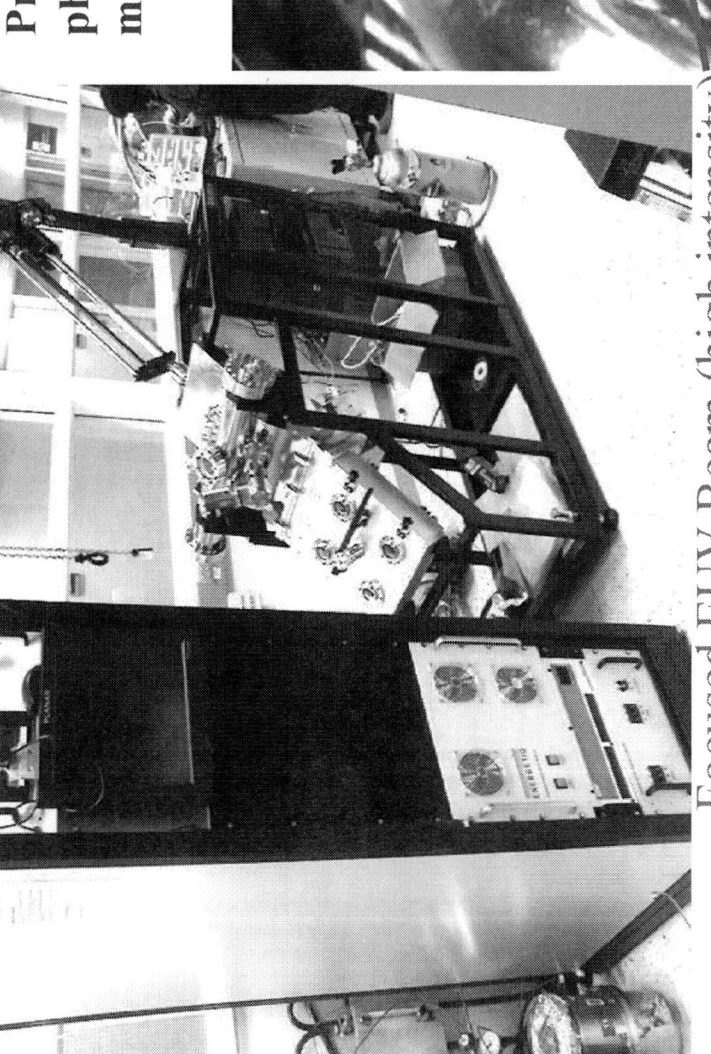

Focused EUV Beam (high intensity)

Visible light from EUV radiation on a phosphor screen

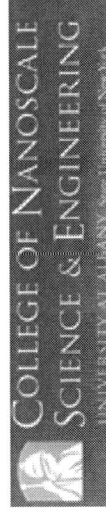

COLLEGE OF NANOSCALE
SCIENCE & ENGINEERING
UNIVERSITY AT ALBANY State University of New York

Witness Plate and Control Mirror Results
(Old Chamber, no collector)

Dose on mirror (J/cm2)

Reflectivity Loss dR/R

- After Chamber Clean Pre Control
- After Chamber Clean Post Control
- After Chamber Clean Resist A
- Post Control
- High outgassing resist
- Pre Control - Bad filter

Prior to chamber clean
After chamber clean

- Before chamber cleaning, there were large reflectivity losses and a wide spread in results
- After chamber cleaning, the results were improved
- The effect of the resist was subtle compared to chamber effects

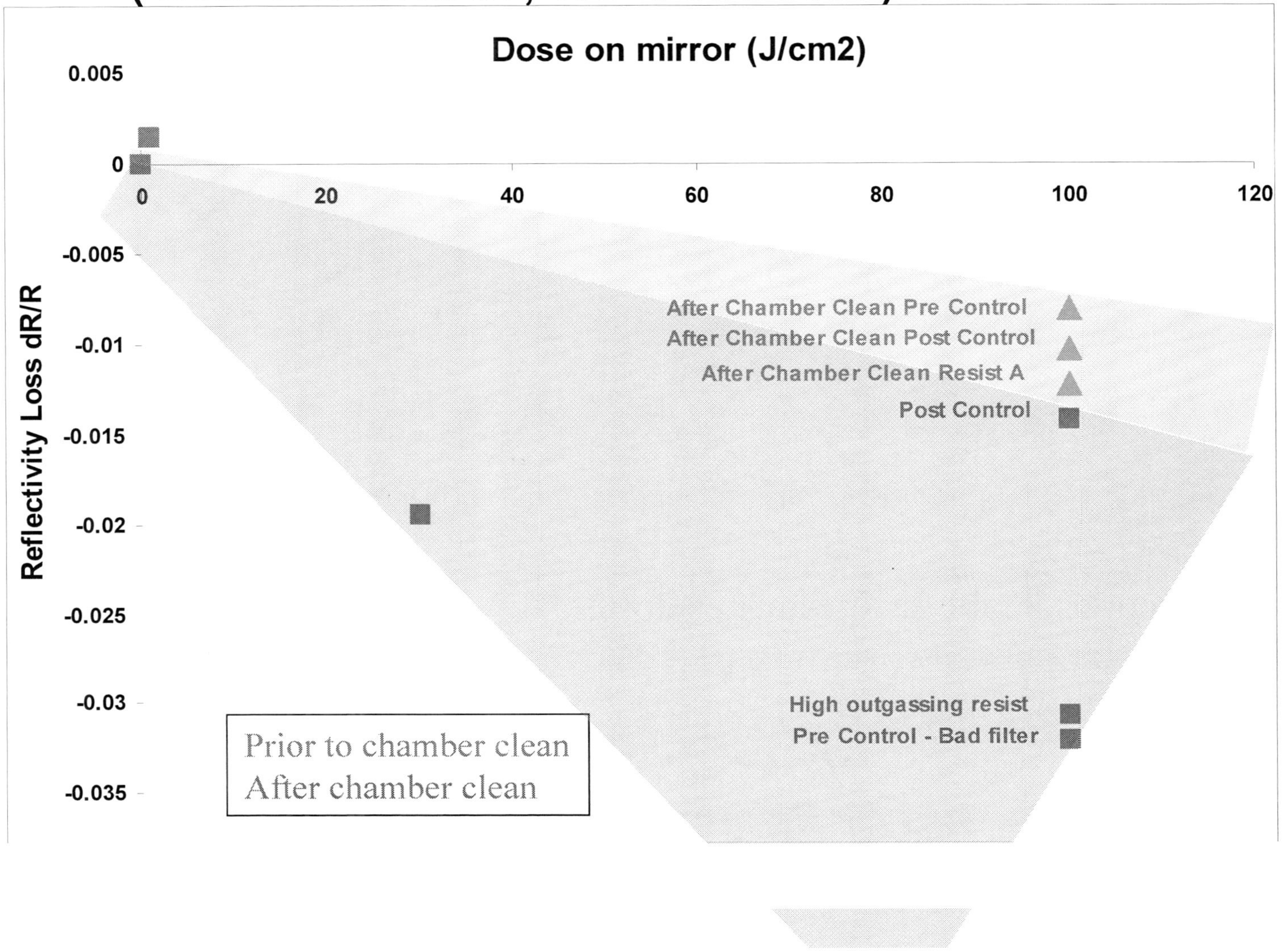

COLLEGE OF NANOSCALE
SCIENCE & ENGINEERING
UNIVERSITY AT ALBANY State University of New York

After Exposure and Reflectivity Loss, XPS Shows Primarily Carbon

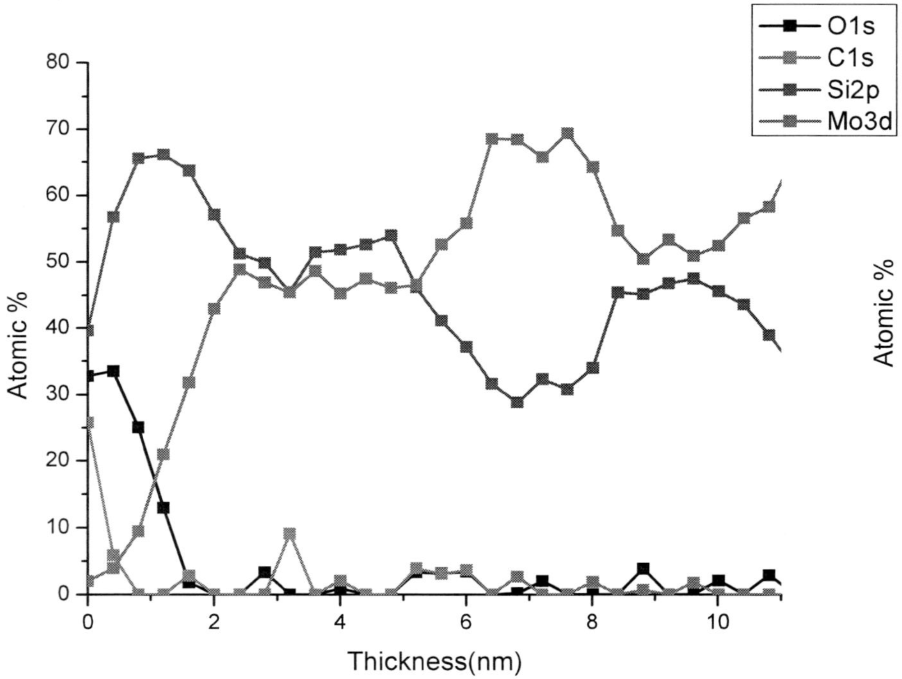

Exposed Witness Plate
Shows primarily an increase in carbon

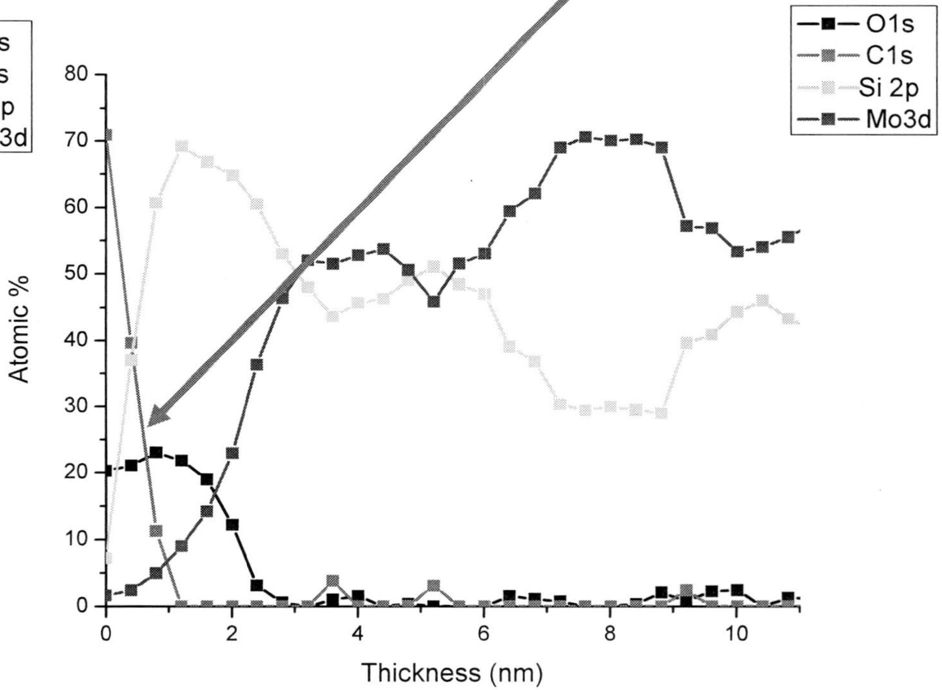

Unexposed Witness Plate

XPS with sputtering to look at materials through sample thickness

Resist Outgassing Due to Vacuum (not exposure)
How long to pump prior to moving wafer near optics?

Typical quartz crystal microbalance (QCM) outgassing test

PMMA resist

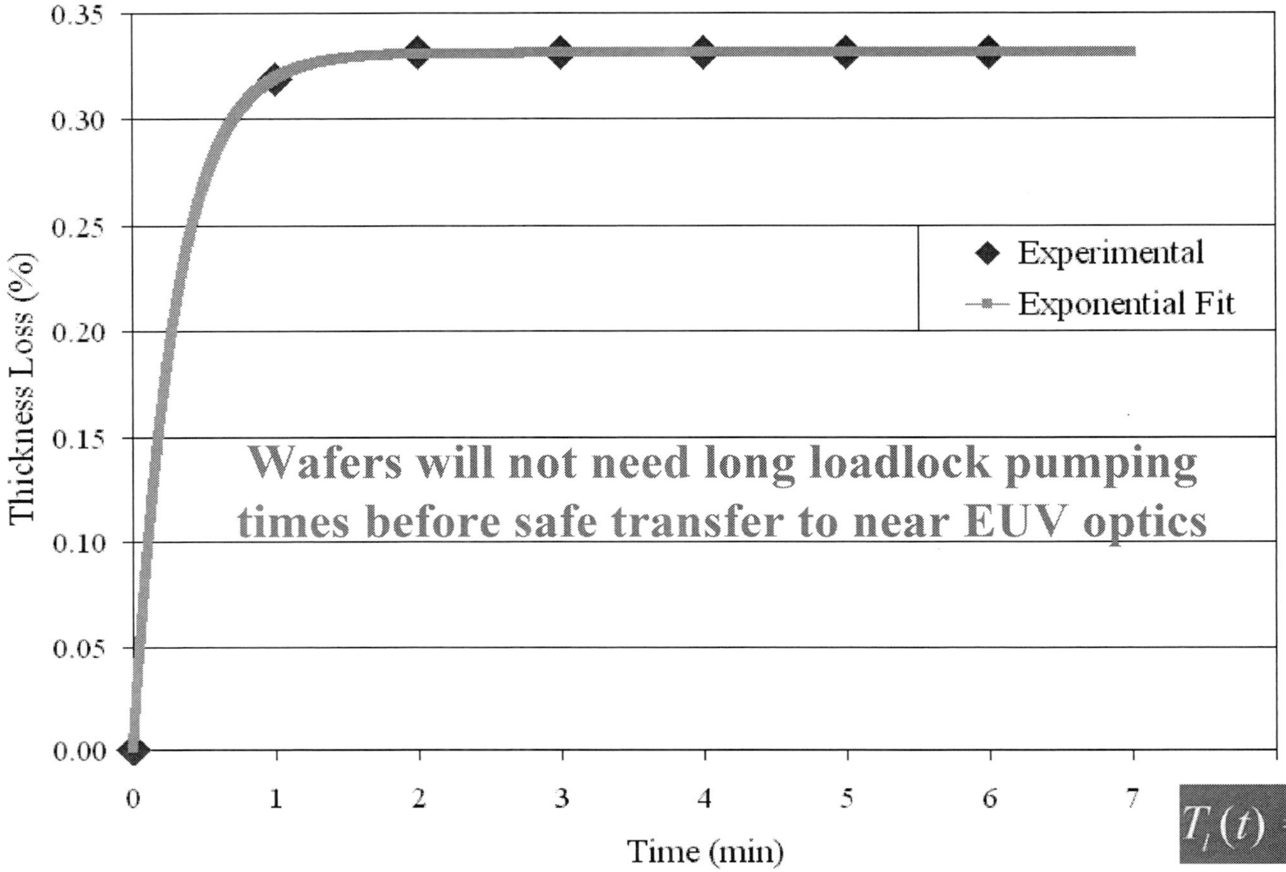

Legend:
- ◆ Experimental
- — Exponential Fit

Wafers will not need long loadlock pumping times before safe transfer to near EUV optics

Y-axis: Thickness Loss (%)
X-axis: Time (min)

$$T_i(t) = T_{i,final}(1 - e^{-t/\tau})$$

Mass Spectrometer Calibration

- Use capacitance gauge in small volume for injection of species
  - Gauge is species independent!
- Use known effective aperture size connecting this volume to the main chamber (after careful measurement of actual aperture size using a spinning rotor gauge)
  - Then with pressure and aperture size known, number of molecules injected in the chamber is known for any species used
- Then measure mass spectrometer sensitivity, counts per second (CPS) for a known number of molecules, Q, entering the chamber
- $Q = C_2 * CPS$
  - Direct measurement between mass spectrometer sensitivity and number of species injected or outgassed

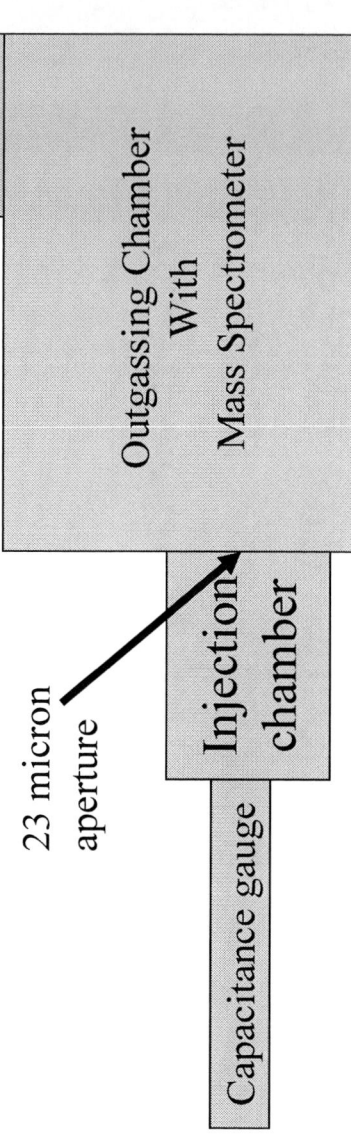

Sample outgassing result
(Resist courtesy of Prof. Robert Brainard)

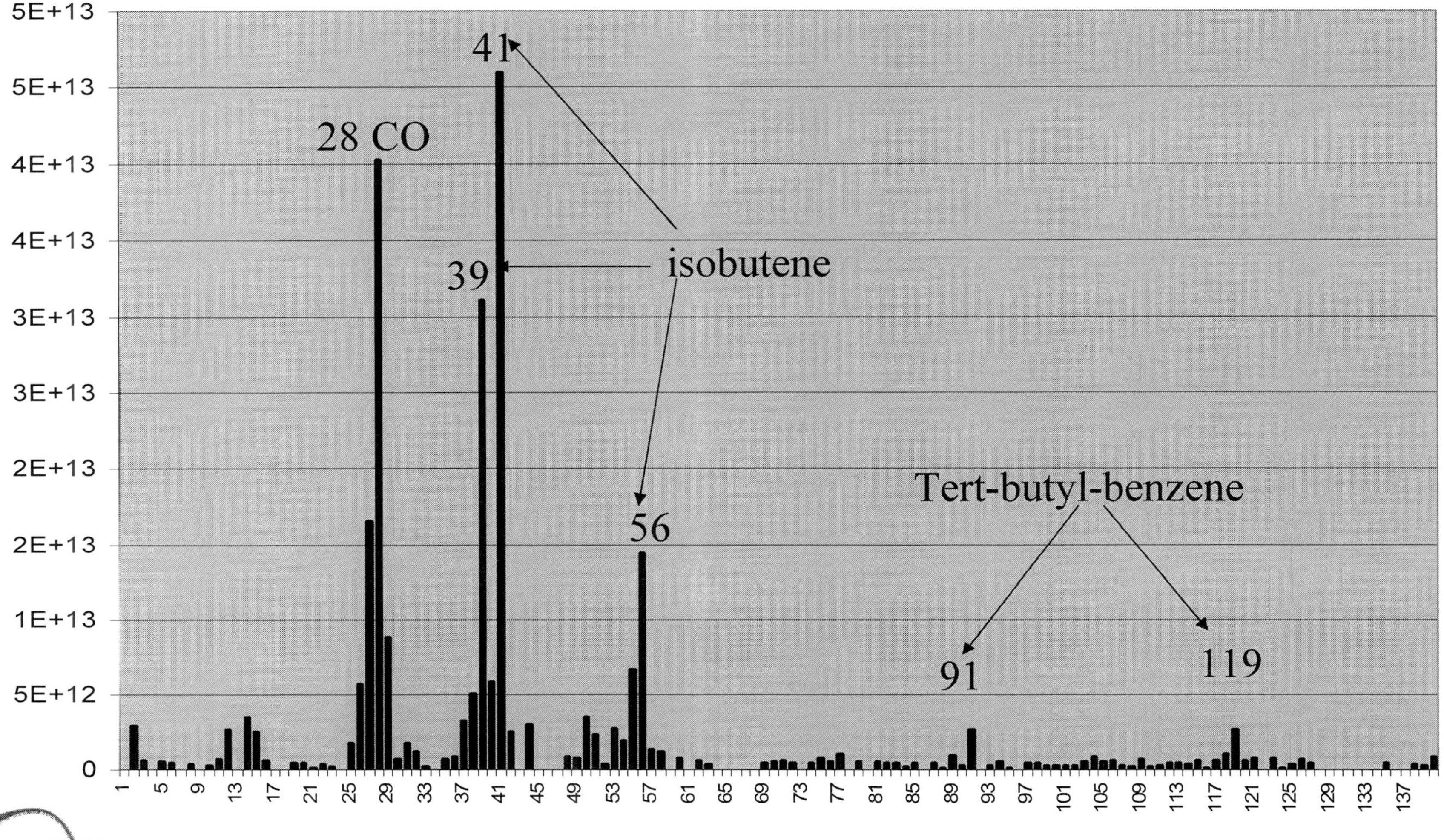

New Chamber Outgassing Measurement Repeatability

Resist	Result	Test Date	(Max-Min)/Min
D	2.2×10^{14}	2/21/2008	25%
	1.8×10^{14}	4/10/2008	
E	3.4×10^{14}	2/21/2008	39%
	2.5×10^{14}	4/10/2008	
F	3.4×10^{14}	2/21/2008	10%
	3.7×10^{14}	4/10/2008	
G	4.5×10^{15}	2/22/2008	15%
	3.9×10^{15}	3/27/2008	
	4.2×10^{15}	3/27/2008	
H	2.9×10^{15}	3/26/2008	30%
	3.7×10^{15}	4/30/2008	
I	2.0×10^{14}	5/8/2008	29%
	2.5×10^{14}	5/9/2008	
J	2.5×10^{14}	5/8/2008	38%
	2.1×10^{14}	5/9/2008	

So far, appears more repeatable than old chamber with old gauges

Contamination Studies of Injected Species

COLLEGE OF NANOSCALE
SCIENCE & ENGINEERING
UNIVERSITY AT ALBANY State University of New York

- We have directly injected a few species at high concentrations of approximately 1×10^{-6} Torr
  - Benzene (known to outgas from resist)
  - Tert-butanol (known to outgas from resist)
  - Diphenyl Sulfide (sulfur compound, possible outgassing component)
- Then, we exposed a mirror to >30 J/cm^2 in these high hydrocarbon environments
- At these high pressures and modest doses, we can not measure reflectivity loss above the measurement accuracy

We have yet to identify any of the outgassed species from resist that contribute significantly to optics contamination!

COLLEGE OF NANOSCALE
SCIENCE & ENGINEERING
UNIVERSITY AT ALBANY State University of New York

Even large hydrocarbons are unlikely to go undetected since molecules fragment during ionization

Two examples of large molecules and their spectra

4-propyl-heptadecane, $C_{20}H_{42}$, 282 AMU   Decyl Disulfide, $C_{20}H_{42}S_2$, 346 AMU

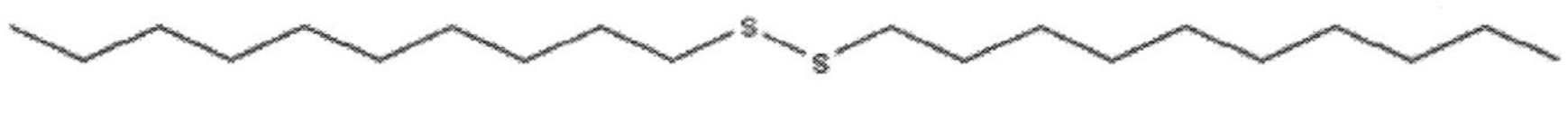

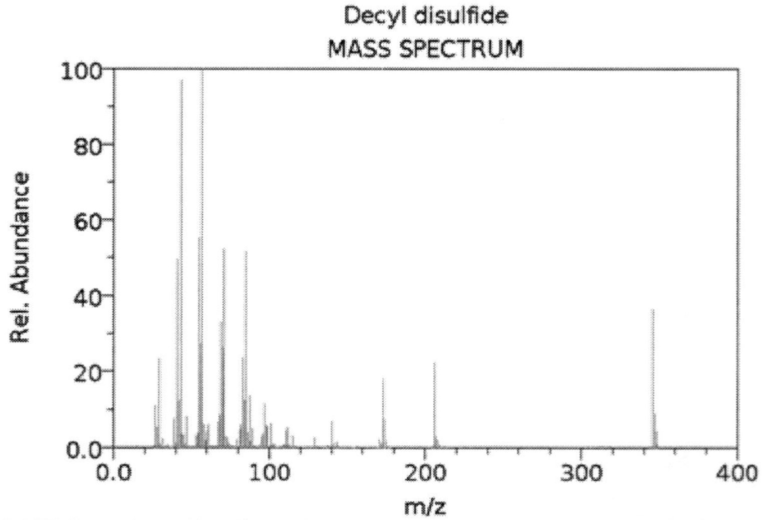

Decyl disulfide
MASS SPECTRUM

Rel. Abundance

m/z

NIST Chemistry WebBook (http://webbook.nist.gov/chemistry)

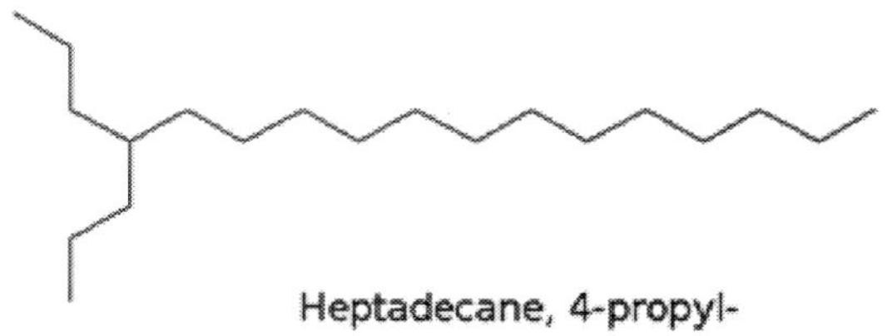

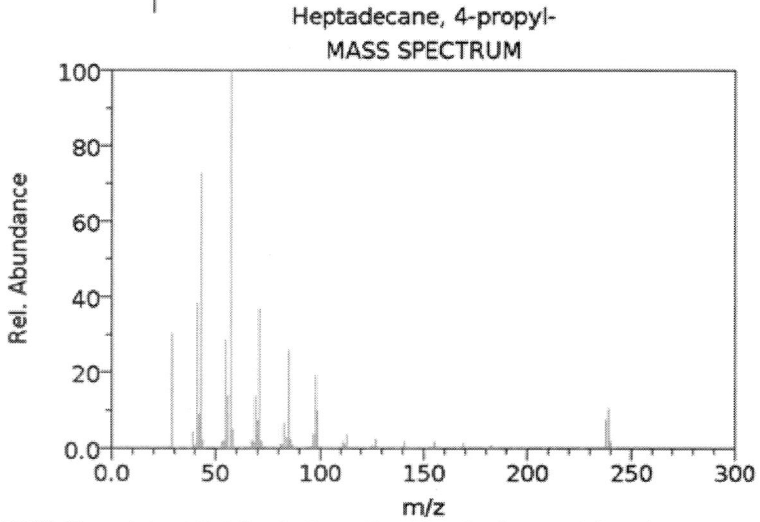

Heptadecane, 4-propyl-
MASS SPECTRUM

Rel. Abundance

m/z

NIST Chemistry WebBook (http://webbook.nist.gov/chemistry)

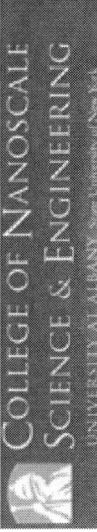

SEMATECH

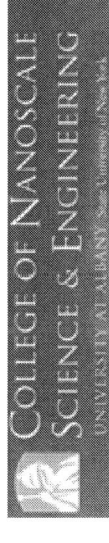

COLLEGE OF NANOSCALE
SCIENCE & ENGINEERING
UNIVERSITY AT ALBANY State University of New York

Summary

- The new chamber has been installed and handles 300 mm wafers for large area exposures

- Outgassing measurements of photoresists using the mass spectrometer are routine

- The new calibration method including a capacitance manometer and a spinning rotor gauge improved repeatability of the outgassing measurements

- The single shell collector has been installed and provides ~100-200 mW/cm^2 intensity (compared to ~ 2mW/cm^2 in direct beam)

- High power outgassing and witness plate exposures will be performed shortly

Experiment of contamination generation by EUV irradiation with the use of high-mass hydrocarbon gas

Masahito Niibe and Keigo Koida
Laboratory of Advanced Science and Technology for Industry (LASTI), University of Hyogo

Background

Contamination of EUVL optics

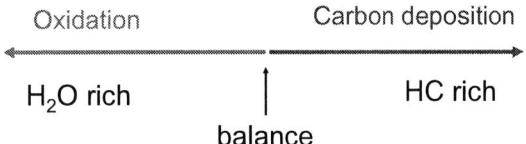

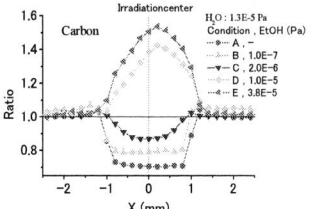

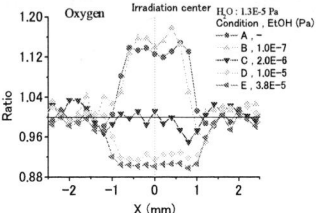

Control of carbon deposition and oxidation by balancing amounts of H_2O and EtOH [3].

Vacuum atmosphere control

Low-mass hydrocarbon (HC) gases, Ex. CH_3OH [1], C_2H_5OH [2, 3], $(CH_3)_2CHOH$ [4]

[1]. S.B. Hill et al., Proc. SPIE, vol. 6151, 61510F (2006).
[2]. L. E. Klebanoff et al., J. Vac. Sci. Technol. A22, 425 (2004).
[3]. Y. Kakutani et al., Proc. SPIE, vol. 6517, 651731 (2007).
[4]. K. Koida et al., Proc. SPIE, vol. 6921, 69213C (2008).

Real exposure system

Out gas from resist, wiring materials, stages etc. → high-mass HC gas

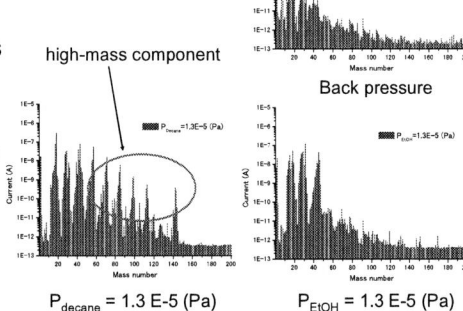

Back pressure

Purpose

Examine the influence of **high-mass HC gas** onto surface contamination

Experimentals

High-mass hydrocarbon gas : **n-decane** ($C_{10}H_{22}$, M=142.28, bp:174 C)

1. Carbon deposition by EUV irradiation
2. Control of carbon deposition by water (H_2O) vapor
3. Remove of deposited carbon by water and EUV irradiation

P_{decane} = 1.3 E-5 (Pa) P_{EtOH} = 1.3 E-5 (Pa)

Mass spectra for BG, EtOH and decane atmosphere

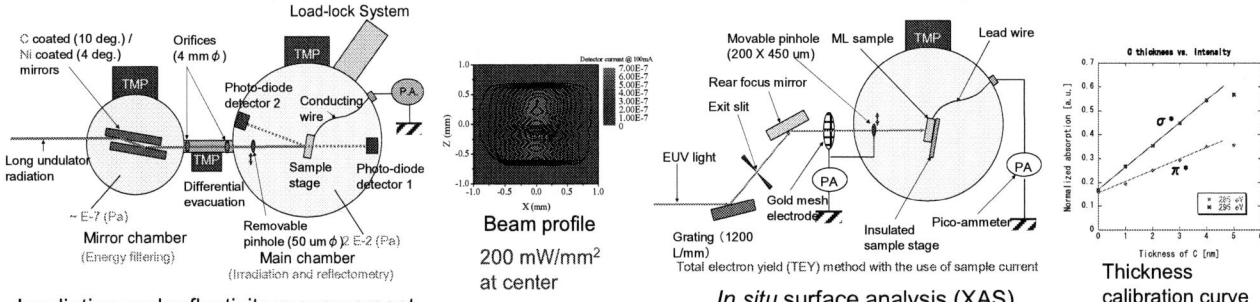

Irradiation and reflectivity measurement

Beam profile
200 mW/mm² at center

In situ surface analysis (XAS)

Thickness calibration curve

Results

1. Carbon deposition by EUV irradiation

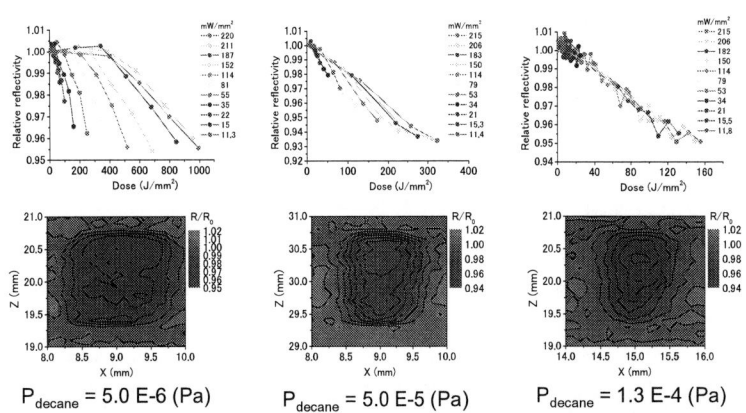

P_{decane} = 5.0 E-6 (Pa) P_{decane} = 5.0 E-5 (Pa) P_{decane} = 1.3 E-4 (Pa)

Reflectivity change at the various EUV intensity points in n-decane introduced atmospheres and reflectivity map after EUV irradiation

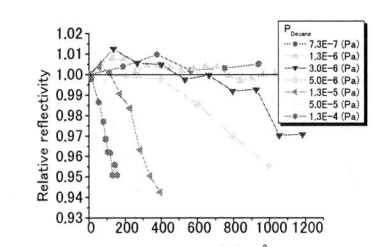

Sequential reflectivity change at irrad. center

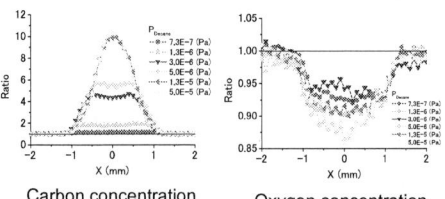

Carbon concentration   Oxygen concentration

Surface analysis by XAS technique

2. Control of carbon deposition by water (H₂O) vapor

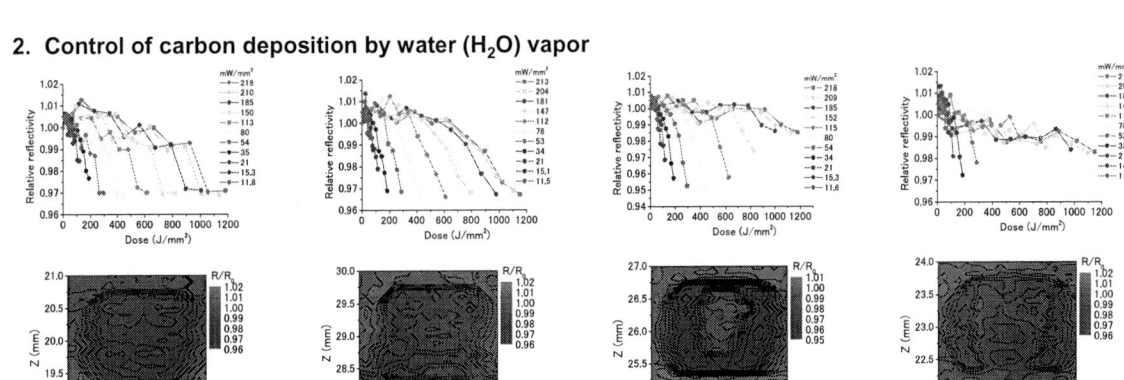

P_{decane} = 3.0 E-6 (Pa)

P_{decane} = 3.0 E-6 (Pa) + P_{H2O} = 1.3 E-5 (Pa)

P_{decane} = 3.0 E-6 (Pa) + P_{H2O} = 5.0 E-5 (Pa)

P_{decane} = 3.0 E-6 (Pa) + P_{H2O} = 1.3 E-4 (Pa)

Reflectivity change at the various EUV intensity points in n-decane and water vapor introduced atmospheres and reflectivity map after EUV irradiation

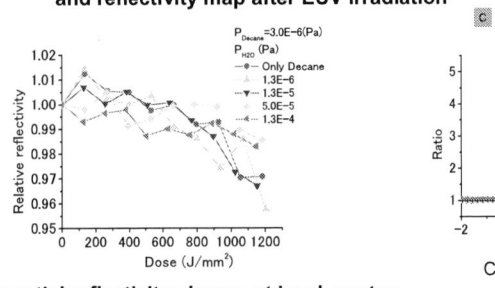

Sequential reflectivity change at irrad. center

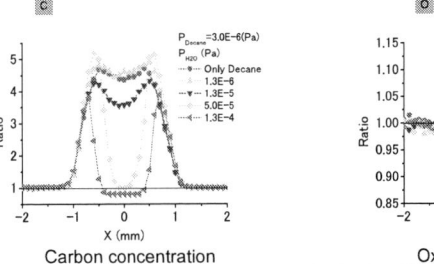

Carbon concentration

Oxygen concentration

Surface analysis by XAS technique

3. Remove of deposited carbon by water and EUV irradiation

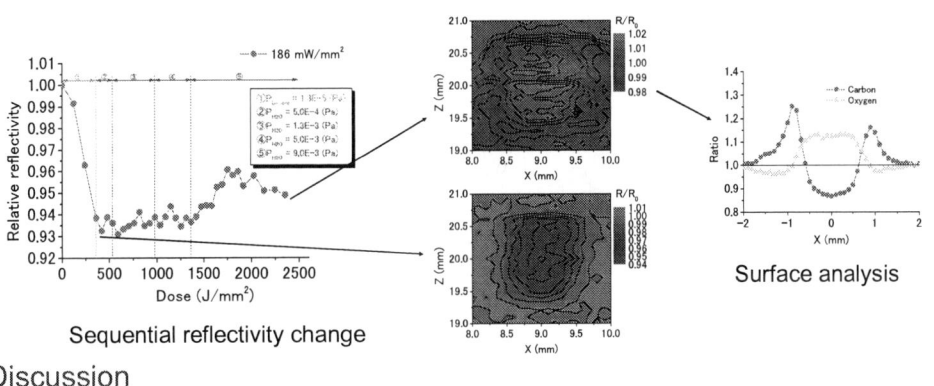

Sequential reflectivity change

Surface analysis

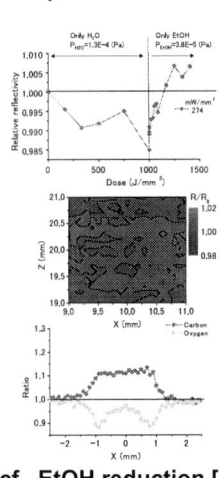

cf. EtOH reduction [4]

Discussion

The origin of non-linear behavior is not clear.

Possibilities 1. Photon induced desorption of HC molecule., 2. Difference of adsorption rate of 1st layer and furthers. 3. ......

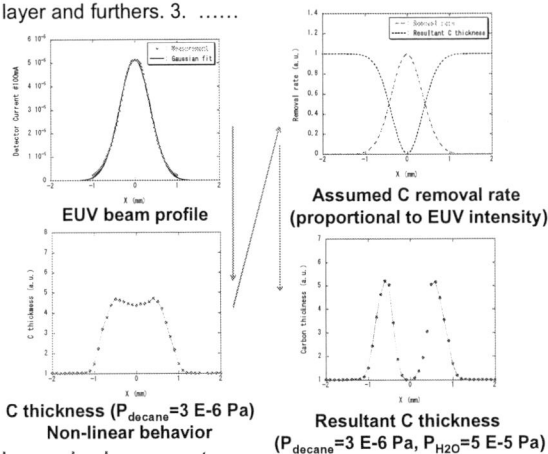

EUV beam profile

Assumed C removal rate (proportional to EUV intensity)

C thickness (P_{decane}=3 E-6 Pa) **Non-linear behavior**

Resultant C thickness (P_{decane}=3 E-6 Pa, P_{H2O}=5 E-5 Pa)

Summary

1. There is a strong EUV intensity dependence (non-linearity) of carbon deposition rate at lower decane pressure. The dependency disappears at P_{decane} > 1.3 E-4 (Pa).

2. Carbon deposition was partially mitigated by water vapor introduction. However, carbon was still deposited at the lower EUV intensity area. (doughnut pattern)

3. Once deposited carbon can partially removed by EUV irradiation in a presence of high water vapor pressure (~E-2 Pa). The removal rate was not uniform.

4. Further study would be needed for deposition, control and removal of carbon contamination which was generated by **high-mass HC gas** and related to **non-linear phenomena**.

Acknowledgements

The authors would like to thank the contamination study team of Nikon and Canon for their helpful discussion.

A part of this work was performed by using the contamination evaluation equipment provided from NEDO.

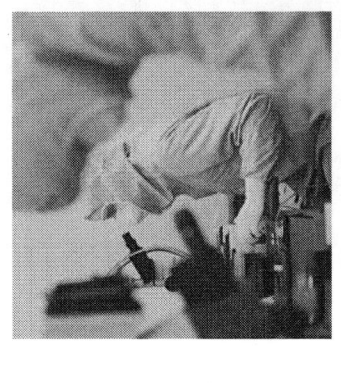
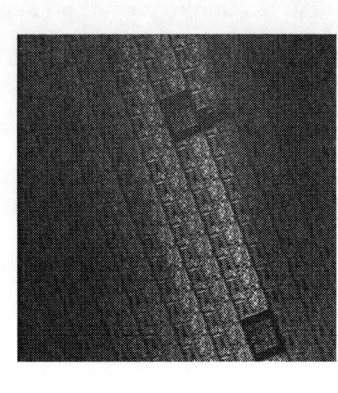
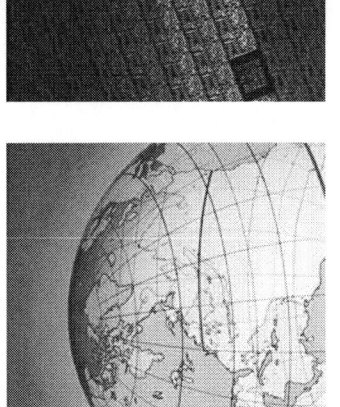

Accelerating the next technology revolution

EUVL Optics Contamination from Resist Outgassing; Status Overview

Kevin Orvek[1], Greg Denbeaux[2],
Alin Antohe[2], Rashi Garg[2],
Chimaobi Mbanaso[2]

[1] SEMATECH
[2] CNSE

INTRODUCTION

- For five years, the semiconductor industry has been testing EUVL resist outgassing levels out of concern for potential contamination of exposure tool optics.

- Many techniques have been used to measure resist outgassing, and early correlation tests showed widely varying values from the different tests and sites.

- The industry is now re-evaluating the resist outgas issue with the advent of full-field tools coming on line and the rapidly increasing number of resist samples that will be required to accelerate EUVL resist development.

- The key issues are whether
  - Resist outgassing is a primary contributor to optics contamination
  - Resist outgas testing is providing us any real information
  - Resist outgas testing should be continued

HISTORY – EUV Optics Contamination and Resist Outgassing

- 2002 - Carbon contamination reported in illuminator of EUV Engineering Test Stand at Sandia Laboratory [1] - attributed to trace organics in system.

- 2003 - Intel Corporation, concerned about photoresist, established a test methodology for their EUV micro-exposure tool (MET) [2]

  – Specification: hydrocarbon outgas [CxHy] $\leq 6.5E+13$ molecules/cm^2

  – Later adopted by SEMATECH for LBNL, Albany MET tools

- 2004 - ASML preliminary spec [CxHy] $\leq 4.7 E+13$ molecules/(cm^{2}sec) [3]

- 2006 - Nikon preliminary spec [CxHy] $\leq 7.0 E+12$ molecules/(cm^{2}sec) [4]

- 2006 - SEMATECH/Intel raises spec to [CxHy] $\leq 6.5E+14$ molecules/cm^2

- 2006 - ASML adopts new test procedure based upon witness damage [5]

  – Spec allows < 2.0% reflectivity loss to witness sample near resist

- 2007 - SEMATECH [6] and Intel Corporation [7] report carbon contamination in the illuminator optics in the MET tools

Overview of Two Approaches to Resist Outgas Measurement

⋏ Two approaches are currently used to set specifications on resist outgassing:

1. Count the number of outgas molecules and set a specified limit.

   • e.g., SEMATECH spec of total hydrocarbons based on RGA counting

   • Can be reported as total molecules or as an outgas rate

2. Determine how much "damage" (reflectivity loss) the resist contributes to a witness sample and set an amount that is permissible.

   • e.g., ASML witness plate testing

⋏ In the end, both approaches should be based upon some reasoning of how much outgassing is allowed in order to meet throughput and optics lifetime requirements in EUVL scanners.

   — Should take into account in situ protection and/or in situ cleaning techniques if they are available on the scanner

1653

Counting Molecules Approach:

How Intel/SEMATECH Specification was Determined

MET primary mirror M2 assumed to be the optic at greatest risk from resist outgas products (only optical lens surface with a direct line-of-sight to the resist).

2006 – current Spec

- Same

- 64 fields/wafer, 36 wafers/week, 40 weeks/year, 3 years.

- 20% of all outgas molecules hit M2.

- 85% of all molecules that hit stick to M2.

- Each molecule that sticks generates 5 carbon atoms.

$\leq$ 6.5E+14 molecules/cm^2

2003/2004/2005 Spec

- M2s 'permitted' to acquire 1 monolayer of carbon after 3 years (about 0.3% reflectivity loss)

- Usage: 64 fields/wafer, 30 wafers/week, 48 weeks/year, 3 years.

- Surface impingement: 100% of all outgas molecules hit M2.

- Surface sticking probability: 100% of all molecules that hit stick to M2.

- Molecular disassociation: Each molecule that sticks generates 10 carbon atoms.

$\leq$ 6.5E+13 molecules/cm^2

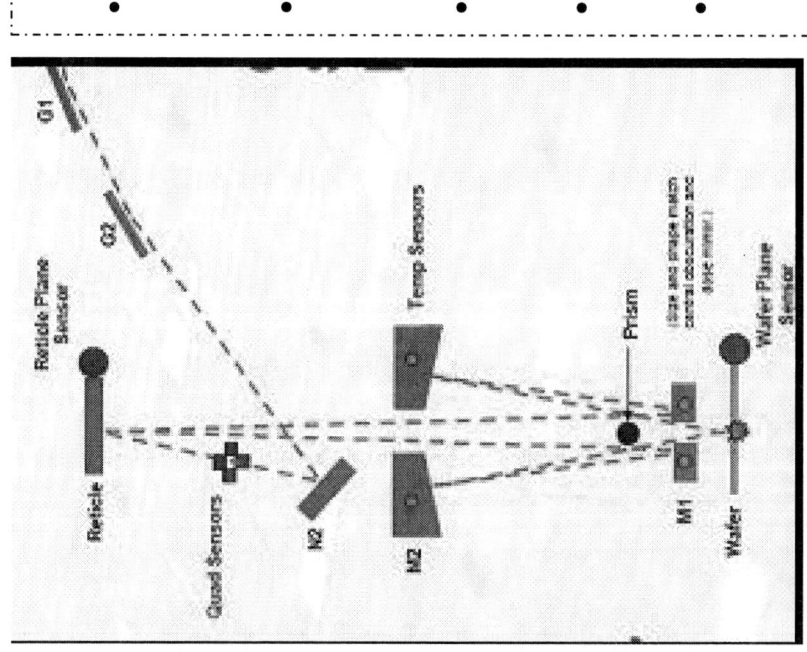

SEMATECH/Intel Approach

△ Still <u>very</u> conservative assumptions in current resist outgas spec level.

△ Yet <u>most resists</u> tested in 2007 passed SEMATECH/Intel outgassing limit.

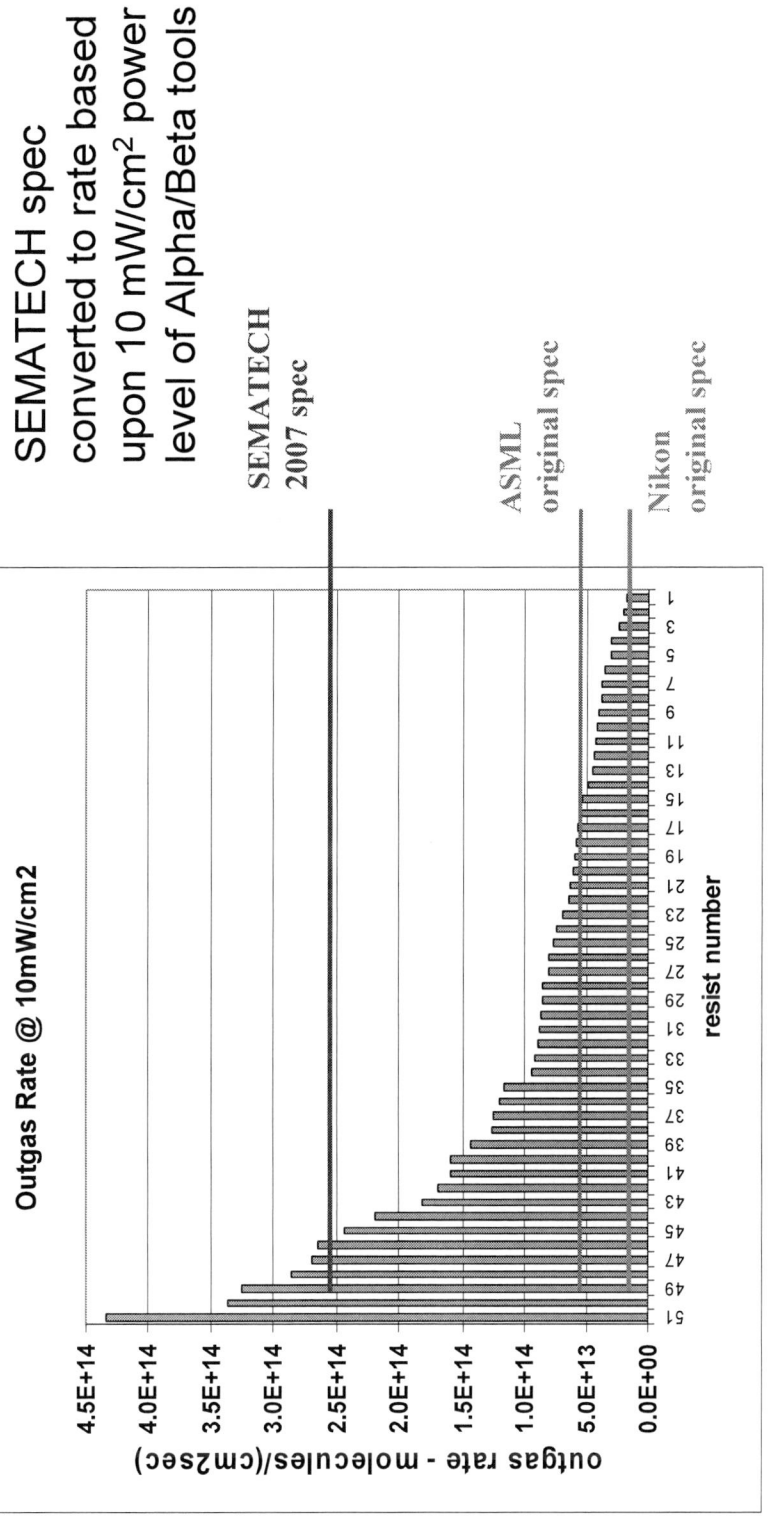

SEMATECH spec converted to rate based upon 10 mW/cm^2 power level of Alpha/Beta tools

SEMATECH 2007 spec

ASML original spec

Nikon original spec

Outgas Rate @ 10mW/cm2

- 119 tested out of 224+ used on MET tools in 2007, 51 commercial shown

- Sometimes we allowed resists on MET tools even if they failed our spec.

- How have we done on protecting the MET tools?

Optics Contamination on SEMATECH's MET Tool

⚑ SEMATECH has swapped illuminator mirror optics twice due to contamination

⚑ With each swap, total system power has been restored to ≥ original value

  ▪ **Intel has experienced same response**

⚑ No sign of significant accumulation of carbon on MET primary optics

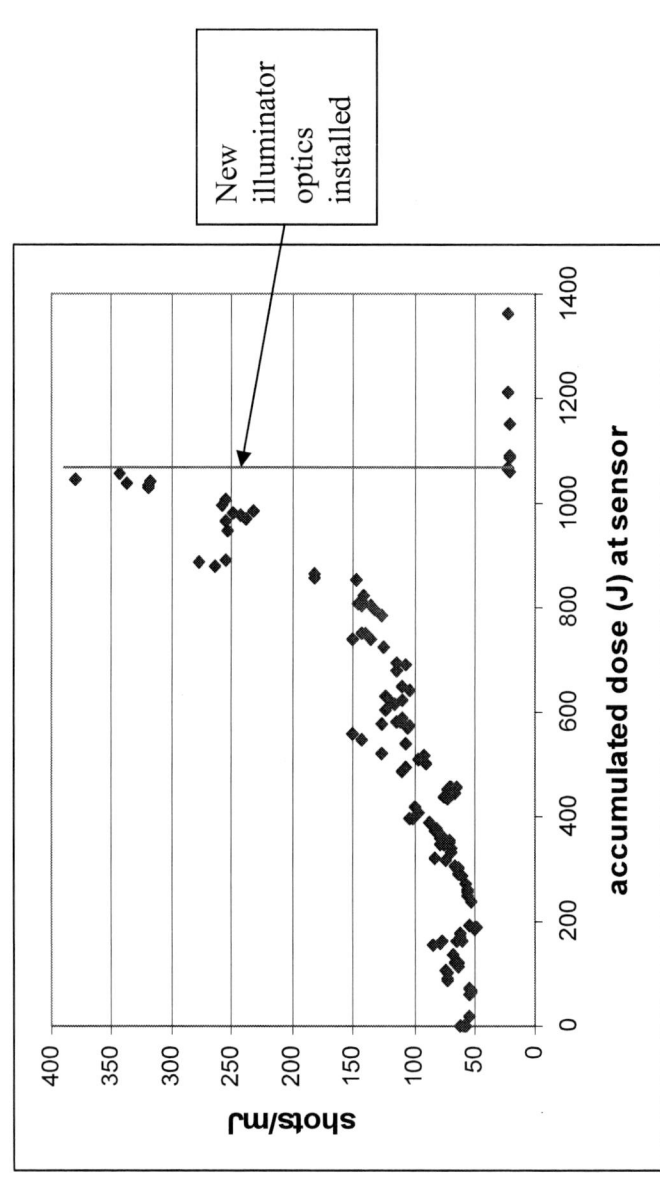

New illuminator optics installed

Shots required per 1 mJ of dose as a function of accumulated dose

It looks like Sandia was correct in 2002; cause is residual organics

Determining Damage Approach:
ASML's Witness Plate Test Methodology

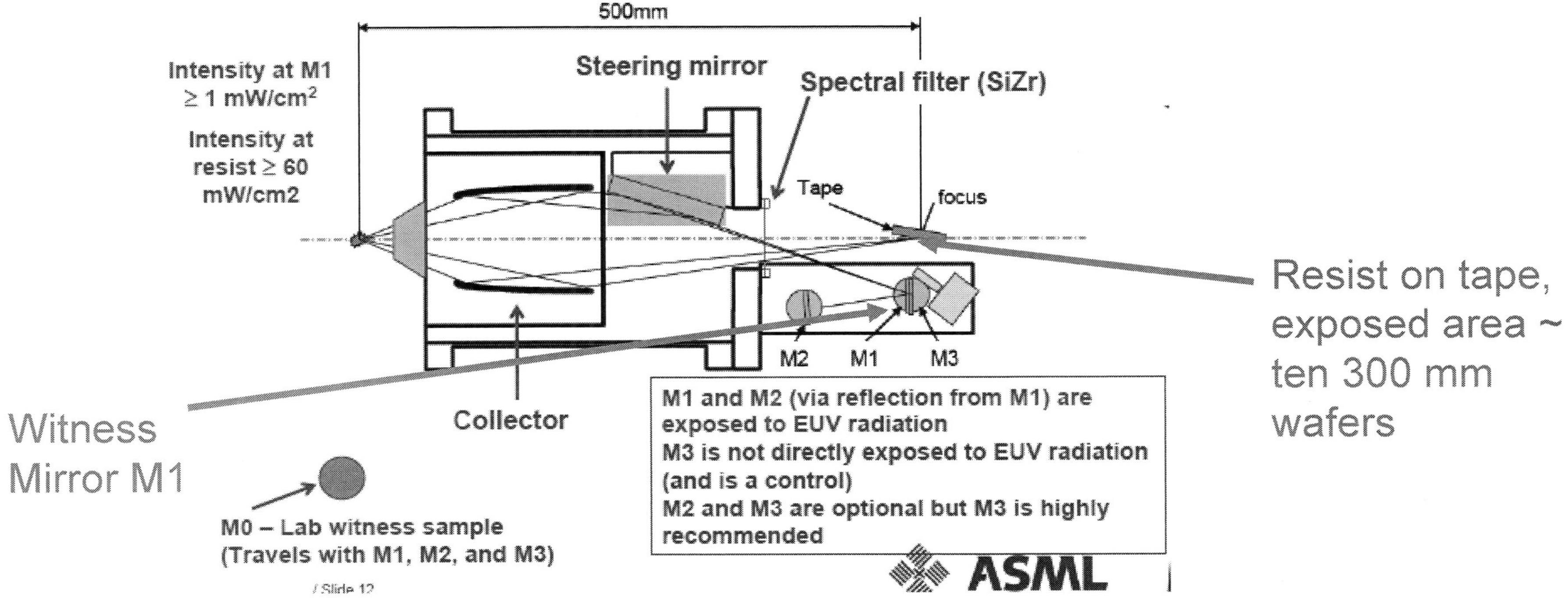

Resist on tape, exposed area ~ ten 300 mm wafers

Spectral filter (SiZr)

focus

Steering mirror

Tape

500mm

M2   M1   M3

Collector

Intensity at M1 ≥ 1 mW/cm²

Intensity at resist ≥ 60 mW/cm2

Witness Mirror M1

M0 – Lab witness sample (Travels with M1, M2, and M3)

M1 and M2 (via reflection from M1) are exposed to EUV radiation
M3 is not directly exposed to EUV radiation (and is a control)
M2 and M3 are optional but M3 is highly recommended

ASML

N. Harned, "Mirror Reflectivity Loss and Resist Outgassing Rates," IEUVI Optics Contamination TWG, February 28, 2008

• Witness mirror M1 allowed to have reflectivity loss just below 2%!!!

• That's a huge amount of carbon allowed (~2 nm) due to ASML outgas mitigation capability on ADT tools. Can we get that much carbon from resist outgassing?

How Much Contamination is Due to Resist?

➤ Our witness plate testing shows contamination due to resist is swamped by contamination due to the chamber

Residual Organics Again !!

	Pre-test, chamber no resist	Test with resist	Post-test, chamber no resist	Average chamber contribution	Resist contribution (subtract chamber)
% Reflectivity Change	- 0.8%	- 1.3%	- 1.1%	- 0.95%	- 0.35%

➤ What if we swamp the chamber with typical resist outgas species?

- Directly inject at high concentrations of ~ 1×10^{-6} Torr
  - Benzene
  - Tert-butanol
  - Diphenyl Sulfide (sulfur compound)

Contamination Studies of Injected 'Resist' Outgas Species – continued

Chosen species for injection and exposure of mirrors to measure contamination

Contaminants	Formula	Structure	Molecular weight (amu)	Boiling point (°C)
Benzene	C_6H_6		78	80.1
Tert-butanol	$C_4H_{10}O$		74	82
Diphenyl Sulfide	$C_{12}H_{10}S$		186	296.2

Sulfur containing – feared from 193 nm days

Contamination Studies of Injected 'Resist' Outgas Species - continued

Reflectivity results due to contamination from these species

Chamber conditions	Chamber pressure (Torr)	Exposure time (hours)	Total dose (J/cm²)	Number of pulses (millions)	Reflectivity change (ΔR/R) %
Clean Chamber (background)	2.5×10^{-8}	8	29	36	- 0.35
Benzene	1×10^{-6}	8	29	36	- 0.35
Tert-Butanol	3×10^{-6}	8	11.5	36	+ 0.09
Diphenyl Sulfide	1×10^{-6}	4.2	15	19	- 0.1
Diphenyl Sulfide	1×10^{-6}	3.6	13	16	+ 0.23
Diphenyl Sulfide	1×10^{-6}	2.9	42	13	- 0.1

➢ No significant reflectivity loss for these species at these pressures & doses

➢ All injected species ΔR/R ≤ Chamber Background Contribution !

Residual Organics Again !!

We have yet to identify any of the outgassed species from resist that contribute significantly to optics contamination in current power witness-type tests!

Comparison of Injected Test Conditions to Outgassing from Resist Wafers

- Benzene is one of the common resist outgassing components

- From our outgassing measurements, a typical resist outgasses 5×10^{13} molecules/cm^2 of benzene

- In our 8-hour, 36 million pulse, 29 J/cm^2 exposure, we require 2.8×10^{20} molecules of benzene in a chamber with a pumping speed of 300 liters/second to keep the pressure at 1×10^{-6} Torr

- Equivalent to Benzene outgas from _8000_ resist wafers sitting in the chamber during the testing

- **Some outgas species are clearly not contributing to carbon contamination in witness-type testing at these power levels!**

- Only data presented to date showing significant witness plate contamination by resist outgas species is NIST data [8] using high concentrations and $\geq 2{,}000$ J/cm^2 dose.

Can Anything Contribute Significant Contamination to Moderate Witness Plate Tests?

Materials Tested with Witness Plate	Filmetrix carbon thickness measured
Apiezon vacuum grease	No signal above background
Heated Apiezon vacuum grease	No signal above background
Neoprene	No signal above background
Heated neoprene	No signal above background
Room temperature carbon conductive tape	No signal above background
Heated (~ 100's of degrees C) carbon conductive tape	2 to 3 nm per exposure hour

⯈ Tests done with EUV and broadband radiation (more likely to contaminate) total power density 0.1 to 1.0mW/cm^2.

⯈ Only 1 material (heated Carbon tape) found to contribute witness plate contamination consistent with 'failure' on ASML type test.

⯈ Witness plate testing at these power levels does not show any problems with most organics.

Discussion – Current Scanners

▲ No data exists to date implicating resist outgassing in any tool optical contamination

▲ No data exists showing resist contamination significantly contributes to witness plate contamination above residual chamber organic contributions at current small-field and full-field power levels

▲ All tests to date indicate residual hydrocarbons are the dominant source of contamination at current power levels

▲ No justification has been shown for continuing resist outgas testing of any kind for small-field and current full-field low power scanners

▲ Significant resources are required to judge which resists should be tested, to arrange for testing, to perform the tests, and to report the results

SEMATECH is discontinuing resist outgas testing for our MET tools with conventional PAG resists.

Discussion – Future Scanners

⋏ Future scanners will need to improve orders of magnitude in residual organics before resist outgassing should be considered as any significant threat.

⋏ Future scanners will hopefully have in situ carbon cleaning techniques; the presence of such capability should be fully comprehended in any budget analysis of allowable hydrocarbons from all sources.

⋏ If in situ cleaning is successful on future scanners, then super-cleanliness is not required, and resist outgassing is not of concern.

⋏ If we need super-clean scanners, then developing techniques to qualify resists will be very difficult, as they need to meet the following requirements:

- Cheap
- Fast
- Accurate

REFERENCES

1. L. Klebanoff, et al., "Environmental data from the engineering test stand," Proc. SPIE, Vol. 4688, 310 (2002).

2. H. Cao, et al., "Quantification of EUV Resist Outgassing," SEMATECH 3rd International EUVL Symposium, Poster session, November 2004.

3. 2004 SEMATECH Resist Advisory Group meeting

4. T. Aoki, K. Murakami, "Acceptable Photoresist Outgassing," IEUVI Resist TWG meeting, October 2006.

5. B. Wolschrijn et al., "New Method for Resist Outgas Qualification," IEUVI Resist TWG meeting, October 2006.

6. A. Wuest, "Optics and Mask Contamination in SEMATECH EUV Micro-Exposure Tools," IEUVI Optics Contamination and Lifetime TWG meeting, March 2007

7. M. Chandhok, "Intel Update Contamination of EUV MET mirrors," IEUVI Optics Contamination and Lifetime TWG meeting, March 2007

8. S. Hill et al., "EUV Resist Outgassing: Data from NIST," IEUVI Optics Contamination TWG meeting, February 2008.

Moisture and Hydrocarbon Management for EUVL tools: Ultra High Vacuum and Purge Gas purification Solutions

Andrea Conte, Cristian Landoni, Paolo Manini
SAES' Getters S.p.A., Lainate (MI), ITALY

Larry Rabellino, Sarah Riddle
SAES' Pure Gas Inc., San Luis Obispo, CA 9340, USA

Abstract

In the present work, the relevant gas sorption properties of Non Evaporable getter (NEG) pumps and purging gas purification solutions for Extreme Ultraviolet Lithography (EUVL) tools are described. Experimental results obtained both purifying typical purge gases (CDA, N_2, H_2) at Part-per-Trillion levels and pumping water and selected organic compounds relevant for EUVL will be presented. These results indicate that the combined use of NEG pumps with state-of-the-art purge gas purification solutions could be effectively adopted to mitigate the carbon/oxygen contamination in UHV systems, improving EUVL optics lifetime and efficiency.

Introduction

Extreme Ultraviolet Lithography is one of the most promising technologies for high volume manufacturing lithography at the 32nm half pitch node and beyond.

One of EUVL technology development hurdles is related with achieving ultra high vacuum level (UHV) during tool operation keeping hydrocarbon and water partial pressures to extremely low levels.

Moisture and Hydrocarbon impurities, either outgassing from the EUV Tool chamber or directly coming from chamber's gas purge ports, interact with the sensitive optics and mirror surfaces and, with EUV radiation, decompose to generate carbon-based films and oxides. These effects are detrimental to the optics performance, significantly reducing the lifetime and increasing the cost of ownership.

Several manufacturers and end-users are evaluating the possibility of purging the EUV chamber with ultra high purity gases and using Non Evaporable Getter (NEG) materials to mitigate the chamber material outgassing.

General properties of NEG pumps and Gas Purifiers, as well as the specific results obtained in pumping water and hydrocarbons under EUV tool typical working conditions will be discussed and reviewed in this study.

What is a NEG Pump

Non-Evaporable Getters are chemical pumps that absorb active gases, such as H_2O, CO, CO_2, O_2 and N_2, while forming a solid solution in the presence of hydrogen.

A NEG pump captures gases by means of a chemical reaction between a chemically active metal or alloy (the getter) and a chemically active gas, thus forming a solid compound with low vapour pressure. The active gases are therefore permanently removed from the sealed device. A getter does not pump noble gases as they do not chemically react.

EUVL Purge Gases Purification Solutions

Critical lenses and optics assembly protection with continuous gas purge is a common strategy for leading edge lithography applications already for 193 nm technologies. In the EUVL world, gas purge solutions are typically implemented not only during chamber pressurization but are also common during critical tool elements manufacturing (such as mirrors, chamber elements etc). Gas purification solutions need to be customized for extreme and aggressive removal of optics detrimental impurities such as H_2O, Hydrocarbons, Refractory compounds (hydrocarbon containing one or more hetero-atoms), Inorganic Acids and Bases at pptv levels. Pressure drop performance and particle filtration are also parameters to carefully consider.

Purity Performances

A dilution manifold for this experiment was used to allow the mixing of certified sources of H_2O and toluene. All the impurities were injected at the same time and the purifier efficiency was tested for several days at a flow rate of 5 slpm and a pressure of 25 psig.

Analytical evaluation of the impurities was carried out as follows:
- H_2O: Inlet AMETEK 5900; Outlet SENSAR TOF-2000 (APIMS)
- Toluene: Inlet NIRA Automatic TD-GC-FID; Outlet CollectTorr Sampling system

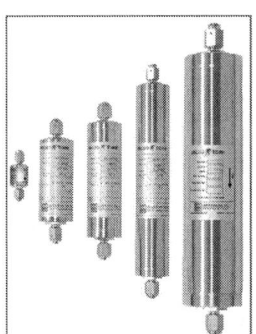
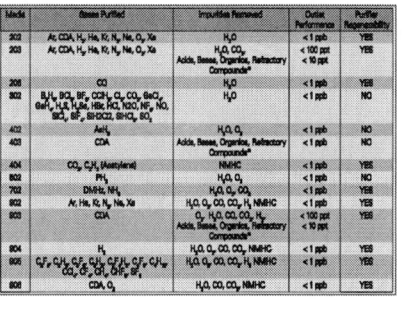
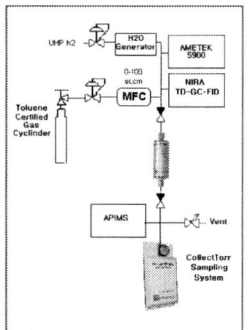
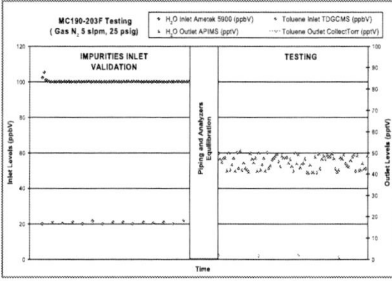

NEG Pumps Possible Application in EUV Lithography

The test system used to characterize the getters is a standard UHV test bench for dynamic sorption tests according to ASTM F798-97. Test gases were CO, H_2, N_2, H_2O, $C_{10}H_{22}$ (decane), C_7H_8 (toluene) and $C_5H_8O_2$(Metyl- metacrylate, MMA).

The getter pump was exposed to the test gas at room temperature, after being activated under standard conditions (500 °C x 1 hour).

- Assuming a toluene partial pressure of 10^{-8}Torr, the SAES CapaciTorr- D 2000 pump can effectively sorb it for ~2000 hours of continuous operation. After that, the getter cartridge can be reactivated and the sorption process could start again. In the case of MMA, the operation times before re-activation is longer, ~5000 hours
- Operation time and pump speed can be increased designing customized configurations. A larger base flange (e.g. CF 200) allows the use of a bigger getter cartridge with correspondingly increased speed and capacity.
- Since the getter pump does not need power to operate after activation, it can be located as close as possible to the source of contamination and the optics.
- The test system used to characterize the getters is a standard UHV test bench for dynamic sorption tests according to ASTM F798-97.

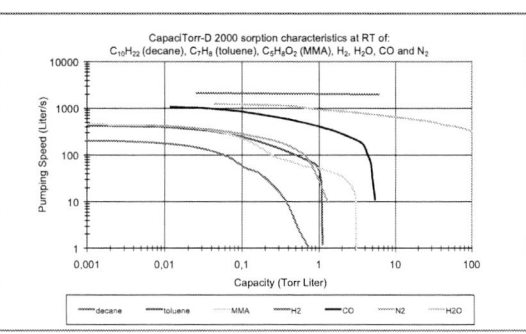

Conclusions

- Purge gas purification solutions are already available for EUVL. Detrimental impurities such as H_2O, Hydrocarbons, Refractory compounds, Inorganic Acids, Bases can be effectively removed down to pptv levels.
- Getter pumps are effective to remove active gases, such as H_2O, H_2, CO, CO_2, O_2 and N_2 as well as several hydrocarbons and volatiles relevant for EUVL process conditions, like Toluene, Decane and MMA.
- Significant gas capacities can be achieved even at room temeperature operation.

we support your innovation

spg@saes-group.com   www.saesgetters.com

Characterization of new EUV stable silicon photodiodes

F. Scholze[*], C. Laubis[*], F. Sarubbi[†], Lis K. Nanver[†], S.N. Nihtianov[†]
[*]PTB, [†]TU Delft, [‡]ASML

Abstract

Development of EUV lithography equipment has triggered a growing interest in EUV radiation detection. Several types of sensors are needed for evaluating and optimizing the imaging performance. The radiation-sensitive surface of the sensors is exposed to high photon flux doses and is affected by hydrocarbons contamination. Consequently, quite aggressive cleaning is required. Therefore, ruggedness to high photon flux and aggressive environments is a key feature of these EUV sensors alongside extreme requirements for stability, reliability, high and spatially uniform responsivity, large dynamic range, and low noise, i.e. low dark current. To meet these requirements, a new EUV photodiode technology is presently being developed and optimized for the requirements of EUV lithography systems at TU Delft in cooperation with ASML. PTB has long experience in the characterization of detectors using synchrotron radiation. This paper presents a study of the EUV performance of p+n photodiodes fabricated by using a novel doping technology. We present measurements of spatial homogeneity, spectral responsivity and high-dose irradiation stability for these new devices. Regarding the combination of high spectral responsivity and irradiation stability, they are already in the present state of development superior to other commercially available detectors.

Conclusion

A new approach[1] was developed at TU Delft in cooperation with ASML for the production of planar diffused silicon p-n diodes for EUV detection to overcome the shortages of state-of-the-art detector technology. PTB performed first studies of the at-wavelength performance of these detectors.

It is shown that the technology is capable of achieving the same nearly ideal responsivity as the best state-of-the-art detectors. The technology also offers the capability to coat the diodes with arbitrary top layers to tailor the efficiency for the particular application without any interference to the responsivity of the underlying photosensitive structure. Due to their almost ideal efficiency, the diodes also show good spatial homogeneity. A particular challenge for EUV photo sensors is the stability under irradiation. Here also the benchmark of the best state-of-the-art detectors is already met.

Summarizing, the development of pure boron-doped photodiodes is proven to be a promising approach toward the improvement of EUV photo sensor performance.

Photodiode structures

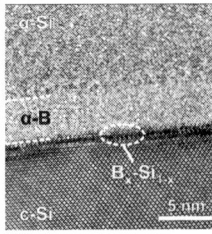 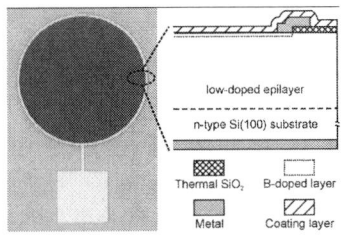

HRTEM image of a B-layer formed after a 2.5 min B₂H₆ exposure at 700 ºC. The sample has been covered with PVD α-Si for TEM analysis.

Image and cross-section of a boron-doped EUV photodiode[1].

Spectral responsivity

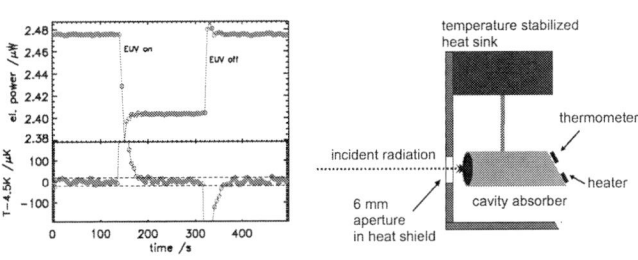

Detector calibration at PTB[2,3] is based on the use of an electrical substitution radiometer[4]. The radiation is totally absorbed by an absorber whose temperature is controlled by an electrical heater. Using on/off cycles for the radiation, the incident radiant power is obtained as the difference in electrical heating power.

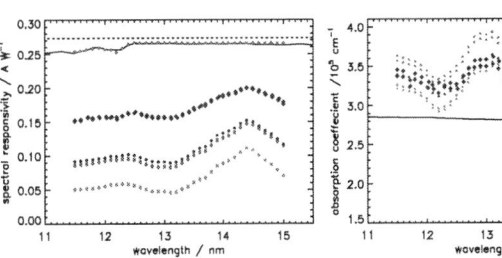

Spectral responsivity of boron-doped silicon diodes. The highest responsivity is for a diode with no additional top layer. The lower responsivity values are for top layers of 15 nm, 30 nm, and 50 nm AlN. The dashed black line shows the theoretical limit[6] for silicon responsivity of 1/3.66 A/W and the solid black line is the best state-of-the-art silicon detector with a very thin nitrided oxide passivation.

Absorption coefficient derived from the responsivity shown left. The data were normalized to the uncoated device and the nominal coating thickness. The good agreement without any artifacts at the Si L-edge proofs the good process reliability and that there is no interference of the coating with the performance of the diodes. Tabulated absorption data from Henke are shown for comparison (black line).

EUV irradiation stability

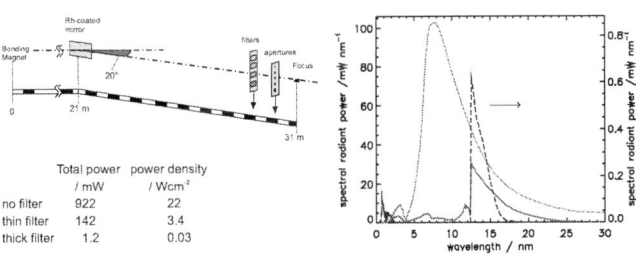

	Total power / mW	power density / Wcm⁻²
no filter	922	22
thin filter	142	3.4
thick filter	1.2	0.03

Irradiation testing is performed at a dedicated beamline[7]. The radiation from a bending magnet is focused by an elliptical mirror with 10° grazing angle. The mirror cuts the white synchrotron radiation distribution at about 5 nm. The total power in the beam is about 1 W, with a power density of 22 W/cm² at a position 360 mm behind the focus. The short wavelength radiation can be further suppressed using Si-based thin film filters resulting in the spectral distributions shown.

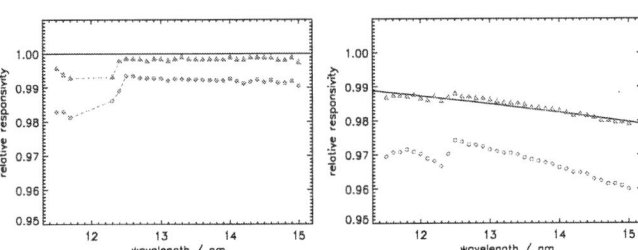

Left: Responsivity of a boron-doped photodiode after irradiation with 200 kJ/cm² normalized to the initial responsivity. The circles represent a diode actually exposed to the radiation while the triangles are a witness diode which was placed under the same vacuum and atmosphere conditions but was shadowed during exposure. No significant change in responsivity is observed for the witness diode.

Right: The same is shown for a state-of-the-art EUV photodiode. Here, also the witness device suffered some reflectivity loss, which might be due to susceptibility of the surface coating to carbon contamination. There is, however, no effect at the silicon L-edge for the witness device indicating that the sensitive region of the diode did not suffer. The solid line represents the effect of 2.5 nm carbon contamination.

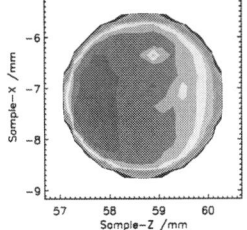 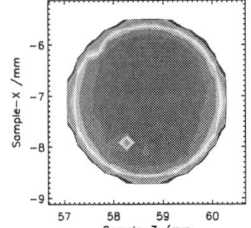

Spatial homogeneity of an early prototype of a boron-doped EUV diode, left, and a commercial state-of-the-art diode, right. The color steps are 0.25 % relative responsivity.

References

1. F. Sarubbi, L. K. Nanver and T. L. M. Scholtes, CVD delta-doped boron surface layers for ultra-shallow junction formation, ECS Transactions, vol. 3, no. 2, pp. 35-44, November 2006.
2. R. Klein, C. Laubis, R. Müller, F. Scholze, G. Ulm, The EUV metrology program of PTB, Microelectronic Engineering 83, 707-709 (2006)
3. F. Scholze, J. Tümmler, G. Ulm, High-accuracy radiometry in the EUV range at the PTB soft X-ray radiometry beamline, Metrologia 40, S224-S228 (2003)
4. H. Rabus, V. Persch, G. Ulm, Synchrotron-Radiation Operated Cryogenic Electrical-Substitution Radiometer as High-Accuracy Primary Detector Standard in the Ultraviolet, Vacuum Ultraviolet and Soft X-ray Spectral Ranges, Appl. Opt. 36, 5421-5440 (1997)
5. F. Scholze, R. Klein, R. Müller, Characterization of detectors for extreme UV radiation, Metrologia 43, S6-S10 (2006)
6. F. Scholze, H. Rabus, and G. Ulm, Mean energy required to produce an electron-hole pair in silicon for photons of energies between 50 and 1500 eV, J. Appl. Phys. 84, 2926-2939 (1998)
7. F. Scholze, R. Klein, T. Bock, Irradiation Stability of Silicon Photodiodes for Extreme-Ultraviolet Radiation Appl. Opt. 42, 5621-5626 (2003)

contact: frank.scholze@ptb.de

2008 EUVL Symposium

EXTREME ULTRAVIOLET LITHOGRAPHY

The Comprehensive Cost for the Mainstream NGL and Simplified Extreme Ultraviolet Lithography Method

Wynn.L.Bear,
Xiangwen,Xiong * .

ABSTRACTS

The next generation lithography is decided by economy and practicality. They included the extreme ultraviolet lithography, Nano-Imprint Lithography, or multiple-e-beam direct write, etc. Now, along with the development of the polarized illumination, high-index materials, solid-immersion mask, double exposure and double patterning, special with the solid immersion on mask supports up to k1 $\geq$ 0.3 at 1.6 NA and more, the immersion 193 nm wavelength lithography is leading to 32 nm half pitch, even 22. INTEL says it won't be using EUV lithography to introduce 22 nm technologies into its chips; it will push EUVL at 16? [1], the Intel now was afraid for the past countless mistakes and errors or not? Furthermore, we concluded the EUVL, a simplify EUVL will show its most advantage in 22 nm node, but not be the double patterning immersion lithography. At the same time, it will become the mainstream next generation lithography technology for 22 nm node and less. We proposed new theories and methods for the mainstream next generation lithography at 22 nm half pitch beyond in the paper. It is a general mechanism to analyze and judge of the mainstream next generation lithography technology. Our current works afford a new guide and strategy of Extreme Ultraviolet Lithography research and its future development. 3 specific descriptions based on our experience for "Moore's Law", and a mathematical model & technical conditions for the semiconductor factory production line for the mainstream next generation lithography technology were built. Base on such the mathematical models and technical conditions, a clear conclusion is emerged that the EUVL be the best choice for the mainstream lithography, and then we are given an only direction and guiding principle for EUVL that is to develop the multiple Catadioptric and composite double reflection EUV Lithography.

Cost is the biggest problem of standard extreme ultraviolet lithography. Based on our Multiple EUV Lithography method and Composite EUV Lithography method, we provide a general cost mechanism. Our current work affords a new strategy of the cost control. At the same time, we have

---

2008 International Symposium on Extreme Ultraviolet Lithography, SEMATECH.

September 28 - October 1, 2008

Lake Tahoe, California, USA.

Wynn.L.bear, Wynn Bear International Cutting-edge High-Tech Institute, CA, USA

Xiangwen,Xiong * , Wynn Bear International advanced Business Machines Company, China.

Tel: (028) 38902-0677, Fax: (028) 38902-0677, Email: wynnbear@gmail.com, wynnbear.inc@gmail.com.

(Wynn.bear is the English name, Xiangwen, xiong is the pinying name).

created mathematical models and their calculation formulas. The models and formulas show that EUV lithography has a great cost advantage over the other two kinds of NGL.

INTRODUCTION

The tendency of modern IC product development is greater size, thinner line width, higher accuracy, higher efficiency and lower cost, etc. Along with integration rate enhancement, the lithography technology of the most essential manufacture process in chip fabrication is facing increasingly difficult problems. The key lithography technology to solve these questions being the 193 nm immersion, extreme ultraviolet lithography, Nano-Imprint Lithography, or multiple-e-beam direct write, etc [2], [3]. Everyone has their own technology that is the most promising for one of the next generation lithography technologies. Nevertheless, Nano- Imprint future is uncertain, EUVL is difficulties, and e-beam lithography is difficult to support such a huge burden. Now, the entire NGL industry is fragmented, chaotic. Some scholars, and even heavyweight experts trying to give a demonstration for the mainstream next generation lithography technology, it seems that the conclusion give by them may be the laughing stock [4], [5], [6], [7].

The Nanoimprint claimed that it has currently demonstrated its ability for semiconductor manufacturing by offering resolution below sub-10nm feature size, high throughput and low cost. There are a few varieties in nanoimprint, which be loosely divided into hot embossing, UV-cured imprinting, micro-contact printing, and room temperature imprint lithography, etc. Based on mechanical micro-replicating technique, the Nanoimprint had avoided the diffracting limitation of traditional optical lithography and can replicate patterns with 6 nm feature sizes[8], [9], [10], [11], [12]. Furthermore, also, the nanoimprint is used in MEMS and other applications to reproduce nanostructure. In the next generation of lithography (NGL), the multiple-e-beam direct write also claimed that it is one of the first-selected schemes for modern IC product manufacturing for its technique is simple and the cost is relatively low[4], [5], [6]. It is the maskless lithography. It seems that has an advantage after the contrastively analyzing with the nanoimprint. Although the shape and energy of the electron beam and the material and depth of substrate will make the proximity effect of Electron Beam Lithography, and a tremendous amount of computation, but it seems that no one will believe that it will be an insurmountable problem [13], [14]. Using in low energy e-beam exposure, it not only obtain high resolution images, but also be unrestricted in a high vacuum.

The EUV lithography technology is the world's most fortunate and most unfortunate. That it is fortunate because it is successfully, that it is unfortunate because it just had the rotten luck of being born in the wrong century. The same should be true of immersion lithography technology.

Talk of the immersion lithography; it was invented by Japanese in the 1980s. It has been unknown until today's "pheasant changed phoenix" -- well-known. Now, 193 nm immersion lithography technologies are widely accepted as the first choice for getting to 45 nm nodes and less. Combining with double exposure, it would be even extended to 32 nm node and 22nm with the cost increasing and yield decreasing [15], [16], [17], and [18].

The Immersion lithography improves critical dimension uniformity (CDU) as well as avoiding the necessity for strong resolution enhancement techniques (RET) as compared with dry lithography. Thus it is possible to significantly reduce the burden of optical proximity correction (OPC) work with immersion lithography [19], [20], [21].

Talk of the resolution enhancement techniques (RET); first we must talk that what is the limited of half pitch? The traditional 193 nm ArF lithography technology was considered to cannot be the technology node of wavelength value and less due to diffraction effect and NA limited, etc. The Raleigh criterion gives a resolution of a diffraction limited lithographic process. It is shown in the following equation. Where λ is the wavelength of laser, NA is the numerical aperture of the lens in the projection tool exposure system, and k1 is a process dependent adjustment factor, R is the resolution. The k1 factor is controlled by a variety of things including the photoresist performance and tool issues such as lens aberrations, etc.

$$R = k1\lambda / (NA)$$

The Resolution Enhancement Technology (RET) are very useful for the optical lithography, like the off-axis illumination (OAI) [25], optical proximity effect correction (OPC) [26], phase shift mask (PSM), etc[27], [28], [29]. These technologies greatly improve the resolution of lithography. In other words, it greatly enhanced the ability to focus the beam, as they mainly use the 193 nm wavelength of the laser. Now, many manufacturers use the 193 nm wavelengths (ArF) laser to get 45 nm points, lines and other patterns, even is the 32 [30].

What is the Immersion lithography? It uses some kind of fluid filling the space between the bottom surface of the last lens and wafer to enlarge the numerical aperture of system. We need to analyze and control the flare, and other to achieve good lithography performances when the immersion liquid and high NA being adopted. The 193nm ArF immersion lithography equipment with double patterning technology Display one's skills to the full technology for half pitch process in the 45, 32, and 22 nm node chips, and resolution enhancement techniques (RET) have become increasingly important. We will require innovations in high-index fluids, lenses and resists to get immersion lithography to the 32 nm node and beyond. , it seems that the immersion lithography technology opened a bright window for the global semiconductor technology roadmap, but the high prices let people daunting for its price almost is the same with Boeing 737 aircraft. At present, it is estimated that need 3 to 5 billion dollars to build one of the most advanced chip plants.

The Mask of the prices for the 193nm ArF dry and immersion lithography are as follows:

90 nm node = 1.5 -2.0 million U.S. dollars,
65 nm node = 3-4 million U.S. dollars,
45 nm node = 6-7 million U.S. dollars.
32 nm node and beyond $\geqslant$ 10 million U.S. dollars (DP).

Though the immersion lithography has advantage in the numerical aperture of optics by an influencing factor of the refractive index of the liquid filled into the space between the bottom lens and wafer, but its comprehensive cost is still relatively expensive.

The Lithography is so important that American Defense Advanced Research Projects Agency (DARPA) had said "the United States must maintain 2 leading generation". To this end, the world's other countries, organizations and groups in the manufacture of integrated circuits to invest a great deal of manpower, resources and finance to ensure them in modern integrated circuits manufactured on the leading position. Although the lithography technology in the manufacture of integrated circuits in the history obtained a glorious success for semiconductors, particularly in the silicon-based semiconductor manufacturers, but the world lithography industry has experienced major setbacks in recent years. Therefore, from the DUV to the VUV (193 nm KrF laser lithography), or 157 nm F2 laser to EUVL, it reflects the history of a world lithography. As previously mentioned, in the DUV and VUV lithography, the mainstream of the world lithography technology is very stable; the development of the world semiconductor industry is also fast and healthy. However, as advocated by Intel the 157 nm lithography failure and the EUVL delay, the lithography industry chaotic situation was dramatically exacerbated. Now, all kinds of new lithography technology has been proposed, the related papers even have been published in the Science magazines, authority Physics journal, and authority nano-technology journal, etc. The related inventors and founder are almost world celebrities or scholars. The Modern lithography and next generation lithography technology industry have also emerged with several super powerful groups. They include the Extreme Ultraviolet Lithography (EUVL), Nano-Imprint Lithography (NIL), e-beam lithography (parallel, multiple or single), interferometric lithography, ion-beam lithography, drop-ondemand nano-inkjet lithography , Dip-pen lithography, scanning array lithography, near-field/evanescent lithography, plasmonic nanowriters, shade Lithography, etc.

Perhaps few people disillusion that "manufacturing is the King" in the semiconductor industries and semiconductor manufacturing is more complex than the design. Take CPU of the INTEL and AMD for example, Intel's products always occupied most of the market share, while AMD is difficult; despite AMD's design is the most leading in most of the time. This is because in the semiconductor industry has a "Moore's Law," which is often talked as the landmark by the industry.

What is the Moore's Law? Moore's law describes an important trend in the history of IC manufacturing: that the number of transistors that can be inexpensively placed on an integrated circuit is increasing exponentially, doubling approximately every two year [22], [23]. Almost every measure of the capabilities of digital electronic devices is linked to Moore's law: processing speed, memory capacity, even the resolution of digital cameras. All of these are improving at exponential rates as well. This has dramatically increased the usefulness of digital electronics in nearly every segment of the world economy. Moore's law describes this driving force of technological and social change in the late 20th and early 21st centuries [24].

What are the best Describes Moore's law's Contributions and Achievements? Here is a simple analogy is that now from San Francisco to New York will only need about 15 seconds if the vehicle speed in the same period also able to increase with the speed of the Moore's law and its prices will be lower.

Doubling the number of transistors the benefits is summarized as follows: expensive is faster, smaller, and less. There are three major technical factors makes it possible to Moore's Law is that reduced feature size, yield improvement and integration density increases. Up to now, the contribution of the feature size reduction is the largest. The feature size reduction depends on the improvement of advanced lithography. So, the judgment and prediction of the mainstream next generation lithography is life-and-death. It relates to all the factories hundreds of billions of dollars in equipment investment, and entire semiconductor industry more than 5~20 trillion dollars the Rise and fall of the market in the next ten-years.

SPECIFIC DESCRIPTIONS OF THE MOORE'S LAW

Moore's Law contain the complex feature size reduction, yield improvement and integration density increases rather than simply Doubling of the number of transistors every two years. The number, surface at area and half pitch of the IC are the most important factor. Based on our research and analysis, they should be described as the following three ways:

1) Words's Descriptions:

Descriptions 01 of the Moore's Law:
Doubling the number of transistors, doubling surface area of the chip, half pitch unchanged every two years.
Descriptions 02 of the Moore's Law:
Doubling the number of transistors, the surface area increase 70%, half pitch to reduce 30% every two years.
Descriptions 03 of the Moore's Law:
Doubling the number of transistors, the surface area unchanged, half pitch to reduce 30% every two years.

2) Mathematical Formulas 'S Descriptions:

The indefinite integral formula is as follows:

$$\int f(x)dx = F(x) + C$$

Note that: $\int f(x)dx$ -- primitive function, $\int$ - integral sign, $f(x)$ - integrand, x - integral variable.

In order to ably direct expression, we transformed into elementary mathematical expression, and the Mathematical formula 01 for the Descriptions 01 of the Moore's Law is as follows:

$$|T_1| = \begin{cases} DN_i = n \times (1+1)^{(i-1)} \\ DS_i = s \times (1+1)^{(i-1)} \\ HP_i = h \end{cases} \quad (1)$$

DNi is the Doubling the number of transistors every two years, DSi is doubling surface area of the chip every two years, HPi is unchanged half pitch, "n" is the number of transistors in the first year, "s" is the surface area of the chip in the first year, "i" is the years, "h" is unchanged half pitch parameter values.

The Mathematical formula 02 for the Descriptions 02 of the Moore's Law is as follows:

$$|T_2| = \begin{cases} DN_i = n \times (1+1)^{(i-1)} \\ DS_i = s \times (1+70\%)^{(i-1)} \\ HP_i = h \times (1-30\%)^{(i-1)} \end{cases} \quad (2)$$

DNi is the Doubling the number of transistors every two years, DSi is the surface area of chip increase 70% every two years, HPi is half pitch to reduce 30% every two years, "n" is the number of transistors in the first two year, "s" is the surface area of the chip in the first two year, "i" is the years, "h" is half pitch parameter values in the first two year.

The Mathematical formula 03 for the Descriptions 03 of the Moore's Law is as follows:

$$|T_3| = \begin{cases} DN_i = n \times (1+1)^{(i-1)} \\ DS_i = s \\ HP_i = h \times (1-30\%)^{(i-1)} \end{cases} \quad (3)$$

DNi is the Doubling the number of transistors every two years, DSi is unchanged surface area of the chip, HPi is half pitch to reduce 30% every two years, "n" is the number of transistors in the first year, "s" is the surface area of the chip in the first year, "i" is the year, "h" is half pitch parameter values in the first year. (In the first two years, we assume that i = 1; in the second one "two years", i = 2; in the third one "two years", i = 3; in the fourth "two years", i = 4; in the fifth "two years", i = 5.)

We expand all the Mathematical formula (1), Mathematical formula (2), and Mathematical formula (3), the results are as follows:

$$|T_1| = \begin{cases} DN_i = n \times (1+1)^{(i-1)} \Rightarrow \begin{cases} DN_1 = n \\ DN_2 = n \times 2 = 2n \\ DN_3 = n \times 2^2 = 4n \\ DN_4 = n \times 2^3 = 8n \\ DN_5 = n \times 2^4 = 16n \end{cases} \\ DS_i = s \times (1+1)^{(i-1)} \Rightarrow \begin{cases} DS_1 = s \\ DS_2 = s \times 2 = 2s \\ DS_3 = s \times 2^2 = 4s \\ DS_4 = s \times 2^3 = 8s \\ DS_5 = s \times 2^4 = 16s \end{cases} \quad (4) \\ HP_i = h \Rightarrow \begin{cases} HP_1 = h \\ HP_2 = h \\ HP_3 = h \\ HP_4 = h \\ HP_5 = h \end{cases} \end{cases}$$

$$|T_2| = \begin{cases} DN_i = n \times (1+1)^{(i-1)} \Rightarrow \begin{cases} DN_1 = n \\ DN_2 = n \times 2 = 2n \\ DN_3 = n \times 2^2 = 4n \\ DN_4 = n \times 2^3 = 8n \\ DN_5 = n \times 2^4 = 16n \end{cases} \\ DS_i = s \times (1+70\%)^{(i-1)} \Rightarrow \begin{cases} DS_1 = s \\ DS_2 = s \times 1.7 = 1.7s \\ DS_3 = s \times 1.7^2 = 2.89s \\ DS_4 = s \times 1.7^3 = 4.91s \\ DS_5 = s \times 1.7^4 = 8.35s \end{cases} \quad (5) \\ HP_i = h \times (1-30\%)^{(i-1)} \Rightarrow \begin{cases} HP_1 = h \\ HP_2 = h \times 0.7 = 0.7h \\ HP_3 = h \times 0.7^2 = 0.49h \\ HP_4 = h \times 0.7^3 = 0.34h \\ HP_5 = h \times 0.7^4 = 0.24h \end{cases} \end{cases}$$

The Extension of the Mathematical Formula (1)   **The Extension of the Mathematical Formula (2)**

$$|T_3| = \begin{cases} DN_i = n \times (1+1)^{(i-1)} \Rightarrow \begin{cases} DN_1 = n \\ DN_2 = n \times 2 = 2n \\ DN_3 = n \times 2^2 = 4n \\ DN_4 = n \times 2^3 = 8n \\ DN_5 = n \times 2^4 = 16n \end{cases} \\[2em] DS_i = S \Rightarrow \begin{cases} DS_1 = S \\ DS_2 = S \\ DS_3 = S \\ DS_4 = S \\ DS_5 = S \end{cases} \\[2em] HP_i = h \times (1-30\%)^{(i-1)} \Rightarrow \begin{cases} HP_1 = h \\ HP_2 = h \times 0.7 = 0.7\,h \\ HP_3 = h \times 0.7^2 = 0.49\,h \\ HP_4 = h \times 0.7^3 = 0.34\,h \\ HP_5 = h \times 0.7^4 = 0.24\,h \end{cases} \end{cases} \quad (6)$$

HP_i	DS_i	DN_i	DRAWING
h	s	n	
h	2s	2m	
h	4s	4m	
h	8s	8m	
h	16s	16m	

The Extension of the Mathematical Formula (3)   **Figure-1**

3) The Descriptions of the Schematic Table and Drawing

The schematic table and drawing 07 for the descriptions 01 and the mathematical formula 1 is shown in figure-1. The schematic table and drawing 08 for the descriptions 02 and the mathematical formula 02 is shown in figure-2. The schematic table and drawing 09 for the descriptions 03 and the mathematical formula 03 is shown in figure-3.

HP_i	DS_i	DN_i	DRAWING
h	s	n	
0.7 h	1.7 s	2 m	
0.49 h	2.89s	4 m	
0.34 h	4.91s	8 m	
0.24 h	8.35s	16m	

HP_i	DS_i	DN_i	DRAWING
h	s	n	or
0.7 h	s	2 m	
0.49 h	s	4 m	
0.34 h	s	8 m	
0.24 h	s	16m	

Figure-2   **Figure-3**

Based on above clear description, the Mathematical formula 03, and equivalent Mathematical formula 06 & Table 9 is the best description for the Moore's Law, the others are metamorphic. Half pitch of IC chip is to reduce 30% every two years is the core for Moore's Law.

THE TECHNICAL CONDITIONS OF THE NEXT GENERATION LITHOGRAPHY FOR THE PRACTICAL PRODUCTION LINE OF THE FACTORY

According to our observation, the yield requirement of the actual factory of the production line is $\geqq$ 100 wph, equipment life in general is $\geqq$ 10 years. Therefore, combined with our previous calculation formula, we come to the technical conditions listed as follows:

The technical conditions for "Doubling the number of transistors, doubling surface area of the chip, half pitch unchanged every two year" is as follows:

$$F(x_4) = |T_4| = \begin{cases} DN_i = n \times (1+1)^{(i-1)} \\ DS_i = s \times (1+1)^{(i-1)} \\ HP_i = h \\ i \geq 5 \text{ (5 Every Two Years)} \\ Y \geq 100 \text{ wph (Wafer Output of Per Hour)} \\ L \geq 10 \text{ Years (Duration Service of Equipment)} \end{cases} \quad (10)$$

The technical conditions for "Doubling the number of transistors, the surface area increase 70%, and half pitch to reduce 30% every two year" is as follows:

$$F(x_5) = |T_5| = \begin{cases} DN_i = n \times (1+1)^{(i-1)} \\ DS_i = s \times (1+70\%)^{(i-1)} \\ HP_i = h \times (1-30\%)^{(i-1)} \\ i \geq 5 \text{ (5 Every Two Years)} \\ Y \geq 100 \text{ wph (Wafer Output of Per Hour)} \\ L \geq 10 \text{ Years (Duration Service of Equipment)} \end{cases} \quad (11)$$

The technical conditions for "Doubling the number of transistors, the surface area unchanged, half pitch to reduce 30% every two year" is as follows:

$$F(x_6) = |T_6| = \begin{cases} DN_i = n \times (1+1)^{(i-1)} \\ DS_i = s \\ HP_i = h \times (1-30\%)^{(i-1)} \\ i \geq 5 \ (5 \ \text{Every Two Years}) \\ Y \geq 100 \ \text{wph} \ (\text{Wafer Output of Per Hour}) \\ L \geq 10 \ \text{Years} \ (\text{Duration Service of Equipment}) \end{cases} \quad (12)$$

It is common requirements and general industry rules for "the equipment life in general is $\geqq 10$ years". The "yield requirement of the actual factory of the production line is $\geqq 100$ wph" means that "wafer per hour" $\geqq 100$. Note that also the equipment life means that it needs to maintain and Catch up Moore's Law 10 years, at least.

The Mathematical formula is as follows:

$$F(x_7) = |T_7| = \begin{cases} \text{Wph} = \dfrac{\text{Batch size}}{\text{Process time}} \\ \text{Process time} = \text{Critical Path Time} + \\ \qquad\qquad \text{Waiting time} + \text{Hold time} \\ \text{Critical Path Time} = \text{Preparation time} + \\ \qquad\qquad \text{Actual process time} + \text{Transfer time} \end{cases} \quad (13)$$

The Motion Frequency of the worktable of the lithography machine

It is important for lithography. The general Yield =100-180 wph; Now, diameter of the wafer is 300MM in Silicon-based semiconductor manufacturing; one wafer produce 400 – 500 Die. The Motion Frequency of the worktable reached hundreds of moves per minute, the highest being 7-9 Times per second. The Dies in the wafer is shown in figure-four.

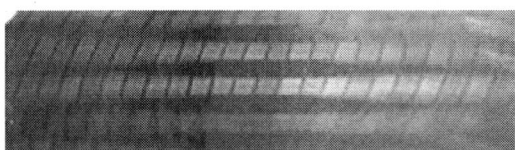

Figure-four

The k1 of the lithography

It is crucial important for the lithography process. It expresses the difficulty level of the process of the factory. The Mathematical Formula is as follows:

$$k1 = R/NA$$

Note that the "R" is the wavelength of the lithography wave; the "NA" is the numerical aperture.

THE MANUFACTURING PRINCIPLE AND ANALYZING FOR EUVL, NIL and MEBDW LITHOGRAPHIES

The popular candidate will handle features smaller than 22 nm and can be extendable for a few more generations are the extreme ultraviolet lithography (EUVL), Nano-Imprint Lithography (NIL), or multiple-e-beam direct write (MEBDW), etc. The figure-five is the 193i transmitted exposure system; the figure-six is the EUVL reflex exposure system.

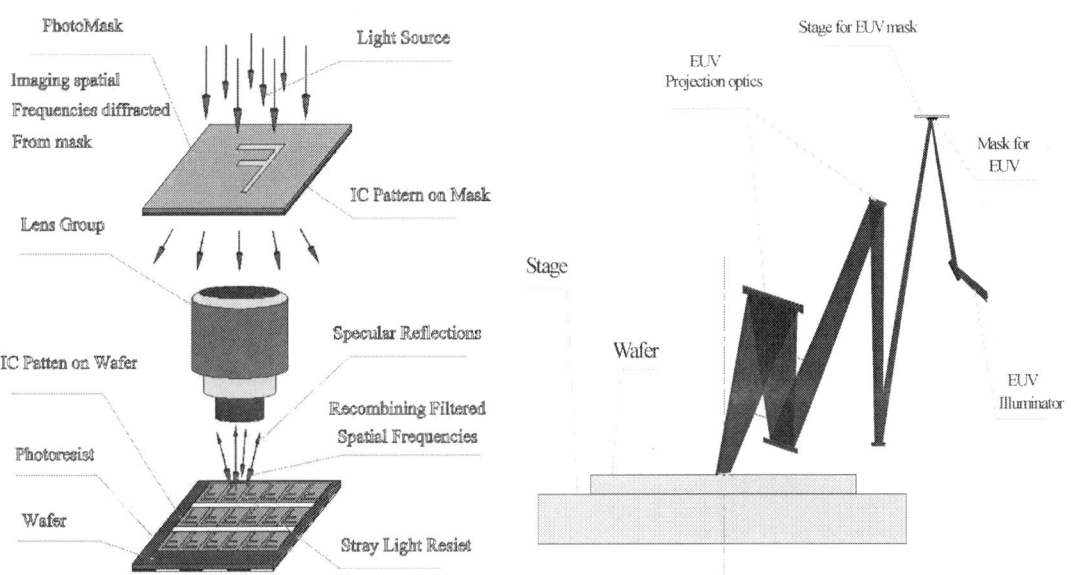

Figure-five   **Figure-six**

The nanoimprint technology is used an intimately contacting mould (1: 1) to replicate nanometer-size features. It is shown in figure-seven.

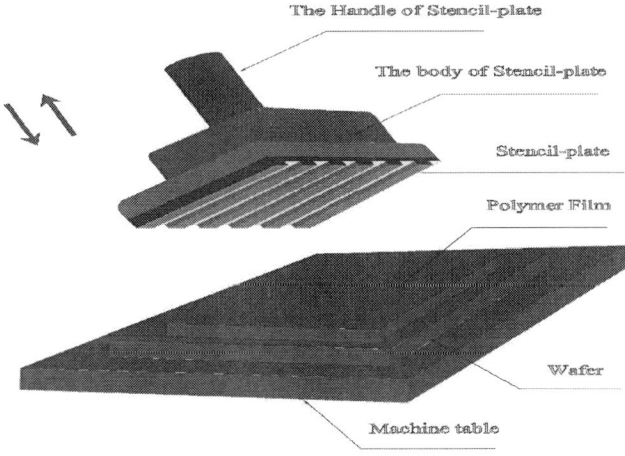

Figure-seven

The Latest Research Progress of the multiple-e-beam direct write (MEBDW) technology declares that it is an advanced, cost-effective and universal variable shaped beam (1: 1) lithography system for direct writing and mask making for both production as well as advanced research applications.

Moore's Law and Economy

It must maintain and Catch up Moore's Law 10 years at least and have economy if every Lithography technology wants to become the next generation lithography. Based on the clear description of the Mathematical formula 03, and equivalent Mathematical formula 06 & the figure-3, the core for Moore's Law is that half pitch of IC chip is to reduce 30% every two year.

We assume the extreme ultraviolet lithography, Nano-Imprint Lithography, or multiple-e-beam direct write all are maintained "reduce 30% of half pitch". How about their doubling the number of transistors and Economy?

The EUVL is non-contact photographic projection of the optical microlithography; it only needs to change the mask. The NIL only needs to change the stencil-plate. MEBDW need doubling lifting the calculation tasks of the host computer, and at the same time doubling lifting the executive ability of the Electron Guns.

Now, the Yield of the EUVL is $\geqq$ 100 wph. The NIL is 20-30 wph. The MEBDW is 30-50 wph (for reference only). Generally speaking, the Yield of the EUVL is easy reach to more, such as 150-180 wph with the power increase in light source(for reference only).

The lowest Yield proportion of EUVL, MEBDW and NIL is about   4 : 2 : 1 .
The maximum Yield proportion of EUVL, MEBDW and NIL is about   8 : 2 : 1 .

The lowest selling price proportion of EUVL, MEBDW and NIL is about   8 : 4 : 1 .
The maximum Yield proportion of EUVL, MEBDW and NIL is about   20 : 10 : 1 .(for reference only)

Based on what the Author personal opinion and experiences, the approximate technical difficulty proportion of EUVL, MEBDW and NIL is as follows (for reference only):
:

The lowest technical difficulty proportion of EUVL, MEBDW and NIL is about   7 : 2 : 5 .
The maximum Yield proportion of EUVL, MEBDW and NIL is about   12 : 2 : 7 .

These data will change with the development and change of the technology, market and product, etc.

Based on what the Author personal opinion and experiences, the rate of finished product and rate of the good product of EUVL, MEBDW and NIL are acceptable, but NIL need to give enough sufficient contexts to justify that it contents for the requirements. Other lithographies need to justify that it contents for the requirements of the Moore's Law, and be use in the production line of the actual factory but not the research test.

WHOLE ANALYZING AND JUDGMENT TO THE EUVL, NIL AND MEBDW

The Lithography technology, as the bellwether of the modern semiconductor technology, the status of the importance of evaluation not is overemphasized if you no matter how you praise it. As one of the most challenging things to NGL, we creatively based on Moore's Law to make the reference judgement.

General Comment

Although the Nano-Imprint (NIL) future is uncertain, extreme ultraviolet lithography (EUVL) is difficulties, and e-beam lithography (MEBDW) is difficult to support such a great burden, the EUVL will be the best choice for future 22 nm technology nodes and beyond. However, the standard EUVL will not be able to catch up with the pace of chip design, in particular the multi-core design of the CPU, etc; Therefore, no matter now or in the future from the point of view, it is imperative to using the Simplified EUV lithography.

Detail Basis and Proof

Although the entire owner and the backer of the extreme ultraviolet lithography (EUVL), Nano-Imprint Lithography (NIL), or multiple-e-beam direct write (MEBDW) have that their own technology that is the most promising to be the best next generation lithography technology. However, we regret to say that the majority of cases, they are extreme, short-sighted, including many world-class scholars and experts. At least on the issue of Moore's Law, they have mentioned it, but in the actual process of judgments seems to have forgotten it.

The future 22 nm technology nodes and beyond for NGL is not only the experimental study, but must use to the actual batch production process and lines of the factory.

Remember that in the experiment wrong, we have the chance come back again. However, when you are wrong in the actual production line, you should not assume that competitors will give you the opportunity to correct. If we have an error judgment after the equipments installed on the production line, the consequences are difficult to imagine. It relates to all the factories incalculable dollars in equipment investment and the fate of a company, and entire semiconductor industry astronomical dollars market in the future, even influencing on the related industries. The Detail Basis and Proof are as follows:

The Multiple-E-Beam Direct Write (MEBDW) Lithography

The MEBDW lithography, on the surface of this technology looks very promising as the best NGL; actually, it is only a superficial conclusion that easy to be exposed. Moore's Law is "Doubling the number of transistors, the surface area unchanged, and half pitch to reduce 30% every two year". The MEBDW reach to the "Doubling the number of transistors" almost need doubling lifting the calculation tasks of the host computer and doubling lifting the executive ability of the Electron Guns. Doubling is much bigger data than the existing system, can this be practically achieved based no current science and technology? Even if you have an unexpected new technology barely able to achieve it, at this time the users will not agree with you for they face to Re-Purchase new equipments every two years. In nature, it is only equivalent to the high-speed multi-axis machine tools of the general manufacturing, rather than Mould. From the point of view of the semiconductor, the manufacturing of transistors by using it is one by one; from the view of lithography, the manufacturing of lines is one by one still. In addition, they will be used again or redistribute whether the stencil-plates of the Nano-Imprint (NIL) Lithography and the Masks of the Extreme Ultraviolet Lithography (EUVL) Lithography, but the Multiple-E-Beam Direct Write (MEBDW) cannot.

Take a CPU as an example, it's transistor counts is 500 million, it will be 16000 million after 10 years with Moore's Law. How can the firm supporter to MEBDW achieve such a huge jump?

The Nano-Imprint (NIL) Lithography

About the NIL, they have always stressed that the cost is low, but they seem to forget that it needs to 5-8 sets of equipments to keep up with the yield of one set of existing optical equipment. It means that the NIL need the same of 5-8 sets of stencil-plates and machines; now, is it cheaper? At the same time, it needs the factory production line re-layout, and adds 5-8 skilled workers and maintenance staff like the MEBDW lithography. It also needs the alignment and overlay control of X-axis, Y-axis and Z-axis that the modern technology almost cannot be achieved, because it is mechanical contact moving rather than the non-contact optical exposure. They need 0.5 – 3nm precision in the future lithography. The 0.5 – 3nm precision is very hard to reach in the optical Lithography that only need the alignment and overlay control of X-axis and Y-axis. This doubt cannot be quickly eliminated unless the technical miracles appear, or the NIL can avoid these questions. The Manufacturing will be very hard for the stencil-plate of 1:1. Even the photomask of 4: 1 is very difficult in optical Lithography, note that the photomask is only about 5nm plating. It will be an enormous challenge for its half pitch to reduce 30% every two years. But so far as situation at present, it is hard for the NIL to catch up with the Moore's Law.

The Extreme Ultraviolet Lithography (EUVL) Lithography

The EUVL is difficulties for it used x-ray, the short wavelength & strong penetrated EUV. Please be acutely aware that EUVL problems, by their very nature, are more difficult than another Lithography. However, it is the most perfect for maintain the Moore's Law. It catches up the Moore's Law only need to replace the Mask in the future 10 years at least. It has many currently problems, these challenges include progress in key critical technology issues such as Demonstration of reliable high-power source, collector modules, simultaneously meeting to resist resolution, sensitivity and line edge roughness specifications and fabricating defect-free masks, etc. However, we developed the simplify EUVL to completely solve the problem.

THE SIMPLIFY EUV LITHOGRAPHY.

Like an F-16 fighter in comparison to an F-15 or an F-35 fighter in comparison to an F-22, the simplify EUVL is the use of idea for CPU design to catch up the Moore's Law with multi-core method. An advanced multi-core CPU are no means a few single-core CPU with a combination & connectivity, but be the seamless integration of internal functional units.

The Simplified EUVL method is used the seamless integration of internal functional systems to obtain enough power, economy and practicality, so that it is an absolute dominance the mainstream lithography technology.

The more detailed data for cost comparison are shown in Table – 1 to Table – 5(All the data for reference only, the data in the Table – 1 is come from [4], and the data for DP in the Table – 1 is come from [4] also). Table – 1 to Table – 5 is taking a CPU as an example in technology node 65, 45, 32, 22 and 16 nm. We assumed that the transistor count of the CPU in the 65 nm node is 250, 000, 000, based on the Moore's Law, the transistor count of the CPU in the 45 nm node is 500, 000, 000; in the 32 nm node is 1, 000, 000, 000; the 22 nm is 2, 000, 000, 000, the 16 nm is 4, 000, 000, 000.

In the Table – 1 to Table – 5, SP is the single patterning lithography, DP is the double patterning, I is the immersion lithography, EUV is the Extreme Ultraviolet lithography, S-EUV is the Simplified Extreme Ultraviolet lithography, MEBDW is the multiple electron beam direct write systems for lithography, NIL is the nanoimprint lithography.

The Table – 1 is shown a choice for 65 nm node is the single patterning lithography and the multiple electron beam direct write systems. Because our goal is the EUV lithography and Simplified EUVL, so we have used anti-EUV data to show impartiality.

Cost Comparison 01					(Technology node: 65nm ; Transistor Count: 250, 000, 000)				
	H_2O 1mm SP	H_2O 1mm DP	H_2O 1mm I	H_2O 1mm DP + I	EUV 50M/100	S-EUV 180M	MEB DW 20M/20	MEB DW 20M/40	NIL 20M/20
Expo tool cost / 1mm SP tool	1.0	X	X	X	X	X	0.62	0.62	X
Track cost / 1mm SP tool	0.13	X	X	X	X	X	0.06	0.10	X
Raw throughput (wph)	145	X	X	X	X	X	20	40	X
Mask cost/layer / 1mm SP mask	1.0	X	X	X	X	X	/	/	X
Exposure+material /layer/1mm	1.0	X	X	X	X	X	2.67	1.42	X
Error Coefficient	/						/	/	
Attention	SP and MEBDW can be for 65 nm node								

Table – 1

The Table – 2 is shown a choice for 45 nm node is the double patterning, immersion lithography and MEBDW. On a comparison to the table – 1, the raw throughput of the MEBDW fell by almost half, it has shown that it be no advantage.

Cost Comparison 02	(Technology node: 45nm ; Transistor Count: 500, 000, 000)								
	H_2O lmm SP	H_2O lmm DP	H_2O lmm I	H_2O lmm DP + I	EUV 50M/100	S-EUV 180M	MEB DW 20M/20	MEB DW 20M/40	NIL 20M/20
Expo tool cost / lmm SP tool	X	1.19	1.45	X	X	X	0.62	0.62	X
Track cost / lmm SP tool	X	0.15	0.14	X	X	X	0.06	0.10	X
Raw throughput (wph)	X	100	135	X	X	X	11	22	X
Mask cost/layer / lmm SP mask	X	2.0	1.0	X	X	X	/	/	X
Exposure+material /layer/lmm	X	2.19	1.2	X	X	X	2.9	1.56	X
Error Coefficient		10 %	10 %				10 %	10 %	
Attention	DP, I, and MEBDW can be for 45 nm node.								

Table – 2

The Table – 3 is shown a choice for 32 nm node is the double patterning immersion lithography, Simplified Extreme Ultraviolet lithography and MEBDW. On a comparison to the table – 2, the raw throughput of the MEBDW fell by almost half again, it is failure. The double patterning immersion lithography and Simplified EUVL has shown it's own advantage.

Cost Comparison 03	(Technology node: 32nm ; Transistor Count: 1, 000, 000, 000)							
	H_2O lmm I	H_2O lmm DP	H_2O lmm DP + I	EUV 50M/100	S-EUV 180M	MEB DW 20M/20	MEB DW 20M/40	NIL 20M/20
Expo tool cost / lmm SP tool	X	X	1.45	X	1.4	0.62	0.62	X
Track cost / lmm SP tool	X	X	0.17	X	0.13	0.06	0.10	X
Raw throughput (wph)	X	X	90	X	180	6	12	X
Mask cost/layer / lmm SP mask	X	X	2.5	X	1.5	/	/	X
Exposure+material /layer/lmm	X	X	2.4	X	1.9	3.2	1.72	X
Error Coefficient			10%		10%	10%	10%	
Attention	DP+I, S-EUV and MEBDW can be for 32 nm node.							

Table – 3

The Table – 4 is shown a choice for 22 nm node is the double patterning immersion lithography, Extreme Ultraviolet lithography, Simplified Extreme Ultraviolet lithography and MEBDW. On a comparison to the table – 3, the MEBDW must be eliminated. The Extreme Ultraviolet lithography, Simplified Extreme Ultraviolet lithography has shown its own advantages, the double patterning immersion lithography is failure.

Cost Comparison 04				(Technology node: 22nm ; Transistor Count: 2, 000, 000, 000)				
	H_2O lmm SP	H_2O lmm DP	H_2O lmm DP + I	EUV 50M/100	S-EUV 180M	MEB DW 20M/20	MEB DW 20M/40	NIL 20M/20
Expo tool cost / lmm SP tool	X	X	1.45	1.54	1.4	0.62	0.62	X
Track cost / lmm SP tool	X	X	0.17	0.13	0.13	0.06	0.10	X
Raw throughput (wph)	X	X	90	100	180	3.5	7	X
Mask cost/layer / lmm SP mask	X	X	3.5	1.5	2.0	/	/	X
Exposure+material /layer/lmm	X	X	2.65	2.1	2.1	3.5	1.9	X
Error Coefficient			10%	10%	10%	10%	10%	
Attention	DP+I, EUV, S-EUV, MEBDW can be for 22 nm node.							

Table – 4

The Table – 5 is shown a choice for 16 nm node is the Extreme Ultraviolet lithography and the simplified Extreme Ultraviolet lithography, their costs basically is the same.

Cost Comparison 05				(Technology node: 16nm ; Transistor Count: 4, 000, 000, 000)				
	H_2O lmm I	H_2O lmm DP	H_2O lmm DP + I	EUV 50M/100	S-EUV 180M	MEB DW 20M/20	MEB DW 20M/40	NIL 20M/20
Expo tool cost / lmm SP tool	X	X	X	1.54	1.5	0.62	0.62	X
Track cost / lmm SP tool	X	X	X	0.13	0.13	0.06	0.10	X
Raw throughput (wph)	X	X	X	150	180	2	4	X
Mask cost/layer / lmm SP mask	X	X	X	2.0	2.5	/	/	X
Exposure+material /layer/lmm	X	X	X	2.3	2.3	3.8	2.1	X
Error Coefficient				10%	10%	10%	10%	
Attention	EUV, S-EUV, and MEBDW can be for 16 nm node.							

Table - 5

CONCLUSIONS, CHALLENGES, AND DIRECTIONS FOR FUTURE RESEARCH

It is the needs of future, age and industry to find out the mainstream next generation lithography technology for humankind. The MEBDW must eliminate the concerns of users about equipment investment. The NIL must eliminate the concerns of users for its technological challenge. The EUVL must eliminate the concerns of users for the technological difficulties. The EUVL Belongs to the optical lithography still, so people have the confidence to solve the technical problems. All of us in the field of optical lithography have several decades' years of rich experience, and has the very complete industrial chain. The users have the confidence to the EUVL, and expect it to enter the market.

ACKNOWLEDGMENT

First I thank myself, because this active research is using Wynn. Bear's personal funds. Another acknowledgment pleases sees my homepages. We thank Prof. Bingkun,Zhou, the President of the Chinese Optical Society.

REFERENCES

[1], http://www.semiconductor.net/blog/870000487/post/510025451.html?q=intel%2C+22nm

[2] Onkels E D. SPIE27th, Annual international symposium and education program on microlithography. 2002, 4691: 203.

[3] R. Pelzer, Paul Kettner etc. Full Wafer Replication of Nanometer Features. [J]. Micro-and Nanotechnology.2005, 34(2) :256~259 .

[4], http://spiedl.aip.org/journals/doc/JMMMGF-ft/vol_6/iss_4/040101_1.html

[5], http://spiedl.aip.org/journals/doc/JMMMGF-ft/vol_6/iss_1/010101_1.html

[6], http://spiedl.aip.org/journals/doc/JMMMGF-ft/vol_5/iss_4/040101_1.html

[7], Burn J. Lin. Sober view on extreme ultraviolet lithography. J. Microlith., Microfab., Microsyst. 5, 033005 (2006).

[8], HEYDERMAN L J,SCHIFT H,DAVID C,et al. Flow behaviour of thin polymer films used for hot embossing lithography[J] .Microelectron Eng, 2000, 54 (3-4) :229-245.

[9] CHOU Y,KRAUSS R. Imprint lithography with sub-10 nm feature size and high throughput .J. Microelectronic Engineering. 1997, 35 :237-240 .

[10] VRATZOV B,FUCHS A,KURZ H. Large scale ultraviolet-based nanoimprint lithography .J.Vac.Sci. Technol.B. 2003, 21(6) :2760-2764 .

[11] Resnick DJ,Dauksher WJ. I imprint lithography:lab curios-ity or the real NGL .SPIE Microlithography Conference. 2003, :12 .

[12] Y. Chen, E. Roy. Soft nanoimprint lithography .2005, 1 :283~288

[13], Aniguchi J,Tokano Y,Miyamoto I,et al. Diamond nano-imprint lithography .Nanotechnology, 2002, 13 :592 .

[13] Chang T HP,Wilson AD,Speth AJ,et al. Electronbeamlithography for complex high density devices[A] .Proceedings of the 6th International Conference on Elec-tron and Ion BeamScience and Technology. San Fran-cisco: The Electro Chemical Society, 1974, .

[14] Mihir Parikh. Correction to proxi mity effects in electronbeamlithography I[J] .Thery J App

Phys, 1979, 50 (6) :4371-4377 .

[15],BURNJ L. The ending of optical lithography andthe prospectsof its successors[J] .Microelectronic Engineering, 2006, 83 (4) :604-613 .

[16] Takahito Chibana,Hitoshi Nakano,Hideo Hata et al. Development status of a193-nm immersion exposure tool[C] .SPIE, 2006, 6154 :61541V .

[17] VONGEHR M,PREDEHLP,HASINGER G. Development of ametrologyto characterize EUVoptics at 13.5 nm[J] .Proceedingsof SPIE, 2006, 6187 (09) :1-9 .

[18] TAKAYOSHI N,AKIRA K. Formation factors of watermark forimmersion lithography[J] .Japanese Journal of AppliedPhysics, 2006, 45 (6) :5383-5387 .

[19] J.Mulkens et.al. Defects, overlay and focus performance improvements with five generations of immersion expo- sure tools[C] .SPIE6520. 2007, .

[20] J.W.de Klerk et. al. Performance of a 1.35NA ArF im- mersion lithography system for 40nm applications [C] .SPIE 6520. 2007, .

[21] E.van Setten et al. Pushing the boundary: low-k1 exten- sion by polarized illumination[C] .SPIE 6520. 2007, .

[22], http://www.intel.com/technology/mooreslaw/

[23], Gordon E. Moore. Cramming more components onto integrated circuits. Electronics, Volume 38, Number 8, April 19, 1965

[24], http://en.wikipedia.org/wiki/Moore's_law

[25], Kamon K, Miyanmoto T, Myoi Y, et al. Photolithography system using annular illumination. Jpn J Appl Phys, 1991,30 (Part 1): 3012.

[26], sudaka K , Kawahira H, Sugawara M, et al. Practical optical proximity effect correction adopting process latitude consideration. Jpn J Appl Phys, 1995, 34(Part 1, 12B): 6552.

[27], Levinson M D, Viswanathan N S, Simpson R A. Improving resolution in Photolithography with a phase shift mask. IEEE Trans Electron Devices, 1982, ED-29: 1828.

[28], Smith B W, Butt S, Alam Z, et al. Attenuated phase shift masks materials for 248 and 193nm lithography. . J Vac   Sci Technol, 1996, B14(6): 3719.

[29], Liu H Y, Karklin L, Wang Y T, et al. The application of phase shifting mask to 140nm gate patterning (II): mask tolerances. Proc. SPIE, 1998, 3334:2

[30], http://www.intel.com/technology/architecture-silicon/45nm-core2/.

AUTHOR INDEX

Agovic, K. 1119
Ahn, B.-S. 646, 721
Allred, B. 1479
Amano, T. 1565, 1593
Amemiya, M. 1564, 1631
Anazawa, T. 1611
Anderson, C. N. 312, 426, 487, 1389, 1390, 1500
Antohe, A. 602, 1634, 1650
Aoyama, H. 1174, 1593
Aquila, A. 1560
Aramaki, F. 1565
Arisawa, Y. 1593
Auželyte, V. 363
Ayers, T. 1461
Bacuita, T. 1433
Banine, V. 1018, 1055, 1481
Bear, W. L. 1418, 1449, 1668
Bender, M. 602
Bergeson, S. 1479
Bernath, R. 956
Bianucci, G. 1041
Bollanti, S. 1391
Borisov, V. M. 996
Borisova, G. N. 996
Brainard, R. 1529
Brandt, D. C. 890
Brunner, T. 1227
Burkhardt, M. 1227
Caudillo, R. 1433
Cha, B. 663
Chiba, H. 1275
Cho, H. K. 646, 721, 848
Cho, W. 721
Choi, S. 1201
Cloutier, P. 1529
Conte, A. 1666
Corthout, M. 935
Cummings, K. 199, 291
Cunado, J. 956
De Groote, F. P. J. 1119
Denbeaux, G. 602, 1634, 1650
Deng, Y. 426, 1227
Denham, P. 312, 1560
Di Lazzaro, P. 1391
Echigo, M. 1499
Ehm, D. 1055
Eichenlaub, S. 663
Ellwi, S. 956
Endo, A. 916
Engelstad, R. 689
Fan, Y. 602, 1634
Flora, F. 1391
Fulford, B. 956
Fumar-Pici, A. 291, 602
Gallatin, G. M. 487

Garg, R. 602, 1634, 1650
Gargini, P. 136
George, S. A. 312, 956, 1530, 1560
Gielissen, K. 1481
Glushkov, D. 1481
Goethals, A. M. 742, 1242
Goldberg, K. 312, 602, 721, 867, 1563, 1633
Goldstein, M. 1321, 1416, 1480
Goo, D. 1201
Goodwin, F. 520, 602, 1417, 1480
Grove, D. 1461
Gullikson, E. 583, 623, 1560
Gustafson, D. 1461
Hada, H. 1501, 1513
Hagiwara, R. 1565
Han, H. 646, 721
Han, S. I. 291
Han, W. 71, 1201
Hansen, S. 291
Harned, N. 199, 1055
Hartley, J. 689, 1417
Hartman, R. 1, 97
Hasegawa, T. 264
Hay, N. 956
Hayes, A. 709
Hazelton, A. J. 520
He, L. 1155
Henderson, C. L. 376
Henderson, I. 956
Hirai, Y. 1542
Hirayama, T. 1501, 1513
Hoef, B. 312
Holfeld, C. 602
Honda, T. 264
Hotta, K. 1462
House, M. 663
Hudyma, R. 1321, 1341, 1416
Hughes, G. 520
Huh, S. 602, 646, 867
Huijbregtse, J. 1055
Huli, L. 291
Hultermans, B. 199
Ikuta, Y. 563
Irie, M. 1513
Ishihara, S. 1583
Itani, T. 335, 1499, 1531, 1542, 1561
Ivanov, A. S. 996
Iwai, T. 1513
Iwamoto, F. 742
Iwasaki, T. 805
Iwashita, J. 1513
Jak, M. 1018
Jonckheere, R. 742
Jones, G. 312
Kadaksham, A. J. 663

AUTHOR INDEX

Kai, T. 1542
Kamo, T. 1174
Kamptaprasad, R. 956
Kaneyama, K. 335
Kang, I.-Y. 721
Kashiwamura, T. 1531
Kawamura, D. 335, 1542
Kearney, P. 709
Kemp, D. 1530, 1560
Kemp, J. 1479
Kempen, A. 1055
Kessels, B. 291
Khristoforov, O. B. 996
Kim, D. 646, 721, 848
Kim, G. 721
Kim, H. 646, 721
Kim, I. 1201
Kim, S.-S. 646, 721, 848
Kinoshita, H. 848, 1501, 1562
Kirukhin, Y. B. 996
Kishimoto, J. 848
Kluenkov, E. 1018
Knight, L. 1479
Koay, C.-S. 1227
Kobayashi, S. 335, 1561
Koh, C. 312, 398, 1417
Koida, K. 1648
Komori, H. 1482
Kondo, H. 1275
Kools, J. 1041
Koops, R. 1119
Koster, N. B. 1119
Kozawa, T. 505
Kuerz, P. 1300
La Fontaine, B. 312, 602, 867, 1227
Landoni, C. 1666
Langer, E. 602
Laubis, C. 1667
Lawson, R. 376
Lee, C.-T. 376
Lee, D. 646, 721, 848
Lee, J. 1201
Lee, S. 721, 1583
Levinson, H. J. 21
Liang, T. 583, 829
Lin, C. C. 709
Litt, L. C. 520
Loginova, E. 1529
Lok, S. 199
Lopatin, A. 1018
Lopez, J. 1479
Luchin, V. 1018
Lyons, A. 689
Lytle, W. M. 1077
Ma, A. 312, 398, 829

Madey, T. E. 1141, 1529
Magoshi, S. 1174, 1483
Mallmann, J. 291
Manini, P. 1666
Maruyama, K. 1542
Mbanaso, C. 1634, 1650
McIntyre, G. 1227
Meijerink, M. G. H. 1119
Meiling, H. 199
Mezi, L. 1391
Mimura, T. 1513
Mirkarimi, P. 709
Mishchenko, V. A. 996
Miura, T. 230
Miyakawa, R. 312, 1632, 1633
Mizokoshi, H. 1462
Mizuno, H. 291, 1227
Mochi, I. 867, 1563
Montgomery, W. 312
Moors, R. 1055
Mori, H. 1483
Mori, I. 1, 183, 551
Morishima, H. 264
Murakami, K. 1275
Murra, D. 1391
Nakamura, J. 505
Nanver, L. K. 1667
Naulleau, P. 10, 124, 312, 398, 426, 487, 602, 867, 1321, 1367, 1389, 1390, 1416, 1500, 1530, 1560, 1563, 1632, 1633
Neumann, M. J. 1077
Niakoula, D. 312, 1560
Nihtianov, S. N. 1667
Niibe, M. 1648
Nishihara, K. 975
Nishimura, N. 1564, 1583
Nishiyama, I. 551, 1583, 1611
Nishiyama, Y. 1565, 1611
Nomura, K. 1275
Oguro, D. 1499
Ohara, L. 291
Oizumi, H. 335, 1499, 1531, 1561, 1611
Okamoto, K. 505
Okoroanyanwu, U. 602
Omata, K. 829
Onodera, J. 1501
Orvek, K. 689, 1634, 1650
Oshino, T. 1275
Ota, K. 1564, 1583, 1631
Owada, T. 1531
Park, C. 1201
Park, J. 312, 398, 1201, 1201
Park, S.-J. 829
Partlow, M. 1461
Peters, J. H. 602

AUTHOR INDEX

Petrillo, K................................291
Pierson, B................................199
Pierson, W................................291
Prokofiev, A. V...........................996
Rabellino, L..............................1666
Raghunathan, S...........................1417
Raju, R..................................1077
Randive, R...............................709
Rastegar, A..............................663
Reiss, I.................................709
Rekawa, S..........................1530, 1560
Ren, L...................................1417
Richardson, M............................956
Riddle, S................................1666
Roberts, J...............................1433
Roller, J................................312
Ronse, K.................................742
Rossinger, M.............................602
Routh, R.................................291
Ruzic, D. N..............................1077
Sahoo, P.................................363
Saidani, M...............................363
Salamassi, F.............................623
Salashchenko, N..........................1018
Salmaso, G...............................1041
Sanche, L................................1529
Santillan, J. J..........................1561
Sarubbi, F...............................1667
Sasaki, T................................1562
Sato, H..................................1462
Schietinger, C...........................1461
Scholze, F...............................1667
Schwarz, M...............................1341
Seo, H.-S.......................646, 721, 848
Shevelko, A..............................1479
Shigemura, H..............551, 1565, 1593
Shiina, K................................1565
Shimizu, M...............................1542
Shimokawa, T.............................1542
Shin, H..................................1077
Shiono, D...........................1501, 1513
Shirai, S............................1174, 1483
Shroff, Y. A.............................623
Sidelnikov, Y............................1481
Sjmaenok, L..............................1018
Sligte, E. te............................1055
Soer, W.............................1018, 1481
Sohn, J..................................689
Solak, H. H..............................363
Spiller, E...............................709
Sporre, J................................1077
Stepanenko, N............................742
Storm, A.................................1055
Suga, O.............551, 805, 1564, 1565, 1583, 1593,
        1611, 1631

Sugiyama, T..............................709
Szilagyi, J..............................956
Tagawa, S................................505
Taguchi, T...........551, 1564, 1583, 1631
Takenoshita, K...........................956
Tamura, T................................829
Tanaka, H...........................1174, 1483
Tanaka, T...........................805, 1593
Tanaka, Y...........................1174, 1483
Tawarayama, K.......................1174, 1483
Tchikoulaeva, A..........................774
Teramoto, Y..............................1462
Terasawa, T.................551, 805, 1565
Thomas, M................................1341
Tolbert, L. M............................376
Tomie, T.................................805
Torre, A.................................1391
Trogisch, S..............................602
Tsybin, N................................1018
Uzawa, S.................................264
Van Der Mullen, J........................1481
Van Herpen, M............................1481
Van Ingen-Schenau, K.....................291
Van Kampen, M............................1055
Van Setten, E............................199
Vandentop, G.............................1433
Vanneer, R...............................1055
Verberk, R...............................1055
Vinokhodov, A. Y.........................996
Wagner, C................................199
Wallow, T................291, 312, 426, 602
Warisawa, S..............................1583
Watanabe, T.............848, 1501, 1562
Watso, R.................................291
Weber, A.................................363
Wieringa, F. P...........................1119
Wood, O............10, 291, 602, 1227, 1367
Wüest, A...........520, 602, 1529, 1634
Wurm, S..........1, 312, 398, 1367, 1417, 1480
Xiong, X...............1418, 1449, 1668
Yakshinskiy, B......................1141, 1529
Yakunin, A...............................1018
Yamamoto, T..............................1583
Yamane, T................................805
Yamatani, D..............................1462
Yan, P.-Y................................623
Yankulin, L.........................602, 1634
Yasaka, A................................1565
Yeo, J.-H................................1201
Yokokoji, O..............................1562
Yokoyama, T..............................1462
Yomogita, A..............................1531
Yoshioka, M..............................935
Yun, H.............................663, 709
Zakhor, A................................1632

AUTHOR INDEX

Zalkind, S. 1141
Zhang, G. 829, 1563
Zimmerman, J...............................1155
Zocchi, F. E...............................1041

SEMATECH
2706 Montopolis Drive
Austin, Texas 78741

ISBN 978-1-61567-661-3